Introduction to SOLIDWORKS

A Comprehensive Guide with Applications in 3D Printing

Introduction to SOLIDWORKS

A Comprehensive Guide with Applications in 3D Printing

GODFREY C. ONWUBOLU

CRC Press is an imprint of the
Taylor & Francis Group, an **informa** business

CRC Press
Taylor & Francis Group
6000 Broken Sound Parkway NW, Suite 300
Boca Raton, FL 33487-2742

Printed on acid-free paper
Version Date: 20160706

International Standard Book Number-13: 978-1-4987-3119-5 (Pack - Book and Ebook)

Library of Congress Cataloging-in-Publication Data

Names: Onwubolu, Godfrey C., author.
Title: Introduction to SolidWorks : a comprehensive guide with applications in 3D printing / Godfrey C. Onwubolu.
Description: Boca Raton : Taylor & Francis Group, a CRC title, part of the Taylor & Francis imprint, a member of the Taylor & Francis Group, the Academic Division of T&F Informa, plc, [2017] | Includes bibliographical references and index.
Identifiers: LCCN 2016014732 | ISBN 9781498731195 (acid-free paper)
Subjects: LCSH: SolidWorks. | Computer-aided design. | Three-dimensional printing.
Classification: LCC T386.S55 O59 2017 | DDC 620/.00420285--dc23
LC record available at https://lccn.loc.gov/2016014732

Visit the Taylor & Francis Web site at
http://www.taylorandfrancis.com

and the CRC Press Web site at
http://www.crcpress.com

Printed and bound in the United States of America by Sheridan

This book is dedicated entirely to God who did solid work in the creation of everyone and everything in existence and in sustaining life. I owe Him all that I have because all that I have comes from Him.

Contents

Section II Intermediate Engineering Design Principles with SolidWorks

Preface

This textbook, *Introduction to SolidWorks: A Comprehensive Guide with Applications in 3D Printing*, is written to assist students in colleges and universities, designers, engineers, and professionals who are interested in using SolidWorks for practical applications. This textbook pitches at an intermediate level for SolidWorks users, although it has been so organized in such a way that the earlier chapters meet the needs of newcomers.

The textbook is divided into four sections. Section I covers the introductory principles of SolidWorks: simple and advanced-part modeling, assembly modeling, drawing, configuration and design tables, and part modeling with equation-driven curves. Section II covers the intermediate principles of SolidWorks: reverse engineering, top–down design, surface modeling, toolboxes and design libraries, animation, and rendering. Section III covers the practice of SolidWorks. Section IV covers the introduction of how SolidWorks is used as a modeling software for 3D printing. It presents in a simple manner how designers can use CAD software for modeling, then save the 3D model(s) in .STL file that is forwarded to a 3D printer for manufacturing. Applications are in the areas of manufacturing processes, mechanical systems, electromechanical systems, and engineering analysis. The sections on manufacturing processes include the design of molds, sheet metal parts, dies, and weldments. The sections on mechanical systems include the aspects of routing such as piping and tubing, power transmission systems, and mechanism design. The section on engineering analysis covers finite element analysis.

Newcomers to SolidWorks should concentrate on Section I of this textbook. Intermediate users of SolidWorks should move on to Section II of this textbook. Section III presents the new paradigm shift in manufacturing, which is additive manufacturing, showing the connections between computer-aided design and this relatively new technology. Organizing the textbook in this way helps students, designers, engineers, and professionals to decide how to optimize their strategy in covering the contents of the textbook; this is also useful to instructors in planning their delivery strategy depending on their course outlines. The American National Standards Institute and International Organization for Standardization standards have been used in this textbook.

This textbook is written using a hands-on approach in which students can follow the steps that are described in each chapter to model parts, assemble parts, and produce drawings. They create applications on their own with little assistance from their instructors during each teaching session or in the computer laboratory. This textbook has a significant number of pictorial descriptions of the steps that a student should follow. This approach makes it easy for the users of the textbook to work on their own as they use the steps that are described as guides. Instructional support is also provided, including SolidWorks files for all the models, drawings, applications, and answers to end-of-chapter questions.

The principles and exercises presented in Sections I and II have been tested in the SolidWorks courses that the author taught in Ontario institutions in Canada. All the examples in this textbook have been solved by the author. Several practical exercises are included at the end of chapters.

The 2014 version of this textbook has brought a new dimension to SolidWorks by introducing readers to the role of SolidWorks in the relatively new manufacturing paradigm shift, known as additive manufacturing, with one chapter on this topic included:

- Additive Manufacturing—Chapters 35–44

Another enhancement of the 2014 version of this textbook is the complete revision of Chapter 5 (Advanced Part Modeling), Chapter 8 (Assembly Modeling), and Chapter 9 (Part and Assembly Modeling).

Other enhancements include additional end-of-chapter problems. It is hoped that students and instructors will find this current edition more useful than the earlier one.

Additional material is available from the CRC Press website: http://www.crcpress.com/product/isbn/9781498731195.

Bibliography

Bethune, J.D. 2009. *Engineering Design and Graphics with SolidWorks.* Prentice-Hall.

Howard, W.E., Musto, J.C. 2008. *Solid Modeling Using SolidWorks 2008.* McGraw-Hill Ryerson.

Lombard, M. 2009. *SolidWorks 2009.* Wiley Publishing Inc.

Onwubolu, G.C. 2014. *Applied Mechanics with SolidWorks.* Imperial College Press, September.

Planchard, D.C., Planchard, M.P. 2013. *Engineering Design with SolidWorks 2013.* Schroff Development Corporation.

Tickoo, S., Sandeep, D. 2009. *SolidWorks 2009 for Designers.* CADCIM Technologies.

Acknowledgments

I would especially like to thank Cindy Renee Carelli, senior acquisitions editor, and Ashley Segal, project coordinator, CRC Press, for their patience and assistance throughout this book project. Special thanks goes to SolidWorks Inc. for providing the resources that enabled me to get this version of the textbook earlier than planned to the readers. Some academic colleagues at different institutions who evaluated this textbook have given me valuable and encouraging feedback, and their efforts are greatly appreciated. My wife, Ngozi, and our children are greatly appreciated. My wife, Ngozi, shared some very challenging times with me during the early stage of my learning of SolidWorks and when I contemplated writing the first version of this textbook, as well as during the current revision. Without the role that the members of my family played, this textbook project would not have succeeded.

Author

Dr. Godfrey C. Onwubolu teaches *computer-aided design (CAD), machine design,* and *engineering analysis using SolidWorks* as well as *applied mechanics* and *engineering mechanics* where he currently applies SolidWorks very extensively. He currently researches in three areas: (1) CAD, (2) additive manufacturing, and (3) inductive modeling. He holds a BEng degree in mechanical engineering and both an MSc and PhD from Aston University, Birmingham, England, where he first developed a geometric modeling system for his graduate studies. He worked in a number of manufacturing companies in the West Midlands, England, and he was a professor of manufacturing engineering at the University of South Pacific, having taught courses in engineering design and manufacturing for several years. Dr. Onwubolu has published several books with international publishing companies, such as Imperial College Press, Elsevier, and Springer-Verlag, and has published over 140 articles in international journals and refereed international conferences. He is a registered professional engineer in Canada. He is an active member of the American Society of Mechanical Engineers.

SolidWorks Solutions Files

The SolidWorks files for all the chapters are in the publisher's website, with permission given to users as required.

Section I

Introductory Engineering Design Principles with SolidWorks

1

Introduction

Objectives:

When you complete this session, you will have

- Understood the role of SolidWorks within the context of the additive manufacturing framework
- Understood the computer-aided engineering (CAE) framework
- Understood the role of SolidWorks within the context of the CAE framework
- A good background of SolidWorks
- Learnt how to start a SolidWorks session
- Understood SolidWorks user interface
- Learned how to set the document options
- Learned how to set up a good file management
- Learned how to start a new document in SolidWorks
- Learned how to model your first part
- Known about useful SolidWorks resources

SolidWorks within the Context of Additive Manufacturing Framework

A new paradigm shift has been relatively recently experienced in manufacturing in which no material cutting is involved; rather, material adding is involved in manufacturing products. This relatively new way of manufacturing, known as additive manufacturing (AM), is drawing the attention of people because it is a complete shift from the well-known machining process. AM builds parts by adding layers of materials as the process continues, resulting in a complete solid part manufacturing without a single cutting taking place.

The intriguing thing to know is that computer-aided design (CAD) systems are central to AM, as could be seen in the framework for this new manufacturing environment that is shown in Figure 1.1. Active research is taking place in this relatively new area of manufacturing, and several developed economies are showing interest in this new manufacturing shift because it is seen as the *future for manufacturing.* AM has several nicknames such as material adding manufacturing, rapid manufacturing, rapid prototyping, environmentally friendly manufacturing, three-dimensional (3D) printing, and what you see is what you manufacture;

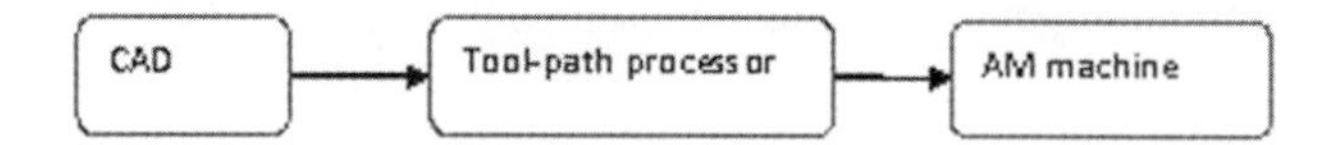

Figure 1.1

AM framework.

it depends from the angle that you see this new technology from. People in the industry seem to refer to it most of the time as 3D printing.

Why is this new manufacturing method referred as listed in the preceding paragraph? It is environmentally friendly because virtually no material scrap arising from manufacturing is emptied into the bin at the end of each day in the manufacturing facility. It is like you are in the office, manufacturing your parts.

Which CAD systems can be used to drive this new manufacturing system? Any of the commercially available ones in the current market qualify, such as SolidWorks, Inventor, computer aided three-dimensional interactive application (CATIA) and Pro/ENGINEER (ProE). If you can model a part or an assembly using any CAD software, you can produce what you model; this is how powerful and how intriguing this relatively new manufacturing environment is. People are intrigued when they see products that are manufactured using AM machines.

The tool-path processor is a software, usually from the owners of the AM machine, for users to create all the pieces of information that are required for producing their parts. The tool-path processor is normally very easy to use.

AM machines are classified according to material classification (machines that manufacture parts from metals, polymers, ceramics, etc.). In this book, the author presents how SolidWorks is used to drive fused deposition modeling (FDM) machines, which manufacture parts from polymers.

But how does the FDM machine work? It is simple in concept but mind-blowing in architecture. Materials are in the form of solid thermoplastics, in filaments that are wound on reels that are known as canisters. The material is fed into the machine where it melts in the heated compartment and flows as fluid into the tool tips (one for model building, the other for support building). The tool tip is moved in x–y directions to deposit the fluid (molten thermoplastic) on a build sheet that is placed on a platen inside the machine. The platen can move in the z-direction (up or down) according to the slice height that the user chooses. Therefore, the FDM machine prints CAD geometries layer by layer until the product is complete. The concept is that simple to explain, but it is intriguing, and this technology has taken manufacturing into another level, previously considered inconceivable. The more interesting thing is that AM can manufacture very complex parts that cannot be manufactured using computer numerically controlled technology. It is amazing.

The author has included this brief introduction to this book on SolidWorks so that the current users of SolidWorks can explore the new manufacturing method of AM, which uses CAD as the driver. At present, SolidWorks (and indeed using any CAD system) should no longer be seen as desktop CAD software, which may not be used to achieve the manufacturing of complex shapes. If you can model any using SolidWorks (and indeed using any CAD system), you can manufacture that part using AM technology.

SolidWorks within the Context of Computer-Aided Engineering Framework

Computer-aided engineering (CAE), which is the performance of engineering tasks or functions with the aid of a computer, has experienced rapid changes over the years. Engineering is a wide-ranging multidisciplinary subject area, and, consequently, so is the subject of CAE. In order for us to cover the area in its entirety, we must examine the ways in which a computer can assist the mechanical, manufacturing, electrical, electronics, chemical, aeronautical, and civil engineer. In this book, we concentrate on the areas of mechanical and manufacturing engineering, which can benefit from CAE functions (Figure 1.2). Recent advances have resulted in concurrent engineering (CE) in which it is no longer necessary to go in a sequential manner from one step to the other within the product development cycle, thereby resulting in shortening the time from design conceptualization to production and shipping to the customer.

The first-generation CAD packages were simply two-dimensional (2D) computer-aided drafting programs, which simply mimicked drafting boards. Several views had to be created as they would be on the drafting boards. Designers had to think in terms of 2D models while having 3D models in mind. During my graduate studies at Aston University, the commonly available CAD packages were of 2D type. Then, I was

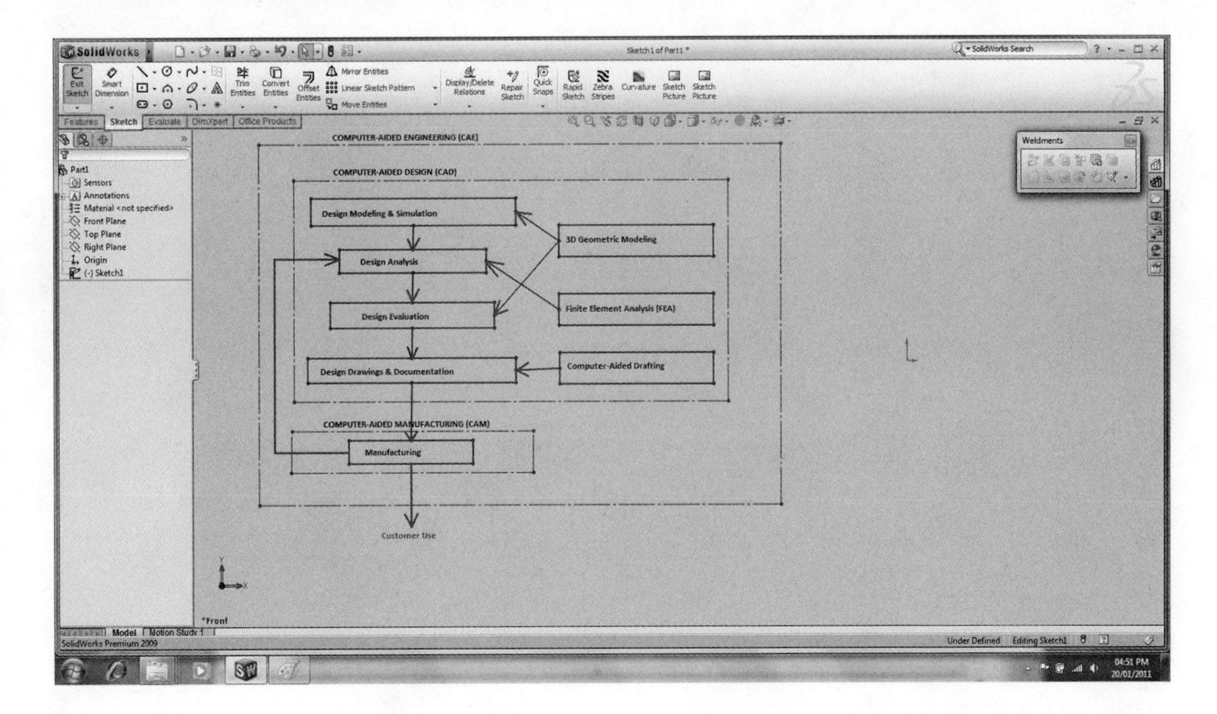

Figure 1.2

CAE framework.

involved in developing an *Engineering Drawing Interpreter*; what that simply means is a program that takes multiple 2D views of an object and interpret them to realize a 3D model as it would appear in reality. First, these 2D CAD packages were very limited in what they could do and tedious to use. There was, therefore, an obvious need for the development of 3D geometric modelers. The development of early 3D geometric modelers will now be briefly reviewed.

3D Wireframes

The development of early 3D geometric modelers commenced with 3D wireframes, which are models that are made up of points and edges in the form of straight lines connecting between appropriate points. There were no well-organized faces until the edges were interpreted to form closed-loop faces, which in turn defined enclosed volumes. Again, at Aston University, I was involved in developing a General Surface Geometric Modeler; the surfaces were simply defined by a mesh of edges that had to be interpreted to form closed loops. Computation of the overall surface areas and the inertial properties of simple/complex general surfaces depended on the interpretation of the meshes describing the surfaces.

Constructive Solid Geometry

In this representation, an object is described in terms of elementary shapes, or primitives. The constructive solid geometry (CSG) representation is based on a two-level scheme. On the second level, bounded primitive volumes are combined by Boolean set operations. The leaf nodes are either primitive leaves, which represent the subsets of 3D Euclidean spaces (solid, primitive shapes that are sized and positioned in space), or transformation leaves, which contain the defining arguments of rigid motions. The branch nodes are operators, which may be a regularized union, an intersection, or a difference or may be rigid motions (Figure 1.3). The solid scheme is the most compact of all known representations, at least for the class of commonly machined parts. The greatest advantage of CSG is that it guarantees the validity of uniqueness of the model: a boundary representation can always be derived in a unique way.

Boundary Representation Scheme

In this representation, a solid is defined by its boundaries. Each surface is *planar* or *sculptured* and bounded by the edges of an adjacent boundary. The boundaries of a solid are usually represented as a union of faces, with each face represented in terms of its boundary (union of edges), together with data that define the surface in which the face lies (Figure 1.4).

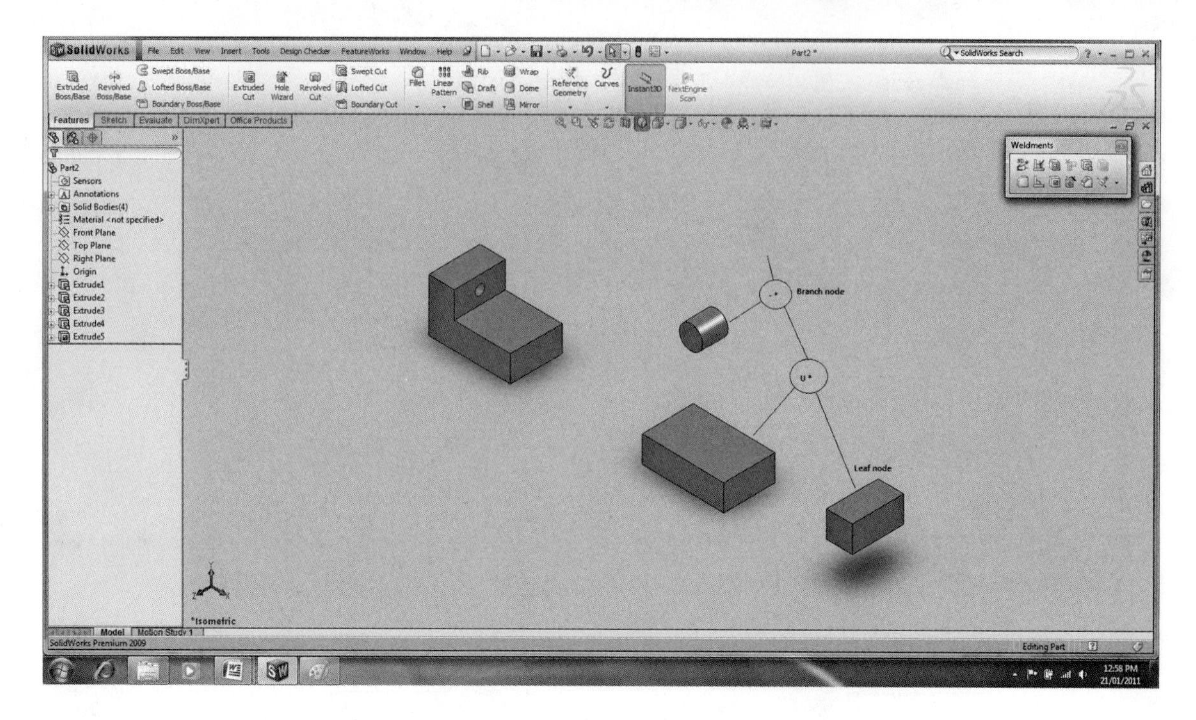

Figure 1.3

CSG.

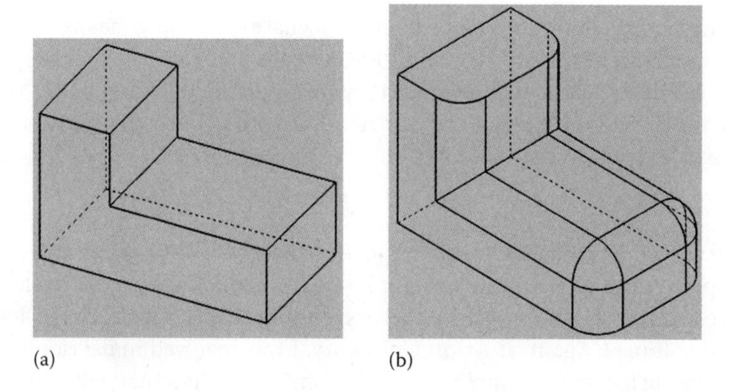

Figure 1.4

Boundary representation (B-Rep) scheme: (a) planar-face B-Rep and (b) curved-face B-Rep.

Feature-Based Parametric Modeling

By the 1980s, a new paradigm called *concurrent engineering* had emerged. With reference to Figure 1.1, it was no longer necessary for the sequential tasks of the left-hand side to be done in that manner. Instead, designers, design engineers, analysts, process planners, manufacturing engineers, and industrial engineers all work together closely right from the start of the product conception until production and shipping to customers. Personnel from different departments could work on their tasks concurrently, thereby supporting teamwork. Any problem encountered at any stage is collectively resolved, resulting in the shortening of time from design conception to product manufacture. This concept is shown on the right-hand side of Figure 1.1. This meant that a more reliable geometric modeling technique had to be used. Feature-based parametric modeling technique was developed to play its role in a CE environment. In parametric modeling, shapes are driven by dimensions. At the leaf nodes are sketches that are fully constrained (fully dimensioned). Features are made up of sketches that are acted upon by some operations such as *Extrude (Add)*, *Cut (Subtract)*, *Sweep*, *Revolve*, and *Loft*. Changing the dimensions of the sketches changes the 3D feature. In a parametric modeling environment, changing the dimension of a sketch automatically changes the shape of the feature, as well as the assembly. Therefore, the bottom–up hierarchy is defined as sketches–features (solid parts)–assembly.

Another interesting observation is that the part, the assembly, and the drawings are all connected parametrically. If any change is made at the part level, the change cascades to the assembly and drawing documents. Different designs can be obtained and evaluated very quickly. Existing design data can be reused to create new designs. Geometric relations, dimensional constraints, and relational parametric equations can be used to capture the design intent. Design tables and configurations are easily generated to realize different design variations. All these advantages make feature-based parametric modeling technique the choice for a CE environment.

Background of SolidWorks

SolidWorks, a registered trademark of the Dassault Systemes Corporation in Concord, MA, is a design automation software package that is used to produce parts, assemblies, and drawings. It is a Windows native 3D solid modeling CAD program that is based on *parametric modeling*. This particular attribute of SolidWorks means that the dimensions of the parts, assemblies, and drawings drive the shapes that are produced. SolidWorks provides an easy-to-use, highest-quality design software for engineers and designers for creating 3D parts, assemblies, and 2D drawings, which are all related. This means that the changes that are made in a part document will affect the assembly and 2D drawing documents. The advantages of parametric modeling are numerous, but some distinct ones include the fact that when a designer realizes that changes need to be made in the dimensions in an assembly document, making such changes in the part document automatically updates the assembly and 2D drawing documents. There is no need to start all over. Parametric modeling is not rigid; it is flexible and makes designing extremely flexible. This attribute of SolidWorks makes it extremely flexible when compared to some other CAD programs.

Starting a SolidWorks Session

There is more than one way to start a SolidWorks session. To start a SolidWorks session, choose Start > All Programs > SolidWorks from the Start menu or double-click on the SolidWorks icon on the desktop of your computer. When the SolidWorks program is first accessed, the window that is displayed is the one that is shown in Figure 1.5. It is more or less a blank page from which the designer then decides what document to open: (a) part, (b) assembly, or (c) drawing.

SolidWorks User Interface

There are basically four menus that are first seen when a new SolidWorks session commences: (1) Menu bar menu, (2) Menu bar toolbar, (3) SolidWorks Help, and (4) SolidWorks Resources. The menus are visible when you move the mouse over or click the SolidWorks logo. You can pin the menus to keep them visible at all times. In order to create a part, an assembly, or a drawing, click File, New from the Menu bar menu or

Figure 1.5

Opening SolidWorks screen.

click New or click Open (Standard toolbar) from the Menu bar toolbar. The Menu bar contains the following.

Menu Bar Toolbar

This is a set of the most frequently used tool buttons from the Standard toolbar, as shown in Figure 1.6. By clicking the down arrow next to a tool button, you can expand it to display a flyout menu with additional functions. When the cursor is moved across the SolidWorks logo, the Menu bar toolbar switches over to the Menu bar menu. The toolbar moves to the right when the menus are pinned.

The available tools are as follows:

- *New* —Creates a new document.
- *Open* —Opens an existing document.
- *Save* —Saves an active document. This lets you access most of the File menu commands from the toolbar. For example, the Save flyout menu includes Save, Save As, and Save All.
- *Print* —Prints an active document.
- *Undo* —Reverses the last action that is taken.
- *Rebuild* —Rebuilds the active part, assembly, or drawing.
- *Options* —Changes the system options and add-ins for SolidWorks.

Menu Bar Menu

The SolidWorks menus are visible when you move the mouse over or click the SolidWorks logo. You can pin the menus to keep them visible at all times. The default menu items for an active document are (a) File, (b) Edit, (c) View, (d) Insert, (e) Tools, (f) Window, (g) Help, and (h) Pin (see Figure 1.7). However, the menu items change depending on which type of document is active.

Task Pane

The Task Pane is displayed when the SolidWorks session commences. The Task Pane contains the following default tabs: (a) SolidWorks Resources , (b) Design Library, (c) File Explorer, (d) SolidWorks Search, (e) View Palette, (f) Appearances/Scene, and (g) Custom Properties.

- SolidWorks Resources
  The SolidWorks Resources tab in the Task Pane includes commands, links, and information.
- Design Library
  The Design Library tab in the Task Pane provides a central location for reusable elements, such as parts, assemblies, and sketches, but not for nonreusable elements, such as SolidWorks

Figure 1.6

Menu bar toolbar.

Figure 1.7

Menu bar menu.

drawings, and text files. It is the gateway to the *3D Content Central* Website. The Design Library abounds with the resources for design, and designers should take time to be familiar with these resources.

- SolidWorks Explorer

  SolidWorks Explorer is a file management tool that is designed to help you perform tasks such as renaming, replacing, and copying SolidWorks files. You can show a document's references, search for documents using a variety of criteria, and list all the places where a document is used. Renamed files are still available to those documents that reference them.
- SolidWorks Search

  SolidWorks Search is used to search key words. The 10 recent searches may be found by clicking the pull-down arrow.
- View Palette

  The View Palette provides the ability to drag and drop the drawing views of an active document, or click the Browse button to locate the desired document.
- Appearances/Scene

  Appearances/Scene is a useful and easy-to-use way of providing a photorealistic rendering of models.
- Help Options

  The SolidWorks flyout menu of Help options is used to access Help, Tutorials, Reference Guide, and other functionalities, as shown in Figure 1.8. Another route for the Help options is through the SolidWorks menu.

So far, what we have considered are the menus that are displayed when a new session of SolidWorks is started. Once a task begins (part modeling, assembly modeling, or drawing), three more menus appear: (1) CommandManager, (2) FeatureManager Design Tree, and (3) Heads-up View toolbar. These are now briefly discussed.

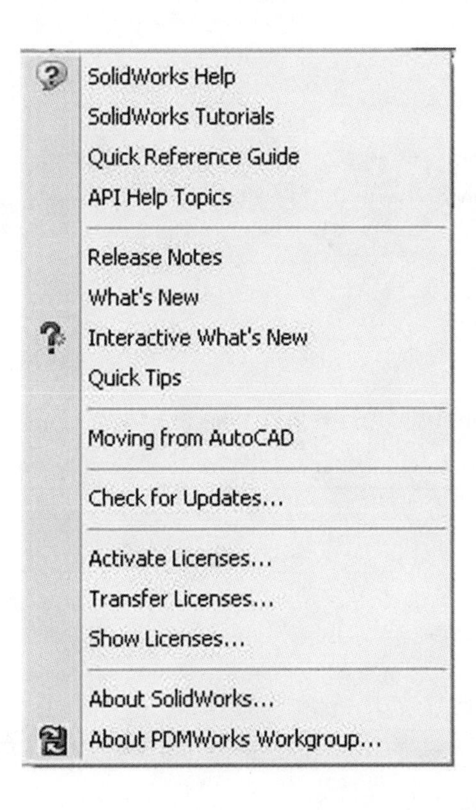

Figure 1.8

Help options.

CommandManager

The CommandManager, which is document dependent, contains most of the tools that you will need to create parts, assemblies, or drawings. The CommandManager is a context-sensitive toolbar that dynamically updates based on the toolbar that you want to access. By default, it has toolbars that are embedded in it based on the document type. For example, the default Part tabs are Features, Sketch, Evaluate, DimXpert, and Office Products. These tabs are illustrated hereafter. The two most widely used categories of tools for part modeling are (1) the feature tools, which are used to create and modify 3D features, and (2) the sketch tools, which are used in creating 2D sketches.

The main *feature tools* used to create and modify 3D features are (a) Extruded Boss/Base, Revolved Boss/Base, Swept Boss/Base, Lofted Boss/Base, and Boundary Boss/Base for adding materials to a part; (b) Extruded Cut, Revolved Cut, Swept Cut, Lofted Cut, and Boundary Cut for removing materials from a part; (c) Fillet, Linear Pattern, and Mirror for operations; (d) specific features such as Rib, Draft, Shell, Wrap, and Dome; and (e) Reference Geometry and Curves, as well as (f) the Instant3D tool (see Figure 1.9).

The *sketch tools* used in creating 2D sketches are (a) Sketch, (b) Smart Dimension, (c) Line, (d) Rectangle, (e) Slot, (f) Circle, (g) Arc, (h) Polygon, (i) Spline, (j) Ellipse, (k) Fillet, (l) Plane, (m) Text, (n) Point, (o) Convert Entities, (p) Offset, (q) Mirror, (r) Linear Pattern, (s) Move, (t) Display/Delete Relations, (u) Repair Sketch, (v) Quick Snaps, and (w) Rapid Sketch (see Figure 1.10).

The *Evaluate tools* are mainly used for *analysis* such as measuring the distances between points on features, mass properties, and section properties. SimulationXpress, FloXpress, DFMXpress, and DriveWorksXpress Wizards are also accessible through the Evaluate tools (see Figure 1.11).

The *DimXpert tools* are mainly for *dimensions* and *tolerance* (see Figure 1.12).

The *Office products toolbar* allows you to activate any add-in application that is included in the SolidWorks Professional or Premium package. *eDrawings* and *Animate* are available using this toolbar, as shown in Figure 1.13.

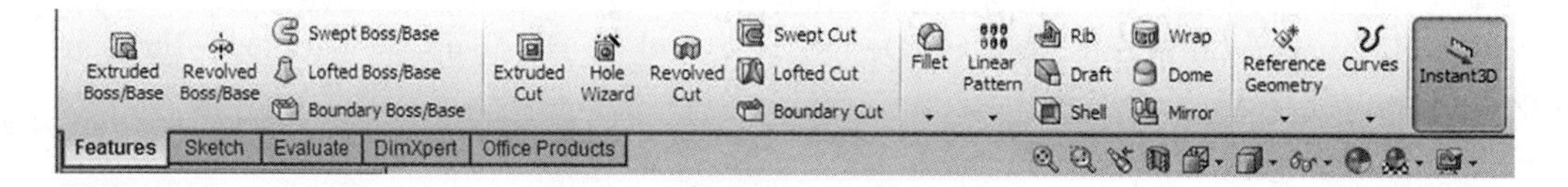

Figure 1.9

Feature tools.

Figure 1.10

Sketch tools.

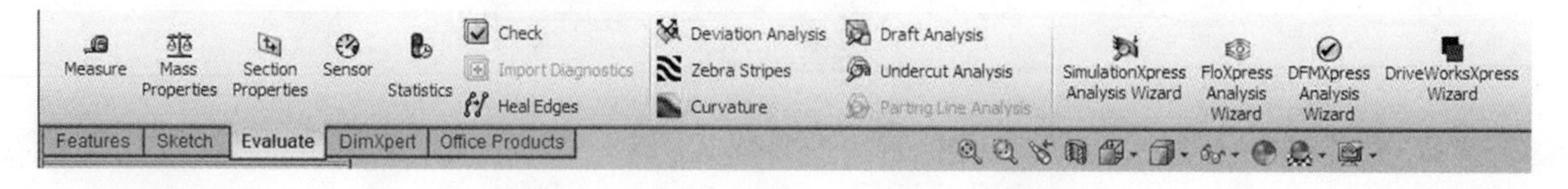

Figure 1.11

Evaluate tools.

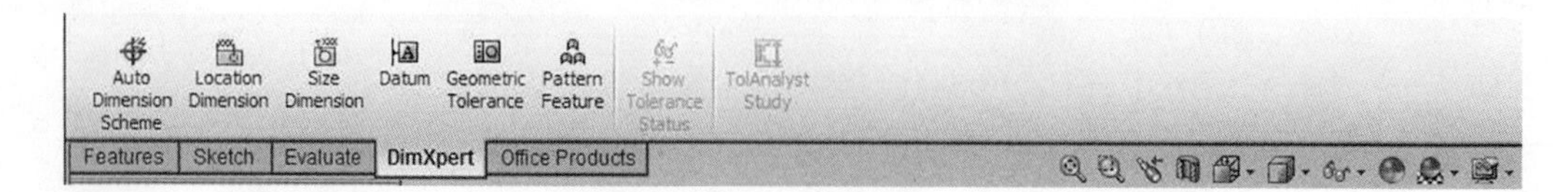

Figure 1.12

DimXpert tools.

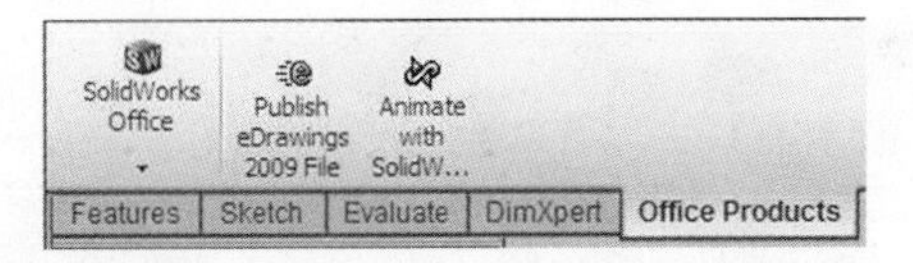

Figure 1.13

Office Product tools.

FeatureManager Design Tree

The FeatureManager Design Tree, which is located on the left side of the SolidWorks Graphics window, summarizes how the part, assembly, or drawing is created. It is necessary to understand the FeatureManager Design Tree in order to be able to troubleshoot a model having some problems. The default tabs of the FeatureManager are the FeatureManager Design Tree, PropertyManager, ConfigurationManager, and DimXpertManager. For example, in the FeatureManager Design Tree shown in Figure 1.14, the standard planes are shown, and the first activity is shown to be *Extrude1* (extruded boss/base). A sketch (or sketches), which defines the extruded part, would normally appear when the Extrude1 is expanded. All the details for designing a part are summarized and stored in the FeatureManager Design Tree. It is the design information storehouse, which fully describes a designed part, assembly, or drawing. It is a design tree showing the relationship between *parents* and *children* in a design context. When information relating to a parent is changed, it automatically affects the children. Every designer using SolidWorks should be very familiar with the semantics and syntax of the FeatureManager Design Tree.

Heads-Up View Toolbar

The Heads-up View toolbar (see Figure 1.15) is a useful tool for the user to have view options during the modeling of parts, assemblies, or drawing. The following views are available: (a) zoom to fit, (b) zoom to area, (c) previous view, (d) section view, (e) view orientation, (f) display style, (g) hide/show items, (h) edit appearance, (i) apply scene, and (j) view setting.

The *view orientation* (top, front, bottom, left, right, and back; isometric, trimetric, and diametric; normal; single view, two view horizontal, two view vertical, four view) and the *display style* (shaded with edge, shaded, hidden lines removed, hidden lines visible, and wireframe) are illustrated in Figure 1.16.

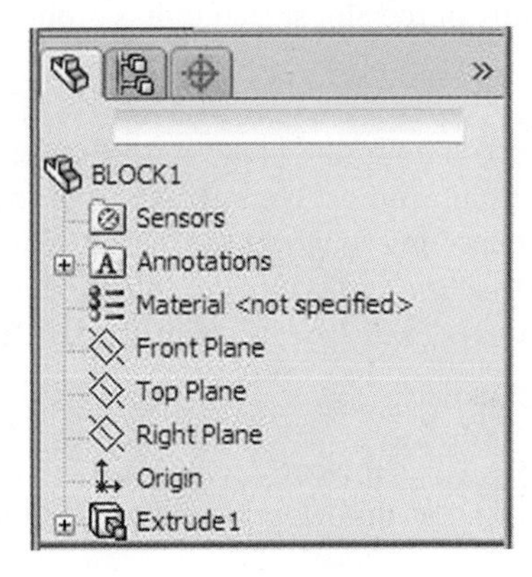

Figure 1.14

FeatureManager Design Tree.

Figure 1.15

Heads-up View toolbar.

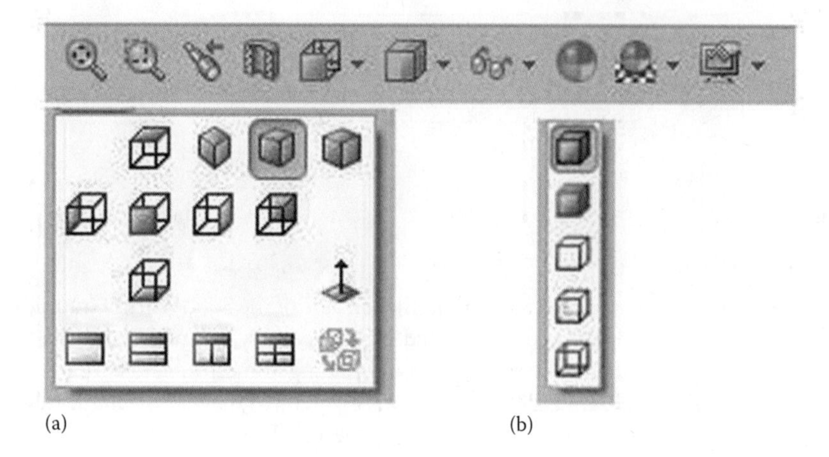

Figure 1.16

(a) View orientation and (b) display style.

Drop-Down Menu

SolidWorks shares the familiar Microsoft® Windows user interface. Users communicate with SolidWorks through drop-down menus (see Figure 1.17), context-sensitive toolbars, consolidated toolbars, or the CommandManager tabs. As the name implies, a drop-down menu drops down other lower-level menus that give more information in terms of design toolbars. A drop-down menu has a black triangular-shaped symbol, which, when clicked, drops other menus. For example, in Figure 1.18, choosing the Features option drops another menu showing the features that are available: fillet/round, chamfer, hole, draft, shell, rib, scale, dome, freeform, shape, deform, indent, flex, wrap, etc.

Right-Click

Right-clicking in the Graphics window on a *model* or in the FeatureManager on a *feature* or *sketch* results in the display of a context-sensitive toolbar, as shown in Figure 1.18.

Consolidated Toolbar

In the CommandManager, similar commands (instructions that inform SolidWorks to perform a task) are grouped together. For example, a family of the Slot sketch tool is grouped together in a single flyout button, as illustrated in Figure 1.19.

System Feedback

SolidWorks provides system feedback by attaching a symbol to the mouse pointer cursor, which indicates what you are selecting or what the system is expecting you to select. Placing your cursor pointer across a model results in system feedback being displayed in the form of a symbol next to the cursor, as illustrated in Figure 1.19.

Setting the Document Options

As has been already mentioned in the "Menu Bar Toolbar" section, the Options tool changes the system options and add-ins for SolidWorks. There are two components of Options: (1) System Options and (2) Document Properties.

System Options

System Options tool allows you to specify the File Locations Options, which contain a list of folders that are referenced during a SolidWorks session. The default templates folder for a new installation on a local drive C:\ is located at: C:\Documents and Settings\All Users\Applications Data\SolidWorks\SolidWorks200x\templates. It is therefore important that you advise SolidWorks software where to find your customized templates, as shown in Figure 1.20.

Document Properties

There are numerous settings that need to be made when you begin a modeling session. Document Properties tool offers users the resources to do this.

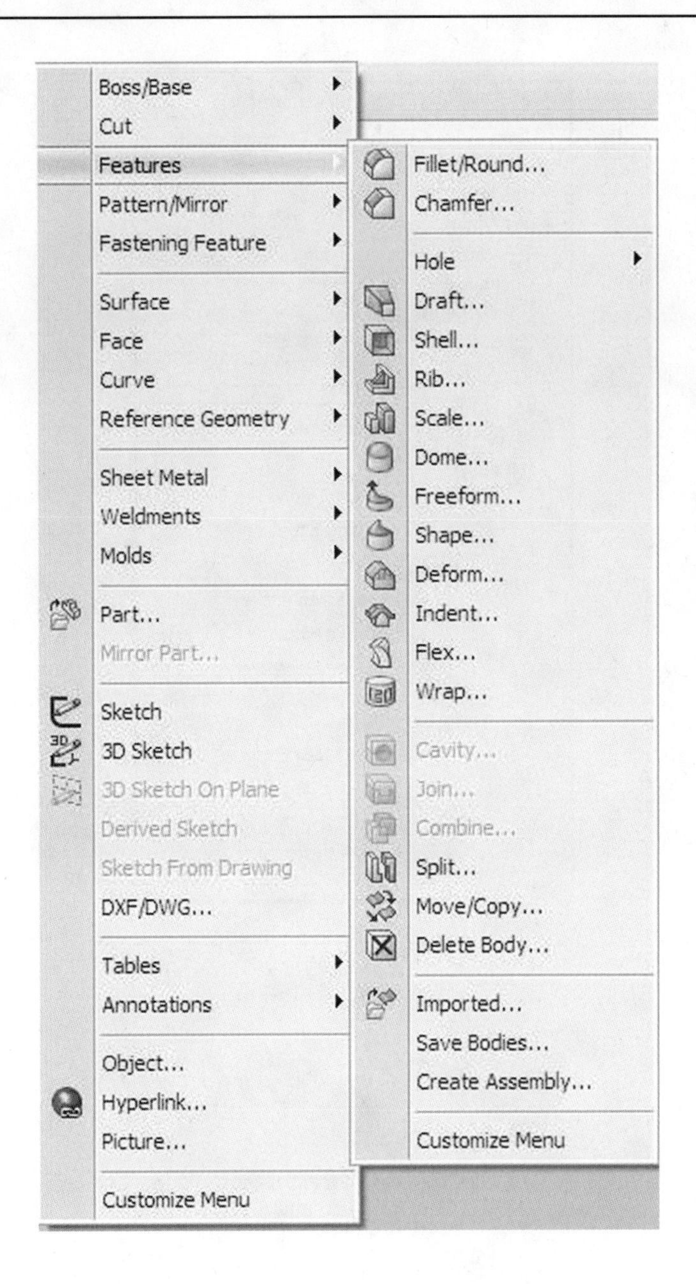

Figure 1.17

Drop-down menus.

Modifying the Drafting Standard

For example, to choose a drafting standard, the Drafting Standard tool is accessed. The standards that are available in this drop-down list are ANSI, ISO, DIN, JIS, BSI, GOST, and GB, as shown in Figure 1.21a and further expanded in Figure 1.21b.

Modifying the Dimension Arrows

The Dimensions option in the document properties allows you to control the parameters that define dimension arrows such as the height, width, and overall length (see Figure 1.22). It also allows you to specify whether dual dimensions are needed, and if so, the precision can also be set. The text font can also be specified.

Modifying the Units Document Properties

The Units Document Properties assists the designer to define the Unit System, Length unit, Angular unit, Density unit, and Force unit of measurement for the Part document (see Figure 1.23). The Decimals option displays the number of decimal places for the Length and Angular units of measurement.

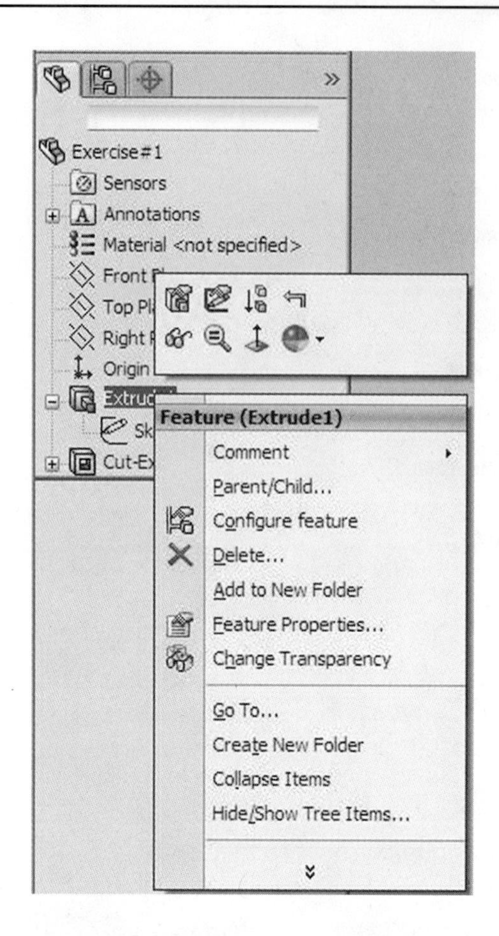

Figure 1.18

Right-clicking options.

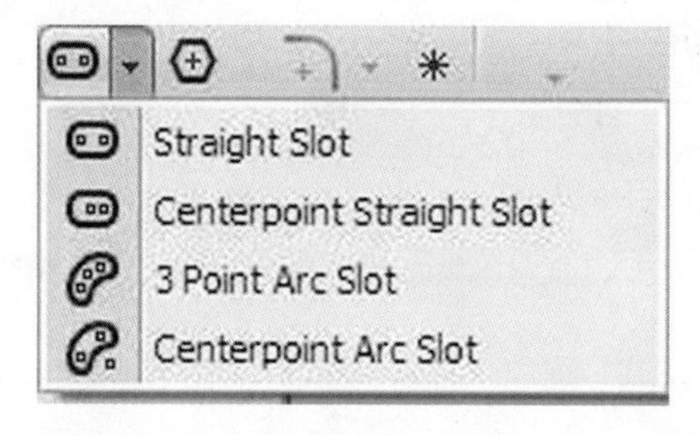

Figure 1.19

Group of the slot sketch.

There are hundreds of document properties. However, the Document Properties that are commonly modified include Drafting Standards, Units, and Decimal Places. For a number of applications, we will need to set up the following:

- ANSI-MM-PART TEMPLATE
  Choose the *millimeter, gram, second (MMGS)*, decimal for length = 0.12, and None for Angle.
- ANSI-INCH-PART TEMPLATE
  Choose the *inch, pound, second (IPS)*, decimal for length = 0.12, and None for Angle.

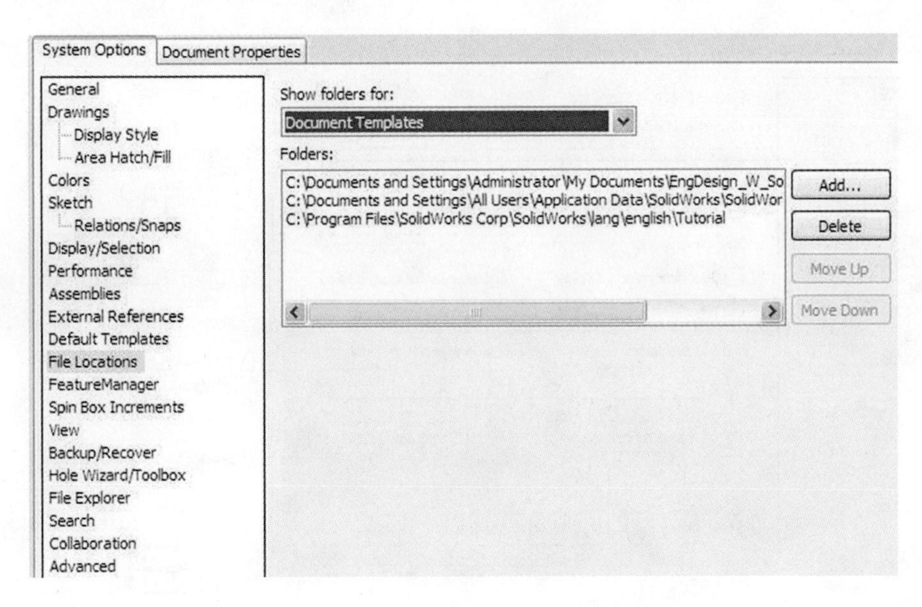

Figure 1.20

System Options for controlling the path for saving document templates.

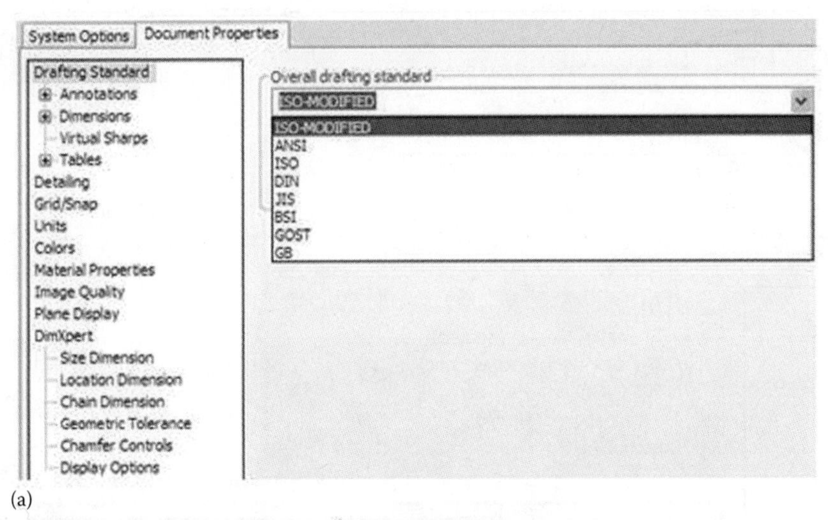

(a)

Drafting Standard Options	Description
ANSI	American National Standards Institute
ISO	International Standard Organization
DIN	Deutsche Institute fur Normumg (Germany)
JIS	Japanese Industry Standard
BSI	British Standards Institution
GOST	Gosndarstuennye State Standard (Russian)
GB	Guo Biao (Chinese)

(b)

Figure 1.21

(a, b) Modify the drafting standard.

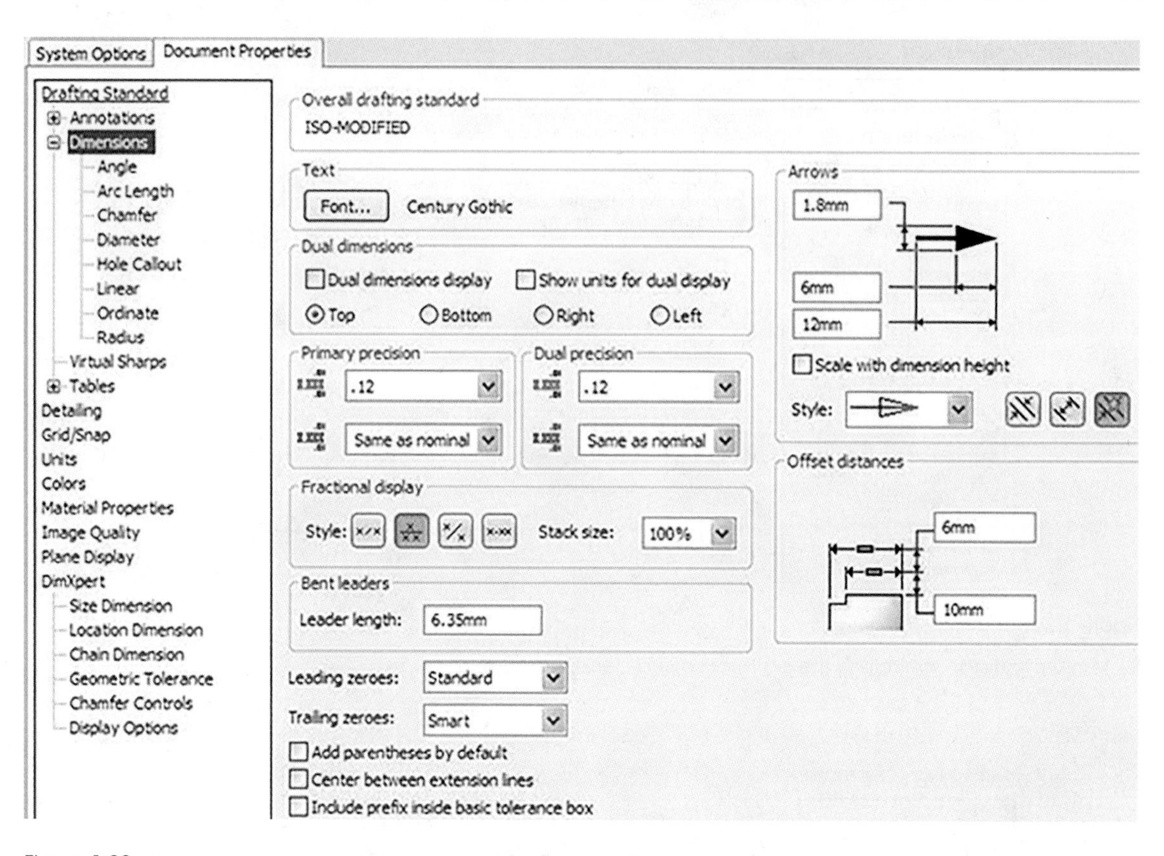

Figure 1.22

Modifying the dimension arrows.

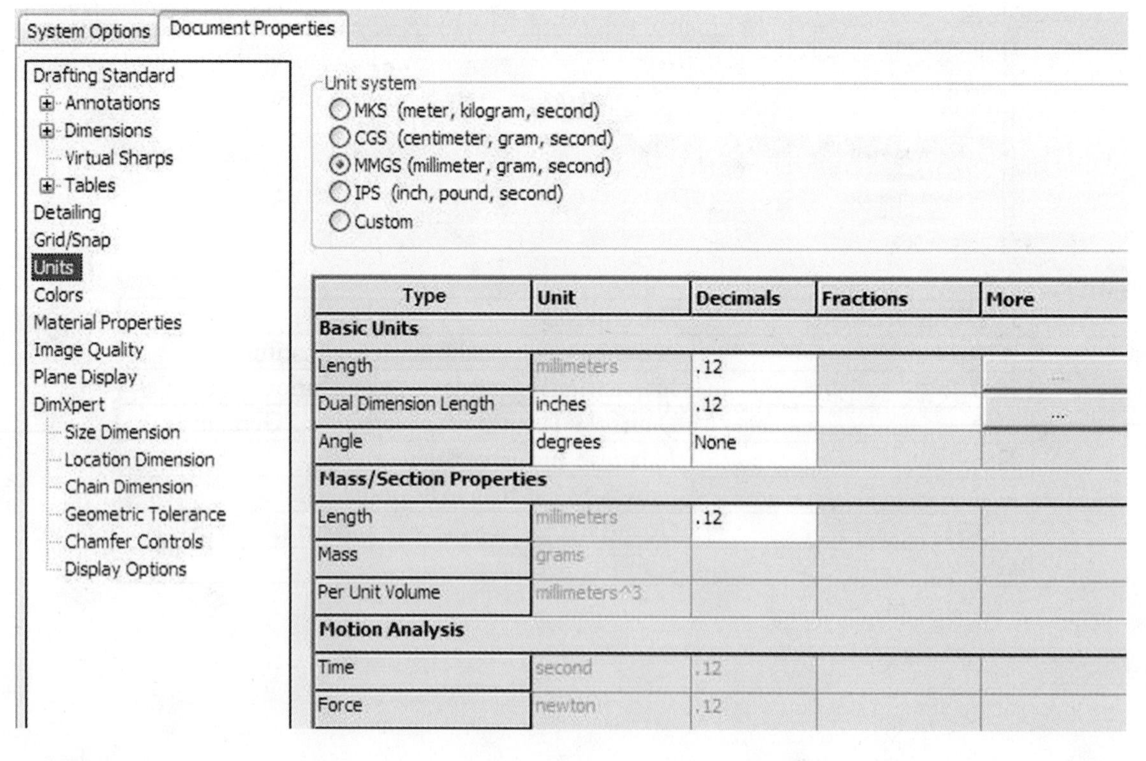

Type	Unit	Decimals	Fractions	More
Basic Units				
Length	millimeters	.12		...
Dual Dimension Length	inches	.12		...
Angle	degrees	None		
Mass/Section Properties				
Length	millimeters	.12		
Mass	grams			
Per Unit Volume	millimeters^3			
Motion Analysis				
Time	second	.12		
Force	newton	.12		

Figure 1.23

Document Properties settings.

File Management

An important aspect of working with SolidWorks is to have a well-organized file management system. Engineers and designers model many parts, and it would be helpful to form the habit of having a well-organized file management system. Generally, in SolidWorks, three types of documents are common: (1) part, (2) assembly, and (3) drawing. Customized templates are needed for these documents; let us decide to save these templates as MY-TEMPLATES. There are components that may be sources, so let us also have a folder for SOURCED-COMPONENTS. We will also decide to save our models in a particular folder under a known path, on our computer's hard drive. Other folders may be created for projects, exercises, etc. It is important to set up a file management system that would help us in a typical SolidWorks class.

Caution Needed during SolidWorks Sessions

During SolidWorks sessions, it is necessary that you save your model from time to time because you may receive a surprise of your computer freezing. When this happens, you might lose information that was not saved before the freezing takes place.

Starting a New Document in SolidWorks

To select a new document (part, assembly, or drawing) in SolidWorks, select the New Document option from the Menu bar. The SolidWorks screen shown in Figure 1.24 should be open at this point. Another route to select the New Document is through the Getting Started rollout of the SolidWorks Resources. The New SolidWorks Document dialog box displayed is shown in Figure 1.24. Click the Advanced button to select the advanced mode. The advanced mode remains selected for all new documents in the current SolidWorks session; the setting is saved when you exit the current session.

- Part

  The *Part* button is chosen by default in the New SolidWorks Document dialog box. Choosing the OK button enables you to start a new part document to create solid models or sheet metal components.
- Assembly

  Choose the Assembly button and then the OK button from the New SolidWorks Document dialog box to start a new assembly document.

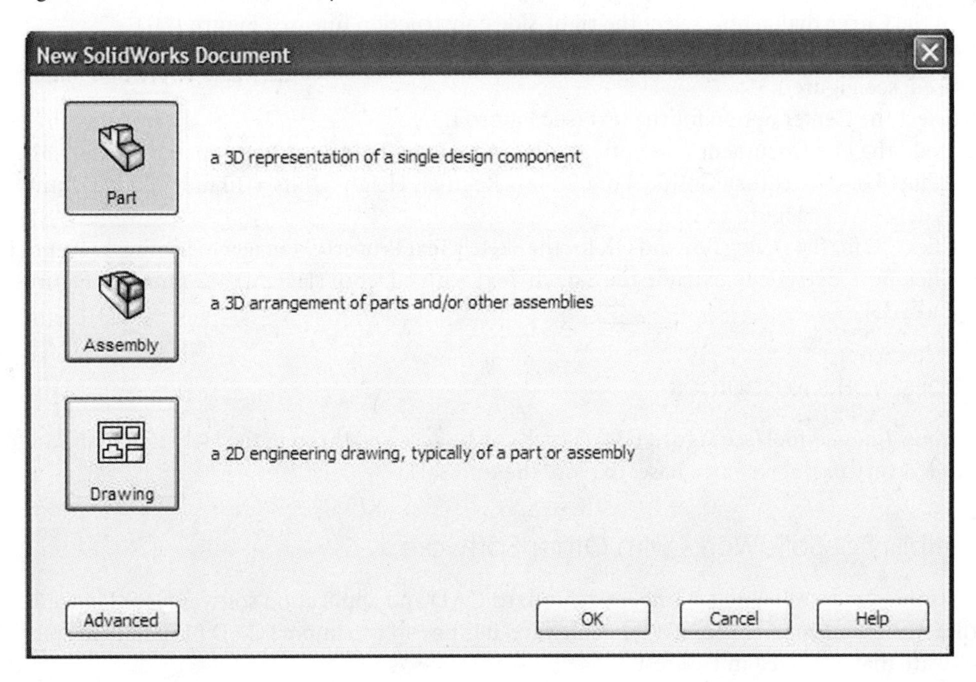

Figure 1.24

New SolidWorks Document dialog box.

- Drawing
  Choose the Drawing button and then the OK button from the New SolidWorks Document dialog box to start a new drawing.

My First Part

After understanding what SolidWorks is all about, let us now create our first part.

Example 1

Create a model (on the right-hand side) whose sketch is shown (on the left-hand side) in Figure 1.25a and b. Ensure that the sketch is fully constrained. Extrude the sketch to a depth of 50 mm, midplane to produce the model. The outer vertical and horizontal edges have a fillet radius = 5 mm. Create your initials on the top, left flat face. Change the extrusion depth and observe the model.

SolidWorks SOLUTION

The following steps are recommended:

1. Create a New document.
2. Ensure that you have set up the following:
   ANSI-MM-PART TEMPLATE: MMGS, decimal for length = 0.12 and None for Angle.
   ANSI-INCH-PART TEMPLATE: IPS, decimal for length = 0.12 and None for Angle.
3. Choose the Front Plane.
4. Starting from the origin, sketch an L-shape, 75 mm × 65 mm with 30-mm offset.
5. Dimension the sketch (see Figure 1.26).
6. Extrude it 50 mm (see Figure 1.26).
7. Click the Fillet option and select the seven edges with a fillet radius = 10 mm, as shown in Figure 1.27. See the filleted model in Figure 1.28.
8. Select the topmost face, be in sketch mode, and sketch two construction lines 25-mm apart (see Figure 1.29).
9. Click the Sketch Text tool (see Figure 1.30).
10. In the Curve dialog box, select the right-side construction line (see Figure 1.31).
11. In the Text dialog box, write your initials (the acronym for computer-aided design, CAD, is used; see Figure 1.31).
12. Select the Center option for the text (see Figure 1.31).
13. Undo the Use Document font option. (A Choose Font dialog box pops up; see Figure 1.31).
14. Adjust Font = Century Gothic, Font Style = Regular, Height Units = 10 mm, Height Points = 28 as desired (see Figure 1.31).
15. Click OK for the Dialog Box and OK for the Sketch Text PropertyManager to finish (see Figure 1.31).
16. Click Boss Extrude to extrude the Sketch Text with a Depth (Height) = 2 mm (see Figures 1.32 and 1.33).

Useful SolidWorks Resources

The *SolidWorks Tutorials* tool (see Figure 1.34), which can be accessed through the Help tool in the SolidWorks Menu Bar, is a very useful resource base. Explore the tutorials.

Compatibility of SolidWorks with Other Software

SolidWorks interfaces well with a number of standard CAD and application software (see Figure 1.35). This means that while working with SolidWorks software, it is possible to import CAD files that are created using other software that are listed in Figure 1.35.

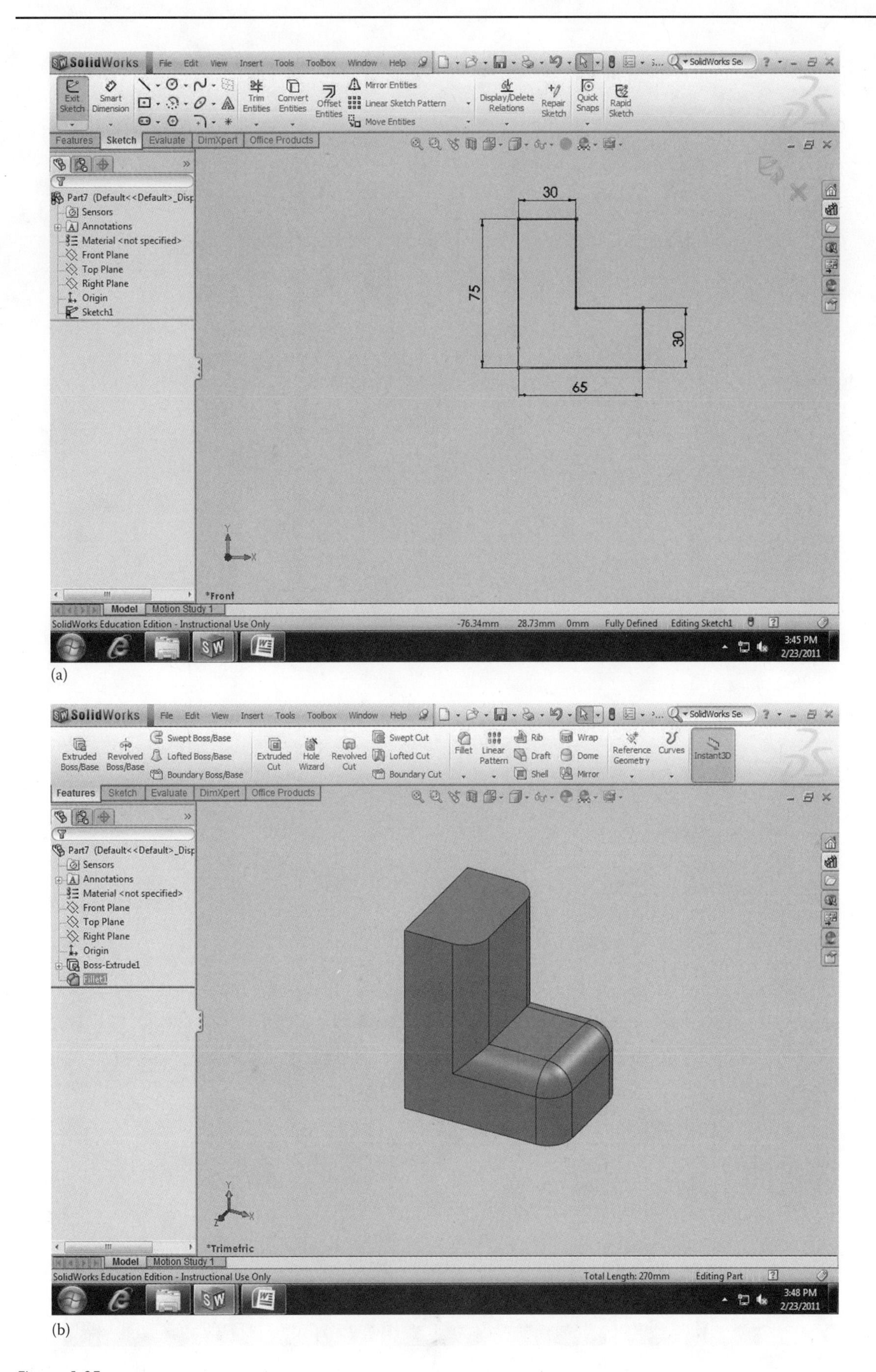

Figure 1.25

(a) Sketch and (b) model.

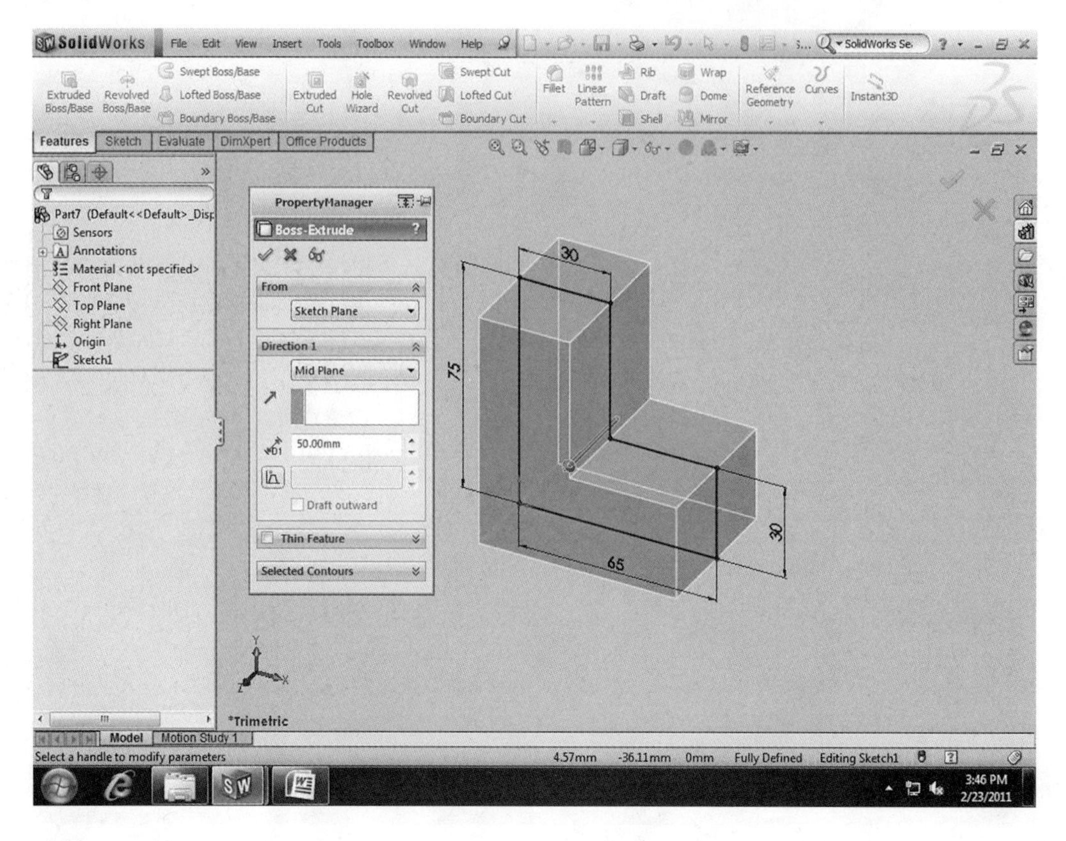

Figure 1.26

Boss Extrude PropertyManager.

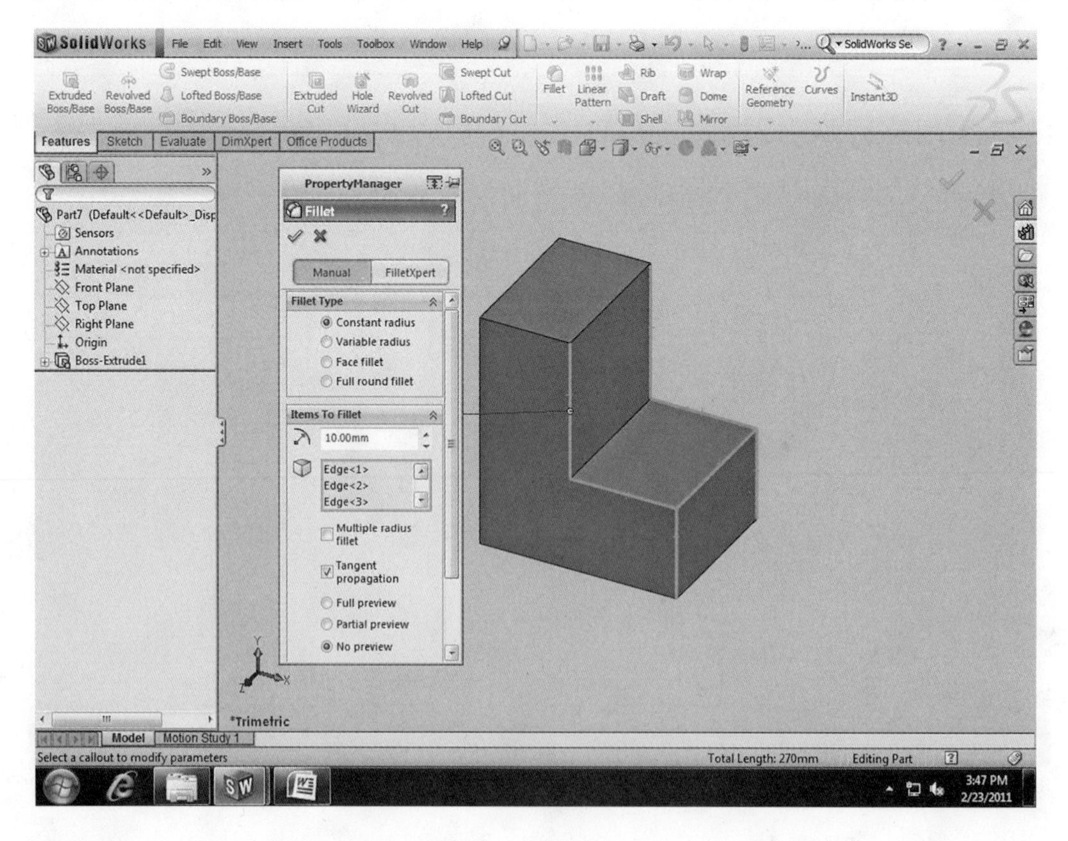

Figure 1.27

Fillet edges.

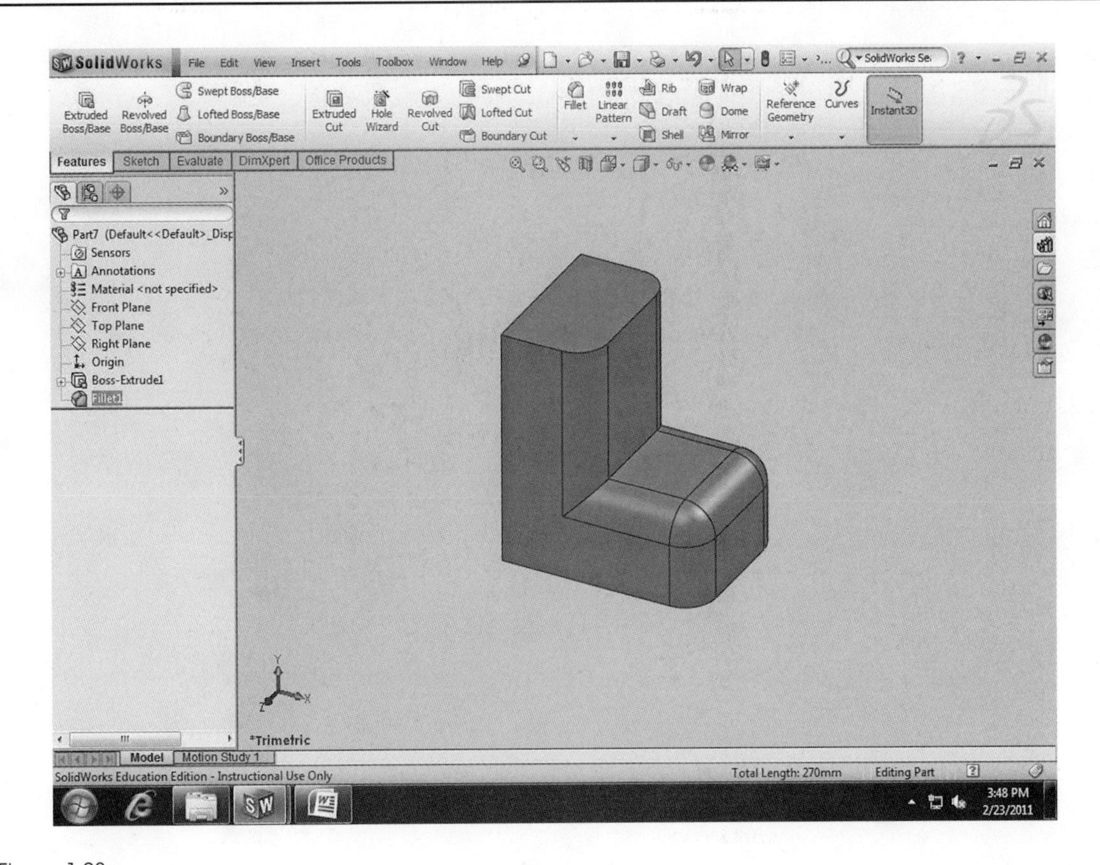

Figure 1.28

Model has fillets.

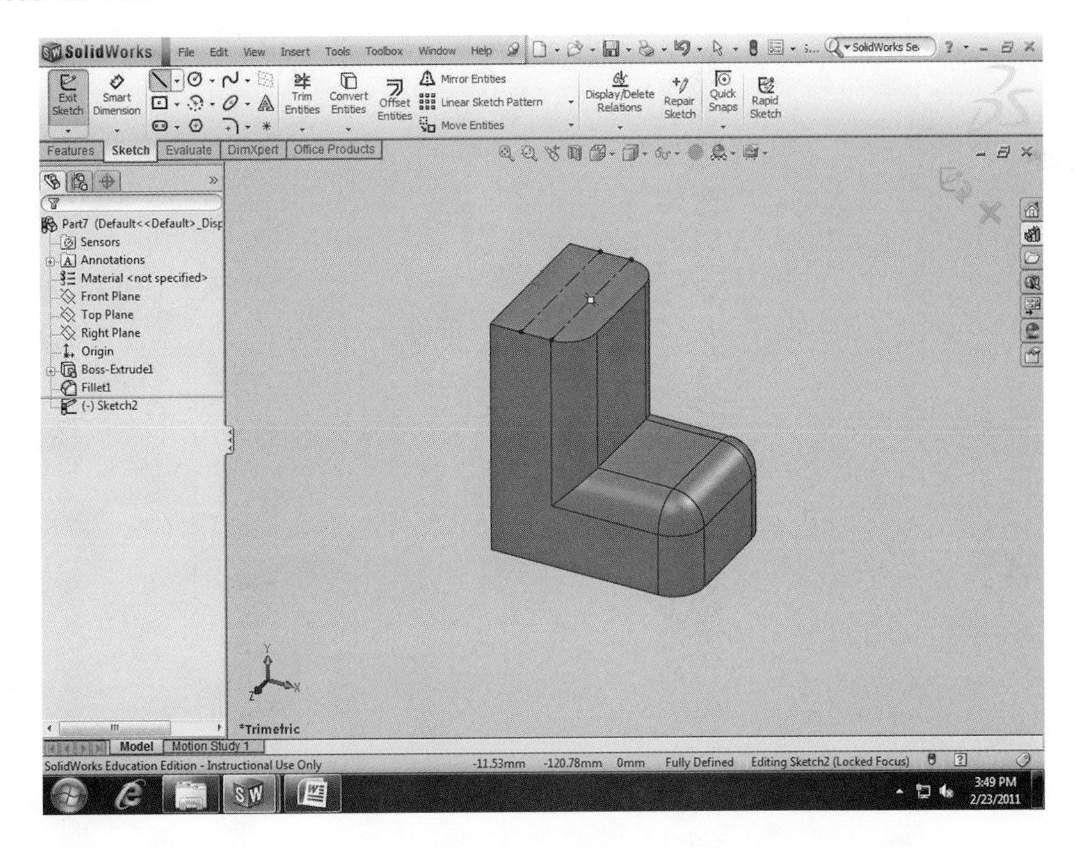

Figure 1.29

Sketch two construction lines.

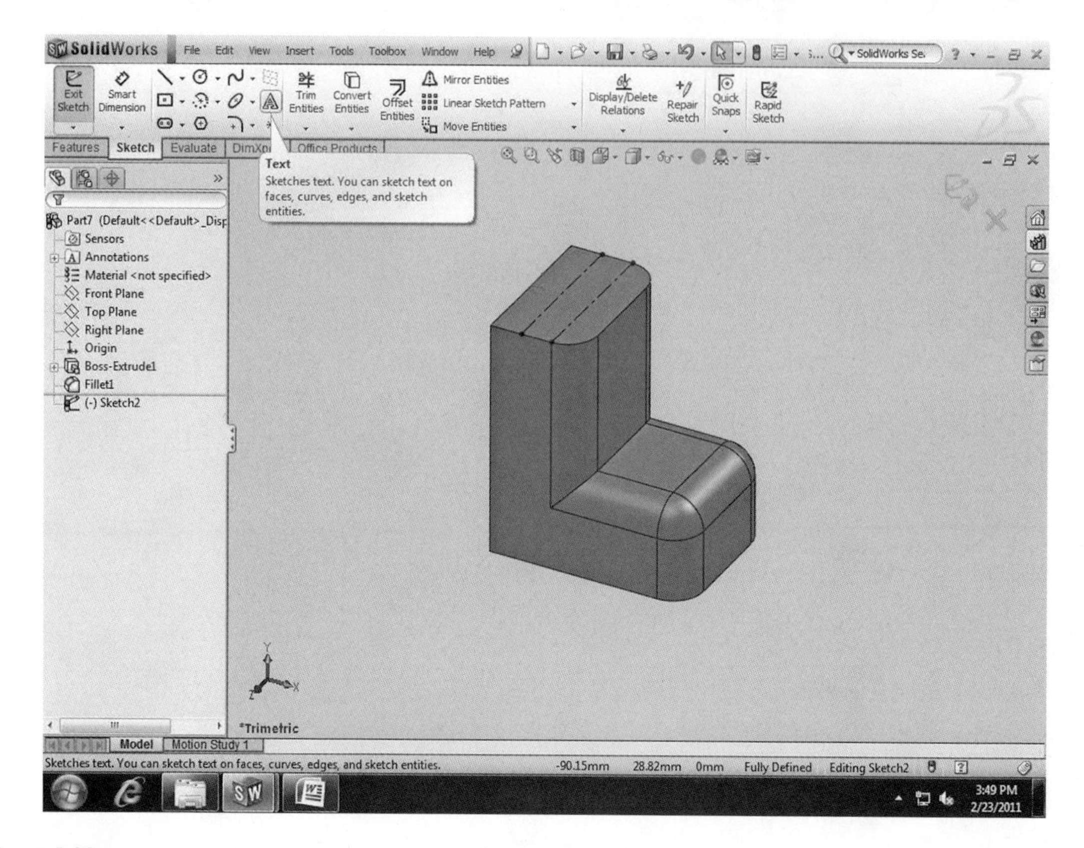

Figure 1.30

Text tool.

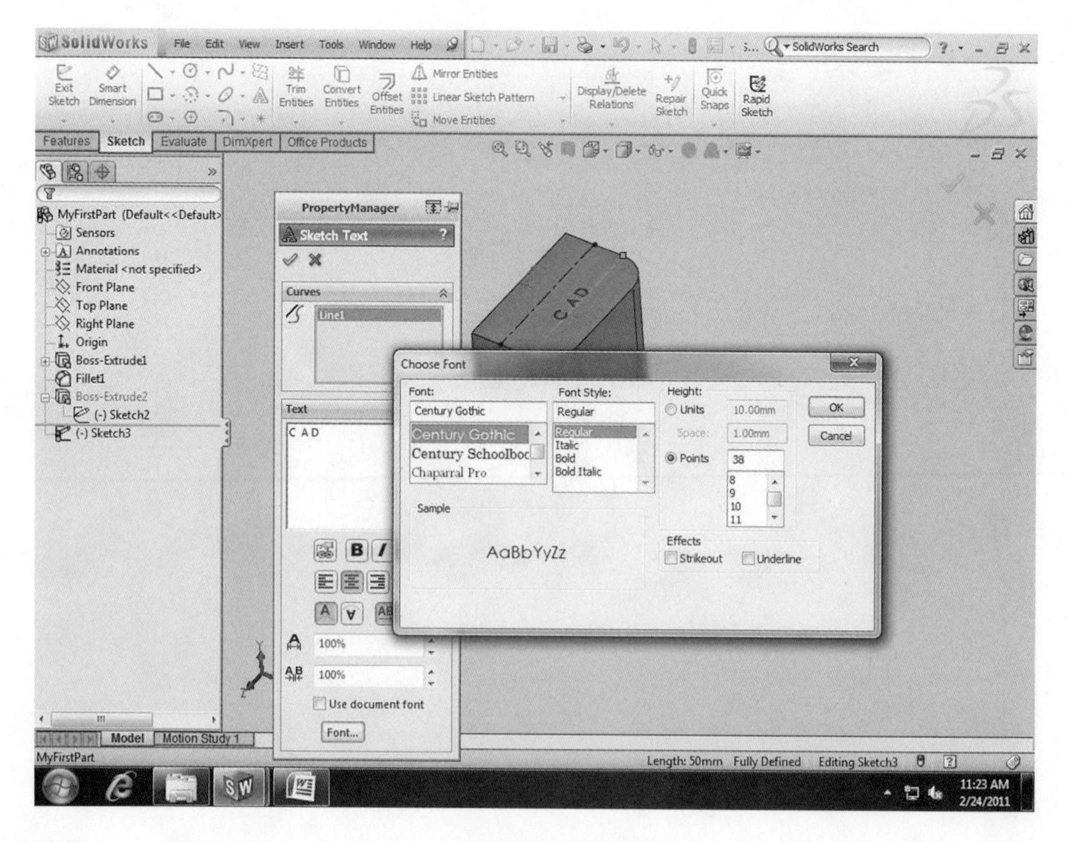

Figure 1.31

Sketch Text PropertyManager.

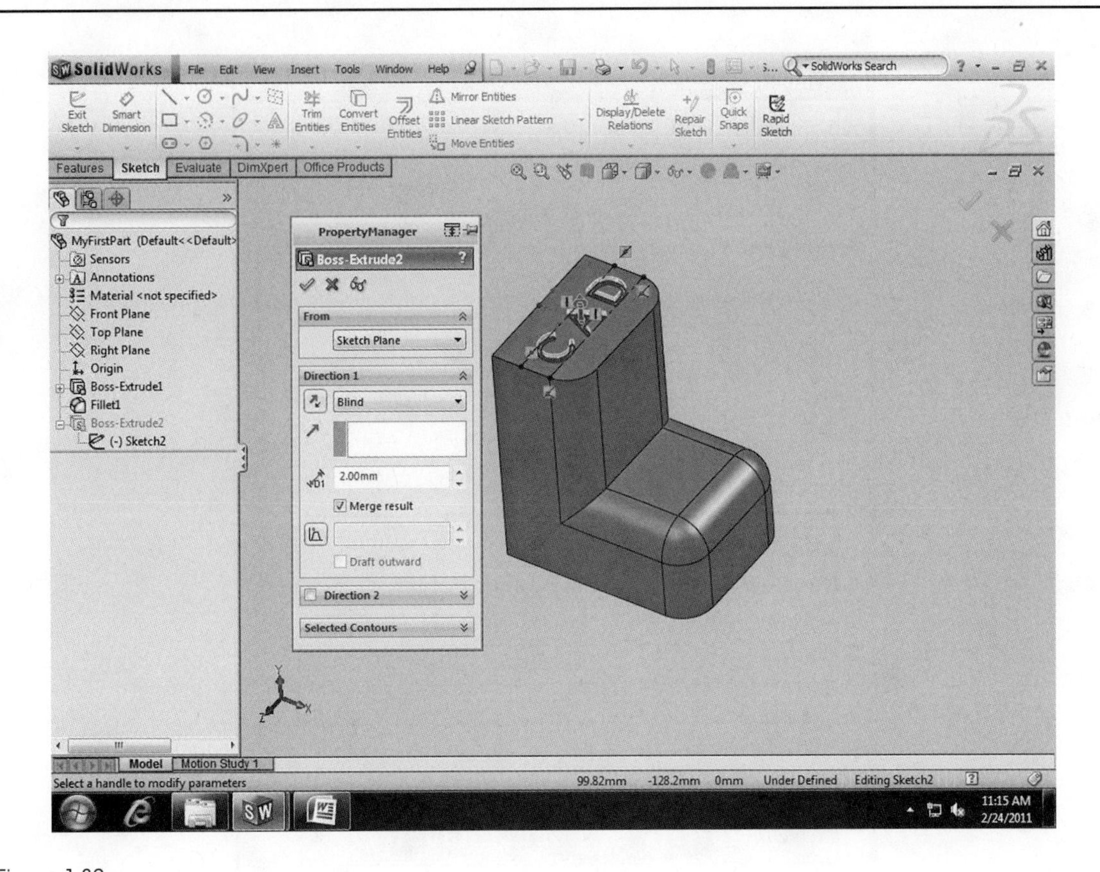

Figure 1.32

Extrude Sketch Text.

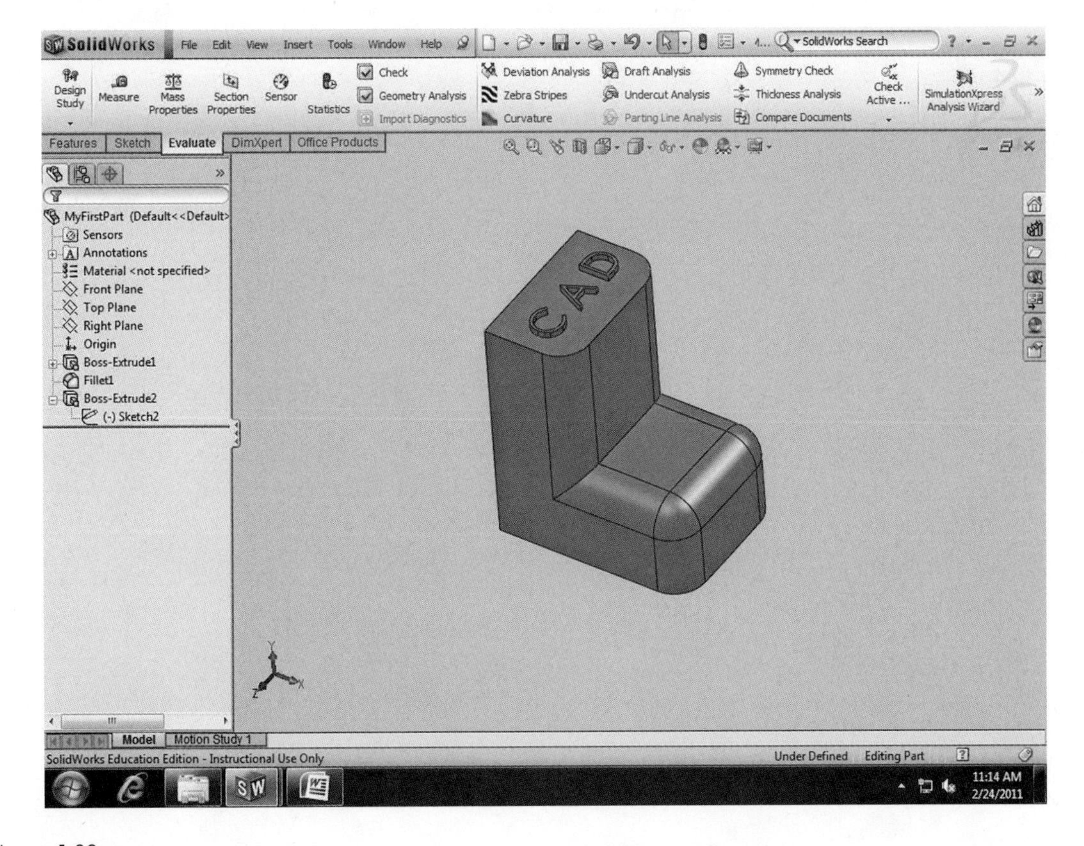

Figure 1.33

Model with initials.

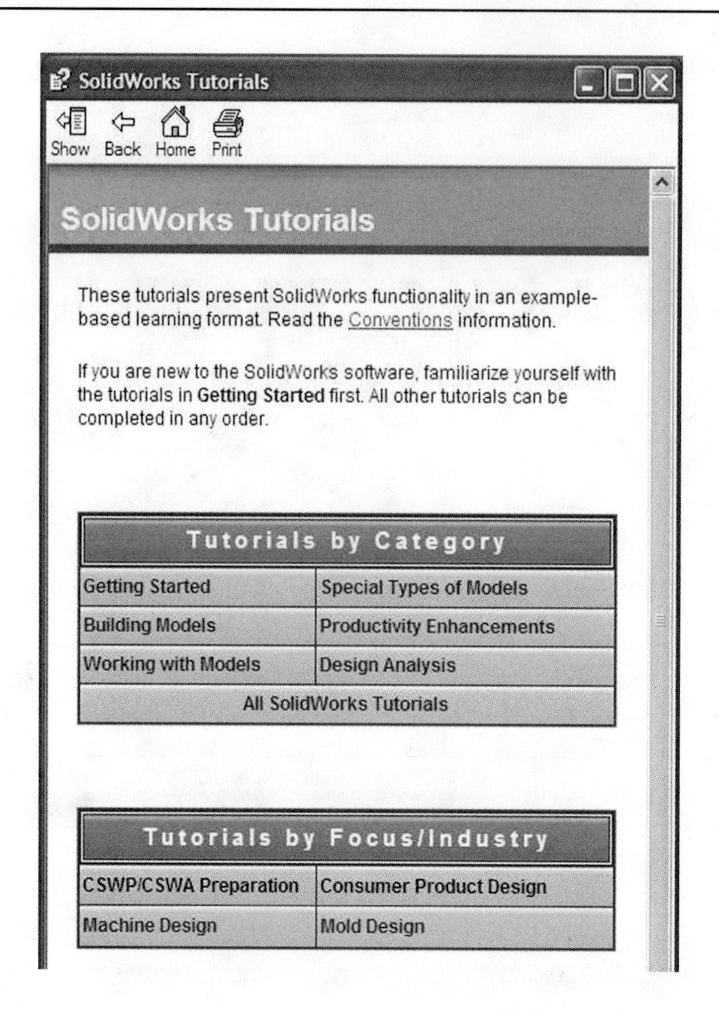

Figure 1.34

SolidWorks Tutorials.

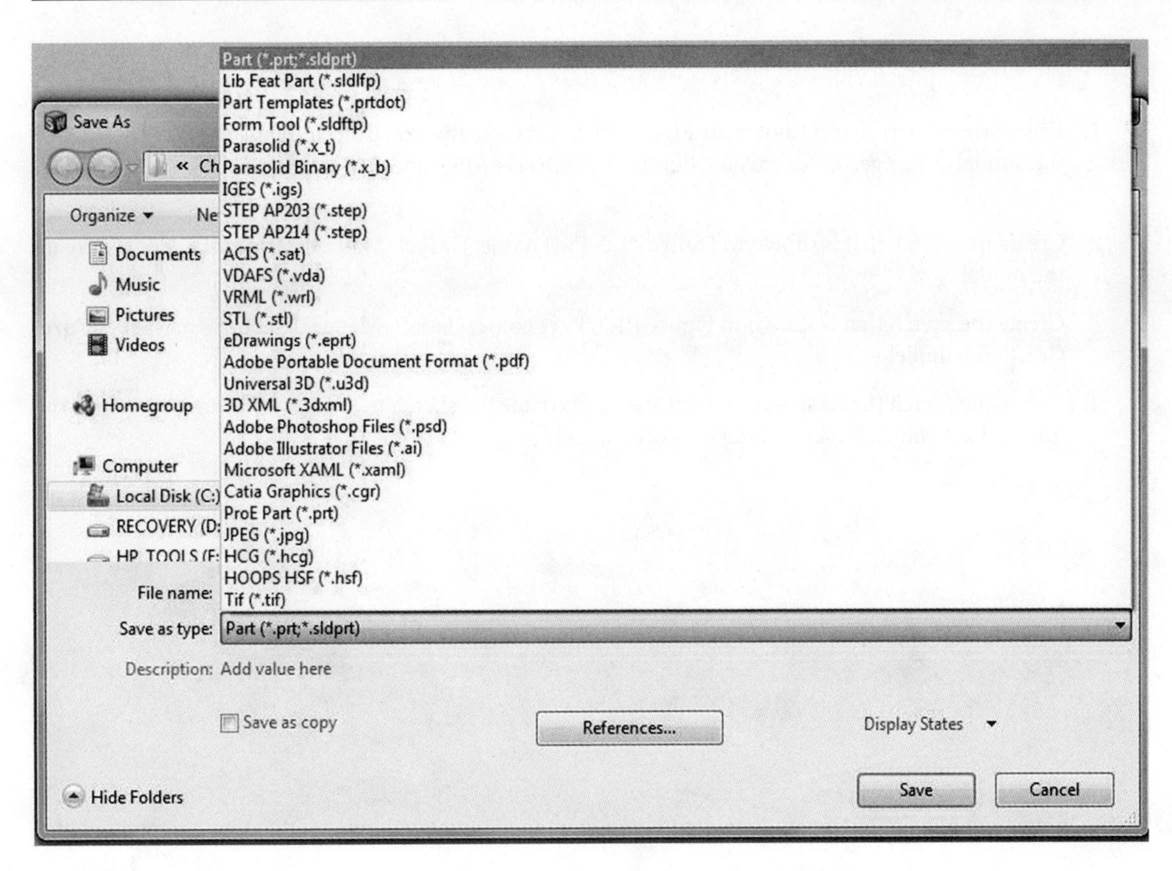

Figure 1.35

SolidWorks interface with a number of standard CAD and application software.

Summary

This chapter gives an overview of the SolidWorks CAD software, which is one of the industry-standard software that are used for the design and analysis of machine parts and assemblies. Industrial designers will also find the software useful. SolidWorks has become very popular with manufacturing industries due to its user-friendliness and cost-effectiveness for small-to-medium enterprises. We have deliberately used a model that is created with the B-Rep scheme of Figure 1.3 as an example to show how easy it is to use SolidWorks, which is a parametrically based design software.

Exercises

1. Create the sketch that is shown in Figure P1.1. Extrude the sketch to a depth of 1 in. to produce the model. Change the extrusion depth and observe the model. Part name: Bracket. Material: Aluminum.
2. Create the sketch that is shown in Figure P1.2. Part name: Gasket. Material: 1-mm thick cork. Create the model.
3. Create the sketch that is shown in Figure P1.3. Part name: Gasket. Material: 0.0625-in. thick bronze. Create the model.
4. Create the sketch that is shown in Figure P1.4. Extrude the sketch to a depth of 5 mm to produce the model. Part Name: Racket. Material: polyvinyl chloride.

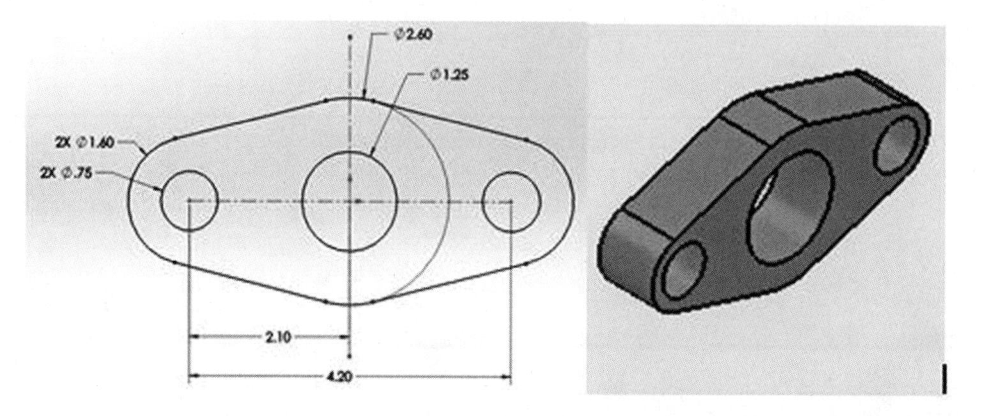

Figure P1.1

Sketch and model.

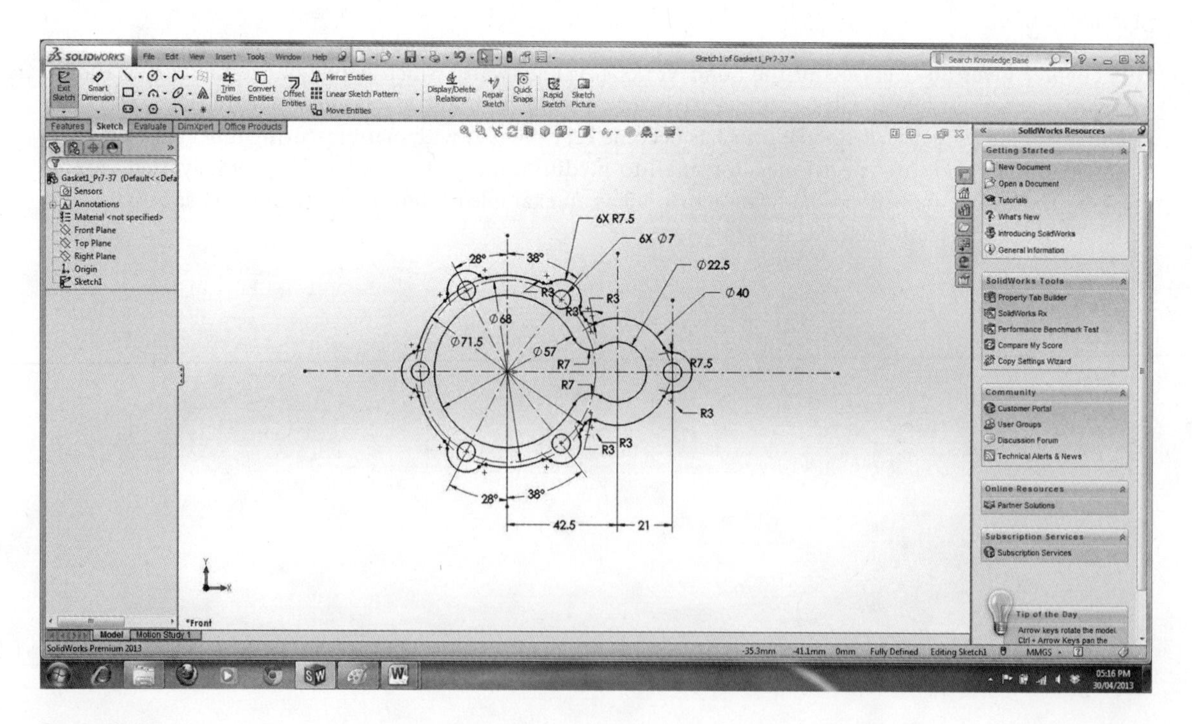

Figure P1.2

Gasket made from cork.

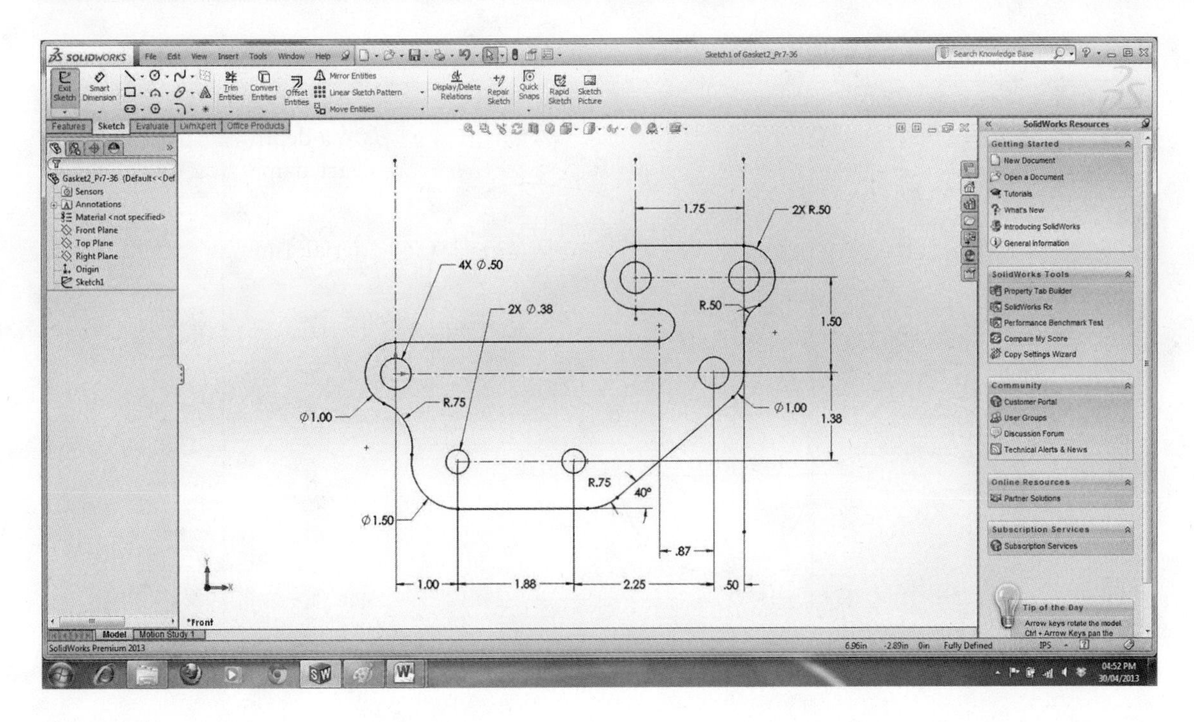

Figure P1.3

Gasket made from bronze.

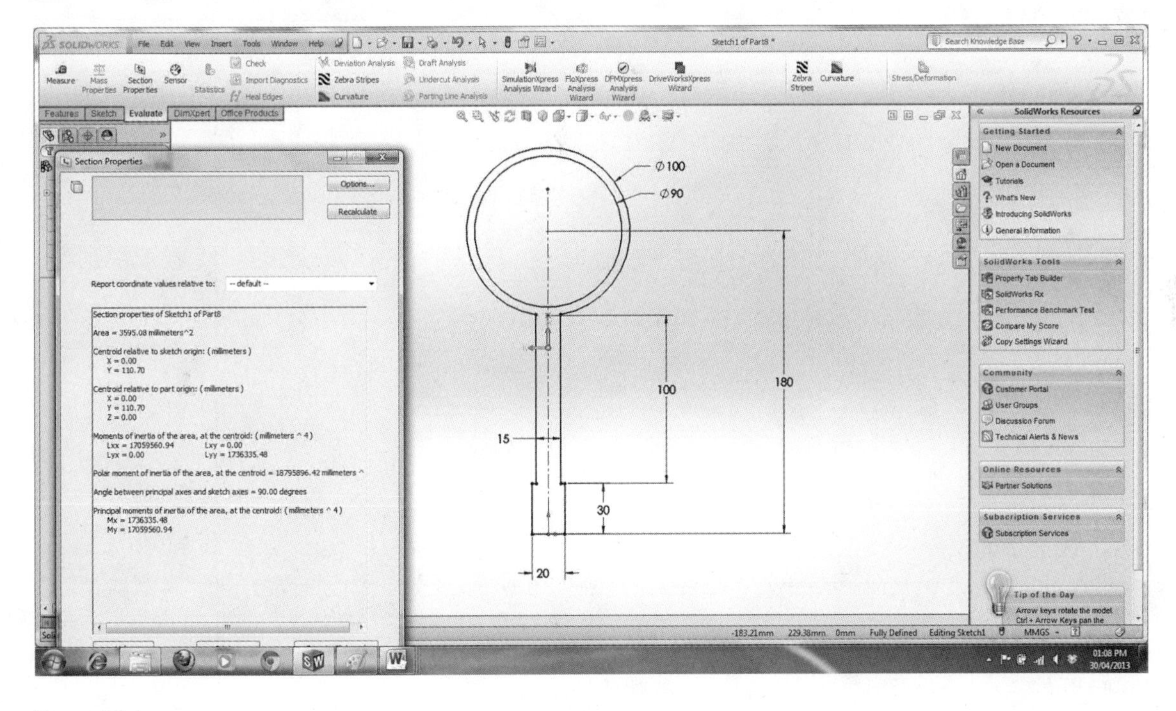

Figure P1.4

Racket.

5. Create the sketch that is shown in Figure P1.5. Part Name: Die stamping. Material: 5-mm thick 1060 Aluminum. Create the model. Determine the mass properties.

6. Create the sketch that is shown in Figure P1.6. Part name: Die stamping. Material: 5-mm thick copper. Create the model. Determine the mass properties.

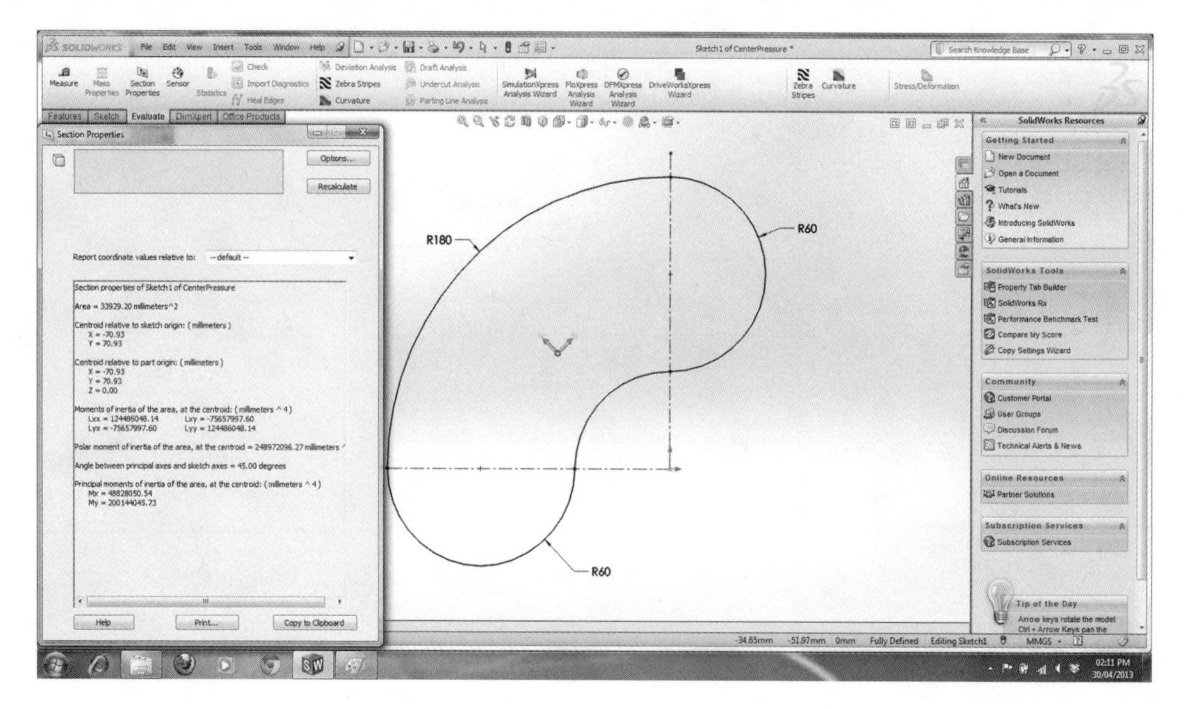

Figure P1.5

Die stamping made from aluminum.

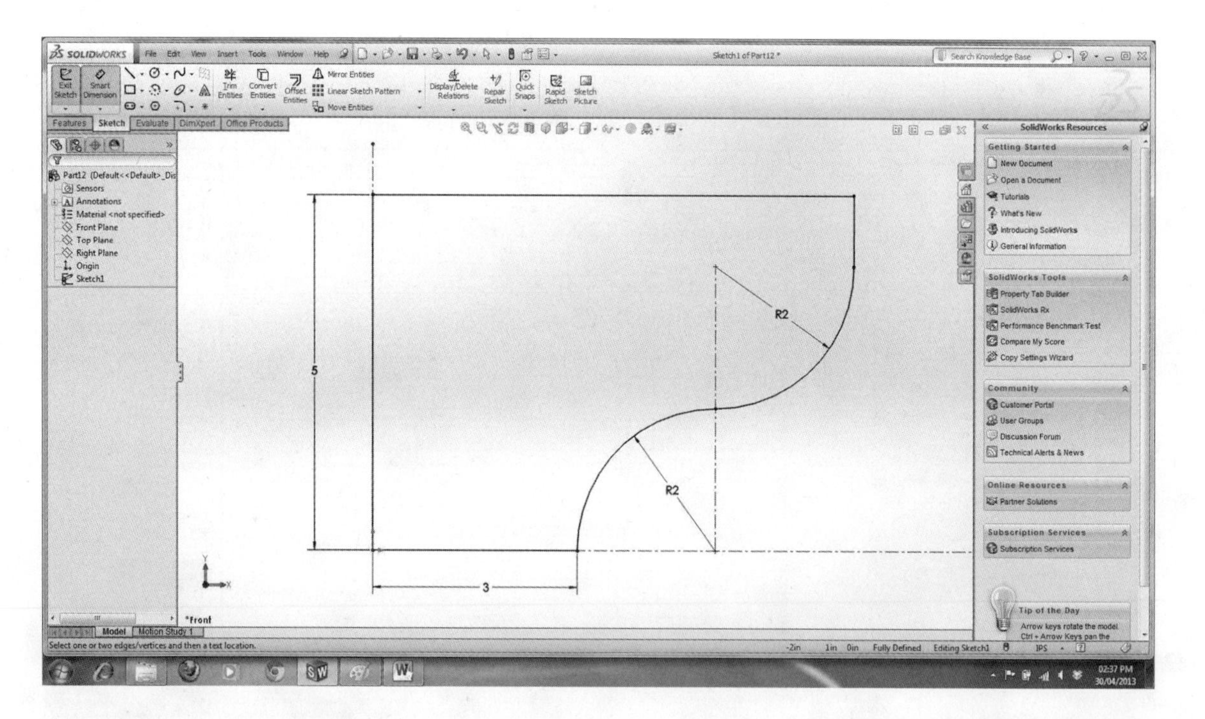

Figure P1.6

Die stamping made from copper.

7. Create the sketch that is shown in Figure P1.7. Part Name: Pivot arm. Material: Aluminum. Create the model: base is 1 high; 4.25 outer diameter (OD) is 1 high from base; 2.13 OD hole is 0.5 high from base; slot is through the base. The dimensions are in inches. Determine the mass properties.
8. Create the sketch that is shown in Figure P1.8. Extrude the sketch to a depth of 15 mm to produce the model. Change the extrusion depth and observe the model.
9. Mirror the part about a plane that coincides with the two vertical rectangular faces.

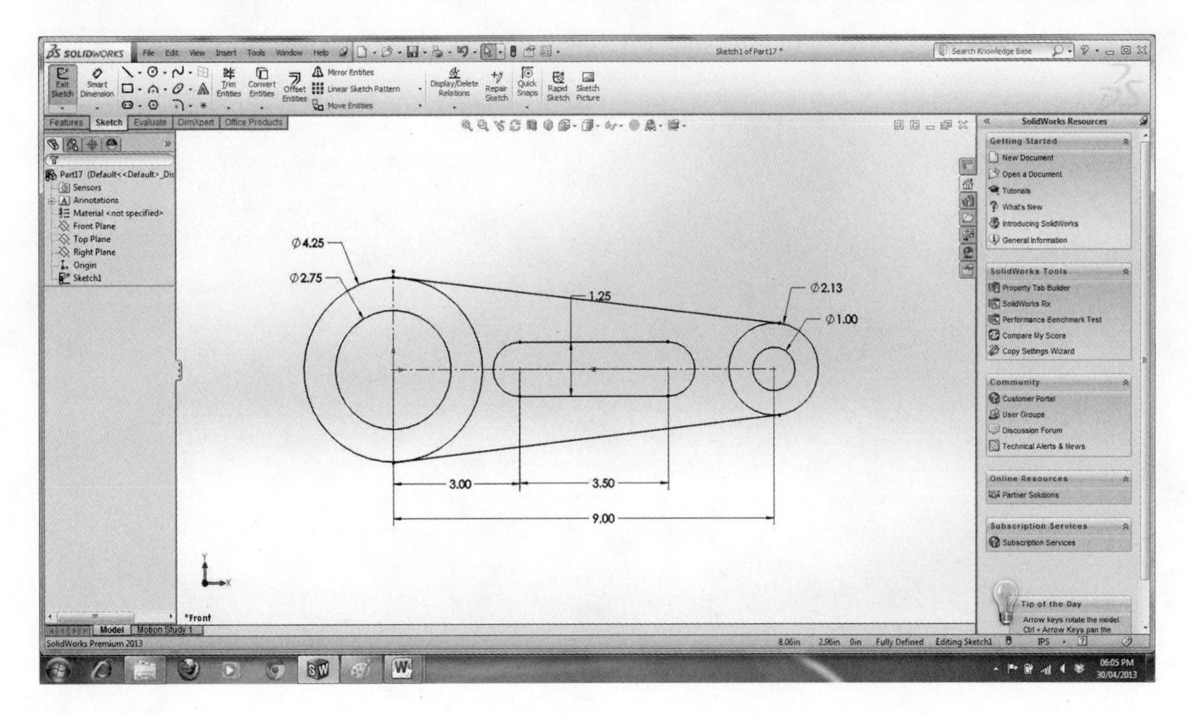

Figure P1.7

Pivot arm.

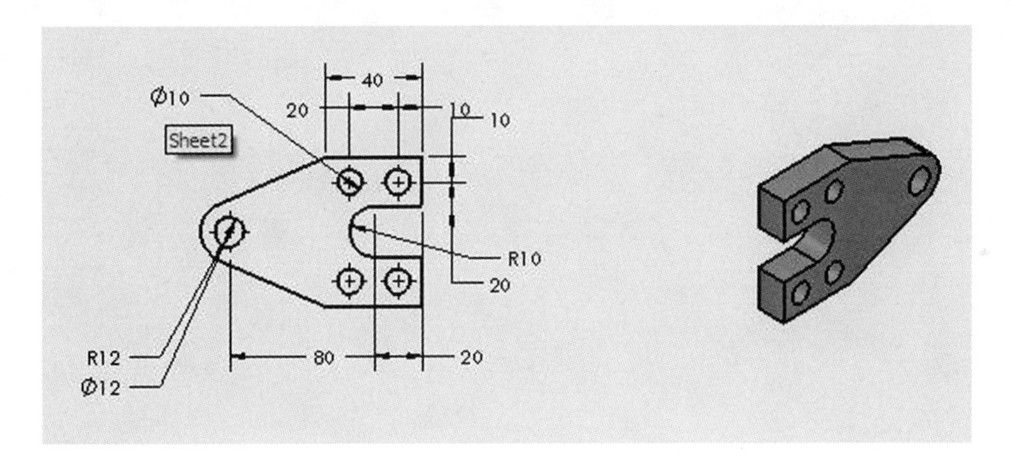

Figure P1.8

Sketch and model.

2

Geometric Construction Tools

Objectives:

When you complete this session, you will have

- Learned about sketch entities in SolidWorks
- Learned about sketch tools in SolidWorks
- Used the sketch tools to modify sketch entities in order to produce parts

Introduction

SolidWorks uses *sketch entities* and *tools* to facilitate the creation of parts. While sketch entities have specific geometries (line, rectangle, parallelogram, slot, polygon, circle, arc, ellipse, parabola, spline, etc.), some of the *sketch tools* (fillet, chamfer, offset, convert entities, intersection curves, trim, extend, split, jog line, constructive geometry, mirror, stretch, move, rotate, scale, copy, pattern, etc.) are used to modify the shapes of sketch entities. Entities are grouped to define a boundary or profile, which are closed regions that are needed for creating extruded or revolved parts. SolidWorks sketch entities and tools can be accessed by clicking the Tools bar.

Sketch entities include *line*, *rectangle* (2-opposite vertices, center, 2-opposite vertices, 3-point corner, 3-point center), *parallelogram*, *slot* (straight, center-point straight, 3-arc, center-point arc), *polygon*, *circle* (center-1-point, 3-points on perimeter), *arc* (center-2-point, tangent, 3-point arc), *ellipse* (full, partial), *parabola*, and *spline*.

Sketch tools include *fillet*, *chamfer*, *offset*, *convert* entities, *intersection curves*, *trim*, *extend*, *split* entities, *jog line*, *construction geometry*, *make path*, *mirror* (straight, dynamic), *stretch* entities, *move* entities, *rotate* entities, *scale* entities, *copy* entities, and *pattern* (linear, circular).

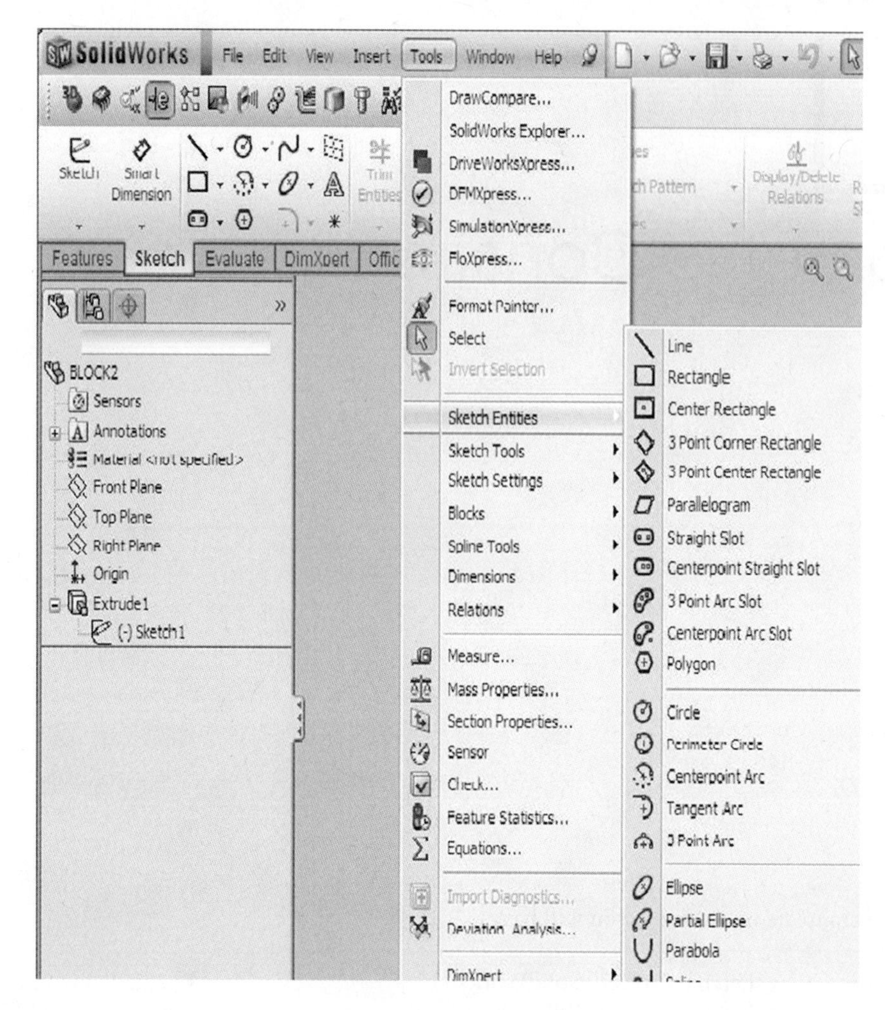

Figure 2.1

Sketch Entities toolbar.

Sketch Entities

Sketch entities are commonly used geometric shapes that are the building blocks for modeling simple and complicated shapes. In general, these entities are grouped into line, rectangle, polygon, slot, arc, circle, ellipse, parabola, and spline. Figure 2.1 shows the *Sketch Entities* toolbar.

For most of the entities' definitions in this section, it assumed that a new *Part* has already been started, and the appropriate *Plane* has already been chosen from the *Feature Manager*. Therefore, a generic format for the sketch entities presented in this chapter is as follows:

1. Start a new Part document.
2. Click Sketch group on the Command Manager.
3. Select (appropriate) Plane from the Feature Manager.

Line

Line is one of the most frequently used sketch entities in part modeling. There are basically two main types of lines: (1) solid and (2) construction lines, as shown in Figure 2.2.

1. Click the Line group on the Command Manager.
2. Select a starting point for the line and click the point.
3. Select other point and extend the line.
4. When the line is complete, right-click the mouse and click the End chain option.

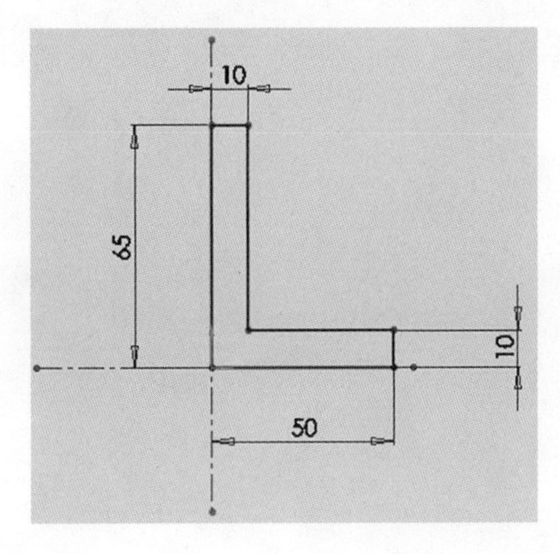

Figure 2.2

Solid and construction lines.

Rectangle

There are a number of options for a rectangle entity: (a) 2-opposite vertices, (b) center, (c) 2-opposite vertices, (d) 3-point corner, and (e) 3-point center, as shown in Figure 2.3. When a user is more interested in centering the rectangle about the origin of the graphics screen, then the *center rectangle* is the one to go for. Let us illustrate the center rectangle.

1. Click the Rectangle group on the Command Manager.
2. Select a center and click the point.
3. Drag the cursor away from the center point to create a rectangle.
4. When the rectangle is complete, click the OK check mark on the PropertyManager to complete the rectangle construction.

Parallelogram

Parallelogram is considered a special type of rectangle; hence, it is grouped as a rectangle type, as shown in Figure 2.4.

1. Click the Rectangle group on the Command Manager.
2. Select and click any two horizontal points.
3. Drag the cursor away from the second point to create a parallelogram.
4. When the parallelogram is complete, click the OK check mark on the PropertyManager to complete the parallelogram construction.

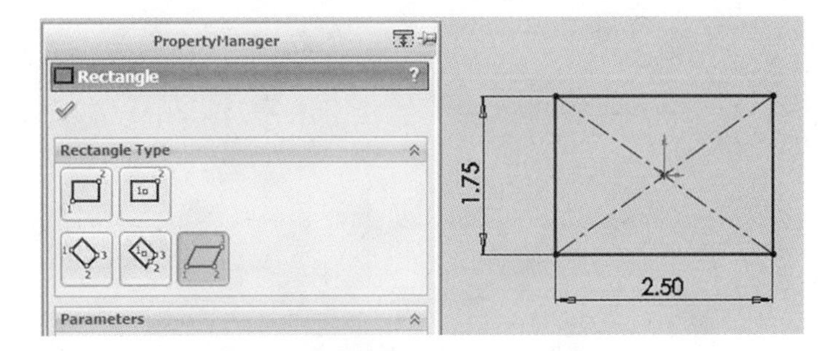

Figure 2.3

Rectangle entities.

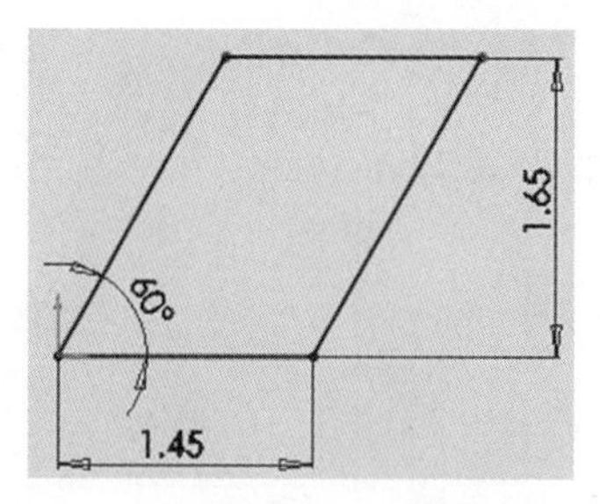

Figure 2.4

Parallelogram entities.

Slot

There are a number of options for a slot entity: (a) straight, (b) center-point straight, (c) 3-arc, and (d) center-point arc, as shown in Figure 2.5.

1. Click the Slot group on the Command Manager.
2. Select and click any two horizontal points.
3. Drag the cursor to the top or bottom of two points for the third point, to get the width of the slot.
4. When the slot is complete, click the OK check mark on the PropertyManager to complete sketching the slot.

Polygon

A polygon has *n*-number of sides as shown in Figure 2.6. To sketch a polygon,

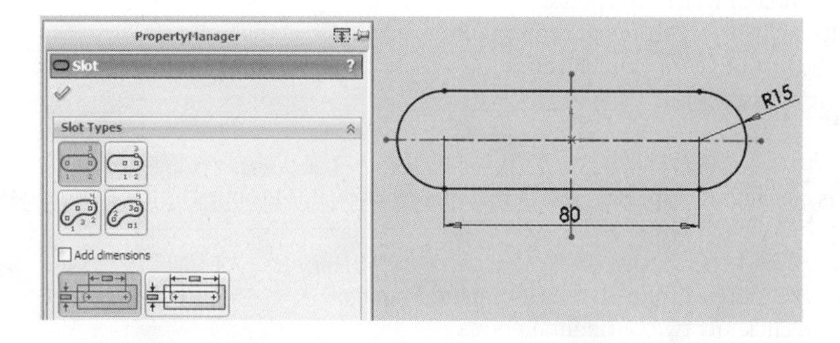

Figure 2.5

Slot entities.

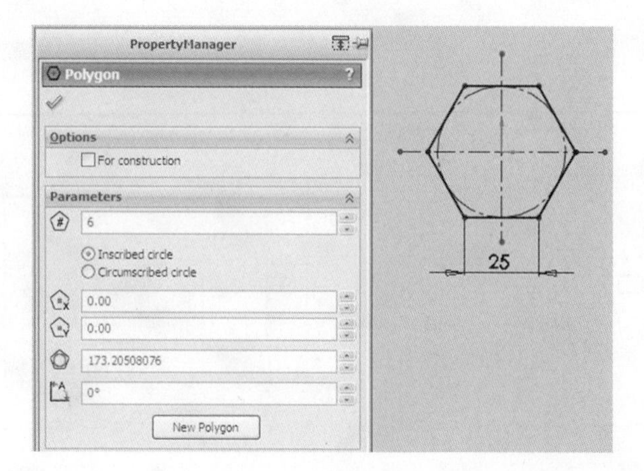

Figure 2.6

Polygon entities.

1. Click the Polygon group on the Command Manager.
2. Define the number of sides as six (hexagon); check Inscribed circle.
3. Click the OK check mark on the PropertyManager to complete sketching the polygon.

Circle

There are generally two ways of defining a circle: (1) center-1-point and (2) 3-points on perimeter. See Figure 2.7.

1. Click the Circle group on the Command Manager.
2. Select and click a point.
3. Drag the cursor away from the center point to create a circle.
4. When the circle is complete, click the OK check mark on the PropertyManager.

Arc

There are generally three ways of defining an arc: (1) center-2-point, (2) tangent, and (3) 3-point arc. See Figure 2.8.

When the center point and two other points are known, the *center-2-point* should be used. When two lines are already sketched, and the two endpoints of an arc are to coincide with the endpoints of the two lines, then *tangent* options should be used. For generally an arc where the center point is not initially known, then the *3-point arc* option should be used. For example, in Figure 2.9, only two points (endpoints of the

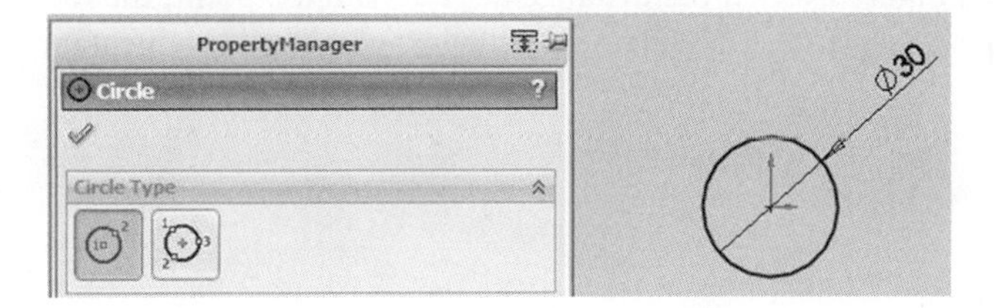

Figure 2.7

Circle entities.

Figure 2.8

Arc entities.

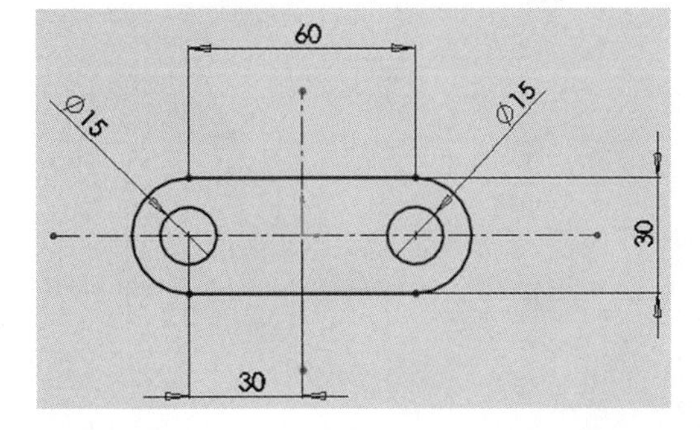

Figure 2.9

Arc entities for sketch definitions.

lines) are defined for the tangent arc. However, the center and the two endpoints of the lines are needed when the *center-2-point arc* option is used, as in Figure 2.9.

Ellipse

An ellipse has a major diameter corresponding to its major axis and a minor diameter corresponding to its minor axis, as shown in Figure 2.10. Two types are available: (1) full and (2) partial.

To create an ellipse,

1. Click Ellipse on the Sketch toolbar, or Tools, Sketch Entities, Ellipse.
2. Click in the graphics area to place the center of the ellipse.
3. Drag and click to set the major axis of the ellipse.
4. Drag and click again to set the minor axis of the ellipse.

An ellipse can be used to model a rugby ball by following these steps:

1. Trim half of the ellipse.
2. Sketch a line to close the shape.
3. Revolve the shape through the centerline.

Parabola

A parabola is a loci of points in which the distance between a fixed point, the focus, and a fixed line, the diretrix, are always equal. See Figure 2.11. It is a well-known geometry in elementary mathematics. To create a parabola,

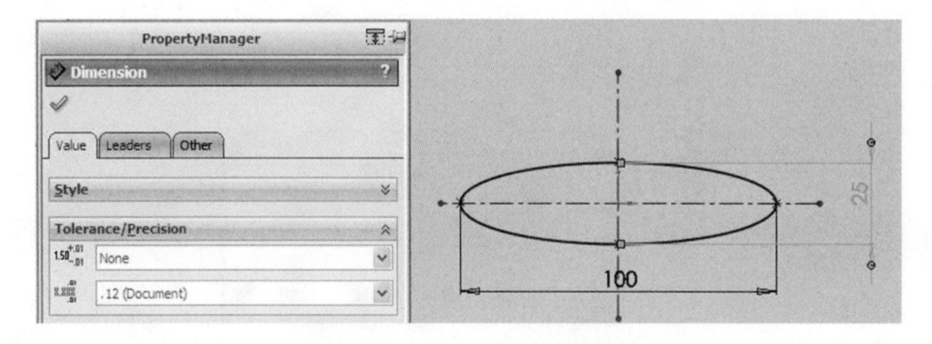

Figure 2.10

Ellipse entities.

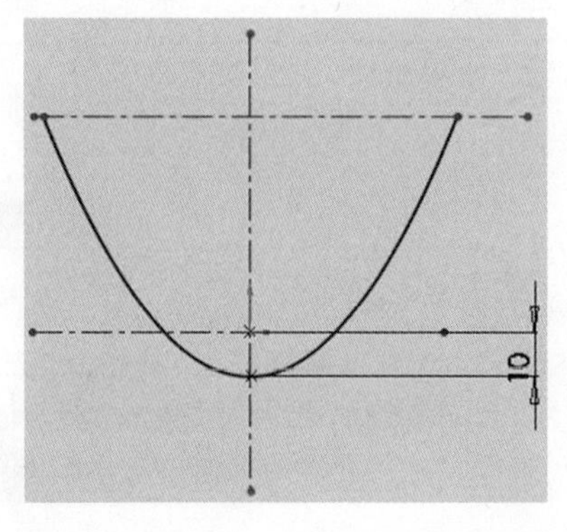

Figure 2.11

Parabola entity.

1. Click Parabola (Sketch toolbar) or Tools, Sketch Entities, parabola.
2. Click to place the focus of the parabola and drag to enlarge the parabola. The parabola is outlined.
3. Click on the parabola and drag to define the extent of the curve.

A parabola can be used to model a satellite dish by following these steps (see Figure 2.12):

1. Trim half of the parabola.
2. Sketch a horizontal line at the top and a vertical line through the focus.
3. Revolve the shape through the vertical centerline.
4. Shell the object with thickness of 1 mm.

Spline

Splines are geometries for modeling complex, general-form curves or surfaces, as shown in Figure 2.13. Understanding the concept of splines is challenging and could be found in advanced computer-aided design documentations. Free-form surfaces are extensively used in the automobile industry. The construction of aircraft, turbomachinery, and automobiles, and recent applications in the advertising and animation industries, has become a driving force behind the research into the techniques for surface design. Splines have been heavily used in the automotive and aircraft industries, and early work in the area began in those sectors for the modeling of automotive bodies and fuselages.

Free-form surfaces are of different forms: (a) quadratic or (b) cubic spline (nonuniform rational B-spline) curve. To sketch a quadratic or cubic spline, specify the following:

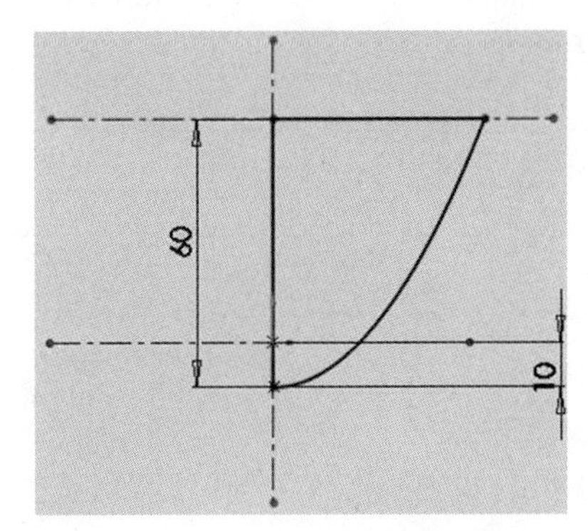

Figure 2.12

Half-parabola used for modeling satellite dish.

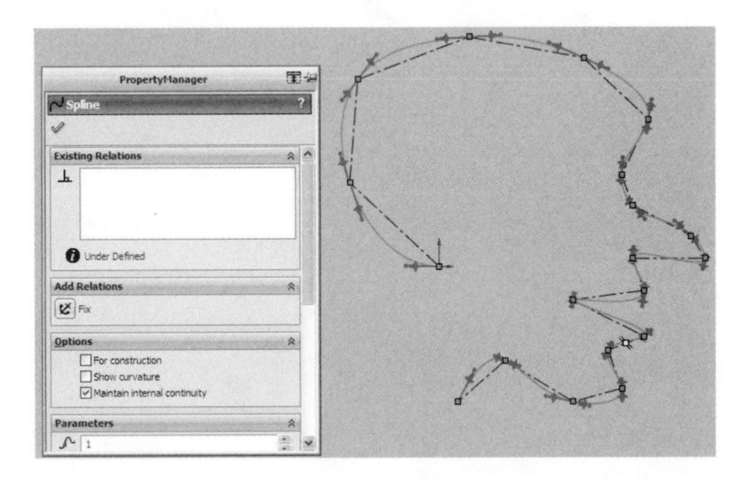

Figure 2.13

Spline entity.

1. First point.
2. Next point (loops so that you can specify as many points as you want).
3. Complete the spline. The spline realized is a single entity.

The shapes modeled using splines can be *controlled*, as shown in Figure 2.13, using controlling polygons.

Parametric B-spline curves and surfaces have several advantages:

- Ability to control the degree of continuity at the joints between adjacent curve segments, and at the borders between surface patches, independent of the order of the segments or the number of control vertices being approximated
- Ability to guarantee that smooth shapes would be generated from smooth data

Sketch Tools

Sketches are two-dimensional in nature, and Figure 2.14 shows the SolidWorks *Sketch Tools* toolbar, which is mainly used for modifying them.

Fillet

The Fillet tool is used to create a fillet between lines as desired in a model in order to remove sharp edges. The sketch shown in Figure 2.15 is filleted at a constant value of 5-mm radius between adjacent lines.

Chamfer

The Chamfer tool is used to create a chamfer in a model. Generally, a chamfer of 45° is common, but other angles could be used. Let us illustrate chamfering by considering an object having all sides horizontal or

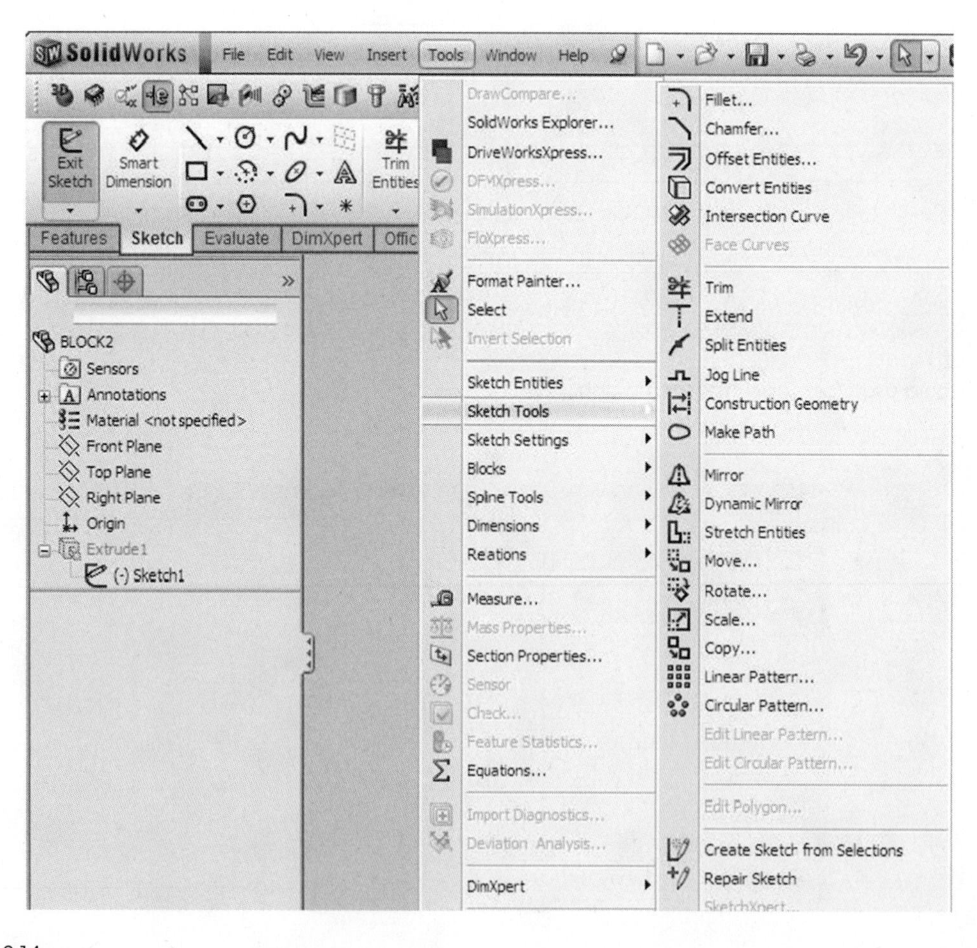

Figure 2.14

Sketch Tools toolbar.

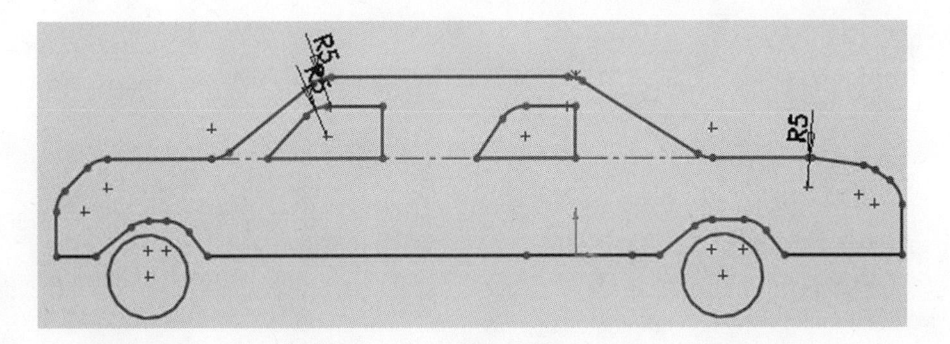

Figure 2.15

Illustrating fillet tool.

vertical. A chamfer is needed between the vertical and horizontal lines to the right of the object at an angle of 45° (see Figure 2.16).

The resulting chamfered object is shown to have a vertical dimension of 30 mm and an angle of 45°, as shown in Figure 2.17.

1. Click the Chamfer group on the Command Manager.
2. Check the Angle-Distance option.
3. For the Distance, click the right vertical line; for the angle, specify 45°.
4. When the chamfer is complete, click the OK check mark on the PropertyManager.

Offset

The *Offset* tool is used to draw entities that are parallel to the existing entities of a sketch, as shown in Figure 2.18. This tool is particularly useful, and it can cut down the time of modeling since all that is needed is the selection of the entities and the definition of the amount of offset.

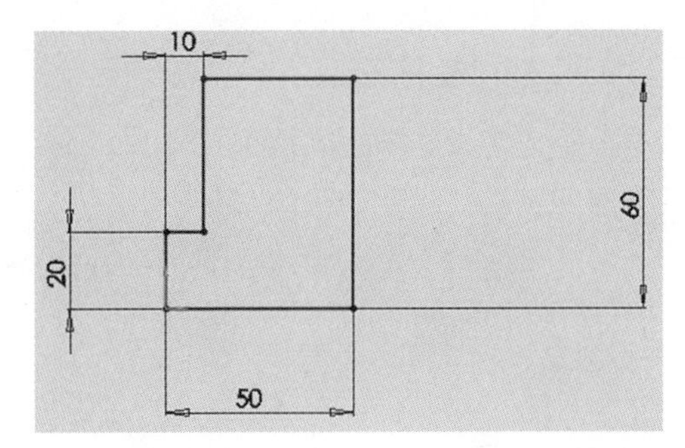

Figure 2.16

Initial geometry to apply Chamfer tool.

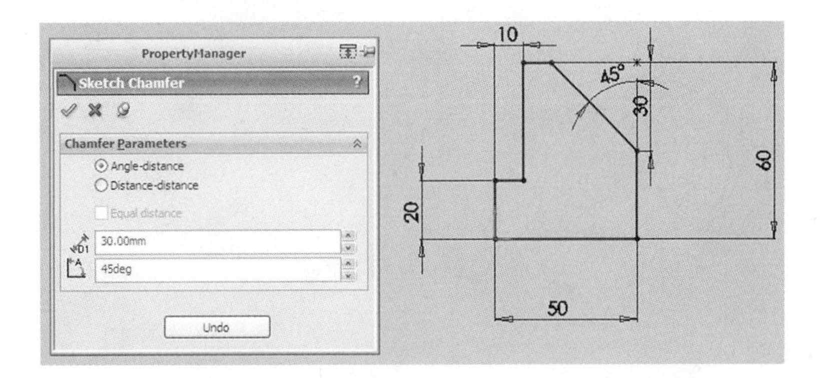

Figure 2.17

Chamfer tool.

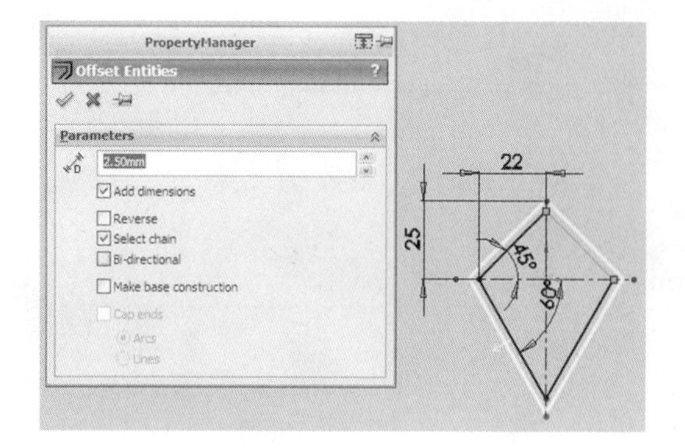

Figure 2.18

Offset tool.

1. Click the Offset group on the Command Manager.
2. Select and click the edges to be offset (see Figure 2.19).
3. Specify the offset that is desired (2.5 mm in this case).
4. When the offset is complete, click the OK check mark on the PropertyManager.

Convert Entities

The *Convert Entities* tool is used to extract the portions of an entity onto a plane and then used for further modeling, as shown in Figure 2.20a and b. For example, a cutout is realized from a circle by using this tool to describe a sector. This sector could then be used to cut through the length or portion of the part being modeled (say, a cylinder).

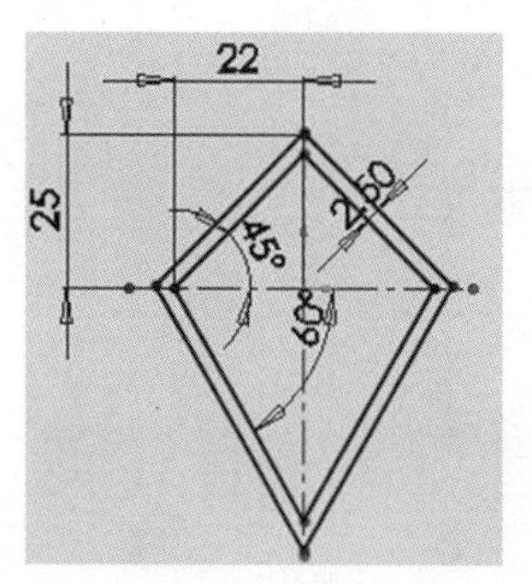

Figure 2.19

Offset geometry realized.

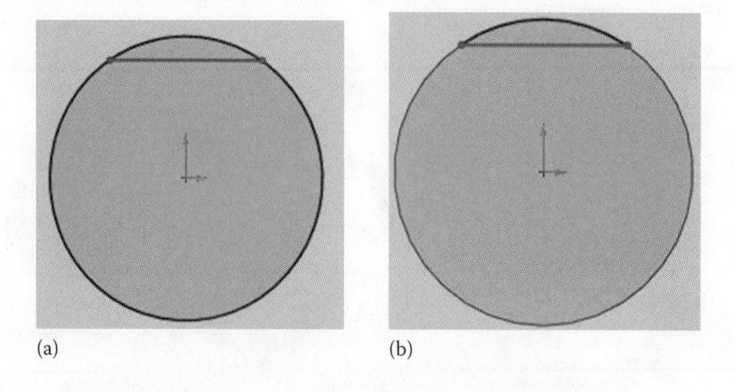

Figure 2.20

Illustrating Convert Entities tool: (a) sketch on a plane and (b) sketch transferred to another parallel plane using Convert Entities tool.

Trim

The *Trim* tool (Figure 2.21) is used to clean up lines or arcs that are not needed, as shown in Figure 2.22.

1. Click the Trim group on the Command Manager.
2. Select the Power Trim option.
3. Hold down and drag the cursor across the entities to be trimmed.
4. When the trim is complete, click the OK check mark on the PropertyManager.

Extend

The *Extend* tool is used to extend an existing line to increase its length. See Figure 2.23. It is found in the Trim Entities group in the *Command Manager.*

1. Click the Extend group on the Command Manager.
2. Click the top and bottom lines to be extended.
3. When the extend task is complete, click the OK check mark on the PropertyManager.

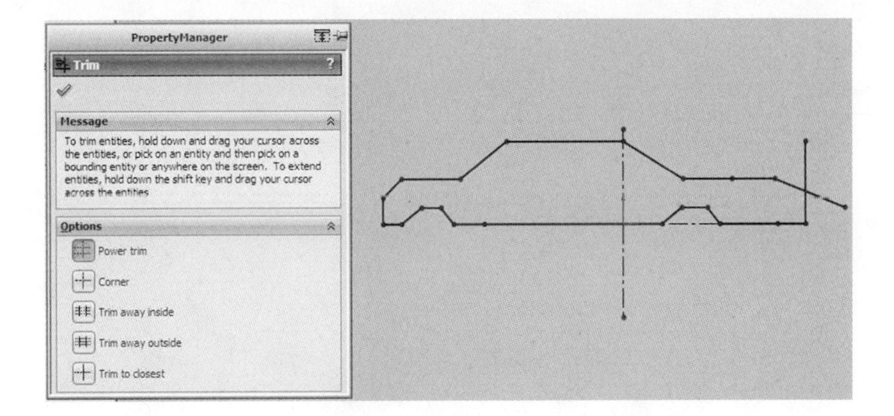

Figure 2.21

Use the Power Trim option.

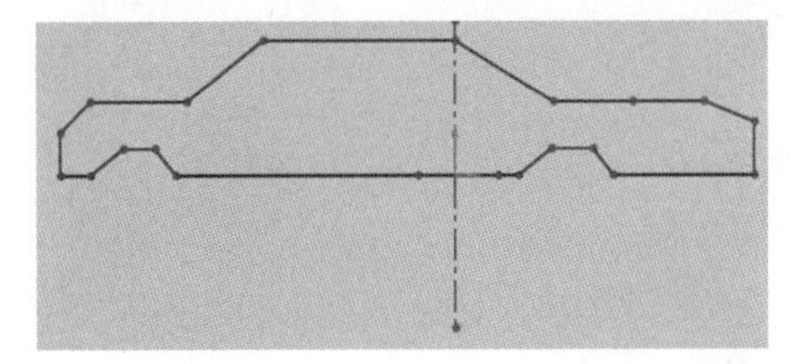

Figure 2.22

Trim Entities tool.

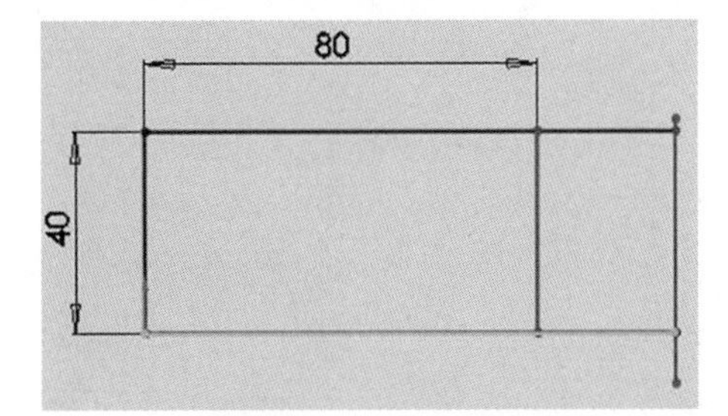

Figure 2.23

Extend Entities tool.

Split Entities

The *Split Entities* tool is found on the Sketching tools. When it is chosen, click on the entity to trim approximately where you want the split. Figure 2.24 shows how this tool is used.

Construction Geometry

It converts a solid line/circle/arc to a construction line/circle/arc.

Mirror

The *Mirror* tool is used to create a mirror image of an entity in a sketch. This tool is used when an axis, usually a line, exists about which an entity is to be mirrored. There are two forms of Mirror tool: (1) simple and (2) dynamic. In the simple form, one-half of the entities are completed and mirrored about an axis, as shown in Figure 2.25. In the dynamic form, as an entity is a sketch, it is dynamically mirrored about the chosen axis. The mirrored entities are shown in Figure 2.26.

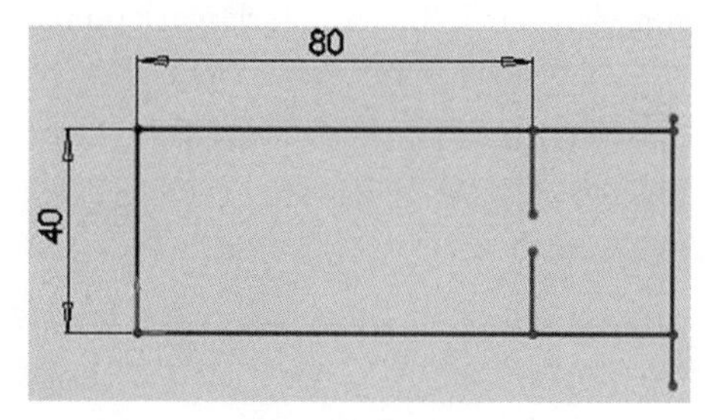

Figure 2.24

Split Entities tool.

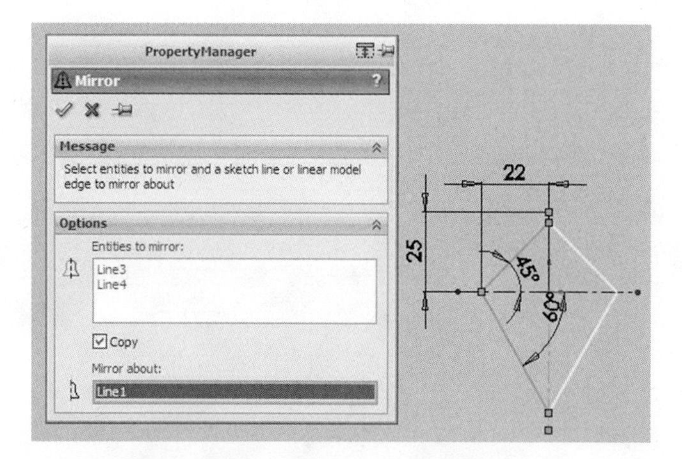

Figure 2.25

Mirror Entities tool.

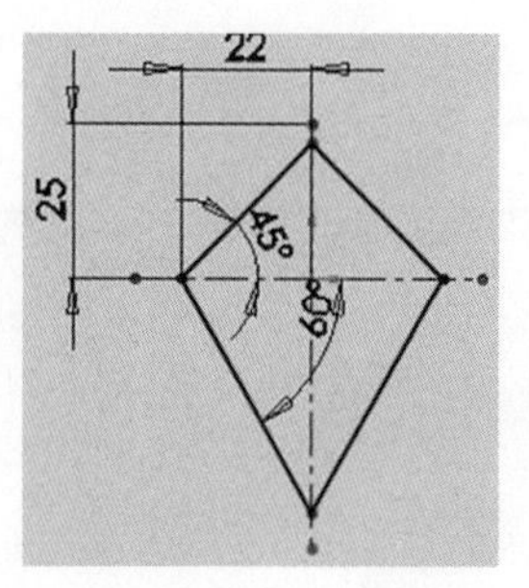

Figure 2.26

Mirrored entities.

Stretch Entities

The *Stretch Entities* tool stretches the entities that are chosen; the value of stretch is based on the datum that is defined, as shown in Figure 2.27. For example, the right-top, side, and bottom lines are stretched about the left vertical line.

1. Click the Stretch group on the Command Manager.
2. Click the lines to be stretched.
3. Click the Base Define by clicking the left vertical line.
4. When the stretch task is complete, click the OK check mark on the PropertyManager.

The bottom line is stretched from 100 to 125 mm, while the top-right line is stretched from 40 to 65 mm. See Figure 2.28 for stretched entities.

Move Entities

The *Move Entities* tool is used to move the entities that are chosen. An individual item or an entire object can be moved, as shown in Figure 2.29.

1. Click the Move group on the Command Manager.
2. Click Entities to Move and window the entities to be moved.
3. Uncheck the Keep Relations.
4. Click any point as a starting point From Point Defined.
5. Drag object to another point and click mouse at the End Point Defined.
6. When the move task is complete, click the OK check mark on the PropertyManager.

Rotate Entities

The *Rotate Entities* tool is used to rotate the entities that are chosen. This tool is illustrated in Figure 2.30.

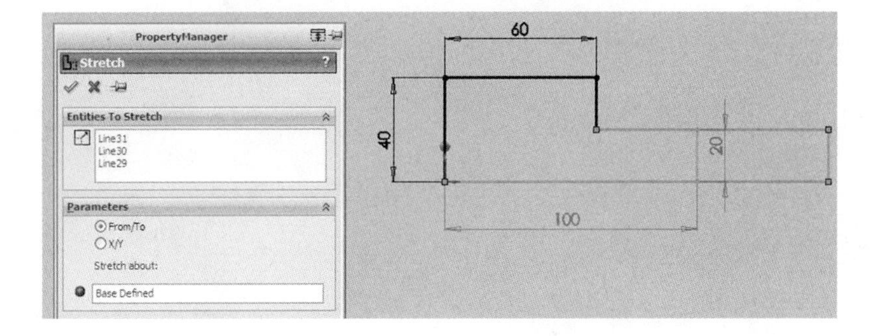

Figure 2.27

Stretch Entities tool.

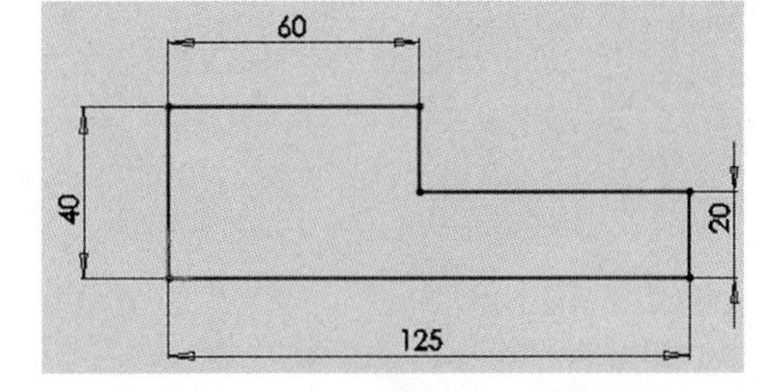

Figure 2.28

Illustrating stretched entities.

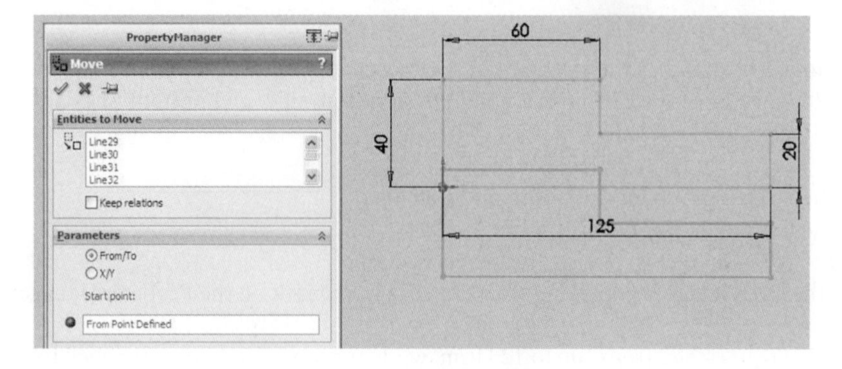

Figure 2.29

Move Entities tool.

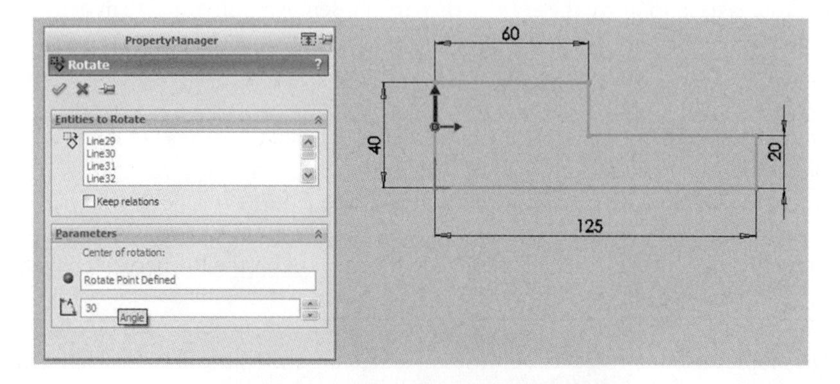

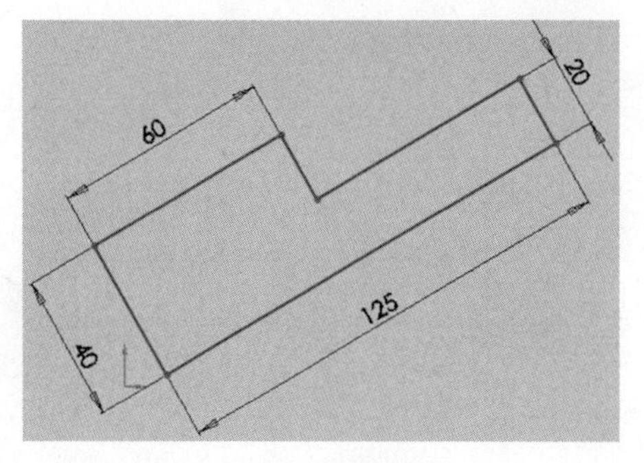

Figure 2.30

Rotate Entities tool.

1. Click the Move group on the Command Manager.
2. Click Entities to Rotate and window the entities to be moved.
3. Uncheck the Keep Relations.
4. Click any point as the Center of rotation.
5. Define the Angle of rotation.
6. When the rotation task is complete, click the OK check mark on the PropertyManager.

Scale Entities

The *Scale Entities* tool is used to scale the entities that are chosen. This tool is illustrated in Figure 2.31.

1. Click the Scale group on the Command Manager.
2. Click Entities to Scale and window the entities to be copied.
3. Click any point as Scale Point Defined.

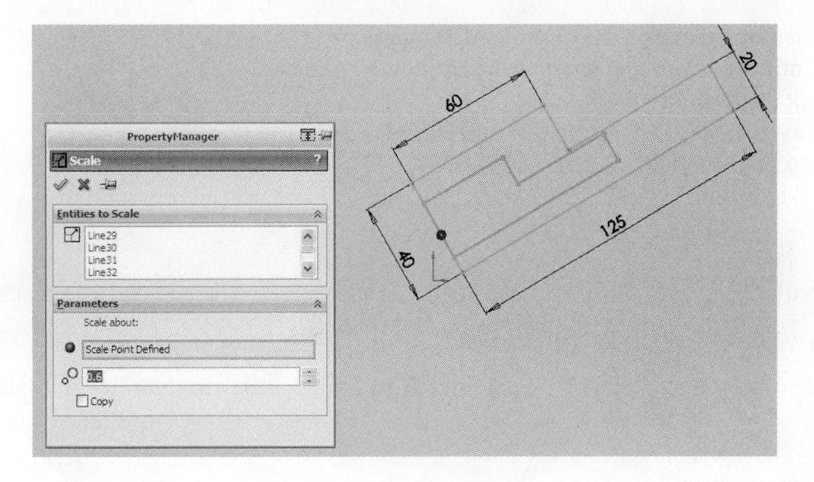

Figure 2.31

Scale Entities tool.

4. Define the scaling value.
5. When the scale task is complete, click the OK check mark on the PropertyManager.

Copy Entities

The *Copy Entities* tool is used to copy the entities that are chosen. An individual item or an entire object can be copied. This tool is illustrated in Figure 2.32.

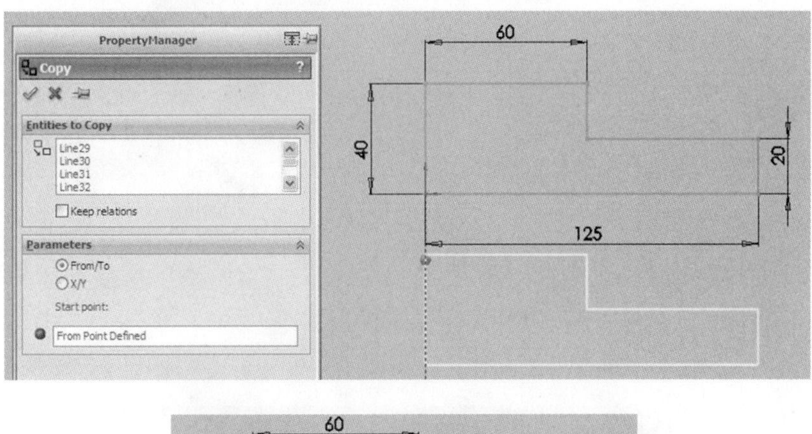

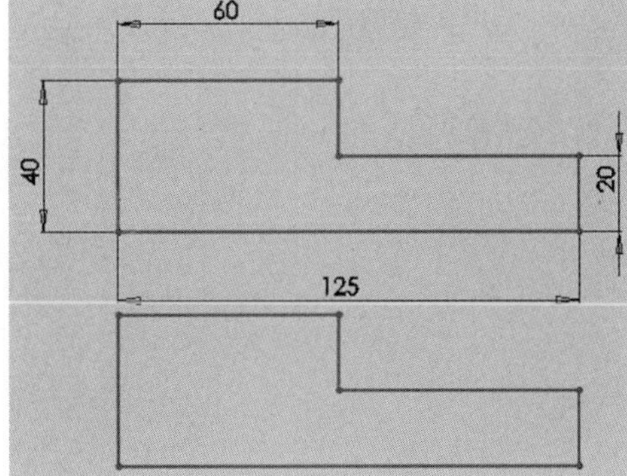

Figure 2.32

Copy Entities tool.

1. Click the Copy group on the Command Manager.
2. Click Entities to Copy and window the entities to be copied.
3. Uncheck the Keep Relations.
4. Click any point as a starting point From Point Defined.
5. Drag object to another point and click mouse at the End Point Defined.
6. When the copy task is complete, click the OK check mark on the PropertyManager.

Pattern

The Linear/Circular Pattern tool is used to create rectangular patterns based on a given model.

Linear Pattern

In linear pattern, the directions for the patterns are linear. Let us consider an example, as shown in Figure 2.33.

1. Sketch a geometry.
2. Sketch the seed pattern.
3. Click the Linear Pattern tool.

   The Linear Pattern Properties Manager appears.

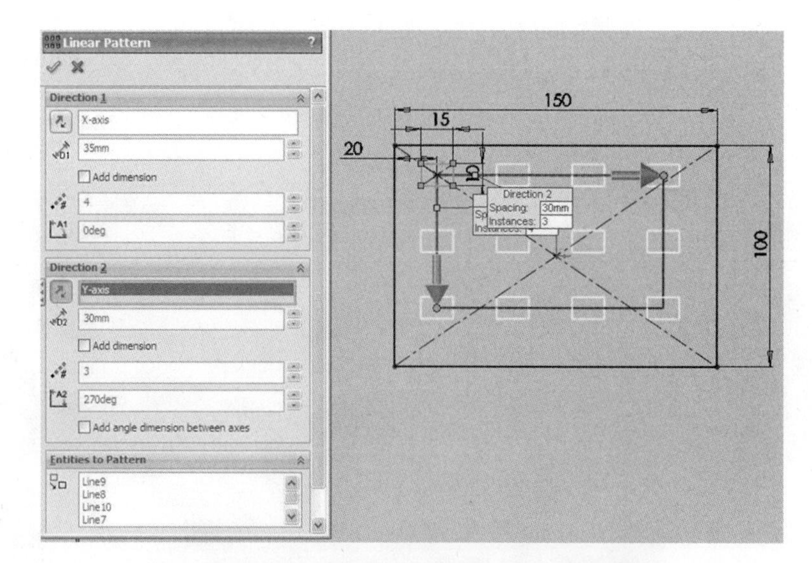

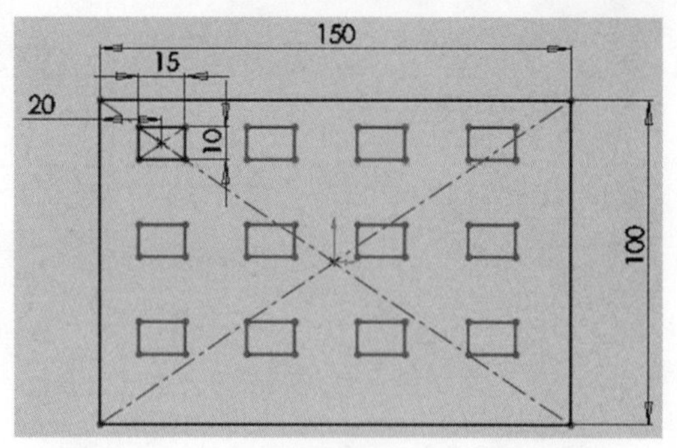

Figure 2.33

Linear Pattern tool and patterns realized.

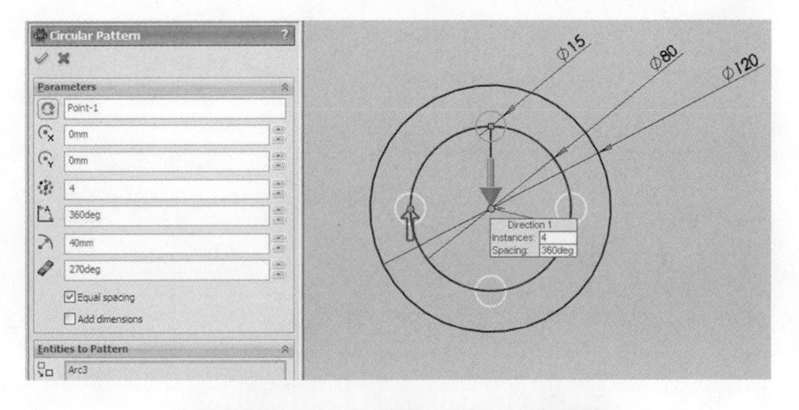

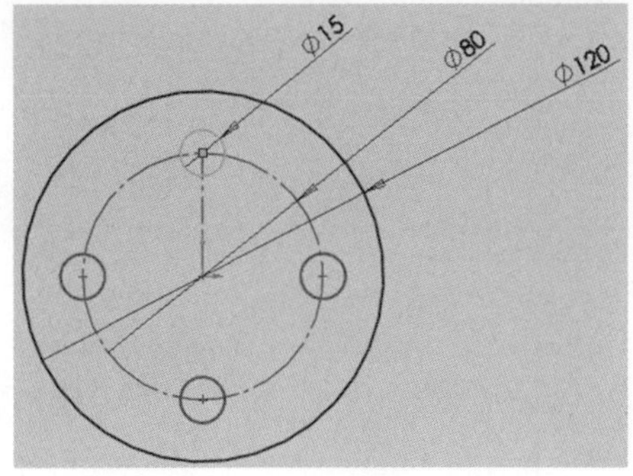

Figure 2.34

Circular Pattern tool and patterns realized.

4. Select the seed pattern.
5. For Direction1>, select a horizontal edge for direction, 35 mm for Distance, and 4 for the number of patterns. For Direction2>, select a vertical edge for direction, 30 mm for Distance, and 3 for the number of patterns.
6. Click OK to complete the patterns.

Circular Pattern

In circular pattern, the patterns are created about an axis in a circular manner, as shown in Figure 2.34.

Summary

This chapter discusses the sketch entities and tools that are used in SolidWorks to facilitate the creation of parts. While sketch entities have specific geometries (line, rectangle, parallelogram, slot, polygon, circle, arc, ellipse, parabola, spline, etc.), some of the sketch tools (fillet, chamfer, offset, convert entities, intersection curves, trim, extend, split, jog line, constructive geometry, mirror, stretch, move, rotate, scale, copy, pattern, etc.) are used to modify the shapes of sketch entities. The sketch entities and the sketch tools have been applied to a number of examples so that users can understand the principles that are involved.

Exercises

1. Use the given dimensions to redraw the given shape in Figure P2.1. Create part models of the objects. Thickness = 0.5 in.

2. Use the given dimensions to redraw the given shape in Figure P2.2; fillet = R10. Create part models of the objects. Thickness = 25 mm.

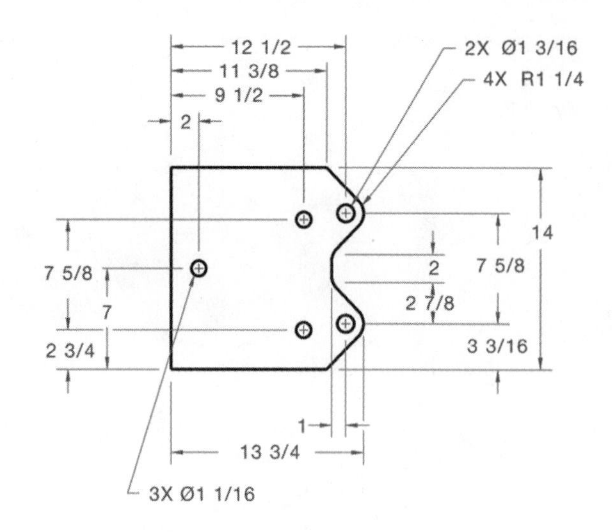

Figure P2.1

Gasket.

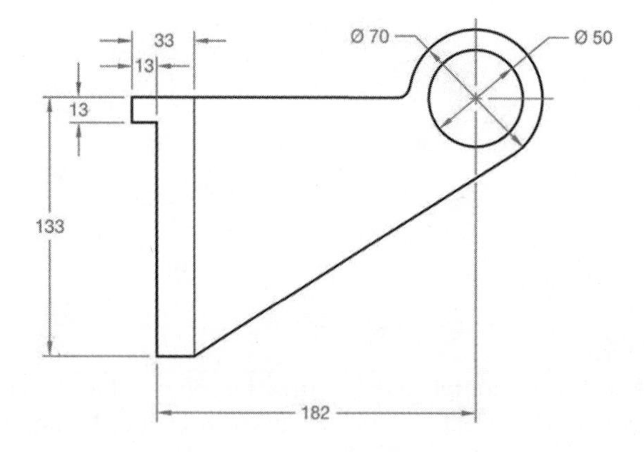

Figure P2.2

Coupler.

3. Use the given dimensions to redraw the given shape in Figure P2.3. Create part models of the objects. Thickness = 1.5 in.
4. Use the given dimensions to redraw the given shape in Figure P2.4. Create part models of the objects. Thickness = 1 in.
5. Use the given dimensions to redraw the given shape in Figure P2.5. Create part models of the objects. Thickness = 2 in.

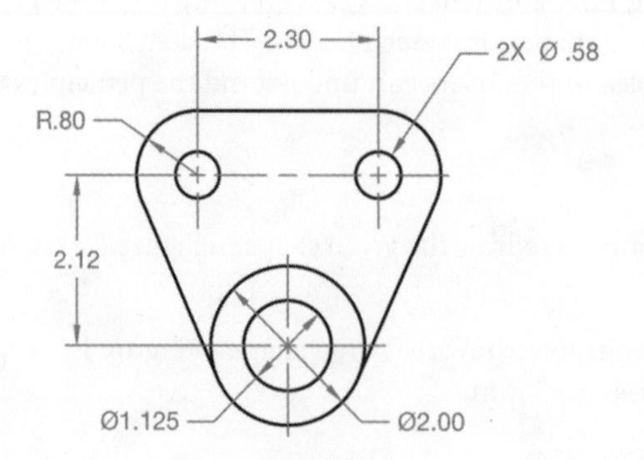

Figure P2.3

Hanger.

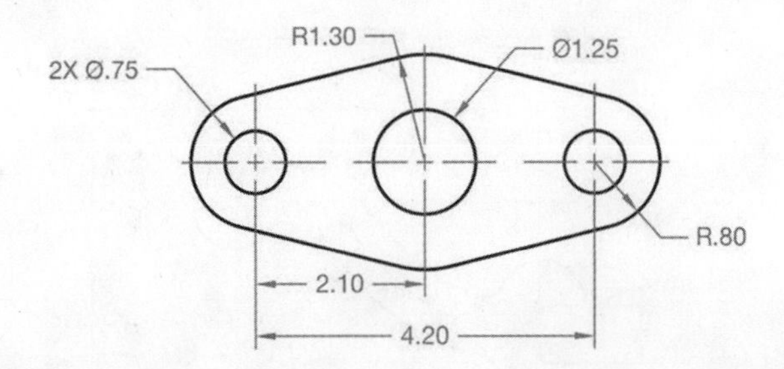

Figure P2.4

Bracket.

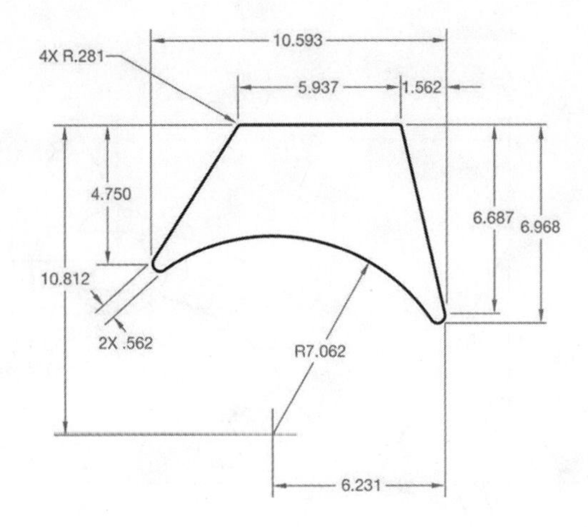

Figure P2.5

Gasket.

6. Use the given dimensions to redraw the given shape in Figure P2.6. Create part models of the objects. Thickness = 30 mm.
7. Use the given dimensions to redraw the given shape in Figure P2.7. Create part models of the objects. Thickness = 0.5 in.
8. Use the given dimensions to redraw the given shape in Figure P2.8. Create part models of the objects. (Be sure to trim part of the 0.625-diameter circle.) Thickness = 0.5 in.
9. Use the given dimensions to redraw the given shape in Figure P2.9. Establish unknown dimensions to your own specifications. Create part models of the objects. (Be sure to trim part of the R1.600 arc.) Thickness = 2 in.

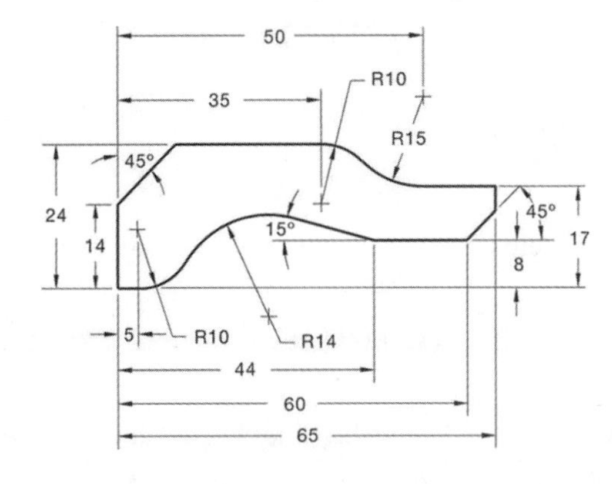

Figure P2.6

Support brace.

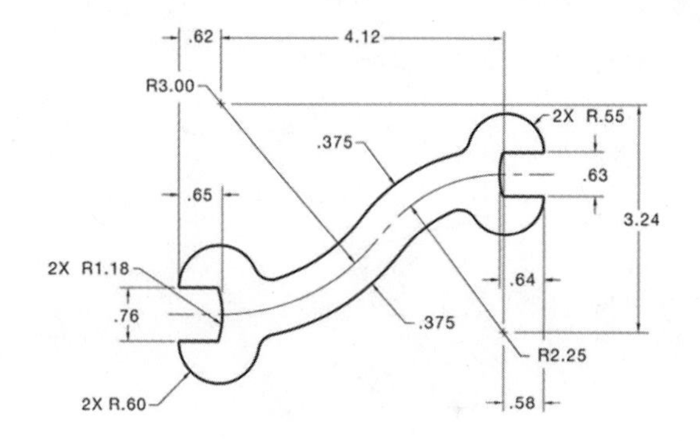

Figure P2.7

Wrench.

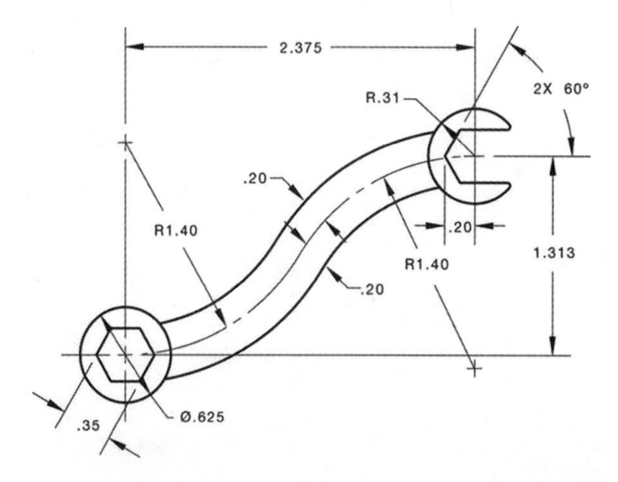

Figure P2.8

Wrench.

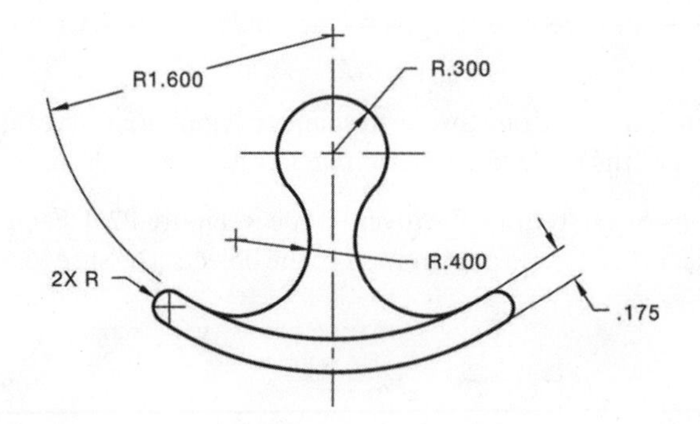

Figure P2.9

Hammer head.

10. Use the information in the ellipse section (refer to Figure 2.10) to create the rugby ball model that is shown in Figure P2.10.
11. Use the information in the parabola section (refer to Figure 2.12) to create the satellite model that is shown in Figure P2.11.
12. Create the sketch that is shown in Figure 2.2 (line entities) and extrude it 10-mm midplane.
13. Create the sketch that is shown in Figure 2.3 (rectangle entities) and extrude it 10 mm.
14. Create the sketch that is shown in Figure 2.4 (parallelogram entities) and extrude it 10 mm.

Figure P2.10

Rugby ball.

Figure P2.11

Satellite dish.

15. Create the sketch that is shown in Figure 2.5 (slot entities) and extrude it 10 mm.
16. Create the sketch that is shown in Figure 2.6 (polygon entities) and extrude it 10 mm.
17. Create the sketch that is shown in Figure 2.9 (arc entities) for sketch definitions.

3

Features

Objectives:

When you complete this session, you will have

- Learned about Features tools in SolidWorks
- Learned how to create three-dimensional (3D) objects using Extrusion Features tools
- Learned how to create 3D objects using Revolved Features tools
- Learned how to create 3D objects using Lofted Features tools
- Learned how to create 3D objects using Swept Features tools
- Learned how to use Modification Features tools
- Learned how to Edit Features and hence be able to troubleshoot models
- Learned how model sketch-driven patterns
- Learned how model curve-driven patterns
- Learned how model table-driven patterns
- Learned how to use reference planes

Introduction

SolidWorks creates three-dimensional (3D) objects based on Features. This chapter introduces *Features* tools (see Figure 3.1) and classifies them into five categories: (1) *extrusion features* (extruded boss/base, draft, dome, rib, and extruded cut), (2) *revolved features* (revolved boss/base, and revolved cut), (3) *lofted features* (lofted boss/base, and lofted cut), (4) *swept features* (swept boss/base and swept cut), and (5) *modification features* (hole wizard, shell, fillet, chamfer, pattern, and mirror). In the first four categories, boss/base or cut models are realized. In the fifth category, the tools for altering the geometry of models are introduced. How to *edit features* work with *reference planes* to create more complex models is also introduced. Advanced *Patterns* (*Sketch Driven, Curve Driven,* and *Table Driven*) tools are also introduced.

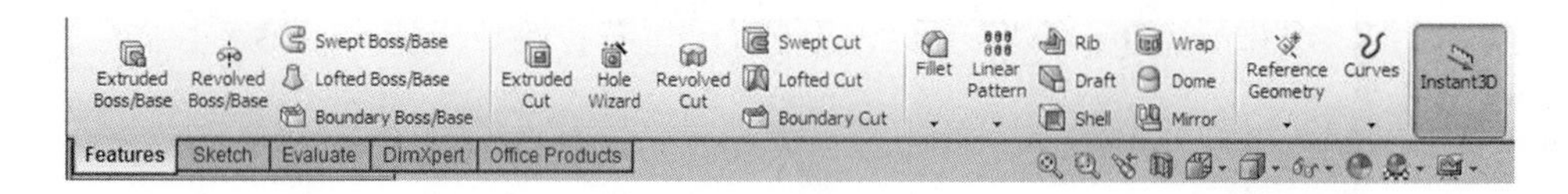

Figure 3.1

SolidWorks *Features* tools.

In illustrating all the features tools presented in this chapter, it is assumed that a new document has been opened for modeling, and the *Document Properties* have been appropriately set to work with an appropriate design standard in millimeters per inch.

Extruded Boss/Base

The extruded boss/base tool is for adding height or thickness to an existing two-dimensional (2D) sketch in order to realize a 3D model. It is an *addition* feature. To use an extruded boss/base tool,

1. Select the front plane and create Sketch1, as shown in Figure 3.2.
2. Click the Features tool.
3. Click the Extruded Boss/Base tool.
   The Extrude Properties Manager appears.
4. Define the extrusion height as 25 mm. A real-time preview will appear (see Figure 3.3).
5. Click OK to complete the extrusion as shown in Figure 3.4.

Draft, Dome, Rib

Draft

The draft tool is used to create a slanted shape in a feature. Let us illustrate the tool with an example (see Figure 3.5).

1. Select the front plane and create Sketch1 (50 mm × 50 mm × 45 mm).
2. Click the Features tool.
3. Click the Draft tool.
   The Draft Properties Manager appears.

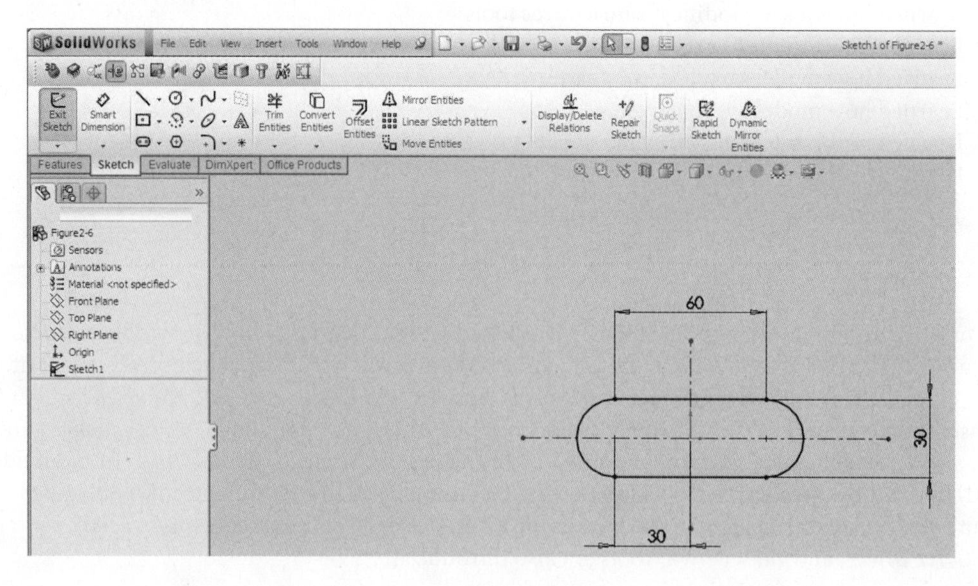

Figure 3.2

Sketch1.

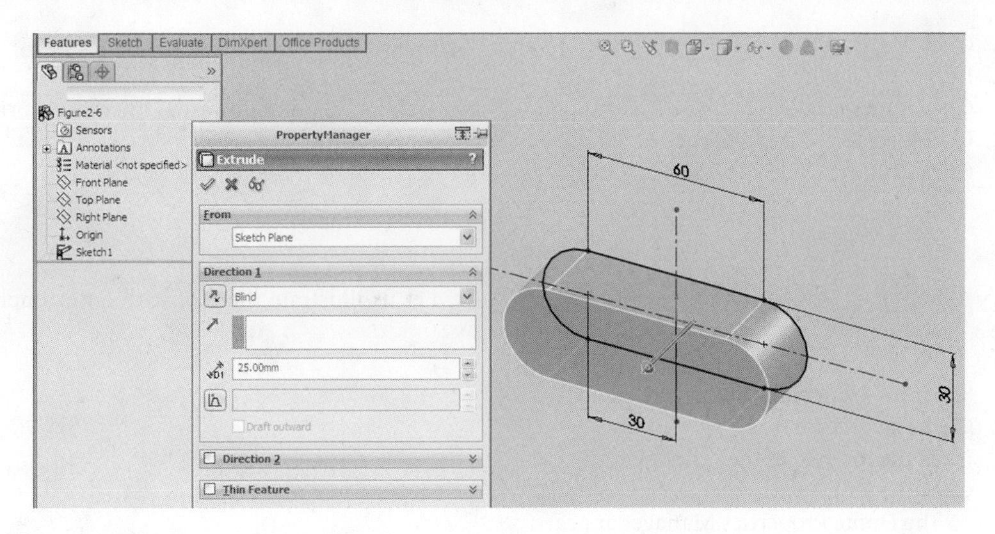

Figure 3.3

Real-time preview of extruded Sketch1.

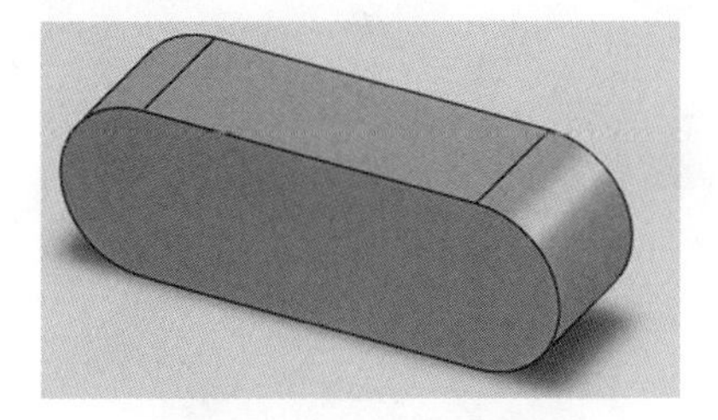

Figure 3.4

Part model obtained through extruded boss/base.

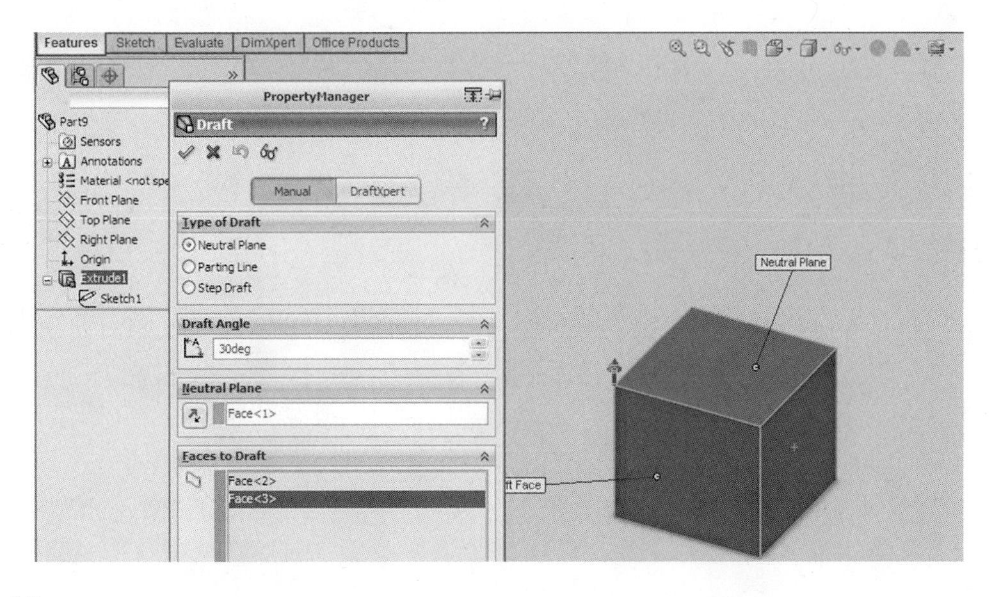

Figure 3.5

Draft tool.

4. Click Manual for Type of Draft. Neutral Plane is chosen as default.
5. Define the Draft Angle as 30°.
6. Click upper face (Face<1>) as Neutral Plane. The arrow should point upward. Otherwise, correct it.
7. Click two adjacent faces (Face<2> and Face<3>) as Faces to Draft.
8. Click OK to complete the process of adding draft to a part model (see Figure 3.6).

Dome

The dome tool is used to add a dome shape to a feature. Let us illustrate the tool with an example (see Figure 3.7).

1. Select the front plane and create Sketch1.
2. Click the Features tool.
3. Click the top face of the cylinder.
4. Click the Dome tool.
   The Dome Properties Manager appears.
5. Define the dome height as 10 mm. A real-time preview will appear.
6. Click OK to complete the process of adding dome to a part model (see Figure 3.8).

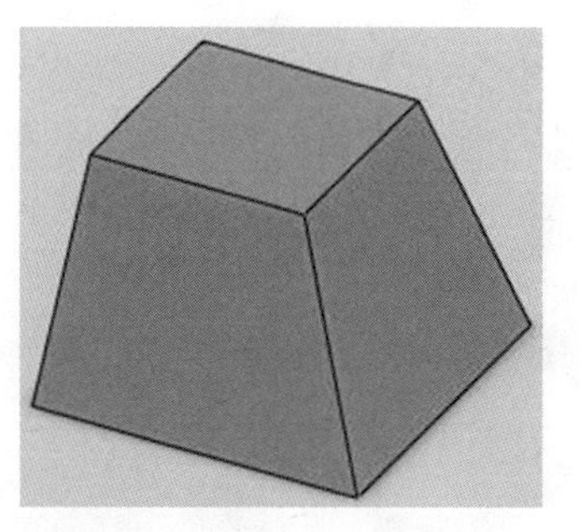

Figure 3.6

Adding draft features to a part model.

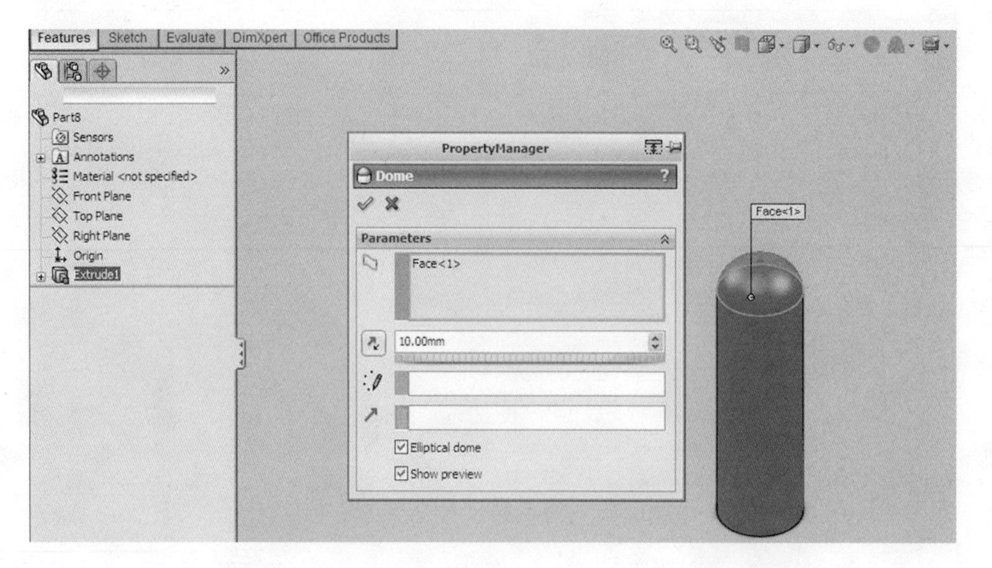

Figure 3.7

Dome tool.

Figure 3.8

Adding dome feature to a part model.

Rib

The rib tool is used to include a rib feature in part. Let there be an existing L-shaped model for which a rib is to be added.

1. Select the front plane (Sketch1) for the L-shape and define a reference plane, midway of the extruded length (see Figure 3.9).
2. Use the reference plane to sketch a slanting line.
3. Click the Features tool.
4. Click the Rib tool.
   The Rib Properties Manager appears (see Figure 3.10).
5. Check *Both Sides* for the Rib Thickness; Check *Parallel to Sketch* for the Extrusion Direction.
6. Set the rib thickness value as 10 mm. A real-time preview will appear.
7. Click OK to complete the rib and hide the reference plane (see Figure 3.11).

Extruded Cut

The extruded cut tool is for removing a portion of existing 2D sketch or part during 3D modeling. It is a *subtraction* feature. Let us illustrate the approach by including two holes, each having a 10-mm diameter located 30 mm from each side of the center of the of the 3D model that is created in Figures 3.2 through 3.4. To use an extruded cut tool,

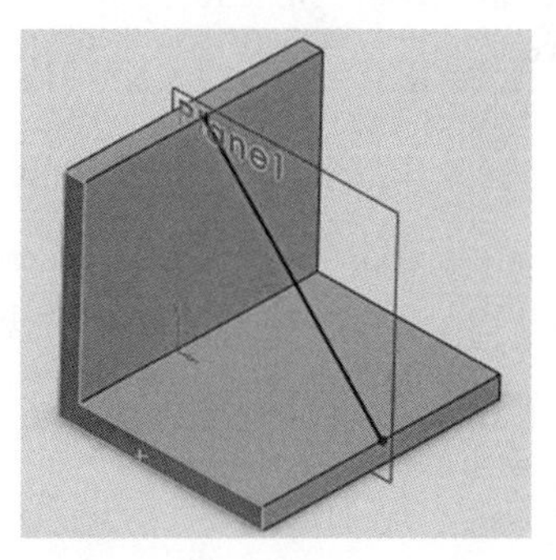

Figure 3.9

Sketching a slanting line on Plane1 of the L-shape to define a rib.

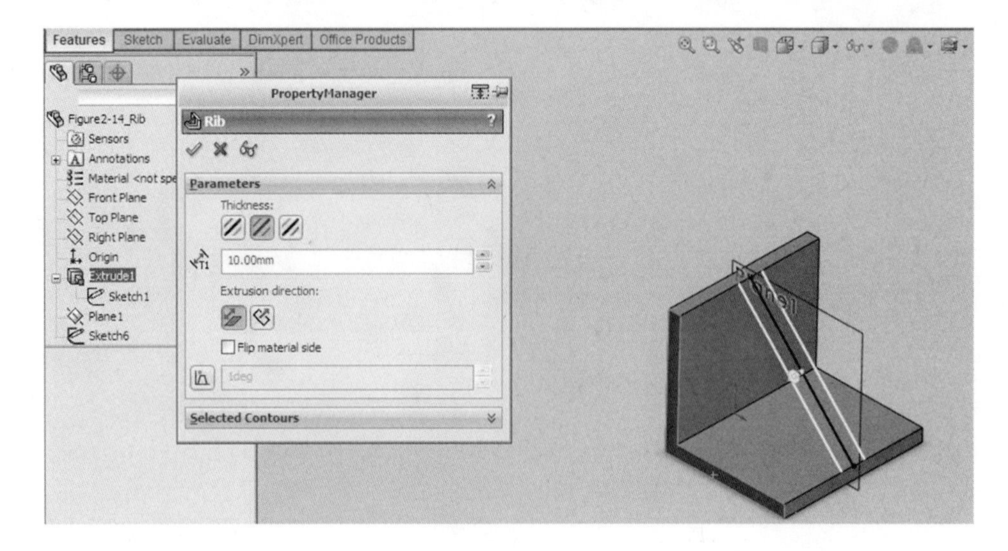

Figure 3.10

Rib Properties Manager.

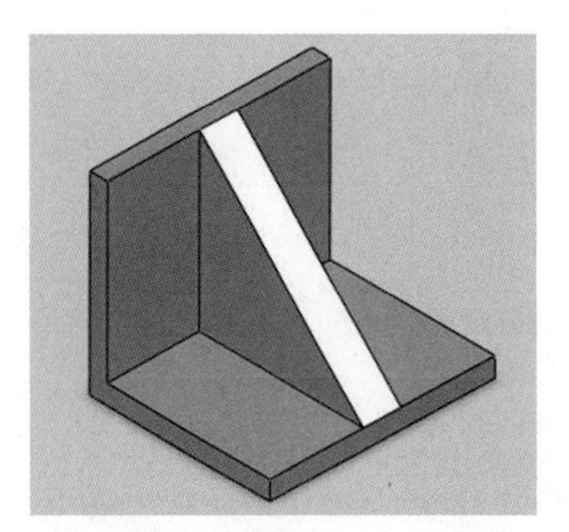

Figure 3.11

Adding rib feature to a part model.

1. Select the front plane and create Sketch2 (circle) to define the holes (see Figure 3.12).
2. Click the Features tool.
3. Click the Extruded Cut tool.
   The Extrude Properties Manager appears (see Figure 3.13).
4. Define the extrude cut distance as *All through*. A real-time preview will appear.
5. Click OK to complete the extrude cut (see Figure 3.14).

Revolved Boss/Base

The revolved boss/base tool rotates a contour about an axis. It is a useful tool when modeling parts that have circular contours. Let us illustrate the revolved boss/base concept as follows:

1. Select the front plane and create Sketch1, as shown in Figure 3.15.
2. Click the Features tool.
3. Click the Revolved Boss/Base tool.
   The Revolve Properties Manager appears (see Figure 3.16).
4. Define the revolved axis (Line1) as the vertical dimension line. A real-time preview will appear.
5. Click OK to complete the revolved part (see Figure 3.17).

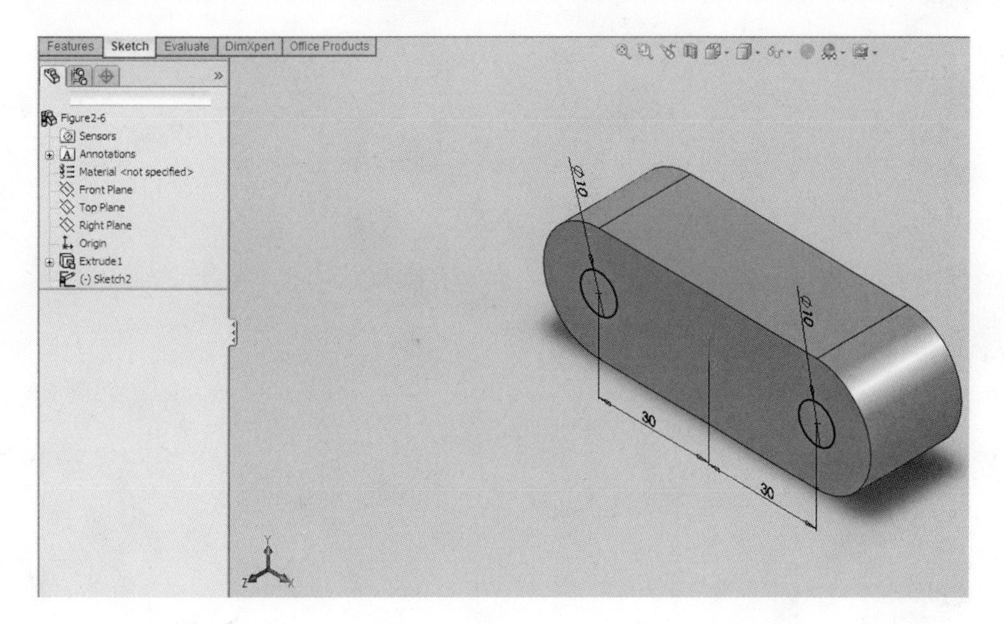

Figure 3.12

Creating circles, Sketch2 to define the holes.

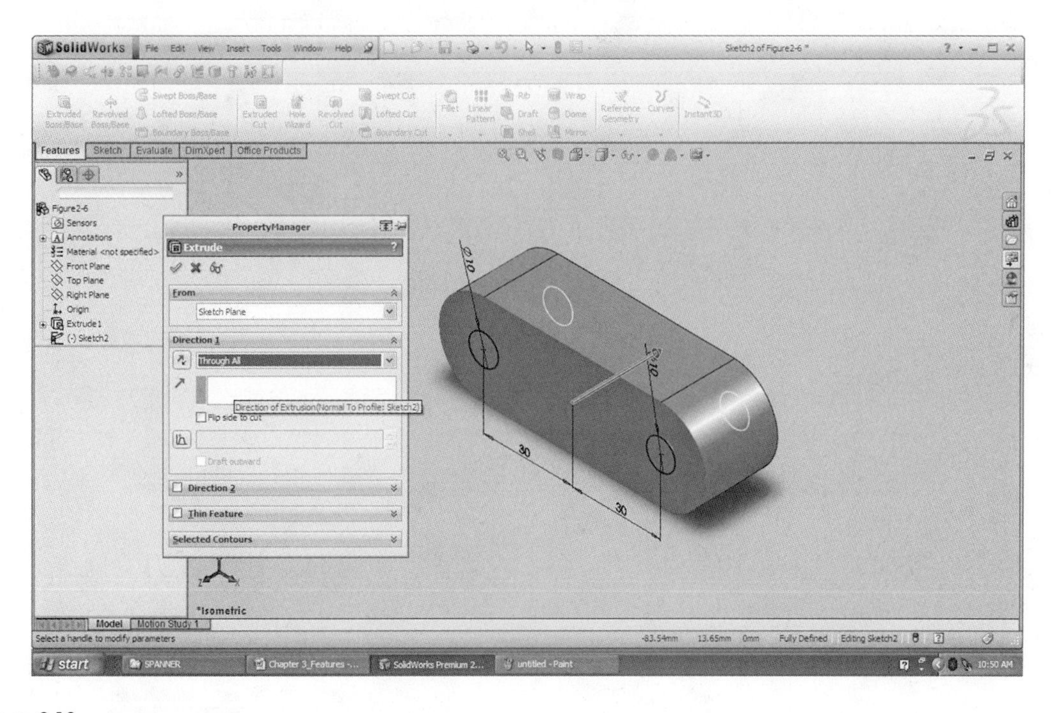

Figure 3.13

Preview for extruding Sketch2.

Creating Four Holes on the Revolved Part

The steps involved in creating four holes are as follows:

1. Sketch four holes, each having a 15-mm diameter, as shown in Figure 3.18.
2. Extrude-cut each hole, Through All, as shown in Figure 3.19. The final revolved part with holes is shown in Figure 3.20.

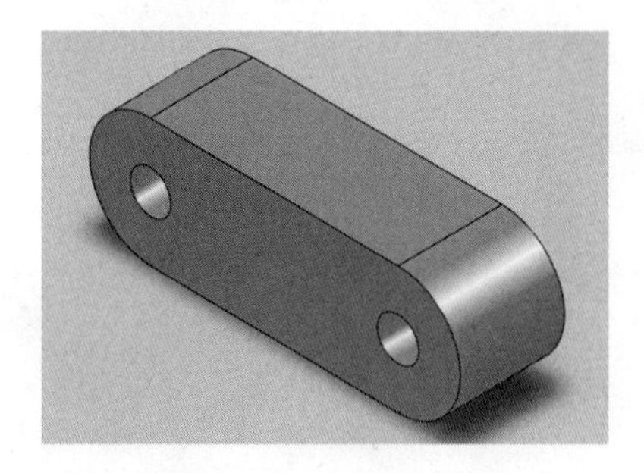

Figure 3.14

Part model obtained through extrude cut.

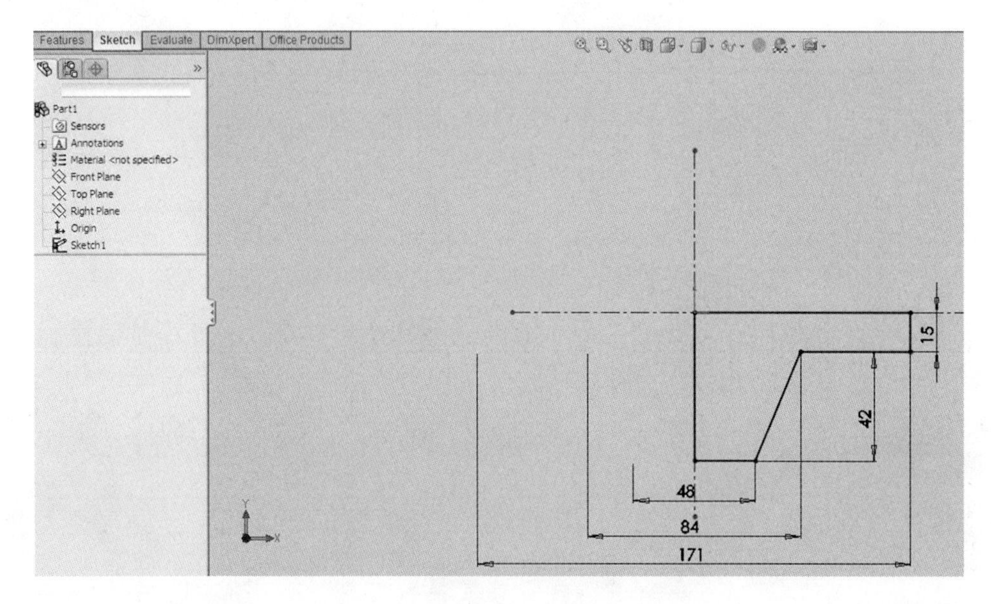

Figure 3.15

Sketch1.

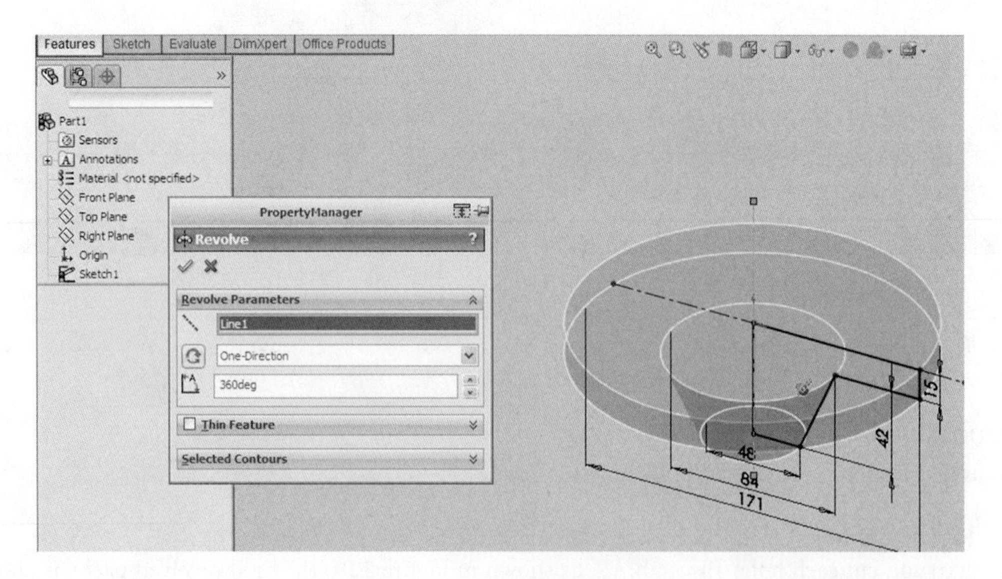

Figure 3.16

Preview.

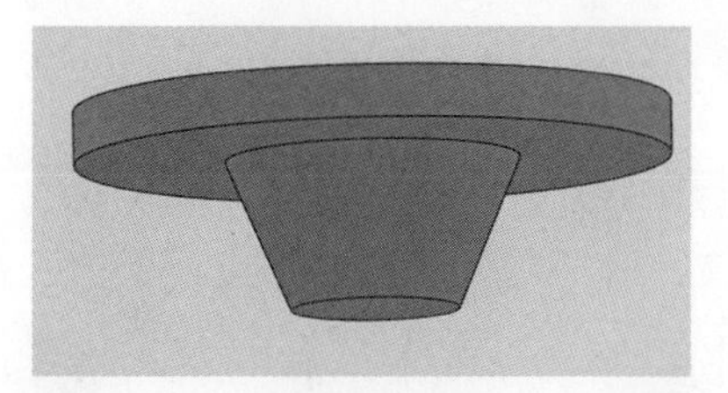

Figure 3.17

Revolved part, Extrude1.

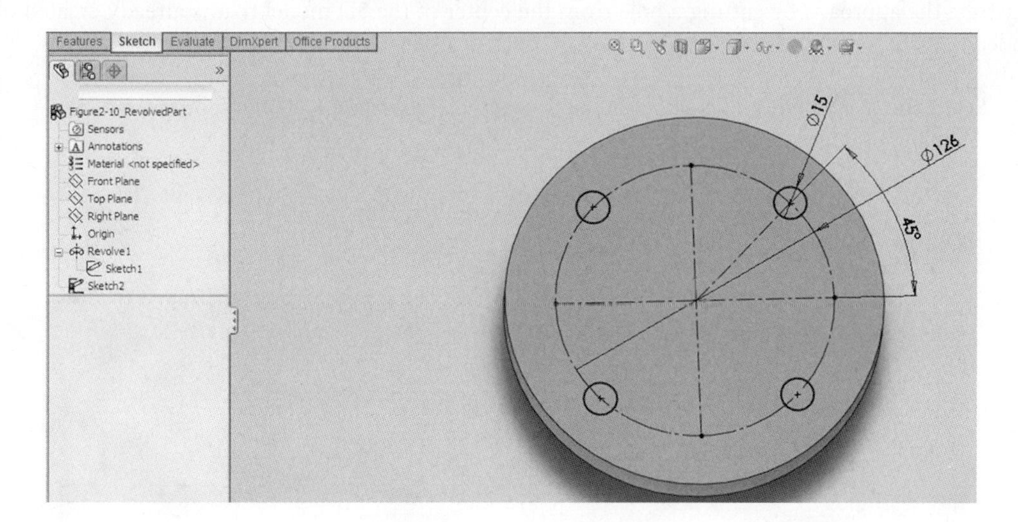

Figure 3.18

Sketches for four holes.

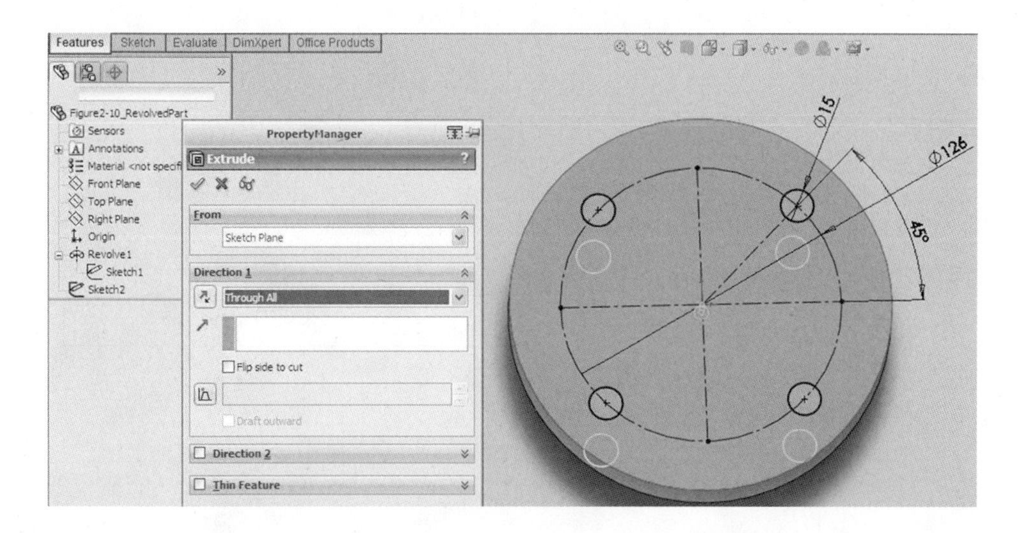

Figure 3.19

Extrude-cut for holes.

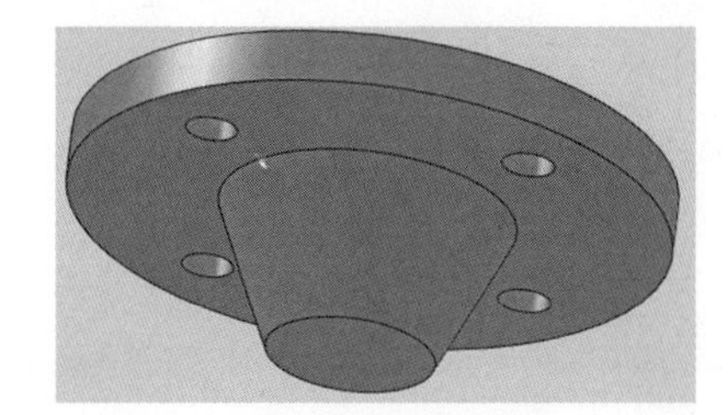

Figure 3.20

Final revolved feature.

Revolved Cut

The revolved cut tool is for removing revolved sections out of 3D objects. It is a subtraction feature. Let us illustrate the approach by cutting a hole from the center of the 3D model that is already created in the Extruded Boss/Base section. To use a revolved cut tool,

1. Select the front plane and create Sketch3 (a rectangle) to define the hole (see Figure 3.21).
2. Click the Features tool.
3. Click the Revolved Cut tool.

   The Revolve Properties Manager appears (see Figure 3.22).

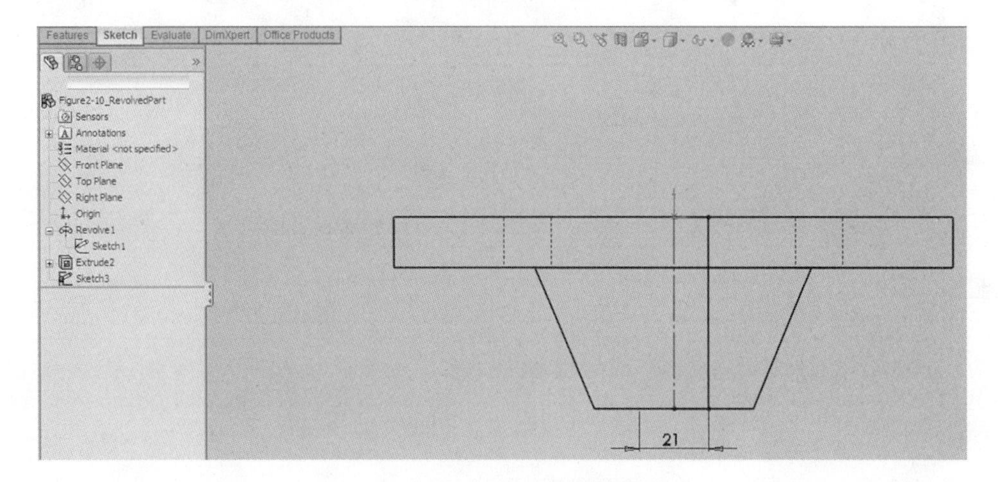

Figure 3.21

Sketch for revolve-cut.

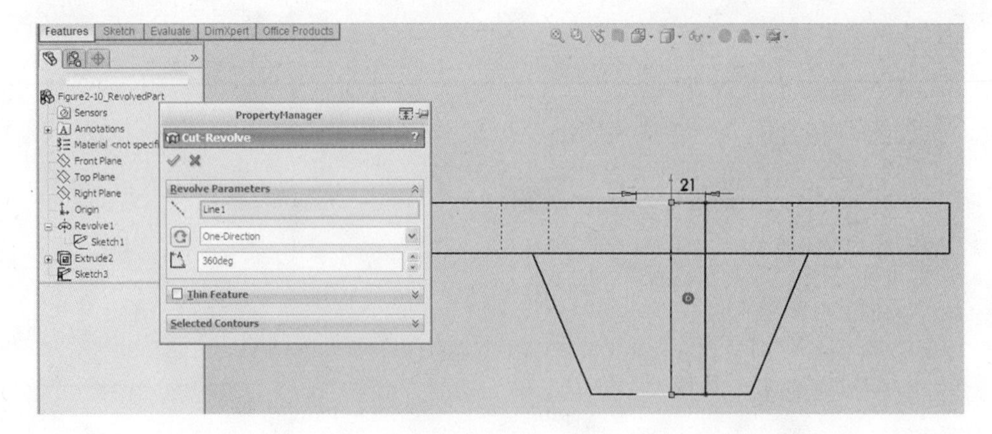

Figure 3.22

Revolve-cut.

4. Define the revolved axis (Line1) as the vertical dimension line. A real-time preview will appear.
5. Click OK to complete the revolved cut (see Figure 3.23).

Although defining a circle and cutting through will still do the same work, the exercise has shown the steps that are required in a creating a revolved cut feature, which may be useful for more complicated modeling.

Lofted Boss/Base

The lofted boss/base tool is used to create a shape between a number of planes, each of which contains a defined shape or geometry. The prerequisite for creating a 3D lofted model is to first sketch the shapes on the different planes. Let us illustrate a lofted boss/base model using a square base, a circle, and an ellipse. There are no restrictions in the shapes that could be used. Let us illustrate the lofted boss/base concept as follows:

1. Select the top plane and create Sketch1 (a square, 90 mm × 90 mm), as shown in Figure 3.24.
2. Exit the sketch mode.
3. Click the Features tool and click the Reference tool. Select the Plane option. The plane box appears; select the top plane.
4. Set the distance between the existing top plane and a new reference plane for 50 mm. Click the OK check mark. A new plane, Plane1, appears, as shown in Figure 3.25.

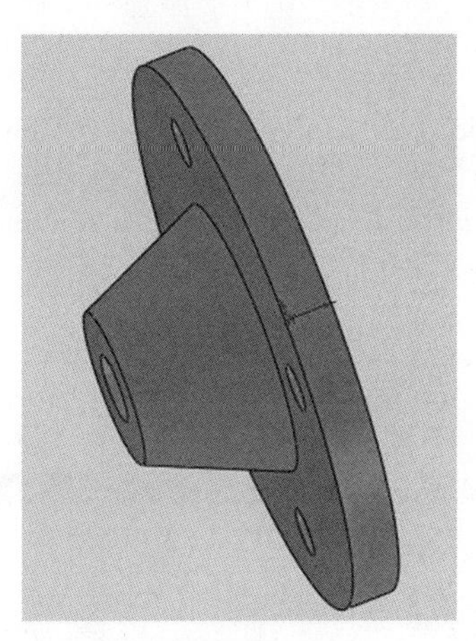

Figure 3.23

Revolved-cut 3D part.

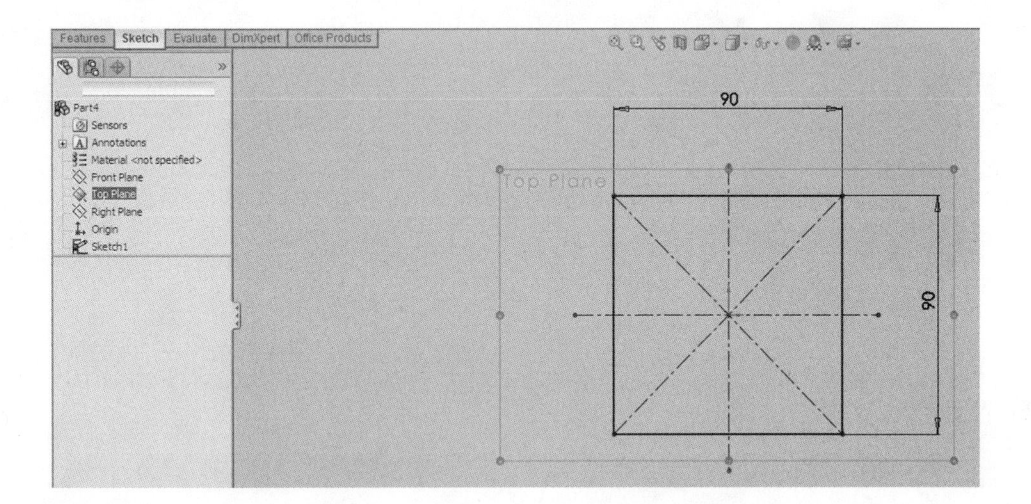

Figure 3.24

Bottom sketch for lofting (Sketch1).

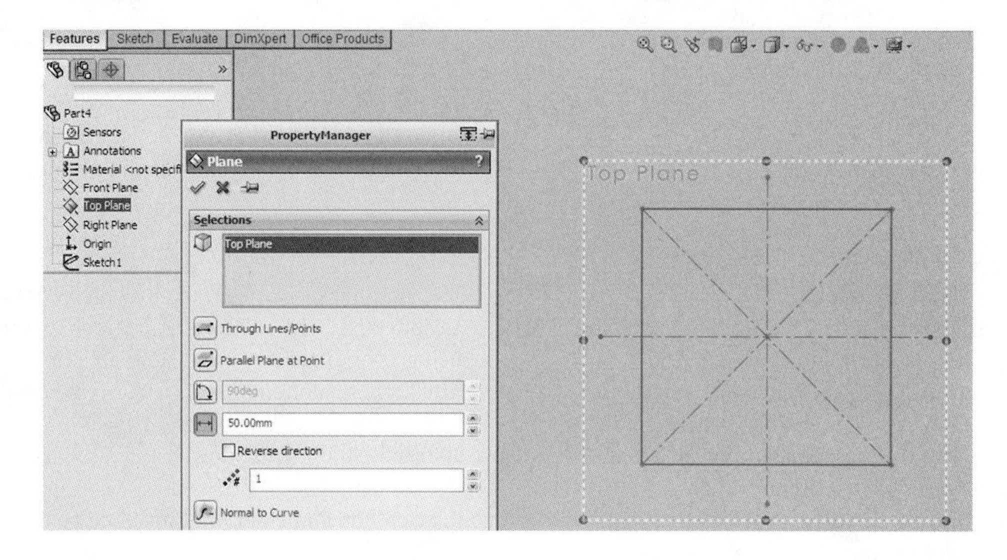

Figure 3.25

Middle sketch for lofting (Sketch2).

5. Create Sketch2 (a circle of diameter, 60 mm) as shown in Figure 3.25.
6. Exit the sketch mode.
7. Click the Features tool and click the Reference tool. Select the Plane option. The plane box appears; select the top plane.
8. Set the distance between Plane1 and a new reference plane for 50 mm. Click the OK check mark. A new plane, Plane2, appears as shown in Figure 3.26.
9. Create Sketch3 (an ellipse of major diameter, 80 mm, and minor diameter, 65 mm) as shown in Figure 3.26.

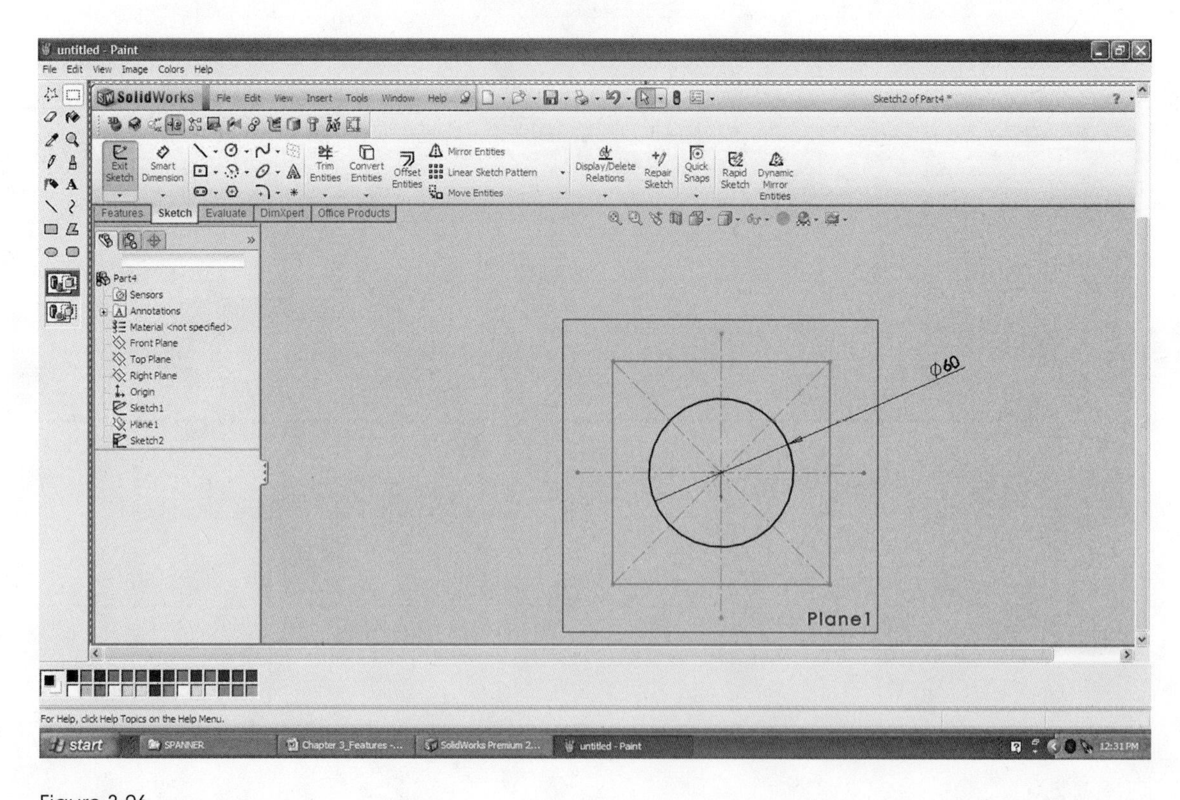

Figure 3.26

Top sketch for lofting (Sketch3).

10. Exit the sketch mode. All sketches appear as shown in Figure 3.27.
11. Click the Lofted Boss/Base tool.
    The Loft Properties Manager appears.
12. Right-click the Profiles box.
13. Click the square, then the circle, and then the ellipse. A real-time preview will appear, as shown in Figure 3.28.
14. Click OK to complete the lofted part. Hide the planes as shown in Figure 3.29.

Lofted Cut

The lofted cut tool (see Figure 3.30) is used to cut a shape between a number of planes, each of which contains a defined shape or geometry. The prerequisite for creating a 3D lofted model is to first sketch the shapes on the different planes. Let us illustrate a lofted cut model using a square base, a circle, and an ellipse, similar to the previous model that is obtained from lofted boss/base but offset into the model.

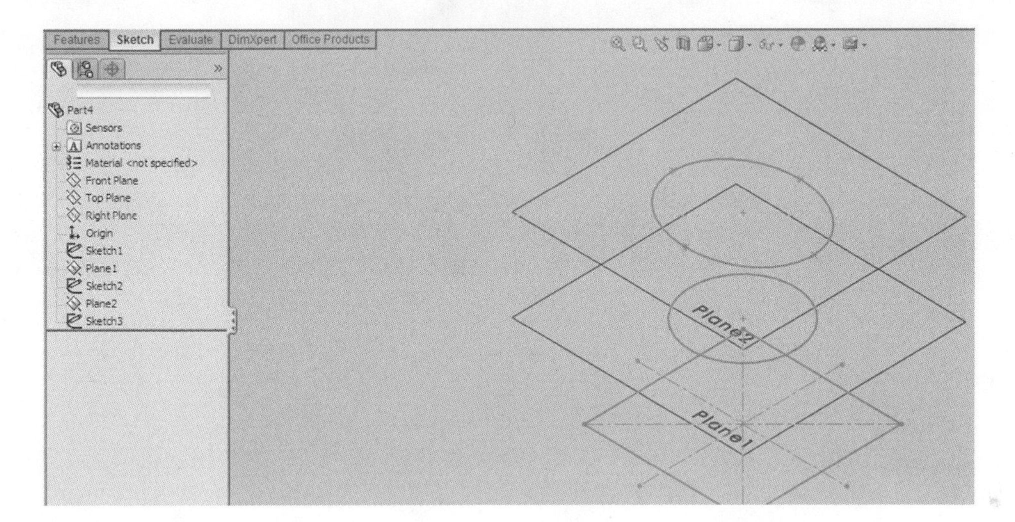

Figure 3.27

Three sketches for lofting.

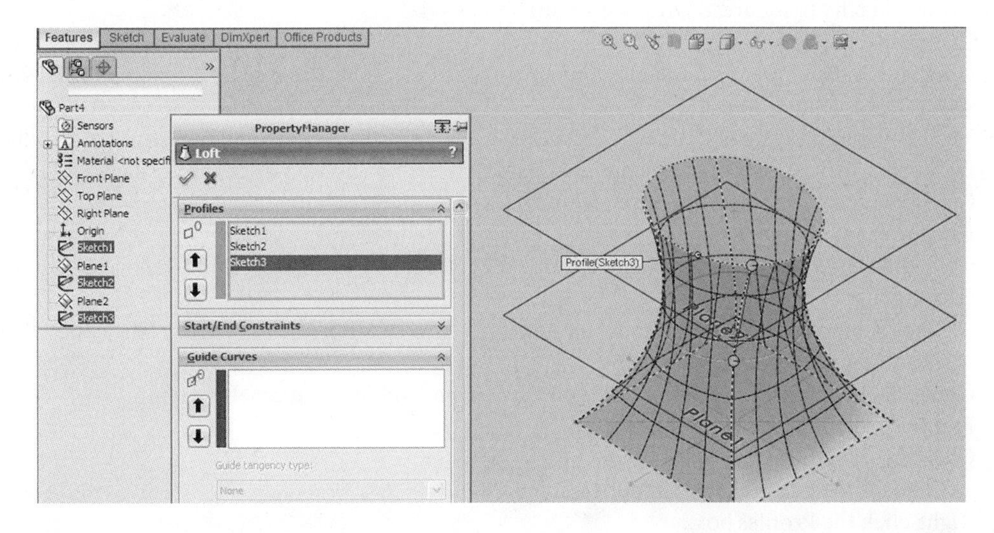

Figure 3.28

Loft based on Sketch1, Sketch2, and Sketch3.

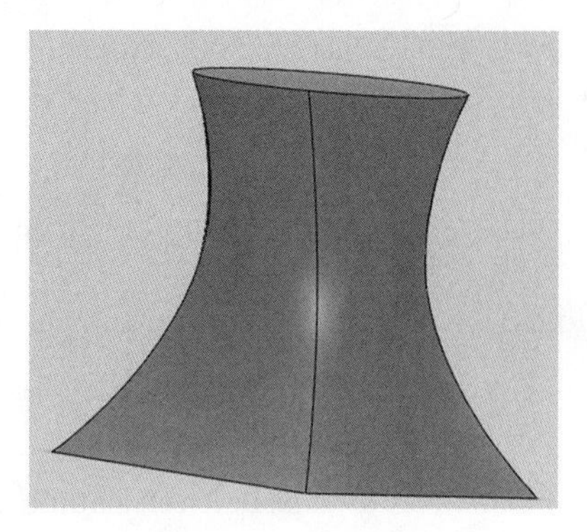

Figure 3.29

3D lofted model.

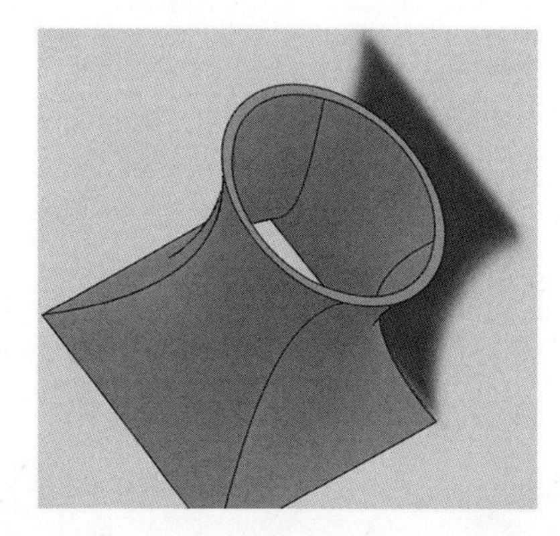

Figure 3.30

Lofted-cut features.

1. Select the top plane. Click the Features tool and click the Reference tool. Select the Plane option. The plane box appears; select the top plane. Set the distance between the existing top plane and a new reference plane for 5 mm further away from the 3D model. Click the OK check mark. A new plane, Plane3, appears.
2. Create Sketch4 (a square, 83 mm × 83 mm).
3. Exit the sketch mode.
4. Click the Features tool and click the Reference tool. Select the Plane option. The plane box appears; select the top plane.
5. Set the distance between the existing Plane1 and a new reference plane for 5 mm below. Click the OK check mark. A new plane, Plane4, appears.
6. Create Sketch5 (a circle of diameter, 57 mm).
7. Exit the sketch mode.
8. Click the Features tool and click the Reference tool. Select the Plane option. The plane box appears; select the top plane.
9. Set the distance between Plane2 and a new reference plane for 5 mm above. Click the OK check mark. A new plane, Plane5, appears.
10. Create Sketch3 (an ellipse of major diameter, 67 mm, and minor diameter, 60 mm).
11. Exit the sketch mode.
12. Click the Lofted Cut tool.

    The Loft Properties Manager appears.
13. Right-click the Profiles box.
14. Click the new square, then the new circle, and then the new ellipse. A real-time preview will appear.
15. Click OK to complete the lofted-cut part, which is now hollow. Hide the planes.

Swept Boss/Base

The swept boss/base tool is used to sweep a profile through a path (arc, spline, etc.). As with lofting, there have to be existing shapes. The prerequisite for creating a 3D swept model is to first sketch the shapes on the different perpendicular planes. In the illustration presented here, a hexagon is the *profile*, while a spline is the *path*, as shown in Figure 3.31.

1. Select the top plane. Create Sketch1 (a hexagon with sides, 0.5 in.).
2. Exit the sketch mode.
3. Select the front plane. Create Sketch2 (a spline with one end coinciding with the origin of the hexagon).
4. Exit the sketch mode.
5. Click the Swept Cut tool.

   The Swept Properties Manager appears (see Figure 3.32).

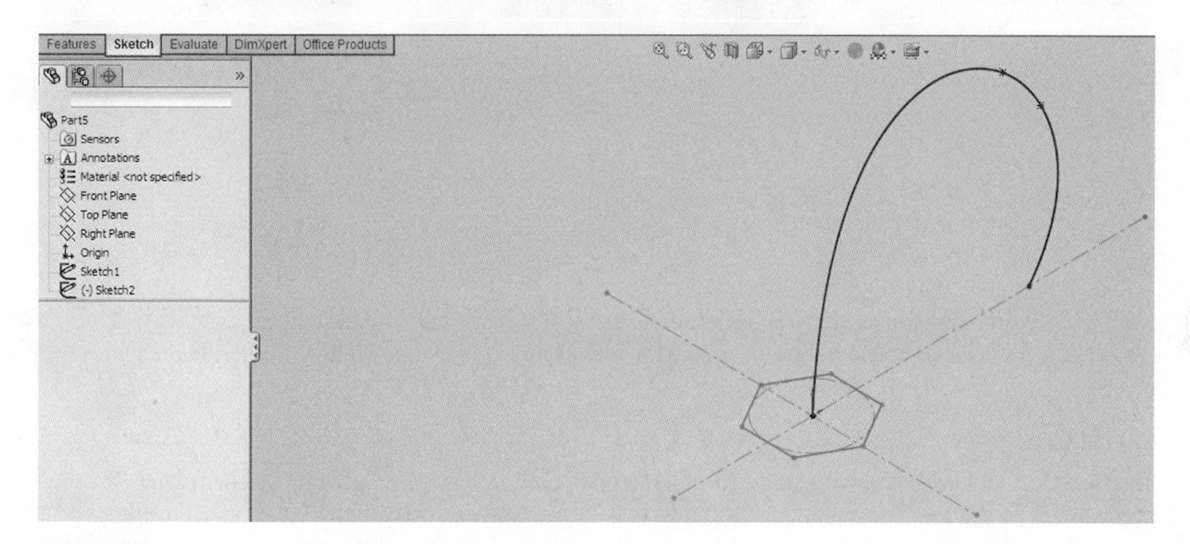

Figure 3.31

Profile (hexagon) and path (spline).

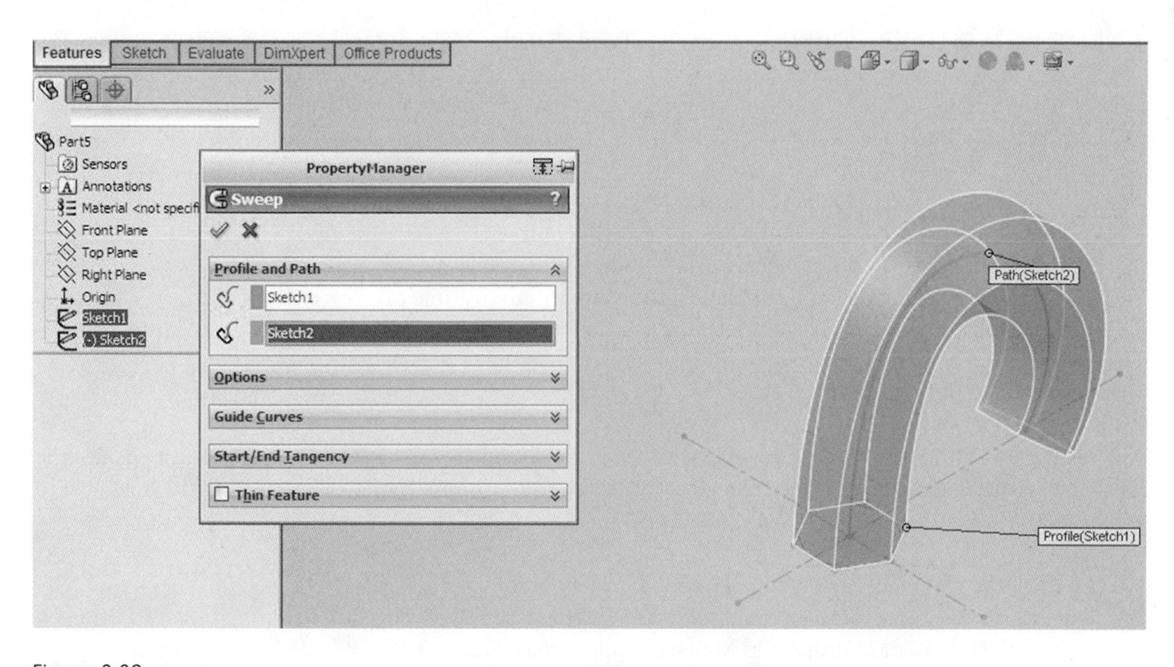

Figure 3.32

Preview.

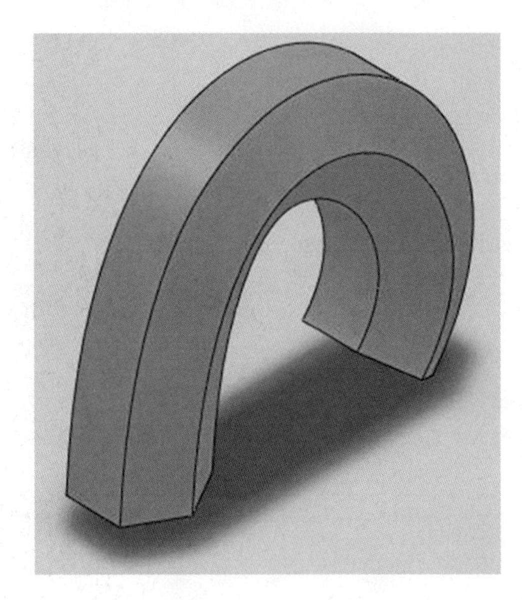

Figure 3.33

3D swept model.

6. Right-click the Profile and Path box.
7. Click the hexagon as profile and the spline as path. A real-time preview will appear.
8. Click OK to complete the swept cut part, which is now hollow (see Figure 3.33). Hide the planes.

Swept Cut

The swept cut tool is used to cut a shape (profile) in one plane along a given path in another plane. The prerequisite for creating a 3D swept model is to first sketch the shapes on the different planes. Let us illustrate a swept-cut model using a hexagon as the profile, while a spline is the path, similar to the previous model that is obtained from swept boss/base (see Figure 3.33). In this case, the new hexagon is offset to be of a smaller size than the one that is previously used.

1. Select the top plane. Click the Features tool and click the Reference tool. Select the Plane option. The plane box appears; select the top plane. Set the distance to be zero so that the new plane is collinear with the existing top plane. Click the OK check mark. A new plane, Plane3, appears.
2. Create Sketch2 (Sketch5) (a hexagon with sides offset inward by 0.05 in.) (see Figure 3.34).
3. Exit the sketch mode.
4. Click the Swept Cut tool.

   The Swept Properties Manager appears (see Figure 3.35).
5. Right-click the Profile and Path box.
6. Click the hexagon as profile and the spline as path. A real-time preview will appear.
7. Click OK to complete the swept cut part, which is now hollow. Hide the planes (see Figure 3.36).

Hole Wizard

The hole wizard tool is used to add hole(s) to an existing 3D model, as shown in Figure 3.37. Let us use the revolved boss/base model as an example; we want to add four 10-mm diameter holes using the hole wizard tool.

1. Click the front face having four 15-mm diameter holes.
2. Click the Features tool and click the Hole Wizard tool.

   The Hole Wizard Properties Manager appears (see Figure 3.38).

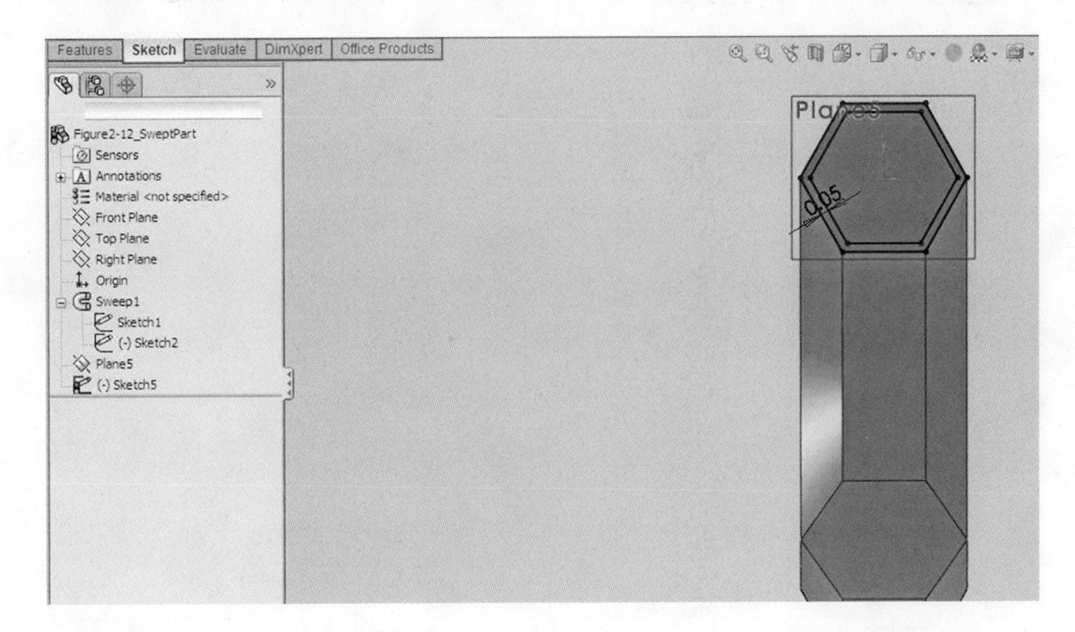

Figure 3.34

Sketch for cutting.

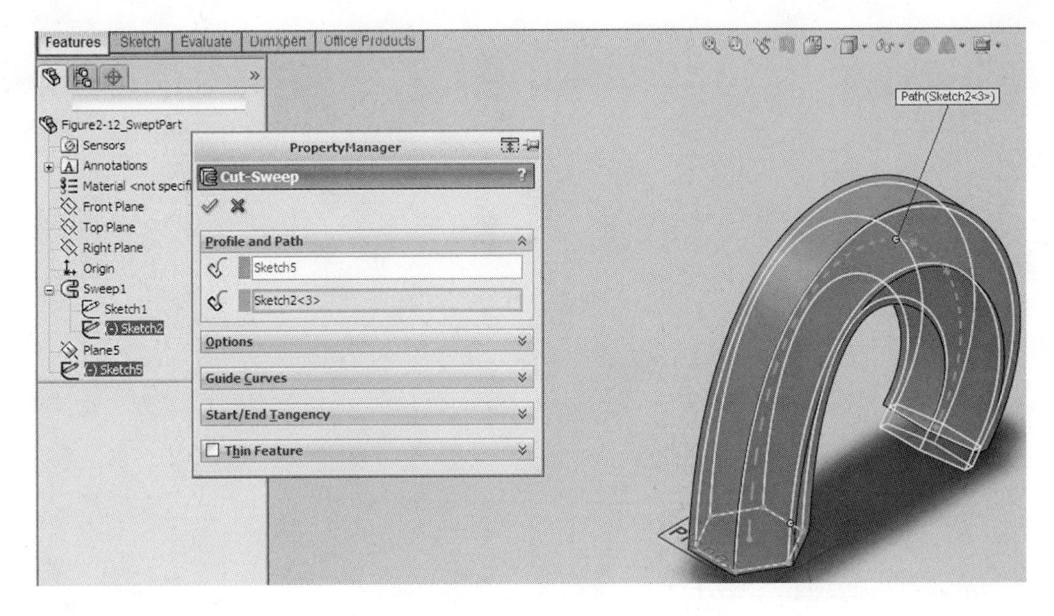

Figure 3.35

Preview.

3. Click the Type button for the hole required. A clear hole was selected for our illustration.
4. Select the ANSI Metric for the Standard unit and 10 mm for the Size.
5. Click the Position button.
6. Click an approximate center point location for the hole.
7. Click the OK check mark. A dialog box will appear (see Figure 3.39). Use the Smart Dimension tool to locate the center of the hole (see Figure 3.40).
8. Click OK in the dialog box.
9. The hole will be added to the model (see Figure 3.41).

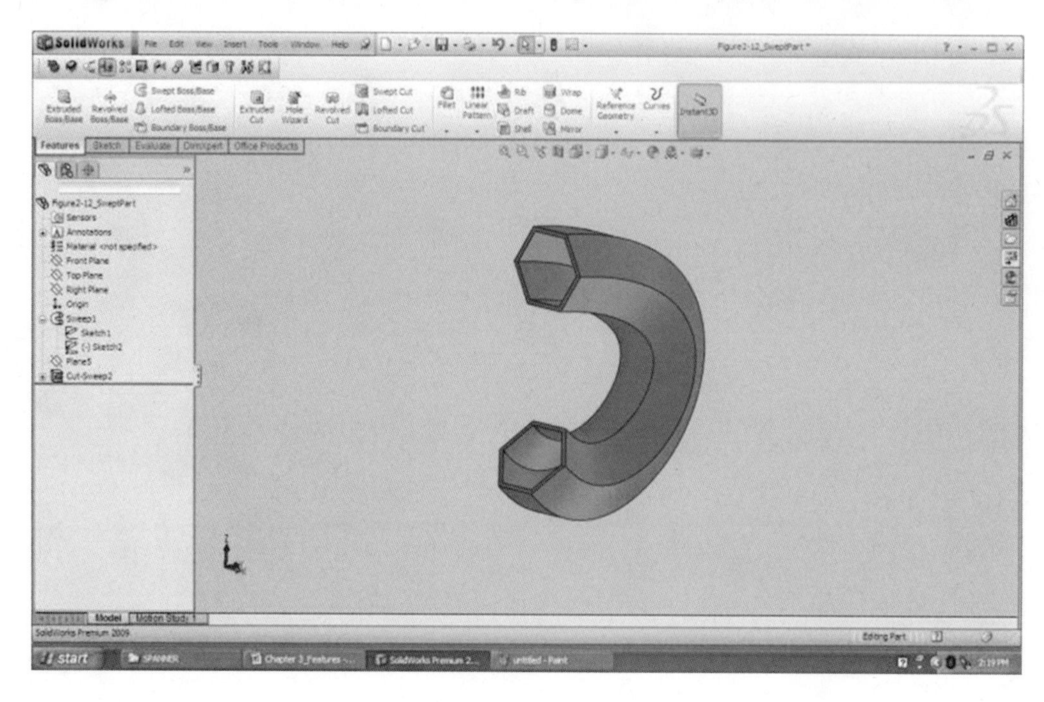

Figure 3.36

3D swept-cut model.

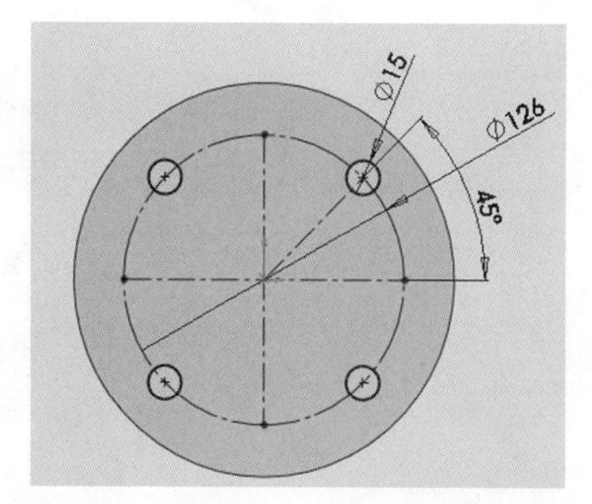

Figure 3.37

An existing model to which hole wizard is to be used to add extra holes.

Shell

The shell tool is used to create a hollow shape in an existing 3D model, transforming it into a hollow object having uniform thickness. Let us use the lofted model for illustrating shell by applying the shell tool on the square base for a value of 5-mm thickness.

1. Select the square base (see Figure 3.42).
2. Click the Shell tool.
   The Shell Properties Manager appears (see Figure 3.43).
3. Apply 2.5-mm thickness.
4. Click OK to complete the shell part, which is now hollow, as shown in Figure 3.44.

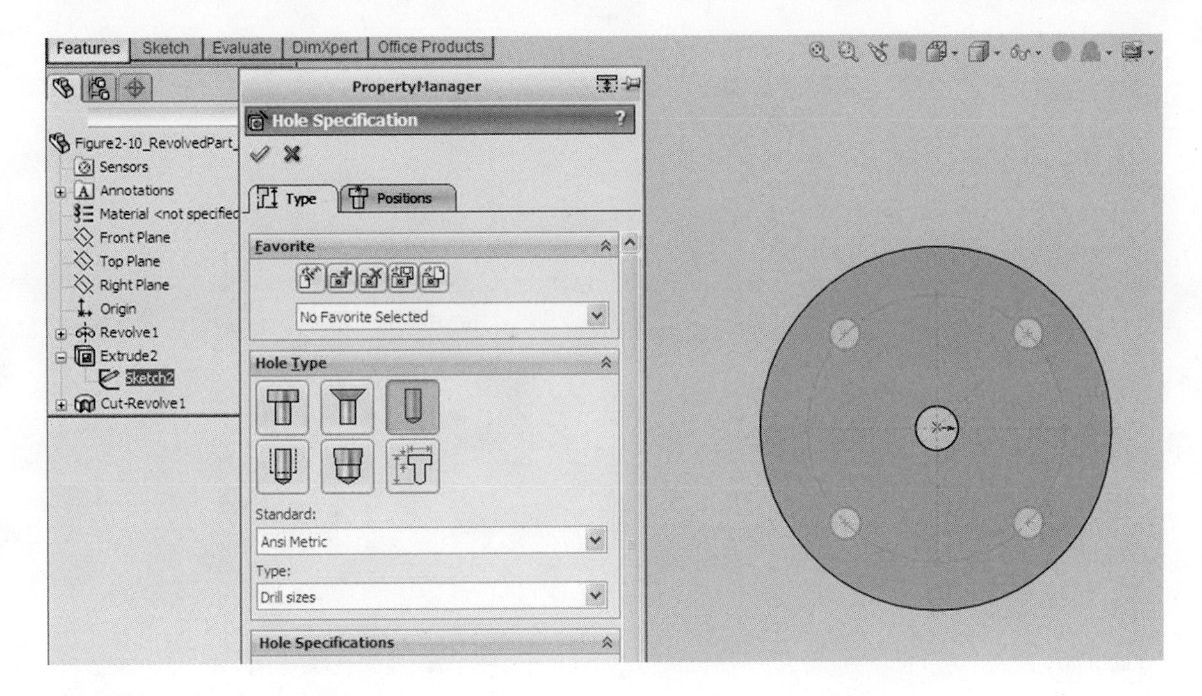

Figure 3.38

Hole Wizard used for additional central hole.

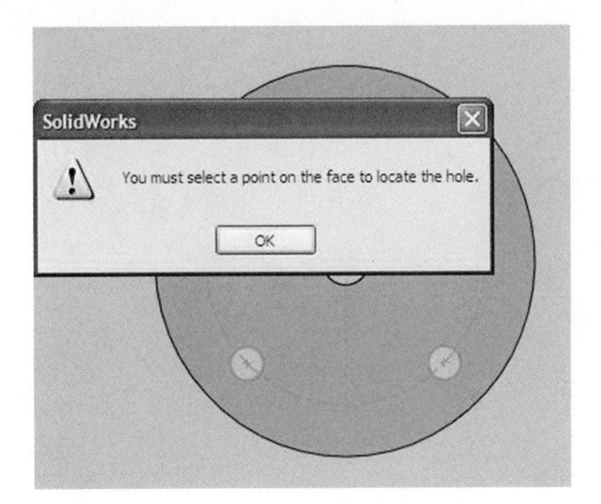

Figure 3.39

Hole Wizard Message.

Fillet Tool

A fillet is a rounded corner. There are usually four ways of defining a fillet: (1) by constant radius, (2) by variable radius, (3) by face, or (4) by full round, as shown in Figures 3.45 through 3.48.

Defining a Constant Radius Fillet

Defining a fillet using a constant radius is illustrated in Figure 3.45.

1. Click the Fillet tool.
   The Fillet Properties Manager appears.

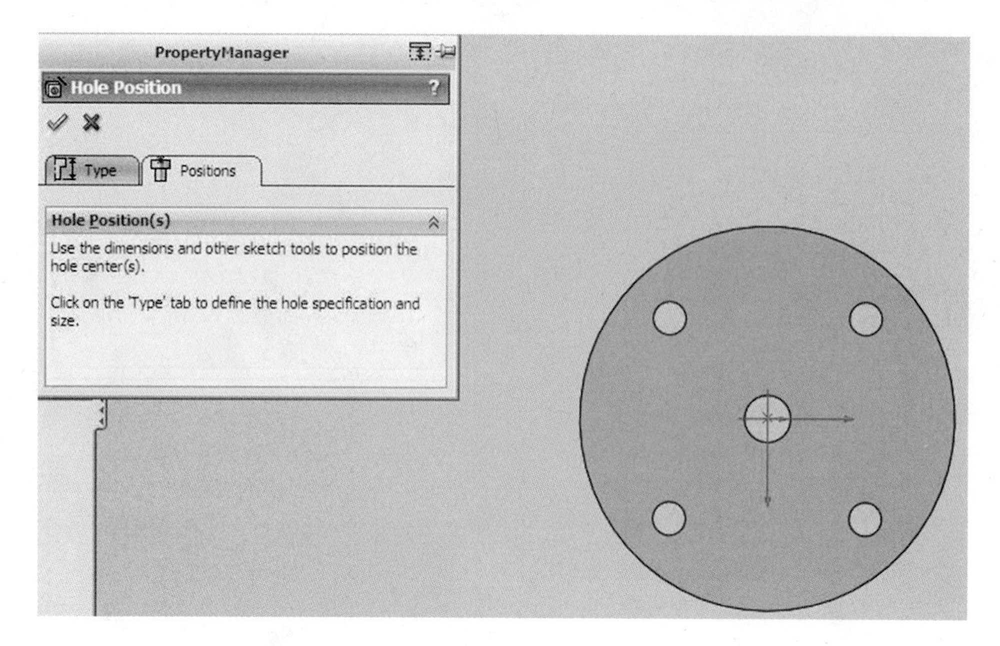

Figure 3.40

Position for central hole.

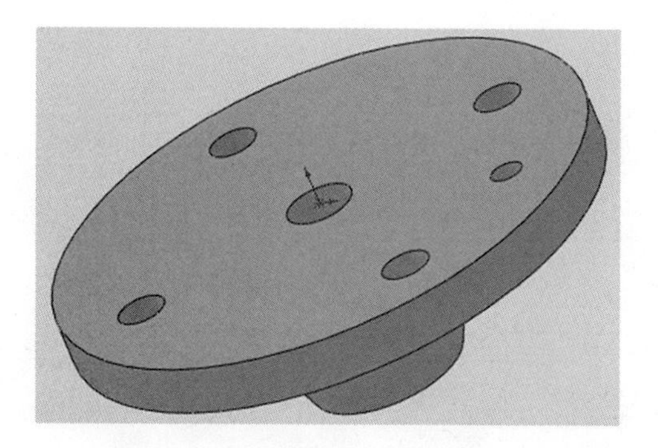

Figure 3.41

Hole created using Hole Wizard tool.

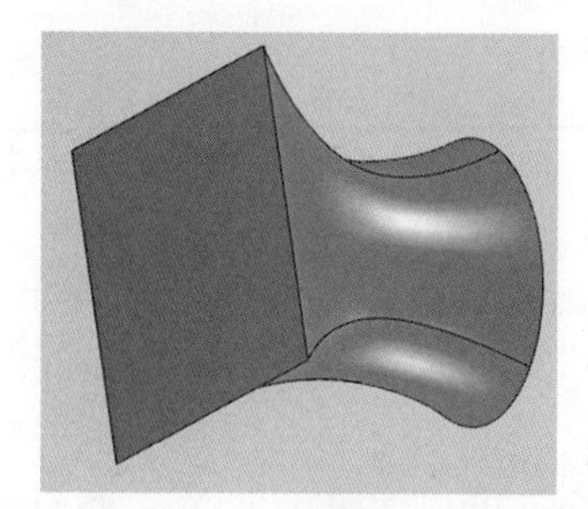

Figure 3.42

Bottom is chosen for shelling.

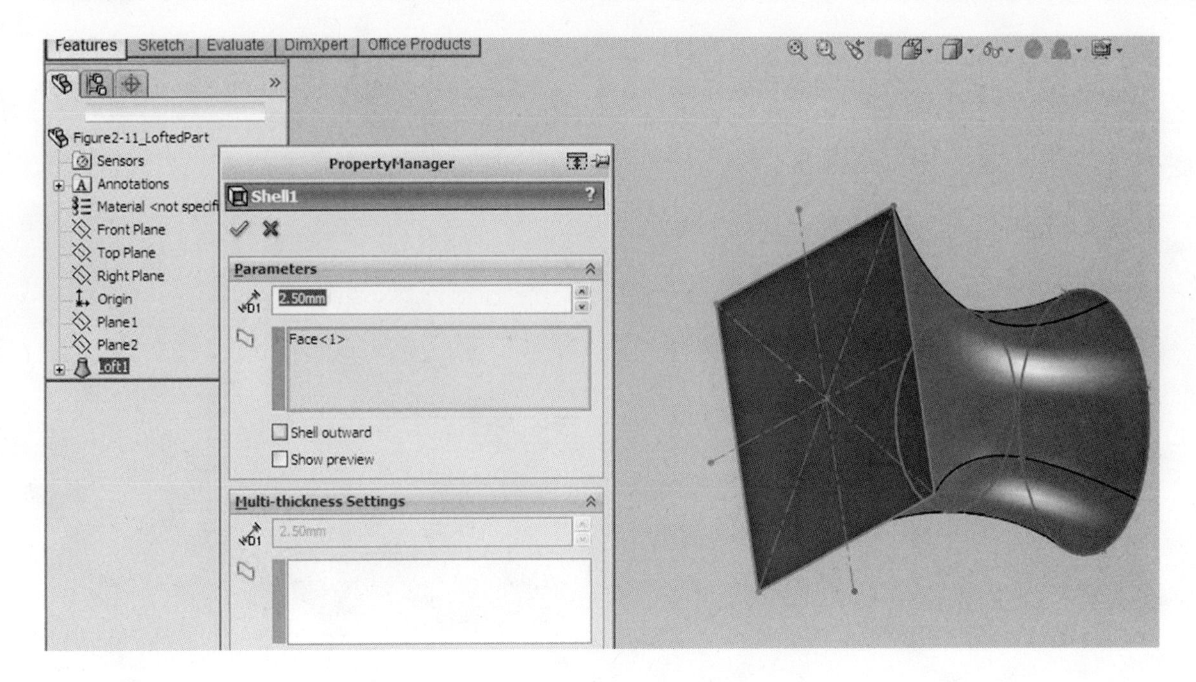

Figure 3.43

Shell thickness defined.

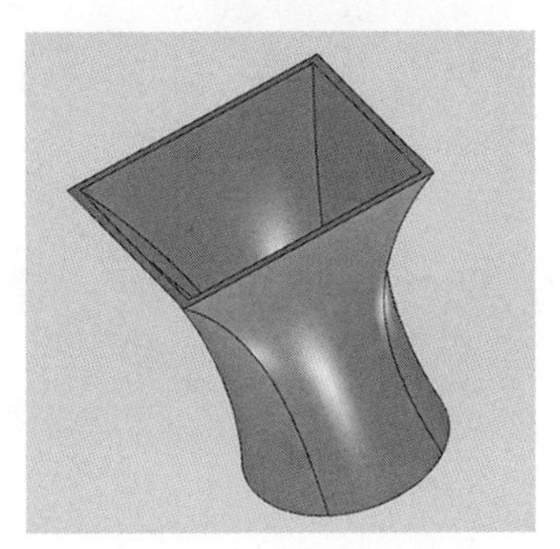

Figure 3.44

Shell tool for creating swept-cut features.

2. Click the Constant Radius option.
3. Apply 5-mm radius. A real-time preview will appear.
4. Click OK to complete the filleted part.

Defining a Variable Radius Fillet

Defining a fillet using a variable radius is illustrated in Figure 3.46.

1. Click the Fillet tool.
   The Fillet Properties Manager appears.
2. Click the Variable Radius option.
3. Click the edge to apply the fillet.
   Two boxes appear on the screen, one at each end of the edge to accept the values.

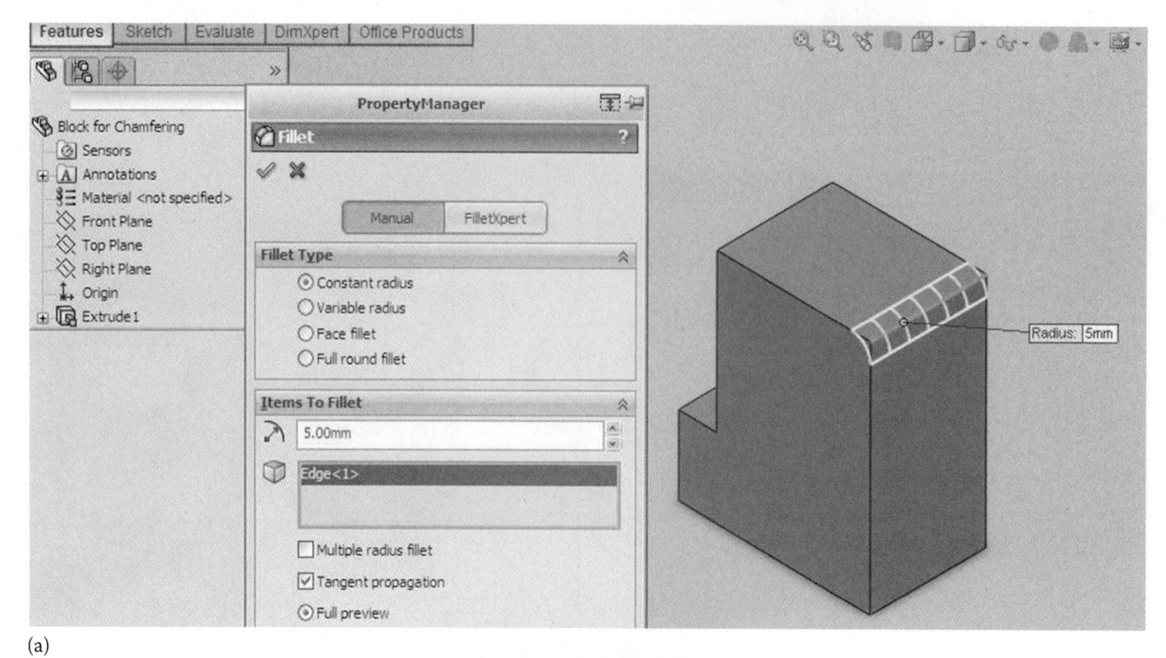

(a)

(b)

Figure 3.45

Fillet tool for creating constant radius fillets: (a) preview and (b) fillet part.

4. Apply 2.5 mm to the word Unassigned in the first box.
5. Apply 5 mm to the word Unassigned in the second box. A real-time preview will appear.
6. Click OK to complete the filleted part.

Defining a Face Fillet

Defining a fillet using the face is illustrated in Figure 3.46.

1. Click the Fillet tool.
   The Fillet Properties Manager appears.
2. Click the Face fillet option.
   Two boxes appear in Items to Fillet box for defining the two faces of the fillet.
3. Apply 5-mm radius.
4. Define Face1 by clicking on a face as shown.
5. Click the second box in the Items to Fillet box and define Face2 by clicking on the face as shown. A real-time preview will appear.
6. Click OK to complete the filleted part.

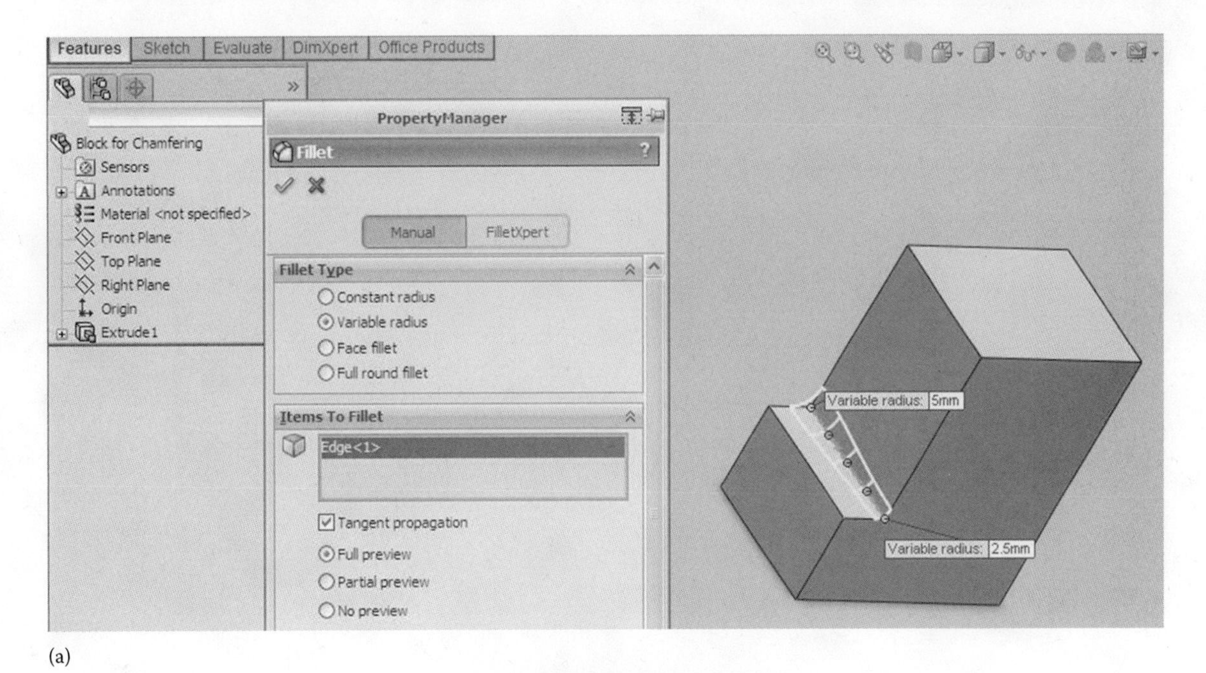

(a)

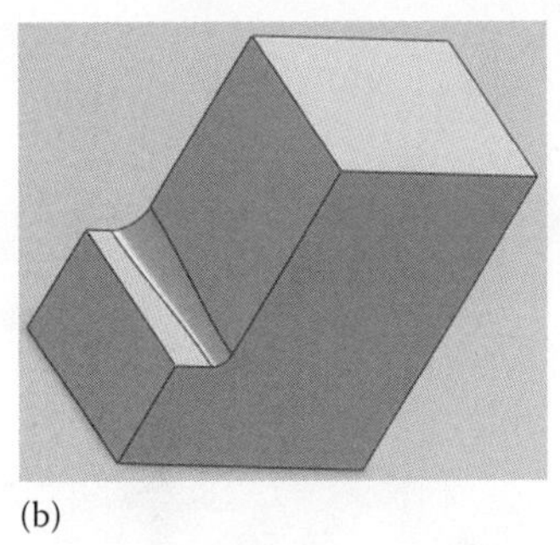

(b)

Figure 3.46

Fillet tool for creating variable radius fillets: (a) preview and (b) fillet part.

Defining a Full Round Fillet

Defining a fillet using the face is illustrated in Figure 3.48.

1. Click the Full round fillet tool.
   The Fillet Properties Manager appears.
2. Click the Full round fillet option.
   Three boxes appear in Items to Fillet box for defining the three faces of the fillet.
3. Define Face1 by clicking on a face as shown.
4. Click the second box in the Items to Fillet box and define Face2 by clicking on the face as shown.
5. Click the third box in the Items to Fillet box and define Face3 by clicking on the face as shown. A real-time preview will appear.
6. Click OK to complete the filleted part.

Chamfer Tool

A chamfer is a slanted surface that is added to the corner of a part. Chamfers are usually manufactured at 45°, but any other angle may be used. There are usually three ways of defining a chamfer: (1) an angle and a distance (2.5 × 45°), (2) by two distances (2.5 × 2.5), or (3) by a vertex, as shown in Figures 3.49 through 3.51.

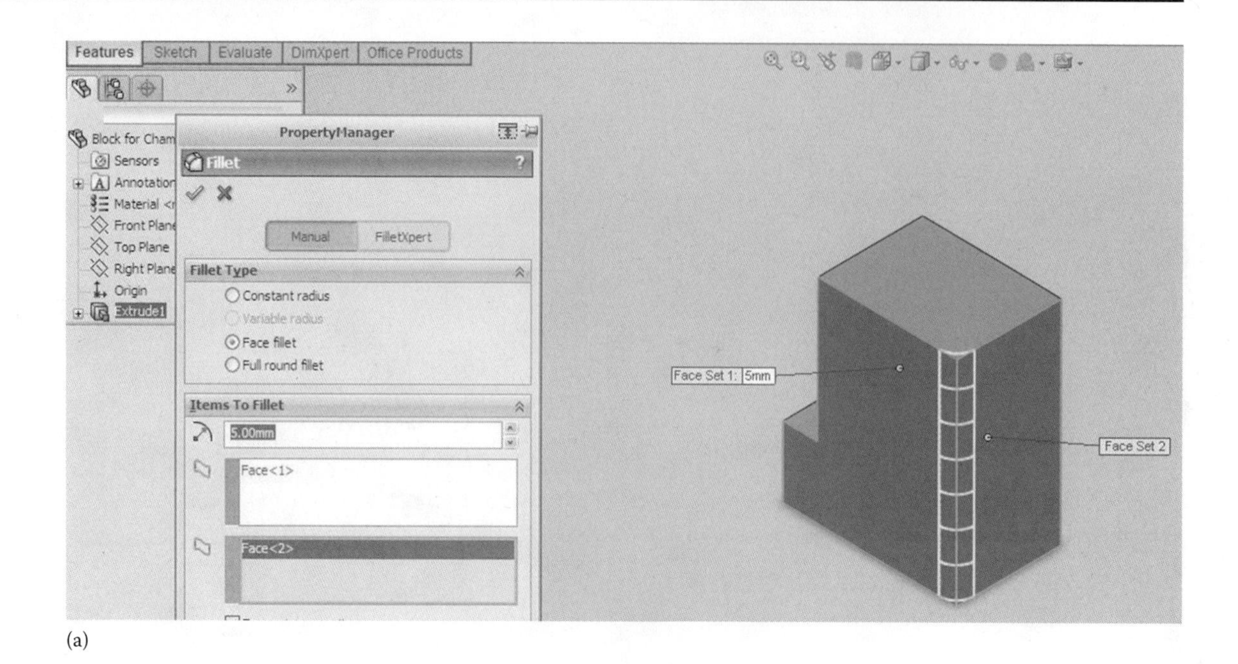

(a)

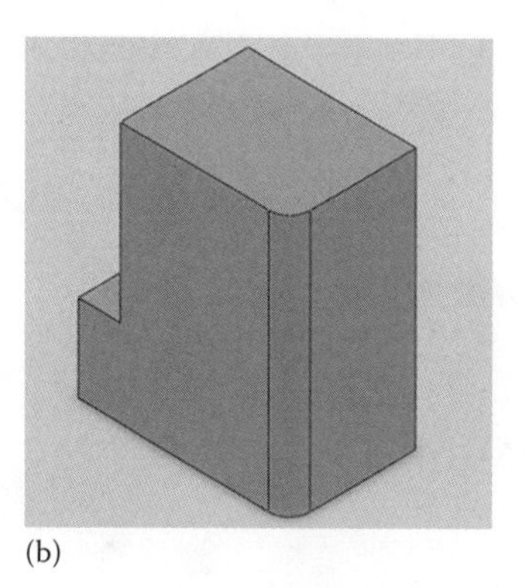
(b)

Figure 3.47

Fillet tool for creating variable radius fillets: (a) preview and (b) fillet part.

Defining a Chamfer Using an Angle and a Distance

Defining a chamfer using an angle and distance is illustrated in Figure 3.49.

1. Click the Chamfer tool.
   The Chamfer Properties Manager appears.
2. Click the Angle-distance tool.
3. Apply 2.5-mm distance and accept the default angle, 45°. A real-time preview will appear.
4. Click OK to complete the chamfered part.

Defining a Chamfer Using Two Distances

Defining a chamfer using two distances is illustrated in Figure 3.50.

1. Click the Chamfer tool.
   The Chamfer Properties Manager appears.
2. Click the Distance-distance tool.
3. Apply 2.5-mm distances each. A real-time preview will appear.
4. Click OK to complete the chamfered part.

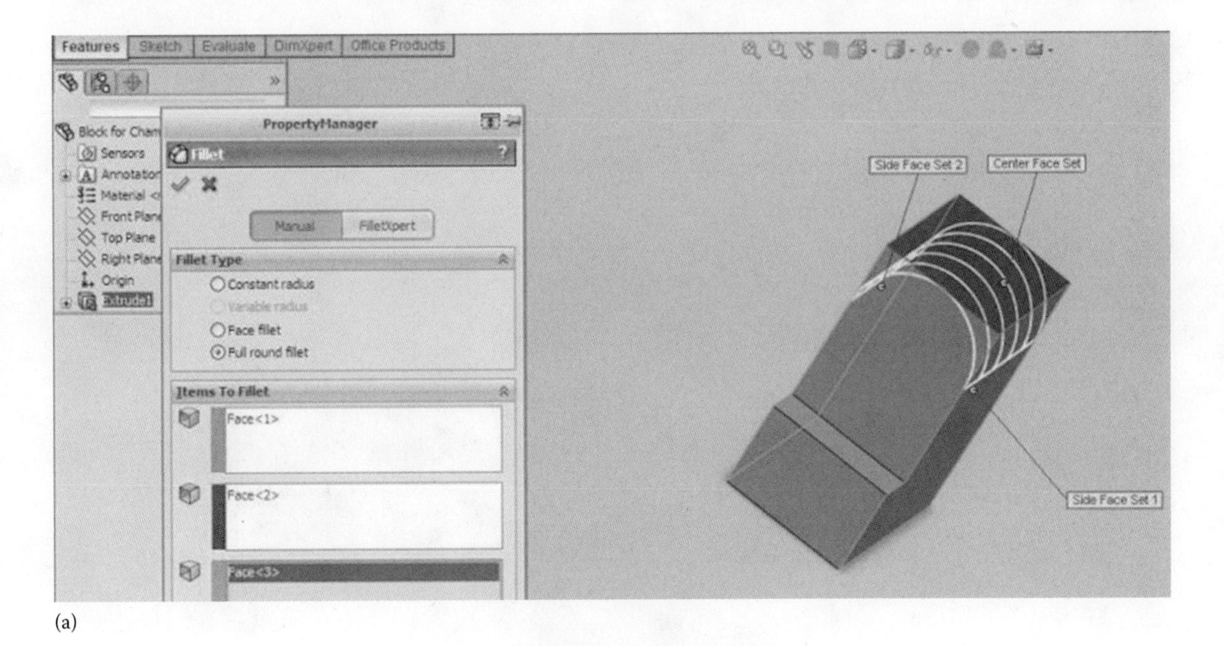

(a)

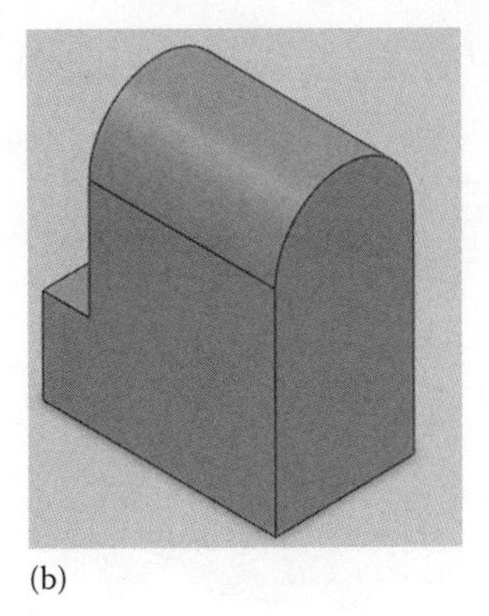

(b)

Figure 3.48

Fillet tool for creating full round fillets: (a) preview and (b) fillet part.

Defining a Chamfer Using a Vertex

Defining a chamfer using a vertex is illustrated in Figure 3.51.

1. Click the Chamfer tool.
   The Chamfer Properties Manager appears.
2. Click the Vertex tool. Three distance boxes will appear.
3. Define the three distances. Although 2.5 mm is used, any other value could be used.
4. A real-time preview will appear.
5. Click OK to complete the chamfered part.

Linear Pattern

The linear pattern tool is used to create rectangular patterns based on a given model. Let us consider an example.

1. Sketch a rectangle, 100 mm × 50 mm and extrude it 5 mm.
2. Click the bottom and sketch one small rectangle, 5 mm × 5 mm; extrude it 50 mm.

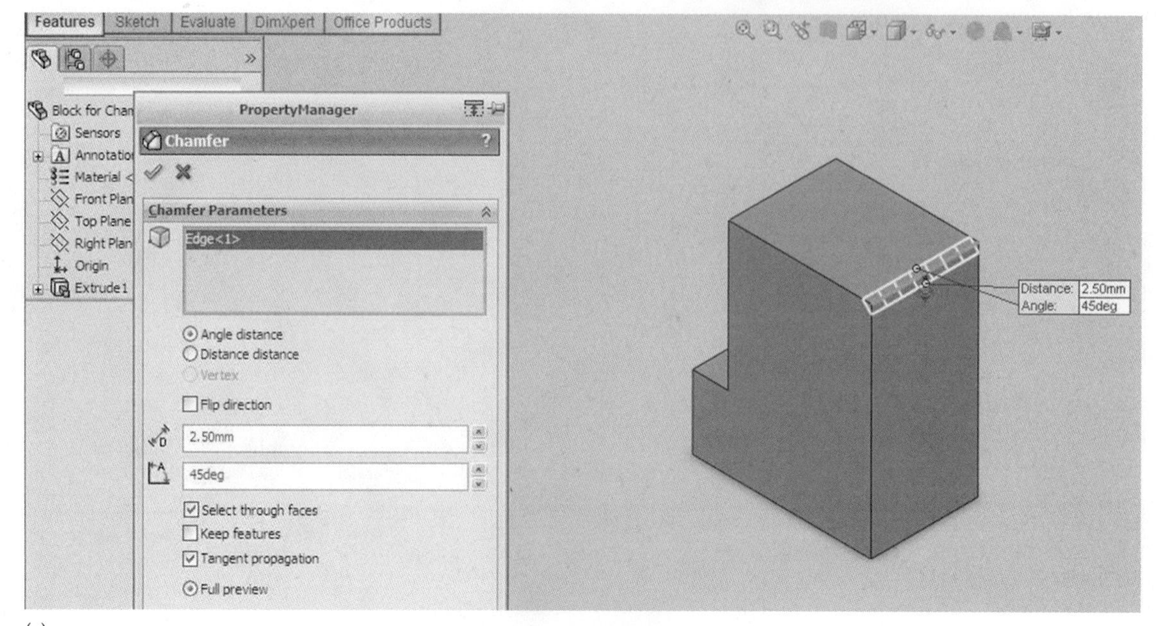

(a)

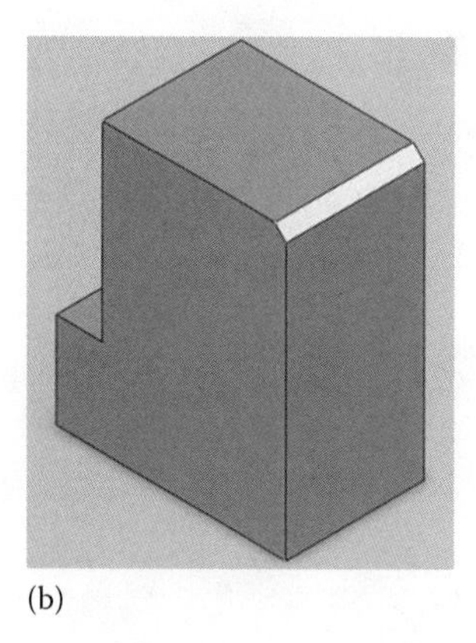

(b)

Figure 3.49

Chamfer tool for creating chamfers defined using an angle and distance: (a) real-time preview and (b) chamfered part.

3. Click the Linear Pattern tool.
   The Linear Pattern Properties Manager appears.
4. Select the 5 mm × 5 mm × 50 mm feature from the Feature Manager for the Feature to Pattern.
5. For Direction1> select a horizontal edge for direction, 70 mm for Distance, and 2 for number of patterns. For Direction2>, select a vertical edge for direction, 36 mm for Distance, and 2 for number of patterns (see Figure 3.52).
6. Click OK to complete the patterns. Observe the object: a table, as shown in Figure 3.53.

Circular Pattern

The pattern tool is used to create a circular pattern on a 3D model about an origin. The 10-mm diameter hole created using the Hole Wizard is now to be duplicated into four patterns in a circular manner.

1. Click the Circular Pattern tool.
   The Circular Pattern Properties Manager appears (see Figure 3.54).
2. Select the 10-mm diameter hole that is created using the Hole Wizard from the Feature Manager for the Feature to Pattern.

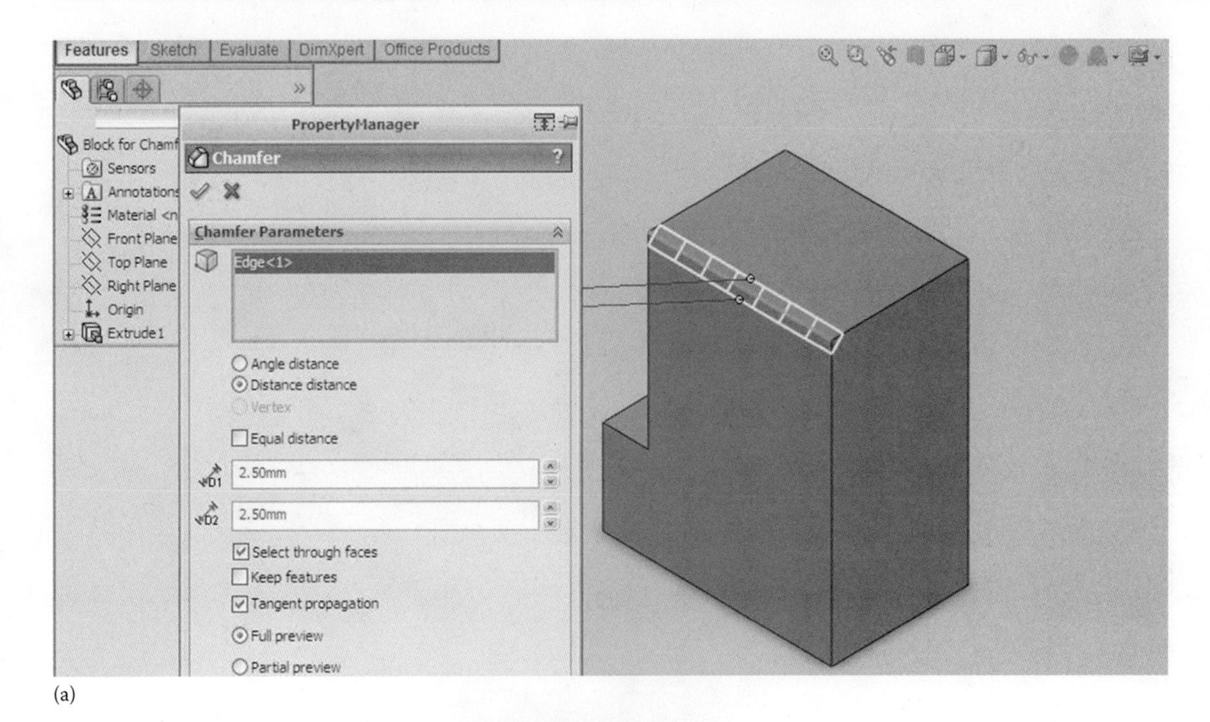

(a)

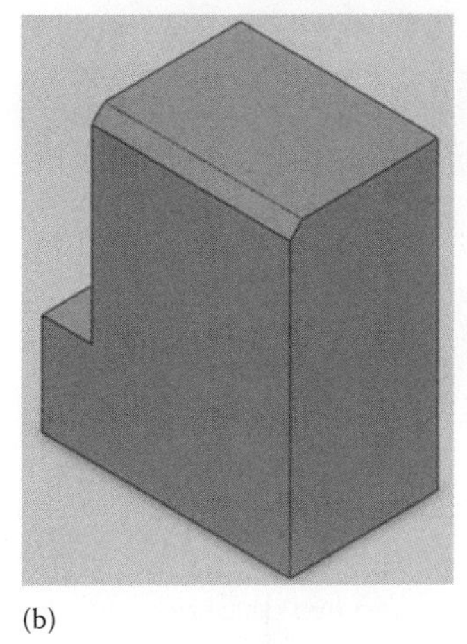

(b)

Figure 3.50

Chamfer tool for creating chamfers defined using two distances: (a) a real-time preview and (b) chamfered part.

3. Click View > Temporary Axis to activate the temporary axes.
4. Select the axis through the origin of the model (Axis<1>) about which to pattern.
5. Define number of patterns.
6. Click OK to complete the patterns (see Figure 3.55).

Mirror

The mirror tool is used to create mirror images of features (see Figure 3.56). The axes for mirroring have to be defined. For example, if only one hole in the upper-right side of the object shown here was initially defined, the bottom-right hole can be obtained by mirroring about the horizontal centerline. Using these two holes, the upper-left and bottom-left holes can be obtained by mirroring about the vertical centerline to obtain Figure 3.57.

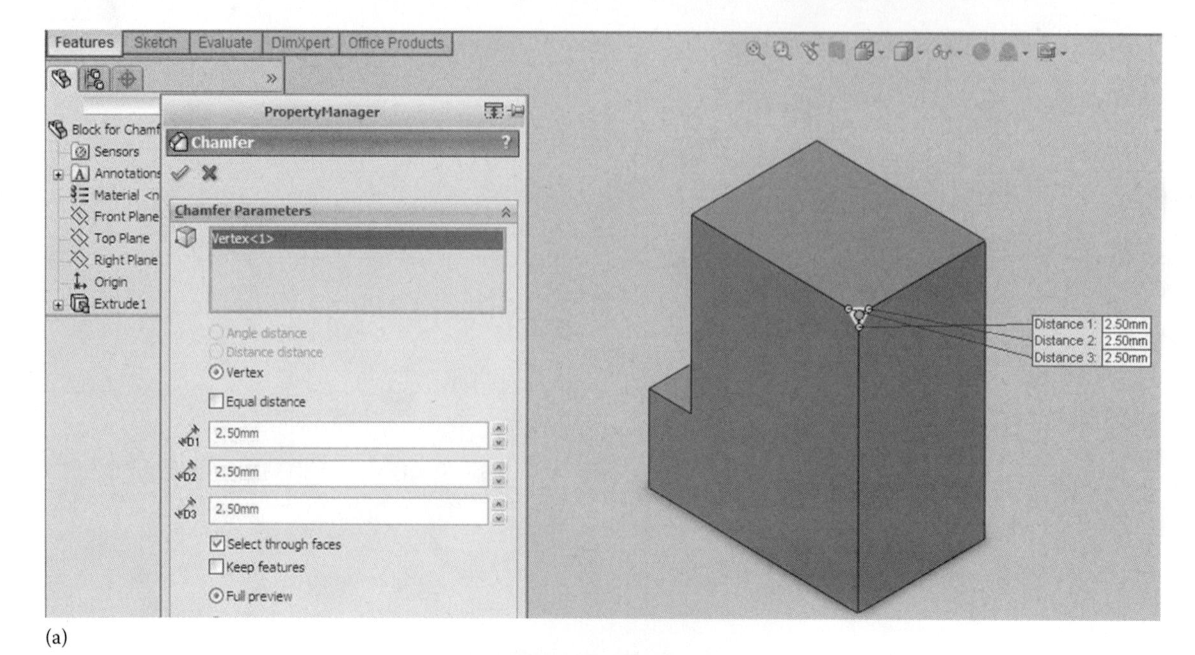

(a)

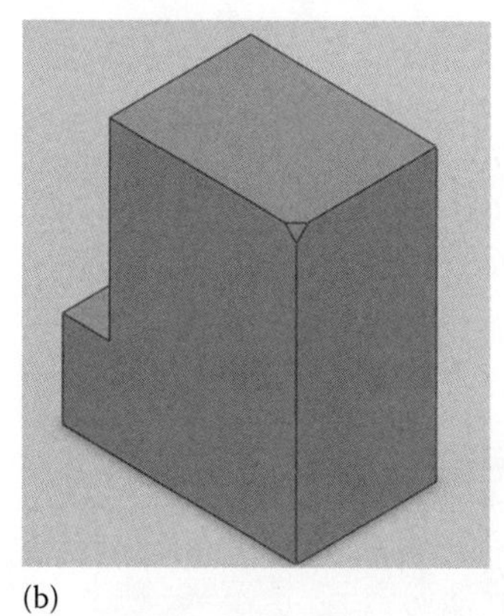

(b)

Figure 3.51

Chamfer tool for creating chamfers defined using two distances: (a) a real-time preview and (b) chamfered part.

Editing Features

SolidWorks *edit features* tools are also extremely powerful. The key to the edit features tools is the extent to which you understand the *FeatureManager design tree*. The design tree gives us the history of how the components of a part were modeled, as shown in Figure 3.58. Since features make up a part, and sketches make up a feature, we can easily edit a part by simply editing the features or the sketches. It is that easy, but you need to try it out for different parts and master the process for troubleshooting your parts. Let us edit the sketch of lofted model and see how this will affect our target shape.

1. Right-click Sketch2 in the Feature Manager.
2. Click Edit Sketch.
3. Change the dimension of the middle sketch (circle) from 60 to 100 mm (see Figure 3.59).
4. Click Exit Sketch. The model bulges outward as shown in Figure 3.60.

SolidWorks edit features tools are powerful for editing models via the design tree. By merely changing the dimension of the middle sketch (circle) in the middle plane, a completely different shape is realized. The designer can change a number of dimensions to realize a suite of different designs. This is a useful tool for industrial designers.

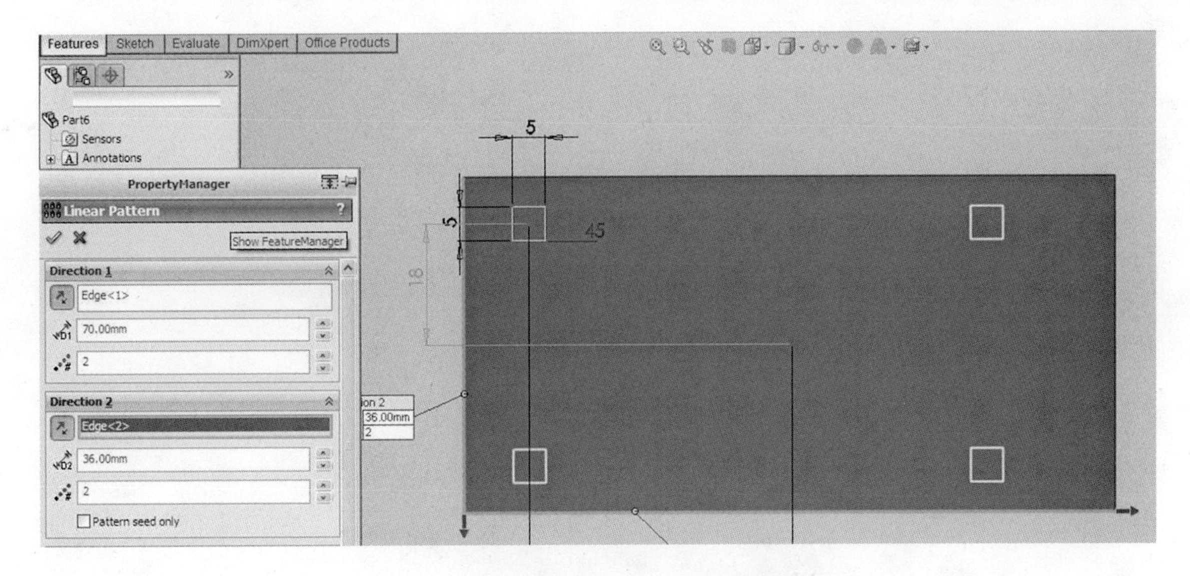

Figure 3.52

Defining distances and number of patterns in Direction 1 and Direction 2.

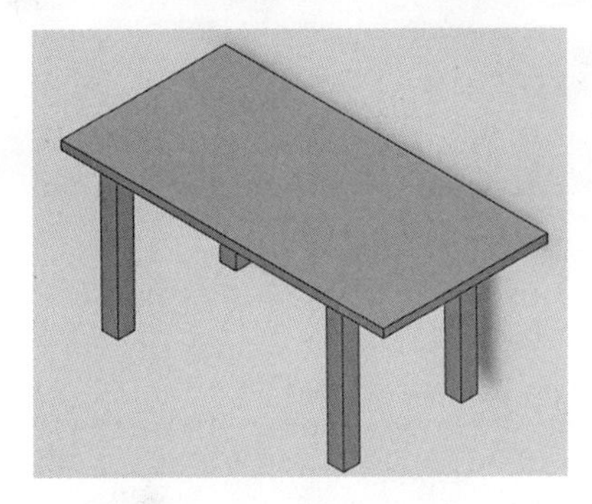

Figure 3.53

3D model using linear pattern tool.

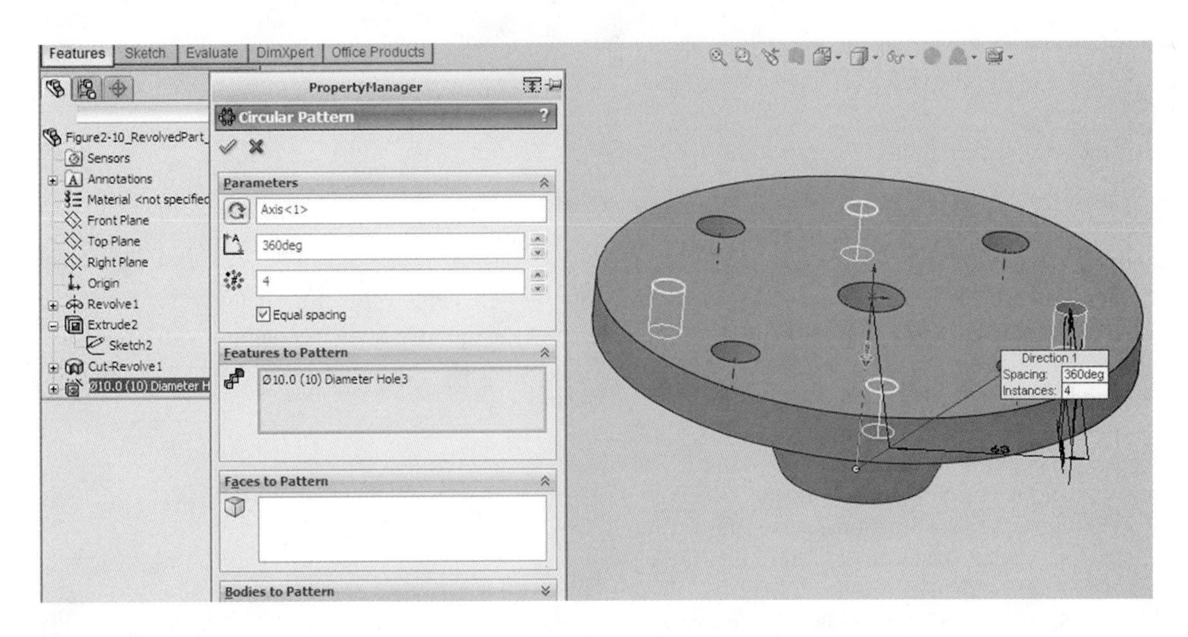

Figure 3.54

A cylindrical hole to be added to an existing 3D model.

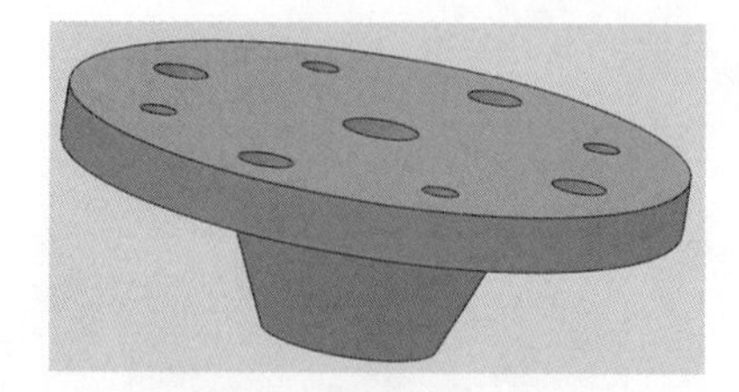

Figure 3.55

3D model using circular pattern tool.

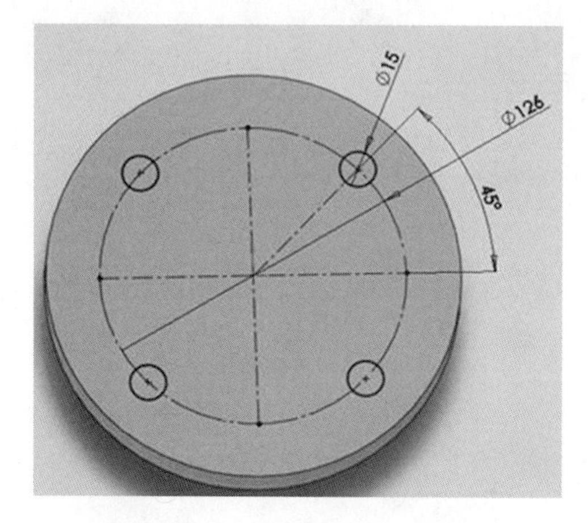

Figure 3.56

3D model illustrating mirror tool.

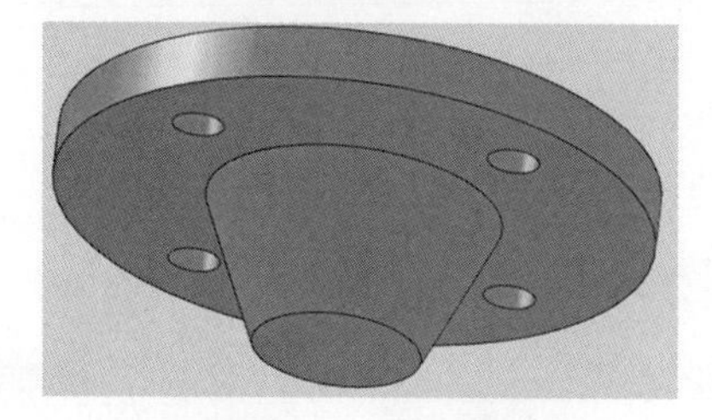

Figure 3.57

3D model with hole features obtained using mirror tool.

Tutorials

Tutorial 3.1: Simple Part 1

Create the model that is shown in Figure 3.61. The sketch of the model is shown in Figure 3.62. Create the sketch and fully dimension it. Extrude the model to the depth of 1.25 in. Determine the center of mass for the model for 1060 Alloy material. Note the origin of the model.

SolidWorks Solution

1. Open a New SolidWorks part document.
2. Set the document properties for the model, with decimal places = 2.
3. Create Sketch1 as shown in Figure 3.62.
4. Click the Features tool.
5. Click the Extruded Boss/Base tool.

   The Extrude Properties Manager appears.

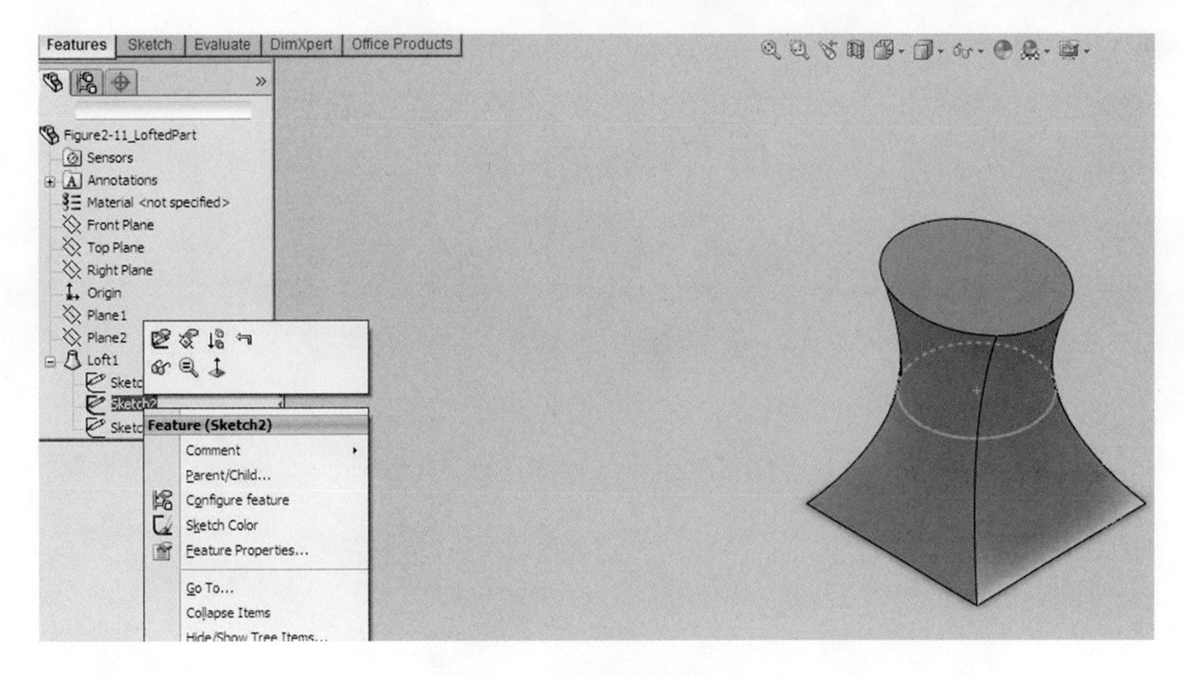

Figure 3.58

Design tree 3D lofted model.

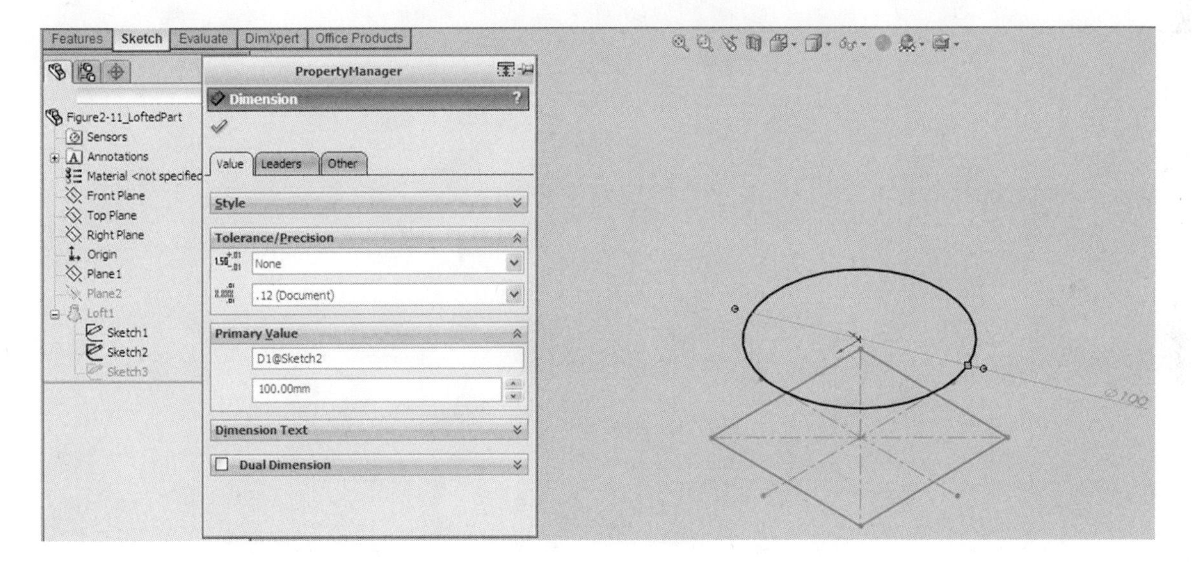

Figure 3.59

Editing middle sketch (Sketch2).

6. Define the extrusion depth as 1.25 in. A real-time preview will appear.
7. Click OK to complete the extrusion, Extrude1, as shown in Figure 3.61.
8. Assign 1060 Alloy material to the part that is modeled.
9. Calculate the mass properties of the part (see Figure 3.63).

Tutorial 3.2: Simple Part 2

Create the model that is shown in Figure 3.64. The sketch of the model is shown in Figure 3.65. Create the sketch and fully dimension it. Extrude the model to the depth of 1.25 in. Determine the center of mass for the model for 1060 Alloy material. Note the origin of the model.

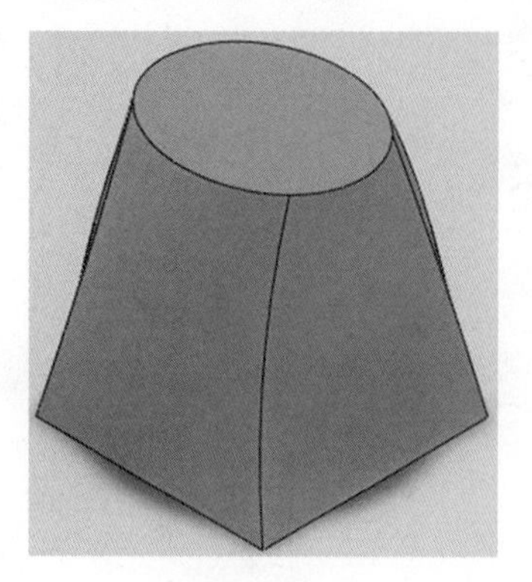

Figure 3.60

Edited model.

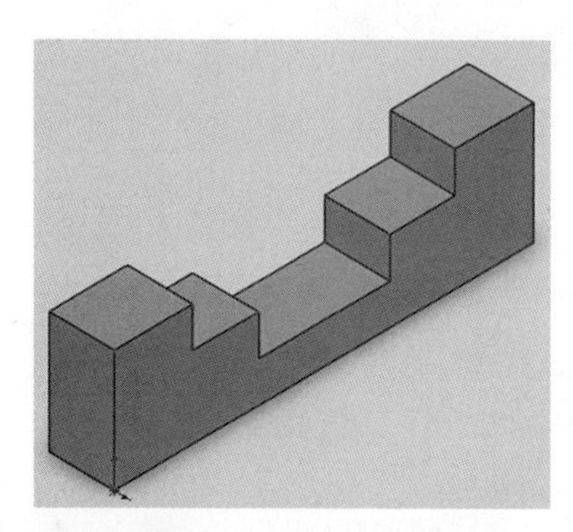

Figure 3.61

Model for Tutorial 3.1.

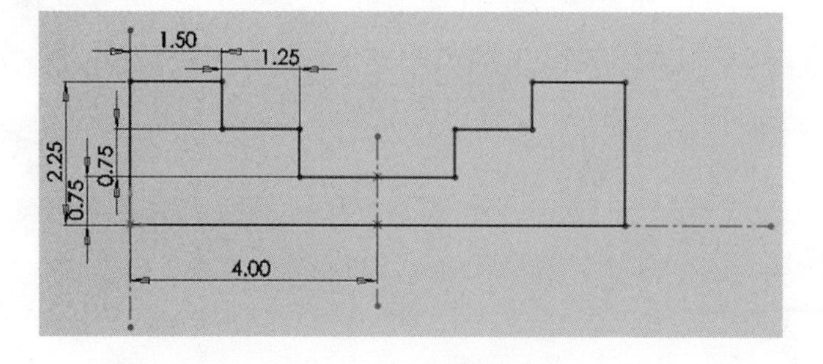

Figure 3.62

Sketch for Tutorial 3.1.

SolidWorks Solution

1. Open a New SolidWorks part document.
2. Set the document properties for the model, with decimal places = 2.
3. Create Sketch1 as shown in Figure 3.65.
4. Click the Features tool.
5. Click the Extruded Boss/Base tool.
   The Extrude Properties Manager appears.
6. Define the extrusion depth as 1.25 in. A real-time preview will appear.
7. Click OK to complete the extrusion, Extrude1, as shown in Figure 3.64.
8. Assign 1060 Alloy material to the part that is modeled.
9. Calculate the mass properties of the part (see Figure 3.66).

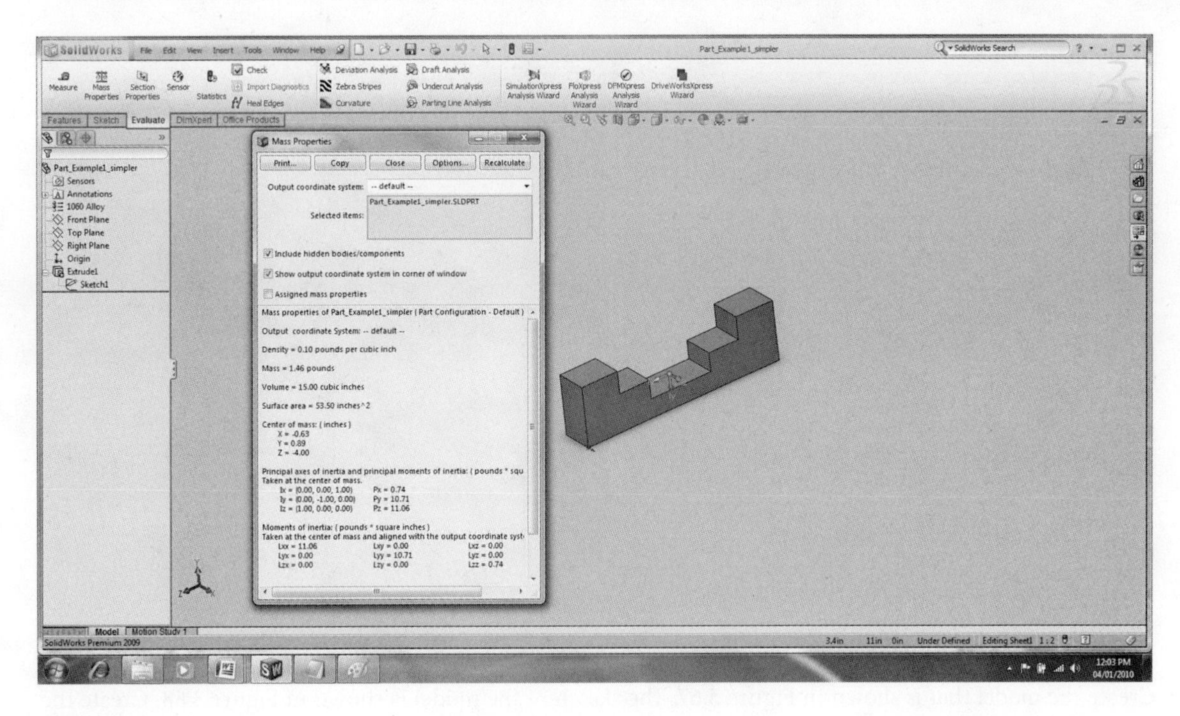

Figure 3.63

Mass properties of the part for Tutorial 3.1.

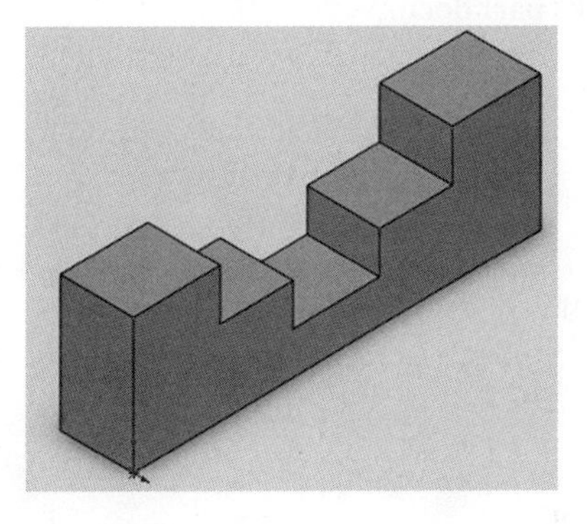

Figure 3.64

Model for Tutorial 3.2.

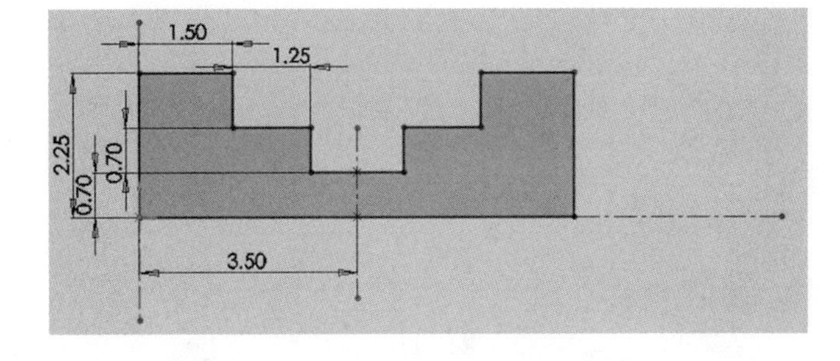

Figure 3.65

Sketch for Tutorial 3.2.

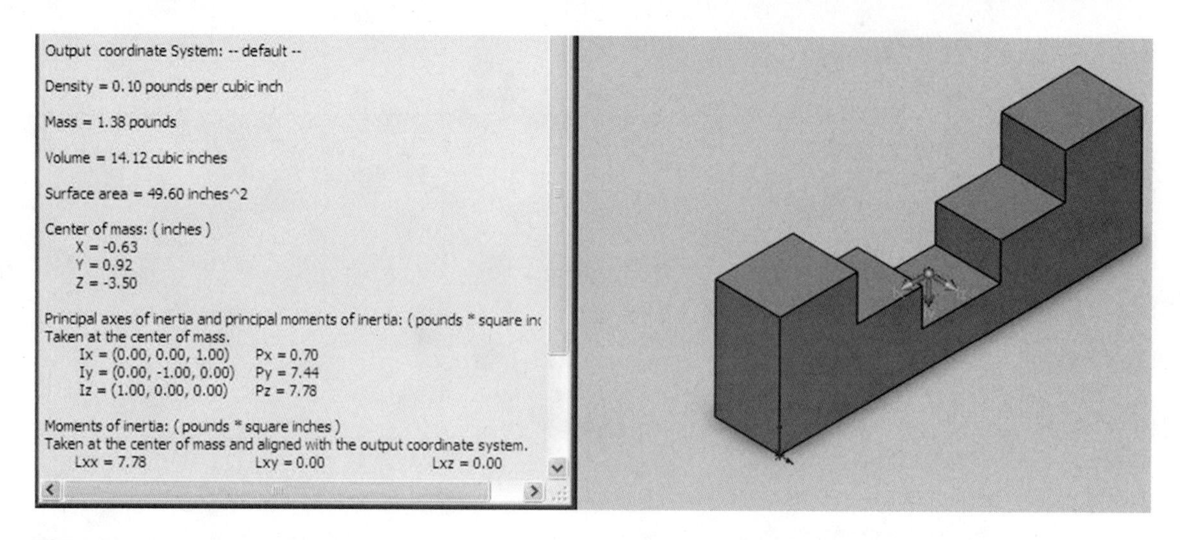

Figure 3.66

Mass properties of the part for Tutorial 3.2.

Tutorial 3.3: Simple Part 3

Create the model that is shown in Figure 3.67. The sketch of the model is shown in Figure 3.68. Create the sketch and fully dimension it. Extrude the model to the depth of 1.25 in. Determine the center of mass for the model for 1060 Alloy material. Note the origin of the model.

SolidWorks Solution

1. Open a New SolidWorks part document.
2. Set the document properties for the model, with decimal places = 2.

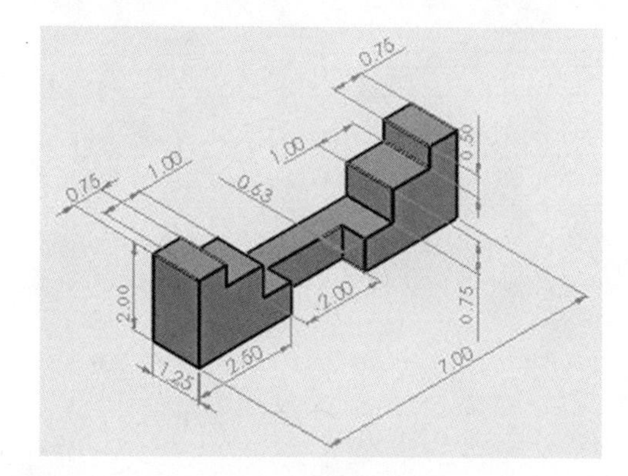

Figure 3.67

Model for Tutorial 3.3.

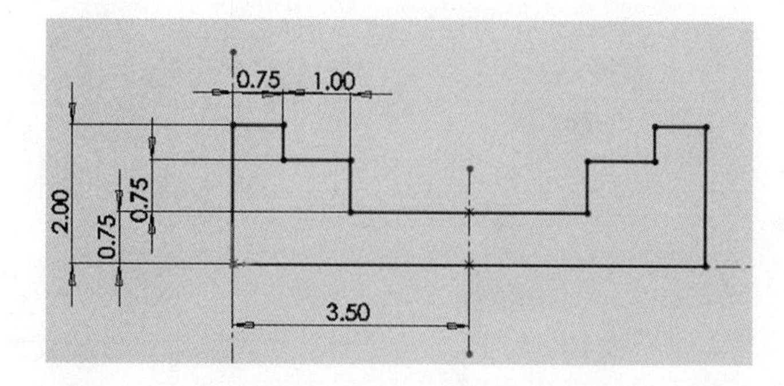

Figure 3.68

Sketch for Tutorial 3.3.

3. Create Sketch1 as shown in Figure 3.68.
4. Click the Features tool.
5. Click the Extruded Boss/Base tool.
   The Extrude Properties Manager appears.
6. Define the extrusion depth as 1.25 in. A real-time preview will appear.
7. Click OK to complete the first extrusion, Extrude1, as shown in Figure 3.69.
8. Create Sketch2 2.00 in. × 0.625 mm on top of the inner face, as shown in Figure 3.69.
9. Extrude-cut Sketch2 All Through, as shown in Figure 3.70, to obtain the final model (see Figure 3.71).
10. Assign 1060 Alloy material to the part that is modeled.
11. Calculate the mass properties of the part (see Figure 3.71).

Tutorial 3.4: Simple Part 4

Repeat Tutorial 3.3 with the material being brass.

1. Right-click the material, 1060 Alloy, on the FeatureManager.
2. Click Edit Material.
3. Select Brass and apply its material properties.
4. Calculate the mass properties of the part (see Figure 3.72).

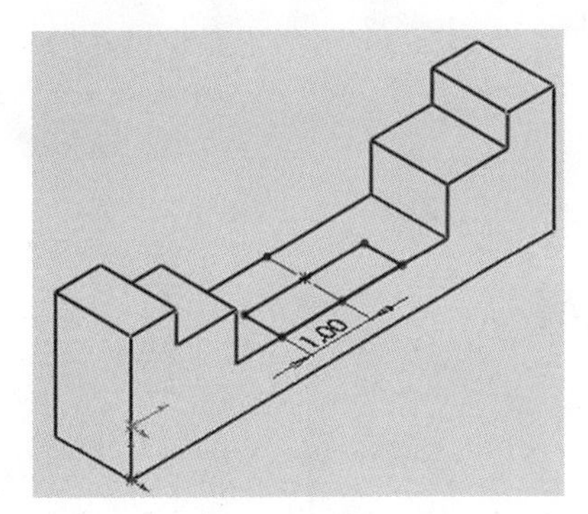

Figure 3.69

Intermediate solution.

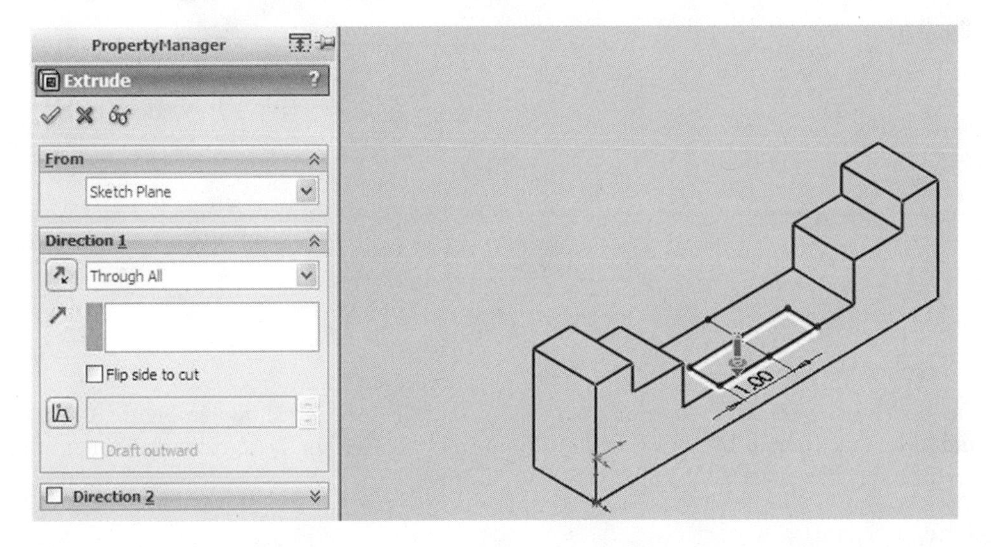

Figure 3.70

Extrude-cut of Sketch2 All Through to obtain the model shown.

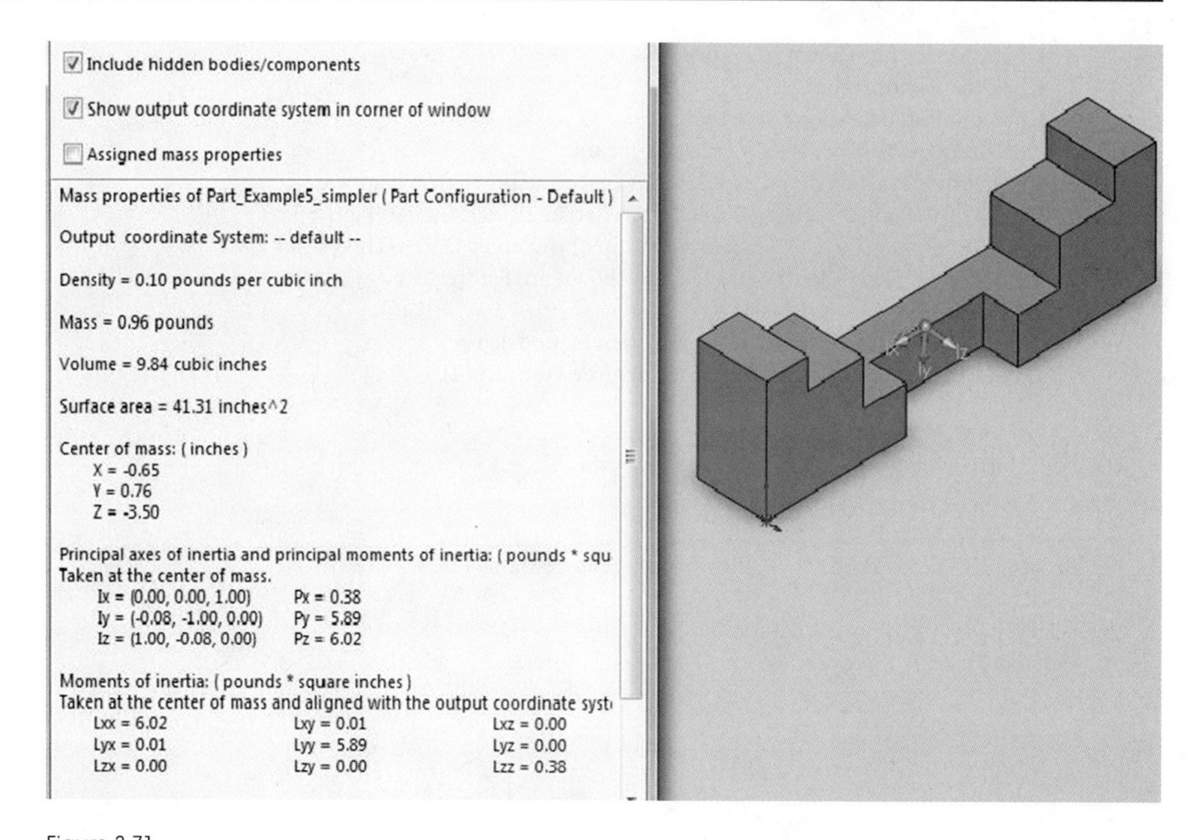

Figure 3.71

Final model and mass properties for Tutorial 3.3.

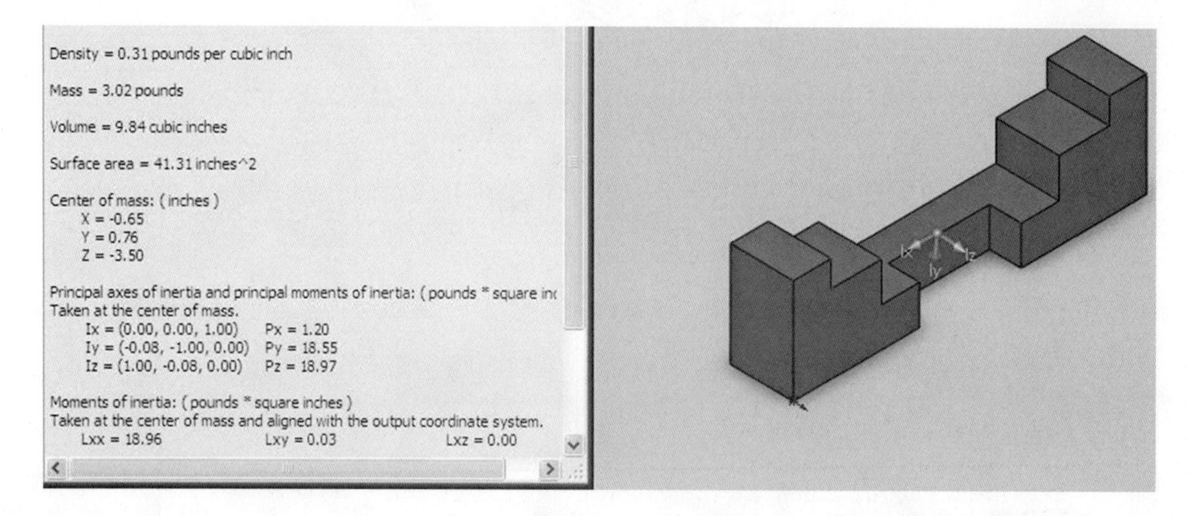

Figure 3.72

Model and mass properties for Tutorial 3.4.

Tutorial 3.5: Simple Part 5

Create the model that is shown in Figure 3.73. The sketch of the model is shown in Figure 3.74. Create the sketch and fully dimension it. Extrude the model to the depth of 1.25 in. Determine the center of mass for the model for 1060 Alloy material. Note the origin of the model.

SolidWorks Solution

1. Open a New SolidWorks part document.
2. Set the document properties for the model, with decimal places = 2.

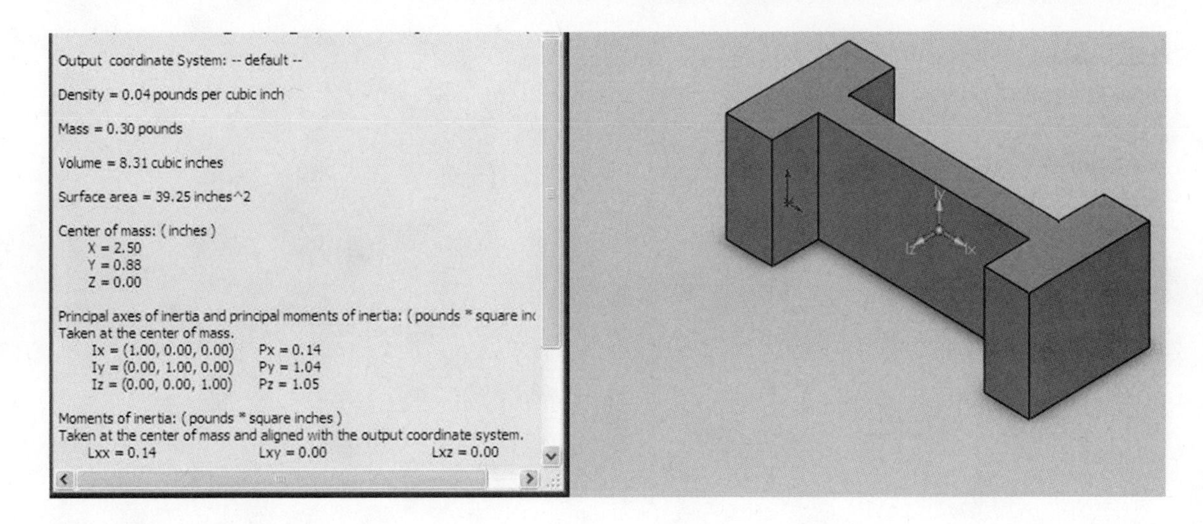

Figure 3.73

Model for Tutorial 3.5.

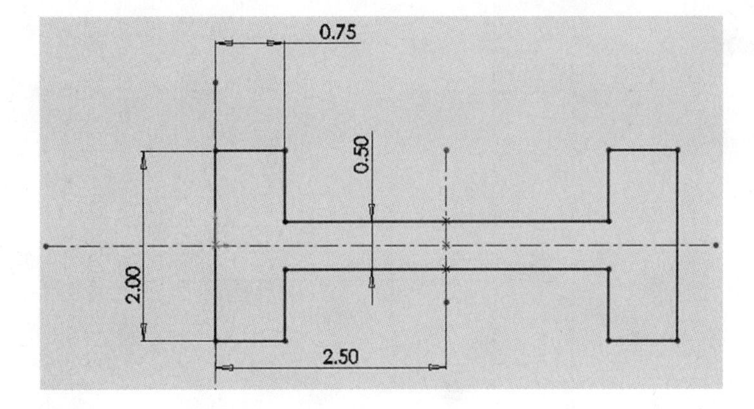

Figure 3.74

Sketch for Tutorial 3.5.

3. Create Sketch1 as shown in Figure 3.74.
4. Click the Features tool.
5. Click the Extruded Boss/Base tool.

   The Extrude Properties Manager appears.
6. Define the extrusion depth as 1.25 in. A real-time preview will appear.
7. Click OK to complete the extrusion, Extrude1, as shown in Figure 3.73.
8. Assign 1060 Alloy material to the part that is modeled.
9. Calculate the mass properties of the part (see Figure 3.75).

Tutorial 3.6: Simple Part 6

Repeat Tutorial 3.5 with the height of sketch being 3.00 in. (see Figure 3.76) and the material being brass.

1. Right-click the material, 1060 Alloy, on the FeatureManager.
2. Click Edit Material.
3. Select Copper and apply its material properties.
4. Calculate the mass properties of the part (see Figure 3.77).

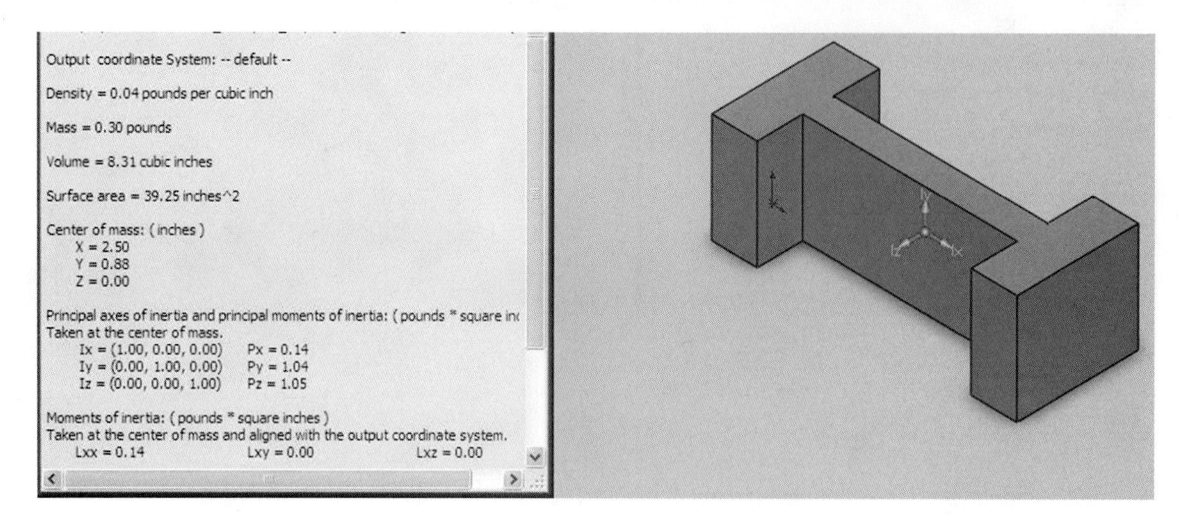

Figure 3.75

Model and mass properties for Tutorial 3.5.

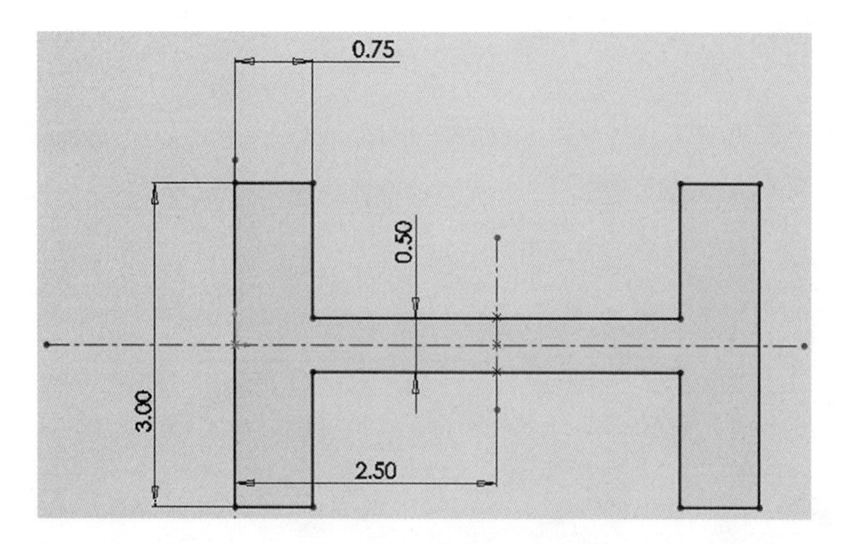

Figure 3.76

Sketch for Tutorial 3.6.

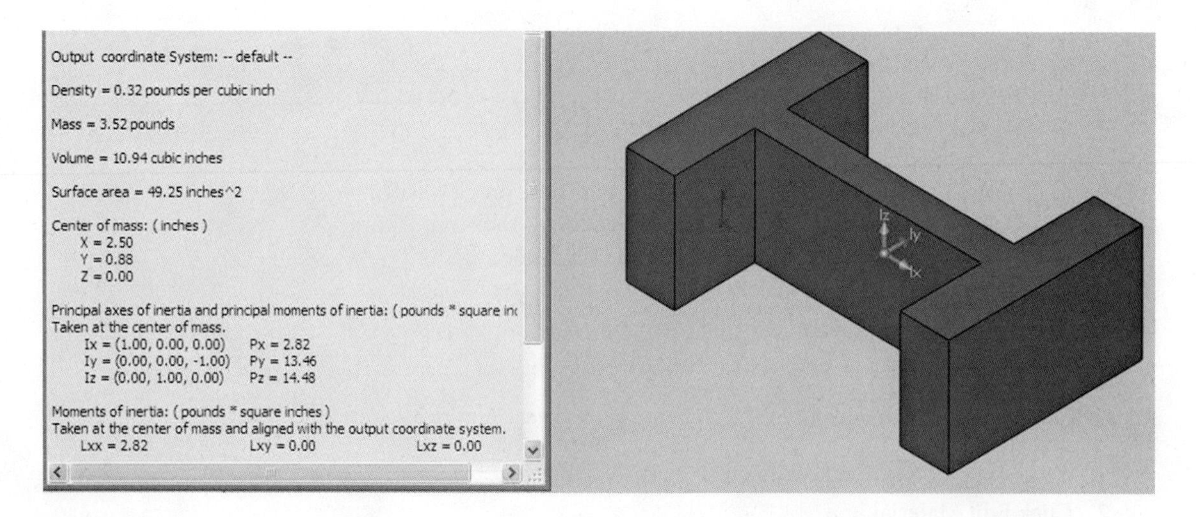

Figure 3.77

Model and mass properties for Tutorial 3.6.

Tutorial 3.7: Simple Part 7

Create the model that is shown in Figure 3.78. Create the sketch and fully dimension it. Determine the center of mass for the model for brass material. Note the origin of the model.

SolidWorks Solution

1. Open a New SolidWorks part document.
2. Set the document properties for the model, with decimal places = 2.
3. Select the top plane; create Sketch1 with dimensions 100-mm long, 60-mm wide, and 10-mm high to realize Extrude1.
4. Select the front plane; create Sketch2, 100-mm long and 30-mm high (40 mm – 10 mm) (30 mm × 10 mm cutout) (see Figure 3.79).
5. Click the Features tool.
6. Click the Extruded Boss/Base tool.
   The Extrude Properties Manager appears.
7. Define the extrusion depth as 20 mm. A real-time preview will appear.
8. Click OK to obtain the extrusion, Extrude2, as shown in Figure 3.80.
9. Select the Right Plane; create Sketch3, 40-mm long and 30-mm high (20 mm × 20 mm cutout) (see Figure 3.81).
10. Click the Features tool.
11. Click the Extruded Boss/Base tool.
    The Extrude Properties Manager appears.
12. Define the extrusion depth as 20 mm. A real-time preview will appear.
13. Click OK to obtain the extrusion, Extrude3, as shown in Figure 3.82.
14. Assign Brass material to the part that is modeled.
15. Calculate the mass properties of the part (see Figure 3.83).

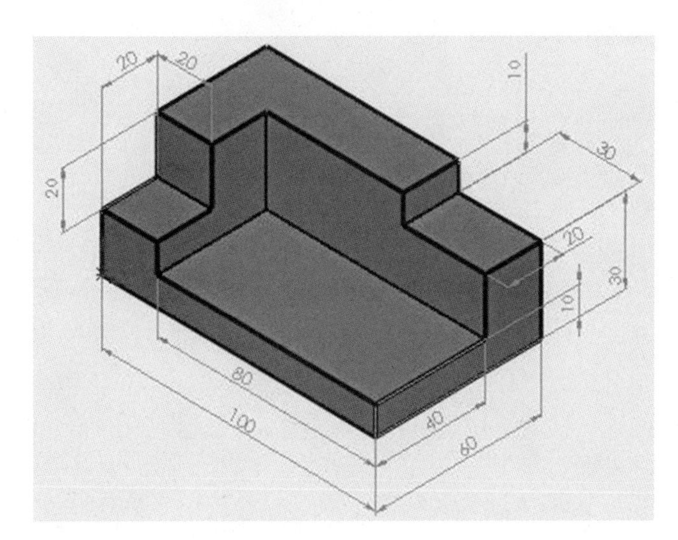

Figure 3.78

Model for Tutorial 3.7.

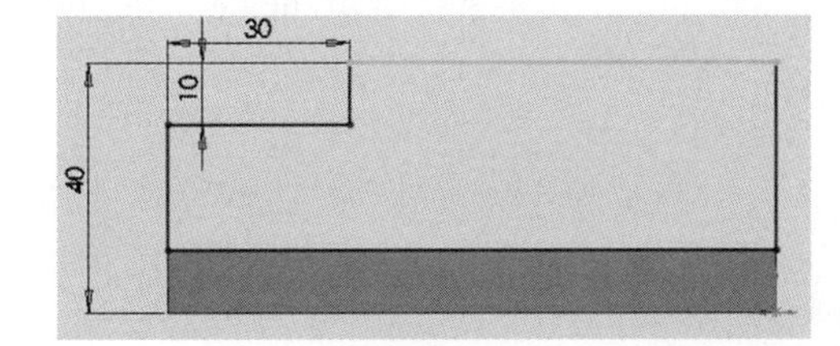

Figure 3.79

Sketch2 with 30 mm × 10 mm cutout.

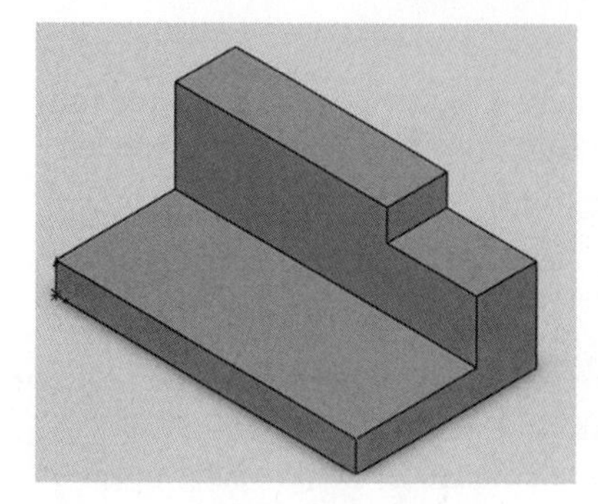

Figure 3.80

Sketch2 extruded 20 mm.

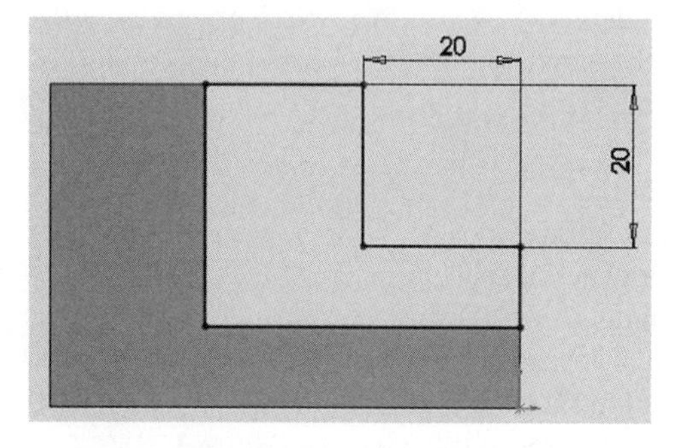

Figure 3.81

Sketch3 with 20 mm x 20 mm cutout.

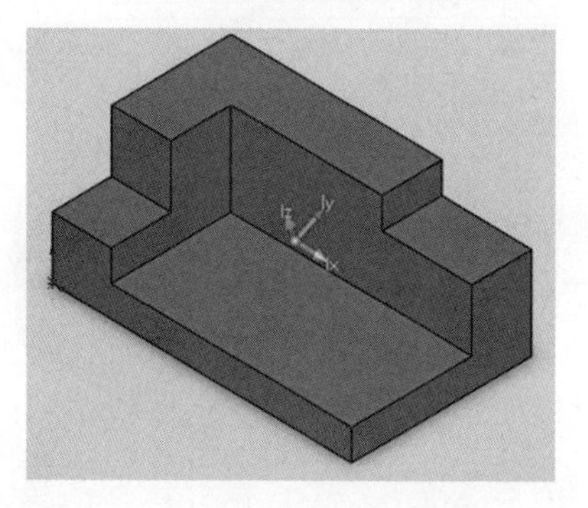

Figure 3.82

Part completed.

Tutorial 3.8: Simple Part 8

Repeat Tutorial 3.7 with a fillet radius of 5 mm, as shown in the model in Figure 3.84. Determine the mass properties.

SolidWorks Solution

1. Click Fillet.
2. Select Constant radius for Fillet Type.
3. Choose Fillet radius, 5 mm.
4. Select the four edges.
5. Click OK to obtain the part as shown in Figure 3.85.
6. Calculate the mass properties of the part (see Figure 3.86).

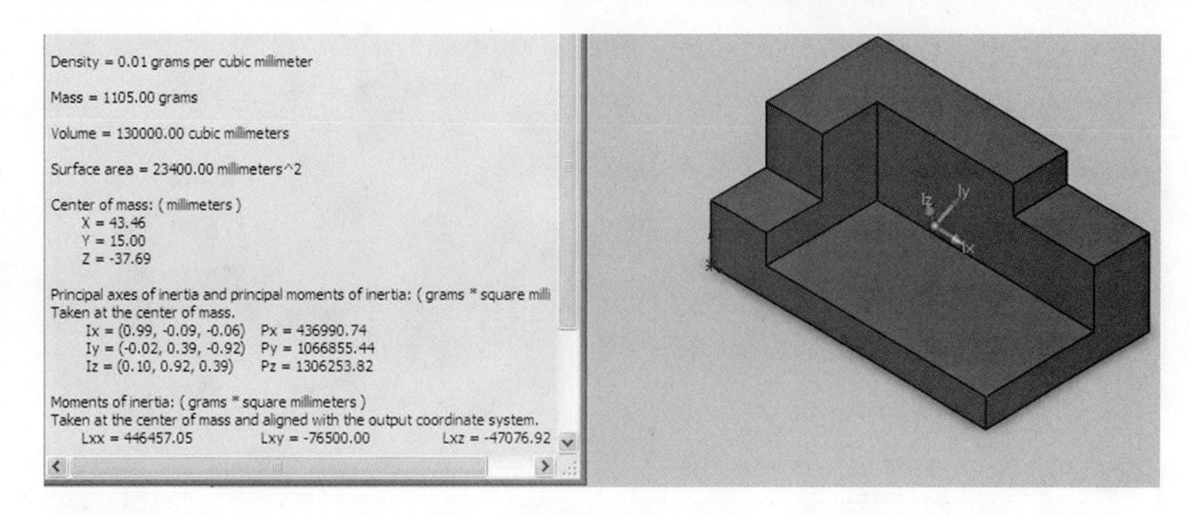

Density = 0.01 grams per cubic millimeter
Mass = 1105.00 grams
Volume = 130000.00 cubic millimeters
Surface area = 23400.00 millimeters^2
Center of mass: (millimeters)
X = 43.46
Y = 15.00
Z = -37.69
Principal axes of inertia and principal moments of inertia: (grams * square milli
Taken at the center of mass.
Ix = (0.99, -0.09, -0.06) Px = 436990.74
Iy = (-0.02, 0.39, -0.92) Py = 1066855.44
Iz = (0.10, 0.92, 0.39) Pz = 1306253.82
Moments of inertia: (grams * square millimeters)
Taken at the center of mass and aligned with the output coordinate system.
Lxx = 446457.05 Lxy = -76500.00 Lxz = -47076.92

Figure 3.83

Model and mass properties for Tutorial 3.7.

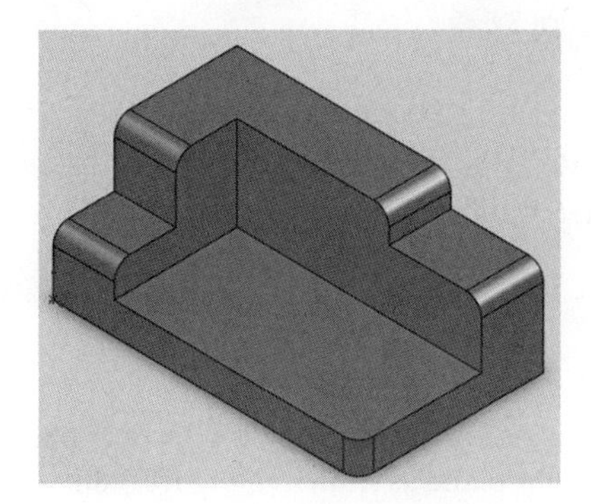

Figure 3.84

Part in Tutorial 3.7 with fillet radius of 5 mm.

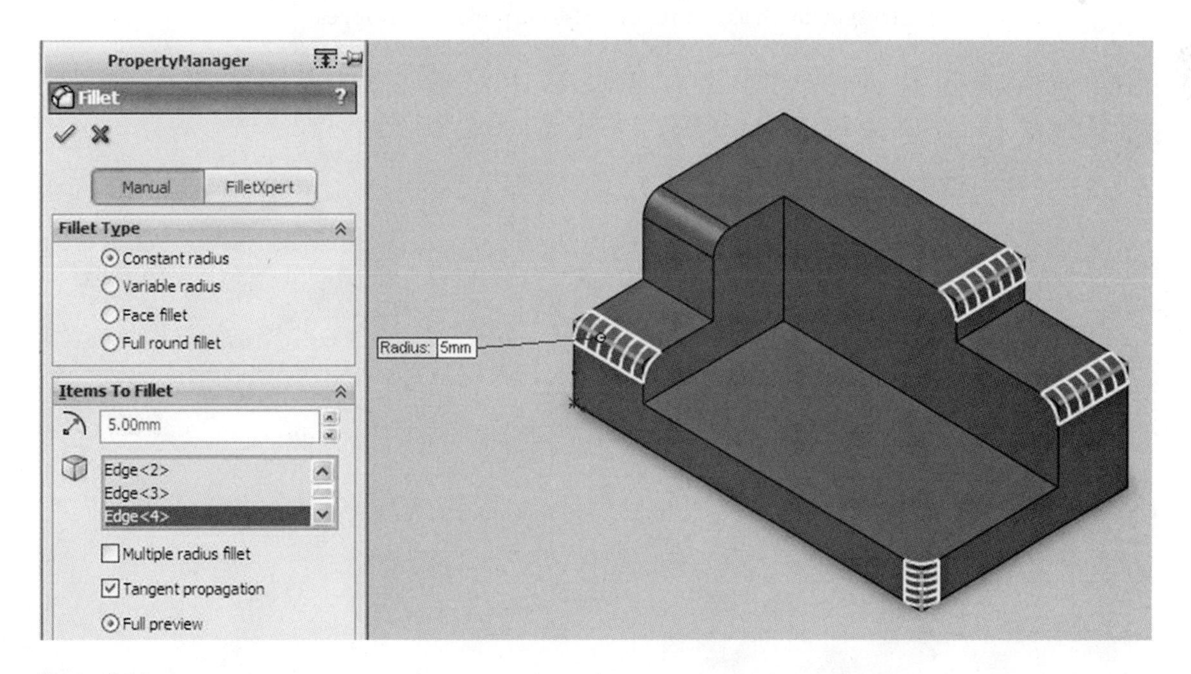

Figure 3.85

Preview of part having fillet radius of 5 mm for some edges.

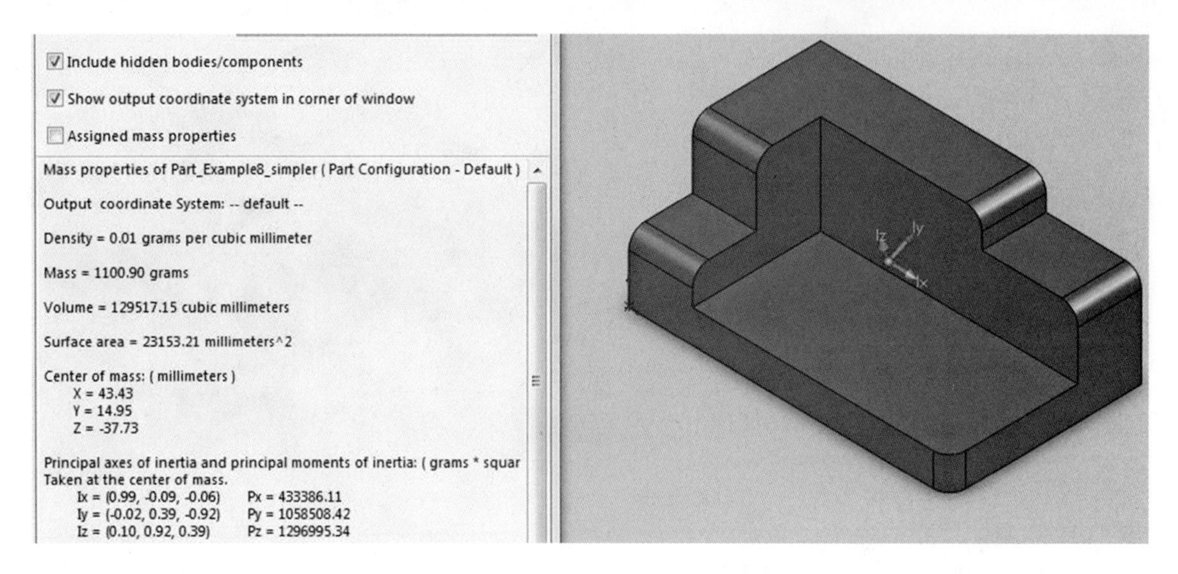

Figure 3.86

Model and mass properties for Tutorial 3.8.

Tutorial 3.9: Simple Part 9

Create the model that is shown in Figure 3.87 for which the sketch is also described. Create the sketch and fully dimension it. Note the symmetry and origin of the model. Determine the center of mass for the model for 1060 Alloy material.

SolidWorks Solution

1. Open a New SolidWorks part document.
2. Set the document properties for the model, with decimal places = 2.
3. Create Sketch1 on the front plane as shown in Figure 3.88.
4. Click the Features tool.
5. Click the Extruded Boss/Base tool.
   The Extrude Properties Manager appears.
6. Define the extrusion depth as 50 mm. A real-time preview will appear.
7. Click OK to complete the partial part, Extrude1, as shown in Figure 3.89.
8. Create Sketch2 on right side of partial model (see Figure 3.90).

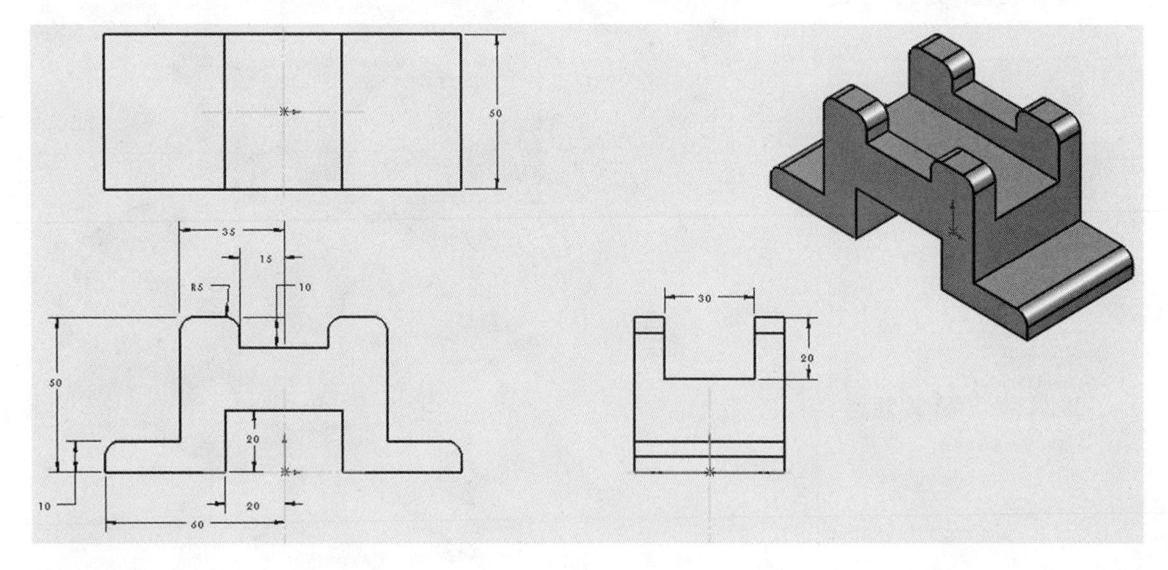

Figure 3.87

Model and sketch describing it.

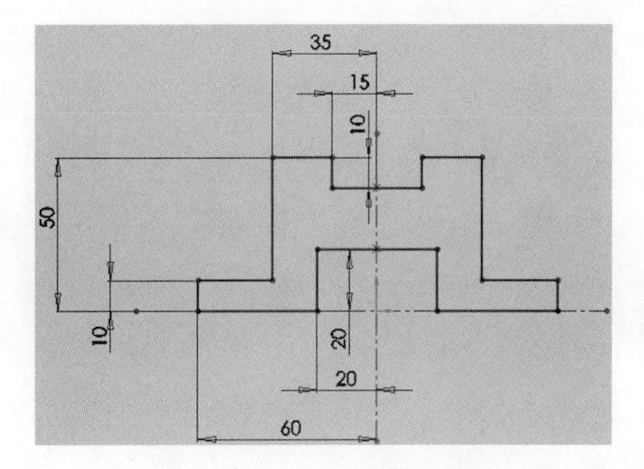

Figure 3.88

Sketch1.

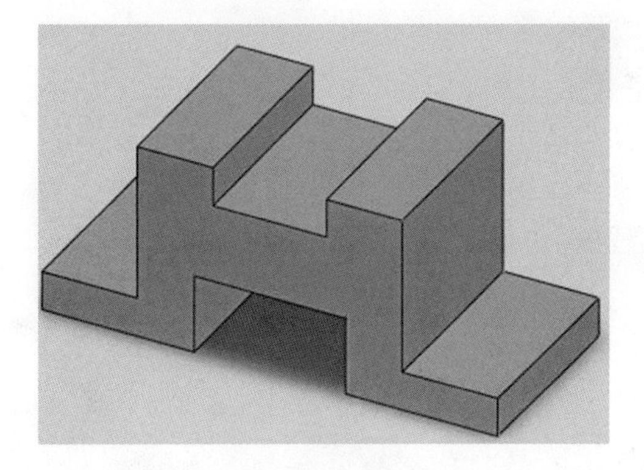

Figure 3.89

Sketch1 extruded 50 mm.

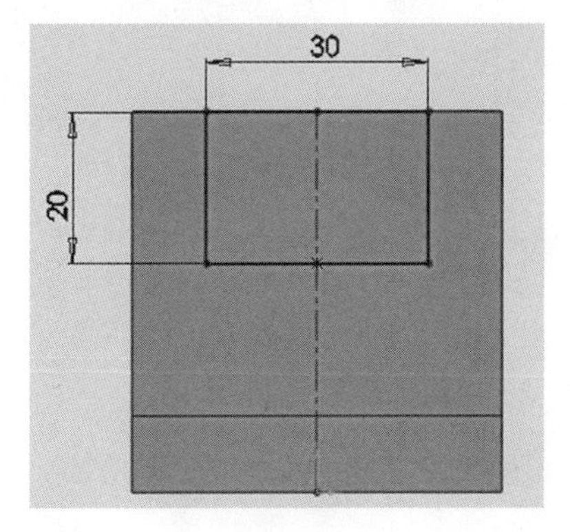

Figure 3.90

Sketch2 on right side of partial model.

9. Extrude-cut Sketch2 through 70 mm (Figure 3.91).
10. Click Fillet.
11. Select Constant radius for Fillet Type.
12. Choose Fillet radius, 5 mm.
13. Select the four edges on the top and two edges on the sides of the part (see Figure 3.92).
14. Assign 1060 Alloy material to the part that is modeled.
15. Calculate the mass properties of the part (see Figure 3.93).

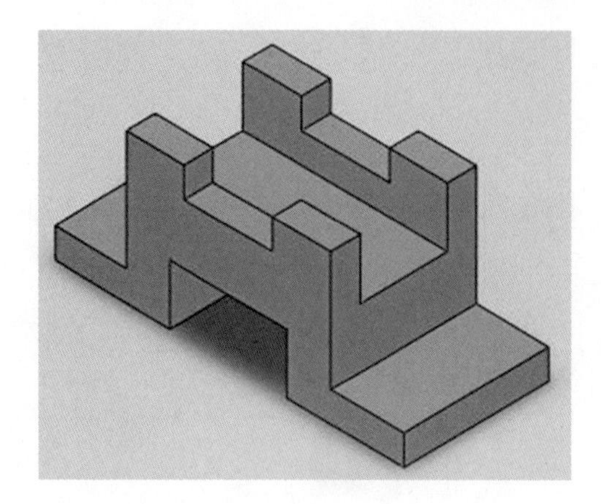

Figure 3.91

Sketch2 extrude-cut 70 mm.

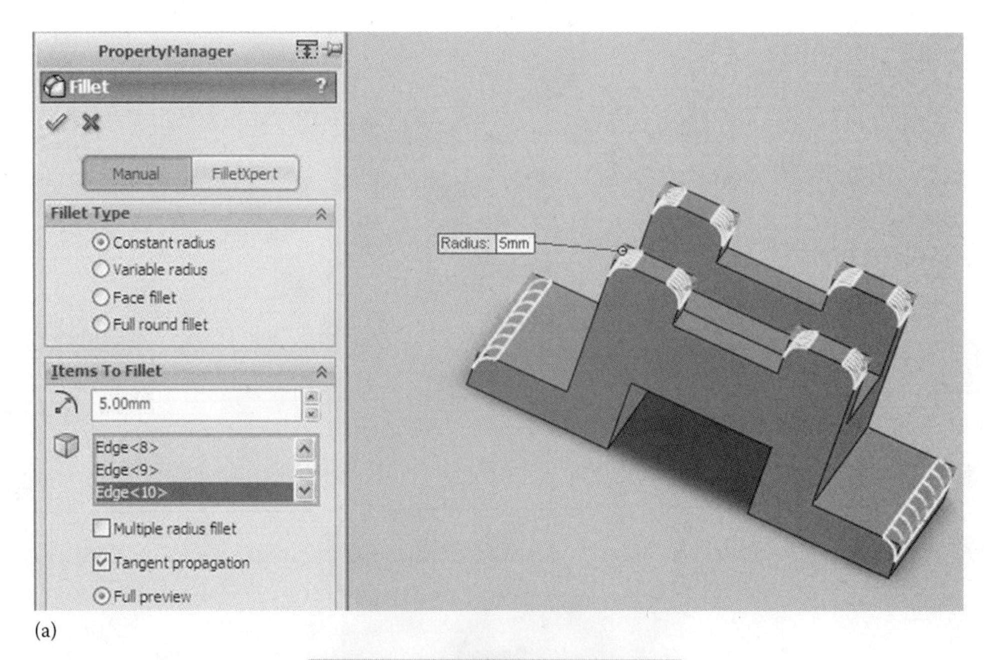

(a)

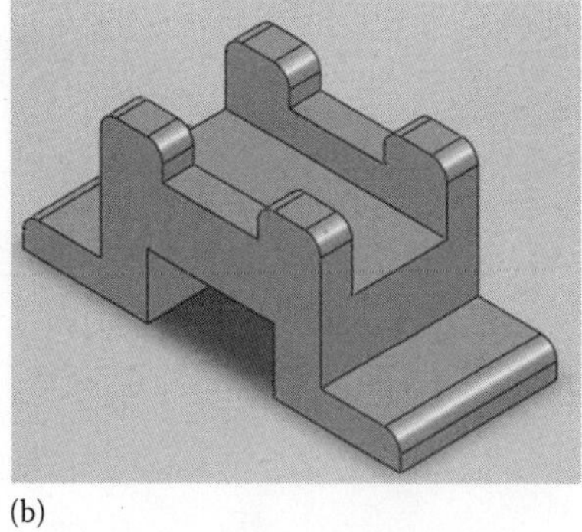

(b)

Figure 3.92

Part with fillets: (a) preview of filleting and (b) part.

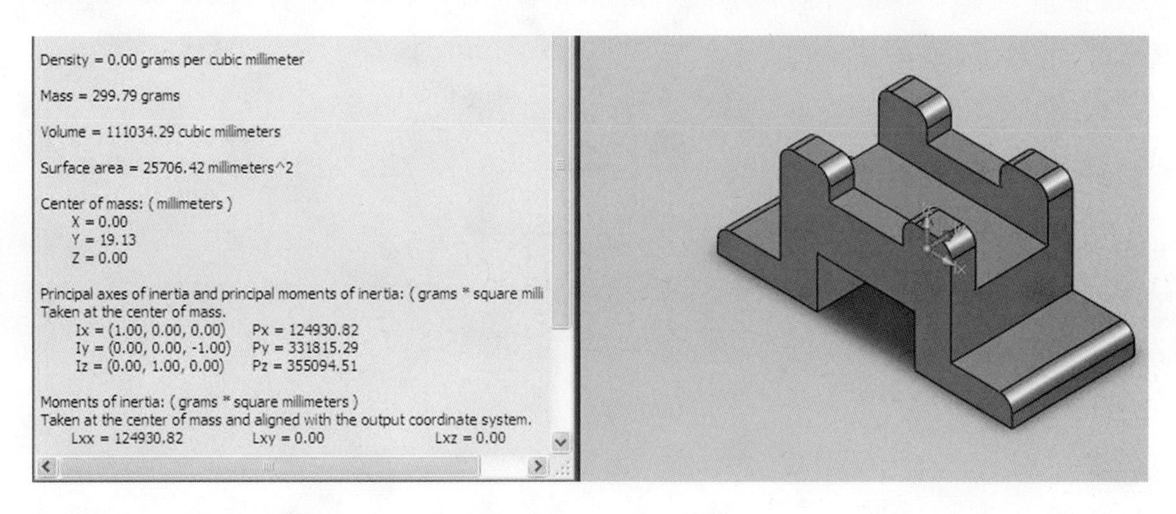

Density = 0.00 grams per cubic millimeter
Mass = 299.79 grams
Volume = 111034.29 cubic millimeters
Surface area = 25706.42 millimeters^2
Center of mass: (millimeters)
X = 0.00
Y = 19.13
Z = 0.00
Principal axes of inertia and principal moments of inertia: (grams * square milli
Taken at the center of mass.
Ix = (1.00, 0.00, 0.00) Px = 124930.82
Iy = (0.00, 0.00, -1.00) Py = 331815.29
Iz = (0.00, 1.00, 0.00) Pz = 355094.51
Moments of inertia: (grams * square millimeters)
Taken at the center of mass and aligned with the output coordinate system.
Lxx = 124930.82 Lxy = 0.00 Lxz = 0.00

Figure 3.93

Model and mass properties for Tutorial 3.8.

Tutorial 3.10: Simple Part 10

Create the model that is shown in Figure 3.94 for which the sketch is also described. Create the sketch and fully dimension it. Note the symmetry and origin of the model. Determine the center of mass for the model for 1060 Alloy material.

SolidWorks Solution

1. Open a New SolidWorks part document.
2. Set the document properties for the model, with decimal places = 2.
3. Create Sketch1 as shown in Figure 3.95.
4. Click the Features tool.
5. Click the Extruded Boss/Base tool.
   The Extrude Properties Manager appears.
6. Define the extrusion depth as 90 mm. A real-time preview will appear.

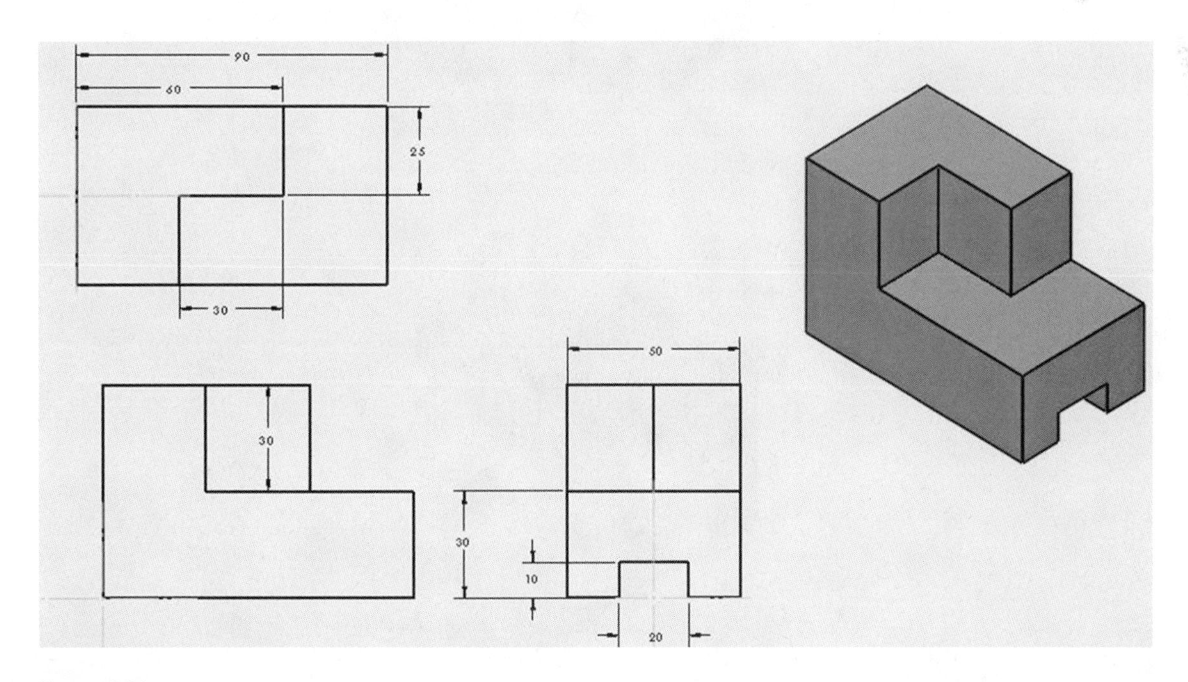

Figure 3.94

Model and sketch describing it.

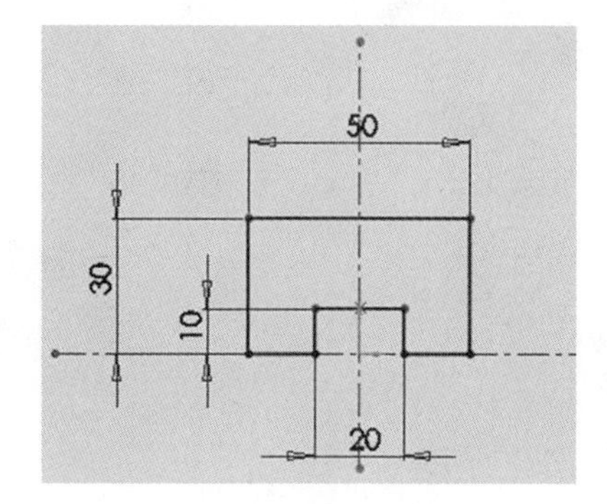

Figure 3.95

Sketch1.

7. Click OK to complete the partial part, Extrude1, as shown in Figure 3.96.
8. Create Sketch2 (L-shaped) on top of the partial model (see Figure 3.101).
9. Extrude Sketch2 through 30 mm to give the completed model in Figure 3.97.
10. Click Fillet.
11. Select Constant radius for Fillet Type.
12. Choose Fillet radius, 5 mm.
13. Select the edges of the part (see Figure 3.98).

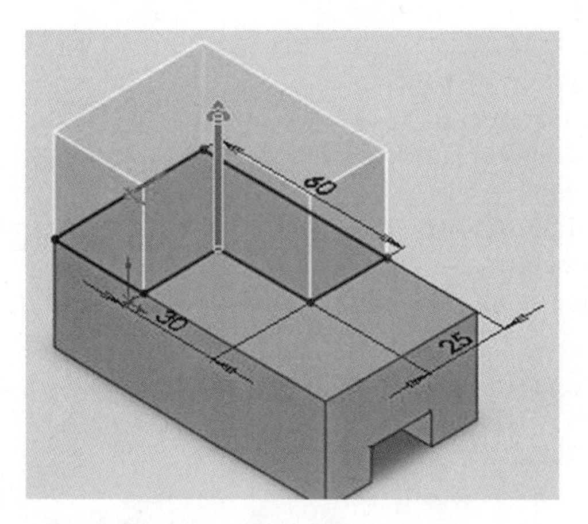

Figure 3.96

Preview of incomplete model.

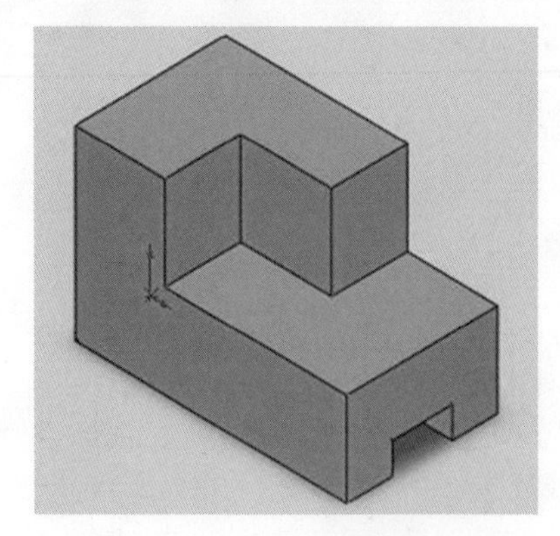

Figure 3.97

Model and mass properties.

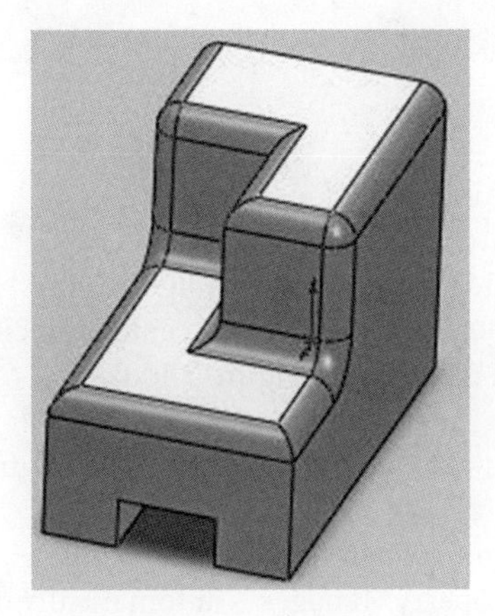

Figure 3.98

Model and mass properties for Tutorial 3.10.

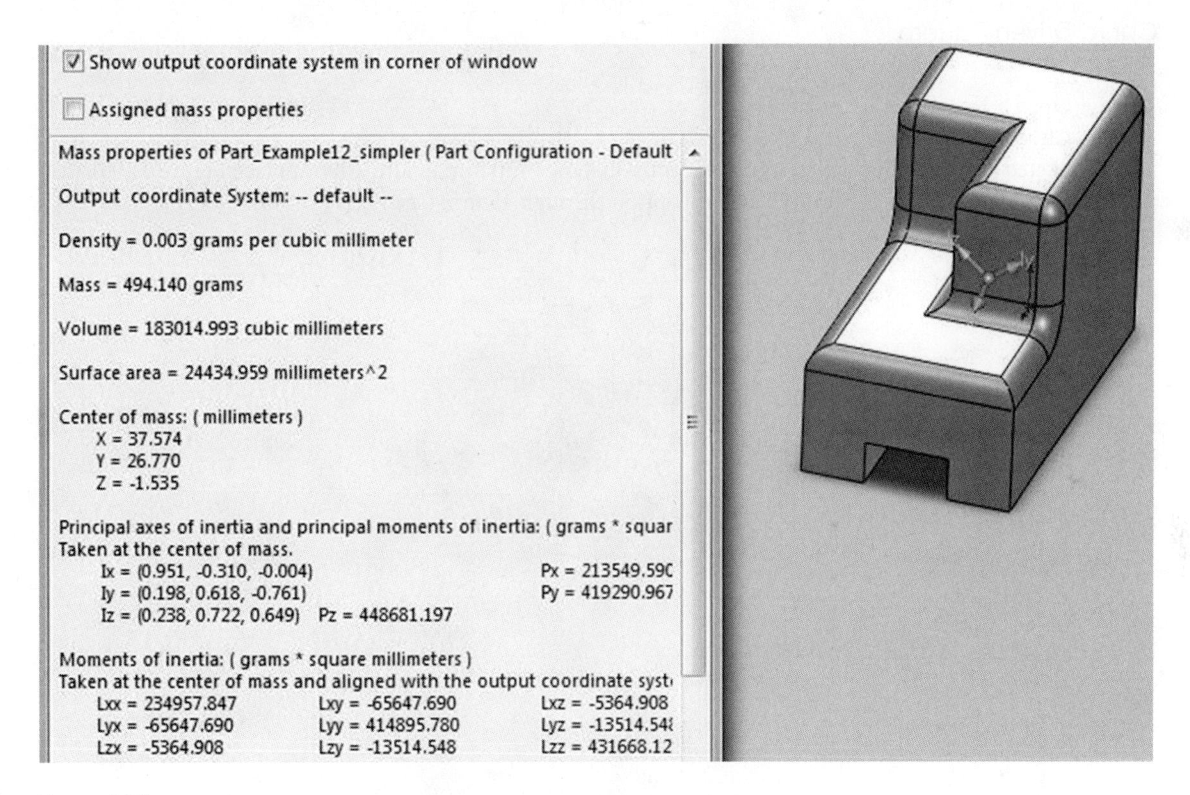

Figure 3.99

Model and mass properties for Tutorial 3.10.

14. Assign 1060 Alloy material to the part that is modeled.
15. Calculate the mass properties of the part (see Figure 3.99).

Patterns—Advanced Methods

The patterns covered so far are quite basic: linear and circular patterns. In some cases, the sketch- and curve-driven pattern approaches are more appropriate. These will be discussed together with table-driven patterns. The pros and cons of these approaches will be obvious as they are discussed.

Sketch-Driven Pattern

1. Open a New Part document.
2. Select the top plane for sketching.
3. Sketch a circle, Sketch1, diameter 150 mm.
4. Extrude Sketch1 using the Mid Plane option through 25 mm.
5. Sketch a circle, Sketch2, diameter 10 mm with its center coinciding with that of Sketch1.
6. Extrude-cut Sketch2 through 25 mm (see Figure 3.100).
7. Click the top of the feature.
8. Sketch several points, Sketch4 at locations of interest (see Figure 3.101).
9. Exit Sketch.
10. Click Feature > Linear Pattern > Sketch Driven Pattern (see Figure 3.102).
    The Sketch Driven Pattern PropertyManager is automatically displayed; see Figure 3.103.
11. In the Selections rollout, click Sketch4 (for the points that are created) (see Figure 3.103).
12. In the Features to Pattern rollout, click Extrude2 (see Figure 3.103).
13. Click OK. (The completed patterns are shown in Figure 3.104.)

Curve-Driven Pattern

1. Open a New Part document.
2. Choose the top plane for sketching.
3. Sketch a slot profile, Sketch1, center-center distance 150 mm, width 100 mm (see Figure 3.105).
4. Extrude Sketch1 using the Mid Plane option through 15 mm (see Figure 3.106).

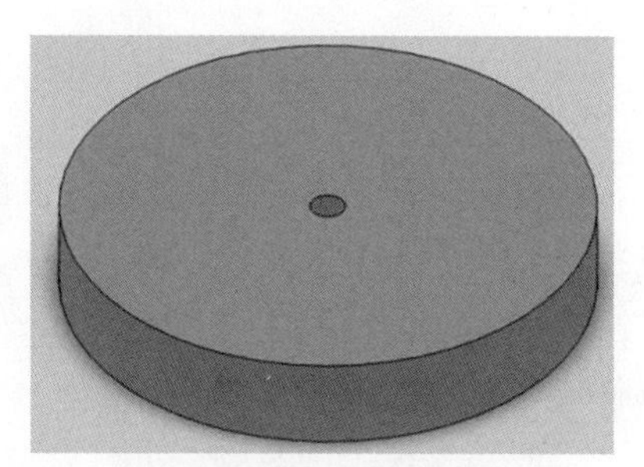

Figure 3.100

Feature to be patterned at the center of model.

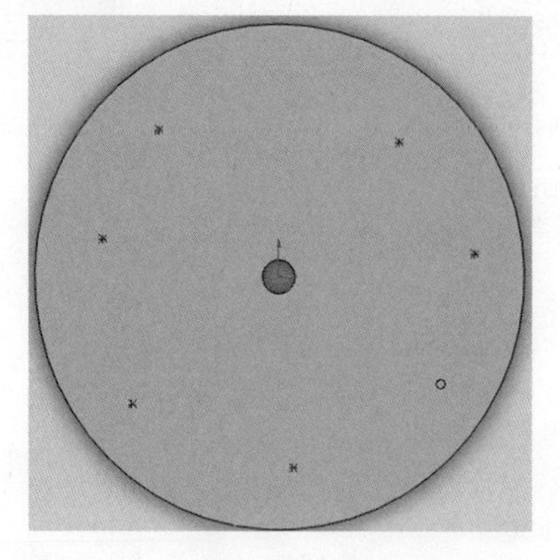

Figure 3.101

Points to drive pattern.

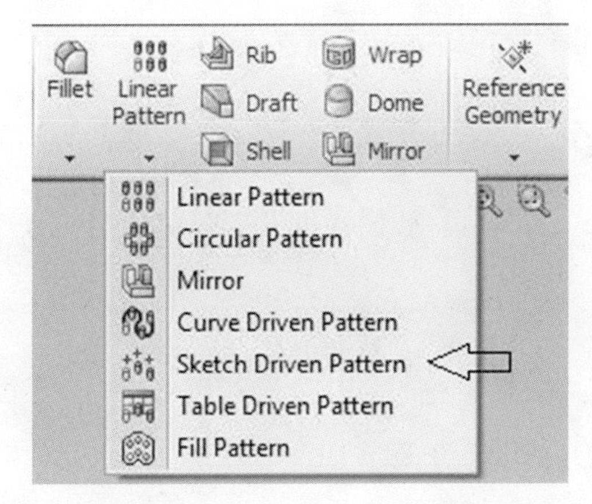

Figure 3.102

Sketch Driven Pattern option.

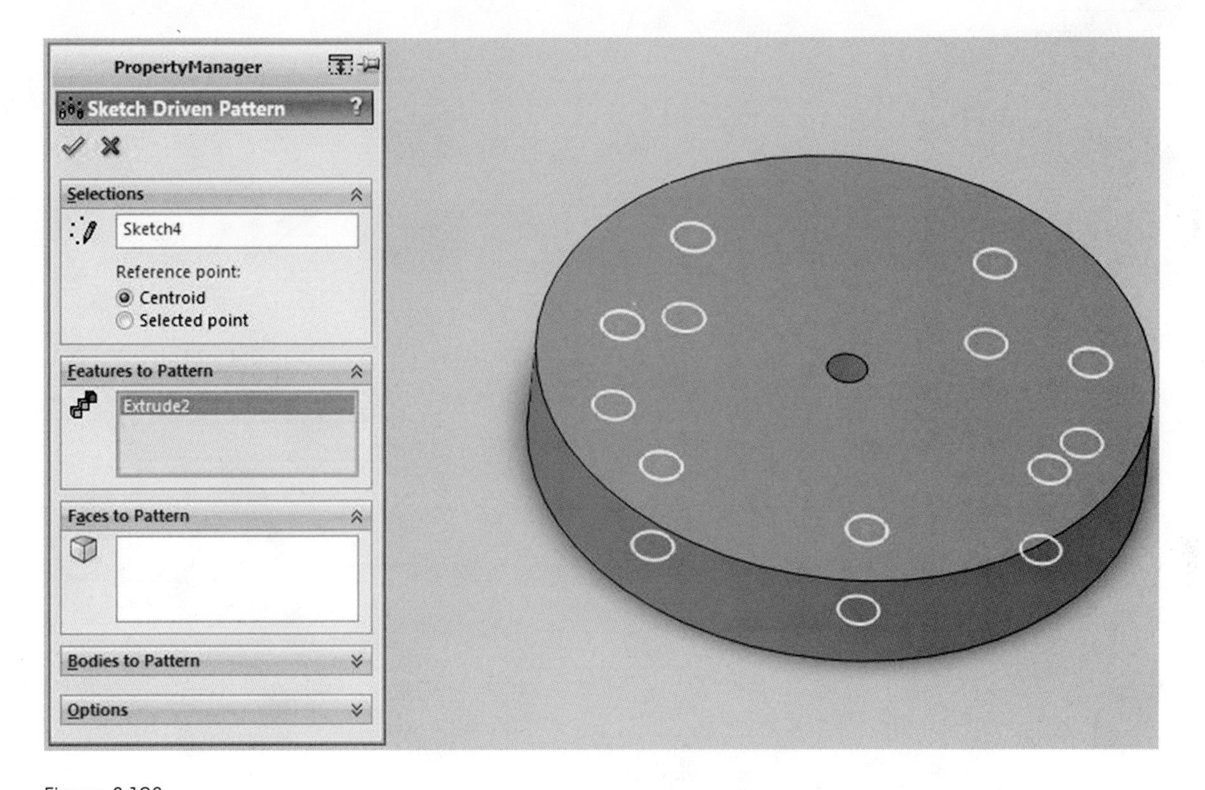

Figure 3.103

Sketch Driven Pattern PropertyManager.

Case 1: Closed Curve

5. Click the top face of the Extruded base.
6. Click the Sketch command option.
7. Click any edge, followed by holding down the Ctrl key and clicking subsequent edges.
8. Click Convert Entities to extract the edges.
9. Click the Offset Entities command option.
10. Set the Offset Distance in the Parameters rollout to be 15 mm and check the Reverse option. (See Figure 3.107 for the extracted closed curve.)
11. Exit Sketch.

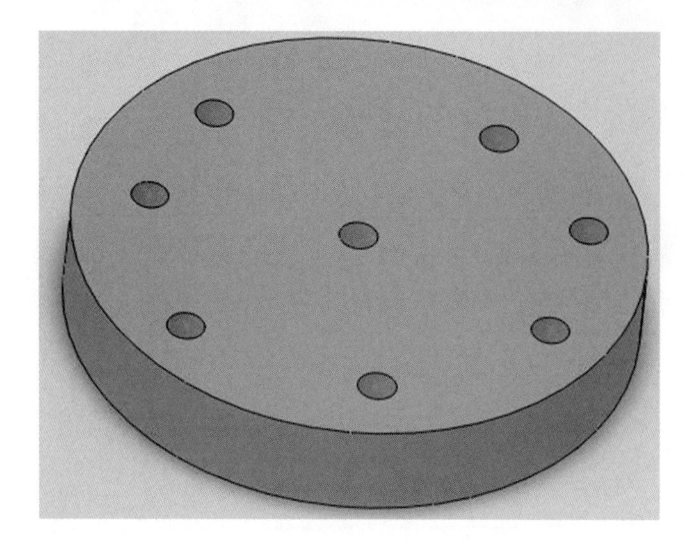

Figure 3.104

The resulting sketch-driven pattern feature.

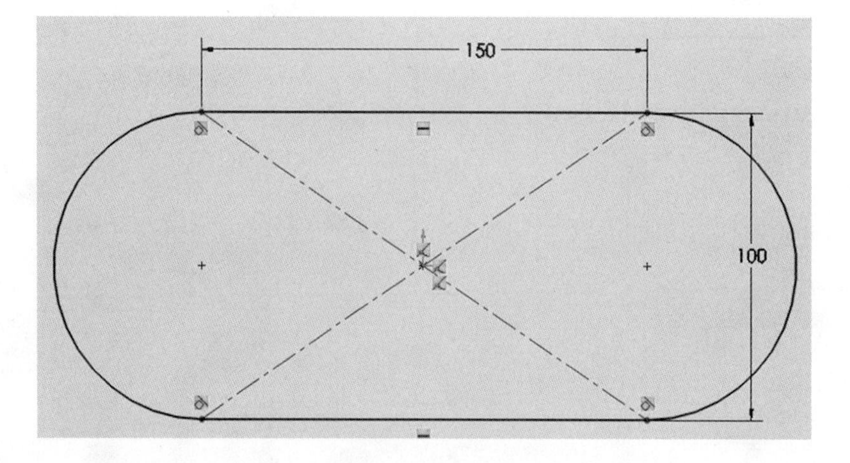

Figure 3.105

Base profile.

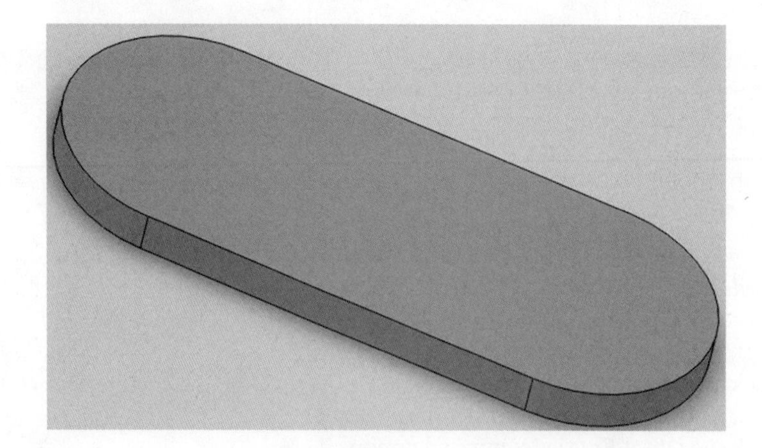

Figure 3.106

Extruded base.

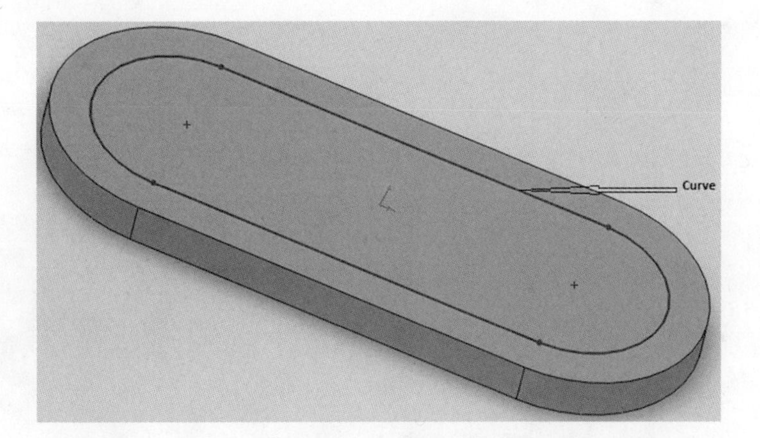

Figure 3.107

Closed curve to drive pattern.

Create a Seed Feature

12. Select the top face of the base.
13. Sketch a circle, Sketch3, diameter 10 mm with center passing through the curve.
14. Extrude Sketch3 through 15 mm (see Figure 3.108).
15. Click Feature > Linear Pattern > Curve Driven Pattern (see Figure 3.109).
    The Curve Driven Pattern PropertyManager automatically displayed; see Figure 3.109.
16. In the Direction1 rollout, select Sketch2 as the Pattern Direction.
17. For the Number of Instances, set it to 10.
18. For the Spacing, set it to 50.
19. In the Features To Pattern rollout, select Extrude2.
20. Click OK. (See the model with pattern in Figure 3.110.)

Mirror the Resulting Curve-Driven Pattern Feature

1. Click the Mirror command option.
2. For the Mirror Face/Plane, select the top plane.
3. For the Features To Mirror, select CrvPattern1.
4. Click OK. (See the model with mirrored pattern in Figure 3.111.)

Case 2: Open Curve

1. Click the top face of the Extruded base (see Figure 3.111 for the previous model).
2. Click the Sketch command option.
3. Sketch a Spline, Sketch2 (see Figure 3.112).

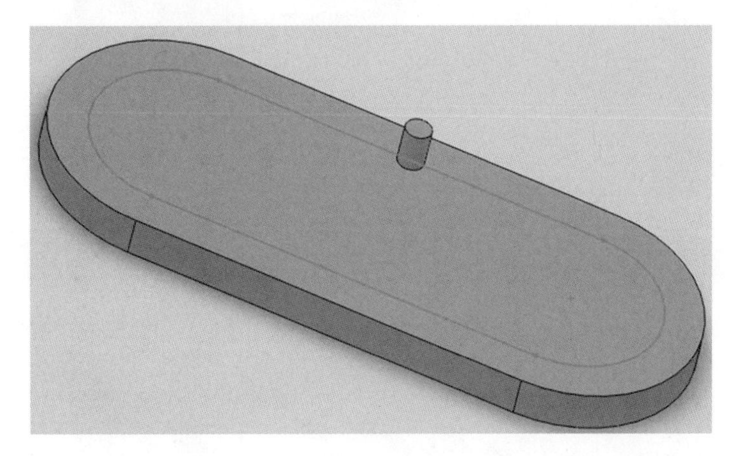

Figure 3.108

Seed feature created on the curve for curve-driven pattern.

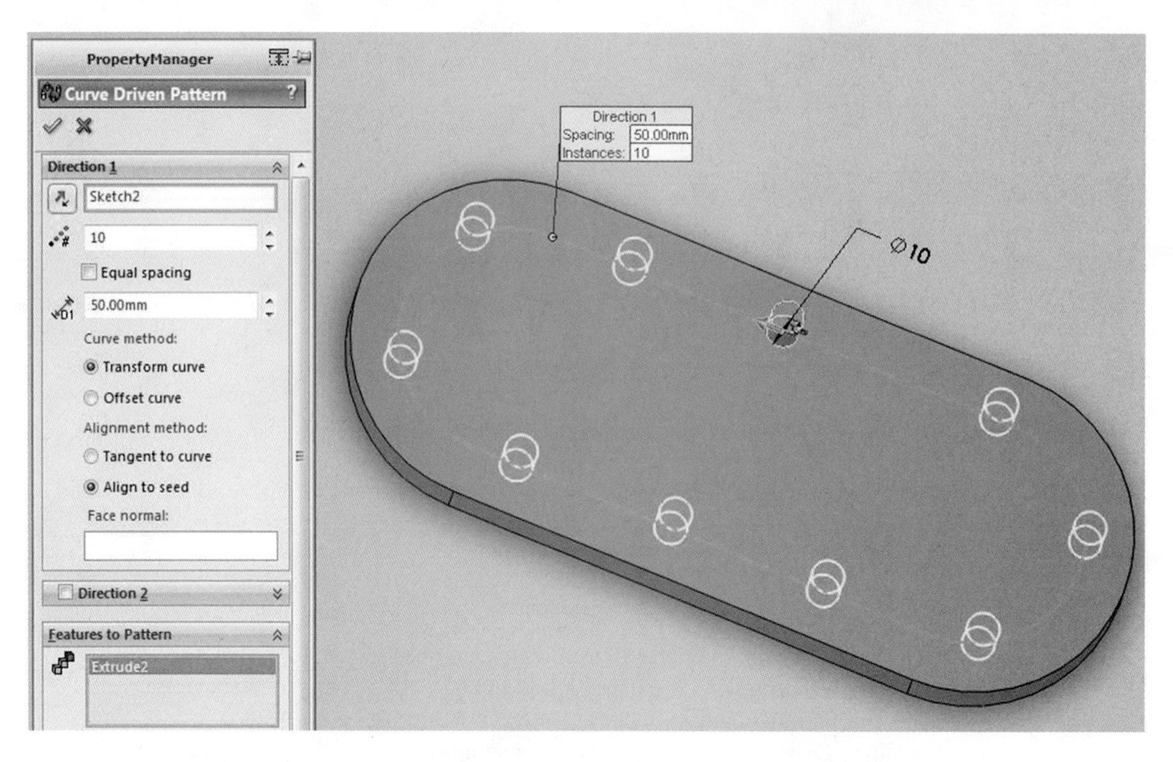

Figure 3.109

Curve-driven pattern PropertyManager.

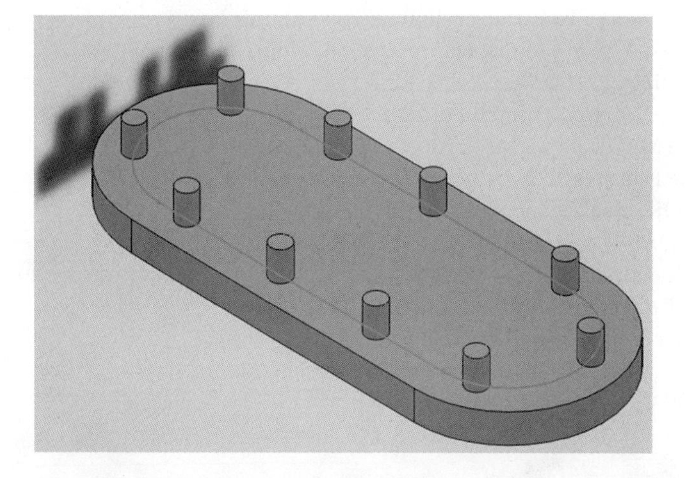

Figure 3.110

Resulting curve-driven pattern feature.

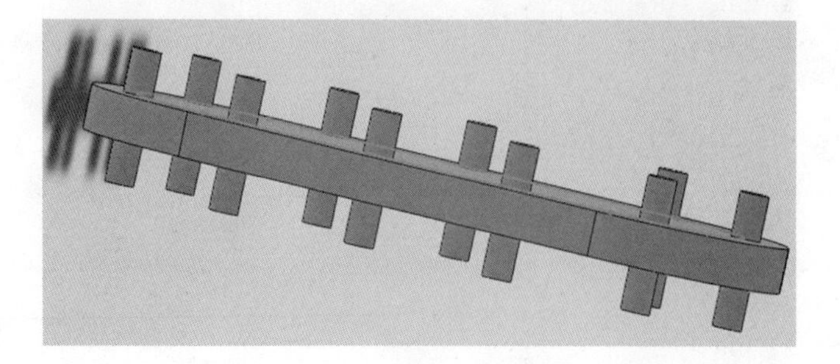

Figure 3.111

Mirrored pattern feature.

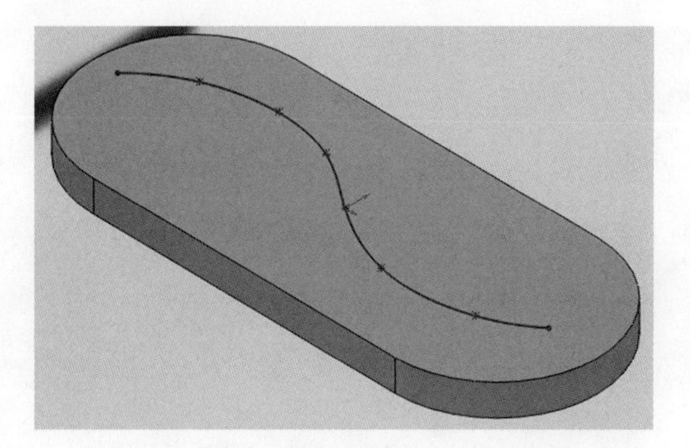

Figure 3.112

Sketch2 for Direction1 pattern-driven curve.

4. Exit Sketch mode.
   Repeat for Line1 (see Figure 3.113).

Create a Seed Feature

5. Select the top face of the base.
6. Sketch a circle, Sketch3, diameter 10 mm with the center passing through the curve.
7. Extrude Sketch3 through 15 mm (see Figure 3.114).
8. Click Feature > Linear Pattern > Curve Driven Pattern (see Figure 3.109).
   The Curve Driven Pattern PropertyManager is automatically displayed; see Figure 3.114.
9. In the Direction1 rollout, select Sketch2 as the Pattern Direction.
10. For the Number of Instances, set it to 5.
11. For the Spacing, set it to 50.
12. In the Direction2 rollout, select Line1@Sketch4 as the Pattern Direction.
13. For the Number of Instances, set it to 2.
14. For the Spacing, set it to 15.
15. In the Features To Pattern rollout, select Extrude2.
16. Click OK. (See the model with pattern in Figure 3.115.)

Mirror the Resulting Curve-Driven Pattern Feature

1. Click the Mirror command option.
2. For the Mirror Face/Plane, select the top plane.
3. For the Features To Mirror, select CrvPattern6.
4. Click OK. (See the Mirror PropertyManager and the mirrored pattern in Figures 3.116 and 3.117, respectively.)

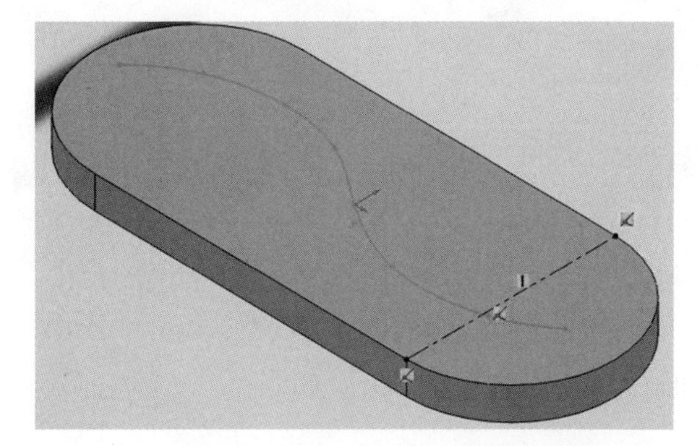

Figure 3.113

Line1 for Direction2 pattern-driven curve.

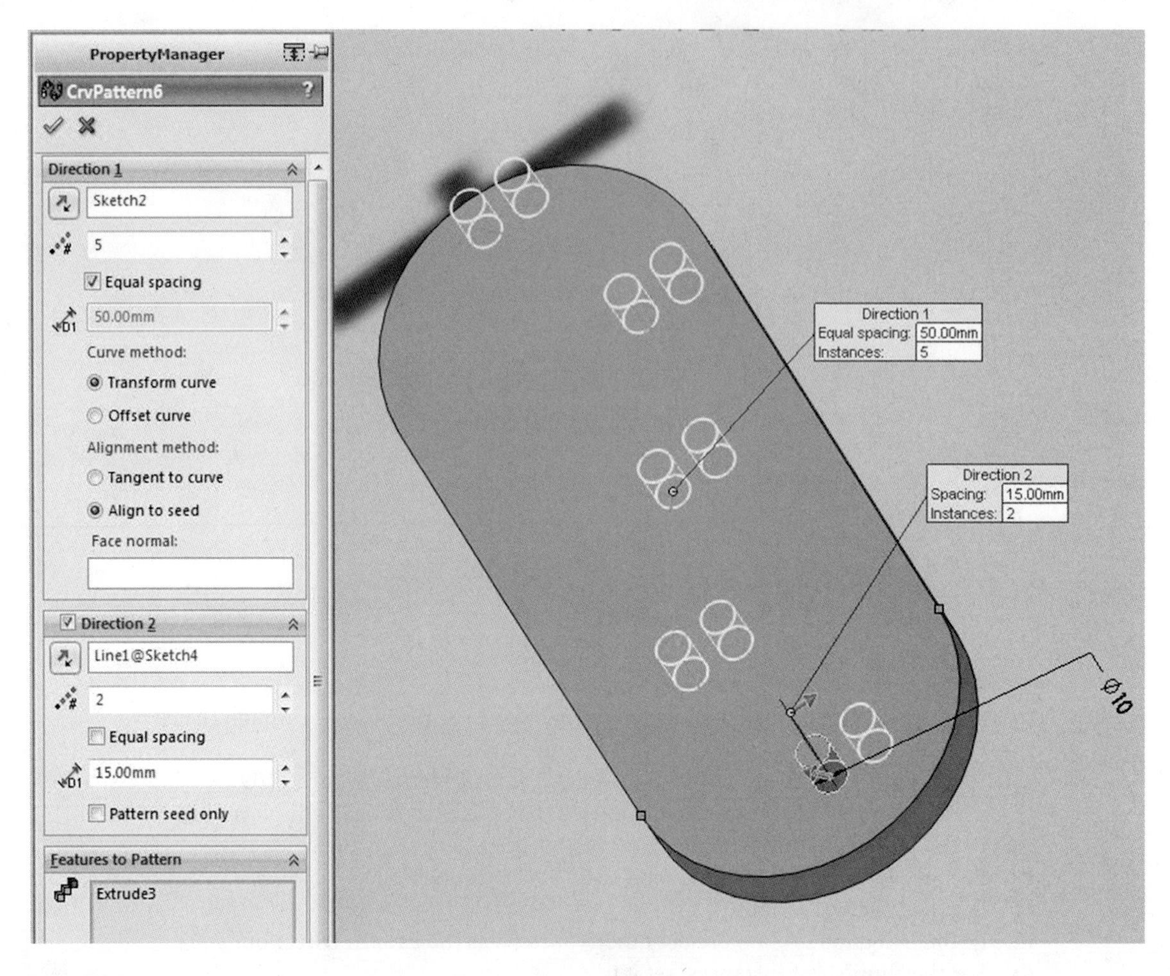

Figure 3.114

Curve-driven pattern PropertyManager.

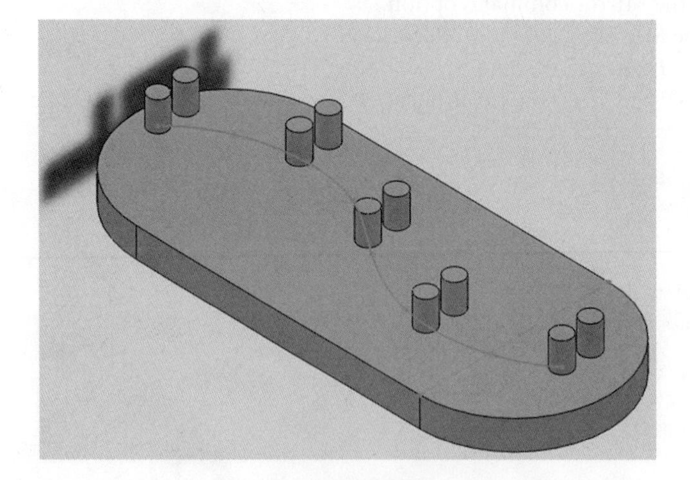

Figure 3.115

Resulting curve-driven pattern feature..

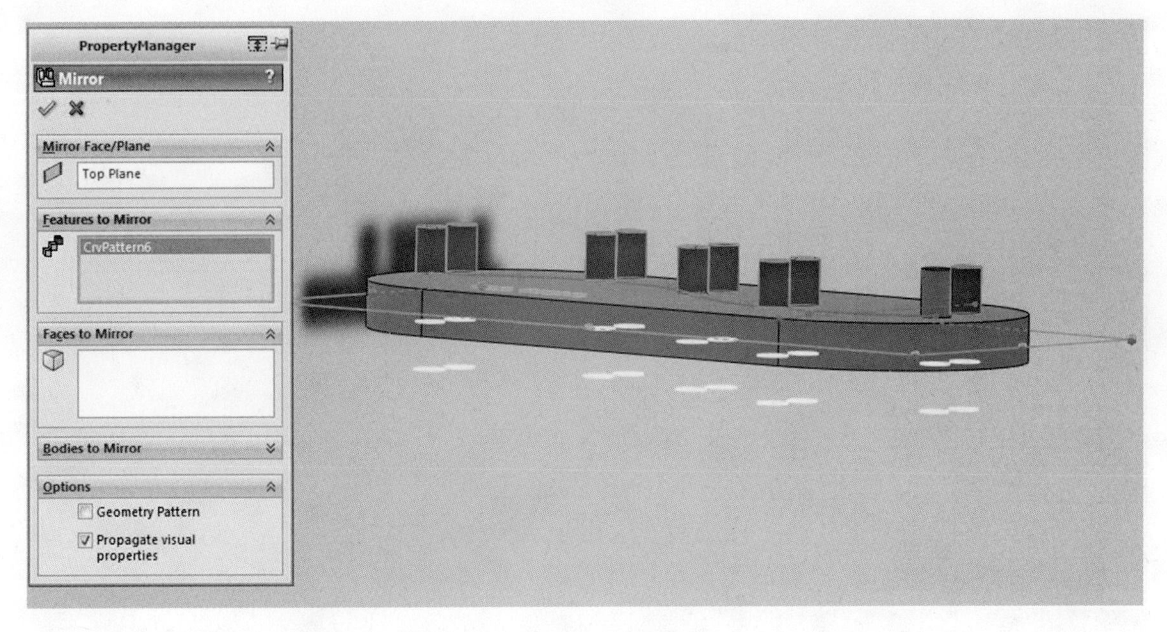

Figure 3.116

Mirror PropertyManager.

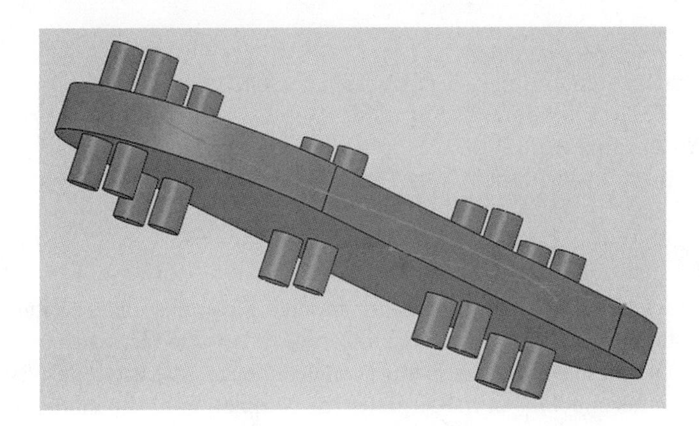

Figure 3.117

Mirrored pattern feature.

Table-Driven Pattern

1. Open a New Part document.
2. Select the top plane for sketching.
3. Sketch a rectangle, Sketch1, with a length of 150 mm and a width of 100 mm.
4. Extrude Sketch1 using the Mid Plane option through 10 mm.
5. Sketch a circle, Sketch2, diameter 5 mm with its center 10 mm × 10 mm from one corner of Sketch1.
6. Extrude-cut Sketch2 using the Mid Plane option through 10 mm (see Figure 3.118).
7. Create a coordinate system, Coordinate System1.
8. Click Feature > Linear Pattern > Table Driven Pattern (see Figure 3.109).

   The Table Driven Pattern PropertyManager is automatically displayed; see Figure 3.118.

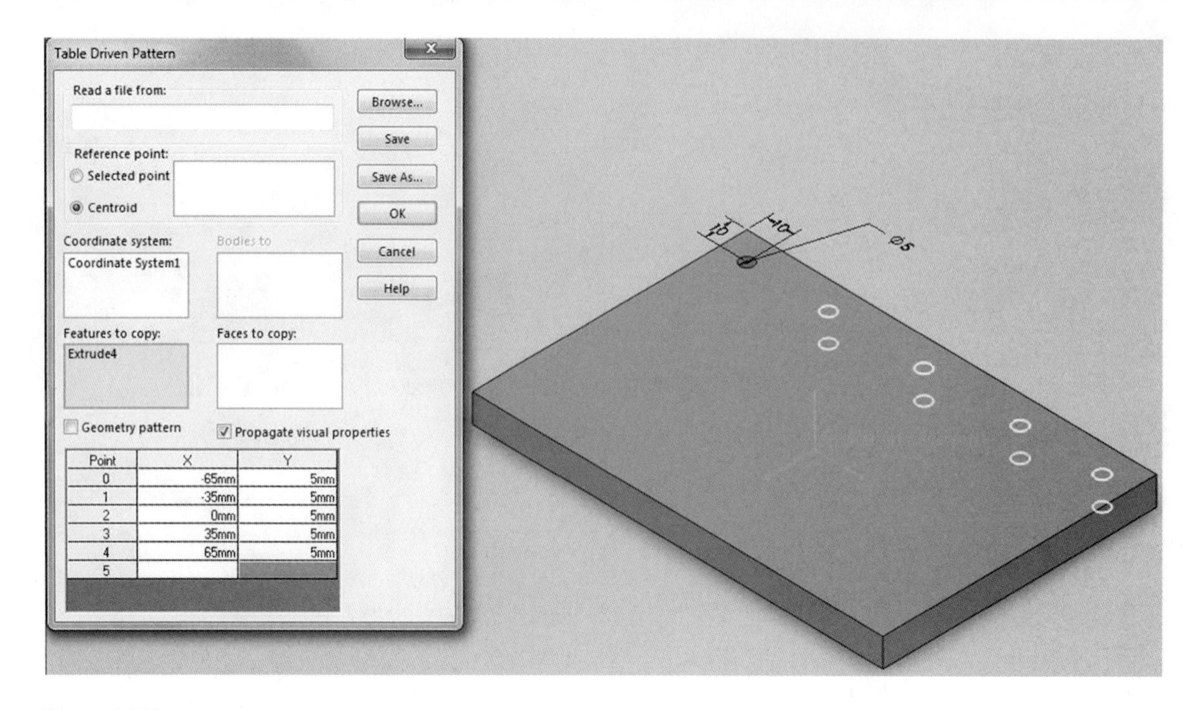

Figure 3.118

Table-driven pattern PropertyManager.

9. In the Coordinate System rollout, select Coordinate System1 from the FeatureManager.
10. In the Features To Copy rollout, select Extrude4.
11. Enter the X, Y Coordinates in the Table for driving the pattern.
12. Click OK. (See the model with pattern in Figure 3.118.)

Reference Planes

The Reference Planes tool is a very powerful tool for modeling complex objects. Reference planes can be defined relative to the standard planes (top, right, and left) or relative to faces on a model from which other geometries are sketched to complete the part being modeled. These were used for lofting, as could be seen. They were also used in other models but hidden after the completion of the modeling exercise. Reference planes are very useful and should be generously used.

The *Plane PropertyManager*, which is new in *SolidWorks 2010*, and newer versions provide the following selections: (a) *First Reference*, (b) *Second Reference*, and (c) *Third Reference*.

Select a face, a plane, or an edge to create a reference plane. The following options are available:

- *Reference Entities:* Displays the selected planes either from the FeatureManager or from the Graphics window.
- *Parallel:* Creates a plane through a point parallel to a plane or a face.
- *Perpendicular:* Creates a plane through a point perpendicular to a plane or a face.
- *Coincident:* Creates a plane through a point coincident to a plane or a face.
- *Angle:* Creates a plane through an edge, axis, or sketch line at an angle to a face or plane. Enter the angle value in the angle box.
- *Distance:* Creates a plane parallel to a plane or a face, which is offset by a specified distance. Enter the *Offset distance* value in the *distance* box.
- *Reverse direction:* Reverses the direction of the angle if needed.
- *Number of Plane to Create:* Displays the selected number of planes to be created.
- *Mid Plane:* Planes are equally spaced on both sides.

Tutorial 3.1: Planes: Using Three Vertices

1. Open a New SolidWorks Part document.
2. Select the front plane.
3. Create a rectangle (100-mm long and 10-mm high) and Extrude (50 mm, mid-plane).
4. Save extruded part.
5. Click Reference Geometry > Plane. (The Plane PropertyManager is displayed.)
6. Click the three vertices (Vertex<1>, Vertex<2>, and Vertex<3>) for the First Reference, Second Reference, and Third Reference, respectively (see Figure 3.119).
7. Click OK to complete the plane definition process.

Tutorial 3.2: Planes: Angled Reference Plane

1. Open a New SolidWorks Part document.
2. Open the extruded part that is previously created in Tutorial 3.1 for Planes.
3. Click Reference Geometry > Plane. (The Plane PropertyManager is displayed.)
4. Click Right Plane from the FeatureManager for First Reference.
5. Click the At Angle button and enter 45° for Angle.
6. Click the front vertical edge of Extrude1 as Second Reference (see Figure 3.120).
7. Click OK to complete the plane definition process.

Tutorial 3.3: Planes: Parallel Planes to Top Plane

1. Open a New SolidWorks Part document.
2. Click Reference Geometry > Plane. (The Plane PropertyManager is displayed; see Figure 3.121.)
3. Click top plane from the FeatureManager for First Reference.
4. Click the Distance button and enter 25 for Distance.
5. Click the Number of Planes button and enter 3 for Number of Planes.
6. Click OK to complete the plane definition process (see Figure 3.122).

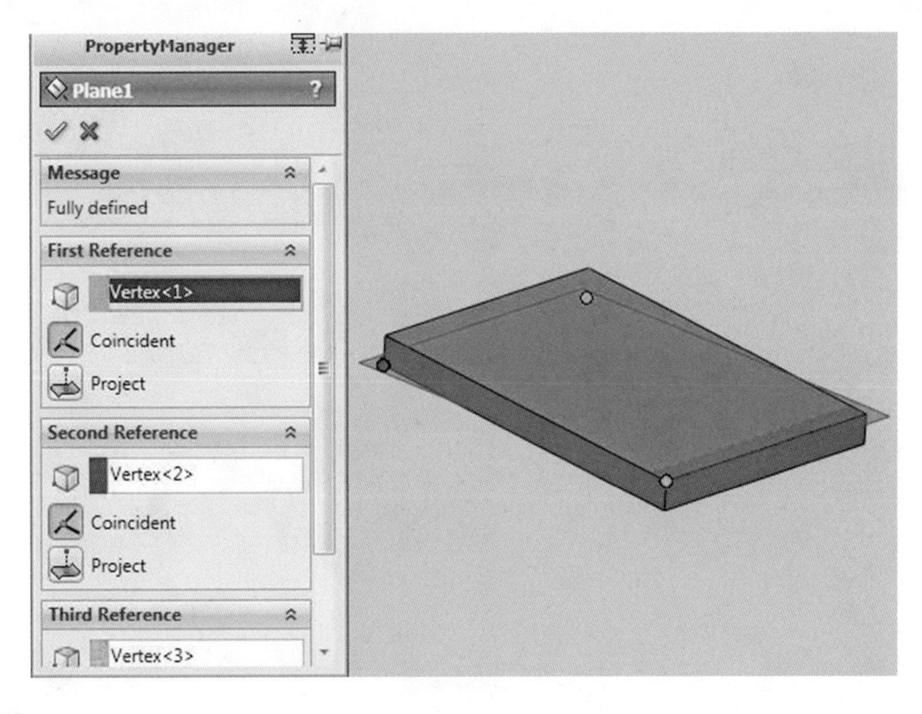

Figure 3.119

Plane defined using three vertices.

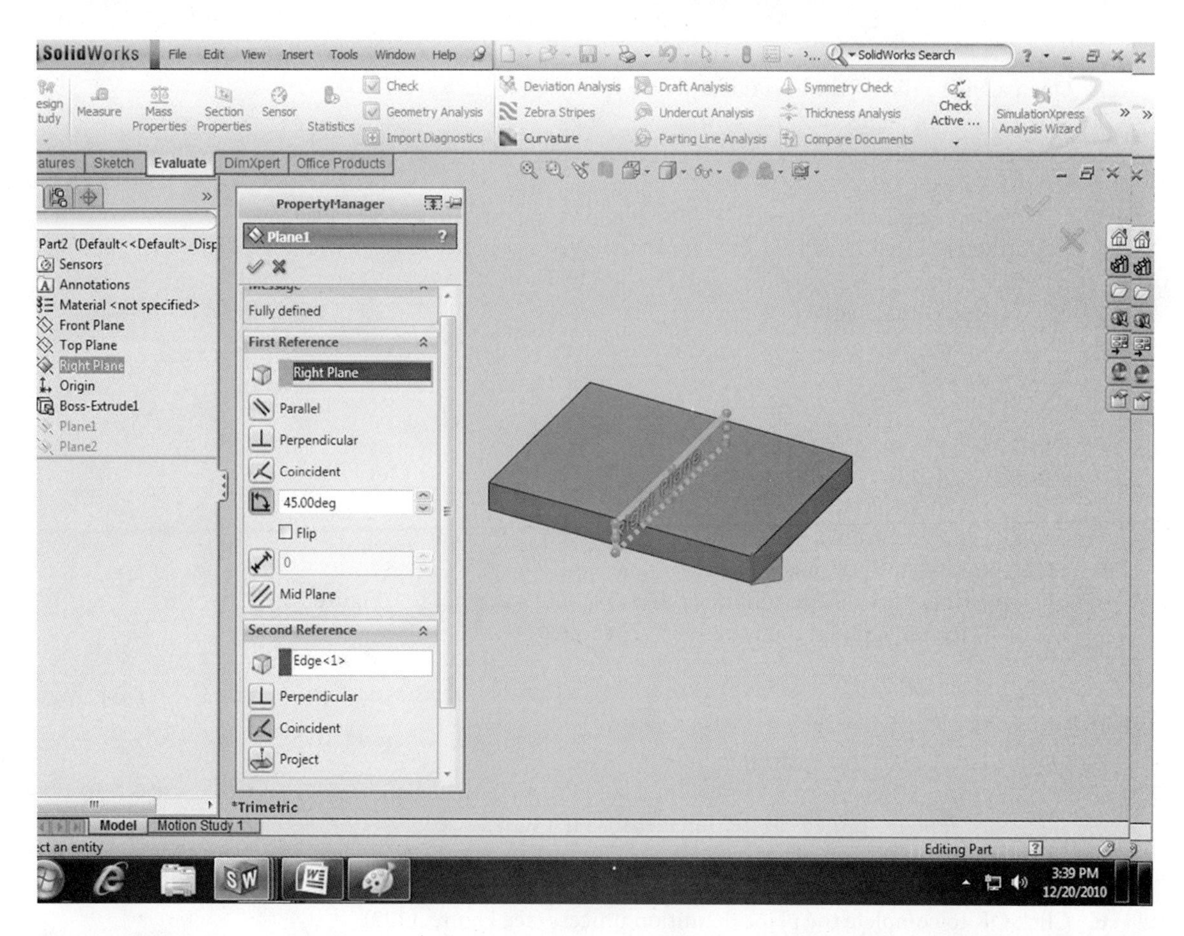

Figure 3.120

Angled reference plane.

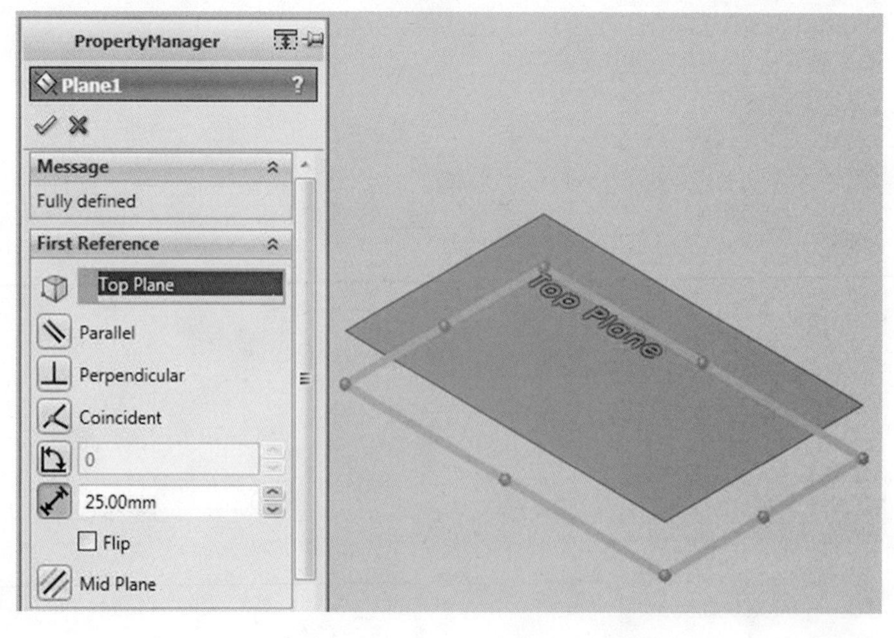

Figure 3.121

Parallel planes.

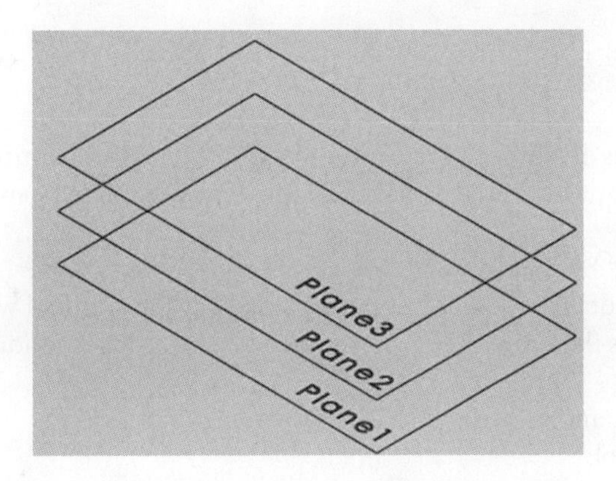

Figure 3.122

Three planes defined.

Tutorial 3.4: Planes: Inclined Plane

1. Open a New SolidWorks Part document.
2. Open the extruded part that is previously created in Tutorial 3.1 for Planes.
3. Create a Sketch, which is a line, and exit sketch.
4. Click Reference Geometry > Plane. (The Plane PropertyManager is displayed; see Figure 3.123.)
5. Click the line (Line4@Sketch2) from the FeatureManager for First Reference.
6. Click the top face (Face<1>) of the parts as Second Reference.
7. Click the At Angle button and enter 45° for Angle.
8. Click OK to complete the plane definition process (see Figure 3.123).

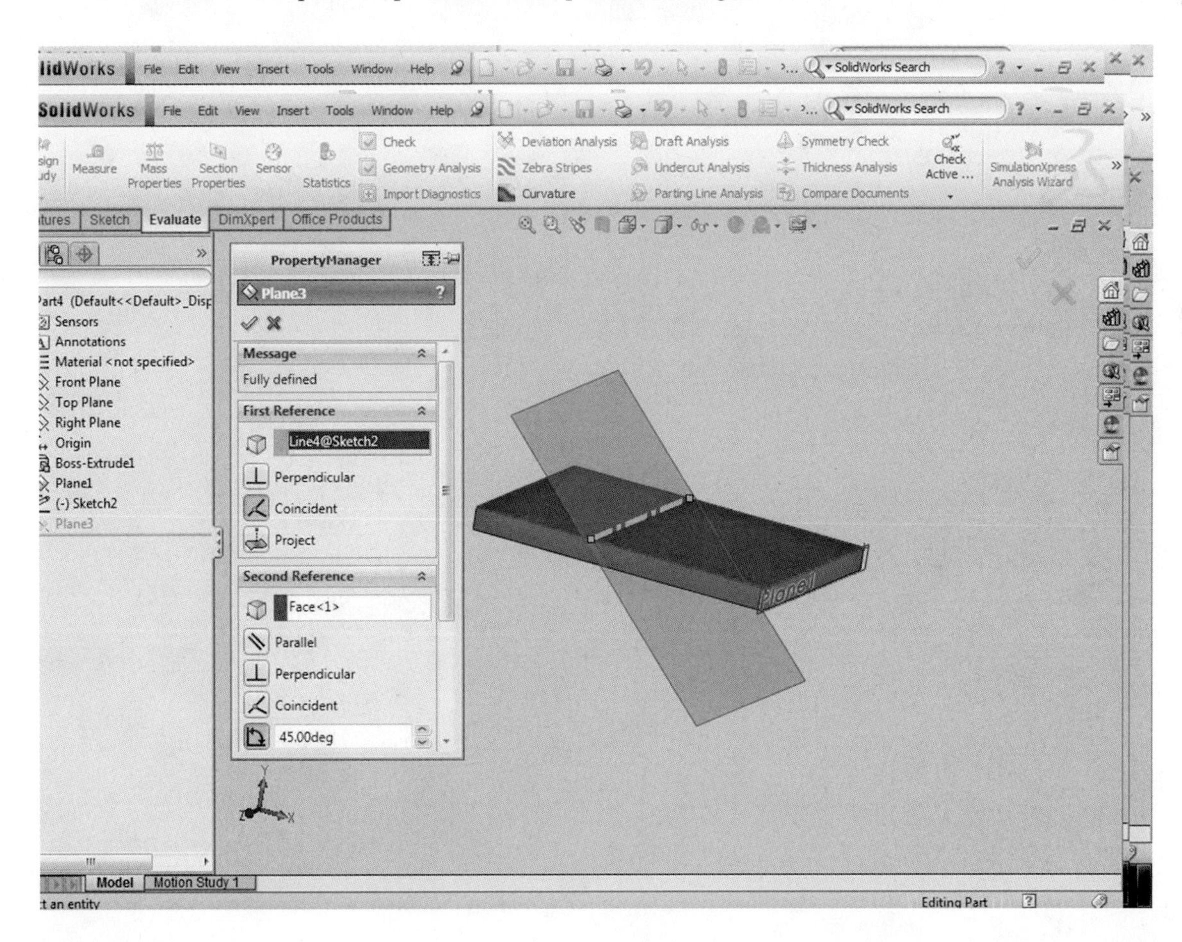

Figure 3.123

Inclined plane.

Creating Patterns

Types of Pattern

There are basically three types of pattern: (1) linear-, (2) circular-, and (3) feature-based patterns. Linear and circular patterns are most frequently used in part design. These are treated here.

Creating the Revolved Boss/Base

The revolved boss/base tool rotates a contour about an axis. It is a useful tool when modeling the parts that have circular contours. Let us illustrate the revolved boss/base concept as follows.

1. Select the front plane and create Sketch1, as shown in Figure 3.124a.
2. Click the Features tool.

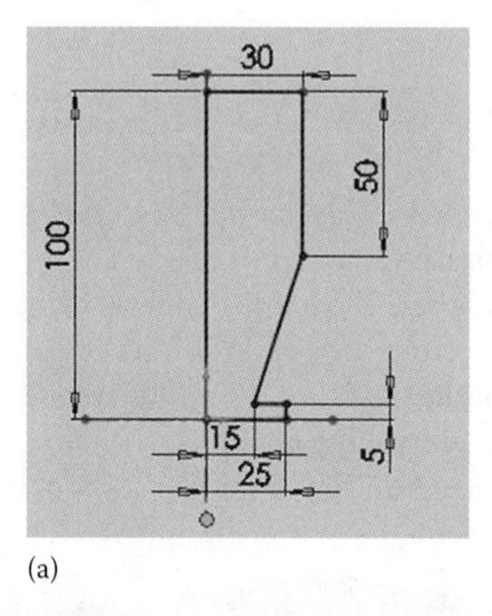

(a)

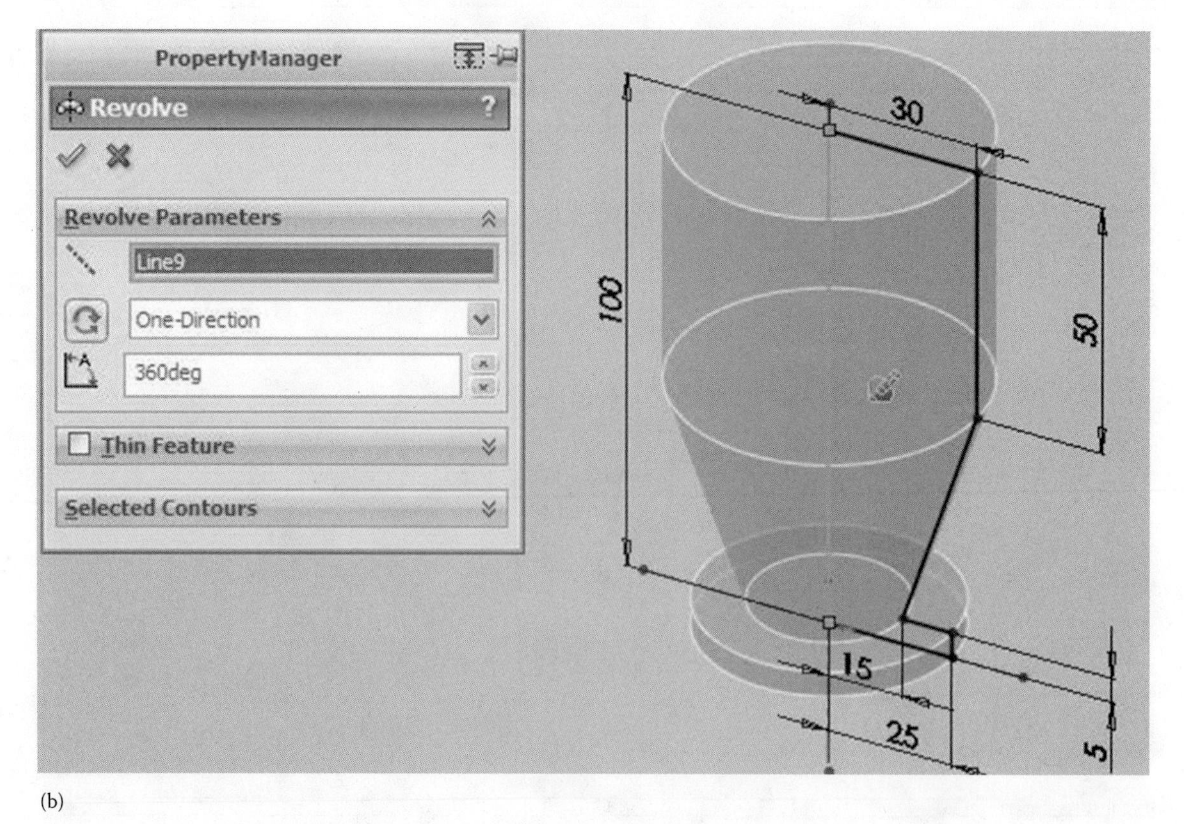

(b)

Figure 3.124

(a) Sketch1 and (b) Revolve Preview.

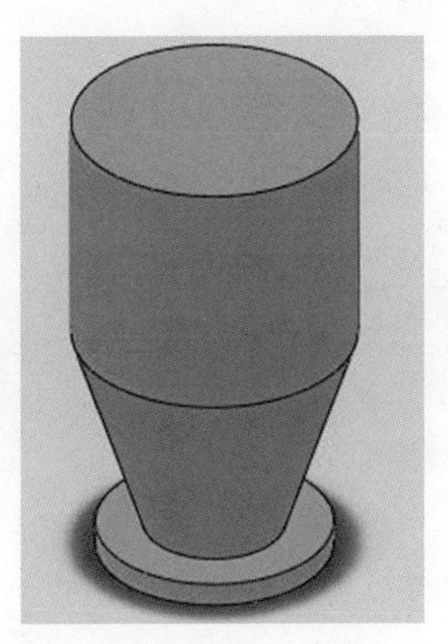

Figure 3.125

Revolve1.

3. Click the Revolved Boss/Base tool.
   The Revolve Properties Manager appears.
4. Define the revolved axis (Line1) as the vertical dimension line. A real-time preview will appear (see Figure 3.124b).
5. Click OK to complete the revolved part without fillet (see Figure 3.125).
6. Fillet with default radius of 2.5 mm (see Figure 3.126) to complete the revolved part.

Shelling

The next step is to shell the part from the top view.

7. Select the top face of Sketch1 as shown in Figure 3.127.
   Click the Shell tool. The Revolve Properties Manager appears.
8. Choose the shell thickness as 3 mm.
9. Click OK to complete the shelled part (see Figure 3.128).

Extruding a Thin Feature

10. Select the thin top plane of the shelled part.
11. Click the Convert Entities tool and click the outer diameter of the top face.
    The circle having a diameter equal to the outer diameter of the top face is extracted (see Figure 3.129).
12. Click the Offset Entities tool.
13. Choose the Offset value of 1 mm.
14. Click Reverse to offset the edge to the inside.
    A second offset circle is extracted.
15. Click OK to complete extraction.
16. Click Extrude-cut, choose Blind for a value of 5 mm.
17. Click OK to complete (see Figure 3.130).

Creating a Slot

18. Create a Reference Plane, Plane1, 30 mm from the front plane. See Figure 3.131.
19. Sketch a vertical centerline on Plane1. See Figure 3.132.
20. Create a slot Sketch on Plane1. See Figure 3.133.
21. Insert dimensions on the slot. See Figure 3.134.
22. Extrude the slot Sketch. See Figure 3.135.

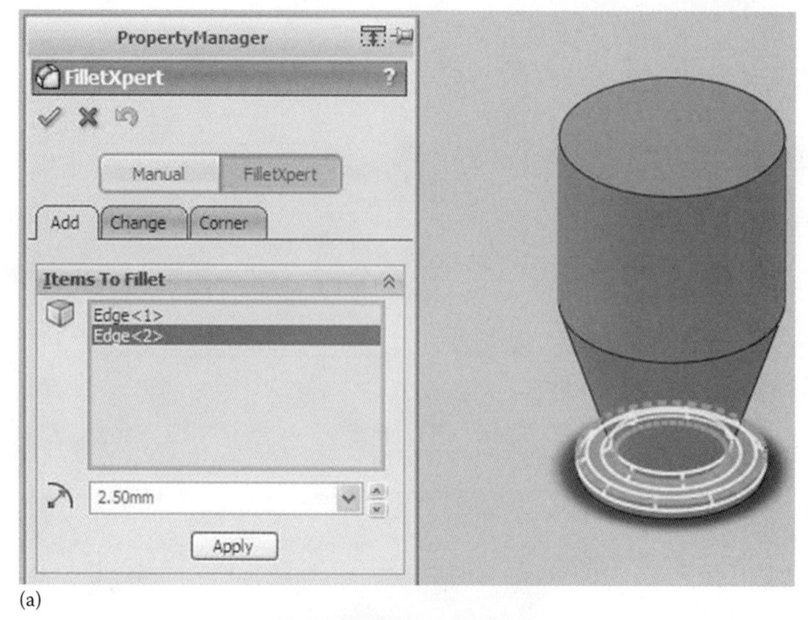

(a)

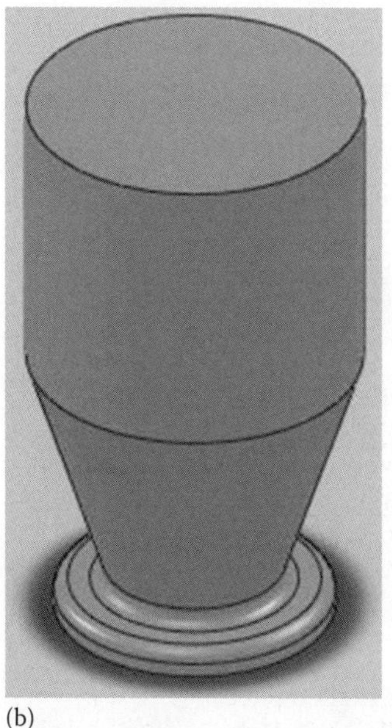

(b)

Figure 3.126

Revolved boss/base tool for creating revolved features: (a) filleting and (b) final revolved feature.

Creating the Linear Pattern

A linear pattern is now created using a temporary axis as the direction in which to create the linear pattern. In this case, we supply the spacing and set the number of instances that is desired.

1. Click the Linear Pattern toolbar. See Figure 3.136.
2. Click View > Temporary Axes.
3. Choose Direction 1 as Temporary Axis.
4. Set Spacing to 10 mm.
5. Set Number of instances to 4.
6. Click Extrude2 as the Features to Pattern from the FeatureManager.
7. Click OK.
8. Hide Plane1.

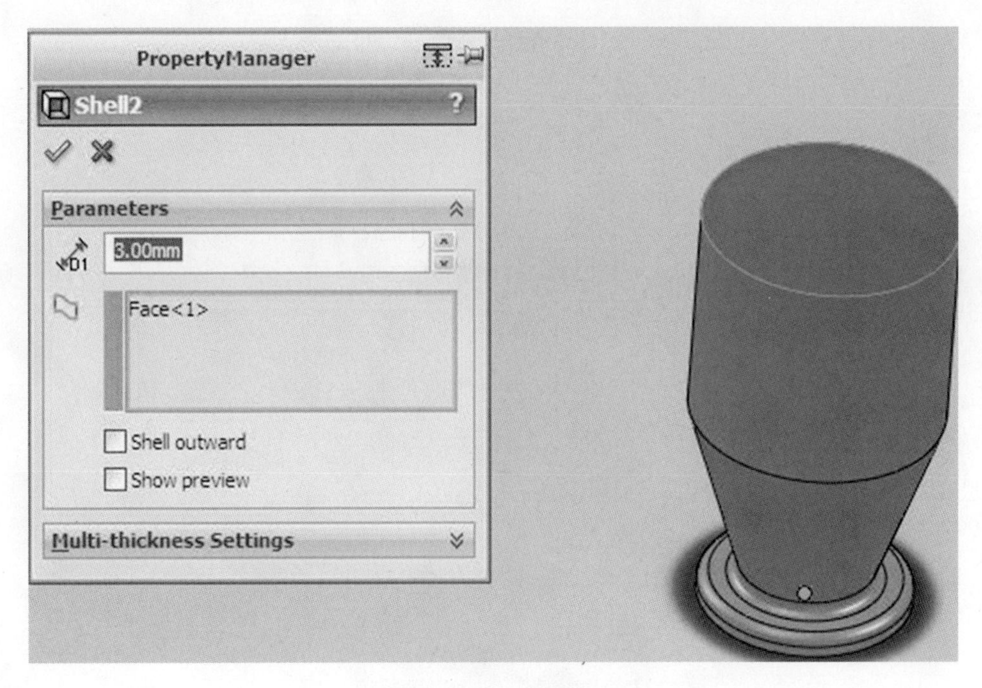

Figure 3.127

Top face of Sketch1 selected for shelling.

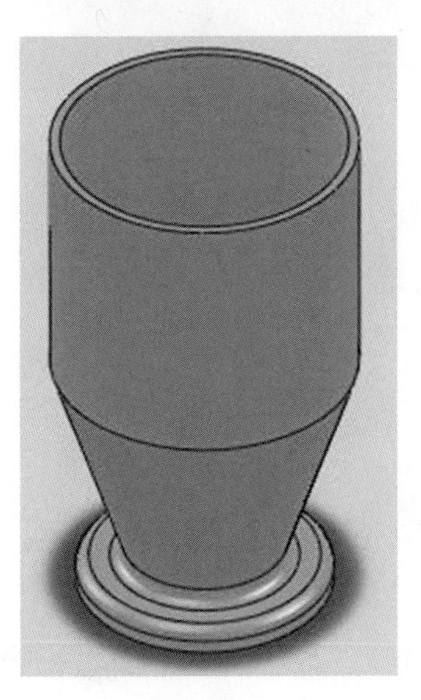

Figure 3.128

Creating revolved features with shelling operation included.

Creating a Circular Pattern of a Linear Pattern

A circular pattern of the linear pattern is now created using a temporary axis as the axis of revolution. In this case, we supply the total angle of revolution as 360° and set the number of instances that is desired.

1. Click the Circular Pattern toolbar (see Figure 3.137).
2. Choose Rotation Axis as Temporary Axis, Axis<1>.
3. Accept the default Total Angle as 360°.

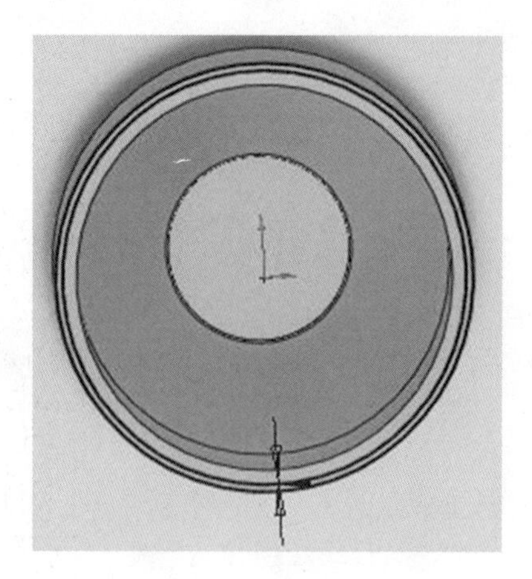

Figure 3.129

Extracting two circles.

Figure 3.130

Extrude-cut.

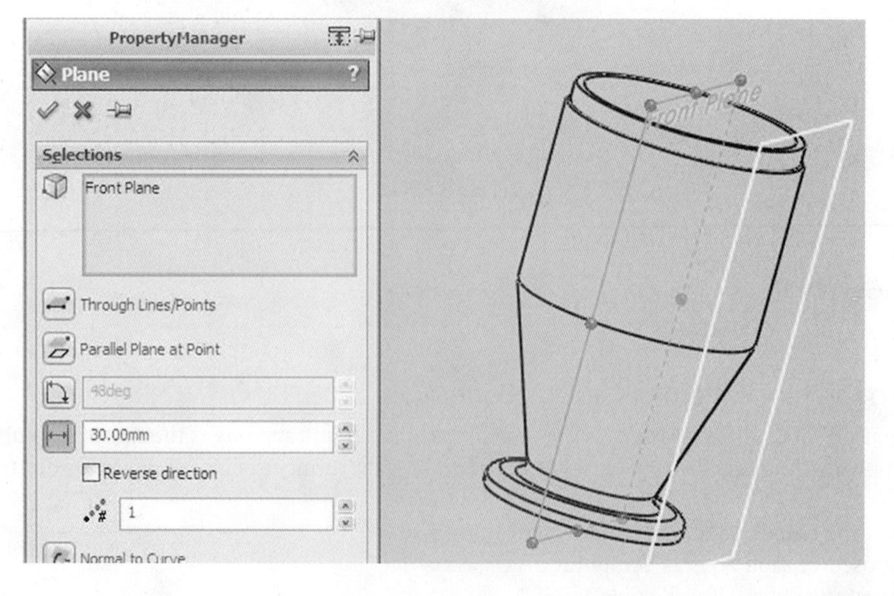

Figure 3.131

Creating a plane, Plane1.

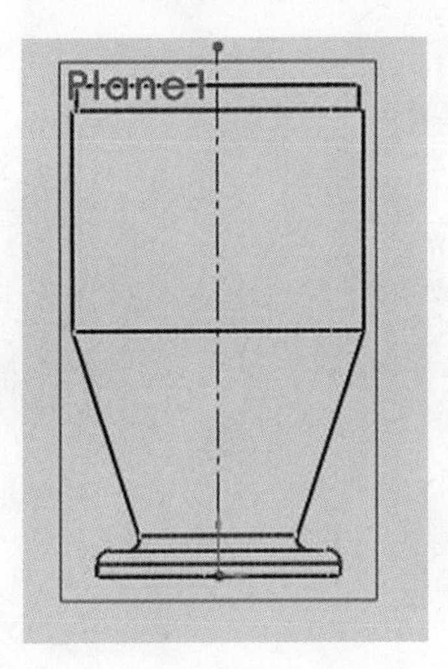

Figure 3.132

Vertical centerline on Plane1.

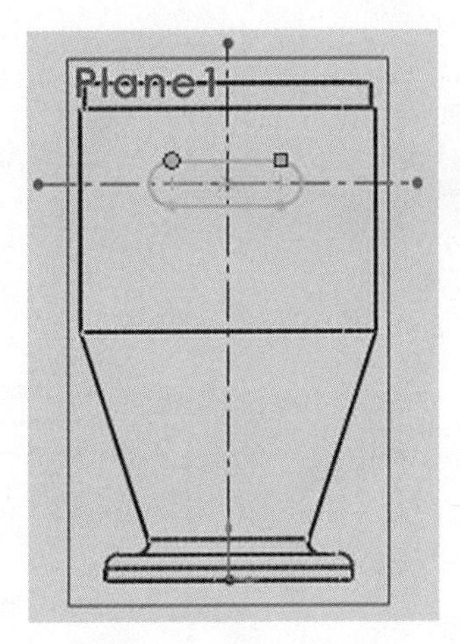

Figure 3.133

Create a slot sketch on Plane1.

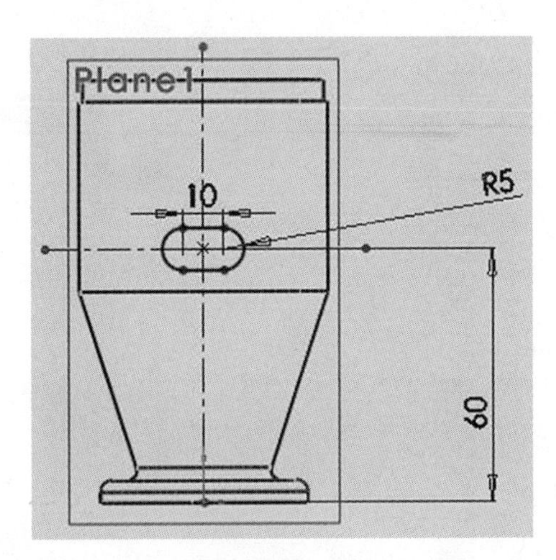

Figure 3.134

Insert dimensions on the slot.

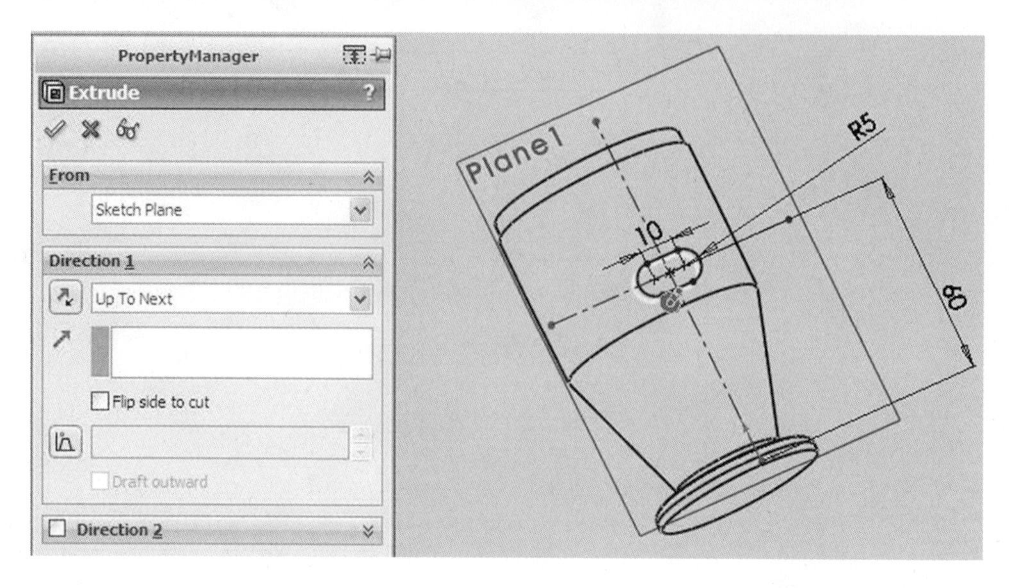

Figure 3.135

Creating a slot feature on the part.

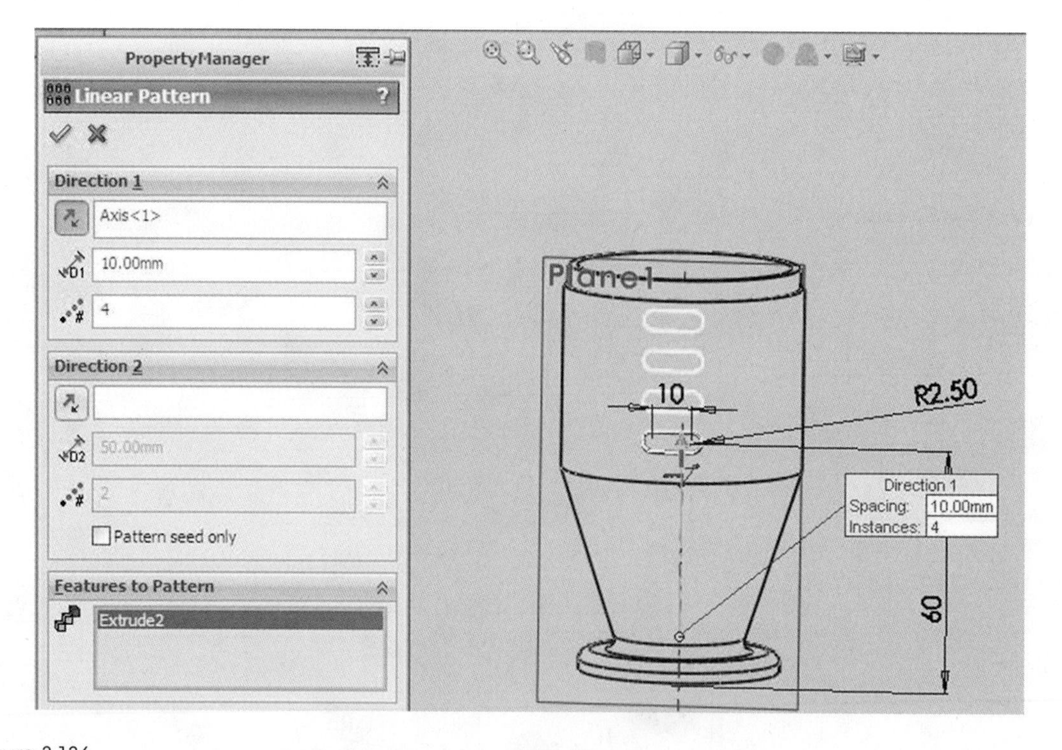

Figure 3.136

Linear pattern created on the part.

4. Set the Number of instances to 3.
5. Click LPattern2 as the Features to Pattern from the FeatureManager.
6. Click OK to obtain the completed model (see Figure 3.138).

Summary

This chapter discusses in detail the features that are used in SolidWorks for creating 3D objects. These Features tools are classified into five categories: (1) extrusion features (extruded boss/base, draft, dome,

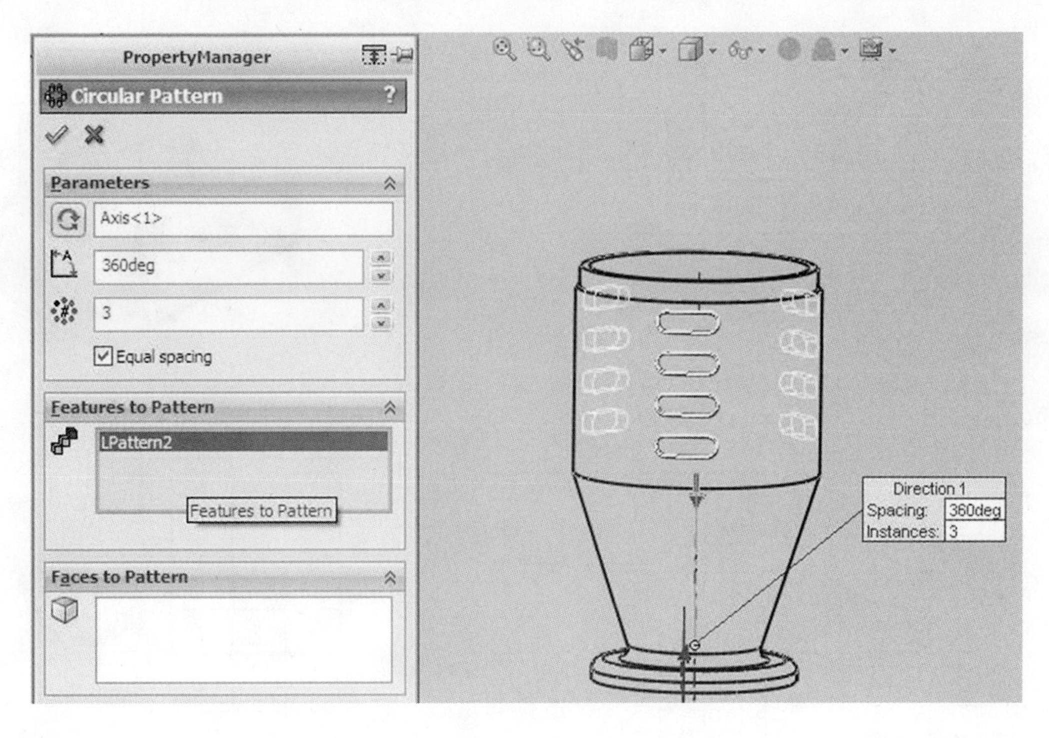

Figure 3.137

Preview of the circular pattern.

Figure 3.138

Finished part.

rib, and extruded cut), (2) revolved features (revolved boss/base and revolved cut), (3) lofted features (lofted boss/base and lofted cut), (4) swept features (swept boss/base and swept cut), and (5) modification features (hole wizard, shell, fillet, chamfer, pattern, and mirror). In the first four categories, boss/base or cut models are realized. Due to its importance, the principles of creating reference planes in order to create more complex models are also introduced. Advanced Patterns (Sketch Driven, Curve Driven, and Table Driven) tools are also introduced.

Exercises

1. Redraw the 3D solid model in Figure P3.1 based on the given dimensions.

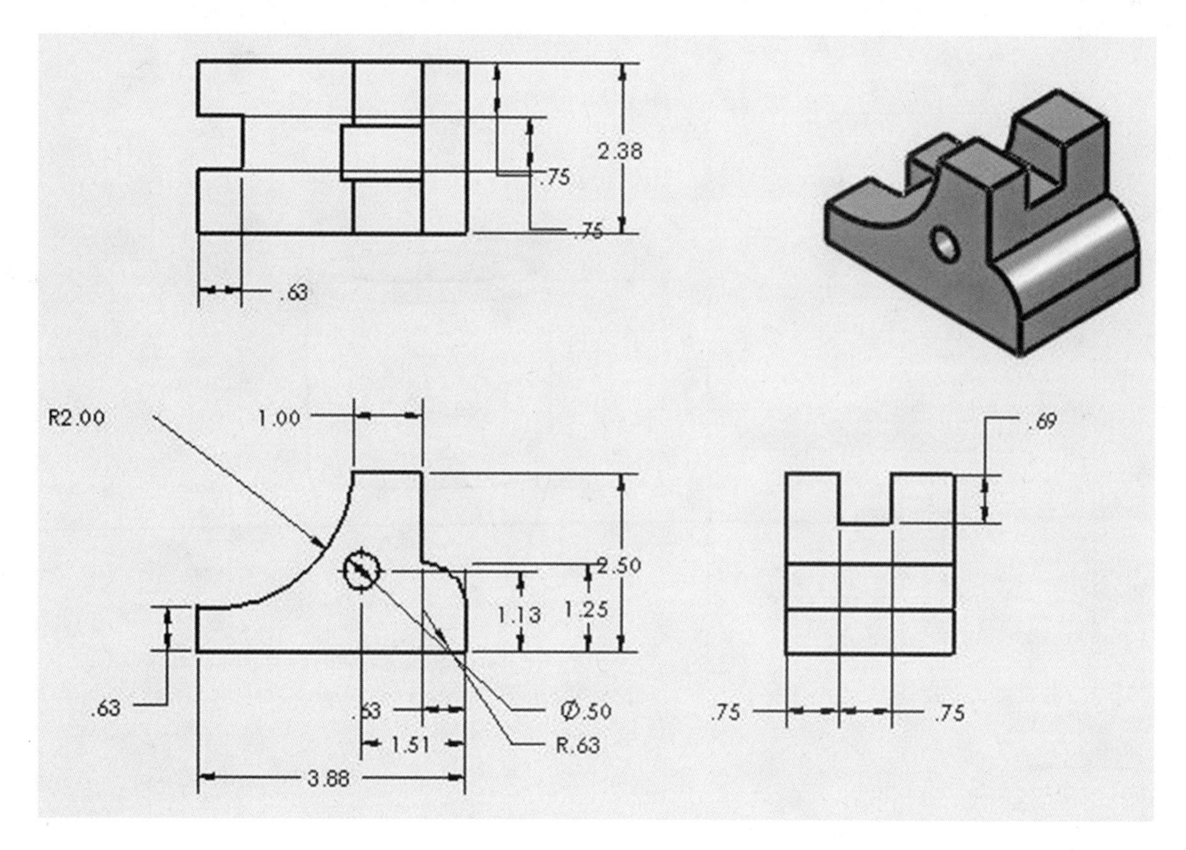

Figure P3.1

Bracket.

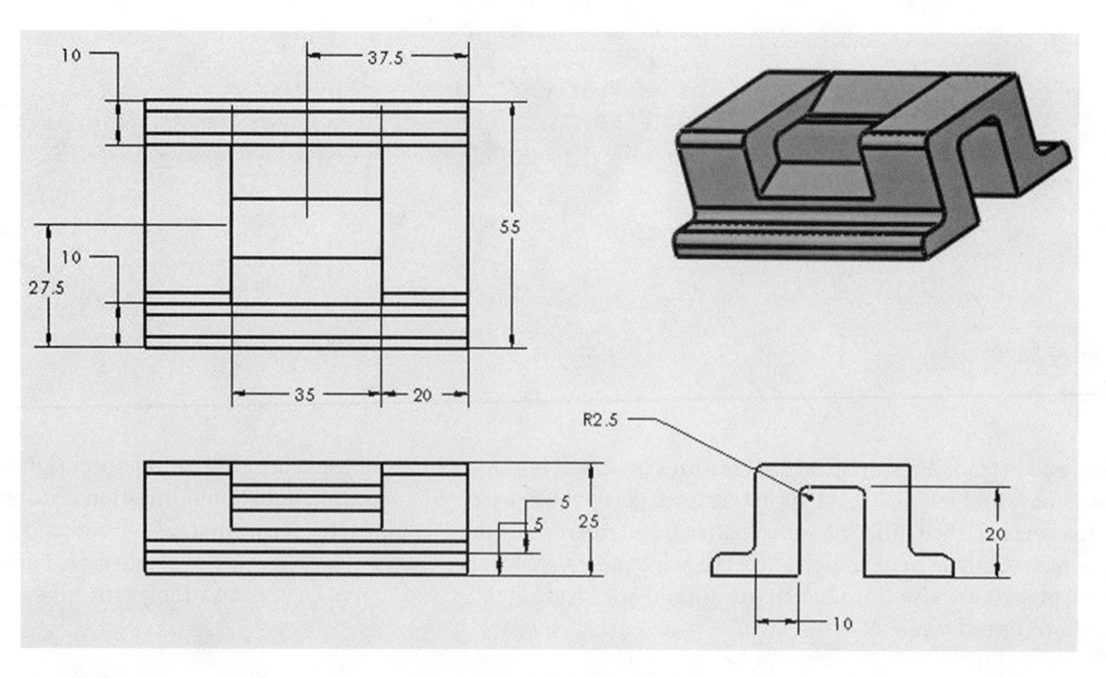

Figure P3.2

Bracket with a cut-out feature.

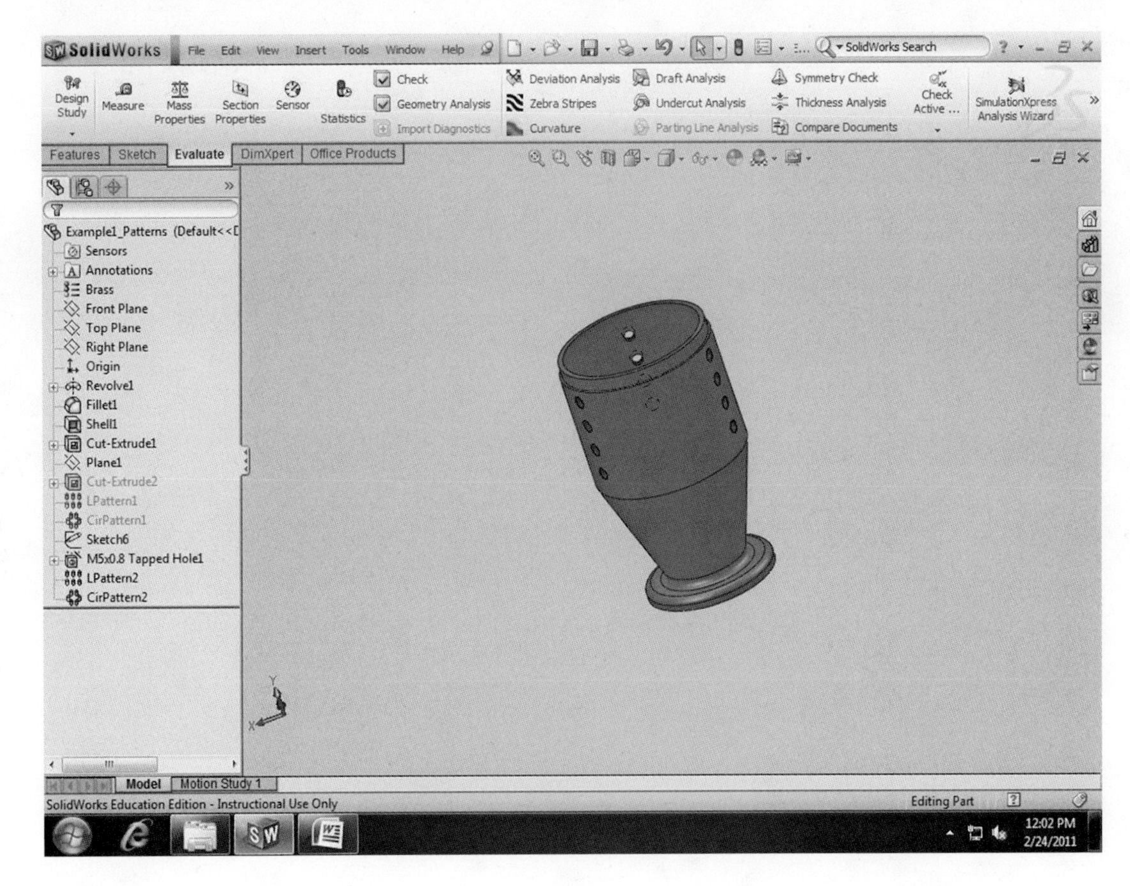

Figure P3.3

Vase.

2. Redraw the 3D solid model in Figure P3.2 based on the given dimensions.
3. Repeat Tutorial 3.9 (Simple Part 9) and mirror the part that is produced about the top plane to obtain a new model. Determine the mass properties of the model that is created. Is the mass twice that obtained in Tutorial 3.9? Explain your answer.
4. For the model having patterns shown in Figure P3.3, suppress the slot patterns. Create a seed tapped hole (M5 × 0.8), 40 mm from the top lip using the Hole Wizard tool. Follow the same procedure that is used to create the patterns that are shown in Figure P3.3. Apply Brass as the material and determine the mass properties for the model.

Part Modeling—CSWA Preparations

Objectives:

When you complete this session, you will have

- Understood the level of expertise in two-dimensional computer-aided design as it applies to engineering
- Learned how to utilize sketch tools, mirror, and draft for part modeling
- Learned how to model three-dimensional parts using Extrusion, Revolved, and Features tools, as well as using reference planes
- Solve part modeling problems that are required for Certified SolidWorks Associates

Introduction

This goal of this chapter is to assist readers to achieve the level of expertise in two-dimensional (2D) computer-aided design (CAD) as it applies to engineering and to help them pass the Certified SolidWorks Associate (CSWA), which is synonymous with competence in SolidWorks at foundation and apprentice level of 2D CAD design and engineering practices and principles. A number of tutorials are presented in this chapter to achieve this goal.

Tutorial 4.1a: Widget

Build the widget part that is shown in Figure 4.1. The lines and the arcs are tangential. Determine the center of mass for this part, if it is made from 1060 Aluminum.

1. Open a New SolidWorks part document.
2. Sketch a horizontal centerline through the origin.
3. Sketch a slanting line on one side of the centerline.
4. Click Mirror Tool.
5. Mirror the slanting line about the horizontal centerline.
6. Take the center of the 30-mm radius arc to be 60 mm to the right of the origin and the center of the 50-mm radius arc to be 40 mm to the left of the origin.

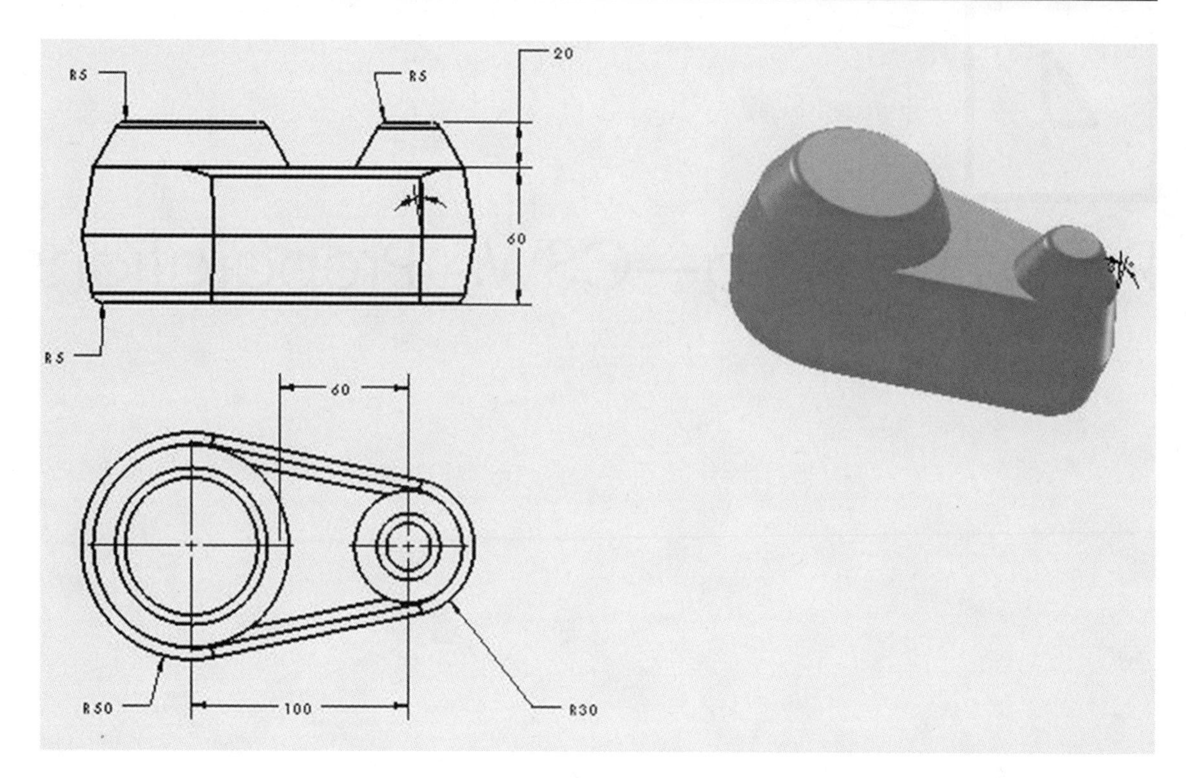

Figure 4.1

Widget.

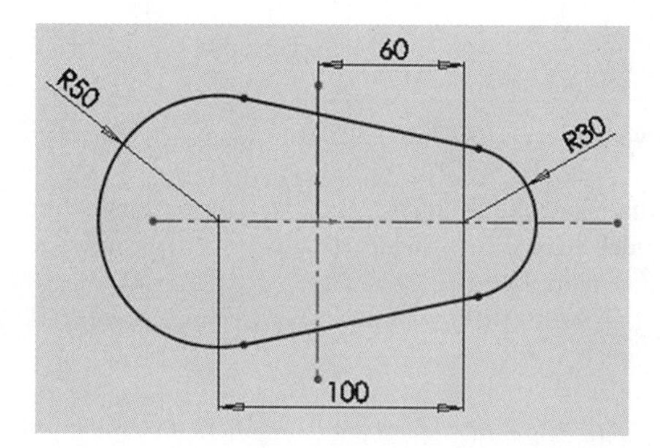

Figure 4.2

Sketch1.

7. Add a Tangent Relation between each of the lines and the two arcs. Trim any dangling edge. See Figure 4.2 for the sketch, Sketch1.
8. Click Extruded Boss/Base.
9. Set Type to Mid Plane, Depth to 60.00 mm, and Draft to 10°. (Ensure that Draft outward is cleared, if it is active; see Figure 4.3.)
10. Click OK.

Add Bosses

11. Sketch two circles using two existing centers of the existing arcs. (Highlight the arcs to awake the centers.)
12. Add Coradial Relation between the larger circle and the larger arc (see Figure 4.4).
13. Add Coradial Relation between the smaller circle and the smaller arc.
14. Click Extruded Boss/Base.
15. Set Type to Blind, Depth to 20 mm, and Draft to 30°. (Ensure that Draft outward is cleared, if it is active; see Figure 4.5.)
16. Click OK. (The modeled part is shown in Figure 4.6.)

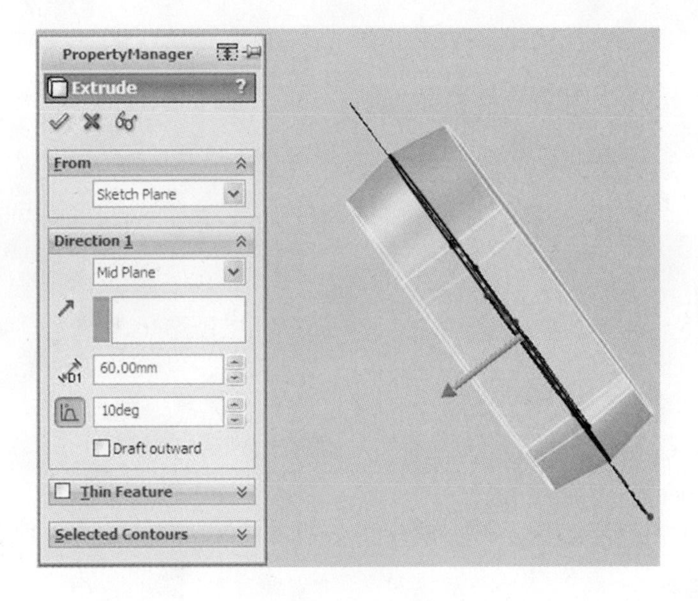

Figure 4.3

Extrude1.

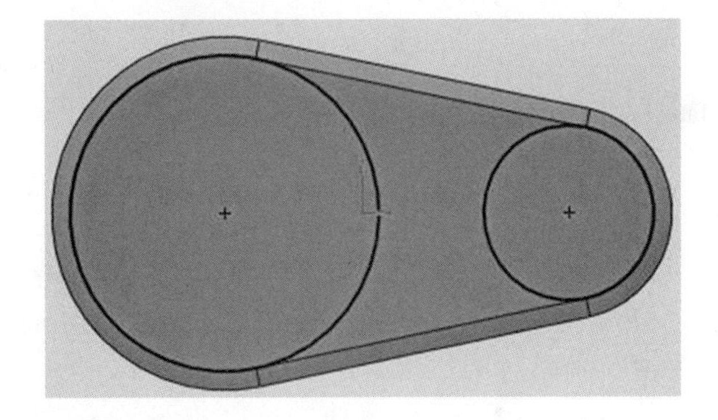

Figure 4.4

Sketches for second extrude.

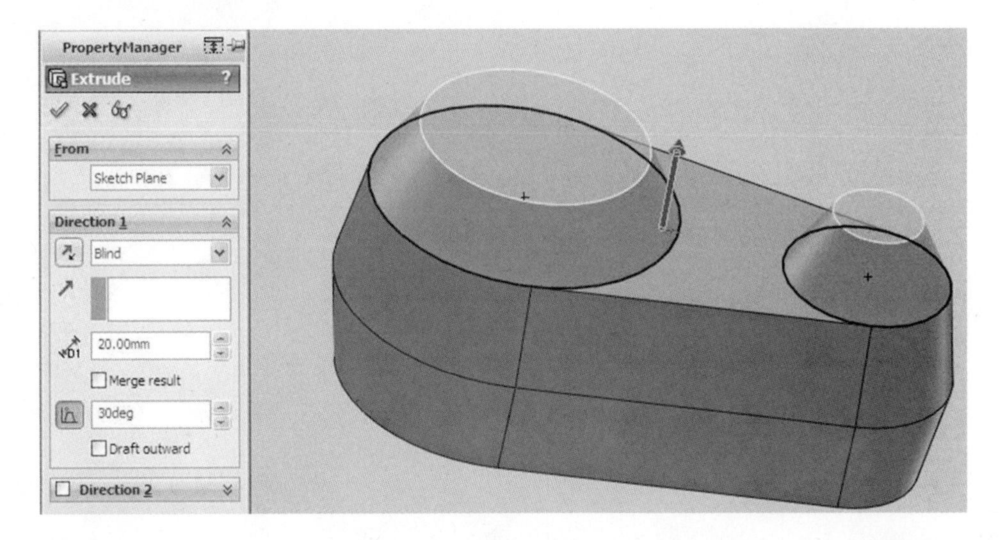

Figure 4.5

Preview.

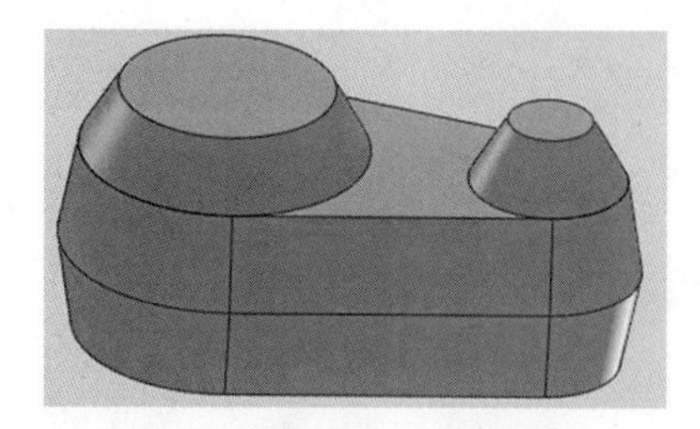

Figure 4.6

Completed part.

Center of mass

1. Right-click Material > Edit Material in FeatureManager.
2. Choose 1060 Alloy and Apply.
3. Click Evaluate > Mass Properties. (See mass calculation output in Figure 4.7.)

Tutorial 4.1b: Widget

Repeat Tutorial 4.1a if the ends of the two slanting lines and the centers of the arcs are collinear. (They have vertical relations.)

Modeling

Similar to Tutorial 4.1a with the appropriate relations applied.

Center of mass

1. Right-click Material > Edit Material in FeatureManager.
2. Choose 1060 Alloy and Apply.
3. Click Evaluate > Mass Properties. (See mass calculation output in Figure 4.8.)

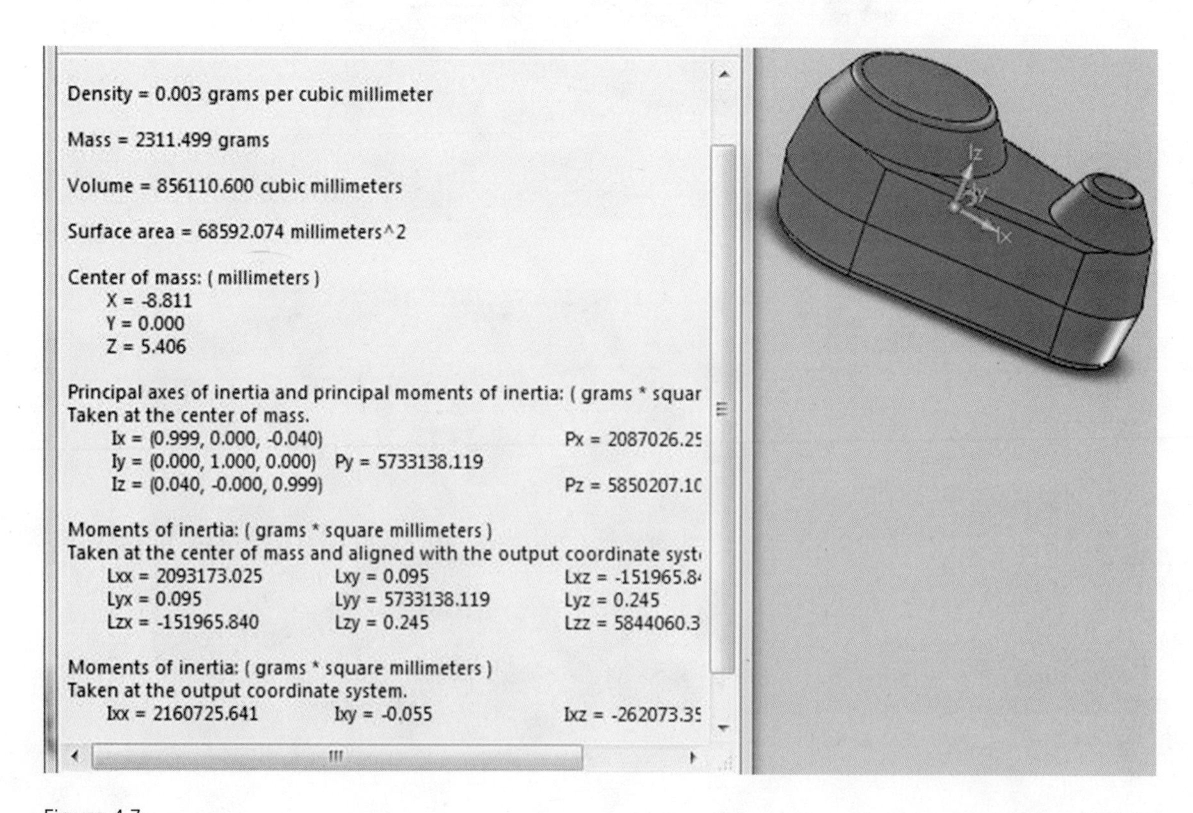

Figure 4.7

Part with mass properties.

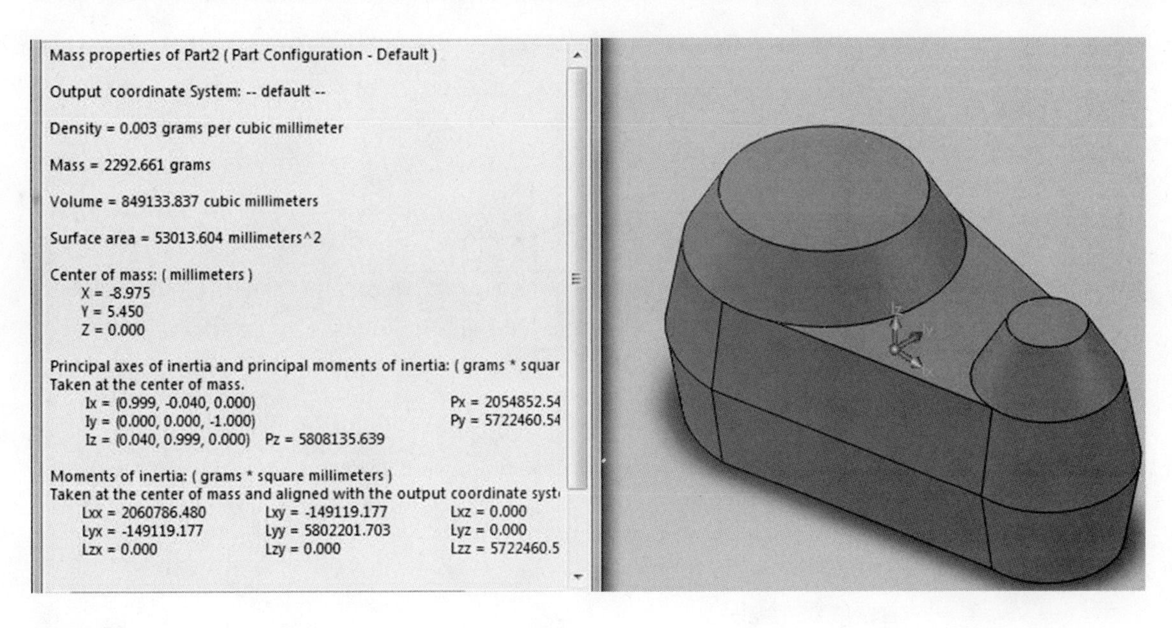

Figure 4.8

Part with mass properties.

Tutorial 4.2

Build the part that is shown in Figure 4.9. Determine the center of mass for this part, if it is made from cast alloy steel with a density of 0.0073 g/mm^3.

1. Create a New part in SolidWorks.
2. Set the document properties for the model.
3. Create Sketch1, which is the profile for Extrude1 (see Figure 4.10).
4. Create the Extrude Base feature. Extrude depth = 5 in. Extrude1 is the Base feature.
5. Create 3 Countersunk holes using Hole Wizard tool (see Figure 4.11).
6. Sketch semicircle R1.5 and a line to give a sector (see Figure 4.12) and Extrude Up To slanting surface.
7. Mirror using left-hand-side rectangular surface (Figure 4.13).

The mass properties are shown in Figure 4.14.

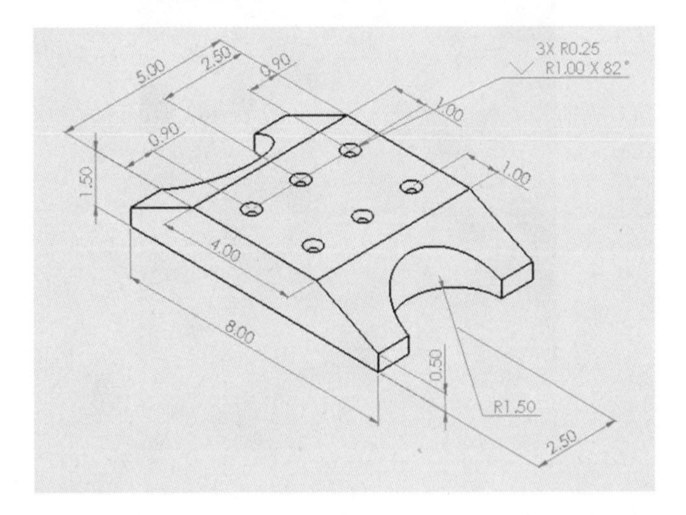

Figure 4.9

Part for modeling. (From David C. Planchard & Marie P. Planchard, Official Certified SolidWorks Associates (CSWA), SDC Publications, 2009.)

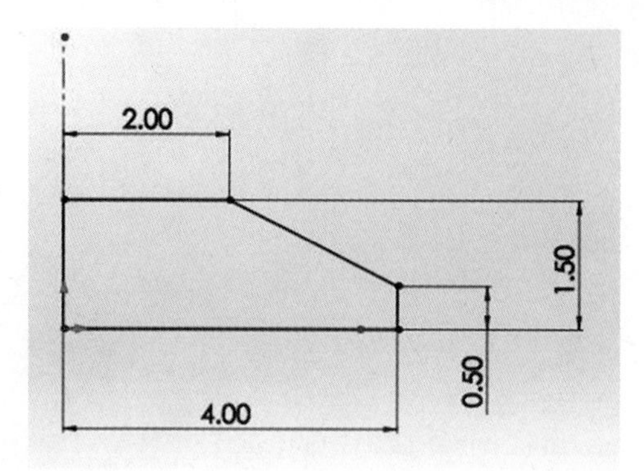

Figure 4.10

Sketch1. (From David C. Planchard & Marie P. Planchard, Official Certified SolidWorks Associates (CSWA), SDC Publications, 2009.)

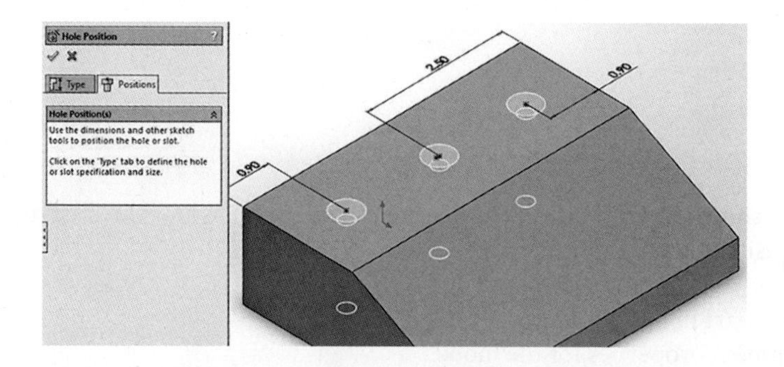

Figure 4.11

Sketch2. (From David C. Planchard & Marie P. Planchard, Official Certified SolidWorks Associates (CSWA), SDC Publications, 2009.)

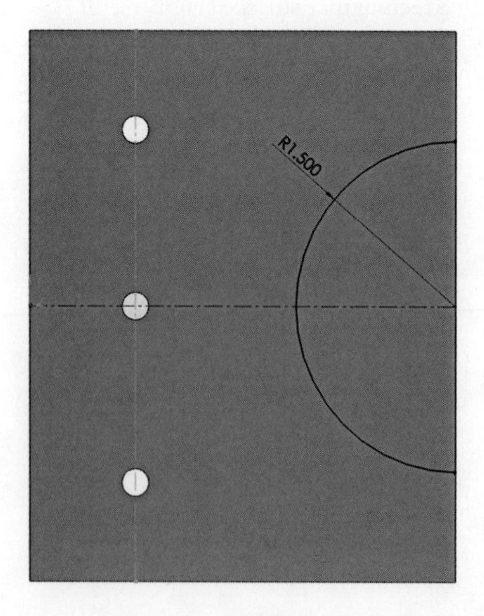

Figure 4.12

Preview for the revolved part. (From David C. Planchard & Marie P. Planchard, Official Certified SolidWorks Associates (CSWA), SDC Publications, 2009.)

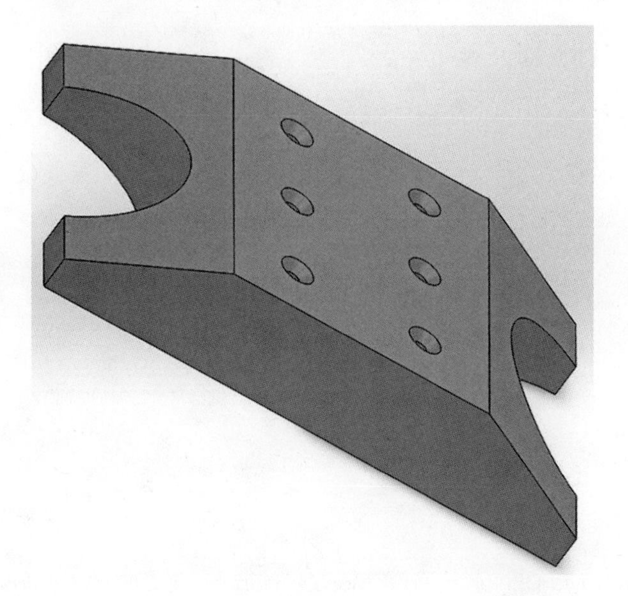

Figure 4.13

Revolved part. (From David C. Planchard & Marie P. Planchard, Official Certified SolidWorks Associates (CSWA), SDC Publications, 2009.)

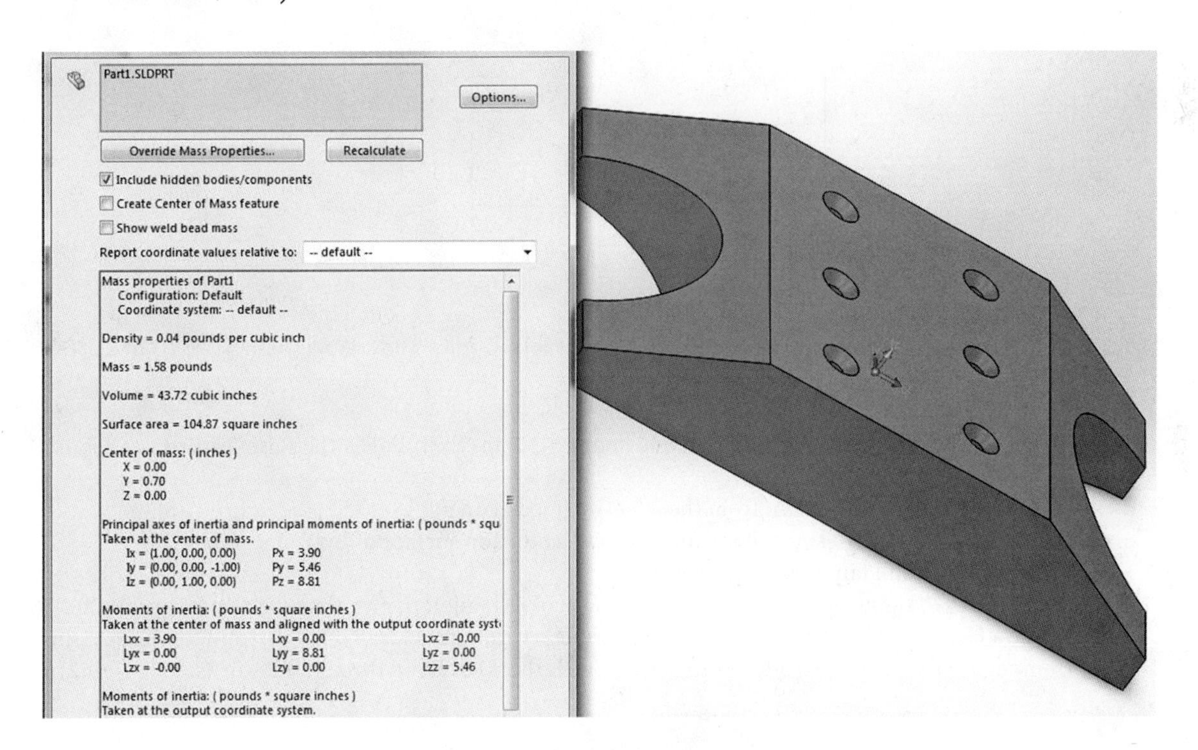

Figure 4.14

Mass properties for the part. (From David C. Planchard & Marie P. Planchard, Official Certified SolidWorks Associates (CSWA), SDC Publications, 2009.)

Tutorial 4.3a

Build the part that is shown in Figure 4.15. Determine the center of mass for this part, if it is made from 6061 Alloy with a density of 0.097 lb./in.^{3}.

1. Create a New part in SolidWorks.
2. Set the document properties for the model.
3. Create Sketch1. Select the Top Plane as the Sketch Plane (see Figure 4.16).

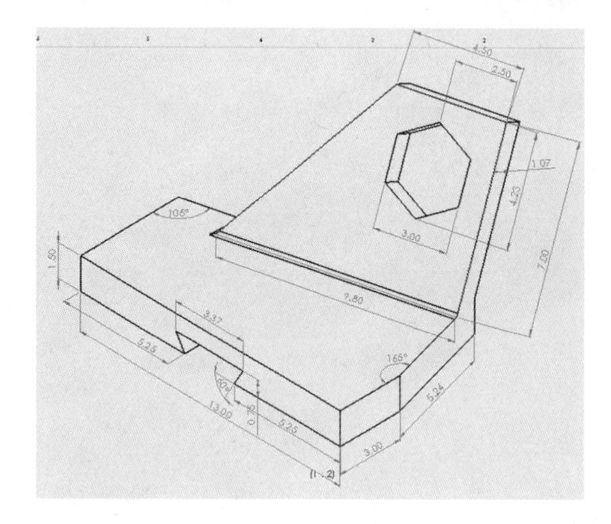

Figure 4.15

Part for modeling. (From David C. Planchard & Marie P. Planchard, Official Certified SolidWorks Associates (CSWA), SDC Publications, 2009.)

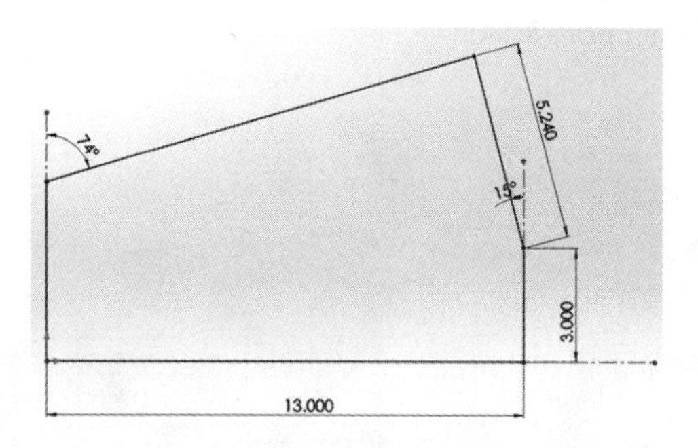

Figure 4.16

Sketch1. (From David C. Planchard & Marie P. Planchard, Official Certified SolidWorks Associates (CSWA), SDC Publications, 2009.)

4. Create the Extrude Base feature. Extrude depth = 1.5 in. Extrude1 is the Base feature (see Figure 4.17).
5. Create Sketch2, a line 1.24-in. from the edge, inclined at an angle of 15° (see Figure 4.18).
6. Create Plane1, passing Sketch at an angle of 120° to the top surface of Extrude1.
7. Create Sketch3 on Plane1 (see Figure 4.19).
8. Extrude Up To right-top vertex of the top-surface of Extrude1, to give Extrude2 (see Figure 4.20).
9. Create a Hexagon, Sketch4 on front face of Extrude2, with center located 4.5-in. from top, and 2.5-in. from edge of Extrude2; Extrude Up To Surface (at the back) or Blind through 1.24-in. (see Figure 4.21).

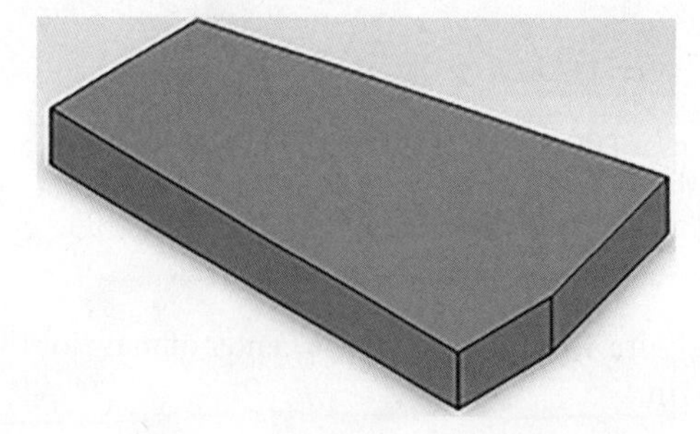

Figure 4.17

Extrude1. (From David C. Planchard & Marie P. Planchard, Official Certified SolidWorks Associates (CSWA), SDC Publications, 2009.)

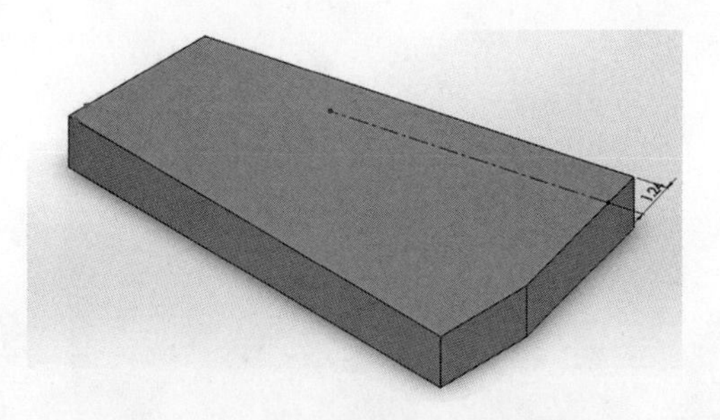

Figure 4.18

Sketch2. (From David C. Planchard & Marie P. Planchard, Official Certified SolidWorks Associates (CSWA), SDC Publications, 2009.)

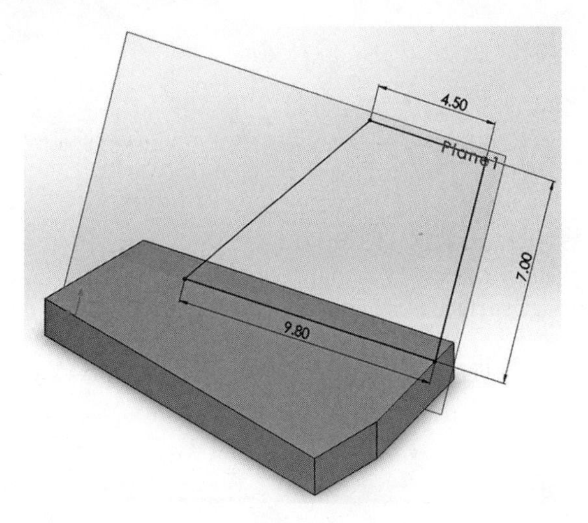

Figure 4.19

Sketch3. (From David C. Planchard & Marie P. Planchard, Official Certified SolidWorks Associates (CSWA), SDC Publications, 2009.)

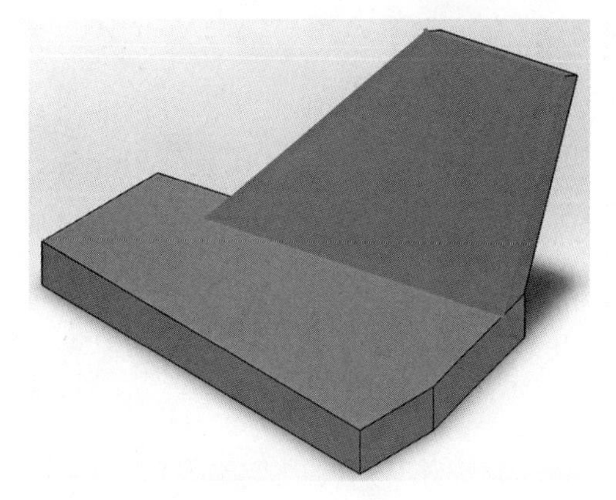

Figure 4.20

Extrude2. (From David C. Planchard & Marie P. Planchard, Official Certified SolidWorks Associates (CSWA), SDC Publications, 2009.)

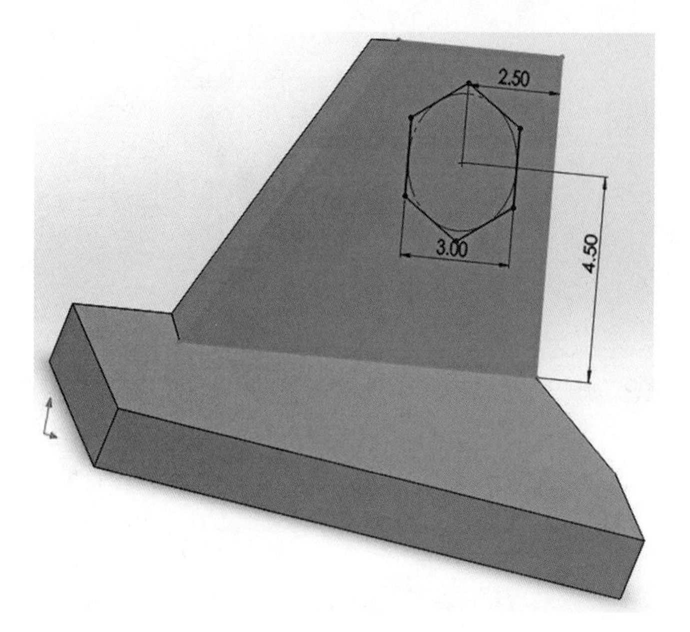

Figure 4.21

Sketch4. (From David C. Planchard & Marie P. Planchard, Official Certified SolidWorks Associates (CSWA), SDC Publications, 2009.)

10. Create Fillets of 0.25-in. radius (see Figure 4.22).
11. Create Sketch5 on the face that is 13-in. long (see Figure 4.23).
12. Extrude Cut Up To back surface (see Figure 4.24).

The completed part is shown in Figure 4.24, while the material properties are shown in Figure 4.25.

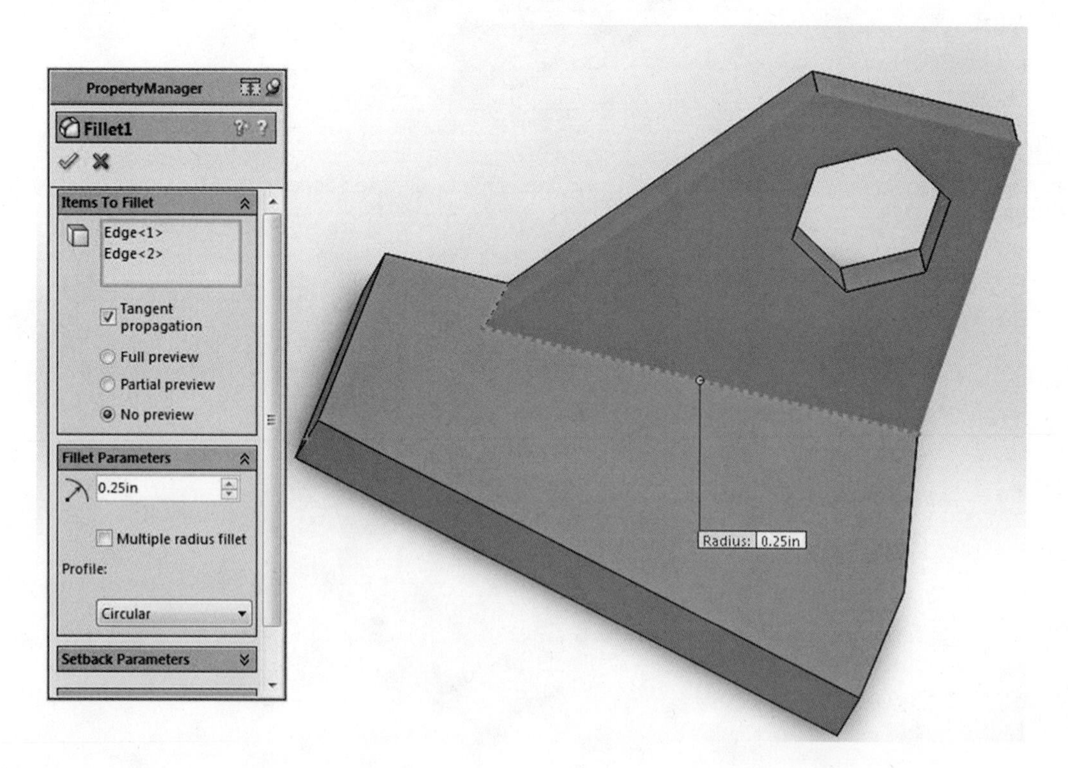

Figure 4.22

Fillet additions. (From David C. Planchard & Marie P. Planchard, Official Certified SolidWorks Associates (CSWA), SDC Publications, 2009.)

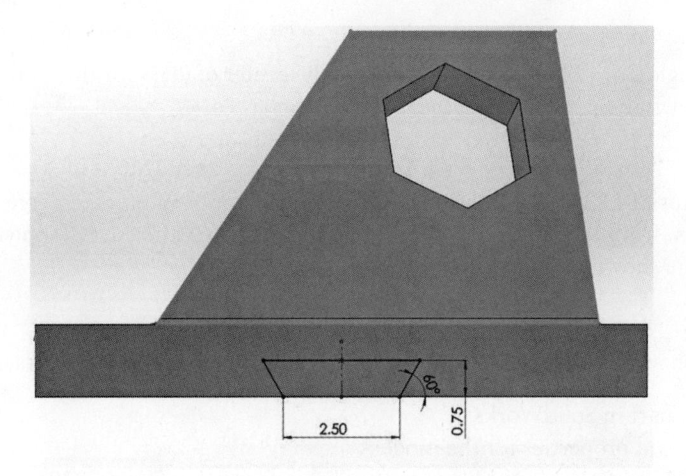

Figure 4.23

Sketch5. (From David C. Planchard & Marie P. Planchard, Official Certified SolidWorks Associates (CSWA), SDC Publications, 2009.)

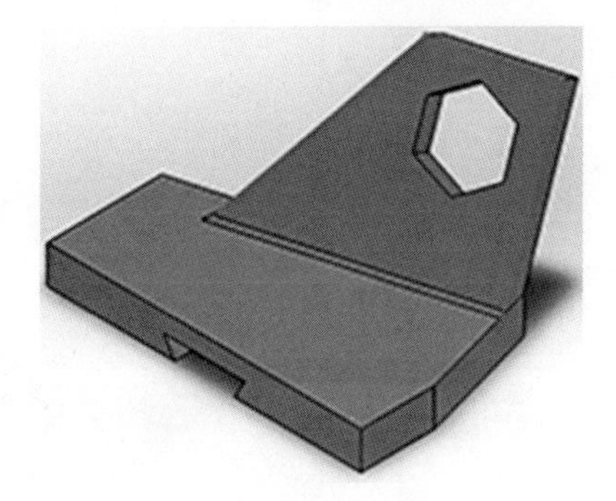

Figure 4.24

Completed part. (From David C. Planchard & Marie P. Planchard, Official Certified SolidWorks Associates (CSWA), SDC Publications, 2009.)

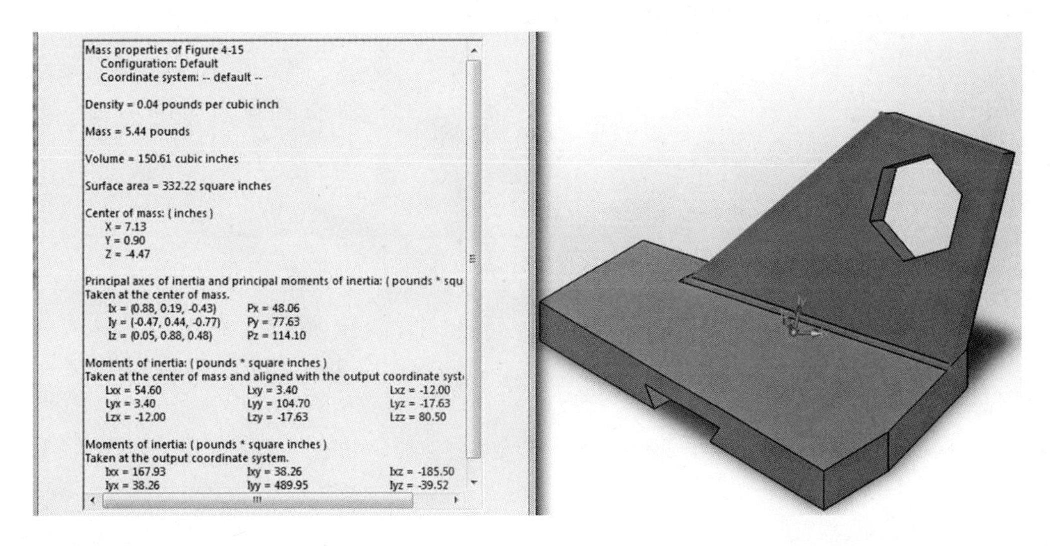

Figure 4.25

Material properties. (From David C. Planchard & Marie P. Planchard, Official Certified SolidWorks Associates (CSWA), SDC Publications, 2009.)

Tutorial 4.3b

Modify the Plane1 angle from 120° to 150°. Determine the center of mass for this part.

The solution is shown in Figure 4.26 for the material properties.

Tutorial 4.4

Build the part for Tutorial 4.3b. Modify the part by sketching two, 1-in. diameter circles on top of Extrude1, located 2-in. and 3-in. from the edge, and then extruding through. The part is shown in Figure 4.27. The material properties are shown in Figure 4.28.

Tutorial 4.5a

Build the part that is shown in Figure 4.29. All fillets are equal: 0.1-in. Determine the center of mass for this part.

1. Create a New part in SolidWorks.
2. Set the document properties for the model.
3. Create Sketch1. Select the Front Plane as the Sketch Plane (see Figure 4.30).
4. Create the Extrude Base feature. Extrude depth = 5.0-in. Extrude1 is the Base feature (see Figure 4.31).
5. Create Sketch2, a circle, 3.0-in. on slanted face, 2.5-in. from the bottom of the slanted face (see Figure 4.32).

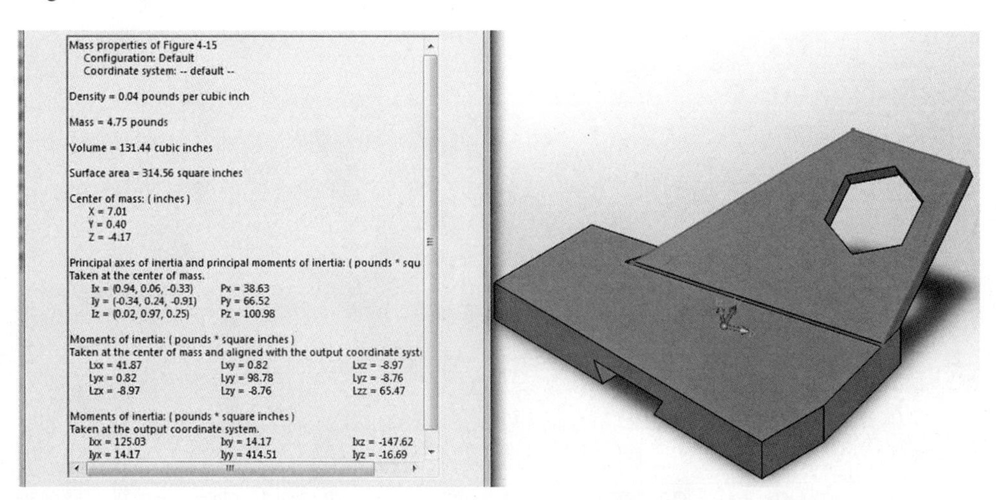

Figure 4.26

Material properties for modified part. (From David C. Planchard & Marie P. Planchard, Official Certified SolidWorks Associates (CSWA), SDC Publications, 2009.)

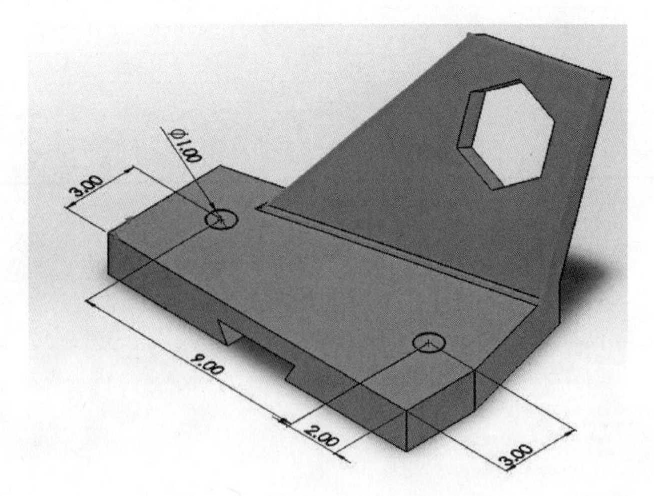

Figure 4.27

Modeled part. (From David C. Planchard & Marie P. Planchard, Official Certified SolidWorks Associates (CSWA), SDC Publications, 2009.)

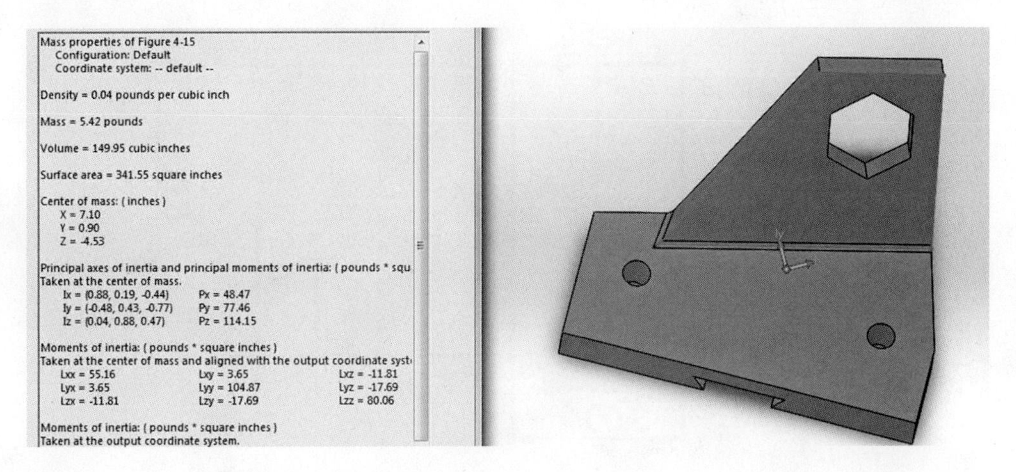

Figure 4.28

Material properties for modified part. (From David C. Planchard & Marie P. Planchard, Official Certified SolidWorks Associates (CSWA), SDC Publications, 2009.)

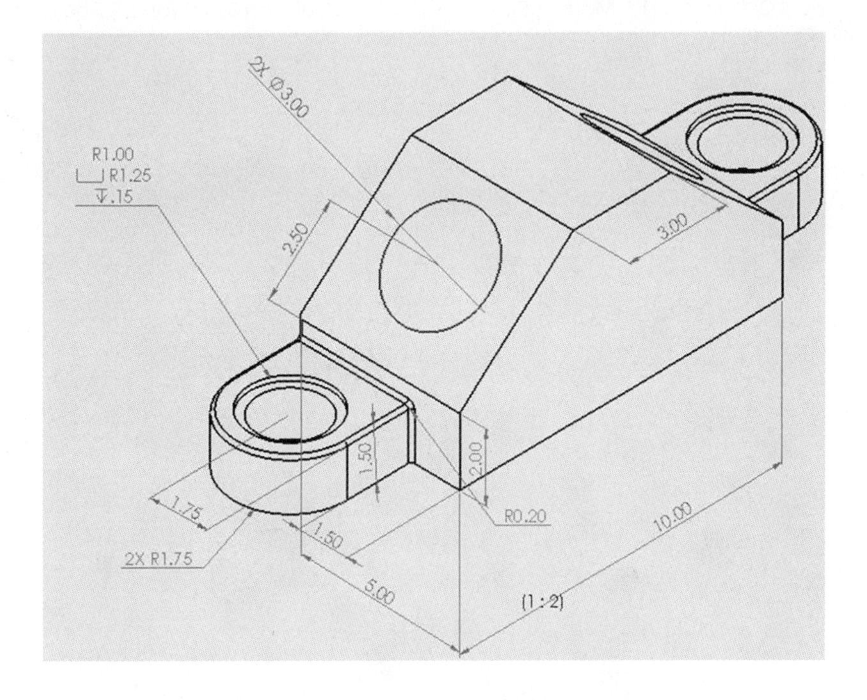

Figure 4.29

Part for modeling. (From David C. Planchard & Marie P. Planchard, Official Certified SolidWorks Associates (CSWA), SDC Publications, 2009.)

6. Extrude Cut Up To bottom-surface. Mirror about Front Plane (see Figure 4.33).
7. Create Sketch3, a profile of two vertical lines, an arc at their endpoints using Tangent Arc tool, and a circle. Select the bottom of Extrude1 as the Sketch Plane. The end of the lines from where they meet the semi-circular arc is 7.0-in. from the center of the bottom face (see Figure 4.33).
8. Extrude Sketch3 1.5-in. from bottom-surface upward, to give Extrude2 (see Figure 4.34).
9. Mirror about Front Plane (see Figure 4.35).
10. Create Fillets of 0.10-in. radius (see Figure 4.36).
11. Create Sketch4, a circle of diameter 2.50-in. on top of Extrude2 (see Figure 4.37).
12. Extrude Cut 0.15-in. downward (see Figure 4.38).
13. Mirror Extrude Cut about Front Plane (see Figure 4.39).

The solution is shown in Figure 4.40, while Figure 4.41 shows the material properties.

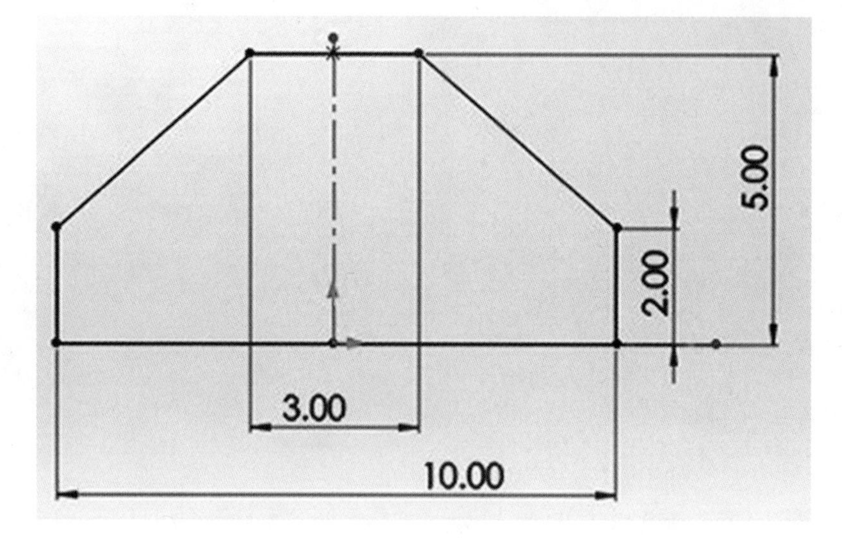

Figure 4.30

Sketch1. (From David C. Planchard & Marie P. Planchard, Official Certified SolidWorks Associates (CSWA), SDC Publications, 2009.)

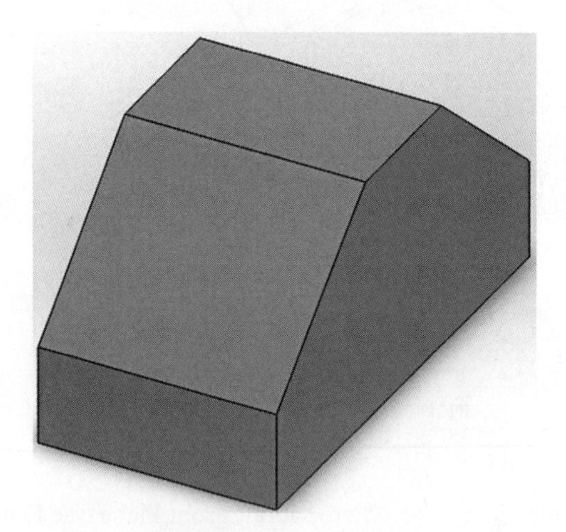

Figure 4.31

Extrude1. (From David C. Planchard & Marie P. Planchard, Official Certified SolidWorks Associates (CSWA), SDC Publications, 2009.)

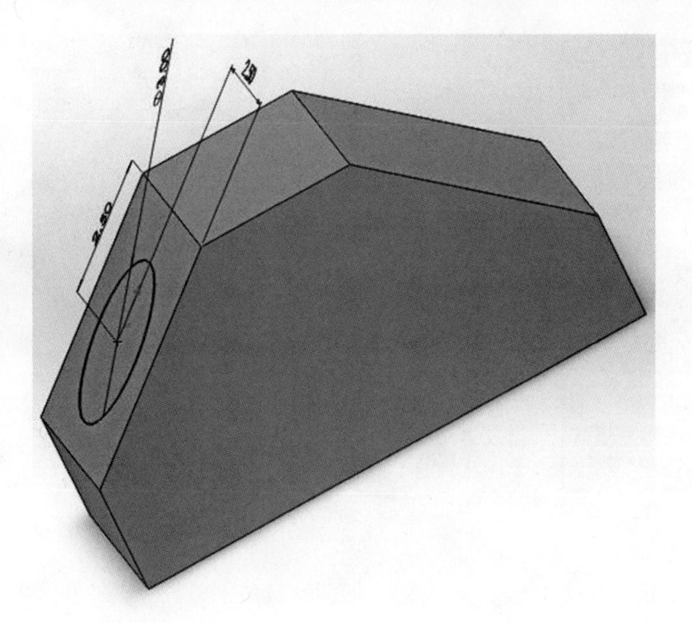

Figure 4.32

Sketch2 and Extrude Cut1. (From David C. Planchard & Marie P. Planchard, Official Certified SolidWorks Associates (CSWA), SDC Publications, 2009.)

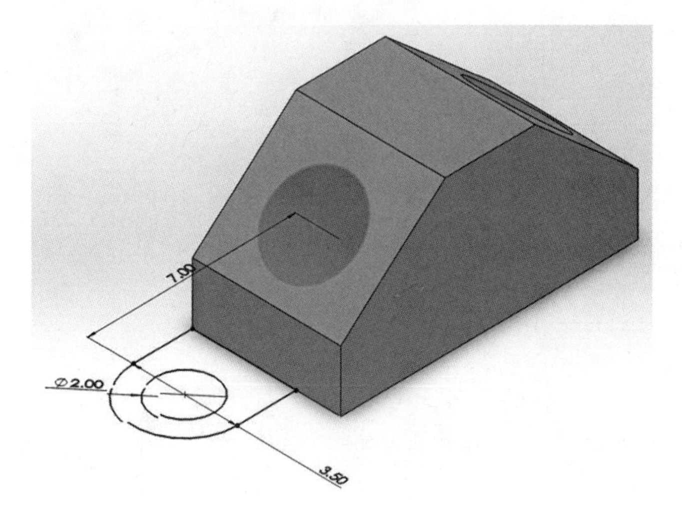

Figure 4.33

Sketch3. (From David C. Planchard & Marie P. Planchard, Official Certified SolidWorks Associates (CSWA), SDC Publications, 2009.)

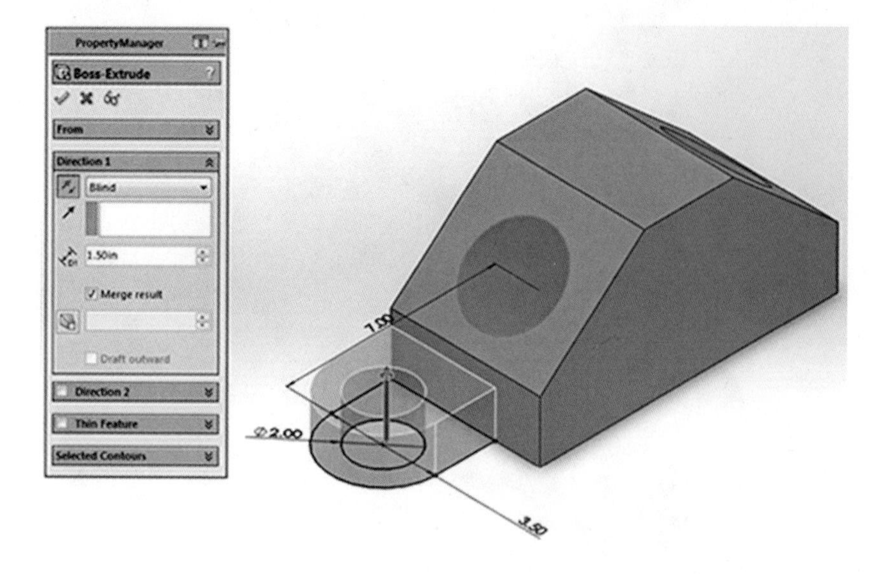

Figure 4.34

Boss feature, Extrude2. (From David C. Planchard & Marie P. Planchard, Official Certified SolidWorks Associates (CSWA), SDC Publications, 2009.)

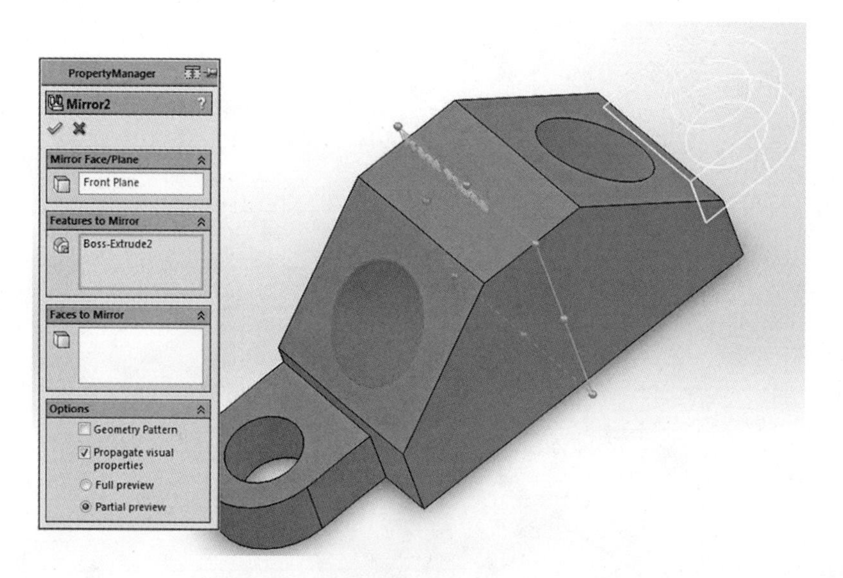

Figure 4.35

Mirror of Extrude2. (From David C. Planchard & Marie P. Planchard, Official Certified SolidWorks Associates (CSWA), SDC Publications, 2009.)

Tutorial 4.5b

Modify the circle on slanted face from dimeter of 3.0-in. to 1.5-in. Determine the center of mass for this part. The material properties are shown in Figure 4.42.

Tutorial 4.6a

Build the part that is shown in Figure 4.43. Determine the center of mass for this part.

1. Create a New part in SolidWorks.
2. Set the document properties for the model.
3. Create Sketch1. Select the Right Plane as the Sketch Plane (see Figure 4.44).

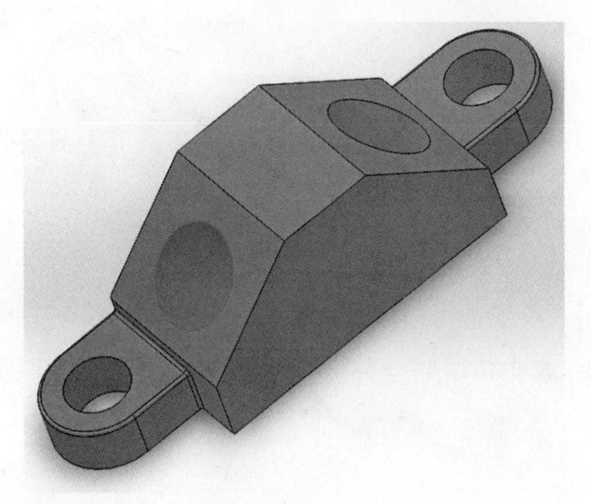

Figure 4.36

Fillets on Extrude2. (From David C. Planchard & Marie P. Planchard, Official Certified SolidWorks Associates (CSWA), SDC Publications, 2009.)

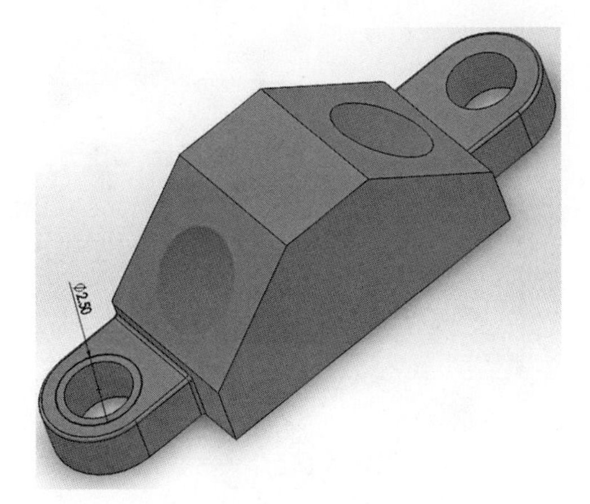

Figure 4.37

Sketch4. (From David C. Planchard & Marie P. Planchard, Official Certified SolidWorks Associates (CSWA), SDC Publications, 2009.)

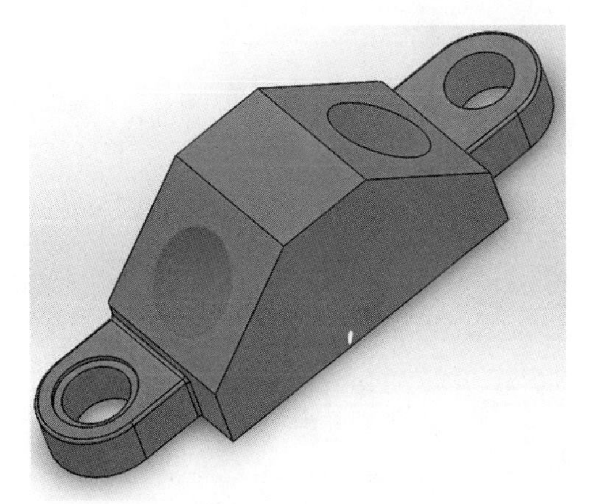

Figure 4.38

Second Extrude Cut. (From David C. Planchard & Marie P. Planchard, Official Certified SolidWorks Associates (CSWA), SDC Publications, 2009.)

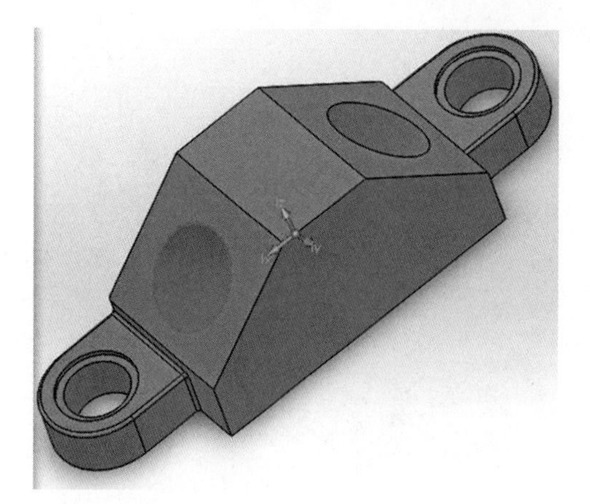

Figure 4.39

Completed part. (From David C. Planchard & Marie P. Planchard, Official Certified SolidWorks Associates (CSWA), SDC Publications, 2009.)

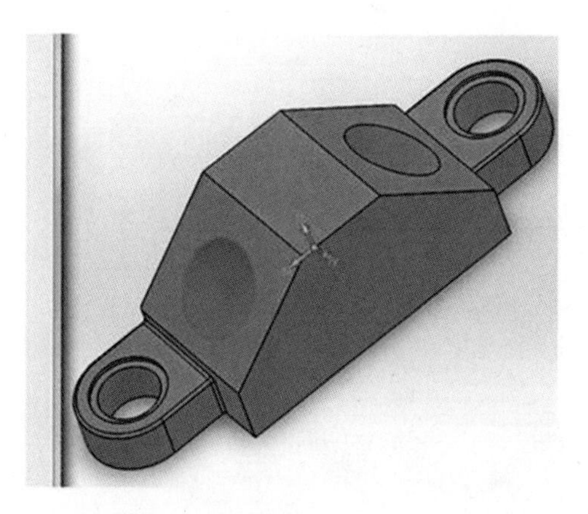

Figure 4.40

Material properties. (From David C. Planchard & Marie P. Planchard, Official Certified SolidWorks Associates (CSWA), SDC Publications, 2009.)

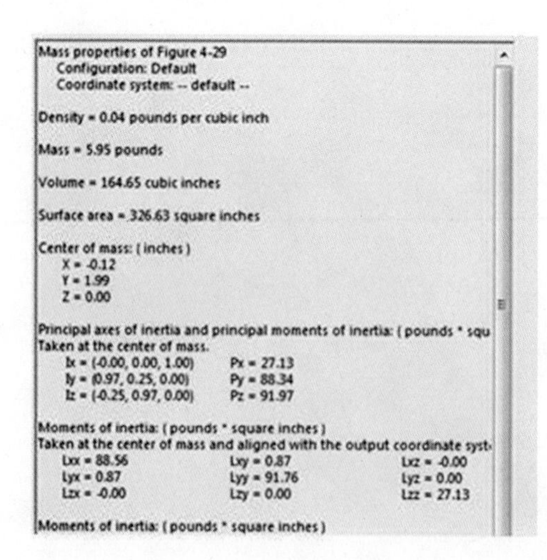

Figure 4.41

Modified part. (From David C. Planchard & Marie P. Planchard, Official Certified SolidWorks Associates (CSWA), SDC Publications, 2009.)

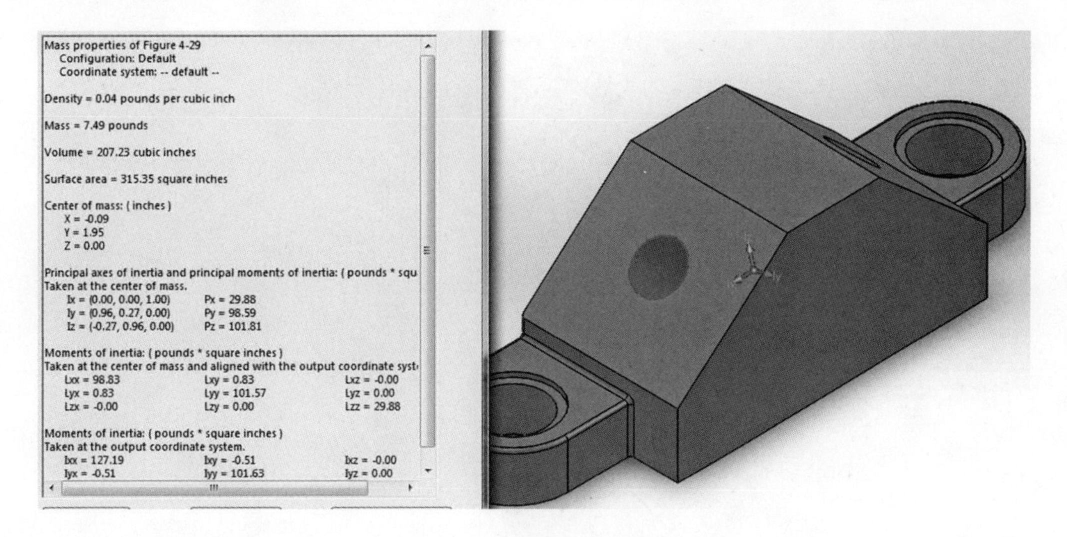

Figure 4.42

Material properties. (From David C. Planchard & Marie P. Planchard, Official Certified SolidWorks Associates (CSWA), SDC Publications, 2009.)

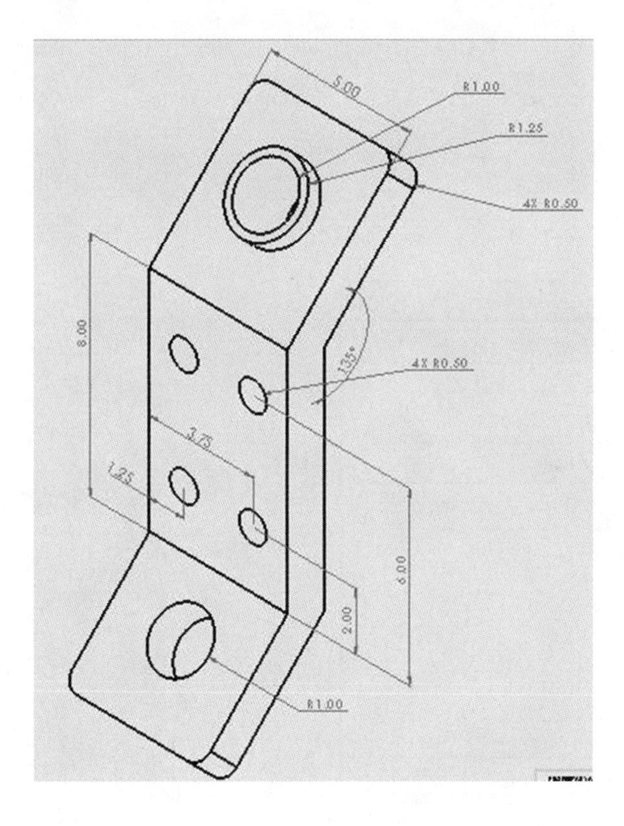

Figure 4.43

Part for modeling. (From David C. Planchard & Marie P. Planchard, Official Certified SolidWorks Associates (CSWA), SDC Publications, 2009.)

4. Create the Extrude-Thin1 feature. Apply Symmetry. Select Mid Plane for End Condition in Direction1. Extrude depth = 5.0 in. Thickness = 1.25 in. Extrude1 is the Base feature (see Figure 4.45).
5. Create Plane1. Select top inclined face of Extrude1. Plane1 is at a distance, 0.625-in. outward (see Figure 4.46).
6. Create Sketch2 on top of inclined face of Extrude1, diameters 2.0-in. and 2.5-in. respectively (see Figure 4.47).
7. Extrude 2.7-in. Blind, to give Extrude2 (see Figure 4.48).

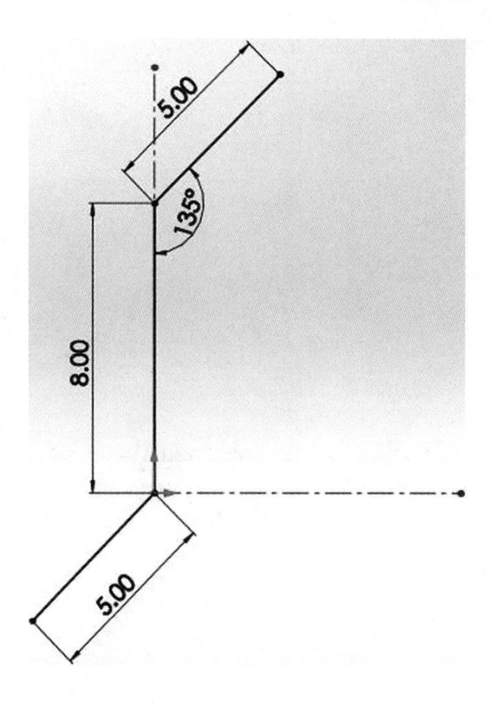

Figure 4.44

Sketch1. (From David C. Planchard & Marie P. Planchard, Official Certified SolidWorks Associates (CSWA), SDC Publications, 2009.)

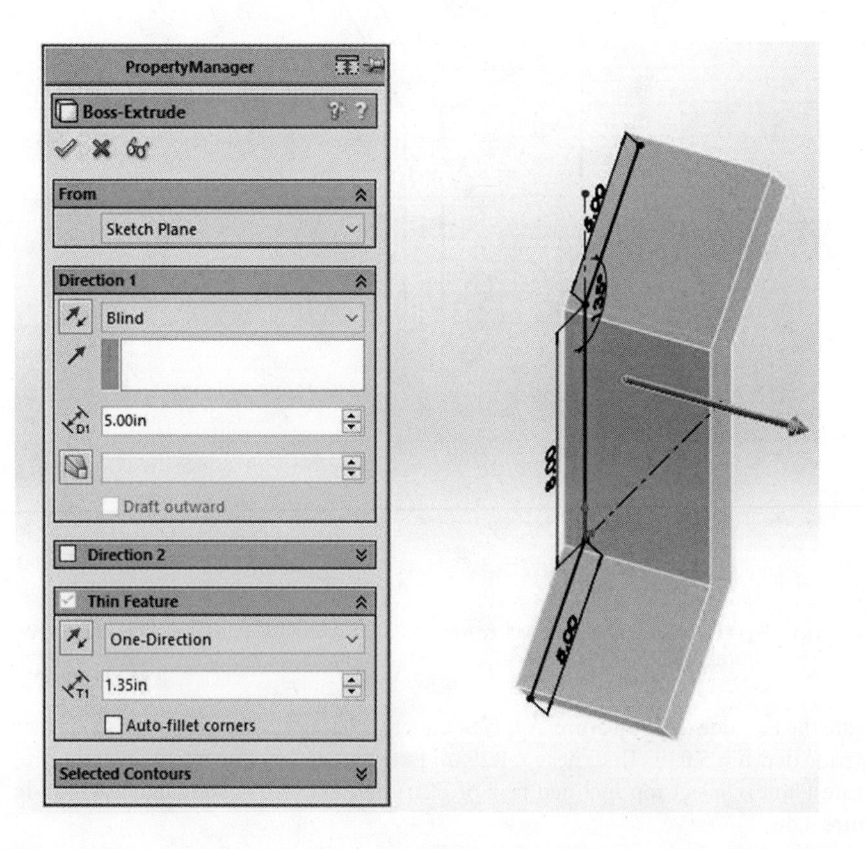

Figure 4.45

Extrude-Thin1. (From David C. Planchard & Marie P. Planchard, Official Certified SolidWorks Associates (CSWA), SDC Publications, 2009.)

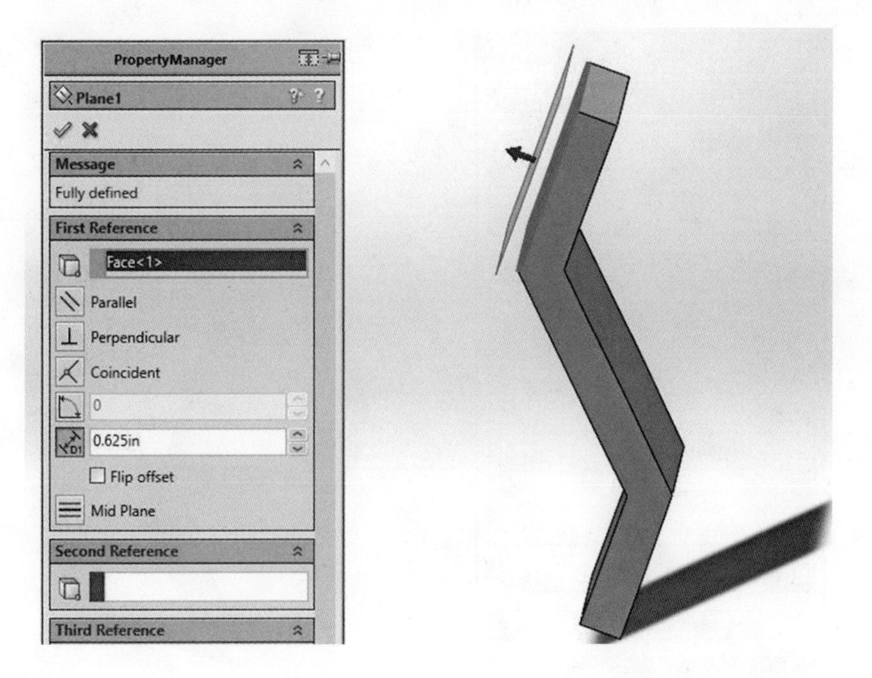

Figure 4.46

Plane1. (From David C. Planchard & Marie P. Planchard, Official Certified SolidWorks Associates (CSWA), SDC Publications, 2009.)

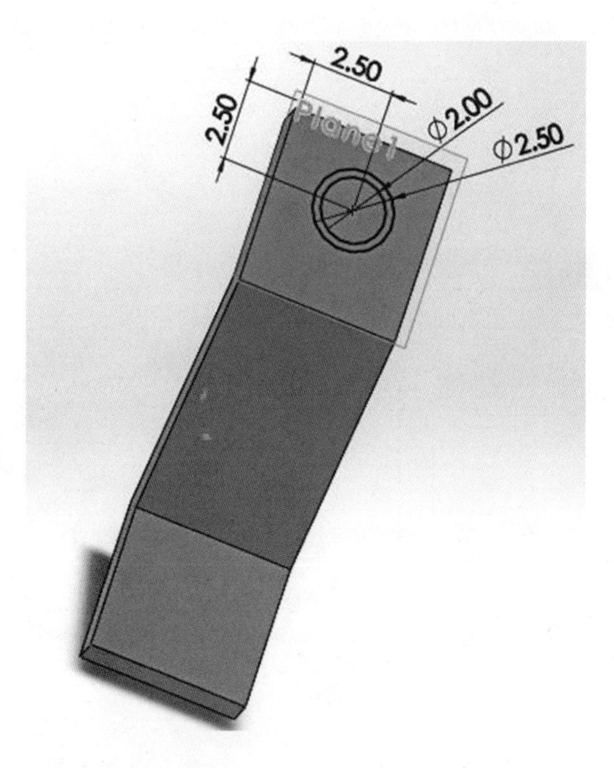

Figure 4.47

Sketch2. (From David C. Planchard & Marie P. Planchard, Official Certified SolidWorks Associates (CSWA), SDC Publications, 2009.)

8. Click top of Surface on Extrude2. Use Convert Entity tool to convert circular edge into a Circle (see Figure 4.49).
9. Extrude-Cut the circle Up To Surface, the bottom surface of Extrude2 (see Figure 4.50).
10. Create Sketch3 on bottom of inclined face of Extrude1, diameter 2.0-in. (see Figure 4.51).
11. Extrude Cut Up To Surface, using bottom of inclined surface as basis (see Figure 4.52).

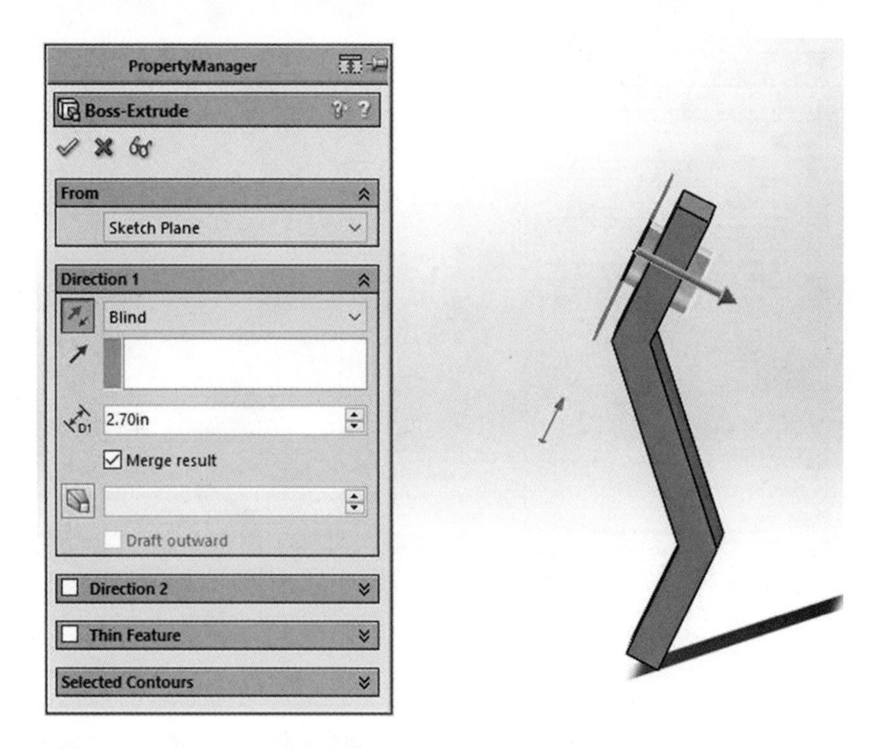

Figure 4.48

Extrude2. (From David C. Planchard & Marie P. Planchard, Official Certified SolidWorks Associates (CSWA), SDC Publications, 2009.)

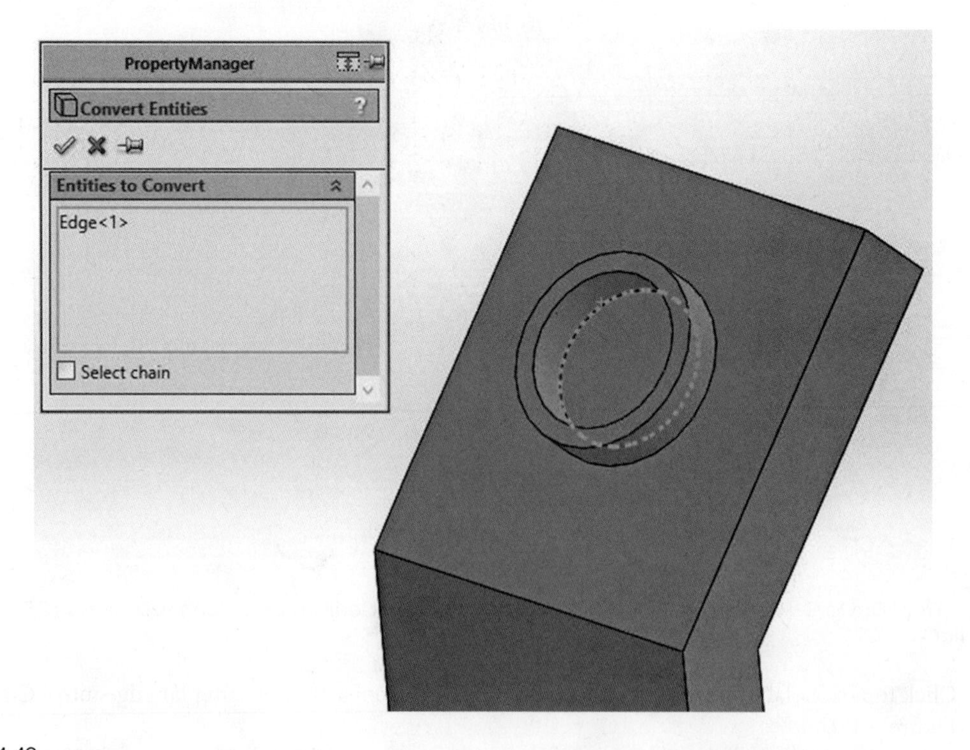

Figure 4.49

First Extrude Cut. (From David C. Planchard & Marie P. Planchard, Official Certified SolidWorks Associates (CSWA), SDC Publications, 2009.)

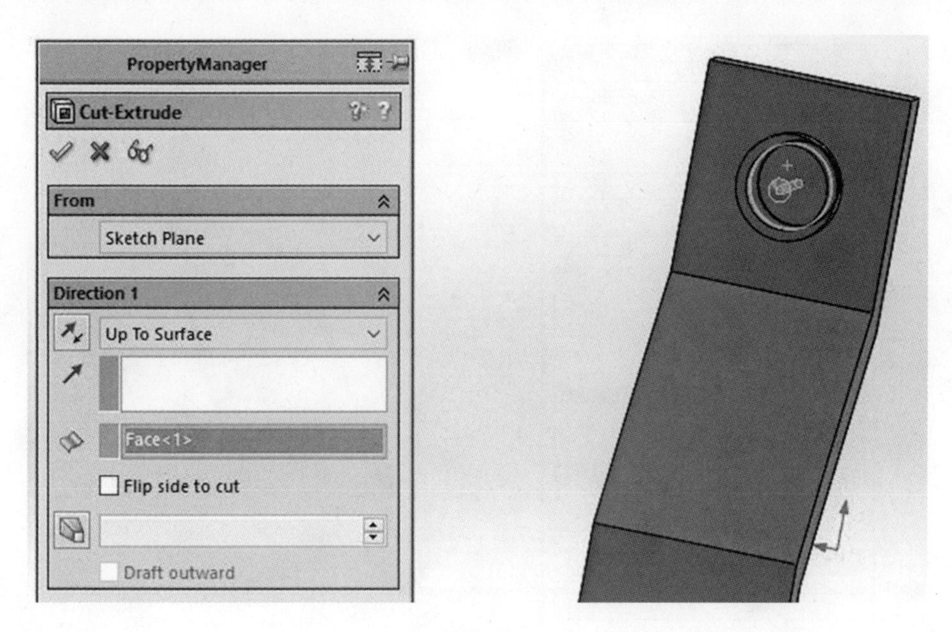

Figure 4.50

Extrude Cut2 and fillets. (From David C. Planchard & Marie P. Planchard, Official Certified SolidWorks Associates (CSWA), SDC Publications, 2009.)

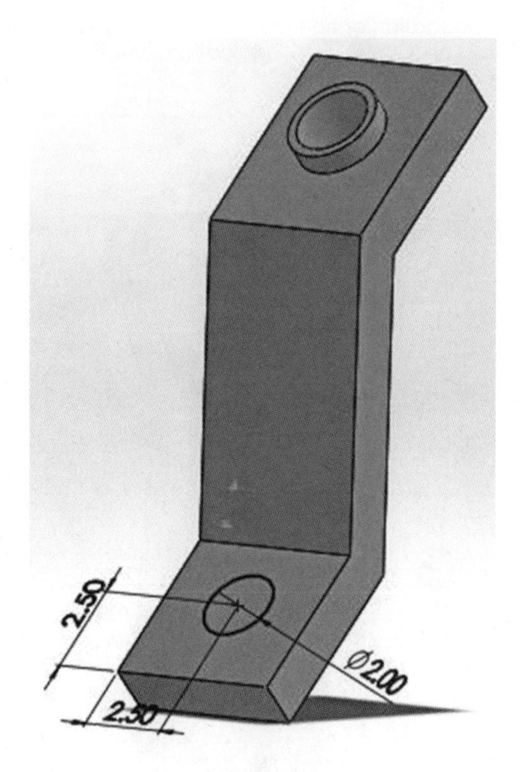

Figure 4.51

More 12-mm radius fillets. (From David C. Planchard & Marie P. Planchard, Official Certified SolidWorks Associates (CSWA), SDC Publications, 2009.)

12. Create Sketch4 on vertical face of Extrude1, 4 circles of diameter 1.0-in. (see Figure 4.53).
13. Create Fillets of 1.0-in. Extrude-Cut Up To Surface using back of vertical face as basis (see Figure 4.54).

Figure 4.54 also shows the material properties.

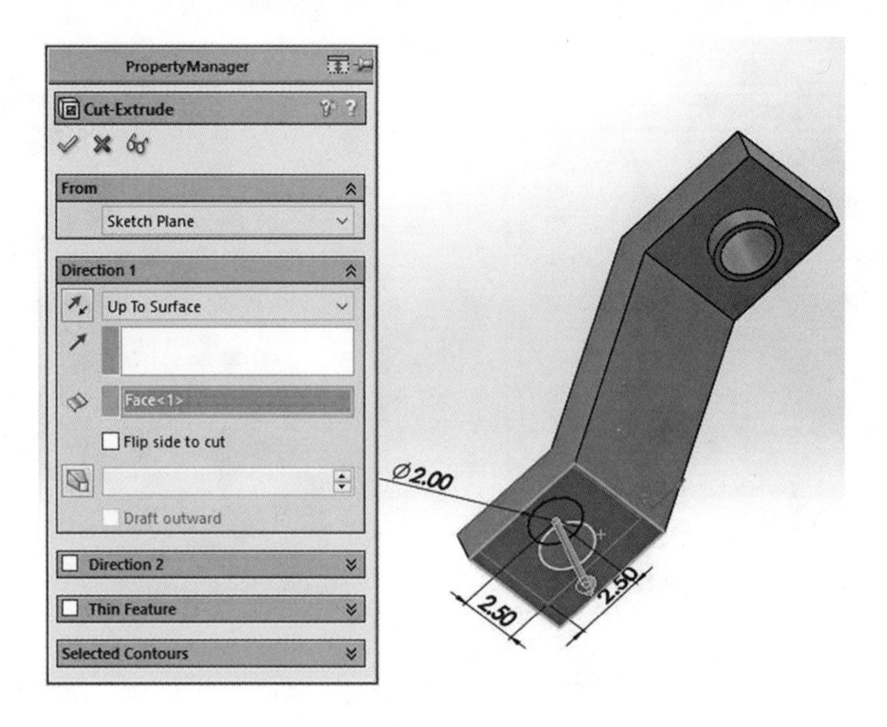

Figure 4.52

Fillets of 4-mm radius, Tangent propagation option chosen. (From David C. Planchard & Marie P. Planchard, Official Certified SolidWorks Associates (CSWA), SDC Publications, 2009.)

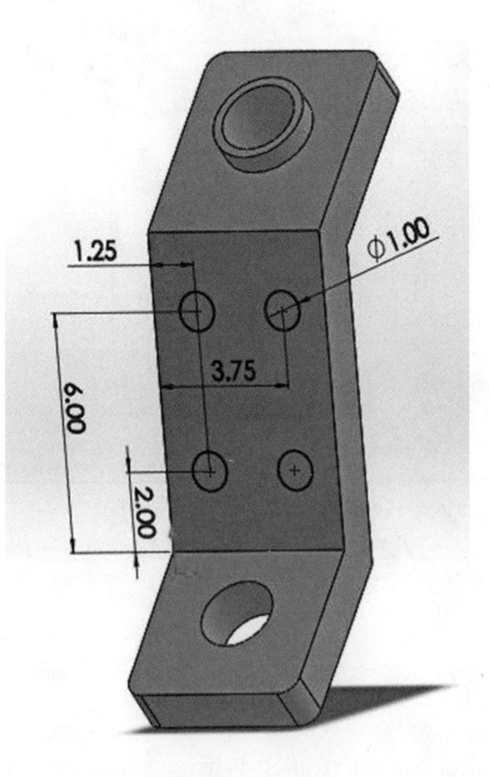

Figure 4.53

Completed part. (From David C. Planchard & Marie P. Planchard, Official Certified SolidWorks Associates (CSWA), SDC Publications, 2009.)

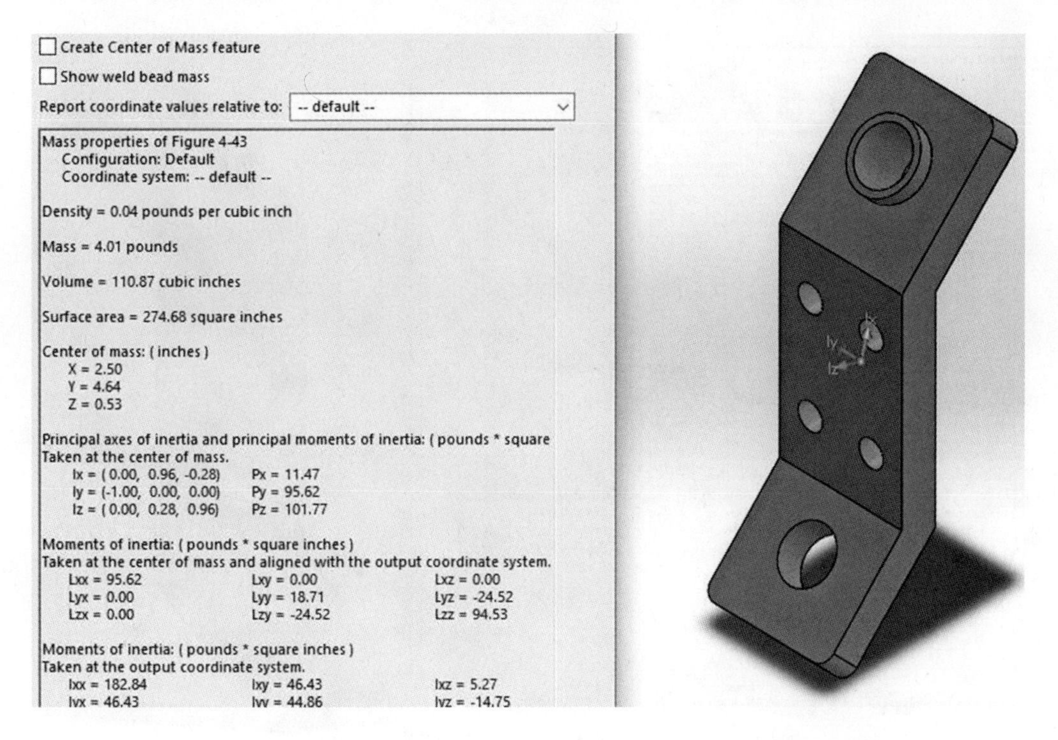

Figure 4.54

Material properties. (From David C. Planchard & Marie P. Planchard, Official Certified SolidWorks Associates (CSWA), SDC Publications, 2009.)

Tutorial 4.6b

Modify the part by suppressing the hole at the bottom of the inclined surface. Recalculate the location of the center of mass with reference to the origin.

The part and material properties are shown in Figure 4.55.

Tutorial 4.7a

Build the part that is shown in Figure 4.56. Determine the center of mass for this part.

1. Create a New part in SolidWorks.
2. Set the document properties for the model.
3. Create Sketch1. Select the Top Plane as the Sketch Plane (see Figure 4.57).
4. Create the Extrude Base feature. Select Mid Plane in Direction1. Extrude depth = 2 in. Caution: Extrude direction is downward. Extrude1 is the Base feature.
5. Create Plane1. Select the midpoints of the top face of Extrude1 to add a construction line. Plane1 is defined using the construction line and the top face of Extrude1, with an Angle = 60° (see Figure 4.58).
6. Create Sketch2, made up of one horizontal line on Extrude1, two lines on Plane1 and an arc using Tangent Arc tool. Select Plane1 as the Sketch Plane (see Figure 4.59).
7. Create the Extrude Boss feature. Select Up To Vertex for End Condition in Direction1. Select the top right vertex of Extrude1. Extrude2 is the Base Boss. (See Figure 4.60 for Preview and Figure 4.61 for Extrude2.)
8. Create Sketch3, made of two circles, diameters 6.0-in. and 5.0-in. respectively. Select the back angle of Extrude2 as the Sketch Plane (see Figure 4.62).
9. Create the Extrude Base feature. Select Blind Condition in Direction1. Extrude depth = 6.93 in., to give Extrude2 (see Figure 4.63).
10. Select foremost face of Extrude2. Create Sketch4, two circles of diameters 6.0-in. and 6.40-in. respectively (see Figure 4.64).
11. Create the Extrude feature. Select Blind of 1.0-in. for End Condition in Direction1 (see Figure 4.65).
12. Create Sketch5, a profile for the Rib feature. Select the Right Plane as the Sketch Plane. Apply a Parallel relation (see Figure 4.66).

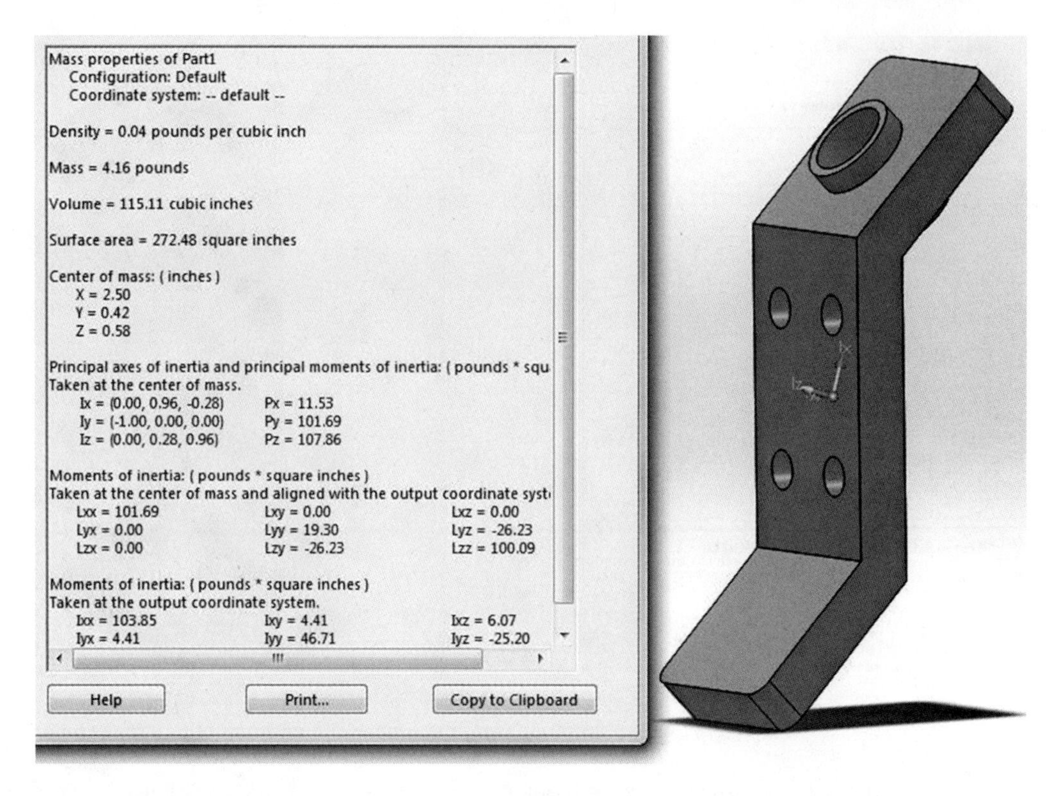

Figure 4.55

Modeled part and material properties. (From David C. Planchard & Marie P. Planchard, Official Certified SolidWorks Associates (CSWA), SDC Publications, 2009.)

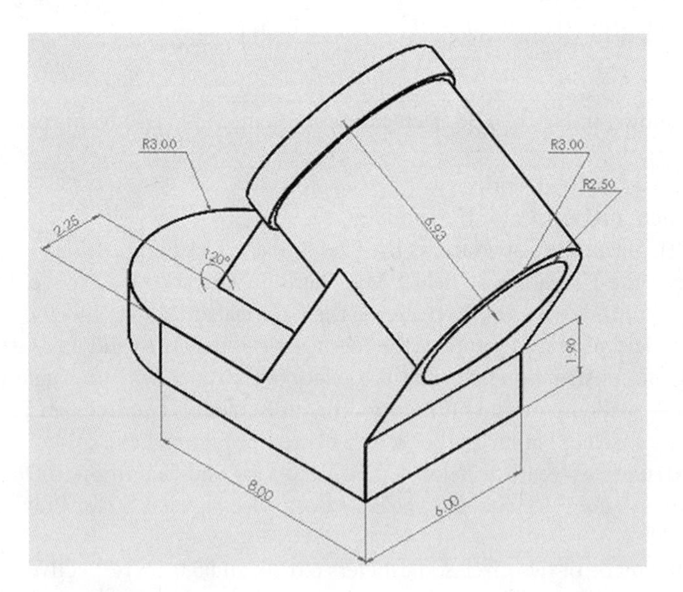

Figure 4.56

Part for modeling. (From David C. Planchard & Marie P. Planchard, Official Certified SolidWorks Associates (CSWA), SDC Publications, 2009.)

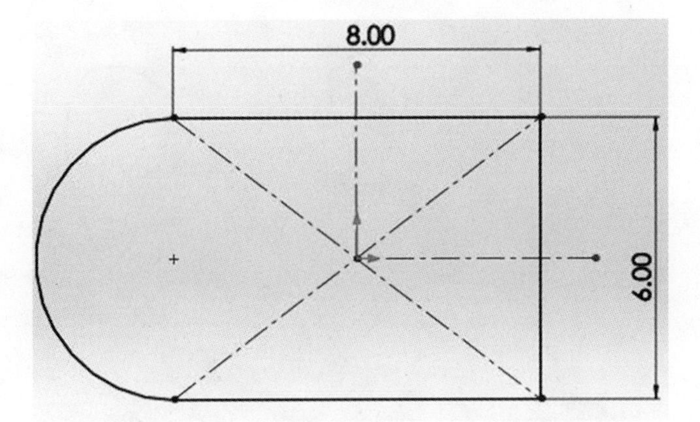

Figure 4.57

Extrude Base. (From David C. Planchard & Marie P. Planchard, Official Certified SolidWorks Associates (CSWA), SDC Publications, 2009.)

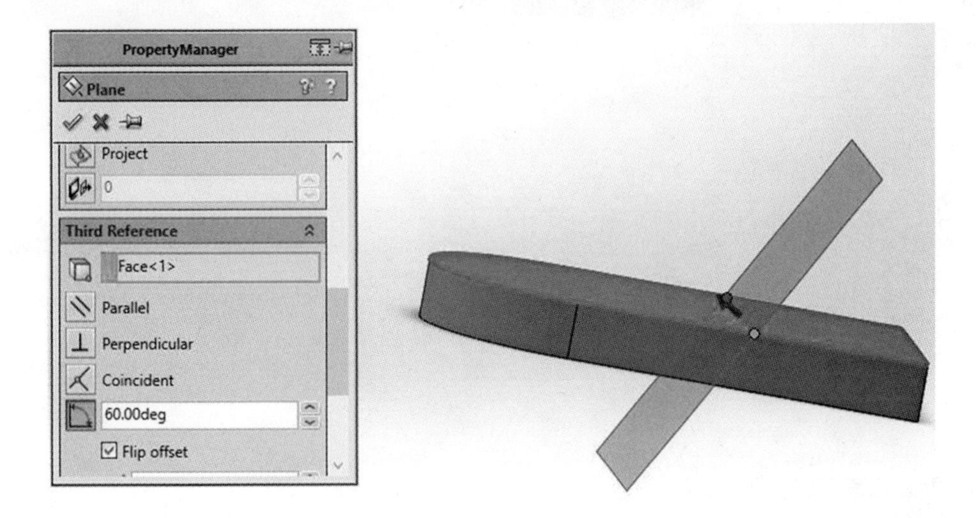

Figure 4.58

Plane1. (From David C. Planchard & Marie P. Planchard, Official Certified SolidWorks Associates (CSWA), SDC Publications, 2009.)

13. Create the Rib feature. Use Insert > Features > Rib Thickness = 1.5 in. (See Figure 4.67 for the Rib feature preview. The completed part and the material assignment are shown in Figures 4.68 and 4.69, respectively.)
14. Calculate the mass properties. Figure 4.70 shows the results.

Tutorial 4.7b

Modify the Rib1 feature from 1.00 to 1.25 in.; modify the depth of Extrude2 from 3.00 to 3.25 in. Modify the Plane1 angle from 48° to 30°. Modify the material from 6061 Alloy to Copper. Recalculate the location of the center of mass with reference to the origin.

	Modify from	Modify to
Rib1 feature	1.00 in.	1.25 in.
Depth of Extrude2	3.00 in.	3.25 in.
Sketch1 angle	48°	30°
Material	6061 Alloy	Copper

The part and the material properties are shown in Figure 4.70.

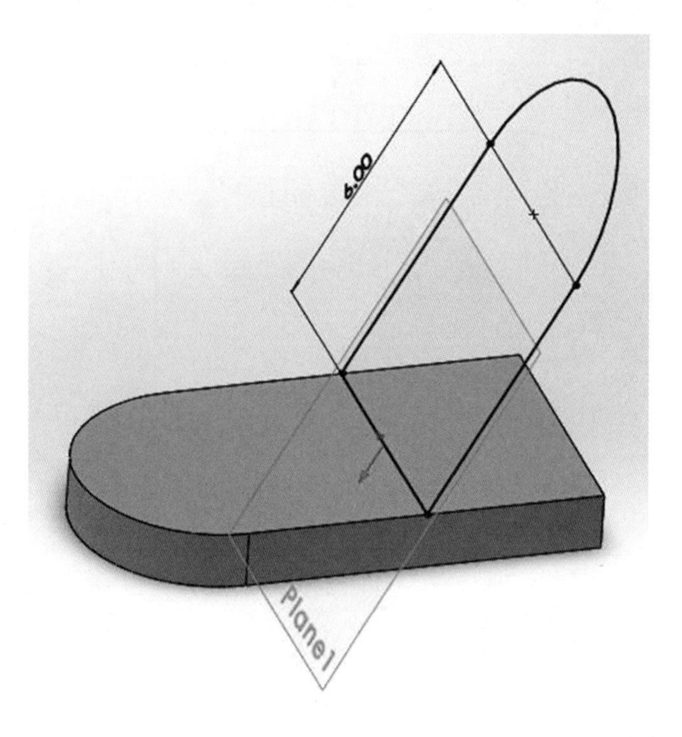

Figure 4.59

Plane1. (From David C. Planchard & Marie P. Planchard, Official Certified SolidWorks Associates (CSWA), SDC Publications, 2009.)

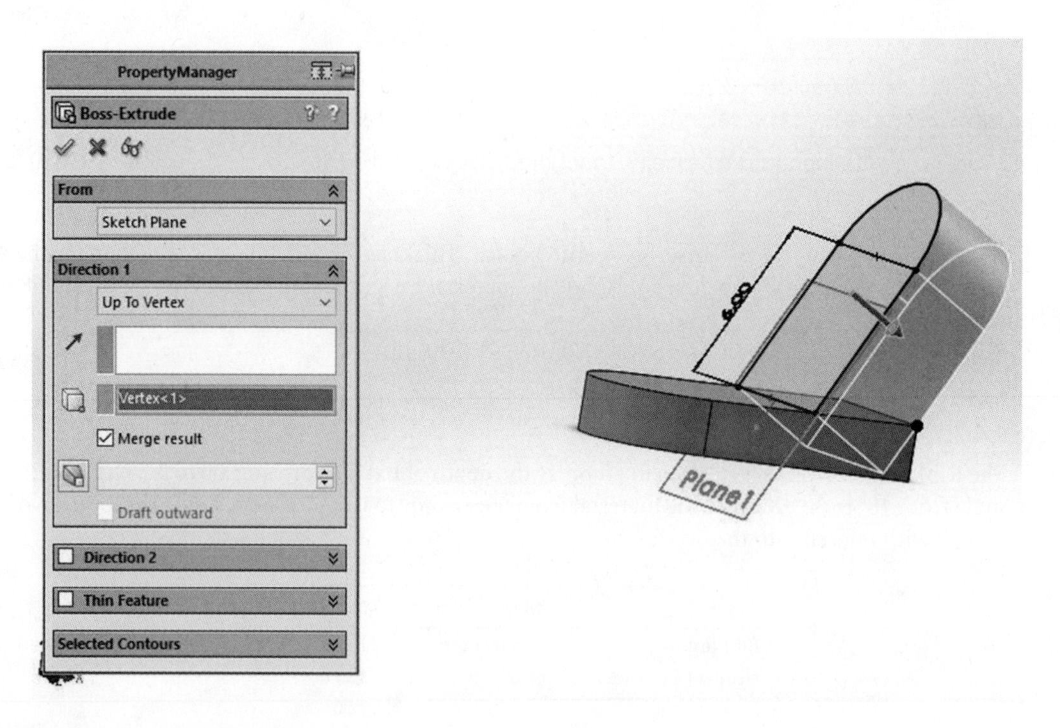

Figure 4.60

Preview of Extrude2. (From David C. Planchard & Marie P. Planchard, Official Certified SolidWorks Associates (CSWA), SDC Publications, 2009.)

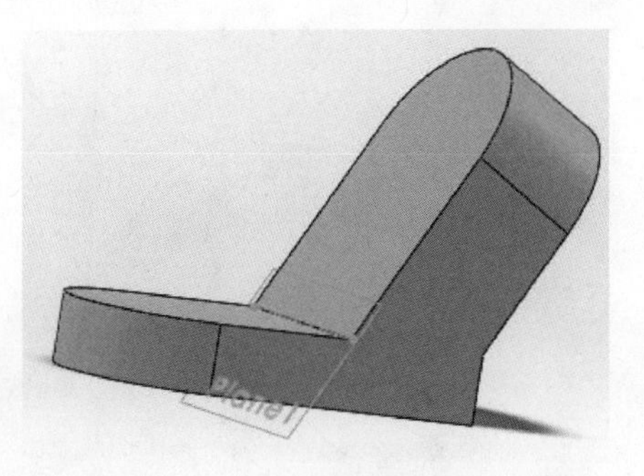

Figure 4.61

Extrude2. (From David C. Planchard & Marie P. Planchard, Official Certified SolidWorks Associates (CSWA), SDC Publications, 2009.)

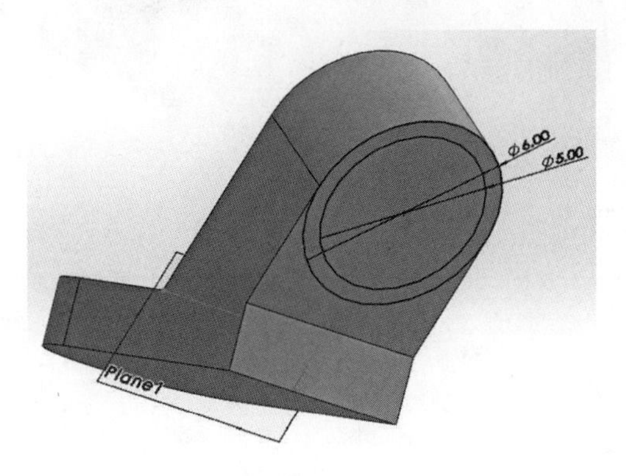

Figure 4.62

Sketch3. (From David C. Planchard & Marie P. Planchard, Official Certified SolidWorks Associates (CSWA), SDC Publications, 2009.)

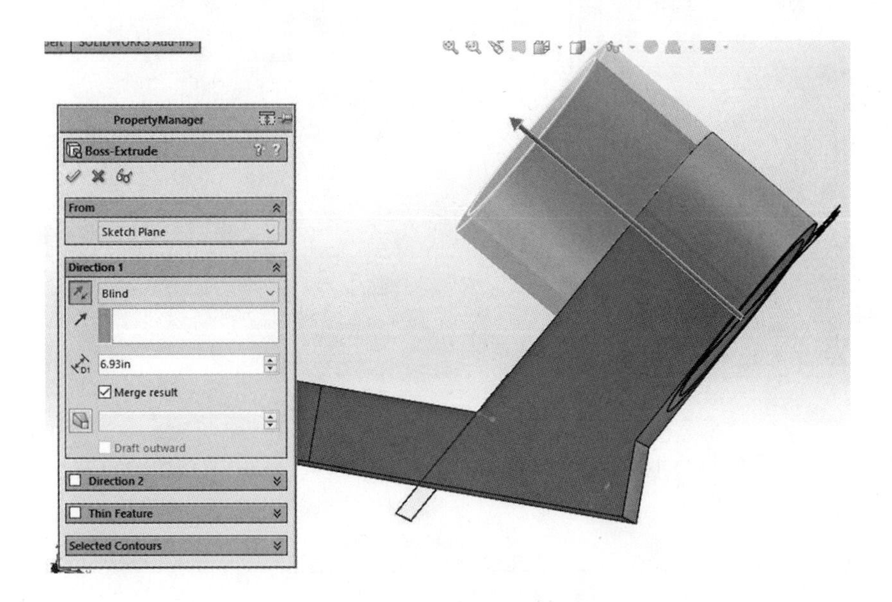

Figure 4.63

Extrude3. (From David C. Planchard & Marie P. Planchard, Official Certified SolidWorks Associates (CSWA), SDC Publications, 2009.)

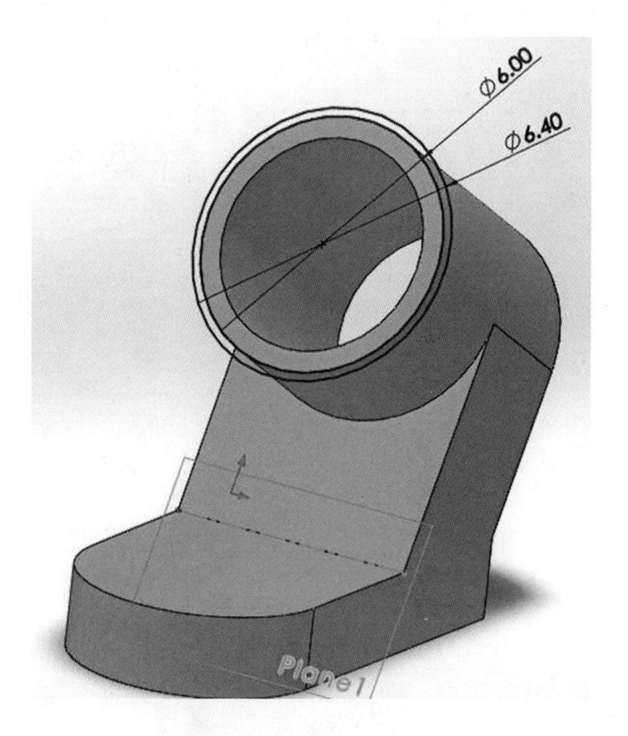

Figure 4.64

Sketch4. (From David C. Planchard & Marie P. Planchard, Official Certified SolidWorks Associates (CSWA), SDC Publications, 2009.)

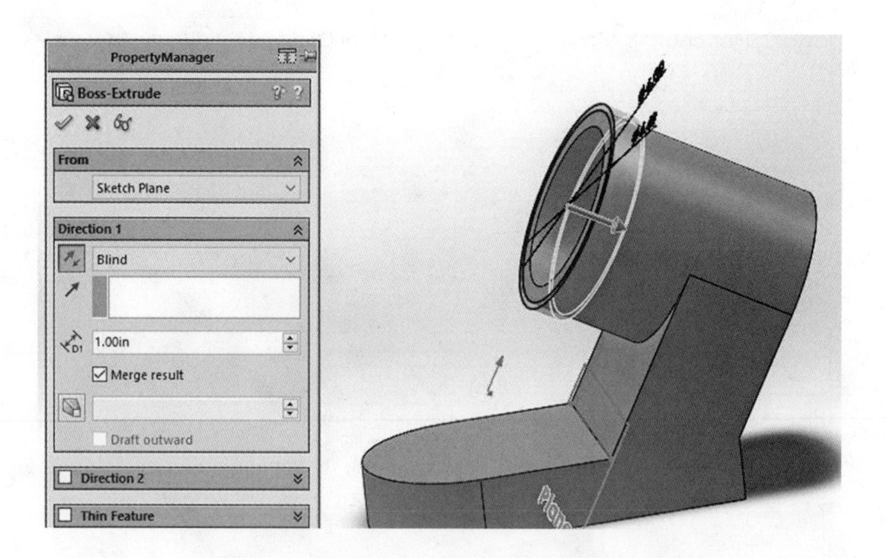

Figure 4.65

Extrude Cut. (From David C. Planchard & Marie P. Planchard, Official Certified SolidWorks Associates (CSWA), SDC Publications, 2009.)

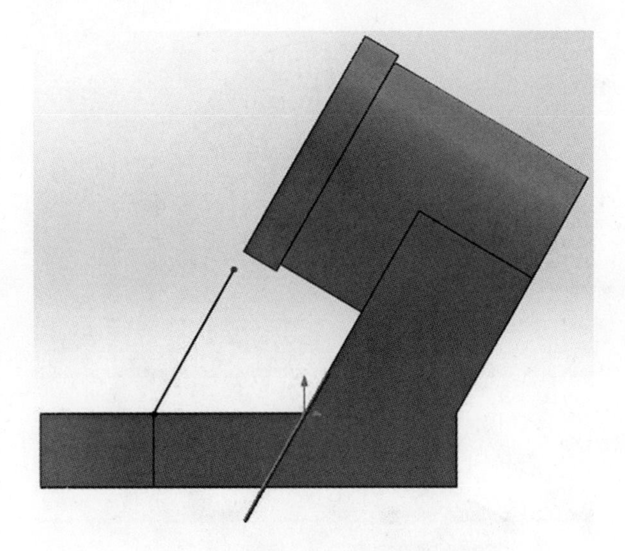

Figure 4.66

Sketch5. (From David C. Planchard & Marie P. Planchard, Official Certified SolidWorks Associates (CSWA), SDC Publications, 2009.)

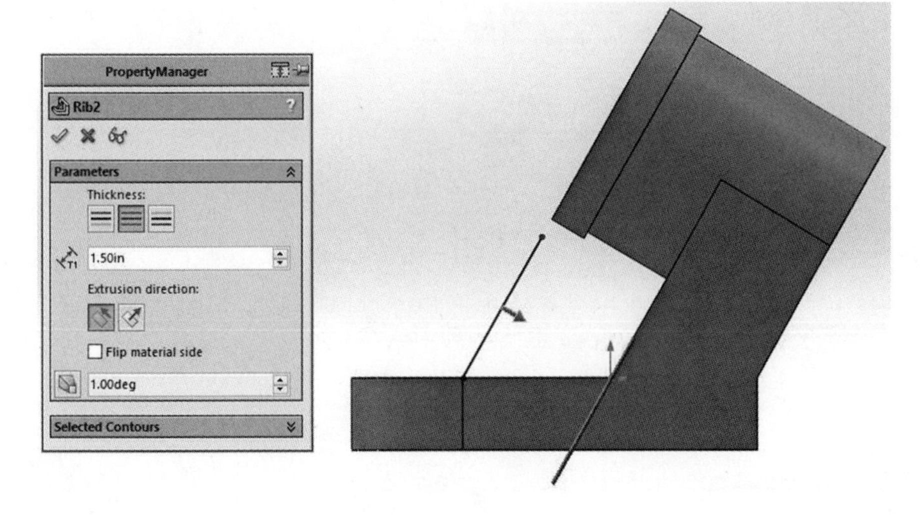

Figure 4.67

Rib feature preview. (From David C. Planchard & Marie P. Planchard, Official Certified SolidWorks Associates (CSWA), SDC Publications, 2009.)

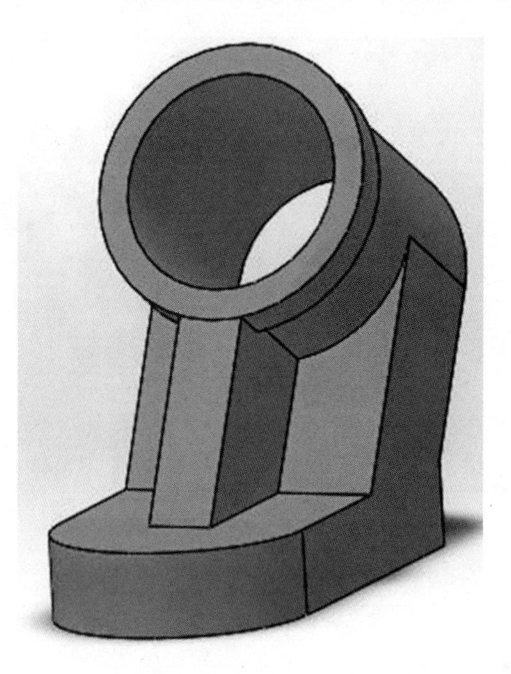

Figure 4.68

Completed part. (From David C. Planchard & Marie P. Planchard, Official Certified SolidWorks Associates (CSWA), SDC Publications, 2009.)

Mass properties of Part1
Configuration: Default
Coordinate system: -- default --

Density = 0.04 pounds per cubic inch

Mass = 9.30 pounds

Volume = 257.42 cubic inches

Surface area = 478.62 square inches

Center of mass: (inches)
X = 0.59
Y = 1.29
Z = 0.00

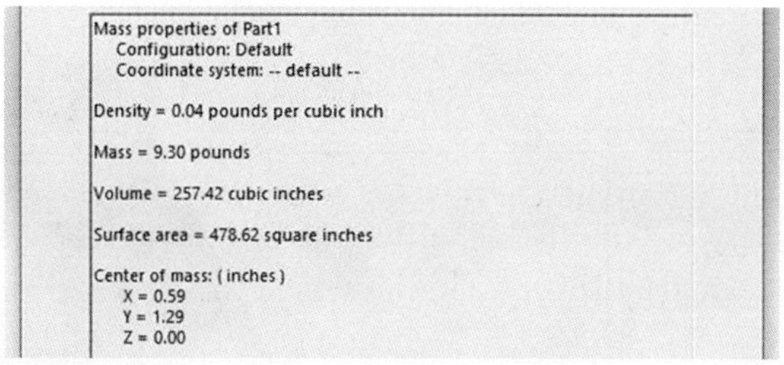

Figure 4.69

Mass properties. (From David C. Planchard & Marie P. Planchard, Official Certified SolidWorks Associates (CSWA), SDC Publications, 2009.)

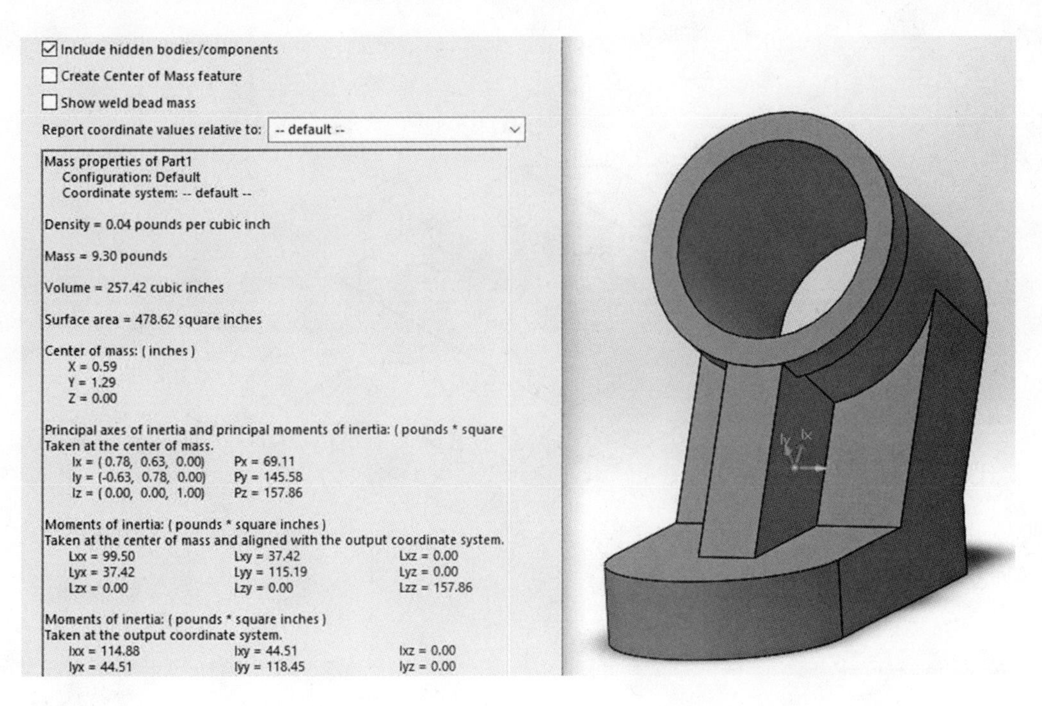

Figure 4.70

Modeled part and material properties. (From David C. Planchard & Marie P. Planchard, Official Certified SolidWorks Associates (CSWA), SDC Publications, 2009.)

Summary

This chapter discusses part modeling at the level of CSWA, and most of the examples are similar to the materials that are covered in the CSWA examinations. Students preparing for this exam will find this chapter helpful. Instructors should consider covering this chapter before moving to the next chapter, which pitches at a higher level.

Exercises

1. Referring to the part in Tutorial 4.2 (see Figure P4.1), create a mirror of the part about the bottom face. Compute the material properties of the part.
2. Referring to the part in Tutorial 4.3a (see Figure P4.2), change the angles of the bottom feature from 105-deg to 100-deg, and 165 (or 15)-deg to 175 (or 5)-deg. Compute the material properties of the part.
3. Referring to the part in Tutorial 4.5a (see Figure P4.3), create a mirror of the part about the right vertical face. Compute the material properties of the part.
4. Referring to the part in Tutorial 4.6a (see Figure P4.4), create a mirror of the part about the right vertical face. Compute the material properties of the part.
5. Referring to the part in Tutorial 4.7a (see Figure P4.5), change the dimension of the hollow feature from 6.93-in. to 6.0-in. Compute the material properties of the part.

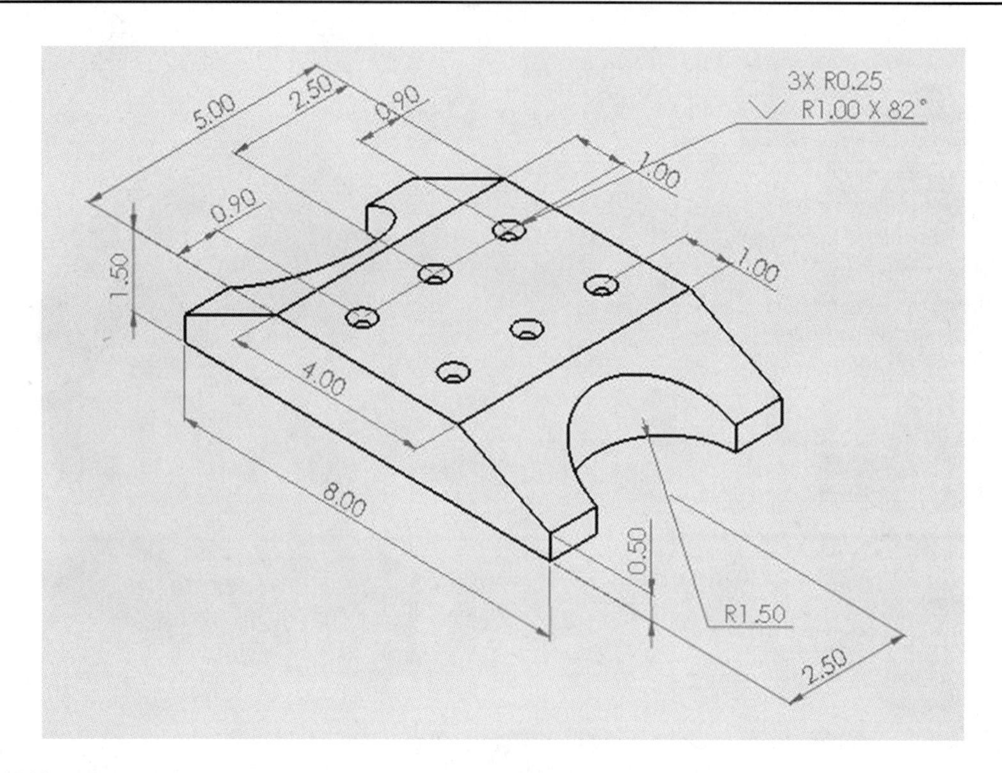

Figure P4.1

Support base. (From David C. Planchard & Marie P. Planchard, Official Certified SolidWorks Associates (CSWA), SDC Publications, 2009.)

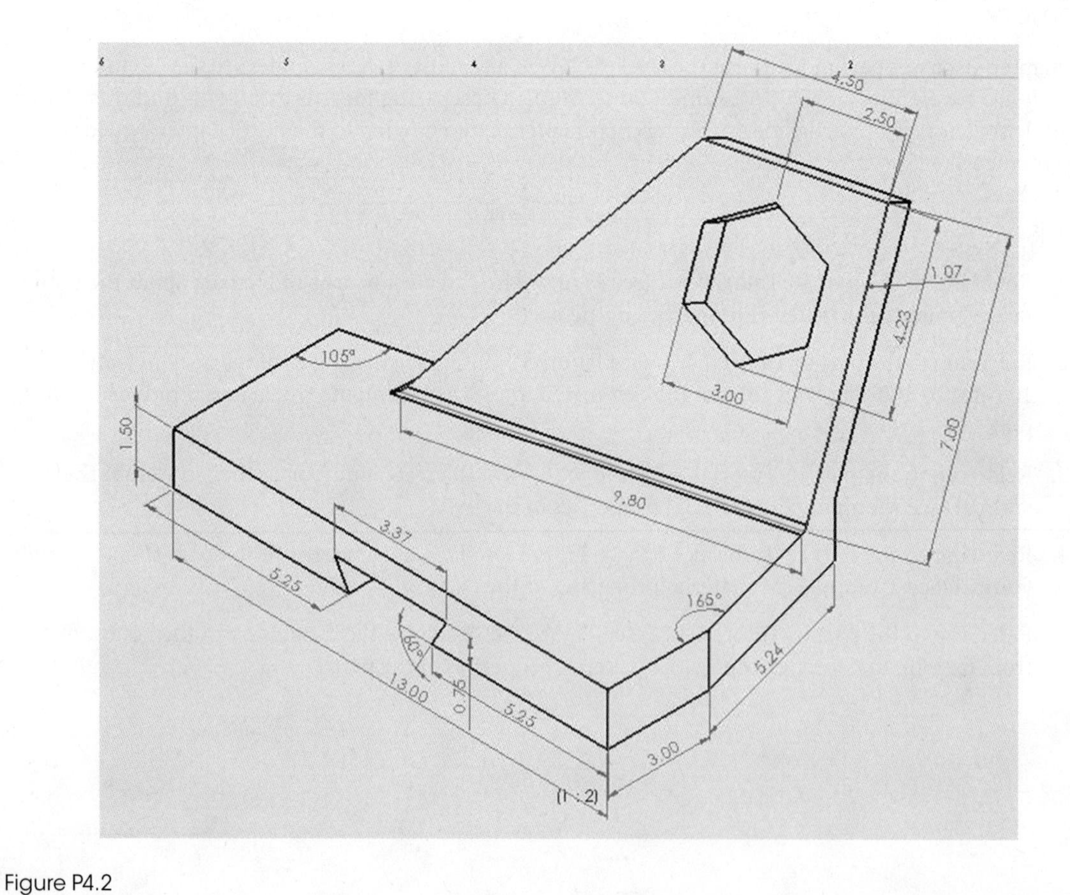

Figure P4.2

Slide bracket. (From David C. Planchard & Marie P. Planchard, Official Certified SolidWorks Associates (CSWA), SDC Publications, 2009.)

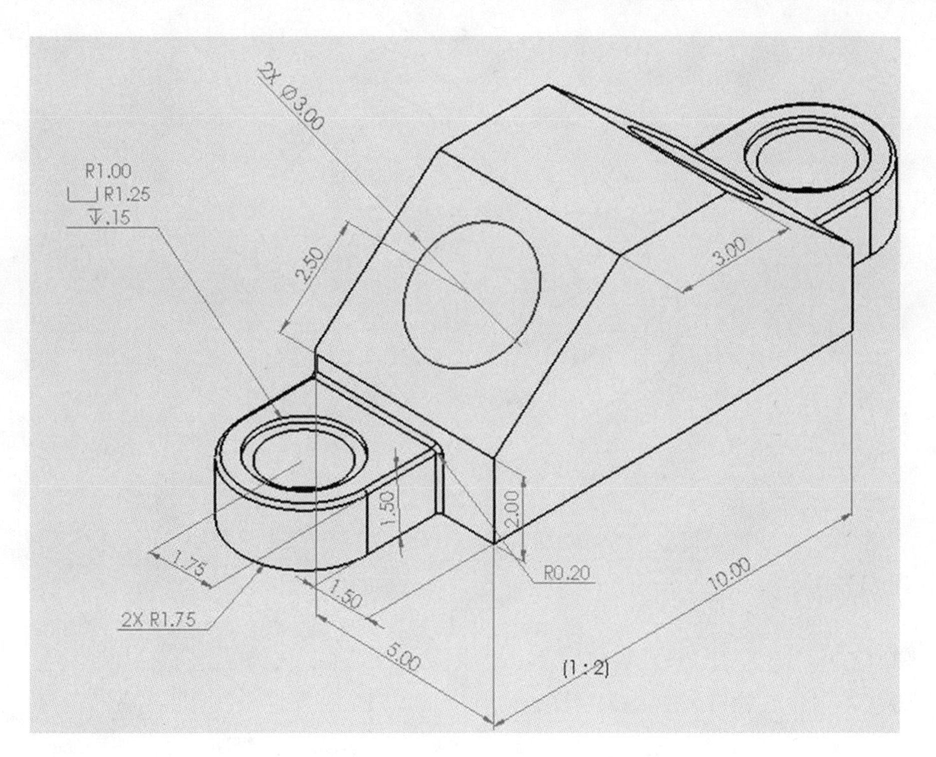

Figure P4.3

Inclined slide bracket. (From David C. Planchard & Marie P. Planchard, Official Certified SolidWorks Associates (CSWA), SDC Publications, 2009.)

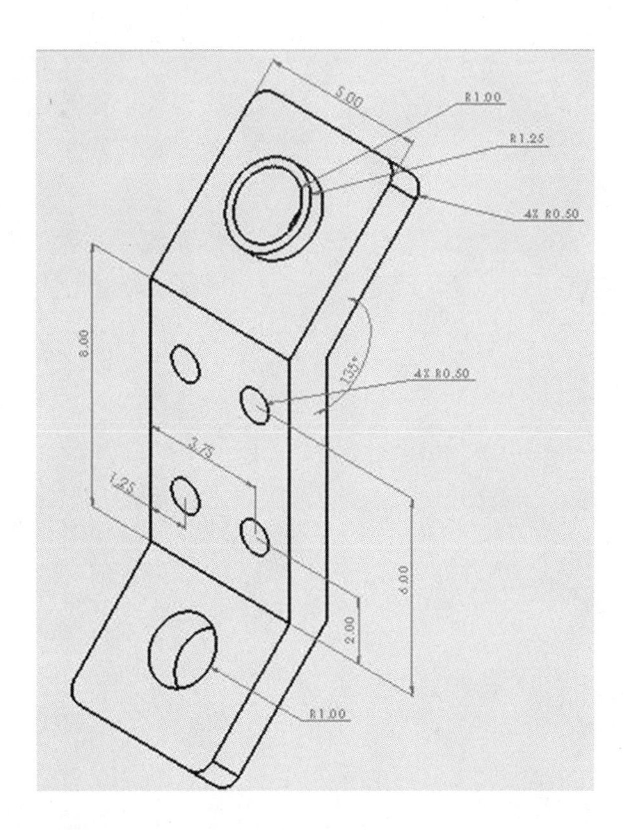

Figure P4.4

Cylinder support. (From David C. Planchard & Marie P. Planchard, Official Certified SolidWorks Associates (CSWA), SDC Publications, 2009.)

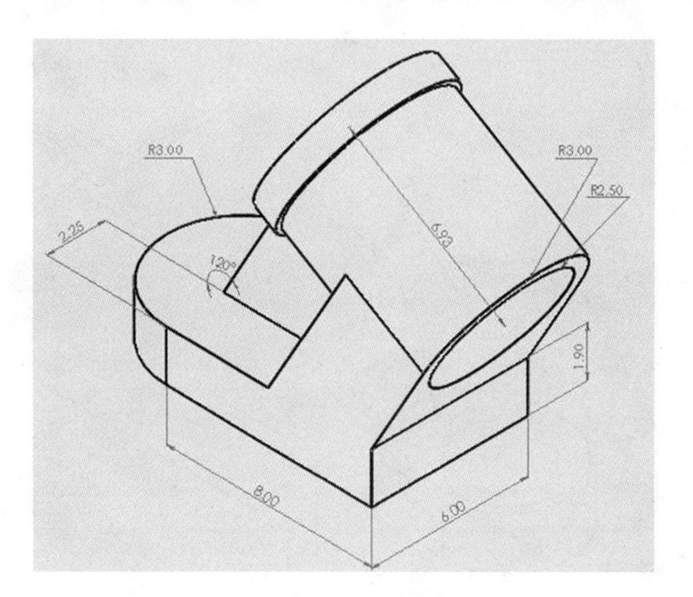

Figure P4.5

Cylinder support.

5

Advanced Part Modeling—CSWA Preparations

Objectives:

When you complete this session, you will have

- Learned about creating and manipulating a model coordinate system
- Learned how to model advanced 3D parts using Extrusion, Revolved, Lofted, Swept, Modification, Edit Features, and Features tools, as well as using reference planes
- Learned how to apply advanced modeling tools

Introduction

There are two differences between advanced part modeling and part modeling: (1) the complexity of the sketches that are involved and (2) the number of dimensions and geometric relations and increased number of features. This means that the modeler has to be able to deal with a significant amount of information in order to complete an advanced part modeling. Another difference between advanced part modeling and part modeling is the complexity of features that are involved in the modeling. It is important to keep in mind the design intent. Take advantage of symmetric features and build in relations that would shorten design time. Maintenance of future design depends to a great extent on how well the current design has been carried out. In the first part of this chapter, we present some *advanced part modeling tutorials* at the Certified SolidWorks Associate (CSWA) level.

Advanced part modeling is one of the five categories of the CSWA examination that is taken from the SolidWorks Website (http:/www.solidworks.com/cswa). There is one question in this chapter that is taken from the CSWA category.

In providing a SolidWorks solution to the problems or tutorials, we will first create a part based on a dimensioned drawing with the origin and coordinate axes annotated. We will then create the part, apply the correct material properties, and retrieve the mass and center of mass information. When building the model, we will ensure that it is oriented relative to the default coordinate system in the manner that is indicated in the problem. Knowledge of *reference plane* is essential in solving most CSWA problems. Characteristic of CSWA exam questions, care should be taken to locate the origin of the part, applying the material properties, and retrieving the mass properties. Creating reference axes to match those in the problem statement is essential when retrieving the center of mass of the model(s).

Advanced Part Modeling Tutorials

Tutorial 5.1: Block with Hook

Create the model that is shown in Figure 5.1. Determine the overall mass and volume of the part and the center of mass for the model.

1. Create a New part in SolidWorks.
2. Set the document properties for the model.
3. Create Sketch1. Select the Front Plane as the Sketch Plane (see Figure 5.2).
4. Create the Extrude Base feature. Extrude depth = 2.5-in. Mid Plane. Extrude1 is the Base feature (see Figure 5.3).
5. Create Sketch2. Select the foremost face of Extrude1 as the Sketch Plane. Sketch a Slot (see Figure 5.4).
6. Create the first Extrude Cut feature. Extrude-cut feature using Through All for End Condition (see Figure 5.5).
7. Create Sketch3. Select the top face of Extrude1 as the Sketch Plane. Sketch the profile that is shown; maintain a surface begins to curve, the 0.75-in being for each side (see Figure 5.6).
8. Extrude Cut Up To Surface, using the bottom surface as basis (see Figure 5.7).
9. Create Sketch4, a Circle of diameter 1.75-in. on top-right surface of Extrude1 (see Figure 5.8).

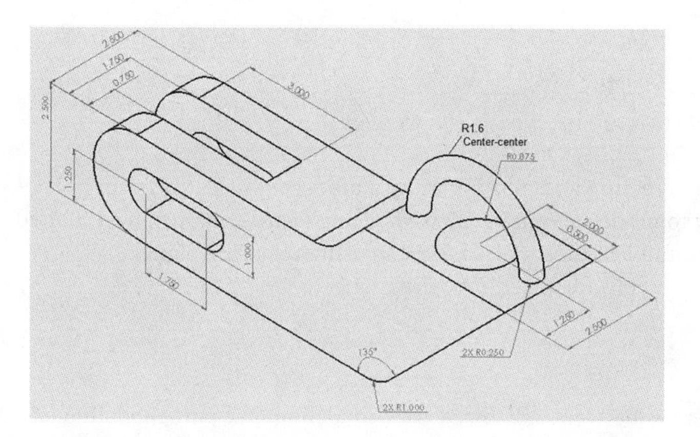

Figure 5.1

Model to be built. (From David C. Planchard & Marie P. Planchard, Official Certified SolidWorks Associates (CSWA), SDC Publications, 2009.)

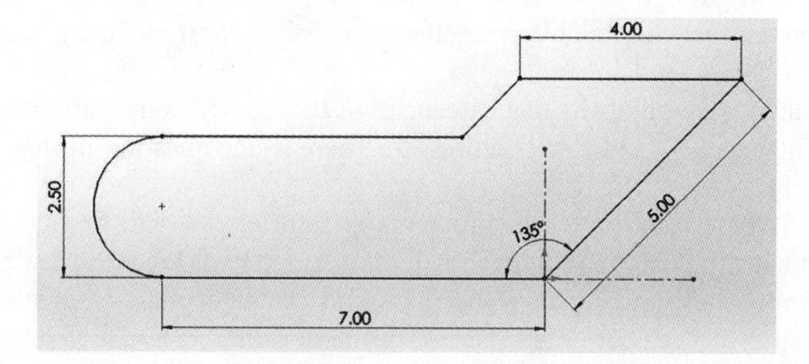

Figure 5.2

Sketch1. (From David C. Planchard & Marie P. Planchard, Official Certified SolidWorks Associates (CSWA), SDC Publications, 2009.)

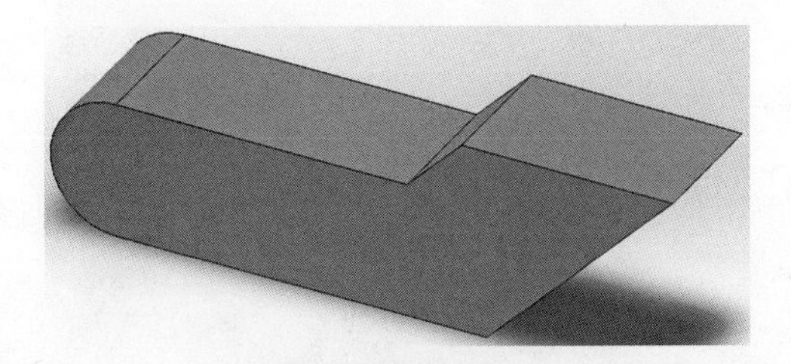

Figure 5.3

Extrude1. (From David C. Planchard & Marie P. Planchard, Official Certified SolidWorks Associates (CSWA), SDC Publications, 2009.)

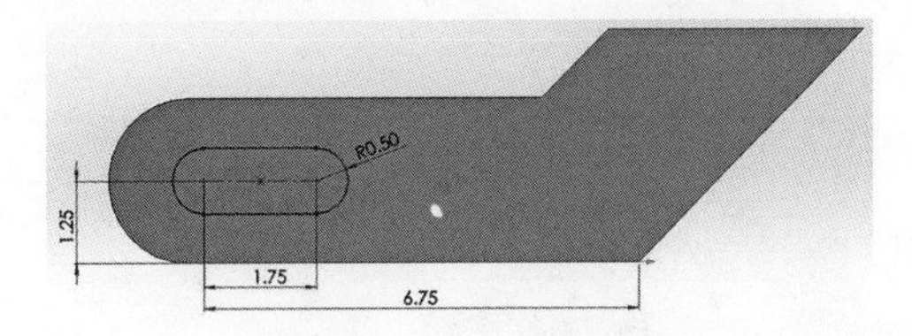

Figure 5.4

Sketch2. (From David C. Planchard & Marie P. Planchard, Official Certified SolidWorks Associates (CSWA), SDC Publications, 2009.)

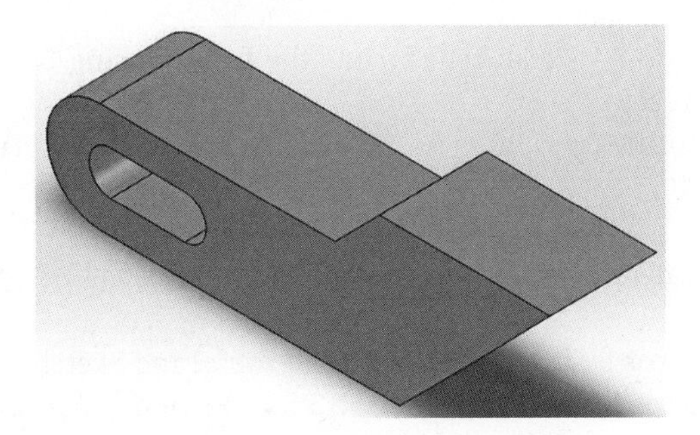

Figure 5.5

Extrude Cut1. (From David C. Planchard & Marie P. Planchard, Official Certified SolidWorks Associates (CSWA), SDC Publications, 2009.)

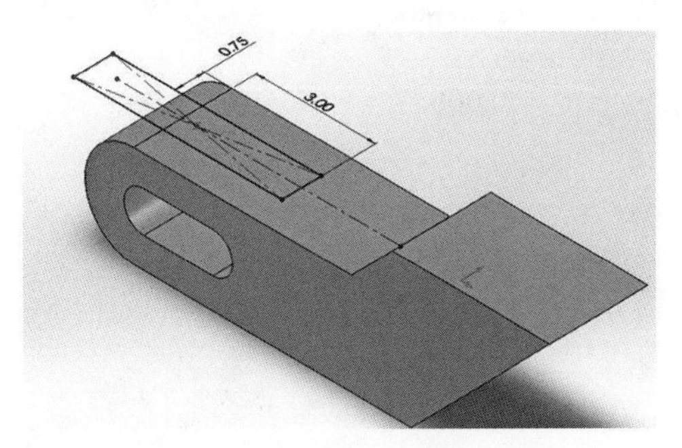

Figure 5.6

Sketch of profile on top of Extrude1. (From David C. Planchard & Marie P. Planchard, Official Certified SolidWorks Associates (CSWA), SDC Publications, 2009.)

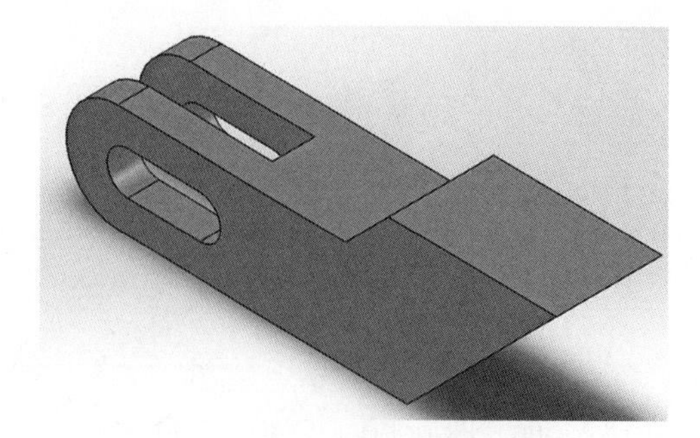

Figure 5.7

Extrude2 of sketched profile. (From David C. Planchard & Marie P. Planchard, Official Certified SolidWorks Associates (CSWA), SDC Publications, 2009.)

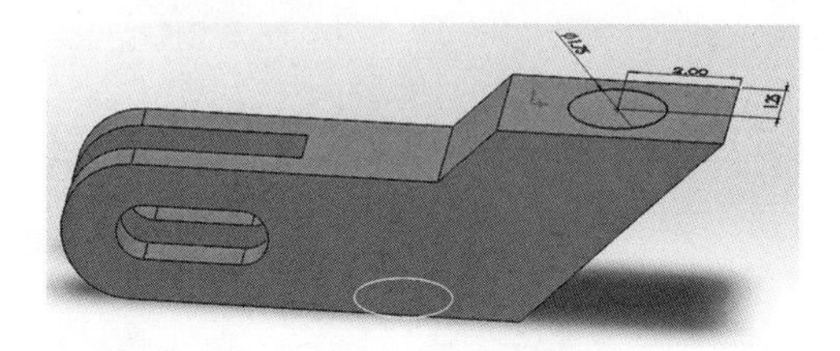

Figure 5.8

Sketch of profile on top of Extrude2. (From David C. Planchard & Marie P. Planchard, Official Certified SolidWorks Associates (CSWA), SDC Publications, 2009.)

10. Create the third Extrude Boss feature. Extrude the feature using Up To Surface, with the bottom face as basis (see Figure 5.9).
11. Create Sketch5. Select the top-right face of Extrude1 as the Sketch Plane. Sketch a circle, diameter 0.50-in. (see Figure 5.10).
12. Exit Sketch5.
13. Using Front Plane, create Sketch6, an arc, radius 1.60-in. (see Figure 5.11).
14. Exit Sketch6.
15. Create the first Swept Boss feature. Sketch5 is the Profile, and Sketch6 is the Path (see Figure 5.12).
16. Create Fillets of Radius of 1.0-in. (see Figure 5.13 for the modeled part).

The mass properties are given as follows:

- Mass properties of housing (Part Configuration-Default)
- Density = 0.003 grams per cubic millimeter
- Mass = 37416.191 grams
- Volume = 13857848.552 cubic millimeters
- Surface area = 542149.374 millimeters^2
- Center of mass: (millimeters) X = 0.000; Y = 70.936; Z = 0.000

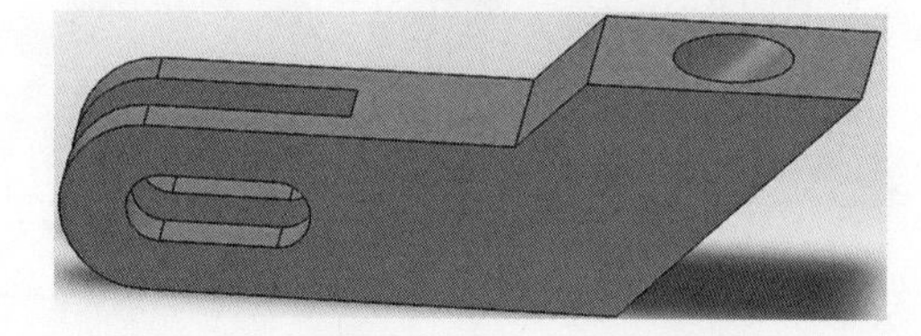

Figure 5.9

Extrude3 of sketched profile. (From David C. Planchard & Marie P. Planchard, Official Certified SolidWorks Associates (CSWA), SDC Publications, 2009.)

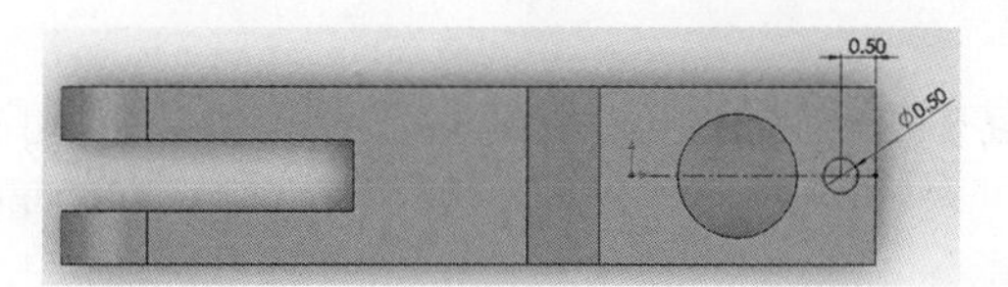

Figure 5.10

Sketch5: a circle on top of Extrude2. (From David C. Planchard & Marie P. Planchard, Official Certified SolidWorks Associates (CSWA), SDC Publications, 2009.)

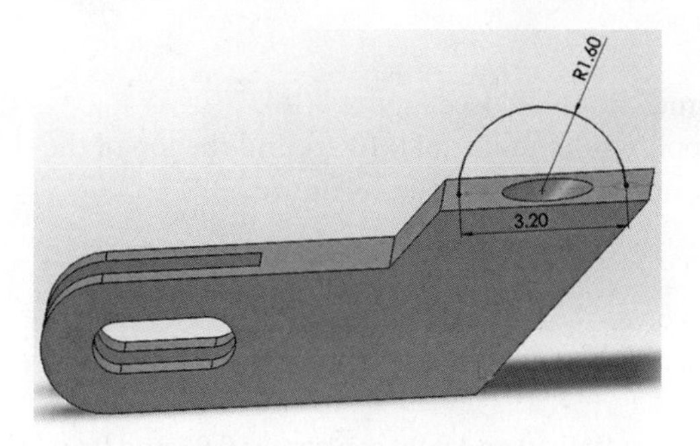

Figure 5.11

Sketch6: a circle on Plane1 on top of Extrude2. (From David C. Planchard & Marie P. Planchard, Official Certified SolidWorks Associates (CSWA), SDC Publications, 2009.)

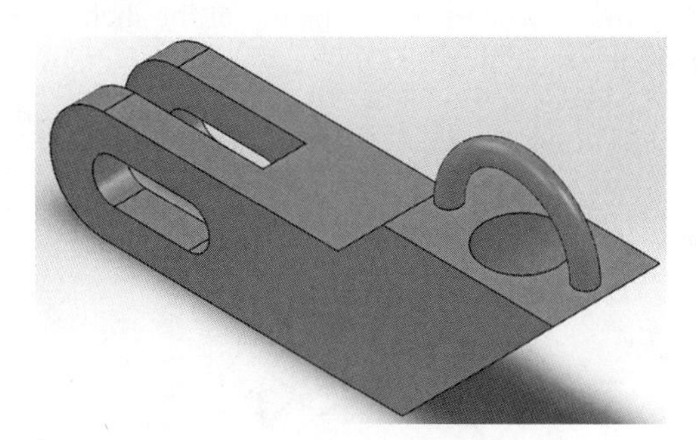

Figure 5.12

Swept feature. (From David C. Planchard & Marie P. Planchard, Official Certified SolidWorks Associates (CSWA), SDC Publications, 2009.)

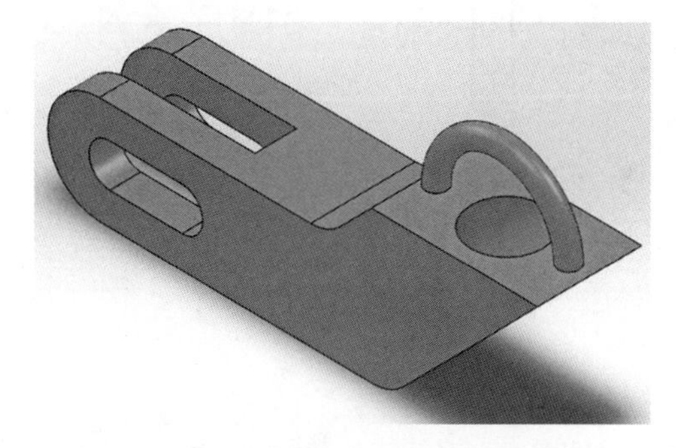

Figure 5.13

Model of part. (From David C. Planchard & Marie P. Planchard, Official Certified SolidWorks Associates (CSWA), SDC Publications, 2009.)

Tutorial 5.2: Housing

Create the housing model that is shown in Figure 5.14. Determine the overall mass and volume of the part, and the center of mass for the model for 1060 Alloy material. Note the origin of the model. Units: inch, pound, second (IPS).

SolidWorks Solution

1. Create a New part in SolidWorks.
2. Set the document properties for the model.
3. Select the Top Plane and create Sketch1, 3.50 × 3.50 (see Figure 5.15).
4. Extrude.

Shell

1. Select lower face and click Shell (see Figure 5.16).
2. Click Fillet and apply a fillet radius of 0.10 around the top of the shelled object.
3. Select Front Plane and create Sketch2 (see Figure 5.17).

Revolve

1. Click Features/Revolve.
2. Select vertical Construction Line as the axis to revolve (see Figure 5.18).
3. Create Plane1.
4. Sketch two circles of radii 0.60 and 0.90 at a center 0.95 and 0.60 from the edges (see Figure 5.19).
5. Click Features/Extrude.
6. Extrude Up To Surface, selecting rightmost face, Face<1> (see Figure 5.20).
7. Fillet Inside Edges and the bottom of cylinder features.
8. Create a circle at the bottom of Revolved feature (see Figure 5.21).
9. Click Extrude Cut.
10. Select Up To Surface, and choose inside of the top of the shelled feature (see Figure 5.22).
11. Click OK to finish model (see Figure 5.23).

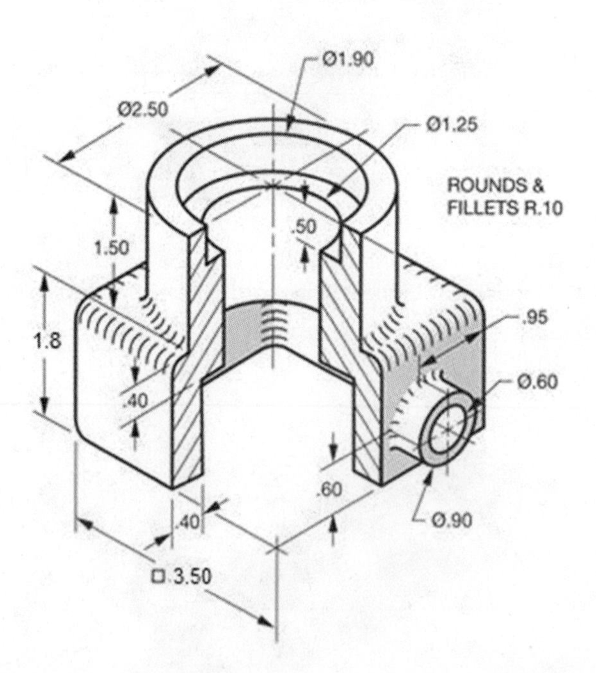

Figure 5.14

Housing model to be built. (From JENSEN/HELSEL/ESPIN. Interpreting Engineering Drawings, 6E. © 2012 Nelson Education Ltd. Reproduced by permission. http://www.cengage.com/permissions.)

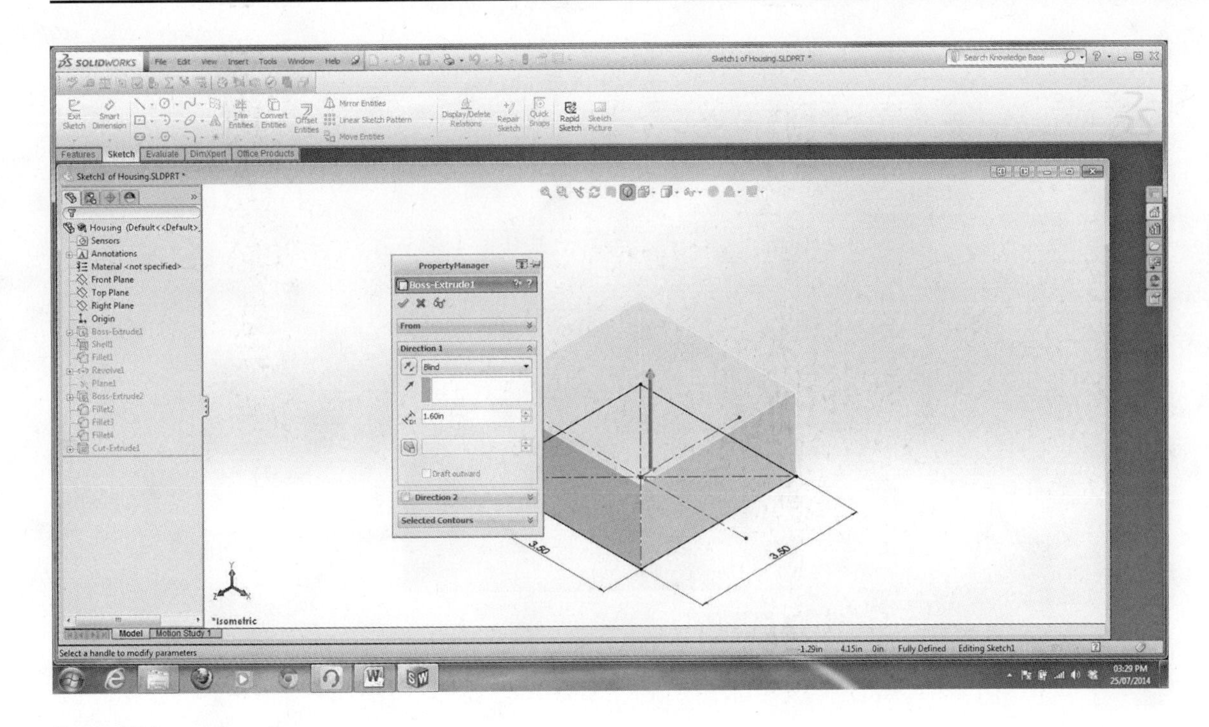

Figure 5.15

Sketch1, 3.50 × 3.50.

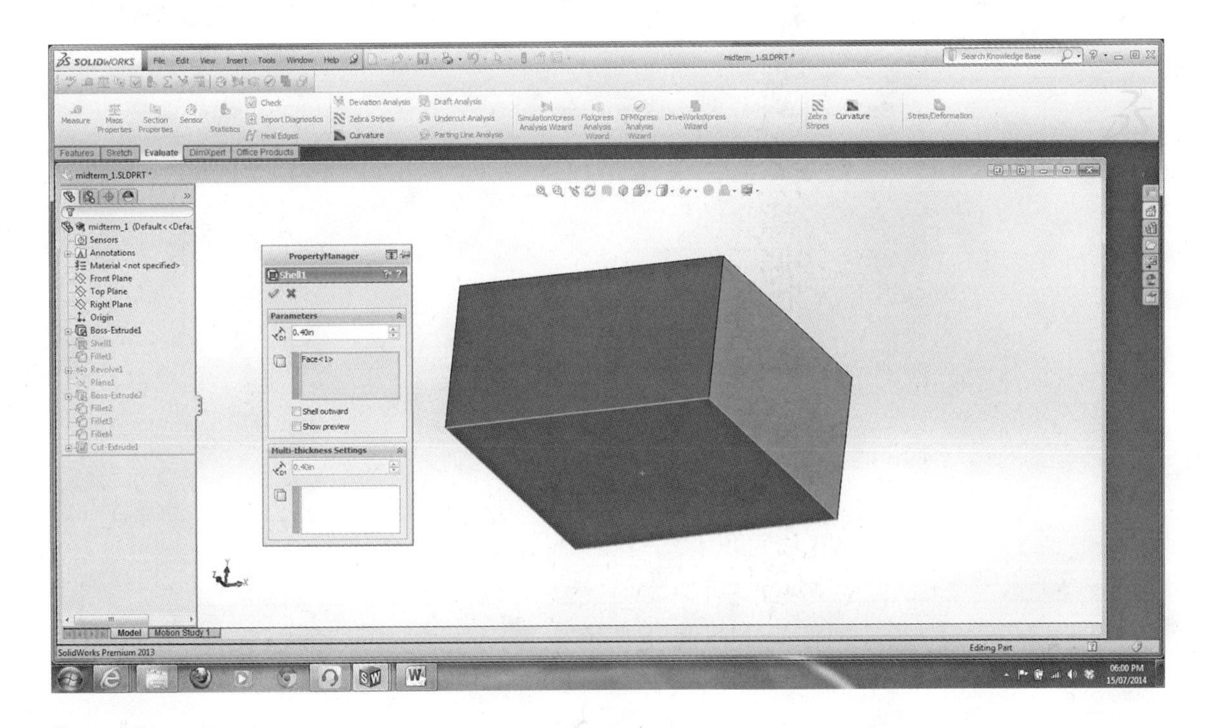

Figure 5.16

Shelling the solid.

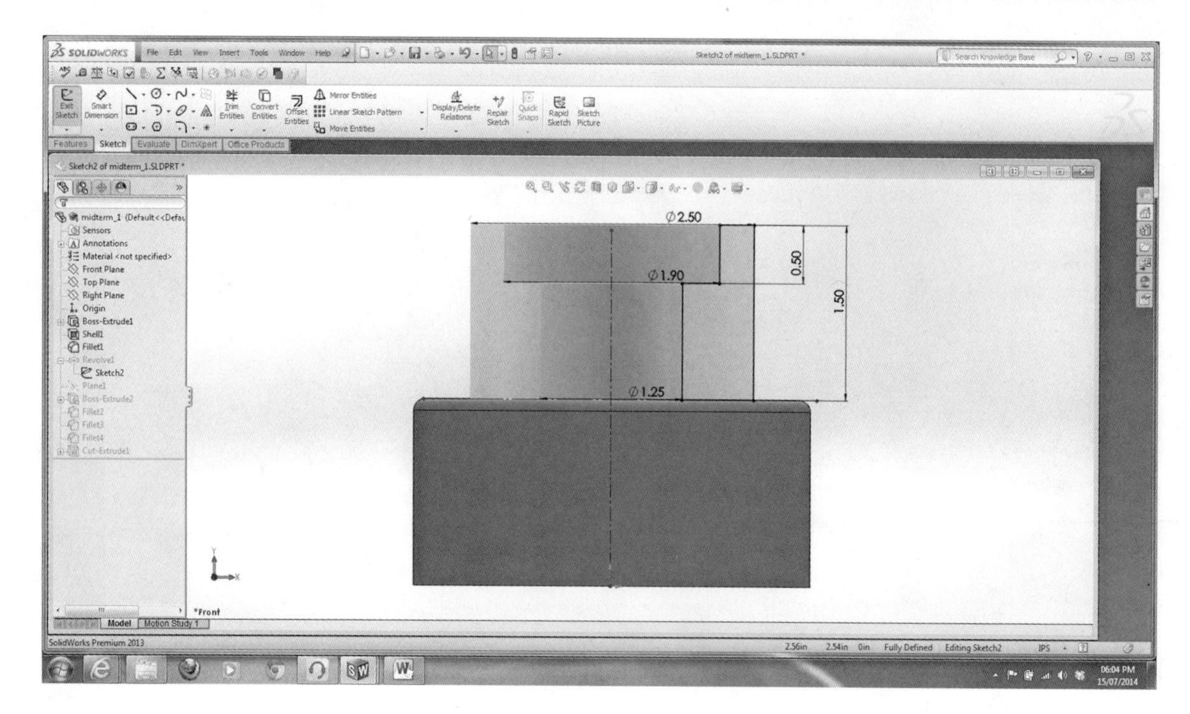

Figure 5.17

Sketch2.

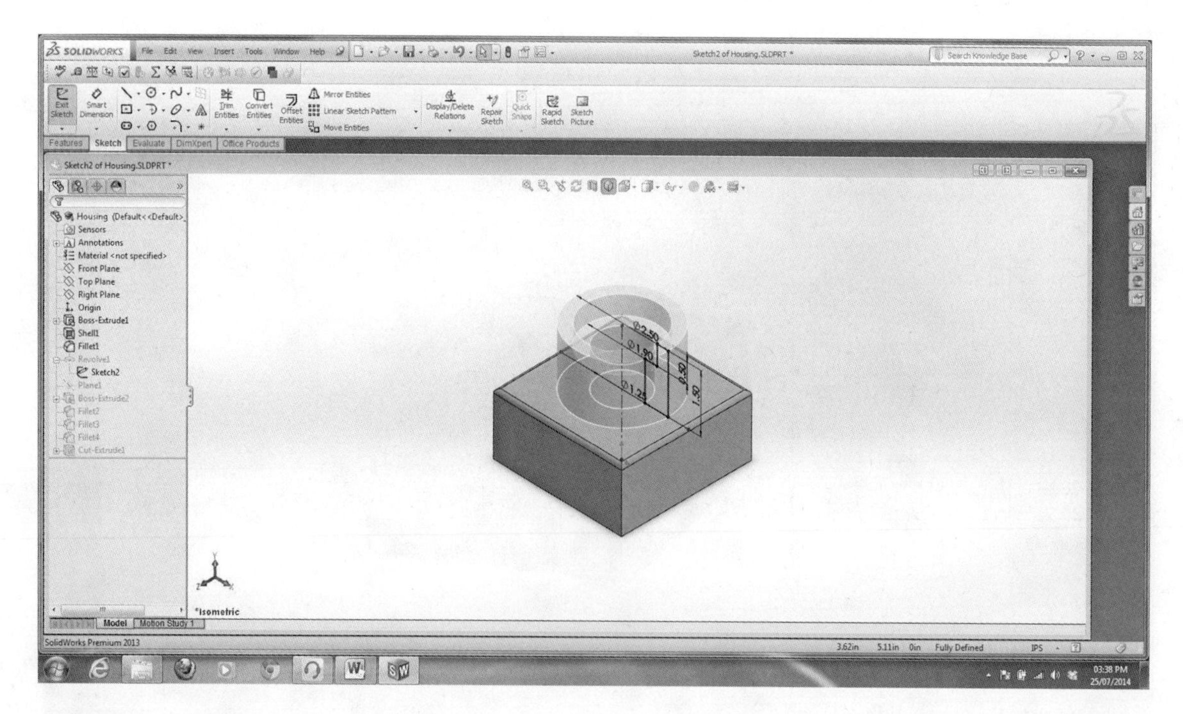

Figure 5.18

Revolve operations.

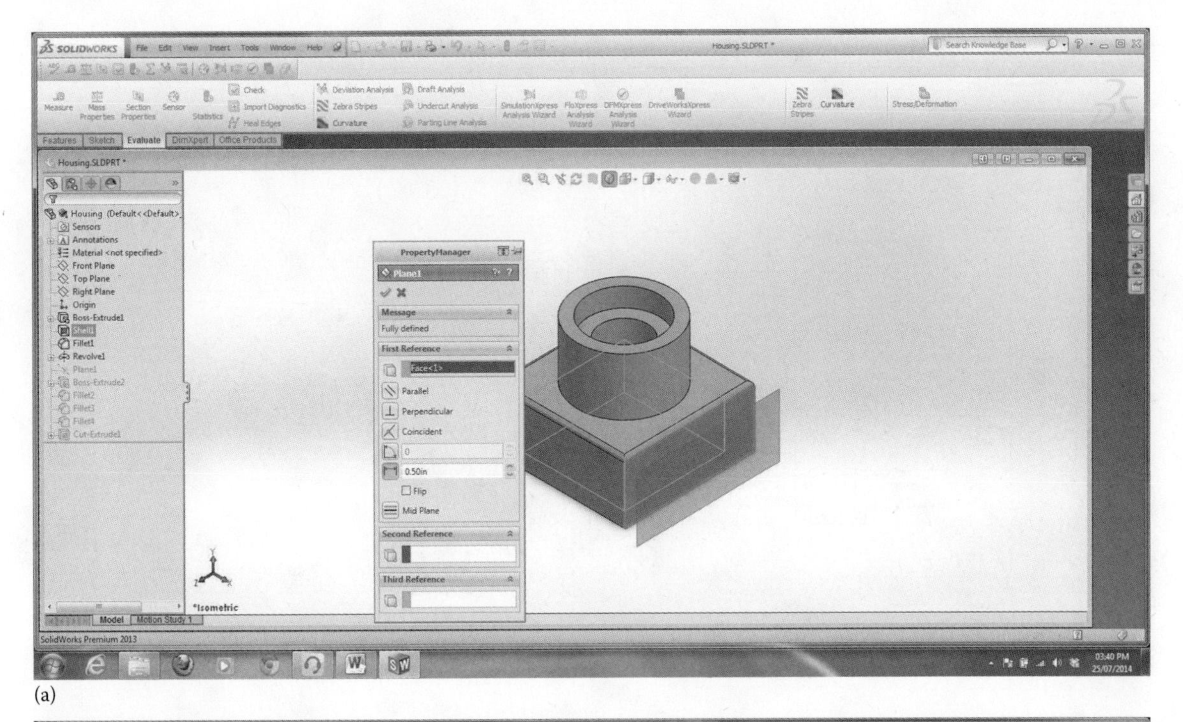

(a)

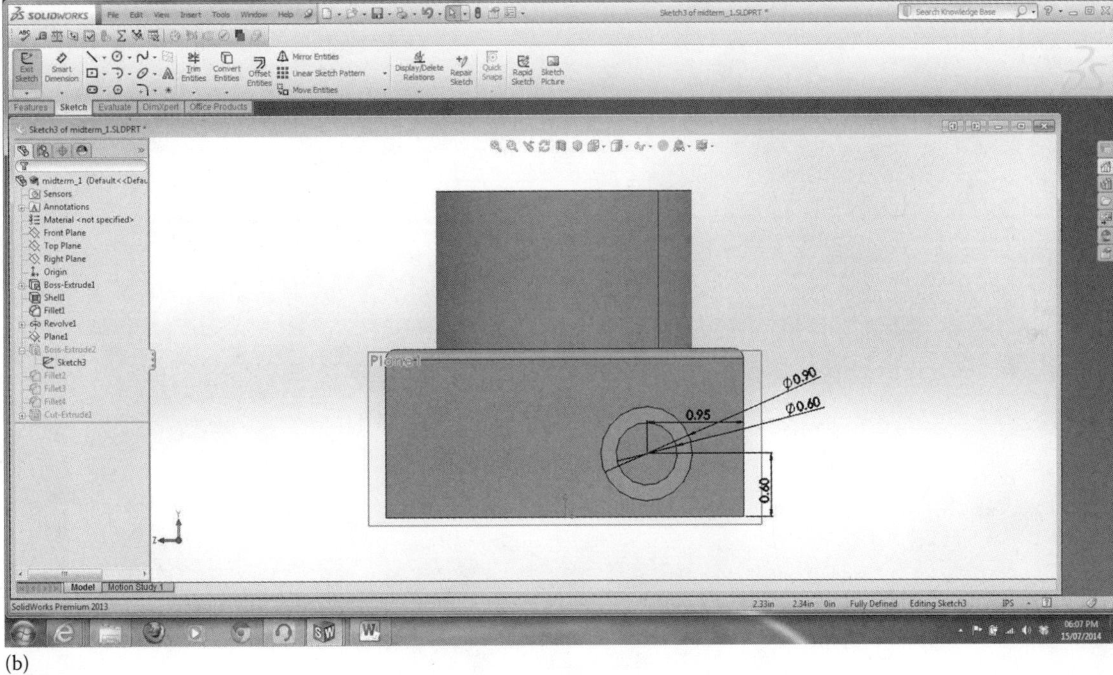

(b)

Figure 5.19

Creating Plane1 (a) and sketching circle (b).

The mass properties are given as follows:

- Mass properties of housing (Part Configuration-Default)
- Density = 0.1 pounds per cubic inch
- Mass = 1.5 pounds
- Volume = 15.3 cubic inches
- Surface area = 78.8 square inches
- Center of mass: (inches) X = 0.0; Y = 1.3; Z = −0.0

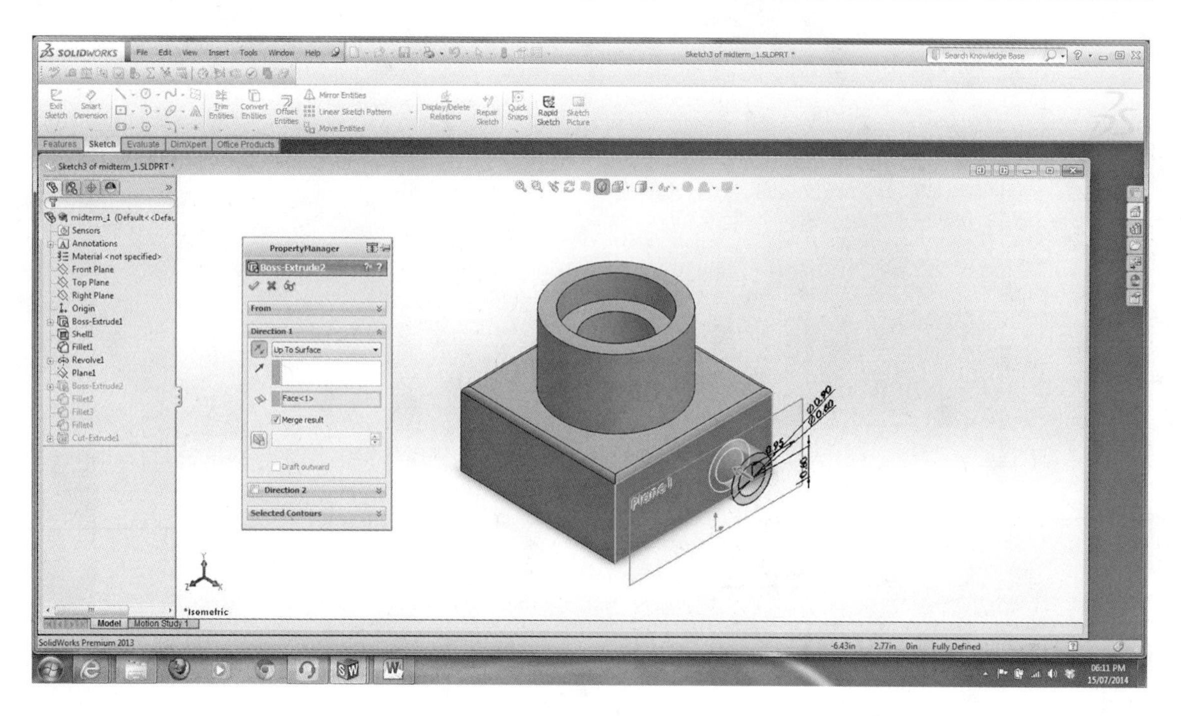

Figure 5.20

Boss-Extrude2.

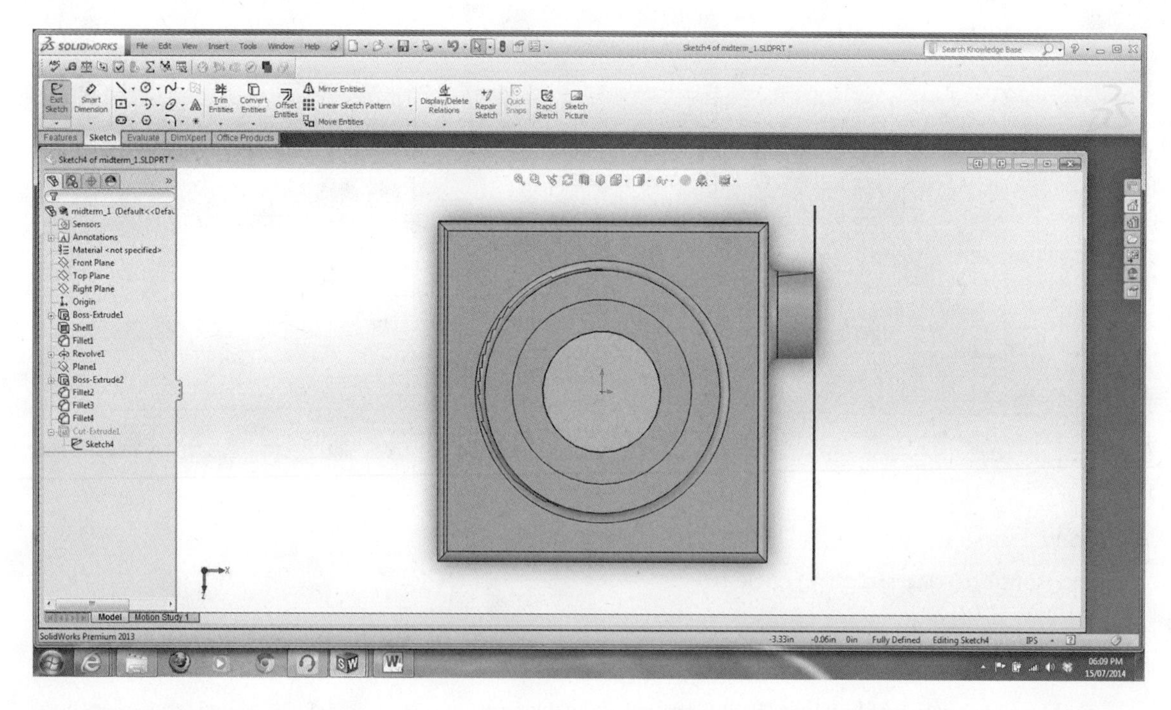

Figure 5.21

Creating circle at the bottom of Revolved feature.

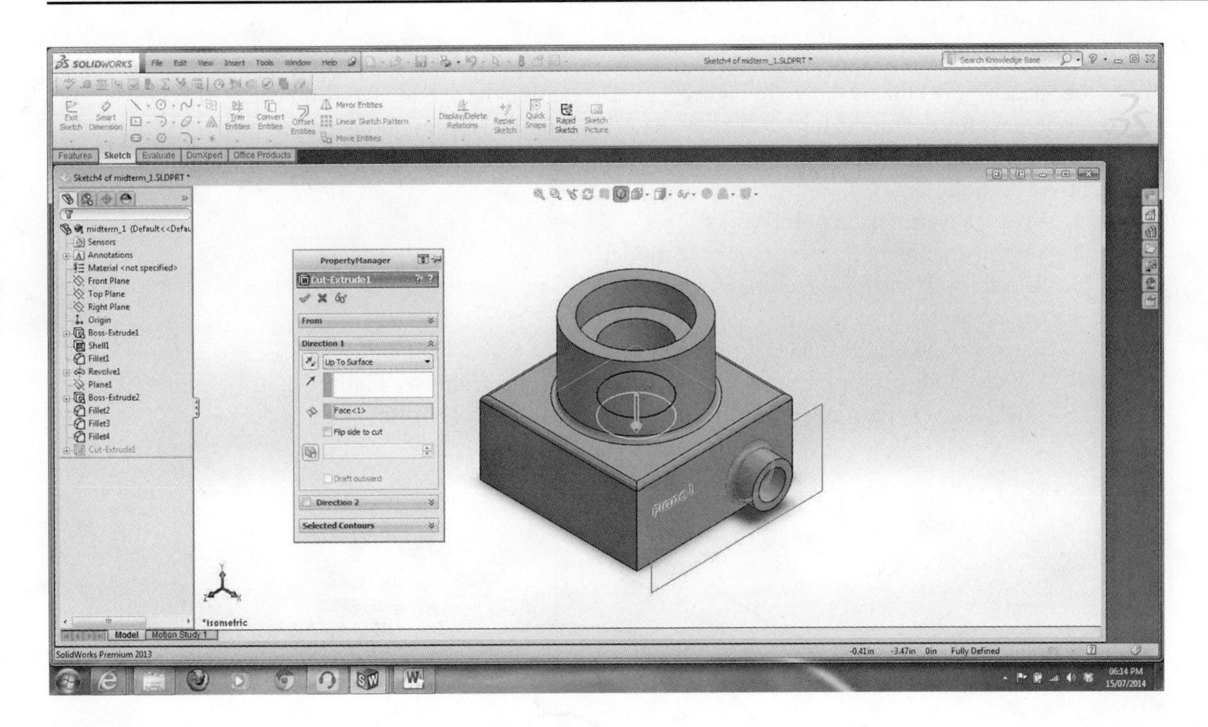

Figure 5.22

Extrude-Cut of circular feature.

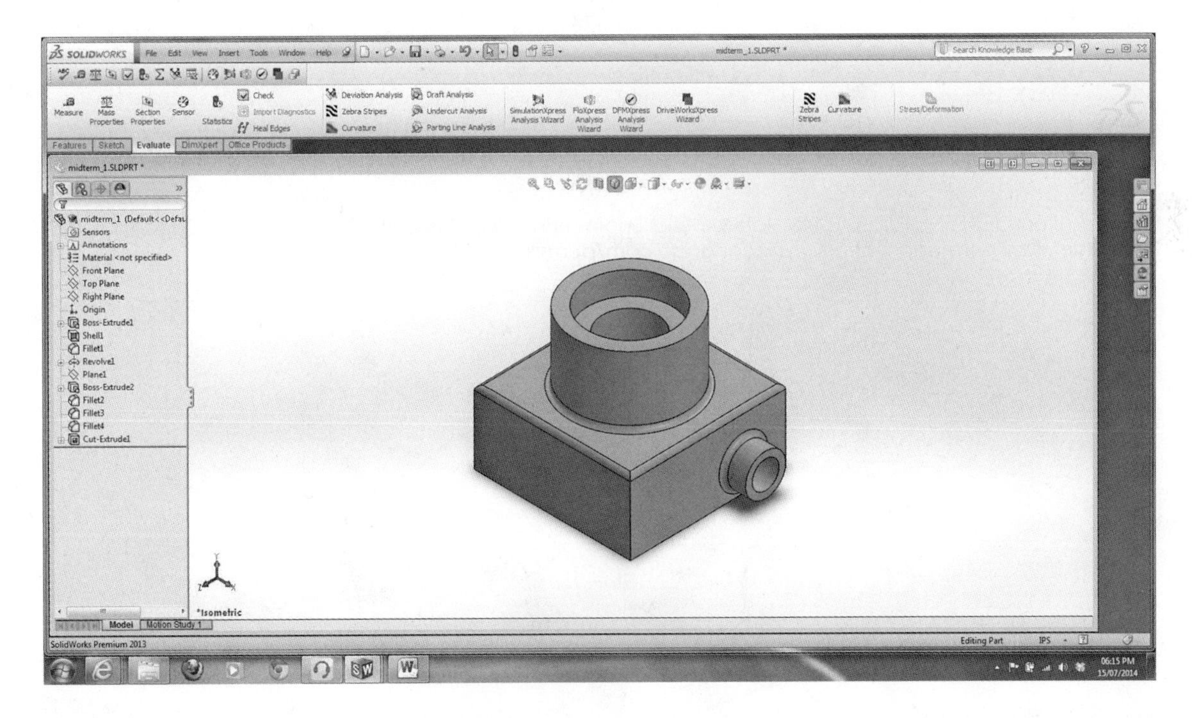

Figure 5.23

Final model.

Tutorial 5.3: Spacer

Create the Spacer model that is shown in Figure 5.24. Determine the overall mass and volume of the part and the center of mass for the model for 6061 Alloy material. Note the origin of the model.

SolidWorks Solution

1. Create a New Part in SolidWorks.
2. Set the document Properties for the model.
3. Select Top Plane for sketching.
4. Create Sketch1 (see Figure 5.25).
5. Extrude 0.59 Mid Plane (see Figure 5.26).
6. Sketch a Line 3.15 from the center of the circle on the right (see Figure 5.27).
7. Click Convert Entities and capture the edges to the left of the line.

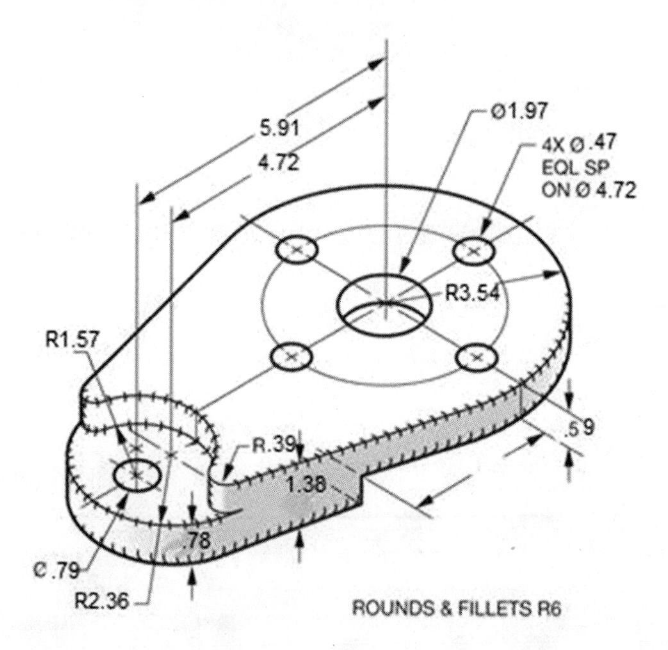

Figure 5.24

Spacer. (From JENSEN/HELSEL/ESPIN. Interpreting Engineering Drawings, 6E. © 2012 Nelson Education Ltd. Reproduced by permission. http://www.cengage.com/permissions.)

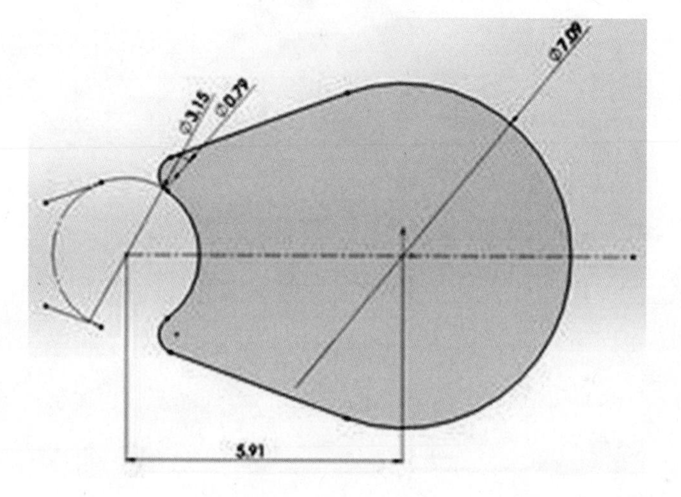

Figure 5.25

Sketch1 for Spacer.

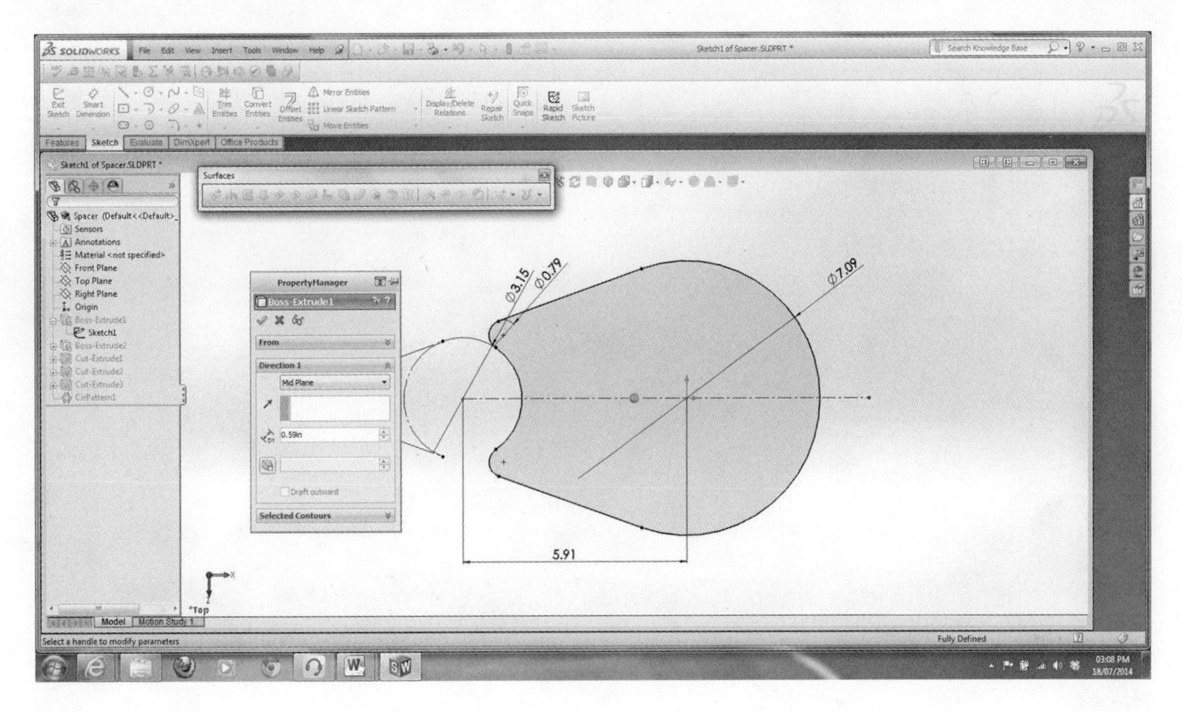

Figure 5.26

Extrude operation.

8. Extrude 0.78 downward.
9. Sketch a circle, diameter 0.79, at the center of extreme-left arc.
10. Extrude-Cut up to the bottom surface (see Figure 5.28).
11. Sketch a circle, diameter 1.97, at the center of extreme-right circle (see Figure 5.29).
12. Extrude-Cut up to the bottom surface.
13. Sketch a circle, diameter 4.72, using the Construction Geometry mode at the center of the extreme-right circle (see Figure 5.30).
14. Sketch another circle, diameter 0.47, on the previous circle.
15. Extrude-Cut up to the bottom surface.
16. Click Circular Pattern.
17. Create four instances of the small circle (see Figure 5.31).
18. Click OK to complete model (see Figure 5.32).

Click Evaluate > Mass Properties to determine the weight of the model as 3.42 lb.

Density = 0.10 pounds per cubic inch
Mass = 3.42 pounds
Volume = 35.10 cubic inches
Surface area = 131.86 square inches
Center of mass: (inches) X = −1.70; Y = −0.18; Z = 0.00

Tutorial 5.4: Oarlock Socket

Create the oarlock socket model that is shown in Figure 5.33. Determine the overall mass and volume of the part and the center of mass for the model for 6061 Alloy material. Note the origin of the model. Units: IPS.

SolidWorks Solution

1. Create a New part in SolidWorks.
2. Set the document properties for the model.
3. Select Right Plane for sketching.
4. Create Sketch1 (Figure 5.34).
5. Extrude Mid Plane 4.7.
6. Create Rounds of 0.47 on four corners (Figure 5.35).

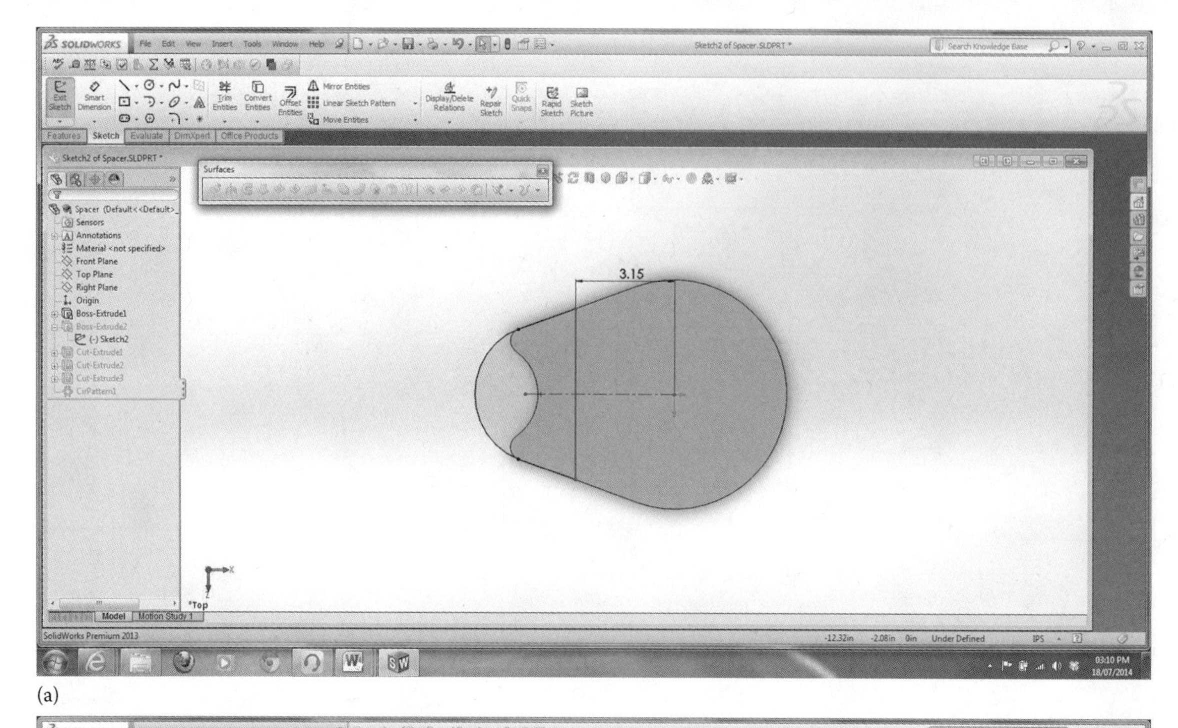

(a)

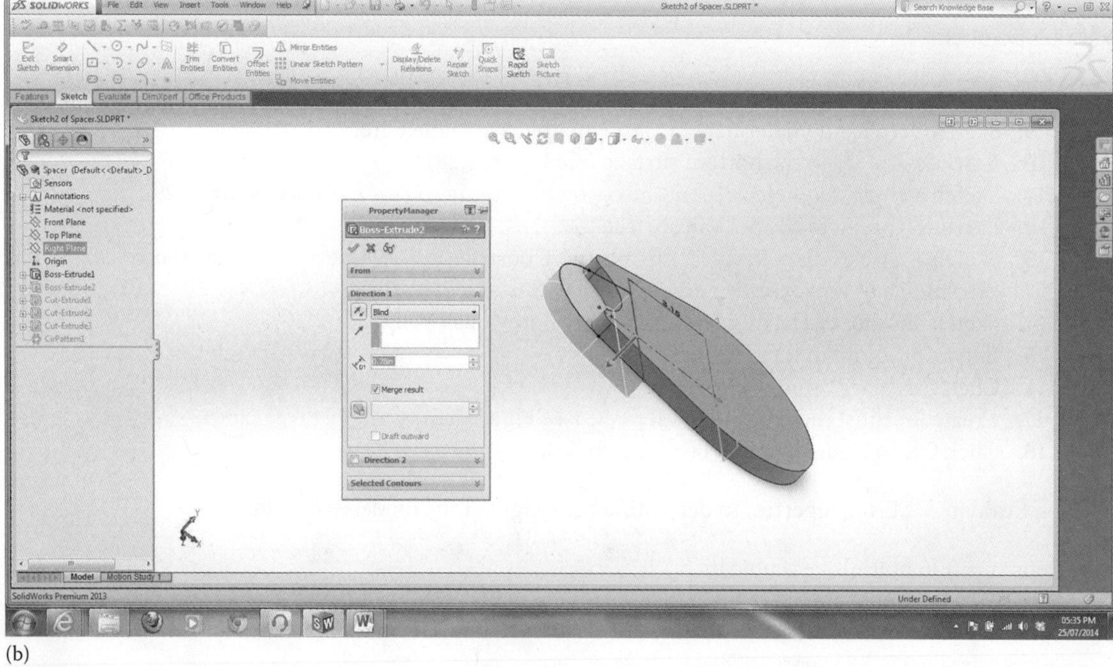

(b)

Figure 5.27

Reference line for lower feature (a) and extruding features to the left (b).

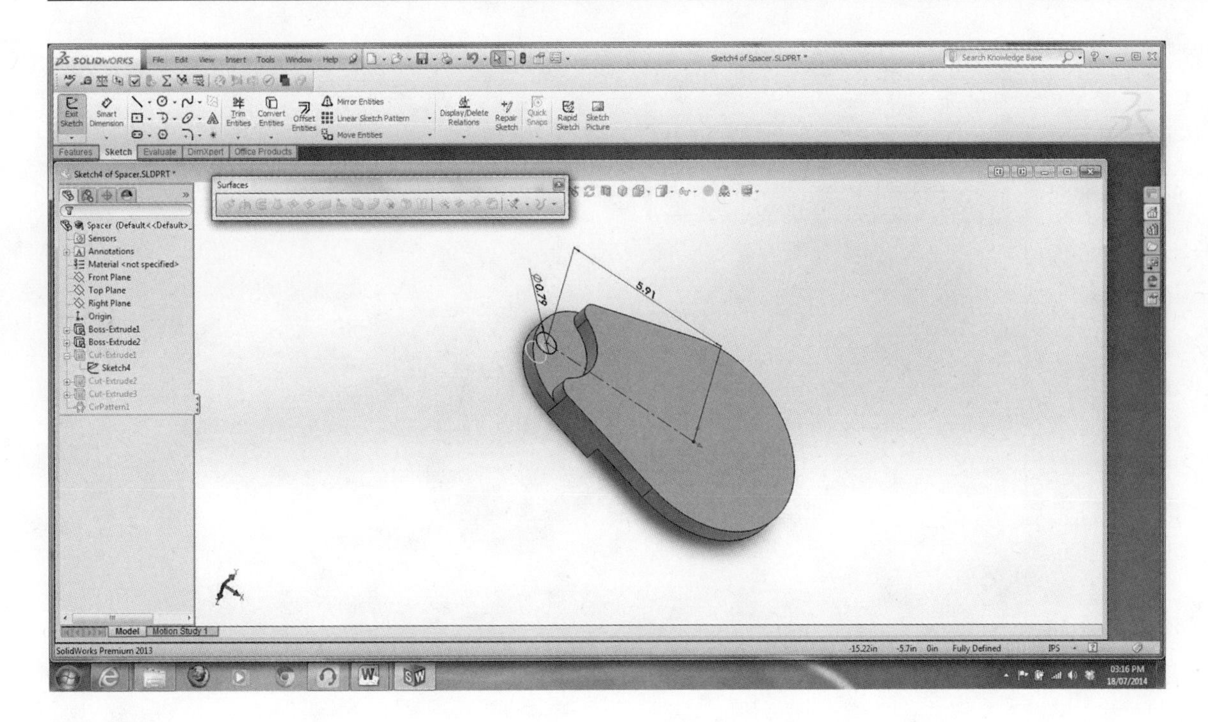

Figure 5.28

Extrude-Cut of hole feature.

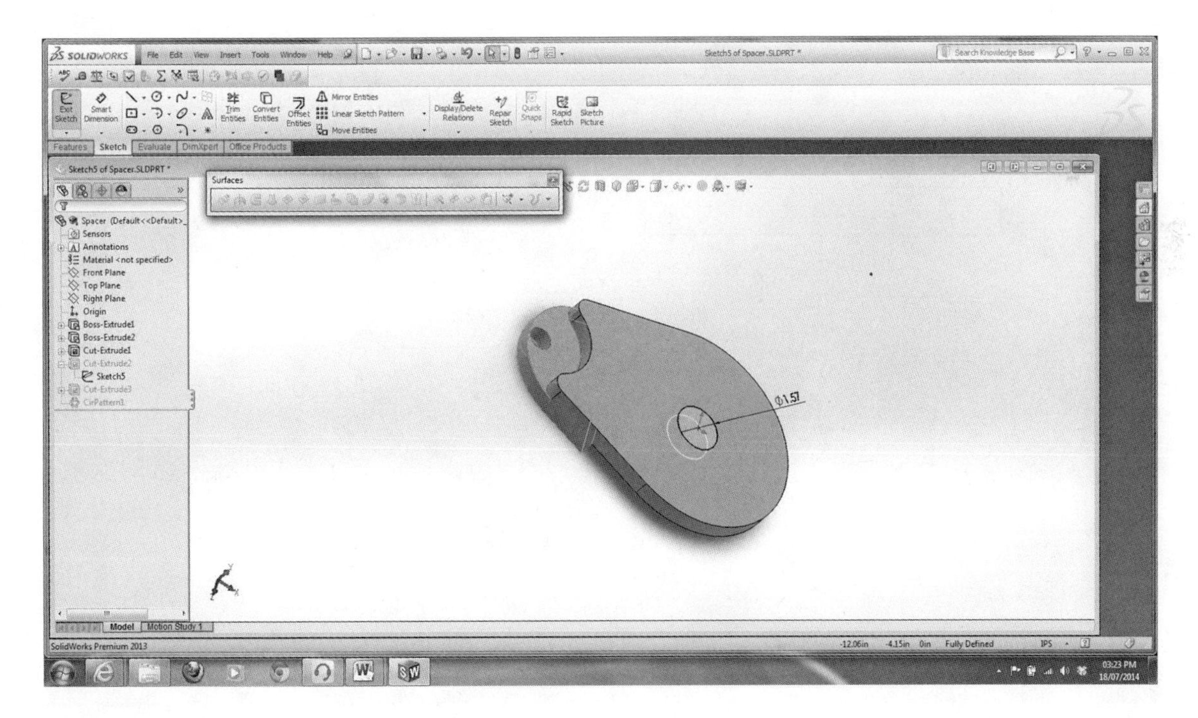

Figure 5.29

Extrude-Cut of hole at the top.

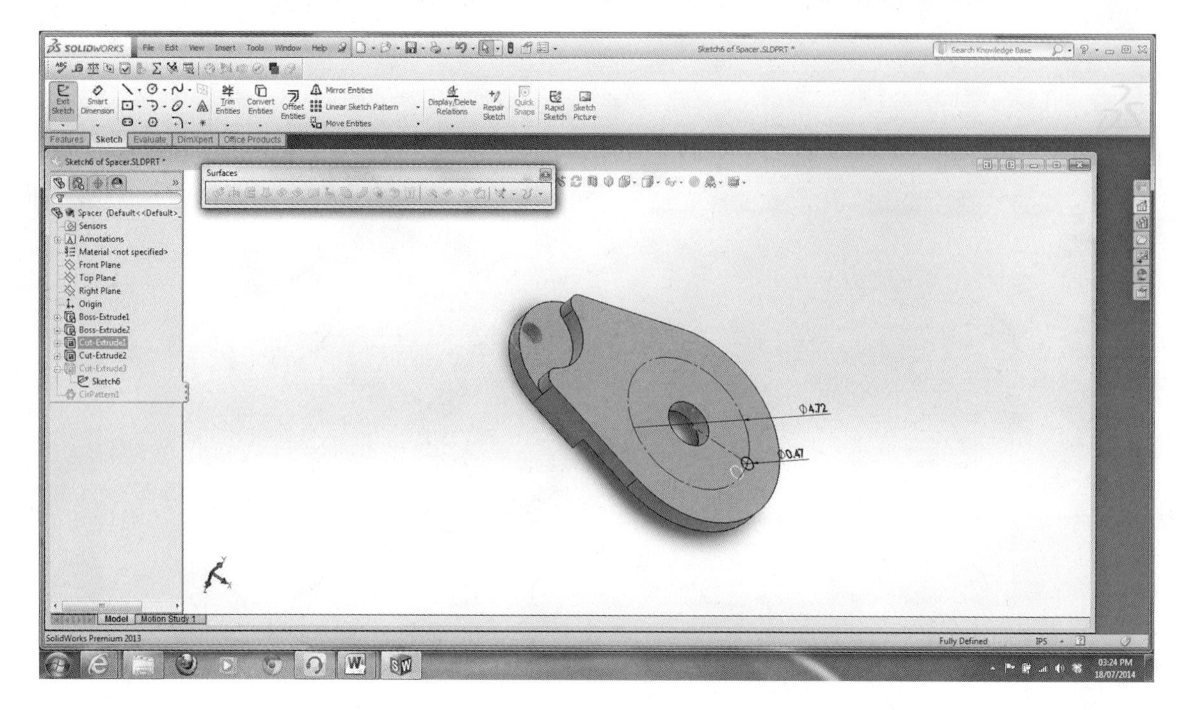

Figure 5.30

Reference circle and sketch for holes at top face.

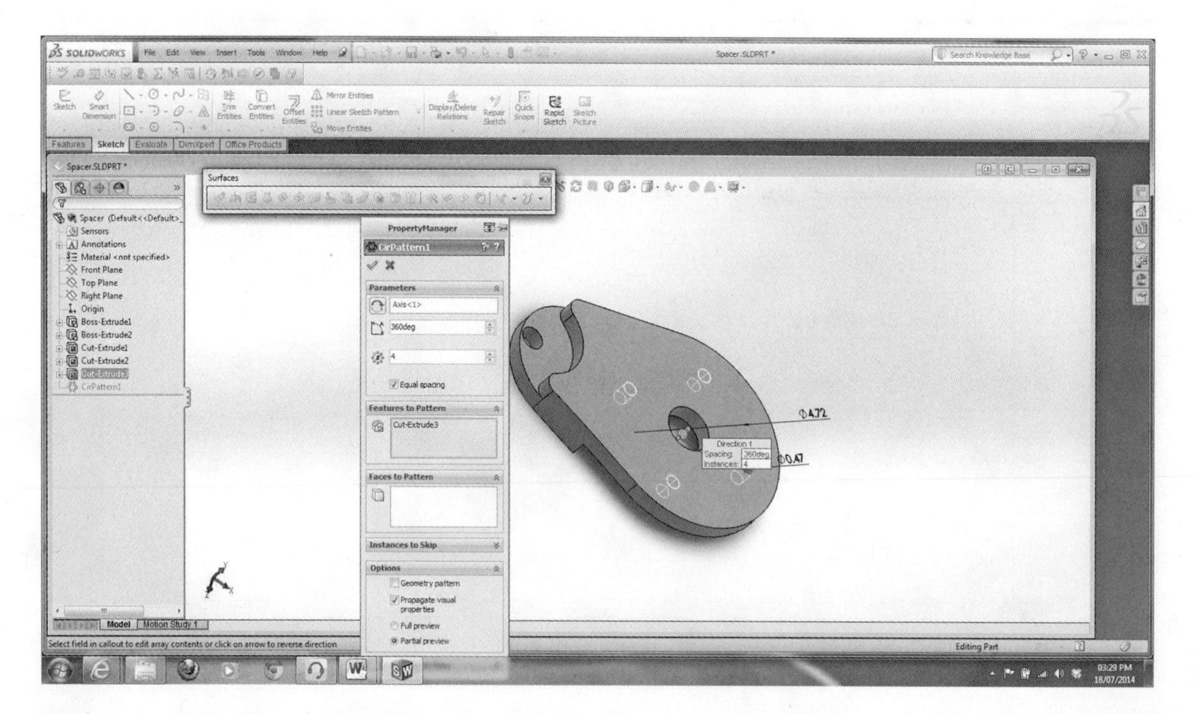

Figure 5.31

Circular pattern option.

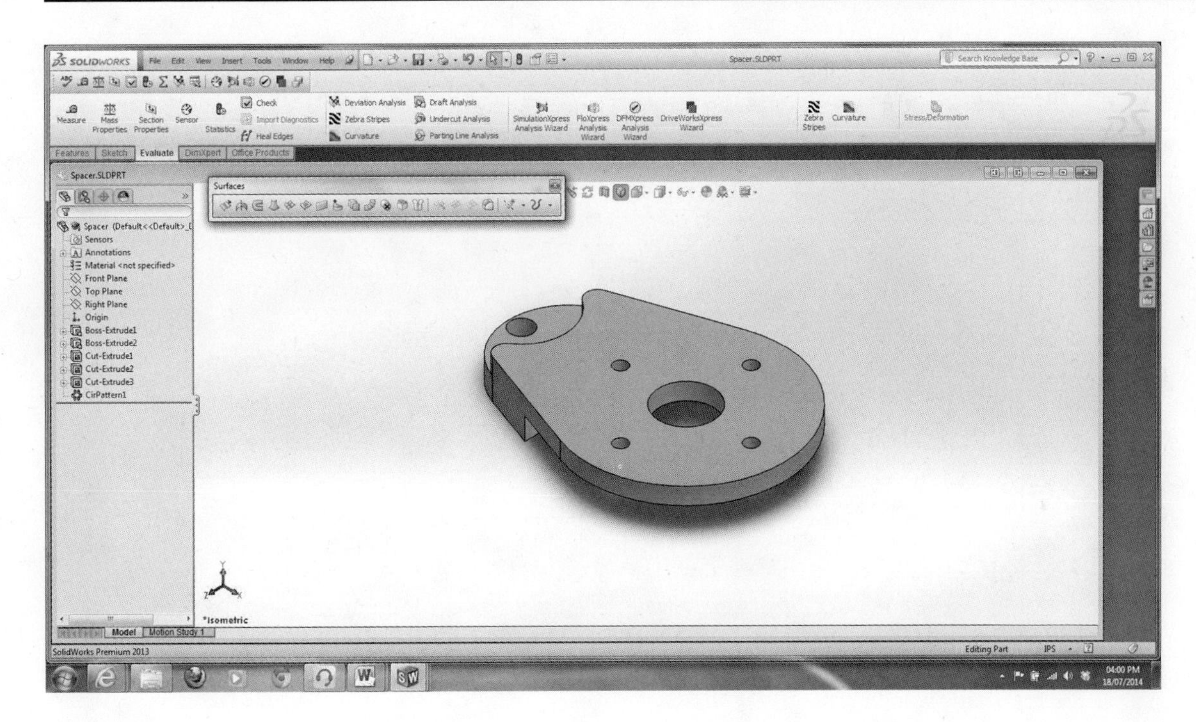

Figure 5.32

Final spacer model.

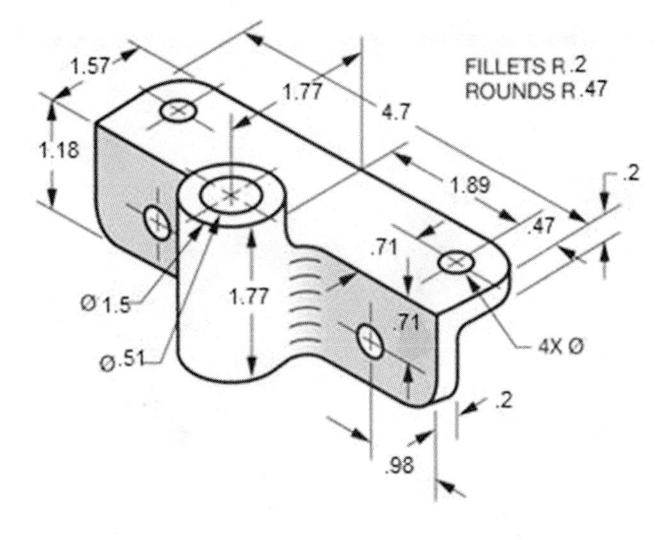

Figure 5.33

Oarlock socket model. (From JENSEN/HELSEL/ESPIN. Interpreting Engineering Drawings, 6E. © 2012 Nelson Education Ltd. Reproduced by permission. http://www.cengage.com/permissions.)

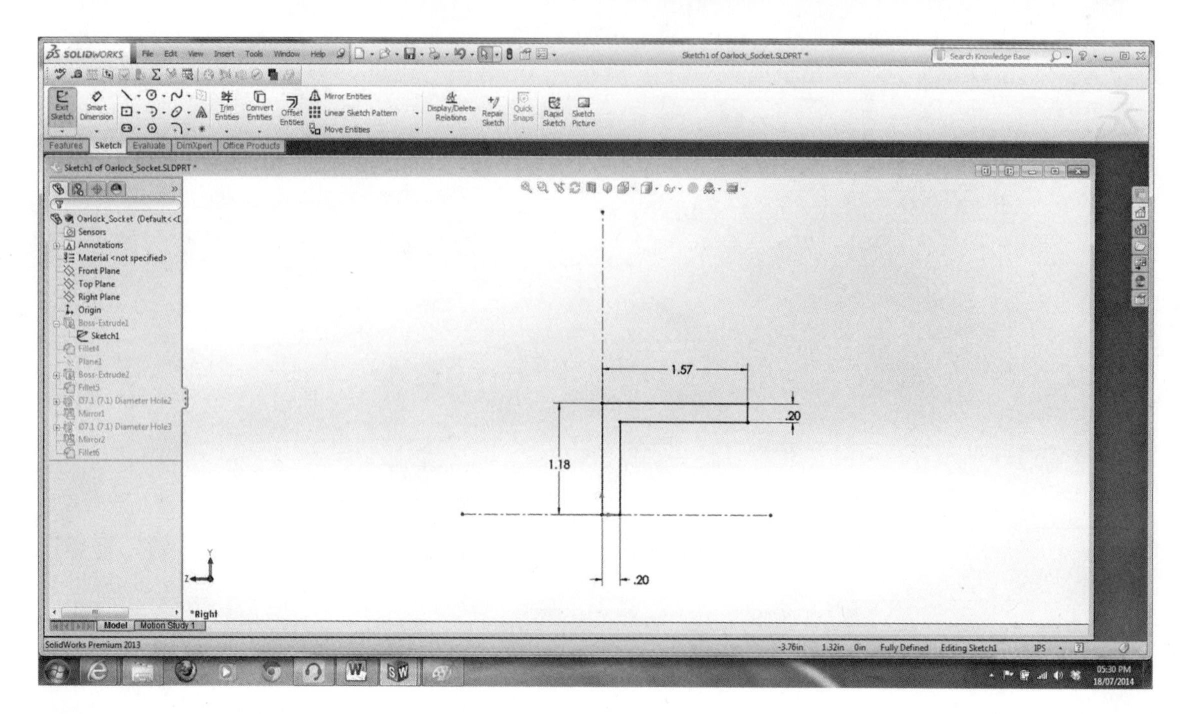

Figure 5.34

Sketch1.

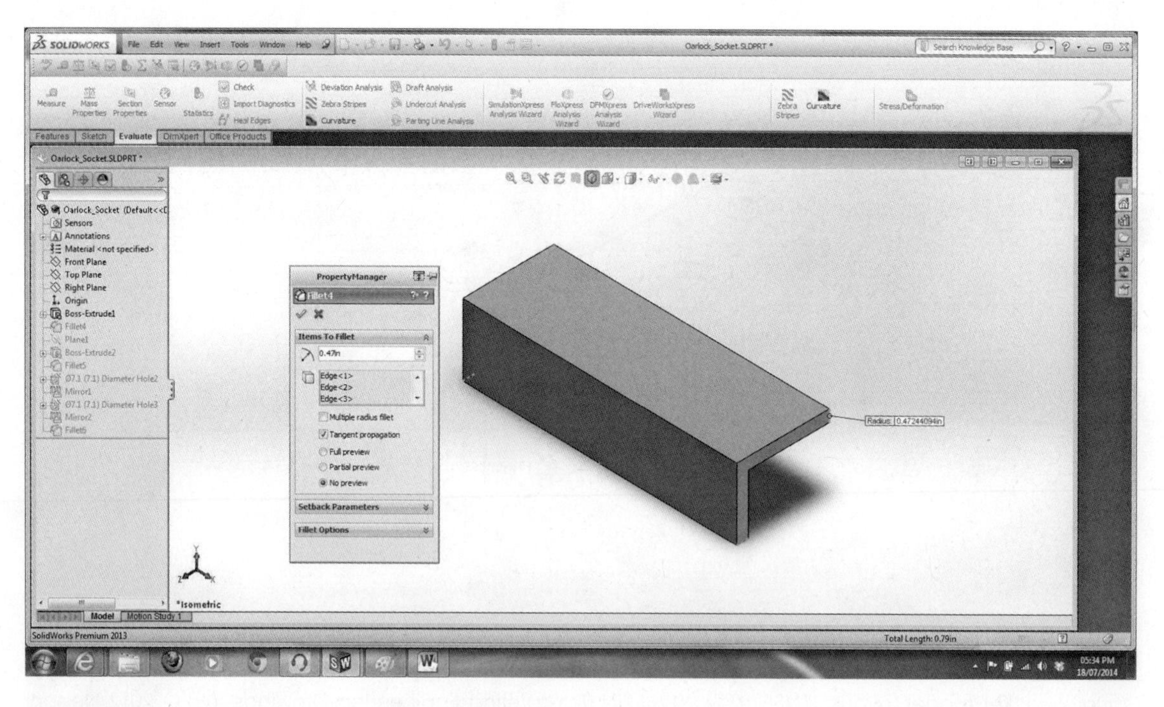

Figure 5.35

Defining fillet radius.

7. Create a Plane 0.5905 from top of partly finished object, midway or top-left edge (Figure 5.36).
8. Sketch two circles, radii 1.50, 0.51, respectively, with the center coinciding with the left edge at the top of the object (Figure 5.37).
9. Extrude Up To Surface corresponding to the lower surface on the L-shape (Figure 5.38).
10. Fillet using radius of 0.2 where the Cylinder meets the L-shape (Figure 5.39).
11. Create a Hole using Hole Wizard tool for a diameter of 0.28 (9/32) (Figure 5.40).

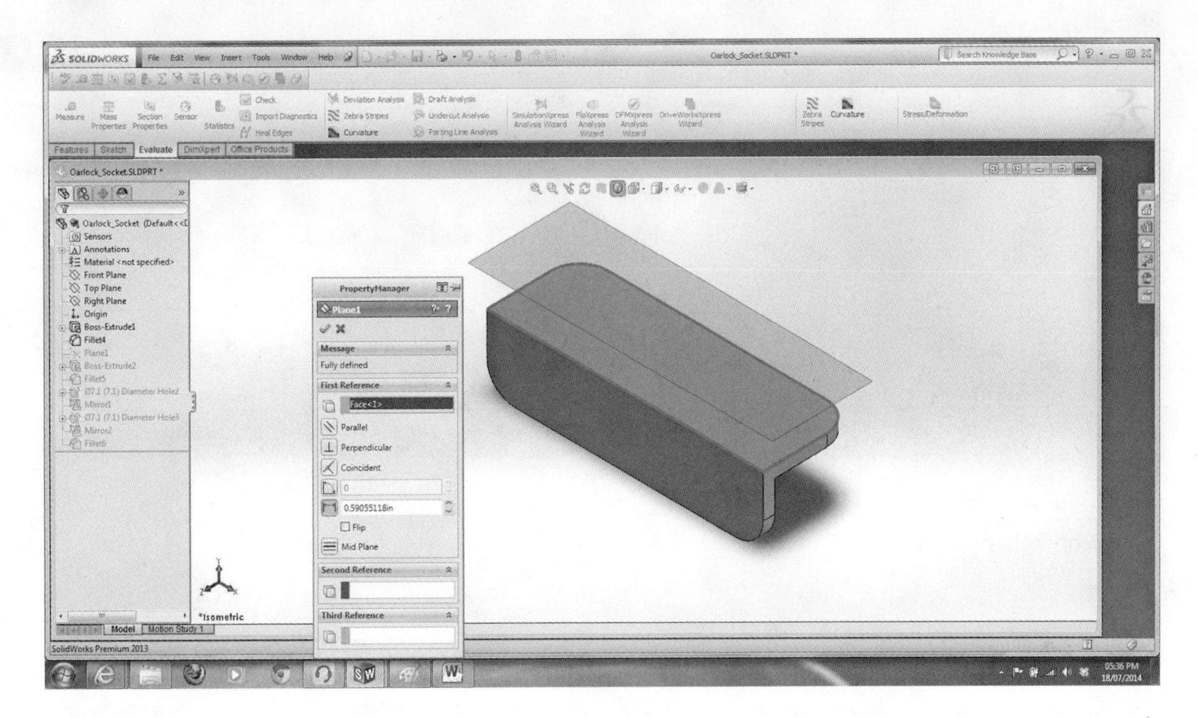

Figure 5.36

Defining Plane1.

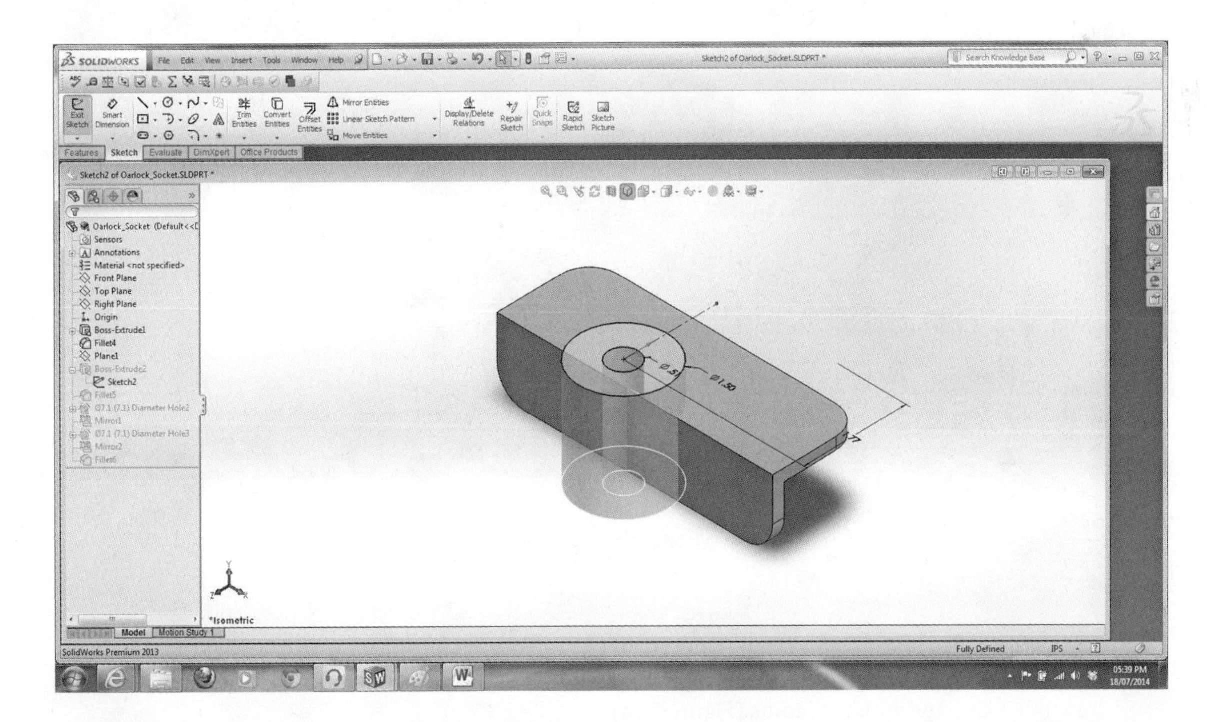

Figure 5.37

Two circles sketched on Plane1.

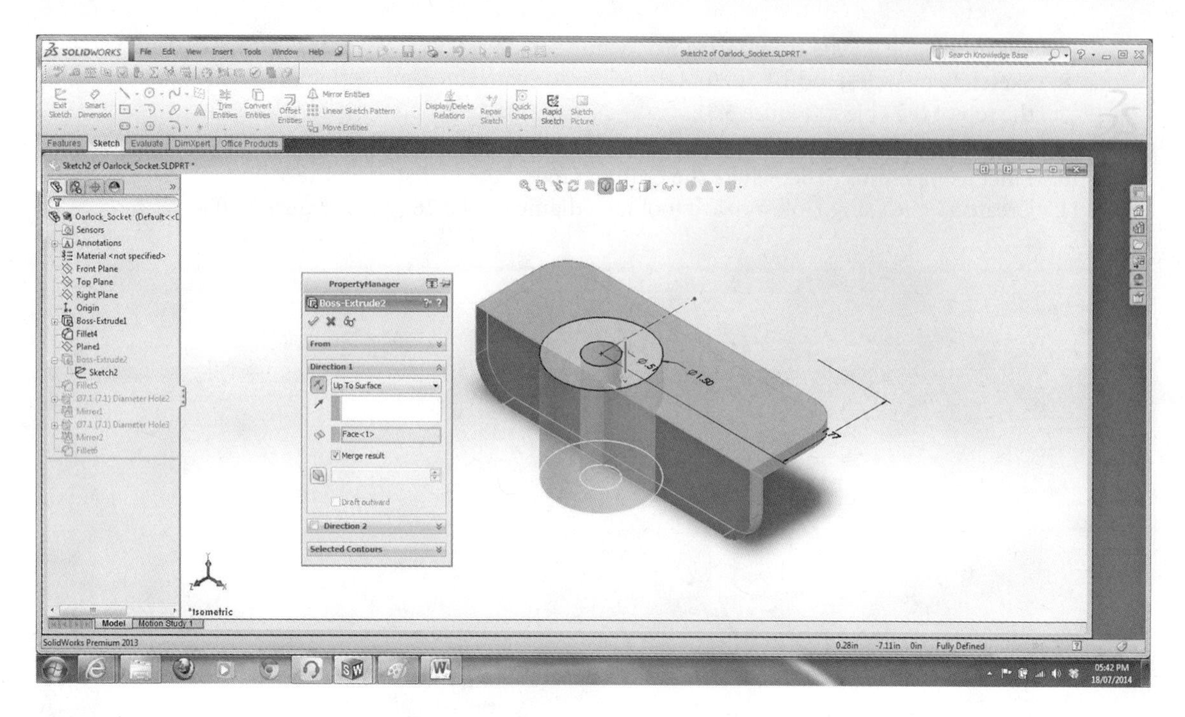

Figure 5.38

Extrude operation.

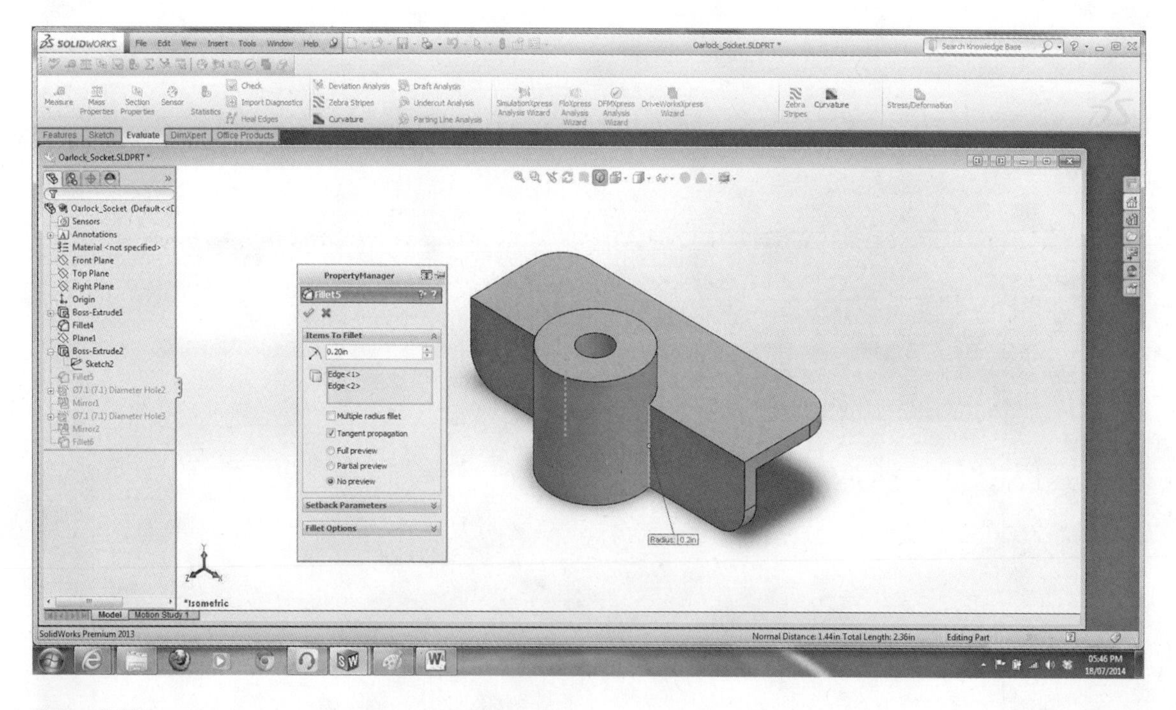

Figure 5.39

Filleting.

12. Mirror the Hole about the Right Plane (Figure 5.41).
13. Repeat the Hole and Mirror about the Right Plane (Figures 5.42 and 5.43).
14. Fillet L-shape with 0.2 radius (Figure 5.44).
15. Click OK to complete model (Figure 5.45).
16. Click Evaluate > Properties to give weight = 0.17 lb, and volume = 4.70 cub in.

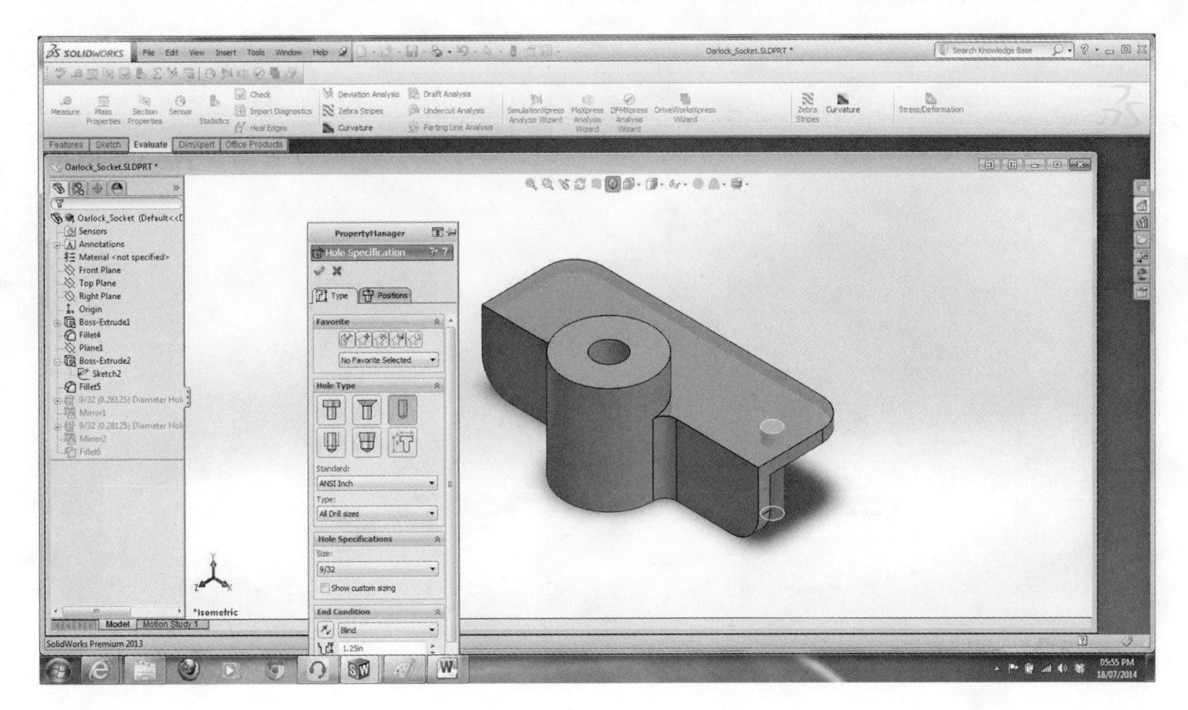

Figure 5.40

Using Hole Wizard to define a hole.

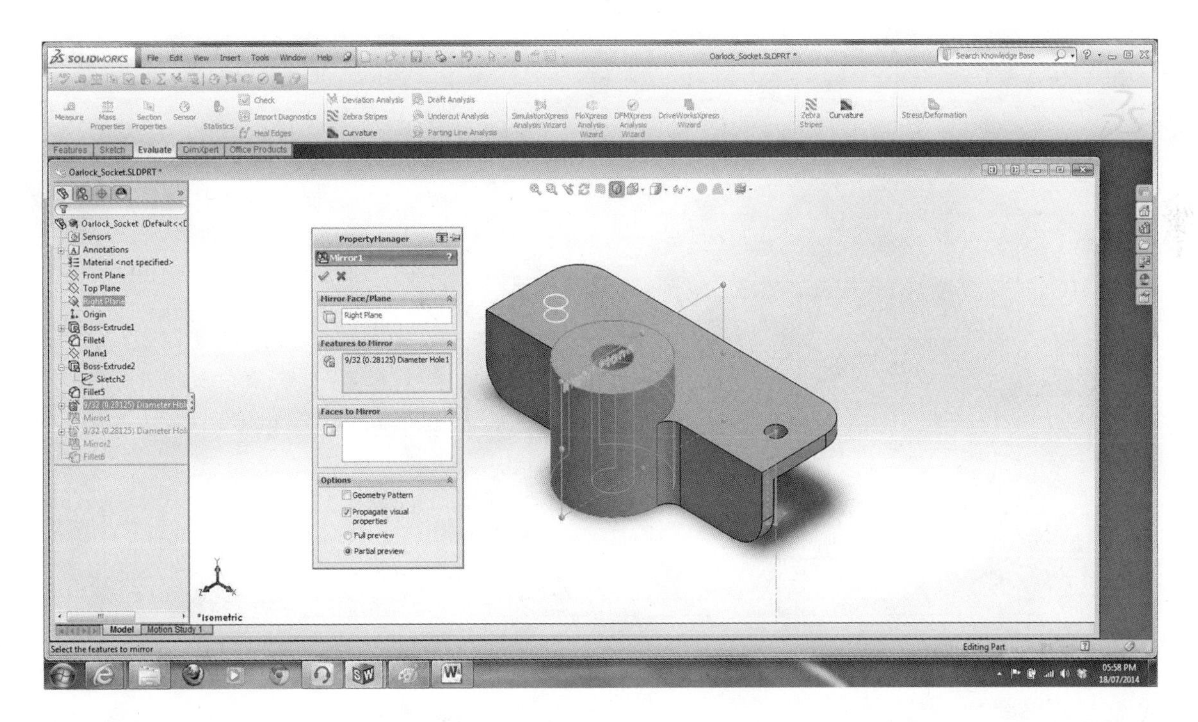

Figure 5.41

Mirroring the hole.

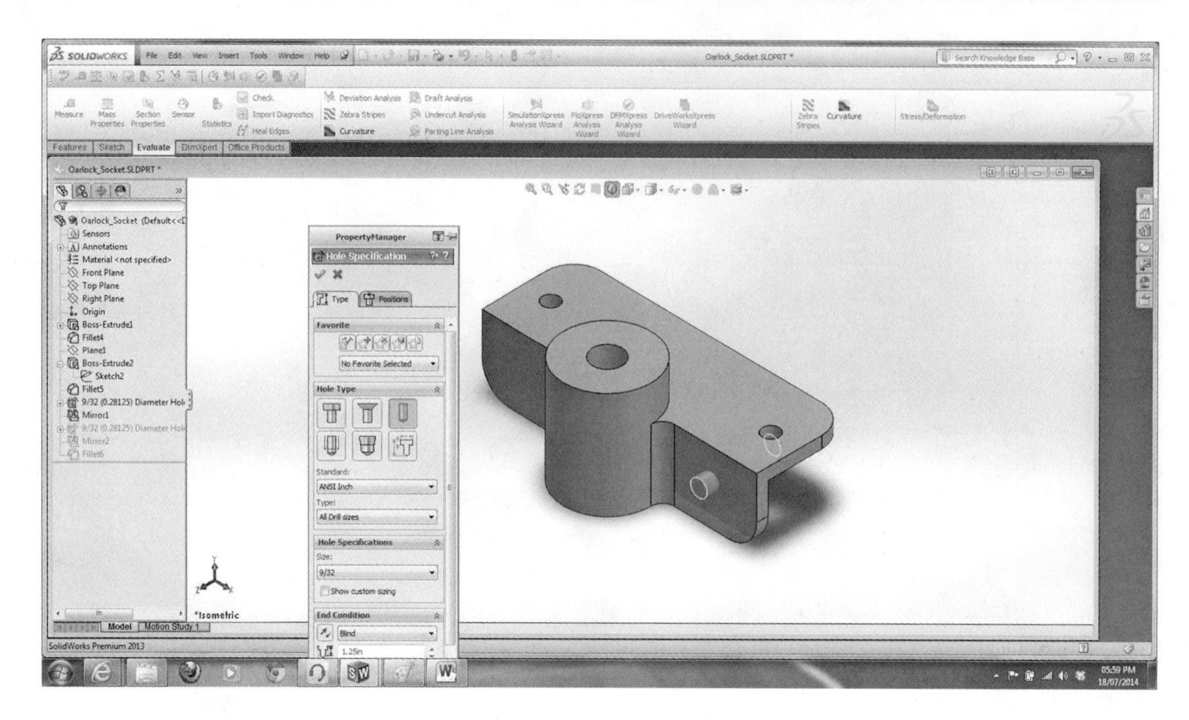

Figure 5.42

Hole Wizard used for creating another hole.

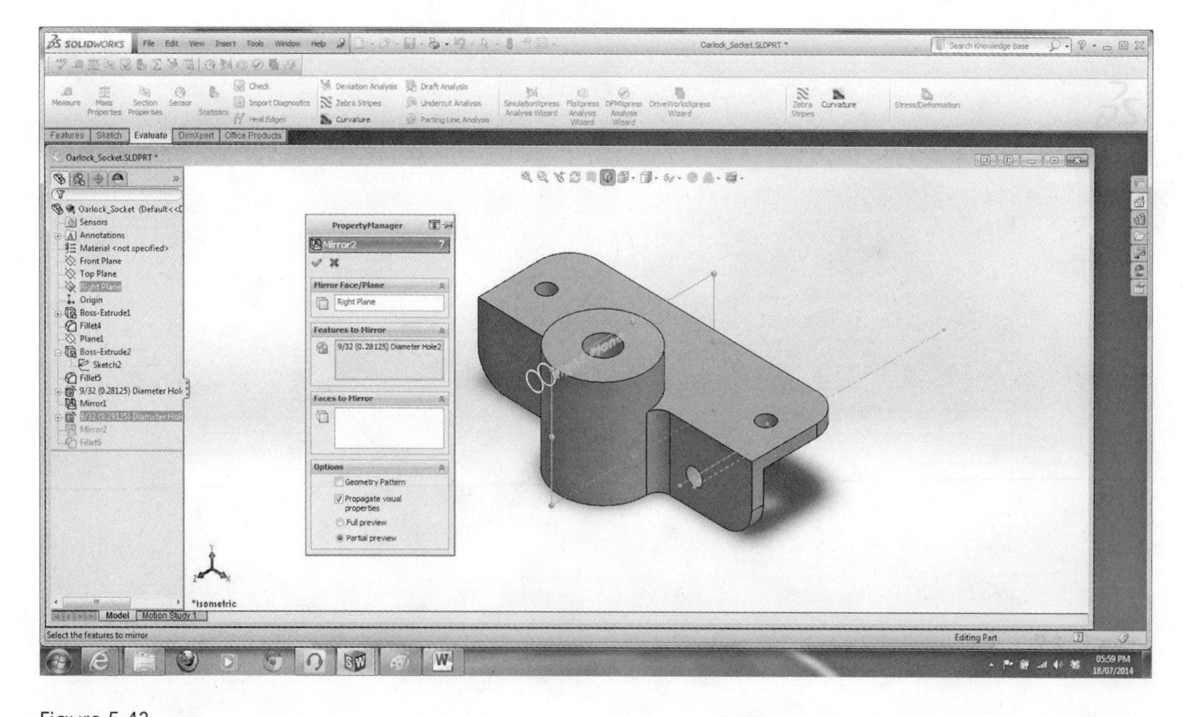

Figure 5.43

Use of Hole Wizard tool.

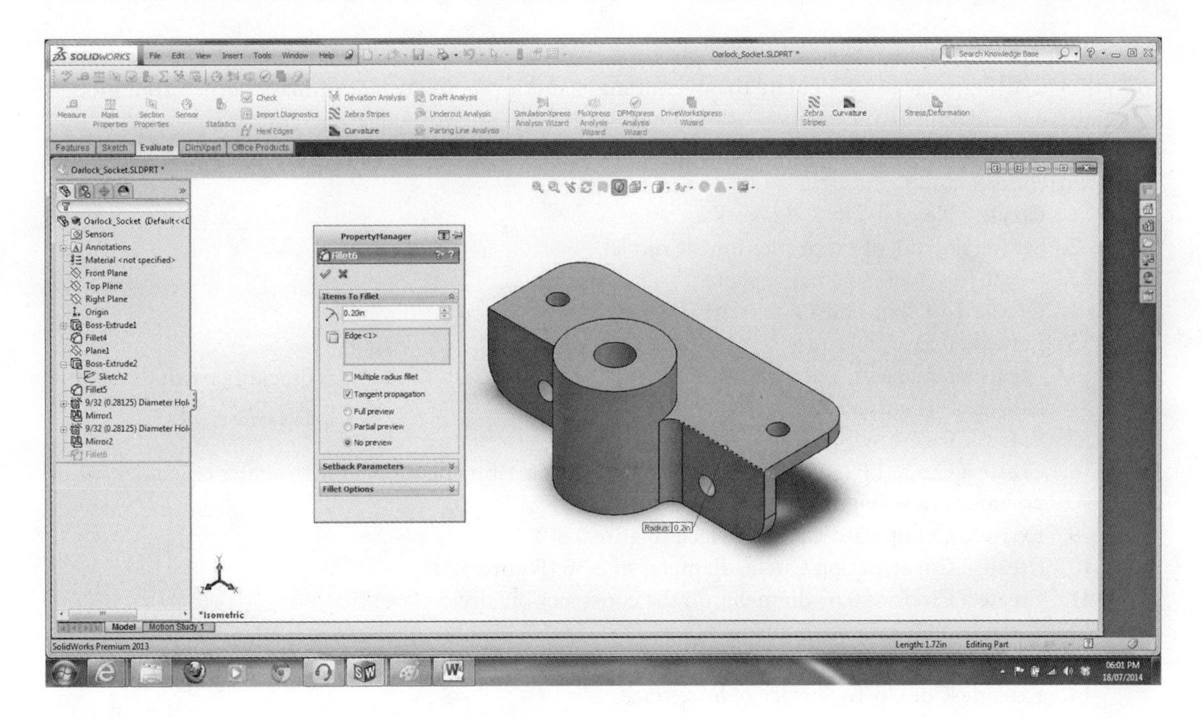

Figure 5.44

Fillet operation.

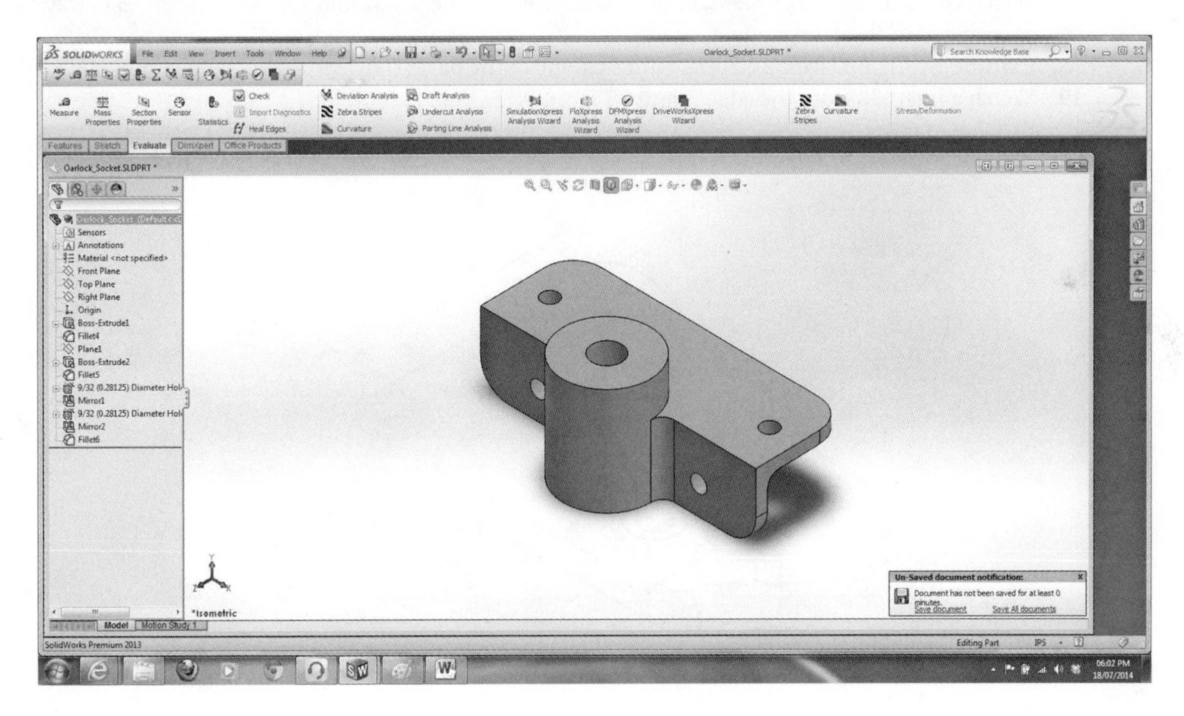

Figure 5.45

Final oarlock model.

Click Evaluate > Mass Properties to determine the weight of the model as 0.46 lb.

Density = 0.10 pounds per cubic inch
Mass = 0.46 pounds
Volume = 4.70 cubic inches
Surface area = 35.50 square inches
Center of mass: (inches) X = 0.00; Y = 0.89; Z = −0.12

Tutorial 5.5: Link

Build this part (Figure 5.46). Calculate the overall mass and locate the center of mass of the given model. Material: Copper. Units: IPS. Density = 0.3215 lb./in.^{3}. All holes are through all.

SolidWorks Solution

1. Create a New part in SolidWorks.
2. Set the document properties for the model.
3. Select Top Plane to create a sketch.
4. On the Top Plane, create Sketch1 (Figure 5.47).
5. Extrude 0.51 Upward.
6. Create a circle with a diameter of 1.5 at the bottom of part, the center coinciding with that of the half-circle (Figure 5.48).
7. Extrude 0.47 downward.
8. Create a Circle 1.02 diameter the bottom of the part-finished model with center coincident with previous circle (Figure 5.49).
9. Extrude Cut upward Up To Surface (Figure 5.50).
10. Create a Construction Circle, diameter of 2.36 (Figure 5.51).
11. Create a Circle of 0.63 diameter on the construction circle.
12. Extrude 0.12 upward (Figure 5.52).
13. Create a Circle of diameter 0.33 at top of boss (Figure 5.53).
14. Extrude-Cut Up To Surface of lower face.
15. Create Circular Pattern of four instances of boss (Figure 5.54).
16. Create Plane1 0.39 from the right side of the model that is completed so far (Figure 5.55).
17. Sketch two circles (Figure 5.56) on the plane with diameters 2.52 and 1.89 and center 3.54 away, as specified in the design intent.
18. Extrude Up To Surface coinciding with extreme of the part (Figure 5.57).
19. Sketch a line 1.38 from the edge (Figure 5.58).

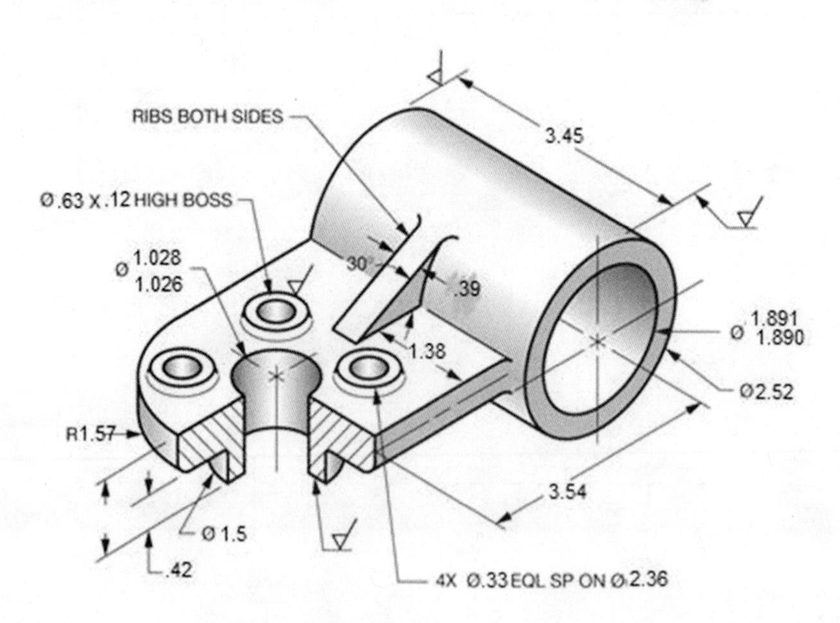

Figure 5.46

Link model. (From JENSEN/HELSEL/ESPIN. Interpreting Engineering Drawings, 6E. © 2012 Nelson Education Ltd. Reproduced by permission. http://www.cengage.com/permissions.)

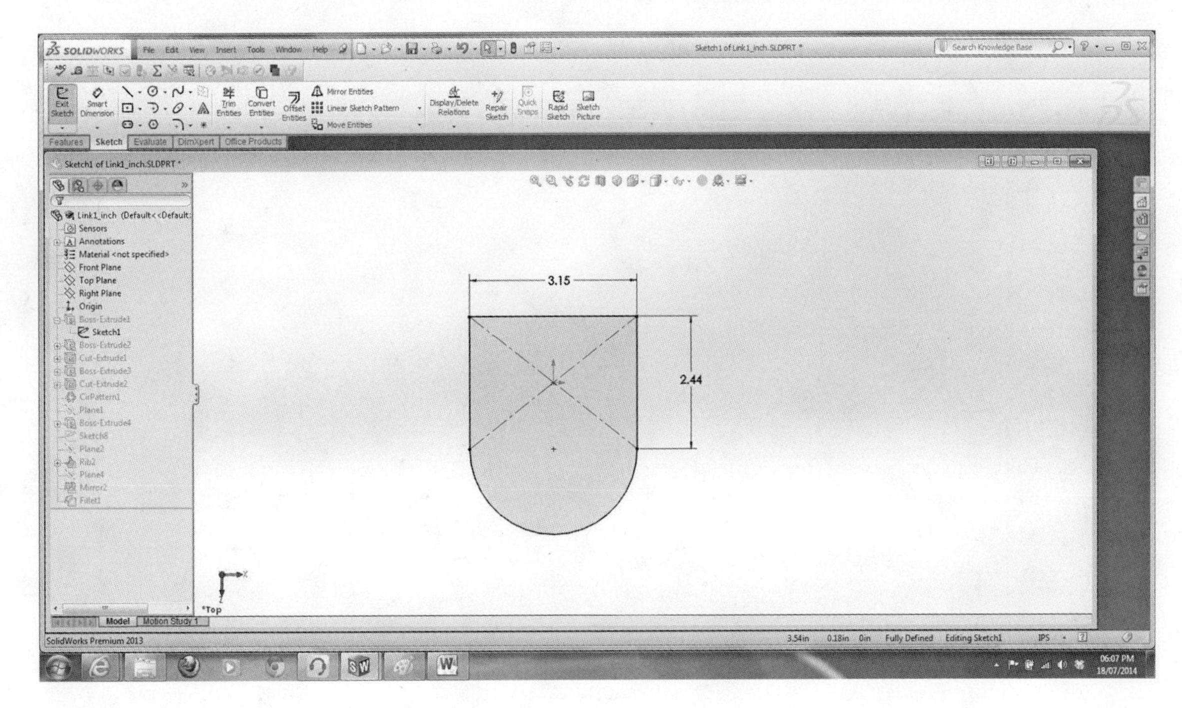

Figure 5.47

Sketch1.

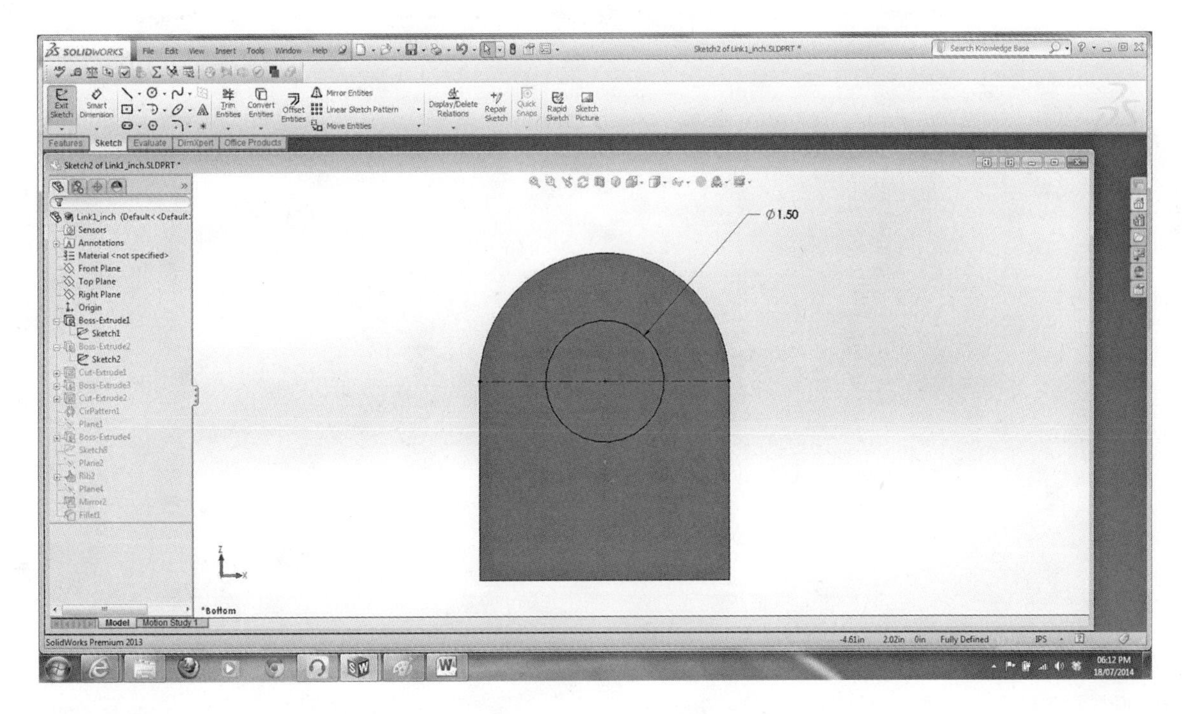

Figure 5.48

Sketch2.

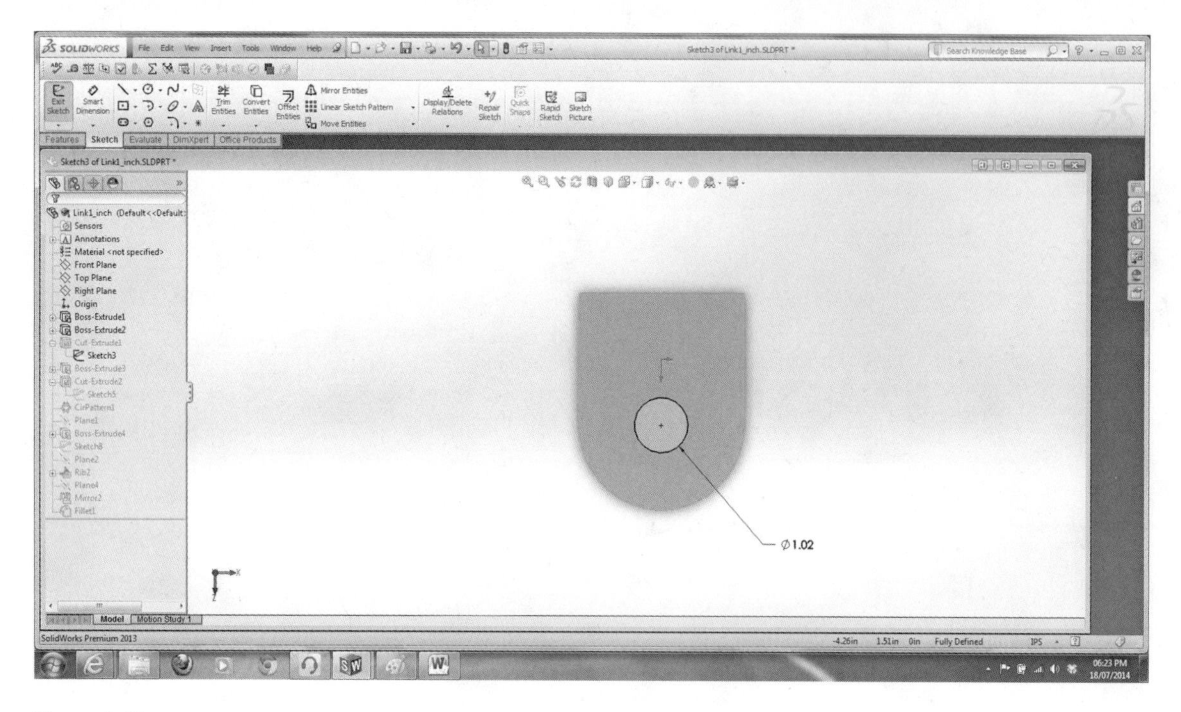

Figure 5.49

Circle for realizing a feature.

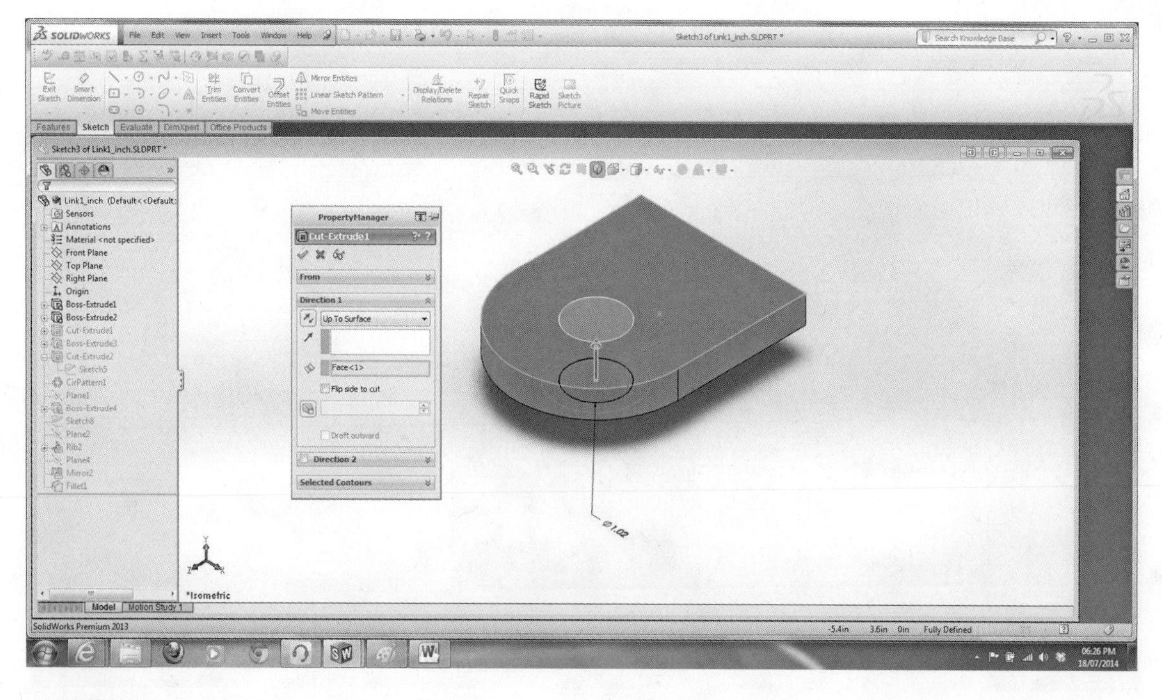

Figure 5.50

Extrude-Cut.

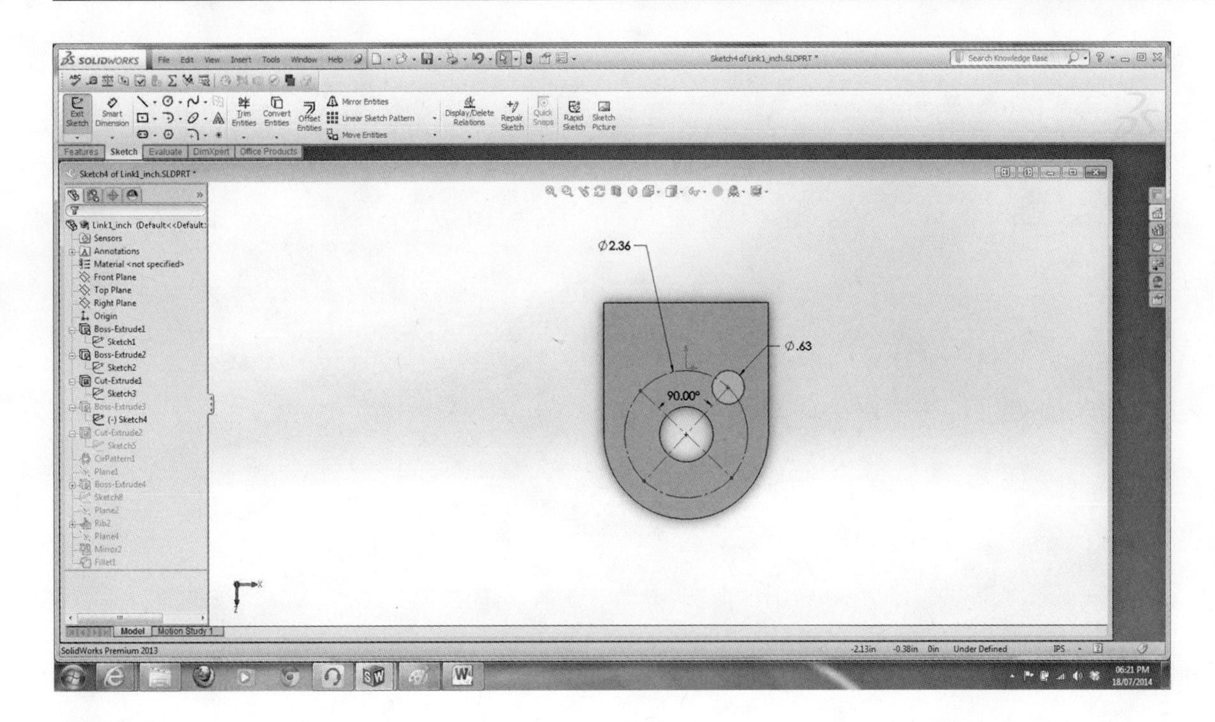

Figure 5.51

Reference circle and Sketch.

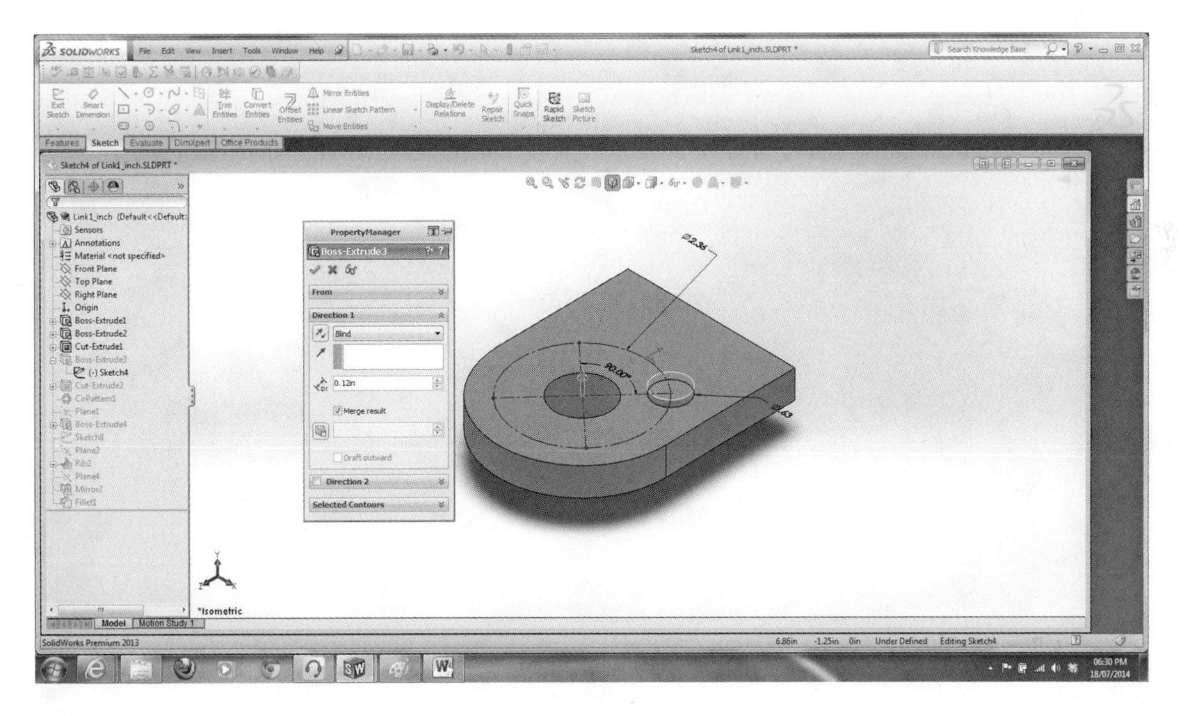

Figure 5.52

Boss created.

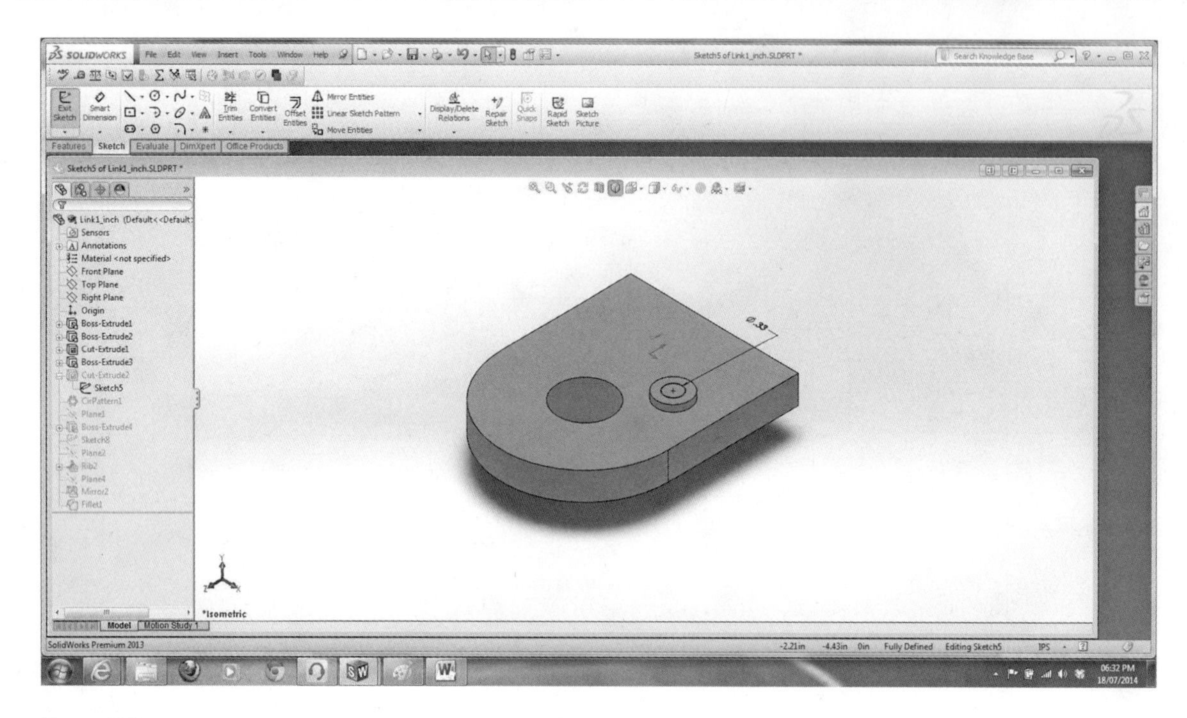

Figure 5.53

Circle for Extrude-Cut operation.

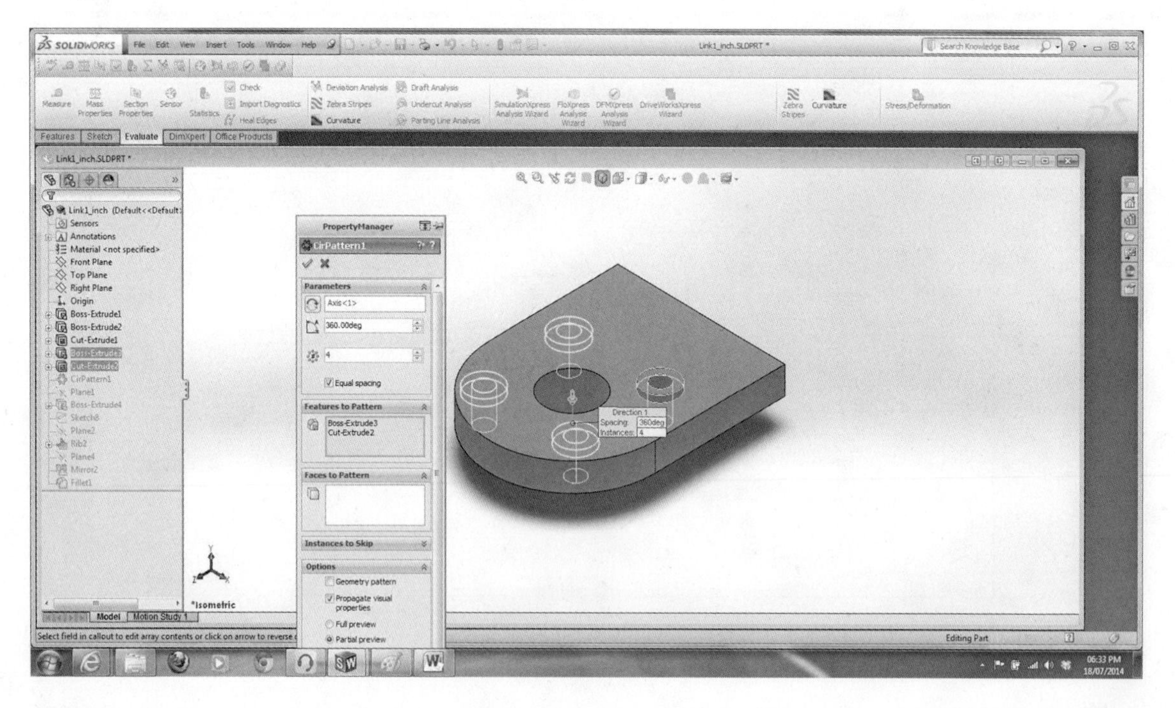

Figure 5.54

Circular Pattern operation.

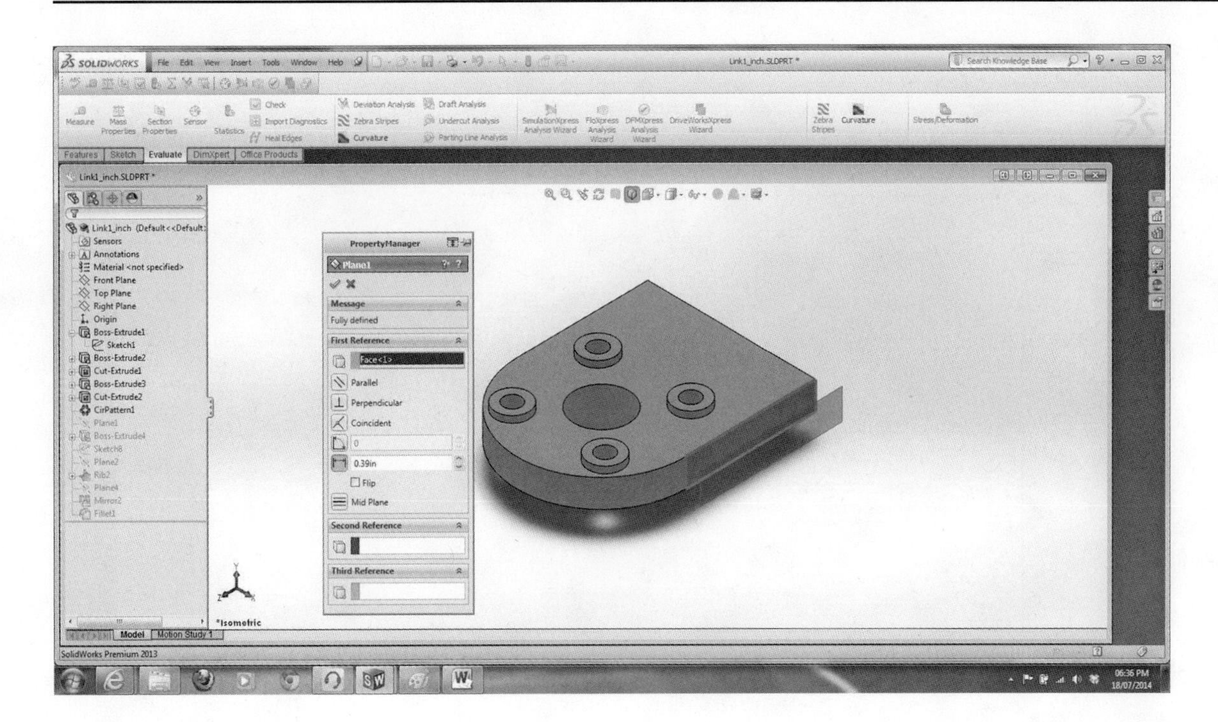

Figure 5.55

Plane1 created.

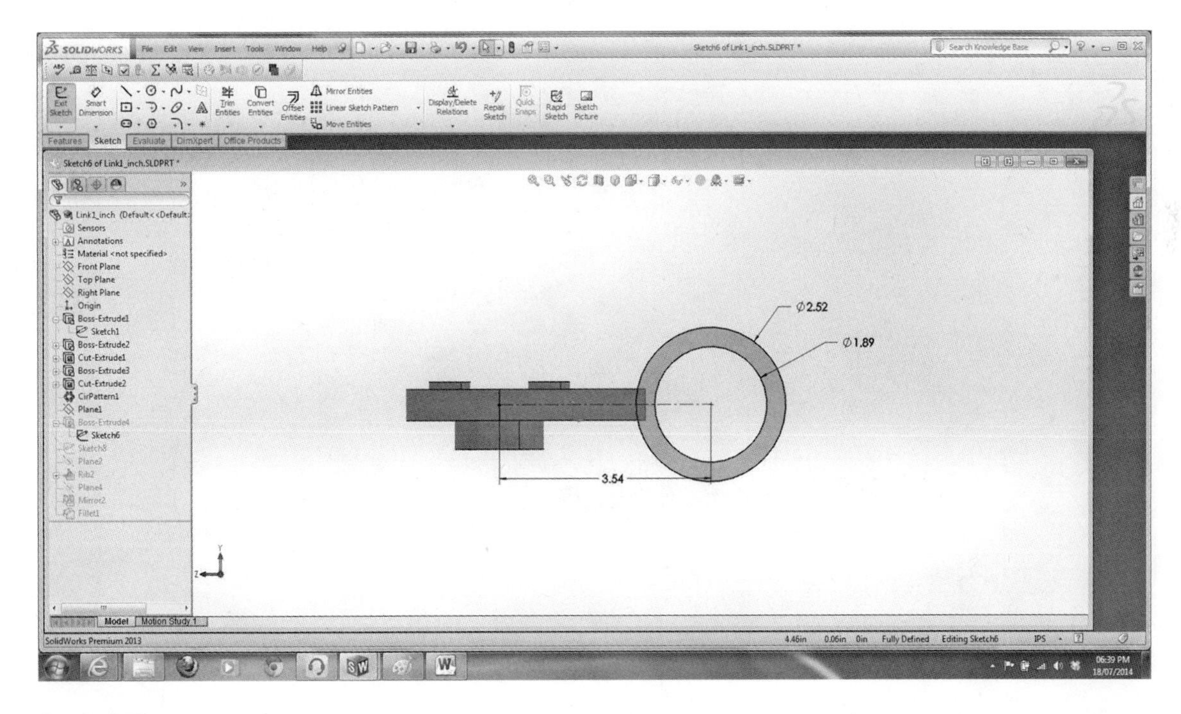

Figure 5.56

Two circles created.

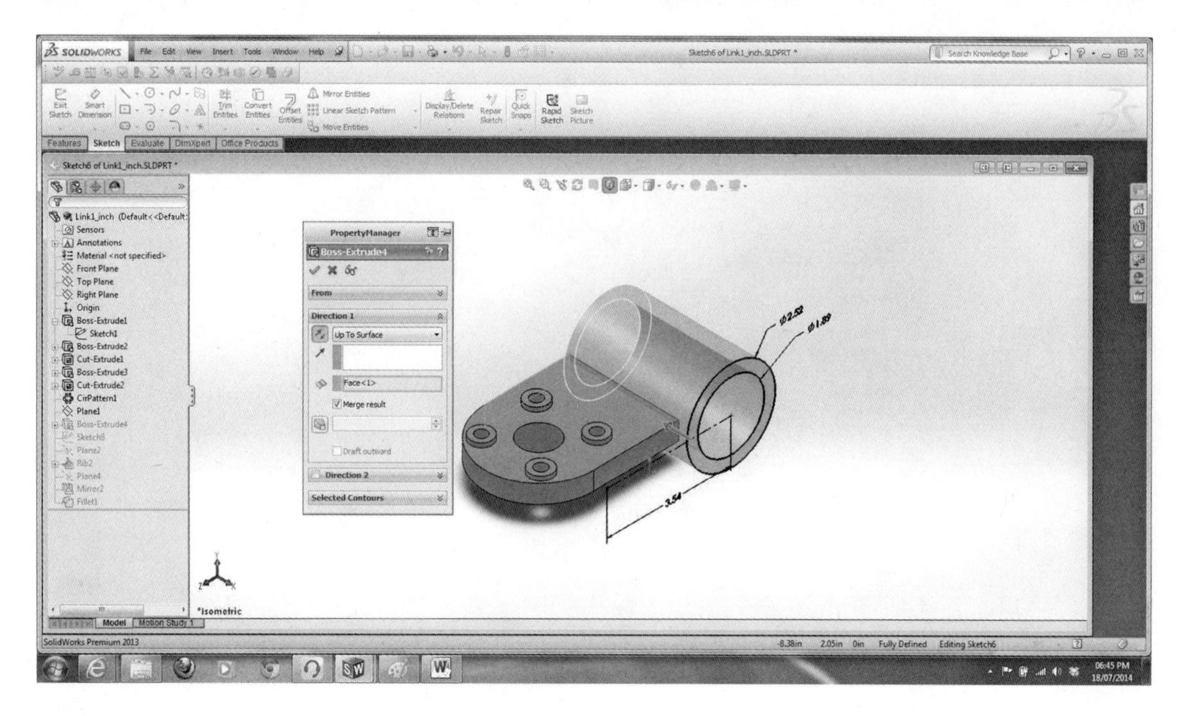

Figure 5.57

Extruding the two circles to create a hollow feature.

Figure 5.58

Reference line.

20. Create Plane2 coincident with this line using the following conditions (Figure 5.59):
    a. First Reference is one endpoint of the line.
    b. Second Reference is the other endpoint of the line.
    c. Third Reference is Coincident condition.
21. Sketch an inclined line, 30° to the horizontal (Figure 5.60).
22. Click Insert/Rib.
23. Choose Thickness Both Side at a value of 0.39 (Figure 5.61).
24. Create a Plane3 0.2759 from the Top Plane (Figure 5.62).
25. Mirror the Rib about Plane3 (Figure 5.63).
26. Fillet the bosses, radius 0.05 (Figure 5.64).
27. Click OK to complete the model (Figure 5.65).

Click Evaluate > Mass Properties to determine the weight of 4.36 lb.

Density = 0.32 pounds per cubic inch
Mass = 4.36 pounds
Volume = 13.56 cubic inches
Surface area = 86.26 square inches
Center of mass: (inches) X = 0.11; Y = 0.24; Z = −1.04

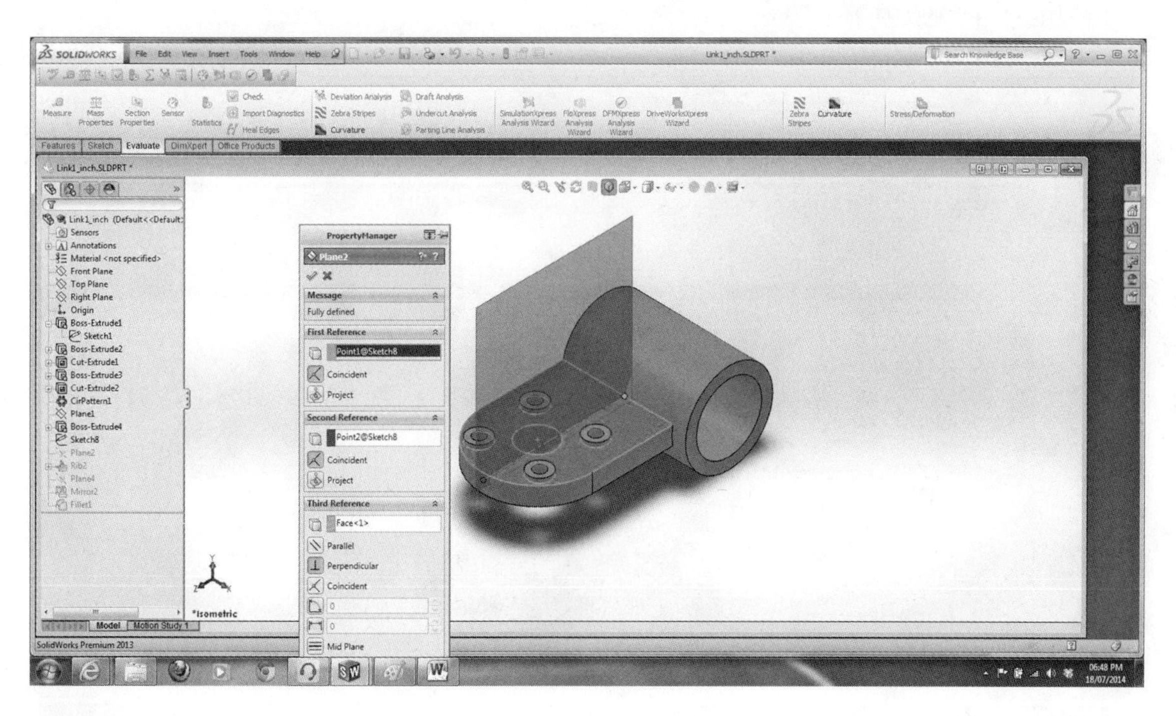

Figure 5.59

Creating Plane2.

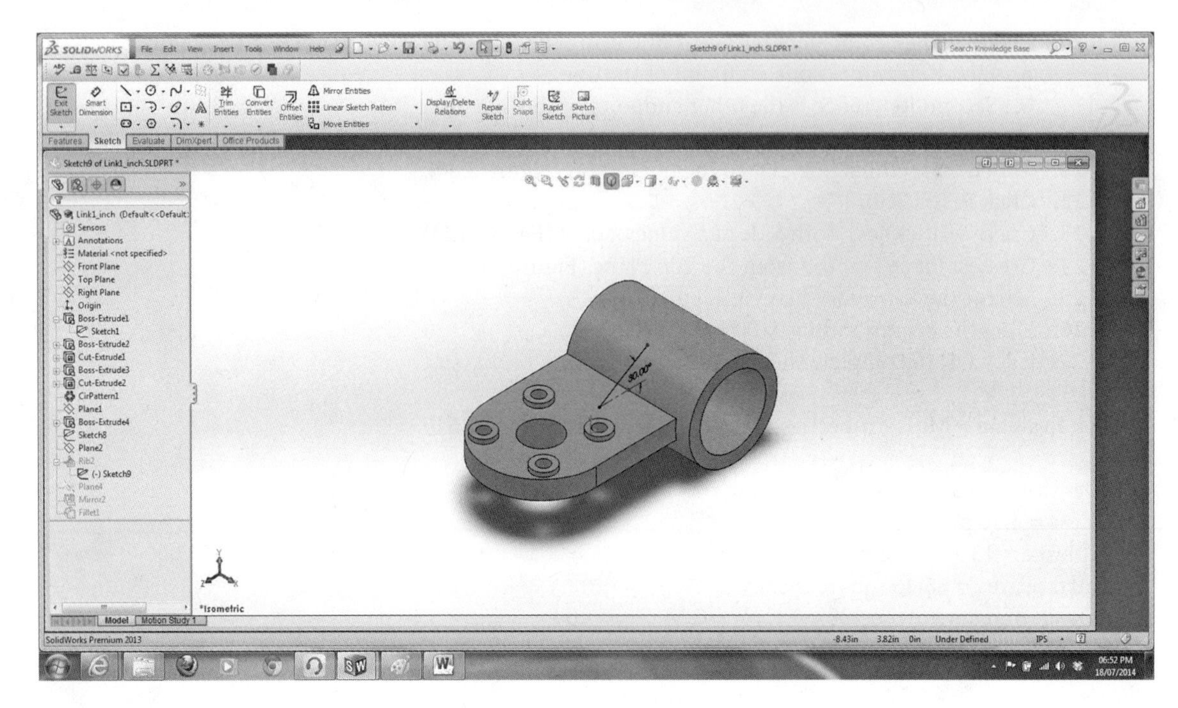

Figure 5.60

Inclined line for defining a rib.

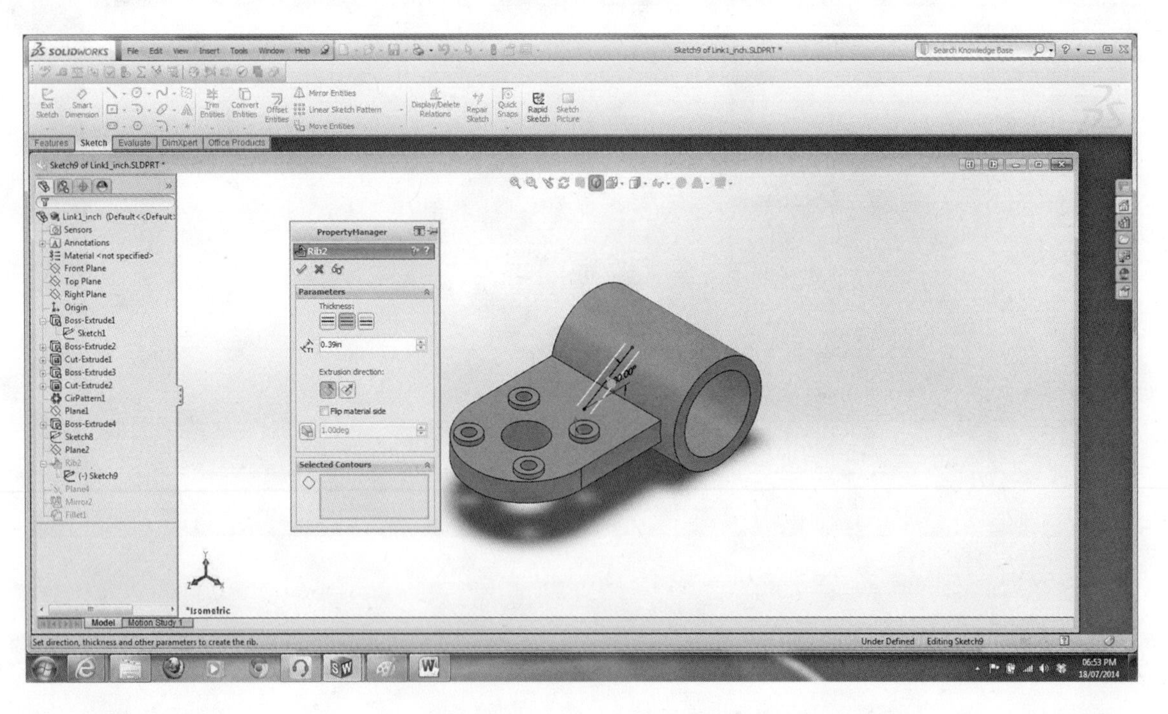

Figure 5.61

Rib created.

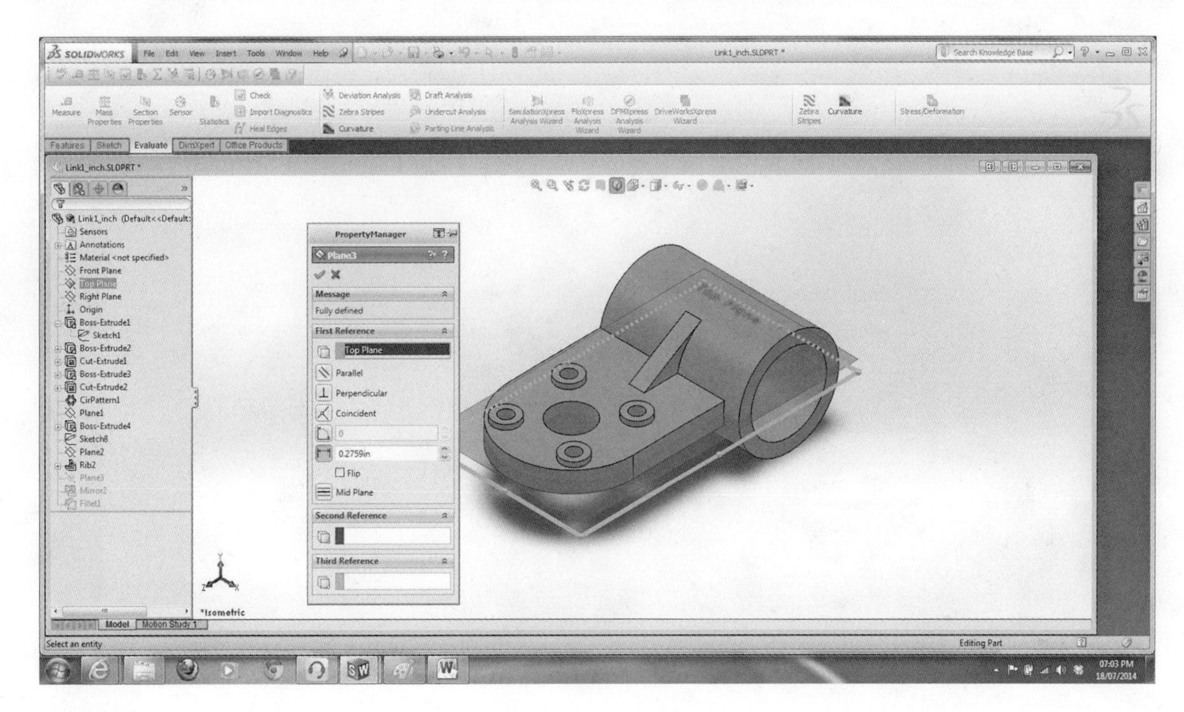

Figure 5.62

Plane3.

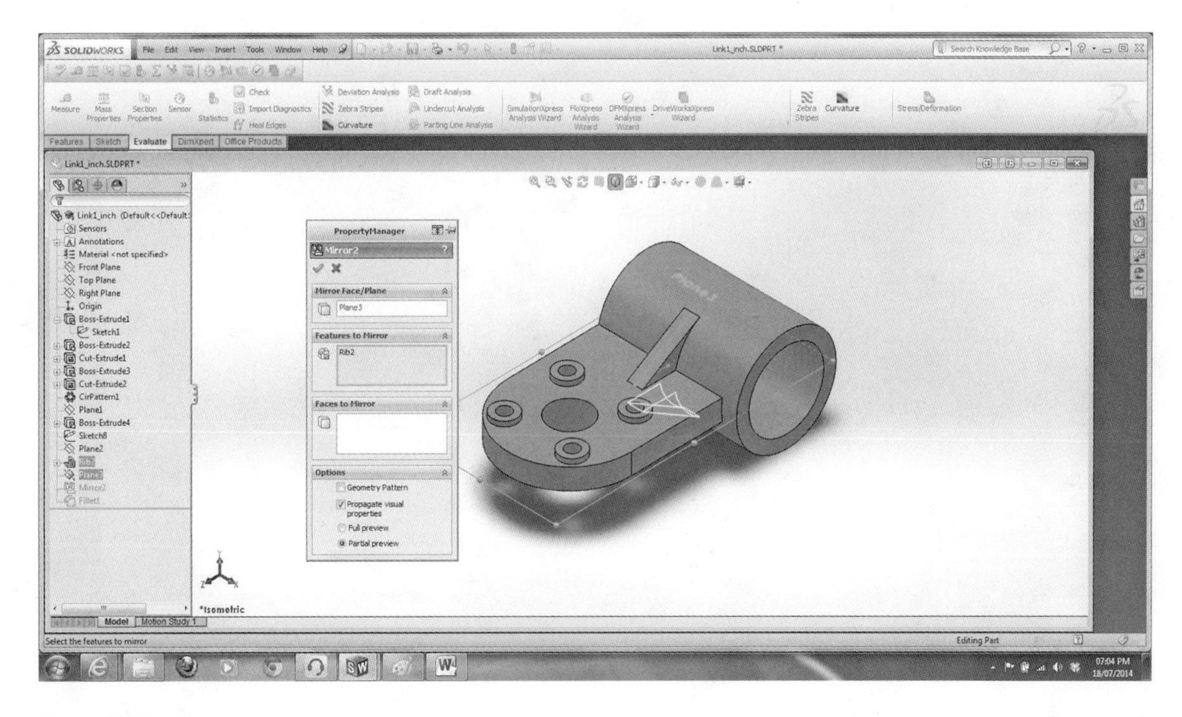

Figure 5.63

Mirrored rib.

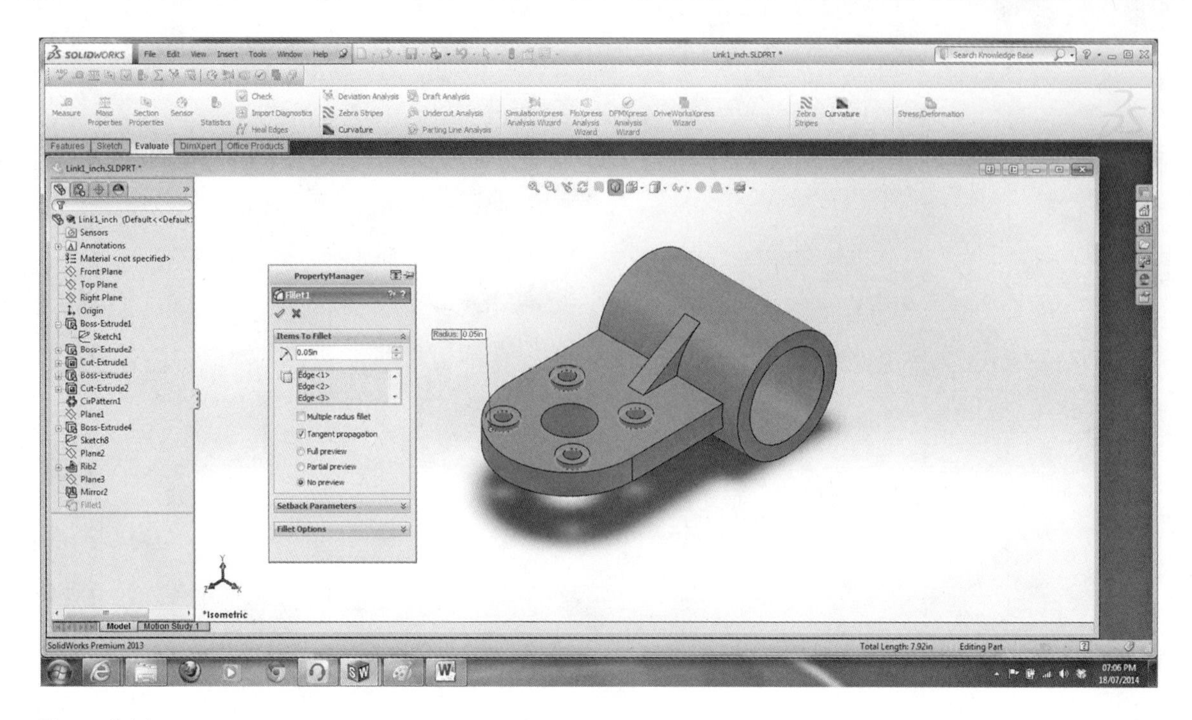

Figure 5.64

Rib is mirrored.

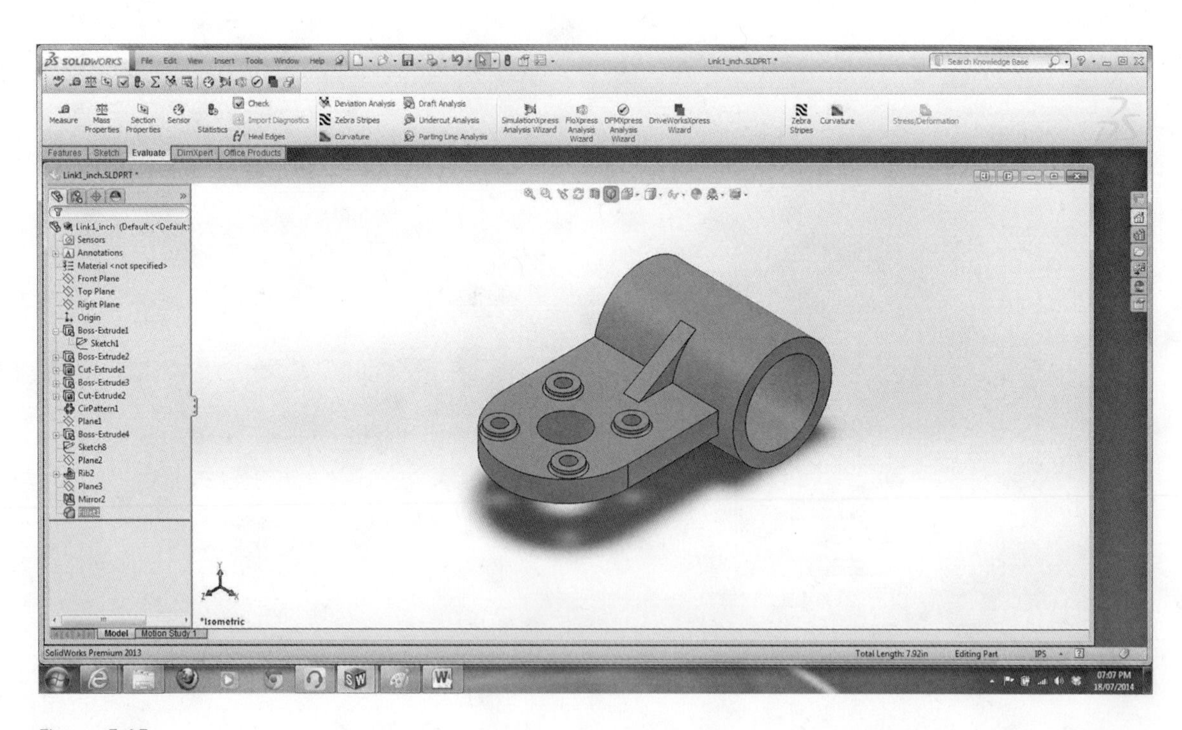

Figure 5.65

Link1 model.

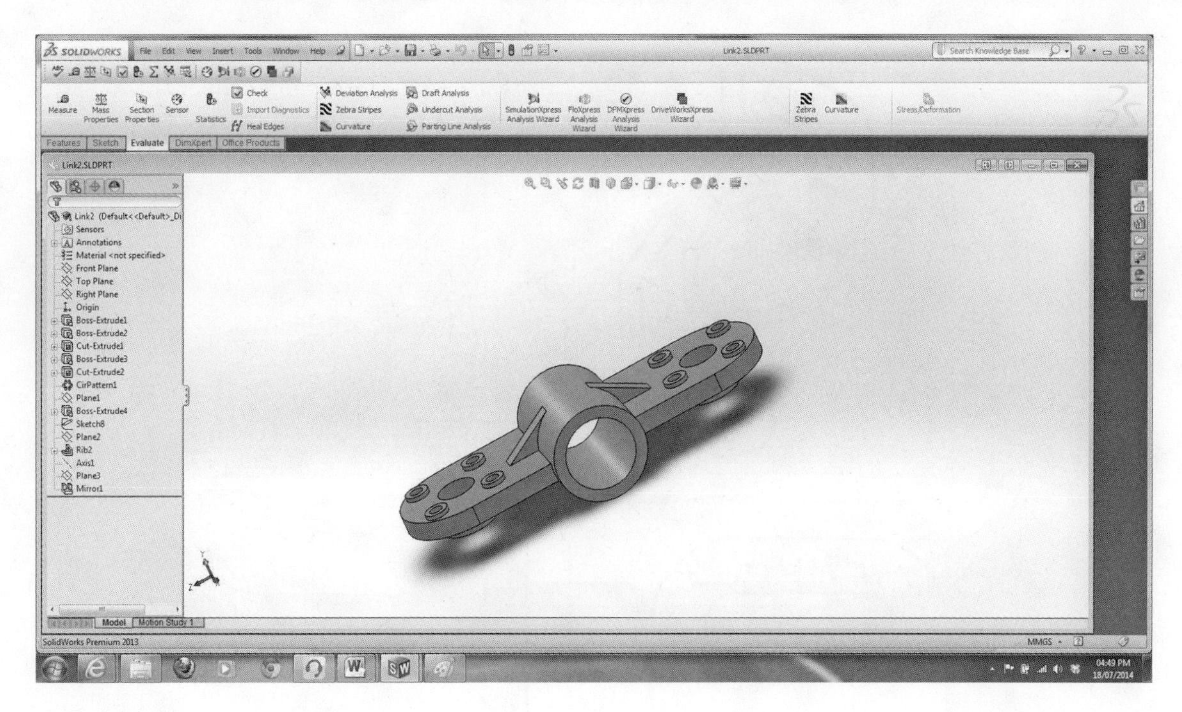

Figure 5.66

Double link model.

Tutorial 5.6: Double Link

Build this part (Figure 5.66) using the Tutorial 5.3 as the starting point. (Mirror the first-half about the Front Plane.) Calculate the overall mass and locate the center of mass of the given model. Material: Copper. Units: IPS. Density = 0.3215 lb./in.^{3}. All holes are through all.

SolidWorks Solution

1. Create a New part in SolidWorks.
2. Set the document properties for the model.
3. Mirror features except from the cylinder about the Front Plane.

Click Evaluate > Properties to determine the weight and volume as follows:

Density = 0.32 pounds per cubic inch
Mass = 6.16 pounds
Volume = 19.16 cubic inches
Surface area = 118.94 square inches
Center of mass: (inches) X = 0.08; Y = 0.24; Z = −2.32

Tutorial 5.7: Inclined Stop

Build the part in Figure 5.67. The broken out details are given as well as the 3D model view on the top-left corner. Calculate the overall mass and locate the center of mass of the given model. Material: 6061 Alloy material. Units: IPS.

SolidWorks Solution

1. Create a New part in SolidWorks.
2. Set the document properties for the model.
3. Select Front Plane for sketching.
4. Create Sketch1 on the Front Plane (Figure 5.68).
5. Extrude 3.5 Mid Plane (Figure 5.69).
6. Create a Slot feature in the upper half as shown in Figure 5.70.

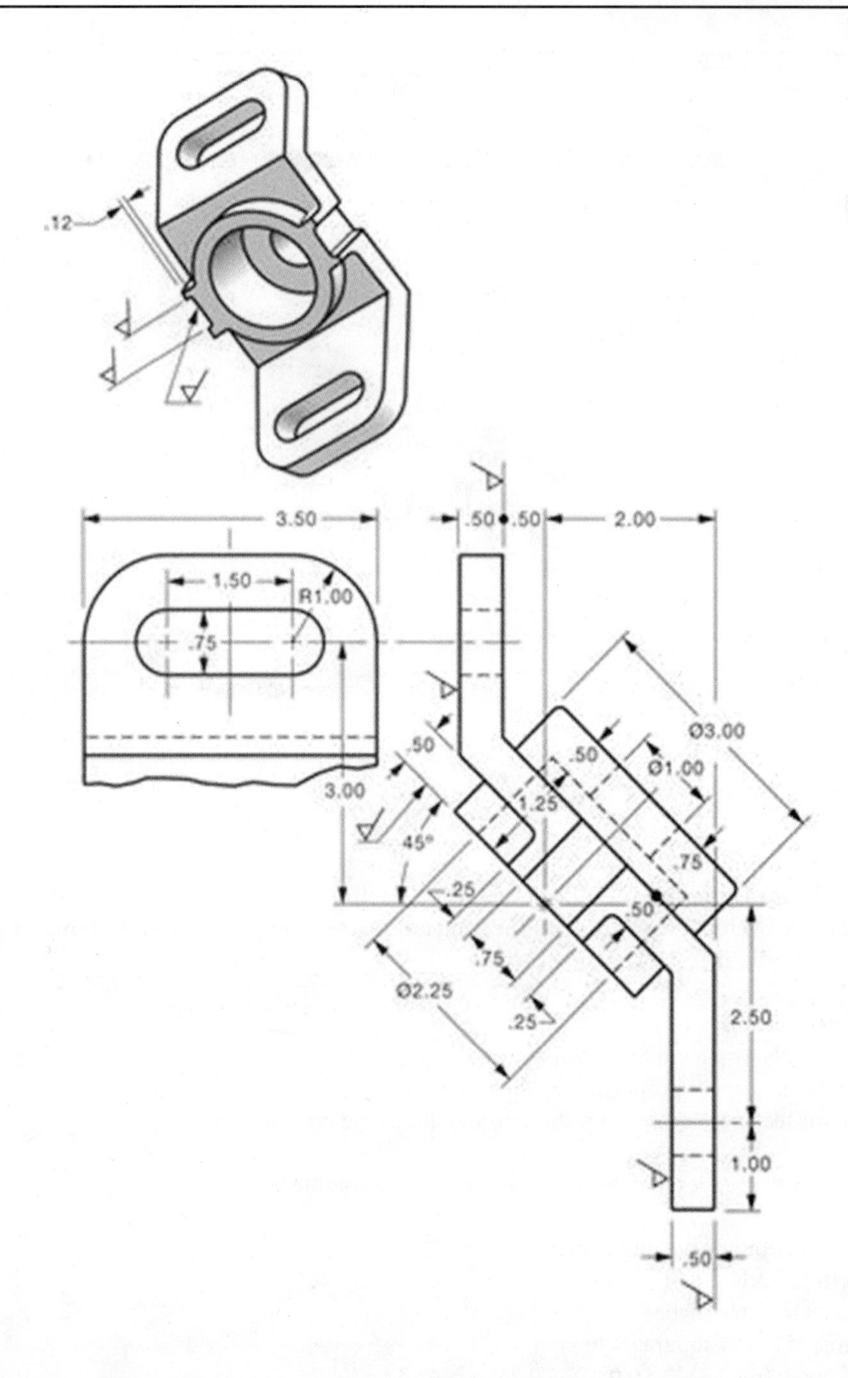

Figure 5.67

Inclined Stop. (From JENSEN/HELSEL/ESPIN. Interpreting Engineering Drawings, 6E. © 2012 Nelson Education Ltd. Reproduced by permission. http://www.cengage.com/permissions.)

7. Extrude-Cut Up To Surface using the back face (Figure 5.71).
8. Create a Slot feature in the lower half as shown in Figure 5.72.
9. Extrude Up To Surface using the back face (Figure 5.73).
10. Fillet with 1.00 radius on four corners (Figure 5.74).
11. Sketch Circle of diameter 3.00 (Figure 5.75).
12. Extrude 0.75 upward (Figure 5.76).
13. Create Plane1.5 from the bottom of uncompleted Part (Figure 5.77).
14. On Plane1, sketch two Circles of diameters 2.25 and 3.00, respectively (Figure 5.78).
15. Extrude Up Surface of bottom of inclined plane to the left (Figure 5.79).
16. Click the Circle in the interior (bottom of inclined plane to the left).

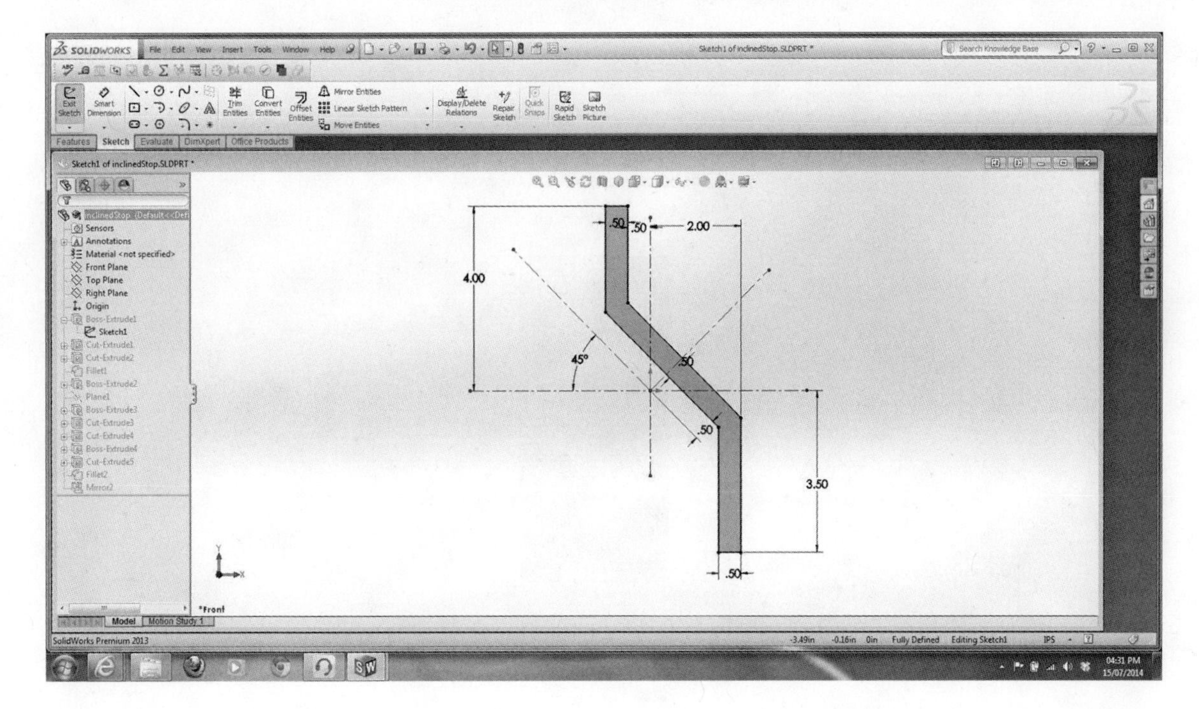

Figure 5.68

Sketch1.

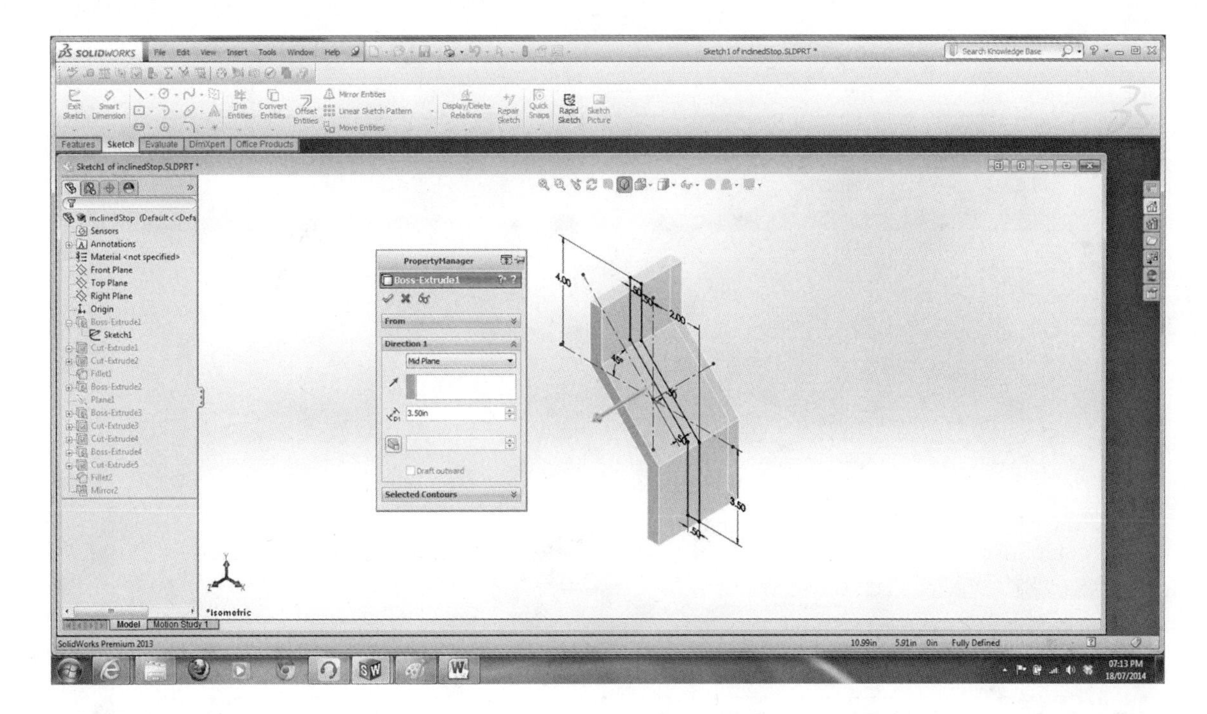

Figure 5.69

Extrude operation.

17. Be in Sketch mode and click Convert Entities to extract the Circle (Figure 5.80).
18. Extrude-Cut 0.75 further northwest of the uncompleted part.
19. Sketch a Circle on top of Circular feature as shown in Figure 5.81.
20. Extrude-Cut 0.5 into the circular portion (Figure 5.82).
21. From the Bottom Face, sketch the Profile that is shown in Figure 5.83.

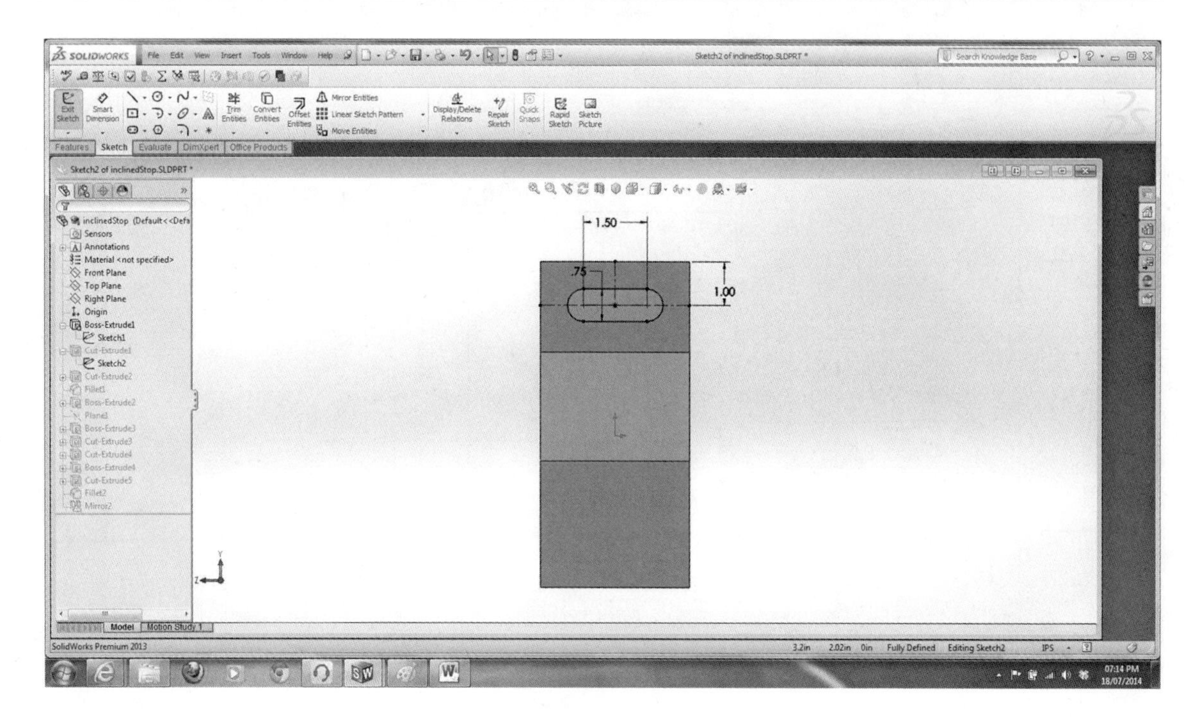

Figure 5.70

Slot feature.

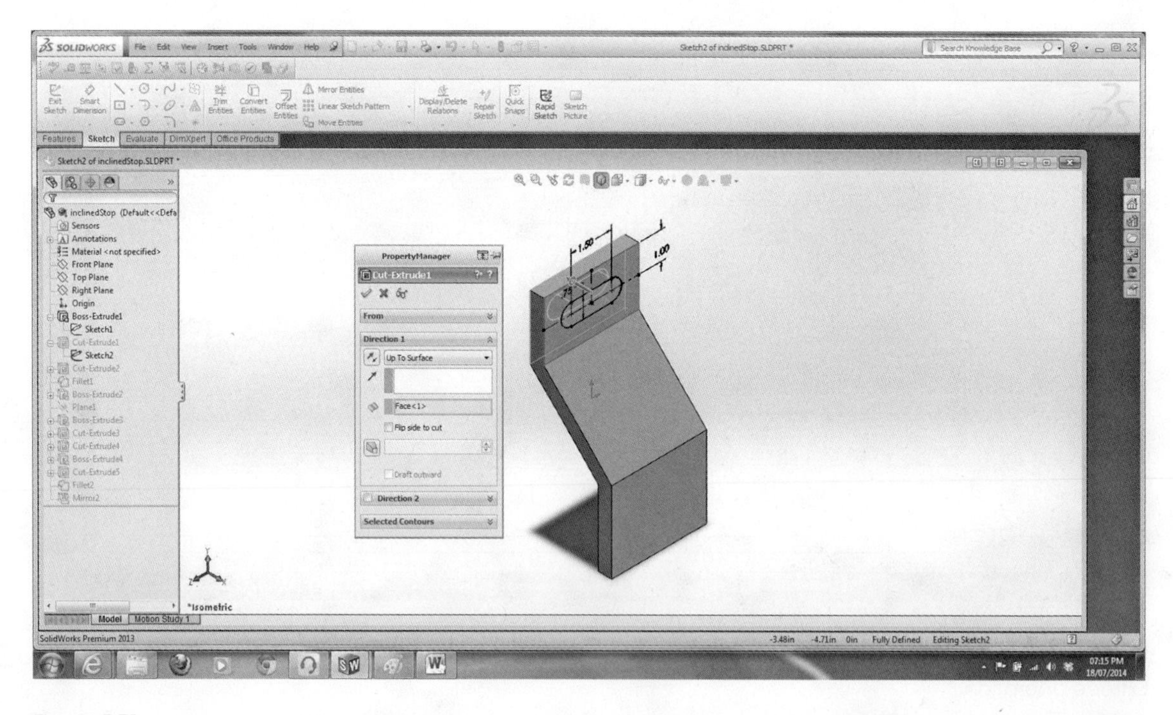

Figure 5.71

Extrude-Cut operation.

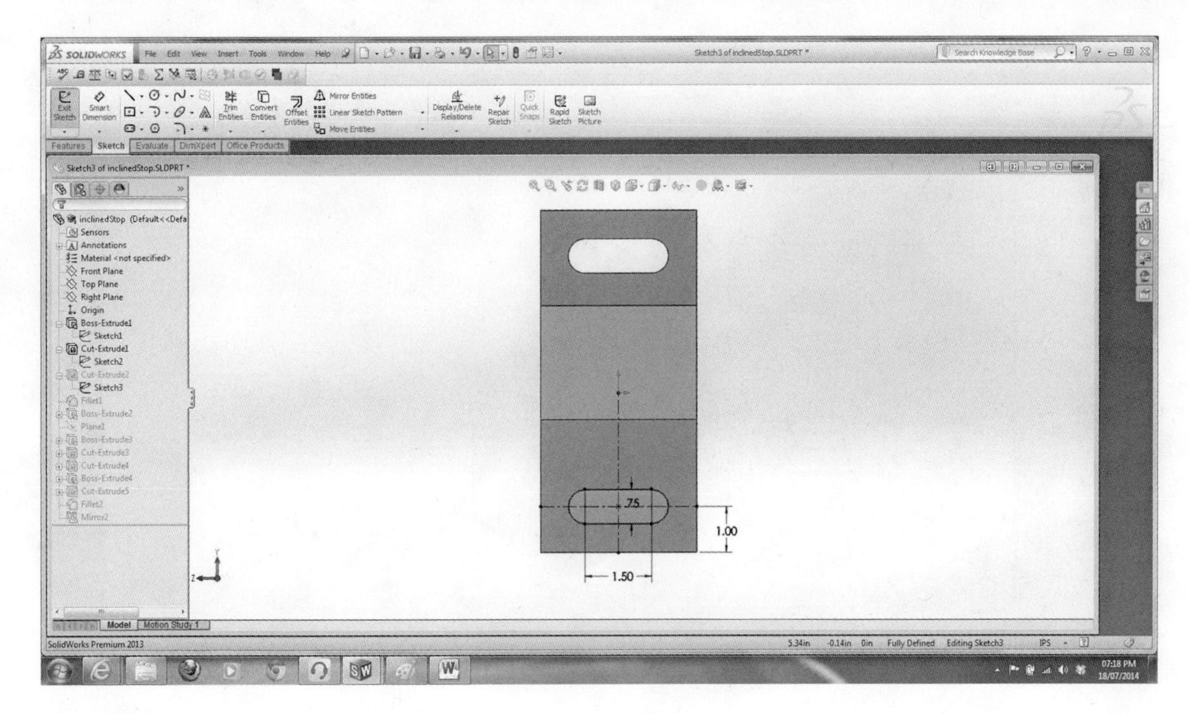

Figure 5.72

Slot feature.

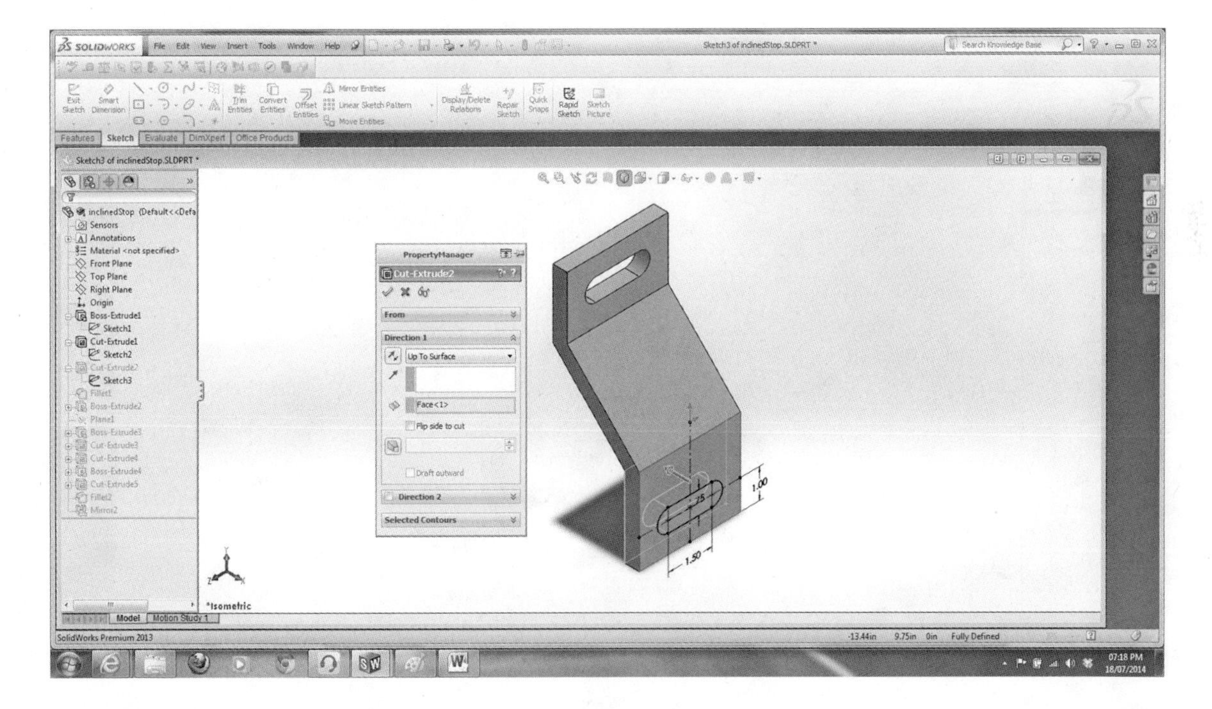

Figure 5.73

Another extrude-cut operation.

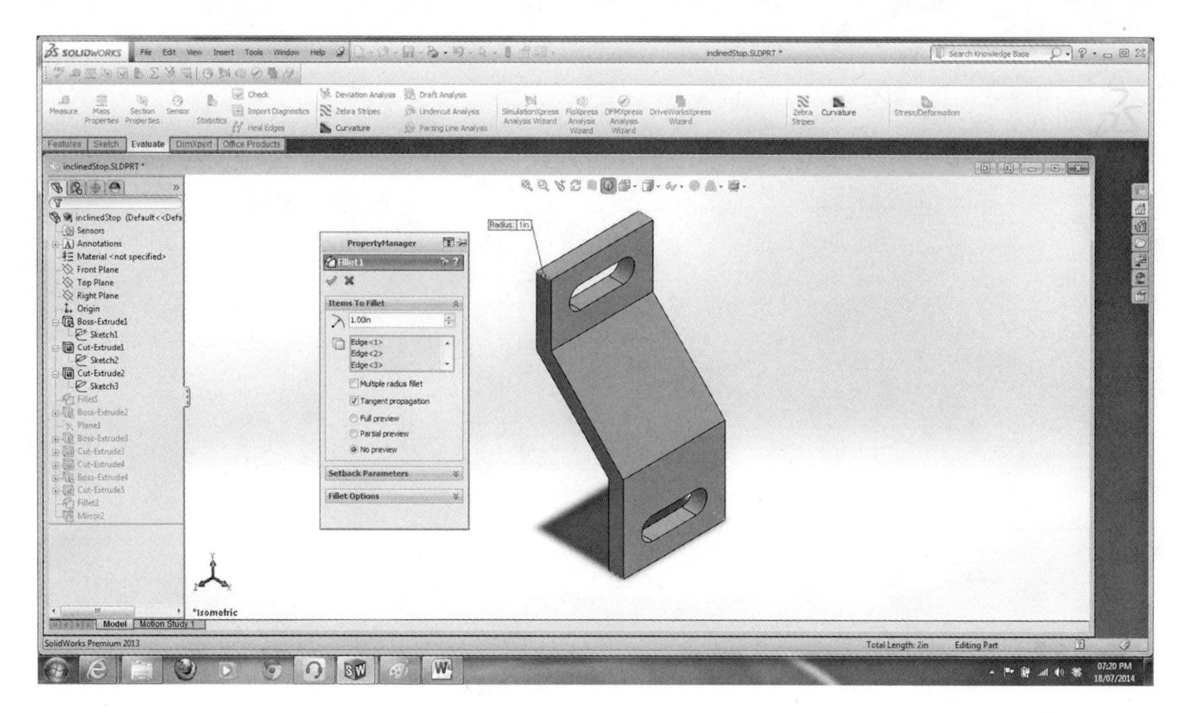

Figure 5.74

Fillet operation.

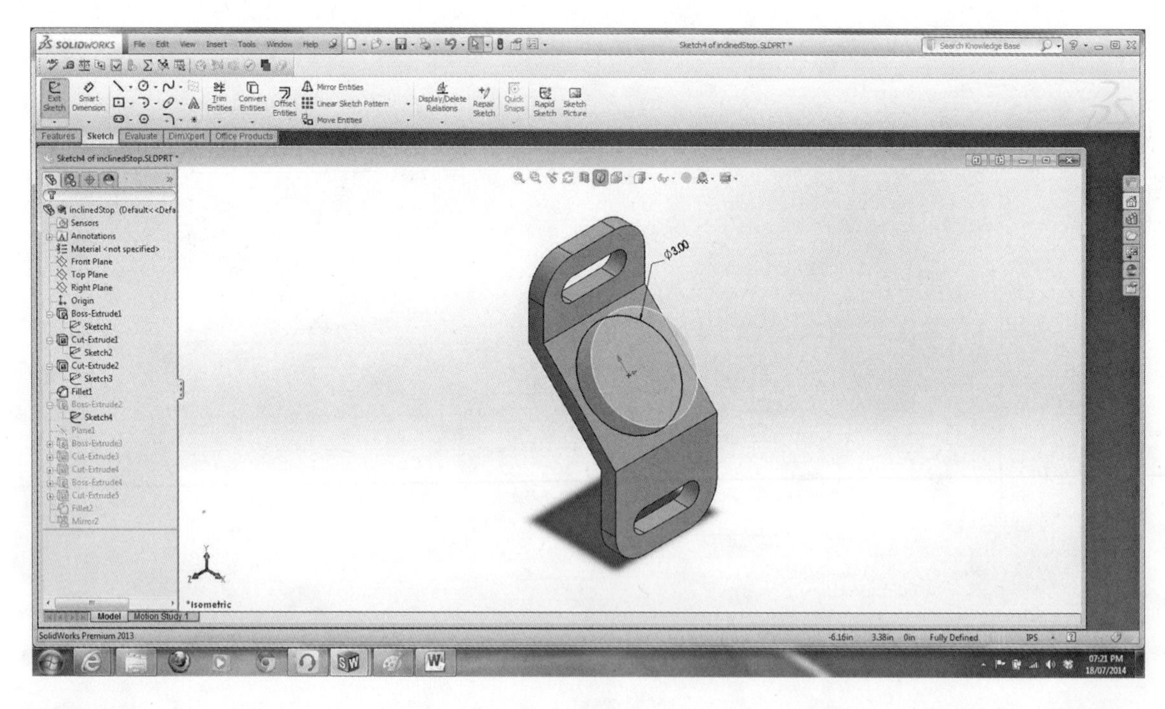

Figure 5.75

Circular feature on top inclined plane.

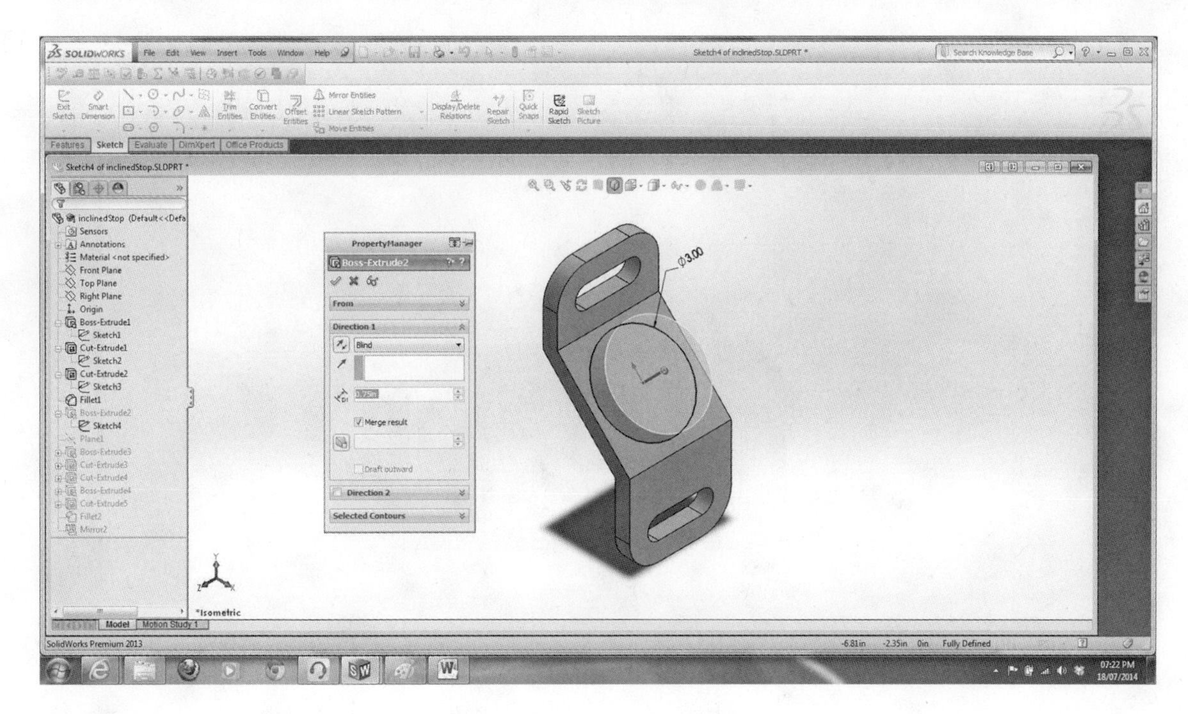

Figure 5.76

Extruding the circle.

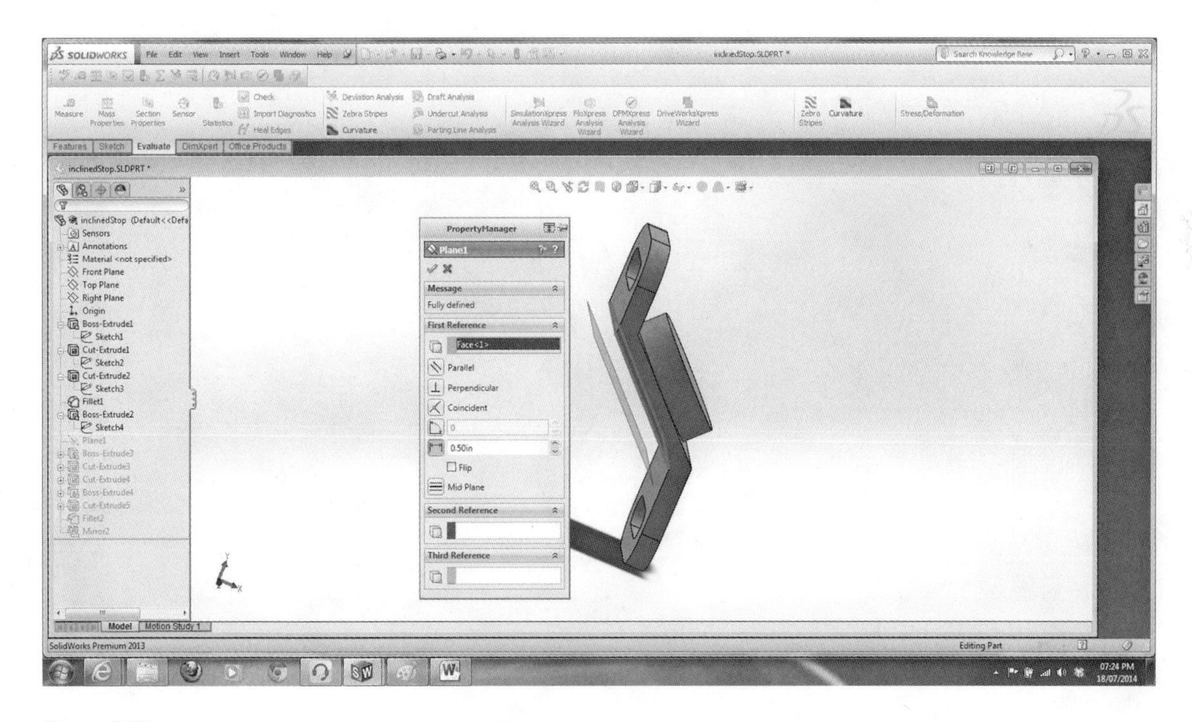

Figure 5.77

Plane1 away from bottom of inclined face.

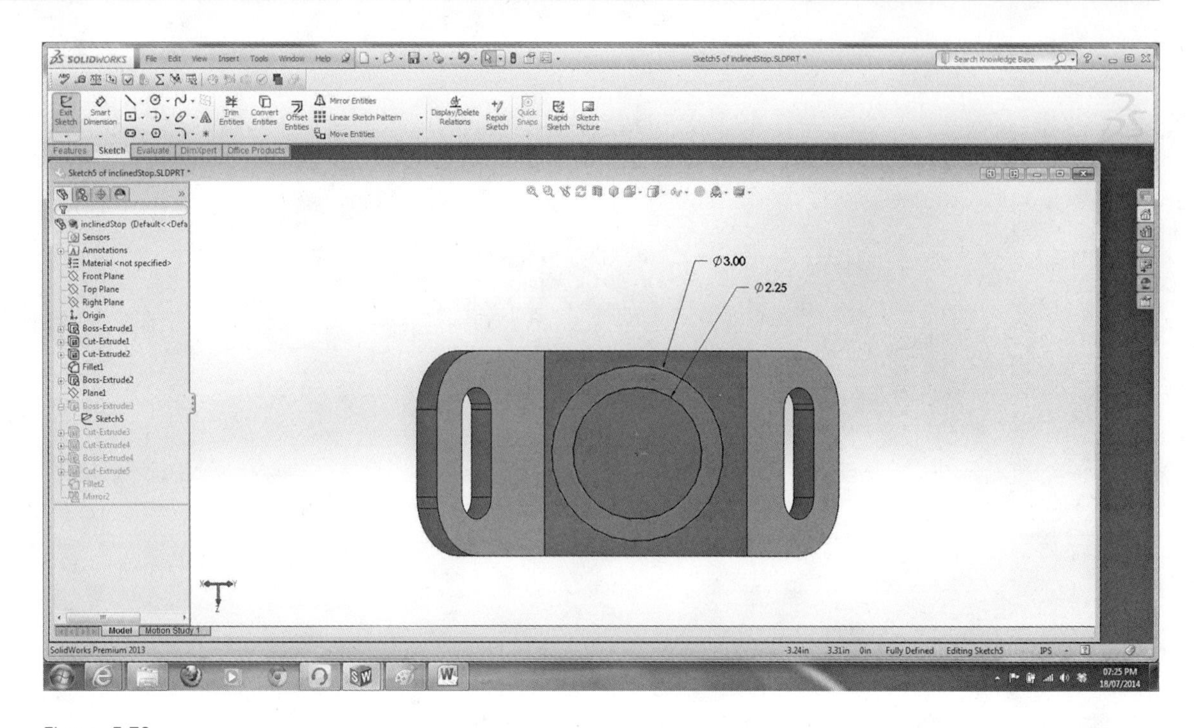

Figure 5.78

Two circles defined on Plane1.

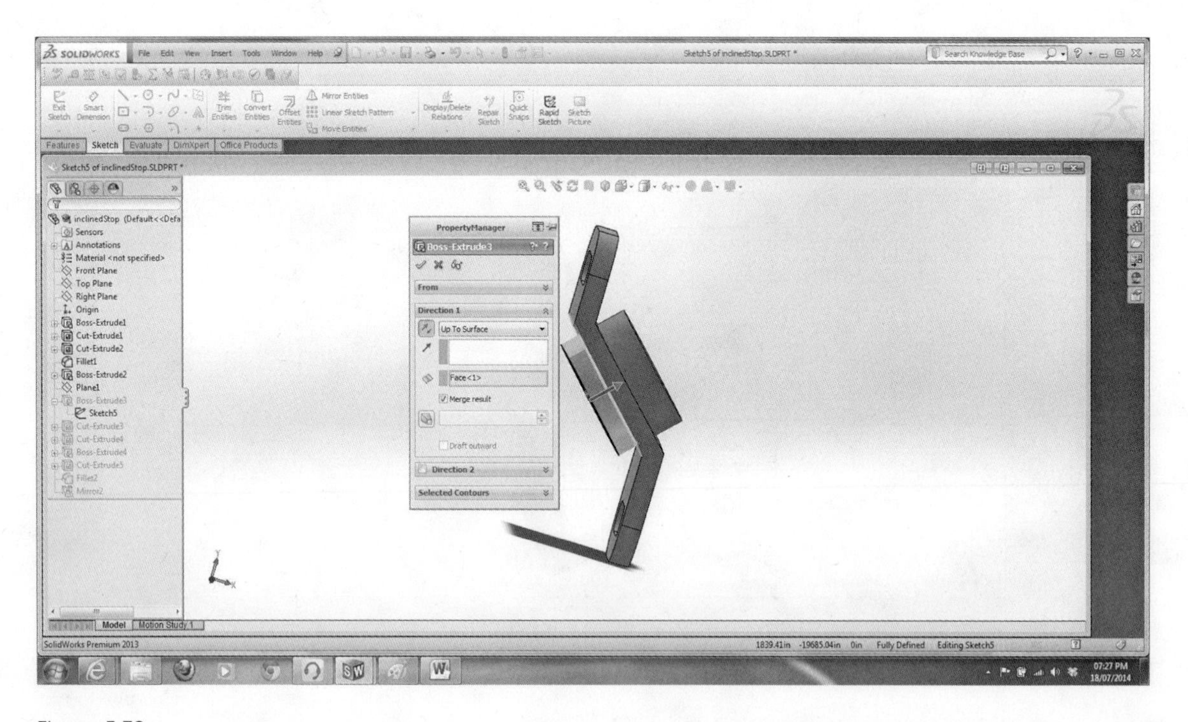

Figure 5.79

Extrude Cut operation.

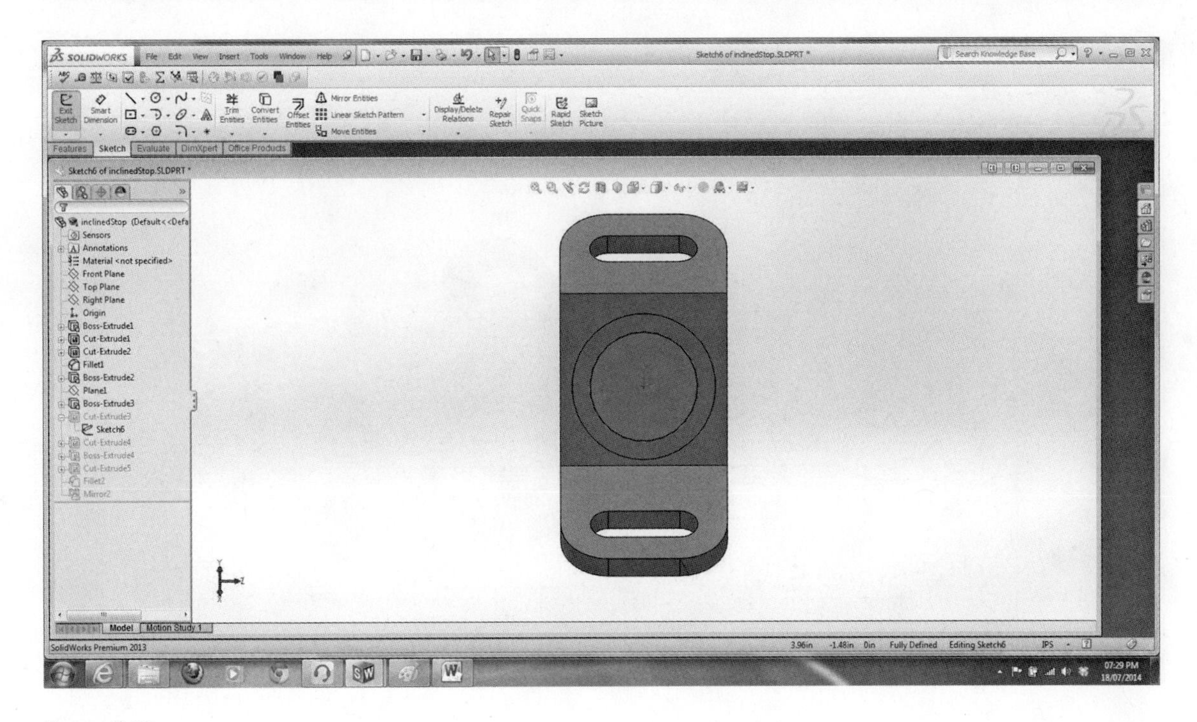

Figure 5.80

Circular feature extracted.

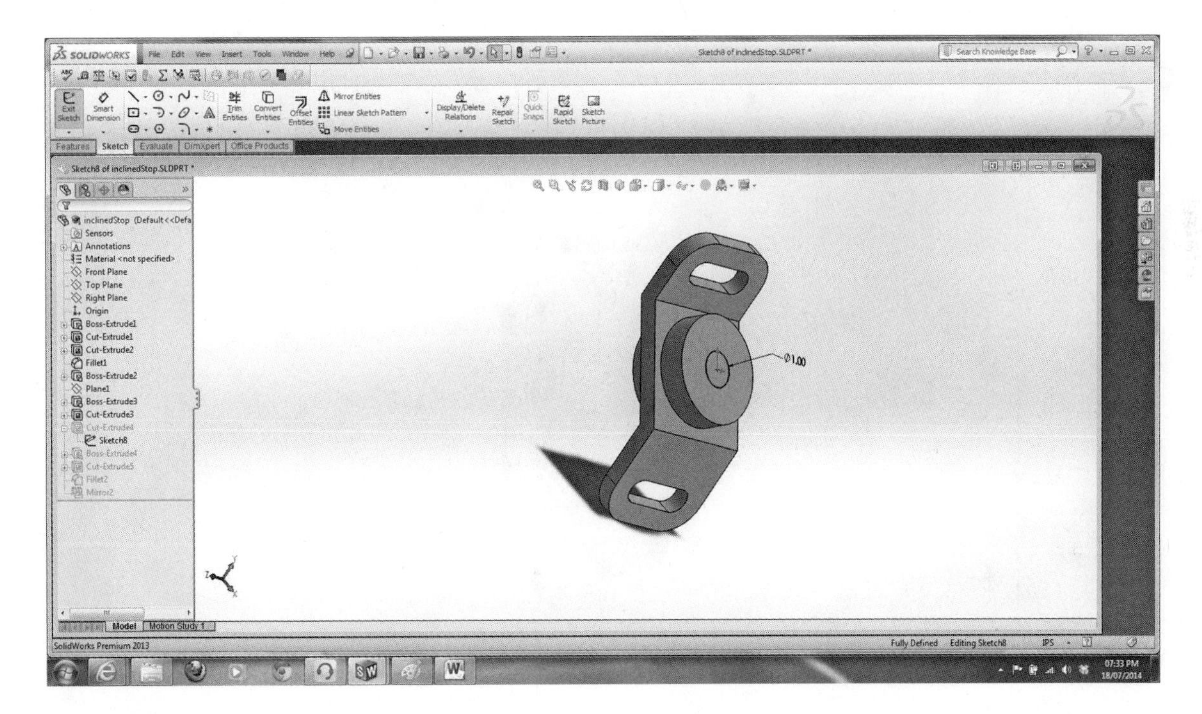

Figure 5.81

Circle for Extrude-Cut operation.

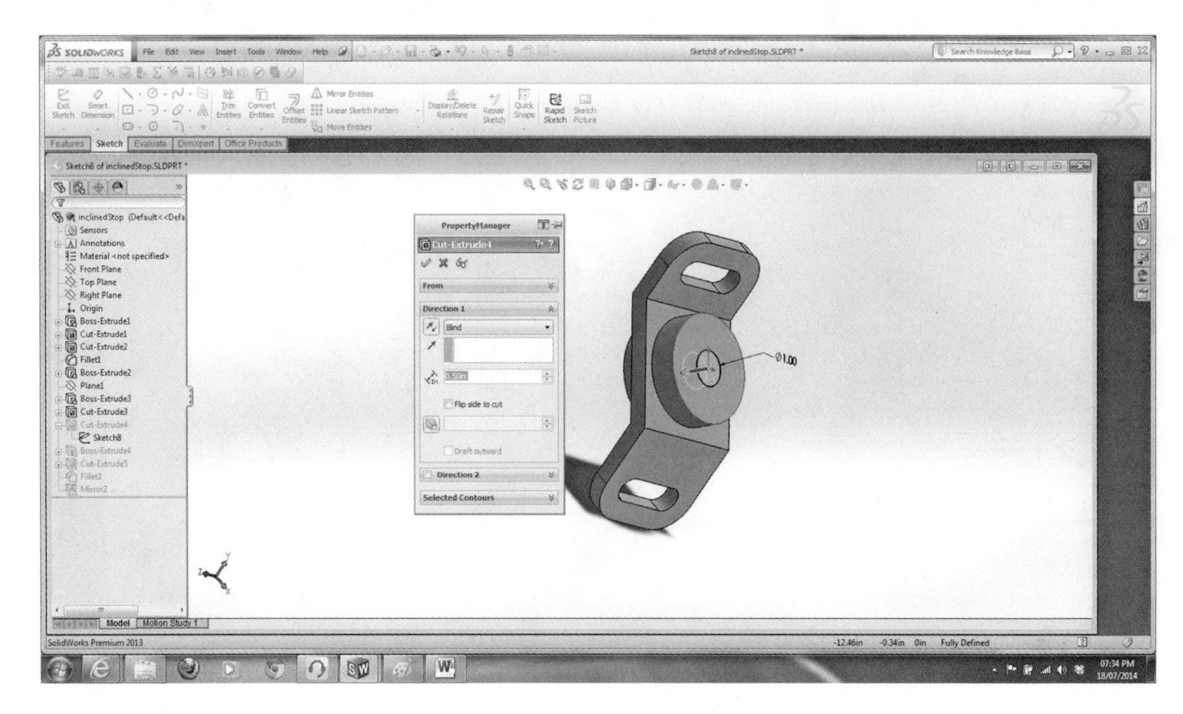

Figure 5.82

Extrude-Cut to realize a blind hole.

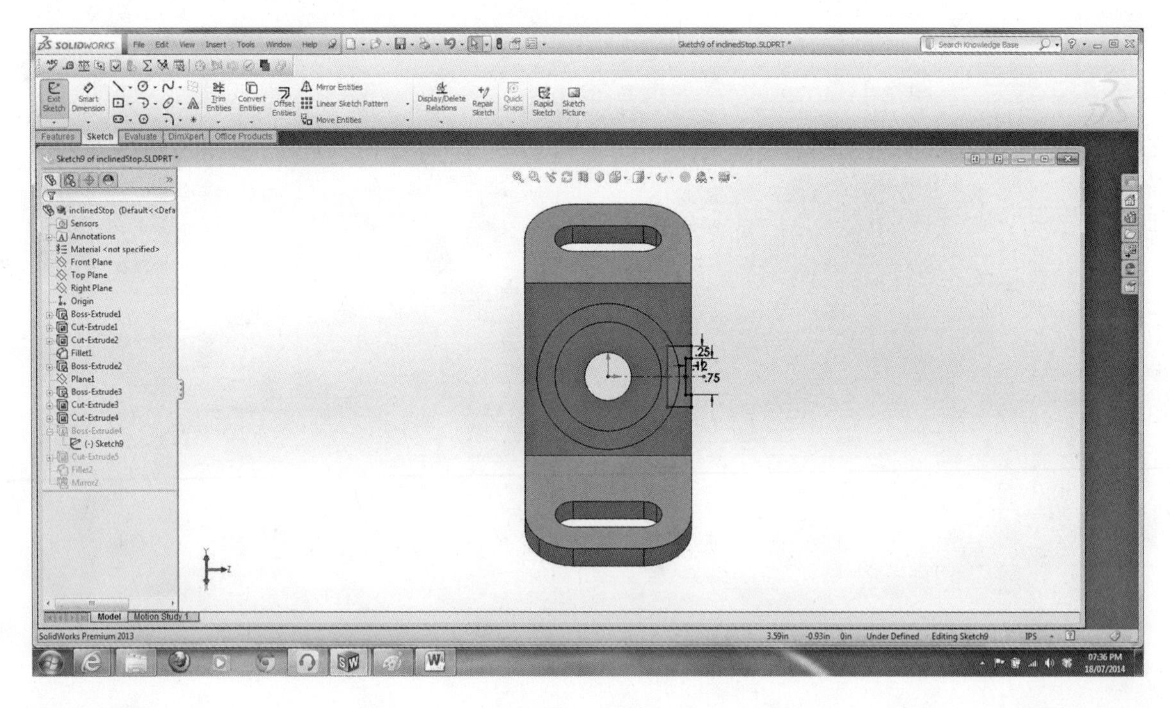

Figure 5.83

Sketching profile.

22. Extrude Up To Surface of Circular feature (Figure 5.84).
23. Use Convert Entities tool to pick up portion that is left after the Extrude operation (Figure 5.85).
24. Extrude Cut Up To Face (Figure 5.86).
25. Add Fillets of radius 0.01 (Figure 5.87).
26. Mirror the last three operations: (1) Boss-Extrude, (2) Cut-Extrude, and (3) Fillet2, about Front Plane (Figure 5.88).
27. The final model is shown in Figure 5.89.

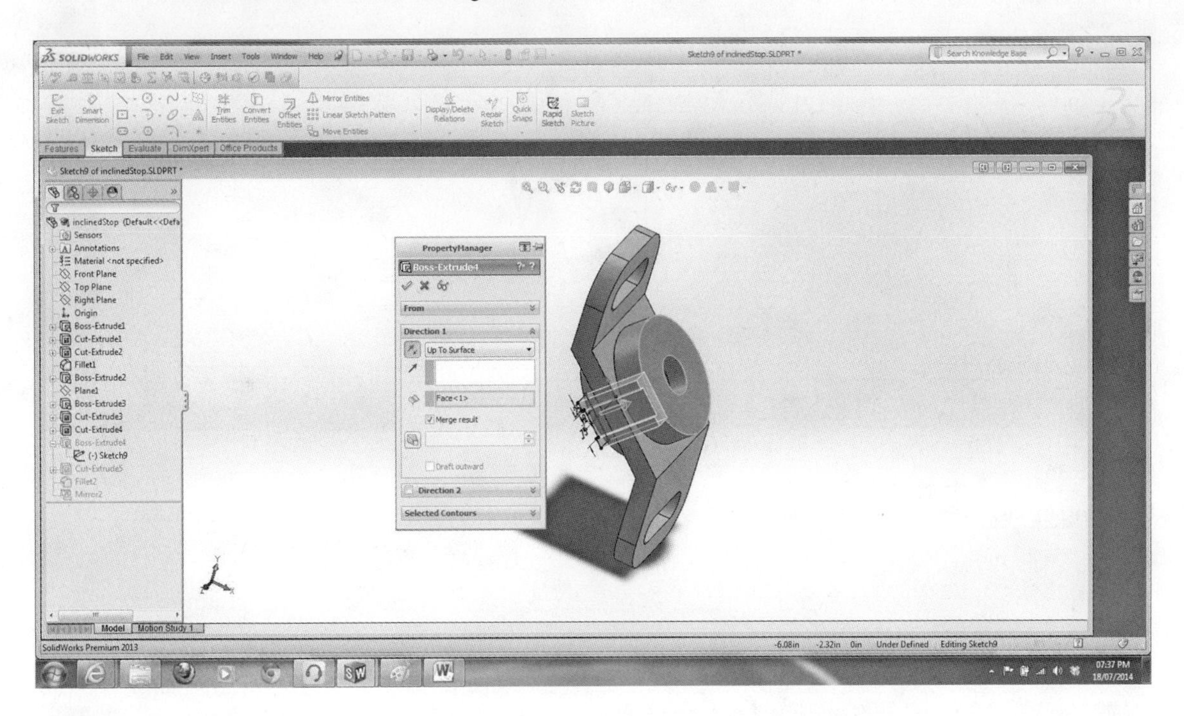

Figure 5.84

Boss-Extrude.

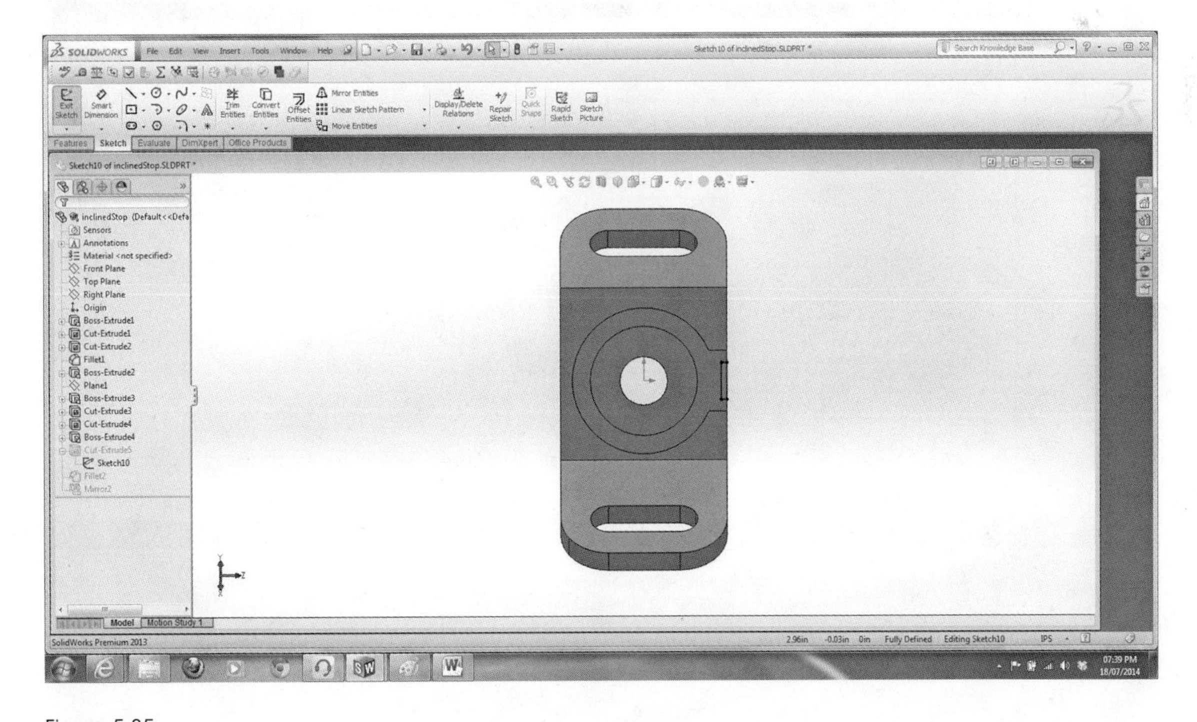

Figure 5.85

Sketch for cleaning up the model.

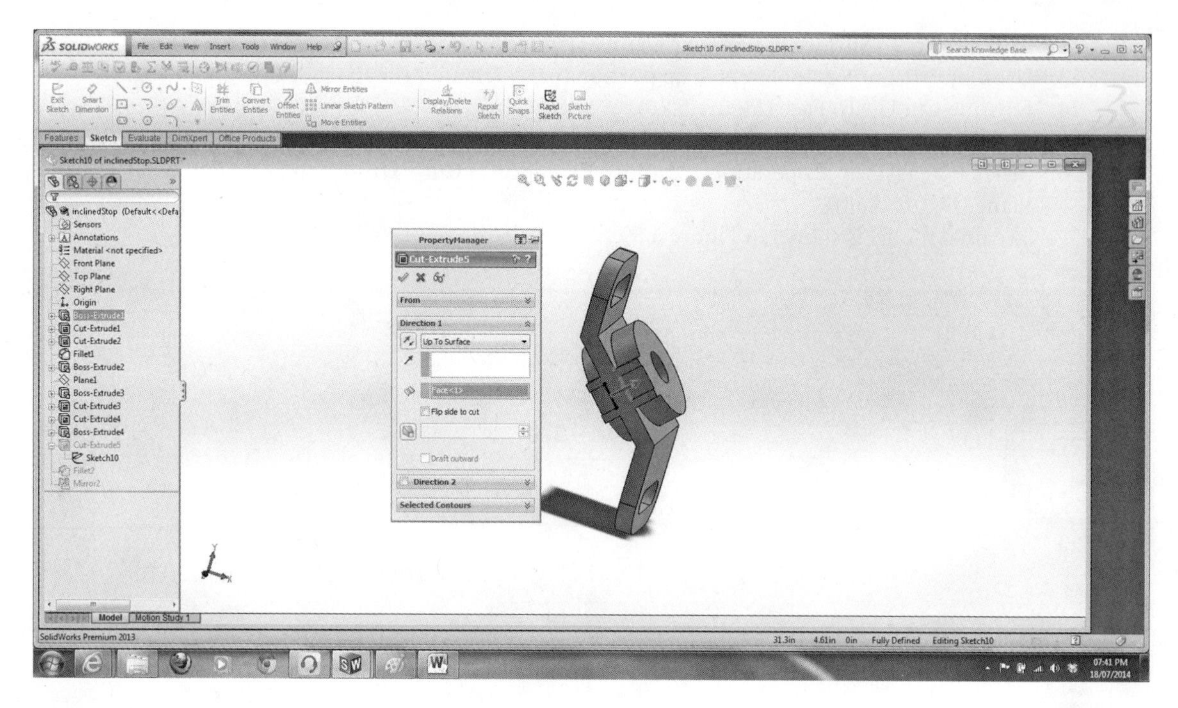

Figure 5.86

Cleaning up using Extrude-Cut operation.

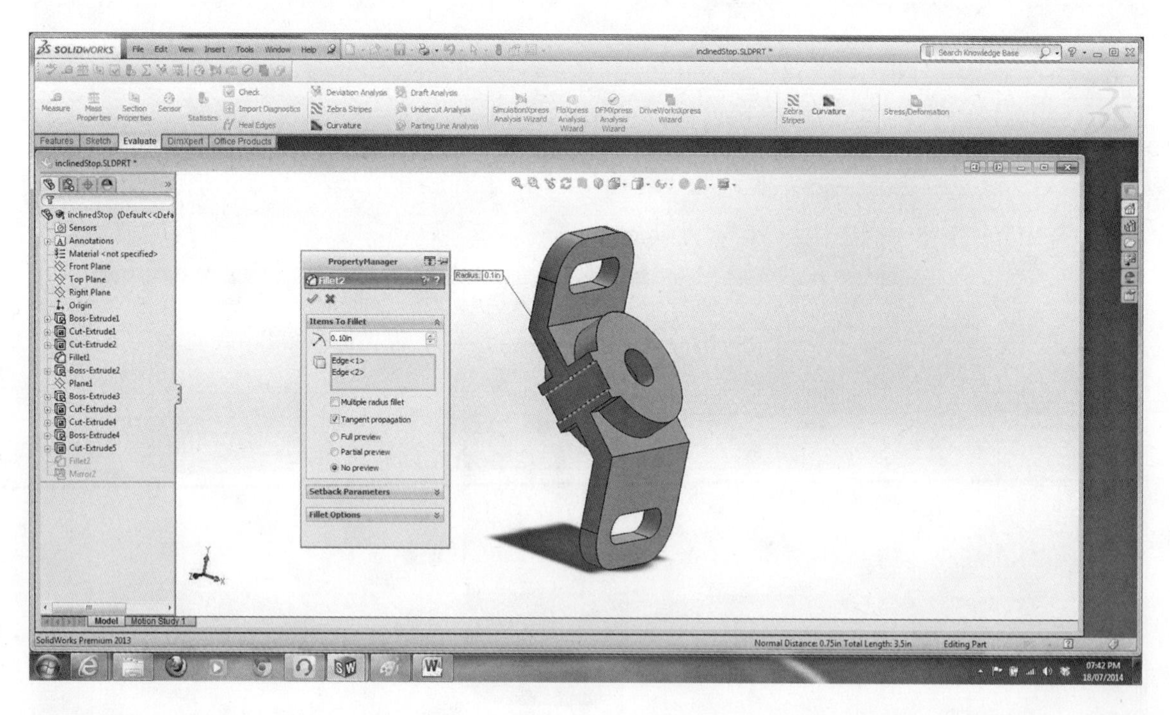

Figure 5.87

Filleting.

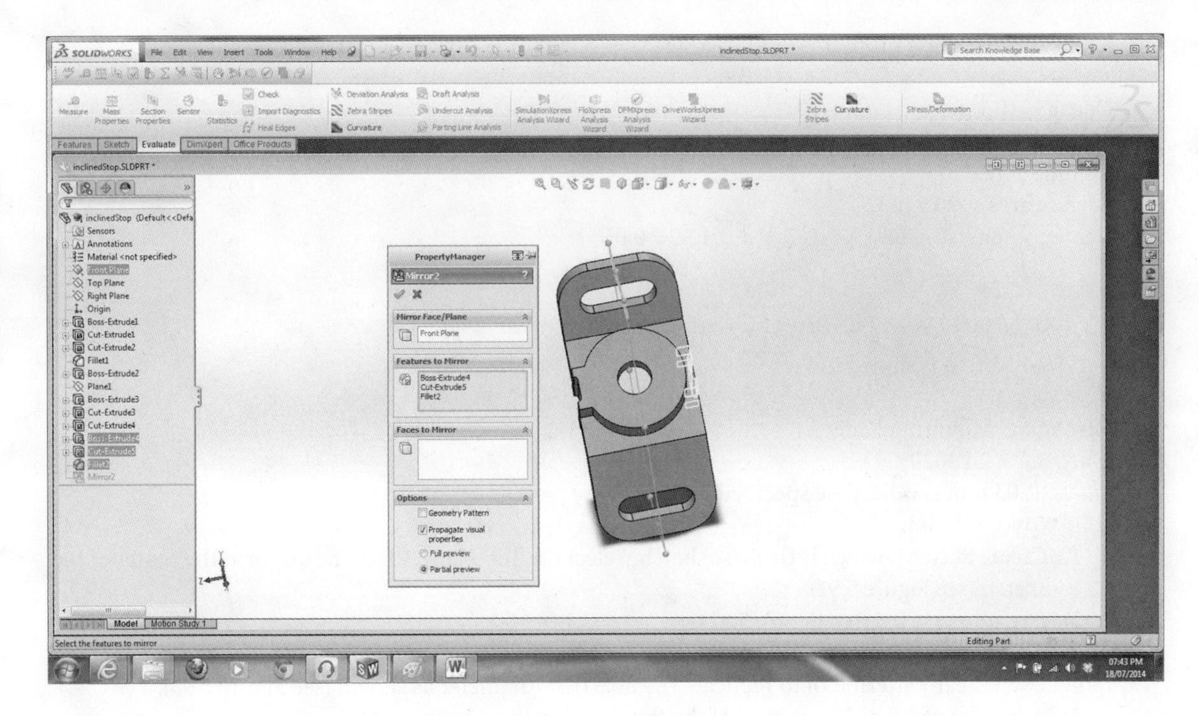

Figure 5.88

Mirror operation.

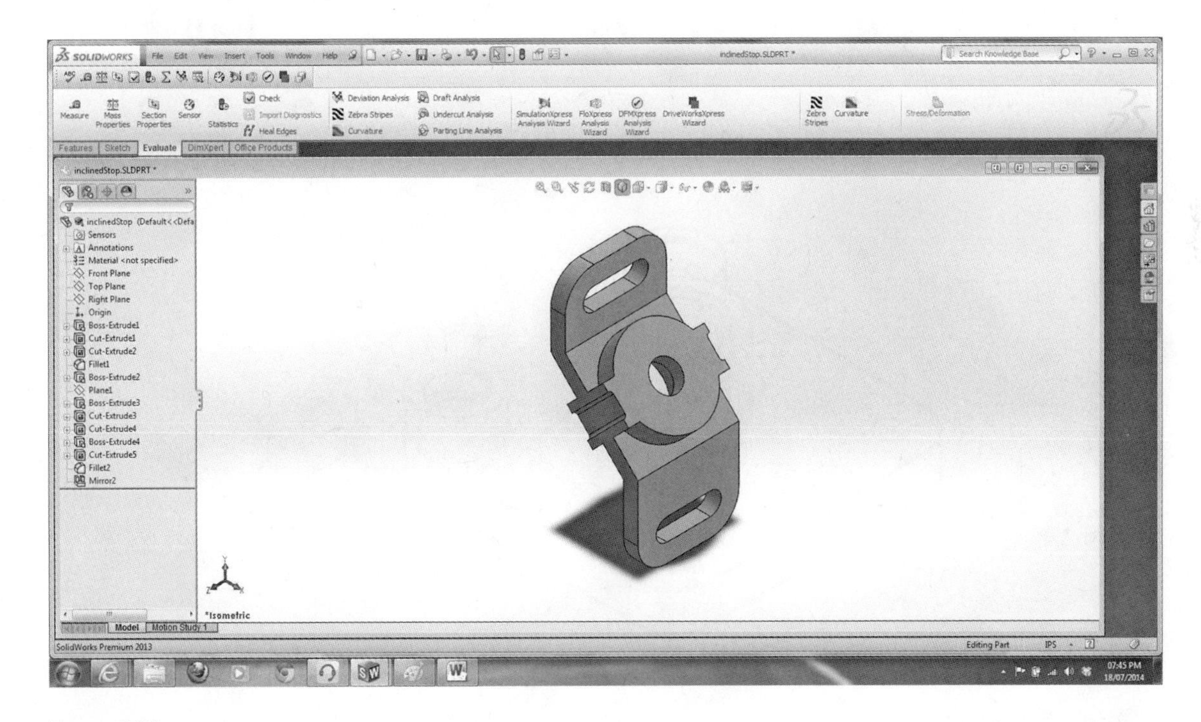

Figure 5.89

Final model of inclined plane.

Click Evaluate > Mass Properties to determine the weight of 1.7 lb. and 17.00 in.^{3}

Density = 0.1 lb./in.^{3}
Mass = 1.7 lb.
Volume = 17.0 in.^{3}
Surface area = 90.9 in.^{2}
Center of mass: (inches) X = 0.7; Y = 0.4; Z = 0.0

Tutorial 5.8: Model with Notched Offset Section View

Create the model that is shown in Figure 5.90.

Part name: Mounting Plate
Material: AISI 1020
Fillets: R.03 unless otherwise specified.
SolidWorks Solution

1. Create Sketch1, which is the base sketch. Select the Top Plane. Place the origin at the center of the sketch (see Figure 5.91).
2. Extrude base 0.5 (see Figure 5.92).
3. Create a hole, 0.38-diameter as shown (see Figure 5.92).
4. Use Linear Pattern tool to replicate the hole 0.38-diameter as shown (see Figure 5.93).
5. Add fillet of 0.5 radius at each of the four corners (see Figure 5.94).
6. Create boss 1.0 diameter height of 0.5 from top of base (see Figure 5.95).

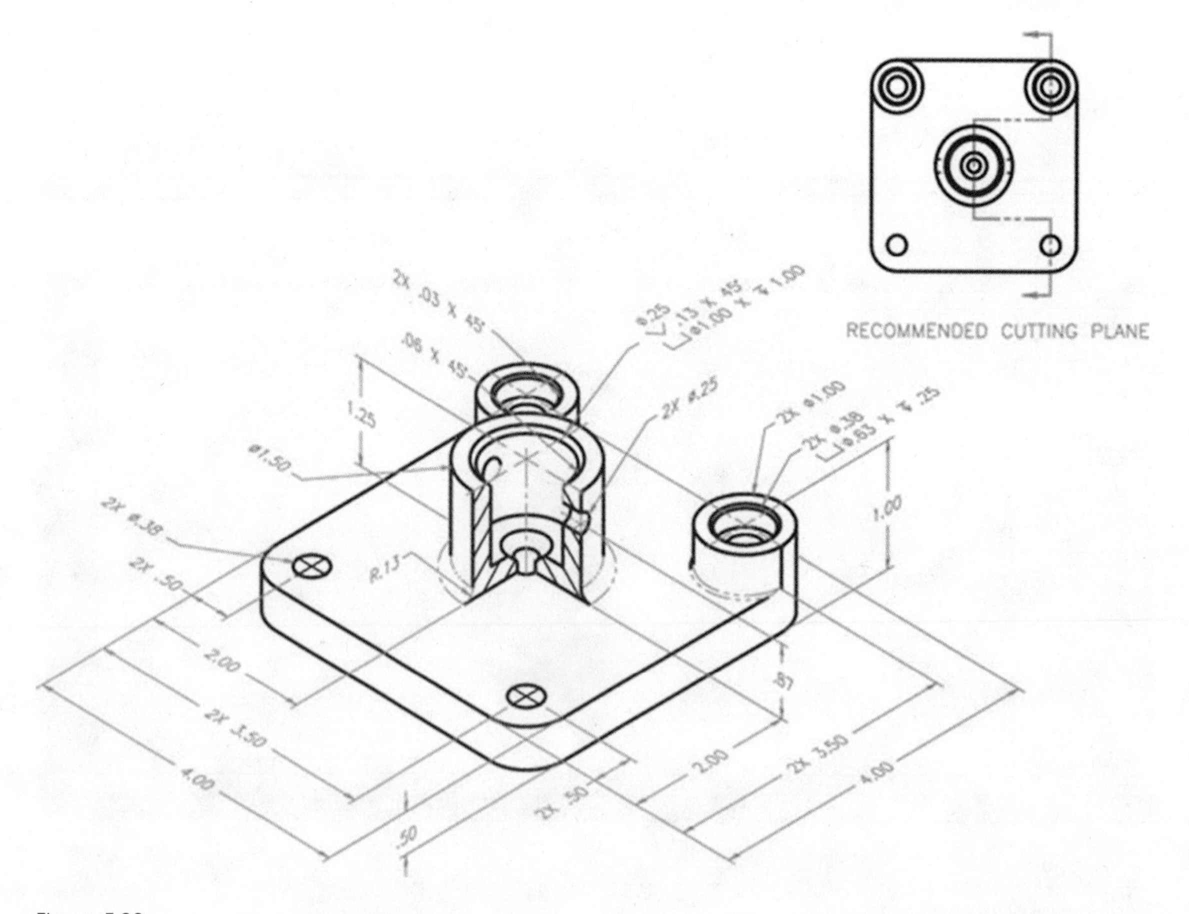

Figure 5.90

Model description. (From JENSEN/HELSEL/ESPIN. Interpreting Engineering Drawings, 6E. © 2012 Nelson Education Ltd. Reproduced by permission. http://www.cengage.com/permissions; From MADSEN. *ENGINEERING DRAW/DESIGN IML*, 1E. © 1991 Delmar Learning, a part of Cengage Learning, Inc. Reproduced by permission. http://www.cengage.com/permissions.)

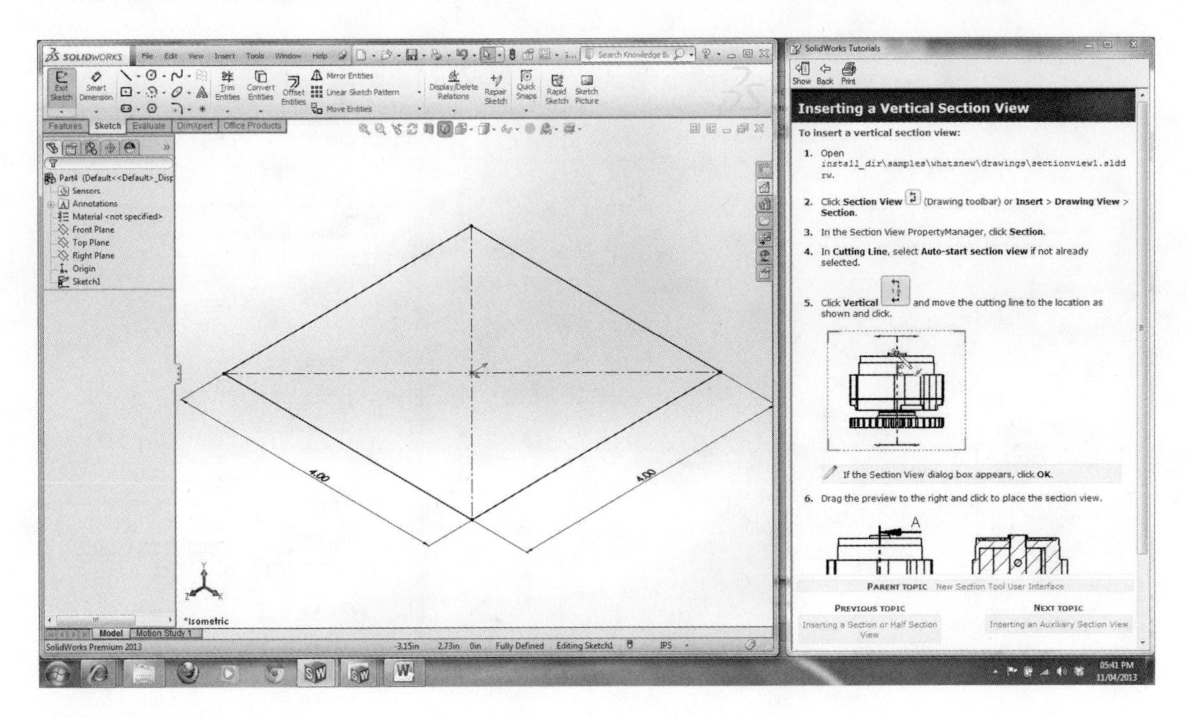

Figure 5.91

Sketch1 for the base of the model.

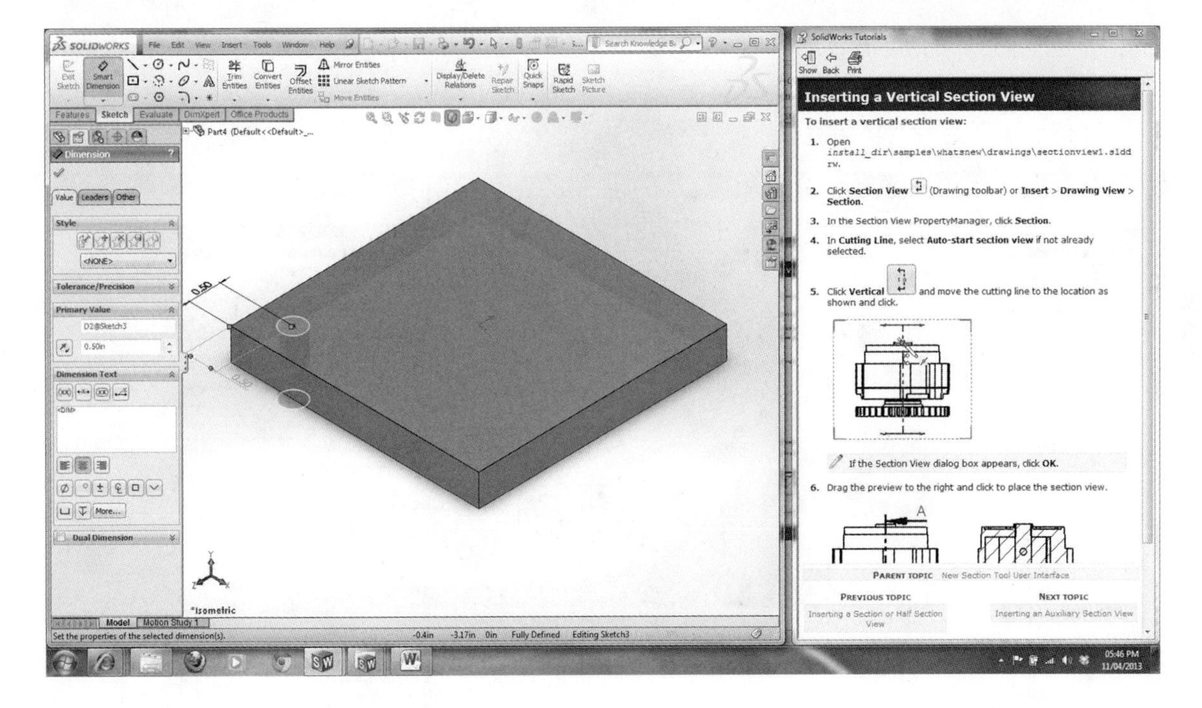

Figure 5.92

Extruded base.

7. Create a circle, 0.63 diameter on top of boss (Figure 5.96).
8. Extrude Cut 0.25 downward to create a countersink (Figure 5.97).
9. Create a circle, 0.38 diameter on top of countersink (Figure 5.98).
10. Extrude Cut Through All (Figure 5.99).
11. Mirror the countersink and through hole about Right Plane (Figure 5.100).
    (See Figure 5.101 for the mirrored features.)

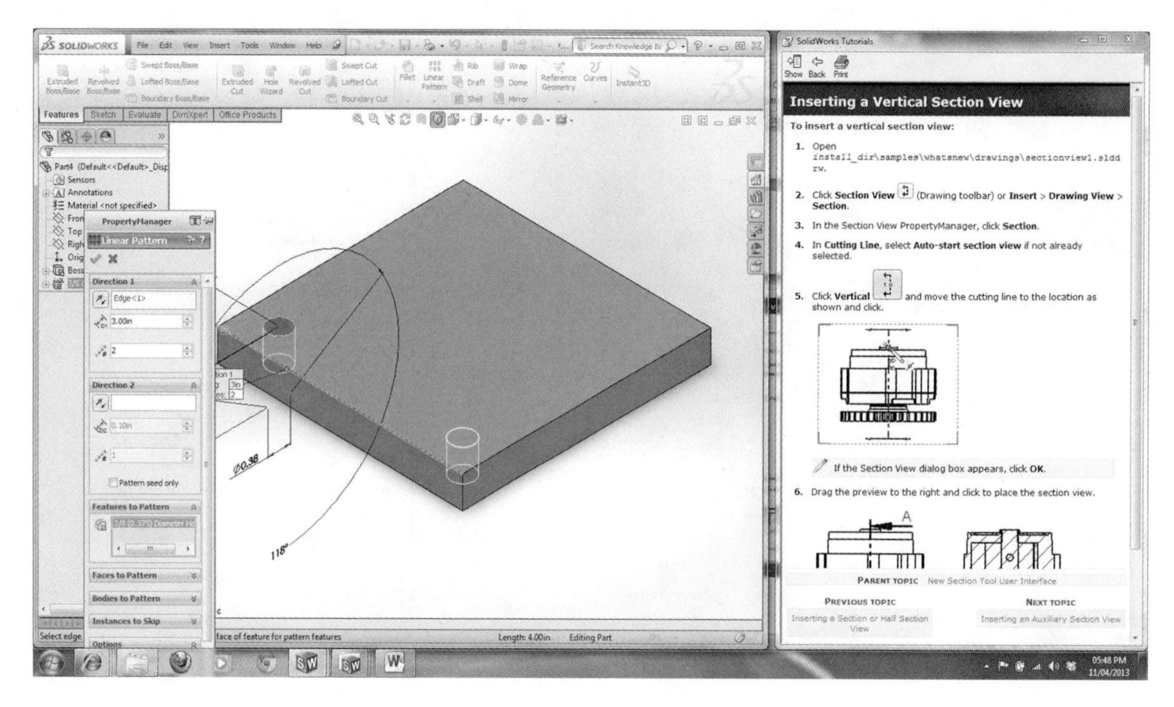

Figure 5.93

Duplicating the hole feature.

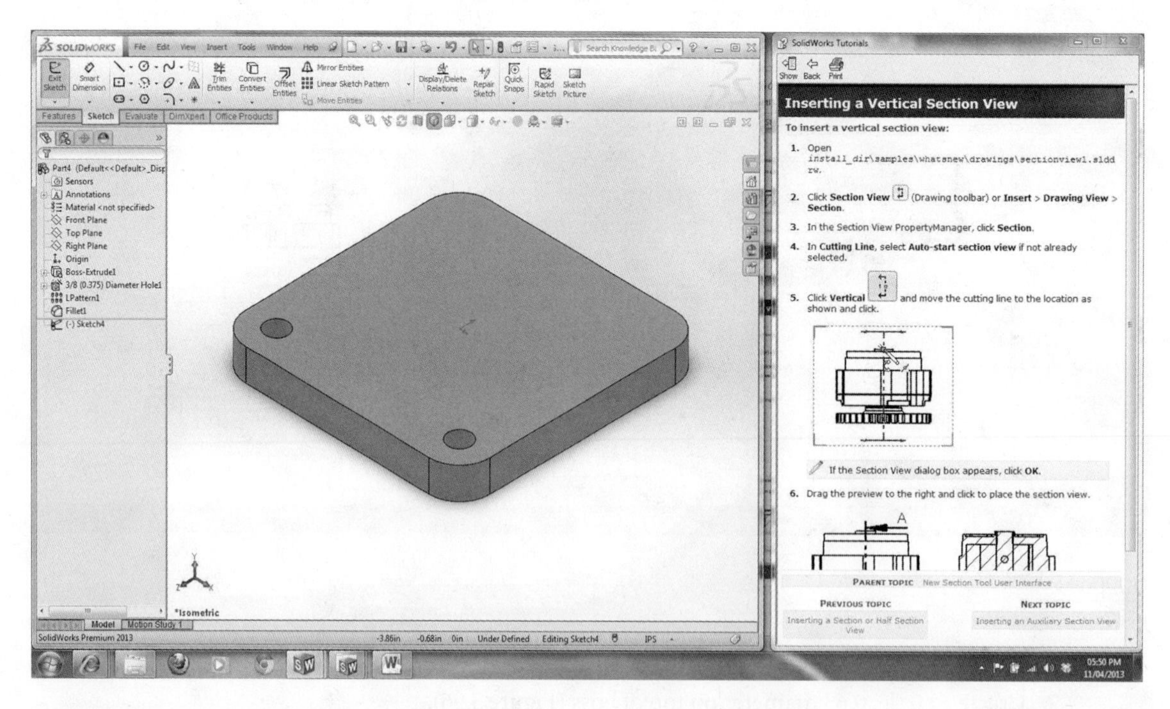

Figure 5.94

Four fillets of 0.5 radius each.

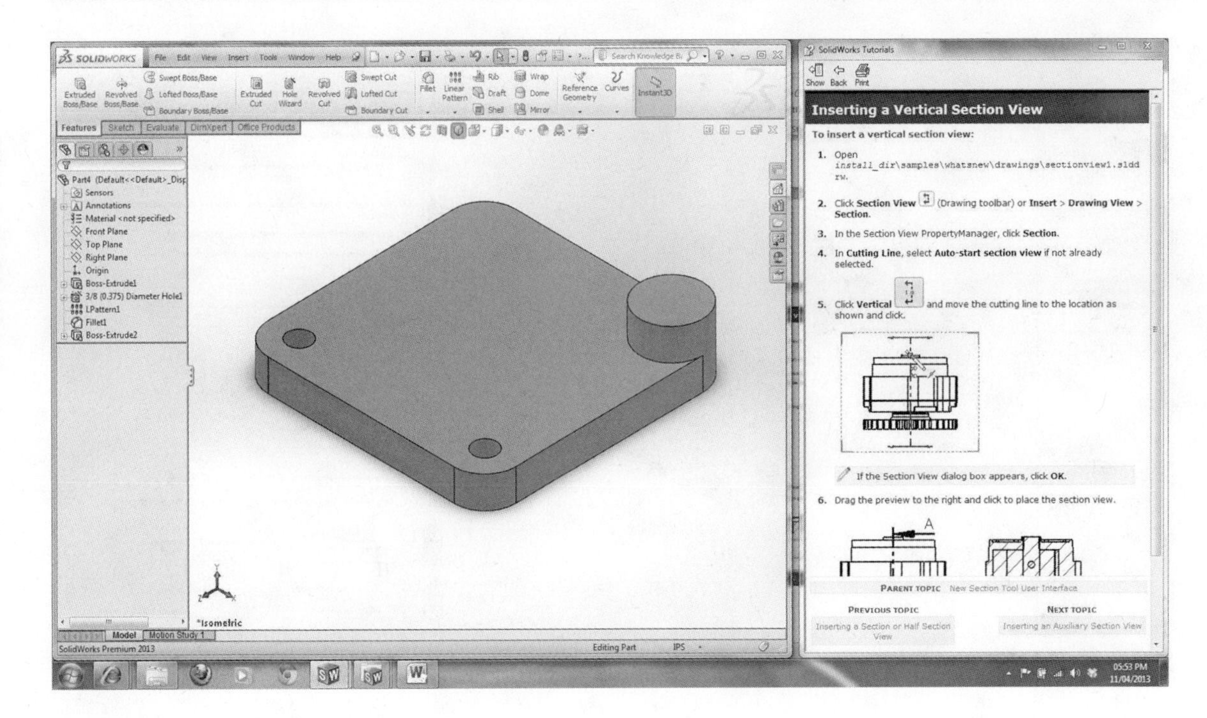

Figure 5.95

Boss added.

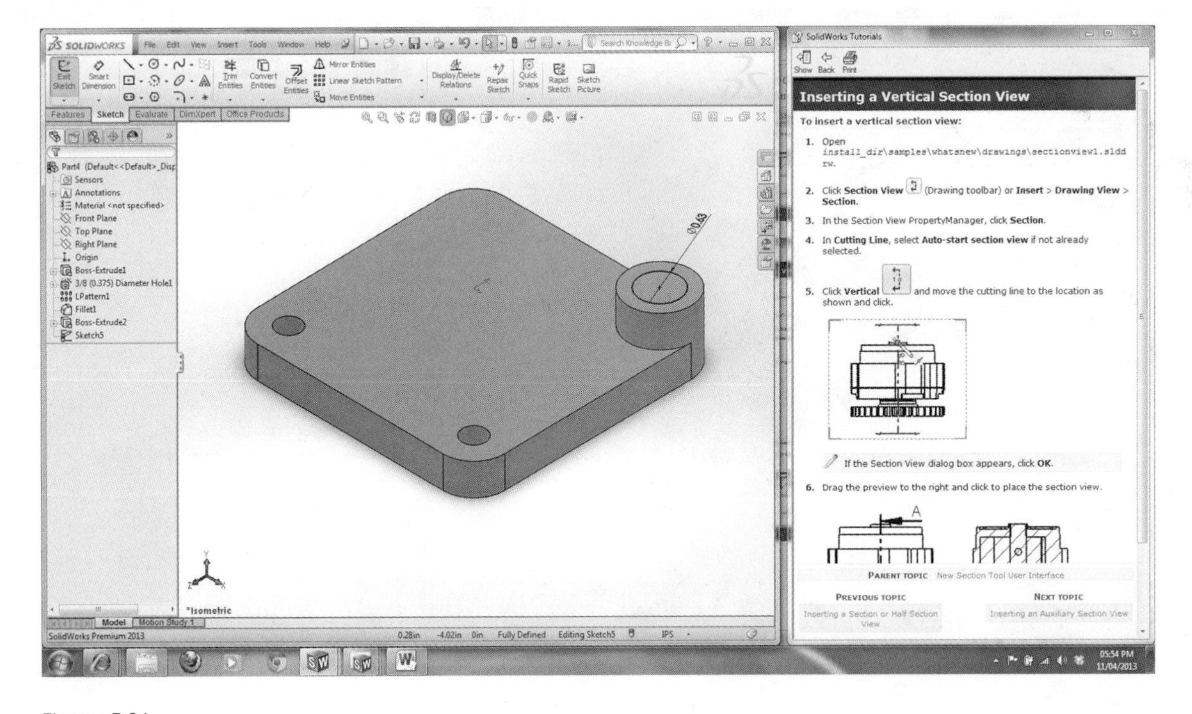

Figure 5.96

Sketch a circle.

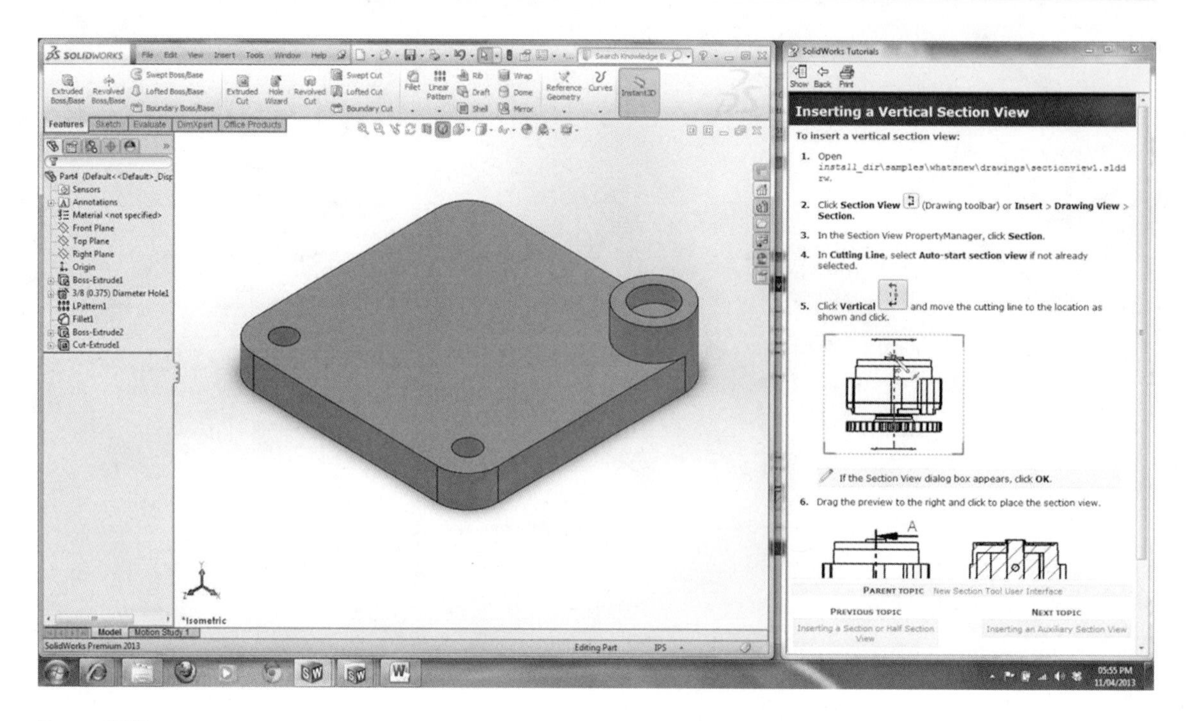

Figure 5.97

Extrude Cut the circle.

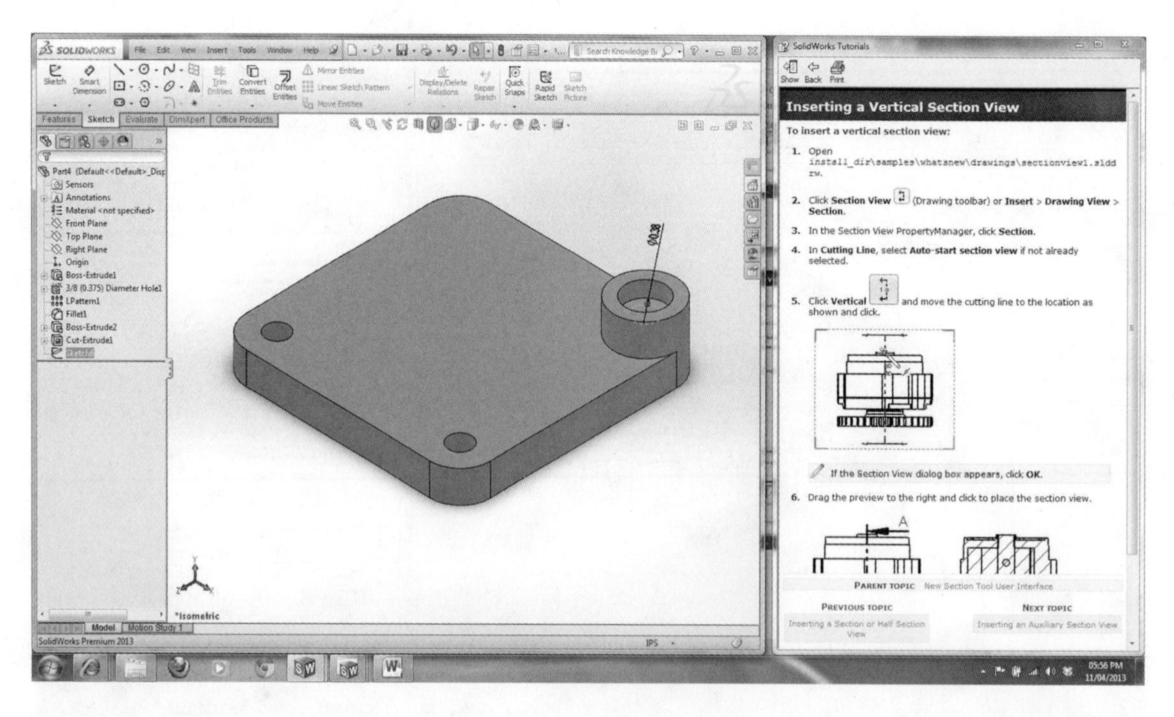

Figure 5.98

Countersink created.

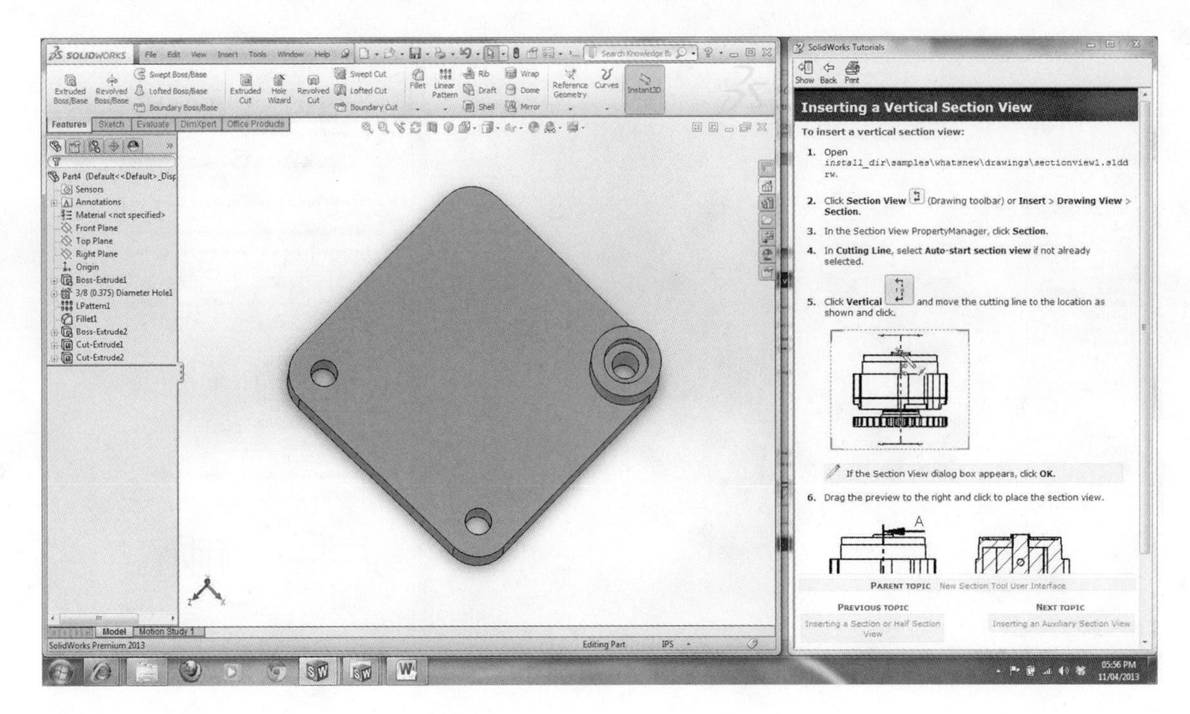

Figure 5.99

Hole right through created.

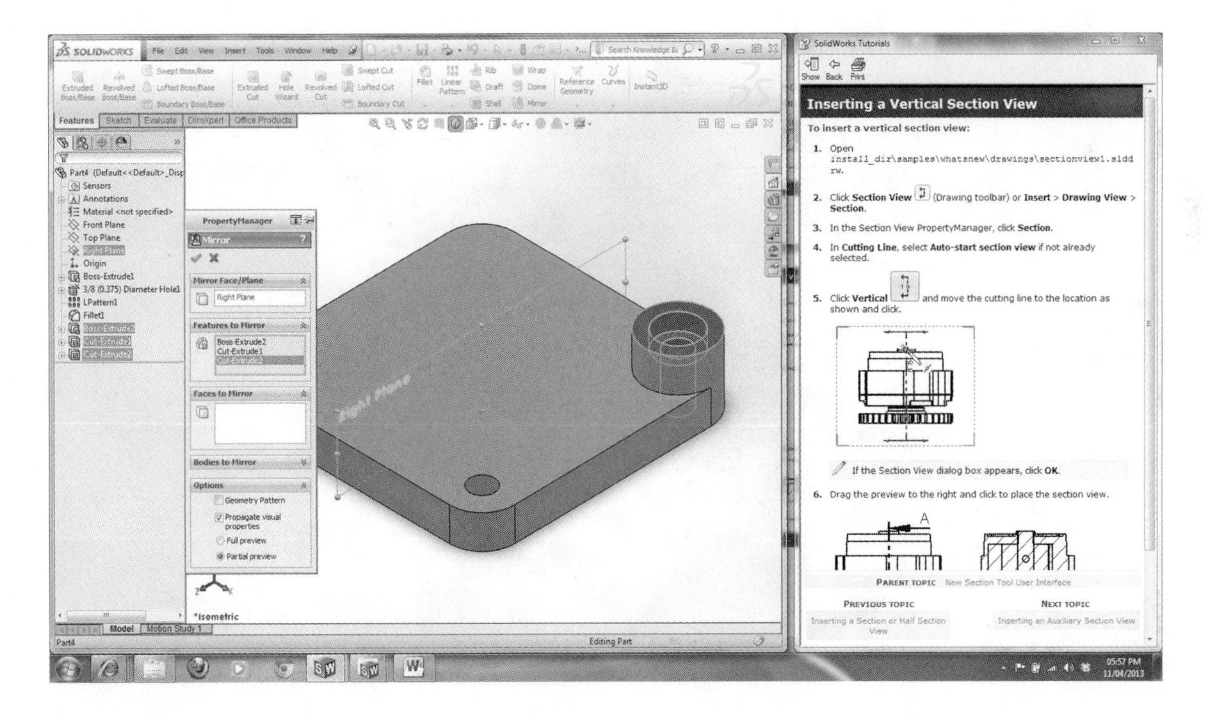

Figure 5.100

Mirror operation.

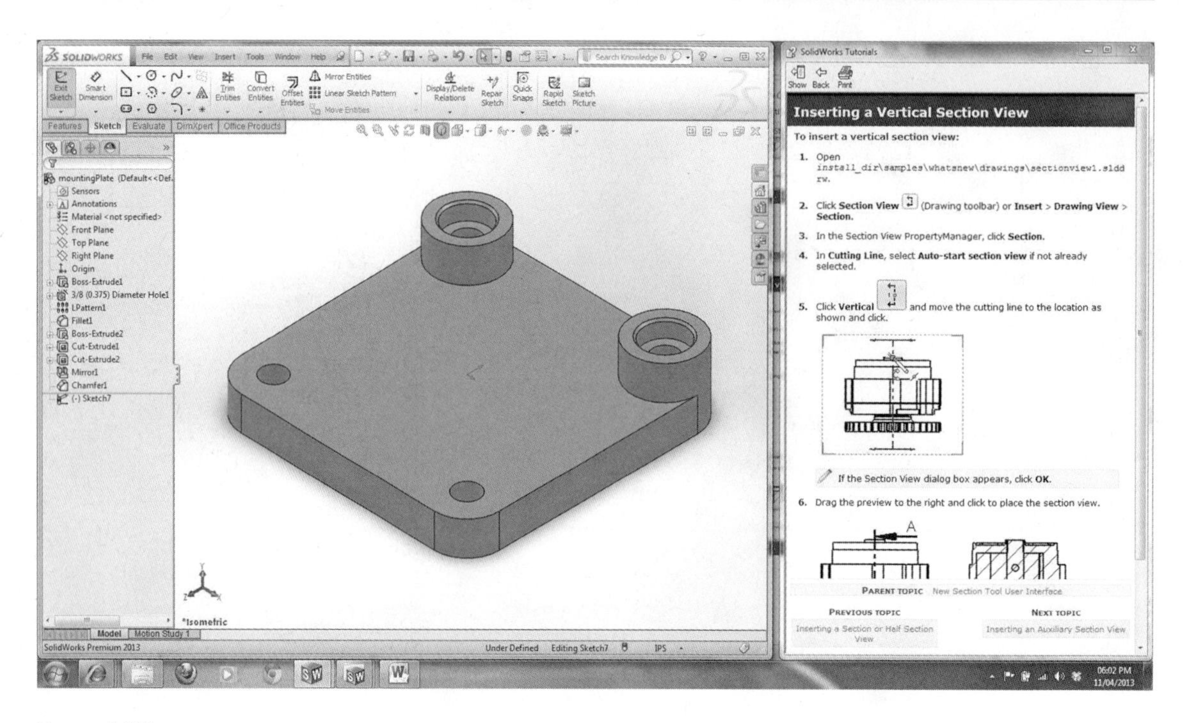

Figure 5.101

Mirrored features.

12. Create central boss, 1.5 diameter, and 1.25 high from top of base (Figure 5.102).
13. Sketch circle, 1 diameter on top of central boss (Figure 5.103).
14. Extrude Cut, 1 downward to create a counter bore (Figure 5.104).
15. Create a circle, 0.25 diameter (Figure 5.105).
16. Extrude Cut right through (Figure 5.106).
17. Add Chamfers (Figures 5.106 and 5.107).
18. Add Material (Figure 5.108).

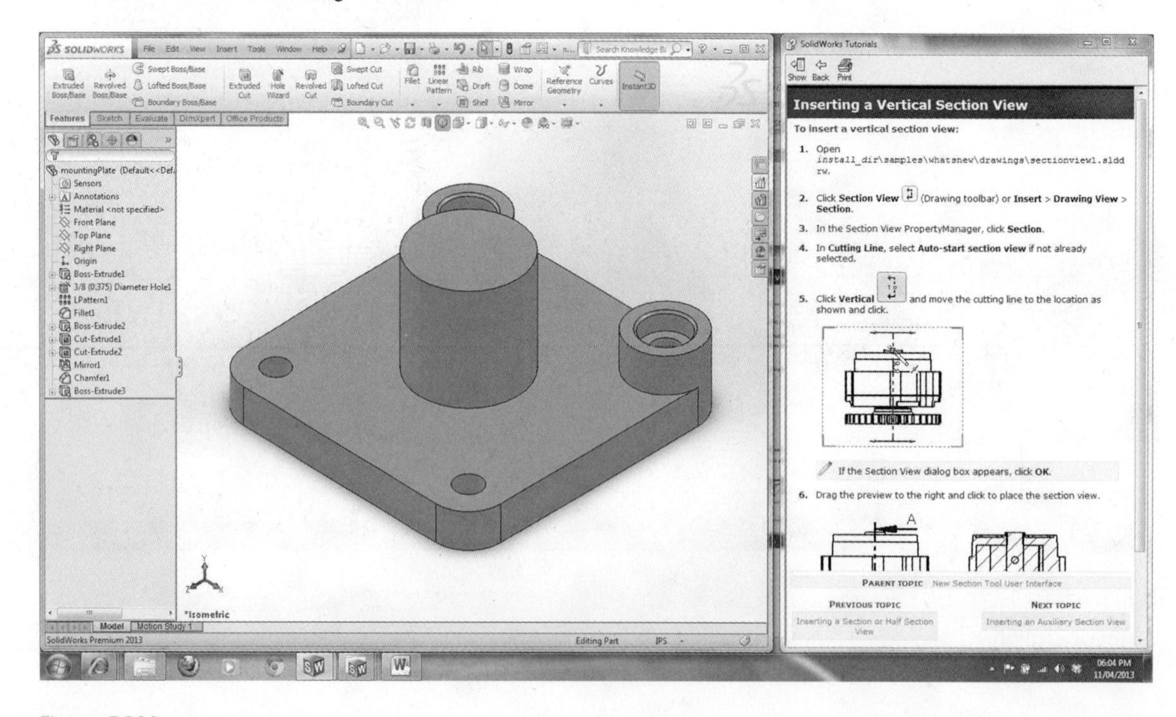

Figure 5.102

Central boss.

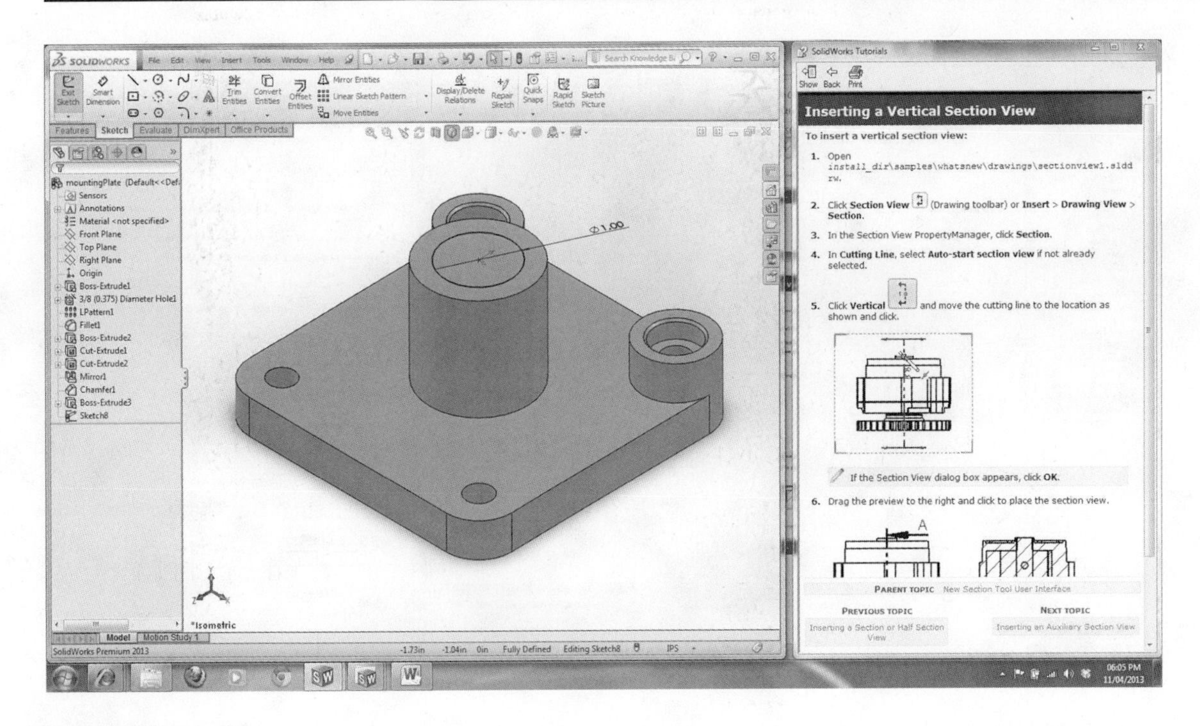

Figure 5.103

Sketch on top of boss.

19. Add Fillet to bottom of central boss (Figure 5.109).
20. Create Plane to coincide with extreme-right vertical face (see Figure 5.110).
21. Be in Sketch mode, and create a circle, 0.25 diameter on the Plane at Normal orientation (see Figure 5.111).
22. Orient model in Isometric view (see Figure 5.112).
23. Extrude Cut to opposite face (see Figure 5.113). (The finished model is shown in Figure 5.114.)

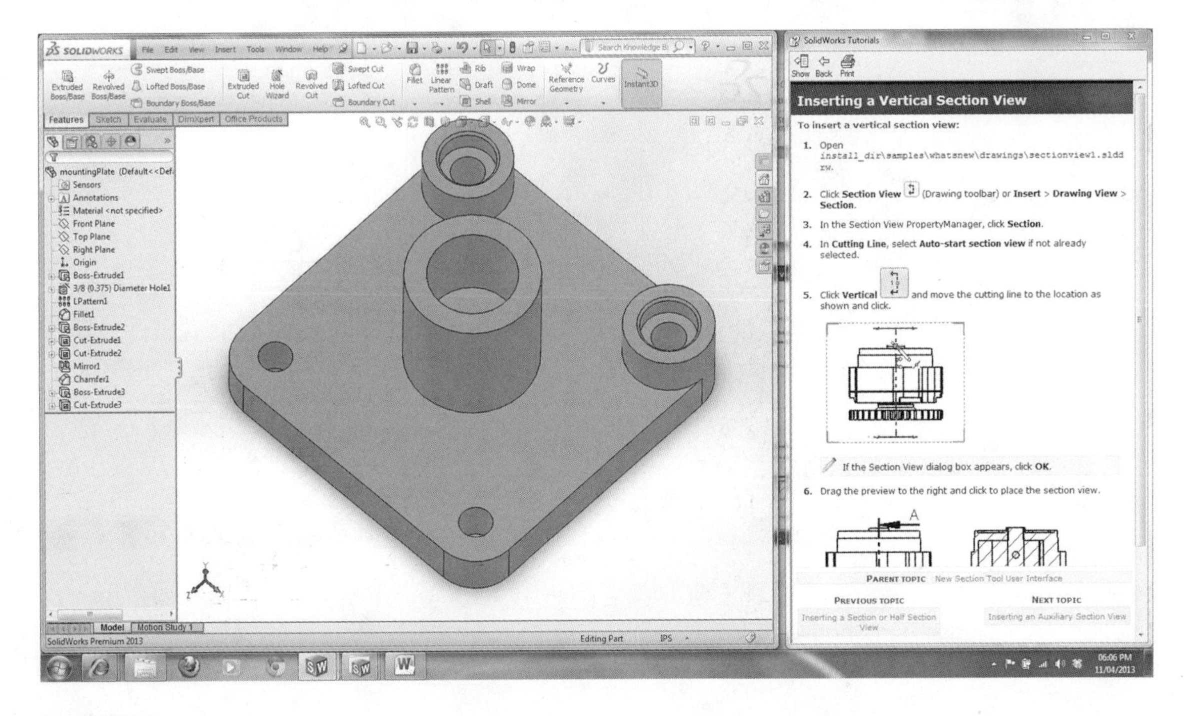

Figure 5.104

Central counterbore.

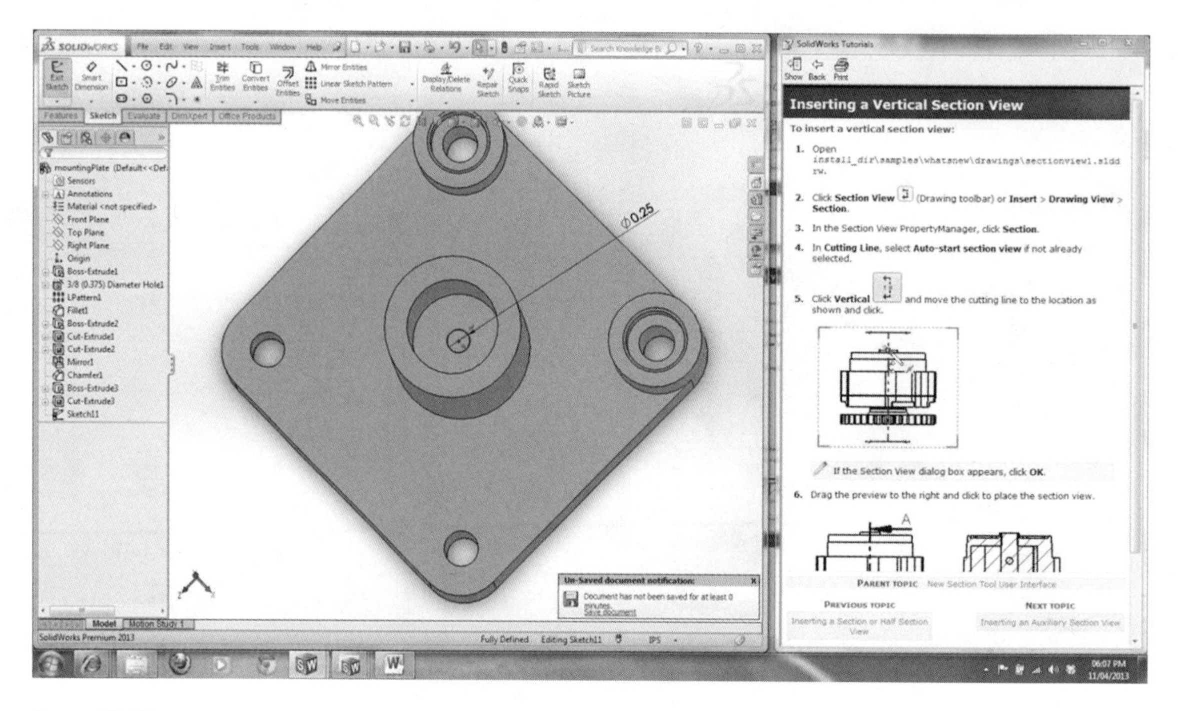

Figure 5.105

Circle for through hole.

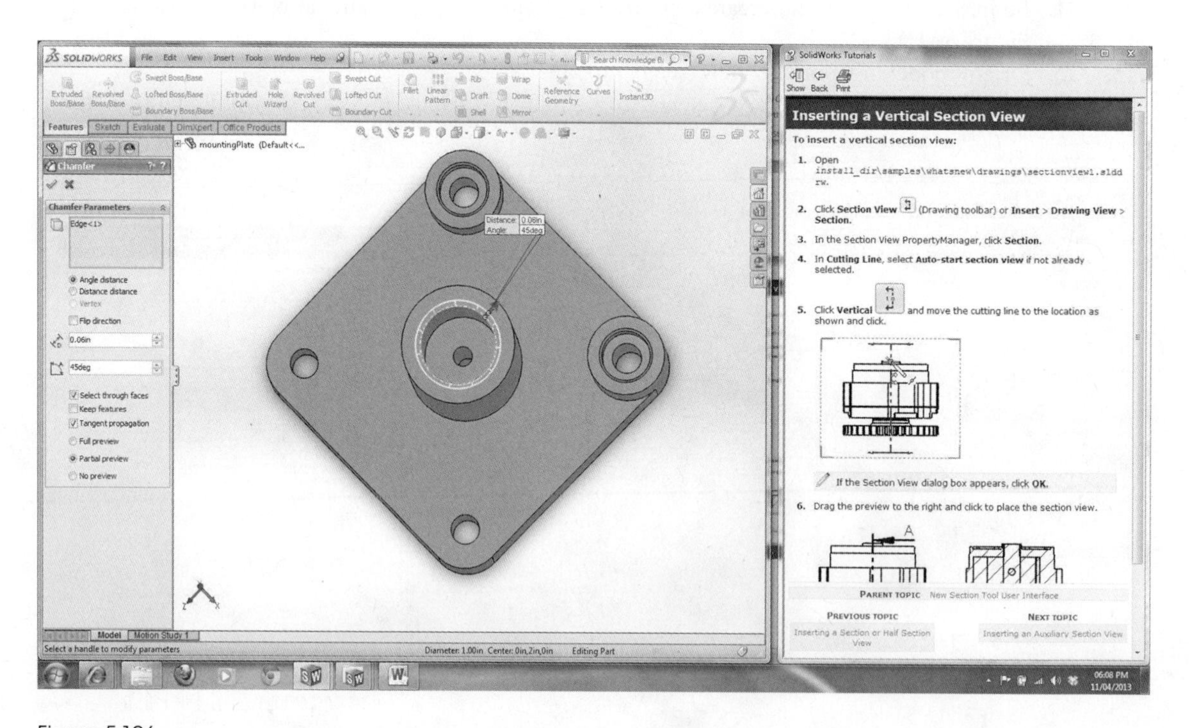

Figure 5.106

Through hole included and Chamfer added.

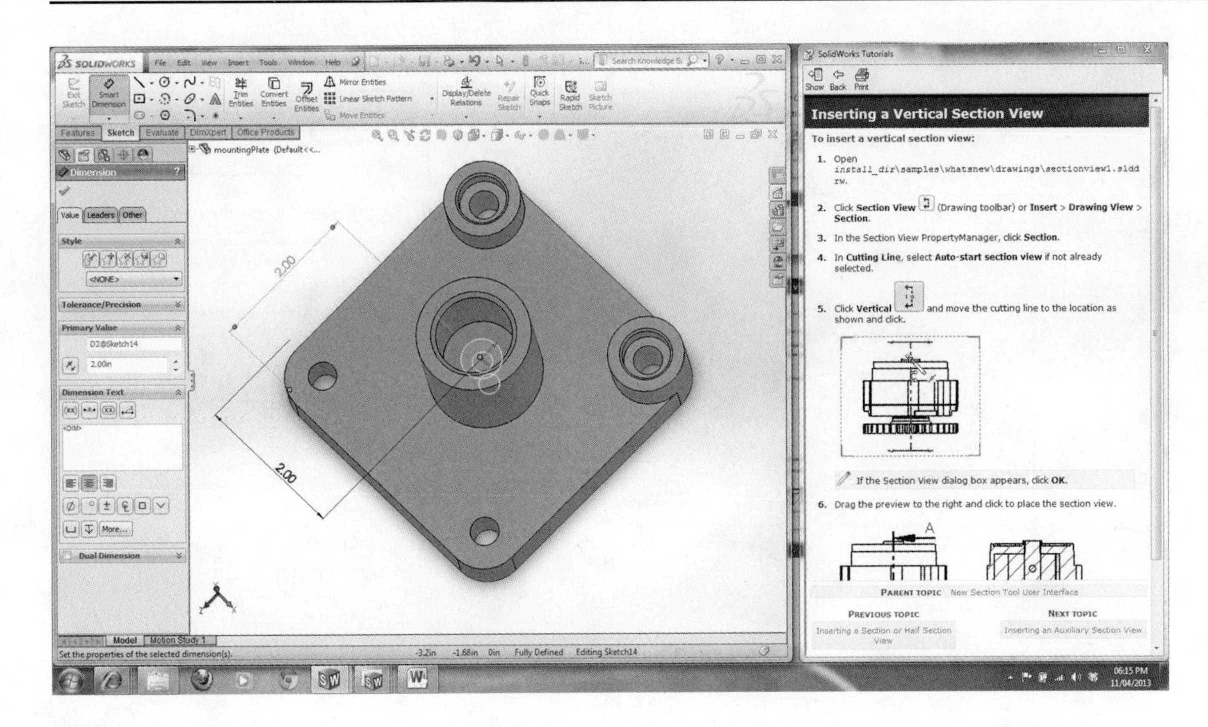

Figure 5.107

Add Chamfer.

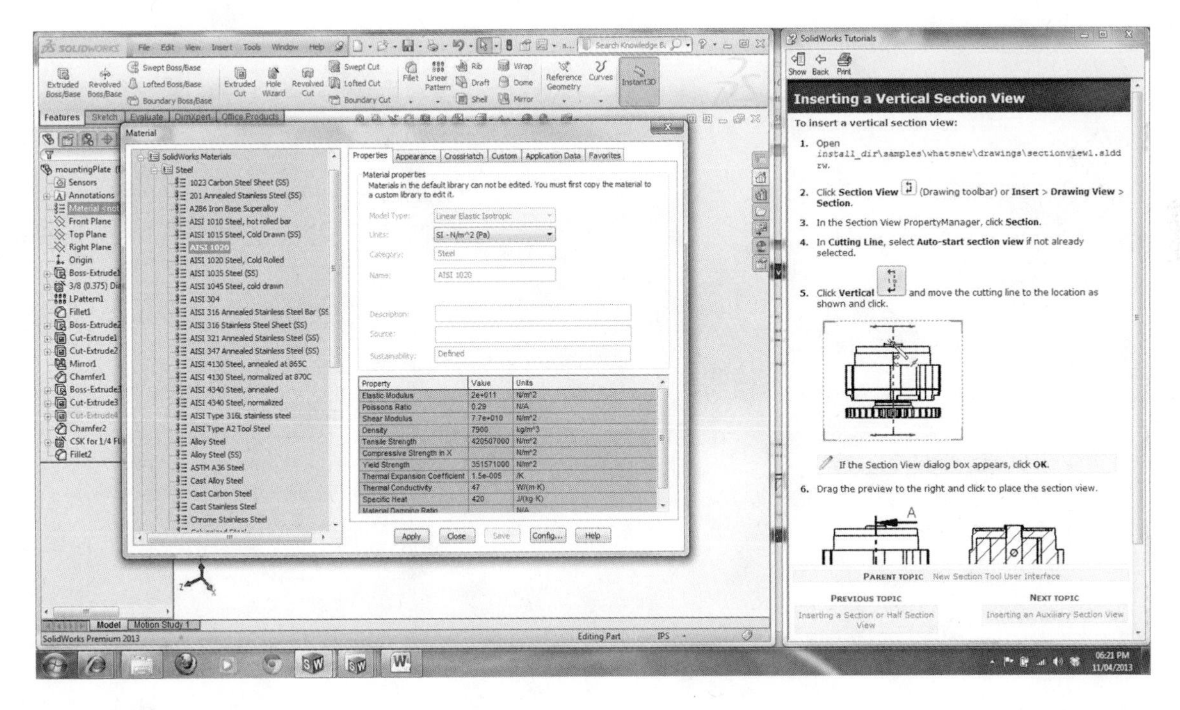

Figure 5.108

Material PropertyManager.

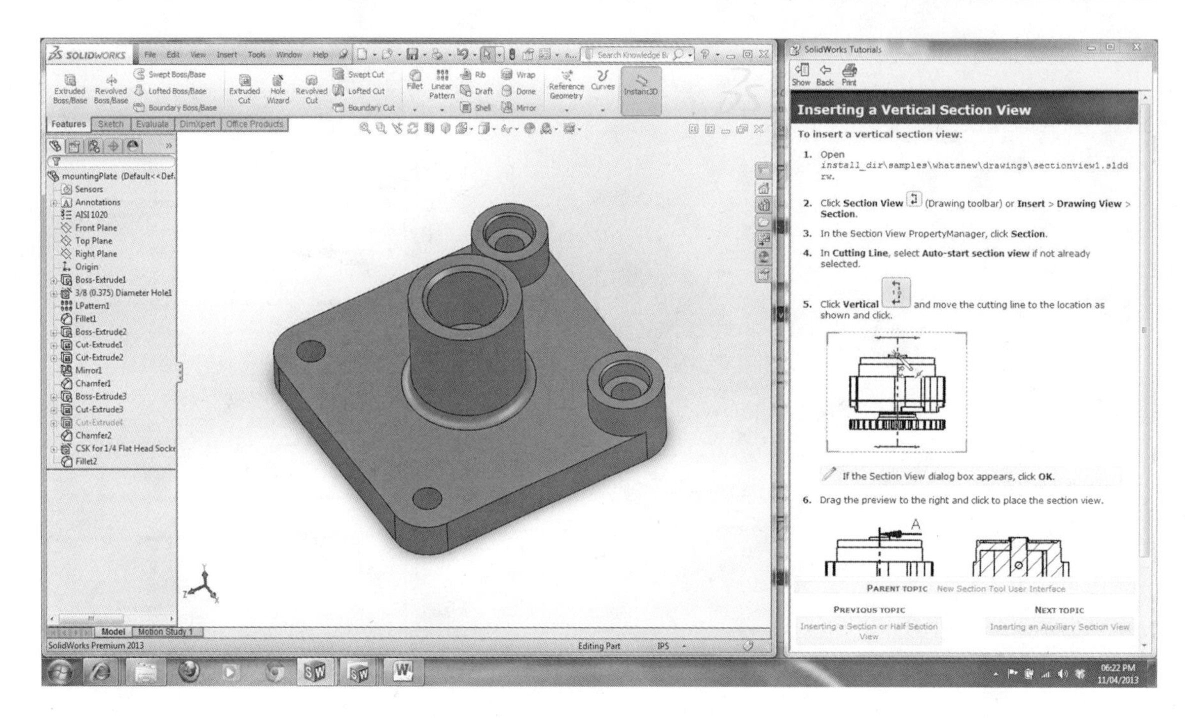

Figure 5.109

Fillet added.

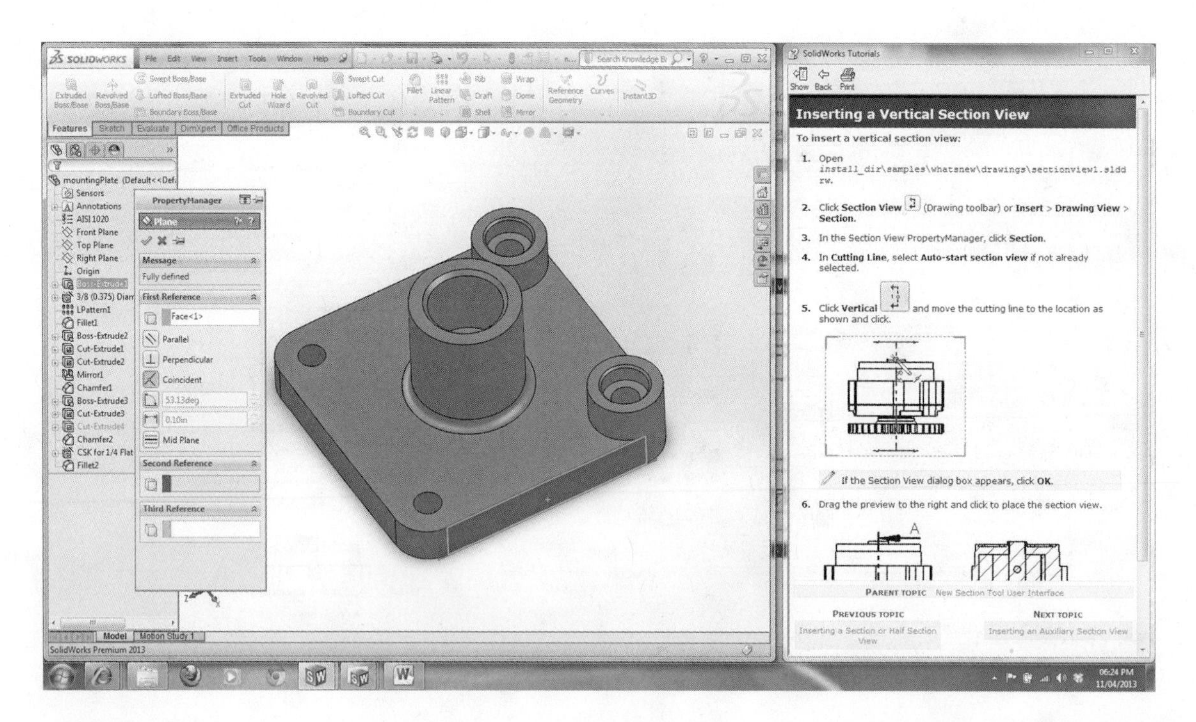

Figure 5.110

Plane1.

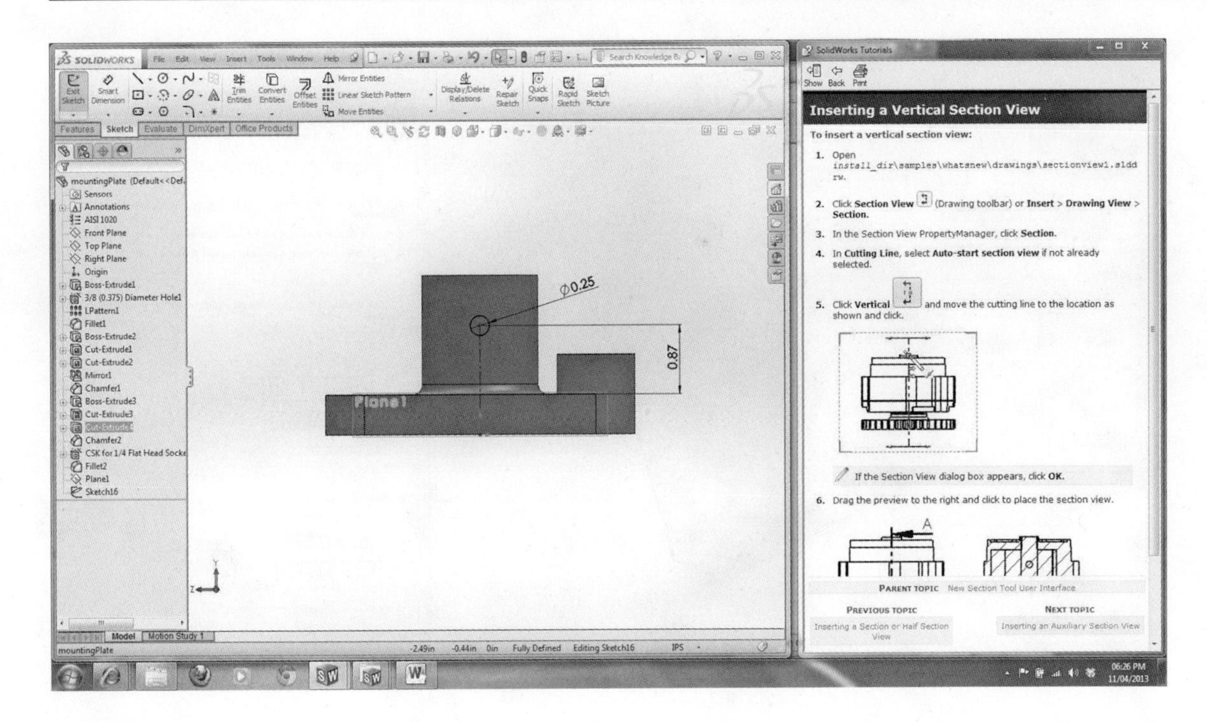

Figure 5.111

Circle on Plane1.

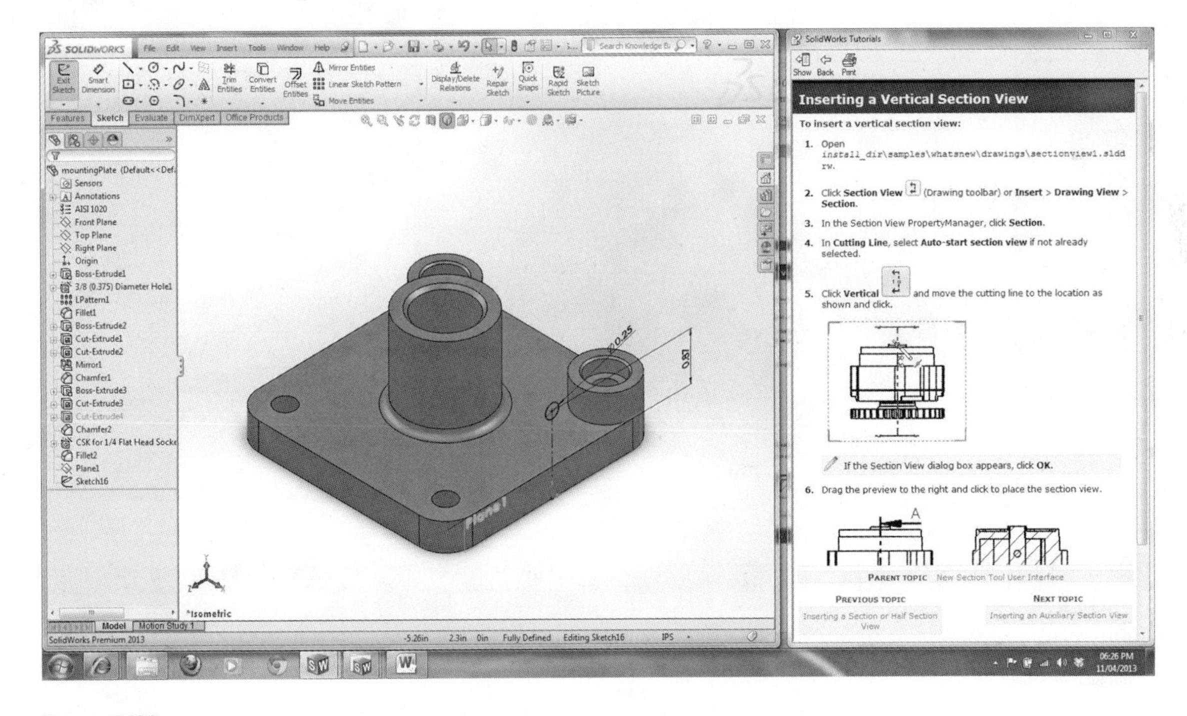

Figure 5.112

Isometric view orientation.

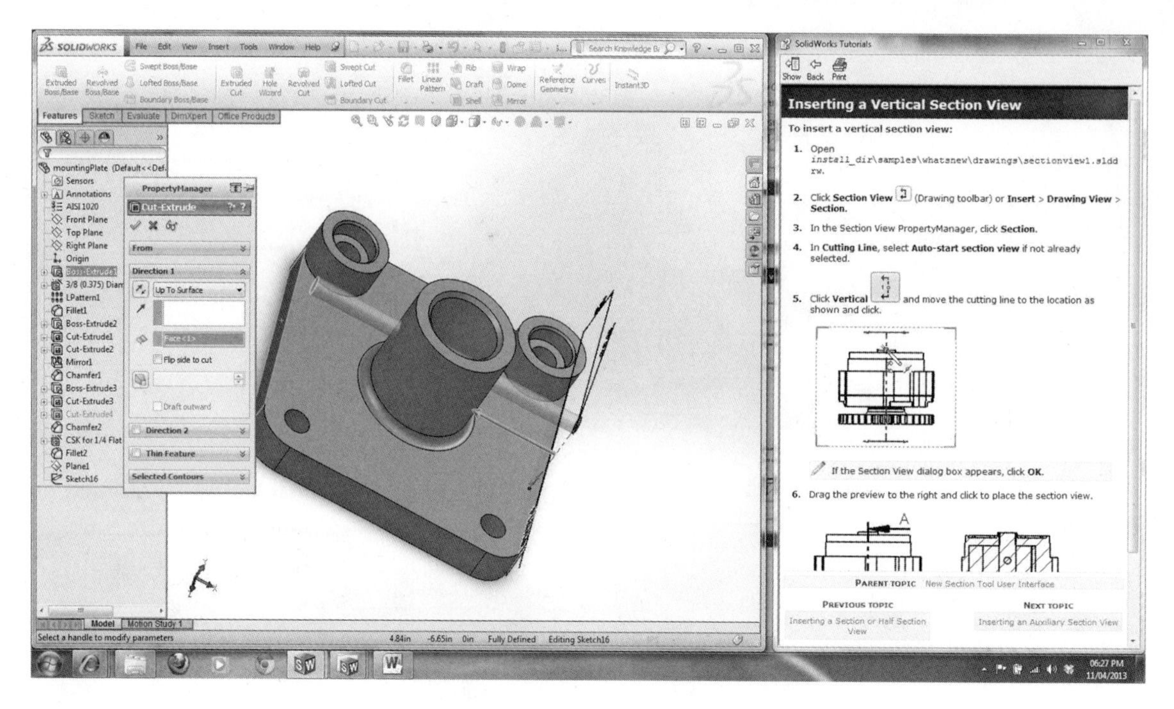

Figure 5.113

Extrude Cut to create hole in central feature.

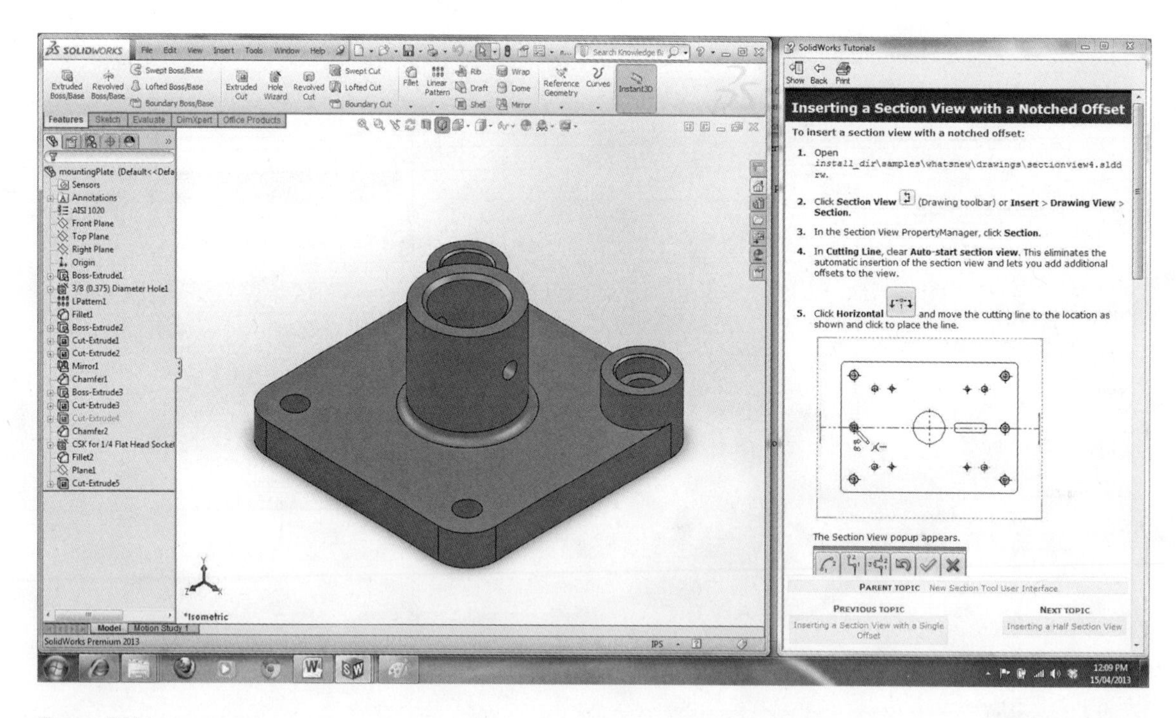

Figure 5.114

Finished model.

Tutorial 5.9: Part Models for Assembly

Create the parts P1 to P8 for the adjustable shaft support shown in Figure 5.115.

> *Material:* As specified per part
> SolidWorks Solution

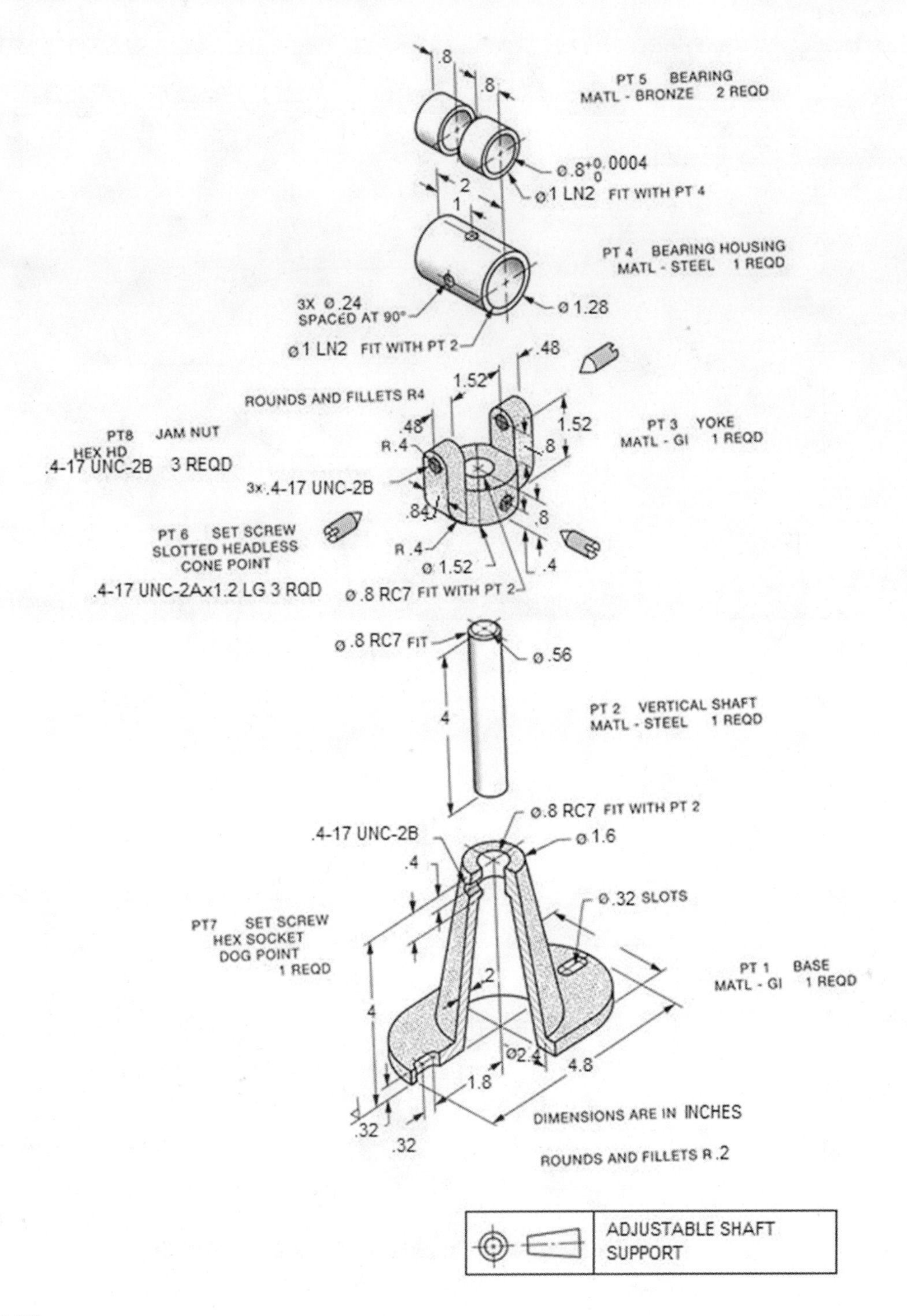

Figure 5.115

Adjustable support shaft details. (From JENSEN/HELSEL/ESPIN. Interpreting Engineering Drawings, 6E. © 2012 Nelson Education Ltd. Reproduced by permission. http://www.cengage.com/permissions.)

PT 1: Base

1. Open a New part in SolidWorks.
2. Set the document properties for the model.
3. Select Top Plane for sketching.
4. Create Sketch1 on the Front Plane (Figure 5.116).
5. Extrude through 8 mid-plane.
6. Select the Front Plane and create Sketch2 that is shown in Figure 5.117.

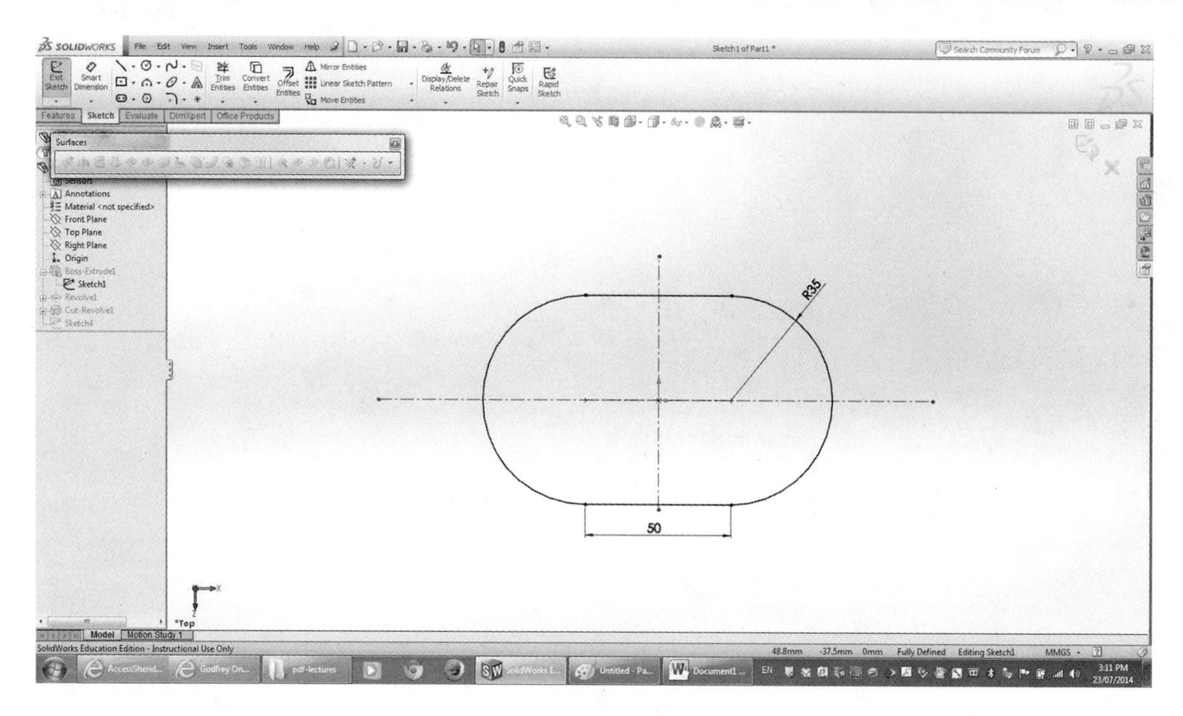

Figure 5.116

Sketch1 on Top Plane.

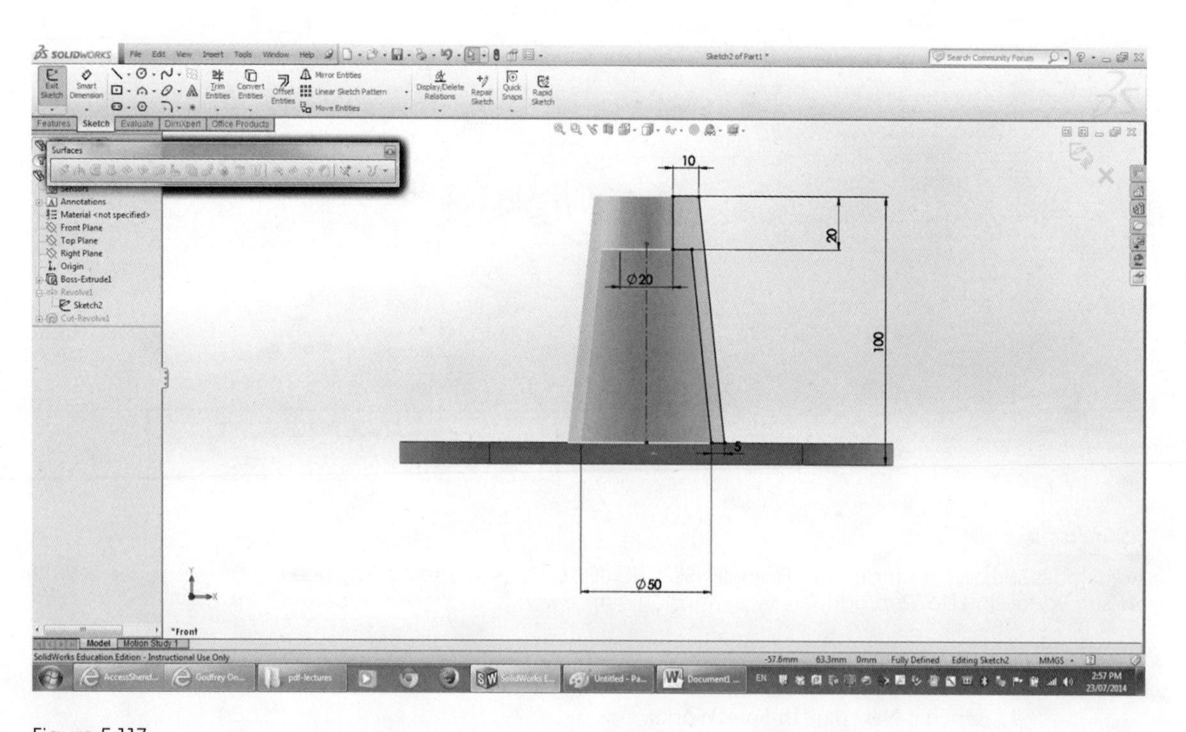

Figure 5.117

Sketch2 in Front Plane.

7. Revolve about the Centerline.
8. Select the Front Plane and create Sketch3 that is shown in Figure 5.118.
9. Revolve-Cut about the Centerline.
10. Select the top of the Base and create Sketch4 that is shown in Figure 5.119.
11. Extrude-Cut through Up To Next Surface.
12. Mirror Extrude-Cut1 about Right Plane (see Figure 5.120 for incomplete part).

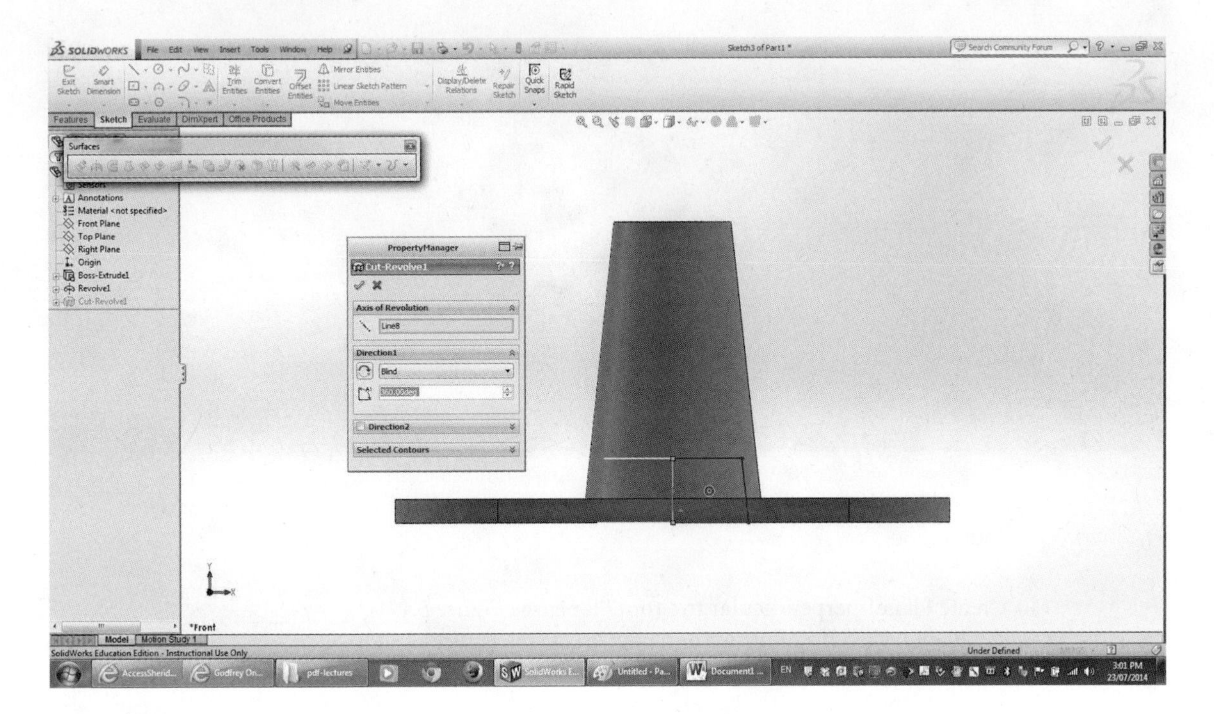

Figure 5.118

Sketch3 Cut-Revolved about centerline.

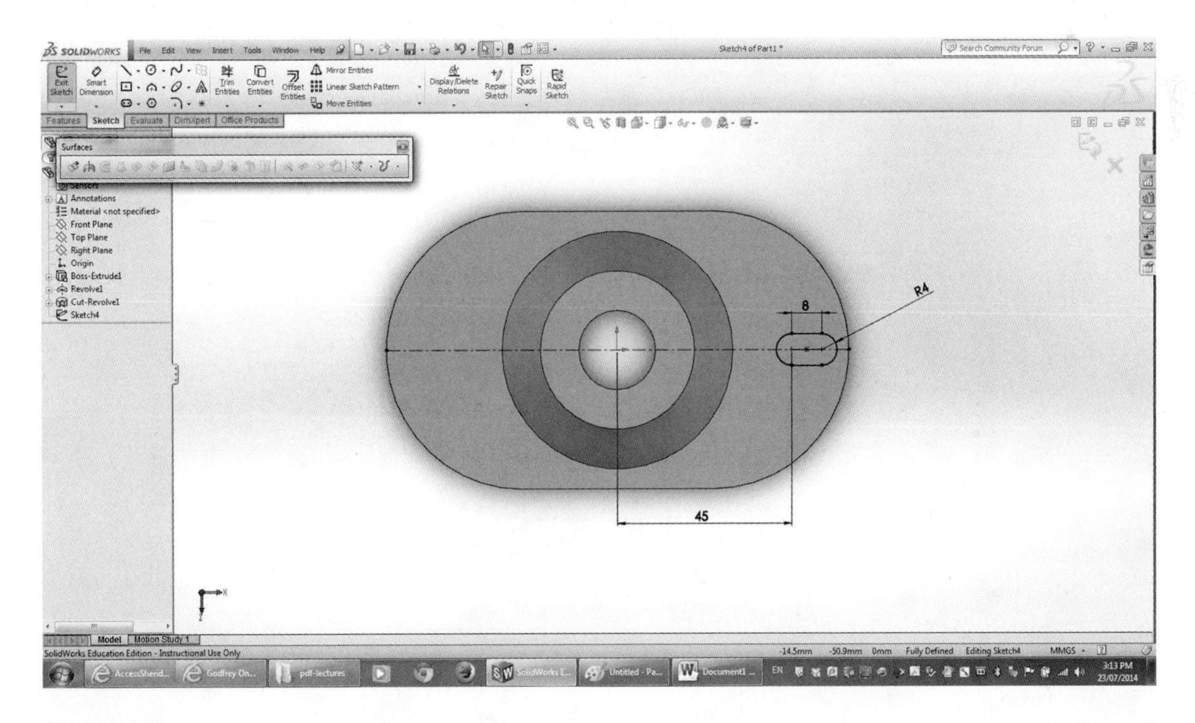

Figure 5.119

Sketch4 on top of the base used for extrude-cut operation.

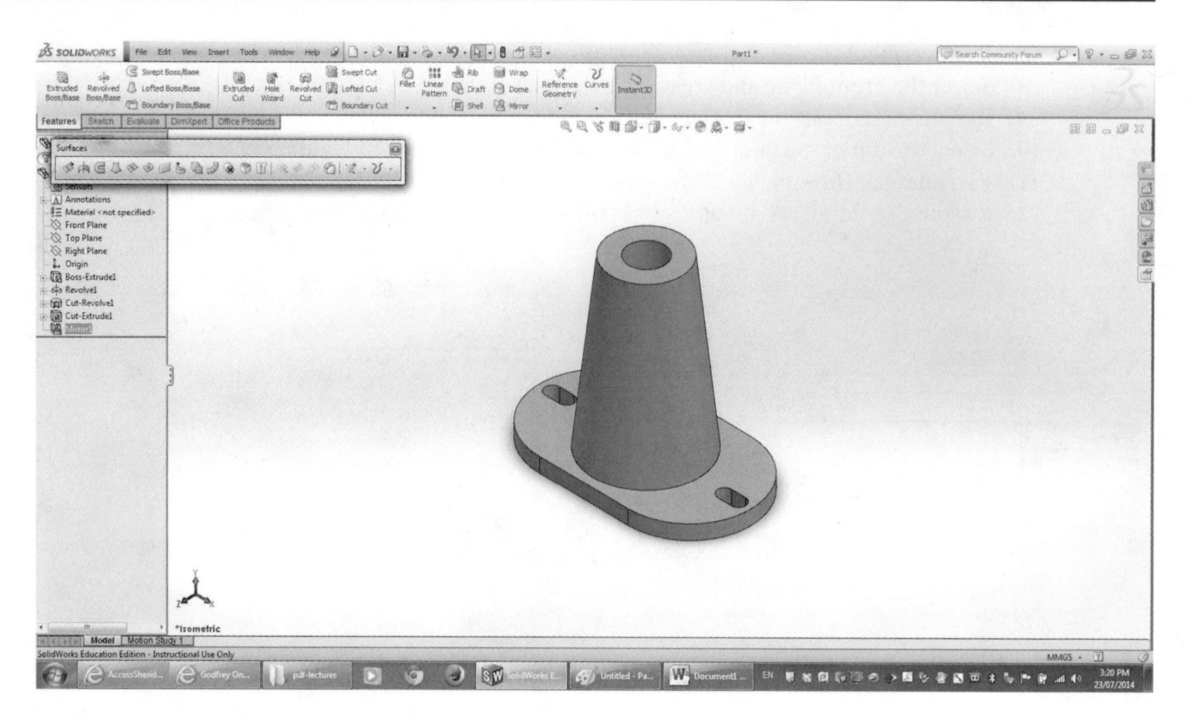

Figure 5.120

PT 1 incomplete part model.

13. Create Plane1 perpendicular to Front Plane (see Figure 5.121).
14. On Plane1, sketch a circle with diameter 0.4 (see Figure 5.122).
15. Extrude-Cut through the thickness of the incomplete part.
16. Create Plane2 similar to Plane1.
17. Extract the 0.4-diameter Circle onto Plane2.
18. Click Features > Curves > Helix and Spiral.
19. Enter the Helix parameters (see Figure 5.123).

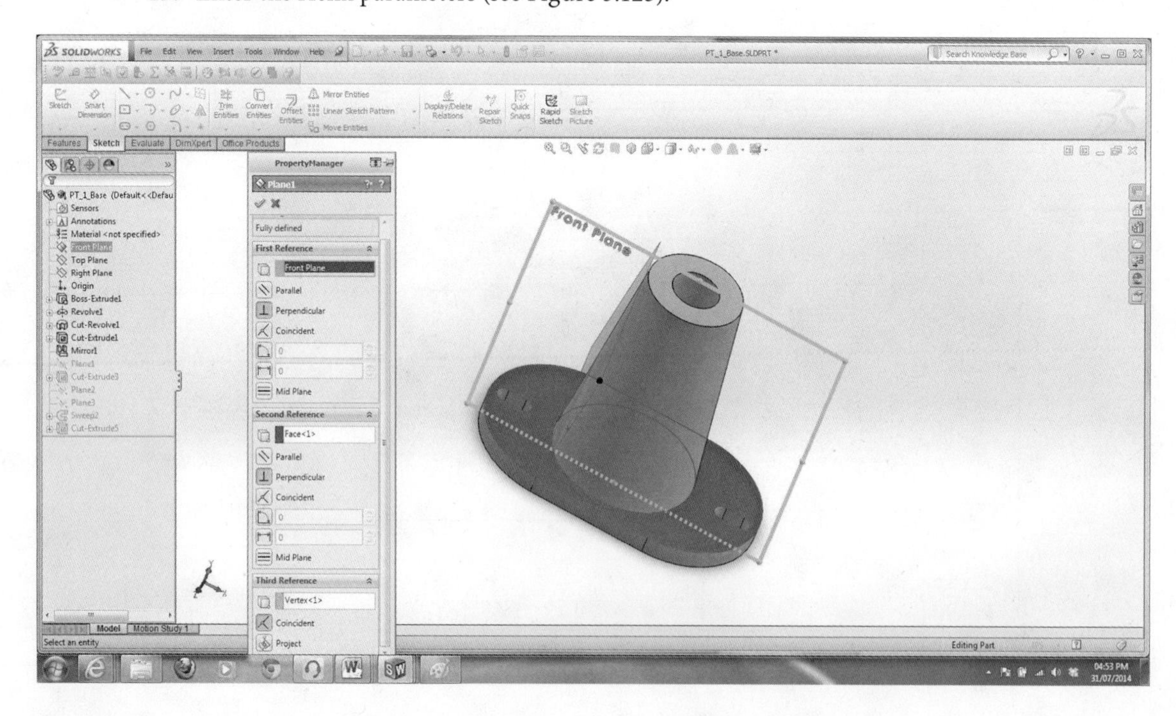

Figure 5.121

Plane1 for creating hole on slanted face.

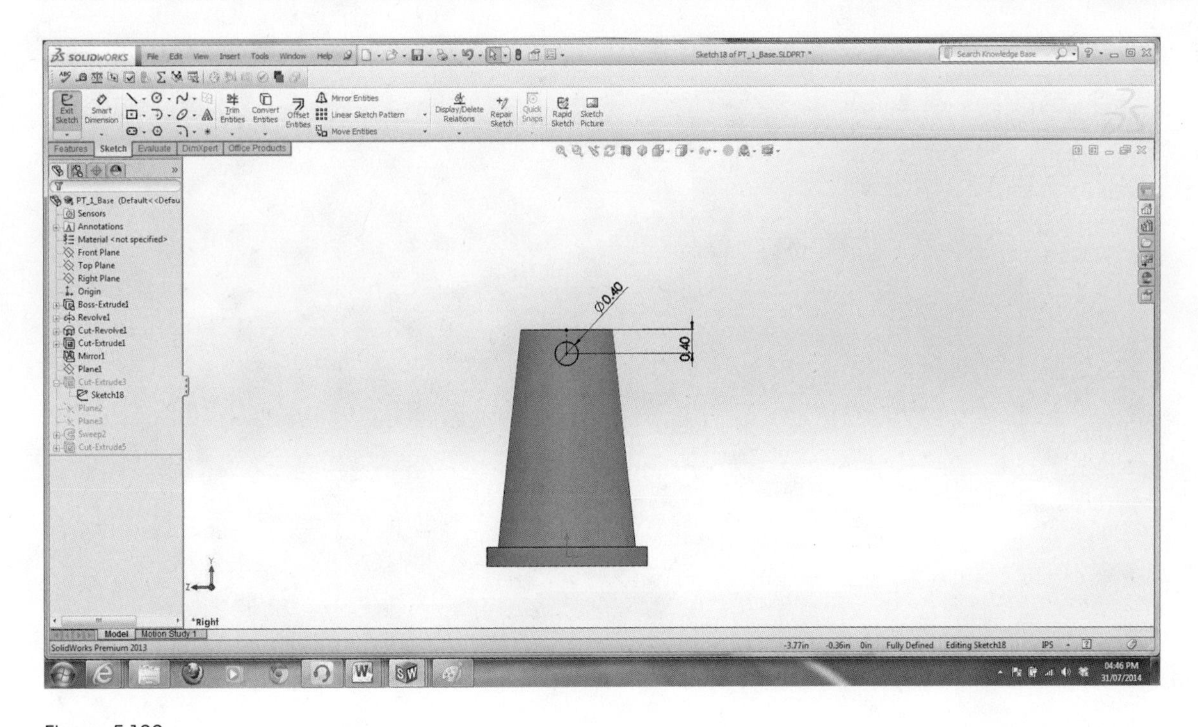

Figure 5.122

Sketch for circle.

20. Create Plane3, and using the Relations tool, align Plane3 to be perpendicular to the endpoint of the Helix.
21. Choose Plane3 while in Sketch mode and sketch the thread Profile (see Figure 5.124).
22. Use Relations tool to Pierce the Profile to the Helix.
23. Click Sweep and carry out the sweep operation (see Figure 5.125).
24. Apply Gray Cast Iron. (See the completed part model in Figure 5.126.)

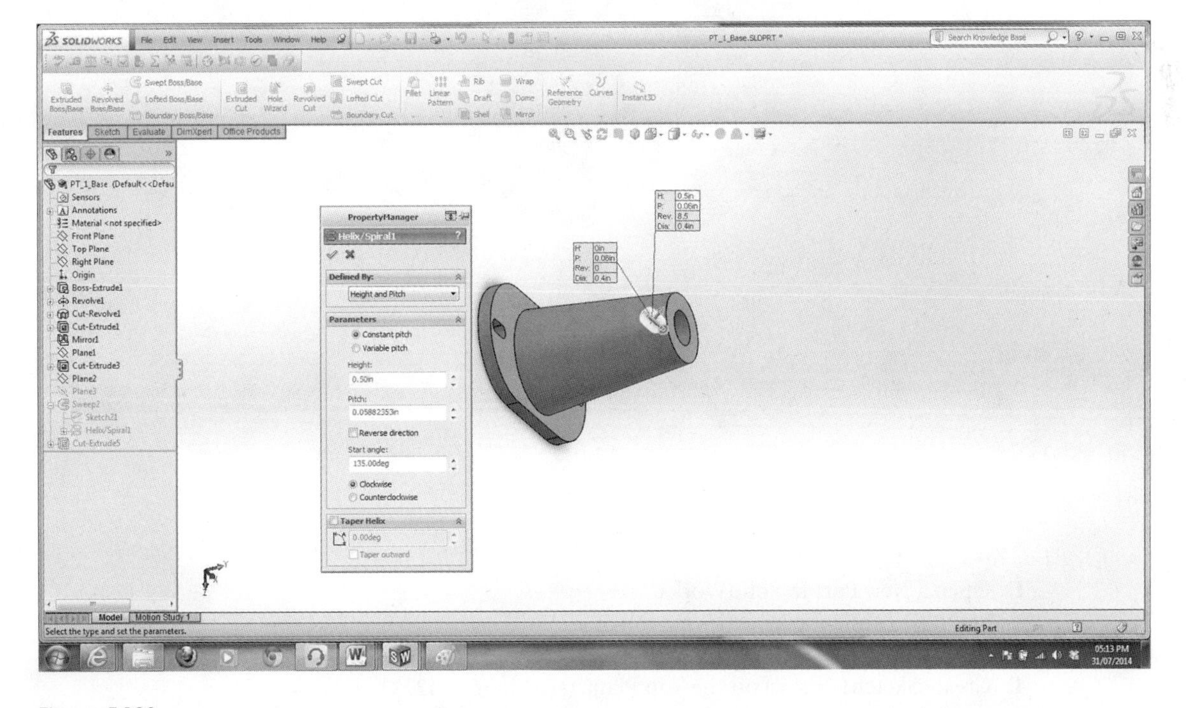

Figure 5.123

Helix parameters.

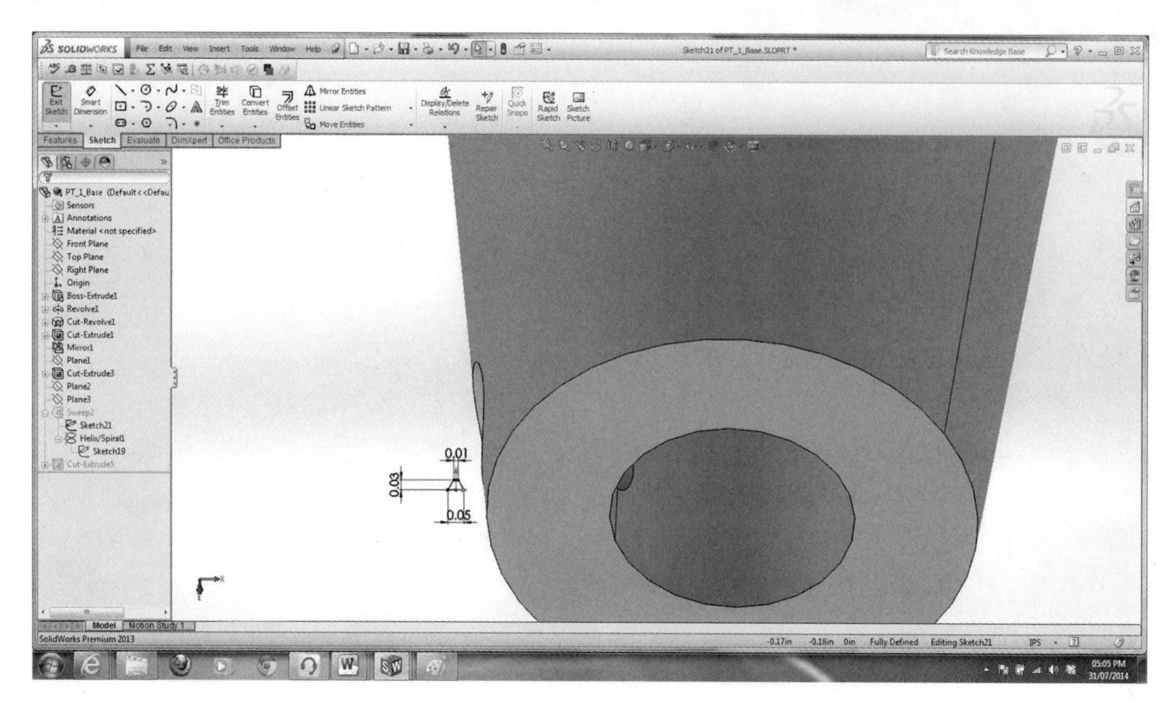

Figure 5.124

Thread profile.

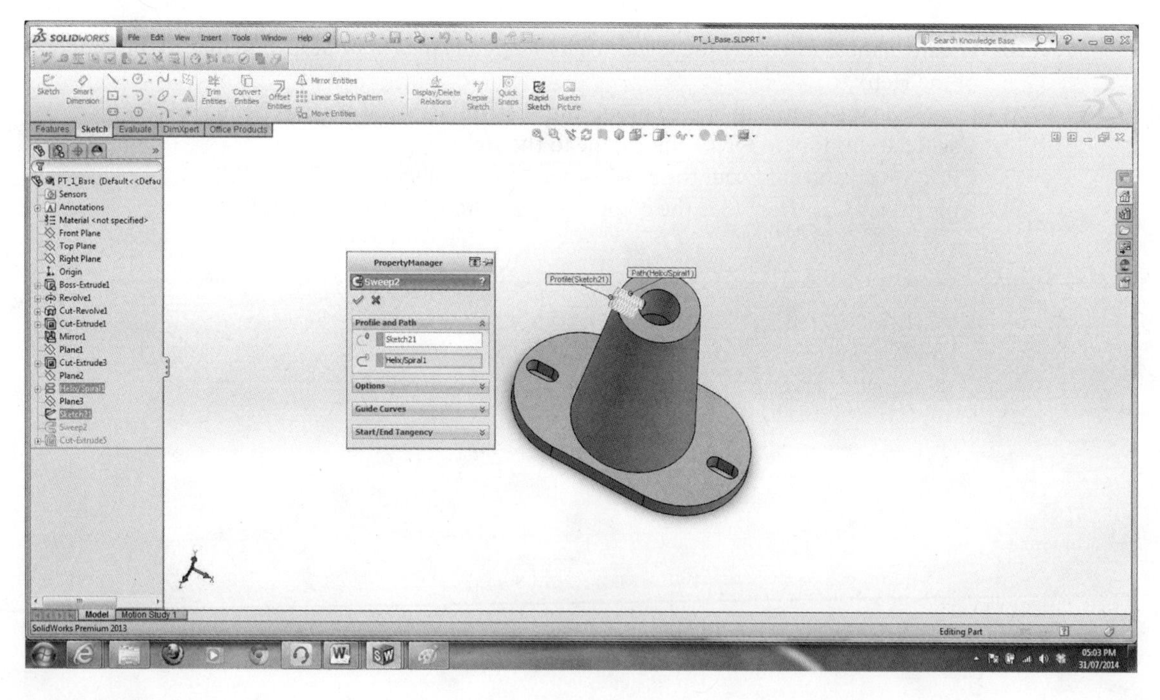

Figure 5.125

Sweep operation.

PT 3: Yoke

1. Open a New Part in SolidWorks.
2. Set the document properties for the model.
3. Select Top Plane for sketching.
4. Create Sketch1 (a slot) on the Top Plane (see Figure 5.127).
5. Extrude through 0.8 mid-plane (see Figure 5.128).
6. Select the top of Extrude1 and sketch a Circle with 0.8 diameter (see Figure 5.129).

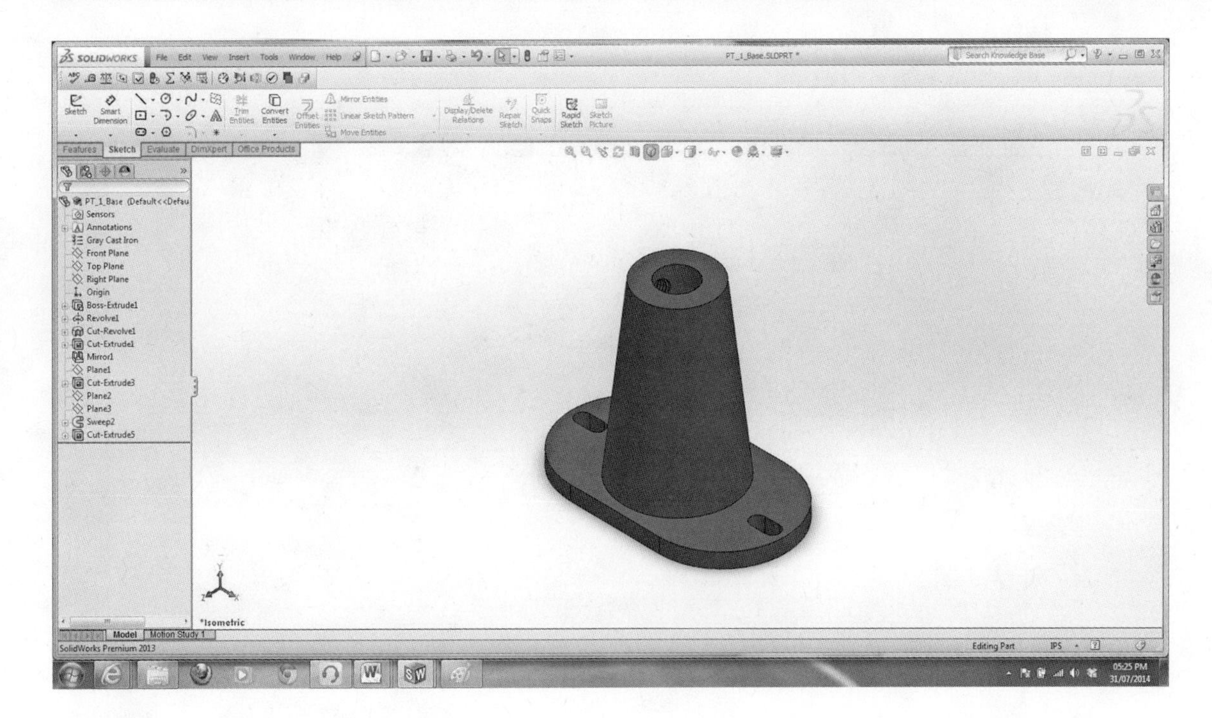

Figure 5.126

Completed part model, PT 1-Base.

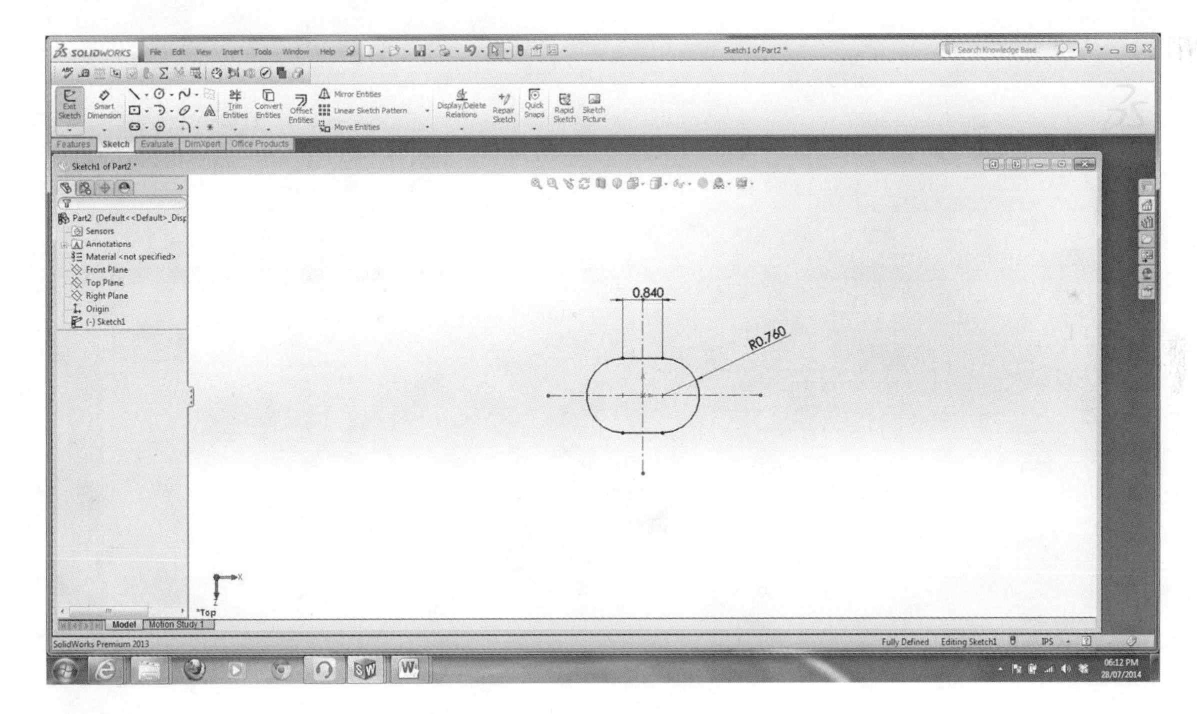

Figure 5.127

Sketch1, a slot for yoke model.

7. Extrude-Cut Up To Next.
8. Create side features by adding the Sketch that is shown in Figure 5.130 on one side of the Slot.
9. Extrude 0.8 outward.
10. Fillet the bottom using the Face Fillet option (see Figure 5.131).
11. Create side Hole, using a Circle of diameter 0.4 (see Figures 5.132 and 5.133).
12. Select the extreme left Face of the feature, be in sketch mode and extract the Circle for base circle for the helix, using Convert Entities tool (see Figure 5.134).

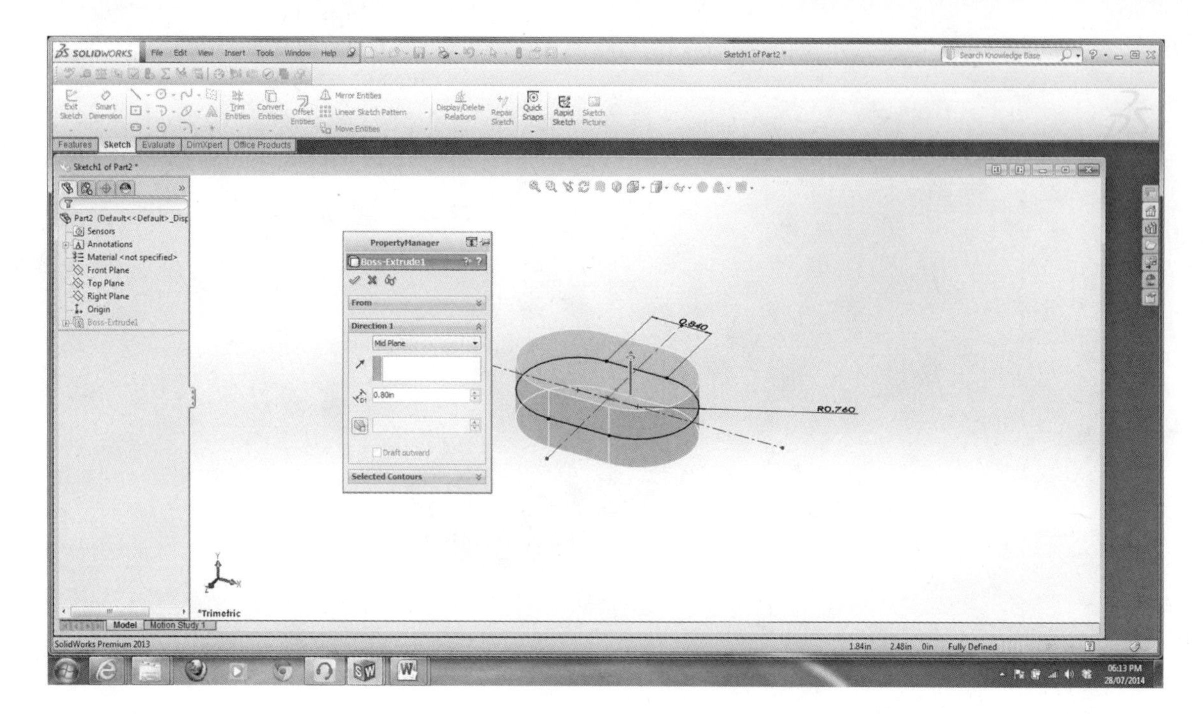

Figure 5.128

Extrusion.

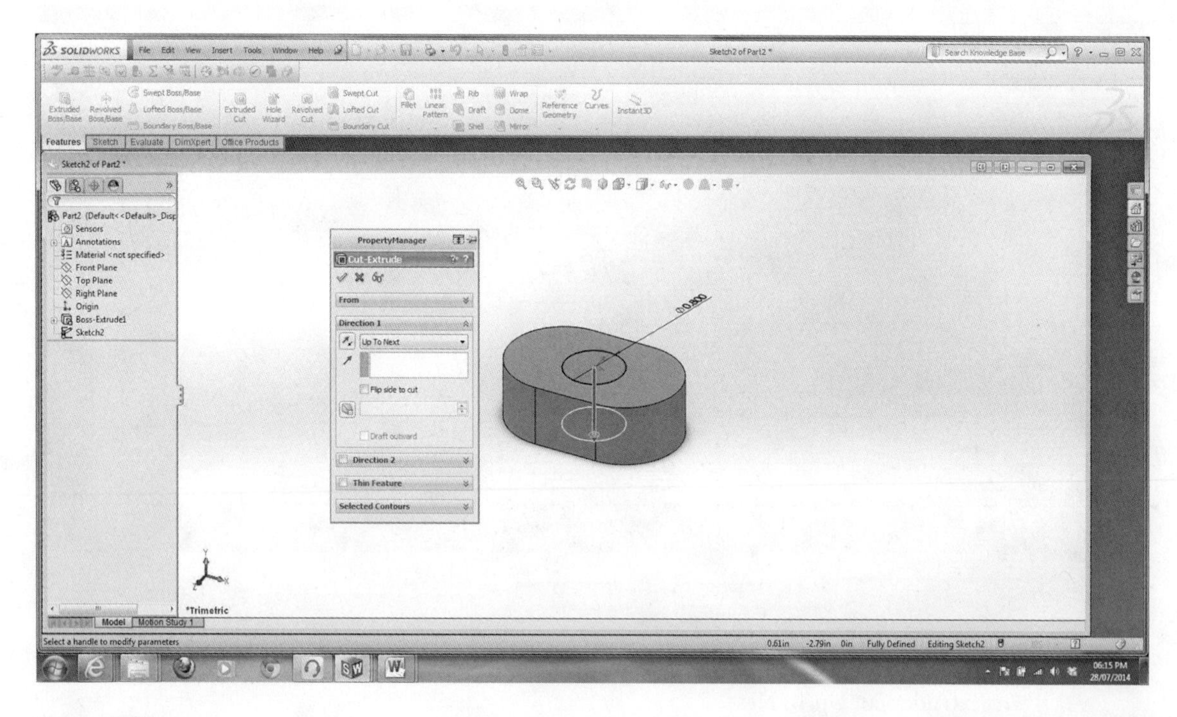

Figure 5.129

Extrude-Cut for circular feature.

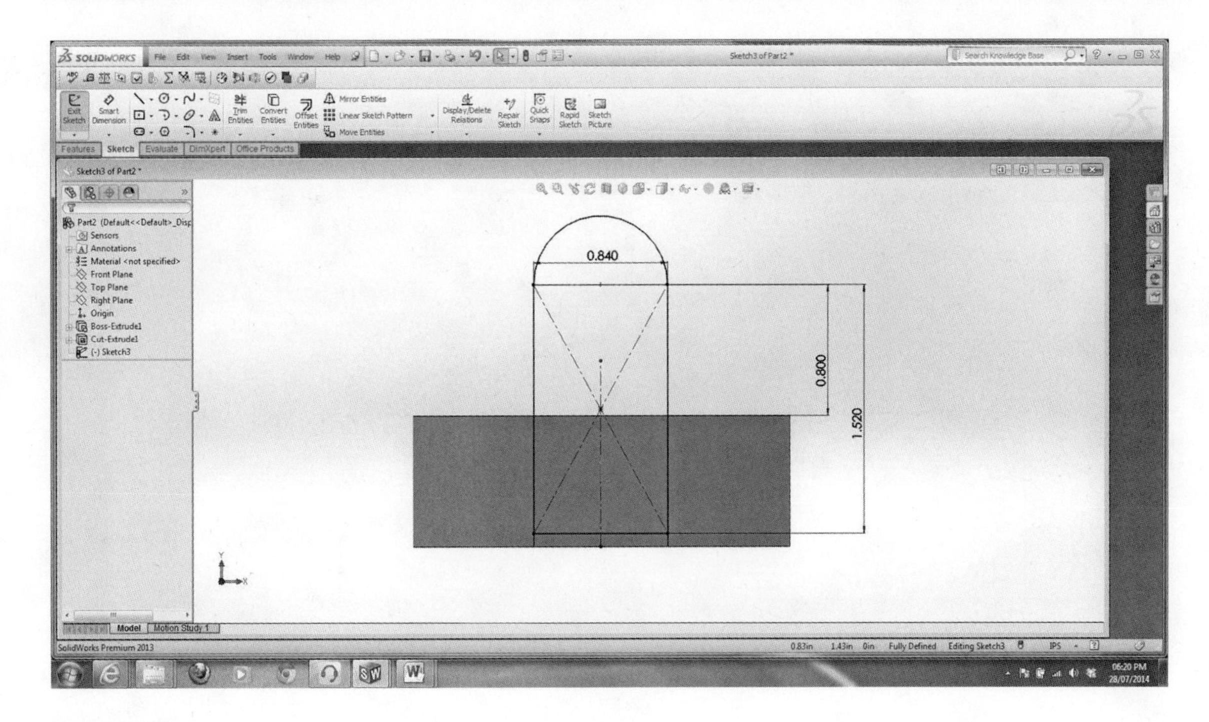

Figure 5.130

Sketch for side feature.

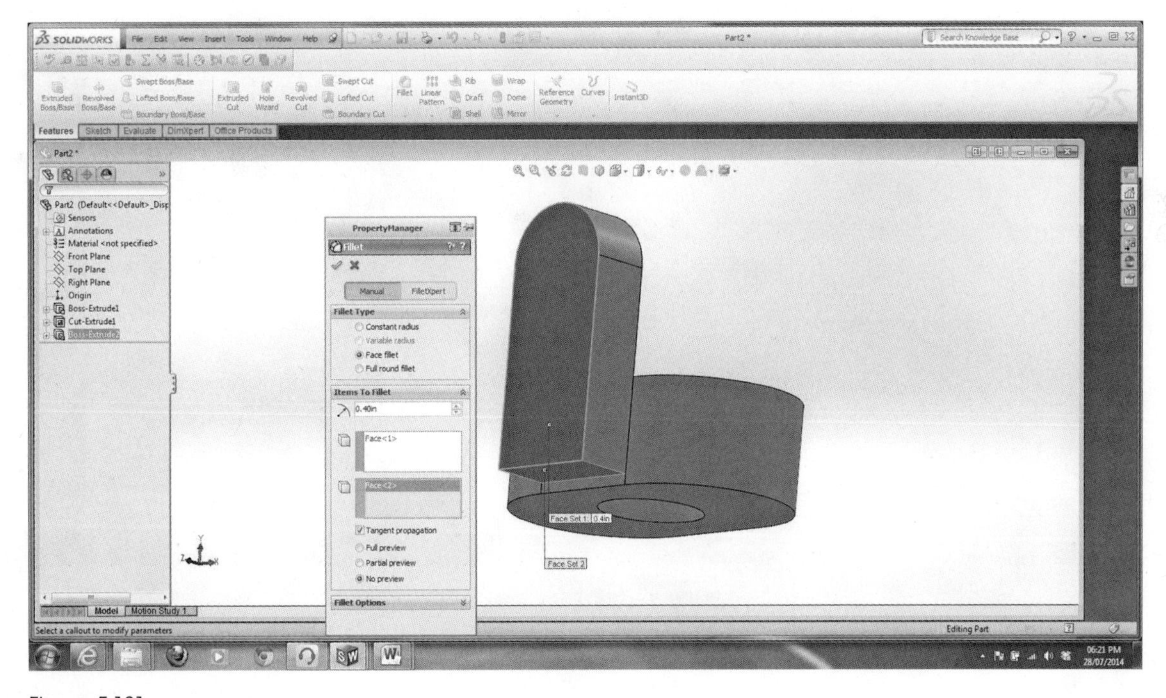

Figure 5.131

Face fillet.

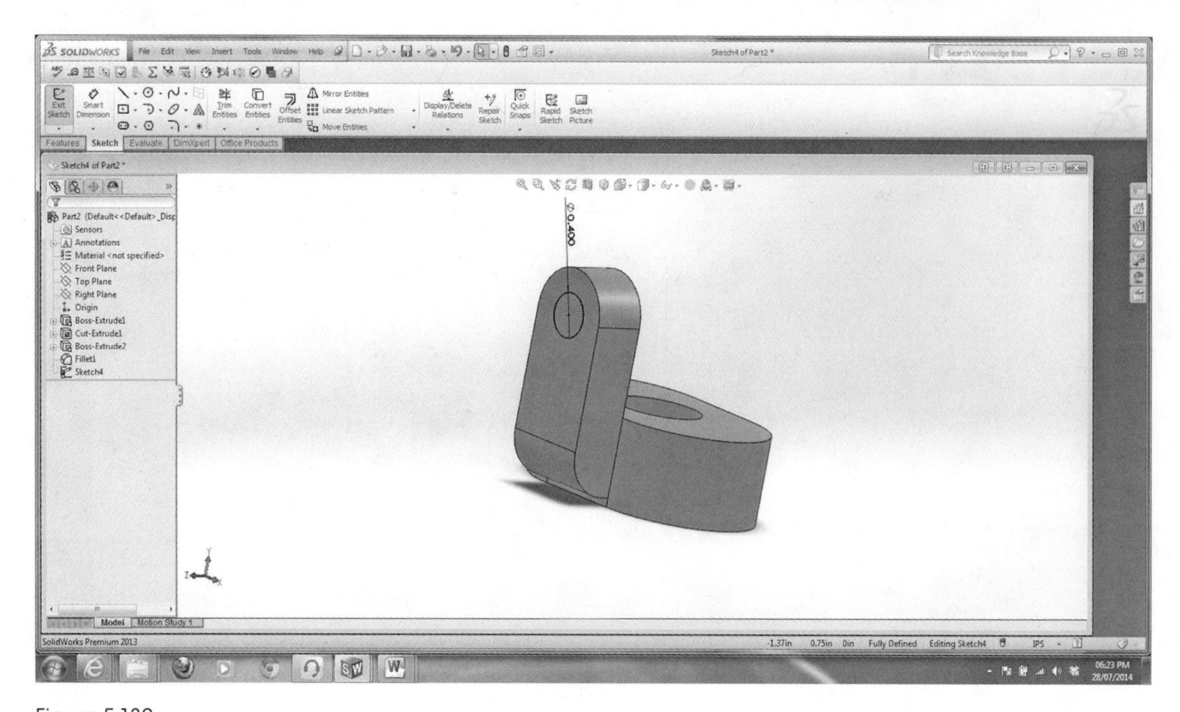

Figure 5.132

Circle for side hole.

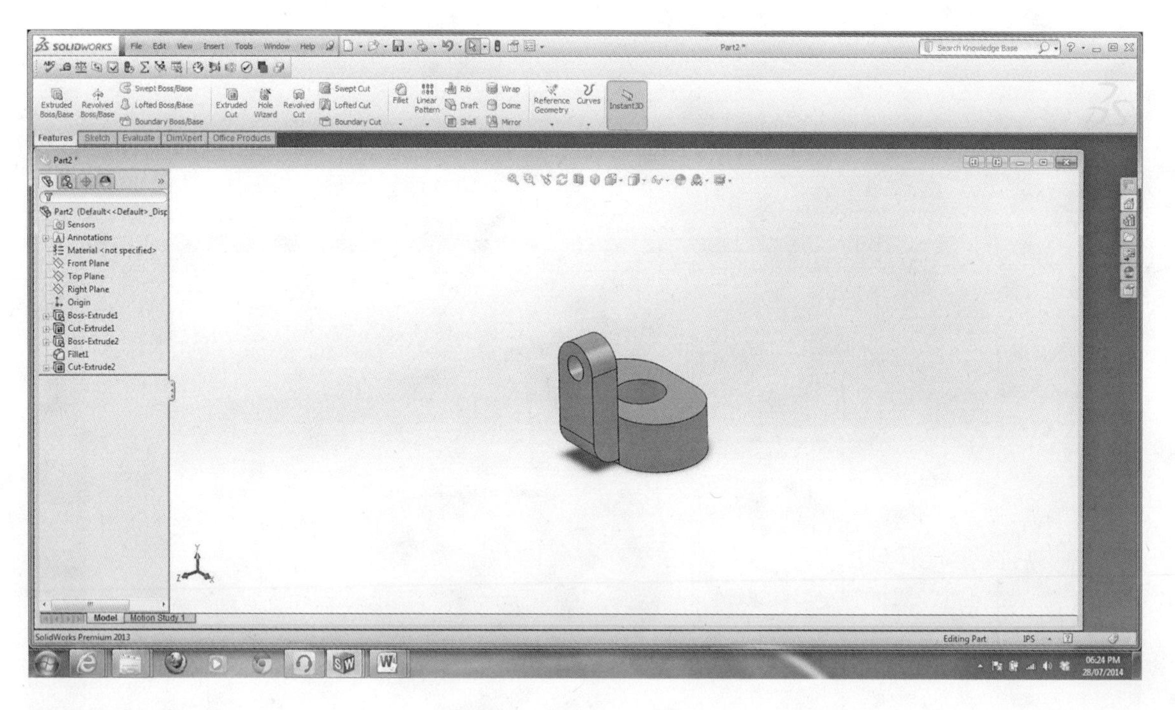

Figure 5.133

Hole created.

13. Proceed as in yoke to create an internal threaded Hole (see Figures 5.135 through 5.141).
14. Click Features > Mirror.
15. Select entities for the left feature as Features to Mirror (see Figures 5.142 and 5.143).
16. Create Plane2 (see Figure 5.144).
17. Create a circle as a base circle for the third internal thread (see Figure 5.145). Figure 5.146 shows the completed model.

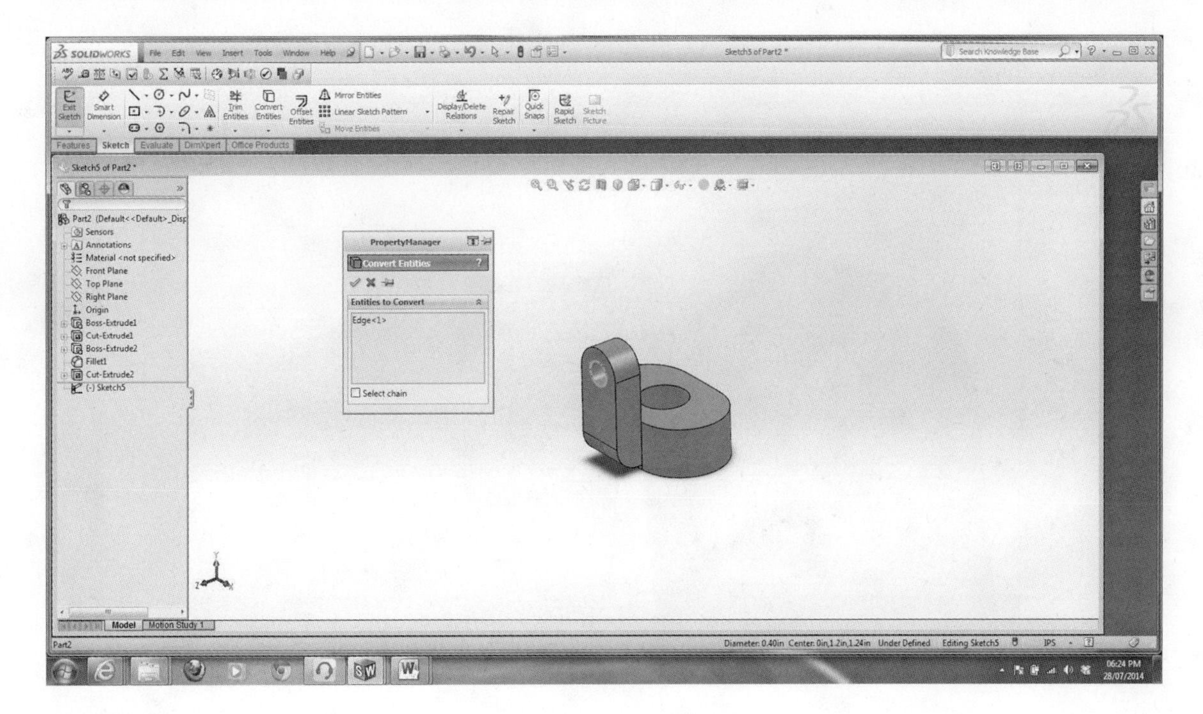

Figure 5.134

Base circle for helix.

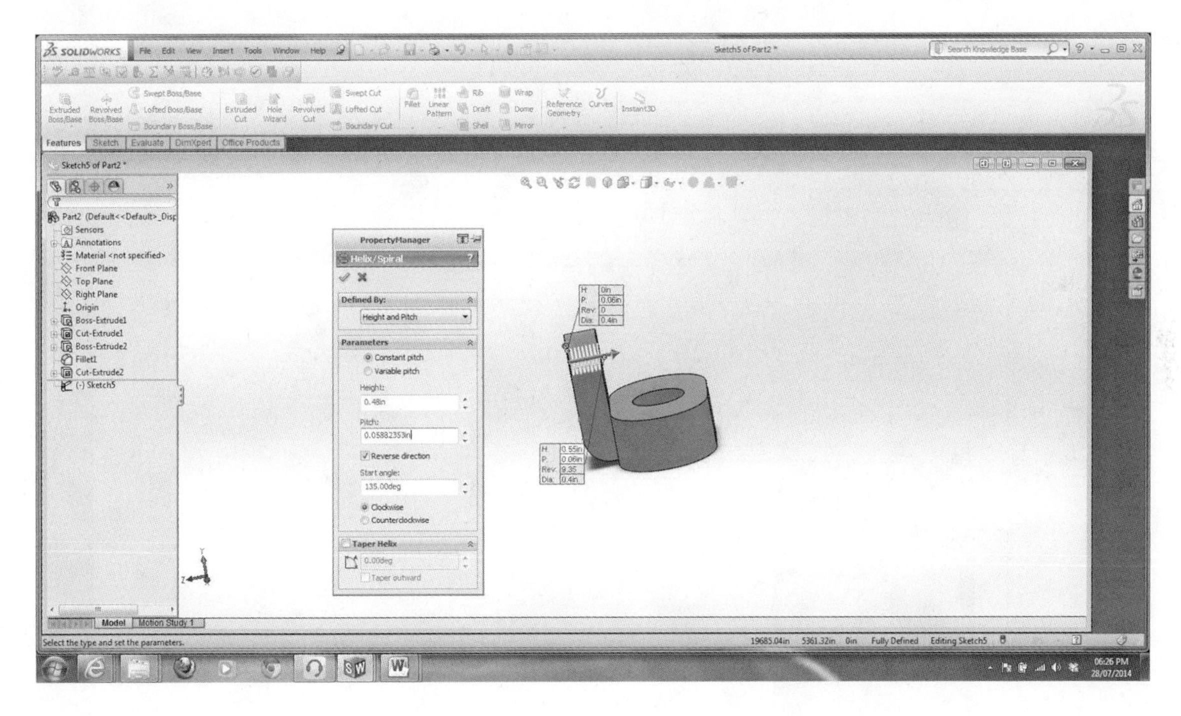

Figure 5.135

Helix parameters defined.

PT 6: Set Screw Slotted Headless Cone Point

1. Open a New Part in SolidWorks.
2. Set the document properties for the model.
3. Select Front Plane for sketching.
4. Create Sketch1 (a slot) on the Front Plane (see Figure 5.147).
5. Revolve about the vertical axis (see Figure 5.148).
6. Create a base circle for helix (see Figures 5.149 and 5.150 for the helix parameter settings).

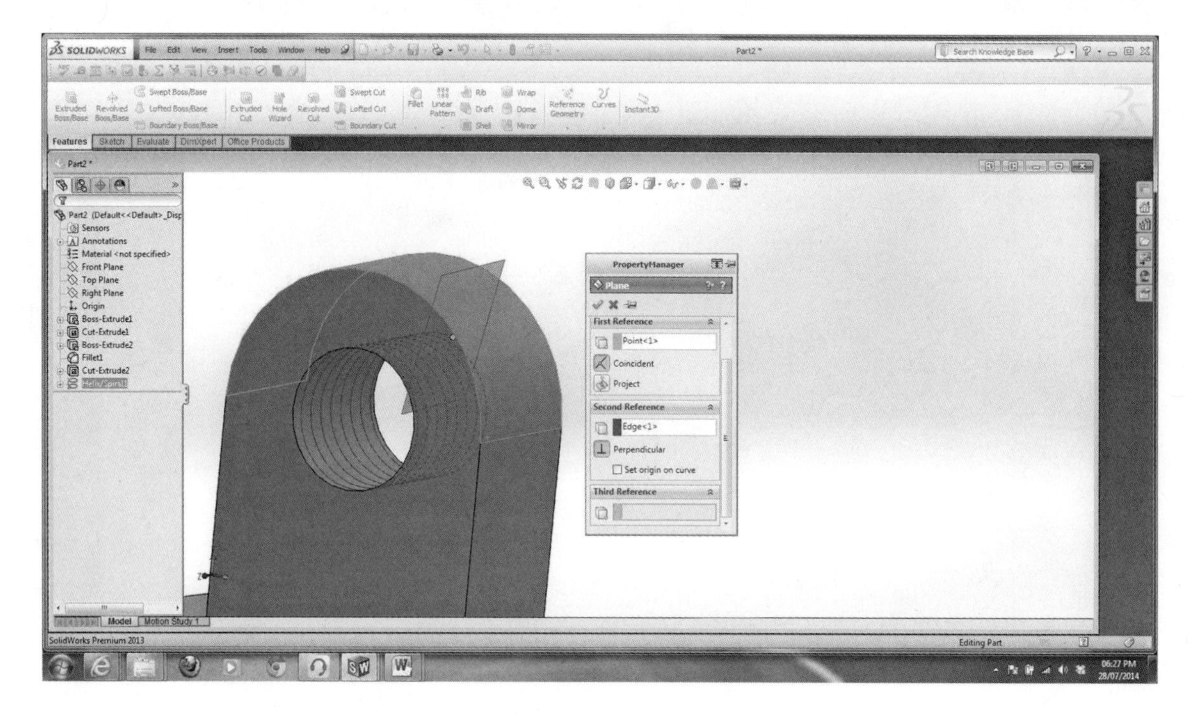

Figure 5.136

Orthogonal plane to helix.

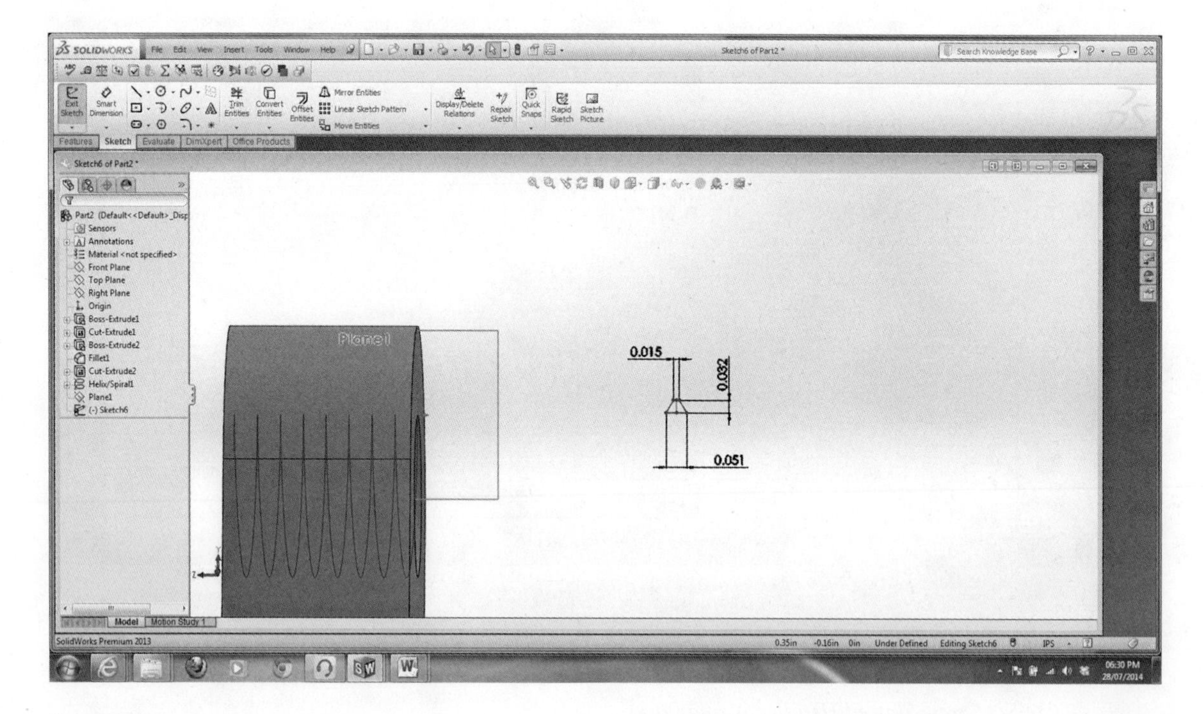

Figure 5.137

Thread profile defined.

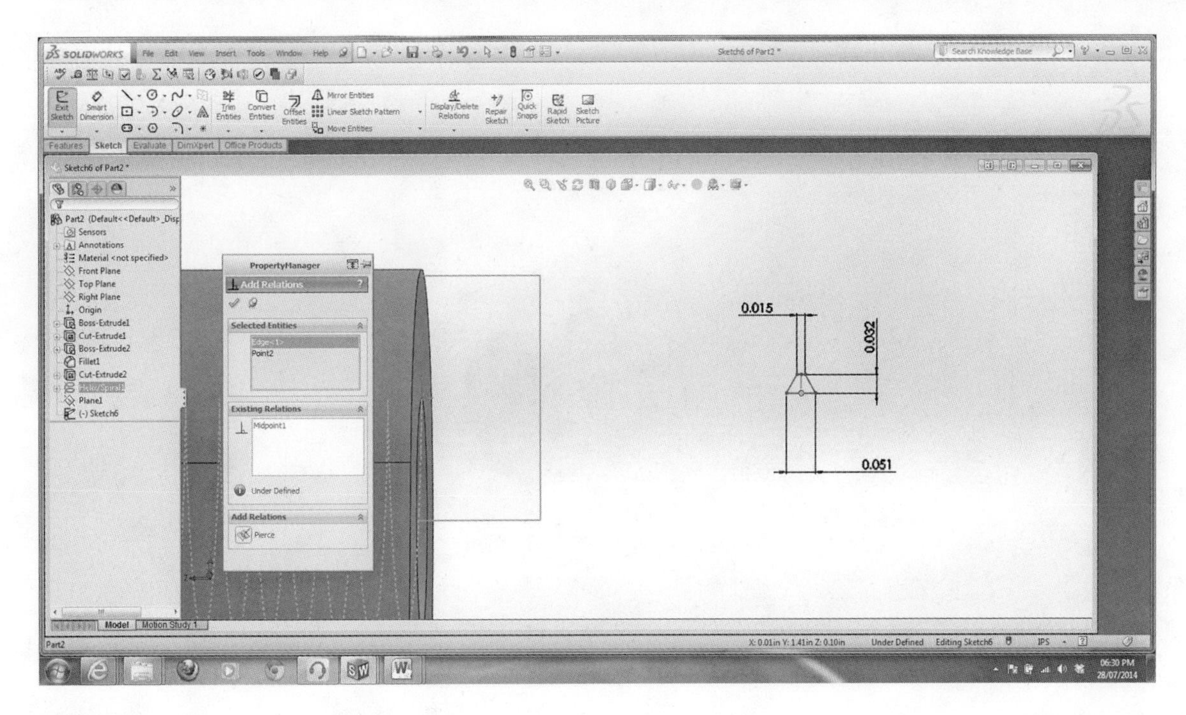

Figure 5.138

Adding relations to pierce the profile with the helix.

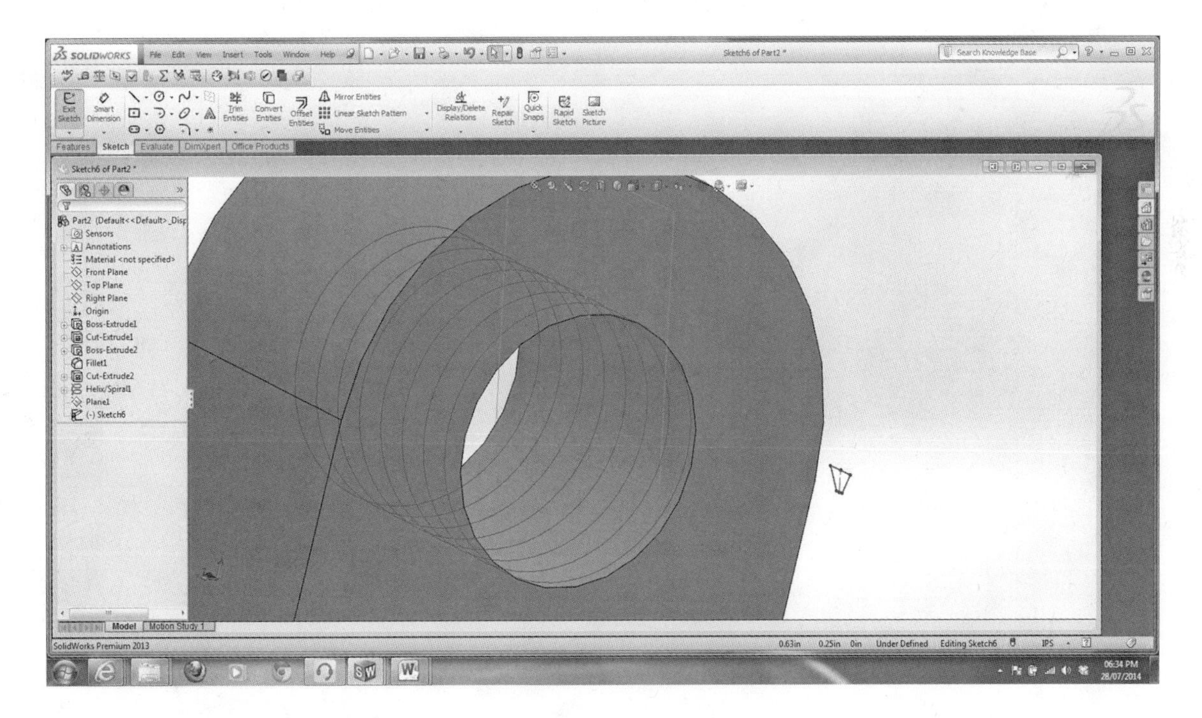

Figure 5.139

Proper orientation of the thread profile.

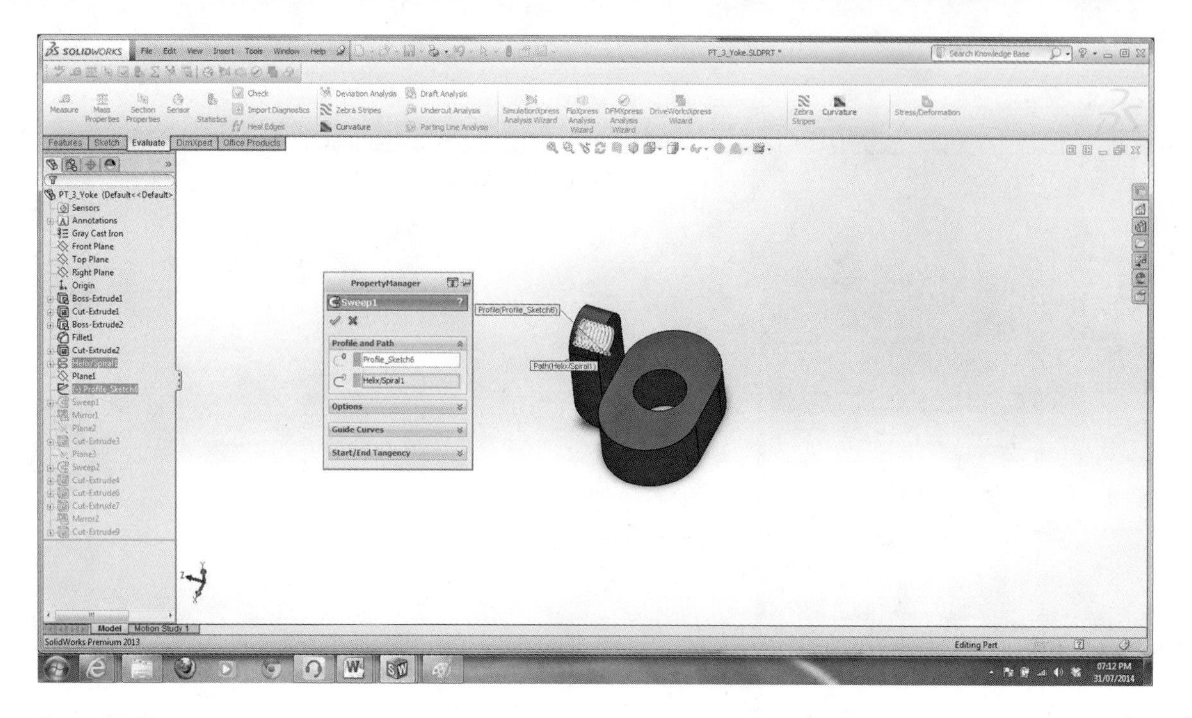

Figure 5.140

Sweep operation.

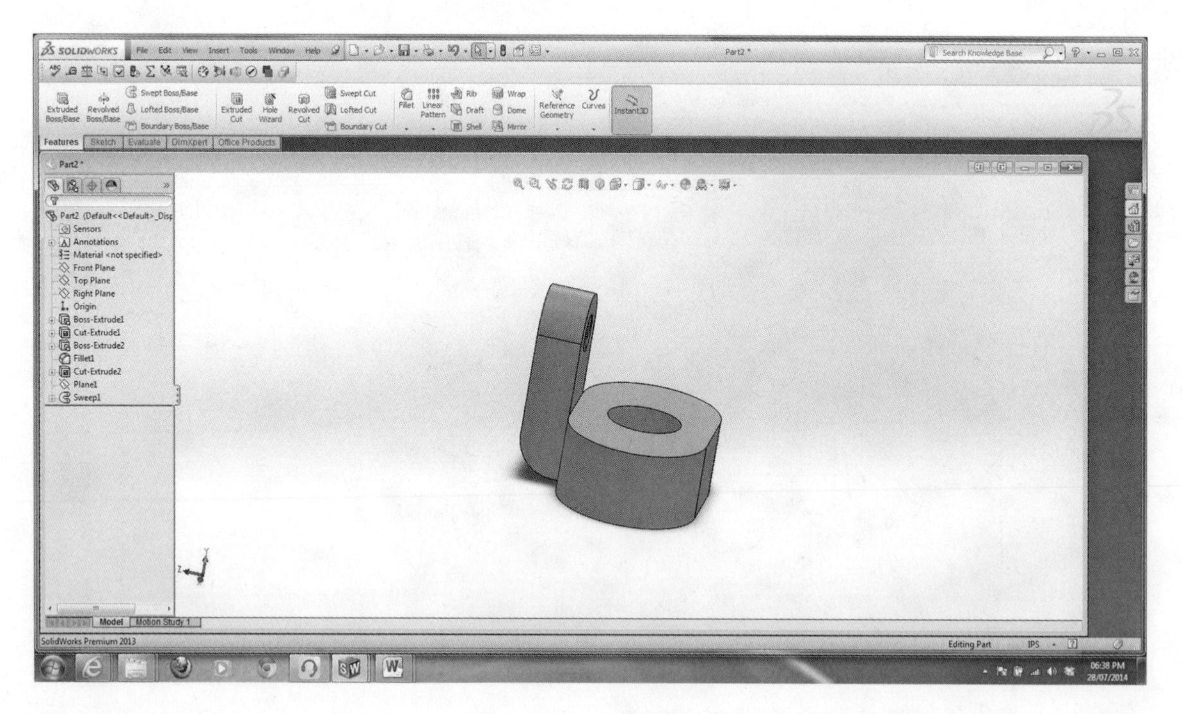

Figure 5.141

Half-left feature.

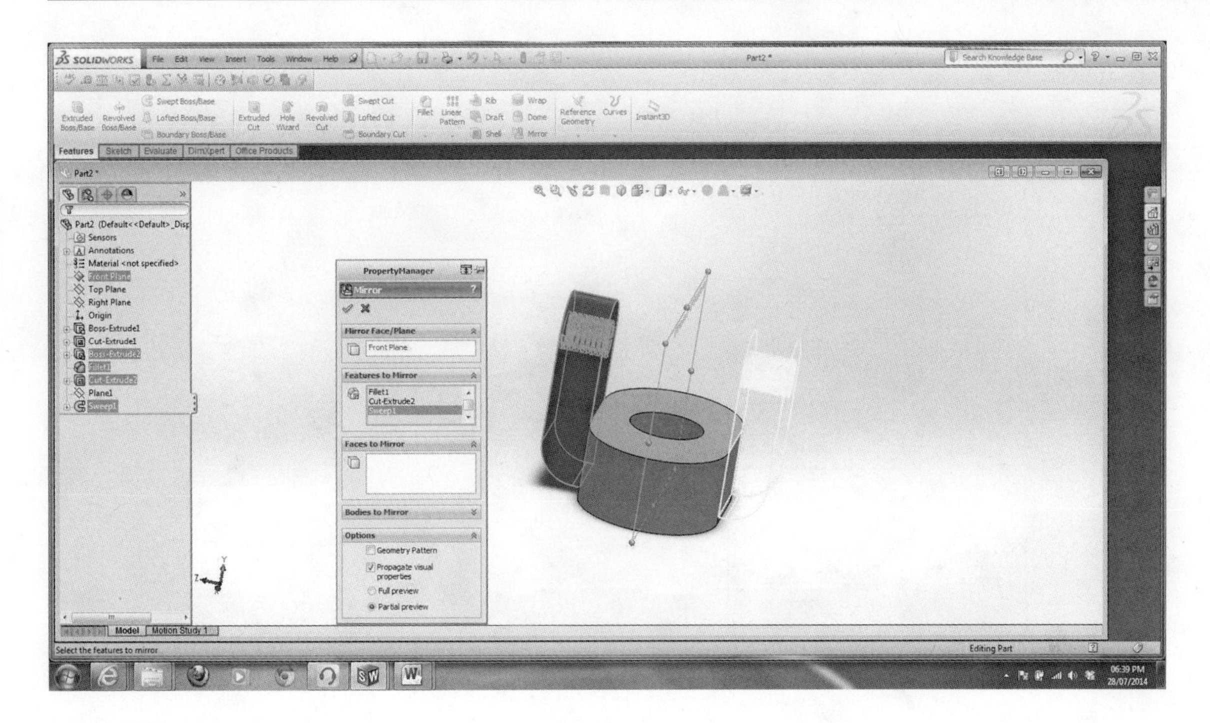

Figure 5.142

Mirror features.

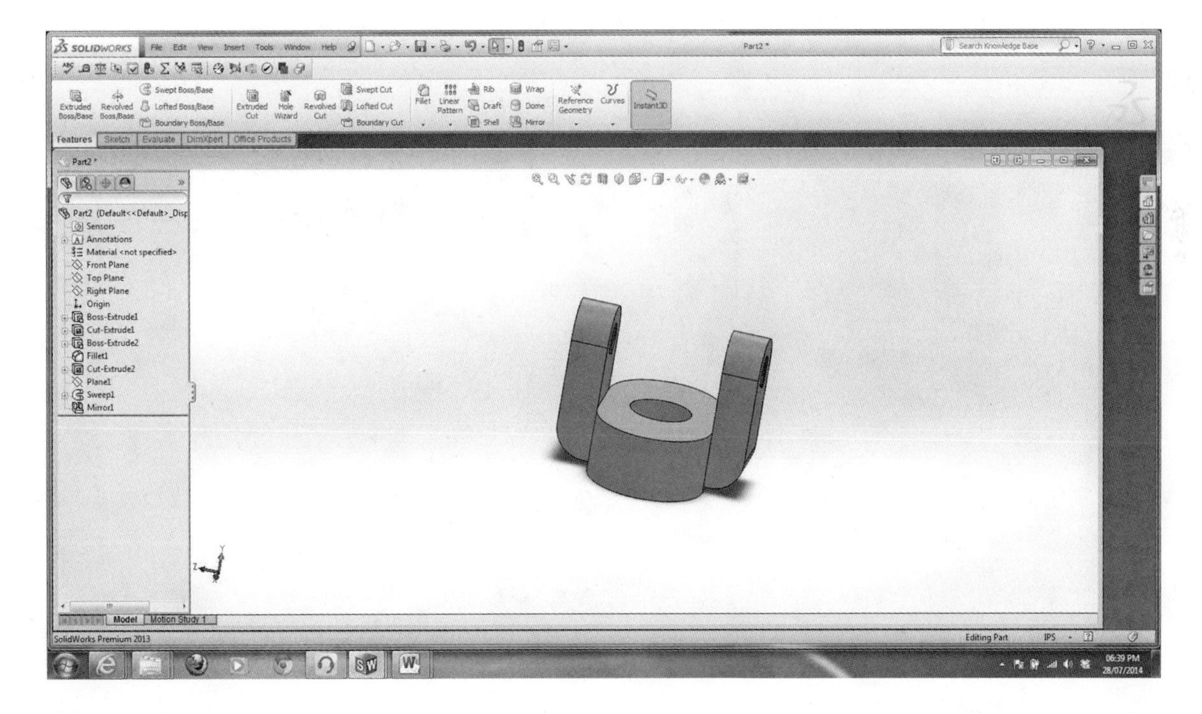

Figure 5.143

Partly completed part.

Figure 5.144

Plane2 tangent to circular face.

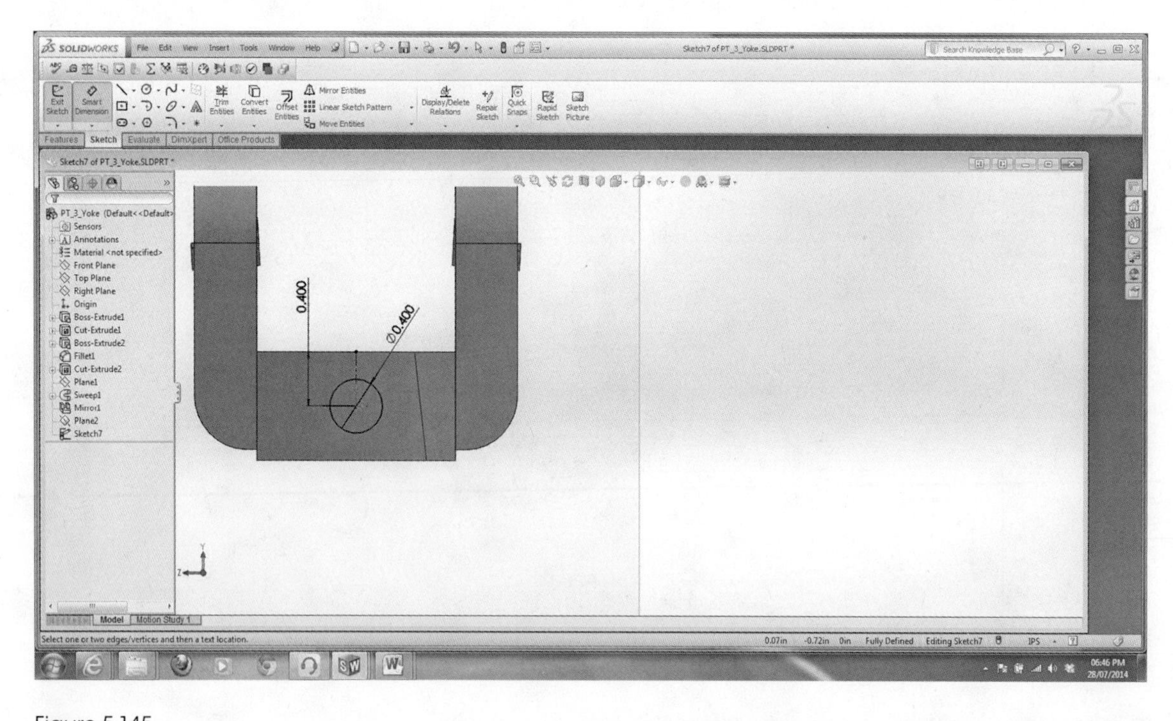

Figure 5.145

Base circle for third thread.

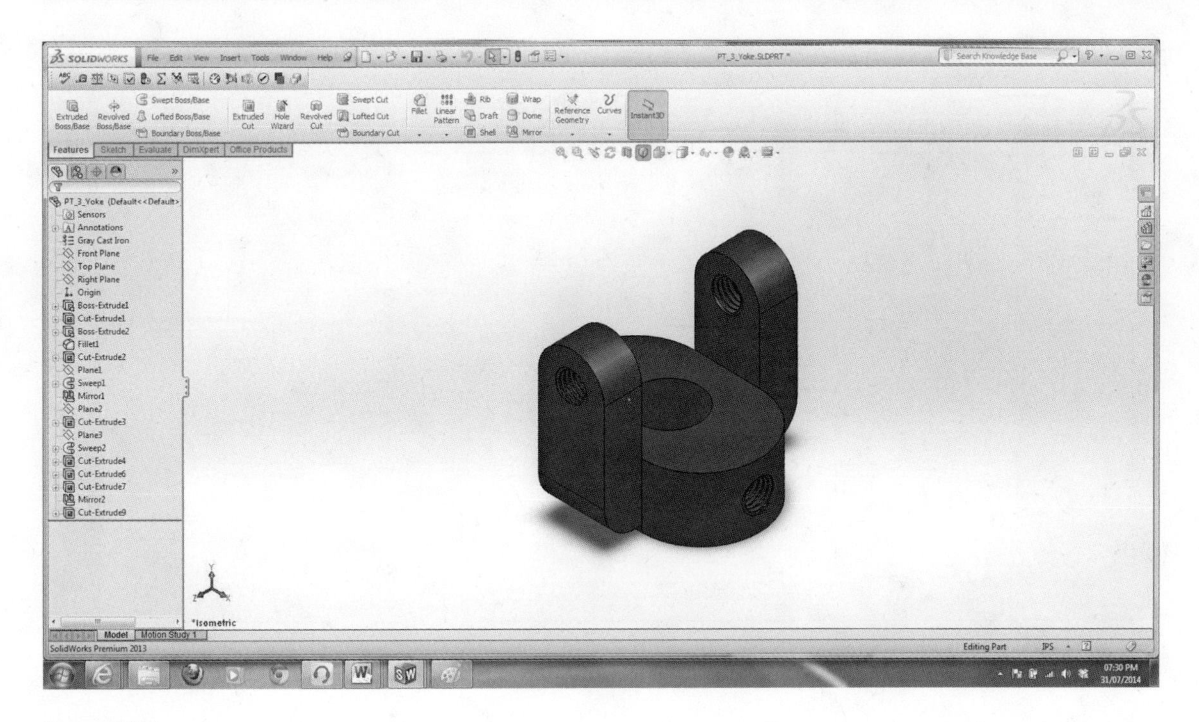

Figure 5.146

Model with cleaned edges.

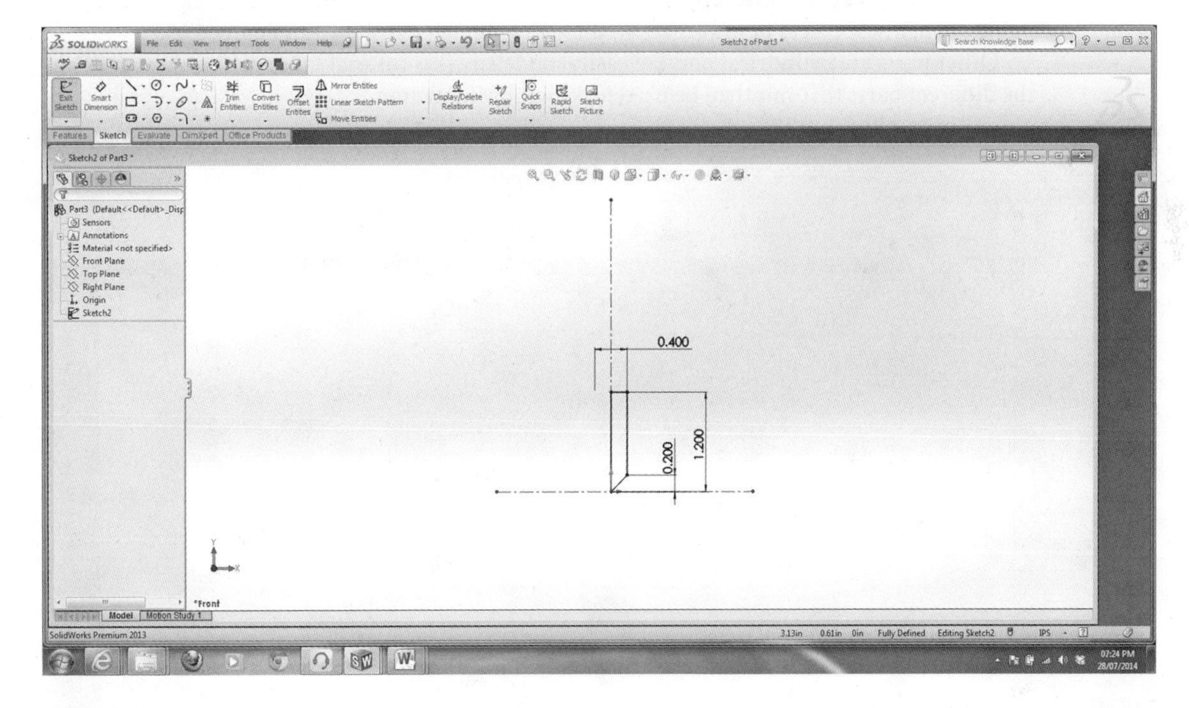

Figure 5.147

Sketch for revolving.

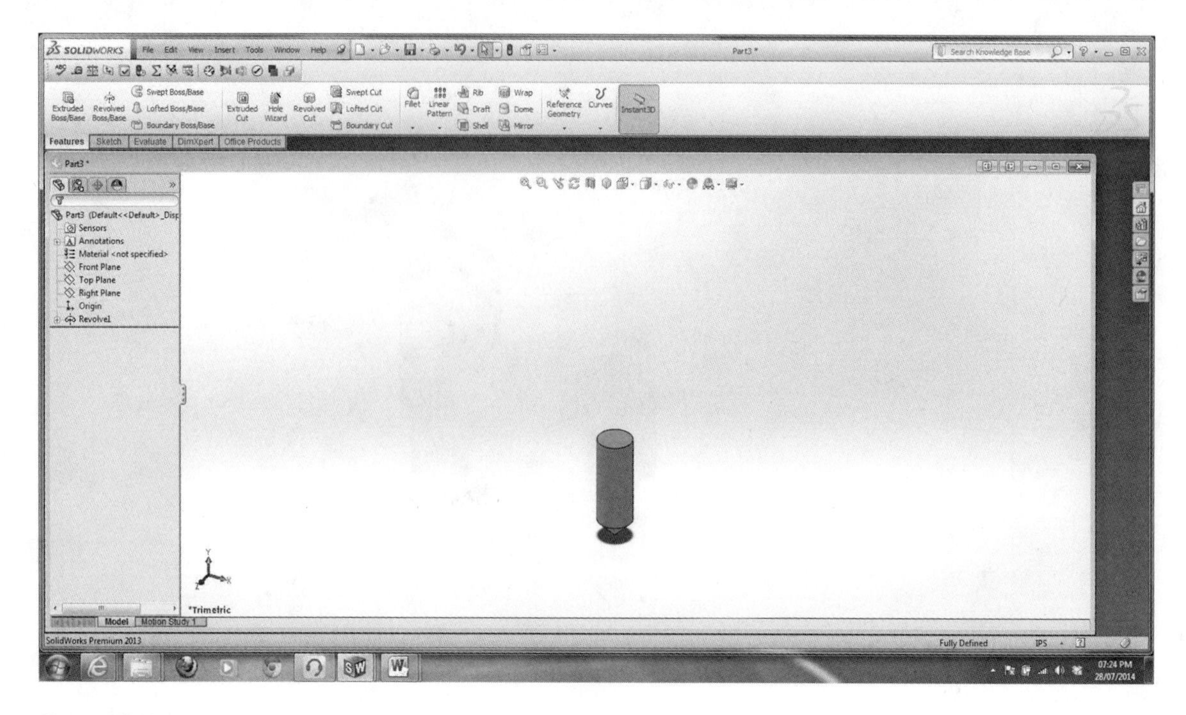

Figure 5.148

Revolved part.

The procedure for creating thread is followed to arrive at the screw of Figure 5.151.

PT 7: Set Screw Hex Socket Dog Point

The data sheet shown in Figure 5.152 is useful for modeling the set screw hex socket dog point. Using these pieces of information, the screw modeled is shown in Figure 5.153.

Other parts are cylindrical and are easily modeled. This tutorial takes us through modeling the different parts that can then be used for assembly modeling.

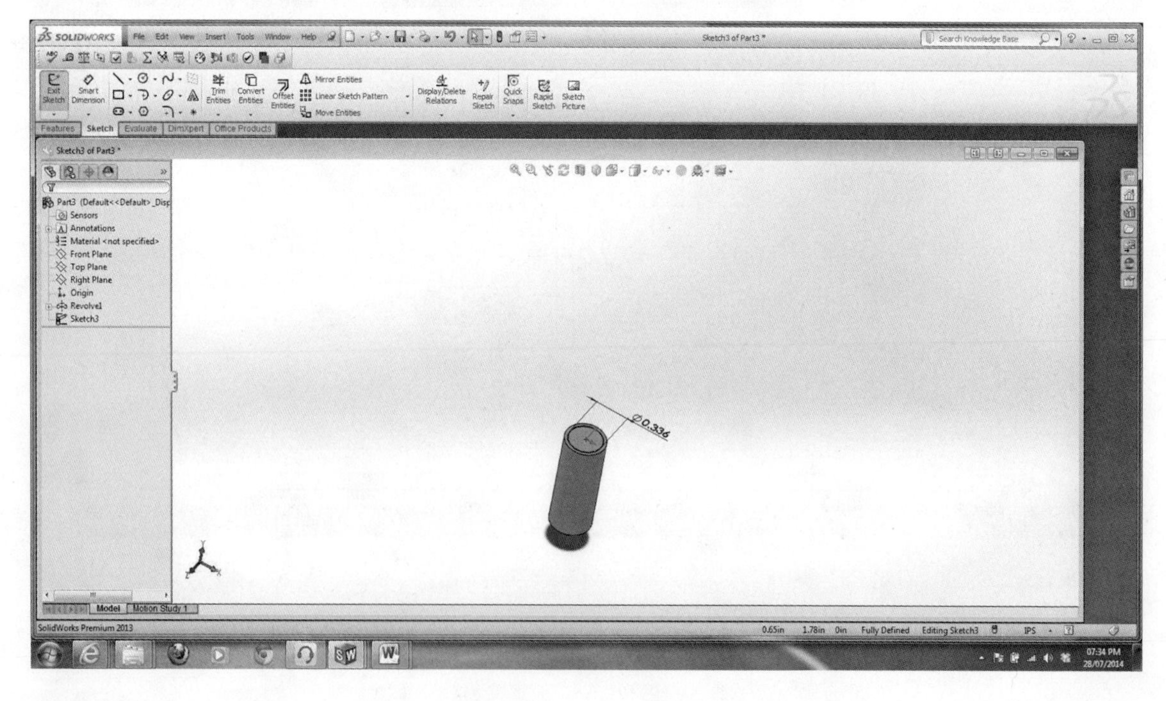

Figure 5.149

Base circle for helix.

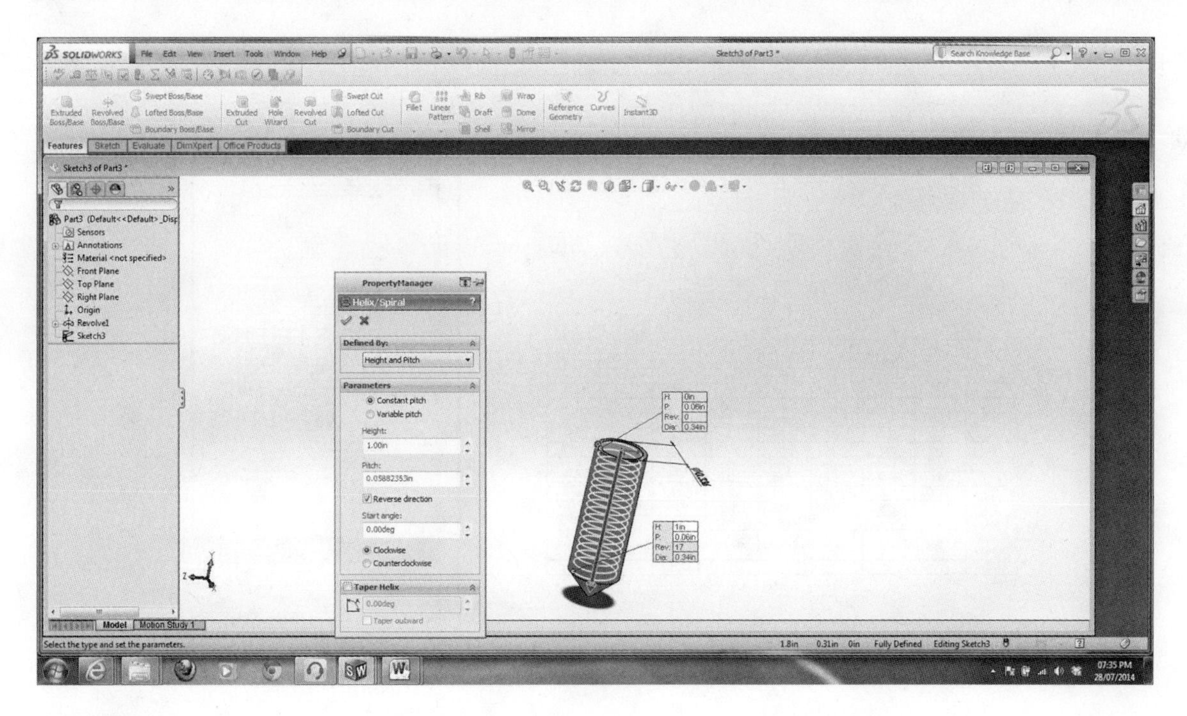

Figure 5.150

Helix parameters.

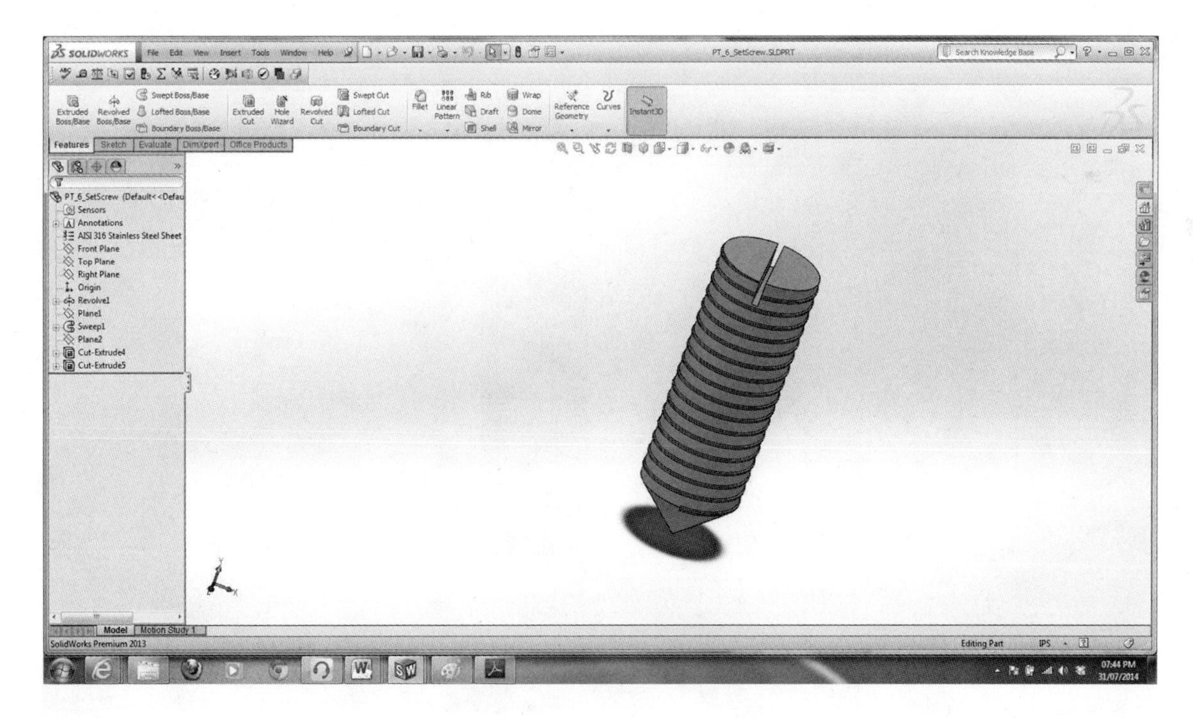

Figure 5.151

Screw model.

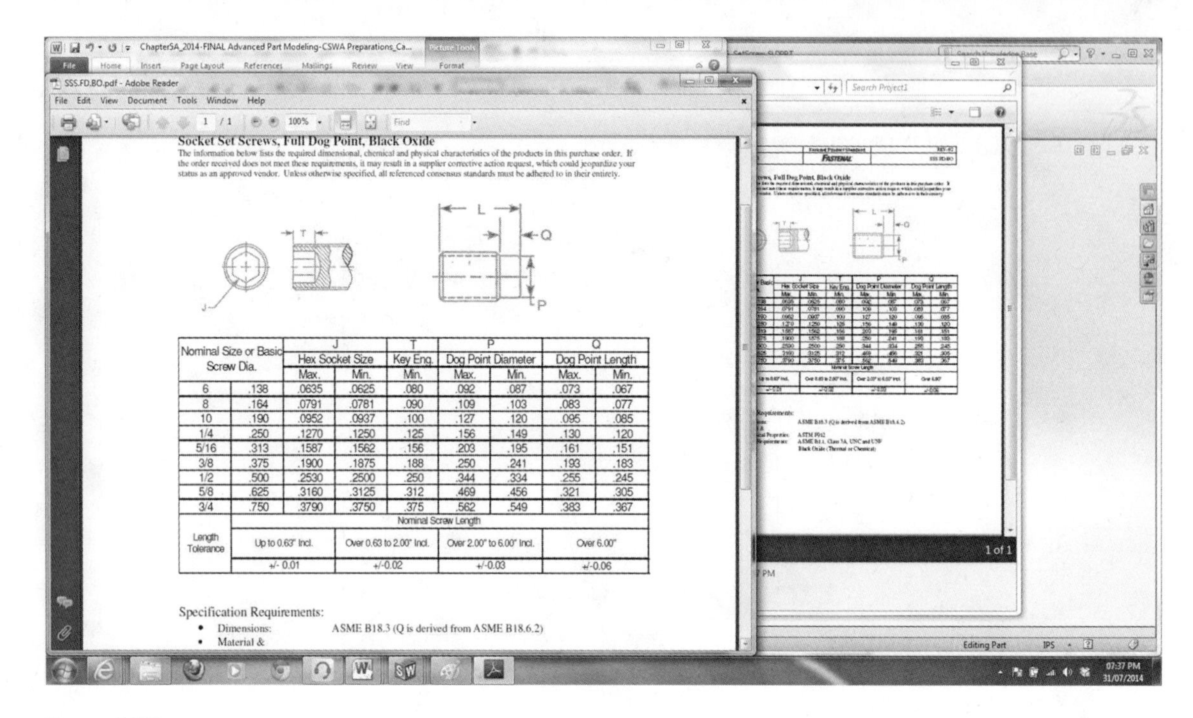

Socket Set Screws, Full Dog Point, Black Oxide

The information below lists the required dimensional, chemical and physical characteristics of the products in this purchase order. If the order received does not meet these requirements, it may result in a supplier corrective action request, which could jeopardize your status as an approved vendor. Unless otherwise specified, all referenced consensus standards must be adhered to in their entirety.

Nominal Size or Basic Screw Dia.		J		T	P		Q	
		Hex Socket Size		Key Eng.	Dog Point Diameter		Dog Point Length	
		Max.	Min.	Min.	Max.	Min.	Max.	Min.
6	.138	.0635	.0625	.080	.092	.087	.073	.067
8	.164	.0791	.0781	.090	.109	.103	.083	.077
10	.190	.0952	.0937	.100	.127	.120	.095	.085
1/4	.250	.1270	.1250	.125	.156	.149	.130	.120
5/16	.313	.1587	.1562	.156	.203	.195	.161	.151
3/8	.375	.1900	.1875	.188	.250	.241	.193	.183
1/2	.500	.2530	.2500	.250	.344	.334	.255	.245
5/8	.625	.3160	.3125	.312	.469	.456	.321	.305
3/4	.750	.3790	.3750	.375	.562	.549	.383	.367
Length Tolerance	Nominal Screw Length							
	Up to 0.63" Incl.		Over 0.63 to 2.00" Incl.		Over 2.00" to 6.00" Incl.		Over 6.00"	
	+/- 0.01		+/-0.02		+/-0.03		+/-0.06	

Specification Requirements:

- Dimensions: ASME B18.3 (Q is derived from ASME B18.6.2)
- Material &

Figure 5.152

Data sheet for screw hex socket dog point.

Figure 5.153

Screw hex socket dog point.

Summary

This chapter discusses part modeling at the levels of CSWA and Certified SolidWorks Professional (CSWP), and most of the examples are similar to the materials that are covered in the CSWA and CSWP examinations. Students preparing for these exams will find this chapter helpful. Instructors should consider covering Chapter 4 before moving to this chapter, which pitches at a higher level. The chapter finishes with the modeling of parts that are used in an assembly modeling.

Exercises

SolidWorks Corporation: CSWP Sample Exam

Certified SolidWorks Professional: Solid Modeling Specialist (CSWP/CORE)

Sample Exam—Part Modeling Portion

These questions are similar to the Parametric Part Modeling portion of the CSWP exam.

Design this part in SolidWorks in Figure P5.1.

Unit system: millimeter, gram, second (MMGS)

Decimal places: 2

Part origin: Arbitrary

Part material: Brass

Material density: 0.0085 g/mm^3

Design note: the part is shelled throughout (single open face as shown).

Question 1:

A = 60, B = 64, C = 140, D = 19

What is the overall mass of the part (in grams)?

Question 2:

A = 50, B = 70, C = 160, D = 23

What is the overall mass of the part (in grams)?

Update part with new features/dimensions in Figure P5.2.

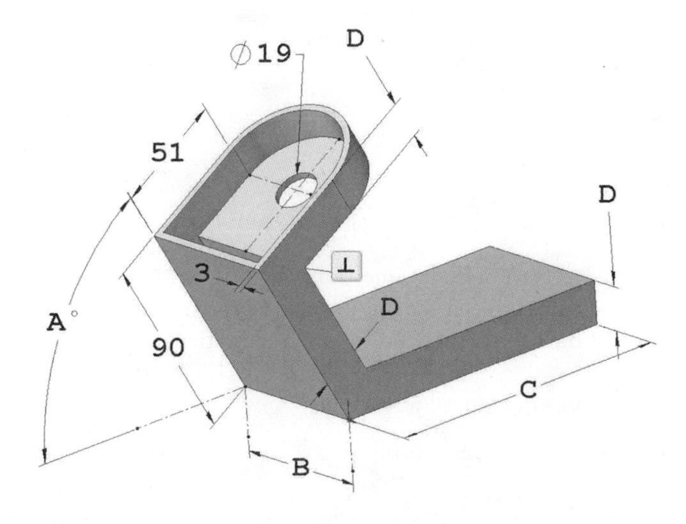

Figure P5.1

Simple cylinder support.

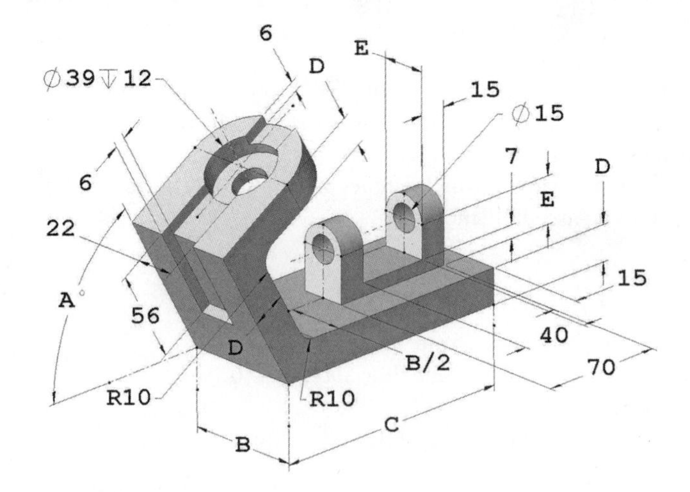

Figure P5.2

Cylinder support.

Unit system: MMGS

Decimal places: 2

Part material: Brass

Material density: 0.0085 g/mm³

Design note: no shell remaining

Question 3:

A = 60, B = 64, C = 140, D = 19, E = 25

What is the overall mass of the part (in grams)?

Question 4:

Build this part in SolidWorks in Figure P5.3.

Material: 6061 Alloy.

What is the overall mass of the part in grams?

Answers

1. 1006.91 grams
2. 1230.82 grams
3. 2859.51 grams
4. 3218.14 grams

Advice: You should be able to answer all four questions correctly within 20–30 min. Read through every question first. This will help you save time and make correct decisions when choosing which sketch plane to use and which sketch profile is best. Avoid sketch fillets in this particular design.

Build this part in SolidWorks in Figure P5.3.

Material: 6061 Alloy. Density = 0.0027 g/mm³

Unit system: MMGS

Decimal places: 2

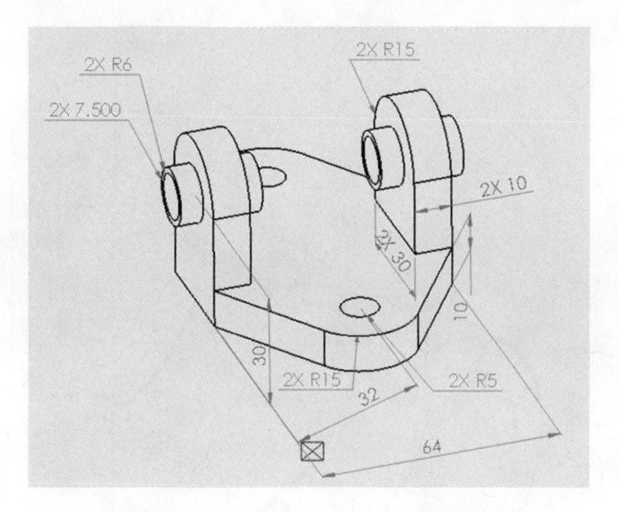

Figure P5.3

Multi-faced Inclined Slide Bracket. (From David C. Planchard & Marie P. Planchard, Official Certified SolidWorks Associates (CSWA), SDC Publications, 2009.)

Part origin: Arbitrary

A = 100

All holes through all, unless otherwise specified

Question 4:

What is the overall mass of the part in grams?

Answers

a. 2040.57

b. 2004.57

c. 102.63

d. 1561.23

Projects

These projects ensure that you know how to read drawings based on American Society of Mechanical Engineers (ASME) standards.
Create the models for Figures P5.4 to P5.7.

Figure P5.4:
Part name: Hydraulic valve cylinder
Material: Phosphor bronze
All fillets and rounds: R1
Figure P5.5:
Part name: Hub
Material: Cast iron
Fillets: R.03 unless otherwise specified
Figure P5.6:
Part name: Slide bracket
Material: AISI 1020
Fillets: R.25 unless otherwise specified

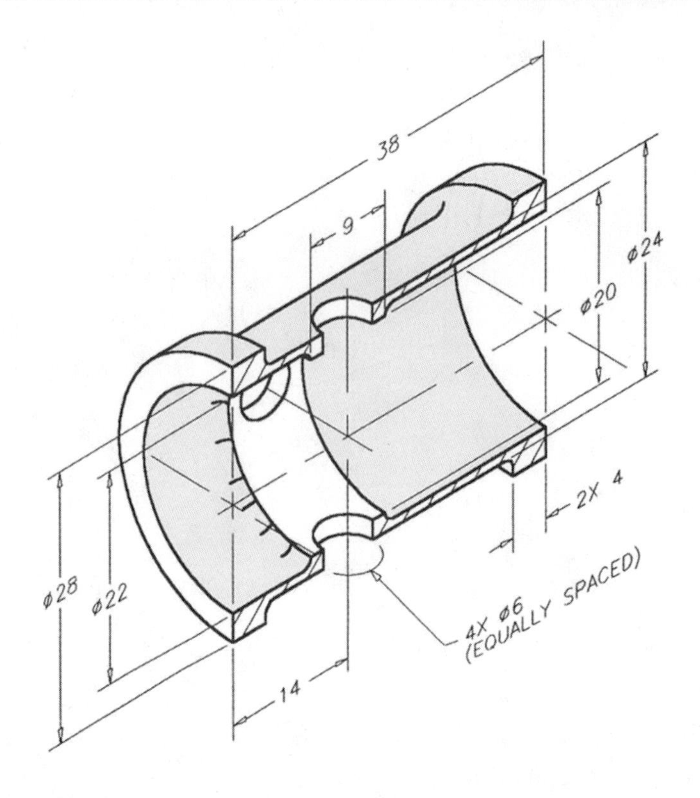

Figure P5.4

Hydraulic valve cylinder. (From Custom Edition, *Engineering Drawing & Design Vol 1*, Nelson Education, 2013; From MADSEN. *ENGINEERING DRAW/DESIGN IML*, 1E. © 1991 Delmar Learning, a part of Cengage Learning, Inc. Reproduced by permission. http://www.cengage.com/permissions.)

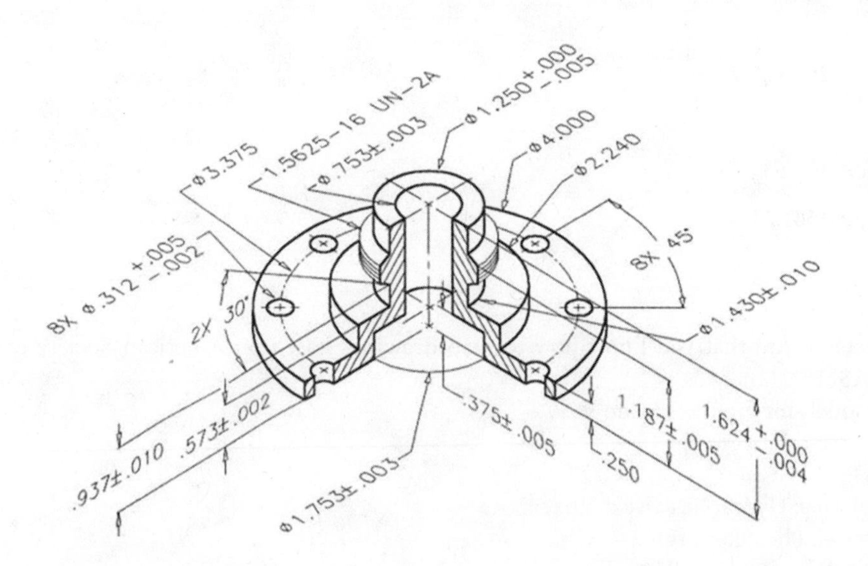

Figure P5.5

Hub. (From Custom Edition, *Engineering Drawing & Design Vol 1*, Nelson Education, 2013; From MADSEN. *ENGINEERING DRAW/DESIGN IML*, 1E. © 1991 Delmar Learning, a part of Cengage Learning, Inc. Reproduced by permission. http://www.cengage.com/permissions.)

Figure P5.7:

Part name: Bracket
Material: AISI 1020
Fillets: R.25 unless otherwise specified

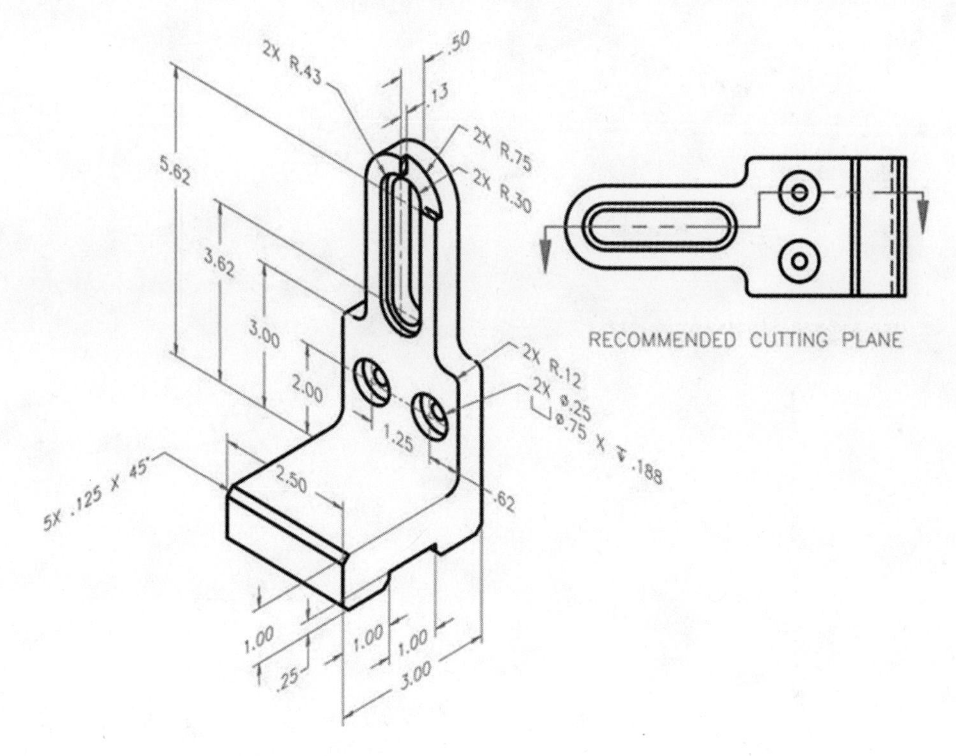

Figure P5.6

Slide bracket. (From Custom Edition, *Engineering Drawing & Design Vol 1,* Nelson Education, 2013.)

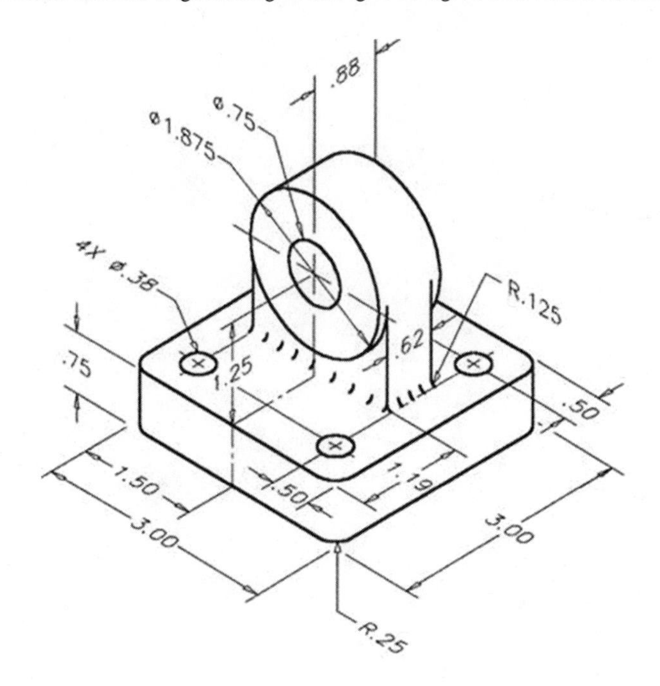

Figure P5.7

Bracket.

Bibliography

Certified SolidWorks Associate (CSWA) Sample Exam Questions. http://www.eng.uwo.ca/designcentre/SWcertification/CSWA/CSWASampleExam2007.pdf.

Fastenal Product Standard: Finished Hex Nuts, Stainless Steel. https://www.fastenal.com/content/product_specifications/FHN.SS.pdf.

6

Creating Revolved, Swept, and Lofted Parts

Objectives:

When you complete this session, you will have

- Learned how to create three-dimensional (3D) objects using Revolved Features tools
- Learned how to use Copy Features tools
- Learned how to create 3D objects using Swept Features tools
- Learned how to use Draft Features tools
- Learned how to create 3D objects using Lofted Features tools
- Learned how to use Circular Pattern Features tools
- Learned how to use reference planes
- Learned how to use Copy Features tools

Revolved Boss/Base

The Revolved Boss/Base tool rotates a contour about an axis. It is a useful tool when modeling parts that have circular contours. Let us illustrate the revolved boss/base concept as follows.

1. Select the front plane and create Sketch1, as shown in Figure 6.1.
2. Click the Features tool.
3. Click the Revolved Boss/Base tool.
   The Revolve Properties Manager appears (see Figure 6.2).
4. Define the revolved axis (Line1) as the vertical dimension line. A real-time preview will appear.
5. Click OK to complete the revolved part (see Figure 6.3).
6. Click the top face of Extrude1.
7. Sketch a construction *Circle* of 126 mm diameter, and four *Circles*, each 15 mm in diameter spaced out at 45° (see Figure 6.4).
8. Select each circle, and click Extrude-cut to create a hole for each of the circle.
9. See Figure 6.5 for the Extrude-Cut preview for the four holes and Figure 6.6 for the final part.

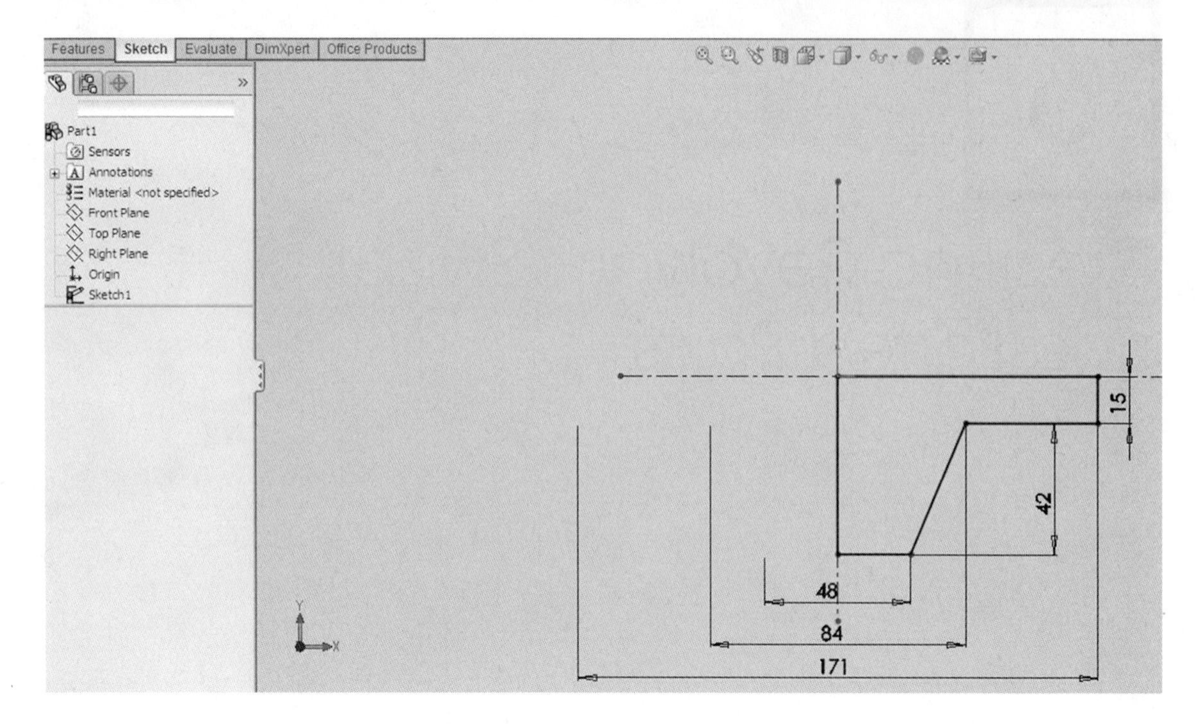

Figure 6.1

Sketch1.

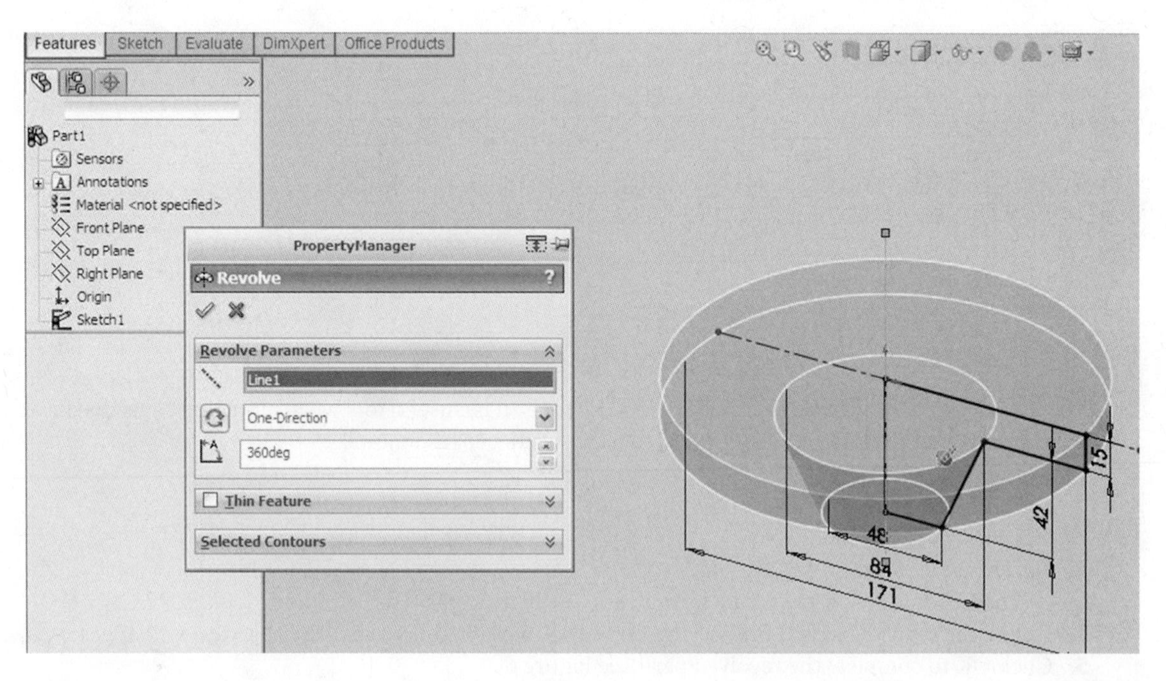

Figure 6.2

Preview.

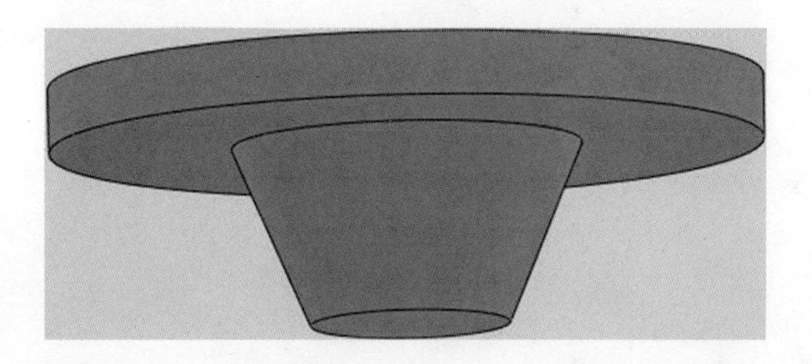

Figure 6.3

Extrude1.

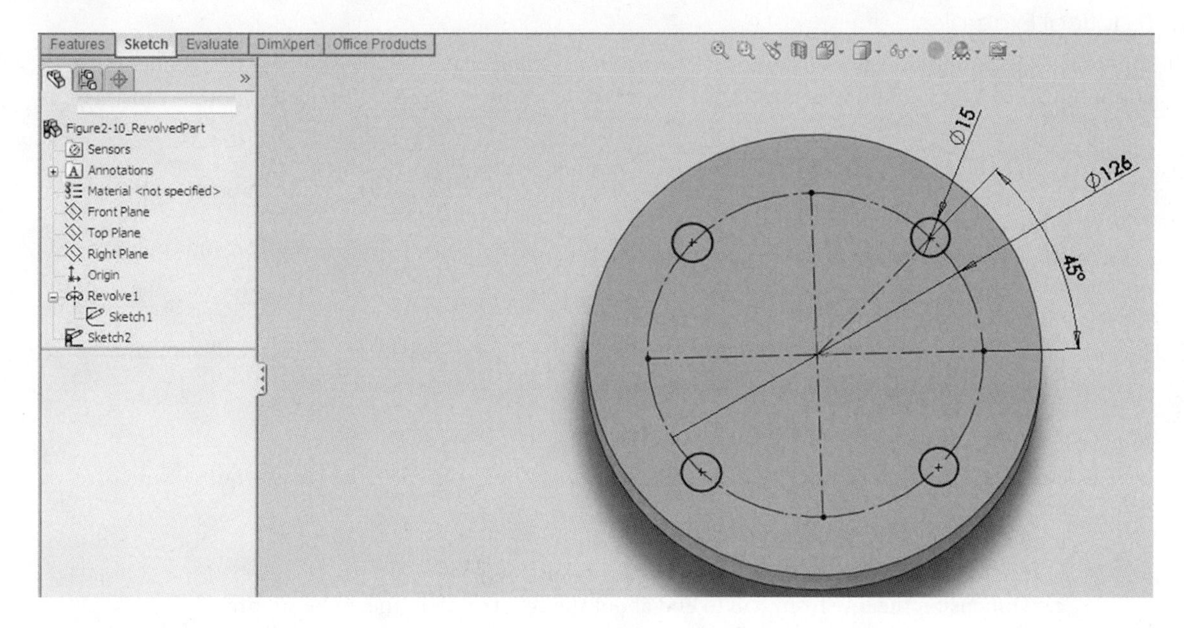

Figure 6.4

Sketches for four holes.

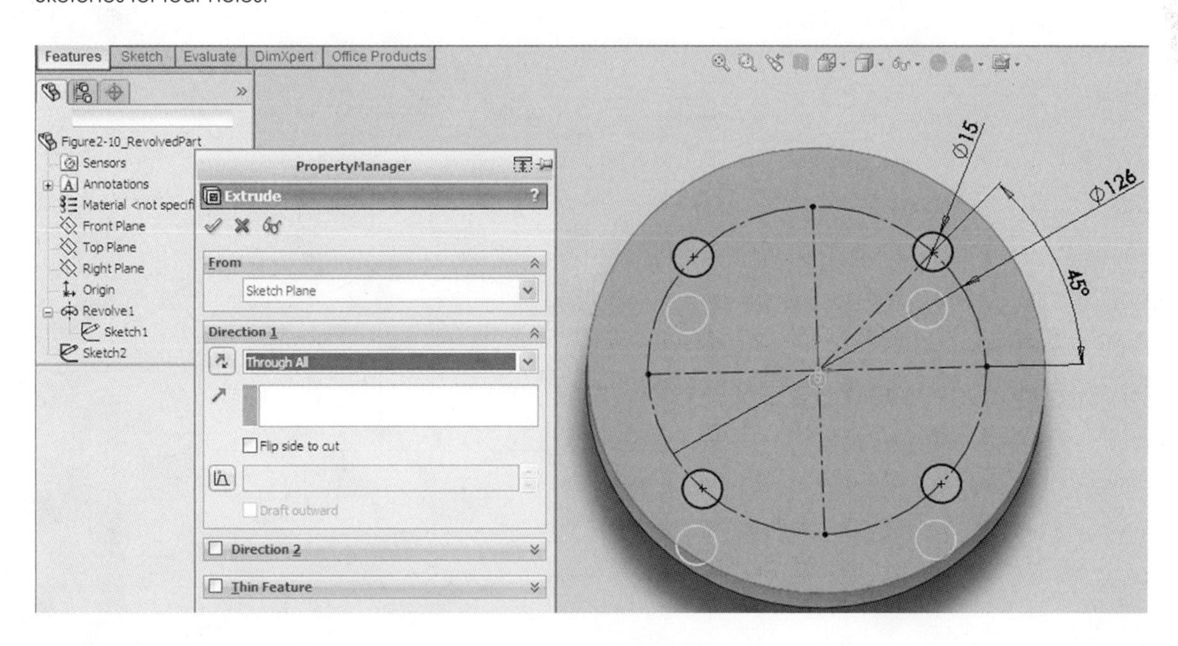

Figure 6.5

Extrude-Cut for holes.

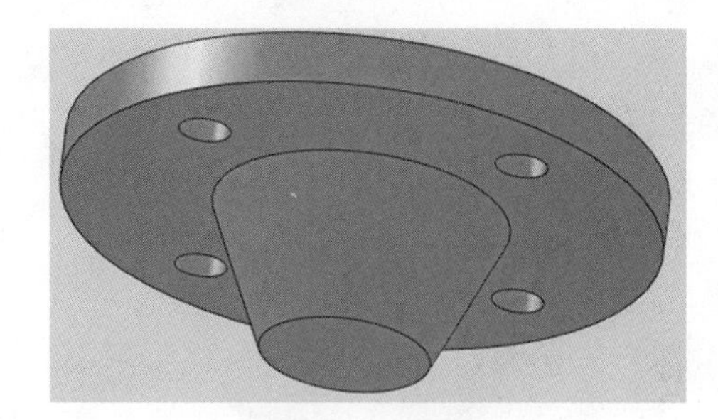

Figure 6.6

Final revolved feature.

Practical Examples

Two practical examples are given based on the principles of revolved parts. One is the engine cylinder, which is commonly used in automobile design, and the other is pulley, which is commonly used in power transmission.

Engine Cylinder

1. Select the front plane and create Sketch1, as shown in Figure 6.7. Define the relations for the slots to ensure that the sketch is fully defined.
2. Click the Features tool.
3. Click the Revolved Boss/Base tool.
   The Revolve Properties Manager appears.
4. Define the revolved axis (Line1) as the vertical centerline. A real-time preview will appear.
5. Click OK to complete the revolved part as shown in Figure 6.8.

Pulley

1. Select the right plane and create Sketch1, as shown in Figure 6.9. The steps required for correctly defining the sketch are as follows:
   a. Sketch a line to the left of the vertical construction line.
   b. Mirror this line about the vertical construction line.
   c. Dimension the line from end to end about the construction line, to be 10 mm.
   d. Sketch the first tooth-profile completely, and define the relations for the three horizontal lines to be equal to 10 mm and the angles to be 60° (see Figure 6.10 for steps a to d).

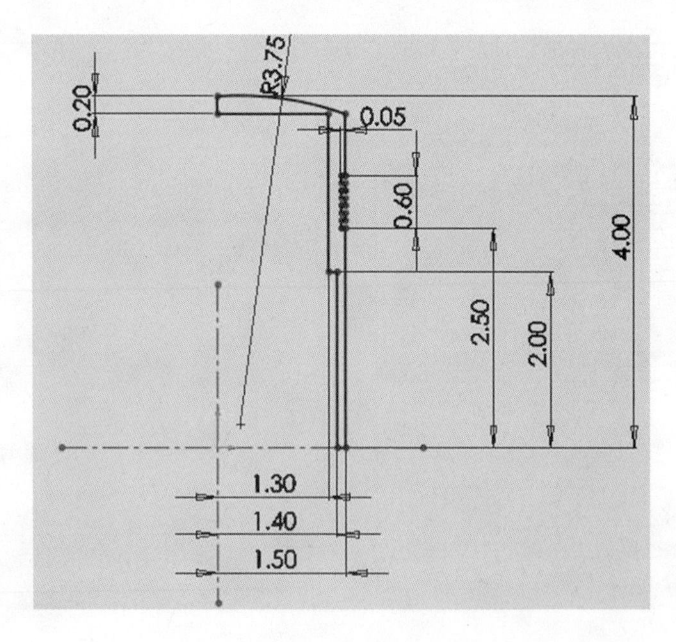

Figure 6.7

Sketch1.

Figure 6.8

Engine cylinder model.

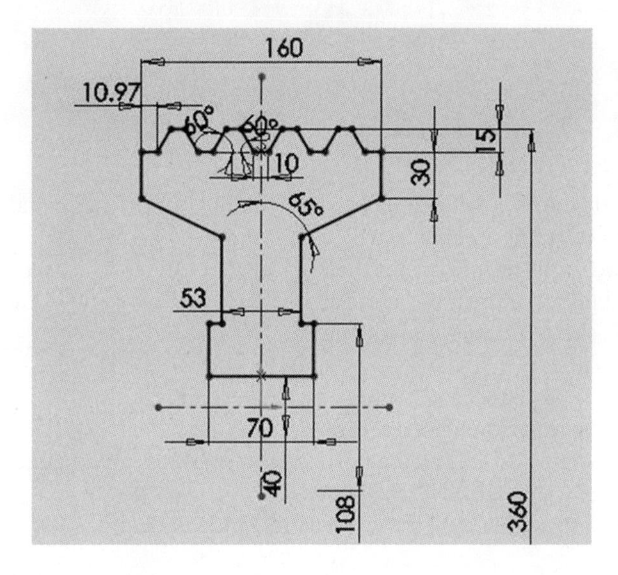

Figure 6.9

Sketch1 required.

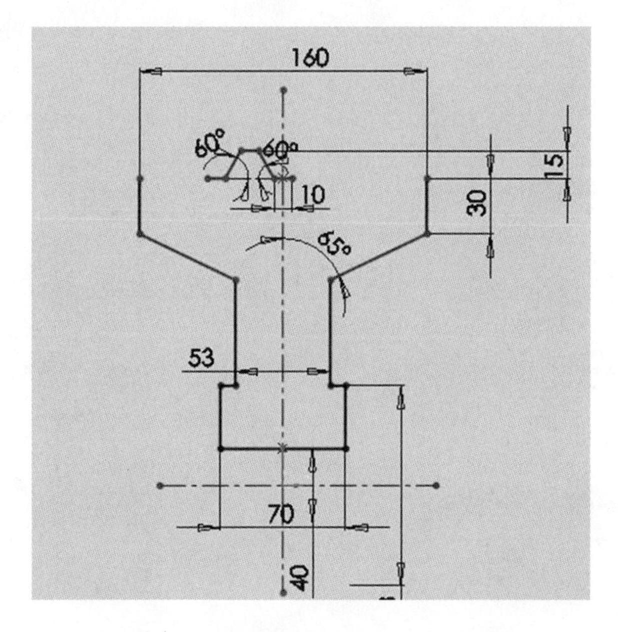

Figure 6.10

Steps a to d.

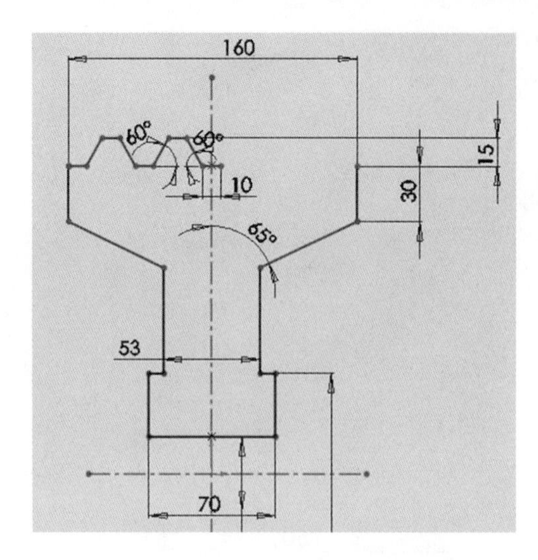

Figure 6.11

Step e.

e. Copy part of the tooth-profile to increase the number of profiles equal to 2 and ensure that the end-condition relations are met (see Figure 6.11 for this step).
f. Mirror the two tooth-profiles about the vertical construction line (see Figure 6.12 for this step).
g. Check that all points at the bottom are collinear, as well as the points at the top of profiles. Also, check that the sketch forms a closed loop.

2. Click the Features tool.
3. Click the Revolved Boss/Base tool.
   The Revolve Properties Manager appears.
4. Define the revolved axis (Line1) as the horizontal centerline. A real-time preview will appear (see Figure 6.13 for steps 2–4).
5. Click OK to complete the revolved part as seen in Figure 6.14.

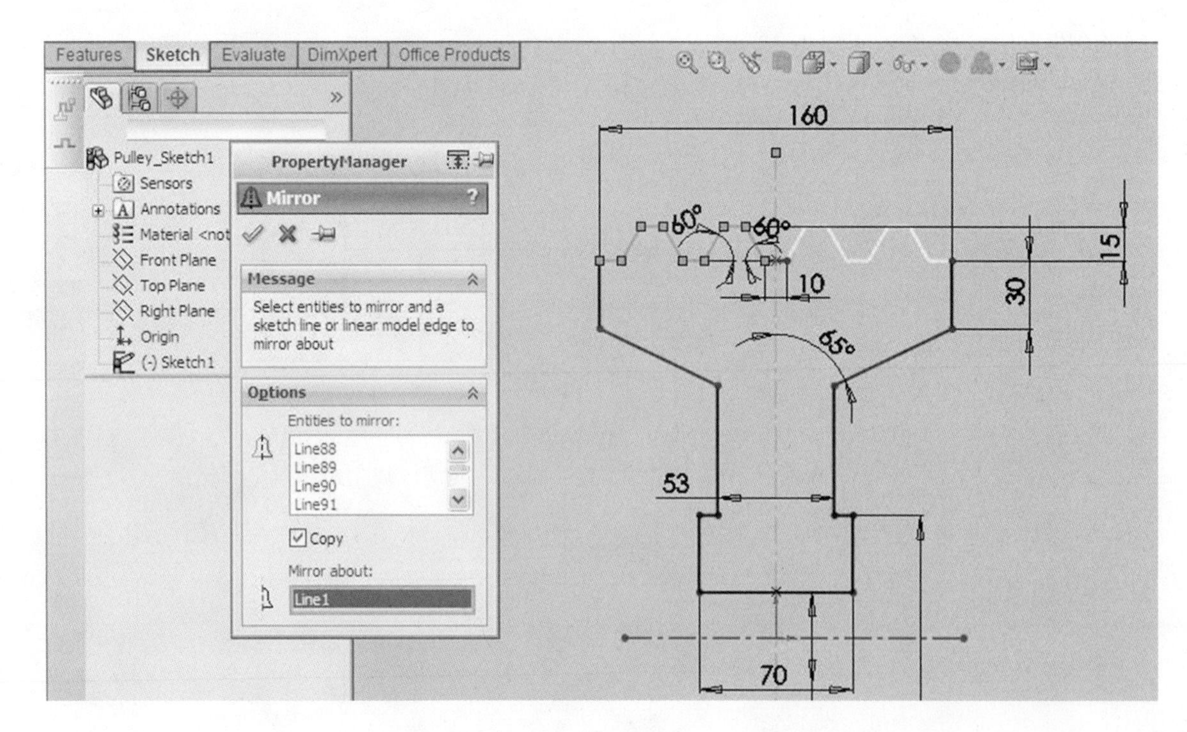

Figure 6.12

Step f.

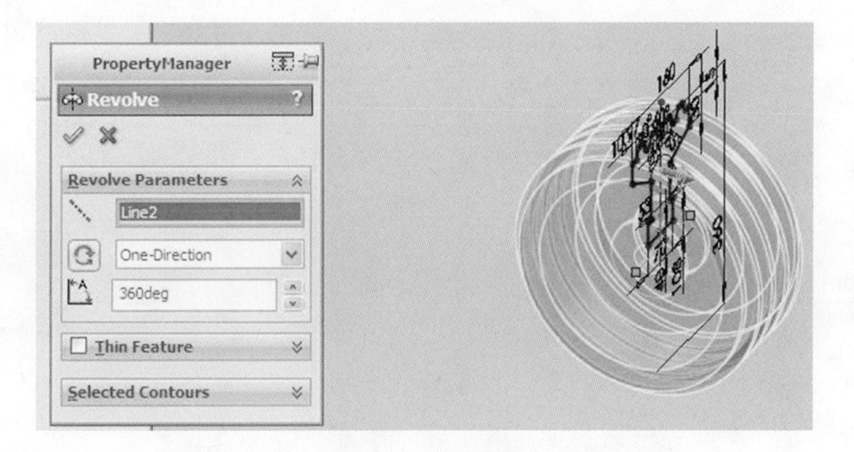

Figure 6.13

Steps 2–4.

Figure 6.14

Pulley model.

Swept Boss/Base

The Swept Boss/Base tool is used to sweep a profile through a path (arc, spline, etc.). As with lofting, there have to be existing shapes. The prerequisite for creating a 3D swept model is to first sketch the shapes on the different perpendicular planes. In the illustration presented here, a hexagon is the *profile*, while a spline is the *path*, as shown in Figure 6.15.

1. Select the Front Plane from the CommandManager.
2. Click Sketch from the CommandManager or from the Context tool bar.
3. Create a Context tool bar (a spline with one end coinciding with the origin; see Figure 6.15).
4. Exit the sketch mode.
5. Click Features > Reference Geometry > Plane from the CommandManager.
6. For the First Reference, select the endpoint of the spline, Context tool bar.
7. For the Second Reference, select the spline, Context tool bar from the graphics window. (SolidWorks enhanced features automatically creates the plane, Plane1.) (Note: this plane is normal to Sketch1 at any point; see Figure 6.15.)
8. Right-click Plane1 and create Sketch2 (a hexagon with sides, 0.5 in.).
9. Exit the sketch mode; see Figure 6.16 for both profile and path.
10. Click the Swept Boss/Base tool.

    The Swept Properties Manager appears (see Figure 6.16).
11. Right-click the Profile and Path box.
12. Click the hexagon as profile and the spline as path. A real-time preview will appear.
13. Click OK to complete the swept cut part, which is now hollow (see Figure 6.17). Hide the planes.

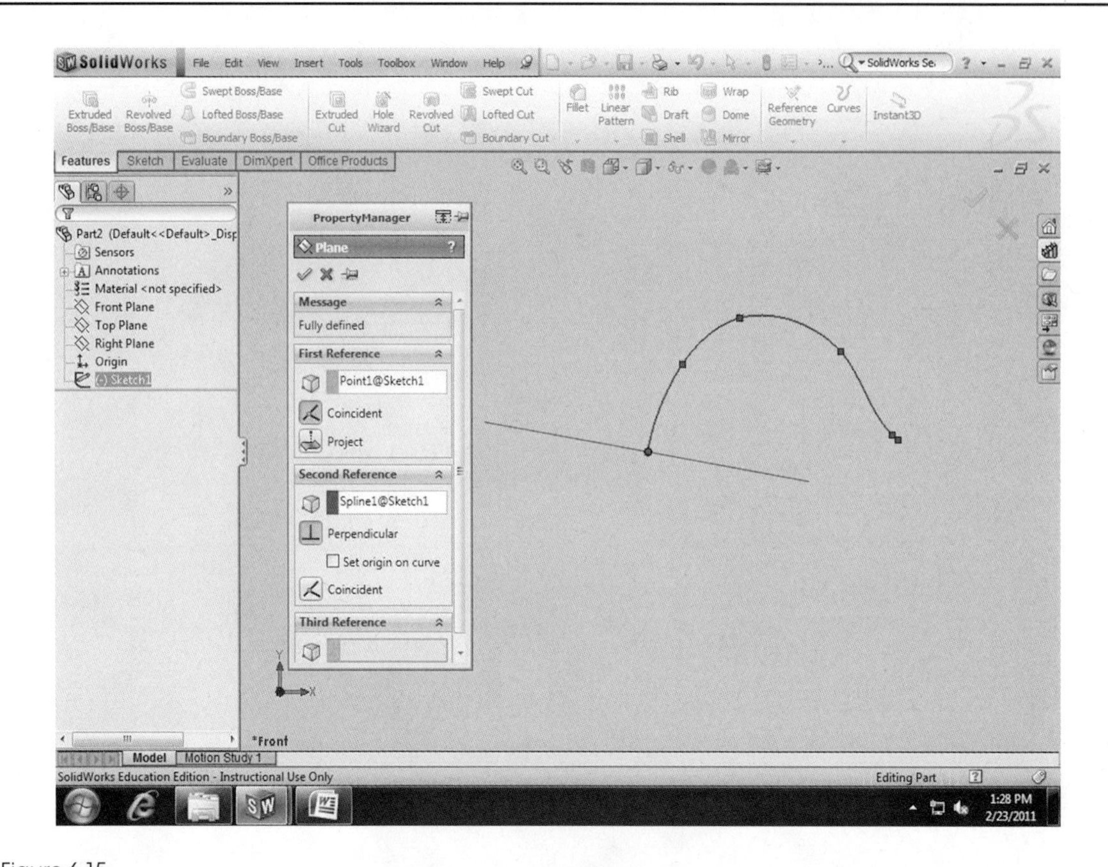

Figure 6.15

Path (Spline) and Plane1 created at the Spline left endpoint. Profile (hexagon) and path (spline).

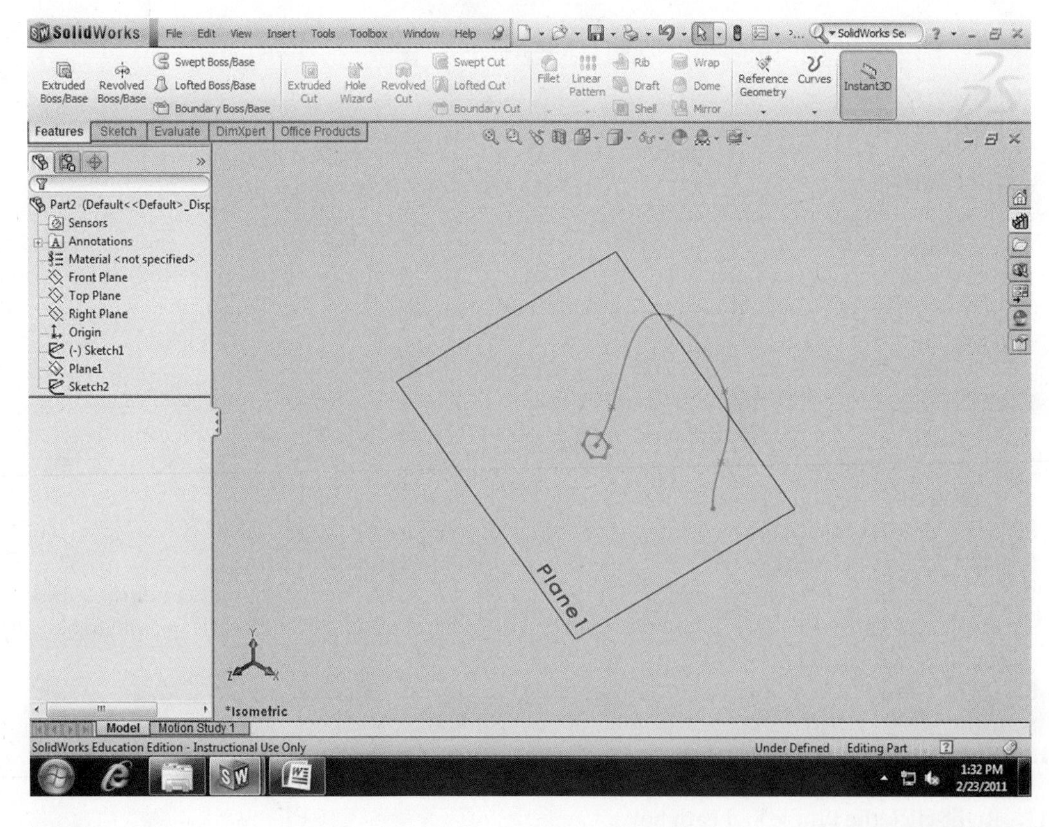

Figure 6.16

Preview.

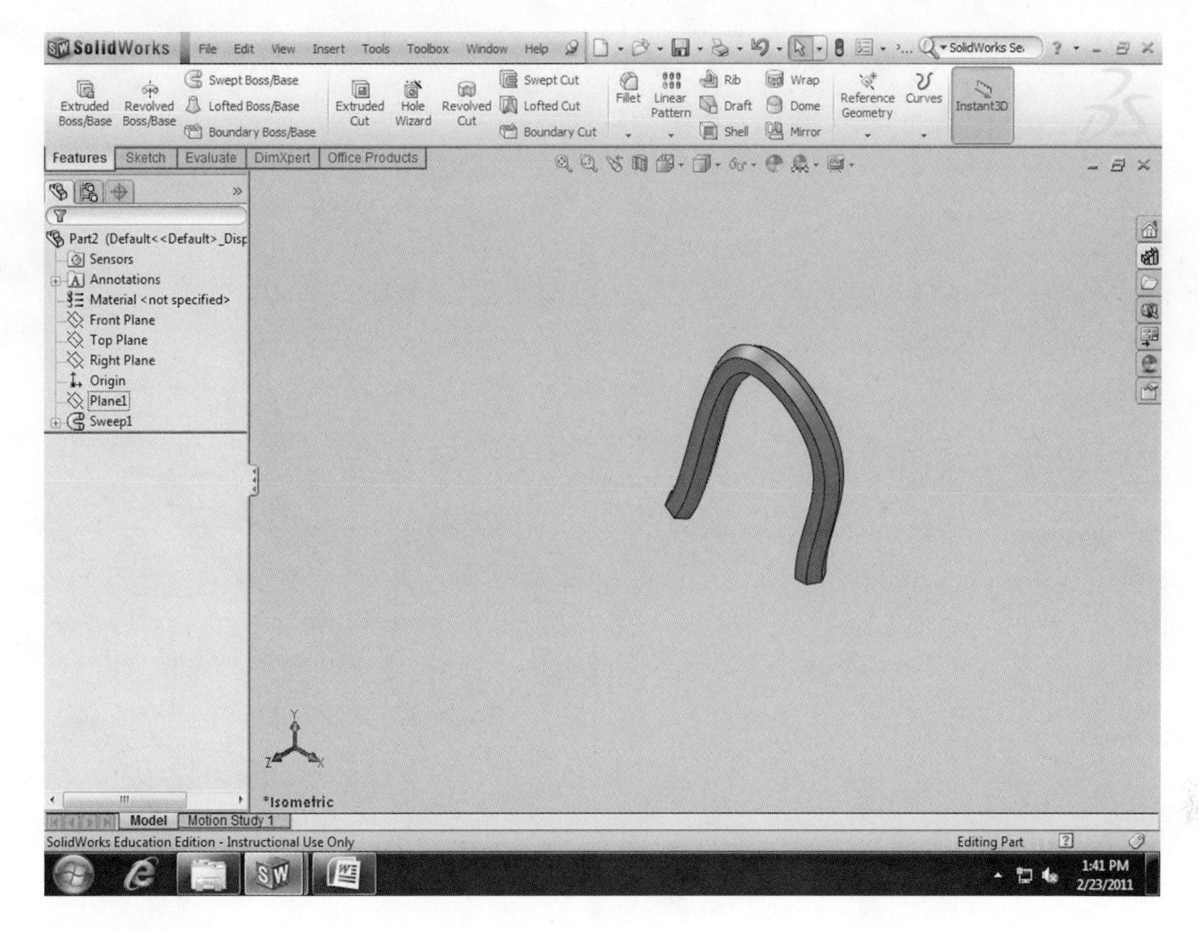

Figure 6.17

3D swept model.

Practical Examples

Among the commonly used swept parts in industrial applications are O-rings, springs, and threaded parts (e.g., nuts and bolts, although these may be archived in the SolidWorks library).

Spring

Springs are modeled using a helix as the path and a circle that is swept through the path as the profile.

1. Select the Top Plane. Create Context tool bar (a circle, radius 0.5 in.) with center at the origin (see Figure 6.18).
2. Select the Isometric orientation.
3. Click on Features > Curves > Helix and Spiral (or Insert > Curves > Helix and Spiral).
   The Helix/Spiral dialog box appears (see Figure 6.19a).
4. Define Helix by Height (1.75 in.); Revolution (6); Starting Angle = 0; and check Clockwise (see Figure 6.19a for values; and see Figure 6.19b for the helix that is created).
5. Click OK to finish helix feature.
6. Click Features > Reference Geometry > Plane from the CommandManager. (The Plane PropertyManager is automatically displayed; see Figure 6.20.)
7. Click the endpoint of the Helix, Context tool bar.
8. Click the Helix, Context tool bar from the FeatureManager. (A plane normal to the Helix at its endpoint is automatically created; see Figure 6.20.)
9. Click OK to accept the plane Plane1 created. (Note: Plane1 is normal to Sketch1 at any point.)
10. Right-click Plane1 and create Sketch2 (a circle, diameter 0.125 in.).
11. Use the Relations tool to make the center of the circle coincident with the origin of the helix, if necessary using pierce relation; click center of circle and helix as Selected Entities and choose Pierce as Relation (see Figure 6.21).

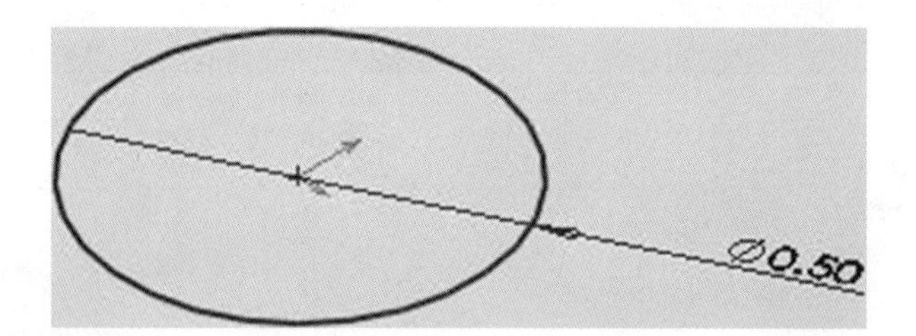

Figure 6.18

Sketch1.

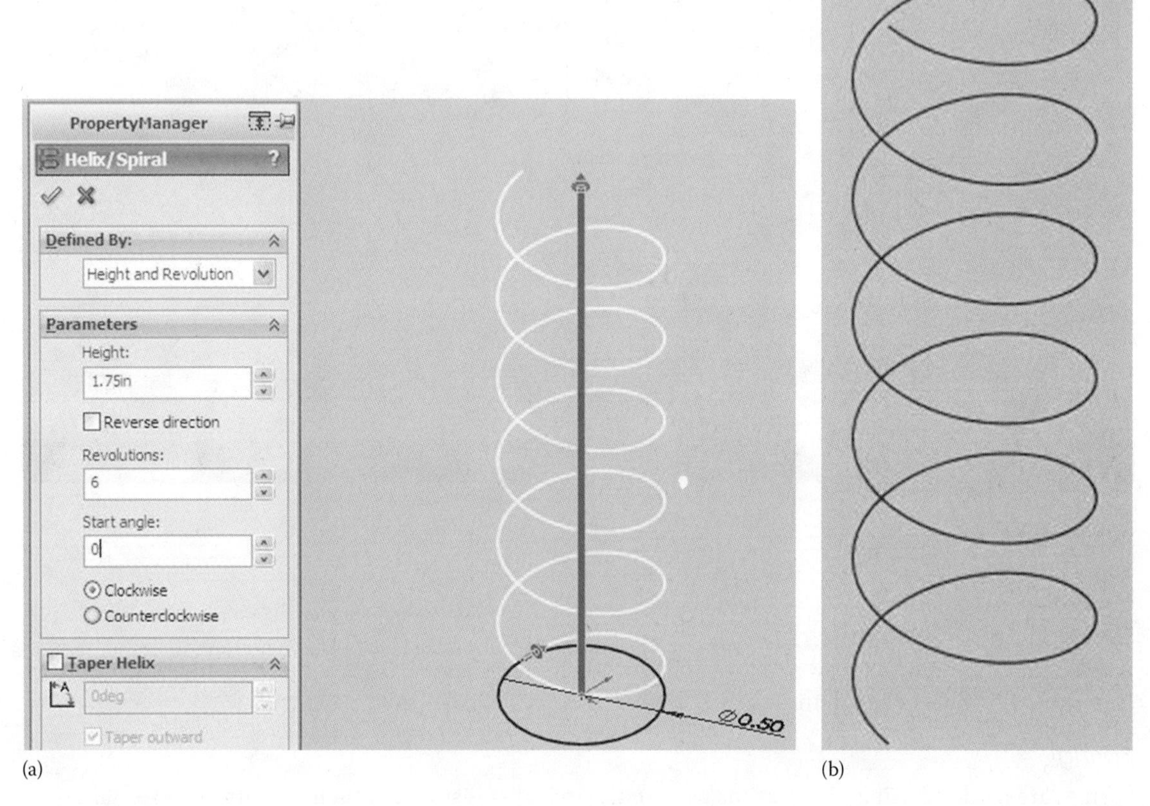

Figure 6.19

Helix/spiral PropertyManager for creating a Helix feature: (a) Helix/Spiral PropertyManagement and (b) Helix.

12. Exit the sketch mode.
13. Click the Swept Boss/Base tool.
    The Swept Properties Manager appears.
14. Right-click the Profile and Path box.
15. Click the Circle as Profile and the Helix as Path. Note: It is preferred to do this through the FeatureManager. A real-time preview will appear (see Figure 6.22a).
16. Click OK to complete the swept cut part, which is now hollow as shown in Figure 6.22b. Hide the planes.

O-Ring

The procedure for modeling an O-ring is similar to that of the spring except that the path is now a circle (say, on the top plane), while the profile remains a circle on another perpendicular plane (front plane). As an illustration, let us model an O-ring with a circular path of diameter 5 in. and a cross section of diameter 0.125 in.

1. Select the Top Plane. Create Context tool bar (a circle, diameter 5 in. and a point) with the center at the origin, as shown in Figure 6.23.
2. Select the Isometric orientation.

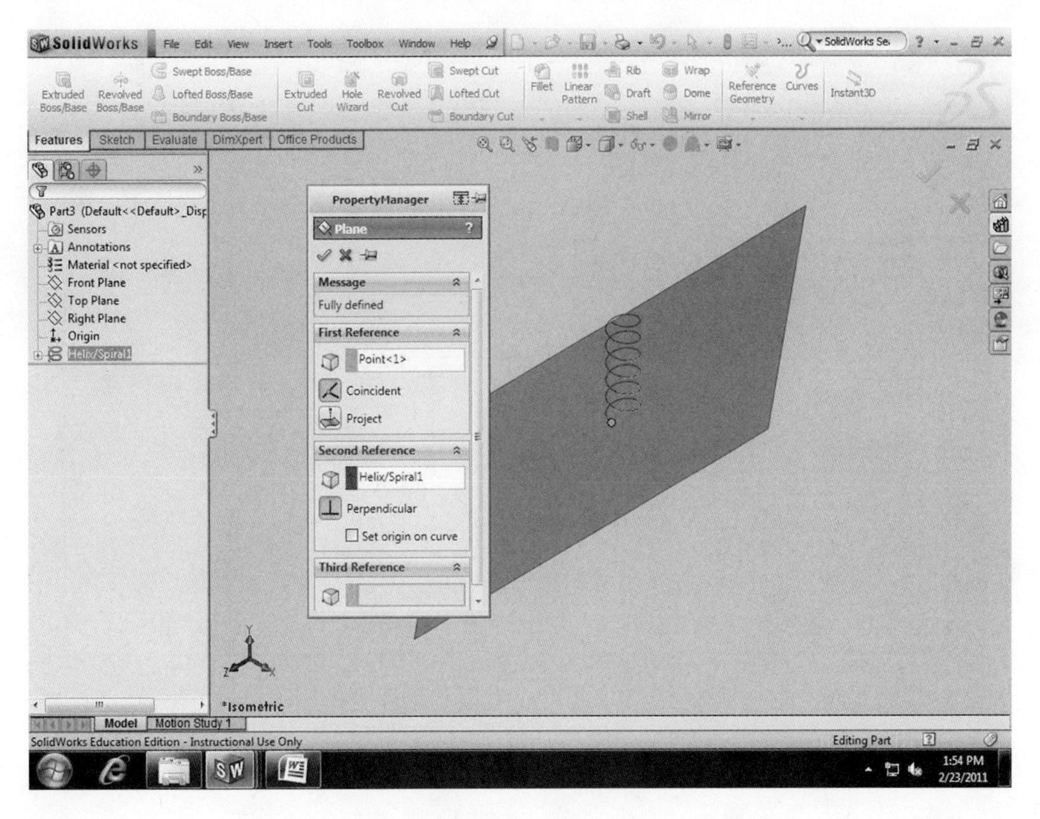

Figure 6.20

Defining normal plane at endpoint of helix.

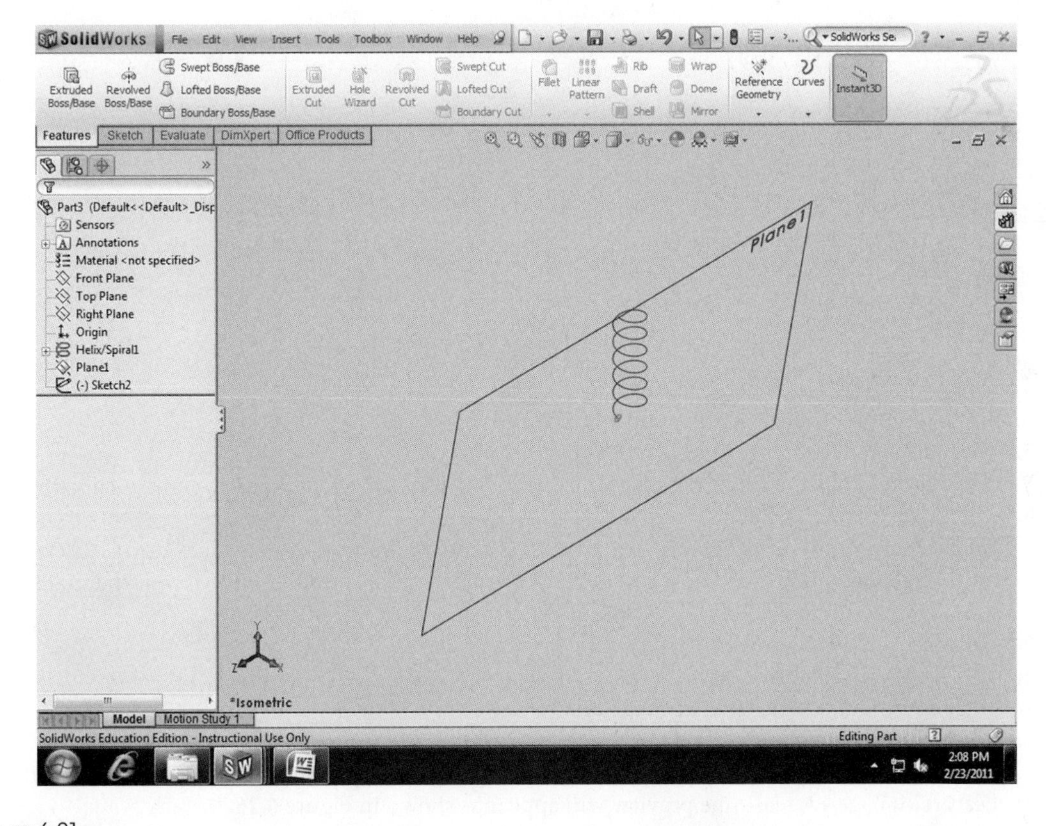

Figure 6.21

Sketch2, a circle on plane normal to helix.

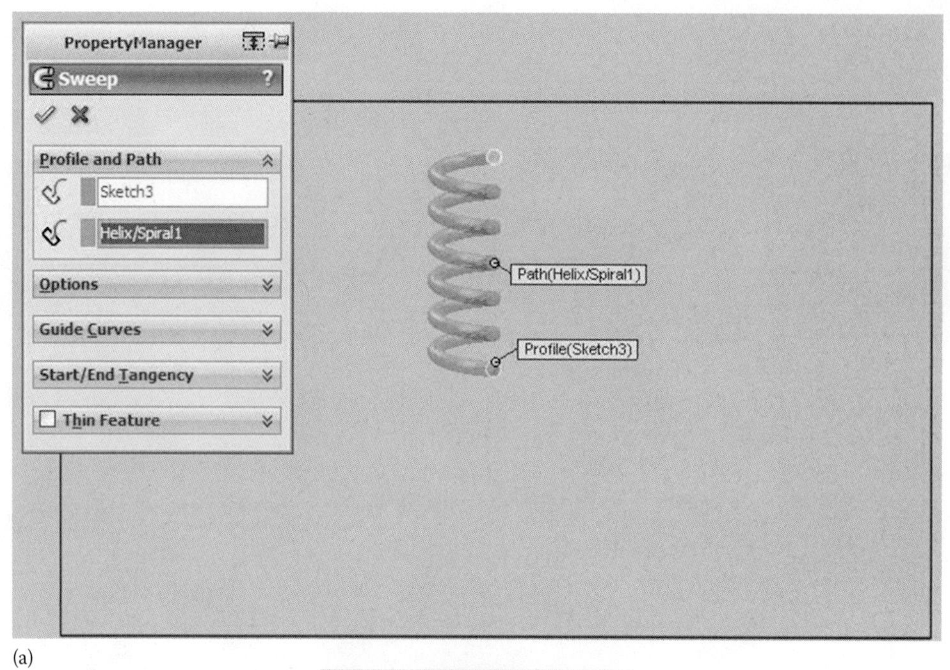

(a)

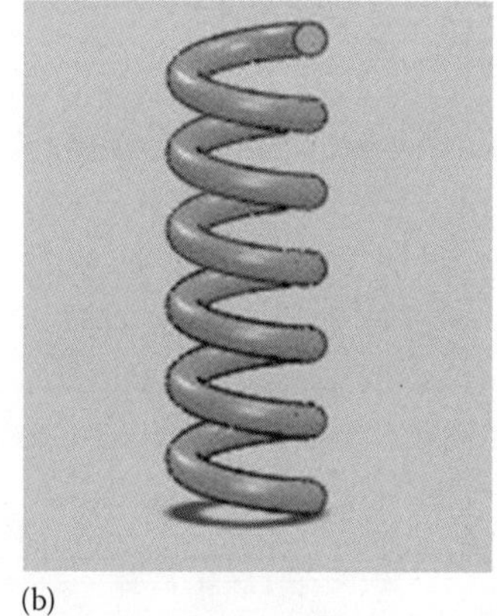

(b)

Figure 6.22

Swept Boss/Base tool for modeling mechanical spring element: (a) preview using path (Helix/Spiral) and profile (circle) and (b) spring element.

3. Exit the Sketch mode.
4. Click Features > Reference Geometry > Plane from the CommandManager.
5. For the First Reference select the Curve and the Perpendicular option.
6. For the Second Reference, select the Point a.
7. Click OK to create the plane Plane1. (Note: Plane1 is normal to Sketch1 at any point; see Figure 6.24.)
8. Right-click Plane1 and create Sketch2 (a circle, diameter 0.125 in.).
9. Add a Pierce relation by clicking the small circle center point. Hold the Ctrl key down. Click on the large circle circumference and release the Ctrl key. Right click Make-Pierce from the dialog box. Click OK (see Figure 6.25).
10. Exit the Sketch mode.
11. Click the Swept Boss/Base tool.

    The Swept Properties Manager appears.
12. Right-click the Profile and Path box.
13. Click the small circle (Sketch2) as profile and the large circle (Context tool bar) as path using the FeatureManager. A real-time preview will appear as shown in Figure 6.26.
14. Click OK to complete the swept cut part, which is now hollow as shown in Figure 6.27. Hide the planes.

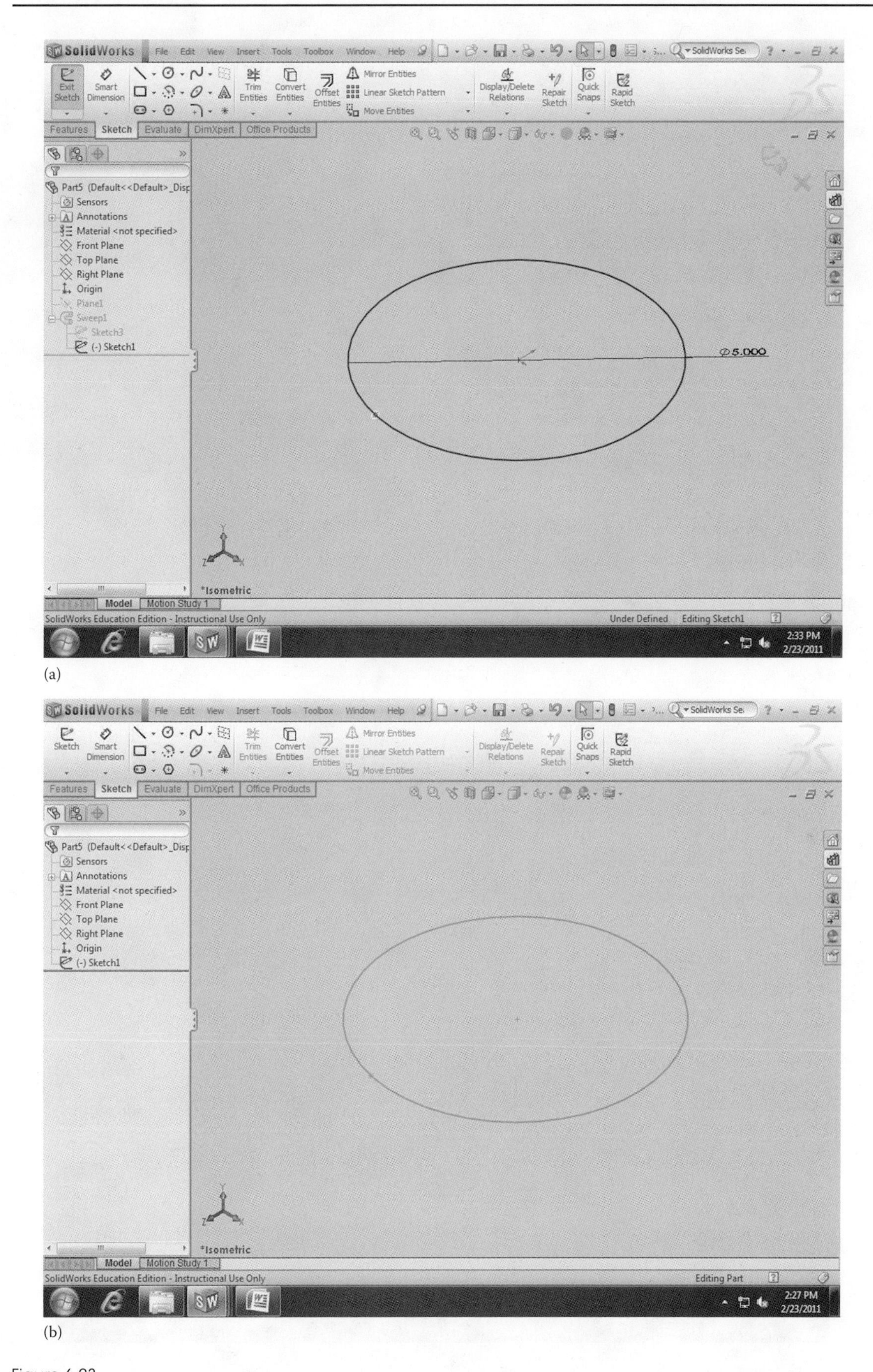

Figure 6.23

Path: (a) path definition and (b) point for defining a plane.

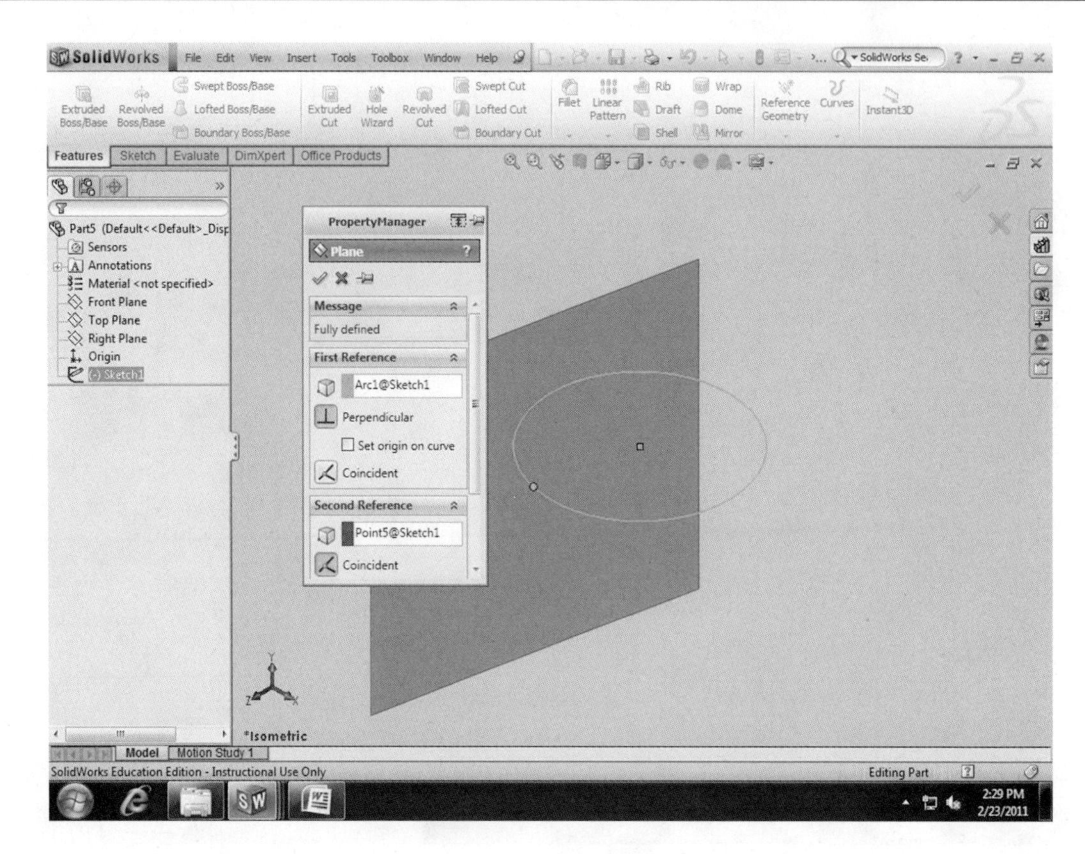

Figure 6.24

Plain1.

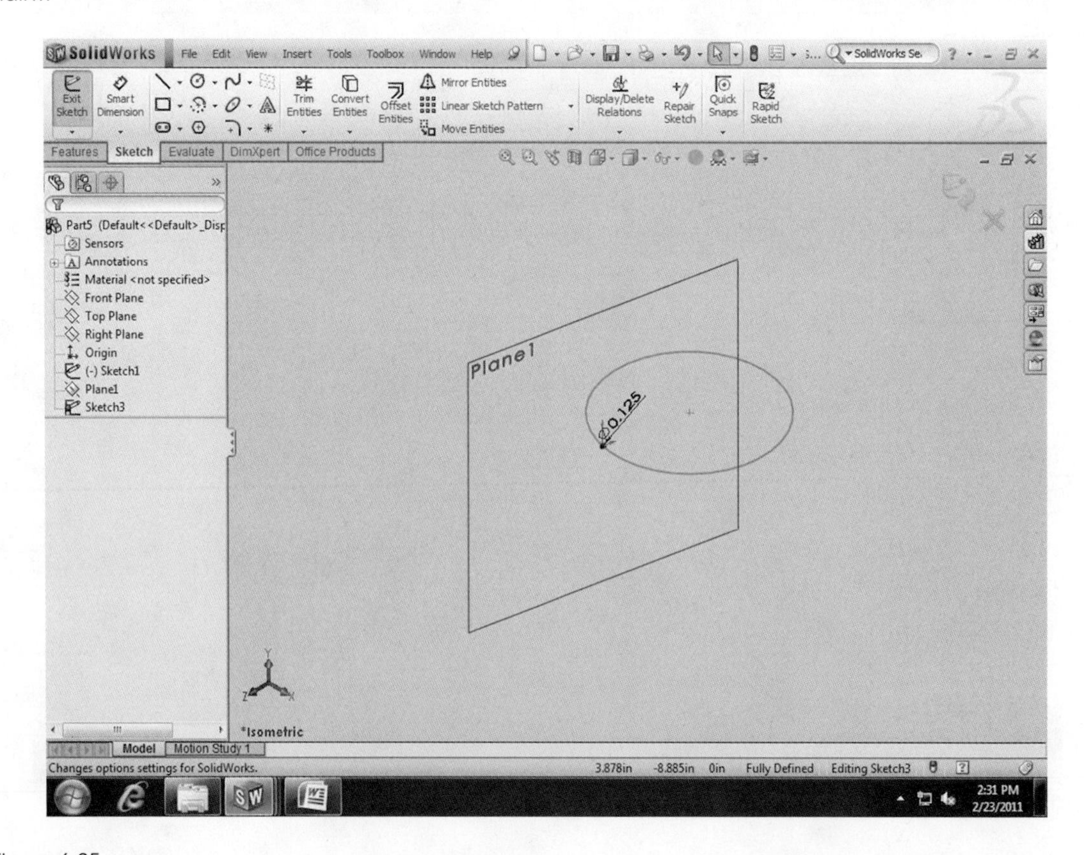

Figure 6.25

Plane definition with Profile (small circle) and path (large circle).

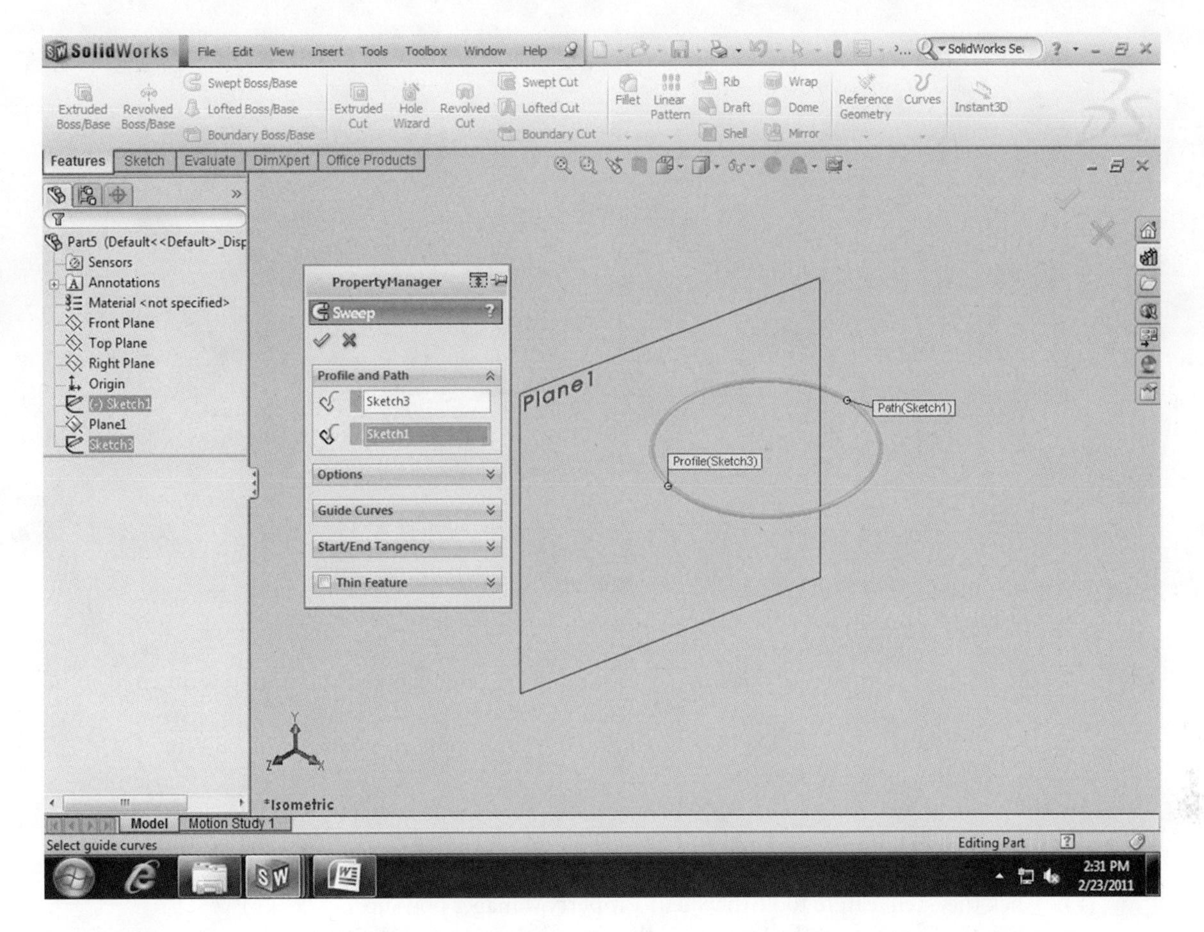

Figure 6.26

Sweep PropertyManager for O-ring preview.

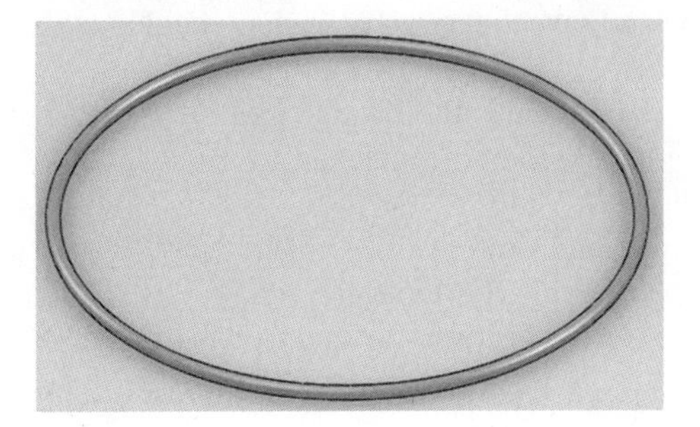

Figure 6.27

O-ring model.

Threaded Cap

Threaded parts are modeled using a helix as the path and a shape of interest swept through the path, as the profile. As a small project, let us first model a cap having a draft angle. Then, we will add a trapezoidal profile as an internal thread for a helical path. The procedure for achieving our objective could be simplified by considering two modules: (1) Cap without internal thread (steps 1–6) and (2) Cap with internal thread (steps 7–11). Let us consider these two modules.

Cap without Internal Thread

1. Select the Front Plane. Create Context tool bar (a circle, radius 5 in.) with center at the origin.
2. Click the Extruded Boss/Base feature tool.

   The Extrude PropertyManager is displayed with Blind as the default End Condition in Direction 1.
3. Enter 1.725 in. as Depth in Direction 1. Click the Draft On/Off button.

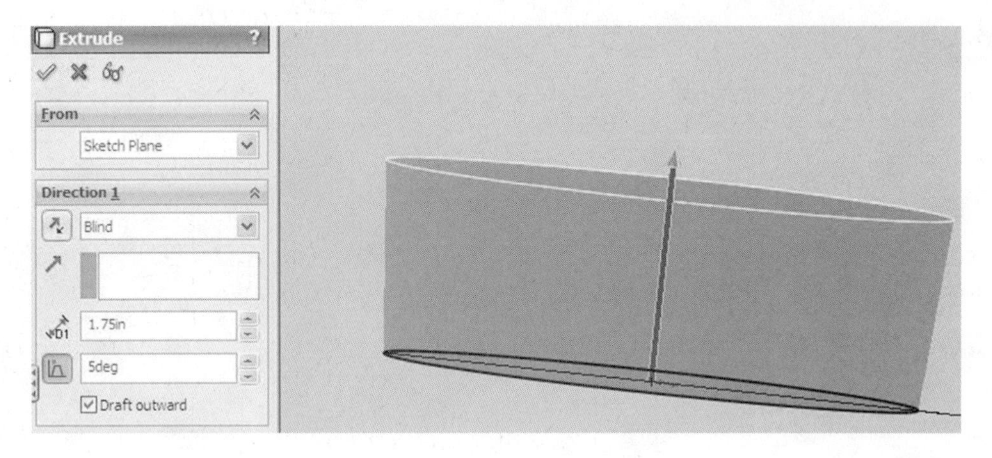

Figure 6.28

Preview of Sketch1 having outward draft and extruded.

4. Enter 5deg as Draft Angle. Click the Draft outward box.
5. Click OK to accept the model (see Figure 6.28).
6. Right-click the front face of the Sketch plane.
7. Click Sketch from the toolbar. Then click the Circle Sketch tool for a circle that is centered at the origin (see Figure 6.29).
8. Using the Smart Dimension tool, dimension the circle to have a diameter of 3.875 in.
9. Click Extrude cut.
10. Enter depth of 0.275 in. for Depth in Direction 1. (See preview in Figure 6.30.)
11. Click Draft On/Off button and enter 5 degrees for Angle, accepting the default settings.
12. Click OK (see Figure 6.31).
13. Click the Shell feature tool. The Shell1 PropertyManager pops up.
14. Click the front face of the Front-Cut, then rotate the part and click the back face.
15. Enter 0.15 in. for shell thickness (see Figure 6.32).
16. Click OK from the Shell1 PropertyManager.
17. Click isometric view and save the part that is shown in Figure 6.33.

Cap with Internal Thread

18. Select the narrow back face. Click the Hidden Lines Removed option.
19. Click Feature > Reference Geometry > Plane.
20. Enter 0.45 in. for Distance.
21. Click the Reverse direction box.
22. Click OK from the plane PropertyManager.

    Plane1 is displayed in the FeatureManager (see Figure 6.34).

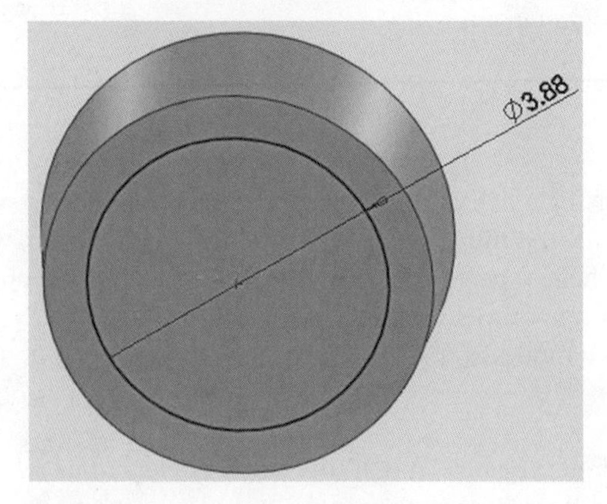

Figure 6.29

Creating Sketch2, a circle for and extrude-cut.

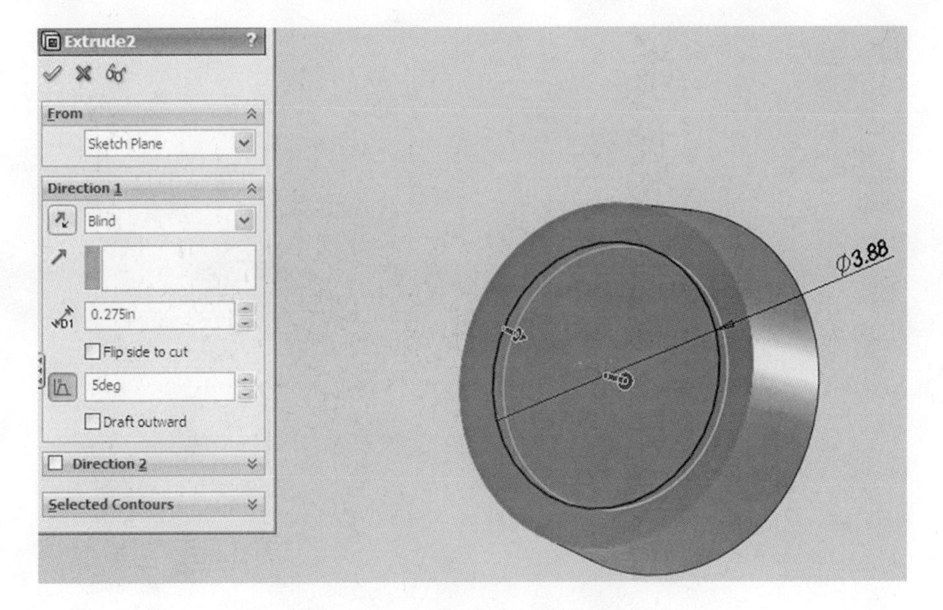

Figure 6.30

Preview for Extrude-cut using Sketch2 (circle).

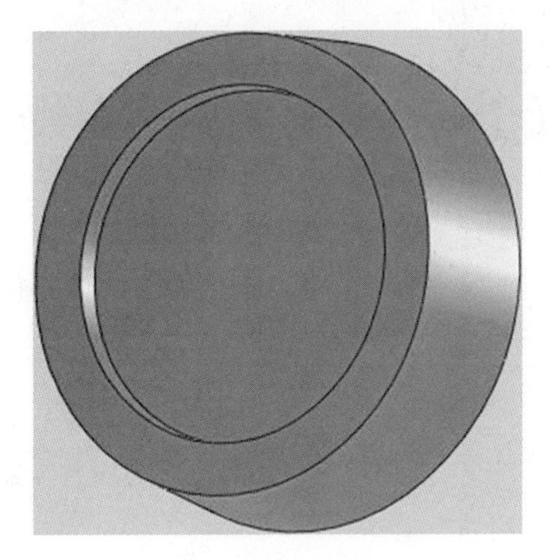

Figure 6.31

Extrude-cut using Sketch2 (circle).

23. Right-click Plane1 from the FeatureManager.
24. Click Sketch from the tool bar.
25. Click the back inside circular edge of the Shell.
26. Click Convert Entities Sketch tool. The circular edge is displayed on Plane1 (see Figure 6.35).
27. Click on Features > Curves > Helix and Spiral (or Insert > Curves > Helix and Spiral).
    The Helix/Spiral dialog box appears.
28. Define Pitch by Pitch (0.125 in.) and Revolution (3) and check Clockwise (see Figure 6.36).
29. Click OK to complete the helix.
30. Create a plane, Plane1 Normal To the spring endpoint: Click the endpoint of the spring and click Feature > Reference Geometry > Plane. Alternatively: Sketch on Right Plane.
    The Helix PropertyManager dialog box appears. Click on the Helix/Spiral1 on the FeatureManager to which this endpoint, <Point1> belongs.
31. Click on Plane1 and create Sketch2 (a circle, diameter 0.125 in. coinciding with the origin of the helix).
32. Exit the sketch mode.
33. Click the Swept Boss/Base tool.
    The Swept Properties Manager appears.

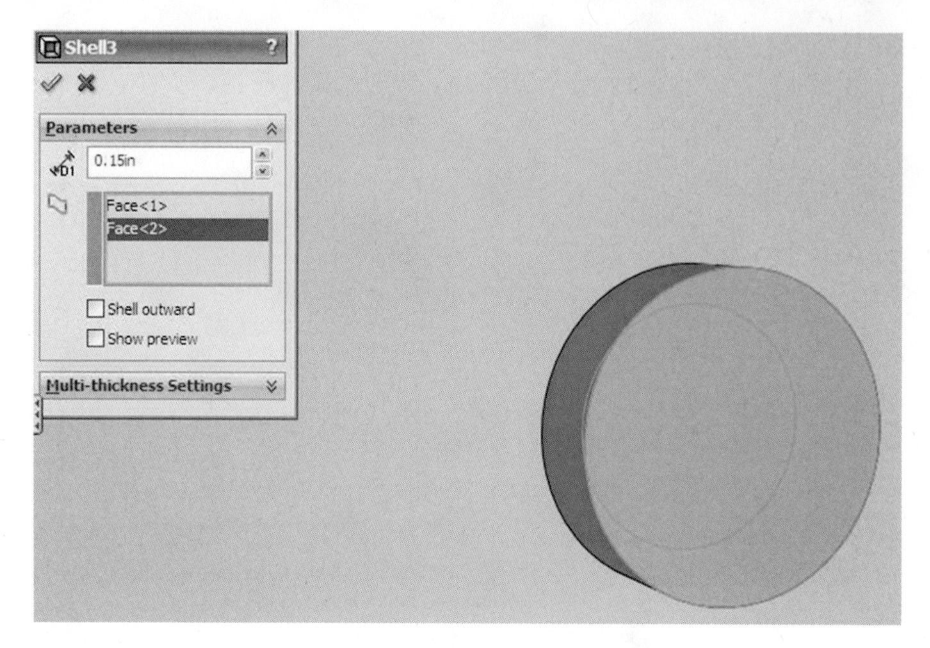

Figure 6.32

Preview for shelling operation.

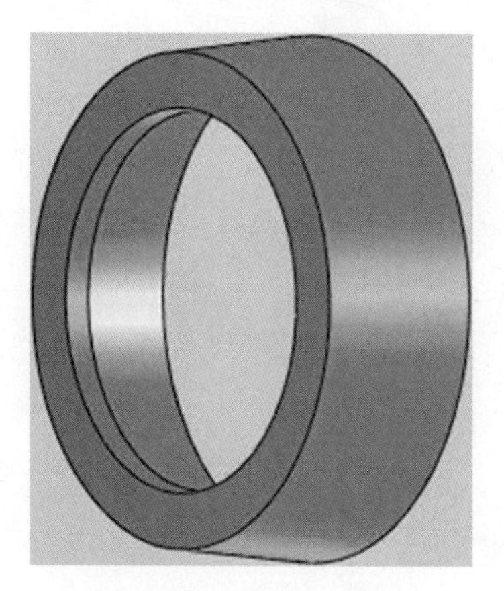

Figure 6.33

Completed part.

34. Right-click the Profile and Path box.
35. Click the circle as profile and the helix as path. Note: It is preferred to do this through the FeatureManager. A real-time preview will appear.
36. Click OK to complete the swept cut part, which is now hollow (see Figure 6.37). Hide the planes.

Lofted Boss/Base

The Lofted Boss/Base tool is used to create a shape between a number of planes, each of which contains a defined shape or geometry. The prerequisite for creating a 3D lofted model is to first sketch the shapes on the different planes. Let us illustrate a lofted boss/base model using a square base, a circle, and an ellipse. There are no restrictions in the shapes that could be used. Let us illustrate the lofted boss/base concept as follows:

1. Select the Top Plane, be in sketch mode and create Context tool bar (a square, 90 mm × 90 mm) as shown in Figure 6.38.
2. Exit the sketch mode.
3. Click the Features tool and click the Reference tool. Select the Plane option. The plane box appears; select the top plane.

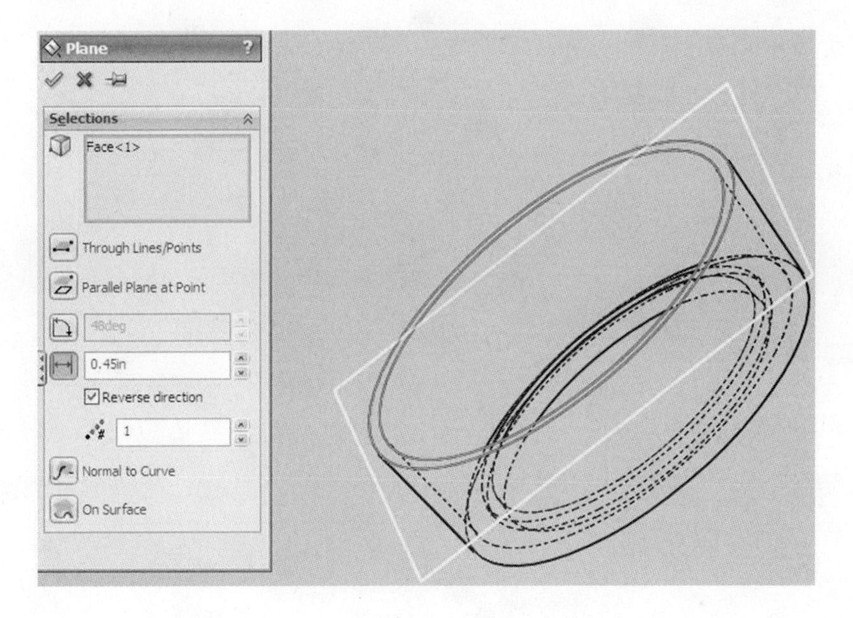

Figure 6.34

Creating Plane1.

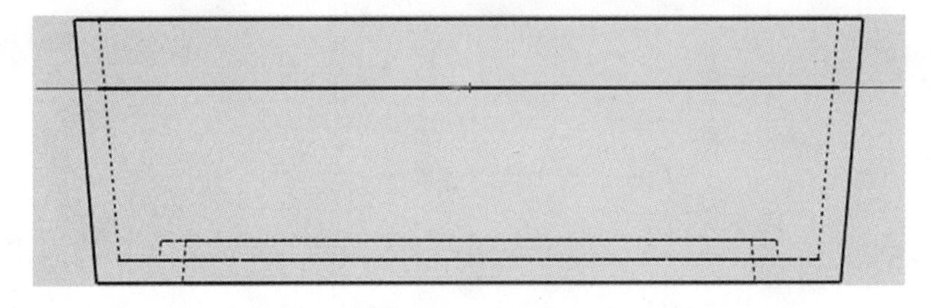

Figure 6.35

Extracting the circular edge using the Convert Entities Sketch tool.

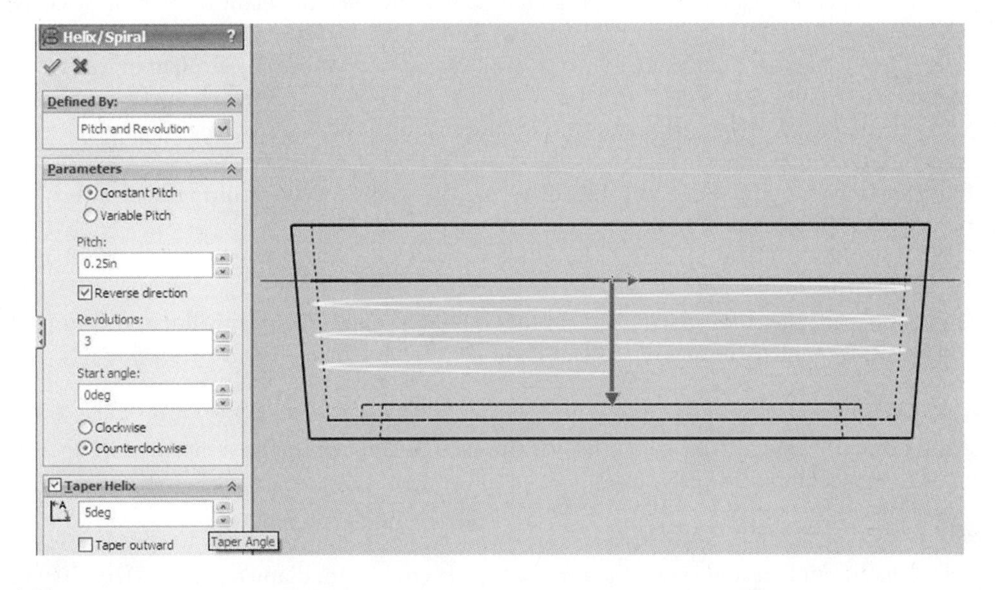

Figure 6.36

Preview of the helix for internal threading.

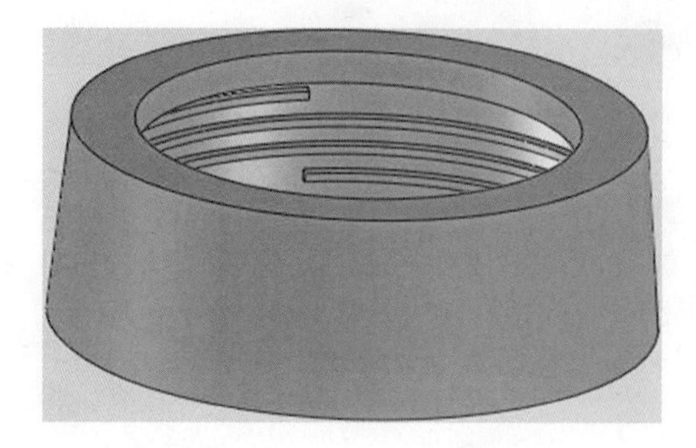

Figure 6.37

Completed part with internal thread.

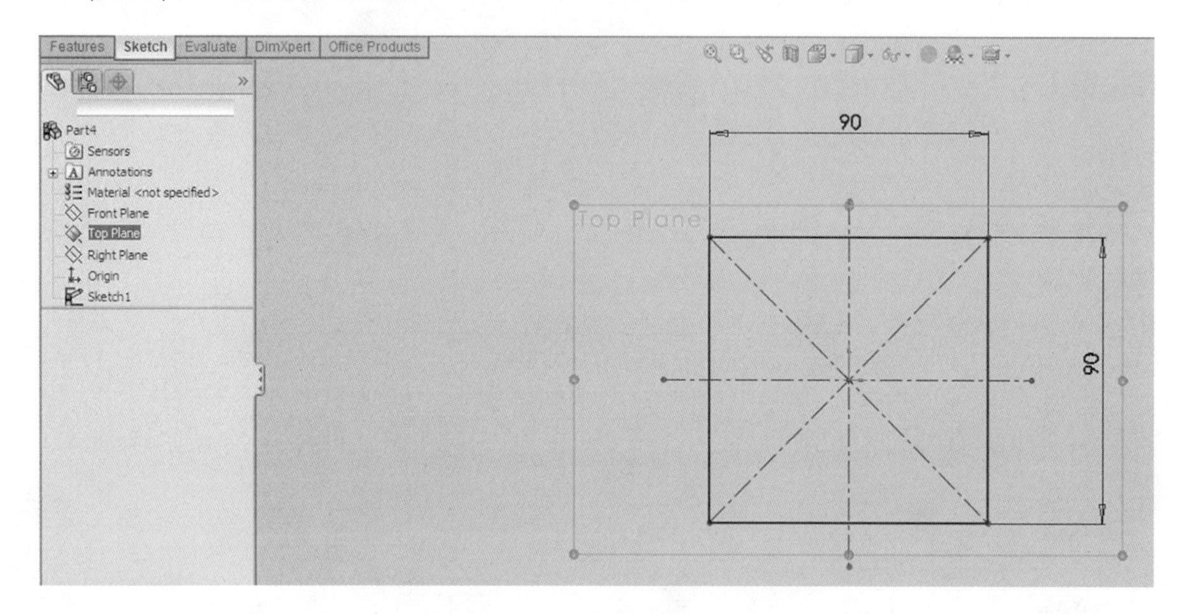

Figure 6.38

Bottom sketch for lofting (Sketch1).

4. Set the distance between the existing Top Plane and a new reference plane for 50 mm. Select Number of Planes = 2. Click OK check mark. Two new planes, Planel and Plane2, appear (see Figure 6.39).
5. Select the Planel, be in sketch mode and create Sketch2 (a Circle of diameter, 60 mm; see Figure 6.40a).
6. Exit the sketch mode.
7. Select the Plane2, be in sketch mode and create Sketch3 (an ellipse of major diameter, 80 mm and minor diameter, 65 mm; see Figure 6.40b).
8. Exit the sketch mode (see Figure 6.41 for the three profiles needed for lofting).
9. Click the Lofted Boss/Base tool.
   The Loft Properties Manager appears.
10. Right-click the Profiles box.
11. Click the Square, then the Circle, and finally the Ellipse. A real-time preview will appear (see Figure 6.42).
12. Click OK to complete the lofted part (see Figure 6.43). Hide the planes.

Practical Examples

Two practical examples are given based on the principles of lofted parts. One example is an impeller that is commonly used in a compressor, turbine design, and the other is an aircraft wing.

Impeller

1. Select the front plane and create *Context tool bar* (a circle 3-in. diameter) and extrude 0.08 in. (see Figure 6.44).
2. Create *Sketch2* (a circle 0.6-in. diameter) and extrude 1.5 in. (see Figure 6.45).

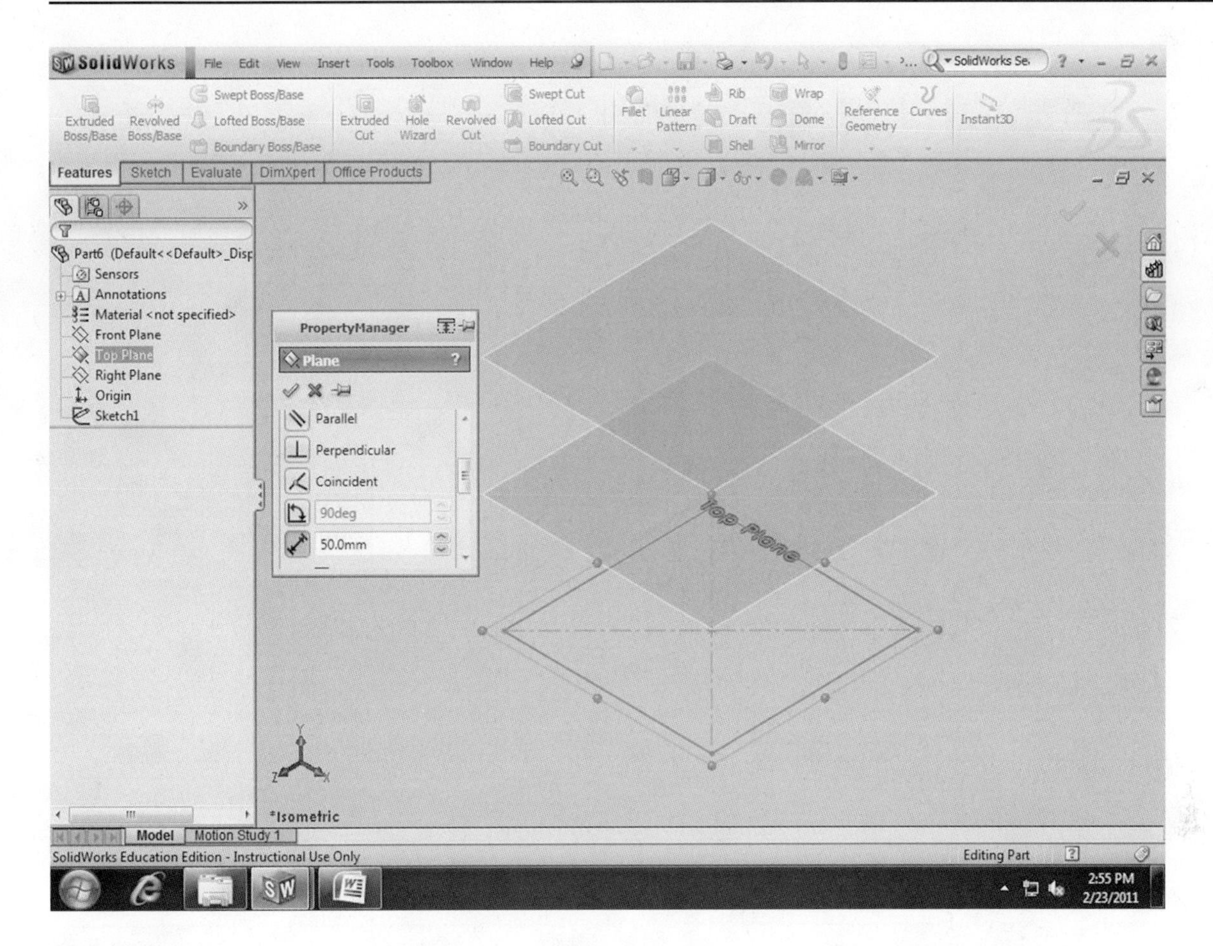

Figure 6.39

Plane1 and Plane2.

3. Create Sketch3 (two arcs, 1.05-in. and 1.03-in. radii, respectively) on top of the extruded face; the center of one of them is (0.75 in., 0.81 in.) away from the center of Sketch1; the center of the other is offset 0.08 in. from the first (see Figure 6.46). Use Convert Entities to extract the concentric circles (3-in. and 0.6-in. diameter, respectively). Trim the circles from Sketch3 (see Figure 6.47).
4. Extrude Sketch3 to 0.6 in. (see Figure 6.48).
5. Define 2 Reference Planes, Plane1 and Plane2 offset 0.68 in. and 0.85 in., respectively, from the Front Plane (see Figure 6.48).
6. Insert Sketch4 on Plane1, selecting all edges to extrude the fin and convert them to entities using Convert-Entities tool. Notice that Plane1 is active (see Figure 6.49).
7. Exit the sketch mode.
8. Insert Sketch5 on Plane2. Create a circle, 1.77 in. diameter (see Figure 6.50).
9. Create Sketch5 with two arcs with their ends touching the circles 1.77 in. diameter and 0.62 in. radius at an angle of 30 degrees; the centers are 0.08 offset from one another (see Figure 6.50). The geometry, Sketch5 is trimmed, as shown in Figure 6.51. The fully defined Sketch5 is shown in Figure 6.52.
10. Return to isometric view.
11. Click the *Features* tool and click the Reference tool. Select the Plane through Vertices option. Create two planes, Plane3 (including the left vertices of Sketch4 and Sketch5 and any other point) and Plane4 (including the right vertices of Sketch4 and Sketch5 and any other point).
12. Sketch spline using 3D Sketch Tool on Plane3 to define Lofting Profile1 and another Lofting Profile2 on Plane4, as shown in Figure 6.53.
13. Create Loft base using Sketch4 and Sketch5 as Profiles, Profile1 and Profile2 as Guide Curves, as shown in Figure 6.54.
14. Click OK and Hide Plane3 and Plane4.

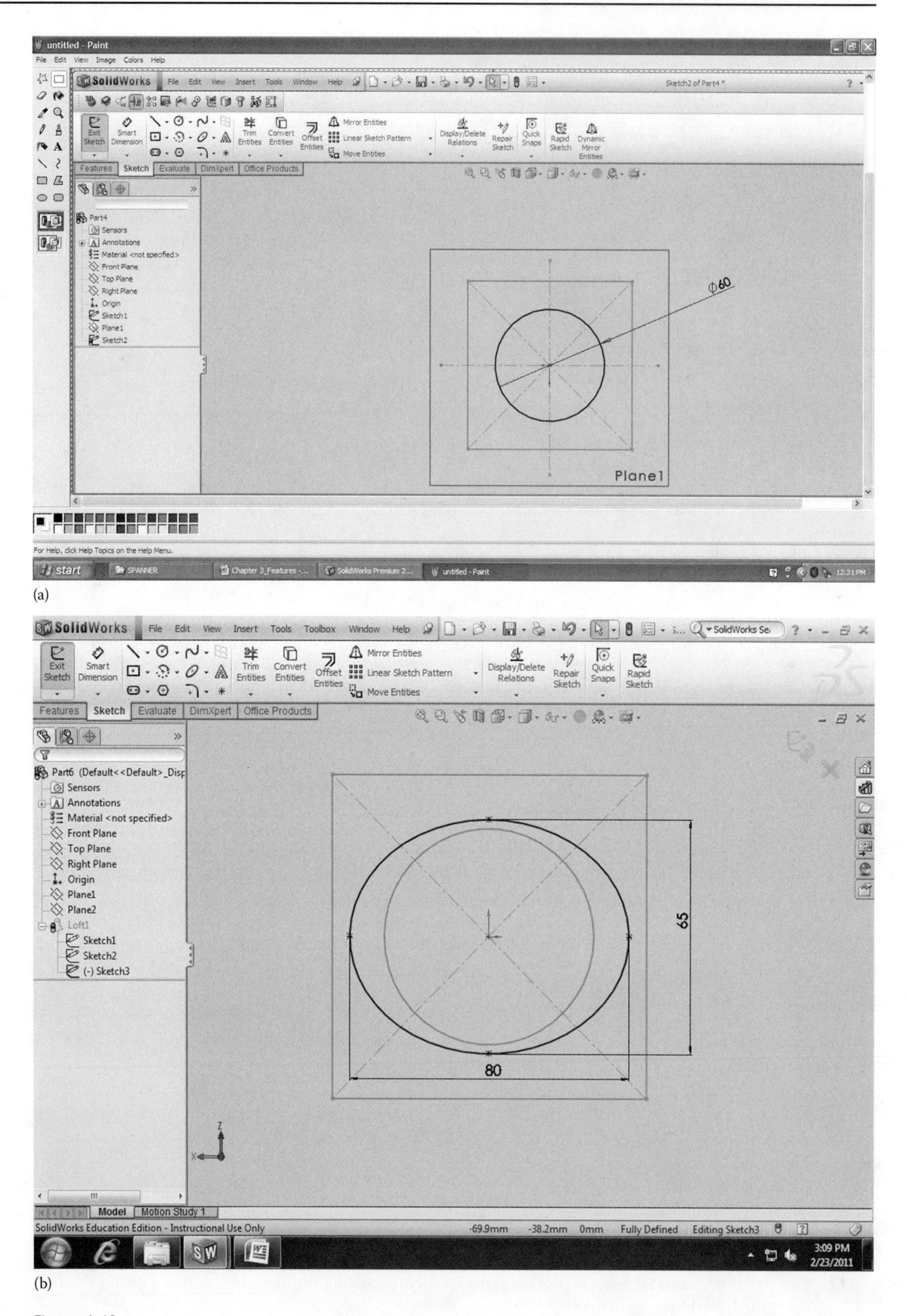

Figure 6.40

Sketch2 and Sketch3 on Plane1 and Plane2, respectively: (a) middle sketch for lofting (Sketch2) and (b) top sketch for lofting (Sketch3).

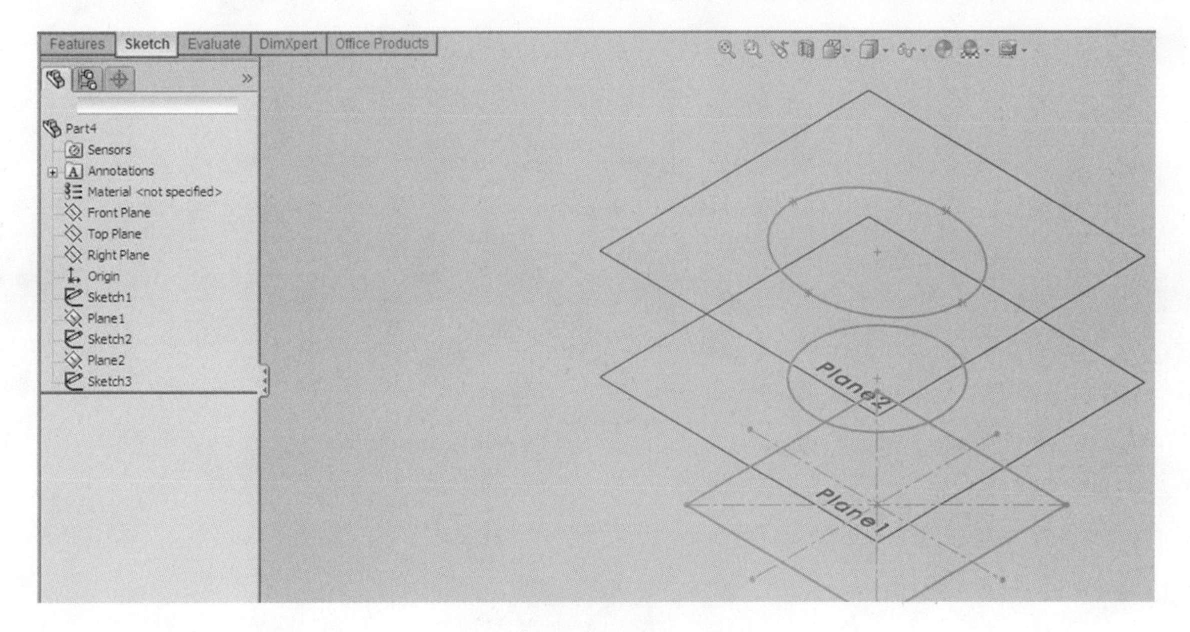

Figure 6.41

Three sketches for lofting.

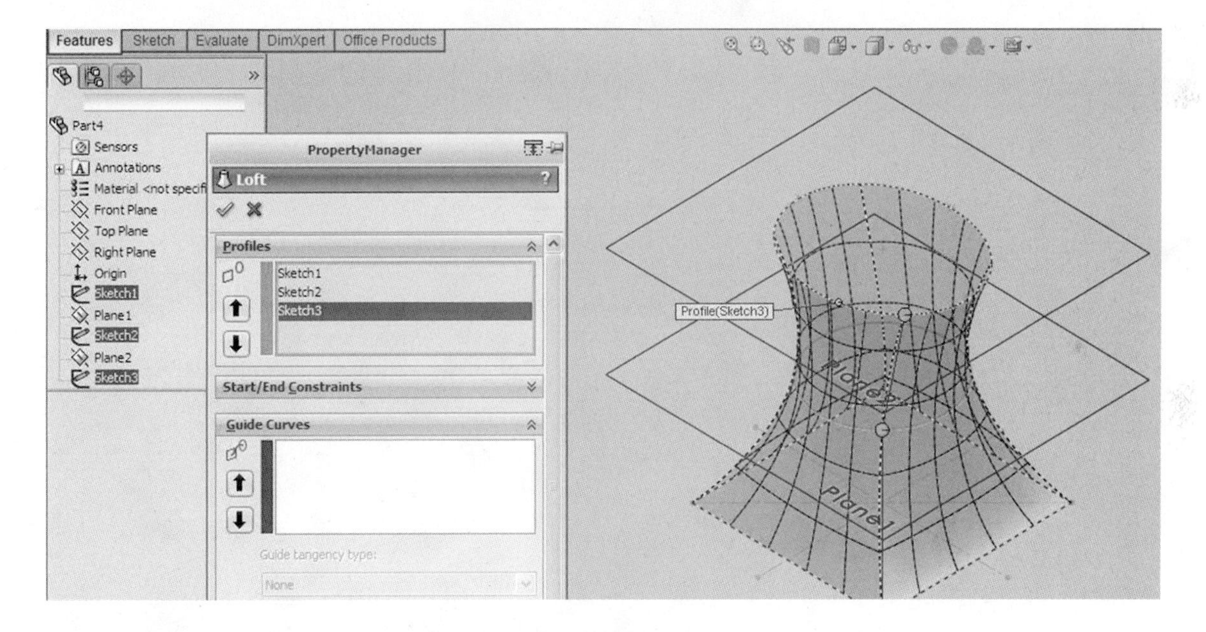

Figure 6.42

Loft based on Sketch1, Sketch2, and Sketch3.

15. Click View > Temporary Axes (axes are displayed; look out for the axis about the cylinder at its top. Note that it may be very small to detect).
16. Click Circular Pattern, temporary axis, Axis<2> about which rotation takes place, 360 degrees for angle, 12 for number of patterns and check Equal Spacing (see Figure 6.55). Click OK. The fully modeled part is shown in Figure 6.56.

Aircraft Wing

1. Select the Right Plane and create Context tool bar as shown in Figure 6.57. A control polygon is used to help in the definition of the sketch.
2. Use Spline Tool to sketch a spline profile with length of 6 in.
3. Create Plane1 10 in. from the Right Plane (see Figure 6.58).

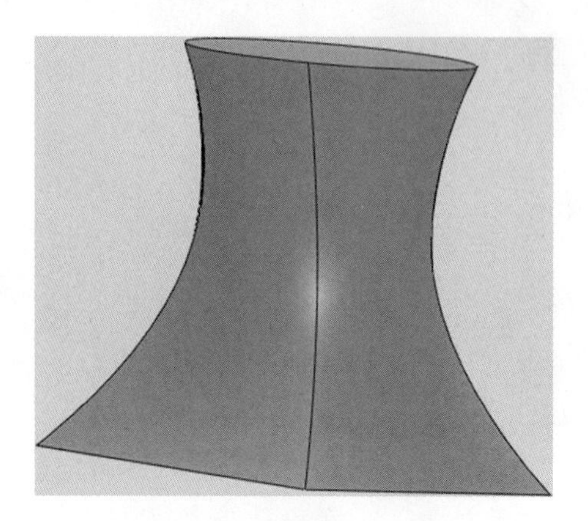

Figure 6.43

3D lofted model.

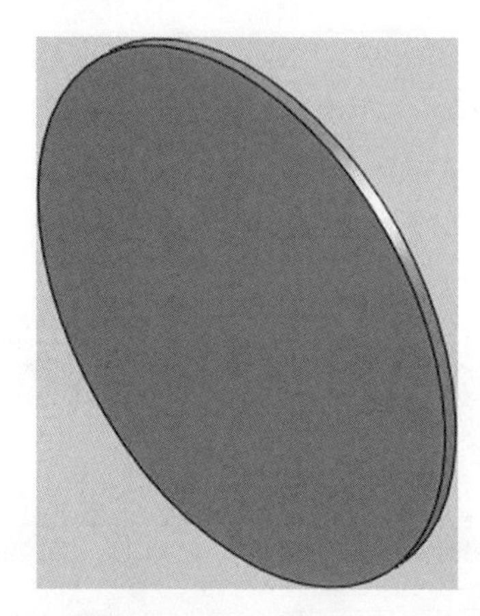

Figure 6.44

Extrude1 from Sketch1.

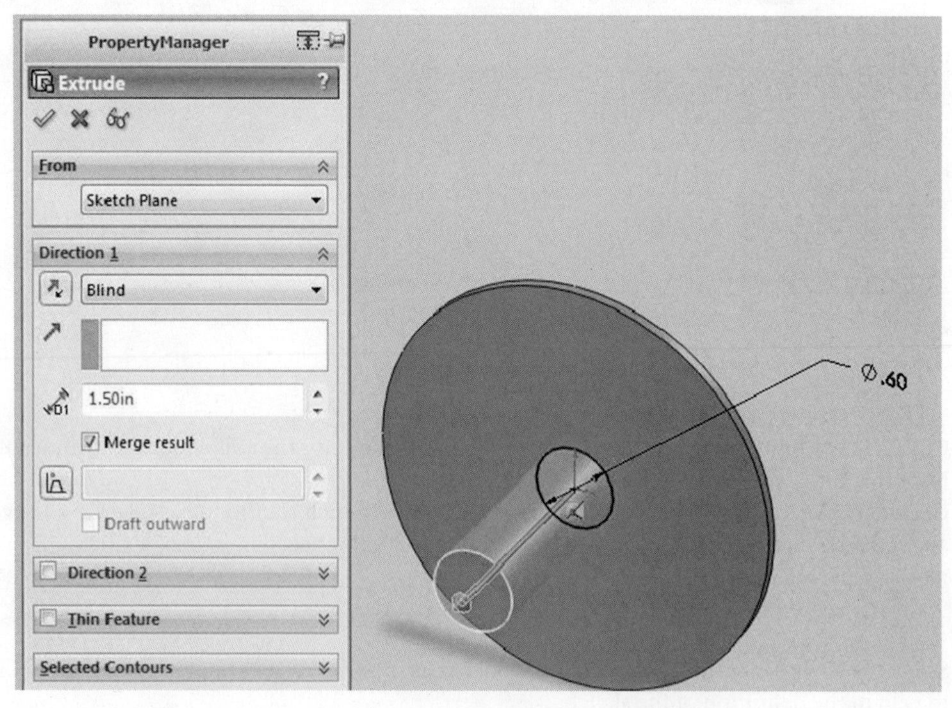

Figure 6.45

Extrude1 from Sketch2.

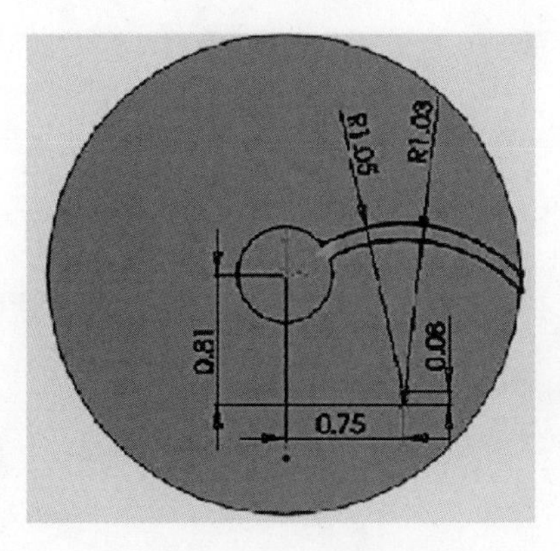

Figure 6.46

Geometric definitions for fin profile on top of Extrude1.

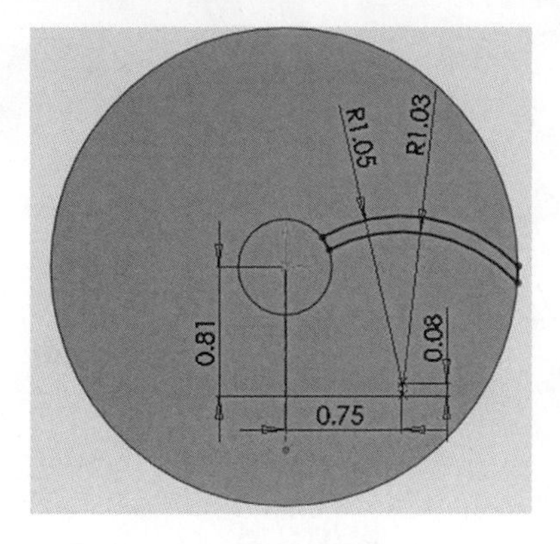

Figure 6.47

Trimmed geometries to fully define fin profile on top of Extrude1.

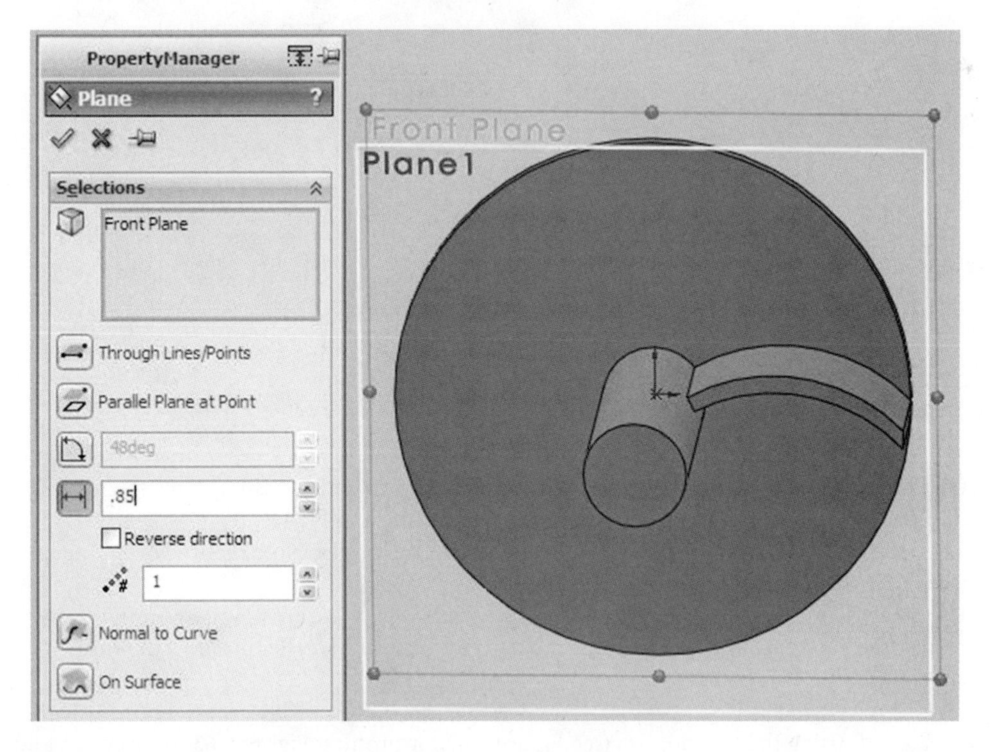

Figure 6.48

Extrude2 for fin profile and creating Plane1 on top of Extrude2.

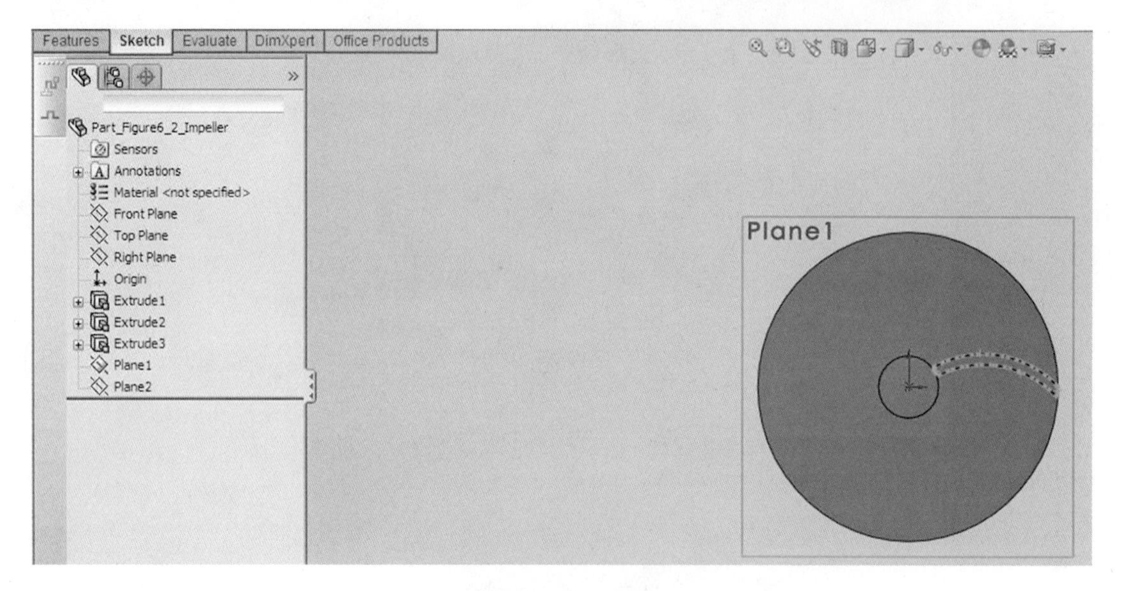

Figure 6.49

Extracting fin profile onto Plane1 from top of Extrude2.

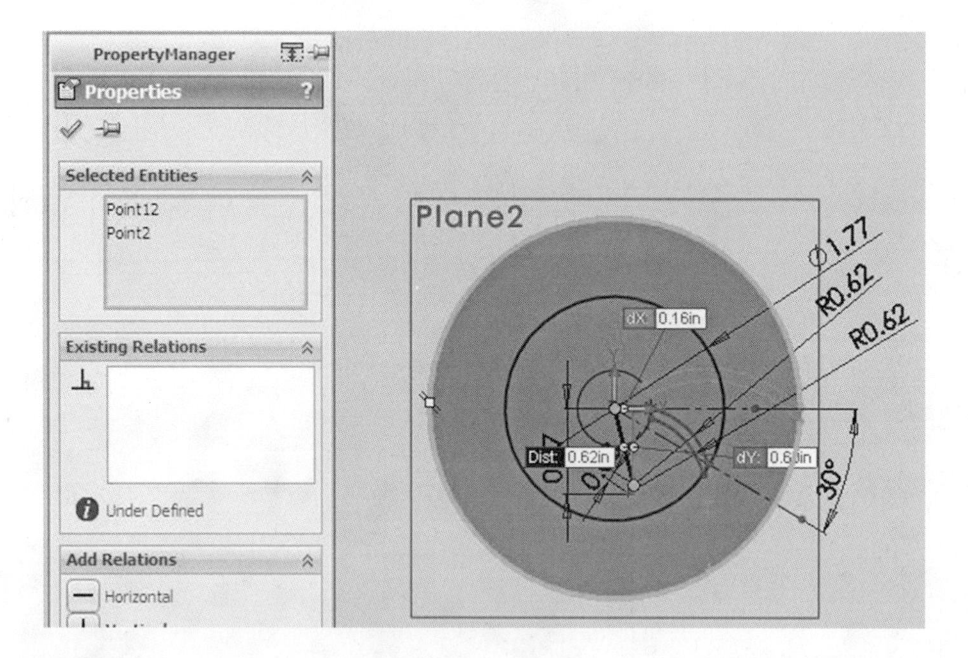

Figure 6.50

Defining geometries for fin profile on Plane2.

4. On Plane1, select all edges of Sketch1 and convert them to entities using Convert-Entities tool. This ensures that we maintain similar profiles on Right Plane and Plane1.
5. Copy the converted entities to start 4 in. away from the origin. Note that the overall length is 6 in. (see Figure 6.59).
6. Use Smart Dimension Tool to change 6 in. to 3.5 in. Note: there are now two similar profiles on two planes (see Figure 6.60). Delete the larger profile to retain the smaller profile if the entities converted are copied as in Figure 6.61.
7. Exit Sketch.
8. Create Loft using the two profiles (see Figure 6.62 for profiles; Figure 6.63 for preview; and Figure 6.64 for completed part).
9. Hide Plane1.

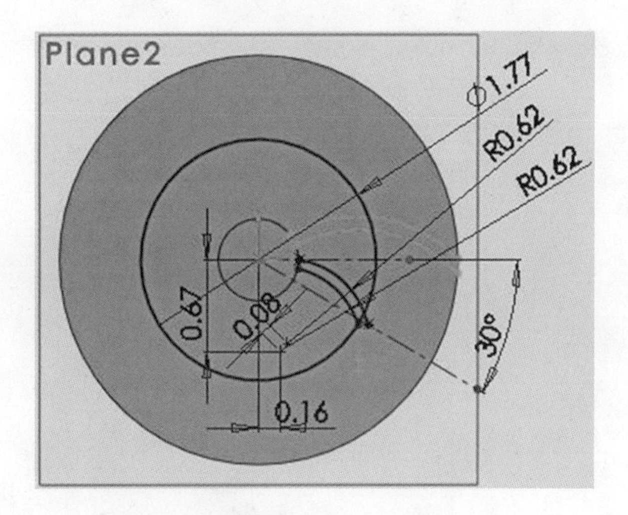

Figure 6.51

Geometries to be trimmed for fin profile on Plane2.

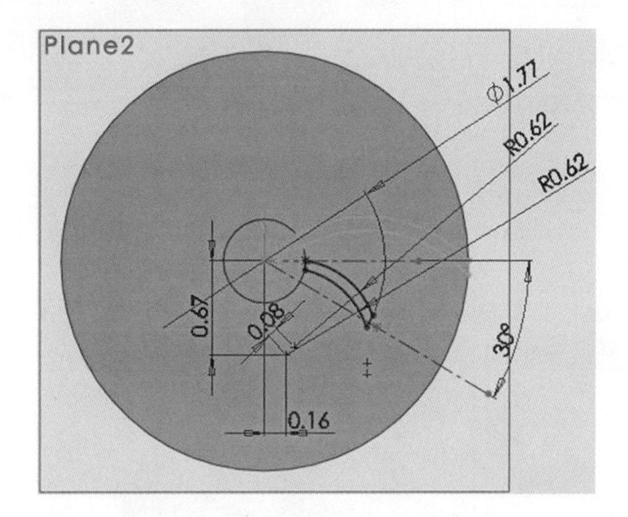

Figure 6.52

Fin profile on Plane2 fully defined.

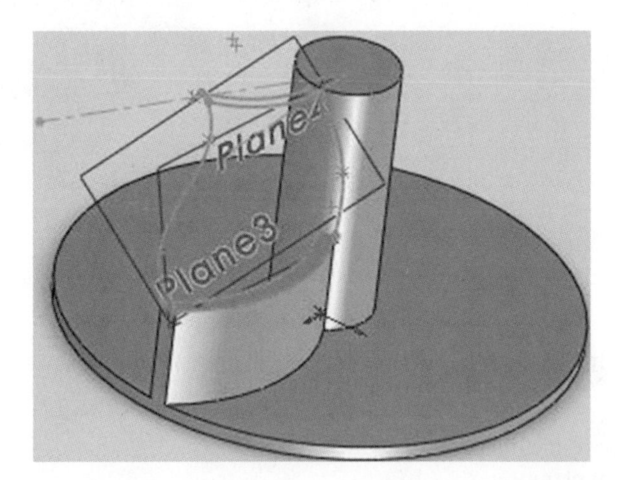

Figure 6.53

Creating Planes Plane3 and Plane4.

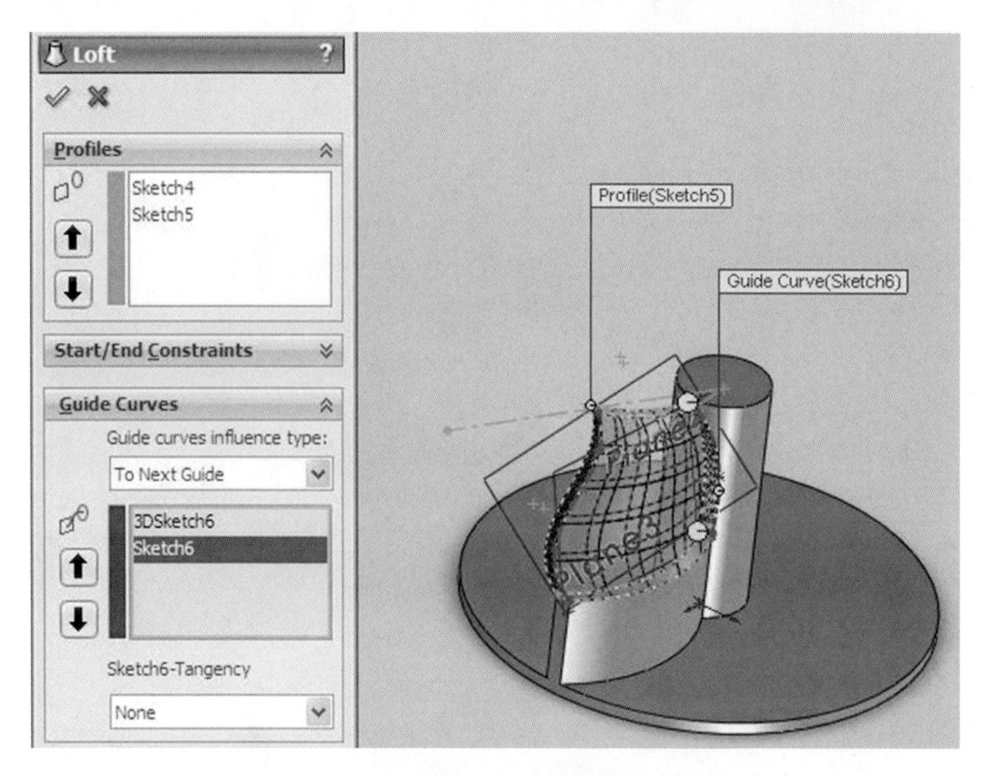

Figure 6.54

Lofted upper fin.

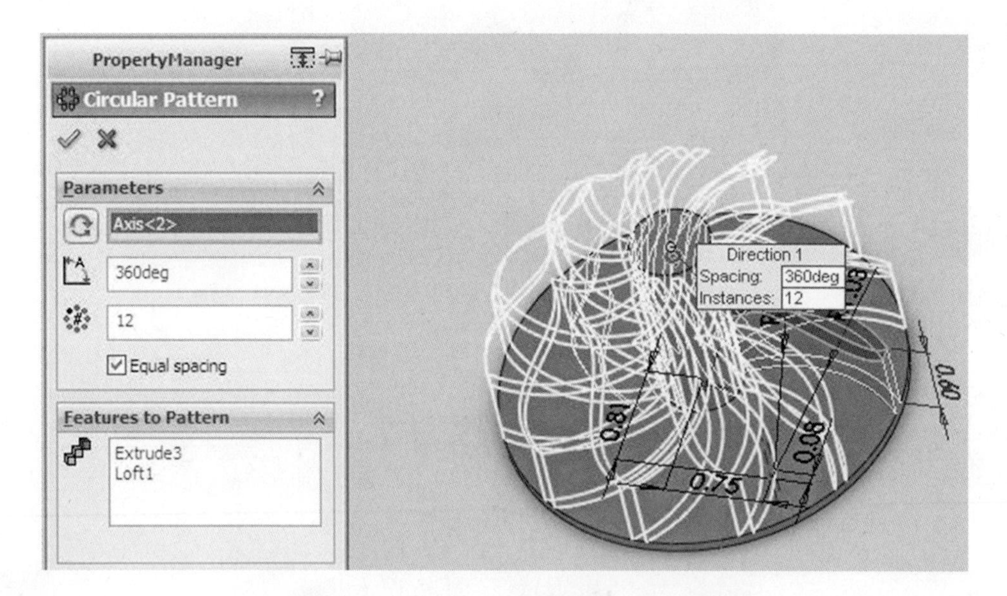

Figure 6.55

Circular pattern of fins.

Practical Swept Feature: Elbow Casting for Steam Valve

Sketch Path for Sweep

1. Open a new part document.
2. Select the front plane.
3. Sketch a quarter-circle, Sketch1, with radius equal to 70 mm (see Figure 6.65).
4. Exit sketch mode.

Figure 6.56

Completed impeller.

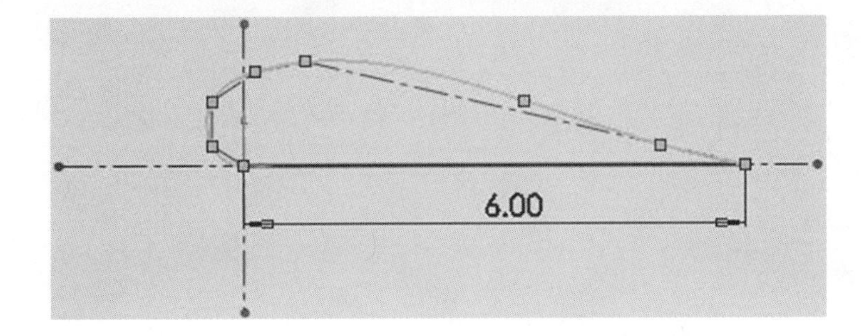

Figure 6.57

Sketch1.

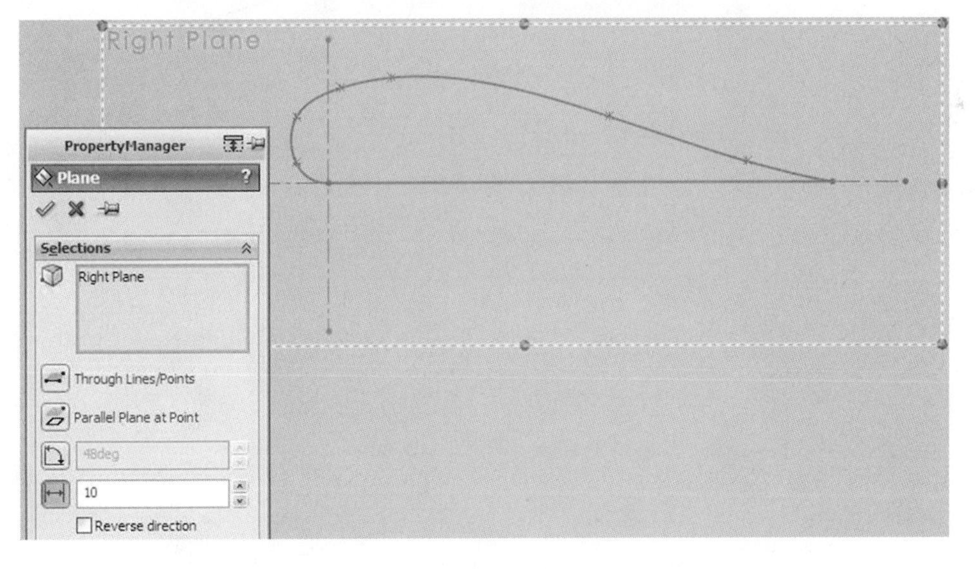

Figure 6.58

Plane1.

Create a Plane Normal to the Endpoint of Sketch1

5. Click the left endpoint of Sketch1.
6. Select the reference geometry icon and click Features > Plane.
7. In the Section rollout, select Sketch1.
8. Select Normal To Curve.
9. Click OK to complete the definition of Plane1.

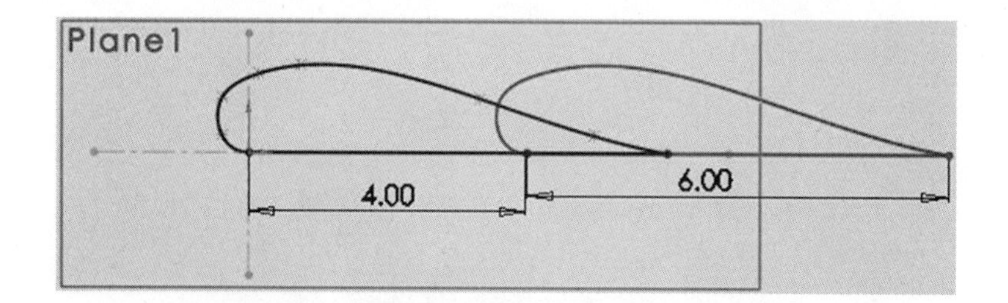

Figure 6.59

Convert entities on Right Plane to entities on Plane1 and copy entities to 4 in. away.

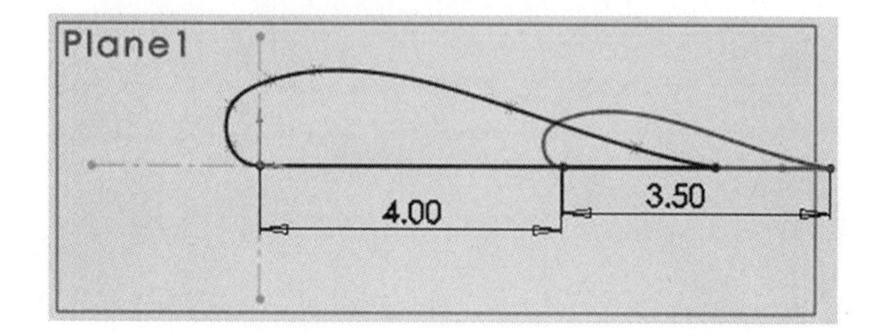

Figure 6.60

Change size of copied profile by resizing 6 in. to 3.5 in.

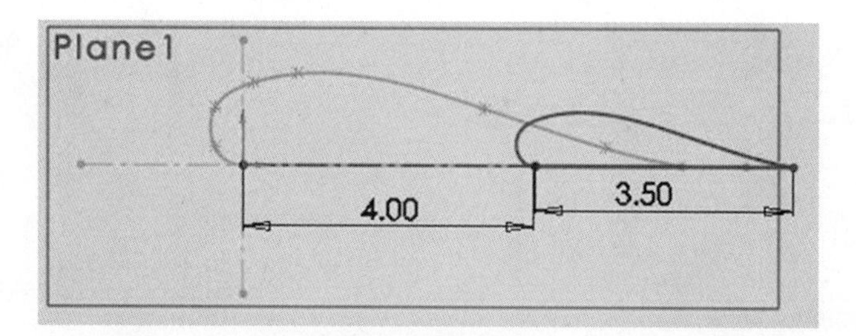

Figure 6.61

Delete larger profile and retain resized profile having length, 3.5 in. and 4 in. from origin.

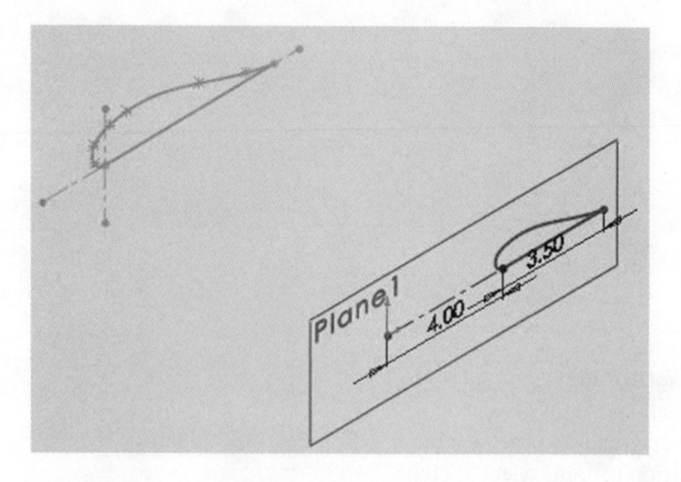

Figure 6.62

Two profiles on two planes (Right Plane and Plane1).

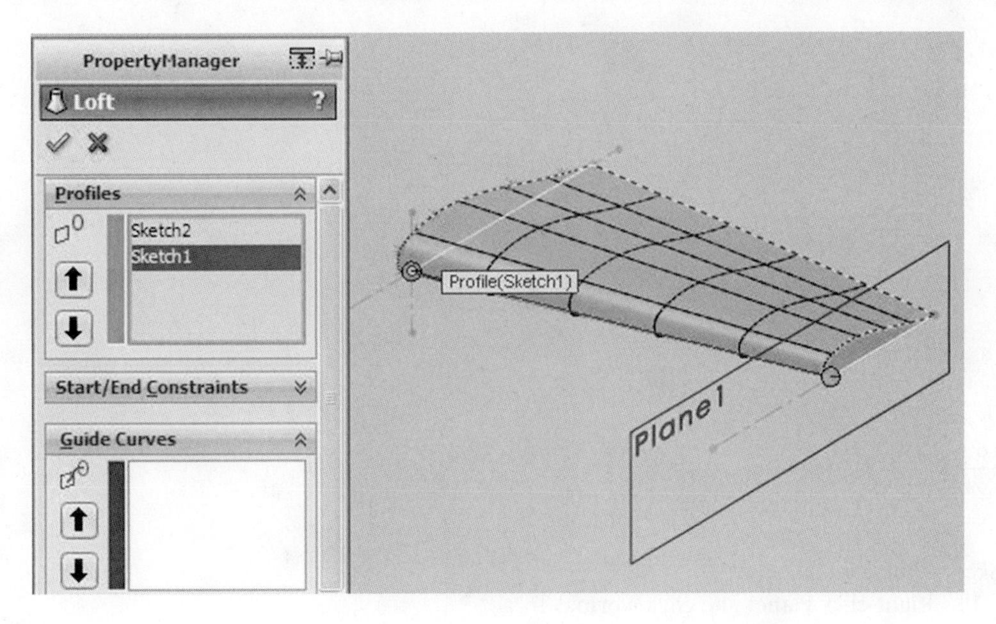

Figure 6.63

Loft using the two profiles.

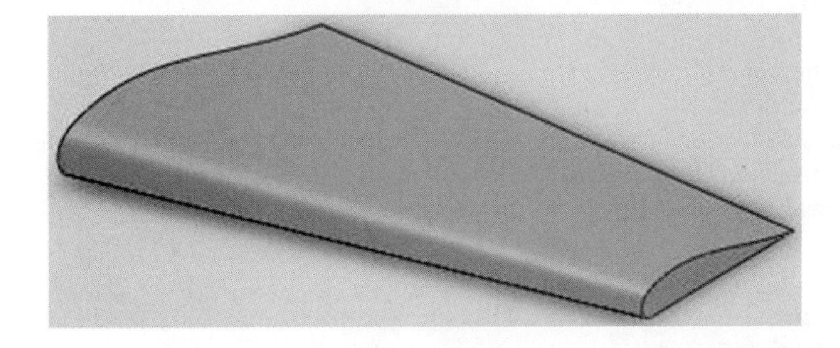

Figure 6.64

Lofted wing.

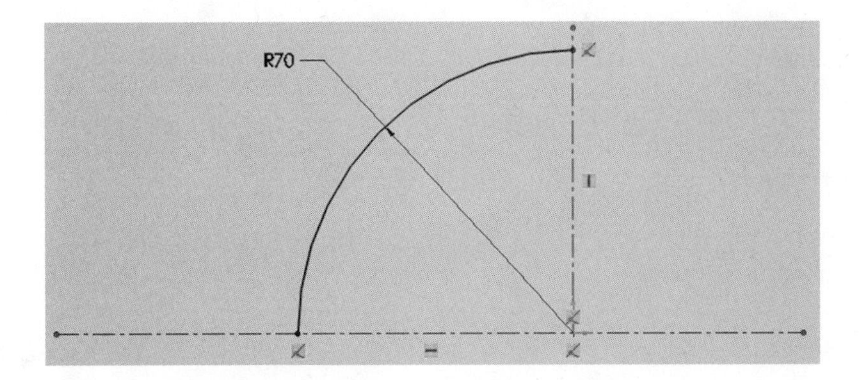

Figure 6.65

Sketch1.

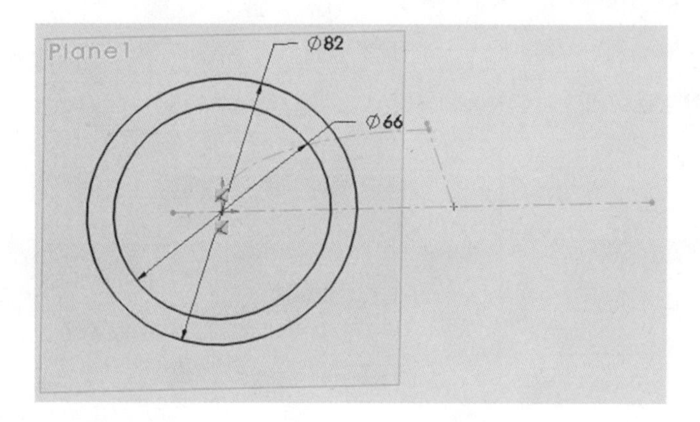

Figure 6.66

Sketch2.

Sketch Profile for Sweep

10. Right-click Plane1 and click Normal To.
11. Sketch two concentric circles, Sketch2, with diameters of 82 and 66 mm, respectively (see Figure 6.66).
12. Exit sketch mode.

Sweep Profile through the Part

13. Click Features > Sweep Boss/Base (see the Sweep PropertyManager in Figure 6.67).
14. In the Sweep PropertyManager, define Sketch2 as the profile and Sketch1 as the path (see Figure 6.67).

Sketch a Line at 45° to the Horizontal (This Line Will Be Used to Create a Plane Normal to Its Endpoint.)

15. Click the front plane.
16. Click Sketch and Normal To.
17. Sketch a dimension line, Sketch3, 152.7-mm long, from the origin at an angle of 45°, and dimension it from the origin (152.7 = (124 – 16)/Cosine(45°)), as shown in Figure 6.68.
18. Exit sketch mode.

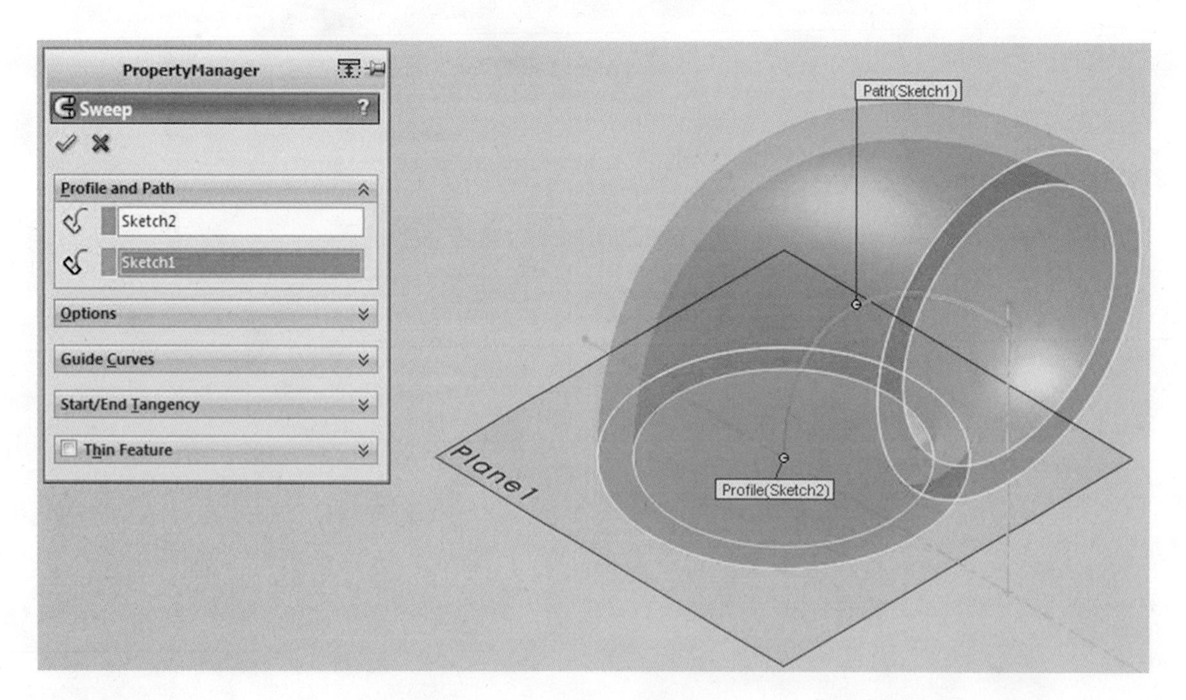

Figure 6.67

Sweep PropertyManager.

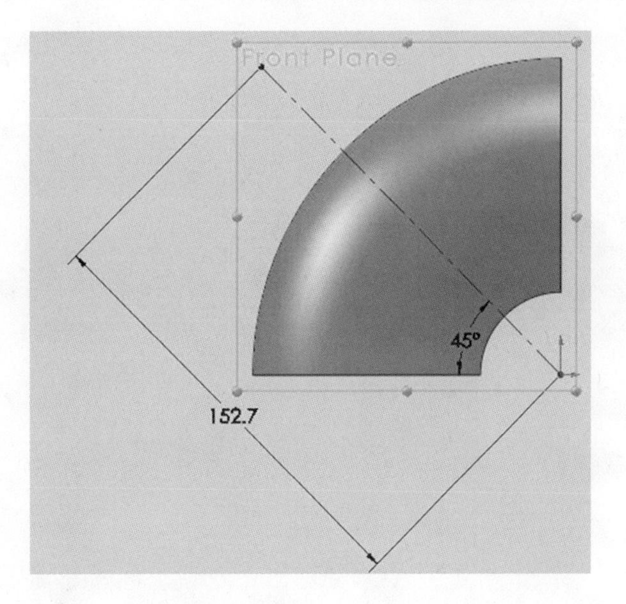

Figure 6.68

Line sketched for creating a plane.

Create a Plane Normal to the Endpoint of Sketch3: Plane2

19. Click the endpoint of Sketch3.
20. Select the reference geometry icon and click Features > Plane.
21. In the Section rollout, select Sketch3.
22. Select Normal To Curve.
23. Click OK to complete the definition of Plane2 (see Figure 6.69).

Create Sketch4 on Plane2

24. Click Plane2 and rotate appropriately, but do not choose Normal To.
25. Create two circles, Sketch4, with diameters of 45 and 58 mm, respectively (see Figure 6.70).

Extrusion of Concentric Circles

26. Click Extrude.
27. In the Extrude PropertyManager, click Up Too Surface for Direction 1, and select the inner surface of the elbow (see Figure 6.71).

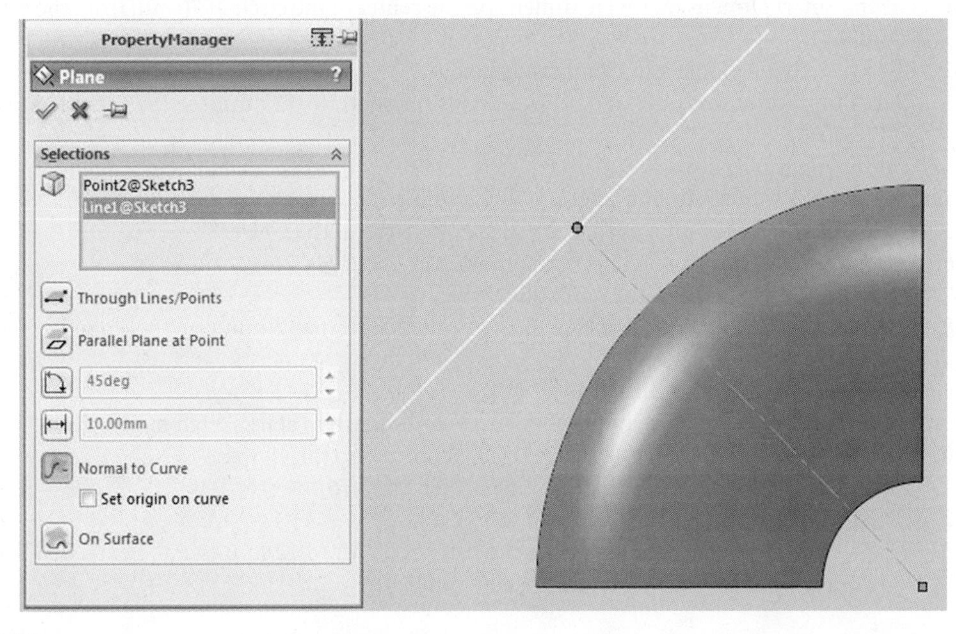

Figure 6.69

Plane created normal to line defined.

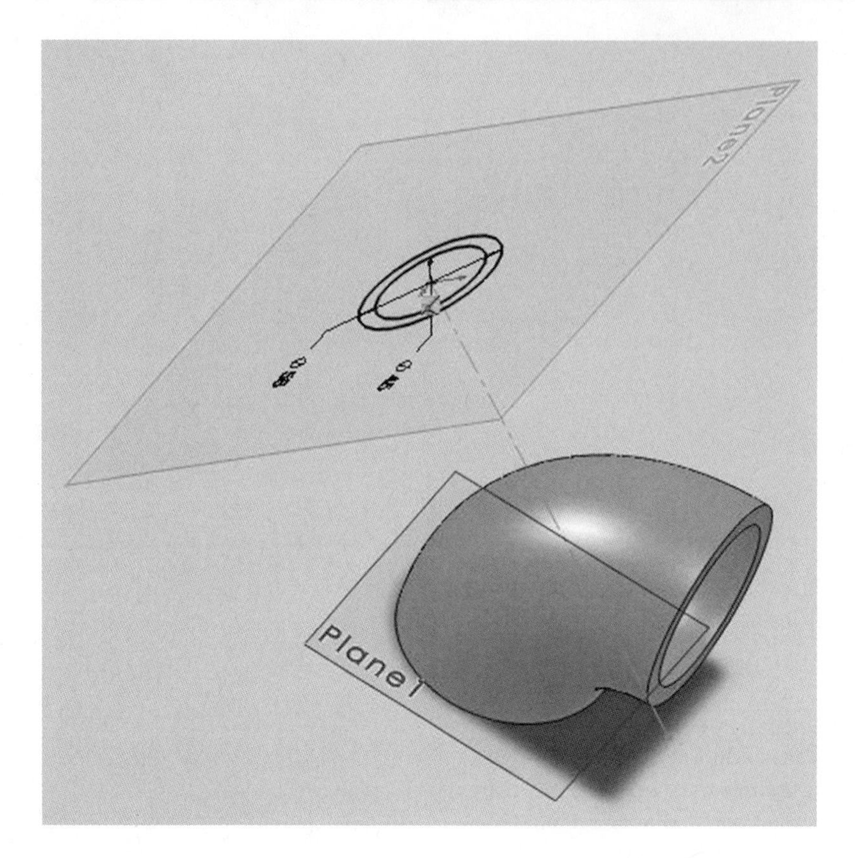

Figure 6.70

Concentric circles sketched on Plane2.

Create Right-Hand Flange for Elbow

28. Click the right-hand face of the elbow and click Sketch, to start sketch mode.
29. Click Normal To; this makes the face selected normal to the viewer.
30. Sketch a circle with a diameter of 128 mm.
31. Use the Smart Dimension tool to dimension the center of the circle at 70 mm from the elbow (see Figure 6.72).
32. Click the Extrude tool from CommandManager.
33. Click Blind for Direction 1 and set the extrusion depth to 12 mm (see the extruded feature in Figure 6.73).

Create Bottom Flange for Elbow

34. Click Planel at the bottom face of the elbow, and click Sketch to start sketch mode.
35. Click Normal To; this makes the face selected normal to the viewer.
36. Sketch a circle with a diameter of 125 mm (see Figure 6.74).
37. Click the Extrude tool from CommandManager.
38. Click Blind for Direction 1 and set the extrusion depth to 12 mm (see the extruded feature in Figure 6.75).

Create Top Flange for Elbow

39. Click Plane2 at the top face of the elbow and click Sketch to start sketch mode.
40. Click Normal To; this makes the face selected normal to the viewer.
41. Sketch a circle with a diameter of 88 mm (see Figure 6.76).
42. Click the Extrude tool from CommandManager.
43. Click Blind for Direction 1 and set the extrusion depth to 12 mm (see the extruded feature in Figure 6.77). The final elbow model is shown in Figure 6.78.

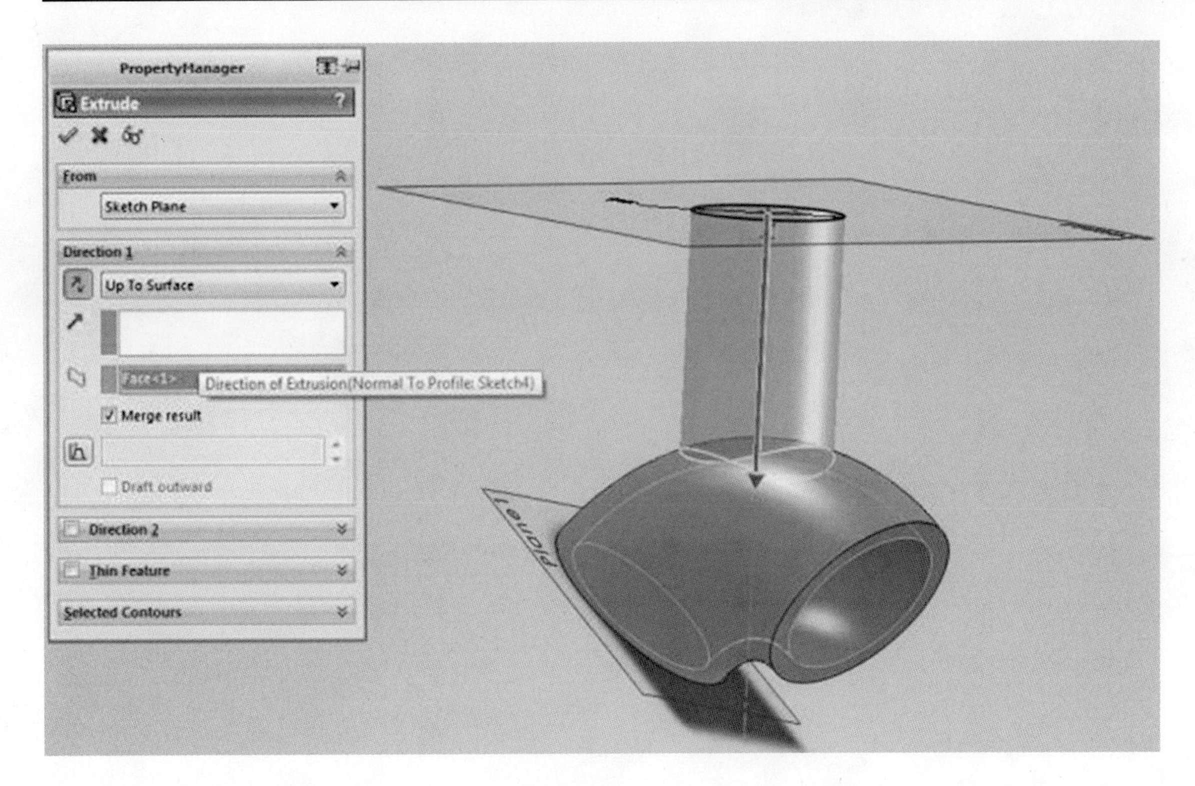

Figure 6.71

Extrude PropertyManager.

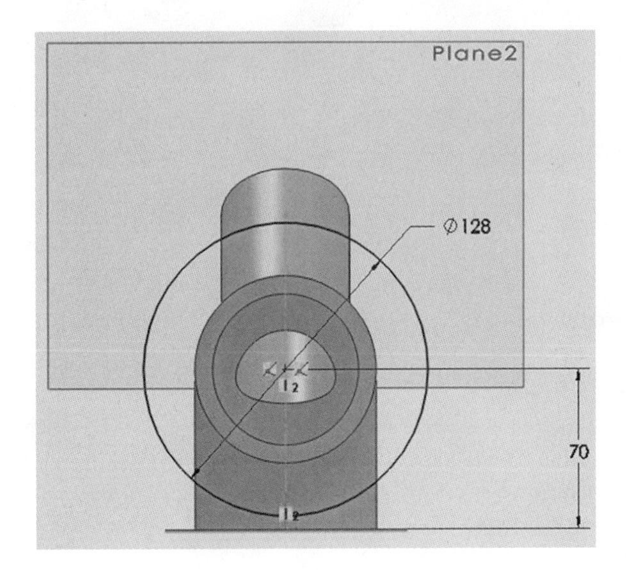

Figure 6.72

Circular profile for right-hand elbow flange.

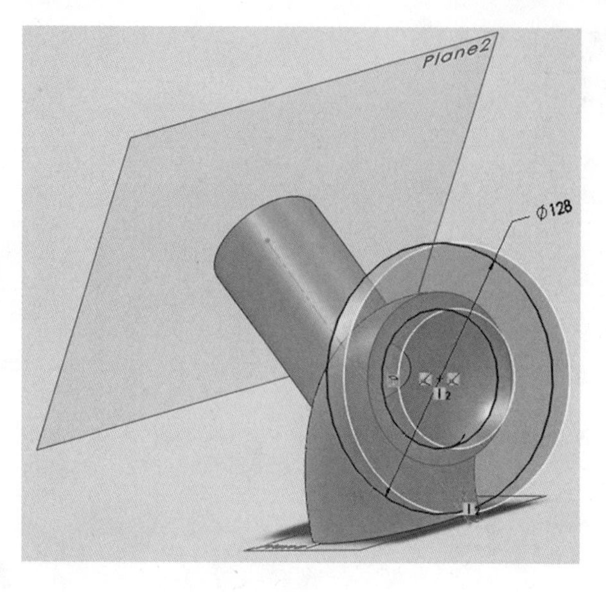

Figure 6.73

Extrusion to complete right-hand elbow flange.

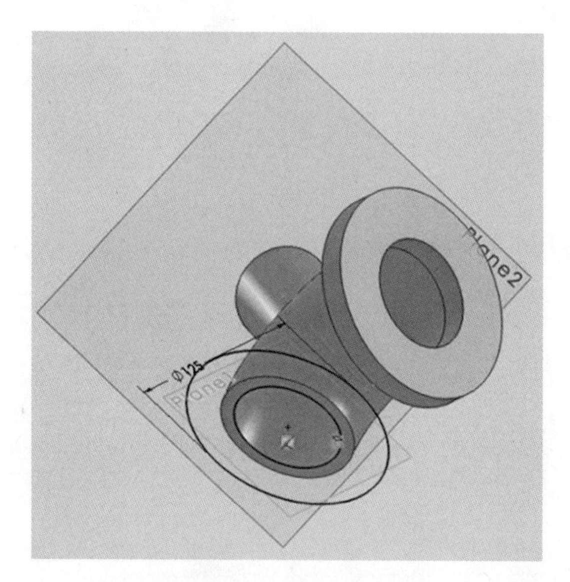

Figure 6.74

Circular profile for bottom elbow flange.

Practical Swept Feature: Lathe Tailstock

Tailstock Ring

1. Select the top plane. Create Sketch1 (a circle of diameter 176 mm) with center at the origin, as shown in Figure 6.79.
2. Select the isometric orientation.
3. Exit sketch mode.
4. From CommandManager, click Features > Reference Geometry > Plane.
5. Select the Normal to Curve option.
6. Click Sketch1.
7. Click OK to create Plane1, normal to Sketch1.
8. Right-click Plane1 and create Sketch2 (a circle of diameter 22 mm).
9. Add a piercing relation by clicking the center of the small circle, then hold the Ctrl key down. Click somewhere on the circumference of the large circle and release the Ctrl key. Right-click Make Pierce from the dialog. Click OK (see Figure 6.80).

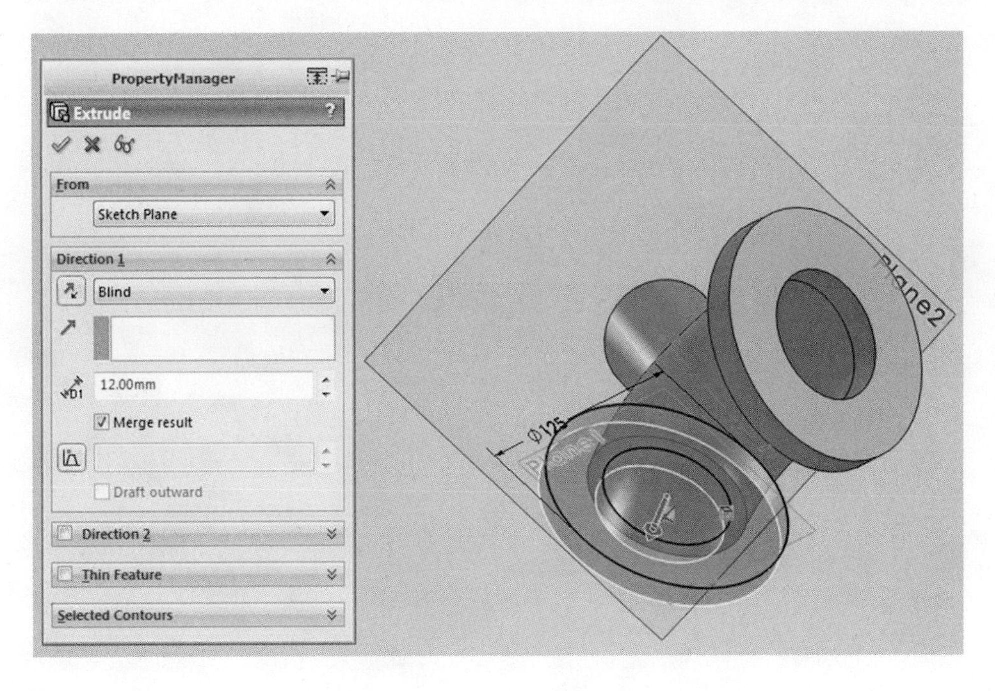

Figure 6.75

Extrusion to complete bottom elbow flange.

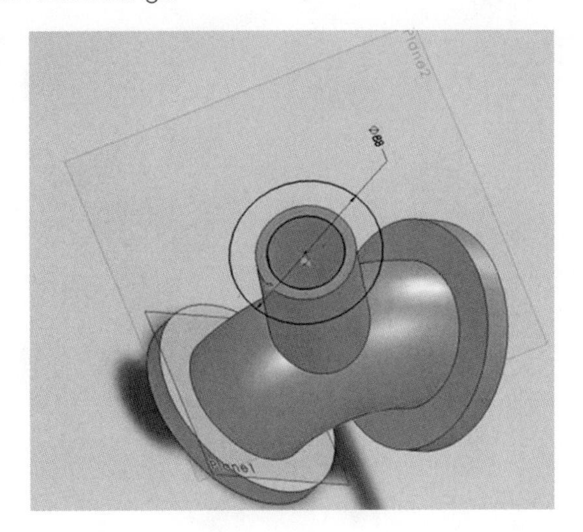

Figure 6.76

Circular profile for top elbow flange.

10. Exit sketch mode.
11. Click the Swept Boss/Base tool. The Sweep PropertyManager appears.
12. Right-click the Profile and Path box.
13. Click the small circle (Sketch2) as the profile and the large circle (Sketch1) as the path using FeatureManager. A real-time preview will appear, as shown in Figure 6.81.
14. Click OK to complete the swept cut part, which is hollow as shown in Figure 6.81. Hide the planes.

Tailstock Central Boss

15. Select the right-hand plane.
16. Create Sketch3, 8-mm offset from the origin, as shown in Figure 6.82.
17. Revolve Sketch3 about the vertical dimension line (see Figure 6.83).

Create Two Planes

18. Select the right-hand plane.
19. Select the reference geometry icon and click Features > Plane.

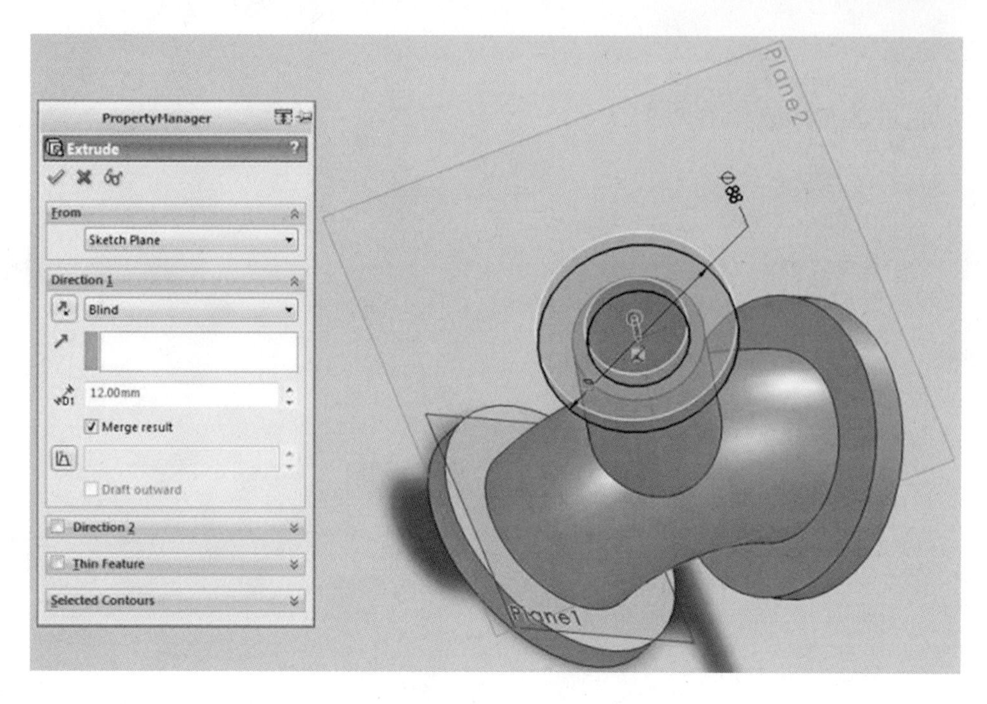

Figure 6.77

Extrusion to complete top elbow flange.

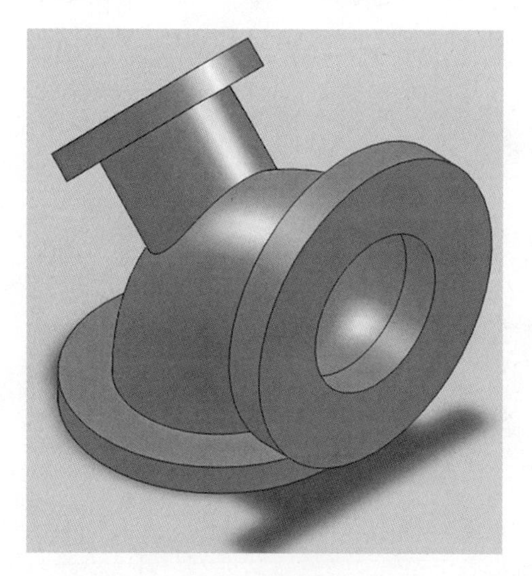

Figure 6.78

Final elbow model.

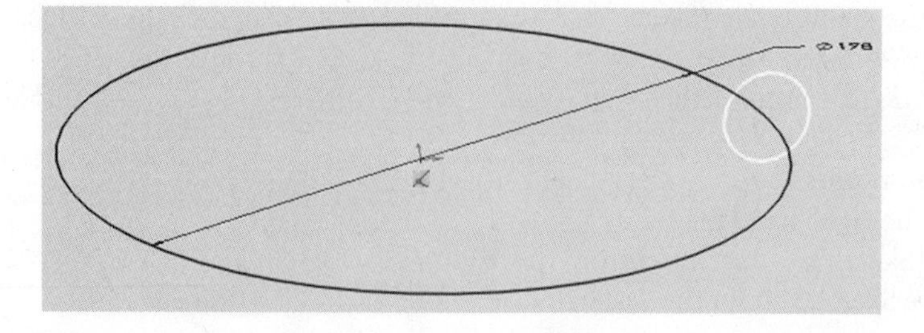

Figure 6.79

Profile (small circle) and path (large circle).

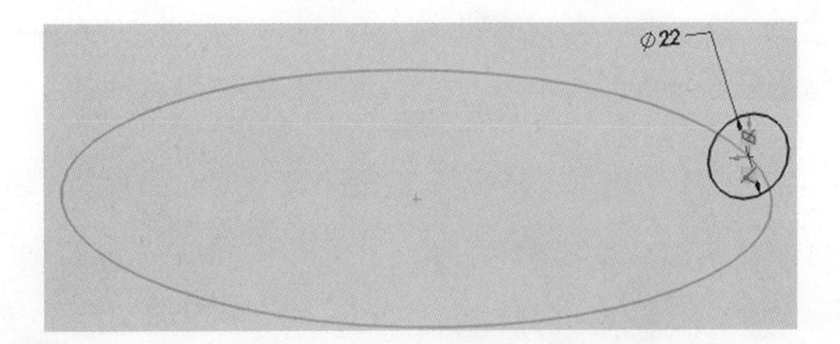

Figure 6.80

Adding a piercing relation.

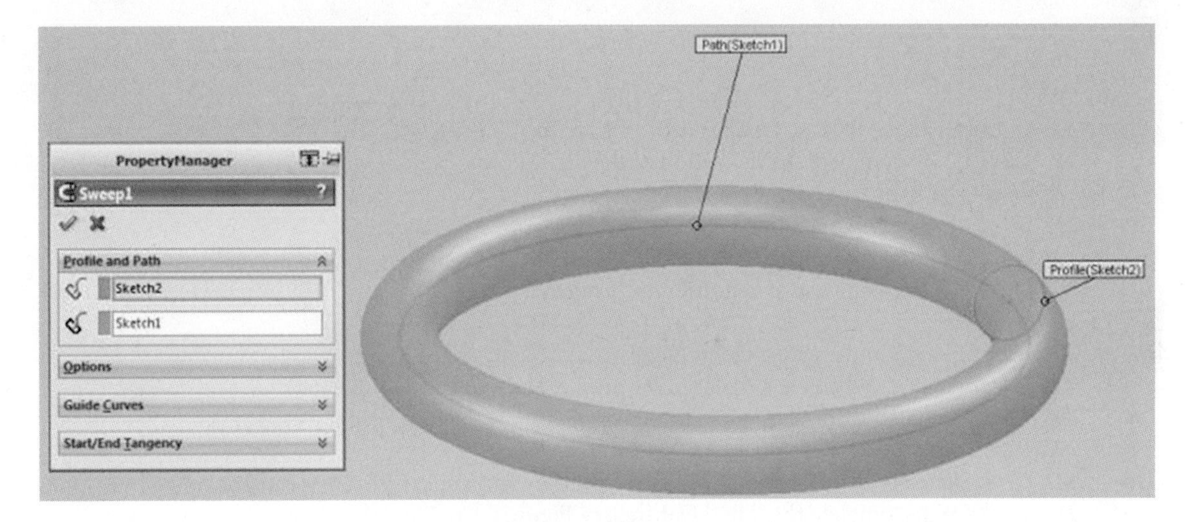

Figure 6.81

Sweep PropertyManager and preview of tailstock ring.

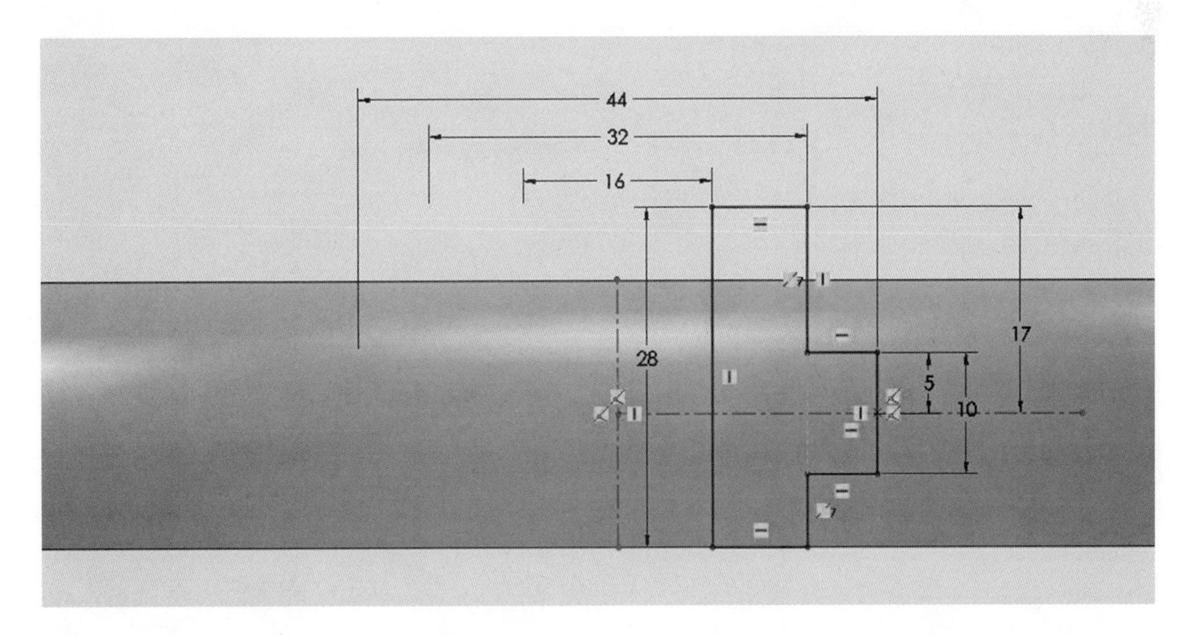

Figure 6.82

Sketch2 to create revolved central boss.

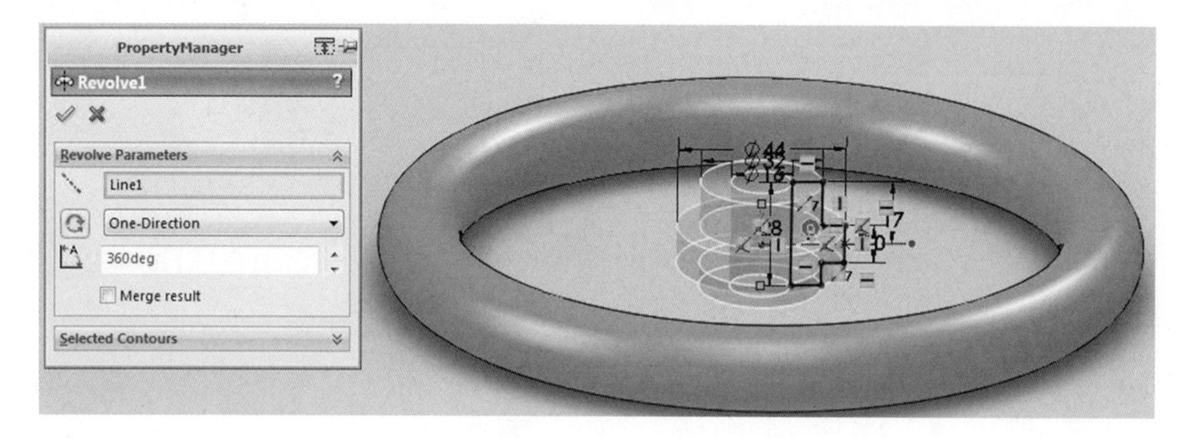

Figure 6.83

PropertyManager and preview of the revolved central boss.

20. In the Distance rollout, set a value of 18 mm (see Figure 6.84).
21. Click OK to complete the definition of the plane.
22. Select the right-hand plane.
23. Select the reference geometry icon and click Features > Plane.
24. In the Distance rollout, set a value of 82 mm (see Figure 6.85).
25. Click OK to complete the definition of the plane.

Lofting

26. Select Plane2, and create a rectangular profile, Sketch4 (see Figure 6.86).
27. Select Plane3, and create a rectangular profile, Sketch5 (see Figure 6.87).
28. Click the Lofted Boss/Base tool. The Loft PropertyManager appears.
29. Right-click the Profiles box.
30. Click Sketch4, then Sketch5. A real-time preview will appear (see Figure 6.88).
31. Click OK to complete the lofted feature (see Figure 6.88). Hide the planes.

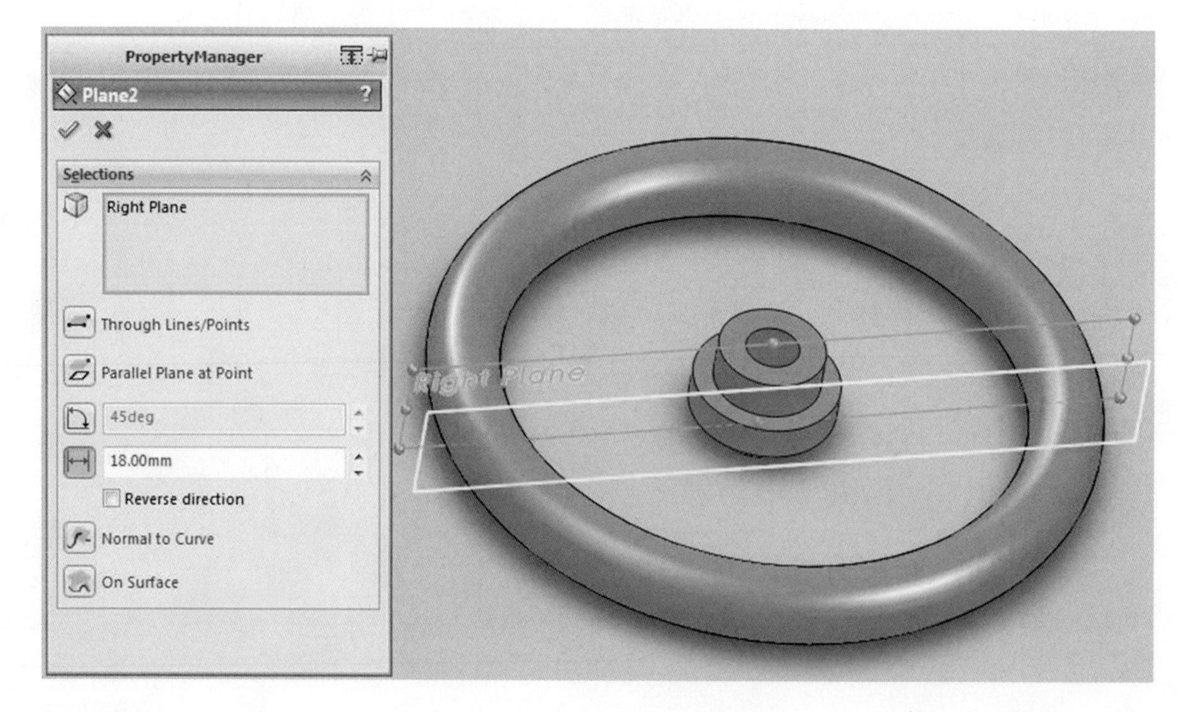

Figure 6.84

Plane2 definition.

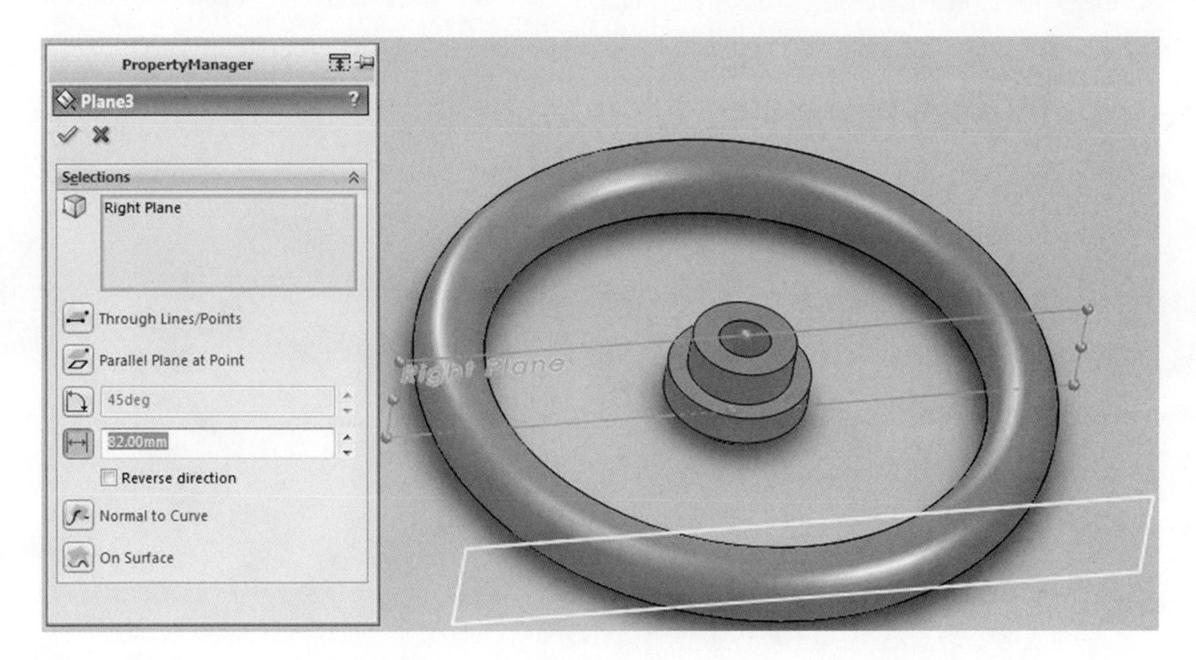

Figure 6.85

Plane3 definition.

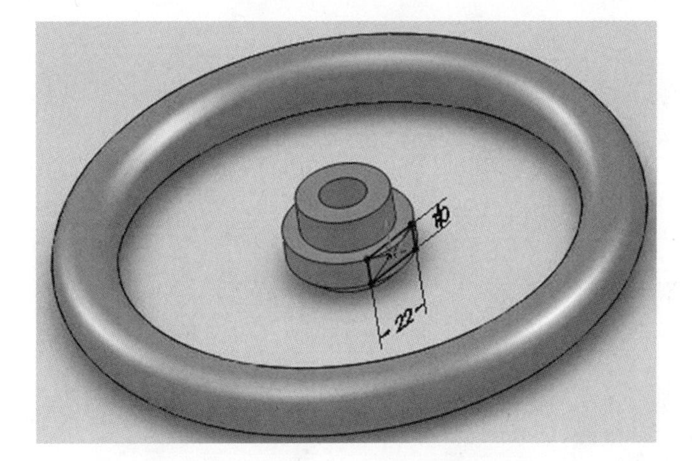

Figure 6.86

Sketch4 on Plane2.

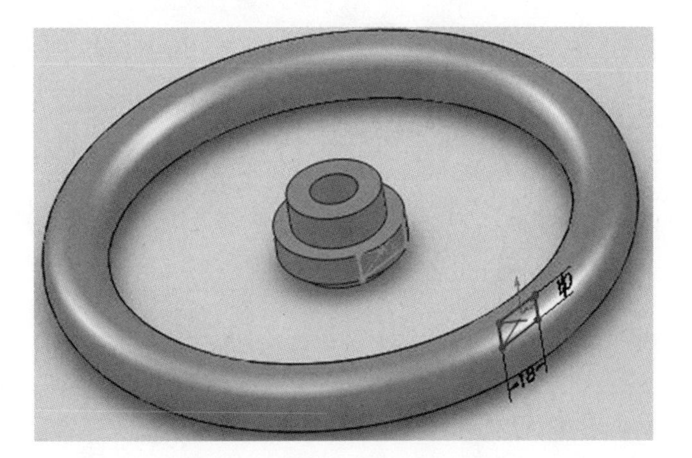

Figure 6.87

Sketch5 on Plane3.

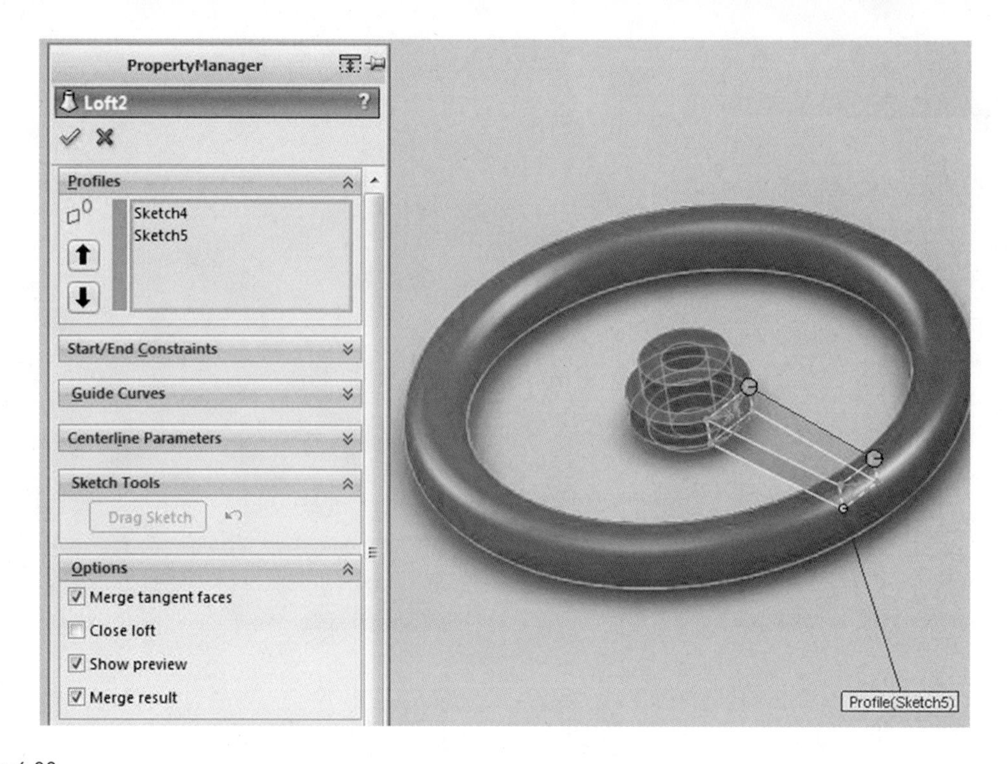

Figure 6.88

Preview of lofted rib.

Create Circular Pattern of Ribs

32. Click Circular Pattern, the Circular Pattern PropertyManager appears (see Figure 6.89).
33. Set the temporary axis, and select it as the axis for the circular pattern.
34. Set the number of instances to 4.
35. Select Loft2 as the Features to Pattern.
36. Add fillets.
37. Click OK to complete the design, as shown in Figure 6.90.

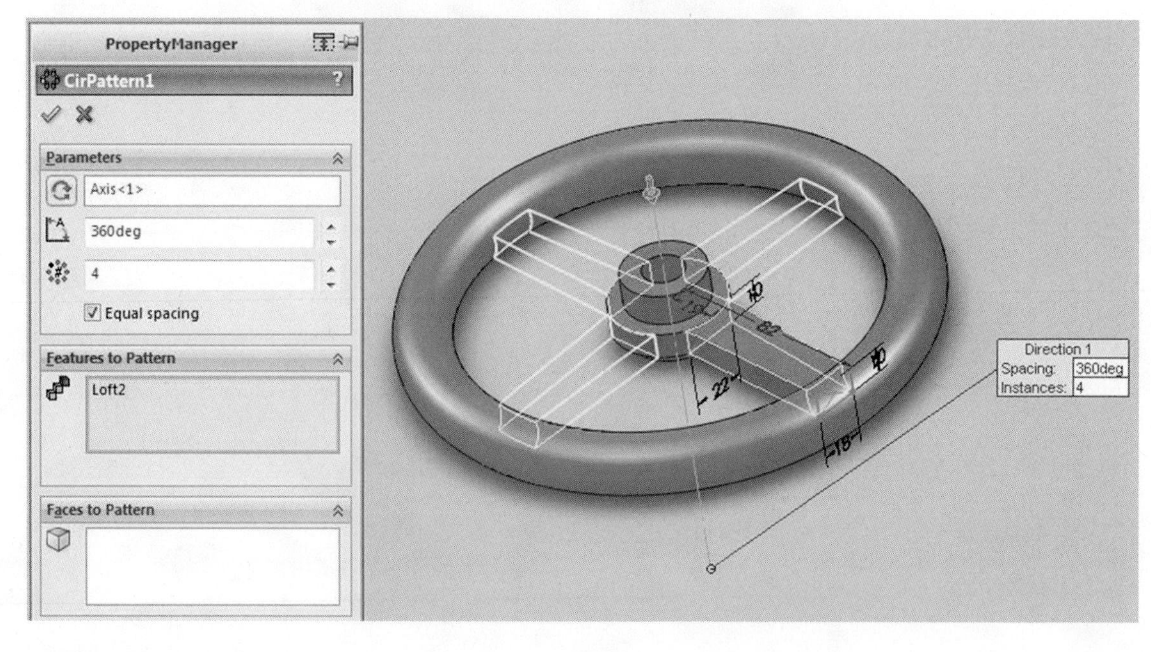

Figure 6.89

Circular Pattern PropertyManager.

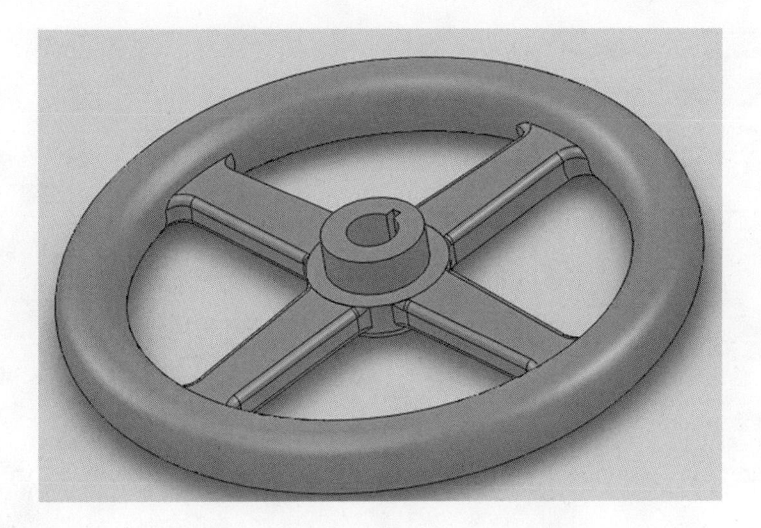

Figure 6.90

Final model of lathe tailstock.

Summary

This chapter discusses the three methods of creating revolved, swept, and lofted features, which are commonly encountered in machine parts. These features are commonly found in automotive assemblies; therefore, a proper understanding of how to model these parts is essential.

Exercises

1. Repeat the design of the tail stock ring of Section 6.5. Create an ellipse with a major diameter of 22 mm and a minor diameter of 10 mm. Create another ellipse with a major diameter of 18 mm and a minor diameter of 10 mm. Loft between these features and use the Circular Pattern tool to replicate the lofted feature four times. Follow the steps that are shown in Figures P6.1 through P6.4.
2. Create a spring with the following parameters:
   a. Base circle for helix, radius 0.5 in.
   b. Helix Height = 1.75 in.
   c. Revolution = 8

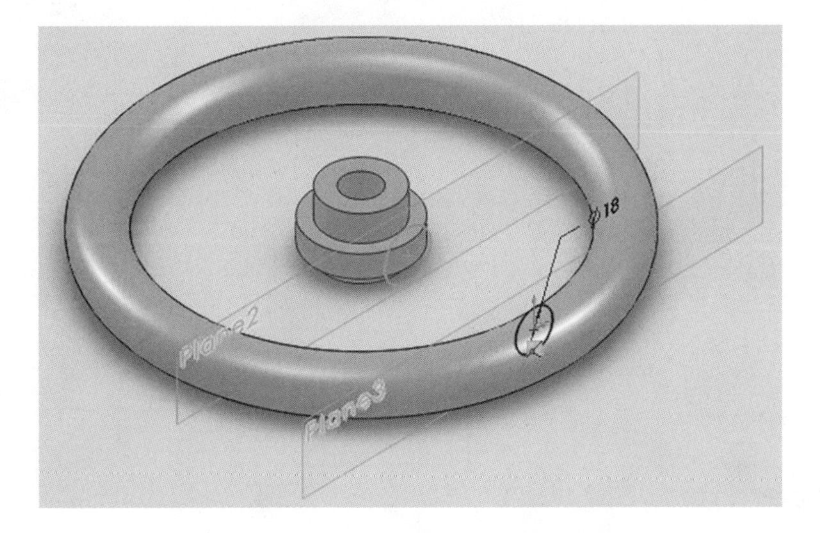

Figure P6.1

Ellipses.

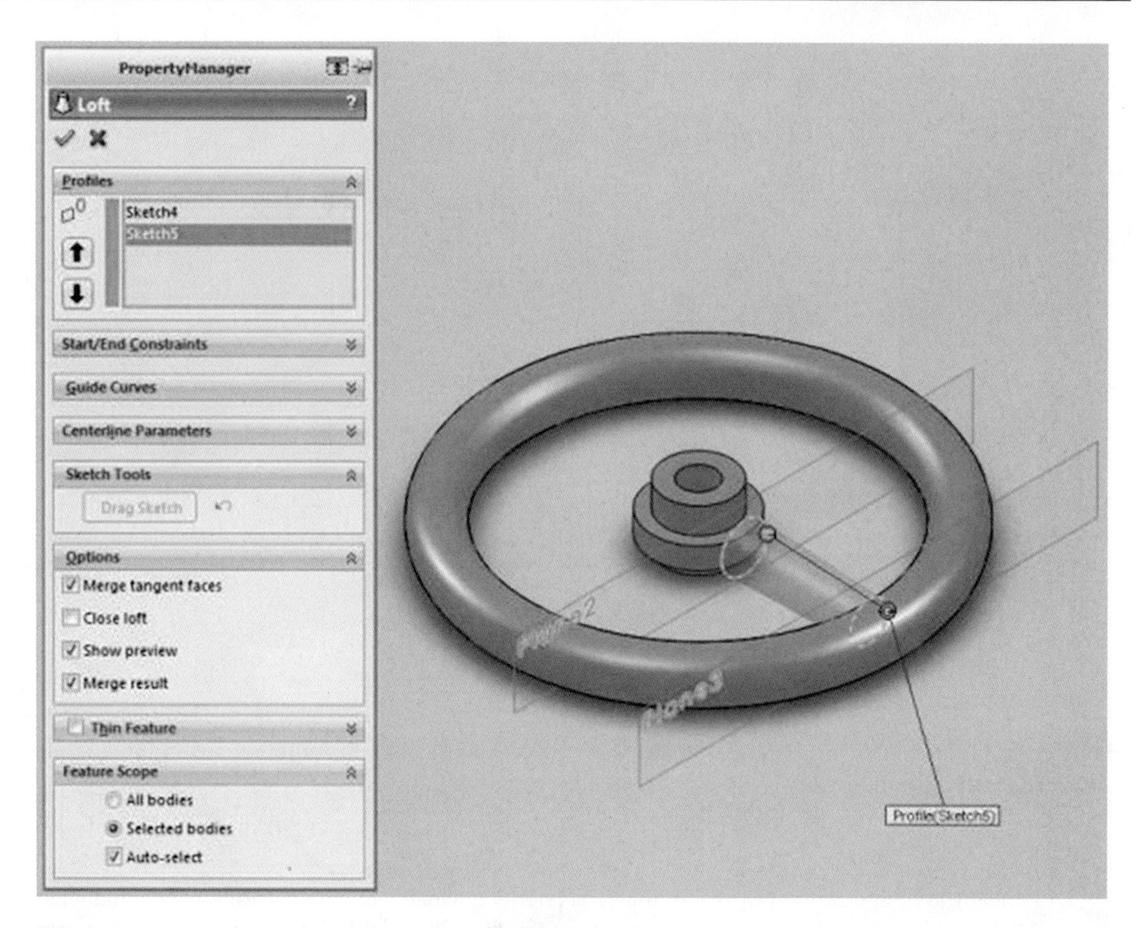

Figure P6.2

Preview of loft.

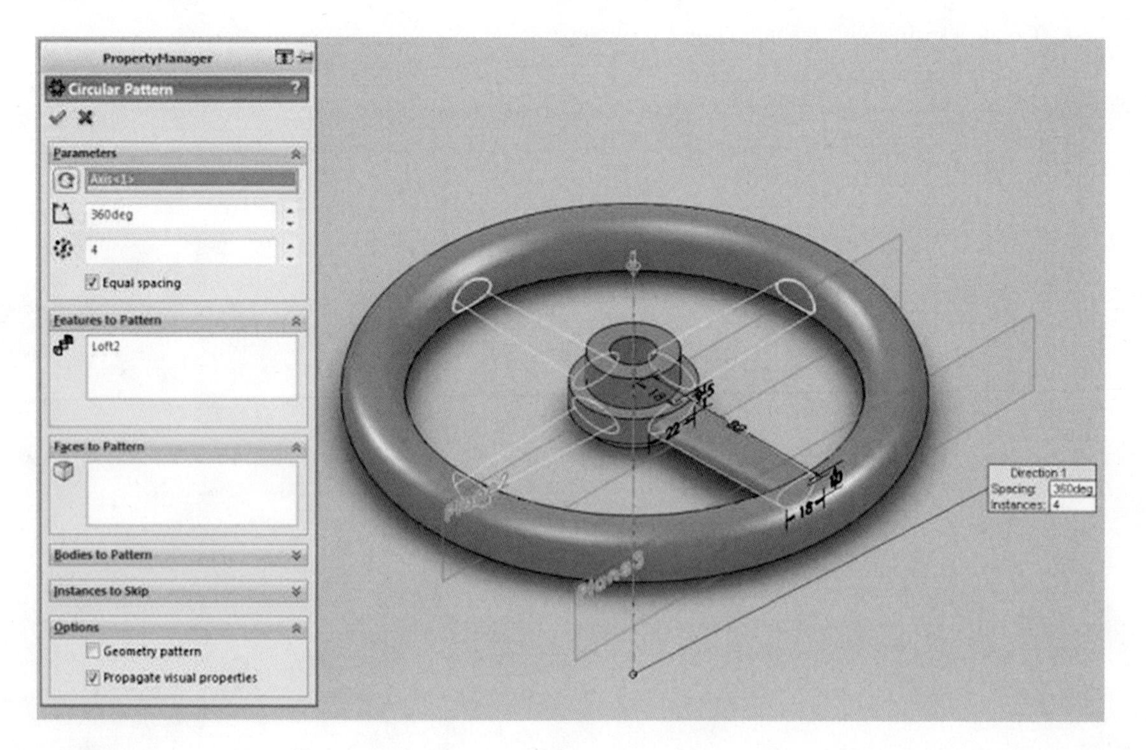

Figure P6.3

Preview of circular pattern.

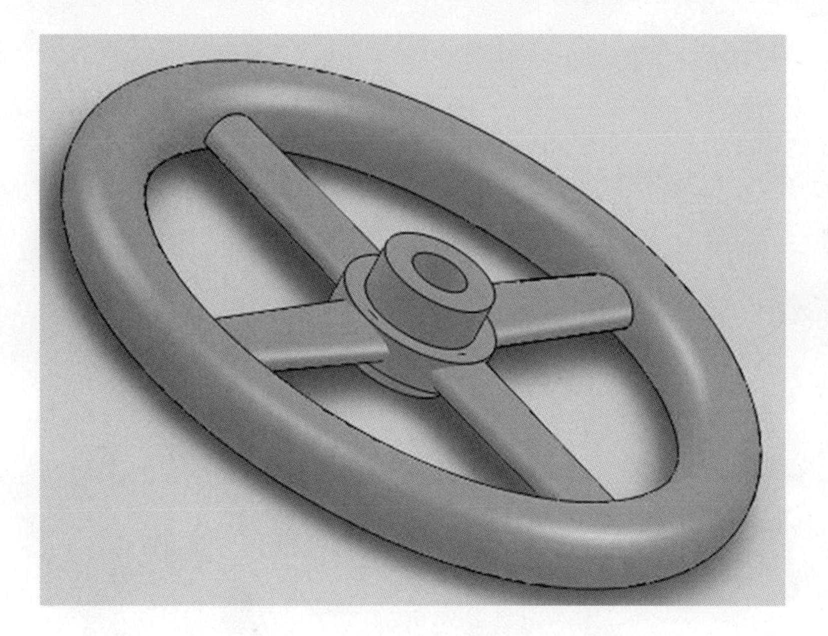

Figure P6.4

Final tailstock.

d. Starting Angle = 270-deg

e. Helix Direction = Anticlockwise

f. Diameter = 0.125 in.

3. Create an O-ring with a circular path of diameter 7.5 in. and a cross section of diameter 0.125 in.

4. Create a lofted boss/base model using a square base, a circle, and an ellipse, with the following parameters:

a. The distance between planes is 75 mm.

b. The square is 100 mm × 100 mm dimension.

c. The circle has a diameter of 70 mm.

d. The ellipse has a major diameter, 90 mm, and a minor diameter, 75 mm.

7

Part Modeling with Equation-Driven Curves

Objectives:

When you complete this session, you will have

- Learned how to use the Equation Driven Curve tool in SolidWorks
- Learned how to create three-dimensional (3D) objects using Equation-Driven Curves
- Learned how to determine the surface areas and volumes of 3D objects that are created using Equation-Driven Curves

Introduction

SolidWorks uses *sketch entities* such as line, rectangle, parallelogram, slot, polygon, circle, arc, ellipse, parabola, spline, etc., as the foundational building blocks for defining three-dimensional (3D) objects. This chapter introduces a new concept of part modeling with the Equation Driven Curve tool in SolidWorks.

Equation-Driven Curves

1. Open a New SolidWorks part document.
2. Choose a *Plane* (such as the Front Plane).
3. Click Sketch to be in sketch mode.
4. Click Spline > Equation Driven Curves (see Figure 7.1).
   The Equation Driven Curves PropertyManager is displayed as in Figure 7.2.
5. In the Enter and equation as a function of x, enter the equation (see Figure 7.2).
6. In the Enter start x value for the equation, enter the start value (see Figure 7.2).
7. In the Enter end x value for the equation, enter the end value (see Figure 7.2).
8. Click OK to complete the procedure.

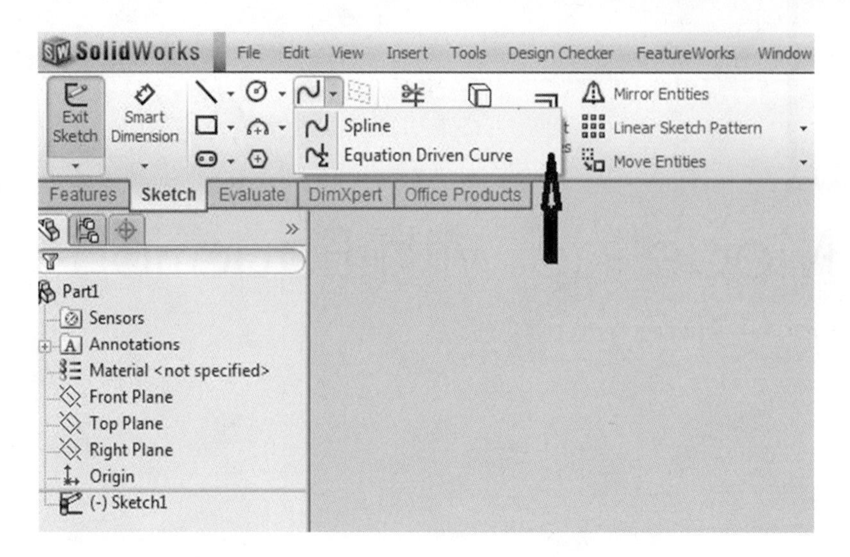

Figure 7.1

Equation Driven Curve tool.

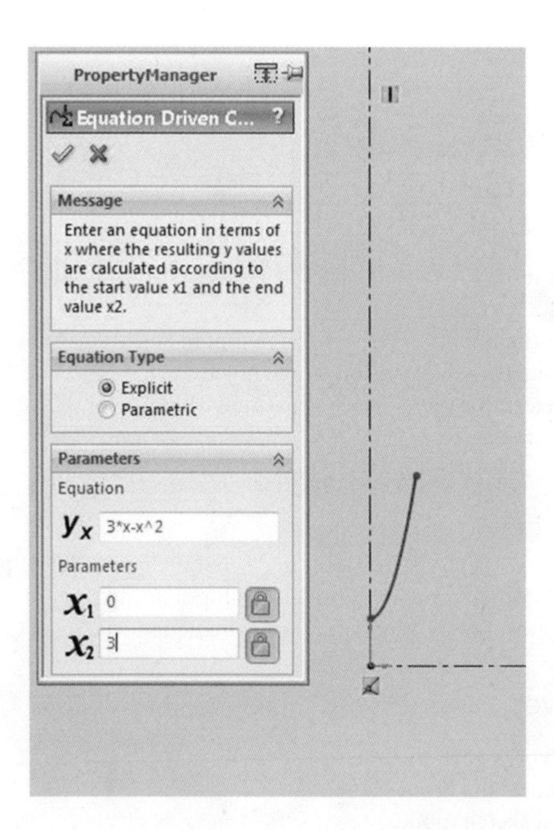

Figure 7.2

Equation Driven Curves PropertyManager.

Problem 1

The problem we are solving is to define the shape that is made up of the first-quadrant area, which is bounded by the curves

$$y = x^2 + 3$$

$$y = 3x - x^2$$

$$x = 0$$

$$x = 3.$$

This problem is from Chapter 31, Section 3, Example 9 of the referenced textbook.

We will first present the integral calculus solution and then follow the methodology that is described in this chapter to present the SolidWorks solution. The shape is then revolved about the y-axis to create a 3D part model.

Integral Calculus Solution for Area

Let

$$f(x) = x^2 + 3$$

$$g(x) = 3x - x^2$$

Then,

$$f(x) - g(x) = (x^2 + 3) - (3x - x^2) = 2x^2 - 3x + 3$$

The area between the curves is given as

$$A = \int_0^3 [f(x) - g(x)]dx = \int_0^3 (2x^2 - 3x + 3)dx = \left[\frac{2x^3}{3} - \frac{3x^2}{2} + 3x\right]_0^3 = 18 - \frac{27}{2} + 9 = 13.5 \text{ sq. units}$$

The next step is to use the SolidWorks Equation Driven Curve tool to define the curves that are represented by the two equations that are used with the integral calculus method to ascertain the accuracy of SolidWorks.

SolidWorks Solution for Area

In this case, we use the Equation Driven Curve tool in SolidWorks to first define the two curves.

1. Open a New SolidWorks part document.
2. Choose the Front Plane.
3. Click Sketch to be in sketch mode (Figure 7.3).

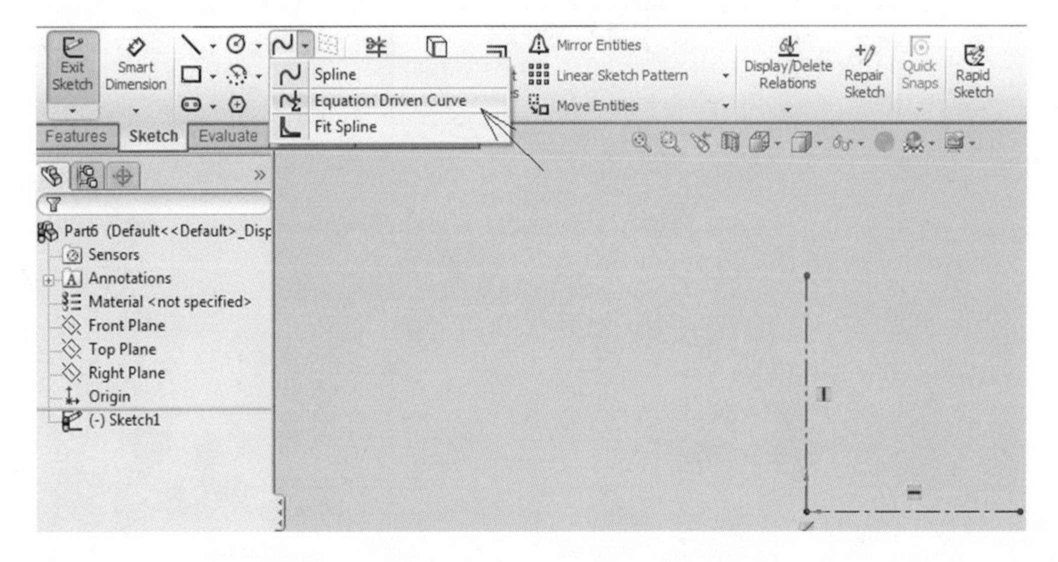

Figure 7.3

Equation Driven Curve is selected from Spline menu.

4. Click Spline > Equation Driven Curves (see Figure 7.3).
   The Equation Driven Curves PropertyManager is displayed as in Figure 7.4.
5. In the Enter and equation as a function of x, enter the equation (see Figure 7.4).
6. In the Enter start x value for the equation, enter the start value (see Figure 7.4).
7. In the Enter end x value for the equation, enter the end value (see Figure 7.4).
8. Click OK to complete the procedure.
9. Repeat steps 1–8 for the second curve (see Figure 7.5).
10. Sketch vertical lines at the left and right ends of the two curves (see Figure 7.6).
11. Click Evaluate > Section Properties (see Figure 7.7 for the solution).

Comparing the values of the area enclosed by the two curves between the upper and lower limits shows that both integral calculus and SolidWorks have the same solution, as shown in Table 7.1.

Part Design Using the SolidWorks Equation-Driven Curves

The part design is obtained by simply revolving the area that is already obtained about the y-axis. Figure 7.8 shows the sectioned and solid views of the revolved shape.

We apply the material property and choose AISI 304 steel (see Figure 7.9), and then compute the mass properties. Figure 7.10 shows the SolidWorks mass properties for the solid that is generated in this study.

Effect of Changing the Axis of Rotation

Here, we present the study of the effect of changing the position of the axis of rotation. By merely offsetting the axis of revolution from the y-axis, we obtain the solid that is generated in Figure 7.11.

Solving this problem using integral calculus could be quite challenging for such a simple solid. Therefore, SolidWorks offers the designer such a flexible tool for determining the solid properties during the design stage.

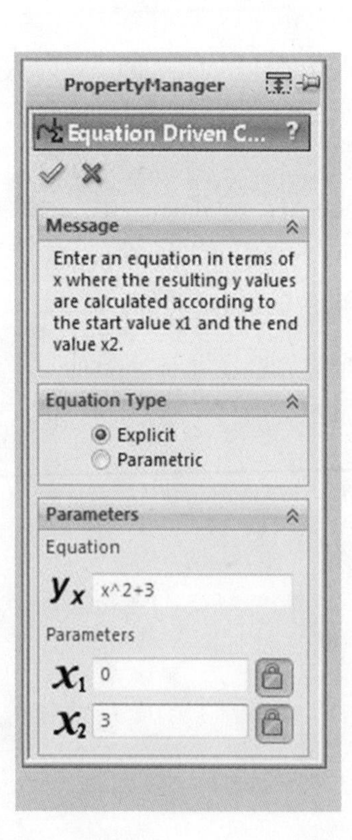

Figure 7.4

Equation 1.

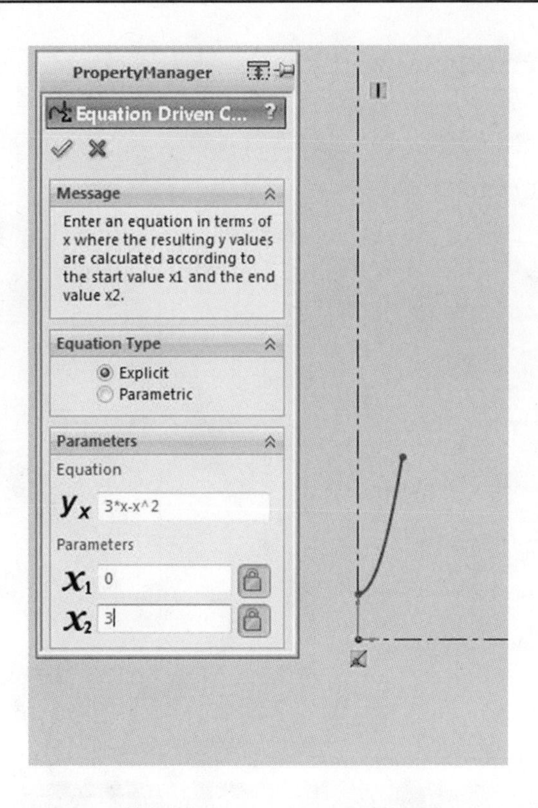

Figure 7.5

Equation 2.

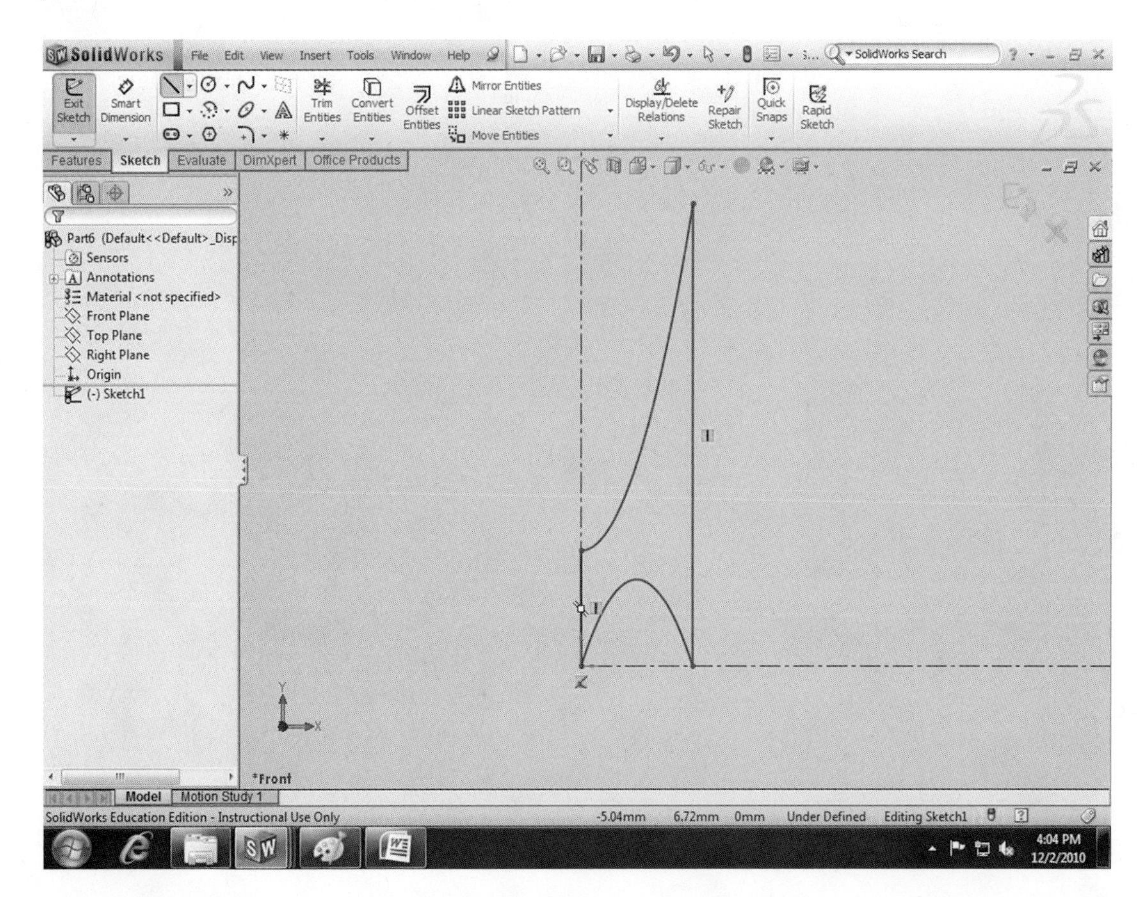

Figure 7.6

Curves from equations 1 and 2.

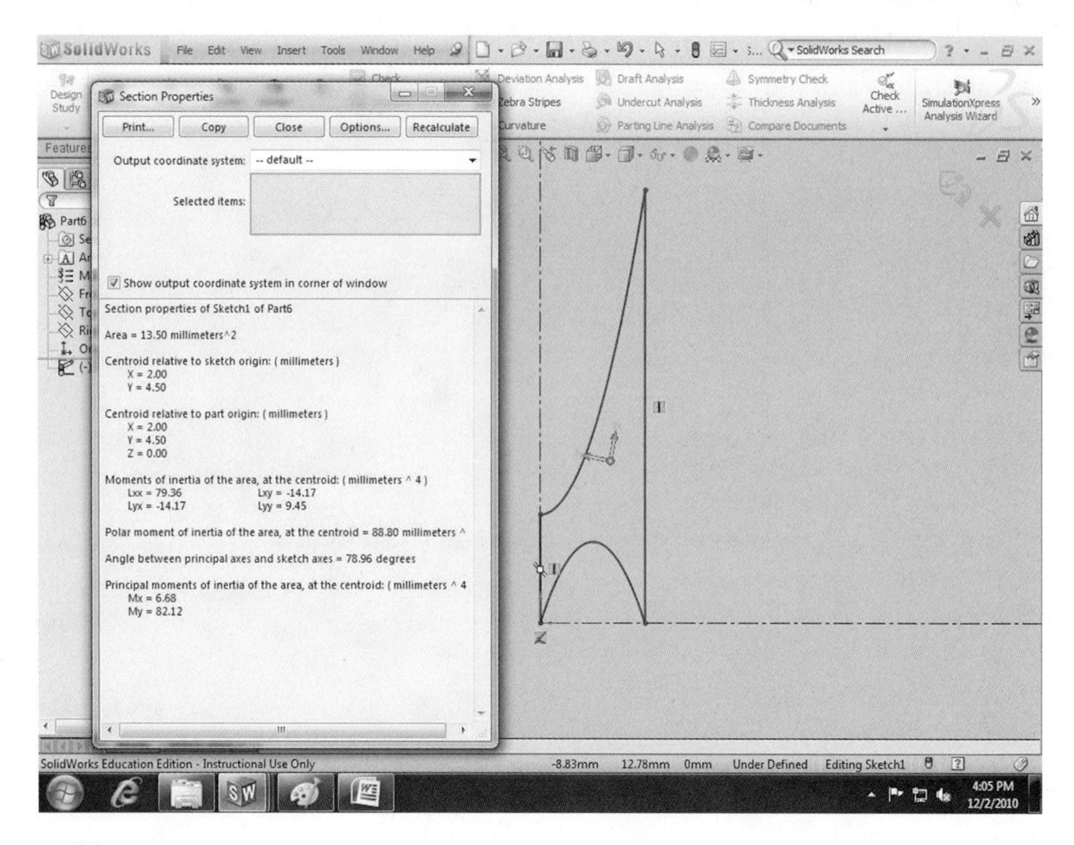

Figure 7.7

Section properties.

Table 7.1 Areas Obtained

Method	Value
Integral calculus	$A = 13.5$ sq. units
SolidWorks	$A = 13.5$ sq. units

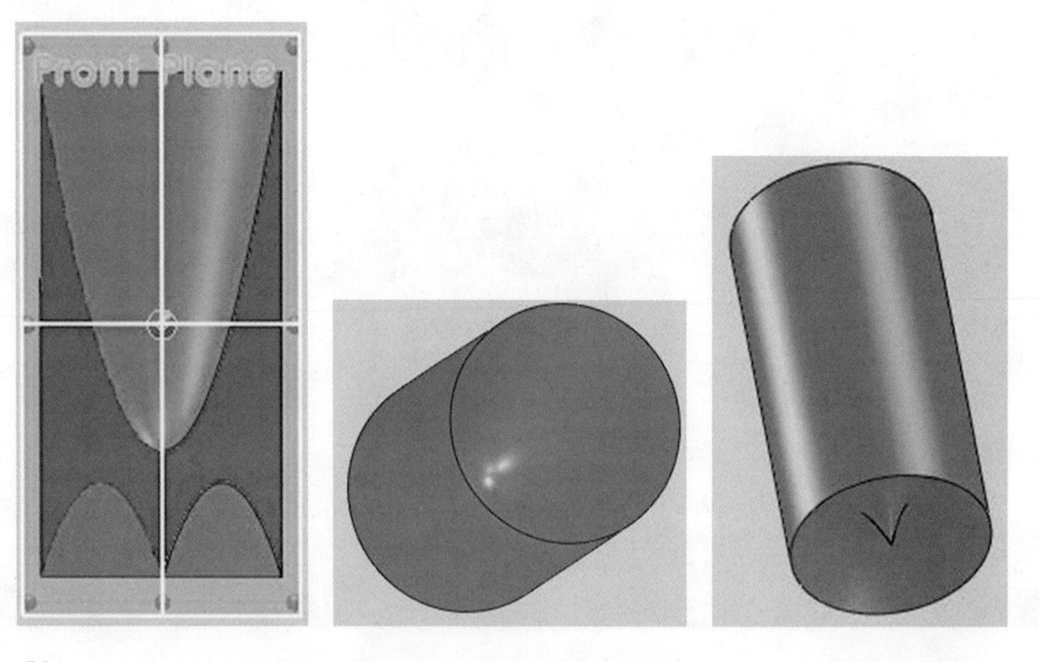

Figure 7.8

Sectioned and solid views of the revolved solid.

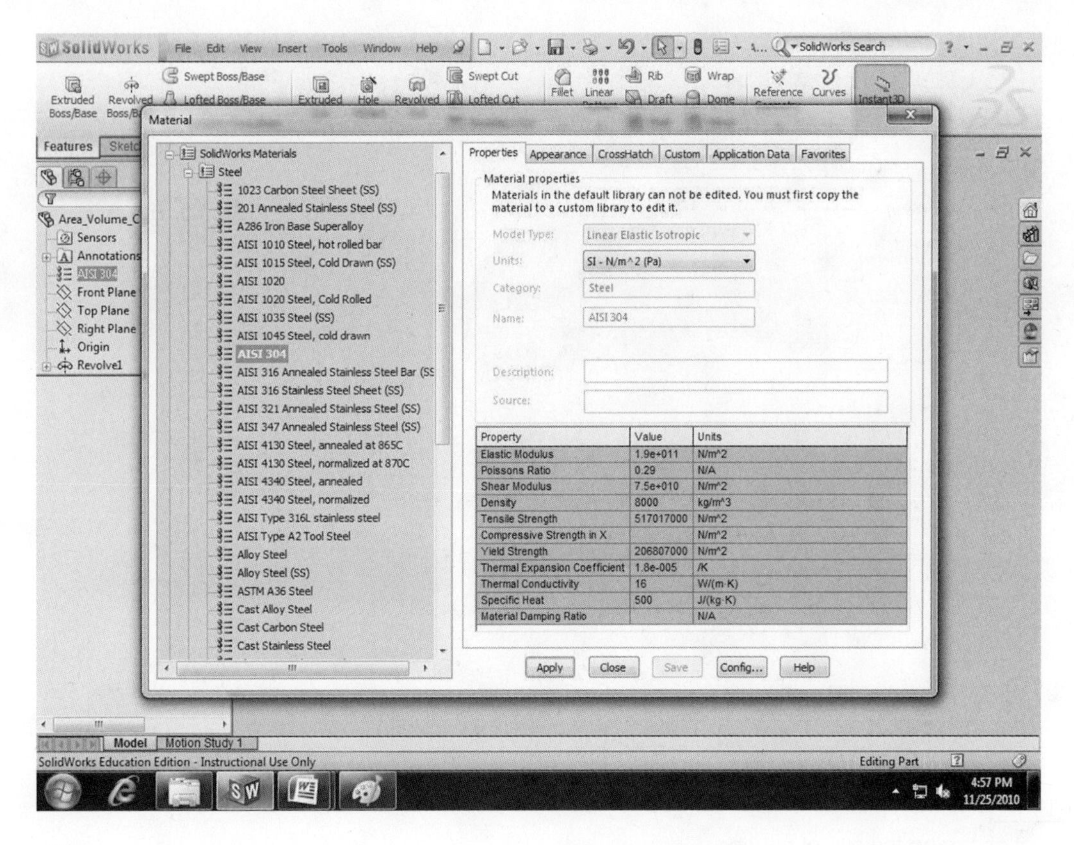

Figure 7.9

Material property: AISI 304 steel is selected.

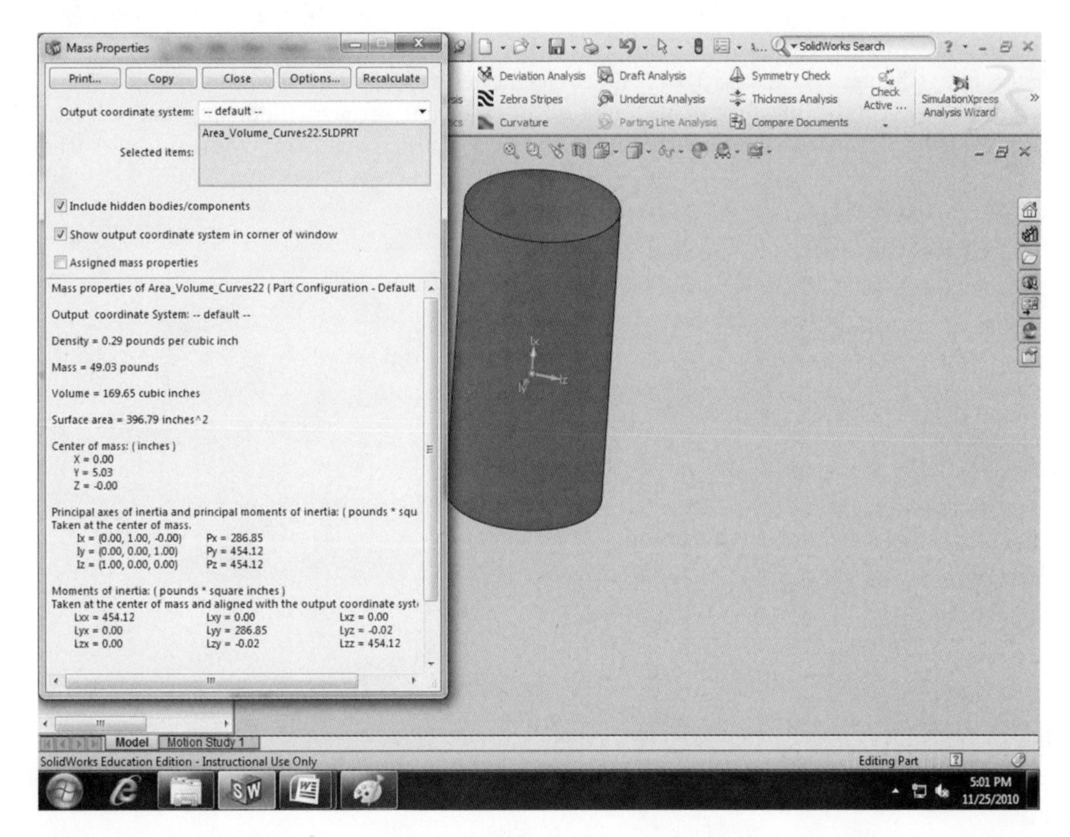

Figure 7.10

SolidWorks mass properties for the solid.

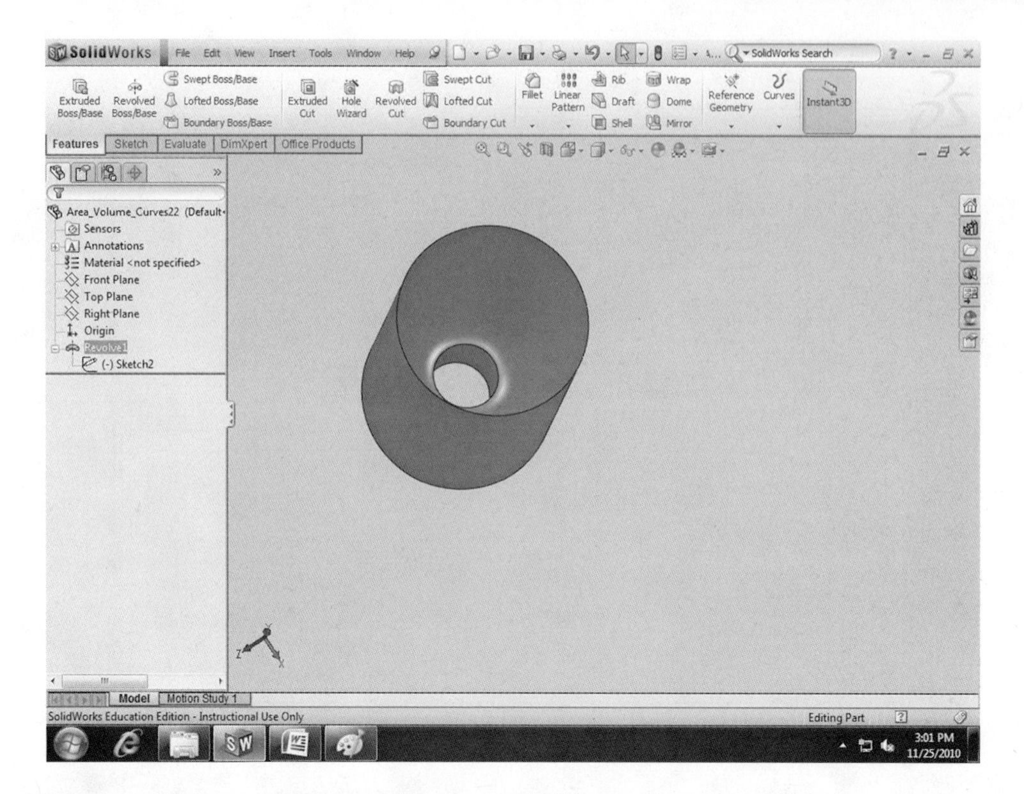

Figure 7.11

Effect of changing the axis of rotation.

Problem 2

The problem we are solving is to find the area that is bounded by the parabolas $y^2 = 4x$ and $x^2 = 4y$. This problem is from Chapter 31, Exercise 3, #22 of the referenced textbook.

We will first present the integral calculus solution and then follow the methodology that is described in this chapter to present the SolidWorks solution. The shape is then extruded 25 mm to create a 3D part model.

Integral Calculus Solution for Area

To find the points of intersection $y^2 = 4x$ with $x^2 = 4y$,

$$\left(\frac{x^2}{4}\right)^2 = 4x$$

or

$$x^4 - 64x = 0$$

leads to $x(x^3 - 64) = 0$; and the solution becomes $x = 0, 4$.

Solving for y:

$$y = \frac{(0)^2}{4} = 0;\ y = \frac{(4)^2}{4} = 4$$

leading to points of intersection (0, 0) and (4, 4).

We integrate the functions taking limits as the values of x (0 and 4):

$$A = \int_0^4 \left[2x^{1/2} - \frac{x^2}{4}\right] dx = \left[\frac{4}{3}x^{3/2} - \frac{x^3}{12}\right]_0^4 = 5.333$$

SolidWorks Solution for Area

In this case, we use the Equation Driven Curve tool in SolidWorks to first define the two curves.

1. Open a New SolidWorks part document.
2. Choose the Front Plane.
3. Click Sketch to be in sketch mode.
4. Click Spline > Equation Driven Curves.
   The Equation Driven Curves PropertyManager is displayed as in Figure 7.12.
5. In the Enter an equation as a function of x, enter the equation (see Figure 7.12).
6. In the Enter start x value for the equation, enter the start value (see Figure 7.12).
7. In the Enter end x value for the equation, enter the end value (see Figure 7.12).
8. Click OK to complete the procedure.
9. Repeat steps 1–8 for the second curve (see Figures 7.13 and 7.14 for curves).
10. Click Evaluate > Section Properties (see Figure 7.15 for the solution).

Comparing the values of the area enclosed by the two curves between the upper and lower limits shows that both integral calculus and SolidWorks have the same solution, as shown in Table 7.2.

Part Design Using the SolidWorks Equation-Driven Curves

The part design is obtained by extruding the area already obtained 25 mm, in Blind mode. Figure 7.16 shows the preview and solid views of the extruded shape.

We apply the material property and choose AISI 304 steel and then compute the mass properties. Figure 7.17 shows the SolidWorks mass properties for the solid that is generated in this study.

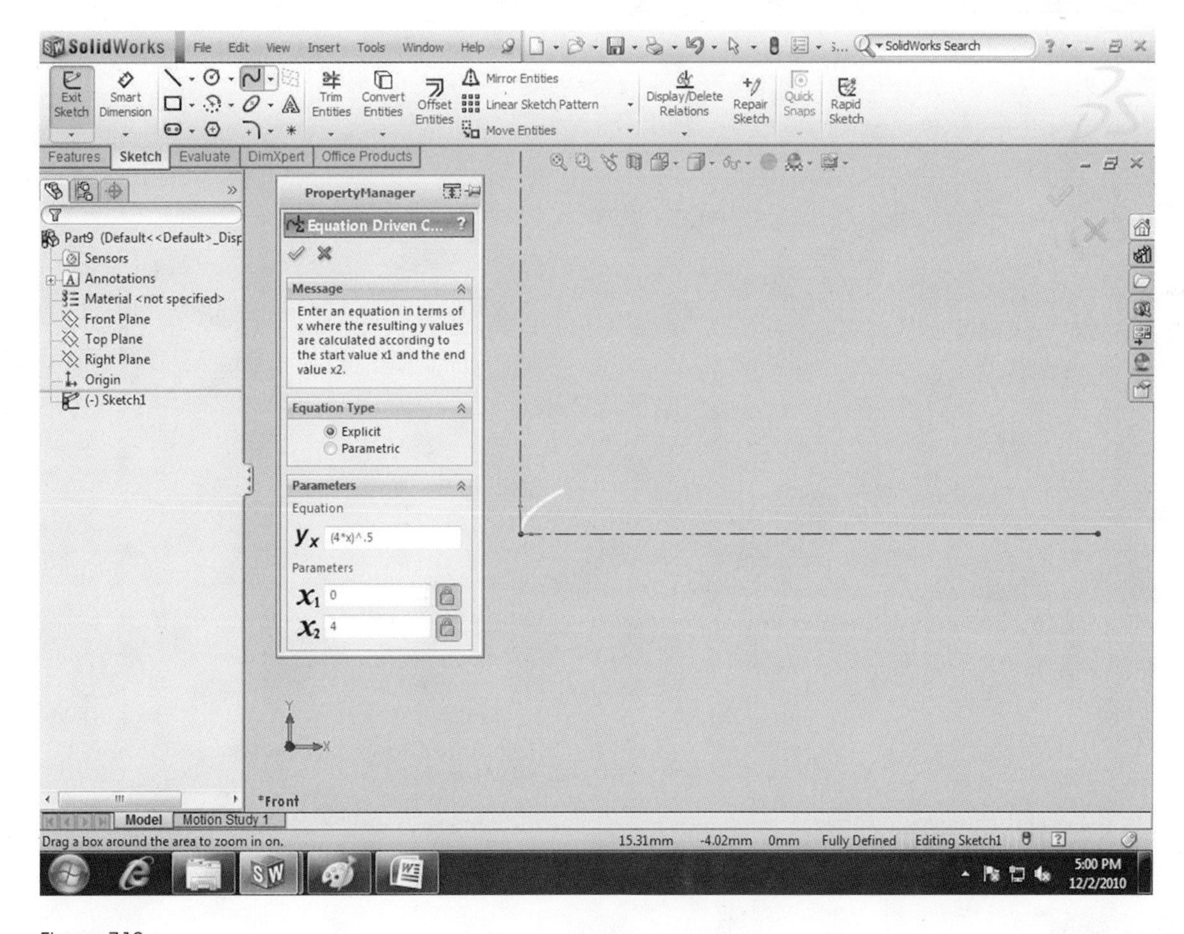

Figure 7.12

Equation 1.

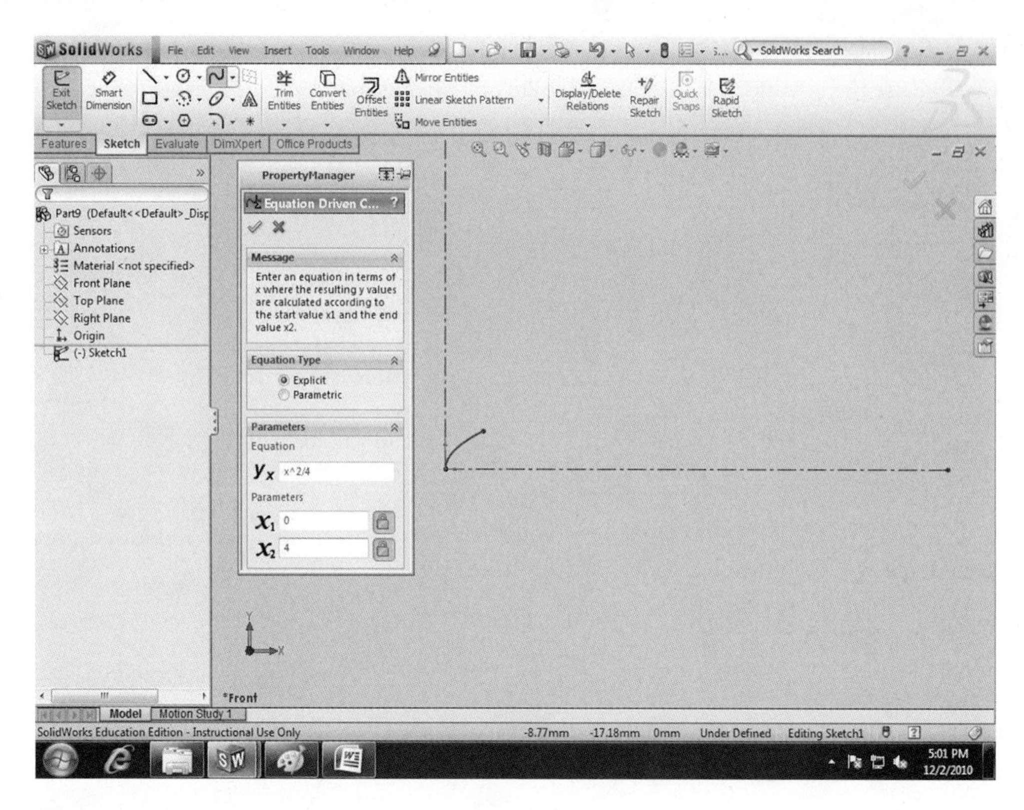

Figure 7.13

Equation 2.

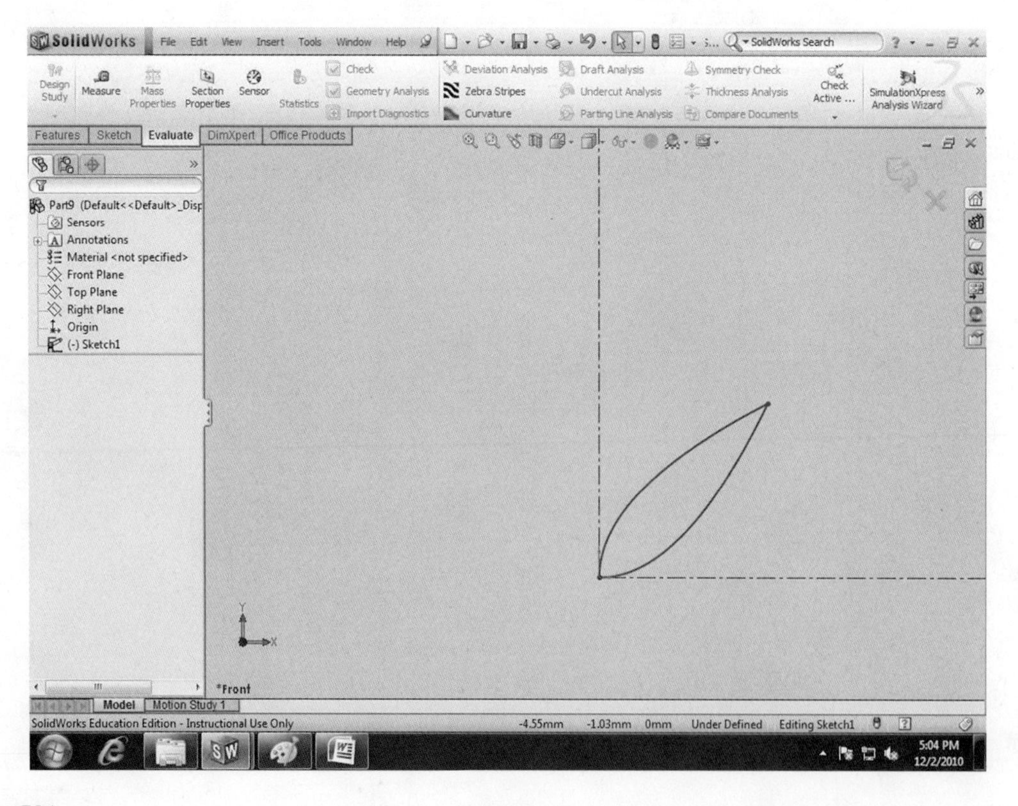

Figure 7.14

Curves are defined.

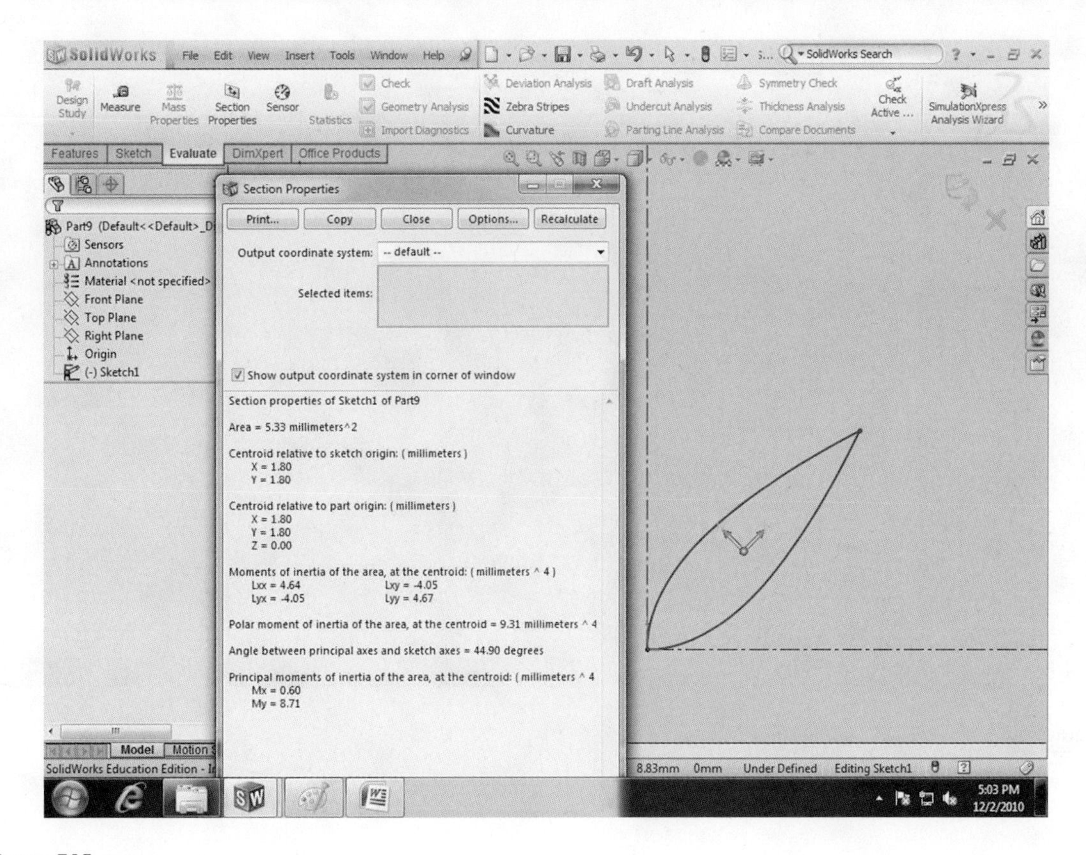

Figure 7.15

Section properties.

Table 7.2 Areas Obtained

Method	Value
Integral calculus	$A = 5.33$ sq. units
SolidWorks	$A = 5.33$ sq. units

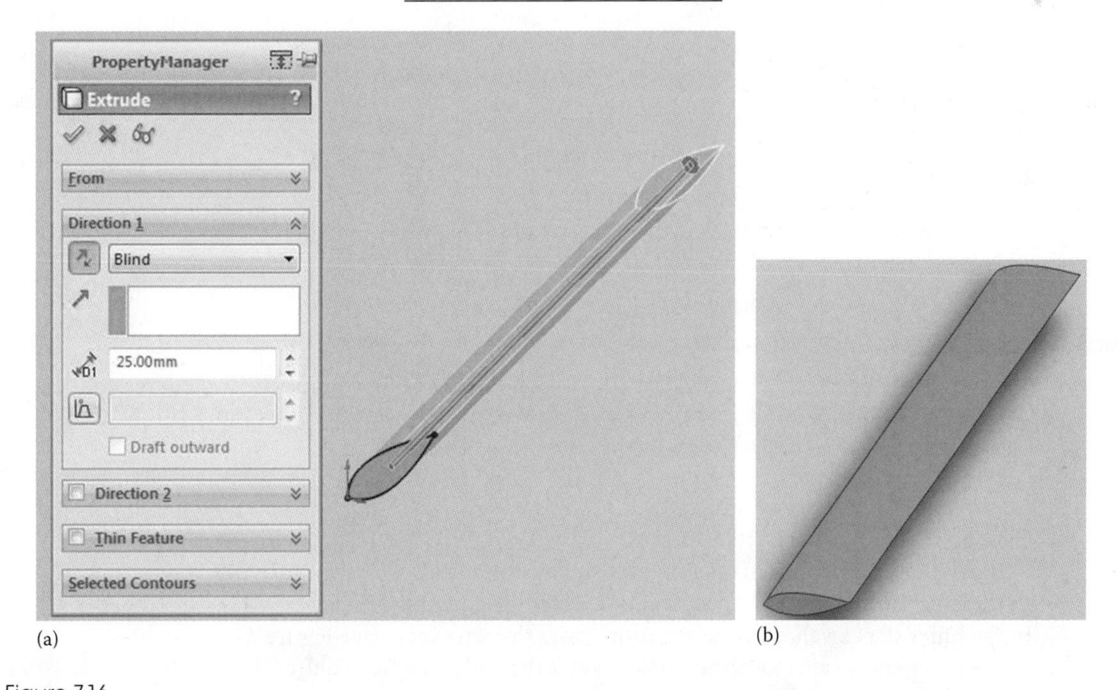

Figure 7.16

Extruded shape: (a) preview and (b) model.

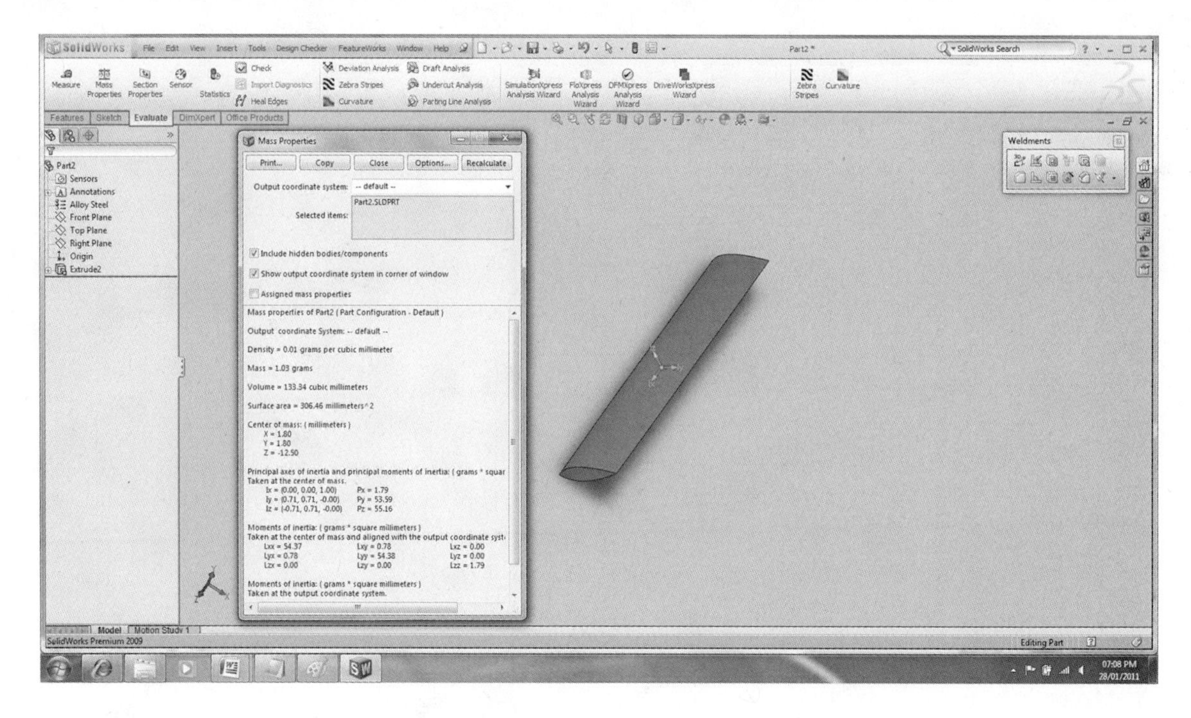

Figure 7.17

Mass properties.

Problem 3

The problem we are solving is to find the volume that is generated when the area bounded by the curve $y = 8/x$, the x-axis, and the lines $x = 1$ and $x = 8$ are rotated about the x-axis.

This problem is from Chapter 31, Section 4, Example 16 of the referenced textbook.

We will first present the integral calculus solution and then follow the methodology that is described in this chapter to present the SolidWorks solution. The shape is then revolved about the x-axis to create a 3D part model.

Integral Calculus Solution for Volume

Since we are rotating the elemental area about the x-axis, the radius of the disc is equal to y, and the thickness dh of the disc is dx.

$$V = \pi\int_a^b y^2\,dy = \pi\int_1^8 \left(\frac{8}{x}\right)^2 dx = 64\pi\int_1^8 x^{-2}\,dx = -64\pi\left[x^{-1}\right]_1^8 = 56\pi\,\text{cub. units}$$

SolidWorks Solution for Area

In this case, we use the Equation Driven Curve tool in SolidWorks to first define the two curves.

1. Open a New SolidWorks part document.
2. Choose the Front Plane.
3. Click Sketch to be in sketch mode.
4. Click Spline > Equation Driven Curves.

   The Equation Driven Curves PropertyManager is displayed as in Figure 7.18.
5. In the Enter an equation as a function of x, enter the equation (see Figure 7.18).
6. In the Enter start x value for the equation, enter the start value (see Figure 7.18).
7. In the Enter end x value for the equation, enter the end value (see Figure 7.18).
8. Click OK to complete the procedure.
9. Sketch two vertical lines and one horizontal line to close the shape (see Figure 7.19).

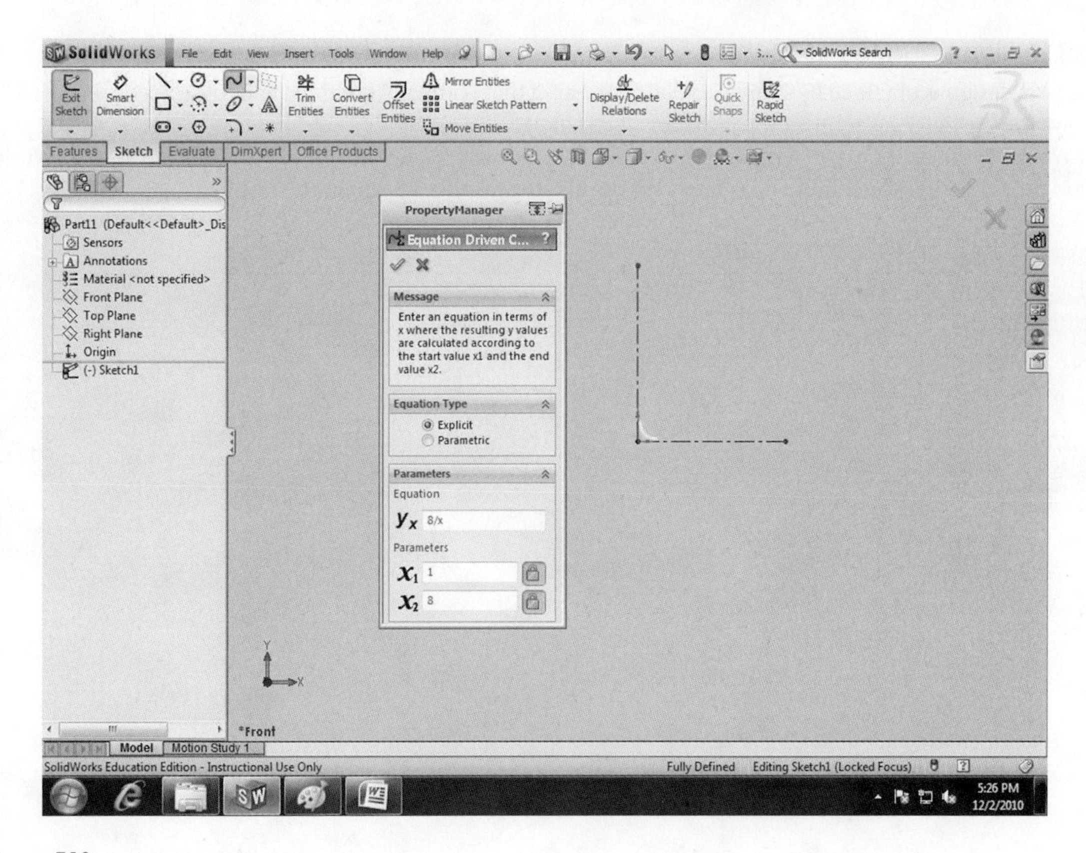

Figure 7.18

Equation for the curve.

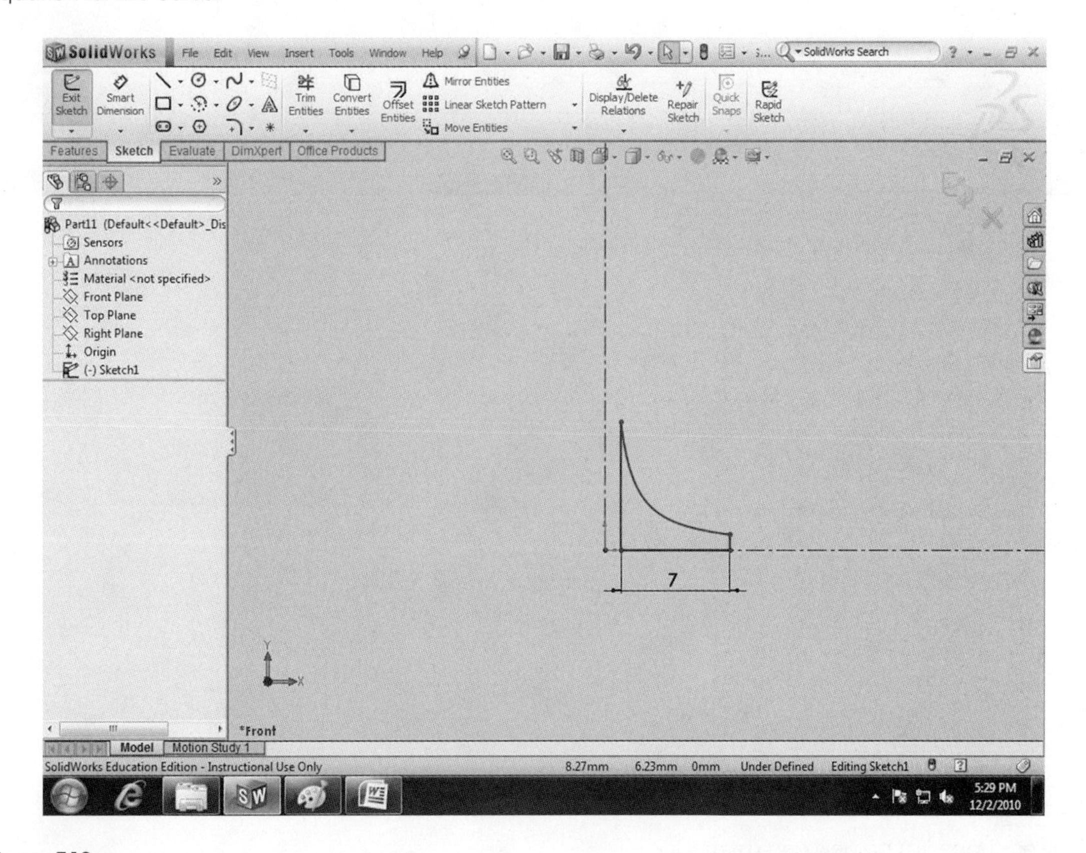

Figure 7.19

Curve is defined; vertical lines and horizontal line included.

Part Design Using the SolidWorks Equation-Driven Curves

The part design is obtained by simply revolving the area that is already obtained about the x-axis. Figure 7.20 shows the preview and solid views of the revolved shape.

We apply the material property and then compute the mass properties. Figure 7.21 shows the SolidWorks mass properties for the solid that is generated in this study. Table 7.3 shows the results.

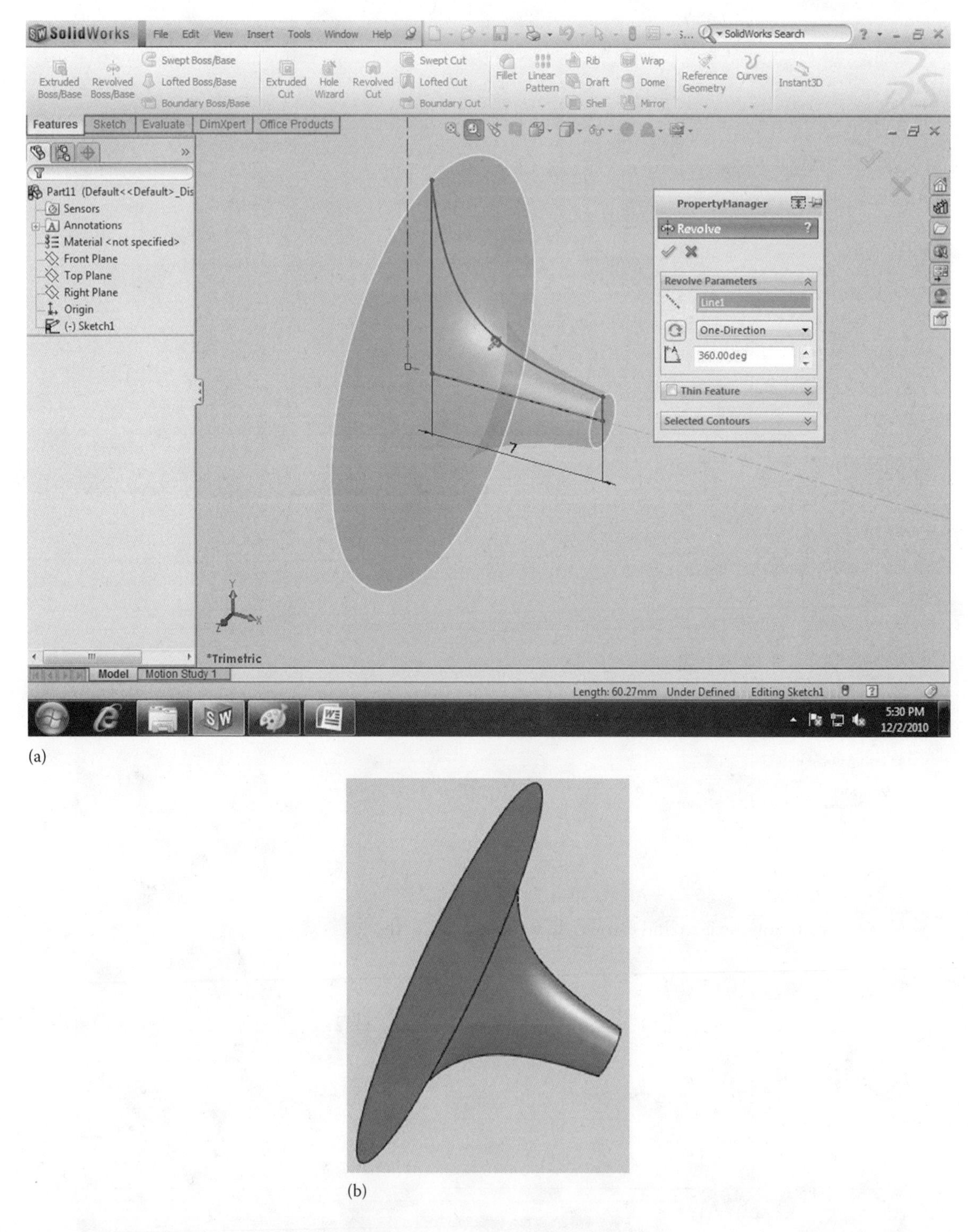

Figure 7.20

Revolved shape: (a) preview and (b) model.

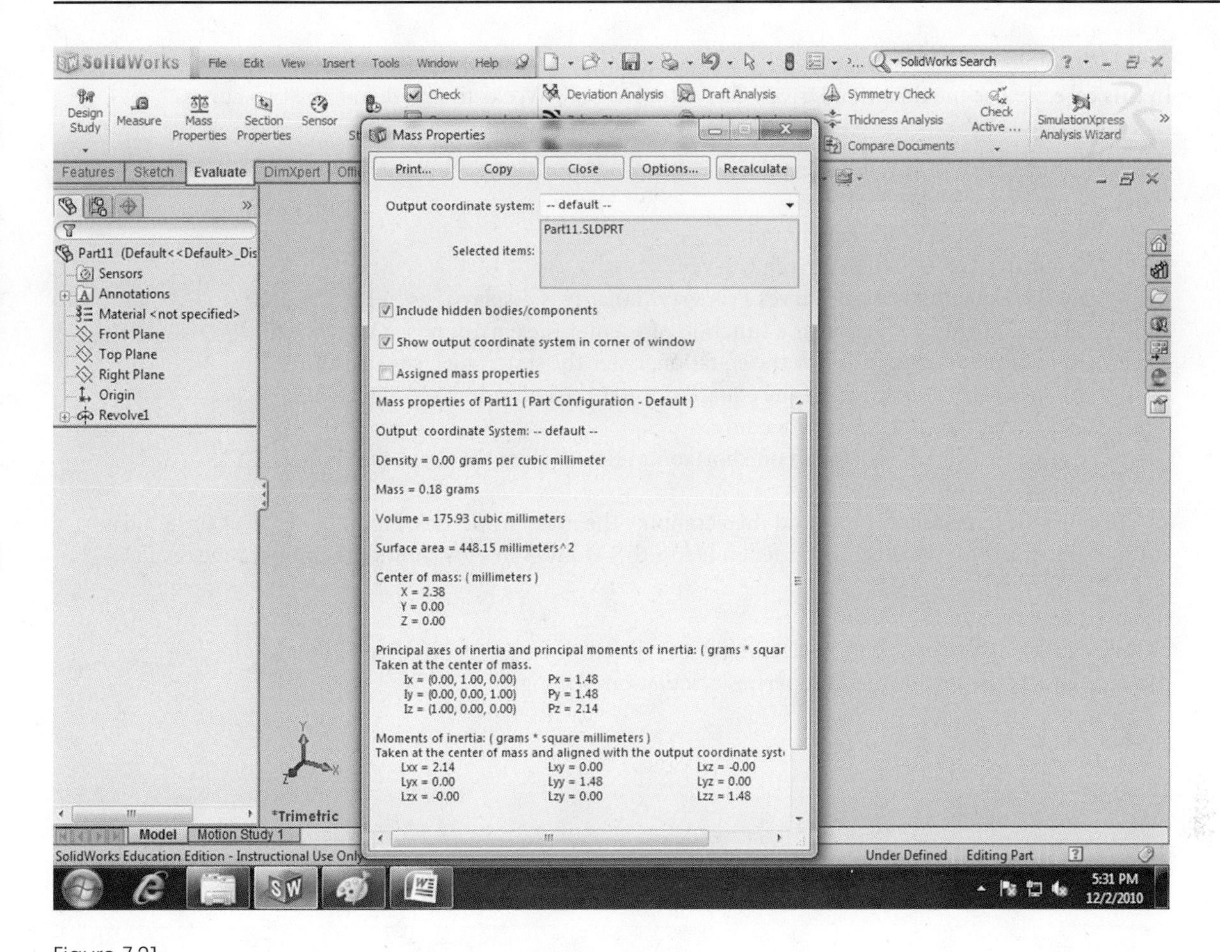

Figure 7.21

Mass properties.

Table 7.3 Volumes Obtained

Method	Value
Integral calculus	$V = 56\,\pi$ sq. units
SolidWorks	$V = 56\,\pi$ sq. units

Problem 4

The problem we are solving is to find the area that is bounded by the curve $y = x^2 + 3$, the y-axis, and the lines $y = 7$ and $y = 12$, which are rotated about the y-axis.

This problem is from Chapter 31, Section 3, Example 14 of the referenced textbook.

We will first present the integral calculus solution and then follow the methodology that is described in this chapter to present the SolidWorks solution. The shape is then revolved about the y-axis to create a 3D part model.

Integral Calculus Solution for Area

$$dA = xdy = (y - 3)^{1/2}\,dy$$

$$A = \int_7^{12} [(y-3)^{1/2}\,dy]dx = \left[\frac{2(y-3)^{3/2}}{3}\right]_7^{12} = \frac{2}{3}[9^{3/2} - 4^{3/2}] = \frac{38}{3} \text{ sq. units}$$

SolidWorks Solution for Area

In this case, we use the Equation Driven Curve tool in SolidWorks to first define the two curves.

1. Open a New SolidWorks part document.
2. Choose the Front Plane.
3. Click Sketch to be in sketch mode.
4. Click Spline > Equation Driven Curves.
   The Equation Driven Curves PropertyManager is displayed as in Figure 7.22.
5. In the Enter and equation as a function of x, enter the equation (see Figure 7.22).
6. In the Enter start x value for the equation, enter the start value (see Figure 7.22).
7. In the Enter end x value for the equation, enter the end value (see Figure 7.2).
8. Click OK to complete the procedure.
9. Sketch two vertical lines and one horizontal line to close the shape (see Figure 7.22).

We apply the material property and then compute the mass properties. Figure 7.23 shows the SolidWorks section properties for the area that is generated in this study. Table 7.4 shows the comparative results.

Revolve about y-Axis

10. Click Revolve option and select the y-axis about which to revolve (see Figure 7.24).
11. Click Evaluate for Mass Properties calculation (see Figure 7.25).

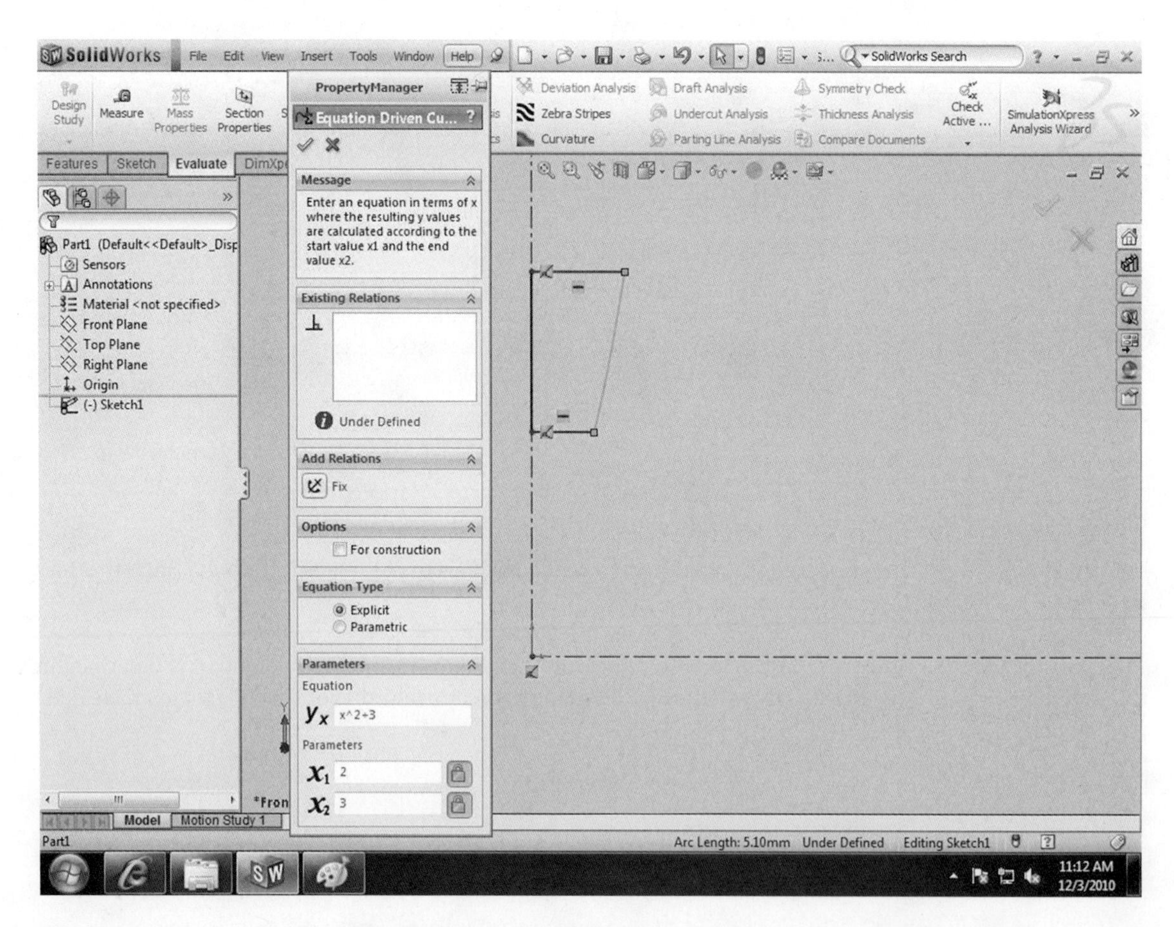

Figure 7.22

Curve is defined; vertical line and horizontal lines included.

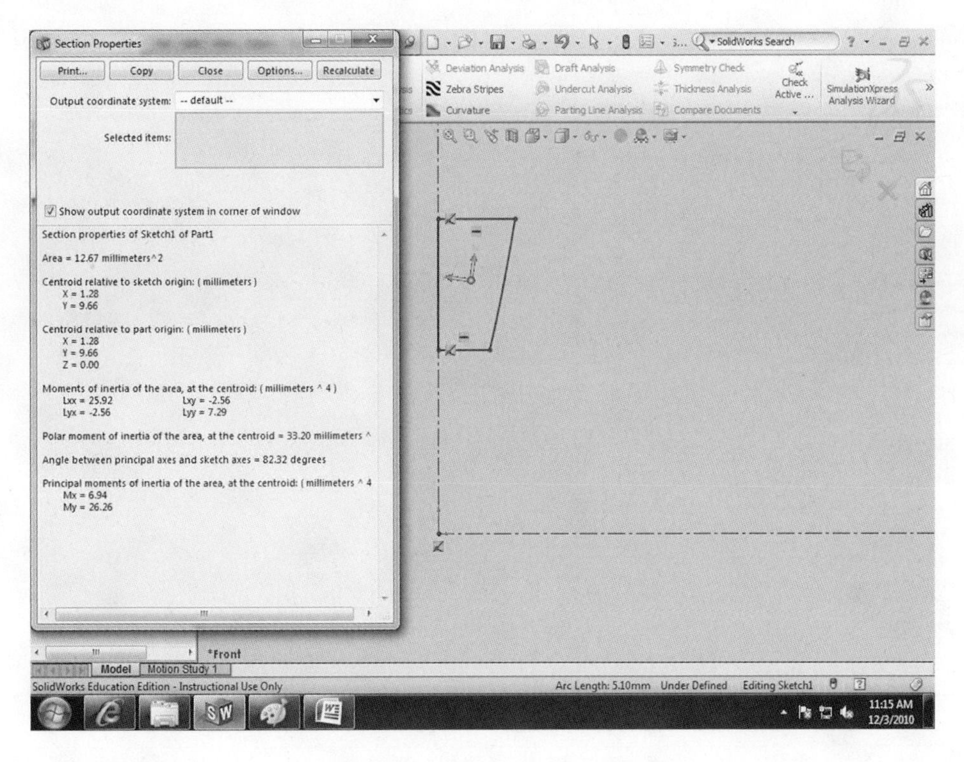

Figure 7.23

Section properties.

Table 7.4 Areas Obtained

Method	Value
Integral calculus	$A = 38/3$ sq. units
SolidWorks	$A = 38/3$ sq. units

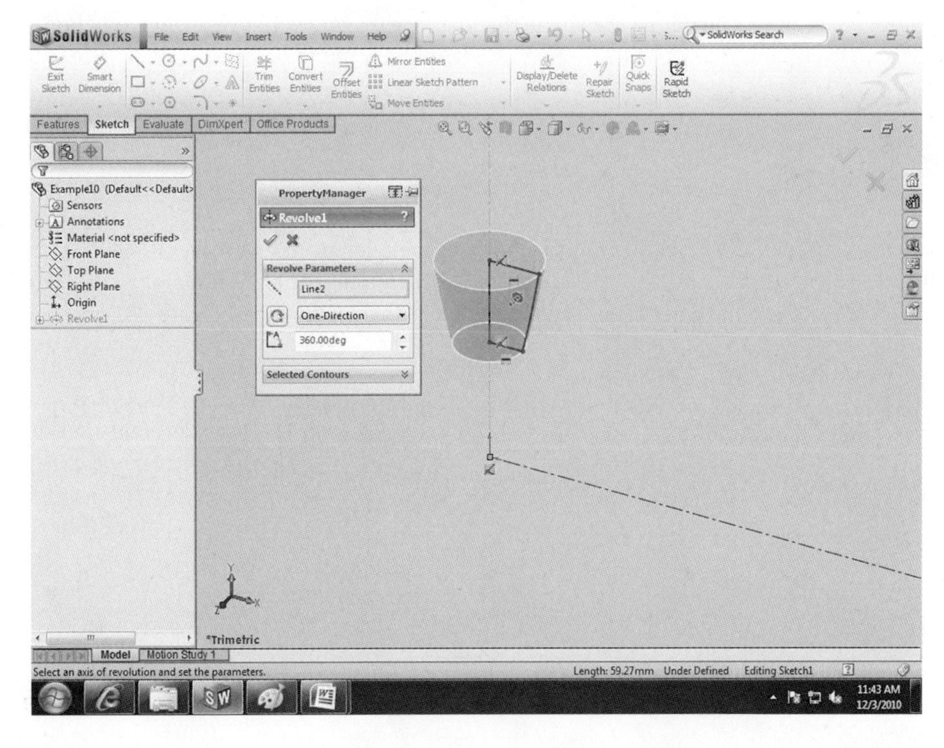

Figure 7.24

Revolve1.

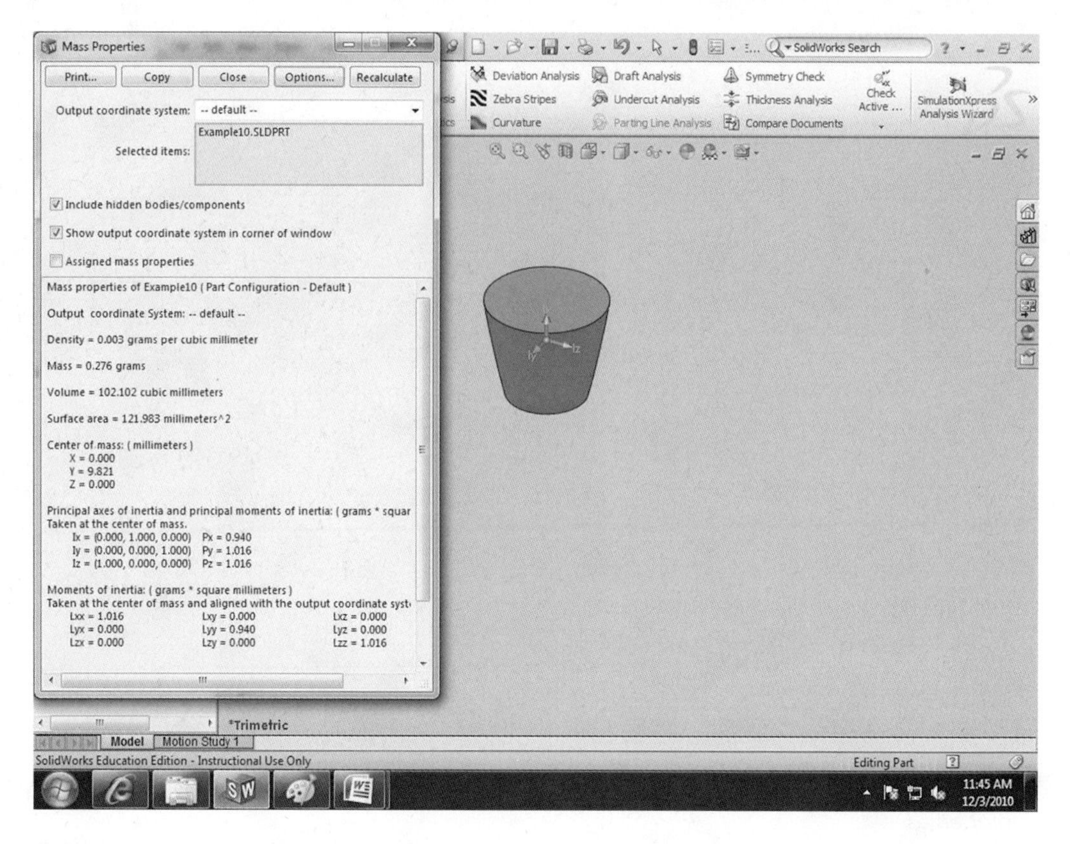

Figure 7.25

Mass properties.

Problem 5

The problem we are solving is to define the shape that is made up of the first-quadrant area bounded by the curves

$$y^2 = 12x$$

$$y^2 = 24x - 36.$$

This problem is from Chapter 31, Section 3, Example 9, of the referenced textbook.

We will first present the integral calculus solution and then follow the methodology that is described in this chapter to present the SolidWorks solution. The shape is then extruded to create a 3D part model.

Integral Calculus Solution for Area

The points of intersection are found by solving the simultaneous equations:

$$12x = 24x - 36$$

Yielding $x = 3$ and $y = \pm\sqrt{12x} = \pm 6$.

Let

$$x_1 = \frac{y^2}{12}$$

$$x_2 = \frac{y^2 + 36}{24}$$

The area between the curves is given as

$$A = \int_0^6 [x_1 - x_2]dx = \int_0^6 \left(\frac{y^2}{24} + \frac{36}{24} - \frac{y^2}{12}\right)dy = \int_0^6 \left(\frac{3}{2} - \frac{y^2}{24}\right)dy = \left[\frac{3y}{2} - \frac{y^3}{72}\right]_0^6 = \frac{3\times 6}{2} - \frac{6^3}{72} = 6 \text{ sq. units}$$

By symmetry, the total area between the two curves is twice this value, or 12 square units.

The next step is to use the SolidWorks Equation Driven Curve tool to define the curves that are represented by the two equations that are used with the integral calculus method to ascertain the accuracy of SolidWorks.

SolidWorks Solution for Area

In this case, we use the Equation Driven Curve tool in SolidWorks to first define the two curves.

1. Open a New SolidWorks part document.
2. Choose the Front Plane.
3. Click Sketch to be in sketch mode.
4. Click Spline > Equation Driven Curves.
   The Equation Driven Curves PropertyManager is displayed as in Figure 7.26.
5. In the Enter and equation as a function of x, enter the equation (see Figure 7.26).
6. In the Enter start x value for the equation, enter the start value (see Figure 7.26).
7. In the Enter end x value for the equation, enter the end value (see Figure 7.26).
8. Click OK to complete the procedure.
9. Repeat steps 1–8 for the second curve (see Figure 7.27).
   Mirror about x-axis
10. Click Mirror.
11. Select the two curves that are defined and click the horizontal construction line to mirror about it (see Figure 7.28).
12. Click Evaluate > Section Properties (see Figure 7.29 for the solution).

Comparing the values of the area enclosed by the two curves between the upper and lower limits shows that both integral calculus and SolidWorks have the same solution, as shown in Table 7.5.

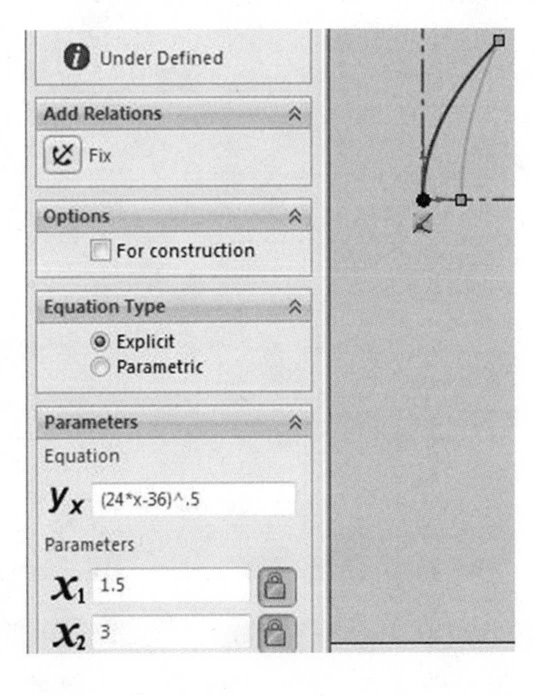

Figure 7.26

Equation 1.

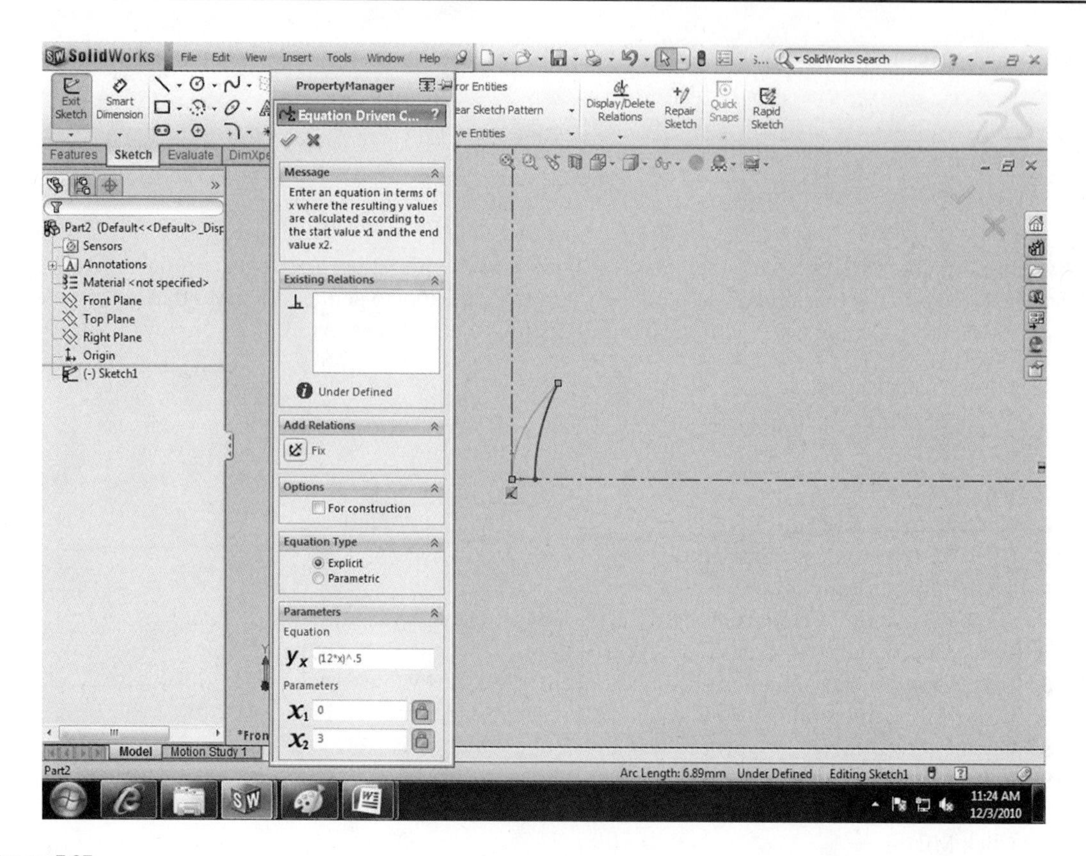

Figure 7.27

Equation 2.

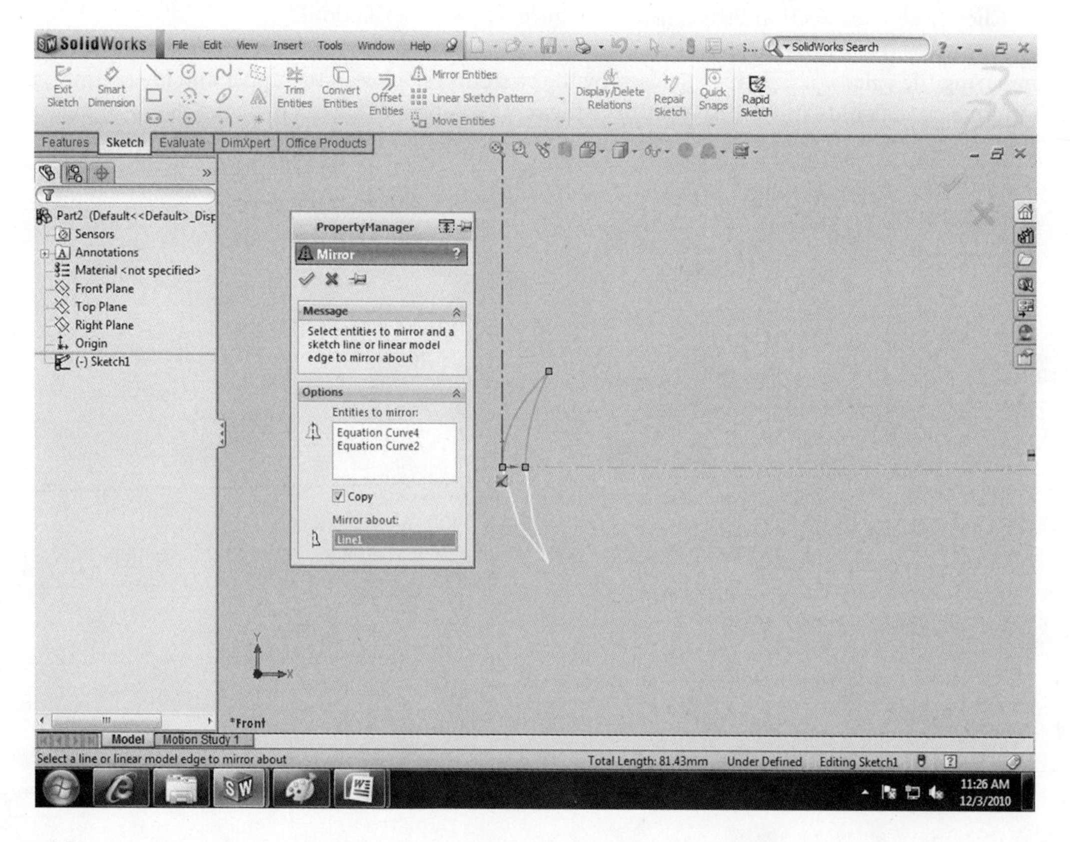

Figure 7.28

Curves are defined and mirrored about the x-axis.

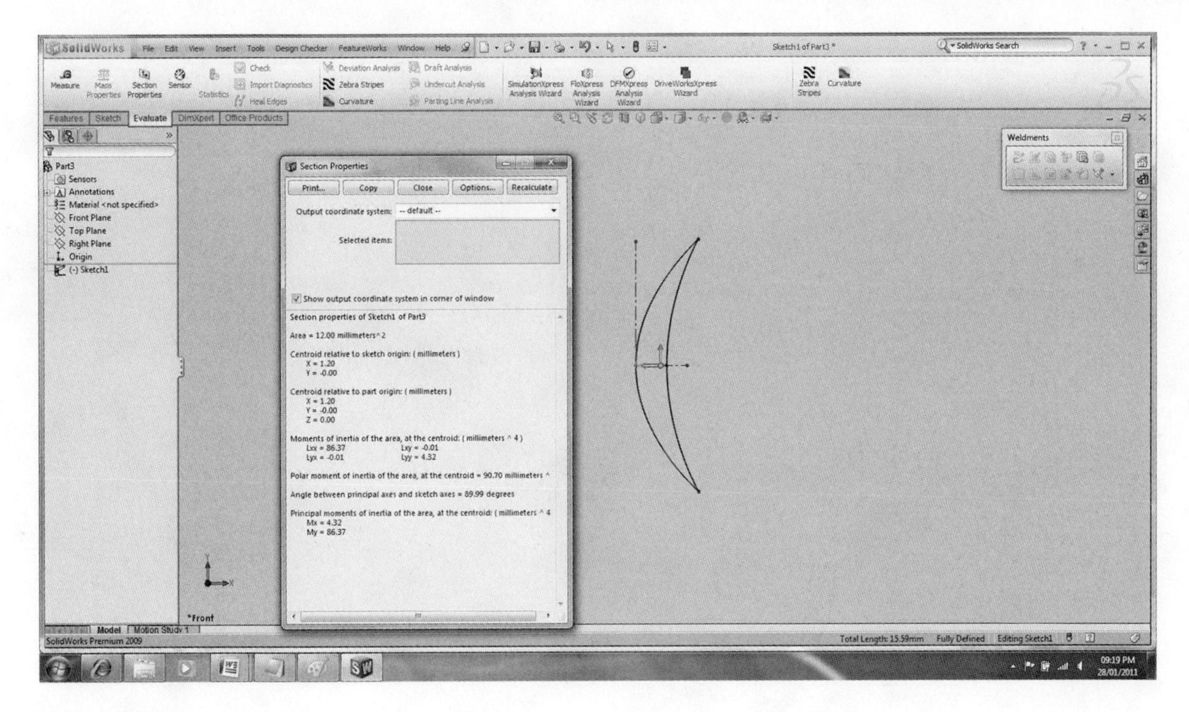

Figure 7.29

Section properties.

Table 7.5 Areas Obtained

Method	Value
Integral calculus	$A = 12$ sq. units
SolidWorks	$A = 12$ sq. units

Part Design Using the SolidWorks Equation-Driven Curves

The part design is obtained by extruding the area already obtained, in Blind mode. Figure 7.30 shows the solid of the extruded shape.

We apply the material property (Figure 7.31) and then compute the mass properties. Figure 7.32 shows the SolidWorks mass properties for the solid that is generated in this study.

Summary

This chapter presents the steps that are involved in using SolidWorks software for the design parts that are bounded by equation-driven curves. The results for known functions show that the areas and volumes obtained using SolidWorks agree with those that are obtained using basic integral methods in mathematics. Integral calculus methods are limited in functionality because general-shaped objects may not have known mathematical definitions, and hence these methods may fail to compute the areas and volumes of such shapes. Numerical methods are used in such cases. This study shows that SolidWorks is a useful design tool that automatically computes the areas and volumes of parts being designed by a designer. We conclude that using SolidWorks to determine the areas and volumes of objects is extremely useful in engineering because these parameters may be required for estimating the costs that are involved in manufactured parts.

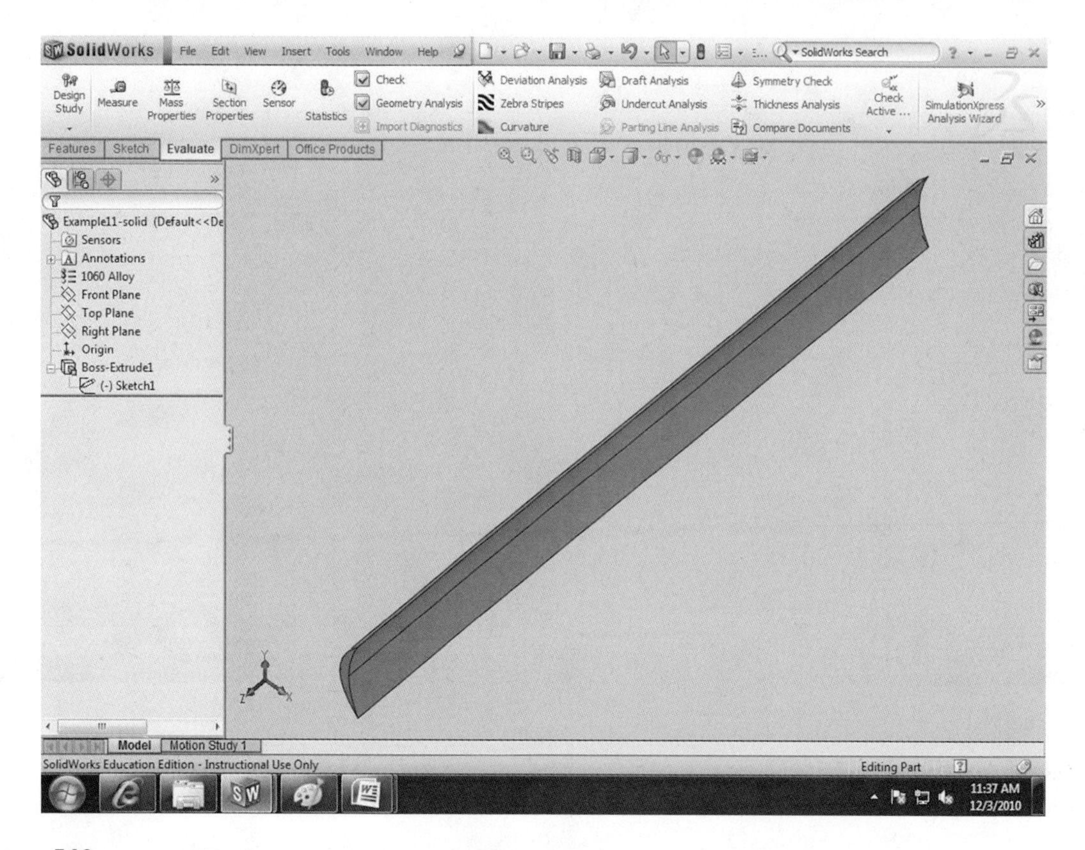

Figure 7.30

Part model.

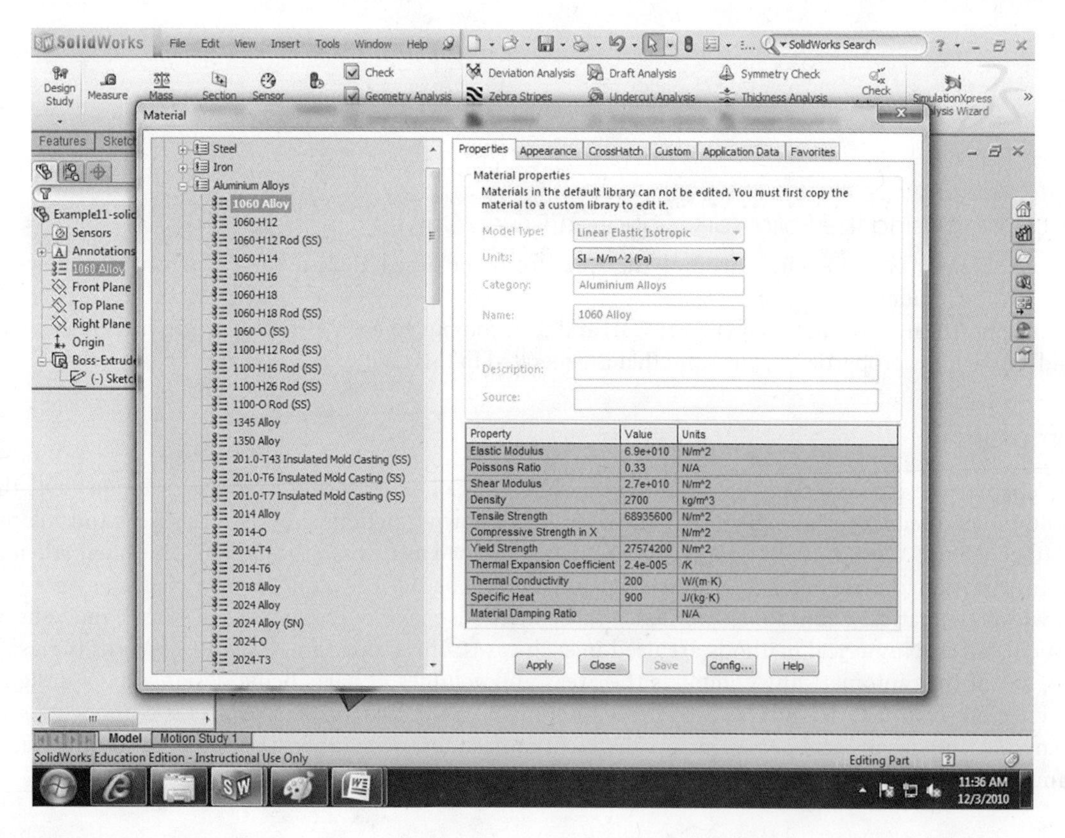

Figure 7.31

Material PropertyManager.

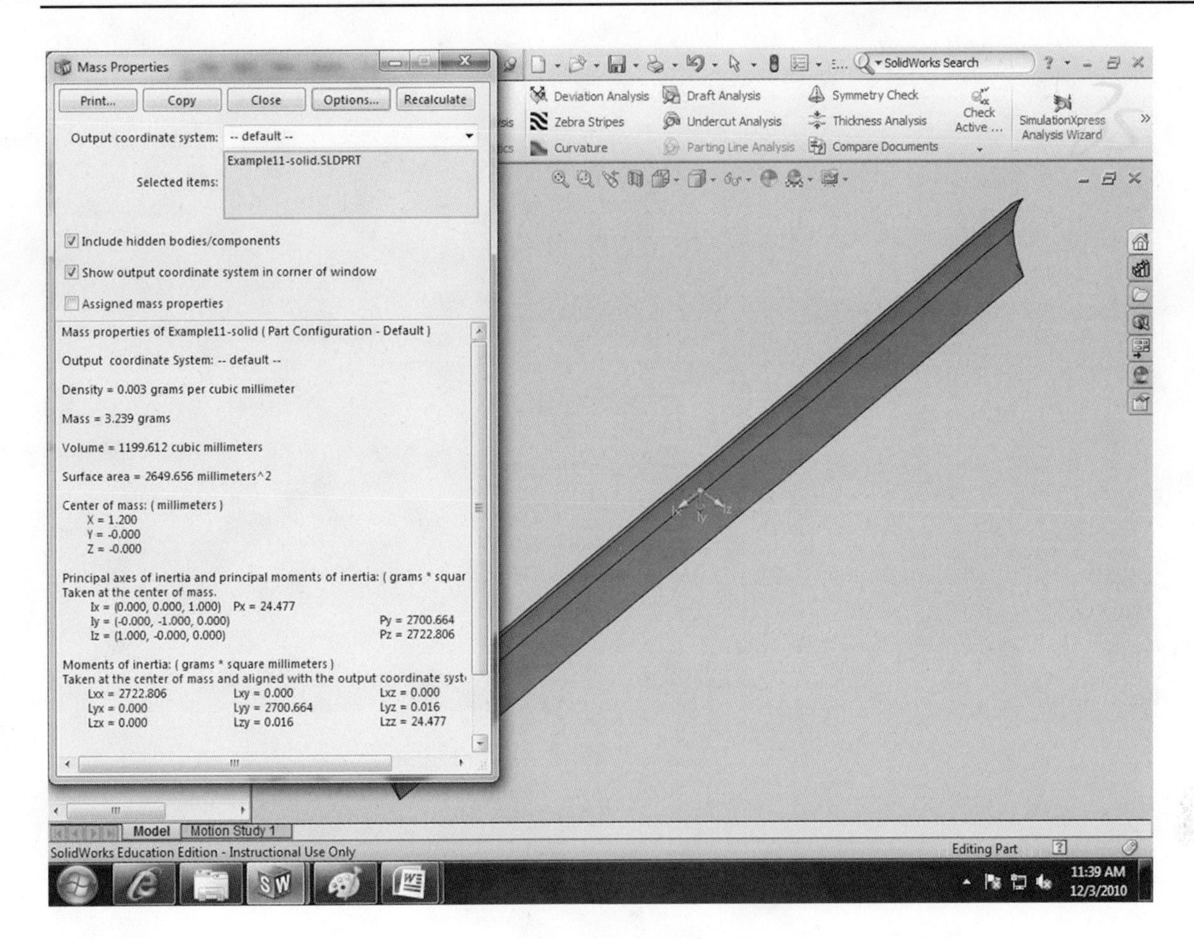

Figure 7.32

Mass properties.

Exercises

1. The problem we are solving is to find the volume that is generated when the first-quadrant area bounded by the curves $y = x^2$ the y-axis, and the lines $y = 1$ and $y = 4$ are rotated about the y-axis. (This problem is from Chapter 31, Section 4, Example 16 of the referenced textbook.) Use the SolidWorks approach that is described in this chapter to determine the volume of the part that is generated. Check your solution with the integral calculus solution that is given as follows:

$$V = \pi \int_a^b x^2\,dy = \pi \int_1^4 y\,dy = \pi \left[\frac{y^2}{2} \right]_1^4 = 7.5\pi \text{ cub. units}$$

2. Find the area that is bounded by the curves $y = \sqrt{x}$ and $y = x - 3$ between $x = 1$ and 4. Revolve the shape that is obtained about the y-axis. Test your knowledge. (See the solutions in Figure P7.1; area = 6.17 square units.)
3. Revolve the shape that is obtained in Question 2 about the left vertical edge in Figure P7.1a.
4. Revolve the shape that is obtained in Question 2 about the right vertical edge in Figure P7.1a.
5. Compare the surface areas and volumes of Questions 2–4.

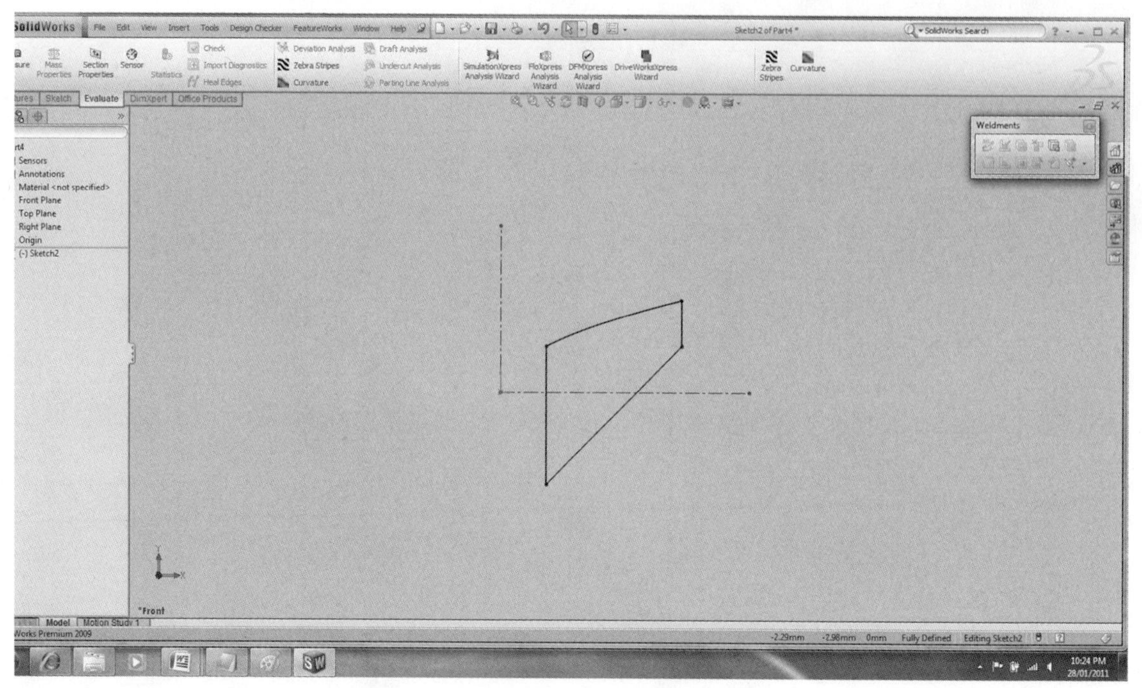

(a)

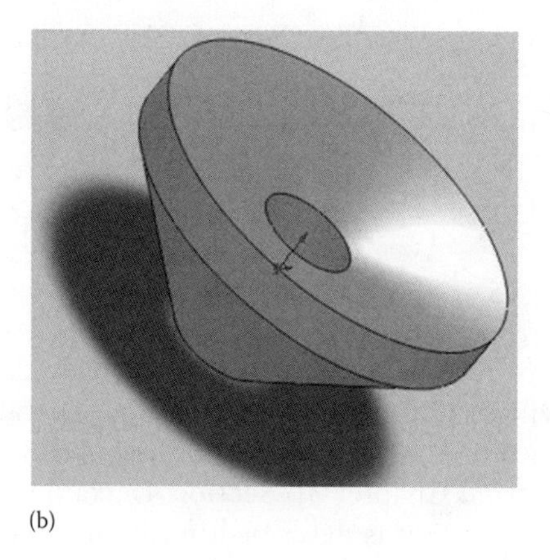

(b)

Figure P7.1

(a) Area between curves. (b) Volume obtained by revolving area about y-axis.

Bibliography

Calter, P.A., Calter, M.A. 2008. *Technical Mathematics with Calculus*, Canadian Edition. Canada: John Wiley & Sons.

8

Assembly Modeling—CSWA Preparations

Objectives:

When you complete this chapter, you will have

- Understood the differences between top–down assembly modeling and bottom–up assembly modeling approaches
- Learned how to apply the bottom–up approach to assembly modeling
- Learned how to mate components in an assembly
- Learned how to assemble parts using the assembly model methodology
- Learned how to use assembly analysis–interference analysis to analyze how good an assembly is and to get a feel of the quality of the parts design
- Learned how to use the exploded view assembly tool
- Learned how to animate the exploded view
- Learned how to handle large assemblies

Introduction

Assembly modeling is the combining of part models into complex, interconnected solid models. There are two well-known approaches to assembly modeling: (1) top–down and (2) bottom–up.

1. *Top–Down Assembly Modeling*

   In the top–down assembly modeling approach, major design requirements are translated into assemblies, subassemblies, and components.
2. *Bottom–Up Assembly Modeling*

   In the bottom–up assembly modeling approach, based on design requirements, components are developed independently and combined into subassemblies and assemblies. The three basic steps in the bottom–up assembly modeling approach are as follows:
   1. Create each component independent of any other component in the assembly.
   2. Insert the components in the assembly.
   3. Mate the components in the assembly as they relate to the physical constraints of the design.

The bottom–up assembly modeling approach is used in this book. Since mating the components in an assembly is the new concept among the three steps that are involved, this is first presented, and then the entire process of the bottom–up assembly modeling approach is applied to typical assembly problems.

Starting the Assembly Mode of SolidWorks

There are two ways to start the assembly mode of SolidWorks: (1) from a new SolidWorks document or (2) from an existing part that we wish to place first in the assembly document.

1. To start the assembly mode of SolidWorks, invoke the New SolidWorks Document dialog box and choose the Assembly button and click OK, as shown in Figure 8.1.
   The *Begin Assembly* PropertyManager pops up, as shown in Figure 8.2. Browse to open the first part to insert.
2. From an existing part document, click the New dialog box, as shown in Figure 8.3. Click Make Assembly from Part/Assembly. The Begin Assembly PropertyManager pops up, as shown in Figure 8.2. Browse to open the first part to insert.

Inserting Components in the Assembly Document

There are several ways of inserting a component in the assembly document of SolidWorks:

1. Click Assembly > Insert Components (see Figure 8.4).
2. Click Insert > Component > Existing Part/Assembly (see Figure 8.5).
3. Insert components using the opened document window (see Figure 8.6).
4. In this case, there are currently opened part documents. Choose Windows > Tile Horizontally or Vertically from the Menu Bar menus. All opened SolidWorks document windows will be tiled accordingly, depending on the option that is chosen (see Figure 8.7).

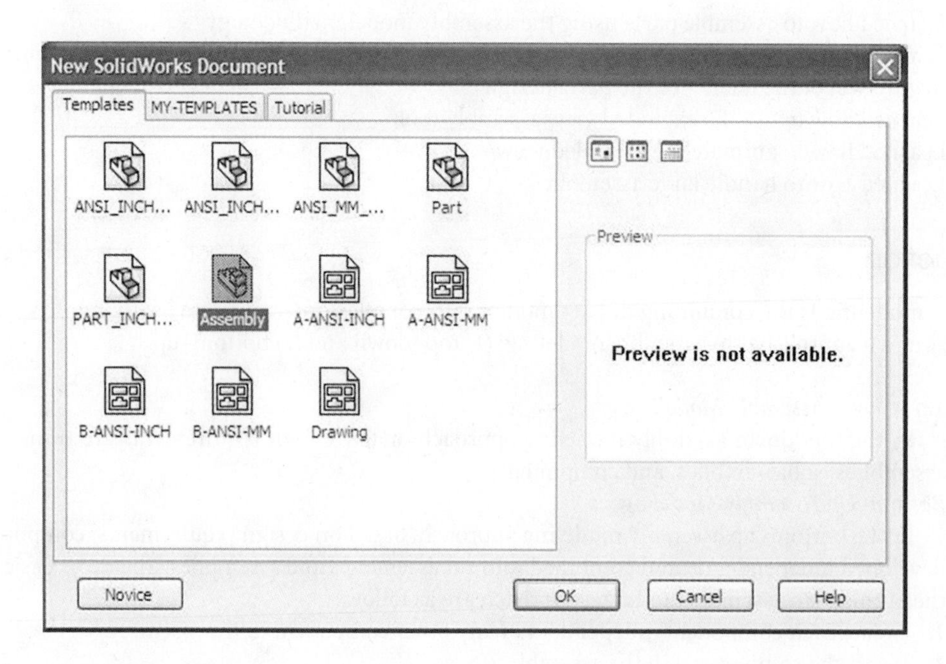

Figure 8.1

The New SolidWorks Document dialog box.

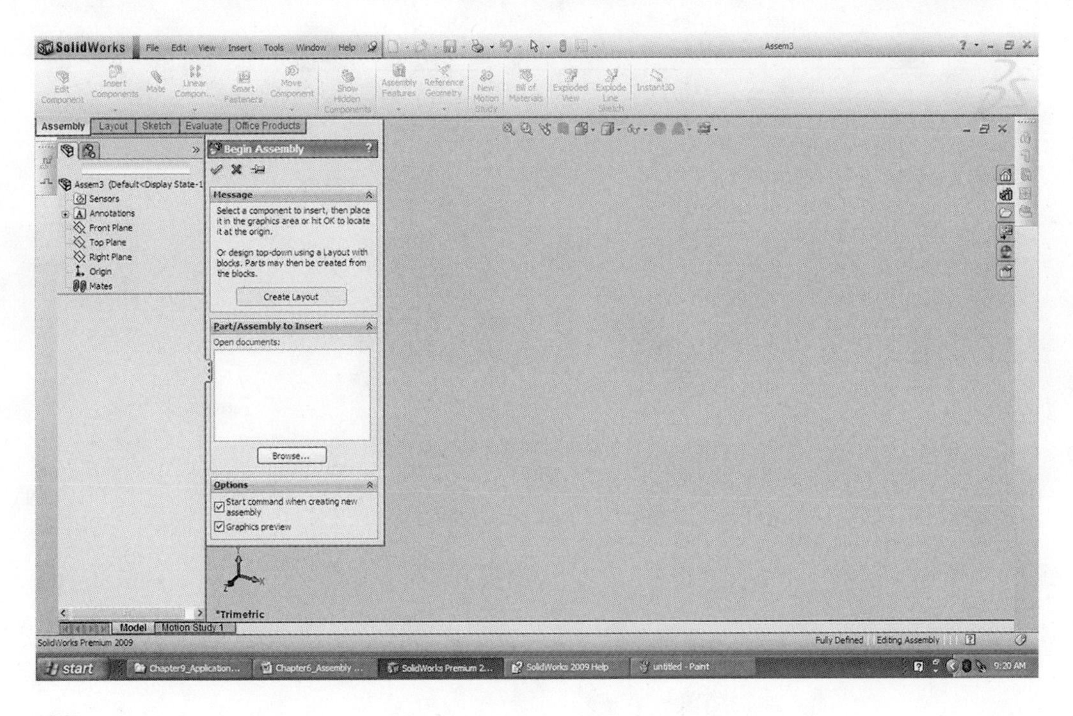

Figure 8.2

The Assembly mode showing the Begin Assembly PropertyManager.

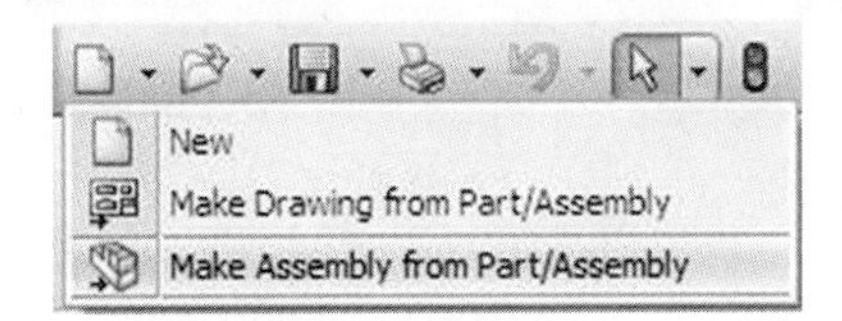

Figure 8.3

Accessing Assembly mode from the New dialog box in an existing part document.

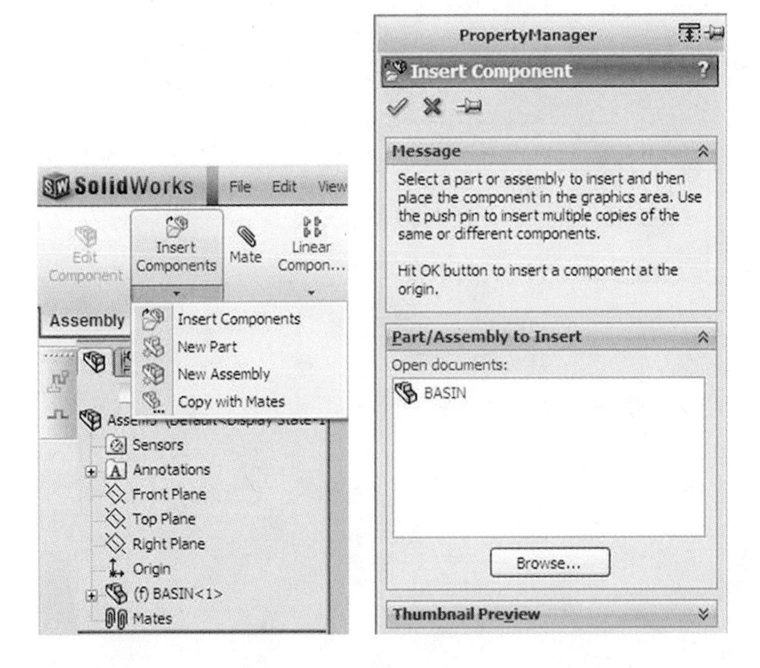

Figure 8.4

Inserting an existing component into an assembly document.

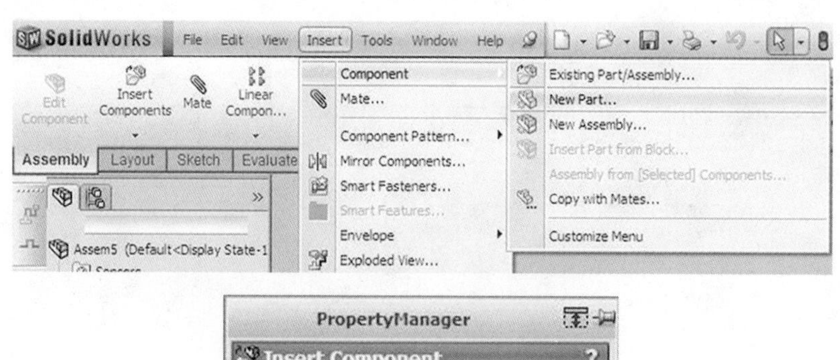

Figure 8.5

Inserting an existing component into an assembly.

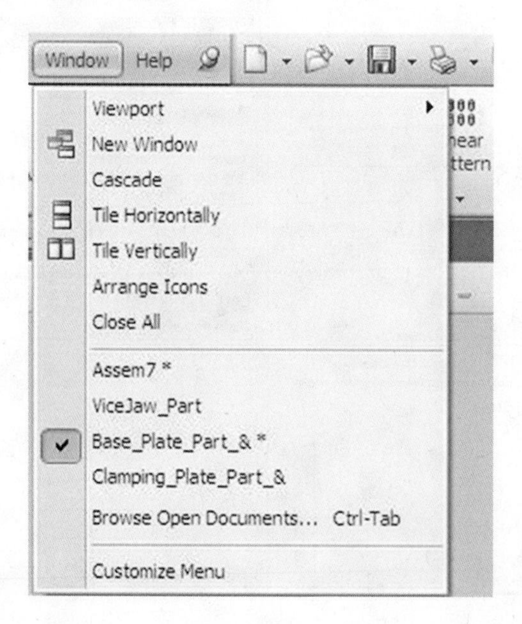

Figure 8.6

Using the opened document window.

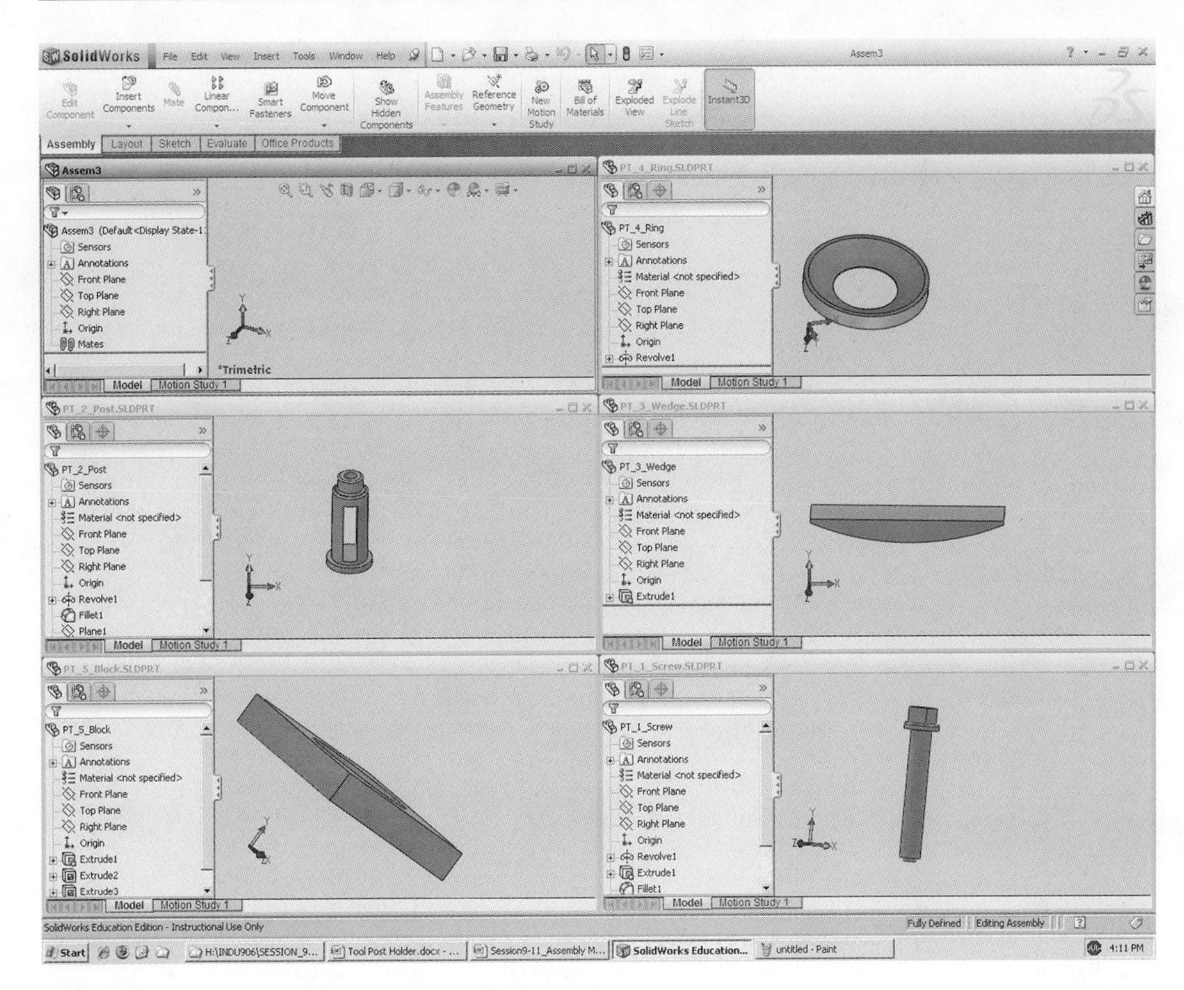

Figure 8.7

Tiled window.

Mates

Mates create geometric relationships between assembly components. There are three types of mates that are available in SolidWorks: (1) standard, (2) advanced, and (3) mechanical mates (see Figure 8.8). Standard mates create geometric relationships, such as coincident, parallel, perpendicular, tangent, concentric, lock, distance, and angle. Each mate is valid for specific combinations of geometry. A comprehensive list of possible mates for different part-types is given here.

Circular or Arc Edge

- *Circular or Arc Edge/Cone*—Coincident, Concentric
- *Circular or Arc Edge/Line*—Concentric
- *Circular or Arc Edge/Cylinder*—Concentric, Coincident
- *Circular or Arc Edge/Plane*—Coincident
- *Circular or Arc Edge/Circular or Arc Edge*—Concentric

Cone

- *Cone/Circular or Arc Edge*—Coincident, Concentric
- *Cone/Cone*—Angle, Coincident, Concentric, Distance, Parallel, Perpendicular
- *Cone/Cylinder*—Angle, Concentric, Parallel, Perpendicular
- *Cone/Extrusion*—Tangent
- *Cone/Line*—Angle, Concentric, Parallel, Perpendicular
- *Cone/Plane*—Tangent
- *Cone/Point*—Coincident, Concentric
- *Cone/Sphere*—Tangent

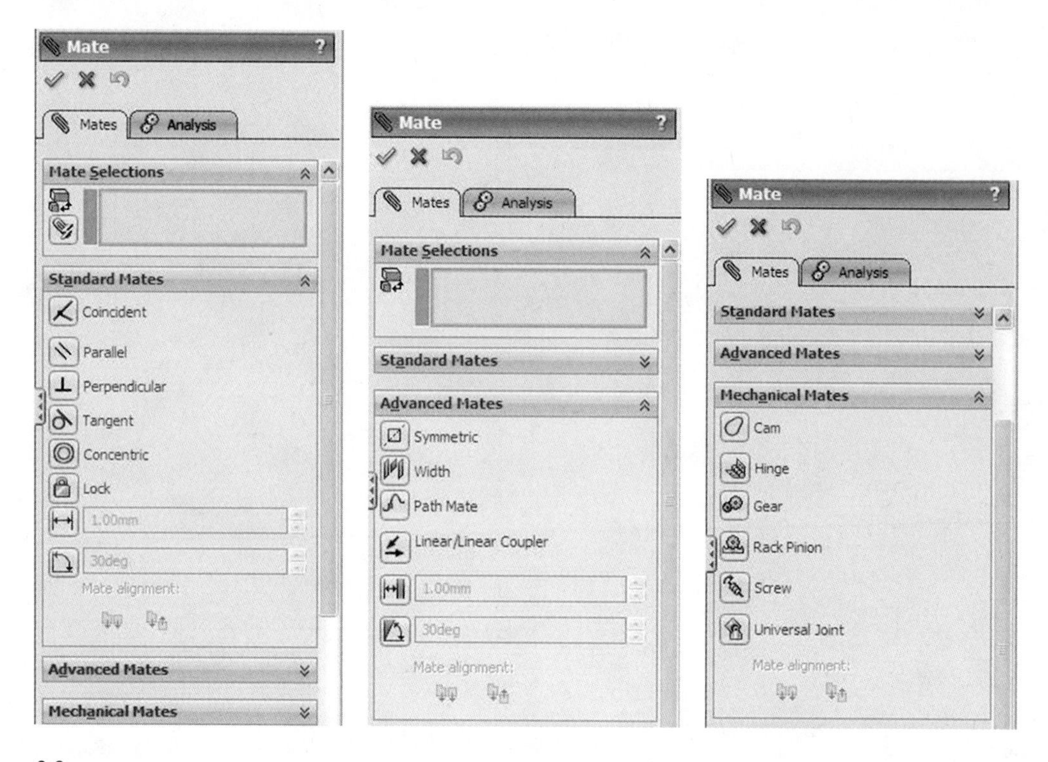

Figure 8.8

Standard, advanced, and mechanical mates in SolidWorks.

Curve

- *Curve/Point*—Coincident, Distance

Cylinder

- *Cylinder/Cone*—Angle, Concentric, Parallel, Perpendicular
- *Cylinder/Cylinder*—Angle, Concentric, Distance, Parallel, Perpendicular, Tangent
- *Cylinder/Extrusion*—Angle, Parallel, Perpendicular, Tangent
- *Cylinder/Line*—Angle, Coincident, Concentric, Distance, Parallel, Perpendicular, Tangent
- *Cylinder/Plane*—Distance, Tangent
- *Cylinder/Point*—Coincident, Concentric, Distance
- *Cylinder/Sphere*—Concentric, Tangent
- *Cylinder/Circular Edge*—Concentric, Coincident
- *Cylinder/Surface*—Tangent

Extrusion

- *Extrusion/Cone*—Angle, Parallel, Perpendicular
- *Extrusion/Cylinder*—Angle, Parallel, Perpendicular, Tangent
- *Extrusion/Extrusion*—Angle, Parallel, Perpendicular
- *Extrusion/Line*—Angle, Parallel, Perpendicular
- *Extrusion/Plane*—Tangent
- *Extrusion/Point*—Coincident

Line

- *Line/Cone*—Angle, Concentric, Parallel, Perpendicular
- *Line/Cylinder*—Angle, Coincident, Concentric, Distance, Parallel, Perpendicular, Tangent
- *Line/Extrusion*—Angle, Parallel, Perpendicular
- *Line/Line*—Angle, Coincident, Distance, Parallel, Perpendicular
- *Line/Plane*—Coincident, Distance, Parallel, Perpendicular
- *Line/Point*—Coincident, Distance

- *Line/Sphere*—Concentric, Distance, Tangent
- *Line/Circular Edge*—Concentric

Plane

- *Plane/Cone*—Tangent
- *Plane/Cylinder*—Distance, Tangent
- *Plane/Extrusion*—Tangent
- *Plane/Line*—Coincident, Distance, Parallel, Perpendicular
- *Plane/Plane*—Angle, Coincident, Distance, Parallel, Perpendicular
- *Plane/Point*—Coincident, Distance
- *Plane/Sphere*—Distance, Tangent
- *Plane/Circular Edge*—Coincident
- *Plane/Surface*—Tangent

Point

- *Point/Cone*—Coincident, Concentric
- *Point/Curve*—Coincident, Distance
- *Point/Cylinder*—Coincident, Concentric, Distance
- *Point/Extrusion*—Coincident
- *Point/Line*—Coincident, Distance
- *Point/Plane*—Coincident, Distance
- *Point/Point*—Coincident, Distance
- *Point/Sphere*—Coincident, Concentric, Distance
- *Point/Surface*—Coincident

Sphere

- *Sphere/Cone*—Tangent
- *Sphere/Cylinder*—Concentric, Tangent
- *Sphere/Line*—Concentric, Distance, Tangent
- *Sphere/Plane*—Distance, Tangent
- *Sphere/Point*—Coincident, Concentric, Distance
- *Sphere/Sphere*—Concentric, Distance, Tangent

Surface

- *Surface/Cylinder*—Tangent
- *Surface/Plane*—Tangent
- *Surface/Point*—Coincident

Assembly Modeling Methodology

Two windows should be open: one for part and another for assembly.

1. *Open Part Window*
   - **a.** Click Open > Path where part is located.
   - **b.** Select Part for Files of Type: for example, Part*.sldprt.
   - **c.** Click View Menu > check Thumbnails.
   - **d.** Double-click Part of interest to be fixed. The Part FeatureManager is displayed.
2. *Open Assembly Window*
   - **a.** Click New.
   - **b.** Double-click Assembly from the default Templates tab.
   - **c.** Double-click Part in the Open documents box.
   - **d.** Click OK from the Begin Assembly PropertyManager.
   - **e.** Click Window > Part > File > Close. (Close the Part and leave only the Assembly open.)
3. *Set Assembly Units*
   - **a.** Click Options > Documents > Units > MMGS (IPS) > OK > Save (to set the Assembly units).

4. *Assemble Parts*

   For each part to be inserted to the fixed part (first in Graphics Window)

   a. Click Assembly > Insert Components > Browse for existing parts.
   b. Double-click the Part from the Path as Part*.sldprt format.
   c. Click a location to position the part.
   d. Click the Mate Assembly Tool and check "Faces" to Mate (). Next Part
5. *Exploded Assembly View Tool*

   For each component to *explode*,

   a. Click the Component.
   b. Drag the Blue Manipulator Handle in the appropriate Direction.
   c. Click Done. Next Part
6. *Animate the Exploded View*
   a. Right-click ExplView1 from the ConfigurationManager.
   b. Click Animate Explode.
   c. Click Stop.
   d. Close the Animator Controller.
   e. Right-click Collapse.
7. *Exploded Assembly View Tool*
   a. Right-click the Component to view in the Graphics Window.
   b. Click Open Part from the shortcut toolbar.
   c. Click Front Plane (or any other Plane).
   d. Click Section View for the Heads-Up View Toolbar.

Project

Let us start presenting an assembly methodology using the adjustable shaft support assembly as an example. The adjustable shaft support has eight parts: (1) PT1-Base, (2) PT2-Vertical shaft, (3) PT3-Yoke, (4) PT4-Bearing housing, (5) PT5-Bearing, (6) PT6-Set screw, (7) PT7-Set screw-hex, and (8) PT8-Jam nut.

Planning the Assembly

There is the need to plan how we are going to go about the assembly of parts in a bottom–up approach. We have to decide which part should be fixed, while other parts are added in a sequential manner. In the tool post holder assembly, the following order applies (see Table 8.1):

Table 8.1 Tool Post Holder Assembly

Order of Assembly	Part	Mate
1	PT1-Base	Fix this
2	PT2-Vertical shaft	Mate with base PT1
3	PT3-Yoke	Mate the bottom part with the top of PT2
4	PT4-Bearing housing	Mate opposite holes with two holes on PT3
5	PT5-Bearing	Mate concentricity with inside ends of PT4
6	PT6-Set screw	Mate with two holes in PT3
7	PT7-Set screw-hex	Mate with one hole in PT3
8	PT8-Jam nut	Mate with screws in PT3

Starting the Assembly Mode of SolidWorks

There are two ways to start the assembly mode of SolidWorks: (1) from a new SolidWorks document or (2) from an existing part that we wish to place first in the assembly document.

1. To start the assembly mode of SolidWorks, invoke the New SolidWorks Document dialog box and choose the Assembly button and click OK, as shown in Figure 8.9.

   The Begin Assembly PropertyManager pops up, as shown in Figure 8.10. Browse to open the first part to insert.

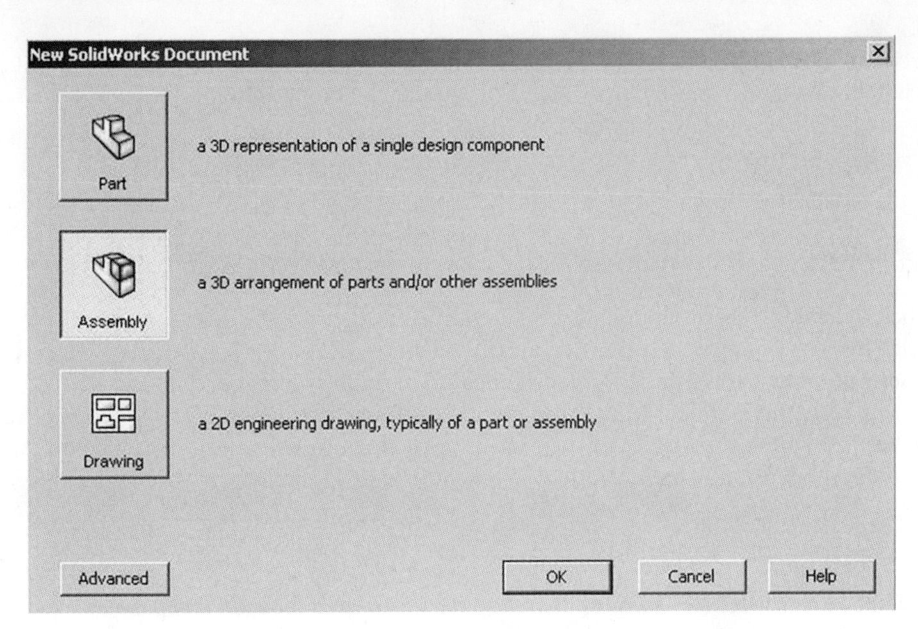

Figure 8.9

The New SolidWorks Document dialog box.

Figure 8.10

The Assembly mode showing the Begin Assembly PropertyManager.

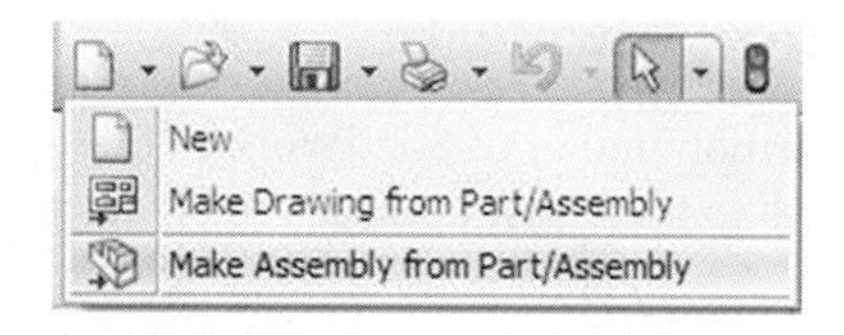

Figure 8.11

Accessing Assembly mode from the New dialog box in an existing part document.

2. From an existing part document, click the New dialog box, as shown in Figure 8.11. Click Make Assembly from Part/Assembly. The Begin Assembly PropertyManager pops up, as shown in Figure 8.4. Browse to open the first part to insert.

Inserting Components in the Assembly Document

There are several ways of inserting a component in the assembly document of SolidWorks:

- Click Assembly > Insert Components (see Figure 8.12).
- Click Insert > Component > Existing Part/Assembly (see Figure 8.13).
- Inserting components using the opened document window (see Figure 8.14).
- In this case, there are currently opened part documents. Choose Windows > Tile Horizontally or Vertically from the Menu Bar menus. All opened SolidWorks document windows will be tiled accordingly, depending on the option that is chosen (see Figure 8.15).

Assembling the Components

Now that all the parts needed for assembly are ready, the first step in the assembly process is to assemble the first two components.

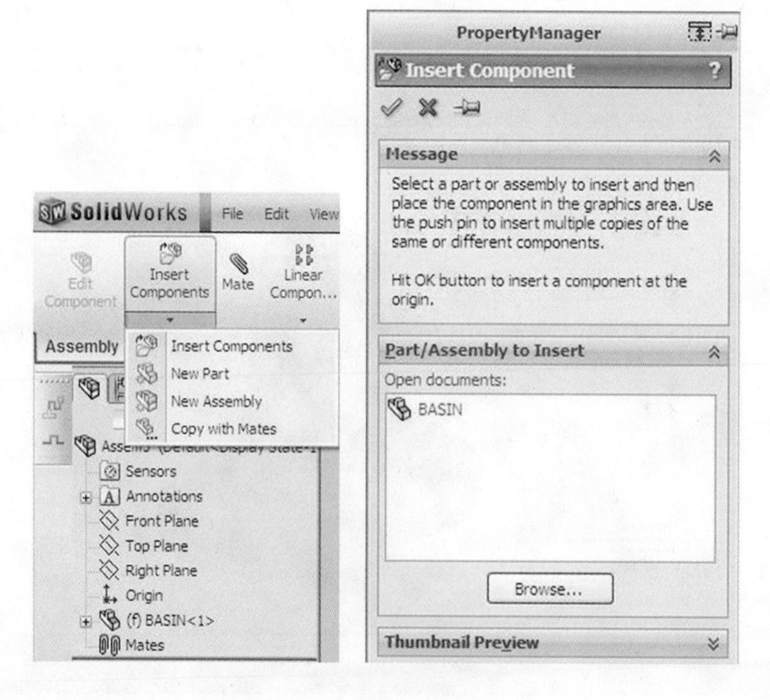

Figure 8.12

Inserting an existing component into an assembly document.

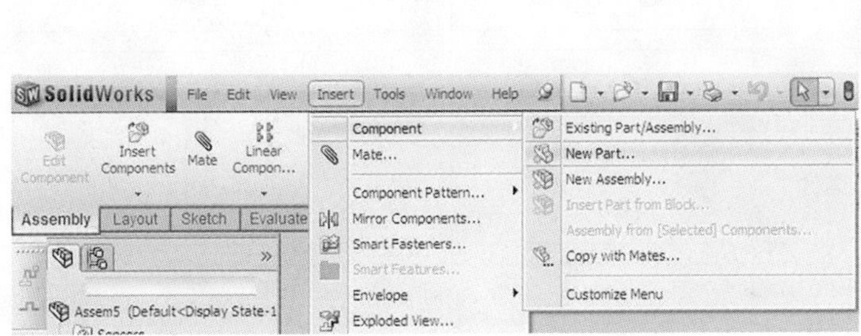

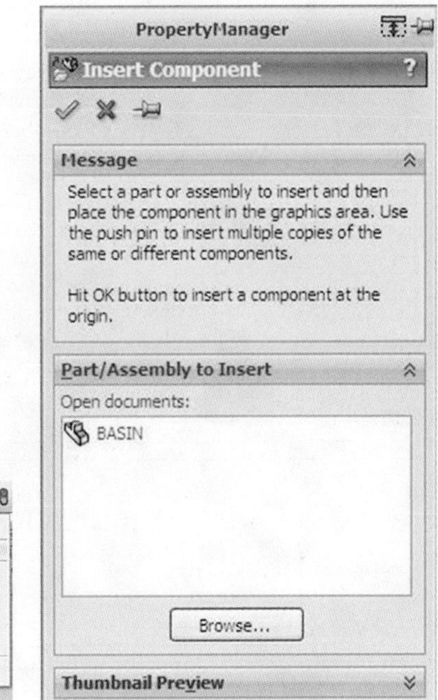

Figure 8.13

Inserting existing component into an assembly.

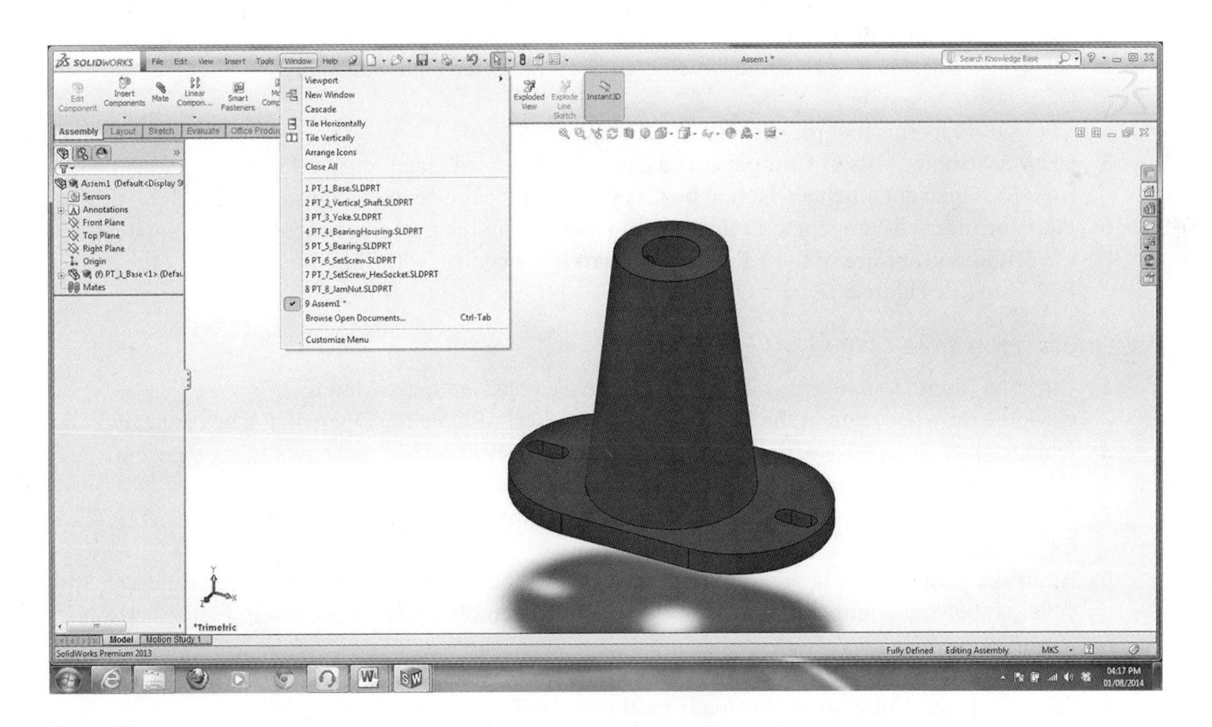

Figure 8.14

Using the opened document.

Figure 8.15

Tiled windows.

Assembling the First Two Components (Base and Vertical Shaft) of the Assembly

Open the first two (or all the) parts (base and vertical shaft) to be assembled. (Note: you are in part mode.)

1. Start a New SolidWorks assembly document.
2. Click on the first part to be fixed, which is the PT1-Base (see Figure 8.16).
3. Click Begin Assembly.
4. Click OK to fix this first part, PT1-Base, at the origin.
5. Click Assembly > Insert Components as already discussed (see Figure 8.18). Select the PT1-Base from the Open documents selection box, as shown in the preview (see Figure 8.17).
6. Place the inserted component in the required location on the first component. The first mate is that the cylindrical surface of the PT2-Vertical shaft is concentric with the inner surface of the hole of PT1-Base (see Figure 8.18).

Assembling the Yoke

1. Click the Insert Components button from the Assemble CommandManager.
2. Click the Browse button in the Part/Assembly to Insert rollout. The Open dialog box appears.
3. Double-click on *PT3-Yoke* from the Open documents selection box, as shown in the preview.
4. Place the inserted component in any location in the drawing area.
5. Apply the Concentric relation that is automatically chosen as default, as shown in Figure 8.19. Click OK to accept it.
6. Apply the Parallel mate between the lower flat face of the base and the upper-left vertical face of the yoke, as shown in Figure 8.20. The result of this subassembly process is shown in Figure 8.21.

Assembling the Bearing Housing

1. Click the Insert Components button from the Assembly CommandManager.
2. Click the Browse button in the Part/Assembly to Insert rollout. The Open dialog box appears.
3. Double-click on PT4-Bearing Housing from the Open documents selection box, as shown in the preview.
4. Click View > Temporary Axes to activate all Axes that are used in modeling cylindrical features.
5. Apply the Parallel mate to the Temporary Axes between one of the three Holes in the Bearing Housing and one of the threaded holes in the Yoke.

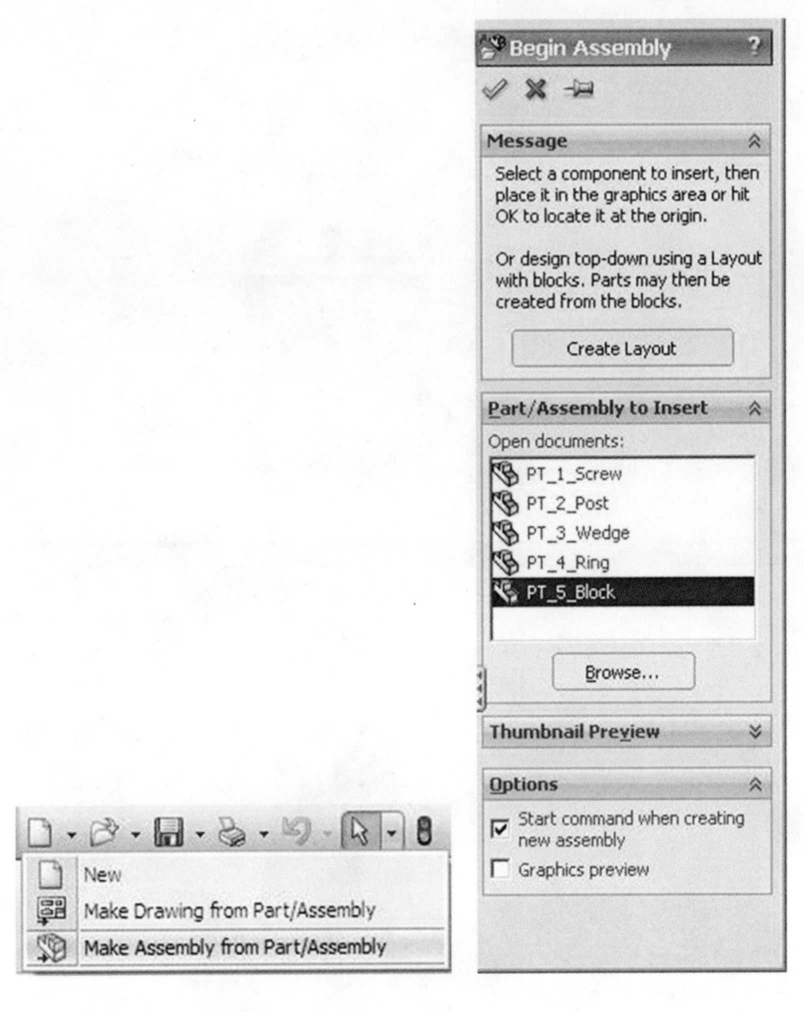

Figure 8.16

Begin assembly with the first part to be fixed.

6. Apply the Parallel mate to the Temporary Axes between the opposite of one of the three Holes in the Bearing Housing and the other of the threaded holes in the Yoke.
7. Apply the Parallel mate between the Top Plane of the Yoke and the Right Plane of the Bearing Housing.
8. Apply the Parallel mate between the Front Plane of the Bearing Housing and the side flat face of the Yoke.
9. Apply the Distance mate between the Front Plane of the Bearing Housing and the inside flat face of the Yoke to be 0.76 in.

See Figure 8.22 for the resulting subassembly.

Assembling the Bearings

1. Click the Insert Components button from the Assemble CommandManager.
2. Click the Browse button in the Part/Assembly to Insert rollout. The Open dialog box appears.
3. Double-click on *PT5-Bearing* from the Open documents selection box, as shown in the preview.
4. Apply the Concentric relation between the outside surface of the Bearing and the inside surface of the Bearing Housing.
5. Apply the Coincident relation between the end of the outside surface of the Bearing and the end of the inside surface of the Bearing Housing.
6. Repeat steps 4 and 5 for the second Bearing.
7. Click OK to accept it.

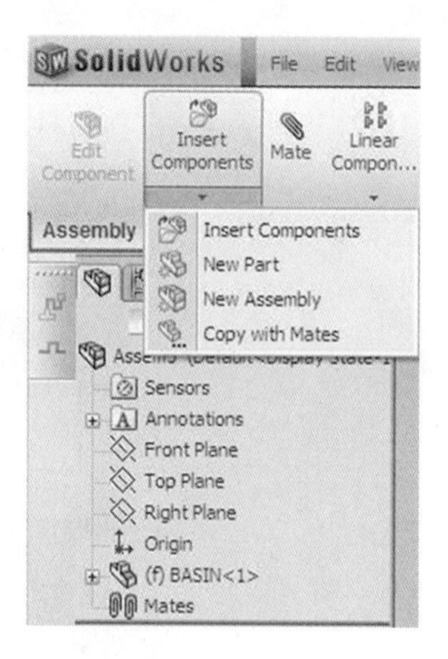

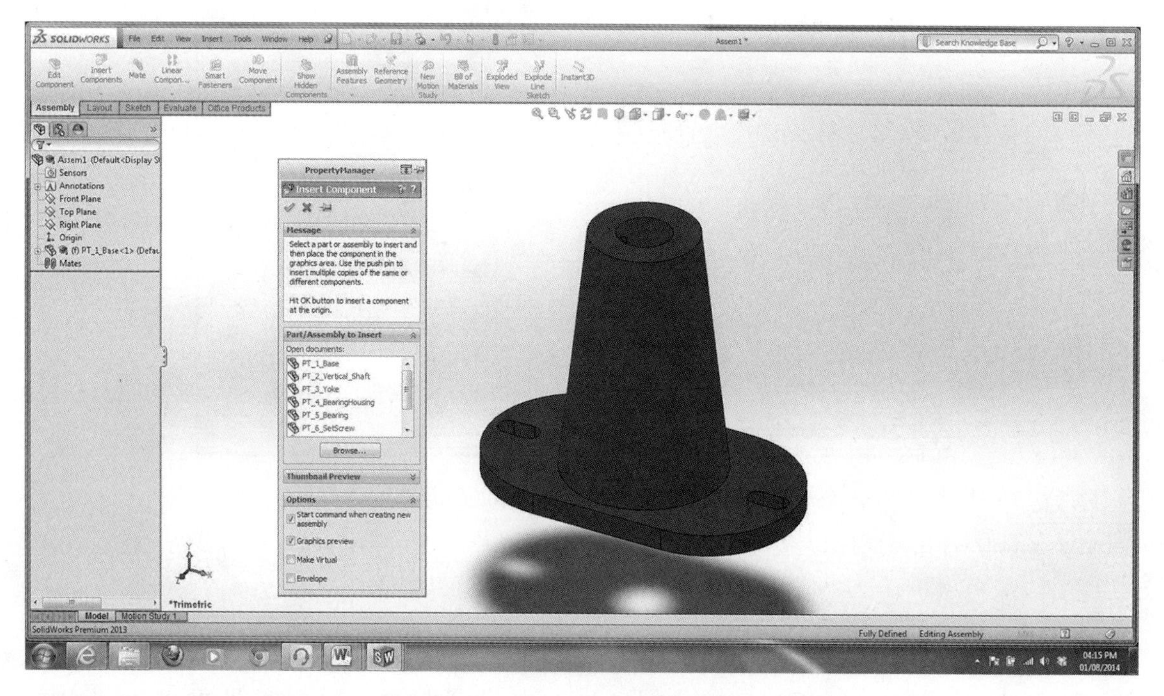

Figure 8.17

Insert a component to the first fixed part in the assembly.

Assembling the Screws and Nuts

1. Click the Insert Components button from the Assemble CommandManager.
2. Click the Browse button in the Part/Assembly to Insert rollout. The Open dialog box appears.
3. Double-click on *PT6-Screw* from the Open documents selection box, as shown in the preview.
4. Apply the Concentric relation between the screw and the threaded hole in the Yoke.
5. Repeat step 4 for all the screws (see Figure 8.23).
6. Repeat steps 2–4 for the Nuts. Click OK to accept it. The final assembly is shown in Figure 8.24.

Assembly Analysis

It is necessary to determine how good the assembly task is by carrying out an interference analysis. An assembly process may deceptively seem alright until an interference analysis is done only to find that there are some issues to be resolved. This step is important because it directs us to where the problems are, which

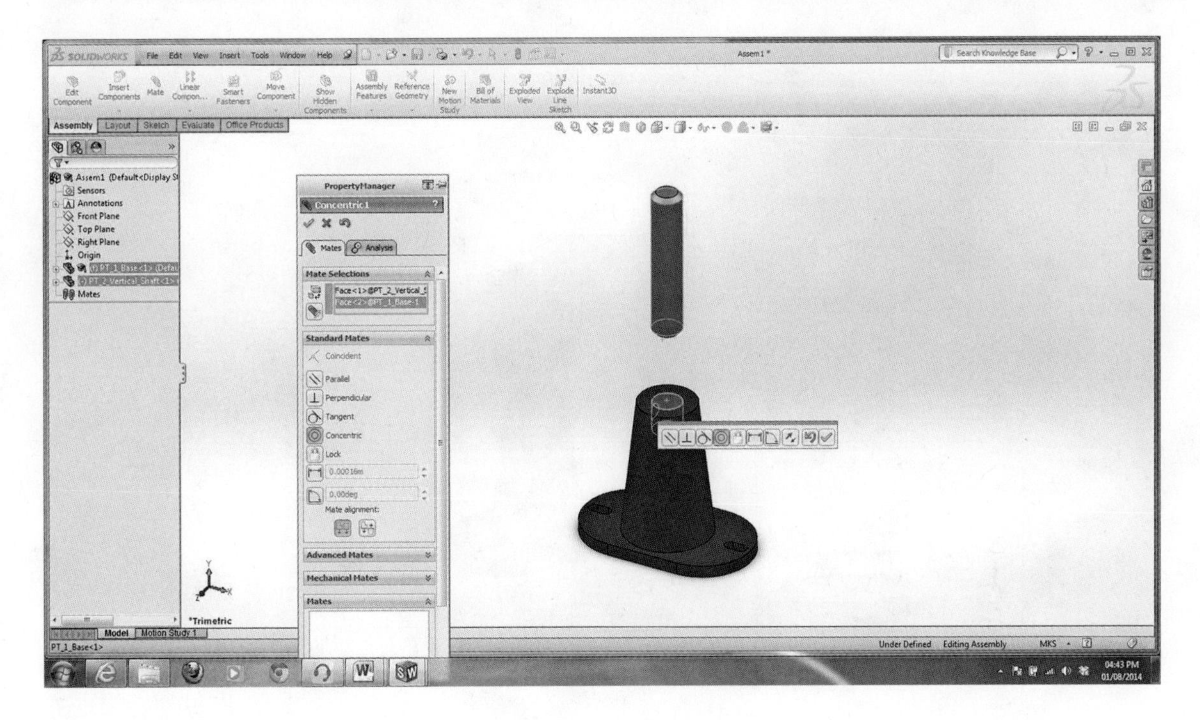

Figure 8.18

First two assembled components: PT-Base and PT2-Vertical shaft.

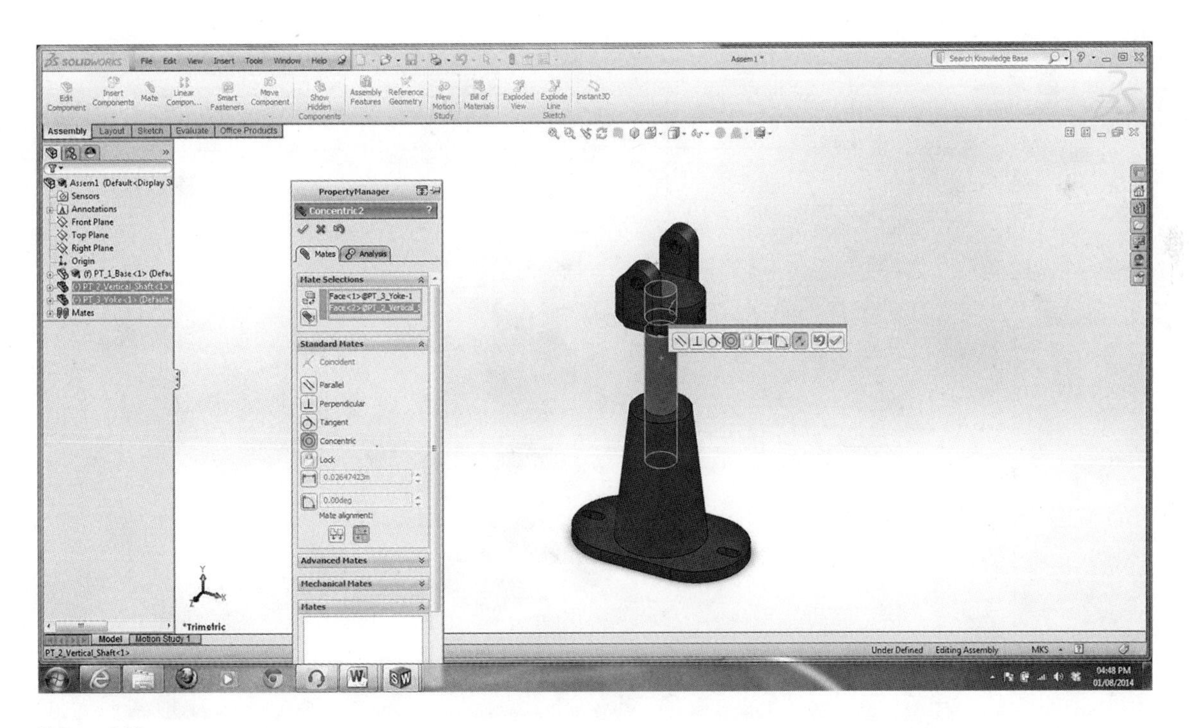

Figure 8.19

Concentric mating between the Base and the Vertical Shaft.

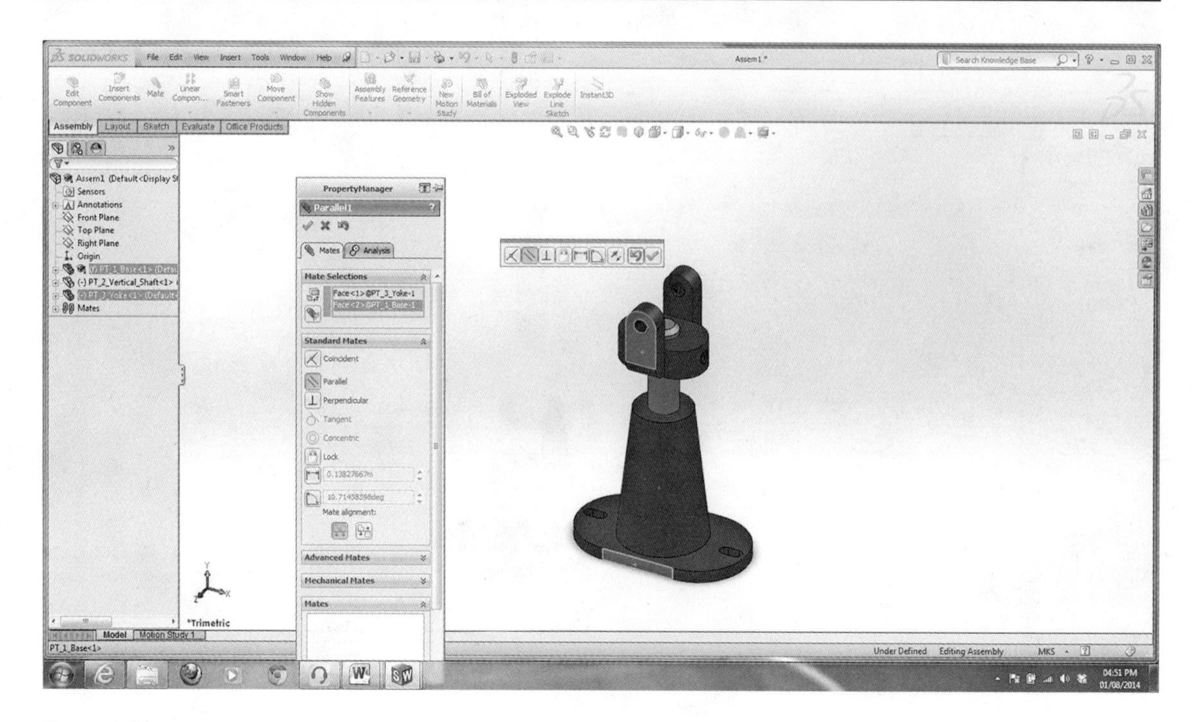

Figure 8.20

Parallel mating between the base and the yoke.

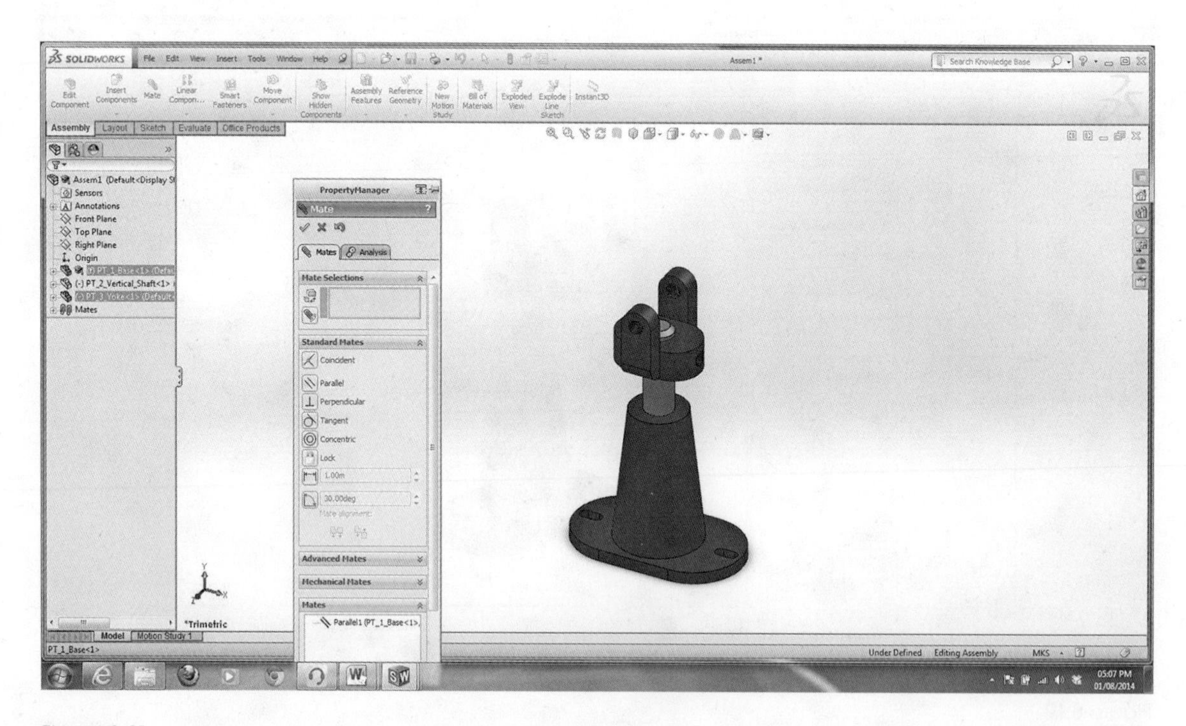

Figure 8.21

Subassembly of first three components: PT1-Base, PT2-Vertical shaft, and PT3-Yoke.

Figure 8.22

Subassembly: PT1-Base, PT2-Vertical Shaft, PT3-Yoke, and PT4-Bearing Housing.

Figure 8.23

Preview of adding PT1-Screw to the assembly.

Figure 8.24

Final Assembly.

need to be fixed before moving over to the expensive phase of machining the individual parts that make up the overall assembly. To analyze your assembly for interference detection, do the following:

1. Click Evaluate > Interference Detection.
2. In the Selected Component, click Calculate (see Figure 8.25 for the outcome).

Exploded View

1. Click the Exploded View Assembly Tool (see Figure 8.26).
2. For each part to explode.
3. Click the part.
4. Drag the blue manipulator handle in the appropriate direction (see Figure 8.27).
5. Click Done.
6. Click Next.
7. Click OK from the Explode Property Manager. (Note: at this point, ExplView1 is automatically found in the ConfigurationManager of the Assembly; see Figure 8.28 for the exploded view.)

To Remove Exploded View,

1. Right-click the Graphics Window.
2. Click Collapse (see Figure 8.29).

Animate Exploded View

1. Right-click the ExplView1 from the ConfigurationManager (see Figure 8.31).
2. Click Animate Explode (see Figure 8.30).
3. Click Stop (see Figure 8.31).
4. Close (X) the Animator Controller.
5. Right-click the Graphics Window.
6. Click Collapse.
7. Save the document.

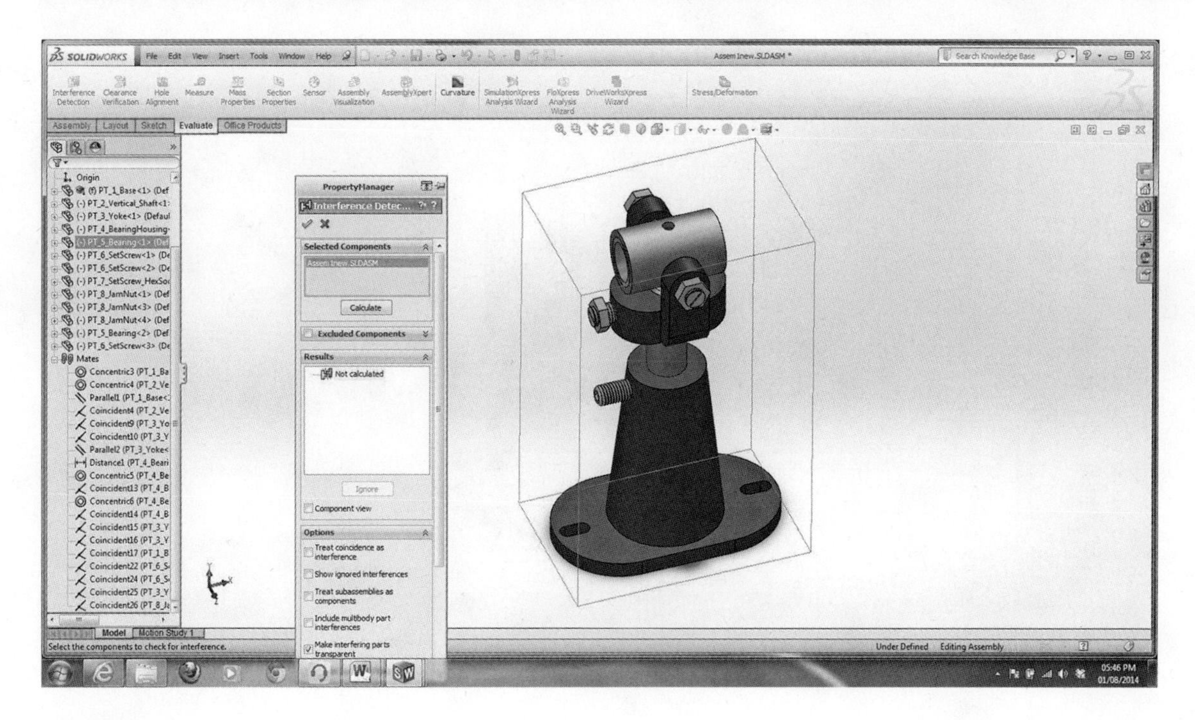

Figure 8.25

Interference detection.

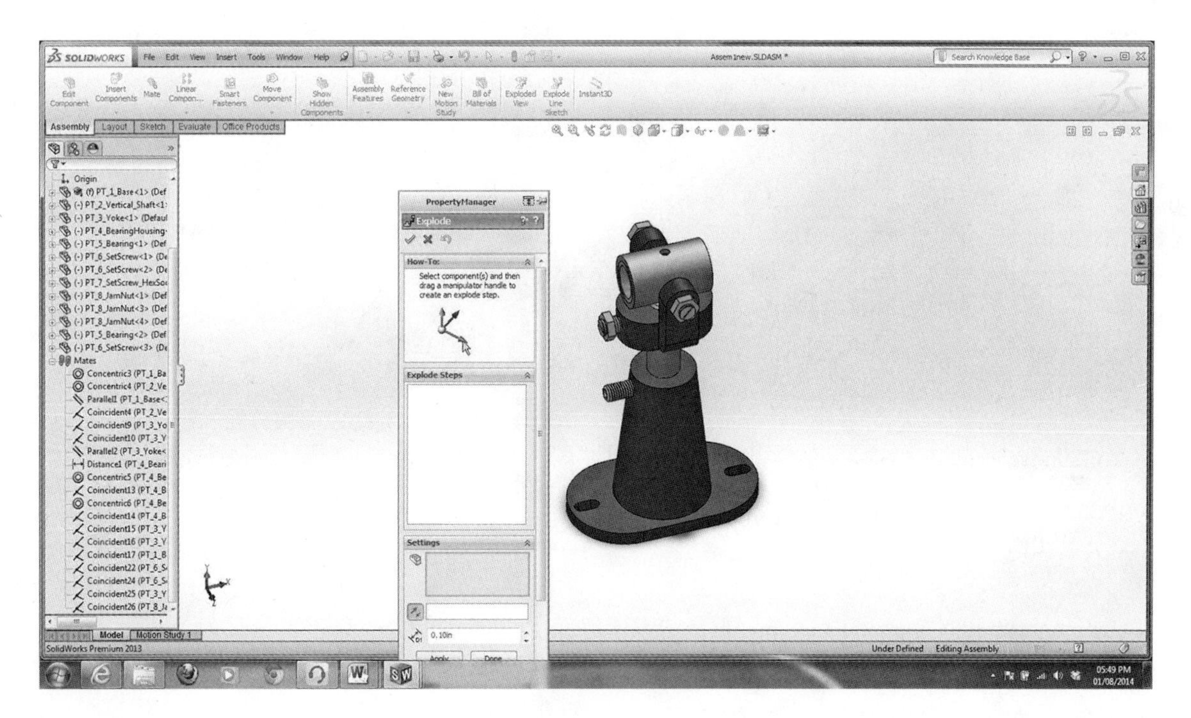

Figure 8.26

Exploded View Assembly tool.

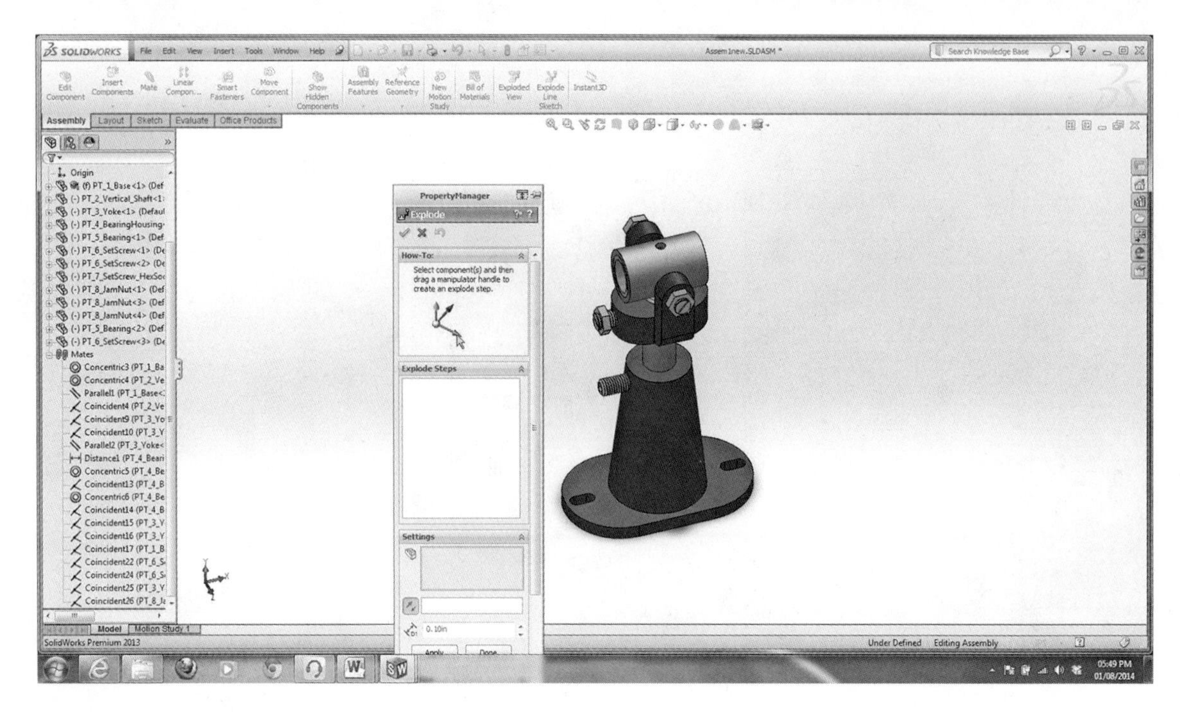

Figure 8.27

Exploding each part using the Triad.

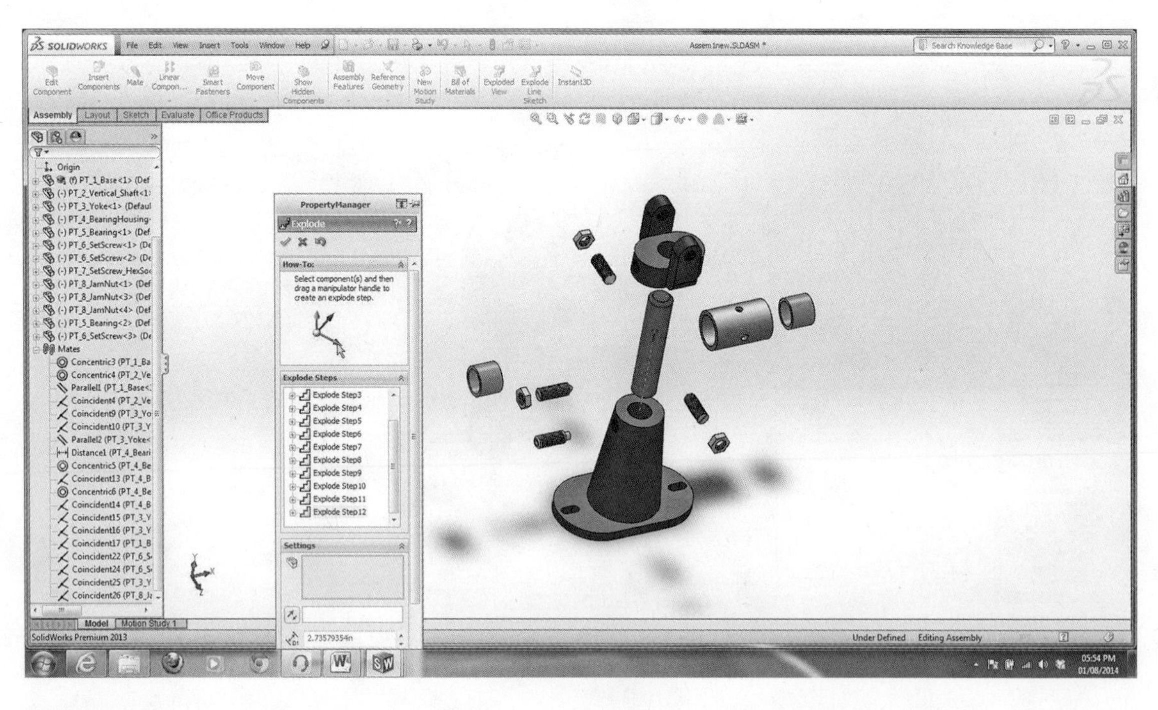

Figure 8.28

Exploded view.

Figure 8.29

Collapsing parts in exploded view.

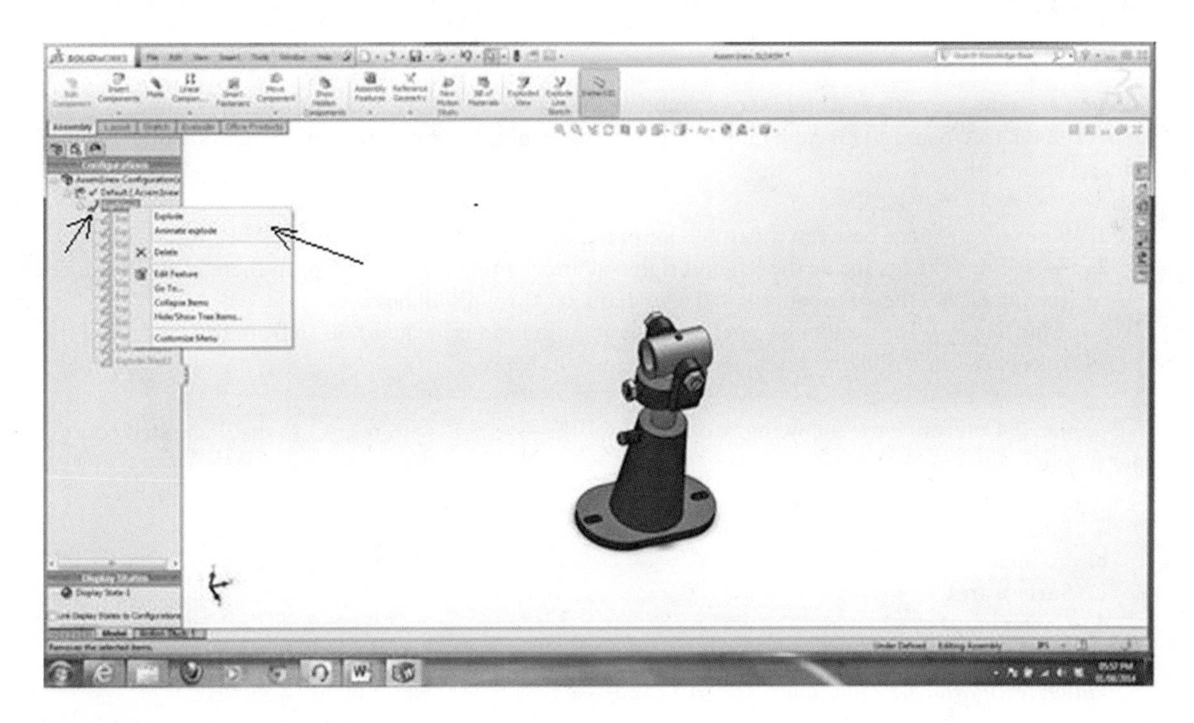

Figure 8.30

Animating the exploded view process.

Figure 8.31

Stopping the animation session.

Tutorials

Tutorial 8.1

Question 1: Model the parts and build the assembly that is shown in Figure 8.32. The assembly contains three parts: (1) Base, (2) Yoke, and (3) Adjusting Pin. (Note the origin. Fully constrain all sketches. Fully mate all parts.)

1. *Base:* The distance between the front face of the Base and the front face of the Yoke = 1.10-in.
2. *Yoke:* The Yoke fits inside the left and right square channels of the Base part; there is no clearance. The top face of the Yoke contains a 0.6-in diameter through-all hole.
3. *Adjusting Pin:* The bottom face of the Adjusting Pin head is located 1.60-in from the top face of the Yoke part. The Adjusting Pin has square shape at the top.

Question 2: Determine the following parameters for the assembly with respect to the illustrated coordinate system in Q1:

a. Mass
b. Volume
c. Surface area
d. Center of mass

Solution to Tutorial 8.1, Question 1: The parts and assembly are shown in Figure 8.33.

Base

1. Create a New part in SolidWorks.
2. Set the document properties for the model.
3. Create Sketch1, which is the base sketch with overall dimensions of 2.0-in × 2.0-in.
4. Create the Extruded Base feature. Extrude1 is the base feature that is obtained by extruding Sketch1 through a depth of 23.0-in.

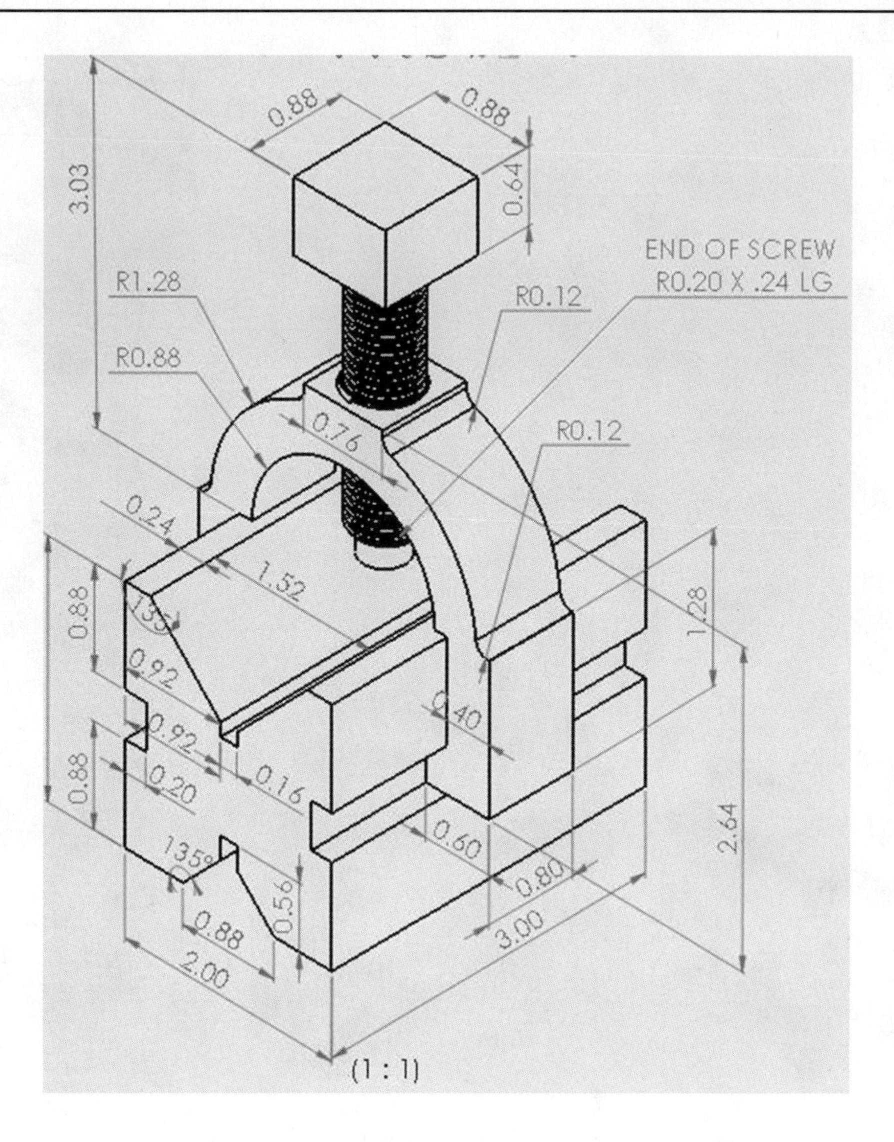

Figure 8.32

Assembly model. (From David C. Planchard & Marie P. Planchard, Official Certified SolidWorks Associates (CSWA), SDC Publications, 2009.)

Yoke

1. Create a New part in SolidWorks.
2. Set the document properties for the model.
3. Create Sketch1, which is the sketch with overall dimensions of 2.80-in × 2.64-in.
4. Create the Extruded Base feature. Extrude1 is the Base feature that is obtained by extruding Sketch1 through a Depth of 0.8-in.

Pin

1. Create a New part in SolidWorks.
2. Set the document properties for the model.
3. Create Sketch1, which is the cylindrical sketch in the form of 0.52-in diameter and 2.16-in long; .6-14UNC-2A thread is created for the length of 2.16-in. Total length of the pin is 2.40-in, the remaining not threaded.
4. The head of the pin is a 0.88-in square, extruded 0.64-in.

   The base, yoke, and pin are shown in Figure 8.33.

Assembly Modeling

1. Create a New assembly in SolidWorks.
2. Click Cancel from the Begin Assembly PropertyManager.

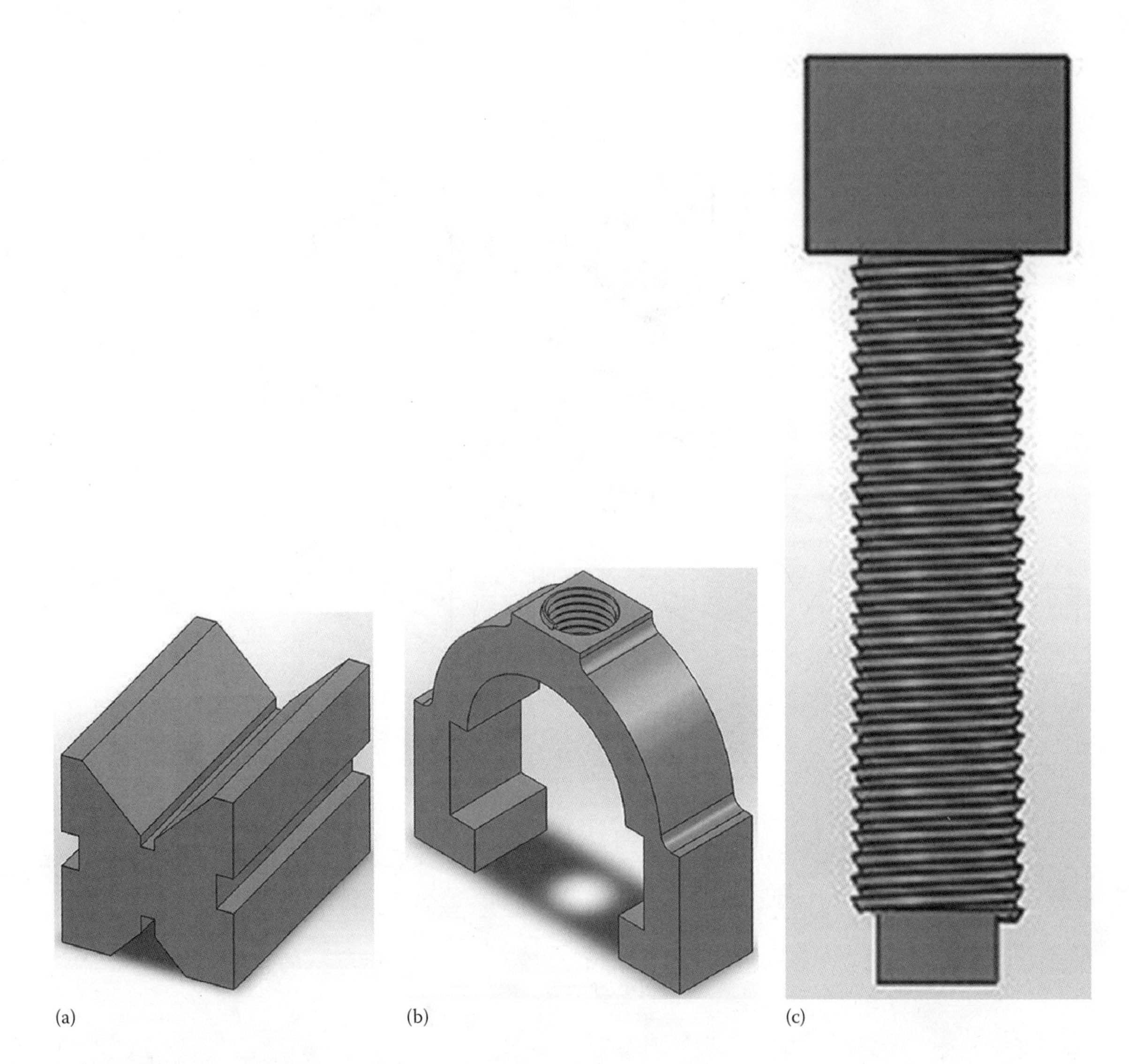

(a) (b) (c)

Figure 8.33

Parts for assembly: (a) base, (b) yoke, and (c) pin. (From David C. Planchard & Marie P. Planchard, Official Certified SolidWorks Associates (CSWA), SDC Publications, 2009.)

3. Set the document properties for the model.
4. Insert the base as the first part in the assembly.
5. Insert the yoke as the second part in the assembly.
6. Insert Parallel and Concentric mates between the upper sides of the base and the notched part of the yoke in the bracket in the assembly.
7. Insert a Distance mate of 1.10-in between the front face of the Base and the front face of the Yoke.
8. Insert the pin as the third part in the assembly.
9. Insert Concentric mate between the pin and the yoke in the assembly.
10. Insert Distance mate of 1.60-in between the bottom face of the Adjusting Pin head and the top face of the Yoke part.

    The base, yoke, and pin are shown in Figure 8.34.

Solution to Tutorial 8.1, Question 2: For the assembly in Q1,

a. Mass: 0.47 pounds
b. Volume: 13.04 cubic inches
c. Surface area: 65.77 square inches
d. Center of mass: (inches): X = –0.00; Y = 1.41; Z = –0.00

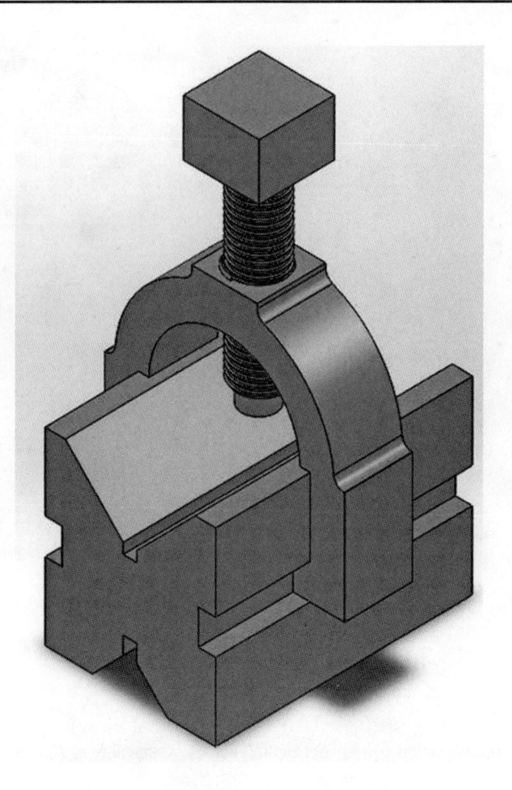

Figure 8.34

Assembly Model solution. (From David C. Planchard & Marie P. Planchard, Official Certified SolidWorks Associates (CSWA), SDC Publications, 2009.)

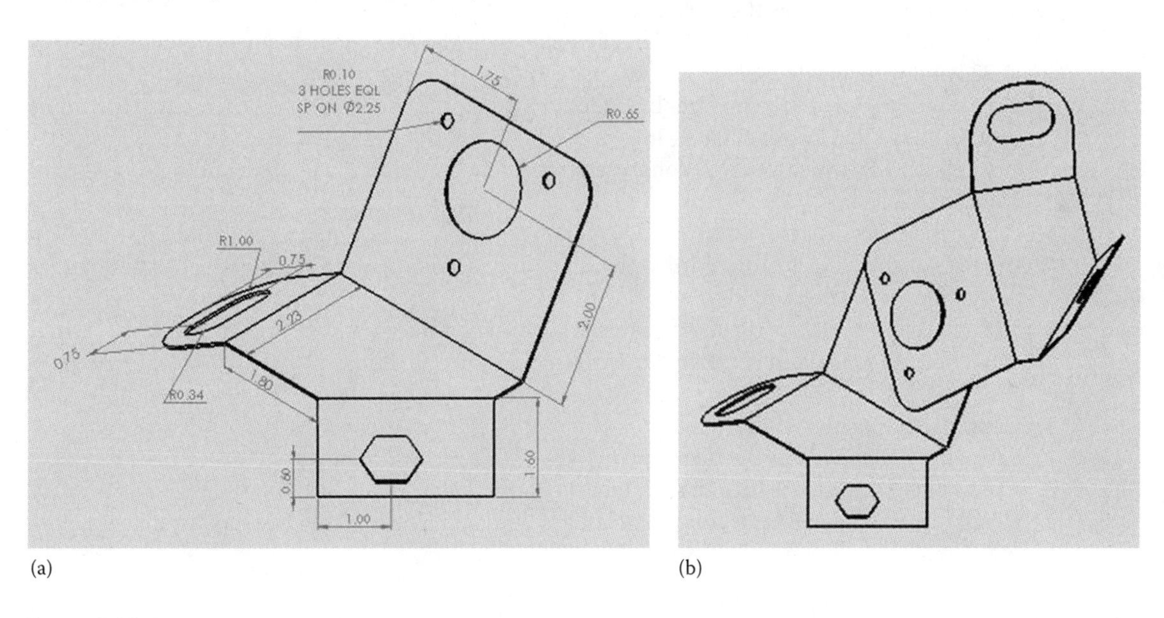

(a) (b)

Figure 8.35

Assembly Model: (a) part with dimensions and (b) assembly of parts. (From David C. Planchard & Marie P. Planchard, Official Certified SolidWorks Associates (CSWA), SDC Publications, 2009.)

Tutorial 8.2

Question 1: The assembly shown in Figure 8.36 contains three machined brackets and two pins. Apply the MMGS unit system. Decimal places: 2. The assembly origin is as shown.

Model the parts. (Note: fully constrain all sketches.)

Question 2: Build the assembly of the parts that are described in Question 1 and shown in Figure 8.35.

Solution to Tutorial 8.2, Question 1: Parts Modeling

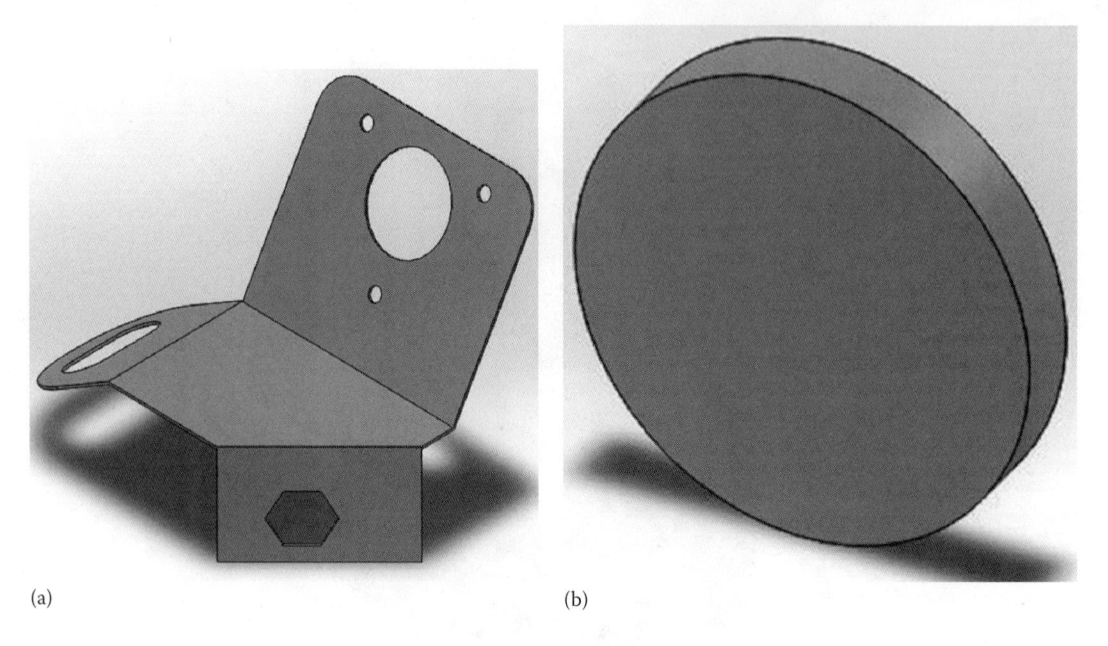

Figure 8.36

Part model for the machined bracket and pin: (a) bracket (right view), (b) pin (front view). (From David C. Planchard & Marie P. Planchard, Official Certified SolidWorks Associates (CSWA), SDC Publications, 2009.)

Bracket

1. Create a New part in SolidWorks.
2. Select the Right Plane as the sketch plane.
3. Create Sketch1, which is the Base (bracket) sketch in the form of a chopped rectangular shape (2.23 × 1.8 with an edge, at an angle of 45-degrees).
4. Create the Extruded Base feature. Extrude1, which is the base feature, is obtained by extruding Sketch1 through a Depth of 0.038-in.
5. Create Sketch2, which is the second sketch in the form of a rectangle, including a hex-shape (2.56-in × 2.40-in).
6. Create the Extrude feature. Extrude2, which is the Extrude feature, thickness of 0.038-in.
7. Create Sketch3, which is the third sketch including a slot; this feature is inclined at an angle of 30-degrees to the horizontal.
8. Create Sketch4, which is the fourth sketch including one large hole, diameter 2.25-in, and three small holes 0.19-in diameter. Extrude with thickness of 0.038-in.

Pin

1. Create a New part in SolidWorks.
2. Select the Right Plane as the sketch plane.
3. Create Sketch1, which is the pin sketch in the form of Hex (.75 ACR FLT).
4. Create the Extruded Base feature. Extrude1 is obtained by extruding Sketch1 through a length of 0.2-in.

The bracket and pin are shown in Figure 8.36.

Solution to Tutorial 8.2, Question 2: Assembly Modeling

1. Create a New assembly in SolidWorks.
2. Click Cancel from the Begin Assembly PropertyManager.
3. Set the document properties for the model.
4. Insert the bracket as the first part in the assembly.
5. Insert the pin as the second part in the assembly.
6. Insert a Concentric mate between the pin and the hole in the vertical part of the bracket in the assembly.

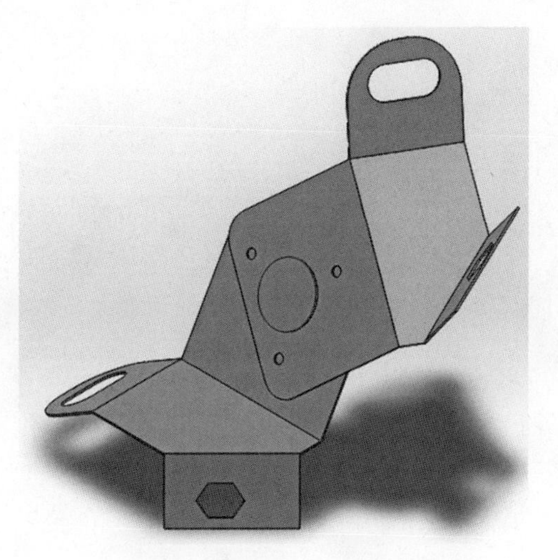

Figure 8.37

Assembly model for the machined brackets and pins. (From David C. Planchard & Marie P. Planchard, Official Certified SolidWorks Associates (CSWA), SDC Publications, 2009.)

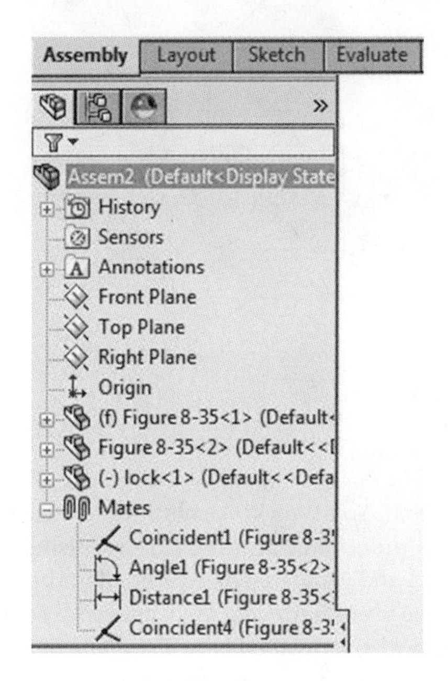

Figure 8.38

FeatureManager information on mating of parts in assembly. (From David C. Planchard & Marie P. Planchard, Official Certified SolidWorks Associates (CSWA), SDC Publications, 2009.)

7. Insert a Distance mate of 0.04-in gap between the brackets.
8. Insert an Angular mate 45° between the edges of the first two brackets.

The assembly of the bracket and pin is shown in Figure 8.37.

The FeatureManager information on mating of parts in assembly is shown in Figure 8.38.

Tutorial 8.3

Repeat Tutorial 8.2 with straight edges for the simple brackets, a Distance mate of 1-mm gap between the brackets, and an Angular mate 45° between the edges of the brackets.

Solution to Tutorial 8.3 The bracket and the pin are shown in Figure 8.39, while the assembly is shown in Figure 8.40.

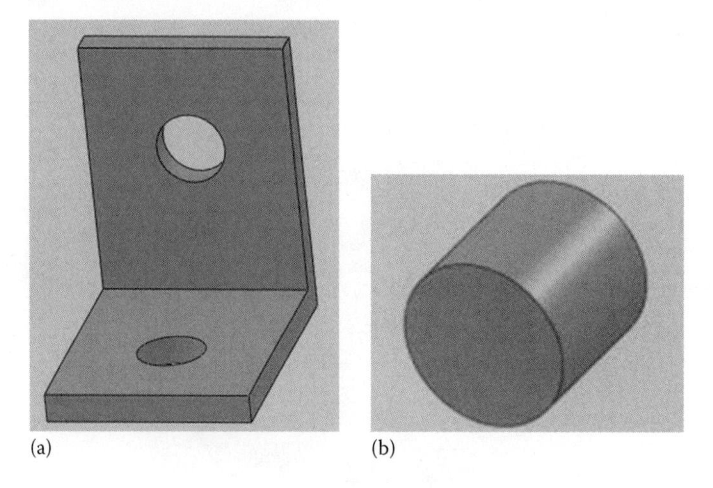

Figure 8.39

Part model for the simple machined bracket and pin: (a) simple bracket and (b) pin for bracket.

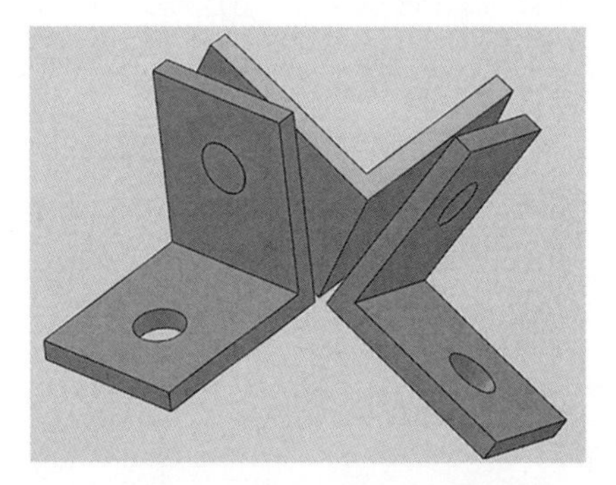

Figure 8.40

Assembly for simple bracket and pins.

Large Assemblies

The controlling display performance in SolidWorks is split into two categories: (1) central processing unit (CPU) processing and (2) graphics processing unit (GPU) processing. This is, in reality, the difference between calculating the parametrics and geometry on one hand and calculating the graphics and display on the other hand. Depending on the user's hardware, drivers, and system maintenance, either of the CPU processing or the GPU processing will be performed better.

There are four ways in which users attempt to minimize the load on the CPU and the GPU in large assemblies. These are (1) suppressed, (2) hidden, (3) lightweight parts, and (4) resolved.

1. *Suppressed*

   When attempting to speed up the performance of an assembly, the biggest impact is made if the user can reduce the load on both the CPU and the GPU. This can be achieved by suppressing a part. When a part is suppressed, it is neither calculated for its parametrics and geometry nor displayed. Consequently, the load on each processor for that part is zero.
2. *Hidden*

   When a part is hidden, the CPU still calculates its parametrics. However, since the part is hidden, it creates no load on the GPU. If the user has a good main processor but a weak video card, then the benefits could be obtained by removing the graphics load from the display by hiding them.
3. *Lightweight parts*

   If a user is interested in showing a part but not calculating any of its parametric relations, then Lightweight parts should be used. Lightweight default settings can be found in Options, on both the assemblies and Performance pages. The user can make parts lightweight through the right mouse button menu.

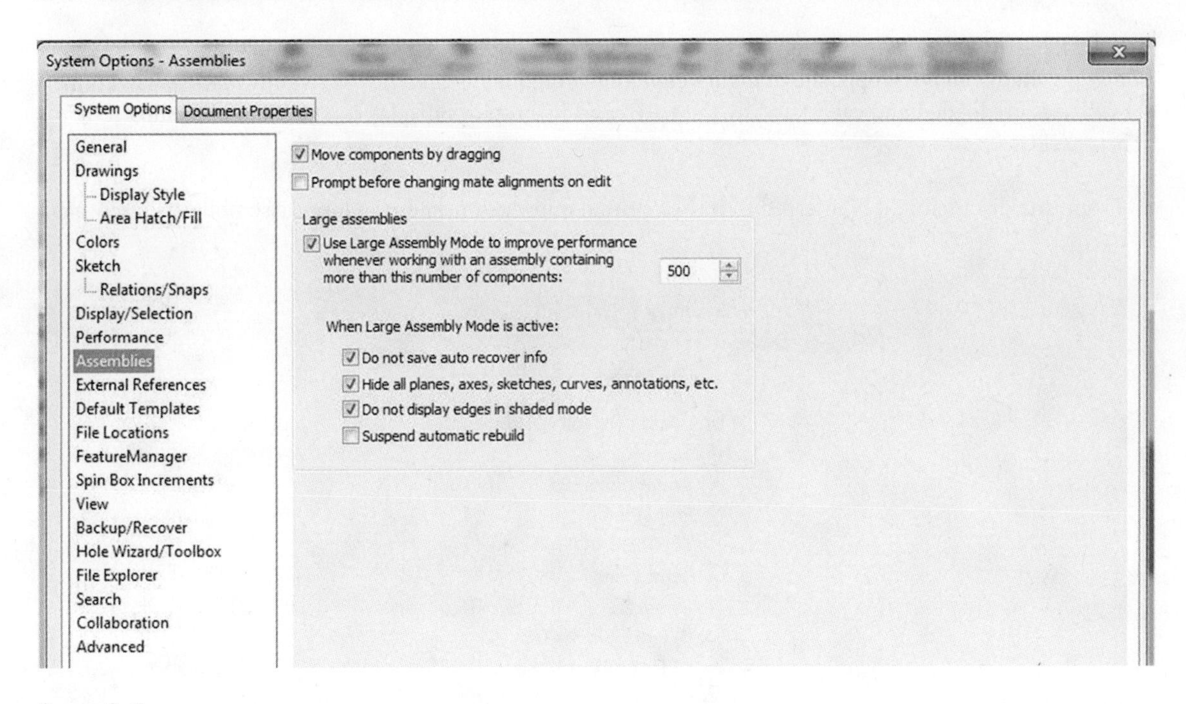

Figure 8.41

System Options.

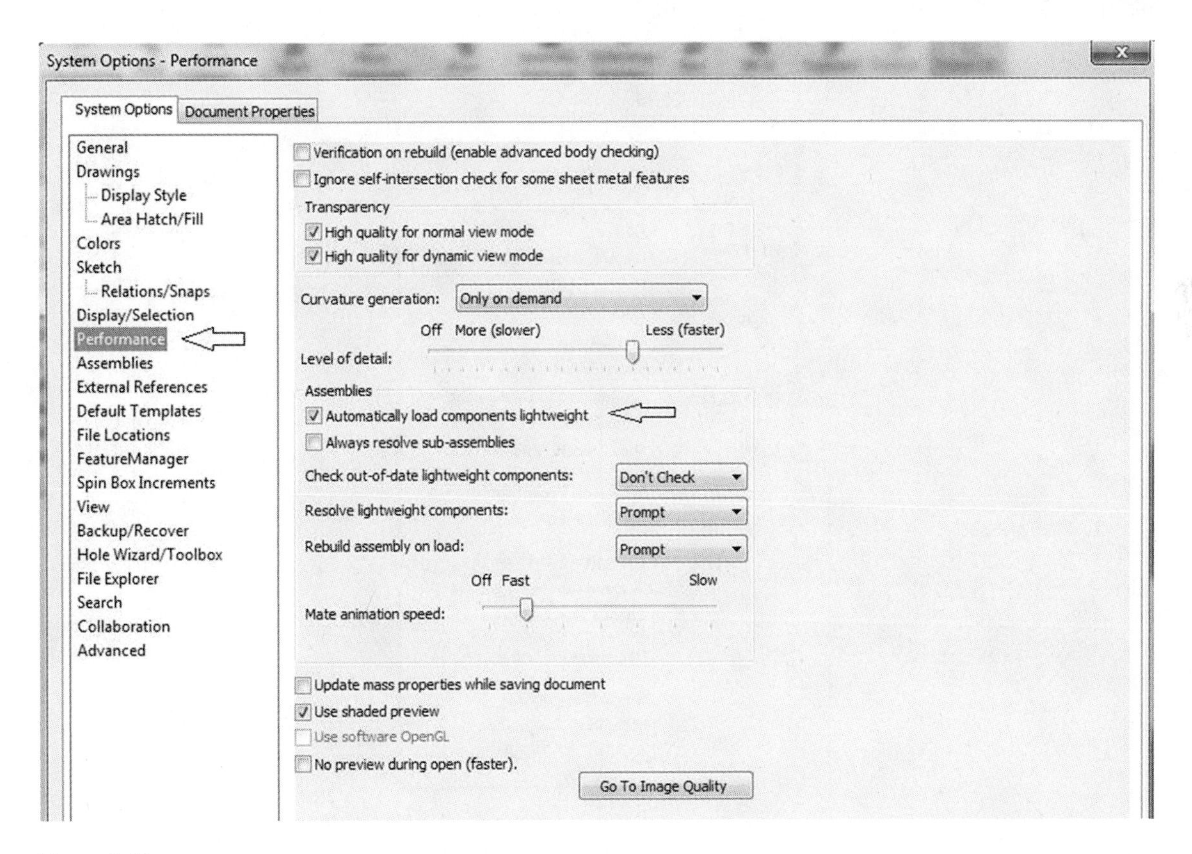

Figure 8.42

Performance setting to automatically load components lightweight.

4. *Resolved*

Resolved is the opposite of Lightweight. Resolved means that the part is fully loaded, its parametrics are loaded and calculated by the CPU, and its graphics display is calculated and shown by the GPU.

From this discussion, it is inferred that the optimal approach to manage large assemblies is *lightweight parts.*

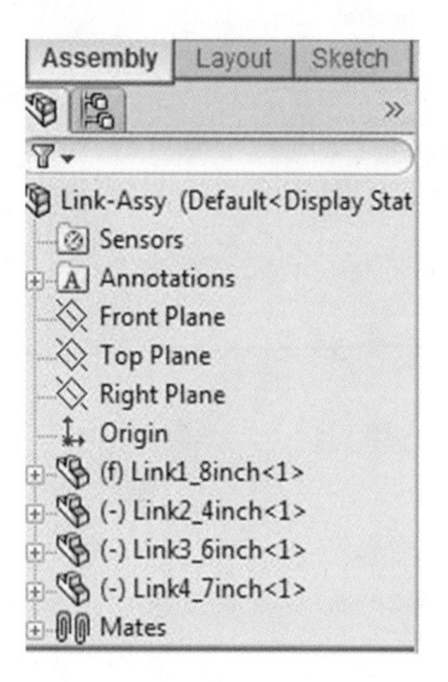

Figure 8.43

Resolved mode.

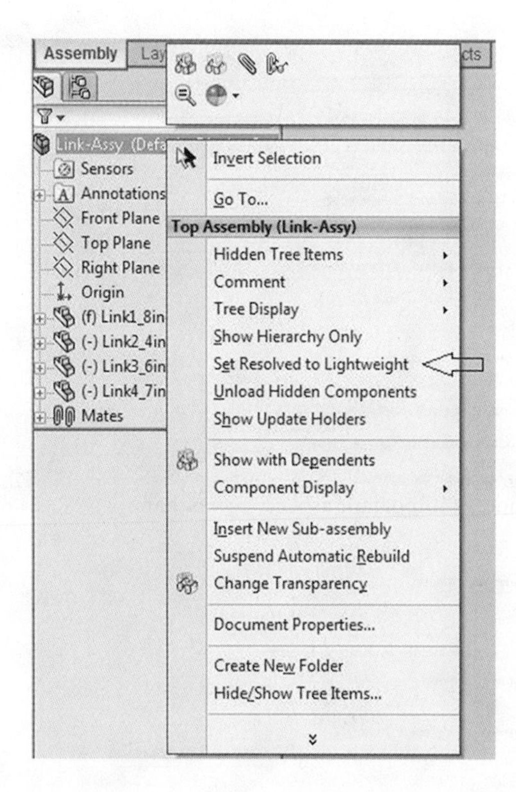

Figure 8.44

Manually setting the topmost resolved feature to Lightweight.

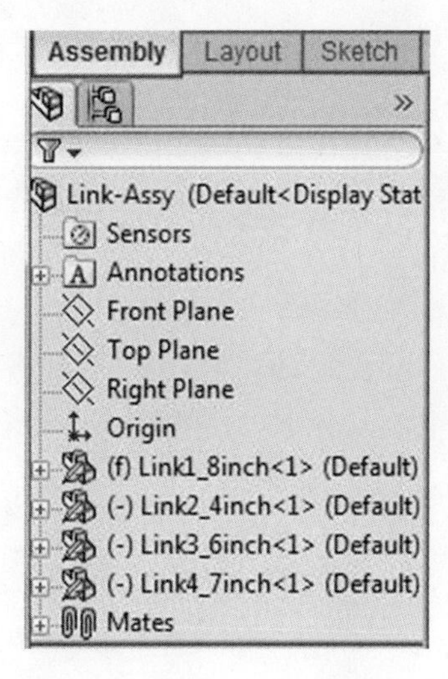

Figure 8.45

Lightweight mode.

Automatic Setting to Lightweight

1. Click Options > System Options > Assemblies (see Figure 8.41). (Note: the number of assembly modes to improve performance is set by default to 500.)
2. Click Options > System Options > Performance (see Figure 8.42).
3. Under Assemblies mode, check the box that is designated. Automatically load components lightweight (see Figure 8.42).
4. Click OK.

Manual Setting to Lightweight

Figure 8.43 shows the Resolved mode. To manually set to Lightweight,

5. Right-click on the topmost resolved feature (see Figure 8.44).
6. Click Set Resolved to Lightweight (see Figure 8.44).

The parts change to Lightweight mode with feathers attached (see Figure 8.45).

Summary

We have gone through the necessary steps for assembling parts in a bottom–up approach. We simply fix a major part and add other parts while ensuring that the mating conditions are met. Thereafter, it is important to analyze how well the parts are modeled by checking for interference detection. To finish up, presenting an exploded view is useful to help the people who are involved in the assembly to figure out how to go about assembling the parts. This information is normally included in the product guide/manual. SolidWorks provides an animation tool to help us visualize how we assembled our parts.

This chapter also covers another important aspect: how to handle large assemblies.

9

Part and Assembly Drawings—CSWA Preparations

Objectives:

When you complete this chapter, you will have

- Understood the differences between part drawing and assembly drawing
- Learned how to apply the bottom–up approach to assembly modeling
- Learned how to create a new drawing template
- Learned how to insert and position views on a drawing
- Learned how to use the exploded view assembly tool
- Learned how to animate the exploded view
- Learned how to insert half-, offset-, and aligned-section views to drawings

Introduction

An engineering drawing is used to fully and clearly define the requirements for engineered items. The main objective of an engineering drawing is to convey all the required information that will allow a manufacturer to produce that component. Engineering drawings are usually created in accordance with the standard conventions for layout, nomenclature, interpretation, appearance, and size.

The previous versions of SolidWorks have been limited to mainly handling mainly full-sectional views of drawings. The enhanced Section View tool now makes creating section views in drawings faster, with simple drag-and-drop placement. Section profiles can now be clicked, dragged, and jogged to modify section profiles on the fly.

This chapter deals first with part and assembly drawing and then follows with describing the steps that are involved in inserting half-, offset-, and aligned-section views to drawings, which are enhancements in the most recent version of SolidWorks. Throughout this chapter, the American Society of Mechanical Engineers standards for views, drawings, and dimensions are used. Although first-angle projection and third-angle projection are briefly discussed, the third-angle projection is used since this is the standard that is mainly used in North America.

Orthographic Projection

An orthographic projection shows the object as it looks from the six sides (front, top, bottom, right, left, or back) and are positioned relative to each other according to the rules of either first- or third-angle projection (see Figure 9.1).

First-angle projection is the International Organization for Standardization (ISO) standard and is primarily used in Europe. The three-dimensional (3D) part is projected into a two-dimensional paper space as if one is looking at an x-ray of the object in which the top view is under the front view and the right view is at the left of the front view (see Figure 9.2).

Third-angle projection is primarily used in the United States and Canada, where it is the default projection system according to the BS 8888 (2006). The left view is placed on the left and the top view on the top.

Drawing Sizes

The sizes of a drawing normally comply with either of the two different standards of ISO (world standard) or US customary, according to Table 9.1.

Creating a SolidWorks Drawing Template

The drawing template is made up of three components: (1) document properties, (2) sheet properties, and (3) title block.

Document Properties

1. Click New.
2. Double-click Drawing.
3. Click Cancel X from Model View Manager. (If the *Start when creating new drawing* is checked, the Model View PropertyManager is selected by default [see Figure 9.3].)
4. Click Options > Document Properties from the Menu bar.

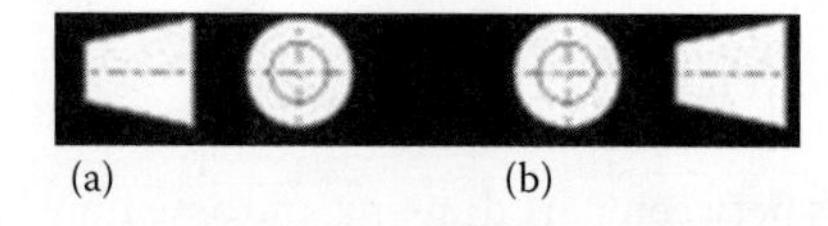

Figure 9.1

Orthographic projections: (a) first angle and (b) third angle.

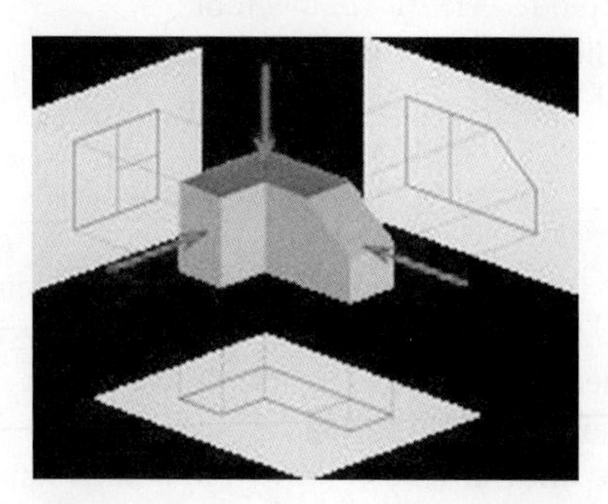

Figure 9.2

First-angle projection of a part.

Table 9.1 ISO (World Standard) or US Customary Drawing Sizes

ISO Drawing Sizes (mm)		US Customary Drawing Sizes	
A4	210 × 297	A	8.5″ × 11″
A3	297 × 420	B	11″ × 17″
A2	420 × 594	C	17″ × 22″
A1	594 × 841	D	22″ × 34″
A0	842 × 1189	E	34″ × 44″

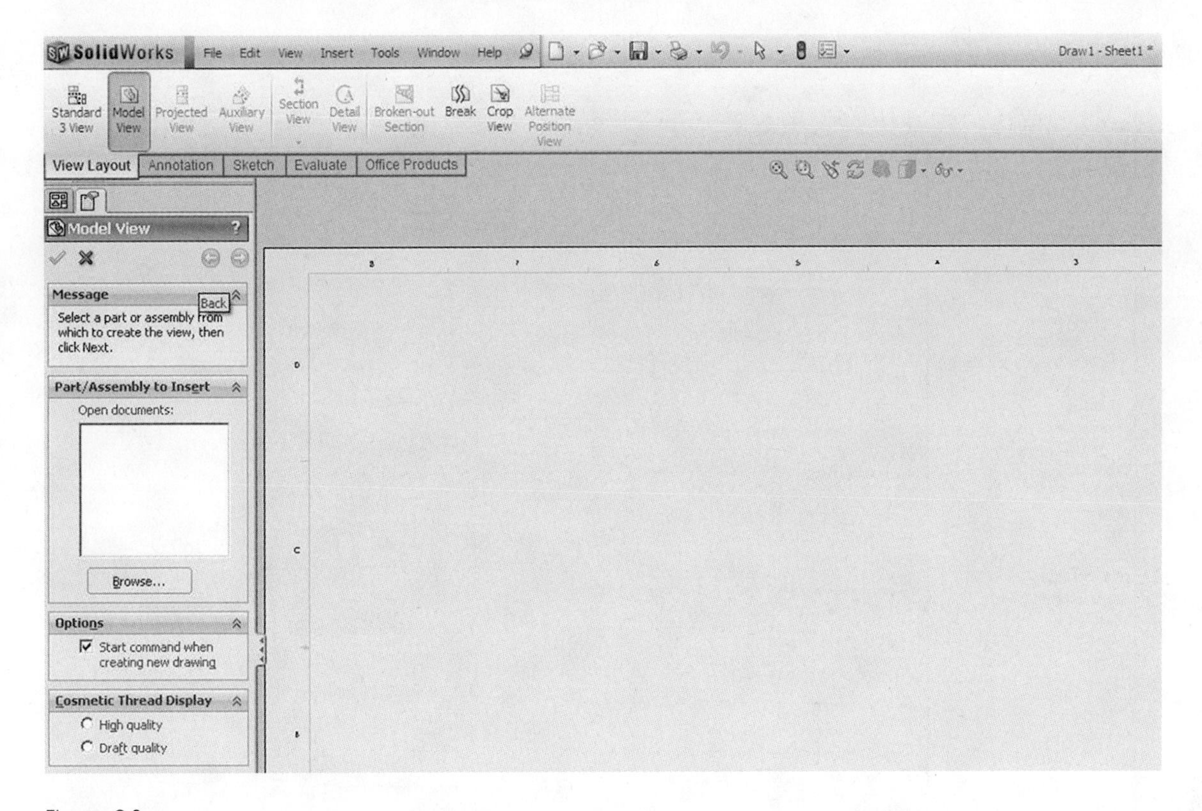

Figure 9.3

Clicking Cancel X from Model View Manager because a template is being created.

5. Click Drafting Standard > ANSI.
6. Click Annotations > Font > Units (enter 3 mm) > OK. (Font-height) (see Figure 9.4).
7. Click Dimensions (enter 1 mm, 3 mm, 6 mm) > OK. (Arrowhead for dimensioning) (see Figure 9.5).
8. Click View Labels > Section (enter 2 mm, 6 mm, 12 mm) > OK (Arrowhead for sectioning) (see Figure 9.6).
9. Click Units > MMGS (select 0.12 basic unit length, None basic unit angle) (see Figure 9.7).
10. Click Layer Properties (select Layers ON). (If not active, right-click the CommandManager and activate it) (see Figure 9.8).
11. Click Systems Options > File Location > Add path and Click OK > OK > OK (Set System Options File Location) (see Figure 9.9).

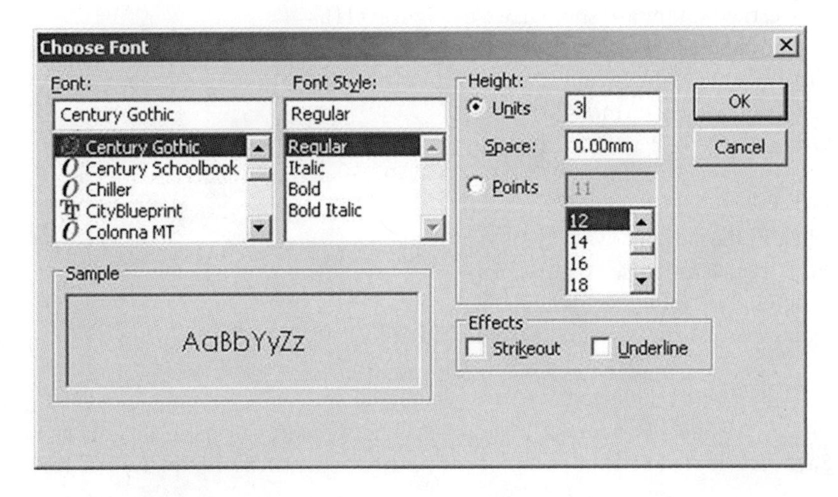

Figure 9.4

Clicking Annotations > Font > Units.

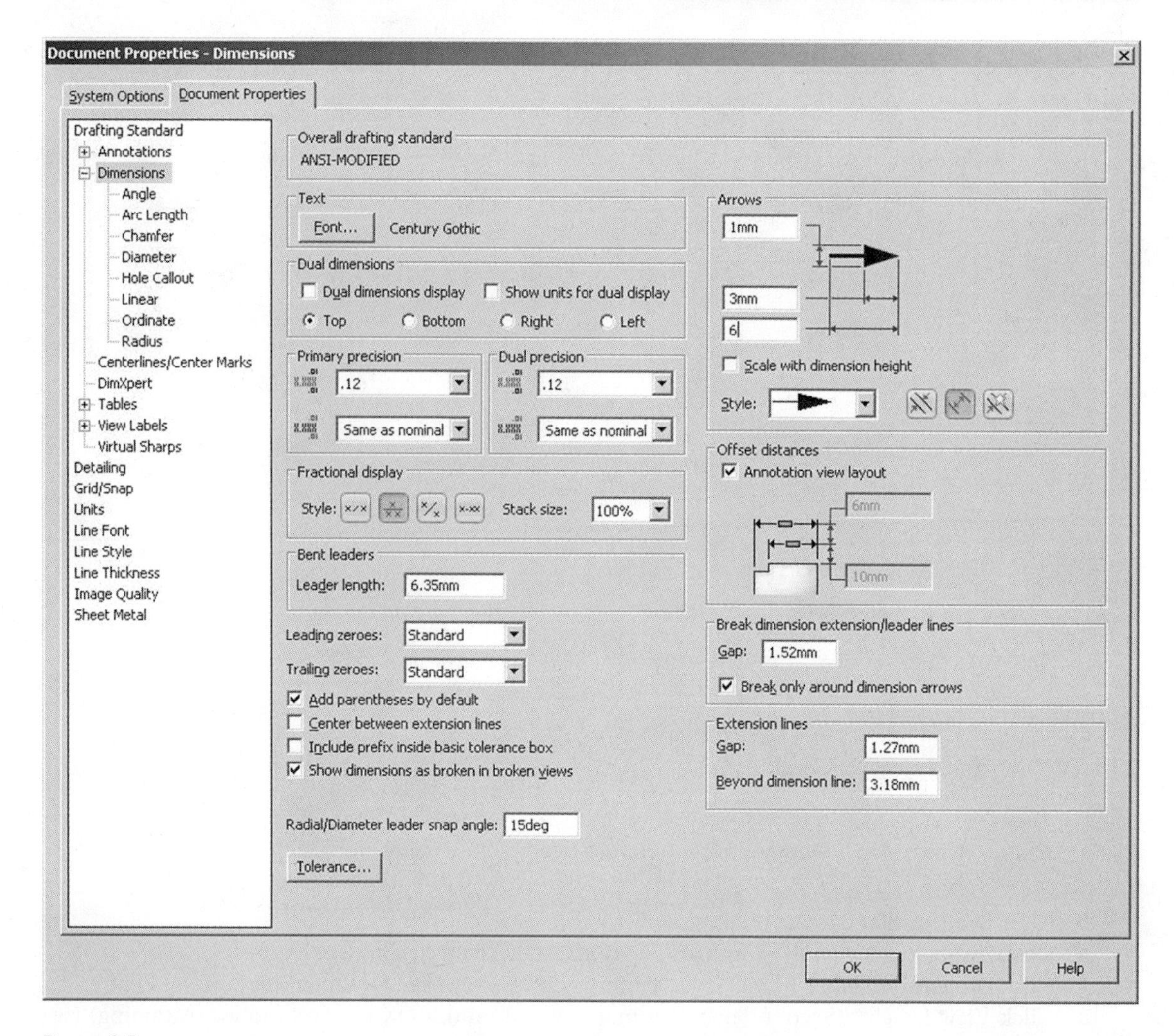

Figure 9.5

Clicking Dimensions.

Sheet Properties

12. Right-click in the Graphics window and select the Properties option (see Figure 9.10).
13. Click Properties > Standard sheet size (see Figure 9.11).
14. Select A-Landscape, A-Portrait, B-Landscape, C-Landscape, D-Landscape, or E-Landscape.
15. Check Third Angle; select Scale 1:1.
16. Uncheck Display sheet format.
17. Click OK.

Title Block

18. Right-click in the Graphics window.
19. Click Edit Sheet Format (see Figure 9.12).
20. Modify Tolerance Note, Angular Tolerance, etc.
21. Create a new Microsoft Logo and paste in Title Block.

Saving the Template

22. Click File > Save Sheet Format > CUSTOM-B.slddrt in path. (Save the Sheet Format.)
23. Click Save As > B-ANSI-MM.drwdot in path (save the Drawing Template) (see Figure 9.13).

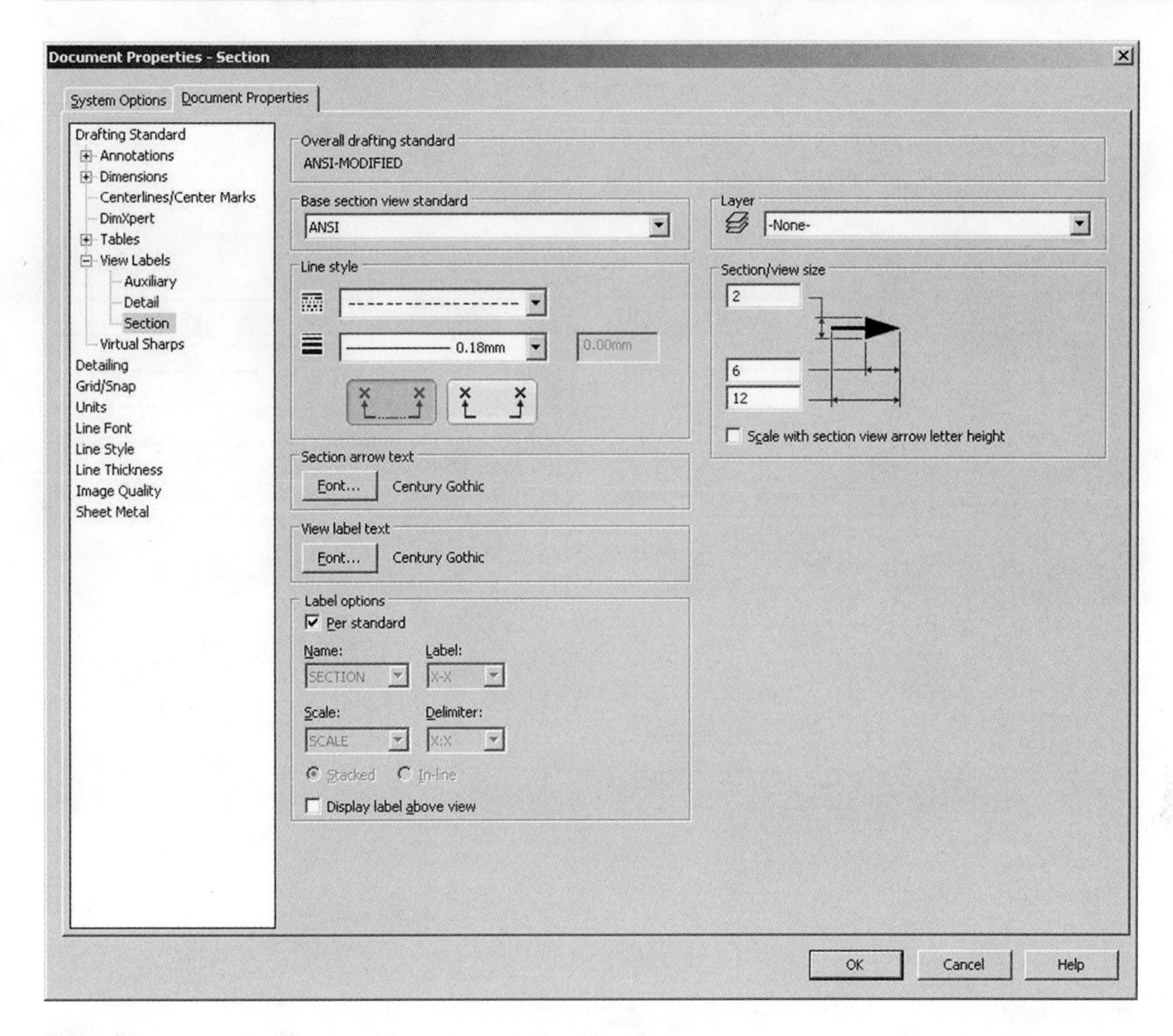

Figure 9.6

Click View Labels > Section.

Part Drawing of Adjustable Shaft Support

This section describes the steps that are required to generate a part drawing. Standard views are first obtained, followed by the exploded view and the bill of materials (BoM). Let us examine some examples.

The assembly shown in Figure 9.14 contains eight machined parts that are already modeled in Chapter 5 and assembled in Chapter 8.

- **a.** Build the assembly of the parts.
- **b.** Produce the Exploded View of assembly, the BoM.
- **c.** Produce the drawing of each of the five parts to include necessary views.

Screw: Standard Views

To display Drawing FeatureManager

1. Click New from the Menu bar.
2. Double-click B-ANSI-MM from MY-TEMPLATE tab.
3. Click Cancel X from the Model View Manager (if this appears).

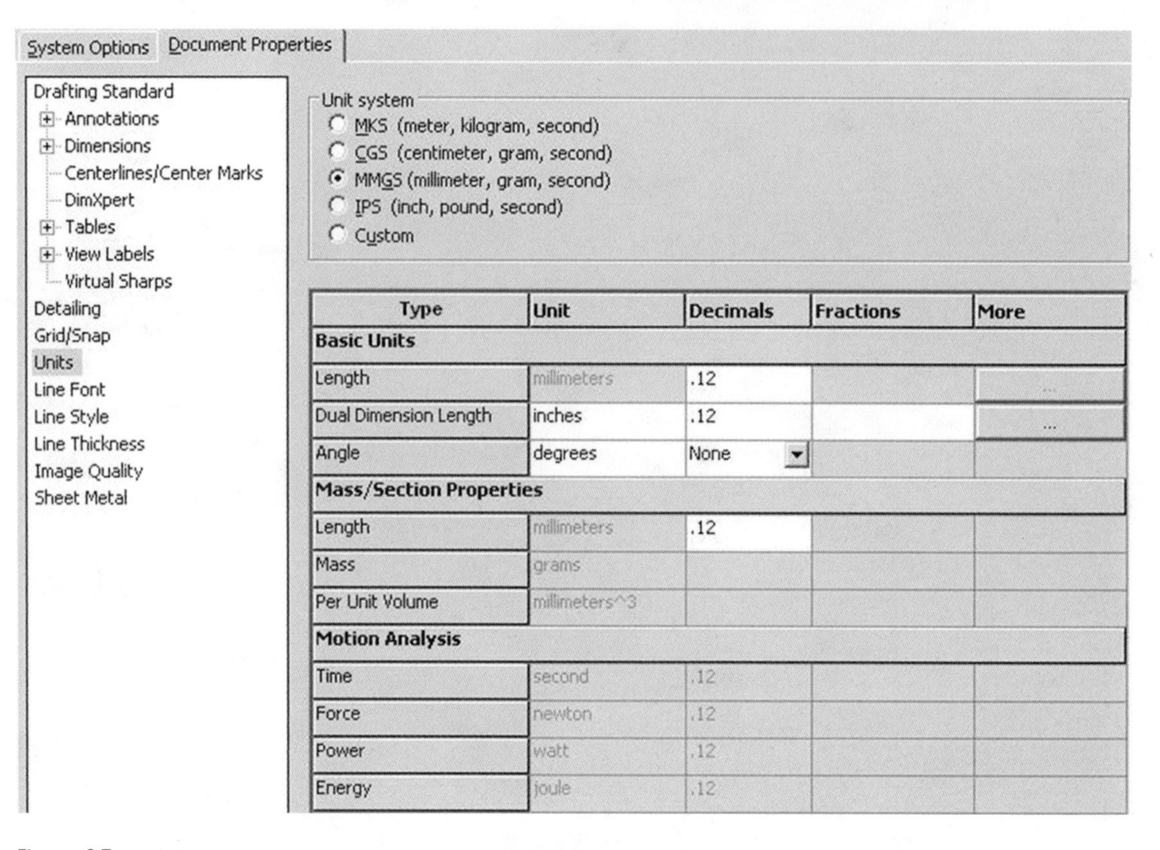

Figure 9.7

Clicking Units > MMGS.

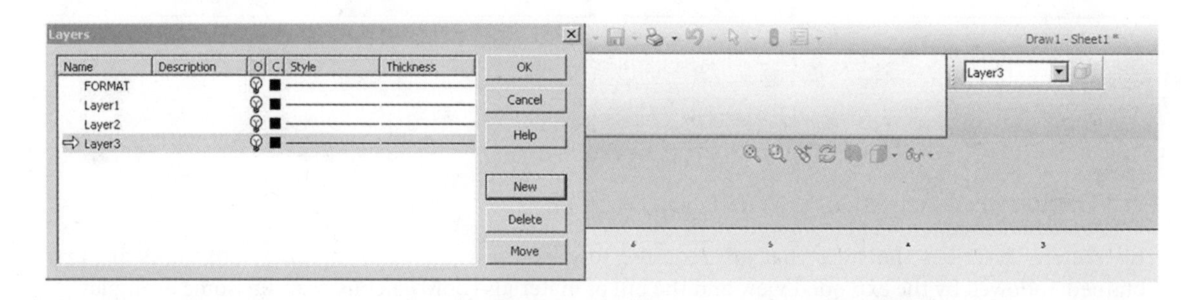

Figure 9.8

Clicking Layer Properties.

To Insert Drawing View and Standard Views

4. Click Model View (see Figure 9.15).
5. Double-click Part from the Part/Assembly to Insert box.
6. Click Multiple Views in the Number of Views rollout.
7. Select Standard views needed: Front and Top, as well as 3D from the Orientation rollout (see Figure 9.15).
8. Click OK.

   The following message may appear, to which you should answer, *Yes: SolidWorks has determined that the following view(s) may need Isometric (True) dimensions instead of standard Projected dimensions. Do you want to switch the view(s) to use Isometric (True) dimensions?*

   (See Figure 9.16 for the query and Figure 9.17 for the views.)
9. Click Save As from the Menu bar (give a name for the drawing) and Save.

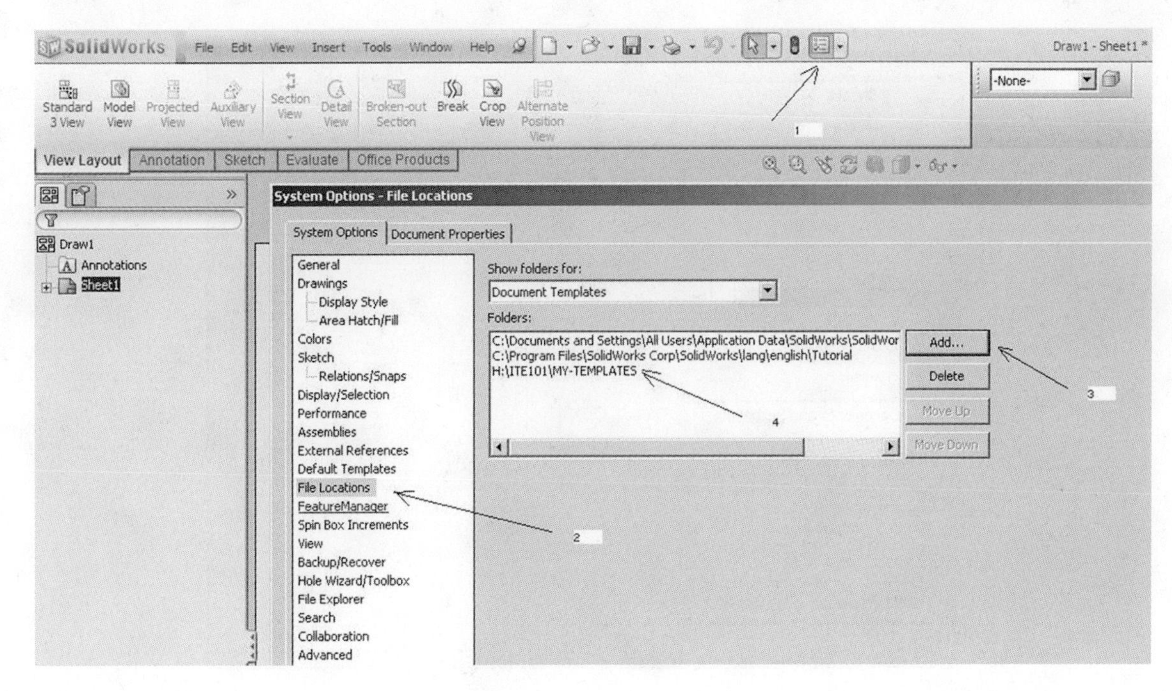

Figure 9.9

Clicking Options > File Location > Add path to set System Options File Location.

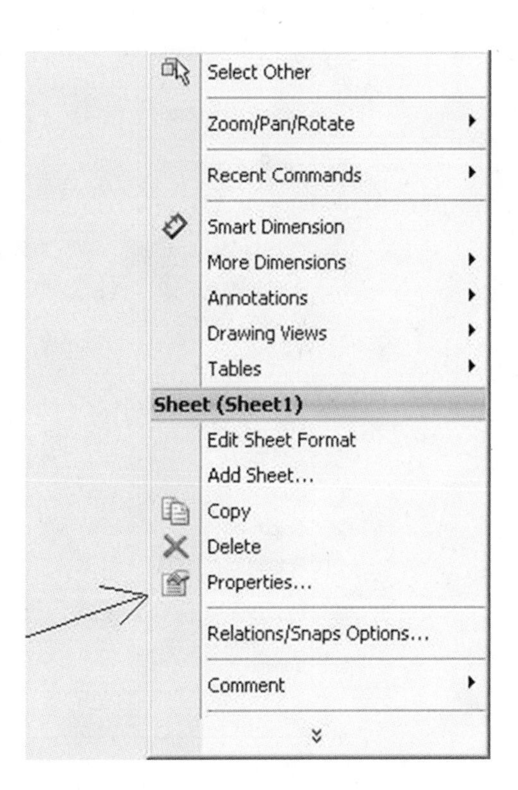

Figure 9.10

Right-clicking in the Graphics window and selecting the Properties option.

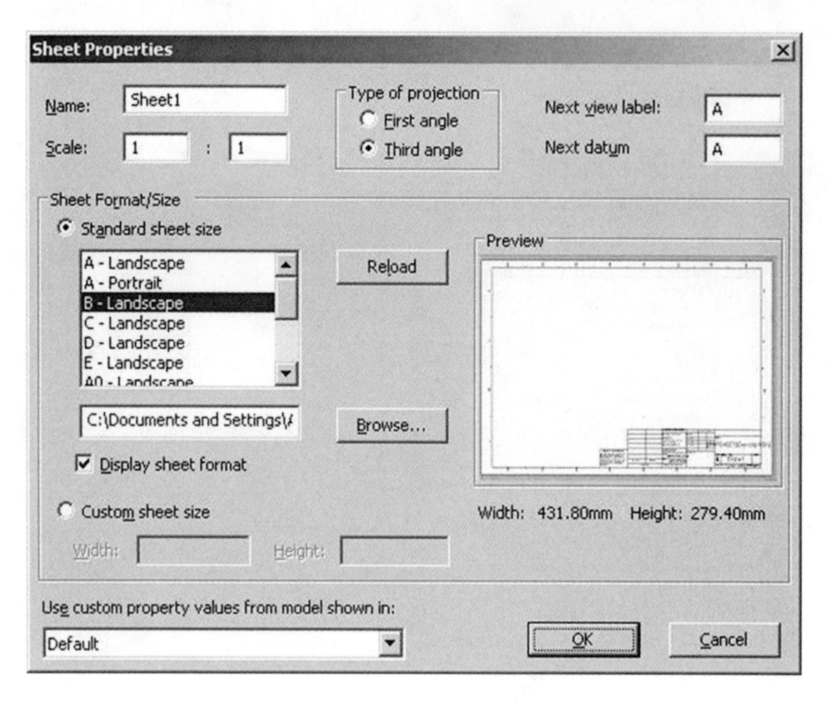

Figure 9.11

Clicking Properties and accessing Standard sheet.

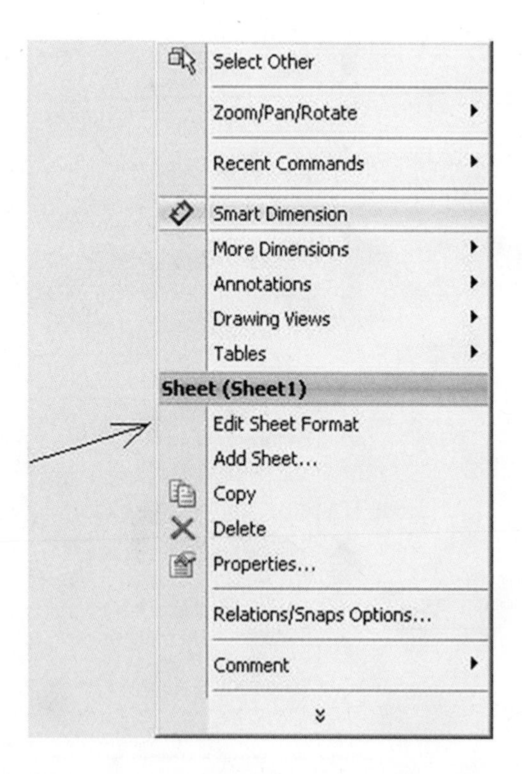

Figure 9.12

Right-clicking in the Graphics window and selecting the Edit Sheet Format option.

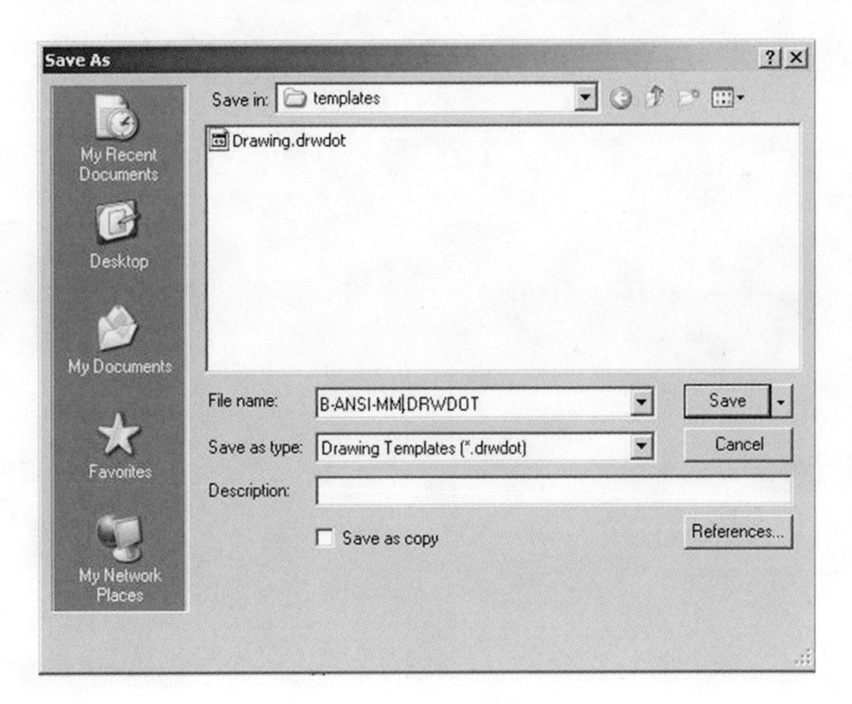

Figure 9.13

Save the drawing template.

Figure 9.14

Adjustable shaft support.

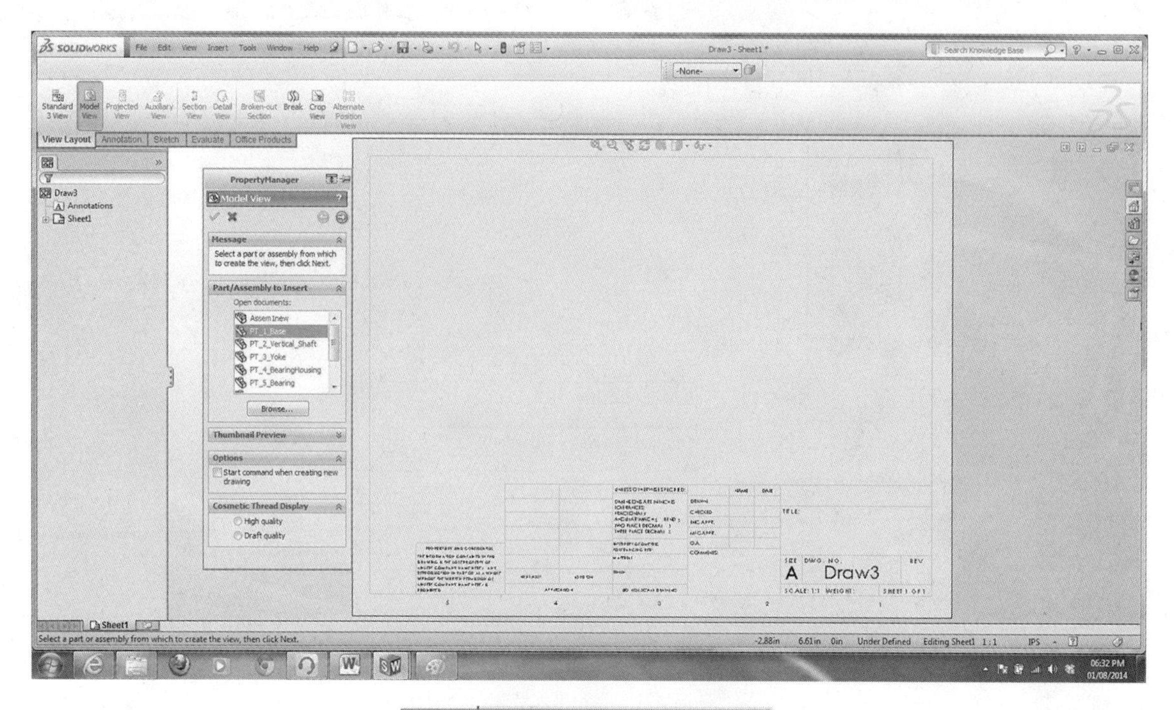

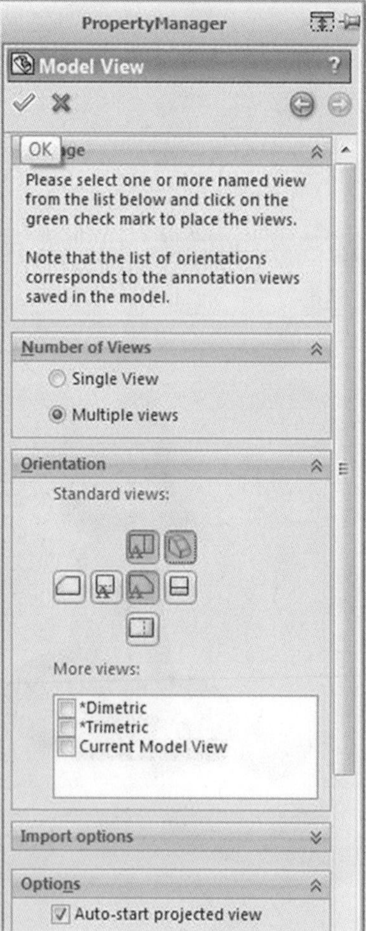

Figure 9.15

Multiple views selected.

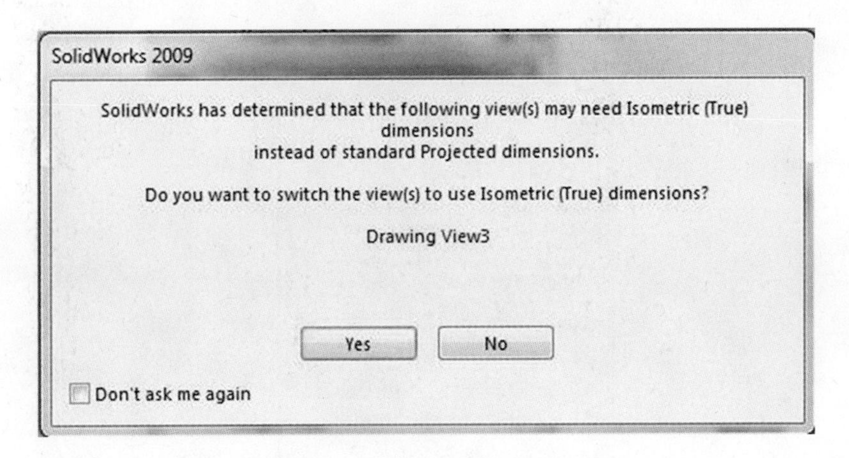
SolidWorks 2009
SolidWorks has determined that the following view(s) may need Isometric (True) dimensions
instead of standard Projected dimensions.
Do you want to switch the view(s) to use Isometric (True) dimensions?
Drawing View3
Yes
No
Don't ask me again

Figure 9.16

SolidWorks query.

Import Entire Dimensions from Source to Destination

1. Click Annotation > Model items.
2. Select Entire model for Source/Destination (see Figure 9.17).
3. Click OK.

Vertical Shaft: Standard Views

To Display Drawing FeatureManager

1. Click New from the Menu bar.
2. Double-click B-ANSI-MM from MY-TEMPLATE tab.
3. Click Cancel X from the Model View Manager (if this appears).

To Insert Drawing View and Standard Views

4. Click Model View (see Figure 9.18).
5. Double-click Part from the Part/Assembly to Insert box.
6. Click Multiple Views in the Number of Views rollout.
7. Select Standard views needed: Front and Top, as well as 3D from the Orientation rollout (see Figure 9.18).
8. Click OK.

   The following message may appear, to which you should answer, *Yes: SolidWorks has determined that the following view(s) may need Isometric (True) dimensions instead of standard Projected dimensions. Do you want to switch the view(s) to use Isometric (True) dimensions?*

   (See Figure 9.19 for the query and Figure 9.20 for the views.)
9. Click Save As from the Menu bar (give a name for the drawing) and Save.

Import Entire Dimensions from Source to Destination

1. Click Annotation > Model items.
2. Select Entire model for Source/Destination (see Figure 9.20).
3. Click OK.

Yoke: Standard Views

To Display Drawing FeatureManager

1. Click New from the Menu bar.
2. Double-click B-ANSI-MM from MY-TEMPLATE tab.
3. Click Cancel X from the Model View Manager (if this appears).

To Insert Drawing View and Standard Views

4. Click Model View (see Figure 9.21).
5. Double-click Part from the Part/Assembly to Insert box.
6. Click Multiple Views in the Number of Views rollout.
7. Select the Standard views that are needed: Front and Top, as well as 3D from the Orientation rollout (see Figure 9.21).

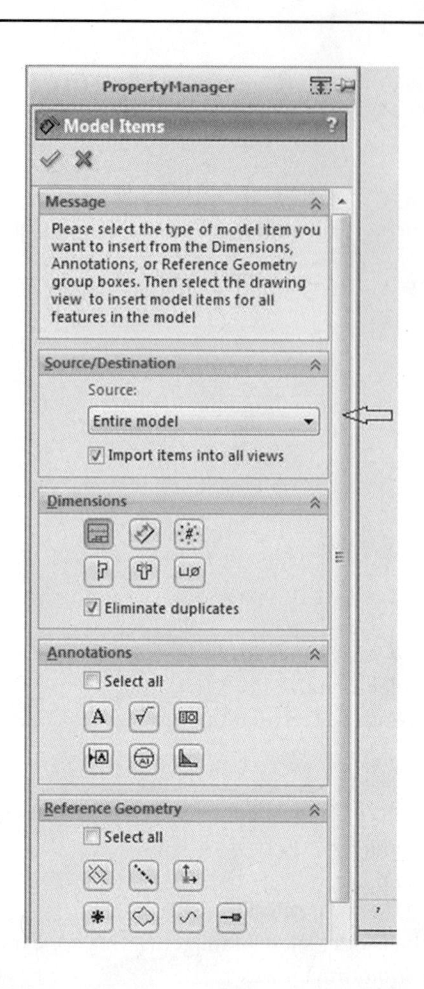

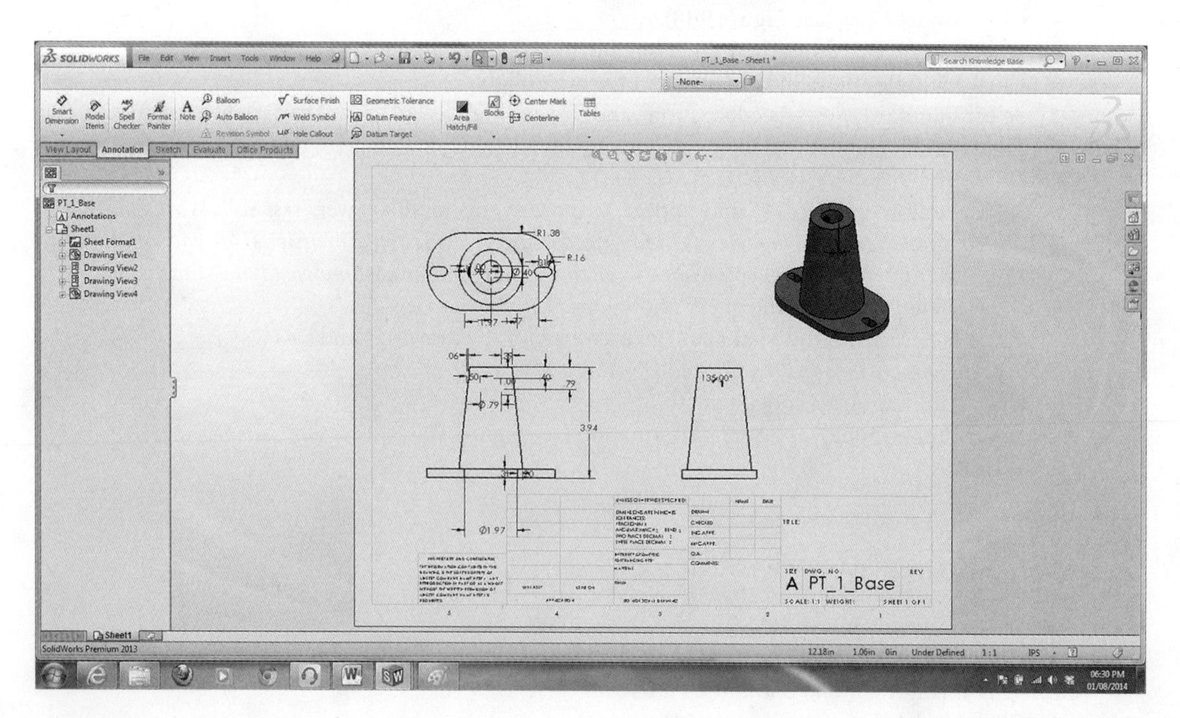

Figure 9.17

Importing dimensions of entire model.

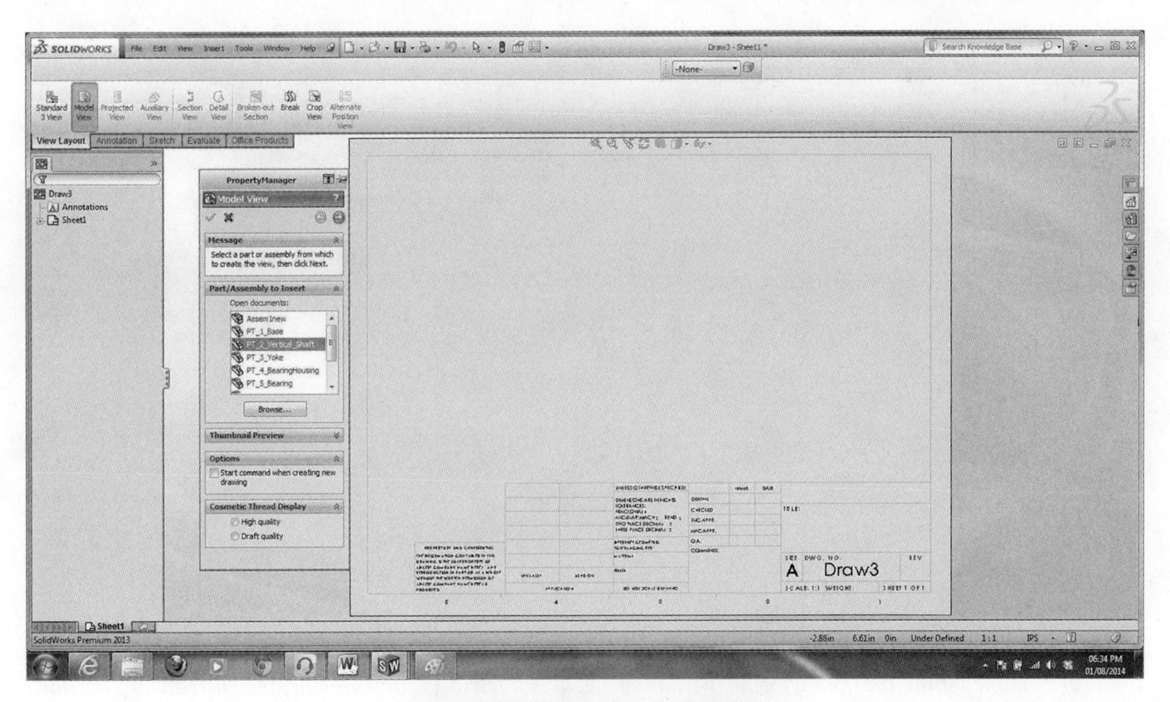

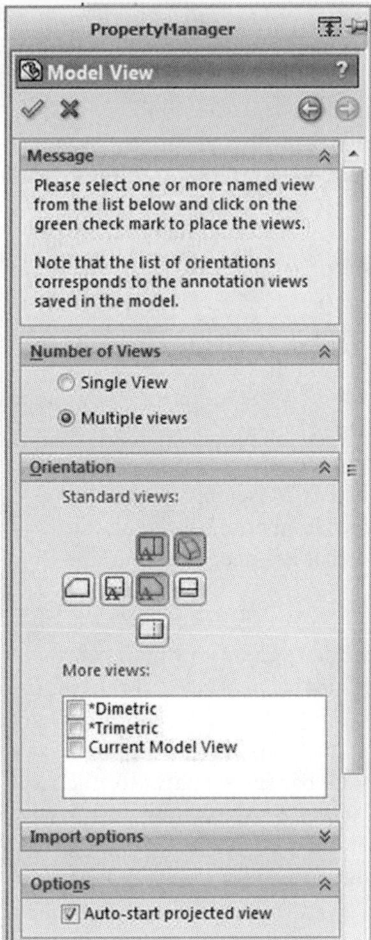

Figure 9.18

Multiple views selected.

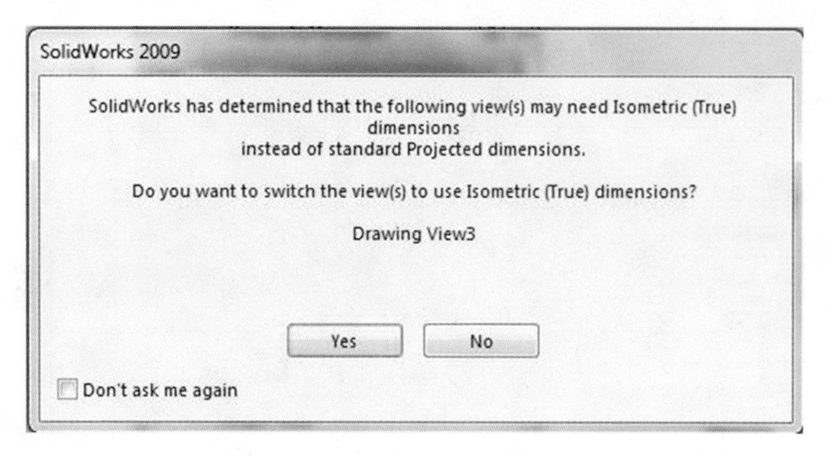
SolidWorks 2009
SolidWorks has determined that the following view(s) may need Isometric (True) dimensions
instead of standard Projected dimensions.
Do you want to switch the view(s) to use Isometric (True) dimensions?
Drawing View3
Yes
No
Don't ask me again

Figure 9.19

SolidWorks query.

8. Click OK.

   The following message may appear, to which you should answer, *Yes: SolidWorks has determined that the following view(s) may need Isometric (True) dimensions instead of standard Projected dimensions. Do you want to switch the view(s) to use Isometric (True) dimensions?*

   (See Figure 9.22 for the query and Figure 9.23 for the views.)
9. Click Save As from the Menu bar (give a name for the drawing) and Save.

Import Entire Dimensions from Source to Destination

1. Click Annotation > Model items.
2. Select Entire model for Source/Destination (see Figure 9.23).
3. Click OK.

Bearing Housing: Standard Views

To Display Drawing FeatureManager

1. Click New from the Menu bar.
2. Double-click B-ANSI-MM from MY-TEMPLATE tab.
3. Click Cancel X from the Model View Manager (if this appears).

To Insert Drawing View and Standard Views

4. Click Model View (see Figure 9.24).
5. Double-click Part from the Part/Assembly to Insert box.
6. Click Multiple Views in the Number of Views rollout.
7. Select the Standard views that are needed: Front and Top, as well as 3D from the Orientation rollout (see Figure 9.24).
8. Click OK.

   The following message may appear, to which you should answer, *Yes: SolidWorks has determined that the following view(s) may need Isometric (True) dimensions instead of standard Projected dimensions. Do you want to switch the view(s) to use Isometric (True) dimensions?*

   (See Figure 9.25 for the query and Figure 9.26 for the views.)
9. Click Save As from the Menu bar (give a name for the drawing) and Save.

Import Entire Dimensions from Source to Destination

1. Click Annotation > Model items.
2. Select Entire model for Source/Destination (see Figure 9.26).
3. Click OK.

Bearing: Standard Views

To Display Drawing FeatureManager

1. Click New from the Menu bar.
2. Double-click B-ANSI-MM from MY-TEMPLATE tab.
3. Click Cancel X from the Model View Manager (if this appears).

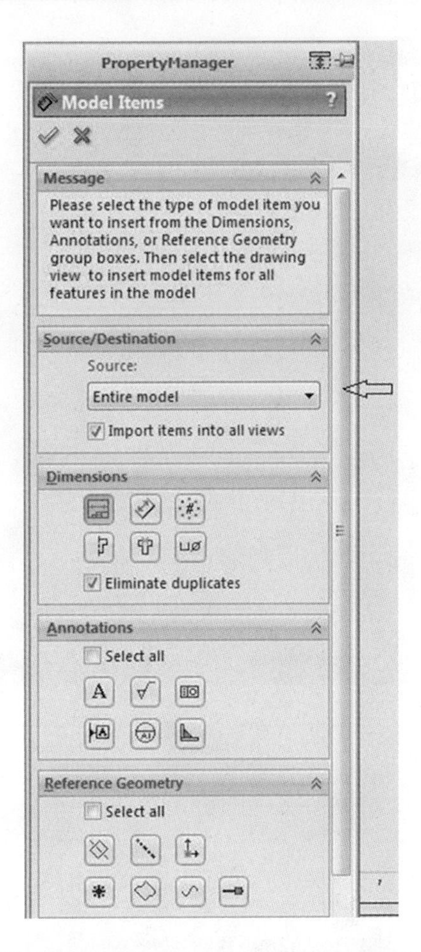

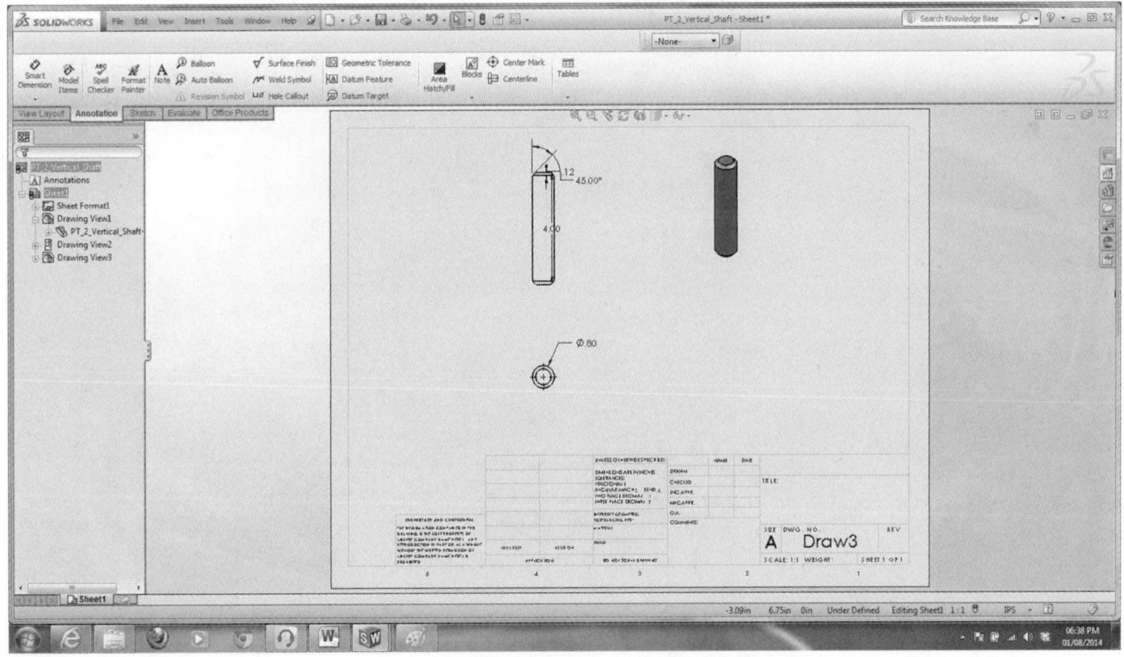

Figure 9.20

Importing dimensions of entire model.

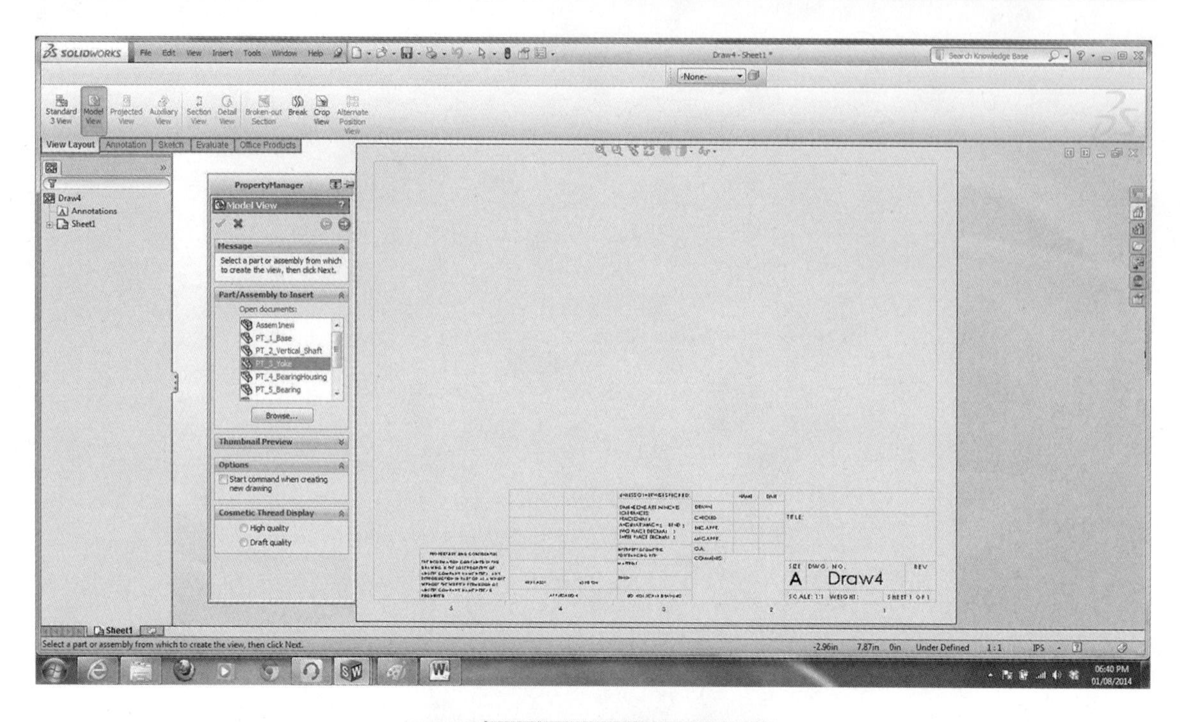

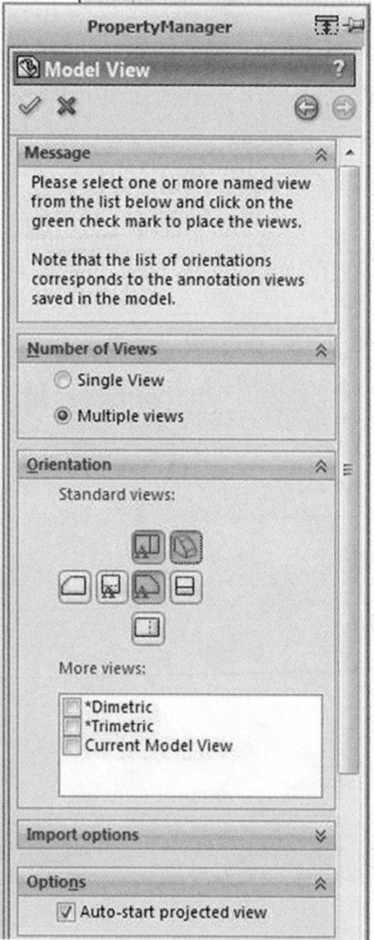

Figure 9.21

Multiple views selected.

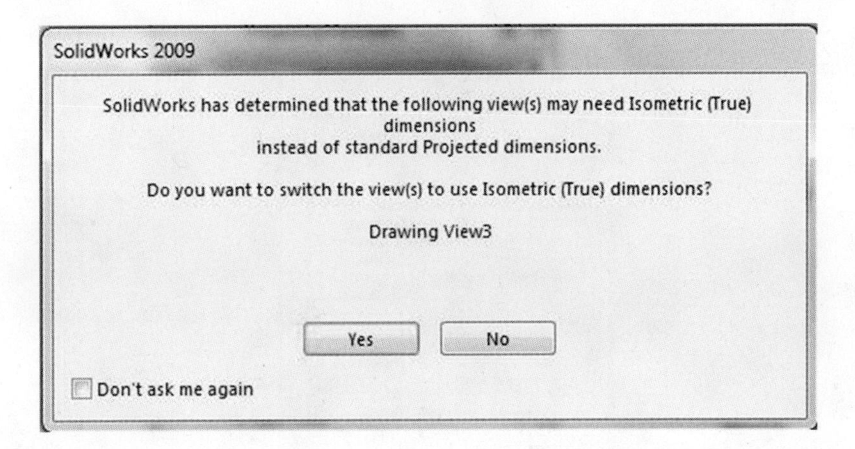
SolidWorks 2009
SolidWorks has determined that the following view(s) may need Isometric (True) dimensions
instead of standard Projected dimensions.
Do you want to switch the view(s) to use Isometric (True) dimensions?
Drawing View3
Yes
No
Don't ask me again

Figure 9.22

SolidWorks query.

To Insert Drawing View and Standard Views

4. Click Model View (see Figure 9.27).
5. Double-click Part from the Part/Assembly to Insert box.
6. Click Multiple Views in the Number of Views rollout.
7. Select Standard views needed: Front and Top as well as 3D from the Orientation rollout (see Figure 9.27).
8. Click OK.

   The following message may appear, to which you should answer, *Yes: SolidWorks has determined that the following view(s) may need Isometric (True) dimensions instead of standard Projected dimensions. Do you want to switch the view(s) to use Isometric (True) dimensions?*

   (See Figure 9.28 for the query and Figure 9.29 for the views.)
9. Click Save As from the Menu bar (give a name for the drawing) and Save.

Import Entire Dimensions from Source to Destination

1. Click Annotation > Model items.
2. Select Entire model for Source/Destination (see Figure 9.29).
3. Click OK.

Assigning Properties through File > Properties

One way of assigning properties to a part is through the File > Properties route (see Figure 9.30).

1. Click File > Properties > Custom (see Figure 9.31).
2. Select CompanyName, Revision, Number, DrawnBy, DrawnDate, Material, Mass, Description and modifying them (see Figures 9.32 and 9.33).
3. Link properties to Title Block.

See Figure 9.34 for the configuration-specific information for the base.
See Figure 9.35 for the configuration-specific information for the vertical shaft.
See Figure 9.36 for the configuration-specific information for the yoke.
See Figure 9.37 for the configuration-specific information for the bearing housing.
See Figure 9.38 for the configuration-specific information for the bearing.

Assigning Properties through Configuration Manager

To assign properties through the configuration manager, the route to follow is shown in Figure 9.39. See Figure 9.40 for the configuration-specific information for the base.

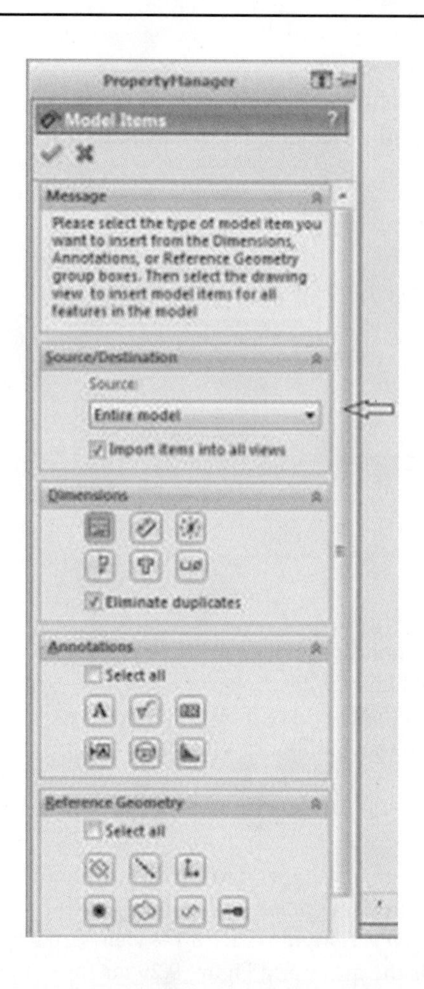

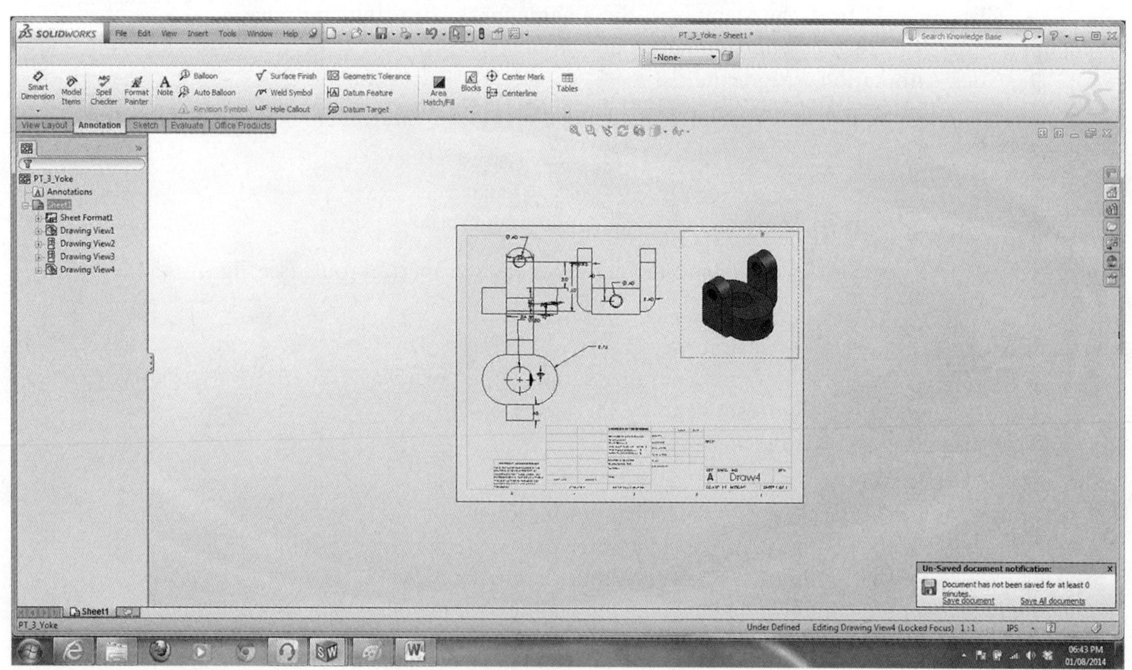

Figure 9.23

Importing dimensions of entire model.

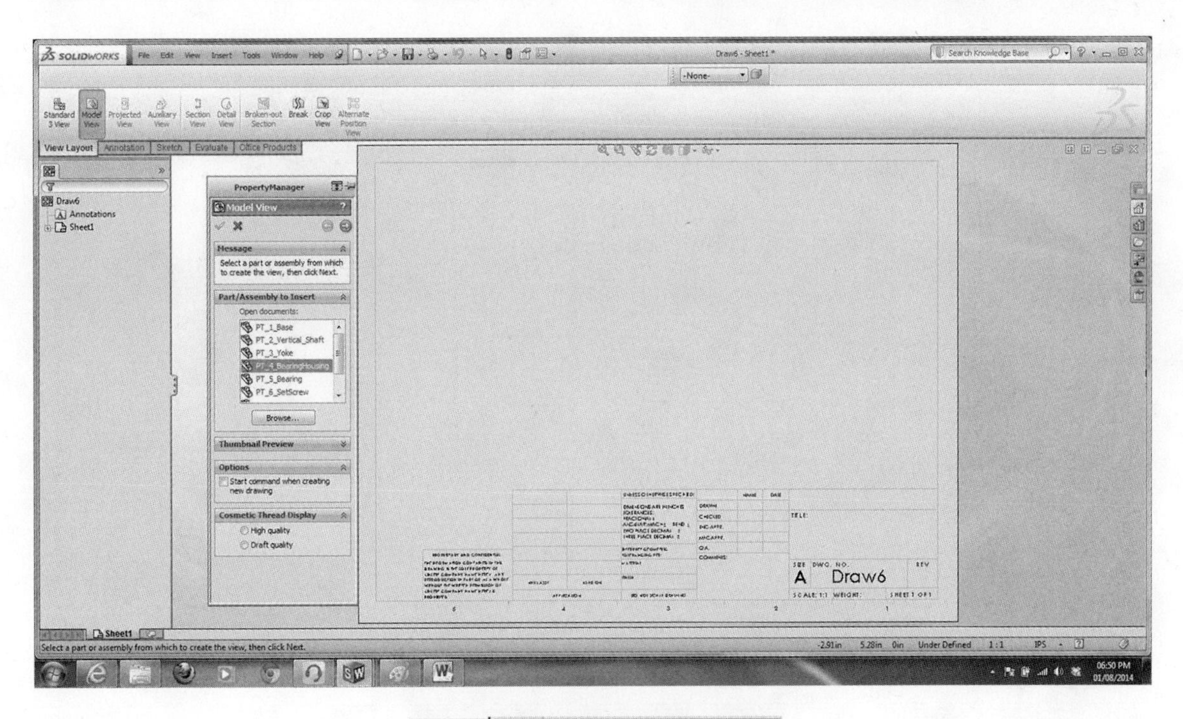

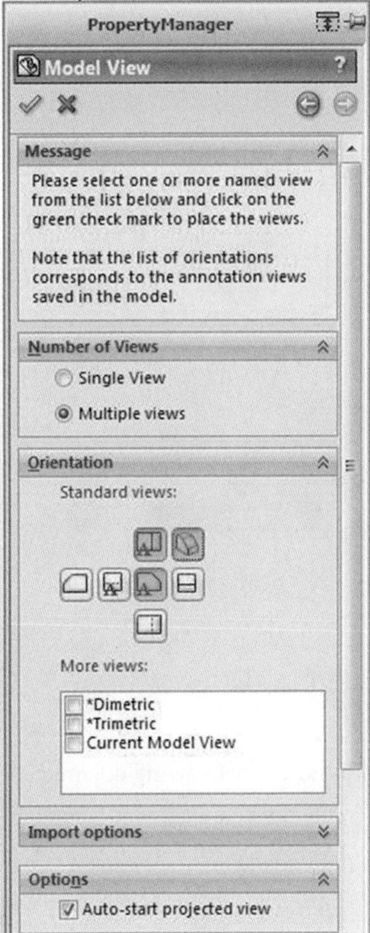

Figure 9.24

Multiple views selected.

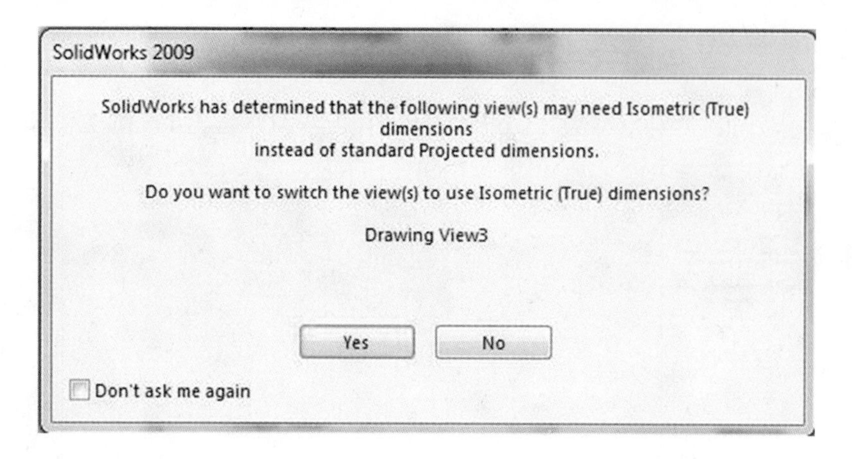
SolidWorks 2009
SolidWorks has determined that the following view(s) may need Isometric (True) dimensions
instead of standard Projected dimensions.
Do you want to switch the view(s) to use Isometric (True) dimensions?
Drawing View3
Yes
No
Don't ask me again

Figure 9.25

SolidWorks query.

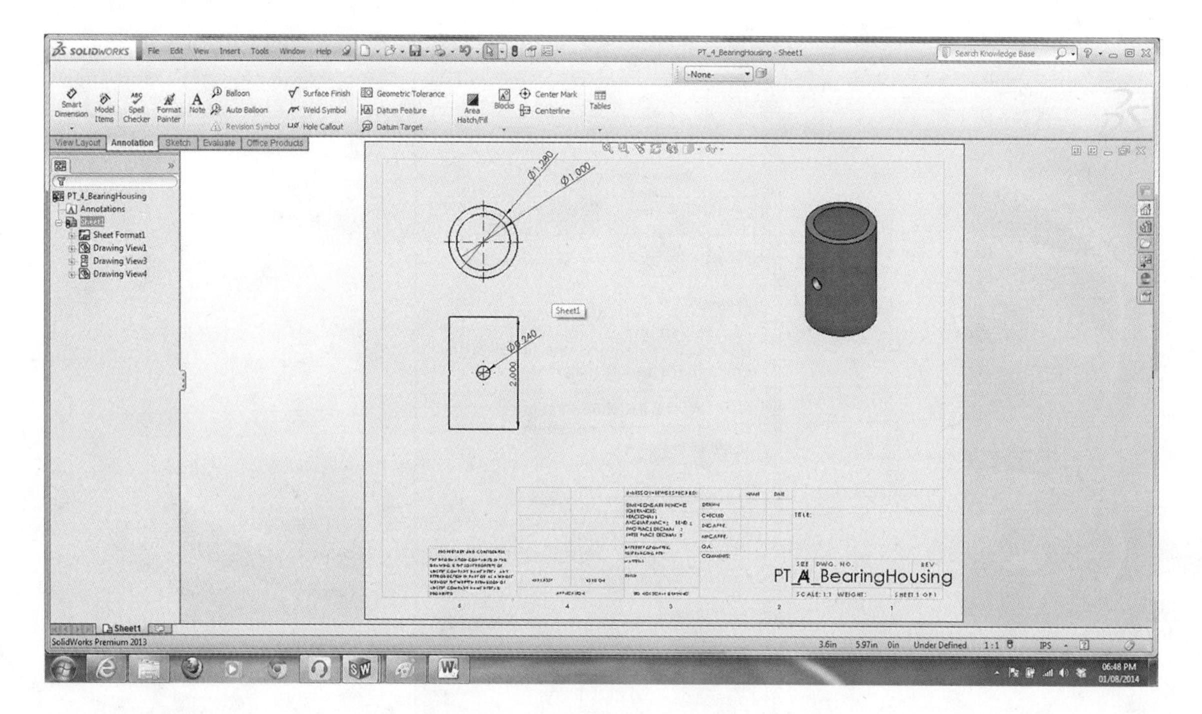
PT_4_BearingHousing

Figure 9.26

Importing dimensions of entire model.

Adjustable Shaft Support Assembly Drawing

Exploded View

Two views must be open (the assembly and a new Drawing document).

Open the Assembly Document Already Completed

1. Click Windows > Close All.
2. Open the Assembly from appropriate folder (see Figure 9.41).

Create a New Drawing

1. Click Make Drawing from Part/Assembly from New (see Figure 9.42).
2. Double-click B-ANSI-MM template from MY-TEMPLATES tab.
3. Click View Palette tool from Task Pane (see Figure 9.43).
4. Click and drag Isometric view from View Palette to Sheet1 (see Figure 9.44).
5. Click OK.

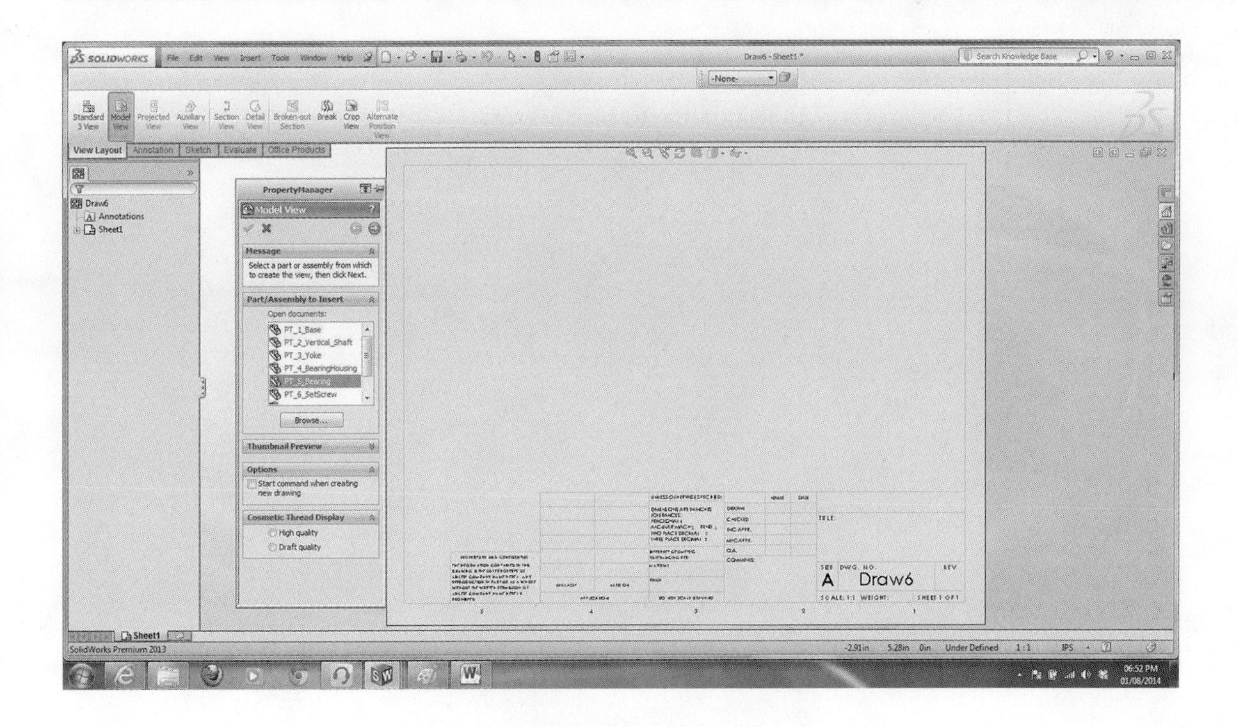

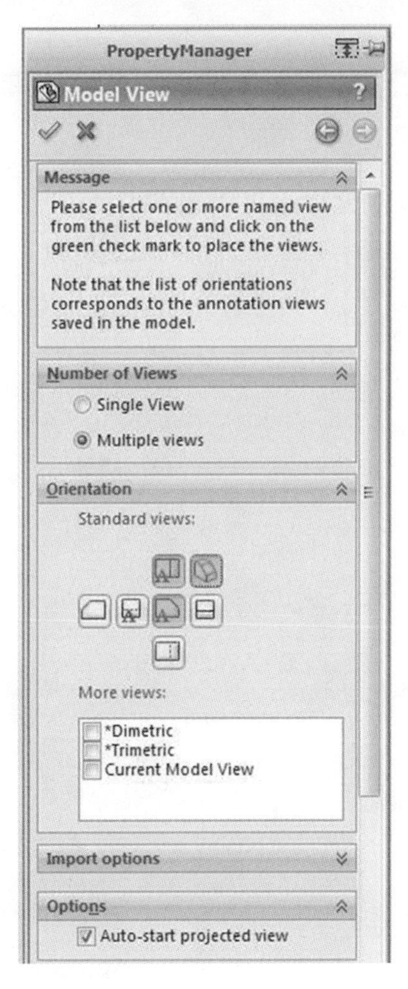

Figure 9.27

Multiple views selected.

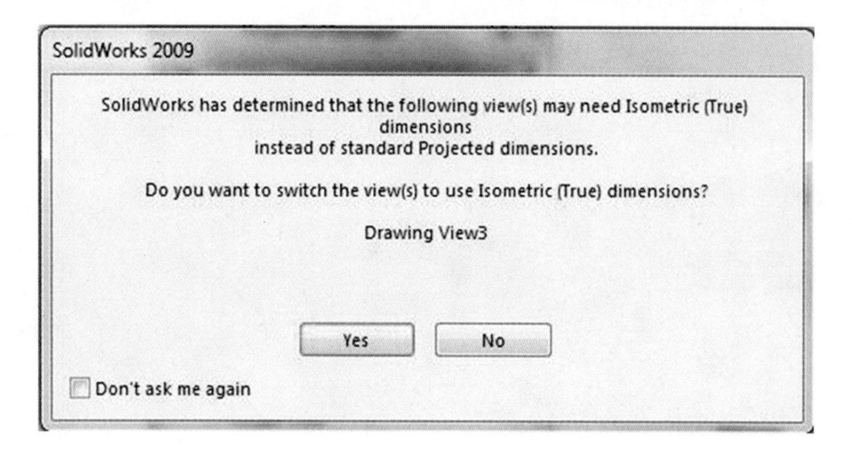

Figure 9.28

SolidWorks query.

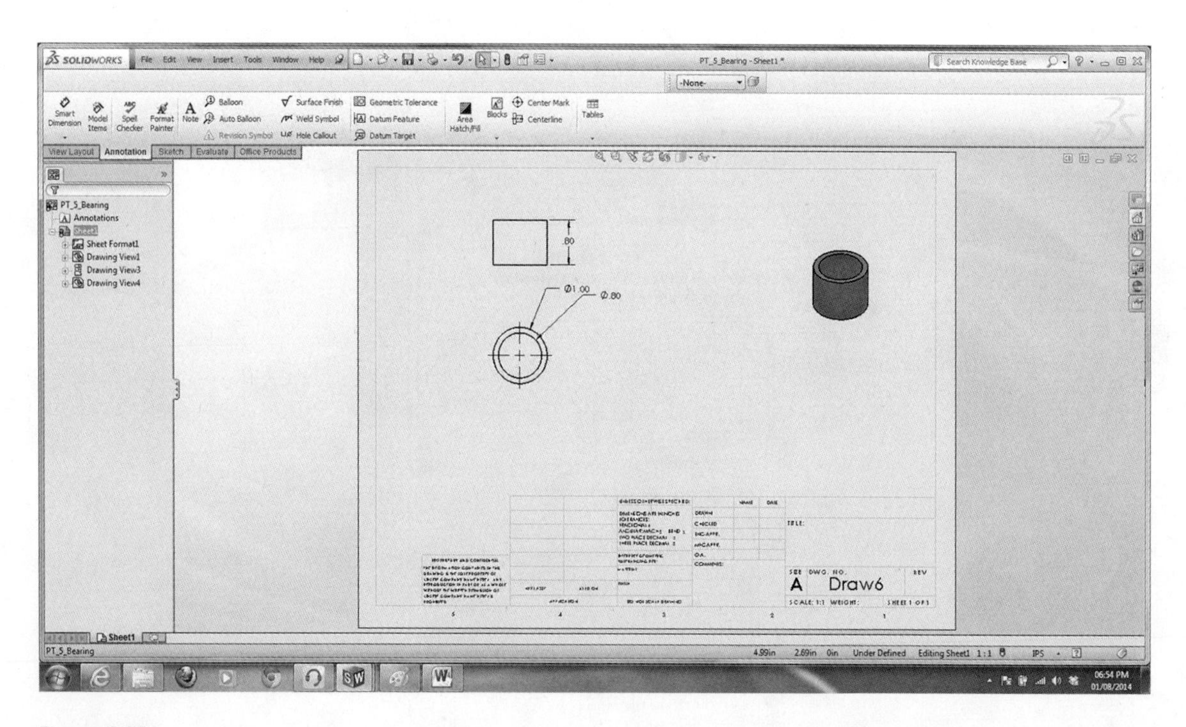

Figure 9.29

Importing dimensions of entire model.

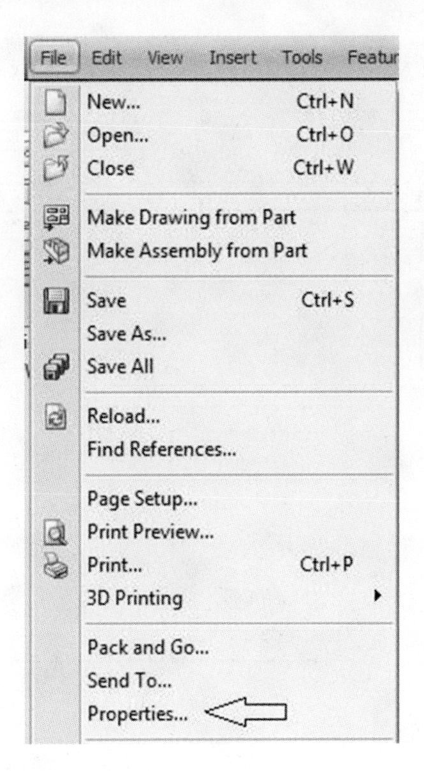

Figure 9.30

Assigning properties to a part: one route.

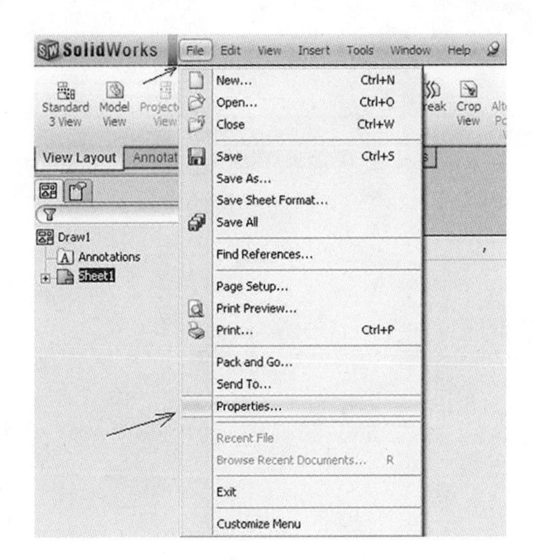

Figure 9.31

Clicking File > Properties.

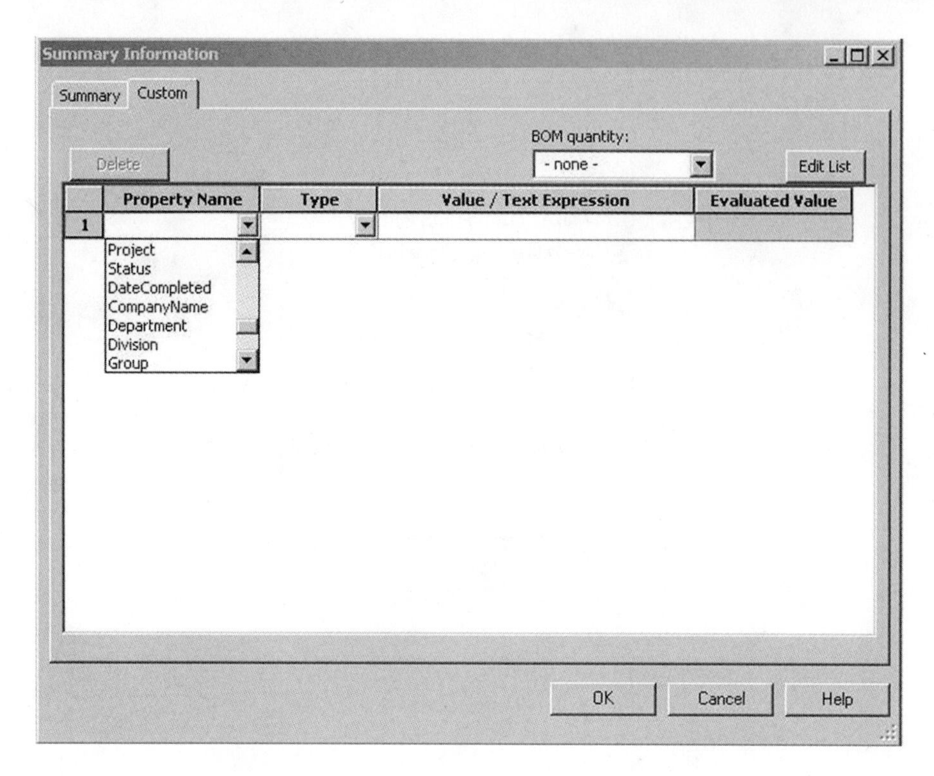

Figure 9.32

Clicking Custom from Properties and selecting Property Name option.

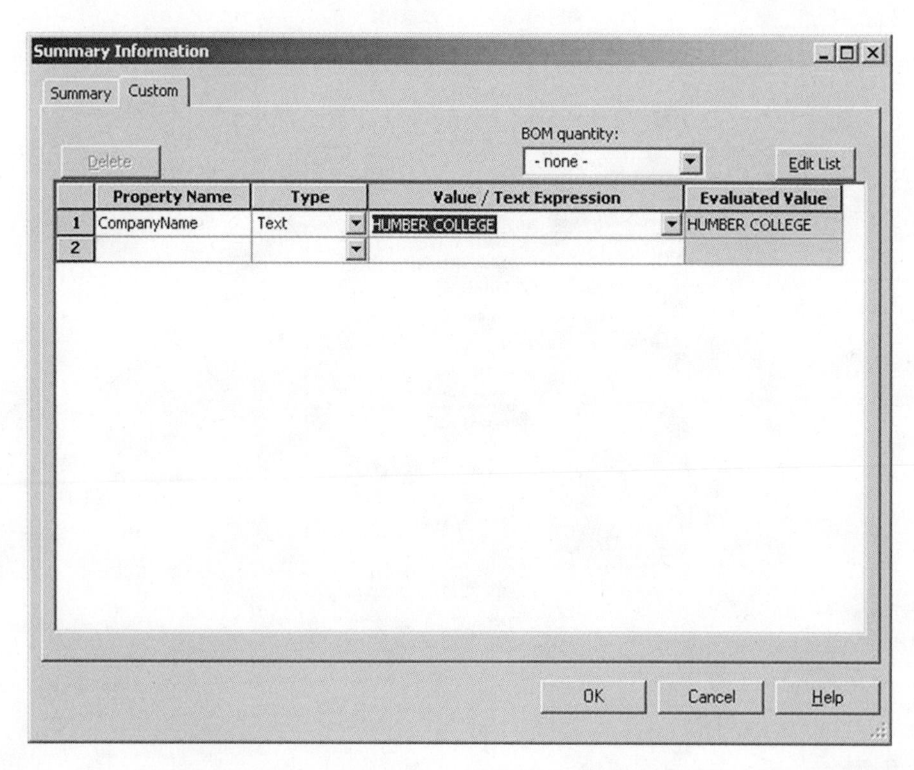

Figure 9.33

Modifying Property Name option.

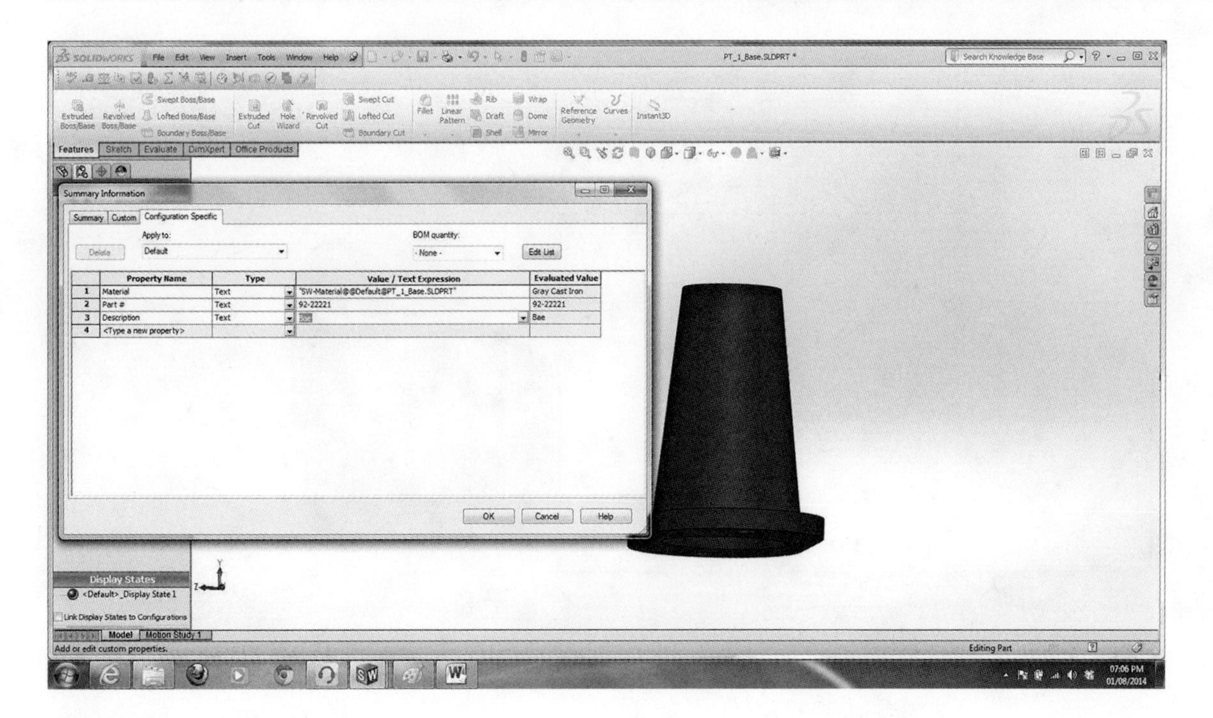

Figure 9.34

Base configuration-specific information.

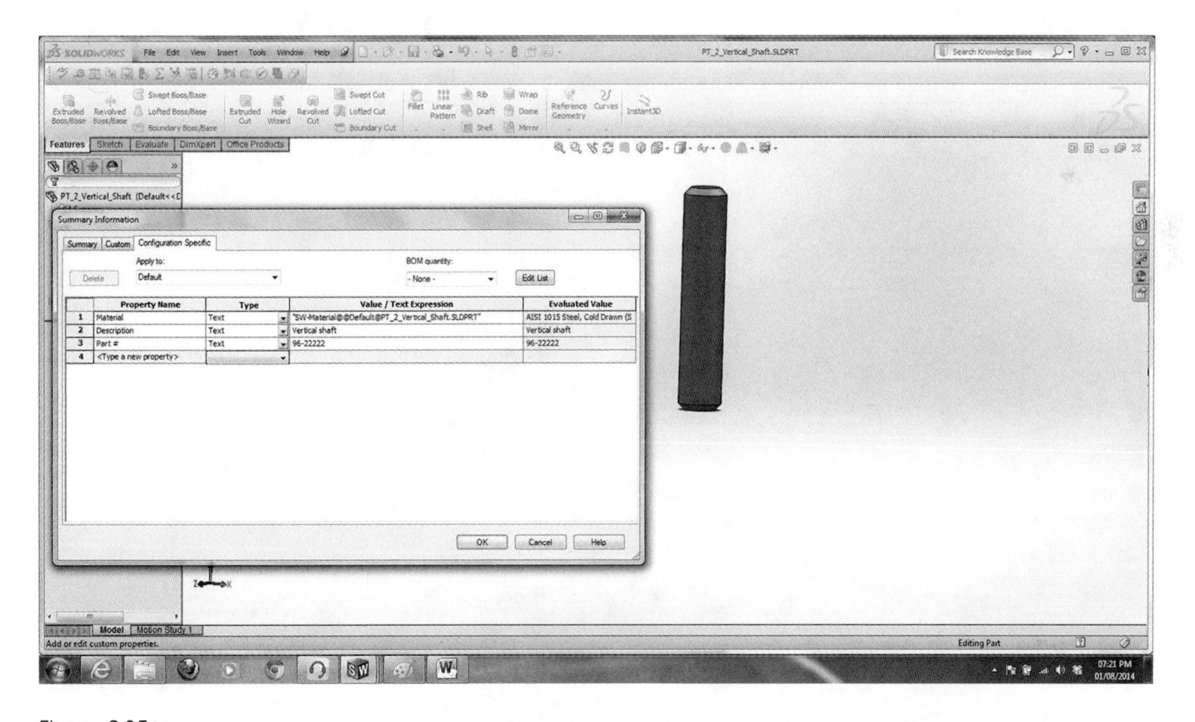

Figure 9.35

Vertical shaft configuration-specific information.

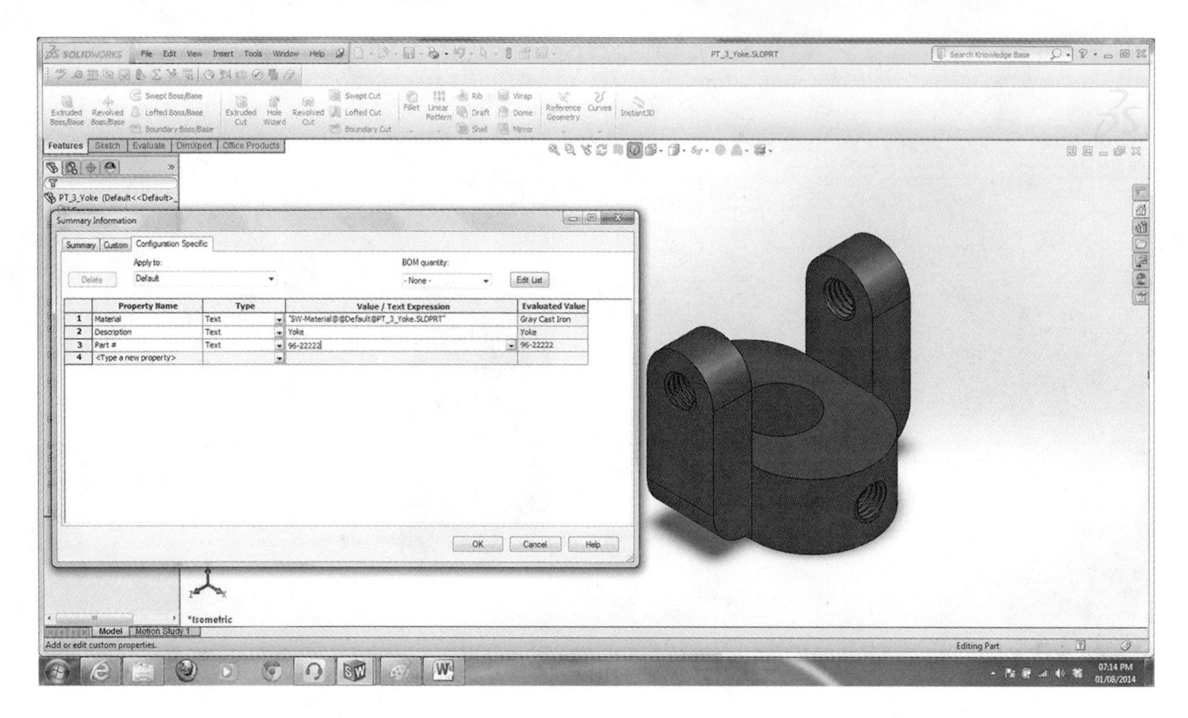

Figure 9.36

Yoke configuration-specific information.

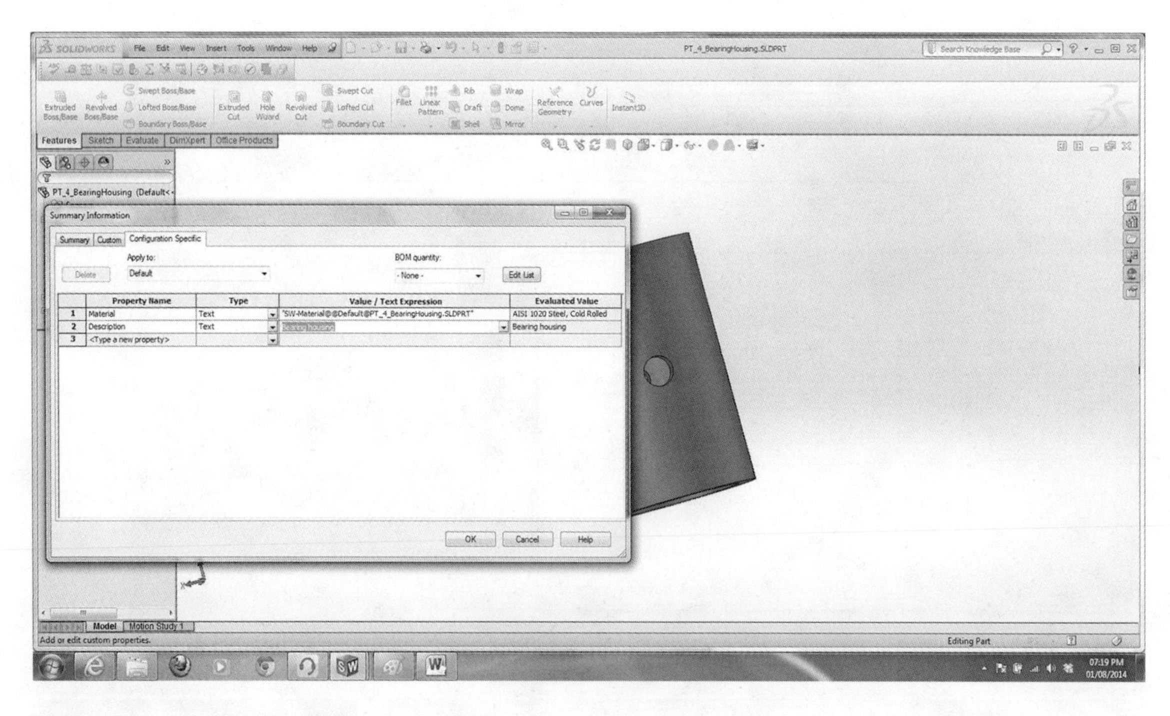

Figure 9.37

Bearing housing configuration-specific information.

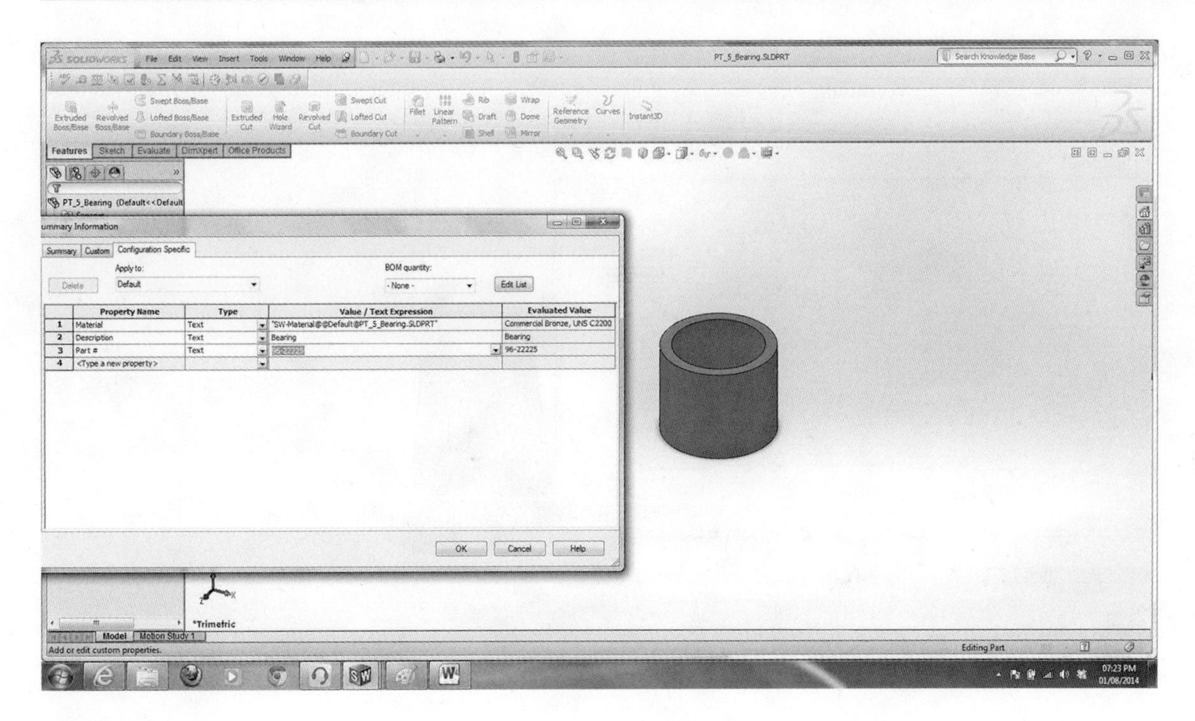

Figure 9.38

Bearing configuration-specific information.

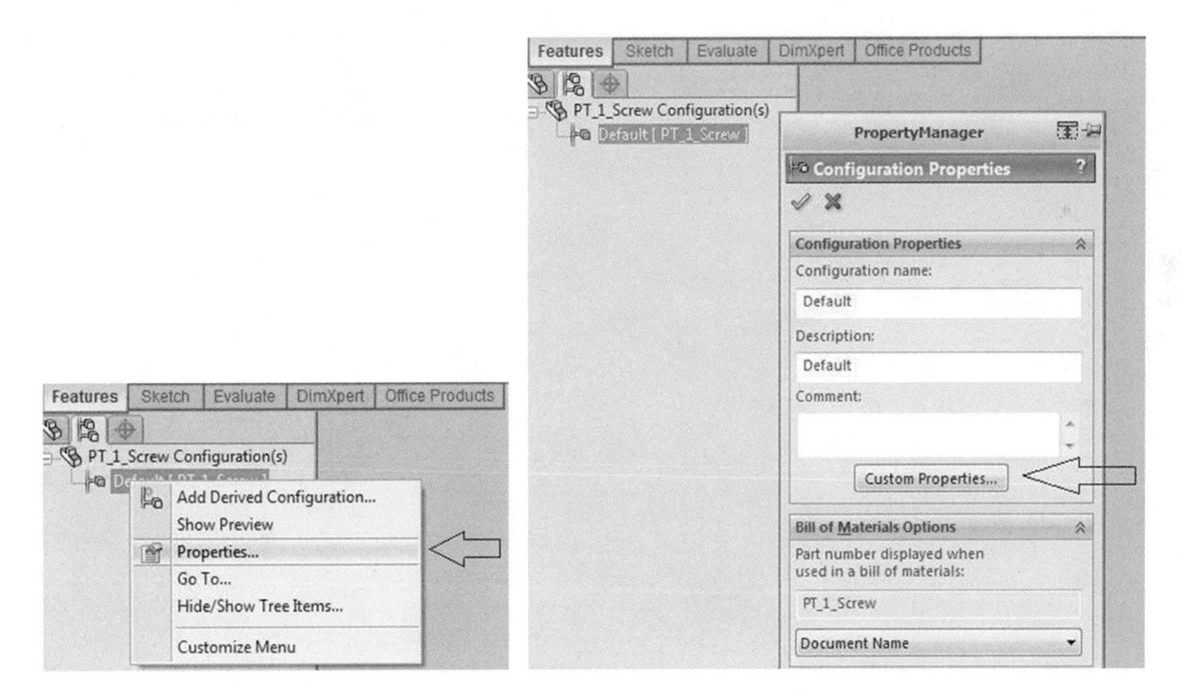

Figure 9.39

Accessing properties configuration for specific information.

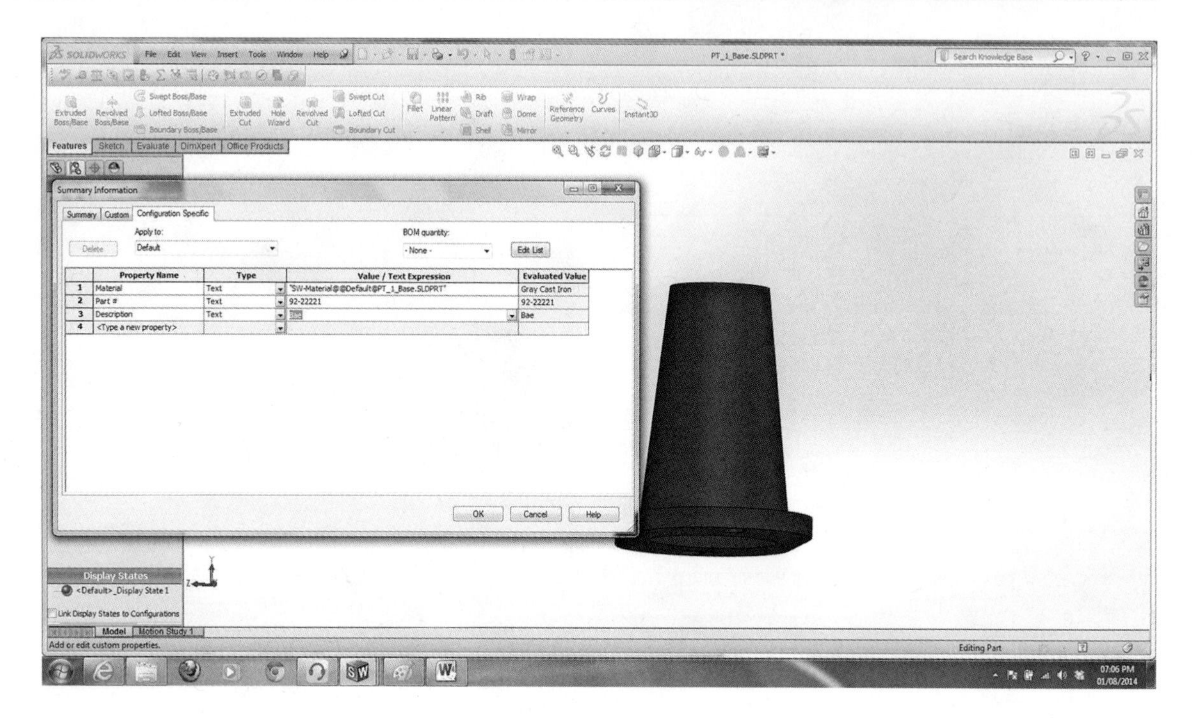

Figure 9.40

Screw configuration-specific information.

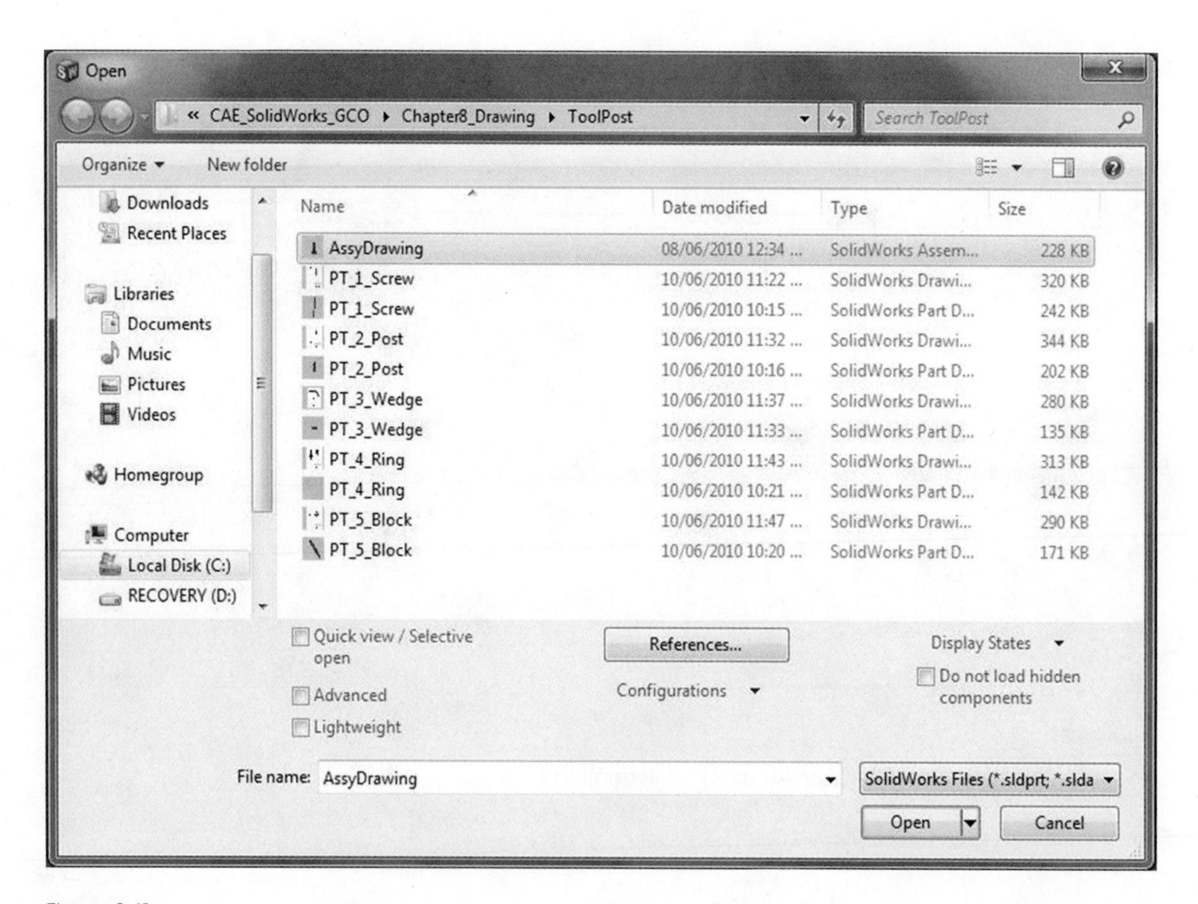

Figure 9.41

Opening the assembly.

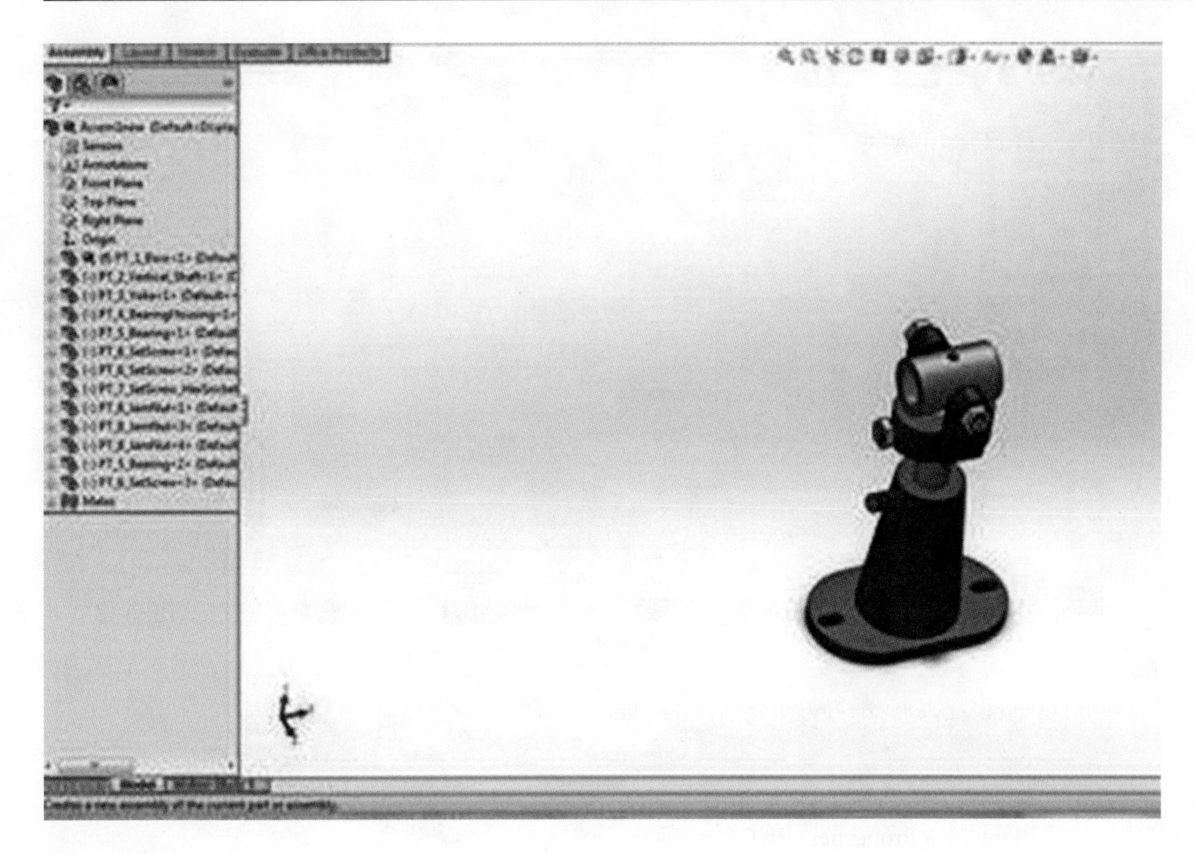

Figure 9.42

Making a drawing from an assembly.

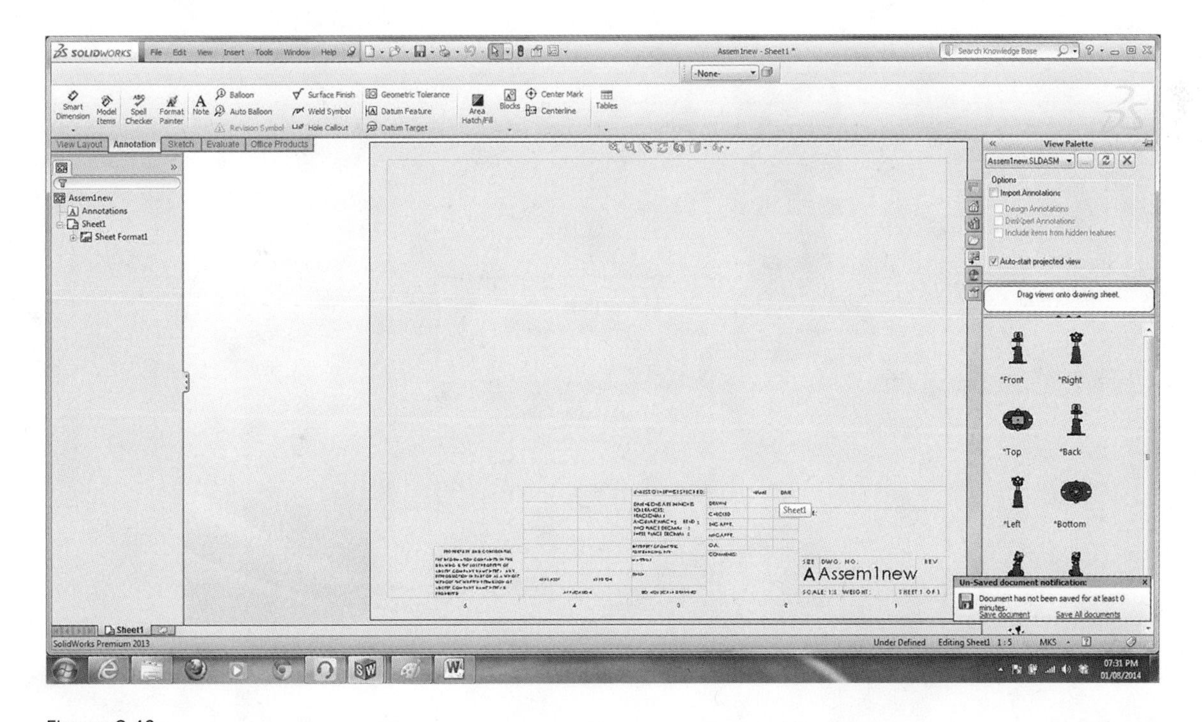

Figure 9.43

View Palette tool.

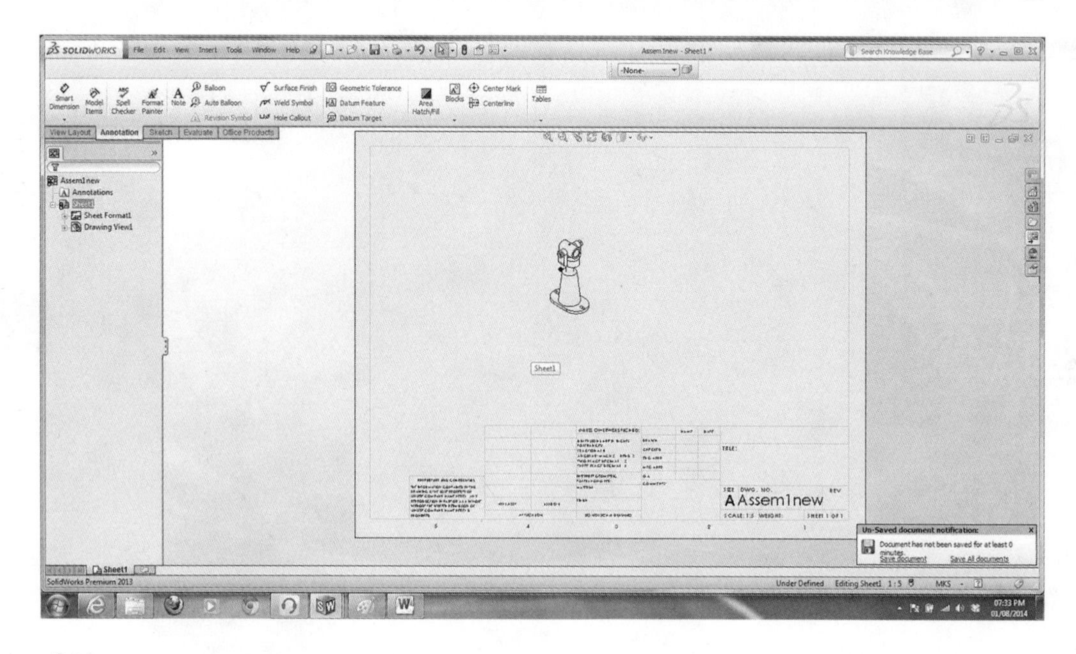

Figure 9.44

Imported assembly by clicking and dragging view from View Palette.

Display the Exploded View

1. Click inside the Isometric view boundary.
2. Right-click Properties (see Figure 9.45).
3. Click Show in exploded state (see Figure 9.46). (Note: the Exploded View, ExplView1, would have already been created in Assembly-Explode View for it to be visible in the exploded state option [see Figure 9.47].)
4. Click OK > OK.
5. Save As Exploded on file. (You need to click Save As to have a copy of the exploded view.)

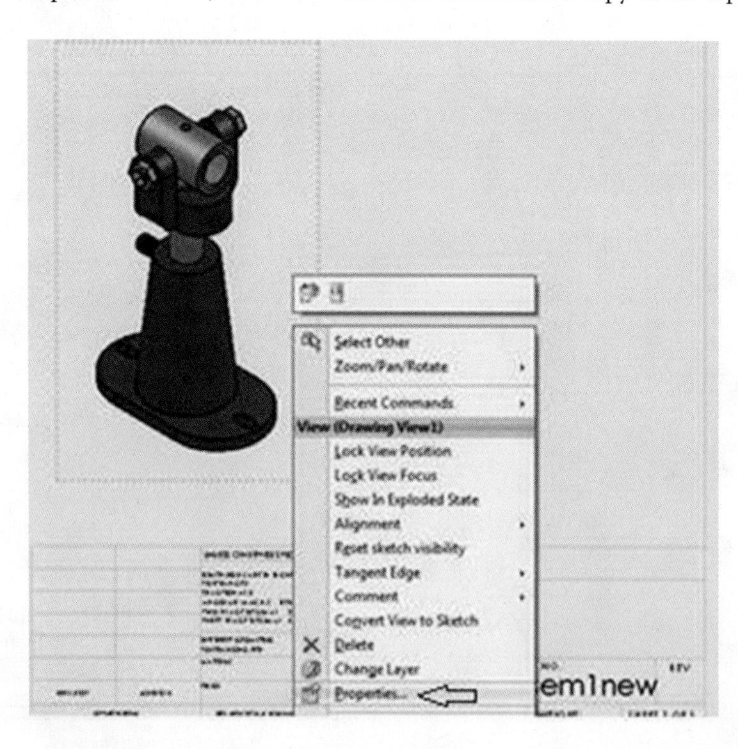

Figure 9.45

Accessing Assembly Properties.

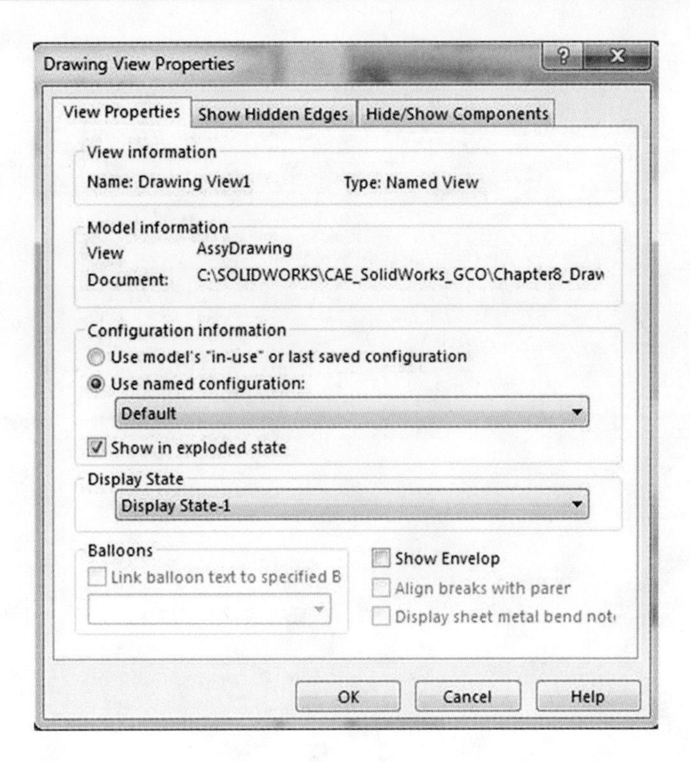

Figure 9.46

Checking Show in exploded state.

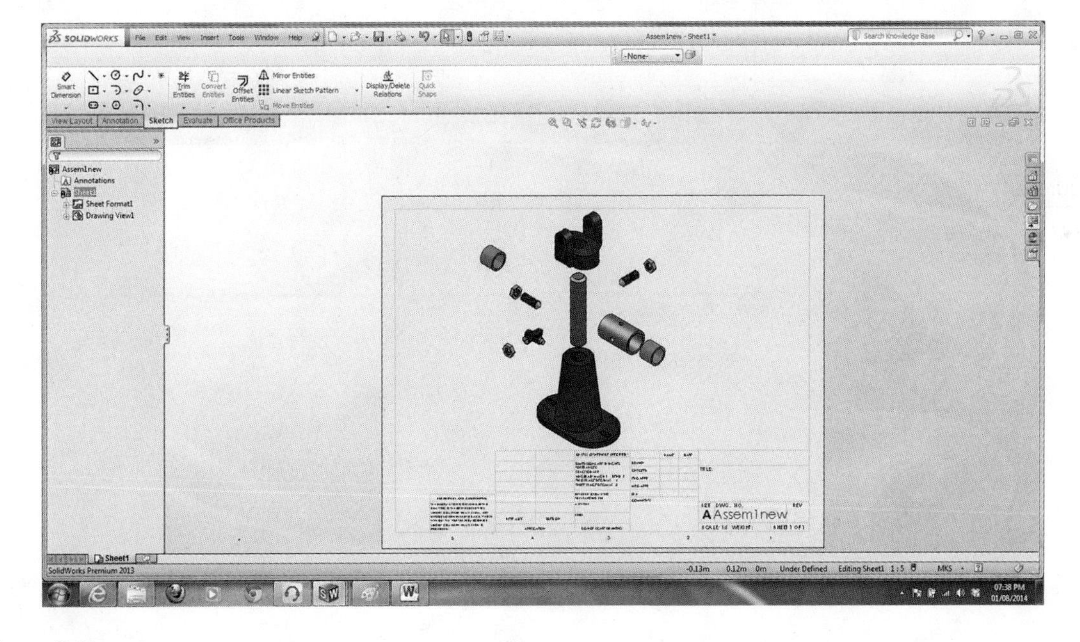

Figure 9.47

Exploded view of tool post assembly.

Balloons

1. Click inside the isometric view boundary.
2. Click Auto Balloon from Annotation toolbar (see Figure 9.48).
3. Click OK.
4. Select Balloon Settings > Select Circular Split Line (see Figure 9.49).
5. Click OK (see Figure 9.50 for the assembly balloon).
6. Save.

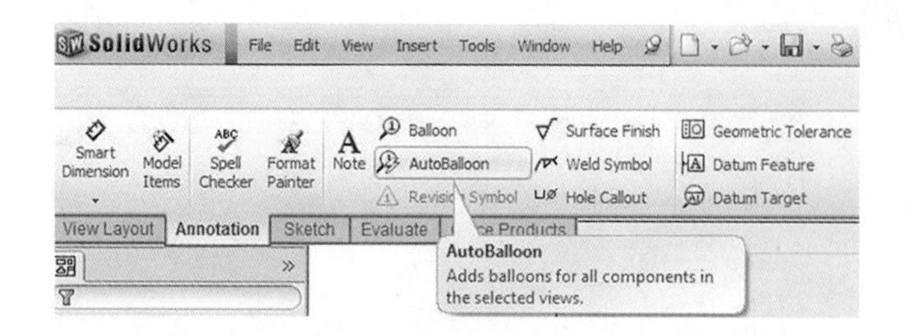

Figure 9.48

Auto Balloon tool.

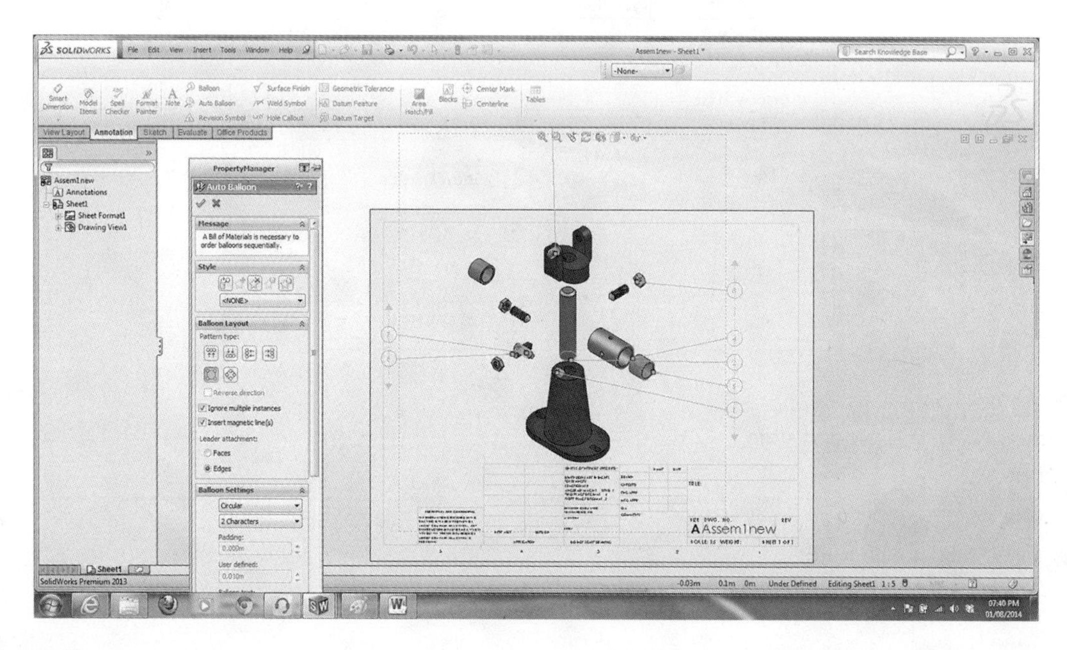

Figure 9.49

Auto Balloon options selected.

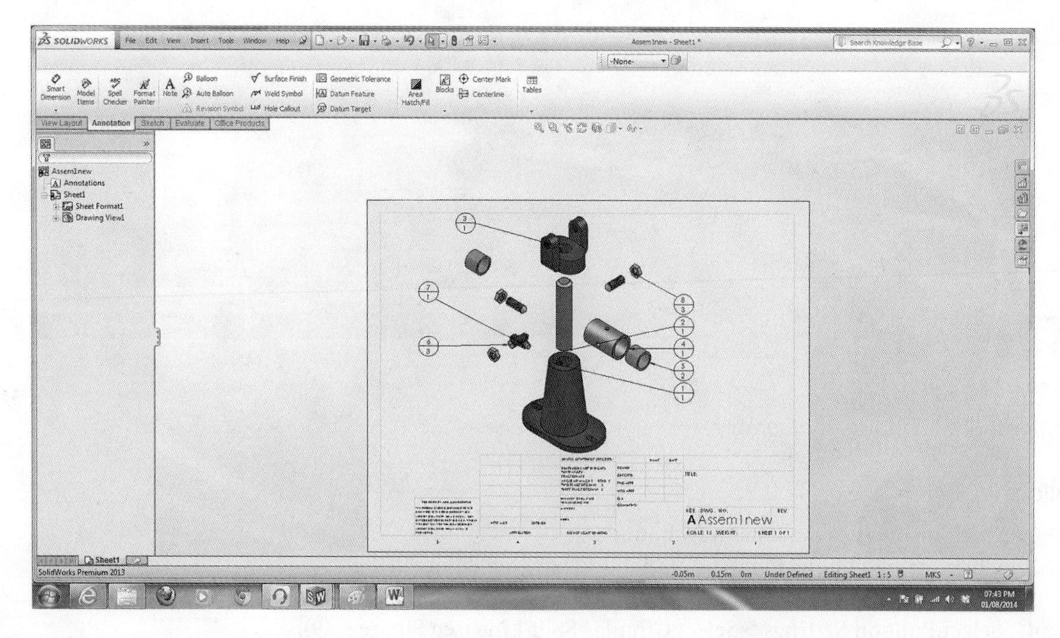

Figure 9.50

Auto Balloon of the assembly.

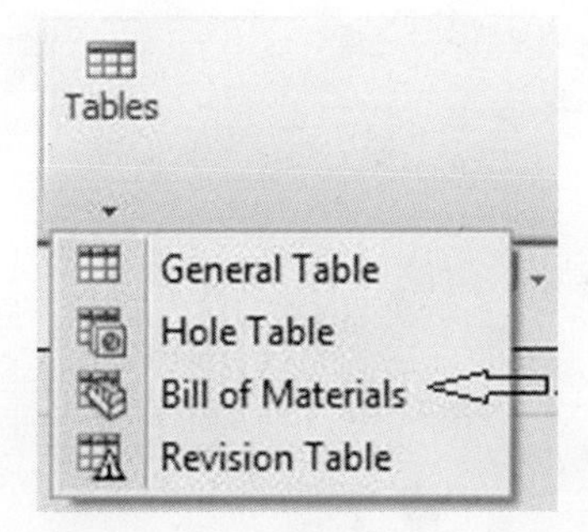

Figure 9.51

BoM option.

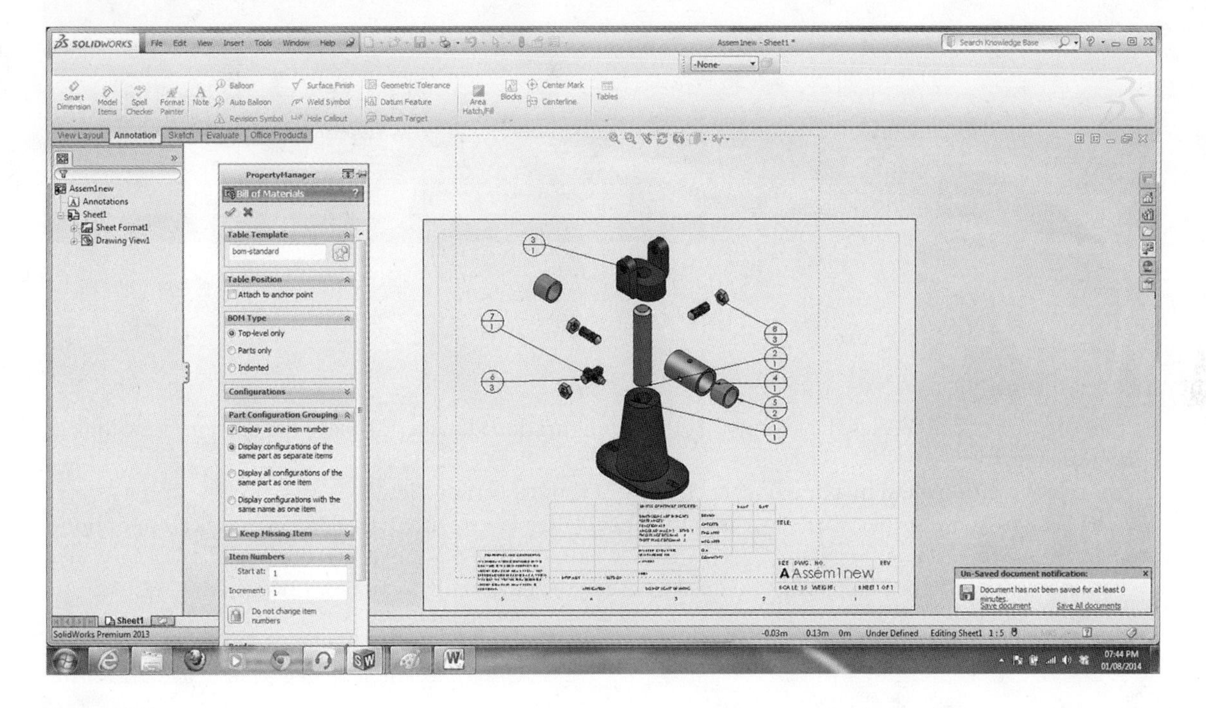

Figure 9.52

BoM PropertyManager.

BoM

1. Click inside the Isometric view boundary (exploded view).
2. Click Annotation > Tables > Bill of Materials (see Figure 9.51 for the BoM option and Figure 9.52 for the PropertyManager).
3. Click a position on any point within the graphics window and click the Return key; the BoM is shown in Figure 9.53.

Inserting Section Views

SolidWorks 2013 now addresses the customers' need to include sectional views other than only the full section, which it has supported previously. This means that the enhanced SolidWorks 2013 will later compete with other computer-aided design (CAD) packages that support the options of section views other than the full section. This chapter discusses the following additional features of sectional views:

- Full-section view
- Half-section view
- Notched offset-section view
- Aligned-section view

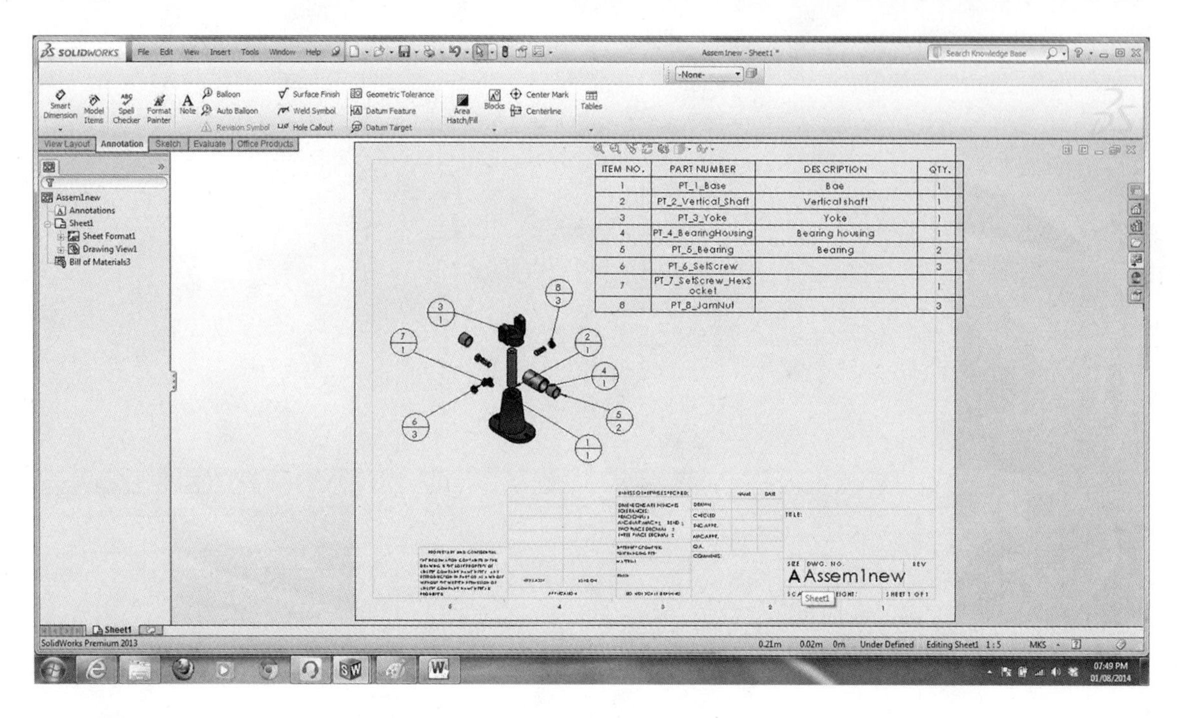

Figure 9.53

BoM of the tool post assembly.

Inserting a Section View

Ensure that you click the View Layout so that the ConfigurationManager displays the Section View tool, as shown in Figure 9.54.

To insert a section view, in a drawing, click View Layout > Section View.

(The Section View PropertyManager is automatically displayed with Half-Section and Section options; see Figure 9.55.)

Let us now discuss the two options that users have to select: (1) half-section or (2) section.

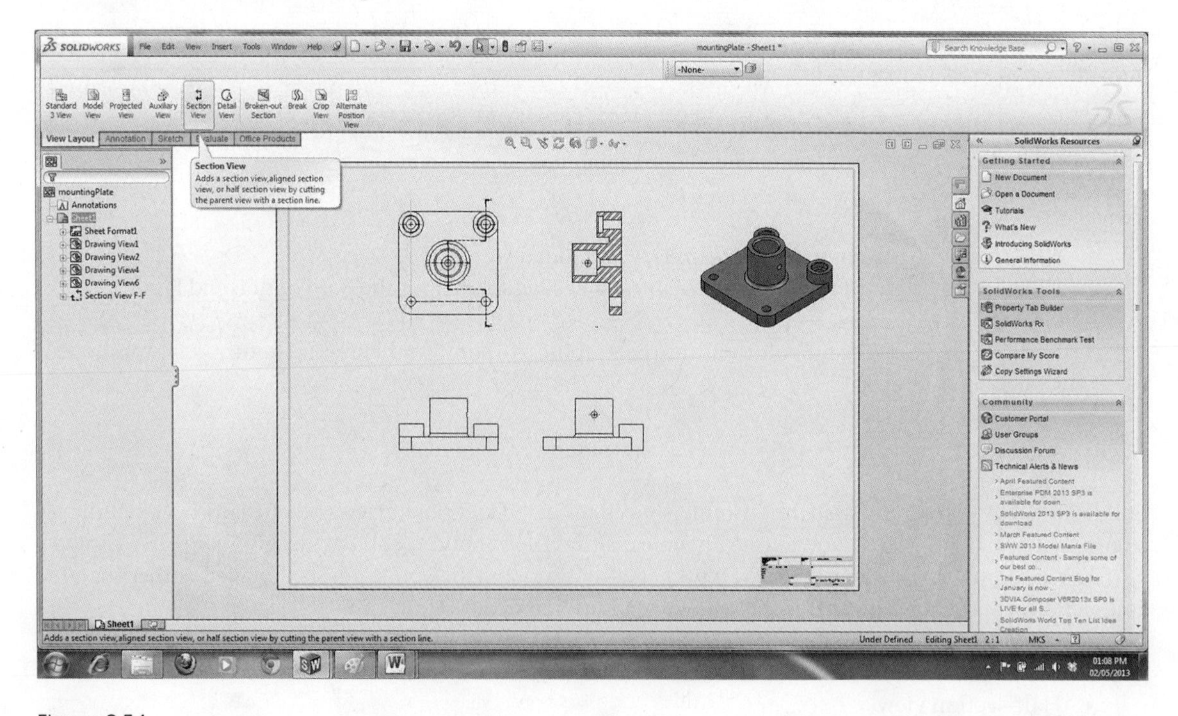

Figure 9.54

View Layout ConfigurationManager showing Section View tool.

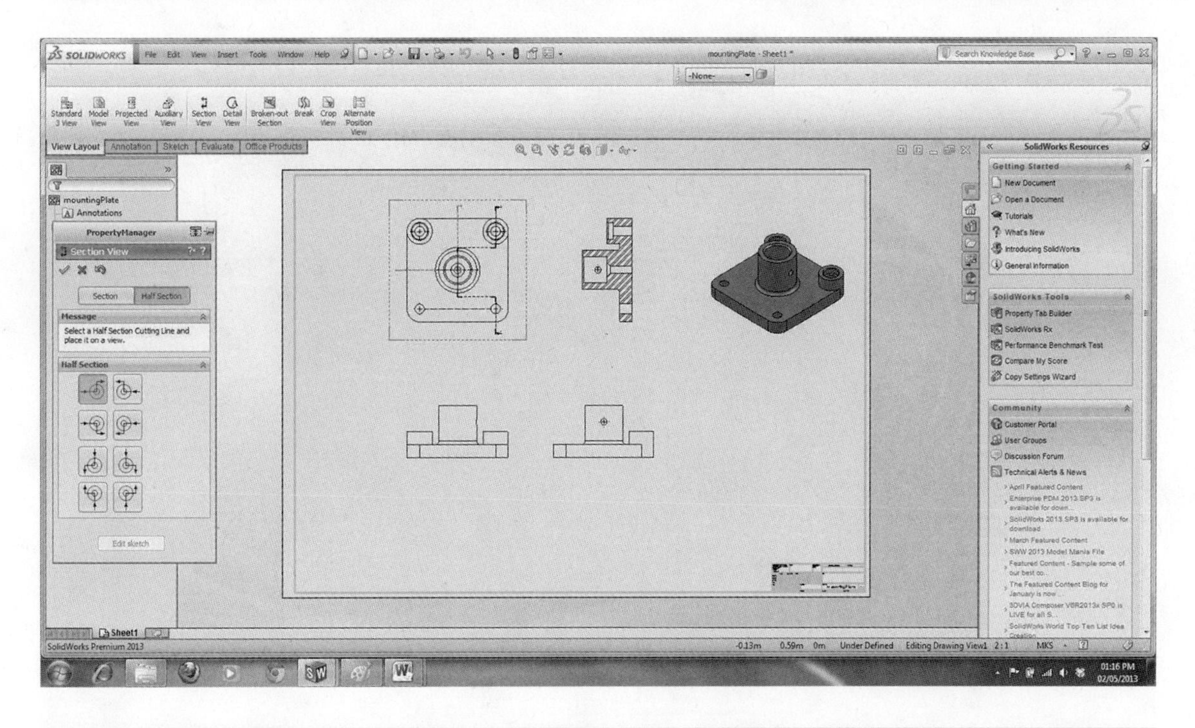

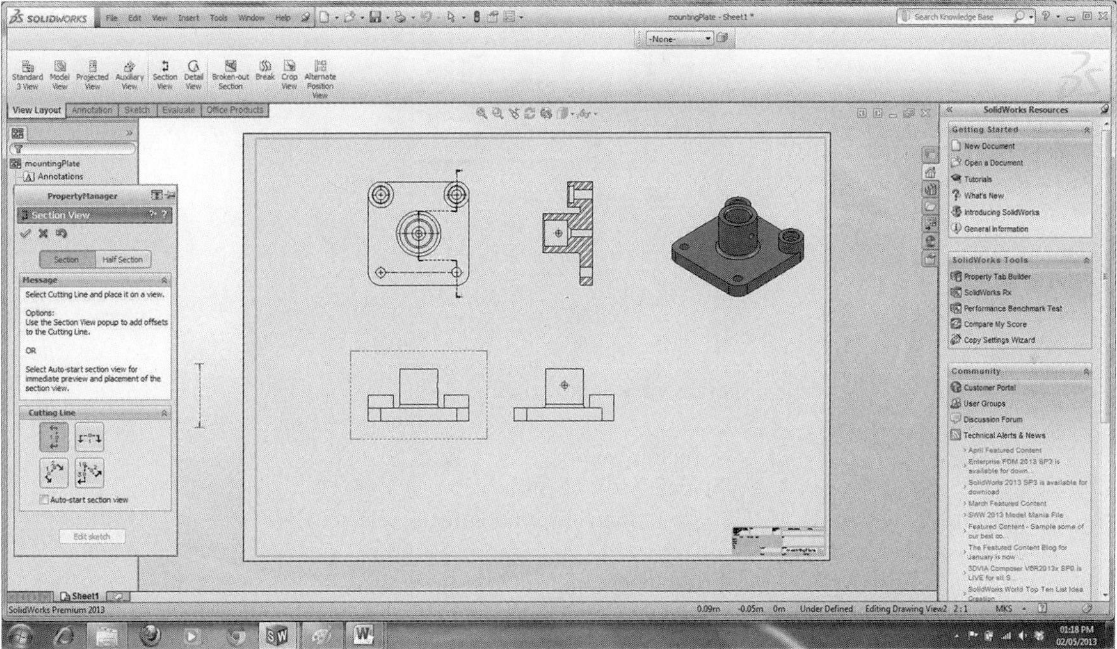

Figure 9.55

Section View PropertyManager.

Half-Section View

When the Half-Section option is selected from the Section View PropertyManager (Figure 9.55), a number of configurations appear from which the user selects (see Figure 9.56).

Select the type of half-section.

Section View

When the Section option is selected from the Section View PropertyManager (Figure 9.55), a number of Cutting Line configurations appear from which the user selects (see Figure 9.57).

Select the cutting line.

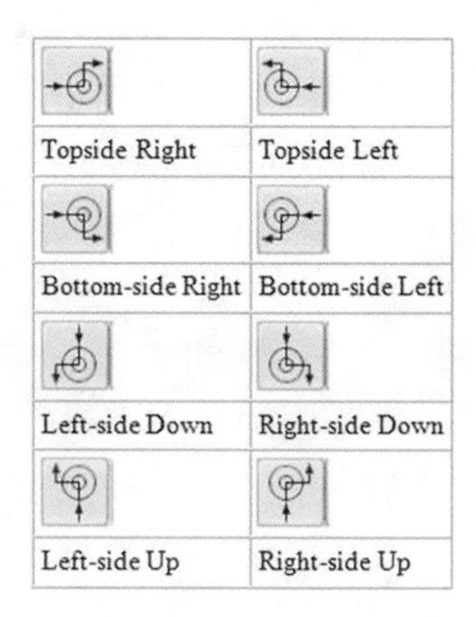

Figure 9.56

Half-section options.

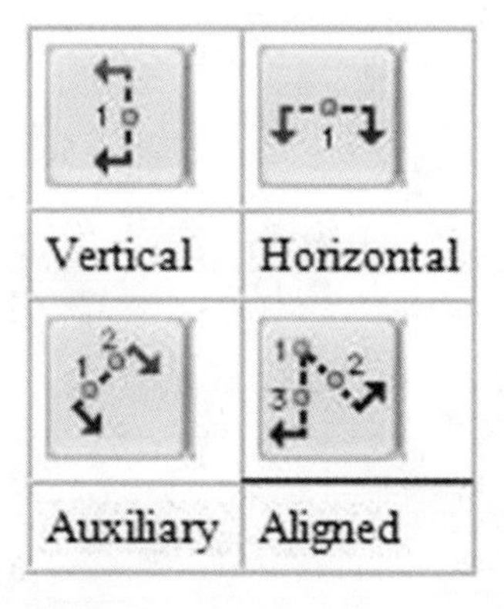

Figure 9.57

Cutting line options.

For example, to insert a vertical section view, follow these steps:

1. Click Section View (Drawing toolbar).
2. In the Section View PropertyManager, click Section.
3. In Cutting Line, select Auto-start section view if not already selected.
4. Click Vertical and move the cutting line to the location as shown and click.
   If the Section View dialog box appears, click OK.
5. Drag the preview to the right and click to place the section view.

After the user selects and places the cutting line in the drawing view, the Section View popup displays (see Figure 9.58).

Use the Section View popup to add offsets to the section view (see Figure 9.59).

If you select Auto-start section view in the Section View PropertyManager, the Section View popup does not appear, allowing you to immediately preview and place the section view in the drawing.

- You can add multiple offsets to a section view. Notch offset may be applied to any cutting line segment. Single offset and Arc offset may only be applied to one of the two outer cutting line segments.
- The cutting line inferences to the drawing's geometry.

Now that we have discussed how to go about sectioning, some examples will now follow in order to clarify the concepts. Sectioning is now made very easy in the enhanced SolidWorks CAD software, which now competes with other CAD software, such as Inventor, for this particular functionality.

Figure 9.58

Section View popup.

Selection	Function	Additional Steps
	Add Arc Offset	Select first point of arc on cutting line, then select second point of arc.
	Add Single Offset	Select first point of offset on cutting line, then select second point of offset.
	Add Notch Offset	Select first point of notch on cutting line, select second point on cutting line for width of notch, then select third point for depth of notch.
	Step Back	
	OK (Add the view)	
	Cancel (Cancel the view)	

Figure 9.59

Using the Section View popup to add offsets to the section view.

Example A: Half-Section

Create a half-section for the model that is shown in Figure 9.60.

Part name: Rod support
Material: 6061-T6 aluminum
Fillets: R.03 unless otherwise specified
SolidWorks Solution

Figure 9.61 shows the partial model dimensions for the necessary sketches, whereas Figure 9.62 shows the solid model. These steps are not part of sectioning but are the prerequisites to produce the model that is required, based on the initial dimensions that are given in Figure 9.60. In other words, these prerequisite steps are useful to users to consolidate their 3D modeling experience. You may skip these steps depending on your level of model capability.

Now, let us proceed to the main steps that are required for the sectioning of this model.

1. 

Click Make Drawing from Part/Assembly.

2. Pin the View Palette.
3. Drag and Drop Views.

To create a half-section view, follow these steps:

1. Highlight the bounding box enclosing the view of interest.
2. Click View Layout > Section View.
3. In the Section View PropertyManager, click Half Section (see Figure 9.63).

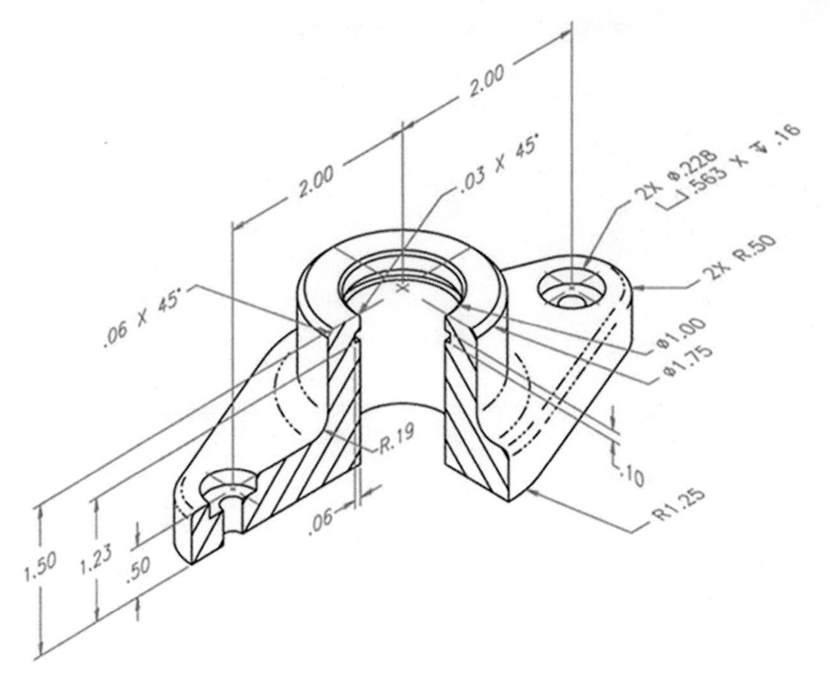

Figure 9.60

Model description for half-sectioning.

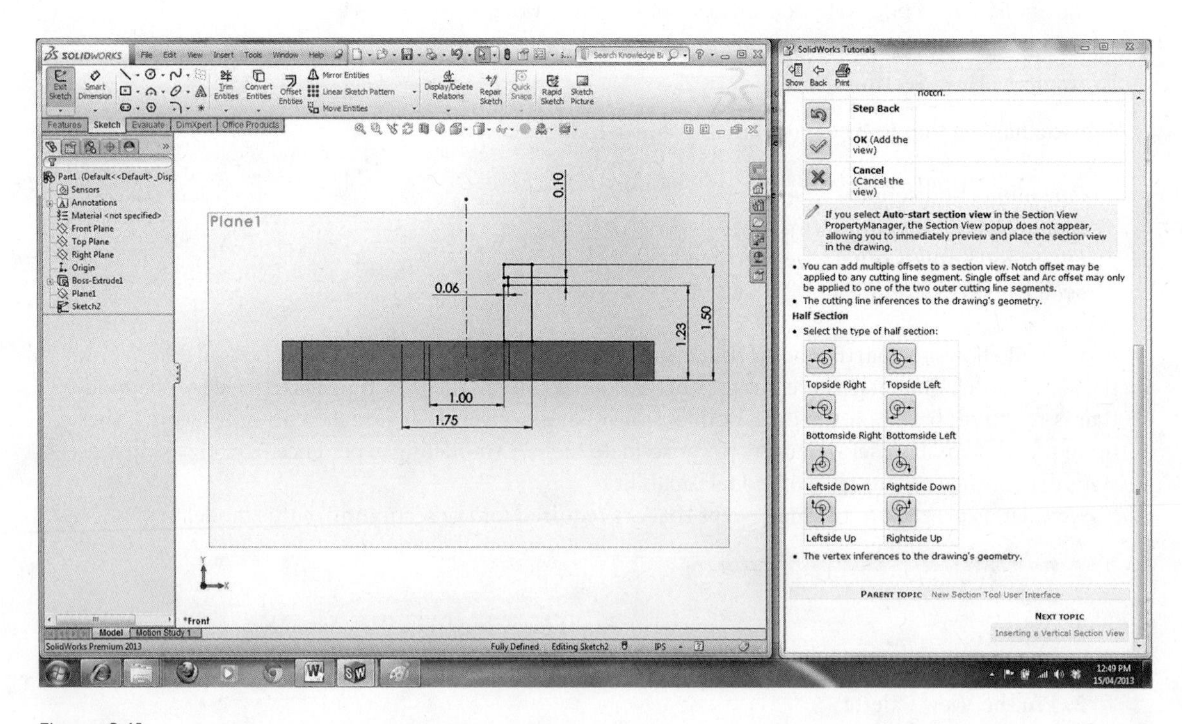

Figure 9.61

Partial model dimensions for the necessary sketches.

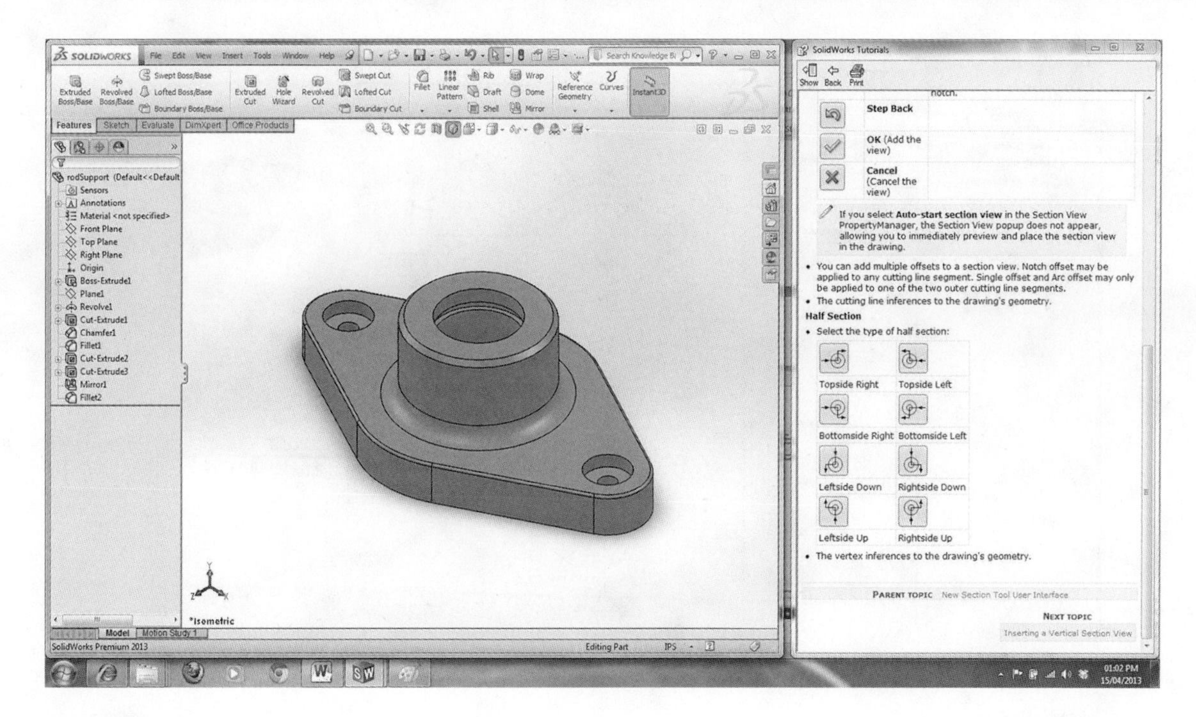

Figure 9.62

Solid model for sectioning.

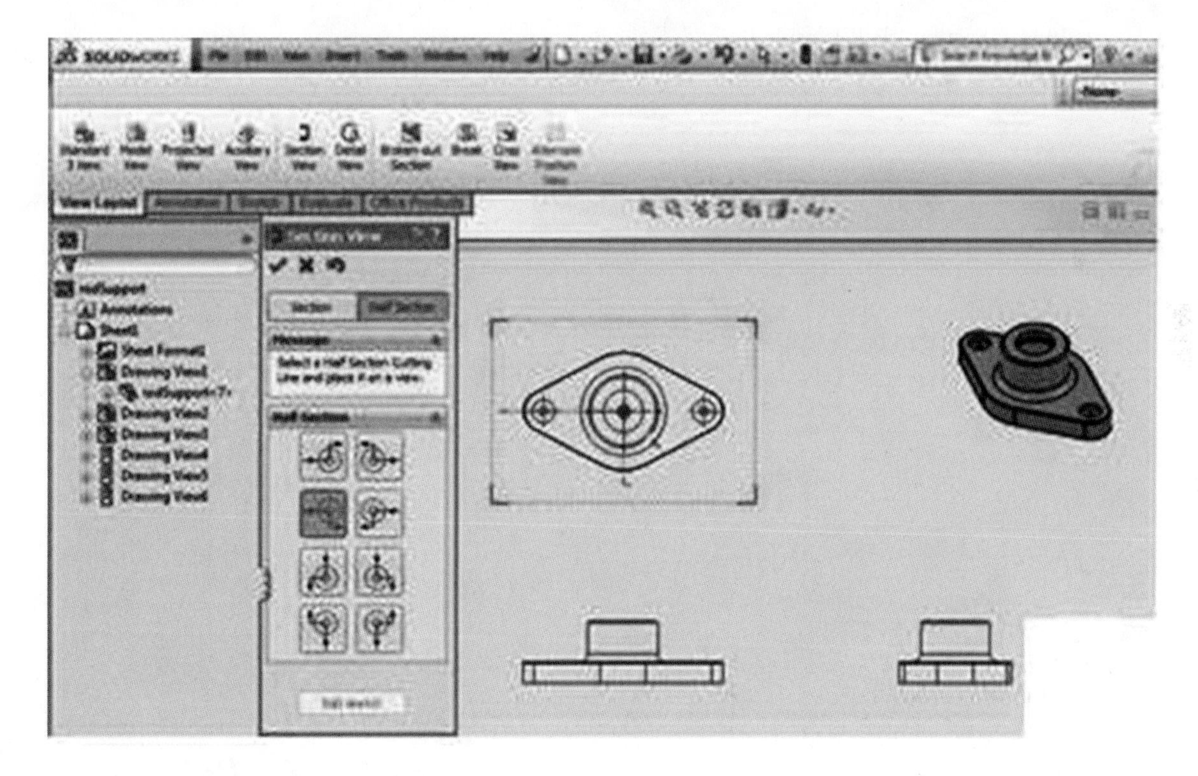

Figure 9.63

Section View option chosen.

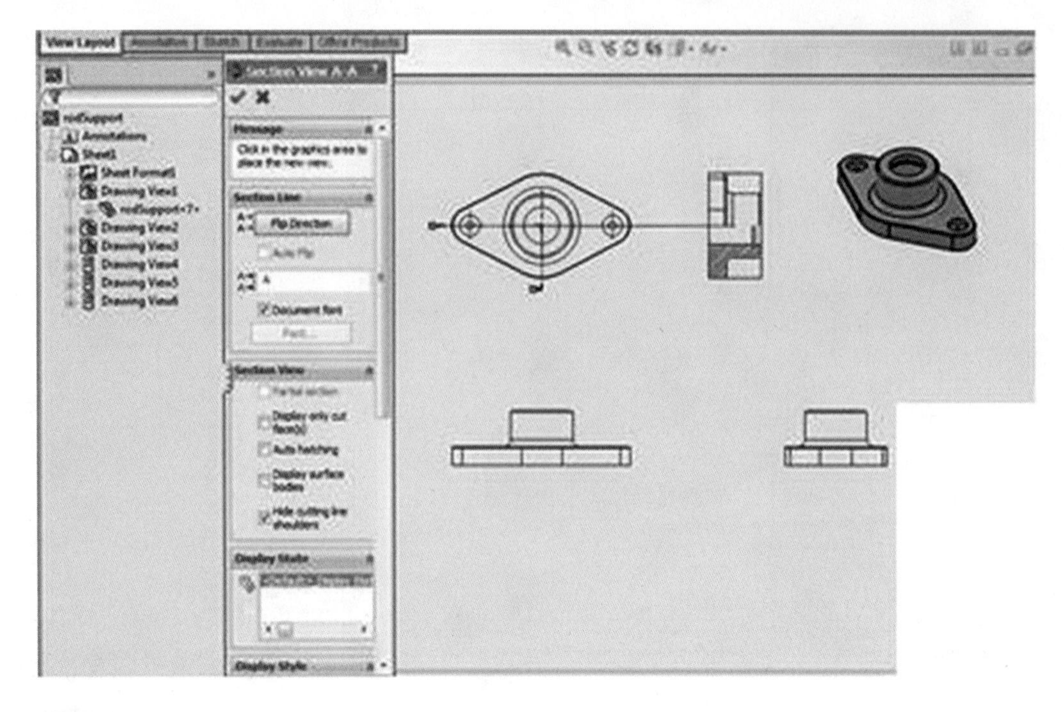

Figure 9.64

Move the pointer to the location as shown and click to place it.

4. In Half Section, click Topside Right.
5. Move the pointer to the location, as shown, and click to place it (see Figure 9.64).
6. Drag the preview to the right and click to place the section view (see Figure 9.65).

Example B: Notched Offset-Section View

Create a half-section for the model that is shown in Figure 9.66.

Part name: Mounting plate
Material: American Iron Steel Institute (AISI) 1020
Fillets: R.03 unless otherwise specified
SolidWorks Solution

Figure 9.67 shows the 3D model.

1. 

Click Make Drawing from Part/Assembly.
2. Pin the View Palette.
3. Drag and Drop Views.

To insert a section view with a notched offset,

1. Click View Layout > Section View.
2. In the Section View PropertyManager, click Section (see Figure 9.68).
3. In Cutting Line, clear Auto-start section view. This eliminates the automatic insertion of the section view and lets you add additional offsets to the view.
4. Click Vertical and move the cutting line to the location, as shown, and click to place the line.

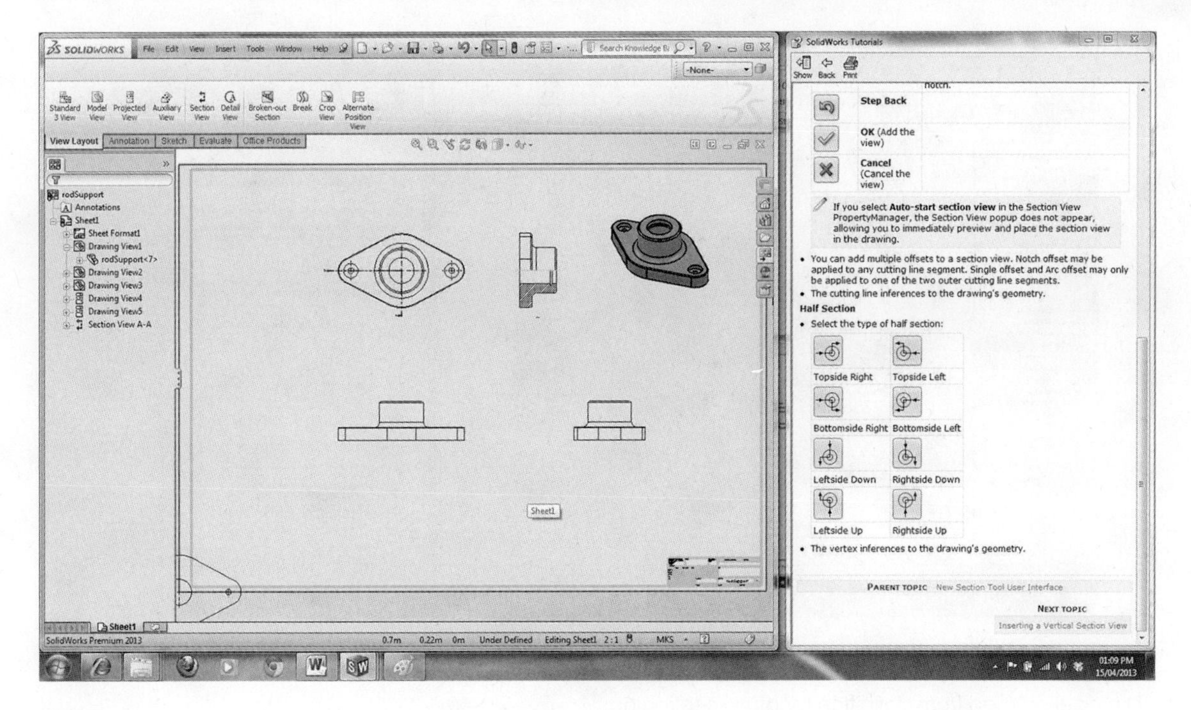

Figure 9.65

Drag the preview to the right and click to place the section view.

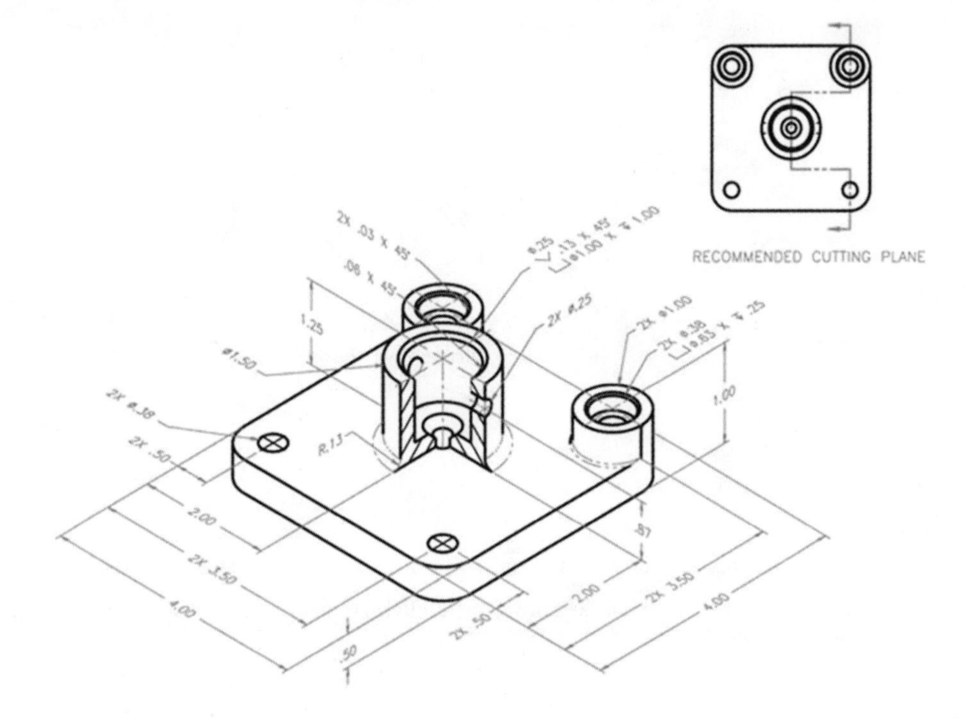

Figure 9.66

Model description for offset sectioning.

5. Click to add a notched offset (see Figure 9.69).
6. Move the pointer to the location, as shown, and click to select the first and second points of the notch. These points must be on the cutting line.
7. Move the pointer to the location, as shown, and click to select the depth of the notch. The Section View popup appears. At this point, you could add additional offsets to the view (see Figure 9.70).

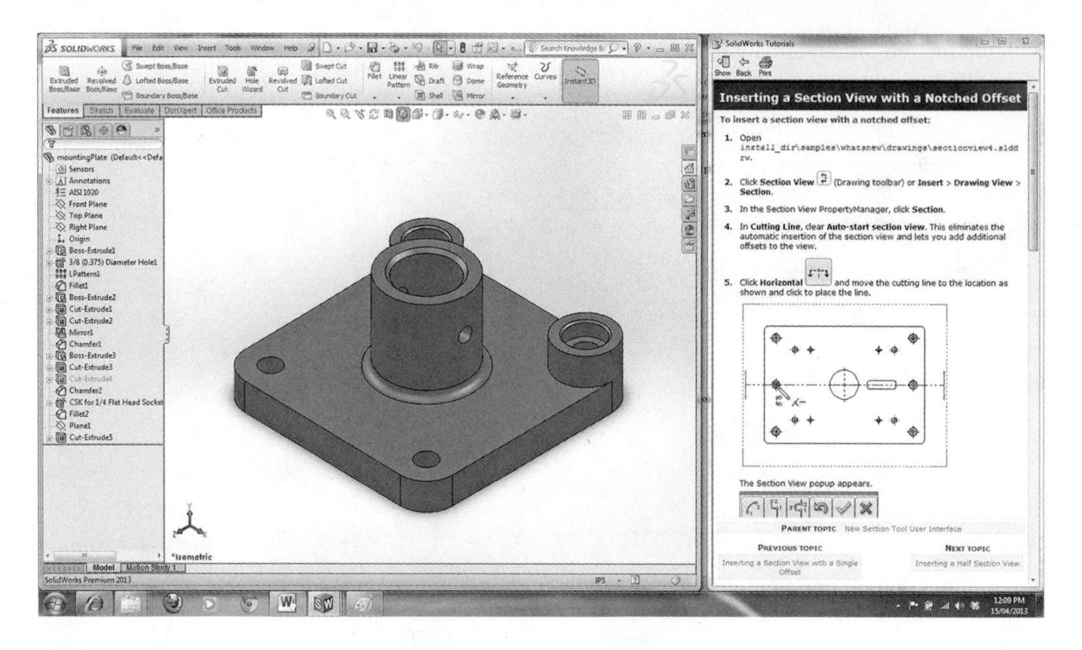

Figure 9.67

3D model created from definition for offset sectioning.

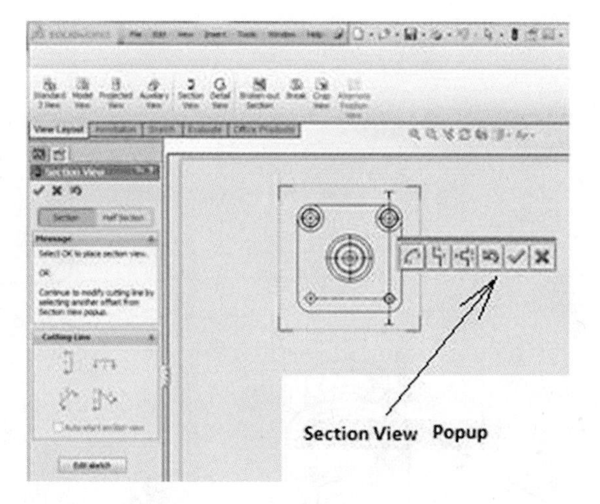

Figure 9.68

The Section View popup appears.

8. Click [checkmark icon] to close the Section View popup.
9. Drag the preview to the location, as shown, and click to place the section view. (Figure 9.71 shows the notched offset.)

Example C: Aligned Section View

Create an aligned section for the model that is shown in Figure 9.72.

Part name: Hub
Material: SAE 3145
Fillets: R.03 unless otherwise specified
SolidWorks Solution

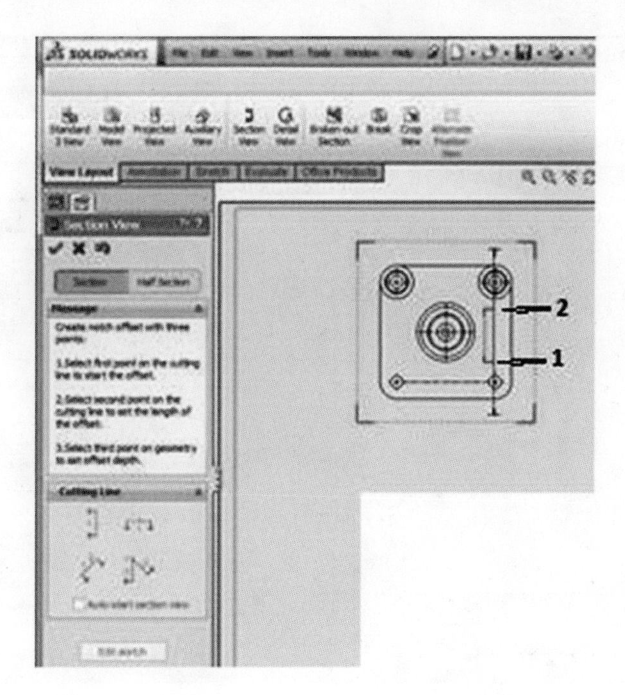

Figure 9.69

Notched offset is added.

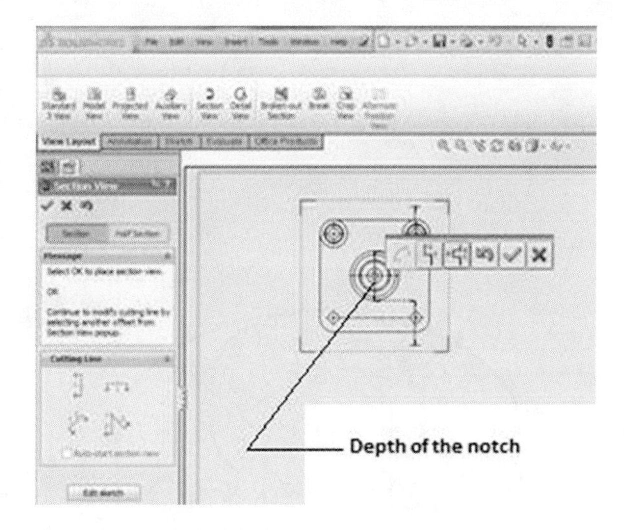

Figure 9.70

Select the depth of the notch.

Figure 9.73 shows the 3D model.

1. Click Make Drawing from Part.
2. Pin the View Palette.
3. Drag and drop the views (see Figure 9.74).
4. Click Drawing View1 bounding box containing the Top View (see Figure 9.75).
5. Click View Layout > Section View.
6. Select Section.

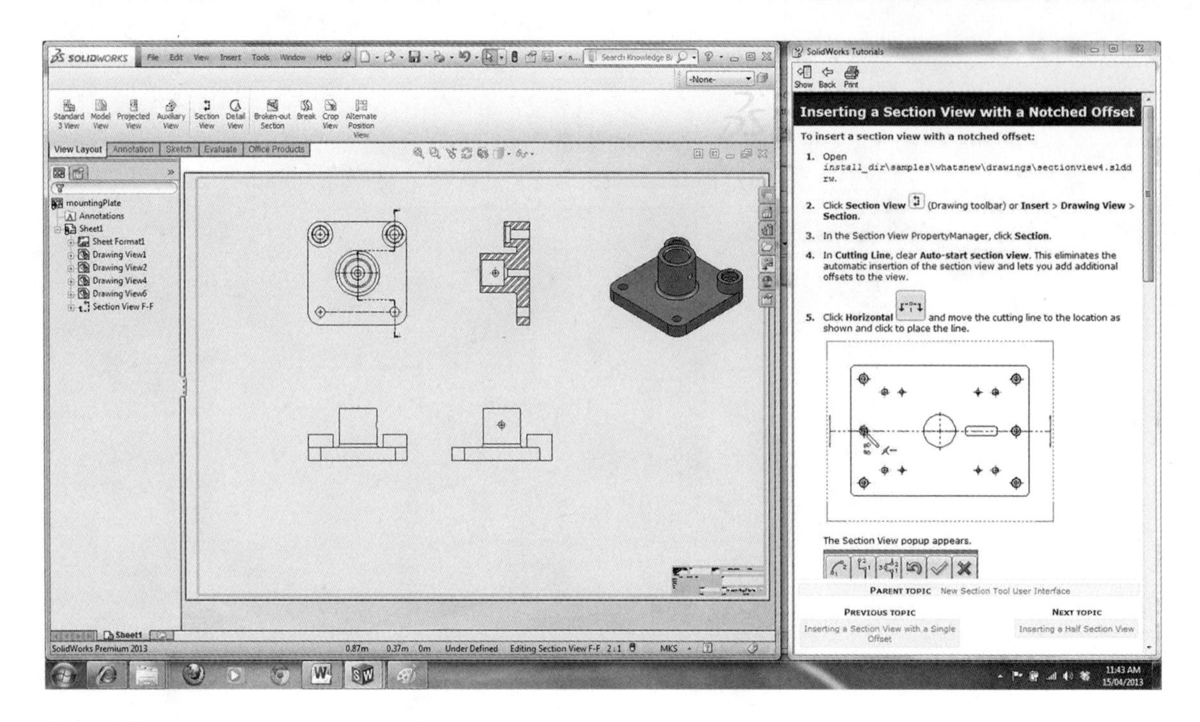

Figure 9.71

Notched offset.

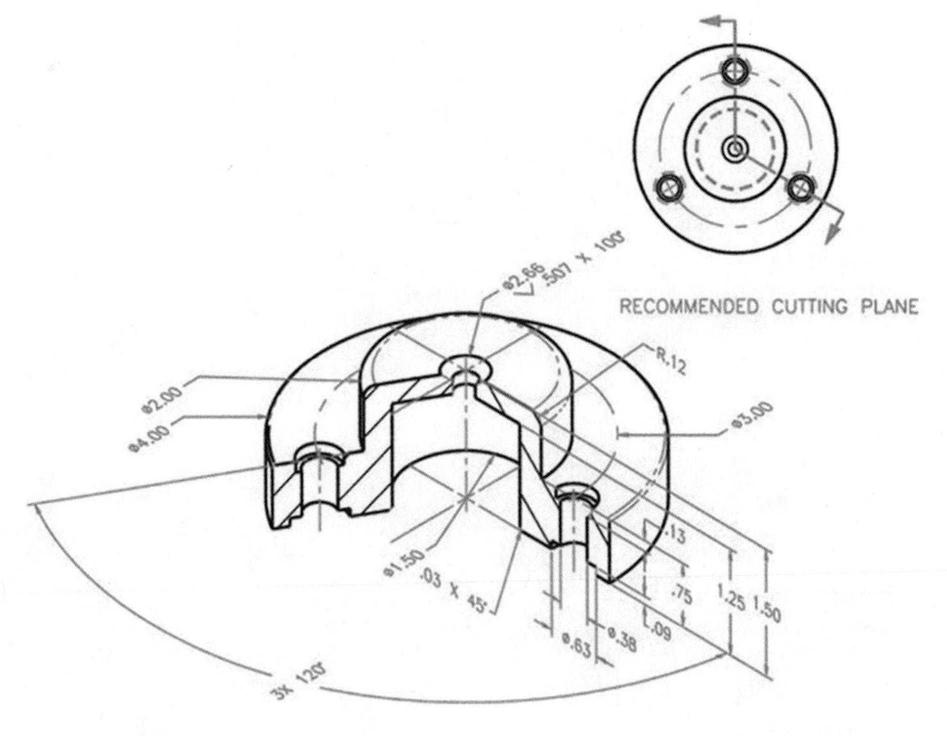

Figure 9.72

Model description for offset sectioning.

7. Click Cutting Line > Aligned (see Figure 9.76a).
8. Select the center of the central circle as 1, the center of the vertical circle as 2, and the center of the bottom-right circle as 3 (see Figure 9.76b).
9. Click OK.
10. Drag Aligned View to the right to complete (see Figure 9.77).

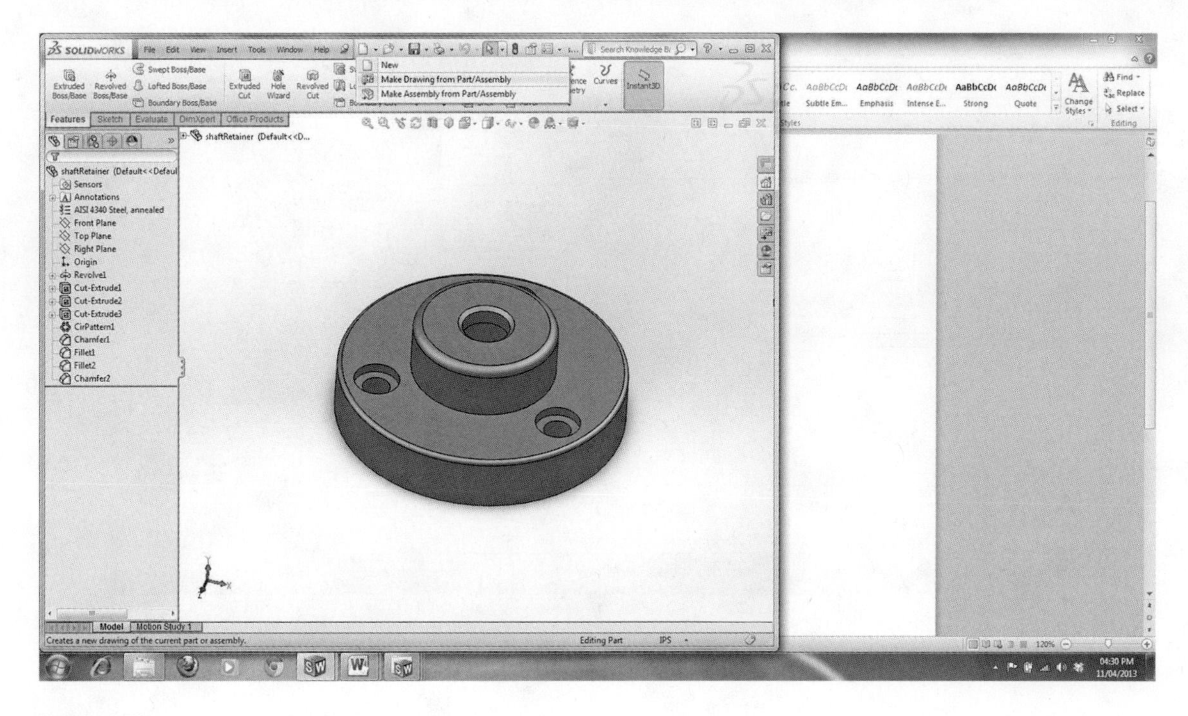

Figure 9.73

3D Model created from definition for aligned sectioning.

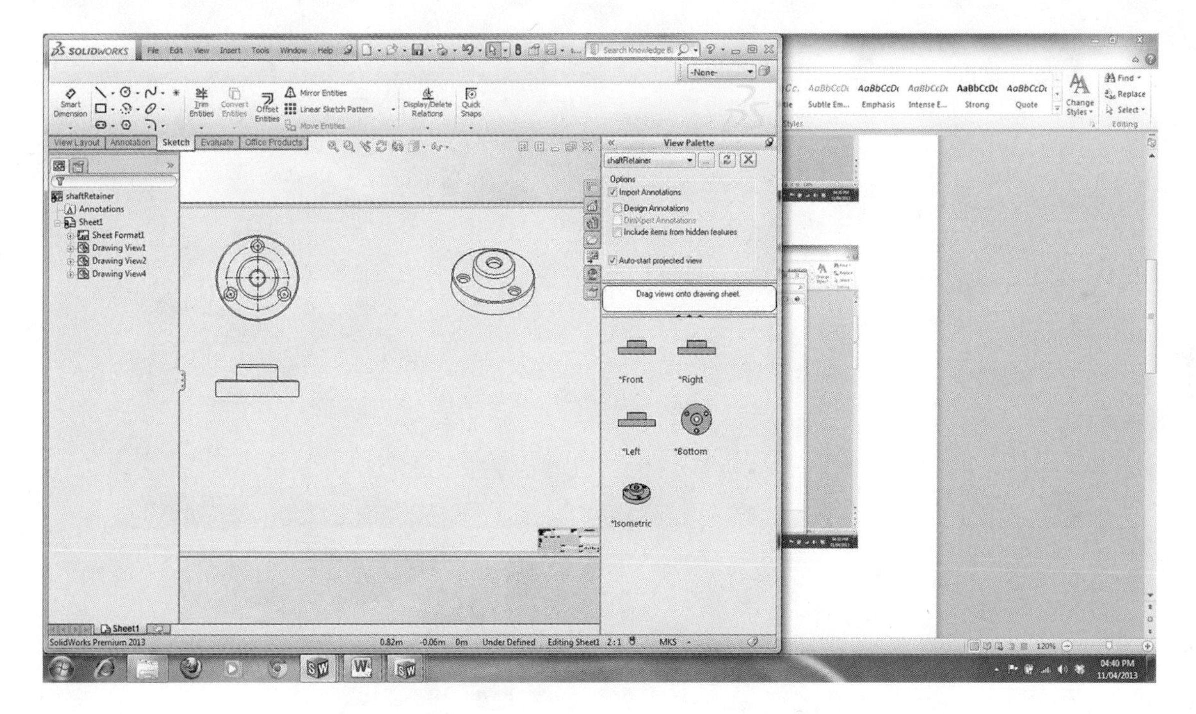

Figure 9.74

Drag and drop the views.

Example D: Full Section

Create a half-section for the model that is shown in Figure 9.60.

Part name: Rod support
Material: 6061-T6 aluminum
Fillets: R.03 unless otherwise specified

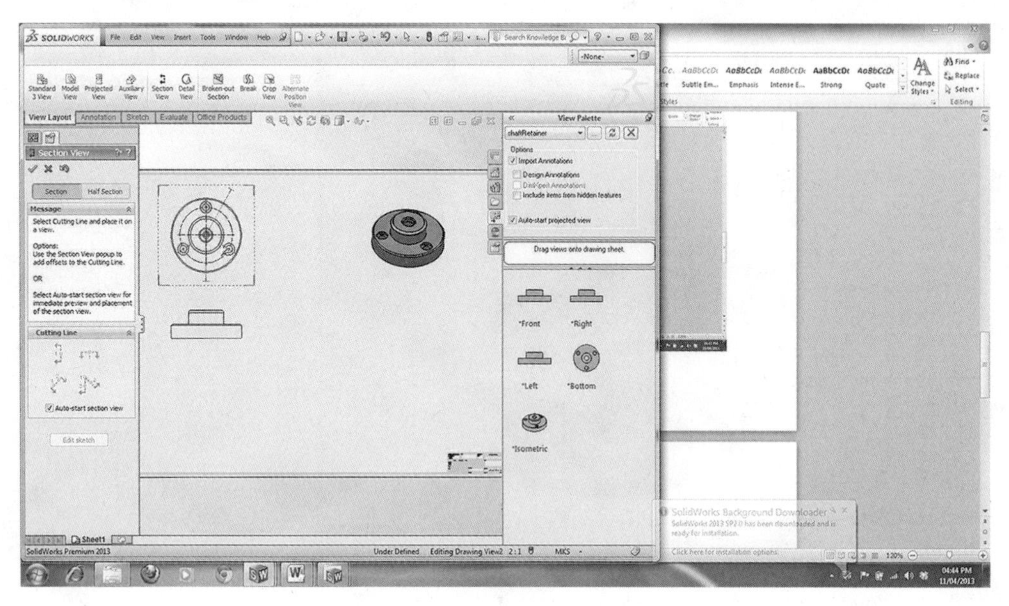

Figure 9.75

Click *Drawing View1* bounding box containing the Top View.

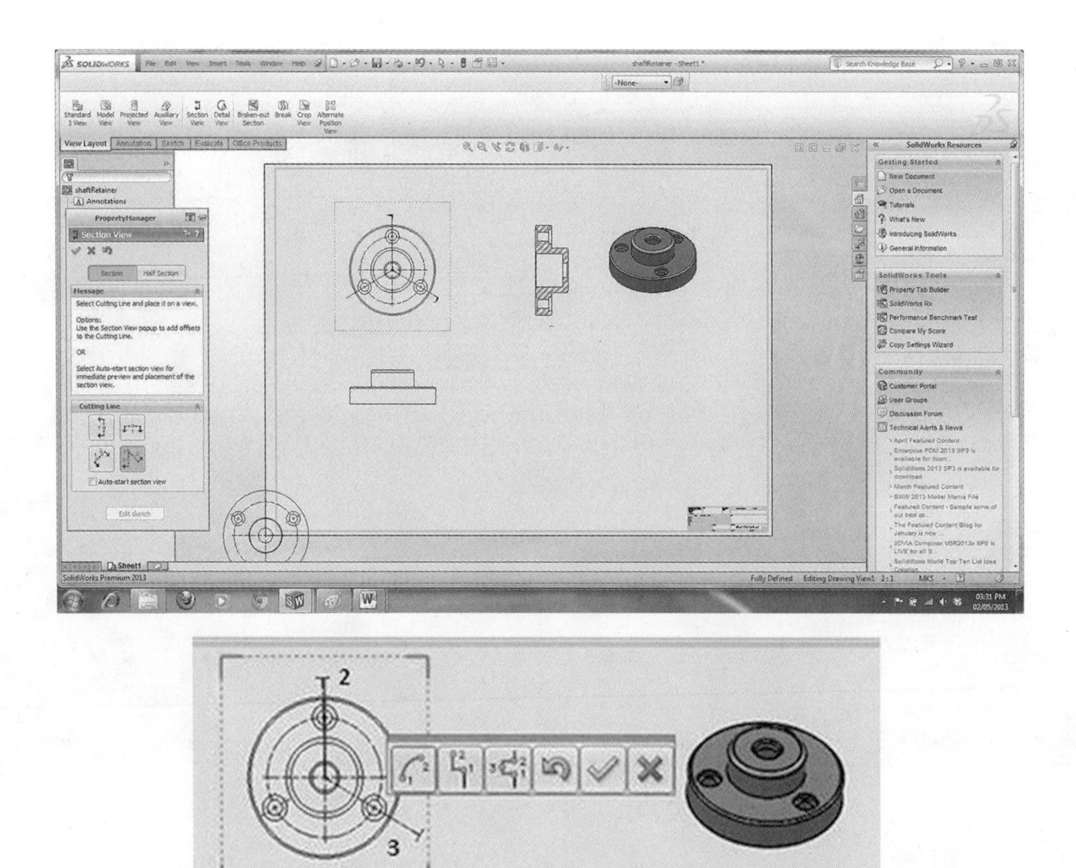

Figure 9.76

Click *Cutting Line* and select *Aligned.*

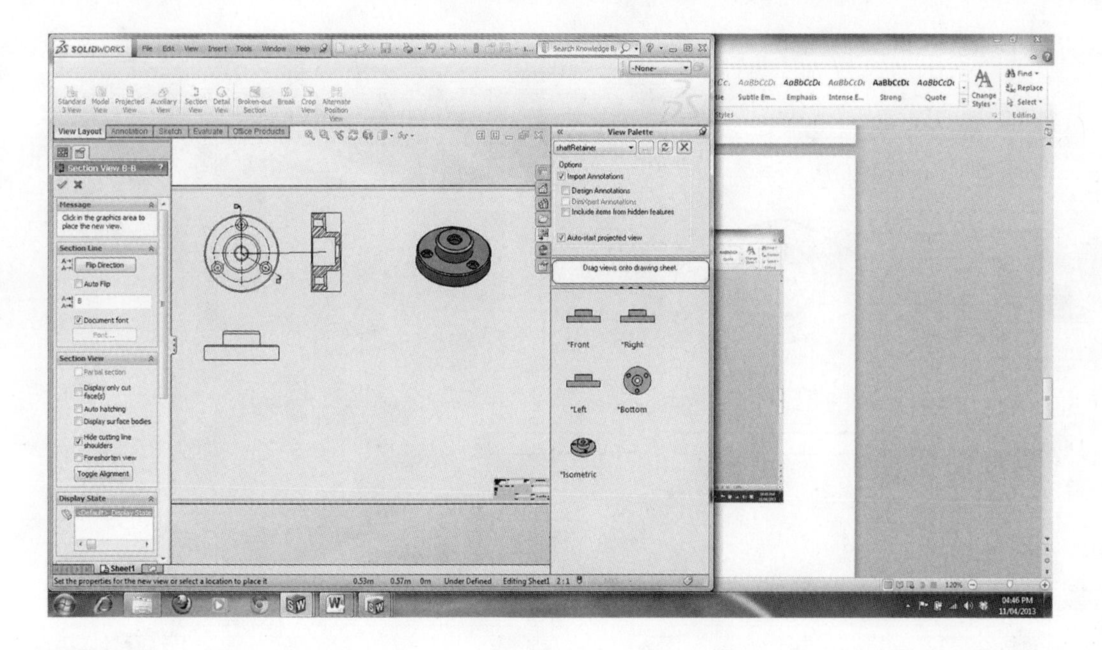

Figure 9.77

Aligned view.

SolidWorks Solution

The solution is very easy using the enhanced SolidWorks 2013 or better.

1. Click Drawing View1 bounding box containing the Top View (see Figure 9.78).
2. Click View Layout > Section View.
3. Select Section.
4. Click Cutting Line > Horizontal (see Figure 9.79).
5. Select *center of central circle.*
6. Click OK.
7. Drag Aligned View to the right to complete (see Figure 9.80).

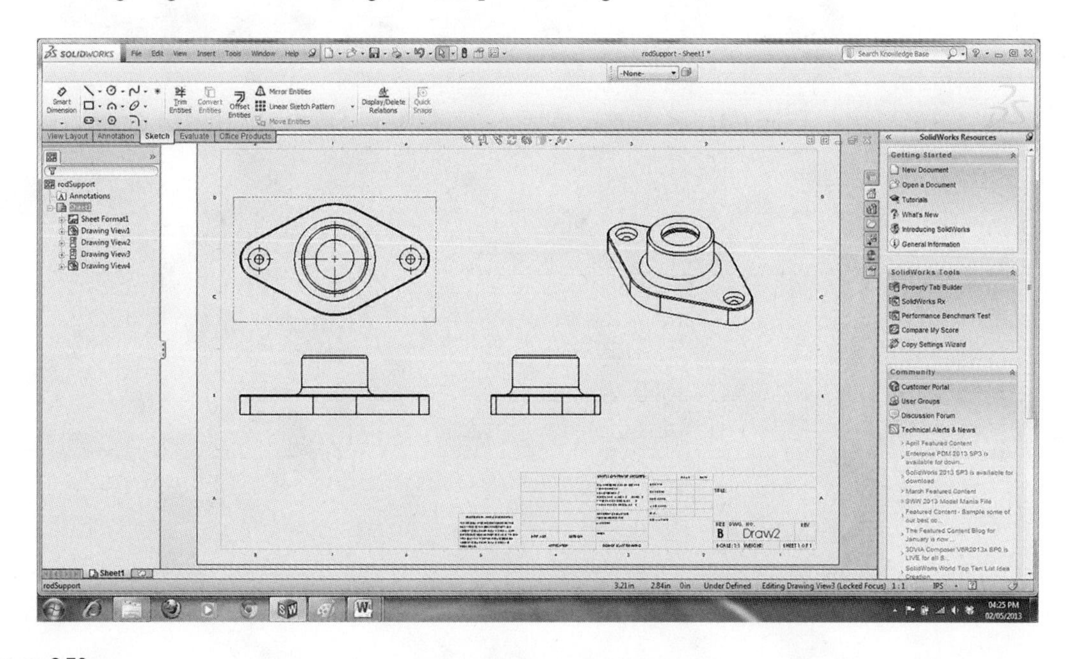

Figure 9.78

Top View highlighted.

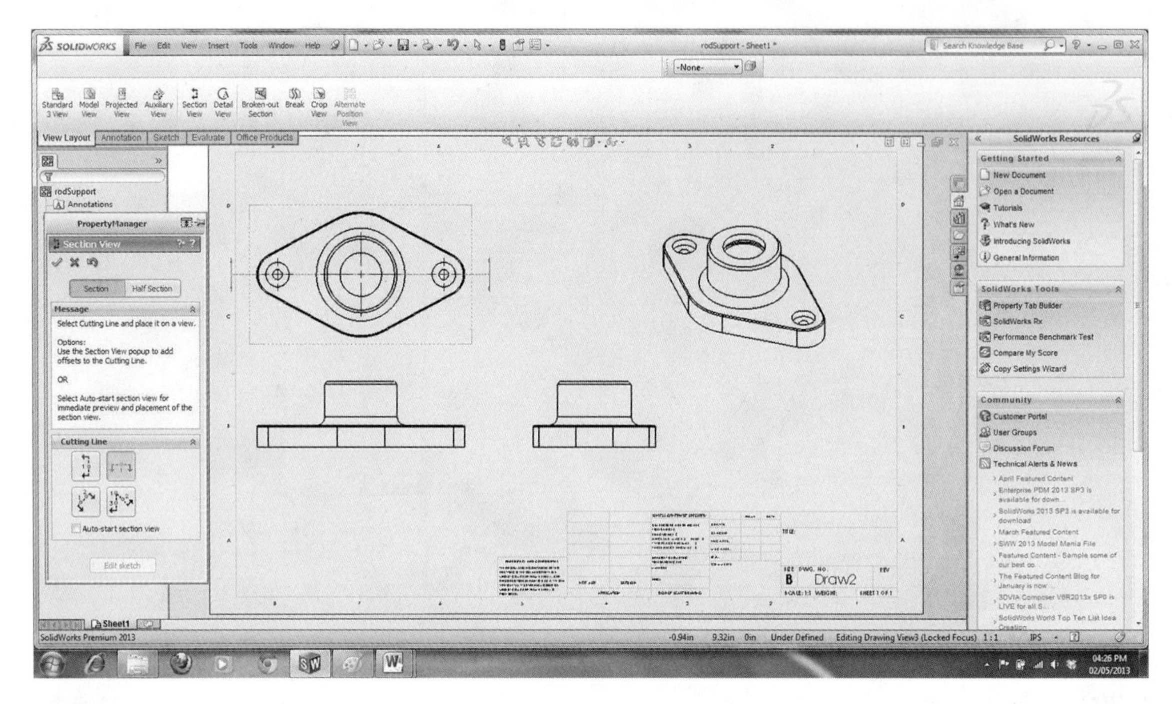

Figure 9.79

Click Cutting Line and select Horizontal.

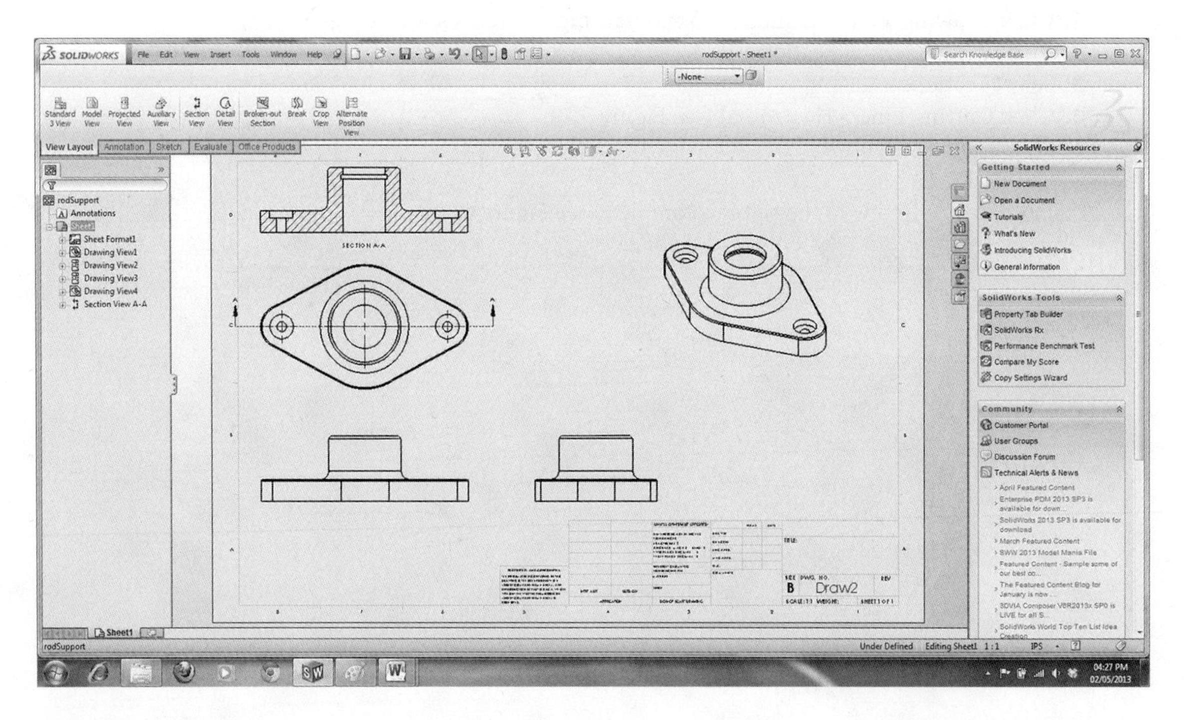

Figure 9.80

Full view.

Summary

The basic principles involved in parts and assembly modeling are discussed in this chapter. This is an important aspect of the CSWA examination. Also, the enhanced capability of SolidWorks 2013 and the better versions for handling the additional features of sectional views are now included in this chapter: (a) half-section view, (b) notched offset-section view, and (c) aligned-section view.

Exercises

1. Produce the drawing of the *bracket* that is described in Tutorial 7.2, which is shown in Figure 9.81, to include the necessary views.
2. Produce the drawing of the bracket that is described in Tutorial 8.3, which is shown in Figure 9.82, to include the necessary views.
3. The assembly shown in Figure 9.83 contains eight machined parts (see Chapters 5, 8, and 9). Produce the following views for each of the eight parts:
    a. Auxiliary view (if necessary)
    b. Sectional view
    c. Detail view
    d. Crop view
    e. Detailed drawing

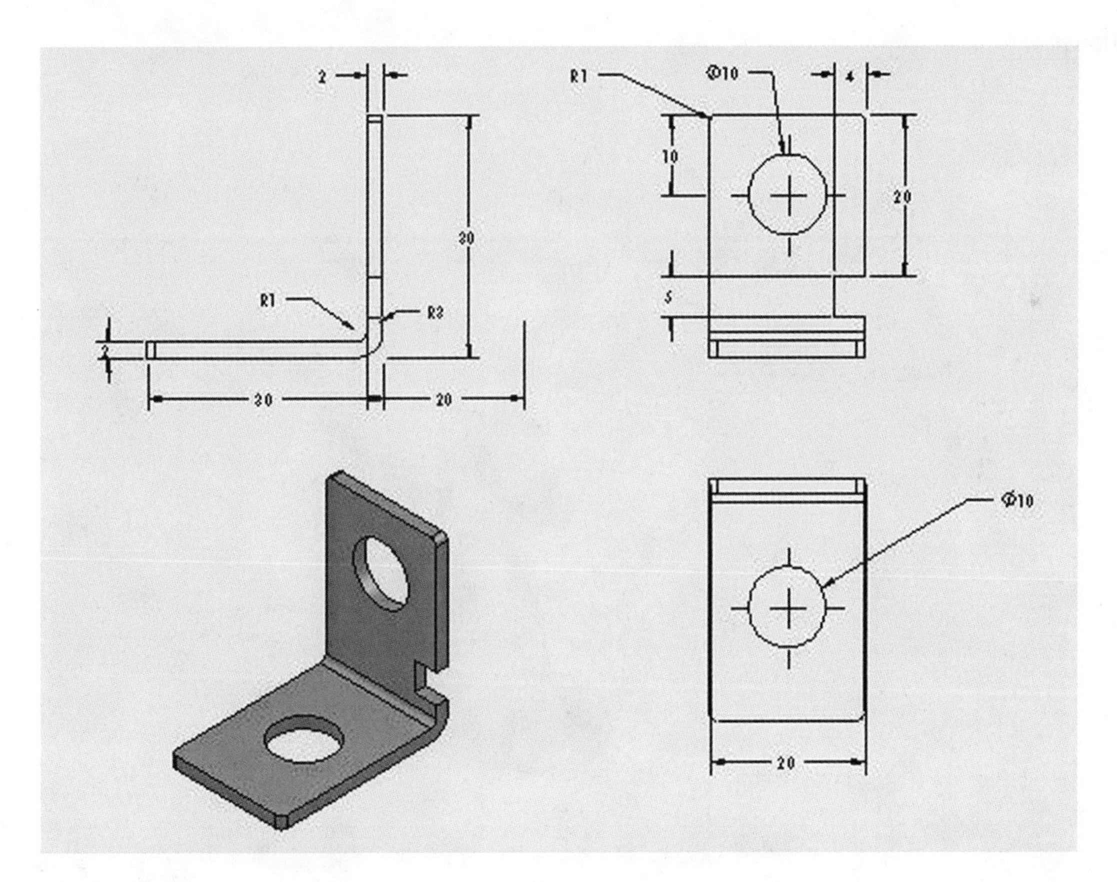

Figure 9.81

Drawing of the bracket described in Tutorial 2.

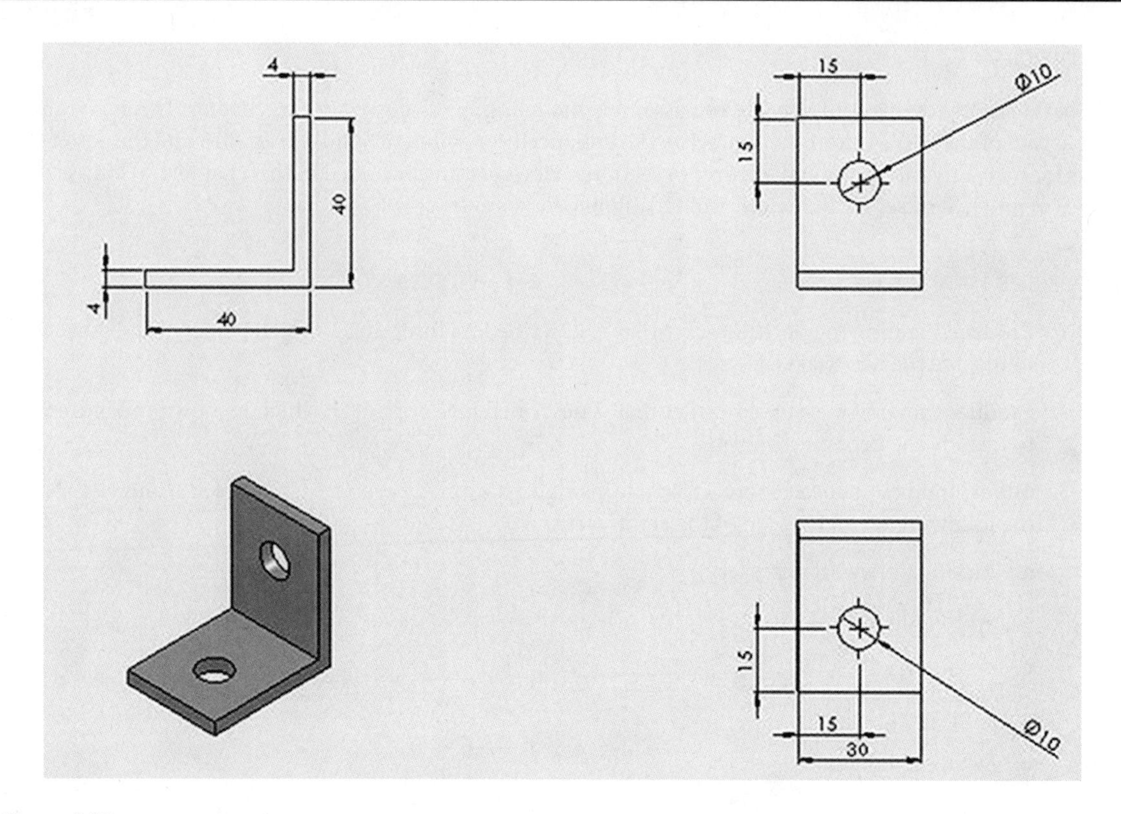

Figure 9.82

Drawing of the simple bracket described in Tutorial 3.

Figure 9.83

Tool post holder.

Projects

Create the sectional views for projects Figures P9.1 to P9.3.

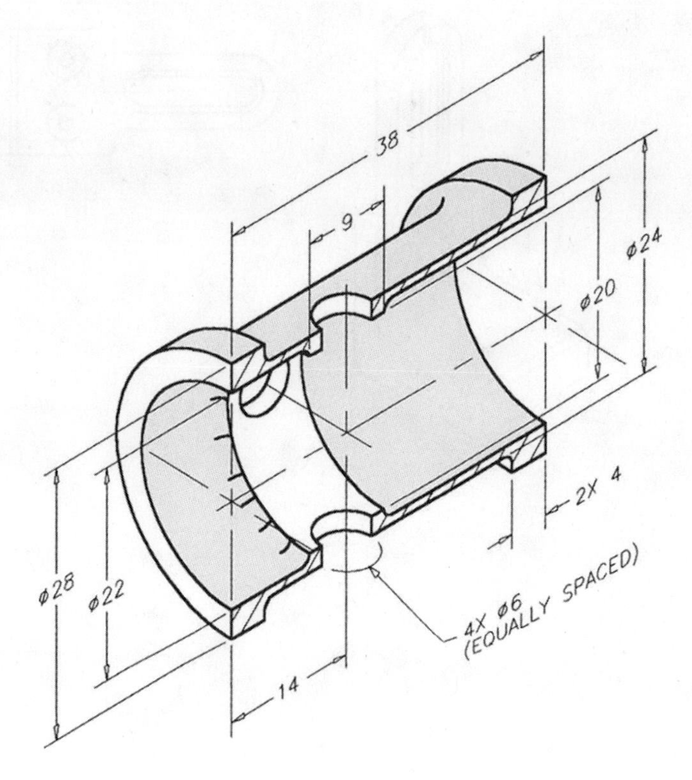

Figure P9.1

Hydraulic valve cylinder.

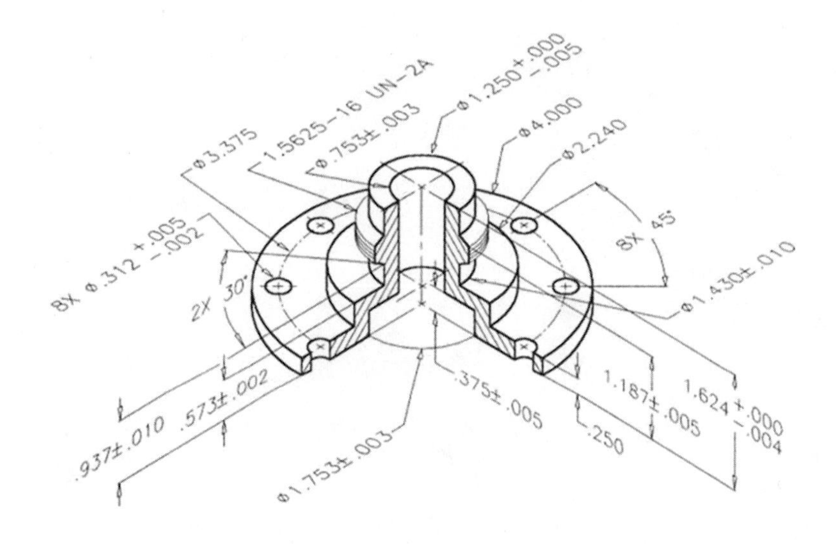

Figure P9.2

Hub.

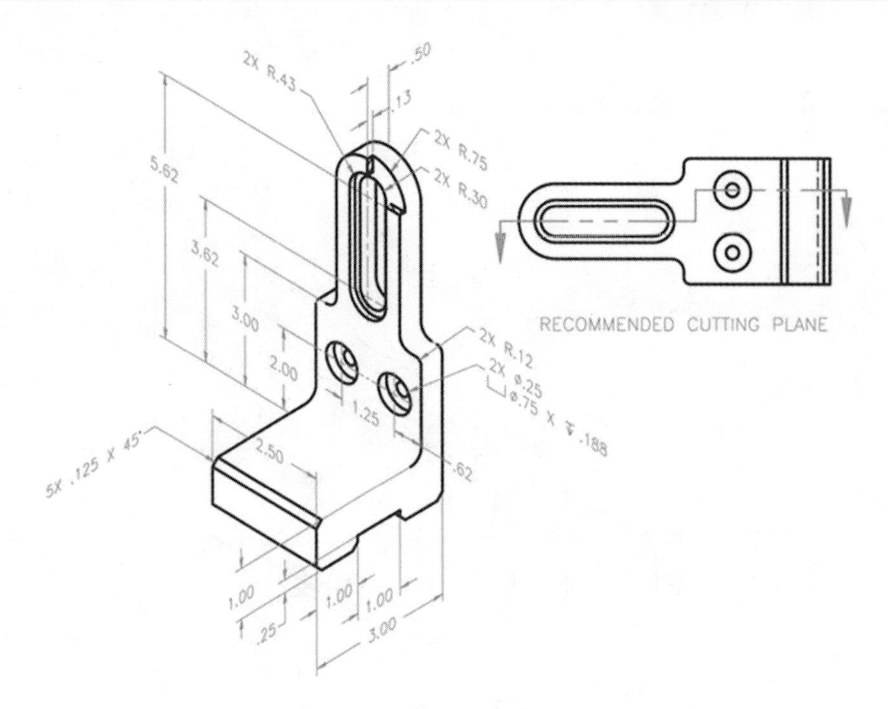

Figure P9.3

Slide bracket.

Section II
Intermediate Engineering Design Principles with SolidWorks

Reverse Engineering Using Auto Trace and FeatureWorks

Objectives:

When you complete this chapter, you will have

- Understood the concept of reverse engineering in a competitive marketplace
- Understood the functionalities of SolidWorks reverse engineering tools
- Import graphic images such as company logos and scanned geometric designs to SolidWorks
- Converted graphic images such as company logos and scanned geometric designs into digital data and traced them for reverse engineering
- Understood the concept of reverse engineering in a competitive marketplace
- Understood the functionalities of SolidWorks reverse engineering tools
- Understood FeatureWorks Add-In and Options
- Imported Geometry from out the SolidWorks environment
- Used Import Diagnostics tool to eliminate gaps between adjacent faces
- Carried out the Automatic Recognition of Features
- Carried out the Interactive Recognition of Features

Reverse Engineering

Engineering is the profession that is involved in the designing, manufacturing, constructing, and maintaining of products, systems, and structures. At a higher level, there are two types of engineering: (1) forward engineering and (2) reverse engineering.

Forward engineering is the traditional process of moving from high-level abstractions and logical designs to the physical implementation of a system. In some situations, there may be a physical part without any technical details, such as drawings and bills of materials, or without engineering data, such as thermal and electrical properties.

The process of duplicating an existing component, subassembly, or product without the aid of drawings, documentation, or a computer model is known as reverse engineering.

Reverse engineering can be viewed as the process of analyzing a system to

- Identify the system's components and their interrelationships
- Create representations of the system in another form or a higher level of abstraction
- Create the physical representation of that system

Reverse engineering is very common in diverse fields such as software engineering, entertainment, automotive, consumer products, microchips, chemicals, electronics, and mechanical designs. For example, when a new machine comes to market, competing manufacturers may buy one machine and disassemble it to learn how it was built and how it works. A chemical company may use reverse engineering to defeat a patent on a competitor's manufacturing process. In civil engineering, bridge and building designs are copied from past successes so that there will be less chance of catastrophic failure. In software engineering, a good source code is often a variation of other good source codes.

In some situations, designers give a shape to their ideas by using clay, plaster, wood, or foam rubber, but a computer-aided design (CAD) model is needed to enable the manufacturing of the part. As products become more organic in shape, designing in CAD may be challenging or impossible. There is no guarantee that the CAD model will be acceptably close to the sculpted model. Reverse engineering provides a solution to this problem because the physical model is the source of information for the CAD model. This is also referred to as the part-to-CAD process.

Another reason for reverse engineering is to compress product development times. In the intensely competitive global market, manufacturers are constantly seeking new ways to shorten lead times to market a new product. Rapid product development (RPD) refers to recently developed technologies and techniques that assist manufacturers and designers in meeting the demands of reduced product development time. For example, injection-molding companies must drastically reduce the tool and die development times. By using reverse engineering, a three-dimensional (3D) product or model can be quickly captured in digital form, remodeled, and exported for rapid prototyping/tooling or rapid manufacturing.

The following are the reasons for reverse engineering a part or a product:

- The original manufacturer of a product no longer produces a product.
- There is inadequate documentation of the original design.
- The original manufacturer no longer exists, but a customer needs the product.
- The original design documentation has been lost or never existed.
- Some bad features of a product need to be designed out. For example, excessive wear might indicate where a product should be improved.
- To strengthen the good features of a product based on the long-term usage of the product.
- To analyze the good and bad features of the competitors' product.
- To explore new avenues to improve product performance and features.
- To gain competitive benchmarking methods to understand the competitor's products and develop better products.
- The original CAD model is not sufficient to support modifications or current manufacturing methods.
- The original supplier is unable or unwilling to provide additional parts.
- The original equipment manufacturers are either unwilling or unable to supply replacement parts or demand inflated costs for sole-source parts.
- To update obsolete materials or antiquated manufacturing processes with more current, less expensive technologies.

Reverse engineering enables the duplication of an existing part by capturing the component's physical dimensions, features, and material properties. Before attempting reverse engineering, a well-planned life-cycle analysis and cost/benefit analysis should be conducted to justify the reverse engineering projects. Reverse engineering is typically cost-effective only if the items to be reverse engineered reflect a high investment or will be reproduced in large quantities. Reverse engineering of a part may be attempted even if it is not cost-effective, if the part is absolutely required, and if it is mission-critical to a system.

Reverse engineering of mechanical parts involves acquiring 3D position data in the point cloud using laser scanners or computed tomography. Representing the geometry of the part in terms of surface points is the first step in creating parametric surface patches. A good polymesh is created from the point cloud using reverse engineering software. The cleaned-up polymesh, nonuniform rational B-spline (NURBS) curves, or NURBS surfaces are exported to CAD packages for the further refinement, analysis, and generation of cutter tool paths for computer-aided manufacturing (CAM). Finally, the CAM produces the physical part.

It can be said that reverse engineering begins with the product and works through the design process in the opposite direction to arrive at a product definition statement. In doing so, it uncovers as much information as possible about the design ideas that were used to produce a particular product.

SolidWorks Reverse Engineering Tools

SolidWorks offers some reverse engineering capabilities:

- *Reverse engineering*—Scan concept sketches or data into SolidWorks using *ScanTo3D* and complete the product design in SolidWorks.
- *Reverse engineering*—Insert a picture into SolidWorks *Sketch Picture* and trace it using the *Auto Trace* tool for reverse engineering. Insert pictures (*.bmp*, *.gif*, *.jpg*, *.jpeg*, *.tif*, or *.wmf*).
- *Feature recognition*—Import non-SolidWorks CAD data, preserve the design intent, and make changes. Increase the value of translated files while reducing the time that is spent rebuilding existing 3D models.

We will describe the Auto Trace tool first and the feature recognition tool later.

Create Auto Trace Tool

The *SolidWorks Auto Trace tool* helps to convert raster data into vector data. Company logos and scanned geometric designs can be converted into SolidWorks digital data using the Auto Trace Add-In tool, which allows users to insert a picture into the SolidWorks Sketch Picture and trace it for reverse engineering. The SolidWorks Sketch Picture PropertyManager enables you to trace outlines or select areas by color to create vector data. This tool creates a sketch that you can save and edit as needed. The Auto Trace tool inserts .bmp, .gif, .jpg, .jpeg, .tif, or .wmf format pictures.

This tool can help convert raster data to vector data. In Tools > Sketch Tools > Sketch Picture, open a document and click the Next icon to select conversion options. Options include the following:

- Trace Settings
- Display Options
- Adjustments

Once you convert the document to vector data, you have a sketch that you can modify, save, and use as a basis for creating a 3D model. The steps involved can be summarized as follows:

1. Open raster data.
2. Trace shape outline.
3. Convert raster data to vector data.
4. Modify sketch.
5. Create 3D model.

1. Open a New Part SolidWorks document as shown in Figure 10.1.
2. Click Add-In (see Figure 10.2).
   The Add-In PropertyManager is displayed (see Figure 10.3).
3. Select Autotrace option to add-in (see Figure 10.3).
4. Click Customize > Commands > Sketch (see Figures 10.2 and 10.4).
5. Drag Sketch Picture to the CommandManager (see Figure 10.5).

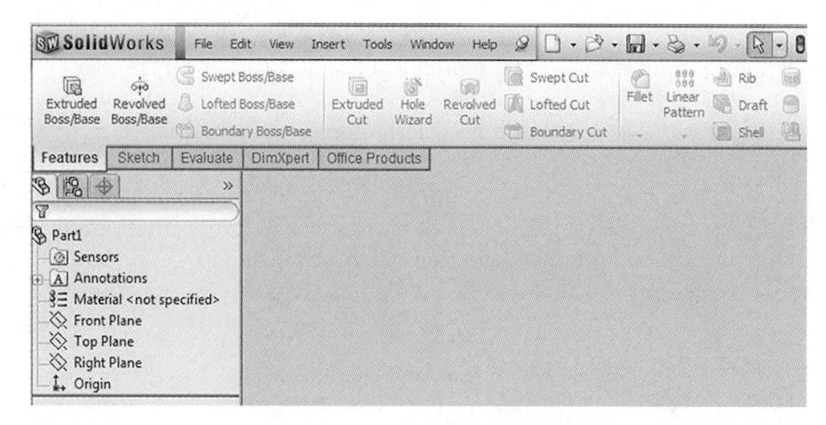

Figure 10.1

A New SolidWorks document.

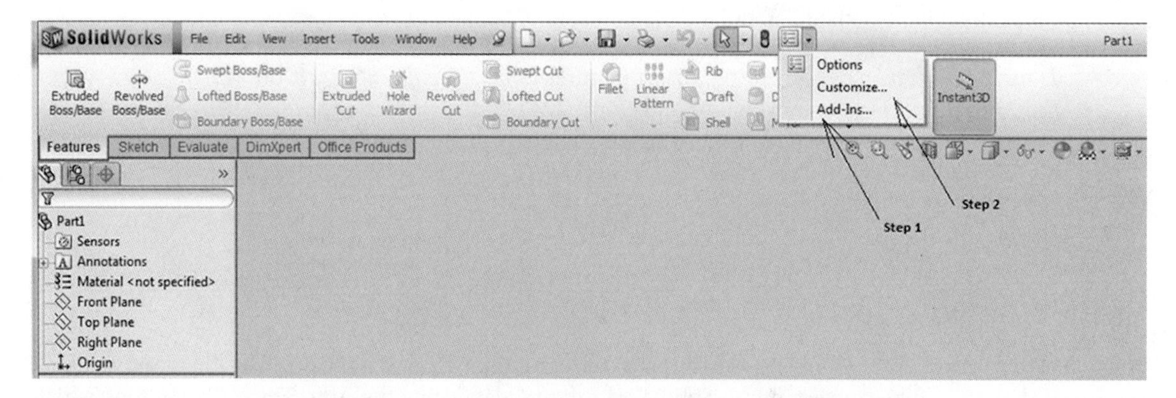

Figure 10.2

Click Add-In from the Feature CommandManager.

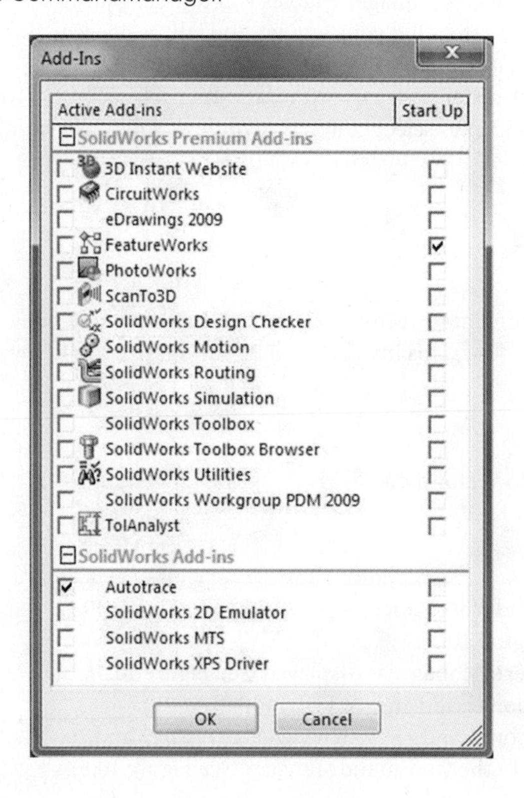

Figure 10.3

Add-In PropertyManager.

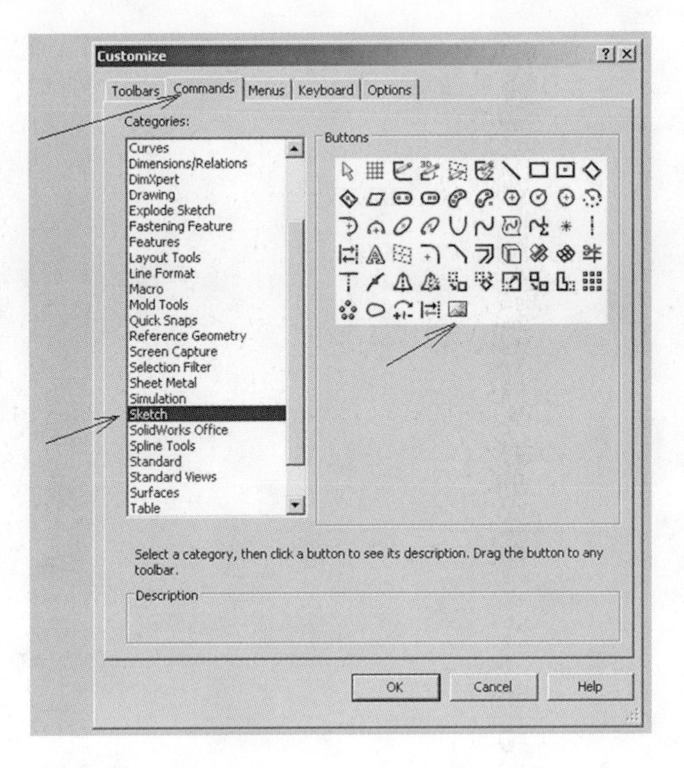

Figure 10.4

Click Customize from the Feature CommandManager to access Sketch.

Figure 10.5

Customizing Sketch Picture.

Methodology for Importing and Extracting Image Features

A four-phase methodology for importing and extracting image features is summarized as follows:

1. *Phase 1:* Import image (by clicking Sketch Picture)
2. *Phase 2:* Image extraction
3. *Phase 3:* Image repair
4. *Phase 4*: Extrude features

Phase 1: Import Image

6. Click Front Plane > Sketch (see Figure 10.6; be in *Sketch mode* to activate Sketch Picture).
7. Click Sketch Picture > SolidWorks Logo.bpm > Open from the appropriate directory.
8. Enter 0 for the first three entries (Origin X Position, Origin Y Position, and Angle, respectively) and 100 mm for the Width (see Figure 10.7 for the resized document).
9. Click Next (arrowhead) to select conversion options (see Figure 10.8). (Sketch Picture Property Manager appears showing conversion options, as in Figure 10.9.)

Phase 2: Image Extraction

10. Press the "F" key to enlarge the image on the screen.
11. Click Use to select the rectangular area in Trace Settings. (See step 1 in Figure 10.10.)

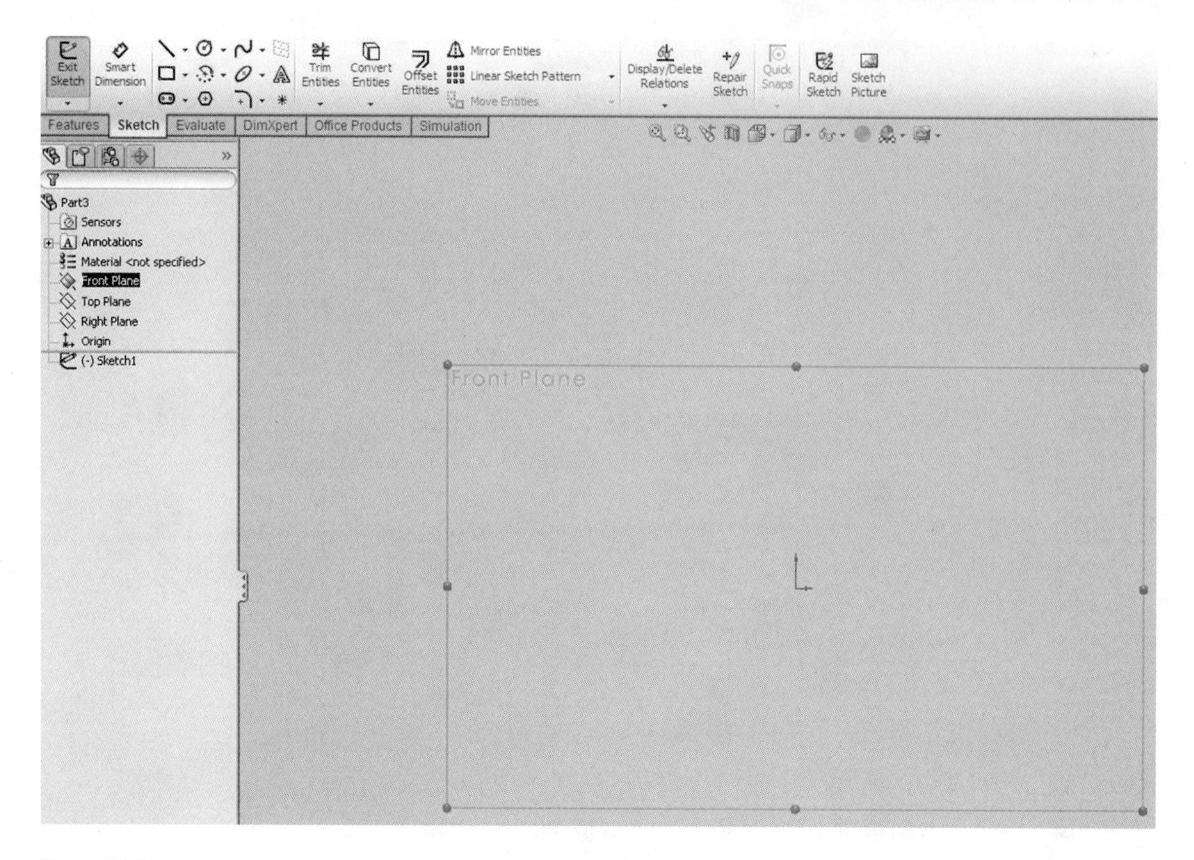

Figure 10.6

Front Plane is chosen.

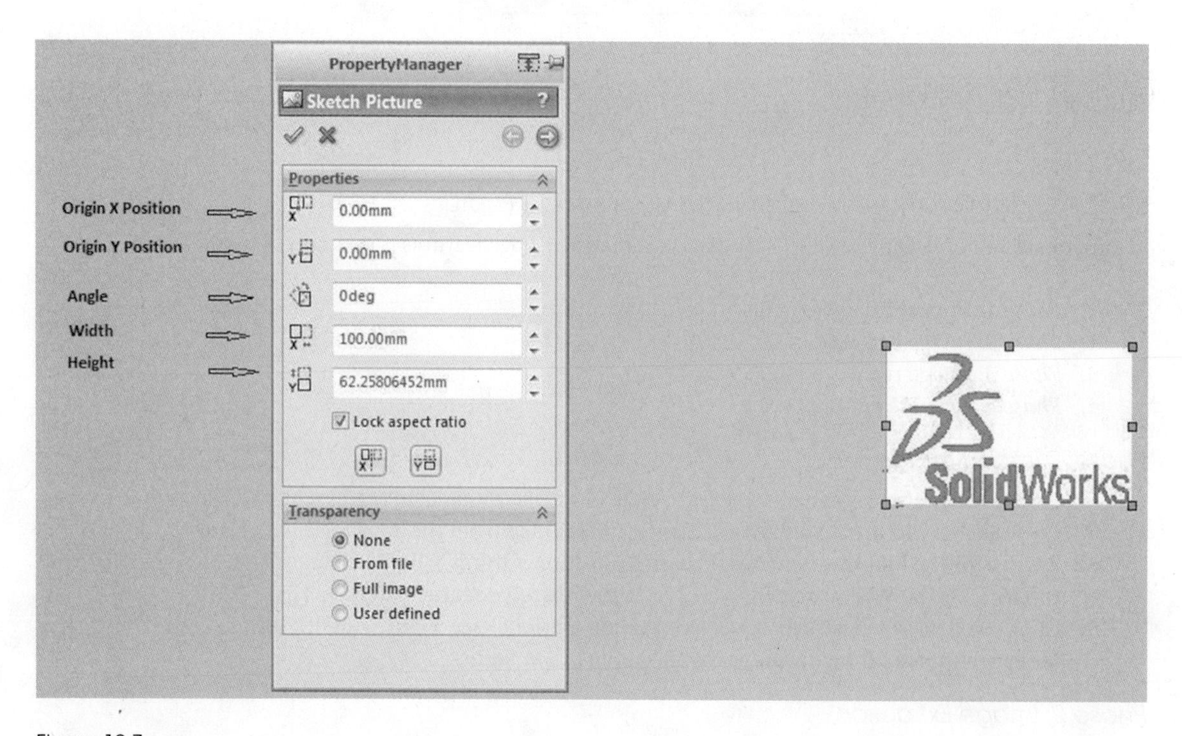

Figure 10.7

Importing a picture to SolidWorks within Sketch Picture environment.

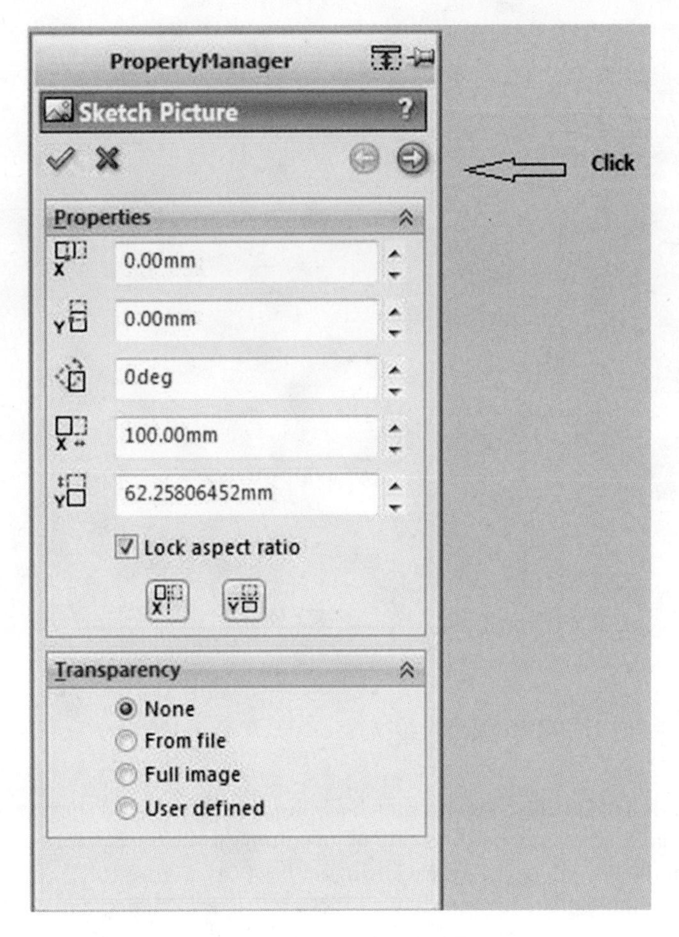

Figure 10.8

Options for Trace Settings, Display Options, and Adjustments.

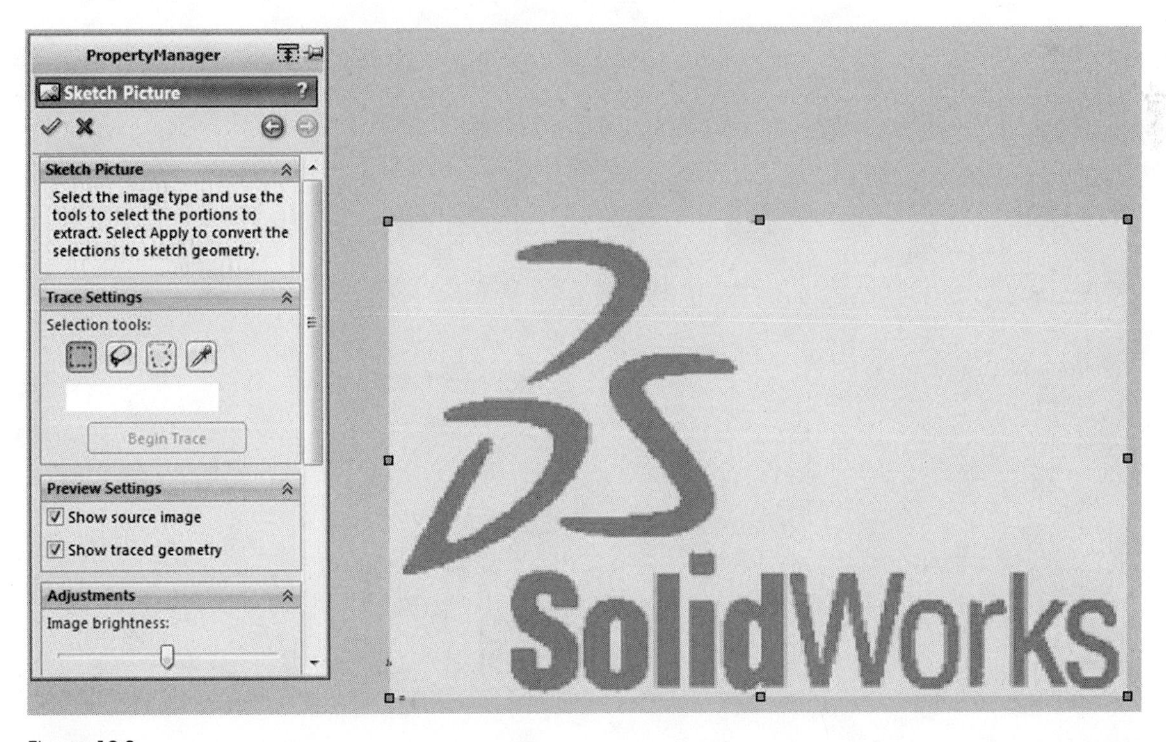

Figure 10.9

Sketch Picture PropertyManager.

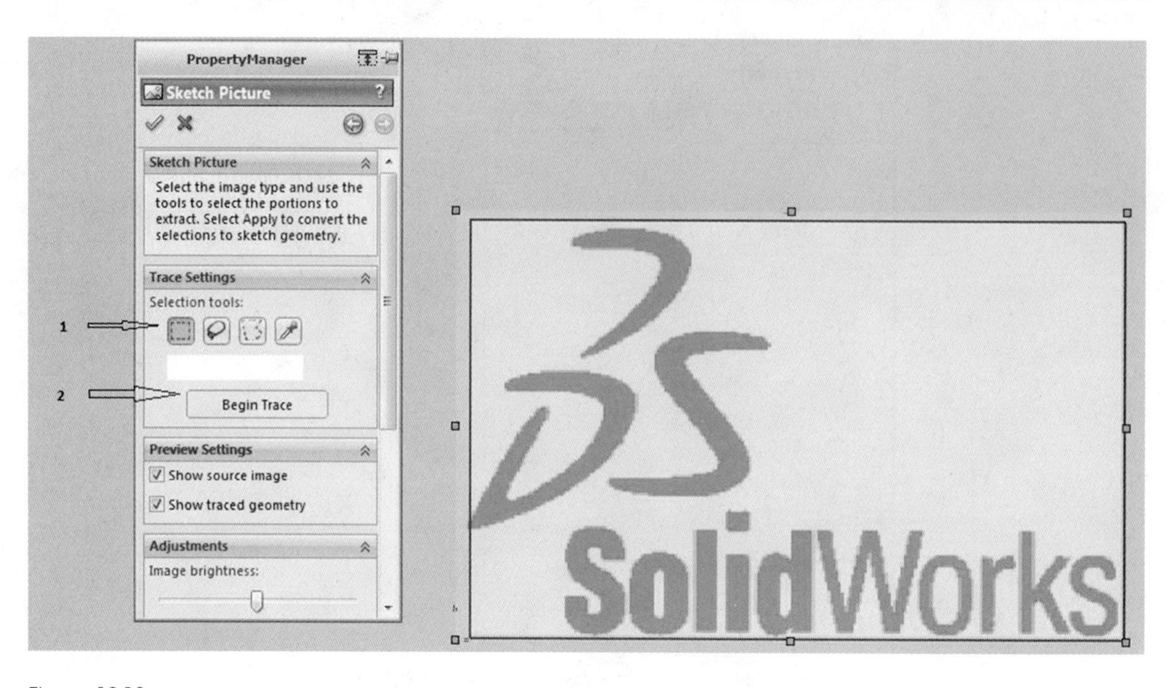

Figure 10.10

Using Selection Tool to extract image automatically.

12. Use the Rectangular Selection Tool to sketch a bounding box around the features (see Figure 10.10).
13. Click Begin Trace in the Trace Settings callout (to automatically trace shape outlines; see Figure 10.10).
14. Click Apply at the bottom of the Trace Settings callout (see Figure 10.11).
15. Click OK on the top-right side of the Sketch Picture PropertyManager (see Figure 10.11).

(The extracted images are shown in Figure 10.12.)

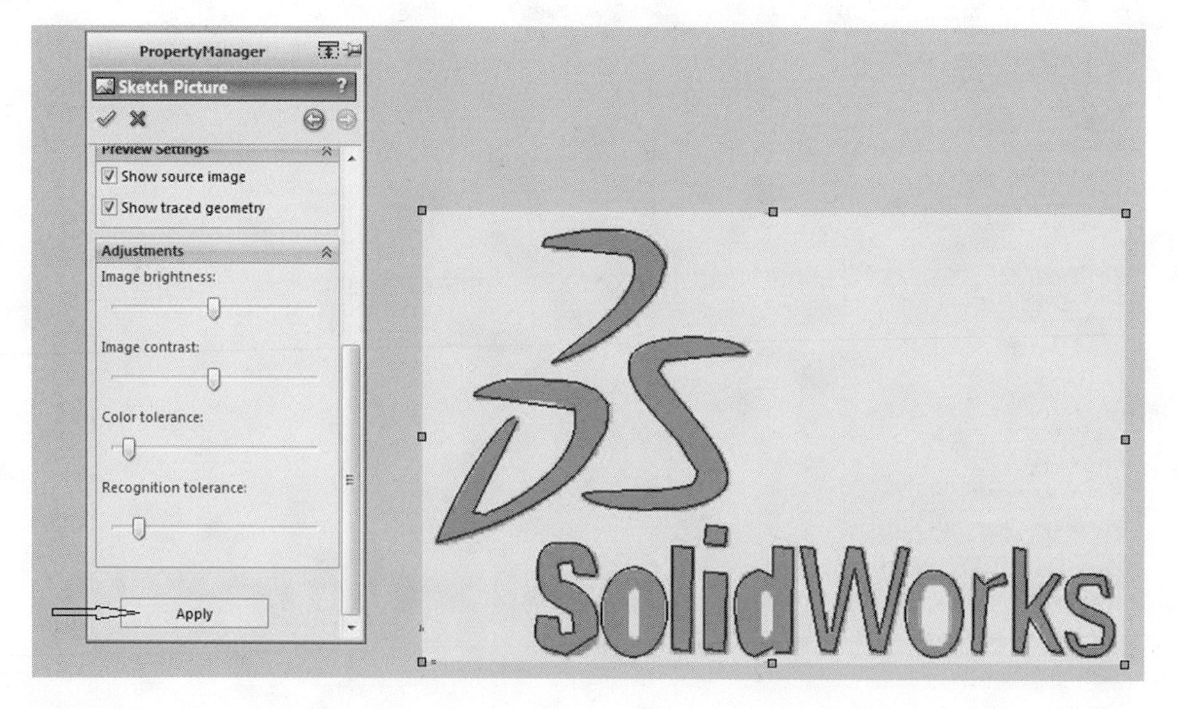

Figure 10.11

Automatic tracing of shape outlines.

Figure 10.12

Extracted images.

Phase 3: Image Repair

Fault Diagnosis

Automatic tracing of shape outlines may result in some missing features, as well as *open contours*, as shown in Figure 10.12. The extracted images have to be repaired. We could use Line or Spline tools to connect open contours and complete the profiles. For example, the words "S," "o," "d," "W," and "D" in italic are not correctly extracted. "S," "W," and "D" have dangling or missing edges. The letters "o" and "d" have their inner profiles missing in the form of "o." The ill-extracted shapes are shown in Figure 10.13. Using the zooming tool, we can zoom into the edge connections because there could be some missing edges (dangling edges).

Fault Repair

Since these shapes are a SolidWorks feature, we could then use Line or Spline tools to correct the missing or ill-extracted edges.

1. Right-click Sketch1 from the FeatureManager (see Figure 10.14).
2. Click Edit Sketch to edit the ill-extracted features.

The corrected shapes are shown in Figure 10.15.

This completes the process of converting the raster data into vector data, which can now be used as a basis for a SolidWorks 3D model.

Phase 4: Extrude Features

To create the SolidWorks 3D model, we follow the normal steps:

1. Click Extruded Boss/Base in the FeatureManager. (The Extrude PropertyManager is displayed, as shown in Figure 10.16.)
2. Accept the Blind option and give a Distance for extrusion as 10 mm. (Figure 10.17 shows the Extruded features.)

Figure 10.13

Ill-extracted shapes.

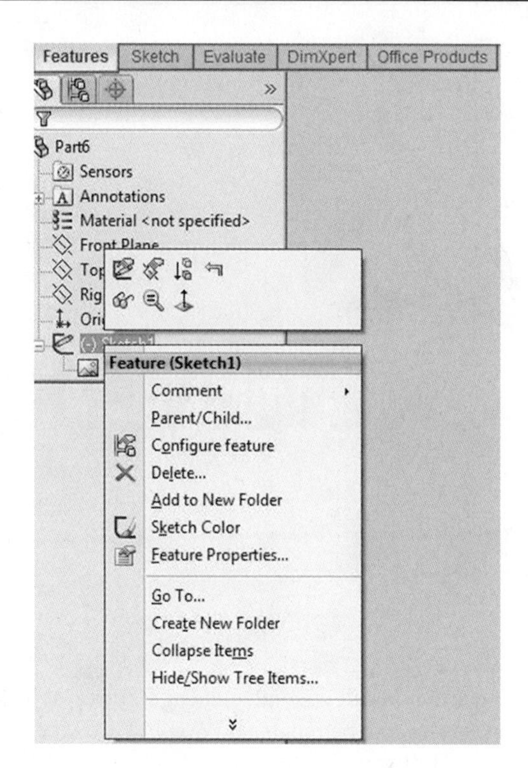

Figure 10.14

Sketch1.

Figure 10.15

Corrected shapes.

Expanding the Solid Bodies folder in the FeatureManager shows that there are 14 solid bodies (see Figure 10.18). It is worthwhile to further investigate the FeatureManager. There is one sketch, Sketch1, that comprises the 14 profiles that are extracted. We notice that the Sketch Picture1 tool icon is attached to Sketch1, showing how the sketch was created.

FeatureWorks Tool

FeatureWorks Product Overview

FeatureWorks® software is a feature recognition software that is fully integrated with SolidWorks 3D CAD software. FeatureWorks software is the first parametric feature recognition solution for CAD users. By applying intelligence to translated 3D CAD files, FeatureWorks brings static 3D data to life, making them ready for use with SolidWorks 3D CAD software. Automatic feature recognition applies intelligence to static

Figure 10.16

Extrude PropertyManager.

Figure 10.17

Extruded features.

geometric data from standard 3D translators. Once feature recognition is complete, the part is fully editable with SolidWorks software. There is additional time saving through the automatic recognition of linear, rectangular, and circular hole patterns (see Figures 10.19 and 10.20).

Ease sharing of 3D models between organizations that use different CAD systems leverages the value of legacy data and makes a complete transition to a SolidWorks solution more quickly. FeatureWorks software reduces the time that is spent rebuilding models, resulting in better product designs, faster development cycles, and lower costs.

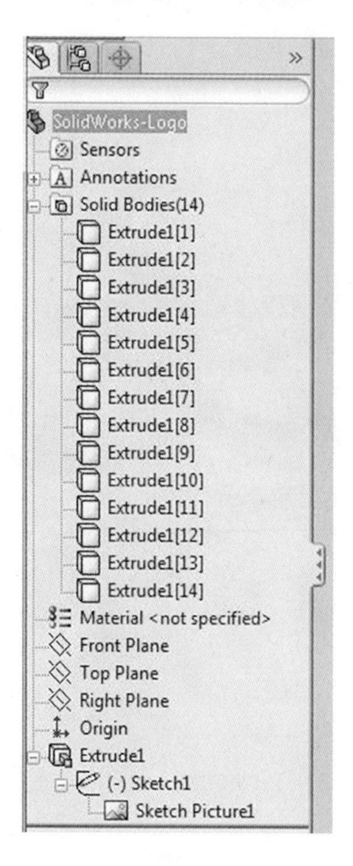

Figure 10.18

Fourteen solid bodies in the Solid Bodies folder.

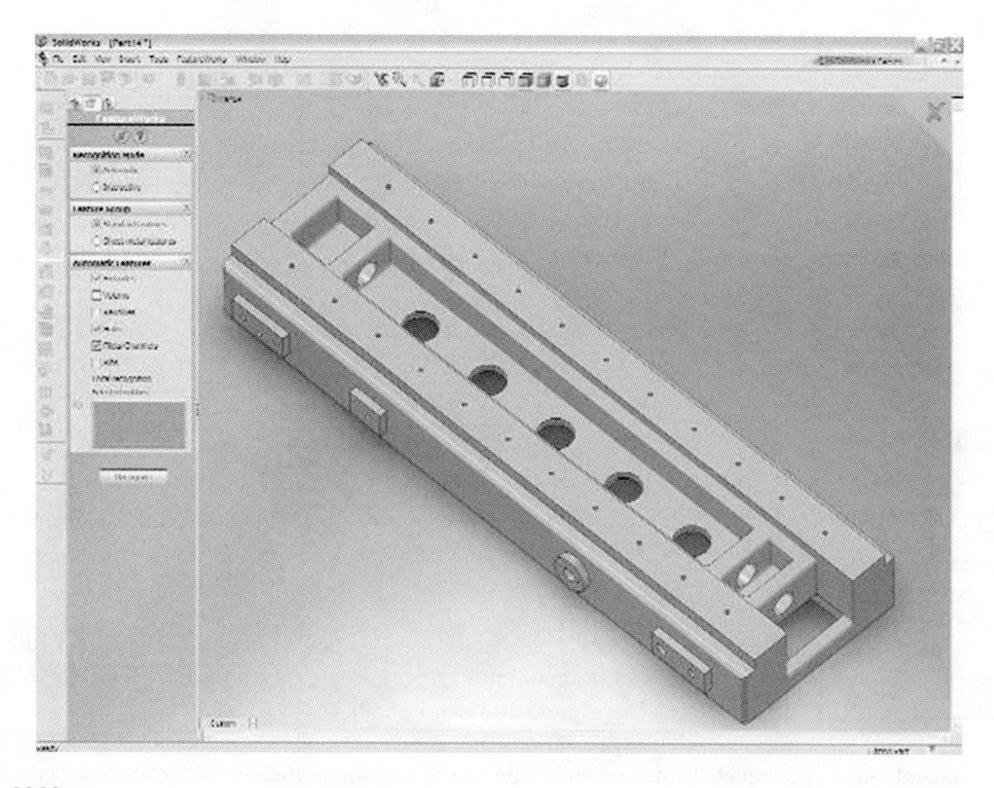

Figure 10.19

Automatic feature recognition.

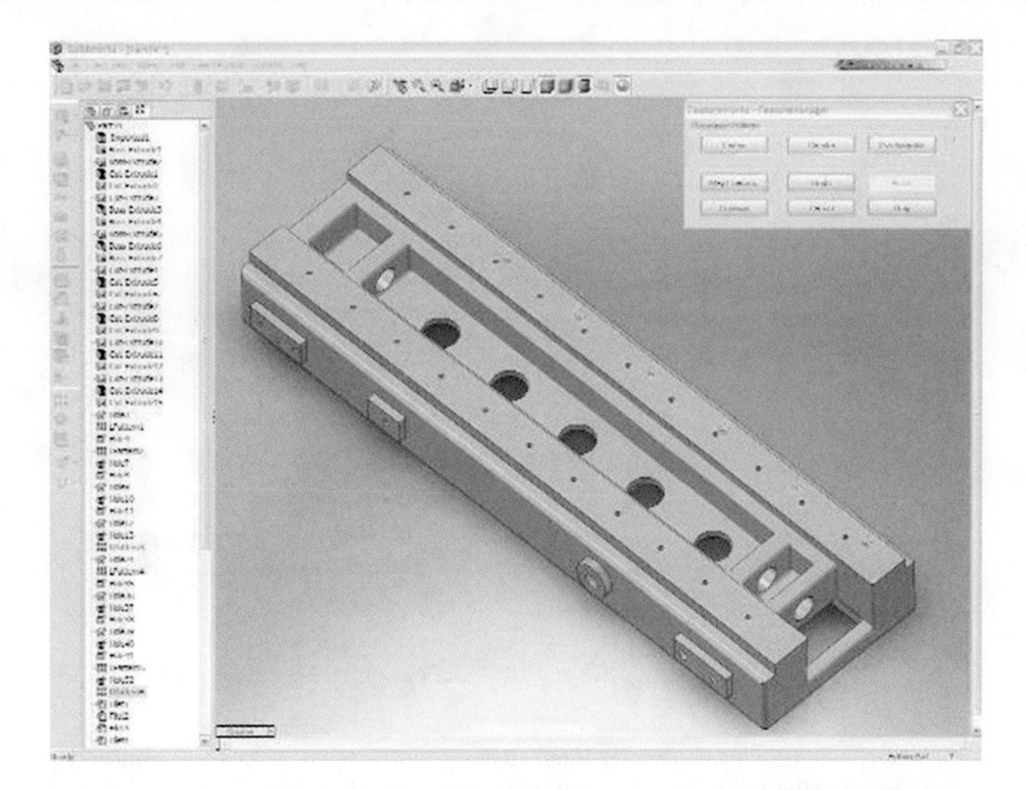

Figure 10.20

Additional time saving through automatic recognition of linear, rectangular, and circular hole patterns.

Why spend hours rebuilding designs? Standard translators allow you to share geometric data with other people and organizations that use different CAD systems, but making changes to these data and performing *what-if* analyses are not so easy. To do so, you typically need to rebuild designs manually, introducing new intelligence or recreating intelligence that was lost during the translation process. Life would be simpler if you could add intelligence to static 3D models more quickly and easily. Now you can, with FeatureWorks software.

FeatureWorks software is the first software of its kind for CAD users. Fully integrated with SolidWorks software, FeatureWorks is the first parametric feature recognition software for CAD users. By recognizing features from the files that are produced by standard translators, FeatureWorks applies intelligence to static geometric data, bringing them back to life and making them ready to use in SolidWorks.

FeatureWorks Software Capabilities

- *Modify the size and location of features easily.*
- *Once the recognition of features—including holes, cuts, chamfers, fillets, extrusions, sheet metal features, ribs, and sketch patterns—is complete, you can easily fine-tune the design using SolidWorks software.*
- *Features recognized by FeatureWorks are fully editable, associative, and parametric, and you can create new features at any time.*
- *FeatureWorks enhances the value of legacy data and enables easier, more productive sharing of 3D models between different CAD systems.*
- *Preserve design intent and maintain quality.* FeatureWorks gives you the flexibility to make changes to static geometric data and helps preserve or introduce a new design intent. For example, a hole originally created as *blind* or *through all* will regain the essential specifications that it may have lost through the translation process, keeping the design intent intact for all downstream modifications and maintaining quality.
- *Choose automatic or interactive methods.* FeatureWorks provides both automatic and interactive feature recognition capabilities. Automatic feature recognition requires no user intervention. The interactive method provides a dialog that lets you control or specify the design intent easily by selecting and clicking on a face or the edges of a cut or boss. The model checker indicates any changes to underlying imported geometry before and after feature recognition. You can perform

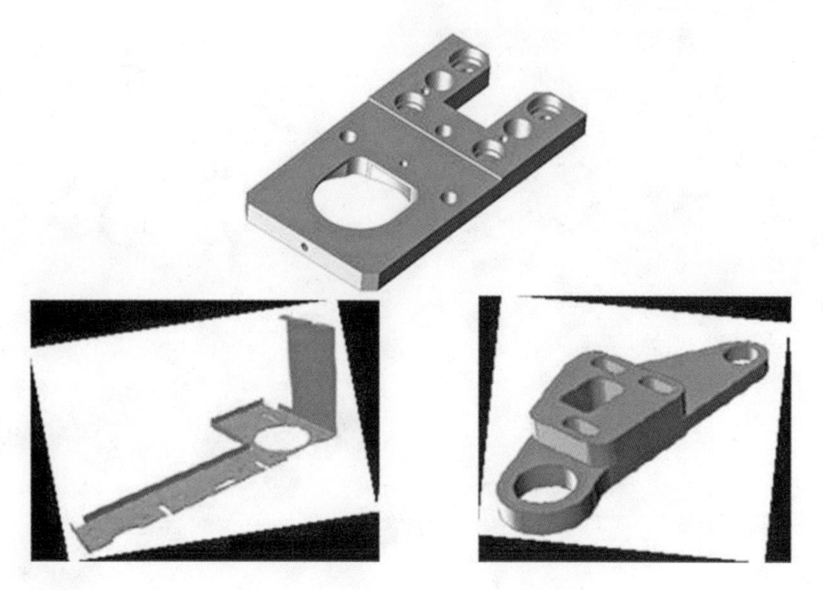

Figure 10.21

Feature recognition of parts from static translated file data.

automatic recognition before or after interactive recognition. All feature recognition steps can be automated using the SolidWorks Application Programming Interface.

- *FeatureWorks requires easy setup and easy use, fully integrated with SolidWorks software.* You can access all the controls for FeatureWorks from the SolidWorks menu bar. The FeatureManager® design tree in SolidWorks software automatically keeps track of the features that are recognized by FeatureWorks, so the entire design process stays consistent and intuitive. FeatureWorks provides the same Windows® look and feel that have made SolidWorks the easiest-to-learn-and-use 3D CAD software available.
- *Save time through parametric feature recognition.* FeatureWorks captures all imported data and recognizes the features from the files that are produced by standard translators such as STEP, IGES, SAT (ACIS®), VDA-FS, and Parasolid® files. Best suited for geometrically regular parts, FeatureWorks recognizes and brings static translated file data to life through the feature recognition of parts that contain holes, cuts, bosses, fillets, chamfers, sketch patterns, curve-driven patterns, sheet metal features, and shells (see Figure 10.21):
  - Extrusion features such as bosses and cuts of the following sketch entities: (a) lines, (b) circles, and (c) circular arcs
  - Revolved features, which are conical or cylindrical
  - Sweep features
  - Hole patterns including linear, rectangular, and circular patterns
  - Any standard hole types, such as simple, tapered, and counterbored
  - Sheet-metal features including edge flange, sketch and bend, hem, and base features
  - Random sketch patterns of features on a plane
  - Shell features such as uniform wall and shell inward only
  - Ribs and draft features
  - Combination of features and imported geometry
  - Constant- and variable-radius fillets
  - Applied features such as chamfers and fillets
  - Multibody part construction

FeatureWorks Add-In

To enable the FeatureWorks application and set the SolidWorks option,

1. Click Tools > Add-Ins (see Figure 10.22).
2. In the dialog box, select FeatureWorks, and then click OK.

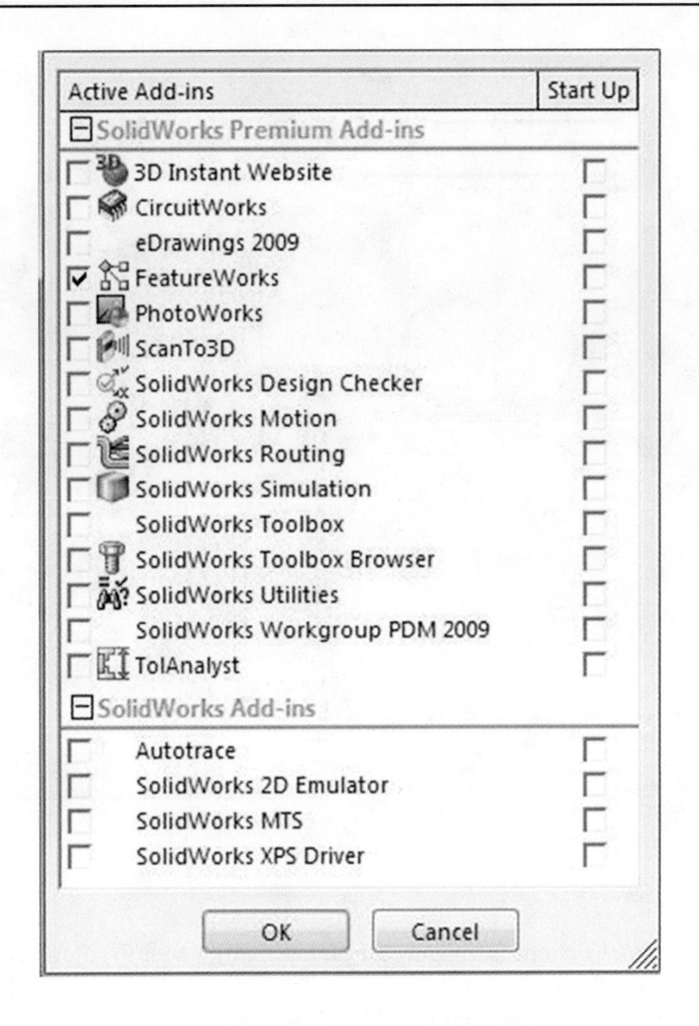

Figure 10.22

SolidWorks Add-Ins.

FeatureWorks Options

1. Click FeatureWorks > Options. (See Figure 10.23 for the FeatureWorks toolbar.)
2. In the dialog box,
   a. For General (see Figure 10.24, top left),
      i. Select Create new file.
      ii. Select Prompt for feature recognition as part opens.

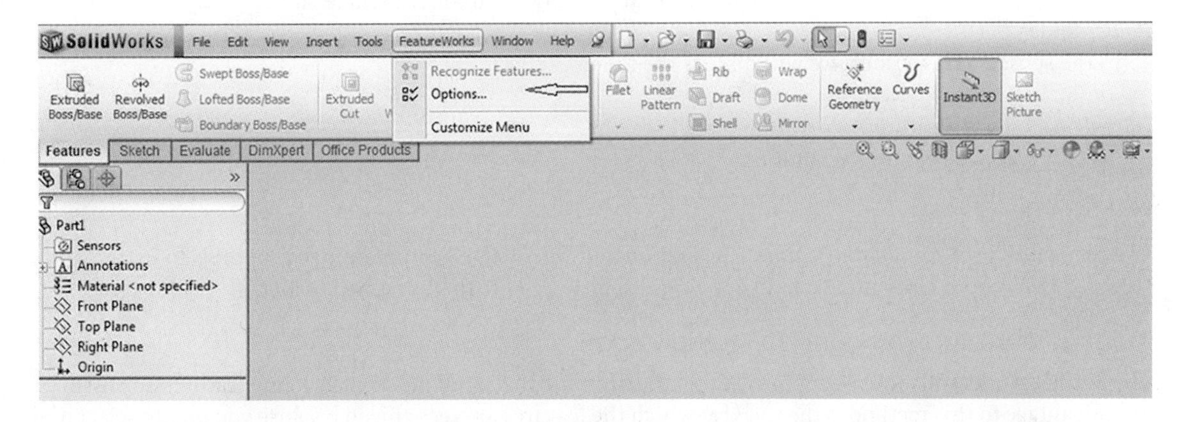

Figure 10.23

SolidWorks FeatureWorks toolbar.

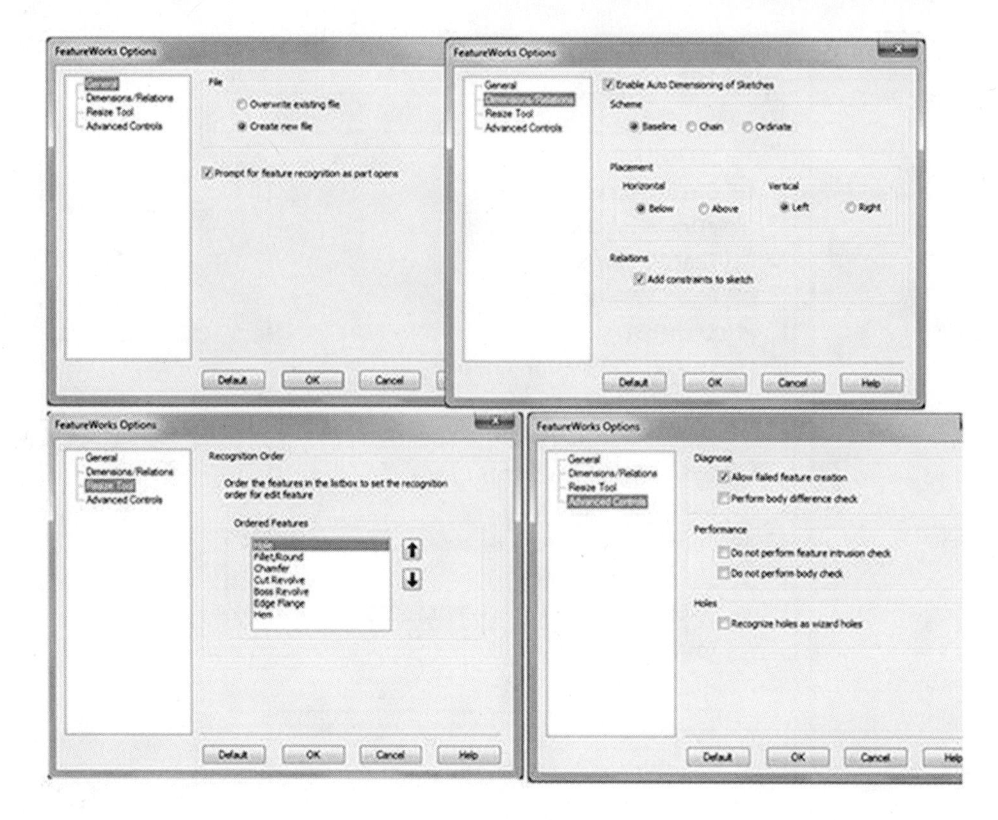

Figure 10.24

SolidWorks FeatureWorks Options.

  b. For Dimensions/Relations (see Figure 10.24, top right), under Relations, select Add constraints to sketch to fully define the sketch.
  c. For Advanced Controls (see Figure 10.24, bottom right),
    i. Under Diagnose, select Allow failed feature creation to allow the creation of features that have rebuild errors.
    ii. Under Performance, clear both check boxes.
3. Click OK.

FeatureWorks PropertyManager

The *FeatureWorks* PropertyManager can be used to set most of the FeatureWorks recognition options. Features can be automatically or interactively recognized. You can use the *step-by-step recognition* with automatic and interactive feature recognition or a combination of these methods. Standard features, or sheet-metal features, can also be recognized. FeatureWorks cannot recognize the entities in the *Surface Bodies* folder.

In the subsequent sections, we will discuss the following:

- Automatic Feature Recognition
- Interactive Feature Recognition

Problem Description

Use the *SolidWorks Automatic Feature Recognition* and *Interactive Feature Recognition* tools to extract and recognize the Initial Graphics Exchange Specification (IGES) file that is named holder.igs.

Automatic Feature Recognition Methodology

The FeatureWorks software attempts to automatically recognize and highlight as many features as possible. The advantage to this method is the speed at which the features are recognized because you do not select the faces or features. If the FeatureWorks software can automatically recognize most or all of the features in your model, then use the Automatic Feature Recognition.

To recognize features automatically,

1. Click Recognize Features (FeatureWorks toolbar) or click FeatureWorks > Recognize Features. The FeatureWorks PropertyManager appears.
2. Under Recognition Mode, click Automatic.
3. Under Feature Type, click one of these options:
   a. *Standard features*
   b. *Sheet metal features*
4. Click Next (arrowhead) to automatically recognize the selected features.
   The Intermediate State PropertyManager appears with the list of Recognized Features.
5. If necessary, you can Find Patterns, Combine Features, or Re-Recognize Features as alternate features.
6. Click to recognize the features. To exit without recognizing the features, click No.
   The features appear in the SolidWorks FeatureManager Design Tree.

Import Geometry

1. Open a file with the filename "HOLDER.IGS" while selecting the files of type IGES (*.igs.*.iges) (see Figure 10.25).
2. Click OK. (See Figure 10.26 for the New SolidWorks Part document PropertyManager.)
3. Click OK. (See Figure 10.27 for the imported geometry.)

Import Diagnostics

Import Diagnostics repairs faulty surfaces, knits repaired surfaces into closed bodies, and makes closed bodies into solids.

When a part is imported, the Import Diagnostics PropertyManager apprears, as shown in Figure 10.27, together with a query that reads, "Do you wish to run Import Diagnostics on this part? Answer Yes if you wish this to run it or No if otherwise."

The Import Diagnostics PropertyManager opens.

If you click Yes, this repair capability is needed because imported surface data often have problems that prevent surfaces from being converted into valid solids. These problems include the following:

- Bad surface geometry
- Bad surface topology (trim curves)
- Adjacent surfaces whose edges are close to each other but do not meet, thus creating gaps between the surfaces

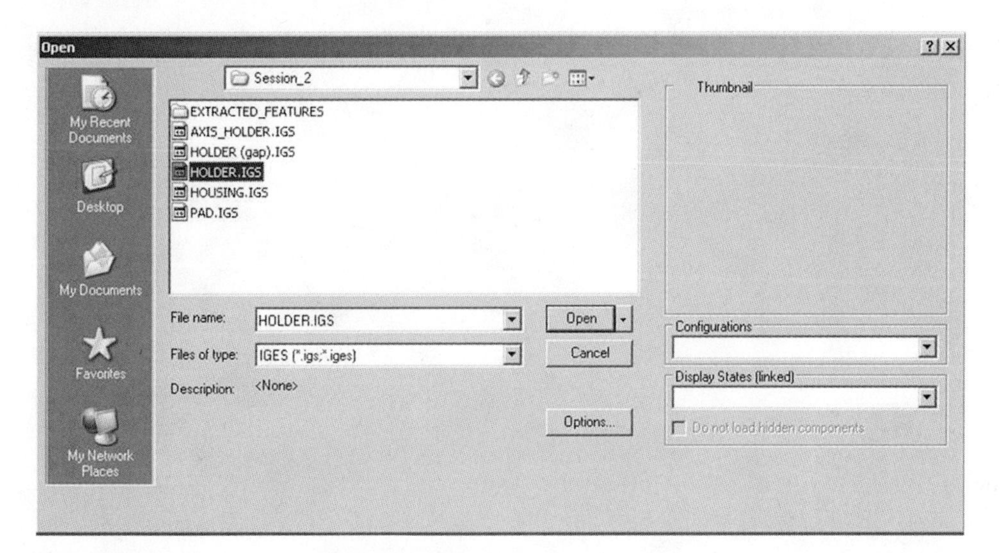

Figure 10.25

Opening a foreign file to SolidWorks (IGES file).

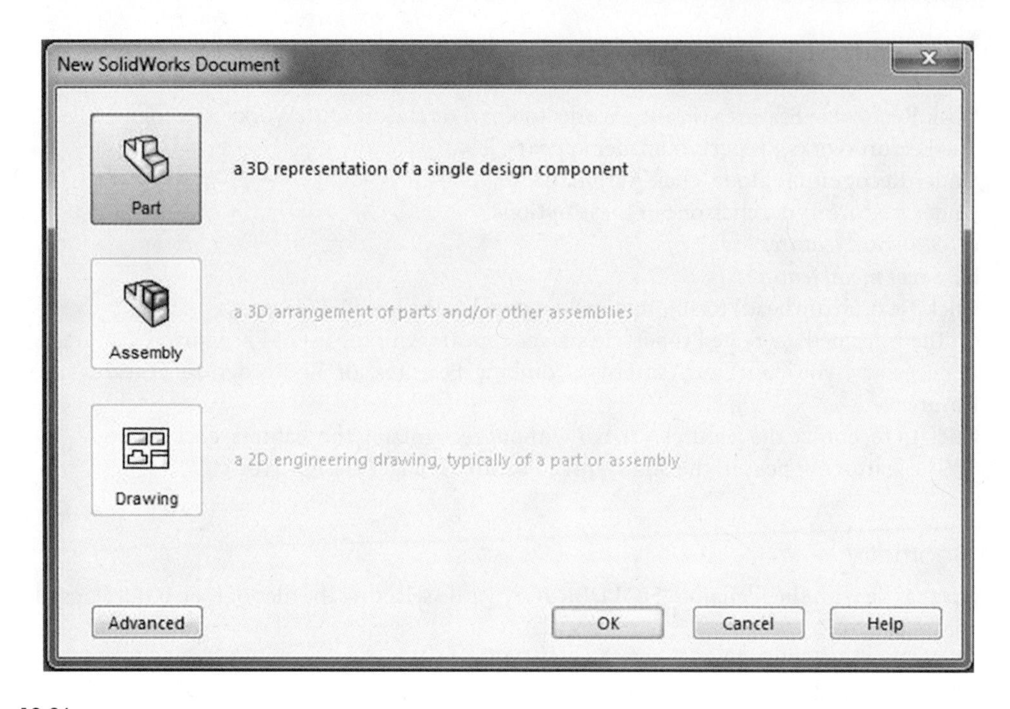

Figure 10.26

New SolidWorks Part document PropertyManager.

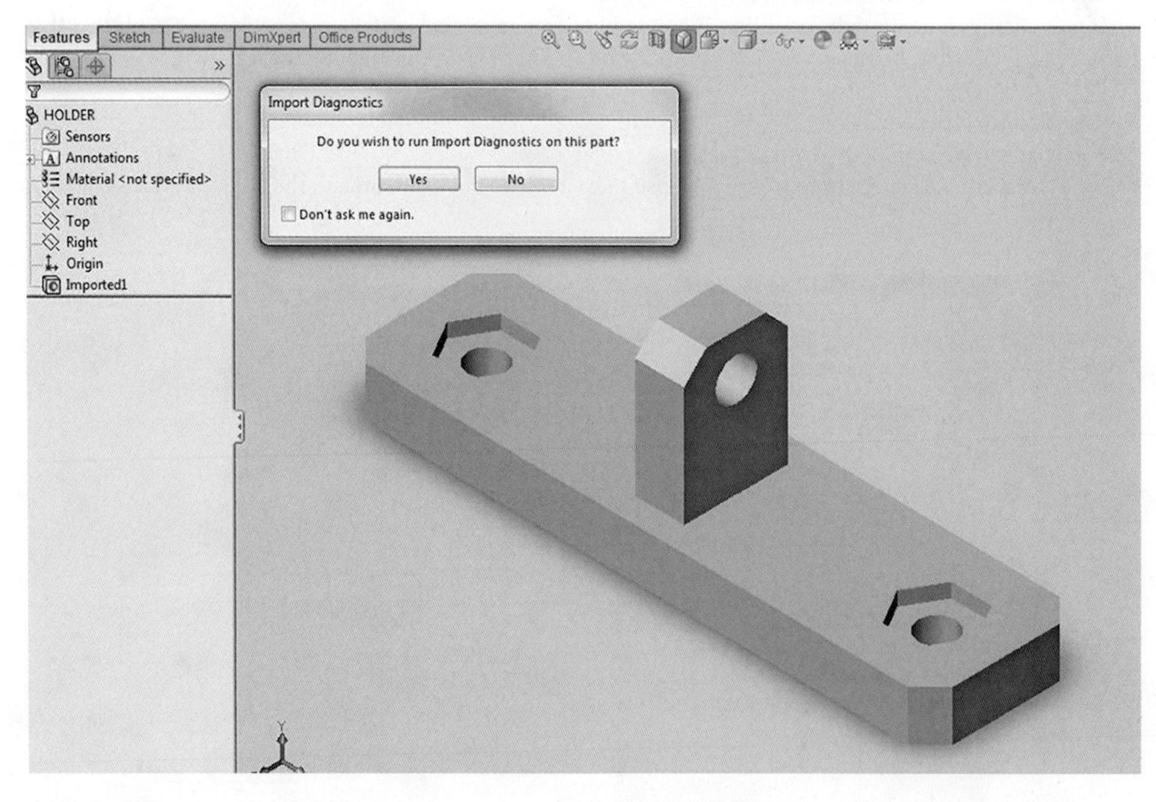

Figure 10.27

Import Diagnostics PropertyManager.

Import Diagnostics finds problems by

- Running the check that is used in Tools > Check,
- Running additional checks, such as overlapping surfaces, and
- Checking for accurate, unsimplified surfaces, which are B-splines that are planar, cylindrical, and so on and therefore can be replaced with equivalent analytic surfaces, improving performance and making the model easier to reference.

Import Diagnostics repairs errors in the geometry (the underlying surface) and topology (the boundaries) of faces by doing one or more of the following:

- Recreating the trim boundaries of a face based on the surrounding geometry (which often fixes overlapping faces).
- Trimming away the defective portions of faces (for cases in which the defective portion is not used in the model).
- Removing the face and using the gap repair algorithm to fill the resulting hole (a last resort).

Import Diagnostics eliminates the gaps between adjacent faces by doing one or more of the following:

- Replacing two close but nonintersecting edges with one tolerant edge.
- Creating a fill surface or lofted surface to fill the gap.
- Extending two adjacent faces into each other to eliminate the gap.

Additional functionality
  - Converts unsimplified surfaces into analytic surfaces.
  - Knits repaired faces into the rest of the surface body, if possible.
  - Converts the body into a solid if the surface body is closed (without gaps). This is done automatically when you click OK in the dialog box.

General approach to using the Import PropertyManager
  - Click Attempt to Heal All.
  - *Repair faces.* Right-click a face in the list and select a command.
  - *Repair gaps.* Right-click a gap in the list and select a command or the Gap Closer tool.

1. Click Yes (if desired) for the imported part, which results in the Import Diagnostic PropertyManager appearing, as shown in Figure 10.28.
2. Click OK to continue (see Figure 10.29 in which the Feature Recognition Query appears).

There is a query that reads, “Do you want to proceed with feature recognition? Answer Yes if you wish this to proceed or No if otherwise.”

Automatic Feature Recognition

1. Click Yes for the query in Figure 10.29 to proceed with feature recognition. (See Figure 10.30 in which the Feature Recognition PropertyManager appears.)
2. Select Automatic for the Selection Mode (see Figure 10.30).
3. Click OK. (See Figure 10.31 for the recognized feature and the FeatureManager.)

Interactive Feature Recognition Methodology

With this methodology, you select the feature type and the entities that make up the feature that you want to recognize.

The advantage to this method is the control that you have over the features that are recognized. For example, you can decide if you want to recognize a cylindrical cut as an extrusion, a revolve, or a hole. Additionally, you can determine the location and complexity of the sketches of your features by the faces and edges that you select.

To recognize features interactively,

1. Click Recognize Features (FeatureWorks toolbar) or click FeatureWorks > Recognize Features.
   The FeatureWorks PropertyManager appears.
2. Under Recognition Mode, click Interactive.

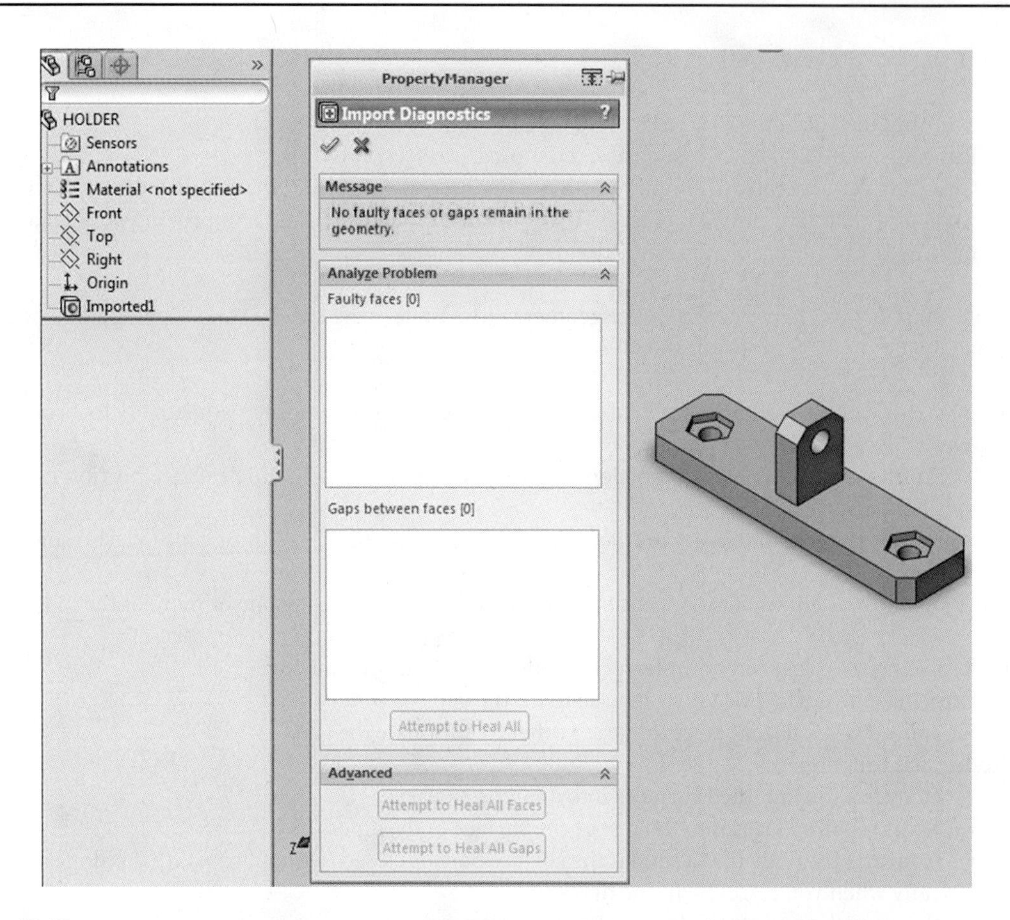

Figure 10.28

Import Diagnostic PropertyManager.

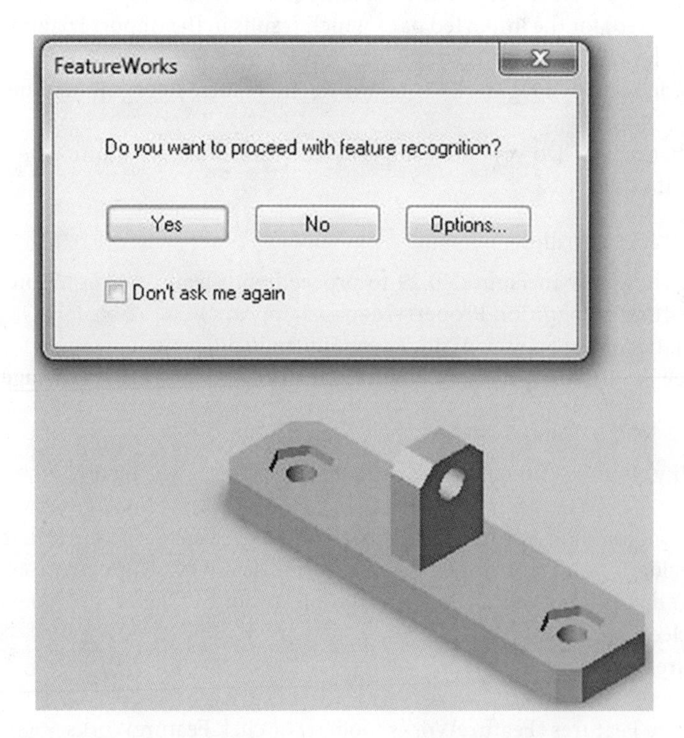

Figure 10.29

FeatureWorks PropertyManager.

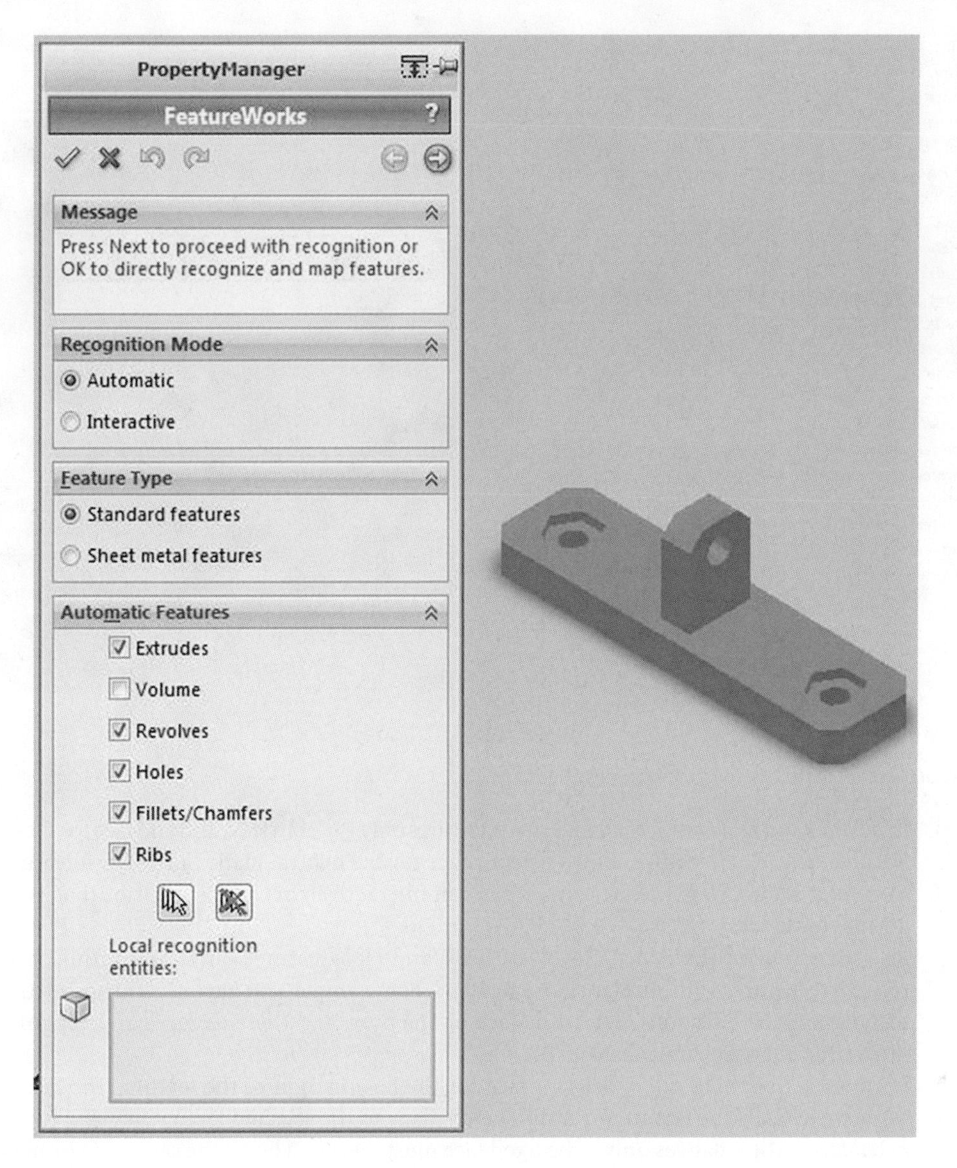

Figure 10.30

FeatureWorks PropertyManager.

3. Under Feature Type, click one of these options:
   a. Standard features
   b. Sheet metal features
4. Under Interactive Features, choose the Feature type.
5. Under Selected entities, select geometry from the graphics area to recognize as the selected Feature type. You do not have to select every face of a feature to recognize it.
6. Select from the following options, which appear depending on the selected feature type:
   a. *Chain chamfer faces.* (Chamfer features only.) Select this check box to have FeatureWorks find additional faces for a chamfer feature that is adjacent to the face that you select.
   b. *Chain fillet faces.* (Fillet features only.) Select this check box to have FeatureWorks find additional fillet faces that are tangent to the fillet face that you select. You can select a chain of fillets with different radii on each edge. Variable radius fillets can be included in a chain of fillets. The selected chain is recognized as a single feature. If the fillet faces do not chain automatically, you may be able to manually select them to create a chain as a single feature.
   c. *Chain revolved faces.* (Revolve and Hole features only.) Select this check box if you want the FeatureWorks software to determine the faces for a revolved feature from a minimum set of faces that you select, and for an illustration of the effect of this check box.

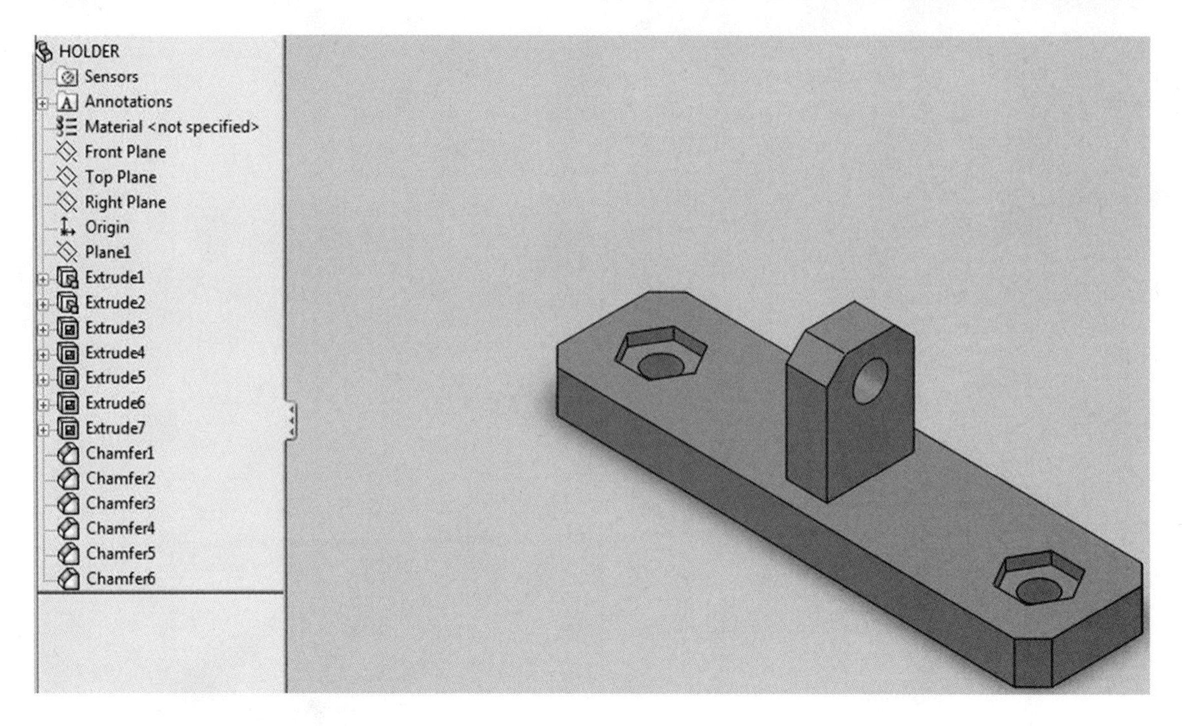

Figure 10.31

Recognized feature and the FeatureManager.

- **d.** *End Face 1 and End Face 2*. (*Loft* and *sweep* features only.) Select a start and end face for these features.
- **e.** *Neutral face*. (Draft features only.) Select a face for the neutral plane of the Draft feature.
- **f.** *Normal to sketch*. (Rib features only.) You can interactively recognize the ribs that are extruded normal to sketch.
- **g.** *Recognize similar*. (Extrudes, Revolves, Ribs, and Hole features only.) Select this check box to recognize features with similar characteristics. For example, you have several boss extrudes with rectangular cross sections. Select the face of one boss, and these features are recognized at the same time but as separate features.
- **h.** *Up to face*. (Extrudes only.) Select a face for the termination of the feature. The FeatureWorks software extends the feature from the sketch plane to the selected face.
- **i.** *Fixed face*. (Miter flanges only.) The fixed face must meet and be on the same side (inner or outer) as the selected miter flange faces.

7. If you want to delete one or more faces, click *Delete Faces*. You can use this prior to feature recognition to get rid of complicated or unwanted geometry. If you delete the faces that are necessary for previously recognized features, FeatureWorks will not be able to recreate those features.
8. Click Recognize to interactively recognize the selected features.
   If successful, the features are removed from the imported body in the graphics area.
9. If you want to undo the recognition of a feature, click Anticlockwise Arrow in the FeatureWorks PropertyManager. Click Clockwise Arrow to recreate the undone feature.
10. Continue to recognize the features of different feature types.
11. Click Next (forward arrowhead) or Recognize.
    The *Intermediate Stage* PropertyManager appears with the list of Recognized Features.
12. Click Next (backward arrowhead) to complete feature recognition and create the new features.

The following steps are common to both Automatic and Interactive Feature Recognition methods: (a) FeatureWorks Add-In, (b) Options, (c) Import non-SolidWorks Geometry, and (d) Import Diagnostics. Therefore, these steps are not repeated here. We will concentrate on the Interactive Feature Recognition method.

1. Select Interactive for the Recognition Mode from the FeatureWorks PropertyManager, which is shown in Figure 10.32, similar to Figure 10.30. (Note: Boss Extrude is the default Feature type.)

Recognize Chamfer features on the Base

2. Select Chamfer from the drop-down button of the Feature type (see Figure 10.33).

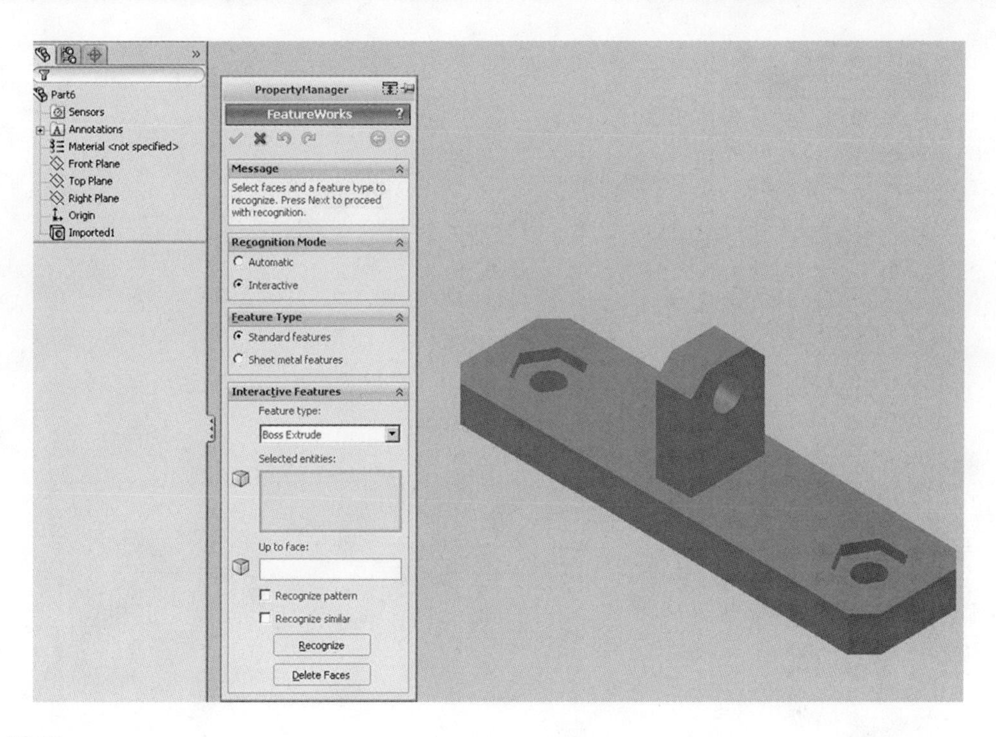

Figure 10.32

FeatureWorks PropertyManager.

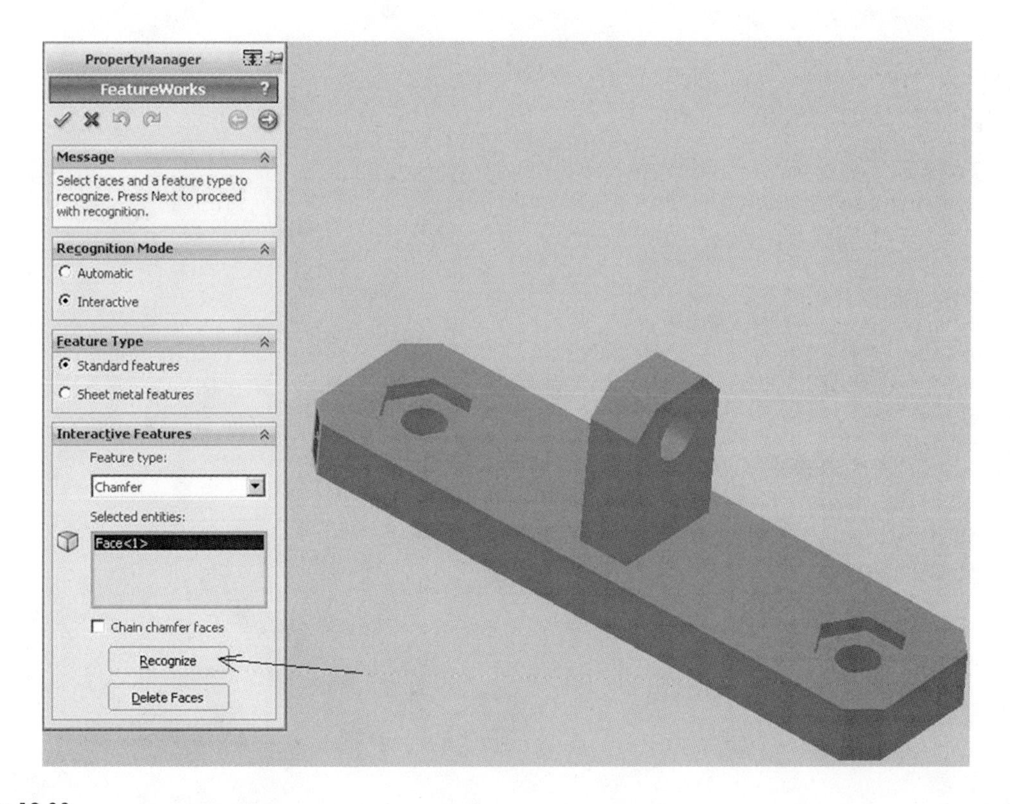

Figure 10.33

Recognition first chamfer feature.

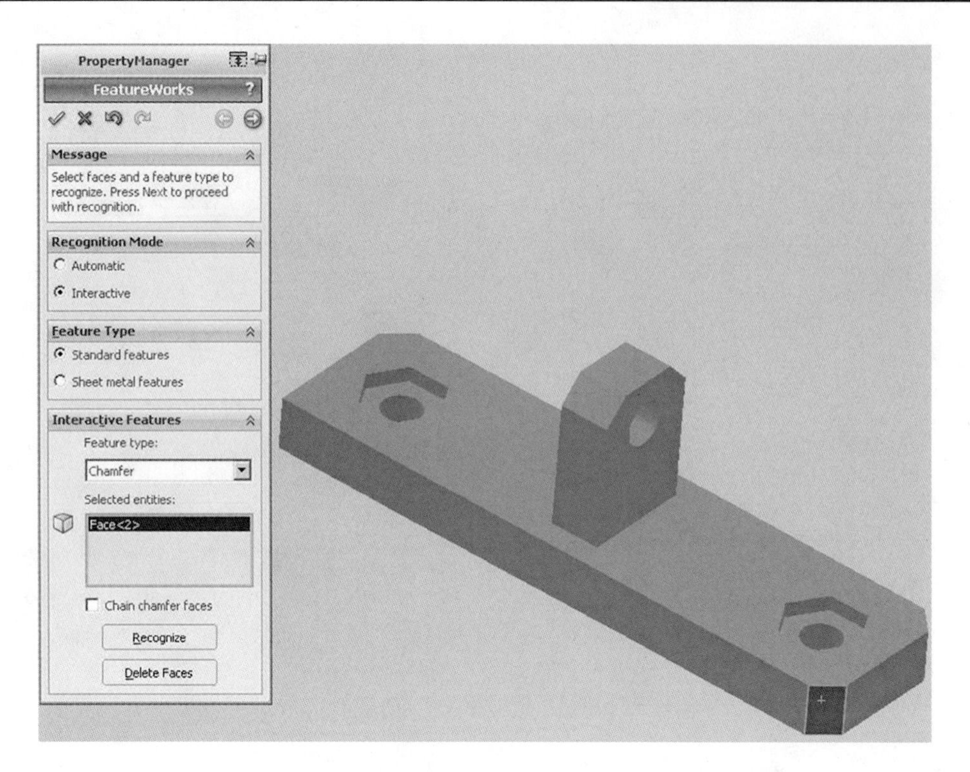

Figure 10.34

Recognition second chamfer feature.

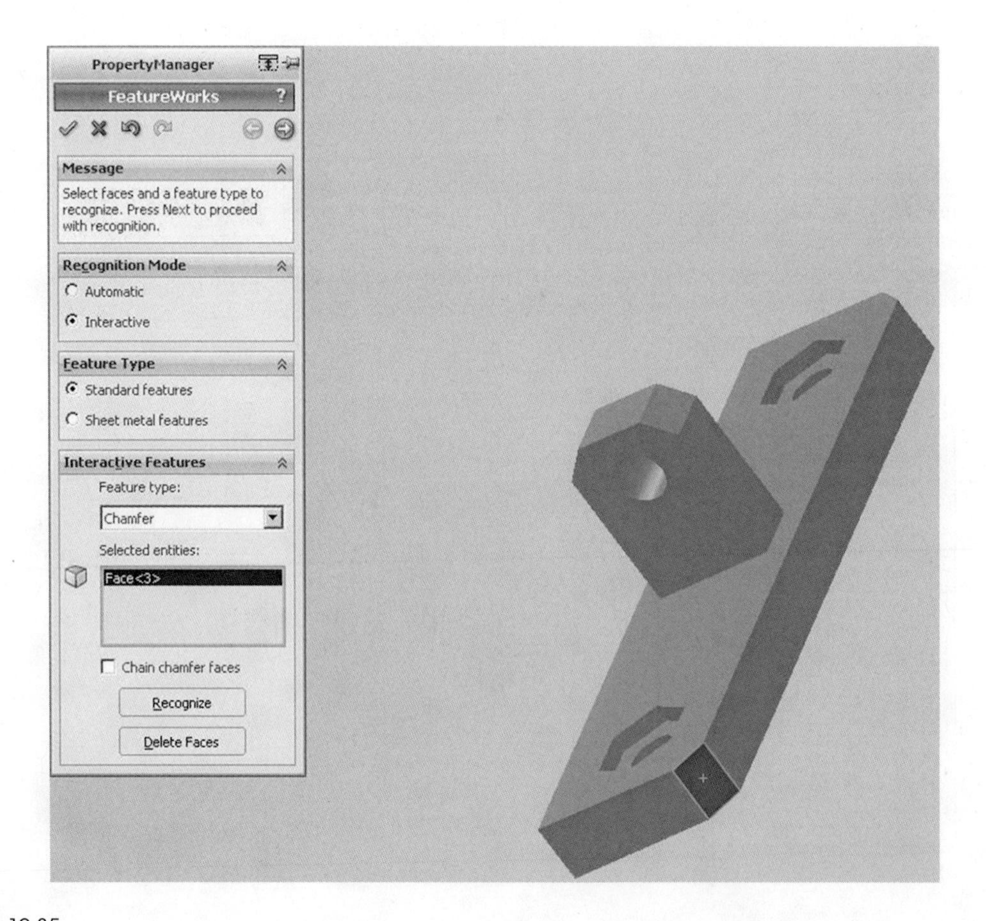

Figure 10.35

Recognition third chamfer feature.

3. Click the face of one chamfer feature (see Figure 10.33).
4. Click the Recognize button (see Figure 10.33).
5. Click the face of the second chamfer feature (see Figure 10.34).
6. Click the Recognize button (see Figure 10.34).
7. Click the face of the third chamfer feature (see Figure 10.35).
8. Click the Recognize button (see Figure 10.35).
9. Click the face of the fourth chamfer feature (see Figure 10.36).
10. Click the Recognize button (see Figure 10.36).

Recognize Cut Extrude Features on the Base and Boss

11. Select Cut Extrude from the drop-down button of the Feature type (see Figure 10.37).
12. Click the face of first cut extrude feature (see Figure 10.37).
13. Click the Recognize button (see Figure 10.37).
14. Click the face of the second cut extrude feature (see Figure 10.38).
15. Click the Recognize button (see Figure 10.38).
16. Click the face of third cut extrude feature (see Figure 10.39).
17. Click the Recognize button (see Figure 10.39).
18. Click the face of fourth cut extrude feature (see Figure 10.40).
19. Click the Recognize button (see Figure 10.40).
20. Click the face of fifth cut extrude feature (see Figure 10.41).
21. Click the Recognize button (see Figure 10.41).

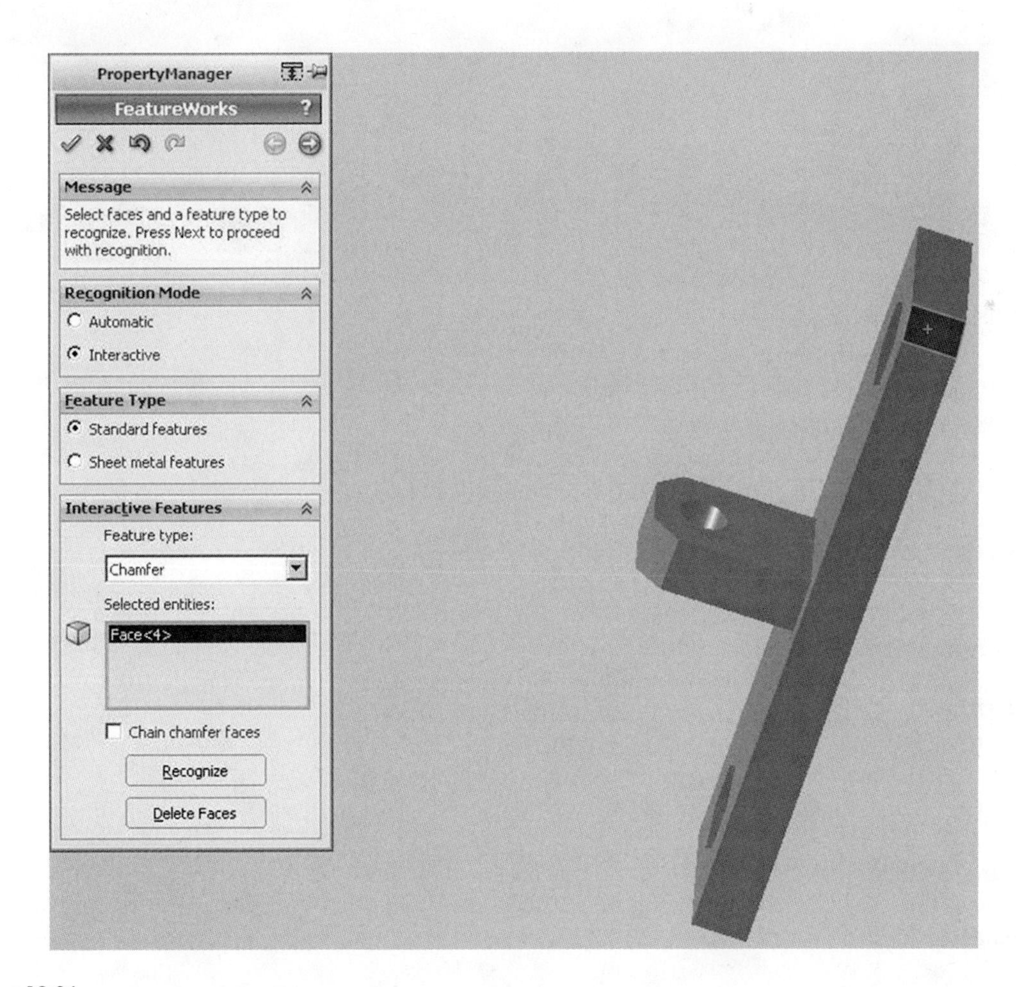

Figure 10.36

Recognition fourth chamfer feature.

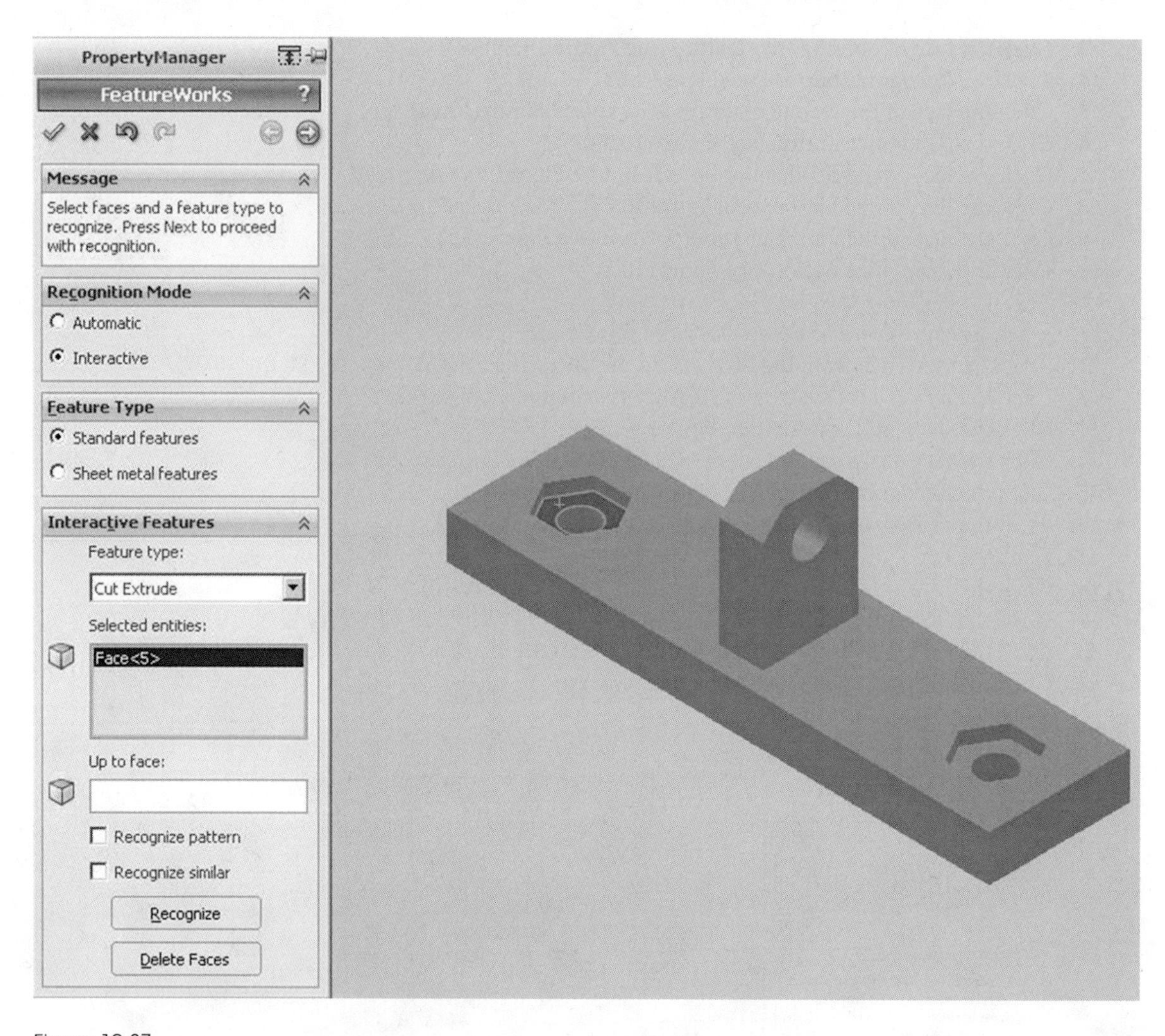

Figure 10.37

Recognition first cut extrude feature.

Recognize Chamfer Features on the Boss

22. Select Chamfer from the drop-down button of the Feature type (see Figure 10.42).
23. Click the face of one chamfer feature (see Figure 10.42).
24. Click the Recognize button (see Figure 10.42).
25. Click the face of the second boss chamfer feature (see Figure 10.43).
26. Click the Recognize button (see Figure 10.43).

Recognize Boss Extrude

27. Select Boss Extrude from the drop-down button of the Feature type (see Figure 10.44).
28. Click the face of top boss extrude feature (see Figure 10.44).
29. Click the Recognize button (see Figure 10.44).
30. Click the face of base boss extrude feature (see Figure 10.45).
31. Click the Recognize button (see Figure 10.45).

Figure 10.46 shows the recognized features that are listed in the FeatureWorks PropertyManager. Figure 10.47 shows the recognized features, while Figure 10.48 shows the SolidWorks FeatureManager having the part renamed.

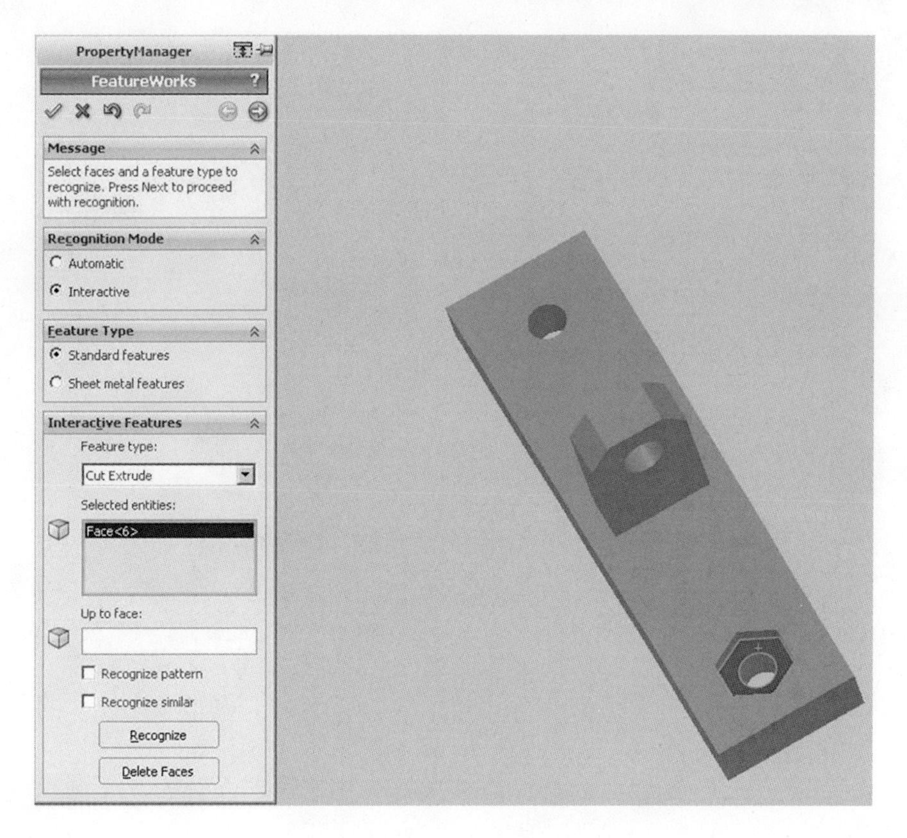

Figure 10.38

Recognition second cut extrude feature.

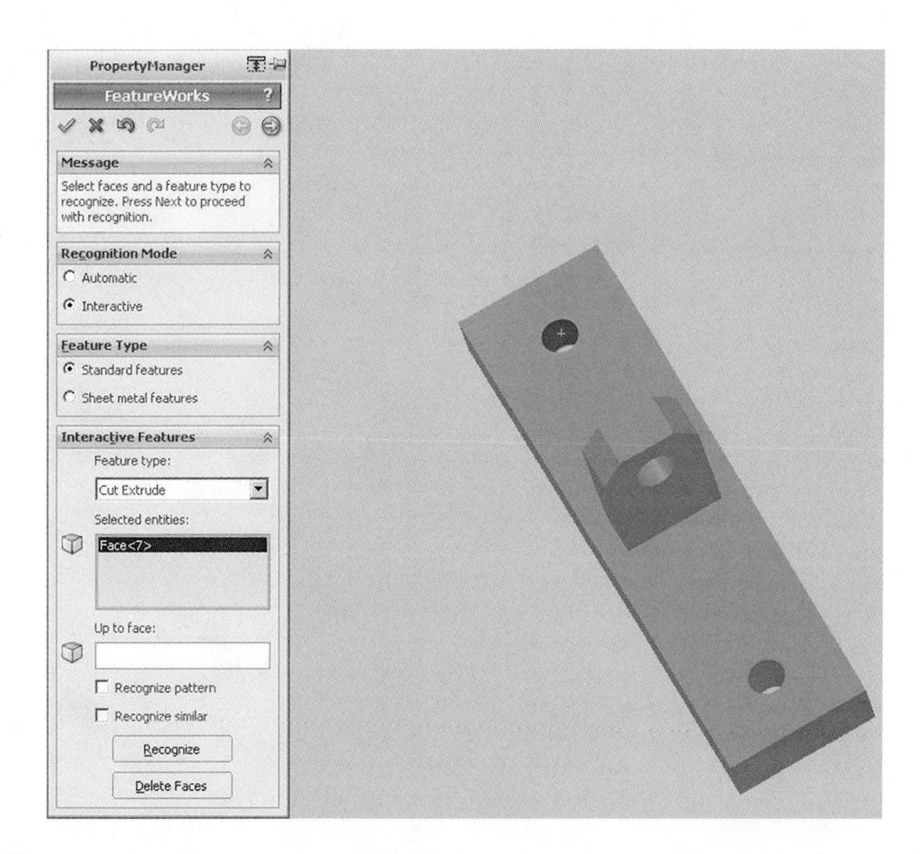

Figure 10.39

Recognition third cut extrude feature.

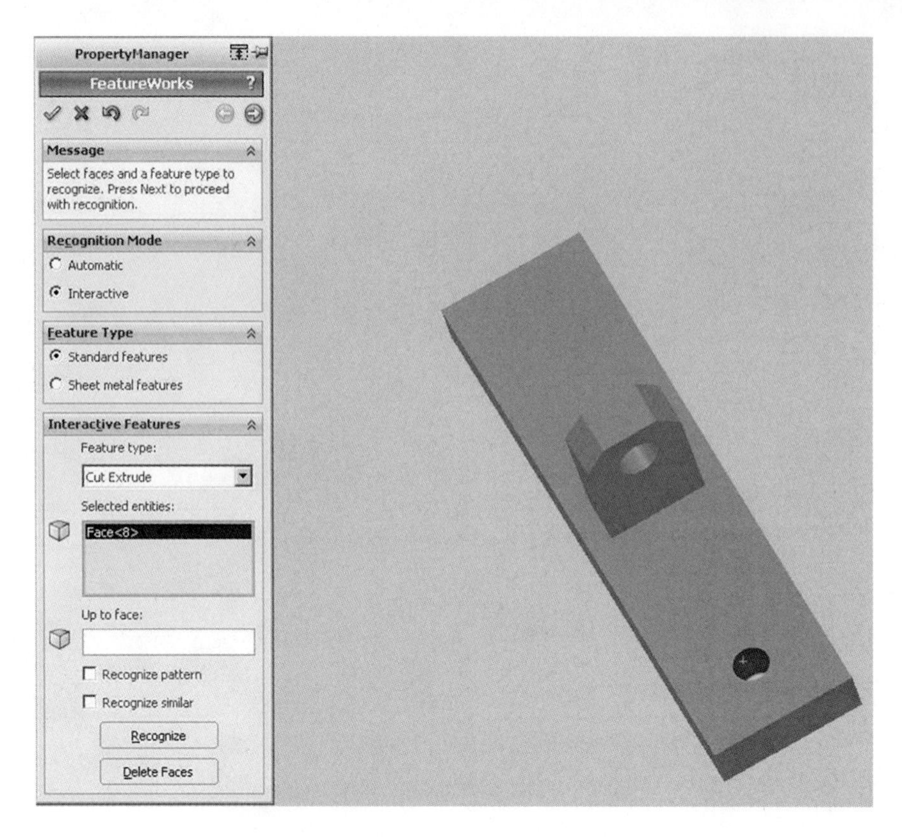

Figure 10.40

Recognition fourth cut extrude feature.

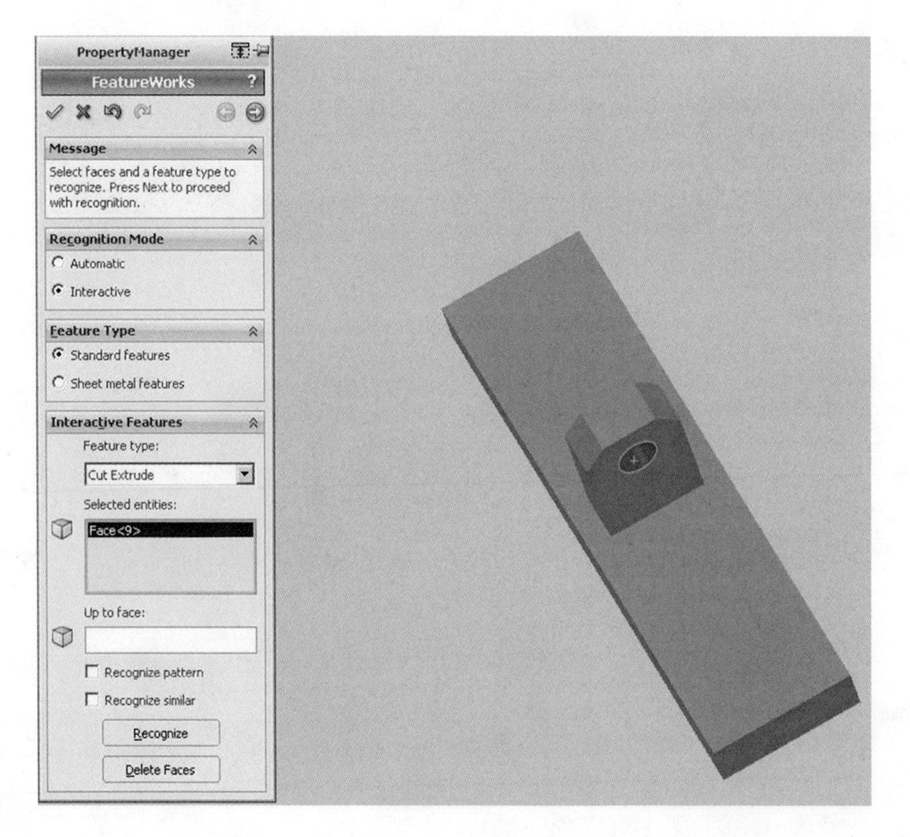

Figure 10.41

Recognition fifth cut extrude feature.

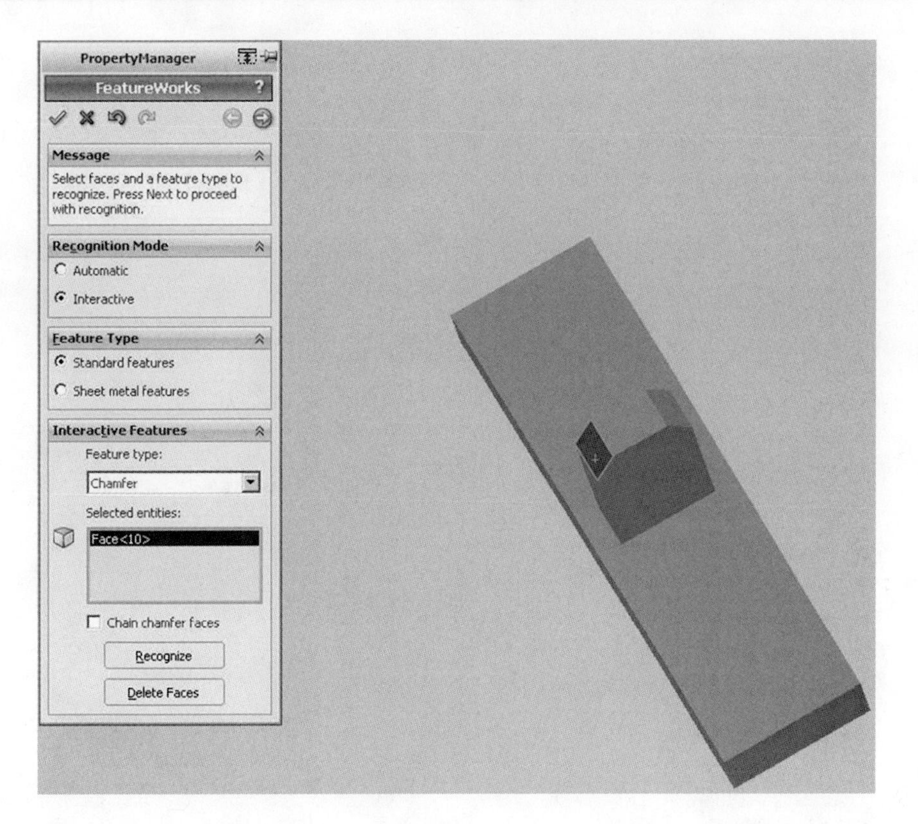

Figure 10.42

Recognition first boss chamfer feature.

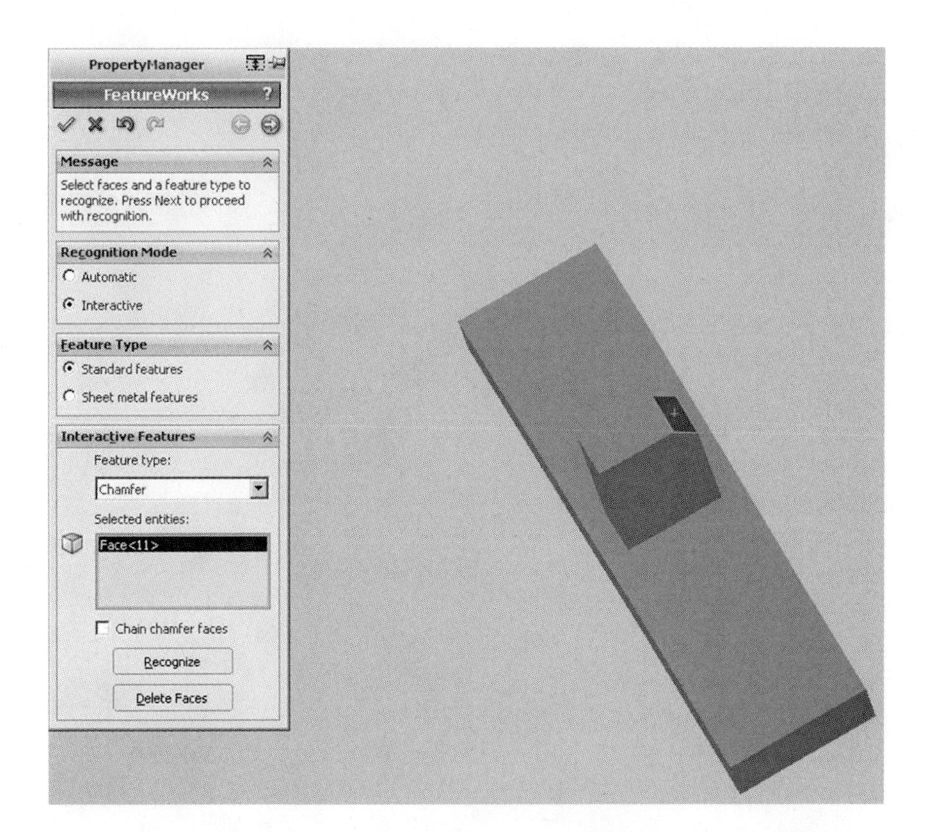

Figure 10.43

Recognition second boss chamfer feature.

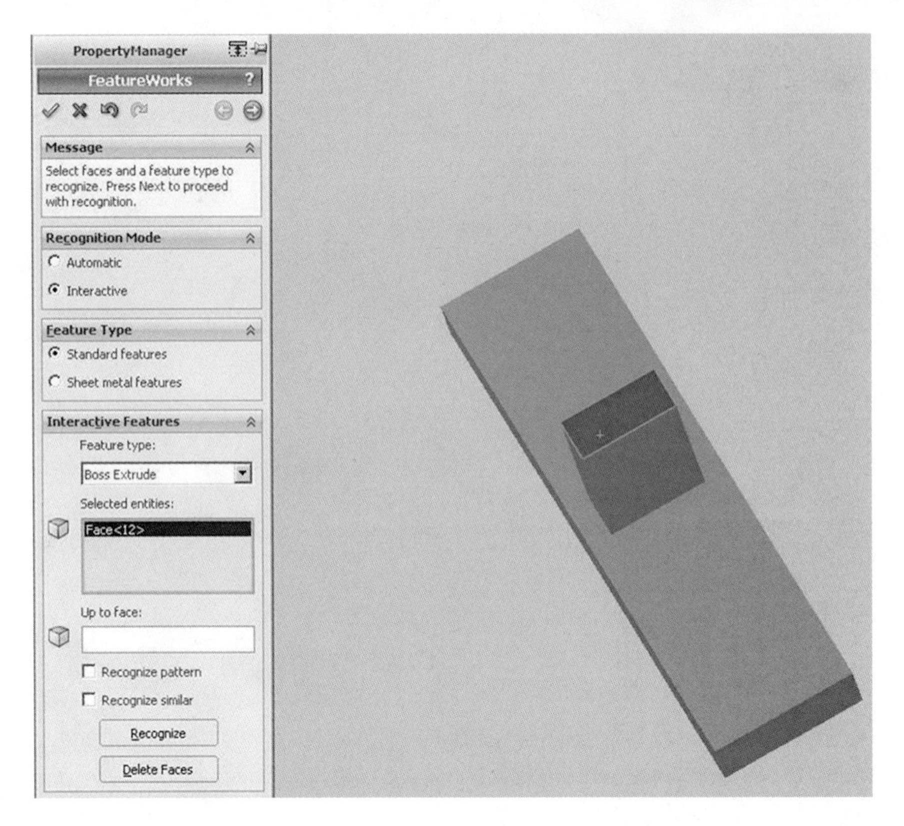

Figure 10.44

Recognition first boss extrude feature.

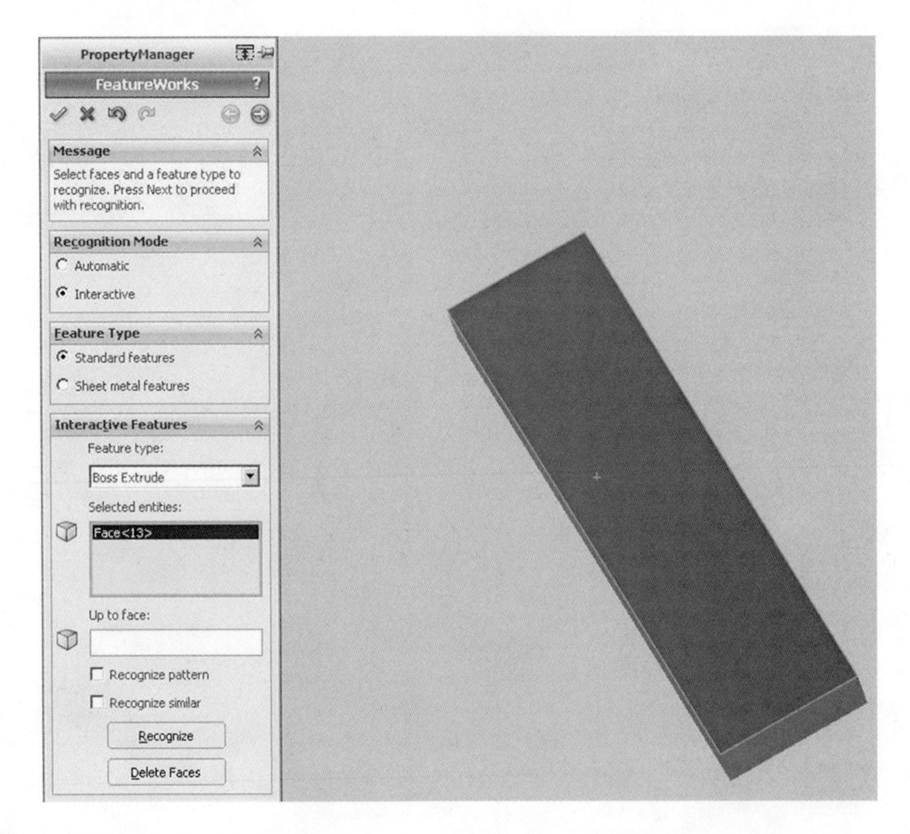

Figure 10.45

Recognition second boss extrude feature.

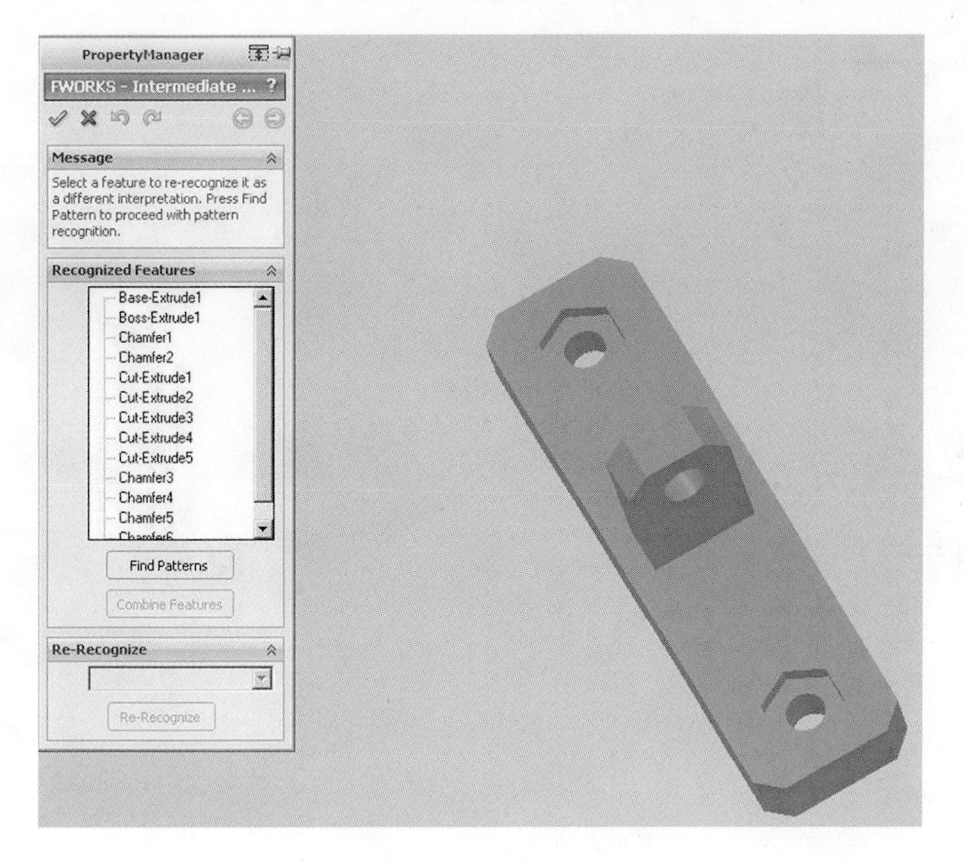

Figure 10.46

Recognized features listed in FeatureWorks PropertyManager.

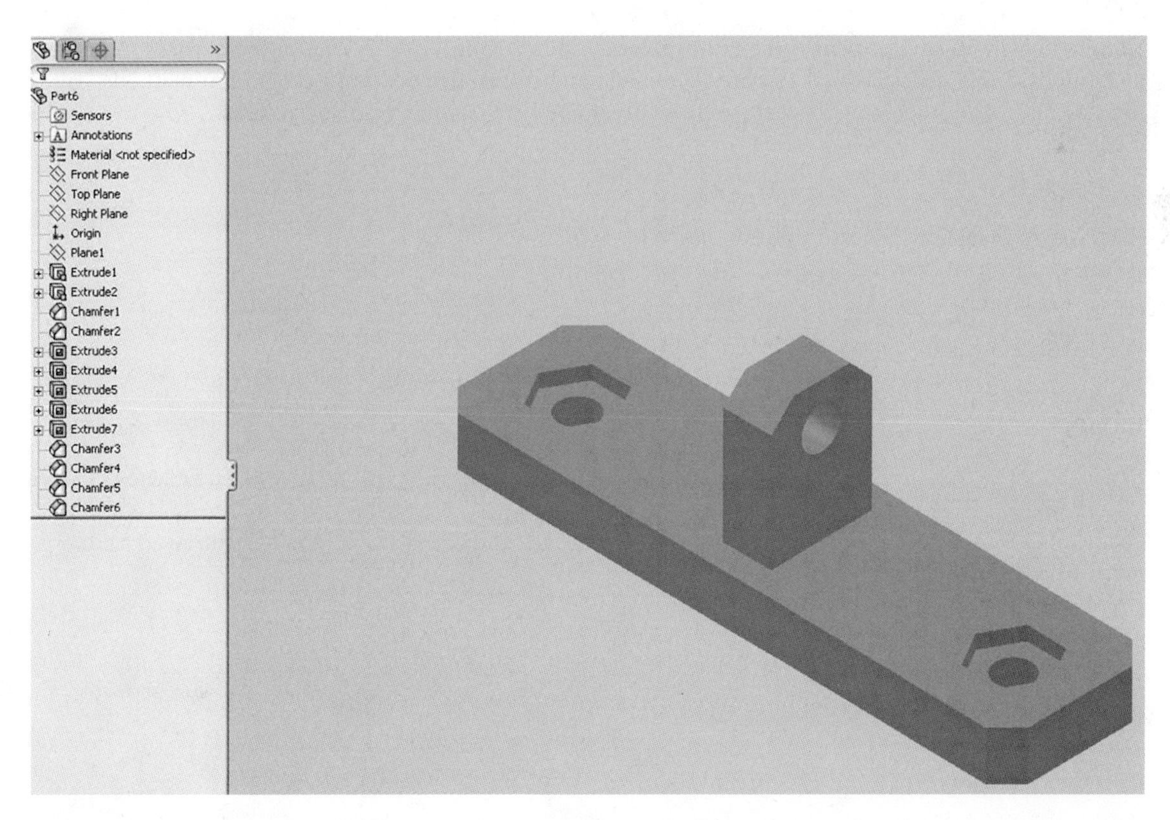

Figure 10.47

SolidWorks FeatureManager showing recognized features.

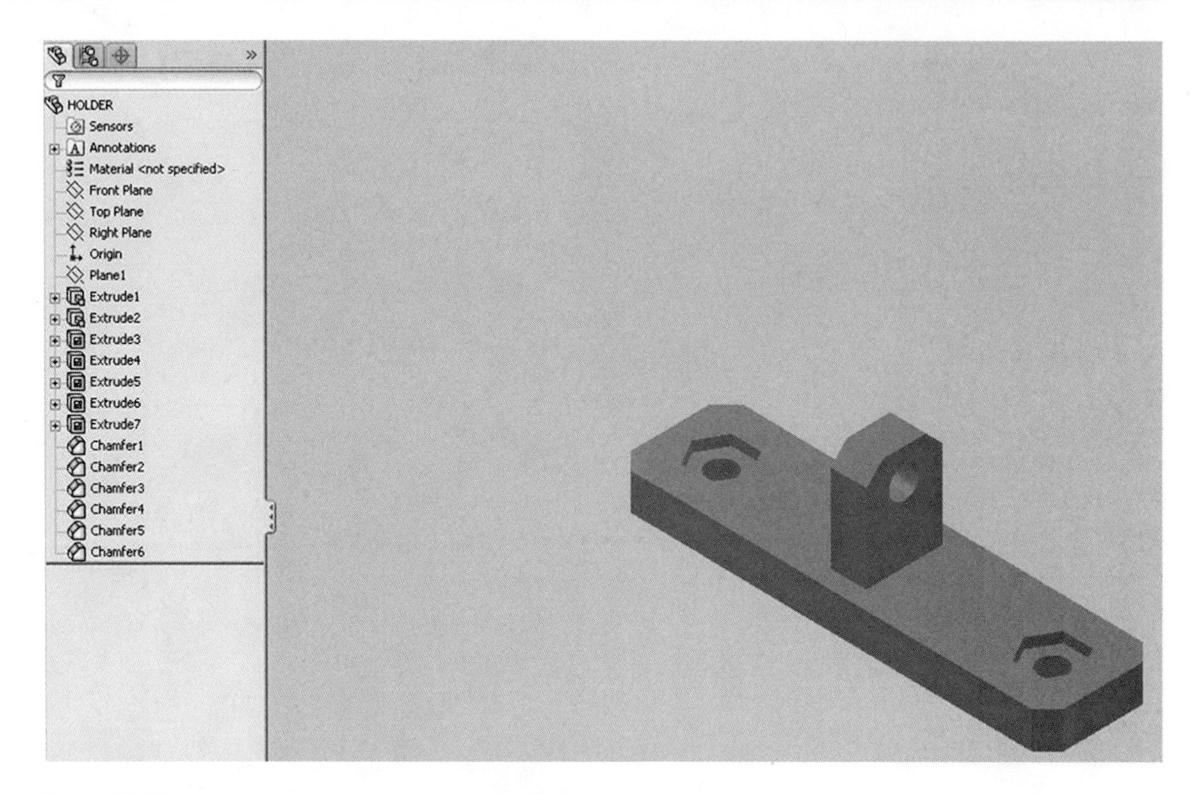

Figure 10.48

SolidWorks FeatureManager showing part renamed.

Summary

In this chapter, we have presented the FeatureWorks Add-In, Options, Import non-SolidWorks Geometry, and Import Diagnostics, as well as both Automatic and Interactive Feature Recognition methods. These tools are useful for importing the features from other CAD environments to SolidWorks.

Exercises

Reconstruct the following features from the available files:

- Housing
- Sheet metal

Bibliography

SolidWorks Help. http://help.solidworks.com/2010/English/SolidWorks/sldworks/SW_Sketch/Sketch_Picture.htm?

solidworksacademic.de. http://www.solidworksacademic.de/solidworks/.../swx_2007_Featureworks.pdf.

Top-Down Design

Objectives:

When you complete this chapter, you will have

- Understood the concept of the top–down design approach
- Used the top–down design approach to develop products from within the assembly

Introduction

The *top–down design approach* is a conceptual approach that is used to develop products from within the assembly. The concept is applied to the design of a cabinet containing electronic parts, as well as to the design of a seal from a lid. Generally, there are two ways in which the top–down design approach could be implemented: (1) designing from layout or (2) designing from part outline. These two approaches will be illustrated through the examples in the following section.

Designing from Layout

Layout of the Cabinet

The layout of the CABINET involves sketching its two-dimensional (2D) profile and dimensioning it. This involves the following steps:

1. Click New from the Menu bar.
2. Click the MY-TEMPLATES tab.
3. Double-click ASM-MM-ANSI.
4. Click Cancel X from the Begin Assembly PropertyManager.
5. Save the model CABINET.
6. Right-click Front Plane from the FeatureManager.
7. Click Sketch from the Context toolbar. The Sketch toolbar is displayed.
8. Sketch a rectangle using the Corner Rectangle option starting from the origin and stretching the top-right corner away from the origin (see Figure 11.1).

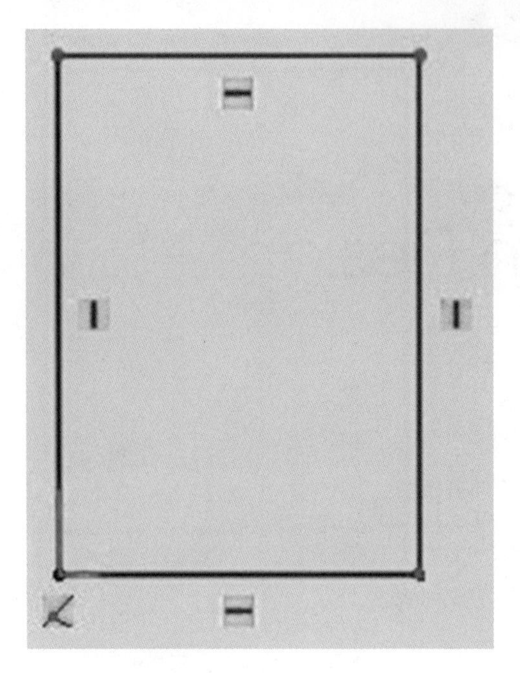

Figure 11.1

Profiles of CABINET.

9. Click Smart Dimensions from the Sketch toolbar.
10. Click one of the horizontal edges and enter the value of 300 as the width (see Figure 11.2).
11. Click one of the vertical edges and enter the value of 400 as the length (see Figure 11.2).

Create Shared Values

Shared Values are linked in a design using Shared Values. The use of Shared Values here is to build design constraints between the layout of the CABINET and the subassemblies of the power supply, the microcontroller, and the cooling unit. Later, we will see how to use Shared Values for configuration management.

12. Right-click the vertical dimension of the CABINET.
13. Click Link Value to open the Shared Values dialog box (see Figure 11.3).

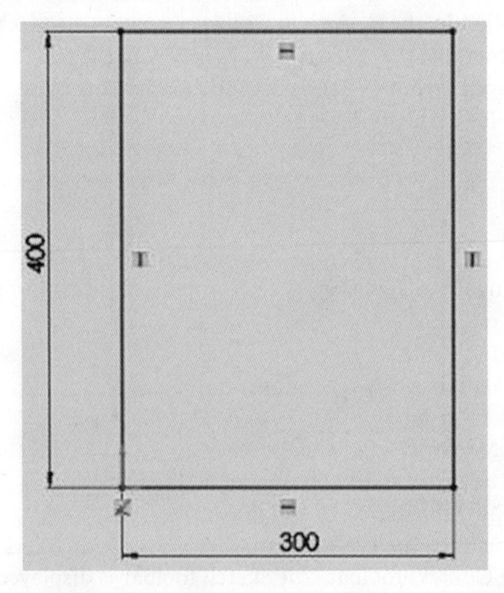

Figure 11.2

Dimensions of CABINET.

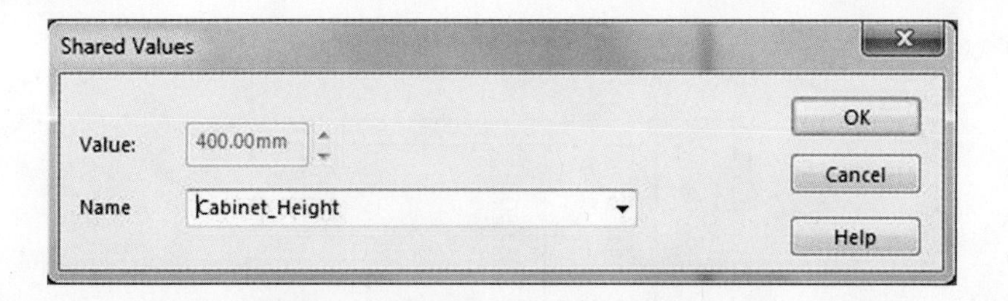

Figure 11.3

Shared Values dialog box.

14. In the Name option, type Cabinet_Height and click OK.
15. Click OK on the Dimension dialog box to complete the process.

Repeat the process for the width of the CABINET.
Notice the dashed (–) shapes that appear beside each of the dimensions (see Figure 11.4).

Sketch the 2D Profile of the Power Supply

16. Click Sketch from the Context toolbar. The Sketch toolbar is displayed.
17. Sketch a rectangle using the Corner Rectangle option starting from a point toward the top left, inside the CABINET and stretching the top-right corner away from the starting point (see Figure 11.5).

Dimension the 2D Profile of the Power Supply

18. Click Smart Dimensions from the Sketch toolbar.
19. Click one of the horizontal edges and enter the value of 75 as the width (see Figure 11.5).
20. Click one of the vertical edges and enter the value of 150 as the length (see Figure 11.5).

Gaps for the Power Supply Profile

21. Click the Smart Dimensions from the Sketch toolbar.
22. Click the top horizontal edges of the power supply profile and the CABINET, and enter the value of 20 as the gap (see Figure 11.5).
23. Click the left (right) vertical edges of the power supply profile and the CABINET, and enter the value of 20 as the gap (see Figure 11.5).

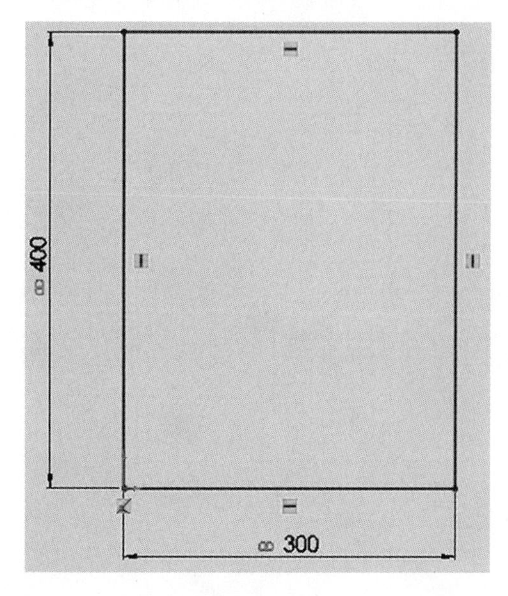

Figure 11.4

Shared Values have dashed (–) signs attached to them.

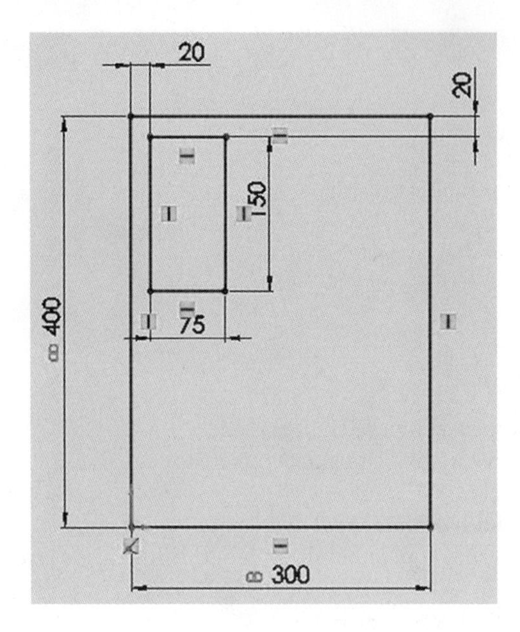

Figure 11.5

Power supply profile, dimensions, and gaps.

Sketch the 2D Profile of the Cooling Unit

24. Click Sketch from the Context toolbar. The Sketch toolbar is displayed.
25. Sketch a rectangle using the Corner Rectangle option starting from a point toward the bottom left, inside the CABINET and stretching the top-right corner away from the starting point (see Figure 11.6).

Dimension the 2D Profile of the Cooling Unit

26. Click Smart Dimensions from the Sketch toolbar.
27. Click one of the horizontal edges and enter the value of 75 as the width (see Figure 11.6).
28. Click one of the vertical edges and enter the value of 100 as the length (see Figure 11.6).

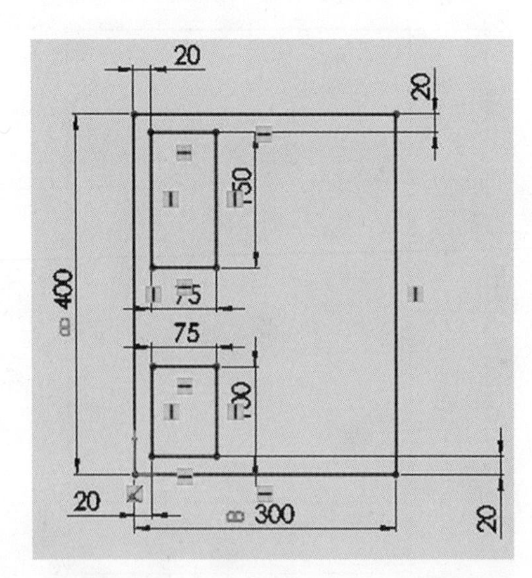

Figure 11.6

Cooling unit profile, dimensions, and gaps.

Gaps for the Cooling Unit

29. Click the Smart Dimensions from the Sketch toolbar.
30. Click the top horizontal edges of the power supply profile and the CABINET, and enter the value of 20 as the gap (see Figure 11.6).
31. Click the left (right) vertical edges of the power supply profile and the CABINET, and enter the value of 20 as the gap (see Figure 11.6).

Sketch the 2D Profile of the Microcontroller

32. Click Sketch from the Context toolbar. The Sketch toolbar is displayed.
33. Sketch a rectangle using the Corner Rectangle option starting from a point toward the bottom right, inside the CABINET and stretching the top-right corner away from the starting point (see Figure 11.7).

Gaps for the Microcontroller

34. Click the Smart Dimensions from the Sketch toolbar.
35. Click the top horizontal edges of the microcontroller profile and the CABINET, and enter the value of 20 as the gap (see Figure 11.7).
36. Click the left (right) vertical edges of the power supply profile and the CABINET, and enter the value of 20 as the gap (see Figure 11.7).
37. Click the right vertical edge of the power supply profile and the left vertical edge of the microcontroller, and enter the value of 30 as the gap (see Figure 11.7).
38. Click Exit Sketch.
39. Rename Sketch1 to Design_Layout. (See Figure 11.8, FeatureManager.)
40. Click Save to save the design layout. (See Figure 11.9 for the design layout.)
41. Right-click the Annotations folder in the FeatureManager (see Figure 11.10).
42. Click Show Feature Dimensions (see Figure 11.10).
43. Right-click any of the gaps of value 20 mm.
44. Click Link Values.
45. Enter Gap as Name in the Shared Values dialog box.
46. Click OK.

Repeat the process for all other 20-mm gap dimensions (or hold the Ctrl key down for the multiple selection process).

Notice the dashed (–) shapes that appear beside each of the dimensions (see Figure 11.11).

If any of the gap values are changed, all gaps change accordingly. There is the need to rebuild the graphics window in order to effect the changes that are made in the new value of the gap.

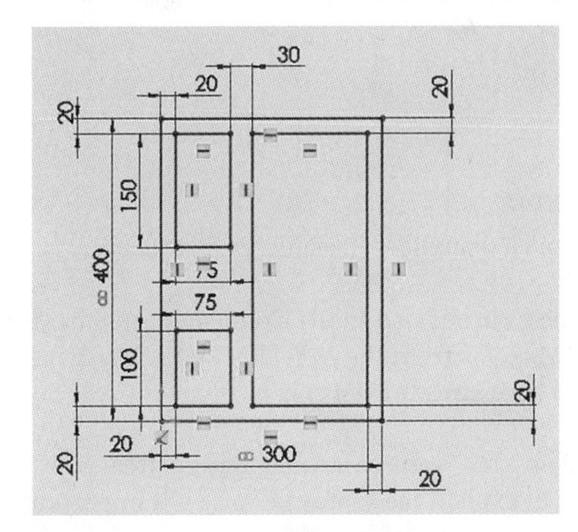

Figure 11.7

Microcontroller profile, dimensions, and gaps.

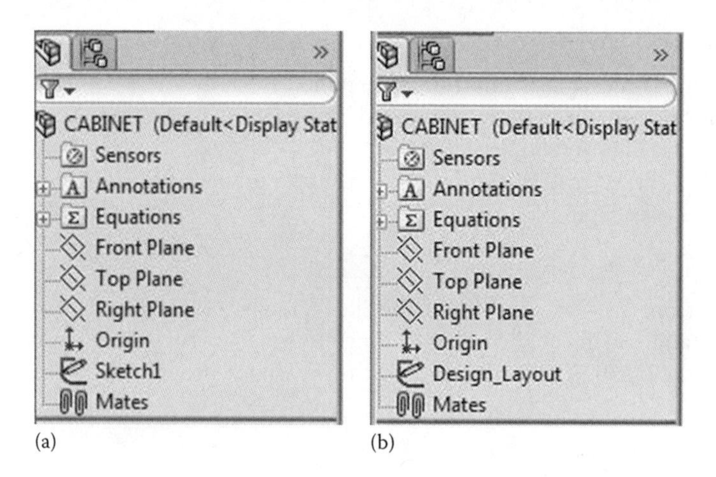

Figure 11.8

Renaming Sketch1 to Design_Layout: (a) before renaming and (b) after renaming.

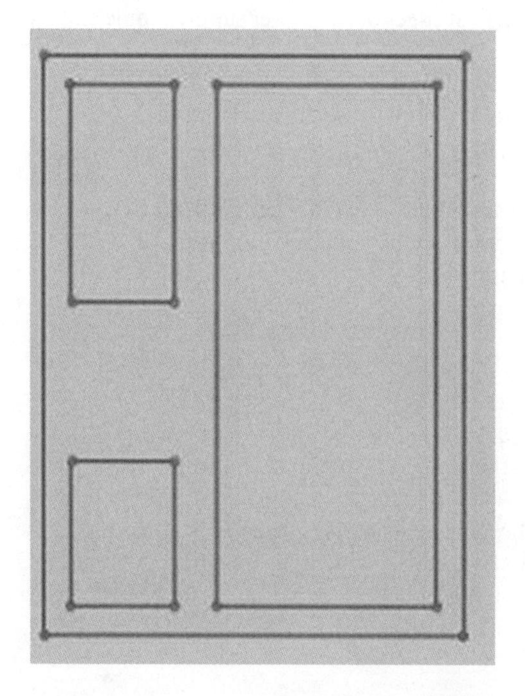

Figure 11.9

Design layout.

Microcontroller—Insert Component

The CABINET assembly is the open document (Default<Display State-1>) at this juncture.

47. Click New Part from the Consolidated Insert Components toolbar (see Figure 11.12).
48. Double-click PART-MM-ANSI from the MY-TEMPLATE tab in the New SolidWorks Document dialog box. (This would have already been created.)

The new part is displayed in the FeatureManager design tree with the new name in the format [Part#^*AssemblyName*]<#>. Let us explain the syntax because it is important to understand this:

- The square bracket shows that the part is a virtual component.
- # is the number of virtual components.
- *AssemblyName* is the assembly name.

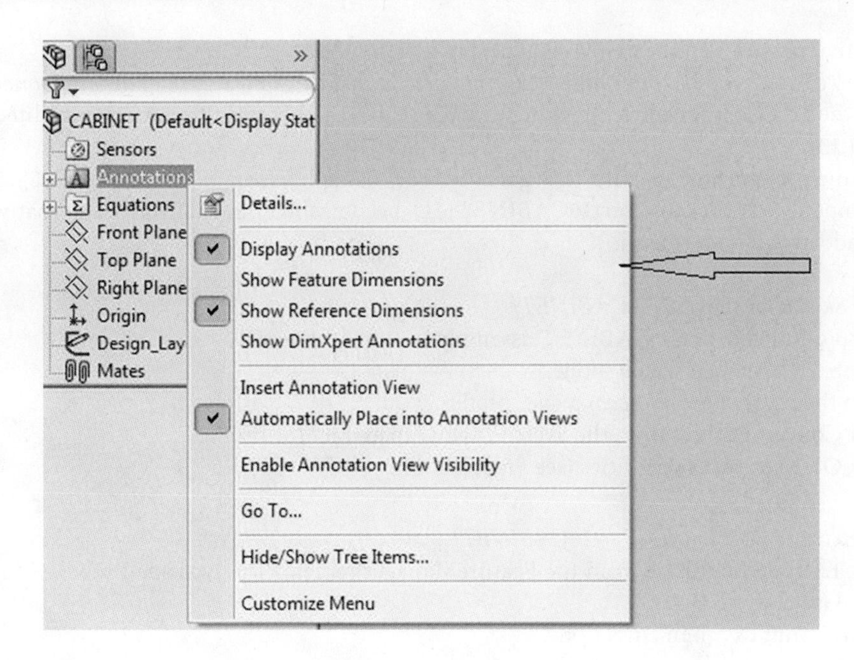

Figure 11.10

Echoing annotations.

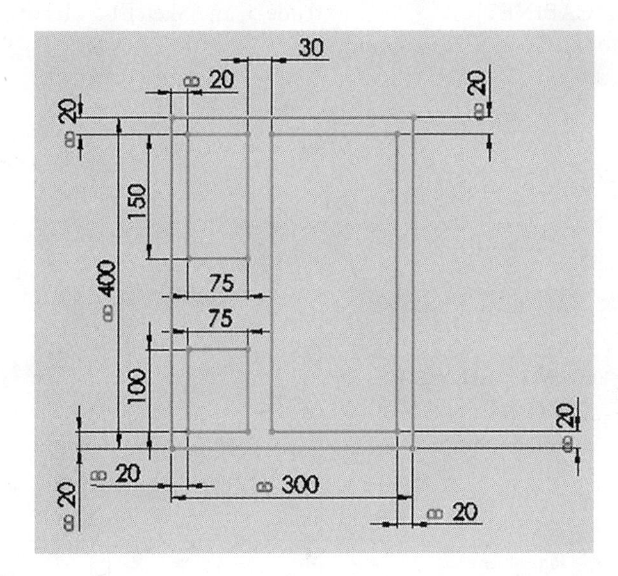

Figure 11.11

Gaps are linked.

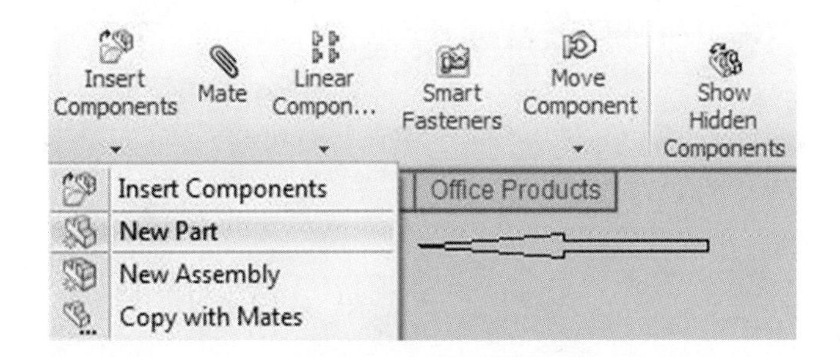

Figure 11.12

Inserting New Part from the Consolidated Insert Components toolbar.

In the current illustration, the first virtual part is [Part1^CABINET]<1>.

The gateway between the assembly and the Component added In-Context is the Edit Component feature. In other words, the Edit Component is the switch between the assembly and the component edited In-Context (see Figure 11.13).

It is extremely important to note this gateway. For example when [Part1^CABINET]<1> is clicked and Edit Component is clicked, [Part1^CABINET]<1> becomes light blue. This means that we are in the Component added In-Context level.

Create the Sketch for the [Part1^CABINET]

49. Click Front Plane in the CABINET assembly FeatureManager.
50. Click Sketch to be in sketch mode.
51. Select the *edges of the microcontroller profile.*
52. Click Convert Entities from the Sketch toolbar to extract the profile.
53. Click OK to extract the entities (see Figure 11.14).

Insert an Extrude Base for the [Part1^CABINET]

54. Click Extrude Boss/Base from the FeatureManager toolbar. The Extrude PropertyManager is displayed (see Figure 11.15).
55. Enter 15-mm Depth in Direction1.
56. Click OK to extrude.
57. Rename Extrude1 to Base Extrude.

Notice now that [Part1^CABINET]<1>->, Base Extrude->, and Sketch1-> all have the "->" symbol indicating External References to the CABINET assembly.

To save the assembly,

58. Click Edit Component. (This returns you to the CABINET assembly.)
59. Click Save.

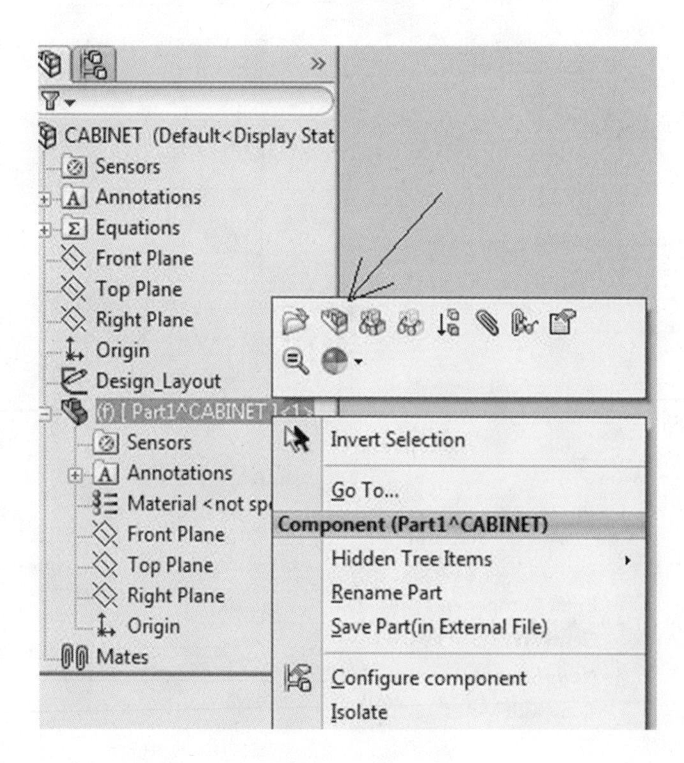

Figure 11.13

Switching to the Component added In-Context level.

Figure 11.14

Creating the sketch for the microcontroller.

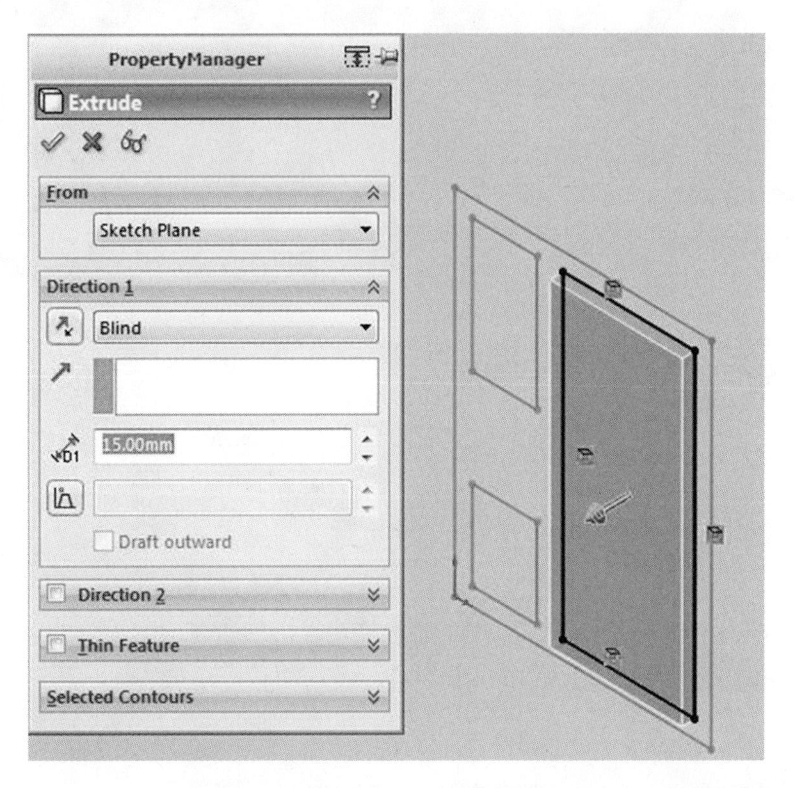

Figure 11.15

Extrude PropertyManager.

To save the [Part1^CABINET]<1>-> to MICROCONTROLLER,

60. Right-click [Part1^CABINET]<1>->. (See Figure 11.16 for the part.)
61. Click Save As.
62. Click OK for "*Resolve Ambiguity*" dialog box.
63. Click OK for the "*Virtual Component*" referring to dialog box.
64. Enter in the File Name box.
65. Close the part.
66. Return to the CABINET assembly.

Cooling Unit—Insert Component

The CABINET assembly is the open document (Default<Display State-1>) at this juncture.

67. Click New Part from the Consolidated Insert Components toolbar (see Figure 11.17).
68. Double-click PART-MM-ANSI from the MY-TEMPLATE tab in the New SolidWorks Document dialog box. (This would have already been created.)

Create the Sketch for the [Part1^CABINET]

69. Click Front Plane in the CABINET assembly FeatureManager.
70. Click Sketch to be in sketch mode.
71. Select the *edges of the cooling unit.*
72. Click Convert Entities from the Sketch toolbar to extract the profile.
73. Click OK to extract the entities (see Figure 11.18).

Figure 11.16

Save the MICROCONTROLLER using Save As option.

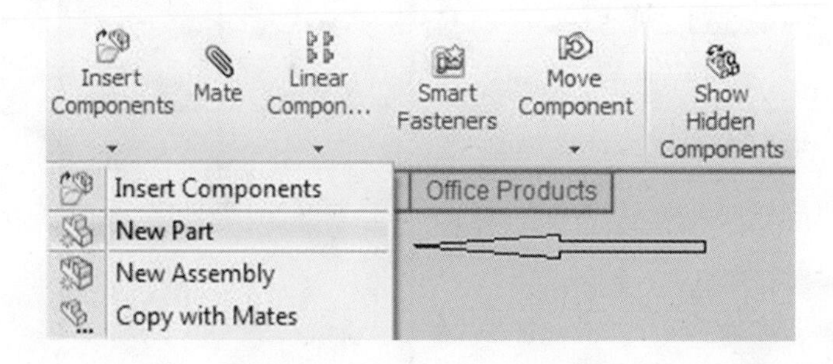

Figure 11.17

Inserting New Part from the Consolidated Insert Components toolbar.

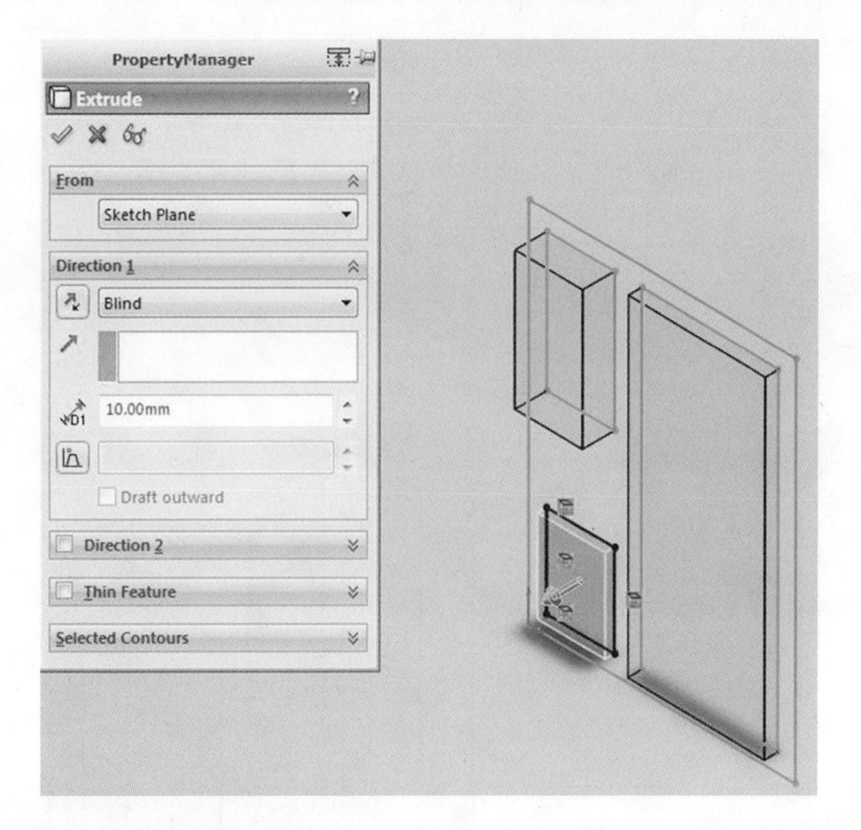

Figure 11.18

Extrude PropertyManager for cooling unit.

Insert an Extrude Base for the [Part1^CABINET]

74. Click Extrude Boss/Base from the FeatureManager toolbar. The Extrude PropertyManager is displayed (see Figure 11.18).
75. Enter 10-mm Depth in Direction1.
76. Click OK to extrude.
77. Rename Extrude1 to Base Extrude.

Notice now that [Part1^CABINET]<1>->, Base Extrude->, and Sketch1-> all have the "->" symbol indicating External References to the CABINET assembly.

To save the assembly,

78. Click Edit Component. (This returns you to the CABINET assembly.)
79. Click Save.

Power Supply—Insert Component

The CABINET assembly is the open document (Default<Display State-1>) at this juncture.

80. Click New Part from the Consolidated Insert Components toolbar (see Figure 11.19).
81. Double-click PART-MM-ANSI from the MY-TEMPLATE tab in the New SolidWorks Document dialog box. (This would have already been created.)

Create the Sketch for the [Part1^CABINET]

82. Click Front Plane in the CABINET assembly FeatureManager.
83. Click Sketch to be in sketch mode.
84. Select the *edges of the power supply.*
85. Click Convert Entities from the Sketch toolbar to extract the profile.
86. Click OK to extract the entities (see Figure 11.20).

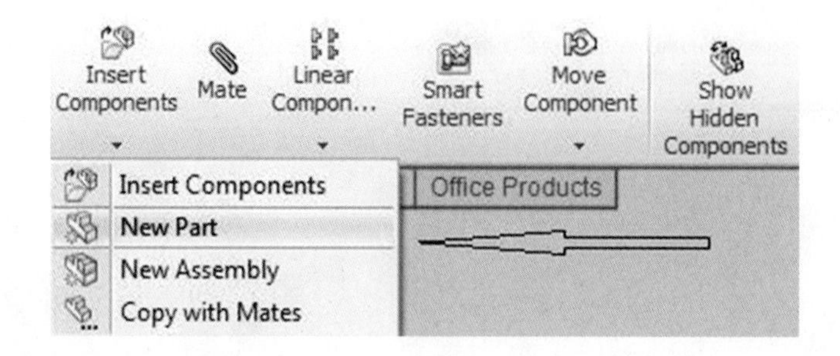

Figure 11.19

Inserting New Part from the Consolidated Insert Components toolbar.

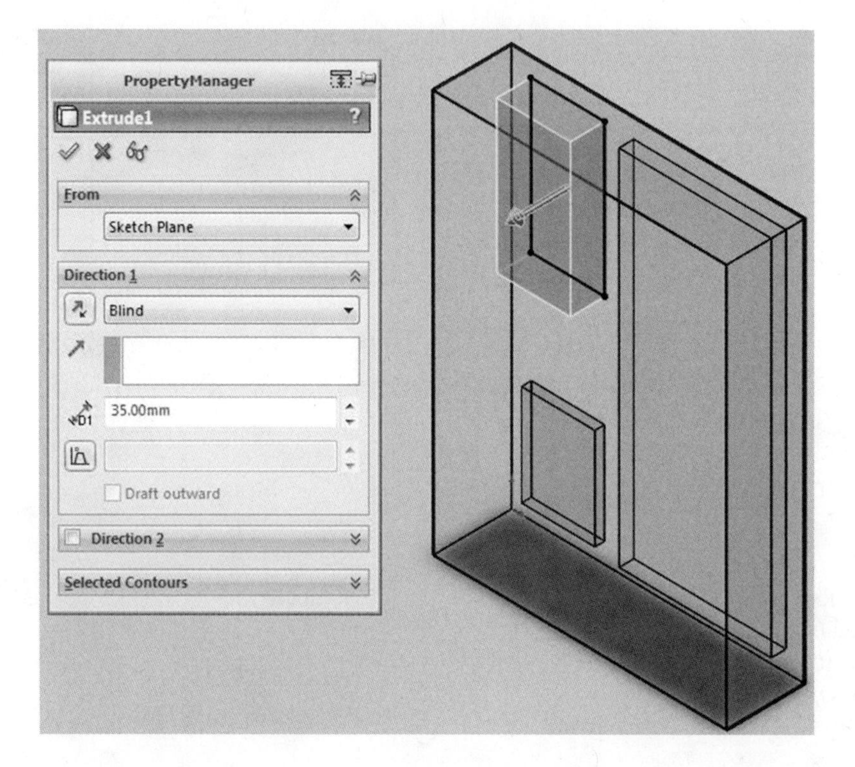

Figure 11.20

Extrude PropertyManager for power supply.

Insert an Extrude Base for the [Part1^CABINET]

87. Click Extrude Boss/Base from the FeatureManager toolbar. The Extrude PropertyManager is displayed (see Figure 11.20).
88. Enter 35-mm Depth in Direction1.
89. Click OK to extrude.
90. Rename Extrude1 to Base Extrude.

Notice now that [Part1^CABINET]<1>->, Base Extrude->, and Sketch1-> all have the "->" symbol indicating External References to the CABINET assembly.

To save the assembly,

91. Click Edit Component. (This returns you to the CABINET assembly.)
92. Click Save.

Housing-Insert Component

The CABINET assembly is the open document (Default<Display State-1>) at this juncture.

93. Click New Part from the Consolidated Insert Components toolbar (see Figure 11.21).

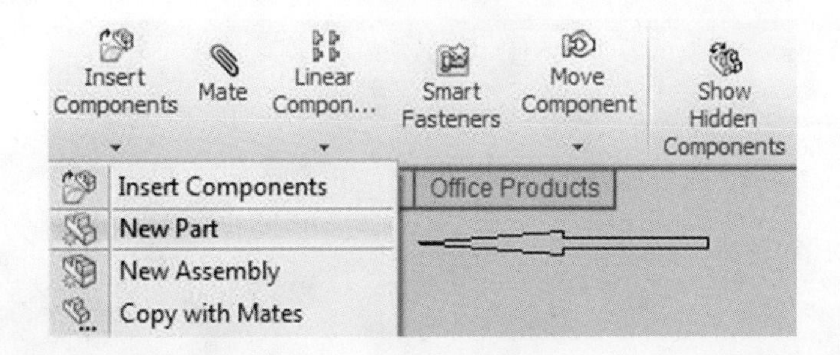

Figure 11.21

Inserting New Part from the Consolidated Insert Components toolbar.

94. Double-click PART-MM-ANSI from the MY-TEMPLATE tab in the New SolidWorks Document dialog box. (This would have already been created.)

Create the Sketch for the [Part1^CABINET]

95. Click Front Plane in the CABINET assembly FeatureManager.
96. Click Sketch to be in sketch mode.
97. Select the *edges of the housing.*
98. Click Convert Entities from the Sketch toolbar to extract the profile.
99. Click OK to extract the entities (see Figure 11.22).

Insert an Extrude Base for the [Part1^CABINET]

100. Click Extrude Boss/Base from the FeatureManager toolbar. The Extrude PropertyManager is displayed (see Figure 11.23).
101. Enter 120-mm Depth in Direction1.

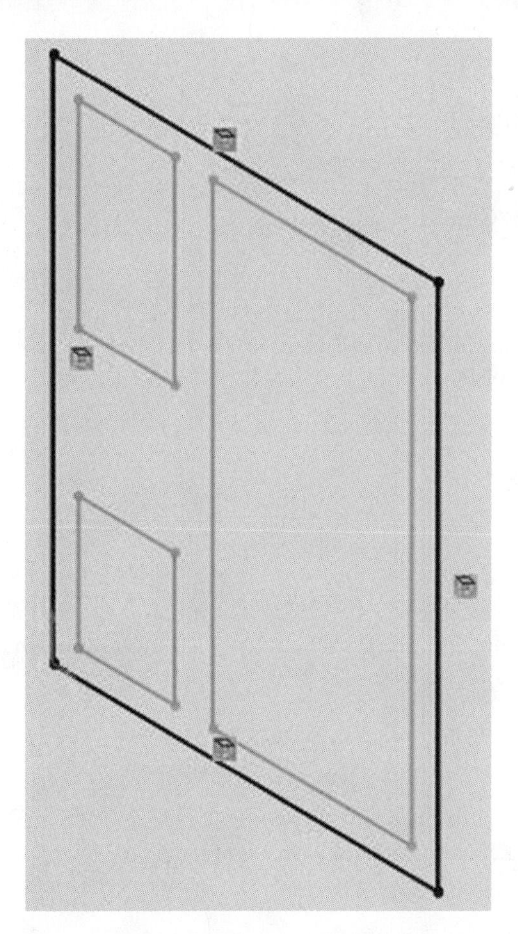

Figure 11.22

Edges of the housing feature extracted.

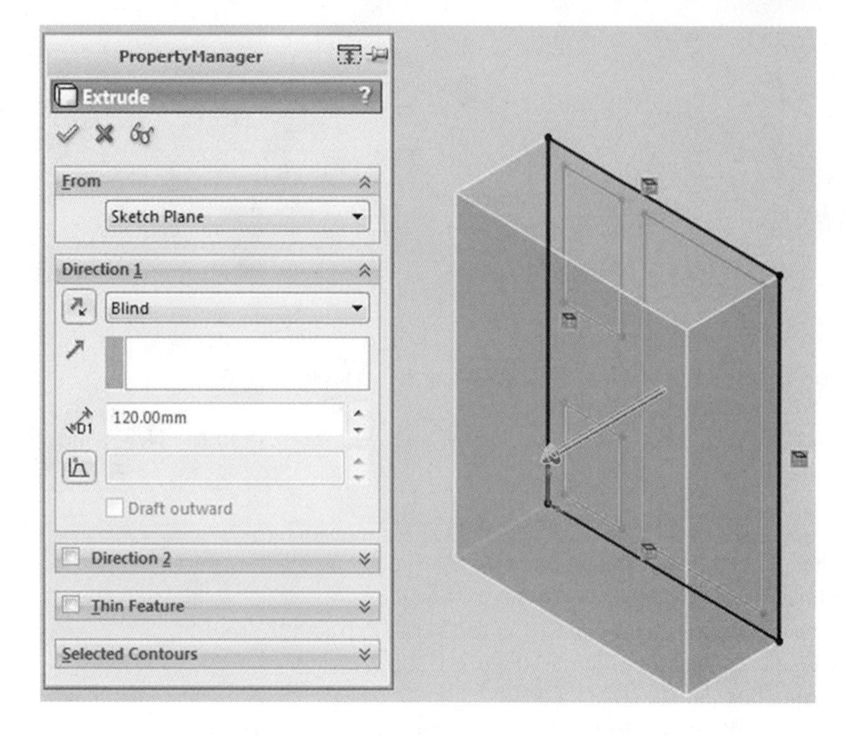

Figure 11.23

Extrude PropertyManager for housing.

102. Click OK to extrude.
103. Rename Extrude1 to Base Extrude.

Notice now that [Part1^CABINET]<1>->, Base Extrude->, and Sketch1-> all have the "->" symbol indicating External References to the CABINET assembly.

Shelling Operation for the Housing Feature
104. Click Feature > Shell in the Feature CommandManager.
105. Click *top face of the housing feature* (Face<1>@HOUSING-1) (see Figure 11. 24).
106. Supply the Thickness of 1.00 mm.

To save the assembly,

1. Click Edit Component. (This returns you to the CABINET assembly.)
2. Click Save.

The FeatureManager details for the steps taken so far are shown in Figure 11.25. Figure 11.26 shows the completed top–down design model, while the labeled components are shown in Figure 11.27.

Designing from Part Outline

Part Model

In the designing from part outline approach, a part exists from which the outline is extracted to create another part in an assembly. Let us commence with a lid (see Figure 11.28) and then create a seal.

Modeling In-Context

1. Click New Assembly document (see Figure 11.29).
2. Click *Lid* as the Open documents in the Part/Assembly to Insert rollout (see Figure 11.29).

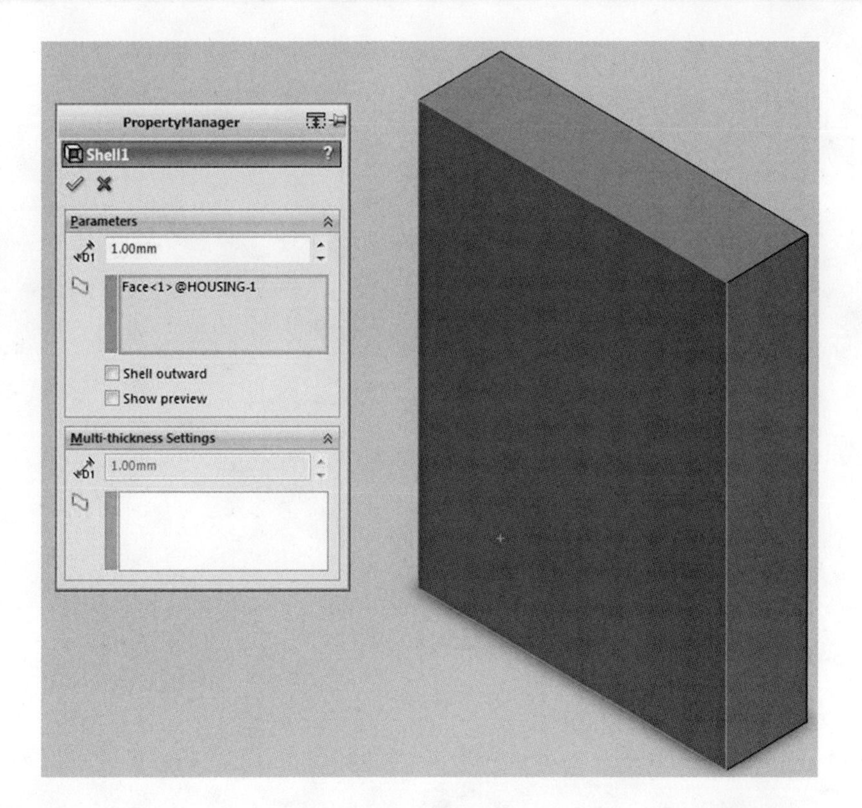

Figure 11.24

Shell PropertyManager.

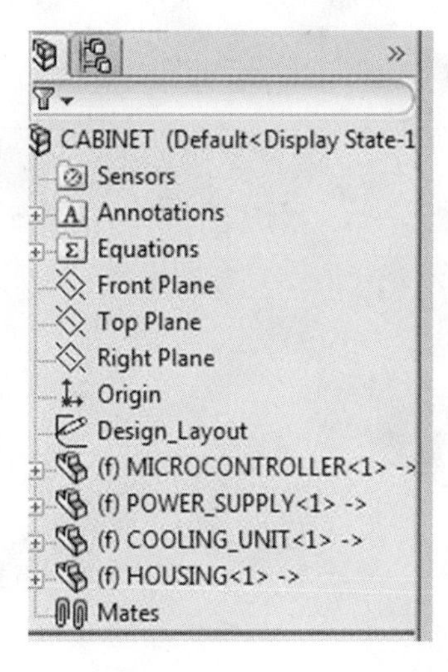

Figure 11.25

FeatureManager details for the steps taken so far.

3. Click OK to insert the Lid.

The Assem1assembly is the open document (Default<Display State-1>) at this juncture.

4. Click New Part from the Consolidated Insert Components toolbar (see Figure 11.30).

The new part is displayed in the FeatureManager design tree with the new name in the format [Part#^*AssemblyName*]<#>. Let us explain the syntax because it is important to understand this.

1. The square bracket shows that the part is a virtual component.
2. # is the number of virtual components.
3. *AssemblyName* is the assembly name.

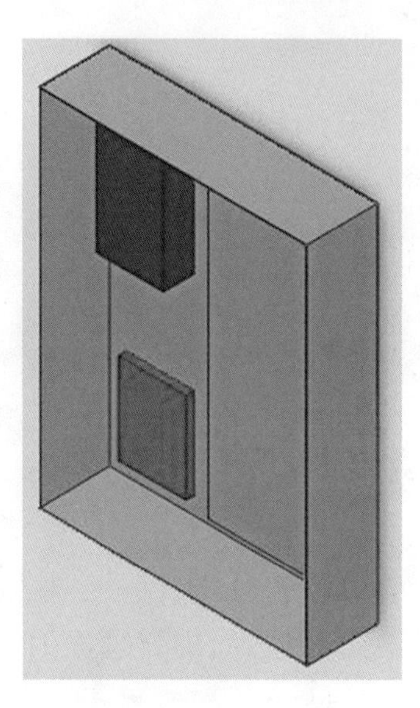

Figure 11.26

Completed top-down design model.

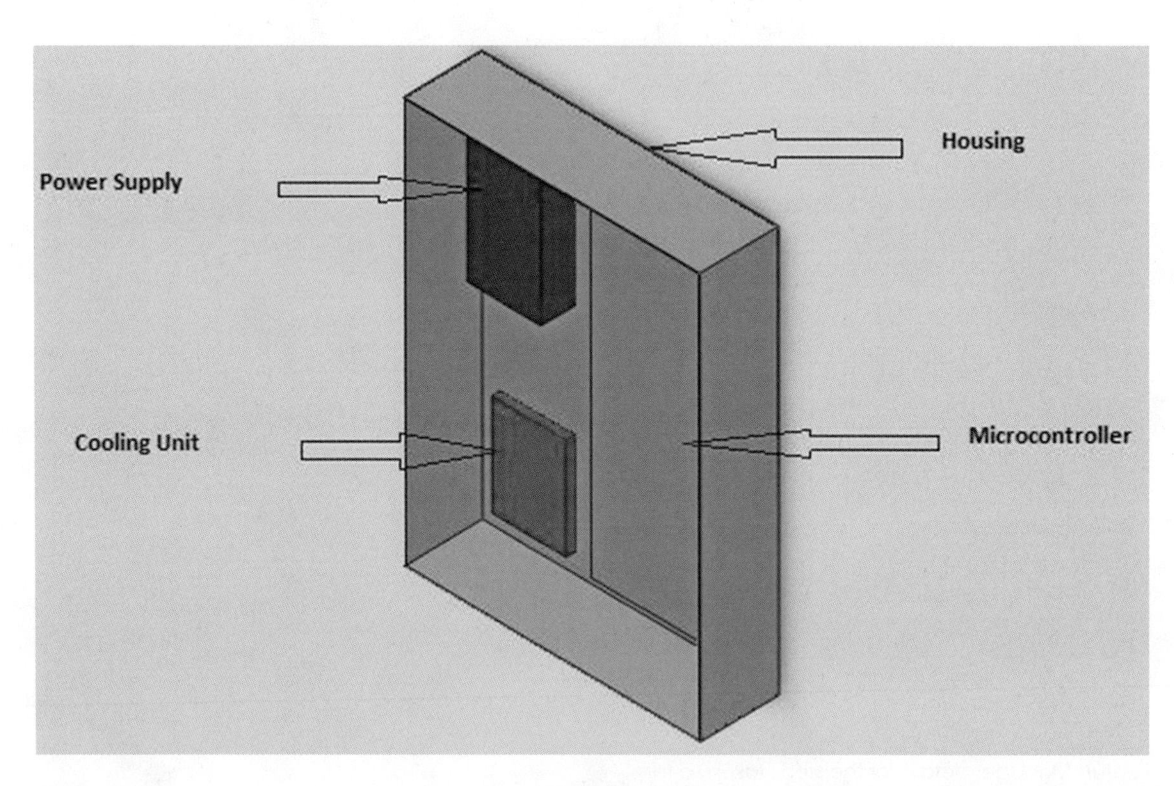

Figure 11.27

Completed top-down design model with labeled components.

In the current illustration, the first virtual part is [Part1^Assem1]<1>->.

The gateway between the assembly and the Component added In-Context is the Edit Component feature. In other words, the Edit Component is the switch between the assembly and the component edited In-Context (see Figure 11.31).

It is extremely important to note this gateway. For example, when [Part1^Assem1]<1>-> is clicked, and Edit Component is clicked, [Part1^Assem1]<1>-> becomes light blue. This means that we are in the *Component added In-Context* level.

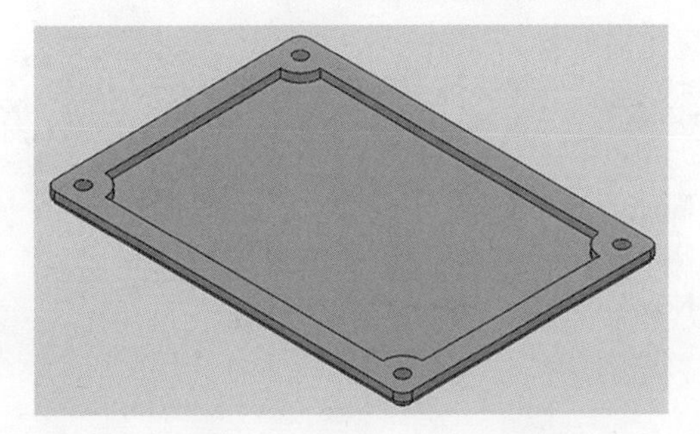

Figure 11.28

Lid model.

Figure 11.29

New Assembly document with the lid opened.

Create the Sketch for the [Part1^Assem1]

5. Click Top Plane in the Assem1 assembly FeatureManager.
6. Click Sketch to be in sketch mode.
7. Select the *edges of the Lid profile.*
8. Click Convert Entities from the Sketch toolbar to extract the profile.
9. Click OK to extract the entities (see Figure 11.32).

Insert an Extrude Base for the [Part1^Assem1]

10. Click Extrude Boss/Base from the FeatureManager toolbar. The Extrude PropertyManager is displayed (see Figure 11.33).
11. Enter 0.1-in. Depth in Direction1.
12. Click OK to extrude.
13. Rename Extrude1 to Base Extrude.

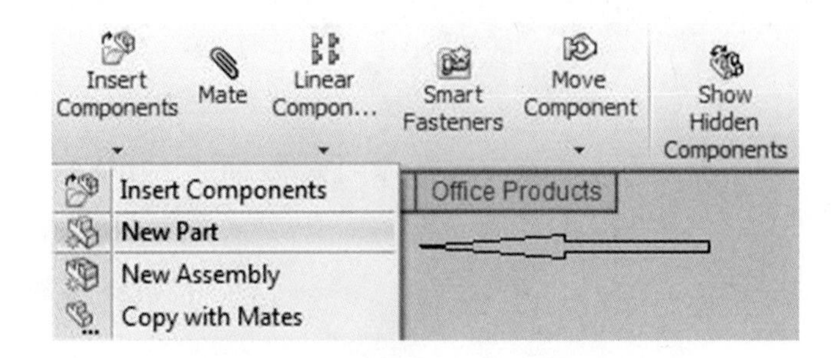

Figure 11.30

Inserting New Part from the Consolidated Insert Components toolbar.

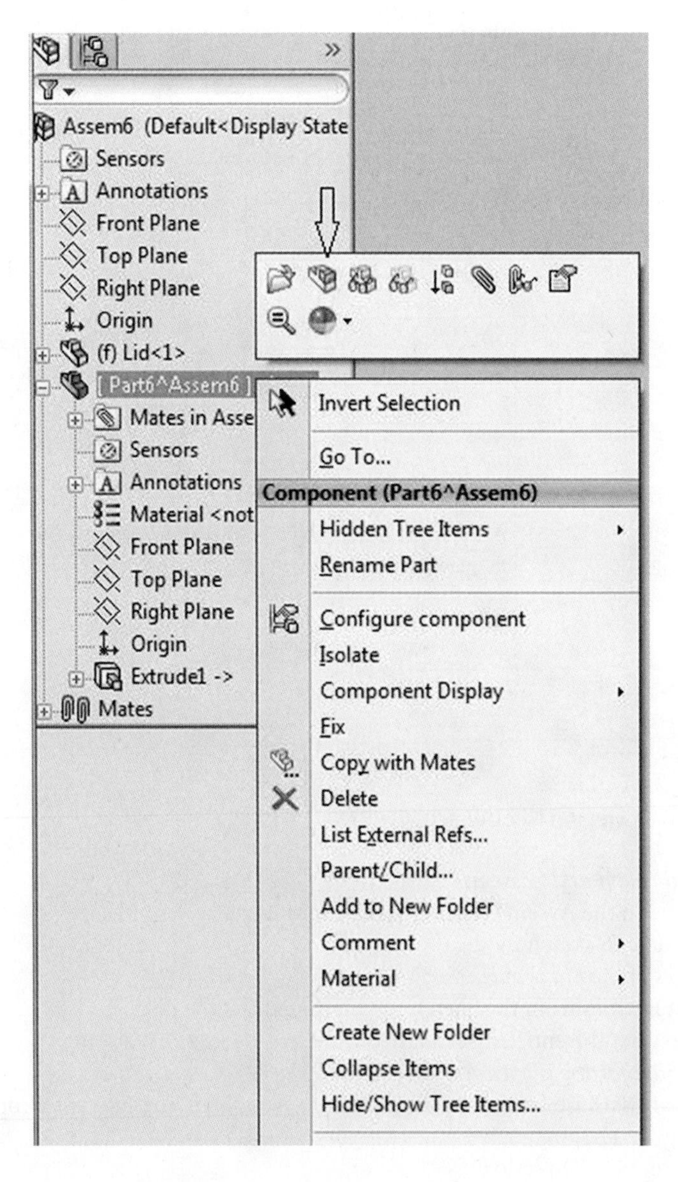

Figure 11.31

Switching to the Component added In-Context level.

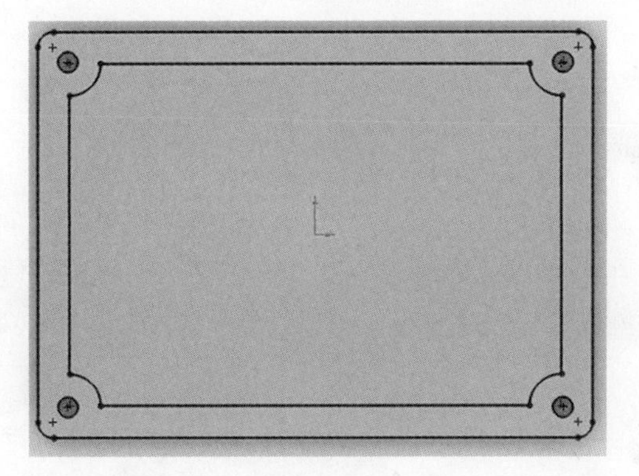

Figure 11.32

In-context sketch for designing the seal.

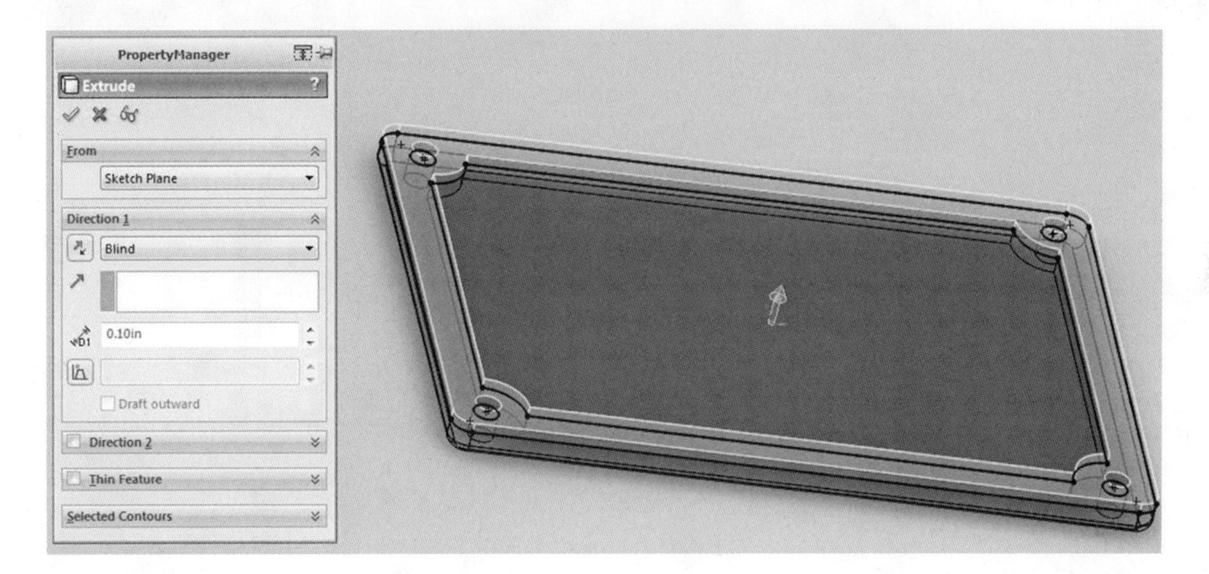

Figure 11.33

Extrude PropertyManager.

Notice now that [Part1^Assem1]<1>->, Base Extrude->, and Sketch1-> all have the "->" symbol indicating External References to the Lid assembly.

Notice that there are now two parts: one of them is Lid<1>; the other is [Part1^Assem1]<1>->.

Color Coding in Design Levels

It is important to understand this nomenclature.

- When [Part1^Assem1]<1>-> and the subnodes are in blue color, we are in part level.
- When [Part1^Assem1]<1>-> and the subnodes are in dark color, we are in assembly level (see Figure 11.34).

Right-click [Part1^Assem1]<1>-> and select Edit to switch from assembly to part level.

Click Edit Component in the CommandManager to switch from part to assembly level.

Saving the New Part

Using the top–down design approach, we have designed a seal from the profile of the lid. The new part now has to be saved.

14. Right-click [Part1^Assem1]<1>-> and select Edit to switch from assembly to part level.
15. Click File > Save As.

    The Resolve Ambiguity window shown in Figure 11.35 automatically appears highlighting the part and awaiting the user to confirm by clicking the OK button.

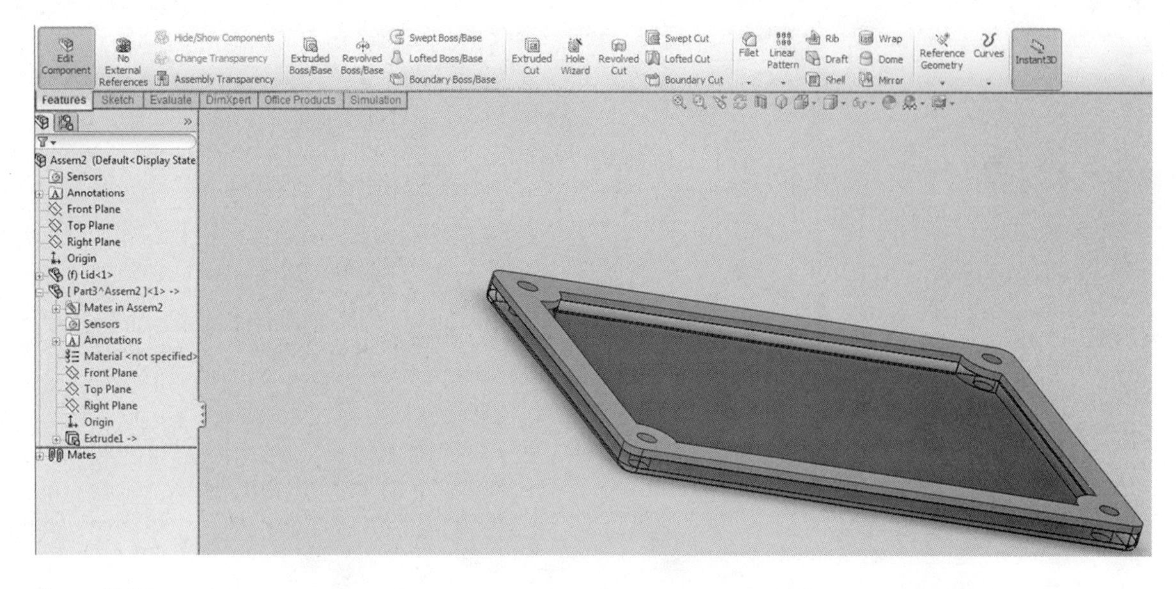

Figure 11.34

Switch mechanism from assembly to part level.

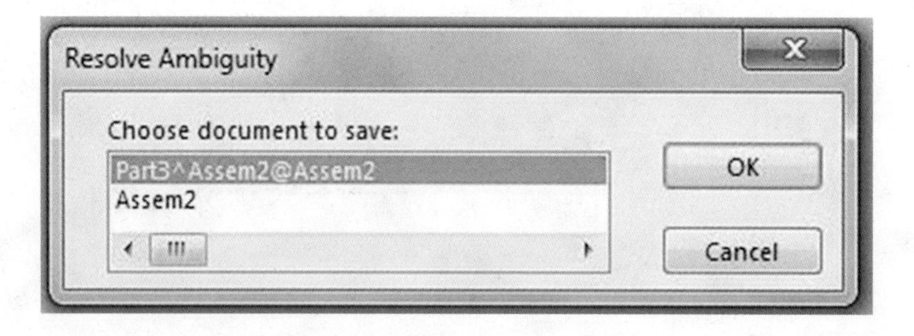

Figure 11.35

Resolve Ambiguity window.

16. Click OK.

Another window appears (see Figure 11.36) displaying the message suggesting the usage of *Save As Copy* option.

17. Click OK.

Another window appears (see Figure 11.37) for saving the document.

18. Browse for the directory to save the new part, with a new name Seal.

19. Click Save.

The AssemblyManager during and after saving the new part is shown in Figure 11.38.

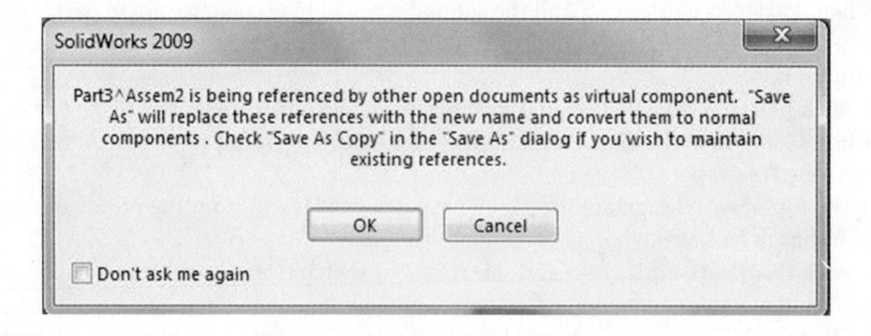

Figure 11.36

Message suggesting the usage of Save As Copy option.

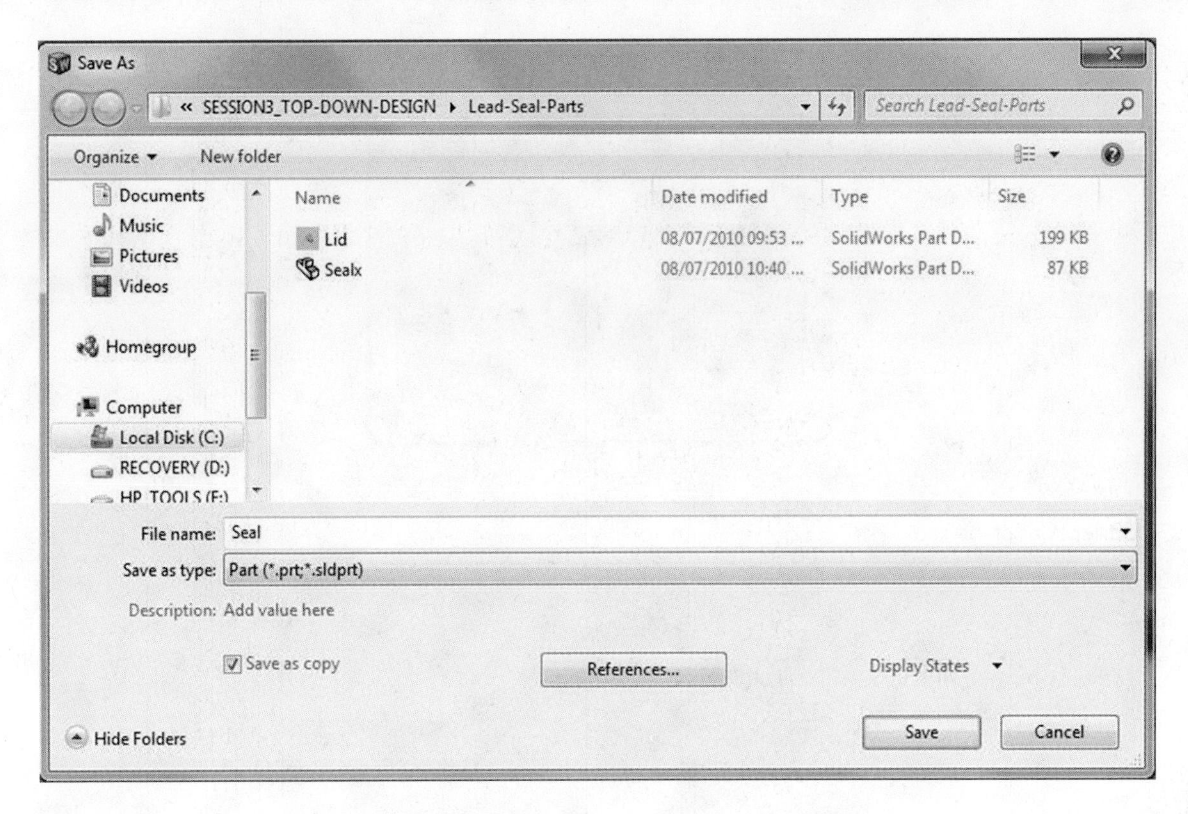

Figure 11.37

Window for saving new part created and the "Save As Copy" option.

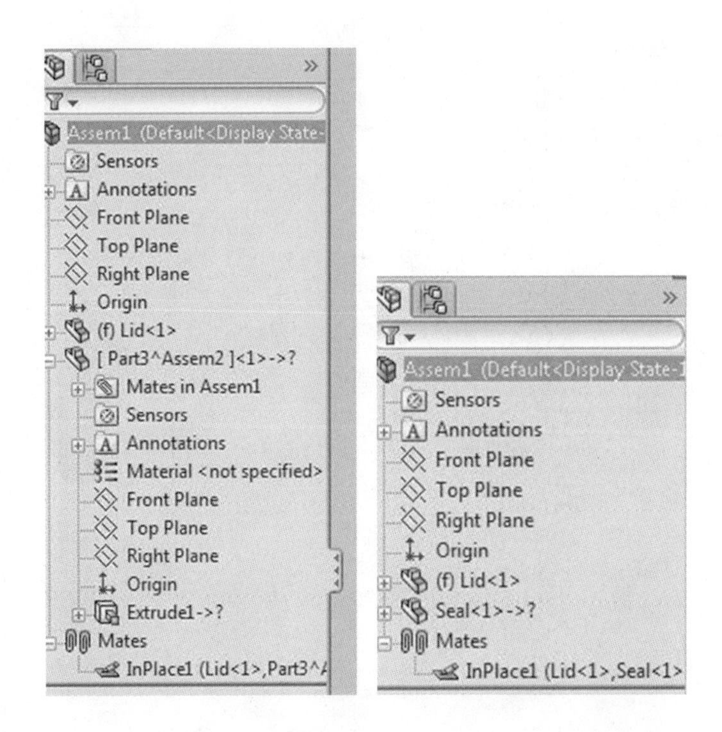

Figure 11.38

AssemblyManager during and after saving the new part created.

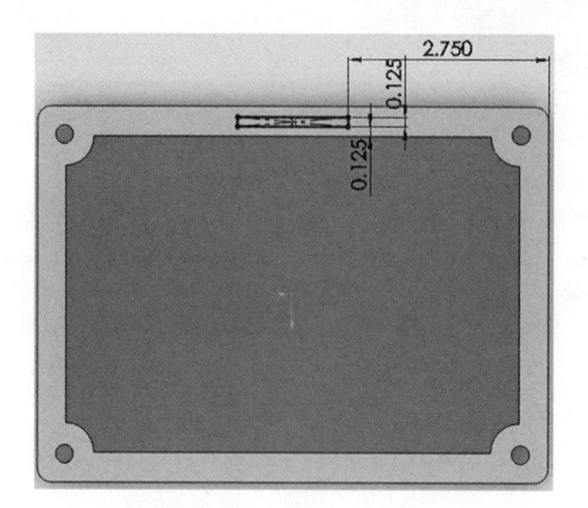

Figure 11.39

Modifications of the seal.

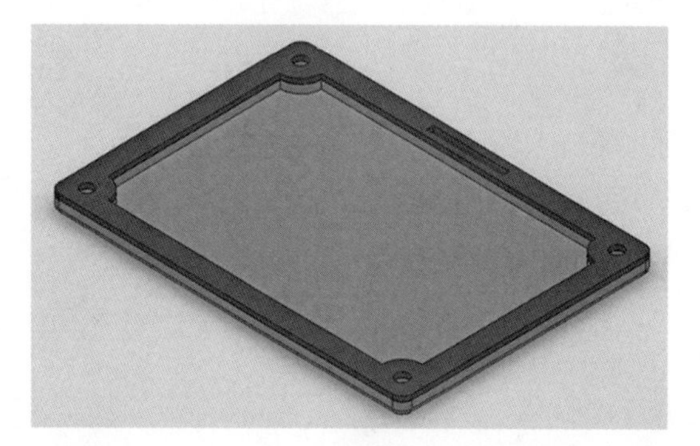

Figure 11.40

Material assignment to the lid and seal.

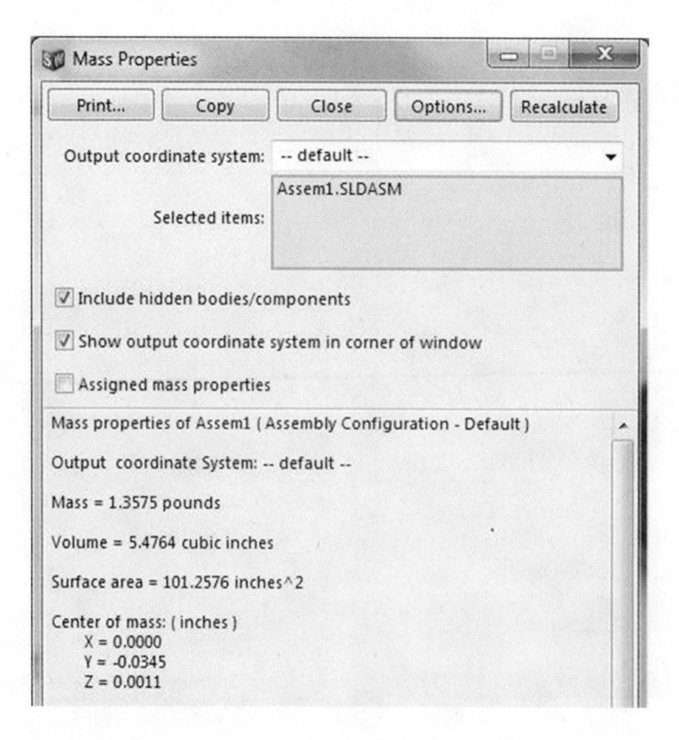

Figure 11.41

Material properties computed.

Further modification of the seal is shown in Figure 11.39, and the assembly of the lid and seal is shown in Figure 11.40. The materials are assigned (see Figure 11.40), and the mass properties are computed (see Figure 11.41).

Mold Design Using Top–Down Approach

In this chapter, a new method is introduced for designing molds based on the top–down paradigm. This is the author's contribution to the mold design using SolidWorks for parts with a specific set of features.

Create Lower Lid from Lid Already Available (Figures 11.42 through 11.44)

1. Open Lid <file name> as active part.
2. Start New SolidWorks Assembly document.
3. Open Lid in *Assembly* environment.
4. Click OK.
5. Save Assembly with a name.
6. Right-click (+)Lid<1> (Default…).
7. Select Edit Part. (The tree for *Lid* turns blue; because we are in part mode.)
8. Check Save As.
9. Check OK for Resolve Ambiguity message.
10. Check OK for Reference as Virtual Part.
11. Save as lidLower<1> (Default…).

Create Mold for lidLower (Figures 11.45 through 11.52)

12. Define a Plane using three points on the part (0.118 from the *flat face* and 0.132 from the *indented face*).
13. Be in Sketch mode.
14. Sketch a *Rectangle* using Center Rectangle tool to enclose lidLower.
15. Dimension the box as 7.5 × 6.5 (inches).
16. Extrude the box inward toward the user through 0.15.
17. Uncheck Merge result to create two Solid Bodies (*Fillet 1 and Boss-Extrude1*).
18. Click Insert > Features > Combine. (The Combine1 PropertyManager appears.)
19. Select Subtract.
20. Expand the SolidBodies and select Boss-Extrude1 as Main Body from the SolidBodies(2).

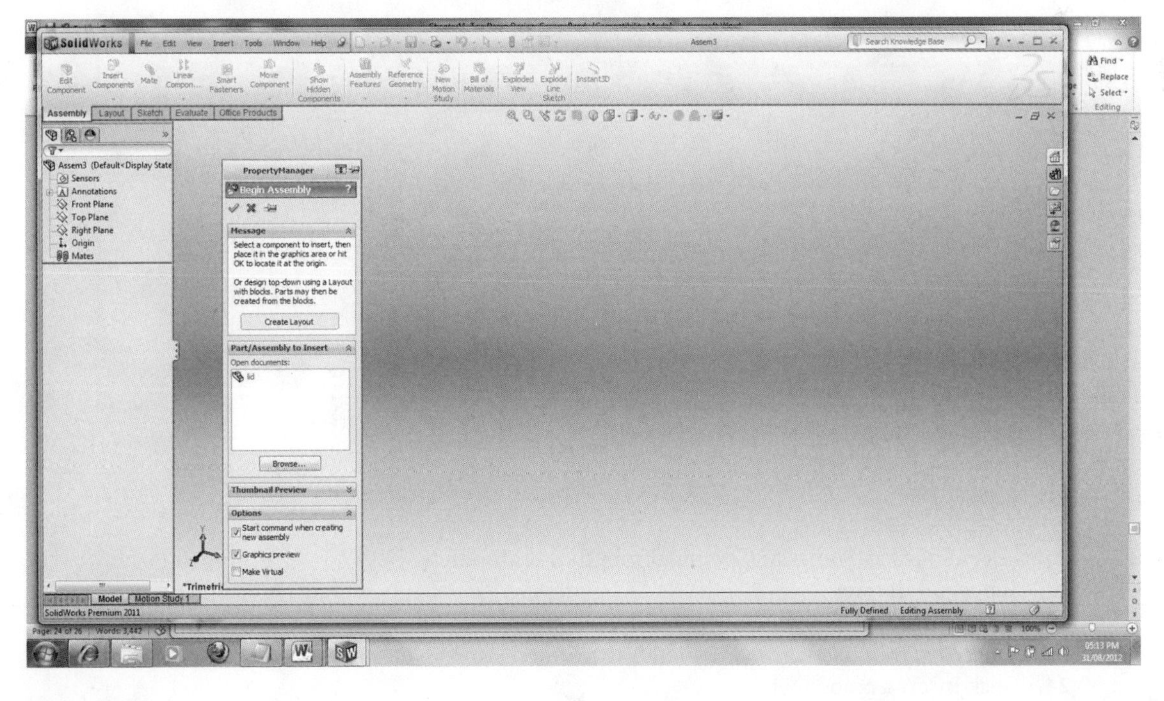

Figure 11.42

Begin Assembly.

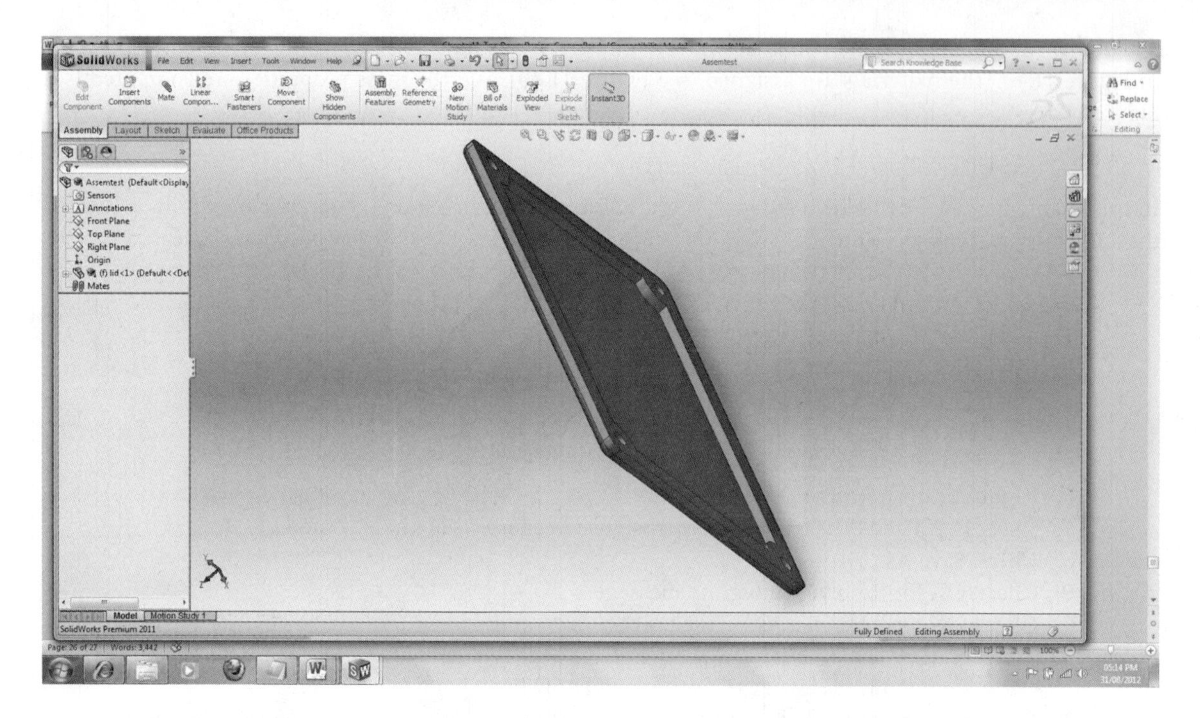

Figure 11.43

Open lid as part.

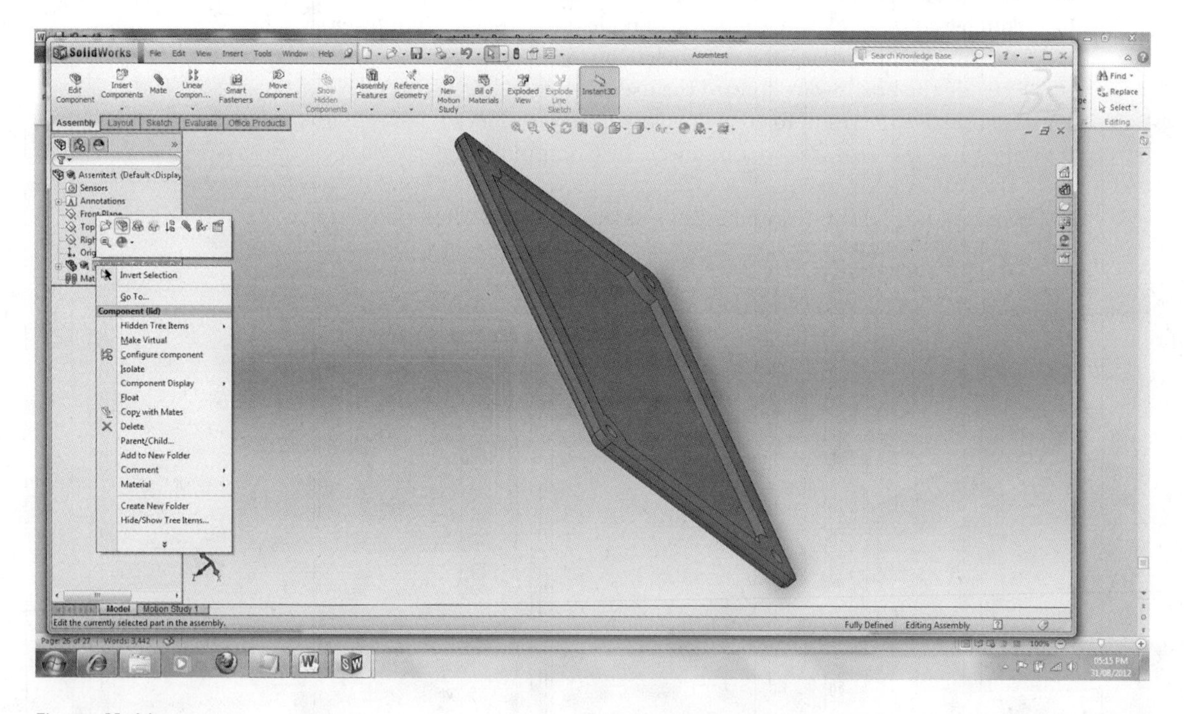

Figure 11.44

Edit Part mode.

21. Select Fillet1 as Bodies to Combine from the SolidBodies(2).
22. Suppress lidLower.
23. Check Edit Component to be in Assembly mode.

Create Upper Lid from Lid Already Available (Figure 11.53)

24. Click Insert Component.
25. Open Lid in *Assembly* environment.
26. Click OK.

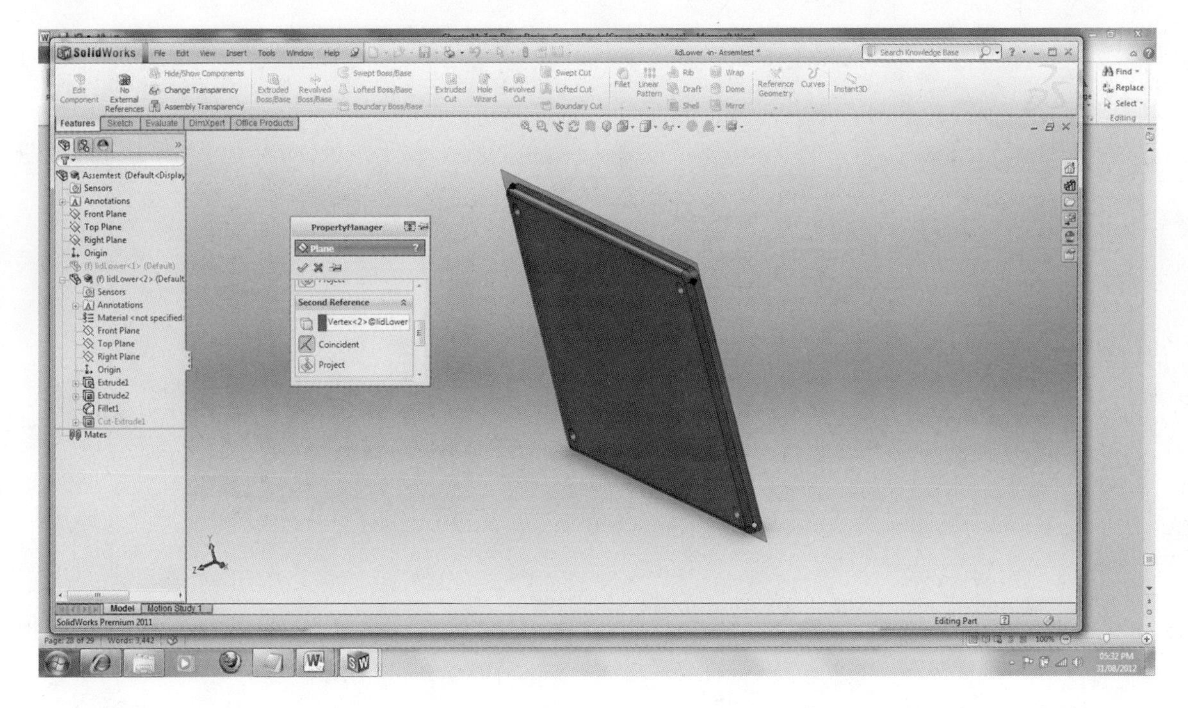

Figure 11.45

Create a plane using three vertices, 0.118 from one face and 0.132 from the other.

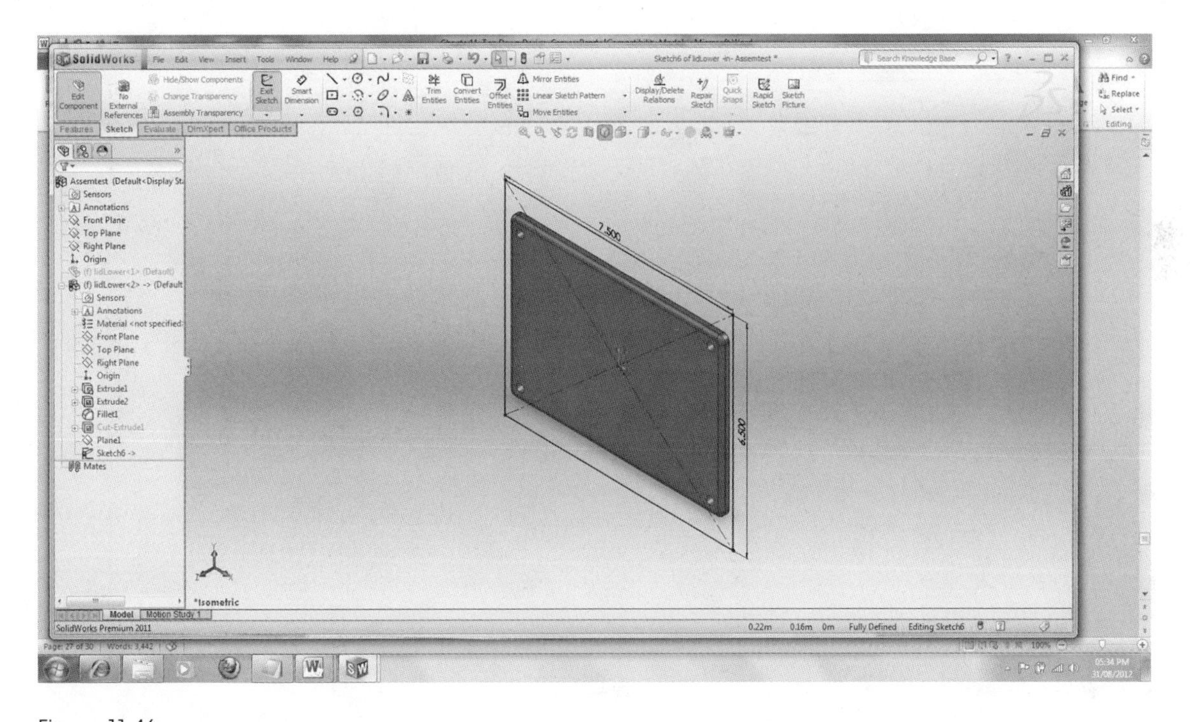

Figure 11.46

Create parting plane 7.5 × 6.5.

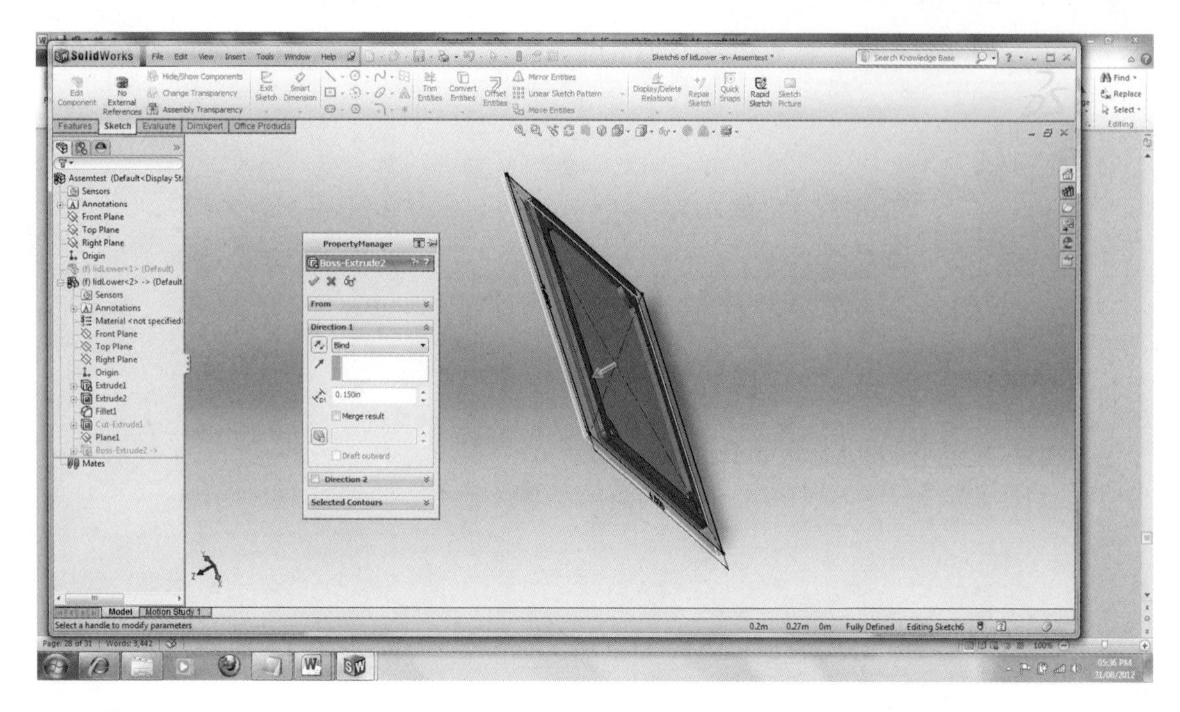

Figure 11.47

Extrude toward the user's view.

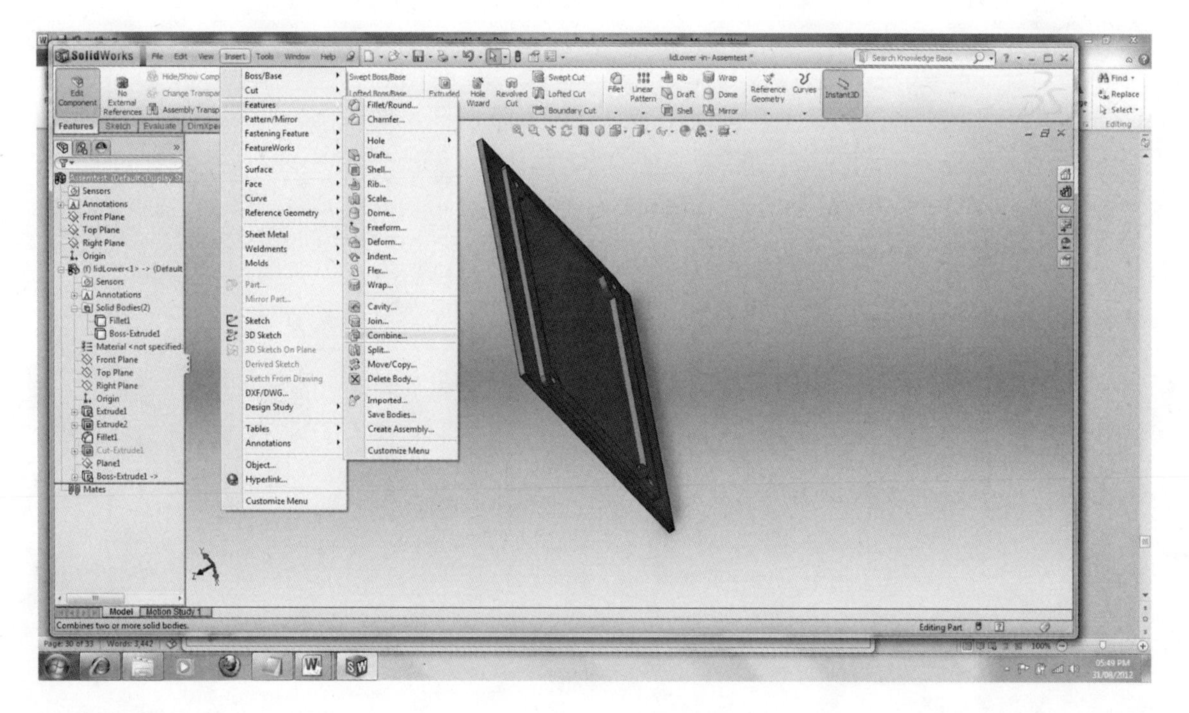

Figure 11.48

Access Boolean operations tool using Insert > Features > Combine.

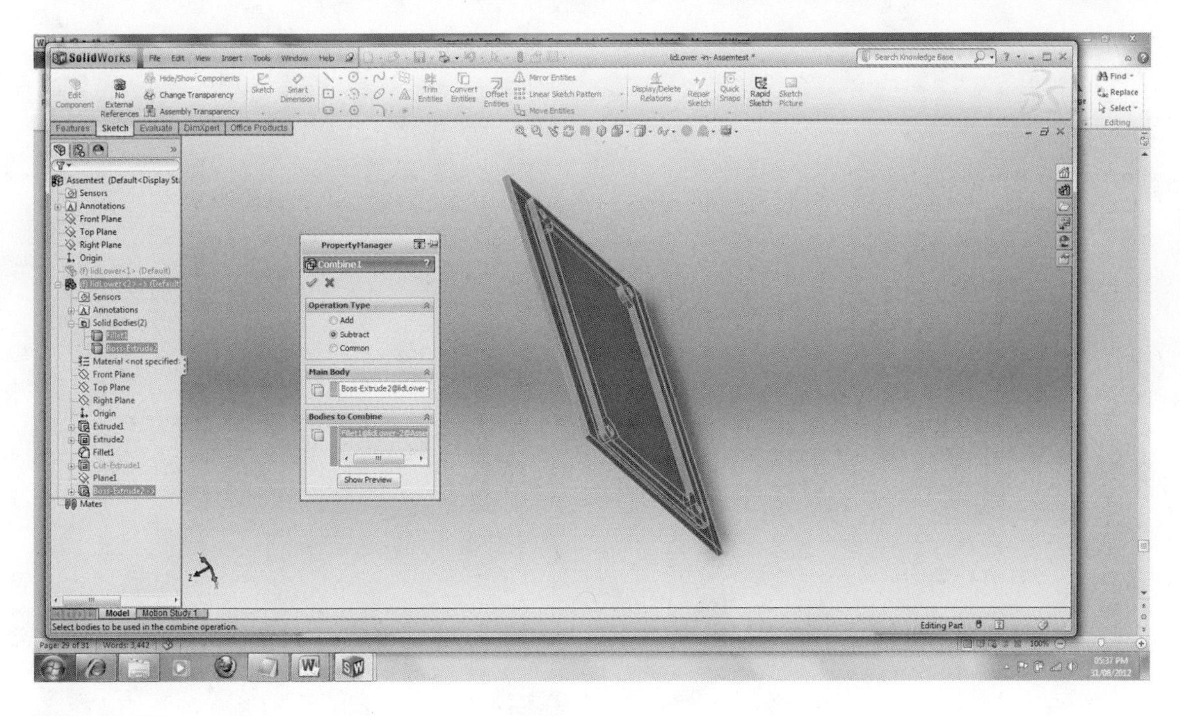

Figure 11.49

Subtract Fillet1 from Boss-Extrude1, using SolidBodies in Boolean operation.

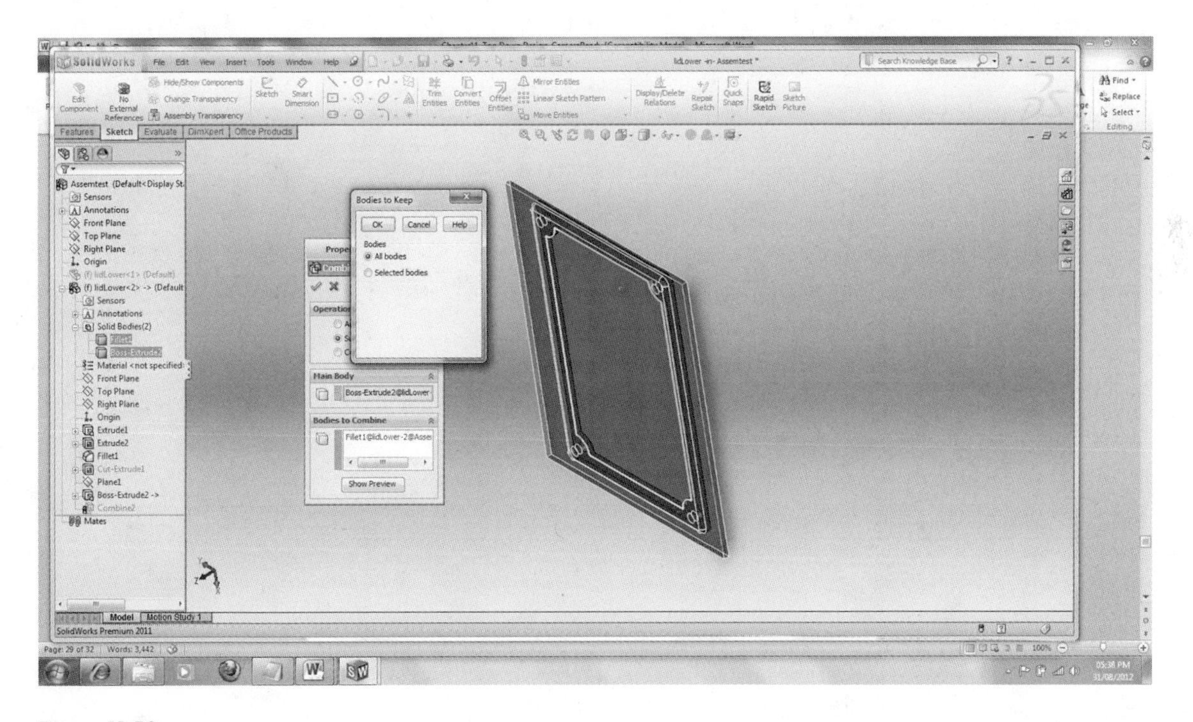

Figure 11.50

Accept All Bodies if prompted on the Bodies to Keep.

Figure 11.51

Lower mold (lidLower).

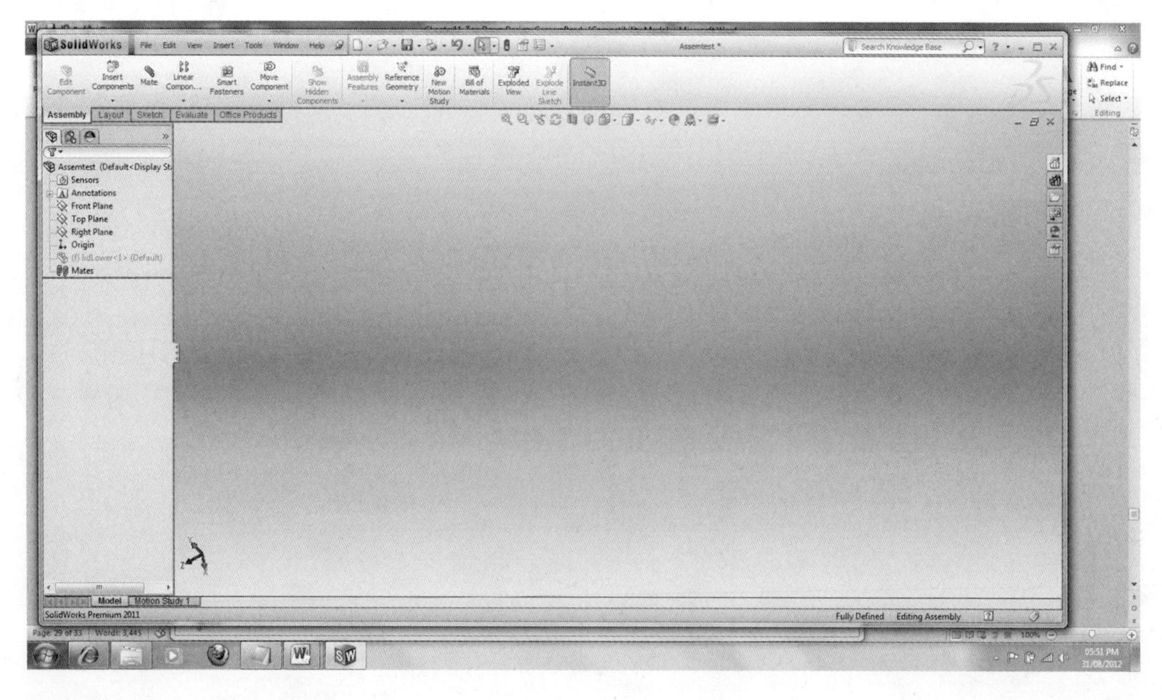

Figure 11.52

Lower mold suppressed.

27. Save Assembly with a name.
28. Right-click (+)Lid<2> (Default…).
29. Select Edit Part.
30. Check Save As.
31. Check OK for Resolve Ambiguity message.
32. Check OK for Reference as Virtual Part.
33. Save as lidUpper<1> (Default…).

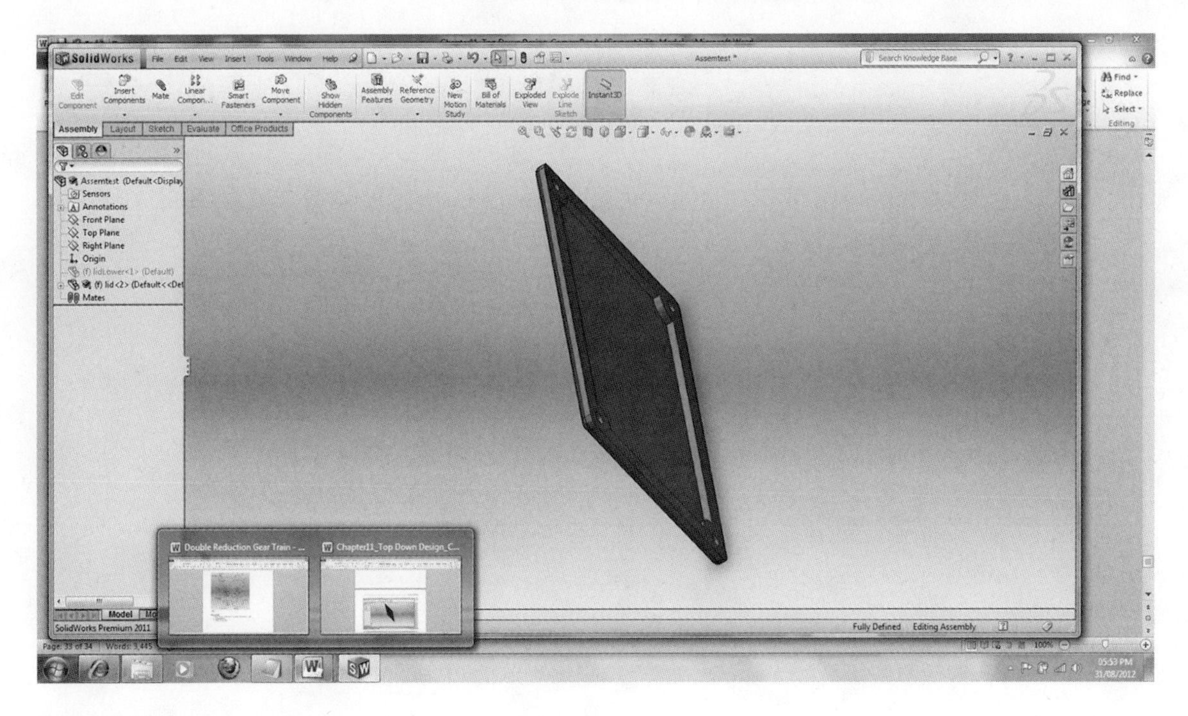

Figure 11.53

Original lid opened and named Upper lid (lidUpper).

Create Mold for lipUpper (Figures 11.54 through 11.57)

34. Define a Plane using three points on the part (0.118 from the *flat face* and 0.132 from the *indented face*).
35. Be in Sketch mode.
36. Sketch a *Rectangle* using Center Rectangle tool to enclose lidUpper.
37. Dimension the box as 7.5 × 6.5 (inches).
38. Extrude the box outward away from the user through 0.20.
39. Uncheck Merge result to create two Solid Bodies (*Fillet 1 and Boss-Extrude1*).

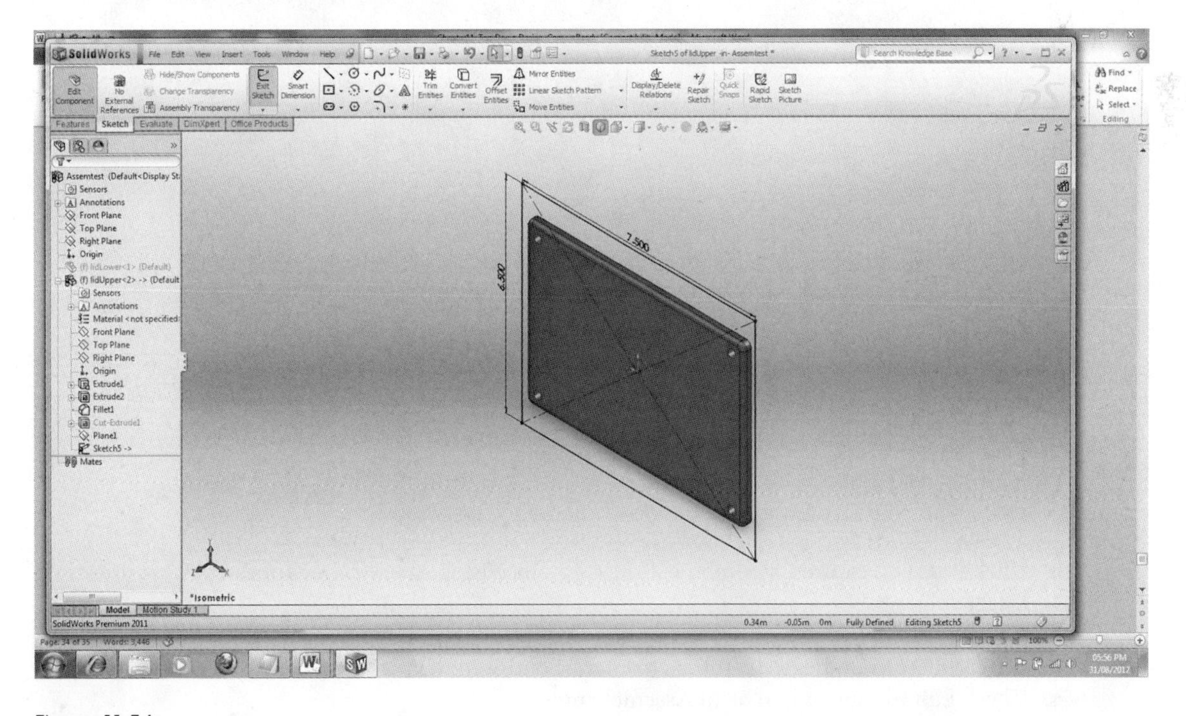

Figure 11.54

Create parting plane 7.5 × 6.5.

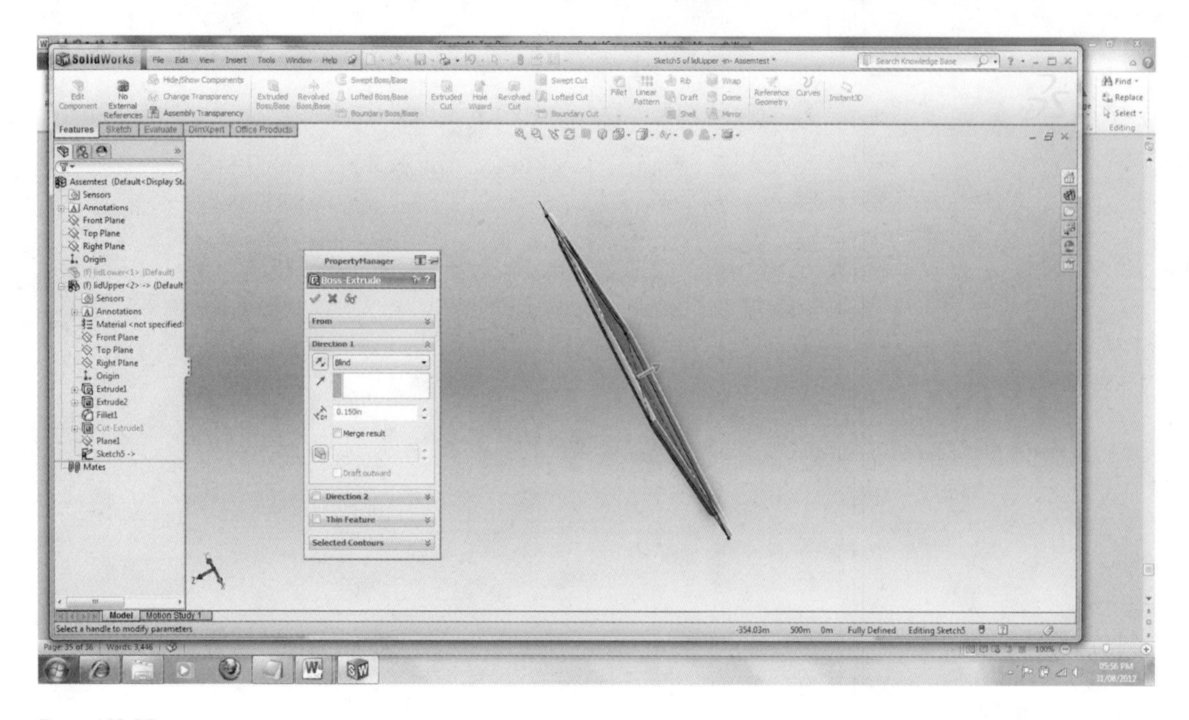

Figure 11.55

Extrude away from the user's view.

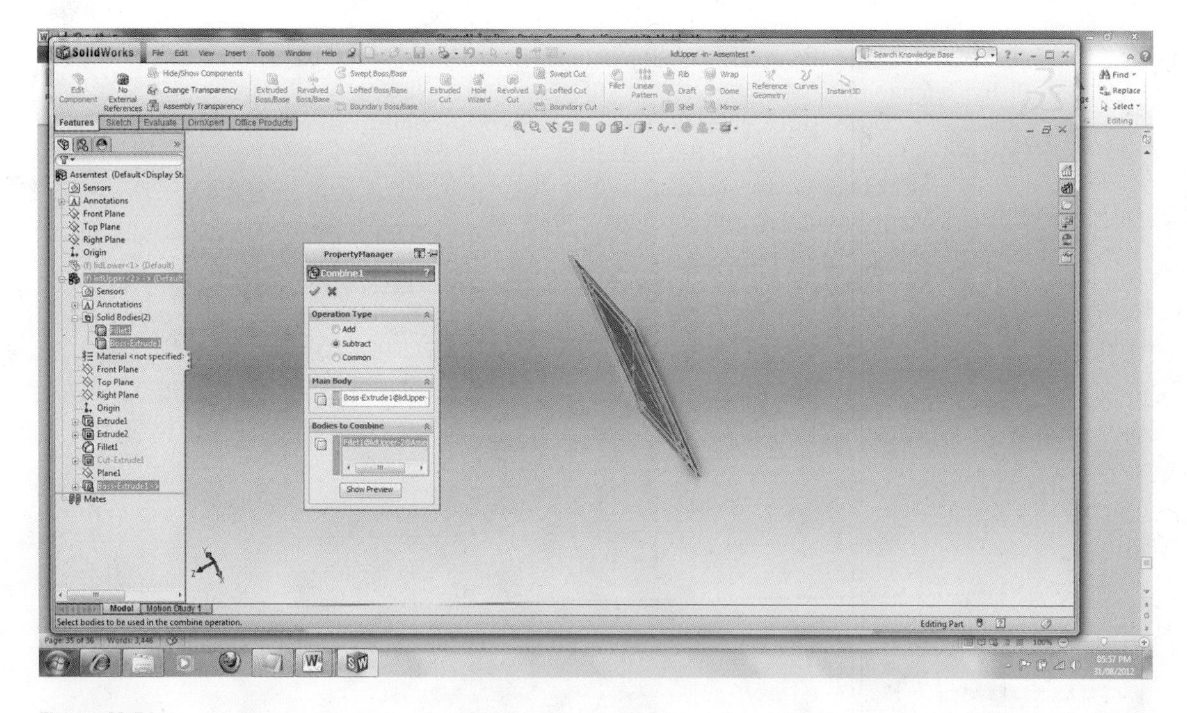

Figure 11.56

Subtract Fillet1 from Boss-Extrude1, using SolidBodies in Boolean operation.

40. Click Insert > Features > Combine. (The Combine1 PropertyManager appears.)
41. Select Subtract.
42. Expand the SolidBodies and select Boss-Extrude1 as Main Body from the SolidBodies(2).
43. Select Fillet1 as Bodies to Combine from the SolidBodies(2).
44. Check Edit Component to be in Assembly mode.

Figure 11.58 shows the complete upper and lower molds.

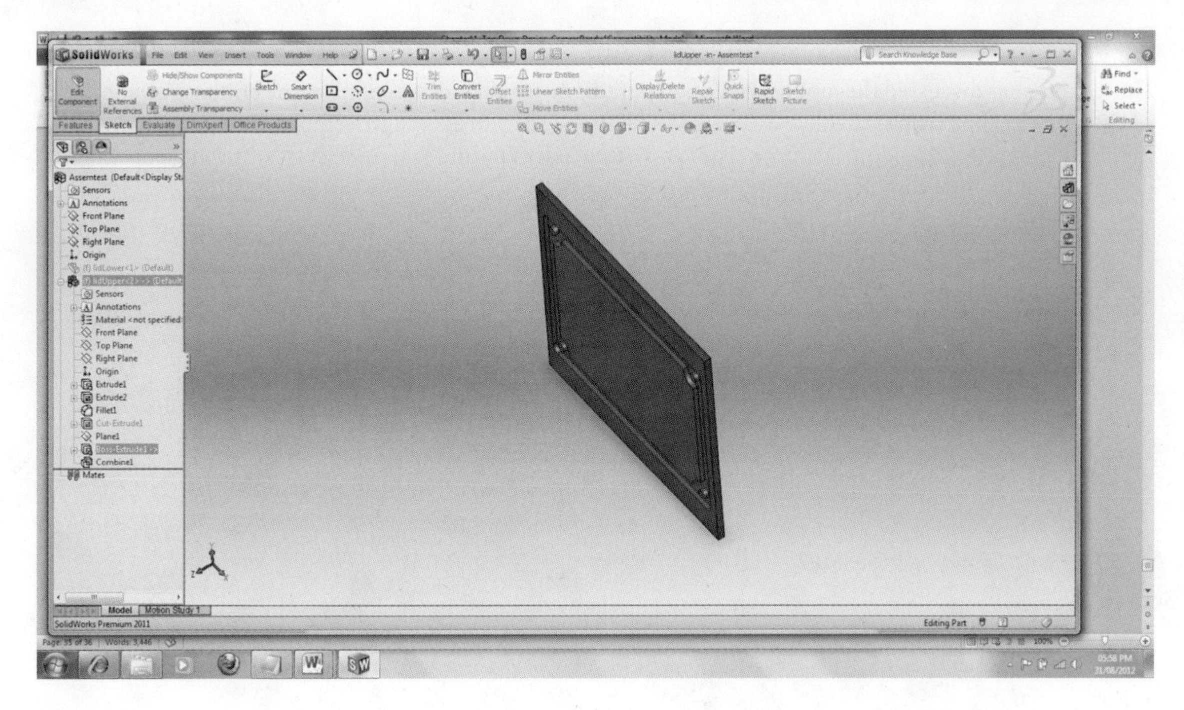

Figure 11.57

Upper mold (lidUpper).

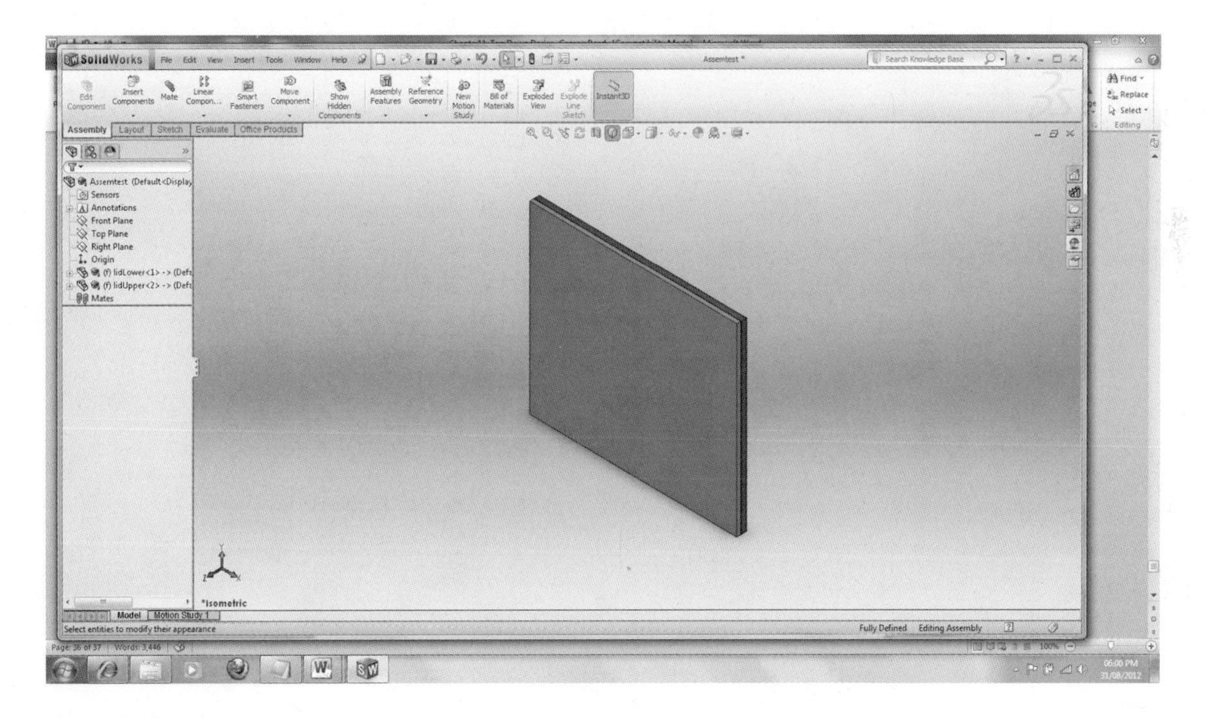

Figure 11.58

Upper and lower molds for the lid model.

Summary

This chapter has discussed the top–down design paradigm, which is different from the bottom–up approach. The top–down approach is useful when parts are being added to an existing assembly. The examples included in the chapter cover the addition of parts to an existing assembly and where we start from a layout and then design the individual components in turn.

Exercises

1. Figure P11.1 shows the model of a ribbed base. You are given the SolidWorks file for this model. Use the top–down design approach to design the 1-mm-thick gasket that is shown.
2. Figure P11.2 shows the model of a connecting rod bushing. You are given the model SolidWorks file. Use the top–down design approach to design the 1-mm-thick gaskets that are shown.

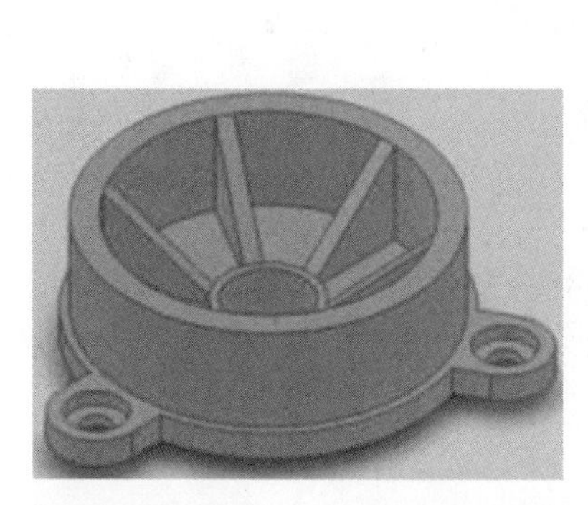
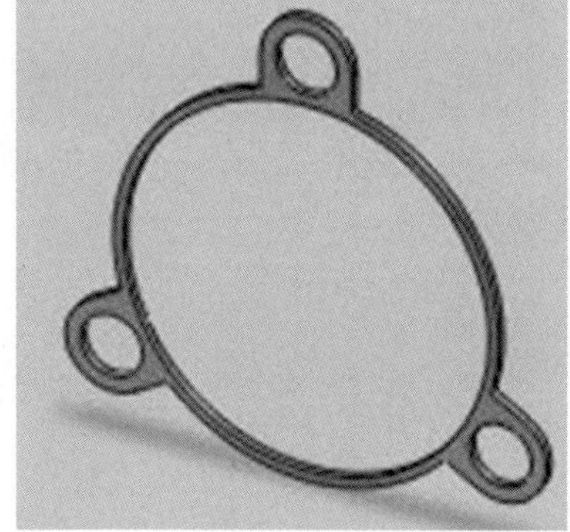

Figure P11.1

Impeller housing.

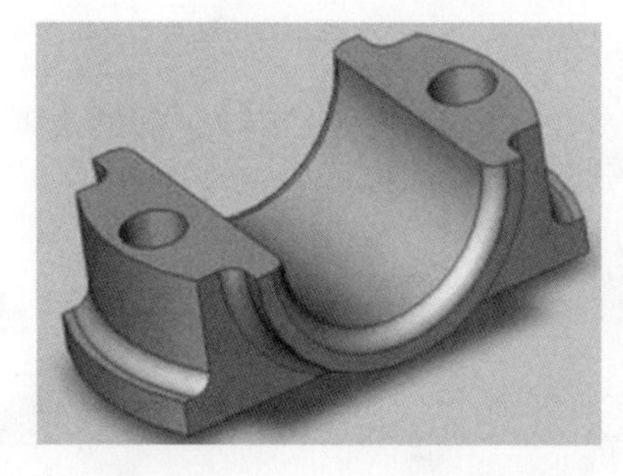
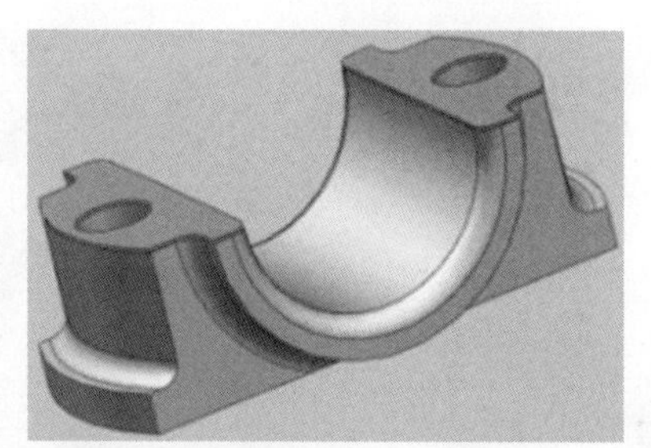

Figure P11.2

Bearing housing.

12

Surface Modeling

Objectives:

When you complete this chapter, you will have learnt the following:

- How to create free-form surfaces
- How to use control polygons for free-form surface design
- How to use control free-form surfaces using control polygons
- How to create extrude surfaces
- How to create boundary surfaces
- How to create lofted surfaces
- How to create revolved surfaces

Generalized Methodology for Free-Form Surface Design

In this section, the author presents a generalized methodology for *free-form surface design*. In SolidWorks, there are at least two ways of creating a free-form design: (1) extracting an existing surface from a three-dimensional (3D) part and creating a free-form surface and (2) creating a free-form surface from scratch. The second method is not popularly used because the users of this method should be fairly conversant with the knowledge of how standard surface modeling methods work. The well-known methods are based on Coons, Bezier, and nonuniform rational B-spline (NURBS); the treatment of these is outside the scope of this course. The SolidWorks free-form surface is based on NURBS.

In this section, the concept of *control polygon* is presented from which free-form surface design is carried out. Control polygon is the key for a flexible and robust free-form surface design methodology.

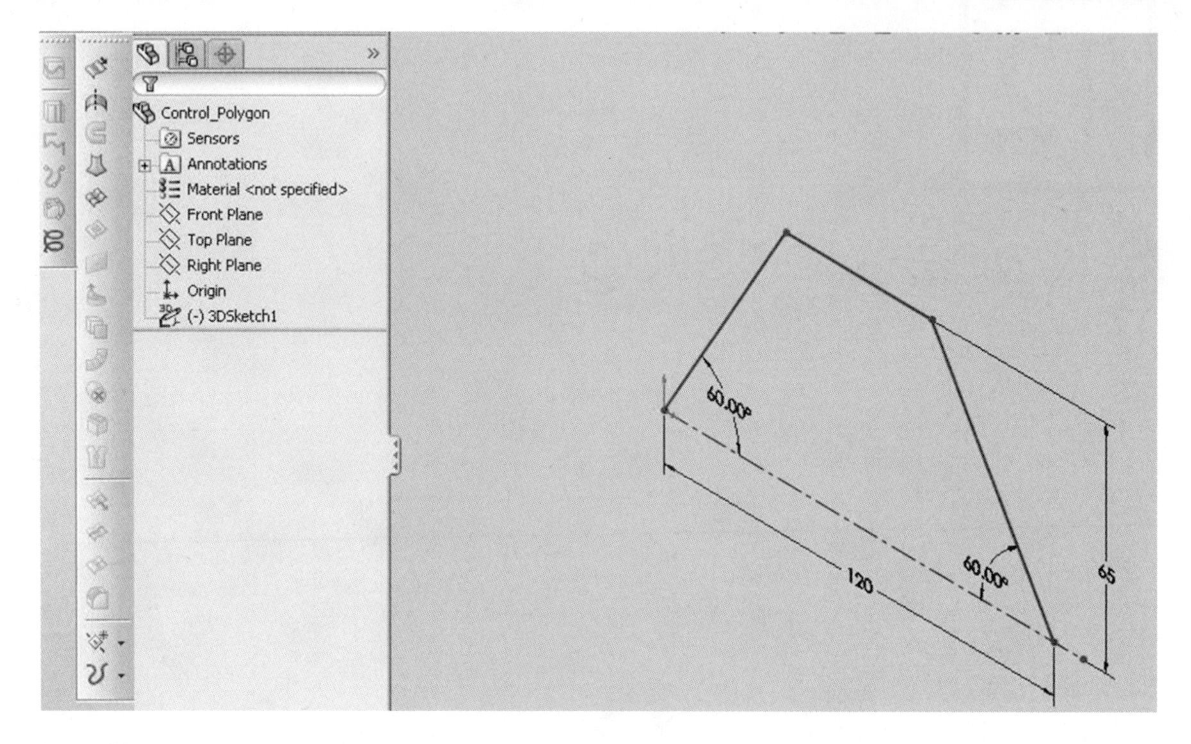

Figure 12.1

Control points on first plane (Front Plane).

Control Polygon

1. Click Insert > 3D Sketch.
2. Click Front Plane.
3. Sketch 3DSketch1 that defines a set of four (or more) control points of interest, as shown in Figure 12.1.
4. Create four (or more) other planes resulting in Plane1, Plane2, Plane3, and Plane4 (see Figure 12.2).
5. Exit 3D Sketch.
6. Select Plane1, Right-click and Click 3D Sketch On Plane.
7. Choose Normal To, to make the plane normal.
8. Sketch lines collinear to the ones on Front Plane. (Choosing Normal To is helpful.)
9. Select Plane2, Right-click and Click 3D Sketch On Plane (Repeat procedure).
10. Select Plane3, Right-click and Click 3D Sketch On Plane (Repeat procedure).
11. Select Plane4, Right-click and Click 3D Sketch On Plane (Repeat procedure).

At this point, our control vertices are ready, as shown in Figure 12.3. There are 20 vertices, four on each plane; there are five planes. Notice that there are five *3DSketches*. This means that we can control each control independently. This is important in *surface design*. For example, *3DSketch1* can be changed independent of *3DSketch2*; *3DSketch2* can be changed independent of *3DSketch3*; *3DSketch3* can be changed independent of *3DSketch4*, etc.

Lofting B-Splines Using Control Polygon

From the control vertices, we can also create B-splines that pass through these vertices, as shown in Figure 12.4.

1. Click 3DSketch1.
2. Click Spline, Click the vertices on 3DSketch1, Click OK, and Exit 3DSketch.
3. Click 3DSketch2.
4. Click Spline, Click the vertices on 3DSketch2, Click OK, and Exit 3DSketch.

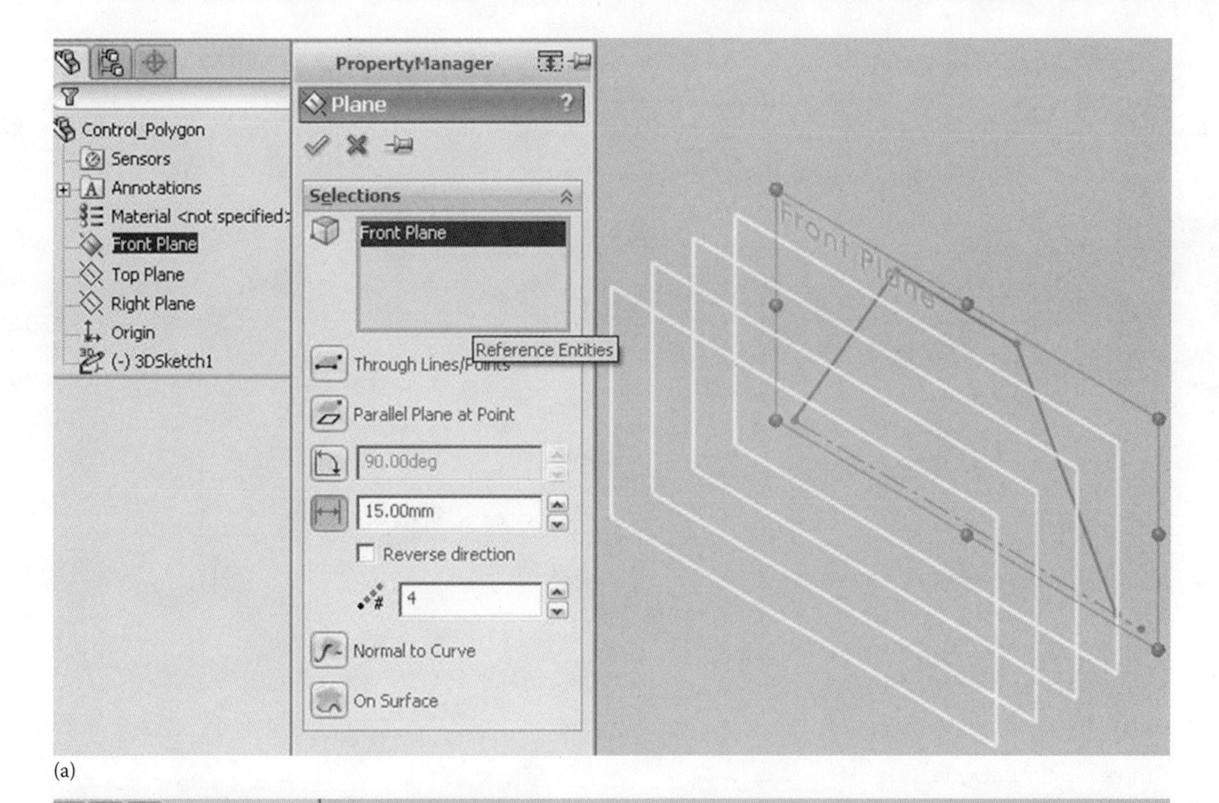

(a)

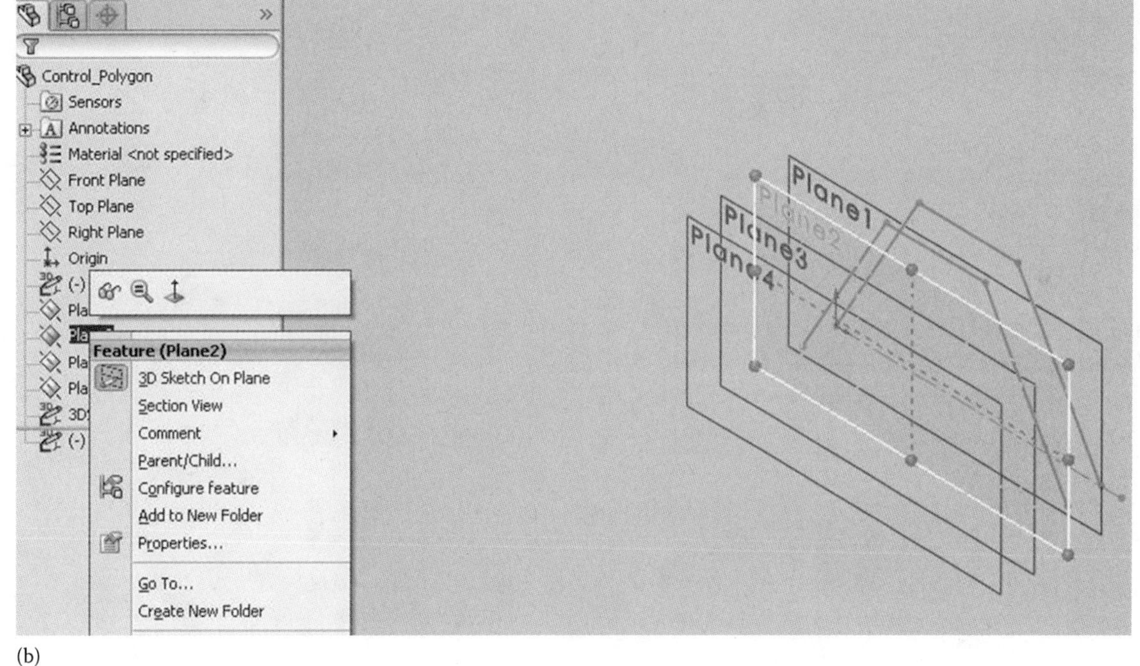

(b)

Figure 12.2

Planes defining control polygon: (a) Plane1 and (b) Front Plane, Plane1,..., Plane4.

5. Click 3DSketch3.
6. Click Spline, Click the vertices on 3DSketch3, Click OK, and Exit 3DSketch.
7. Click 3DSketch4.
8. Click Spline, Click the vertices on 3DSketch4, Click OK, and Exit 3DSketch.

The reason we exit 3DSketch each time is because we are switching a different plane. This is the catch. The lofted B-splines using control polygon are shown in Figure 12.5.

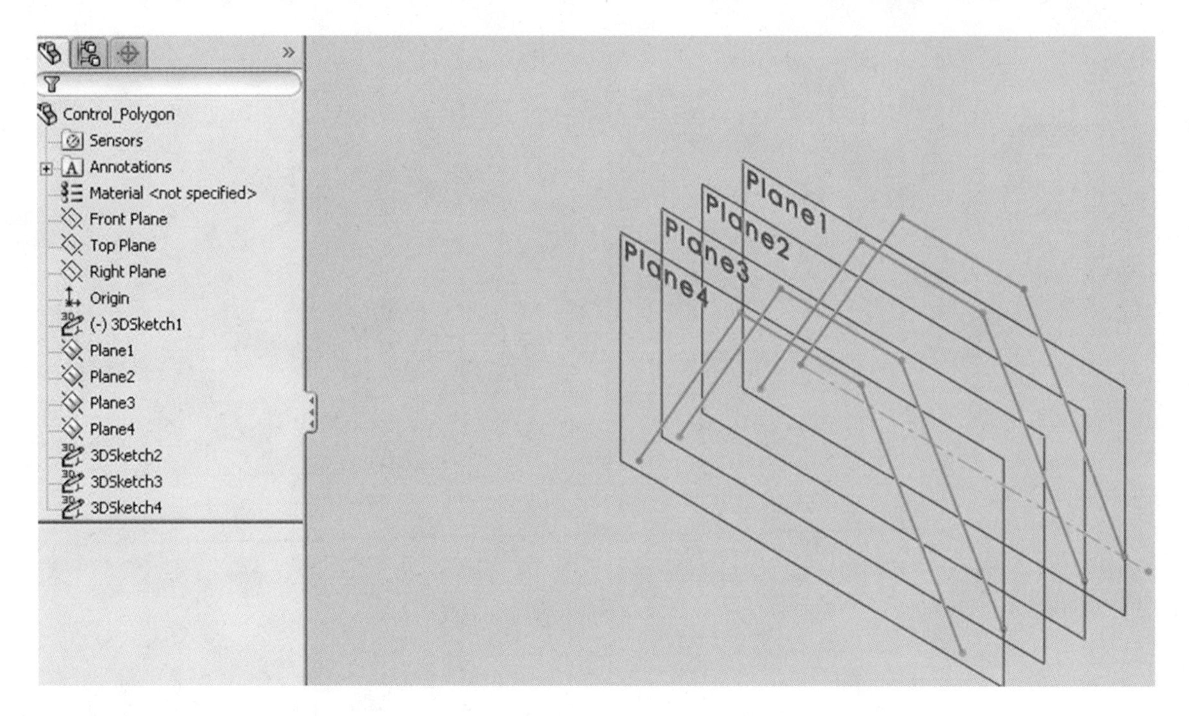

Figure 12.3

Simple control polygon.

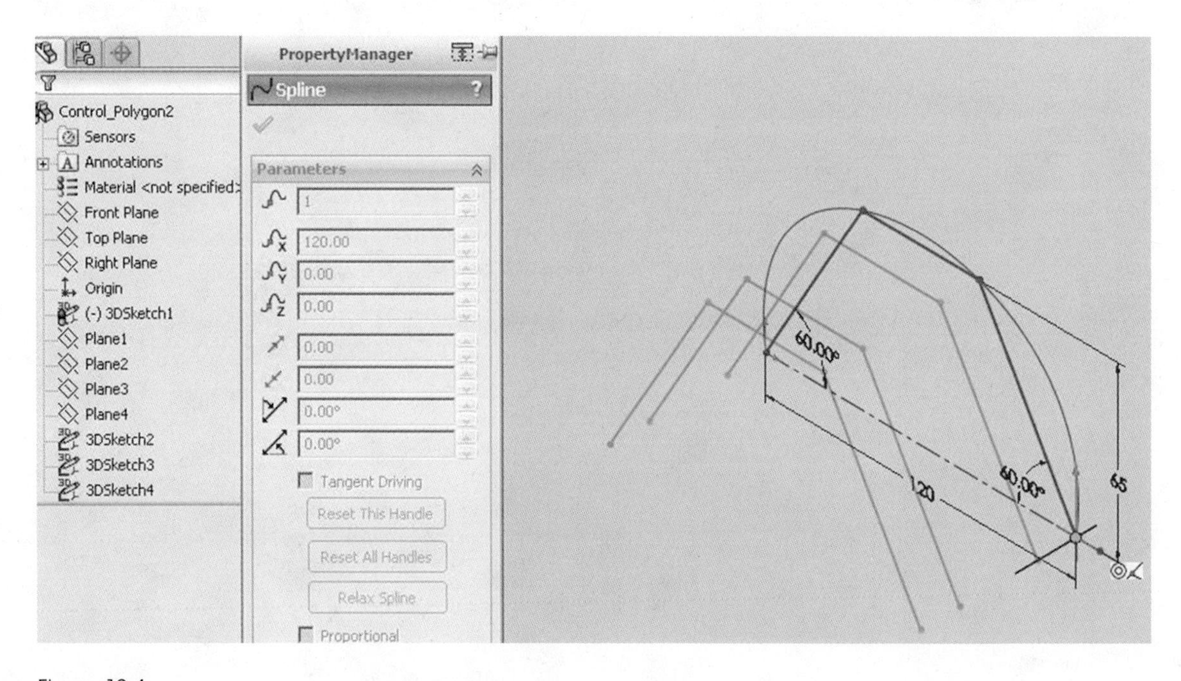

Figure 12.4

First B-spline based on control points on first plane.

Free-Form Surface Design

Now, let us design surfaces based on the Control Polygon that we have created. What it now seems is that we have created a lofting of B-Splines based on the Control Polygon that is realized. We will use the Boundary Surface option with the Control Polygon. If we design from scratch, this is the only option that is active.

1. Click Insert > Surface > Boundary Surface.
2. Click 3DSketch_Spline1 in the Direction1 rollout (see Figure 12.6).

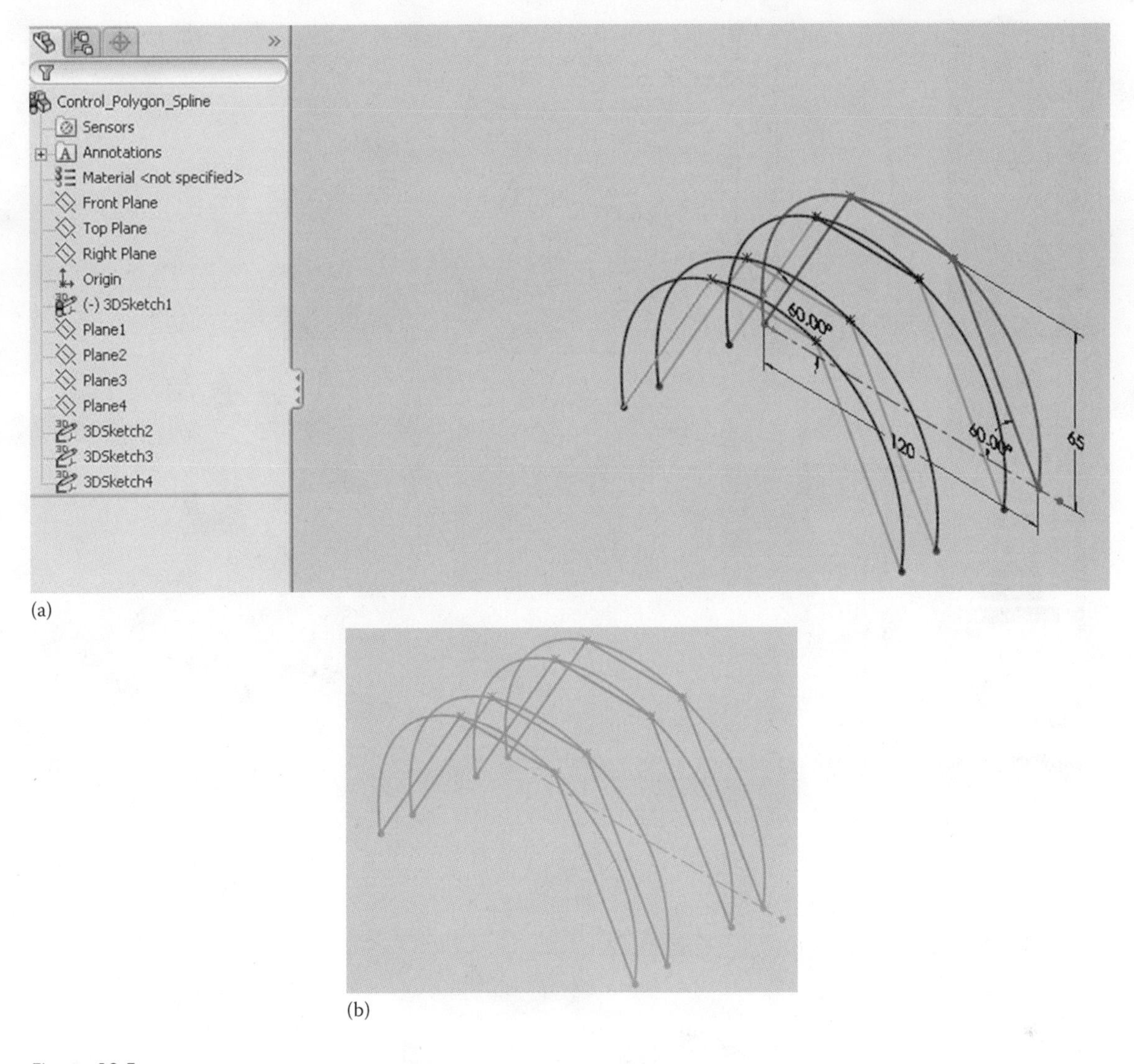

Figure 12.5

B-Spline using Control Polygon: (a) dimensions shown and (b) Exit Sketch.

3. Click 3DSketch_Spline2.
4. Click 3DSketch_Spline3.
5. Click 3DSketch_Spline4.
6. Click 3DSketch_Spline5. (See Figure 12.4 for the preview of the surface.)
7. Click OK. (Figure 12.5 for surface realized.)

We have now created our first free-form surface from scratch, as shown in Figure 12.7. Notice that a new folder, Surface Bodies(1), is now in existence, and the FeatureManager now has a surface definition, Boundary-Surface1, which contains all the 3DSketch_Spline1,..., 3DSketch_Spline5 as well.

Effect of Modifying Control Polygon on Free-Form Surface

If we can modify the control polygon and see the effect of the modification on the free-form surface, that will offer us a flexible way of shape design. As previewed in Figure 12.6, the effect of modifying the control polygon is shown in Figure 12.8. This modification is done through 3DSketch1, not through 3DSketch_Spline1.

Notice that the 3DSketch_Spline1,..., 3DSketch_Spline5 are all bundled and placed under the Boundary-Surface1, which is now created in the FeatureManager. By further modifying the control vertices, as shown in Figure 12.9, we can then have a modified shape (see Figure 12.10).

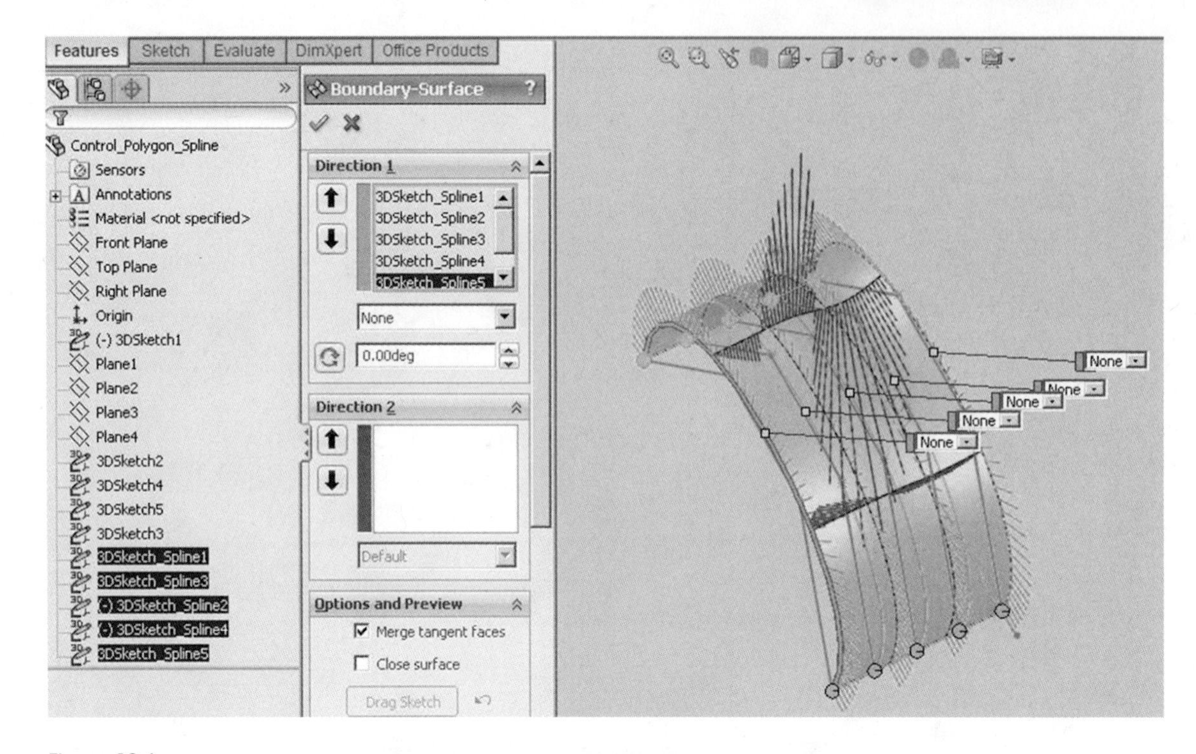

Figure 12.6

Preview of free-form surface using B-Spline-based control polygon.

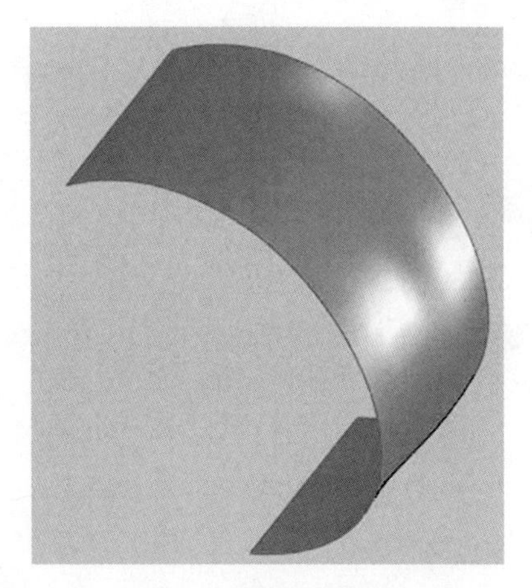

Figure 12.7

Free-form surface using B-Spline-based control polygon.

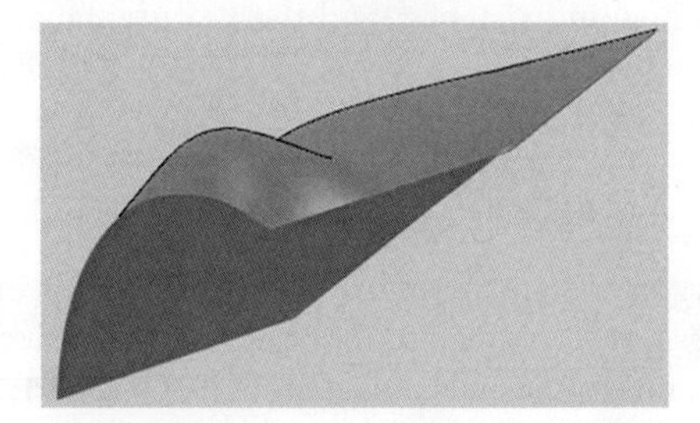

Figure 12.8

Modifying the free-form surface through modifying the control polygon.

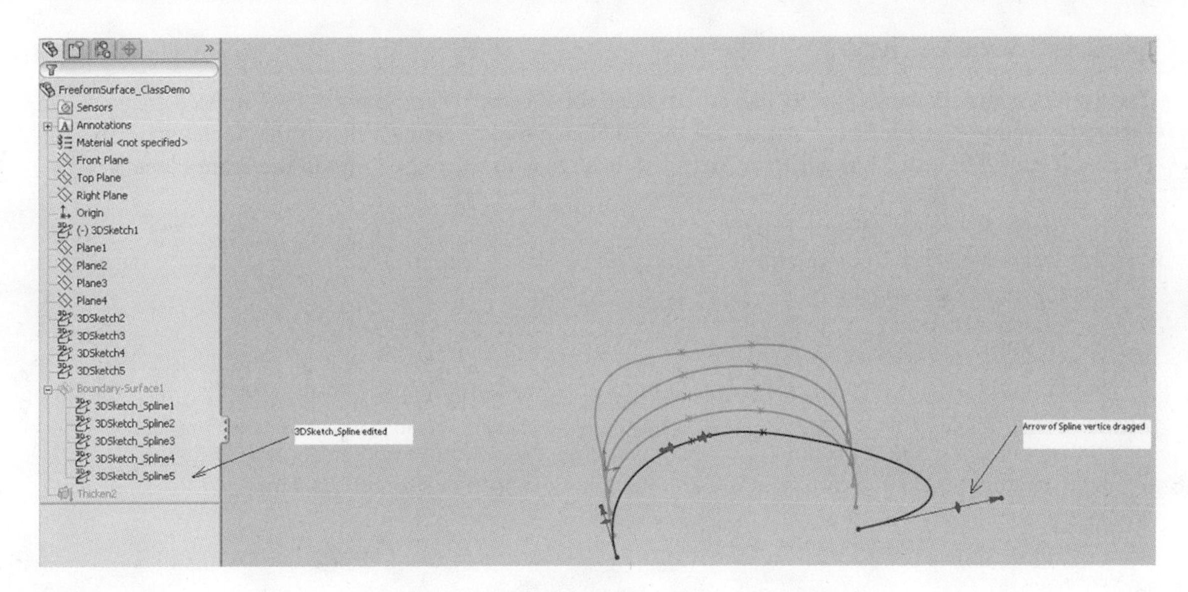

Figure 12.9

Modified control vertices.

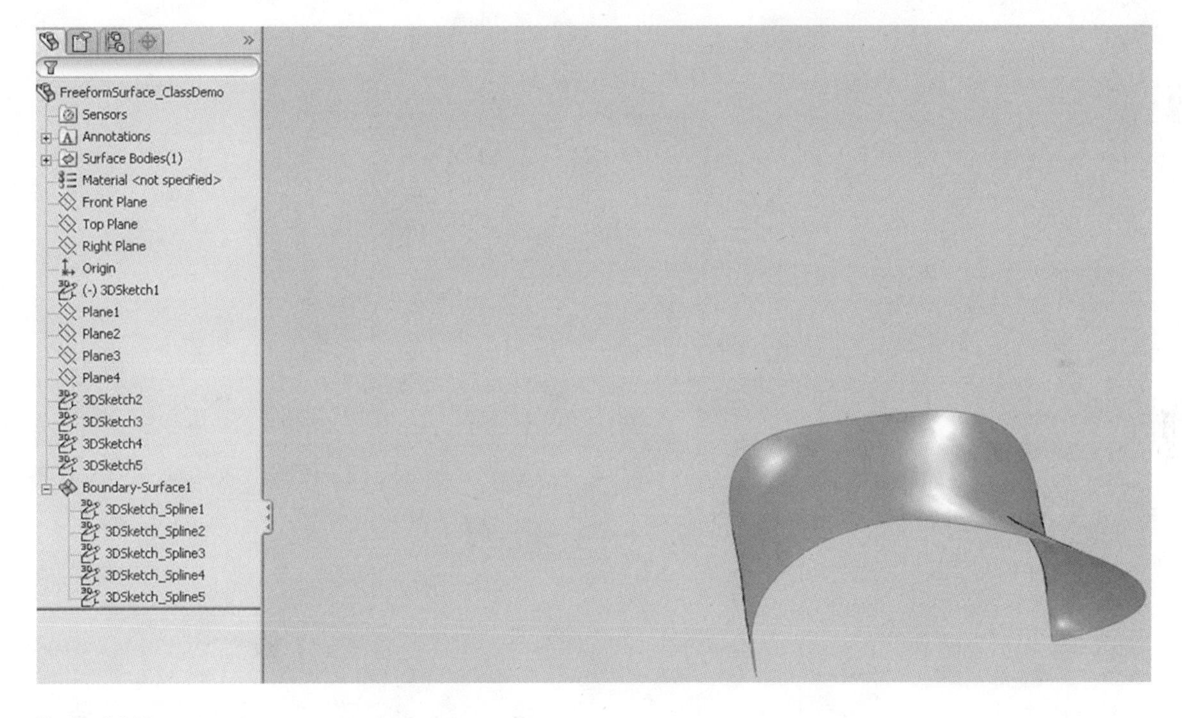

Figure 12.10

Modified shape resulting from modified control vertices.

Notice that when the surface model is complete, Surface Bodies(1) now shows up in the FeatureManager at the top, while Boundary-Surface(1) shows up at the bottom. This now shows that we have Surface Bodies. The number in parenthesis shows the number of Surface Bodies.

In this chapter, we have presented a generalized methodology for free-form surface design. We have also shown that by modifying the Splines obtained using the control vertices by dragging the handle vertices on the Spline vertices, it is possible to obtain a significantly different free-form surface. This generalized approach to free-form design is heavily used in practical applications in the automotive, ship, and aircraft industries for designing car bodies, ship hulls, airplane fuselages, etc.

Extruded Surface: Type I

The Extruded Surface works exactly like an extruded solid, except that the ends of the surface are open. An extrude direction is needed here. Let us use the 3D Sketch tool to define a sketch that is shown in the Top Plane in Figure 12.11, which needs to be extruded through a distance of 65 mm in the Front Plane.

1. Create 3DSketch1 (Figure 12.11).
2. Create 3DSketch2 (a straight line vertical to 3DSketch1 specifying the direction of extrusion, as shown in Figure 12.12).
3. Choose 3DSketch1 in the FeatureManager.
4. Click Insert > Surface > Extrude.
5. From the Selected Contours rollout, highlight the geometry of 3DSketch1 in the Graphics Window (see Figure 12.13).
6. From the Direction 1 rollout, click 3DSketch2. (A Preview pops up; see Figure 12.13.)
7. Give the value of the extrusion in Direction1 in the Distance spinner as 65 mm.

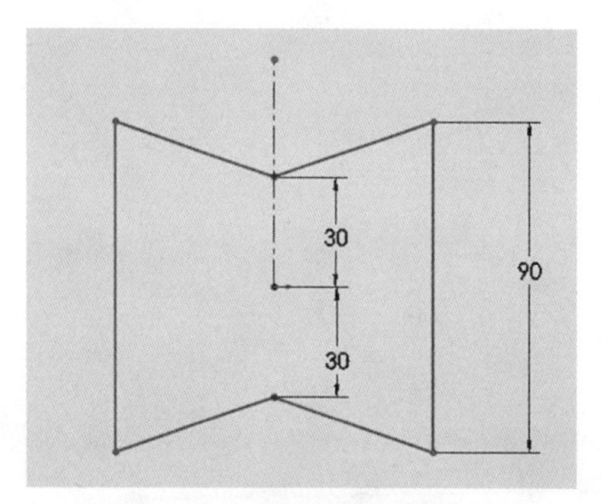

Figure 12.11

Sketch1 for extruding surface.

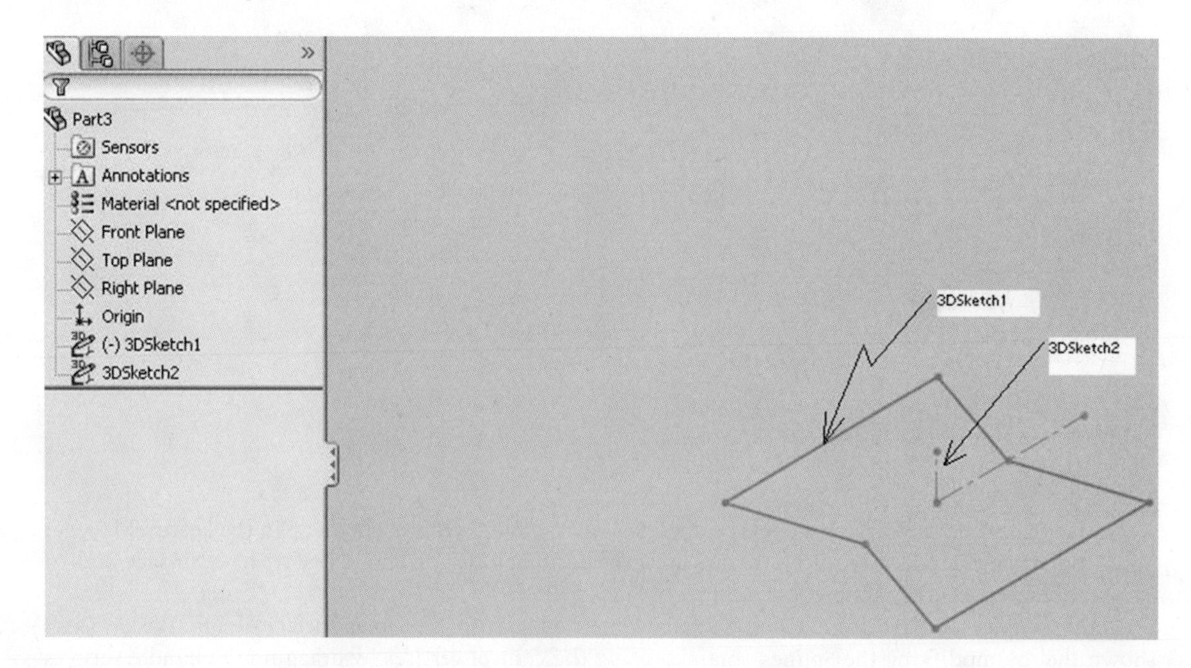

Figure 12.12

Vertical line: direction for extrusion.

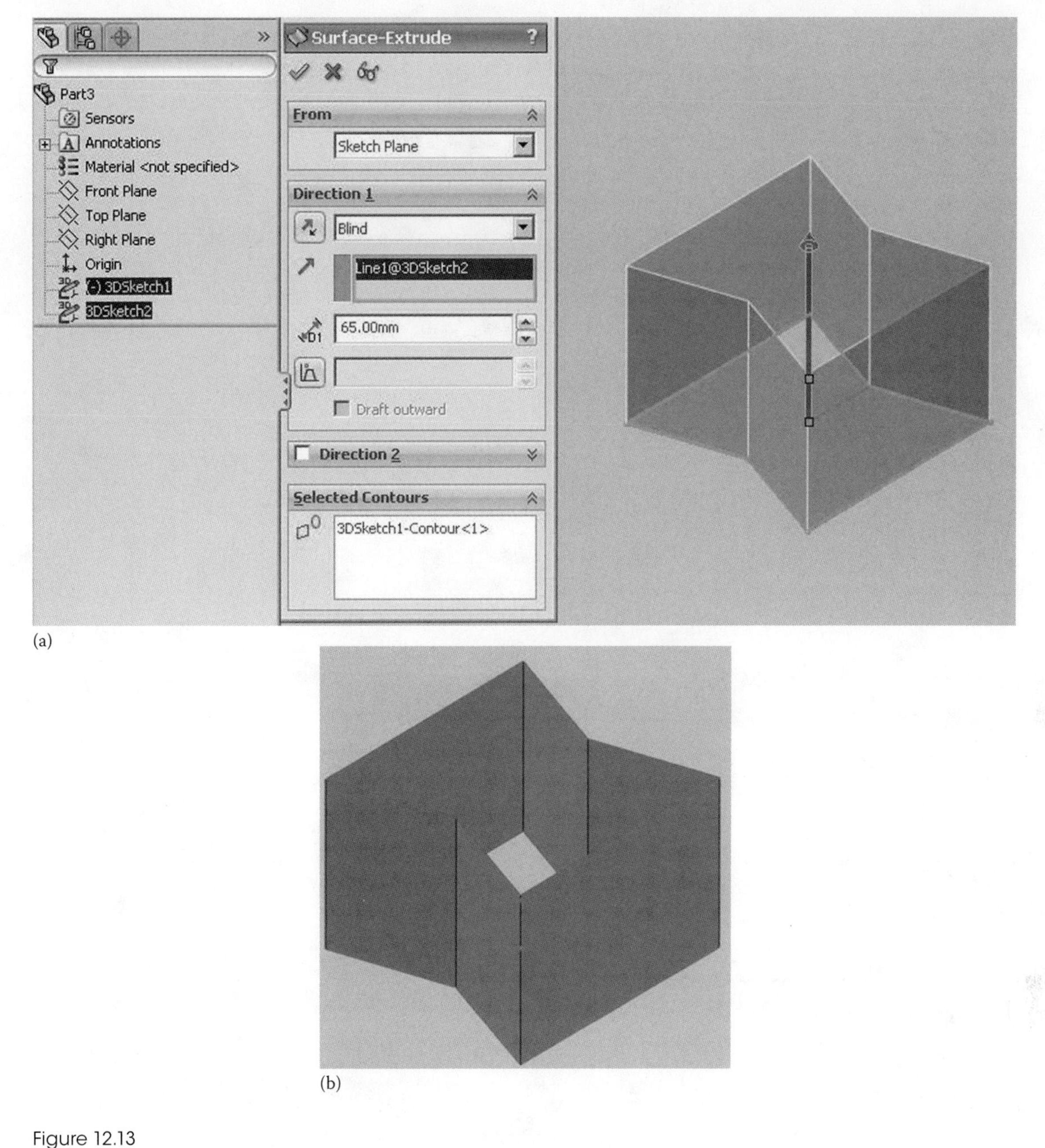

Figure 12.13

Extruded surface: (a) the preview of the extruded surface and (b) the extruded surface.

Revolved Surface: Type I

The *Revolved Surface* works exactly like a revolved solid, except that the ends of the surface are open. A revolve direction is also needed as in solid. Let us use the two-dimensional Sketch tool to define a sketch that is shown in Figure 12.14 in the Front Plane, which needs to be revolved about a vertical construction line or axis. The fillet radius is 1 mm as shown.

1. Click Insert > Surface > Revolve.
2. Select the contour of Sketch1 (Sketch1-Contour<1>) from the Graphics Window.
3. Select the direction, Line1, about which to revolve the contour (see Figure 12.15).
4. Click OK.

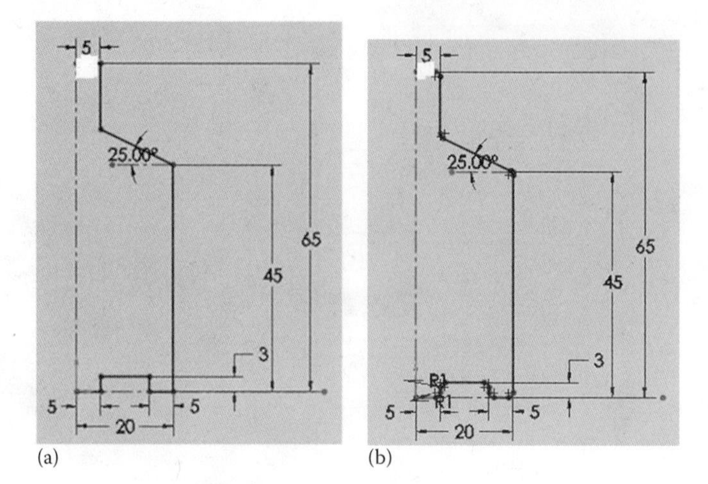

Figure 12.14

Sketch for revolved surface: (a) no fillets and (b) fillets added.

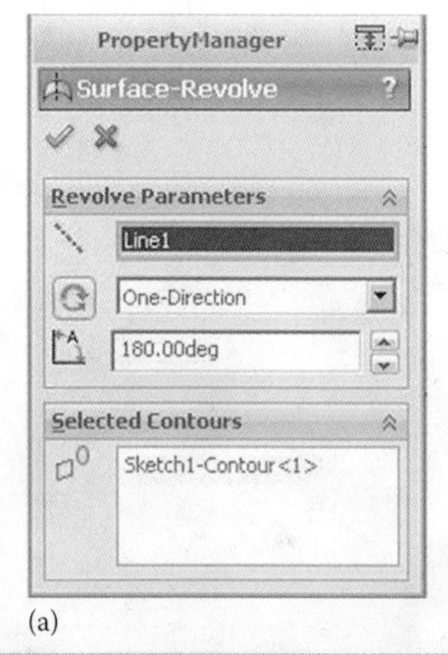

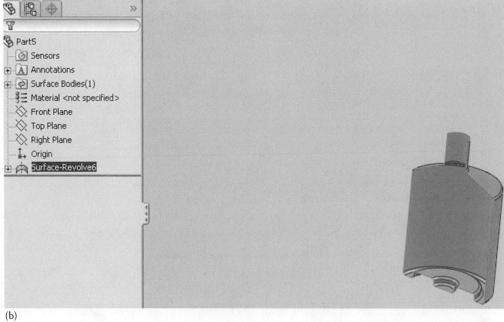

Figure 12.15

Revolved surface: (a) PropertyManager for surface-revolve and (b) revolved surface.

Knitting Multiple Surfaces

When we design different surfaces, they become knitting multiple Surfaces Bodies, just like we have Solid Bodies that are Multiple Bodies. To *sew together* Surfaces Bodies, Knit Surface is used just like a tailor sews pieces of clothes together.

If the knit operation results in a watertight volume, the Fill Surface option turns the volume into a solid.

Thicken Feature

If a surface body that encloses a volume is selected, then the option Create solid from enclosed volume pops up on the Thicken PropertyManager. It could be accessed via the route of

Insert > Boss/Base > Thicken

Let us convert the modified free-form surface that is already designed to a free-form solid. We will use the Thicken Feature option.

1. Click Insert > Boss/Base > Thicken.
2. Choose Boundary-Surface1 in the Thicken Parameter spinner (see Figure 12.16).

The converted solid is shown in Figure 12.17.

Once we have converted the surface model into a solid model using the Thicken Feature, notice that two things happen:

1. The Surface Bodies(1) previously at the top of the FeatureManager, as well as the Boundary-Surface(1) at the bottom, now disappear.
2. The thicken1 feature is added to the bottom of the FeatureManager.

These observations are important: we now only have solids, no more reference to surface bodies.

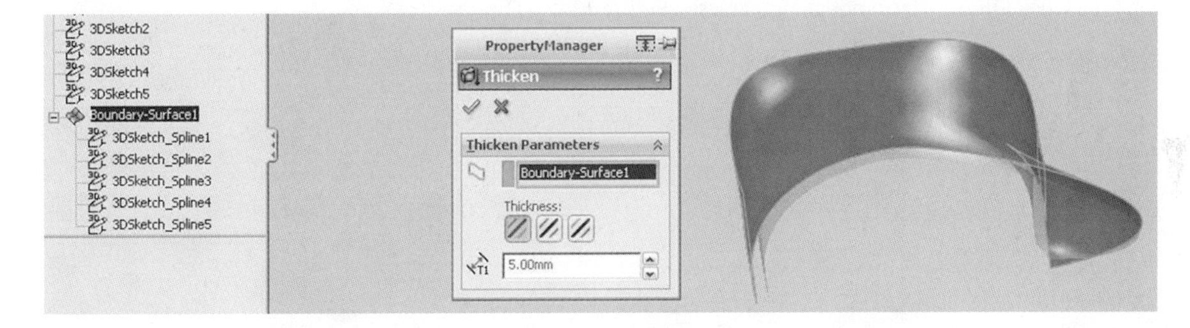

Figure 12.16

Process of converting the surface model into a solid model using Thicken Feature.

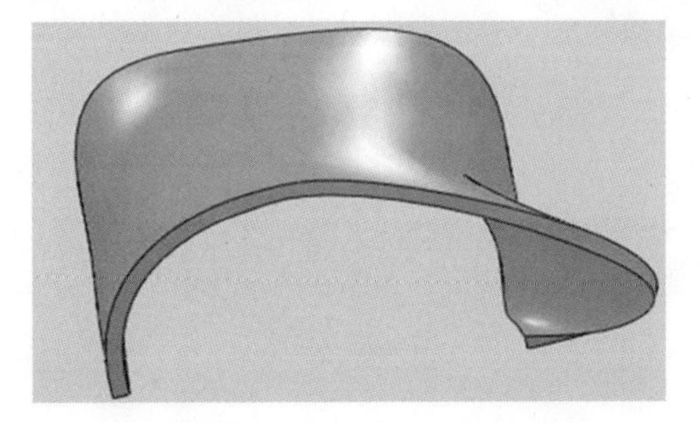

Figure 12.17

Solid model realized using Thicken Feature.

Fill Surface

The *Fill Surface* has an option to merge the fill with a solid or to knit it into a surface body.

Create a Hole

1. Create Plane5, 60 mm from the Top Plane.
2. Create 3DSketch6, a circular profile 15 mm in diameter.
3. Extrude 3DSketch6 Up To the top face of the solid, Thicken1 in the Direction of 3DSketch7 resulting in a hole (see Figure 12.18).

Create a Hole

Let us fill this hole using the Fill Surface option.

1. Click Insert > Surface > Fill Surface. (See Figure 12.19 for preview.)
2. Click the *cutout* in the Graphics Window for the Patch Boundary rollout; this is the Extrude_HoleCut. Edge<1>Contact-50 appears in the Patch Boundary rollout, and the cut is filled at the top (see Figure 12.20). It is amazing. Notice that the solid is not filled; the top surface of the solid is filled. This is the similar principle used in "Shut-off Surfaces" used in Mold design.

The Fill Surface tool is of primary importance in the manufacturing industry. For instance, a company receives an order from another contracting company requesting a job description for filling a hole on the rim of an automobile wheel that is for mass production. A wrong hole was initially drilled. This hole is to be filled and drilled elsewhere on the rim. Redesigning is expensive. How can the company deal with this problem within a short time? The answer is to use the fill surface toolbar that we have described. It is effective and efficient and greatly admired by many designers for what it can achieve.

Extruded Surface: Type II

1. Click Top Plane.
2. Create Sketch1 (see Figure 12.21).
3. Click Insert > Surface > Extrude. (Surface-Extrude1 PropertyManager appears; see Figure 12.22.)
4. In the Direction1 rollout, supply 35 mm for the Depth of extrusion (see Figure 12.22).
5. Click OK. (Surface model appears; see Figure 12.23.)

Revolved Surface: Type II

1. Click Front Plane.
2. Create Sketch1 (see Figure 12.24).

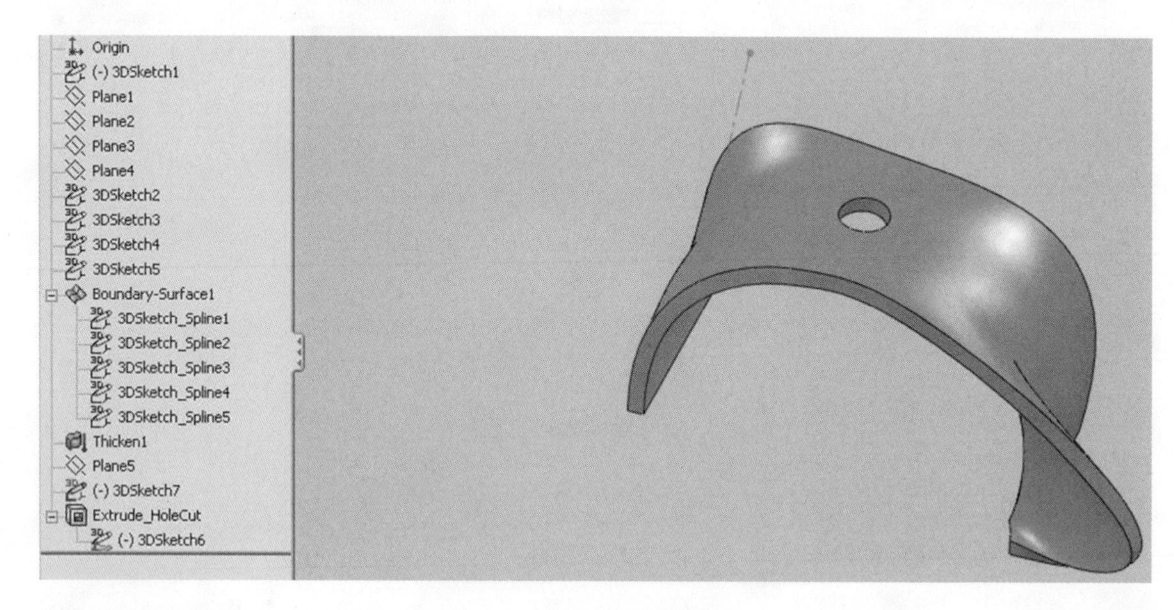

Figure 12.18

Create a hole.

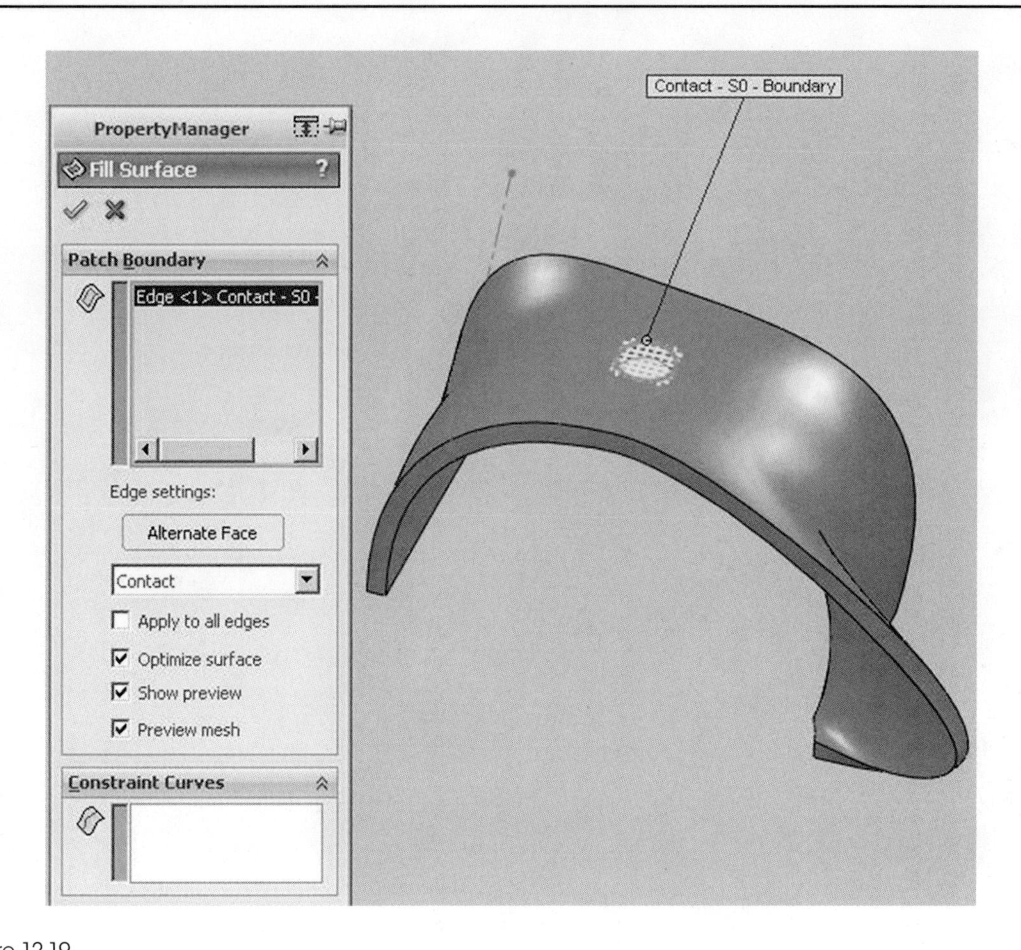

Figure 12.19

Preview for Fill Surface.

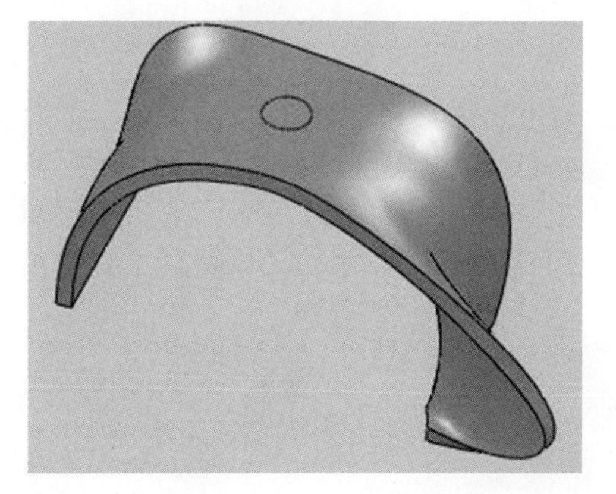

Figure 12.20

Repaired surface.

3. Click Insert > Surface > Revolve. (Surface-Revolve1 PropertyManager appears; see Figure 12.25.)
4. Click OK. (Surface model appears; see Figure 12.26.)

Swept Surface

Rectangular Section

1. Click Front Plane.
2. Create a 3D sketch, 3DSketch1 (see Figure 12.27).
3. Exit Sketch.

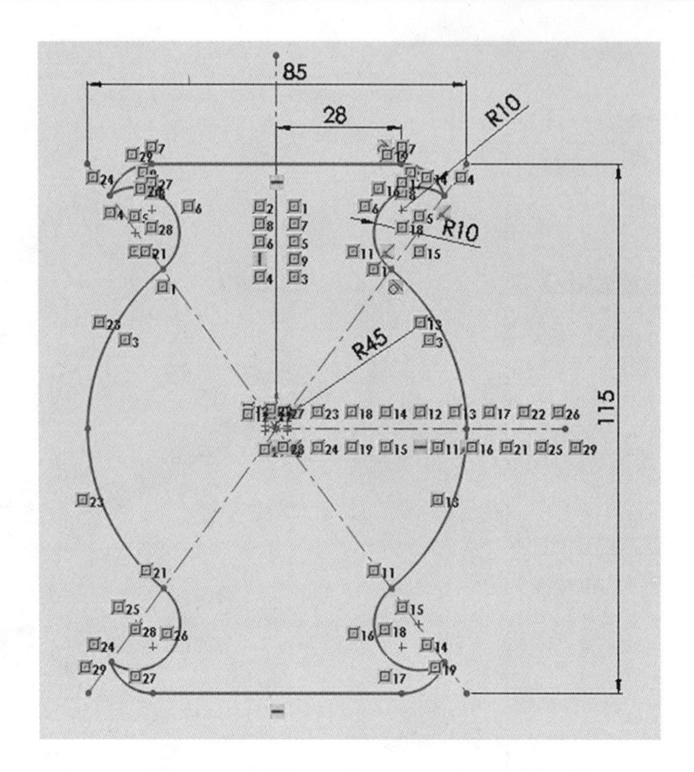

Figure 12.21

Sketch1.

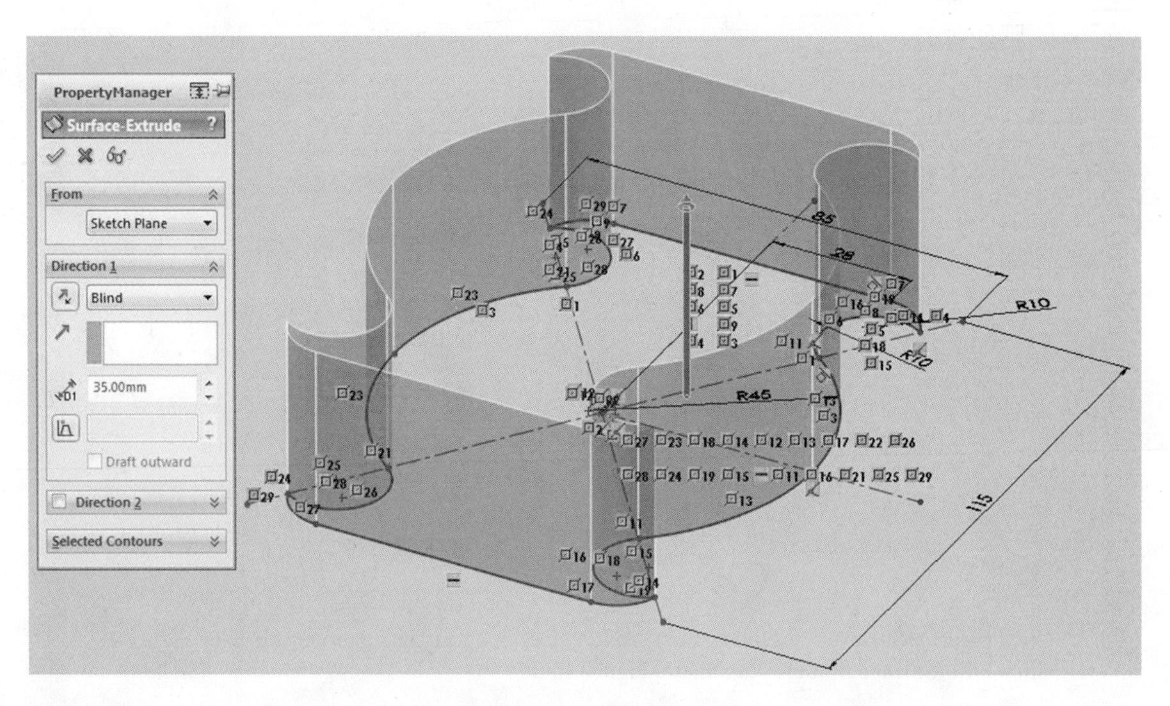

Figure 12.22

Surface-Extrude1 PropertyManager.

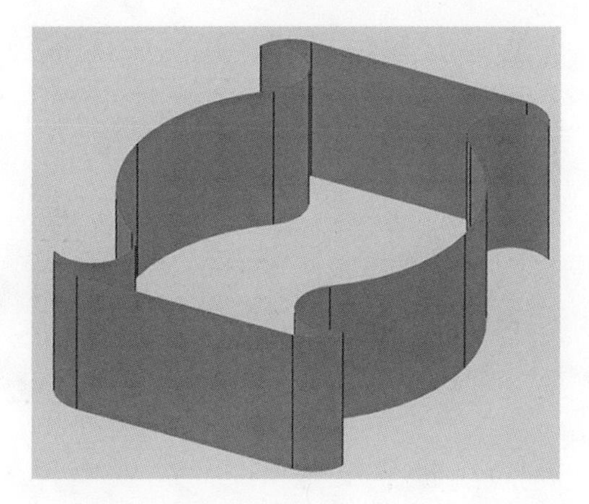

Figure 12.23

Surface model.

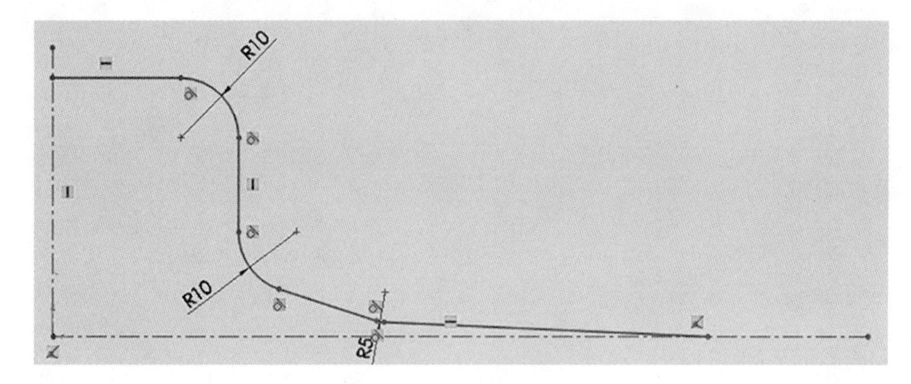

Figure 12.24

Sketch1.

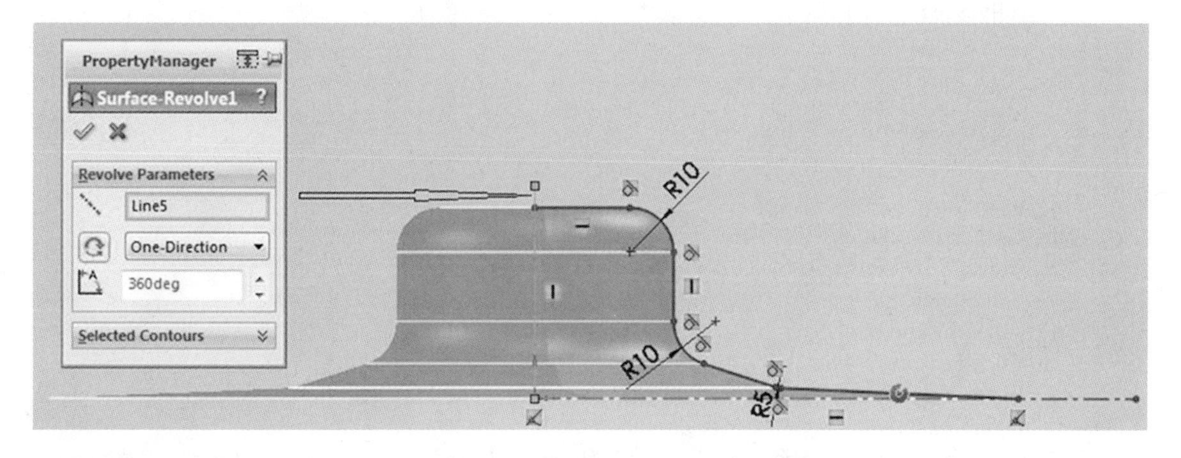

Figure 12.25

Surface-Revolve1 PropertyManager.

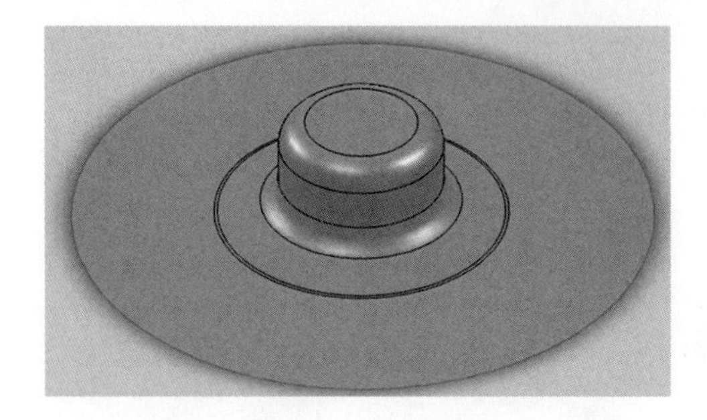

Figure 12.26

Surface model.

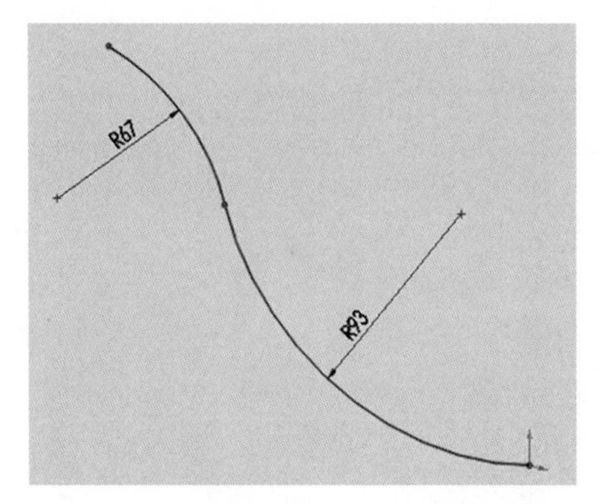

Figure 12.27

Sketch1.

4. Click end of 3DSketch1 (Point9@3DSketch1).
5. Click Features > Plane.
6. In the Selections rollout, click Normal to Curve. (The Plane PropertyManager appears as in Figure 12.28.)
7. Click the arc close to the endpoint that is already selected. (Arc3@3DSketch1 is added automatically; see Figure 12.28.)
8. Exit Sketch.
9. Click Insert > Surface > Sweep (Figure 12.29).
10. The Surface-Sweep PropertyManager appears; see Figure 12.30.
11. Click OK. (Surface model appears; see Figure 12.31.)

Circular Section

1. Click Front Plane.
2. Create a 3D sketch, 3DSketch1 (see Figure 12.32).
3. Exit Sketch.
4. Click the end of 3DSketch1 (Point9@3DSketch1).
5. Click Features > Plane.
6. In the Selections rollout, click Normal to Curve. (The Plane PropertyManager appears as in Figure 12.33.)
7. Click the arc close to the endpoint that is already selected. (Arc3@3DSketch1 is added automatically; see Figure 12.33.)
8. Exit Sketch.
9. Click Insert > Surface > Sweep (Figure 12.34).
10. The Surface-Sweep PropertyManager appears; see Figure 12.35.
11. Click OK. (Surface model appears; see Figure 12.36.)

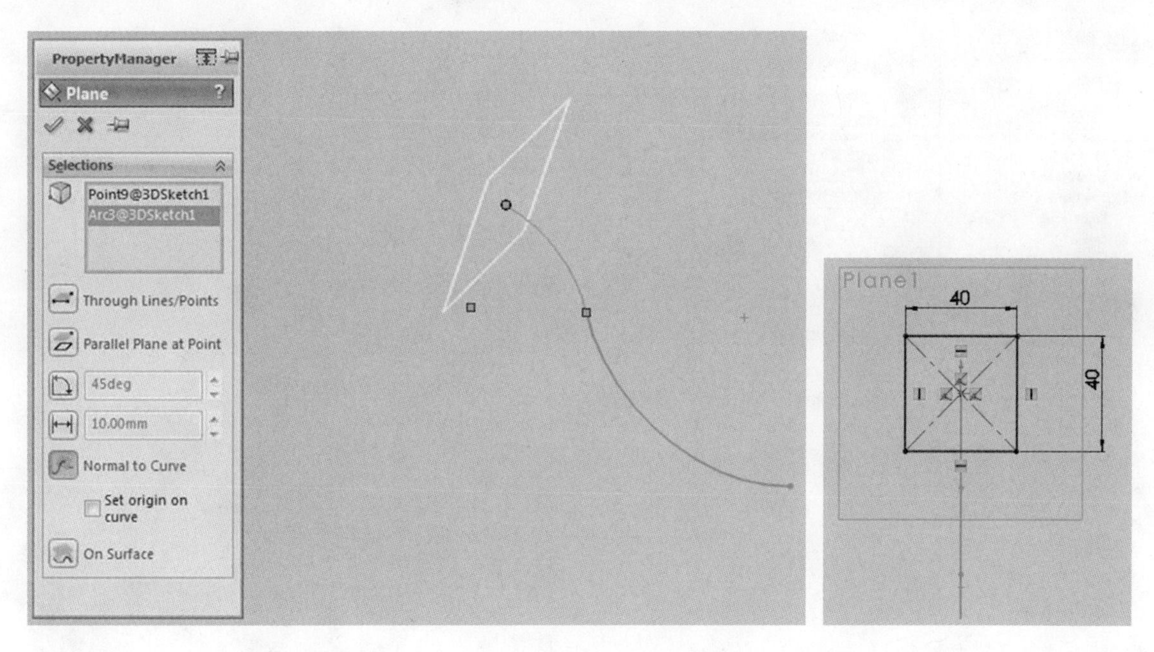

Figure 12.28

Plane PropertyManager.

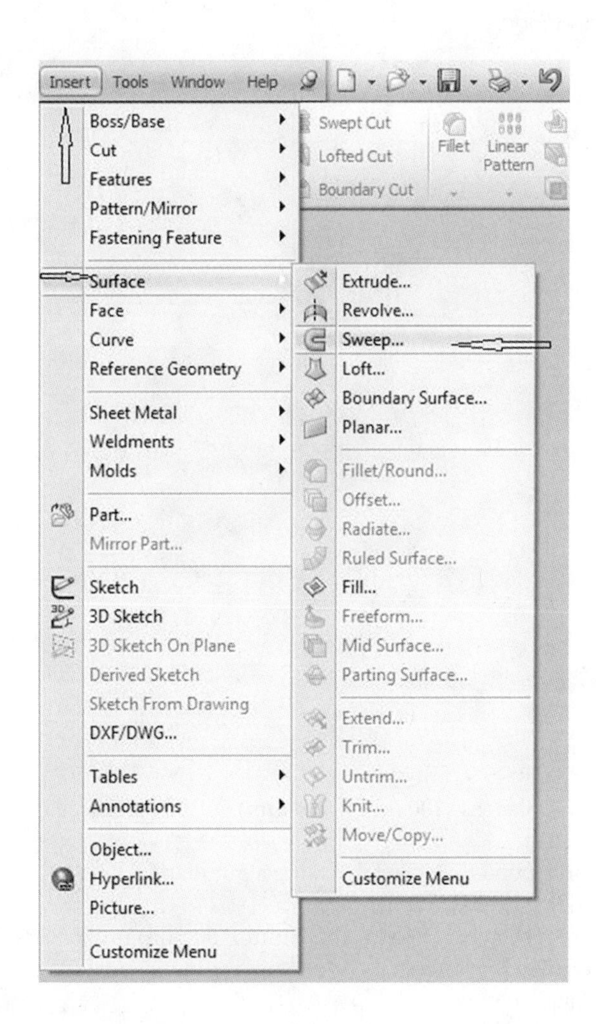

Figure 12.29

Surface sweep option.

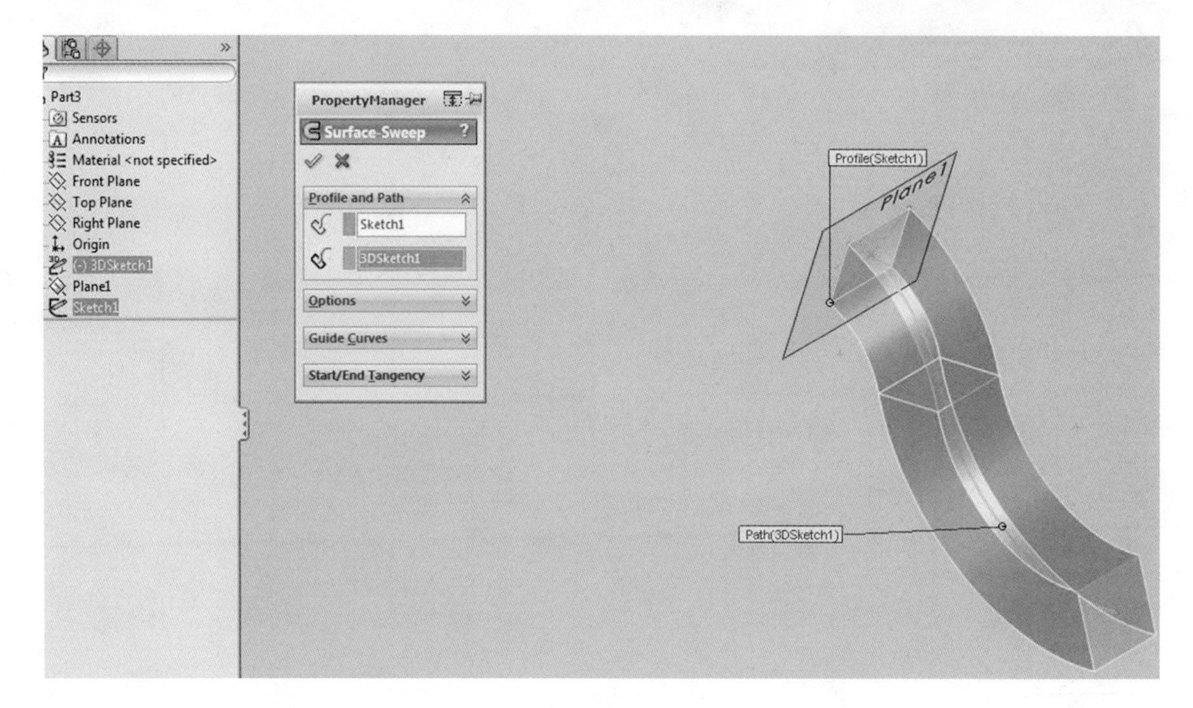

Figure 12.30

Surface sweep PropertyManager.

Figure 12.31

Surface model.

Loft Surface

1. Create a 3D sketch, 3DSketch1 (100 mm × 100 mm), in YZ Plane (see Figure 12.37).
2. Exit sketch, 3DSketch1.
3. Create Plane1 to correspond to 3DSketch1 using a point and a line.
4. Create Plane2 at a distance of 175 mm.
5. On Plane2, create a 3D sketch, 3DSketch2, (50 mm × 50 mm) in YZ Plane (see Figure 12.38).
6. Exit sketch, 3DSketch2.
7. Click Insert > Surface > Loft.
   The Surface-Loft PropertyManager appears; see Figure 12.39.
8. Click OK. (Surface model appears; see Figure 12.40.)

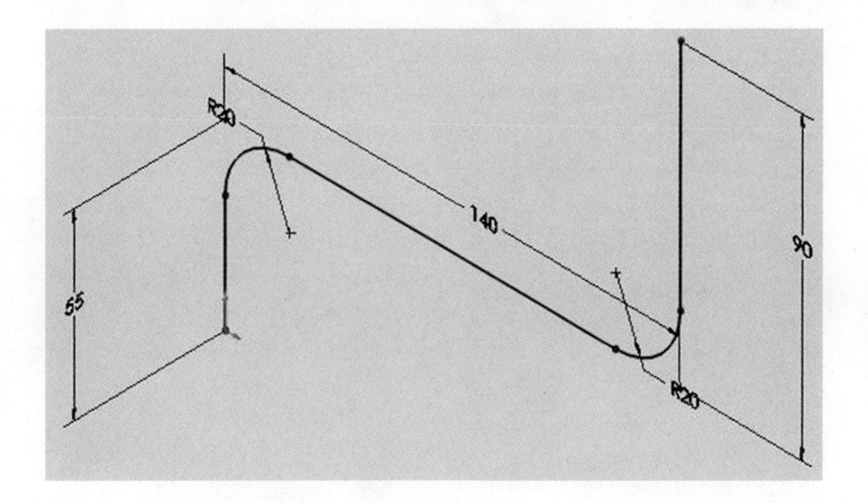

Figure 12.32

Sketch1.

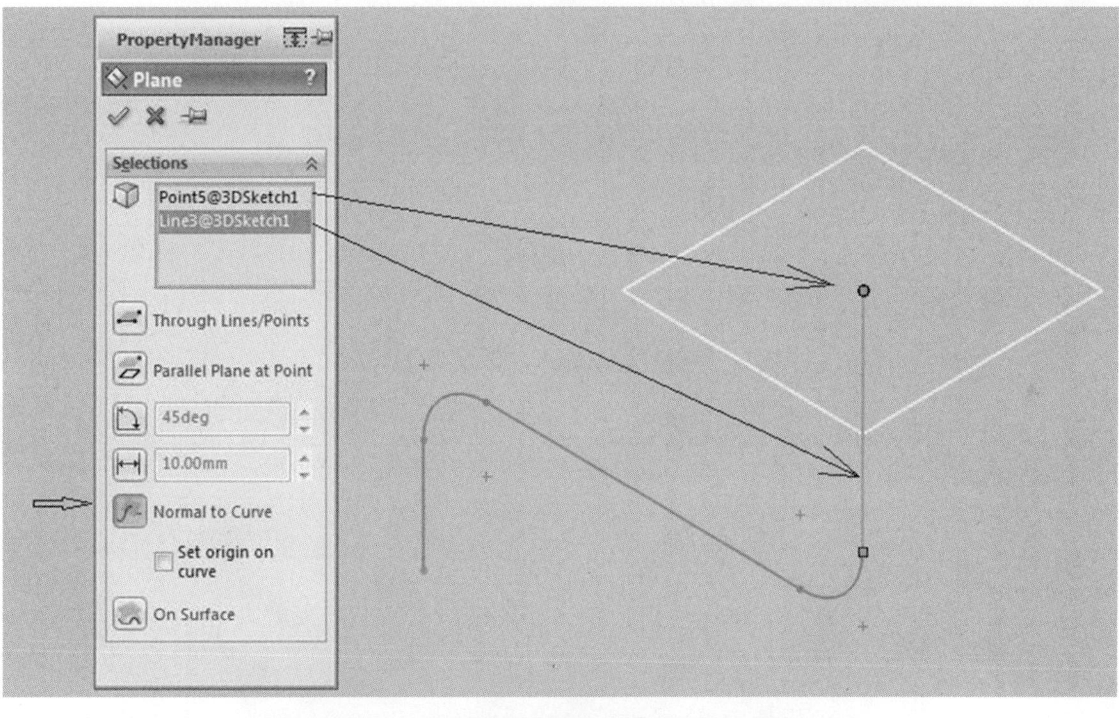

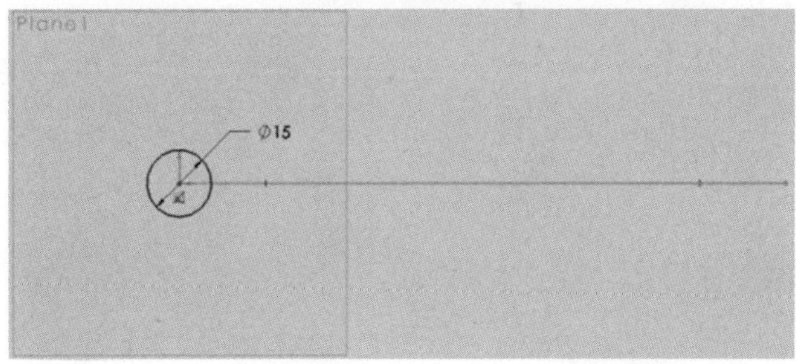

Figure 12.33

Plane PropertyManager.

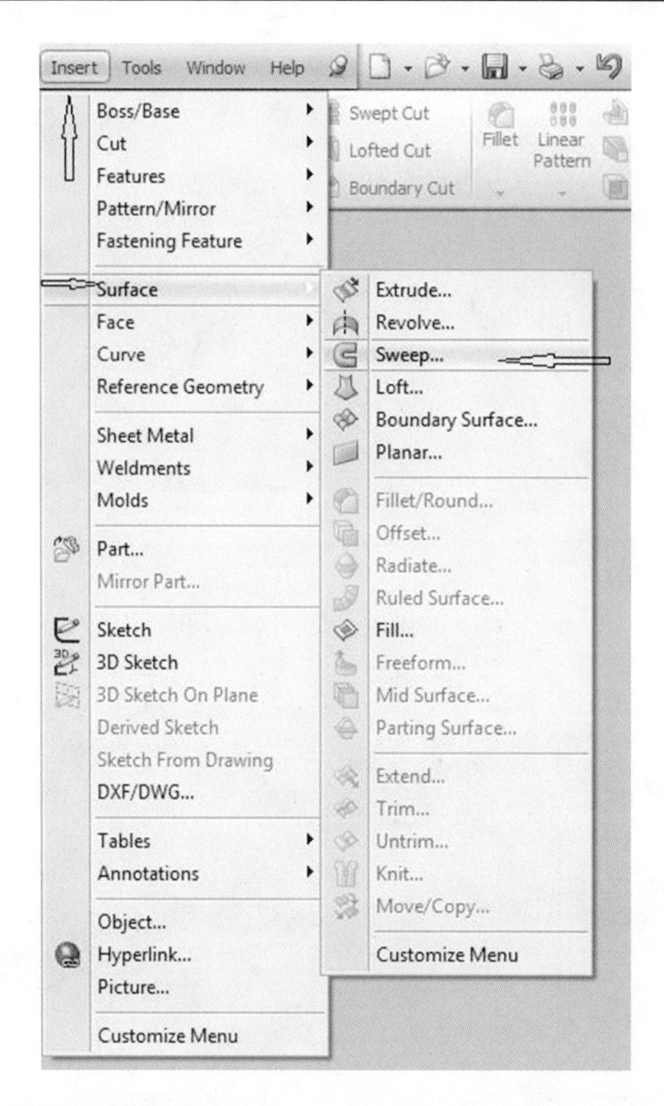

Figure 12.34

Surface sweep option.

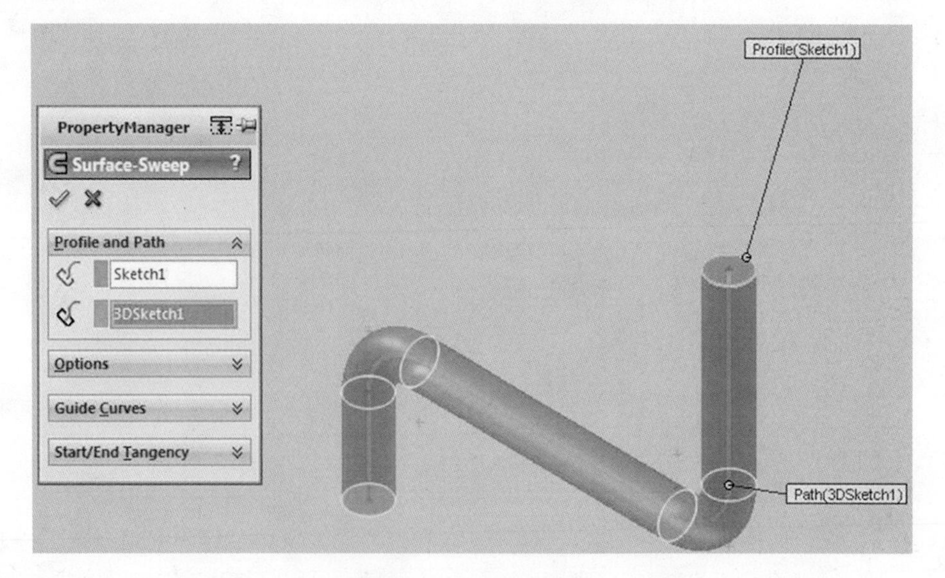

Figure 12.35

Surface sweep PropertyManager.

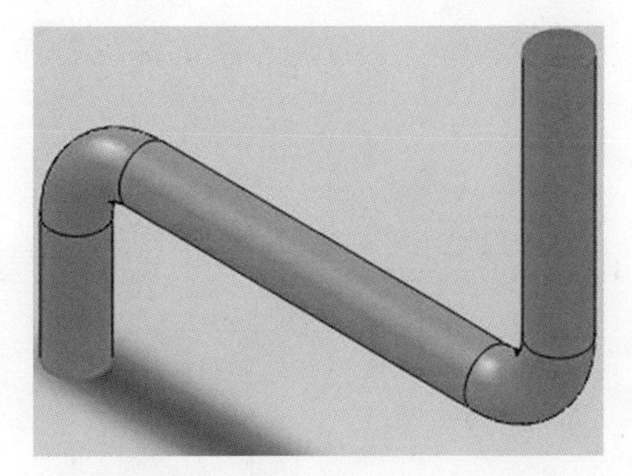

Figure 12.36

Surface model.

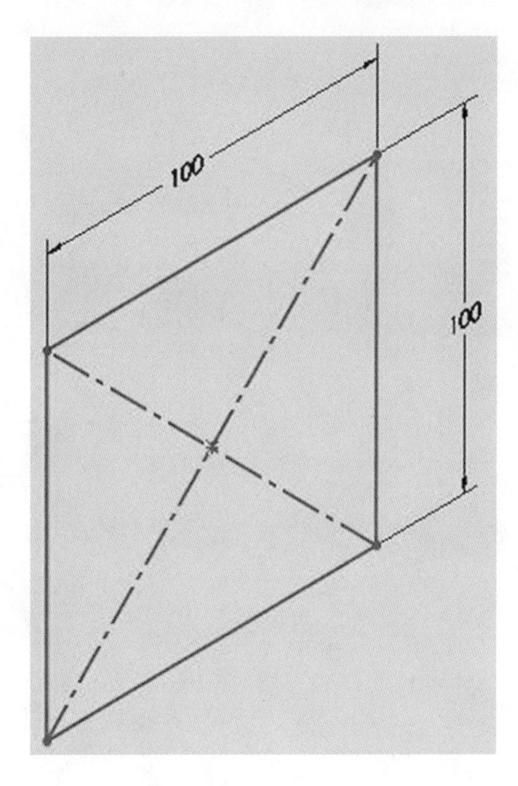

Figure 12.37

3DSketch1 on YZ Plane.

Free-Form Surface Design: Boundary Surface

Control Polygons

1. Start a New SolidWorks Part document.
2. Select the Front Plane.
3. Click Sketch 3DSketch on Plane.
4. Create a 3D sketch, 3DSketch1, on XY Plane (see Figure 12.41).
5. Exit sketch, 3DSketch1.
6. Create Plane1 on the same plane containing 3DSketch1 using a point and a line.
7. Create three more planes: (1) Plane2, (2) Plane3, and (3) Plane4 (see Figure 12.42).
8. Click Plane2 > 3D-Sketch on Plane.
9. Click View Orientation > Normal To.

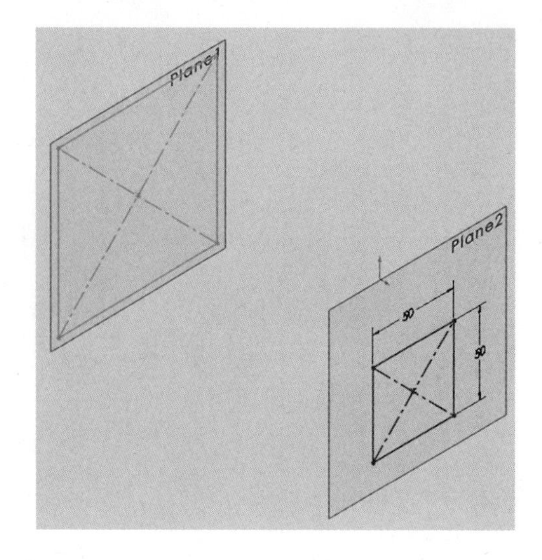

Figure 12.38

3DSketch2 on YZ Plane.

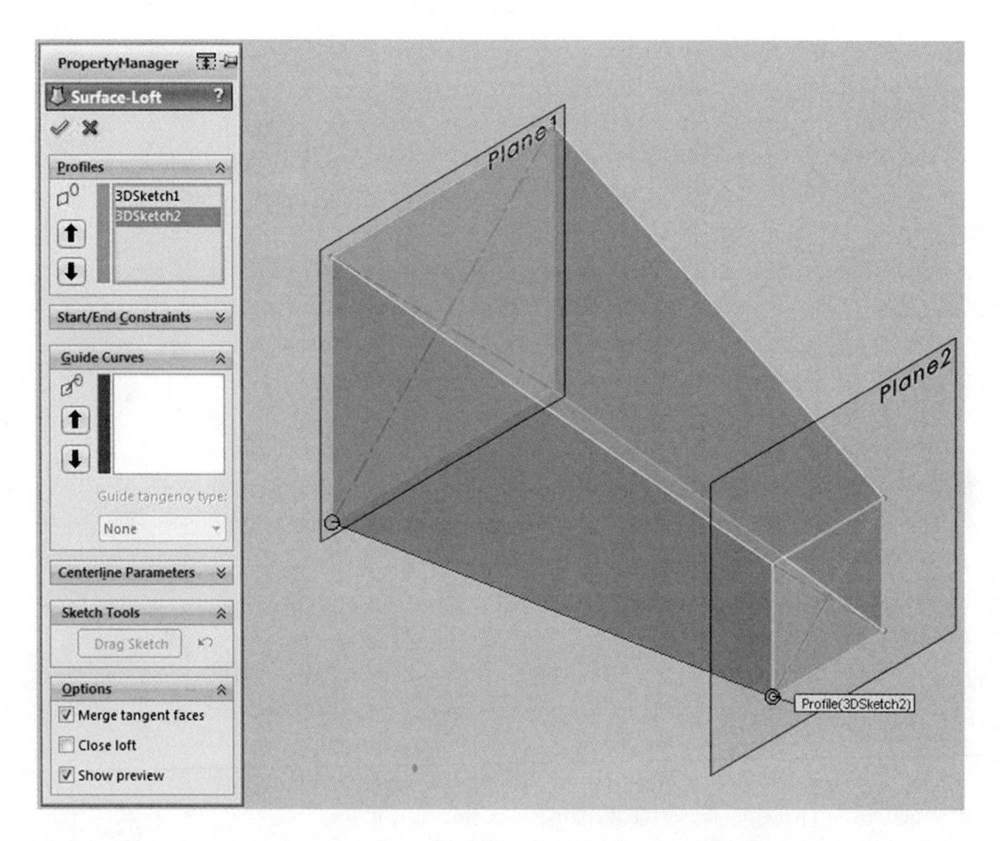

Figure 12.39

Surface-Loft PropertyManager.

10. Create a 3D sketch, 3DSketch2, on XY Plane (see Figure 12.43).
11. Exit sketch, 3DSketch2.
12. Click Plane3 > 3D-Sketch on Plane.
13. Click View Orientation > Normal To.
14. Create a 3D sketch, 3DSketch3, on XY Plane (see Figure 12.43).
15. Exit sketch, 3DSketch3.
16. Click Plane4 > 3D-Sketch on Plane.

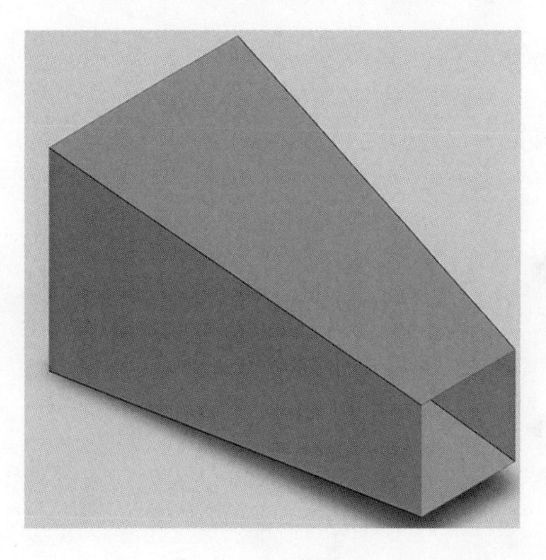

Figure 12.40

Surface model.

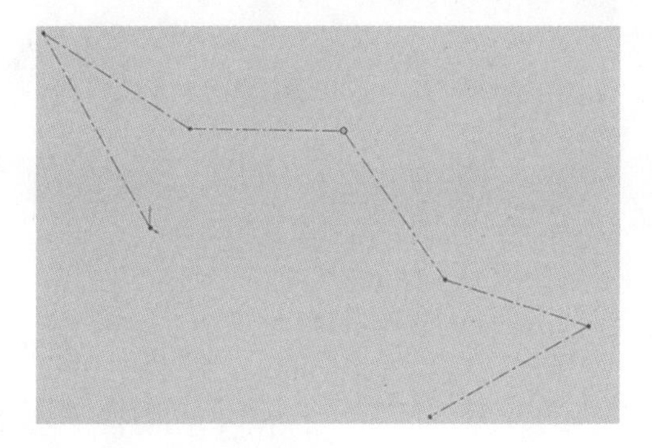

Figure 12.41

Control polygon with control vertices.

17. Click View Orientation > Normal To.
18. Create a 3D sketch, 3DSketch4, on XY Plane (see Figure 12.43).
19. Exit sketch, 3DSketch4.

Modifying Control Polygons

20. Click Plane1 > 3D-Sketch.
21. Modify 3D sketch Vertices as desired. (See Figure 12.44 for modified vertices in Plane1 and Plane4, especially.)

U-direction Control Curves

22. Click Insert > Curve > Curve Through Reference Points (see Figure 12.45).
    The Curve Through PropertyManager automatically appears (see Figure 12.46).
23. Click vertices of the first set of Control Vertices (see Figure 12.46).
24. Click OK.

Repeat for Second Set of Control Vertices

25. Click vertices of the second set of Control Vertices (see Figure 12.47).
26. Click OK.

Repeat for Third Set of Control Vertices

27. Click vertices of the third set of Control Vertices (see Figure 12.48).
28. Click OK.

Repeat for Fourth Set of Control Vertices

29. Click vertices of the fourth set of Control Vertices (see Figure 12.49).
30. Click OK.

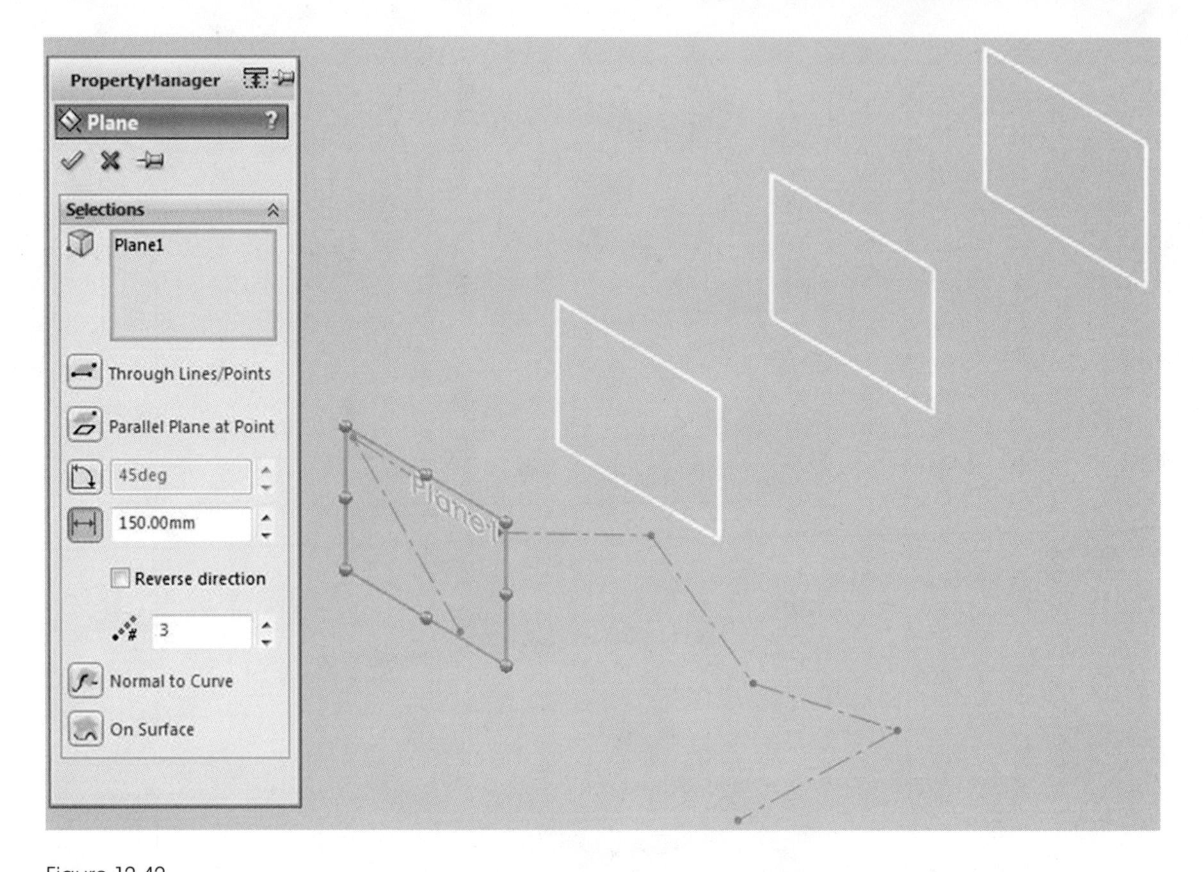

Figure 12.42

Three extra planes created: Plane2, Plane3, and Plane4.

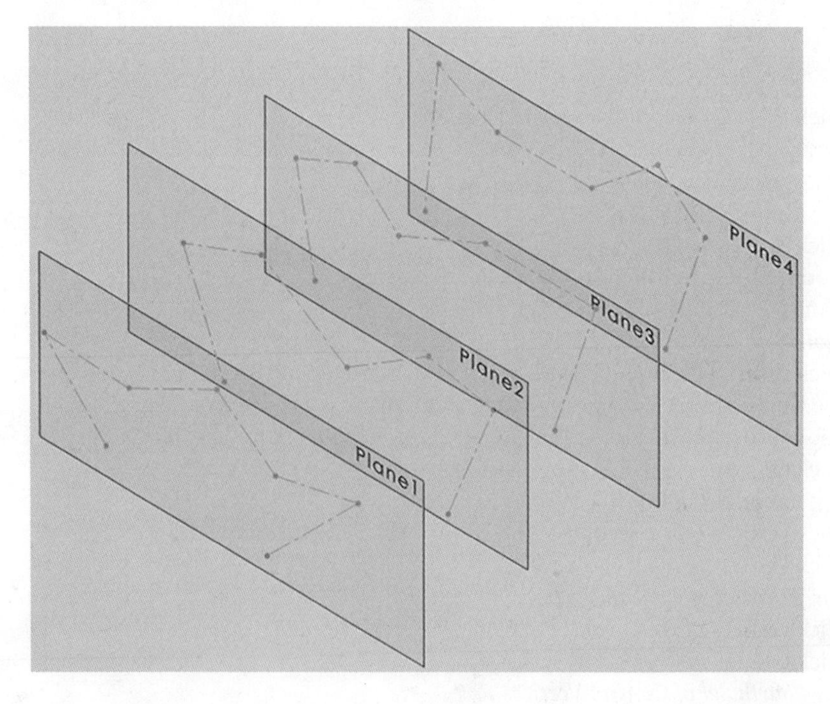

Figure 12.43

Four control polygons.

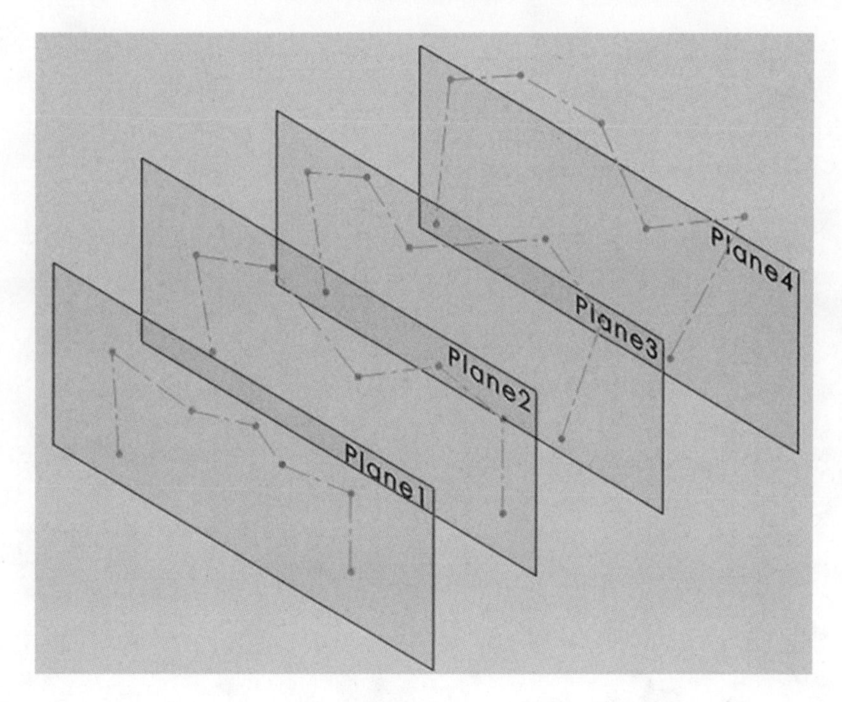

Figure 12.44

Modified control polygons.

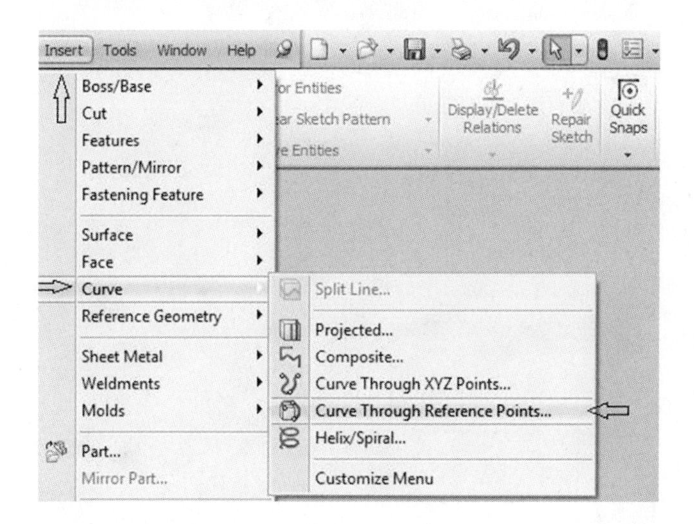

Figure 12.45

Access Spline Curve Through Reference Points tool.

U-direction Control Curves

31. Click Insert > Curve > Curve Through Reference Points.
    The Curve Through PropertyManager automatically appears (see Figure 12.50).
32. Click vertices of the first set of Control Vertices (see Figure 12.50).
33. Click OK.

Repeat for Second Set of Control Vertices

34. Click vertices of the second set of Control Vertices (see Figure 12.51).
35. Click OK.

Repeat for Third Set of Control Vertices

36. Click vertices of the third set of Control Vertices (see Figure 12.52).
37. Click OK.

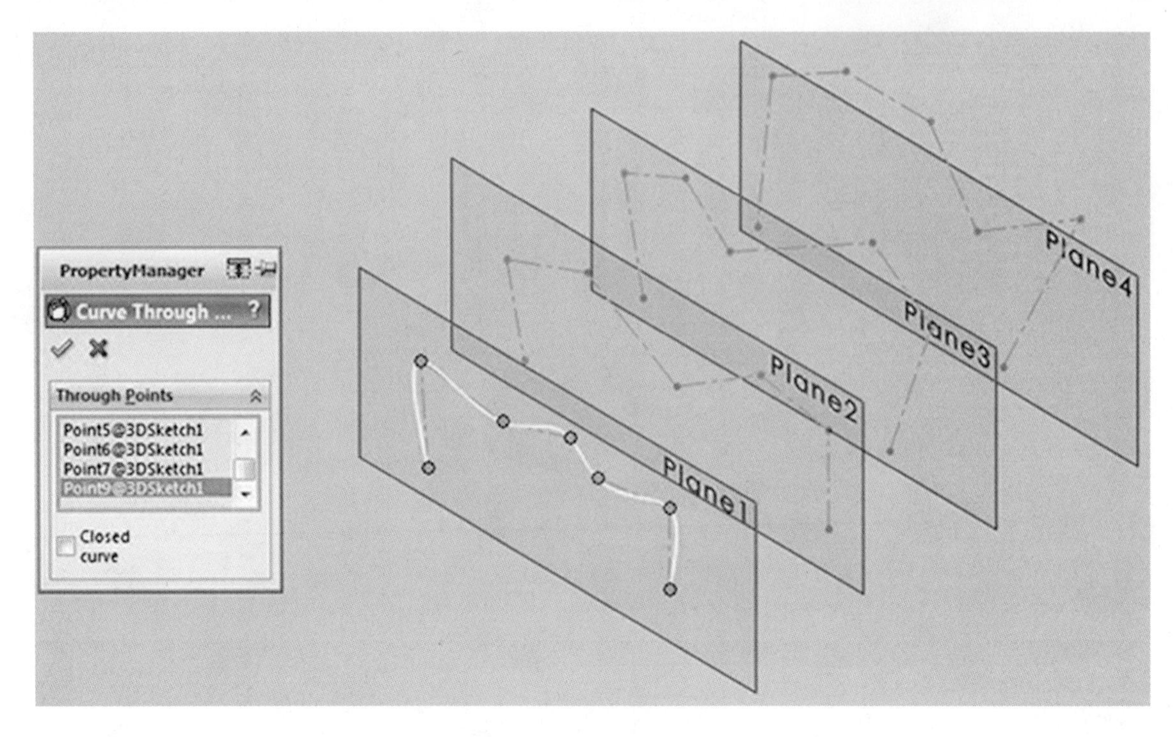

Figure 12.46

Spline Curve Through Reference Points PropertyManager for first U-points.

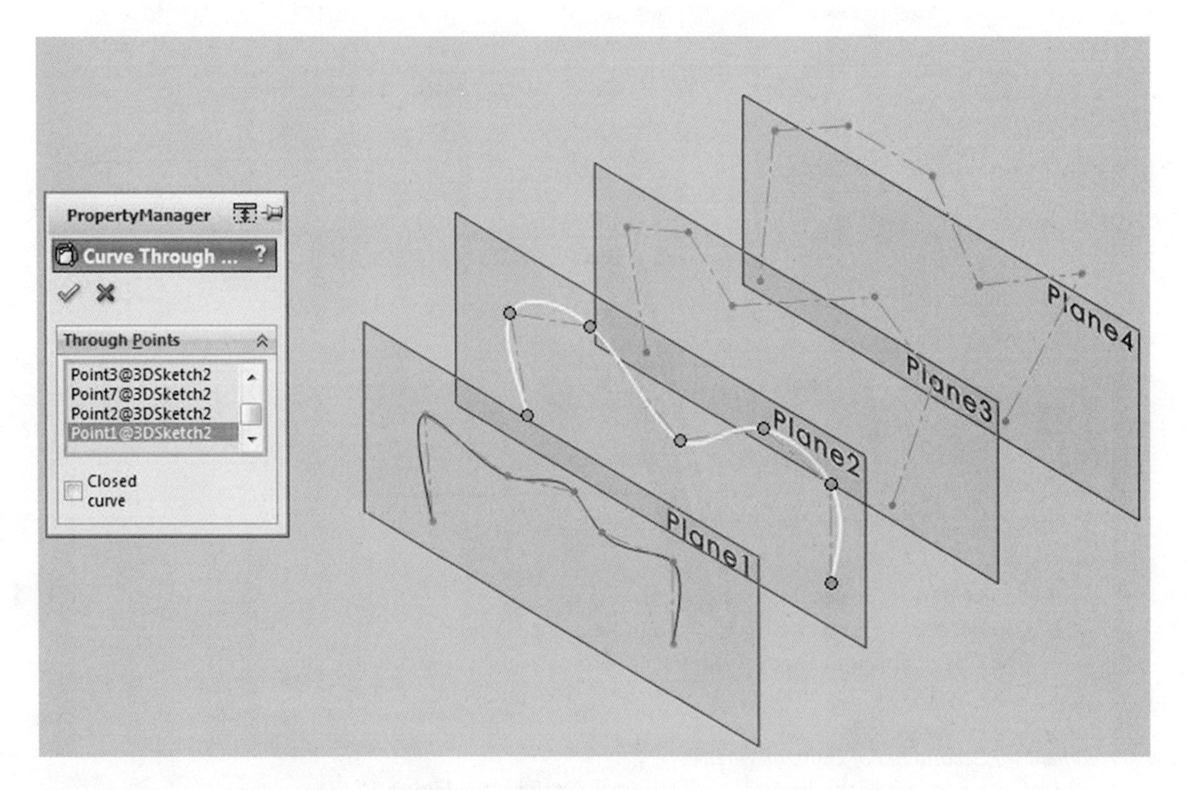

Figure 12.47

Spline Curve Through Reference Points PropertyManager for second U-points.

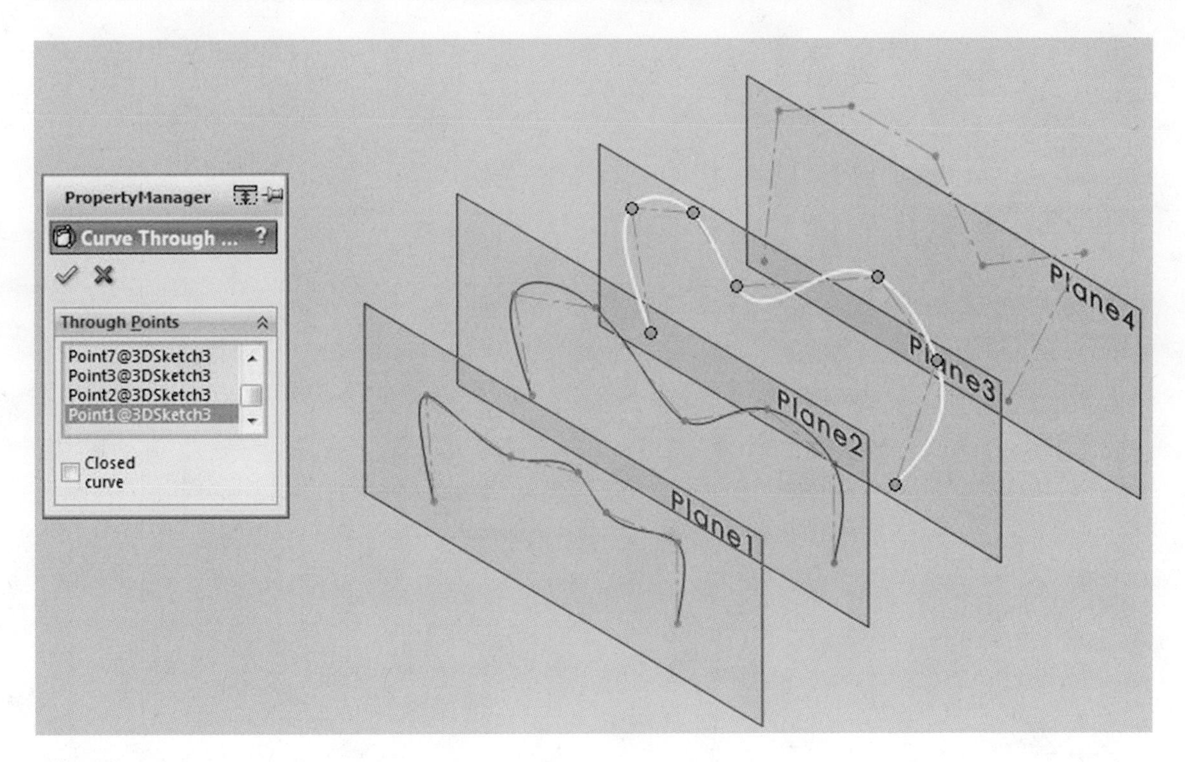

Figure 12.48

Spline Curve Through Reference Points PropertyManager for third U-points.

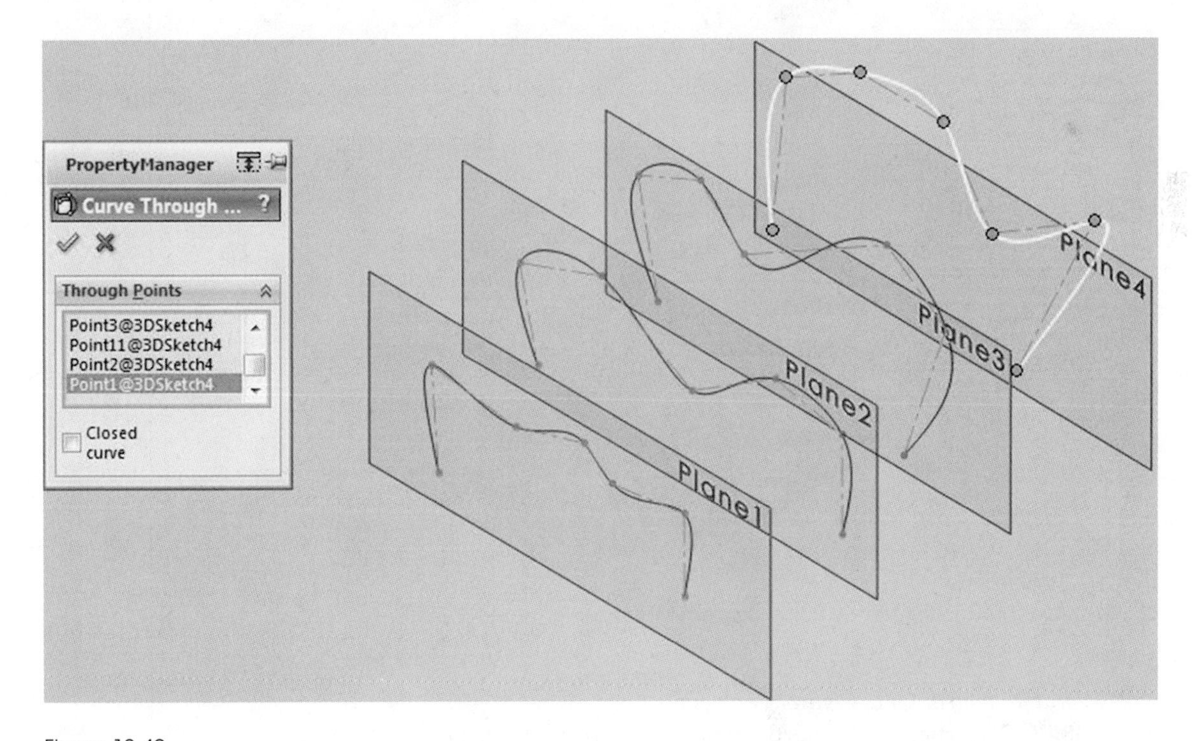

Figure 12.49

Spline Curve Through Reference Points PropertyManager for fourth U-points.

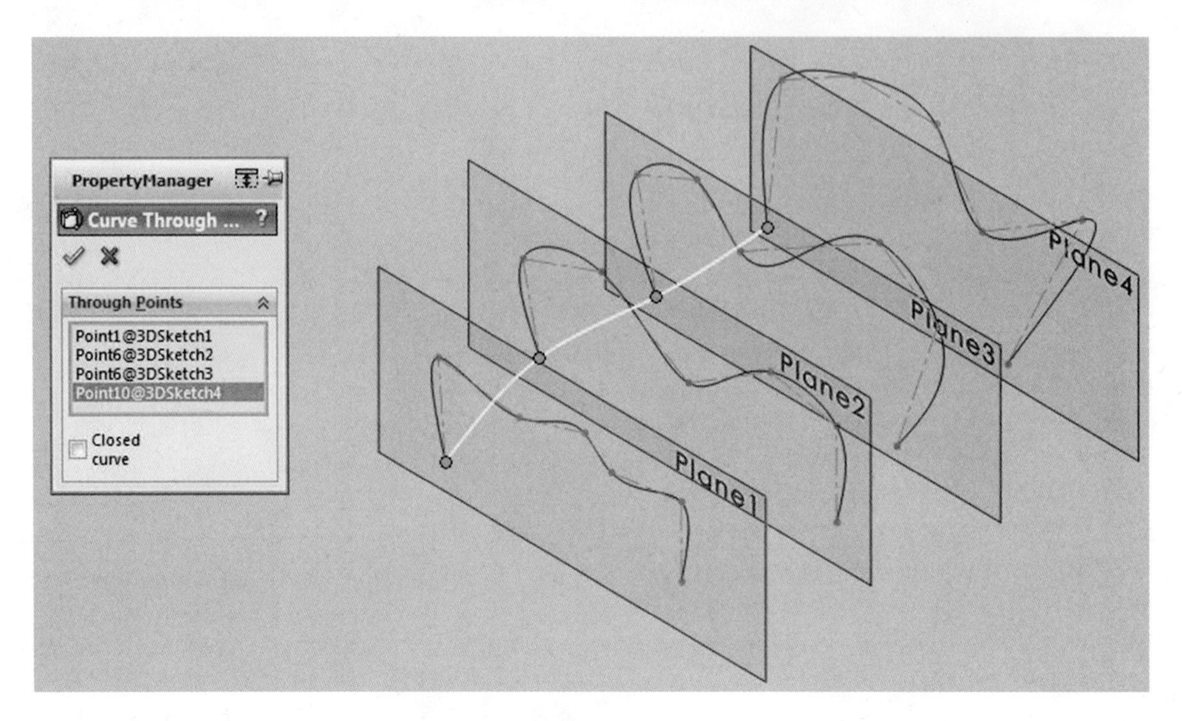

Figure 12.50

Spline Curve Through Reference Points PropertyManager for first V-points.

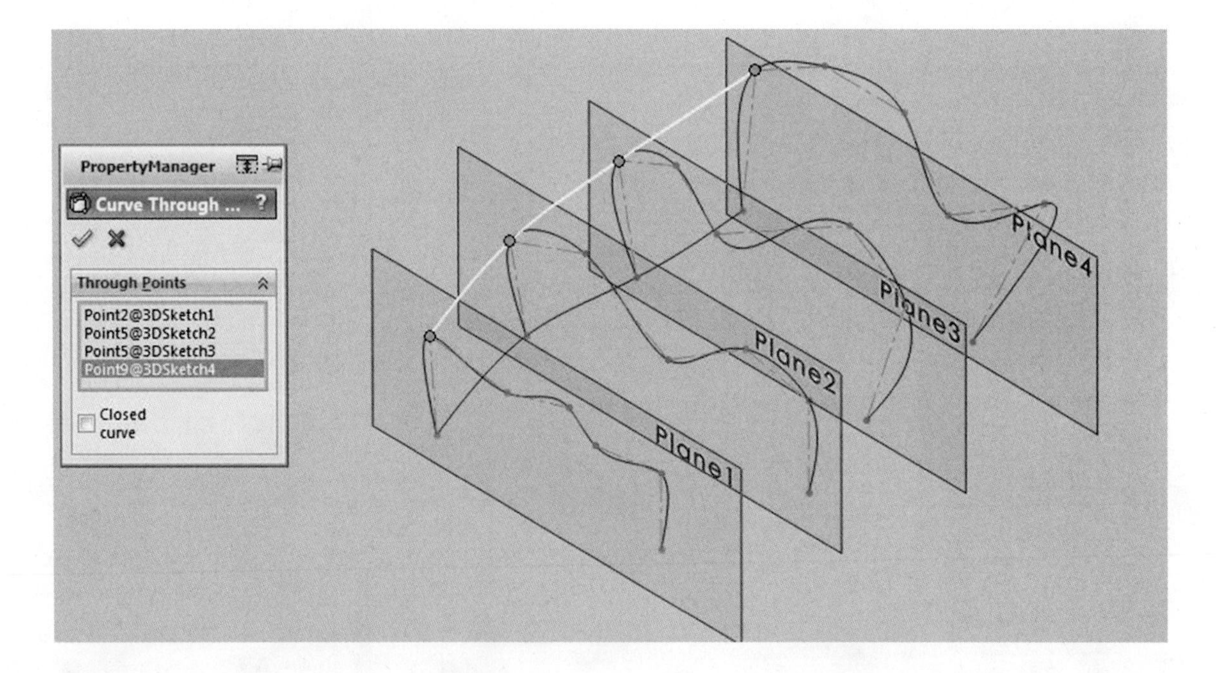

Figure 12.51

Spline Curve Through Reference Points PropertyManager for second V-points.

Repeat for Fourth Set of Control Vertices

38. Click vertices of the fourth set of Control Vertices (see Figure 12.53).

39. Click OK.

Repeat for Second Set of Control Vertices

40. Click vertices of the second set of control vertices (see Figure 12.54).

41. Click OK.

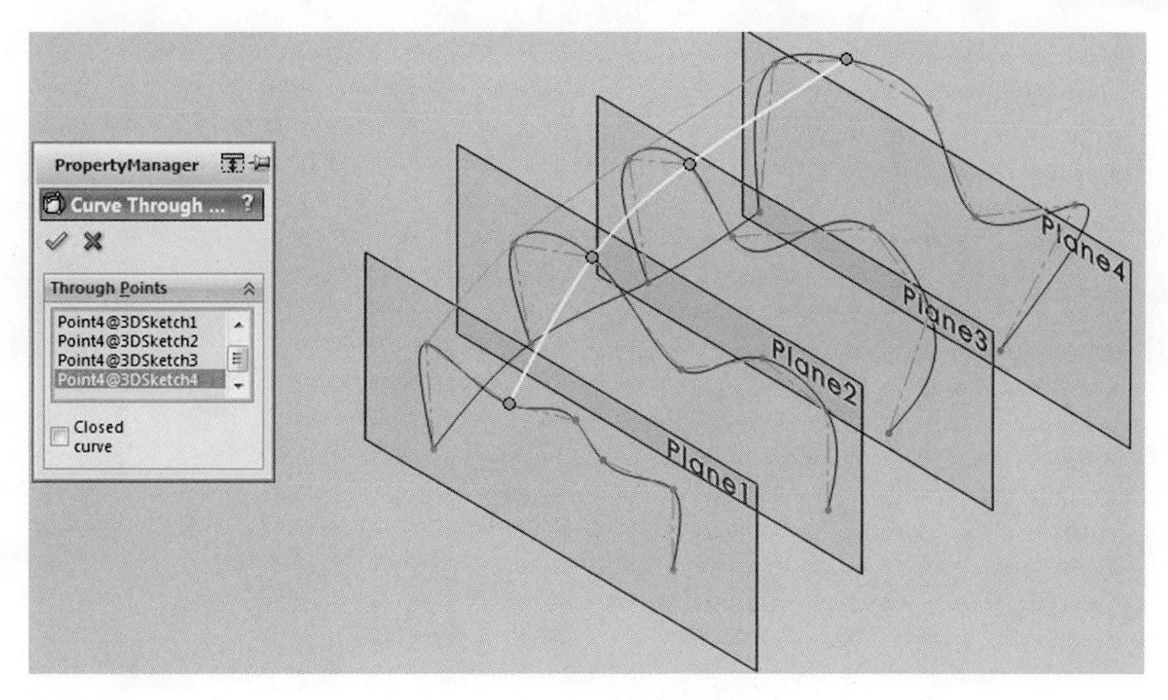

Figure 12.52

Spline Curve Through Reference Points PropertyManager for third V-points.

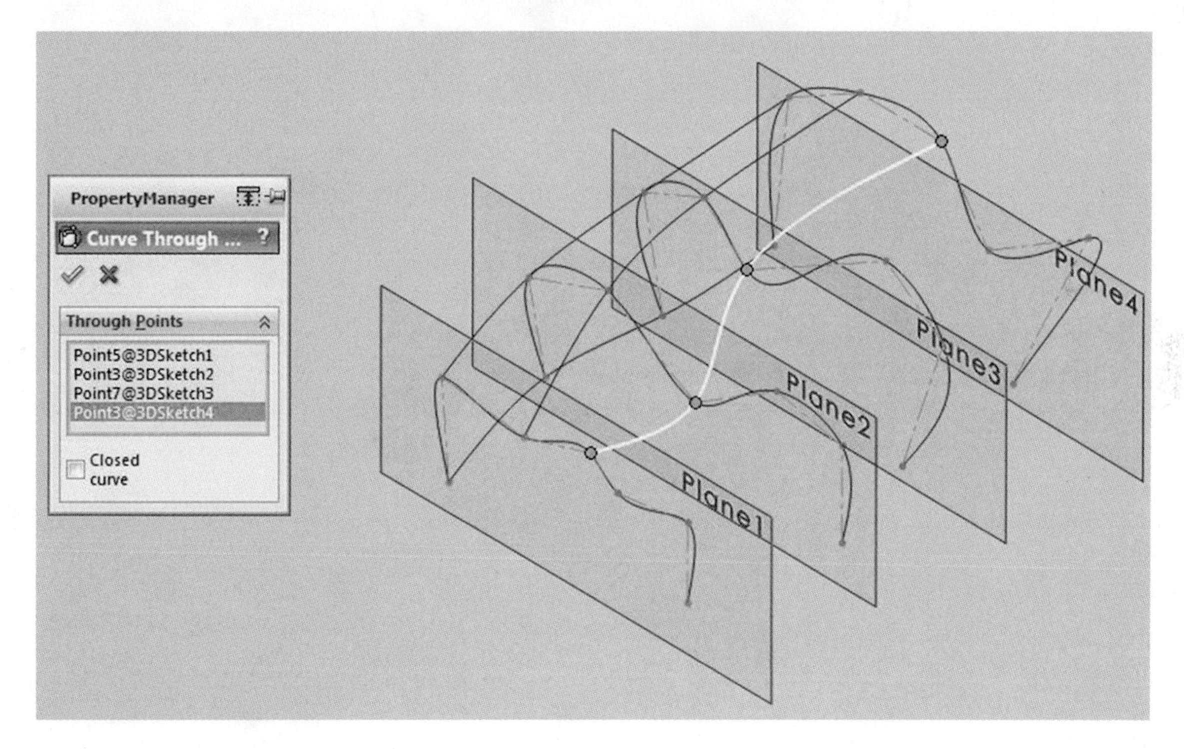

Figure 12.53

Spline Curve Through Reference Points PropertyManager for fourth V-points.

Repeat for Third Set of Control Vertices

42. Click vertices of the third set of control vertices (see Figure 12.55).

43. Click OK.

Repeat for Fourth Set of Control Vertices

44. Click vertices of the fourth set of control vertices (see Figure 12.56).

45. Click OK.

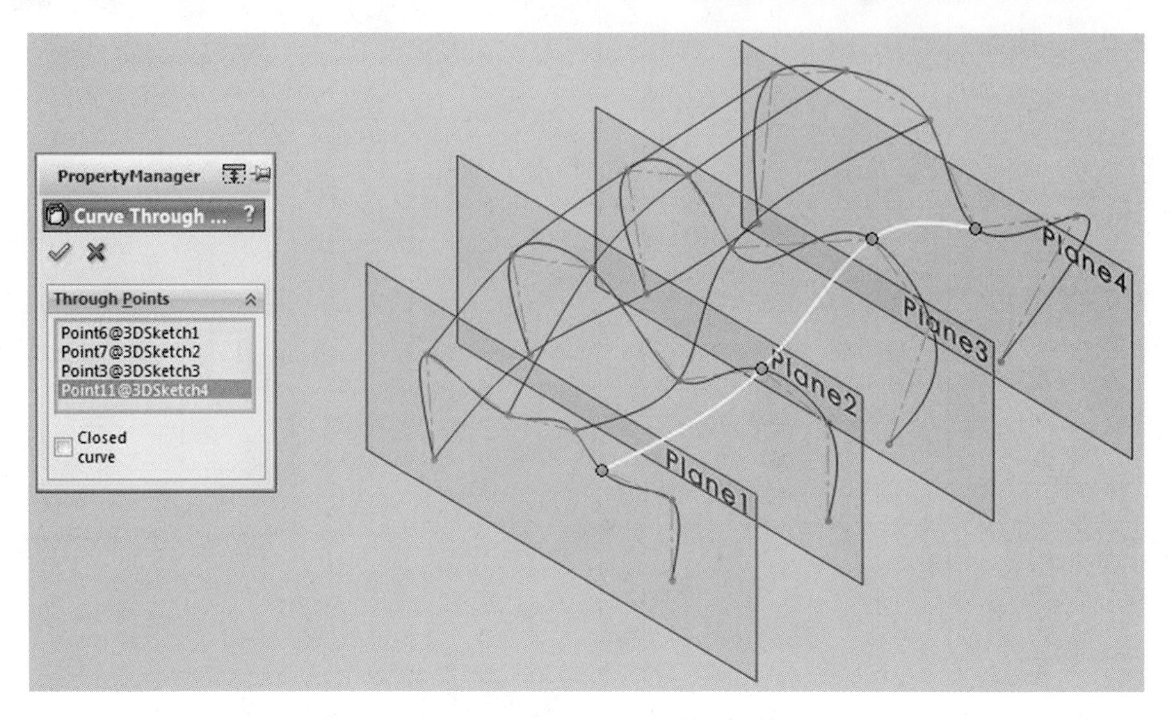

Figure 12.54

Spline Curve Through Reference Points PropertyManager for fifth V-points.

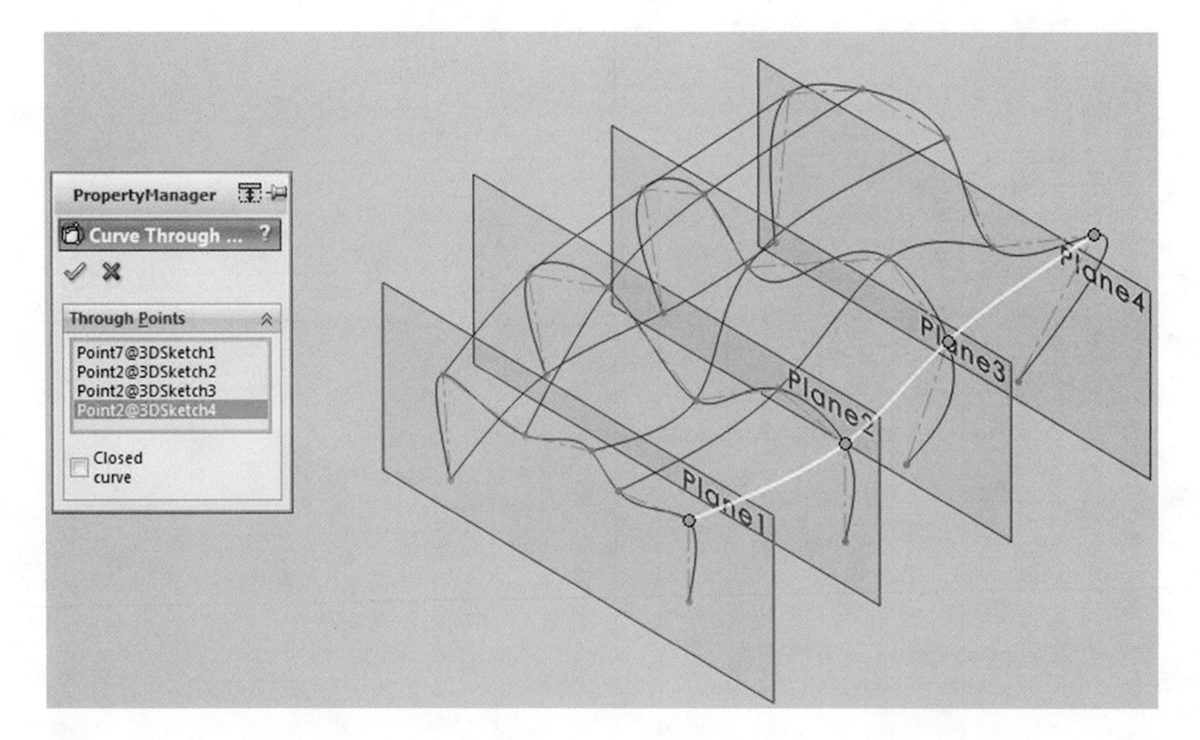

Figure 12.55

Spline Curve Through Reference Points PropertyManager for sixth V-points.

Creating Boundary Surface Using Control Curves

46. Click Insert > Surface > Boundary Surface (see Figure 12.57).
    The Boundary Surface PropertyManager automatically appears (see Figure 12.58).
47. Select Curve1,..,Curve4 for *Dir1 Curves Influence* (see Figure 12.58).
48. Select Curve5,..,Curve11 for *Dir2 Curves Influence* (see Figure 12.58).
    The surface model is shown in Figure 12.58.

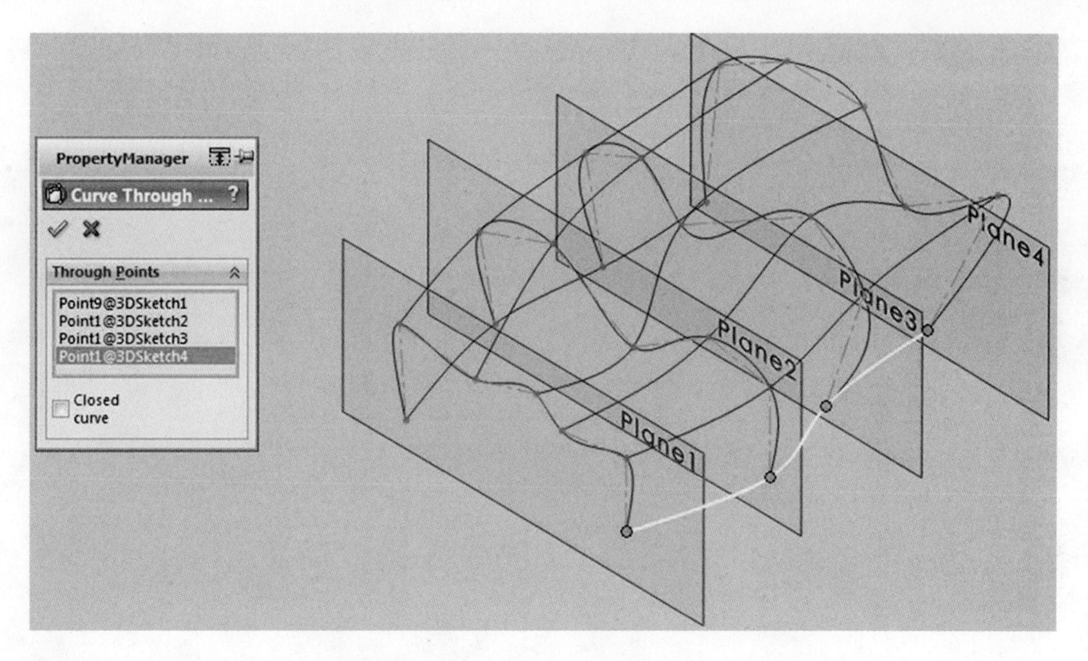

Figure 12.56

Spline Curve Through Reference Points PropertyManager for seventh V-points.

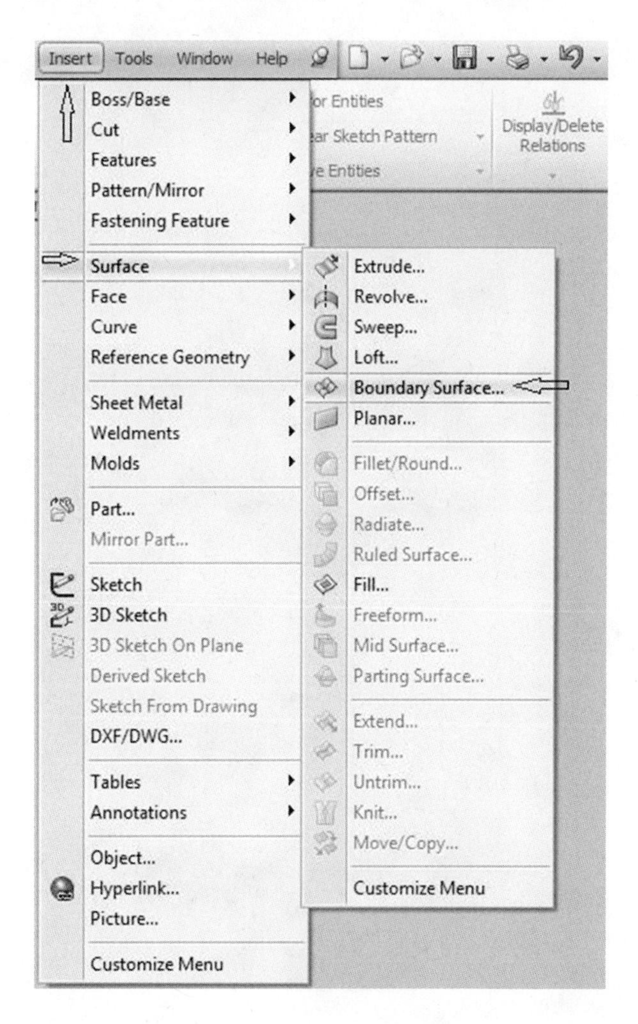

Figure 12.57

Insert Boundary Surface.

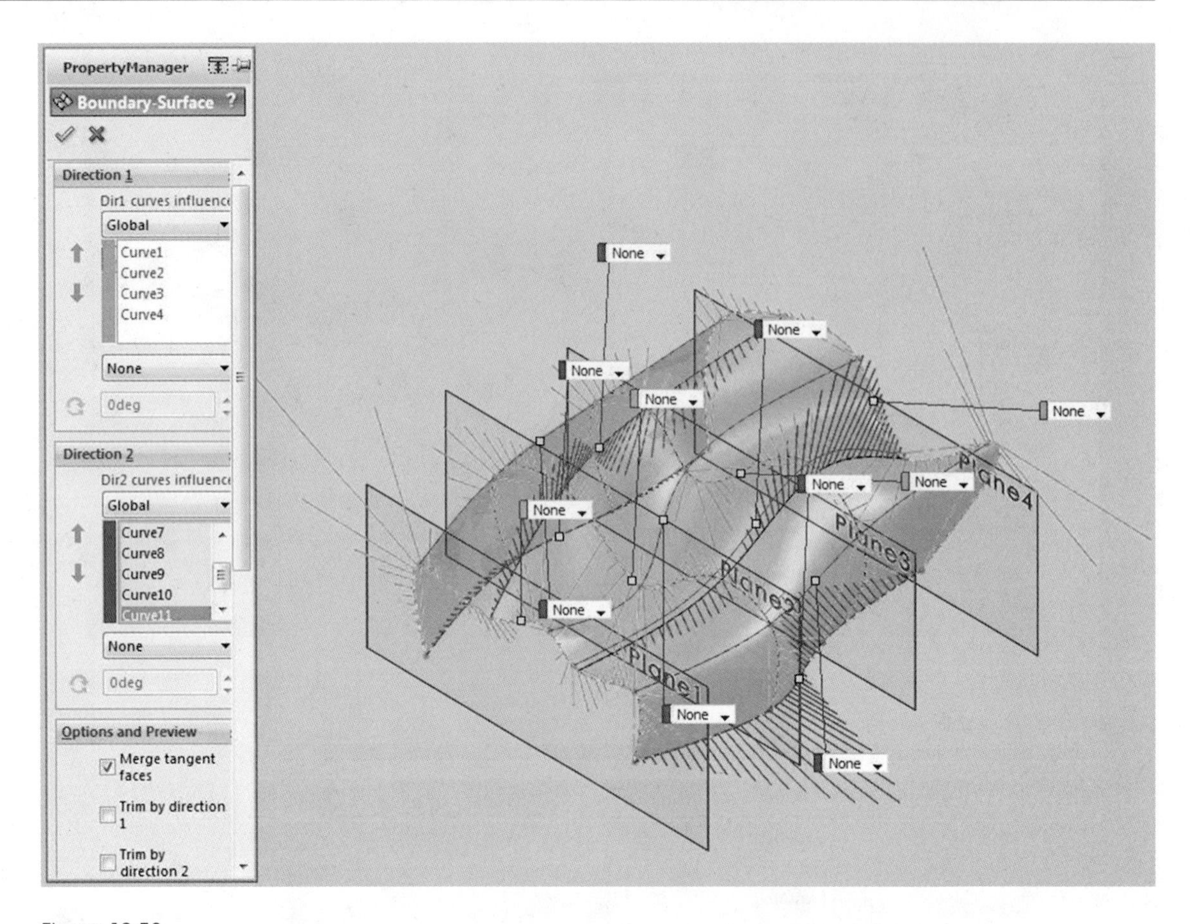

Figure 12.58

Boundary Surface PropertyManager.

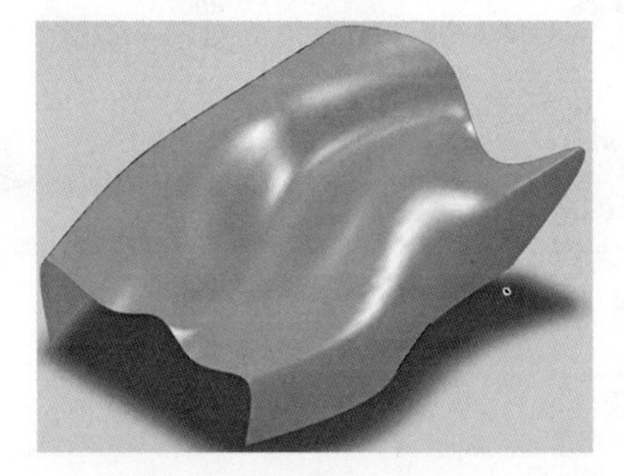

Figure 12.59

Surface model created.

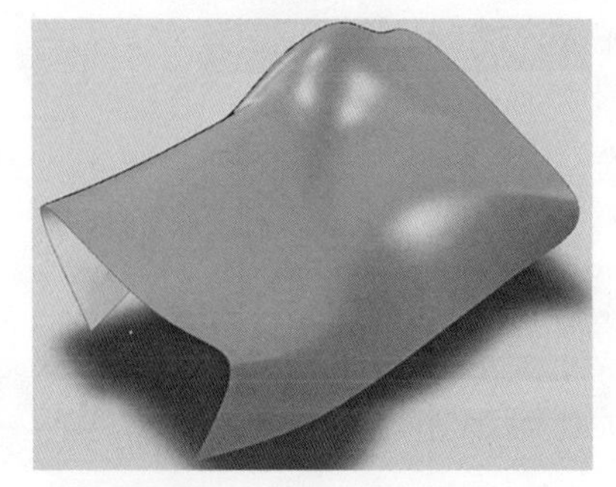

Figure 12.60

Modified surface model created by modifying some control vertices.

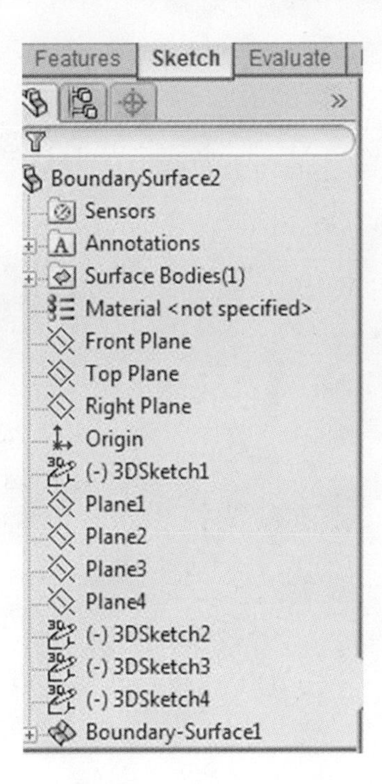

Figure 12.61

FeatureManager for the surface design.

The surface model created is shown in Figure 12.59. By modifying the control vertices, a modified surface model created is shown in Figure 12.60. The FeatureManager for the surface design is shown in Figure 12.61.

Summary

In summary, we have been presented with three ways of realizing solids from surfaces. Try these options. After all, a solid is made up of pieces of surfaces on its boundary; therefore, in general, it is natural to piece together surfaces into a solid. On the other hand, the outer parts of a solid can be extracted for a number of bounding surfaces. We have also presented a very useful tool for filling surfaces. This is a very useful tool in the manufacturing industry.

Exercises

1. Create (drill) a hole, 10-mm diameter anywhere on top of Figure 12.59; then fill the hole.
2. Repeat the process for Figure 12.60.
3. How is the process of filling a hole useful in manufacturing?

13

Toolboxes and Design Libraries

Objectives:

When you complete this chapter, you will have learned the following:

- The usefulness of SolidWorks Toolbox and Design Library
- Modeling and analyzing structural steel using the SolidWorks Toolbox
- Adding Grooves for O-rings and retaining rings to components
- Using the SolidWorks Design Library to add standard mechanical parts (nuts, bolts, screws, etc.) to an assembly
- Using the SolidWorks Toolbox to design gears (spur, helical, and bevel)
- Using the SolidWorks Toolbox to design gearing systems (gears, shafts, bearing selection, etc.)

Introduction

The goals of this chapter are to encourage engineers and designers to take advantage of the design tools that are available in SolidWorks and to extend their understanding of this robust and flexible computer-aided design (CAD) package beyond merely shape design (modeling).

SolidWorks Toolbox Add-Ins

1. Open a New Part document.
2. Click the Options pull-down menu.
3. Select SolidWorks Toolbox *Add-Ins* options (see Figure 13.1).
   The Add-Ins PropertyManager appears as shown in Figure 13.2.
4. Check (select) SolidWorks Toolbox and SolidWorks Toolbox Browser (see Figure 13.2).

How to Use SolidWorks Design Library and Toolbox

SolidWorks software is useful for creating different types of standard mechanical parts. There are two ways of using the Toolbox. One way is to access the Toolbox once it is added in, through the SolidWorks

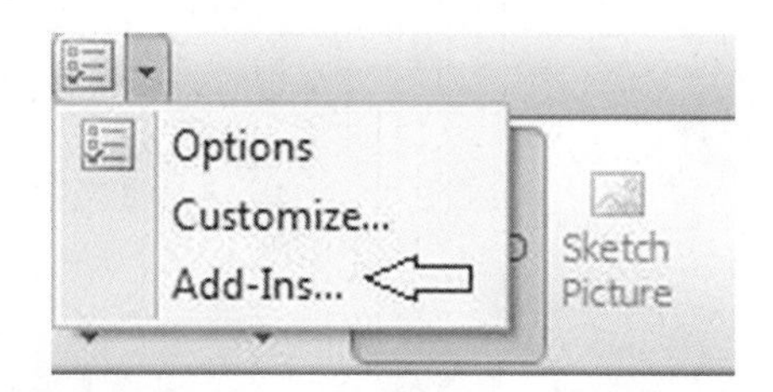

Figure 13.1

Select Add-Ins options.

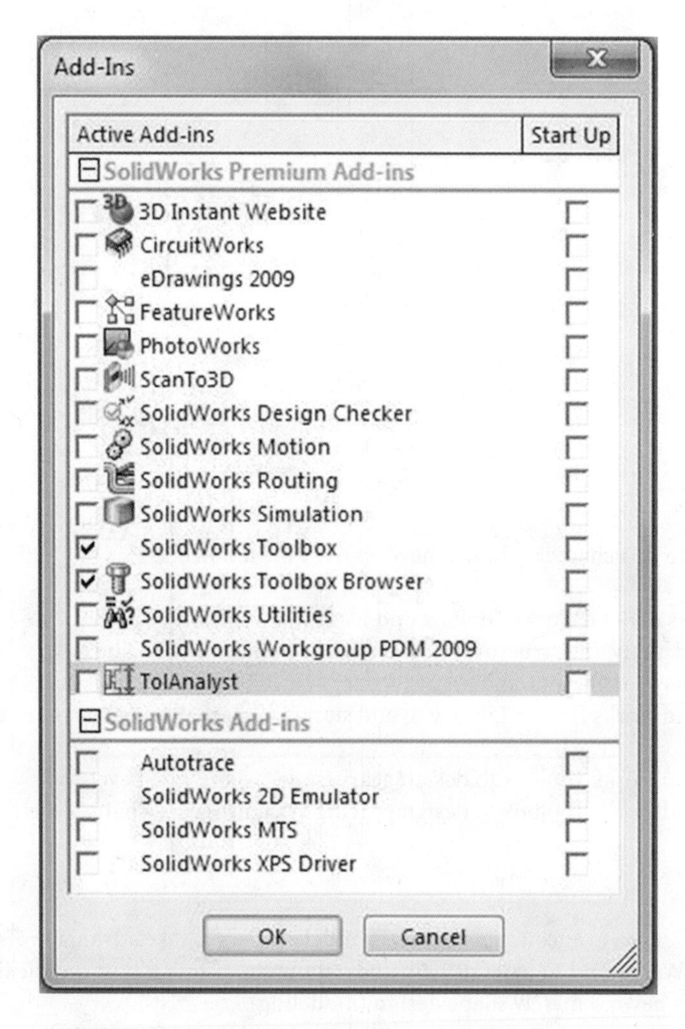

Figure 13.2

SolidWorks Toolbox Add-Ins.

CommandManager (see Figure 13.3) or through the Design Library (to be discussed in the "Creating Standard Parts Using SolidWorks Design Library and Toolbox" section). More features are available to the user when the Toolbox is accessed in the Design Library.

Features Available in the Toolbox via CommandManager

- Structural Steel
- Grooves
- Cams
- Beam Calculator
- Bearing Calculator

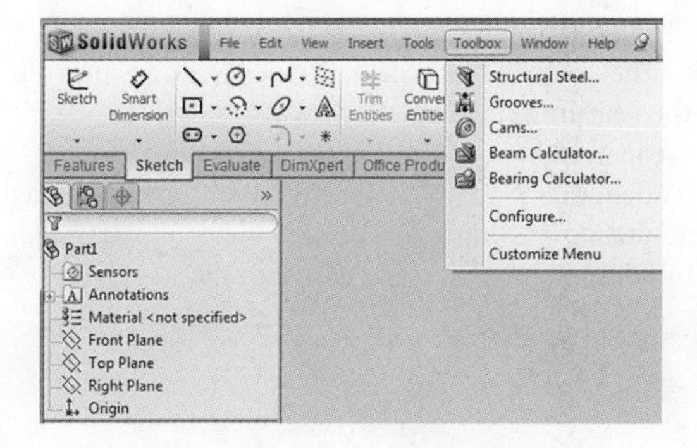

Figure 13.3

SolidWorks Toolbox accessed through the CommandManager.

Features Available in the Toolbox via Design Library

- Bearings
- Bolts and Screws
- Jig Bushings
- Keys
- Nuts
- O-rings
- Pins
- Power Transmission (Chain Sprockets, Gears, Timing Belt Pulleys)
- Retaining Rings
- Structural Members
- Washers

In the subsequent sections, we will discuss some of the SolidWorks Toolbox features that are accessed through the CommandManager and present an overview of the SolidWorks Toolbox features that are accessed through the Design Library.

Structural Steel

1. Open a New Part document.
2. Click Toolbox > Structural Steel (see Figure 13.3).
3. Select Ansi Inch (see Figure 13.4).

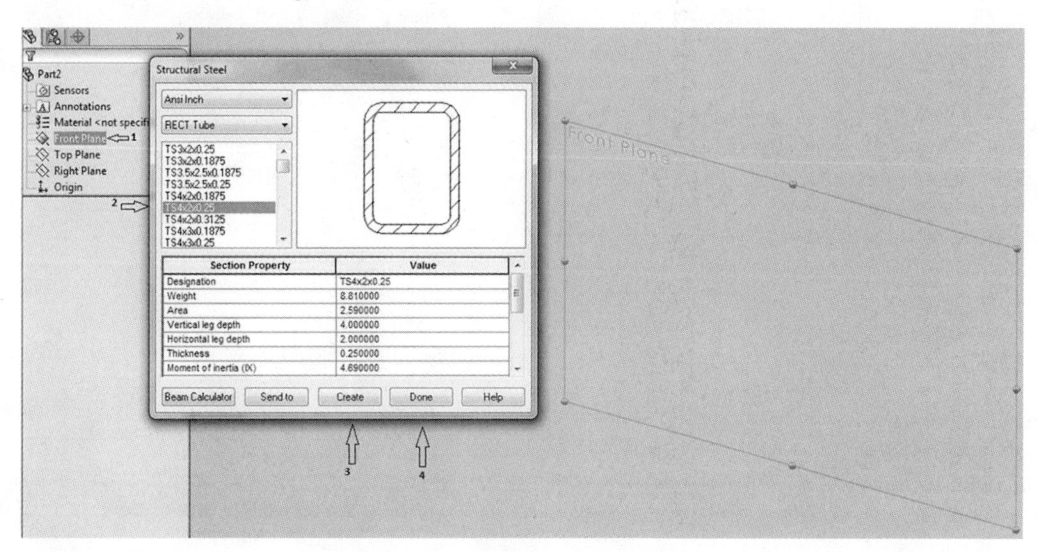

Figure 13.4

Creating TS4x2x0.25 rectangular tube.

4. Select RECT Tube > TS4x2x0.25 (see Figure 13.4).
5. Click Create > Done. (The cross section, Sketch1, is created, as shown in Figure 13.5.)
6. Click Sketch1 from the FeatureManager.
7. Click Edit Sketch to be in *sketch mode* (see Figure 13.6).
8. Click Features > Extruded Boss/Base. (See the Extrude PropertyManager in Figure 13.7.)
9. In the Direction1 rollout, accept Blind and specify 500 for the extrusion Distance.
   The rectangular tube created, *Extrude1*, is shown in Figure 13.8.

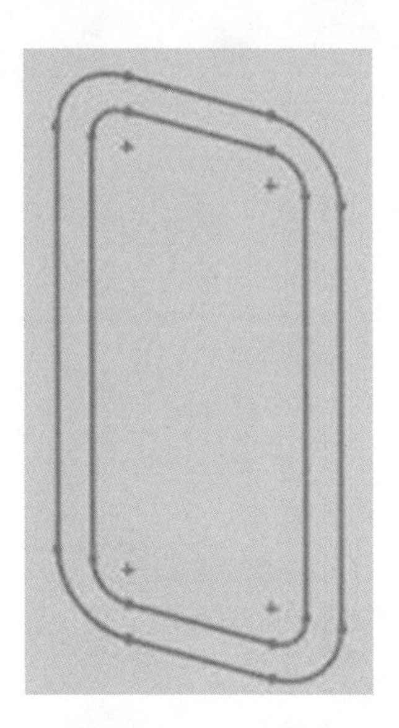

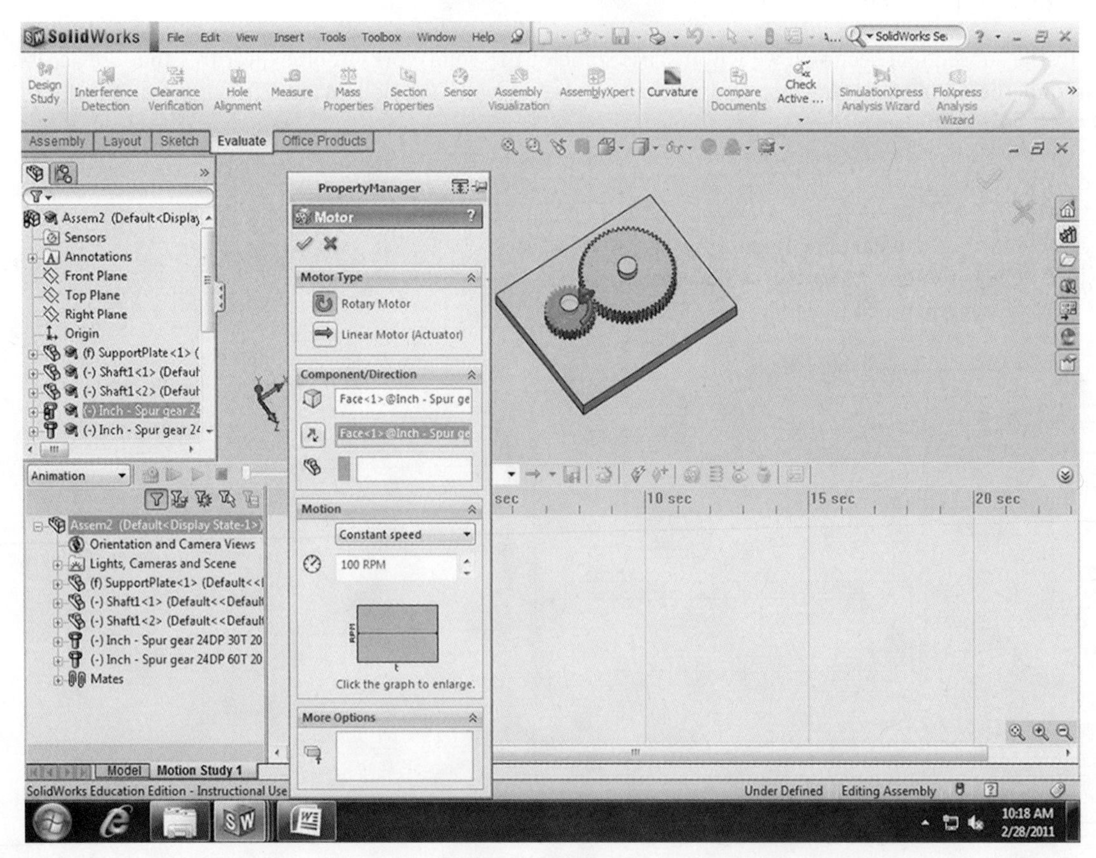

Figure 13.5

Cross section of TS4x2x0.25 rectangular tube.

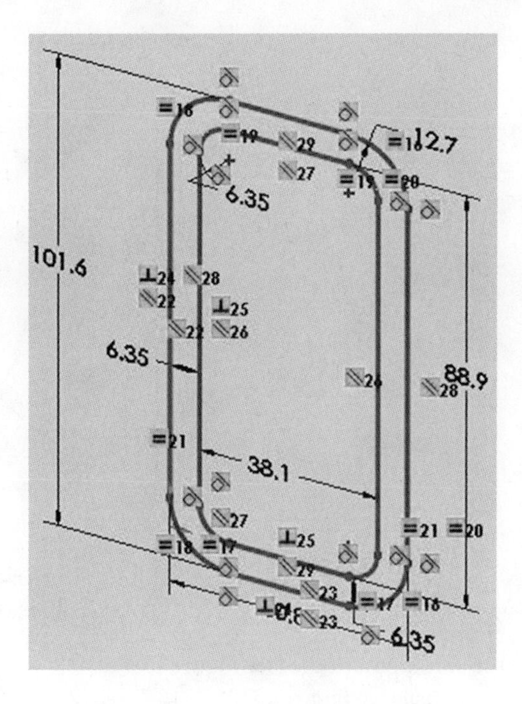

Figure 13.6

Sketch1 in sketch mode.

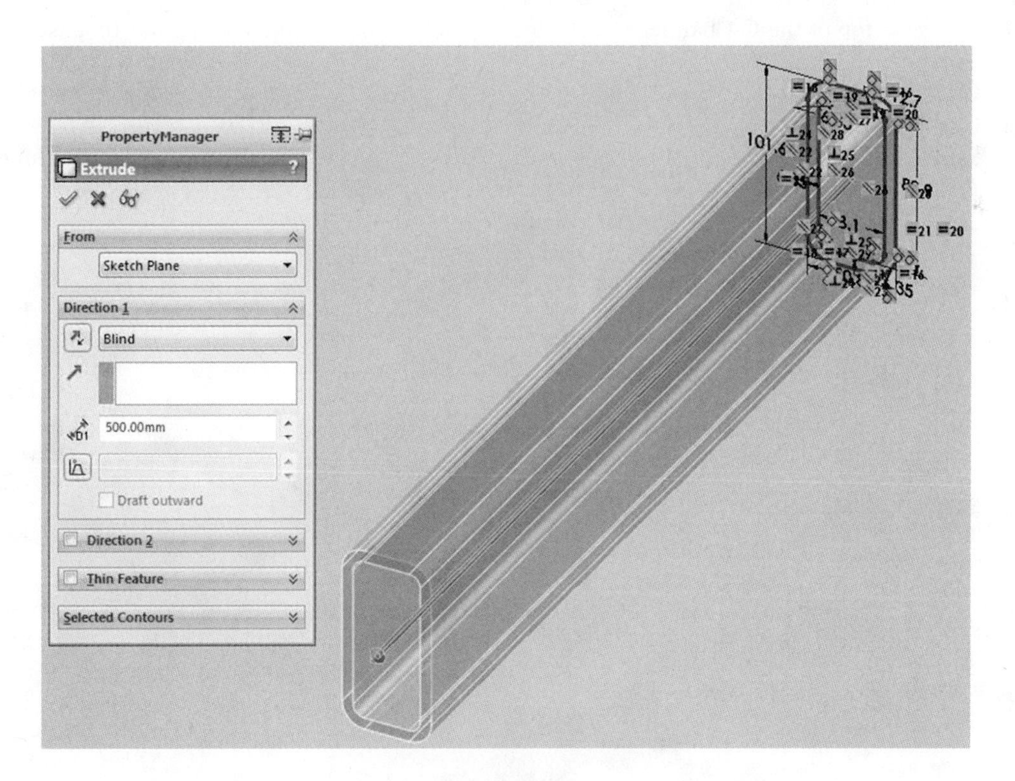

Figure 13.7

Extrude PropertyManager.

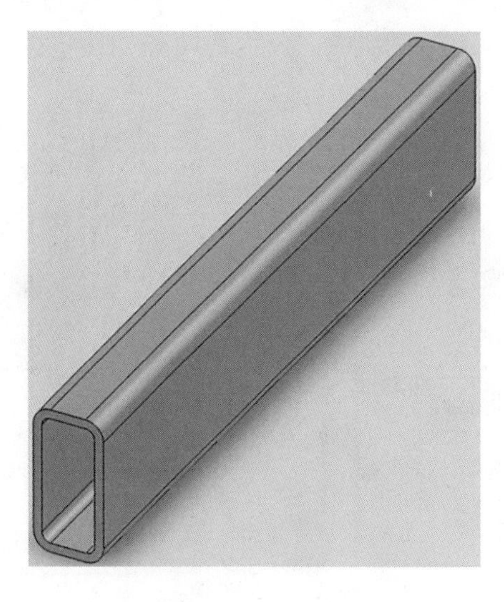

Figure 13.8

TS4x2x0.25 rectangular tube created, Extrude1.

10. Click the front face of the rectangular tube (see Figure 13.9).
11. Click Toolbox > Structural Steel (see Figure 13.9).
12. Select C Channel > C3x4.1 (see Figure 13.9).
13. Click Create > Done. (The cross section, Sketch2, is created, as shown in Figure 13.10.)
14. Click Sketch2 from the FeatureManager.
15. Click Edit Sketch to be in sketch mode (see Figure 13.11).
16. Click Add Relations.
17. Select the top of the C-Channel and the top edge of the hole of the rectangular tube (see Figure 13.11).
18. Click OK to add the relations.
19. Click Add Relations.
20. Select the left edge of the C-Channel and the left edge of the hole of the rectangular tube (see Figure 13.11).

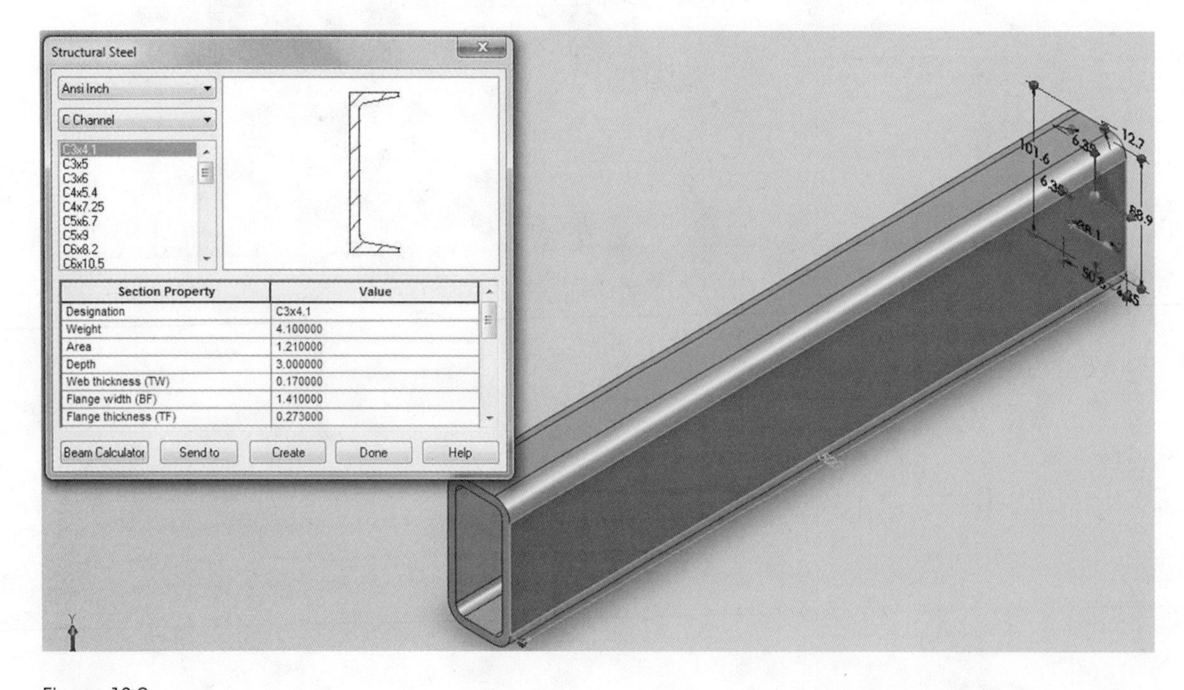

Figure 13.9

Creating C3x4.1 C-Channel.

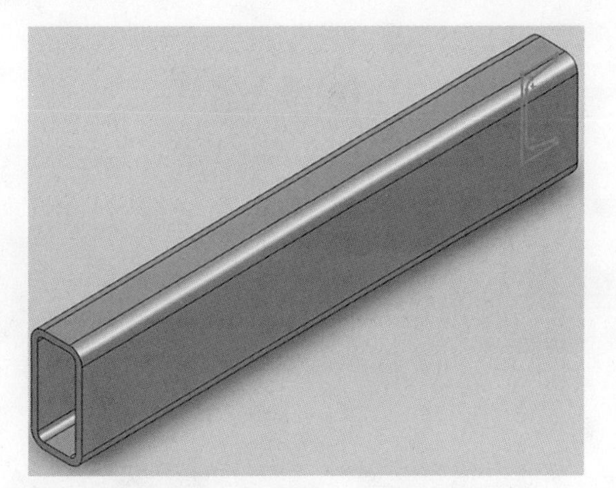

Figure 13.10

Cross section of C3x4.1 C-Channel, Sketch2.

21. Click OK to add the relations. (See Figure 13.12 for the proper positioning of Sketch2.)
22. Click Features > Extruded Boss/Base. (See the Extrude PropertyManager in Figure 13.13.)
23. In the Direction1 rollout, accept Blind and specify 250 for the extrusion Distance.
24. Click OK to complete the process. (See Figure 13.13 for the C-Channel created *Extrude2*.)
25. Create Plane1, 250 mm from one end of the rectangular tube (see Figure 13.14).
26. Click Features > Mirror. (The Mirror PropertyManager is displayed as in Figure 13.14.)
27. Select Plane1 as the Mirror Face/Plane.
28. Select Extrude2 as the Bodies to Mirror.
29. Click OK to complete mirroring.
30. Create Plane2 using any three outer vertices of the mirrored C-channel (see Figure 13.15).
31. Click Plane2 from the FeatureManager.
32. Click Toolbox > Structural Steel (see Figure 13.16).
33. Select C Channel > C3x4.1 (see Figure 13.16).
34. Click Create > Done. (The cross section, Sketch3, is created, as shown in Figure 13.17.)
35. Click Sketch3 from the FeatureManager.
36. Click Edit Sketch to be in sketch mode (see Figure 13.18).
37. Click Add Relations.
38. Select the bottom outer vertex of the S3x5.7 S-beam and the bottom-left (outer) vertex of the mirrored C-channel (see Figure 13.18).
39. Click OK to add the relations.
40. Click Features > Extruded Boss/Base. (See the Extrude PropertyManager in Figure 13.19.)
41. In the Direction1 rollout, select Up To Surface and select *outer face* of C-Channel.
42. Click OK to complete the process. (See Figure 13.20 for the S-beam-created *Extrude3* and Figure 13.21 for the four solid bodies that are created.)

Beam Calculator

The *Beam Calculator* dialog box allows you to perform deflection and stress calculations on structural steel cross sections. The Toolbox has the Beam Calculator for calculating the structural section properties for different available configurations.

The available load options include the following:

- Fixed at one end, loaded at the other end
- Fixed at one end, uniformly loaded
- Supported at both ends, loaded in middle
- Supported at both ends, uniformly loaded
- Supported at both ends, unsymmetrical load
- Supported at both ends, two symmetrical loads

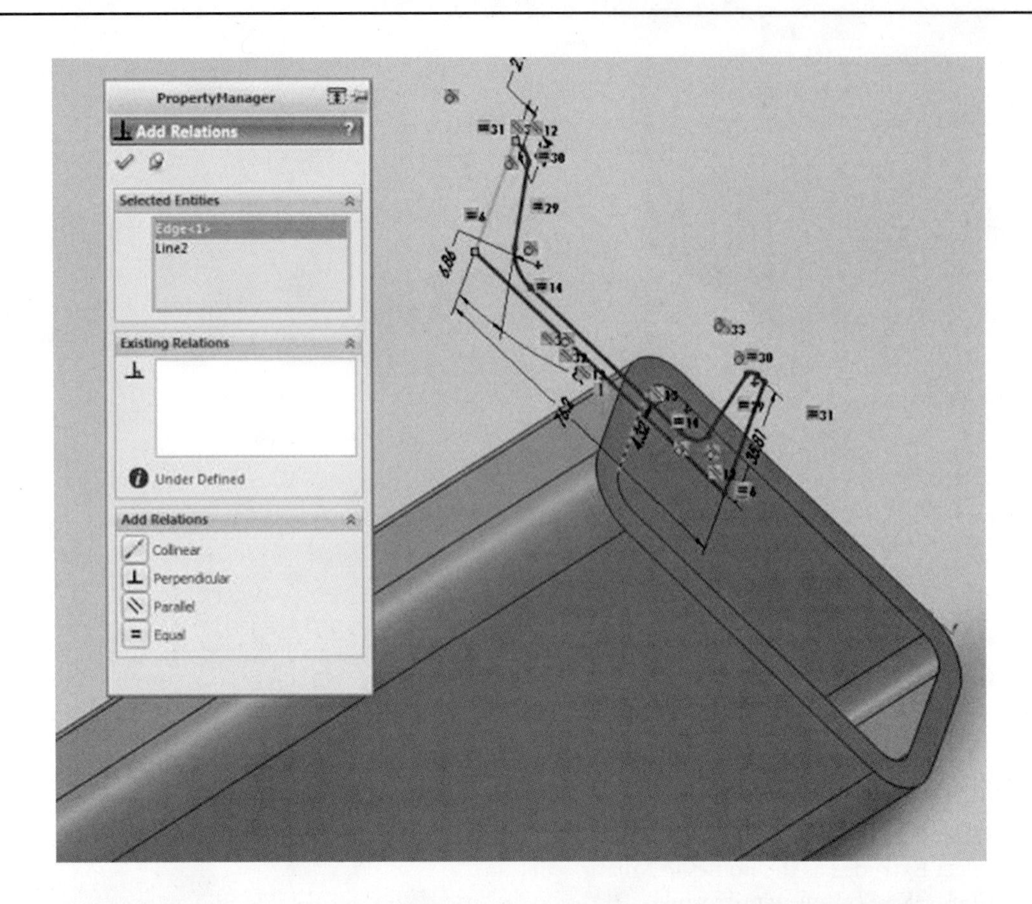

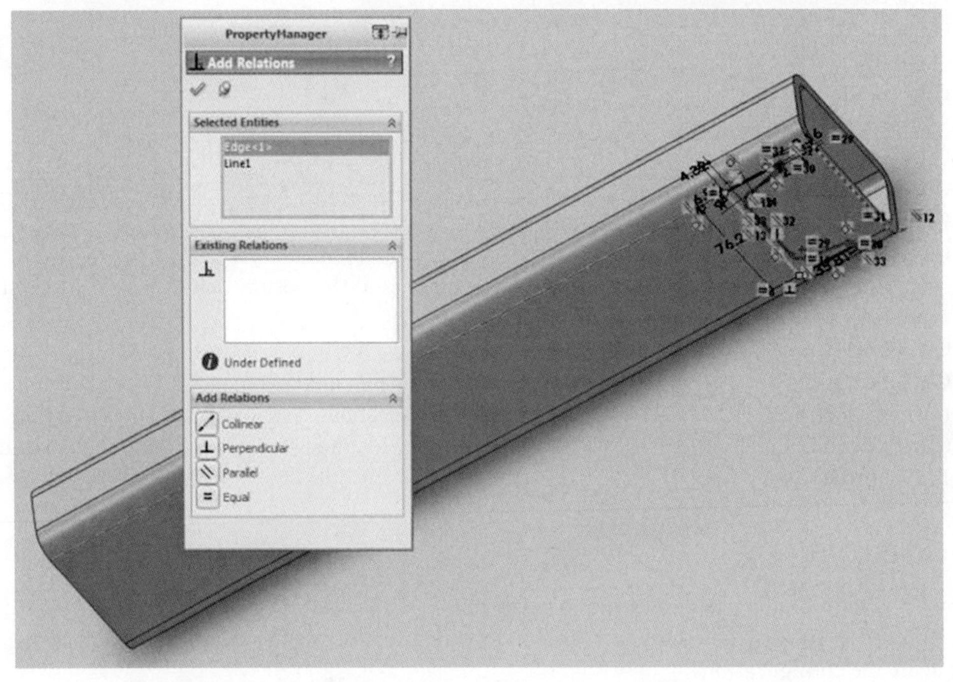

Figure 13.11

Cross section of C3x4.1 C-Channel, Sketch2 in sketch mode.

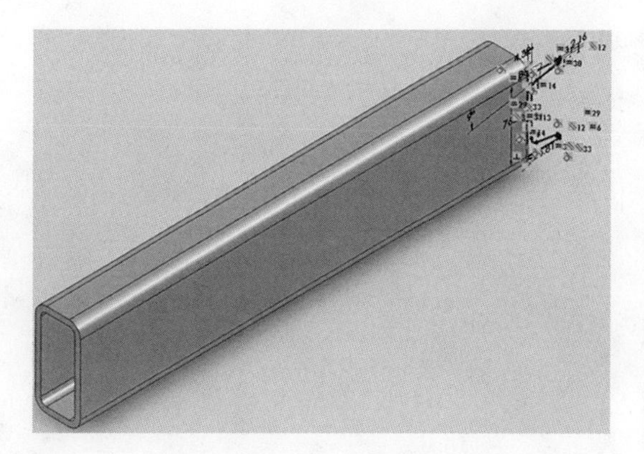

Figure 13.12

Sketch2 placed in correct position using appropriate relations.

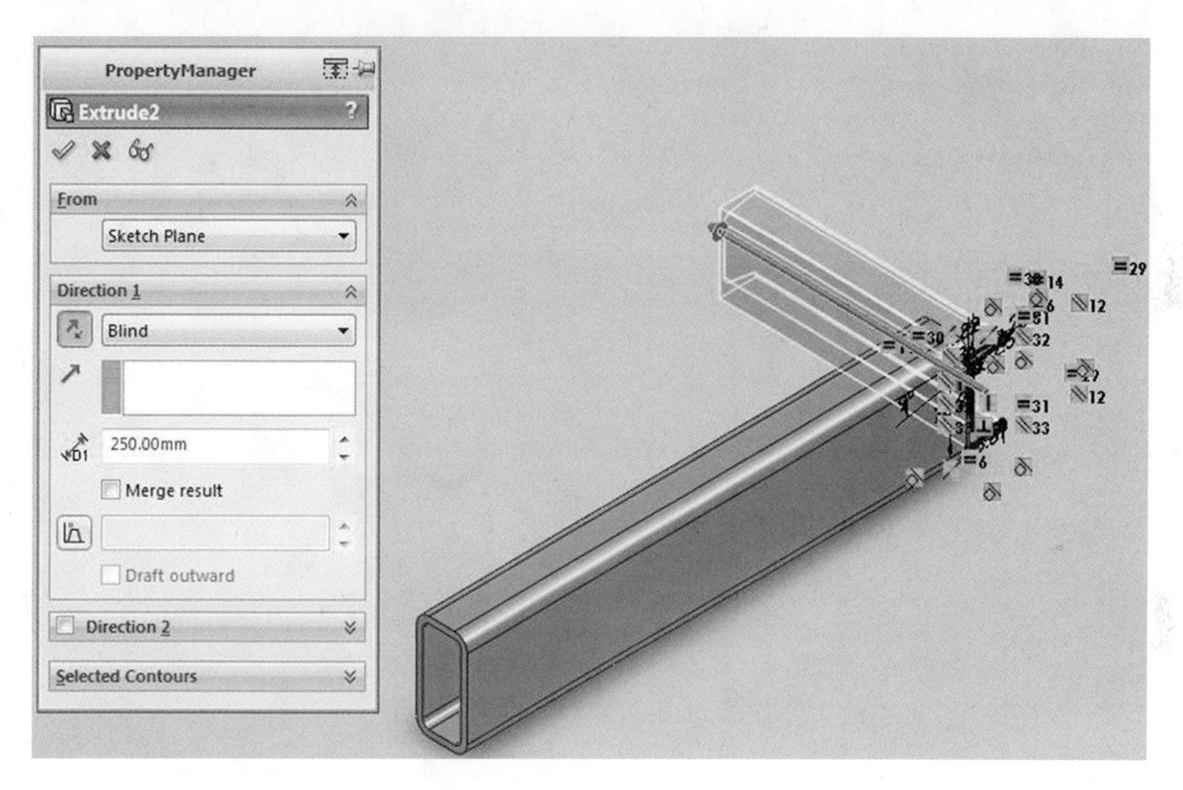

Figure 13.13

Extrude PropertyManager for Extrude2 feature.

To perform a beam calculation,

1. Click Beam Calculator on the SolidWorks Toolbox toolbar, or click Toolbox > Beam Calculator. (See the Beam Calculator PropertyManager in Figure 13.22.)
   The Beam Calculator dialog box appears.
2. Select a Load Type using the slider to the left of the preview window.
3. Under Type of Calculation, select Deflection or Stress.
   The Input area updates to show the boxes that are appropriate for your selection.
4. Click Beams, select a beam in the Structural Steel dialog box, and then click Done to return to the Beam Calculator dialog box (see Figure 13.23).
   Some of the information in the Input area is automatically entered based on the beam that you select.

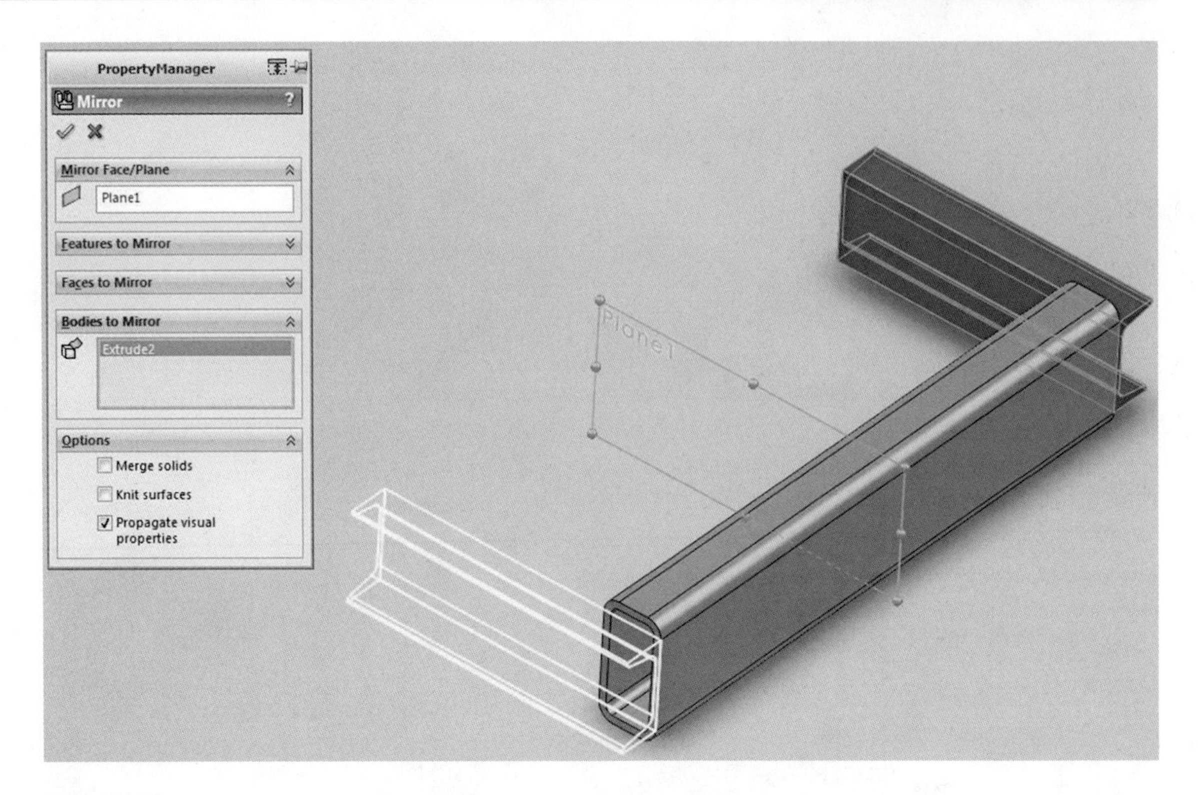

Figure 13.14

Mirror PropertyManager.

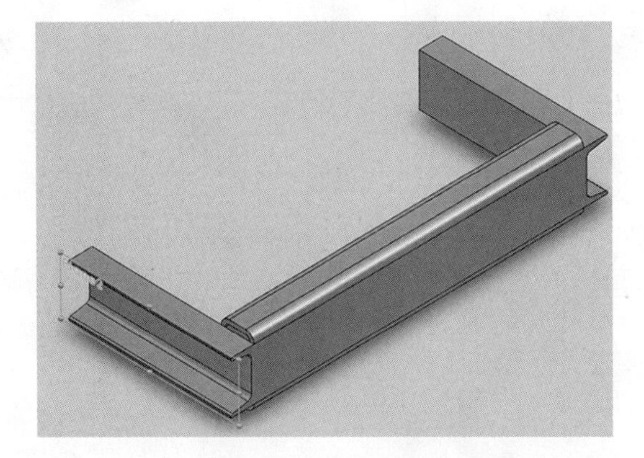

Figure 13.15

Plane2 defined using any three outer vertices of the mirrored C-channel.

5. Select an Axis (X or Y local axis) to determine the value in the Moment of inertia or Section modulus box (see Figure 13.23).
6. Type a value in the rest of the boxes in the Input area *except for the one to be solved*, and click Solve. For example, make sure that there is a value in all of the boxes except for the Deflection box if you are trying to solve for the deflection.

   The Beam Calculator dialog box calculates the remaining value for you.
7. Click Done to close the Beam Calculator dialog box.

   The solutions are displayed in Figure 13.24.

Grooves

Industry standard O-ring grooves can be created on your cylindrical model. The gateway to access the Grooves option in the Toolbox is shown in Figure 13.25.

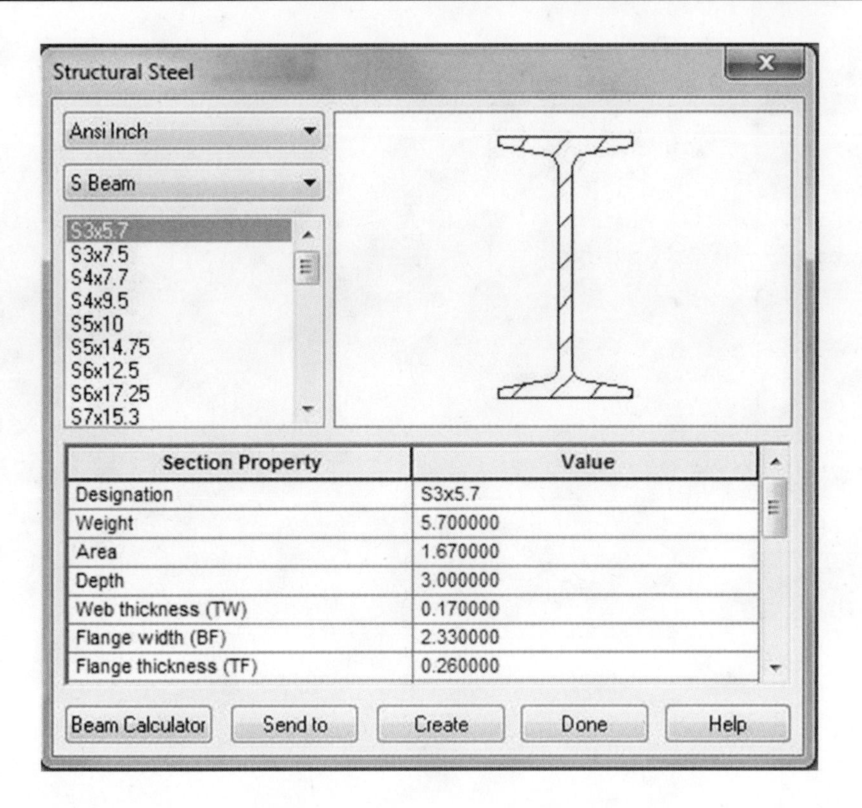

Figure 13.16

Creating S3x5.7 S-beam.

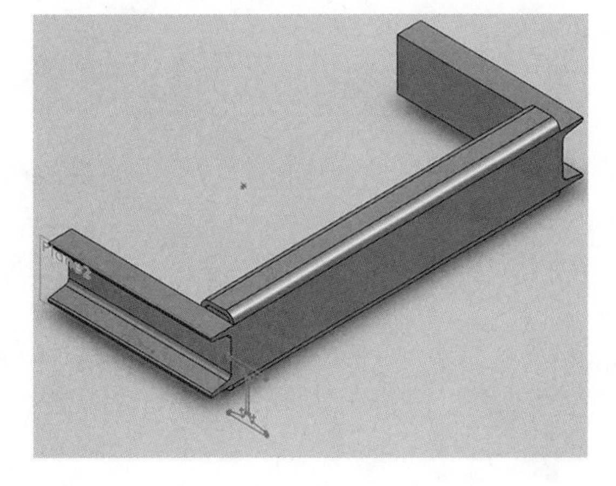

Figure 13.17

Cross section of S3x5.7 S-beam.

To create an O-ring groove,

1. Select a cylindrical face on a part where the groove is to be placed.
   By preselecting a cylindrical face, the SolidWorks software determines the diameter for the groove and suggests the appropriate groove sizes.
2. Click Grooves on the SolidWorks Toolbox toolbar, or click Toolbox > Grooves (see Figure 13.25).
   The Grooves dialog box appears (see Figure 13.26) showing the ANSI Inch and Male Static Groove.
3. On the O-Ring Grooves tab, do the following:
   a. Select a Standard (ANSI Inch), a Groove Type (Male Static Groove), and an available Groove Size (AS 568-338) from the lists on the top left of the tab.
      The fields in the Property and Value columns are updated.

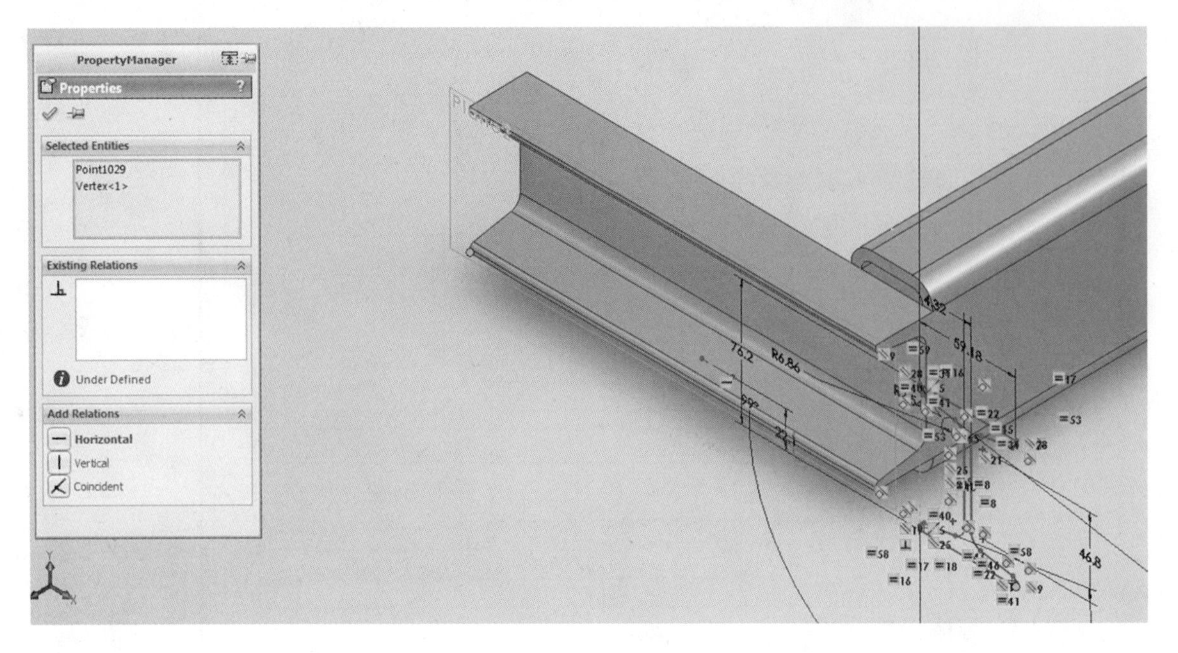

Figure 13.18

Sketch3 placed in correct position using appropriate relations.

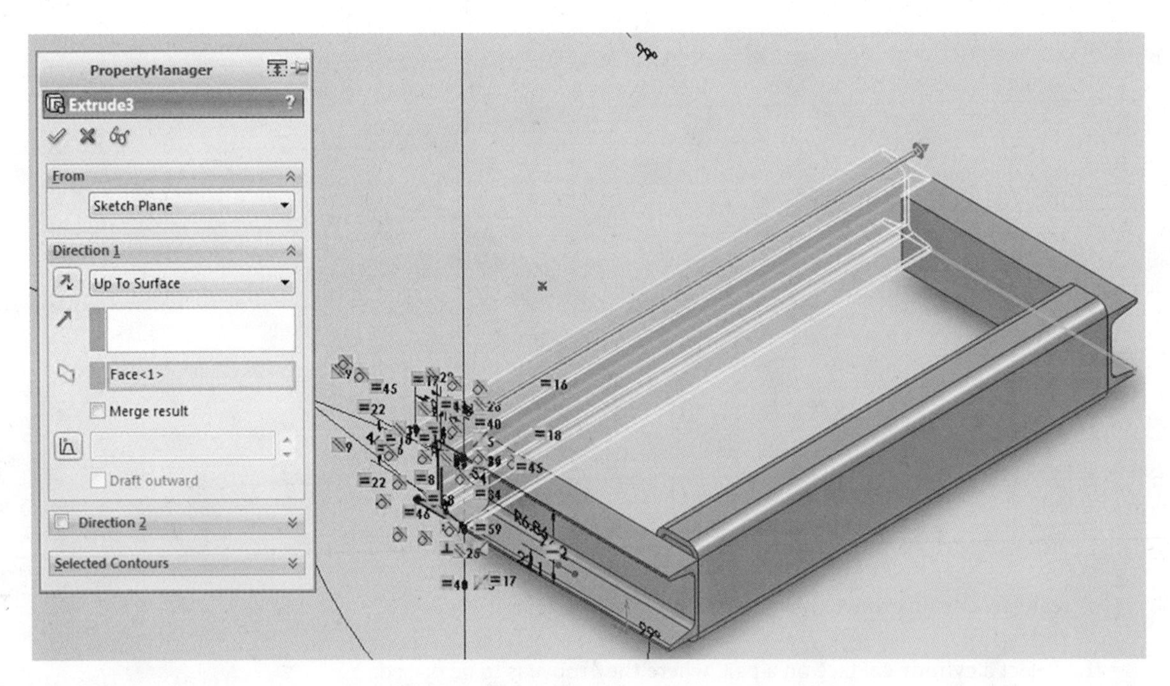

Figure 13.19

Extrude PropertyManager for Extrude3 feature.

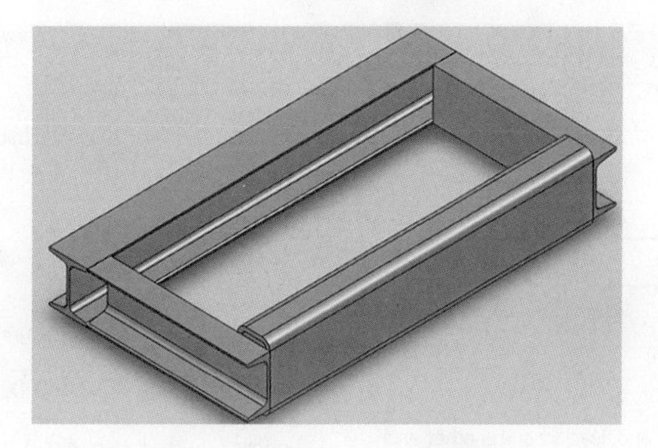

Figure 13.20

Completed structural steel model.

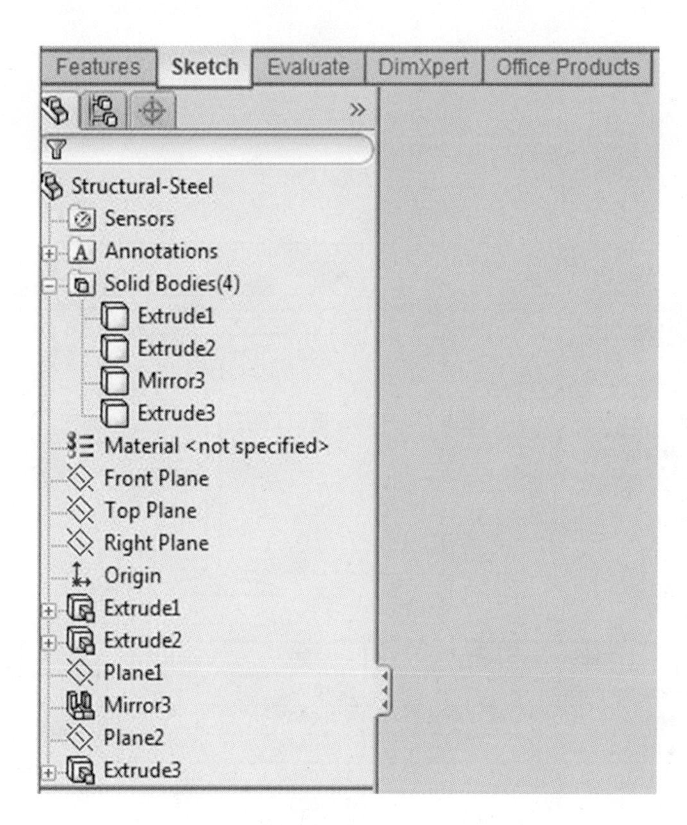

Figure 13.21

FeatureManager showing the four solid bodies created.

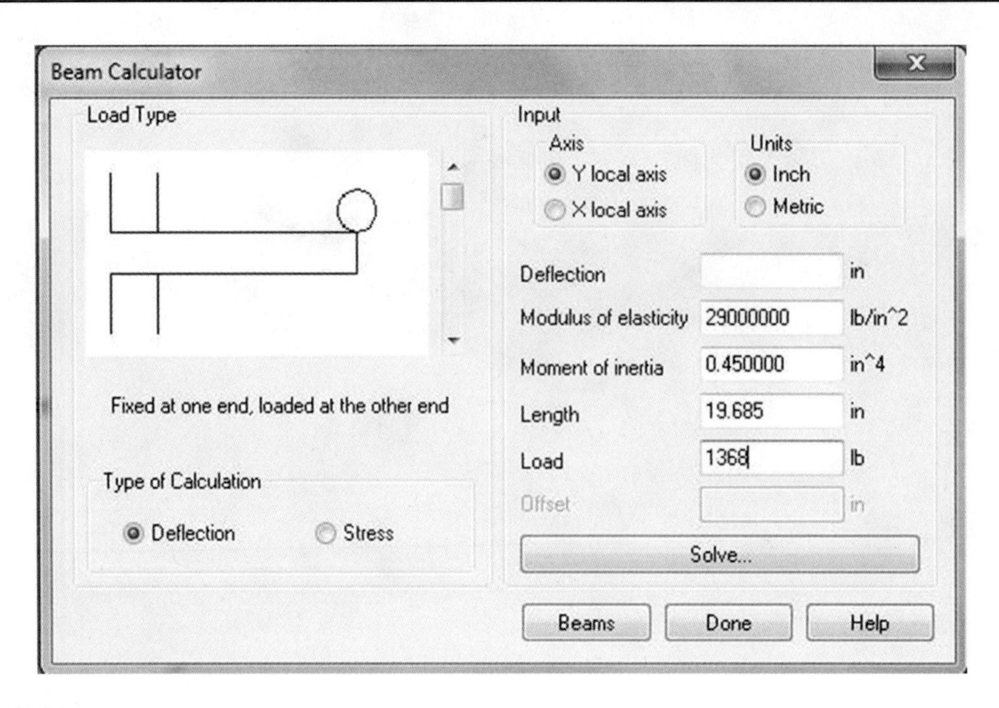

Figure 13.22

Beam Calculator PropertyManager.

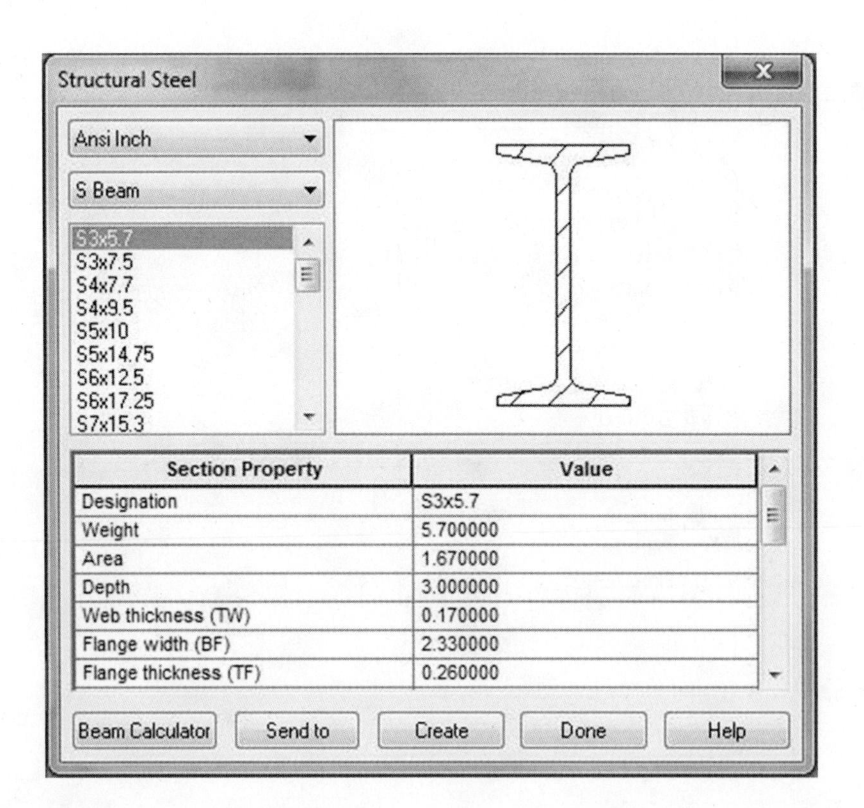

Figure 13.23

Beams available with their properties' values.

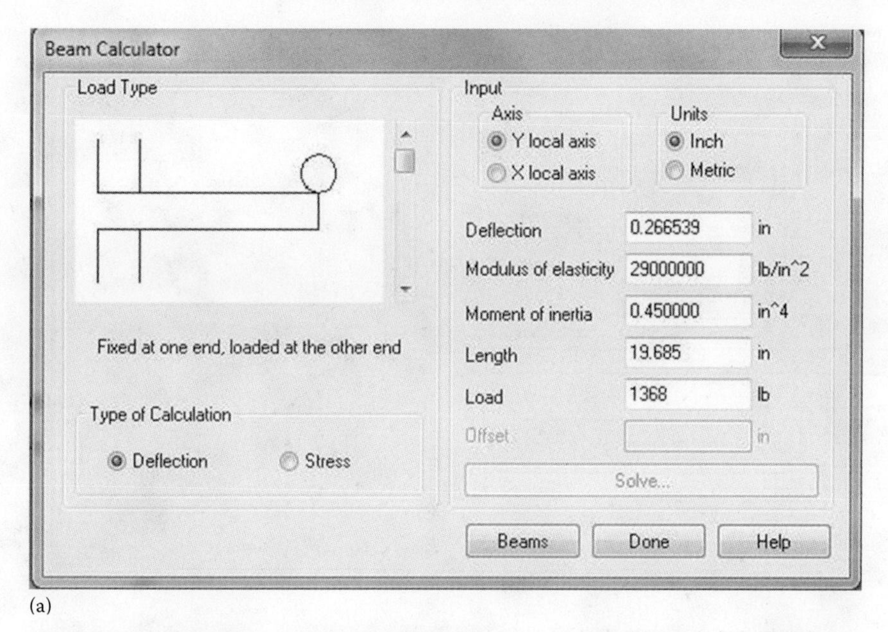

(a)

steel-structure - Notepad

File Edit Format View Help

Structural Steel Properties

Standard:	Ansi Inch
Type:	S Beam
Designation:	S3x5.7
Weight:	5.700000
Area:	1.670000
Depth:	3.000000
Web thickness (TW):	0.170000
Flange width (BF):	2.330000
Flange thickness (TF):	0.260000
Fillet radius (RI):	0.270000
Toe fillet radius (RA):	0.100000
Radius of gyration (RT):	0.570000
Moment of inertia (IX):	2.520000
Section modulus (SX):	1.680000
Radius of gyration (RX):	1.230000
Moment of inertia (IY):	0.450000
Section modulus (SY):	0.390000
Radius of gyration (RY):	0.522000
Torsional constant (J):	0.040000
Plastic modulus (ZX):	1.950000

(b)

Figure 13.24

Solutions generated: (a) deflection output and (b) section properties output to a Notepad file.

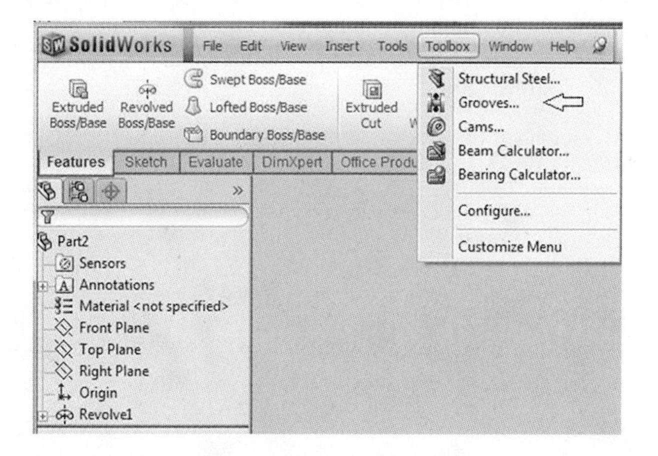

Figure 13.25

Gateway to access the Grooves option in the toolbox.

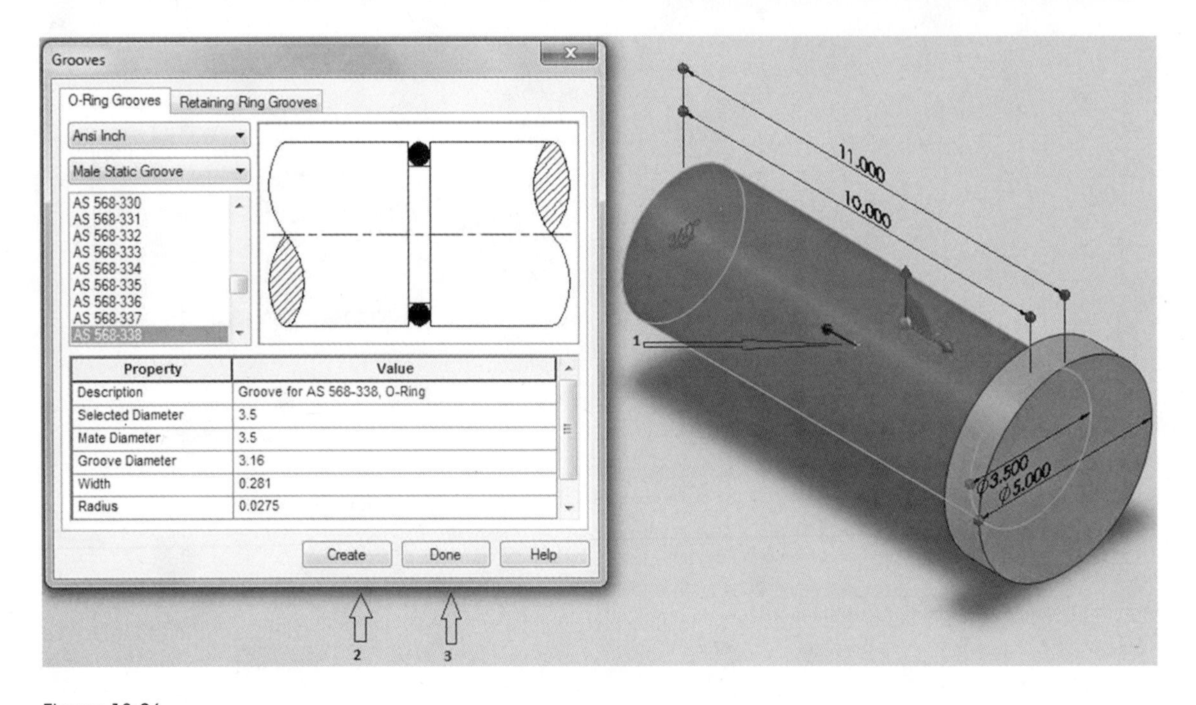

Figure 13.26

Grooves PropertyManager.

b. Notice that the Selected Diameter (3.5) is set for you because you selected a cylindrical face in step 1.
c. Notice the Mate Diameter (3.5). This is a reference value for the diameter of the nongrooved mating part that completes the seal.

4. Notice the values for the Groove Diameter (3.18), Width (0.281), and Radius (0.275).
5. Click Create to add the groove.
   The groove is cut into the model. A feature appears in the FeatureManager design tree with a name (*Groove for AS 568-338, O-Ring1*) that matches the Description.
6. Click Done (OK) to close the dialog box. (See Figure 13.27 for the model with groove.)

The next step is to locate the groove at the exact position that is required. To do this,

1. Right-click Sketch2 in the FeatureManager under Groove for AS 568-338, O-Ring1 Feature tree (see Figure 13.28).
2. Click Edit Sketch to be in sketch mode (see Figure 13.28).
3. Position the part using the Normal To tool (see Figure 13.28).
4. Dimension the *edge of the Groove profile from the top of the part* to locate the desired position (7.5) (see Figure 13.29).

The final part is shown in Figure 13.30.

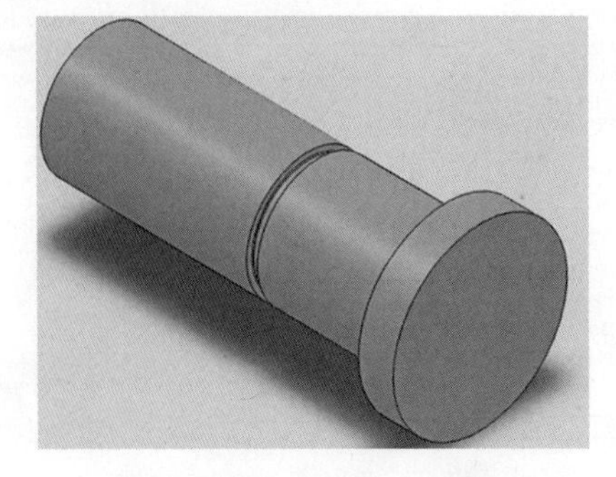

Figure 13.27

Groove created in a model.

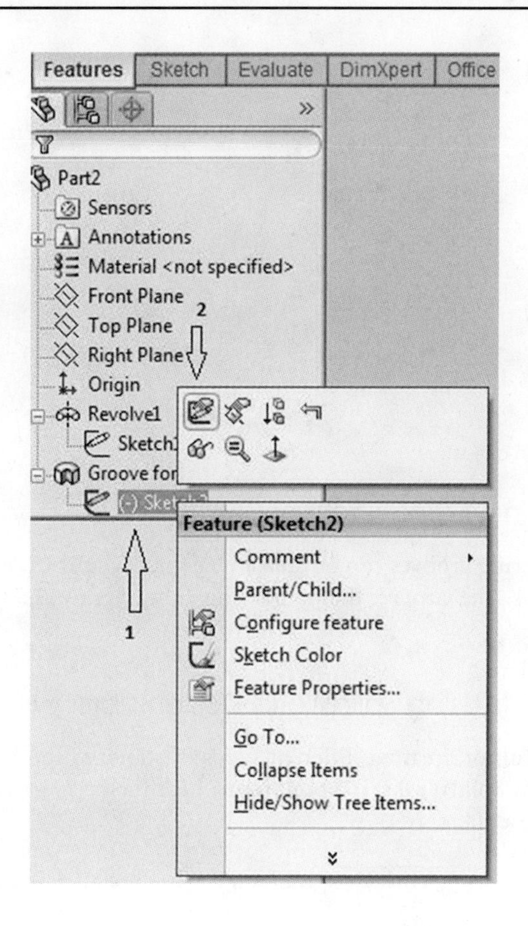

Figure 13.28

Editing Groove location.

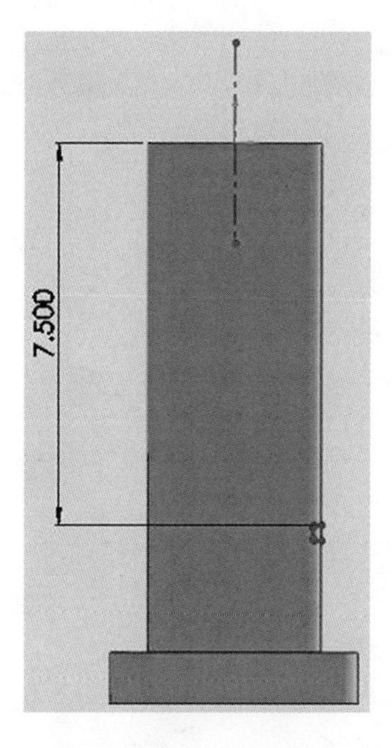

Figure 13.29

Redimensioning Groove position.

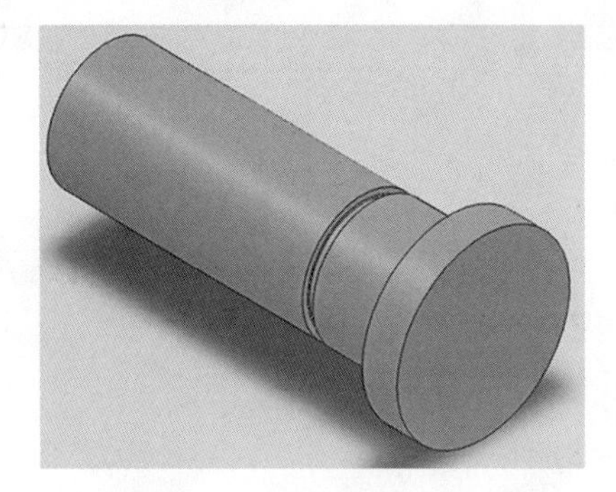

Figure 13.30

Final part with groove position adjusted.

O-Ring Grooves: Retaining Ring Grooves

Industry standard retaining ring grooves can be created on your cylindrical model using exactly the same gateway as in O-ring grooves. The *Grooves* dialog box shown in Figure 13.30 is for both types of grooves (O-ring and retaining ring).

Creating Standard Parts Using SolidWorks Design Library and Toolbox

SolidWorks software is useful for creating different types of standard mechanical parts. As many as 18 standards are available in the SolidWorks Toolbox from which the designer can choose (see Figures 13.16 through 13.32). To access these standards,

click Design Library > Toolbox.

The Toolbox offers quite a number of design tools for bearings, bolts and screws, jig bushings, keys, nuts, O-rings, and power transmission elements (chain sprockets, gears, timing belt pulleys, retaining rings, structural members, and washers), as shown in Figures 13.16 through 13.33. These design tools assist the design engineers in performing their work more effectively and efficiently.

The first step in using the Toolbox in the Design Library is to decide on the standard to use: (a) American National Standards Institute (ANSI) (Inch/Metric), (b) British Standards Institution (BSI), (c) Canadian

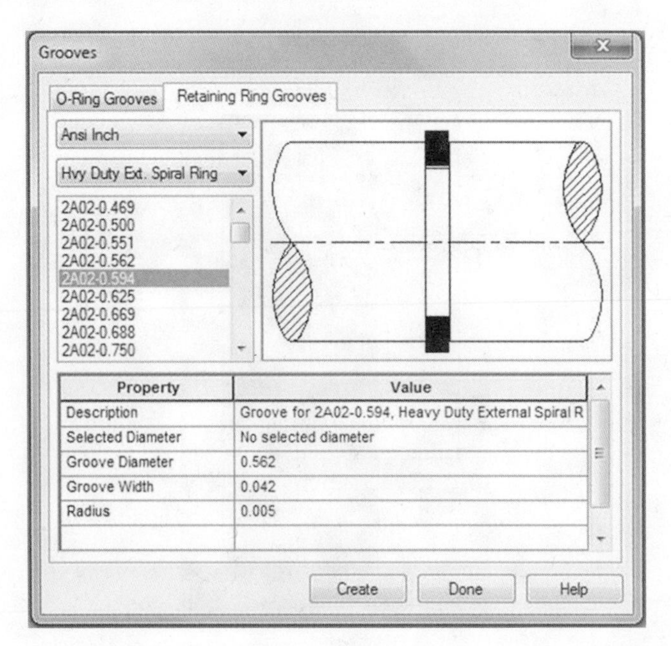

Figure 13.31

Grooves PropertyManager.

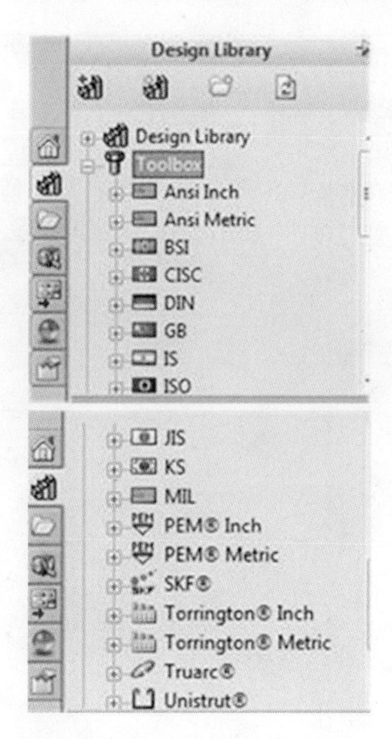

Figure 13.32

SolidWorks Toolbox Standards.

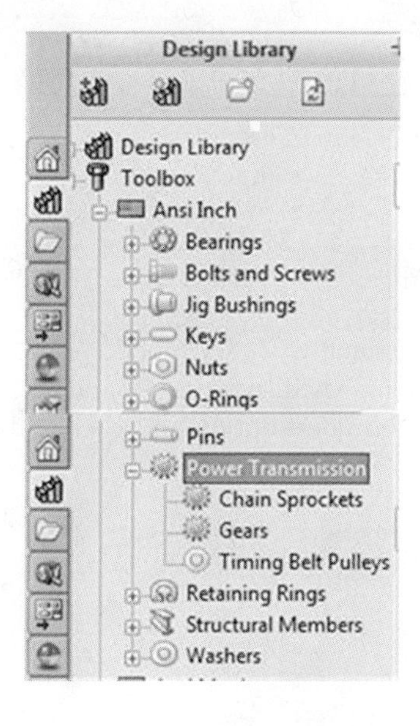

Figure 13.33

Toolbox design tools.

Institute of Steel Construction (CISC), (d) Deutsches Institut für Normung (DIN), (e) Guobiao (GB), (f) International Organization for Standardization (ISO), (g) Japanese Industrial Standards (JIS), (h) Korean Standard (KS), (i) Military Standard (MIL-STD), (j) PEM (Inch/Metric), (k) Svenska Kullagerfabriken AB (SKF), (l) Torrington, (m) (Inch/Metric), (n) Truarc, or (o) Unistrut. The first 10 are national/internal standards, whereas the remaining 5 are registered companies of international reputations that produce specific products.

Adding Set Screws to the Collar of a Shaft

As an illustration, we consider adding set screws to the collar of a shaft. Figure 13.34 shows the Set Screw selection from the Design Library Toolbox and how it is dragged into the subassembly of a shaft and collar.

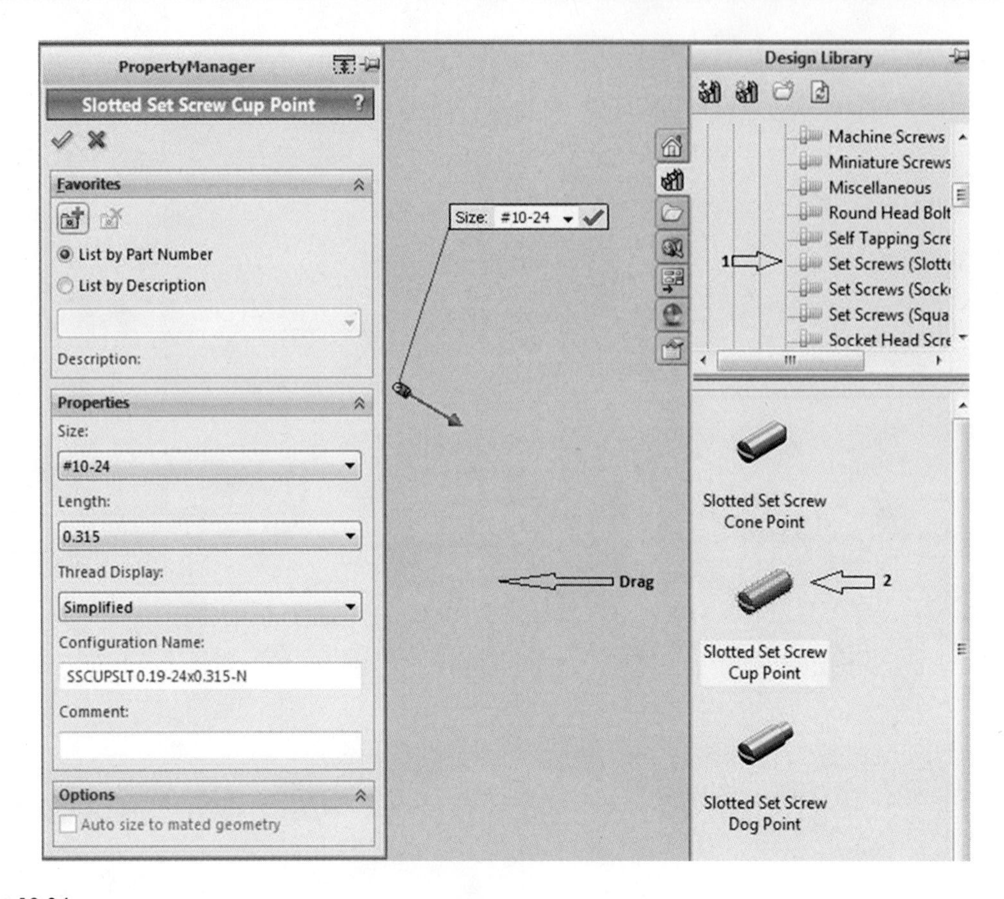

Figure 13.34

Set screw selection from the Design Library Toolbox.

To access set screws, do the following:

1. Open the collar and shaft subassembly.
2. Click Design Library > Toolbox > ANSI Inch > Bolts and Screws > Set Screws (Slotted).
3. Select the Slotted Set Screw Cup Point option and click and drag the set screw into the drawing window (see Figure 13.34).

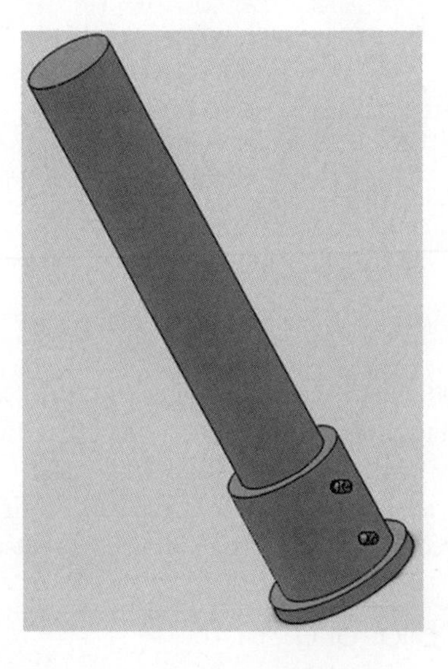

Figure 13.35

Assembly shaft, collar, and set screws.

4. Define the size of the set screw as #10-24 and the length as 0.315.
5. Click OK to complete insertion and specification.
6. Use Mate tool and insert the set screws into the collar and shaft.
7. Save the assembly that is shown in Figure 13.35.

Gear Design Using SolidWorks

Gears are used to transmit rotary motion, torque, and power from one shaft to another. Gears are the most rugged and durable means of power transmission compared to various other means such as belts and chains. They have transmission efficiency as high as 98% but are more expensive than belts and chains.

Types of Gears

The four main types most commonly used in practical application are (1) *spur*, (2) *helical*, (3) *bevel*, and (4) *worm* gears.

1. *Spur gears* are the simplest.
2. *Helical gears*, like spur gears, are cut from a cylindrical blank and have involute teeth. The difference is that their teeth are at some helix angle to the shaft axis. These gears are used to transmit power between parallel and nonparallel shaft.
3. *Bevel gears* are cut on conical blanks to be used to transmit motion between intersecting shafts. The simplest bevel gear type is the straight-tooth bevel gear or the straight bevel gear.
4. *Worm gear set* consists of a worm (similar to a screw) and the worm gear (a special helical gear). The shafts on which the worm and gear are mounted are usually oriented 90° to each other. Worm gearing can be employed to transmit motion between nonparallel nonintersecting shafts.

Generally, spur and helical gears have teeth that are parallel and inclined to the axis of rotation, respectively.

Design Methodology for Gears

Two modes of failure affect gear teeth: (1) *fatigue fracture due to fluctuating bending stress* at the root of the tooth and (2) fatigue (*wear*) of the tooth surface. Both modes of failure must be checked when designing the gears. The shapes and sizes of the teeth are standardized by the American Gear Manufacturers Association (AGMA). The methods of AGMA are widely used in the design and analysis of gear sets. The AGMA approach requires extensive use of charts and graphs together with equations that facilitate application of CAD.

In a gearing system, three tasks have to be performed:

1. Check the gear pairs for the two main modes of failure: the AGMA approach is used.
2. Size the shaft carrying the gears based on gear loading: the AGMA approach is used.
3. Select bearings carrying the shaft based on gear loading: SKF, ANSI, ISO, etc.

SolidWorks Solution Procedure

In this chapter, we consider the most critical steps. The author has worked out the details of the *black box*. Utilization of the AGMA approach for checking the gear pairs for the two main modes of failure is treated as a black box. Utilization of the AGMA approach for sizing the shaft carrying the gears based on gear loading is treated as a black box. The reason for labeling these design decisions as black boxes is due to the fact that the details are not included in this chapter as they are outside the scope of this chapter. The procedure is given to show how SolidWorks or any other CAD design software can be used for gearing system design rather than the mere shape design, which is mainly covered by users.

The general solution procedure involves the following major components (note where SolidWorks could be applied):

1. Sizing of the gearbox based on the design specifications (supports, housing, etc.). (Use SolidWorks.)
2. Check the gear pairs for the two main modes of failure. (AGMA approach is used; treat this as a black box since this is not covered in this chapter; use SolidWorks Simulation as an alternative.)
3. Size the shaft carrying the gears based on gear loading. (AGMA approach is used; treat this as a black box since this is not covered in this chapter; use SolidWorks Simulation as an alternative.)
4. Select bearings carrying the shaft based on gear loading. (Use the SolidWorks Toolbox to select bearings after the initial loading analysis has been completed.)
5. Models gears. (Use SolidWorks.)

6. Assemble Parts (include keyways, keys, etc.). (Use SolidWorks.)
7. Animate gear assemble. (Use SolidWorks.)
8. Create Drawings and BoM. (Use SolidWorks.)

We will now commence SolidWorks gear design procedure by first considering spur gears, then helical gears, and finally, bevel gears. Note that worm gears are not supported in the SolidWorks Design Library/Toolbox. Another observation is that that ANSI Inch supports more elements for power transmission than ANSI mm. Therefore, designers should consider using ANSI Inch where more mechanical elements such as chain sprockets and timing belt pulley are required. It is feasible to convert from ANSI Inch to ANSI mm, if the latter is required.

Spur Gear Design

Let us consider spur gears, which are the simplest of the different classes of gear. In this example, we will use SolidWorks to design the spur gears from the design specifications that are given. Other details for sizing the gearbox, checking that the AGMA specifications are met, are not shown.

Problem Description

A simple gearing system consists of a plate (5 × 4 × 0.5), two shafts each (φ 0.5 × 1.7), and a pair of spur gears that are defined as follows:

Gear 1	Gear 2
Diametral pitch = 24	Diametral pitch = 24
Number of teeth = 30	Number of teeth = 60
Pressure angle = 20-deg	Pressure angle = 20-deg
Face width = 0.5	Face width = 0.5
Hub style = One side	Hub style = One side
Hub diameter = 1.00	Hub diameter = 1.00
Overall length = 0.70	Overall length = 0.70
Nominal shaft diameter = ½	Nominal shaft diameter = ½
Key = None	Key = None

Use SolidWorks for designing the gearing system.

SolidWorks Solution Procedure

The pitch diameter of the pinion is $D_1 = \frac{N_{t1}}{P_d} = \frac{30}{24} = 1.25''$

The pitch diameter of the gear is $D_2 = \frac{N_{t2}}{P_d} = \frac{60}{24} = 2.5''$

The center-to-center distance is $c = \frac{(D_1 + D_2)}{2} = \frac{(1.25 + 2.5)}{2} = 1.875''$

Note that for spur gears, the center-to-center distance calculated agrees with the value that should be used. In helical gears, this may not be the case.

Support

Model the support (see Figure 13.36).

Shaft

Model the shaft (see Figure 13.37).

Pinion

1. Open a New SolidWorks Part document.
2. Click Design Library > Toolbox > ANSI Inch > Power Transmission > Gears.
3. Right-click Spur Gear (see Figure 13.38).
4. Select Create Part. (Note: this is the preferred route that is used in this book; do not *drag and drop* parts from the Design Library.) (See Figure 13.38.)

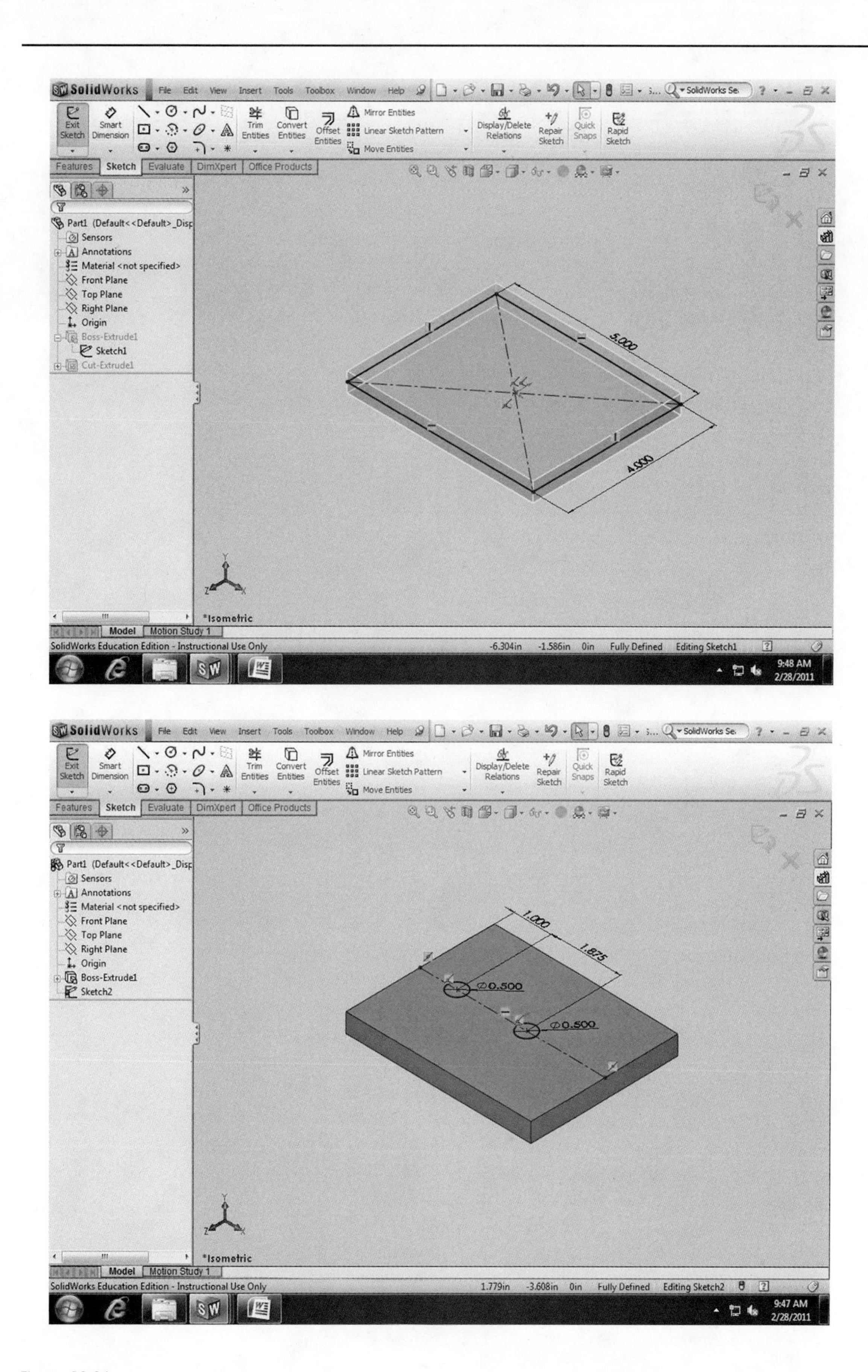

Figure 13.36

Support part.

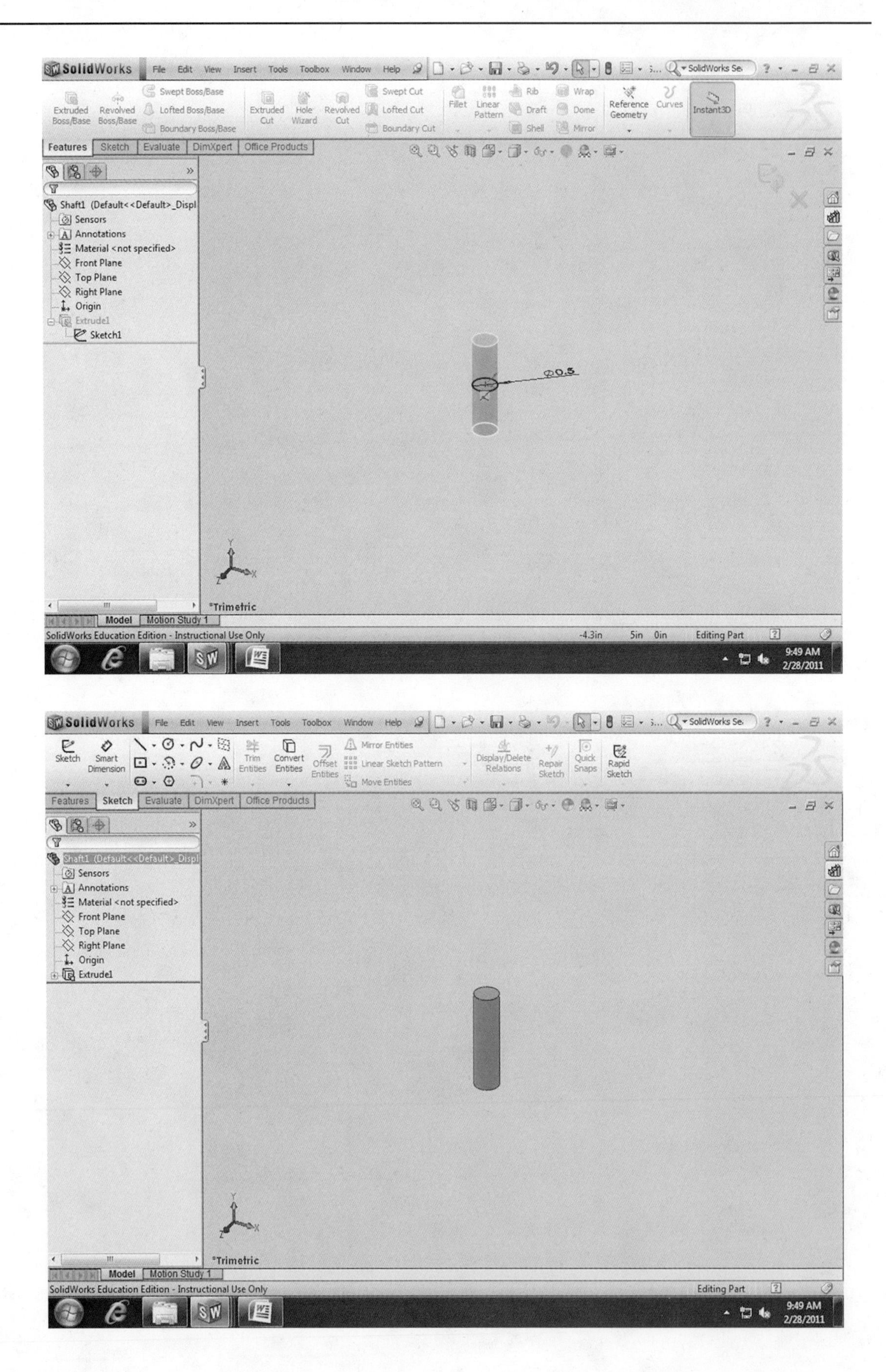

Figure 13.37

Shaft.

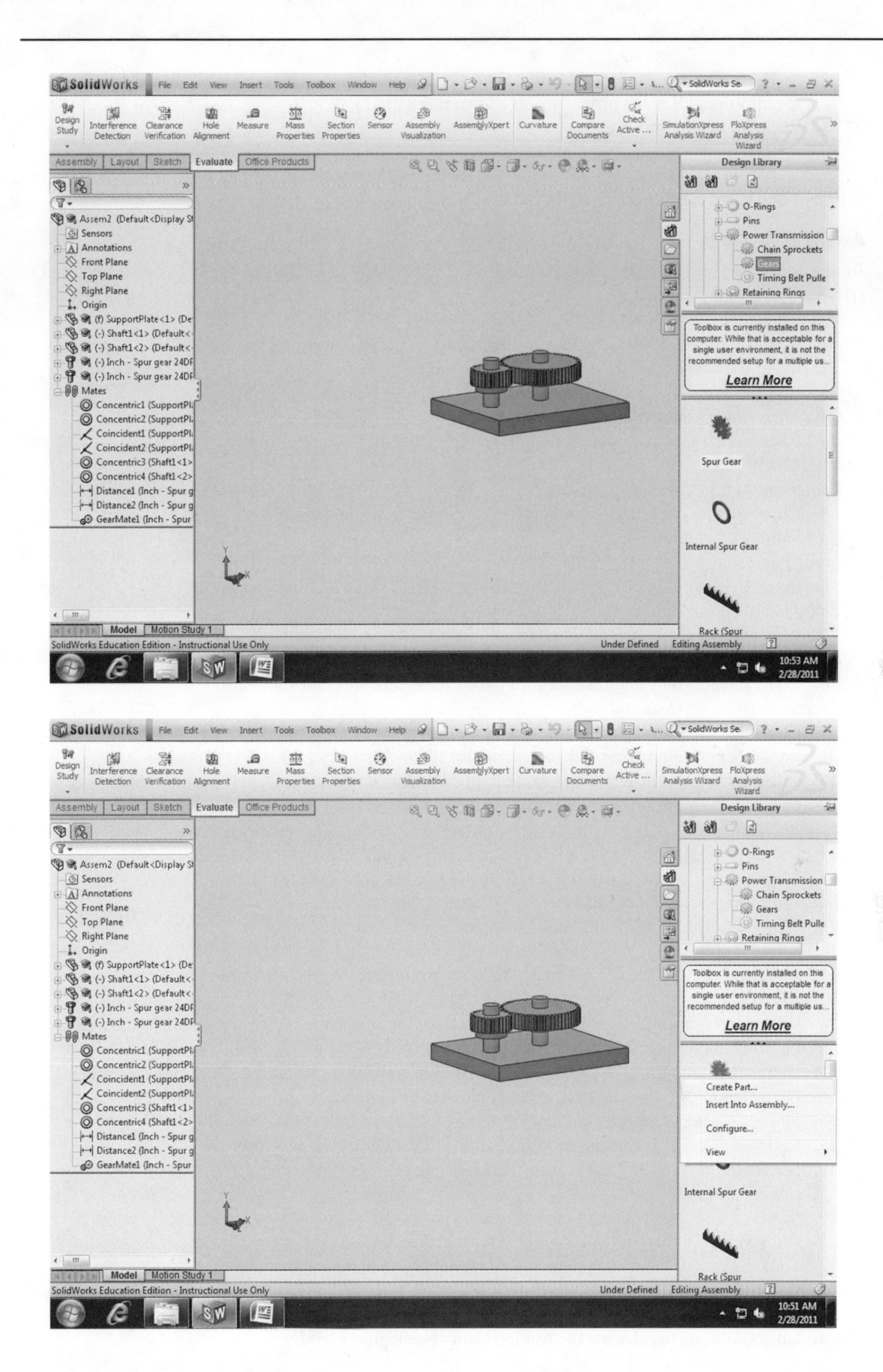

Figure 13.38

Creating Helical gears.

The Spur Gear PropertyManager appears immediately (see Figure 13.39). Based on the design specification, fill in the properties, as shown in Figure 13.39.

5. Click OK when complete.

Gear

Repeat as in pinion. (See Figure 13.40 for the PropertyManager.)

Assembly Modeling

To assemble the parts, the following procedure is given as a guideline (other sequences could be used to arrive at an assembly solution):

1. Open a New SolidWorks Assembly document.
2. Insert the base as the first part (see Figure 13.41).
3. Insert the shaft twice since two are needed (see Figure 13.41).

Concentric Mating for Shafts

4. Click *each shaft* and *each hole*, and apply Concentric Mate condition (see Figures 13.42 and 13.43).

Coincident Mating for Shafts

5. Click *end face of each shaft* and *bottom of support plate*, and apply Coincident Mate condition (see Figures 13.44 and 13.45).

Insert Pinion

6. Insert the gear (or pinion) as the fourth part. (See Figure 13.46 for the pinion.)

Concentric Mate for Pinion and Shaft

7. Click *inside of bore surface of pinion* and *shaft* and apply Concentric Mate (see Figure 13.47).

Insert Gear

8. Insert the gear (or pinion) as the fifth part. (See Figure 13.48 for the gear.)

Concentric Mate for Gear and Shaft

9. Click *inside of bore surface of gear* and *shaft* and apply Concentric Mate (see Figure 13.49).

Distance Mate for Pinion/Gear and Top of Support Plate

10. Click *end face of hub for pinion/gear* and *top face of support plate* and apply Distance Mate with value equal to 0.5 (see Figures 13.50 and 13.51).

Align the Teeth on Pinion and Gear

11. Use Zoom to Area tool to zoom the teeth of pinion and gear in the area of mesh.
12. Align the teeth of Pinion and Gear to ensure that they mesh (see Figure 13.52).

Temporarily Suppress Coincident Mates for Shafts

13. Right-click the Coincident Mates for Shafts.
14. Select Suppress tool to suppress them.
15. Pull the shafts down to expose the bore of the pinion and gear (see Figure 13.53).

Mechanical Mates

16. Click Mates, and for the Mate Selection, select the *inner faces of bore for pinion* and *gears.* (See Figure 13.54 for the GearMate PropertyManager.)
17. Define the Ratio using Number of Teeth for Pinion (= 30) and Gear (= 60), respectively; their Pitch Diameters or Number of Revolutions could also be used (see Figure 13.54).

Unsuppress Coincident Mates for Shafts

18. Right-click the Coincident Mates for Shafts.
19. Select Suppress tool to unsuppress them. (See Figure 13.55 for the model.)

Animation

1. Click Motion Study1 at the bottom left of the Window.
2. For the Type of Study, select Animation (see Figure 13.56).
3. Click the Motor tool to define the motor; select the Rotary Motor, with Pinion chosen as driver (see Figure 13.57).
4. Click Calculate to calculate the Motion Study (see Figure 13.58).

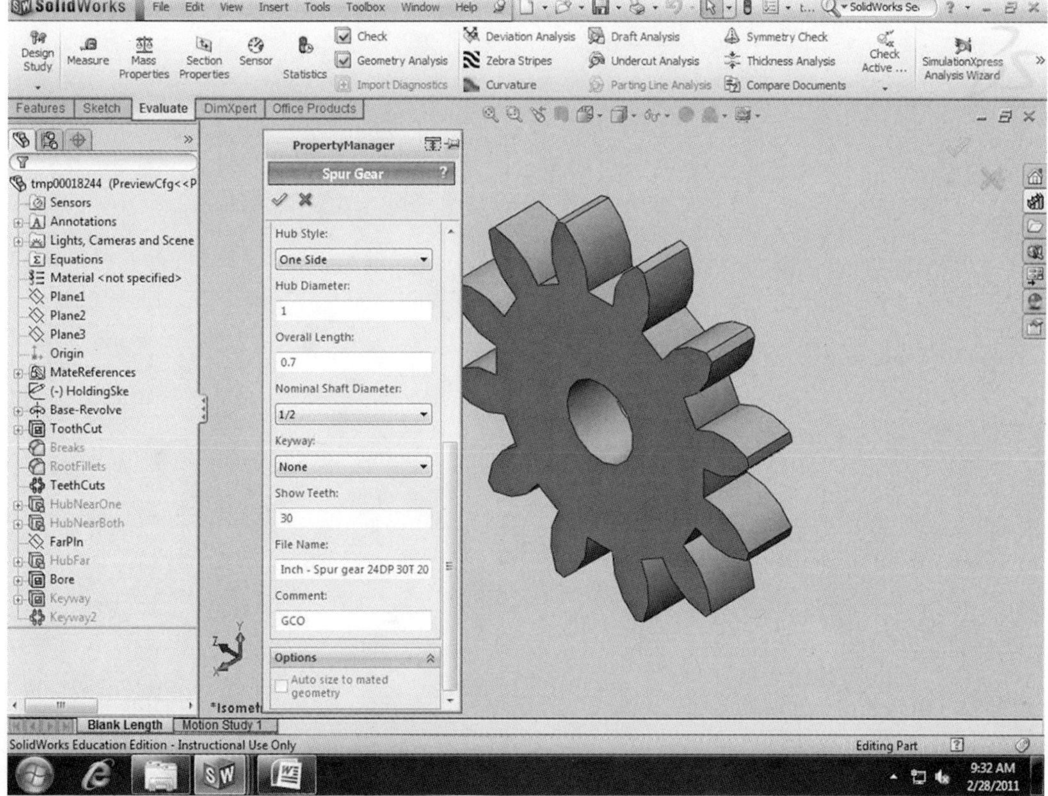

Figure 13.39

Spur gear PropertyManager (pinion defined).

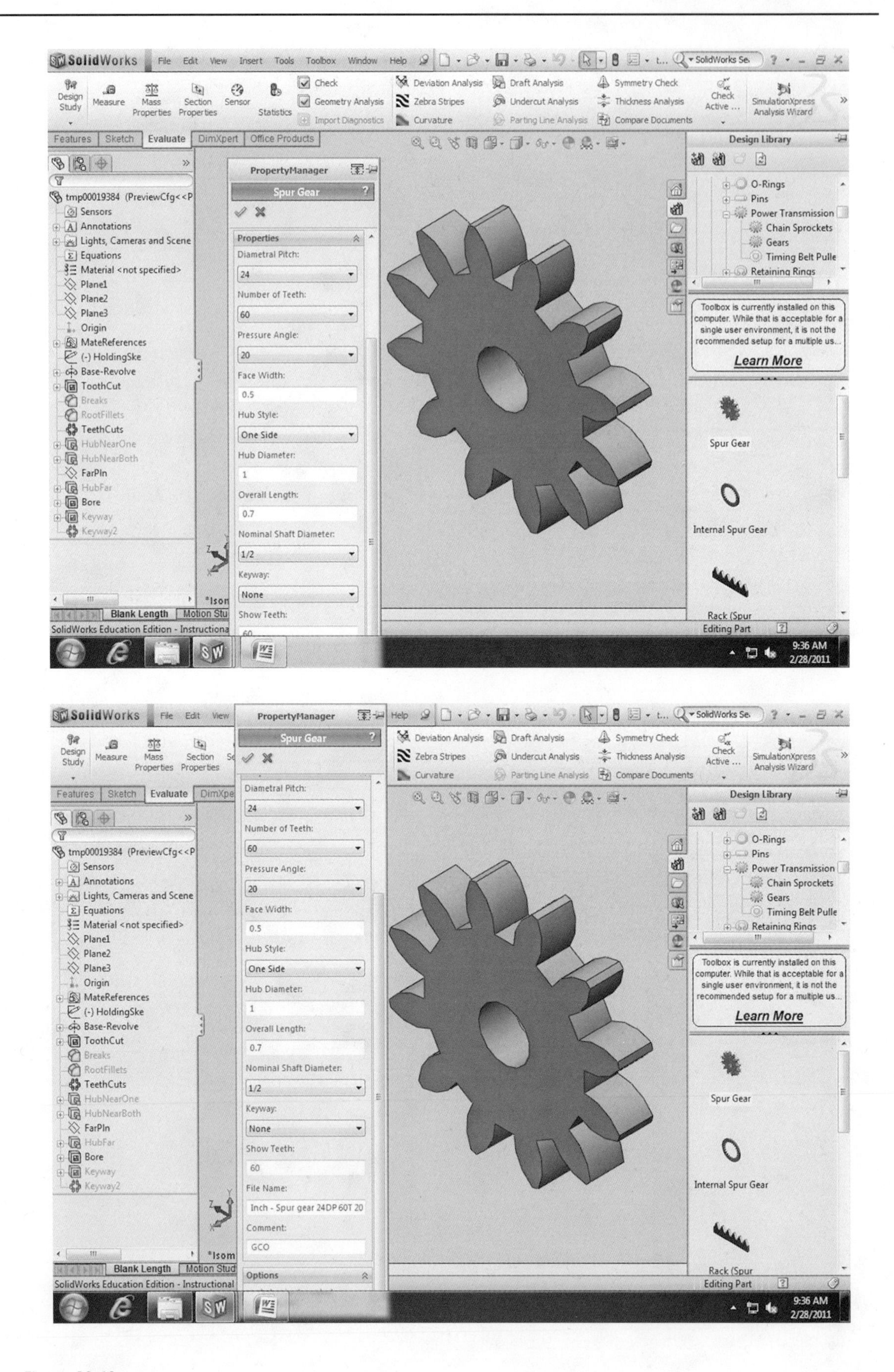

Figure 13.40

Spur gear PropertyManager (gear defined).

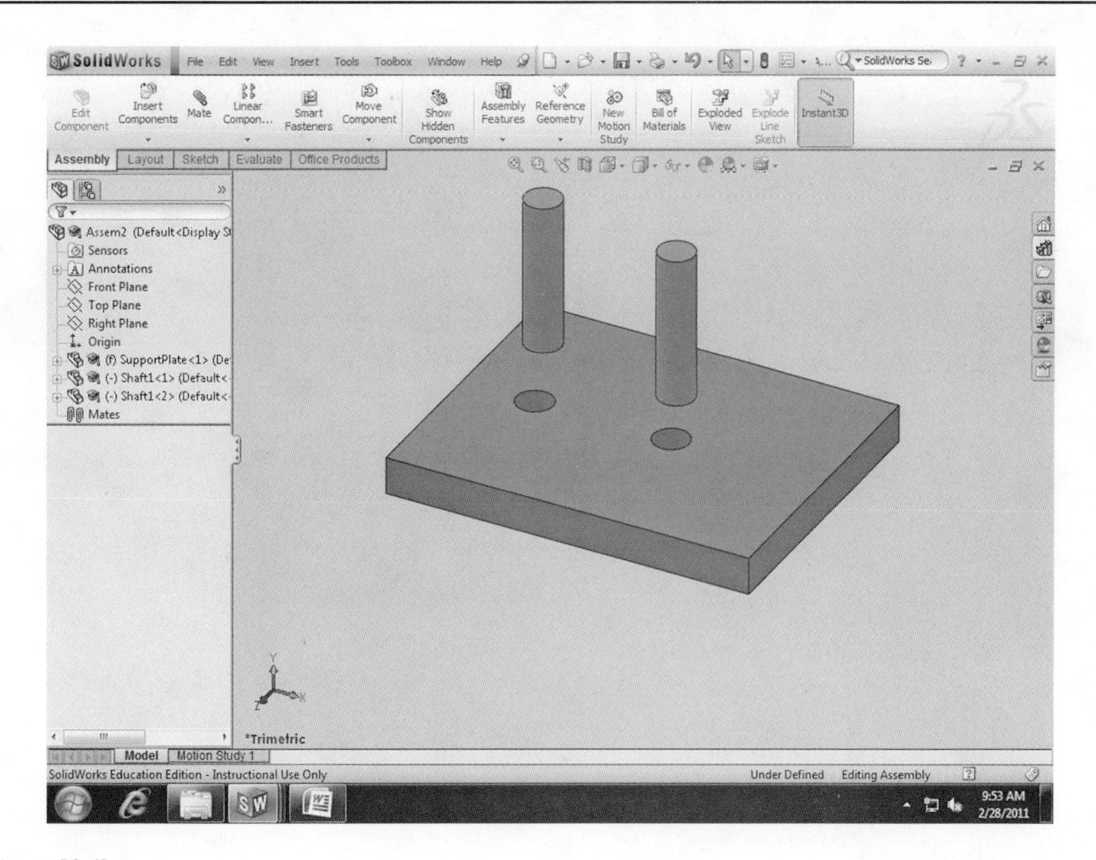

Figure 13.41

Insert support and shafts.

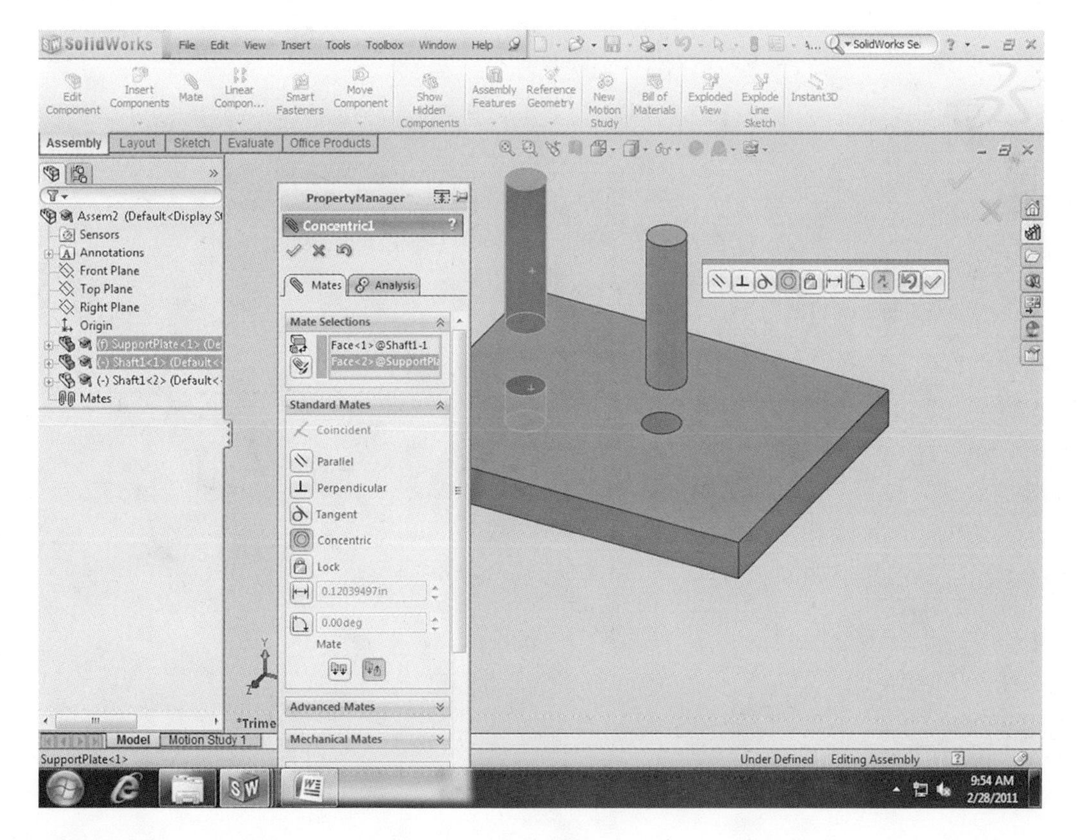

Figure 13.42

Concentric mating for first shaft.

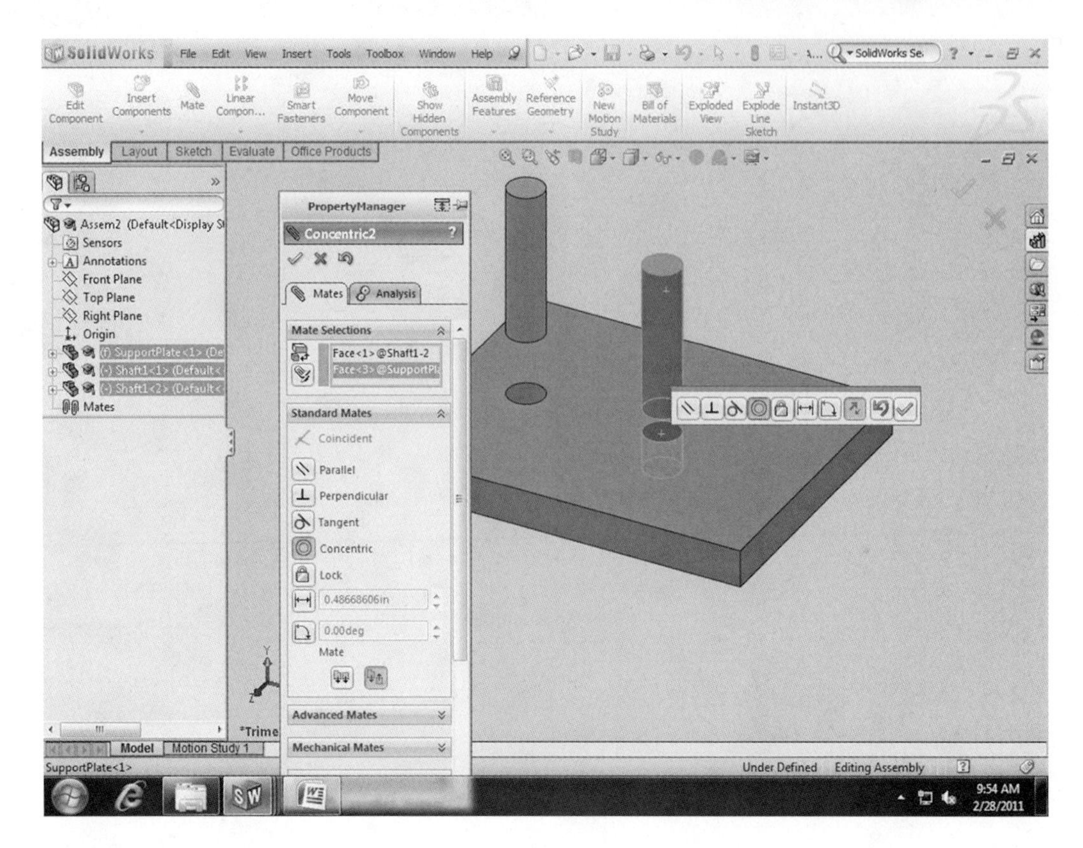

Figure 13.43

Concentric mating for second shaft.

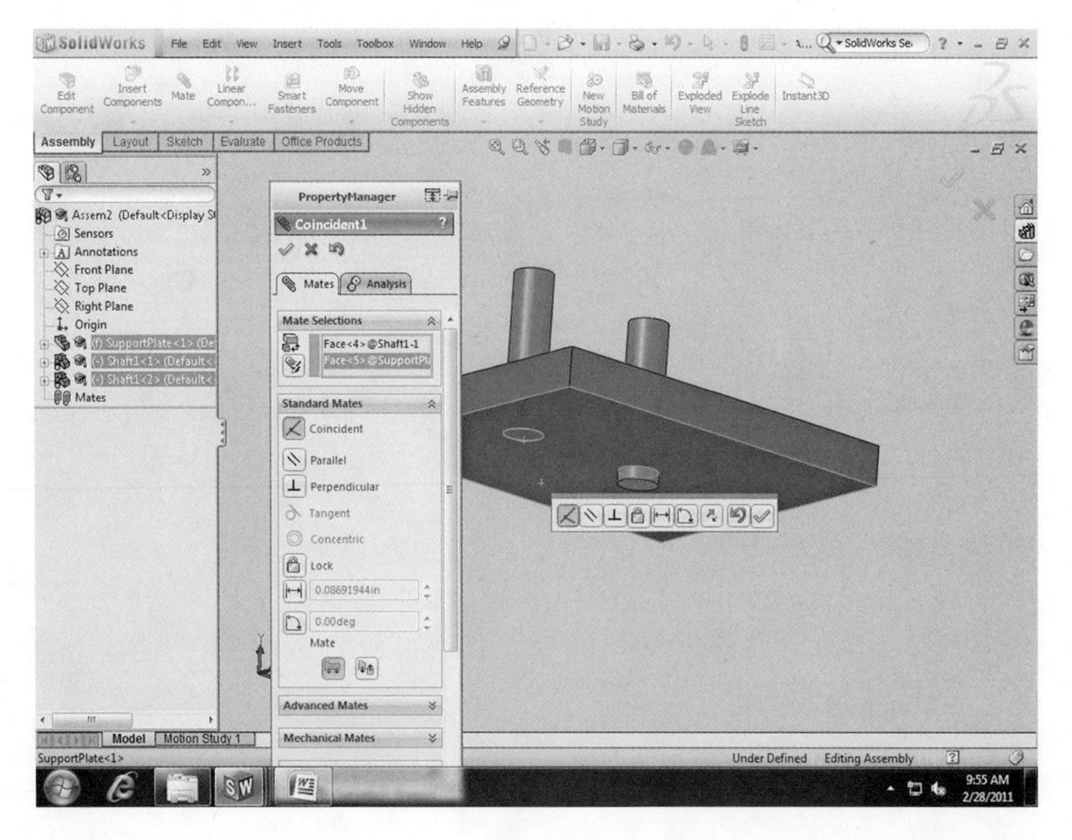

Figure 13.44

Coincident mating for one shaft.

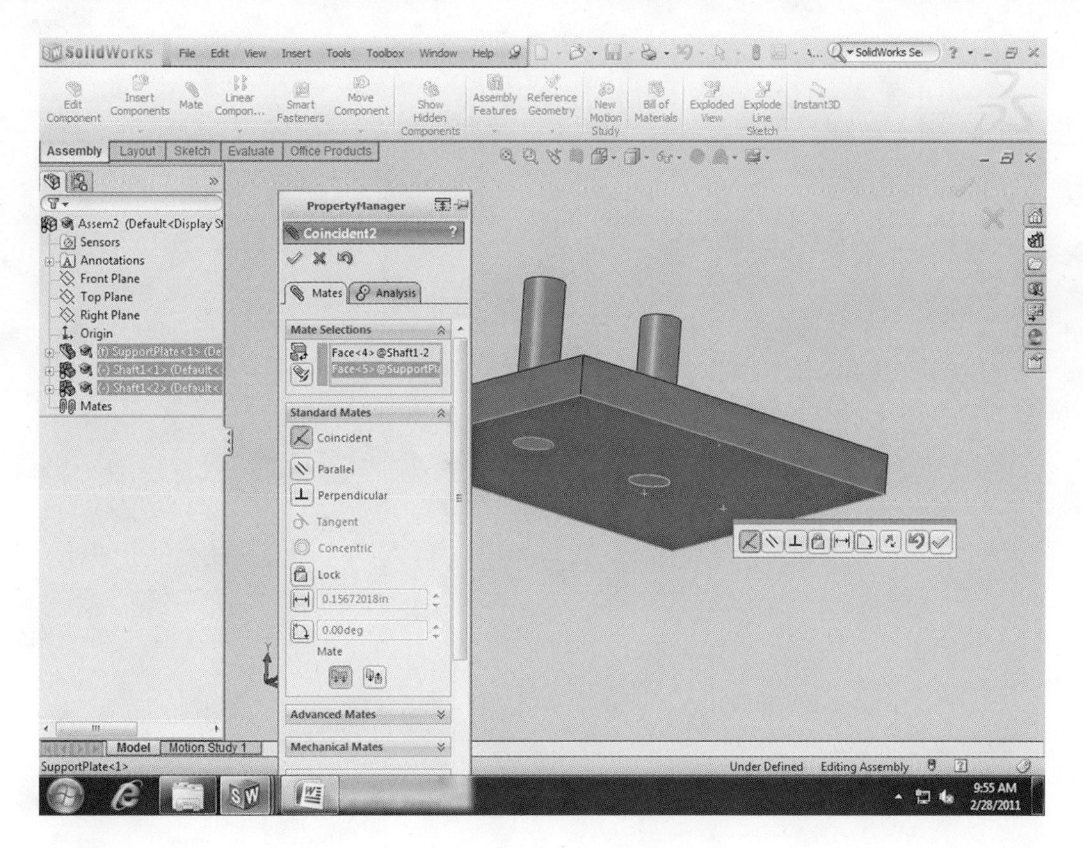

Figure 13.45

Coincident mating for second shaft.

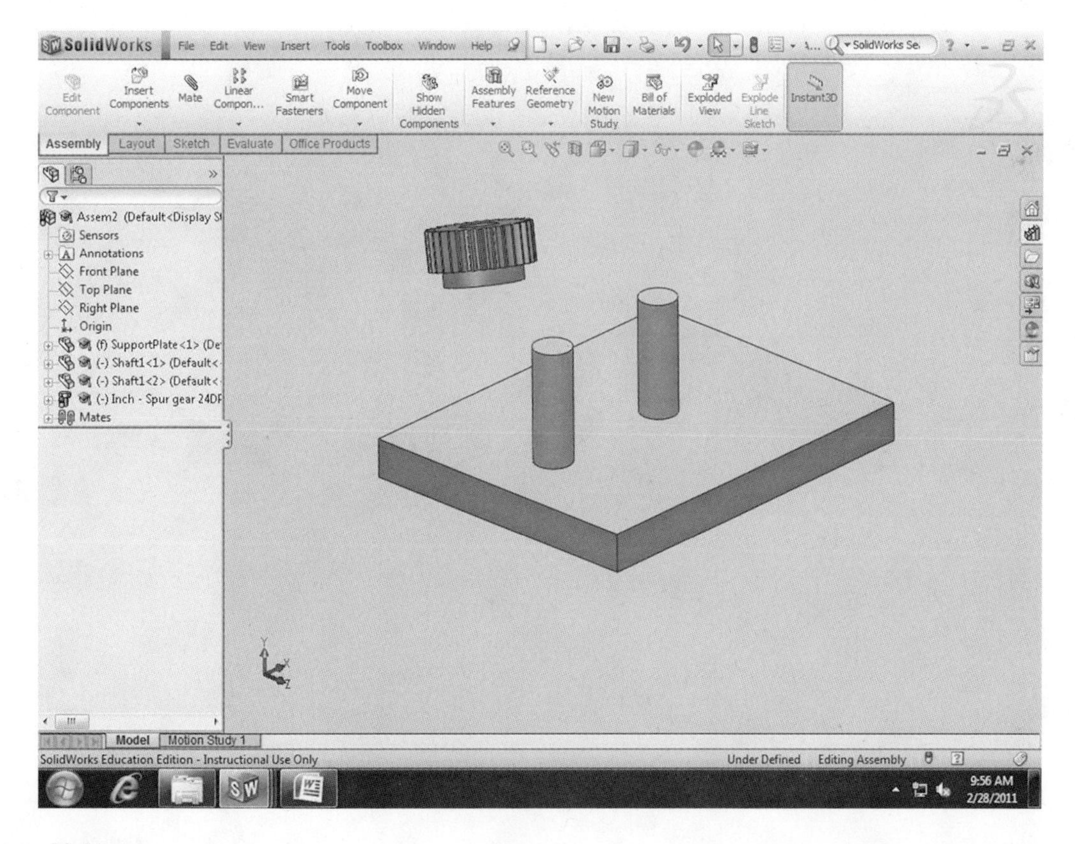

Figure 13.46

Insert the pinion.

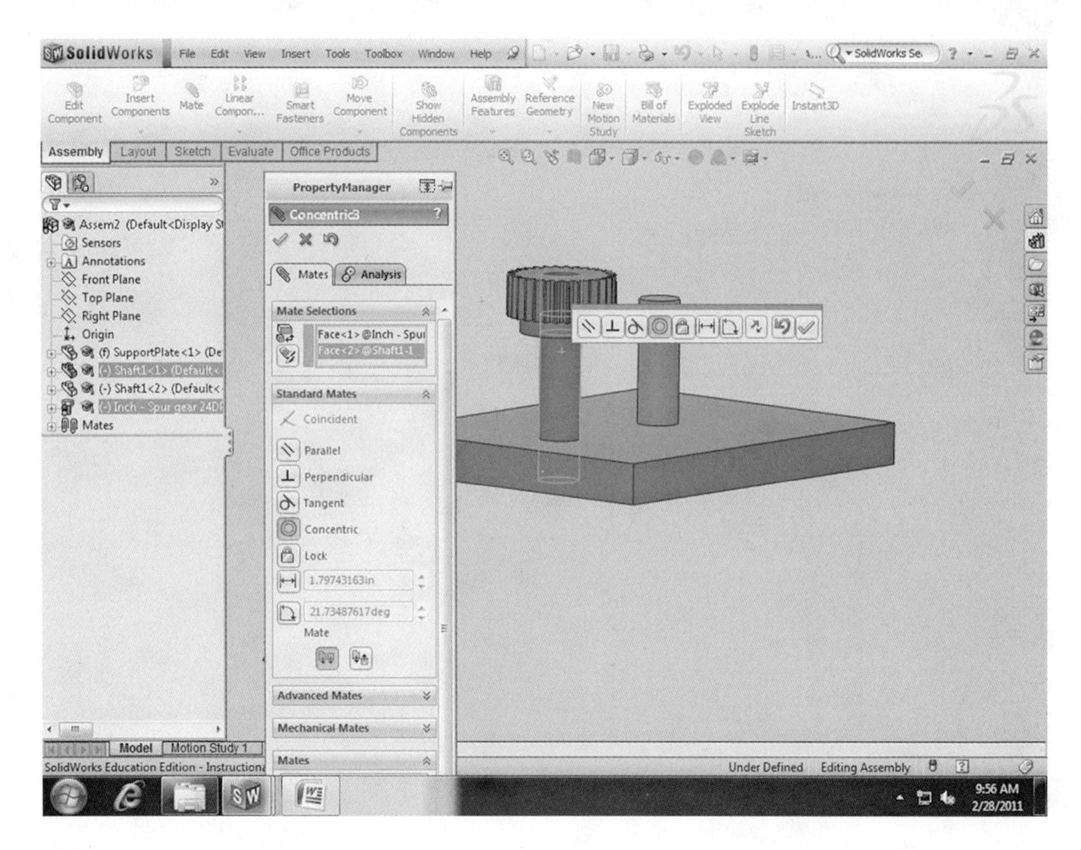

Figure 13.47

Concentric Mate PropertyManager.

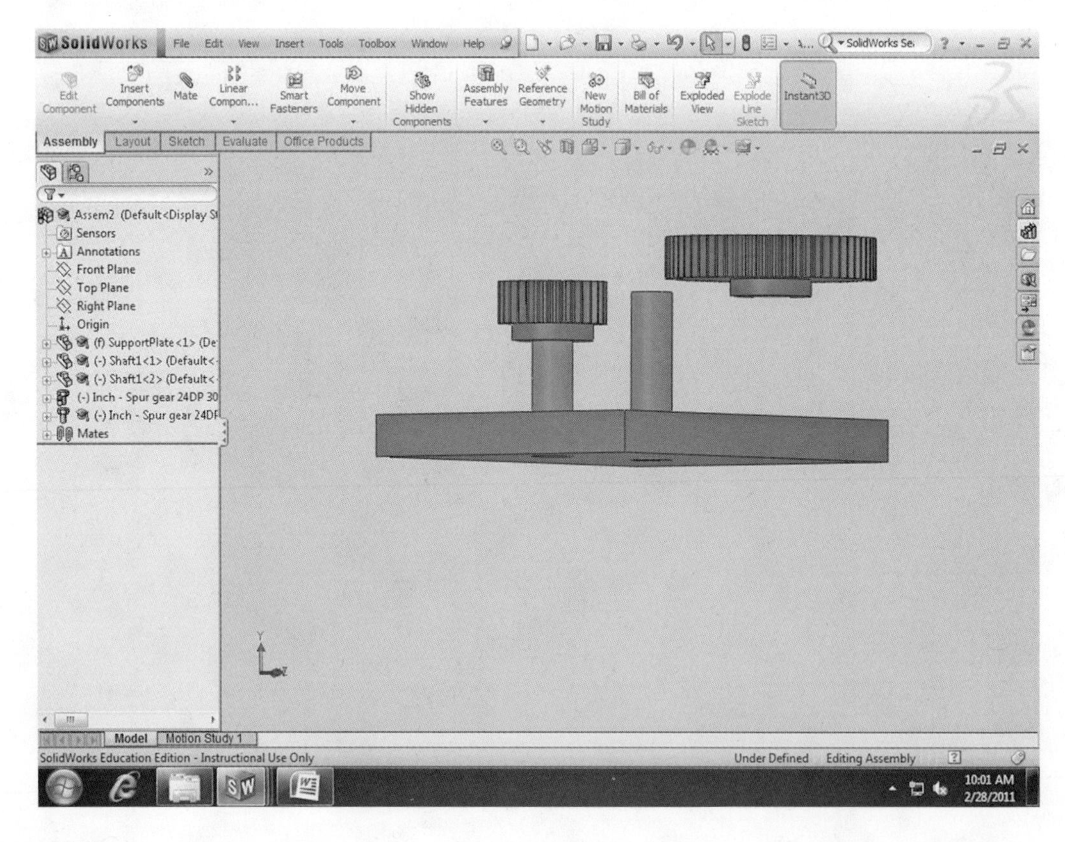

Figure 13.48

Insert the gear.

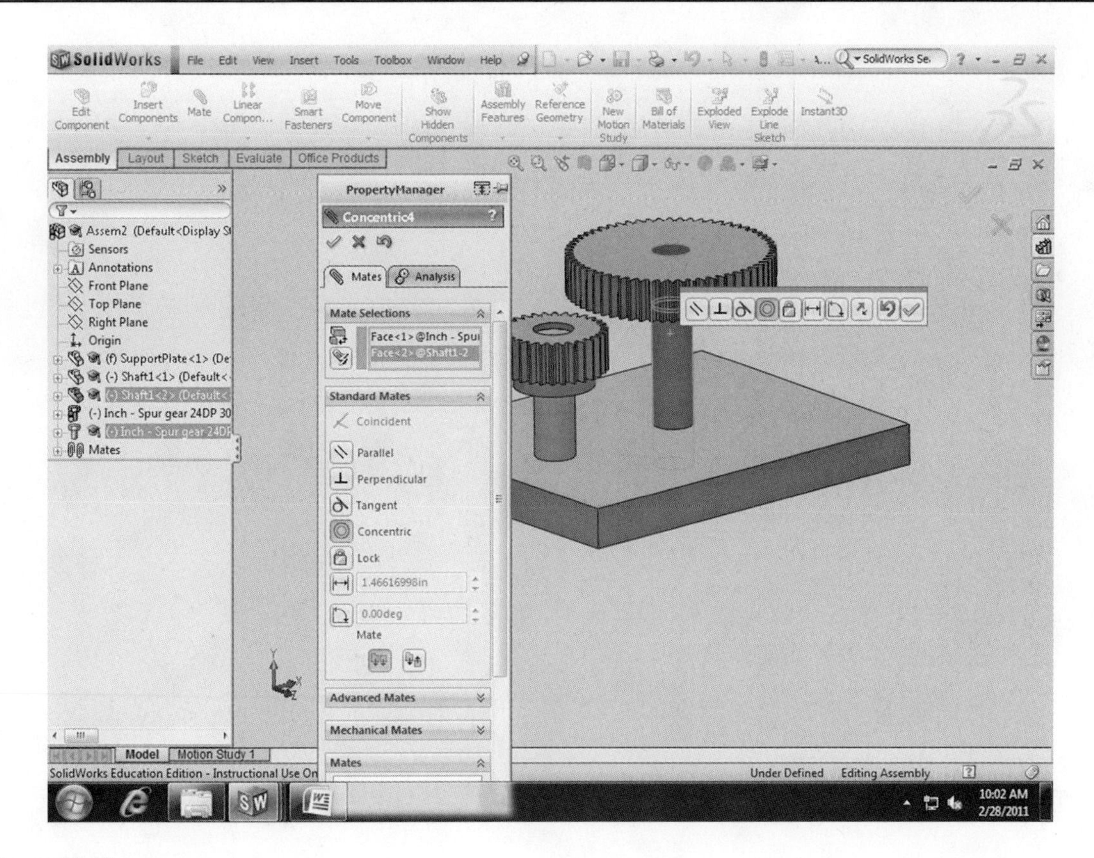

Figure 13.49

Concentric Mate PropertyManager.

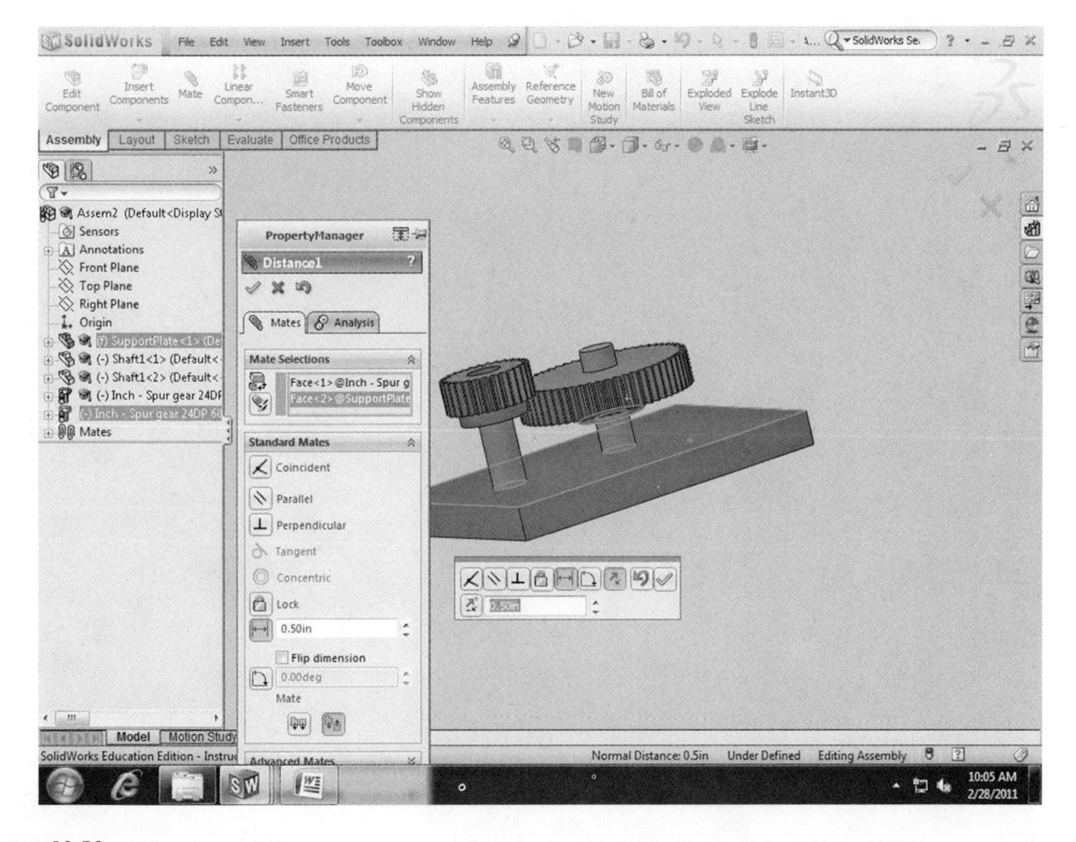

Figure 13.50

Distance Mate PropertyManager.

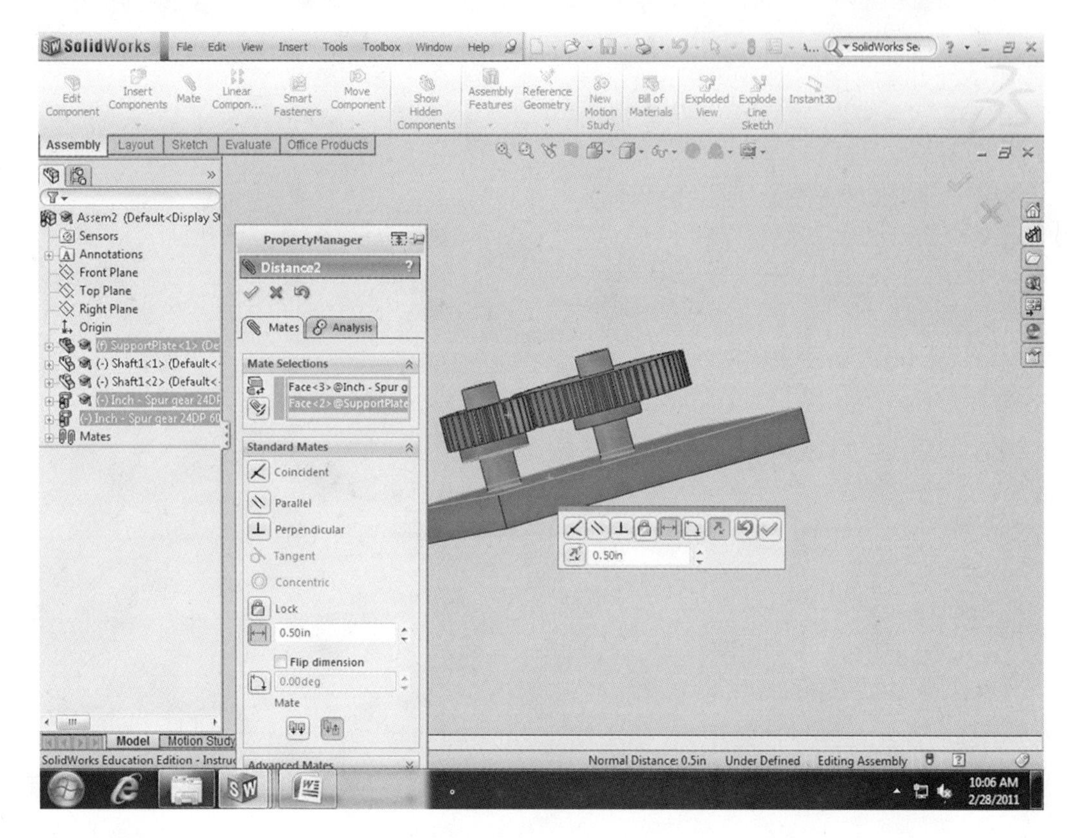

Figure 13.51

Distance Mate PropertyManager.

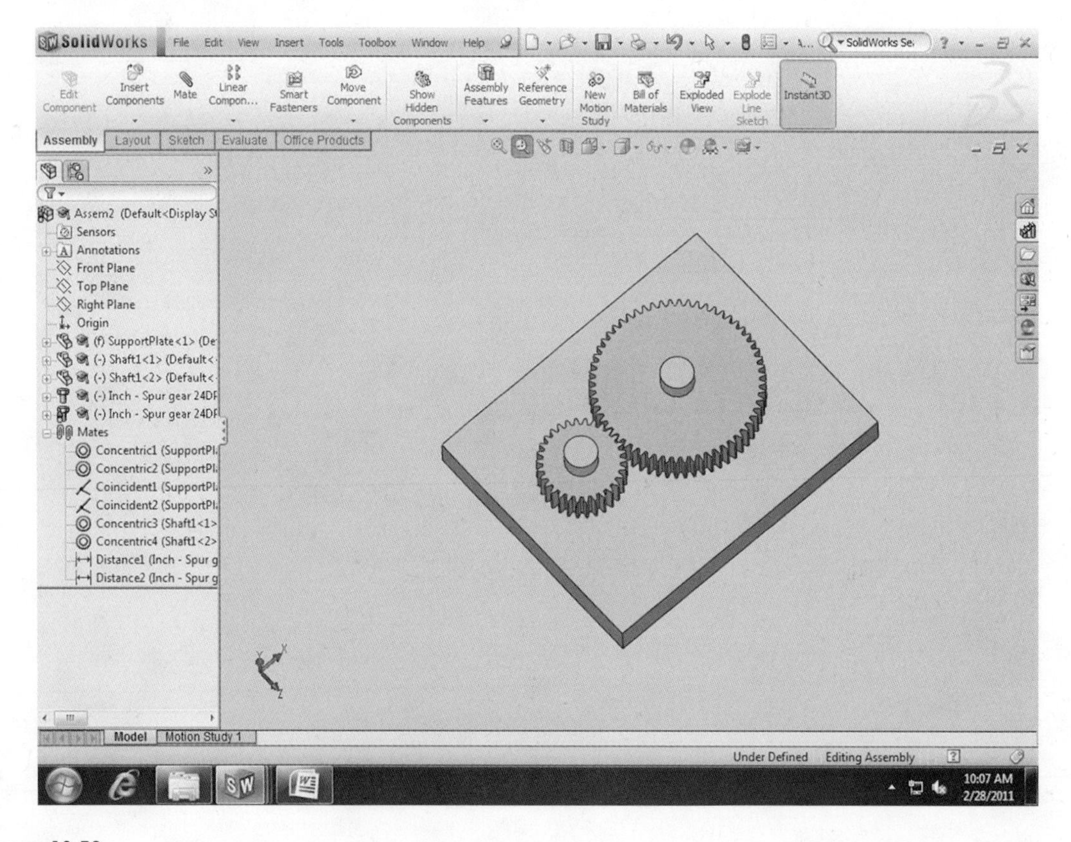

Figure 13.52

Align teeth of Pinion and Gear.

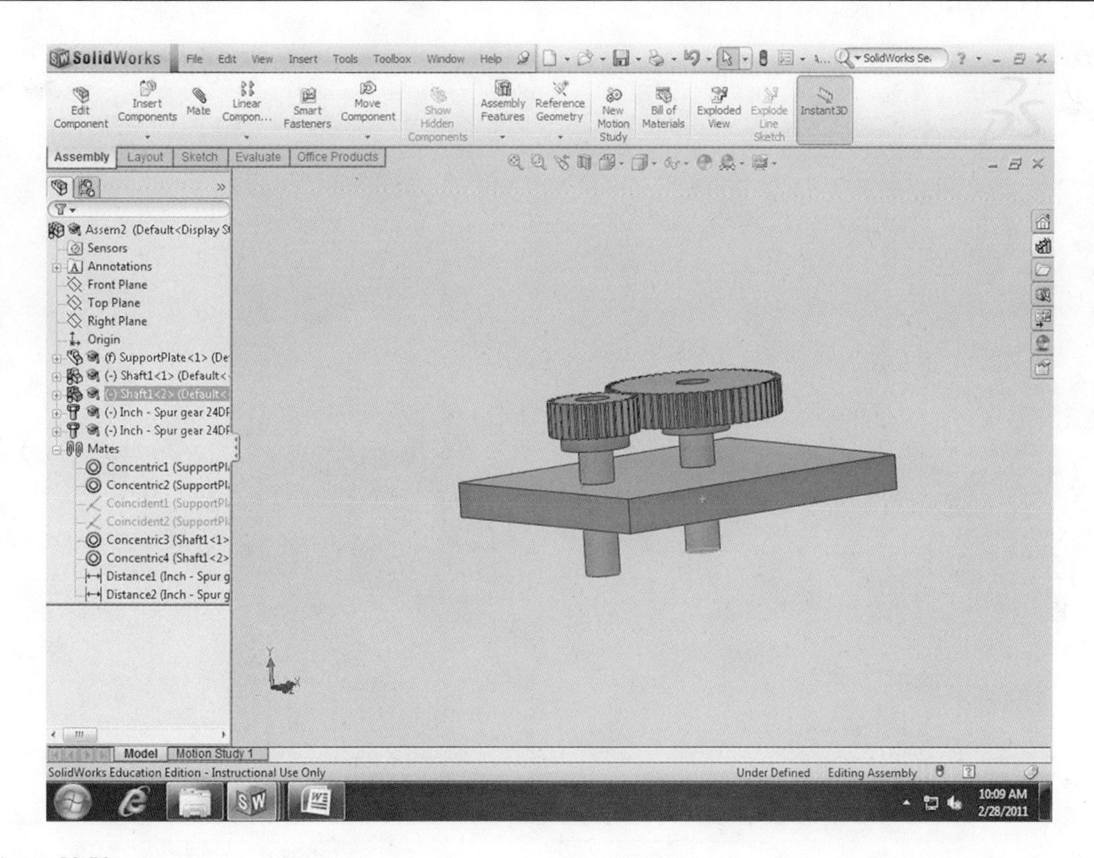

Figure 13.53

Suppress Coincident Mates for shafts and pull shafts.

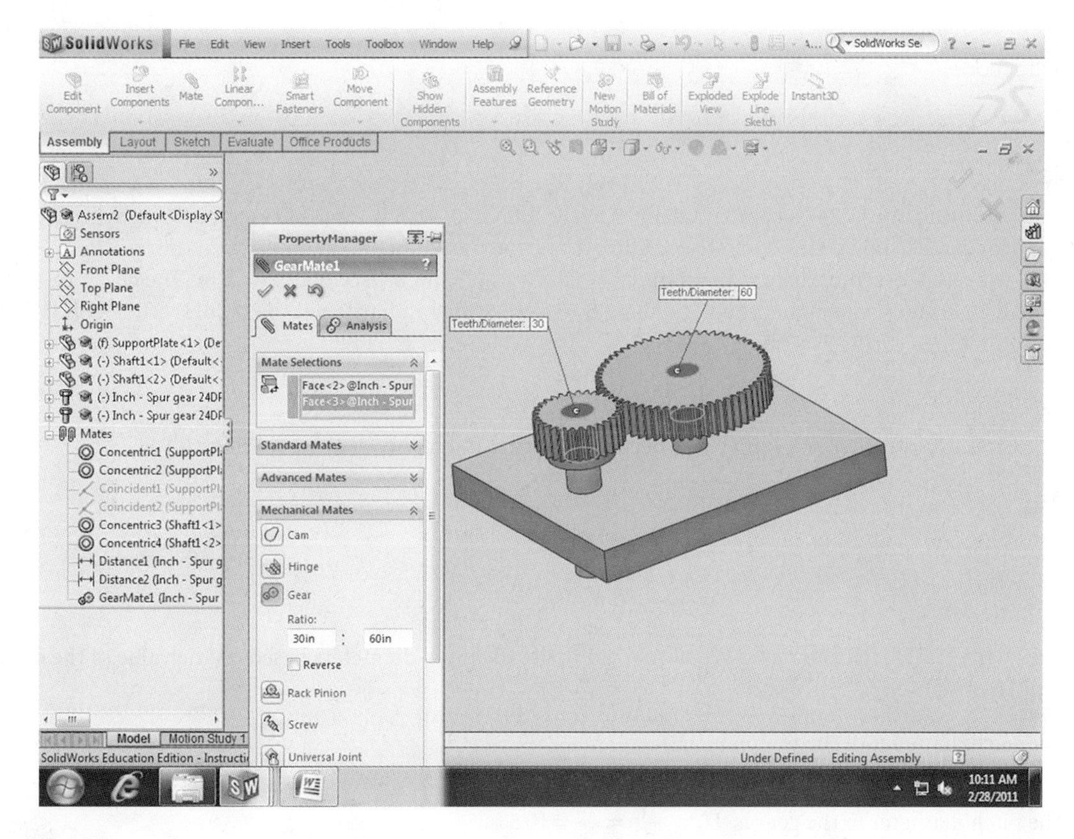

Figure 13.54

GearMate PropertyManager.

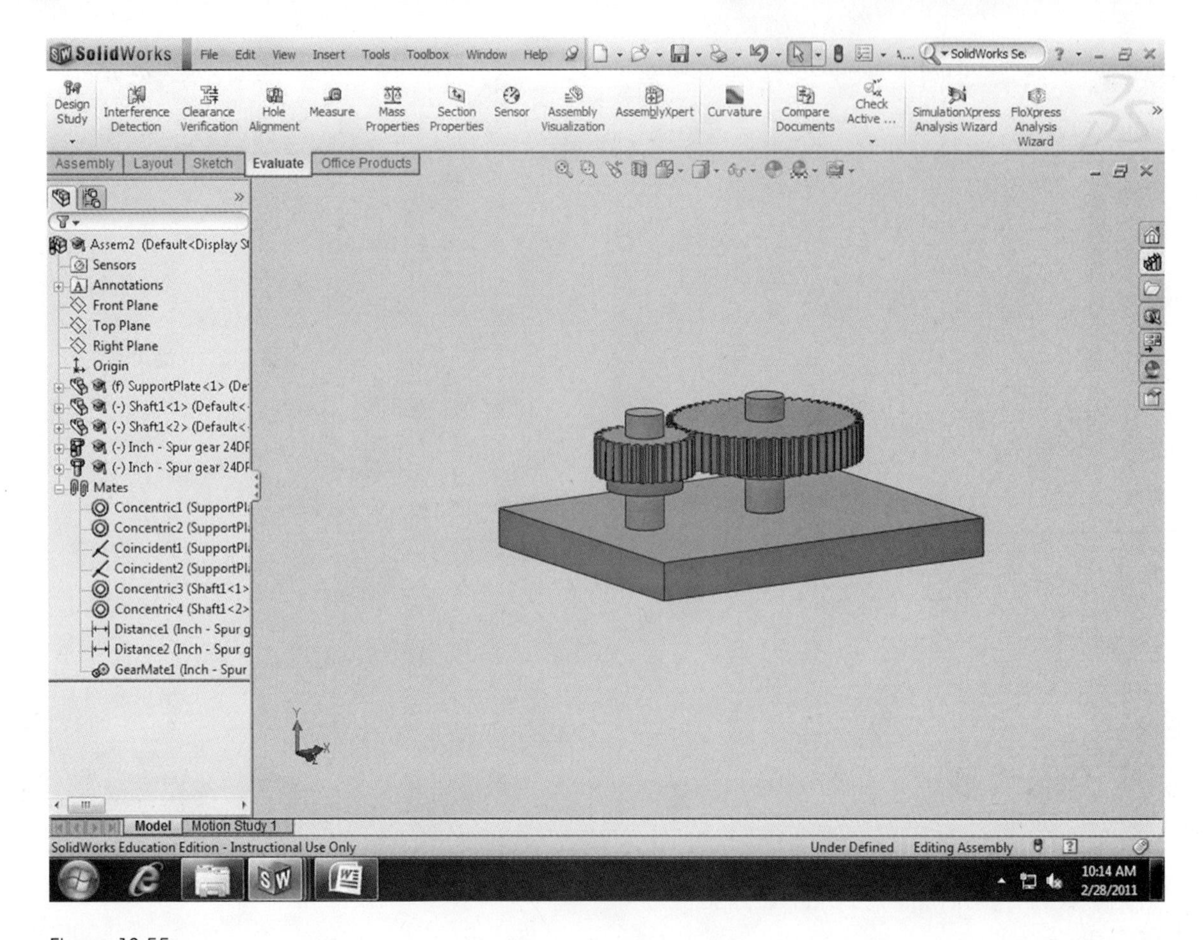

Figure 13.55

Gearing model.

Helical Gear Design

A motor running at n = 2400 rpm and delivering 122-kW (164-hp) power drives a machine by means of a helical gear set, as shown in Figure 13.59. The gears have the following geometric quantities: P_n = 5 in.$^{-1}$, φ = 20°, c (center-to-center distance) = 9 in., N_1 = 30, N_2 = 42, and b (face width) = 2 in. The gears are made from SAE 1045 steel, water-quenched and tempered (WQ&T), and hardened to 200 BHN. Use SolidWorks for designing the gearing system.

Initial Sizing

$P_n = P_d/\text{Cos}\ \psi$. (Note: there is a difference between Normal Pitch and Diametral Pitch.)

$$P_d = \frac{(N_{t1} + N_{t2})}{2c} = \frac{(30+72)}{2(9)} = 4$$

The Diametral Pitch becomes 4. We need to compute the pitch diameters based on the value of the diametral pitch of 4.

The pitch diameter of the pinion is $D_1 = \dfrac{N_{t1}}{P_d} = \dfrac{30}{4} = 7.5''$

The pitch diameter of the gear is $D_2 = \dfrac{N_{t2}}{P_d} = \dfrac{42}{4} = 10.5''$

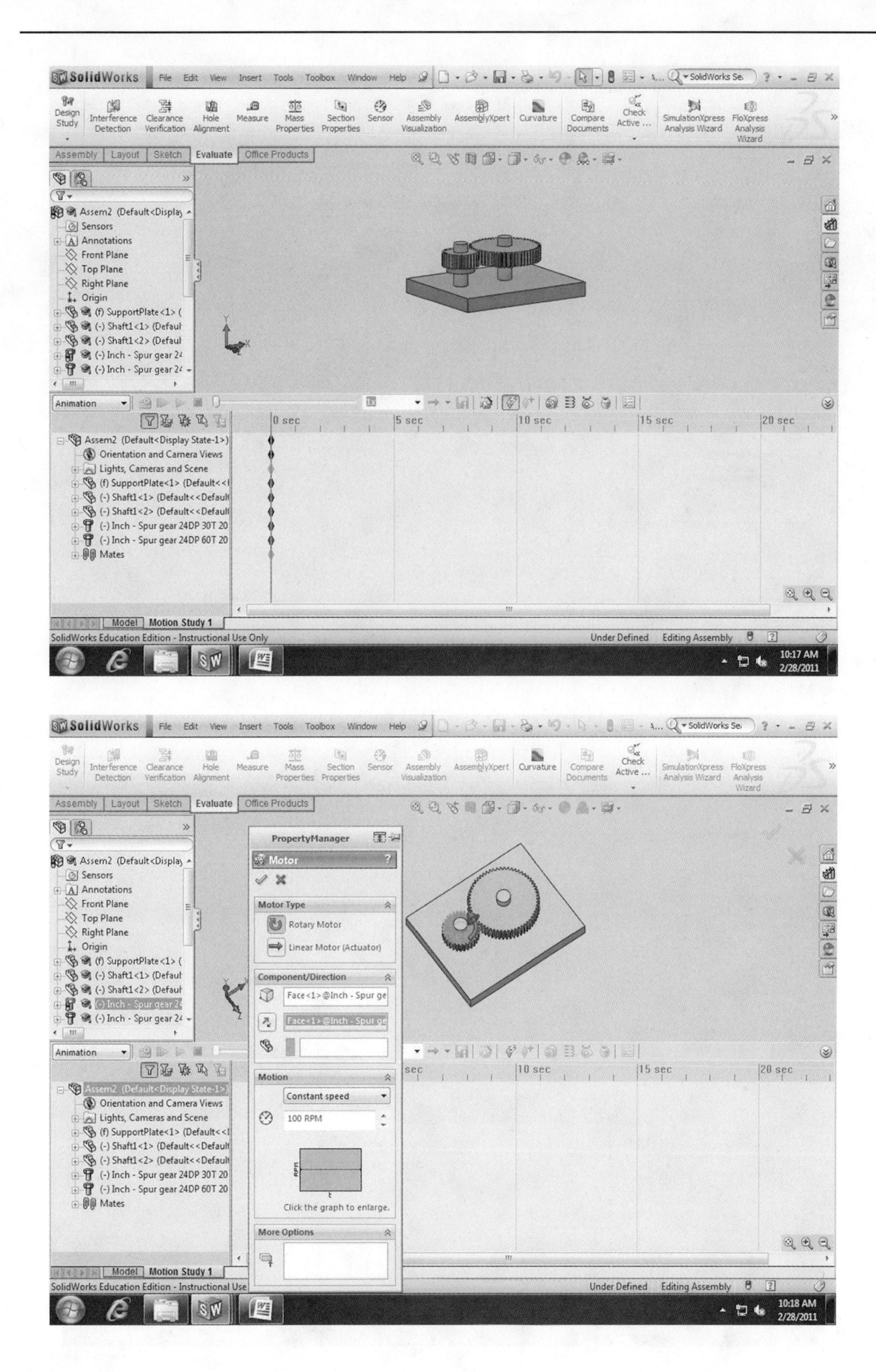

Figure 13.56

Select Type of Study.

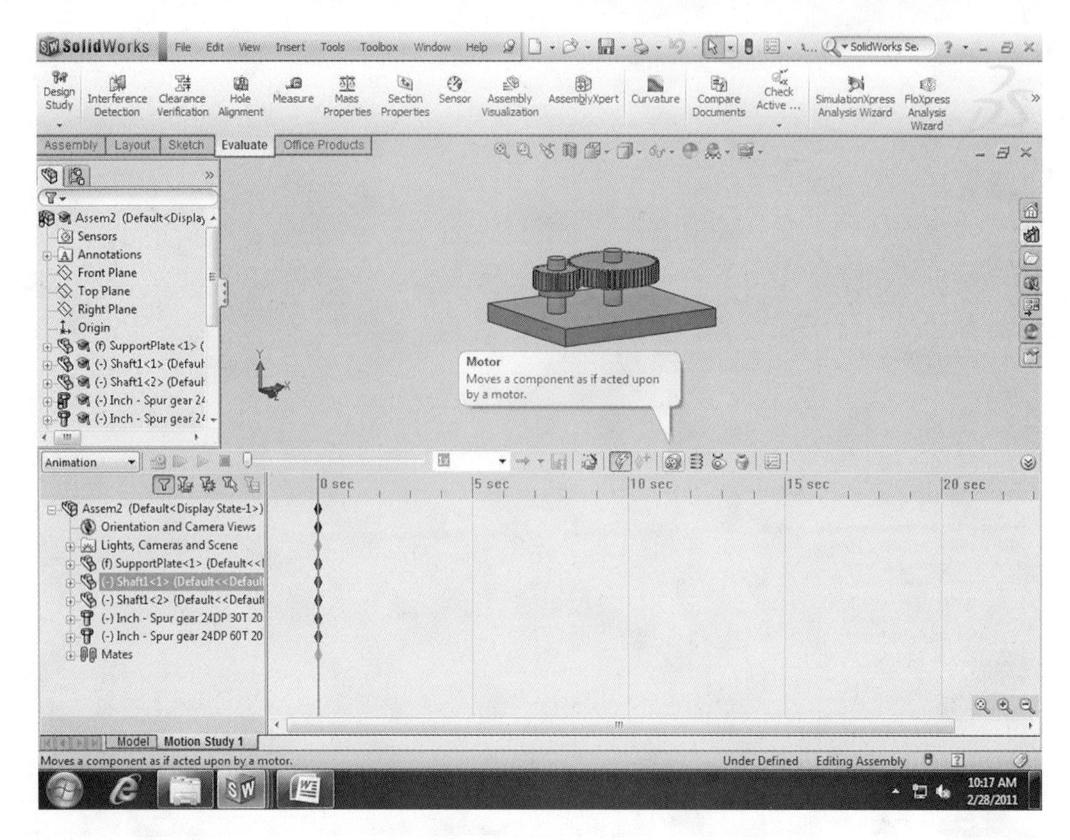

Figure 13.57

Define motor and apply to pinion: Motor tool.

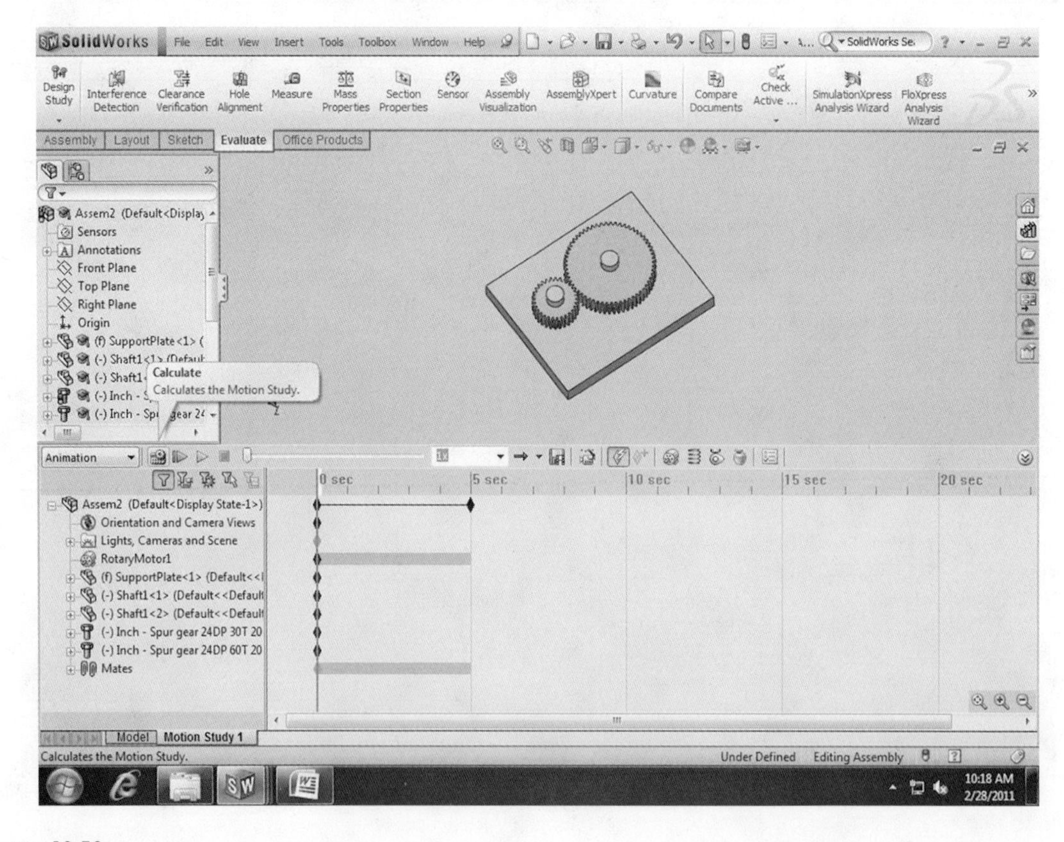

Figure 13.58

Calculate the Motion Study.

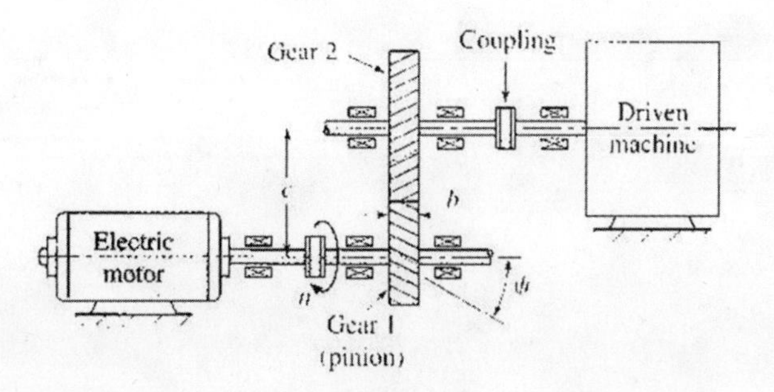

Figure 13.59

Helical gear set. (From A. Ugural, *Mechanical Design: An Integrated Approach*, McGraw Hill, 2004. With permission.)

The center-to-center distance is $c = \frac{(D_1 + D_2)}{2} = \frac{(7.5 + 10.5)}{2} = 9.0''$

Note that this value matches the given value in the problem definition. We need to determine the Helix Angle, given as $\text{Cos}\,\psi_1 = \frac{P_d}{P_n} = \frac{4}{5} = 0.8$

The arccosine of 0.8 is equal to 36.9-deg or 37-deg. But note that due to Virtual Values, our measured diameters from SolidWorks may be different. We will use Imperial units.

Support (Housing) Sizing

Length of support, $L = b + 2b_w + 2\varepsilon_1$
Width of support, $W = d_1 + d_2 + \varepsilon_w$
Thickness of support, $T = \varepsilon_T$

Pinion

1. Open a New SolidWorks Part document.
2. Click Design Library > Toolbox > ANSI Inch > Power Transmission > Gears.
3. Right-click Helical Gear (see Figure 13.60).
4. Select Create Part. (Note: this is the preferred route that is used in this book; do not drag and drop parts from the Design Library.) (See Figure 13.60.)

   The Helical Gear PropertyManager appears immediately (see Figure 13.61). Based on the design specification, fill in the properties, as shown in Figure 13.61.
5. Click OK when complete. (See Figure 13.62 for the pinion.)

Gear

Repeat as in pinion. (See Figure 13.63 for the PropertyManager and Figure 13.64 for the gear.)

Support

Model the support (see Figure 13.65).

Assembly Modeling

To assemble the parts, the following procedure is given as a guideline (other sequences could be used to arrive at an assembly solution):

1. Open a New SolidWorks Assembly document.
2. Insert the support as the first part (see Figure 13.66).
3. Insert the shaft twice since two are needed.

Concentric Mating for Shafts

4. Click *each shaft* and *each hole*, and apply Concentric Mate condition. (See Figure 13.67 for shafts that are mated to support.)

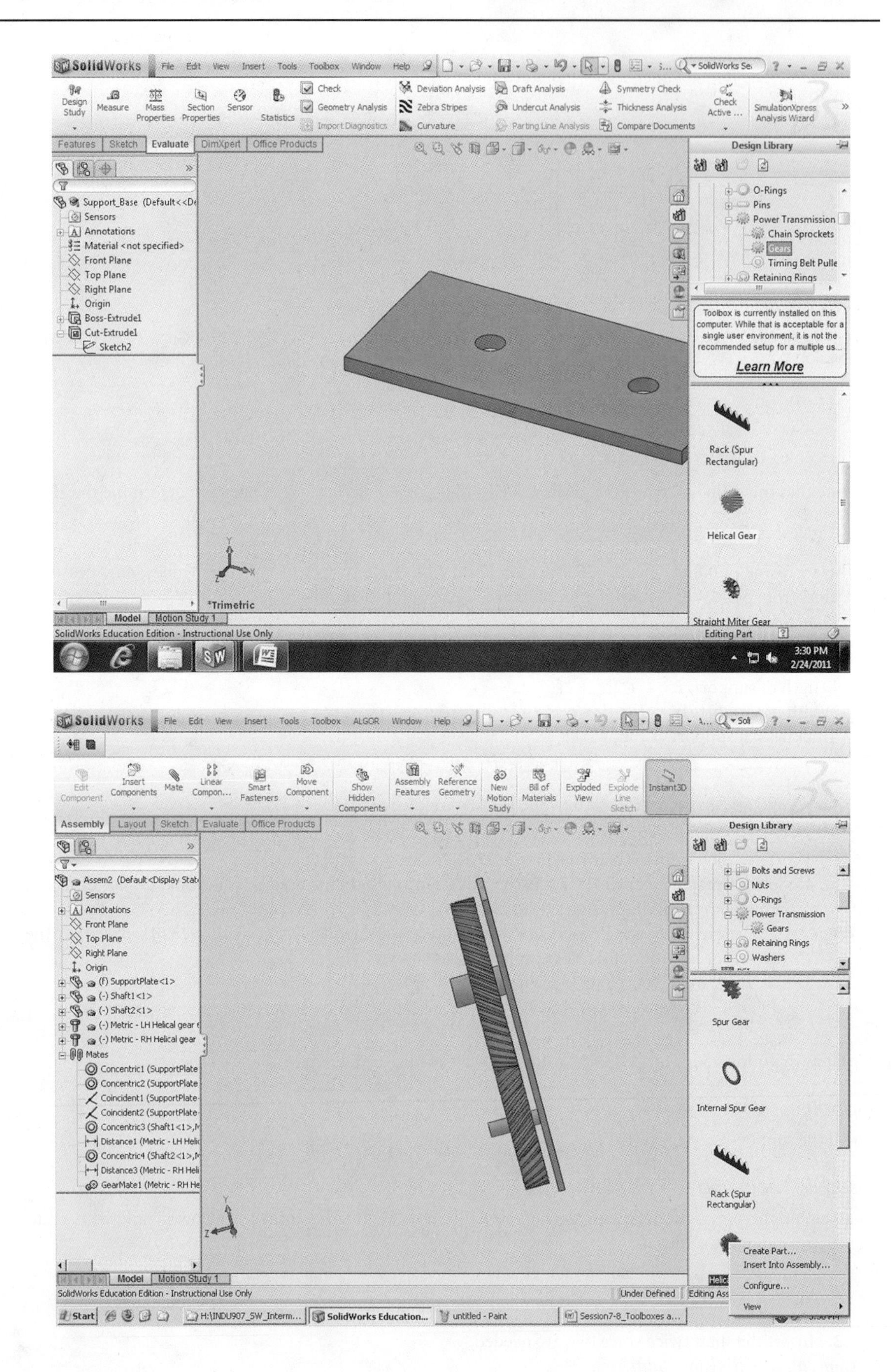

Figure 13.60

Creating Helical gears.

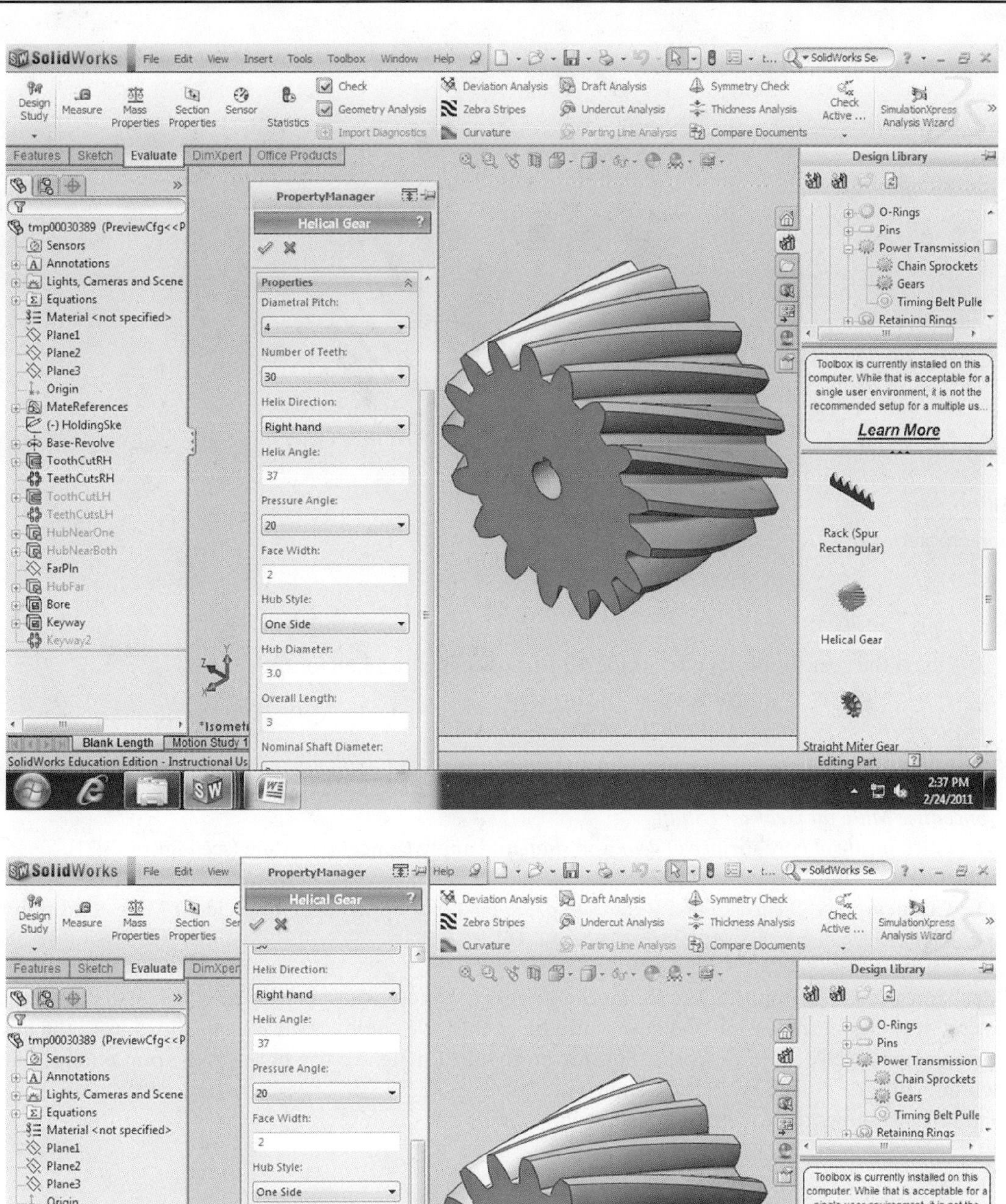

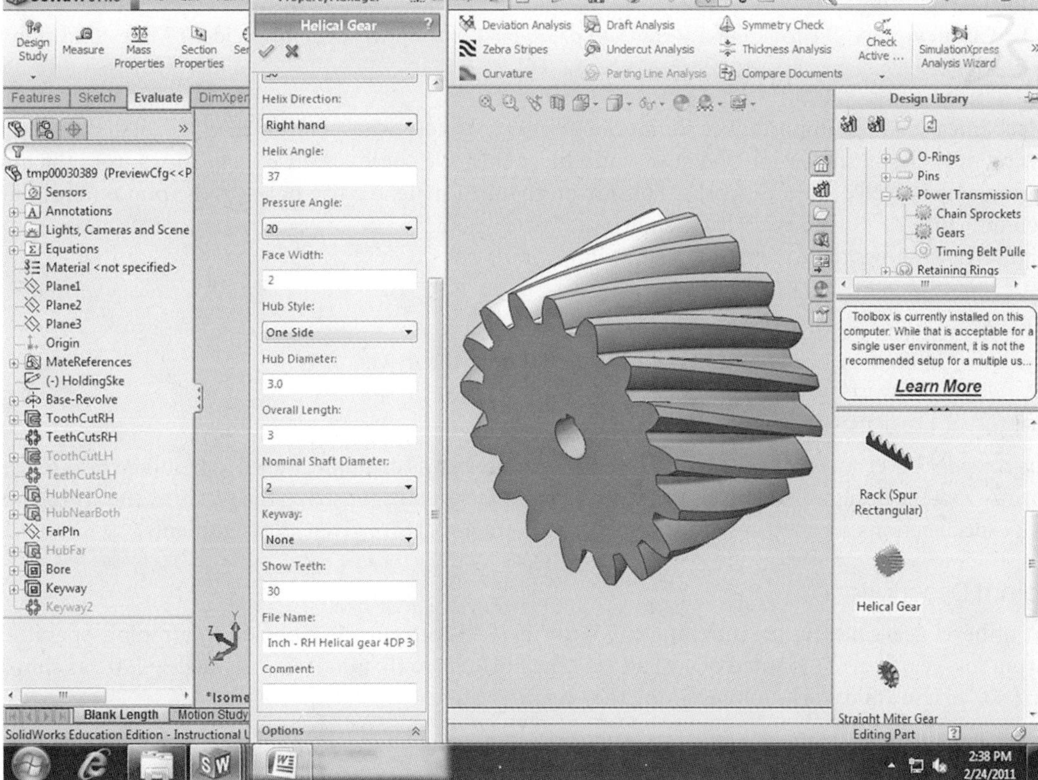

Figure 13.61

Helical gear PropertyManager.

Figure 13.62

Pinion created.

Insert Pinion

1. Insert the gear (or pinion) as the fourth part. (See Figure 13.68 for the pinion.)

Concentric Mate for Pinion and Shaft

1. Click *inside of bore surface of pinion* and *shaft* and apply Concentric Mate (see Figure 13.69).

Insert Pinion

1. Insert the gear (or pinion) as the fifth part. (See Figure 13.70 for the pinion.)

Concentric Mate for Gear and Shaft

1. Click *inside of bore surface of gear* and *shaft* and apply Concentric Mate (see Figure 13.71).

Coincident Mating for Shafts

1. Click *end face of each shaft* and *bottom of support plate*, and apply Coincident Mate condition (see Figure 13.72).

Mechanical mate is important for pinion and gear to rotate when one of them is rotated. Generally, choosing the holes in the pinion and gear will achieve this. We could also use the features from the pinion and gear, respectively; this still works. For our mechanical mate, we use holes in the pinion and gear, as shown in Figure 13.73. The assembly is shown in Figure 13.74.

Figure 13.75 shows the AssemblyManager with details.

Animation

See the steps in Figures 13.76 and 13.77. Pinion is chosen as the driver.

Bevel Gear Design

Let us consider bevel gears as another class of applications. In this example, we will use SolidWorks to design the gears from the design specifications that are given. Other details for sizing the gearbox, checking that the AGMA specifications are met, are not shown. These extra details are given as assignments.

Problem Description

A set of bevel gears having pressure angle $\varphi = 20°$ is used to transmit 20 hp from a pinion operating at 500 rpm to a gear that is mounted on a shaft rotating at 200 rpm that inserts at an angle of 90°, as shown in Figure 13.78. Use SolidWorks for designing the gearing system.

SolidWorks Solution

Pinion

1. Open a New SolidWorks Part document.
2. Click Design Library > Toolbox > ANSI Metric > Power Transmission > Gears.

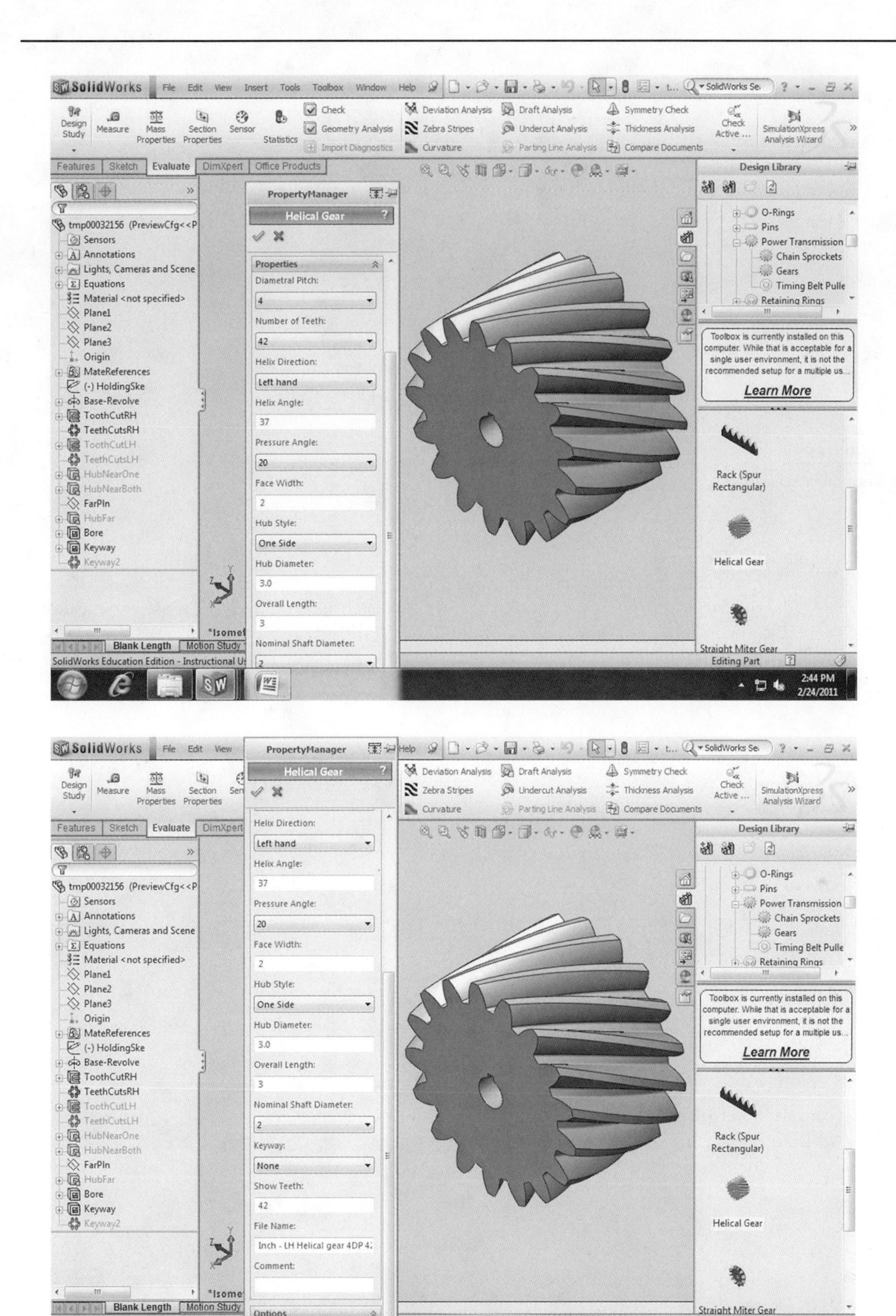

Figure 13.63

Helical gear PropertyManager.

Figure 13.64

Gear created.

3. Right-click Helical Gear.
4. Select Create Part. (Note: this is the preferred route.)

   The Helical Gear PropertyManager appears immediately (see Figure 13.79). Based on the design specification, fill in the properties, as shown in Figure 13.79.
5. Click OK when complete. (See Figure 13.80 for the pinion.)

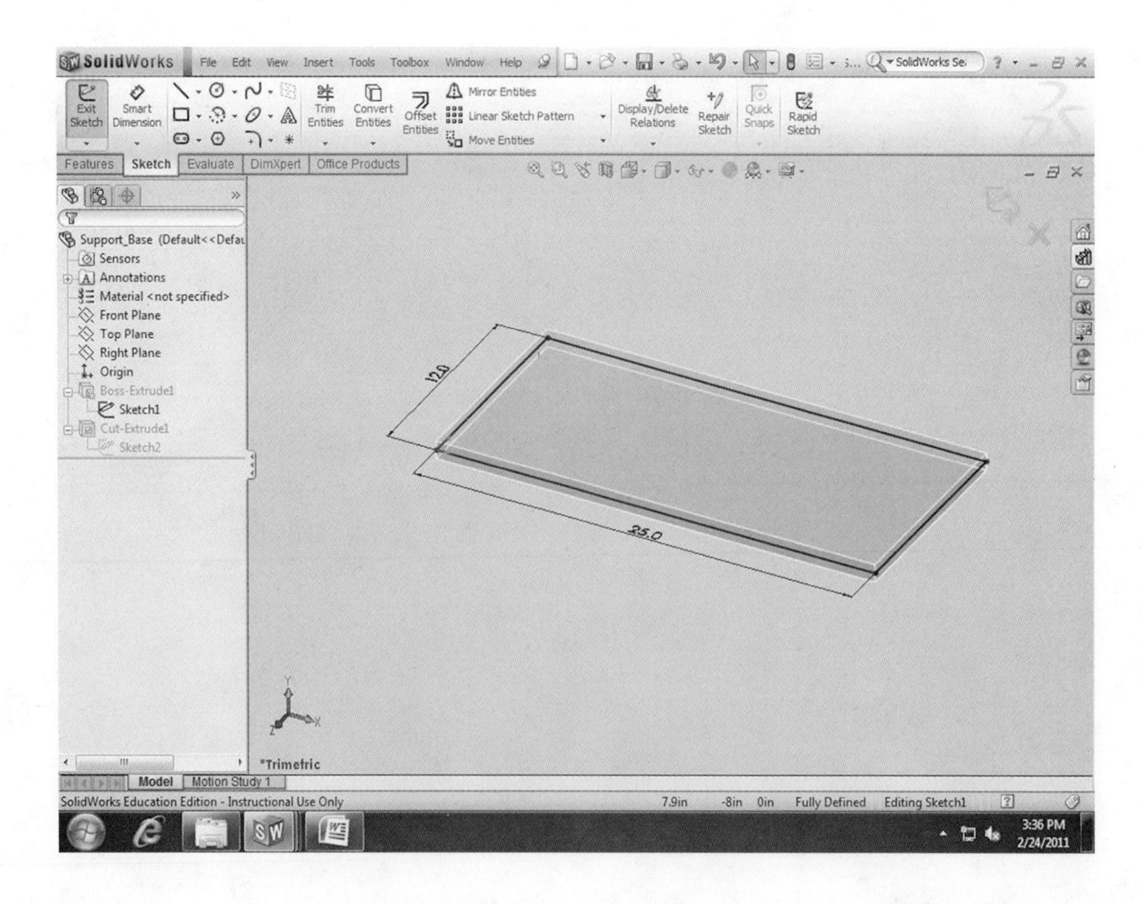

Figure 13.65

Support part.

(Continued)

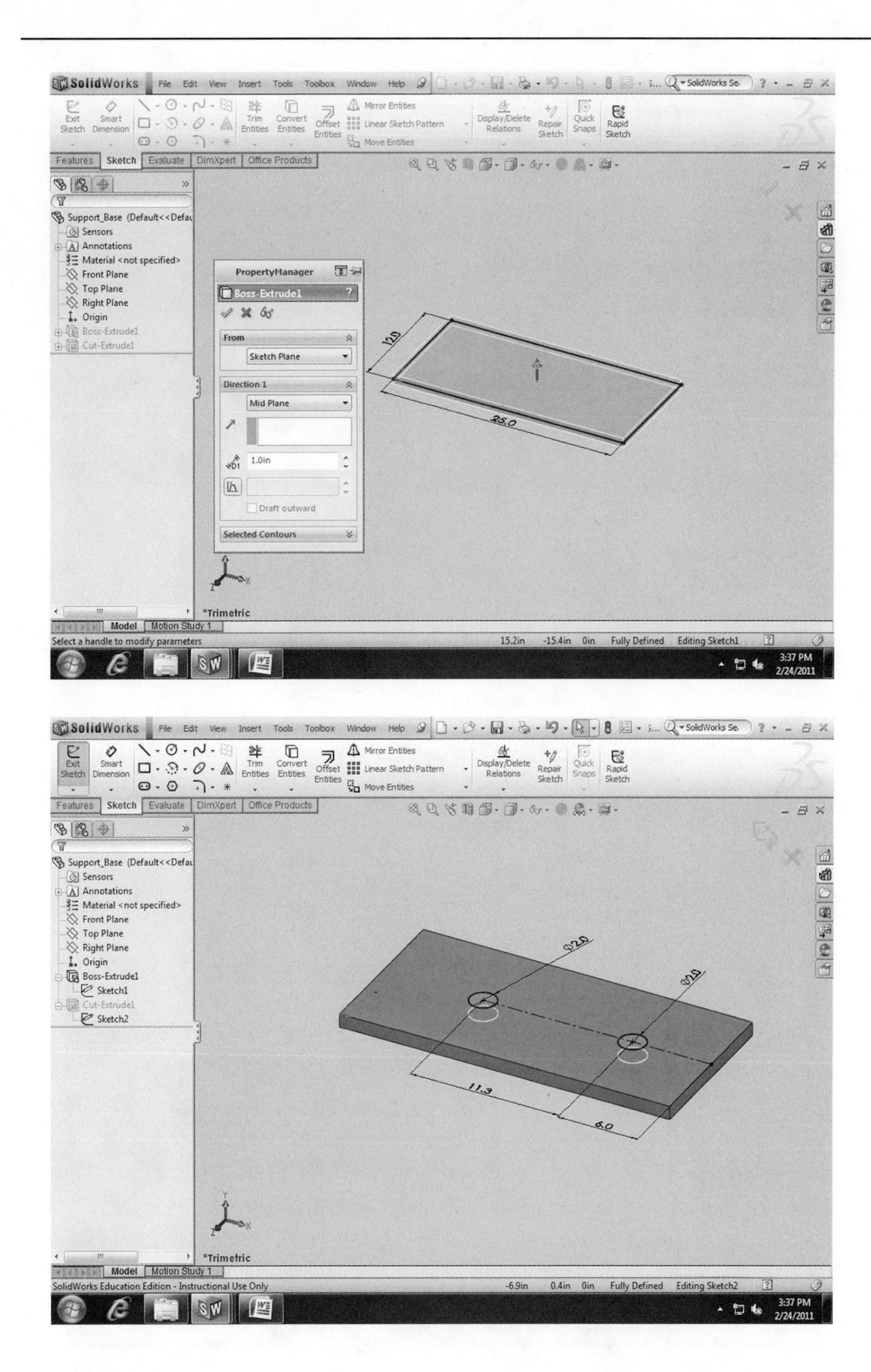

Figure 13.65 (Continued)

Support part.

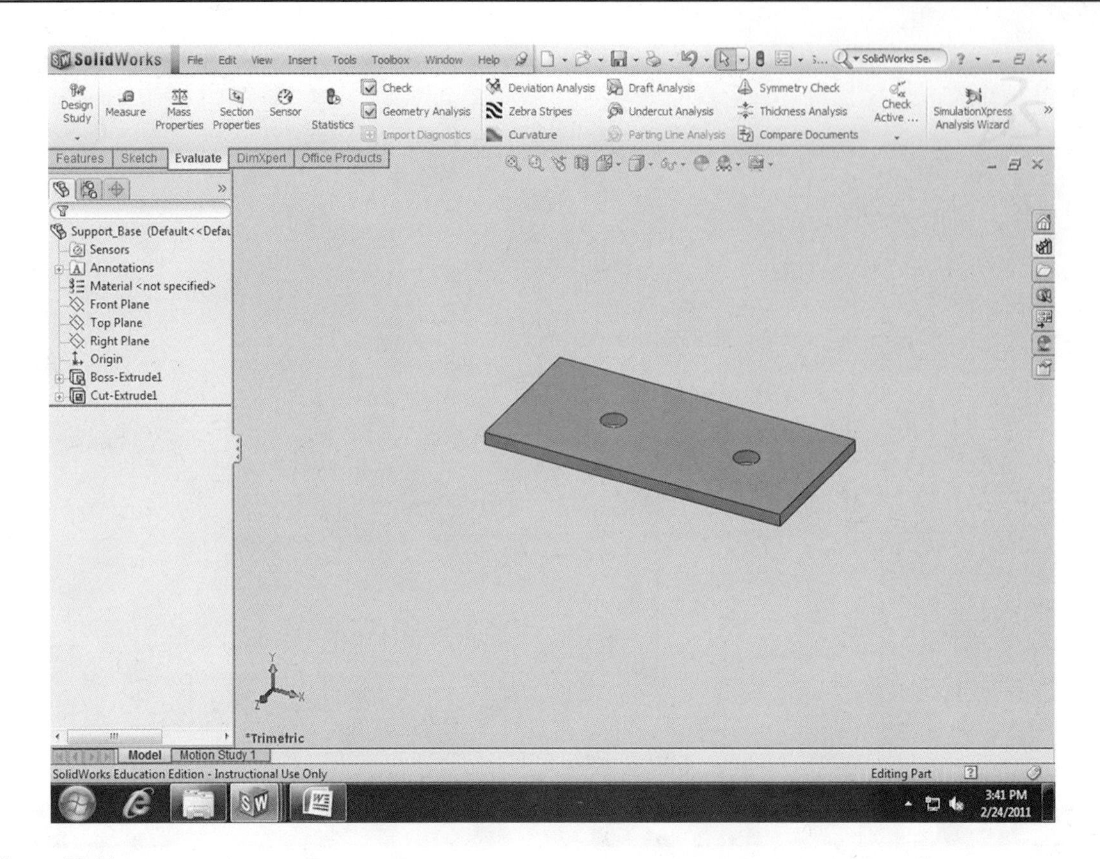

Figure 13.66

Support is the first part inserted.

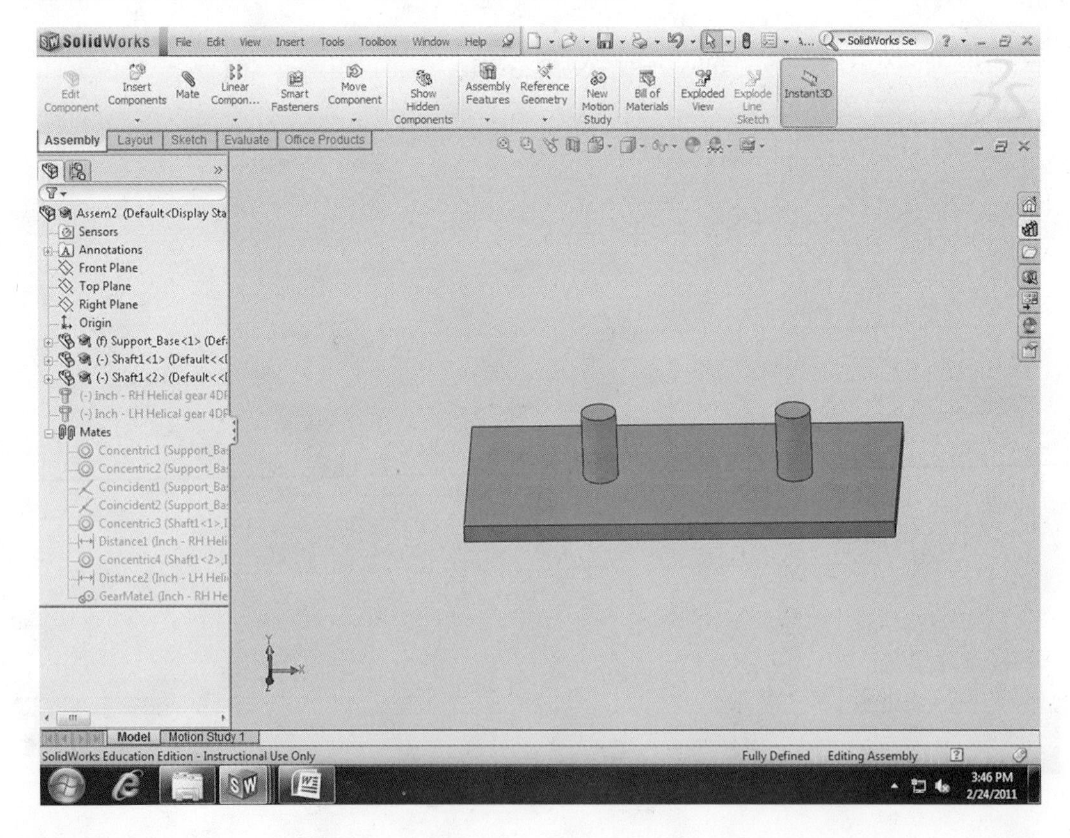

Figure 13.67

Two shafts inserted to the support.

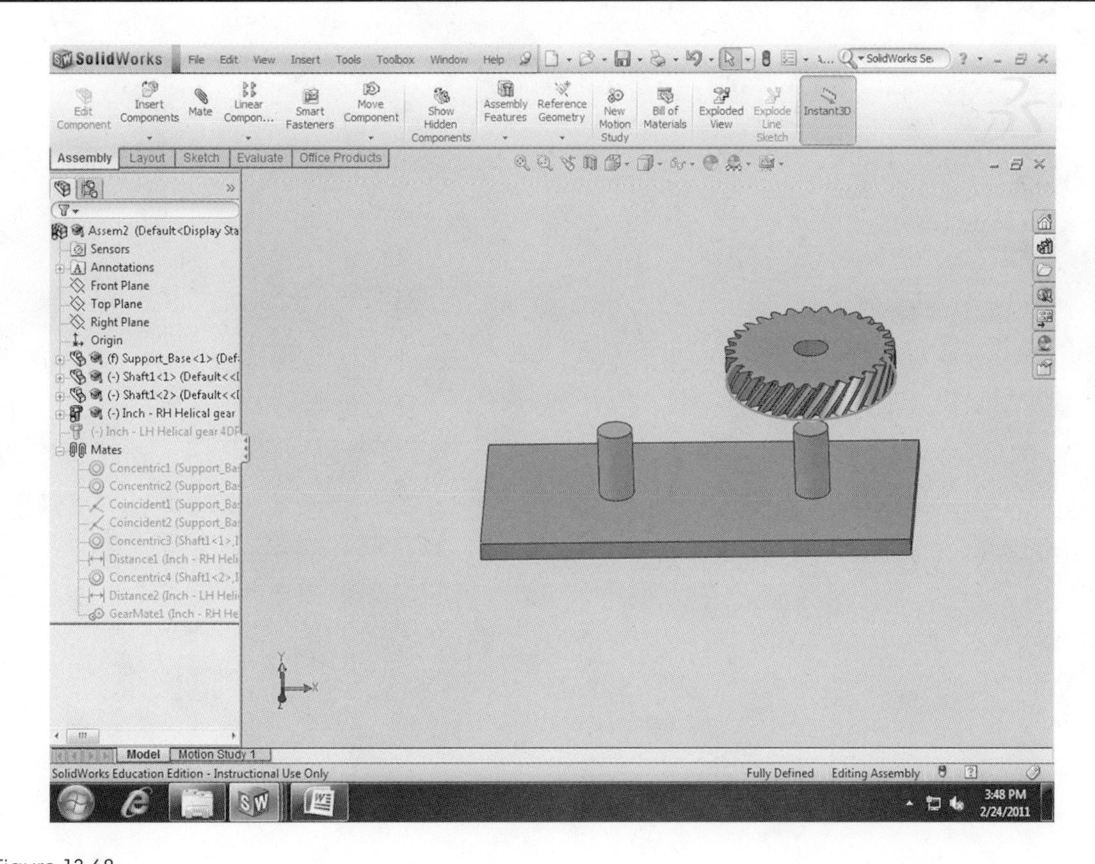

Figure 13.68

Pinion is inserted.

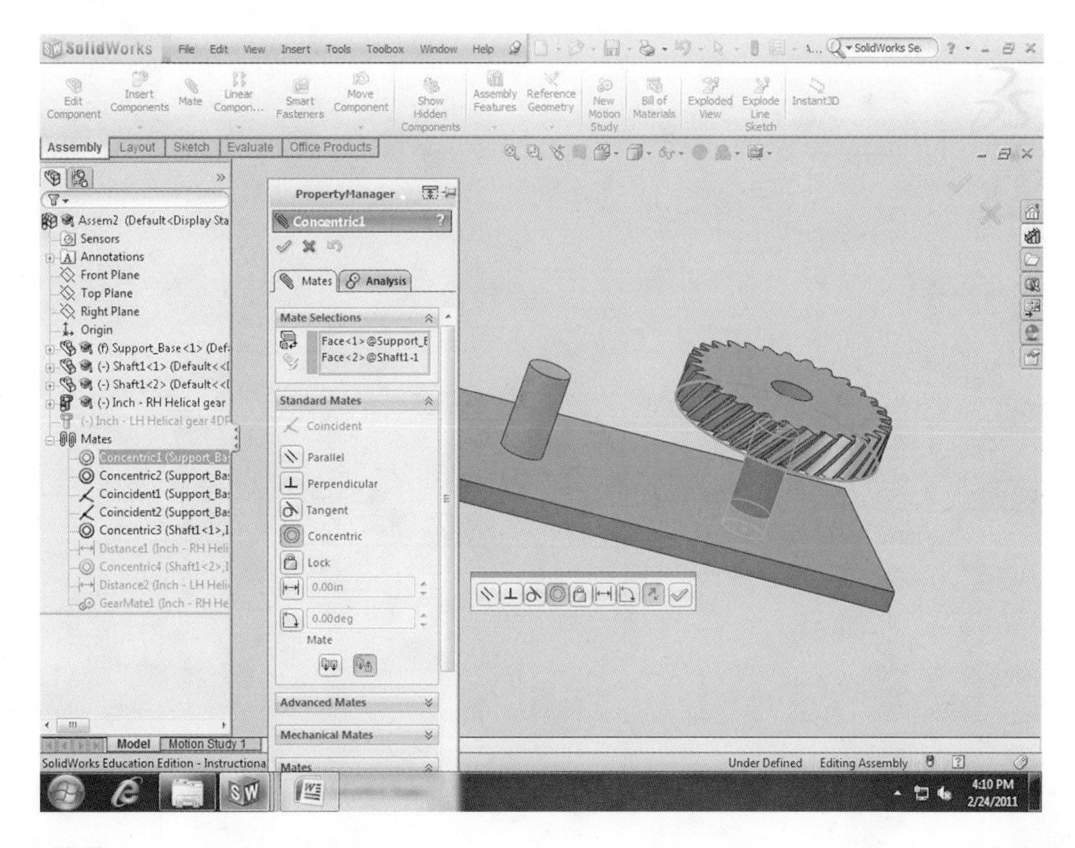

Figure 13.69

Hole in pinion and shaft are made concentric.

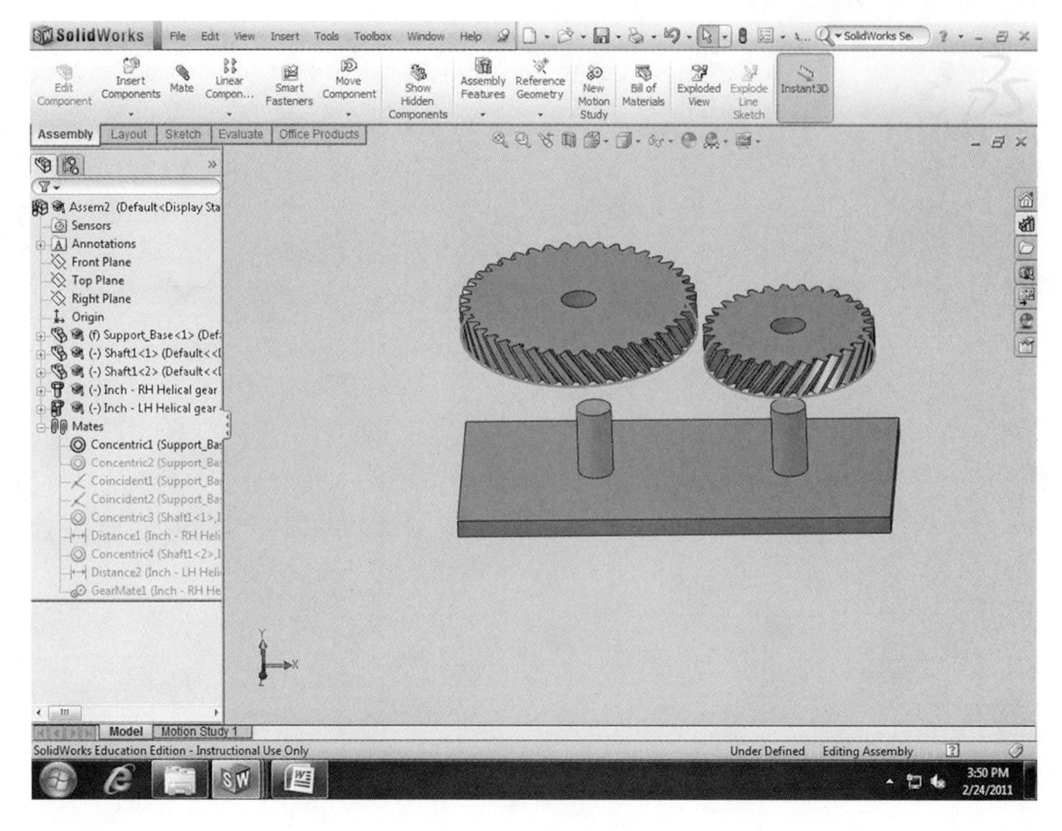

Figure 13.70

Gear is inserted.

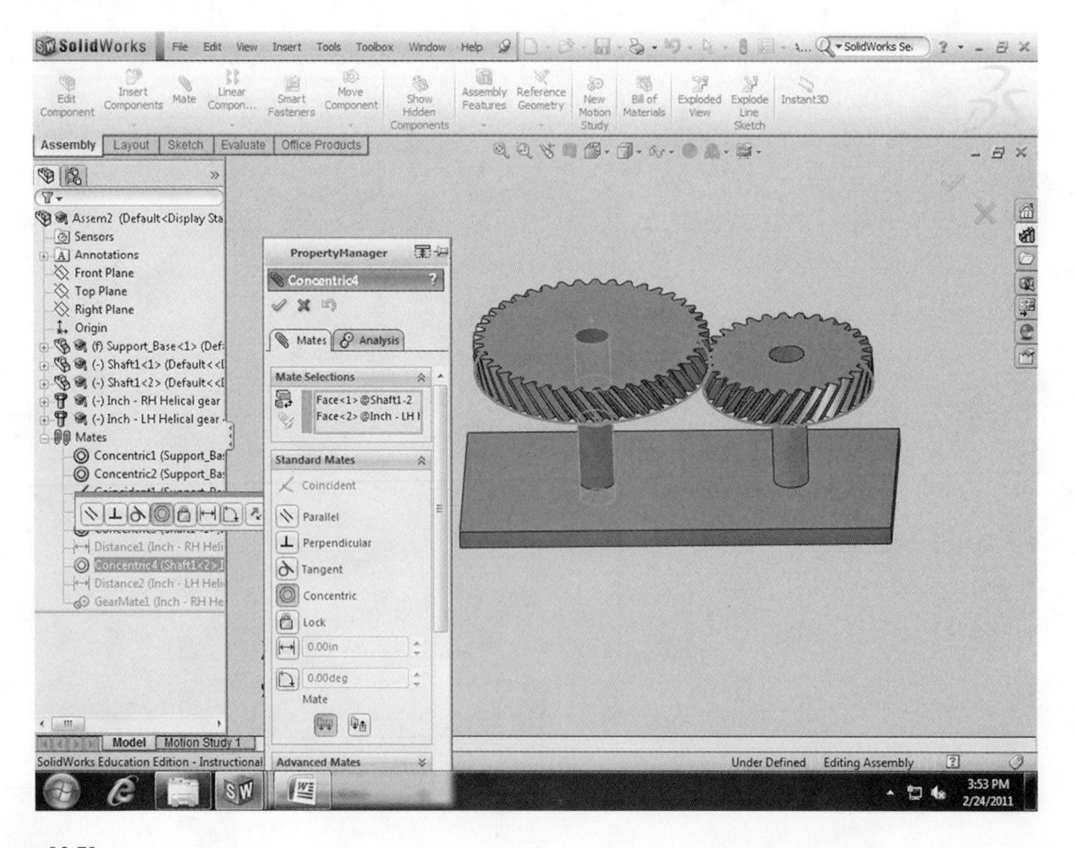

Figure 13.71

Hole in gear and shaft are made concentric.

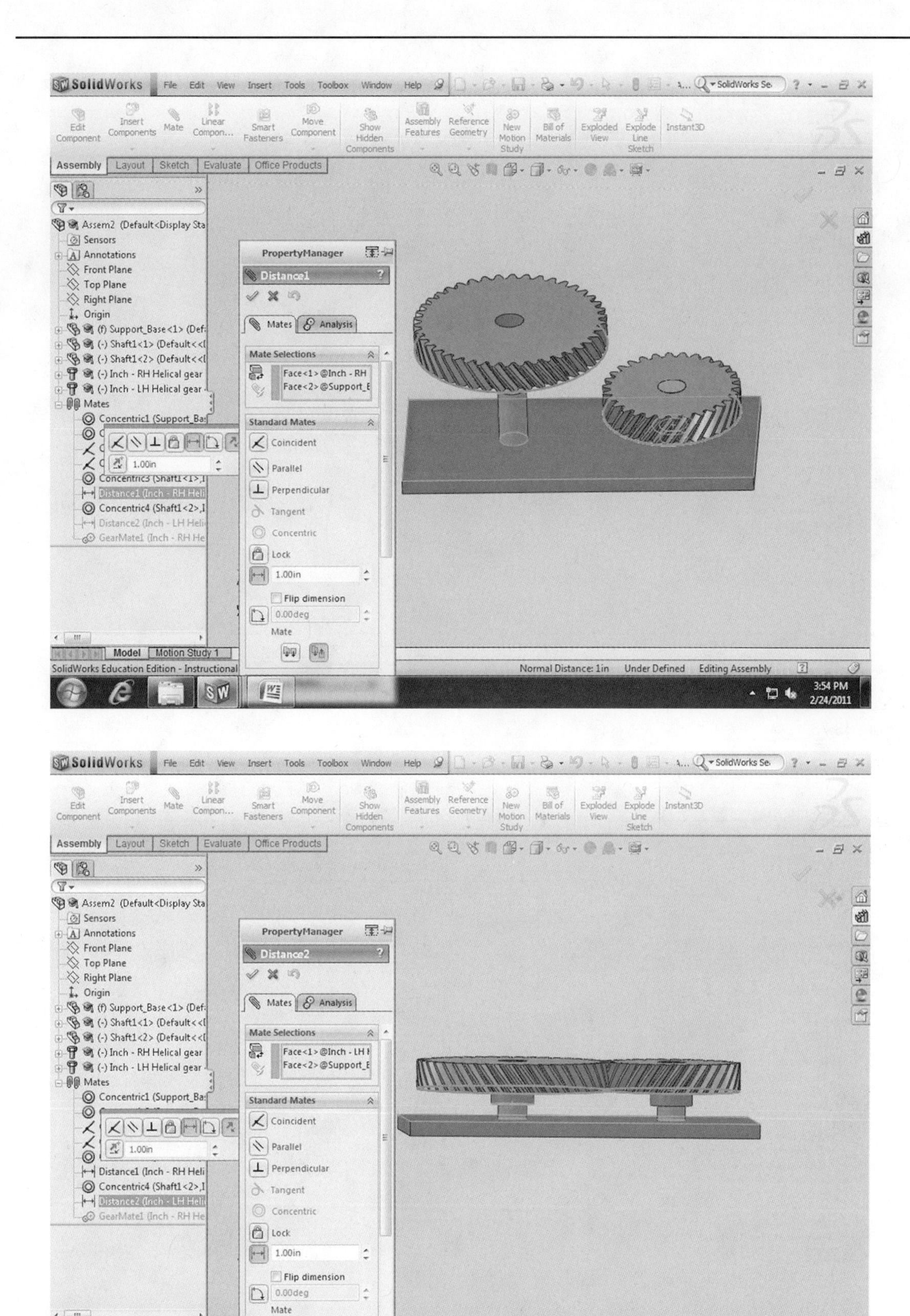

Figure 13.72

Fix distance between the top face of support and bottom face of pinion/gear.

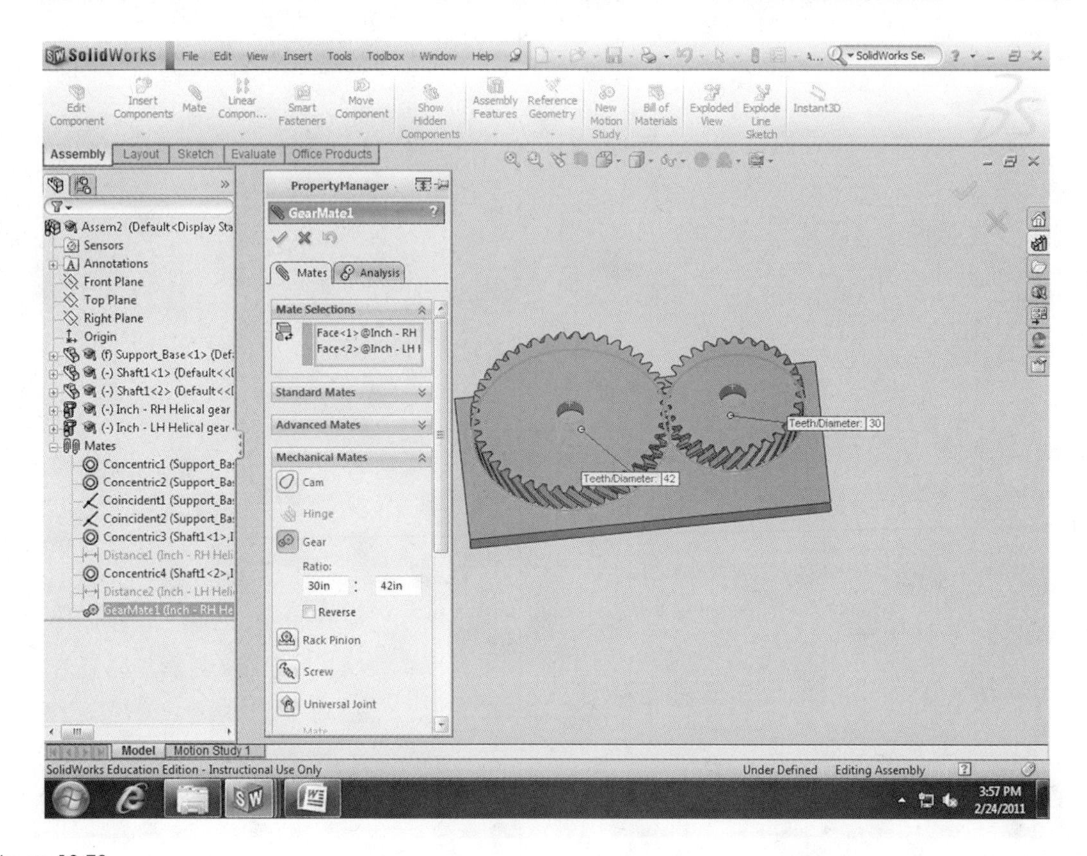

Figure 13.73

Mechanical Mate.

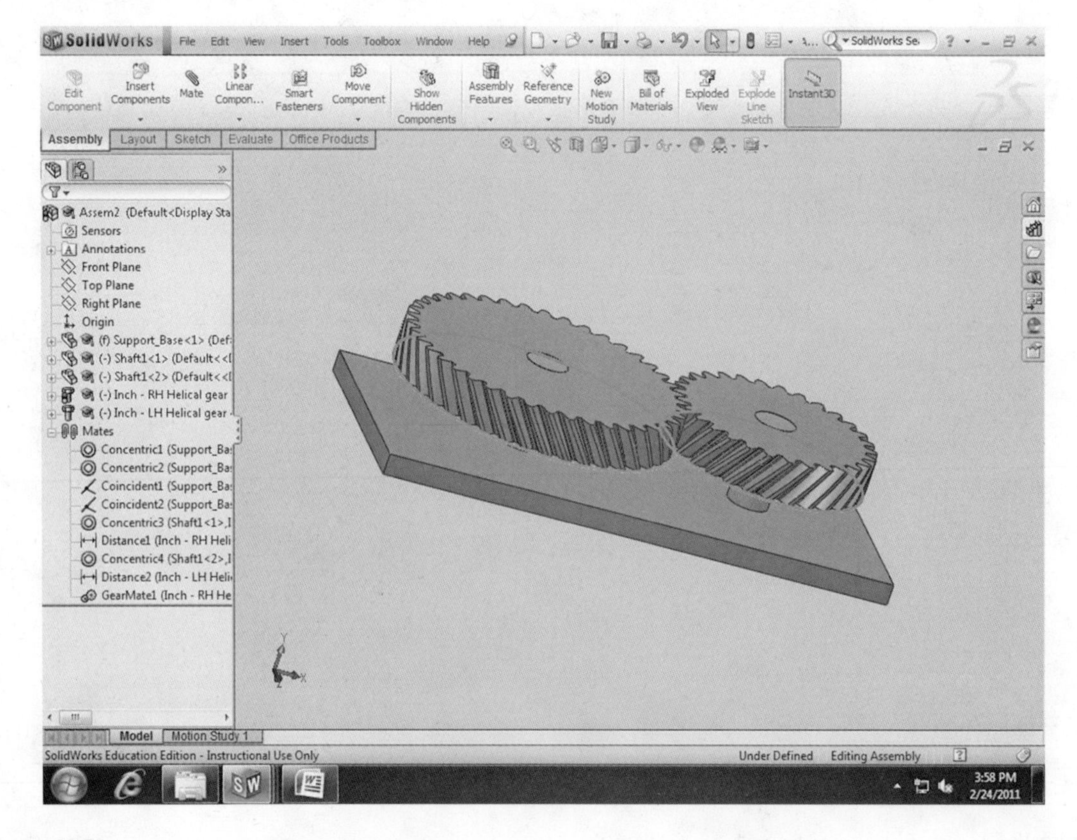

Figure 13.74

Assembly of pinion, gear, shafts and support.

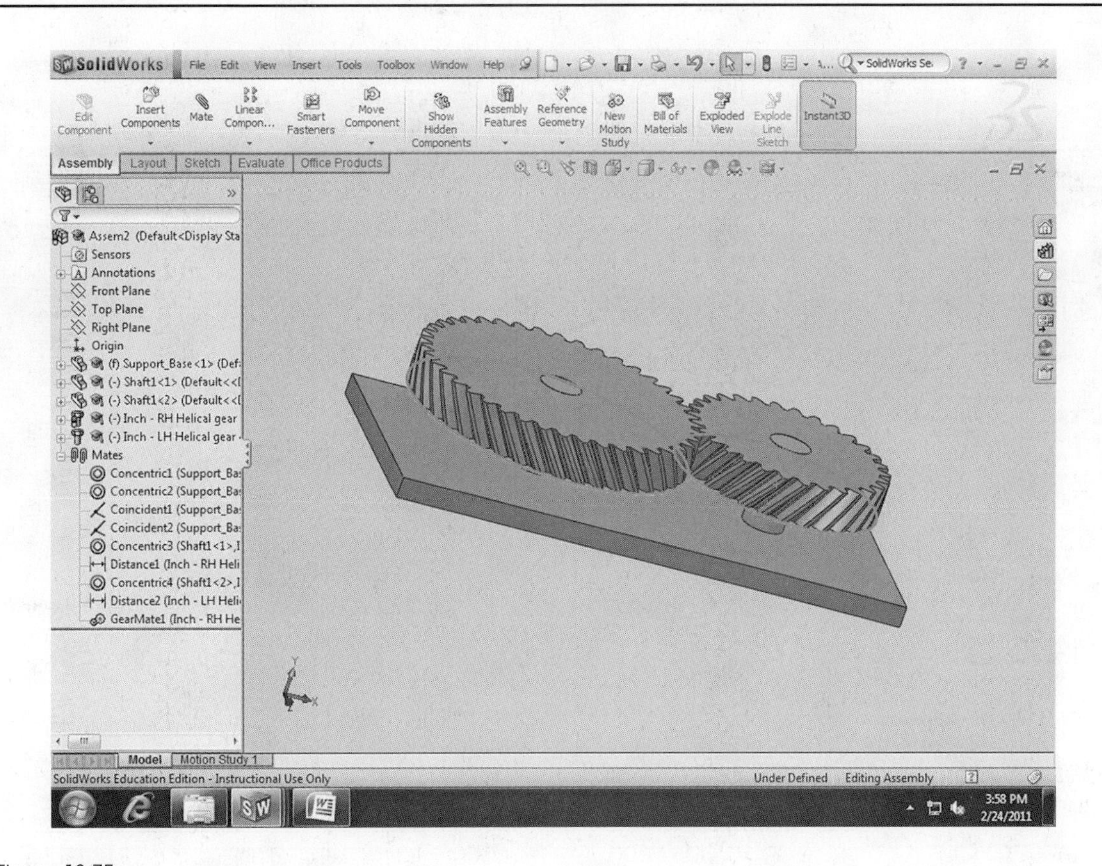

Figure 13.75

AssemblyManager showing details.

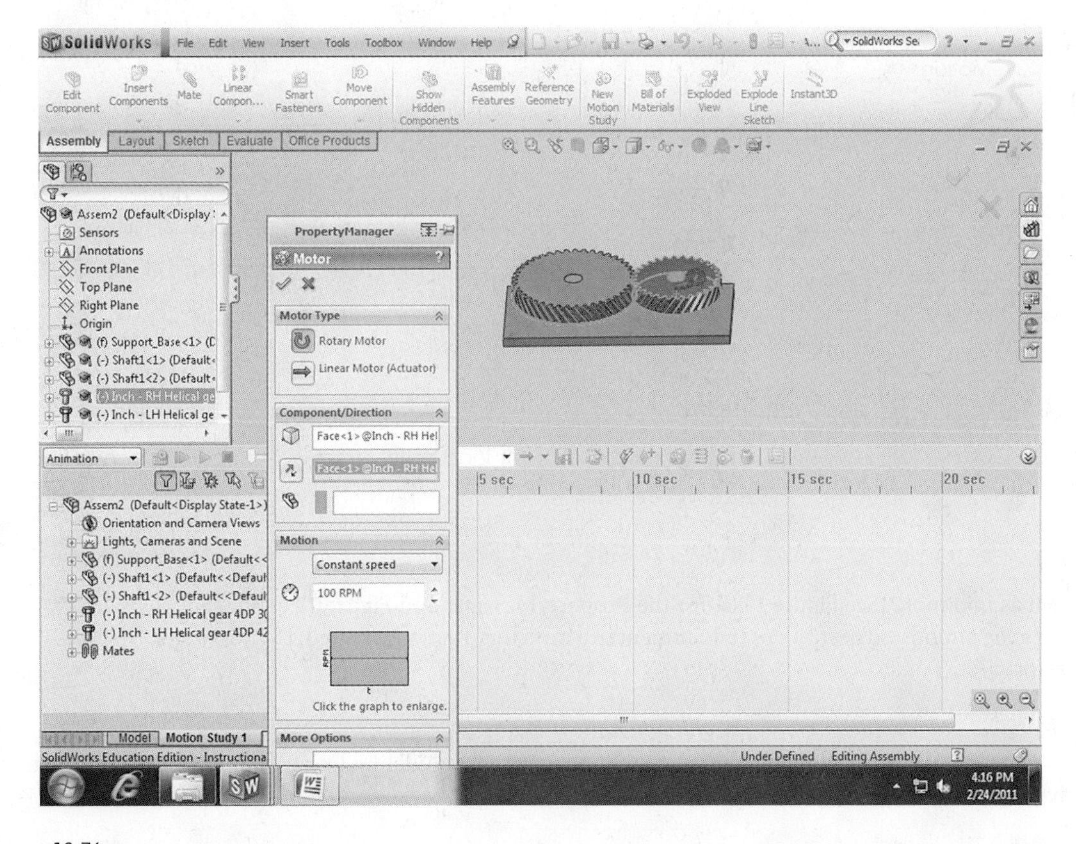

Figure 13.76

Pinion is chosen as driver.

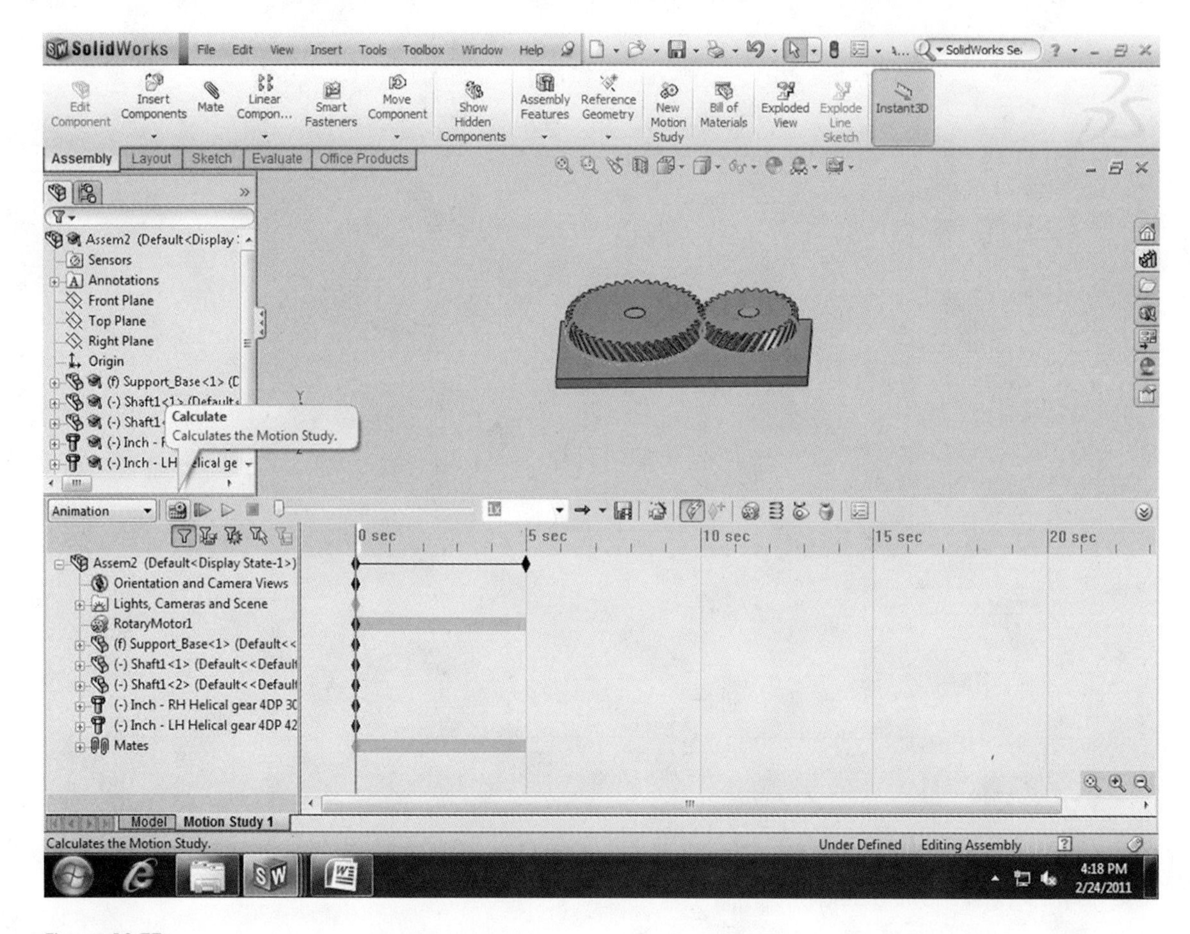

Figure 13.77

Calculate Motion.

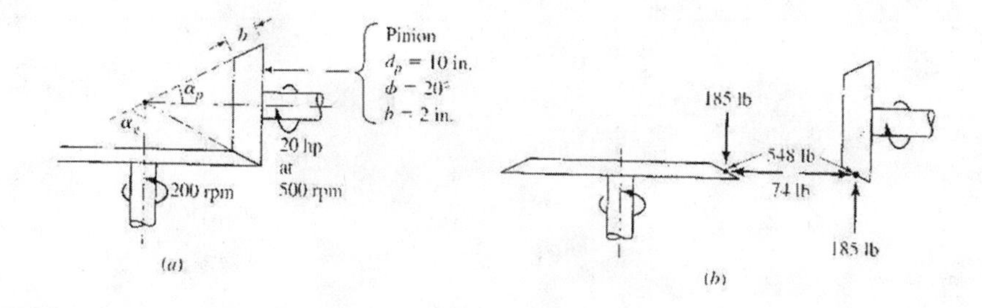

Figure 13.78

Helical gear set: (a) geometric and motion data and (b) load data. (From A. Ugural, *Mechanical Design: An Integrated Approach*, McGraw Hill, 2004. With permission.)

Gear

Repeat as in pinion. (See Figure 13.81 for the PropertyManager and Figure 13.82 for the gear.)

Save the pinion and gear as a part document in a directory that you created. Do not save in the SolidWorks directory.

Assembly

To assemble the parts, the following procedure is given as a guideline (other sequences could be used to arrive at an assembly solution):

1. Open a New SolidWorks Assembly document.
2. Insert a shaft for the pinion (or gear) as the first part (see Figure 13.83).

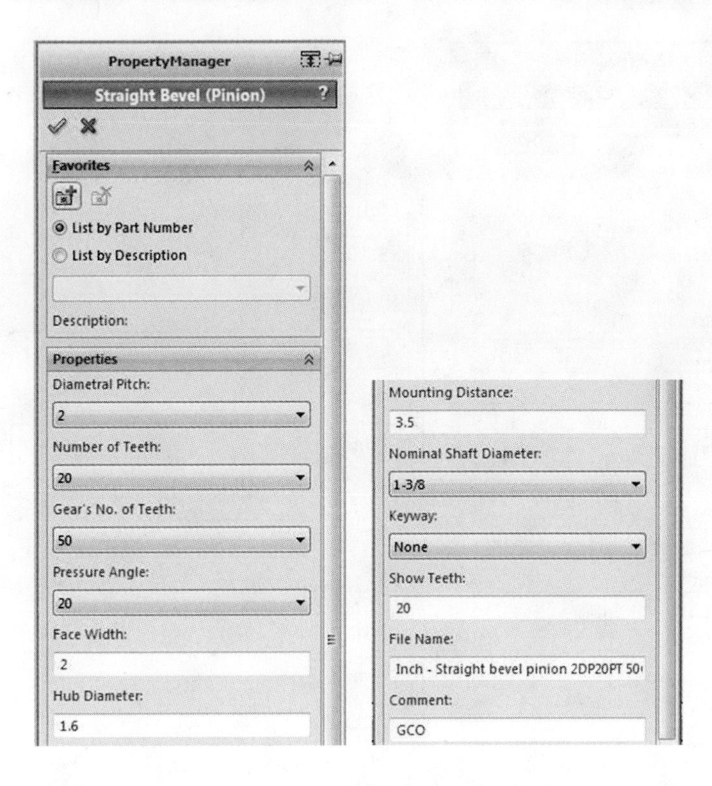

Figure 13.79

Bevel gear PropertyManager for the pinion.

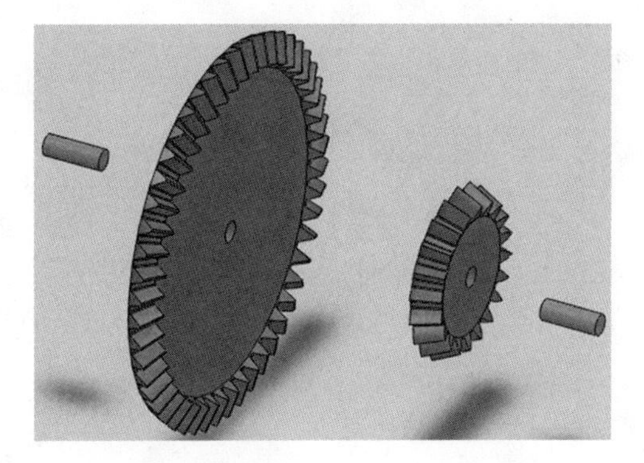

Figure 13.80

Pinion created by SolidWorks Toolbox.

3. Insert the pinion (or gear) as the second part (see Figure 13.83).
4. Insert a shaft for the gear (or pinion) as the third part (see Figure 13.83).
5. Insert the gear (or pinion) as the fourth part (see Figure 13.83).

Mating

6. Mate the shaft and pinion hole as being Concentric (see Figure 13.84).
7. Mate the shaft and the gear hole as being Concentric (see Figure 13.85).
8. Mate the two shafts at 90° to each other. (Alternatively, use the back faces of the gears for mating; check how they are oriented.) (See Figure 13.86.)
9. Mechanical mate is added. (Use the holes in pinion and gear or other features from pinion and gear.) (See Figure 13.87.)
10. Mate end of shaft and face of pinion face/gear face (see Figures 13.88 and 13.89).
11. Mate teeth-to-teeth clearance using distance (see Figure 13.90).

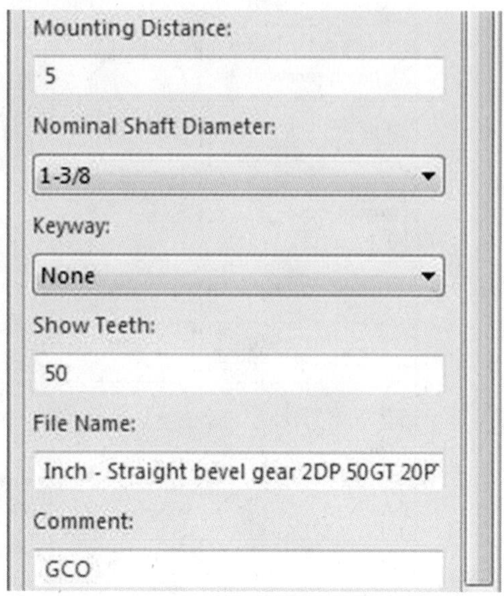

Figure 13.81

Bevel gear PropertyManager for the gear.

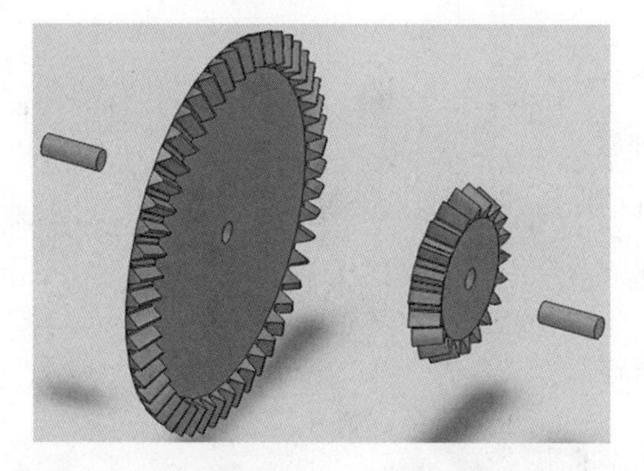

Figure 13.82

Gear created by SolidWorks Toolbox.

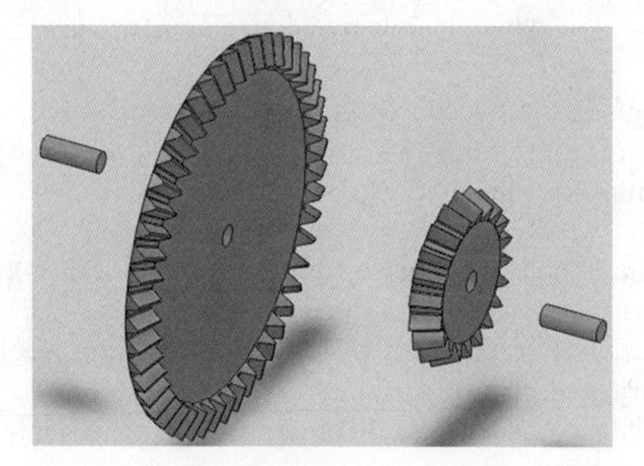

Figure 13.83

Pinion, gear, shaft (twice).

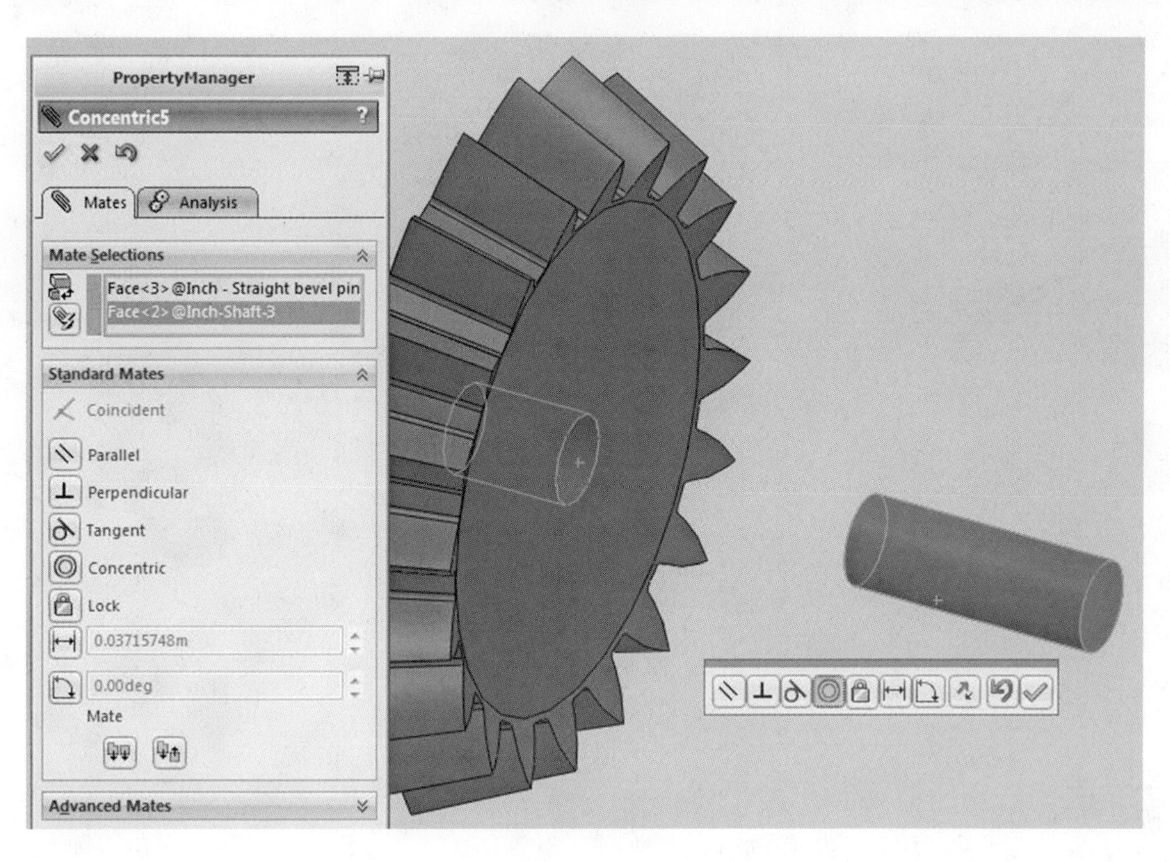

Figure 13.84

Shaft and pinion hole mating.

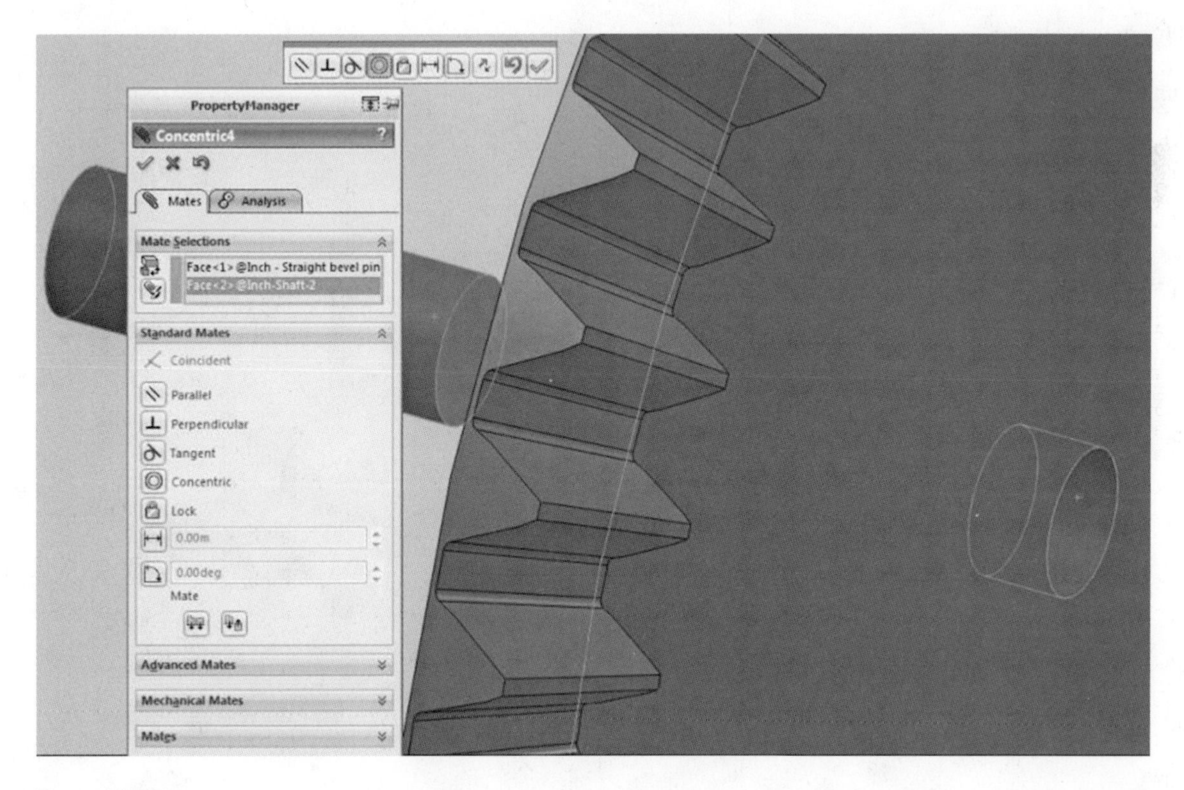

Figure 13.85

Shaft and gear hole mating.

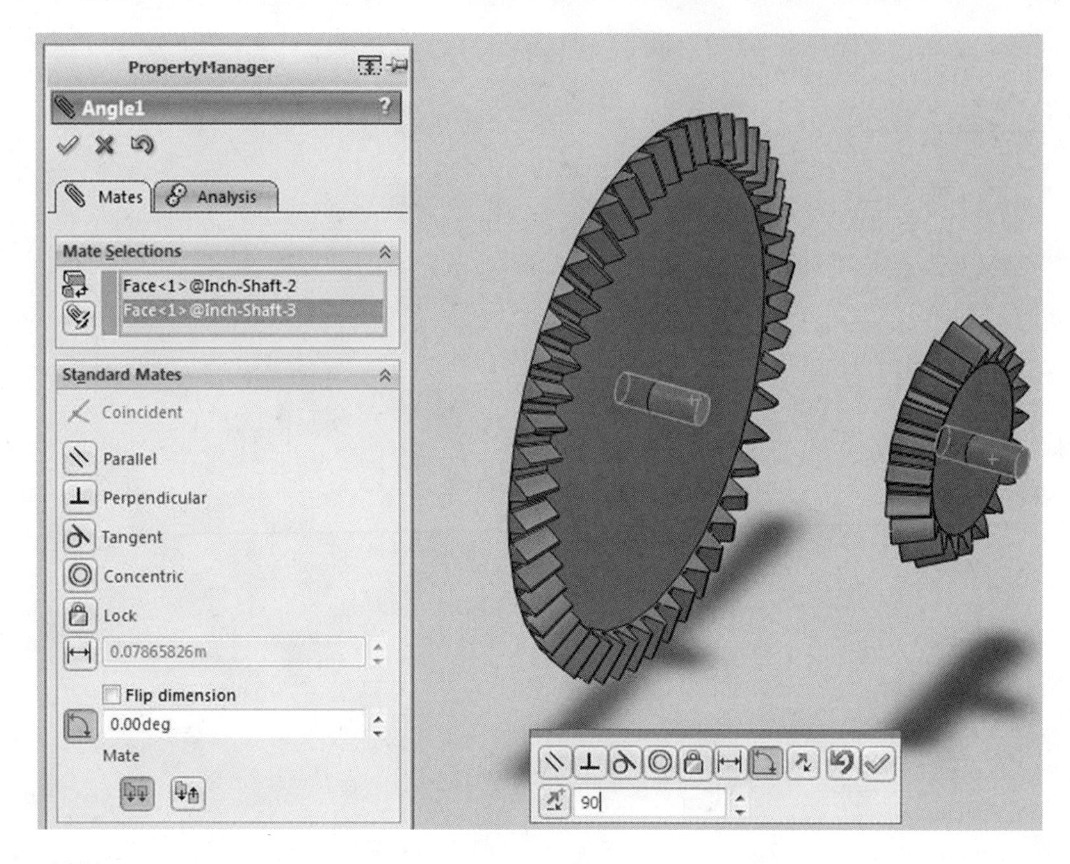

Figure 13.86

Pinion-shaft and gear-shaft mating at 90°. (Alternatively, faces could be used.)

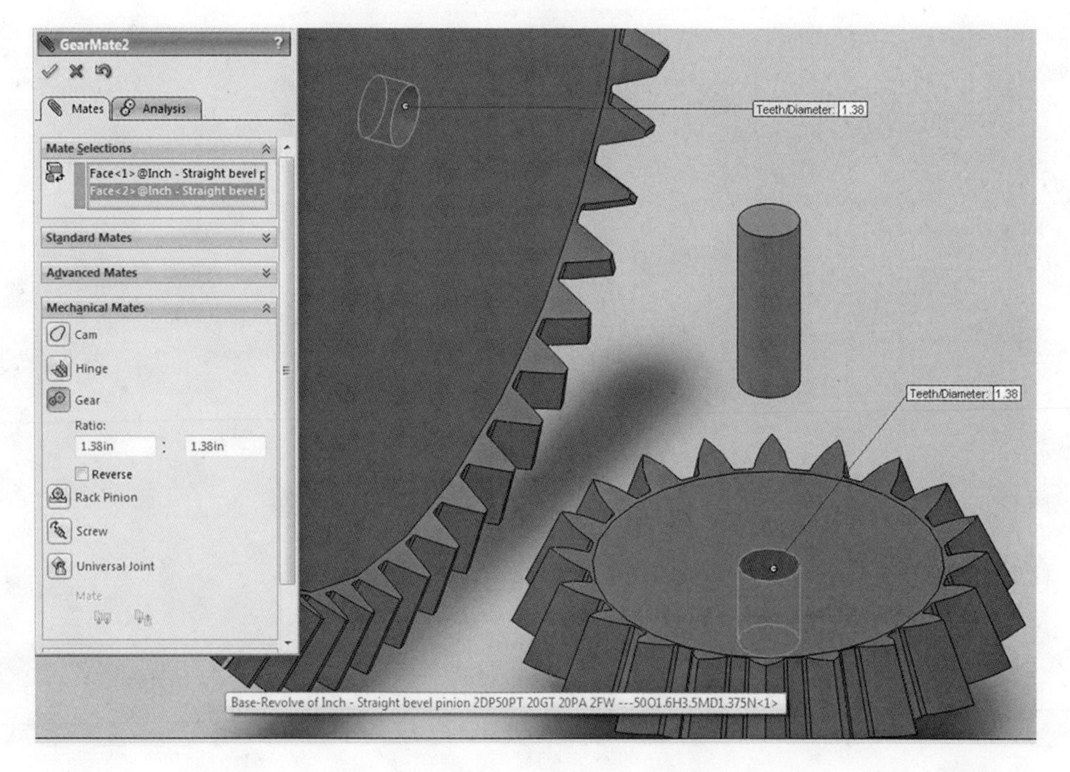

Figure 13.87

Mechanical gear mating.

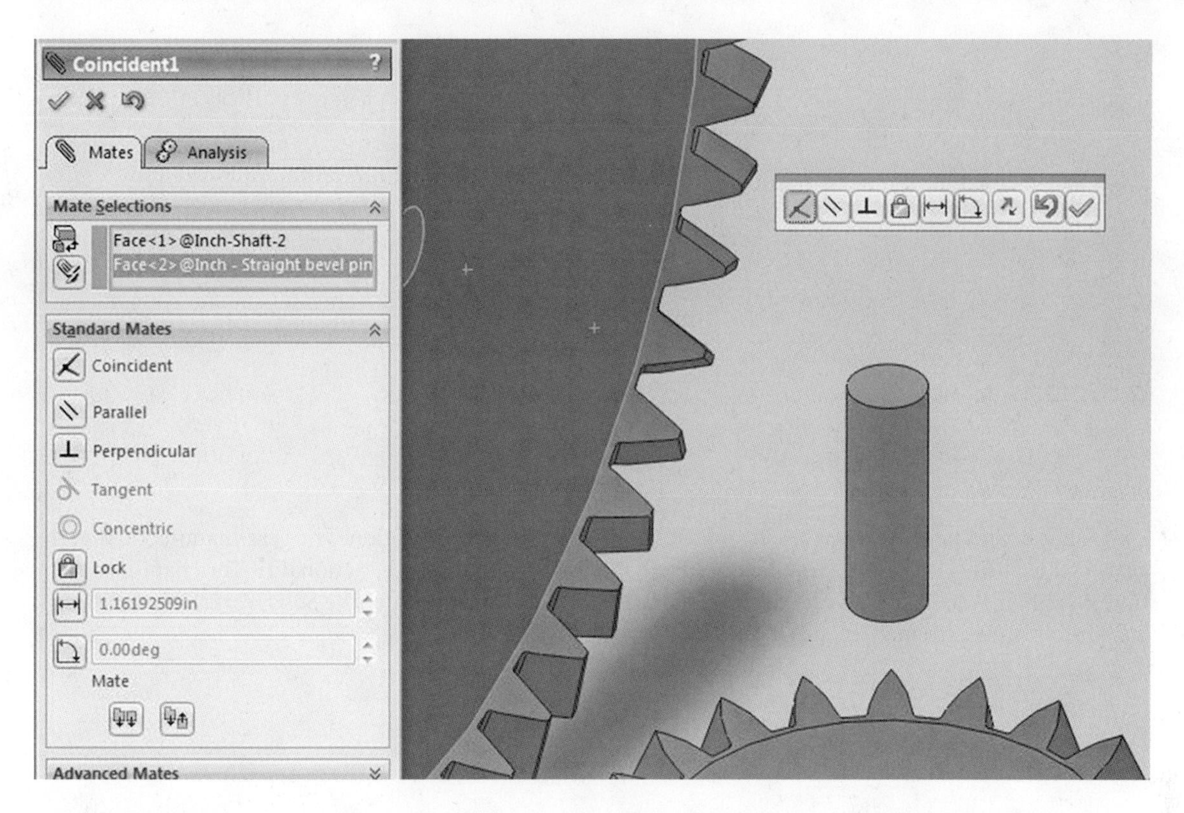

Figure 13.88

Shaft-end and gear face mating.

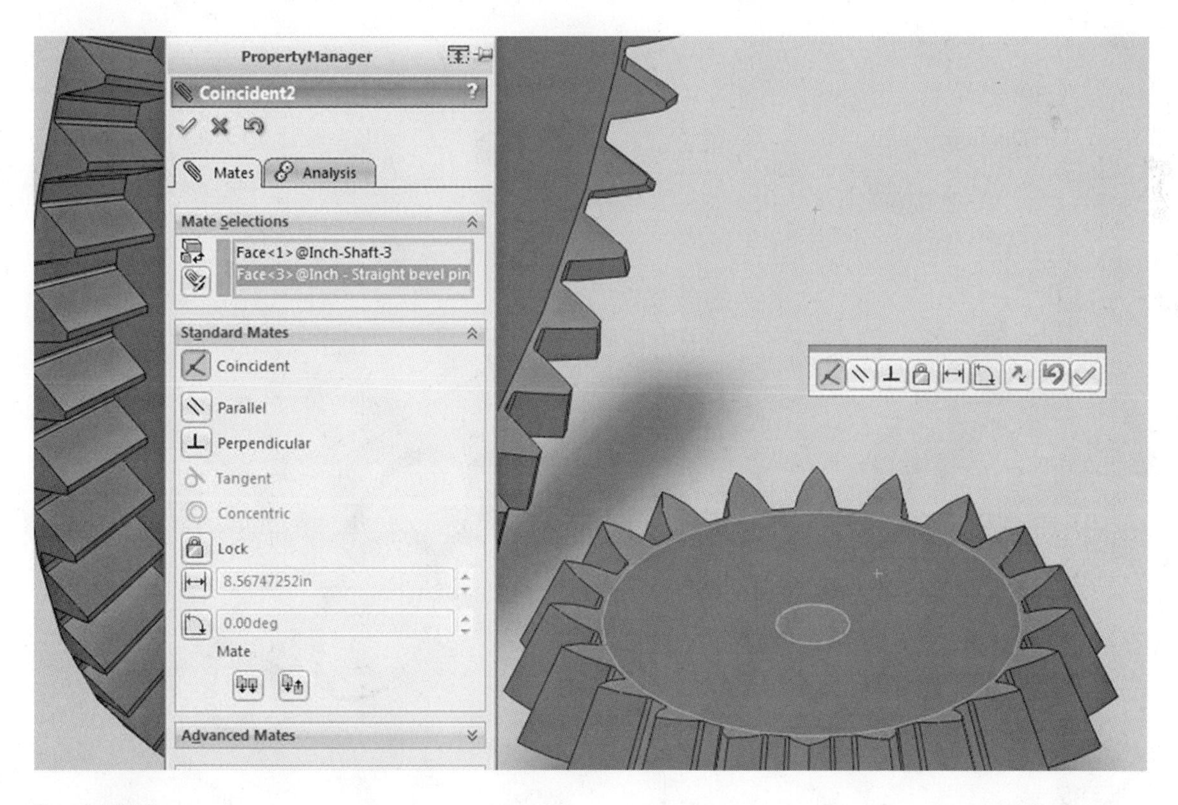

Figure 13.89

Shaft-end and pinion face mating.

Figure 13.90

Teeth alignment for distance mating.

Summary

Spur, helical, and bevel gears have been considered. There is an observation. For gear manufacturing, the gears created using SolidWorks Toolbox need to be revisited for practical functionality. The chapter concentrates on giving the engineers and designers an in-depth understanding of the *SolidWorks Toolbox/Design Library*, which is extremely useful in practice for designers.

Exercises

Project 1 (Measurements in Inches)

1. Create four shafts, 0.375-in. diameter and 3.00-in. long.
2. Model the rectangular support plate that is shown in Figure P13.1. Consider modeling the shaft supports.
3. Access the SolidWorks Design Library and create three Gear 1s and Gear 2s.

   The gears are defined as follows:

   Gear 1:

   Diametral pitch = 12

   Number of teeth = 18

   Pressure angle = 14.5

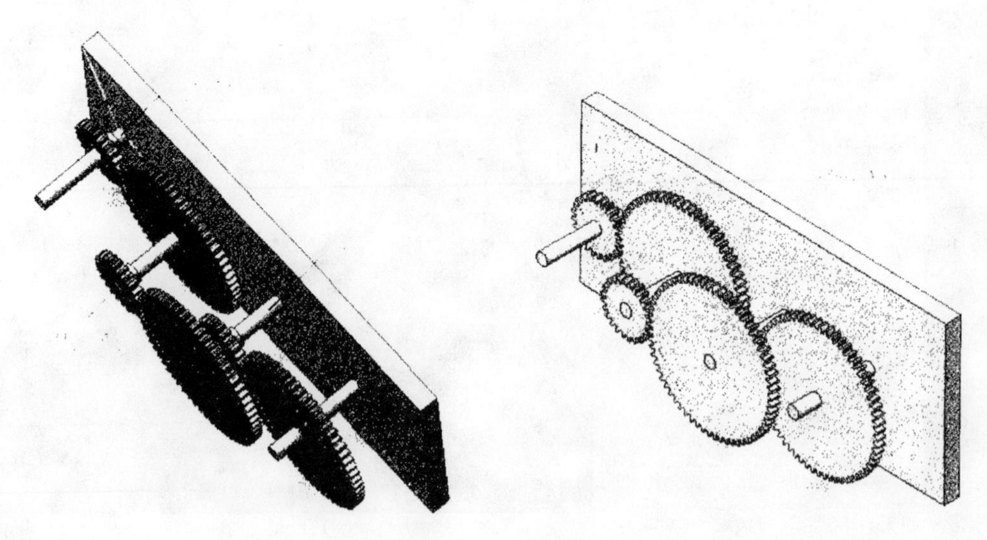

Figure P13.1

Spur gear train.

Face width = 0.25

Hub style = One side

Hub diameter = 0.50

Overall length = 0.50

Nominal shaft diameter = 3/8

Key = None

Gear 2:

Diametral pitch = 12

Number of teeth = 48

Pressure angle = 14.5

Face width = 0.25

Hub style = One side

Hub diameter = 0.50

Overall length = 0.50

Nominal shaft diameter = 3/8

Key = None

4. Add #6-32 threaded holes to each gear hub 0.19 from the top hub surface.
5. Assemble the gears onto the shafts so that Gear 1 and Gear 2 are 0.50 offset from the support plate, and Gear 3 and Gear 4 are 1.25 offset from the support plate.
6. Insert a #6-32 Slotted Set Screw with an Oval Point into each hole.
7. Creating an exploded assembly drawing.
8. Create a bill of materials (BoM).
9. Animate the assembly.

Project 2 (Measurements in Millimeters)
Based on Figure P13.1, define a support plate and shafts that support the following gears. Consider modeling the shaft supports. Use two of each gear.

Parameters:

Plate: 20 thick, a distance of at least 25 beyond the other edge of the gears to the edge of the plate.

Shafts: Diameters that match the bore diameter of the gear; minimum offset between the plate and the gear is 20 or greater.

Gear 1:

Module = 2.0

Number of teeth = 25

Pressure angle = 14.5

Face width = 12

Hub style = One side

Hub diameter = 25

Overall length = 30

Nominal shaft diameter = 20

Key = None

Gear 2:

Diametral pitch = 2.0

Number of teeth = 60

Pressure angle = 14.5

Face width = 12

Hub style = One side

Hub diameter = 30

Overall length = 30

Nominal shaft diameter = 20

Key = None

1. Add M4 threaded holes to each gear hub 12 from the top hub surface.
2. Assemble the gears onto the shafts so that Gear 1 and Gear 2 are 10 offset from the support plate and Gear 3 and Gear 4 are parallel to the ends of the support plate.
3. Insert an M4 Slotted Set Screw with an Oval Point into each hole.
4. Creating an exploded assembly drawing.
5. Create a BoM.
6. Animate the assembly.

Project 3 (Measurements in Millimeters)

Carry out a preliminary design (sizing) for the gearbox of Figure P13.2. The design outcomes are similar to Project 1 (#1–#9). Size the gearbox of Figure P13.2. Consider modeling the shaft supports.

A 0.5-hp, 1725-rpm electric motor drives the input shaft 1 at 95% efficiency. All gears have $\psi = 20°$ pressure angle. Shafts 1 or 2, 3, and 4 are supported by 12-, 19-, and 25-mm bore flanged bearings, respectively. The pinions are made of carburized 55 R_c steel. Gears are Q&T, 180 BHN steel.

Data:

	Module m (mm)	Number of Teeth N	Pitch Diameter d (mm)	Face Width b (mm)
Gear 1 (pinion)	1.3	15	20	14
Gear 2	1.3	60	80	14
Gear 3 (pinion)	1.6	18	28.8	20
Gear 4	1.6	72	115.2	20
Gear 5 (pinion)	2.5	15	37.5	32
Gear 6	2.5	60	150	32

Project 4 (Measurements in Inches)

A motor running at n = 2400 rpm drives a machine by means of a helical gear set, as shown in Figure P13.3. The gears have the following geometric quantities: P_n = 5 in.$^{-1}$, $\psi = 20°$, c = 9 in., N_1 = 30, N_2 = 42, b = 2 in. The gears are made of SAE 1045 steel, WQ&T, and hardened to 200 BHN. Size the gearbox of Figure P13.3. The design outcomes are similar to Project 1 (#1–#9). Consider modeling the couplings and shaft supports.

Project 5 (Measurements in Inches)

A turbine rotates at n = 8000 rpm drives a 250-kW (335-hp) generator at 1000 rpm by means of a helical gear set, as shown in Figure P13.4. The gears have the following geometric quantities: gear set angle $\psi = 30°$,

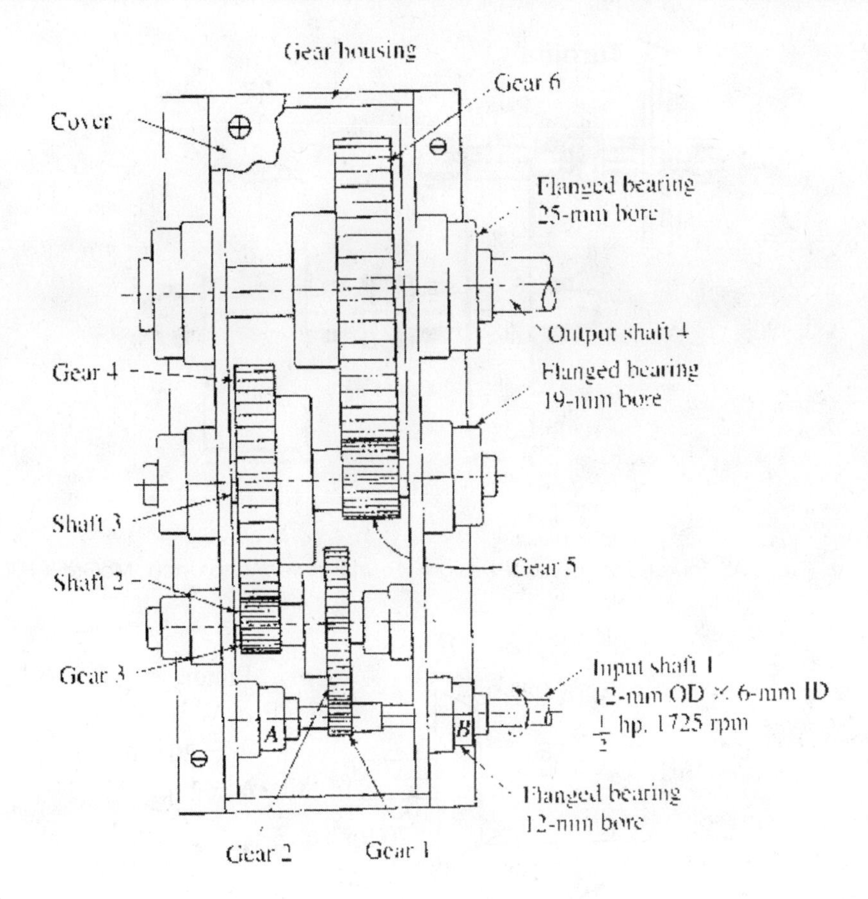

Figure P13.2

Spur gear train. (From A. Ugural, *Mechanical Design: An Integrated Approach*, McGraw Hill, 2004. With permission.)

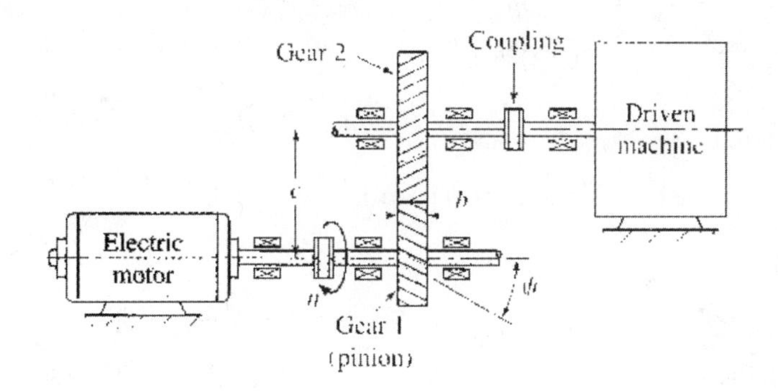

Figure P13.3

Helical gear set. (From A. Ugural, *Mechanical Design: An Integrated Approach*, McGraw Hill, 2004. With permission.)

$\varphi_n = 20°$, $P_n = 10$ in.$^{-1}$, and $b = 8$ in. The pinion is made of steel with 150 BHN, and the gear is cast iron. Size the gearbox of Figure P13.4. The design outcomes are similar to Project 1 (#1–#9). Consider modeling the couplings and shaft supports.

Project 6 (Measurements in Inches)

A set of straight bevel gears having a pressure angle $\psi = 20°$ is to be used to transmit 20 hp from a pinion operating at 500 rpm to a gear that is mounted on a shaft that intersects at an angle of 90°, as shown in Figure P13.5. Size the gearbox of Figure P13.5. The design outcomes are similar to Project 1 (#1–#9). Consider modeling *shaft supports*.

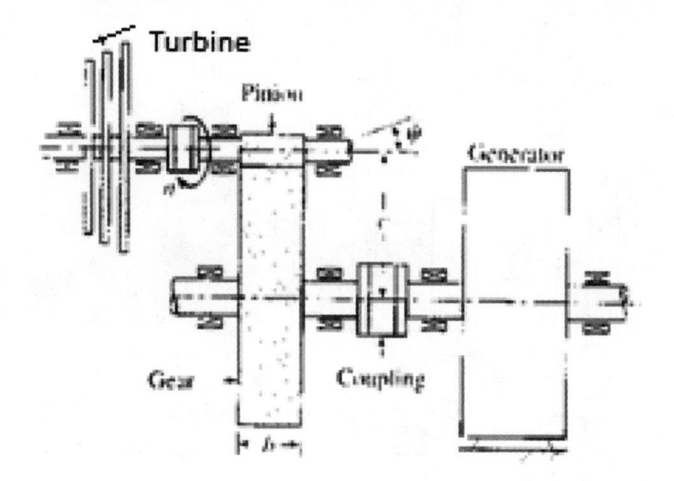

Figure P13.4

Helical gear set. (From A. Ugural, *Mechanical Design: An Integrated Approach*, McGraw Hill, 2004. With permission.)

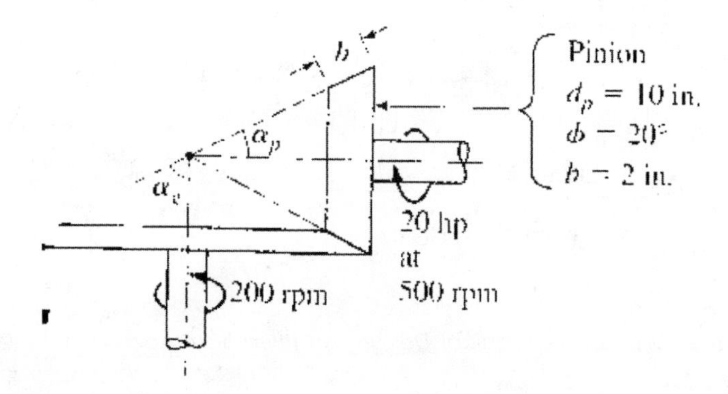

Figure P13.5

Bevel gears. (From A. Ugural, *Mechanical Design: An Integrated Approach*, McGraw Hill, 2004. With permission.)

Basic Gear Kinematics

The nomenclature used in the basic kinematics of gears is given as follows:

P: Diametral pitch
m: Module
p_c: Circular pitch
D: Circular pitch diameter
N_t: Number of teeth
φ: Pressure angle
ψ: Helix angle (for helical gears only)
c: Center-to-center distance

Spur Gears

1. $P_d = \frac{N_t}{D}$: Imperial unit
2. $m = \frac{D}{N_t}$: SI unit
3. $p_c = \frac{\pi D}{N_t}$
4. $P_d \times p_c = \pi$

Helical Gears

1. $P_n = P_d/\text{Cos}\ \psi$
2. $P_a = P_n/\text{Sin}\ \psi = P_n \times \text{Cot}\ \psi$
3. $P_d = \dfrac{N_t}{D}$
4. $c = \dfrac{D_1 + D_2}{2} \Rightarrow P_d = \dfrac{(N_{t1} + N_{t2})}{2c}$

Bevel Gears

1. $D_p = \dfrac{N_{tp}}{P};\quad D_g = \dfrac{N_{tg}}{P}$
2. $\tan\alpha_p = \dfrac{N_{tp}}{N_{tg}};\quad \tan\alpha_g = \dfrac{N_{tg}}{N_{tp}}$
3. $r_s = \dfrac{N_{tg}}{N_{tp}} = \dfrac{N_{tp}}{N_{tg}} = \tan\alpha_p = \text{Cot}\,\alpha_g$

14

Animation with Basic Motion

Objectives:

When you complete this chapter, you will have

- Understood the different types of motion studies
- Modeled linkages
- Animated the linkages using Basic Motion

Different Types of Motion Studies

The *MotionManager Interface* is accessed in the lower left of the graphics window (see Figure 14.1). SolidWorks uses different types of motion studies. We will briefly examine these.

- *Animation*. Animation uses the key frame method in which the software interpolates between the positions that are established by mates, freehand drag, or positioning via Triad or XYZ values.
- *Basic Motion*. Basic Motion uses motors, springs, three-dimensional (3D) contacts, and gravity (*Physical Simulation*); it does not use frames. It includes *Physical Dynamics*, which is the calculation of motion due to collision.
- *Motion Analysis*. Motion Analysis is the highest level of motion study and is an *Add-In* to SolidWorks.

We will model simple linkages, assembly them into a mechanism, and then carry out motion studies using Basic Motion.

Modeling of Linkages

Model Link1

1. Open a New Part document.
2. Create a sketch, Sketch1, which has a rectangular section that is 8-in. long with a half-circular arc 1-in. radius at both ends (see Figure 14.2).
3. Extrude Sketch1 through 0.25 in., Mid Plane (see Figure 14.3).

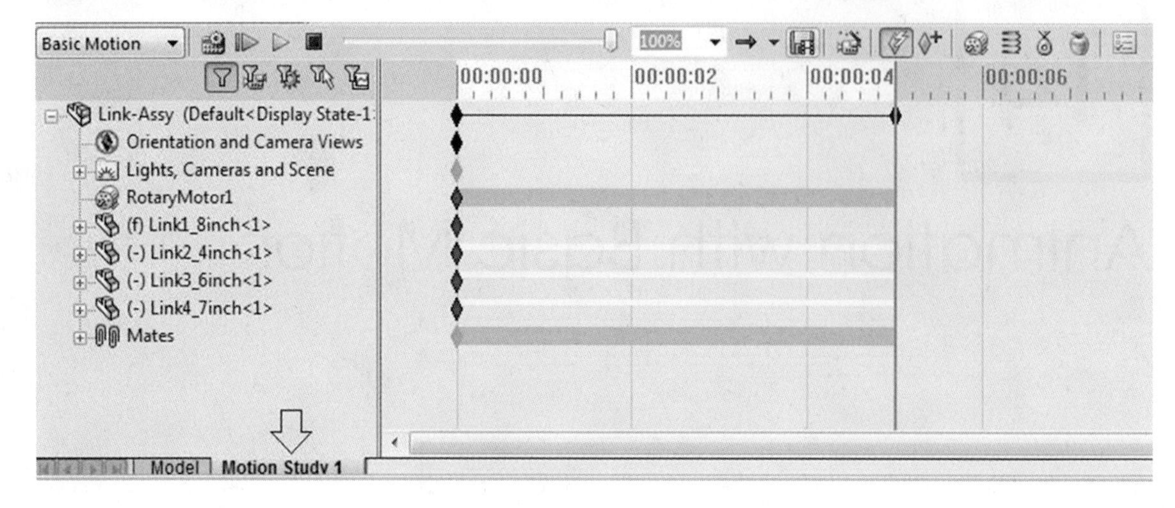

Figure 14.1

MotionManager Interface.

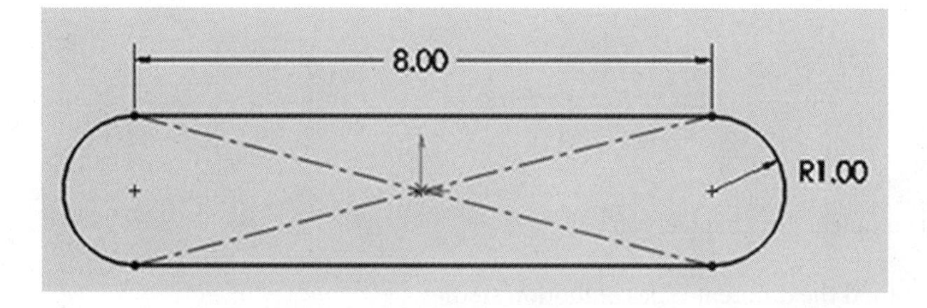

Figure 14.2

Sketch1.

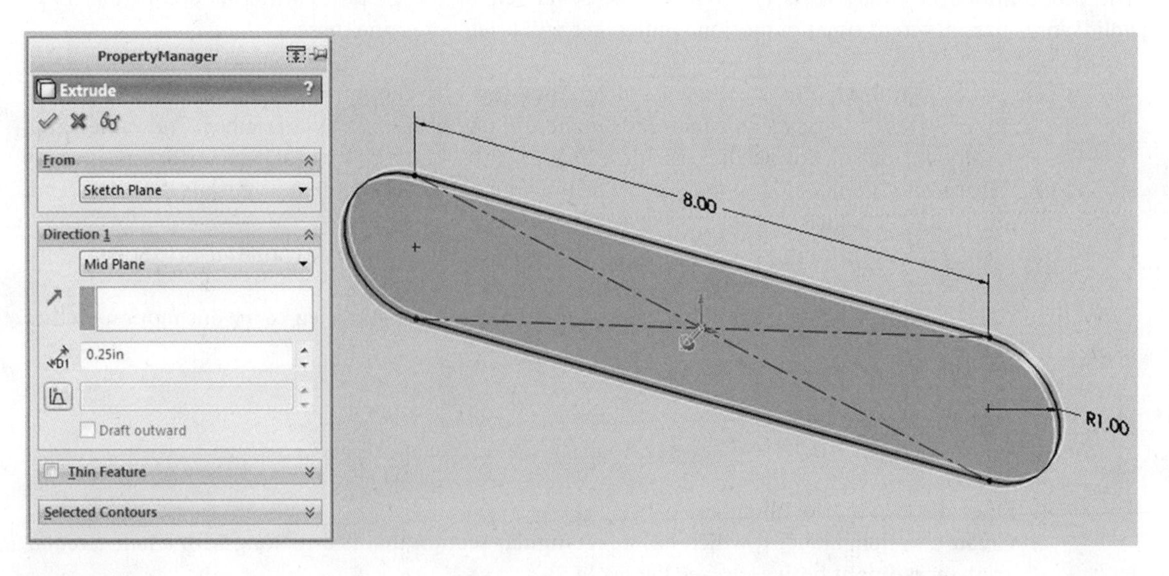

Figure 14.3

Extrude Sketch1.

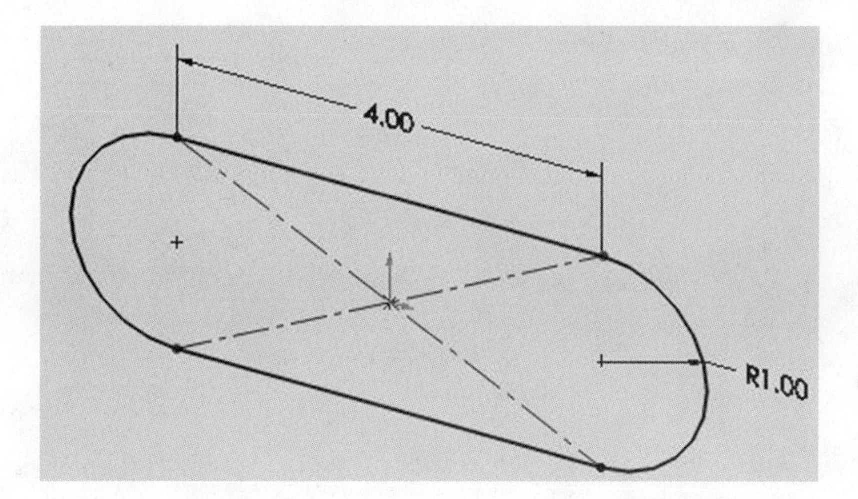

Figure 14.4
Sketch2.

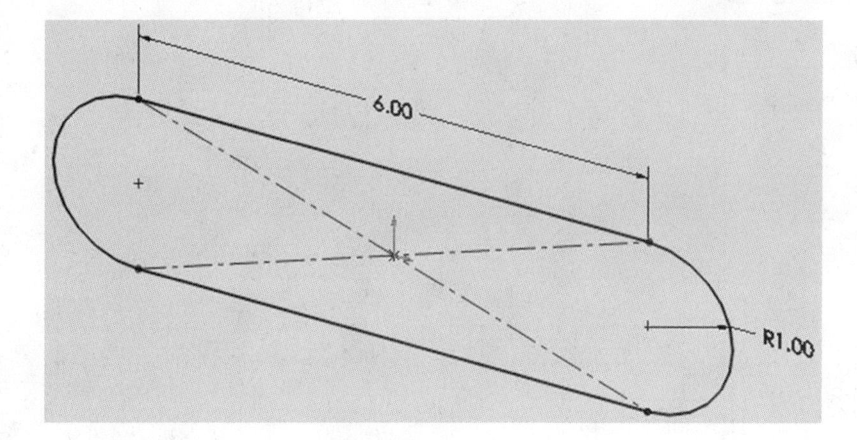

Figure 14.5
Sketch3.

Model Link2

1. Create a sketch, Sketch2, which has a rectangular section that is 4-in. long with a half-circular arc 1-in. radius at both ends (see Figure 14.4).
2. Extrude Sketch1 through 0.25 in., Mid Plane.

Model Link3

1. Create a sketch, Sketch3, which has a rectangular section that is 6-in. long with a half-circular arc 1-in. radius at both ends (see Figure 14.5).
2. Extrude Sketch1 through 0.25 in., Mid Plane.

Model Link4

1. Create a sketch, Sketch4, which has a rectangular section that is 7-in. long with a half-circular arc 1-in. radius at both ends (see Figure 14.6).
2. Extrude Sketch1 through 0.25 in., Mid Plane.

Assembly Modeling of Linkages

1. Open a New Assembly document.
2. Fix the first part, Link1_8inch.
3. Insert Link2_4inch.
4. Create Concentric Mate between Link1_8inch and Link2_4inch (see Figure 14.7).
5. Create Coincident Mate between Link1_8inch and Link2_4inch (see Figure 14.8).

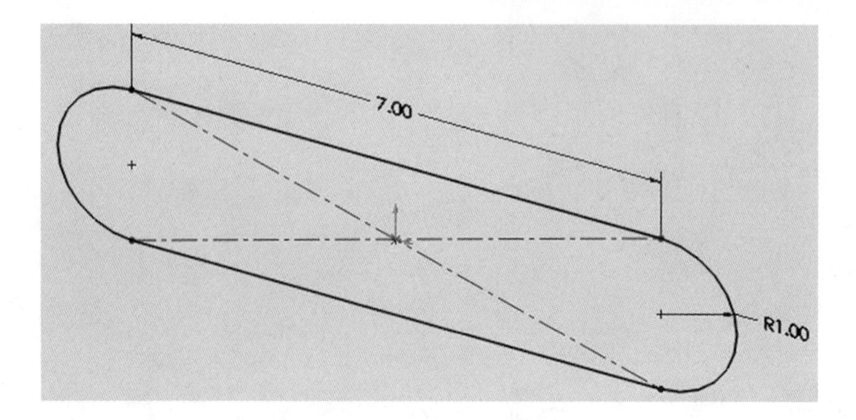

Figure 14.6

Sketch4.

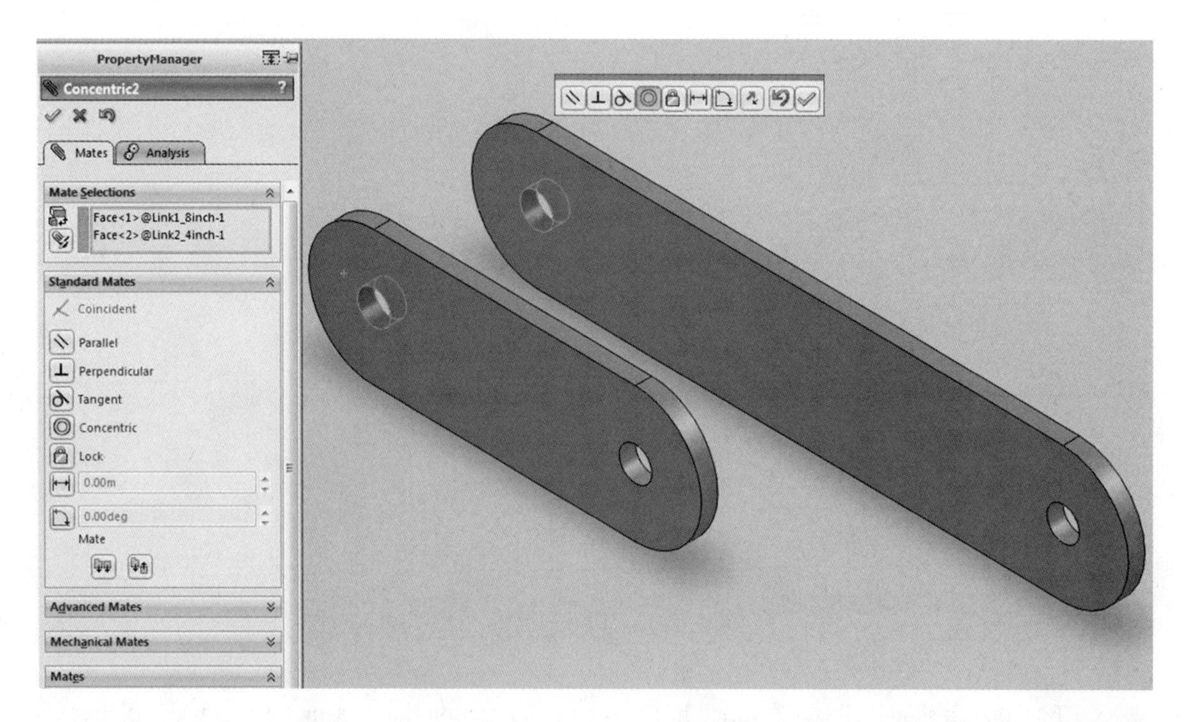

Figure 14.7

Concentric Mate between Link1_8inch and Link2_4inch.

6. Insert Link3_6inch.
7. Create Concentric Mate between Link1_8inch and Link3_6inch (see Figure 14.9).
8. Create Coincident Mate between Link1_8inch and Link3_6inch (see Figure 14.10).
9. Insert Link4_7inch.
10. Create Concentric Mate between Link2_4inch and Link4_7inch (see Figure 14.11).
11. Create Coincident Mate between Link2_4inch and Link4_7inch (see Figure 14.12).
12. Create Concentric Mate between Link2_4inch and Link3_6inch (see Figure 14.13).

Figure 14.14 shows the mechanism that is realized.

MotionManager Interface

The MotionManager Interface is accessed in the lower left of the graphics window (see Figure 14.15). There are four physical simulation elements that are used: (1) *Gravity,* (2) *3D contacts,* (3) *Springs,* and (4) *Motors* (see Figures 14.15 and 14.16 for their PropertyManagers).

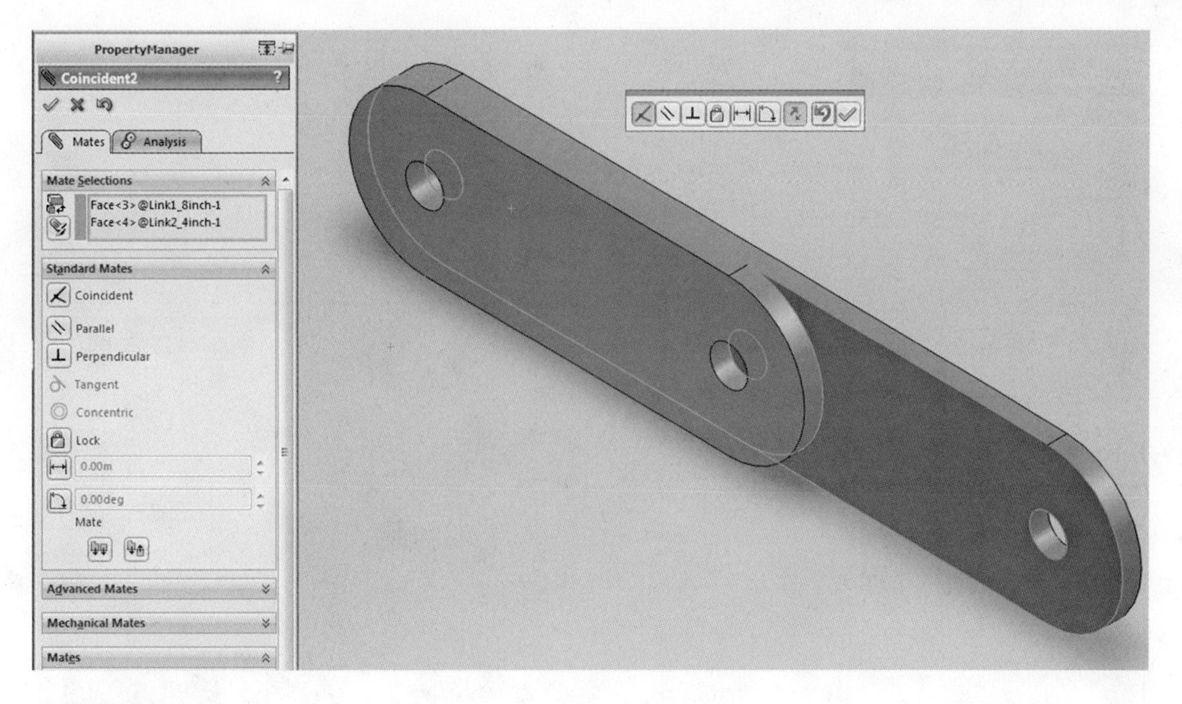

Figure 14.8

Coincident Mate between Link1_8inch and Link2_4inch.

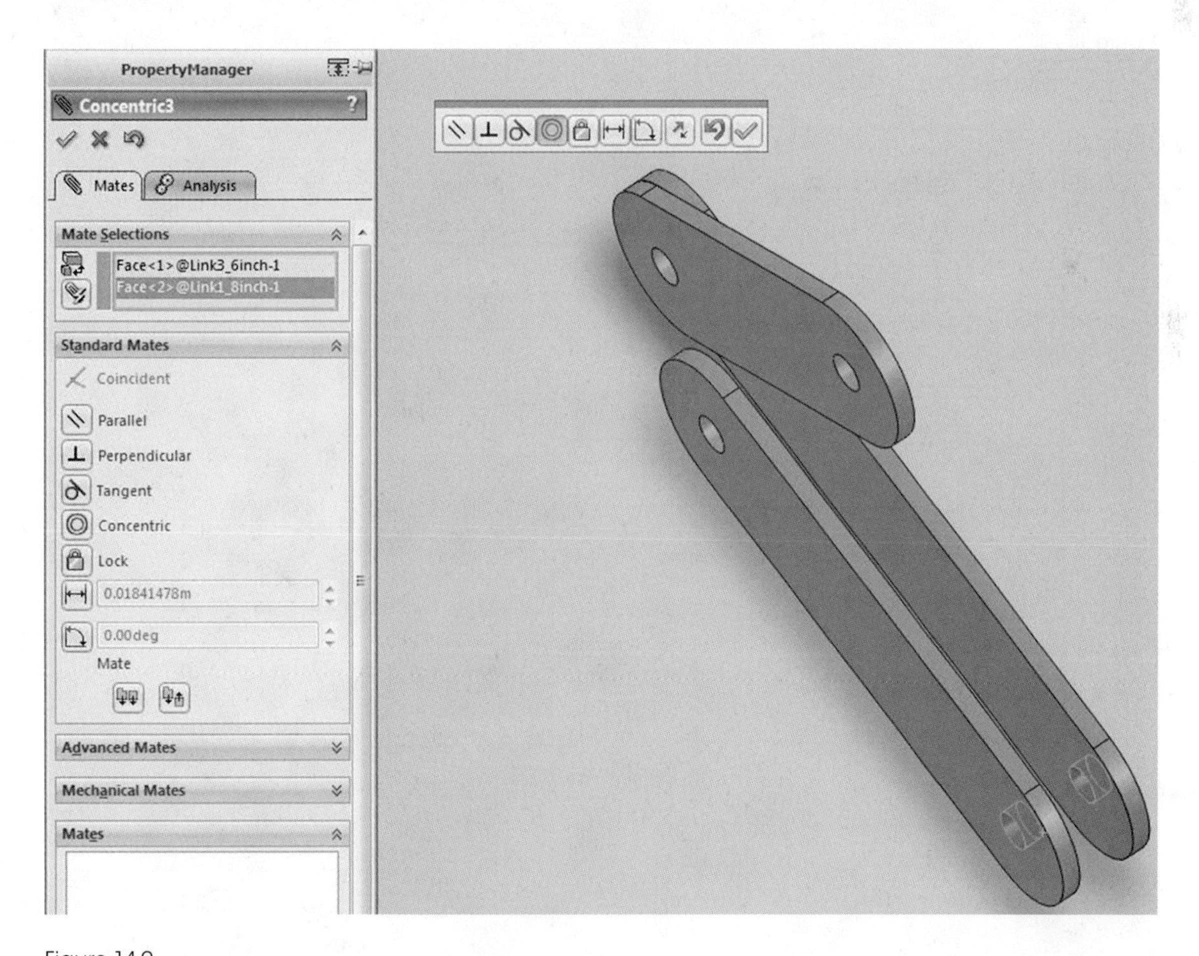

Figure 14.9

Concentric Mate between Link1_8inch and Link2_4inch.

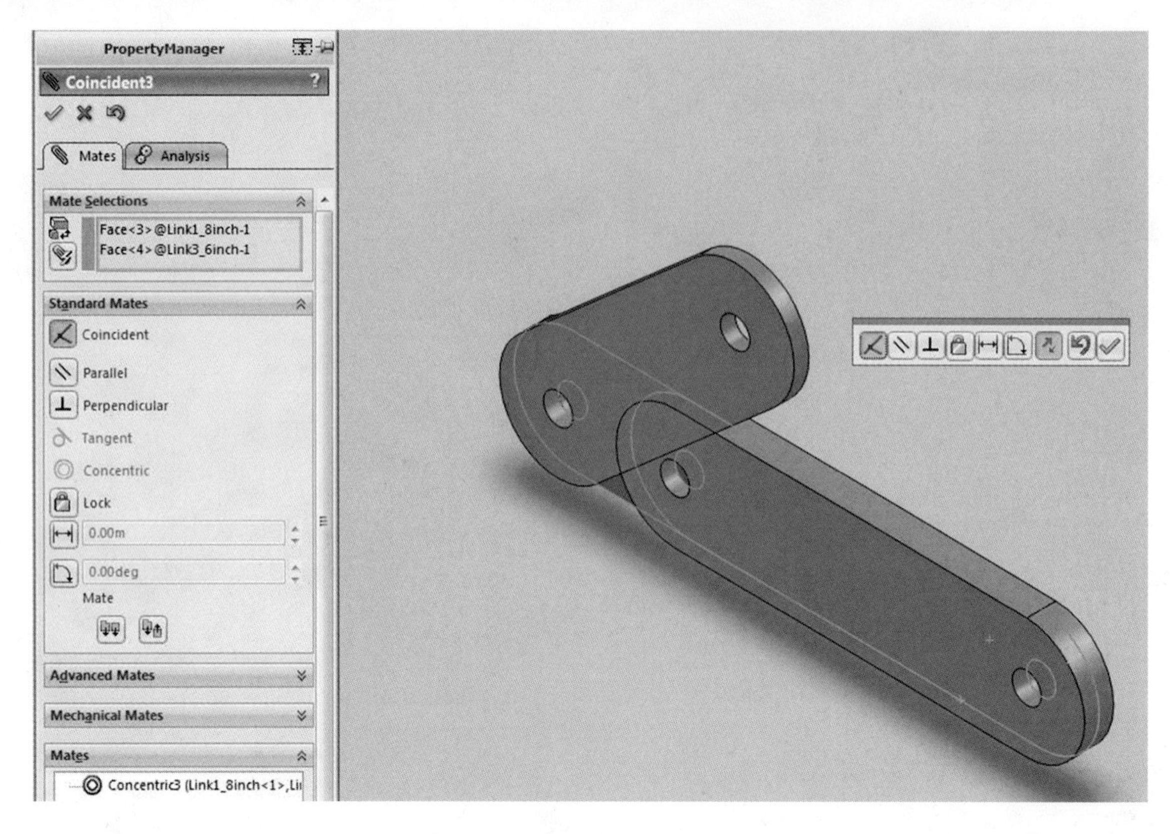

Figure 14.10

Coincident Mate between Link1_8inch and Link2_4inch.

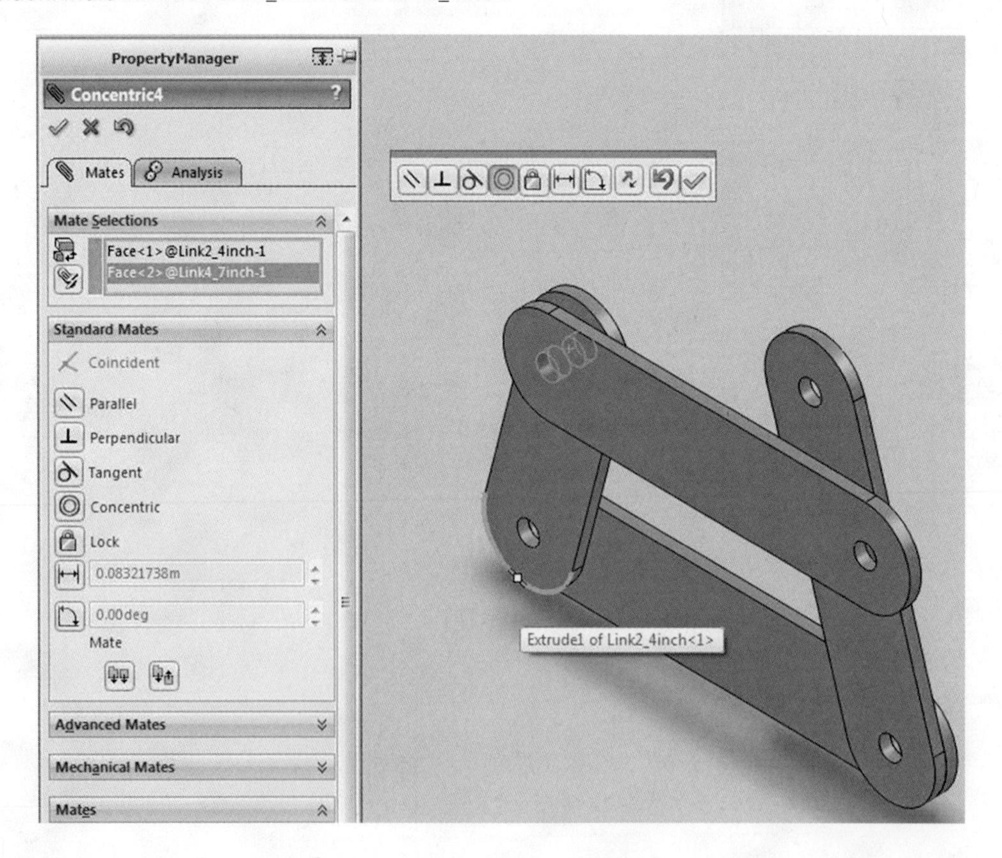

Figure 14.11

Concentric Mate between Link2_4inch and Link4_7inch.

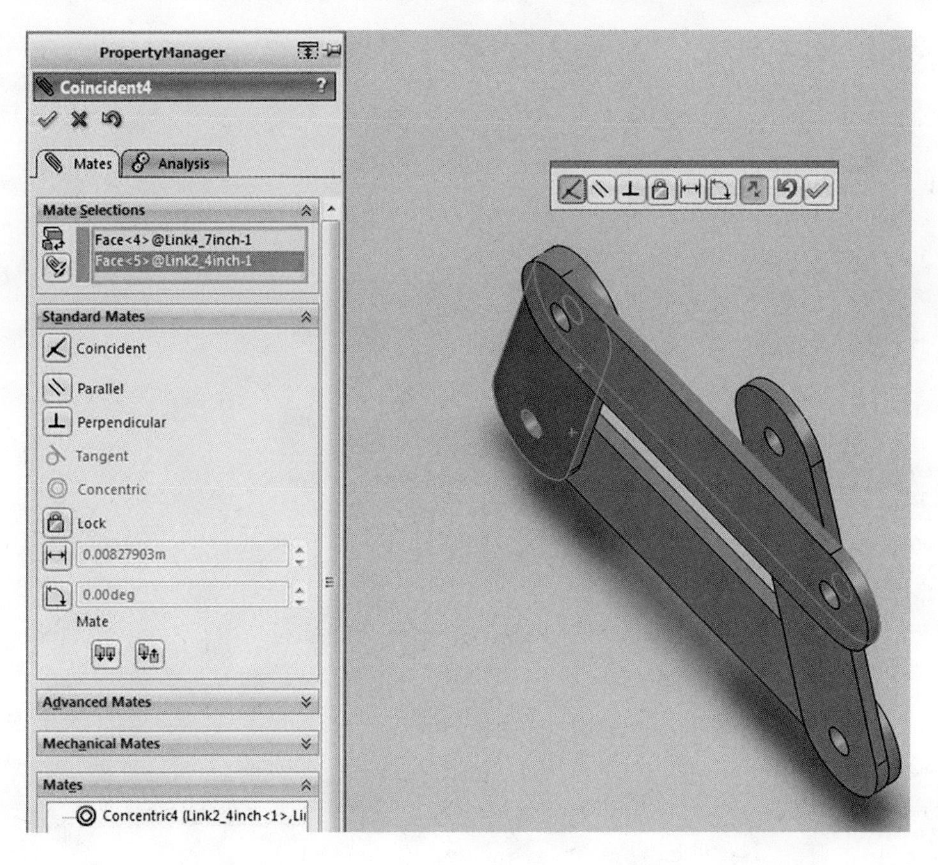

Figure 14.12

Coincident Mate between Link2_4inch and Link4_7inch.

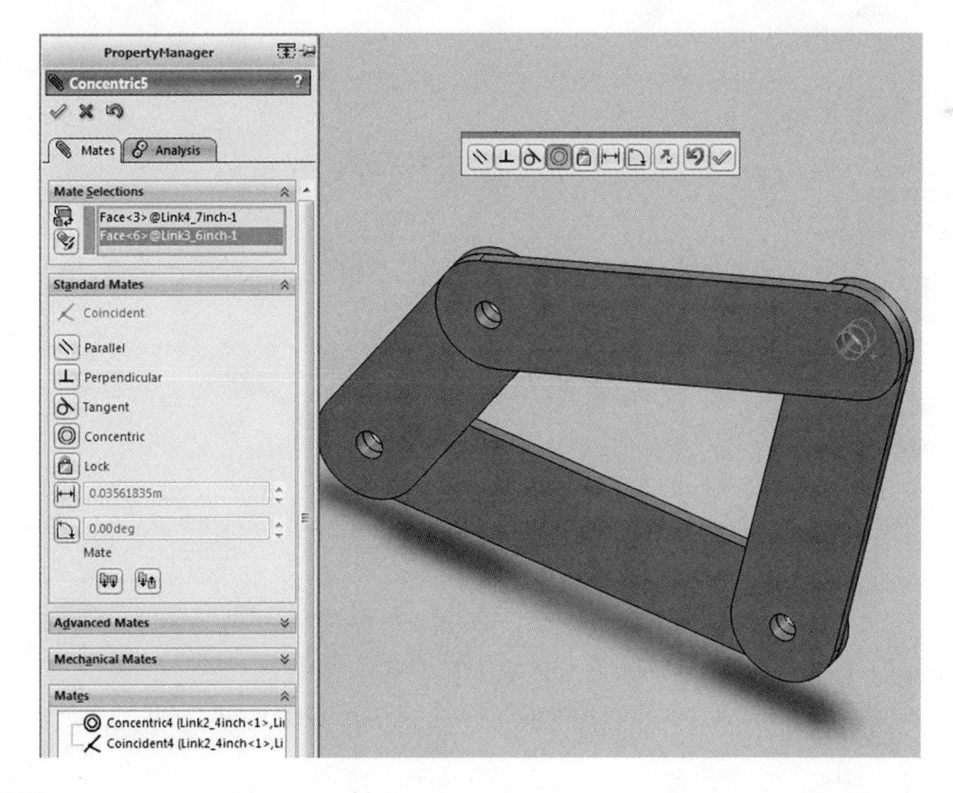

Figure 14.13

Create Concentric Mate between Link2_4inch and Link3_6inch.

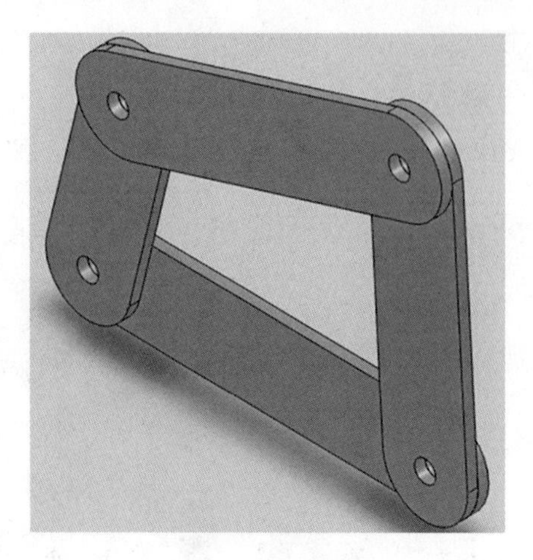

Figure 14.14

Mechanism realized.

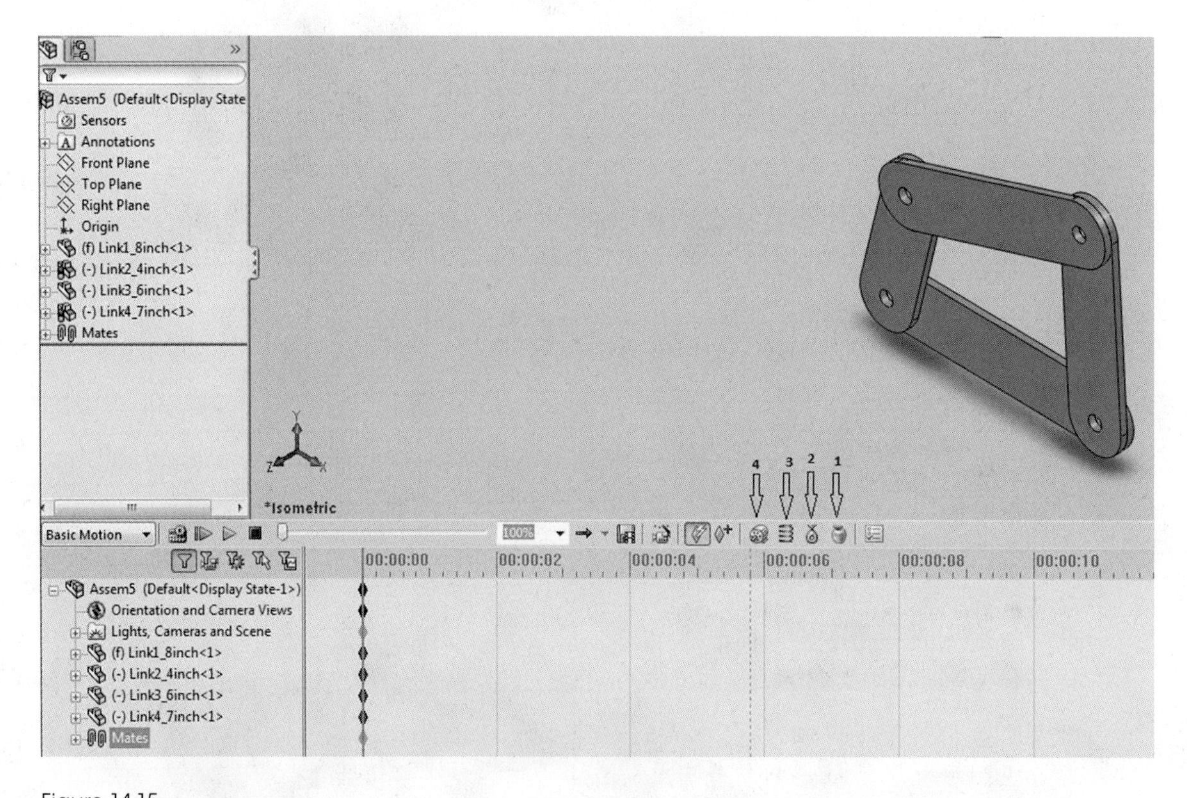

Figure 14.15

MotionManager interface.

Using Motors

1. Click the Motor option. (The Motor PropertyManager is displayed in Figure 14.17.)
2. Select Rotary Motion (see Figure 14.17).
3. Click the face of Link2 for the Component/Direction field (a red arrow showing the direction of rotation is displayed; change the direction if required) (see Figure 14.17).
4. Click OK (see Figure 14.17).
5. Click the Calculate option to *calculate* the motion study (see Figure 14.18).

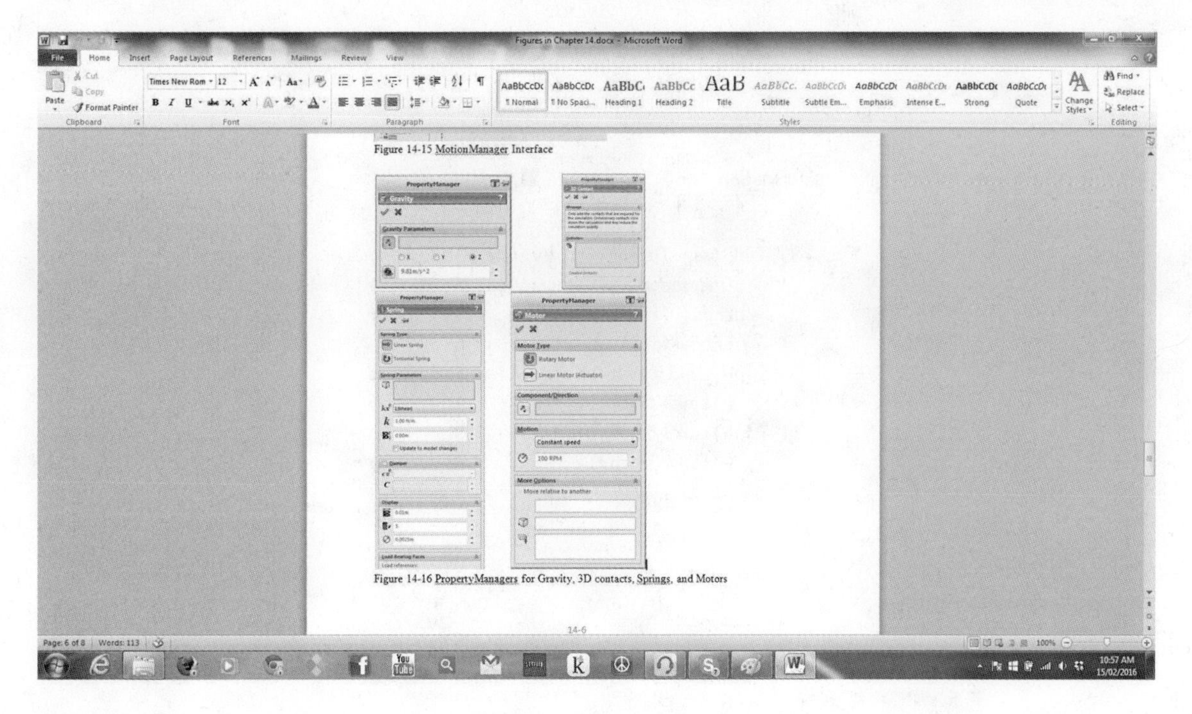

Figure 14.16

PropertyManagers for gravity, 3D contacts, springs, and motors.

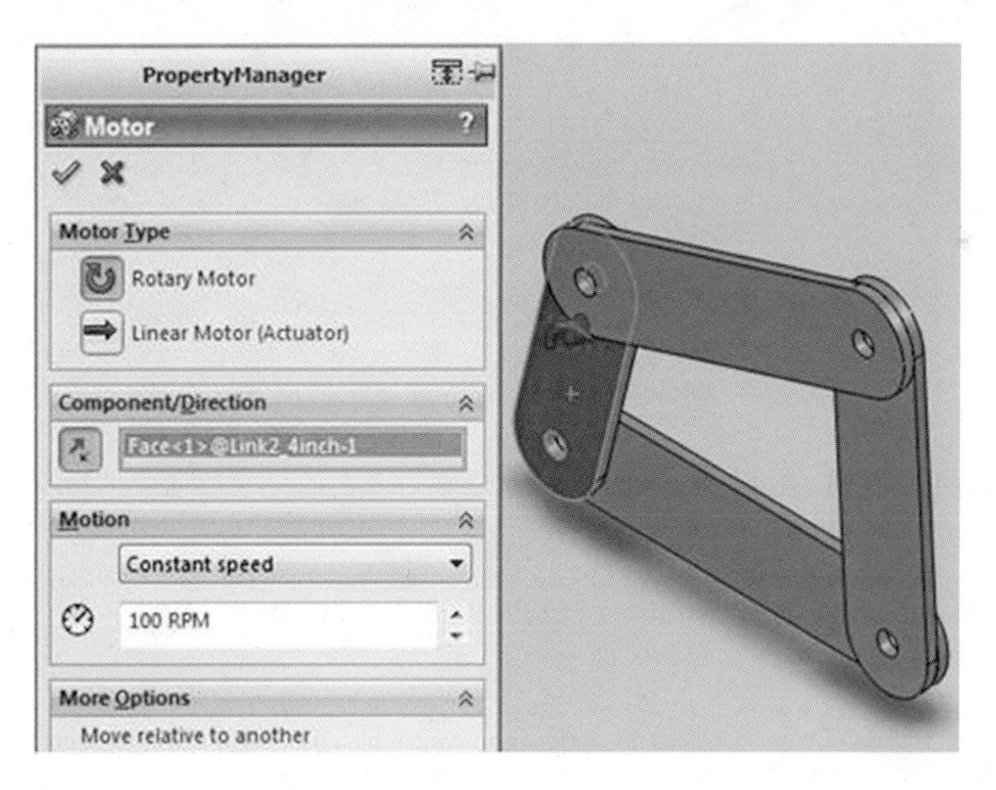

Figure 14.17

Motor PropertyManager.

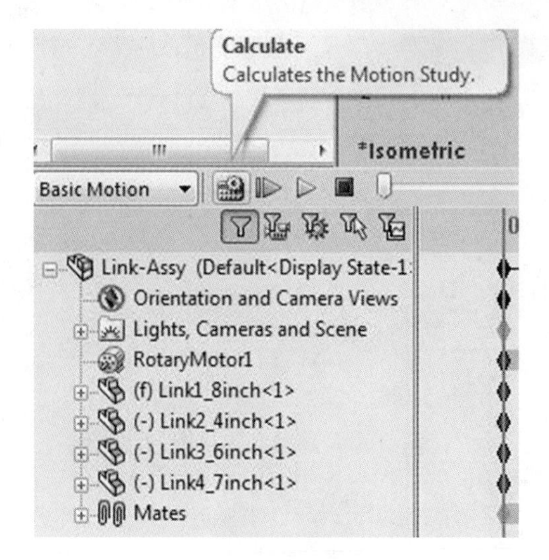

Figure 14.18

Calculate a motion study.

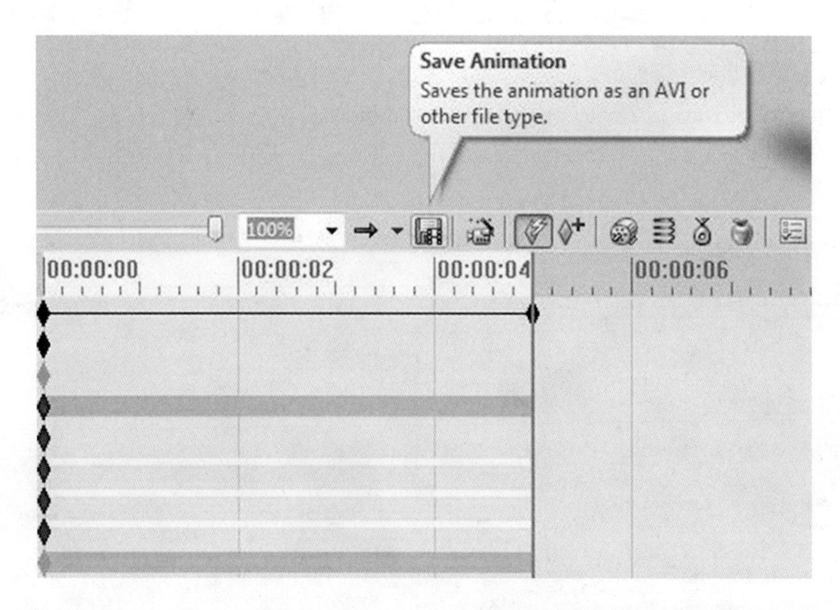

Figure 14.19

Save Animation as an *avi file.

The motion studies can be saved as a Microsoft Audio Video Interleave (*.avi) file by clicking the Save option in the MotionManager Interface (see Figure 14.19).

Summary

Animation with Basic Motion has been discussed. Its limitations have been highlighted. For more meaningful simulations, a premium price has to be paid by using SolidWorks Motion Analysis.

15

Animation with SolidWorks Motion

Objectives:

When you complete this session, you will have learned the following:

- The overview of SolidWorks Motion
- SolidWorks Motion Add-Ins
- The overview of slider–crank assembly that is created in SolidWorks
- Kinematic analysis of the slider–crank mechanism using SolidWorks Motion
- Interference analysis of the slider–crank mechanism using SolidWorks Motion
- Dynamic analysis of the slider–crank mechanism using SolidWorks Motion

Types of Motion Study

Motion studies are graphical simulations of motion for assembly models. Visual properties such as lighting and camera perspective can be incorporated into a motion study. Motion studies do not change an assembly model or its properties. They simulate and animate the motion that the user prescribes for a model. SolidWorks *mates* can be used to restrict the motion of components in an assembly when motion is modeled.

From a motion study, MotionManager, a timeline-based interface can be used, which includes the following motion study tools:

- *Animation* (available in core SolidWorks). Animation can be used to animate the motion of assemblies:
  - Add motors to drive the motion of one or more parts of an assembly.
  - Prescribe the positions of assembly components at various times using set *key points*. Animation uses interpolation to define the motion of assembly components between key points.
- *Basic Motion* (available in core SolidWorks). Basic Motion can be used for approximating the effects of *motors*, *springs*, *contact*, and *gravity* on assemblies. Basic Motion takes mass into account in calculating motion. Basic Motion computation is relatively fast, so you can use this for creating presentation-worthy animations using physics-based simulations.

- *Motion Analysis* (available with the SolidWorks Motion™ add-in to SolidWorks Premium). Motion Analysis can be used for accurately simulating and analyzing the effects of motion elements (including *forces, springs, dampers,* and *friction*) on an assembly. Motion Analysis uses computationally strong kinematic solvers and accounts for material properties, as well as mass and inertia in the computations. Motion Analysis can also be used to plot simulation results for further analysis.

Generally, animation outputs can be posted as Audio Video Interleave (AVI) files; AVI is a multimedia container format that is introduced by Microsoft.

Deciding Which Type of Study to Use

Use Animation to create presentation-worthy animations for motion that does not require accounting for mass or gravity.

Use Basic Motion to create presentation-worthy approximate simulations of motion that account for mass, collisions, or gravity.

Use Motion Analysis to run computationally strong simulations that take the physics of the assembly motion into account. This tool is the most computationally intensive of the three options. The better the user understands the physics of the motion that is required, the better the results that are obtained. Motion Analysis can be used to run impact analysis studies to understand the component response to different types of forces.

Differentiating between Animation, Basic Motion, and Motion Analysis

There are some differences between the animation and motion capabilities in the various SolidWorks modules, namely, Animation, Basic Motion, and Motion Analysis. These differences need to be understood by end users so that they can determine for themselves which is the right one to use. A nice feature is that the interface pretty much stays the same, and changing from one type of output to the other might only take changing the solver type (Animation, Basic Motion, or Motion Analysis). In this subsection, we discuss the classification of motion studies based on the solvers that are used.

- *Animation (or Assembly Motion)* uses the *3D Dimensional Constraint Manager (3D DCM)* from D-Cubed.

  The 3D DCM is commonly used to position the parts in an assembly or mechanism. Fast, fully three-dimensional, nonsequential solving, comprehensive geometry, dimension, and constraint support enable designers to build, modify and animate the most demanding of assemblies and mechanisms efficiently.

  Animation is used to create simple animations that use interpolation to specify the point-to-point motion of parts in assemblies.
- *Basic Motion (or Physical Simulation)* uses *Ageia PhysX.*

  Ageia PhysX is a physics solver that is primarily used in games. It simulates how objects move and react and how they behave. It simulates lifelike motion and interaction. With its Ageia PhysX, SolidWorks Physical Simulation focuses on making the simulation look real.

  Basic Motion is used for approximating the effects of motors, springs, collisions, and gravity on assemblies. Basic Motion takes mass into account in calculating motion. Basic Motion computation is relatively fast, so you can use this for creating presentation-worthy animations using physics-based simulations.
- *Motion Analysis (or SolidWorks Motion)* uses the *ADAMS solver.*

  The Automated Dynamic Analysis of Mechanical Systems (ADAMS) solver can analyze the complex behavior of mechanical assemblies. With this solver, Motion focuses on *accurately* analyzing the forces, torques, contact forces, power consumption, and so on in your mechanism. You can plot any of the resultant kinematic quantities over time (or versus other parameters) in the analysis.

  SolidWorks Motion comes in the top level of our design software, SolidWorks Premium, and is also included in both SolidWorks Simulation Professional and Simulation Premium. You must first turn on the Add-In before using Motion Analysis.

  Motion Analysis is used to accurately simulate and analyze the motion of an assembly while incorporating the effects of Motion Study elements (including forces, springs, dampers, and friction). A Motion Analysis study combines motion study elements with mates in motion calculations. Consequently, motion constraints, material properties, mass, and component contact are included in the SolidWorks Motion kinematic solver calculations. A Motion Analysis study also calculates loads that can be used to define load cases for structural analyses.

Now that the differences between the animation and motion capabilities in the various SolidWorks modules, we will proceed to discuss the SolidWorks Motion overview, SolidWorks Motion Add-In, four-bar linkage mechanism, slider–crank mechanism, and then to the motion analysis of a slider–crank mechanism using SolidWorks Motion.

SolidWorks Motion Overview

SolidWorks Motion is the standard virtual prototyping Add-In package in SolidWorks Premium, SolidWorks Simulation Professional, and SolidWorks Simulation Premium for engineers and designers who are interested in understanding the performance of their assemblies. The most popular virtual prototyping tool for SolidWorks, SolidWorks Motion motion simulator, lets you make sure that your designs will work before you build them.

SolidWorks Motion enables engineers to size motors/actuators, determine power consumption, layout linkages, develop cams, understand gear drives, size springs/dampers, and determine how contacting parts behave.

The result is of motion simulation and analysis is a quantitative reduction in physical prototyping costs and reduced product development time. SolidWorks Motion also provides qualitative benefits such as the ability to consider more designs, risk reduction, and the availability of valuable information early in the design process.

SolidWorks Motion Add-In

1. Open SolidWorks.
2. Open the model file.
3. Click the Add-Ins tool (see Figure 15.1a).
4. Check SolidWorks Motion (see Figure 15.1b).
5. Click OK. (Motion Manager is added; see Figure 15.2.)

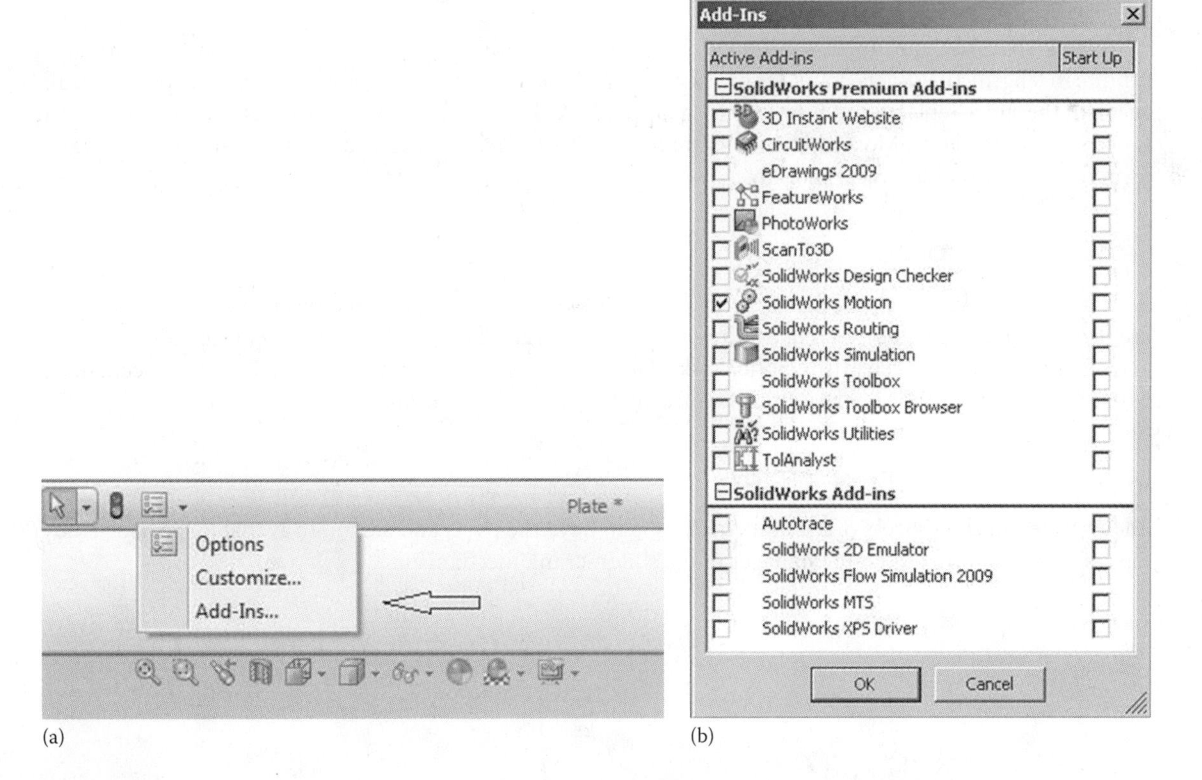

Figure 15.1

SolidWorks Motion Add-In: (a) Add-Ins tool is clicked and (b) SolidWorks Motion tool is clicked.

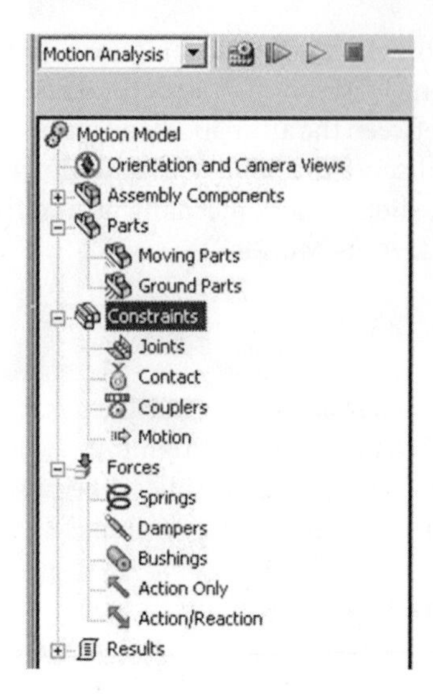

Figure 15.2

SolidWorks Motion Manager.

Four-Bar Linkage Mechanism and Slider–Crank Mechanism

One of the simplest examples of a constrained linkage is the *four-link mechanism*. A variety of useful mechanisms can be formed from a four-link mechanism through slight variations, such as changing the character of the pairs, the proportions of links, etc. Furthermore, many complex link mechanisms are combinations of two or more such mechanisms. The majority of four-link mechanisms fall into one of the following two classes:

1. Four-bar linkage mechanism
2. Slider–crank mechanism

The four-bar mechanism has some special configurations that are created by making one or more links infinite in length. The slider–crank (or crank and slider) mechanism shown in Figure 15.3 is a four-bar linkage with the slider replacing an infinitely long output link. This configuration translates a rotational motion into a translational one. Most mechanisms are driven by motors, and slider–cranks are often used to transform rotary motion into linear motion.

Problem Description

A slider–crank mechanism designed using SolidWorks consists of five parts and one subassembly. They are *bearing, crank, rod, pin, piston,* and *rodandpin* (a subassembly of rod and pin). Figure 15.4 shows the exploded view of the mechanism. SolidWorks Motion is used for the motion analysis of the mechanism, which includes kinematic analysis, interference analysis, and dynamic analysis.

SolidWorks Parts and Assembly

The first step in the process of Motion Analysis is to have the parts and assembly available. For our study, Figure 15.5 shows the AssemblyManager of the slider–crank mechanism.

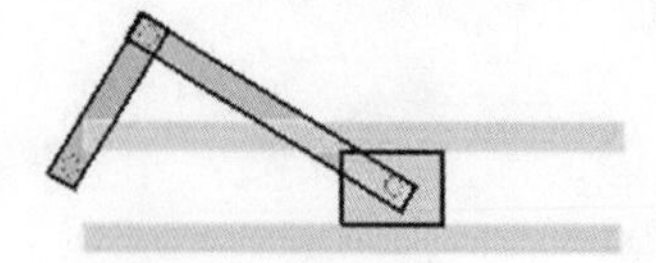

Figure 15.3

Slider–crank mechanism.

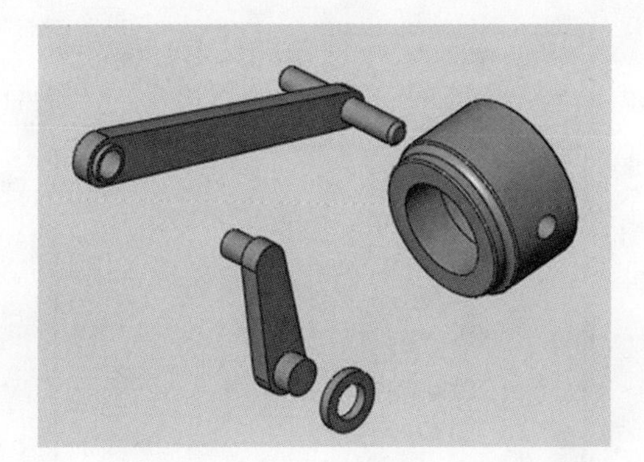

Figure 15.4

Model to be studied using SolidWorks Motion.

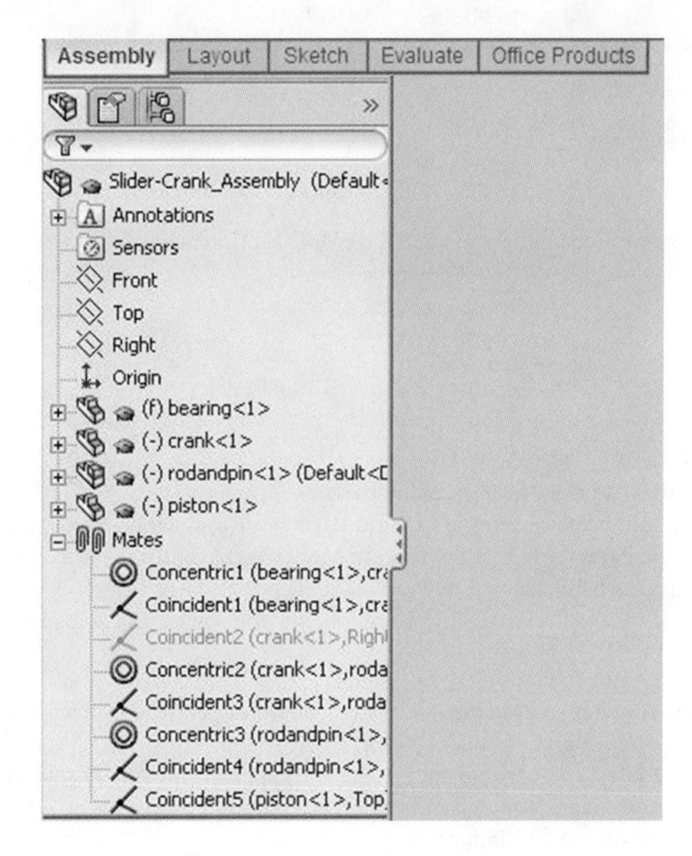

Figure 15.5

AssemblyManager of parts and assembly of model for the motion analysis.

Using SolidWorks Motion

The Motion Model contains five branches (nodes), as shown in Figure 15.6. These are the following:

1. *Assembly Components:* contains all parts and subassembly defining the assembly
2. *Parts:* contains Moving Parts and Ground Parts
3. *Constraints:* contains joints, contact, couplers, motion
4. *Forces:* contains springs, dampers, bushings, Action Only, Action/Reaction
5. *Results*

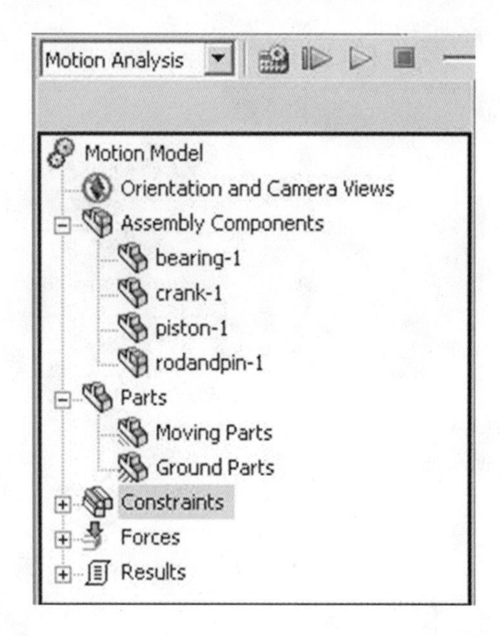

Figure 15.6

Five branches (nodes) of the Motion Model.

Methodology for Motion Analysis Using SolidWorks Motion

The four basic steps for a motion analysis using SolidWorks Motion are as follows:

1. Defining Bodies
2. Driving Joints
3. Turning off Gravity
4. Running Simulation

Defining Bodies

Before defining the bodies, we need to classify the parts as Moving or Grounded. This means that we need decide which part(s) should be placed in the Ground Parts node and which part(s) should be placed in the Moving Parts node of the Parts branch of the Motion Model Manager. In the current example, the bearing will be grounded, whereas other parts will move.

To define the bodies, we require Assembly Components and Parts. The steps involved in defining the bodies are as follows:

1. Expanding the Assembly Components. (There would appear the four components: (1) *bearing-1*, (2) *crank-1*, (3) *piston-1*, and (4) *rodandpin-1*.)
2. Expanding the Parts. (There would appear the two types: (1) *Moving Parts* and (2) *Ground Parts*.)
3. Click *bearing-1* and drag it from Assembly Components and drop it to the Ground Parts in the Parts branch of the Motion Model Manager (see Figure 15.7).
4. Select *crank-1*, *piston-1*, and *rodandpin-1* at the same time using the Shift key, and drag them from Assembly Components and drop them to the Moving Parts in the Parts branch of the Motion Model Manager (see Figure 15.7).

Driving Joints

The constraints branch has Joints, Contact, Couplers, and Motion as submodules (see Figure 15.8). For defining the driving joints, we will examine that Joints could be one of the following types: (a) *concentric*, (b) *revolute*, or (c) *translational* (see Figure 15.9).

1. Right-click the Revolute node and choose Properties (see Figure 15.9).
   The Edit Mate-Defined Joint dialog box is automatically displayed (see Figure 15.10).
2. Choose Velocity for Motion Type, Constant for Function, and enter 360 degrees/sec for Angular Velocity in the Motion tab of the Edit Mate-Defined Joint dialog box (see Figure 15.10).

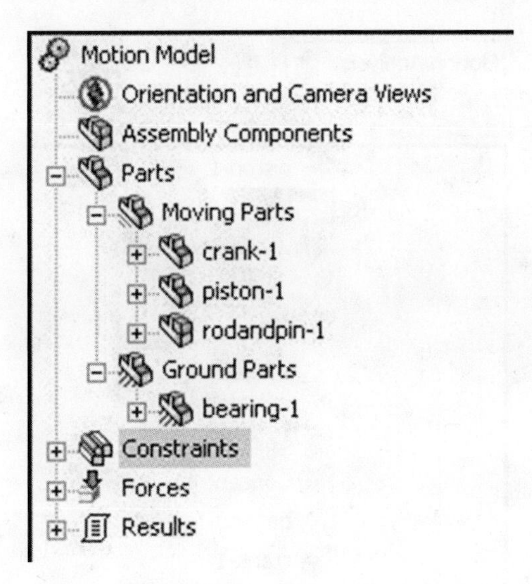

Figure 15.7

Bodies defined.

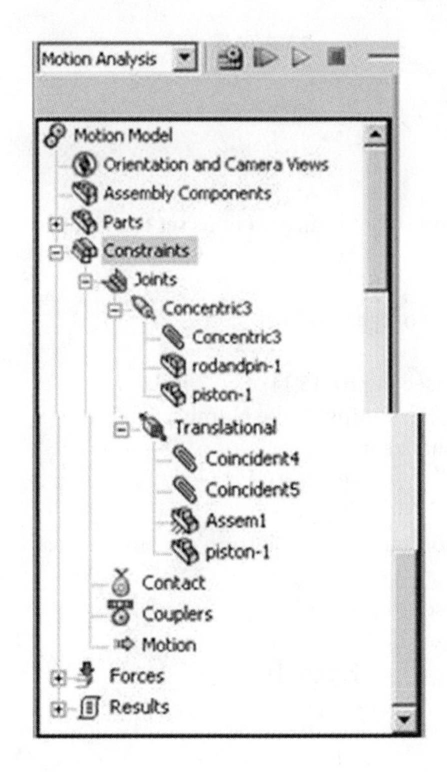

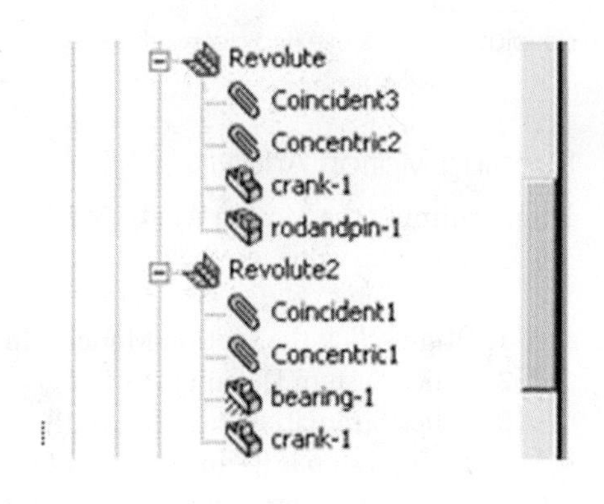

Figure 15.8

Components of Constraints branch and details of Joints.

Turning Off Gravity

There are cases in which the gravity effect is not needed. In order to *turn off Gravity*, do the following:

1. Right-click the Motion ModelManager at the bottom-left side of the window (see Figure 15.11).
2. Click System Defaults. (The SolidWorks Motion Manager is automatically displayed, as shown in Figure 15.12.)
3. In the Simulation Units, select the Force Unit as lbf (see Figure 15.13).
4. Uncheck the Gravity On in the Gravity Parameters options (see Figure 15.13).
5. Click OK.

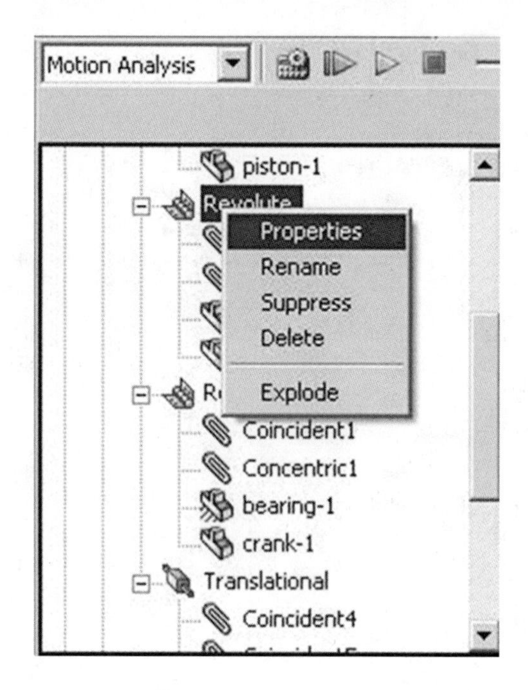

Figure 15.9

Revolute node is used to define the type of motion.

Running Motion Analysis

Before running the Motion Analysis, it may be useful to reset the Number of Frames. Let us set the value to 500. To reset the value, do the following:

1. Right-click the Motion ModelManager at the bottom-left side of the window.
2. Click System Defaults.
3. Click Simulation and change the Number of Frames to 500 (see Figure 15.14).
   This step is useful if we want to visualize what is actually happening. If the Number of Frames is small, the motion may be too fast that we are not able to visualize it.
   To run Motion Analysis,
4. Click the Calculate icon (see Figure 15.15).
   The results are obtained by right-clicking a part or joint of interest and plotting the output(s), as shown in Figure 15.16.
5. Right-click piston-1 > Plot > CM Position > X Component (see Figure 15.16).
6. Right-click piston-1 > Plot > CM Velocity > X Component (see Figure 15.16).
7. Right-click piston-1 > Plot > CM Acceleration > X Component (see Figure 15.16).

Kinematic Analysis

The study setup so far is for the kinematic analysis of the slider–crank mechanism. In this case, no effect of loading is considered; we are only interested in the motion of the machine members without considering the inertial effects.

Figures 15.17 through 15.19 shows the results for position, velocity, and acceleration, respectively, for the piston. Similar results can be obtained for any part of the mechanism that we are interested in.

Figure 15.17 shows the result for position of the piston.

In Figure 15.17, it could be observed that the piston moves from a minimum of 5 in. to a maximum of 11 in., resulting in a stroke of 6 in. (11–5). Initially, the piston is at a position of 7.3 in., progressively increasing within 0.3 sec to 11 in. Then, it progressively decreases until 5 in. for the next 0.5 sec (0.8–0.3). It then moves back to the starting point within the next 3 sec. It could be therefore inferred that the bottom dead center is at a position, 5 in., while the top dead center is at a position, 11 in. There is a dwell at the top dead center (at 0.3 s) for about 0.05 sec. Similarly, there is a dwell at the bottom dead center (at 0.8 s) for about 0.025 sec.

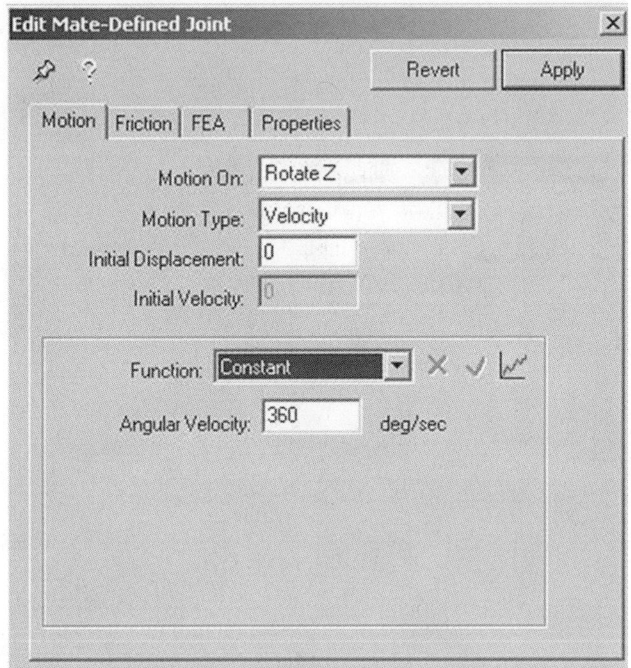

Figure 15.10

Edit Mate-Defined Joint dialog box and definitions for Motion.

Figure 15.18 shows the result for velocity of the piston. The maximum velocity is 19 in./sec when the piston is at a position of 7.3 in., corresponding to 0 sec and 1 sec, respectively. We observed (in Figure 15.17) that there is a dwell at the top dead center (at 0.3 s) for about 0.05 sec. We expect the velocity to be zero in this case. Similarly, we observed (in Figure 15.17) that there is a dwell at the bottom dead center (at 0.8 s) for about 0.025 sec. We also expect the velocity to be zero in this case. Figure 15.18 confirms that the velocity at 0.3 and 0.8 s is zero.

Figure 15.19 shows the result for the acceleration of the piston. There is a basin between 0 and 0.6 s, with constant and minimum acceleration (deceleration) of −86 in/s^2 between 0.25 and 0.35 s, corresponding to

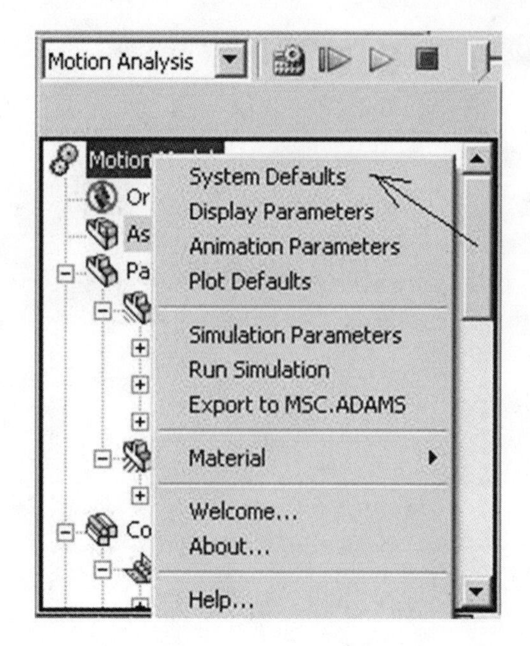

Figure 15.11

Systems Defaults to be accessed.

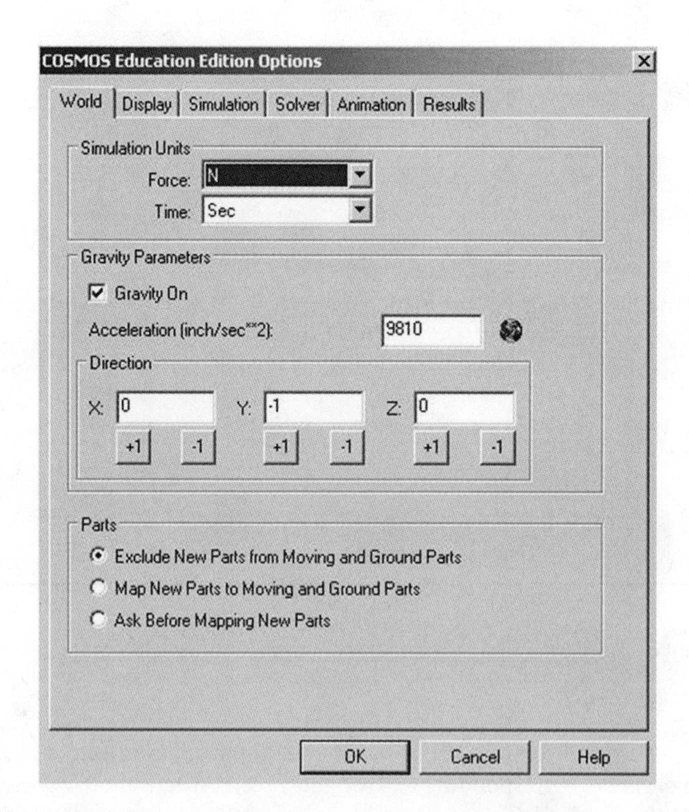

Figure 15.12

Gravity Parameter modified.

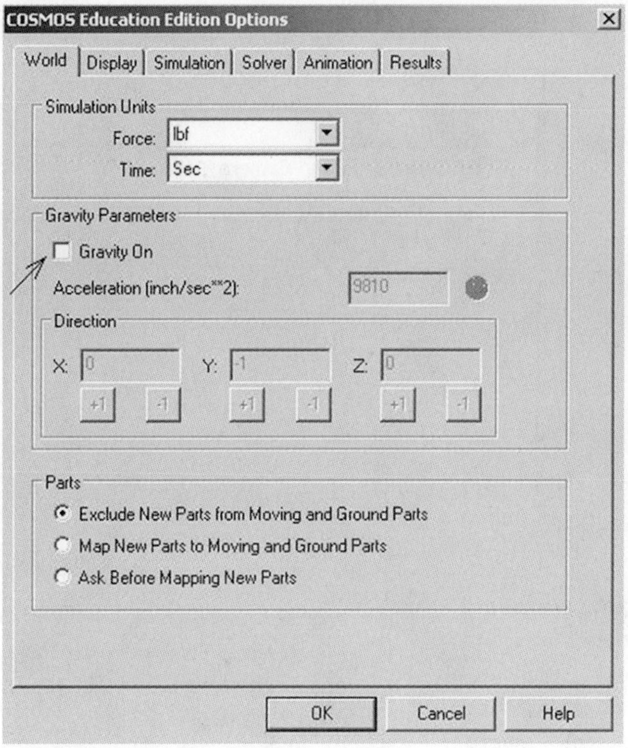

COSMOS Education Edition Options

World | Display | Simulation | Solver | Animation | Results

Simulation Units

Force: lbf

Time: Sec

Gravity Parameters

Gravity On

Acceleration (inch/sec**2): 9810

Direction

X: 0 Y: -1 Z: 0

+1 -1 +1 -1 +1 -1

Parts

Exclude New Parts from Moving and Ground Parts

Map New Parts to Moving and Ground Parts

Ask Before Mapping New Parts

OK Cancel Help

Figure 15.13

Gravity Off.

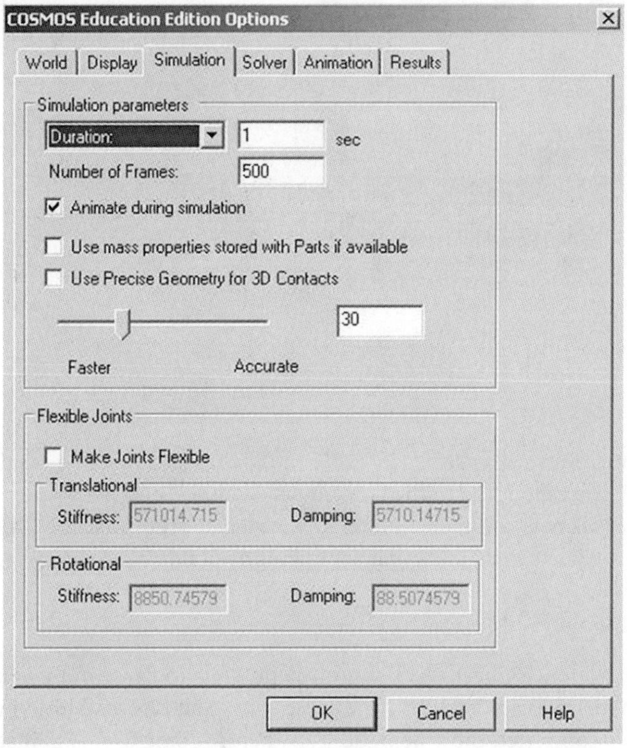

COSMOS Education Edition Options

World | Display | Simulation | Solver | Animation | Results

Simulation parameters

Duration: 1 sec

Number of Frames: 500

Animate during simulation

Use mass properties stored with Parts if available

Use Precise Geometry for 3D Contacts

30

Faster Accurate

Flexible Joints

Make Joints Flexible

Translational

Stiffness: 571014.715 Damping: 5710.14715

Rotational

Stiffness: 8850.74579 Damping: 88.5074579

OK Cancel Help

Figure 15.14

Resetting the Number of Frames.

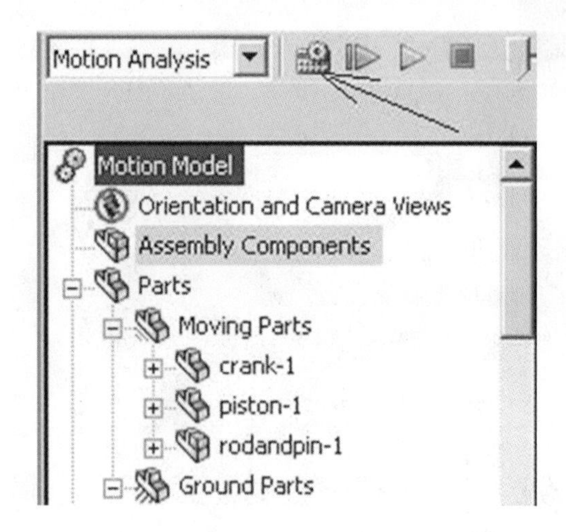

Figure 15.15

Running the Motion Analysis.

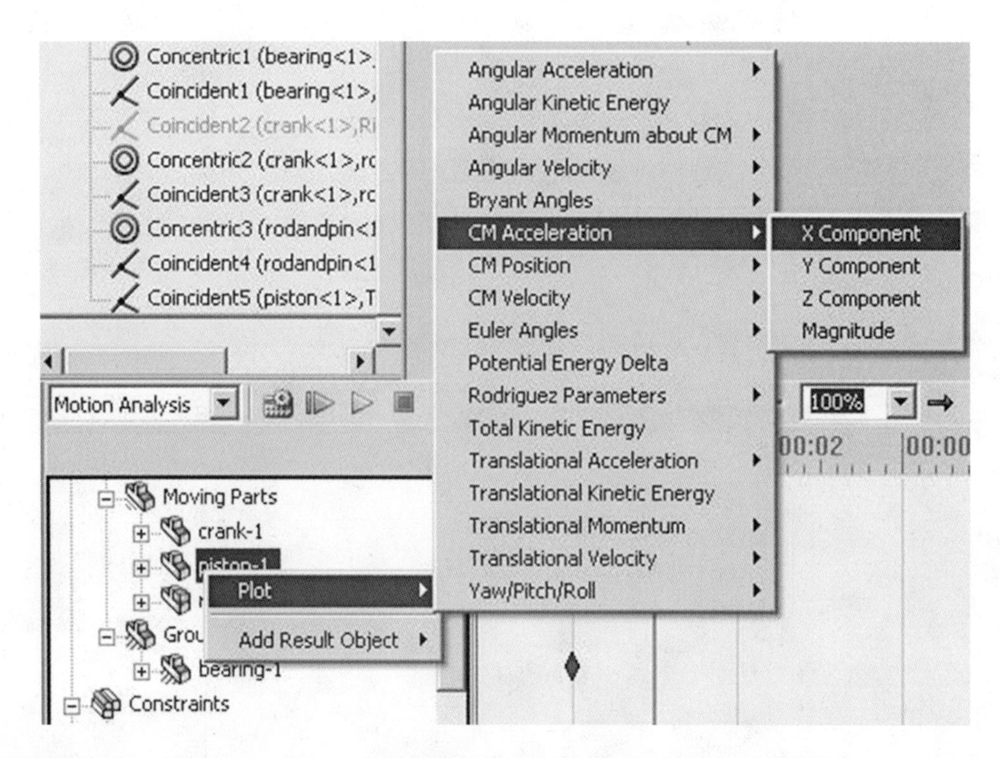

Figure 15.16

Process of generating Results.

the top dead center. There is a similar scenario between 0.8 and 0.85 s with a smaller basin for maximum acceleration of 189 in/s^2, corresponding to the bottom dead center.

Our kinematic analysis is now complete. (Note: to have a new study, use the Calculate icon. SolidWorks will ask you if you want to delete the results that are currently obtained. Answer Yes and continue.)

Interference Check

Our second study is Interference Check. Although it is possible to carry out an interference check in the SolidWorks assembly model, we can do exactly the same using SolidWorks Motion when the parts are moving. Ensure that you have completed the kinematic study before going to this new study. For Interference,

1. Right-click the Motion ModelManager at the bottom-left side of the window. (Notice in Figure 15.20 that the options are now different because there are Results that are available.)

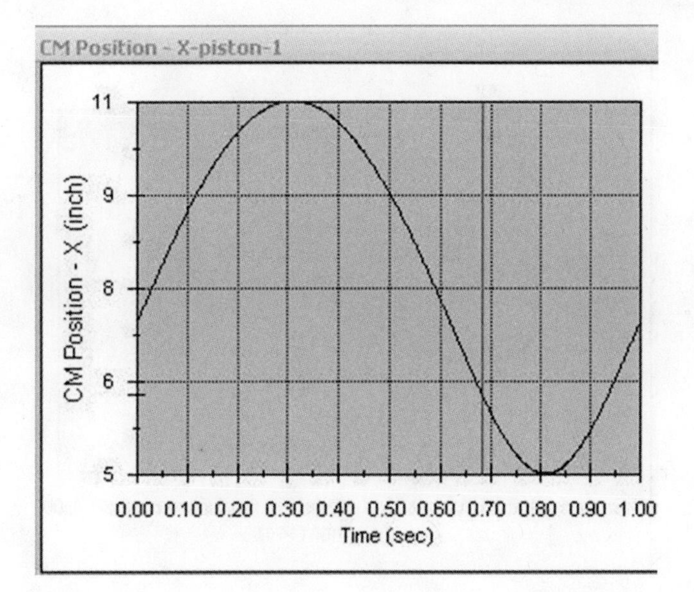

Figure 15.17

Position plot.

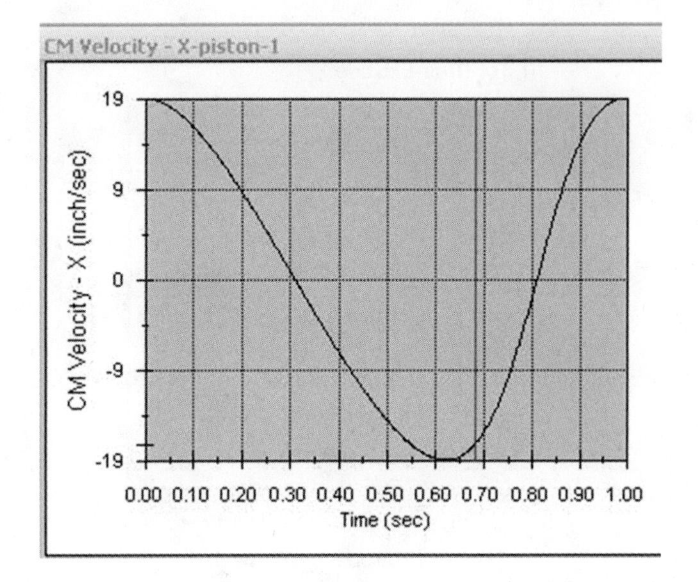

Figure 15.18

Velocity plot.

2. Click Interference Check (see Figure 15.20).
   In the Select Parts to test text field (red in color),
3. Click *each part/subassembly* (piston-1, rodandpin-1, bearing-1, etc.). (See the Find Interferences Over Time PropertyManager in Figure 15.21; we will accept the default values of 1, 501, and 2 for the Start Frame, End Frame, and Increment, respectively.)
4. Click Find Now button. (See Figure 15.22 for the results.)

Dynamic Analysis

Our third study is Dynamic Analysis. Ensure that you delete the current Results before continuing. To delete the current results,

1. Right-click the Motion ModelManager at the bottom-left side of the window.
2. Click Delete Results (see Figure 15.23).

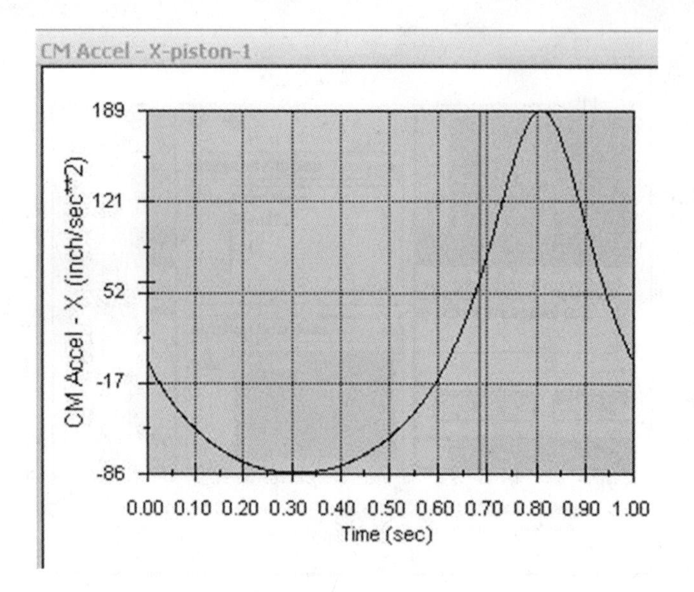

Figure 15.19

Acceleration plot.

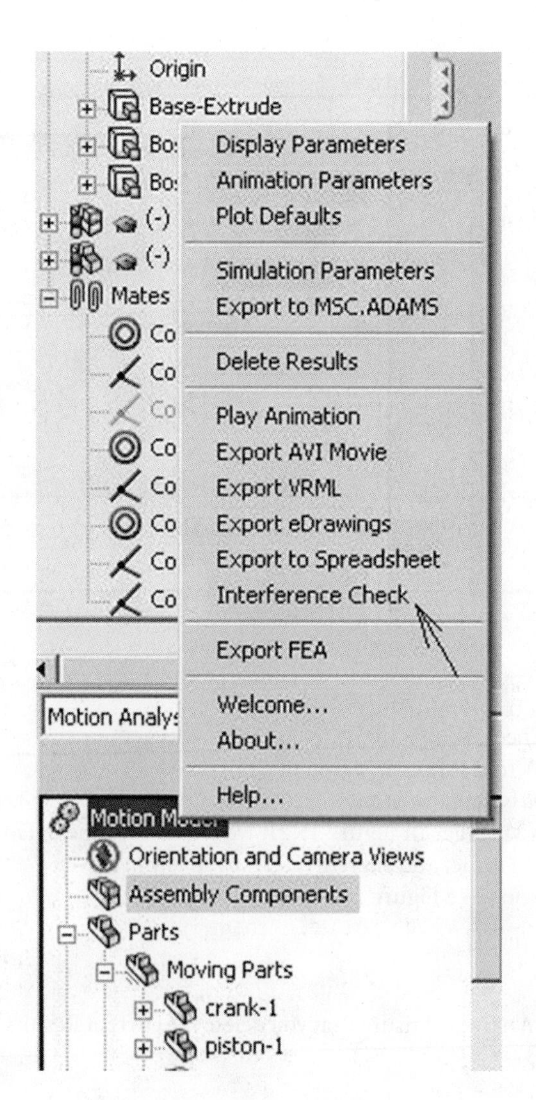

Figure 15.20

Interference Check study selection.

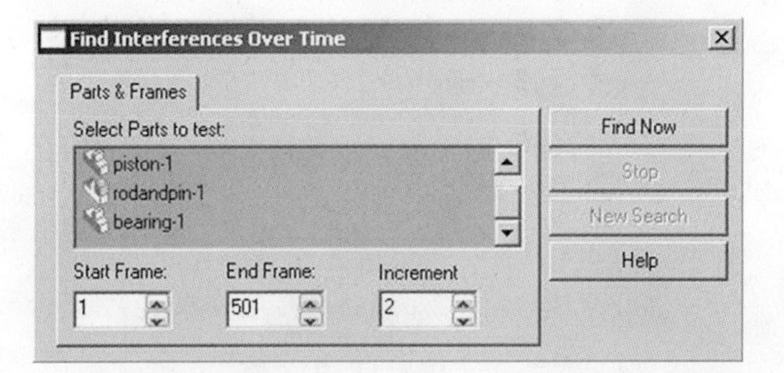

Figure 15.21

Find Interferences Over Time PropertyManager.

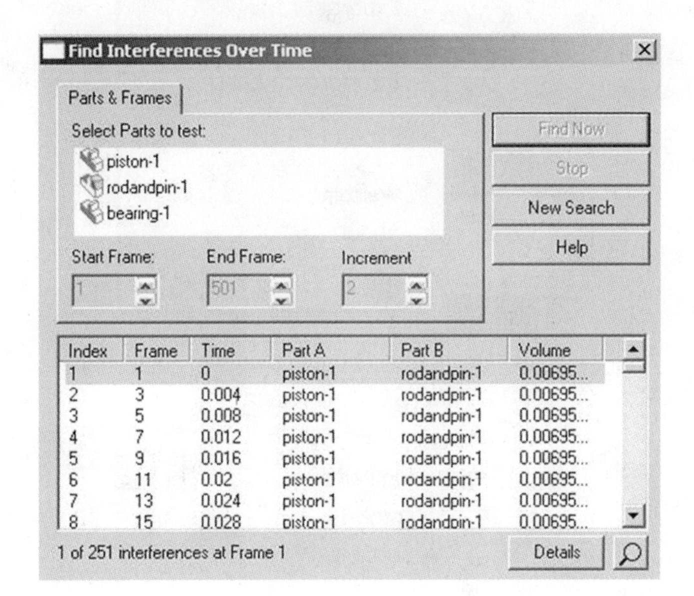

Figure 15.22

Results of interference check.

The steps involved in dynamic analysis are as follows:

1. Freeze the velocity of the driver (at the Joint branch of Constraints).
2. Select the type of Force to apply.
3. Select Component to which Force is Applied (the face to apply the force).
4. Select Reference Component to orient Force (the ground part is the reference).
5. Select Force Function.
6. Run Motion Analysis.

In the dynamic analysis for the slider–crank mechanism, a force of 3 lb. that simulates the engine firing load will be added to the piston. This load will be considered to act along the negative X-direction; hence, a negative value of −3 lbf is given. To begin applying the force, we need to freeze the velocity of the revolute joint first, which is the driver for the slider–crank mechanism.

1. *Freeze the velocity of the driver*

   Freeze the velocity of the driver. (At the Joint branch of Constraints; in our case, we are looking at Revolve.)

   a. Click Revolve > Properties > Motion Type and select "Free" to freeze the velocity (see Figure 15.24).

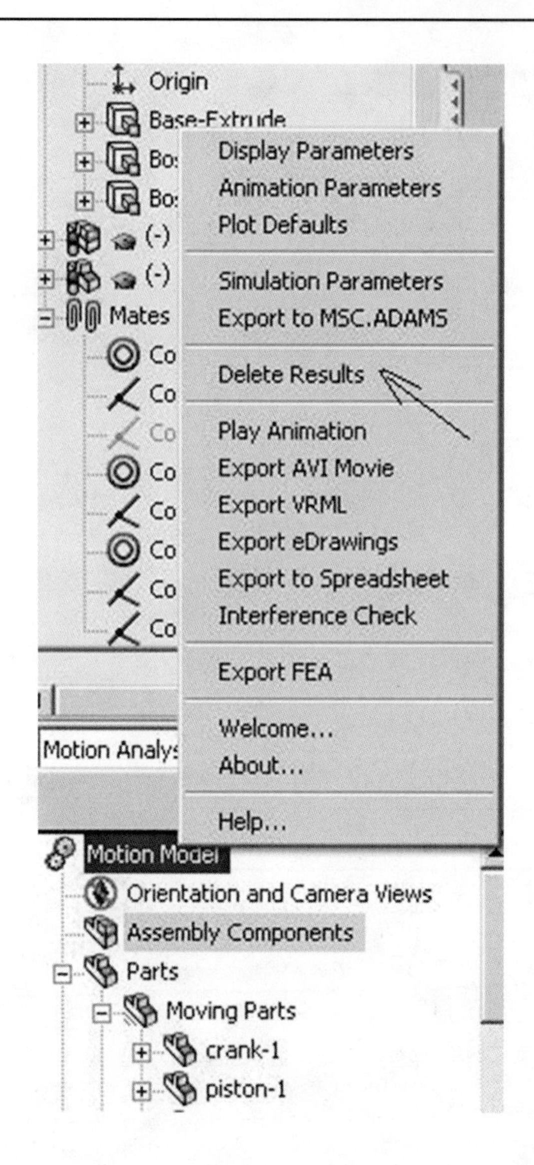

Figure 15.23

Delete current Results.

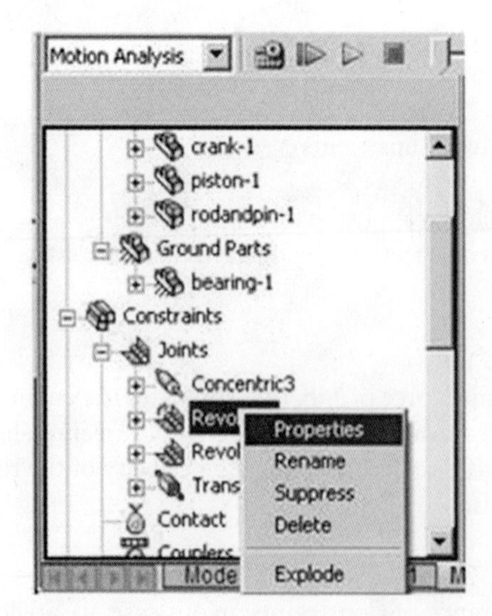

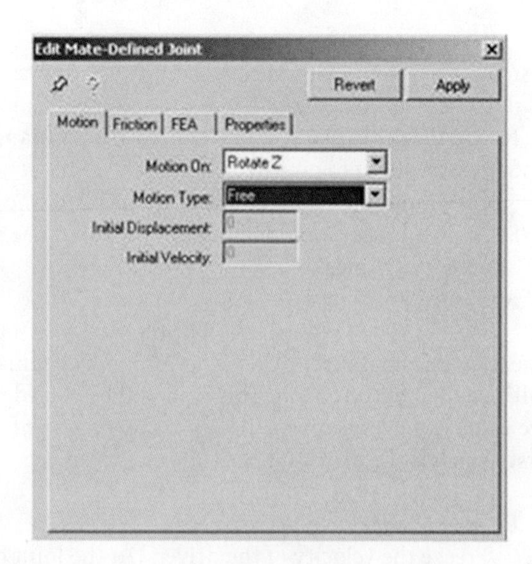

Figure 15.24

Freezing the velocity.

2. *Select the type of Force to apply*

There are a number of types of forces (*springs*, *dampers*, *bushings*, *action-only force*, and *action-only moment*). There is therefore the need to select the type of force to apply. We are interested in Action-Only Force.

a. Click Forces > Add Action-Only Force to Add Forces (see Figure 15.25).

The Insert Action-Only Force PropertyManager (see Figure 15.26).

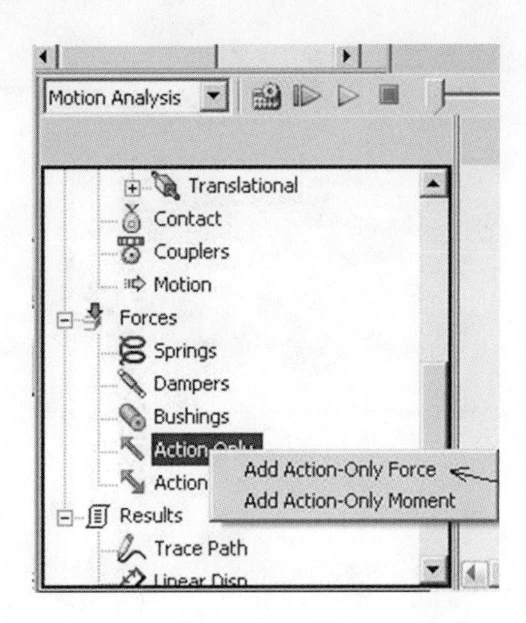

Figure 15.25

Add Force.

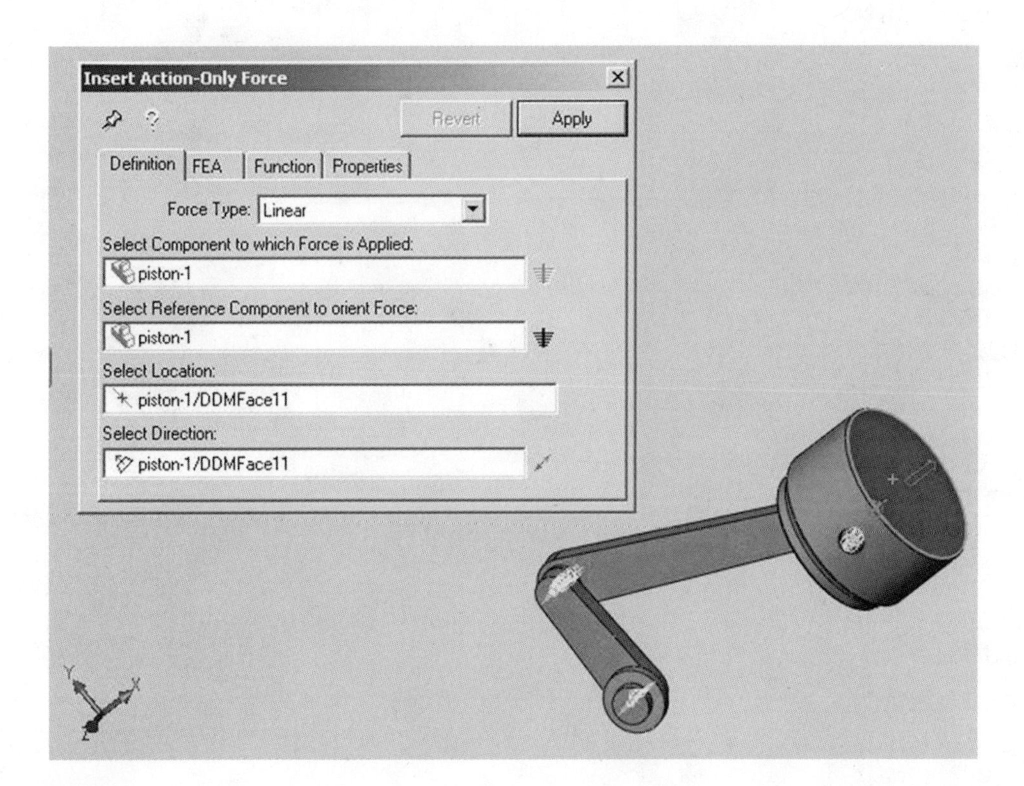

Figure 15.26

Adding force on the face of the piston.

3. **and 4.** *Select Component to which Force is applied and Select Reference Component to orient Force*
   From the Definition option, two fields have to be filled in the Insert Action-Only Force PropertyManager:
   a. Select Component to which Force is Applied (the face to apply the force).
   b. Select Reference Component to orient Force (the ground part is the reference).
5. Select the *End face of the Piston* for the Select Component to which Force is applied field and the *Ground Parts* for the Select Reference Component to orient Force field of the Insert Action-Only Force PropertyManager (see Figure 15.26).

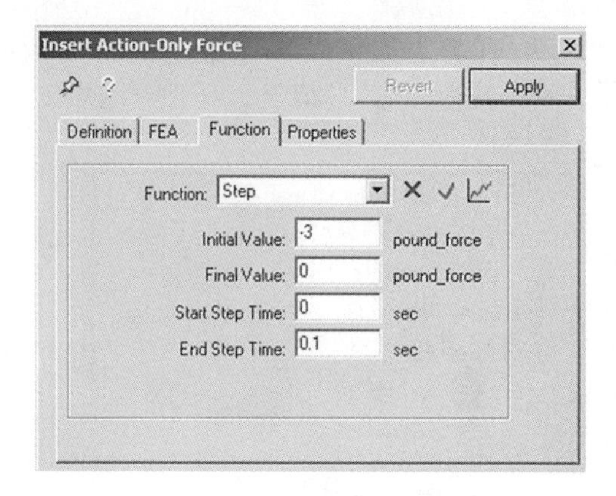

Figure 15.27

Defining the force function.

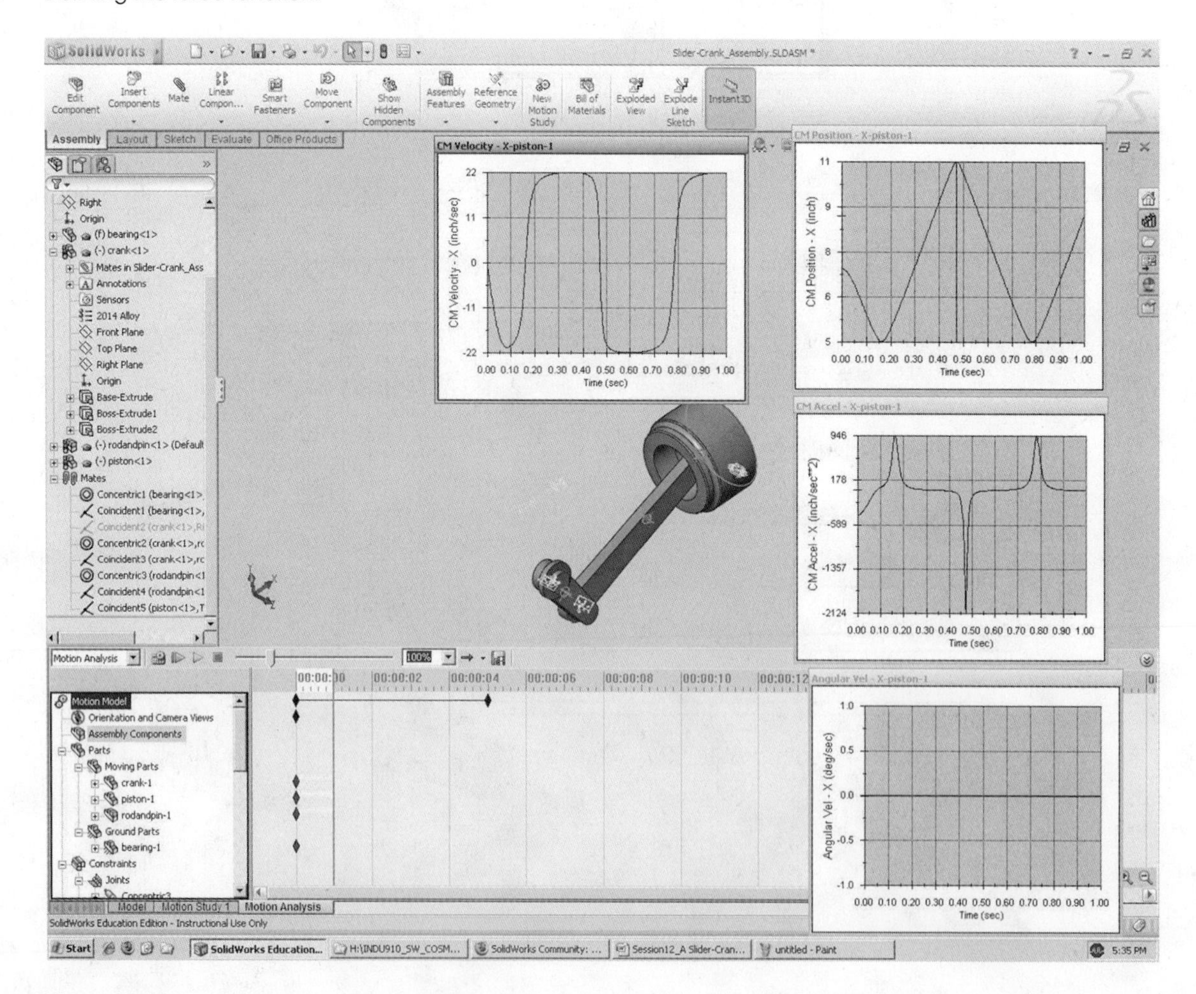

Figure 15.28

Screenshot of position, velocity, and acceleration for dynamic loading.

6. *Select Force Function*

   The force will be defined as a point force (load) acting at the center point on the end face of the piston. Therefore, the force is simulated as a step function of 3 lbf acting along the negative X-direction and applied for 0.1 sec. Notice that the force is applied for a short period. The Type of Function will be Step; the Force has Initial Value = −3 and Final Value = 0; the Start Step Time = 0 and End Step Time = 0.1.

   a. Select the Function Option and choose Step as the Function Type (see Figure 15.27).
   b. For the Force Value (pound_force), supply −3 as the Initial Value of force (lbf) and 0.1 as the End Step Time (Note: accept 0 as Final Value of force (lbf) and 0 as Start Step Time) (see Figure 15.27).
   c. Click *Apply*. (See results in Figure 15.28.)

7. *Run Motion Analysis*

   We are now ready to run the Motion Analysis. To run Motion Analysis,

   a. Click the Calculate icon.

The results shown in Figure 15.28 are automatically generated for the position, velocity, and acceleration of the piston for the dynamic loading. It could be observed that the piston moves in the negative X-direction for 0.15 s and then changes direction moving in the positive direction until 0.45 s when reverses direction until 0.8 in., and moves in the positive X-direction again until 1 s.

Validating the Results

Validating the results obtained using SolidWorks Motion cannot be overemphasized. This is true for using any software. Therefore, we need to check that we are correct by comparing our results with the results from known methods. The results obtained for these motion analyses are consistent with the analytical methods that are found in dynamics textbooks.

Summary

We have presented SolidWorks Motion for solving simple and complex parts or assemblies. The differences between SolidWorks Motion and Animation or Basic Motion have been discussed in detail.

Exercises

1. What is the difference between SolidWorks Motion and Animation or Basic Motion?

2. Are we able to obtain the position, velocity, or acceleration for the slider–crank mechanism using Animation or Basic Motion?

3. Are we able to carry out a dynamic analysis for the slider–crank mechanism using Animation or Basic Motion?

Bibliography

Animation of Crank-Slider Mechanism Using SolidWorks. http://www.me.unlv.edu/~mbt/320/SolidWorks/CrankSliderSWAnimation.htm.

Chang, K.-H. 2008. *Motion Simulation and Mechanism Design with COSMOSMotion 2007*. SDC Publications, Schroff Development Corporation, Mission, Kansas.

Crane, C. EML 2023-Computer Aided Graphics and Design, University of Florida. http://cimar.mae.ufl.edu/~carl/eml2023_spring09/pages/docs/cosmos_motion/CM_Student_Workbook-ENG-2008.pdf.

SOLIDWORKS Forums/MySolidWorks. https://forum.solidworks.com/community/solidworks_simulation/motion_studies/blog/2010/02/08/solvers-used-for-animation-basic-motion-motion-analysis.

Zhang, Y., Finger, S., Behrens, S. Rapid Design through Virtual and Physical Prototyping, Carnegie Mellon University. http://www.cs.cmu.edu/~rapidproto/mechanisms/chpt5.html#fourlink.

16

Rendering

Objectives:

When you complete this chapter, you will have learned the following:

- How to use Zebra Stripes options to render a part or assembly
- How to use Curvature options to render a part or assembly
- How to use the PhotoWorks toolbar to render photorealistic images

Introduction

The *Zebra Stripes* tool allows the user to see small changes in a surface that may be hard to see with a standard display. Zebra Stripes simulate the reflection of long strips of light on a very shiny surface. With Zebra Stripes, the user can easily see wrinkles or defects in a surface and can verify that two adjacent faces are in contact, are tangent, or have a continuous curvature.

The *Curvature* tool allows the user to display a part or assembly with the surfaces rendered in different colors according to the local radius of curvature. Curvature is defined as the reciprocal of the radius (1/radius), in current model units. By default, the greatest curvature value displayed is 1.0000, and the smallest value is 0.0010. As the radius of curvature decreases, the curvature value increases, and the corresponding color changes from black (0.0010) to blue, green, and red (1.0000). As the radius of curvature increases, the curvature value decreases. A planar surface has a curvature value of zero because the radii of flat faces are infinite.

PhotoWorks Studio allows the user to render a model in an existing scene with lights. The user selects one of the studios, and the scene and lights are automatically added. The lights and the scene automatically scale to the size of the model. The floor of the scene positions itself on the bottom of the model relative to the current view orientation.

Surface Model

1. Click New Part document.
2. Create Sketch1 (see Figure 16.1).

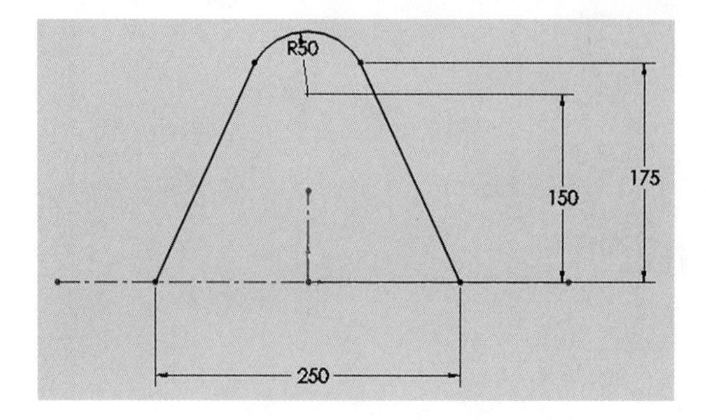

Figure 16.1

Sketch1.

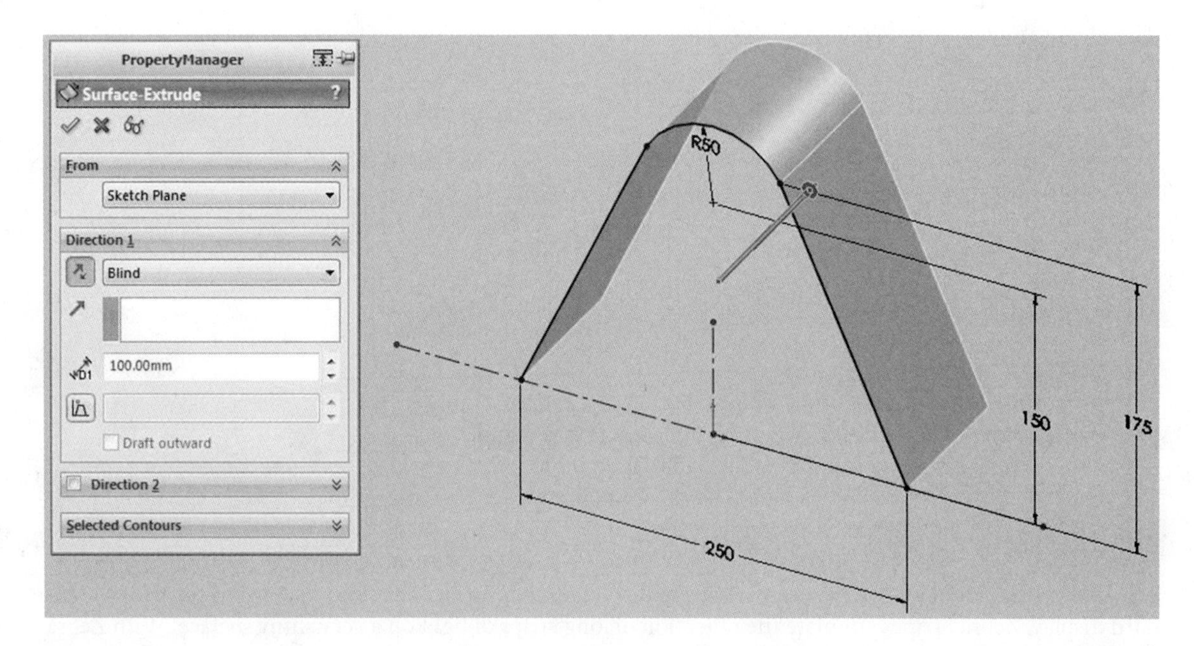

Figure 16.2

Surface-Extrude PropertyManager.

3. Click Insert > Surface > Extrude. (See Figure 16.2 for the Surface-Extrude PropertyManager.)
4. In the Direction1 rollout, set the Extrude Thickness as 100 mm (see Figure 16.2).

Zebra Stripes

Inserting Zebra Stripes Tool

5. Right-click the top-right empty space in the CommandManager.
6. Select the Customize option (see Figure 16.3).
7. In the Customize PropertyManager, click Commands > View (see Figure 16.4).
8. Drag the Zebra button to the CommandManager (see Figure 16.4).

Including Zebra Stripes Effects on a Part

9. With the surface active, click the Zebra Stripes tool from the CommandManager. (See Figure 16.5 for the rendered zebra surface.)

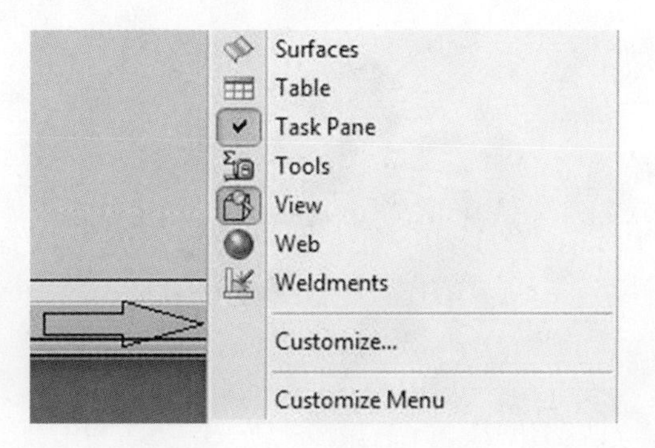

Figure 16.3

Select Customize option.

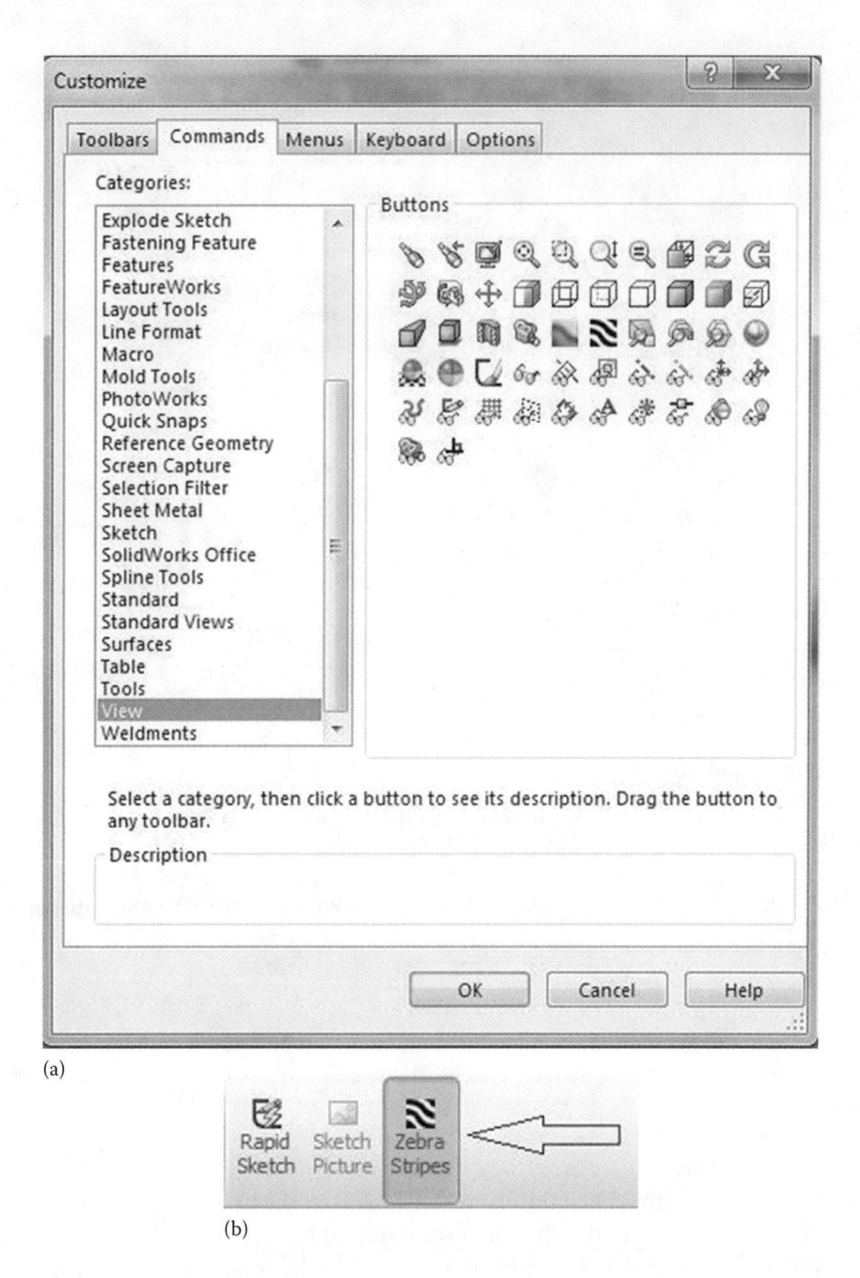

Figure 16.4

Zebra Stripes tool: (a) View Buttons and (b) Zebra Stripes tool.

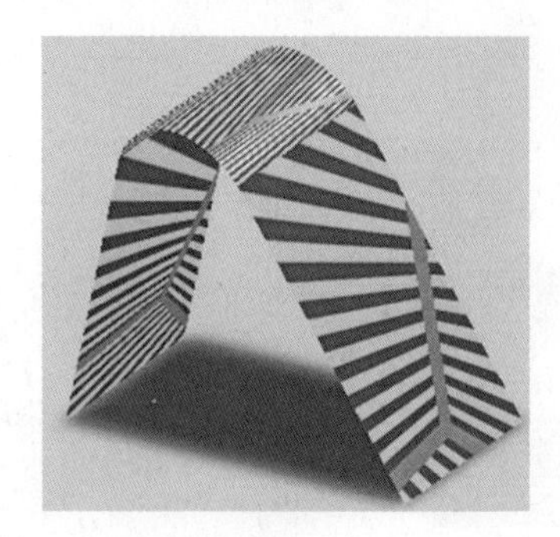

Figure 16.5

Rendered zebra surface.

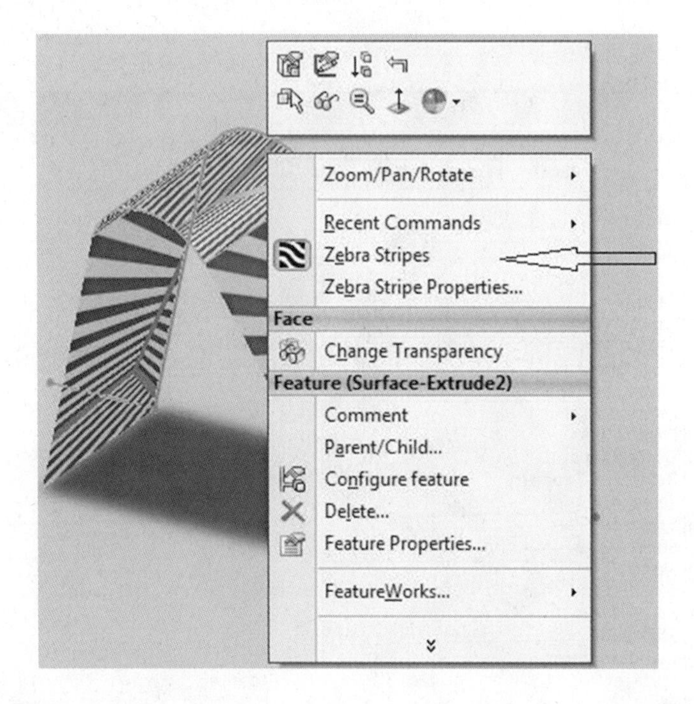

Figure 16.6

Removing Zebra Stripes effects.

Removing Zebra Stripes Effects

10. Right-click the surface having Zebra Stripes effects (see Figure 16.6).
11. Click Zebra Stripes (see Figure 16.6). (The effect disappears.)
    or
12. Click Zebra Stripes from the CommandManager (see Figure 16.6). (The effect disappears.)

Curvature

Inserting Curvature Tool

The procedure is similar to the Zebra Stripes, except that the *Curvature* tool is selected and dragged to the CommandManager.

Including Curvature Effects on a Part

1. With the surface active, click the Curvature tool from the CommandManager. (See Figure 16.7 for the rendered surface with the curvature depicted.)
   Notice: Curvature = 1/Radius of curvature

Removing Curvature Effects

2. Click Curvature from the CommandManager. (The effect disappears.)

Figure 16.7

Rendered surface with the curvature (1/Radius of curvature) depicted.

PhotoWorks Toolbar

Inserting PhotoWorks Tool

1. Click a New Part.
2. Click Add-Ins (see Figure 16.8).
3. Select or check PhotoWorks from the Add-Ins PropertyManager (see Figure 16.9).
4. Click the PhotoWorks Studio tool from the PhotoWorks Studio toolbar (see Figure 16.10). (The PhotoWorks Studio PropertyManager is displayed; see Figure 16.11.)

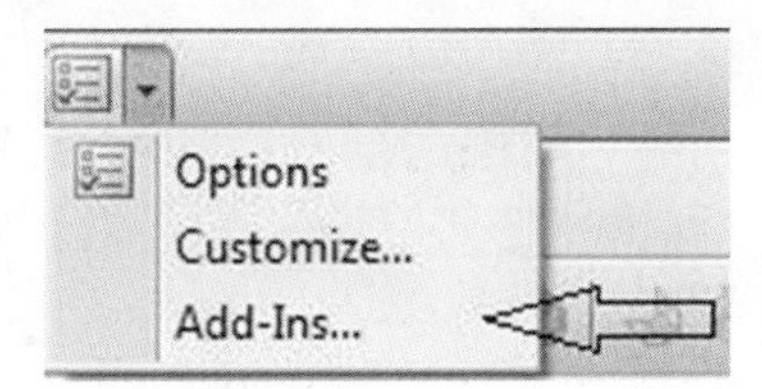

Figure 16.8

Add-Ins options.

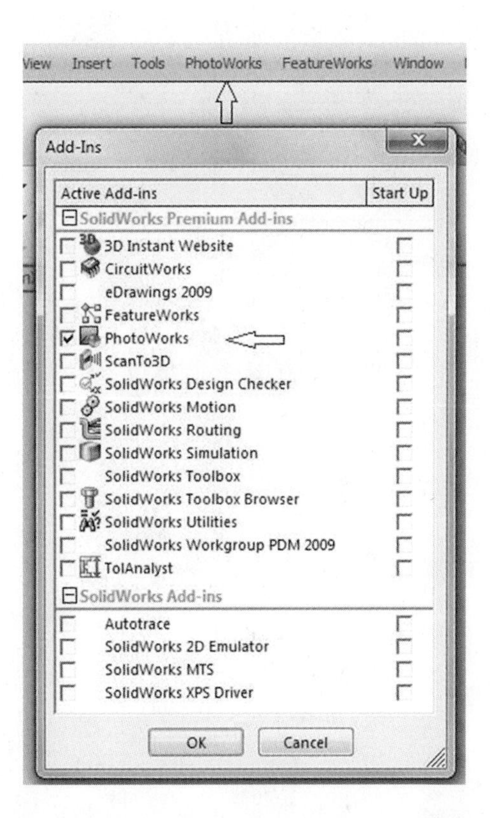

Figure 16.9

PhotoWorks selection from the Add-Ins PropertyManager.

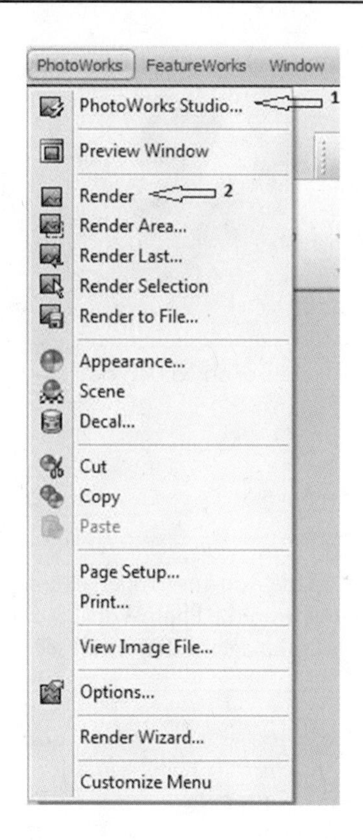

Figure 16.10

PhotoWorks Studio toolbar.

Figure 16.11

Office Space option selection.

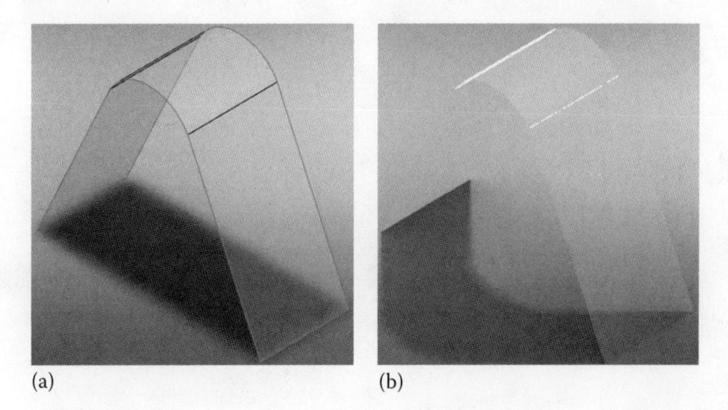

Figure 16.12

Rendered model: (a) before rendering and (b) after rendering.

5. In the Scenery rollout, select Office Space using the drop-down menu (Figure 16.11).
6. Click Render from the PhotoWorks Studio toolbar (see Figure 16.10).
7. Click OK. View the rendered model in the Graphics window (see Figure 16.12).

Summary

This chapter presents Zebra Stripes and Curvature tools, as well as the PhotoWorks Studio, for rendering surface models. The Curvature tool can suggest which area of a surface is under more stress than the others because it is related to the inverse of the radius of curvature. These tools are mainly for aesthetic design and are useful at the design stage for exploring the best surfaces that are required for a product.

Exercises

1. Use the Zebra Stripes tool and PhotoWorks Studio to render the surface that is shown in Figure 12.7.
2. Use the Zebra Stripes tool and PhotoWorks Studio to render the surface that is shown in Figure 12.16.
3. Use the Zebra Stripes tool and PhotoWorks Studio to render the surface that is shown in Figure 12.59.

Section III

Engineering Design Practice with SolidWorks

17

Mold Design

Objectives:

When you complete this chapter, you will have learned the following:

- The terminology and concepts of Draft, Parting Line, Draft Analysis, Shut-Off Surfaces, Parting Surface, and Tooling Split used in mold design
- How to create Parting Line
- How to create Parting Surface
- How to carry out Tooling Split and create cavity and core blocks for injection molds

Mold Design Background

Mold design is a special area of manufacturing processes that considers the process for creating the lower- and upper-half of the mold that is required for producing forged, cast, and plastic parts. Due to its specialty, there is very limited information that is available for a proven methodology using the automated software approach. Mold design is one of the important applications of SolidWorks. The earlier versions of SolidWorks followed a long-winded approach, but the most recent versions of SolidWorks are quite effective in handling a reasonably good number of mold designs. The SolidWorks Mold Tools are available to help users create cavity and core blocks for injection molds. They do not provide libraries for building the entire mold or mold components. The most recent versions of SolidWorks offer significant enhancements in mold design, and these enhancements are covered in this book.

It is a known fact that the Parting Surface in SolidWorks works best on planar parting lines that are convex throughout the part and that mold designers have to create their Parting Surfaces for 70% or more of the involvement in mold design. To put in different words, SolidWorks Mold Tools are not reliable for concave parting lines/surfaces or nonplanar lines/surfaces, which are commonly encountered in complex parts. It is important to note some of the limitations of SolidWorks in mold design. SolidWorks Mold Tools are semi-automated tools. It should helpful to know that creating a moderately complex mold requires some level of manual intervention to get the SolidWorks Mold Tools to deliver reasonably usable results. Several experienced mold designers tend to use different techniques, ranging from cutting away bits of solids, to utilizing surface modeling, to using about 80% Mold Tools techniques, and the rest, manual surfacing.

The remaining part of this chapter presents the application of SolidWorks to mold design. The prerequisite is that the part for which the mold design is to be carried out should have already been modeled. Let us now follow step by step the method that is used in SolidWorks Mold Tools for designing a mold, and we will use the basin, pulley, bowl, plastic cover, and sump parts that we have already modeled.

Mold Design Tools Overview

Mold analysis tools are utilized by the designers of molded plastic parts and by the designers of the mold tools that are used to manufacture those parts. You can create a mold using a sequence of integrated tools that control the mold creation process. You can use these mold tools to analyze and correct deficiencies with either SolidWorks or the imported models of the parts to be molded. Mold tools span from initial analysis to creating the tooling split. The result of the tooling split is a multibody part containing separate bodies for the molded part, the core, and the cavity, plus other optional bodies such as side cores. The multibody part file maintains your design intent in one convenient location. Changes to the molded part are automatically reflected in the tooling bodies.

The following overview from SolidWorks lists the typical mold design tasks and the SolidWorks functions that provide solutions to help users complete those tasks (Table 17.1).

Mold Design Methodology

1. Click Insert > Part > Name*.sldprt.
2. Click Draft > DraftXpert > Add {check Auto paint}.
3. Click Scale > [1.2] {check Uniform Scaling}.
4. Click Parting Line > Pull Direction (Mold Parameter) (Using Temporary Axis; note the direction).
   a. Draft Analysis. (Check "Use for Core/Cavity Split" > OK.)
   b. Pick all lines/curves defining the Parting Line > OK. (An Arrow moves along the Parting Line contour. It should be manually guided.)
5. Click Shut-Off Surfaces > OK. (Clear "Knit option" if advised. If *redundant*, Shut-OFF Surfaces exist, *delete* them.)
6. Click Parting Surface > Top View > Value <1-50> {Check Perpendicular to Pull}.
   a. Check "Knit all surfaces" IF there is a warning.
7. Click TOP PLANE (or other plane) and sketch base (normally rectangular) of TOOLING SPLIT.
   a. Exit Sketch.
8. Check TOOLING SPLIT > PLANE in step 6 and define Heights UP/DOWN from this PLANE, which passes the origin.
   a. Define the height of tooling split upward (mm) from datum.
   b. Define the height of tooling split downward (mm) from datum.
9. Right-click Surface Bodies and *hide* them.
10. Expand Solid Bodies in the FeatureManager.
11. Click Tooling Split [1] > Insert > FeatureManager > Move-Copy.
    a. Move Triad vertical axis in the direction to pull out.
12. Repeat Step 10 for the other Tooling Split [2].

Enhancements in SolidWorks for Mold Design

One area where the most recent versions of SolidWorks significantly differ from their predecessors is in Mold Design. We summarize the new steps for mold design here, for a part, *basin*:

Access Model: **BASIN**

1. Open a New SolidWorks part document. (Note: Mold Tools should be activated.)
2. Set the document properties for the model, with decimal places = 2.
3. Click Insert > Part > BASIN*.sldprt > OK.

Mold Design Procedure

1. Click Draft from the Mold Tools CommandManager (see Figure 17.1).
2. Enter a draft angle and check the Auto part dialog box (see Figure 17.2).

Table 17.1 Mold Tasks and SolidWorks Functions

Tasks	Solutions
When you are not using the models that are built with SolidWorks, import parts into SolidWorks.	Use *Import/Export* tools to import models into SolidWorks from another application. The model geometry in imported parts can include imperfections such as gaps between surfaces. The SolidWorks application includes an *import diagnostic* tool to address these issues.
Determine if a model (imported or built in SolidWorks) includes faces without draft.	Use the *Draft Analysis* tool to examine the faces to ensure sufficient draft. Additional functionality includes the following: • *Face classification*. Display a color-coded count of faces with positive draft and straddle faces. • *Gradual transition*. Display the draft angle as it changes within each face.
Check for undercut areas.	Use the *Undercut Detection* tool to locate trapped areas in a model that prevent ejection from the mold. These areas require a mechanism called a *side core* to produce the undercut relief. Side cores eject from the mold as it is opened.
Scale the model.	Resize the model's geometry with the *Scale* tool to account for the shrink factor when plastic cools. For odd-shaped parts and glass-filled plastic, you can specify nonlinear values.
Select the parting lines from which you create the parting surface.	Generate parting lines with the *Parting Lines* tool, which selects a preferred parting line around the model.
Create shut-off surfaces to prevent leakage between the core and the cavity.	Detect possible sets of holes and automatically shut them off with the *Shut-off Surfaces* tool. The tool creates surfaces to fill the open holes using no fill, tangent fill, contact fill, or a combination of the three. The no-fill option is used to exclude one or more holes though so that you can manually create their shut-off surfaces. You can then create the core and the cavity.
Create the parting surface from which you can create the tooling split. With certain models, use the *Ruled Surface* tool to create interlock surfaces along the edges of the parting surface.	Use the *Parting Surfaces* tool to extrude surfaces form the parting lines that are generated earlier. These surfaces are used to separate the mold cavity geometry from the mold core geometry.
Add interlock surfaces to the model.	Apply these solutions for interlock surfaces: • *Simpler models*. Use the automated option that is part of the *Tooling Split* tool. • *More complex models*. Use the *Ruled Surface* tool to create the interlock surfaces.
Perform tooling split to separate the core and the cavity.	Create the core and the cavity automatically with the *Tooling Split* tool. This tool uses the parting line, shut-off surfaces, and parting surfaces information to create the core and the cavity and allows you to specify the block sizes.
Create side cores, lifters, and trimmed ejector pins.	Use *Core* to extract geometry from the tooling solid to create a core feature. You can also create lifters and trimmed ejector pins.
Display the core and the cavity transparently, enabling you to view the model inside.	Assign different colors to each entity with the *Edit Color* tool. The Edit Color tool also manipulates optical properties such as transparency.
Display the core and the cavity that are separated.	Separate the core and the cavity at a specified distance with the *Move/Copy* Bodies tool.

3. Click the Scale tool and enter the Scale parameter box (see Figure 17.3).
4. Click the Parting Lines tool (see Figure 17.4).
5. When the Parting line PropertyManager appears, click a vertical line for the pull direction; reverse direction if needed (see Figure 17.5).
6. Click the Draft Analysis tool (see Figure 17.5).
7. When the colored model appears, select edges that form a closed loop with the aid of the *GPS* (see Figure 17.6). When the closed loop is successfully selected, a message is displayed (see Figure 17.7).
8. Click the Shut-off Surfaces tool (see Figure 17.8).
9. Click the Parting Surfaces tool (see Figure 17.9).
10. Adjust the Parting Surface Parameter (15 mm or any other larger value; see Figure 17.10).
11. Click the Plane on which the Parting Surface is defined (Top Plane) (see Figure 17.11).

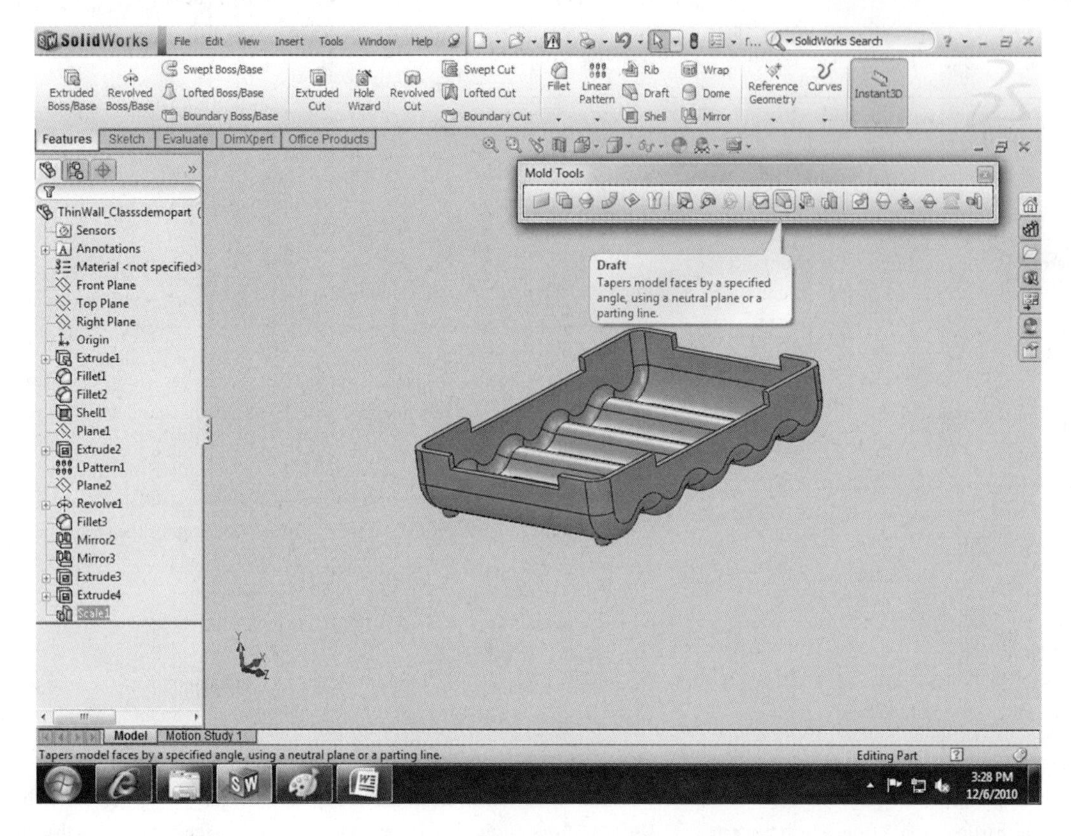

Figure 17.1

Draft tool.

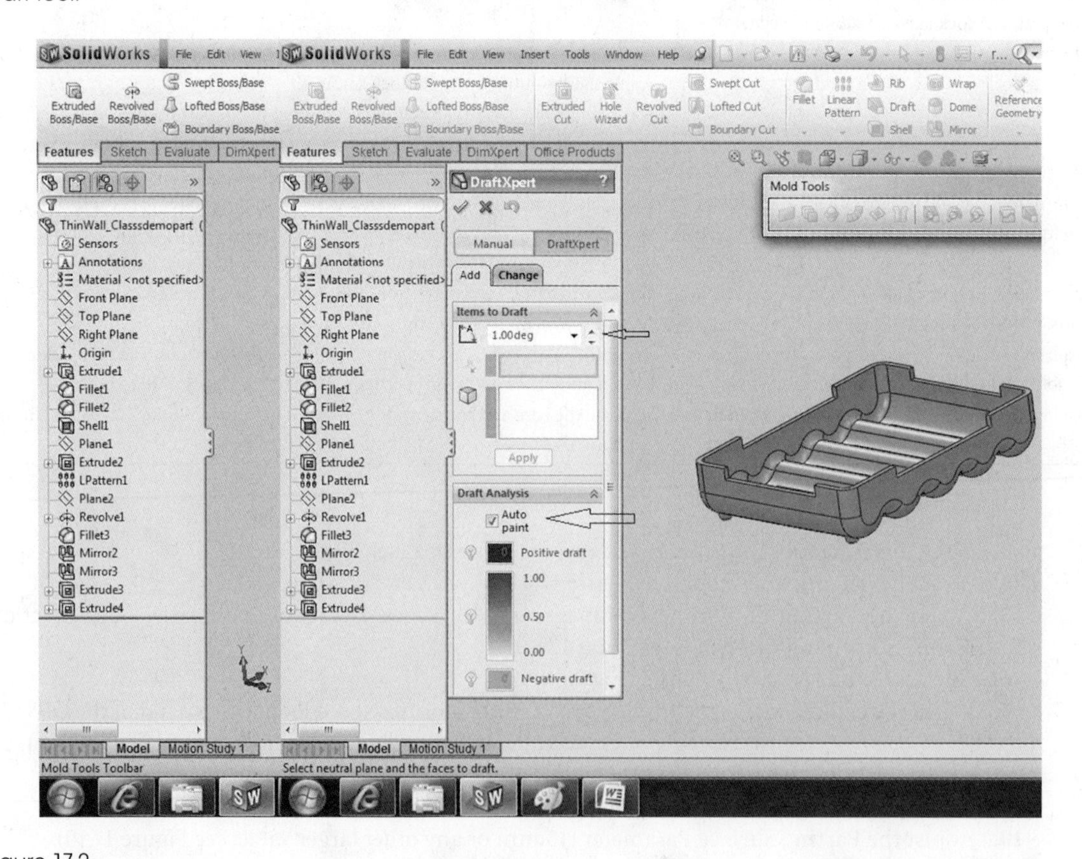

Figure 17.2

Draft angle and Auto part check.

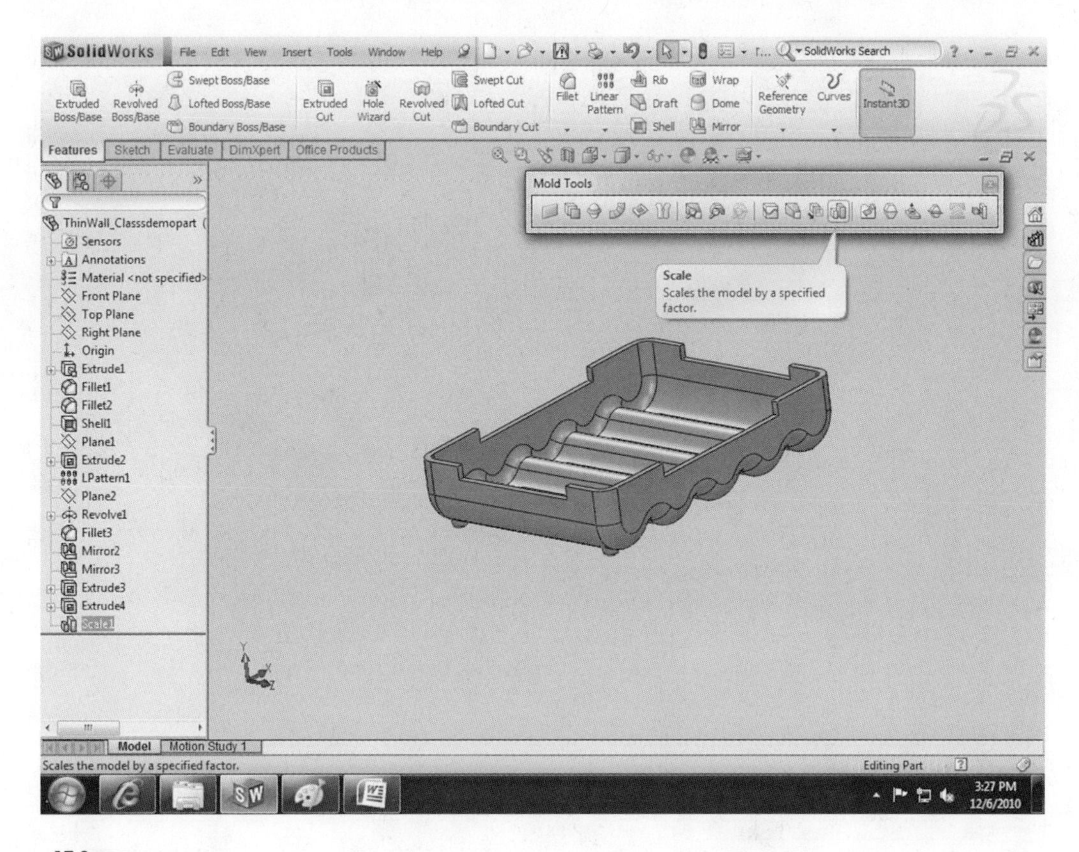

Figure 17.3

Scale factor.

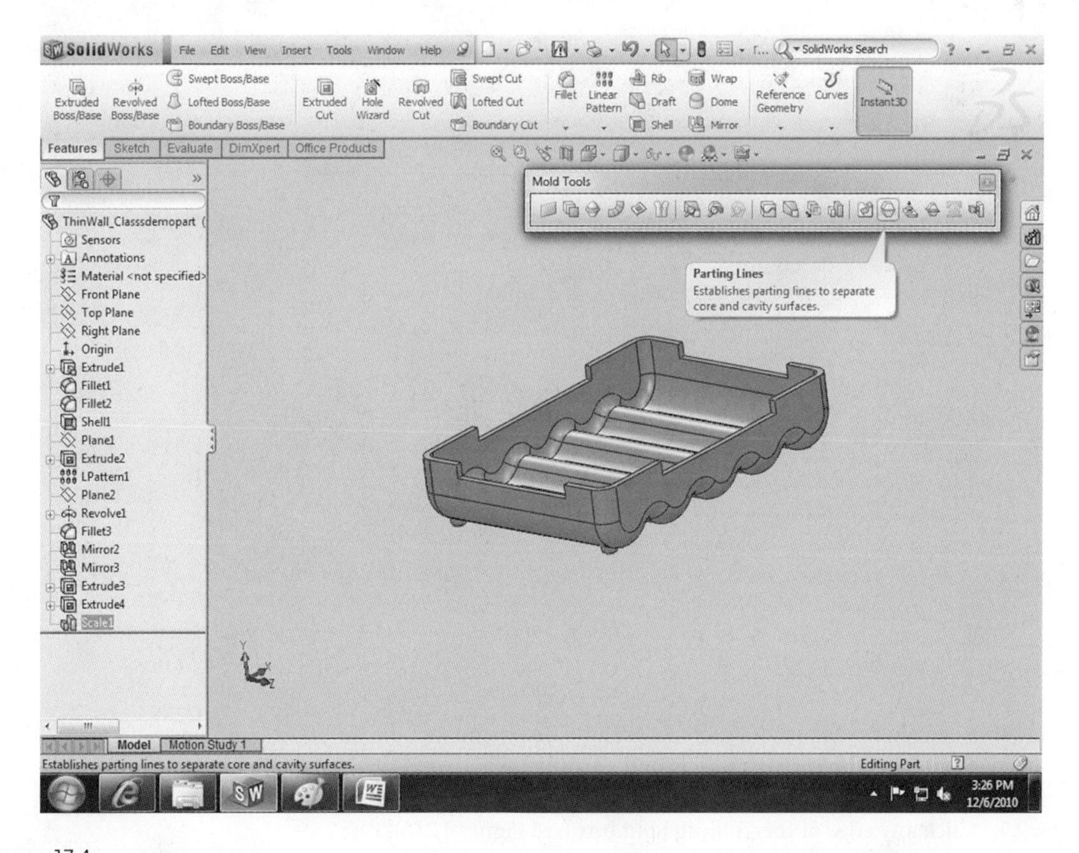

Figure 17.4

Parting lines.

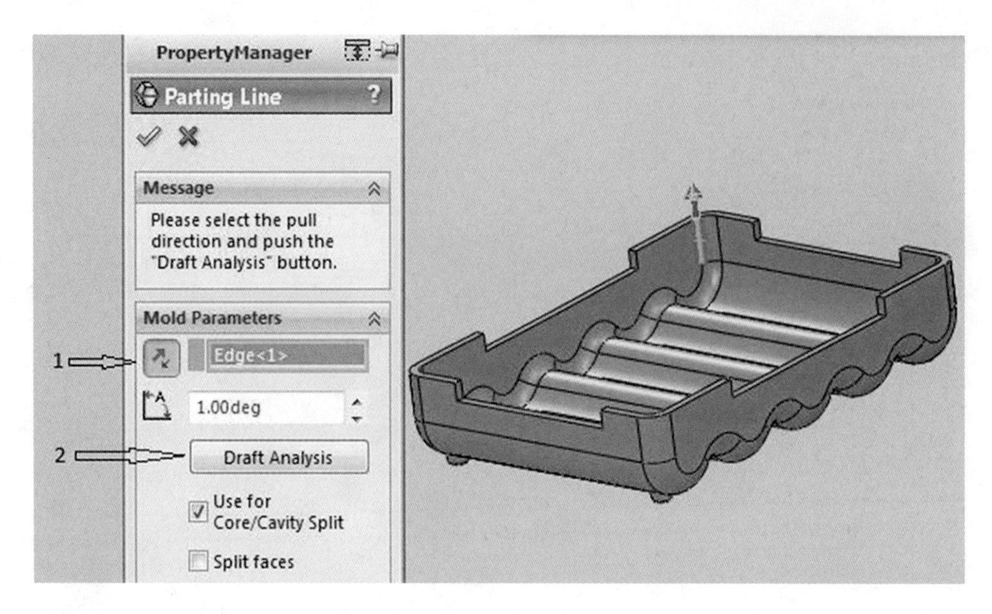

Figure 17.5

Choose Pull direction and Draft Analysis to prepare for defining Parting line.

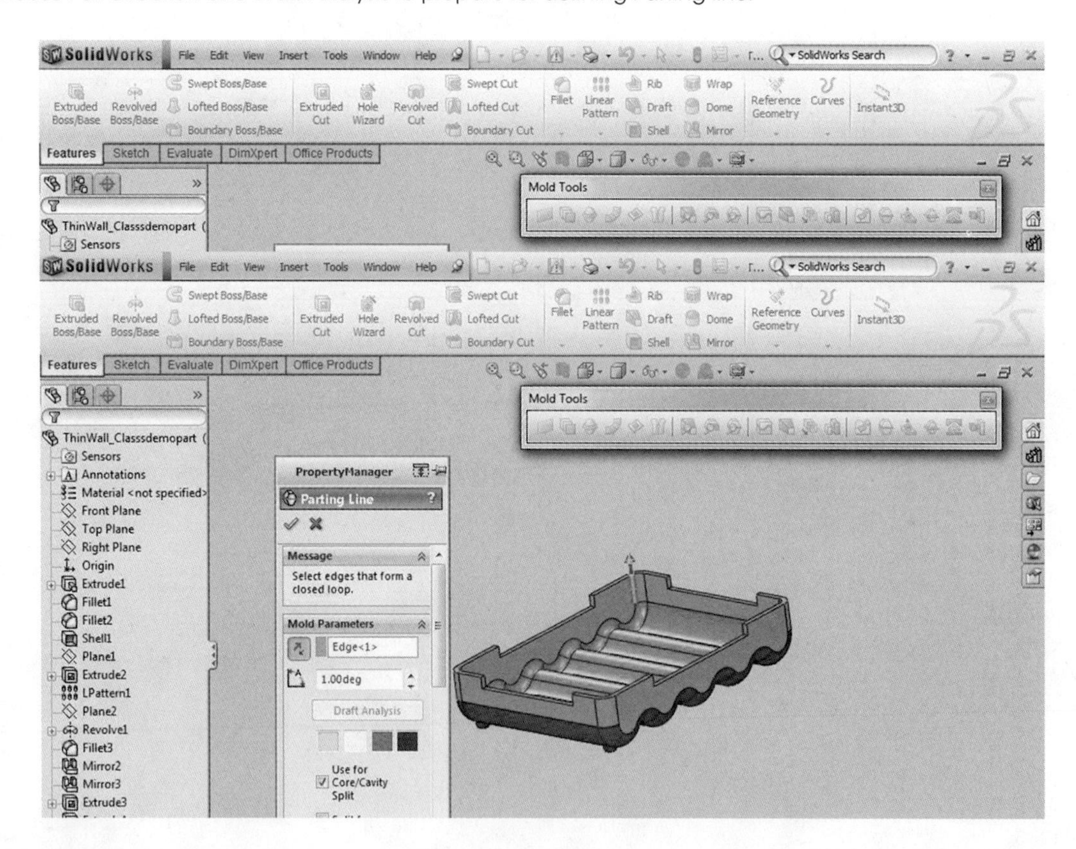

Figure 17.6

Select edges that form a closed loop.

12. Be in Sketch Mode and sketch a rectangle to define Tooling Split box (see Figure 17.12).
13. Exit Sketch Mode (this step is imperative) (see Figure 17.13).
14. Click the Tooling Split icon (see Figure 17.14).
15. Click any edge of the Tooling Split box (see Figure 17.15).
16. When the Tooling Split PropertyManager appears, adjust the Block Size (see Figure 17.16).
17. From the FeatureManager, hide the Surface Bodies.

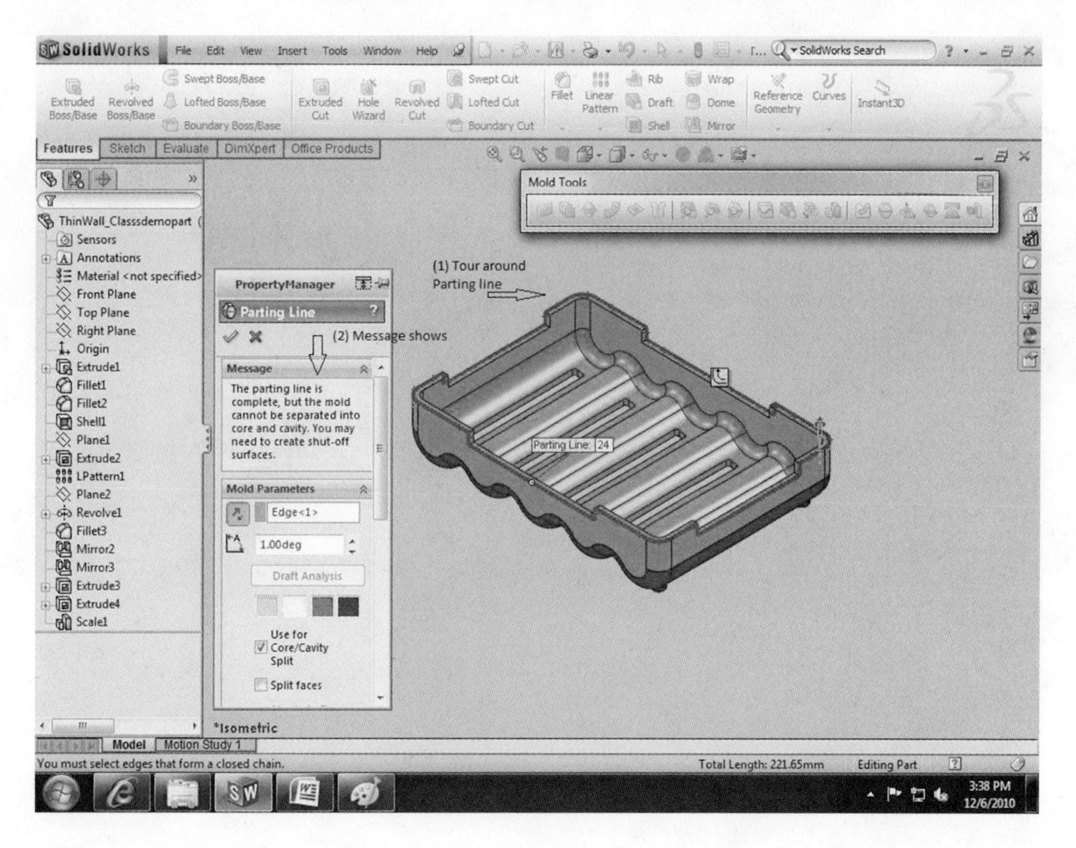

Figure 17.7

Message is displayed to show that parting line is complete.

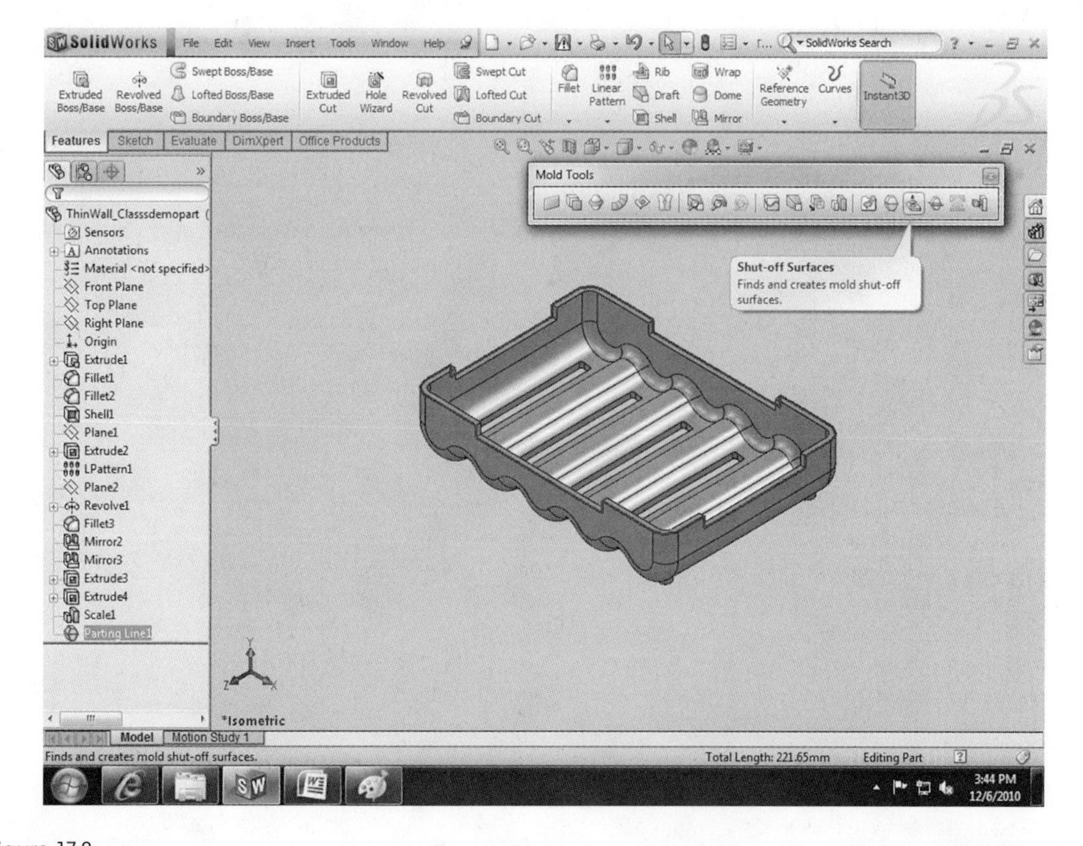

Figure 17.8

Shut-off Surface.

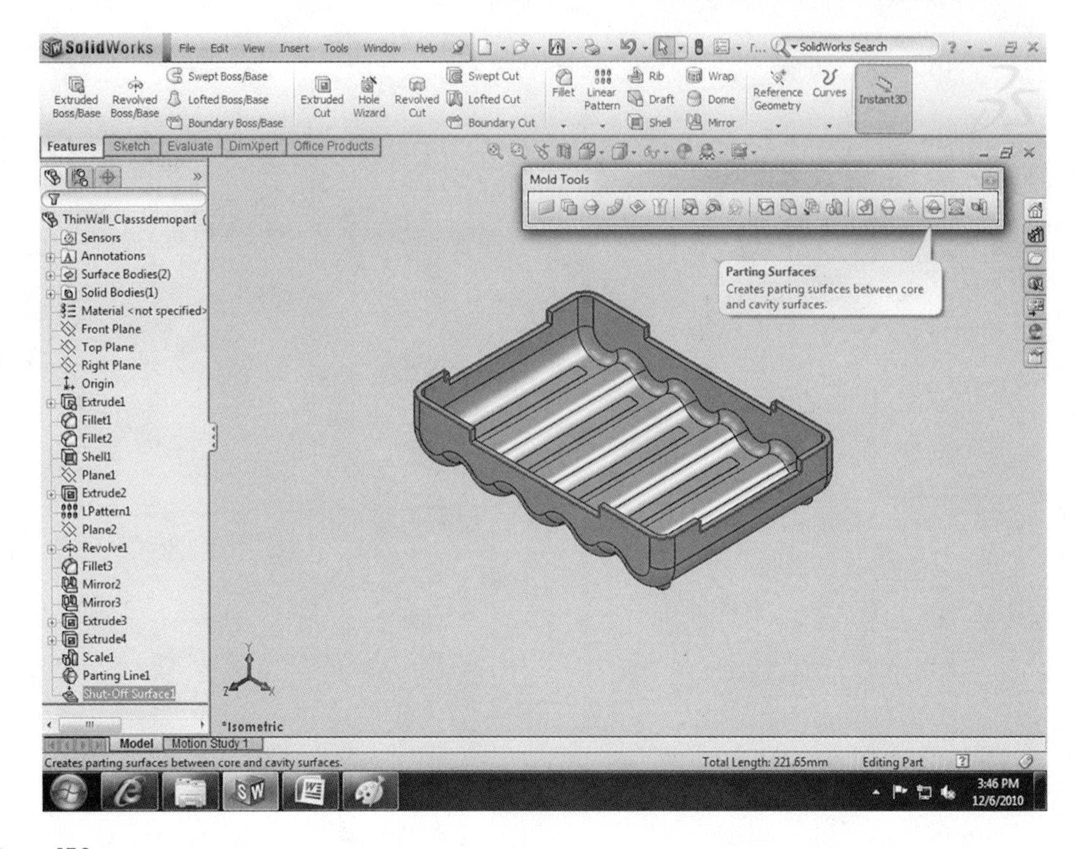

Figure 17.9

Parting Surface.

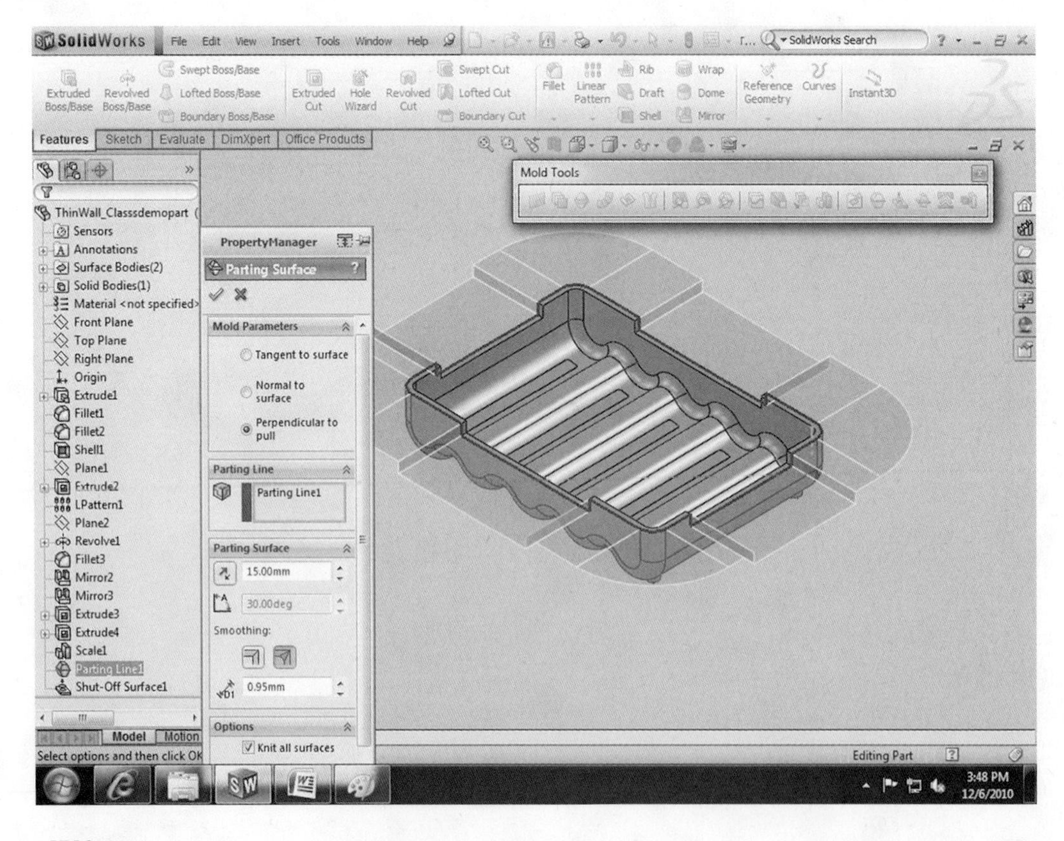

Figure 17.10

Adjust the Parting Surface Parameter to a suitable value (e.g., 15 mm).

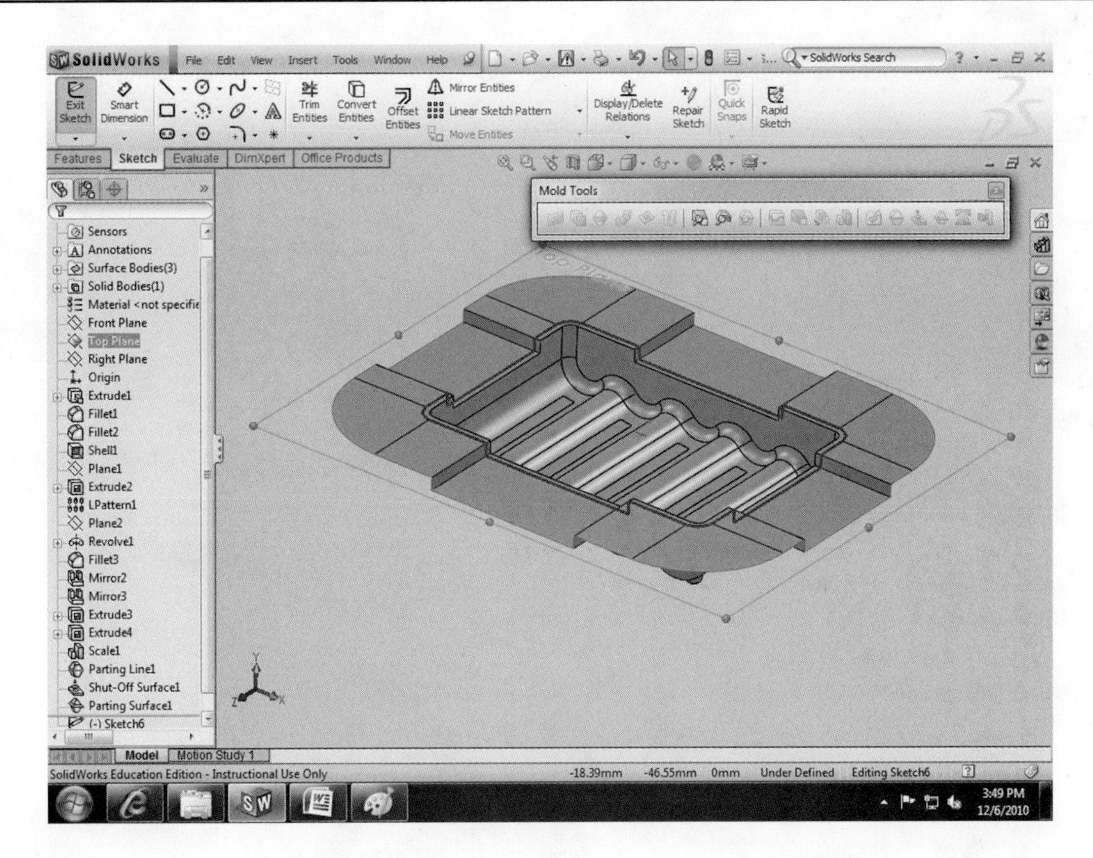

Figure 17.11

Plane defining Parting Surface is selected.

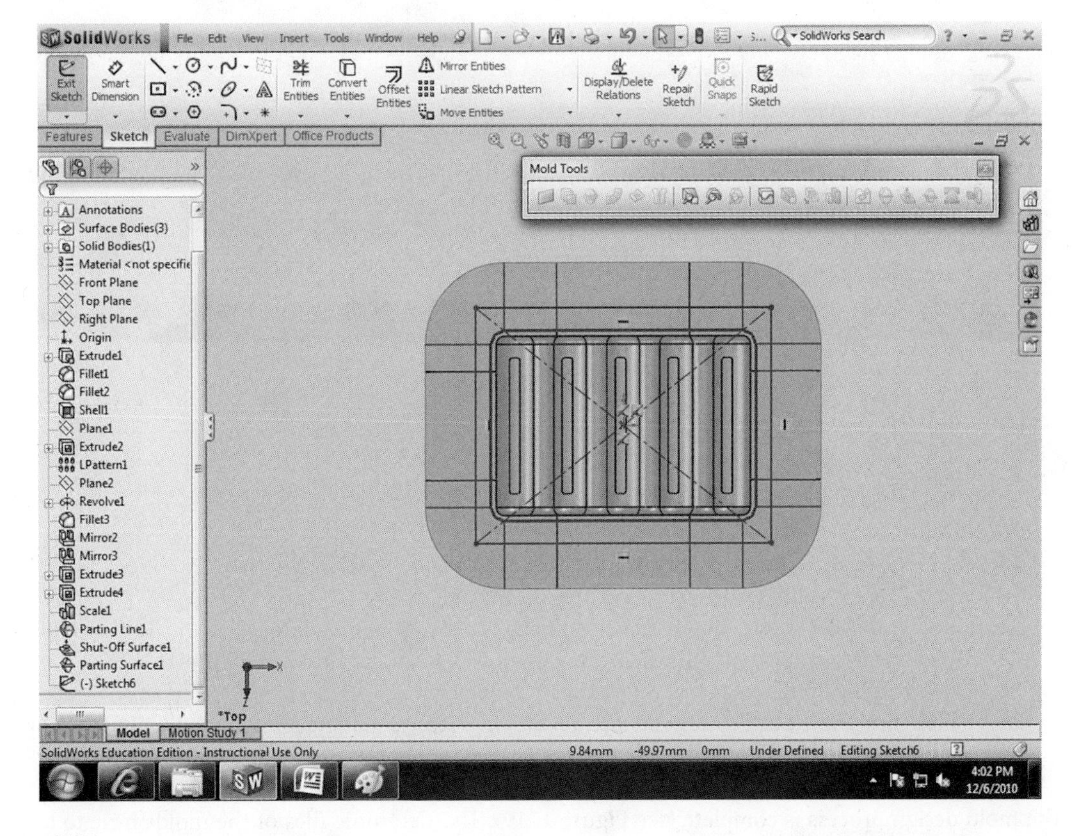

Figure 17.12

Sketch a rectangle to define Tooling Split box.

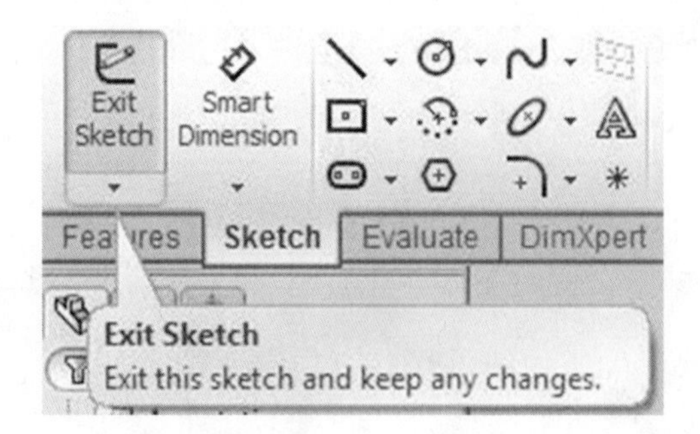

Figure 17.13

Exit Sketch Mode.

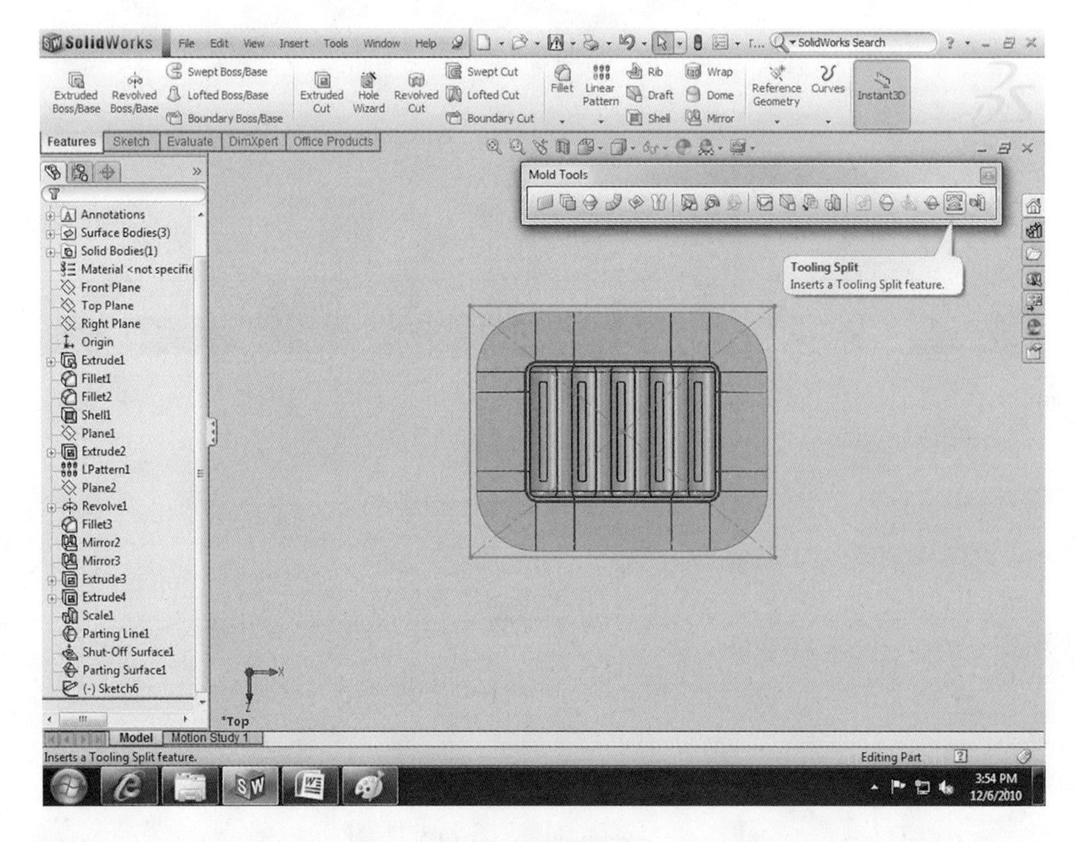

Figure 17.14

Click the Tooling Split icon.

18. From the FeatureManager, hide the Parting Line.
19. From the FeatureManager, expand the Solid Bodies.
20. Click Tooling Split[1] > Insert > Features > Move/Copy.
21. Click Translate/Rotate to activate the Triad. (This step is mandatory for the Triad to appear.) When the Triad appears, pull the Arrow to move part of the Box (see Figure 17.17).
22. Click Tooling Split[2] > Insert > Features > Move/Copy.
23. Click Translate/Rotate to activate the Triad. (This step is mandatory for the Triad to appear.)
24. When the Triad appears, pull the Arrow to move part of the Box (see Figure 17.18).

The mold design process is complete (see Figure 17.19). The electronic files of the molds have to be sent to the manufacturing division for manufacture, while the mold model is retained in the design division for archiving and future revisions.

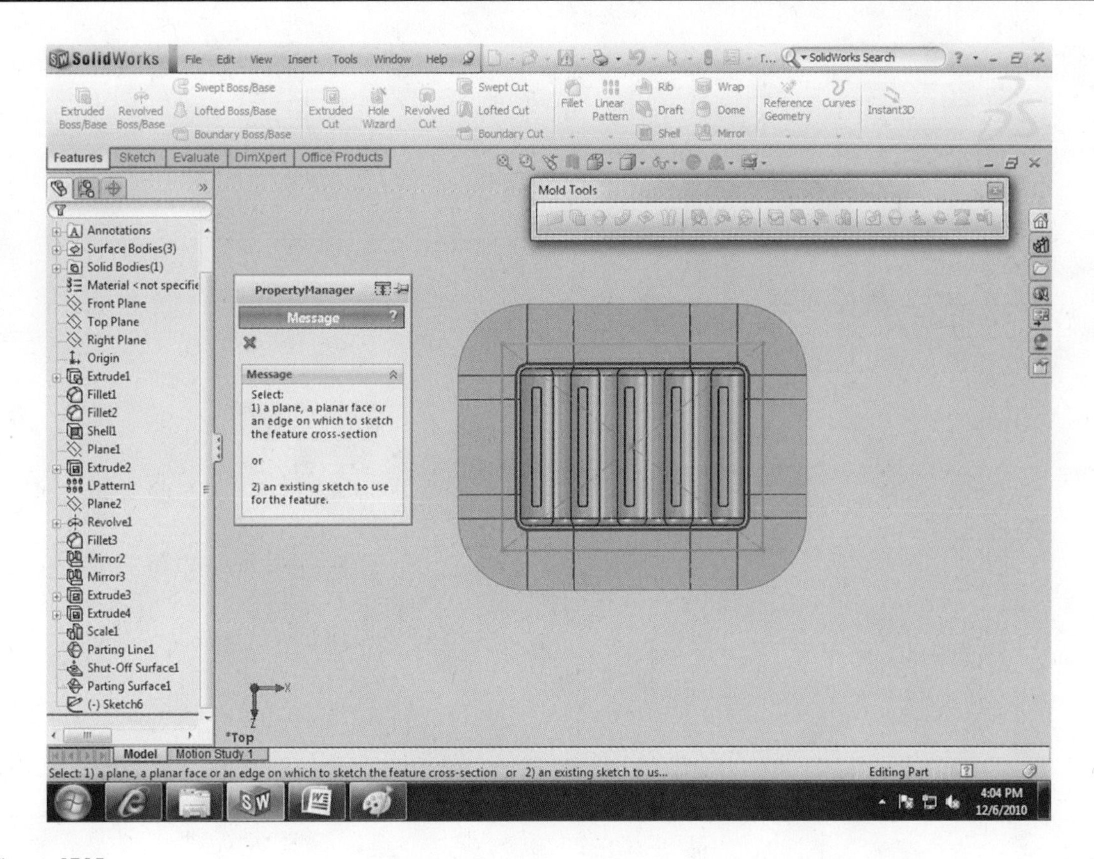

Figure 17.15

Click any edge of the Tooling Split box.

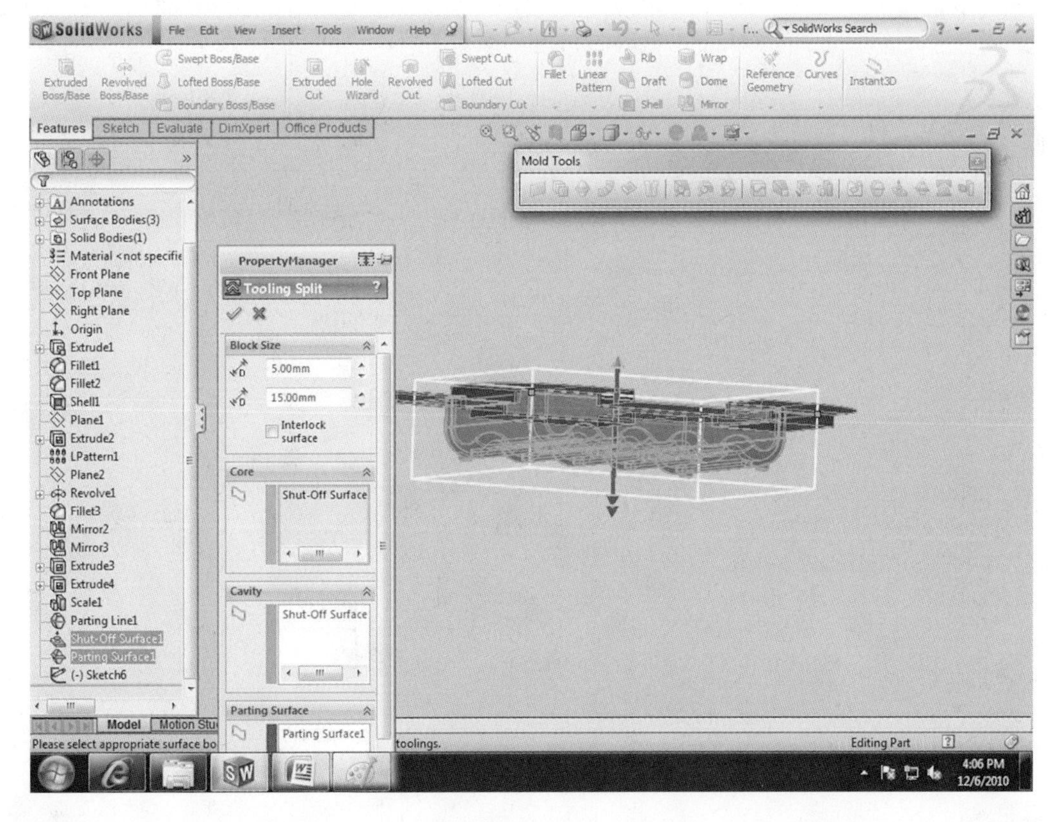

Figure 17.16

Adjust the Block Size from the Tooling Split PropertyManager.

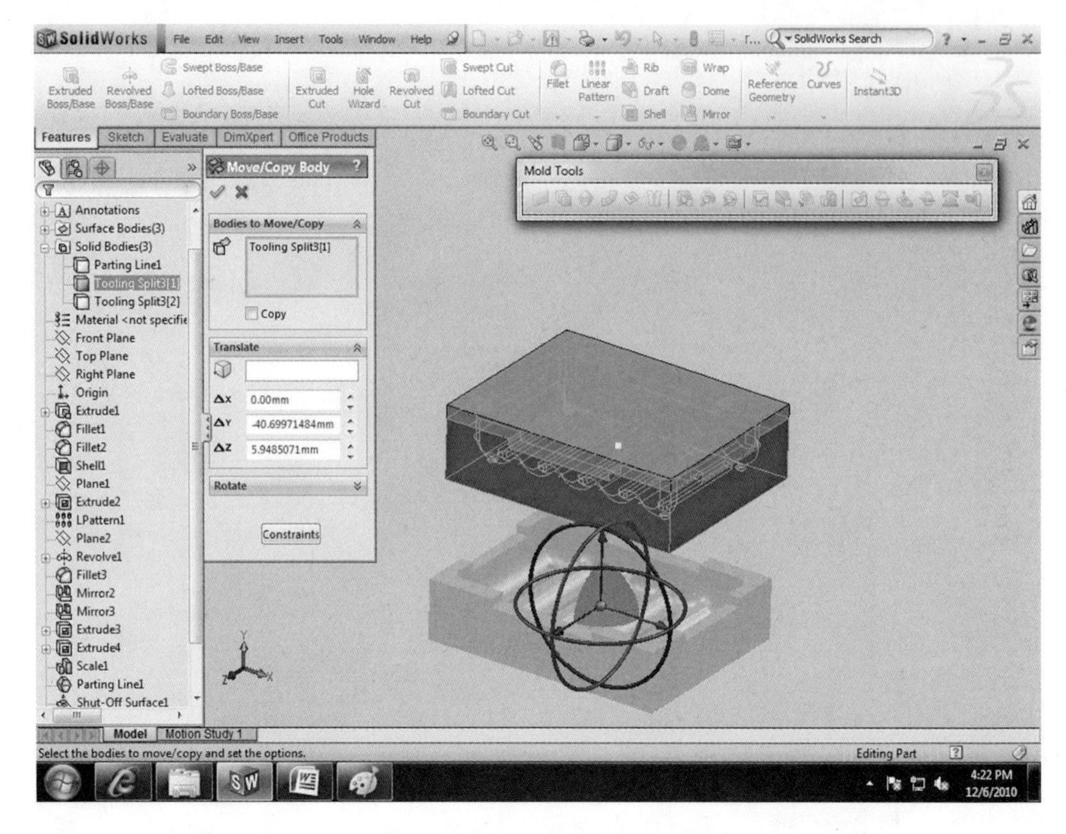

Figure 17.17

Pull Arrow of Triad to move part of the Box.

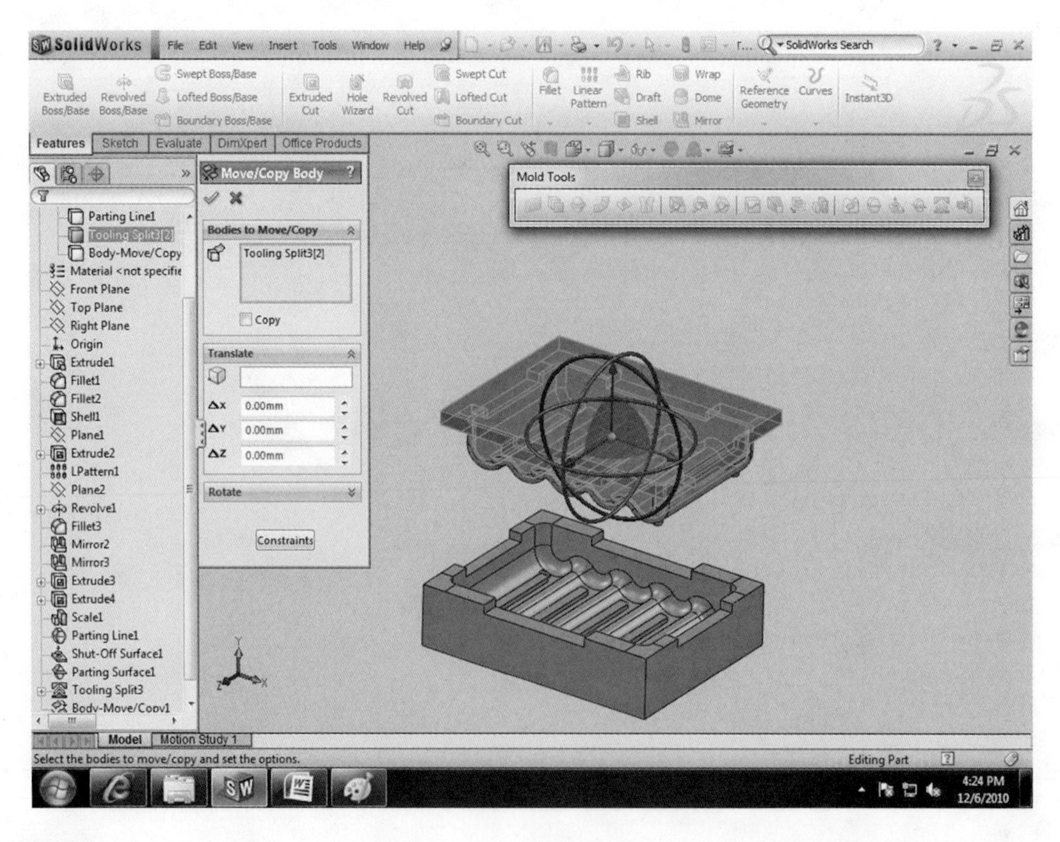

Figure 17.18

Pull Arrow of Triad to move part of the Box.

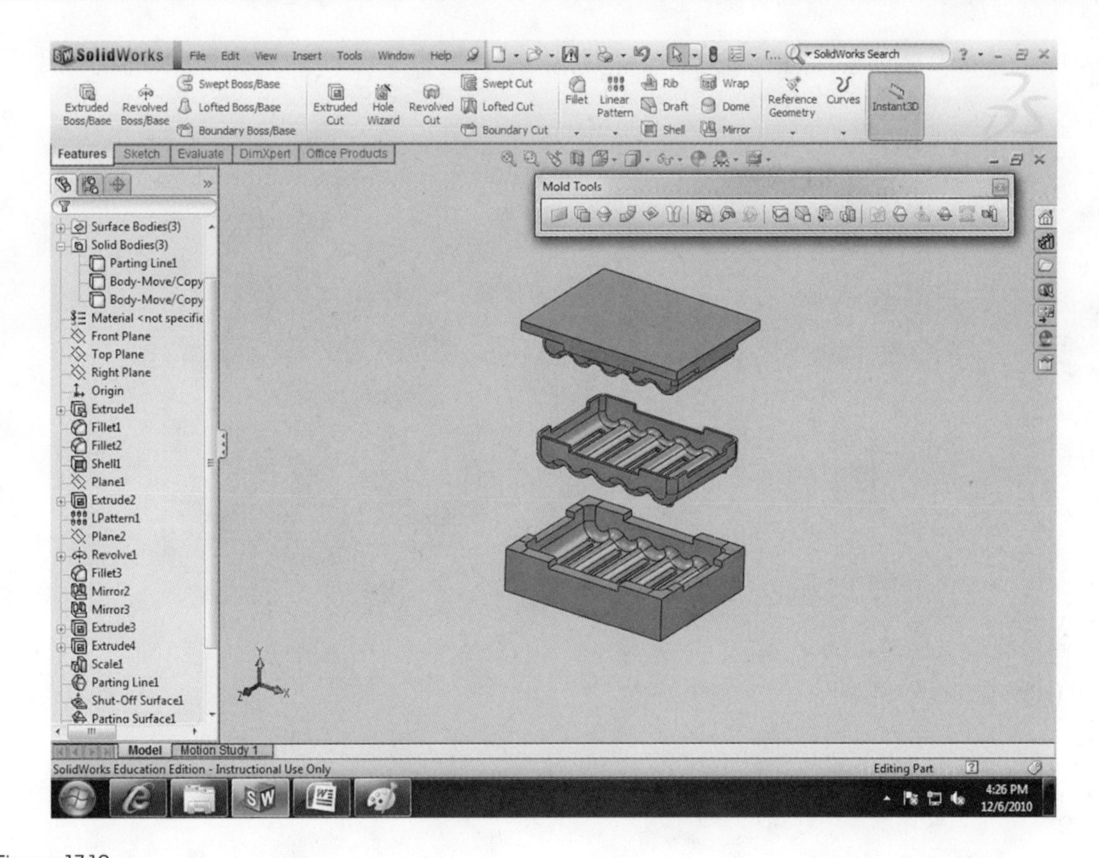

Figure 17.19

Mold design is complete.

Summary

The chapter presents some of the significant enhancements that have been made in the most recent versions of SolidWorks for Mold Design. Such enhancements have made SolidWorks to be very user-friendly, efficient, and effective in mold design. A step-by-step example of the processes involved in mold design is included in this chapter, and following these steps, users should be able to design molds for the similar parts that are encountered in practice.

Exercises

Design the molds for producing the parts that are shown in Figures P17.1 through P17.4; the files are available in the resources archive for this book.

1. Pulley (Figure P17.1)
2. Bowl (Figure P17.2)
3. Plastic cover (Figure P17.3)
4. Sump (Figure P17.4)

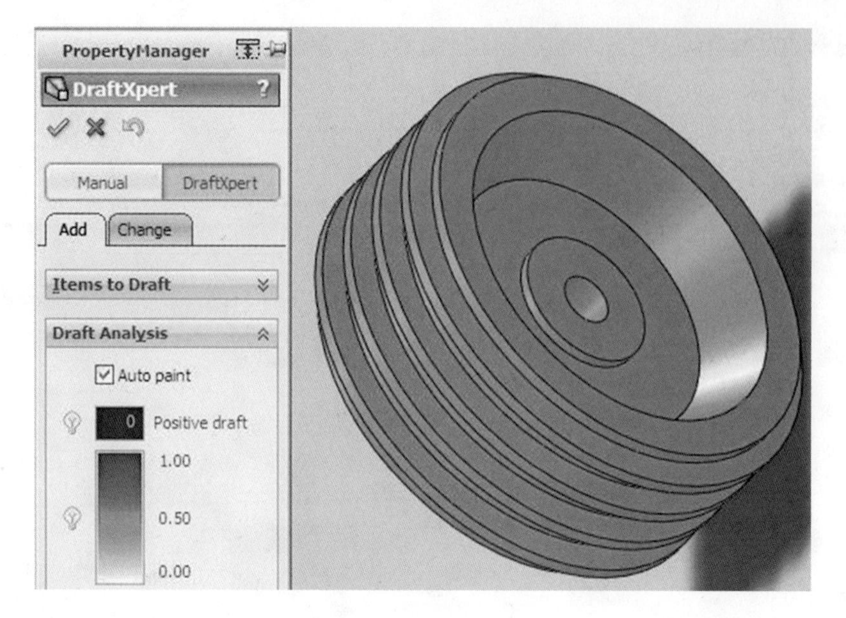

Figure P17.1

Part for which the mold is being designed: pulley.

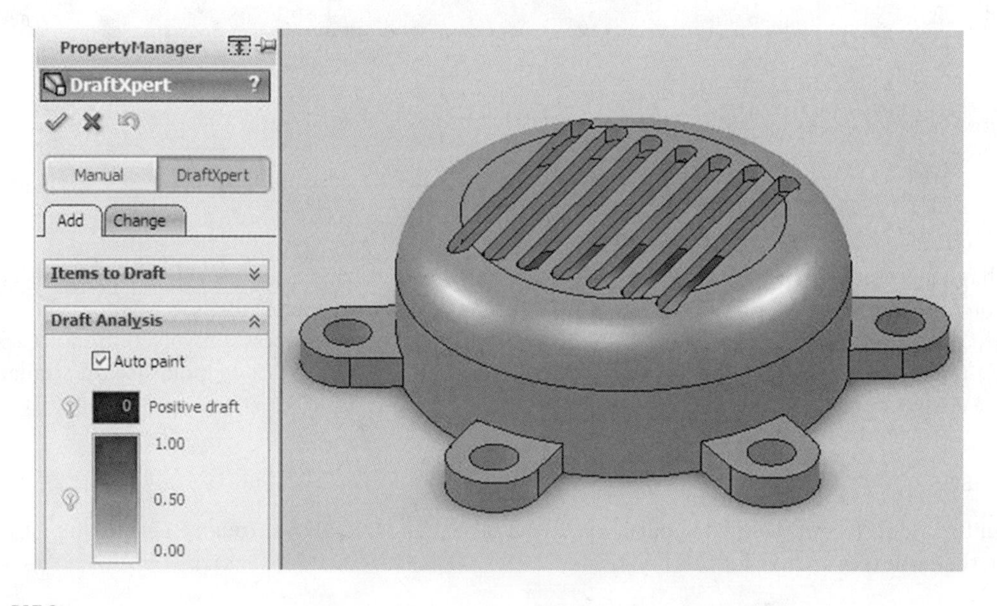

Figure P17.2

Part for which the mold is being designed: bowl.

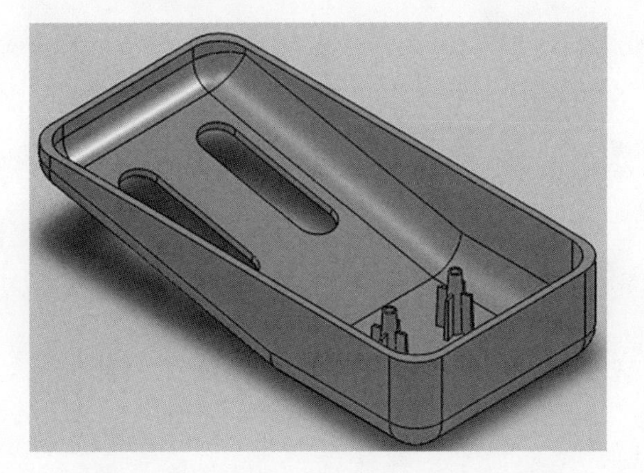

Figure P17.3

Part for which the mold is being designed: plastic cover.

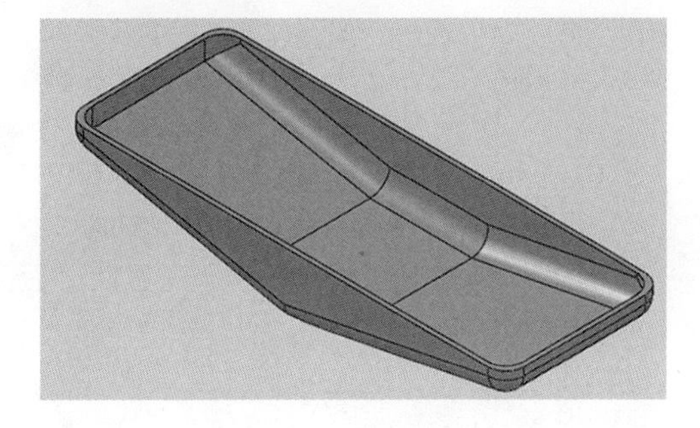

Figure P17.4

Part for which the mold is being designed: sump.

18

Sheet Metal Parts—I

Objectives:

When you complete this chapter, you will have

- Learned the differences between the insert bends and base flange approaches in Sheet Metal Parts
- Learned to create sheet metal parts using the Insert Bends approach
- Learned about Sheet Metal Part tools
- Learned how to insert a Rip feature
- Learned how to insert a Sheet Metal Bend feature
- Learned how to insert a Flange feature
- Learned how to insert a Hem feature
- Learned how to insert a Miter Flange feature
- Learned how to insert a Vent features and make some manufacturing decisions
- Learned to create sheet metal parts using the base flange approach

There are a number of manufacturing functions for which SolidWorks has tools for industrial applications. Some of these manufacturing functions include the mold design, sheet metal work, weldments, hinges, etc. Sheet metal work is one of the several aspects of manufacturing process that requires specific attention.

Sheet Metal Manufacturing Processes

The success of forming is in relation to two things: (1) the flow and (2) stretch of material. As a die forms a shape from a flat sheet of metal, there is a need for the material to move into the shape of the die. The flow of material is controlled through pressure that is applied to the blank and lubrication that is applied to the die or the blank. If the form moves too easily, wrinkles will occur in the part. To correct this, more pressure or less lubrication is applied to the blank to limit the flow of material and cause the material to stretch or thin. If too much pressure is applied, the part will become too thin and break. Drawing metal is the science of finding the correct balance between wrinkles and breaking to achieve a successful part.

There are several methods that are involved in sheet metal work, among which are the following: (a) bending, (b) roll forming, (c) deep drawing, (d) bar drawing, (e) tube drawing, (f) wire drawing, and (g) plastic drawing.

Bending is a common *metalworking* technique to process *sheet metal*, as shown in Figure 18.1. It is usually done by hand on a *box* and a *pan brake*, or industrially on a *brake press* or a *machine brake*. The typical products that are made like this are boxes such as *electrical enclosures* and rectangular *ductwork*. Usually, bending has to overcome both tensile stresses and compressive stresses. When bending is done, the residual stresses make it spring back toward its original position, so we have to overbend the sheet metal keeping in mind the residual stresses. When a sheet metal is bent, it stretches in length. The bend deduction is the amount that the sheet metal will stretch when bent as measured from the outside. A bend has a radius. The term bend radius refers to the inside radius. The bend radius depends upon the dies that are used, the metal properties, and the metal thickness.

In sheet metal parts, knowledge of the following is important:

- *Bending phenomenon* (effect of bend radius, material thickness, and bend angle)
- *Stress relief methods* (rectangular, tear, and obround configurations)

Many software packages refer to the K-factor for bending sheet metal. K-factor is a ratio that represents the location of the neutral sheet with respect to the inside thickness of the sheet metal part. The bend allowance is the length of the arc of the neutral axis between the tangent points of a bend in any material.

$$B_d = 2^*(R + T) - B_a$$

$$B_a = \pi^*(R + K^*T)^*\alpha/180$$

$$K = (180^*B_a)(\pi^*\alpha^*T) - R/T$$

where

B_a = bend allowance
R = inside bend radius
K = K-Factor, which is t/T
T = material thickness
t = distance from inside face to neutral sheet
α = bend angle in degrees (the angle through which the material is bent)

Roll forming is a continuous bending operation in which a long strip of *metal* (typically coiled *steel*) is passed through consecutive sets of rolls, or *stands*, each performing only an incremental part of the bend,

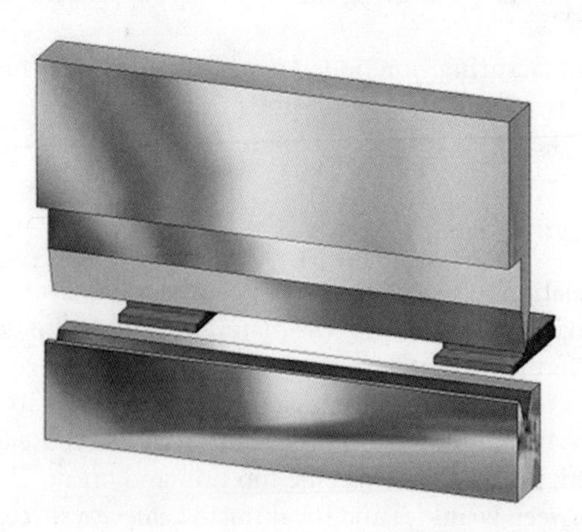

Figure 18.1

Sheet metal bending setup.

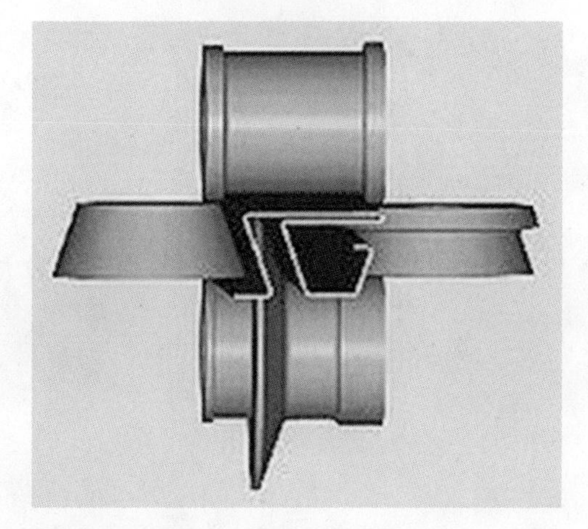

Figure 18.2

Sheet metal roll-forming bending setup.

until the desired cross-section profile is obtained (see Figure 18.2). Roll forming is ideal for producing parts with long lengths or in large quantities.

Deep Drawing

Drawing can also be used to pull metal over a *die* (male mold) to create a specific shape. For example, *stainless steel kitchen sinks* are formed by drawing the stainless steel *sheet metal* stock over a form (the die) in the shape of the sink. *Beverage cans* are formed by drawing *aluminum* stock over can-shaped dies. By comparison, *hydroforming* forces metal into a female mold using pressure.

There are many other manufacturing processes that are applied to sheet metal part design.

There are basically two approaches to sheet metal part design. One is referred to as the *insert bends method*; this is the traditional method. The other is the *base flange method*, which is the more recent one. These approaches are now presented.

Sheet Metal Part Design Methodology Using Insert Bends

Example 1: Sheet Metal Parts

Create a Shell

1. Create a shell. (Note: a shell is not a sheet metal; it is a thin-walled part having constant thickness.)
2. Click the MY-TEMPLATE tab.
3. Double-click ANSI-MM-PART.
4. Click Top Plane (or other appropriate plane).
5. Sketch and dimension the part profile (a rectangle, 350 mm × 450 mm).
6. Click Extruded Boss/Base.
7. Enter a value (100 mm) for Depth in Direction1.
   (In this case, a box [350 mm × 450 mm × 100 mm]; see Figure 18.3).
8. Click OK.
9. Click the front face of the Extrude1 feature.
10. Click Shell for the features toolbar. The Shell1 PropertyManager appears.
11. Click the Shell outward box.
12. Enter 1 mm (or other value) for Thickness.
13. Click OK from the Shell1 PropertyManager.
    Shell1 is displayed in the FeatureManager, and a shelled part is obtained as in Figure 18.3.

As already mentioned in the preceding list, creating a shell is not sheet metal work. A thin-walled part (shell) needs to be available as the starting point for sheet metal part design using the Insert Bends method.

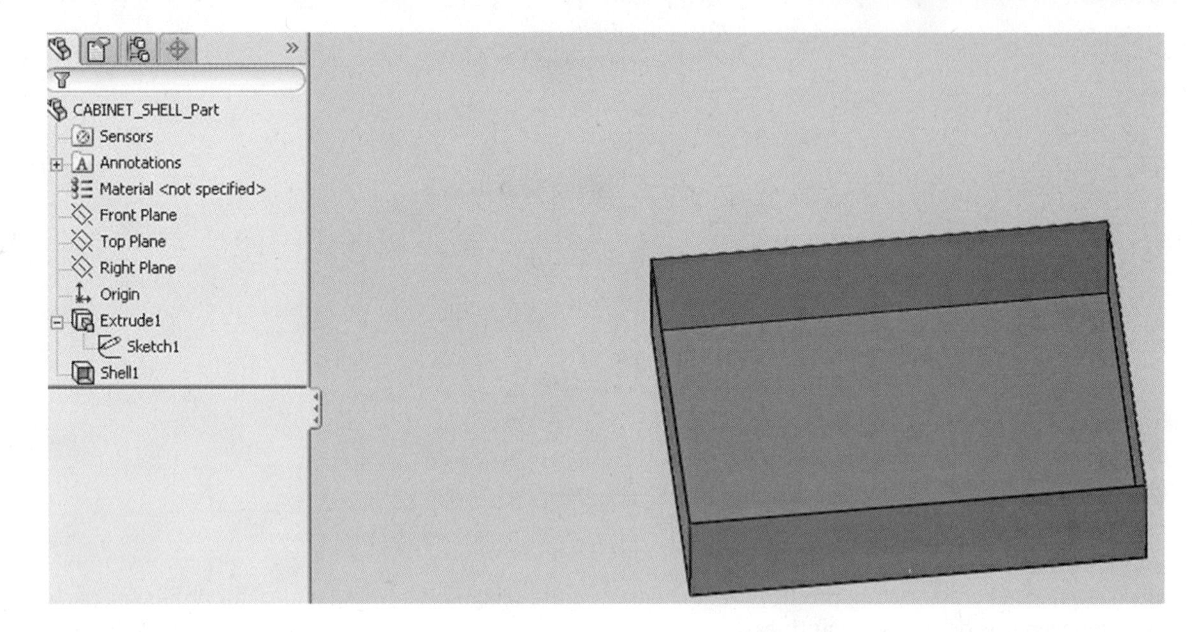

Figure 18.3

Shell1 PropertyManager.

Insert a Rip Feature

A *Rip* feature is normally inserted in a part to create a cut of no thickness along the edges of the Extruded Base feature.

1. Click Rip from the Sheet Metal toolbar. The Rip PropertyManager appears (see Figure 18.4).
2. Click the inside vertical edges. The selected edges are displayed in the Edges to Rip/Rip Parameters box. Accept the default Rip Gap (0.10 mm).
3. Click OK from the Rip PropertyManager. Rip1 appears in the FeatureManager.

The gateway that is used into the Sheet Metal toolbox is the Feature > Insert > Sheet Metal (see Figure 18.5). This route enables the sheet metal tool box when a part (shell) is the starting point.

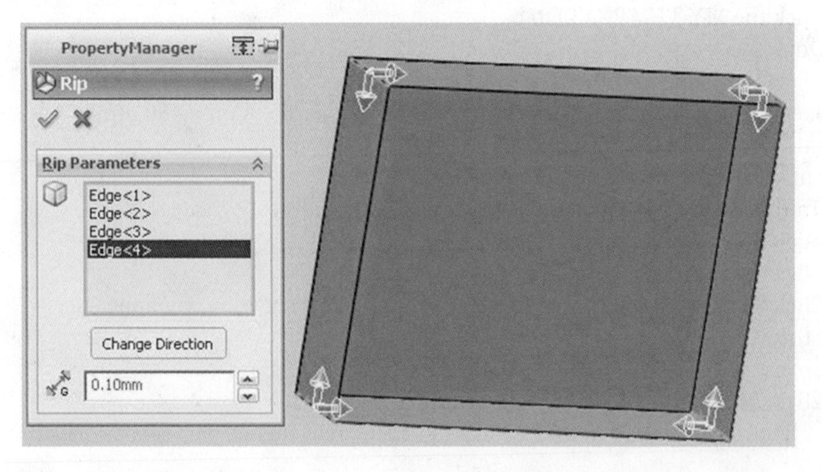

Figure 18.4

Inserting a rip feature.

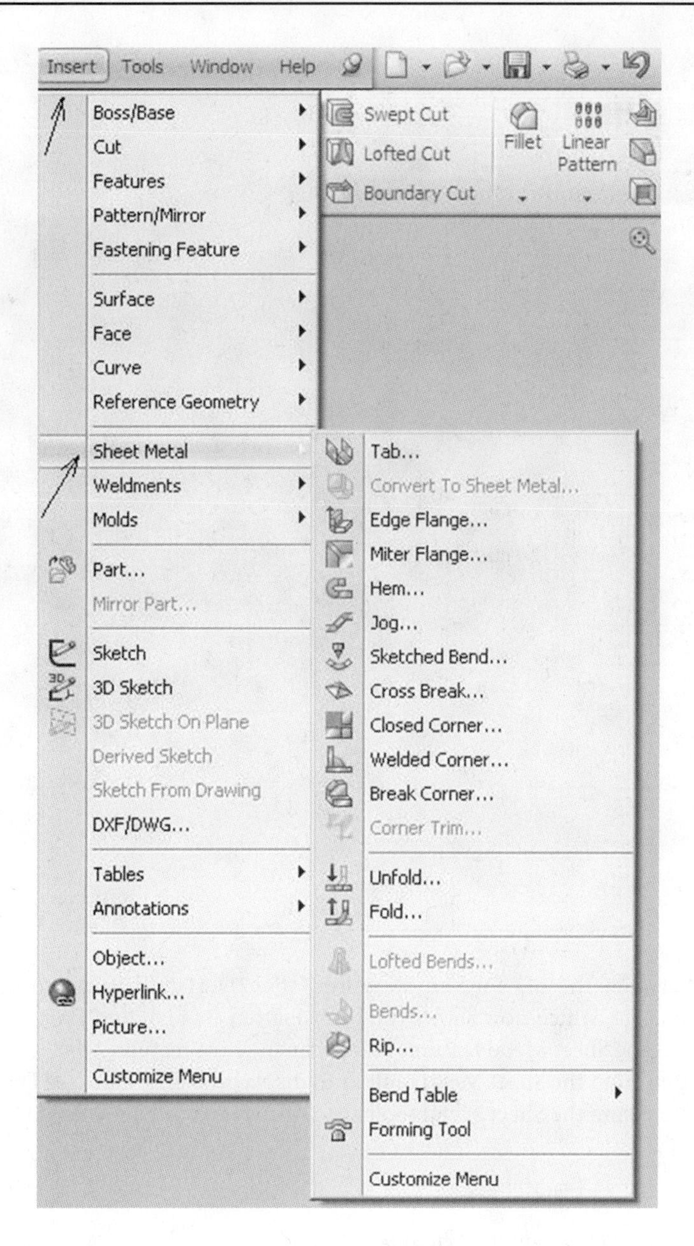

Figure 18.5

Insert Sheet Metal Toolbar gateway.

Insert the Sheet Metal Bends

1. Click the inside bottom face to remain fixed.
2. Click Insert Bends from the Sheet Metal toolbar. The Bends PropertyManager appears (see Figure 18.6). Face<1> is displayed.
3. Enter Bend Radius (2.00 mm).
4. Enter K-factor (0.45 mm).
5. Select Rectangle for Auto Relief Type and value of 0.5 mm.
6. Click OK from the Bends PropertyManager.
7. Click OK to the message "Auto relief cuts were made for one or more bends."
8. Click Isometric view from the Heads-up View toolbar.
9. Click Save.

The part that has been created is in its three-dimensional (3D)-formed state. The two-dimensional (2D) flat (manufactured) state is obtained using the Flatten sheet metal tool. The Flatten tool

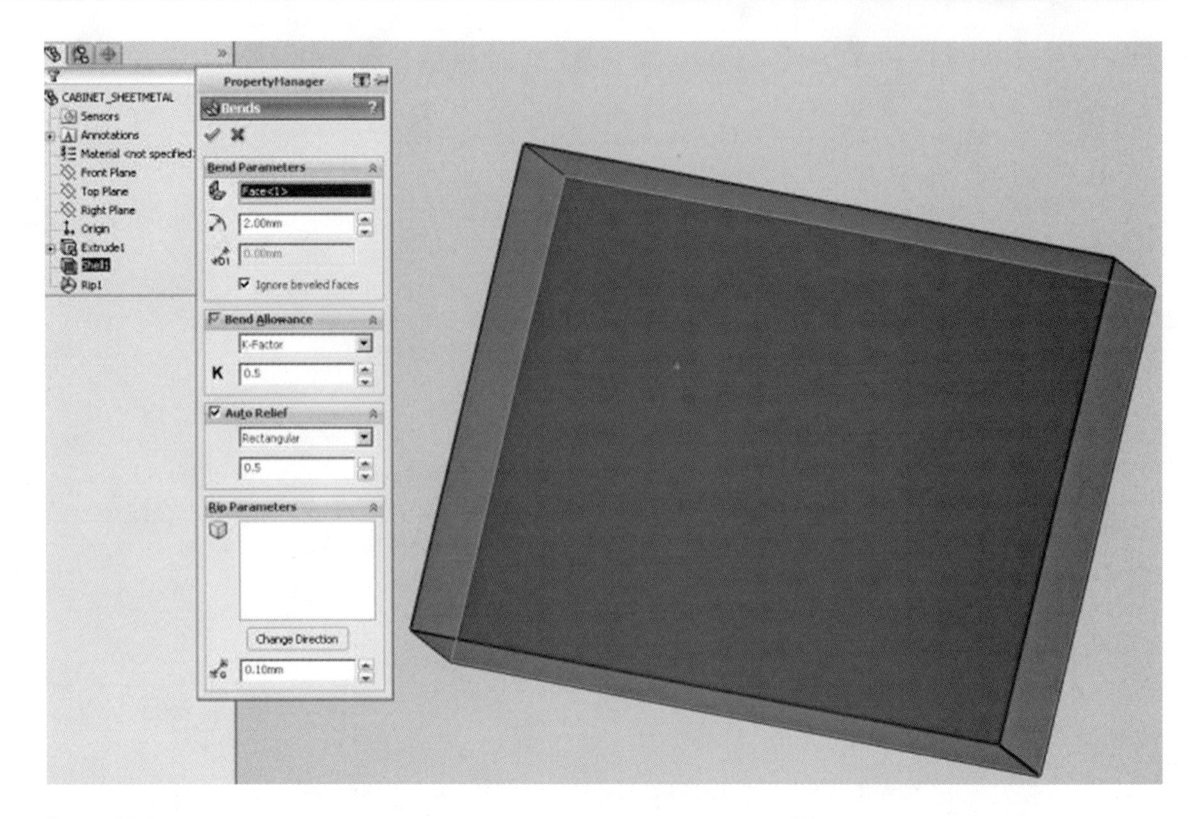

Figure 18.6

Inserting the sheet metal bends.

toggles between the 3D-formed state and the 2D flat (manufactured) state. Notice the FeatureManager, which now shows the Rip1, Flatten-Bends1, and Process-Bends1 features. We observe that the Sheet Metal features start after the Rip1 feature.

10. Click Flatten from the Sheet Metal toolbar to display the Flat State (see Figure 18.7).
11. Click Flatten from the Sheet Metal toolbar to display the fully Formed State (see Figure 18.8).

Insert the Edge Flange Features

Inserting Edge Flange Feature on the Left Side of Cabinet

12. Click the front vertical left edge (Edge<1>) of the cabinet.
13. Click Edge-Flange from the Sheet Metal toolbar (see Figure 18.9).
14. Select Blind from the Flange Length and Enter 25 mm for length. An arrow appears; reverse the direction if needed.
15. Click OK.

Inserting Edge Flange Feature on the Right Side of Cabinet

16. Click the front vertical right edge (Edge<1>) of the cabinet.
17. Click Edge-Flange from the Sheet Metal toolbar (see Figure 18.10).
18. Select Blind from the Flange Length and enter 25 mm for length. An arrow appears; reverse the direction if needed.
19. Click OK.

Insert the Hem Features

Inserting Hem Feature on the Left Side of Cabinet

20. Click the left Edge Flange (Edge-Flange1) of the cabinet.
21. Click Hem from the Sheet Metal toolbar (see Figure 18.11).
22. Select Material Inside.

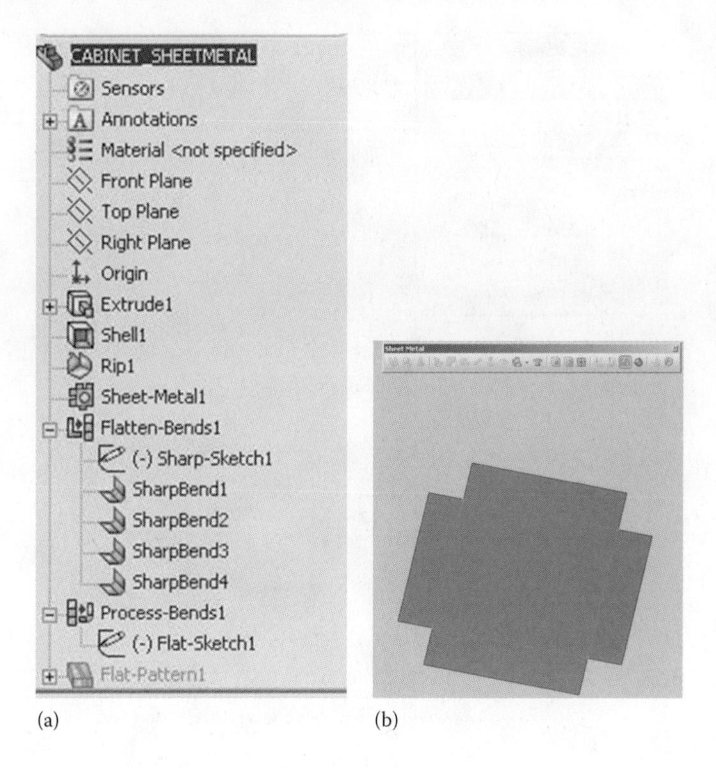

Figure 18.7

The flattened state. (a) FeatureManager shows the Flatten-Bends1. (b) Flat State obtained by clicking Flatten option and Process-Bends1 Sheet Metal features.

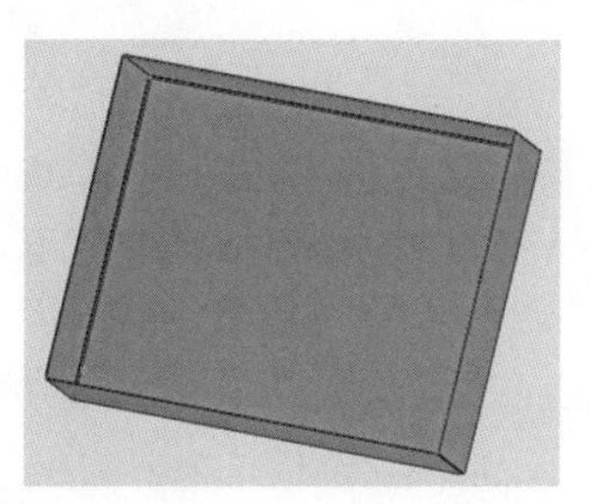

Figure 18.8

The fully Formed State obtained by clicking Flatten option again. (It toggles.)

23. Click the Reverse Direction button. (Zoom to see the effects of choice.)
24. Accept default (Open) or select as required from the Type and Size.
25. Enter 8 mm for Length.
26. Enter 0.1 mm for Gap Distance.
    (Observe the preview.)
27. Click OK.

Inserting Hem Feature on the Right Side of Cabinet

28. Click the right Edge Flange (Edge-Flange1) of the cabinet.
29. Click Hem from the Sheet Metal toolbar.
30. Select Material Inside.
31. Click the Reverse Direction button. (Zoom to see the effects of choice, as shown in Figure 18.12.)
32. Accept default (Open) or select as required from the Type and Size.
33. Enter 8 mm for Length.
34. Enter 0.1 mm for Gap Distance.
35. (Observe the preview.)
36. Click OK. The completed sheet metal cabinet with edge flange and hem features is shown in Figure 18.13.

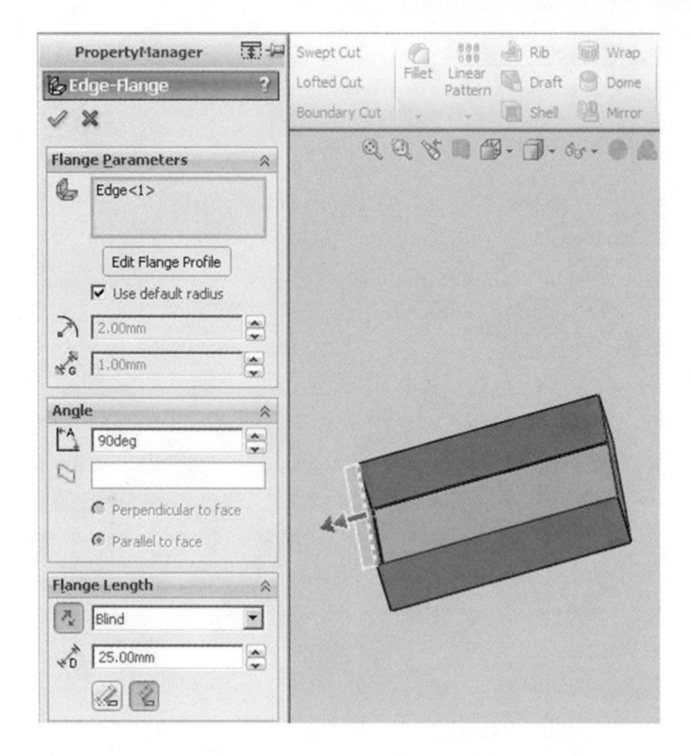

Figure 18.9

Inserting the edge flange features.

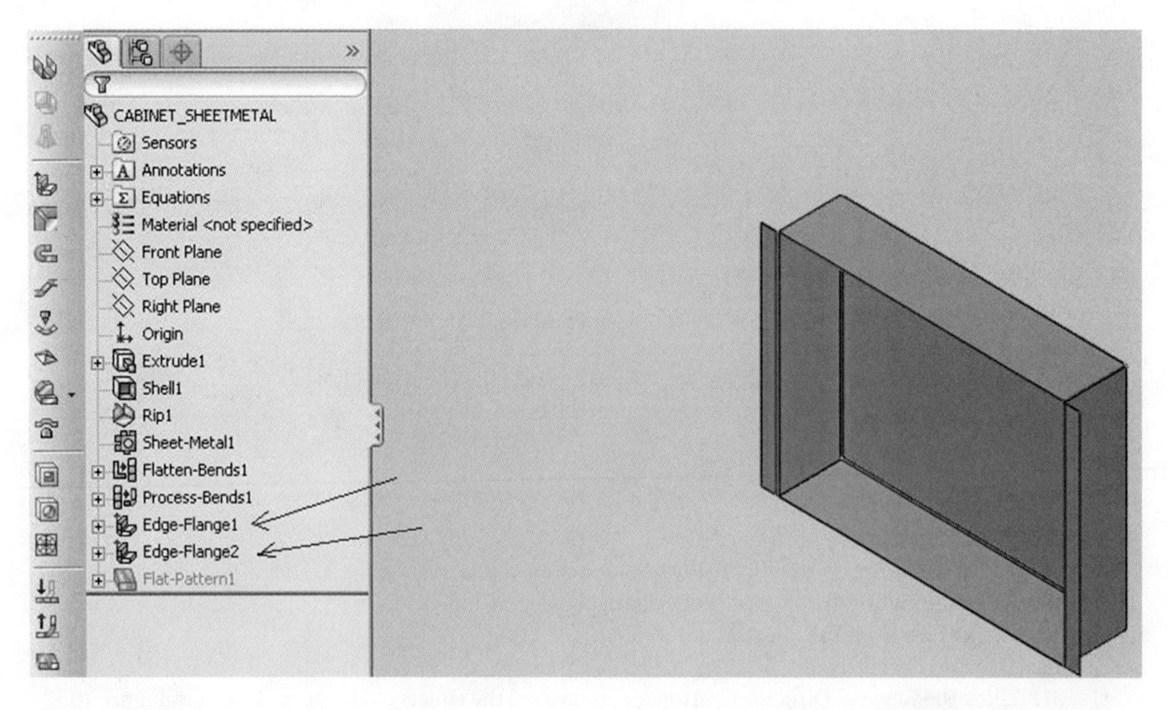

Figure 18.10

Edge Flange features on two sides of the cabinet.

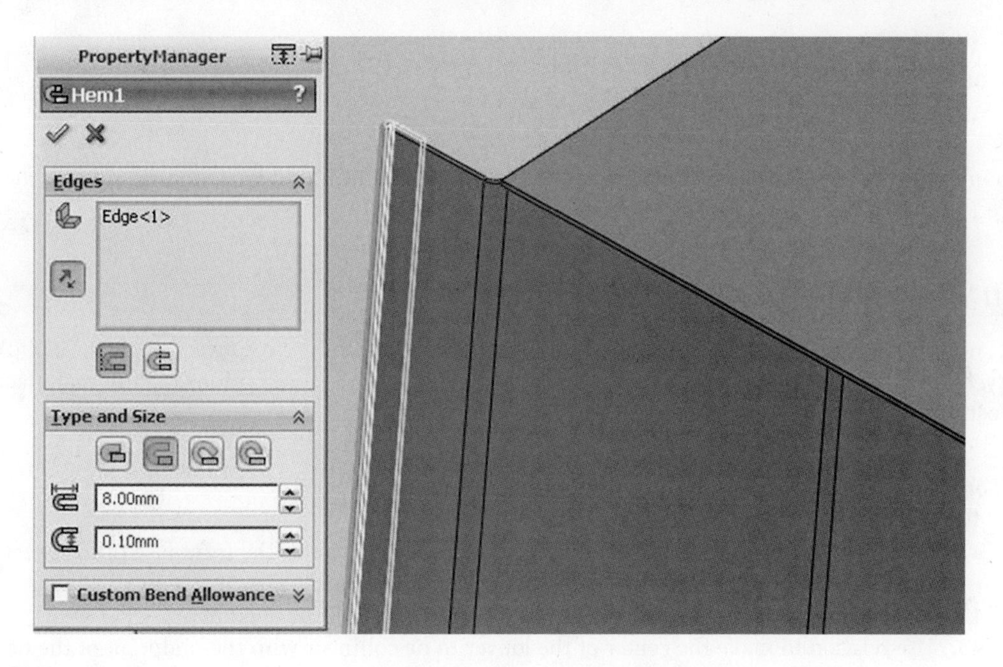

Figure 18.11

Inserting the hem features.

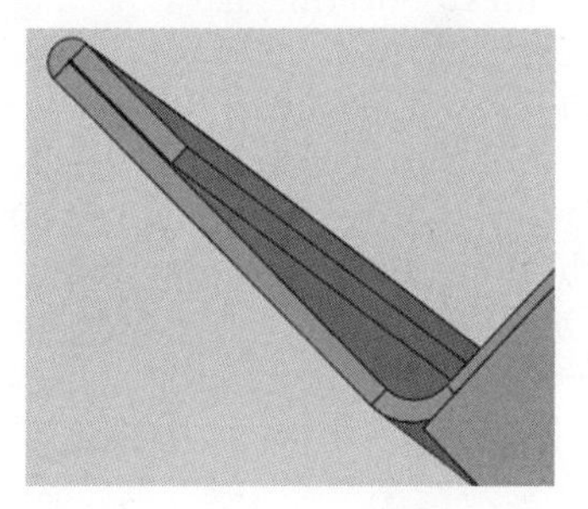

Figure 18.12

A clear view of hem.

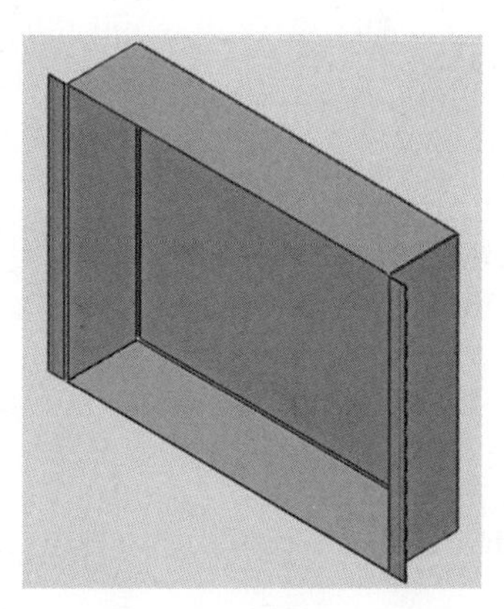

Figure 18.13

Sheet metal cabinet with edge flange and hem features.

37. Rename Hem1 as Hem1_Left.
38. Rename Hem2 as Hem2_Right.
39. Click Save.

Zooming gives a clear view of the hem, as shown in Figure 18.12, and the final cabinet is shown in Figure 18.13.

Insert Ventilation Features

There are several ways of including ventilation features: (a) d-cutout and louvers from the SolidWorks library, (b) holes, and (c) pattern cutouts that are made in-house. All these options require manufacturing and cost considerations.

Louvers

40. Click Design Library > Forming Tools > Louvers.
41. To apply the louver to the sheet metal part, drag the louver to the front face of the sheet metal part. Do *not* drop the forming tool yet until you are done (see Figure 18.14a).
42. To rotate the positioning sketch 90°, click Tools > Sketch Tool > Modify.
43. Use relation to make the center of the louver to be collinear with the midpoint of the height of the cabinet.
44. Dimension this center to be 90 mm from the bottom edge of the cabinet (see Figure 18.14b).
45. Click Finish to complete inserting the louver.
46. Create four patterns of the louver 25-mm apart (see Figure 18.14c).

D-cutout

47. Click Design Library > Features > Sheetmetal.
48. To apply the d-cutout to the sheet metal part, drag the d-cutout to the front face of the sheet metal part. Do *not* drop the forming tool yet until you are done.
49. To rotate the positioning sketch 90°, click Tools > Sketch Tool > Modify.
50. Use relation to make the top edge of the d-cutout to be horizontal.
51. Dimension the center of the d-cutout to be 50 mm from the right edge of the cabinet and 60 mm from the bottom of the cabinet.
52. Click Finish to complete inserting the d-cutout, as shown in Figure 18.15.

Example 2: Sheet Metal Parts

Let us follow exactly the procedure for the Insert Bend to illustrate another example for a channel sheet metal part design.

Extruding Thin Feature

1. Create a U-sketch (200 × 50) and extrude it 100 mm to define a channel, as shown in Figure 18.16.

Inserting Sheet Metal Bends

2. Click the inside bottom face to remain fixed.
3. Click Insert Bends from the Sheet Metal toolbar. The Bends PropertyManager appears (see Figure 18.17). Face<1> is displayed.
4. Enter Bend Radius (2.00 mm).
5. Enter K-factor (0.45 mm).
6. Select Rectangle for Auto Relief Type and a value of 0.5 mm.
7. Click OK from the Bends PropertyManager.
8. Click OK to the message "Auto relief cuts were made for one or more bends."
9. Click Isometric view from the Heads-up View toolbar.
10. Click Save.

Click the Flatten tool to obtain the flattened state, as shown in Figure 18.18.

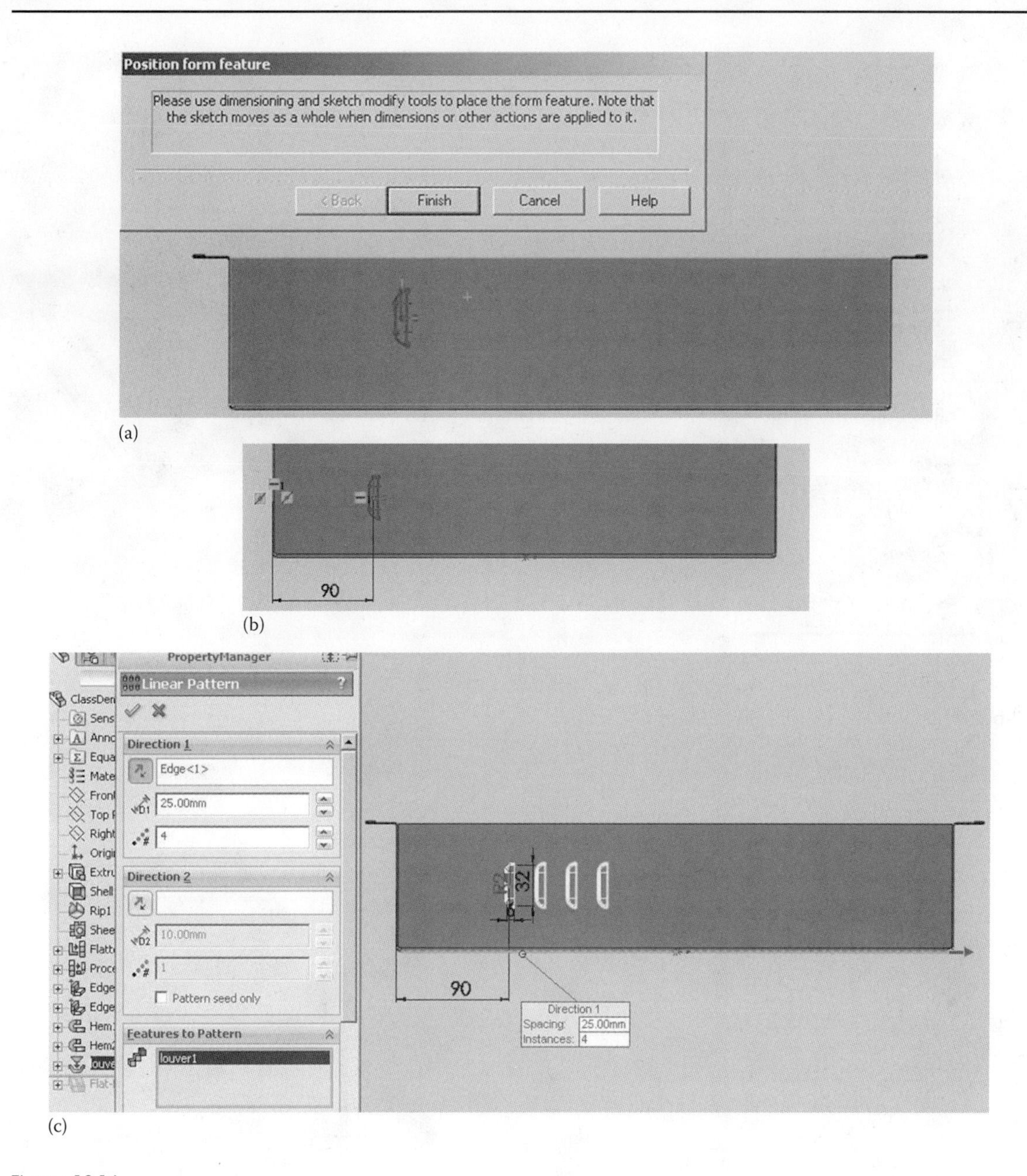

Figure 18.14

Louvers created: (a, b) inserting a louver and (c) linear pattern for several louvers. (You already have the above caption, so that is fine; nothing else).

Inserting the Holes Using Hole Wizard

The Hole Wizard tool is used to create four holes, as shown in Figure 18.19.

Applying the Sheet Metal Forming Tool

11. Click Design Library > Forming Tools > Louvers.

12. To apply the louver to the sheet metal part, drag the louver to the front face of the sheet metal part. Do *not* drop the forming tool yet until you are done (see Figure 18.20a).

13. To rotate the positioning sketch 90°, click Tools > Sketch Tool > Modify (see Figure 18.20b and c).

14. Use relation to make the center of the louver to be collinear with the center of the cabinet (see Figure 18.20d).

15. Dimension this center to be 40 mm from the edge of the cabinet (see Figure 18.20d).

16. Click Finish to complete inserting the louver.

17. Create four patterns of the louver 40-mm apart (see Figure 18.20e and f).
The channel part is shown in Figure 18.20g.

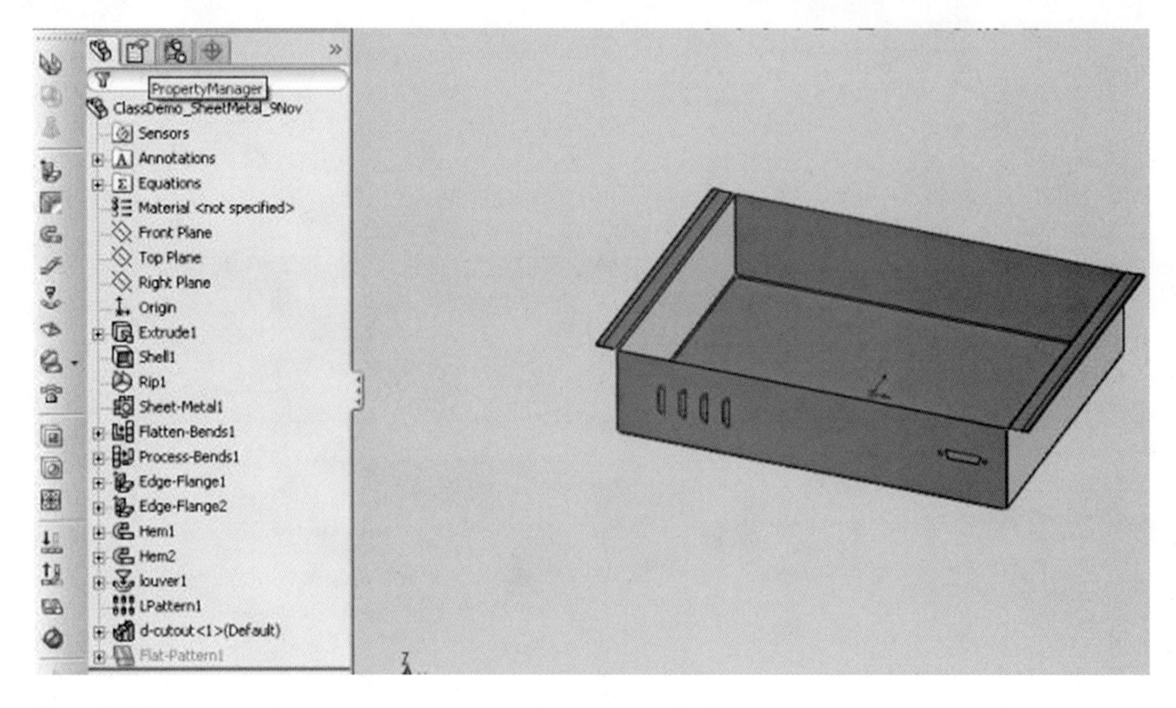

Figure 18.15

D-cutout and louver features in the cabinet.

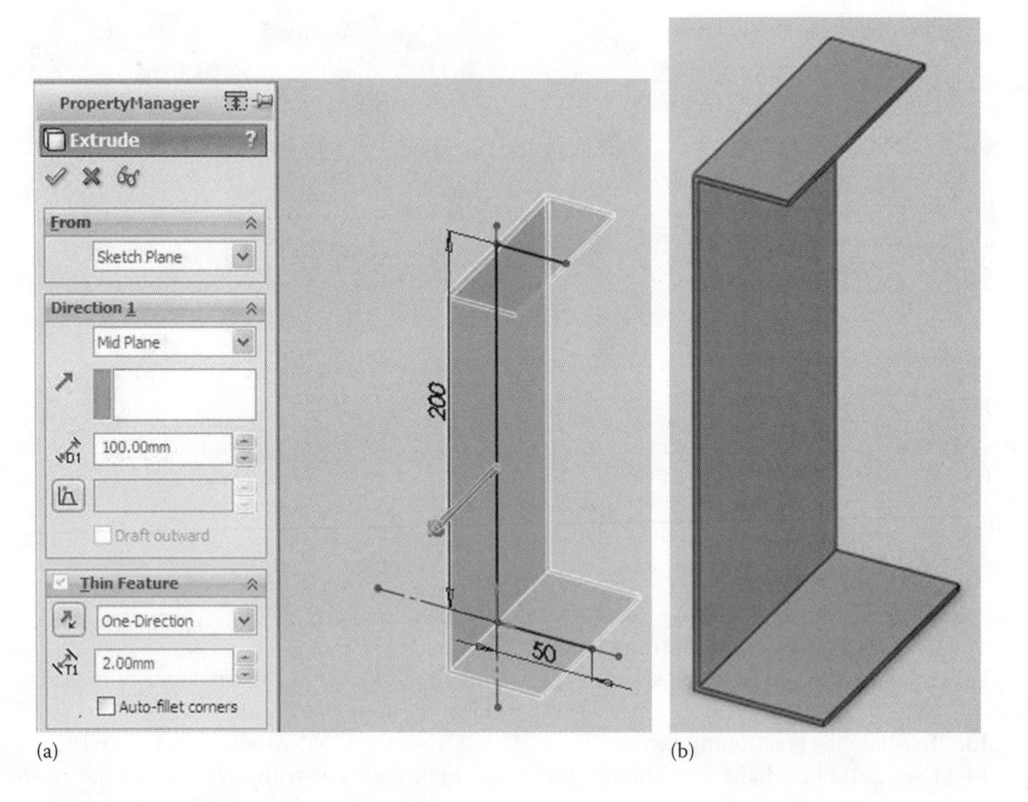

Figure 18.16

U-channel part: (a) preview and (b) part.

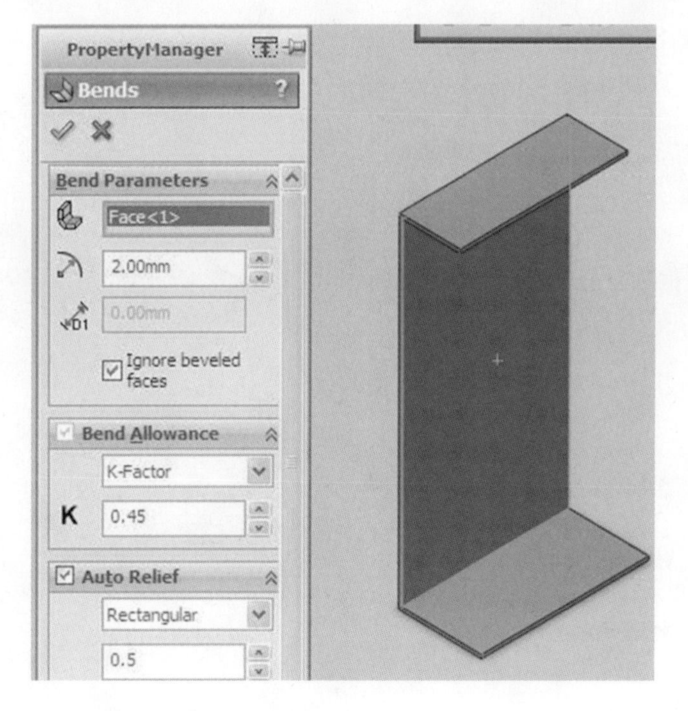

Figure 18.17

Clicking the inside bottom face to remain fixed.

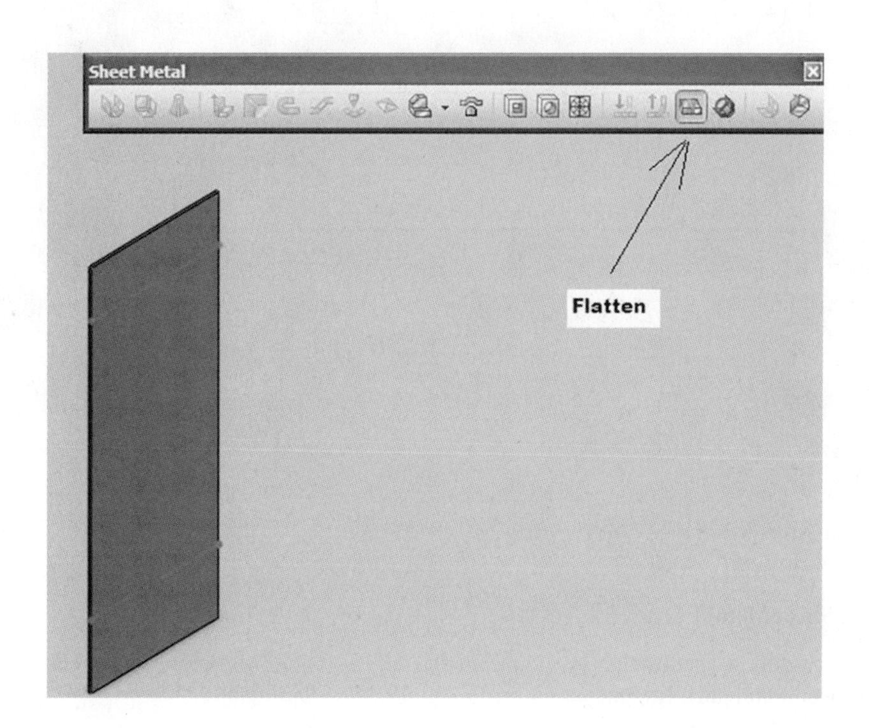

Figure 18.18

Flatten state.

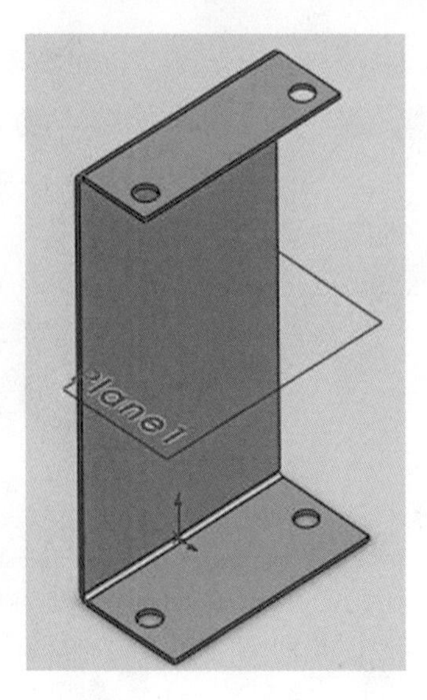

Figure 18.19

Create four holes using the Hole Wizard tool.

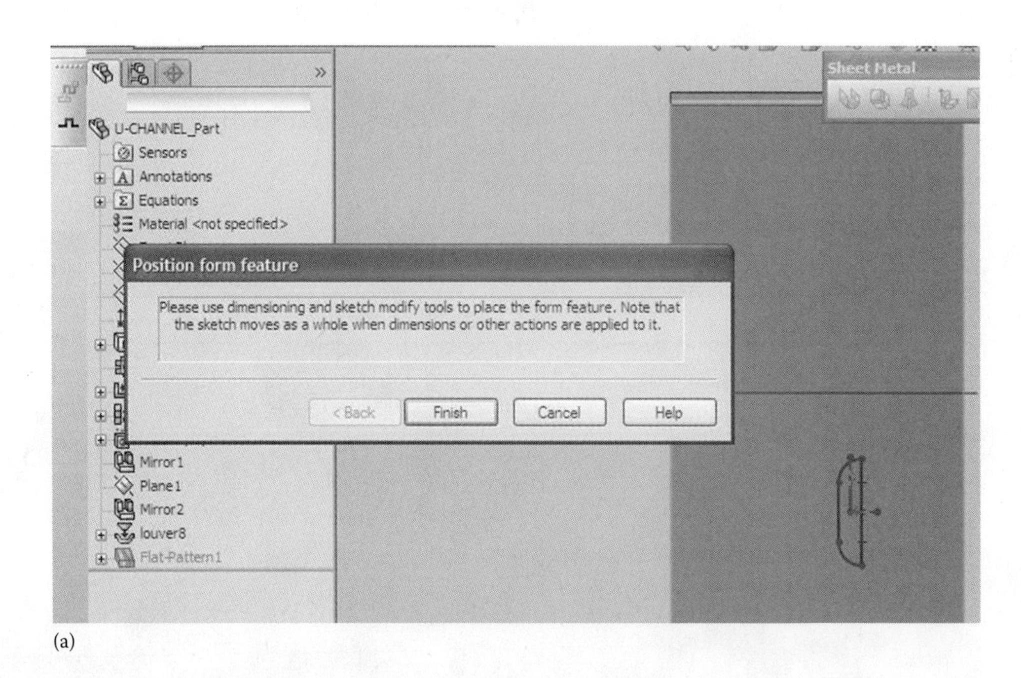

(a)

Figure 18.20

Inserting louvers: (a) a louver dragged from the Design Library. *(Continued)*

Example 3: Sheet Metal Parts

This time, let us follow exactly the procedure for the Insert Bend to illustrate yet another example for a channel sheet metal part design.

Extruding Thin Feature

1. Create on the Right Plane a slanting U-sketch (75 × 60 bent at an angle of 120°) and extrude it 100 mm to define a channel, as shown in Figure 18.21.

Inserting Sheet Metal Bends

2. Click the inside bottom face to remain fixed.

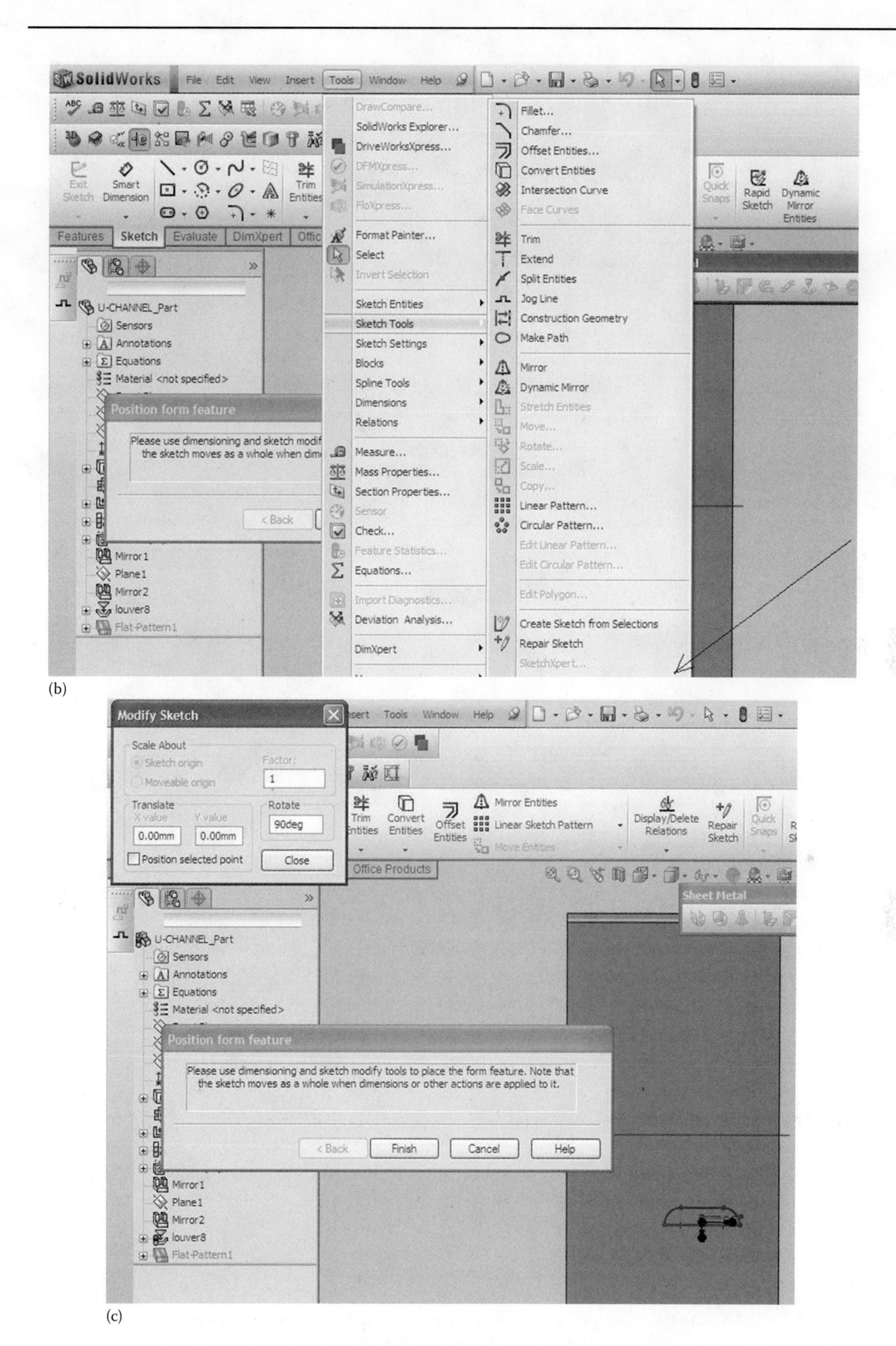

Figure 18.20 (Continued)

Inserting louvers: (b) tool for modifying the orientation of the louver and (c) modified orientation (vertical to horizontal position). *(Continued)*

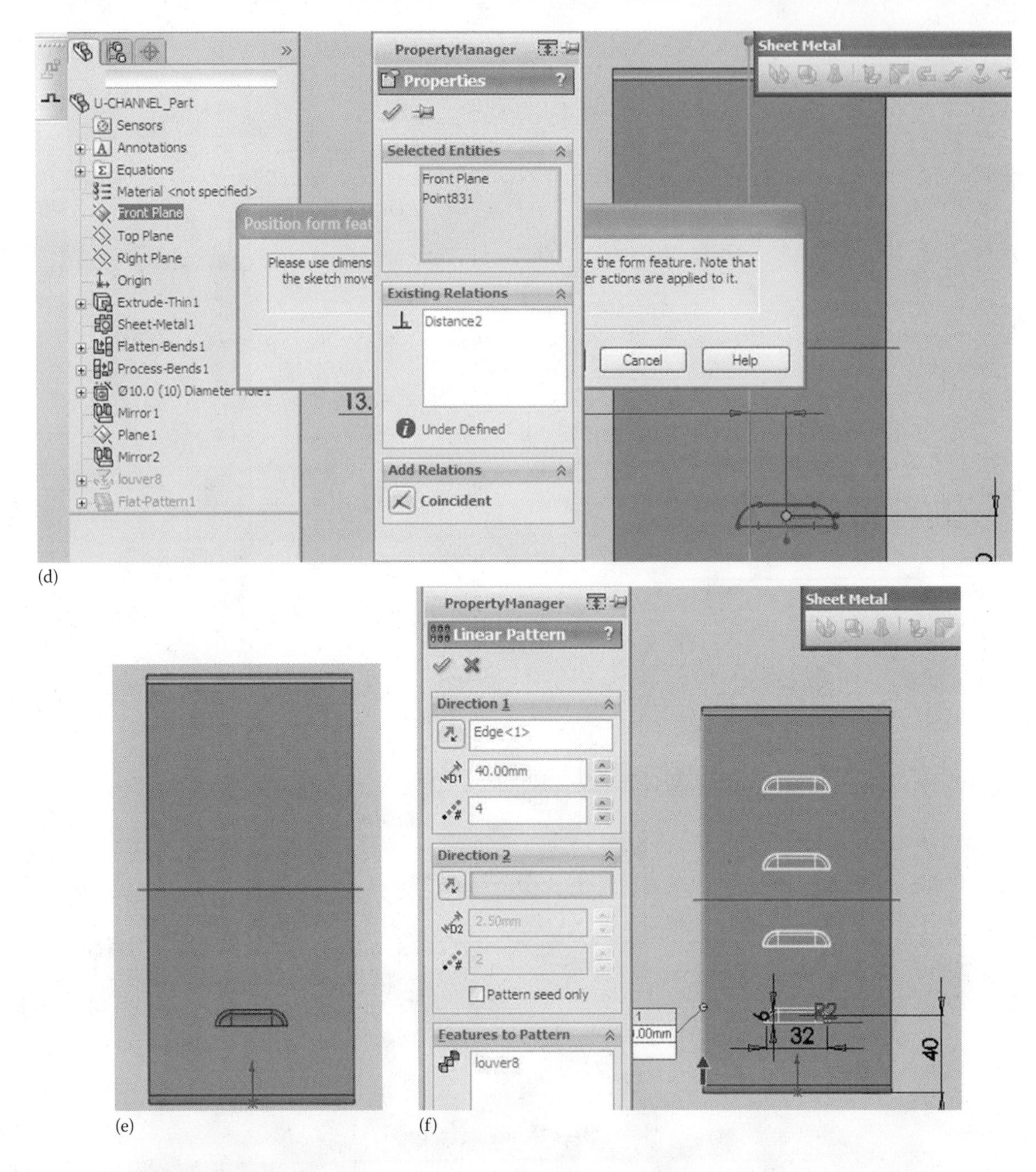

Figure 18.20 (Continued)

Inserting louvers: (d) positioning louver through dimensioning, (e) seed louver for linear pattern and (f) linear pattern for four instances. *(Continued)*

3. Click Insert Bends from the Sheet Metal toolbar. The Bends PropertyManager appears (see Figure 18.22). Face<1> is displayed.
4. Enter Bend Radius (2.00 mm).
5. Enter K-factor (0.45 mm).
6. Select Rectangle for Auto Relief Type and value of 0.5 mm.
7. Click OK from the Bends PropertyManager.
8. Click OK to the message "Auto relief cuts were made for one or more bends."
9. Click Isometric view from the Heads-up View toolbar.
10. Click Save.
11. Click the Flatten tool to obtain the flattened state, as shown in Figure 18.23.
12. Click the Flatten tool again to obtain the 3D-formed state.

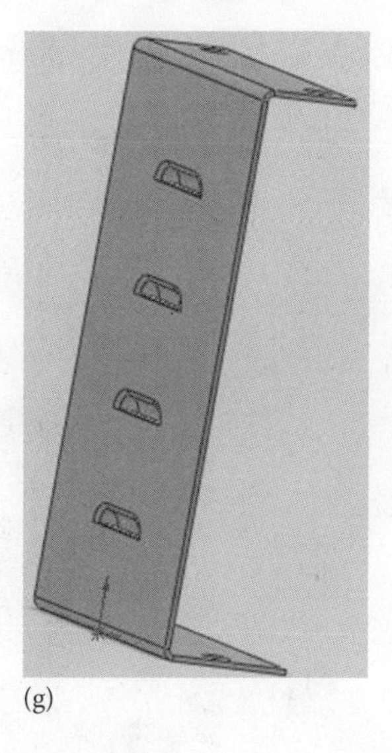

(g)

Figure 18.20 (Continued)

Inserting louvers: (g) channel part.

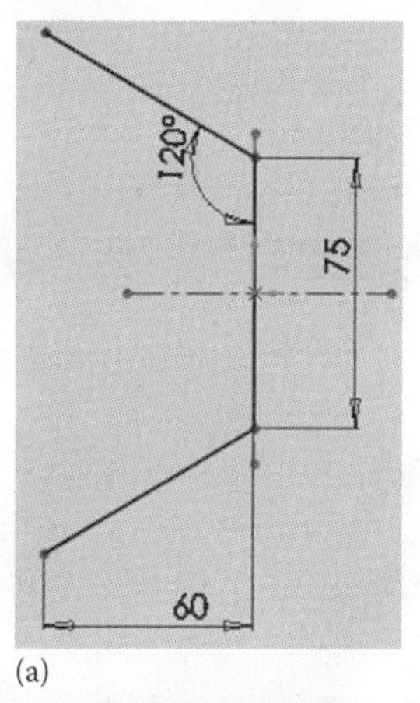

(a)

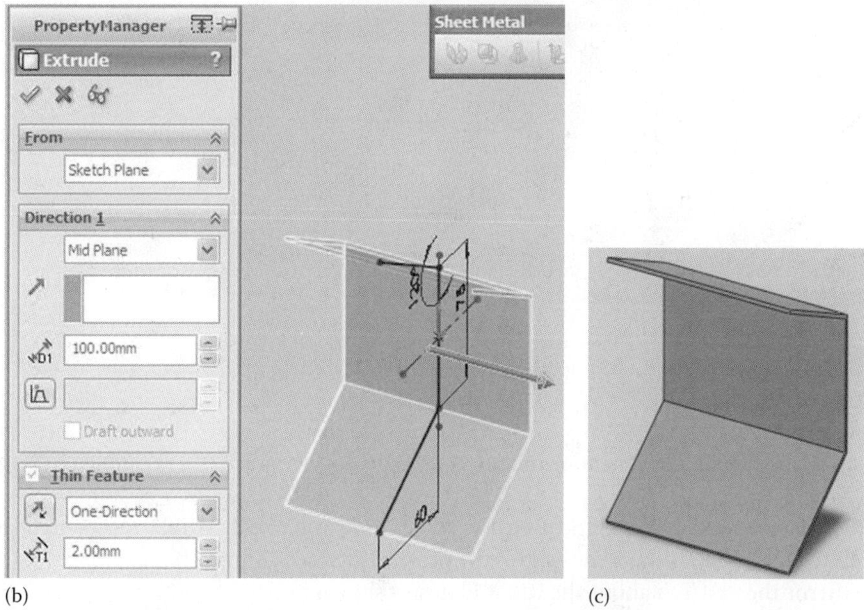

(b) (c)

Figure 18.21

Sketch and part required for sheet metal work: (a) sketch, (b) preview, and (c) part.

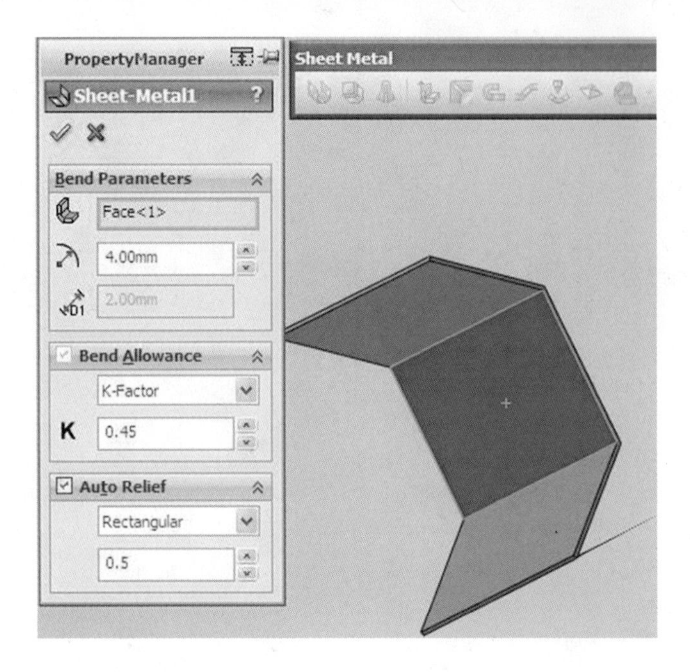

Figure 18.22

Specifying the face to remain fixed during bending.

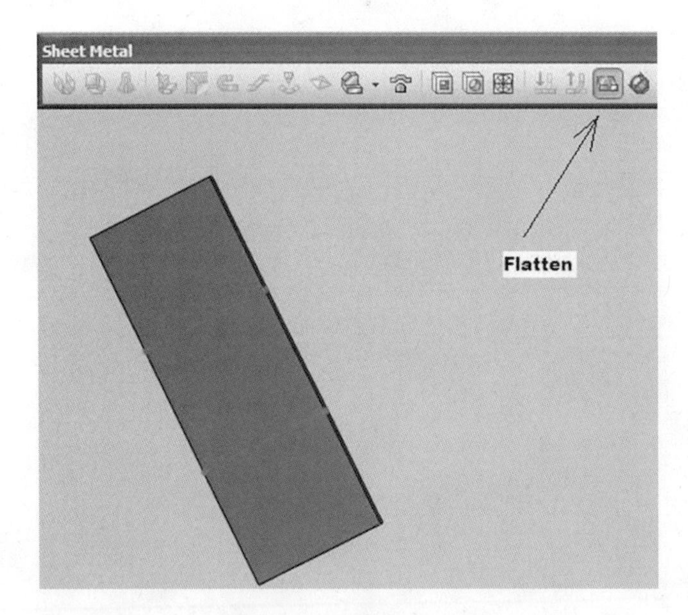

Figure 18.23

The flattened condition.

Inserting a Reference Plane

13. Click the Reference Plane tool and select an edge, as well as a vertex, as shown in Figure 18.24.

Create Tabs

14. Select the Reference Plane and sketch as well as dimension the first Tab, as shown in Figure 18.25.
15. Extrude the Tab toward the rear of the part, as shown in Figure 18.26.
16. Mirror the first Tab about the Right Plane, as shown in Figure 18.27.
17. Mirror the two Tabs about the Top Plane, as shown in Figure 18.28.
    The formed sheet metal is now shown in Figure 18.29.
18. Click the Flatten tool to obtain the flattened state, as shown in Figure 18.30.

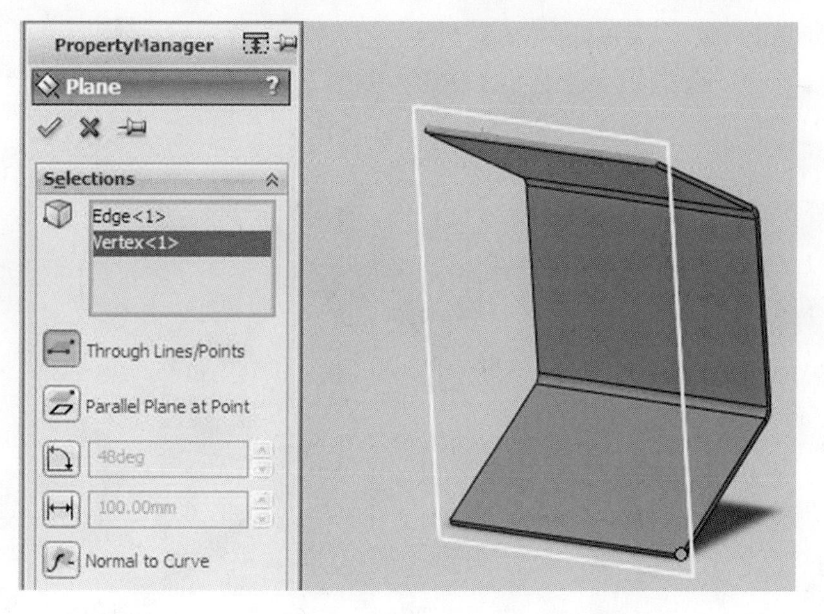

Figure 18.24

Reference plane define.

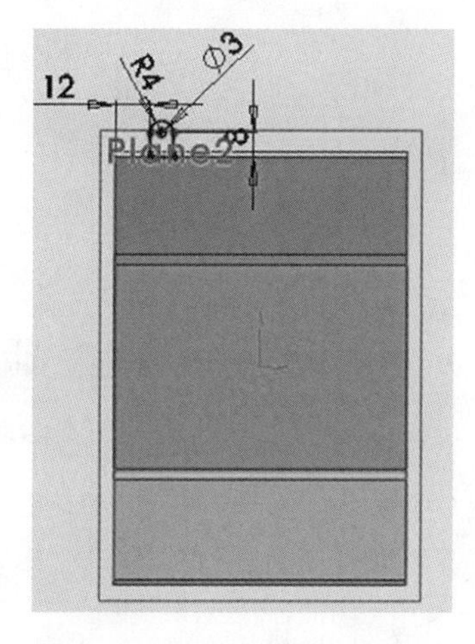

Figure 18.25

Creating sketch for a tab.

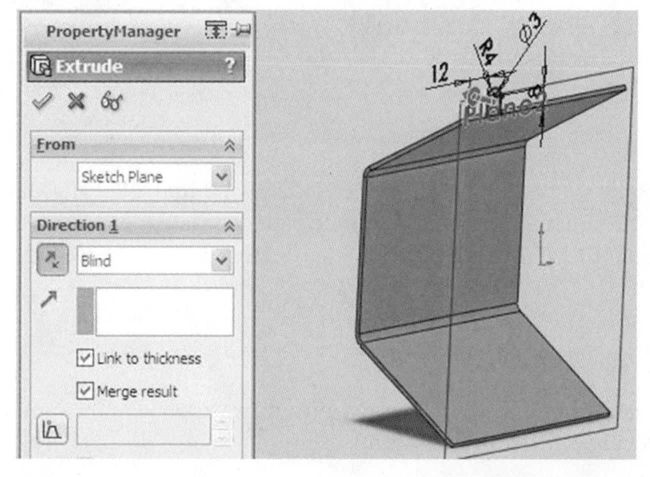

Figure 18.26

Extruding the sketch to form the tab.

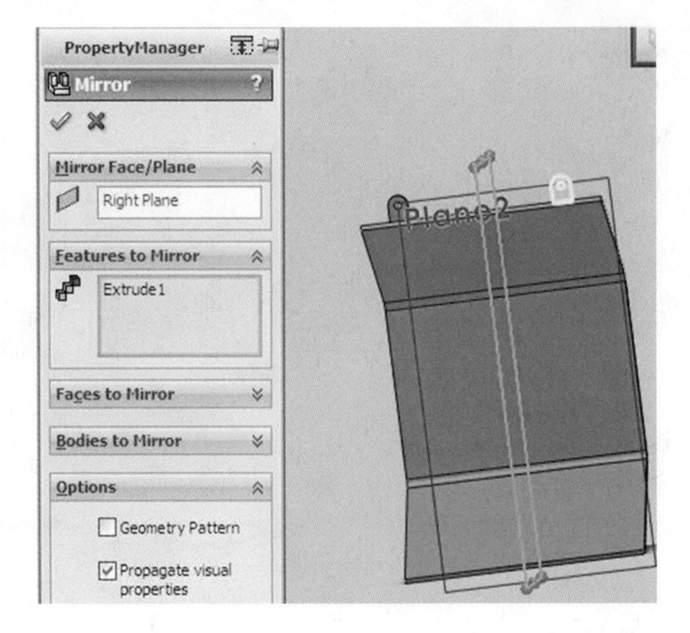

Figure 18.27

Duplication of the tab through mirroring.

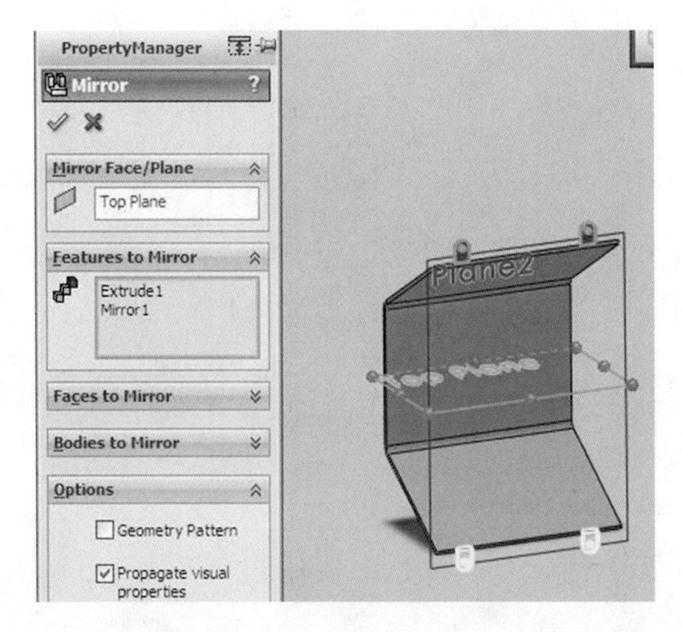

Figure 18.28

Further duplication of the tab through mirroring.

19. Drag the Rollback Bar to just below the Flatten Bend in the FeatureManager.
20. Sketch a circle, 30 mm in diameter, on the flattened state, as shown in Figure 18.30.
21. Drag the Rollback Bar to the end of the FeatureManager to show the cut feature, as shown in Figure 18.31.

Sheet Metal Part Design Methodology Using Base Flange

In this approach, the starting point is a base flange, which, when applied to the cabinet under consideration, is the bottom face (a rectangle, 350 mm × 450 mm).

1. Click the MY-TEMPLATE tab.
2. Double-click ANSI-MM-PART.

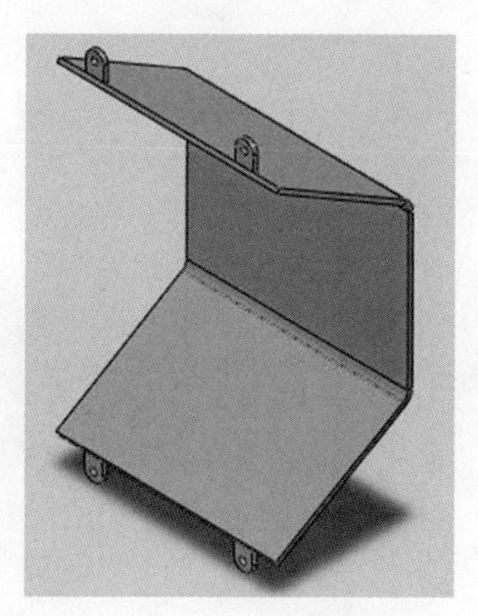

Figure 18.29

Formed sheet metal.

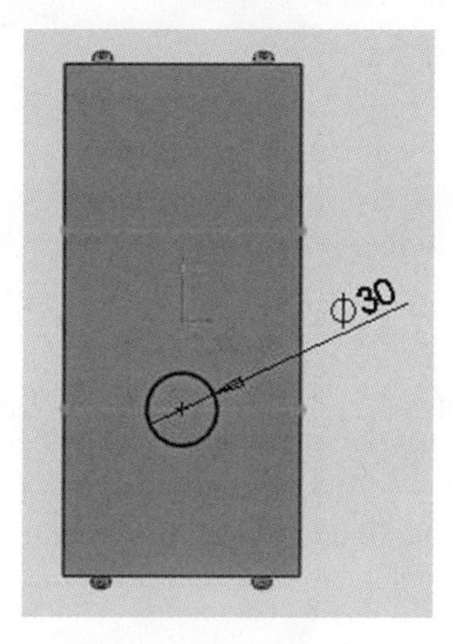

Figure 18.30

Flattened sheet metal.

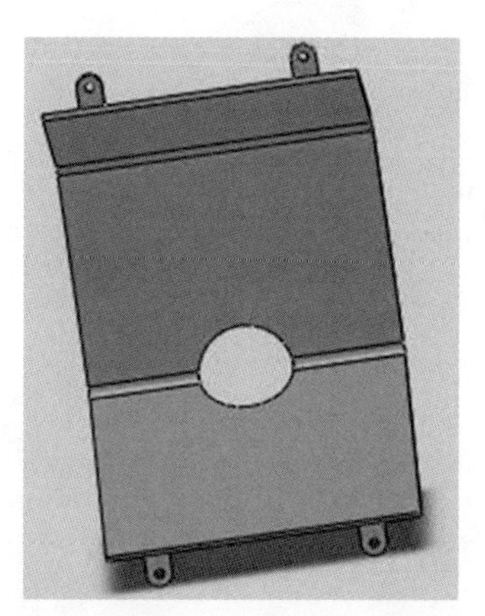

Figure 18.31

Formed sheet metal with cut feature.

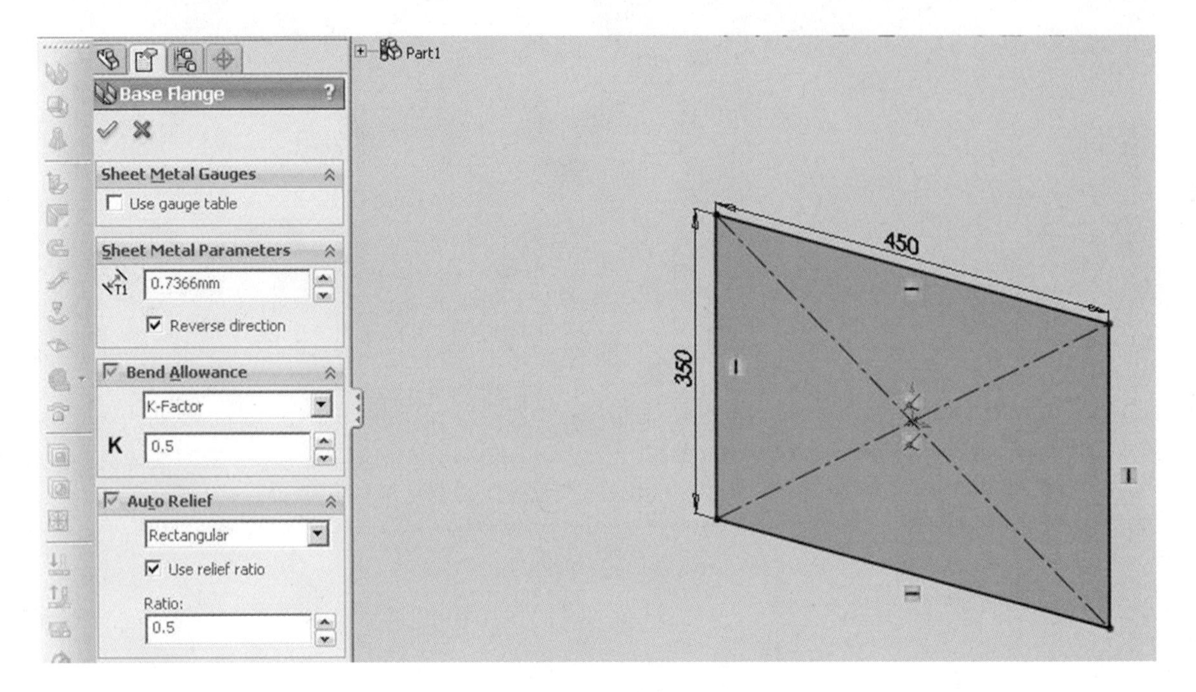

Figure 18.32

Sketch1.

3. Click Front Plane (or other appropriate plane).
4. Sketch and dimension the base flange, Sketch1 (a rectangle, 350 mm × 450 mm).
5. Click Base Flange from the Sheet Metal toolbar. The Base Flange PropertyManager appears (see Figure 18.32).

   Accept the default value of sheet metal thickness. (Notice: there is an option to choose Gauge Table.)

Notice that Base-Flange1 appears after Sheet-Metal1 in the FeatureManager (see Figure 18.33).
The steps involved in creating a Miter Flange are as follows:

a. Click any edge of the Base Flange.
b. Select the Line tool from the Sketch toolbar and sketch the profile for the miter flange.
c. Without exiting sketch mode, click Miter tool and select each edge of the Base Flange to create the Miter feature. (The first one is automatically created on the first edge that is initially selected.) (Note: if you exit the sketch, then it has to be selected for the Miter function.)

Create an Edge Flange

6. Click one of the edges of the Base Flange and then select the Line tool from the Sketch toolbar (see Figure 18.34). (Notice that a perpendicular plane for sketching the line is automatically selected; this is very useful in this approach.)
7. Create an L-sketch with dimensions 100-mm long and 35-mm wide, Sketch2, with a fillet radius of 2.5 mm (see Figure 18.35).
8. Exit sketch.

Create Miter Flange Features

9. Select Sketch2.
10. Click the Miter Flange button on the Sheet Metal toolbar, while the sketch is still active.
11. Select the four edges of the Base Flange, which are not automatically dimensioned to have a height of 100 mm as the height of the Sketch, as shown in Figure 18.36.

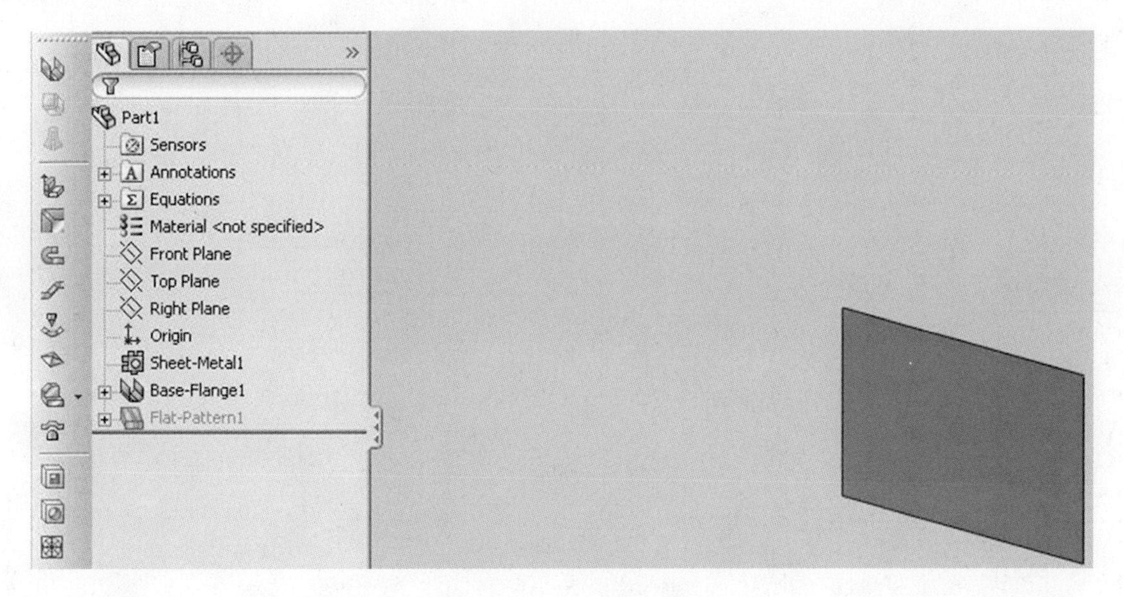

Figure 18.33

Base Flange appears after sheet metal in the FeatureManager.

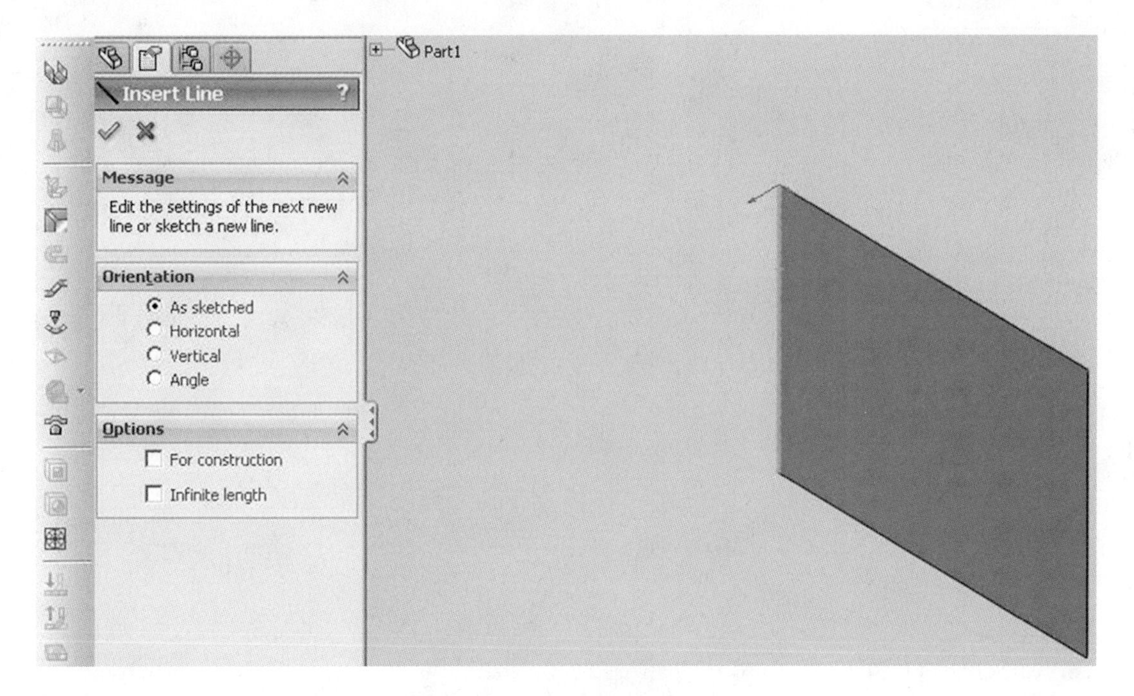

Figure 18.34

Effect of selecting line tool.

12. Click OK from the Miter Flange Property toolbar.
13. Hide Plane1.

Alternatively, still be in sketch mode after the L-shape is sketched and access Miter Flange.

Our design, based on the base flange approach, is complete at this stage, as shown in Figure 18.37, which seems to have more capabilities than the alternative *insert bends* design approach. Other venting features already discussed can now be included.

Figure 18.35

Creating profile sketch.

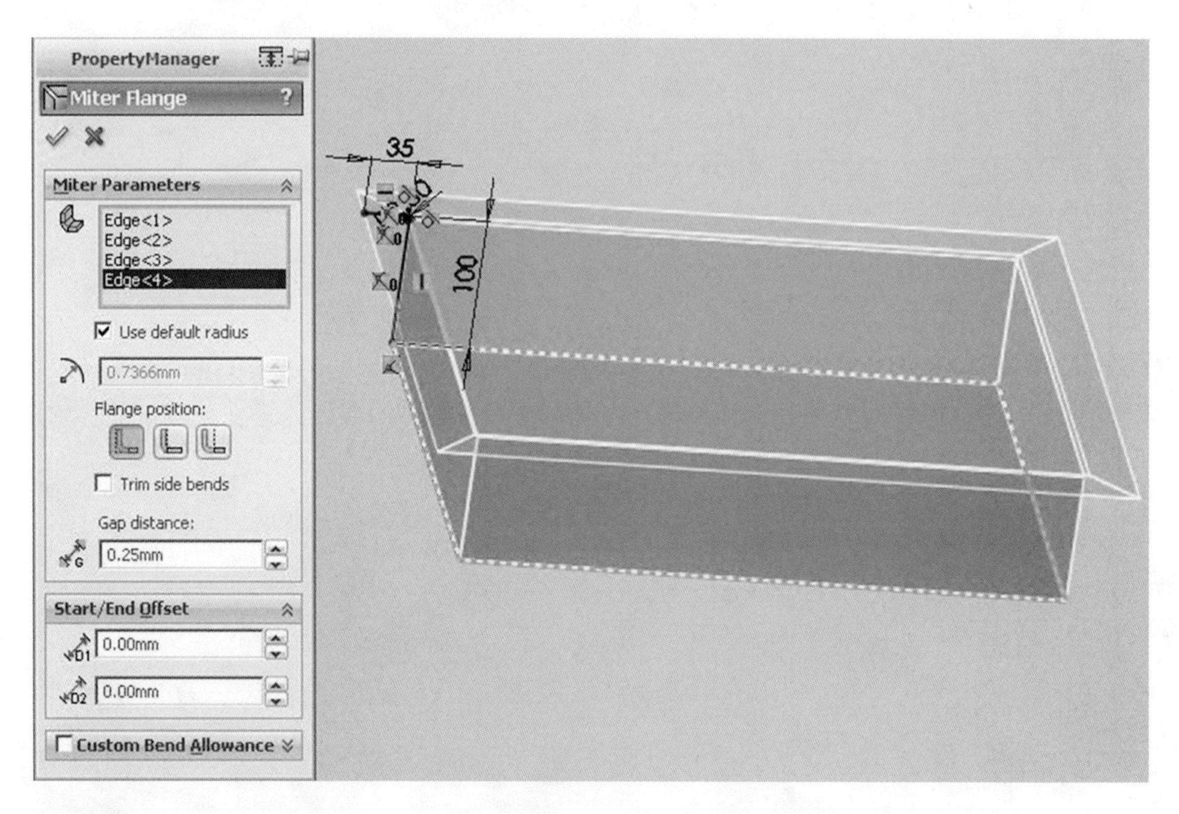

Figure 18.36

Miter flange feature.

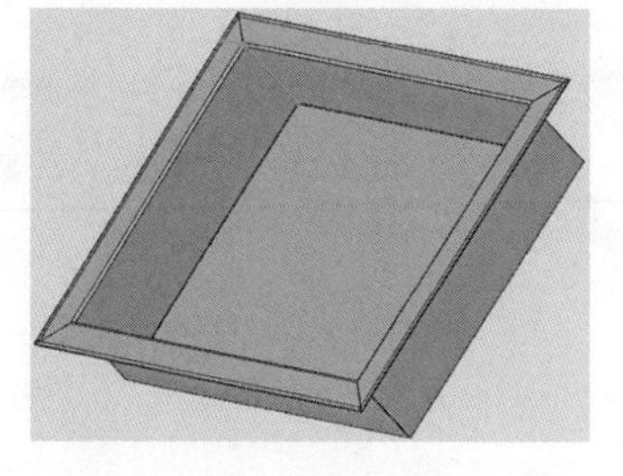

Figure 18.37

Completed part using base flange approach.

Summary

This chapter has discussed the differences between the insert bends and base flange approaches in sheet metal part design and considered some examples. The insert bends approach does not match the natural way that sheet metal design is carried out on the shop floor, whereas the base flange approach does. The next chapter uses the base flange approach to solve a number of problems of practical relevance.

Exercises

1. Repeat the sheet metal part design for Example 1 (refer to Figure 18.15 and Figure P18.1) using the base flange approach.
2. Repeat the sheet metal part design for Example 2 (refer to Figure 18.20 and Figure P18.2) using the base flange approach.
3. Repeat the sheet metal part design for Example 3 (refer to Figure 18.31 and Figure P18.3) using the base flange approach.

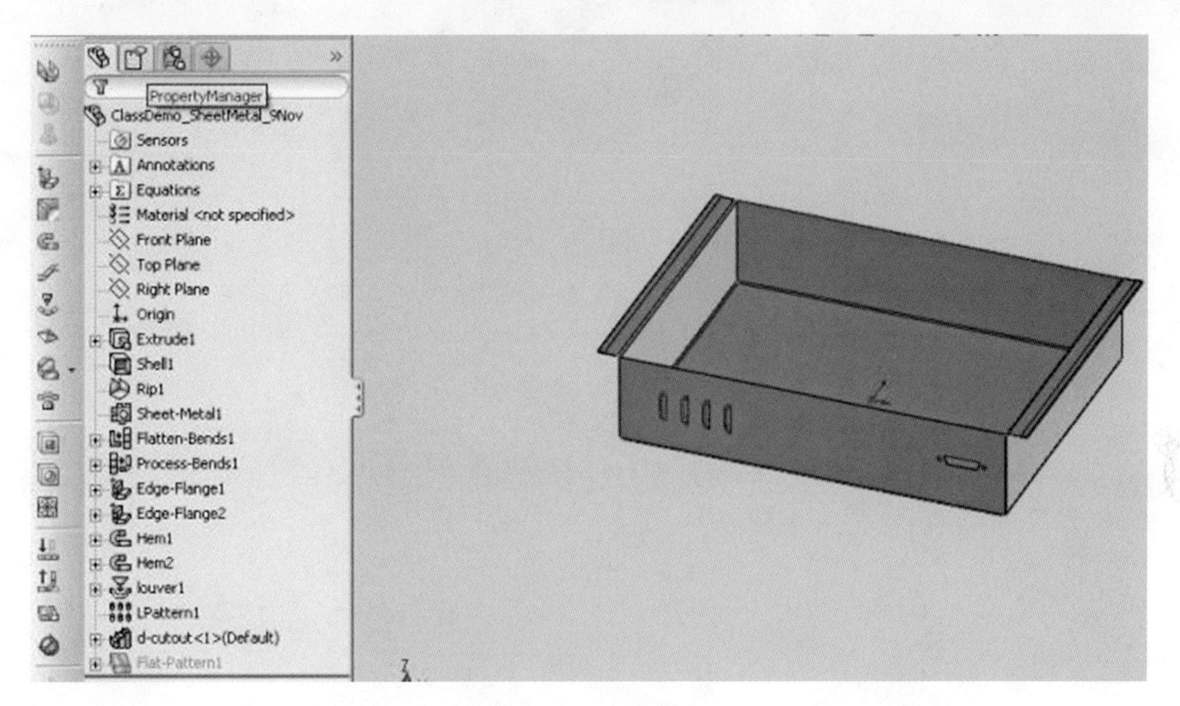

Figure P18.1

Cabinet having D-cutout and louver features.

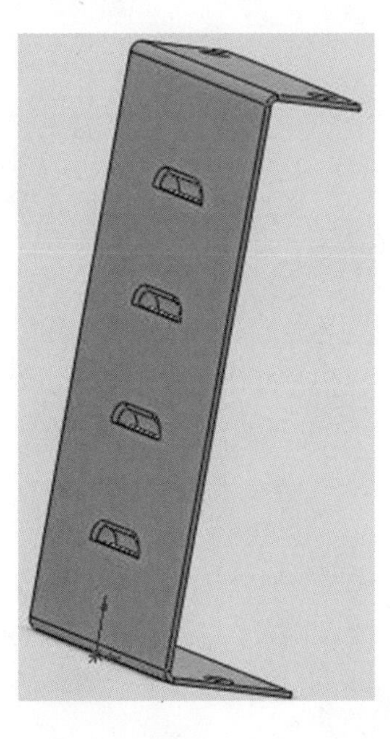

Figure P18.2

U-channel having louver features.

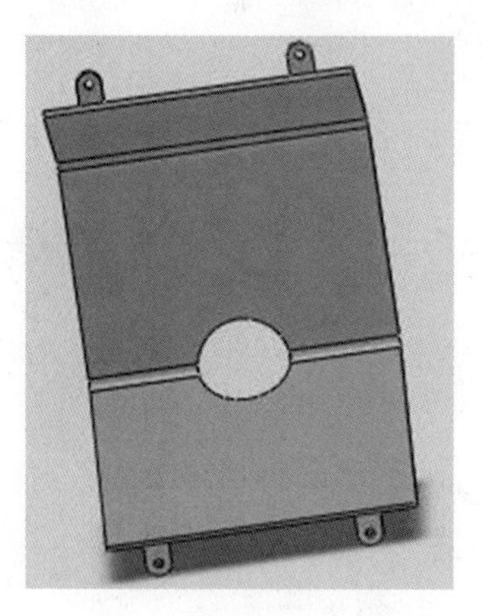

Figure P18.3

Divergent-channel having cut features.

19

Sheet Metal Parts—II

Objectives:

When you complete this chapter, you will have

- Learned to create sheet metal parts using the base flange approach

Comparing Sheet Metal Design Methods

There are three ways to create a sheet metal part:

1. *Convert a solid part to a sheet metal part.*
   You can convert a solid or surface body or an imported part.
2. *Create the part as a sheet metal part using sheet metal-specific features.*
   When you create a part initially out of sheet metal, you use two features: (1) base flange and (2) miter flange. This eliminates extra steps because you create a part as sheet metal from the initial design stage.
3. *Build a part, shell it, and then convert it to sheet metal.*
   If you build a solid and then convert it to sheet metal, you need more features: (a) Base Extrude, (b) Shell, (c) Rip, and (d) Insert Bends. However, there are instances when it is preferable to build a part and then convert it to sheet metal.

Advantages of the Base Flange Approach

The base flange approach has some advantages, which are summarized:

1. A part is created as sheet metal from the initial design stage.
2. The method is quite intuitive and consistent with the manner in which sheet metal work is done in practice.
3. Designing sheet metal using this approach is very flexible and efficient.
4. When some sheet metal tools are used, all the tools necessary for defining sketches are available and could be used.

The remaining parts of this chapter concentrate on the *base flange* approach for sheet metal part design, which is preferred to the *insert bends* approach. Several tutorials are discussed to show how the base flange approach can be applied to a number of sheet metal part designs.

This chapter discusses several tutorials for sheet metal design based on the base flange approach. It is expected that students will become competent in sheet metal part design after covering all the tutorials. Detailed step-by-step procedures have been discussed in such a way that students can reproduce the solutions of the problems being solved. The standard of the sheet metal part design problems discussed in this chapter is at par with the Certified SolidWorks Associate (CSWA) and Certified SolidWorks Professional (CSWP) examinations.

Tutorials on Base Flange Approach for Sheet Metal Design

The following tutorials are discussed in detail in the subsequent subsections of this chapter:

- *Tutorial 19.1:* General Sheet Metal Part
- *Tutorial 19.2:* P1
- *Tutorial 19.3:* P2
- *Tutorial 19.4:* Hanger Support
- *Tutorial 19.5:* Jogged Sheet Metal Part
- *Tutorial 19.6:* Lofted Sheet Metal
- *Tutorial 19.7:* P9
- *Tutorial 19.8:* General Sheet Metal
- *Tutorial 19.9:* Hanger
- *Tutorial 19.10:* CSWP-SMTL

Sheet Metal User Interface

The Sheet Metal User Interface is shown in Figure 19.1, with the definitions of each tool described. The most used tools in the base flange approach are (a) Edge Flange, (b) Sketched Bend, and (c) Flatten.

Tutorial 19.1: General Sheet Metal Part

In this tutorial, we will create the sheet metal part that is shown at the top right, with its views and dimensions shown in Figure 19.2. The flat pattern of the sheet metal part is shown in Figure 19.3.

SolidWorks Solution to Tutorial 19.1

Create the Base Flange

1. Start a New SolidWorks Part document.
2. Select the Top Plane.
3. Be in sketch mode and create Sketch1, 500 mm × 400 mm (see Figure 19.4).
4. Click the Base Flange/Tab option from the Sheet Metal CommandManager (see Figure 19.4).

Sheet Metal

1 2 3 4 5 6 7 8 9 10 11 12 13 14 15 16 17 18 19 20

1	**Base Flange/Tab**		11	Forming Tool
2	Convert to Sheet Metal		12	*Extruded Cut*
3	Lofted-Bend		13	*Simple Hole*
4	**Edge Flange**		14	Vent
5	Miter Flange		15	Unfold
6	Hem		16	Fold
7	Jog		17	Flatten
8	**Sketched Bend**		18	No Bends
9	Cross-Break		19	Insert Bends
10	Corners		20	Rip

Figure 19.1

Sheet Metal User Interface.

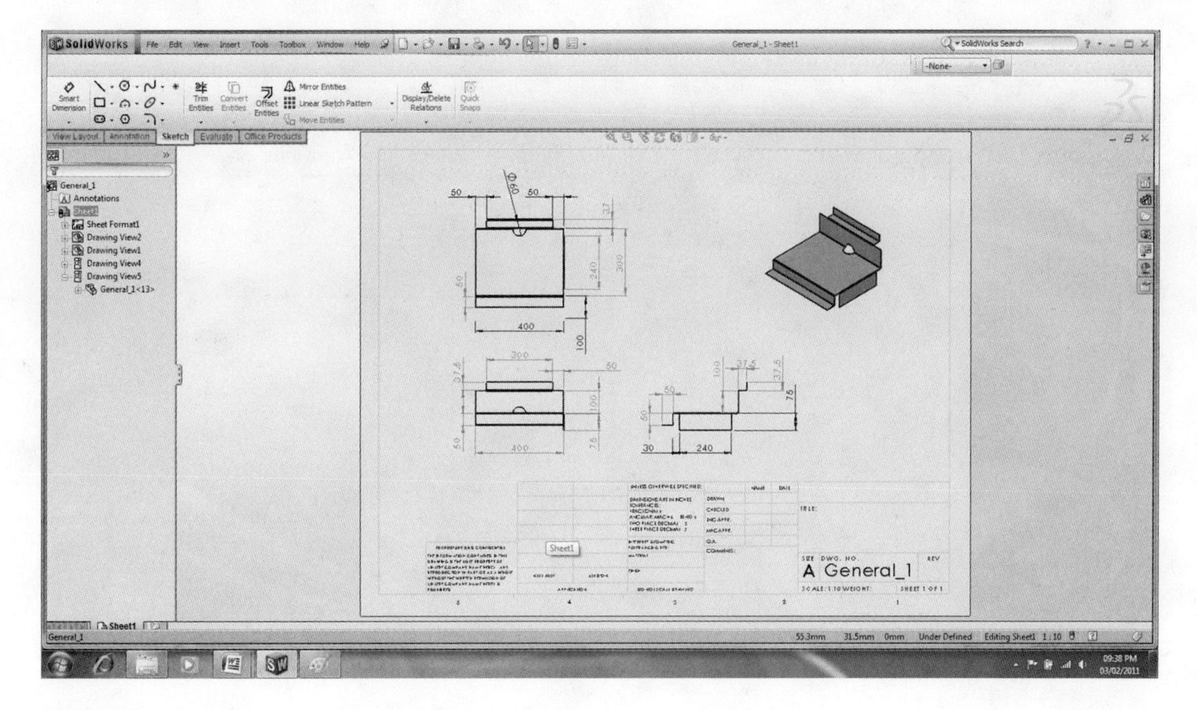

Figure 19.2

Part and drawing views and dimensions for Tutorial 19.1.

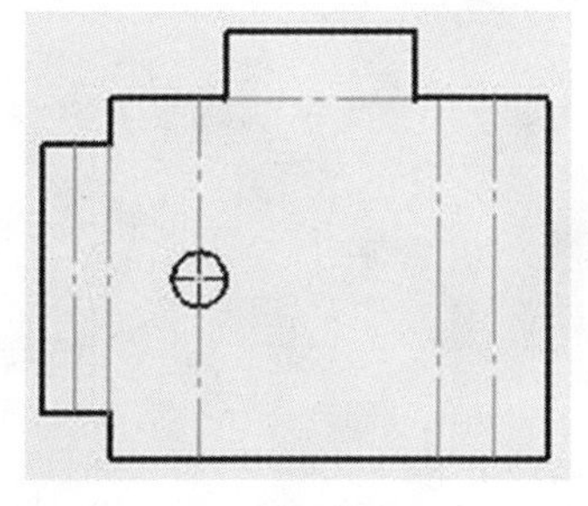

Figure 19.3

Flat pattern of the sheet metal part for Tutorial 19.1.

(Note: The Sheet Metal CommandPropertyManager is automatically displayed; see Figure 19.5.)

5. Set the values of the following parameters: Thickness = 1.5; K-Factor = 0.5; Ratio = 0.5.
6. Click OK.

Create the First Sketched Bend

7. Click top of the Base Flange.
8. Be in sketch mode and create Sketch2 on the Base Flange for the first bending operation, a line 100 mm from one edge (see Figure 19.6).
9. Click the Sketch Bend tool (see Figure 19.7).
10. Accept the default Bend radius of 2 mm and Angle = 90-deg.
11. Click the Fixed Face (Face<1>) to one side of the line (Sketch2) for the Bend Parameters rollout. (The Sketched Bend PropertyManager is automatically displayed; see Figure 19.8.)
12. Click OK to complete the sketched bend operation. (See Figure 19.9 for bend created.)

Create the Second Sketched Bend

13. Click top of the Base Flange.
14. Be in sketch mode and create Sketch3 on the Base Flange for the first bending operation, a line 100 mm from one edge (see Figure 19.9).
15. Click the Sketch Bend tool (see Figure 19.7).

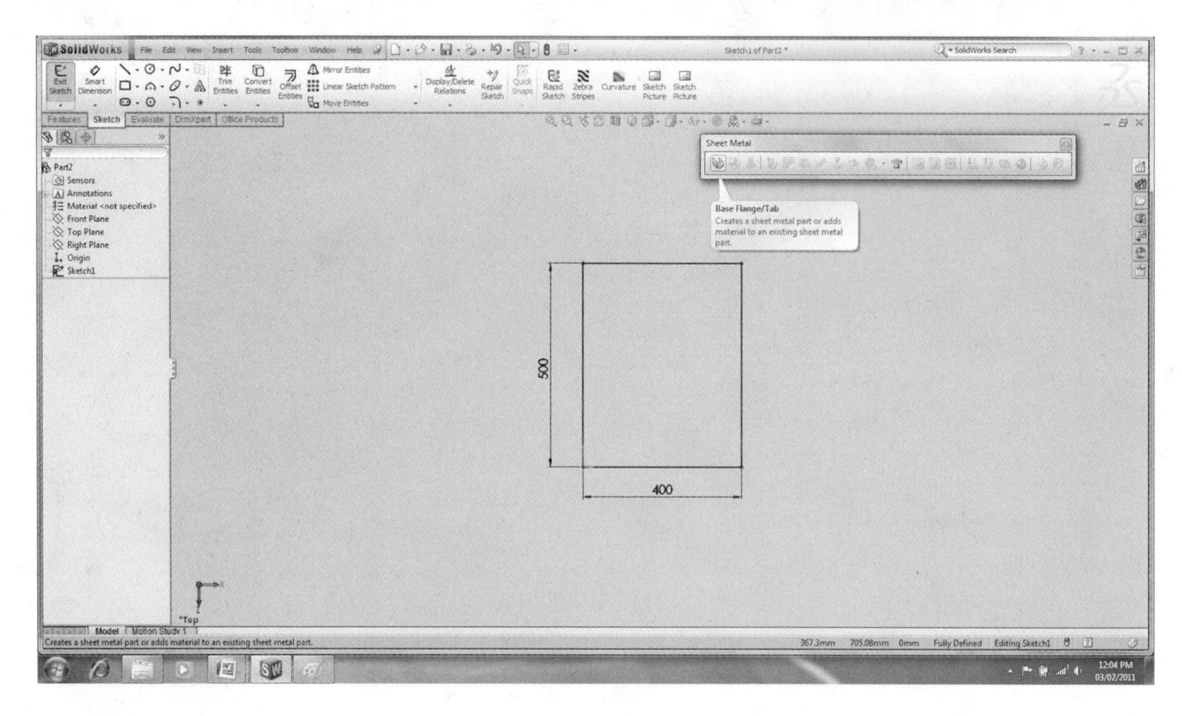

Figure 19.4

Sketch1 and selection of Base Flange/Tab option.

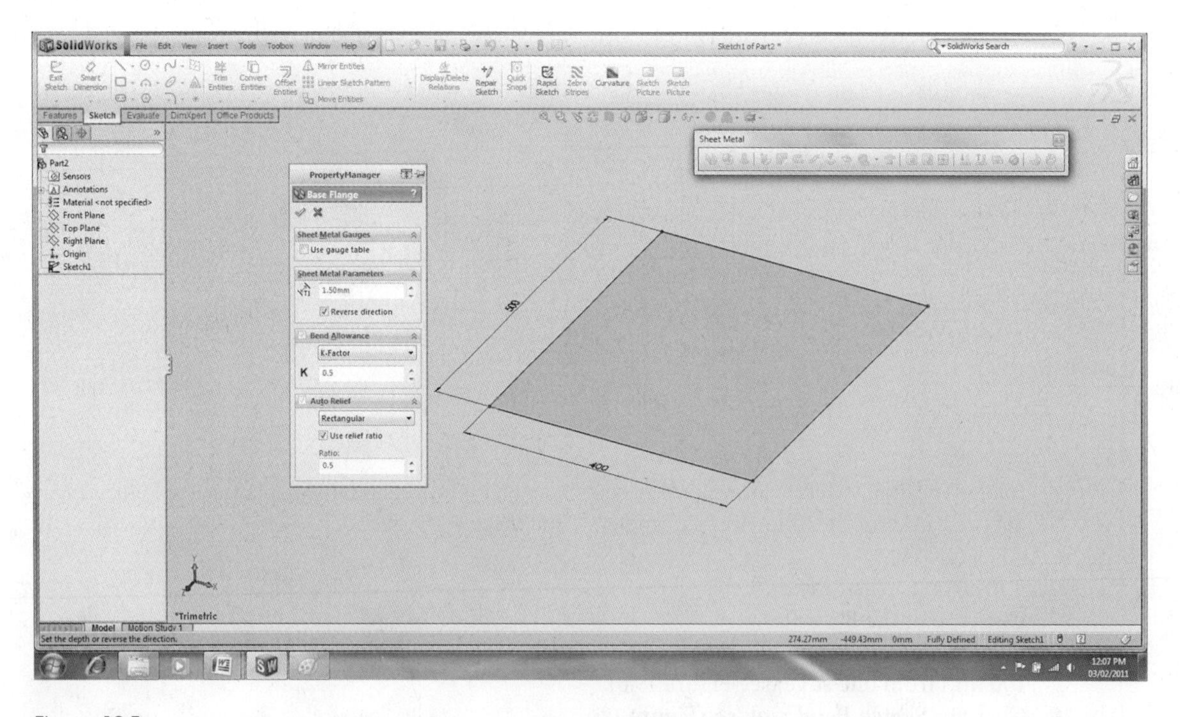

Figure 19.5

Base Flange/Tab PropertyManager.

16. Accept the default Bend radius of 2 mm and Angle = 90-deg.
17. Click the Fixed Face (Face<1>) to one side of the line (Sketch2) for the Bend Parameters rollout. (The Sketched Bend PropertyManager is automatically displayed; see Figure 19.10.)
18. Click OK to complete the sketched bend operation. (See Figure 19.11 for bend created.)

Create the First Edge Flange

19. Select the top edge of the first Sketched Bend (see Figure 19.12).

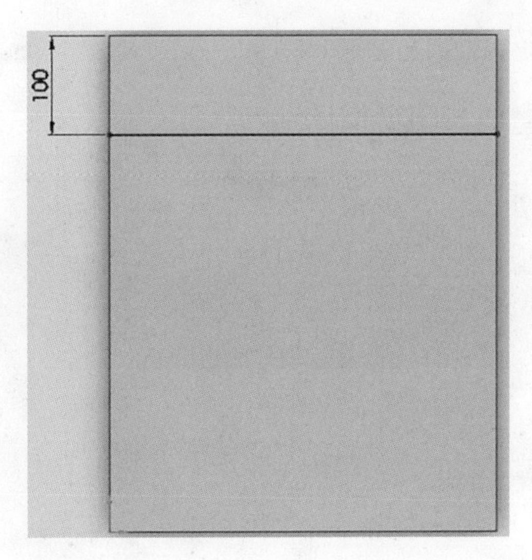

Figure 19.6

Sketch2.

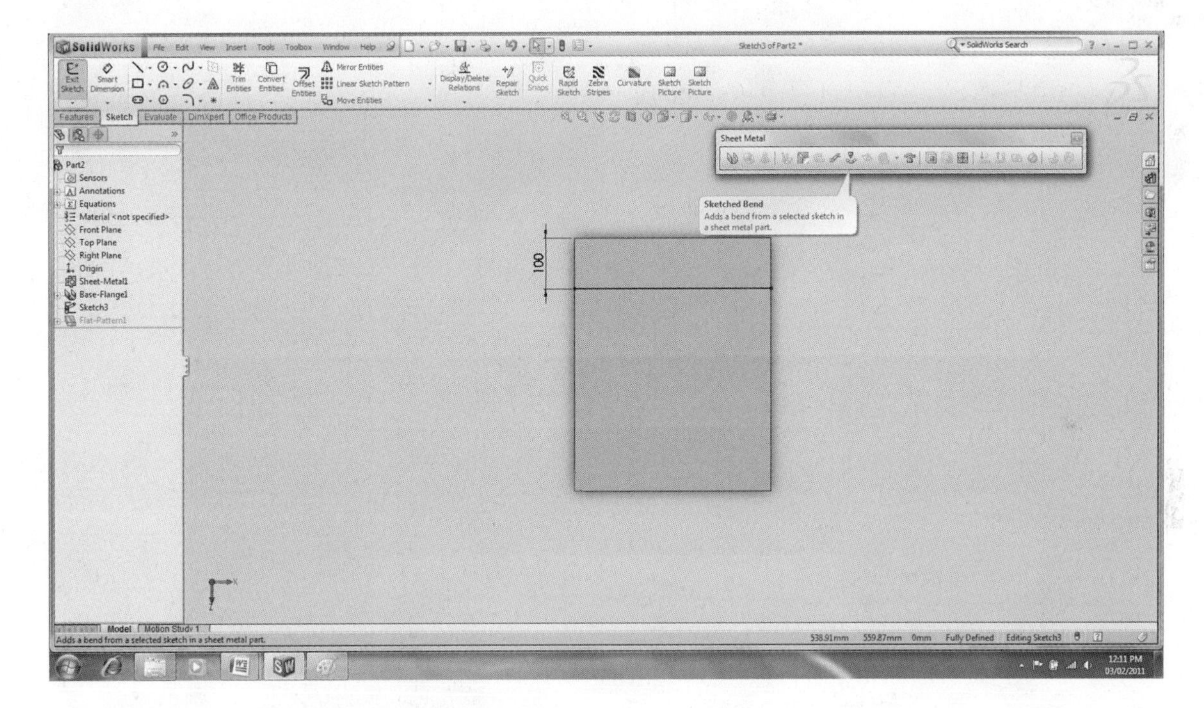

Figure 19.7

Sketched Bend tool is used for bend operation.

20. Set Length of Edge Flange = 75 and Angle = 90-deg (see Figure 19.12).
21. Click the Edit Flange Profile button from the Edge Flange PropertyManager (see Figure 19.12).
22. Click OK if the following message appears: "*The automatic relations inferred by this operation would over define the sketch. They will not be added.*"
23. Adjust the Width of the Edge Flange that is created by dragging the endpoints of the edge to new positions (50 mm from each end) and using sketch tool to add dimensions (see Figure 19.13). (Note: *When the Profile Sketch window appears, do not click Finish until the length of the Edge Flange is fully adjusted.*)
24. Click Finish.

Create the Second Edge Flange

25. Select the top edge of the first Sketched Bend (see Figure 19.14).

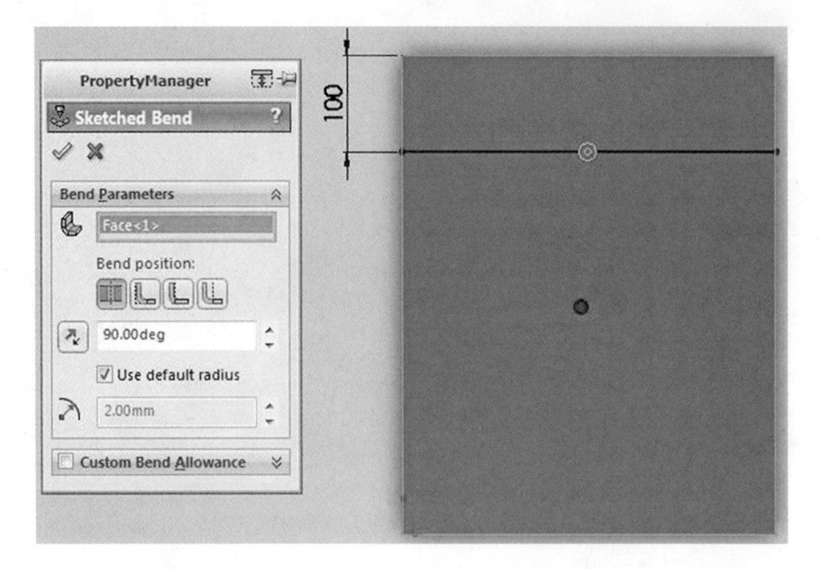

Figure 19.8

Sketched Bend PropertyManager.

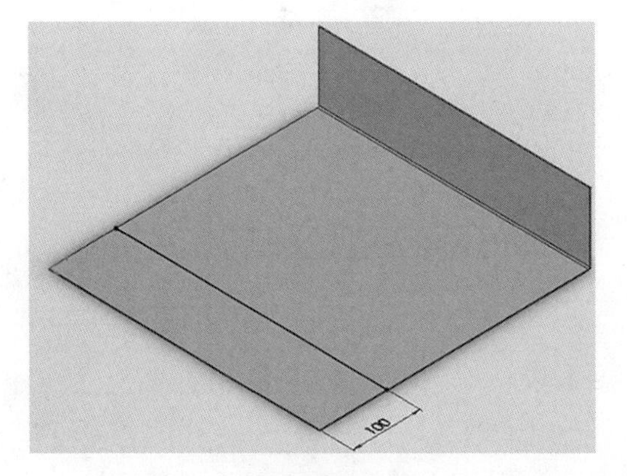

Figure 19.9

Bend created.

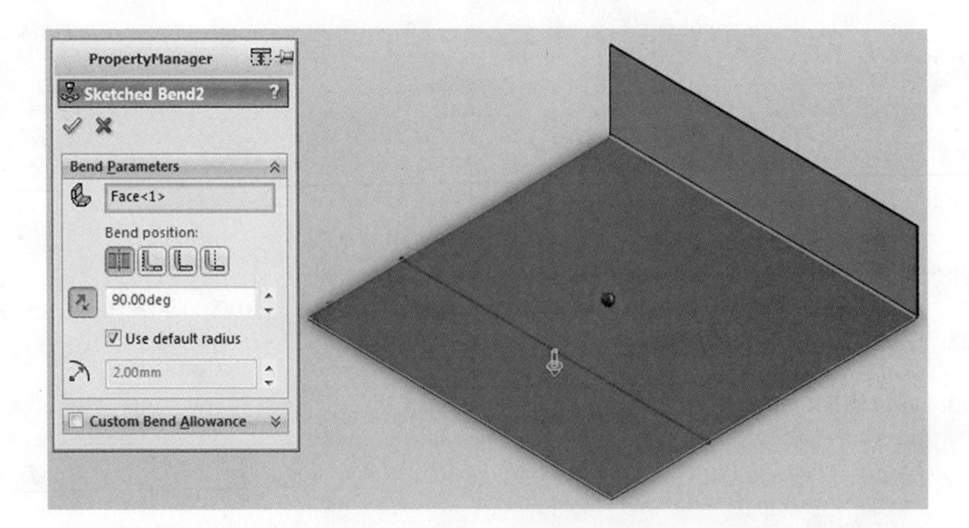

Figure 19.10

Sketched Bend PropertyManager.

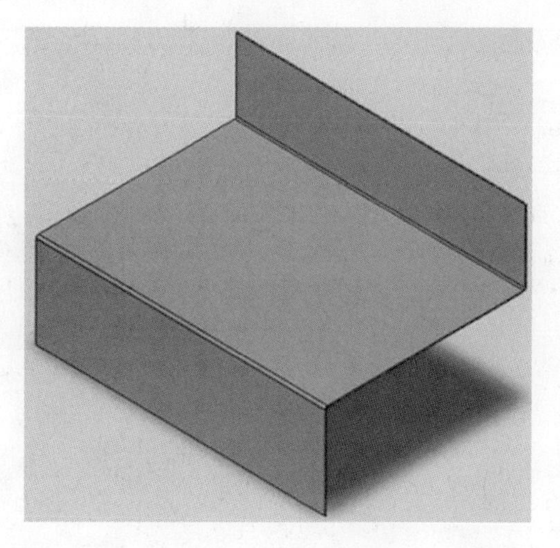

Figure 19.11

Bend created.

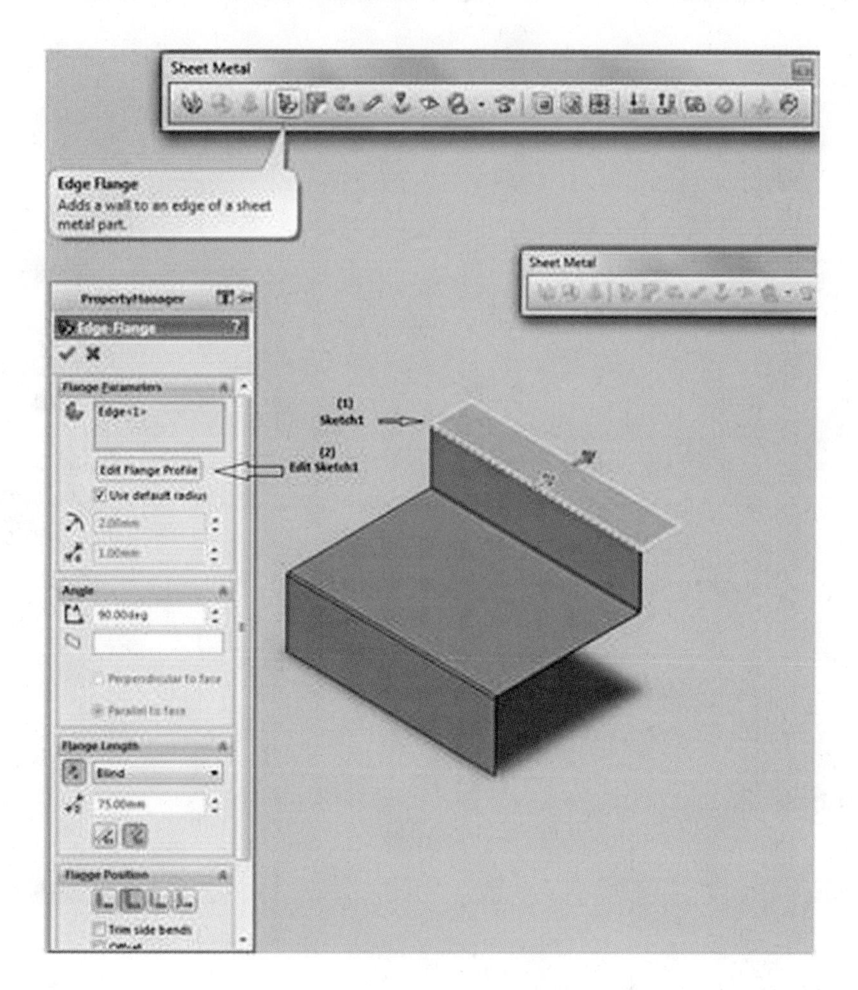

Figure 19.12

PropertyManager for creating Edge Flange.

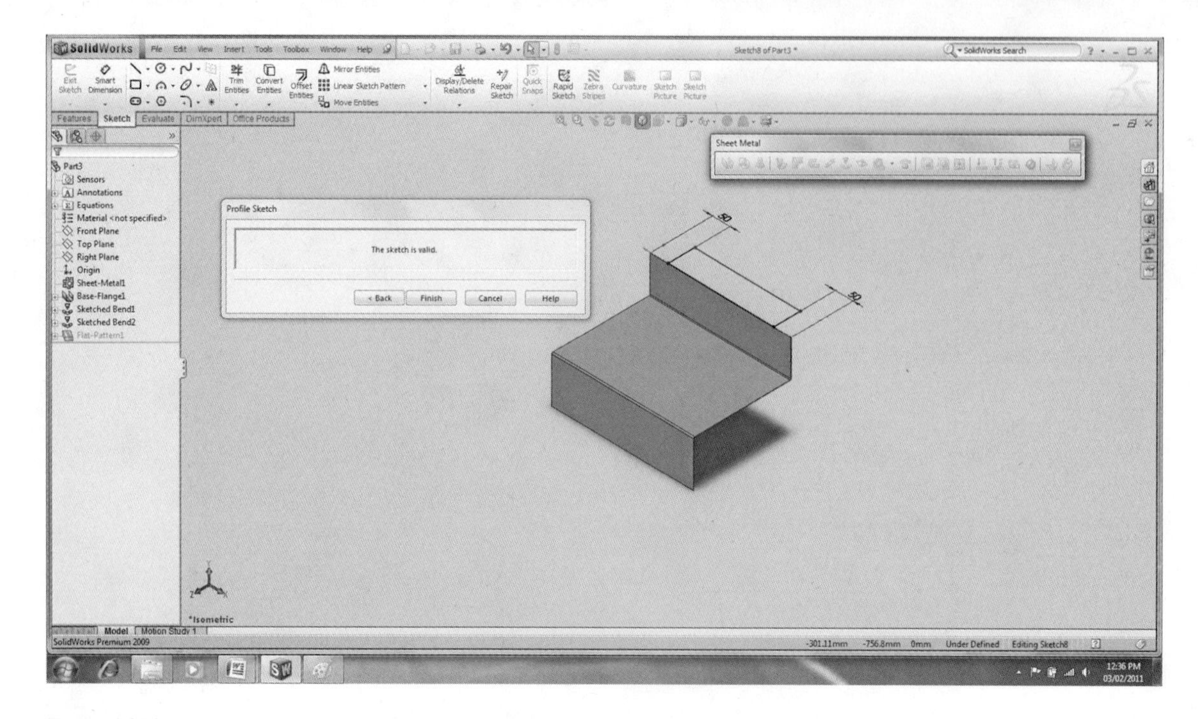

Figure 19.13

Profile Sketch window.

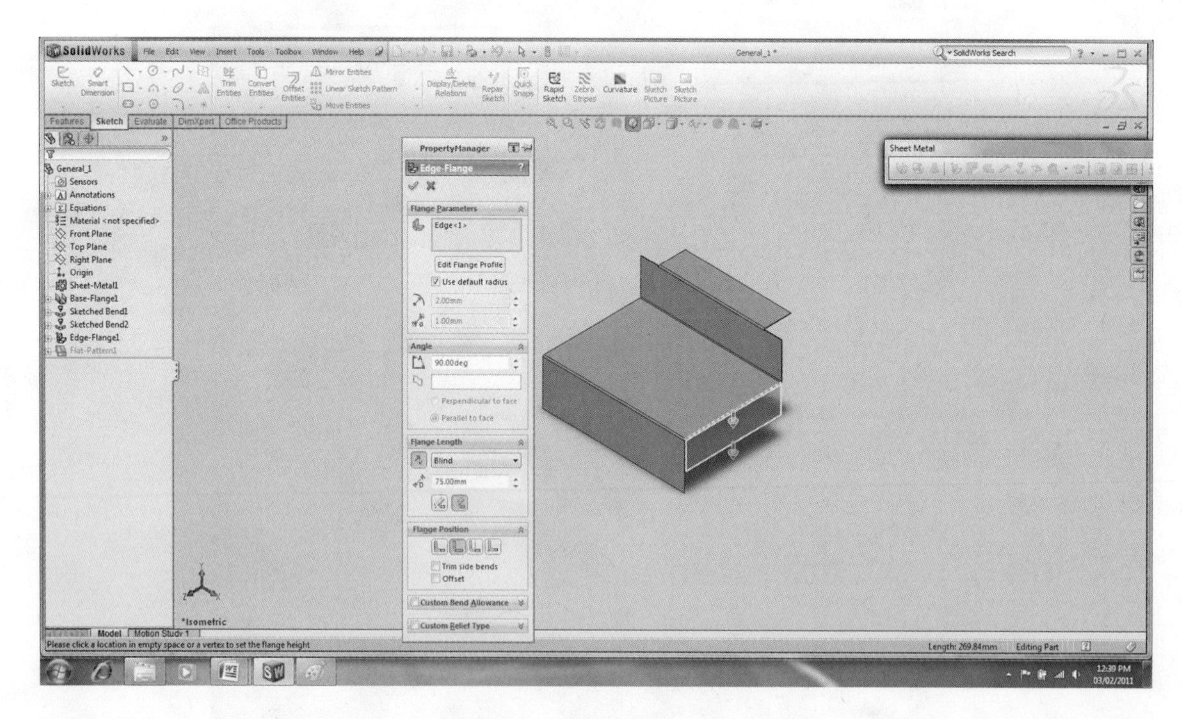

Figure 19.14

PropertyManager for creating Edge Flange.

26. Set Length of Edge Flange = 75 and Angle = 90-deg (see Figure 19.14).
27. Click Edit Flange Profile button from the Edge Flange PropertyManager (see Figure 19.14).
28. Click OK if the following message appears: "*The automatic relations inferred by this operation would over define the sketch. They will not be added.*"
29. Adjust the Width of the Edge Flange that is created by dragging the endpoints of the edge to new positions (30 mm from each end) and using sketch tool to add dimensions (see Figure 19.15). (Note: *When the Profile Sketch window appears, do not click Finish until the length of the Edge Flange is fully adjusted.*)
30. Click Finish. (See Figure 19.16 for the two edge flanges.)

Unfold the First Sketched Bend

31. Click Unfold button. (The Unfold PropertyManager is automatically displayed; see Figure 19.17.)
32. Select Face<1> as the Fixed Face and SketchBend1 as the Bends to unfold (see Figure 19.17).

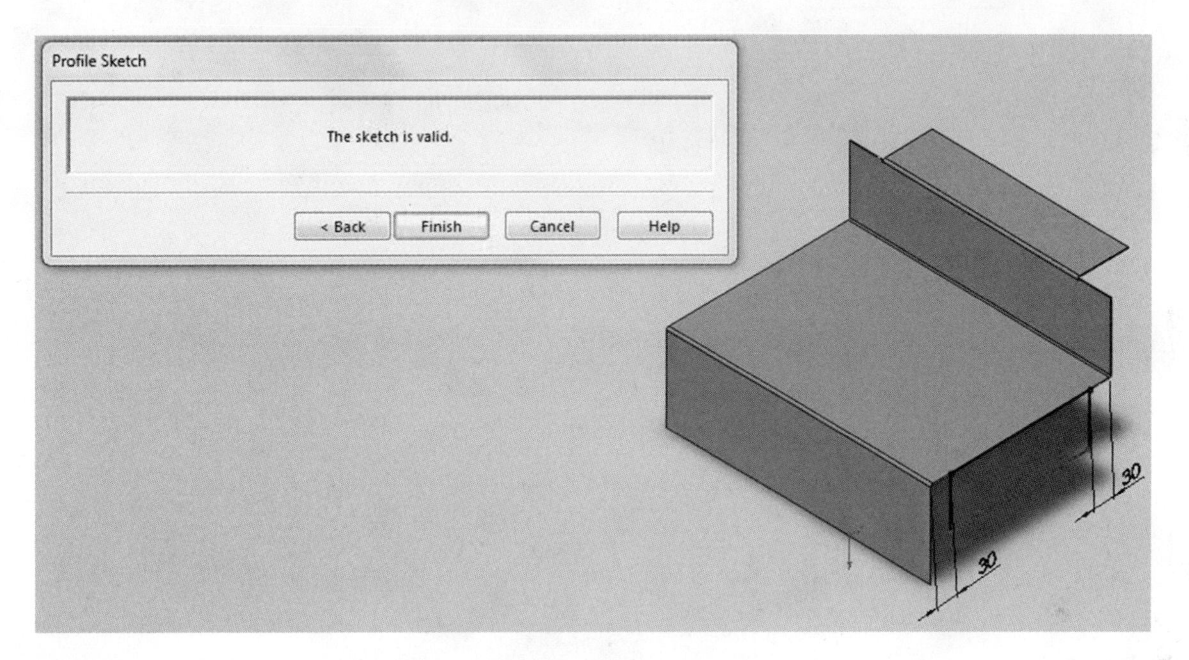

Figure 19.15

Profile Sketch window.

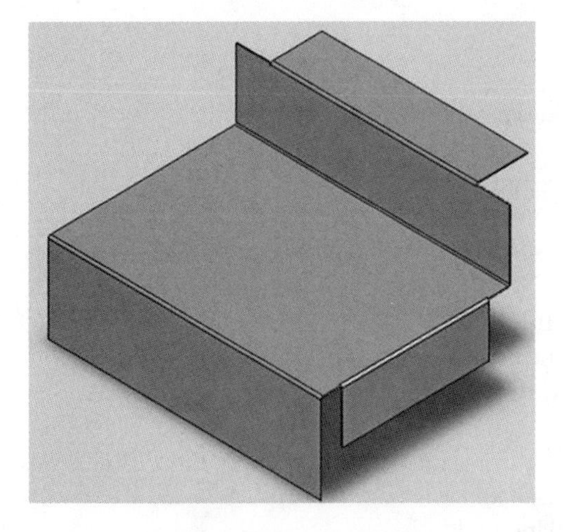

Figure 19.16

Model has the two edge flanges created.

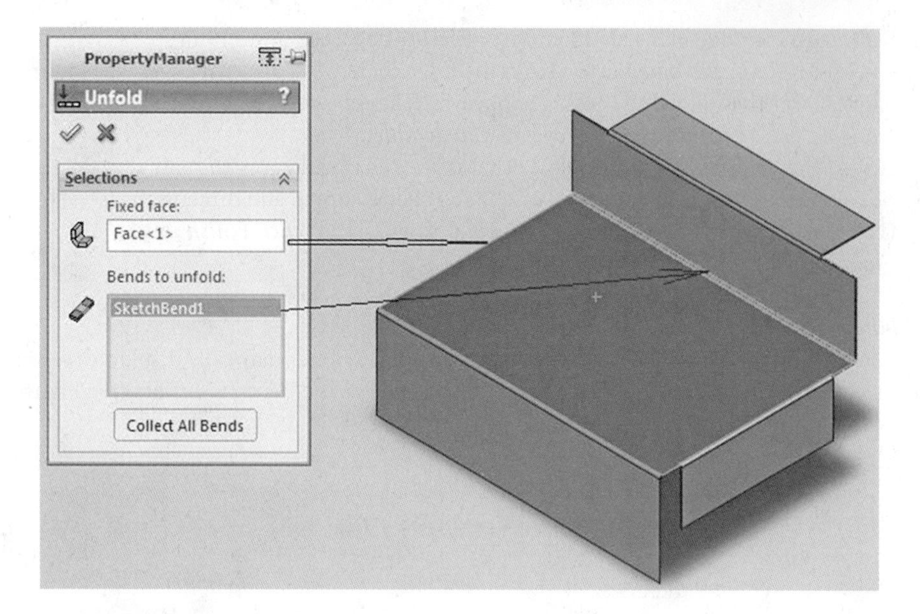

Figure 19.17

Unfold PropertyManager.

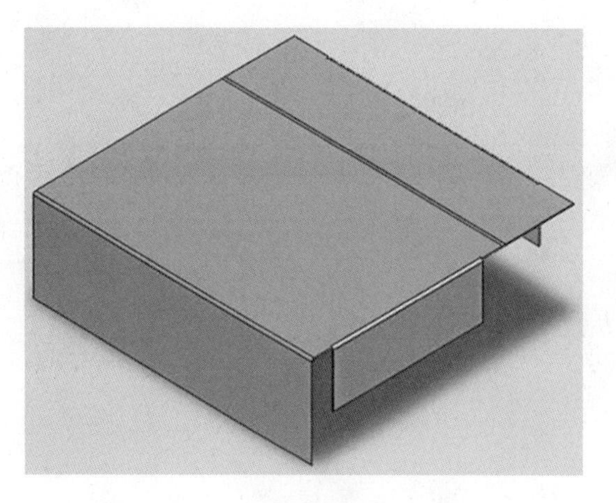

Figure 19.18

Unfolded bend.

33. Click OK to unfold the bend (see Figure 19.18).

Create a Circular Hole Midpoint along the First Sketched Bend

34. Be in sketch mode, and create Sketch6, a Circle having a diameter of 60 mm (see Figure 19.19).
35. Extrude-cut up to Next (see Figure 19.20).

Fold the First Sketched Bend

36. Click Fold button. (The Fold PropertyManager is automatically displayed; see Figure 19.21.)
37. Select Face<1> as the Fixed Face and SketchBend1 as the Bends to Fold (see Figure 19.21).
38. Click OK to fold the bend (see Figure 19.22).

Create the Third Sketched Bend

39. Click bottom Edge Flange.
40. Be in sketch mode and create Sketch7 on the bottom Edge Flange for the third bending operation, a line midpoint from one edge (see Figure 19.23).
41. Click the Sketch Bend tool.
42. Accept the default Bend radius of 2 mm and Angle = 90-deg.

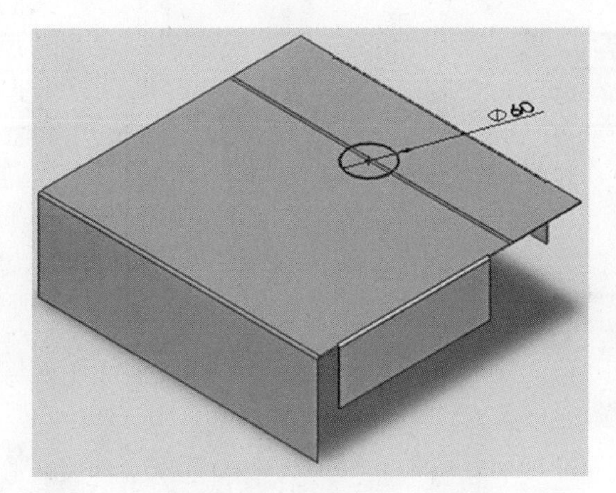

Figure 19.19

Sketch5, a circle.

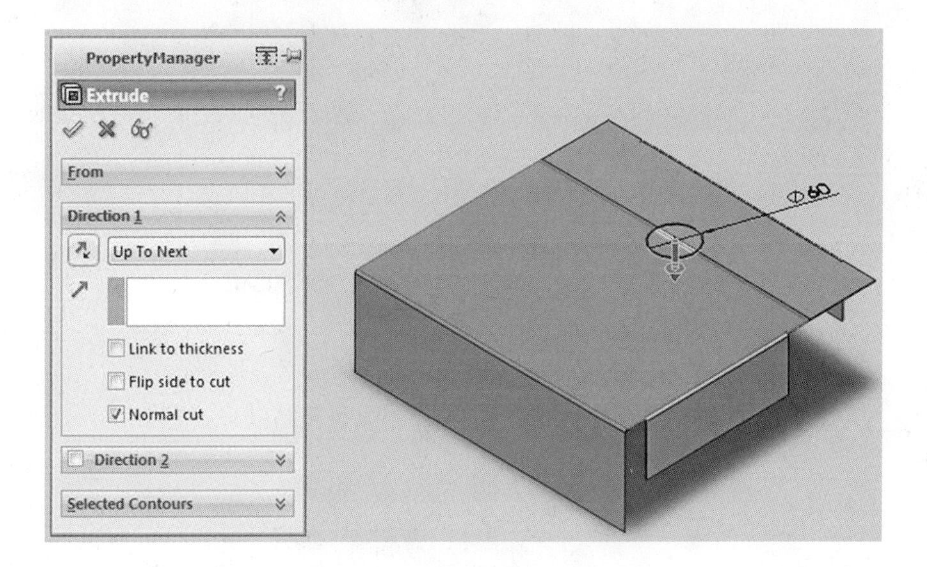

Figure 19.20

Extrude-cut to create a hole.

43. Click the Fixed Face (Face<1>) to one side of the line (Sketch2) for the Bend Parameters rollout (the Sketched Bend PropertyManager is automatically displayed; see Figure 19.23).
44. Click OK to complete the sketched bend operation (see Figure 19.24 for bend created).

Create the Fourth Sketched Bend

45. Click top Edge Flange.
46. Be in sketch mode and create Sketch8 on the bottom Edge Flange for the third bending operation, a line midpoint from one edge (see Figure 19.25).
47. Click the Sketch Bend tool.
48. Accept the default Bend radius of 2 mm and Angle = 90-deg.
49. Click the Fixed Face (Face<1>) to one side of the line (Sketch2) for the Bend Parameters rollout. (The Sketched Bend PropertyManager is automatically displayed; see Figure 19.25.)
50. Click OK to complete the sketched bend operation. (See Figure 19.26 for bend created.)

Create the Flatten State

51. Click the Flatten tool. (The flatten state is as shown in Figure 19.27.)

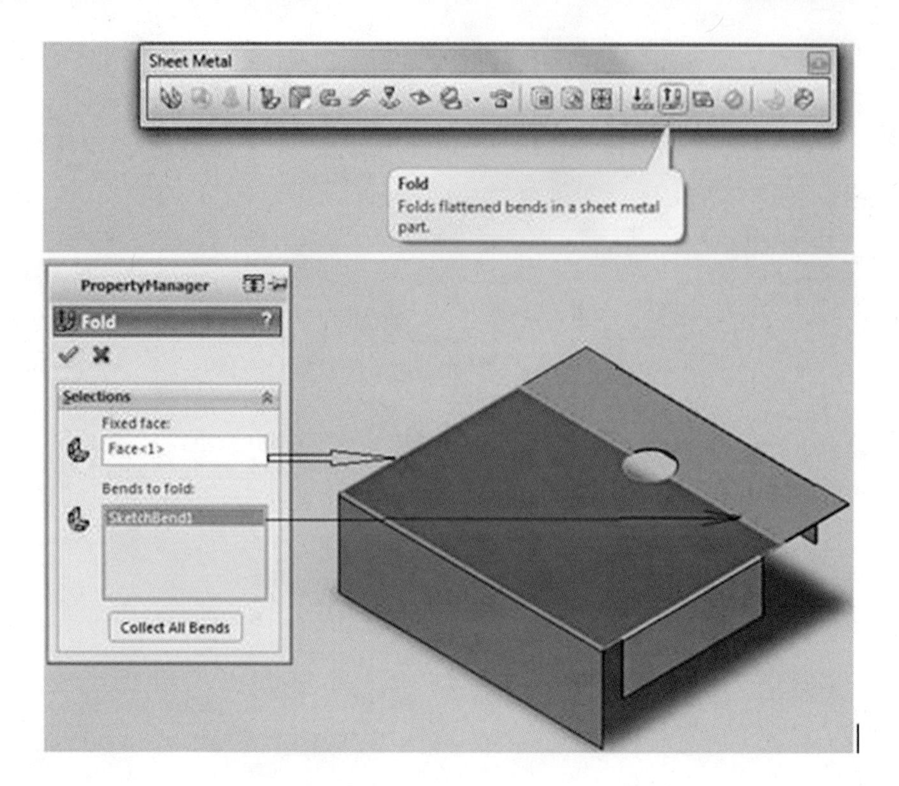

Figure 19.21

Fold PropertyManager.

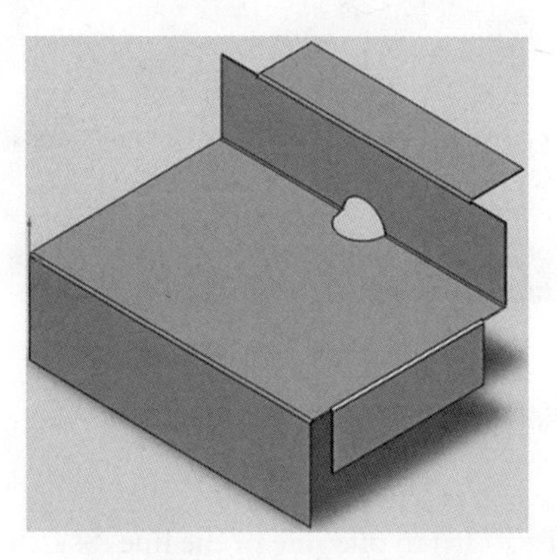

Figure 19.22

Fold bend.

Tutorial 19.2: P1

In this tutorial, we will create the sheet metal part that is shown at the top right, with its views and dimensions shown in Figure 19.28. The flat pattern of the sheet metal part is shown in Figure 19.29.

SolidWorks Solution to Tutorial 19.2

Create the Base Flange

1. Start a New SolidWorks Part document.
2. Select the Top Plane.
3. Be in sketch mode and create Sketch1, 800 mm × 400 mm (see Figure 19.30).
4. Click the Base Flange/Tab option from the Sheet Metal CommandManager (see Figure 19.31).

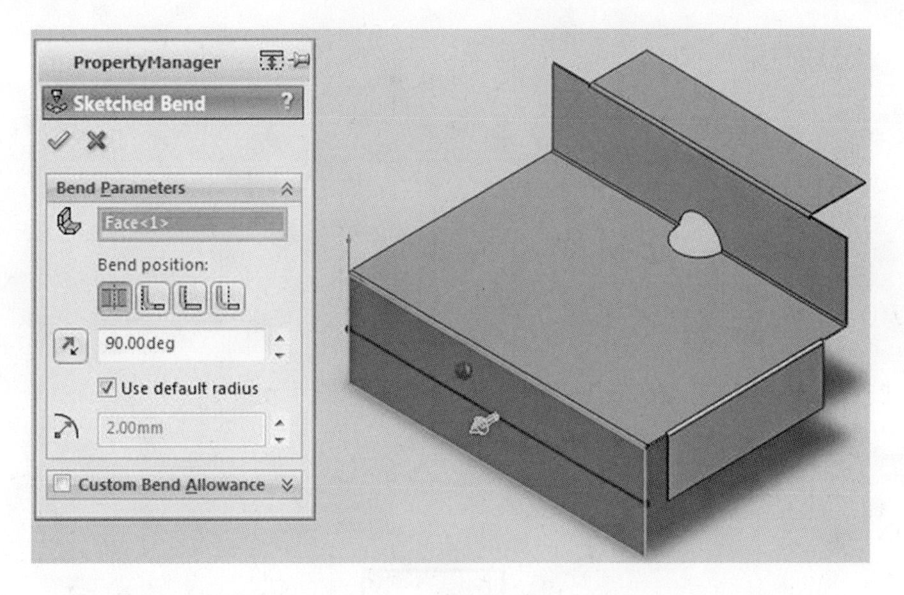

Figure 19.23

Sketched Bend PropertyManager.

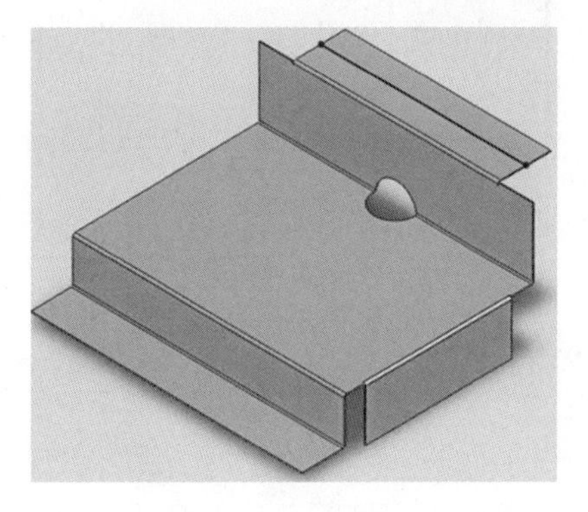

Figure 19.24

Bend created.

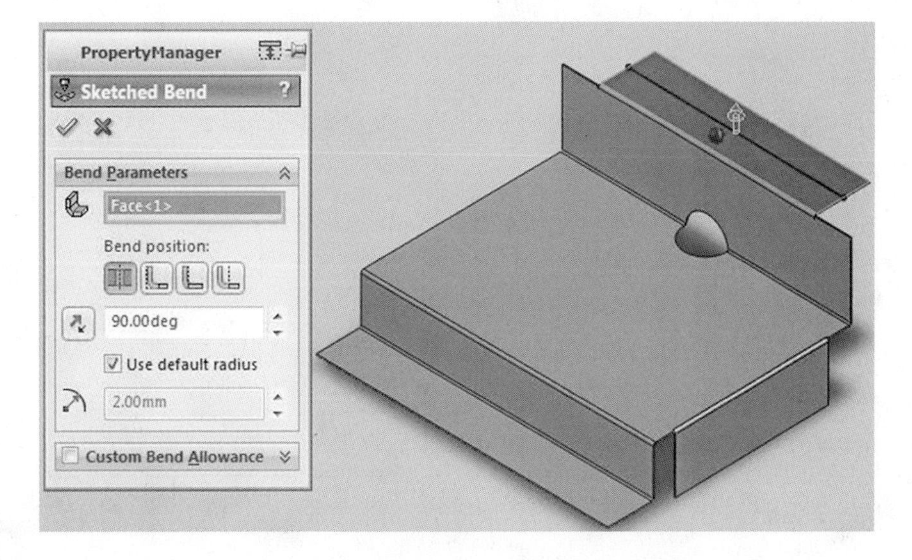

Figure 19.25

Sketched Bend PropertyManager.

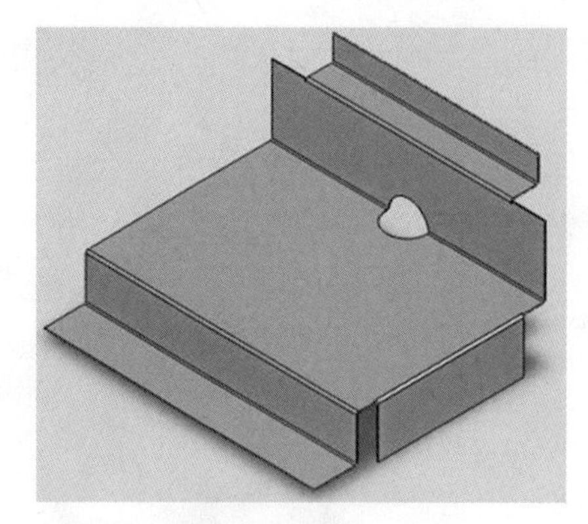

Figure 19.26

Bend created.

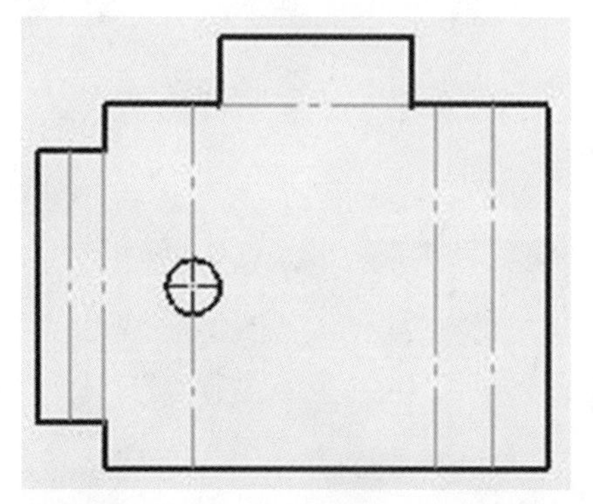

Figure 19.27

Flatten state.

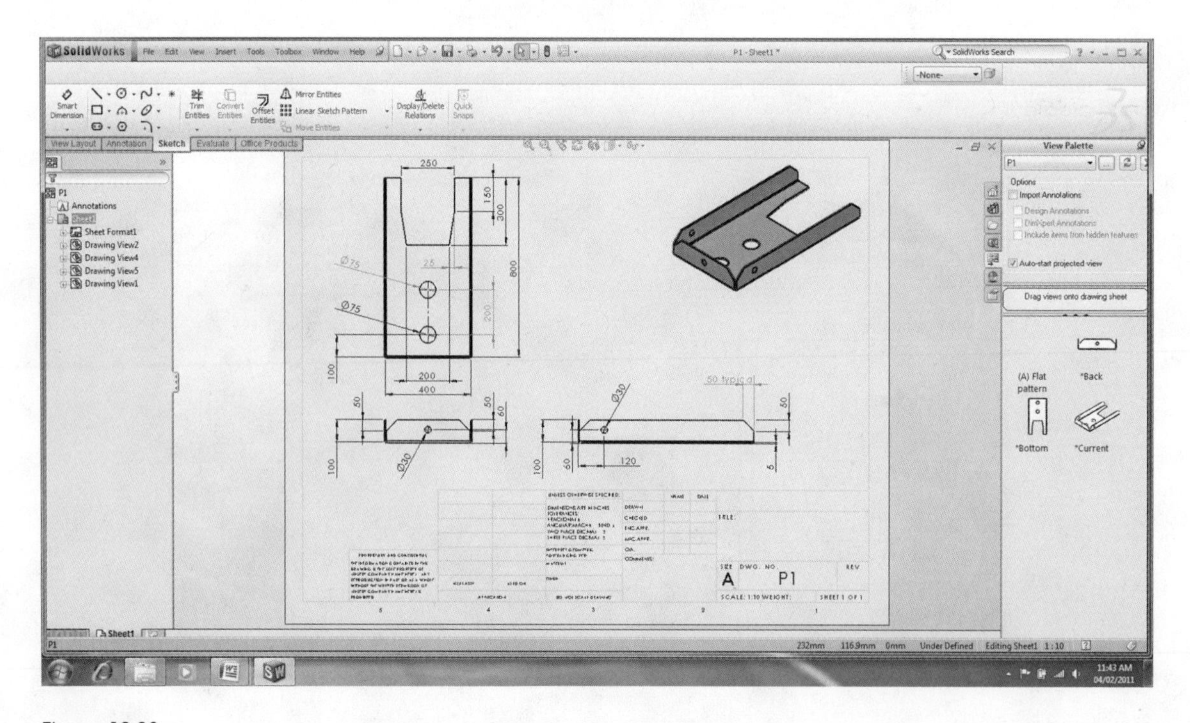

Figure 19.28

Sheet metal part with drawings and dimensions for Tutorial 19.2.

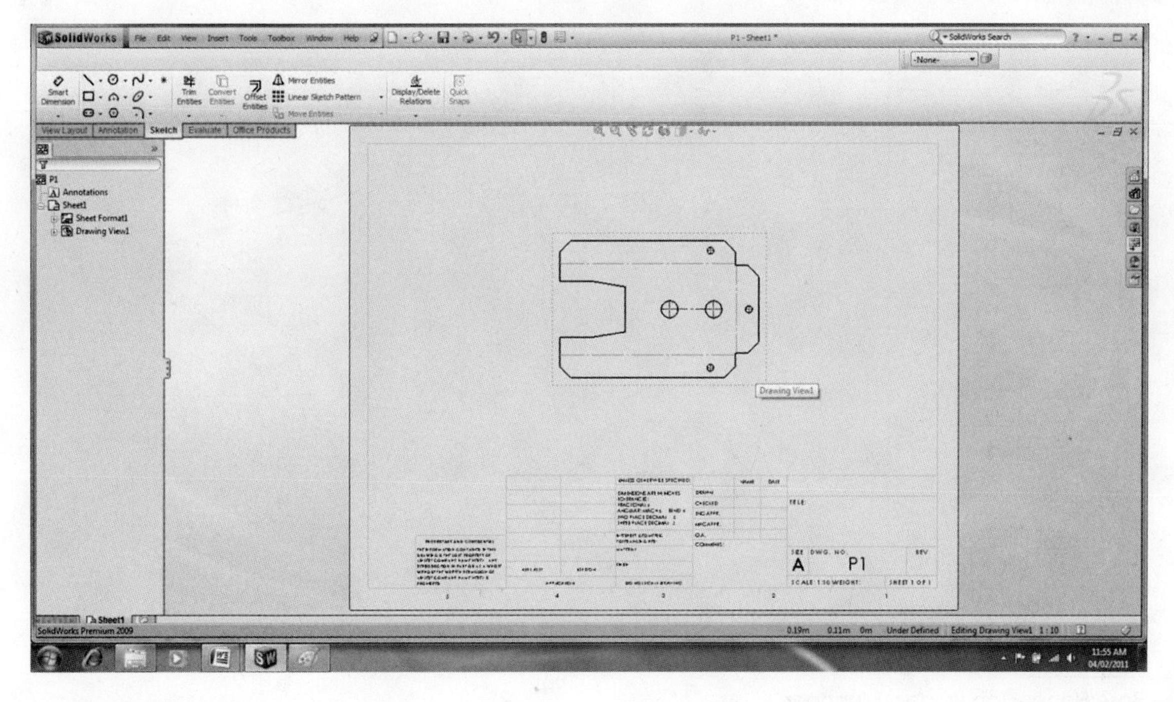

Figure 19.29

Flat pattern of the sheet metal part for Tutorial 19.2.

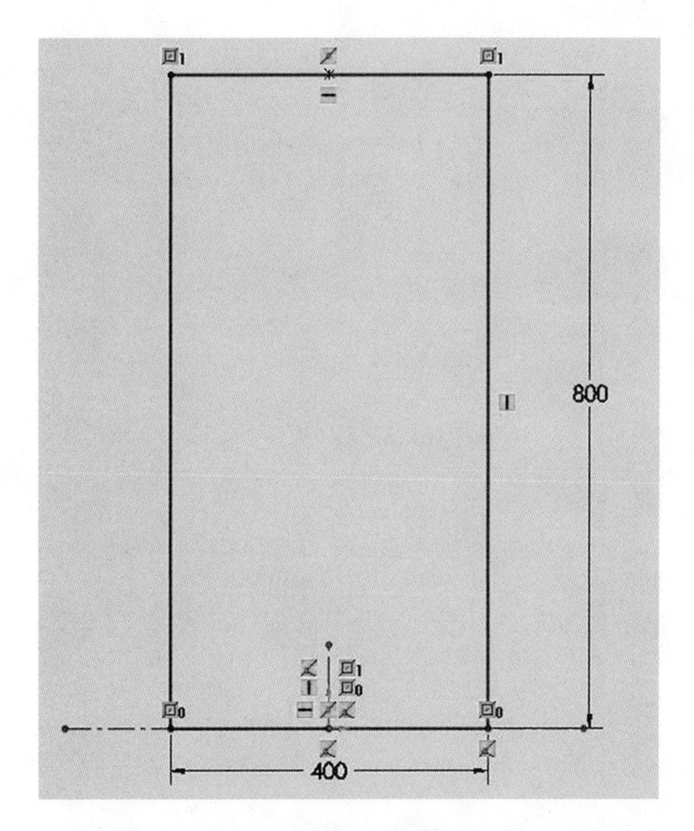

Figure 19.30

Sketch1 and selection of Base Flange/Tab option.

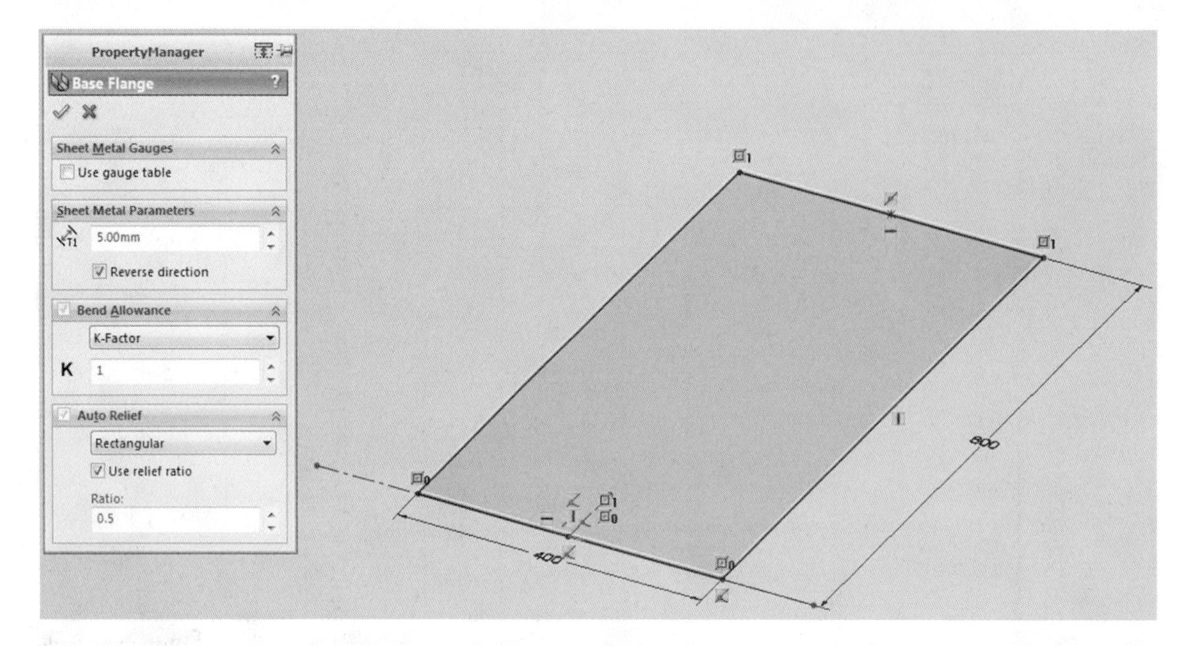

Figure 19.31

Base Flange/Tab PropertyManager.

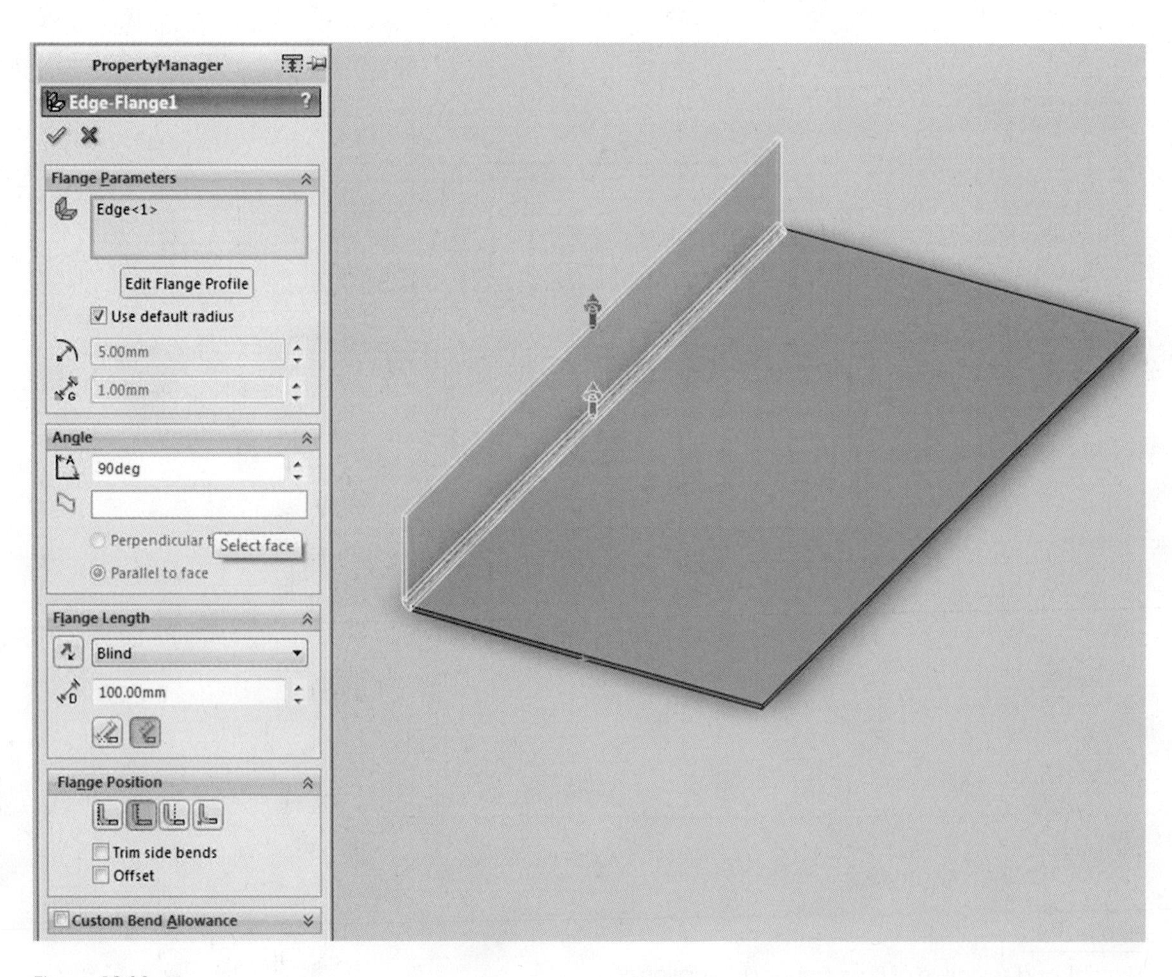

Figure 19.32

PropertyManager for creating first Edge Flange.

5. Set the values of the following parameters: Thickness = 5; K-Factor = 1; Ratio = 0.5.
6. Click OK.

Create the First Edge Flange

7. Select the left edge of the first Base Flange. (See Figure 19.32 for the Edge Flange PropertyManager.)
8. Set Length of Edge Flange = 100 and Angle = 90-deg (see Figure 19.32).
9. Click OK.

Create the Second Edge Flange

10. Select the left edge of the second Base Flange. (See Figure 19.33 for the Edge Flange PropertyManager.)
11. Set Length of Edge Flange = 100 and Angle = 90-deg (see Figure 19.33).
12. Click OK.

Create the Third Edge Flange

13. Select the left edge of the third Base Flange. (See Figure 19.34 for the Edge Flange PropertyManager.)
14. Set Length of Edge Flange = 100 and Angle = 90-deg (see Figure 19.34).
15. Click OK.

Create the First Break Corner

16. Select an Edge Flange. (See Figure 19.35 for the Break Corner PropertyManager.)
17. Select the Edge Flange as the face to break, Face<1> (see Figure 19.35).
18. Set break Distance = 50 (see Figure 19.35).
19. Click OK.

Create the Second Break Corner

20. Select an Edge Flange. (See Figure 19.36 for the Break Corner PropertyManager.)
21. Select the Edge Flange as the face to break, Face<1> (see Figure 19.36).
22. Set break Distance = 50 (see Figure 19.36).
23. Click OK.

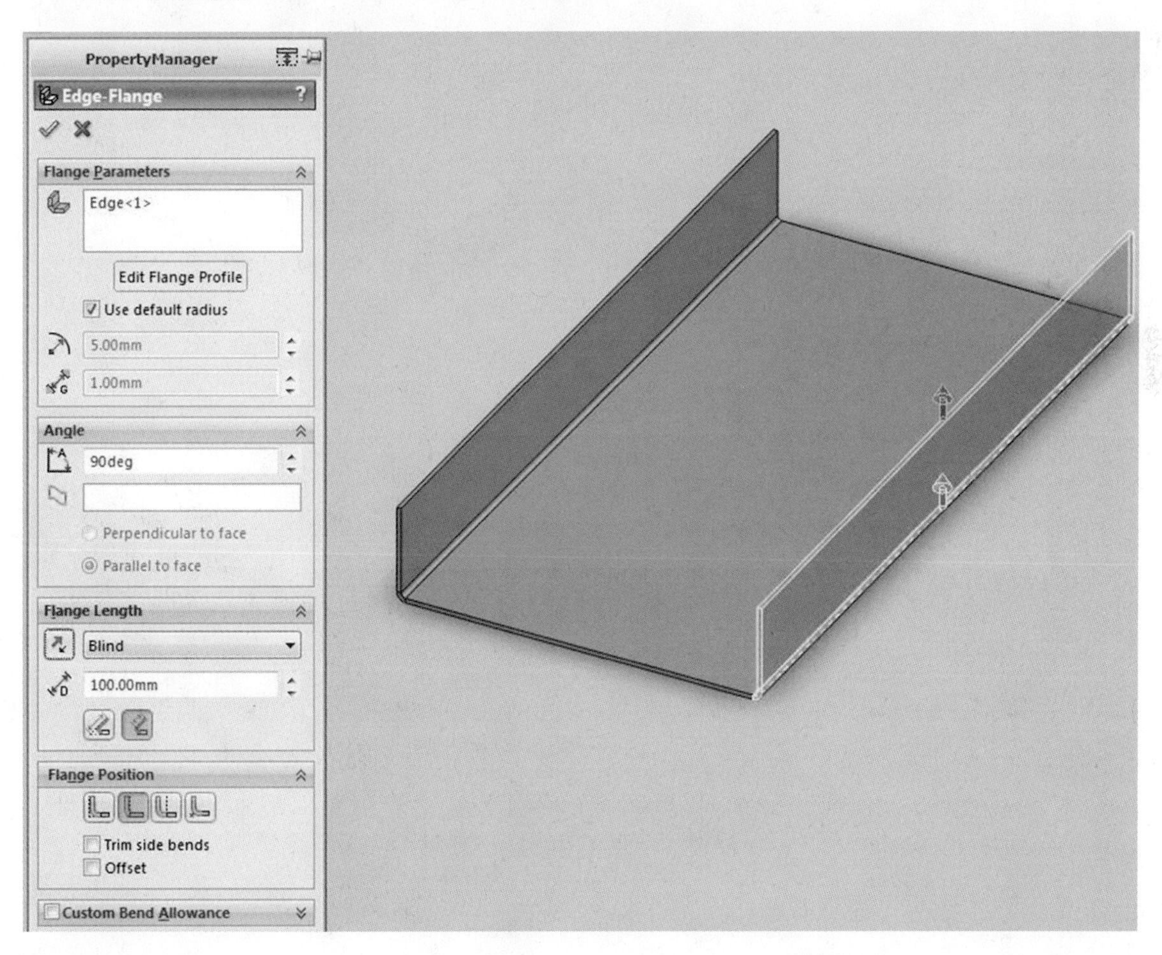

Figure 19.33

PropertyManager for creating second Edge Flange.

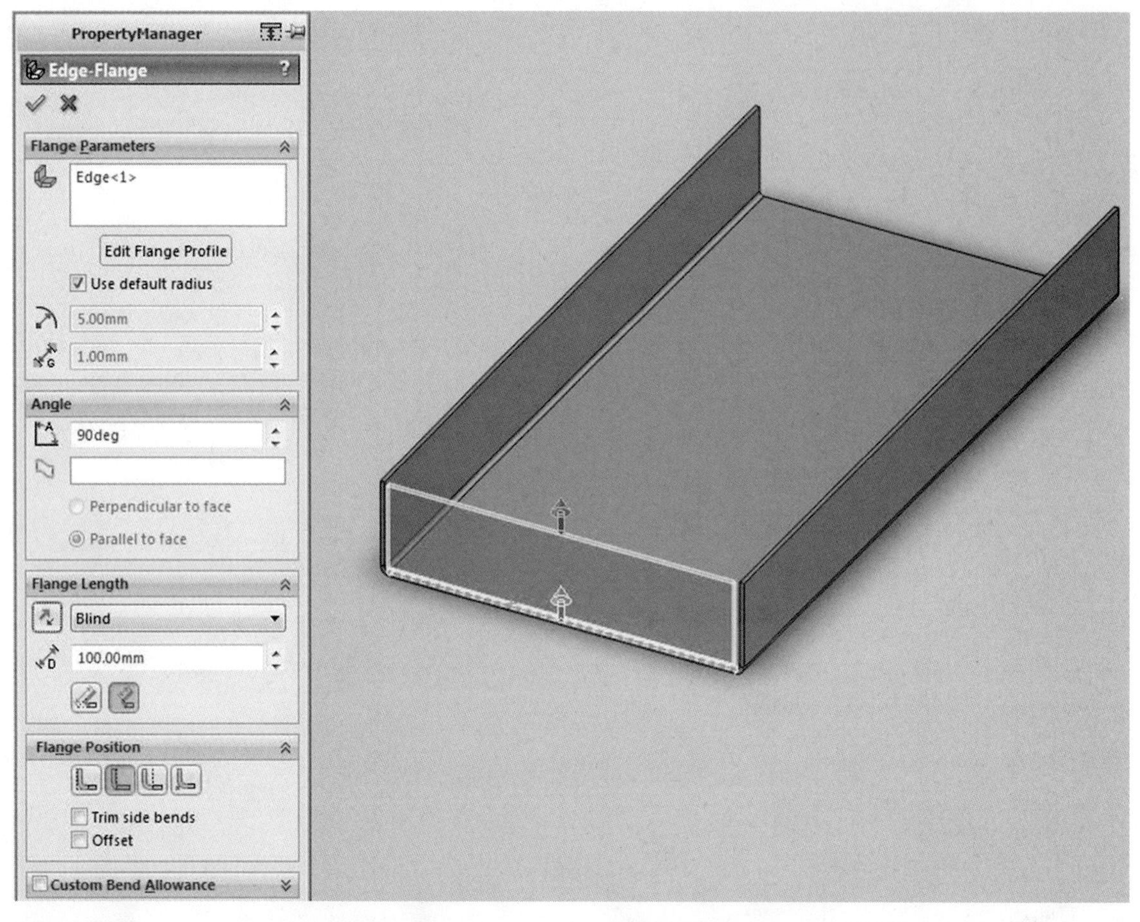

Figure 19.34

PropertyManager for creating third Edge Flange.

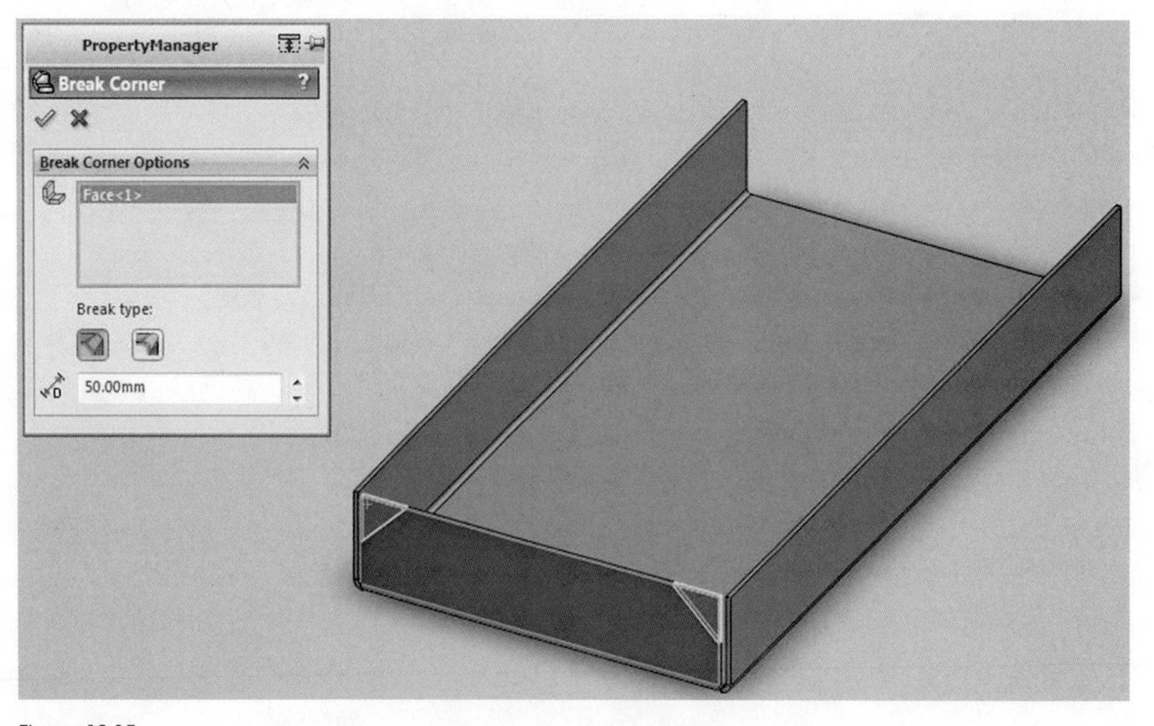

Figure 19.35

PropertyManager for creating first Break Corner.

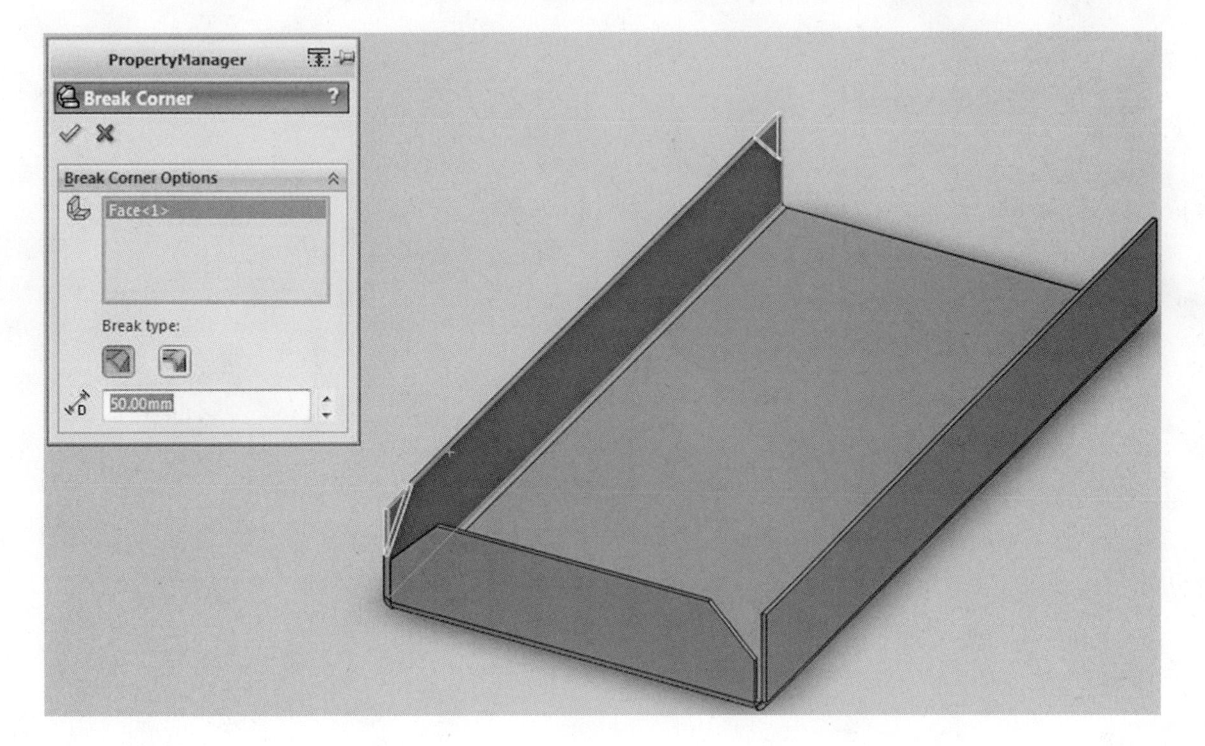

Figure 19.36

Second Break Corner PropertyManager.

Create the Third Break Corner

24. Select an Edge Flange. (See Figure 19.37 for the Break Corner PropertyManager.)
25. Select the Edge Flange as the face to break, Face<1> (see Figure 19.37).
26. Set break Distance = 50 (see Figure 19.37).
27. Click OK.

Create Extrude-Cut Features

28. Create a sketch on top of Base Flange and Extrude-Cut (see Figure 19.38).
29. Create sketches (circles) on top of Base Flange and Extrude-Cut (see Figure 19.39).
30. Create a sketch on end view of Edge Flange and Extrude-Cut (see Figure 19.40).
31. Create a sketch on end view of Edge Flange and Extrude-Cut (see Figure 19.41).
    The completed of the sheet metal part is shown in Figure 19.42.

Tutorial 19.3: P2

In this tutorial, we will create the sheet metal part that is shown at the top right, with its views and dimensions shown in Figure 19.43a. The flat pattern of the sheet metal part is shown in Figure 19.43b.

Create the Base Flange

1. Start a New SolidWorks Part document.
2. Select the Front Plane.
3. Be in sketch mode and create Sketch1, 400 mm × 250 mm (see Figure 19.44).
4. Click the Base Flange/Tab option from the Sheet Metal CommandManager. (Note: The Sheet Metal CommandPropertyManager is automatically displayed; see Figure 19.45.)
5. Set the values of the following parameters: Thickness = 5; K-Factor = 1; Ratio = 0.5.
6. Click OK.

Create the First Edge Flange

7. Select the right edge of the first Sketched Bend (see Figure 19.46).
8. Set Length of Edge Flange = 100 and Angle = 90-deg (see Figure 19.46).
9. Click Edit Flange Profile button from the Edge Flange PropertyManager (see Figure 19.46).
10. Click OK if the following message appears: "*The automatic relations inferred by this operation would over define the sketch. They will not be added.*"

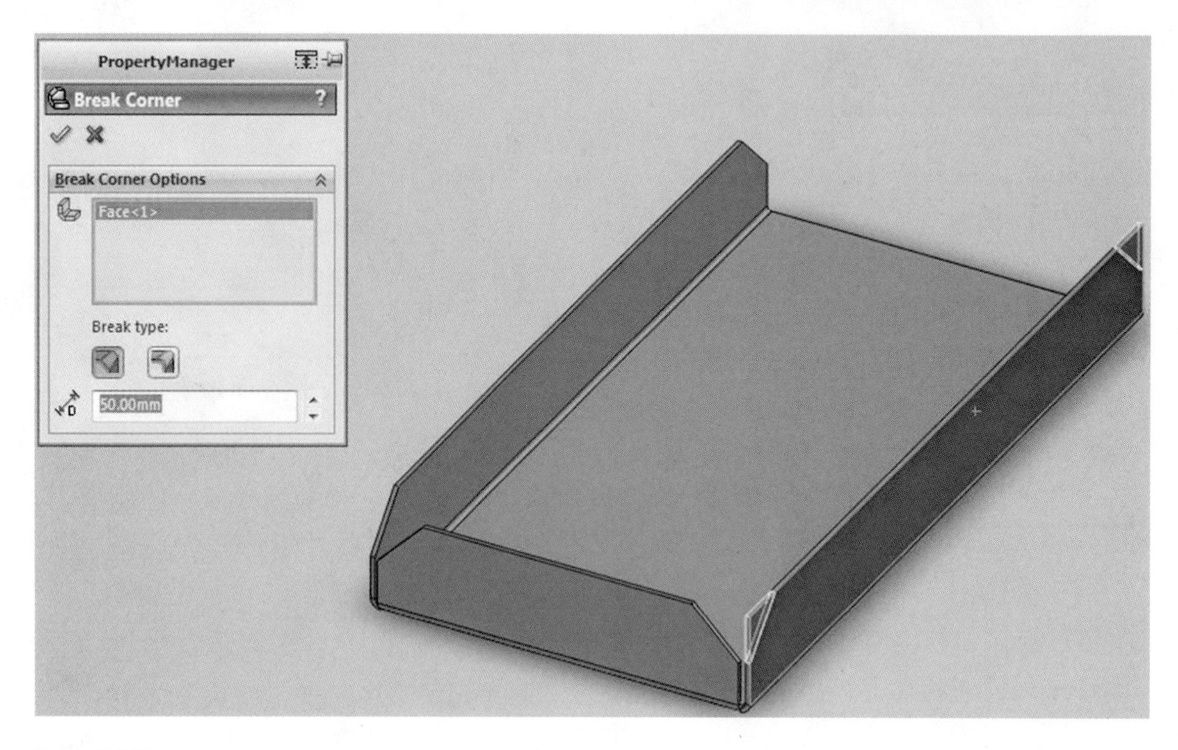

Figure 19.37

Third Break Corner PropertyManager.

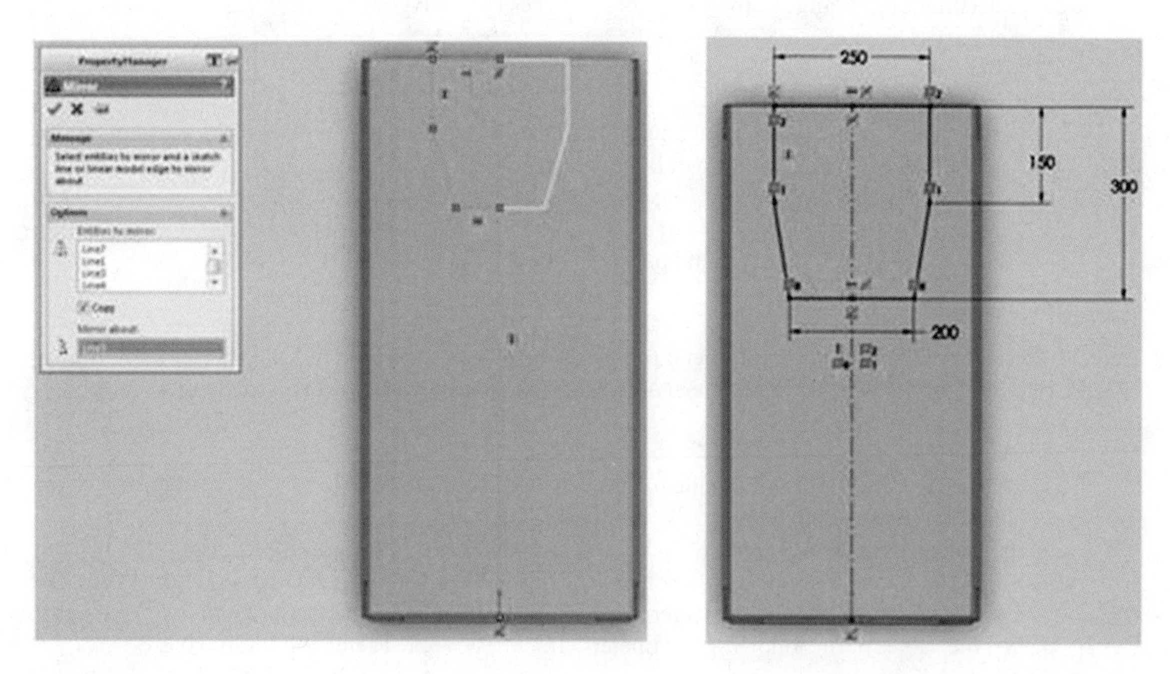

Figure 19.38

Sketch for Extrude-Cut.

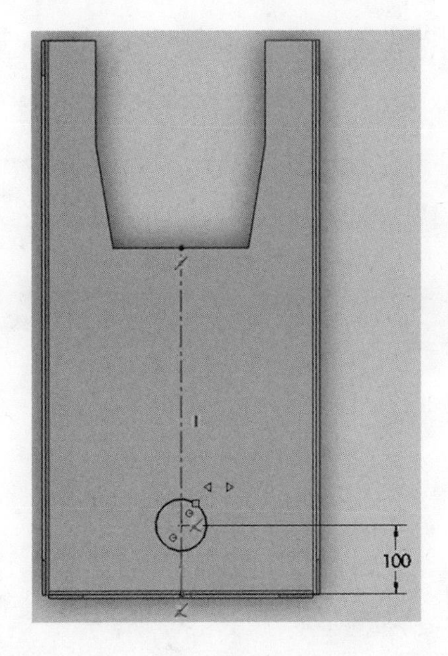

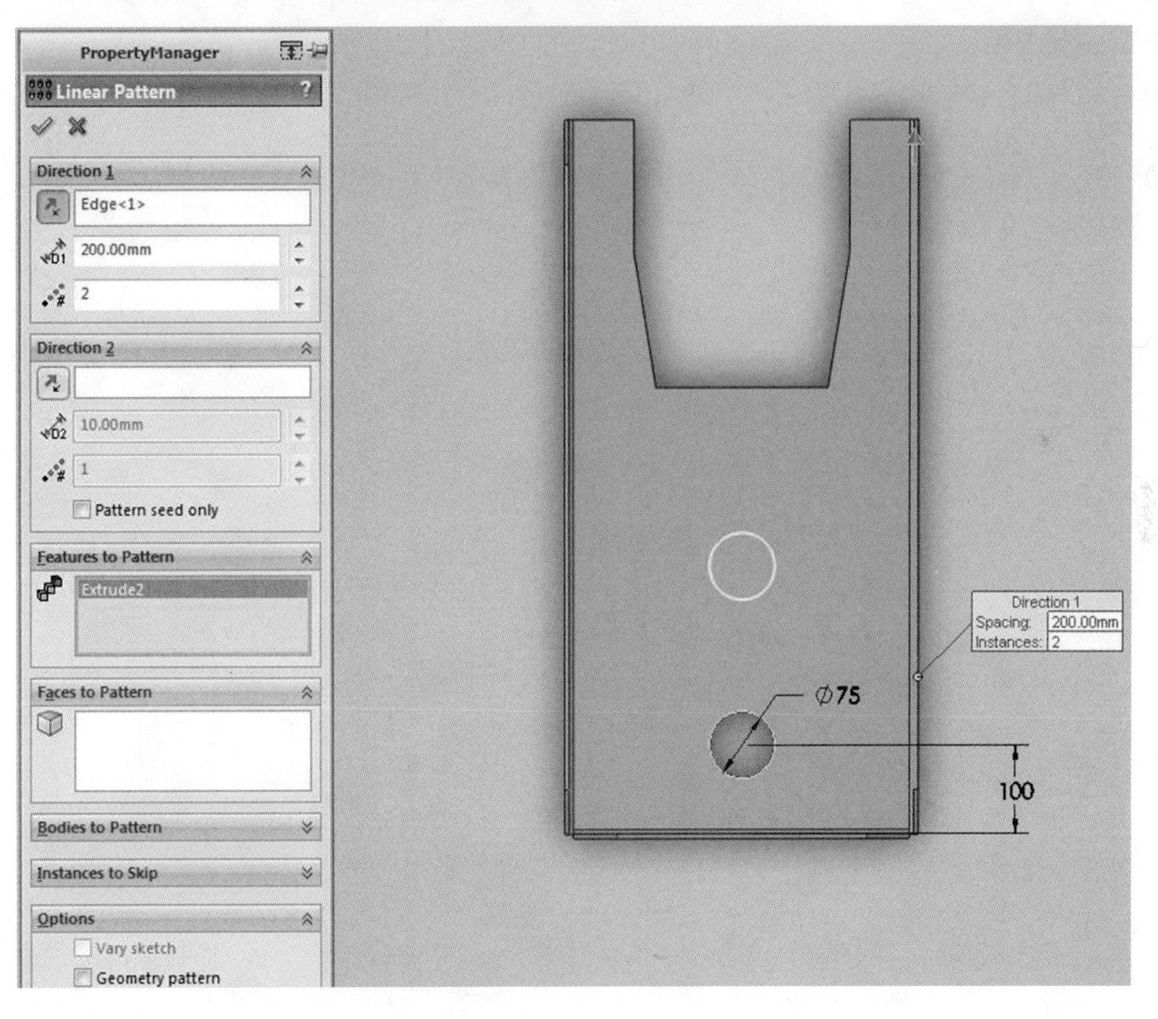

Figure 19.39

Sketches (circles) for Extrude-Cut.

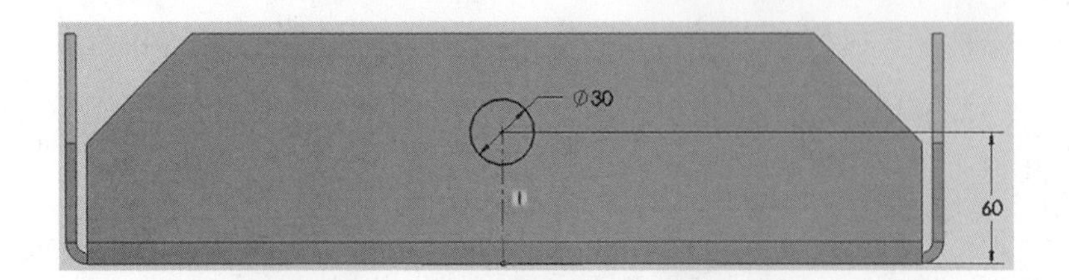

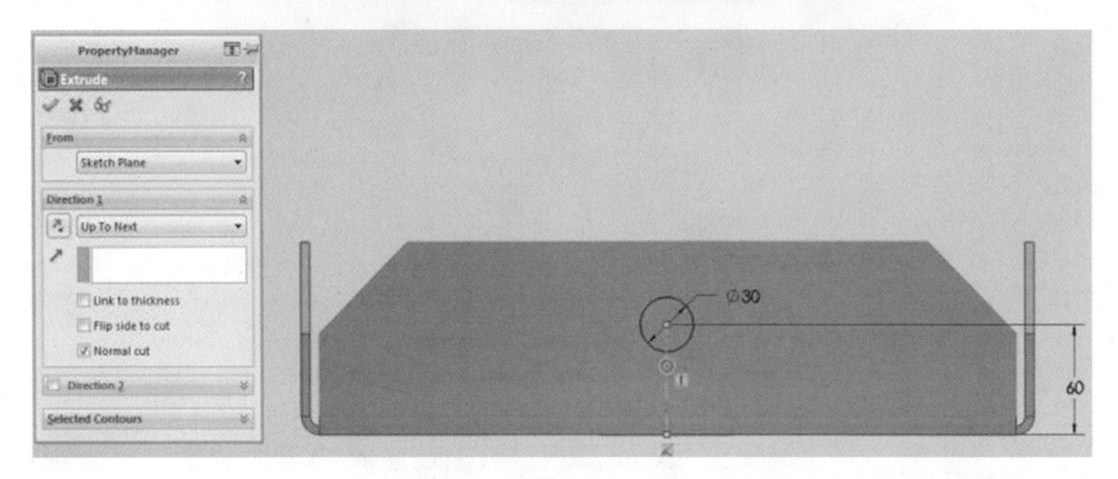

Figure 19.40

Sketch (circle) for Extrude-Cut.

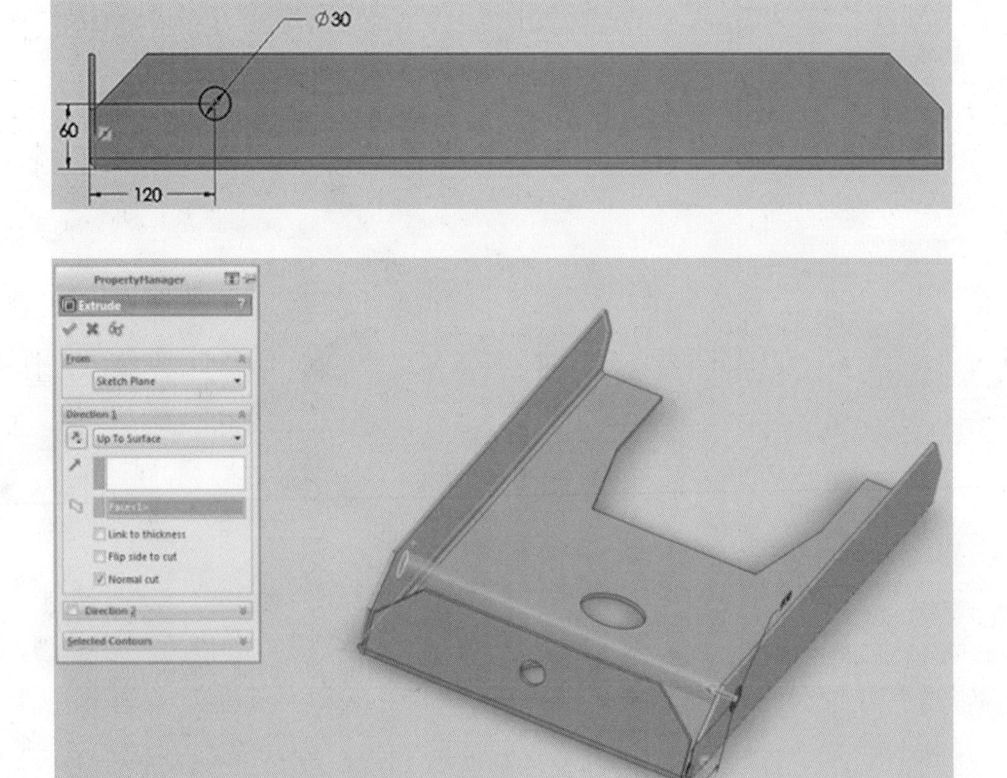

Figure 19.41

Sketch (circle) for Extrude-Cut.

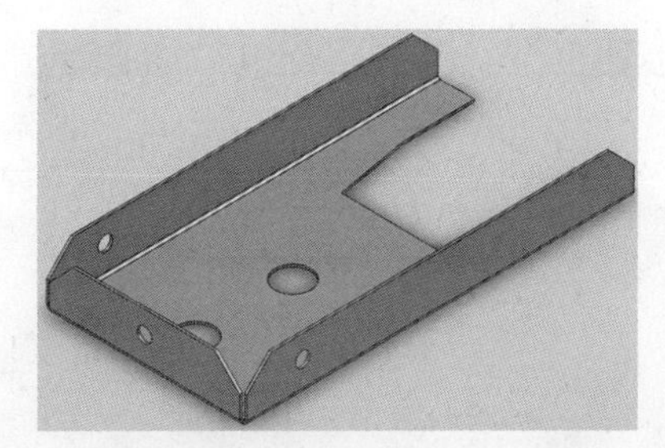

Figure 19.42

Sheet metal part.

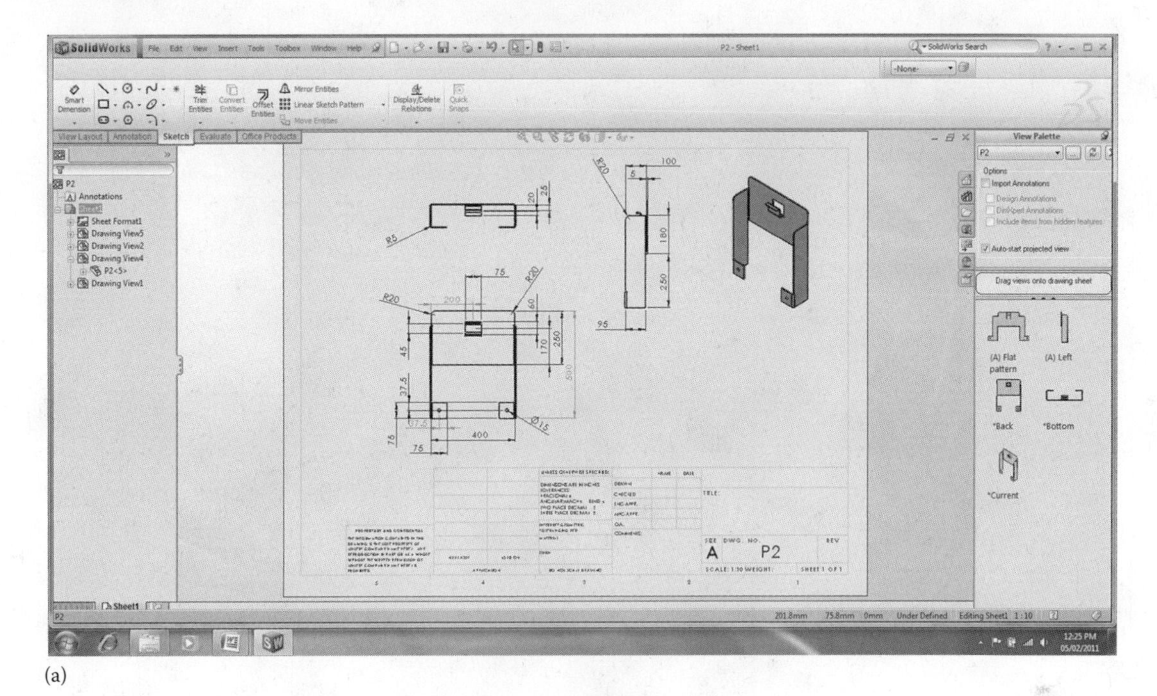

(a)

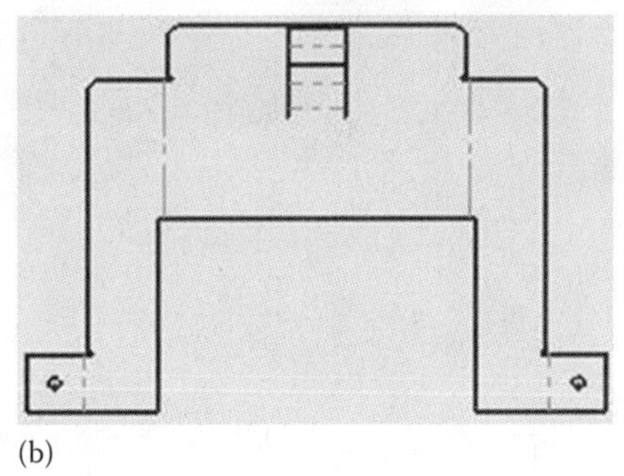

(b)

Figure 19.43

(a) Sheet metal part with drawings and dimensions for Tutorial 19.3 and (b) flat pattern of the sheet metal part for Tutorial 19.3.

11. Adjust the Width of the Edge Flange created by dragging one endpoint of the edge to a new position (180 mm) and using Sketch tool to add dimensions. (Note: *When the Profile Sketch window appears, do not click Finish until the length of the Edge Flange is fully adjusted.*)
12. Click Finish (see Figure 19.47 for the Edge Flange).

Create the First Tab

13. Select the Edge Flange just and sketch a rectangle, 250 mm × 95 mm (see Figure 19.48).
14. Click the Base Tab tool to create a tab (see Figure 19.49).
    (Note: The Tab feature is automatically created as shown in Figure 19.50.)

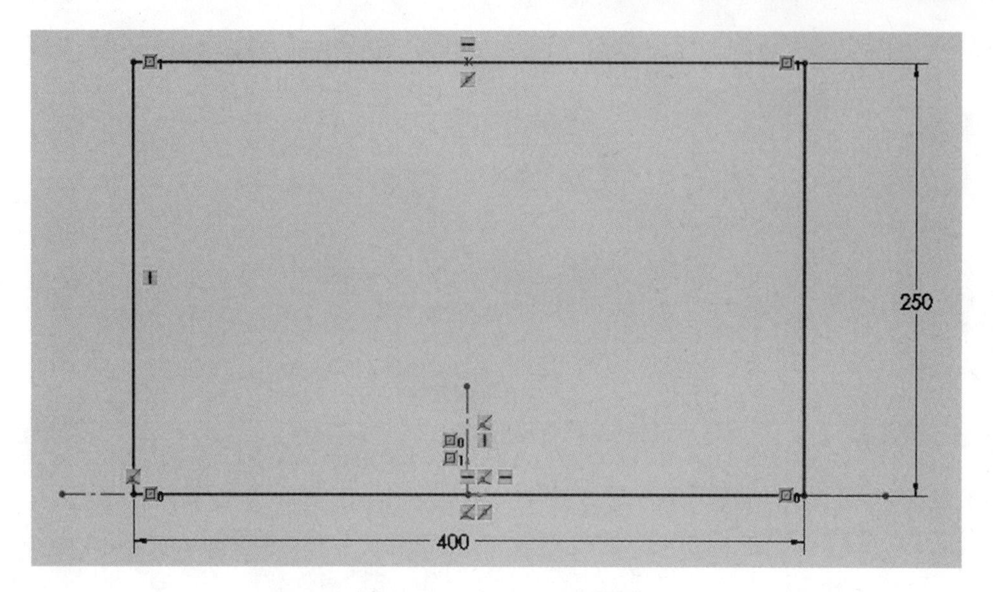

Figure 19.44

Sketch1 and selection of Base Flange/Tab option.

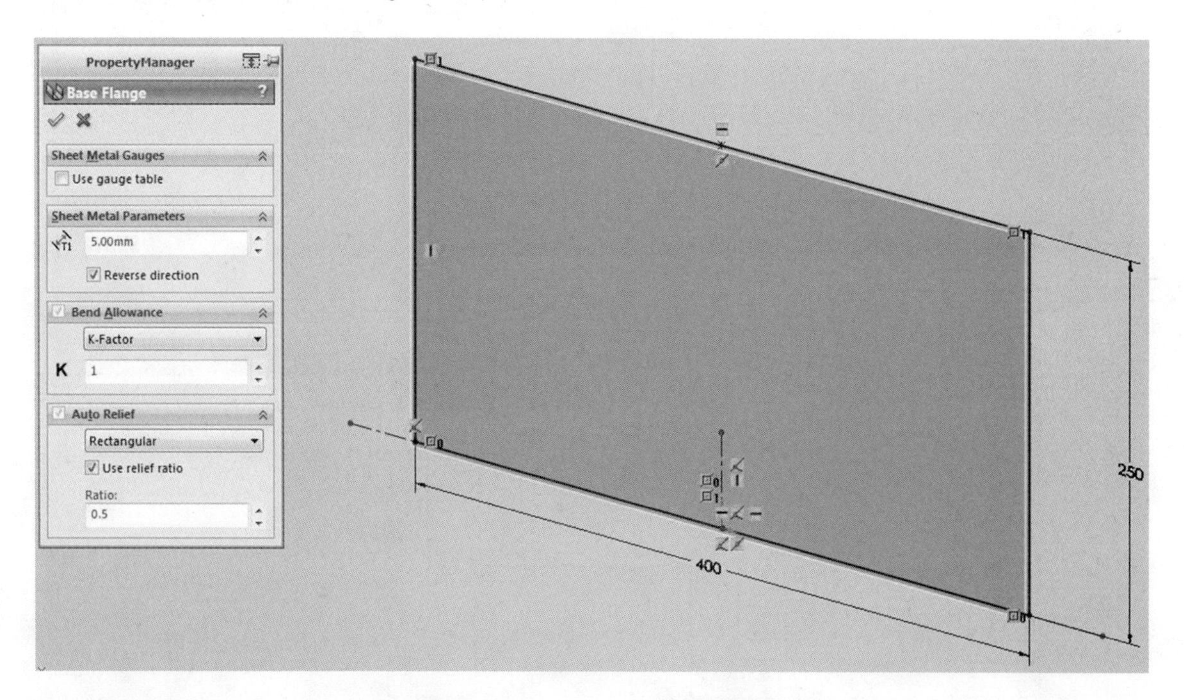

Figure 19.45

Base Flange/Tab PropertyManager.

Create the Second Edge Flange

15. Select the bottom edge of the right Edge Flange (see Figure 19.51).

16. Set Length of Edge Flange = 75 and Angle = 90-deg (see Figure 19.51).

17. Click Edit Flange Profile button from the Edge Flange PropertyManager (see Figure 19.51).

18. Click OK if the following message appears: "*The automatic relations inferred by this operation would over define the sketch. They will not be added.*"

19. Adjust the Width of the Edge Flange created by dragging the endpoints of the edge to new positions (75 mm from each end) and using the Sketch tool to add dimensions.
(Note: *When the Profile Sketch window appears, do not click Finish until the length of the Edge Flange is fully adjusted.*)

20. Click Finish. (See Figure 19.52 for the two edge flanges.)

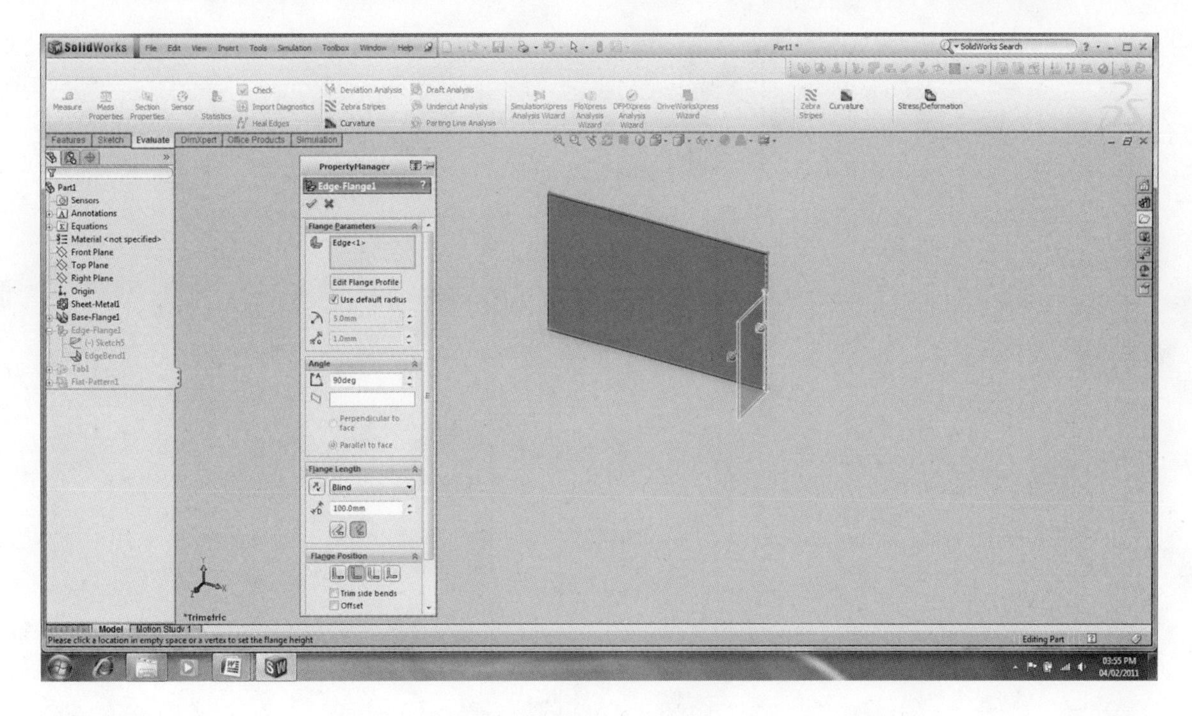

Figure 19.46

PropertyManager for creating Edge Flange.

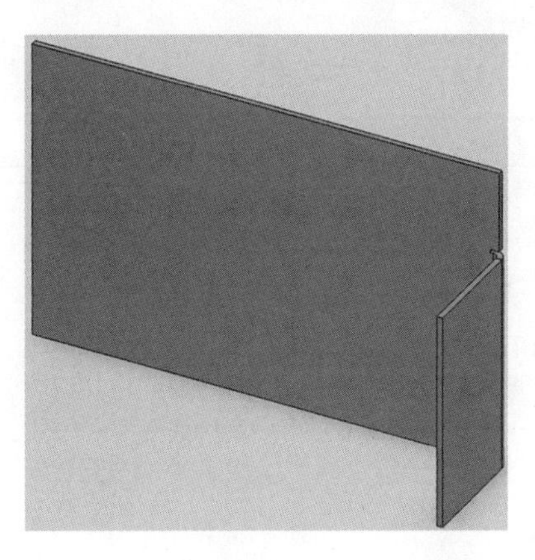

Figure 19.47

Edge Flange created.

Create a Circular Hole

21. Click the face of the second Edge Flange.
22. Be in sketch mode, and create a Sketch, Circle (see Figure 19.52).
23. Click Extrude-Cut and choose Next to create the hole (see Figure 19.53).

Fillet Edges

24. Click Fillet and select corners on the right of the model.
25. Set Radius of fillet = 20 mm.
26. Click OK.

Mirror of Second Edge Flange

27. Click the Mirror tool.
28. For the Mirror Face/Plane, select the Right Plane.
29. For Features to Mirror, select *all the features for the right leg* (see Figure 19.53).
30. Click OK.

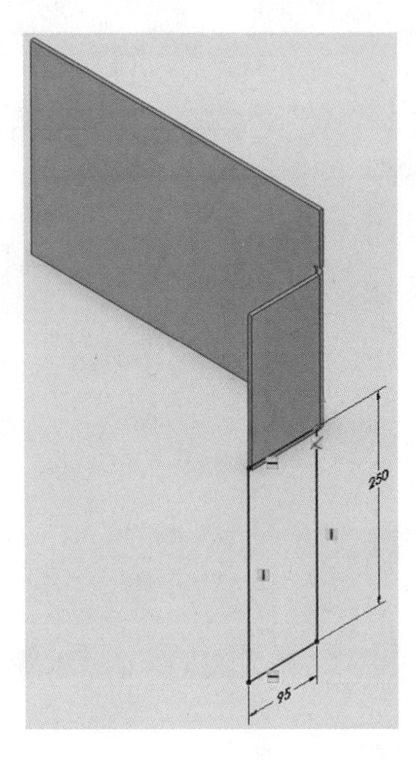

Figure 19.48

Sketch2 create.

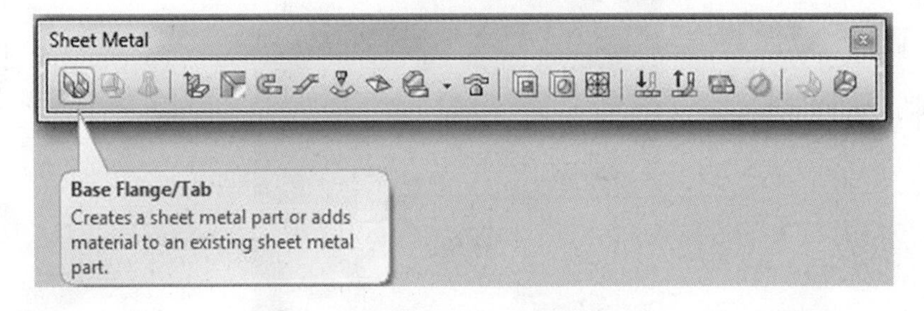

Figure 19.49

Base Tab tool used to create a tab.

Create the Second Extrude-Cut

31. Click the face of the Base Flange.
32. Be in sketch mode, and create a Sketch, Rectangle (see Figure 19.54).
33. Click Extrude-Cut and choose Next to create the slot (see Figure 19.55).

Create First Edge Flange for Rectangular Slot

34. Select the bottom edge of the Rectangular Slot created (see Figure 19.55).
35. Set Length of Edge Flange = 25 and Angle = 90-deg (see Figure 19.55).
36. Click OK.

Create Second Edge Flange for Rectangular Slot

37. Select the edge of the Edge Flange created at the Slot (see Figure 19.56).
38. Set Length of Edge Flange = 45 and Angle = 90-deg (see Figure 19.56).
39. Click OK.

Create Third Edge Flange for Rectangular Slot

40. Select the edge of the second Edge Flange created at the Slot (see Figure 19.57).
41. Set Length of Edge Flange = 20 and Angle = 90-deg (see Figure 19.57).
42. Click OK (see Figure 19.58 for the completed sheet metal model).

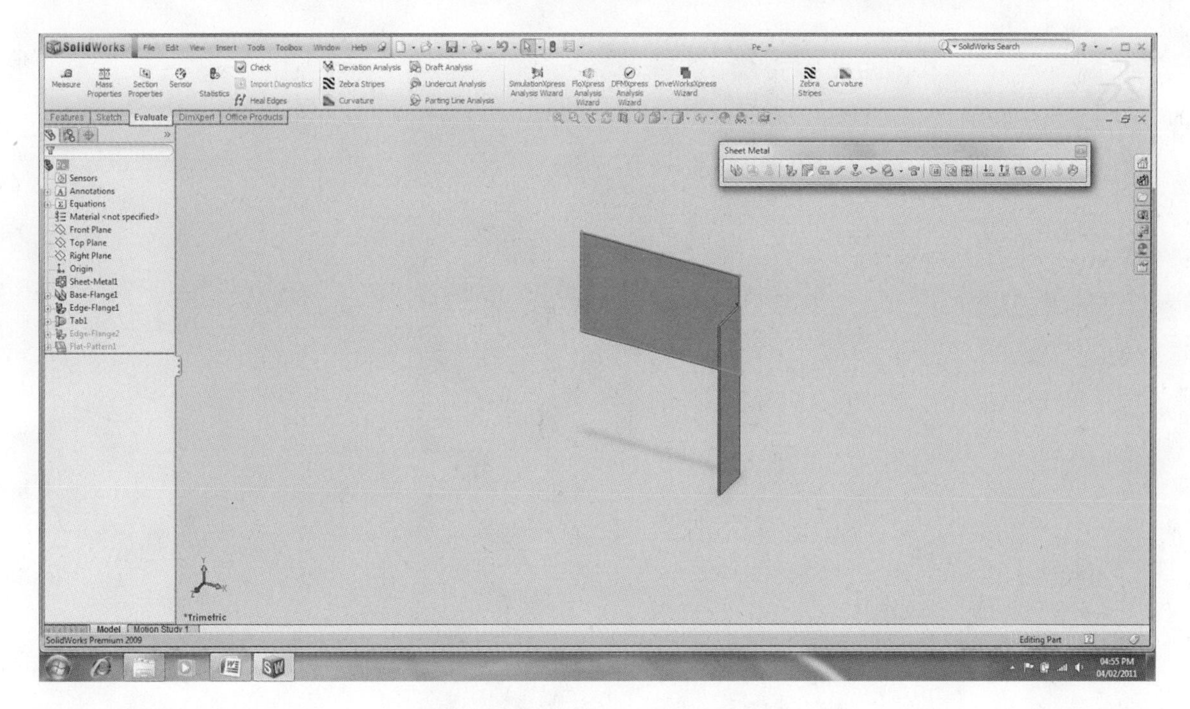

Figure 19.50

Tab tool used to create a tab.

Tutorial 19.4: Hanger Support

In this tutorial, we will create the sheet metal part that is shown at the top right, with its views and dimensions shown in Figure 19.59. The flat pattern of the sheet metal part is shown in Figure 19.60.

SolidWorks Solution to Tutorial 19.4

Create the Base Flange

1. Start a New SolidWorks Part document.
2. Select the Top Plane.
3. Be in sketch mode and create Sketch1 (see Figure 19.61).
4. Click the Base Flange/Tab option from the Sheet Metal CommandManager (see Figure 19.61). (Note: The Sheet Metal CommandPropertyManager is automatically displayed; see Figure 19.62.)
5. Set the values of the following parameters: Thickness = 1; K-Factor = 1; Ratio = 0.5.
6. Click OK.

Create the First Sketched Bend

7. Click top of the Base Flange.
8. Be in sketch mode and create Sketch2 on the Base Flange for the first bending operation, a line 25 mm *from the center of the sketch* (see Figure 19.63).
9. Click the Sketch Bend tool.
10. Accept the default Bend radius of 2 mm and Angle = 90-deg.
11. Click the Fixed Face (Face<1>) to one side of the line (Sketch2) for the Bend Parameters rollout (the Sketched Bend PropertyManager is automatically displayed; see Figure 19.64).
12. Click OK to complete the sketched bend operation (see Figure 19.65 for bend created).

Create the First Sketched Bend

13. Click top of the Base Flange.
14. Be in sketch mode and create Sketch3 on the Base Flange for the first bending operation, a line 25 mm *from the center of the sketch* (see Figure 19.66).
15. Click the Sketch Bend tool.
16. Accept the default Bend radius of 2 mm and Angle = 90-deg.
17. Click the Fixed Face (Face<1>) to one side of the line (Sketch2) for the Bend Parameters rollout. (The Sketched Bend PropertyManager is automatically displayed; see Figure 19.67.)
18. Click OK to complete the sketched bend operation. (See Figure 19.68 for bend created.)

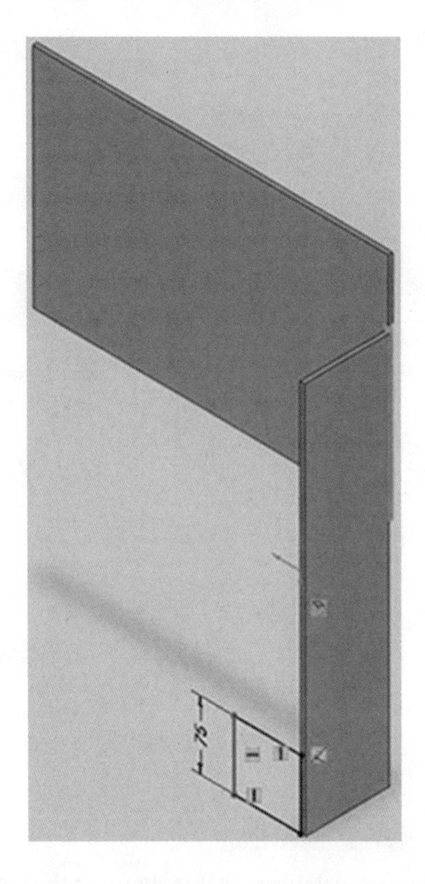

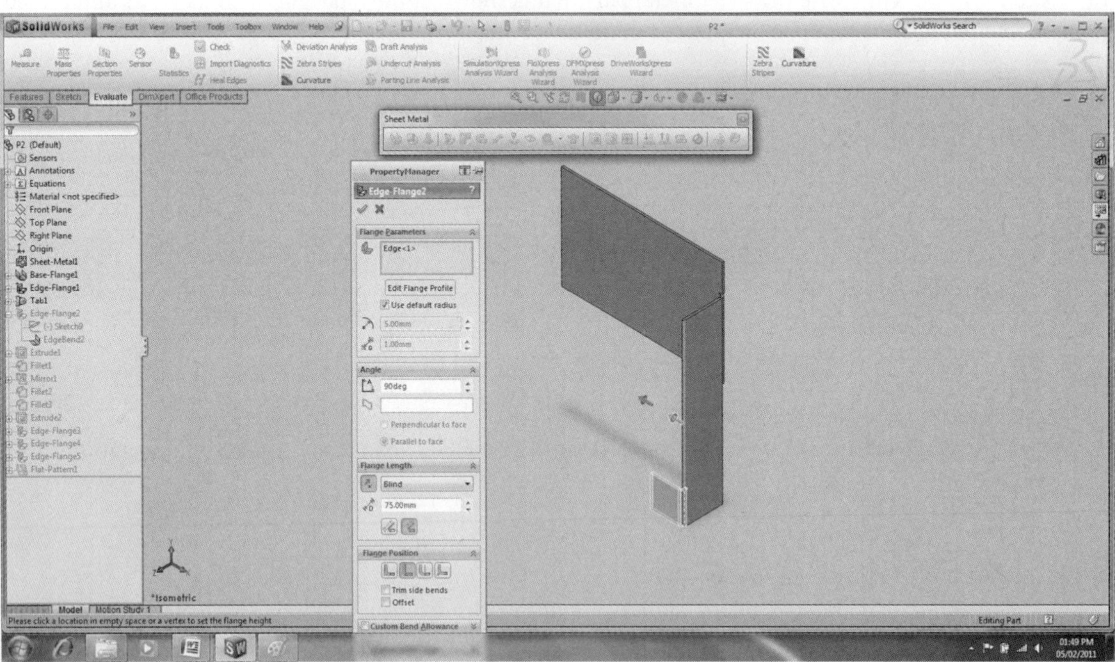

Figure 19.51

Edge Flange PropertyManager.

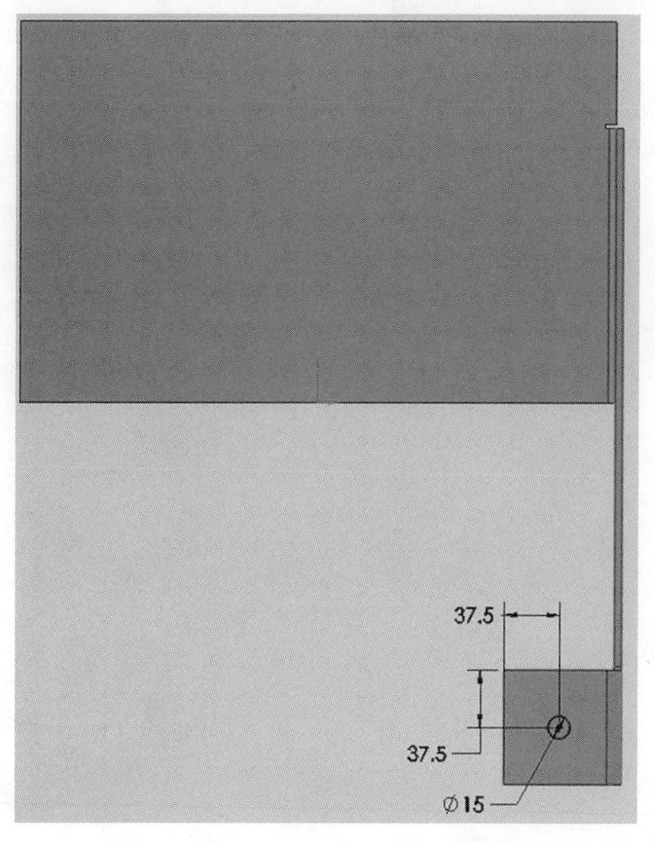

Figure 19.52

Edge Flange PropertyManager.

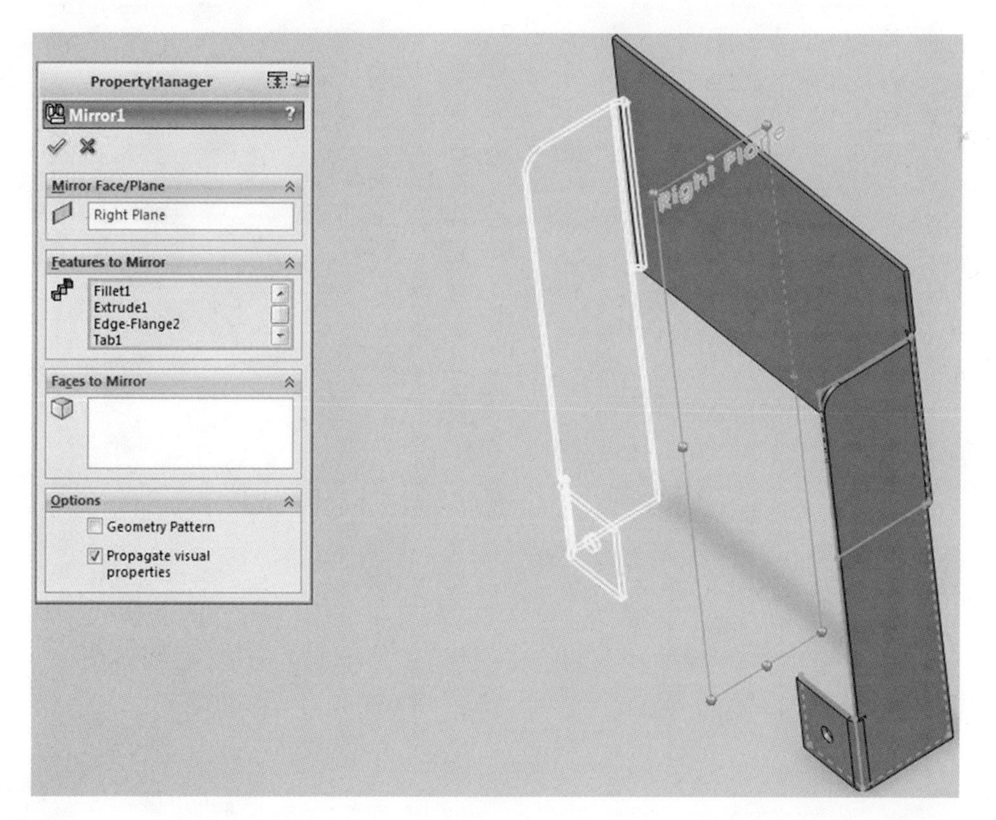

Figure 19.53

Hole created on second edge flange.

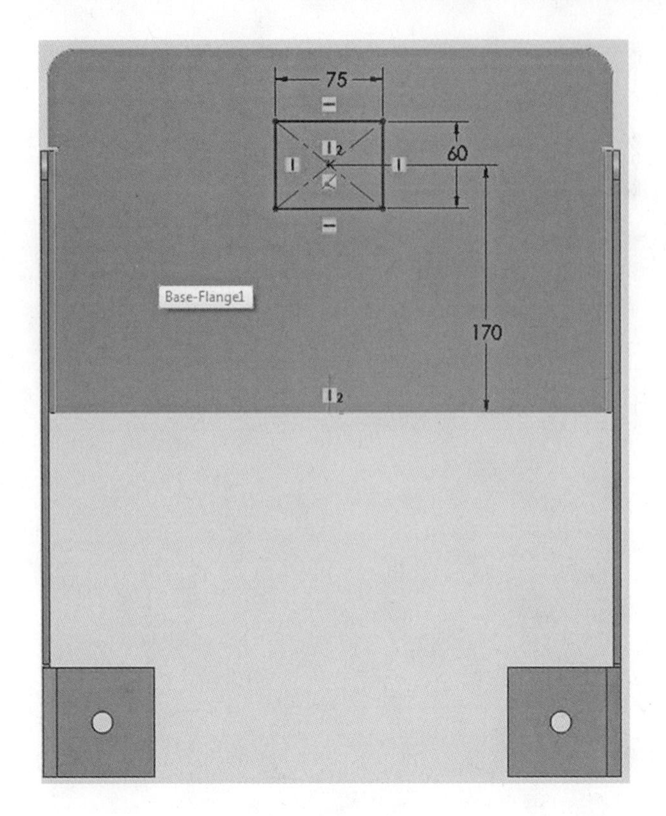

Figure 19.54

Sketch for second Extrude-Cut.

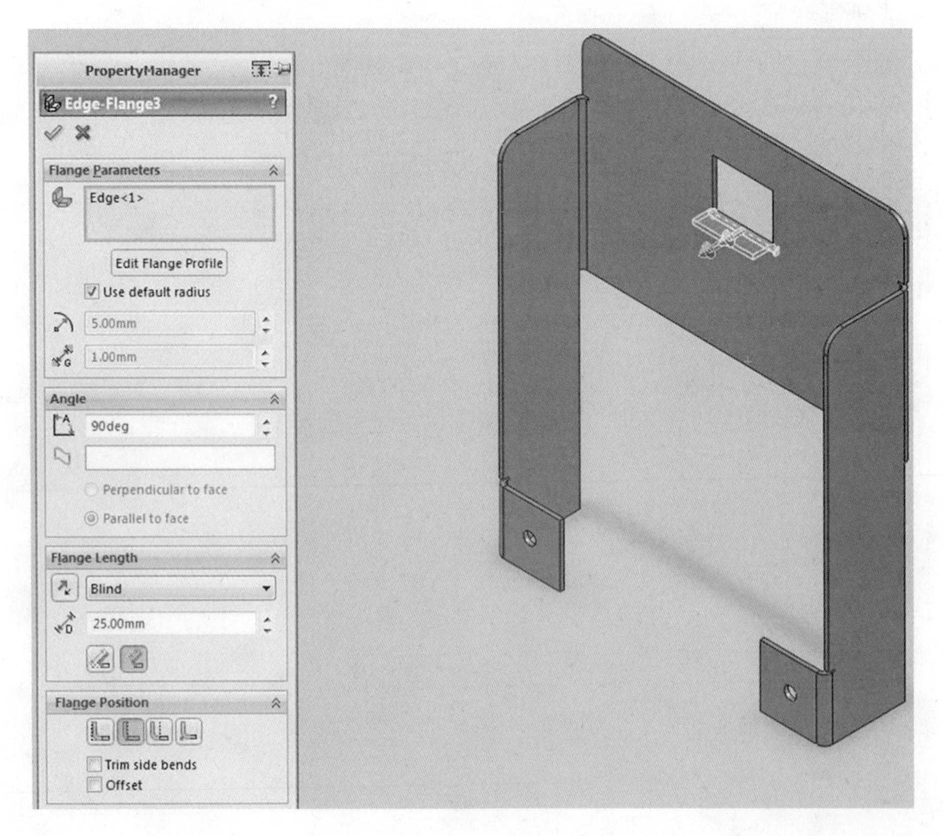

Figure 19.55

First edge flange for rectangular slot.

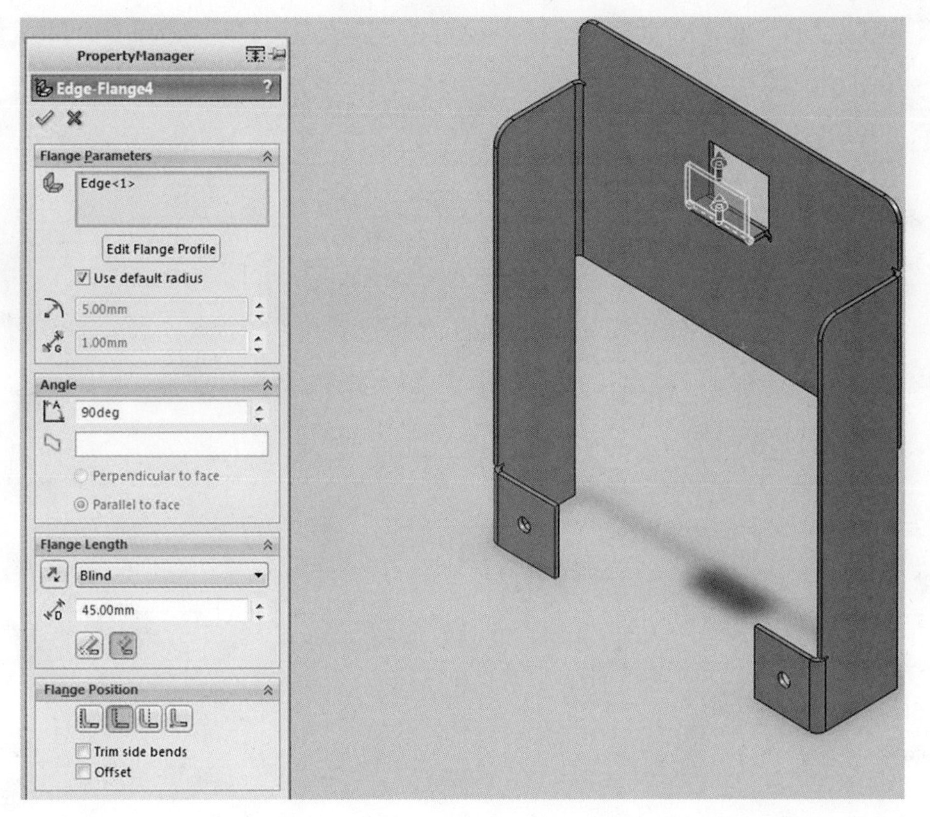

Figure 19.56

Second edge flange for rectangular slot.

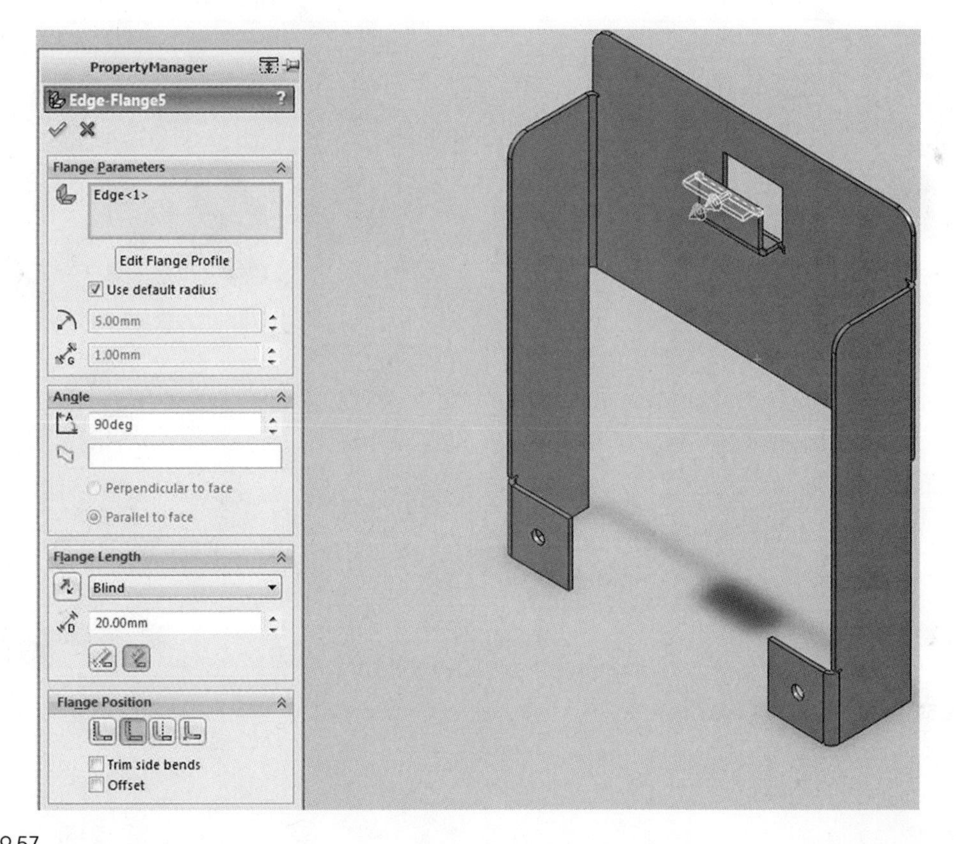

Figure 19.57

Third edge flange for rectangular slot.

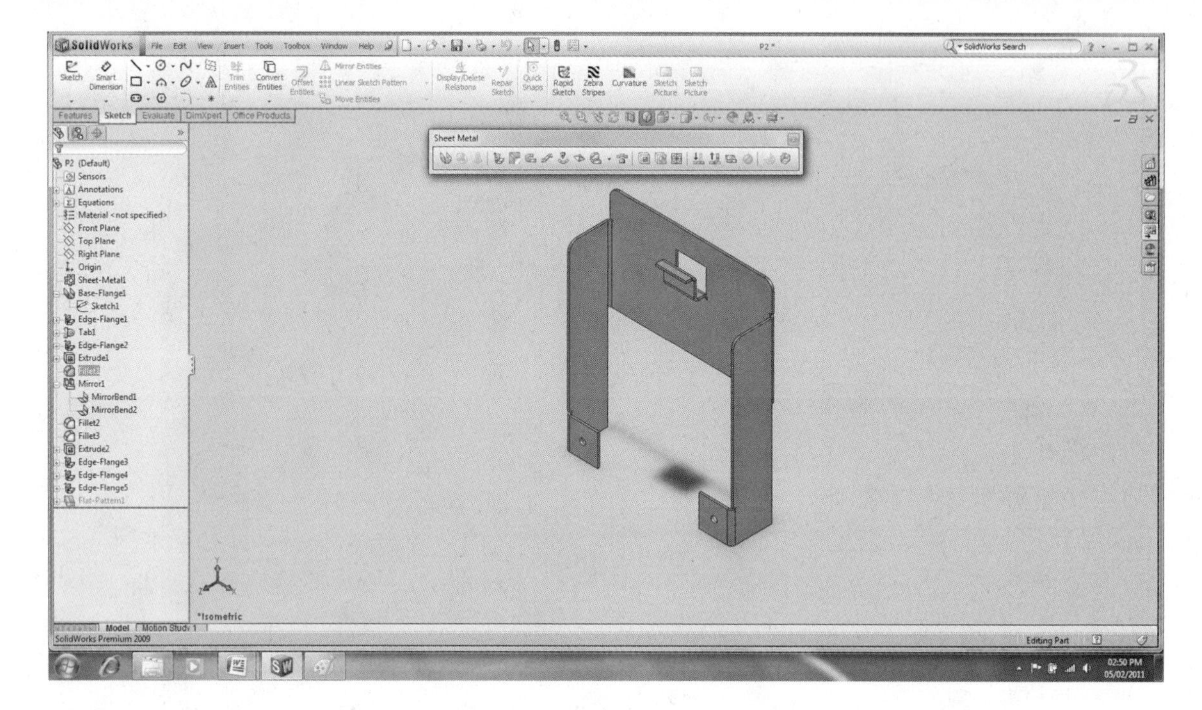

Figure 19.58

Sheet metal part.

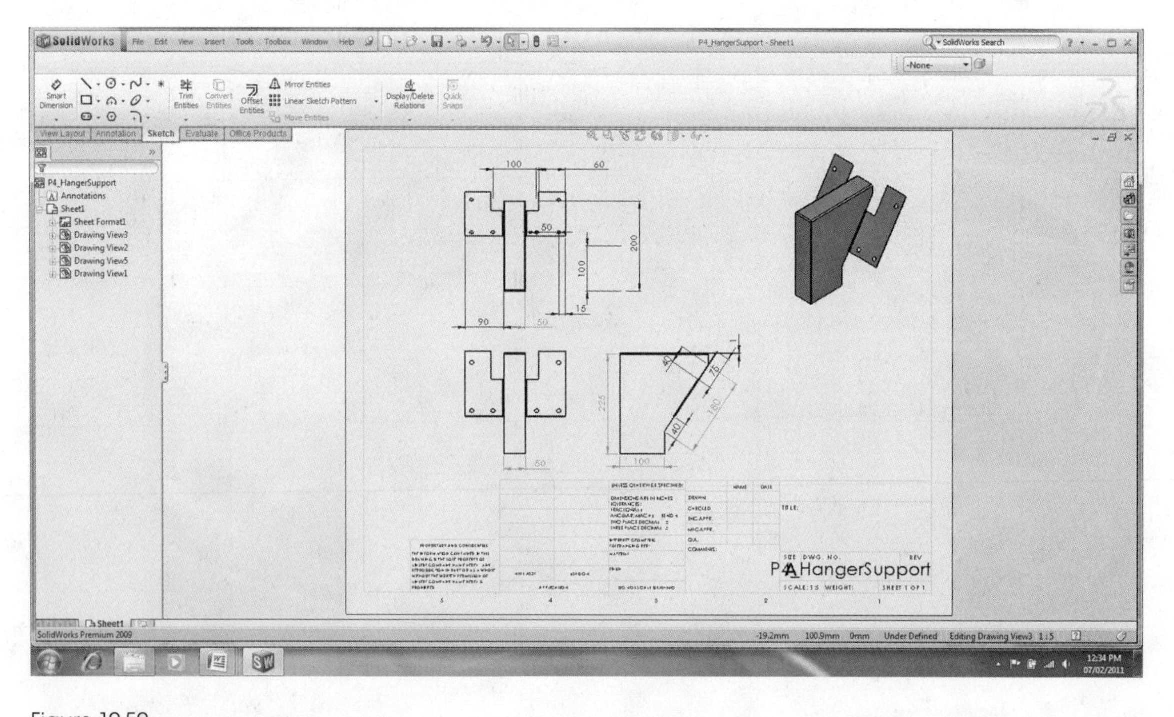

Figure 19.59

Part and drawing views and dimensions for Tutorial 19.4.

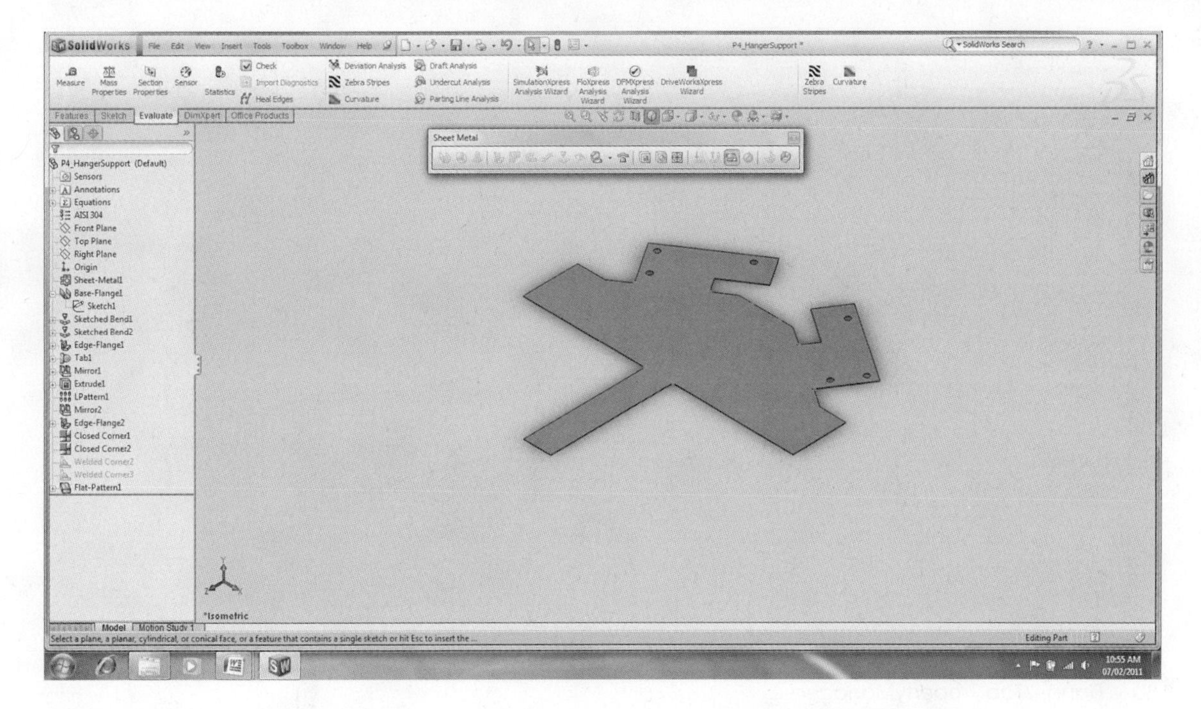

Figure 19.60

Flat pattern of the sheet metal part for Tutorial 19.4.

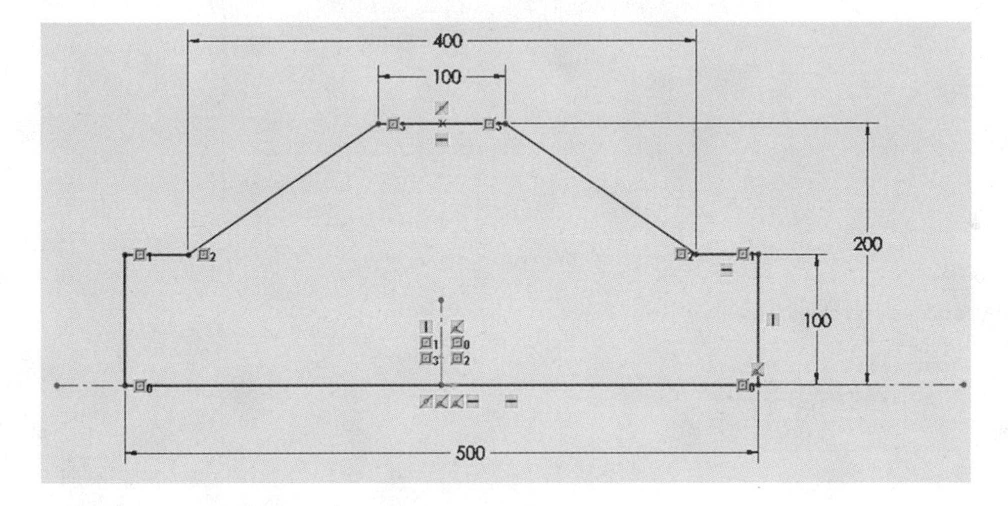

Figure 19.61

Sketch1 and selection of Base Flange/Tab option.

Create the First Edge Flange

19. Select the *right-edge* of the first Sketched Bend (see Figure 19.69).
20. Set Length of Edge Flange = 90 and Angle = 90-deg (see Figure 19.69).
21. Click Edit Flange Profile button from the Edge Flange PropertyManager (see Figure 19.69).
22. Click OK if the following message appears: "*The automatic relations inferred by this operation would over define the sketch. They will not be added.*"
23. Adjust the Width of the Edge Flange created by dragging one endpoint of the edge to a new position (40 mm) and using sketch tool to add dimensions (see Figure 19.70). (Note: *When the Profile Sketch window appears, do not click Finish until the length of the Edge Flange is fully adjusted.*)
24. Click Finish. (See Figure 19.71 for the Edge Flange.)

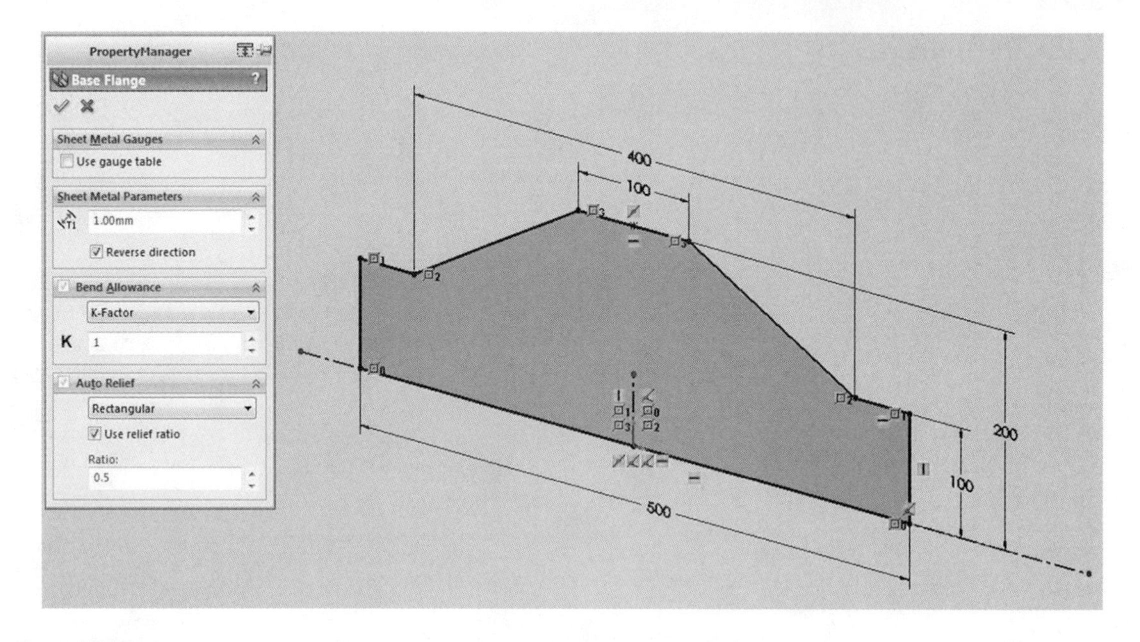

Figure 19.62

Base Flange/Tab PropertyManager.

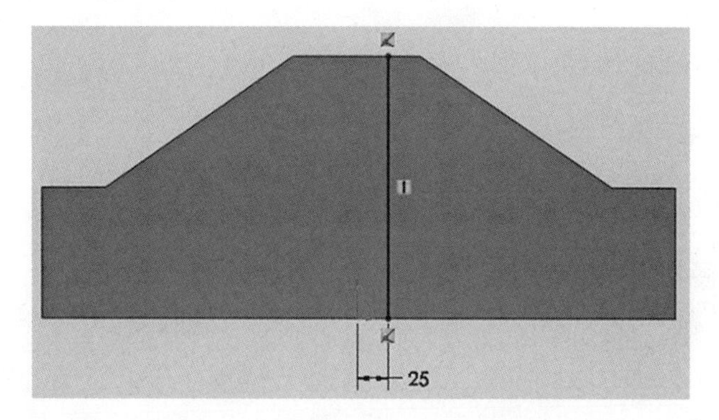

Figure 19.63

Sketch used for bend operation.

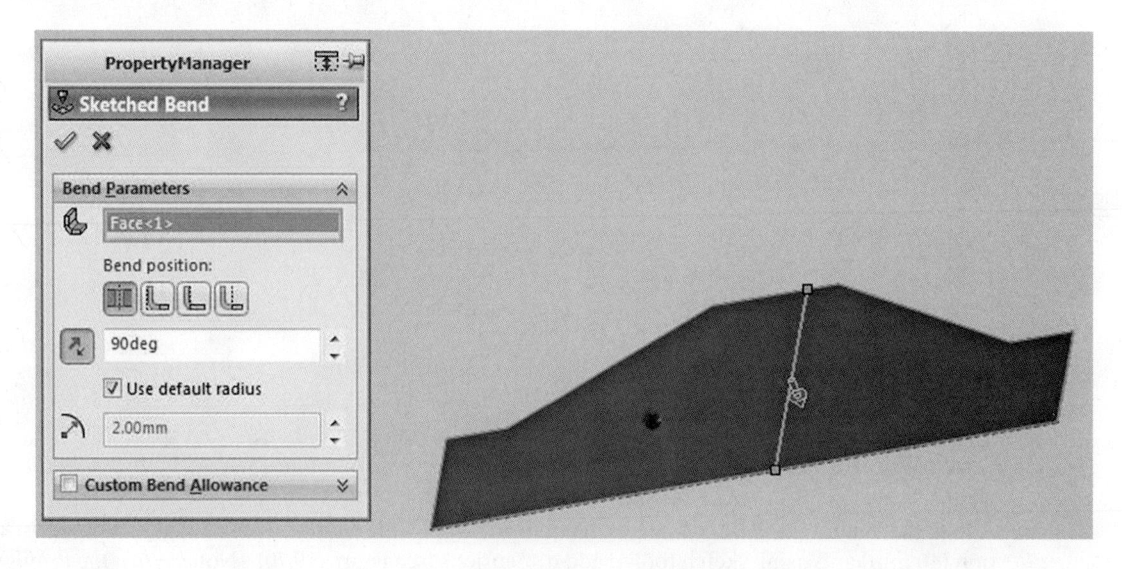

Figure 19.64

Sketched Bend PropertyManager.

Figure 19.65

Bend created.

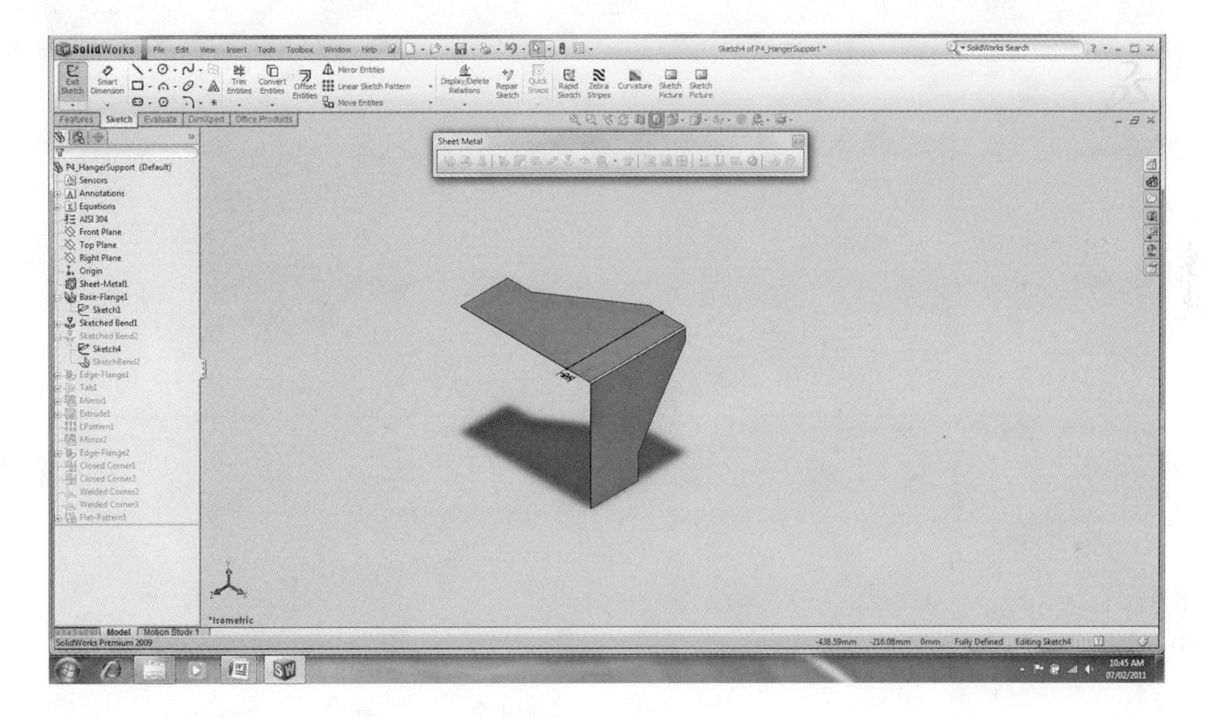

Figure 19.66

Sketch used for bend operation.

First Base Tab Created

25. Click the *right edge flange* that is created and sketch a rectangle, 60 mm × 75 mm (see Figure 19.72a).
26. Click the Base Tab tool. (See Figure 19.72b for the Base Tab created.)

Create the Third Edge Flange

27. Click the Base Flange tool.
28. Select the *top edge* of the Base Flange (see Figure 19.73).
29. Set Length of Edge Flange equal to the length of the vertical *edge* (see Figure 19.73).
30. Click Finish (see Figure 19.74 for the Edge Flange).

Create a Sketch, Circle

31. Click the right edge flange that is created and sketch a Circle, 9-mm diameter (see Figure 19.74).
32. Click the Linear Pattern tool and replicate the circular sketch 2 on x-direction and 2 on y-direction (see Figure 19.75).

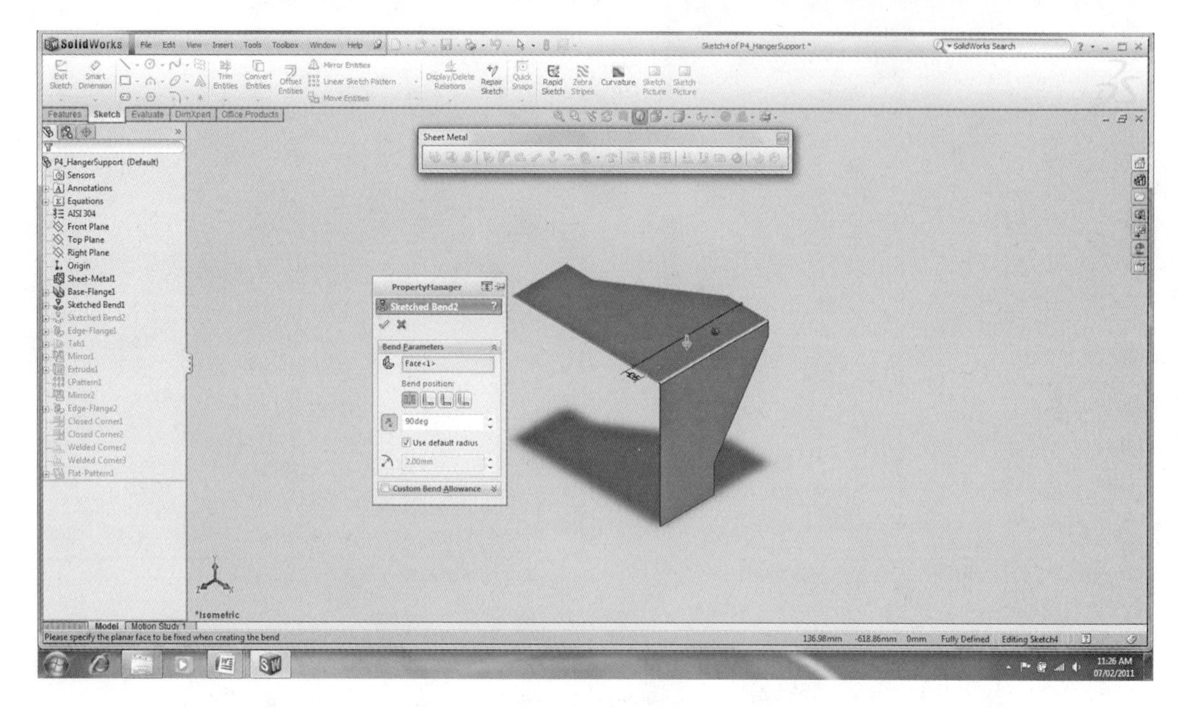

Figure 19.67

Sketched Bend PropertyManager.

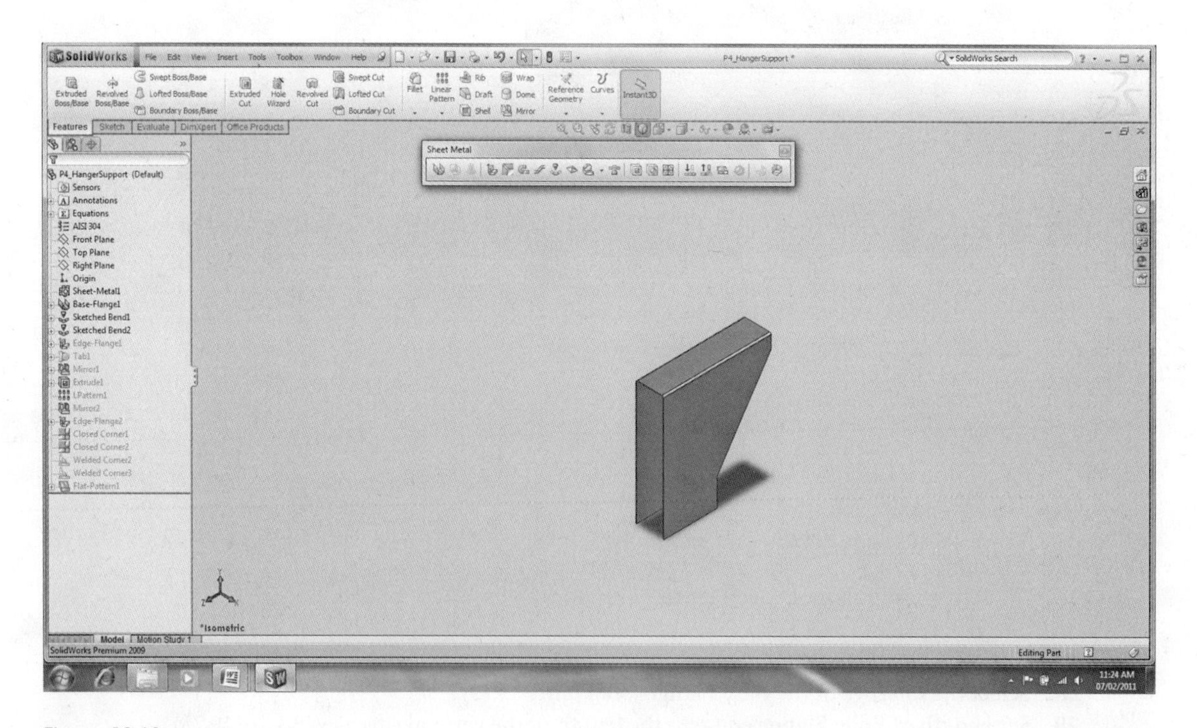

Figure 19.68

Bend created.

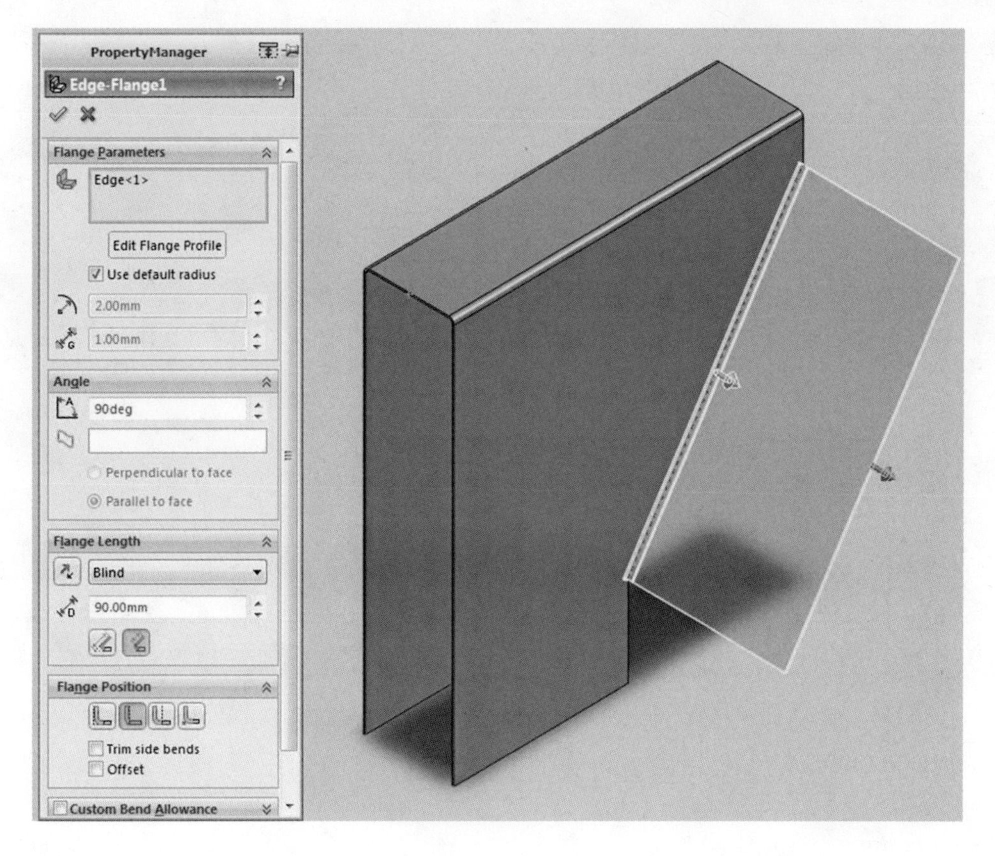

Figure 19.69

Edge Flange PropertyManager.

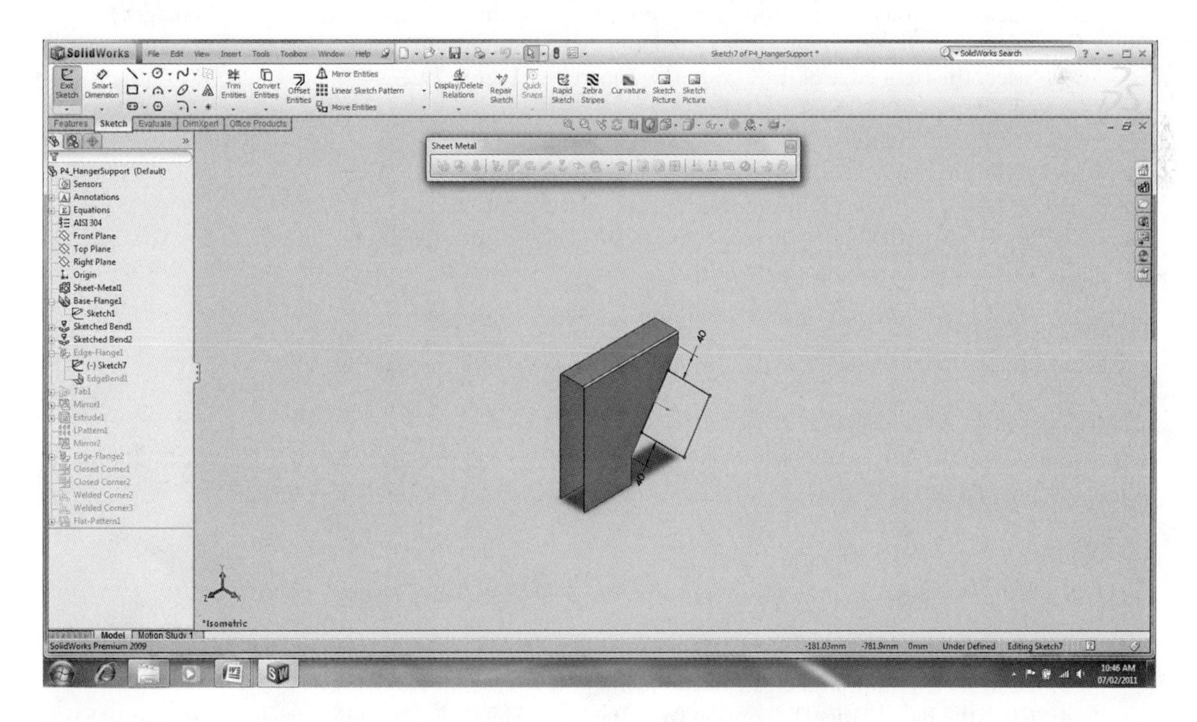

Figure 19.70

Adjusted edge.

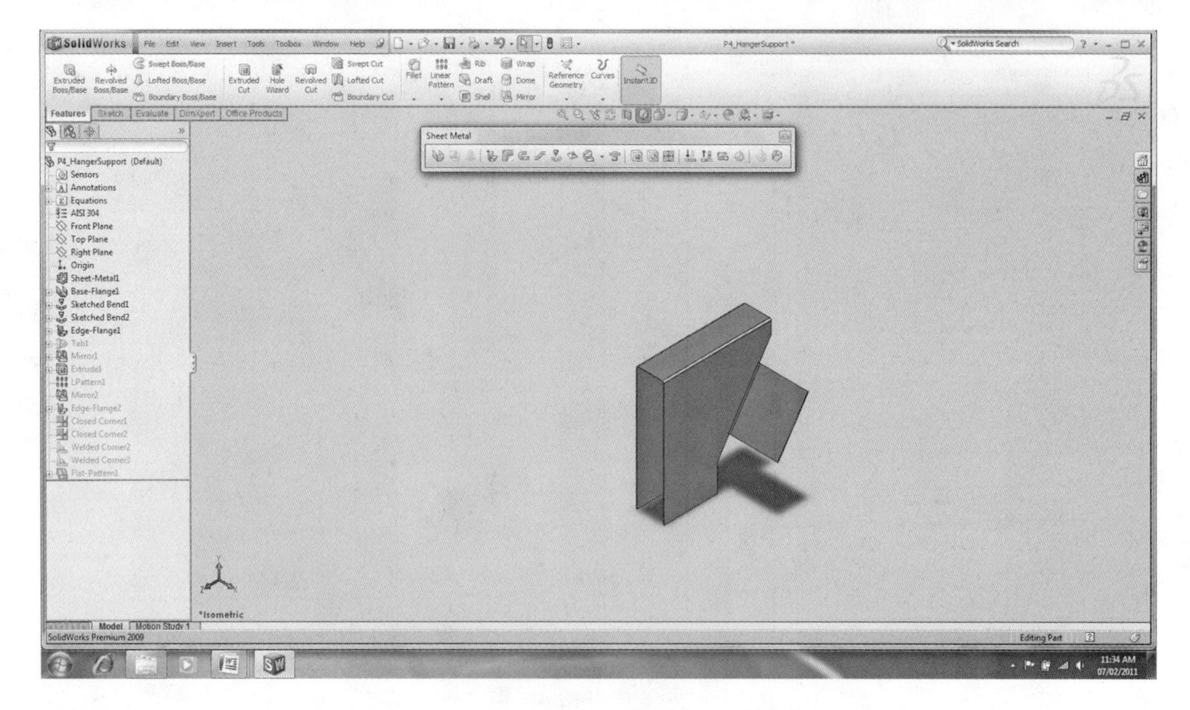

Figure 19.71

First Edge Flange created.

Mirror Features

33. Click the right edge flange that is created and sketch a Mirror Features tool. (See Figure 19.76 for the Mirror Features PropertyManager.)
34. Select all the features for mirror operation (see Figure 19.76).

Closed Corner Operations

35. Click the Closed Corner tool. (See Figure 19.77 for the Closed Corner PropertyManager.)
36. Select the *left edge* of the *Third Edge Flange* (see Figure 19.77).
37. Select the right edge of the *Third Edge Flange* (see Figure 19.77).
38. Click OK to finish the operation.

Welded Corner Operations

39. Click the Welded Corner tool (see Figure 19.78 for the Welded Corner PropertyManager).
40. Select the *left edge* of the *Third Edge Flange* (see Figure 19.78).
41. Select the right edge of the *Third Edge Flange* (see Figure 19.78).
42. Click OK to finish the operation.

Flattened State

Click the Flatten tool. (See Figure 19.79 for the Flattened state of the sheet metal.)

Tutorial 19.5: Jogged Sheet Metal Part

In this tutorial, we will create the sheet metal part that is shown at the top right, with its views and dimensions shown in Figure 19.80. The flat pattern of the sheet metal part is shown in Figure 19.81.

SolidWorks Solution to Tutorial 19.5

Create the Base Flange

1. Start a New SolidWorks Part document.
2. Select the Top Plane.
3. Be in sketch mode and create Sketch1, 100 mm × 200 mm (see Figure 19.82).
4. Click the Base Flange/Tab option from the Sheet Metal CommandManager (see Figure 19.83). (Note: The Sheet Metal CommandPropertyManager is automatically displayed; see Figure 19.83.)
5. Set the values of the following parameters: Thickness = 3.0; K-Factor = 1; Ratio = 0.5.
6. Click OK.

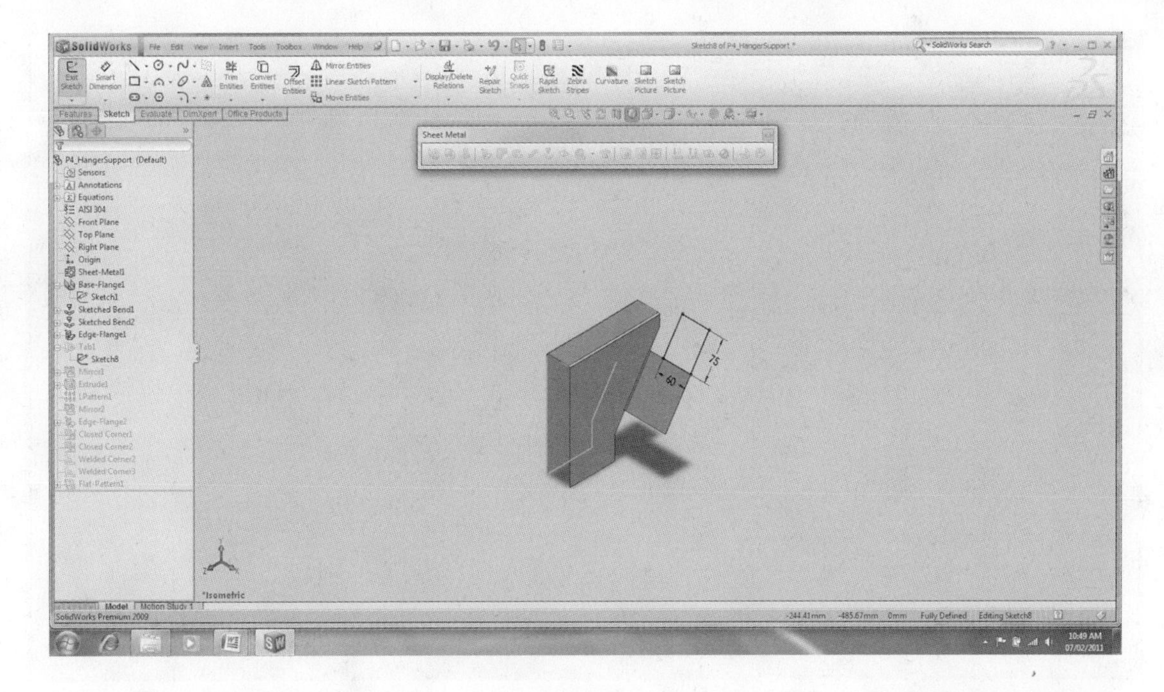

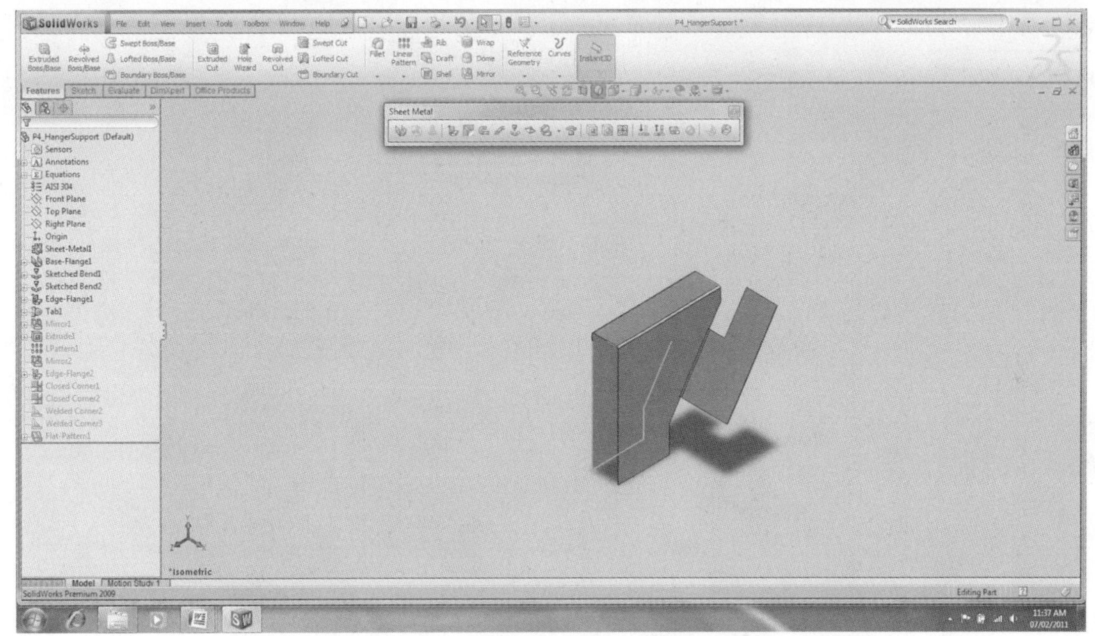

Figure 19.72

First Base Tab created.

Create Sketch for the Extrude-Cut

7. Click the top face of the Base Flange (see Figure 19.84).
8. Be in sketch mode and create Sketch2 (see Figure 19.85).
9. Click Extrude-Cut.
10. For the Distance for Extrude-Cut, choose Up To Next (see Figure 19.86).

Mirror the Extrude-Cut Feature

11. Click Mirror tool to mirror tool.
12. Select the Extrude1 as Feature to Mirror (see Figure 19.87).
13. Select the Front Plane as the Mirror Face (see Figure 19.87).
14. Click OK.

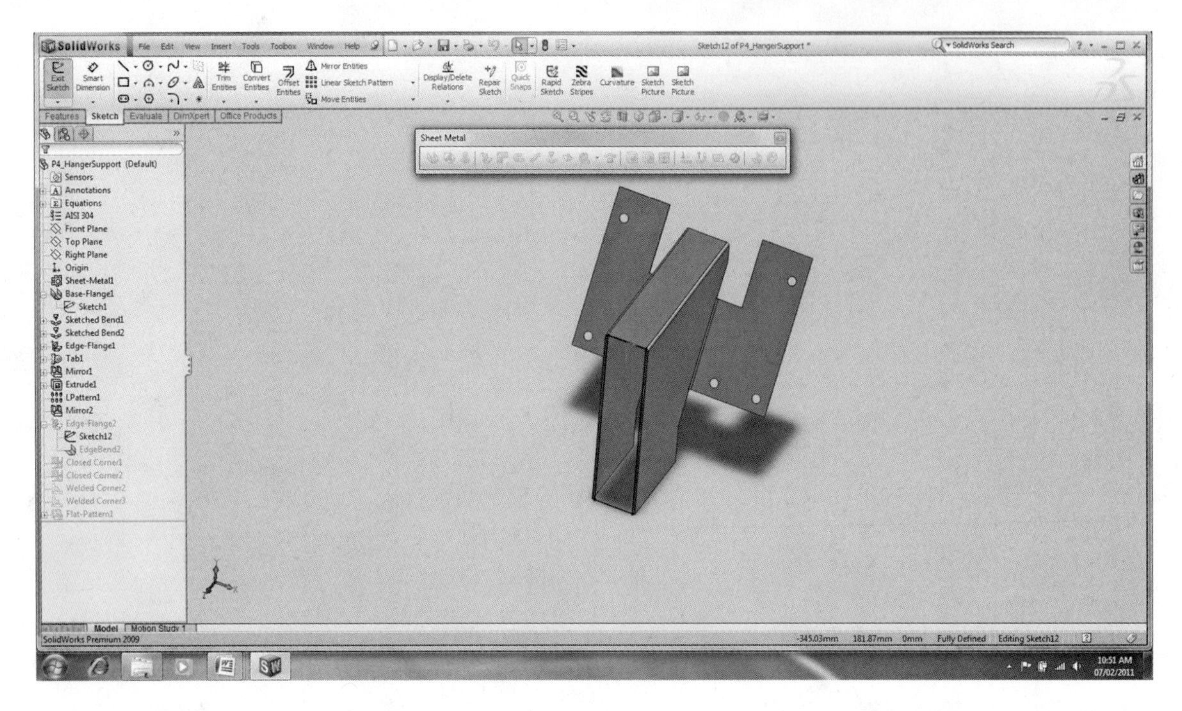

Figure 19.73

Upper horizontal edge chosen for creating third edge flange.

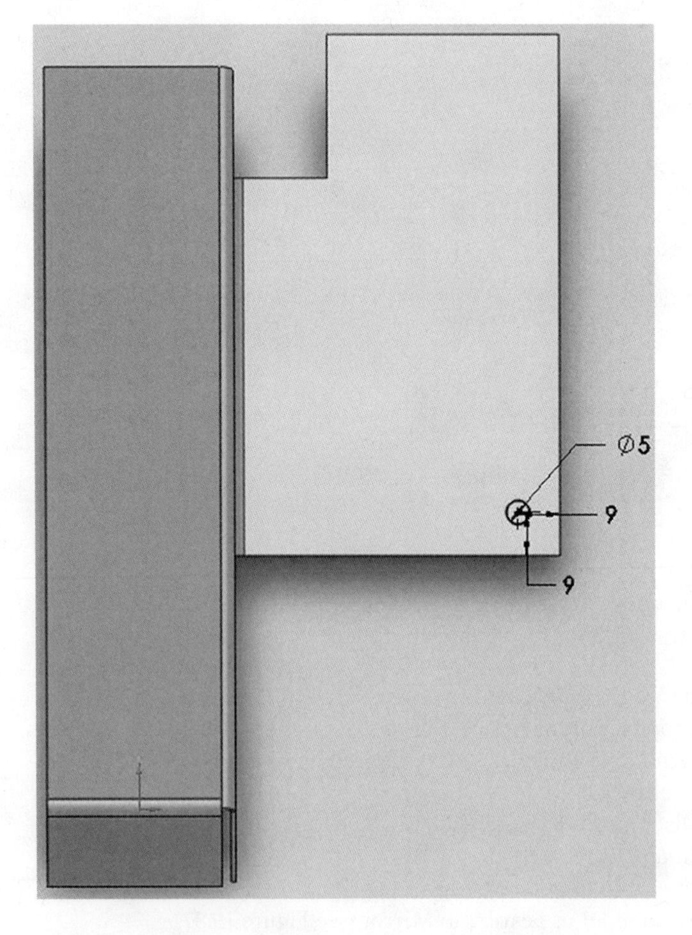

Figure 19.74

Circular sketch on the right edge flange.

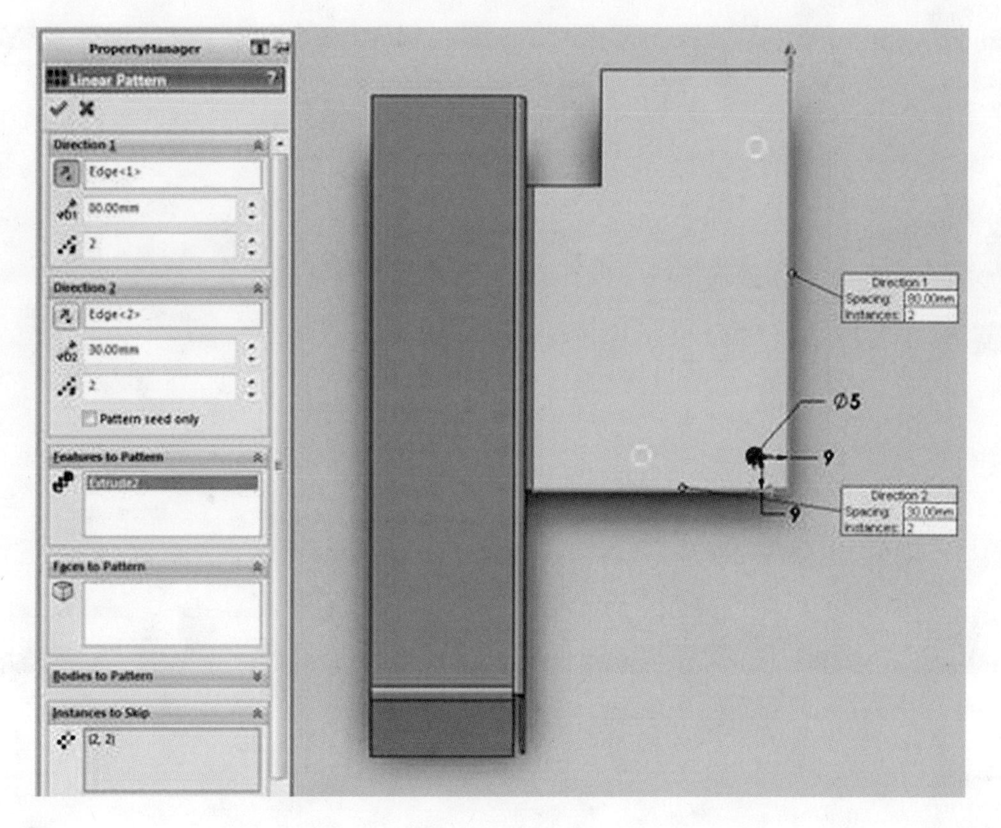

Figure 19.75

Linear patterning of the circular sketch on the right edge flange.

Sketch a Line for Jogging

15. Select the top face of the midfeature (see Figure 19.88).
16. Be in sketch mode and create Sketch3, a line for jogging (see Figure 19.88).

Exit Line Sketch

17. Click Exit to exit Sketch3. (Note: This is extremely important.)

Create a Jog Feature

18. Select Sketch3 that is created.
19. Click Jog (Jog PropertyManager appears, as shown in Figure 19.89).
20. In the Jog Offset rollout, select the Inside Offset for the Dimension Position.
21. *Case 1: Jog Offset of 4.05 mm at an angle of 90 degrees.*
22. Set the Jog Offset (Blind Distance) to 4.05 mm.
23. Reverse the Blind Direction (if jog is in an opposition direction to the one that is desired).
24. Accept the Jog Angle to be 90 degrees.
25. In the Selections rollout, click any point on the surface to the right (or left) of the Jog Line (Sketch) to indicate the Fixed Face (Face<1>). (A Black Sphere appears on the fixed face, and jog is created to the right [or left] of the Jog Line; see Figure 19.86.) (The jog feature is created in Figure 19.90.)

Let us consider another jog feature by changing some of the settings on the Jog PropertyManager.

Case 2: Jog Offset of 35 mm at an Angle of 135°

1. Right-click Jog1 in the FeatureManager.
2. Select Edit Feature from the In-Context Menu that is displayed.
3. Set the Jog Offset to 35 mm at an angle of 135° (see Figure 19.91).
4. Click OK (see Figure 19.92).

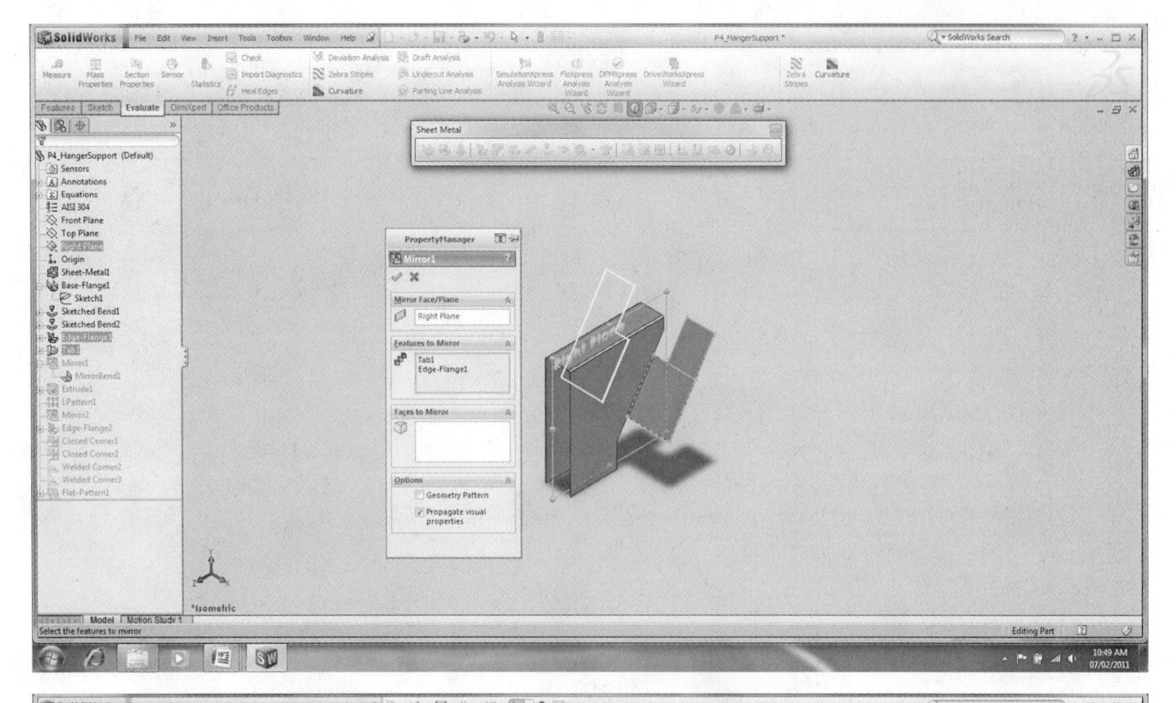

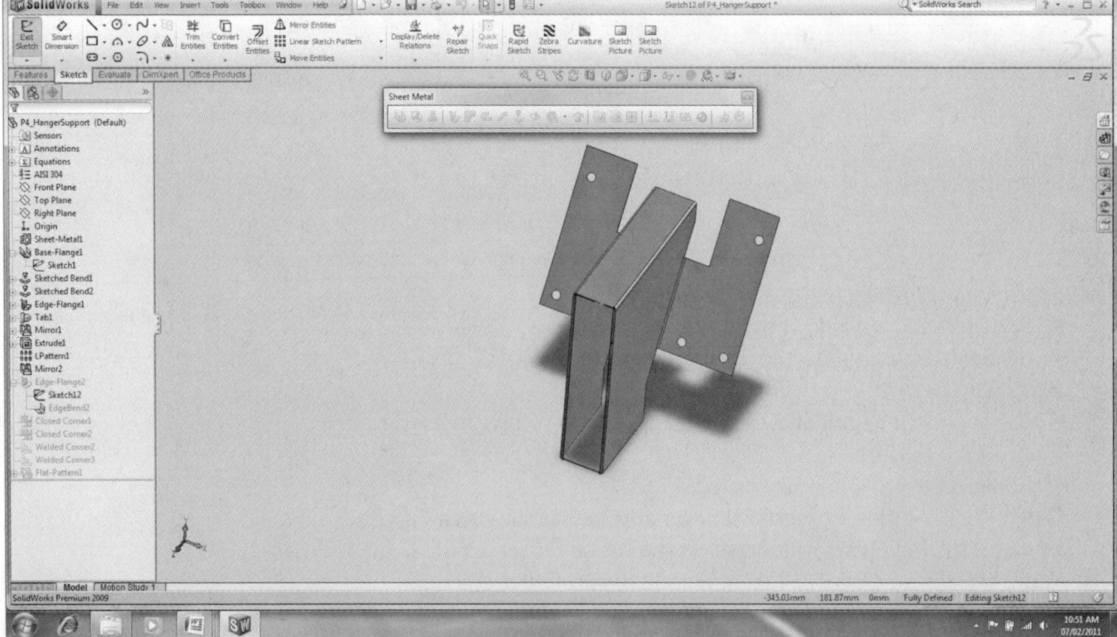

Figure 19.76

Mirror operation for some features created.

Tutorial 19.6: Lofted Sheet Metal

In this tutorial, we will create the sheet metal part that is shown at the top right, with its views and dimensions shown in Figure 19.93.

Create Lofted Boss/Base

1. Select the Front Plane and create Sketch1 (a square, 50 mm × 50 mm), as shown in Figure 19.94a.
2. Create a Split on Sketch1 (0.02-mm width) (see Figure 19.94b).
3. Exit the sketch mode.

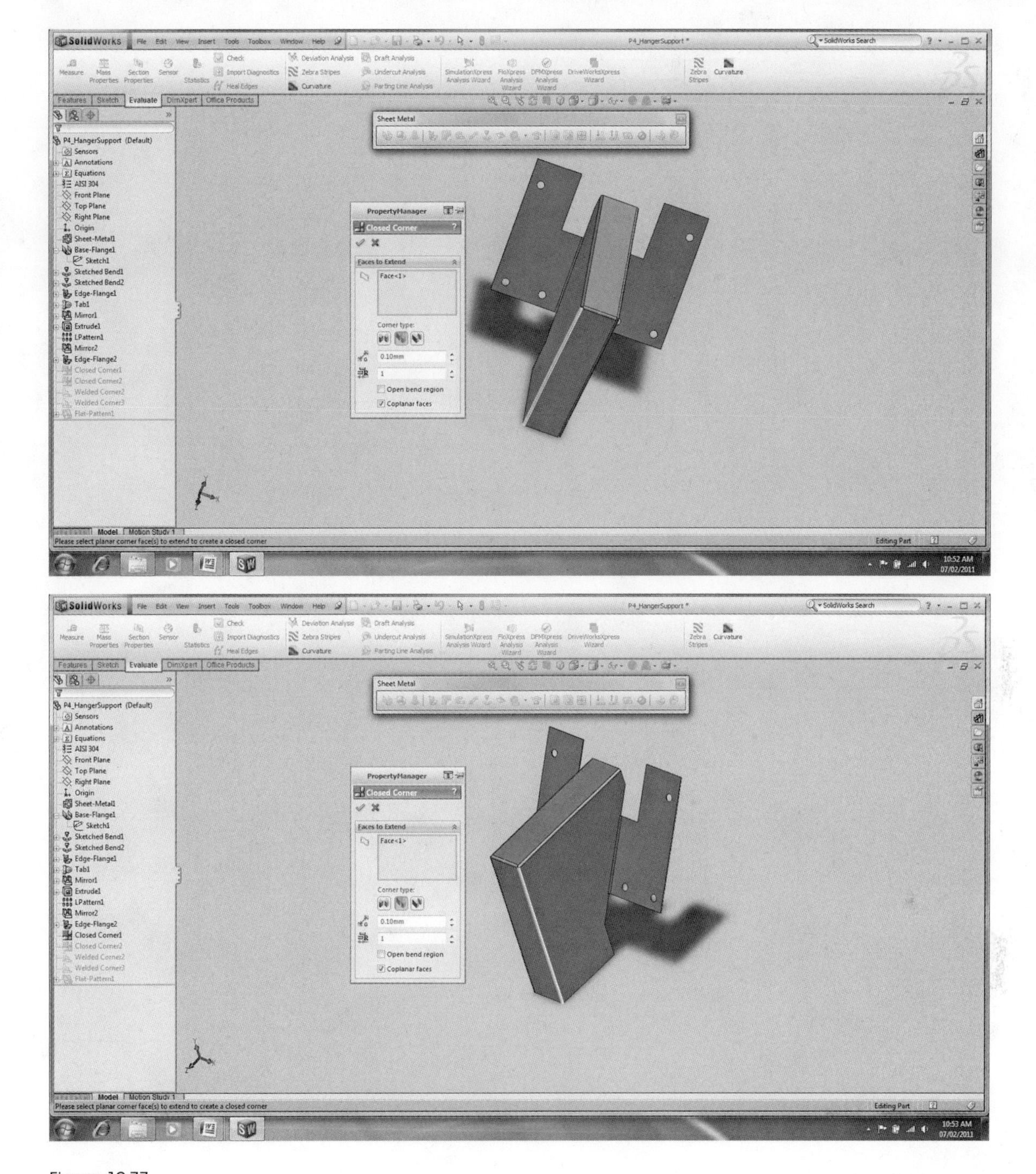

Figure 19.77

Closed corner operations.

4. Click the Features tool and click the Reference tool. Select the Plane option. The plane box appears; select the Front Plane.
5. Set the distance between the existing Front Plane and a new reference plane for 100 mm. Click OK check mark. A new plane, Plane1, appears (see Figure 19.95).
6. Click Plane1 and be in sketch mode.
7. Create Sketch2 (a Circle of diameter, 100 mm) (see Figure 19.96a).
8. Create a Split on Sketch2 (0.02-mm width) (see Figure 19.96b).
9. Exit the sketch mode. (See Figure 19.97 for the two sketches.)
   (Note: It is essential to *break* Sketch1 and Sketch2 Split Entities tool.)

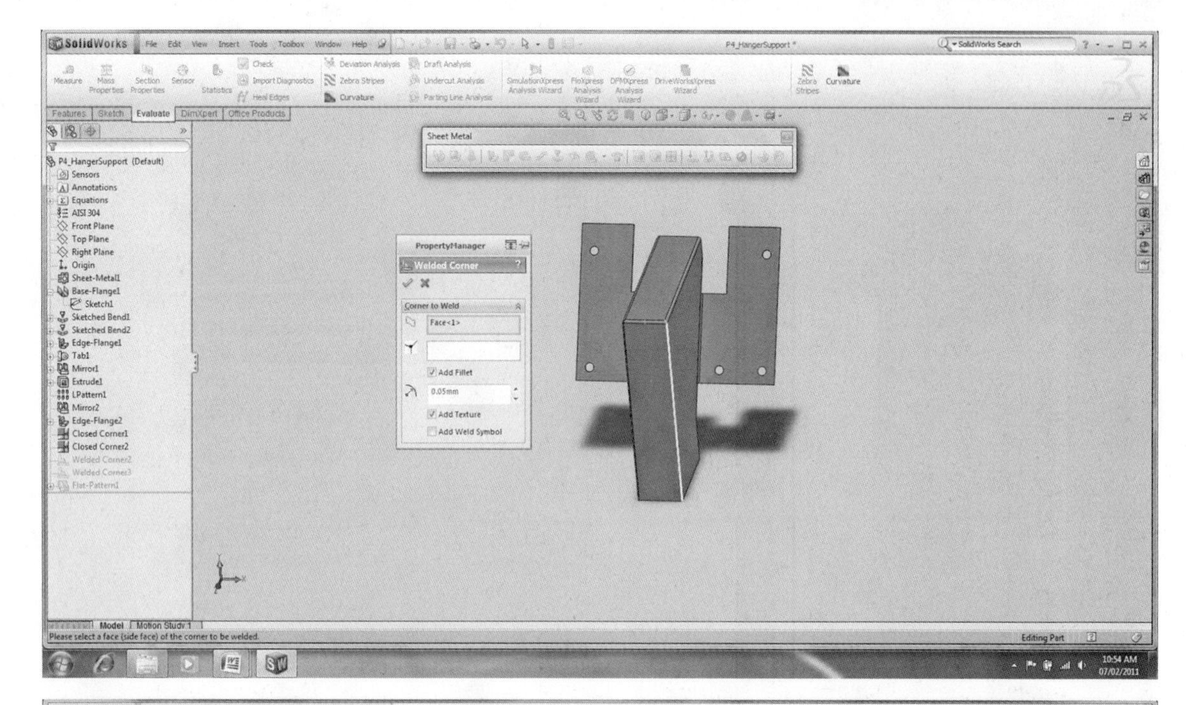

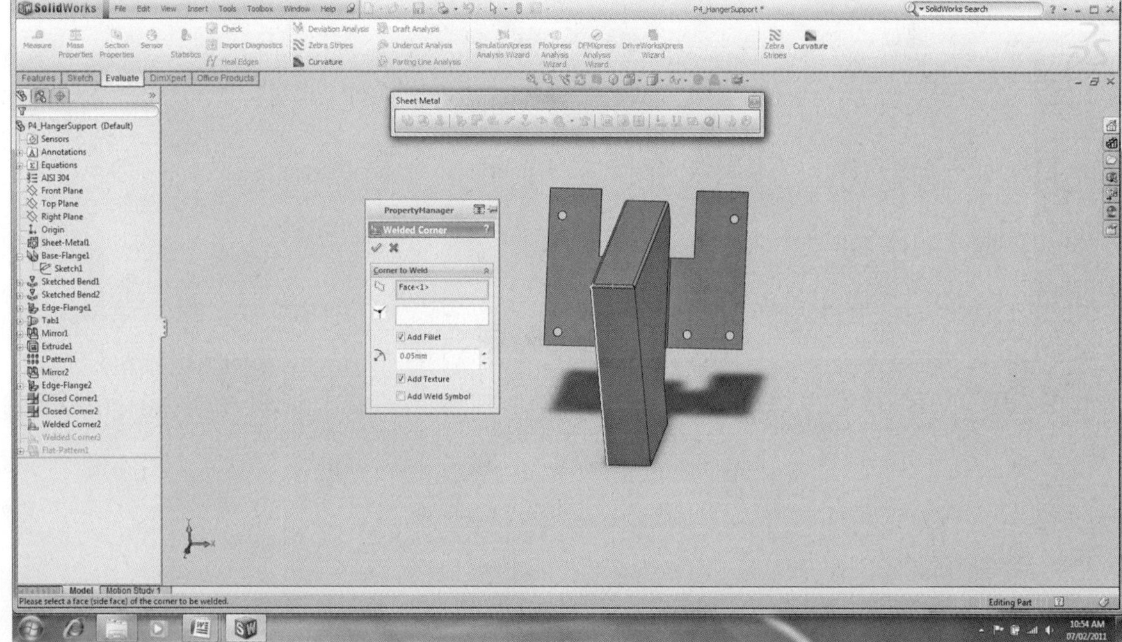

Figure 19.78

Welded corner operations.

Fillet the Square Shape to Remove Sharp Corners (SolidWorks Requires Smooth Corners)

1. Click Fillet tool and select the four edges of the square shape (see Figure 19.98).
2. Select a Fillet Radius of 5 mm.
3. Click OK to complete filleting.

Create Lofted Bend

1. Click the Lofted Bend tool (see Figure 19.99).
   The Lofted Bend Properties Manager appears.
2. Right-click the Profiles box.
3. Click the Square and the Circle. A real-time preview will appear (see Figure 19.100).
4. Click OK to complete the lofted part (see Figure 19.101). Hide the planes.

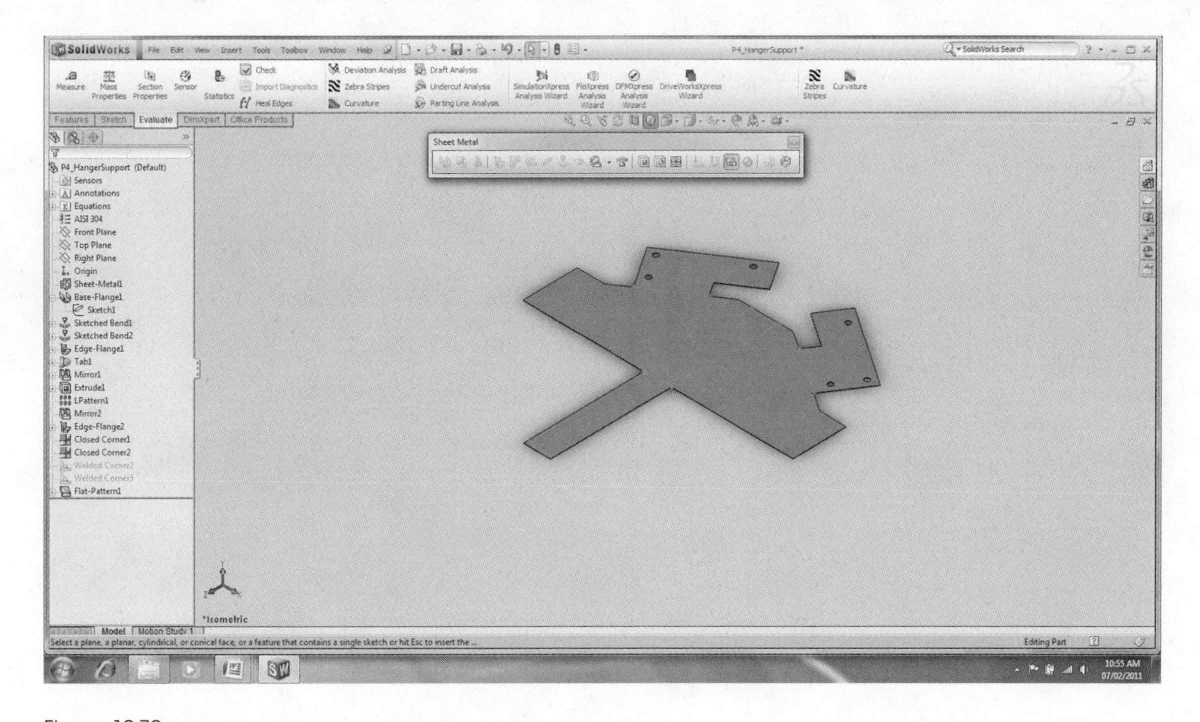

Figure 19.79

Flattened state of the sheet metal.

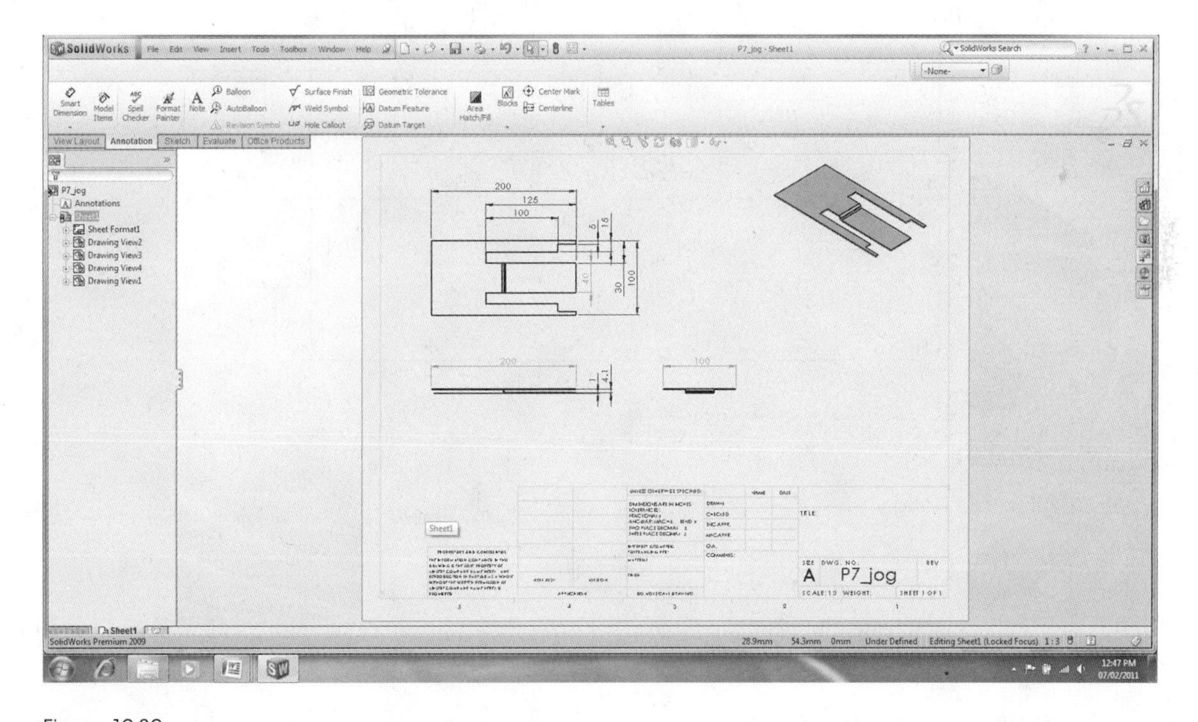

Figure 19.80

Part and drawing views and dimensions for Tutorial 19.5.

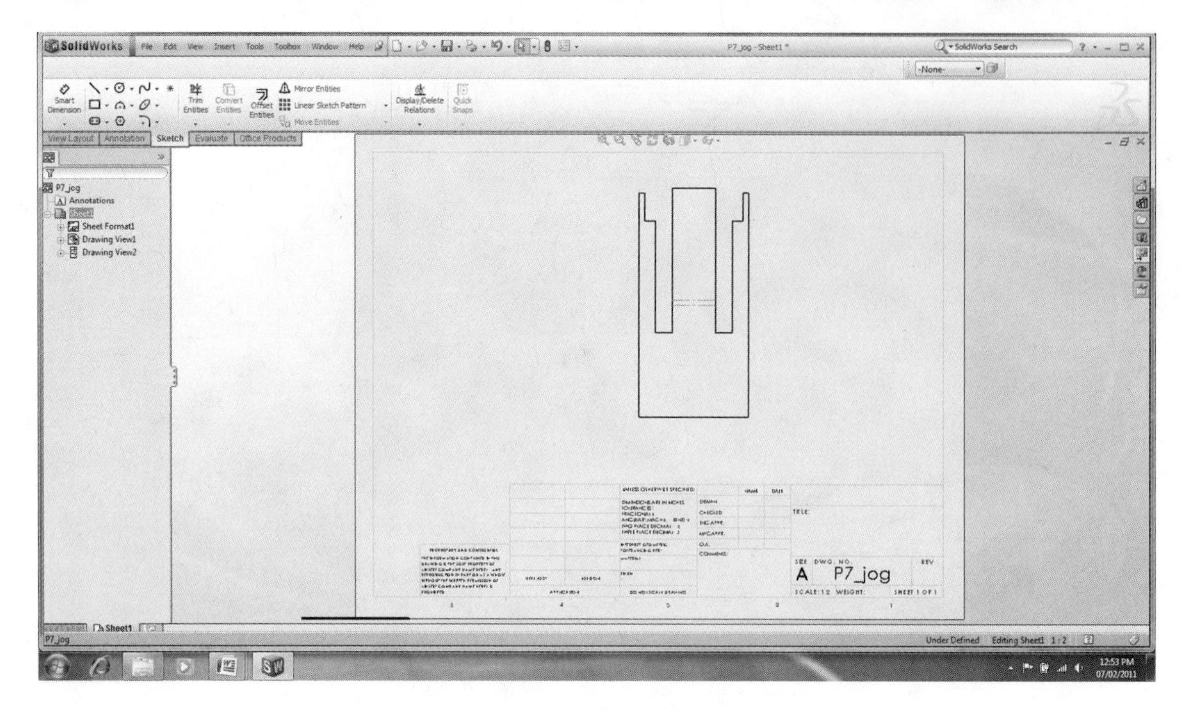

Figure 19.81

Flat pattern of the sheet metal part for Tutorial 19.5.

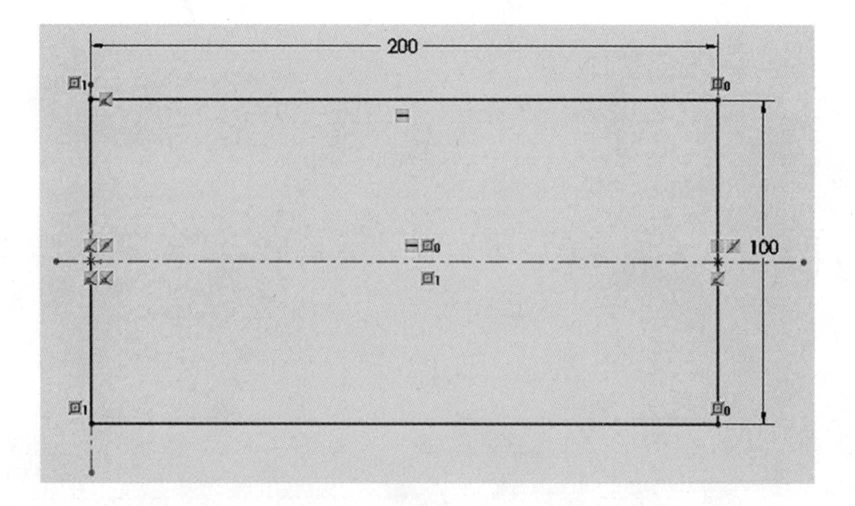

Figure 19.82

Sketch1 for the Base Flange.

Tutorial 19.7: P9

In this tutorial, we will create the sheet metal part that is shown at the top right, with its views and dimensions shown in Figure 19.102. The flat pattern of the sheet metal part is shown in Figure 19.103.

SolidWorks Solution to Tutorial 19.7

Create the Base Flange

1. Start a New SolidWorks Part document.
2. Select the Top Plane.
3. Be in sketch mode and create Sketch1 (see Figure 19.104).
4. Click the Base Flange/Tab option from the Sheet Metal CommandManager. (Note: The Base Flange PropertyManager is automatically displayed; see Figure 19.105.)

(a)

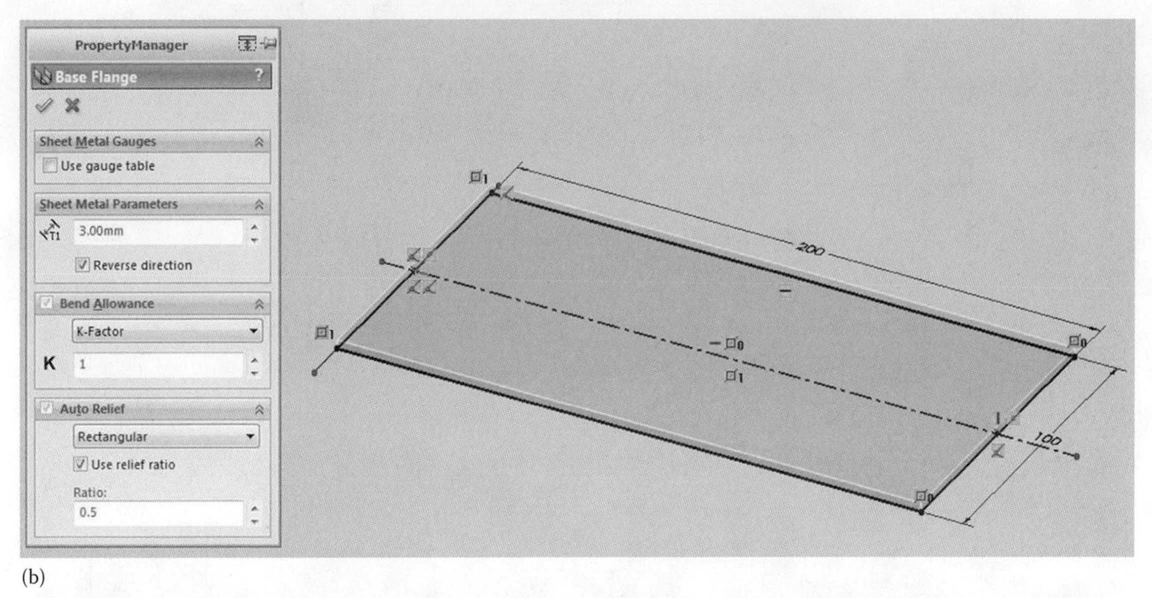

(b)

Figure 19.83

Base Flange: (a) Sheet Metal CommandManager and (b) Base Flange PropertyManager.

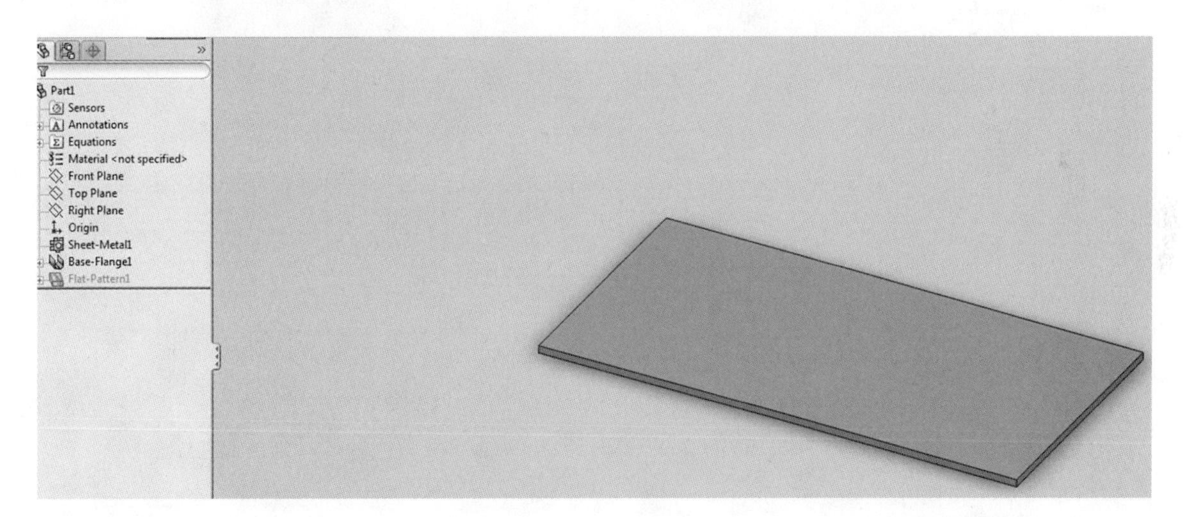

Figure 19.84

Top face of the Base Flange is selected for sketch.

5. Set the values of the following parameters: Thickness = 2; K-Factor = 1; Ratio = 0.5.
6. Click OK. (See Figure 19.106 for the base flange.)

Creating the First Sketched Bend

7. Create a line 200 mm *from the top edge of the base flange* (see Figure 19.107).
8. Click the Sketch Bend tool.
9. Accept the default Bend radius of 2 mm and Angle = 90-deg.
10. Click the Fixed Face (Face<1>) to one side of the line (Sketch2) for the Bend Parameters rollout. (The Sketched Bend PropertyManager is automatically displayed; see Figure 19.108.)
11. Click OK to complete the sketched bend operation. (See Figure 19.109 for bend created.)

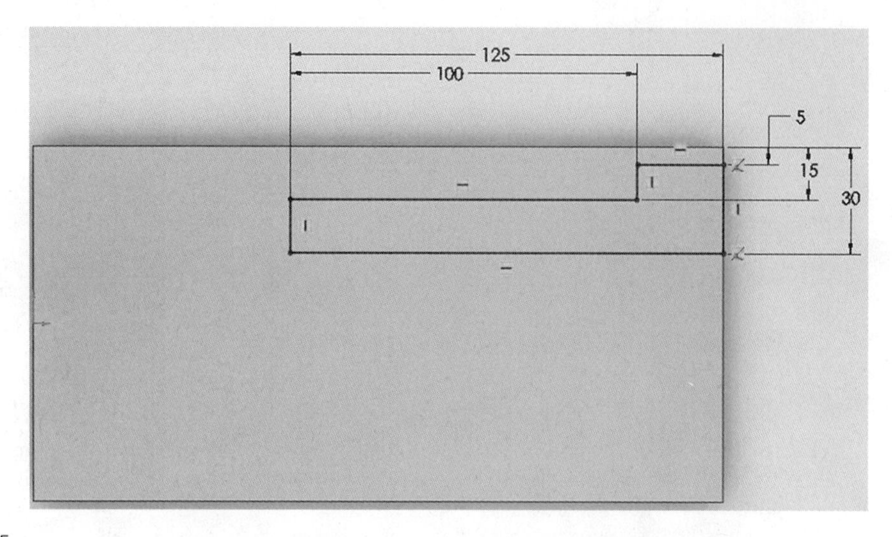

Figure 19.85

Sketch2.

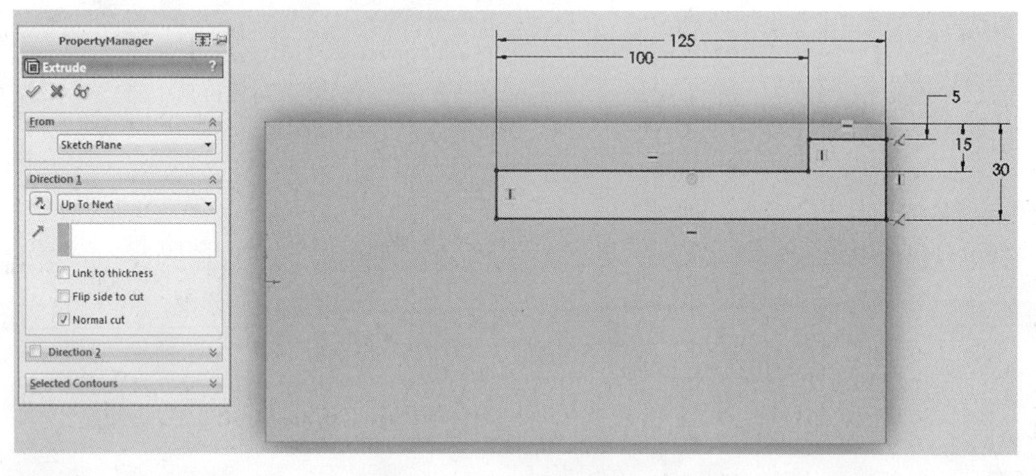

Figure 19.86

Extrude-Cut of Sketch2.

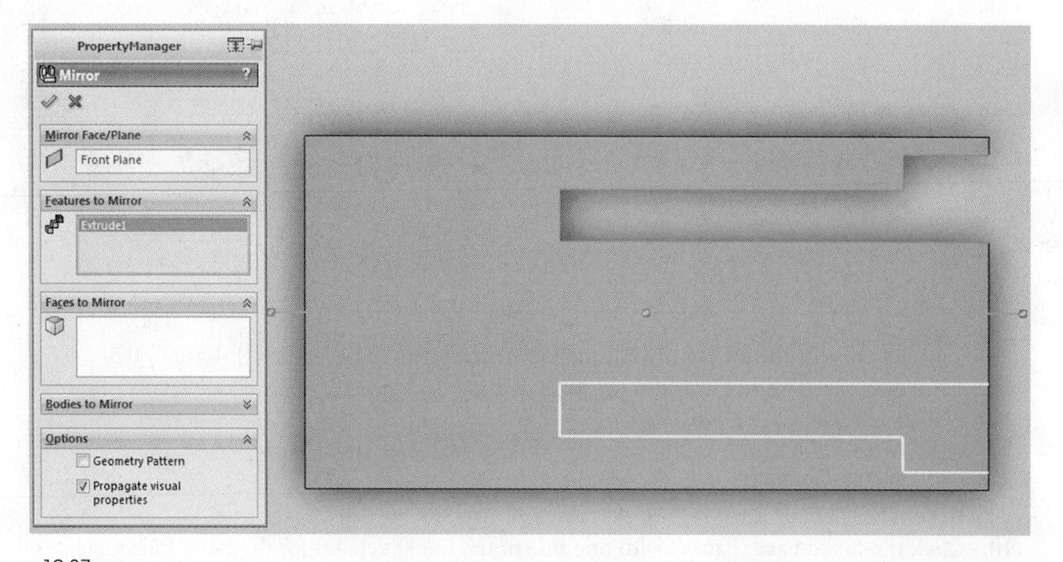

Figure 19.87

Features are mirrored.

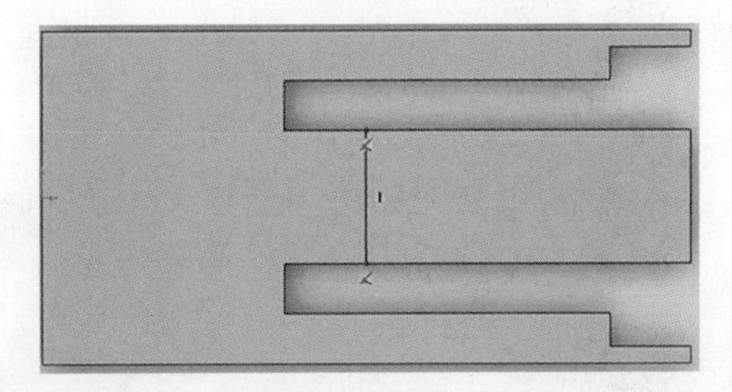

Figure 19.88

Sketch3.

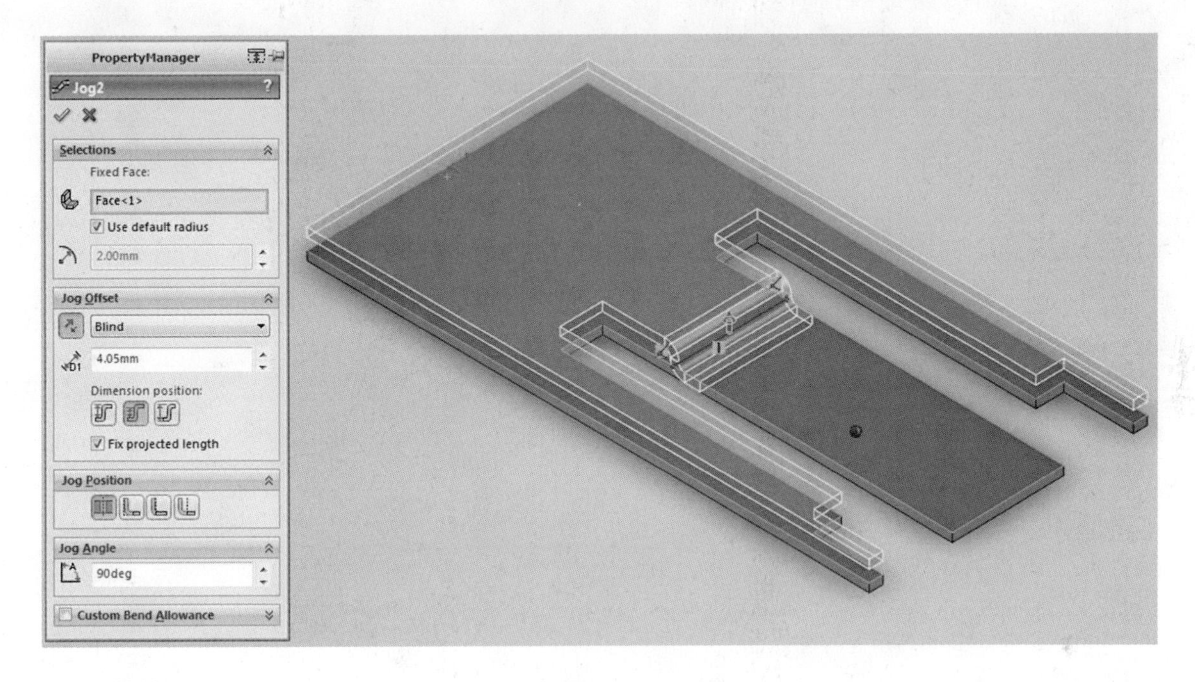

Figure 19.89

Jog PropertyManager.

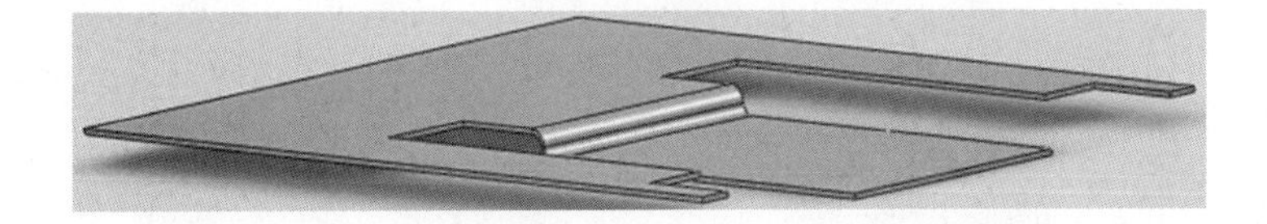

Figure 19.90

Sheet metal with jog feature is created.

Creating the Second Sketched Bend

12. Create a line, collinear with the top of the cut on the first sketch bend feature (see Figure 19.110).
13. Click the Sketch Bend tool.
14. Accept the default Bend radius of 2 mm and Angle = 90-deg.
15. Click the Fixed Face (Face<1>) to one side of the line for the Bend Parameters rollout. (The Sketched Bend PropertyManager is automatically displayed; see Figure 19.110.)
16. Click OK to complete the sketched bend operation. (See Figure 19.111 for bend created.)

Creating the Third Sketched Bend

17. Create a line, midway of the first sketch bend feature (see Figure 19.112).
18. Click the Sketch Bend tool.

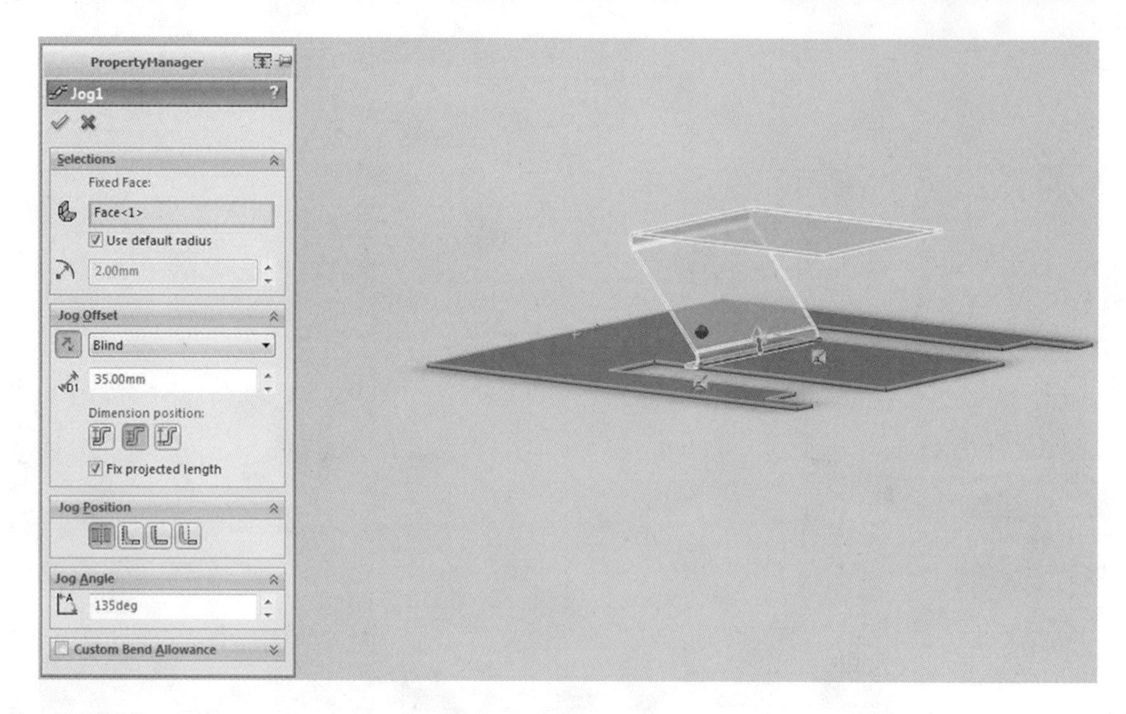

Figure 19.91

Jog PropertyManager.

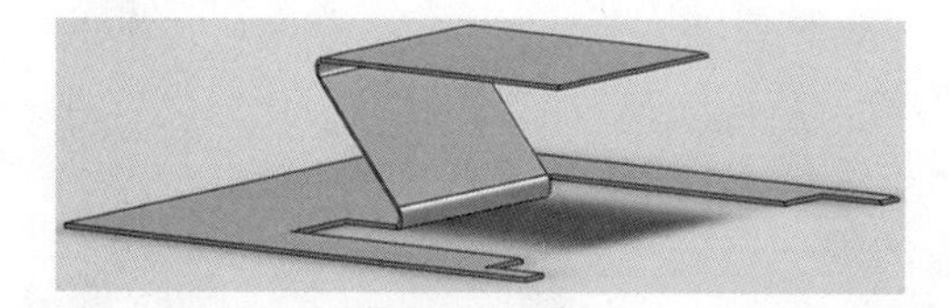

Figure 19.92

Sheet metal with jog feature is created.

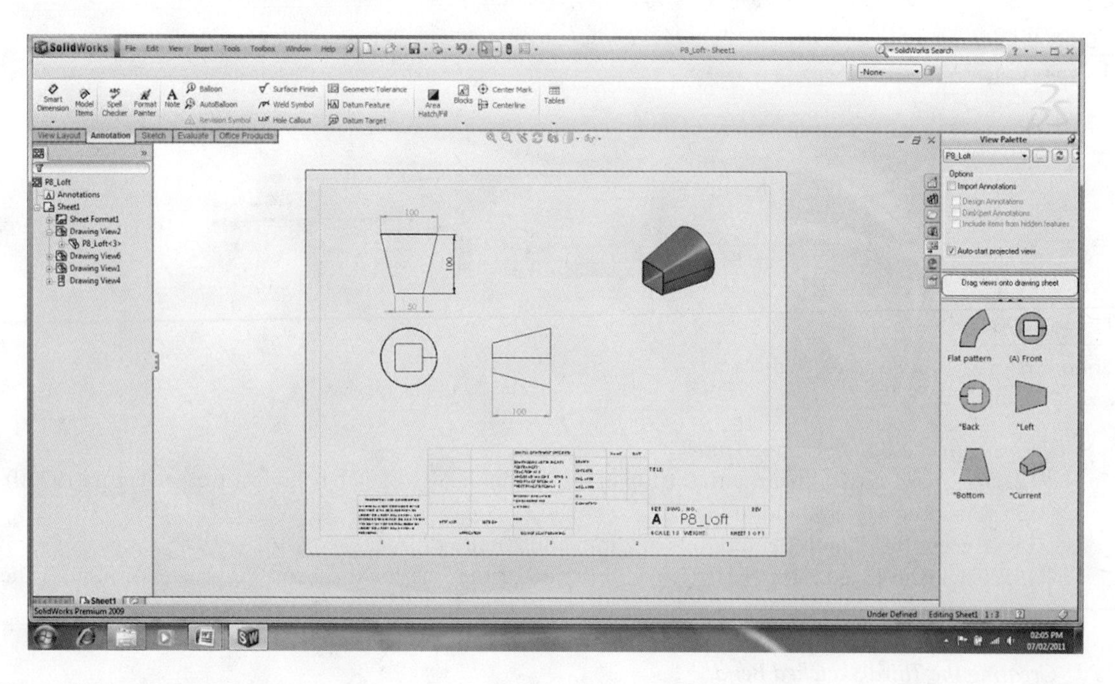

Figure 19.93

Part and drawing views and dimensions for Tutorial 19.6.

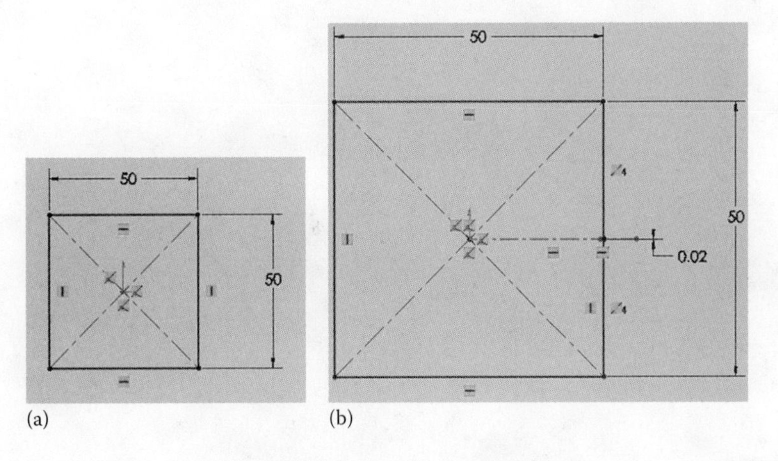

Figure 19.94

Sketch1 for lofting: (a) Sketch1 and (b) Split Sketch1 (0.02-mm width).

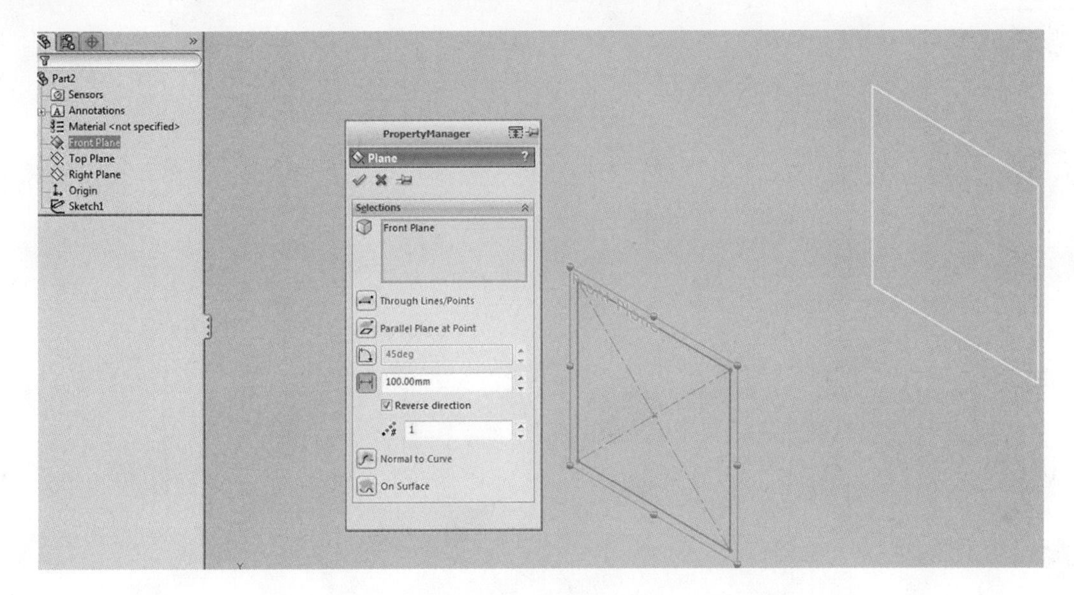

Figure 19.95

Plane1 is created.

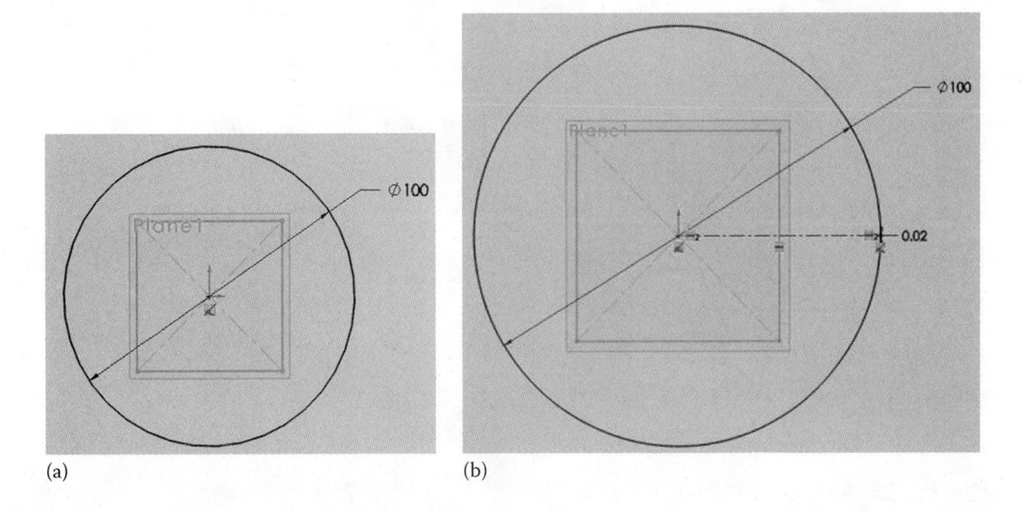

Figure 19.96

Sketch2 for lofting: (a) Sketch2 and (b) Split Sketch2 (0.02-mm width).

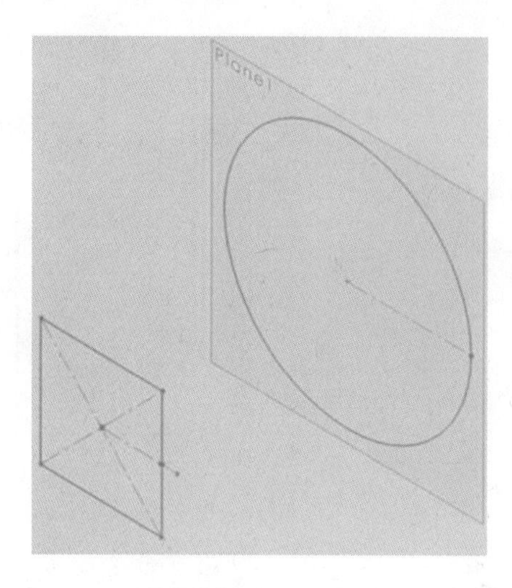

Figure 19.97

Sketches for lofting.

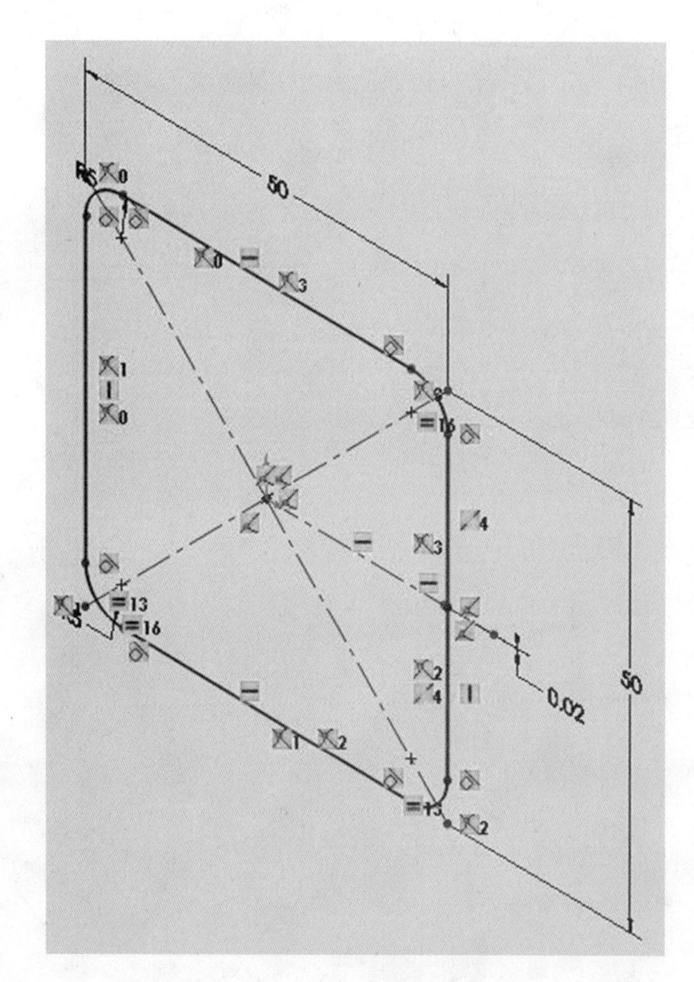

Figure 19.98

Edges of the square feature are filleted.

Figure 19.99

Lofted Bend tool.

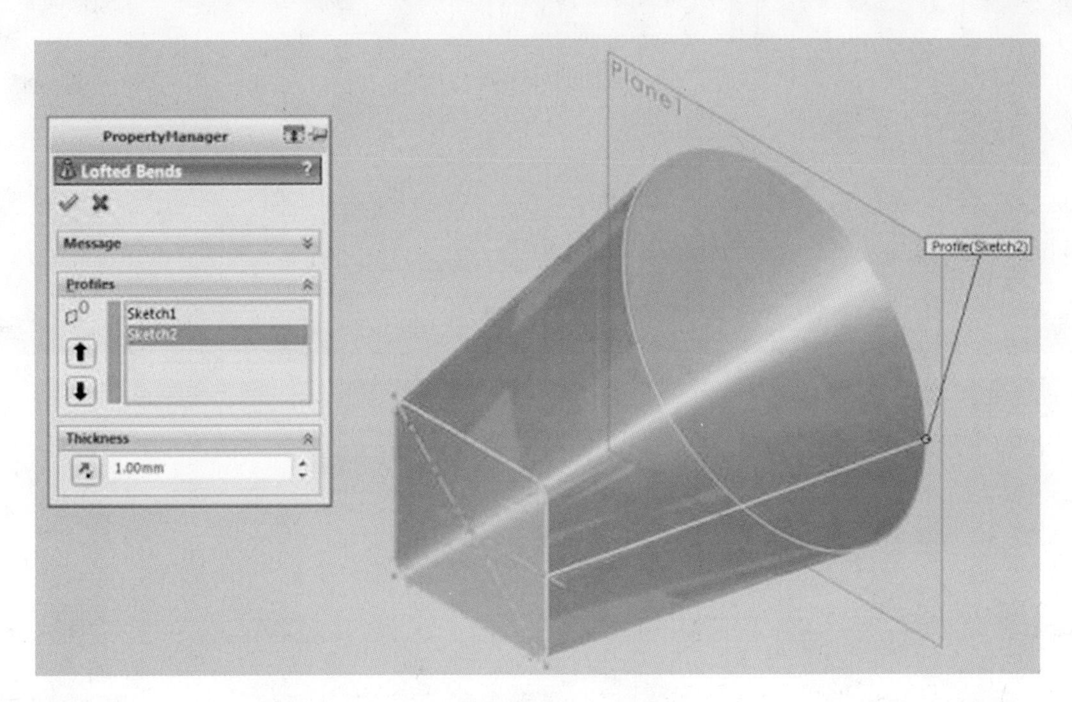

Figure 19.100

Lofted Bend PropertyManager.

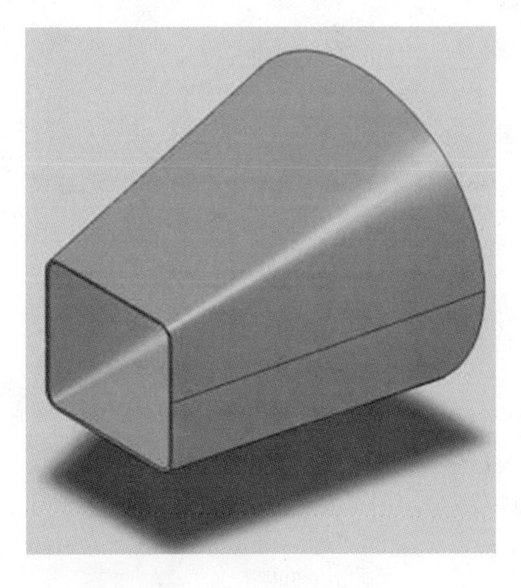

Figure 19.101

Lofted sheet metal.

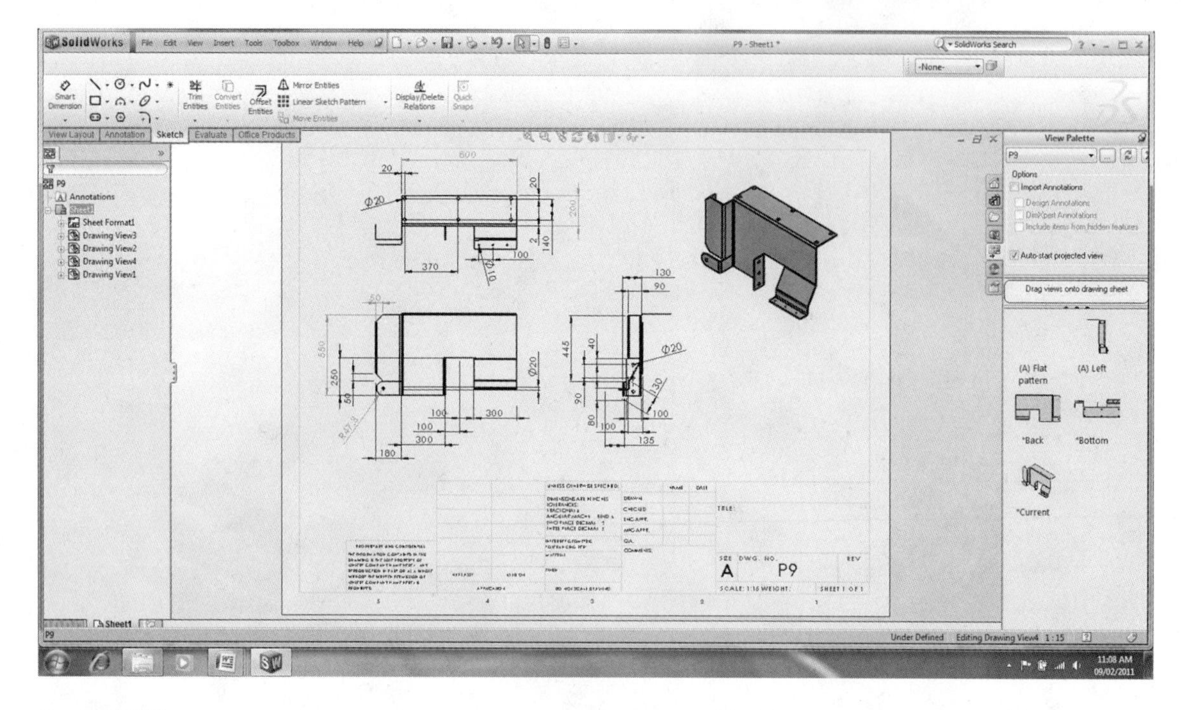

Figure 19.102

Part and drawing views and dimensions for Tutorial 19.7.

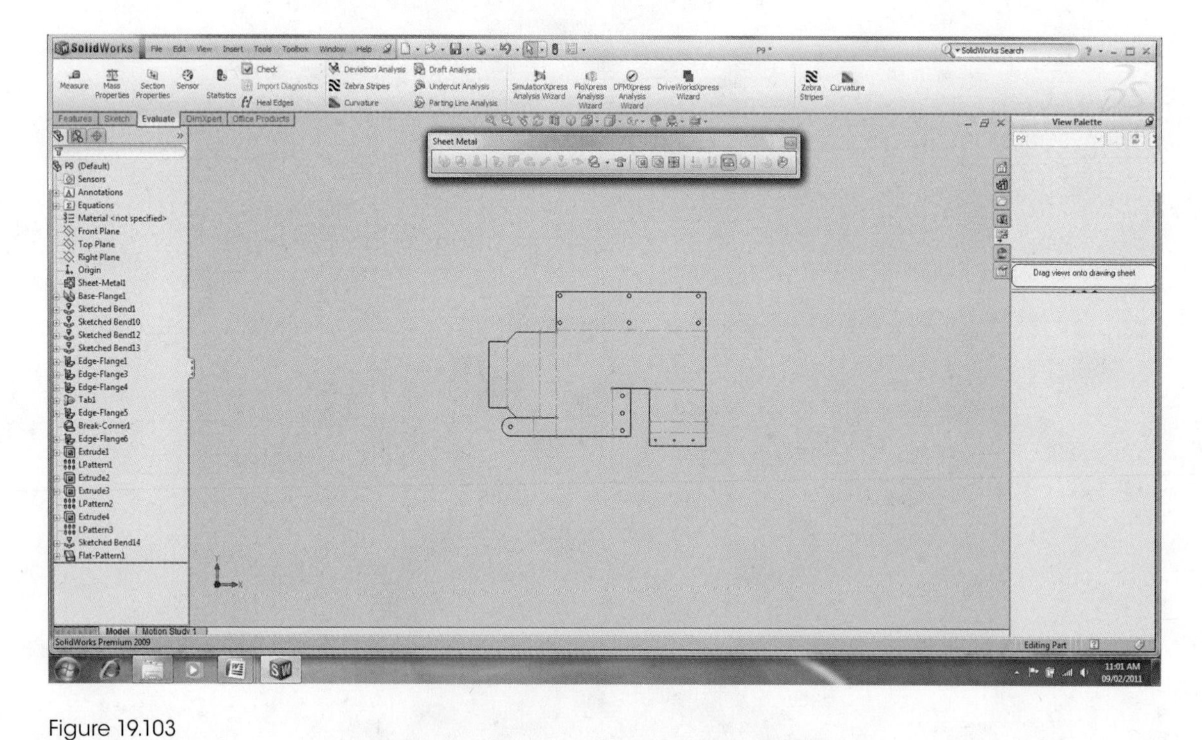

Figure 19.103

Flat pattern of the sheet metal part for Tutorial 19.7.

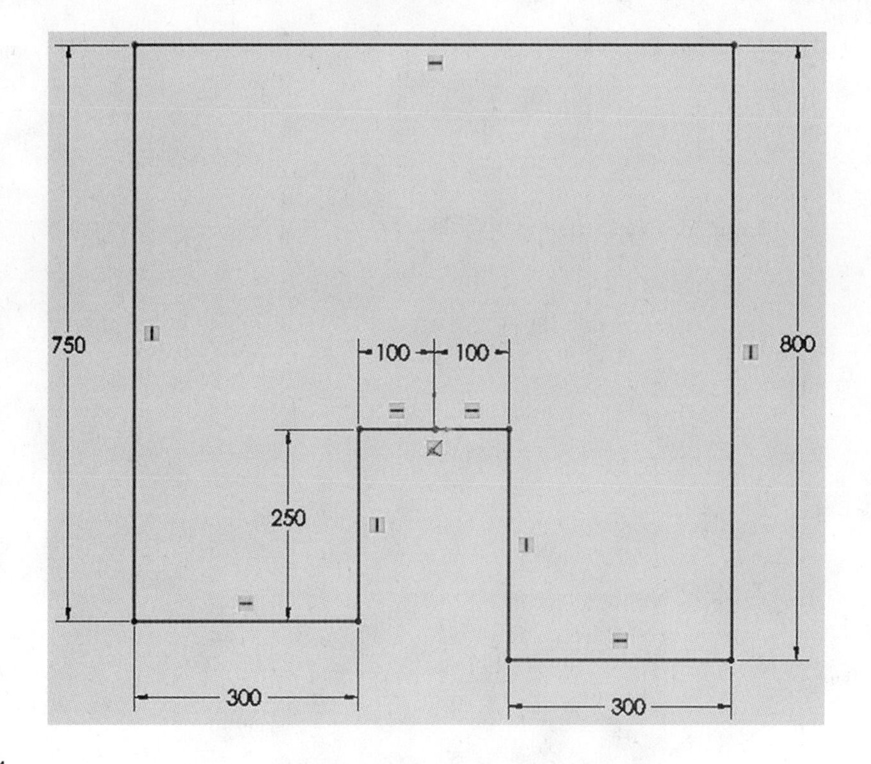

Figure 19.104

Sketch1.

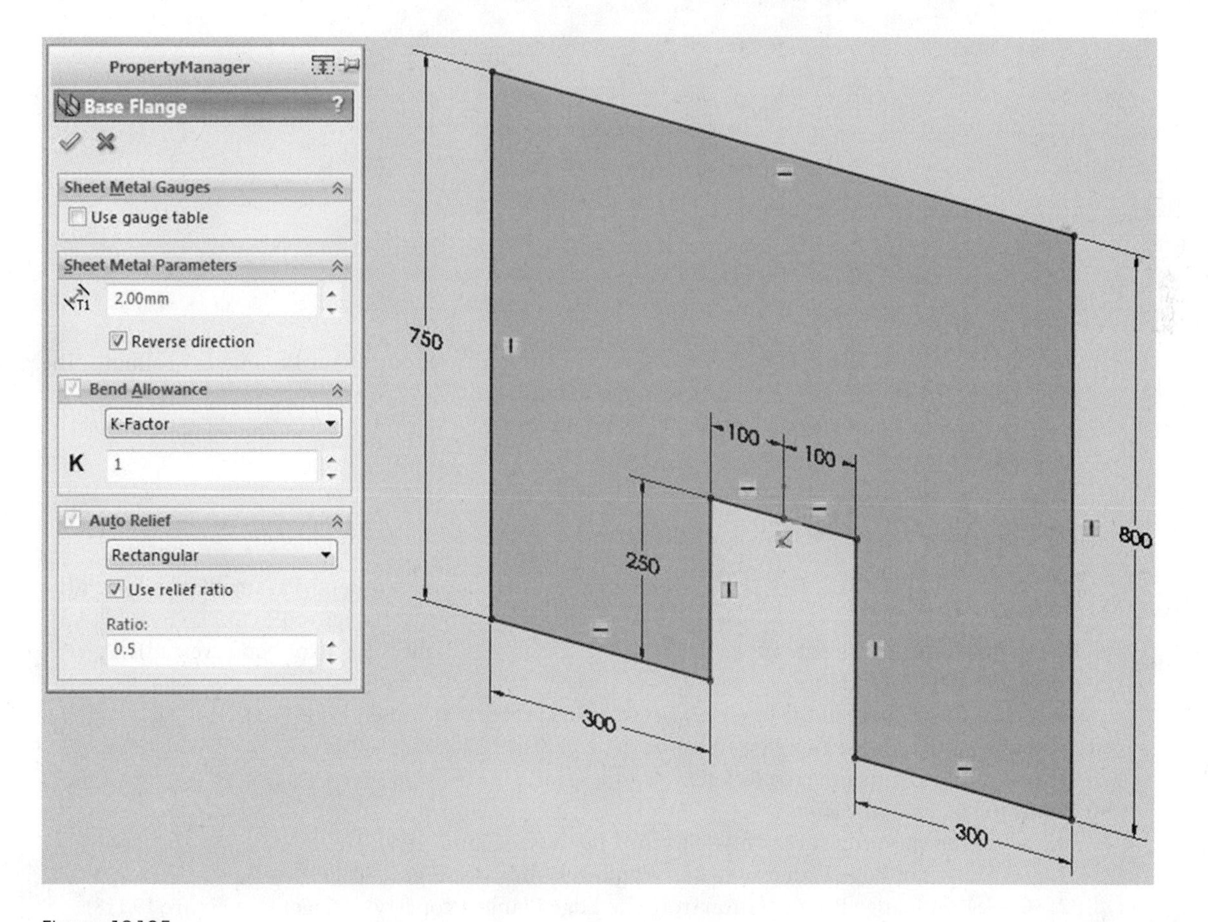

Figure 19.105

Base Flange PropertyManager.

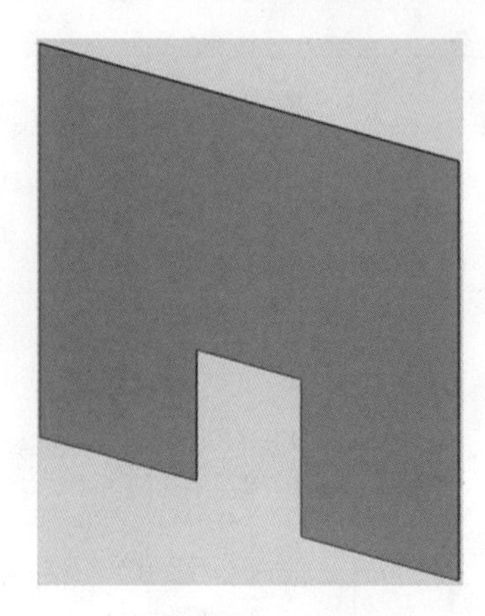

Figure 19.106

Base Flange.

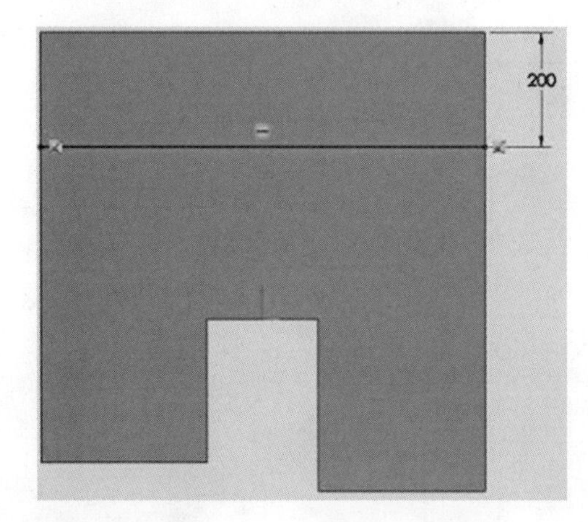

Figure 19.107

Sketch for the first Sketched Bend.

19. Accept the default Bend radius of 2 mm and Angle = 90-deg.
20. Click the Fixed Face (Face<1>) to one side of the line for the Bend Parameters rollout. (The Sketched Bend PropertyManager is automatically displayed; see Figure 19.112.)
21. Click OK to complete the sketched bend operation. (See Figure 19.113 for bend created.)

Creating the Fourth Sketched Bend

22. Create a line, 80 mm from the bottom of the first sketch bend feature (see Figure 19.114).
23. Click the Sketch Bend tool.
24. Accept the default Bend radius of 2 mm and Angle = 90-deg.
25. Click the Fixed Face (Face<1>) to one side of the line for the Bend Parameters rollout. (The Sketched Bend PropertyManager is automatically displayed; see Figure 19.115.)
26. Click OK to complete the sketched bend operation (see Figure 19.116 for bend created).

Creating the First Edge Flange

27. Select the *left edge* of the first cut feature of the first Sketched Bend (see Figure 19.117).
28. Set Length of Edge Flange = 100 and Angle = 90-deg (see Figure 19.117).
29. Click OK to finish the operation.

Creating the Second Edge Flange

30. Select the right edge of the first Sketched Bend (see Figure 19.118).
31. Set Length of Edge Flange = 90 and Angle = 90-deg (see Figure 19.118).
32. Click Edit Flange Profile button from the Edge Flange PropertyManager (see Figure 19.118).
33. Click OK if the following message appears: "*The automatic relations inferred by this operation would over define the sketch. They will not be added.*"

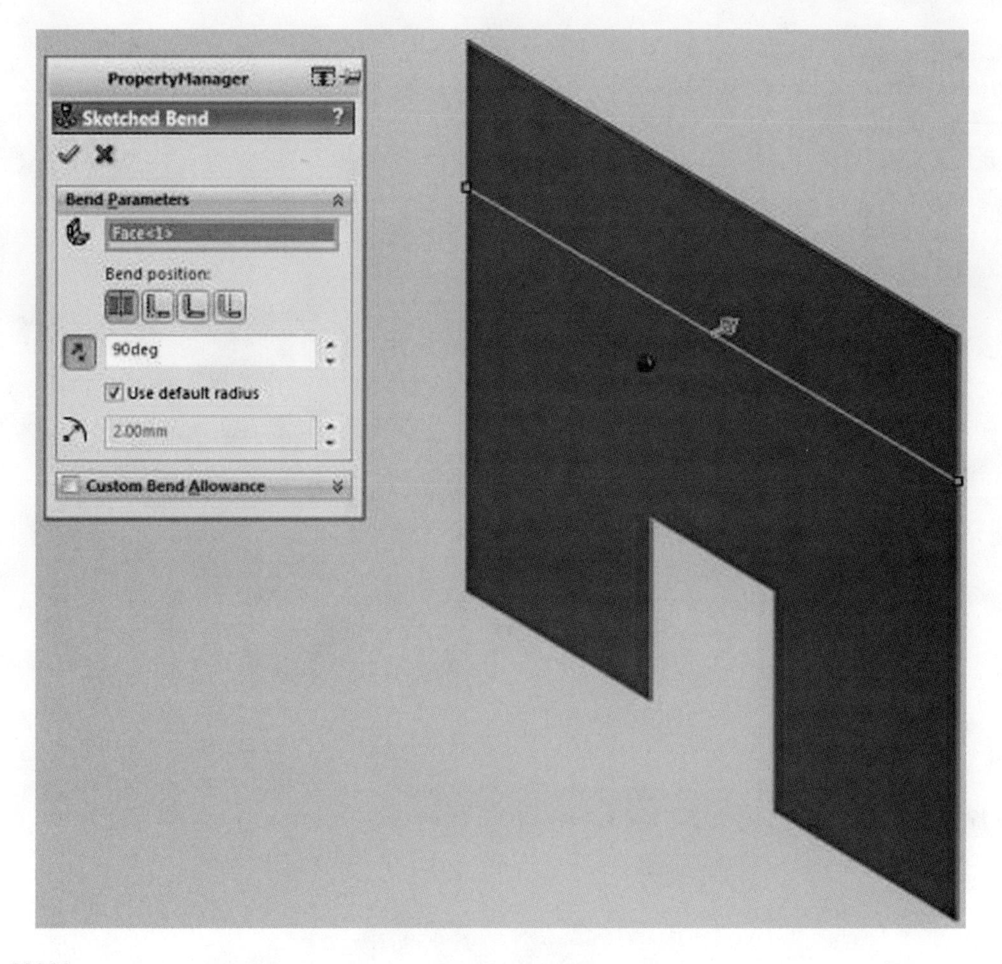

Figure 19.108

Sketched Bend PropertyManager.

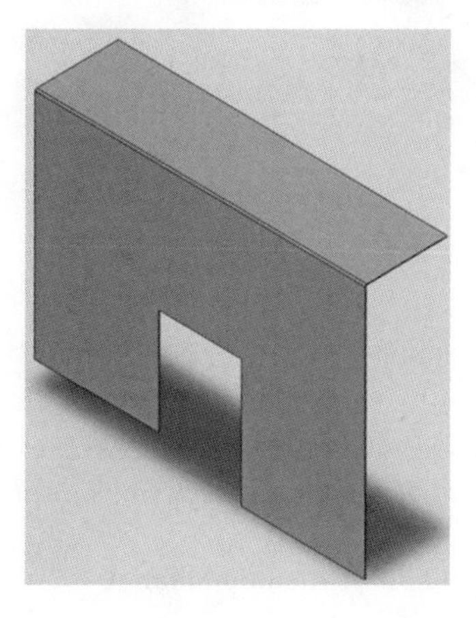

Figure 19.109

First sketched bend created.

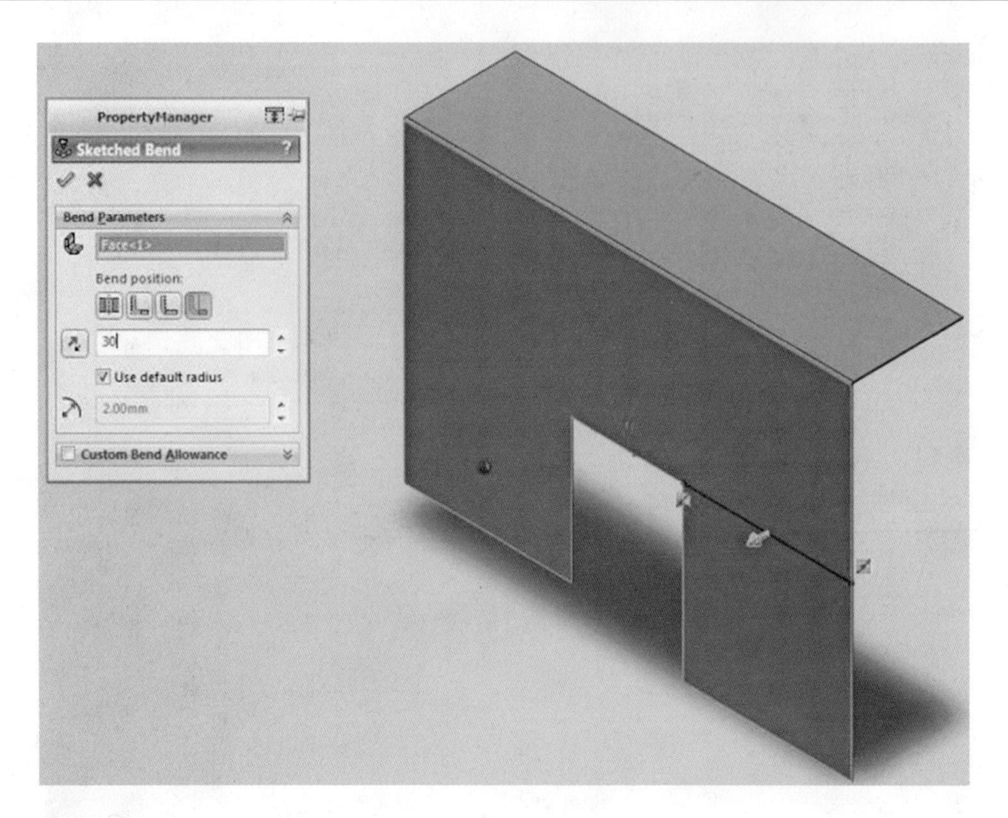

Figure 19.110

Sketch for the second sketch bend.

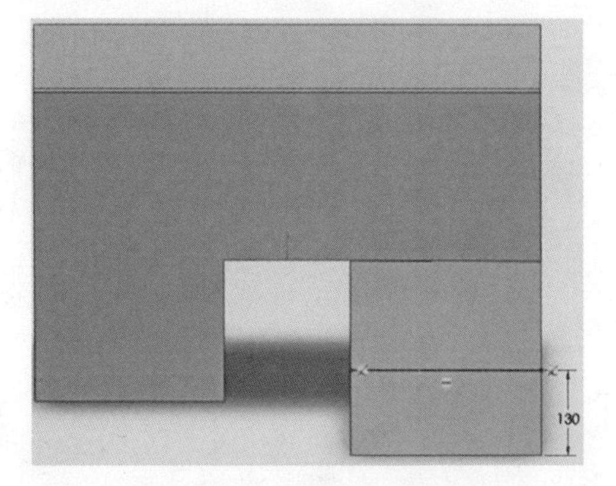

Figure 19.111

Second sketched bend created.

34. Adjust the Width of the Edge Flange created by dragging one endpoint of the edge to a new position (445 mm) and using sketch tool to add dimensions (see Figure 19.119). (Note: *When the Profile Sketch window appears, do not click Finish until the length of the Edge Flange is fully adjusted.*)
35. Click Finish. (See Figure 19.120 for the Edge Flange.)

Creating the Third Edge Flange

36. Select the left-bottom edge of the second edge flange (see Figure 19.121).
37. Set Length of Edge Flange = 130 and Angle = 90-deg (see Figure 19.121).
38. Click OK to finish the operation.

Creating the First Tab

39. Be in sketch mode on the third edge flange and create a sketch for first tab (see Figure 19.122).
40. Click the Base Flange/Tab tool to produce the first tab (see Figure 19.123).

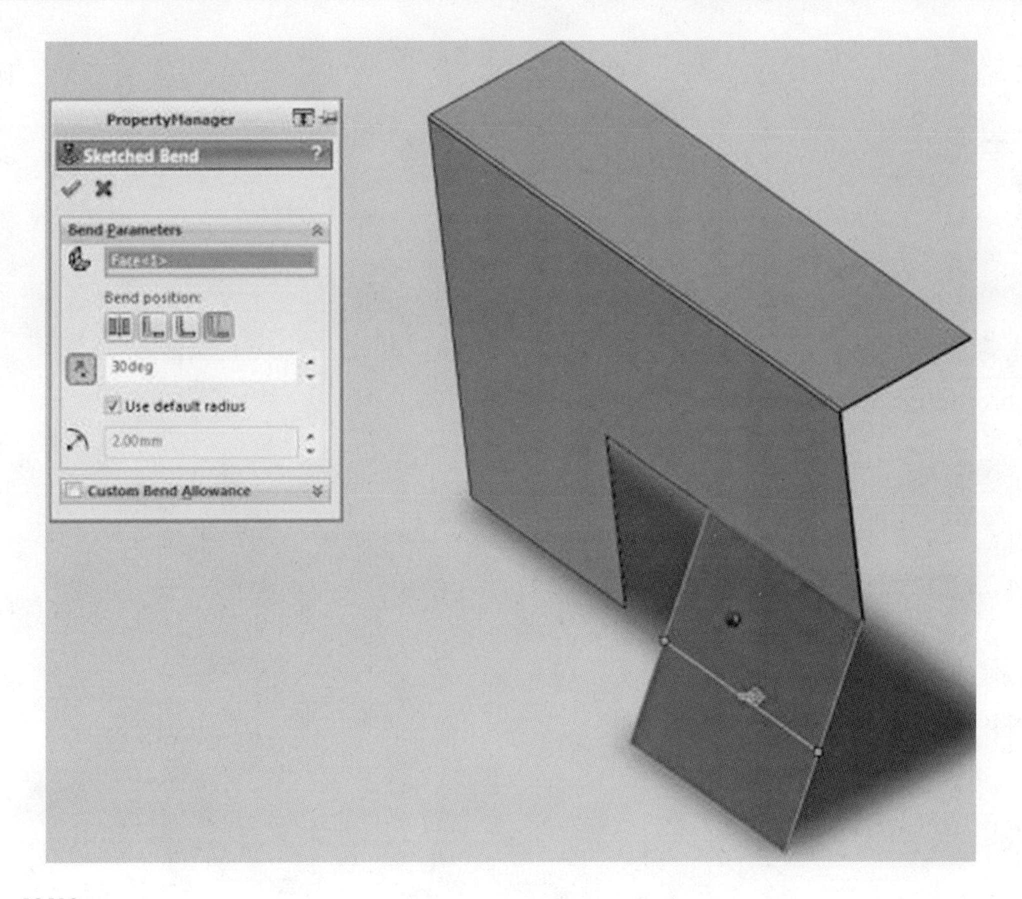

Figure 19.112

Sketch for the third sketch bend.

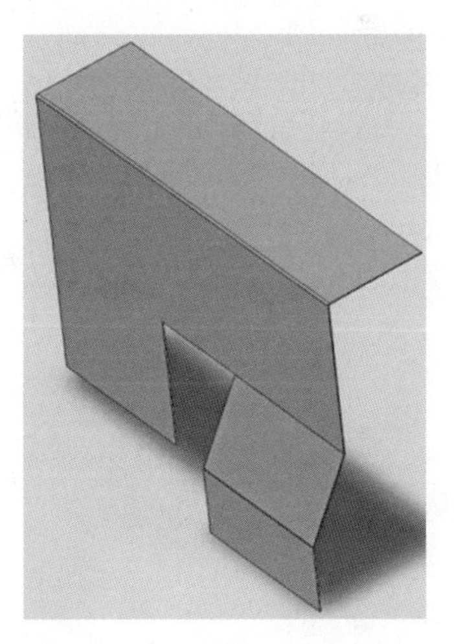

Figure 19.113

Third sketched bend created.

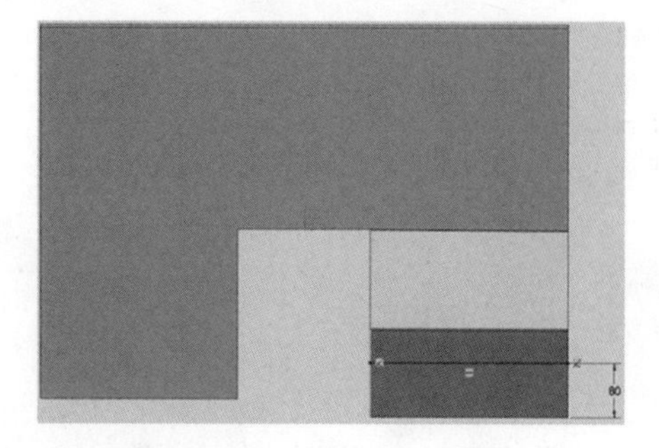

Figure 19.114

Sketch for the fourth sketch bend.

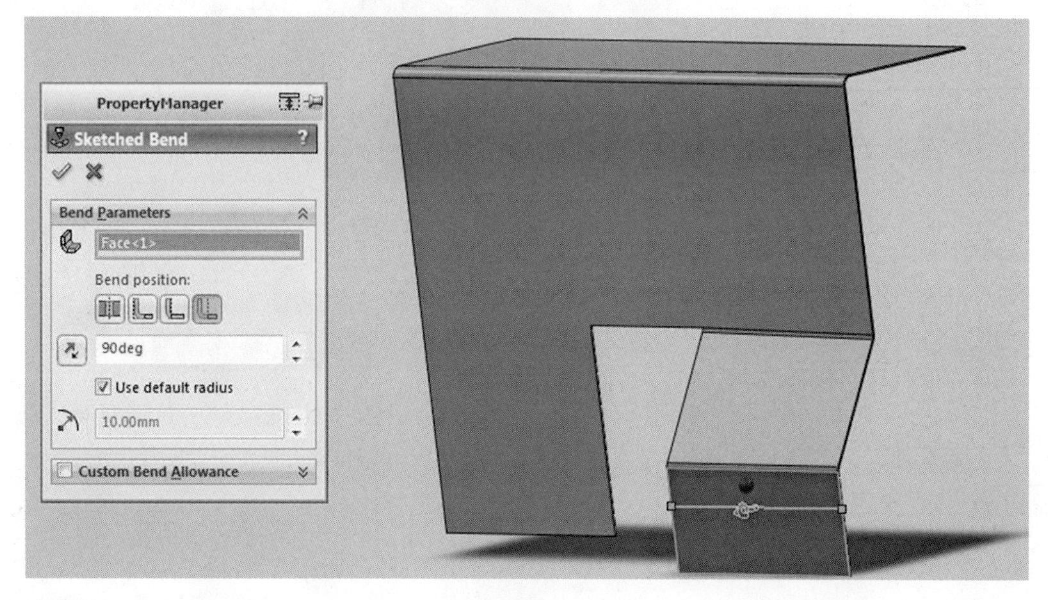

Figure 19.115

Sketched Bend PropertyManager.

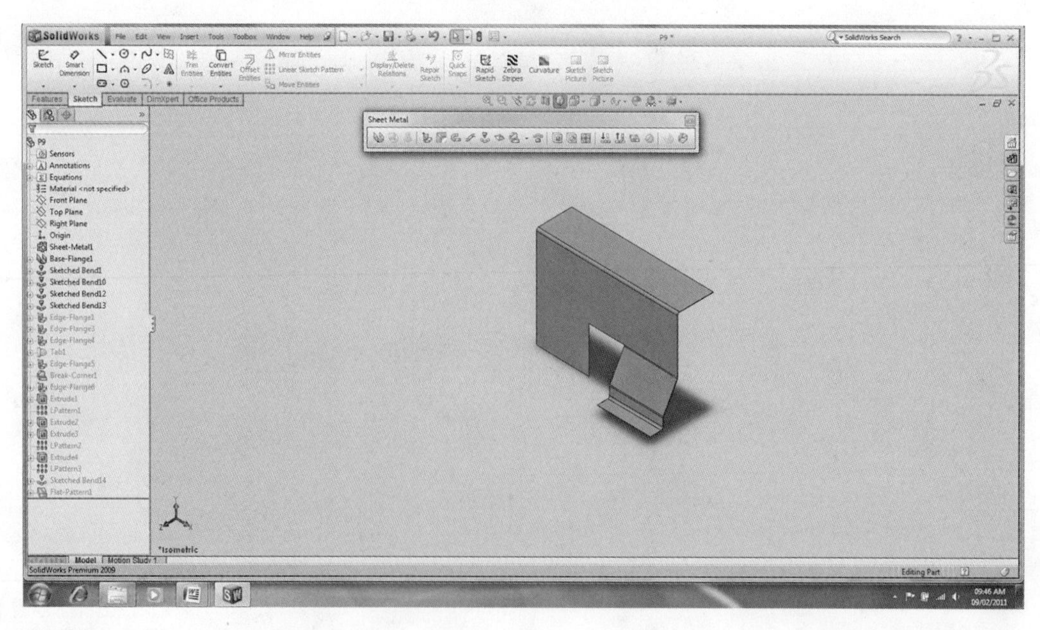

Figure 19.116

Fourth sketched bend created.

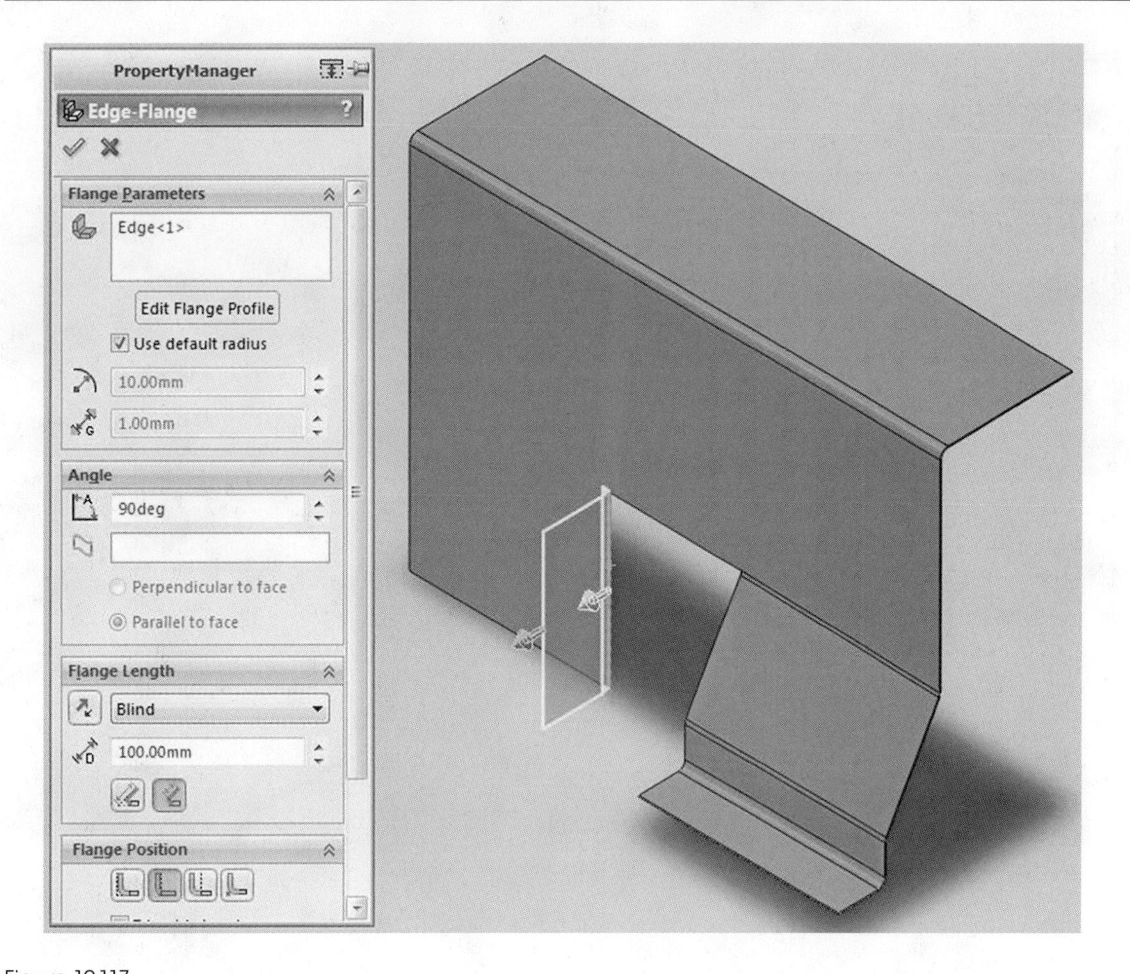

Figure 19.117

Edge Flange PropertyManager.

Creating the Fourth Edge Flange

41. Select the left-outer bottom edge of the third edge flange (see Figure 19.124).

42. Set Length of Edge Flange = 180 and Angle = 90-deg (see Figure 19.124).

43. Click OK to finish the operation.

Creating the First Break Corner

44. Click the Break Corner tool.

45. Select the outer face (Face<1>) of the fourth edge flange (see Figure 19.125).

46. Set Distance = 50 (see Figure 19.125).

47. Click OK to finish the operation.

Creating the Fifth Edge Flange

48. Select the left-outer bottom edge of the third edge flange (see Figure 19.126).

49. Set Length of Edge Flange = 100 and Angle = 90-deg (see Figure 19.126).

50. Click OK to finish the operation.

Creating the Holes of Features

51. Click the top of the Base Flange and be in sketch mode; create a circle 20 mm diameter, located 20 mm × 20 mm from the outer-left vertex (see Figure 19.127).

52. Click Extrude-Cut to create a hole from the circle; choose Up To Next.

53. Create Linear Pattern using the seed hole and select 3 and 2 for Number of features for Direction 1 and Direction 2, respectively (see Figure 19.127).

Figure 19.128 shows a number of holes that are created in the other features of the sheet metal part.

Creating the Second Sketch Bend

54. Create a line 135 mm from the hole of the tab (see Figure 19.129a).

55. Click the Sketch Bend tool.

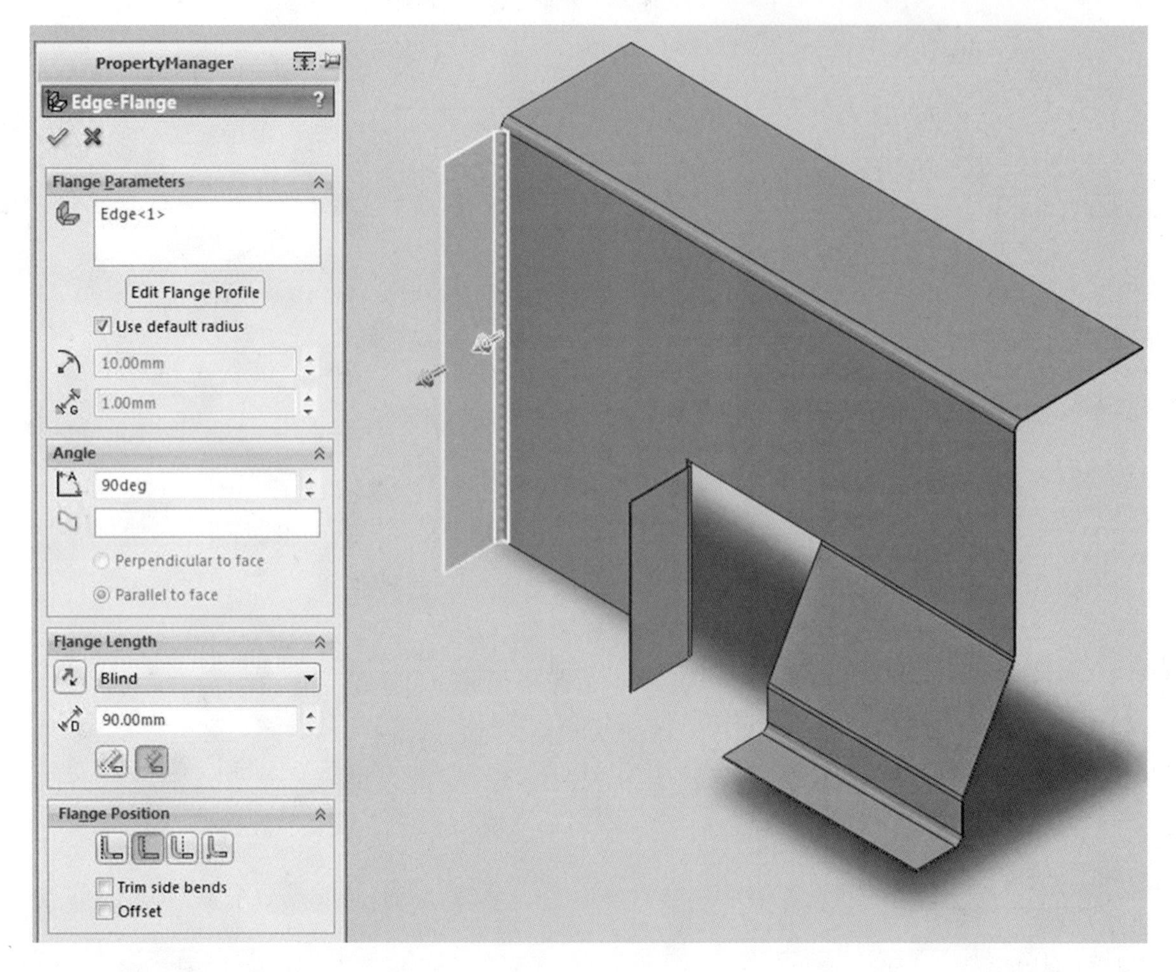

Figure 19.118

Edge Flange PropertyManager.

56. Accept the default Bend radius of 2 mm and Angle = 90-deg.
57. Click the Fixed Face (Face<1>) to one side of the line (Sketch2) for the Bend Parameters rollout. (The Sketched Bend PropertyManager is automatically displayed; see Figure 19.129b.)
58. Click OK to complete the sketched bend operation. (See Figure 19.130 for bend created.)

Tutorial 19.8: General Sheet Metal

In this tutorial, we will create the sheet metal part that is shown at the top right, with its views and dimensions shown in Figure 19.131. The flat pattern of the sheet metal part is shown in Figure 19.132.

SolidWorks Solution to Tutorial 19.8

Create the Base Flange

1. Start a New SolidWorks Part document.
2. Select the Top Plane.
3. Be in sketch mode and create Sketch1 (see Figure 19.133).
4. Click the Base Flange/Tab option from the Sheet Metal CommandManager.
5. Set the values of the following parameters: Thickness = 1; K-Factor = 0.5; Ratio = 0.5.
6. Click OK.

Create the First Edge Flange

7. Select the right edge of the first Sketched Bend (see Figure 19.134).
8. Set Length of Edge Flange = 90 and Angle = 90-deg (see Figure 19.134).
9. Click Edit Flange Profile button from the Edge Flange PropertyManager (see Figure 19.134).

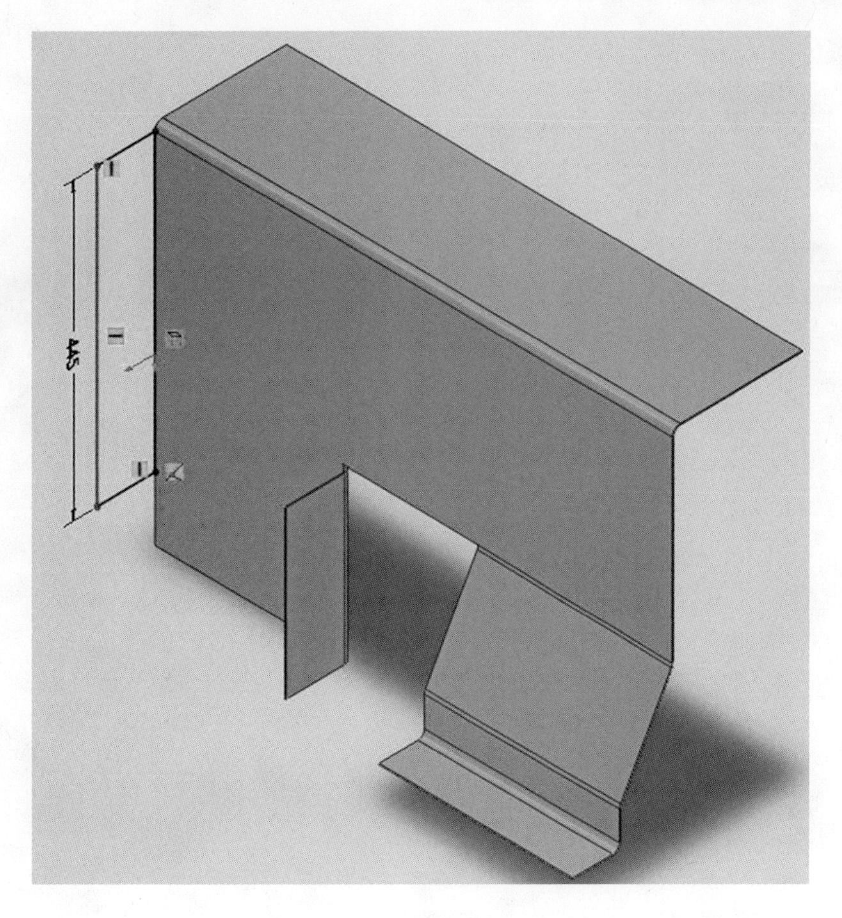

Figure 19.119

Width adjustment for the edge flange.

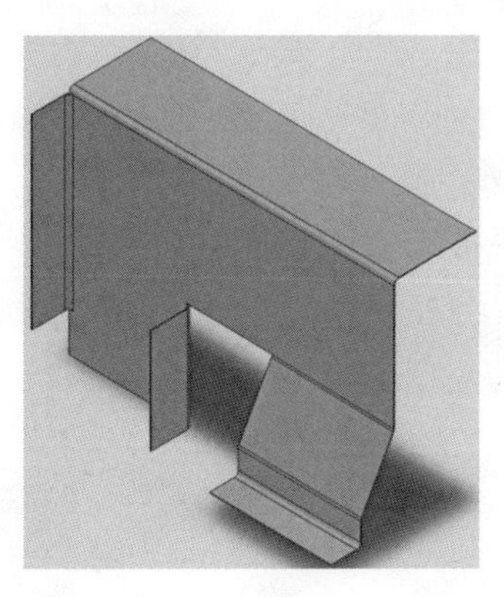

Figure 19.120

Second Edge Flange.

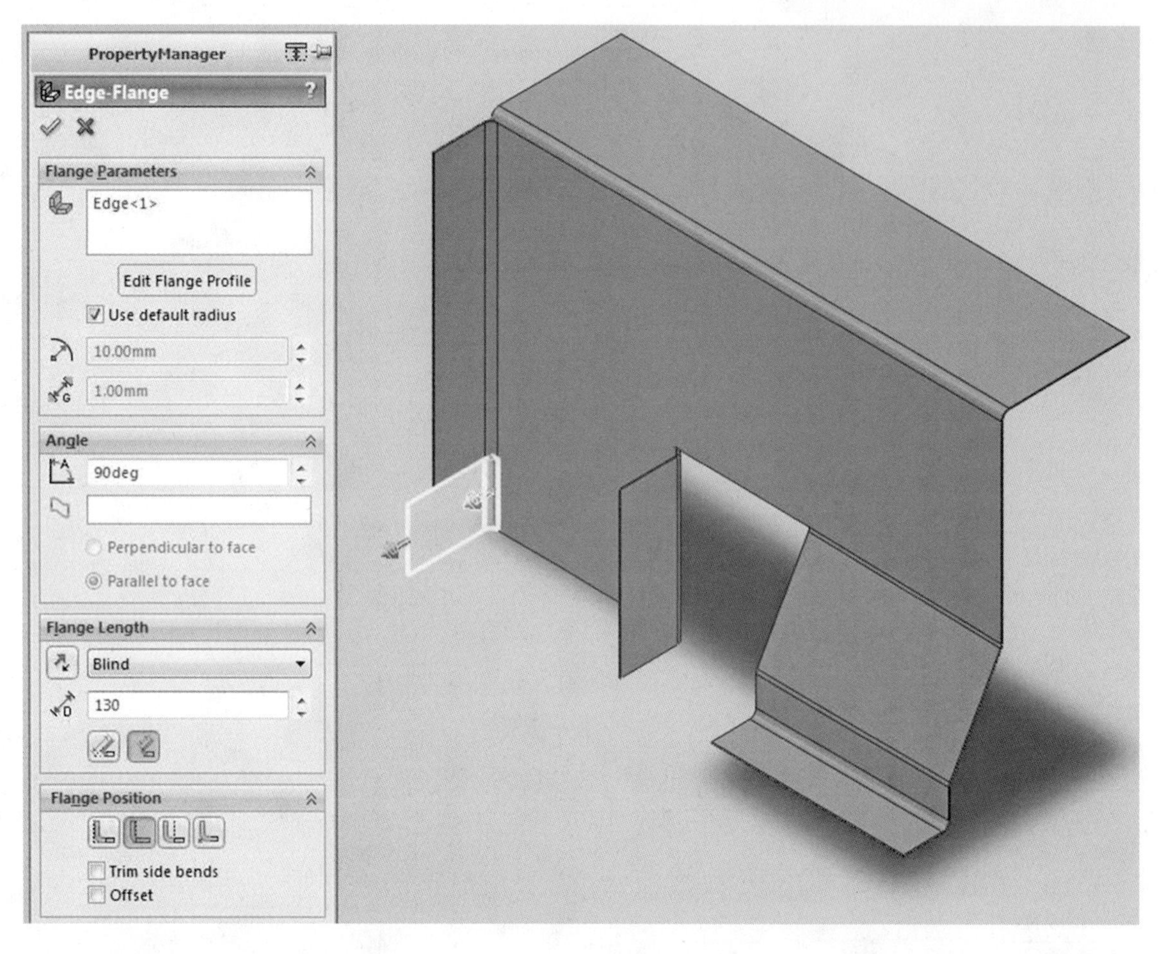

Figure 19.121

Third Edge Flange.

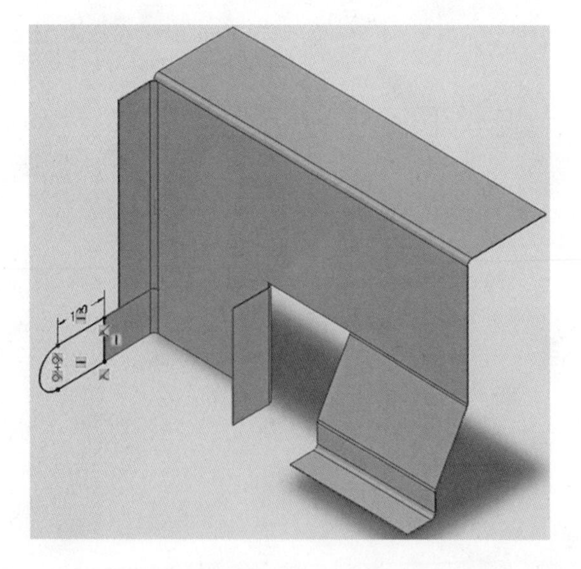

Figure 19.122

Sketch for first Tab.

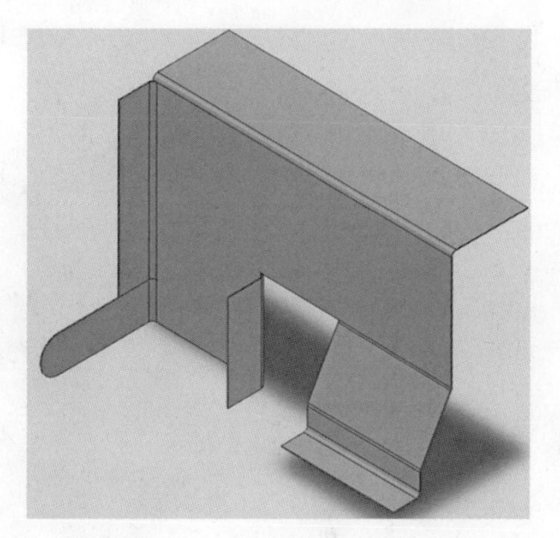

Figure 19.123

First Tab.

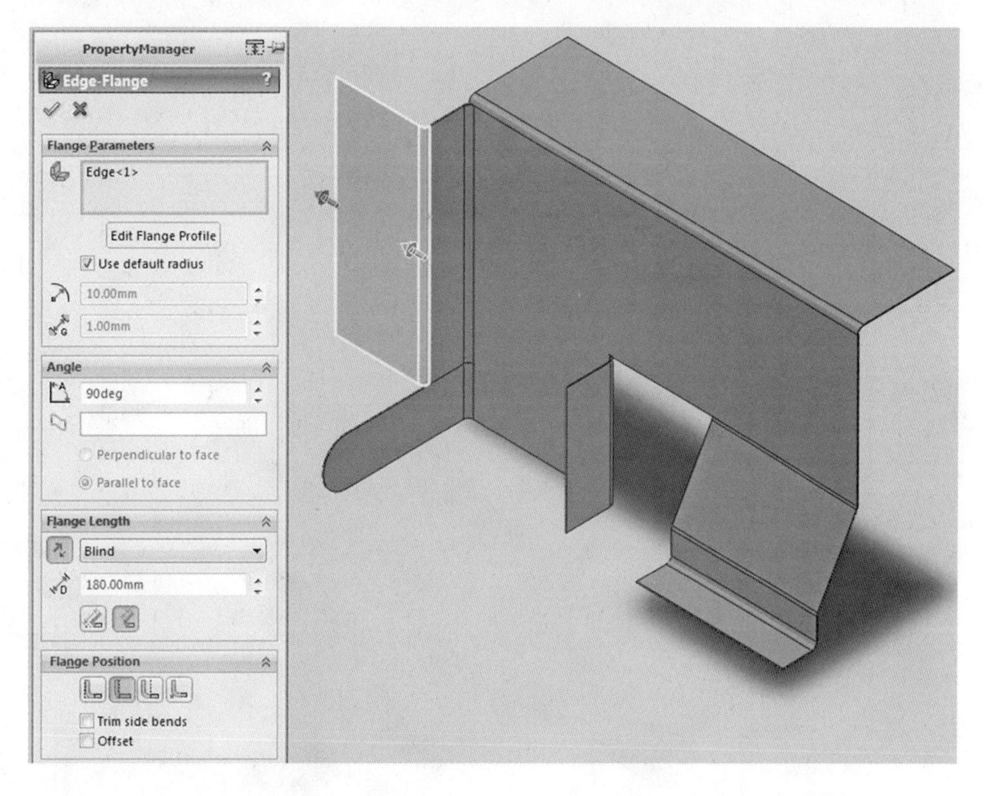

Figure 19.124

Edge Flange PropertyManager.

10. Click OK if the following message appears: "*The automatic relations inferred by this operation would over define the sketch. They will not be added.*"
11. Adjust the Width of the Edge Flange created by dragging one endpoint of the edge to a new position (16 mm) and using sketch tool to add dimensions (see Figure 19.134). (Note: *When the Profile Sketch window appears, do not click Finish until the length of the Edge Flange is fully adjusted.*)
12. Click Finish.

Unfold the First Edge Flange

13. Click the Unfold tool.
14. Click the Fixed Face (Face<1>) and Bends to Unfold (EdgeBend1) (see Figure 19.135).
15. Click OK to unfold the edge.

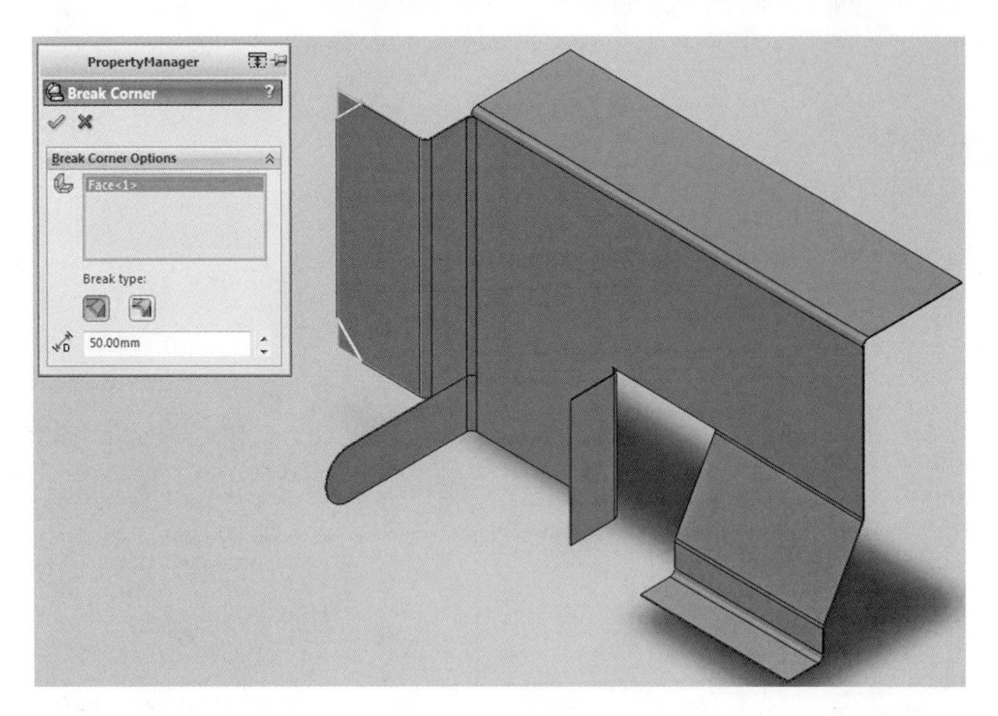

Figure 19.125

Break Corner PropertyManager.

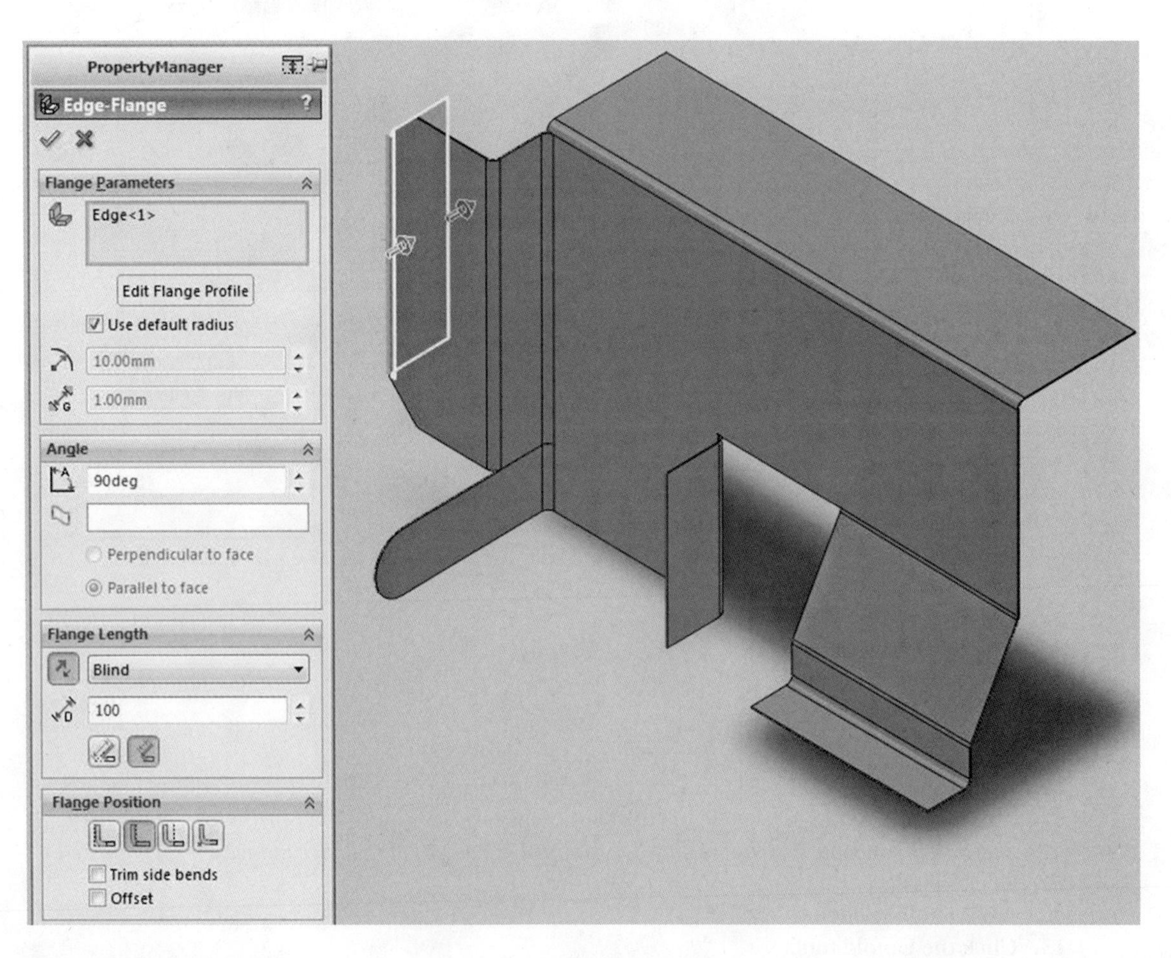

Figure 19.126

Edge Flange PropertyManager.

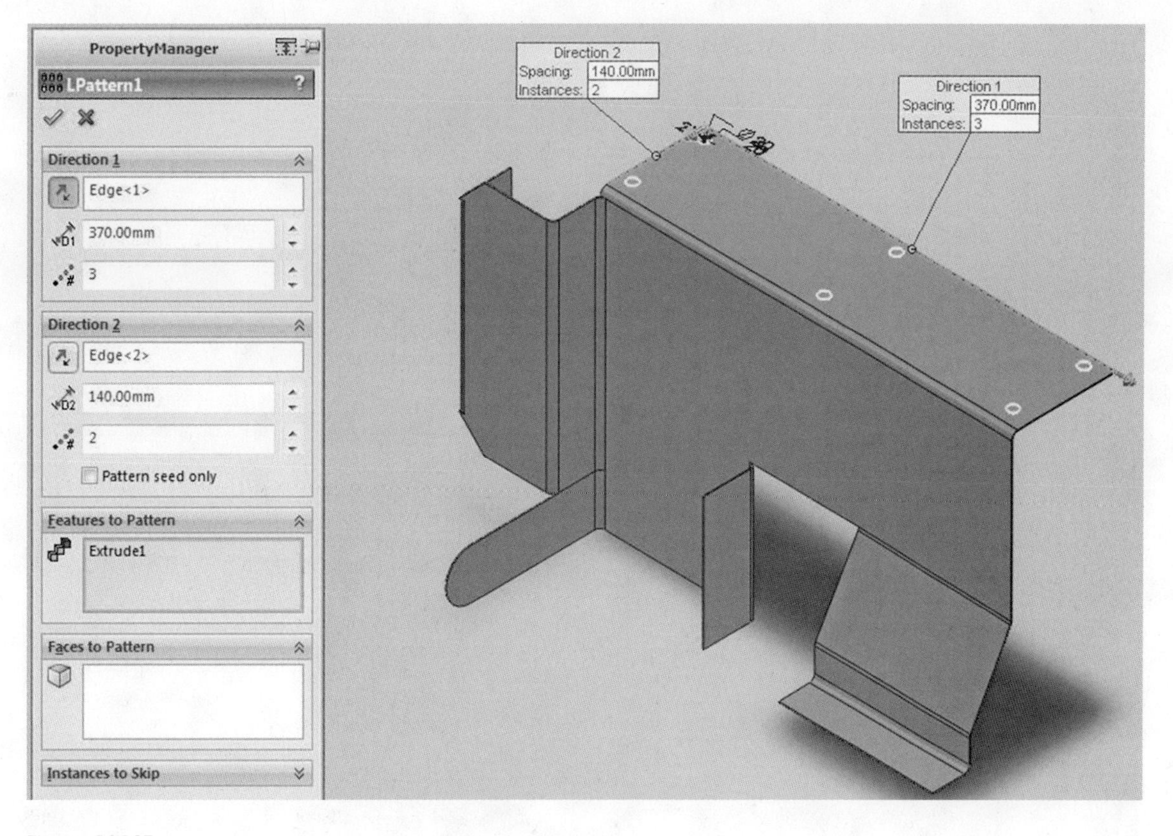

Figure 19.127

Linear Pattern PropertyManager for replicating holes.

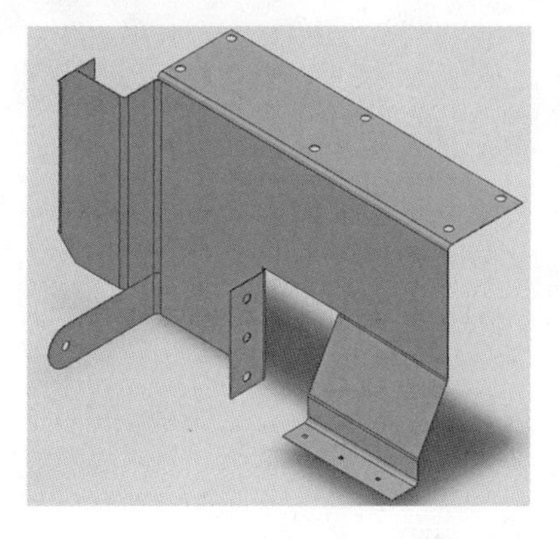

Figure 19.128

Holes created in other features.

Create a Hole along First Edge Flange

16. Click top of the Base Flange and be in sketch mode.
17. Sketch a circle 5 mm in diameter midway of the unfolded bent feature (see Figure 19.136).
18. Extrude-Cut the Circle Up To Next (see Figure 19.136).

Fold the First Edge Flange

1. Click the Fold tool.
2. Click the Fixed Face (Face<1>) and Bends to fold (EdgeBend1) (see Figure 19.137).
3. Click OK to fold the edge.

Inserting Hem Feature on the Right Edge Flange

1. Click the top of the Edge Flange (Edge-Flange1).
2. Click Hem from the Sheet Metal toolbar (see Figure 19.138).

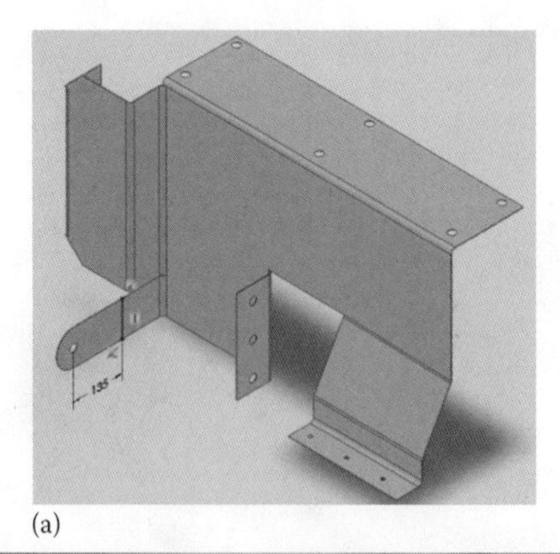

(a)

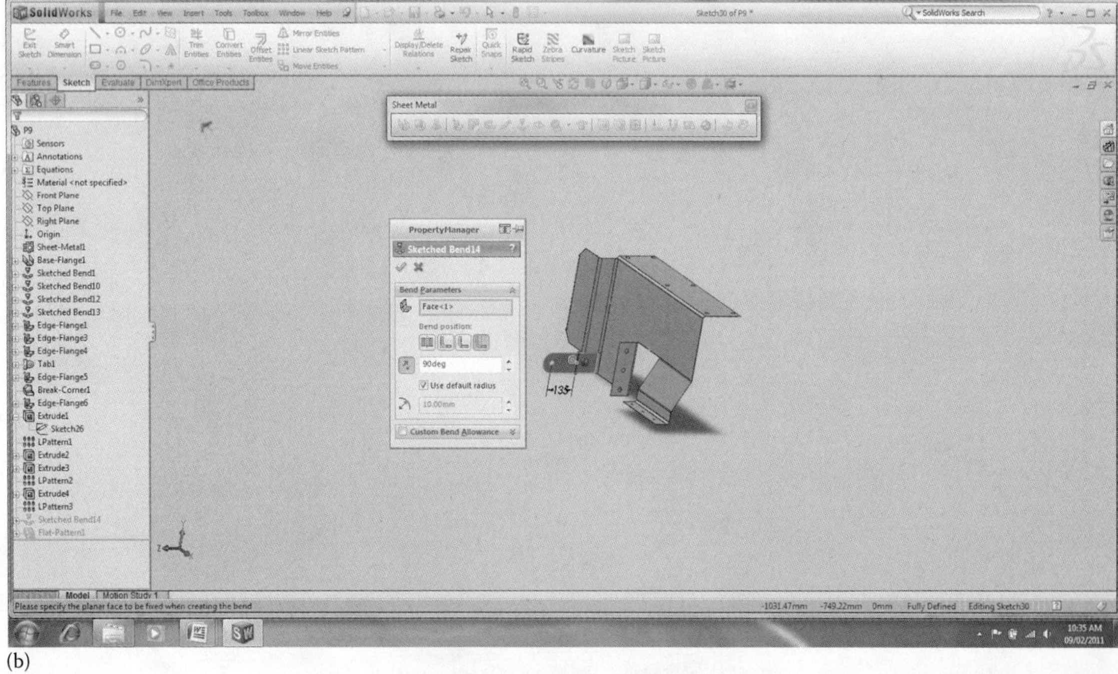

(b)

Figure 19.129

Sketch and Sketched Bend PropertyManager for second bend: (a) Sketch for second bend and (b) Sketched Bend PropertyManager.

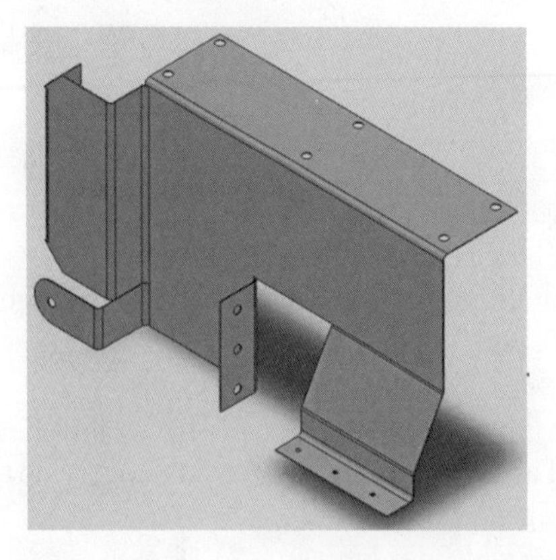

Figure 19.130

Sheet metal part completed.

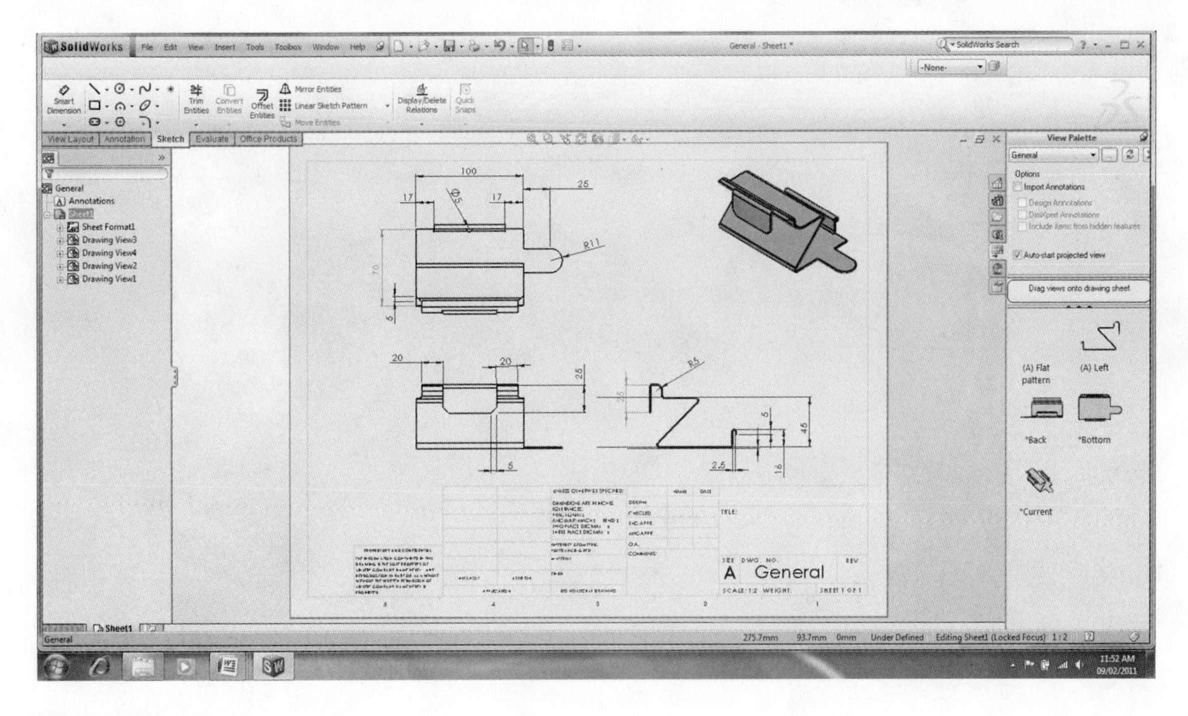

Figure 19.131

Part and drawing views and dimensions for Tutorial 19.8.

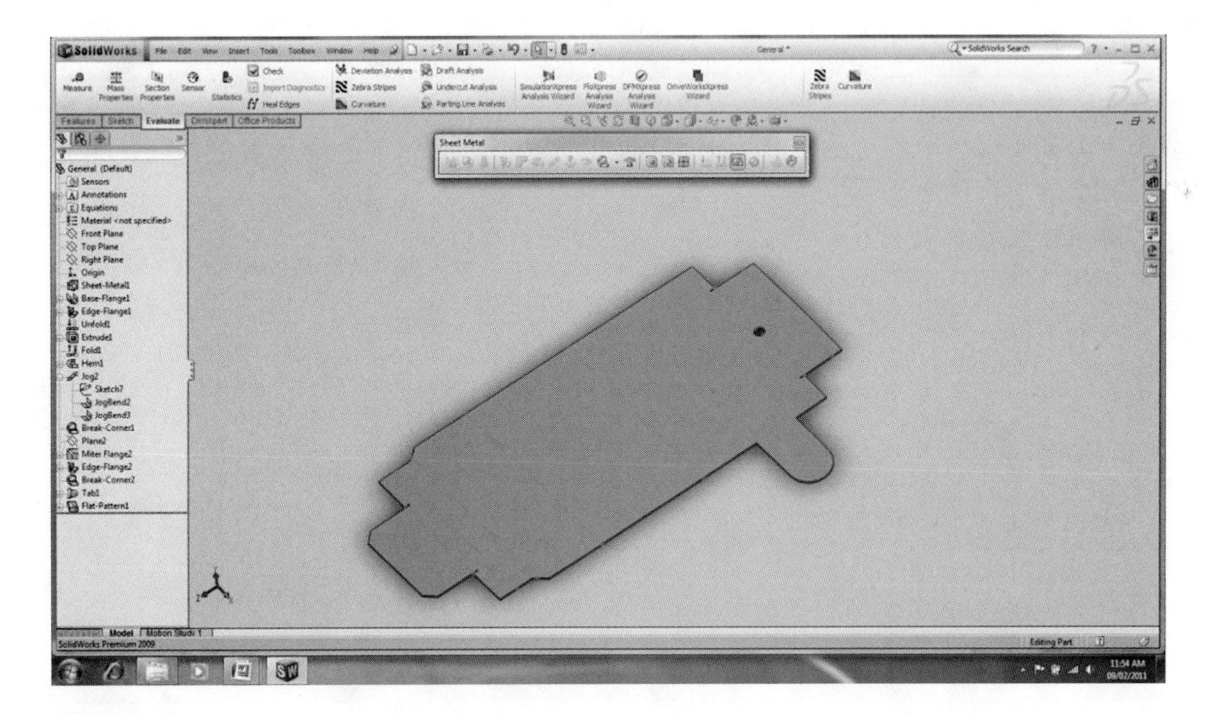

Figure 19.132

Flat pattern of the sheet metal part for Tutorial 19.8.

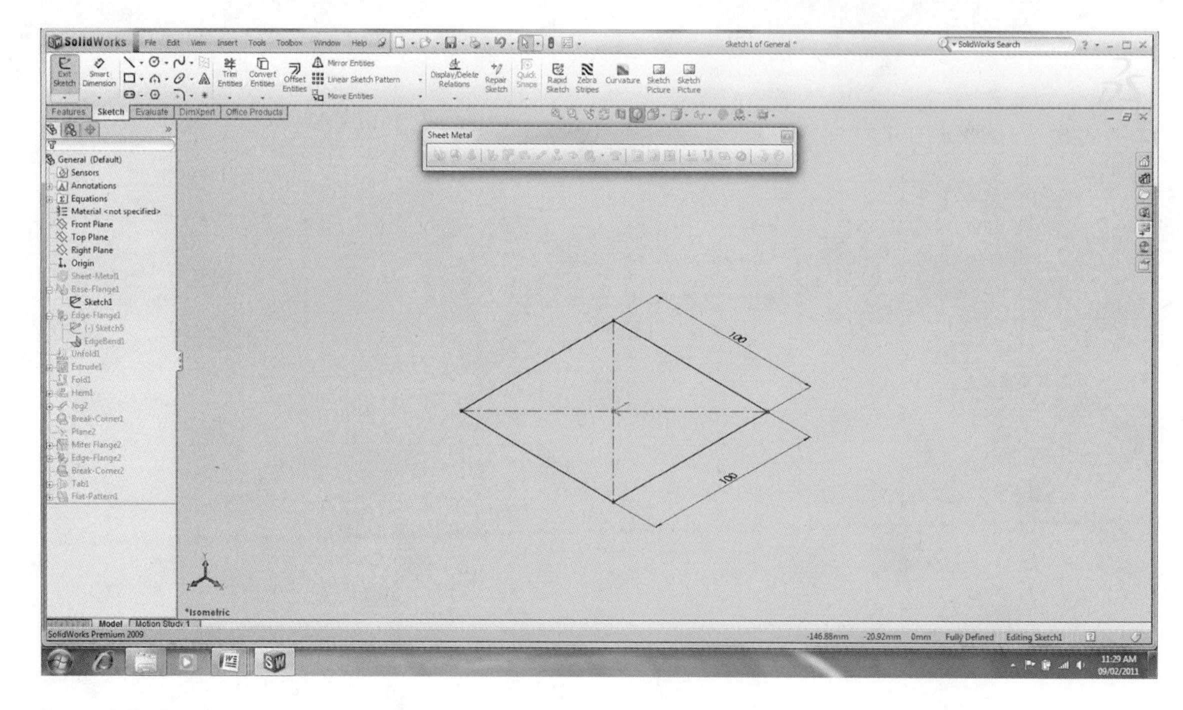

Figure 19.133

Sketch for Base Flange.

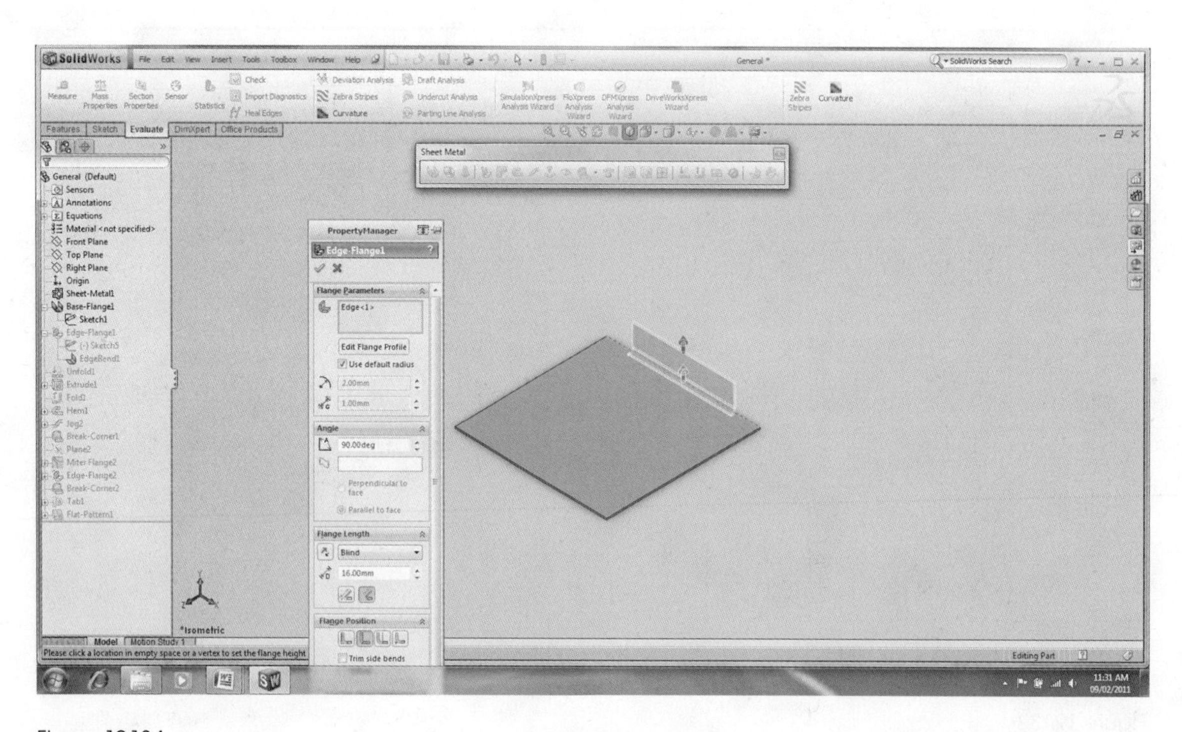

Figure 19.134

Edge Flange PropertyManager.

Figure 19.135

Unfold PropertyManager.

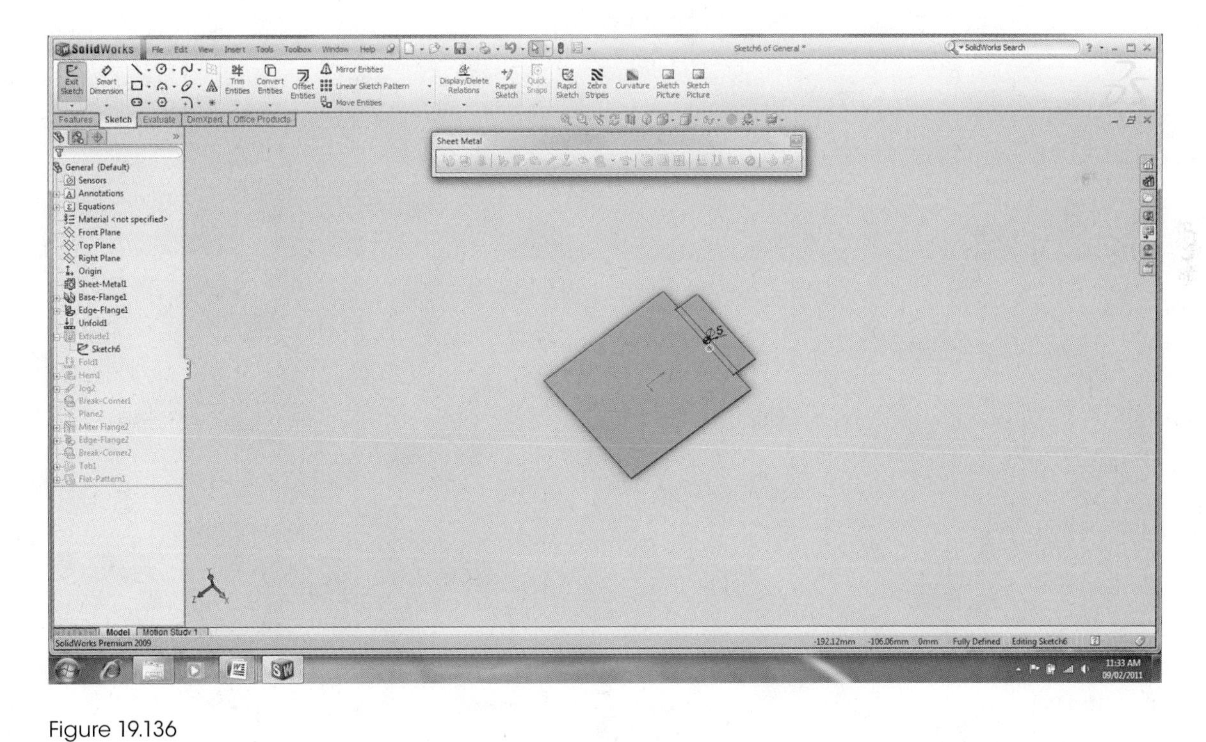

Figure 19.136

Create a hole midway of unfolded bent feature.

Figure 19.137

Unfold PropertyManager.

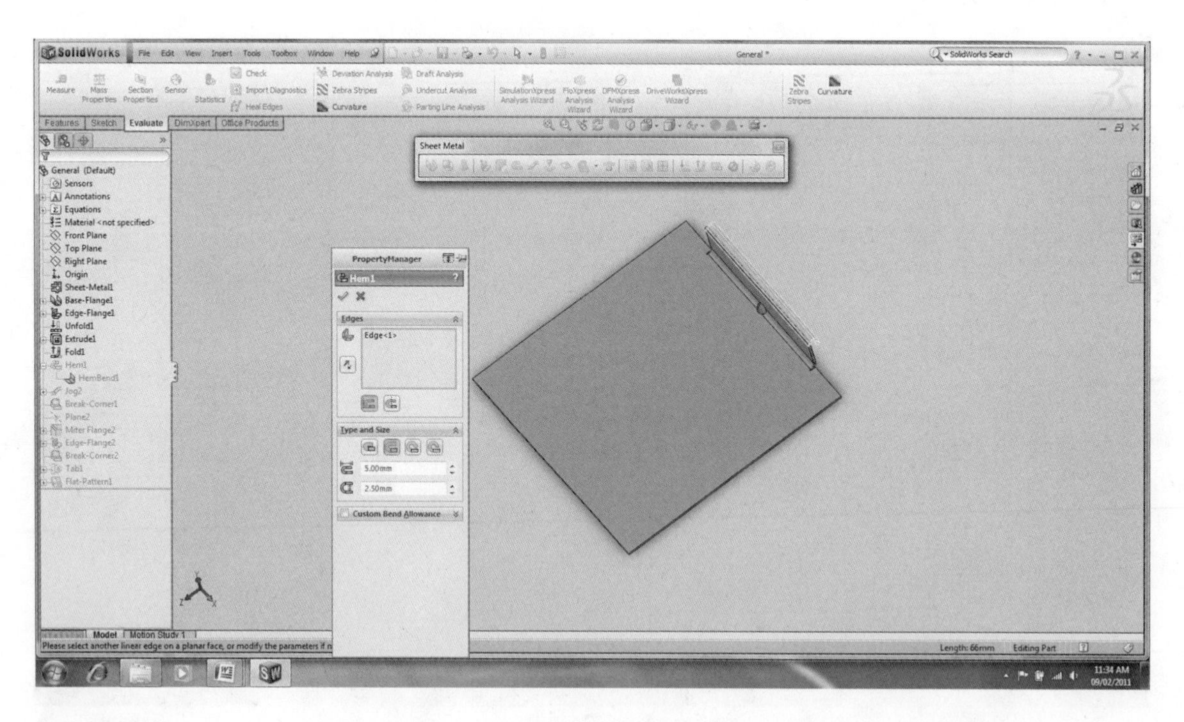

Figure 19.138

Hem PropertyManager.

3. Select Material Inside.
4. Click the Reverse Direction button. (Zoom to see the effects of choice.)
5. Accept default (Open) or select as required from the Type and Size.
6. Enter 5 mm for Length.
7. Enter 2.5 mm for Gap Distance.
   (Observe the preview.)
8. Click OK.

Create a Jog

9. Click the top face of the Base Flange, be in sketch mode and sketch a Line 30 mm from the edge for the Jog Feature (see Figure 19.139).
10. Click the Jog tool (see Figure 19.140).
11. Click the Fixed Face on the right side of the sketch on the Base Flange.
12. Set Jog Offset equal to 45.
13. Set Jog Angle equal to 135.
14. Set OK to finish.

Create Break Corner

15. Select the outer face (Face<1>) of the fourth edge flange (see Figure 19.141).
16. Set Distance = 5 (see Figure 19.141).
17. Click OK to finish the operation.

Create Sketch for Miter Flange

18. Click one of the edges of the Base Flange and then select the Line tool from the Sketch toolbar (see Figure 19.142). (Notice that a perpendicular plane for sketching the line is automatically selected; this is very useful in this approach.)
19. Create an L-sketch, with a fillet radius of 5 mm (see Figure 19.142).
20. Exit sketch.

Create Miter Flange Features

21. Select Sketch2.
22. Click the Miter Flange button on the Sheet Metal toolbar, while the sketch is still active.
23. Select the Jog edge, at the top, as shown in Figure 19.143.
24. Click OK from the Miter Flange Property toolbar.

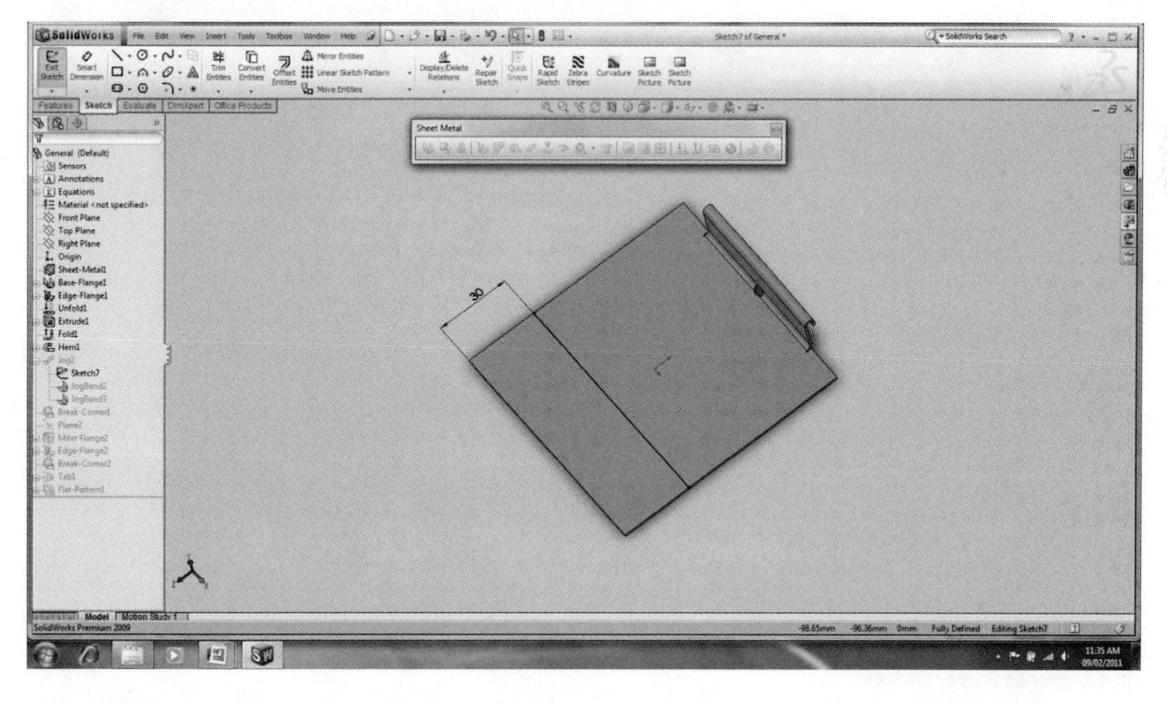

Figure 19.139

Jog Feature.

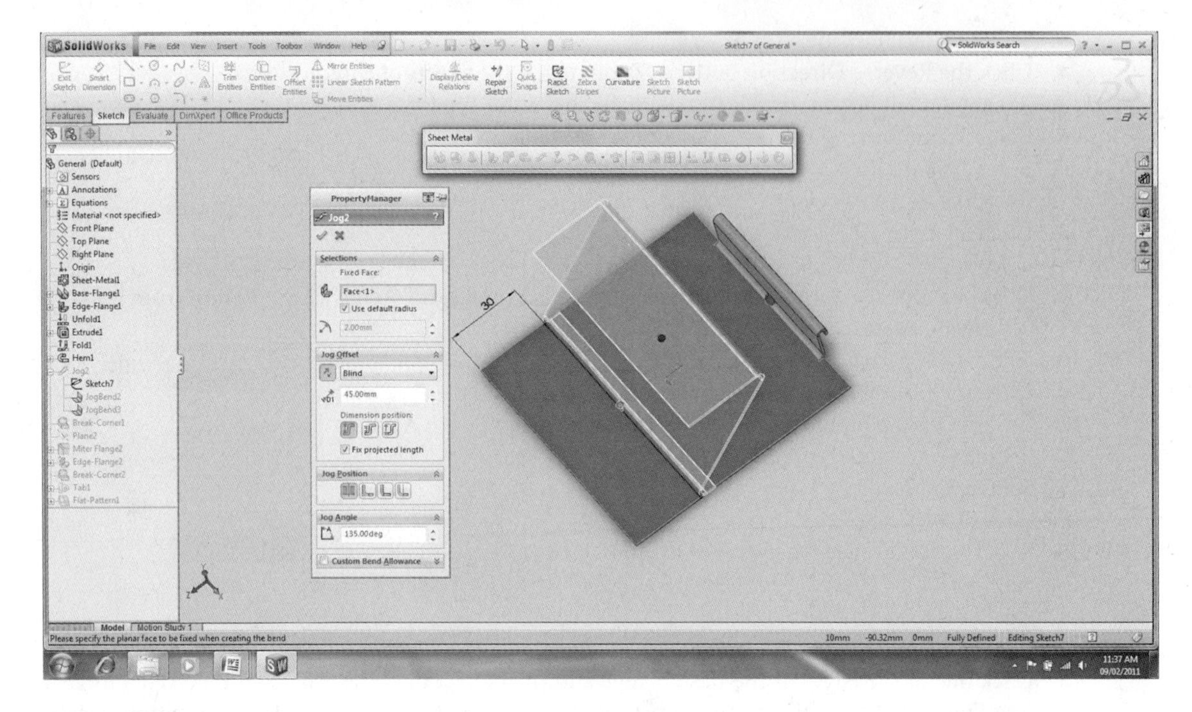

Figure 19.140

Jog PropertyManager.

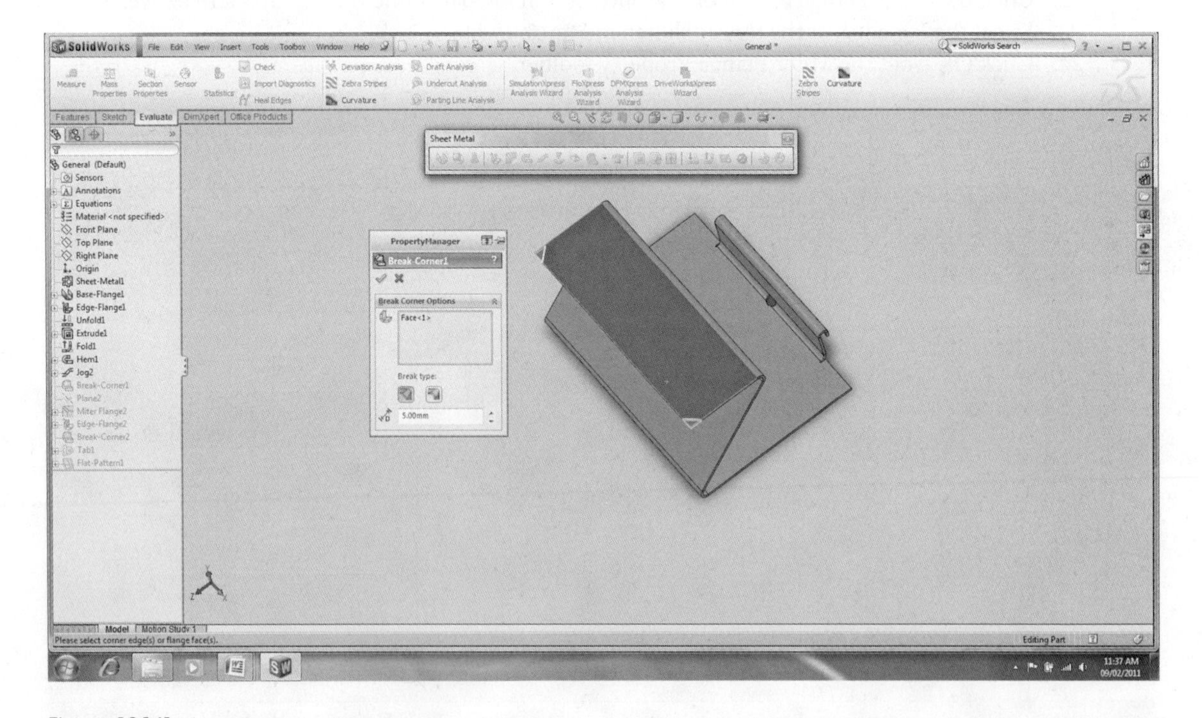

Figure 19.141

Break Corner PropertyManager.

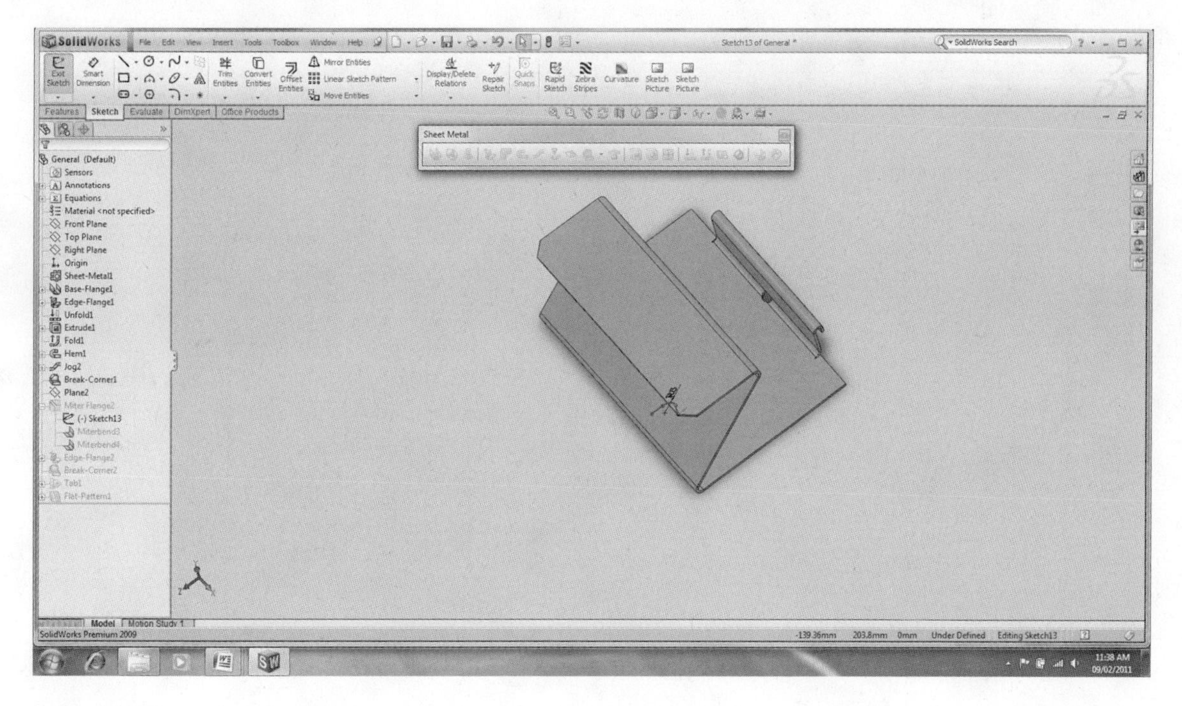

Figure 19.142

Sketch for Miter Flange.

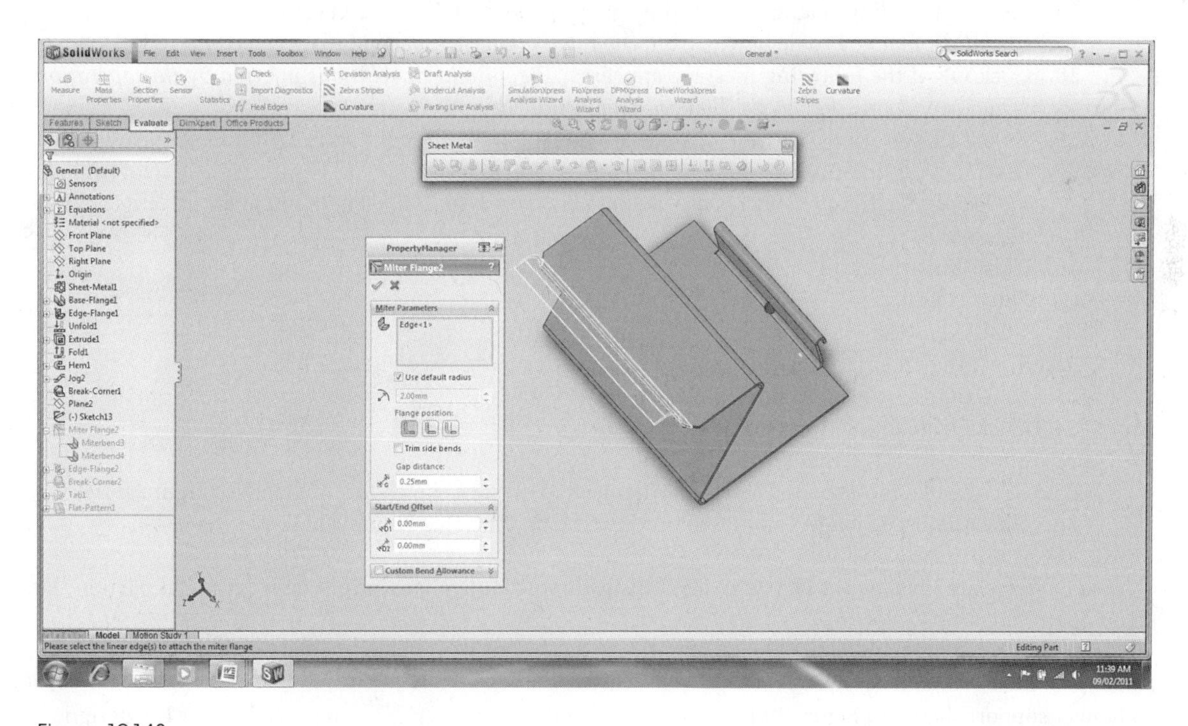

Figure 19.143

Miter Flange PropertyManager.

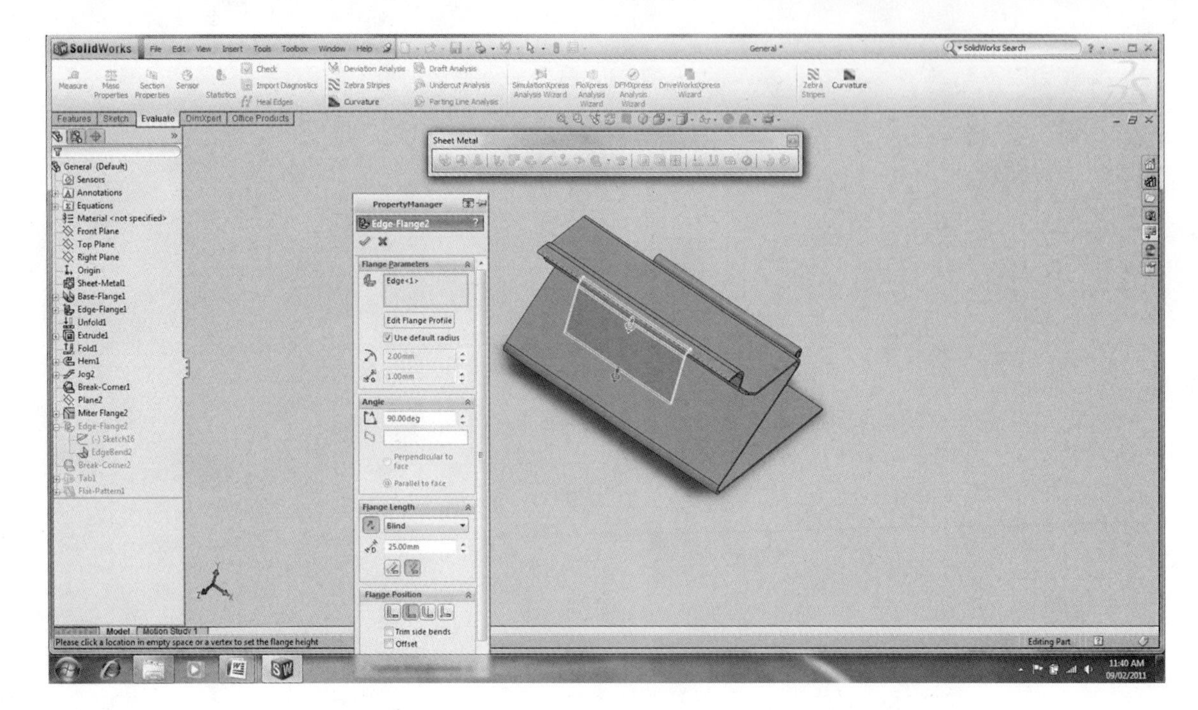

Figure 19.144

Edge Flange PropertyManager.

Create the Second Edge Flange

25. Select the right edge of the first Sketched Bend (see Figure 19.144).
26. Set Length of Edge Flange = 25 and Angle = 90-deg (see Figure 19.144).
27. Click Edit Flange Profile button from the Edge Flange PropertyManager (see Figure 19.144).
28. Click OK if the following message appears: "*The automatic relations inferred by this operation would over define the sketch. They will not be added.*"
29. Adjust the Width of the Edge Flange that is created by dragging one endpoint of the edge to a new position (16 mm) and using sketch tool to add dimensions (see Figure 19.145). (Note: *When the Profile Sketch window appears, do not click Finish until the length of the Edge Flange is fully adjusted.*)
30. Click Finish.

Second Break Corners

31. Click the Break Corner tool.
32. Select the outer face (Face<1>) of the fourth edge flange (see Figure 19.146).
33. Set Distance = 5 (see Figure 19.146).
34. Click OK to finish the operation.

First Tab

35. Be in sketch mode on the third edge flange and create a sketch for first tab (see Figure 19.147).
36. Click the Base Flange/Tab tool to produce the first tab (see Figure 19.148).

The Flatten state is shown in Figure 19.149. This is important for fabricating the sheet metal part on the shop floor.

Tutorial 19.9: Hanger

A hanger support shown in Figure 19.150 is to be modeled. Show a step-by-step procedure for the design.

SolidWorks Sheet Metal Procedure

1. Start a New Part document.
2. Select the Top view.
3. Sketch the base profile, Sketch1 (see Figure 19.151).

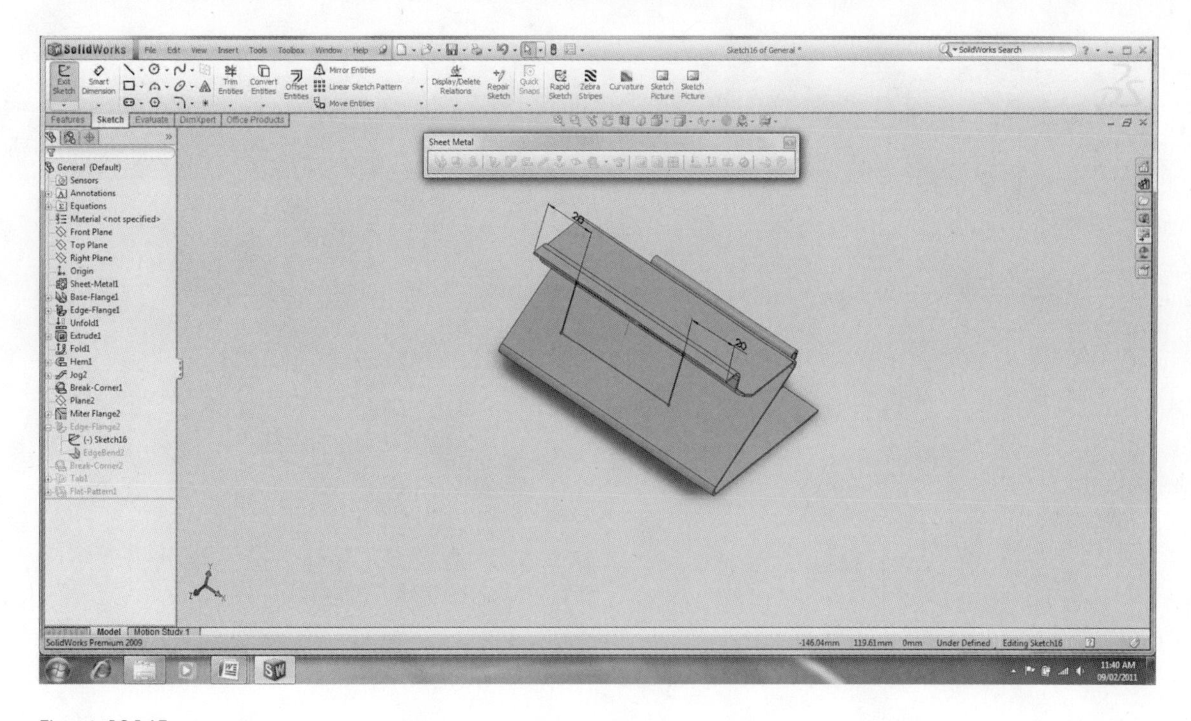

Figure 19.145

Width adjustment for the Edge Flange.

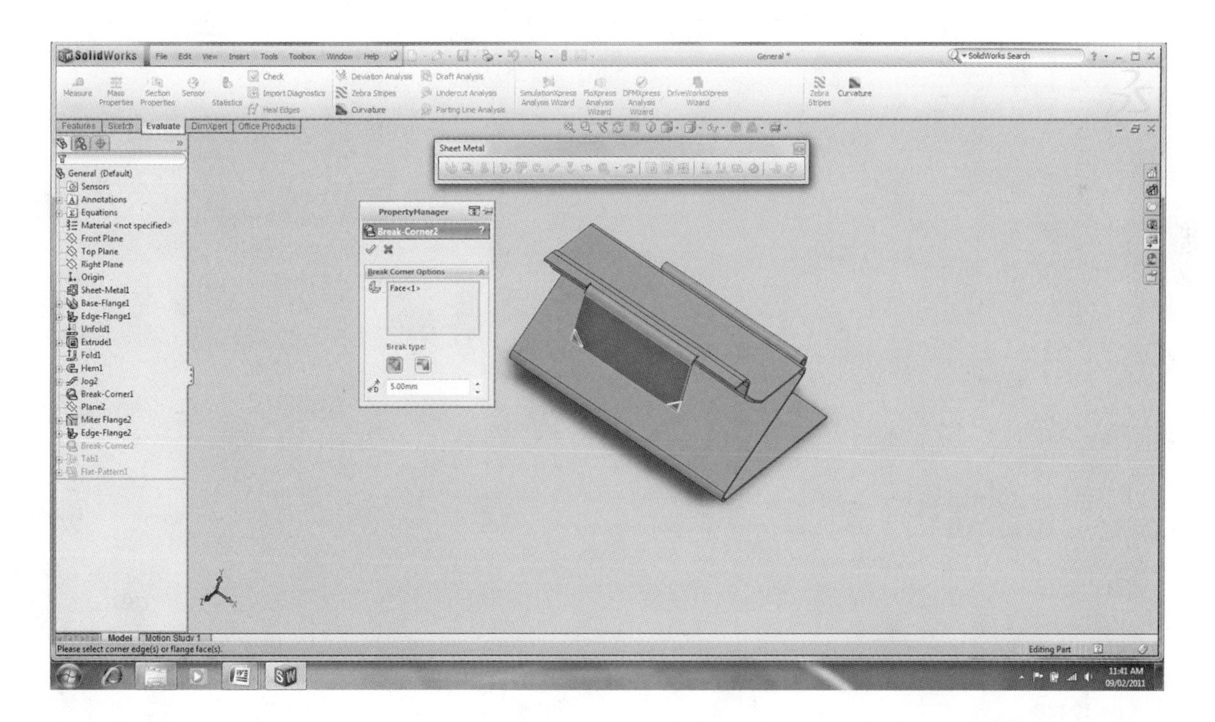

Figure 19.146

Break Corner PropertyManager.

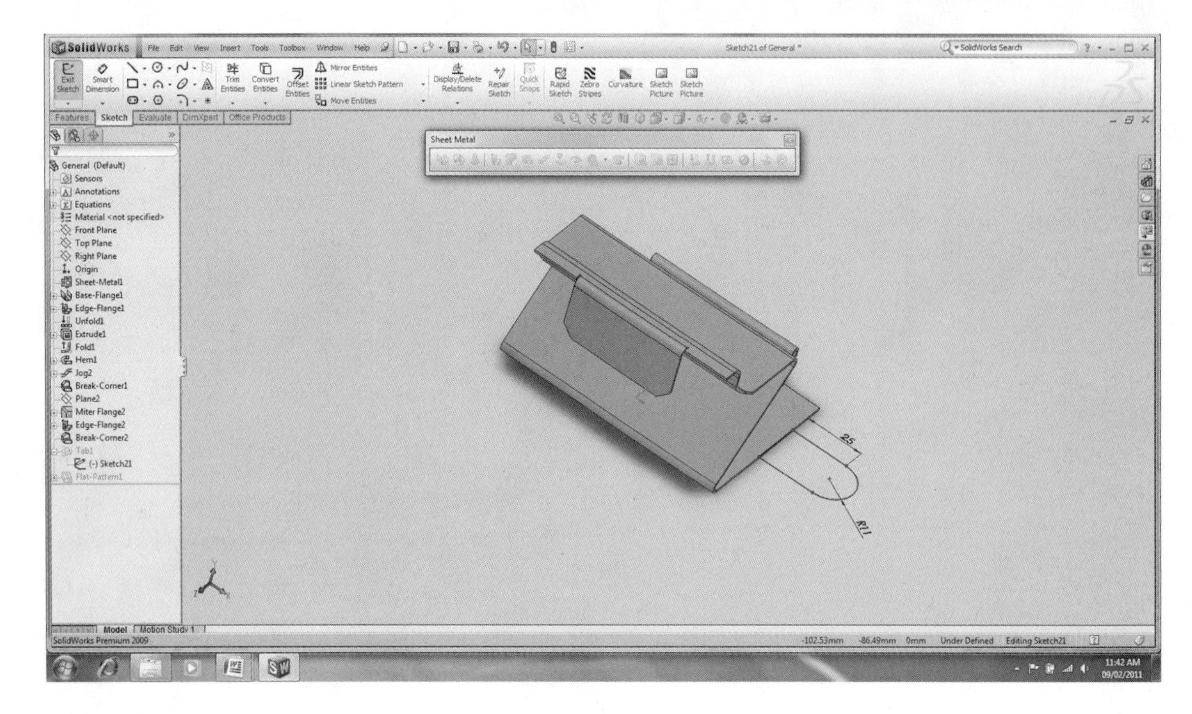

Figure 19.147

Sketch for creating tab feature.

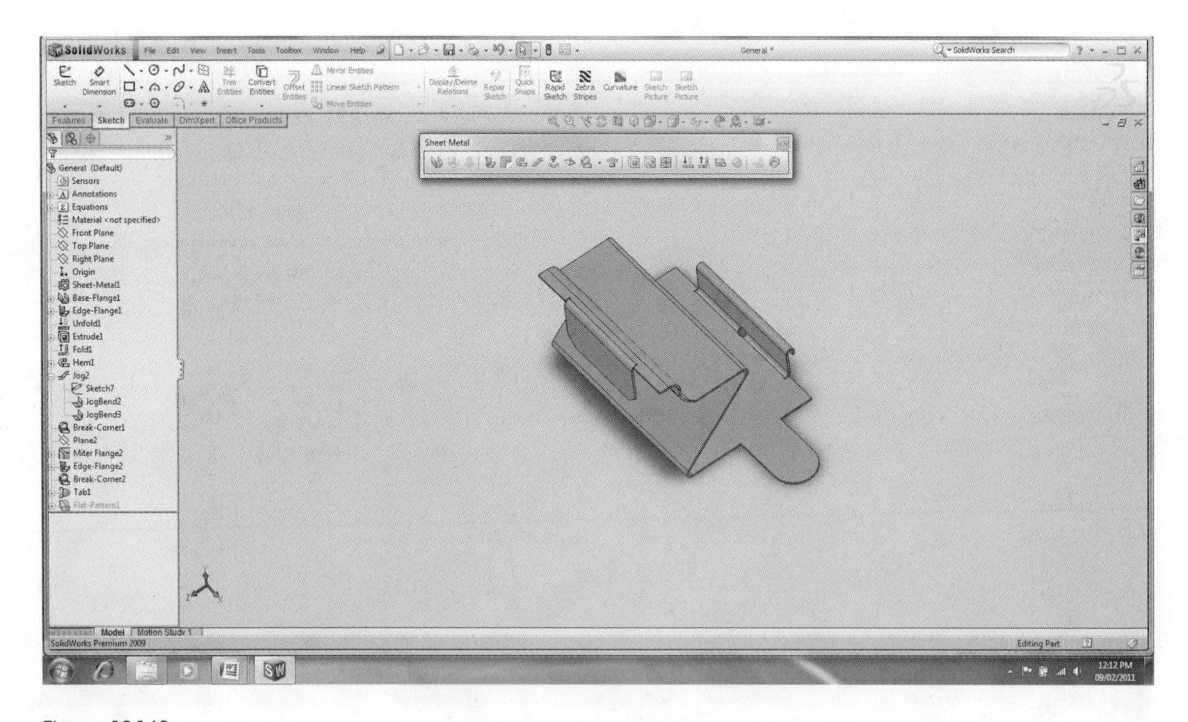

Figure 19.148

Sheet metal part.

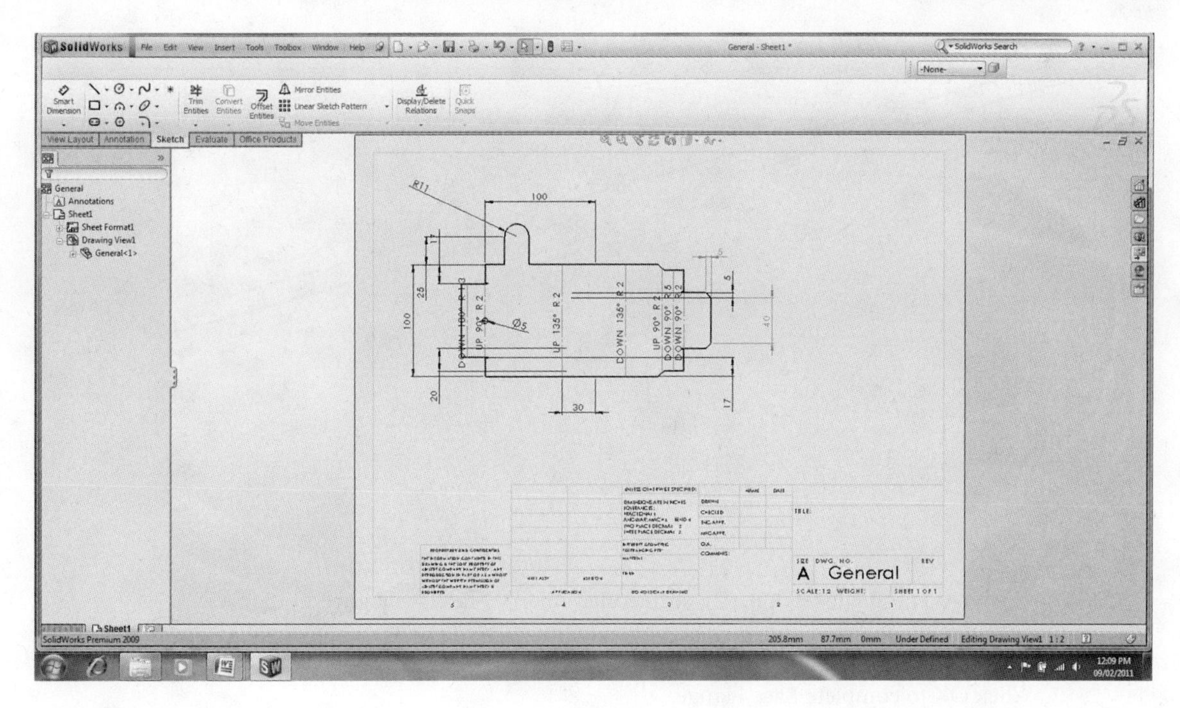

Figure 19.149

Flatten state.

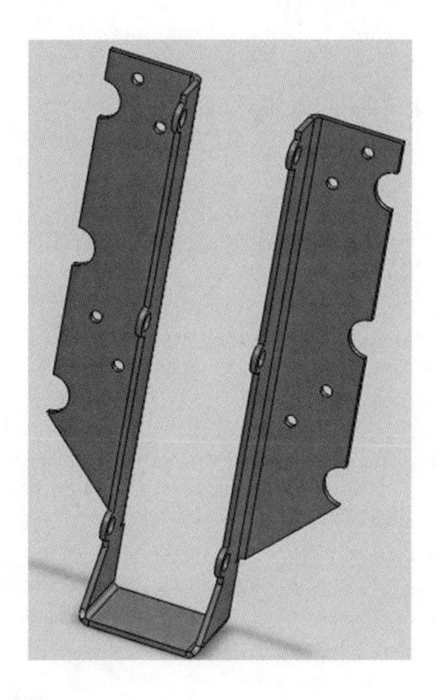

Figure 19.150

Sheet metal model.

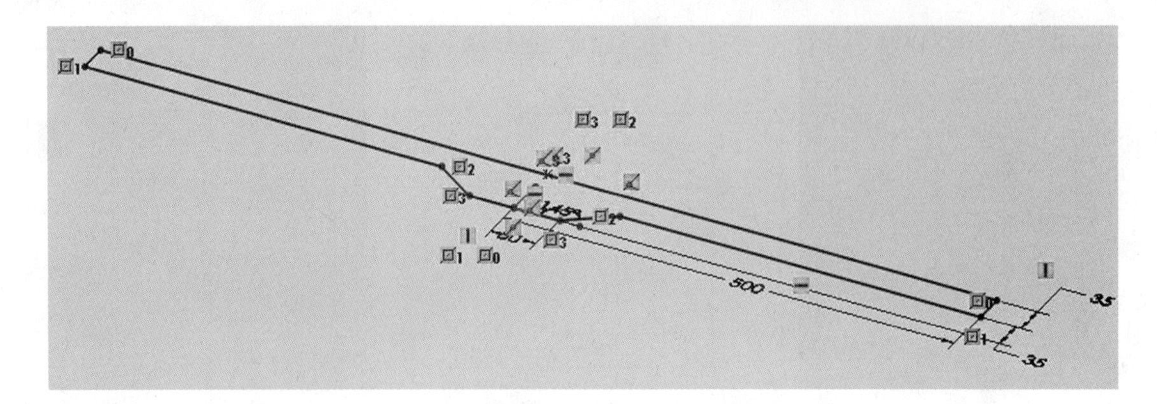

Figure 19.151

Sketch1.

Base Flange

4. With base profile (Sketch1) active, click Base Flange/Tab (see Figure 19.152).
5. Set the Thickness as 5 mm (see Figure 19.152).
6. Accept the K-Factor and Ratio default values (see Figure 19.152).
7. Click OK to complete Base Flange.

First Sketched Bending Operation

8. Sketch a line, Sketch2, on the Base Flange for the first bending operation. (See Figure 19.153 for Sketched Bend PropertyManager.)
9. Click the Sketched Bend tool from the SheetMetal CommandManager.
10. Click the Fixed Face (Face<1>) *to the left of the line* (Sketch2) in the Bend Parameters rollout (see Figure 19.153).
11. Click OK to complete the sketched bend operation. (See Figure 19.150 for Sketched Bend.)

Second Sketched Bending Operation

12. Sketch another line, Sketch2, on the Base Flange for the second bending operation. (See Figure 19.154 for Sketched Bend PropertyManager.)
13. Click the Sketched Bend tool from the SheetMetal CommandManager.
14. Click the Fixed Face (Face<1>) *to the right of the line* (Sketch2) in the Bend Parameters rollout (see Figure 19.154).
15. Click OK to complete the sketched bend operation. (See Figure 19.154 for Sketched Bend.)

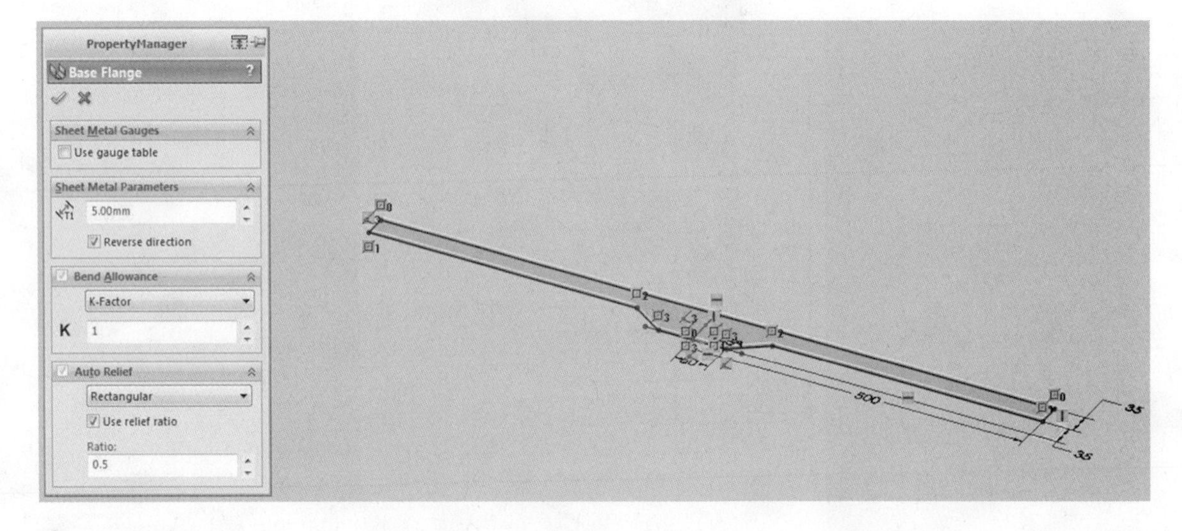

Figure 19.152

Base Flange PropertyManager.

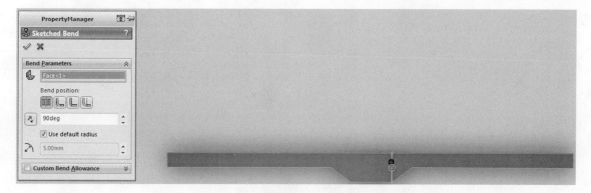

Figure 19.153

Sketched Bend PropertyManager.

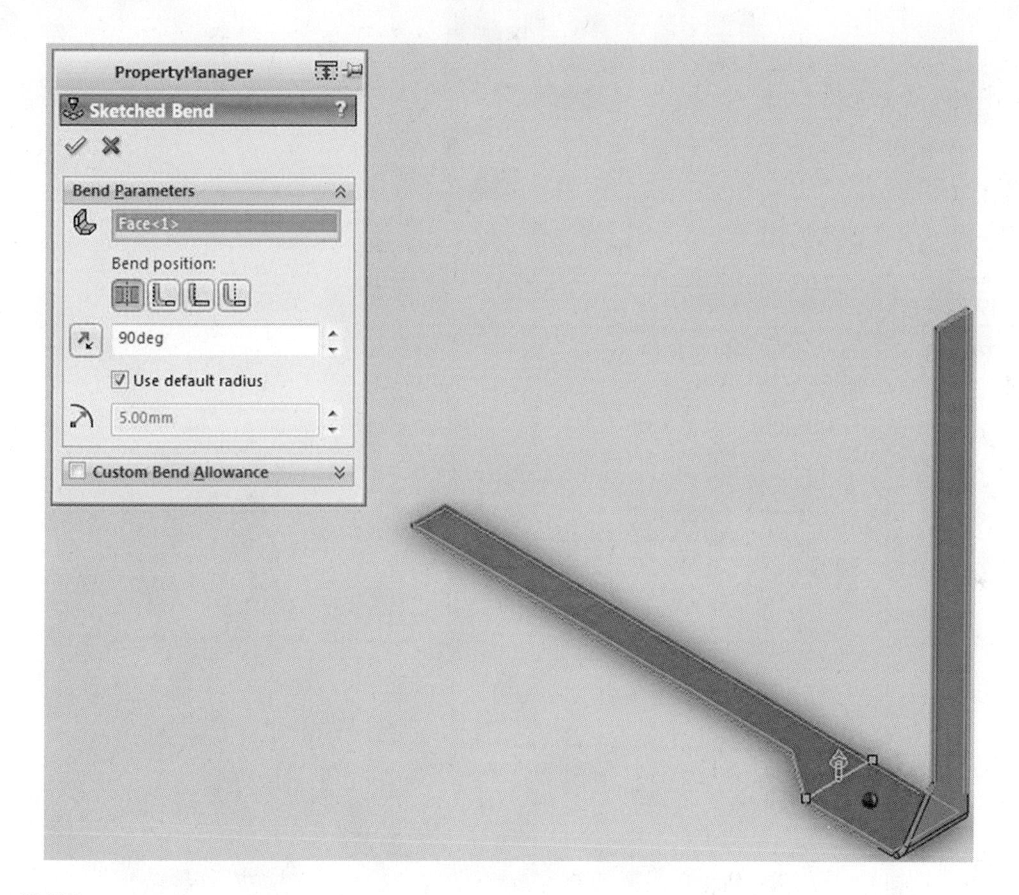

Figure 19.154

Sketched Bend PropertyManager.

Create Circle on the Right Vertical Face

16. Click the right vertical face of the model and use Normal To tool to position it appropriately (see Figure 19.155).
17. Sketch a Circle with diameter of 25 mm on the right vertical face (see Figure 19.155).
18. Use Smart Dimension tool to dimension the diameter as 25 mm, and the center of the circle to be 20 mm from the angular vertex aligned with the edge (see Figure 19.155).

Create an Extrude Feature

19. Click Feature > Extrude. (See Figure 19.156 for the Extrude PropertyManager.)
20. In the Direction1 rollout, select Up To Vertex. (See Figure 19.156 for Vertex at the top.)
21. Click OK to complete this extrusion.

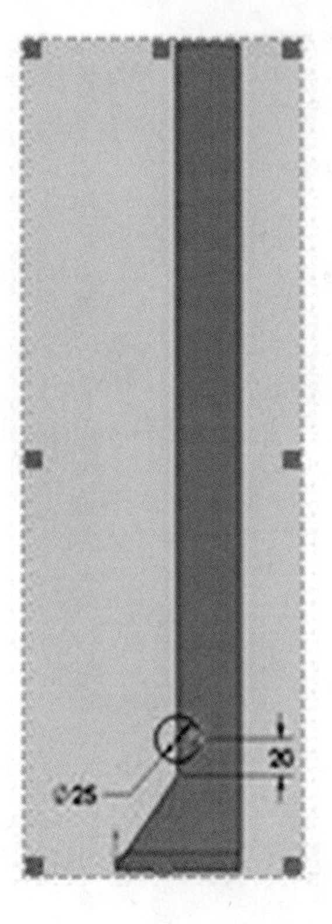

Figure 19.155

Circle sketched on right vertical face of model.

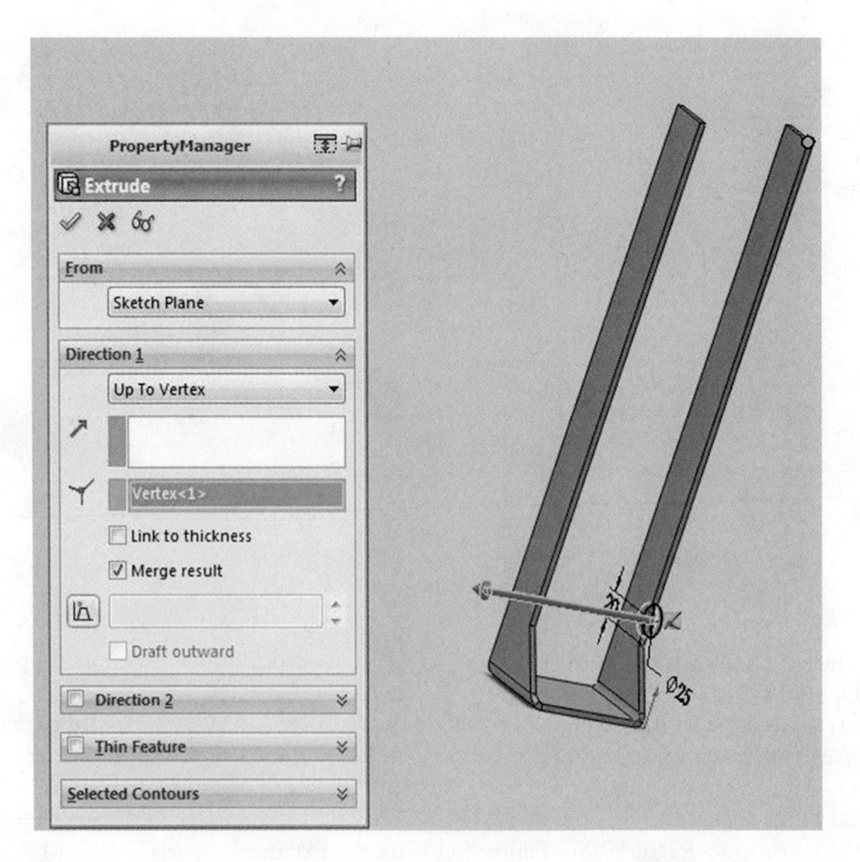

Figure 19.156

Extrude PropertyManager.

Create Another Circle on the Right Vertical Face

22. Sketch another concentric Circle with a diameter of 15 mm (see Figure 19.157).
23. Use the Smart Dimension tool to dimension the diameter as 15 mm (see Figure 19.157).

Create an Extrude Cut Feature

24. Click Feature > Extrude Cut. (See Figure 19.157 for the Extrude PropertyManager.)
25. In the Direction1 rollout, select Up To Vertex. (See Figure 19.157 for Vertex at the top.)
26. Click OK to complete this extrusion.

Linear Pattern

27. Click Features > Linear Pattern. (See Figure 19.158 for the Linear Pattern PropertyManager.)
28. In the Direction1 rollout, click the Edge of the right vertical side of the model (see Figure 19.158).
29. For the Distance between patterns, set it to be 180 mm (see Figure 19.158).
30. For the Number of Instances, set it to be 3 (see Figure 19.158).
31. For the Features To Pattern select Extrude3 and Extrude4 (see Figure 19.158).
32. Click OK to complete the patterns (see Figure 19.158).

Mirror

33. Click Features > Mirror. (See Figure 19.159 for the Mirror PropertyManager.)
34. For the Mirror/Face Plane, select the Right Plane (see Figure 19.159).
35. For the Features Mirror select LPattern2 (see Figure 19.159).
36. Click OK to complete the mirror operation (see Figure 19.159).

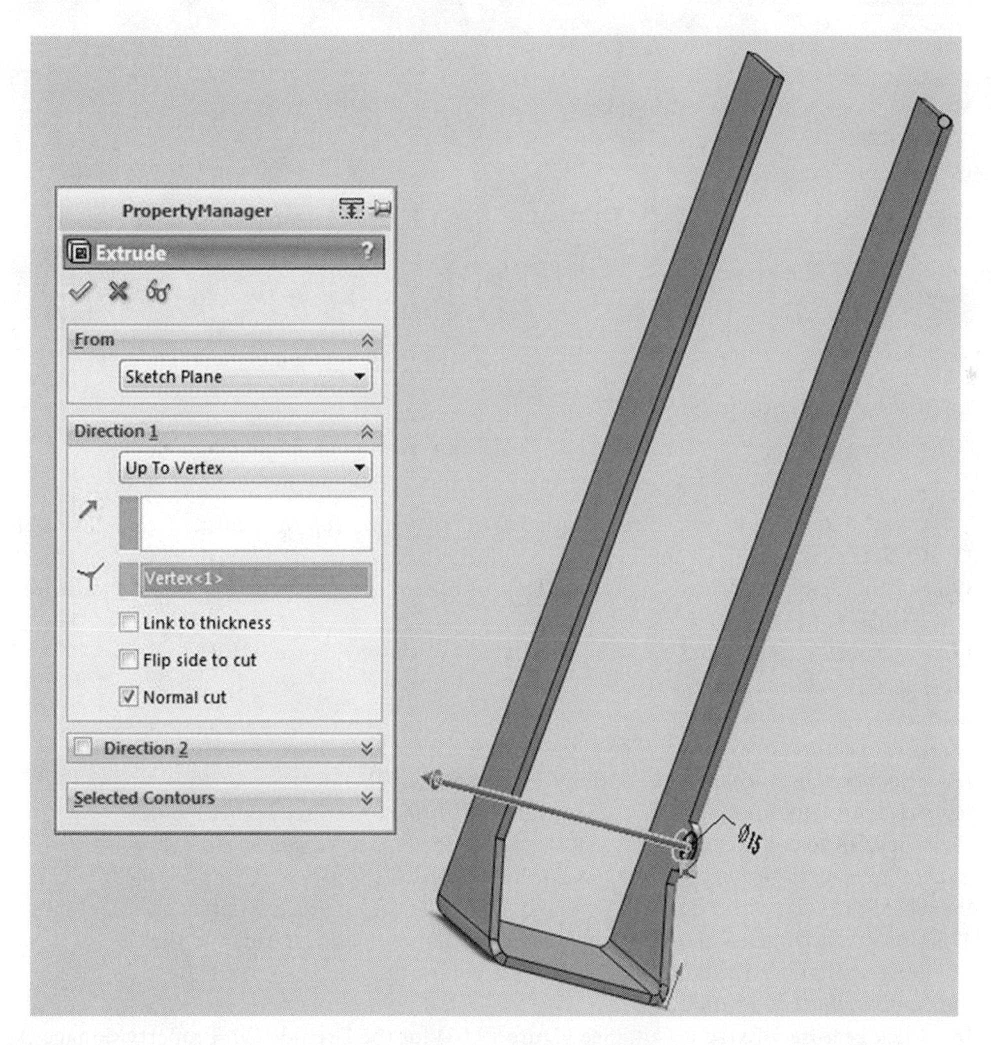

Figure 19.157

Circle created and Extrude Cut PropertyManager.

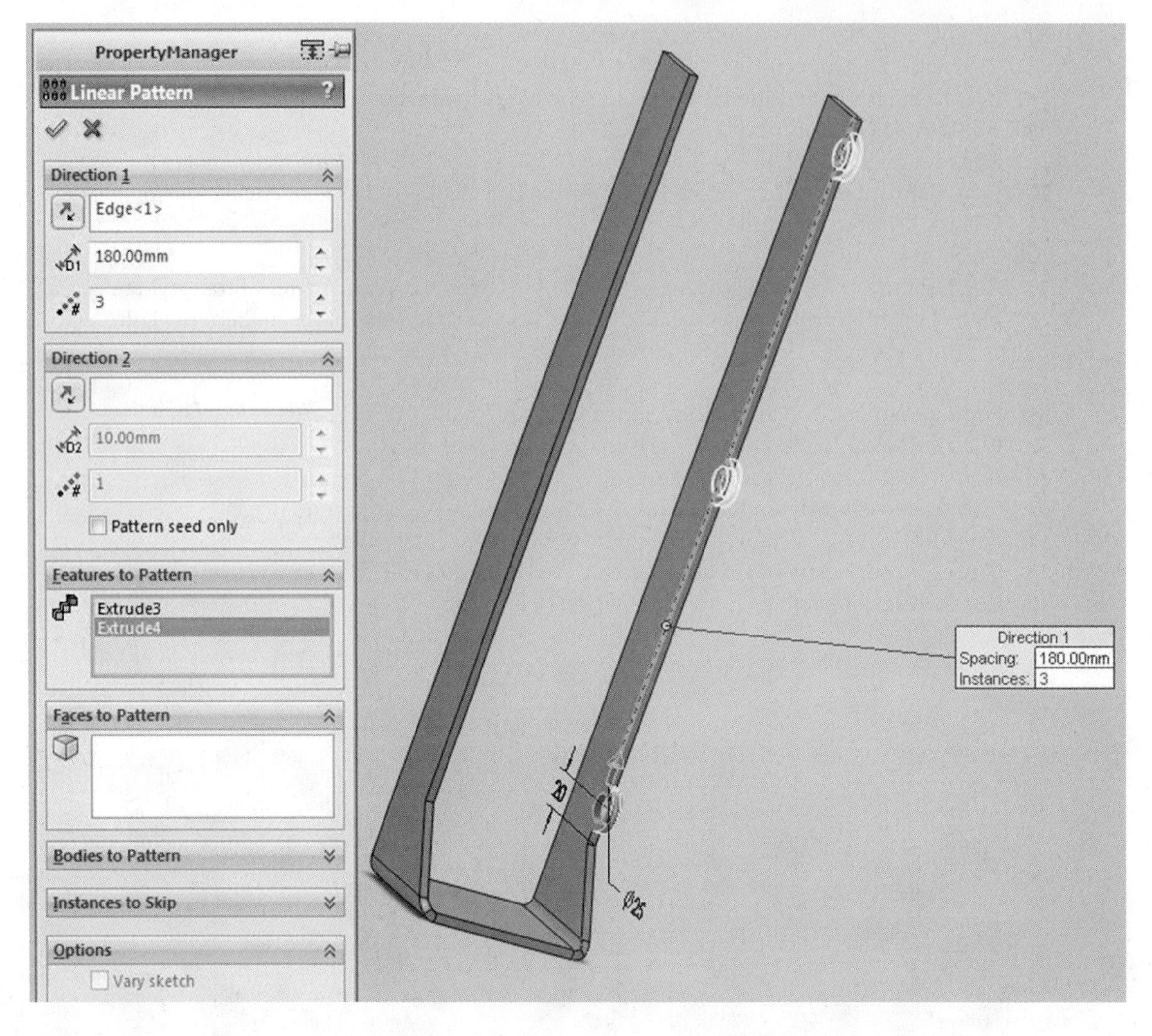

Figure 19.158

Linear Pattern PropertyManager.

Edge Flange

37. Create a Rectangular profile of length 390 mm *from the Top right corner* (see Figure 19.160).
38. Click the Edge-Flange tool (see Figure 19.161).
39. For the Flange Parameter, click the *Edge of the Rectangle common with the model* (see Figure 19.161).
40. For the Flange Length, set it to a value of 75.00 mm (see Figure 19.161).
41. Click OK (see Figure 19.161).

Chamfer

42. Click Features > Chamfer. (See Figure 19.162 for the Chamfer PropertyManager.)
43. Select the Distance distance option (see Figure 19.162).
44. Select an Edge and for Distance set the value to 70.00 mm (see Figure 19.162).
45. Click OK to complete the chamfer operation (see Figure 19.162).

Create a Circle on the Front Face of Right Edge Flange

46. Sketch a Circle with a diameter of 33.4 mm passing through a chamfered edge (see Figure 19.163).
47. Use Smart Dimension tool to dimension the diameter as 33.4 mm and the center, a distance of 148.85 mm from the bottom of model (see Figure 19.163).

Create an Extrude Cut Feature

48. Click Feature > Extrude Cut. (See Figure 19.164 for the Extrude Cut PropertyManager.)
49. In the Direction1 rollout, select Up To Next (see Figure 19.164).
50. Click OK to complete this extrusion.

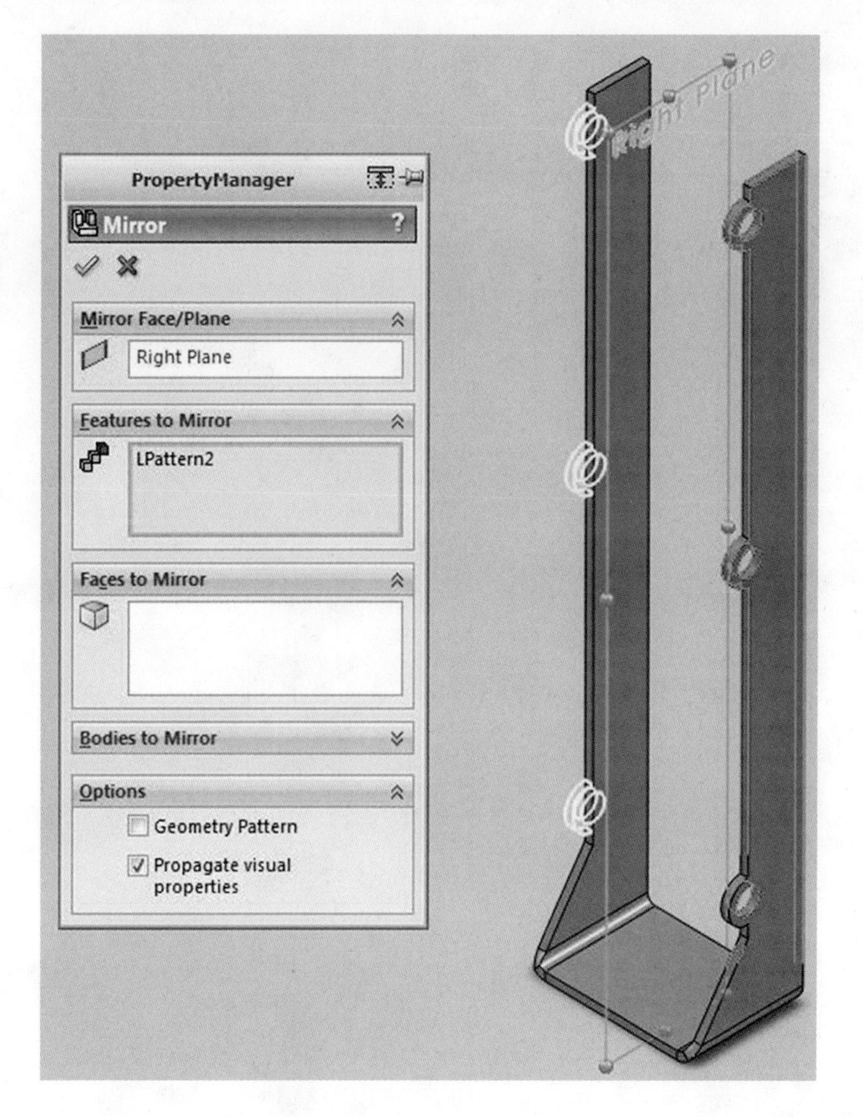

Figure 19.159

Mirror operation.

Linear Pattern

51. Click Features > Linear Pattern. (See Figure 19.165 for the Linear Pattern PropertyManager.)
52. In the Direction1 rollout, click the Edge of the right vertical side of the model (see Figure 19.165).
53. For the Distance between patterns, set it to be 130 mm (see Figure 19.165).
54. For the Number of Instances, set it to be 3 (see Figure 19.165).
55. For the Features To Pattern select Extrude5 (see Figure 19.165).
56. Click OK to complete the patterns (see Figure 19.165).

Create a Circle on the Front Face of Right Edge Flange

57. Sketch two Circles each with a diameter of 10 mm and a center-to-center distance of 35 mm (see Figure 19.166).
58. Use the Smart Dimension tool for the dimensions that are shown (see Figure 19.166).

Create Extrude Cut Feature

59. Click Feature > Extrude Cut. (See Figure 19.167 for the Extrude Cut PropertyManager.)
60. In the Direction1 rollout, select Up To Next (see Figure 19.167).
61. Click OK to complete this extrusion.

Linear Pattern on Right Edge Flange

1. Click Features > Linear Pattern. (See Figure 19.168 for the Linear Pattern PropertyManager.)
2. In the Direction1 rollout, click the Edge of the right vertical side of the model (see Figure 19.168).

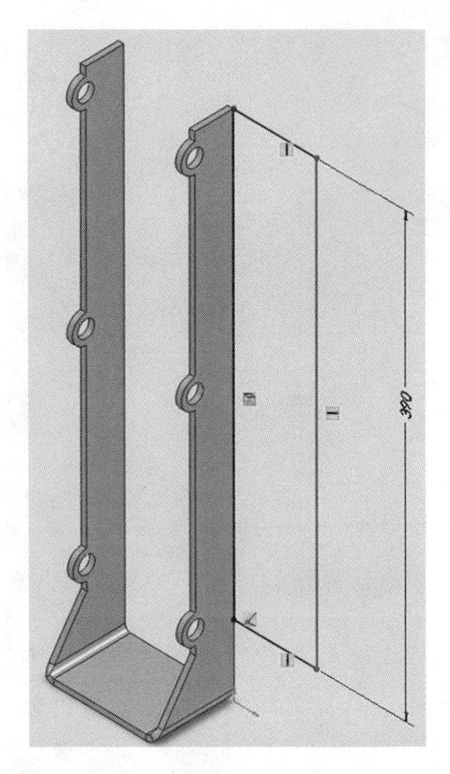

Figure 19.160

Rectangular profile for Edge Flange feature.

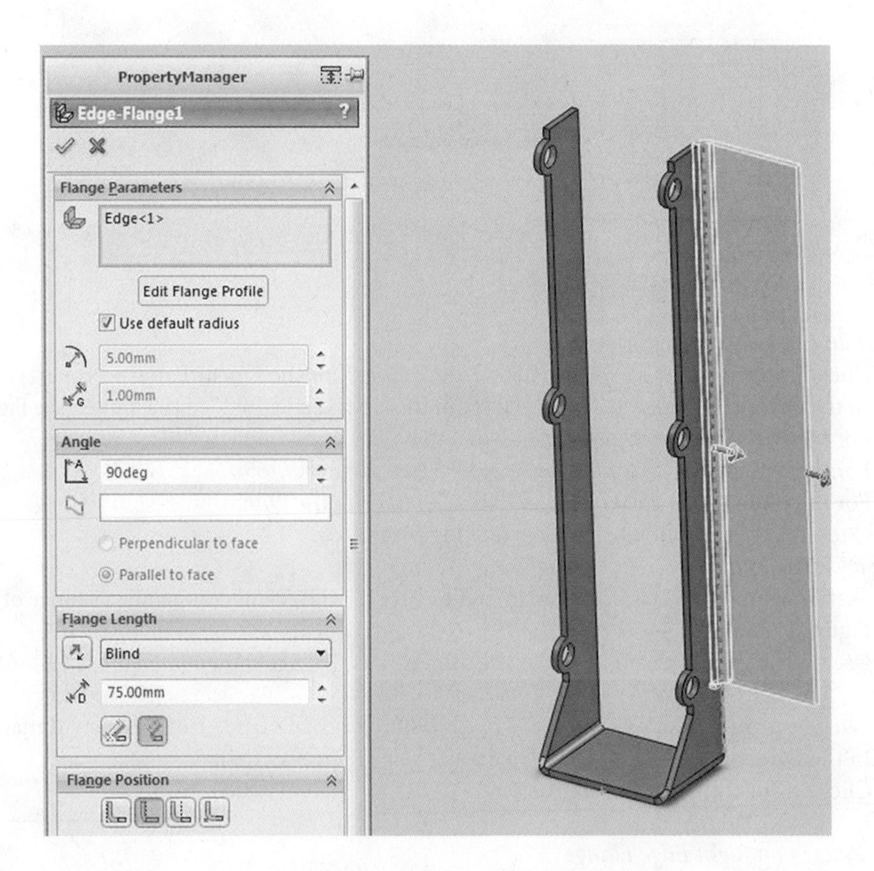

Figure 19.161

Edge Flange PropertyManager.

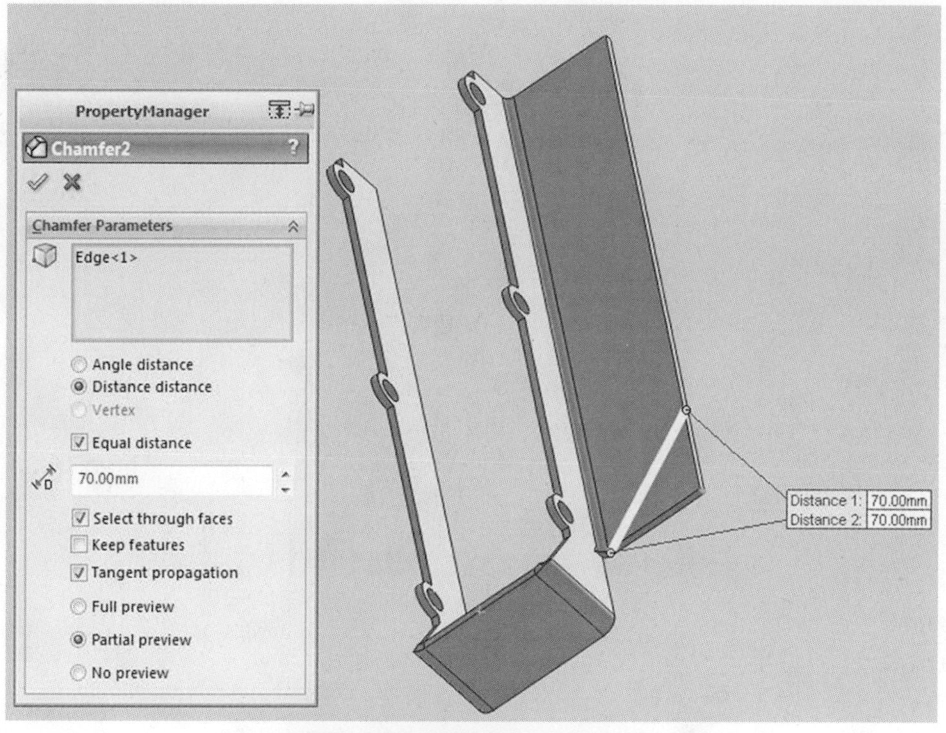

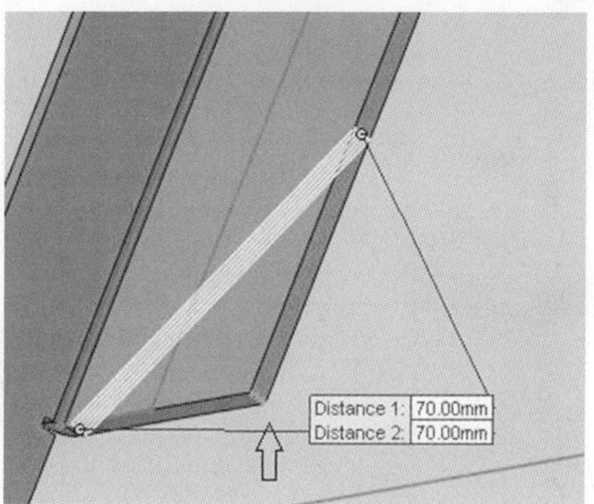

Figure 19.162

Chamfer PropertyManager.

3. For the Distance between patterns, set it to be 210 mm (see Figure 19.168).
4. For the Number of Instances, set it to be 2 (see Figure 19.168).
5. For the Features To Pattern select Extrude6 (see Figure 19.168).
6. Click OK to complete the patterns (see Figure 19.168).

Mirror

1. Click Features > Mirror. (See Figure 19.169 for the Mirror PropertyManager.)
2. For the Mirror/Face Plane, select the Right Plane (see Figure 19.169).
3. For the Features Mirror select LPattern3, Extrude6, LPattern4 (see Figure 19.169).
4. Click OK to complete the mirror operation (see Figure 19.169).

The flattened state for the sheet metal is shown in Figure 19.170.

Tutorial 19.10: CSWP-SMTL

Design the part that is shown in Figure 19.171 using SolidWorks Sheet Metal features. The part material is Aluminum, 1060 Alloy. The sheet metal thickness is 1.20 mm, and the inner bend radius is 1.00 mm. The

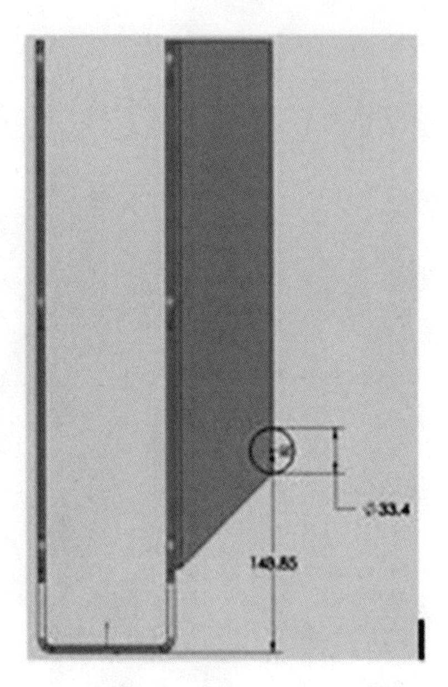

Figure 19.163

Circle on Right Edge Flange.

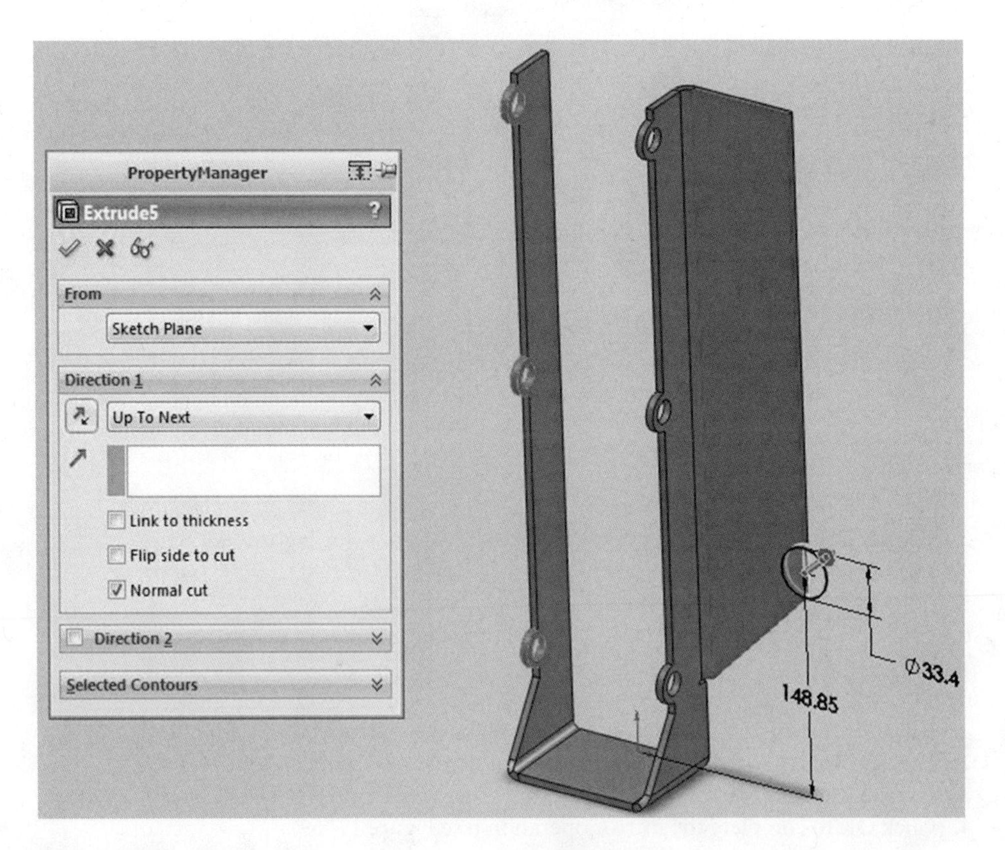

Figure 19.164

Extrude Cut PropertyManager.

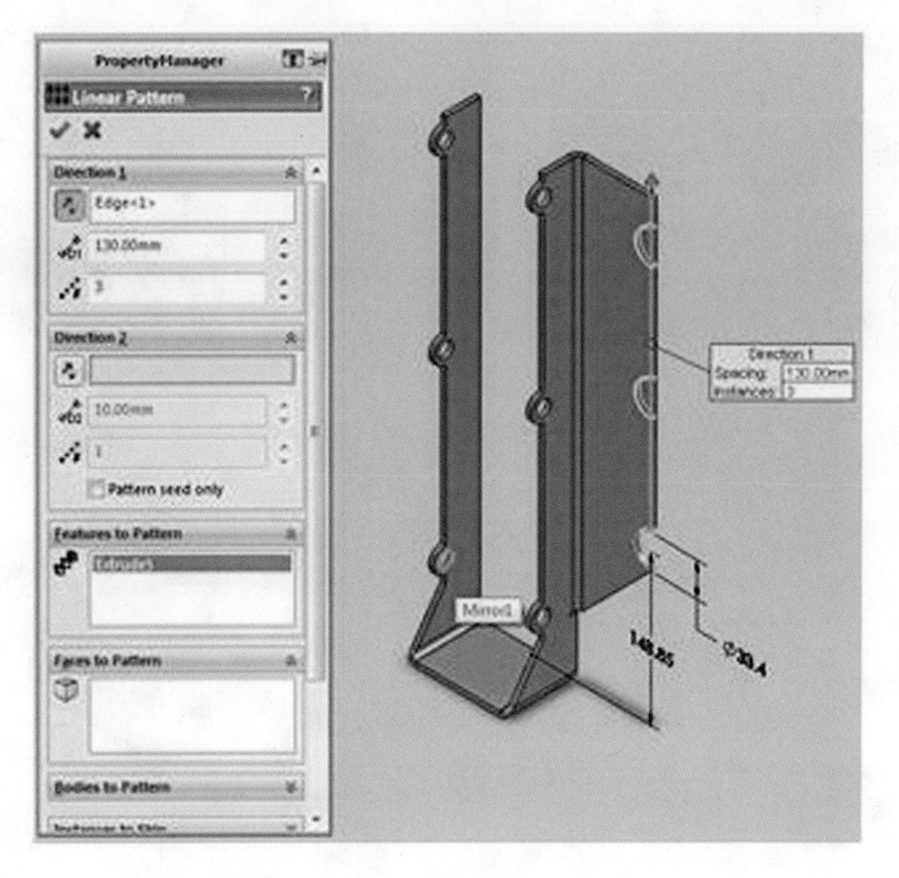

Figure 19.165

Linear Pattern PropertyManager.

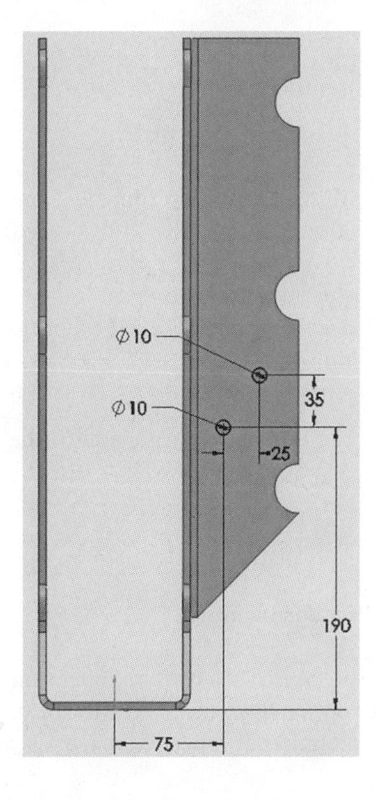

Figure 19.166

Two circles for Extrude Cut.

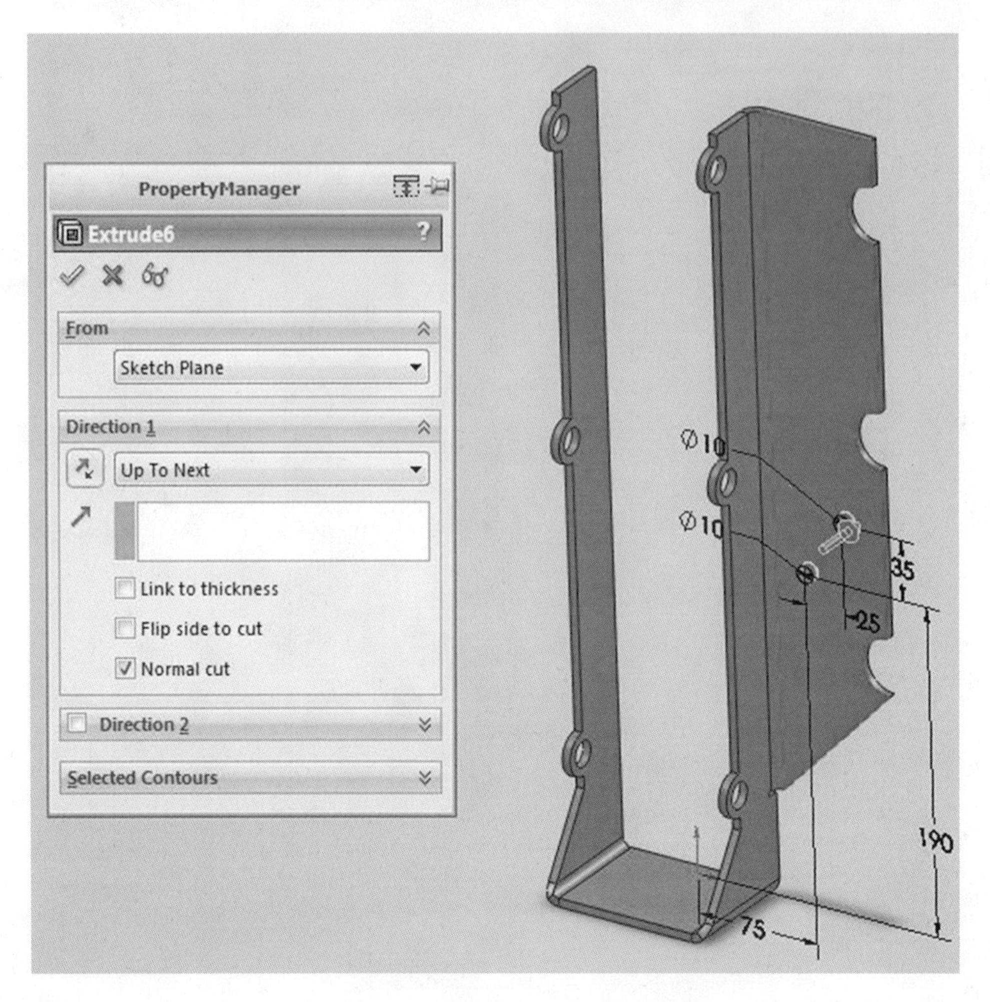

Figure 19.167

Extrude Cut PropertyManager.

part origin is arbitrary. Unit system: millimeter, gram, second. Decimal places: 2. The following dimensions are used for the design: A = 175, B = 100, C = 50, D = 15.

1. What is the overall *mass* of the part (in grams)?
2. What are the overall measured *length* (X) and *width* (Y) for the flattened state?
3. Produce the *Flattened State Drawing* for this part.

Note: These questions are similar to the primary part portion of the Certified SolidWorks Professional-SMTL examination.

SolidWorks Solution

Create Base Flange

1. Start a New SolidWorks Part document.
2. Select the Front Plane.
3. Be in sketch mode and create Sketch1, *A* by *B* in dimension (see Figure 19.172).
4. Click the Base Flange/Tab option from the Sheet Metal CommandManager.
   (Note: The Sheet Metal CommandPropertyManager is automatically displayed; see Figure 19.172.)
5. Set the values of the following parameters: Thickness = 1; K-Factor = 0.5; Ratio = 0.5.
6. Click OK.

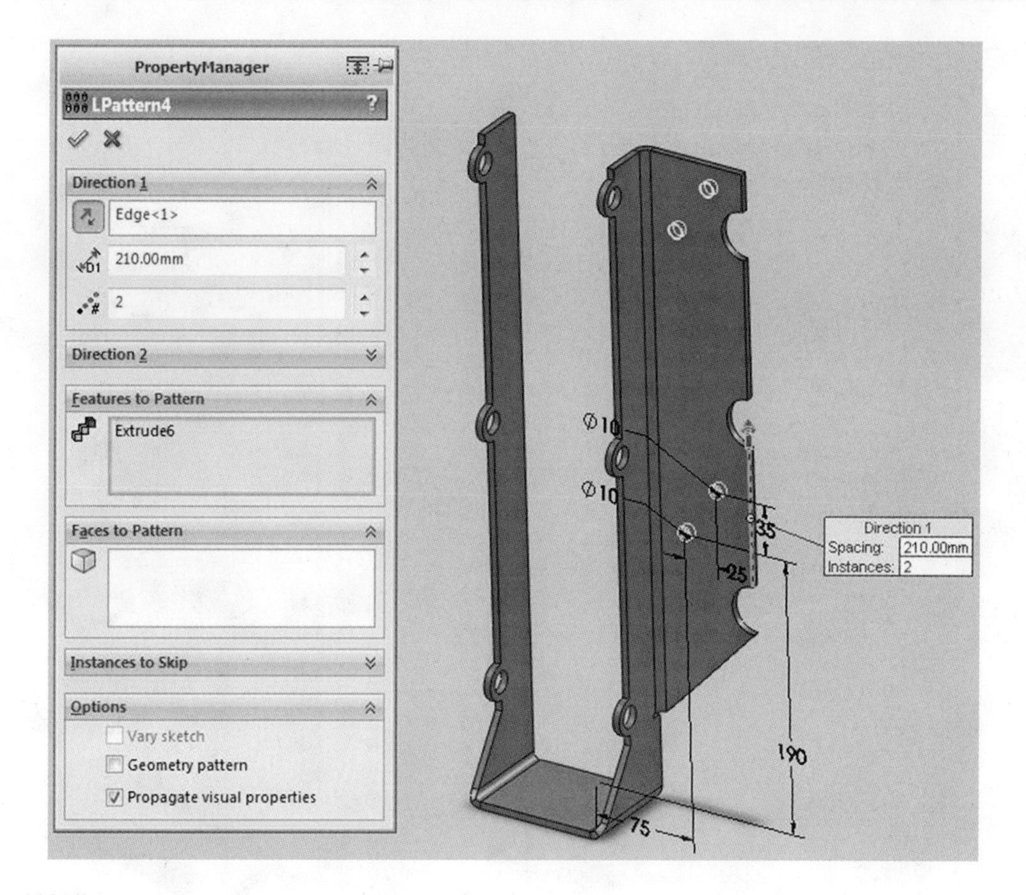

Figure 19.168

Linear Pattern PropertyManager.

Creating the First Edge Flange

7. Select the top edge of the base flange (see Figure 19.173).
8. Set Length of Edge Flange = 50 and Angle = 90-deg in the Edge Flange PropertyManager (see Figure 19.173).
9. Click OK to finish the operation.

Creating the Second Edge Flange

10. Select the bottom edge of the base flange (see Figure 19.174).
11. Set Length of Edge Flange = 50 and Angle = 90-deg in the Edge Flange PropertyManager (see Figure 19.174).
12. Click OK to finish the operation.

The steps involved in creating Miter Flange are as follows:

a. Click any edge of the Base Flange or the Edge Flange where Miter is needed.
b. Select the Line tool from the Sketch toolbar and sketch the profile for the Miter Flange.
c. Without exiting sketch mode, click Miter tool and select each edge of the Base Flange or the Edge Flange where Miter is needed to create the Miter feature. (The first one is automatically created on the first edge that is initially selected.) (Note: If you exit the sketch, then it has to be selected for the Miter function.)

Create the First Miter Flange Feature

13. Click any Edge of the Base Flange or Edge Flange where Miter is needed.
14. Select the Line tool from the Sketch toolbar. (Notice that a perpendicular plane for sketching the line is automatically selected; this is very useful in this approach.)
15. Create a sketch 15-mm wide, Sketch2 (see Figure 19.175).
16. Click the Miter Flange button on the Sheet Metal toolbar, while the sketch is still active.
17. Select the three Edges of the Base Flange/Edge Flange (see Figure 19.175).
18. Click OK from the Miter Flange Property toolbar.

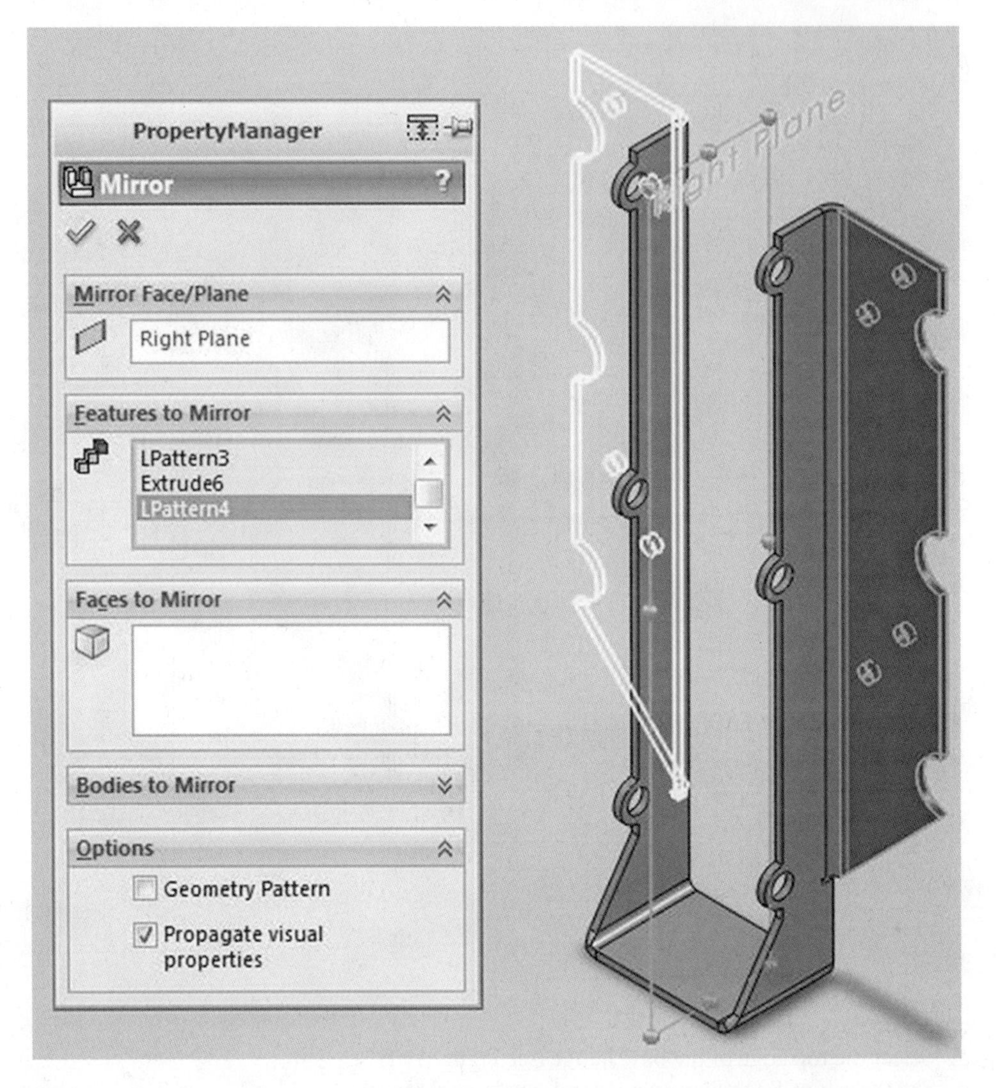

Figure 19.169

For the Mirror PropertyManager.

Create the First Miter Flange Feature

19. Click any Edge of the Base Flange or Edge Flange where Miter is needed.
20. Select the Line tool from the Sketch toolbar. (Notice that a perpendicular plane for sketching the line is automatically selected; this is very useful in this approach.)
21. Create a sketch 15-mm wide, Sketch3 (see Figure 19.176).
22. Click the Miter Flange button on the Sheet Metal toolbar, while the sketch is still active.
23. Select the three Edges of the Base Flange/Edge Flange (see Figure 19.176).
24. Click OK from the Miter Flange Property toolbar.
25. Hide Plane1.

Create Sketch for Extrude Cut

26. While in sketch mode, select the inner face of the Base Flange, and create a partial sketch (see Figure 19.177).
27. Select the partial sketch and Mirror about the horizontal Construction Line (see Figure 19.178).
28. Select the partial sketch and Mirror about the horizontal Construction Line (see Figure 19.179).

Create Extrude Cut

29. Click the Extrude-Cut option.
30. Select the Up To Next for Direction1 (see Figure 19.180).
31. Click OK to complete the extrusion process.

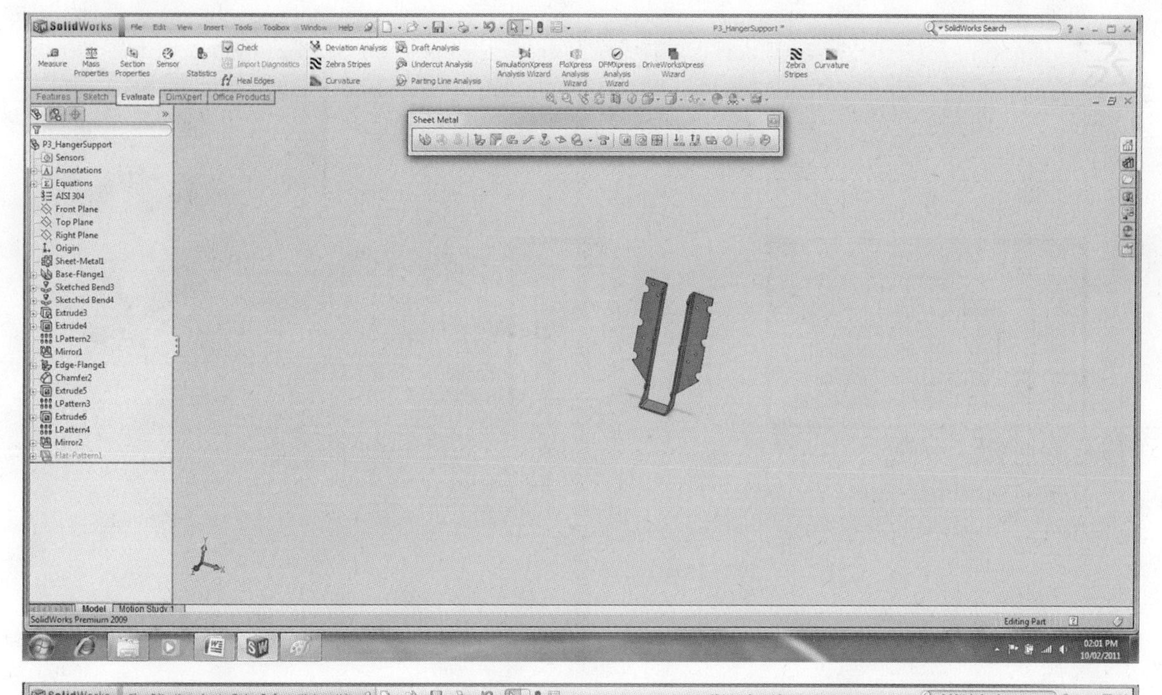

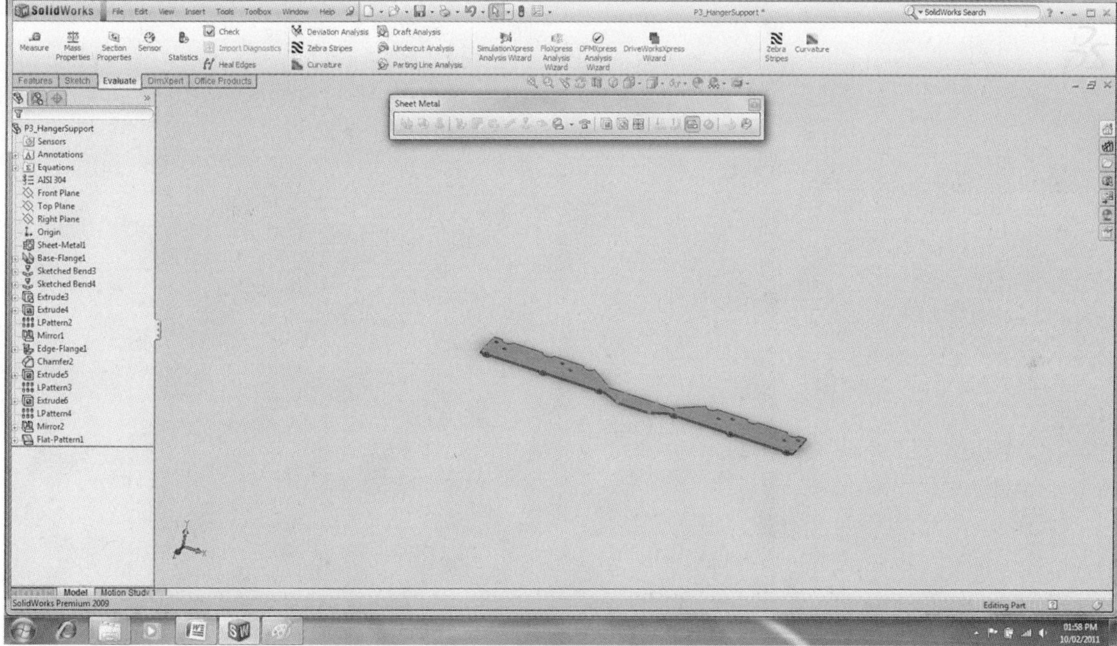

Figure 19.170

Flattened state for the sheet metal.

Create Circular Extrude Cuts

32. Create a circle, 4-mm diameter, located 9 mm × 10 mm from the open end of the sheet metal (see Figure 19.181).
33. Click Extrude-Cut option.
34. Select the Extrude-Cut and Mirror about the Top Plane (see Figure 19.182).
35. Create another circle, 4-mm diameter, located 9 mm from the bottom of the sheet metal (see Figure 19.183).
36. Click Extrude-Cut option.
37. Select the Extrude-Cut and Mirror about the Right Plane (see Figure 19.184).

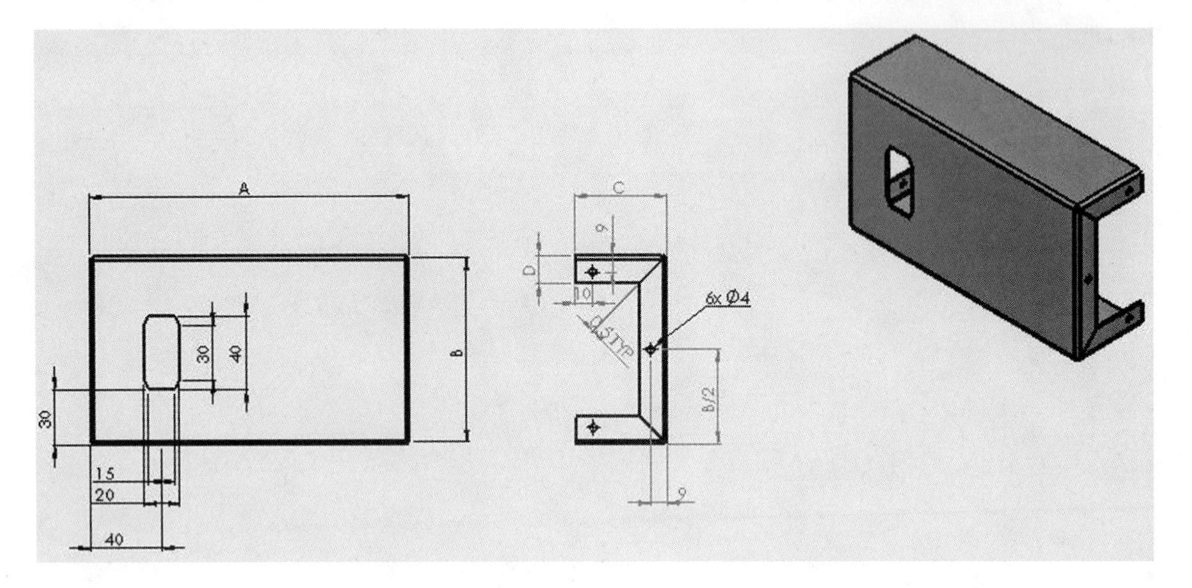

Figure 19.171

Drawings with dimensions and model of sheet metal.

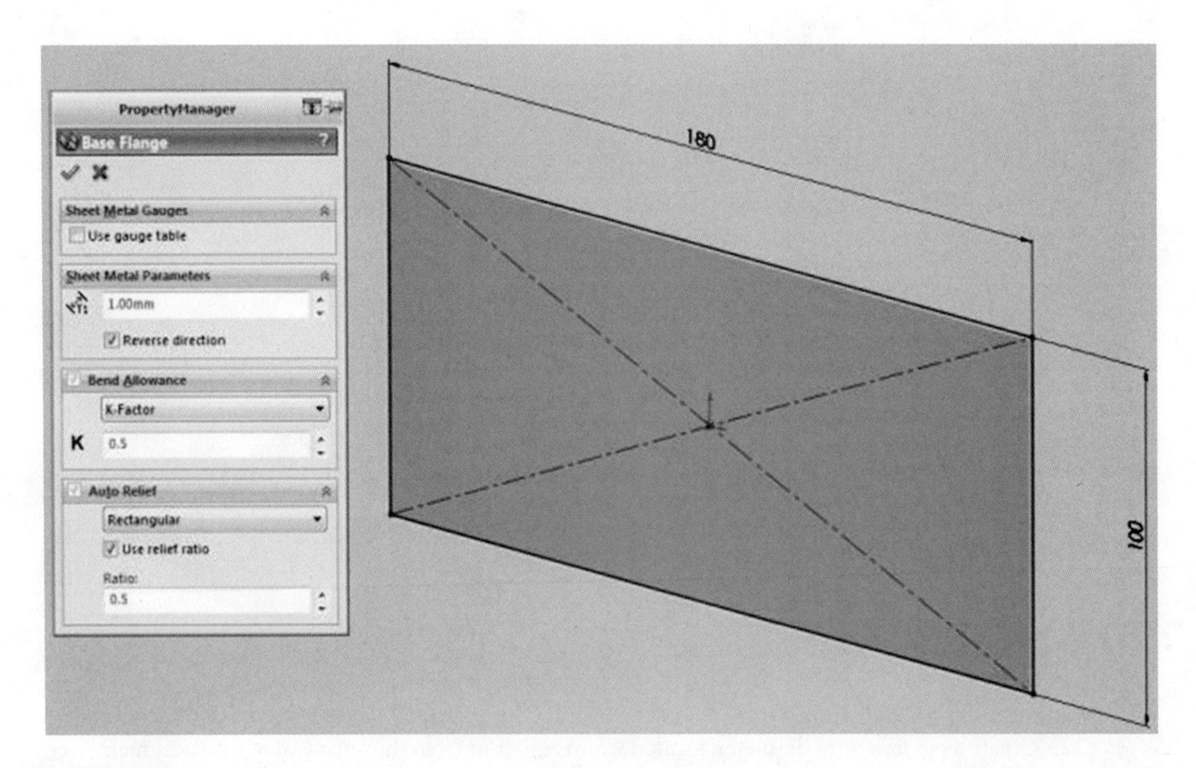

Figure 19.172

Sketch1 and selection of Base Flange/Tab option.

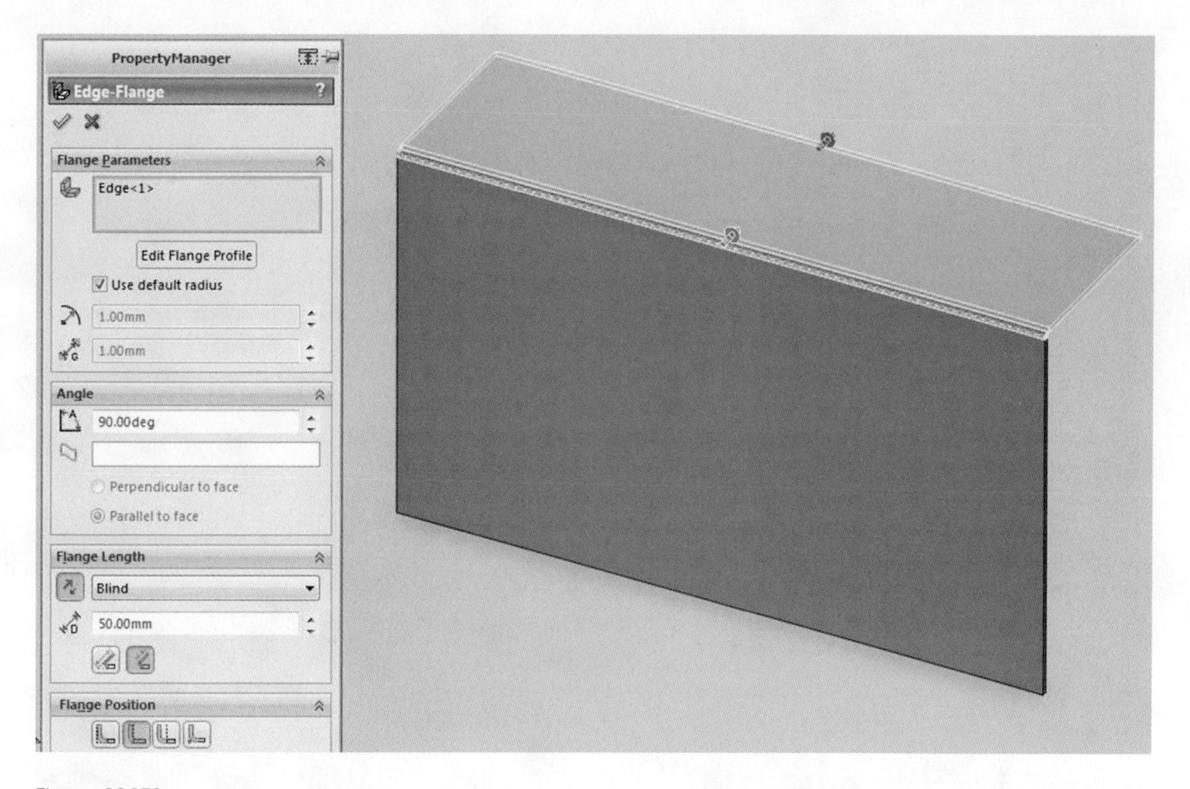

Figure 19.173

Edge Flange PropertyManager.

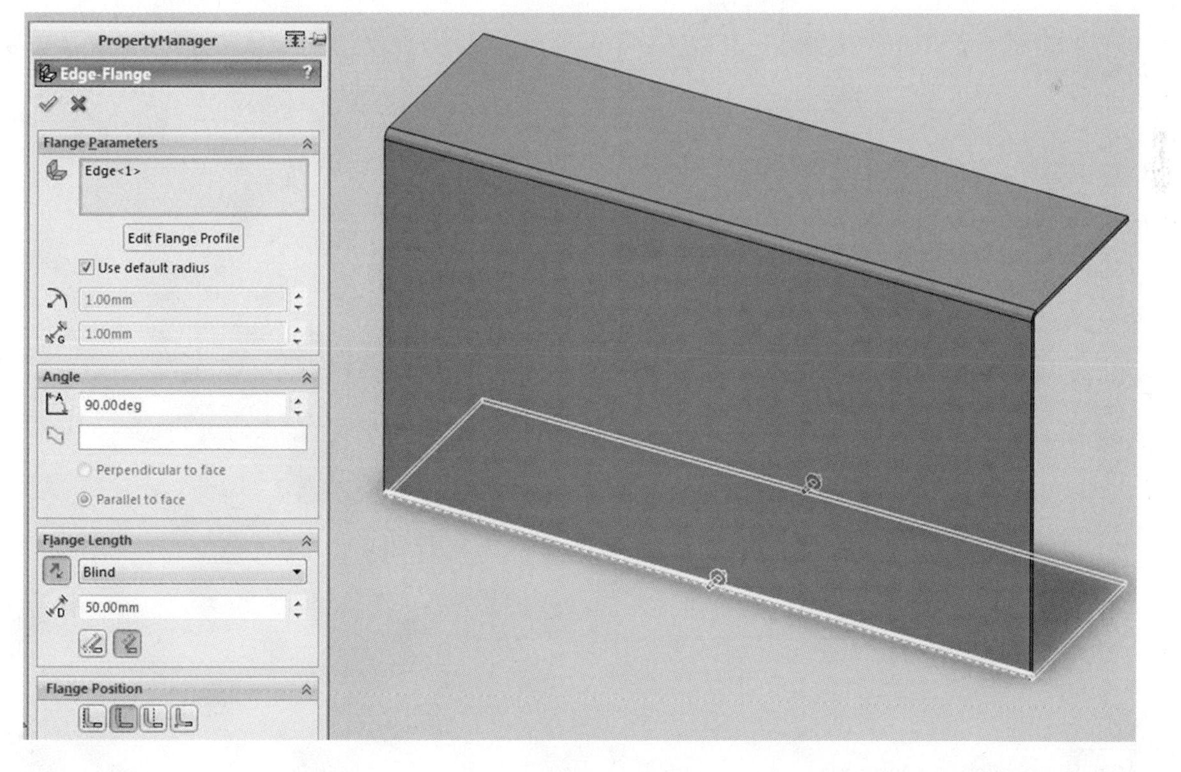

Figure 19.174

Edge Flange PropertyManager.

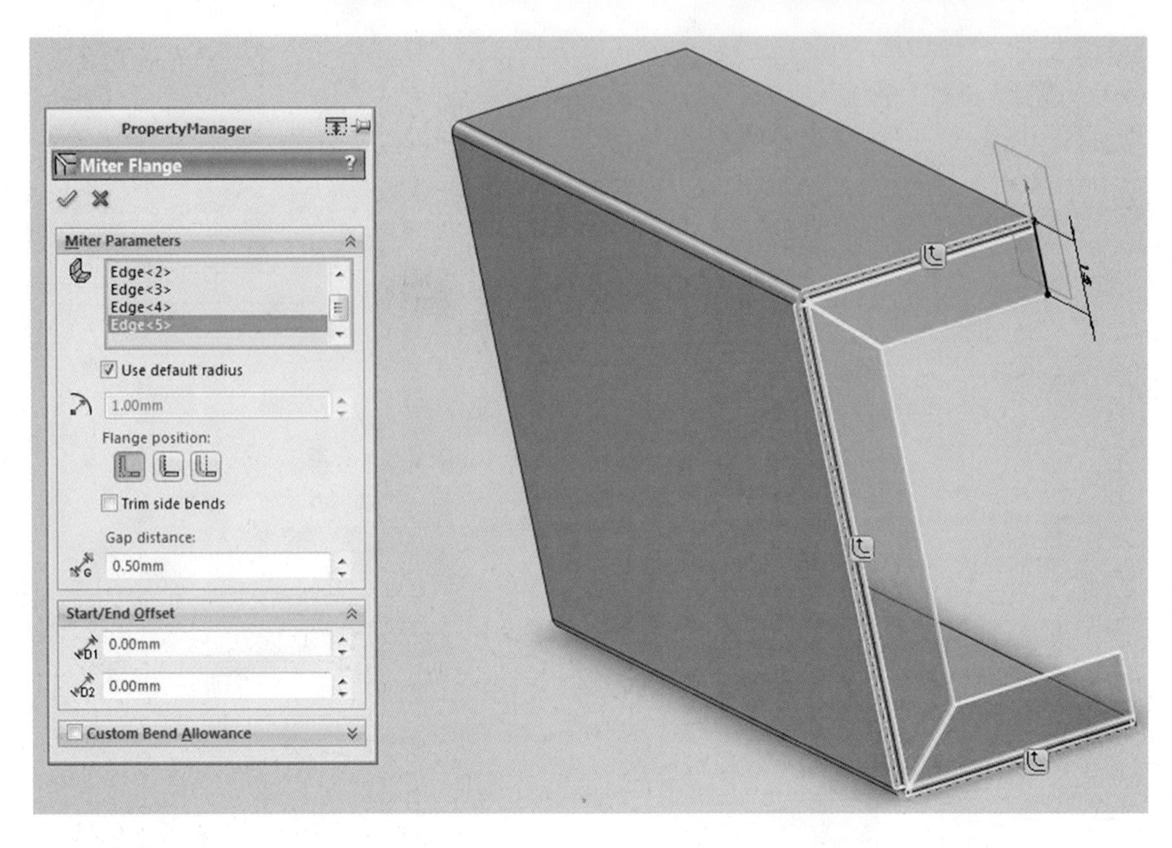

Figure 19.175

Miter Flange PropertyManager.

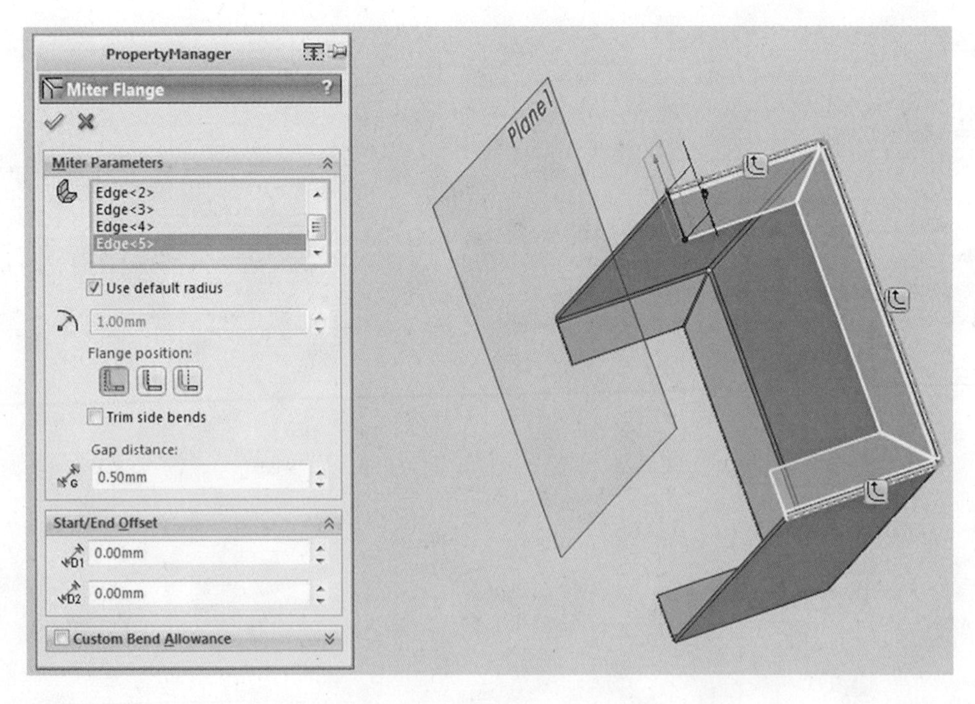

Figure 19.176

Miter Flange PropertyManager.

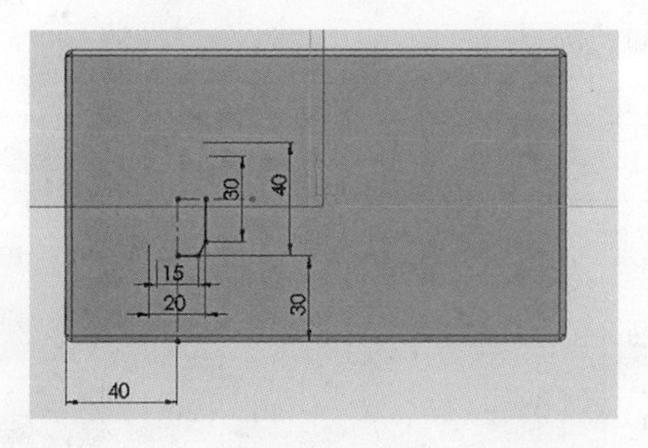

Figure 19.177

Sketch for Extrude-Cut.

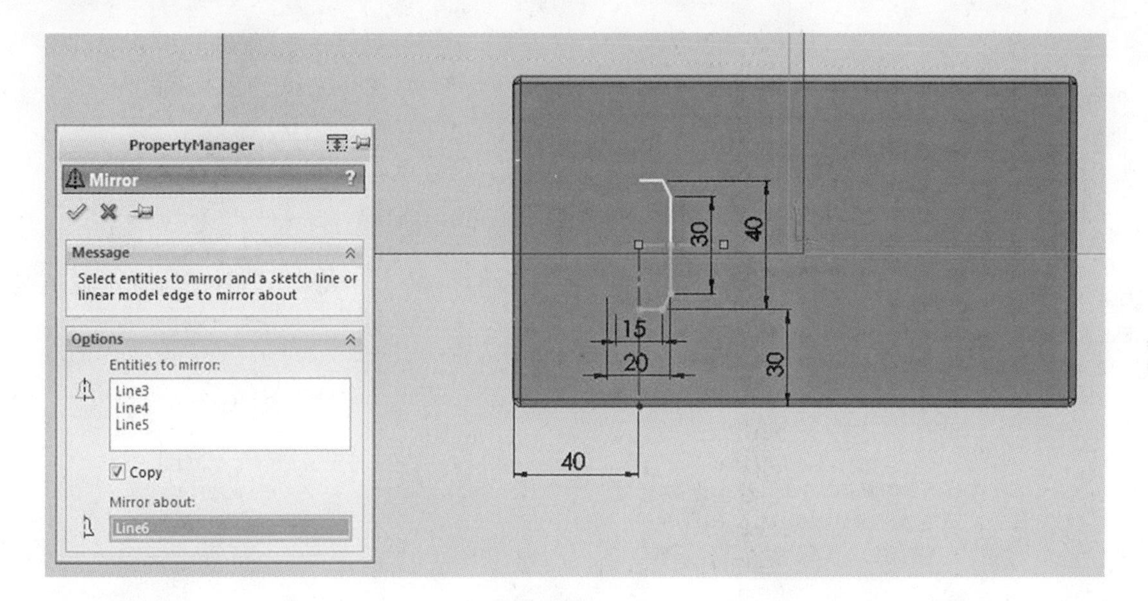

Figure 19.178

Mirror PropertyManager.

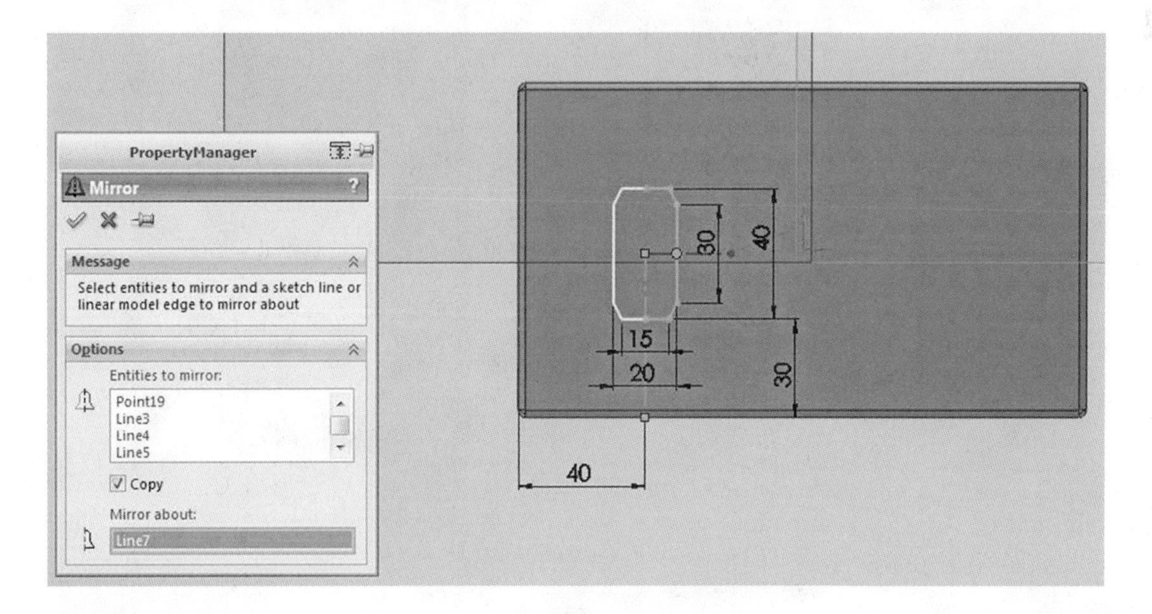

Figure 19.179

Mirror PropertyManager.

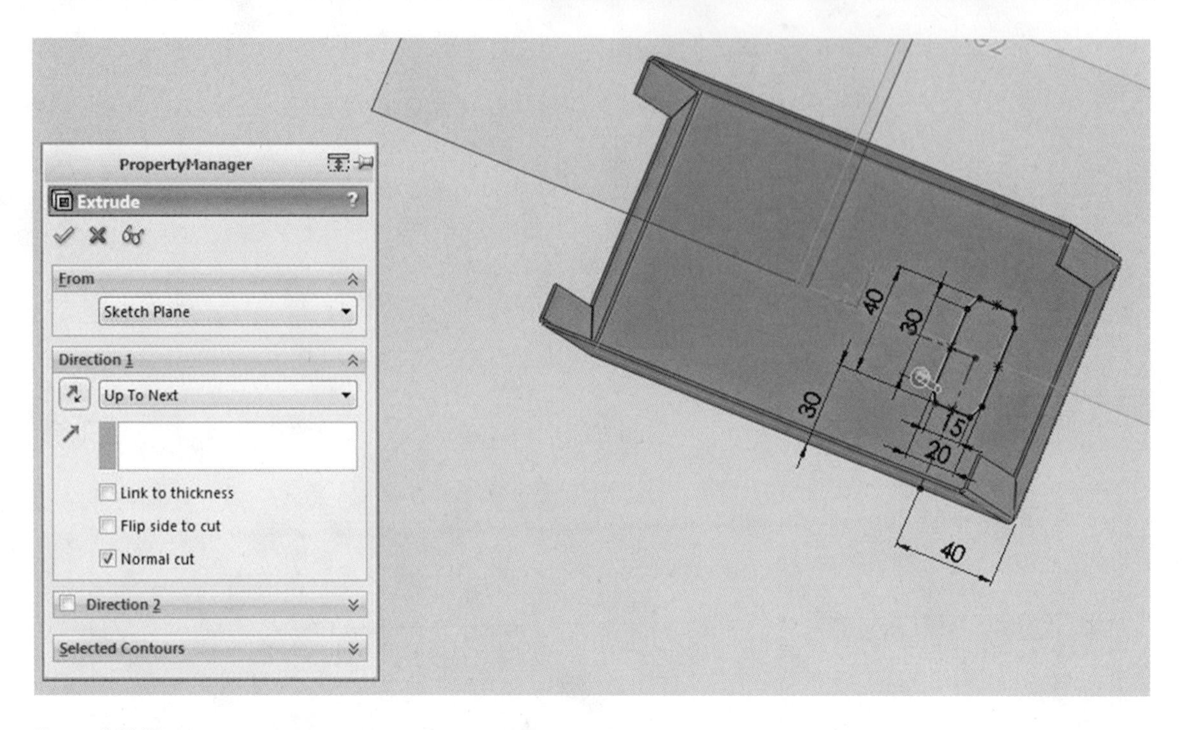

Figure 19.180

Extrude-Cut PropertyManager.

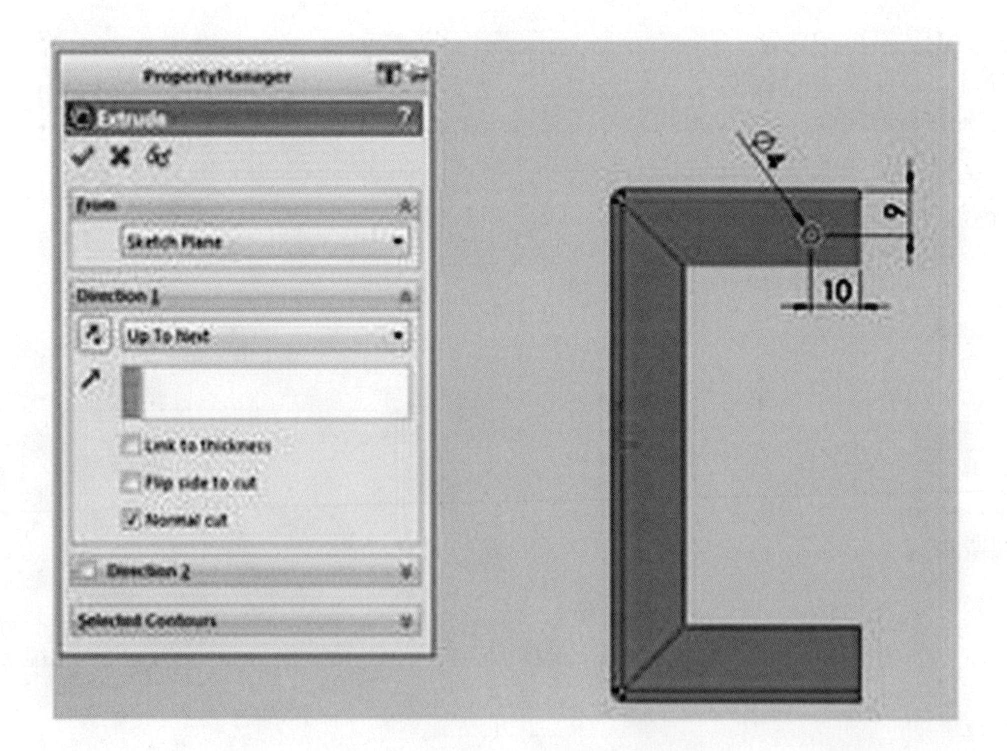

Figure 19.181

Circle created and extrude cut.

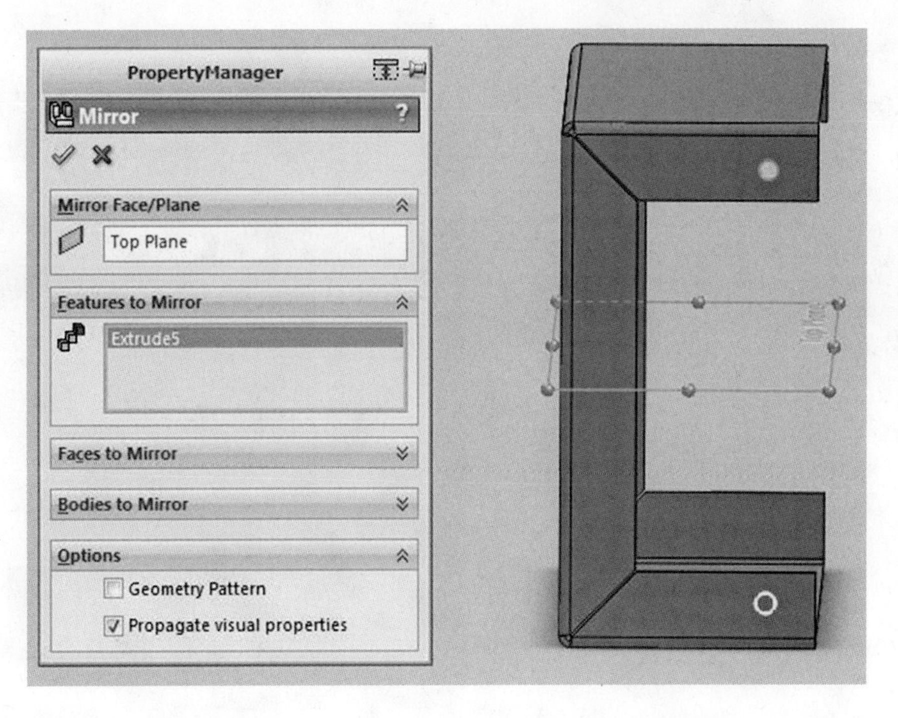

Figure 19.182

Mirror PropertyManager.

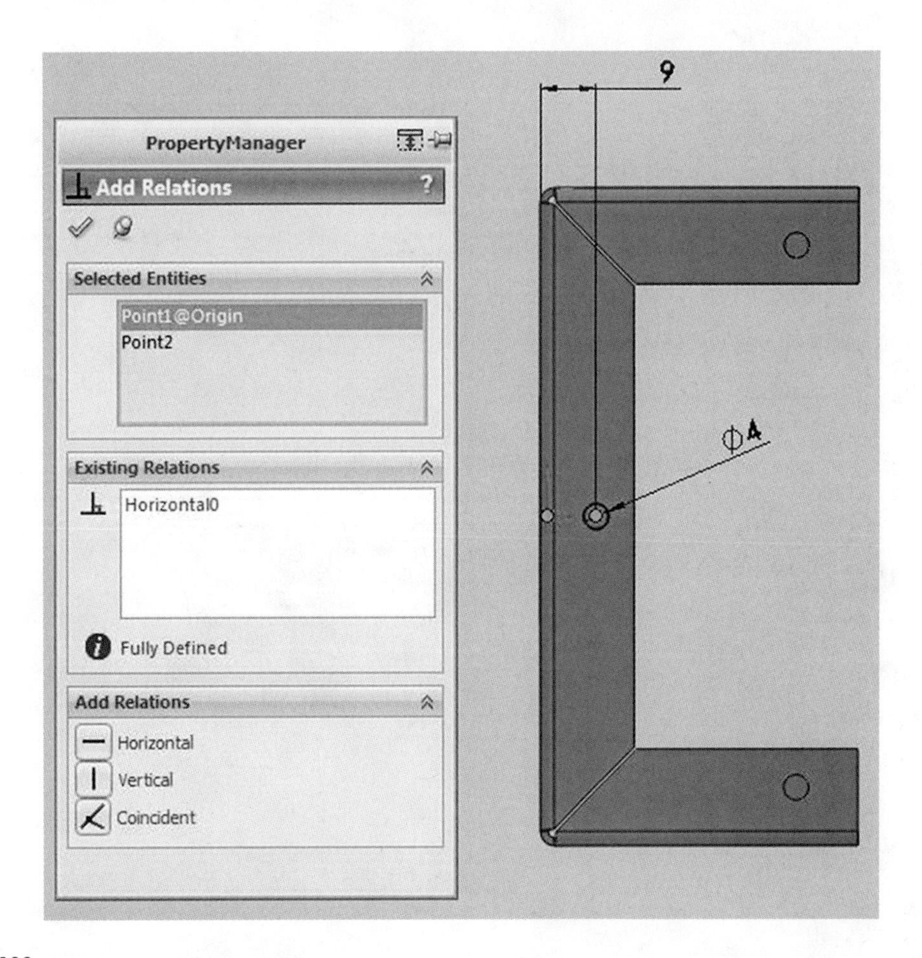

Figure 19.183

Another circle created for extrusion.

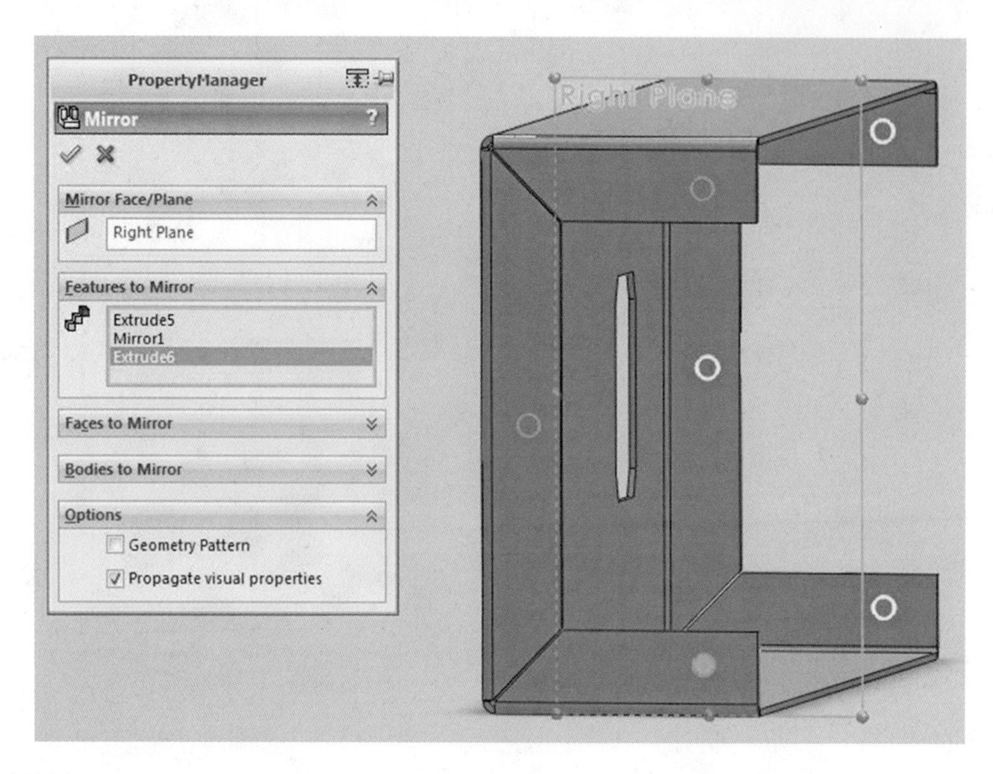

Figure 19.184

Mirror PropertyManager.

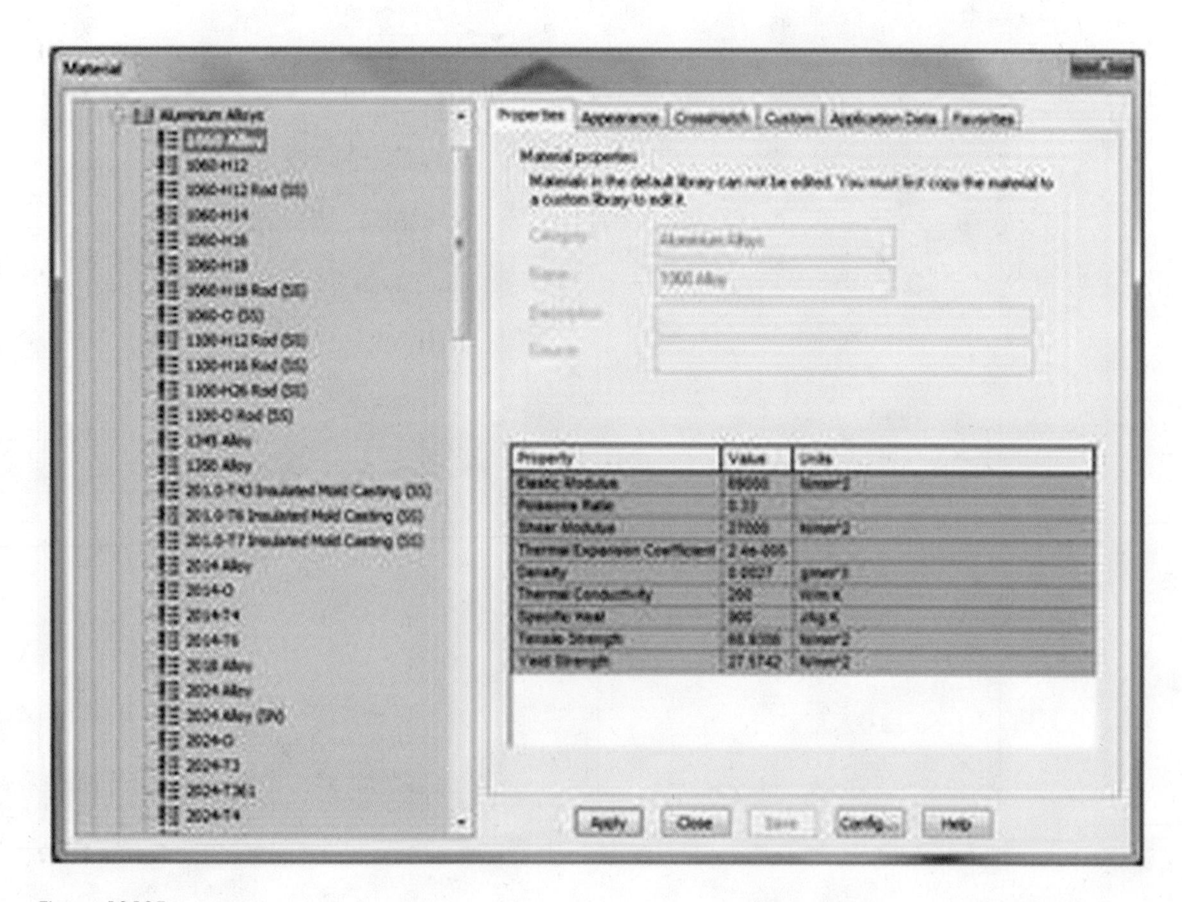

Figure 19.185

Material assignment.

Apply Material

38. Right-click Material <not specified> from the FeatureManager.
39. Select Edit Material.
40. From the Material PropertyManager, select 1060 Alloy from the SolidWorks *Materials Database* (see Figure 19.185).
41. Click Apply to apply the Material.
42. Click Close to complete the process.

Compute Mass Properties

43. Click Evaluate > Mass Properties. (See Figure 19.186 for the values.)

Create Flattened State Drawing

Figure 19.187 shows the flattened state of the sheet metal part.

The overall measured *length* (X) and *width* (Y) for the flattened state are as follows:

$$X = 206.23 \text{ mm}; \quad Y = 201.03 \text{ mm}$$

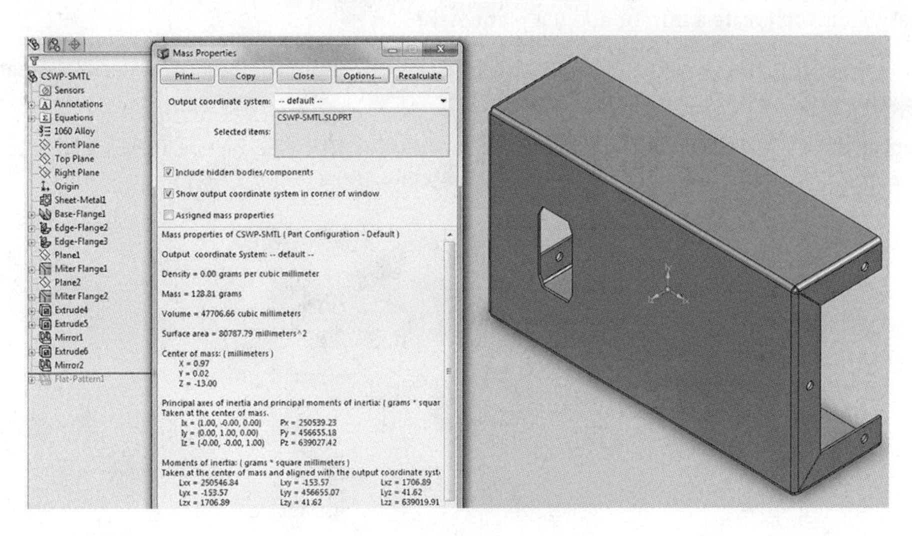

Figure 19.186

Mass properties computation.

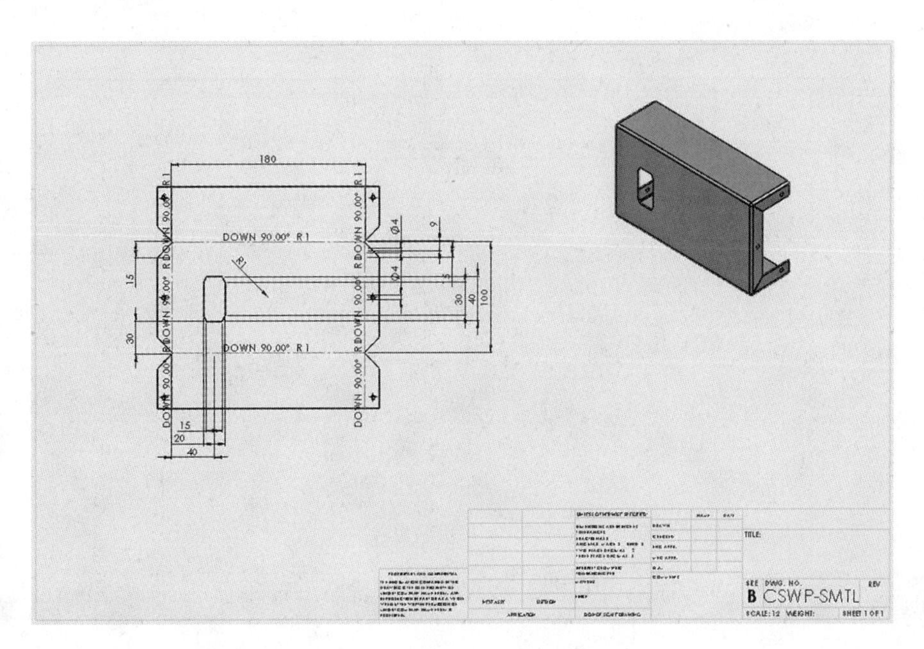

Figure 19.187

Flattened state.

Summary

This chapter has presented in much detail the design of sheet metal parts based on the base flange approach. Ten tutorials have been presented, including questions, which are typical of the CSWA examination. It is expected that if students work through these tutorials, they will significantly enhance their proficiency in sheet metal part design.

Exercises

1. Apply 1060 Alloy material from the SolidWorks to the sheet metal in Tutorial 19.1 (see Figure P19.1) and determine the mass.
2. Apply 1060 Alloy material from the SolidWorks to the sheet metal in Tutorial 19.1 (see Figure P19.2) and determine the mass.
3. For the sheet metal in Tutorial 19.2 (see Figure P19.3) apply 1060 Alloy material and determine the mass. Can you create a mirror about a plane A-B?
4. For the sheet metal in Tutorial 19.8 (see Figure P19.4), apply 1060 Alloy material as material, and determine the mass. Create part drawing.
5. Repeat the CSWP-SMTL sheet metal part in Tutorial 19.9 (see Figure P19.5) with the following dimensions that are used for the design: A = 180, B = 100, C = 50, D = 15.

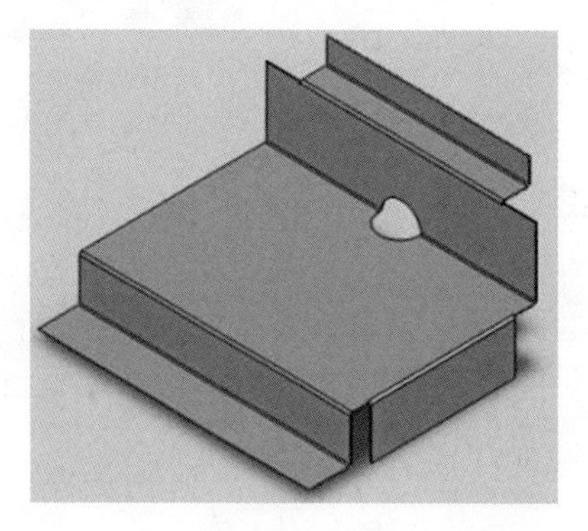

Figure P19.1

Bend created.

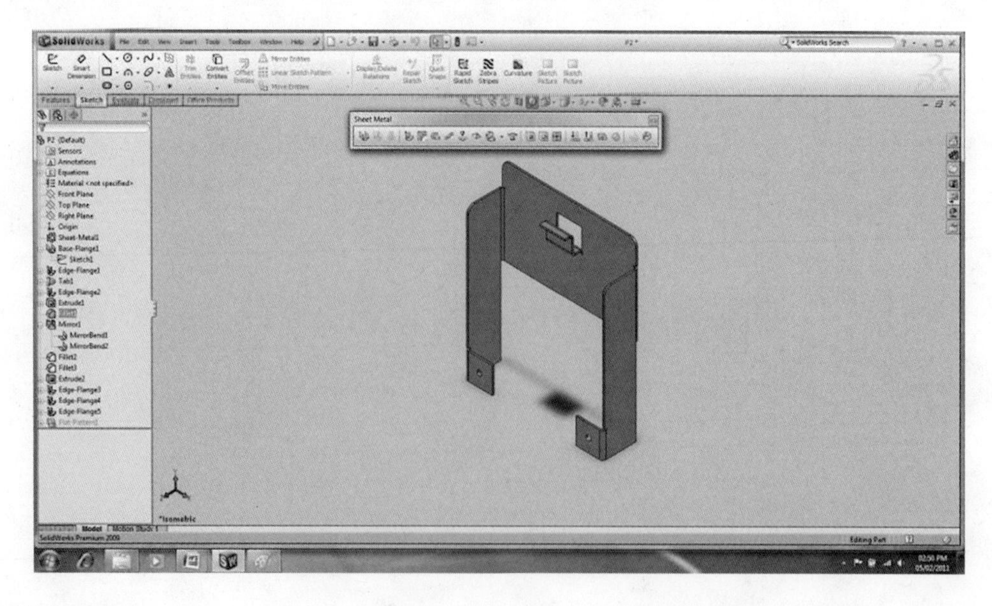

Figure P19.2

Sheet metal part.

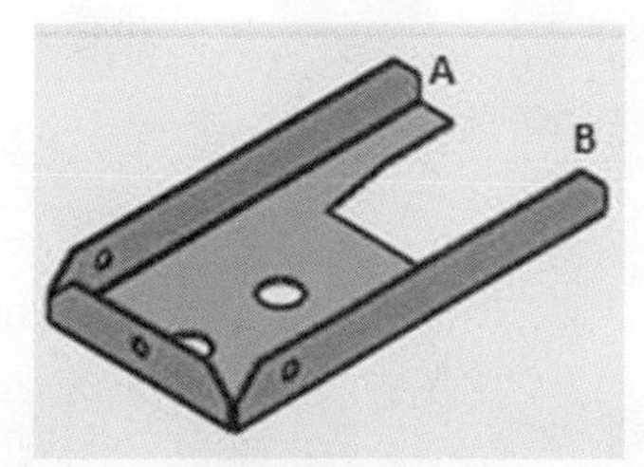

Figure P19.3

Bracket.

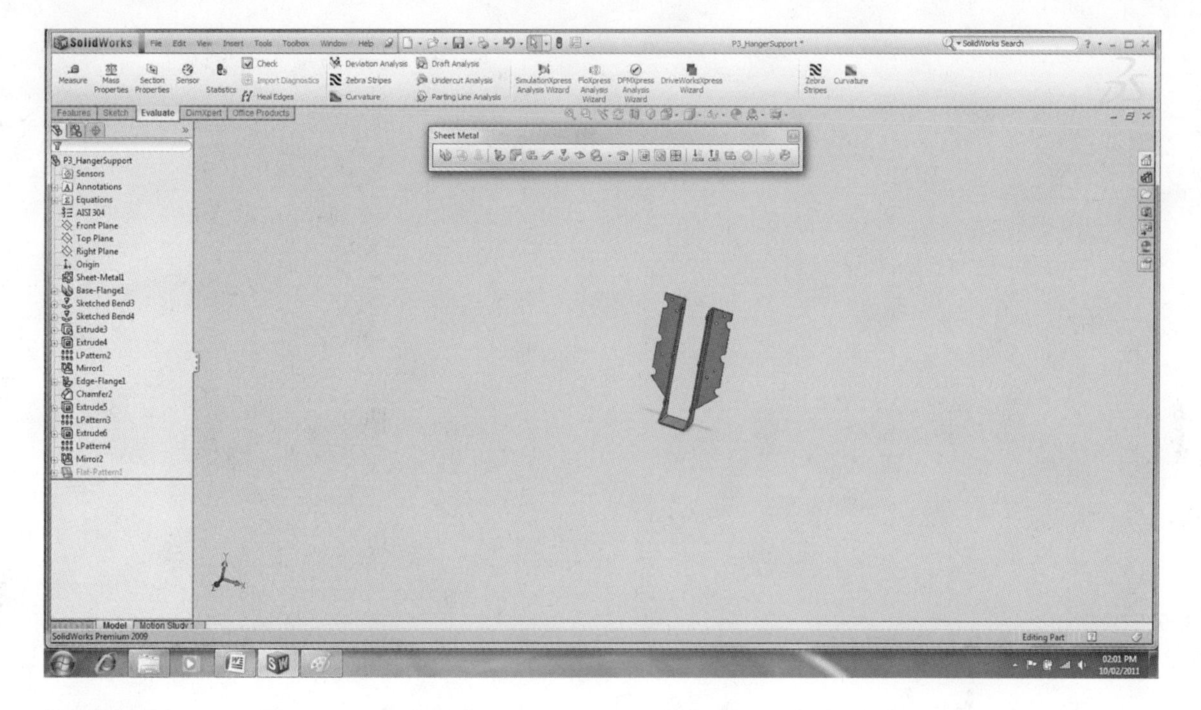

Figure P19.4

Flattened state for the sheet metal.

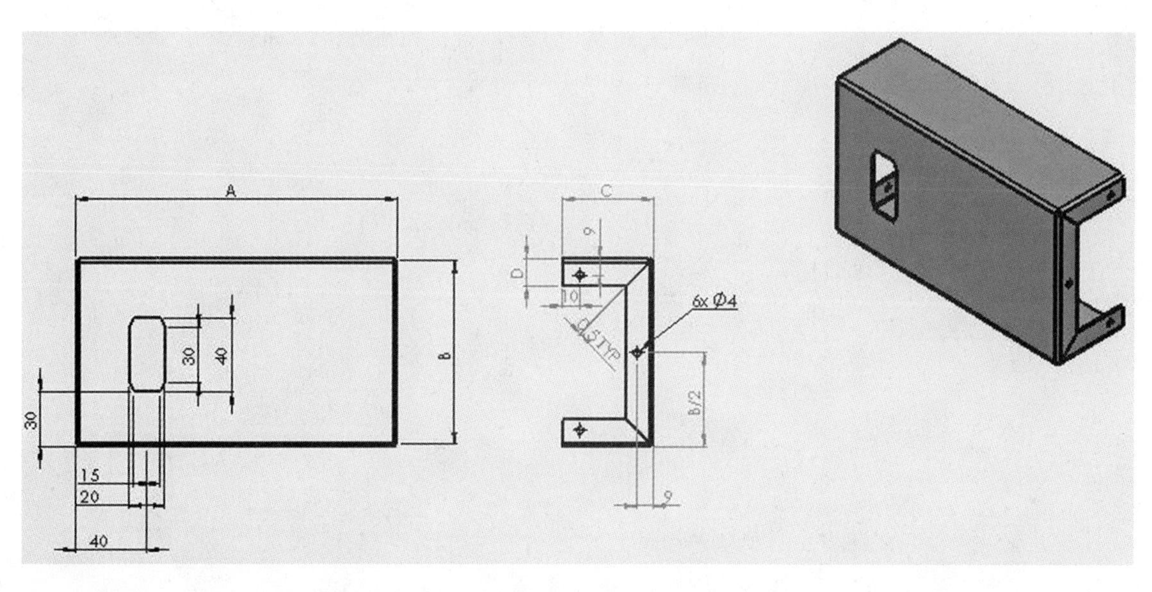

Figure P19.5

Drawings with dimensions and model of sheet metal.

20

Weldments

Objectives:

When you complete this chapter, you will have learned the following:

- About Weldment tools
- How to define the basic weldment framework using 3DSketch
- How to insert a Structural Member
- How to insert a Fillet Bead feature
- How to insert a Trim/Extend feature
- How to insert an End Cap feature
- How to insert a Gusset feature
- How to apply Weldment to part design

Introduction

Weldments play a significant role in joining structural members and hence is of vital importance in the civil engineering discipline. Also, weldments constitute one of the several aspects of manufacturing process that requires specific attention. We now consider weldments. Weldment functionality enables you to design a weldment structure as a single multibody part. The basic approach is to use two-dimensional and three-dimensional (3D) sketches to define the basic framework and then create structural members containing the groups of sketch segments. Features that can be added include gussets, end caps, etc., using tools on the *Weldments toolbar.* This chapter also considers creating weldments from a part design.

Creating Parts with a 3D Sketch

Since the paths of weldments are mainly in 3D, a good grasp of the 3DSketch tool is a prerequisite for creating weldments. From the main menu, select the 3DSketch tool (see Figure 20.1) by clicking on the Sketch tab of the CommandManager. Select the Line Tool. Note that beside the cursor, the plane that the line will be displayed.

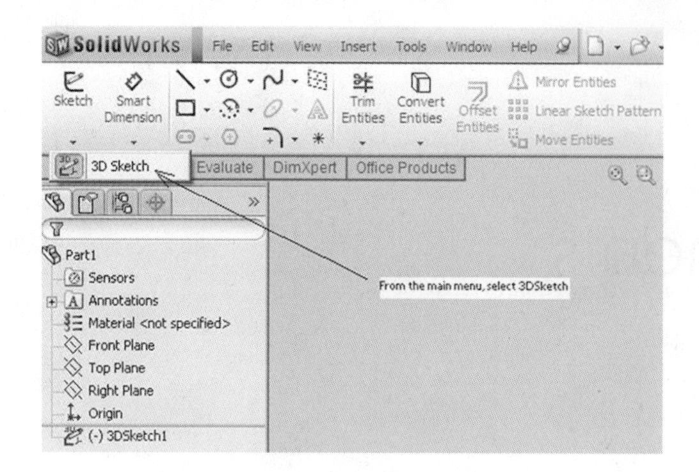

Figure 20.1

3DSketch tool.

Tutorials

Tutorial 20.1: Bent Tube

Model a bent tube with a 0.5-in. outer diameter and a 0.375-in. inner diameter to join the ends of two tubes, as shown in Figure 20.2. Follow the route that is shown by the centerlines in Figure 20.2, and add 1-in. radius fillets to the corners.

Solution

1. Select the 3DSketch tool by clicking on the Sketch tab of the CommandManager.
2. Select the Line Tool. Note that beside the cursor, the plane that the line will be displayed.
3. Press the Tab key until when the plane selected is YZ.
4. Sketch a line 12-in. long from the origin along Y.
5. Sketch a line 8-in. long along Z.
6. Press the Tab key until when the plane selected is ZX.
7. Sketch a line 10-in. long along X.
8. 3DSketch1 is complete (see Figure 20.3).
9. Exit 3DSketch.
10. Click the top vertex of the vertical line, which is 12-in. long.

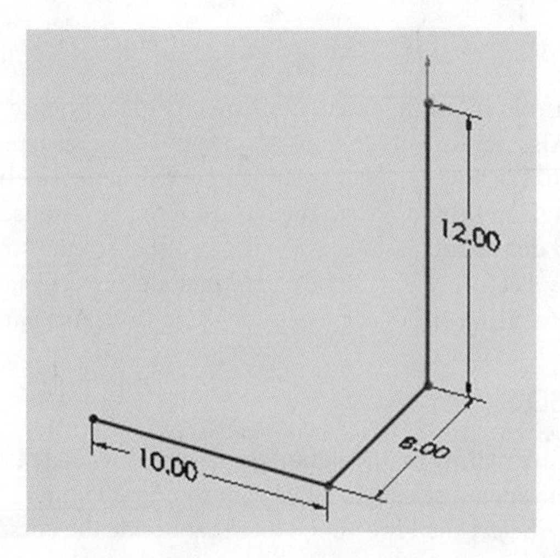

Figure 20.2

Specified route.

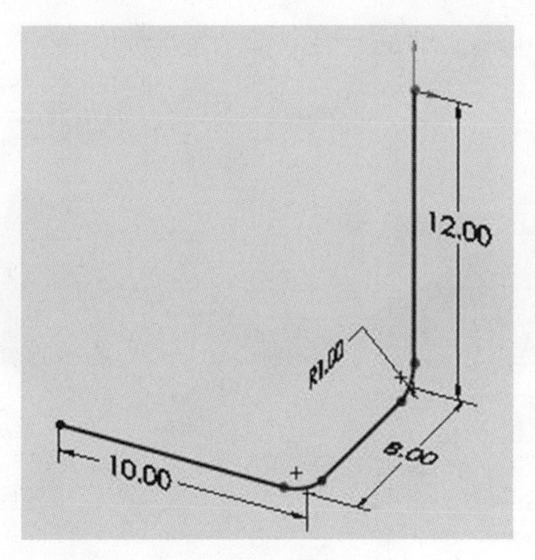

Figure 20.3

3DSketch of route.

11. Click Features > Reference Geometry.
12. Select top vertex and the vertical line that is connected to it, for Selections.
13. Select Normal to Curve to complete defining Plane1.
14. Select Plane1.
15. Right-click and select Sketch tool to sketch the two circles with 0.5-in. outer diameter and 0.375-in. inner diameter, as Sketch1; dimension them (see Figure 20.4).
16. Exit Sketch.
17. Click Features > Sweep.
18. Select Sketch1 as the Profile and 3DSketch1 as the Path (see Figure 20.5).
19. Click OK.

Tutorial 20.2: Bent Tube

Model a bent tube with a 0.5-in. outer diameter and a 0.375-in. inner diameter to join the ends of two tubes, as shown in Figure 20.6. Follow the route that is shown by the centerlines in Figure 20.6, and add 1-in. radius fillets to the corners.

Solution

1. Select the 3DSketch tool by clicking on the Sketch tab of the CommandManager.
2. Select the Line Tool. Note that beside the cursor, the plane that the line will be displayed.
3. Press the Tab key until when the plane selected is YZ.
4. Sketch a line 3-in. long from the origin along Y.
5. Press the Tab key until when the plane selected is ZX.
6. Sketch a line 3-in. long along X.
7. Join vertical and horizontal lines.
8. 3DSketch1 is complete (see Figure 20.6).
9. Exit 3DSketch.
10. Click the left vertex of the horizontal line, which is 3-in. long.
11. Click Features > Reference Geometry.
12. Select top vertex and the vertical line that is connected to it, for Selections.
13. Select Normal to Curve to complete defining Plane1.
14. Select Plane1.
15. Right-click and select the Sketch tool to sketch the two circles with 0.5-in. outer diameter and 0.375-in. inner diameter, as Sketch1; dimension them (see Figure 20.7).
16. Exit Sketch.

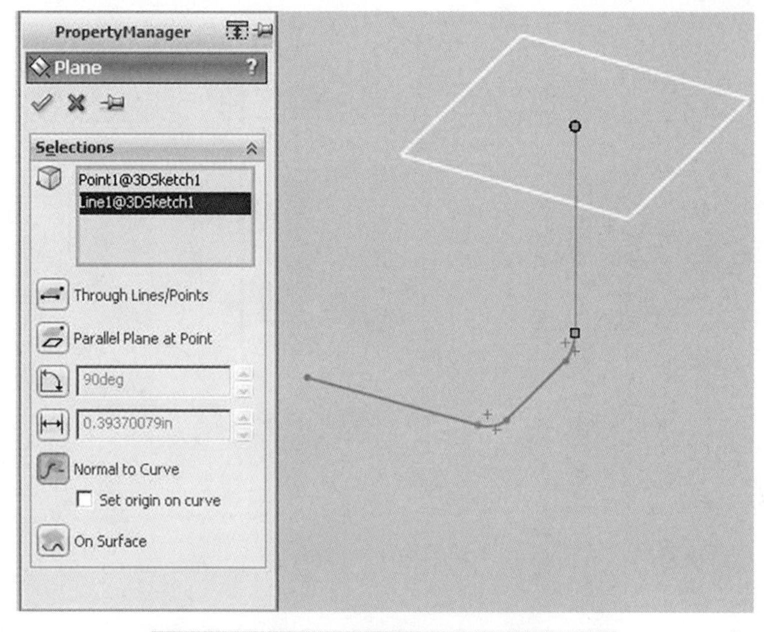

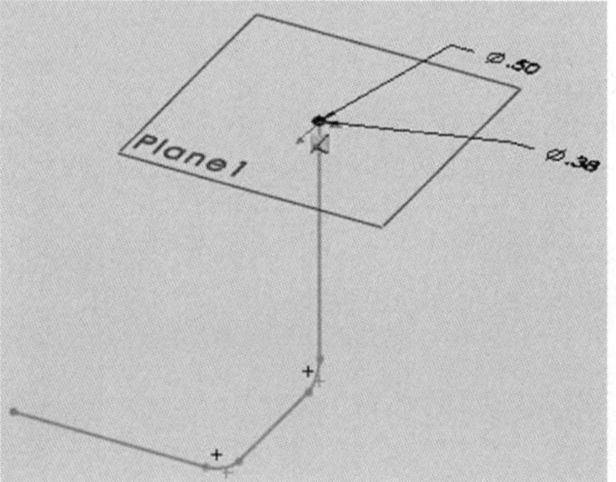

Figure 20.4

Normal plane and sketches for two circles.

Tutorial 20.3: Handlebar Tube

Model a handlebar tube with a 25.4-mm outer diameter and a 23.4-mm inner diameter to join the ends of two tubes, as shown in Figure 20.8. Follow the route that is shown by the centerlines in Figure 20.8 and add fillets to the corners as specified.

Solution

1. Select the 3DSketch tool by clicking on the Sketch tab of the CommandManager.
2. Select the Line Tool. Note that beside the cursor, the plane that the line will be displayed.
3. Press the Tab key until when the plane selected is YZ.
4. Sketch a line 160-mm long from the origin along Z.
5. Sketch a line 120-mm long from the origin along Y.
6. Sketch a line 80-mm long from the origin along –Z.
7. Press the Tab key until when the plane selected is ZX.
8. Sketch a line 200-mm long along –X.
9. Add Fillets to the corners as specified.
10. 3DSketch1 is complete (see Figure 20.9).
11. Exit 3DSketch.
12. Click the end of the 160-mm-long line.
13. Click Features > Reference Geometry.
14. Select top vertex and the vertical line that is connected to it, for Selections.

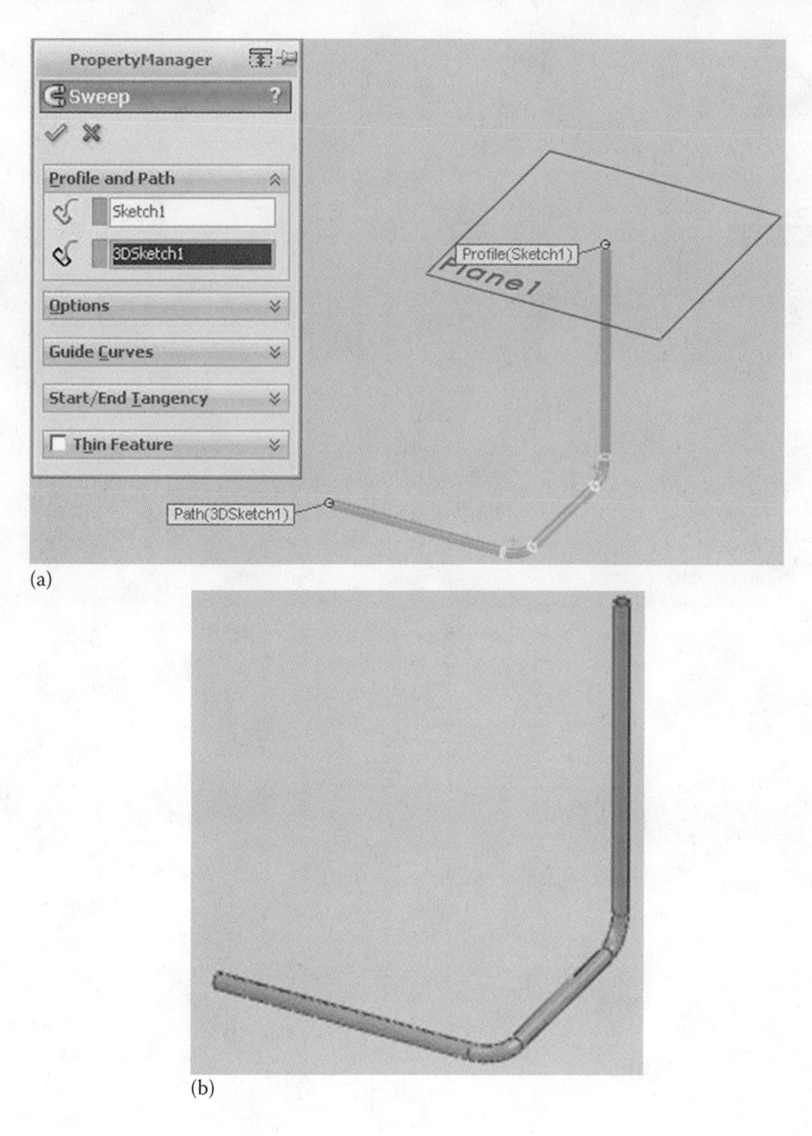

Figure 20.5

Part for Exercise 1: (a) preview and (b) part.

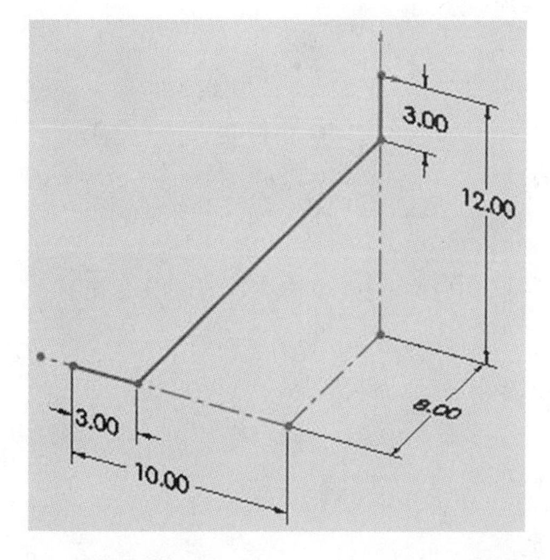

Figure 20.6

Path for a bent tube model.

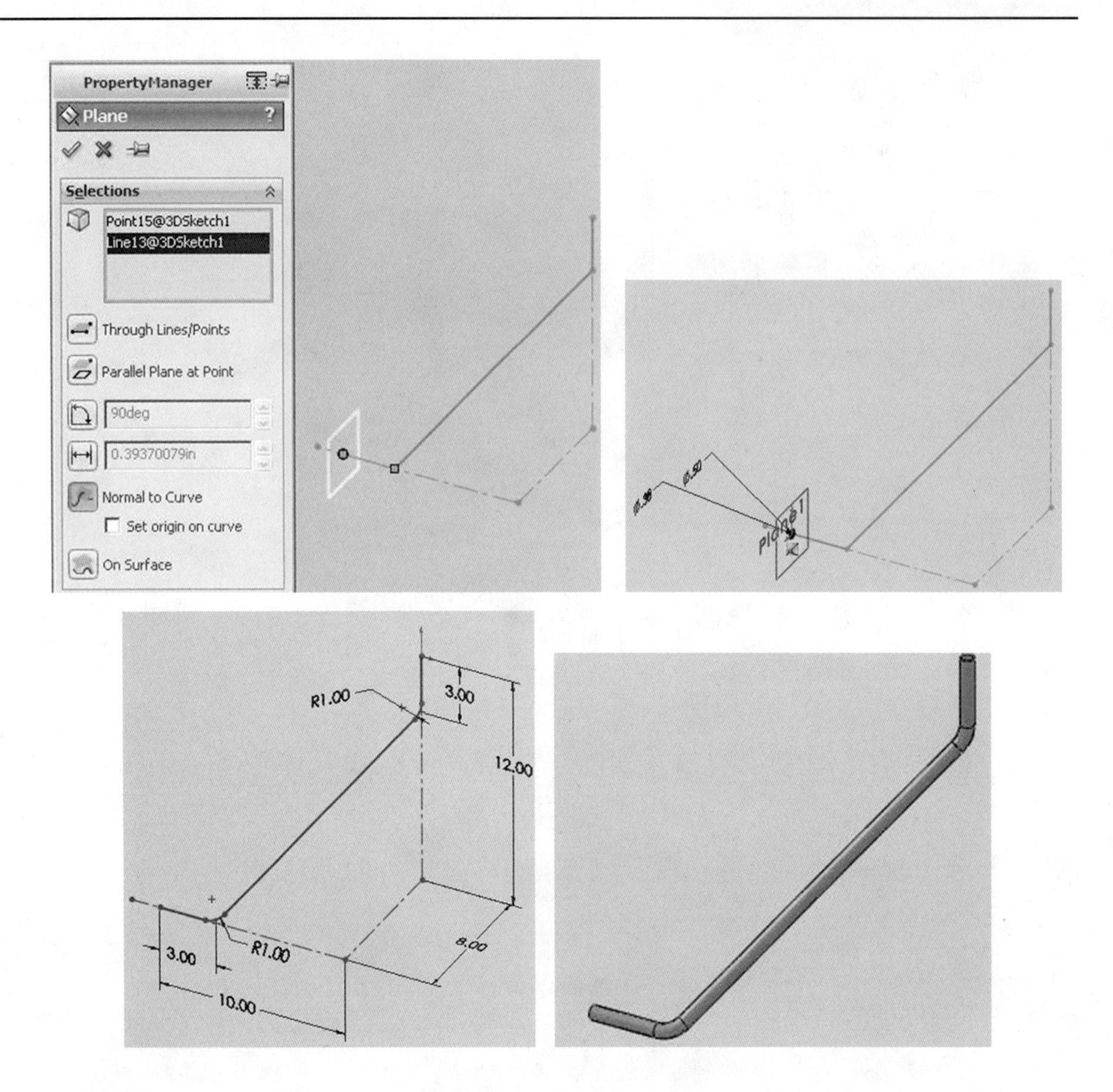

Figure 20.7

Part for Exercise 1.

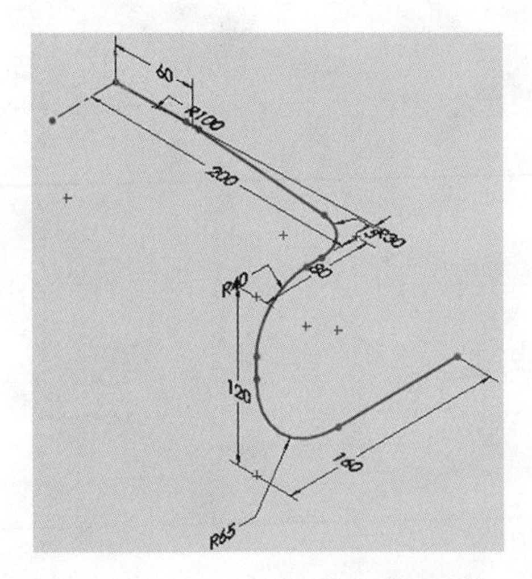

Figure 20.8

Specified route.

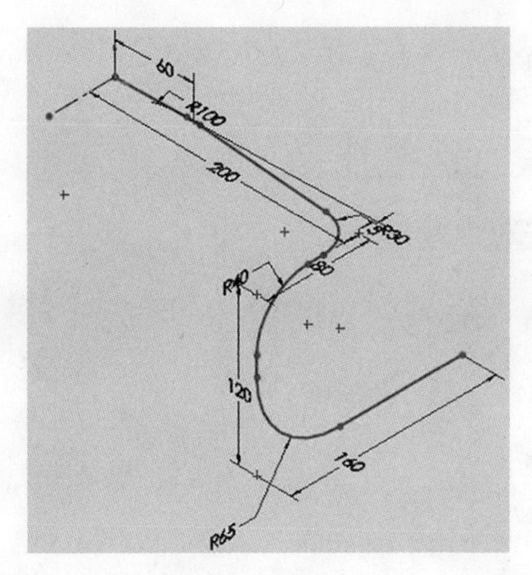

Figure 20.9

3DSketch of route.

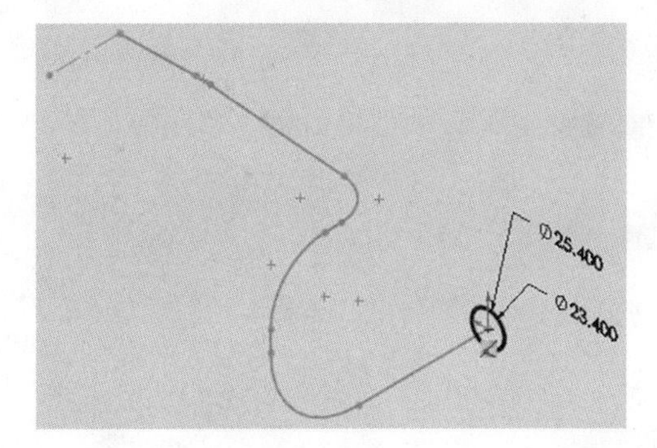

Figure 20.10

Normal plane and sketches for two circles.

15. Select Normal to Curve to complete defining Plane1.
16. Select Plane1.
17. Right-click and select the Sketch tool to sketch the two circles with 25.4-mm outer diameter and 23.4-mm inner diameter, as Sketch1; dimension them (see Figure 20.10).
18. Exit Sketch.

Figure 20.11 shows the preview, while Figure 20.12 shows the final model.

Now that we have a good grasp of the 3DSketch tool, which is a prerequisite for creating weldments, it becomes logical to present the Weldments toolbar.

Weldments Toolbar

The Weldments toolbar provides tools for creating weldment parts. The main tools are as follows:

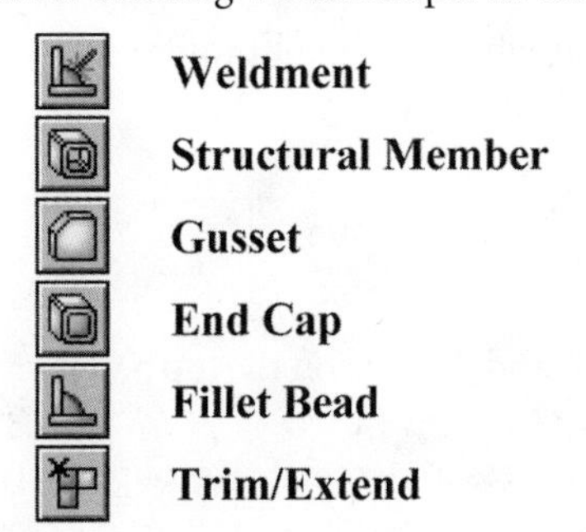

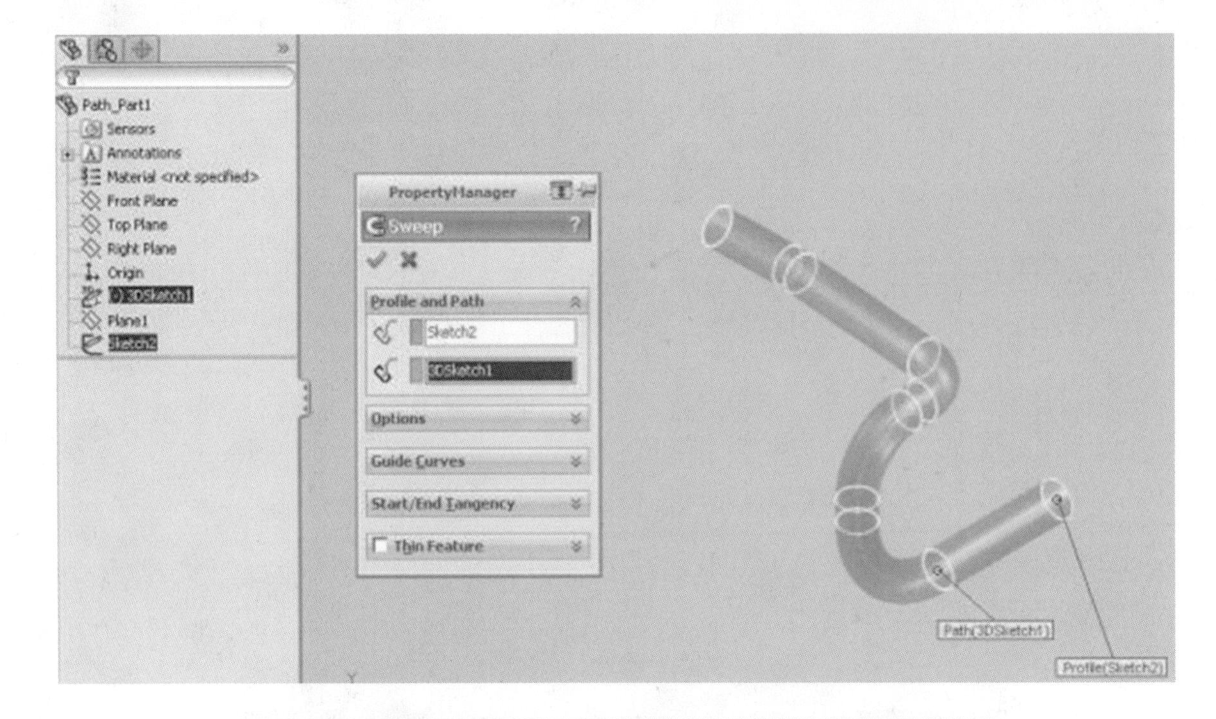

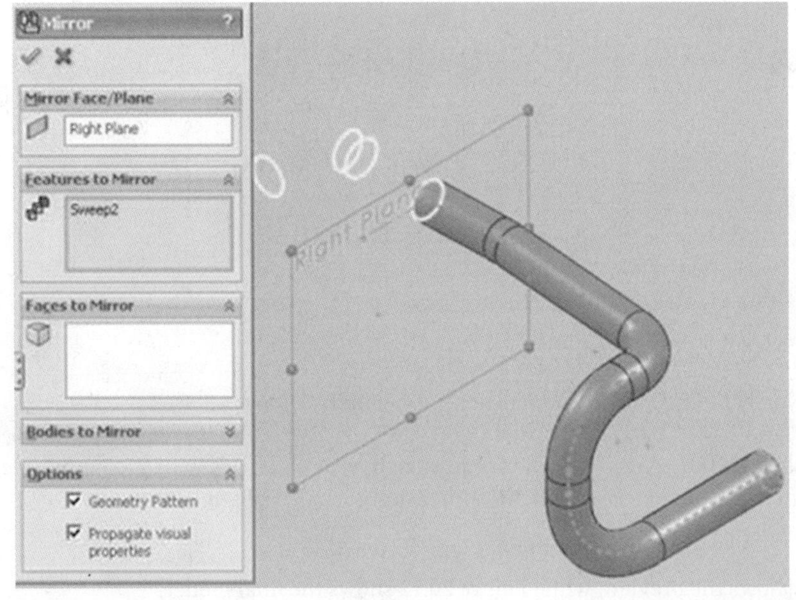

Figure 20.11

Preview of model.

Structural Member

The first step is to define the paths for which different profiles will be created. The paths are created using 3DSketch.

1. Click the MY-TEMPLATE tab.
2. Double-click ANSI-MM-PART.
3. Click Top Plane.
4. Using 3DSketch mode, sketch and dimension the top path (a rectangle, 20 in. × 12 in.). This is in the X–Y plane (see Figure 20.13).
5. Exit 3DSketch mode. This is necessary to create a group of structural members.

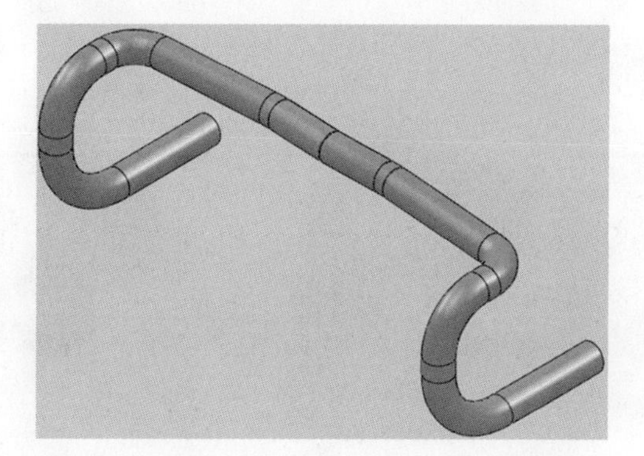

Figure 20.12

Model of handlebars.

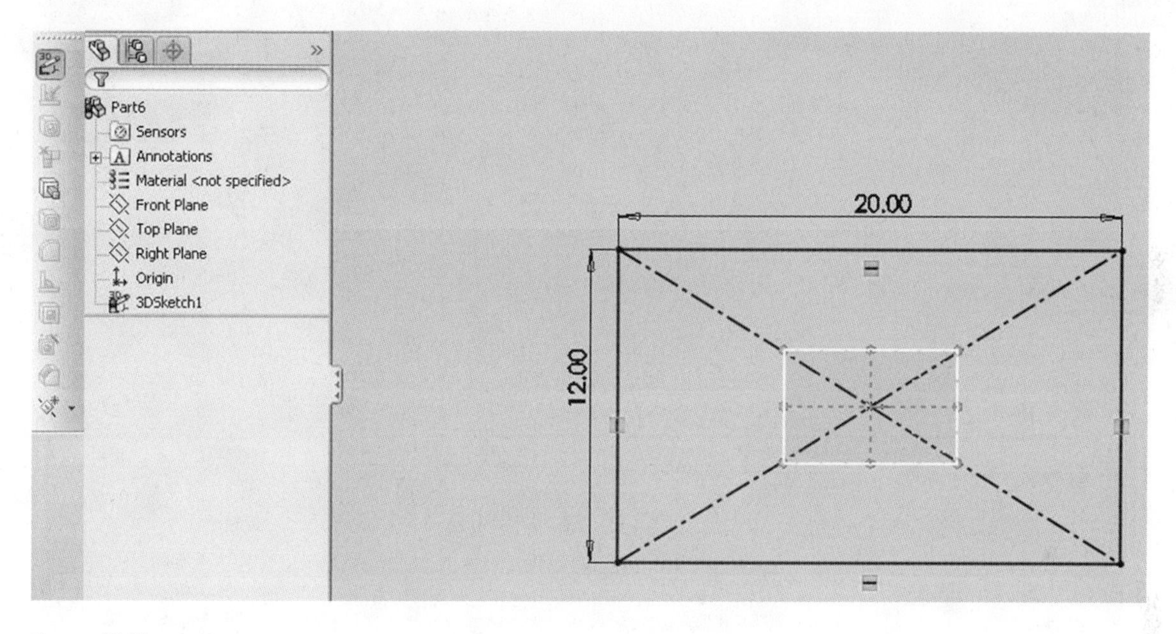

Figure 20.13

Top path (a rectangle, 20 in. × 12 in.) in the X–Y plane.

6. Click Top Plane.
7. Click 3DSketch in CommandFeature.
8. Pick one of the vertices of the Top rectangular feature and Sketch a line. (Switch to YZ plane by pressing Tab key.)
9. Sketch the other legs and use relations to make all four legs parallel and equal.
10. Dimension one of them to be 10 in.
11. Add a relation that one of the legs and the top members are perpendicular (90°) (see Figures 20.14 and 20.15). The complete sketch is shown in Figure 20.15.
12. Exit 3DSketch.
13. Click Structural Member toolbar.
14. In the Selection rollout, specify the profile of the structural member by selecting the following (see Figure 20.16):
    a. *Standard.* Select ansi inch.
    b. *Type.* Select a Profile Type, square tube.
    c. *Size.* Select a Profile, such as 2 × 2 × 0.25.
15. Select all four members of the Top elements as Group1.

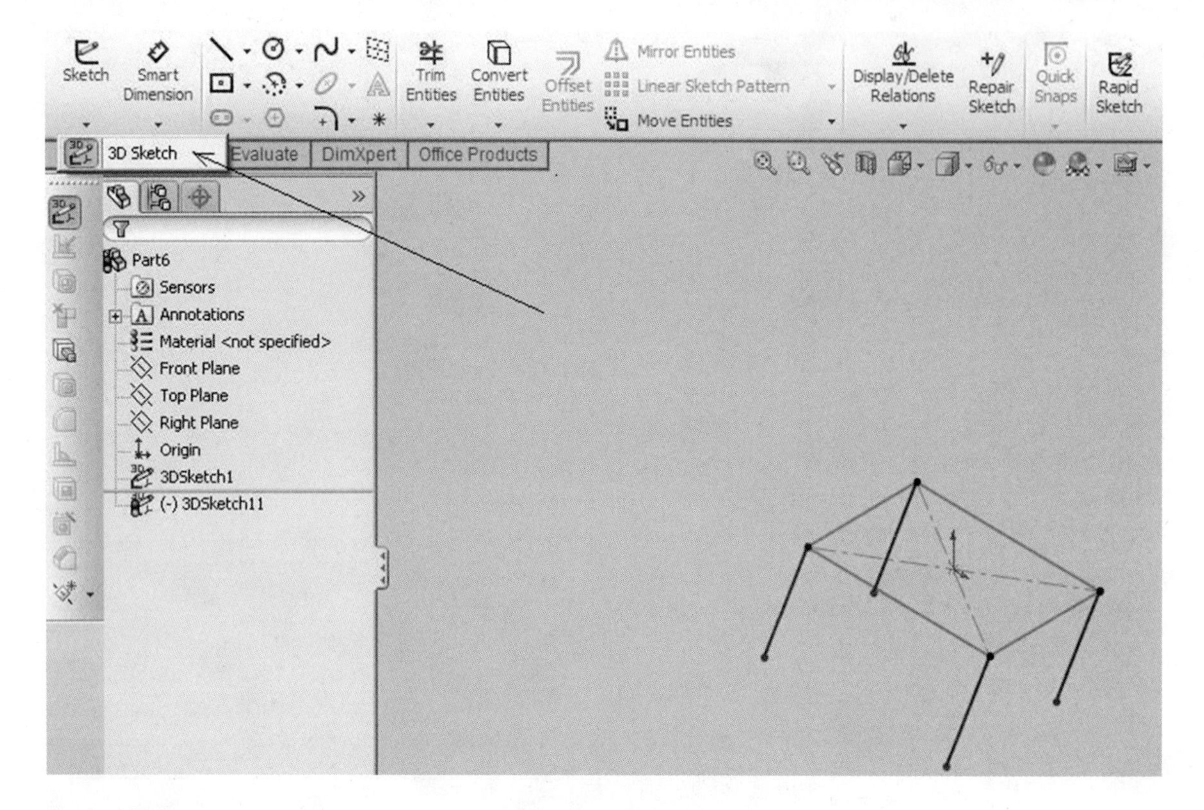

Figure 20.14

3DSketch in CommandFeature necessary for creating weldments.

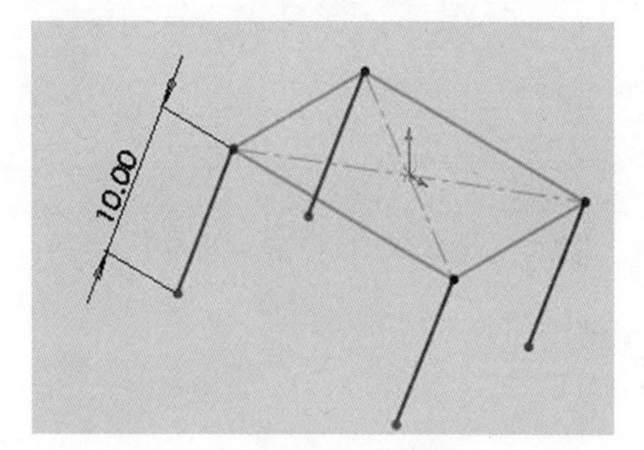

Figure 20.15

Add relations between all legs and top path.

16. Click New Group button and select all 4 legs as Group2.
17. Click OK.
18. In the Settings rollout, select Apply corner treatment and click End Meter.

Expanding the Cut list shows that there are eight multibodies; these are also included in the Structural Member1 if it is expanded, as shown in Figure 20.17.

Trimming the Structural Members

There are cases in which structural members have to be trimmed. For example, if instead of defining Group1 and Group2 at the same time of defining the structural members, they are defined as two different structural members, then the need for trimming may arise. This scenario is shown in Figure 20.18. Notice that there

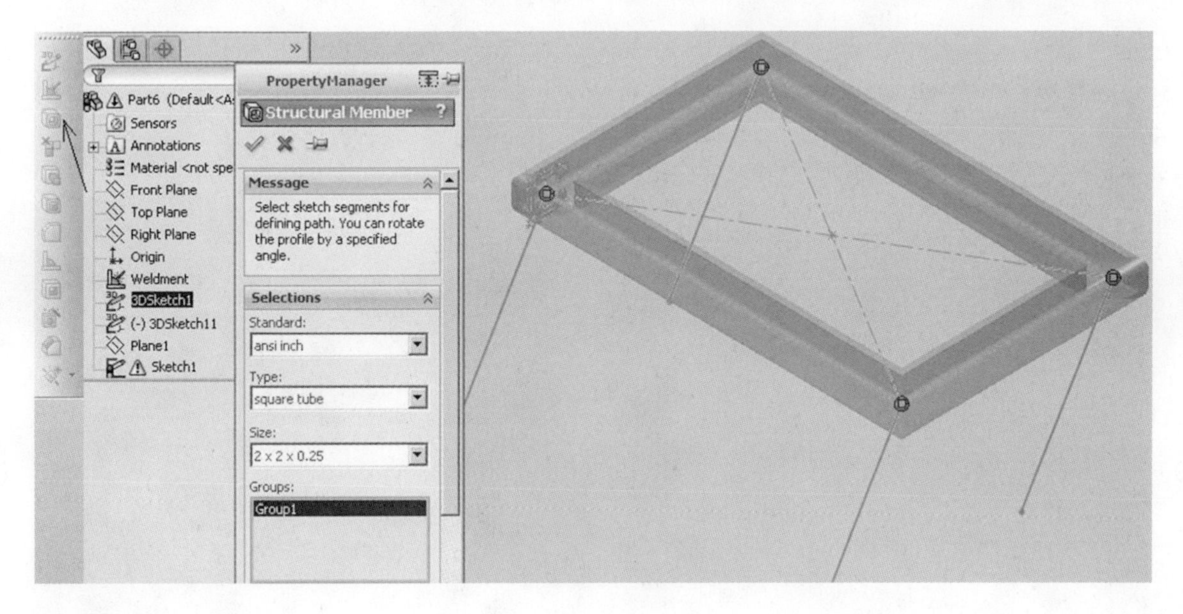

Figure 20.16

Structural Member tool.

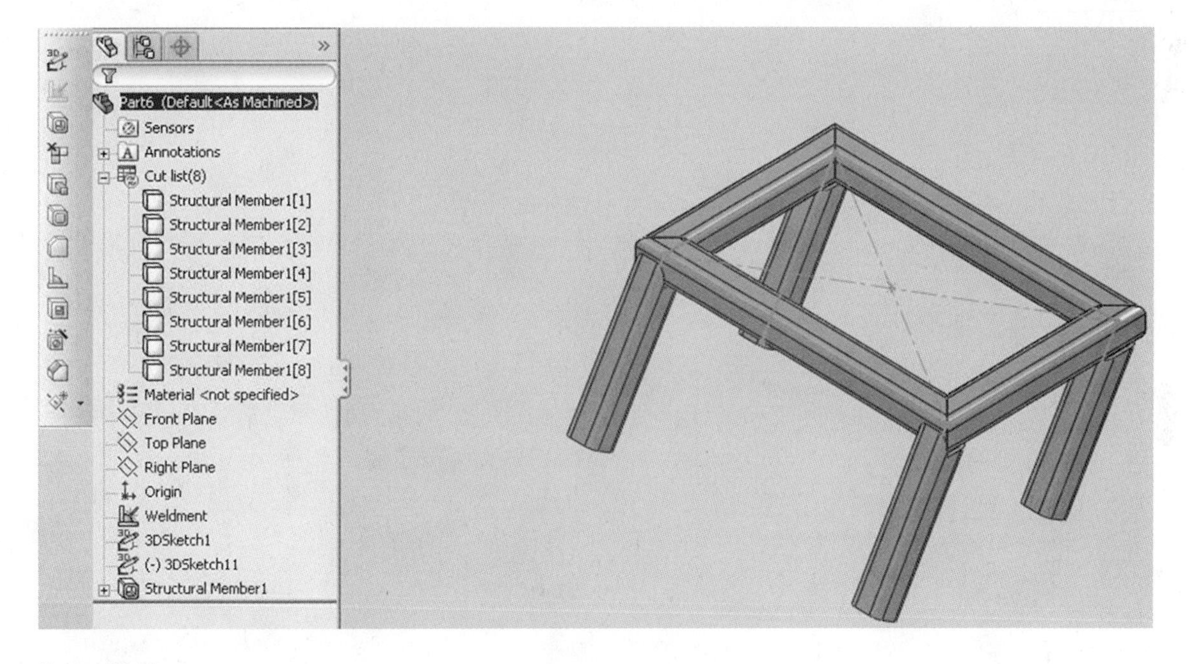

Figure 20.17

Cut list for eight multibodies' structural members.

are Structural Member1 and Structural Member2. However, trimming is now needed because some solid bodies are entering into other solid bodies, which is not allowed in practice.

19. To trim, click Trim/Extend.
20. Accept Corner Type as End Trim (see Figure 20.19).
21. Select one of the vertical legs as the Bodies to be Trimmed.
22. Select the Bodies radio button for Trimming Boundary.
23. Click the bounding surfaces.

Repeat the process for all other legs.

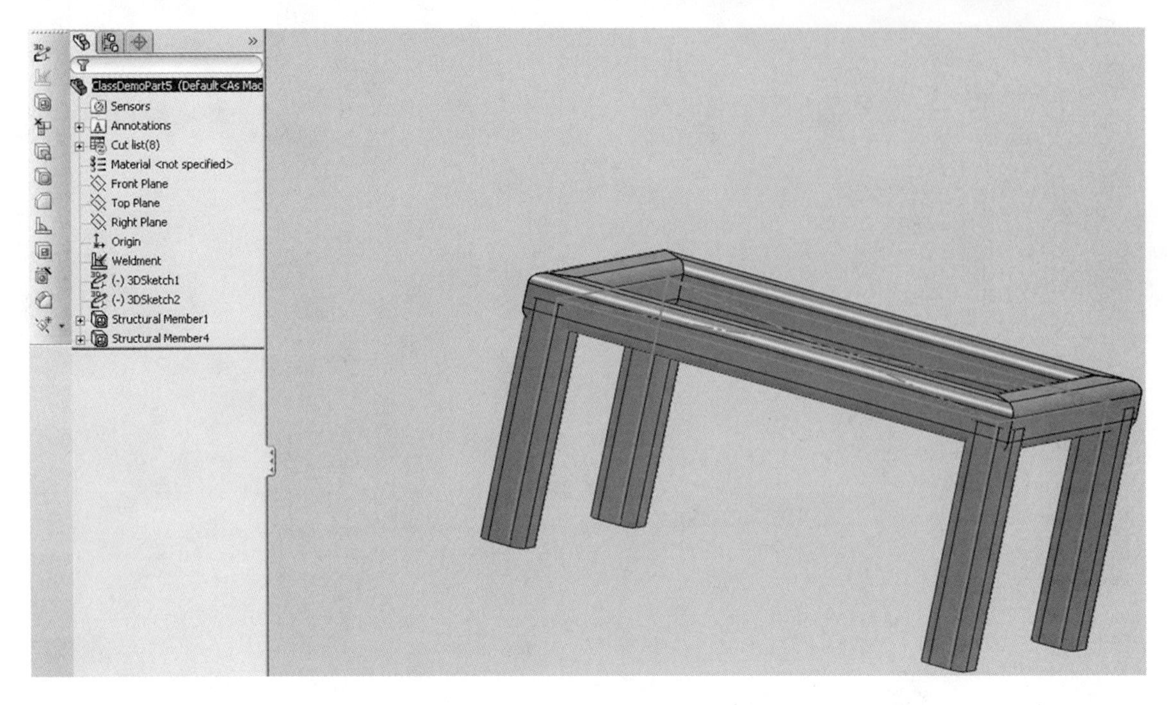

Figure 20.18

Structural Member requiring trimming operation.

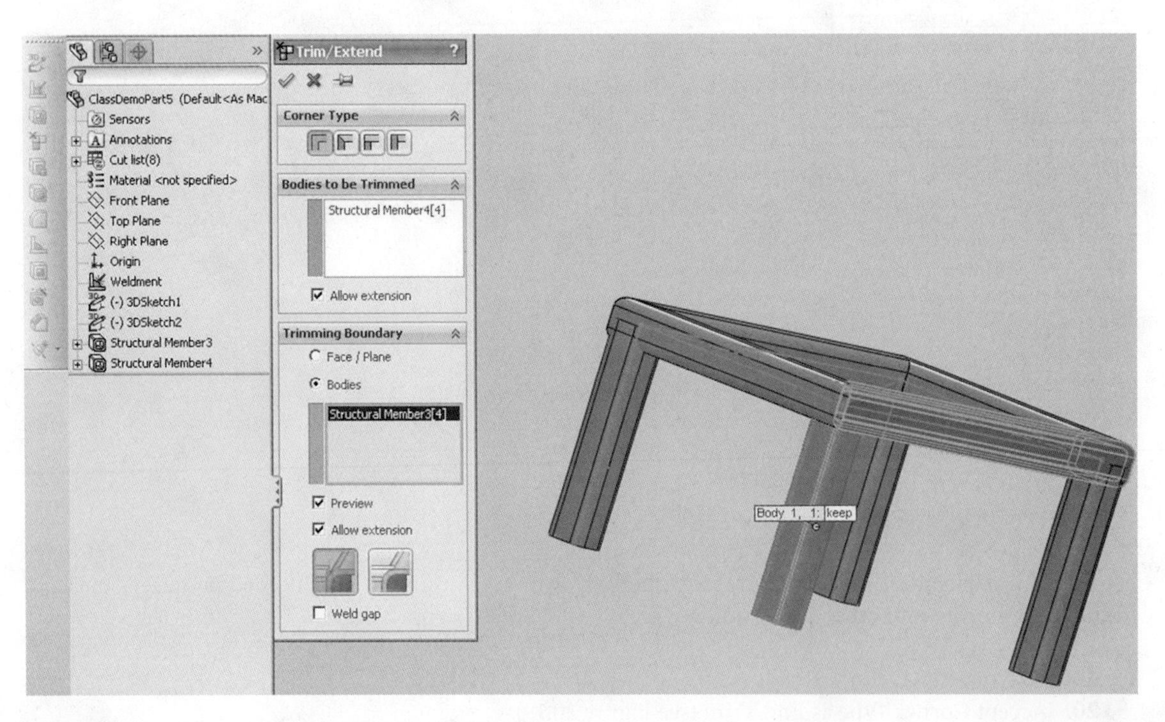

Figure 20.19

Trim/Extend option.

Adding End Caps to Structural Members

Let us add end caps to close the open ends of the segments of the structural members of interest. In this example, the bottom ends of the four legs need to be capped, so they should be selected one at a time (see Figure 20.20).

24. Click the End Cap option of the Weldments toolbar.
25. Click the face (strip) that defines the cross section of the leg structure as the Parameter.
26. Select Thickness direction to Inward to make the end cap flush with the original extent of the structure.
27. Set the thickness to 0.25 in.
28. Select Use thickness ratio.
29. Set Thickness Ratio to 0.65. (Try other values and observe the preview.) The capped legs are shown in Figure 20.21.

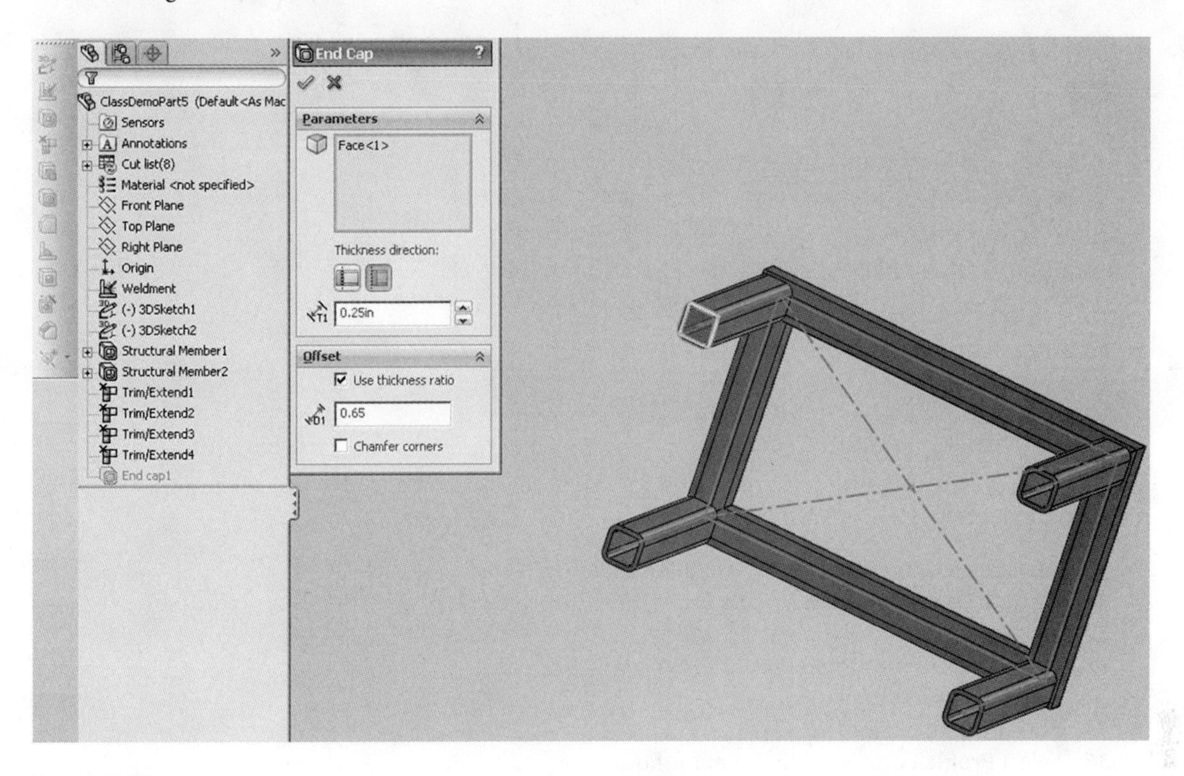

Figure 20.20

End cap option.

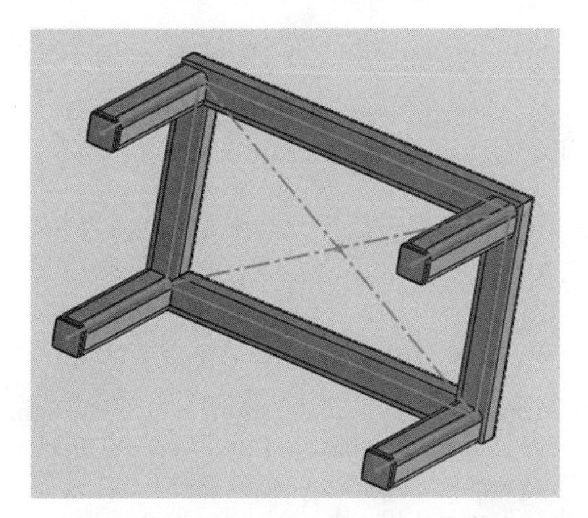

Figure 20.21

Structural Members with end caps.

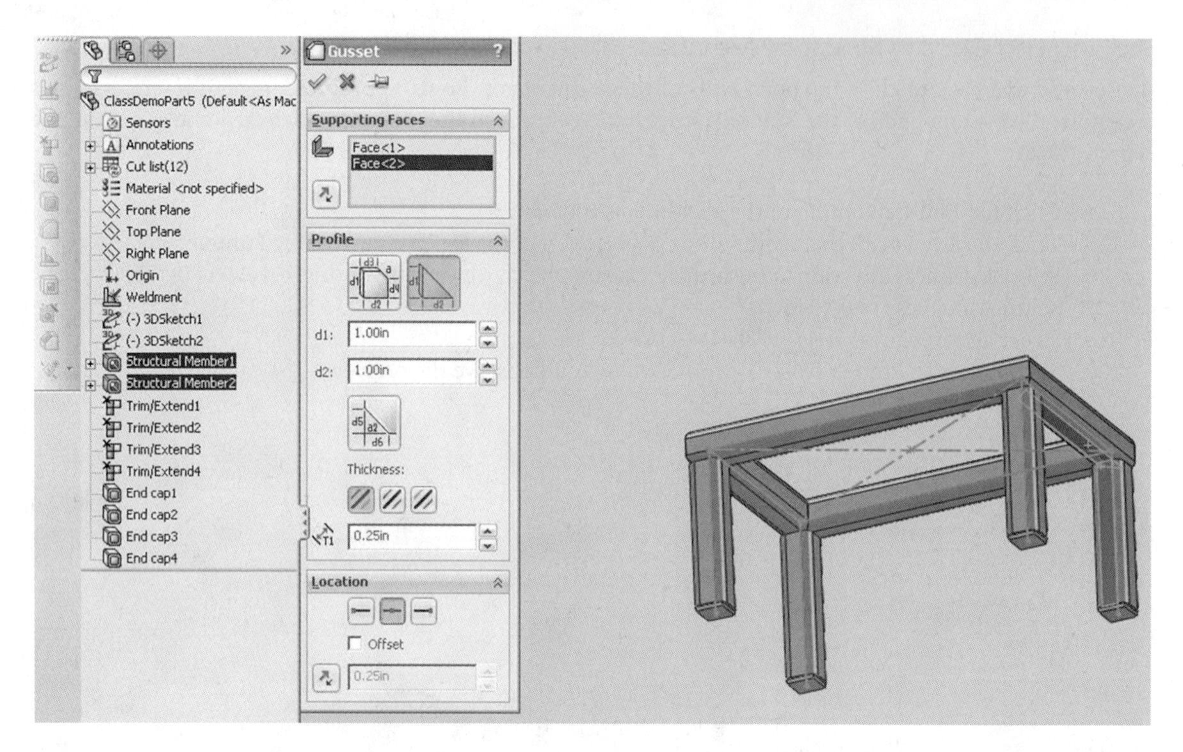

Figure 20.22

Gusset option.

Adding Gussets to Structural Members

Gussets features, which may be triangular or polygonal profiles, can be added to two adjoining planar faces. Let us add gussets to the corners of the four legs.

30. Click the Gusset option of the Weldments toolbar.
31. Click the two faces (Face<1> and Face<2>) as shown for the Supporting Faces (see Figure 20.22).
32. Under Profile, click Triangular Profile.
33. Set Profile Distance1 and Profile Distance2 to 1 in. each.
34. Select Thickness to inner Side.
35. Set Gusset thickness to 0.25 in.
36. Select Profile Locates at Mid Point for Location.

Repeat steps 30–36 for each leg.

Adding Fillet Beads to Structural Members

Fillet beads features can be added to two adjoining planar faces. Let us add fillet beads to the corners of the four legs.

37. Click the Fillet Bead option of the Weldments toolbar.
38. Click Gusset face.
39. Click the two faces (Face<1> and Face<2>) as shown for the Supporting Faces (see Figure 20.23).

Repeat for each leg.

Weldment of Parts

In this section, it is demonstrated that weldment can be applied to part design. We will first create a part that is made up of an inverted channel shape as the base and two vertical brackets. Figure 20.24 shows *Sketch1* for the base, while Figure 20.25 shows the extruded base of Sketch1.

A rectangular sketch, Sketch2, 70 mm × 10 mm, which is offset 15 mm from the horizontal top edges, is used to create a slot, as shown in Figures 20.26 and 20.27.

Extrude-cut is mirrored about the center of the base to create two slots.

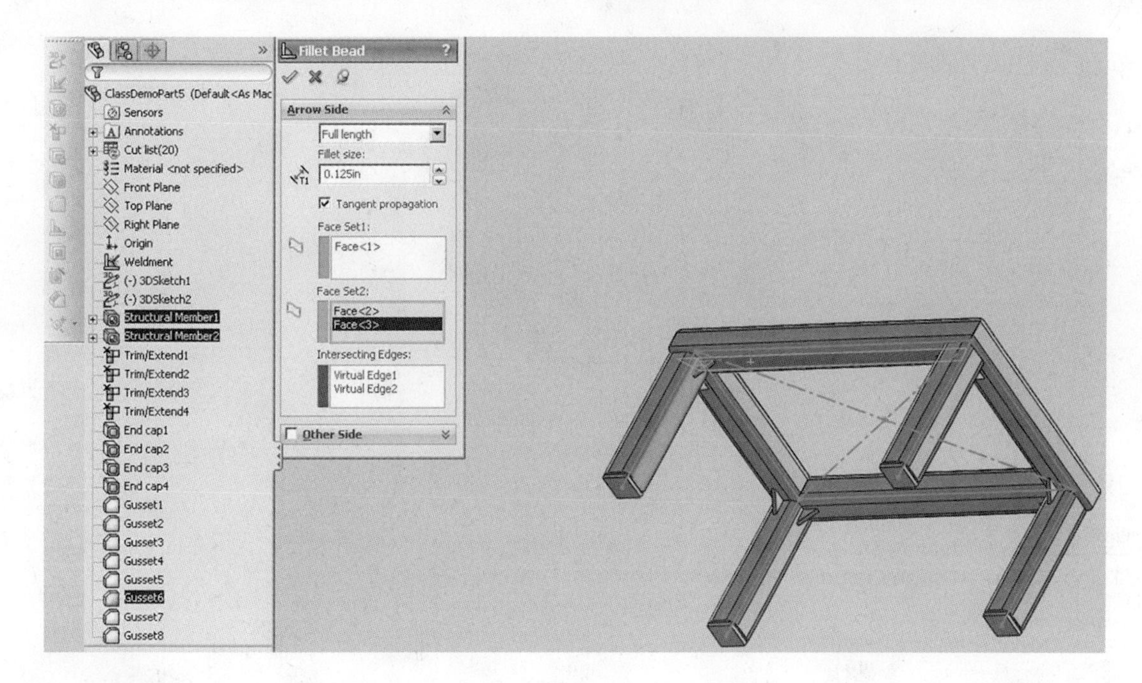

Figure 20.23

Fillet bead option.

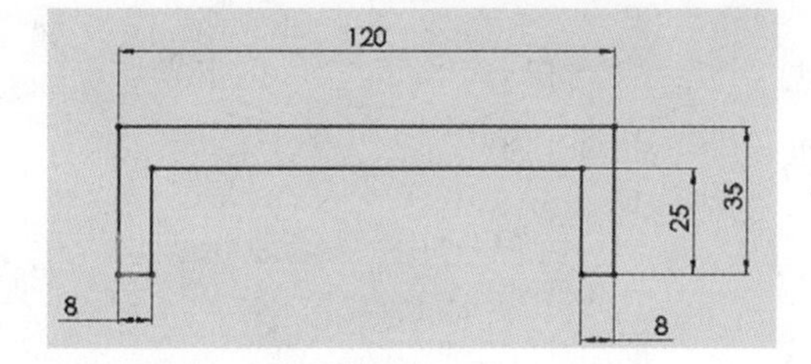

Figure 20.24

Sketch1 for the base.

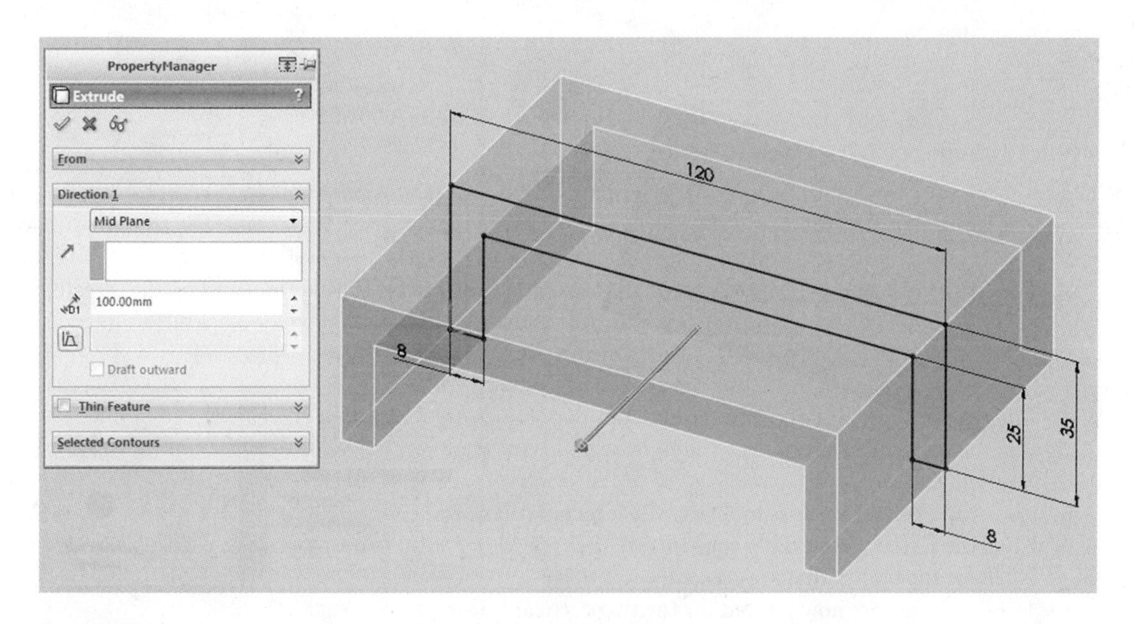

Figure 20.25

Extruded base of Sketch1.

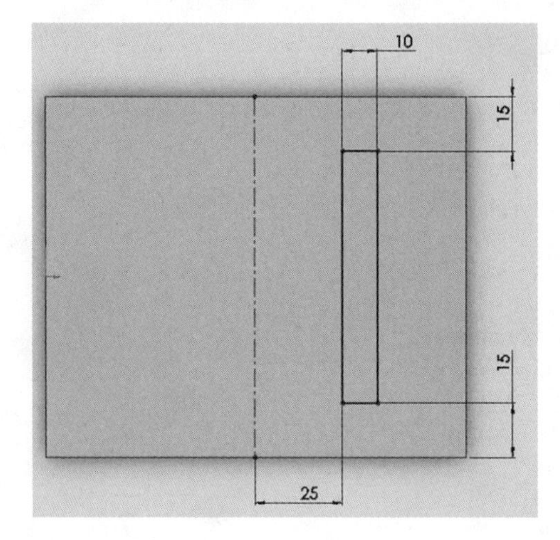

Figure 20.26

Sketch2 for slot through the base.

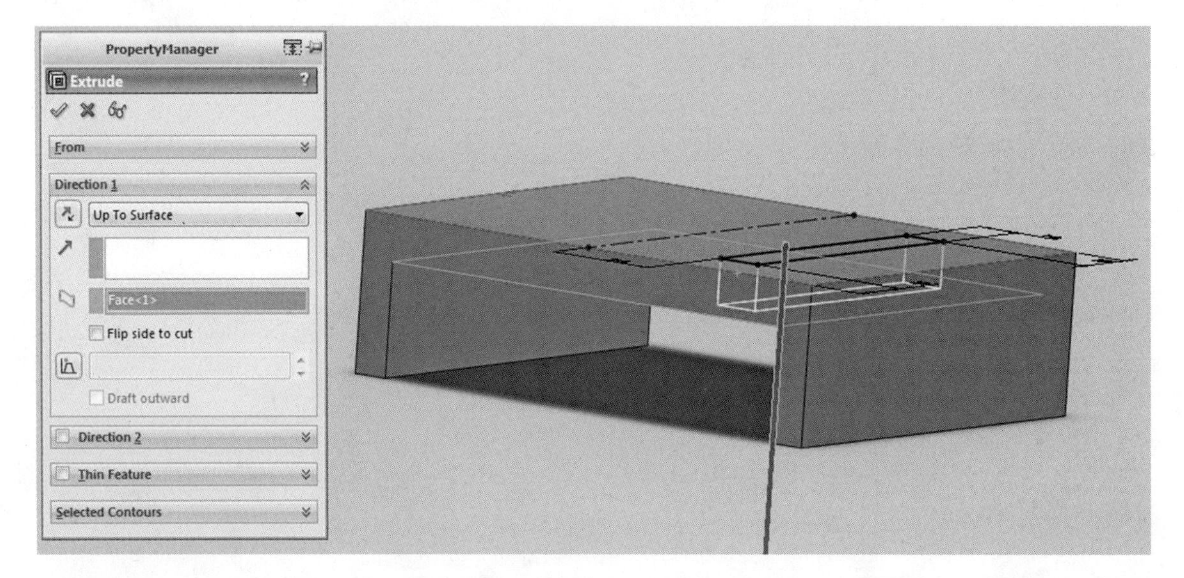

Figure 20.27

Extrude-cut for the slot.

Assembly versus Solid Bodies

When an assembly of parts is created, the Weldment tool does not work. However, when Solid Bodies are created, the Weldment tool functions appropriately. The point made here is important: use solid bodies for the weldments of parts. The principle is now illustrated.

The edges of the Extrude-cut features are extracted using the Convert Entities tool and used to create side features by extrusion. The Merge Results options are unchecked during the extrude operation in order to create solid bodies.

1. Click the bottom face of the horizontal upper portion of the base that is created.
2. Be in sketch mode and select the *four edges of the right slot (Extrude-cut feature)* that is already created.
3. Click the Convert Entities tool to create a rectangular profile.
4. Click the Extrude Boss/Base tool; ensure that the Merge Results options are unchecked. See Figure 20.28 for the bodies that are created.

   Fillet beads are now created for the two vertical features.
5. Click Weldment tool to enable the weldment environment (see Figure 20.29).
6. Click Fillet Bead tool and select one face of the left-side feature, as well as the top of the base.

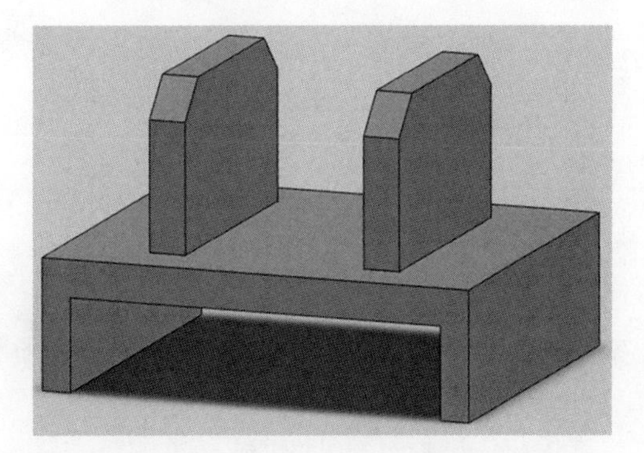

Figure 20.28

Base and side features as three solid bodies.

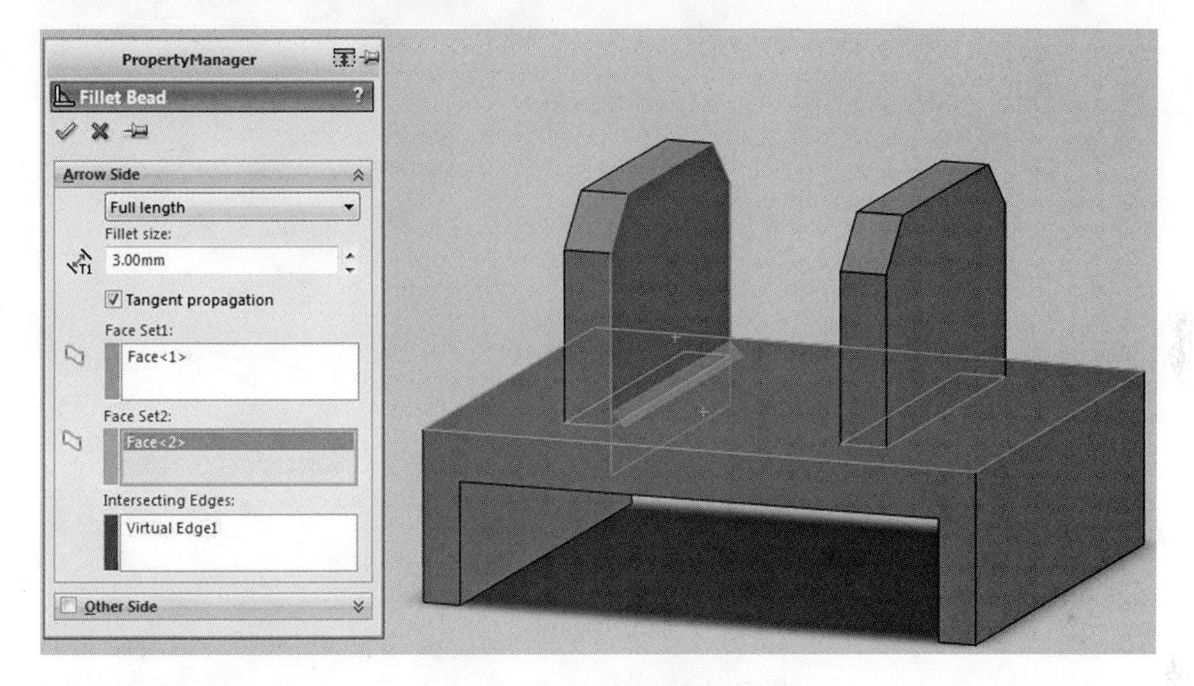

Figure 20.29

First fillet bead is created for the top-left vertical feature.

7. Click Fillet Bead tool and select one face of the left-side feature, as well as the top of the base (see Figure 20.30).
8. Click Fillet Bead tool and select one face of the left-side feature, as well as the top of the base (see Figure 20.31).
9. Click Fillet Bead tool and select one face of the left-side feature, as well as the top of the base (see Figure 20.32).
10. Click Fillet Bead tool and select one face of the right-side feature, as well as the top of the base (see Figure 20.33).
11. Click Fillet Bead tool and select one face of the right-side feature, as well as the top of the base (see Figure 20.34).
12. Click Fillet Bead tool and select one face of the right-side feature, as well as the top of the base (see Figure 20.35).
13. Click Fillet Bead tool and select one face of the right-side feature, as well as the top of the base (see Figure 20.36).

The completed weldment model is shown in Figure 20.37. It is important to note that there are 11 members in the weldment Cut list (see Figure 20.38); *three* from the part model and *four* each from the fillet beads around the *two* vertical features (3 + 4 + 4 = 11).

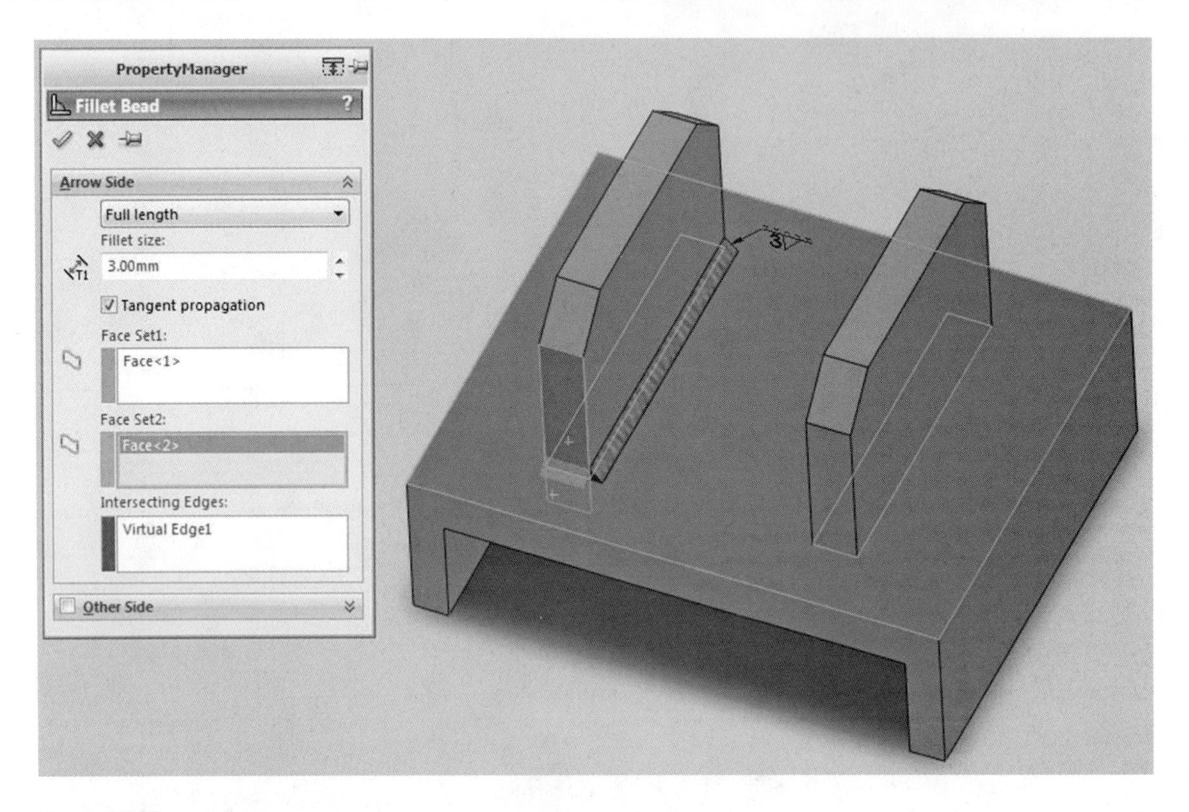

Figure 20.30

Second fillet bead is created for the top-left vertical feature.

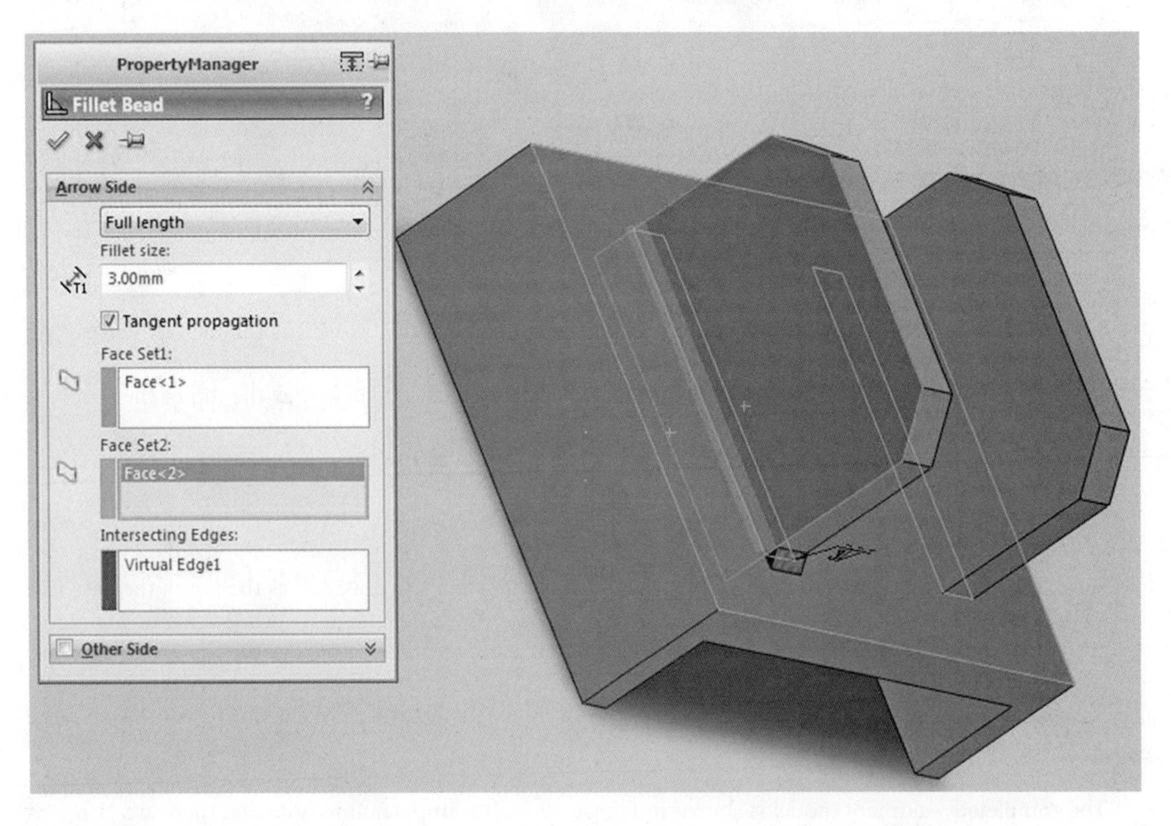

Figure 20.31

Third fillet bead is created for the top-left vertical feature.

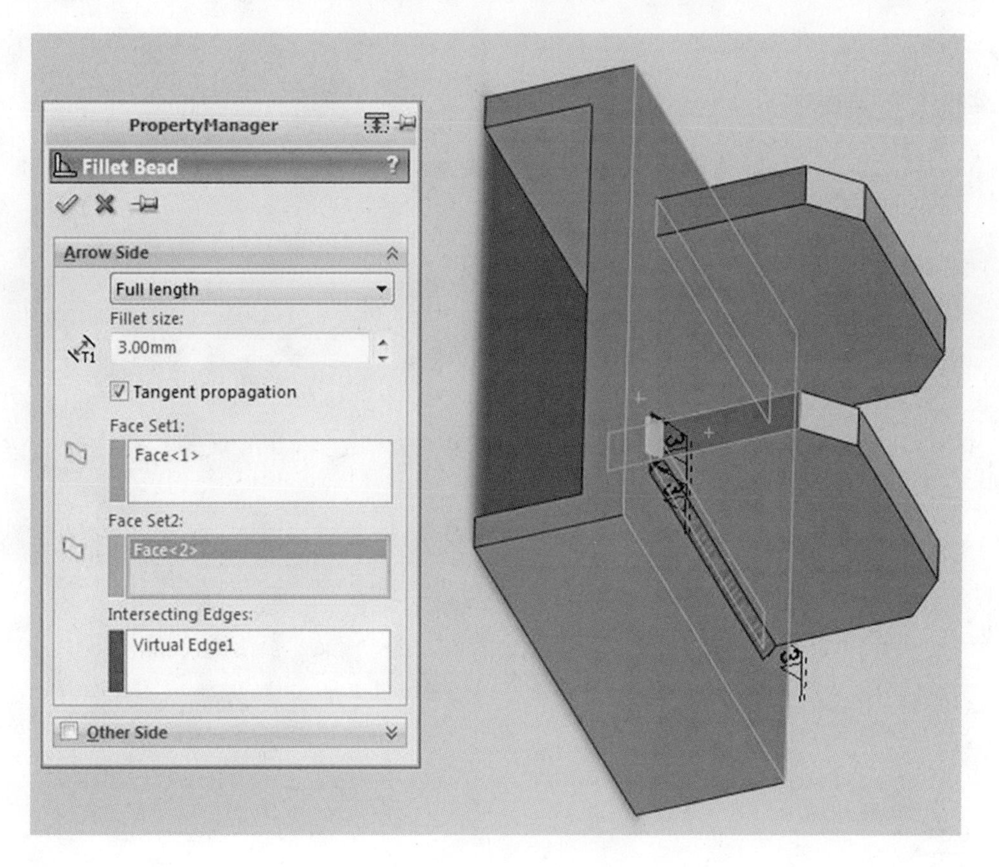

Figure 20.32

Fourth fillet bead is created for the top-left vertical feature.

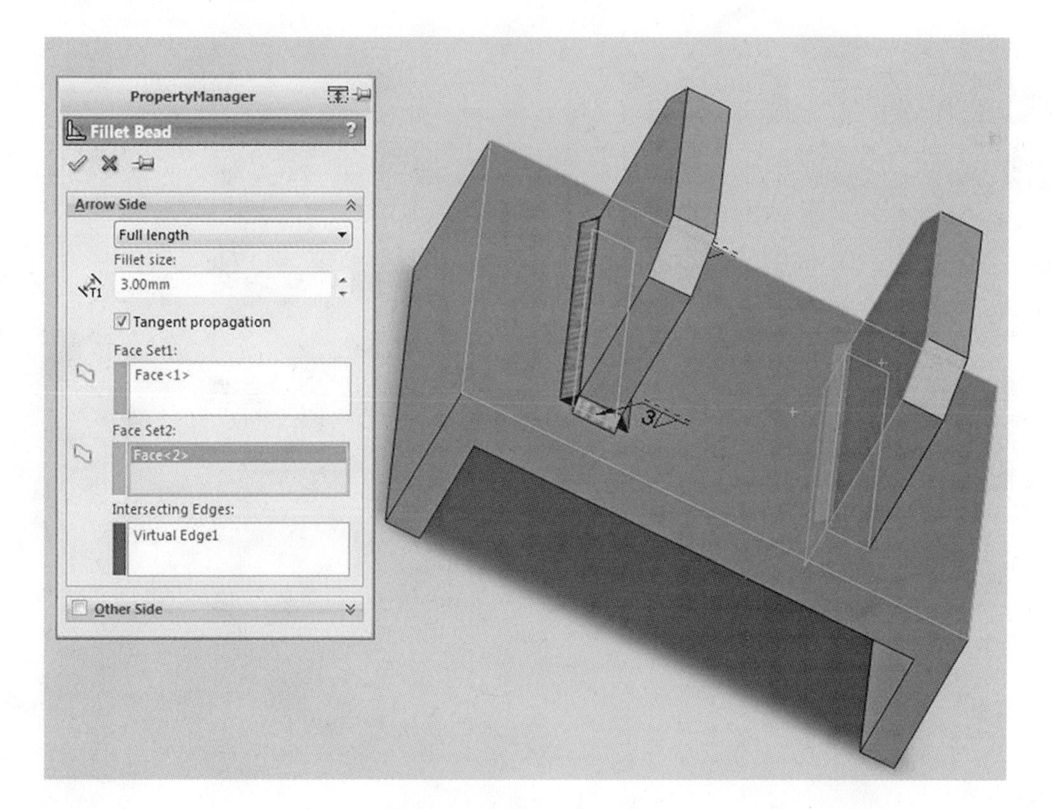

Figure 20.33

First fillet bead is created for the top-right vertical feature.

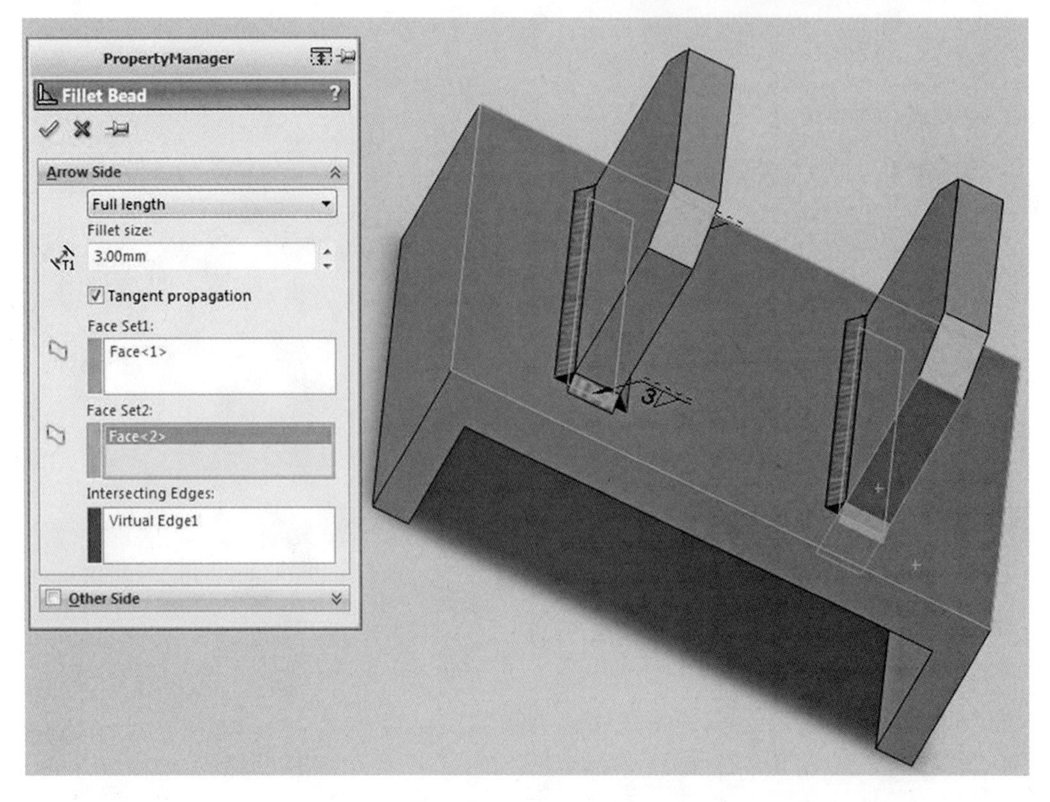

Figure 20.34

Second fillet bead is created for the top-right vertical feature.

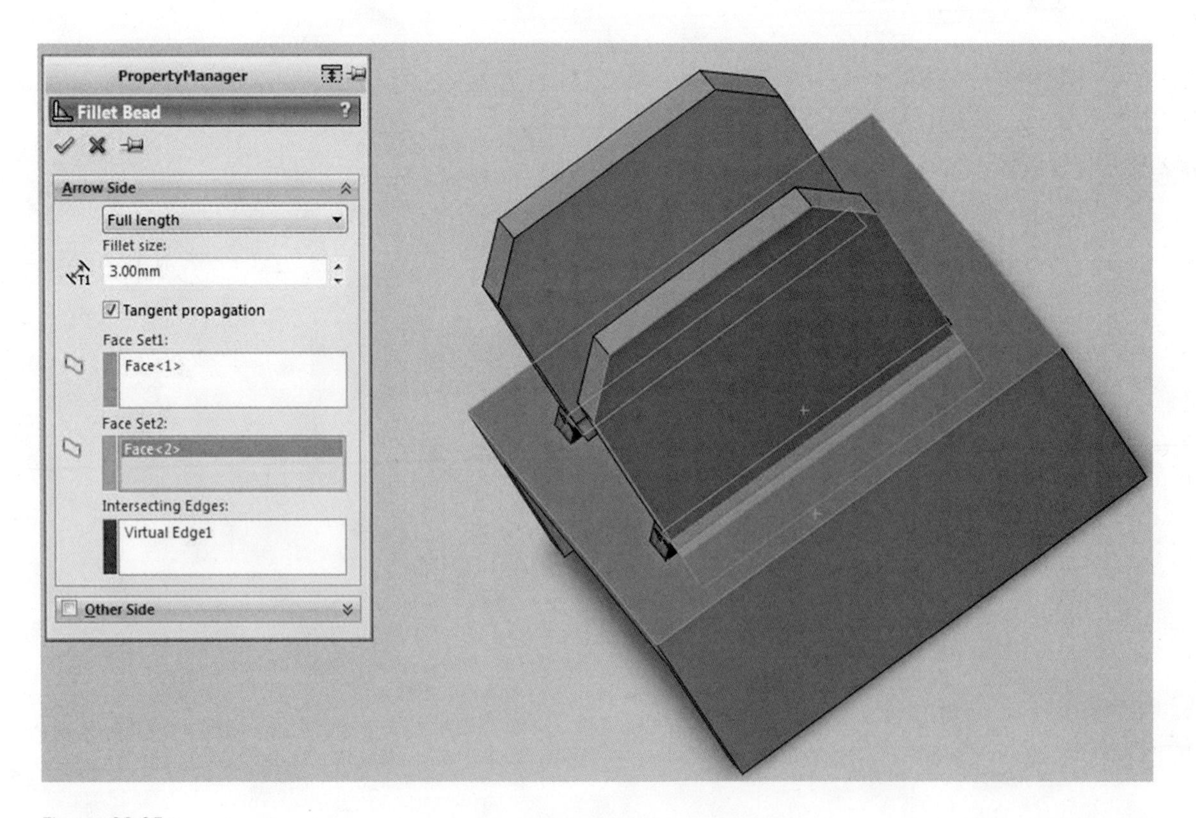

Figure 20.35

Third fillet bead is created for the top-right vertical feature.

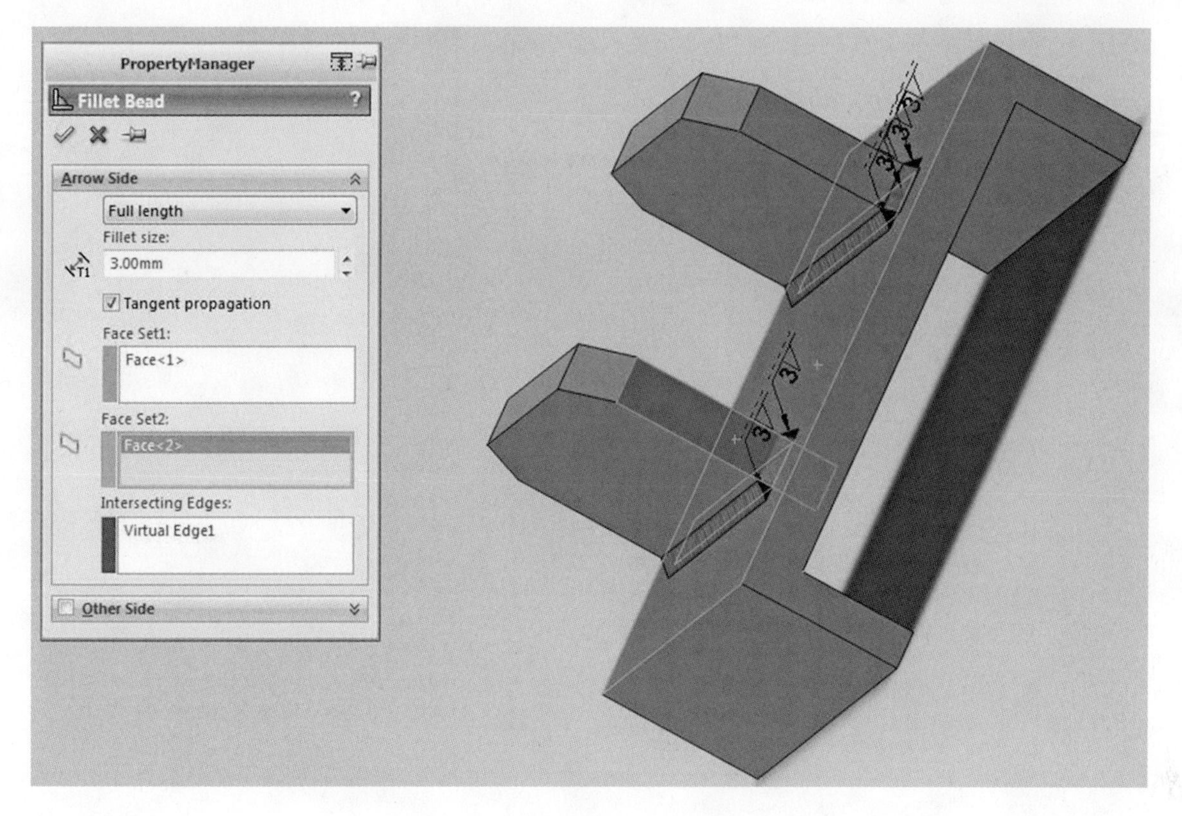

Figure 20.36

Fourth fillet bead is created for the top-right vertical feature.

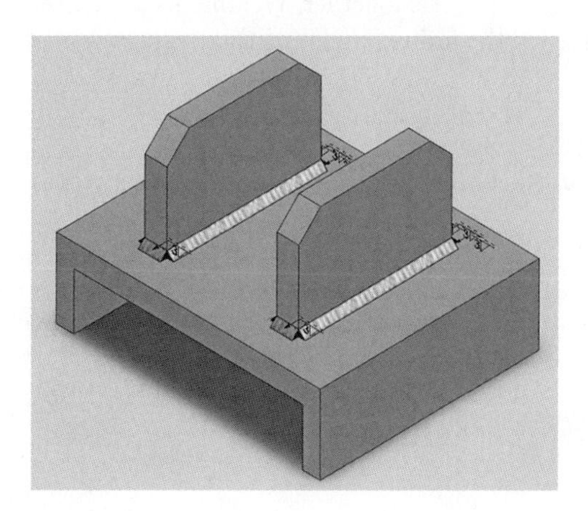

Figure 20.37

Completed weldment of base and two side features.

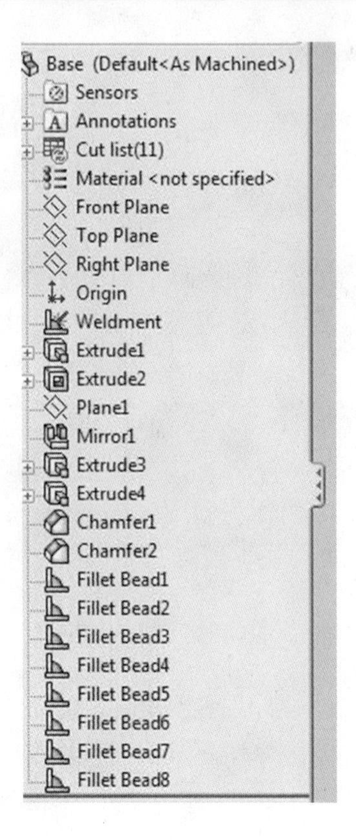

Figure 20.38

FeatureManager for designed weldment.

Summary

In this chapter, we have described the methods for creating weldments from defined paths using the 3DSketch tool and creating weldments from part design. These are two approaches. The first approach is considered the choice for modeling trusses. The second approach is the choice when we are modeling solid parts and intend welding some of the parts together. Weldments are extremely useful in modeling trusses and beams for stress analysis using the finite element method.

Exercises

1. Create the truss structure that is shown in Figure P20.1. The material is steel pipe, NPS 1½″, Schedule 40.
2. Create the truss structure that is shown in Figure P20.2. The material is steel pipe, NPS 1½″, Schedule 40.
3. Create the truss structure that is shown in Figure P20.3. The material is steel pipe, NPS 1½″, Schedule 40.

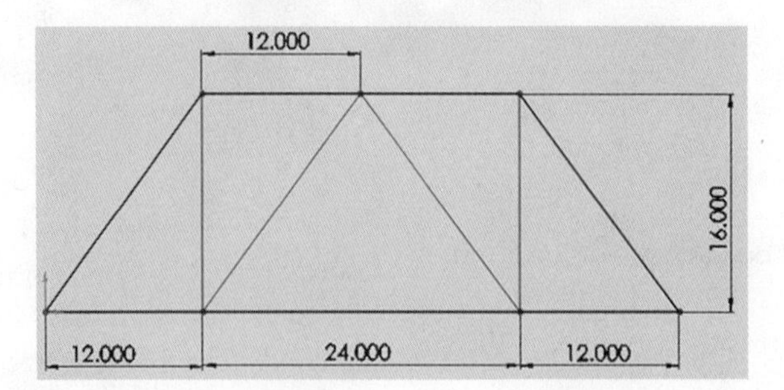

Figure P20.1

Truss 1.

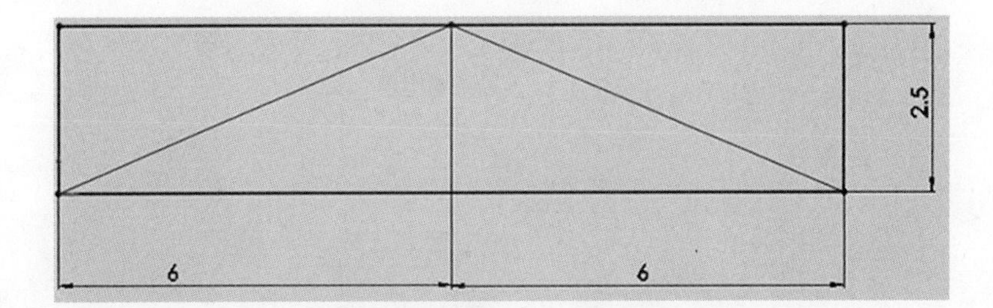

Figure P20.2

Truss 2.

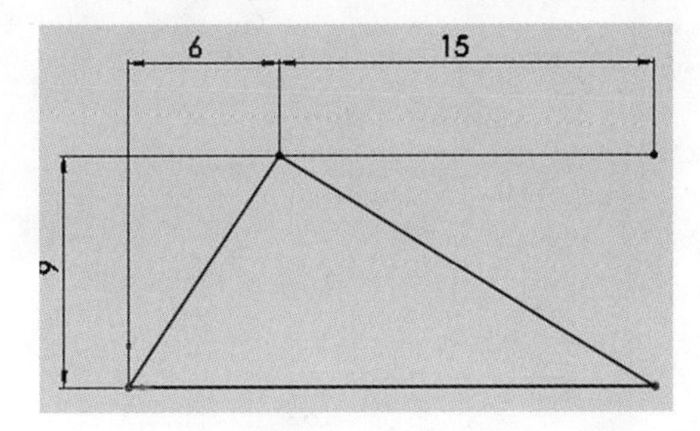

Figure P20.3

Truss 3.

4. Create the truss structure that is shown in Figure P20.4. The material is steel pipe, NPS 1½″, Schedule 40.
5. Create the truss structure that is shown in Figure P20.5. The material is steel pipe, NPS 1½″, Schedule 40.
6. Create the truss structure that is shown in Figure P20.6. The material is steel pipe, NPS 1½″, Schedule 40.
7. Create the truss structure that is shown in Figure P20.7. The material is steel pipe, NPS 1½″, Schedule 40.
8. Create the truss structure that is shown in Figure P20.8. The material is steel pipe, NPS 1½″, Schedule 40.
9. Create the truss structure that is shown in Figure P20.9. The material is steel pipe, NPS 1½″, Schedule 40.
10. Create a plate, 120-mm-×-100-mm-×-10-mm thick, placed on top of the two vertical brackets that are shown in Figure P20.10. Create fillet beads to hold the plate to the top faces of the vertical brackets. Refer to the "Weldment of Parts" section.

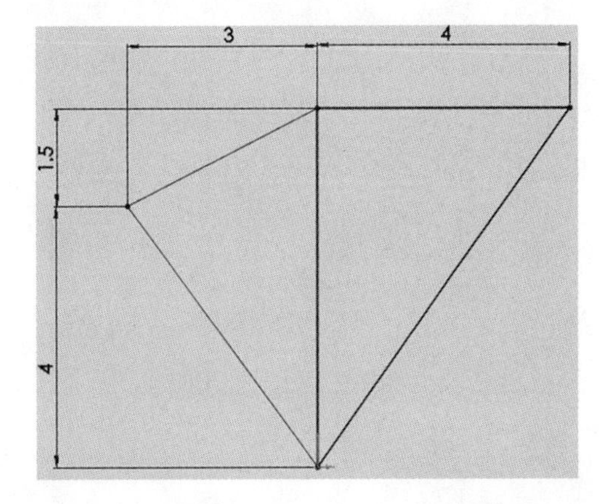

Figure P20.4

Truss 4.

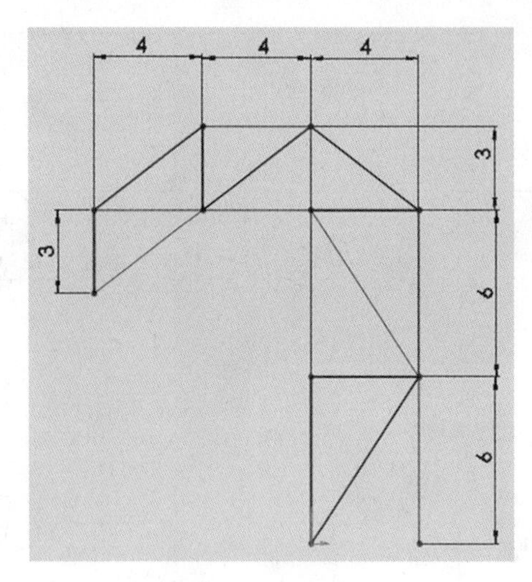

Figure P20.5

Truss 5.

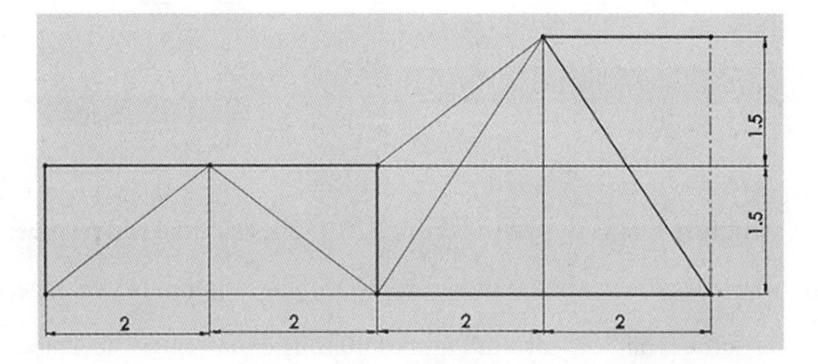

Figure P20.6

Truss 6.

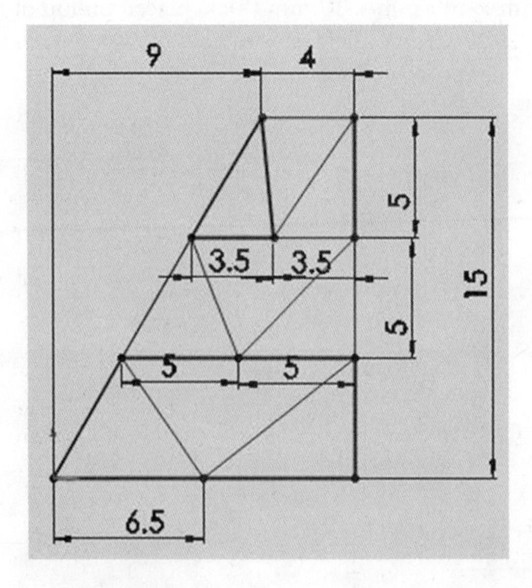

Figure P20.7

Truss 7.

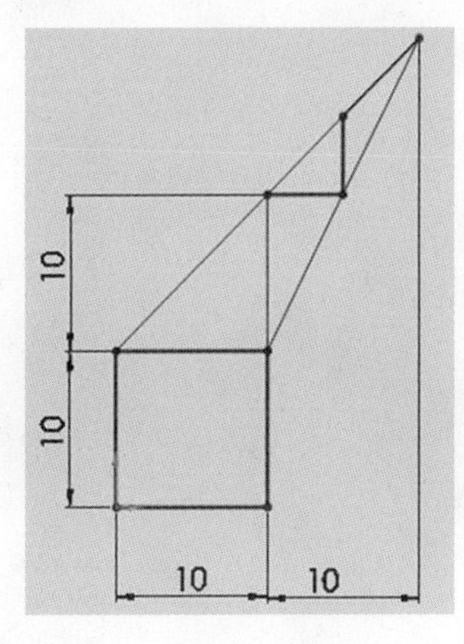

Figure P20.8

Truss 8.

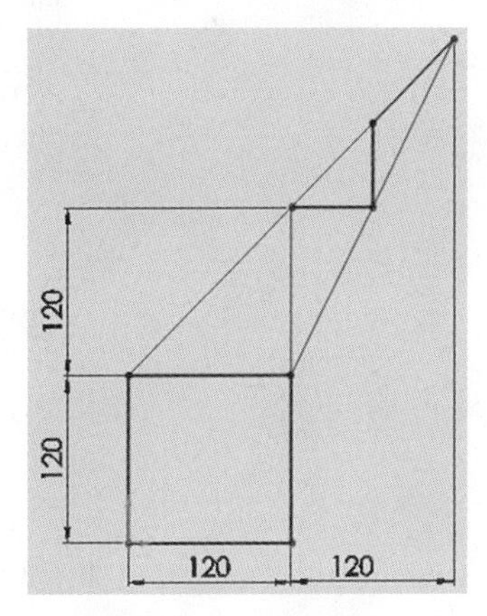

Figure P20.9

Truss 9.

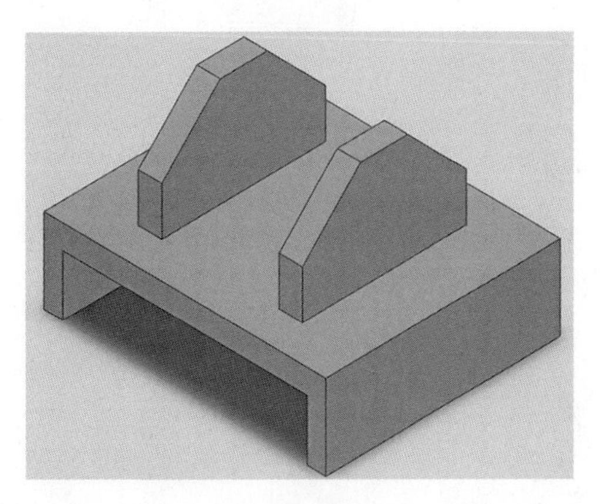

Figure P20.10

Assembly of base and side features.

21

Routings in Piping and Tubing

Objectives:

When you complete this chapter, you will have learned

- How to customize routing templates
- How to add parts to routing library
- How to start a route
- How to create the route for pipes and tubes

Introduction

SolidWorks Routing enables designers to create a special type of subassembly that builds a path of pipes, tubes, or electrical cables between components. SolidWorks Routing also includes harness flattening and detailing capabilities so that designers can develop two-dimensional harness manufacturing drawings from three-dimensional (3D) electrical route assemblies.

A route subassembly is always a component of a top-level assembly. When you insert certain components into an assembly, a route subassembly is created automatically. Unlike other types of subassemblies, you do not create a route assembly in its own window and then insert it as a component in the higher-level assembly.

You model the route by creating a 3D sketch of the centerline of the route path. The SolidWorks software generates the pipe, tube, or cable along the centerline.

The SolidWorks software makes extensive use of design tables to create and modify the configurations of route components. The configurations are distinguished by different dimensions and properties. If you are unfamiliar with these concepts, see the "Design Tables" section.

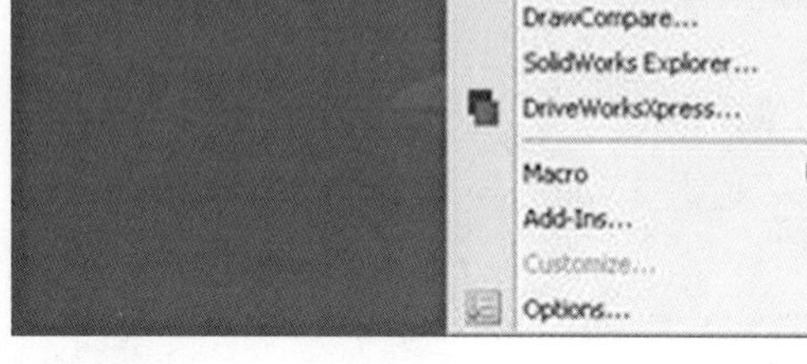

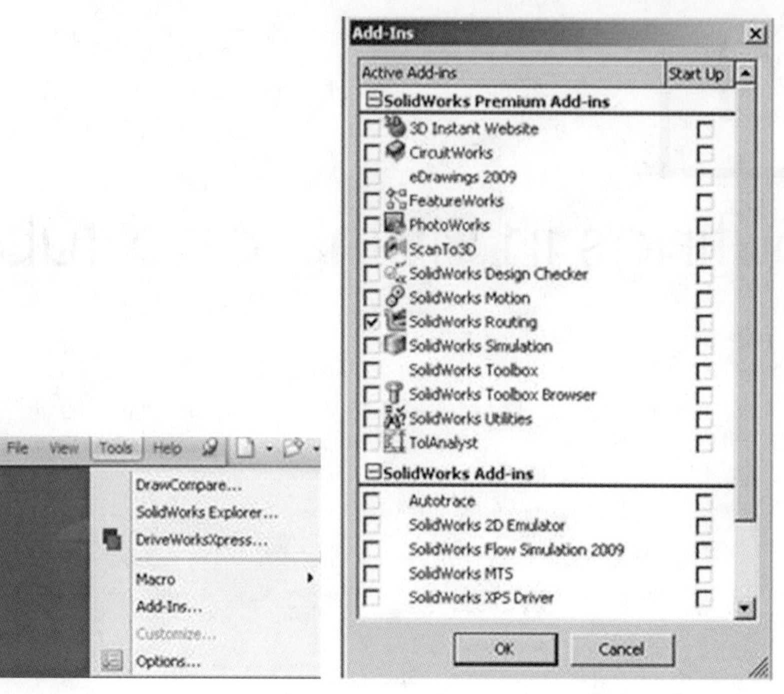

Figure 21.1

SolidWorks Routing add-ins.

Activating the SolidWorks Routing Add-Ins

The starting point for utilizing the routing tool is to activate the SolidWorks Routing add-ins. To activate SolidWorks Routing,

1. Click Tools > Adds-Ins.
2. Select SolidWorks Routing to activate it for the current session in the Active Add-Ins.
3. Click OK to add SolidWorks Routing (see Figure 21.1).

Some Backgrounds

It should be noted that a route is always a component of a top-level assembly. This means that when we insert components into an assembly, a route subassembly is created automatically. There is a way of switching from the subassembly to the top-level assembly. We will consider this later.

Customizing Routing Templates

Now create a custom routing template and set its units to inches. If your organization's policy allows, you could save the custom template in the default template location, but for our current purpose, you save it in a new folder that you create.

1. In Windows Explorer, create a folder on your local drive called H:\MyRouting.
2. In SolidWorks, click Open.
3. In the Open dialog box:
   a. For Look in, browse to your default template location (typically C:\Documents and Settings\All Users\Application Data\SolidWorks\SolidWorks<*version*>\templates).
   If your default template location is different, browse to that location.
   b. In File of type, select Template (*.prtdot;*.asmdot;*.drwdot).
   c. Select routeAssembly.asmdot.
   d. Click Open.

Now save a copy of the template and change some settings in it.

1. Click File, Save As.
2. In the Save As dialog box:
   a. For Save in, browse to H:\MyRouting.
   b. For File name, type MyRouteAssembly.
   c. For Save as type, select Assembly Templates (*.asmdot).
   d. Click Save.
3. Click Options.
4. In the dialog box:
   a. On the Document Properties tab, select Units.
   b. Under Unit system, select IPS (inch, pound, second).
   c. Click OK.
5. Click Save (Standard toolbar).

Adding Parts to the Routing Library

The Routing Library contains parts (such as flanges, fittings, and pipes) for you to use in routes that you create. By default, the Routing Library is located in a folder named *routing* in the Design Library. You can add components to existing folders in the Routing Library, or create new folders. You must have write access to your Design Library to create folders and add parts.

Create a new folder in the Routing Library and add an *assembly fitting without assembly connection points (acp)* assembly.

1. In the Task Pane,
   a. Click the Design Library tab.
   b. Browse to Design Library\routing\assembly fittings.
2. At the top of the Task Pane, click Create New Folder.
3. Type MyLibrary for the folder name and press Enter.

Let us consider some illustrations on routing.

Illustration 1

Now, add the assembly fitting without acp.sldasm to the MyLibrary folder (see Figure 21.2).

1. At the top of the Task Pane, click Add to Library.
2. In the PropertyManager,
   a. For Items to Add, select assembly fitting without acp.sldasm at the top of the flyout FeatureManager Design Tree.
   b. Under Save To, make sure that the MyLibrary folder is selected under Design Library folder.
   c. Click OK.
   The part is added to the Routing Library, and is available for selection when you create a route.
3. Close the part.

Starting a Route

Add some pipes and tube routes to an assembly.

1. Browse to C:\Documents and Settings\All Users\Application Data\SolidWorks\SolidWorks <*version*>\ design library\routing\assembly fittings\MyLibrary.
2. Save the assembly as MyAssyFitting.sldasm.
   The assembly normally will already contain some fittings that need to be connected by pipe or tube routes.

Start the first route by dragging a flange into the assembly. Figure 21.2 shows a preview of the Design Library Assembly Fitting. You can use tools on the View toolbar to zoom, rotate, and pan the model view to facilitate working with the model.

1. Click Start by Drag/Drop (Piping toolbar [see Figure 21.3]).
   The Design Library opens to the piping section of the Routing Library.

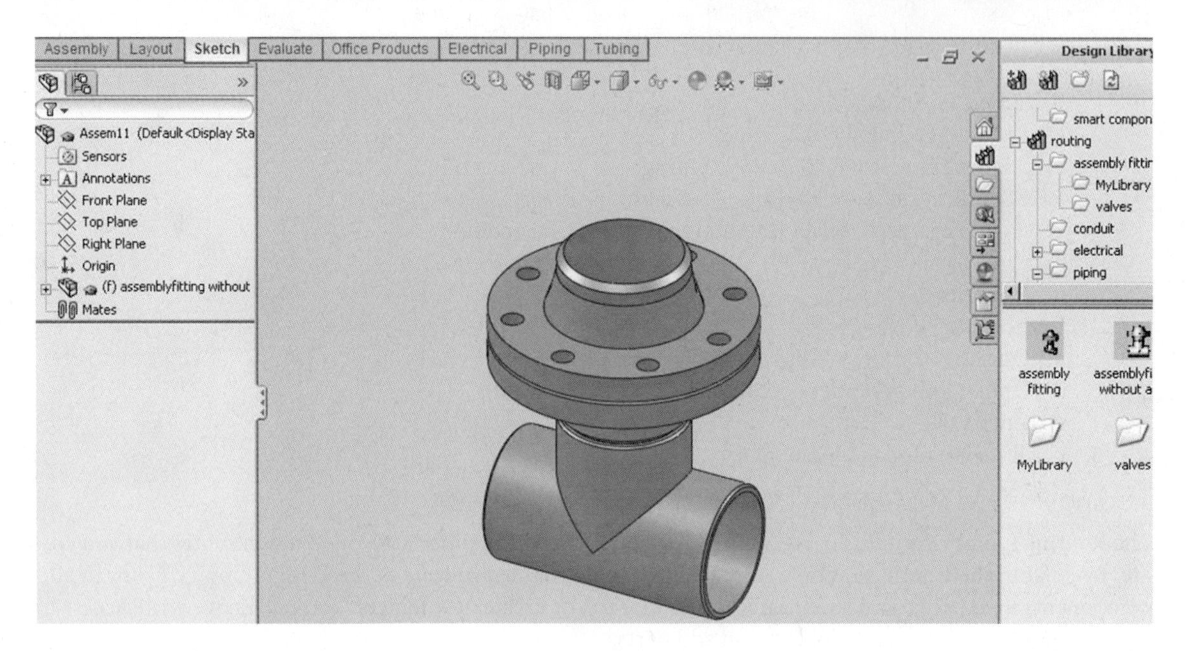

Figure 21.2

Assembly fitting without acp.sldasm.

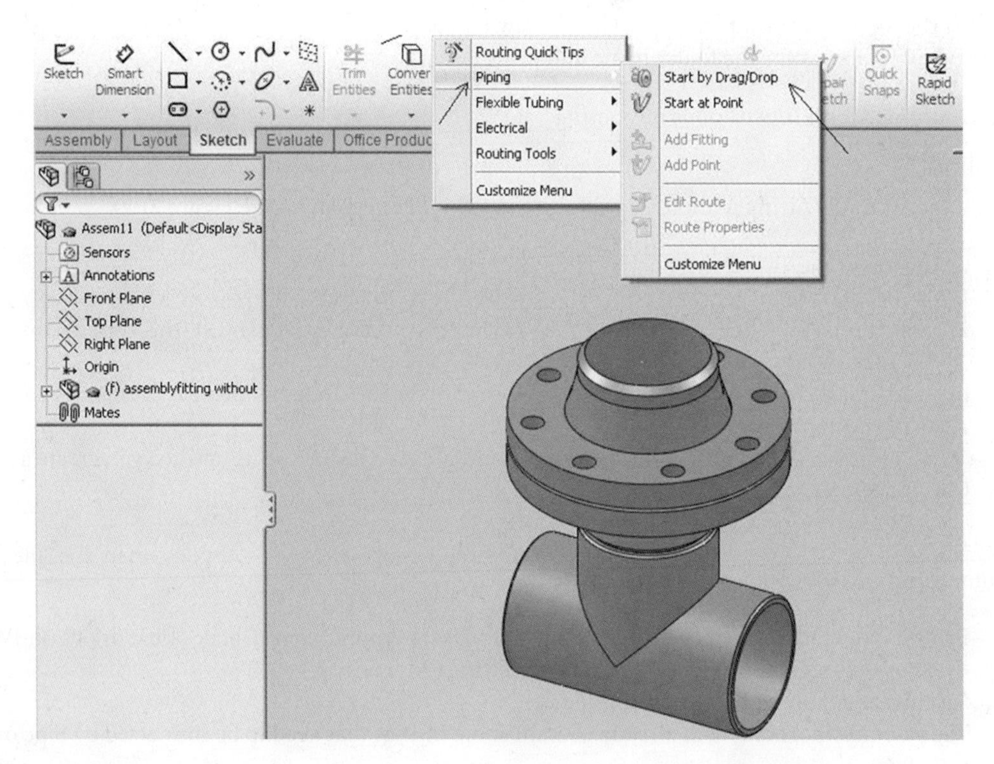

Figure 21.3

Drag/Drop option.

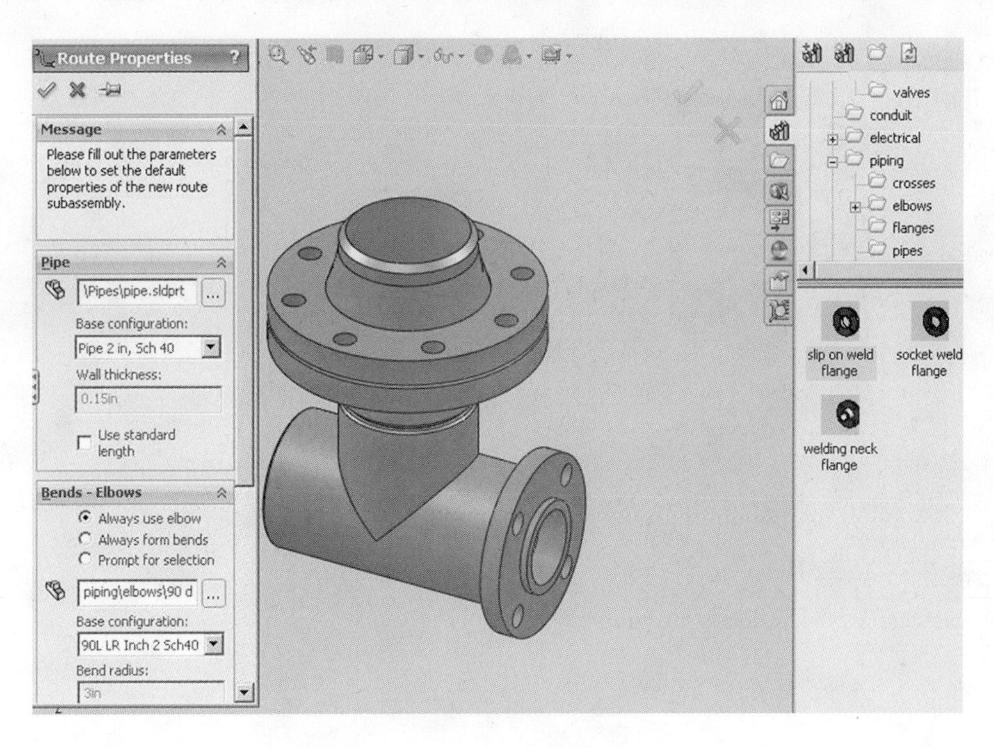

Figure 21.4

Flange for dragging/dropping on the right of window.

2. In the lower panel, double-click the flanges folder.
3. Drag slip on weld flange.sldprt from the library to the flange face on the regulator.
4. Drop the flange when it snaps into place (see Figure 21.4).
5. In the dialog box,
   a. Select Slip On Flange 150-NPS2.
   b. Click OK.

The Route Properties PropertyManager appears.

In the *Route Properties* PropertyManager, you specify the properties of the route you are about to create. Some of the items you can specify include the following:

- Which pipe or tube parts to use
- Whether to use elbows or bends

For this session, use the default settings.

1. Click OK.
2. If a message asks if you want to turn off the option Update component names when documents are replaced, click Yes.

The following happens:

- A 3D sketch opens in a new route subassembly.
- The new route subassembly is created as a virtual component, and appears in the FeatureManager design tree as [Pipe1-Assem1].
- A stub of pipe appears, extending from the flange that you just placed, as shown in Figure 21.5.

Creating the Route

Start creating the segments of the route.

1. Drag the endpoints of the stub (see Figure 21.6) to increase the pipe length, as shown in Figure 21.6. You do not need to be exact.

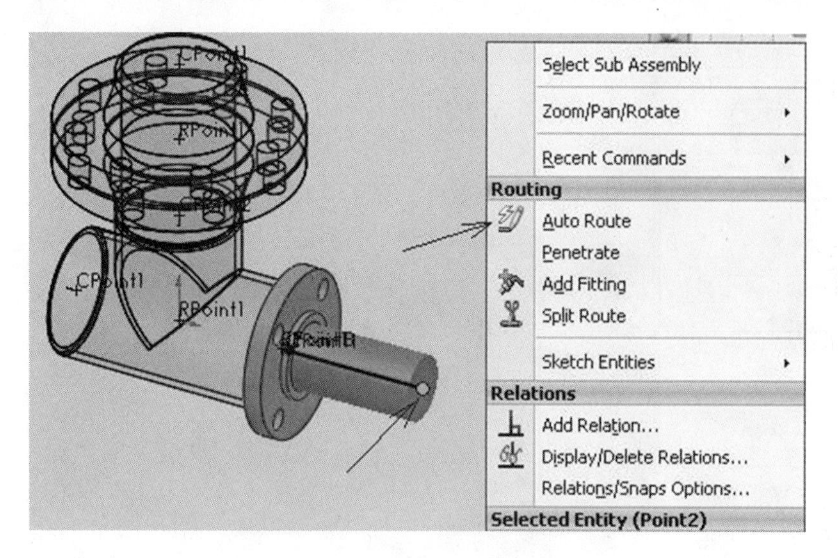

Figure 21.5

Starting a route: a pipe is added to the flange.

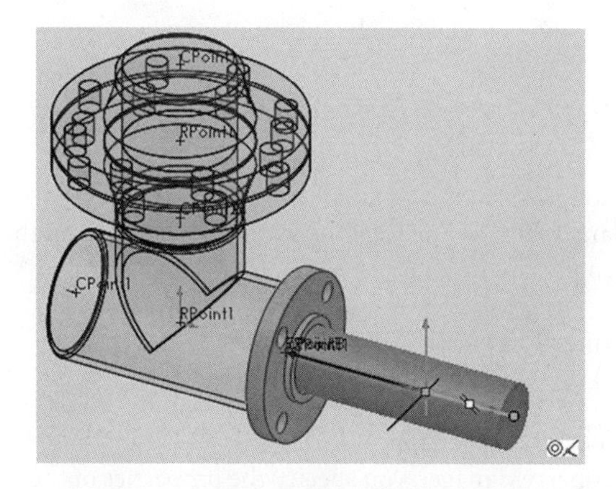

Figure 21.6

Pipe lengthens.

2. If a message appears about not adding automatic relations, click OK. (The software is trying to add sketch relations to the weldment that is behind the pipe but determines that the relations would overdefine the route sketch.)

   Now, add the horizontal flange to the route so that you can connect the pipe to it.
3. Zoom to the horizontal flange.

   On the View menu, make sure that Routing Points is *selected* and Hide All Types is *cleared.*
4. Move the pointer over the connection point (CPoint1) in the center of the flange.

   The connection point is highlighted.
5. Right-click CPoint1 and select Add to Route.

A stub of pipe extends from the flange.

Illustration 2

Let us switch to assembly fitting.sldasm. The subassembly has a Tee and flanges (see Figure 21.7). In this case, we can commence our routing beginning with three pipes. We will do the following:

1. Click CPoint1 and extend the pipe to the right (see Figure 21.8).
2. Click CPoint2 and extend the pipe to the left (see Figure 21.8).

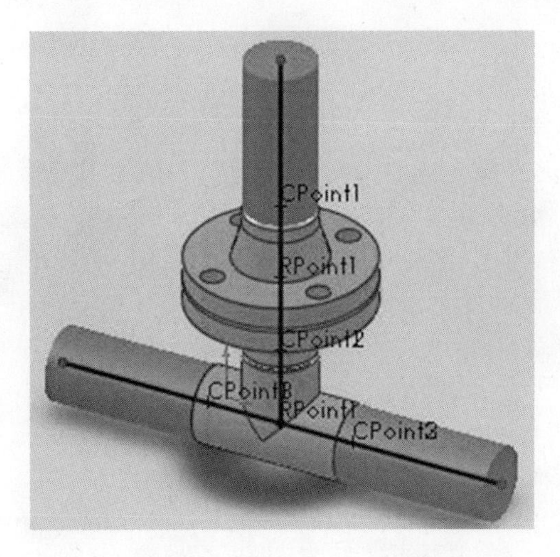

Figure 21.7

The Assembly fitting.sldasm.

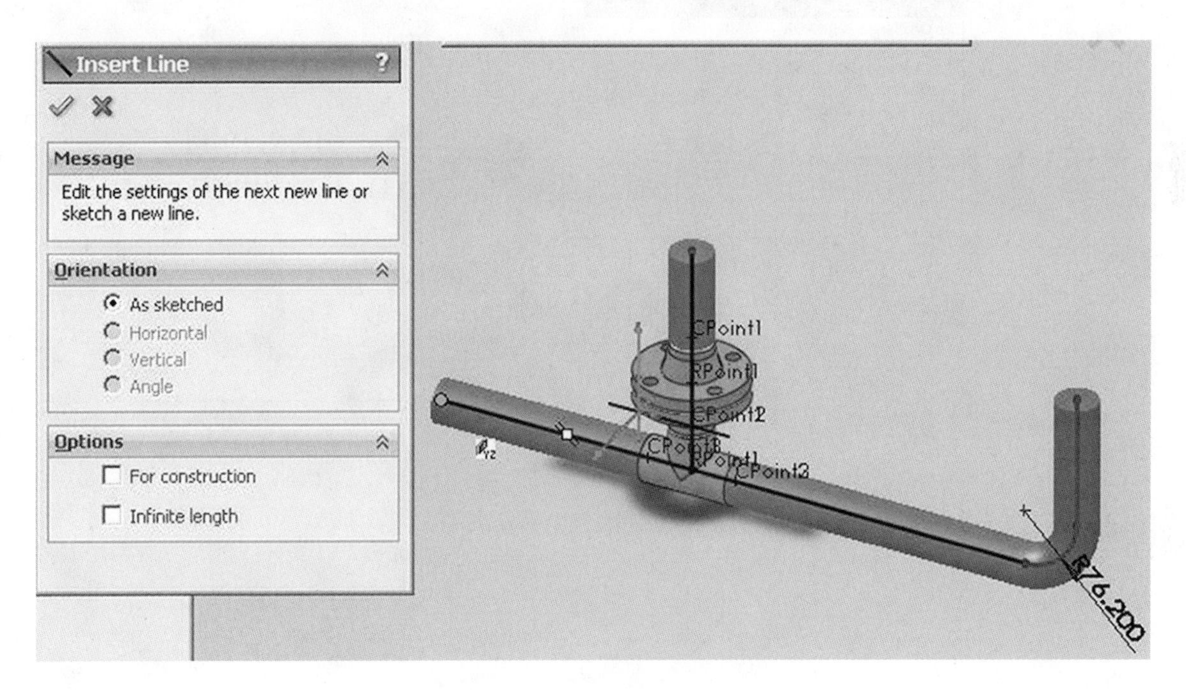

Figure 21.8

Extending right-hand pipe vertically using 3DSketch tool.

3. Click CPoint1 on the right, and insert a line using 3DSketch along Y on the XY plane (see Figure 21.9).
4. Click CPoint2 on the left, and insert a line using 3DSketch along Z on the YZ plane (see Figure 21.9).

   We will now add a pair of flanges on the vertical pipe and add an extra pipe above the pair of flanges.
5. Click the end (CPoint1) of the vertical pipe, and drag slip on flange to the point (CPoint1). Note that a pipe is automatically added to the pair of flanges (see Figure 21.10).
6. Click the end (CPoint1) of the vertical pipe, and extend it (see Figure 21.11).
7. Click CPoint1, and insert a line using 3DSketch along Z on the YZ plane (see Figure 21.11).
8. Click the end (CPoint1) of the vertical pipe, and drag slip on flange to the point (CPoint1), as shown in Figure 21.12.

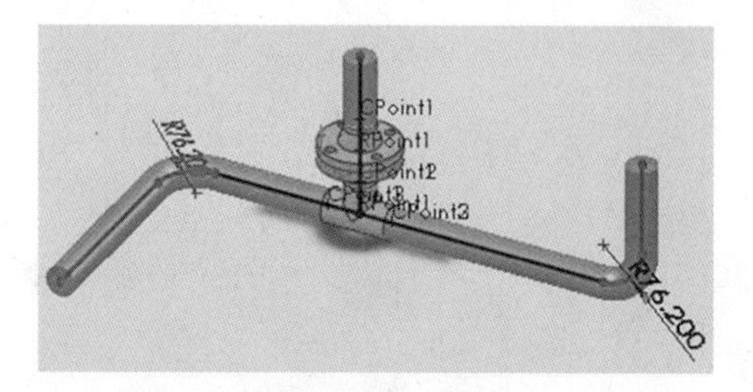

Figure 21.9

Extending left-hand pipe horizontally using 3DSketch tool.

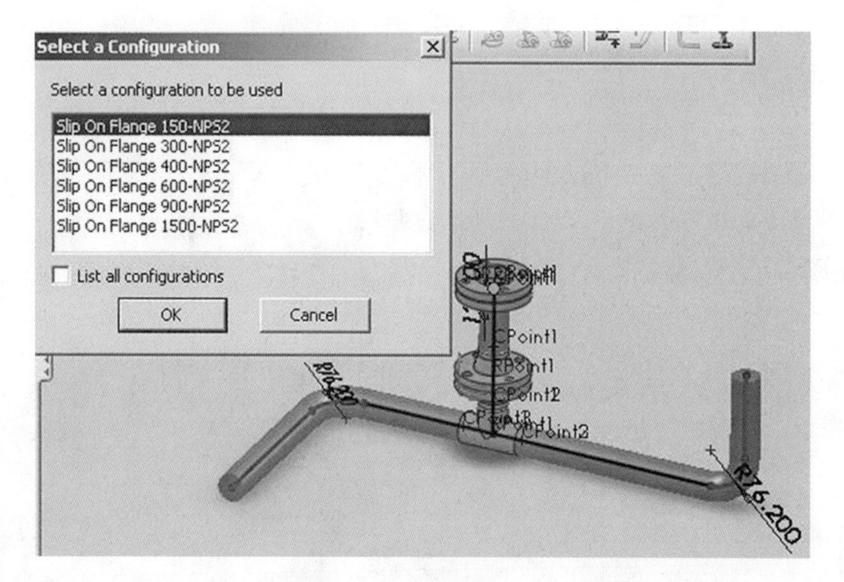

Figure 21.10

Dragging slip from Library unto flange at the top of the third vertical pipe.

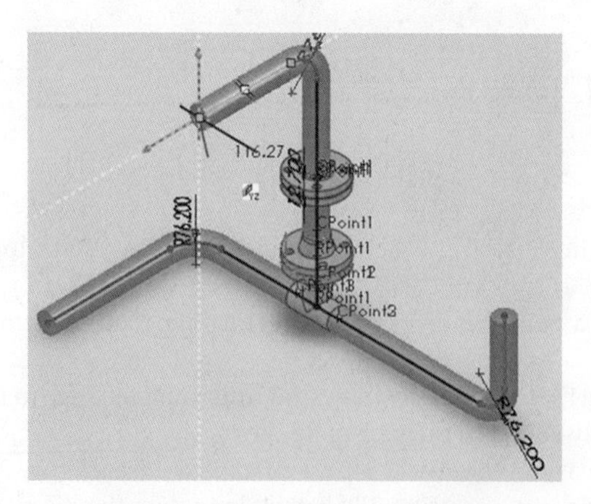

Figure 21.11

Extending third pipe vertically and along Z-direction on the YZ plane.

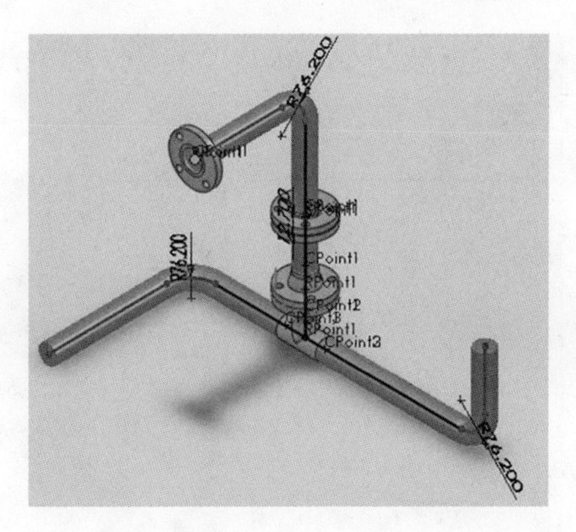

Figure 21.12

Dragging slip from Library unto flange third vertical pipe.

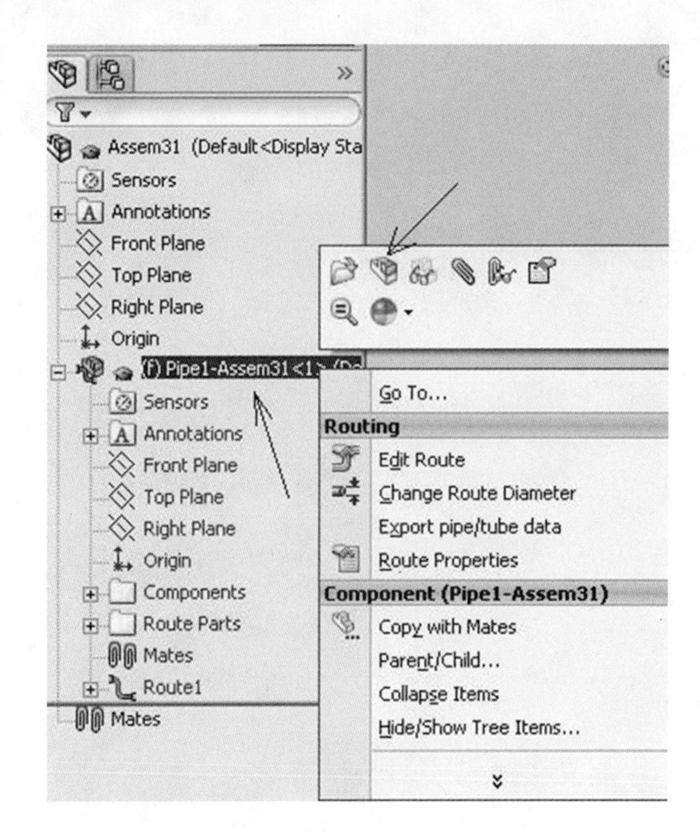

Figure 21.13

Switching to the top-level assembly.

It should be noted that so far, we are creating a subassembly in a top-level assembly. The subassembly is in blue color. To move to the top-level assembly, click Edit Assembly. This concept is shown in Figure 21.13. The routing solution is shown in Figure 21.14. The FeatureManager contains the details of the routing operations.

Route Drawing

When the routing exercise is complete, a new drawing document is opened for presenting the drawing of the route. The same procedure for drawing parts applies here, with the bill of materials added. This is left as an exercise.

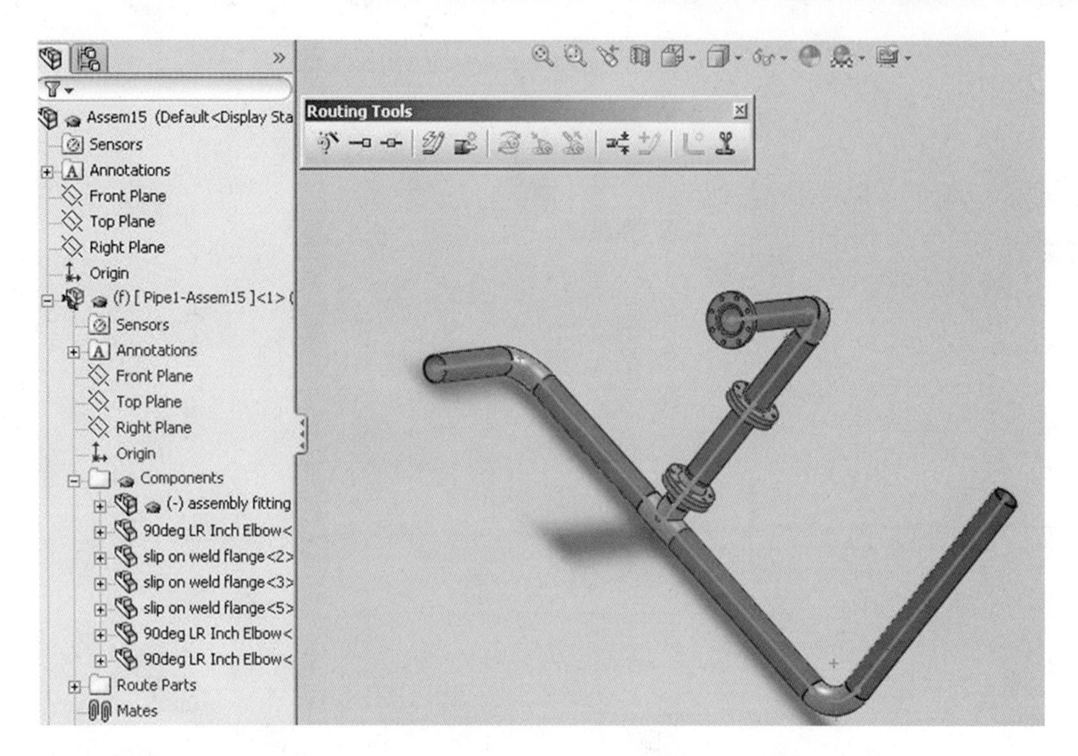

Figure 21.14

The routing problem is done.

22

Power Transmission Elements

Objectives:

When you complete this chapter, you will have learned

- The available Design Tools for power transmission systems
- How to size Supporting Plates and Shafts for power transmission systems
- How to model an assembly of the Pinion and Gear (Spur-type) transmission system
- How to model an assembly of the Rack and Pinion transmission system
- How to model an assembly of the Belt and Pulley transmission system
- How to model an assembly of the Chain and Sprocket transmission system
- How to model an assembly of the Bevel Gear transmission system

Power is generated from an engine, a motor, or a windmill. The power is then transmitted to a mechanism to perform some function. Transmission is generally through the use of elements such as gears, pulleys, and chains. For example, power is generated in the engine of an automobile and then transmitted to the wheels via the gear box. These power transmission elements are considered in this chapter. We discuss how SolidWorks software is used to model these transmission elements.

Gears and Power Transmission

A *gear* is a rotating machine part having cut *teeth*, or *cogs*, which *mesh* with another toothed part in order to transmit torque. Two or more gears working in tandem are called a *transmission* and can produce a mechanical advantage through a gear ratio and thus may be considered a simple machine. Geared devices can change the speed, magnitude, and direction of a power source. The most common situation is for a gear to mesh with another gear; however, a gear can also mesh a nonrotating toothed part, called a rack, thereby producing translation instead of rotation.

When two gears of unequal number of teeth are combined, a mechanical advantage is produced, with both the rotational speeds and the torques of the two gears differing in a simple relationship.

In transmissions that offer multiple gear ratios, such as bicycles and cars, the term gear, as in *first gear*, refers to a gear ratio rather than an actual physical gear. The term is used to describe similar devices even when the gear ratio is continuous rather than discrete, or when the device does not actually contain any gears, as in a continuously variable transmission.

The gears in a transmission are analogous to the wheels in a pulley. An advantage of gears is that the teeth of a gear prevent slipping.

The remaining sections of this chapter discuss the methodology for the assembly modeling of pinion and gear, rack and pinion, belt and pulley, and chain and sprocket transmission systems.

Spur Gears

Spur gears or *straight-cut gears* are the simplest type of gears. They consist of a cylinder or disk, with the teeth projecting radially, and although they are not straight-sided in form, the edge of each tooth is straight and aligned parallel to the axis of rotation. These gears can be meshed together correctly only if they are fitted to parallel axles.

Creating Gears Using SolidWorks

SolidWorks software is useful for creating different types of gears such as spur, bevel, and helical gears. As many as 18 standards are available in the SolidWorks Toolbox from which the designer can choose (see Figure 22.1). To access these standards, click Design Library > Toolbox.

Note: You must have the Toolbox license in order to access Toolbox Add-Ins. First, click the Options tab to make changes in the settings for SolidWorks:

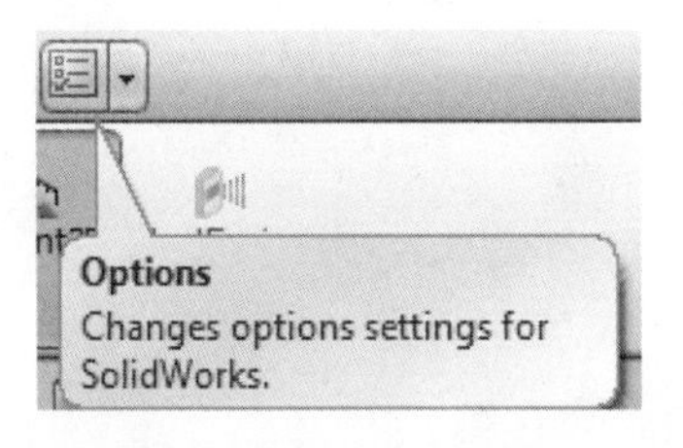

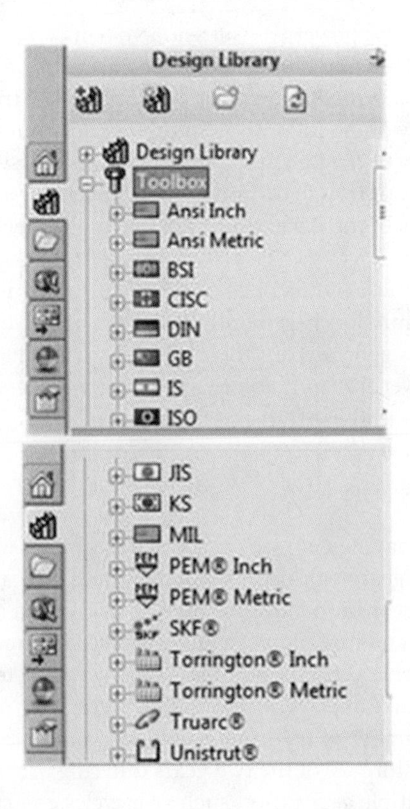

Figure 22.1

SolidWorks Toolbox Standards.

Select the Add-Ins option and tick off the following options:

- SolidWorks Toolbox
- SolidWorks Toolbox Browser

The Toolbox offers quite a number of design tools for bearings, bolts and screws, jig bushings, keys, nuts, O-rings, and power transmission elements (chain sprockets, gears, timing belt pulleys, retaining rings, structural members, and washers), as shown in Figure 22.2. These design tools assist the design engineer performs his/her work more effectively and efficiently. In this chapter, we discuss the use SolidWorks for modeling power transmission elements.

Problem Description

A gear system consists of the pinion (Gear 1) and gear (Gear 2), and the specifications relating to their geometry are given as follows:

Gear 1: Pinion
Diametral pitch = 24
Number of teeth = 30
Face thickness = 0.5
Bore = 0.5 diameter
Hub = 1.0 diameter
Pressure angle = 20 deg.

Gear 2: Gear
Diametral pitch = 24
Number of teeth = 60
Face thickness = 0.5
Bore = 0.5 diameter
Hub = 1.0 diameter
Pressure angle = 20 deg.

Figure 22.2

Toolbox design tools.

Using SolidWorks, create an assembly model consisting of the two spur gears, two pins and a support plate, and animate the movement of the gears.

Support Plate Sizing

The pitch diameters for the pinion and gear are

$$D_p = 30/24 = 1.25''$$

$$D_g = 60/24 = 2.50''$$

The support plate dimensions are given as follows:

$$W = (D_g + 2\times 0.5)$$

$$L = [(D_p + 0.5) + (D_g + D_p)/2 + W/2]$$

leading to $W = 3.5$ in. and $L = 5.375$ in., respectively. Based on these dimensions, the support plate is created using SolidWorks.

Gear Assembly Modeling Using SolidWorks

The gear assembly modeling consists of three main steps: (1) modeling of the support plate; (2) modeling of the pin; and (3) assembly of the support plate, pin, and gears.

Support Plate Model

1. Create a New Part document.
2. Select the Top Plane.
3. Sketch a rectangular profile, Sketch1, of 5.4 in. × 3.5 in. (see Figure 22.3).
4. Extrude Sketch1 through 0.5 in. to get Extrude1.
5. Create a circle, Sketch2, on top of Extrude1 at a point (1.75, 1.00) (see Figure 22.4).
6. Extrude-cut this circle through the thickness, to get Extrude2.

Create a Linear Pattern of Extrude2

7. Click Features > Linear Pattern. (The Linear Pattern PropertyManager is displayed as shown in Figure 22.5.)
8. Click a horizontal edge (Edge <1>) as Direction1.
9. Set the Number of Instances equal to 2.
10. Select Extrude2 as the Features to Pattern.
11. Click OK.
12. Save the document.

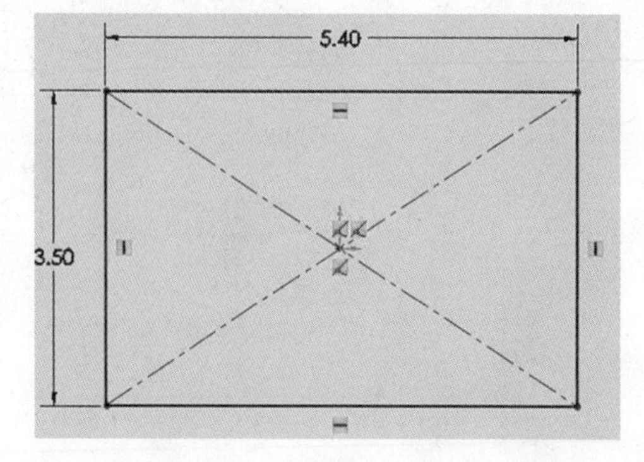

Figure 22.3

Sketch1.

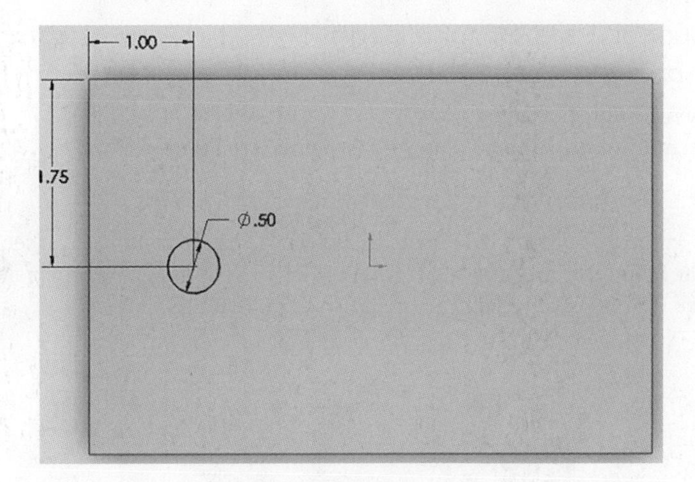

Figure 22.4

Sketch2.

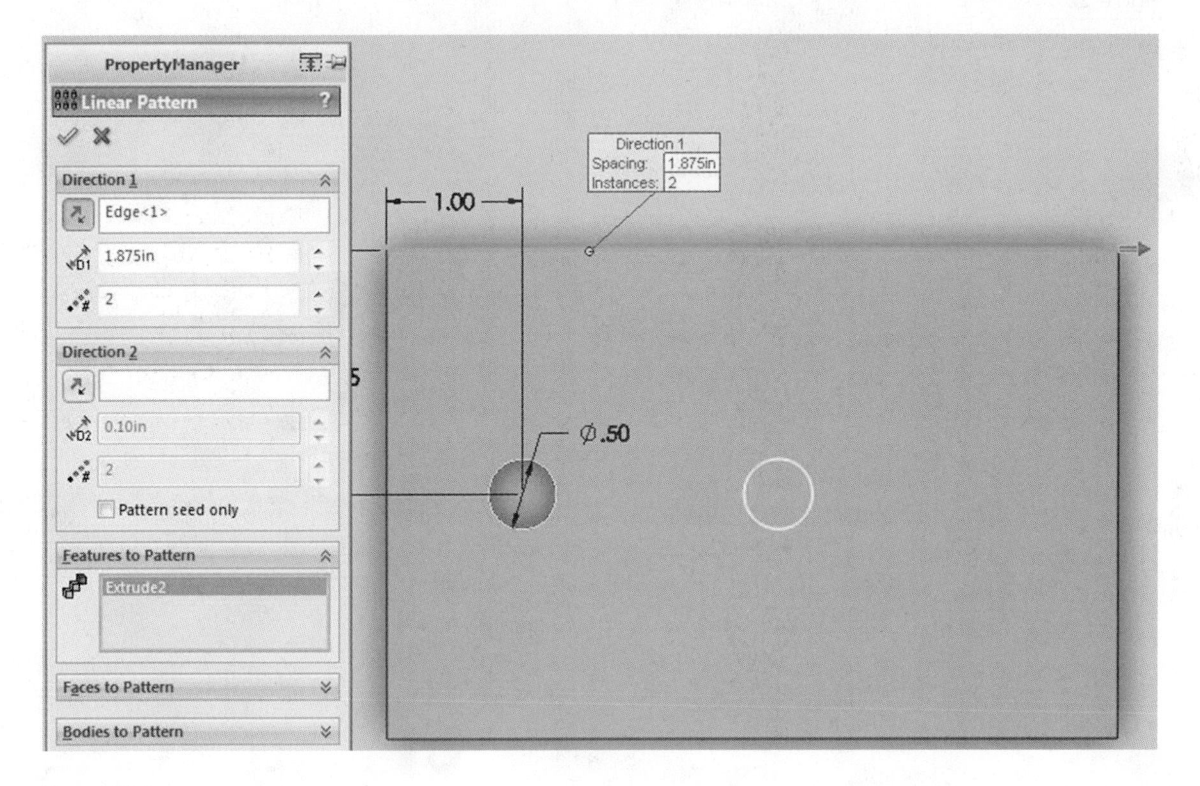

Figure 22.5

Linear Pattern PropertyManager.

Pin Model

1. Create a New Part document.
2. Select the Top Plane.
3. Sketch a circular profile Sketch1 of 0.5-in. diameter.
4. Extrude Sketch1 through 1.5 in. to get Extrude1.
5. Click OK.
6. Save the document.

Assembly of the Support Plate, Pin, and Gears

Assembly of the Support Plate and Two Pins

1. Create a New Assembly document.
2. Create an assembly of the support plate and two pins (see Figure 22.6).

Adding Gears to the Assembly of the Support Plate and Two Pins

3. Click Design Library Toolbox > ANSI Inch > Power Transmission > Gears (see Figure 22.7).
4. Right-click the Spur Gear tool and select Create Part. (The Spur Gear PropertyManager is automatically displayed for the pinion and gear, as shown in Figure 22.8.)

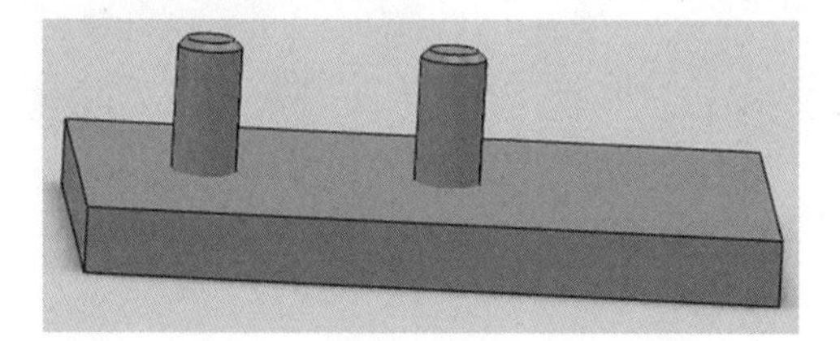

Figure 22.6

Assembly of the support plate and two pins.

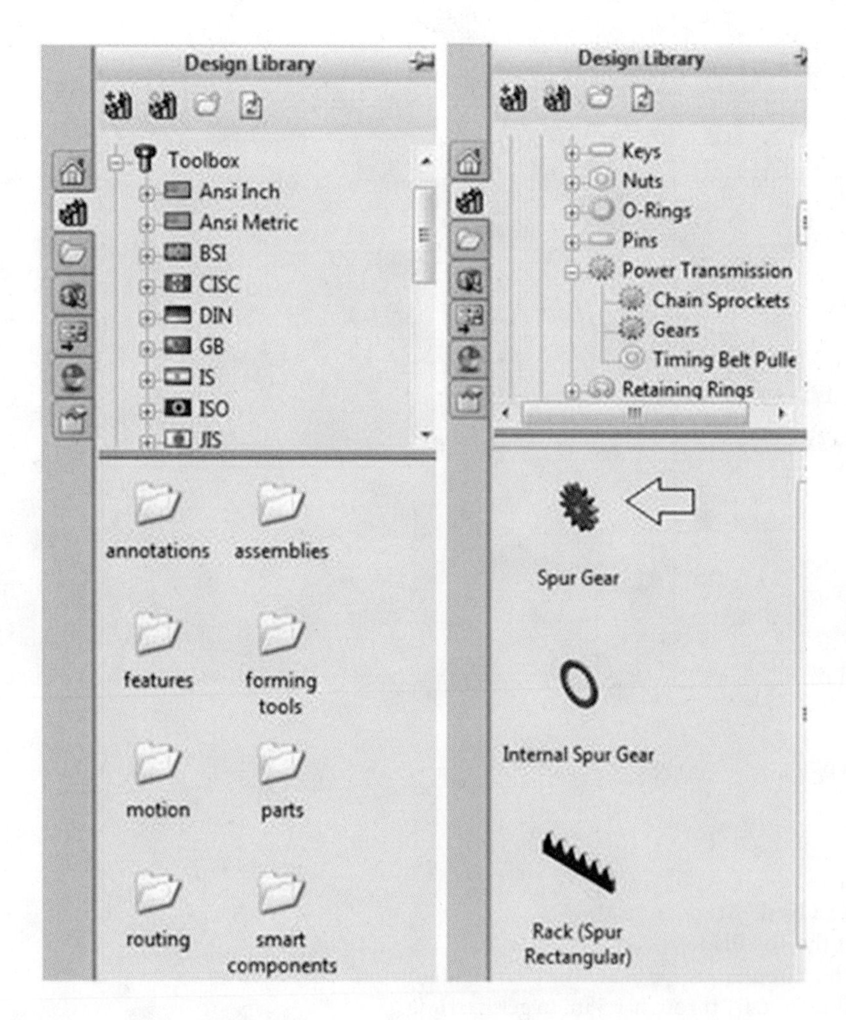

Figure 22.7

Toolbox access for Gears in the SolidWorks Design Library.

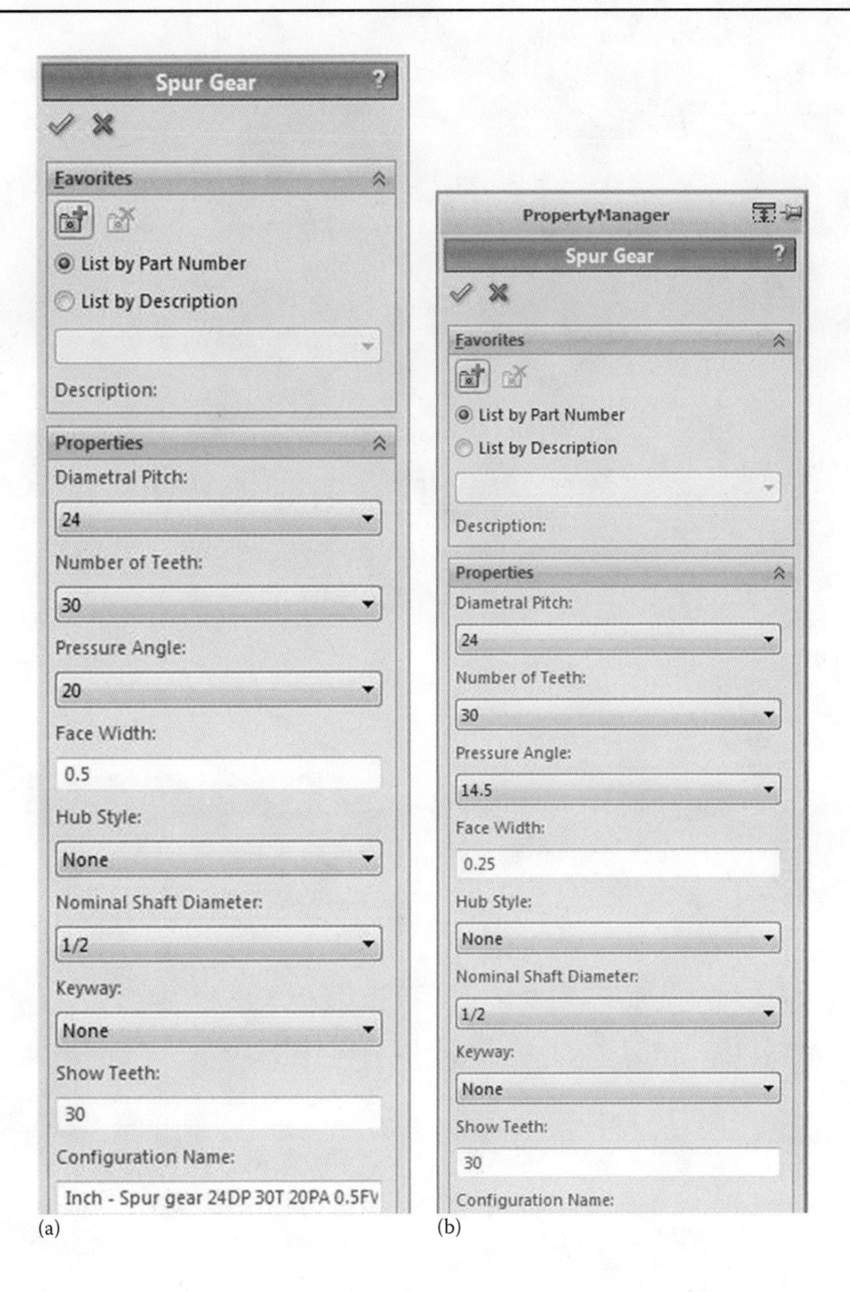

Figure 22.8

Spur Gear PropertyManager showing (a) pinion and (b) gear data.

5. Enter the data for pinion and gear as shown in Figure 22.8.
6. Click OK.
7. Click any point on the graphics drawing area. (A duplicate of the gear appears; see Figure 22.9.)
8. Click Cancel in the Insert Components PropertyManager.

Ordinary Mating

9. Rotate the individual pinion and gear.
10. Click Mate.
11. Select the hole in the pinion and the outer surface of the pin and apply Concentric mate.
12. Select the top of the pinion and the top surface of the support and apply Distance mate of 1.0 (see Figure 22.10).

    Repeat for the gear (see Figure 22.11).

    Repeat the process (see Figure 22.12 for the meshed pinion and gear).

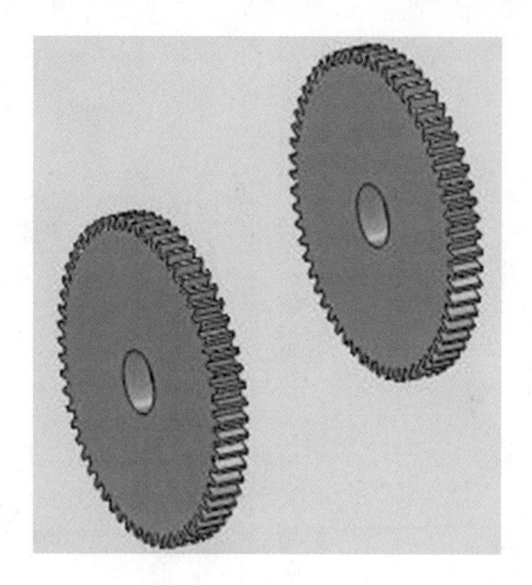

Figure 22.9

Duplicate spur gears created.

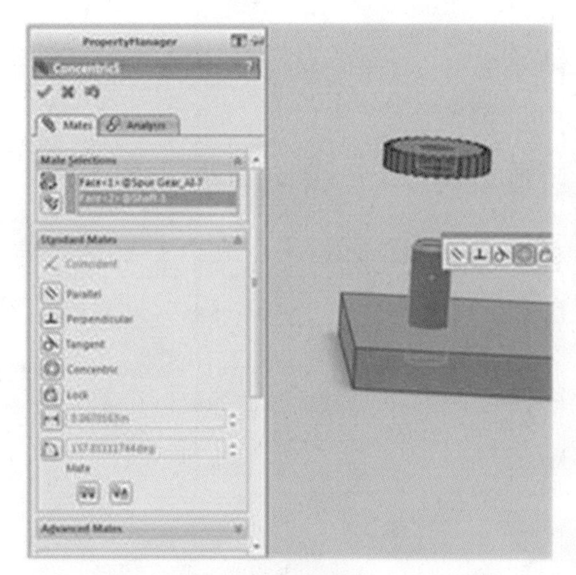

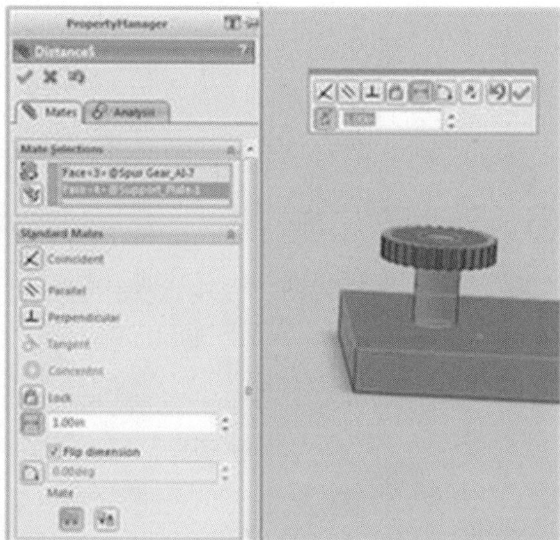

Figure 22.10

Mating conditions for pinion.

Mechanical Mating

13. Click Mate > Mechanical Mates. (See Figure 22.13 for the GearMate PropertyManager.)
14. Click Gear (see Figure 22.13).
15. Select the *inner surface of the pinion hole* and the *inner surface of the gear hole* (see Figure 22.13).
16. Enter the Ratios as 0.5 and 1 (see Figure 22.13).
17. Click OK (see Figure 22.13).

Animation

18. Click the Motion Study1 icon at the bottom-right corner of the graphics window.
19. Click the Motor icon (see Figure 22.14).
20. Click the top of the pinion as the Component (see Figure 22.14).
21. Set the Motion Speed as required (3000 rpm) (see Figure 22.14).

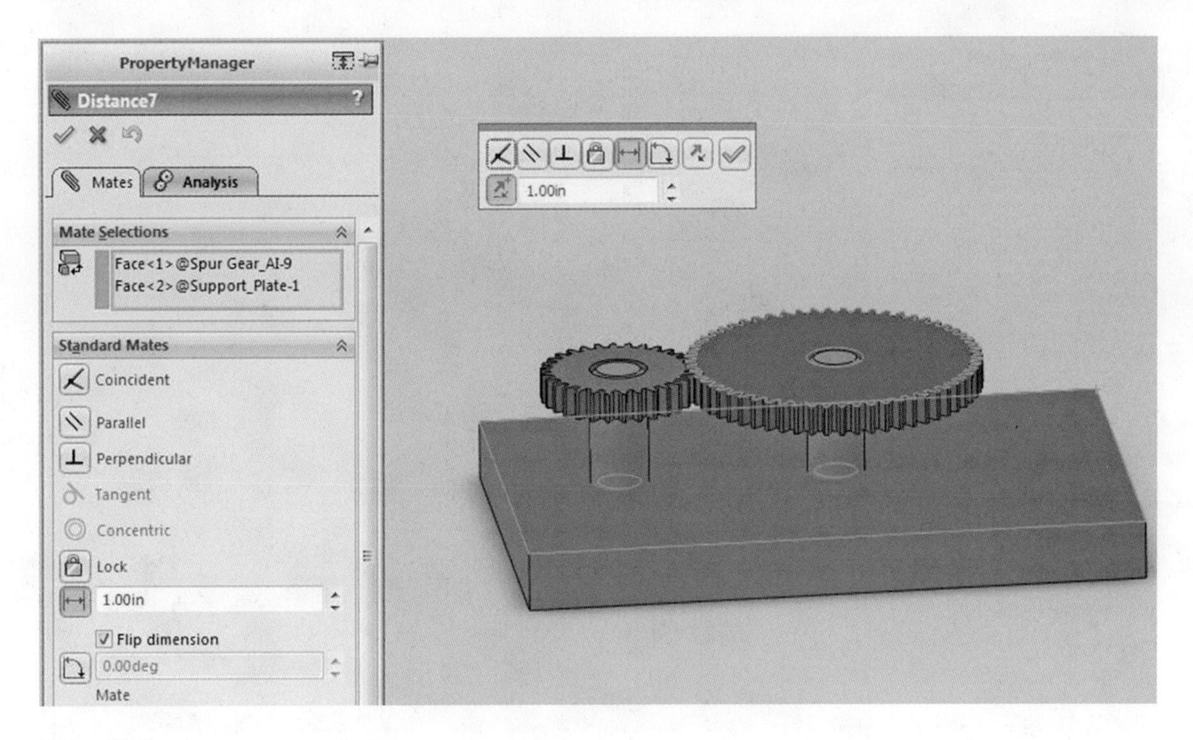

Figure 22.11

Mating conditions for gear.

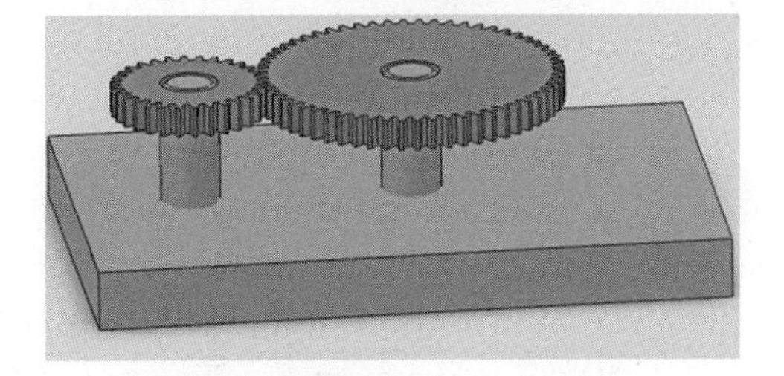

Figure 22.12

Meshed pinion and gear.

Rack and Pinion Gears

A rack is a toothed bar or rod that can be thought of as a sector gear with an infinitely large radius of curvature. Torque can be converted to linear force by meshing a rack with a pinion: the pinion turns; the rack moves in a straight line. Such a mechanism is used in automobiles to convert the rotation of the steering wheel into the left-to-right motion of the tie rod(s). Racks are also featured in the theory of gear geometry, where, for instance, the tooth shape of an interchangeable set of gears may be specified for the rack (infinite radius), and the tooth shapes for the gears of particular actual radii are then derived from that. The rack and the pinion gear type is employed in a rack railway.

Problem Description

A rack-and-pinion system consists of the rack and pinion and the specifications (Imperial units are used) relating to their geometry are given as follows:

Rack
Diametral pitch = 24
Face width = 0.25
Pitch height = 1.5
Length = 5
Pressure angle = 14.5 deg.

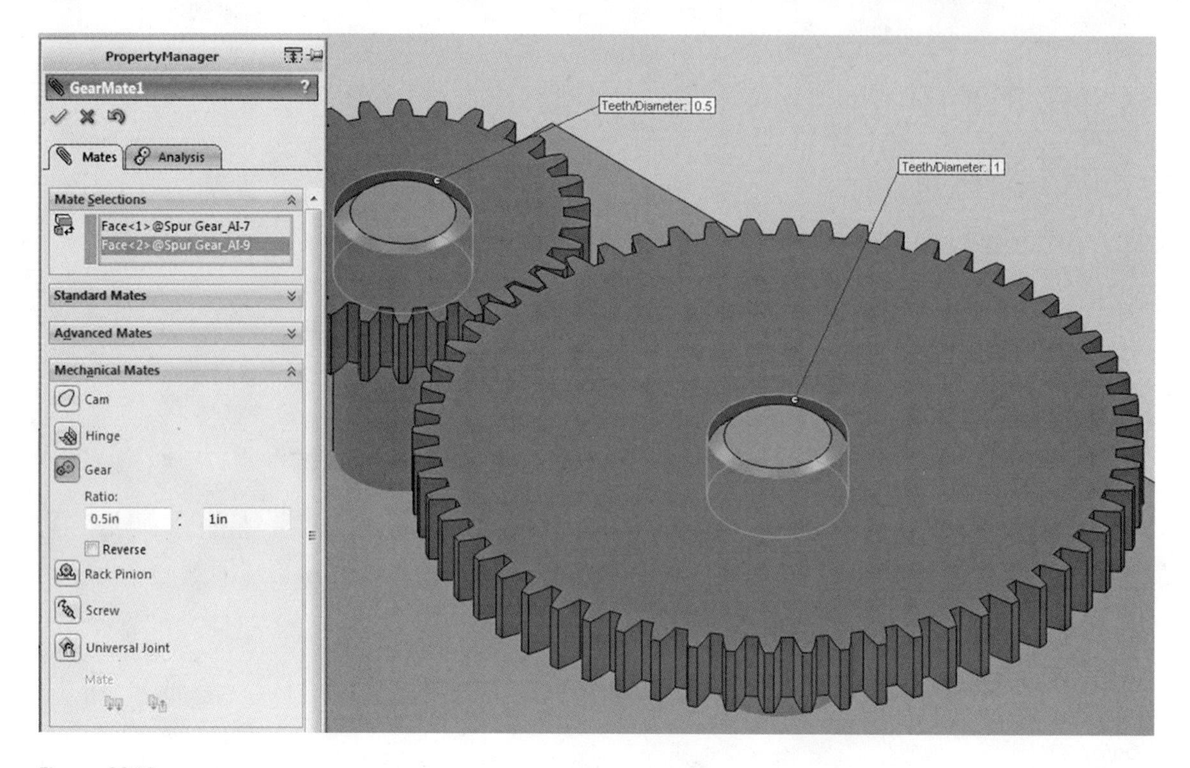

Figure 22.13

Mechanical mating conditions.

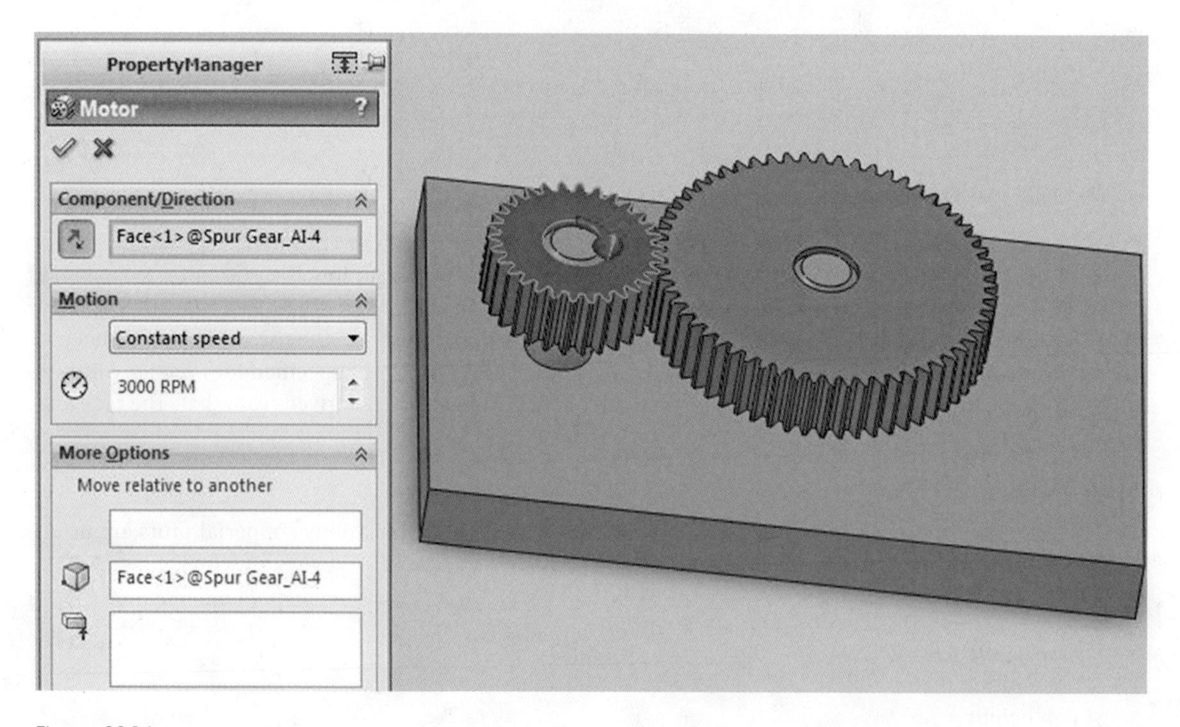

Figure 22.14

Motion of pinion and gear.

Pinion
Diametral pitch = 24
Number of teeth = 20
Face width = 0.25
Hub style: One side
Hub diameter = 1.0
Overall length = 1.0
Nominal shaft diameter = 1/2
Pressure angle = 14.5 deg.

Shaft
Shaft diameter = 1/2
Shaft length = 2.25

Using SolidWorks, create an assembly model consisting of the rack, pinion, and shaft, and animate the movement of the gears.

Gear Assembly Modeling Using SolidWorks

The gear assembly modeling consists of four main steps: (1) insert the shaft, (2) insert the pinion from the Design Library and assemble it to the shaft to form a subassembly, (3) insert the rack from Design Library, and (4) the assembly of the subassembly and the rack.

Insert Shaft in Assembly

1. Create a New Assembly document.
2. Click Assembly > Insert Components.
3. Insert Shaft (see Figure 22.15).

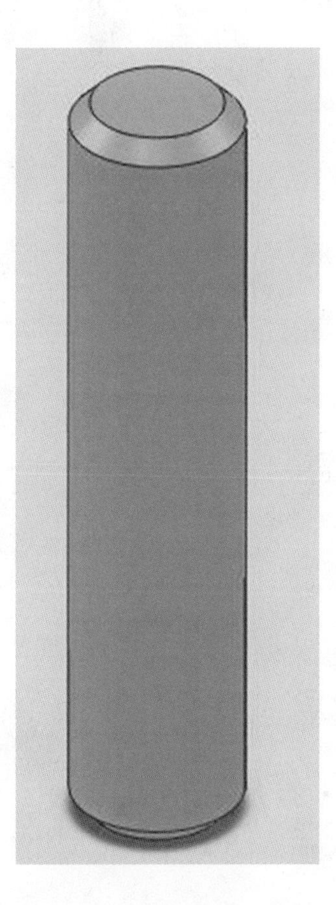

Figure 22.15

Shaft inserted first in the assembly.

Subassembly of Shaft and Inserted Pinion from Design Library

4. Click Design Library Toolbox > ANSI Inch > Power Transmission > Gears (see Figure 22.16).
5. Right-click the Spur Gear tool and select Create Part. (The Spur Gear PropertyManager is automatically displayed for the pinion and gear, as shown in Figure 22.17.)
6. Enter the data for spur gear (pinion), as shown in Figure 22.17.
7. Click OK.
8. Rotate the pinion appropriately.
9. Click Mate.
10. Select the hole in the pinion and the outer surface of the shaft and apply Concentric mate (see Figure 22.18).
11. Select the top of the pinion and the top surface of the shaft and apply Coincident mate (see Figure 22.19).

Insert Rack in Assembly

12. Click Design Library Toolbox > ANSI Inch > Power Transmission > Gears (see Figure 22.16).
13. Right-click the Rack (Spur Rectangular) tool and select Create Part. (The Rack PropertyManager is automatically displayed for the pinion and the gear, as shown in Figure 22.20.)
14. Enter the data for the pinion and the gear, as shown in Figure 22.20.
15. Click OK.

Ordinary Mating

16. Click Mate.
17. Select the top of the pinion and the outer flat surface of the rack and apply Parallel mate (see Figure 22.21).

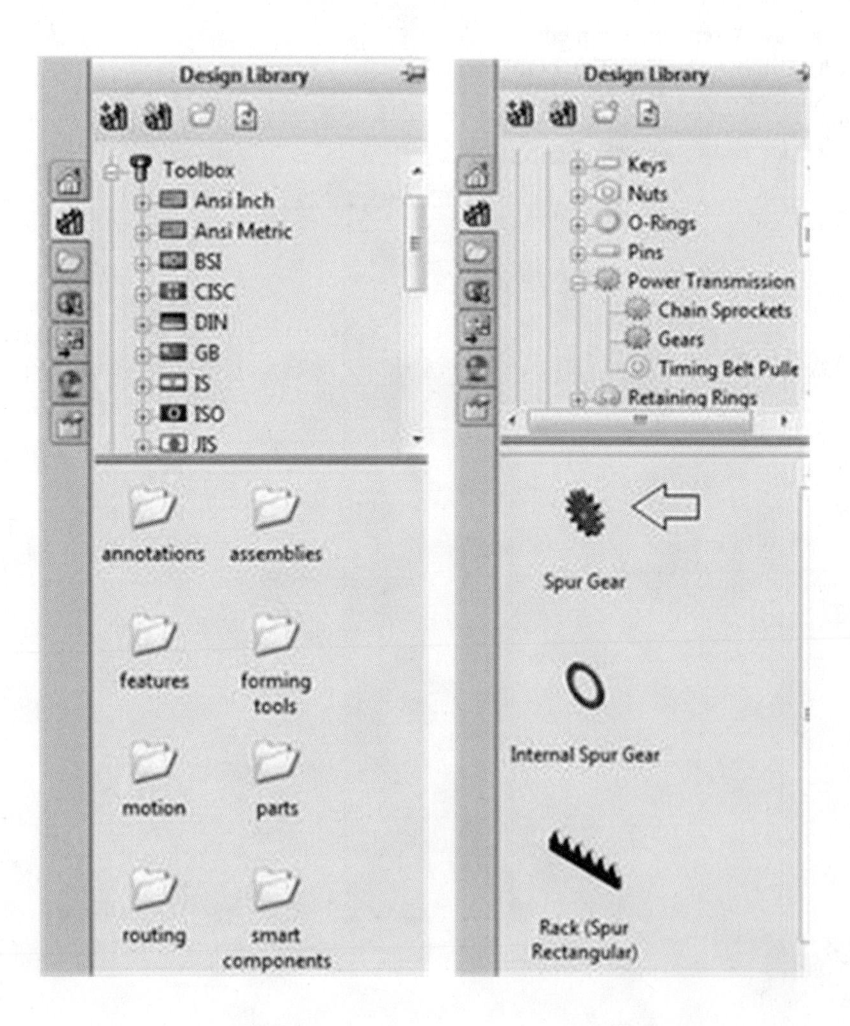

Figure 22.16

Toolbox access for Gears in the SolidWorks Design Library.

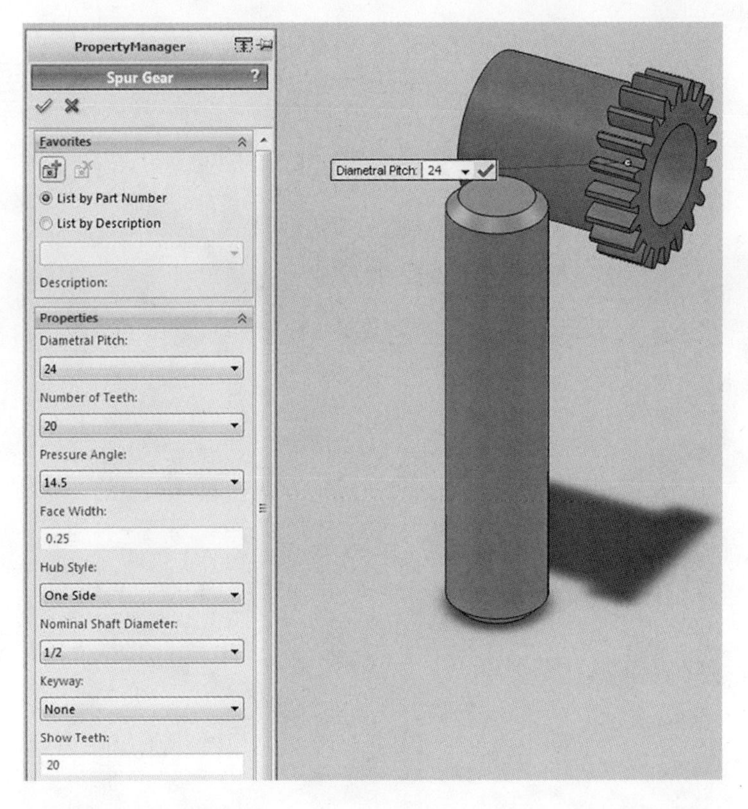

Figure 22.17

Spur Gear PropertyManager.

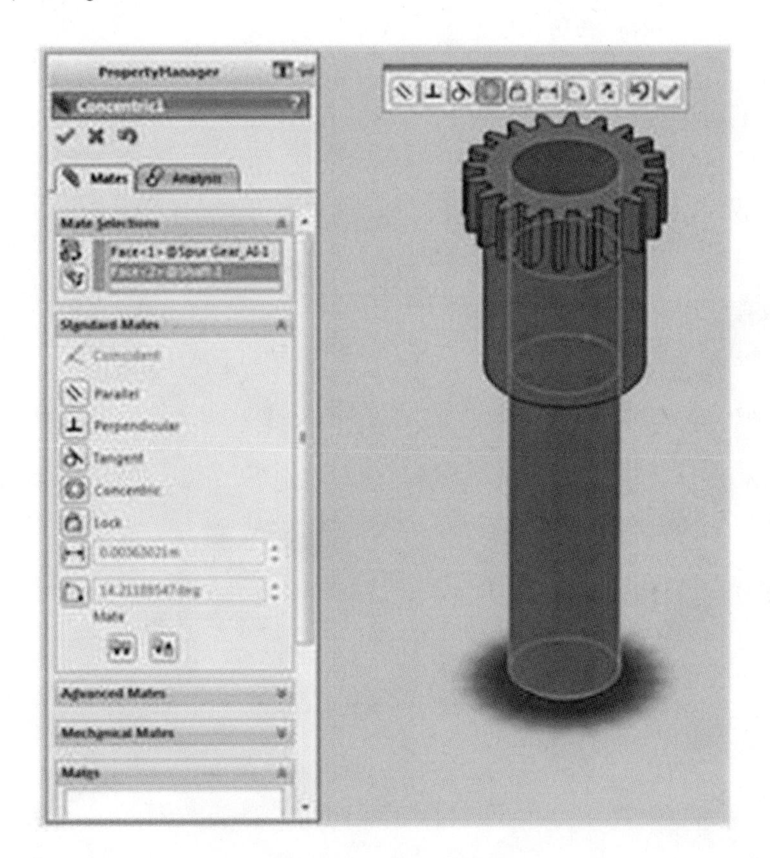

Figure 22.18

Concentric mating condition.

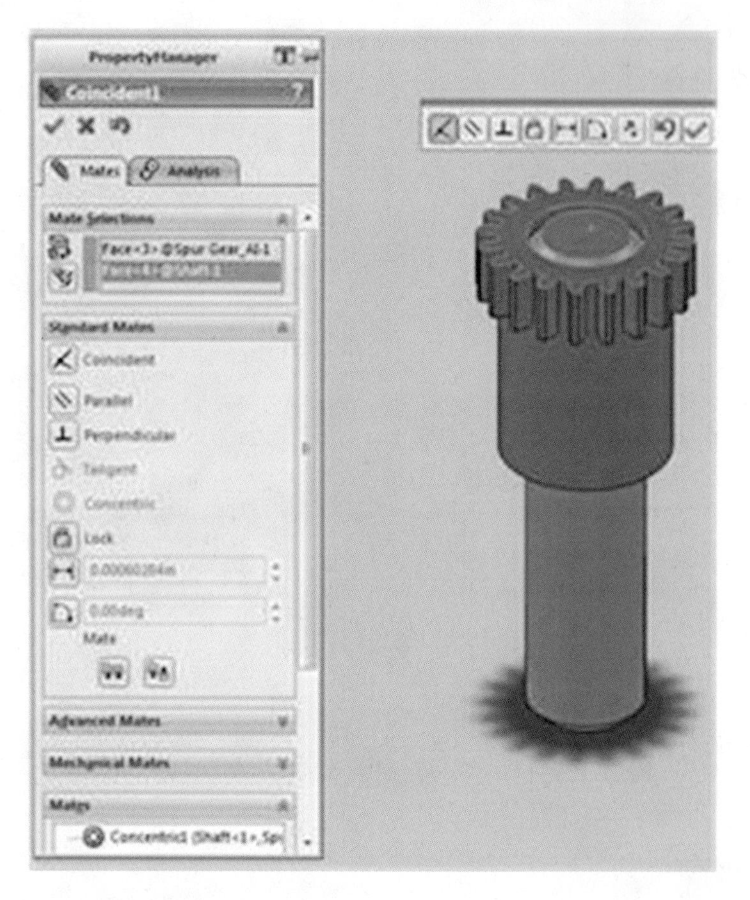

Figure 22.19

Coincident mating condition.

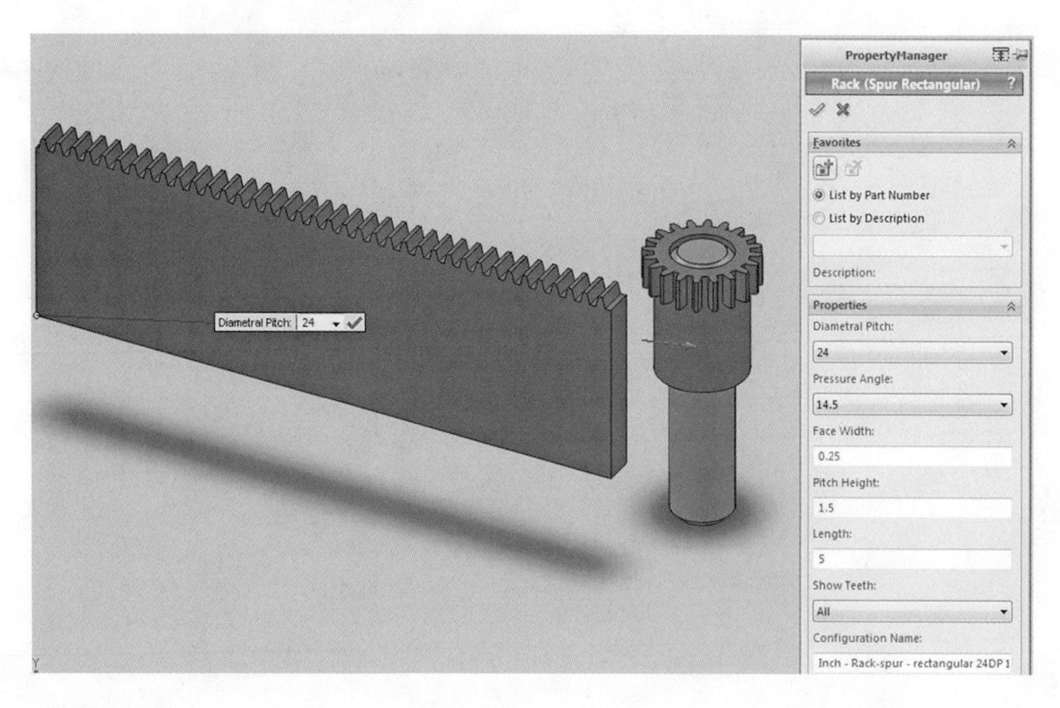

Figure 22.20

Rack PropertyManager.

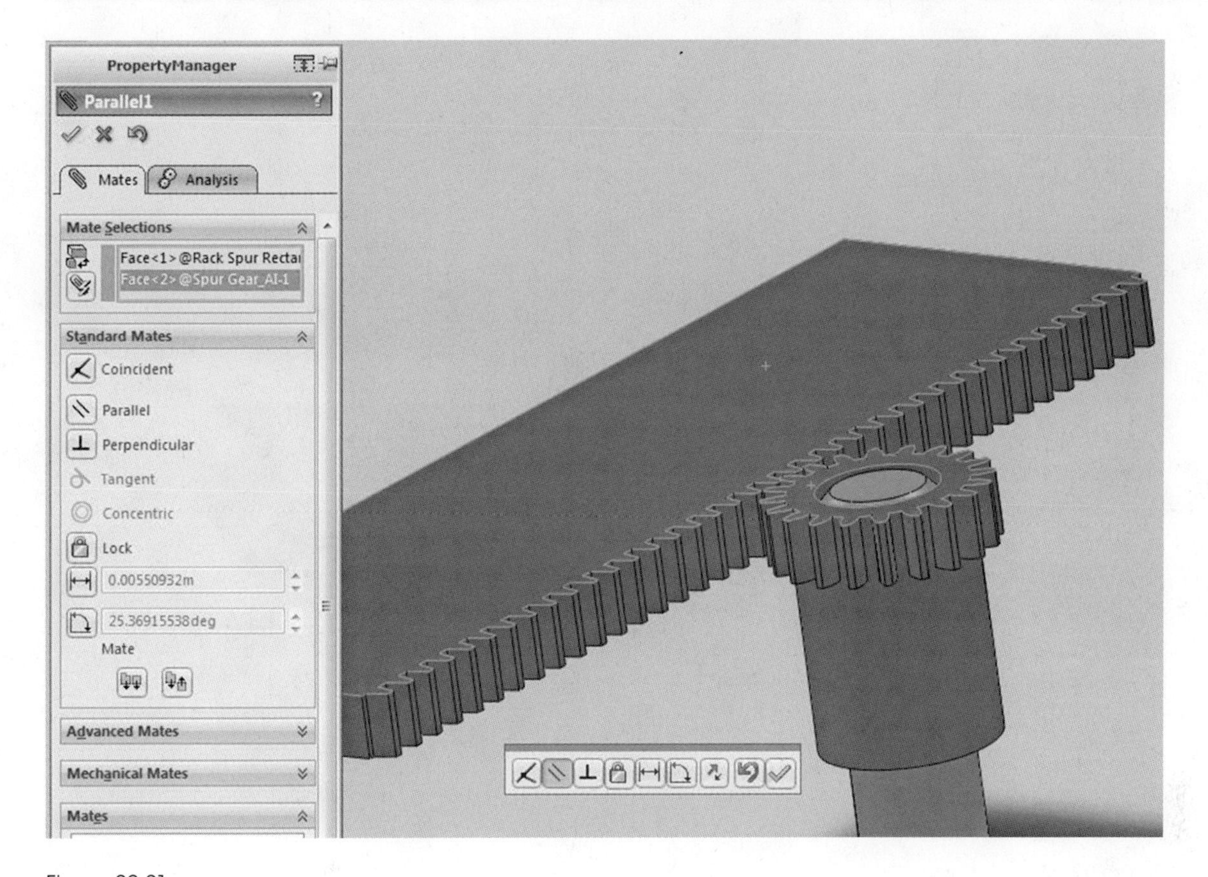

Figure 22.21

Parallel mate condition.

18. Select top of the pinion and the outer flat surface of the rack and apply Coincident mate (see Figure 22.22).
19. Select the bottom tooth of the rack and the top tooth of the pinion and apply Tangent mate (see Figure 22.23).

Mechanical Mating

20. Click Mate > Mechanical Mates. (See Figure 22.24 for the Rack Pinion PropertyManager.)
21. Click Rack Pinion (see Figure 22.24).
22. Select the *top edge of the rack* and the *side face of the pinion* (see Figure 22.24).
23. Click OK. (See Figure 22.25 for completed assembly.)

The FeatureManager is shown in Figure 22.26. At this stage, it would be necessary to save the document under a preferred filename.

Animation

24. Click the Motion Study1 icon at the bottom-right corner of the graphics window.
25. Click the Motor icon (see Figure 22.27).
26. Click the pinion as the Component (see Figure 22.27).
27. Set the Motion Speed as required (100 rpm) (see Figure 22.27).
28. Click OK. (See Figure 22.28 for the position of the pinion at the end of simulation.)

Belts and Pulleys

Belts are the cheapest utility for power transmission between shafts that may not be axially aligned. Power transmission is achieved by specially designed belts and pulleys. The demands on a belt drive transmission system are large, and this has led to many variations on the theme. They run smoothly and with little noise,

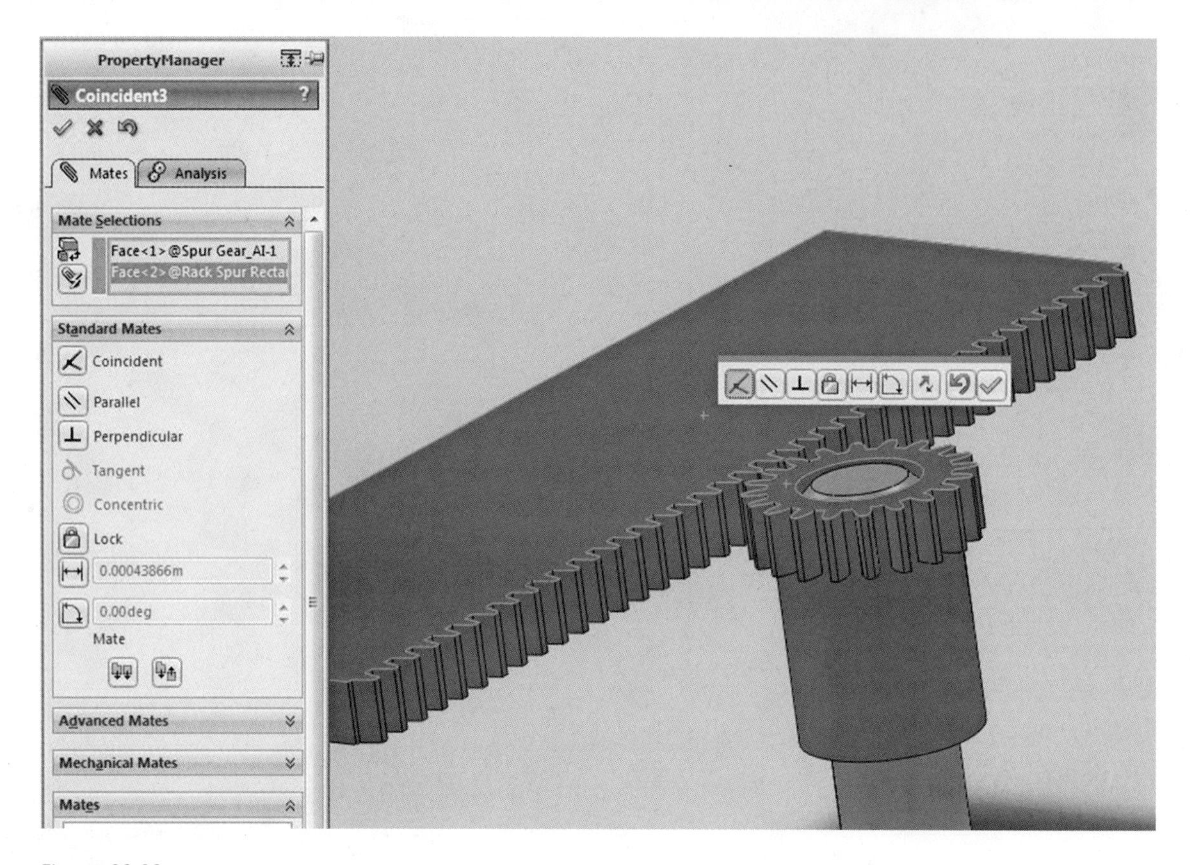

Figure 22.22

Coincident mate condition.

Figure 22.23

Tangent mate condition.

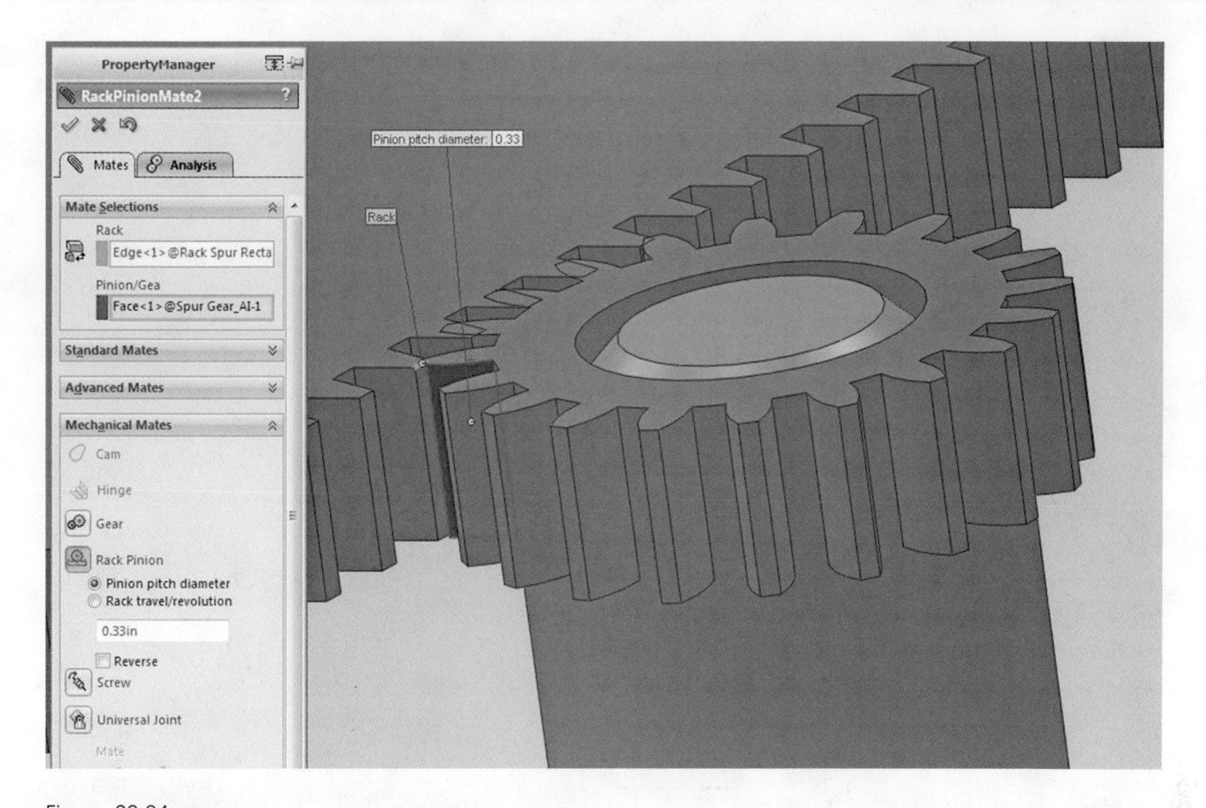

Figure 22.24

Mechanical mate condition.

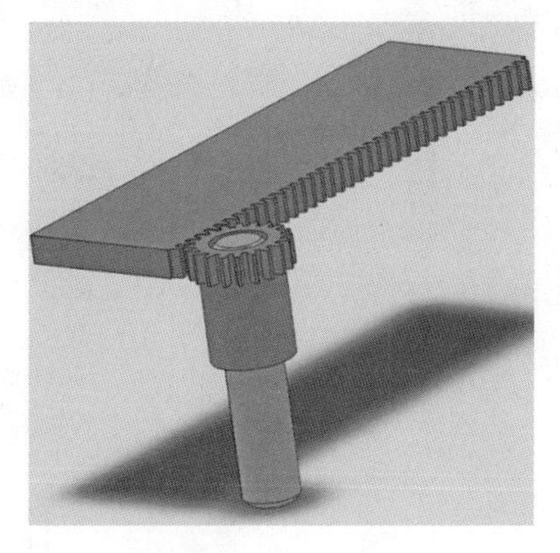

Figure 22.25

Completed rack-pinion assembly.

and with cushion motor and bearings against load changes, albeit with less strength than gears or chains. However, improvements in belt engineering allow the use of belts in systems that only formerly allowed chains or gears.

A *belt* is a loop of flexible material that is used to link two or more rotating shafts mechanically. Belts may be used as a source of motion, to transmit power efficiently, or to track relative movement. Belts are looped over pulleys. In a two-pulley system, either the belt can drive the pulleys in the same direction or the belt may be crossed, so that the direction of the shafts is opposite. As a source of motion, a conveyor belt is one application where the belt is adapted to continually carry a load between two points.

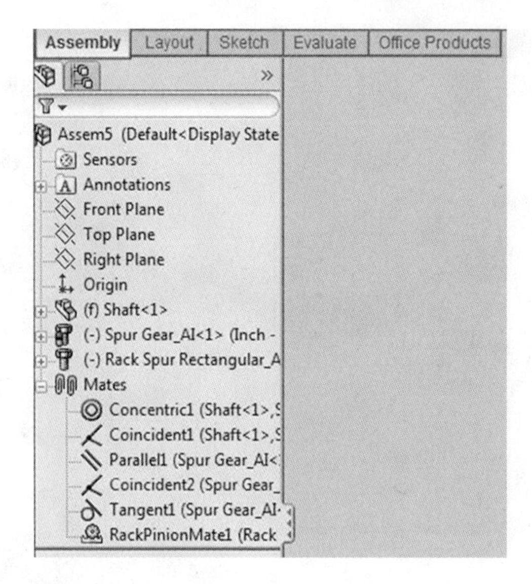

Figure 22.26

FeatureManager showing Design Tree.

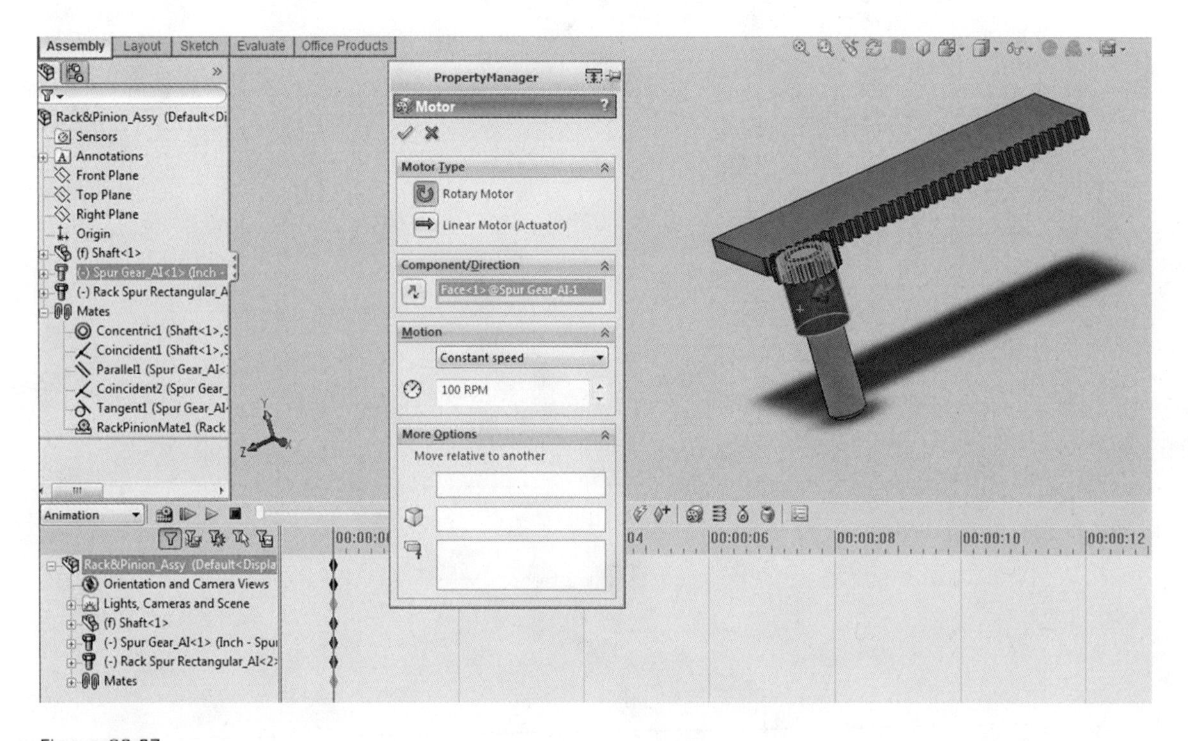

Figure 22.27

Motion of gear system.

Problem Description

A belt-and-pulley system consists of the support plate and shaft, pulley, and belt, and the specifications relating to their geometry are given as follows:

Support Plate

Length = 6.00

Width = 4.00

Thickness = 0.5

Holes: 0.375 diameter-2 holes, one at (1.00, 2.00) diametrically positioned; the other is located 3.00 in. away to the right (see Figure 22.29).

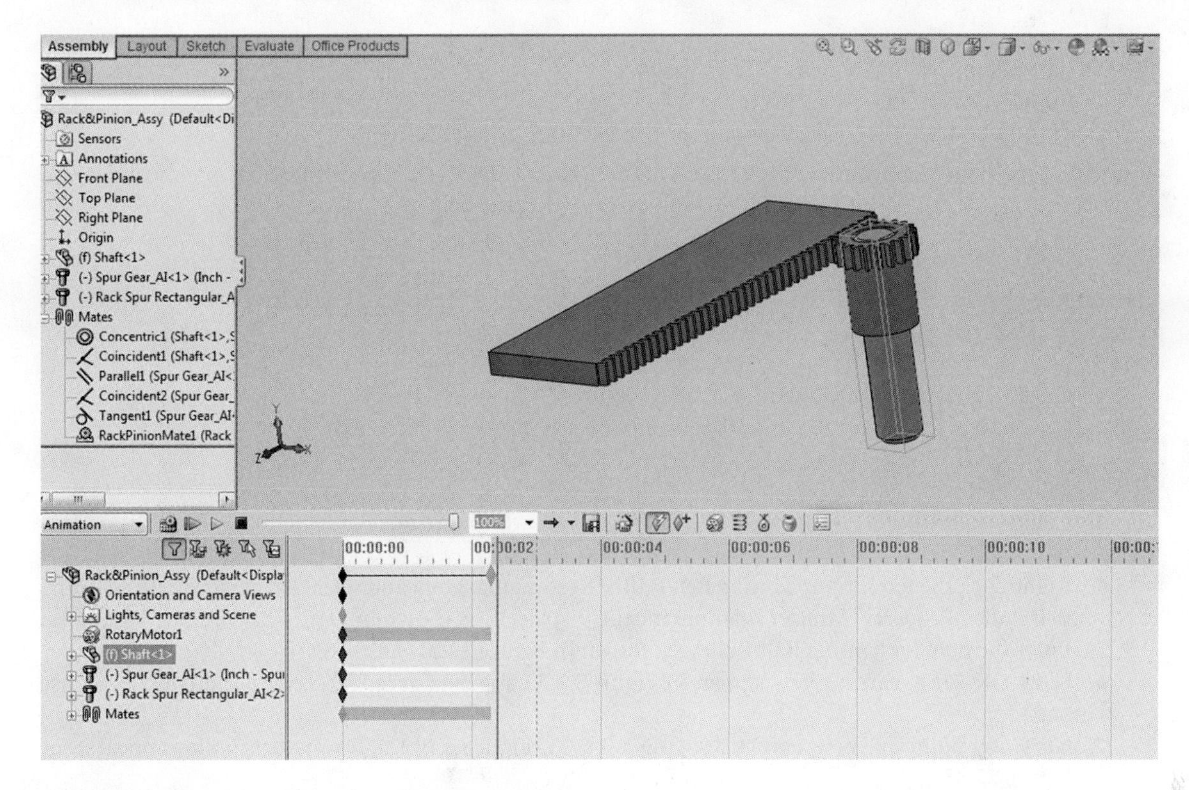

Figure 22.28

Position of the pinion at the end of simulation.

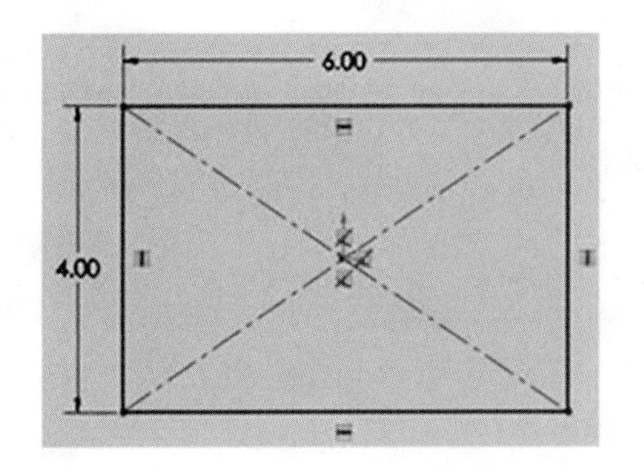

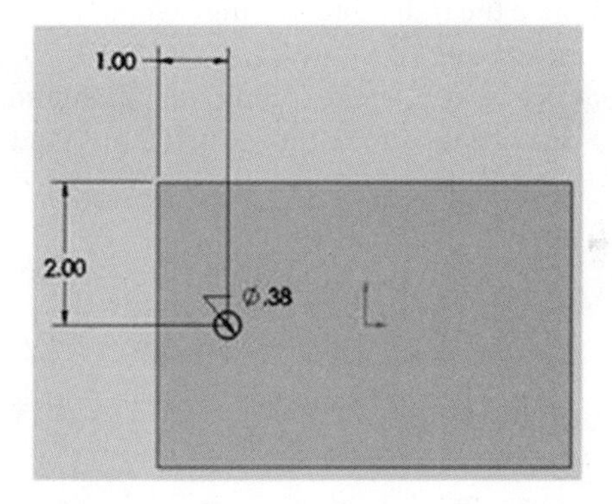

Figure 22.29

Support plate dimensions.

Shaft
Length = 1.75
Diameter: 0.375

Pulley
Belt pitch = (0.200)—XL
Belt width = 0.38
Pulley style: Flanged
Number of grooves = 20
Hub diameter = 0.375
Overall length = 0.5
Nominal shaft diameter = 3/8
Keyway: None

Using SolidWorks, create an assembly model consisting of the support plate and shaft, pulley, and belt, and animate the movement of the belt and the pulley.

Belt and Pulley Assembly Modeling Using SolidWorks

The belt and pulley assembly modeling consists of three main steps: (1) inserting the subassembly of the support plate and shaft, (2) inserting the pulley, and (3) inserting the belt.

Subassembly of Support Plate and Pin

The support plane and the shaft are similar to the one that is used for the spur gear except for the specifications, so modeling details will not be given in this section.

1. Create a New Assembly document.
2. Create an assembly of the support plate and two shafts (see Figure 22.6).

Inserting Timing Belt Pulley

3. Click Design Library Toolbox > ANSI Inch > Power Transmission > Timing Belt Pulley (see Figure 22.30).
4. Right-click the Rack (Spur Timing Belt Pulley Rectangular) tool and select Create Part. (The Timing Belt Pulley PropertyManager is automatically displayed, as shown in Figure 22.30.)
5. Enter the data for Timing Belt Pulley, as shown in Figure 22.31.
6. Click OK. (If a warning box shown in Figure 22.32 appears, accept the error and click the Close box.)
7. Click any point on the graphics drawing area. (A duplicate of the timing belt pulley appears; see Figure 22.33.)
8. Click Cancel in the Insert Components PropertyManager.

Mating

9. Rotate the individual timing belt pulley.
10. Click Mate.
11. Select the hole in the timing belt pulley and the outer surface of the shaft and apply Concentric mate.
12. Select the top of the timing belt pulley and the top surface of the shaft and apply Coincident.

    Repeat for the second timing belt pulley/shaft pair. (See Figure 22.34 for subassembly.)

Figure 22.30

Toolbox access for Timing Belt Pulley in the SolidWorks Design Library.

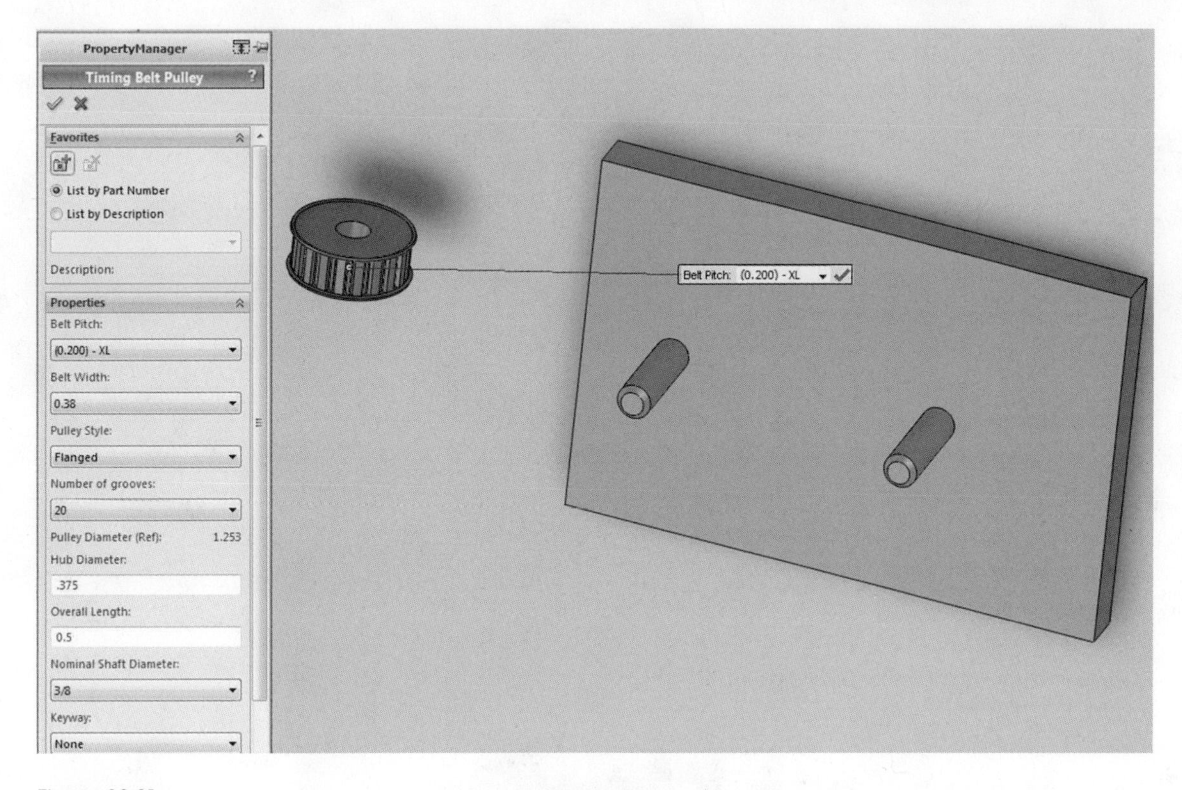

Figure 22.31

Timing Belt Pulley PropertyManager.

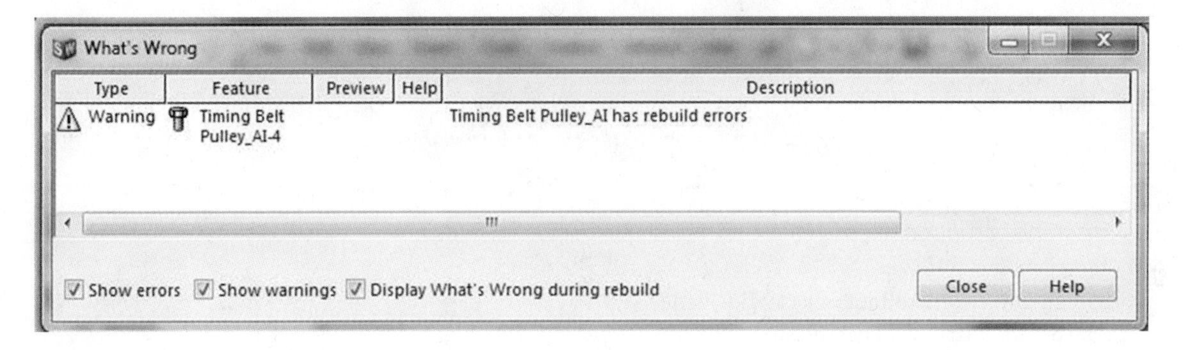

Figure 22.32

Warning box.

Inserting the Belt

13. Click Insert > Assembly Feature > Belt/Chain. (See Figure 22.35 for the Belt/Chain insertion tool.)
14. Click *one top surface on the left pulley's tooth* and *a second top surface on the right pulley's tooth* to define the Belt Members (see Figure 22.36).
15. In the Properties rollout, check the Field for Use Belt Thickness and assign a value of 0.14 (see Figure 22.36).
16. In the Properties rollout, check the Fields for Engage Belt, and Create Belt Part.

    A belt feature is created as shown in Figure 22.37.
17. Expand Belt1 in the FeatureManager.
18. Right-click [Belt1^...]<1>->? and select Edit Part option (see Figure 22.38).
19. Click *any segment of the belt feature* (see Figure 22.39).
20. Click the Sketch icon to be in Sketch mode (see Figure 22.39).
21. Click Features > Extrude Boss/Base. (The Extrude PropertyManager is displayed, as shown in Figure 22.40.)

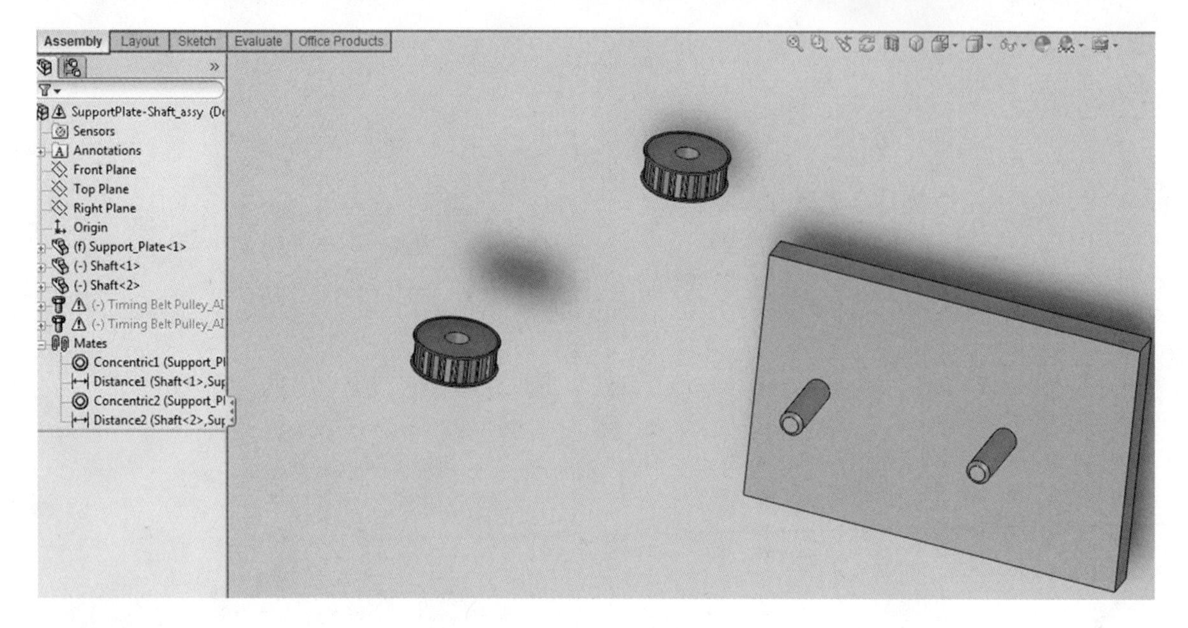

Figure 22.33

Duplicate of the timing belt pulley created.

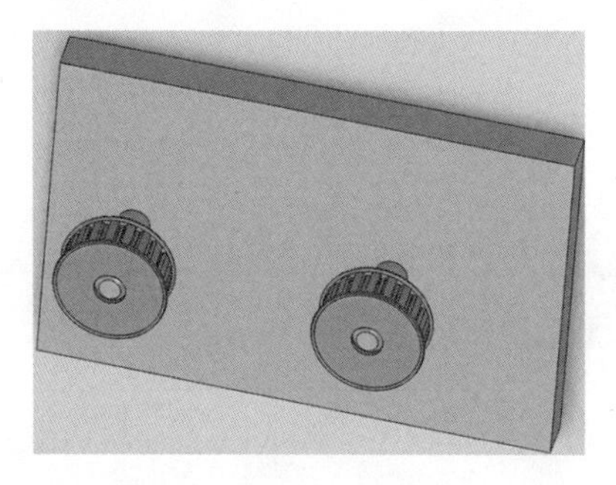

Figure 22.34

Subassembly of support plate, shafts, and timing belt pulleys.

22. In Direction1 rollout, select Mid-Plane.
23. Set the extrusion Distance to be 0.42.
24. Check the box for Thin Feature and set the Thickness to 0.14.
25. Click OK.
26. Click Edit Component to exit editing the assembly.

Animation

27. Click Motion Study1 icon at the bottom-right corner of the graphics window.
28. Click the Motor icon (see Figure 22.41).
29. Click one of the pulleys as the Component (see Figure 22.41).
30. Set the Motion Speed as required (100 rpm) (see Figure 22.41).
31. Click OK. (See Figure 22.28 for the position of the pinion at the end of simulation.)

Chain Drive: Chains and Sprockets

Chain drive is a way of transmitting mechanical power from one place to another. It is often used to convey power to the wheels of a vehicle, particularly bicycles and motorcycles. It is also used in a wide variety of machines besides vehicles. Most often, the power is conveyed by a roller chain, known as the *drive chain* or *transmission chain*, passing over a sprocket gear, with the teeth of the gear meshing with the holes in the links of the chain. The gear is turned, and this pulls the chain putting mechanical force into the system.

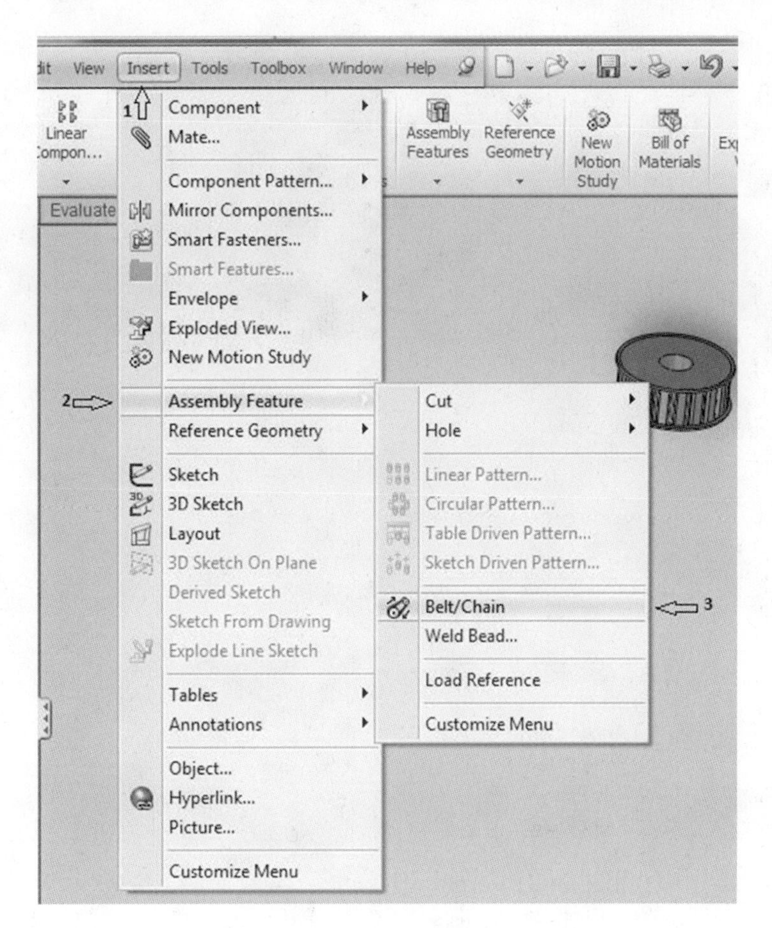

Figure 22.35

Belt/Chain insertion tool.

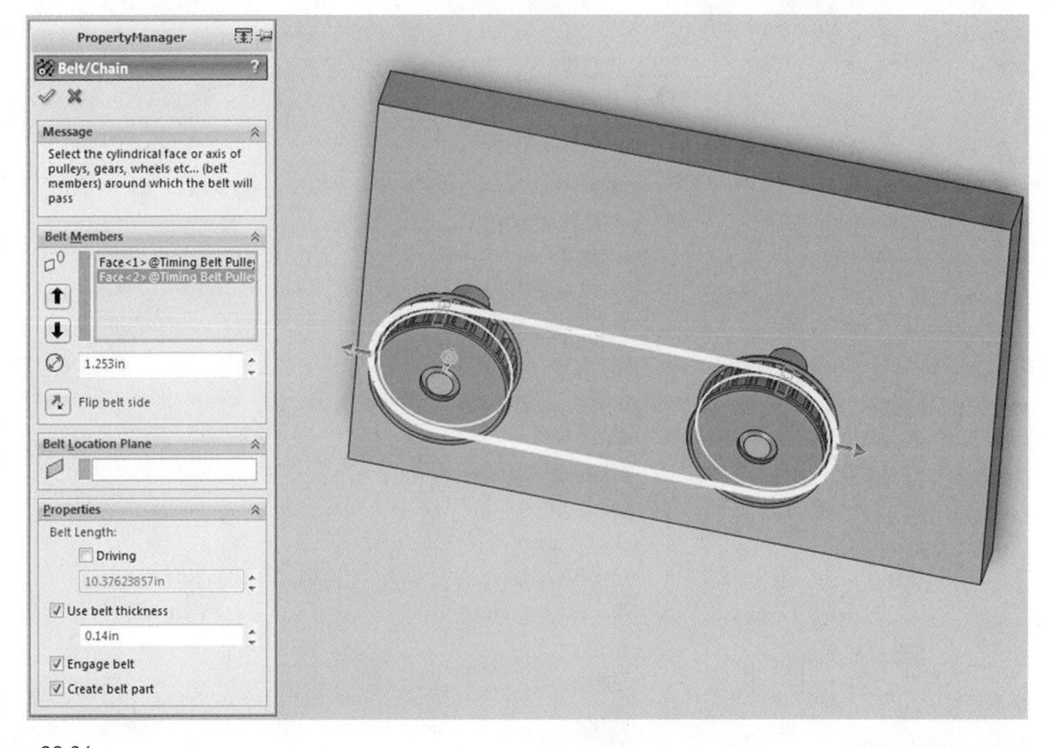

Figure 22.36

Belt Members defined.

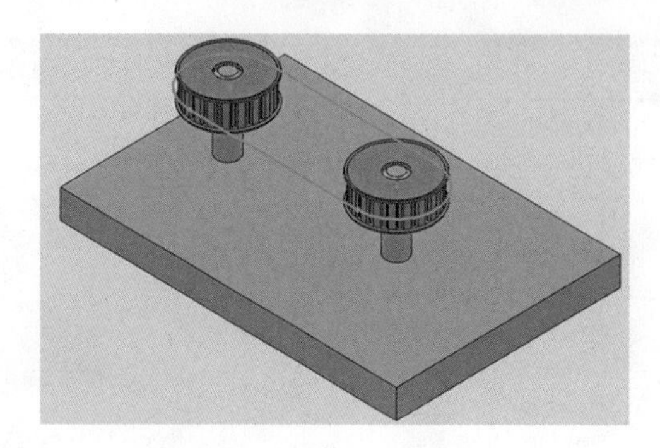

Figure 22.37

A belt feature is created.

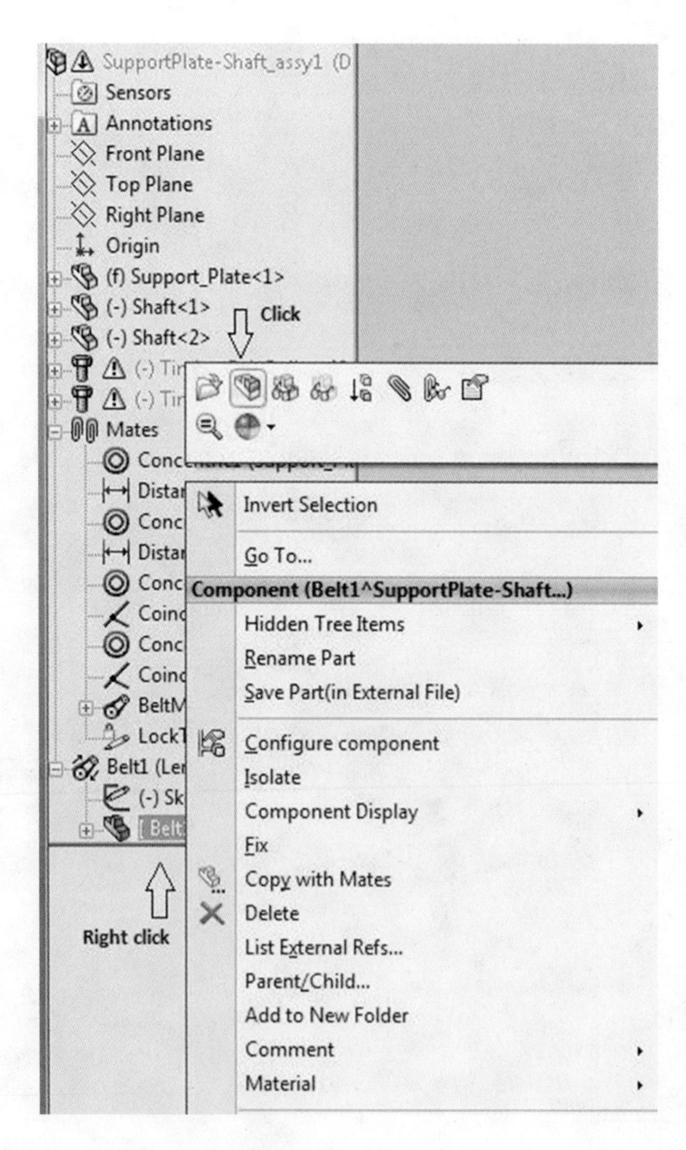

Figure 22.38

In-Context modeling of the belt.

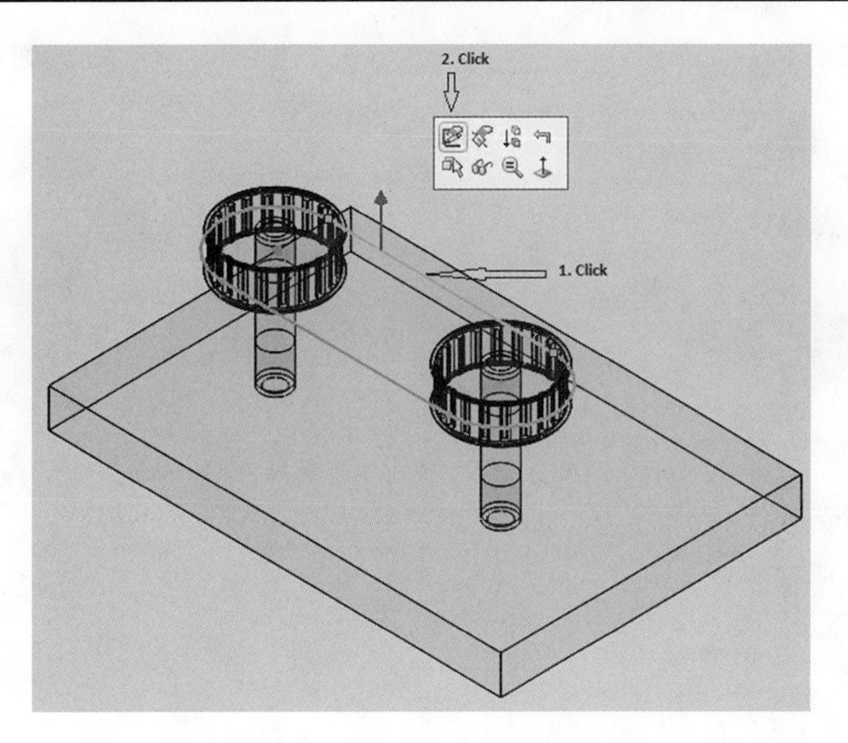

Figure 22.39

Modeling the belt in In-Context mode.

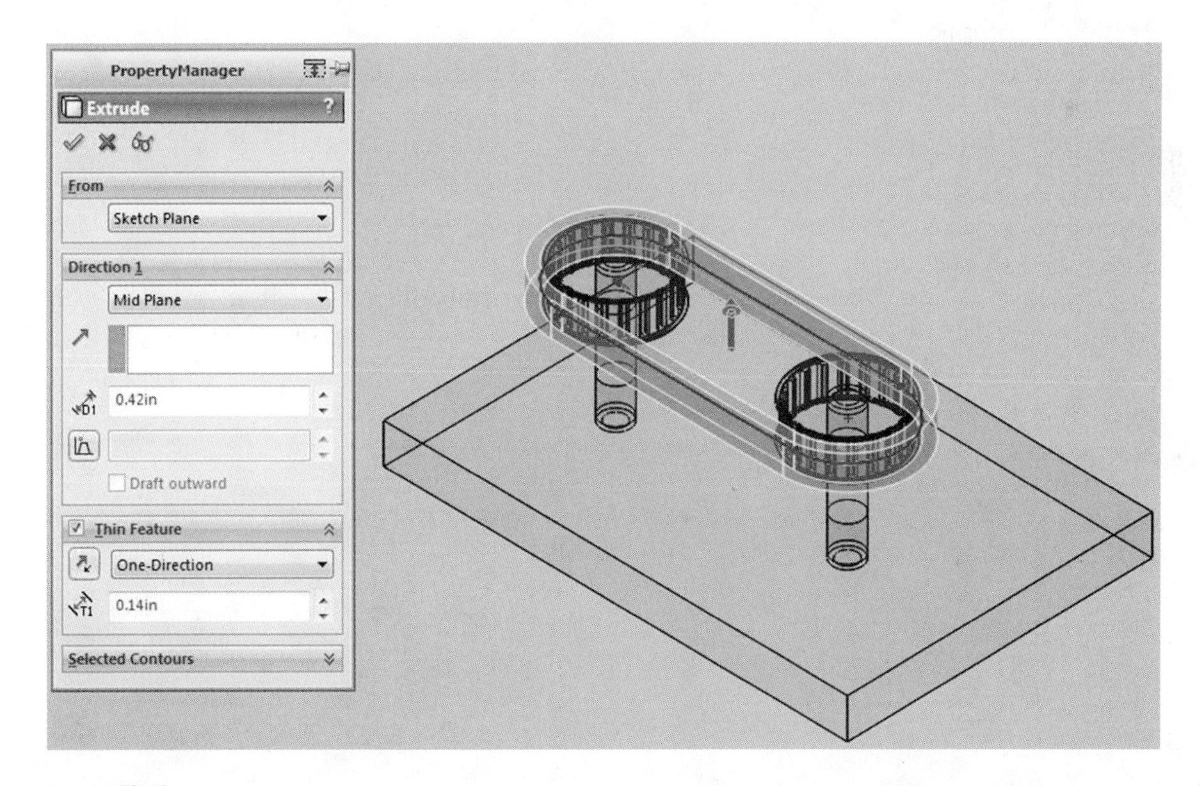

Figure 22.40

Extrude PropertyManager.

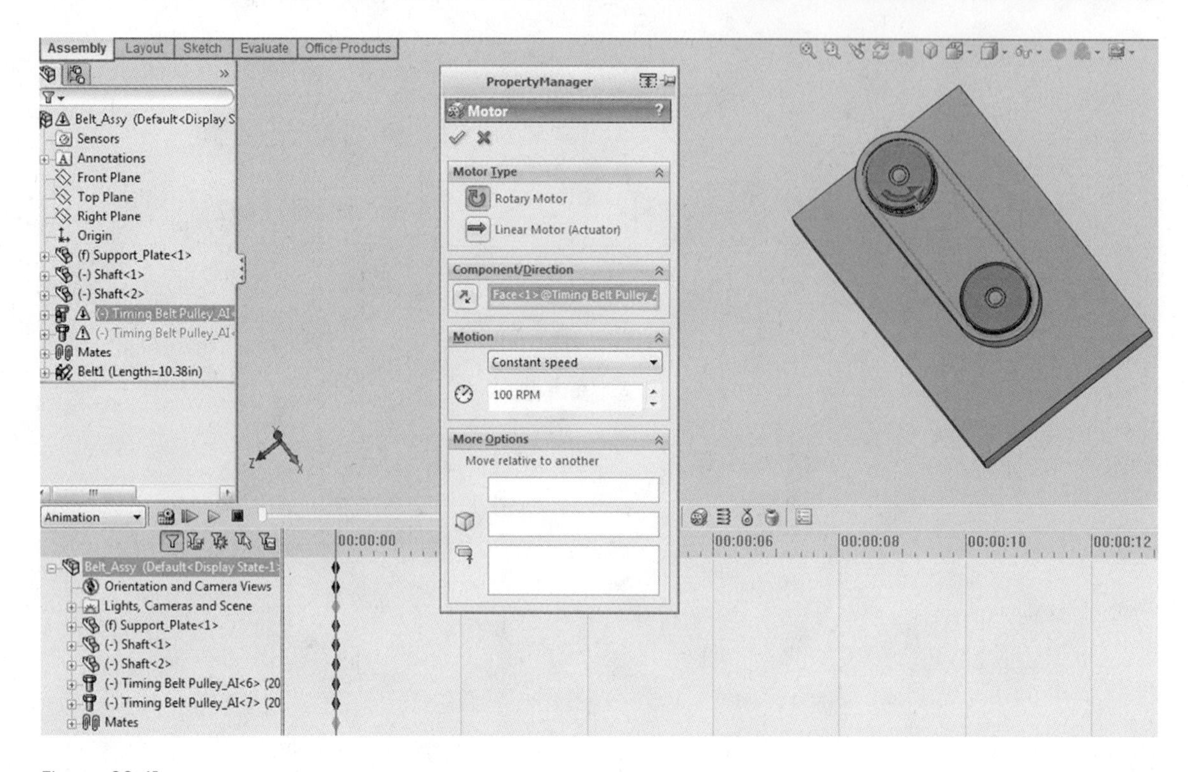

Figure 22.41

Motion PropertyManager.

SolidWorks creates chains and sprockets in a manner that is similar to that which is used for creating belts and pulleys.

Problem Description

A chain-and-sprocket system consists of the support plate and shaft, sprocket, and chain and the specifications relating to their geometry are given as follows:

Support Plate
Length = 20.00
Width = 10.00
Thickness = 1.00
Holes: 1.00 diameter-2 holes, one at (4.00, 5.00) diametrically positioned; the other is located 3.00 in. away to the right (see Figure 22.42).

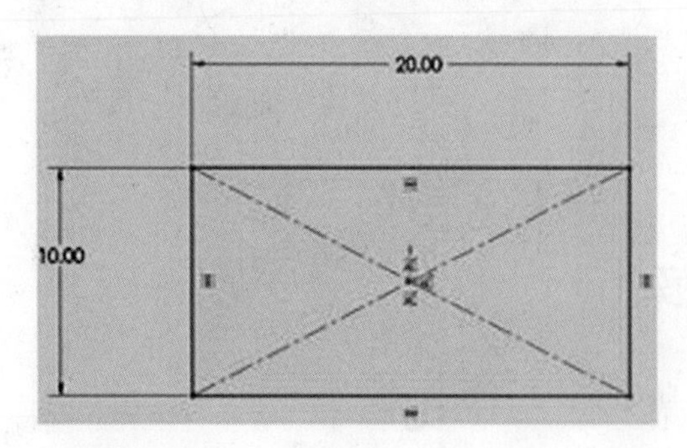

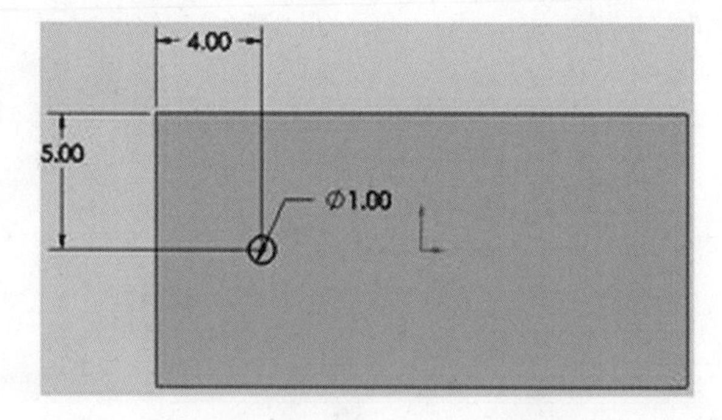

Figure 22.42

Support plate dimensions.

Shaft
Length = 4.00
Diameter: 1.00

Sprocket
Chain number = SC610
Number of teeth = 24
Belt width = 0.38
Pulley style: None
Nominal shaft diameter = 1
Keyway: None

Using SolidWorks, create an assembly model consisting of the support plate and shaft, sprocket, and chain, and animate the movement of the chain and sprocket.

Chain and Sprocket Assembly Modeling Using SolidWorks

The chain and sprocket assembly modeling consists of three main steps: (1) inserting the subassembly of the support plate and shaft, (2) inserting sprocket, and (3) inserting the chain.

Subassembly of Support Plate and Pin

The support plane and shaft are similar to the one that is used for the spur gear except for the specifications, so the modeling details will not be given in this section.

1. Start a New Assembly document.
2. Open SupportPlate-Shaft_assy, which is a 20 × 10 × 1 plate with 1 inch-dia-2 holes and a shaft, 1inch-dia-4 inch long (an assembly of the support plate and two shafts [see Figure 22.6]).

Inserting Timing Belt Pulley

3. Click Design Library Toolbox > ANSI Inch > Power Transmission > Chain Sprockets (see Figure 22.43a).
4. Right-click the Silent Larger Sprocket tool and select Create Part. (The Belt1 PropertyManager is automatically displayed, as shown in Figure 22.43b.)

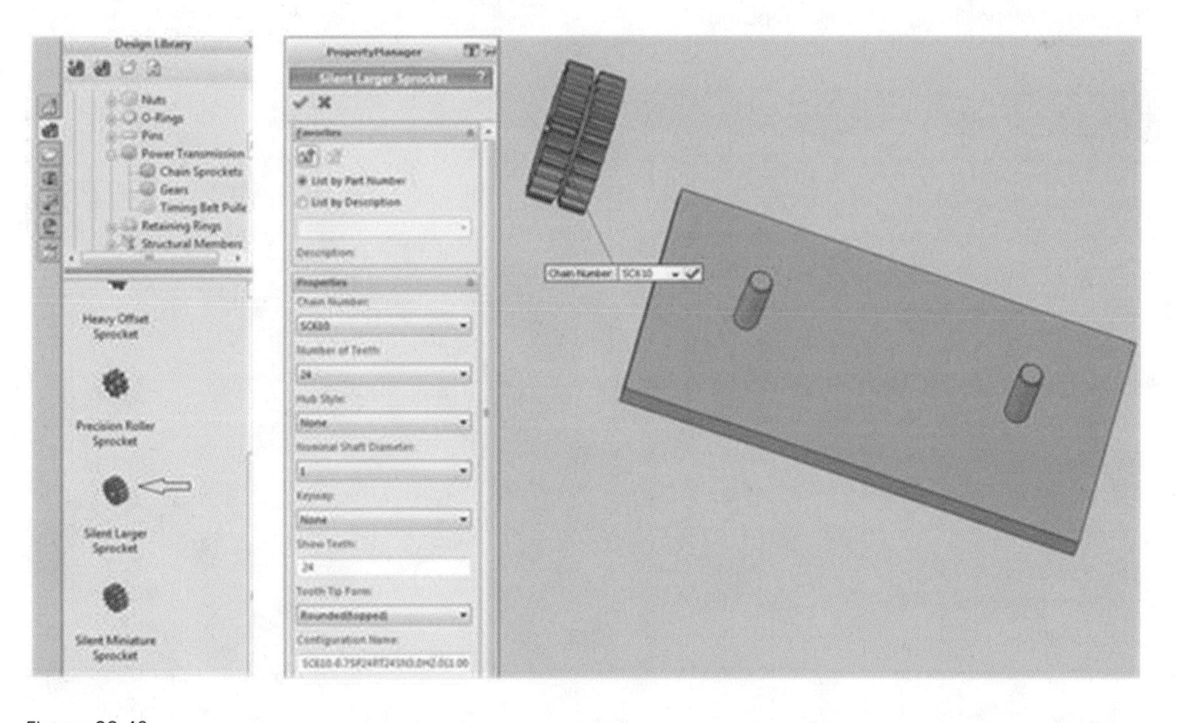

Figure 22.43

Chain Sprockets tool and Sprocket PropertyManager.

5. Enter the data for Timing Belt Pulley as shown in Figure 22.43b.
6. Click any point on the graphics drawing area. (A duplicate of the timing belt pulley appears; see Figure 22.44.)
7. Click Cancel in the Insert Components PropertyManager.
8. Click any point on the graphics drawing area. (A duplicate of the sprocket appears; see Figure 22.44.)
9. Click Cancel in the Insert Components PropertyManager.

Mating

10. Rotate the individual sprocket.
11. Click Mate.
12. Select the hole in the sprocket and the outer surface of the shaft and apply Concentric mate.
13. Select the top of the sprocket and the top surface of the shaft and apply Coincident (see Figure 22.45).
    Repeat for the second sprocket/shaft pair. (See Figure 22.46 for subassembly.)

Inserting the Chain

14. Click Insert > Assembly Feature > Belt/Chain (see Figure 22.47 for the Belt/Chain insertion tool).
15. Click *one top surface on the left sprocket's tooth* and *a second top surface on the right sprocket's tooth* to define the Belt Members (see Figure 22.48).

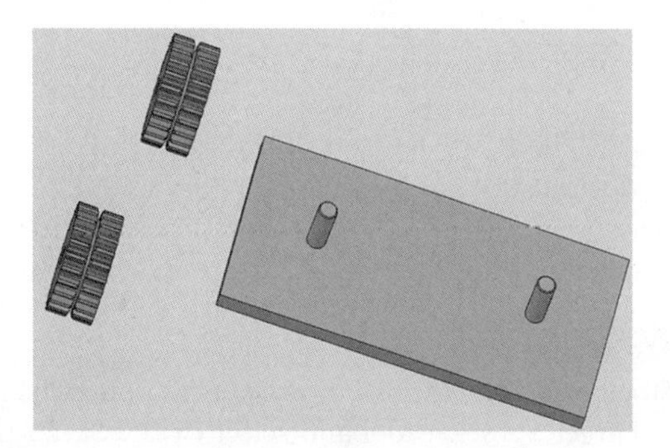

Figure 22.44

Duplicate of the timing belt pulley created.

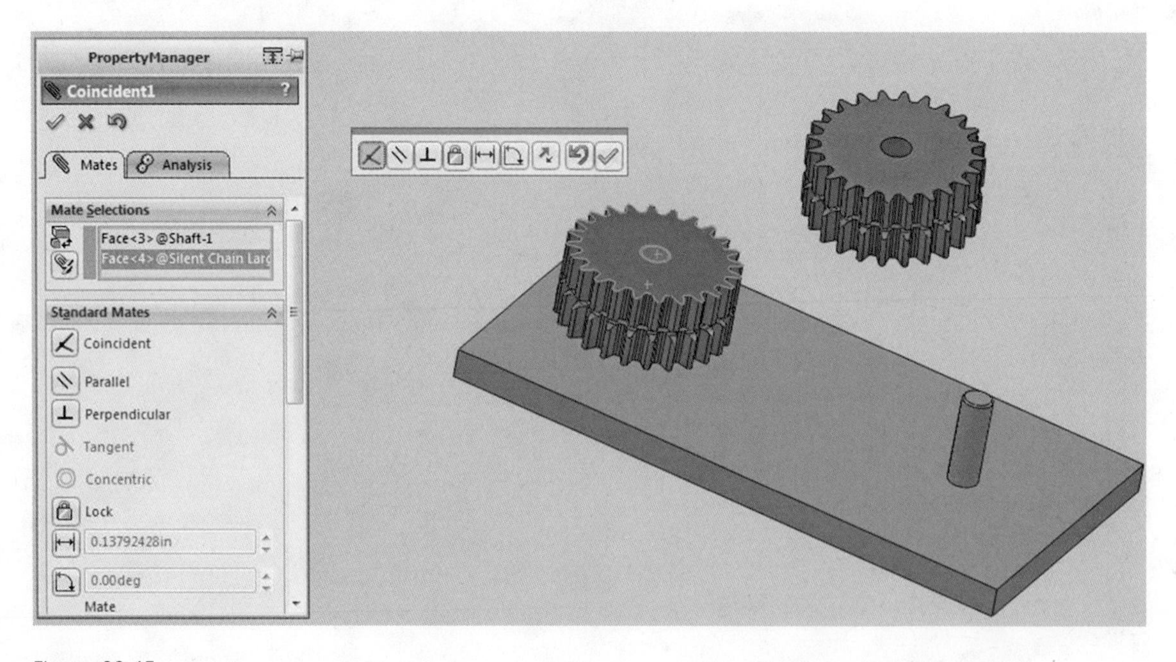

Figure 22.45

Mating conditions.

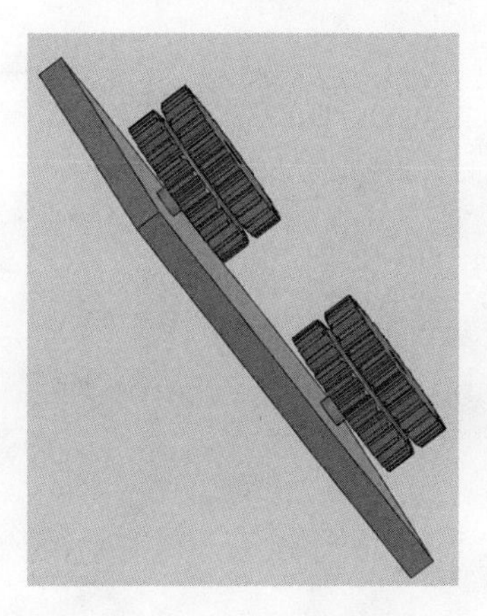

Figure 22.46

Subassembly of support plate, shafts, and sprockets.

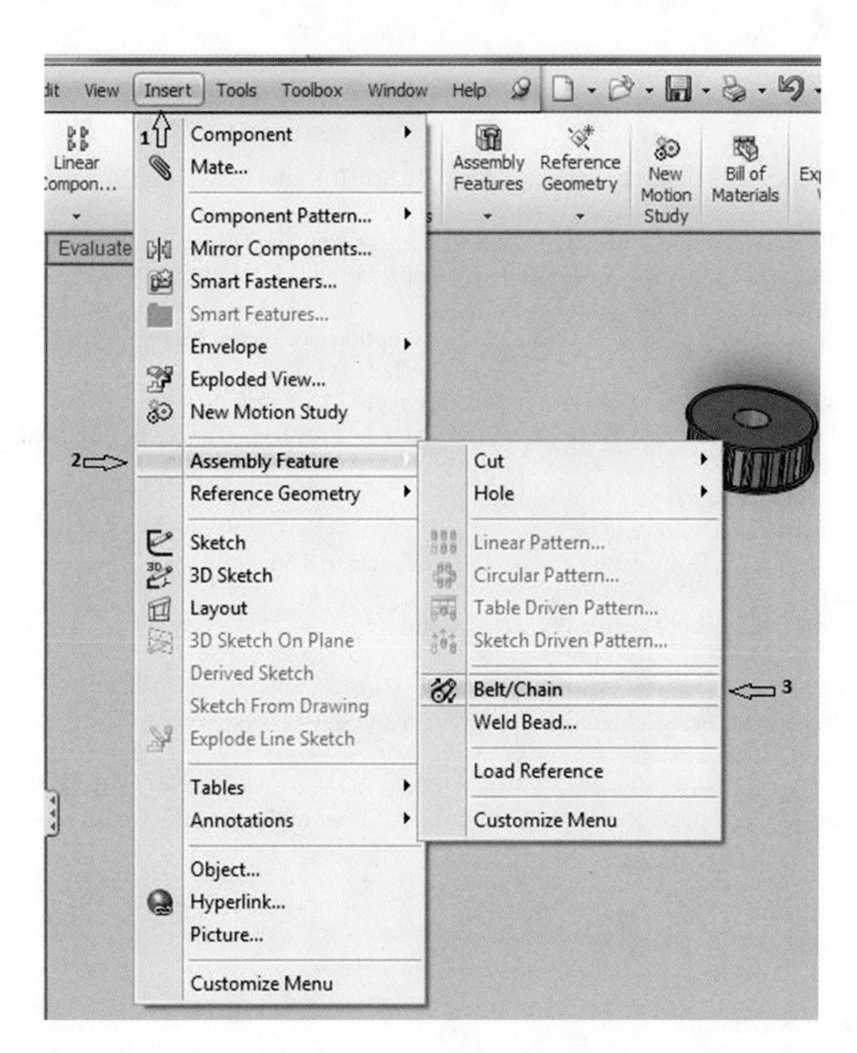

Figure 22.47

Belt/Chain insertion tool.

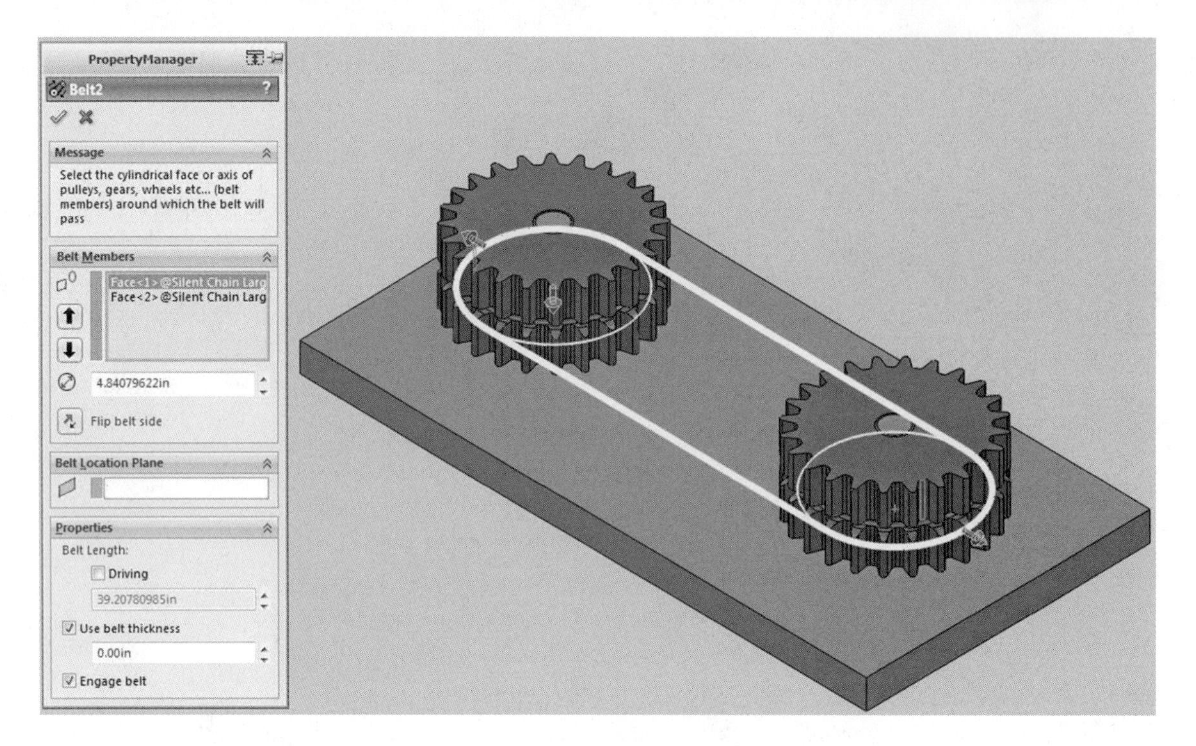

Figure 22.48

Belt Members defined.

16. In the Properties rollout, check the Field for Use Belt Thickness and assign a value of 0.5 (see Figure 22.48).
17. In the Properties rollout, check the Fields for Engage Belt, and Create Belt Part.
    A belt feature is created, as shown in Figure 22.48.
18. Expand Belt1 in the FeatureManager.
19. Right-click [Belt1^...]<1>->? and select Edit Part option (see Figure 22.49).
20. Click *any segment of the belt feature* (see Figure 22.50).
21. Click the Sketch icon to be in Sketch mode (see Figure 22.50).
22. Click Features > Extrude Boss/Base. (The Extrude PropertyManager is displayed, as shown in Figure 22.50.)
23. In Direction1 rollout, select Mid-Plane.
24. Set the extrusion Distance to be 1.00.
25. Check the box for Thin Feature and set the Thickness to 0.50.
26. Click OK.
27. Click Edit Component to exit editing the assembly.

Figure 22.51 shows the chain and sprocket assembly model.
Figure 22.52 shows the FeatureManager for the chain and sprocket assembly model.

Animation

28. Click the Motion Study1 icon at the bottom-right corner of the graphics window.
29. Click the Motor icon.
30. Click one of the sprockets as the Component.
31. Set the Motion Speed as required (100 rpm).
32. Click OK.

Bevel Gear Box Design

Before discussing the gear box design for the bevel gears, let us briefly discuss how to use the Design Library Toolbox (which must be available through the SolidWorks Options). The first illustration is the creation of

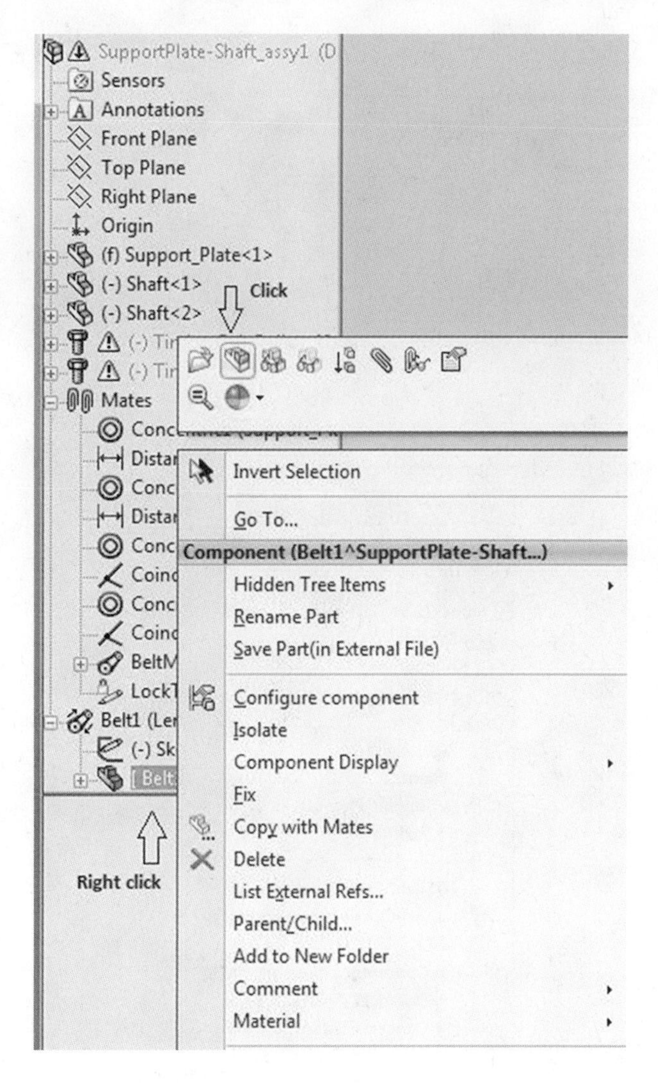

Figure 22.49

In-Context modeling of the belt.

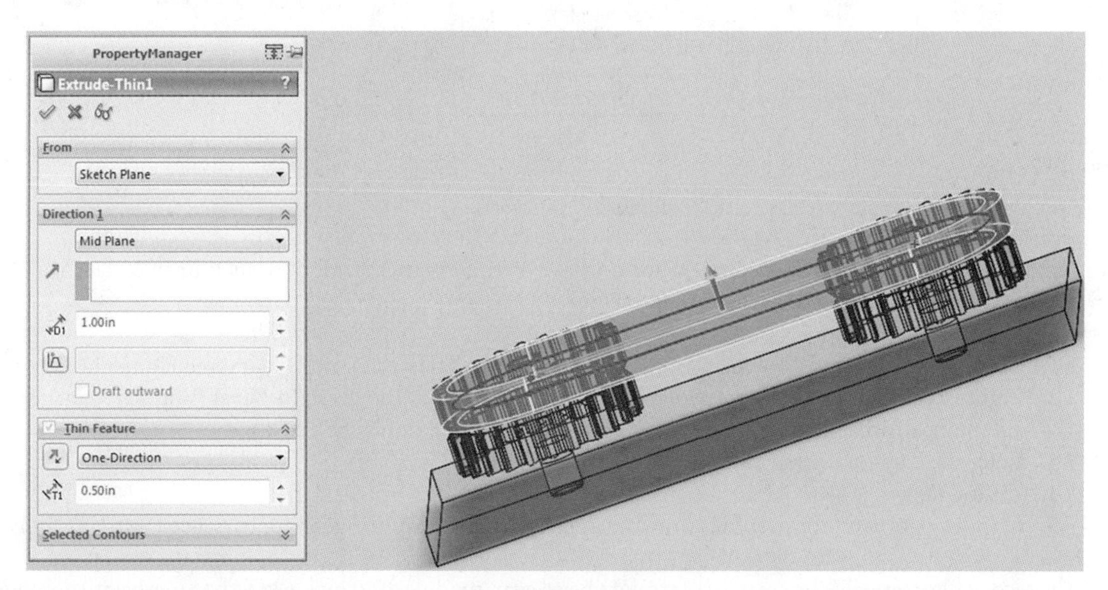

Figure 22.50

Extrude PropertyManager.

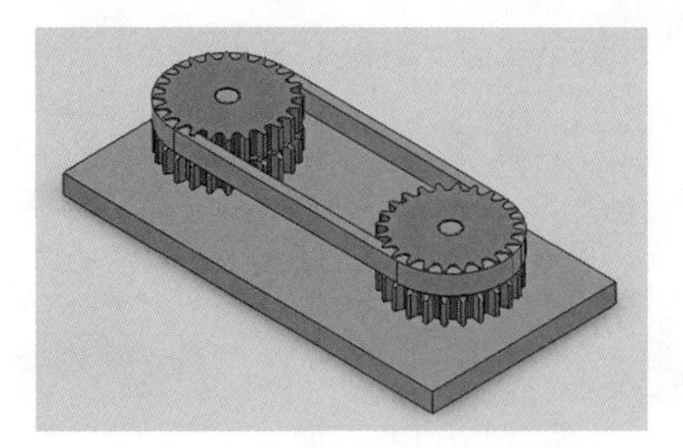

Figure 22.51

Chain and sprocket assembly model.

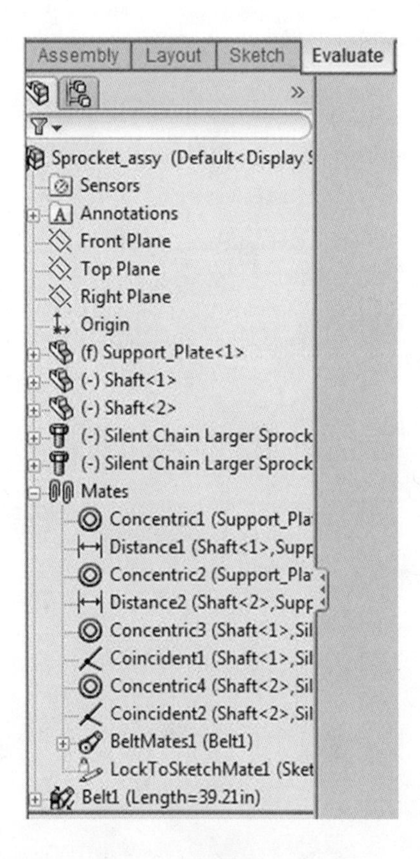

Figure 22.52

FeatureManager for the chain and sprocket assembly model.

straight bevel gear: a module of 10 mm, 25 teeth to mesh with another straight bevel gear of 63 teeth, a 20° pressure angle, and a face width of 51 mm (see Figures 22.53 and 22.54).

1. Click Design Library Toolbox > ANSI Metric > Power Transmission > Gears (see Figure 22.7).
2. Right-click the Straight Bevel tool and select Create Part. (The Straight Bevel PropertyManager is automatically displayed for the pinion and gear, as shown in Figure 22.53.)
3. Enter the data for pinion and gear as shown in Figure 22.53.
4. Click OK.
5. Click any point on the graphics drawing area. (A duplicate of the gear appears; see Figure 22.54.)

The second illustration is the creation of straight bevel gear: module of 10 mm, 63 teeth to mesh with another straight bevel gear of 25 teeth, 20° pressure angle, and face width of 51 mm (see Figures 22.55 and 22.56).

Let us now design the gear box in the subsequent subsections.

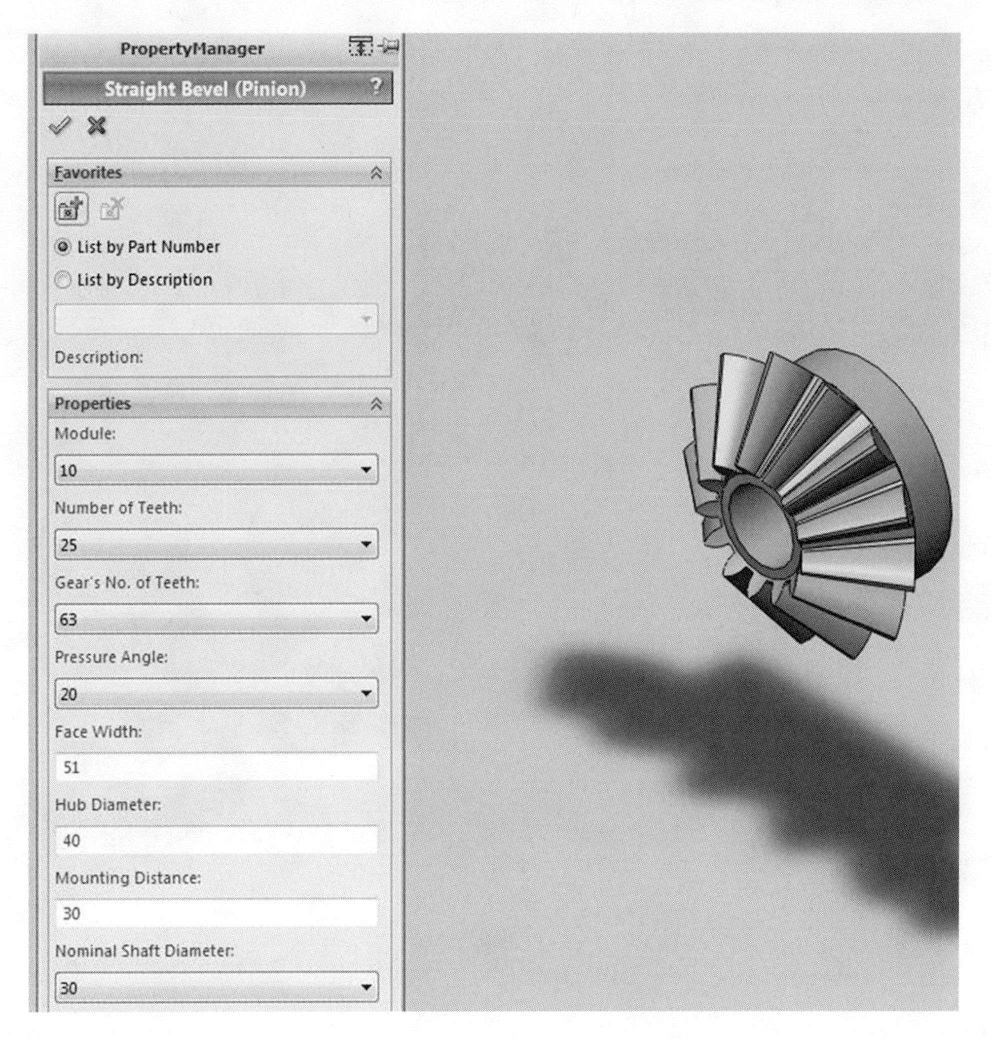

Figure 22.53

Preview in Straight Bevel PropertyManager.

Figure 22.54

Straight Bevel created from specifications.

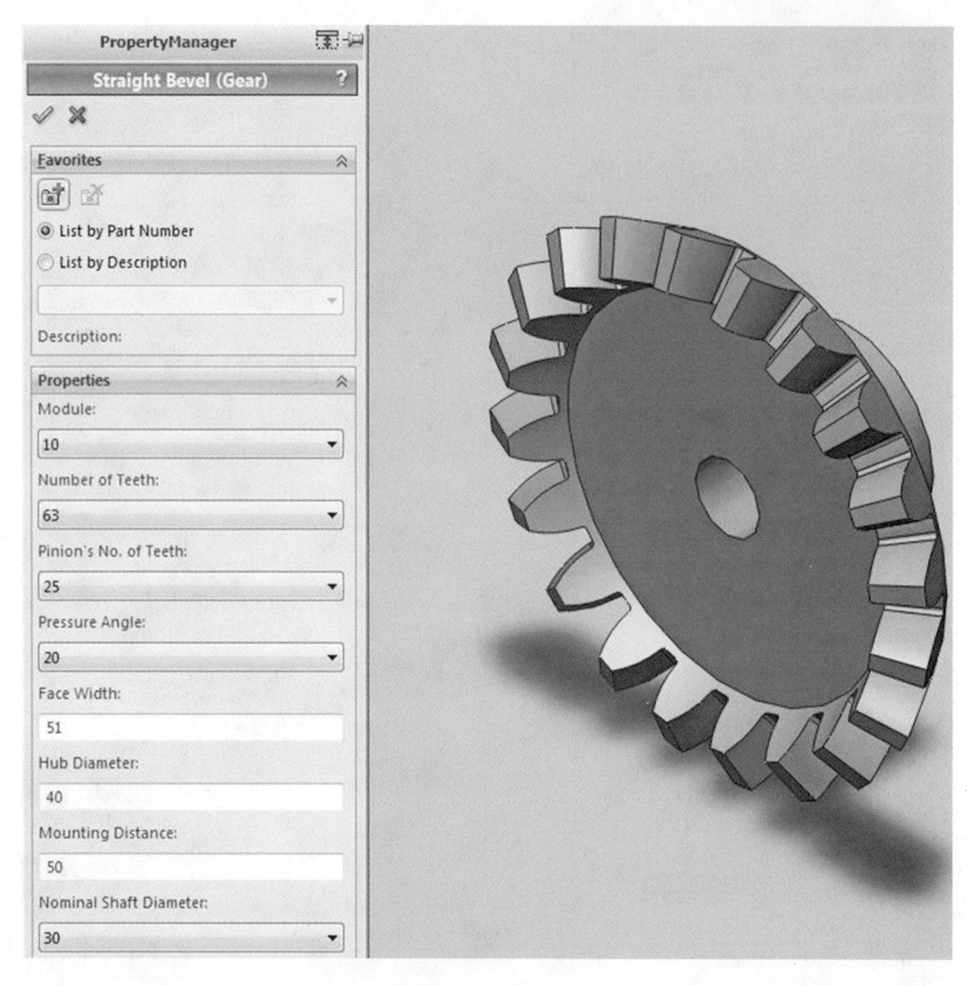

Figure 22.55

Preview in Straight Bevel PropertyManager.

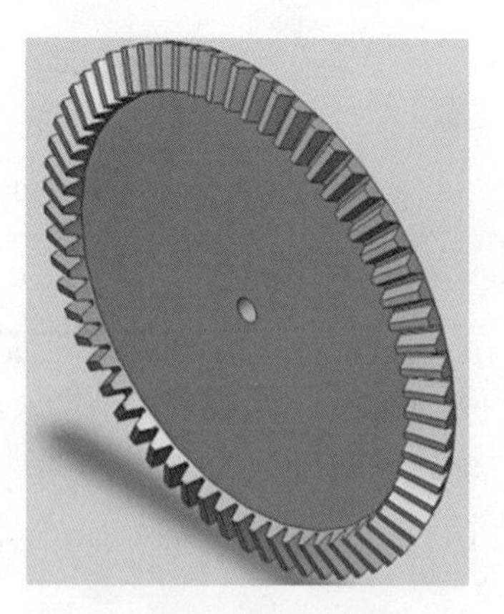

Figure 22.56

Straight Bevel created from specifications.

Problem Description

Design a gear box consisting of an input straight bevel gear having 20 teeth and an input straight bevel gear having 50 teeth. The diametral pitch is 2, a 20° pressure angle, and a face width of 2. The nominal shaft diameter is 1-3/8. (Note: Imperial units are used.)

Input

Pinion

1. Click Design Library Toolbox > ANSI Metric > Power Transmission > Gears (see Figure 22.7).
2. Right-click the Straight Bevel tool and select Create Part. (The Straight Bevel PropertyManager is automatically displayed.)
3. Enter the data for pinion and gear as shown in Figure 22.57.
4. Click OK.
5. Click any point on the graphics drawing area. (A duplicate of the gear appears.)

Output

Gear

1. Click Design Library Toolbox > ANSI Metric > Power Transmission > Gears (see Figure 22.7).
2. Right-click the Straight Bevel tool and select Create Part. (The Straight Bevel PropertyManager is automatically displayed.)
3. Enter the data for pinion and gear as shown in Figure 22.58.
4. Click OK.
5. Click any point on the graphics drawing area. (A duplicate of the gear appears.)

Assembly Modeling

1. Open a New Assembly SolidWorks document.
2. Import the input (pinion) and output (gear) bevel gears, and shafts into the graphics area for assembly (see Figure 22.59).
3. Create Concentric mate for a shaft and the *input* (pinion) bevel gear (Figure 22.60).
4. Create Concentric mate for a shaft and the *output* (gear) bevel gear (Figure 22.61).
5. Apply Concentric mates between the shafts of the bevel gears (see Figure 22.62).
6. Click the Mate > Mechanical Mates > Gear and select the *inside faces hubs of both bevel gears*. (The GearMate PropertyManager is automatically displayed; see Figure 22.63.)
7. Enter 1.38 as the gear Ratio (see Figure 22.63).
8. Ensure that gear teeth are manually aligned to enhance meshing (see Figure 22.64).
9. Apply Coincident mate between the end face of the shaft and the faces of bevel gears (see Figures 22.65 and 22.66).

Figure 22.67 shows the gear box complete without housing.

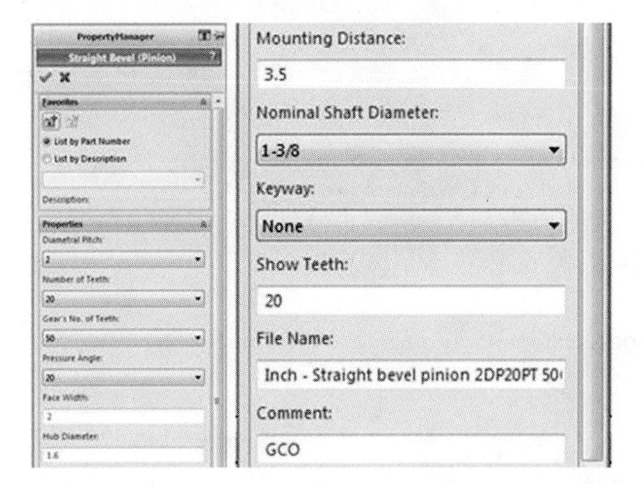

Figure 22.57

Input Straight Bevel specifications.

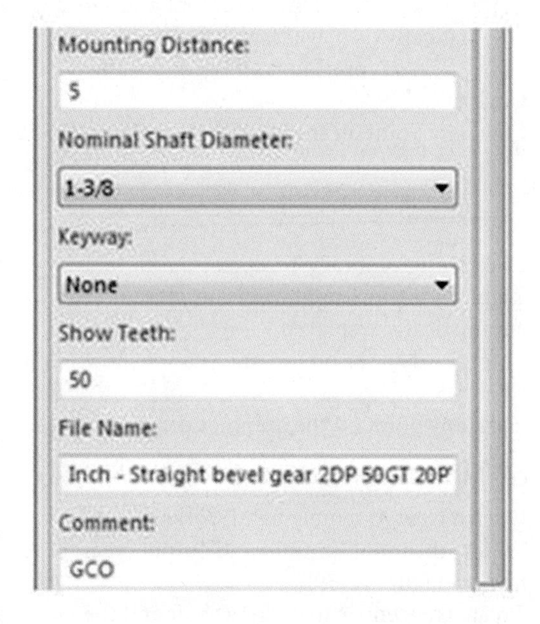

Figure 22.58

Output Straight Bevel specifications.

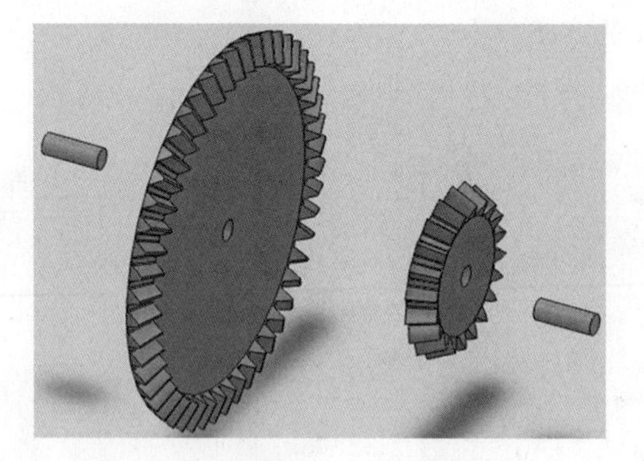

Figure 22.59

Input and output bevel gears and shafts in graphics area for assembly.

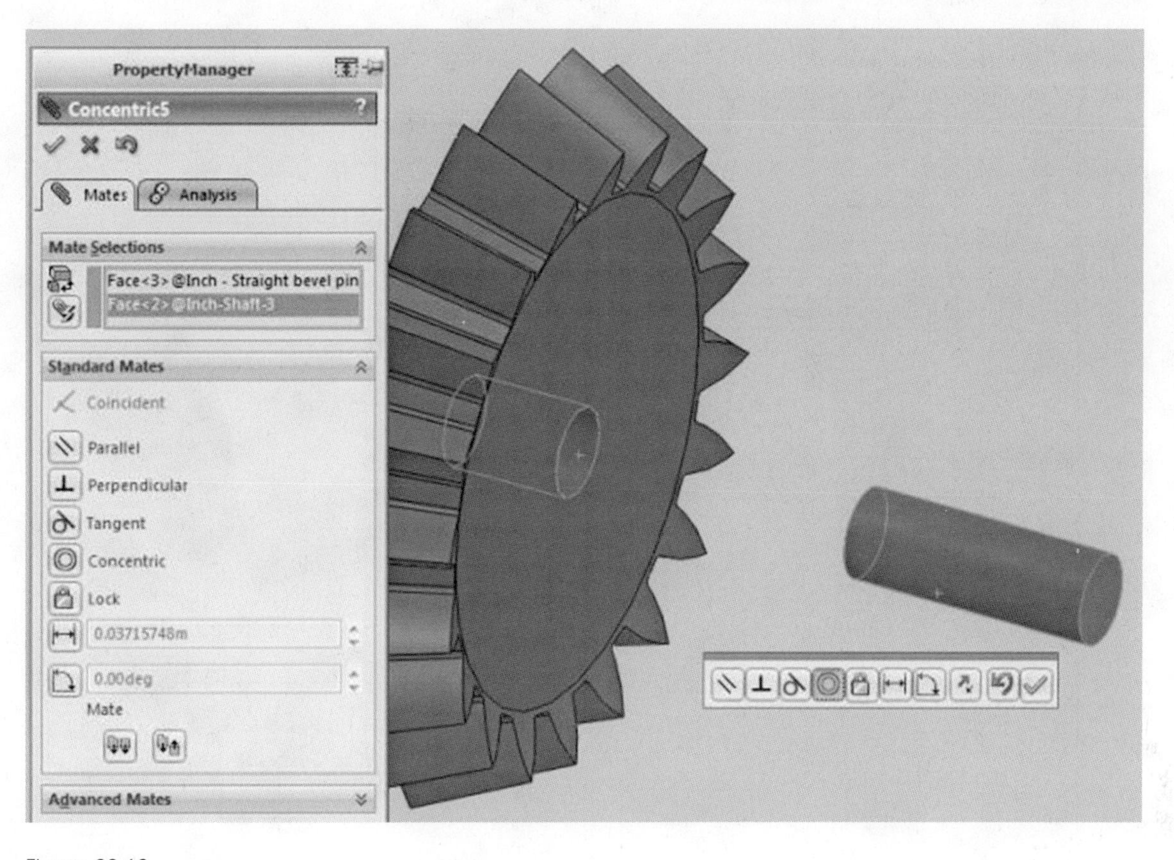

Figure 22.60

Concentric mate for shaft and input bevel gear.

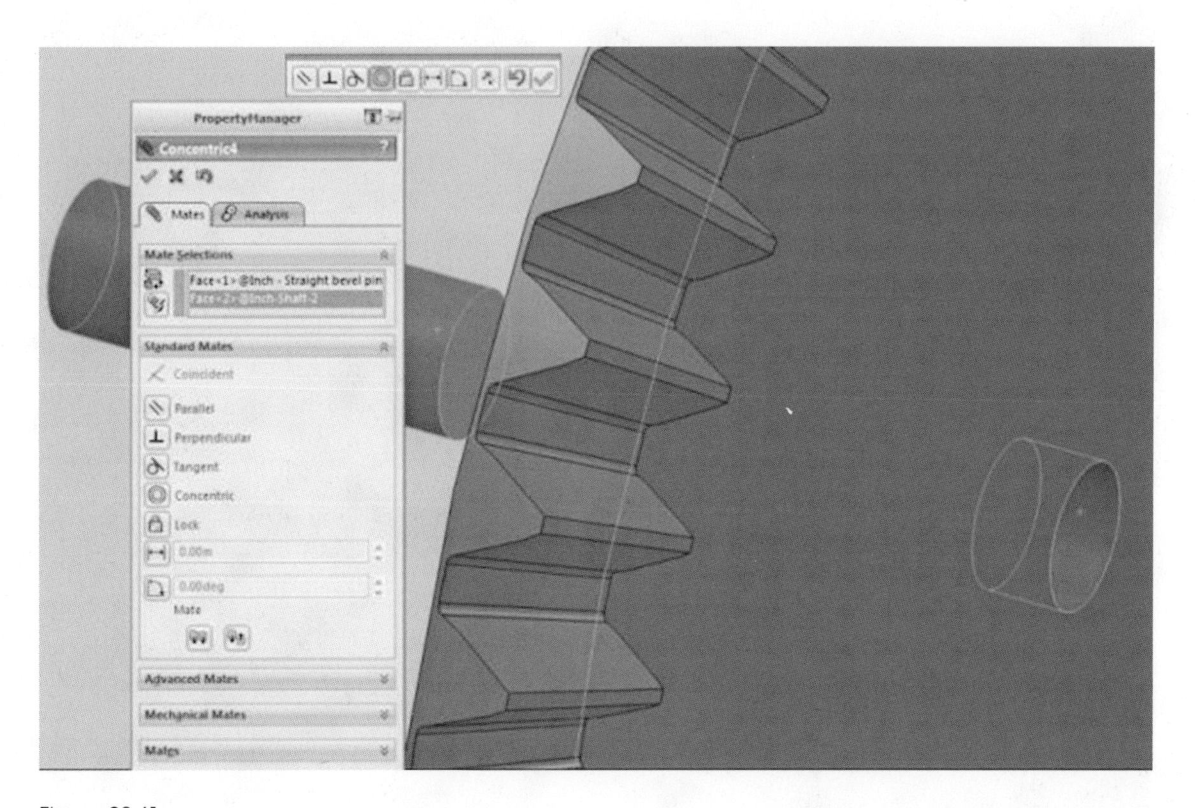

Figure 22.61

Concentric mate for shaft and output bevel gear.

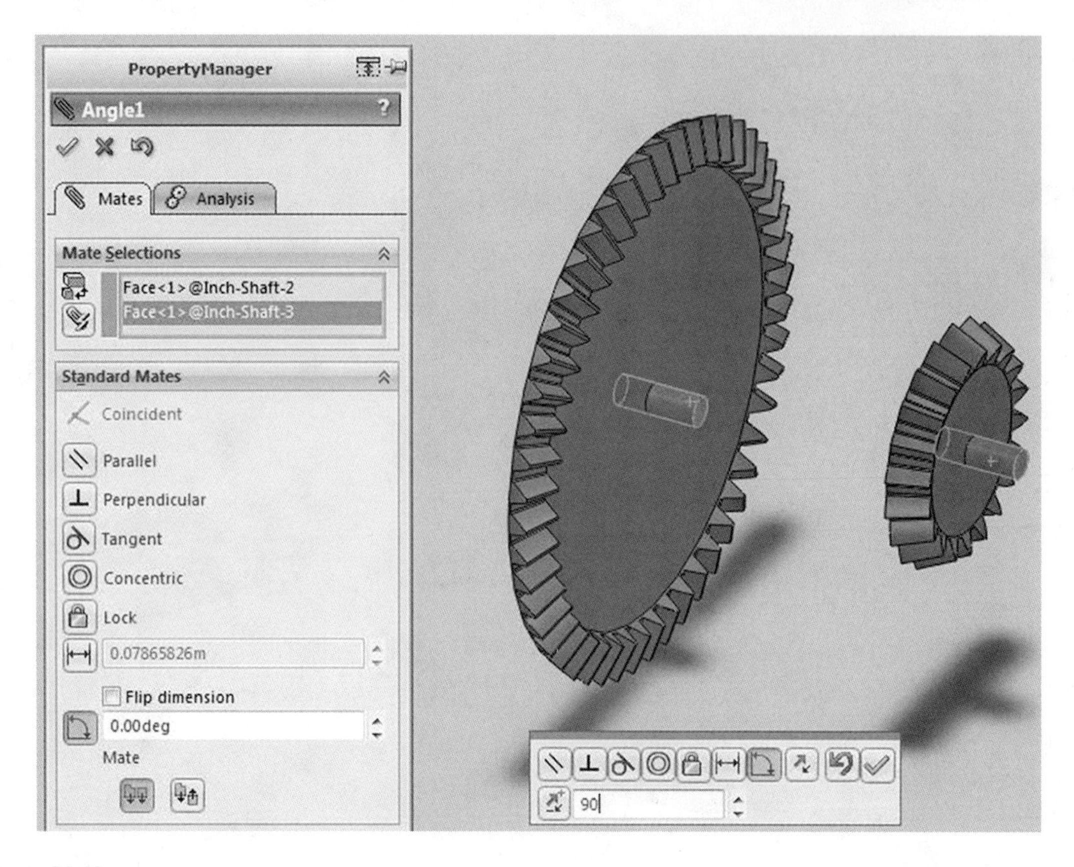

Figure 22.62

Bevel gear shaft for input and output pairs.

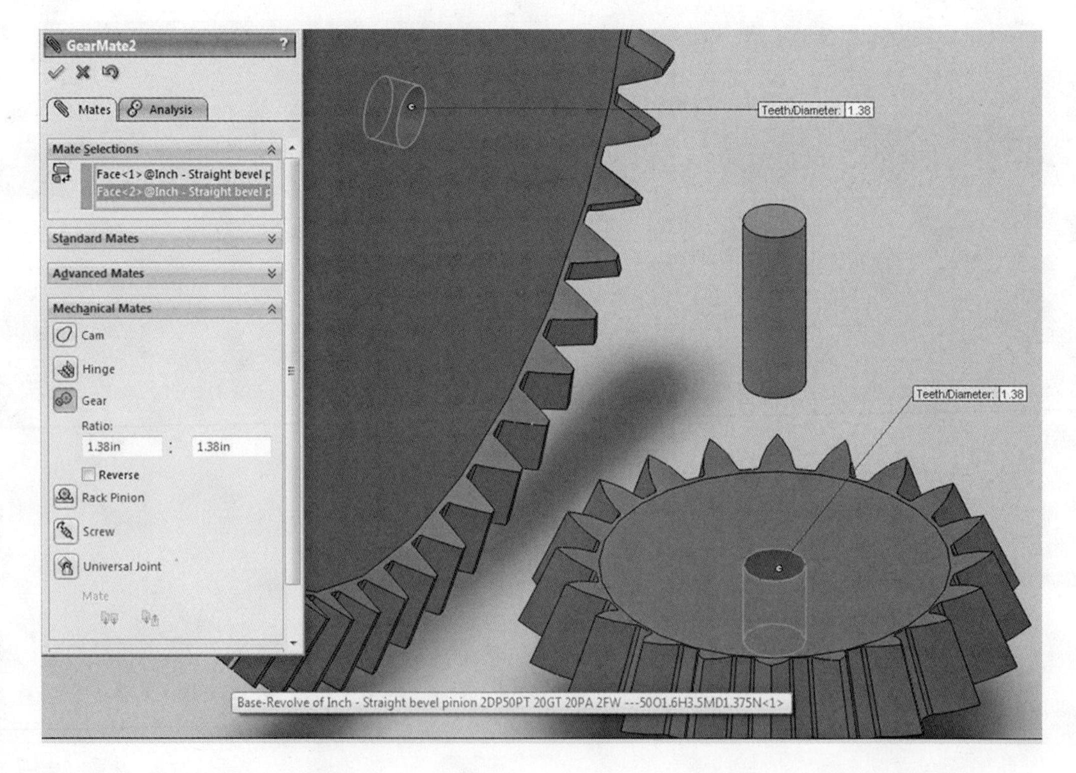

Figure 22.63

GearMate PropertyManager.

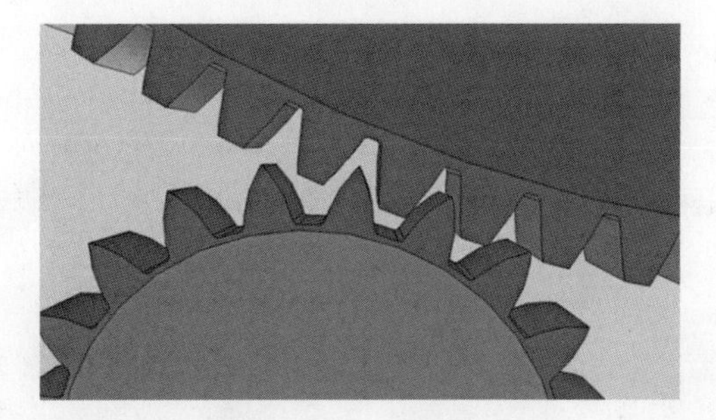

Figure 22.64

Manual alignment of gear teeth.

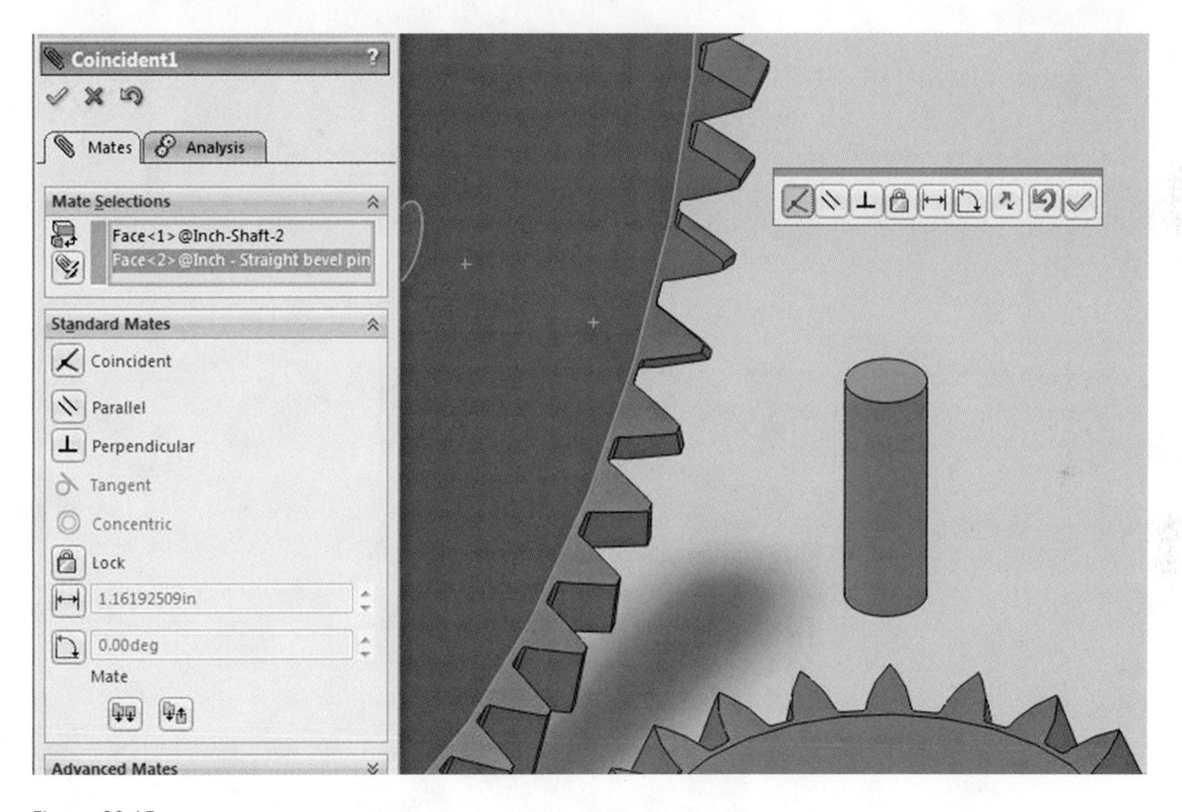

Figure 22.65

End face of shaft is coincident with face of bevel gear.

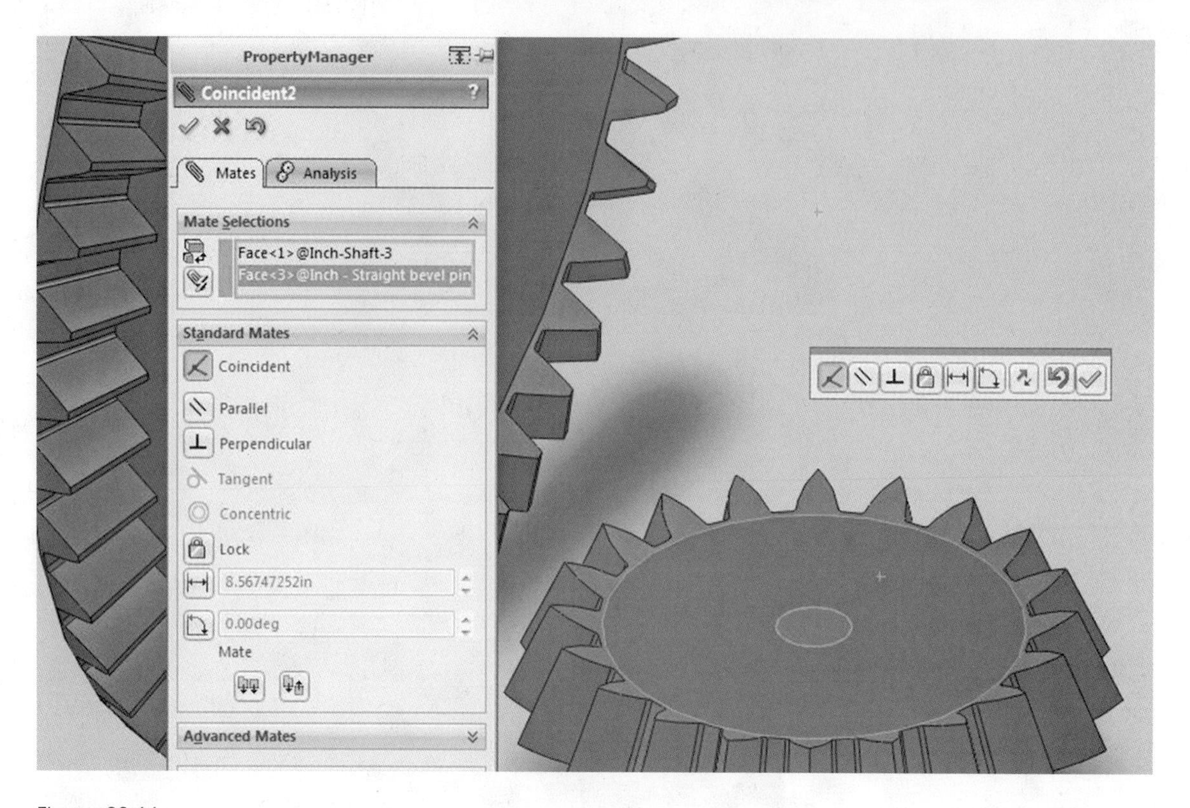

Figure 22.66

End face of shaft is coincident with face of bevel gear.

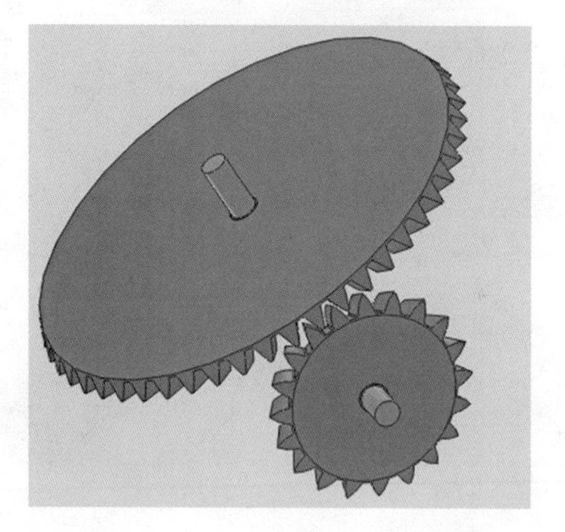

Figure 22.67

Gear box complete (without housing).

Summary

This chapter discusses in detail the design procedures for pinion and gear (spur-type), rack and pinion, belt and pulley, chain and sprocket, and bevel gear transmission systems. Several examples are given to make the procedures clear. The power transmission systems are common in mechanical drive applications. It is important that students attempt to answer the questions that are given at the end of this chapter in order to consolidate their understanding of the materials that are presented.

Exercises

1. A gear system consists of the pinion (Gear 1) and gear (Gear 2), and the specifications relating to their geometry are given as follows:

Gear 1: Pinion	Gear 2: Gear
Diametral pitch = 24	Diametral pitch = 24
Number of teeth = 24	Number of teeth = 48
Face thickness = 0.5	Face thickness = 0.5
Bore = 0.5 diameter	Bore = 0.5 diameter
Hub = 1.0 diameter	Hub = 1.0 diameter
Pressure angle = 20 deg.	Pressure angle = 20 deg.

Using SolidWorks, create an assembly model consisting of the two spur gears, two pins, and a support plate, and animate the movement of the gears.

2. A gear system consists of the pinion (Gear 1) and gear (Gear 2), and the specifications relating to their geometry are given as follows:

Gear 1: Pinion	Gear 2: Gear
Diametral pitch = 30	Diametral pitch = 30
Number of teeth = 36	Number of teeth = 72
Face thickness = 0.5	Face thickness = 0.5
Bore = 0.5 diameter	Bore = 0.5 diameter
Hub = 1.0 diameter	Hub = 1.0 diameter
Pressure angle = 20 deg.	Pressure angle = 20 deg.

Using SolidWorks, create an assembly model consisting of the two spur gears, two pins, and a support plate, and animate the movement of the gears.

3. Design a gear system consisting of an input straight bevel gear having 25 teeth and an input straight bevel gear having 63 teeth. The module is 10 mm, a 20-degree pressure angle, and a face width of 51 mm. The nominal shaft diameter is 30 mm (see Figures 22.53 through 22.56).

23

Cam Design

Objectives:

When you complete this chapter, you will have

- Learned how to design cams
- Understood the relationship between cams and followers

Introduction

A *cam* is a mechanical device having a profile or groove machined on it, which gives an irregular or special motion to a *follower*. The type of follower and its motion depend on the shape of the profile or groove.

Types of Cams

Cams fall into two main classes: (1) *radial (edge* or *plate) cams* and (2) *cylindrical cams*. The follower of a radial cam reciprocates or oscillates in a plane perpendicular to the cam axis, whereas with a cylindrical cam, the follower moves parallel to the cam axis.

Types of Followers

The *knife edge* or *point follower* is the simplest type of follower. It is not often used as it wears rapidly, but it has the advantage that the cam profile can have any shape.

With the *roller follower*, the rate of wear is reduced, but the profile of the cam must not have any concave portions with a radius that is smaller than the roller radius.

The *flat follower* is sometimes used, but the cam profile must have no concave portions.

Creating Cams Using the Traditional Method

The traditional method of creating cam profiles is to define a displacement diagram and then transfer the displacement diagram to a *base circle*.

What is a base circle? A base circle is a circle passing through the nearest approach of the follower to the cam center.

Creating Cams in SolidWorks

SolidWorks creates cams by utilizing existing cam templates. There are templates for *circular* and *linear* cams and *internal* and *external* cams. The templates allow the designer to work directly on the cam profile and eliminate the need for a displacement diagram. The SolidWorks Toolbox must be available before we can access the Cams Tool. Therefore, the first step is to Add-In the Toolbox.

Problem Definition

Let us illustrate how to create a cam in SolidWorks with a circular cam having a 4.00-in. base circle and a profile that rises 0.5 in. over 90° using harmonic motion, dwells for 180°, falls 0.50 in. over 45° using harmonic motions, and dwells for 45°.

SolidWorks Toolbox Add-Ins

1. Open SolidWorks.
2. Open the model file.
3. Click the Add-Ins tool (see Figure 23.1).
4. Check SolidWorks Toolbox and SolidWorks Toolbox Browser (see Figure 23.2).
5. Click OK. (The Toolbox tool is added.)

To Access the SolidWorks Cams Tool

To access the Cams tool,

1. Create a new part document.
2. Select the front plane.
3. Click the toolbox icon from the top of the CommandManager.
4. Click the Cams tool from the menu (see Figure 23.3). The Cam-Circular toolbox will be automatically displayed (see Figure 23.4).

The Cam-Circular toolbox has three tabs: (1) Setup, (2) Motion, and (3) Creation.

Cam-Circular Setup

1. Click the List button on the Setup tab of the Cam-Circular toolbox (see the left side of Figure 23.4). The Favorites dialog will be displayed (see the right side of Figure 23.4). This dialog lists the cam templates that can be used to create different cams.
2. Select Sample 2—Inch Circular.
3. Click Load.

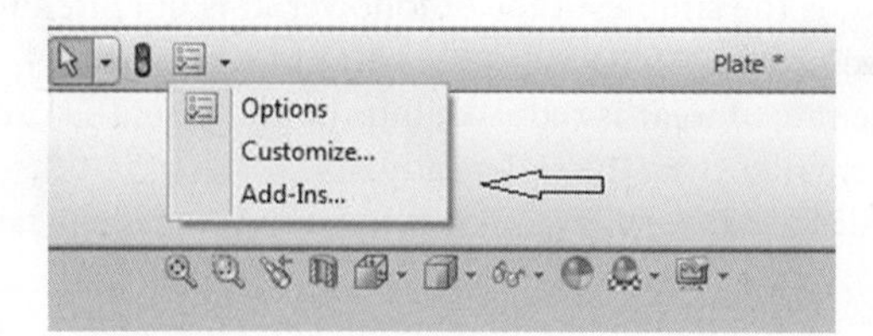

Figure 23.1

Add-Ins tool.

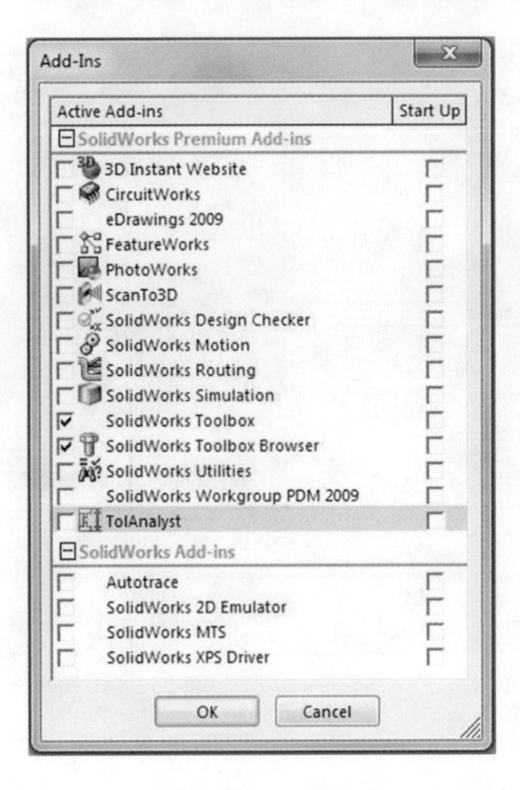

Figure 23.2

SolidWorks Add-Ins PropertyManager.

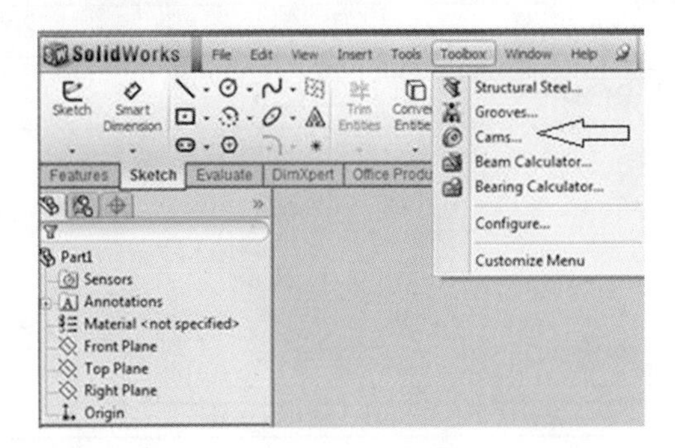

Figure 23.3

Cams tool.

4. Expand the Follower Type from the pull-down menu and set the type as Translating. The PropertyManager for the Circular-Cam, Translating-Follower, appears (see Figure 23.5).
5. Define the Properties that are required for the Setup of the cam.

The base circle of the cam has a diameter of 4.00 in. (radius of 2.00 in.). and the follower has a diameter of 0.50 in. (radius of 0.25 in.). This means that the Starting Radius is 2.25 in. (2.00 + 0.25). Therefore, in the Property rollout on the Setup tab. enter the following:

Units: Inch
Cam Type: Circular
Follower Type: Translating
Follower Diameter: 0.50
Starting Radius: 2.25
Starting Angle: 0
Rotation Direction: Clockwise

Figure 23.4

Cam-Circular Setup.

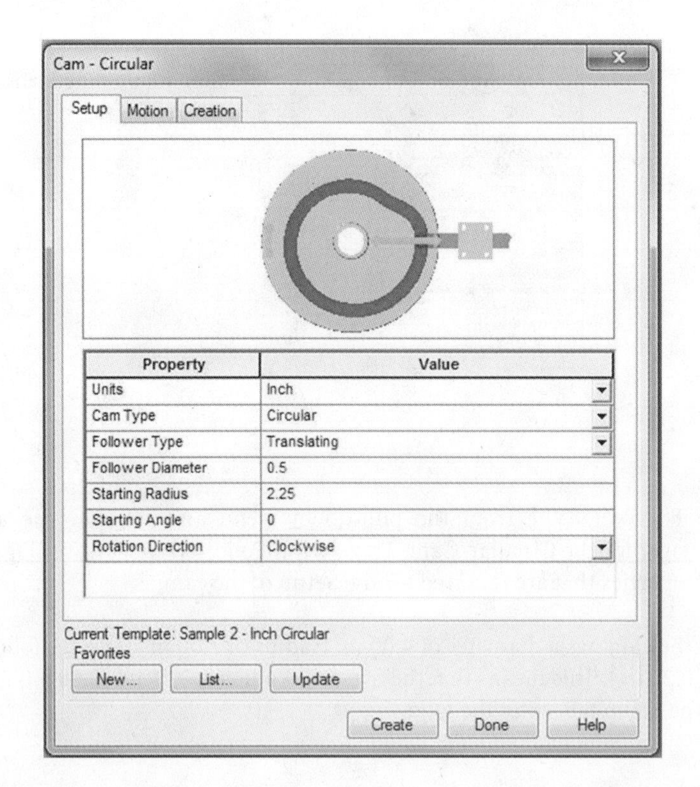

Figure 23.5

Setup for Cam-Circular.

Cam-Circular Motion

From the problem definition, the motion of the circular cam has four sectors. It rises 0.5 in. over 90° using harmonic motion, dwells for 180°, falls 0.50 in. over 45° using harmonic motions, and dwells for 45°. This information is utilized to define the motion.

6. Select the Motion tab in the Cam-Circular toolbox.
7. Click Add to display the Motion Creation Details dialog (see Figure 23.6).
8. In the Motion Creation Details dialog, enter the first sector by setting the Motion Type to Harmonic, the Ending Radius to 2.75 in. (2.25 + 0.5), and Degrees Motion to 90°.
9. Click OK.
10. Click Add.
11. For the second sector, set the Motion Type to Dwell and the Degrees Motion to 180°. (See the upper dialog in Figure 23.7.)
12. Click OK.
13. Click Add.
14. For the third sector, set the Motion Type to Harmonic, the Ending Radius to 2.25 in., and the Degrees Motion to 45°. (See the middle dialog in Figure 23.7.)
15. Click OK.
16. Click Add.
17. For the final sector, set the Motion Type to Dwell and the Degrees Motion to 45°. (See the lower dialog in Figure 23.7.)
18. Click OK.

Figure 23.8 shows the motion for the different sectors of the cam.

Cam-Circular Creation

19. Select the Creation tab.
20. Modify the default settings (see the upper dialog in Figure 23.9) as follows. (See the lower dialog in Figure 23.9.) Set the Blank Outside Diameter to 6.00 in., Thickness to 0.5 in., and Thru Hole Diameter to 1.5 in.
21. Set Track Type & Depth to Thru. (The Blind condition requires a value.)

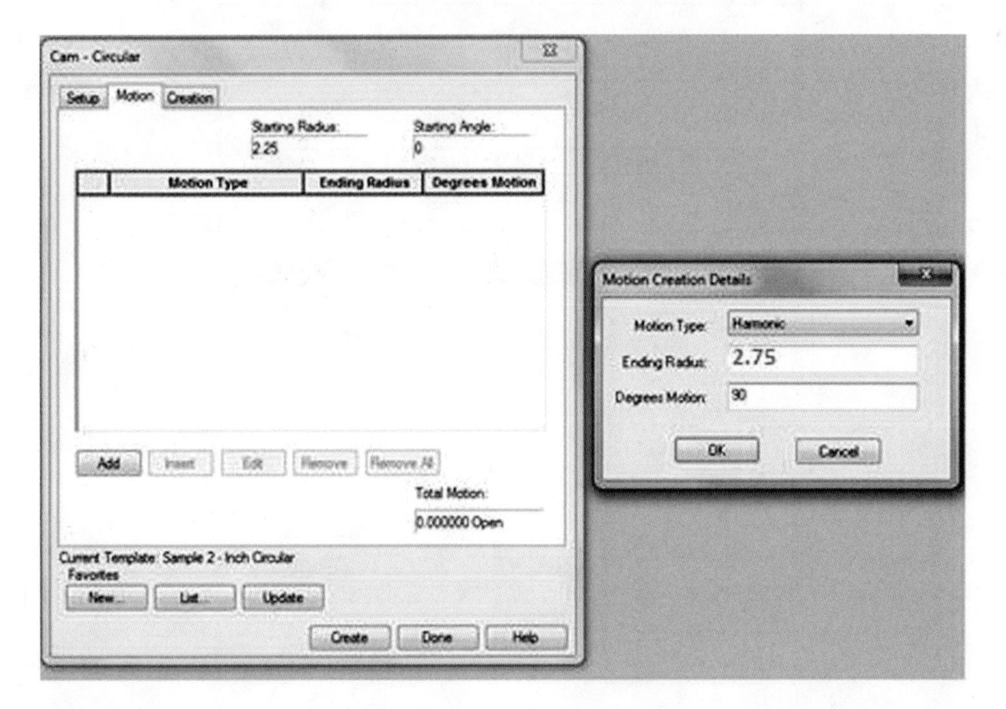

Figure 23.6

Motion Creation Details dialog.

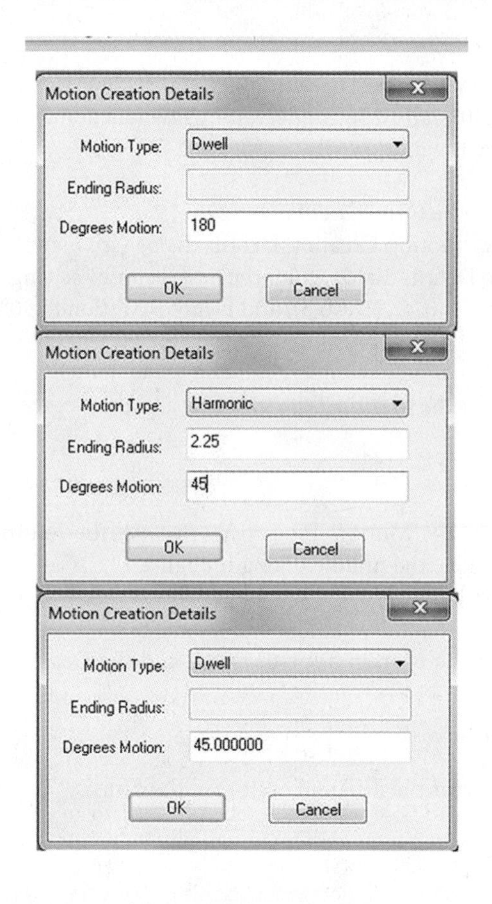

Figure 23.7

Motion Creation Details.

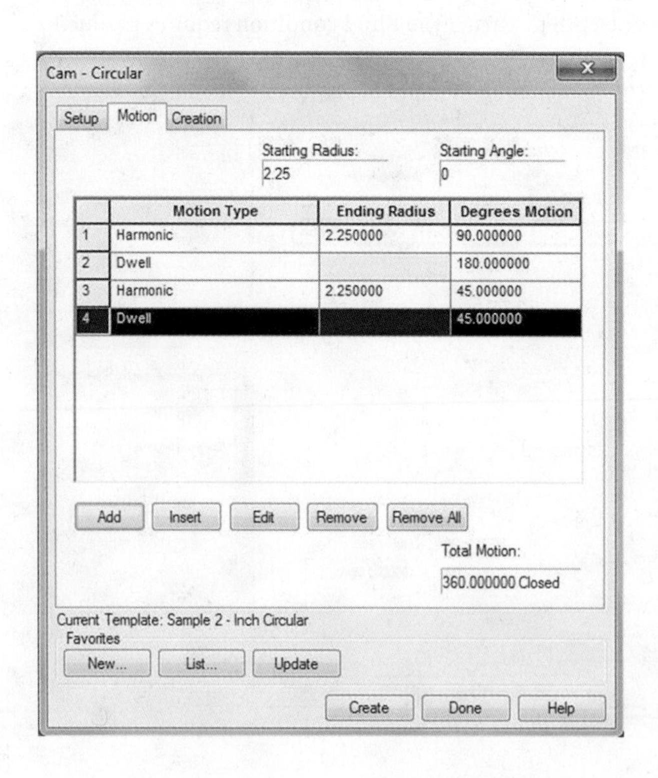

Figure 23.8

Motion summary.

Figure 23.9

Creation tab for Cam-Circular.

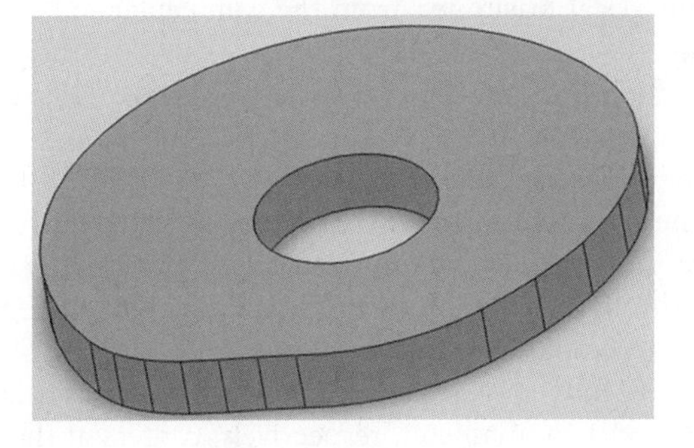

Figure 23.10

Final cam model.

22. Set Track Surfaces to Inner.
23. Click Create.
24. Click Done.

The final cam model is displayed in Figure 23.10.

Cam Model Modification

The cam model in Figure 23.10 fails to mate with the cam follower because of the discontinuities of the face. This is what SolidWorks generates. Therefore, some modifications have been made to make the assembly of the parts feasible. The steps given here are required to be followed by users to obtained feasible cam profiles other than the solutions that SolidWorks provide:

1. Design a cam model based on given specifications.
2. Use the Convert Entities tool, while in *Sketch mode* on one of the faces of the cam, to extract the edges.
3. Use Spline tool to create a spline through the edges converted into entities; add short intervals where the shape of the spline deviates from the cam profile. You may explore other approaches.

Figure 23.11

Modified cam profile.

4. Add Construction line(s) <at least one> from the cam center to one point on the profile. (This is necessary for copying.)
5. Click Spline>Construction-line and *Copy* <alternatively, use CTLR>c>.
6. Open a New Part document.
7. Select Top view and be in *Sketch mode.*
8. Paste the copied Spline and centerline from the previous document.
9. Move both the Spline and *Construction line* until origin matches the end of the construction line.
10. Extrude cam profile through 0.5-in. (or as originally specified). See Figure 23.11 for the modified cam profile; it is smooth compared to that of Figure 23.10.
11. Sketch a 0.5-in. circle and Extrude-Cut to form hole in cam (or as originally specified).
12. Add two Circles, 0.5- and 1-in. diameter, respectively, centers at the origin, and extrude through 0.5-in. to create the hub (or as originally specified).

Creating a Hub

There are two ways to create a hub: (1) manually and (2) by using the Cam-Circular dialog.

In the manual method, the cam model is first obtained, as shown in Figure 23.10. Then, using the Sketch tool, a circle concentric with the circle that defines the shaft is sketched, and extruded, as shown in Figure 23.12. This is not an efficient method compared to the second method, using the Cam-Circular dialog.

Creating a Hub Using the Cam-Circular Dialog

This is easily achieved by defining the following fields in Figure 23.13:

Near Hub Diameter = 1.5 in.
Near Hub Length = 0.75 in.

The hub is created as shown in Figure 23.14.

Creating a Hole for a Key Using the Hole Wizard

The Hole Wizard can be used to create a hole for a key (see Figure 23.14).

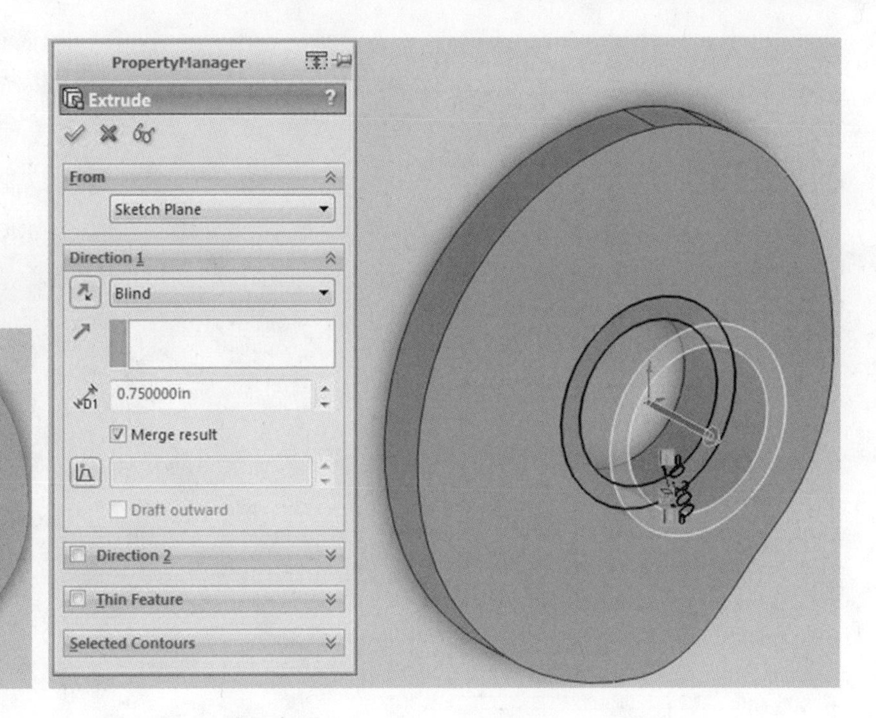

Figure 23.12

Manual method of creating a hub.

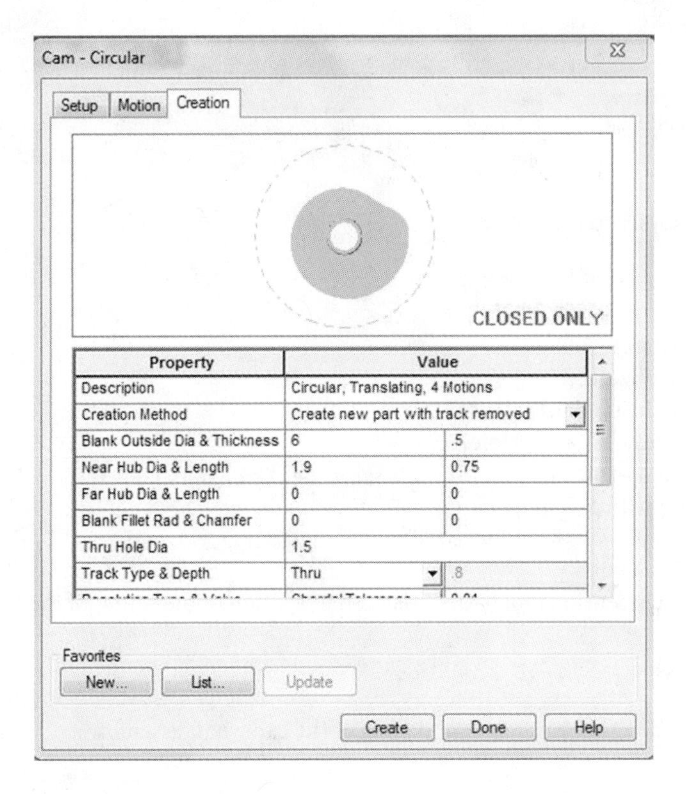

Figure 23.13

Hub definition in Creation tab for Cam-Circular.

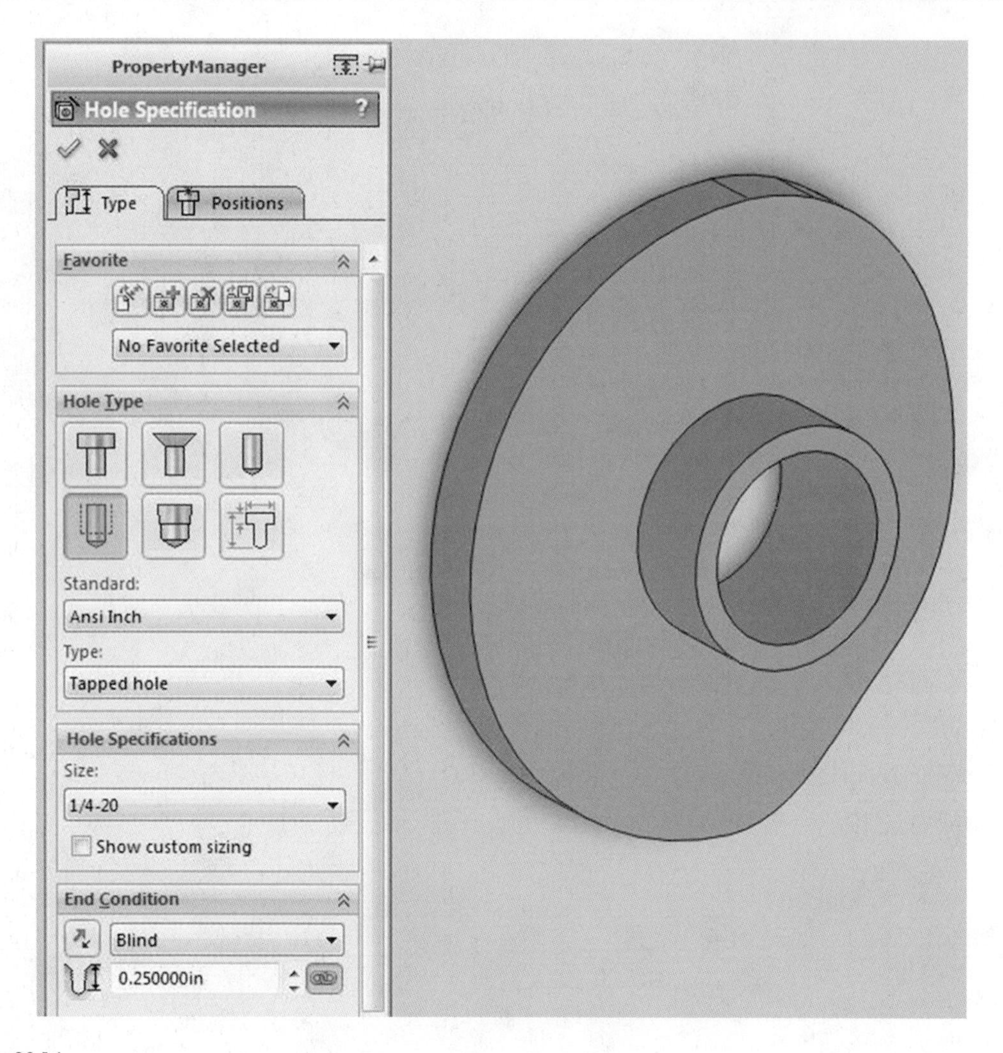

Figure 23.14

Creating a hole for a key using the Hole Wizard.

Type

1. Click the Type tab.
2. Click the Hole Wizard.
3. Select ANSI Inch Standard.
4. For Type, select the Tapered Hole.
5. For Size, select 1/4-20.

Position

6. Click the Position tab.
7. Use the Smart Dimension tool to dimension the position of the center of the hole (see Figure 23.15).

Cam Shaft Assembly

The following sections design the other components of the cam shaft assembly.

Spring

The "Spring" section in Chapter 6 gives the steps that are involved in spring design.

The helix path diameter is 0.5 in., and the diameter of the circular profile is 0.125 in. The other specifications for the helix are shown in Figure 23.16.

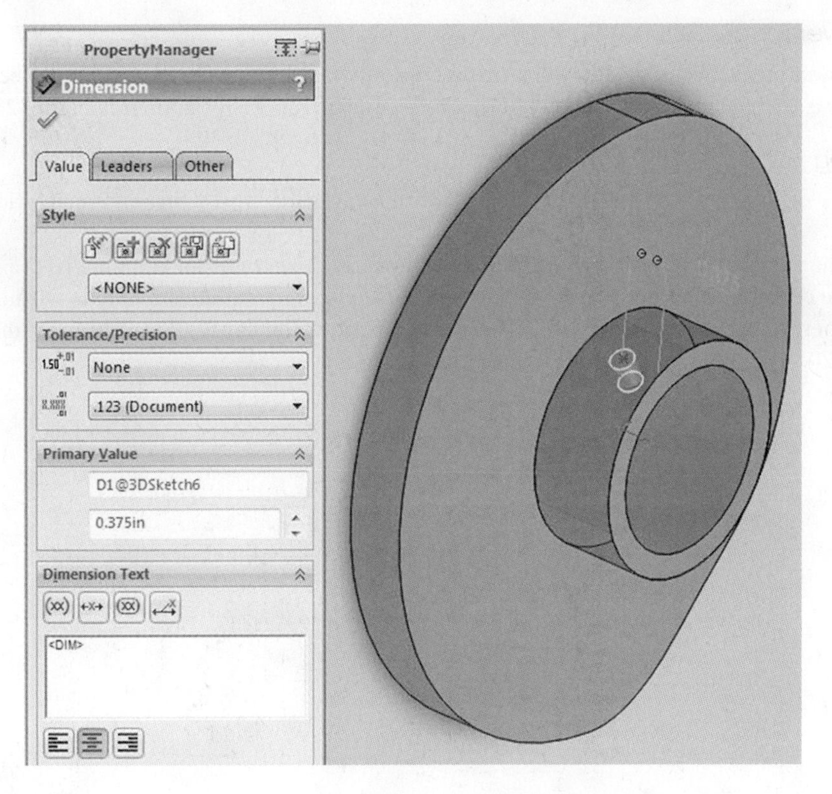

Figure 23.15

Hole location defined using the Smart Dimension tool.

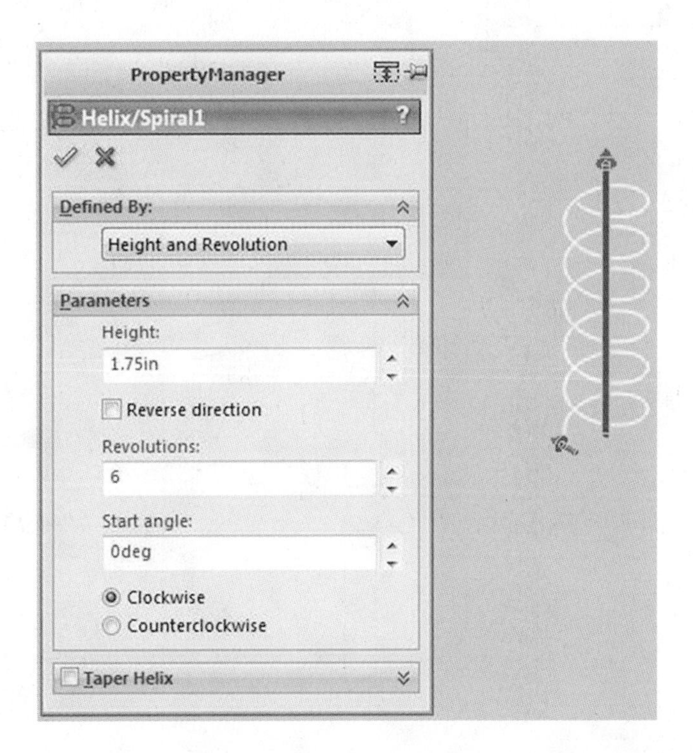

Figure 23.16

Specifications for the helix.

Cam Follower Bracket

The cam follower bracket is modeled by creating a U-profile 1.00 in. × 0.875 in. × 1.00 in. (see Figure 23.17). The thickness is 0.13 in., and it is extruded in the Mid Plane direction by 0.5 in. (see Figure 23.18).

Face Filleting

The two arms of the U, Face<1> and Face<2>, are filleting using the Tangent Propagation option (see Figure 23.19).

Holes of diameter 0.25 in. are cut into the top and sides. The center of the hole in the top face coincides with the center of that face, while the centers of the holes in the sides coincide with the center that is used to generate the fillets, that is, 0.25 in. from each side. The cam follower bracket model is shown in Figure 23.20.

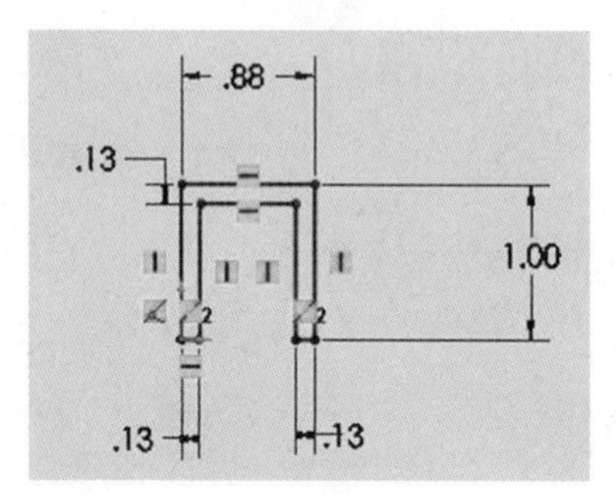

Figure 23.17

Profile for cam follower bracket.

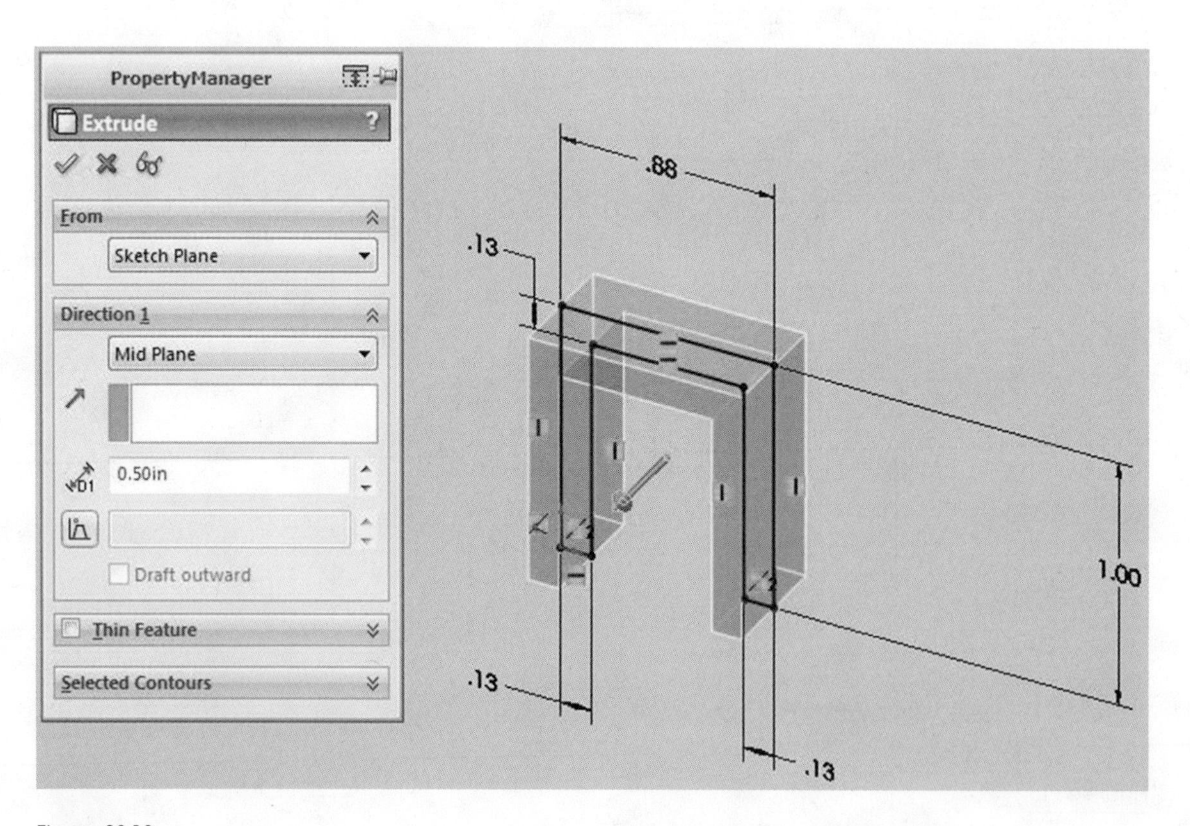

Figure 23.18

Extruding the profile.

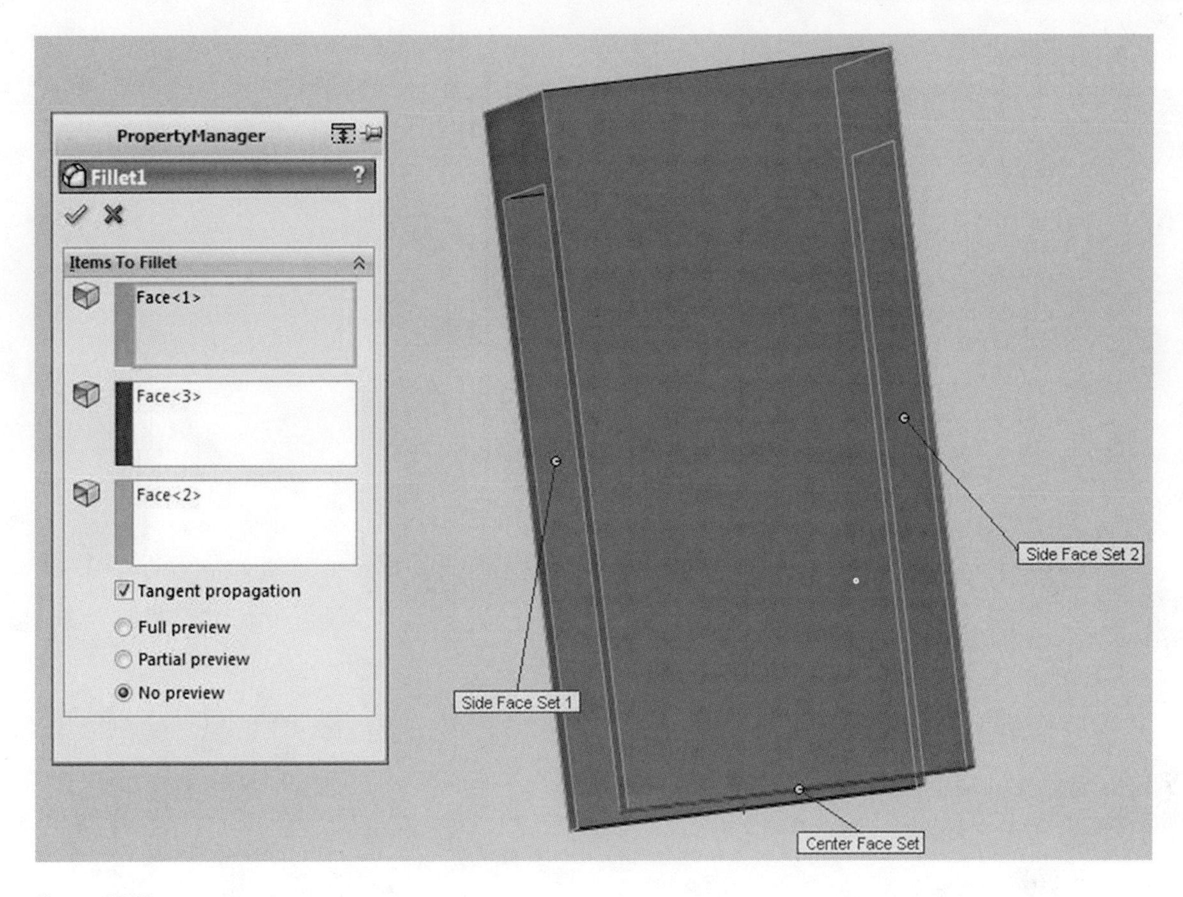

Figure 23.19

Face filleting the model.

Figure 23.20

Cam follower bracket model.

Cam Bracket

The cam bracket is modeled by creating a U-profile 3.00 in. × 8.5 in. × 4.00 in. (see Figure 23.21). The thickness is 0.25 in., and it is extruded in the Mid Plane direction by 6.00 in. (see Figure 23.22).

Roller

The roller is modeled by creating a circle of diameter 0.50 in., and the hole is a concentric circle of diameter 0.25 in. Both circles are extruded through 0.5 in. (see Figure 23.23).

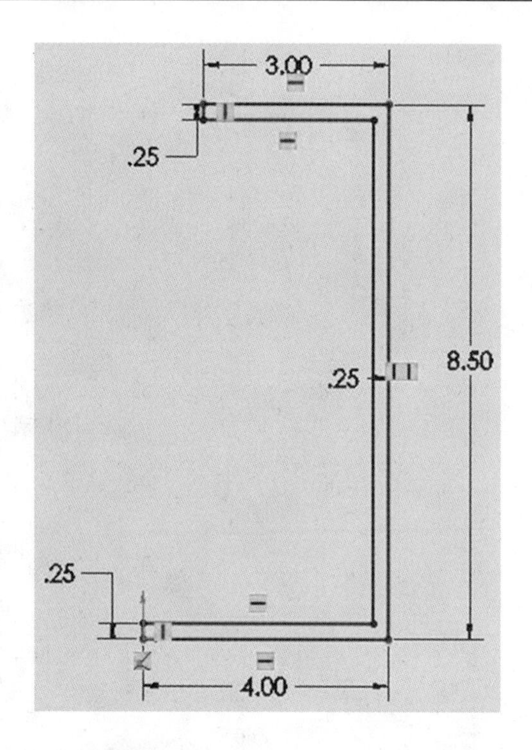

Figure 23.21

Profile for cam bracket.

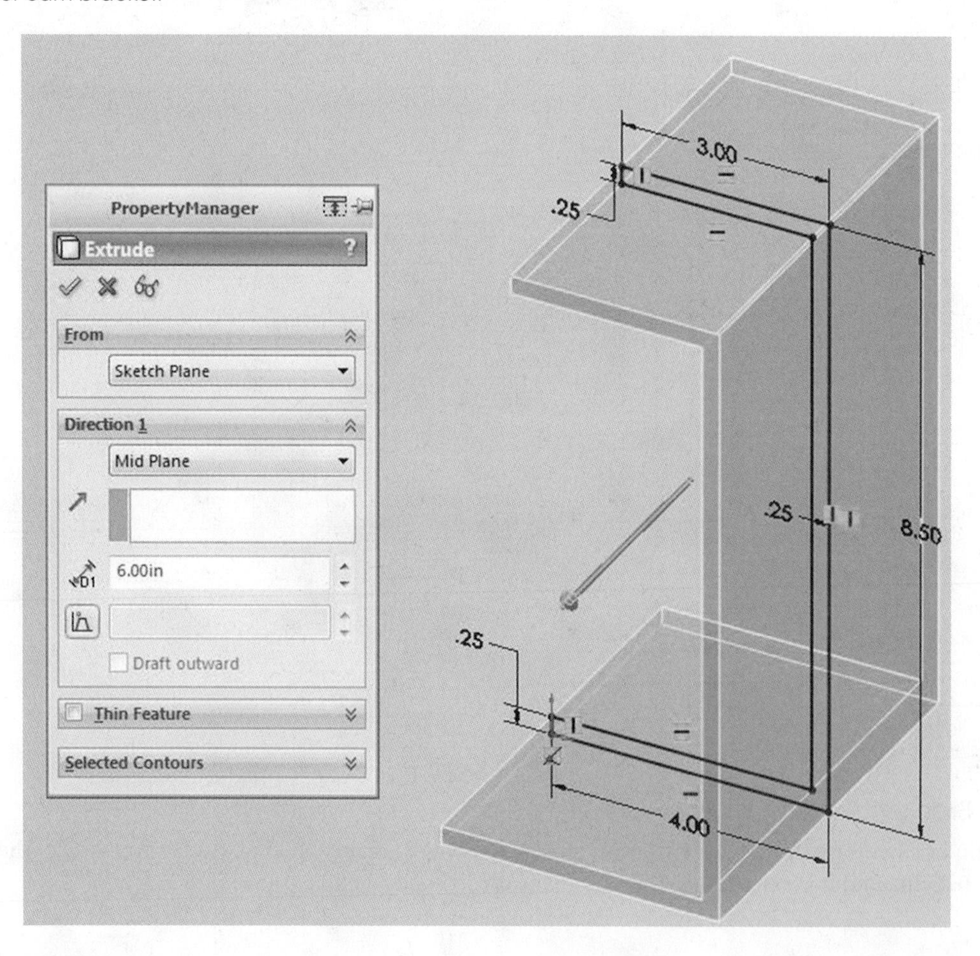

Figure 23.22

Extrusion of profile for cam bracket.

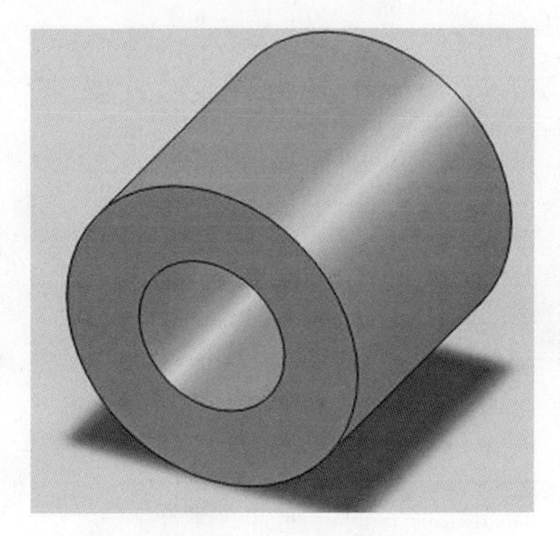

Figure 23.23

Roller model.

Cam Shaft

The cam shaft is modeled by creating a circle of diameter 1.5 in. and extruded in the Mid Plane direction by 2.5 in. (see Figure 23.24).

Handle

The handle is modeled by creating a circle of diameter 0.25 in. and extruded in the Mid Plane direction by 2.75 in. (see Figure 23.25).

Pin

The pin is modeled by creating a circle of diameter 0.25 in. and extruded in the Mid Plane direction by 0.875 in. (see Figure 23.26).

Assembly of Cam Shaft Components

The assembly of all the parts designed in this chapter is shown in Figure 23.27. Note that the cam follower in Figure 23.27 will not mate correctly due to the file of the cam surface (patches) but the modified cam follower in Figure 23.27 will mate correctly due to its smooth profile. This is the author's contribution to SolidWorks cam modeling. Try it out and you will have the experience for yourself.

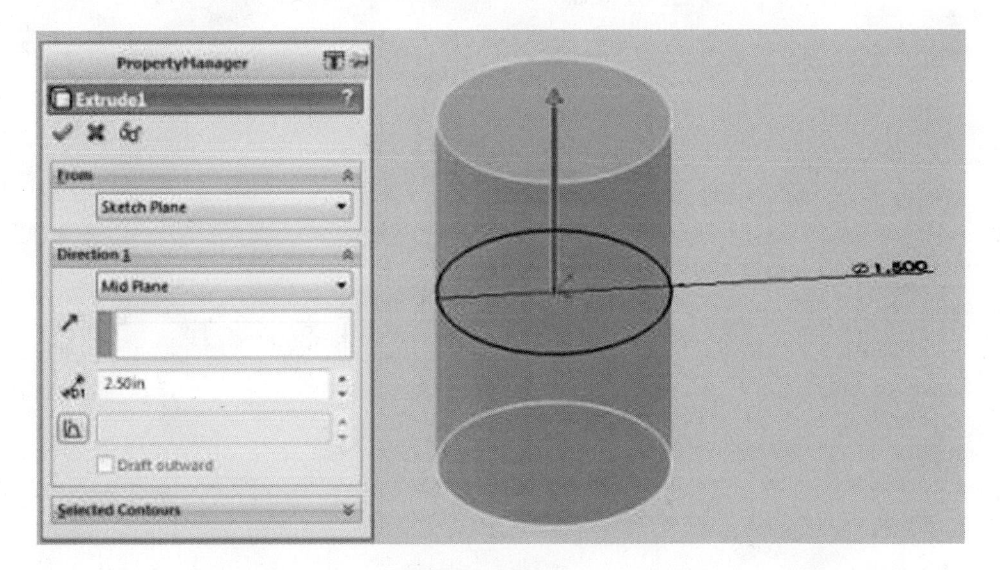

Figure 23.24

Cam shaft model.

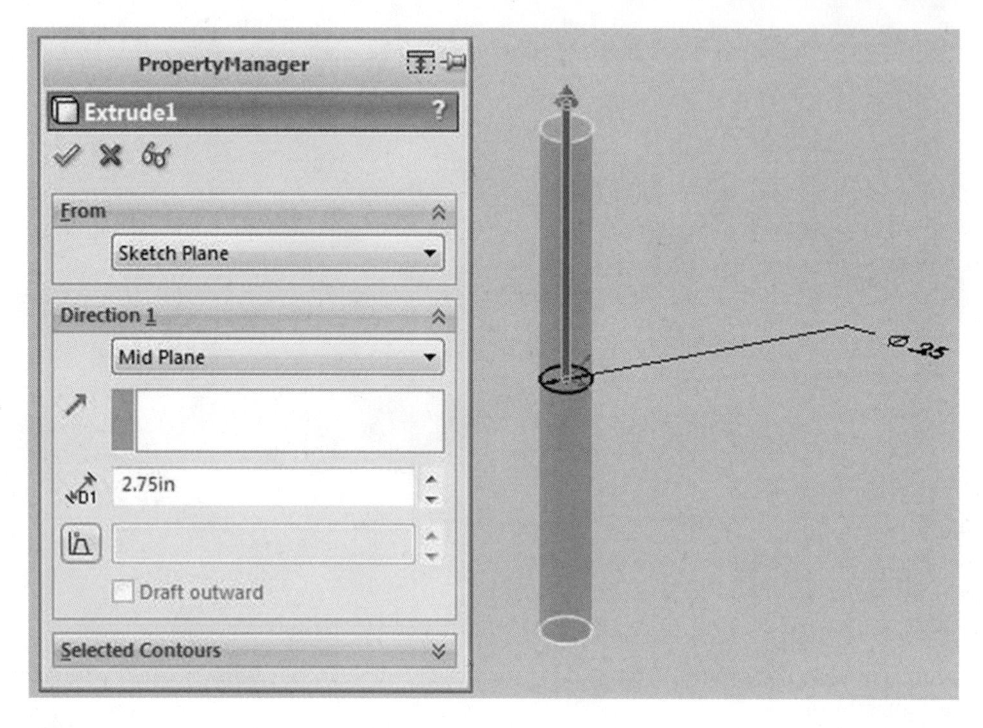

Figure 23.25

Handle model.

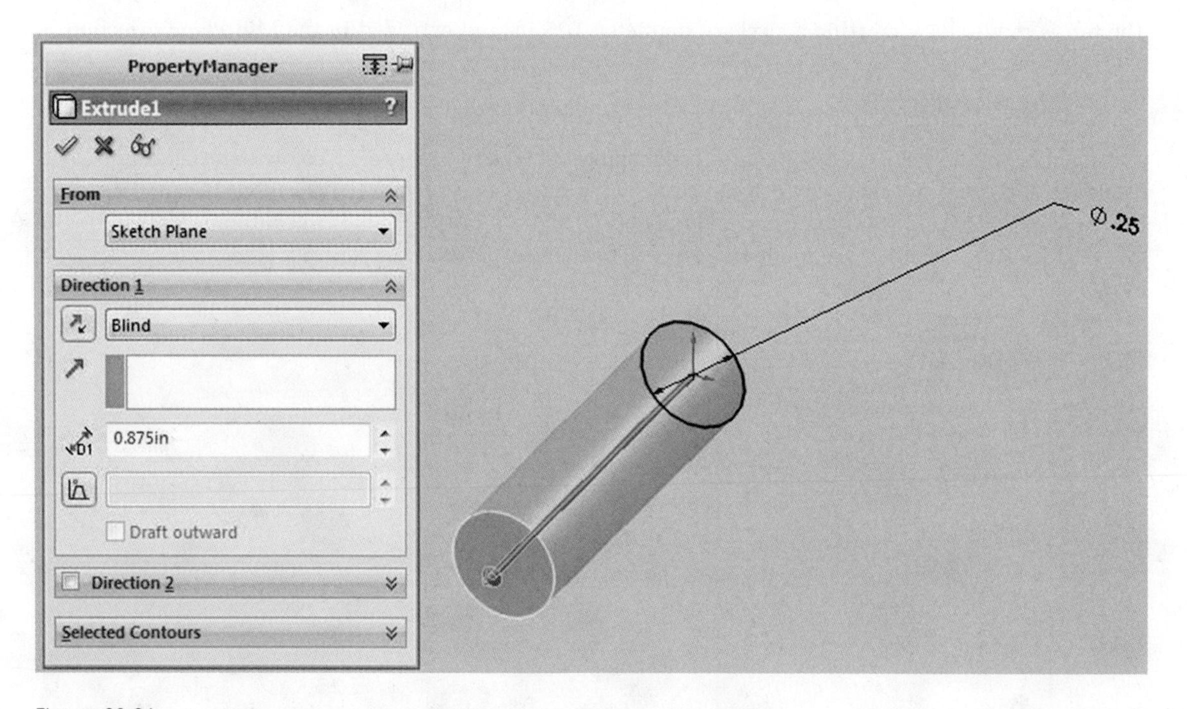

Figure 23.26

Pin model.

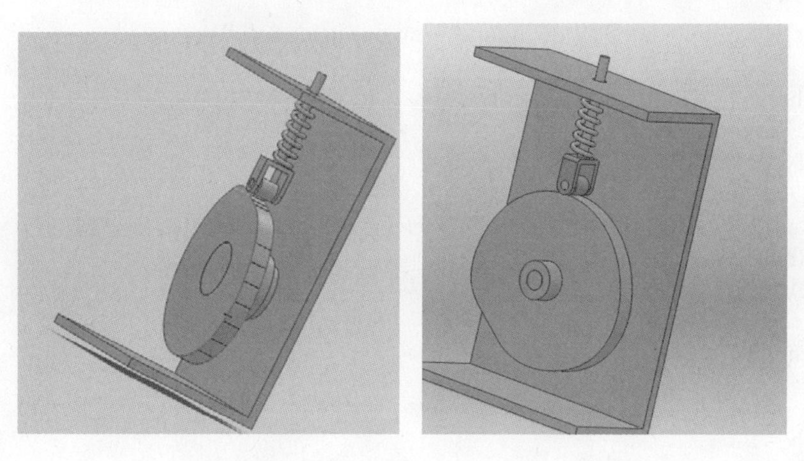

Figure 23.27

Assembly of cam and follower.

Exercises

1. Assemble the follower bracket, roller, handle, and pin. Name this the cam subassembly.
2. Assemble the cam subassembly, cam, cam shaft, and spring. Name this the cam assembly.

    All files for this assembly are available in the textbook's resources database.

Project 1: Model the cam with the following specifications using SolidWorks:

Units: Inches
Cam Type: Circular
Follower Type: Translating
Follower Diameter = 0.5
Starting Radius = 2.00 [= (Base Circle Diameter + Follower Diameter)/2]
Starting Angle = 0°
Rotation Direction = Clockwise

Cam Motion:
Dwell 45°
Rise 0.375, Modified Trapezoidal Motion, 90°
Dwell 90°
Fall 0.375, Modified Trapezoidal Motion, 90°
Dwell 45°

Others:
Blank Outside Dia. = 5.00
Thickness = 0.50
The hub has an outer diameter of 1.00 in. and extends 0.50 in. from the surface of the cam.
The cam bore is 0.50-diameter

Create a keyway in both the cam and the cam shaft that will accept a 0.375 × 0.375 × 0.500-in square key. Is this keyway sizing feasible?

Make any modifications/adjustments to the cam-bracket and the length of the spring in order to accommodate the cam.

You are supplied parts and subassembly to complete the cam design.

Assemble the cam that you have modeled together with the subassembly and other parts given to you to complete a cam system.

Project 2: Model the cam with the following specifications using SolidWorks:

Units: Inches
Cam Type: Circular
Follower Type: Translating
Follower Diameter = 0.375
Starting Radius = 1.4375 [= (Base Circle Diameter + Follower Diameter)/2]
Starting Angle = 0°
Rotation Direction = Clockwise

Cam Motion:
Dwell 45°
Rise 0.375, Harmonic Motion, 135°
Dwell 90°
Fall 0.375, Harmonic Motion, 90°

Others:
Blank Outside Dia. = 4.00
Thickness = 0.375
The hub has an outer diameter of 1.00 in., and extends 0.50 in. from the surface of the cam.
The cam bore is 0.50-diameter

Create a keyway in both the cam and the cam shaft that will accept a 0.375 × 0.375 × 0.500-in. square key. Is this keyway sizing feasible?

Make any modifications/adjustments to the cam-bracket and the length of the spring in order to accommodate the cam.

You are supplied parts and subassembly to complete the cam design.

Assemble the cam that you have modeled together with the subassembly and other parts given to you to complete a cam system.

24

Mechanism Design Using Blocks

Objectives:

When you complete this chapter, you will have

- Created sketches of a mechanism
- Saved the sketches created as block files
- Inserted a block into the layout environment
- Applied relations to the blocks
- Converted blocks into parts

Introduction

A block is a set of entities that are grouped together as a single entity. Blocks are used to create complex mechanisms as sketches and to check their functionality before being developed into complex three-dimensional (3D) models.

Blocks Toolbar

The Blocks toolbar, shown in Figure 24.1, is used to control the sketched entities of the blocks.

Problem Description

The different views of the parts of a reciprocating mechanism with required dimensions are shown in Figures 24.2 through 24.4. We will create the sketches and save them as SolidWorks blocks. We will convert the blocks to parts and assemble the reciprocating mechanism.

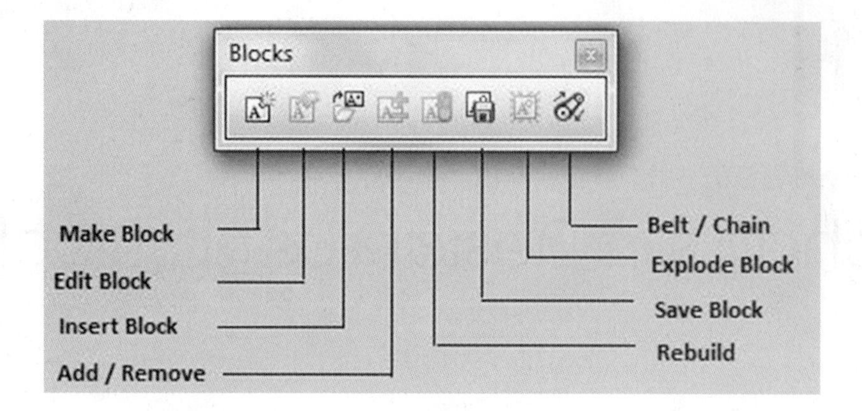

Figure 24.1

The Blocks toolbar.

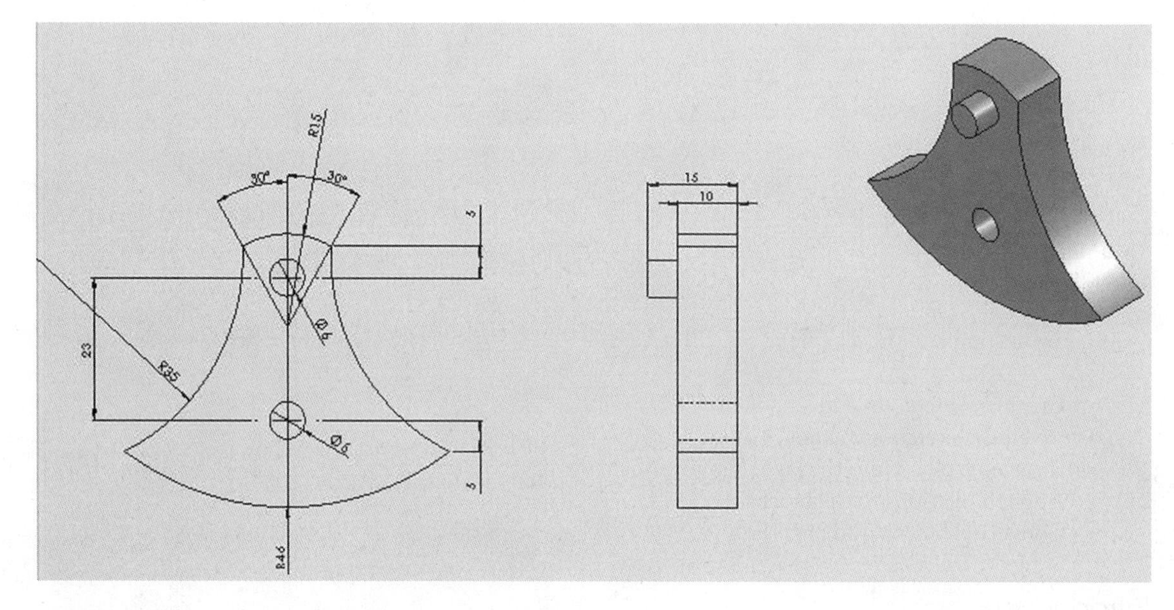

Figure 24.2

Crank front view, right view, and part.

Creating Sketches of a Mechanism

Crank

1. Start a new SolidWorks part document.
2. Select the front plane.
3. Sketch the crank profile that is shown in Figure 24.5.

Piston Rod

4. Start a new SolidWorks part document.
5. Select the front plane.
6. Sketch the piston rod profile that is shown in Figure 24.6.

Piston Tank

7. Start a new SolidWorks part document.
8. Select the front plane.
9. Sketch the piston tank profile that is shown in Figure 24.7.

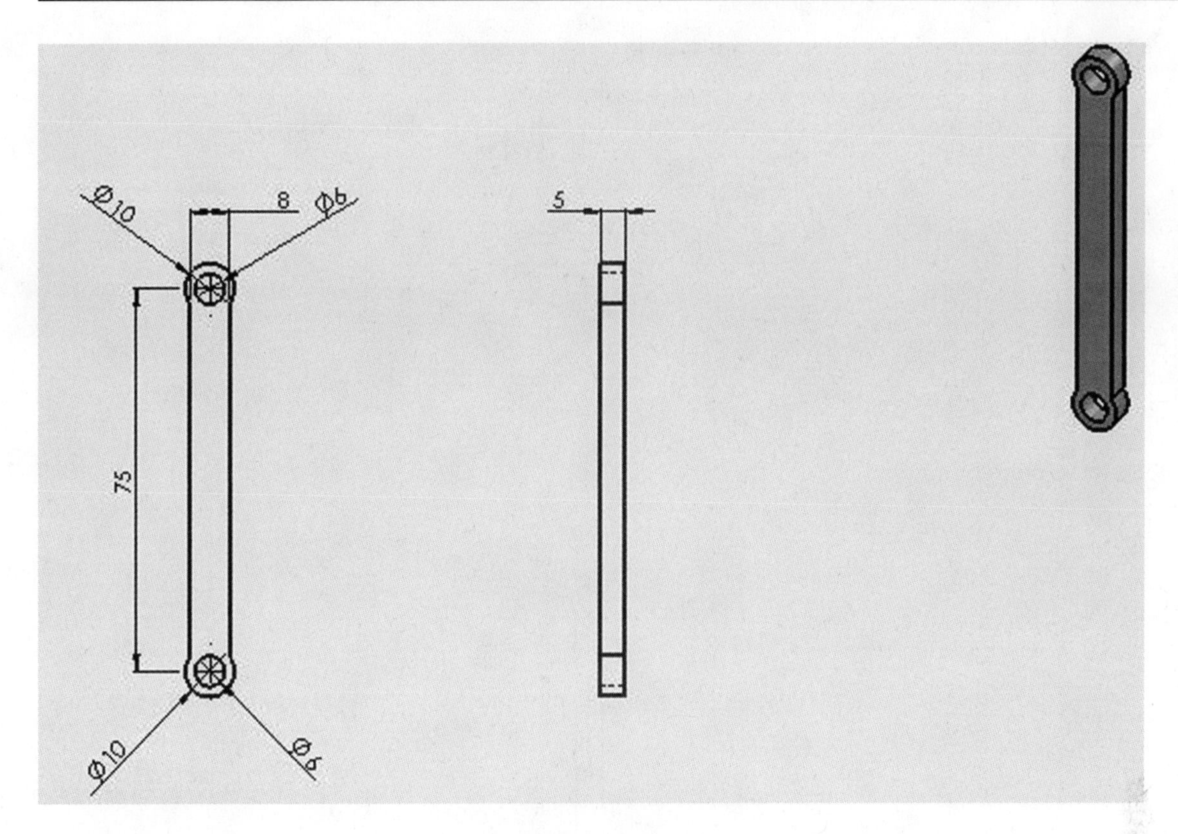

Figure 24.3

Piston rod front view, right view, and part.

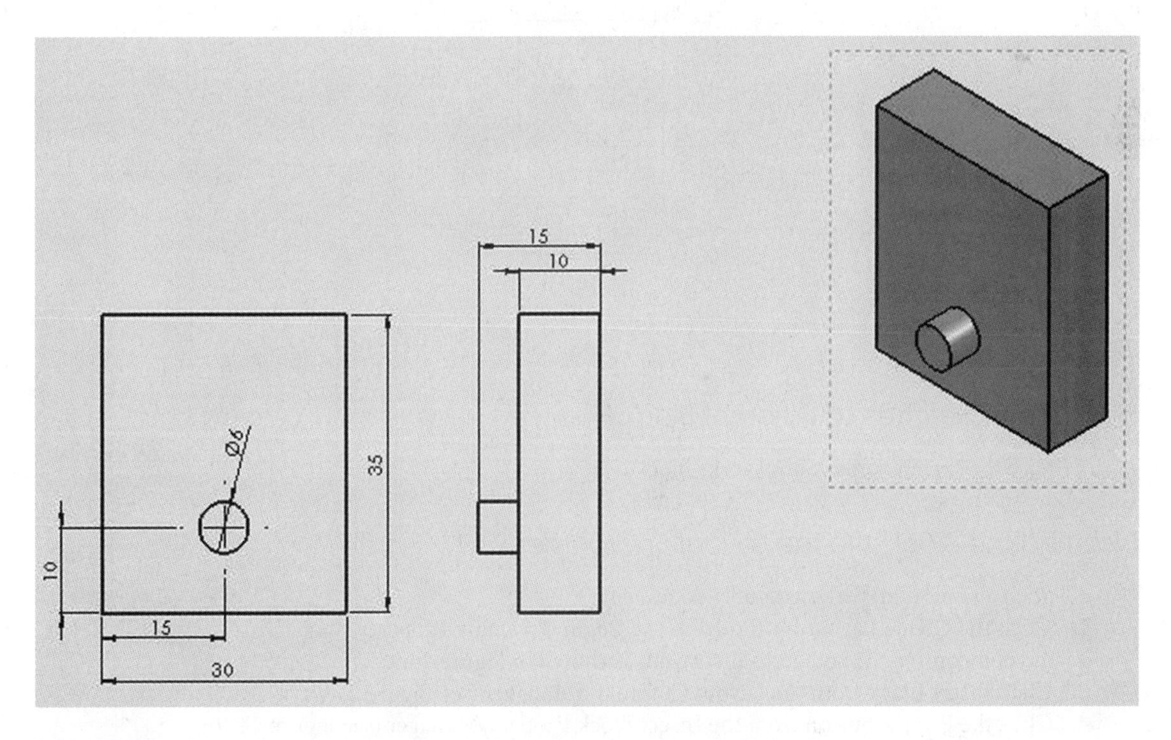

Figure 24.4

Piston tank front view, right view, and part.

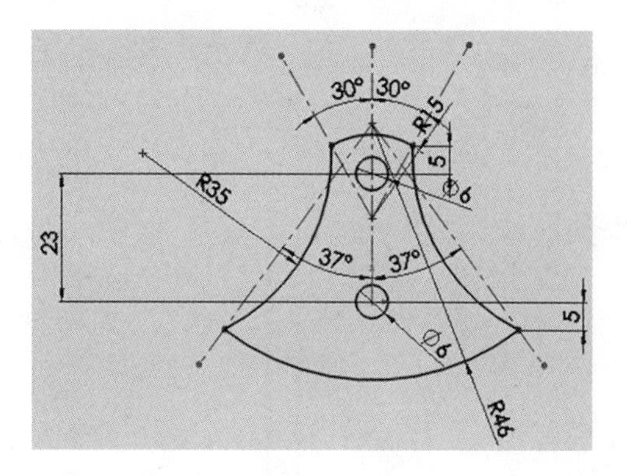

Figure 24.5

Sketch of crank.

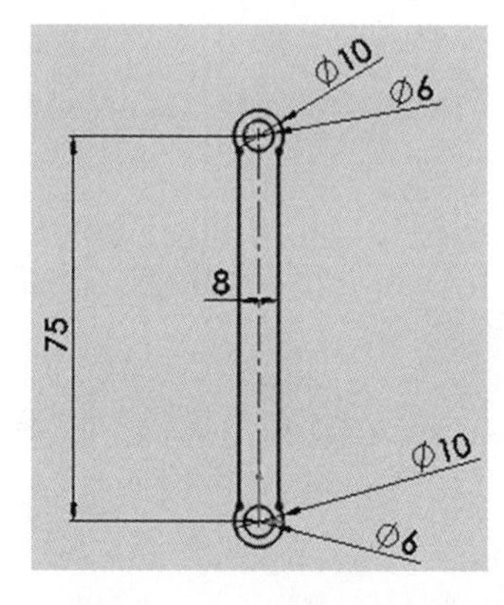

Figure 24.6

Sketch of piston rod.

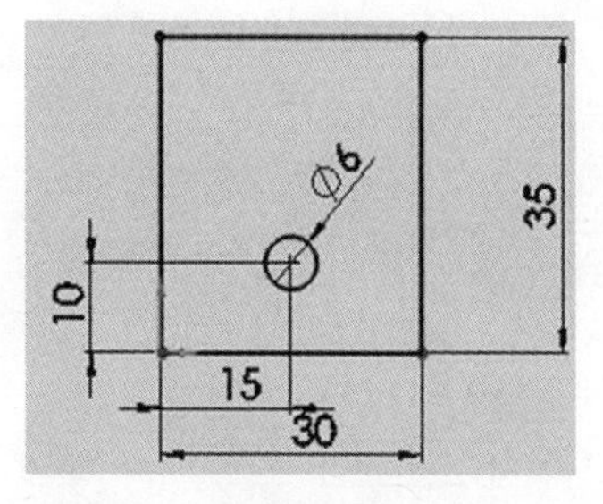

Figure 24.7

Sketch of piston tank.

Saving the Sketches as Different Block Files

1. Save the sketch files as SolidWorks blocks.

Inserting the Block into the Layout Environment

1. Start a new SolidWorks assembly document.
2. Click the Create Layout button from the Begin Assembly PropertyManager (see Figure 24.8). The Layout CommandManager is displayed, as shown in Figure 24.9.
3. Click Insert Block from the Layout CommandManager (see Figure 24.9).
4. Click the Browse button from the Insert Block PropertyManager (see Figure 24.10).
5. Select the crank, piston rod, and piston tank blocks from the file (see Figure 24.10).
6. Click OK.

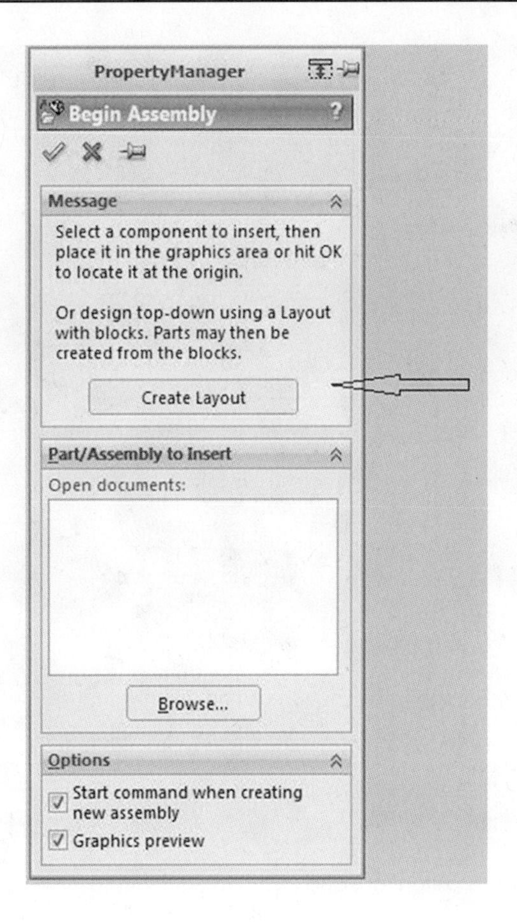

Figure 24.8

Begin Assembly PropertyManager.

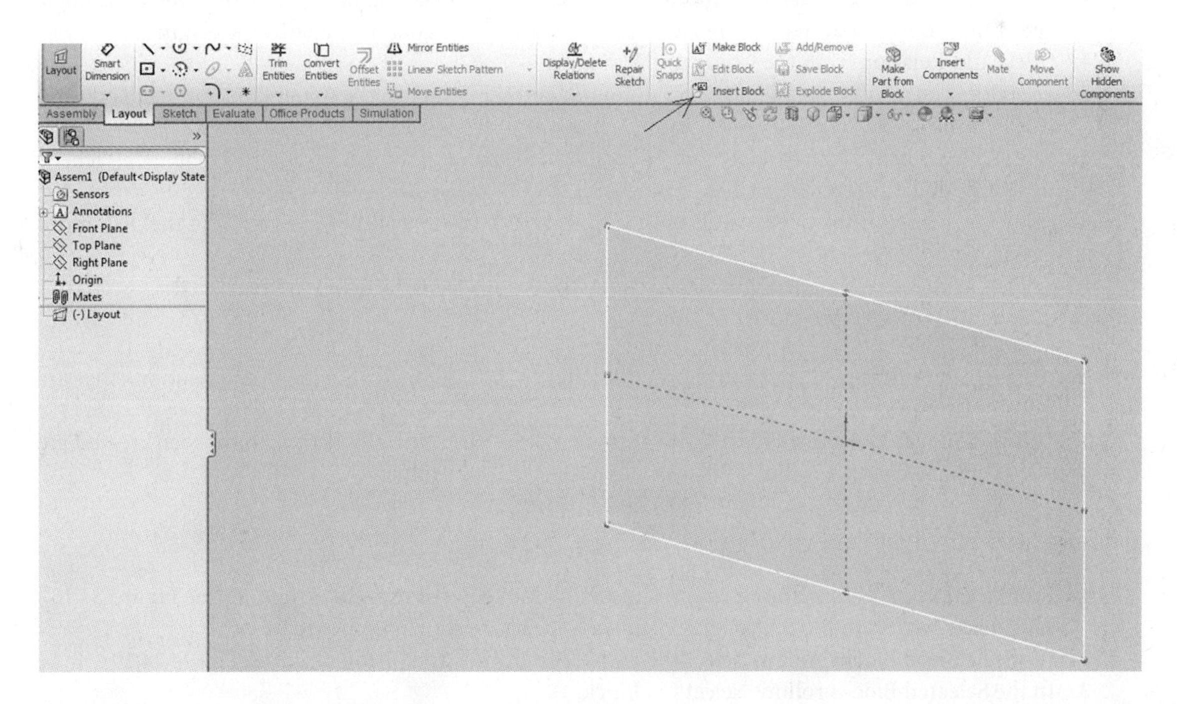

Figure 24.9

Layout CommandManager.

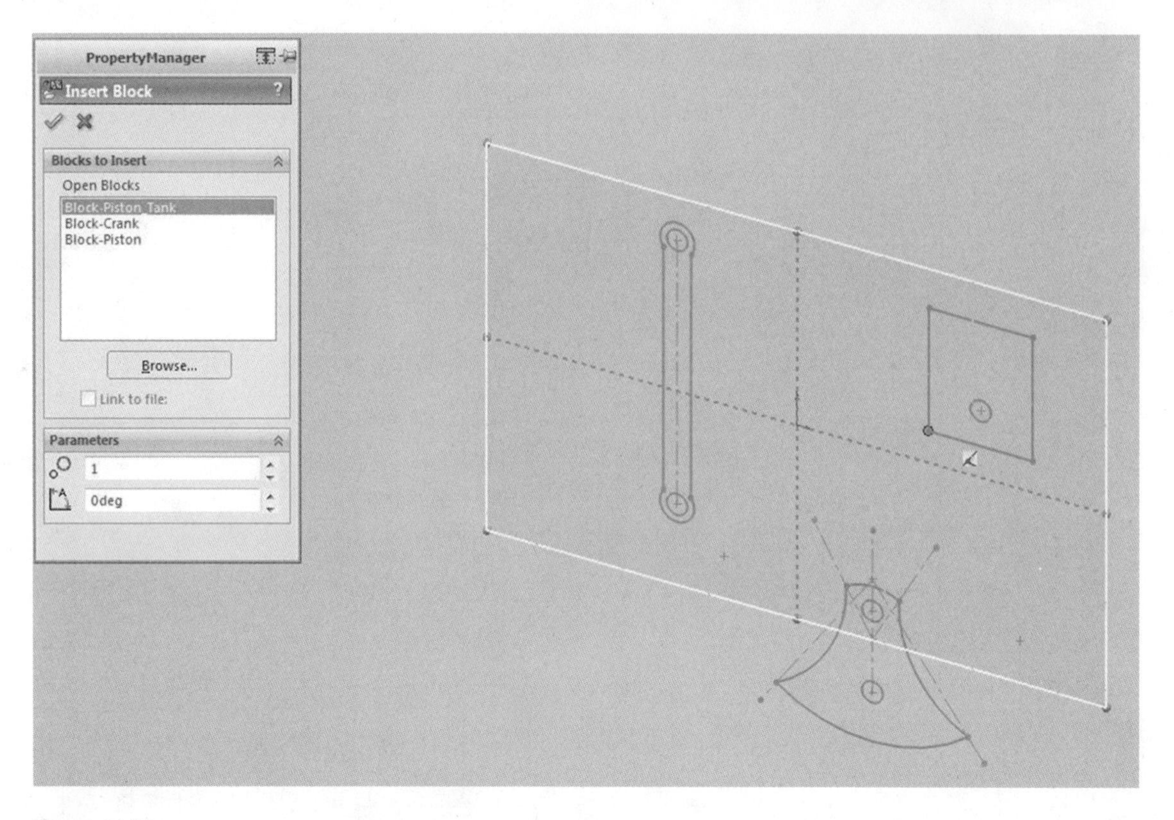

Figure 24.10

Opening the blocks for the crank, piston rod, and piston tank.

Applying Relations to the Blocks

1. Click Add Relation from the CommandManager (see Figure 24.11, arrow labeled 1). The Add Relations PropertyManager appears.
2. Select the origin and the center of the lower circle of the crank (see Figure 24.11, arrows labeled 2 and 3).
3. Select the Coincident relation. Figure 24.12 shows the crank that is located at the origin.
4. Click Add Relation from the CommandManager.
5. Select the center of the circle on the piston tank and the center of the upper circle of the piston rod (see Figure 24.12).
6. Select the Coincident relation. Figure 24.13 shows the piston tank and piston rod that are attached.
7. Click Add Relation from the CommandManager.
8. Select the center of the upper circle of the crank and the center of the lower circle of the piston rod (see Figure 24.14).
9. Select the Vertical relation. Figure 24.15 shows the crank that is vertically attached to the piston rod/piston tank (in a reciprocating relation).

Now, when the crank is rotated, the piston tank will reciprocate vertically due to the relations that are defined (see Figure 24.16).

The FeatureManager looks like Figure 24.17 at this juncture. The three blocks that have been created are displayed as (1) Block-Crank, (2) Block-Piston, and (3) Block-Piston-tank.

Converting Blocks into Parts

1. From the Layout CommandManager, choose the Make Part From Block option (see Figure 24.18). The Make Part From Block PropertyManager appears, as shown in Figure 24.19.
2. In the Selected Blocks rollout, select the crank, piston rod, and piston tank (see Figure 24.19).
3. In the Selected Blocks rollout, select On Block.
4. Click OK three times, each time the new SolidWorks document dialog is displayed and Part is selected by default (see Figure 24.20).

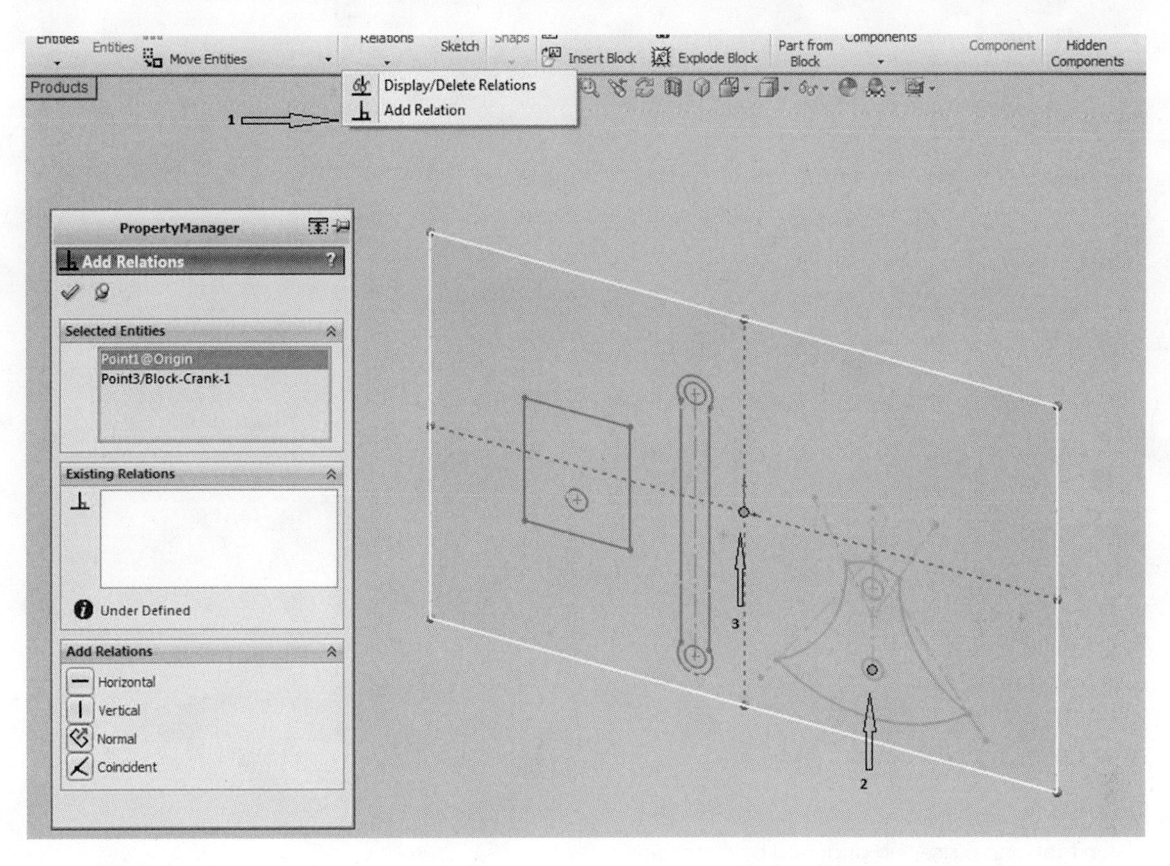

Figure 24.11

Add Relations PropertyManager for the crank.

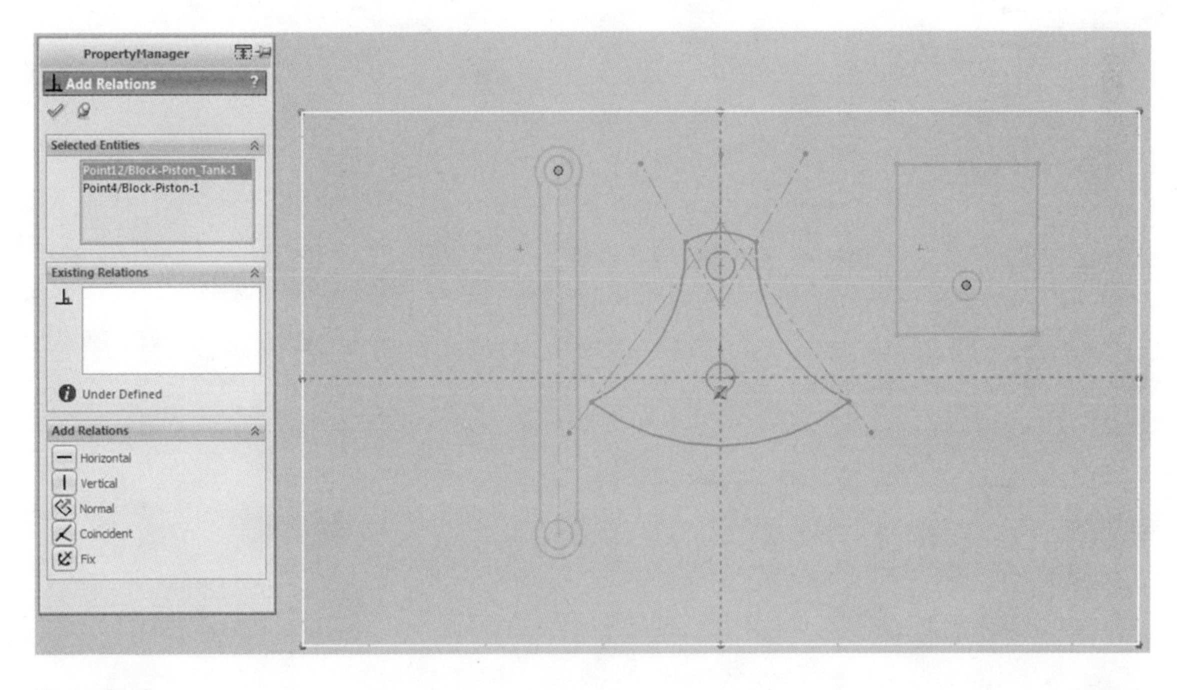

Figure 24.12

Add Relations PropertyManager for piston rod and piston tank.

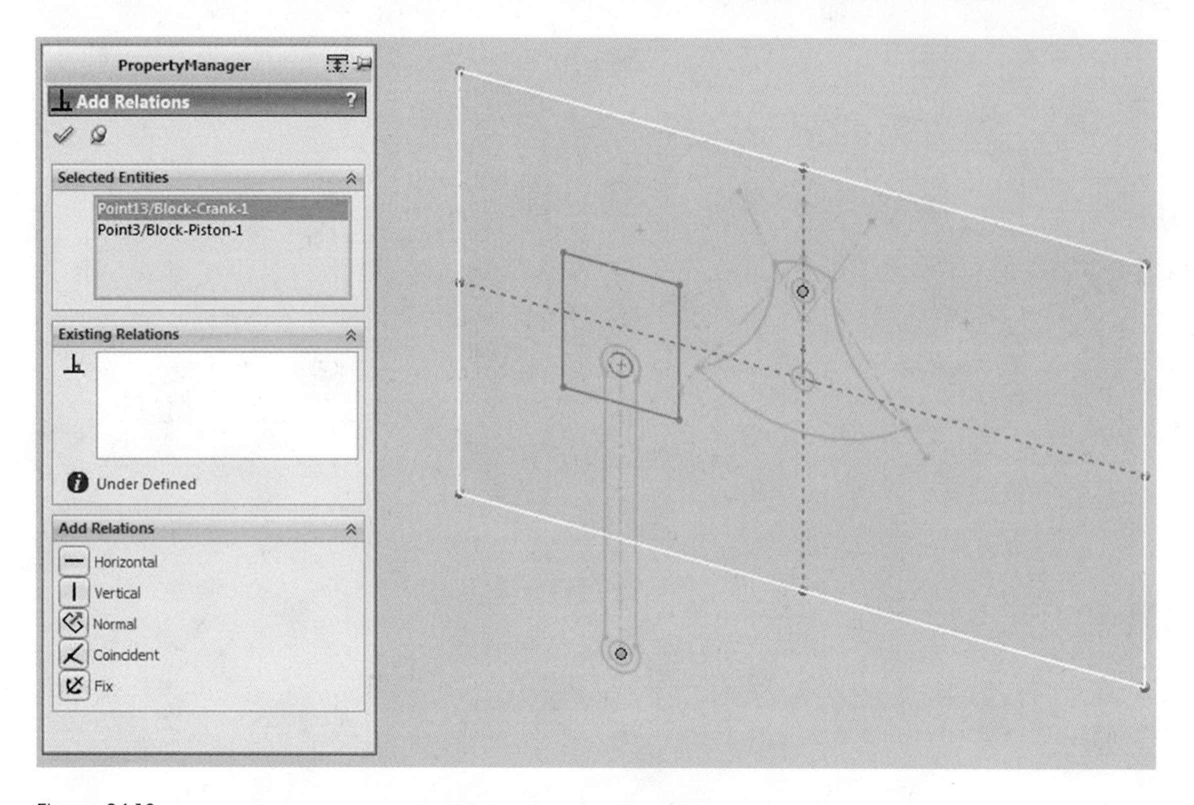

Figure 24.13

Piston tank and piston rod attached by a coincident relation.

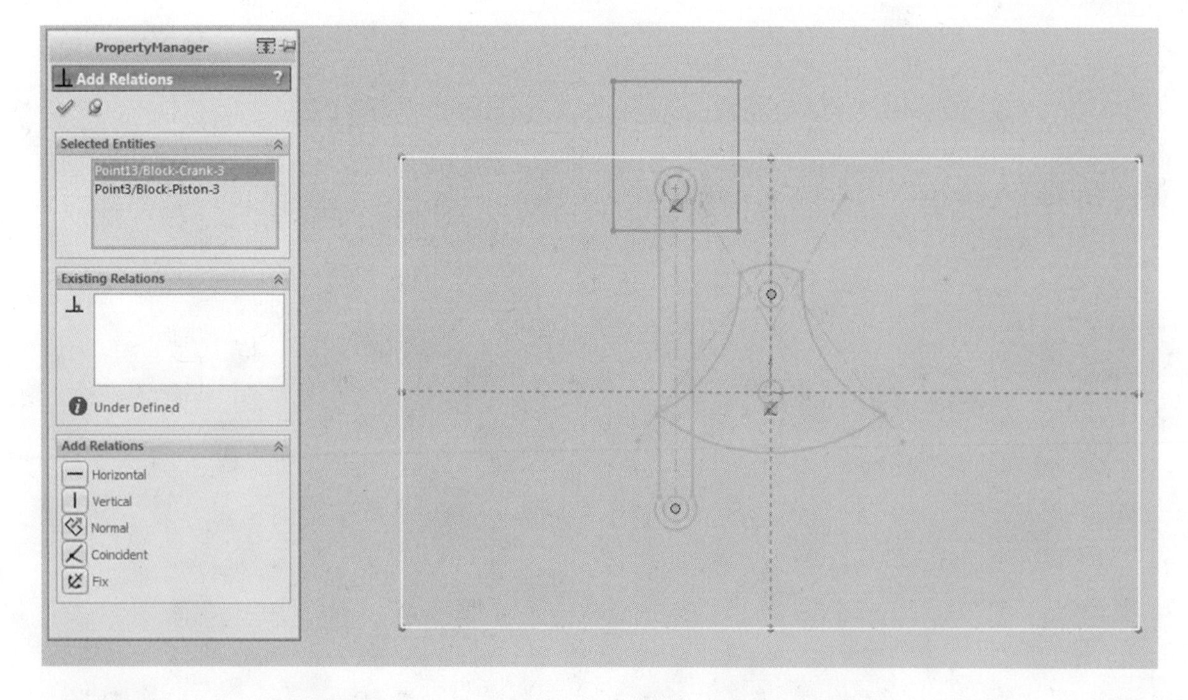

Figure 24.14

Add Relations for crank and piston rod.

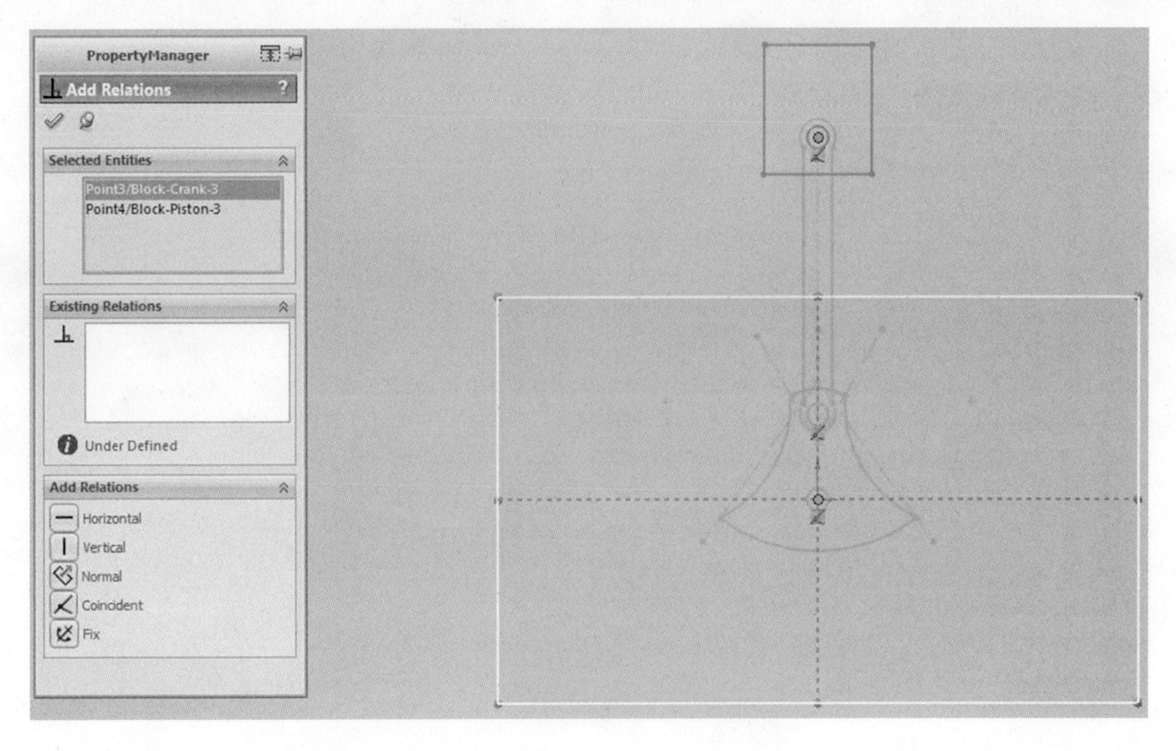

Figure 24.15

Crank vertically attached to piston rod/piston tank.

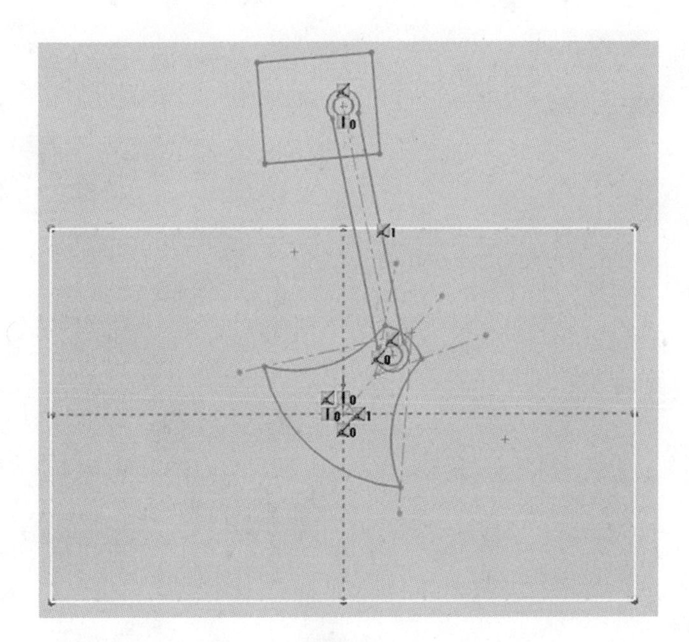

Figure 24.16

Crank, piston rod, and piston tank properly connected.

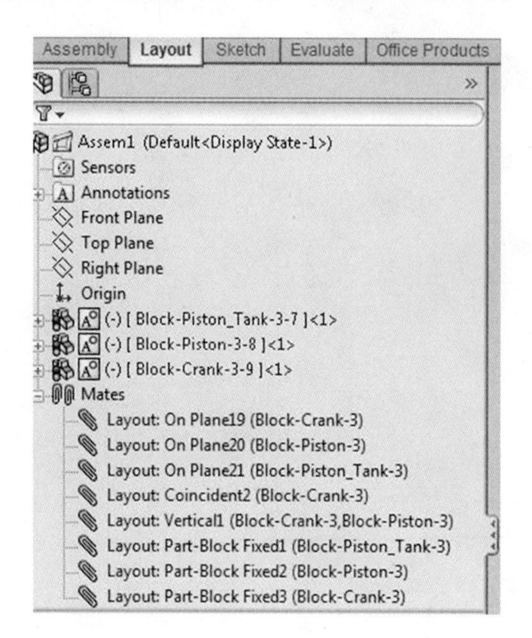

Figure 24.17

FeatureManager.

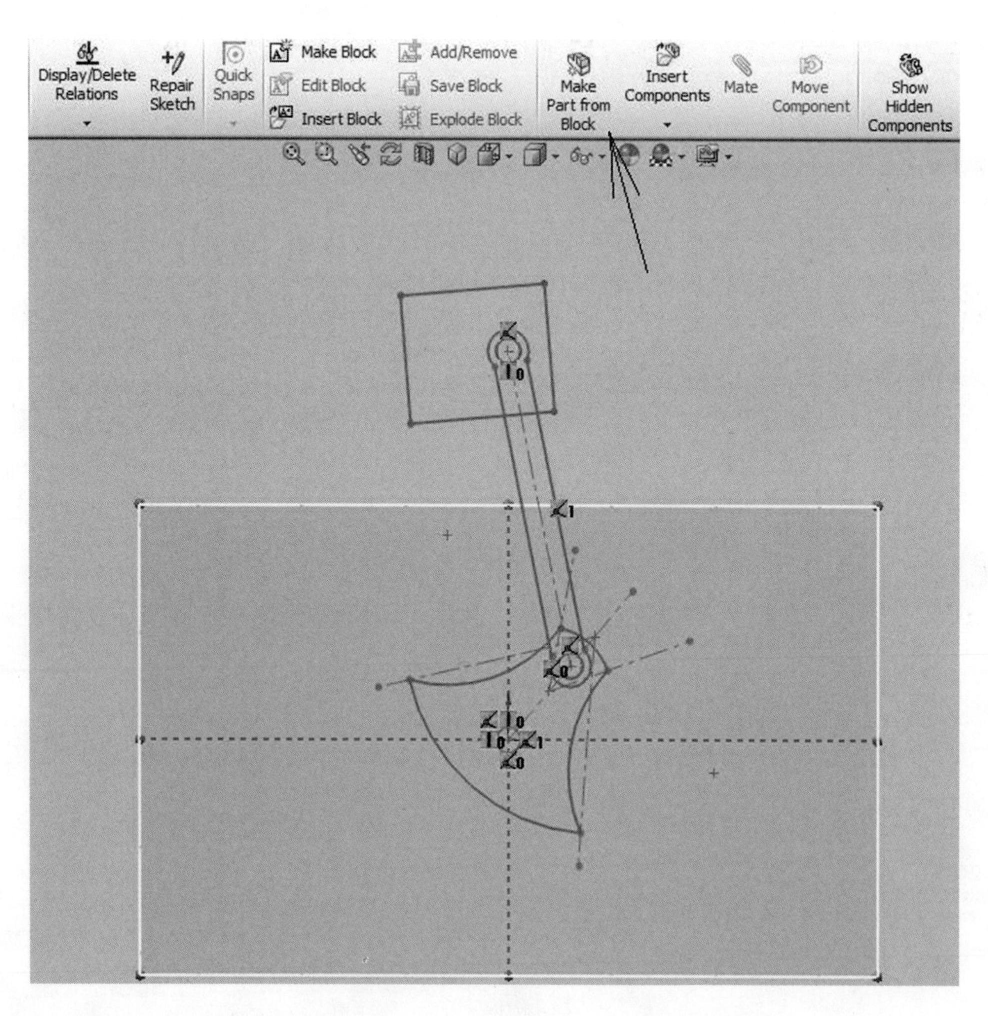

Figure 24.18

Make Part from Block option.

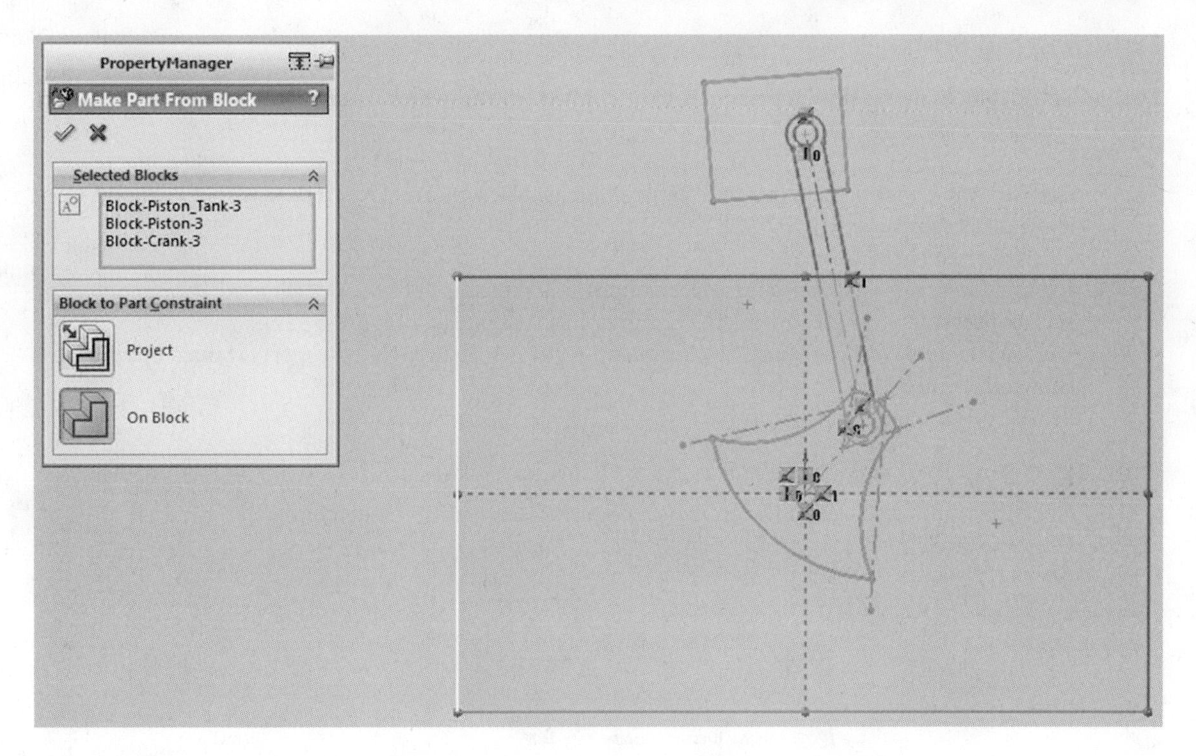

Figure 24.19

Make Part from Block PropertyManager.

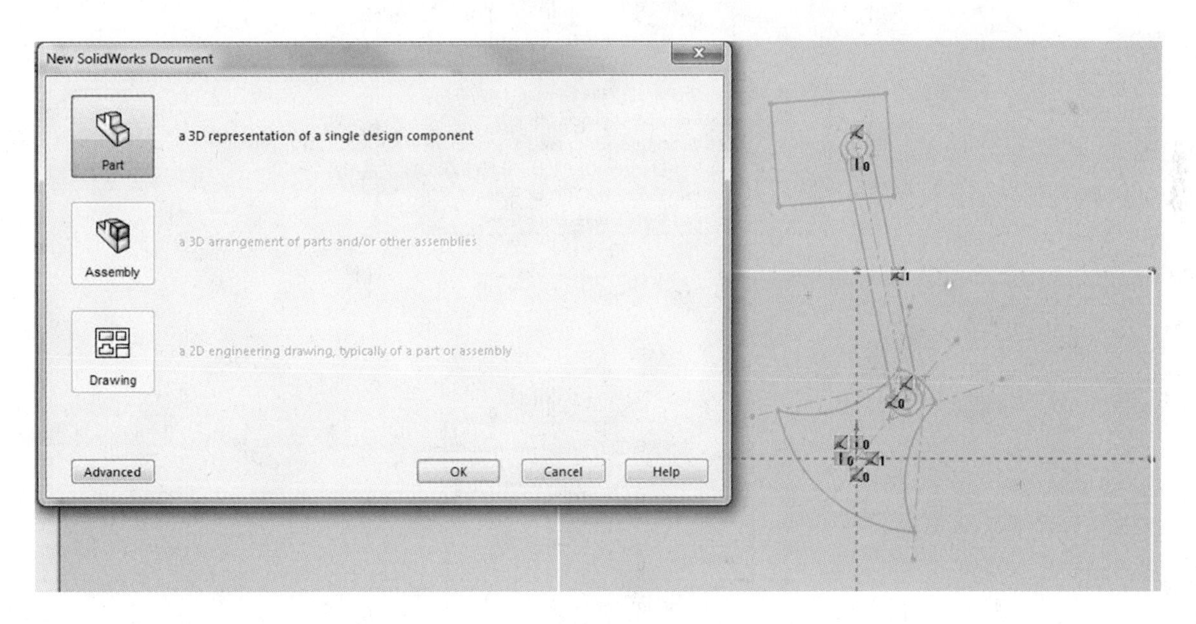

Figure 24.20

Click OK for the new SolidWorks document displayed.

Extruding the Parts

Designing with blocks shares similar principles to in-context editing, which is used in the top–down approach.

Crank

1. Click on the crank block from the FeatureManager Design Tree. A window with in-context icons appears, as shown in Figure 24.21.
2. Click the edit part icon.
3. Expand the crank block tree and select Sketch1. If a sketch is not selected, there will be a message on the PropertyManager prompting the designer to select one (see Figure 24.22).
4. Click Extrude Base/Boss from the CommandManager. The Extrude PropertyManager appears, as shown in Figure 24.23.
5. Set the extrusion depth to 10 mm.

Creating a Boss to Coincide with the Upper Circle on the Crank

6. Click the right-hand side of the crank (see Figure 24.24) and start sketch mode.
7. Click the circle.

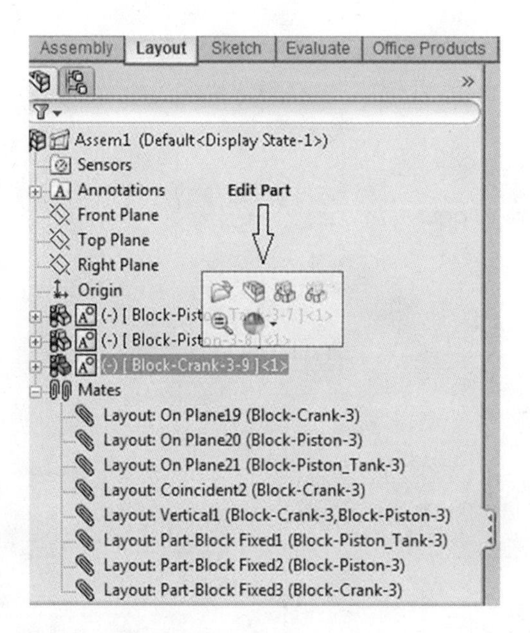

Figure 24.21

In-context editing.

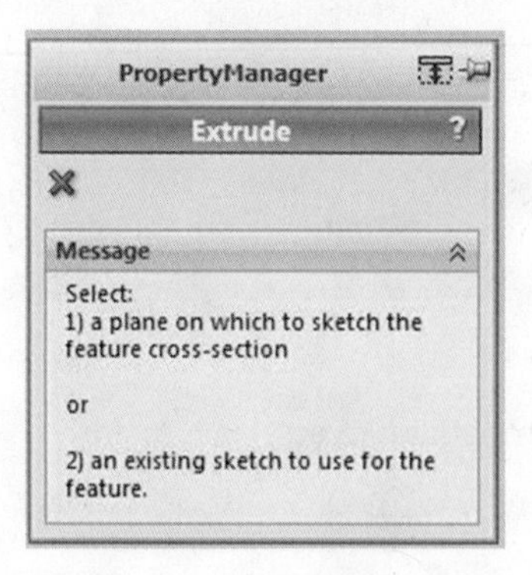

Figure 24.22

Extrude PropertyManager prompting user to select a sketch.

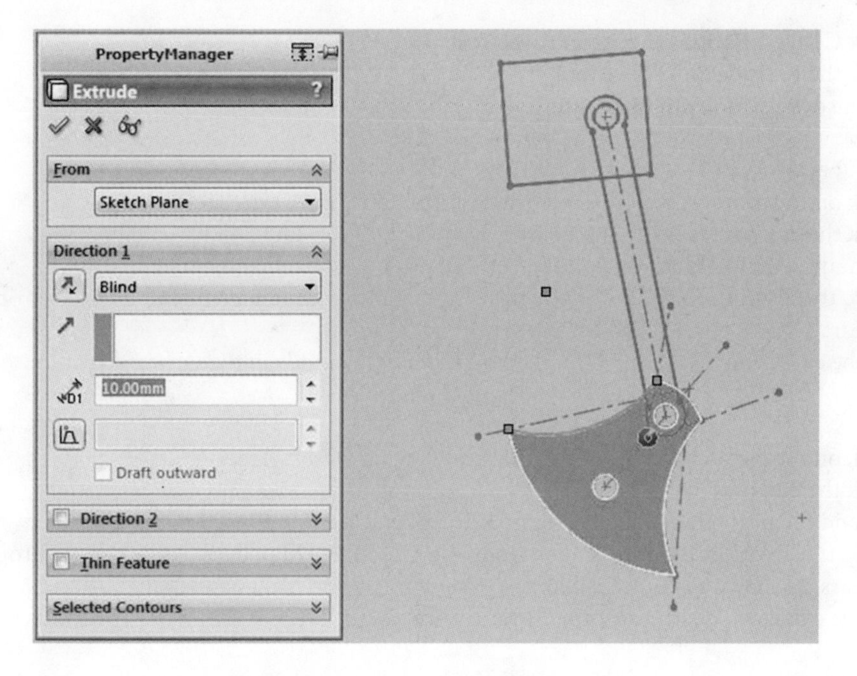

Figure 24.23

Extrude PropertyManager for the crank.

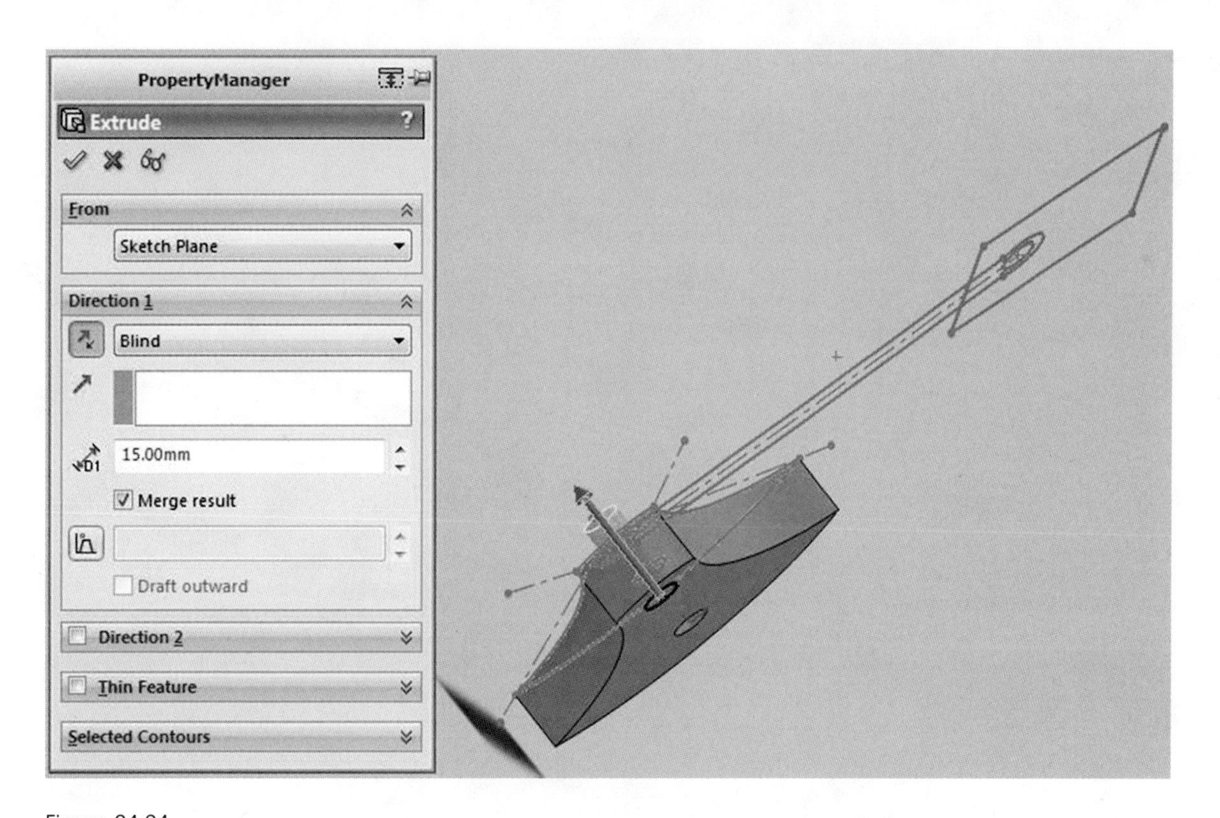

Figure 24.24

Extrude PropertyManager to create a boss.

8. Click Convert Entities to extract this circle.
9. Click the Extrude Base/Bose tool.
10. Set the extrusion depth to 15 mm.

Piston Rod

11. Click on the piston rod block from the FeatureManager Design Tree.
12. Click the edit part icon.
13. Expand the piston block tree and select Sketch1.
14. Click the Extrude Base/Boss tool from the CommandManager. The Extrude PropertyManager appears, as shown in Figure 24.25.
15. Set the extrusion depth to 5 mm. Reverse Direction 1 if it is in the wrong direction.

Piston Tank

16. Click on the piston block from the FeatureManager Design Tree.
17. Click the edit part icon.
18. Expand the piston tank block tree and select Sketch1.
19. Click the Extrude Base/Boss tool from the CommandManager. The Extrude PropertyManager appears, as shown in Figure 24.26.
20. Set the extrusion depth to 5 mm. Reverse Direction 1 if it is in the wrong direction.

The final reciprocating mechanism model is shown in Figure 24.27. As the crank is rotated, the piston rod and the piston tank will reciprocate along the vertical axis due to the relations that are defined at the block level (not at the assembly level).

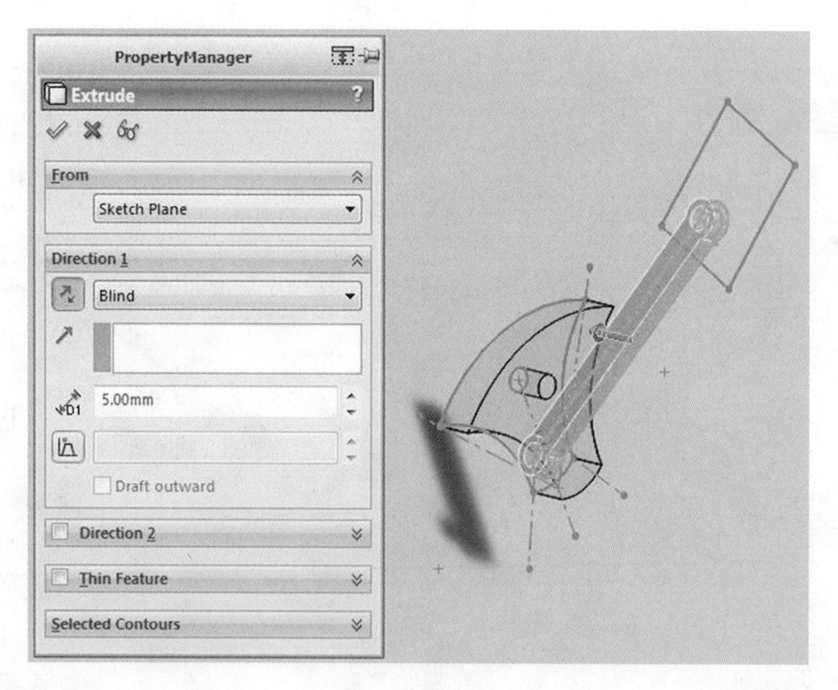

Figure 24.25

Extrude PropertyManager for the piston rod.

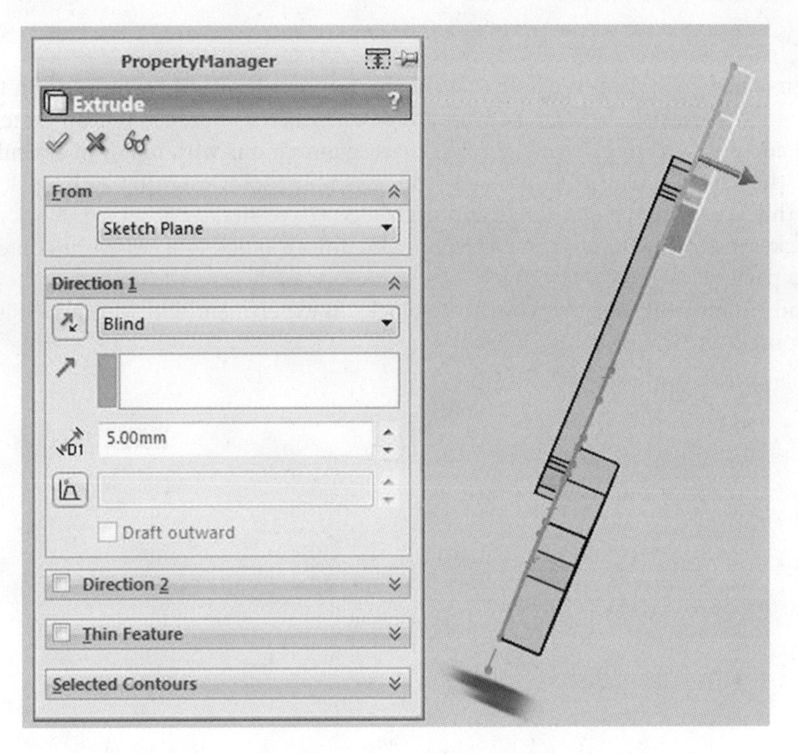

Figure 24.26

Extrude PropertyManager for the piston tank.

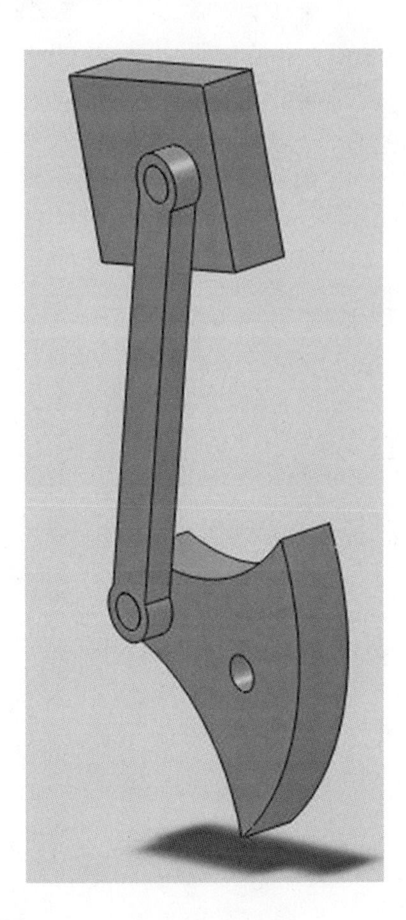

Figure 24.27

Final reciprocating mechanism model.

Summary

The block design is now complete. We note that the mating conditions were imposed on the blocks rather than on the assembly of parts. This is the advantage of using blocks because we can sketch in two dimensional (2D) and add relations to 2D sketches, which are synonymous with mates in assembly mode. When blocks are used, the Layout tool is used instead of the Assembly tool. Designing with blocks is similar to in-context editing that is used in the top–down approach, as we have already observed. Block design creates an assembly of blocks (instead of parts), and a sketch that defines a block is accessed in-context and extruded to convert it to a part.

Industrial and engineering designers often use blocks to experiment with various designs in 2D before committing resources to 3D design.

25

Die Design

Objectives:

When you complete this chapter, you will have

- Understood the scope of die design
- Understood the components of a die set
- Designed a die holder of a die set
- Designed a punch holder of a die set
- Designed a die block
- Designed a blanking punch
- Designed a punch plate
- Designed a pilot
- Designed a stripper
- Designed a back gauge
- Designed a front spacer
- Designed an automatic stop
- Assembled a design set

Scope of Die Design

Die design in a general sense either involves designing an entire press tool with all components taken together or, in a limited way, involves designing that particular component which is machined to receive the blank.

Components of a Die Set

Die sets are made by several manufacturers, and they may be purchased in a great variety of shapes and sizes. The major components of a die set are the die holder (the lower part of the die) and the punch holder (the upper part of the die) (see Figure 25.1). The *punch shank* A is clamped in the ram of the press, which is reciprocated up and down by a crank. In operation, the *punch holder* B moves up and down with the ram. *Bushings* C, pressed into the punch holder, slide on *guide posts* D to maintain the precise alignment of

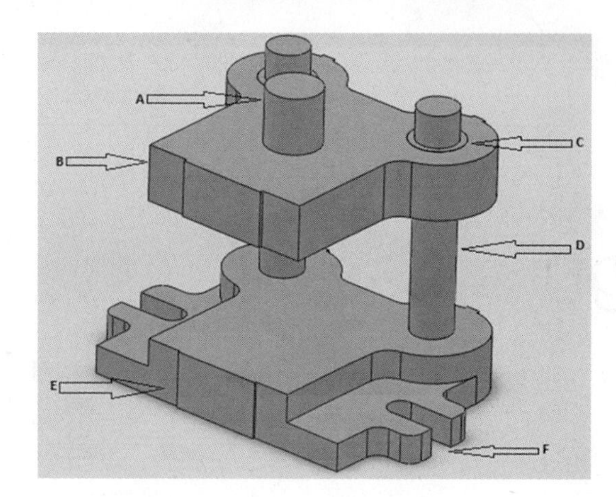

Figure 25.1

A typical die set.

cutting members of the die. The *die holder* E is clamped to the bolster plate (a thick steel plate that is fastened to the press frame) of the press by bolts passing through slots F.

The other components of a die set include *die block*, *blanking punch*, *punch plate*, *pilot*, *stripper*, *back gauge*, *front spacer*, and *automatic stop*. Each of these has to be designed, and all components have to be assembled to cut a particular *scrap strip*.

Pierce and Blank Die

The pierce and blank die is one of the most basic die sets to be studied. The steps involved include the following:

1. Carefully study the part print since the information given on it provides many clues for solving the design problem.
2. Design a scrap strip as a guide for laying out the view of the actual dies.
3. Design the parts of the die set and assemble them to have complete die set for producing the part.

In a pierce and blank die operation, the die pierces two holes at the first station, and then the pat is blanked out at the second station. The material from which the blanks are removed is cold-rolled steel strip.

Part Design

In manual method, the part drawing is the starting point for die design. Using computer-aided design (CAD) such as SolidWorks, the part design, as well as the part drawing, is fairly easy and straightforward.

1. Start a New SolidWorks Part document.
2. Click the Front Plane.
3. Figure 25.2 shows the basic profile for the part.
4. Using the Straight Slot tool, a slot is sketched with the center–center being 1.5 in., while the arc Radius is 9/16 in. (0.563 in.).
5. Extrude the basic profile 0.0625 in., as shown in Figure 25.3. This is the blanked feature.
6. Sketch a Circle with its center coinciding with the end arcs, 0.563-in. diameter (see Figure 25.4).
7. Extrude-cut the Circle in order to define a hole (see Figure 25.4).
8. Mirror the hole so as to have the pierced holes (see Figure 25.5); the blanked strip is shown in Figure 25.6.

Scrap Strip

A scrap strip is normally designed to guide the laying out of the views of the actual die. Manually, this is sketched using pencil and paper. Using SolidWorks CAD software, the scrap strip is produced, as shown in Figure 25.7, but we have only included dimensions to fully define the scrap strip layout.

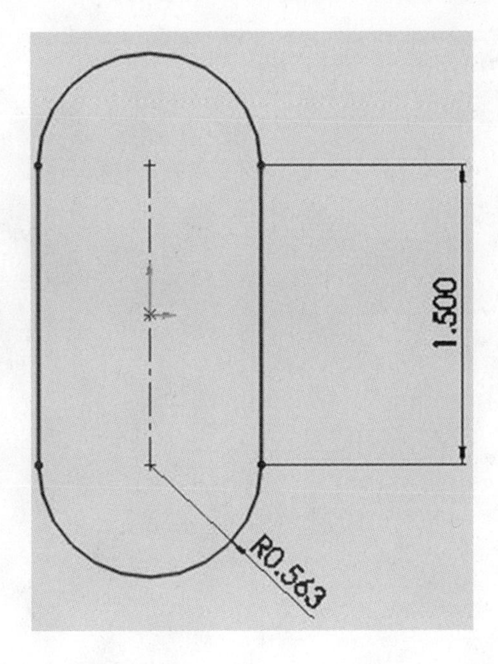

Figure 25.2

Basic profile for the part.

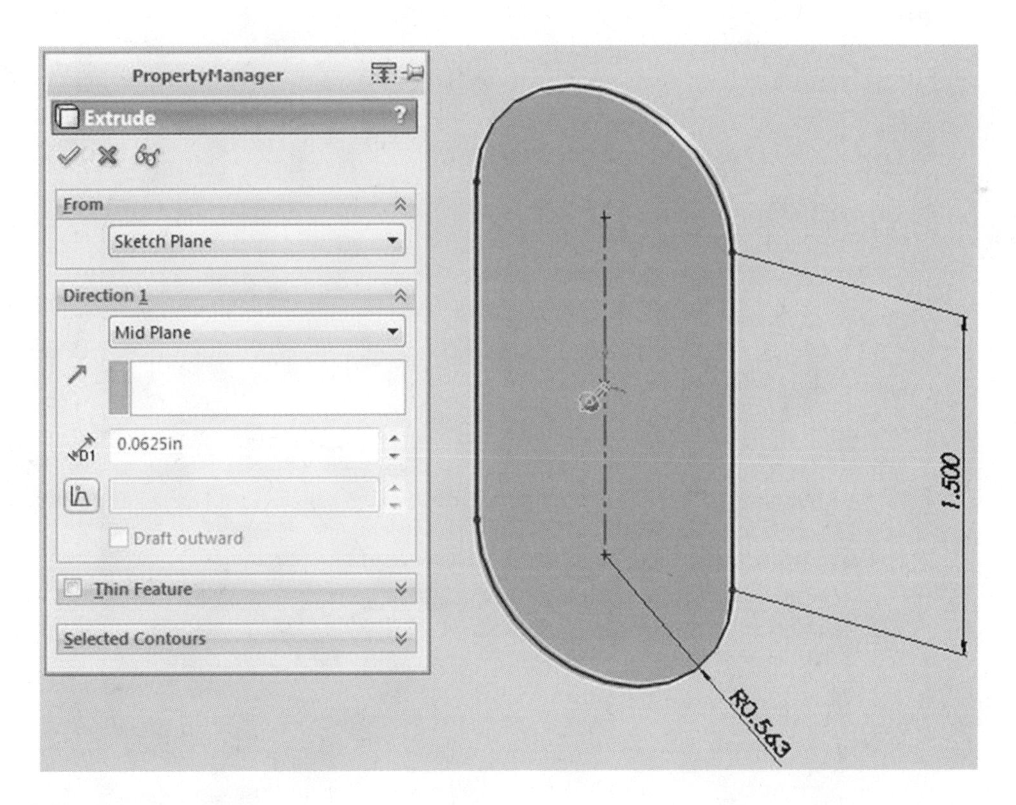

Figure 25.3

Basic profile extruded 0.0625 in.

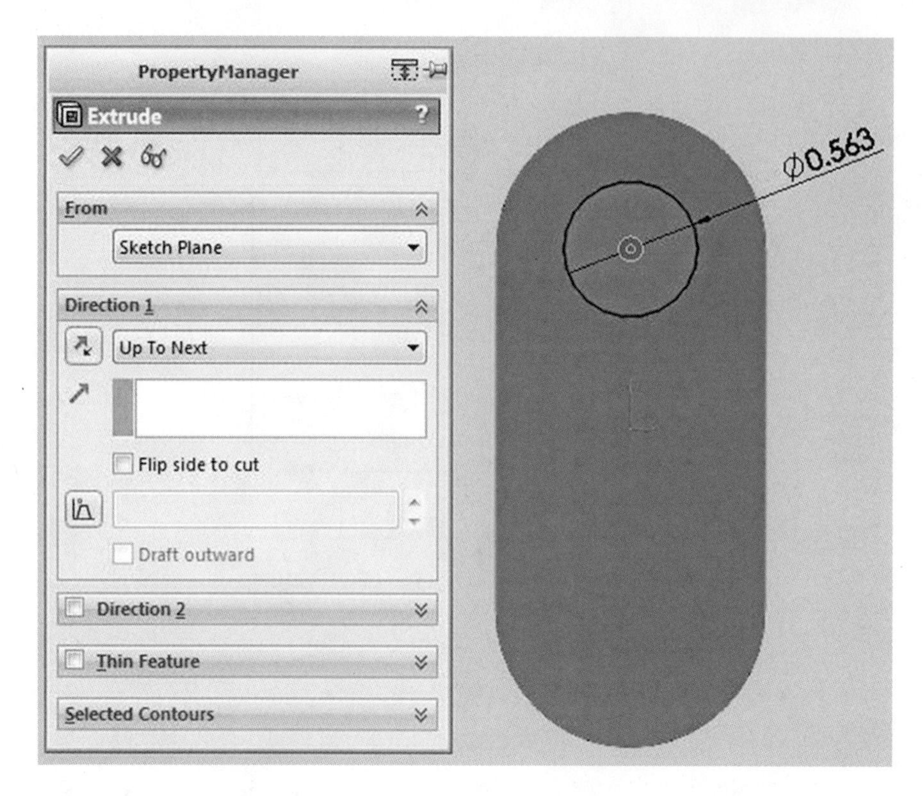

Figure 25.4

Extrude-cut for the circle to define a hole.

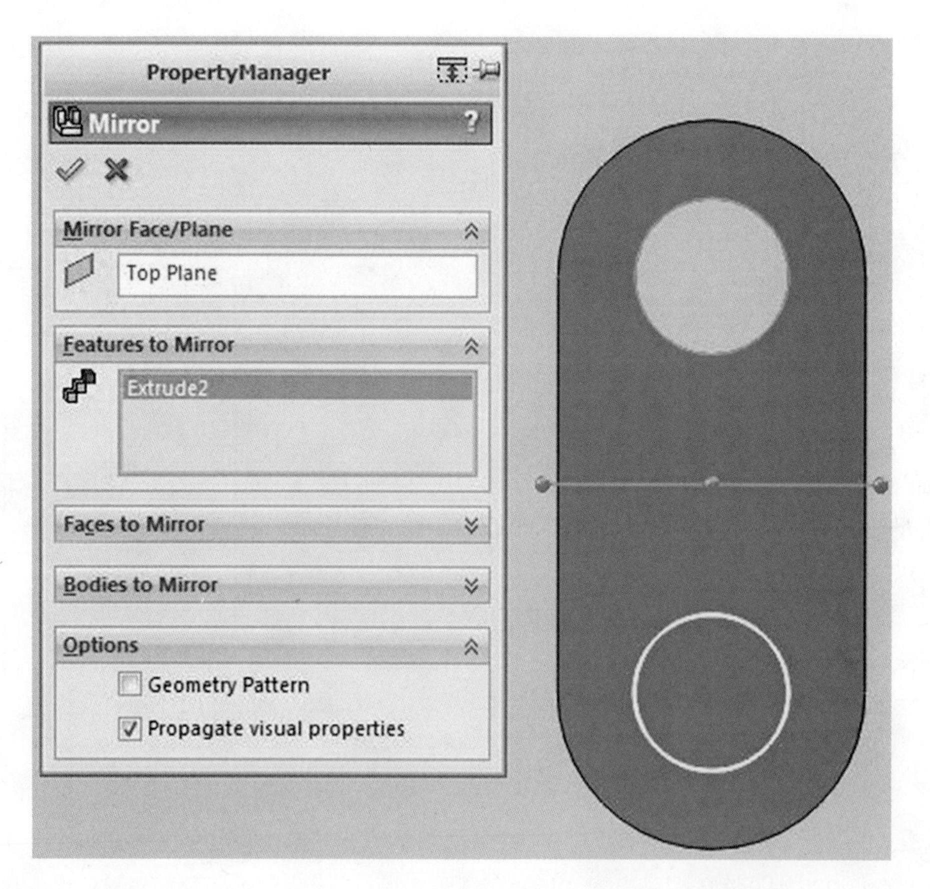

Figure 25.5

Mirrored hole.

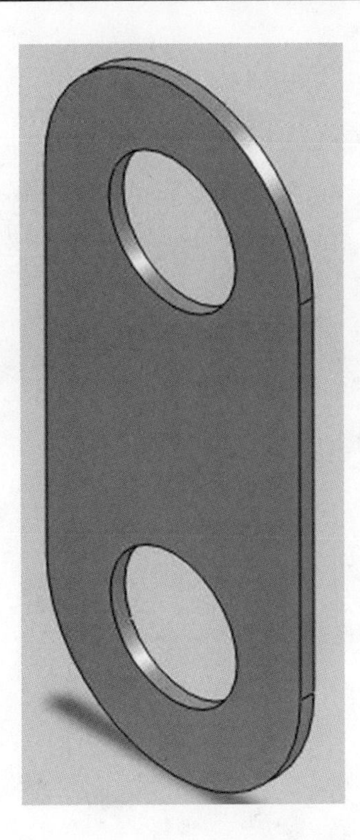

Figure 25.6

Pierced holes and blanked strip.

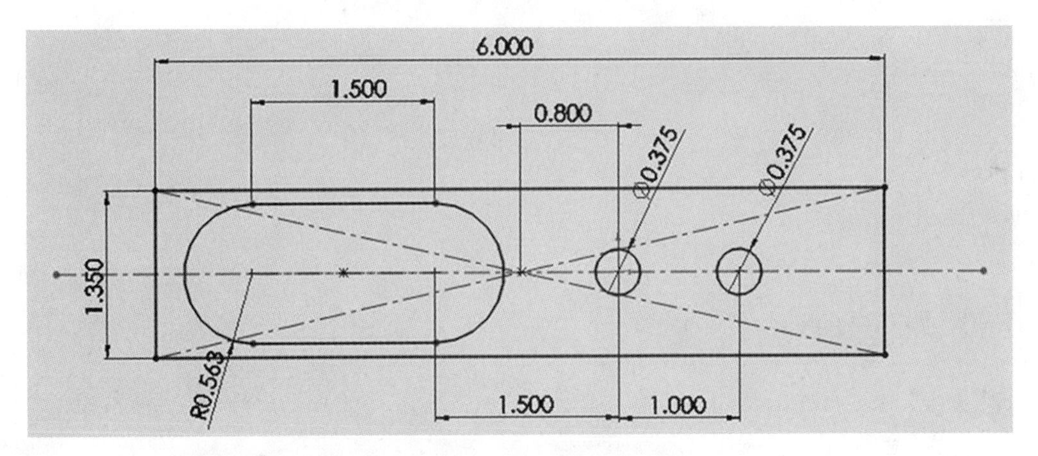

Figure 25.7

Scrap strip layout using SolidWorks CAD software.

Design of Die Holder of a Die Set

Designing the die holder of a die set is one of the major tasks in die design. Standard dimensions are available from manufacturers' websites.

1. Create the *basic profile* of a Die Set Die Holder on the Top Plane, as shown in Figure 25.8.
2. Extrude the basic die holder sketch through 0.502 in., as shown in Figure 25.9.
3. Extrude an additional portion through a value 0.5 in., as shown in Figure 25.10.
4. Create one hole, 0.75 in. in diameter.
5. Mirror the hole about the Right Plane, as shown in Figure 25.11, to realize Figure 25.12.

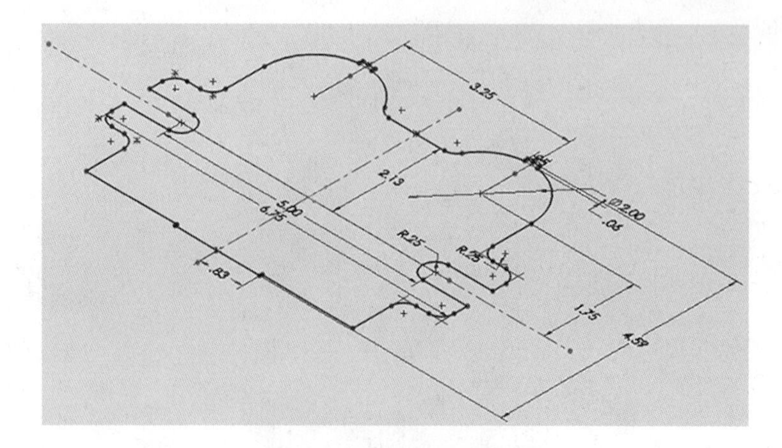

Figure 25.8

Sketch for die holder of a die set.

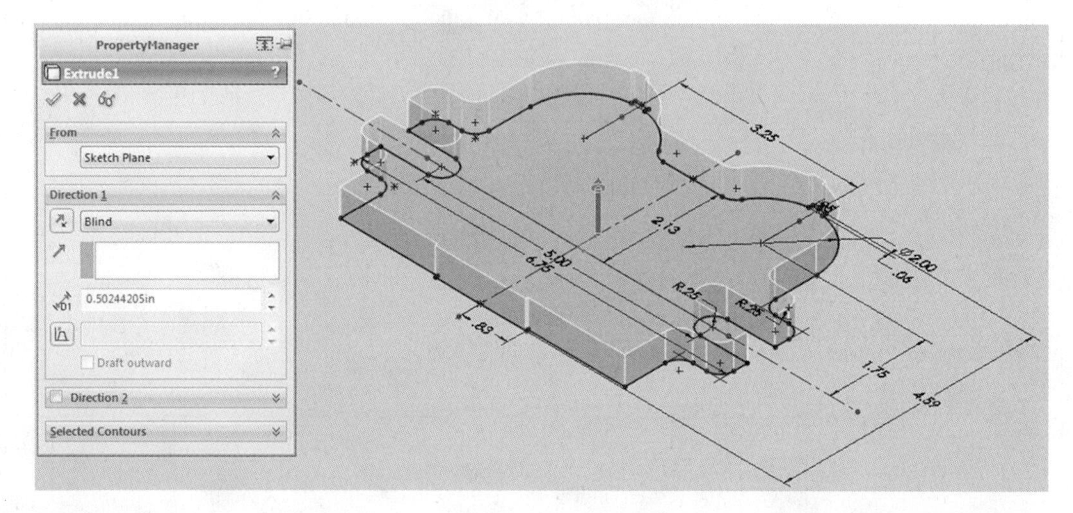

Figure 25.9

Extrude PropertyManager.

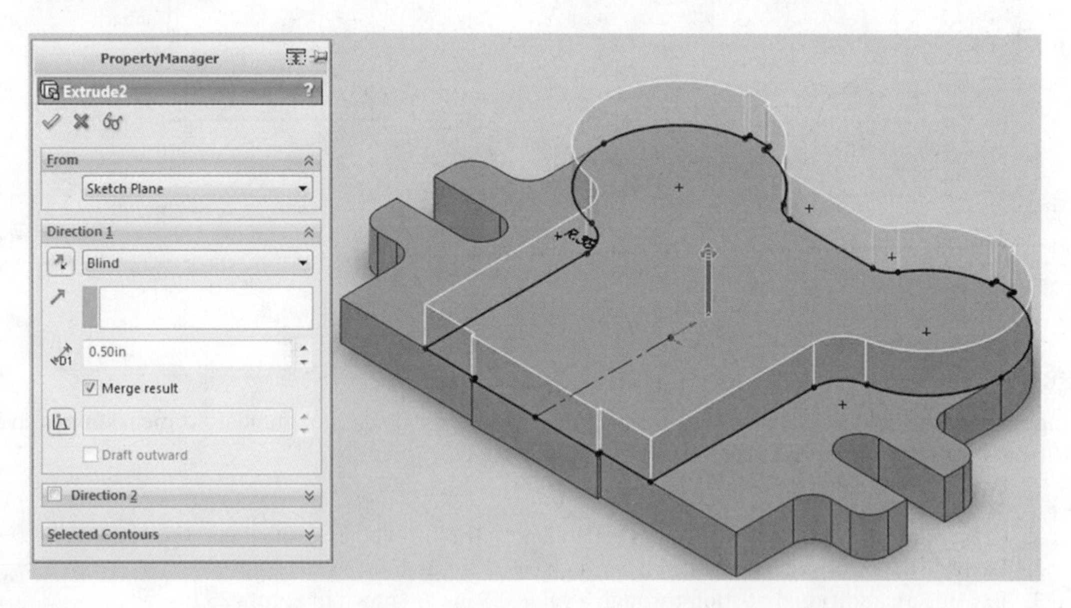

Figure 25.10

Extrusion of portion of upper surface.

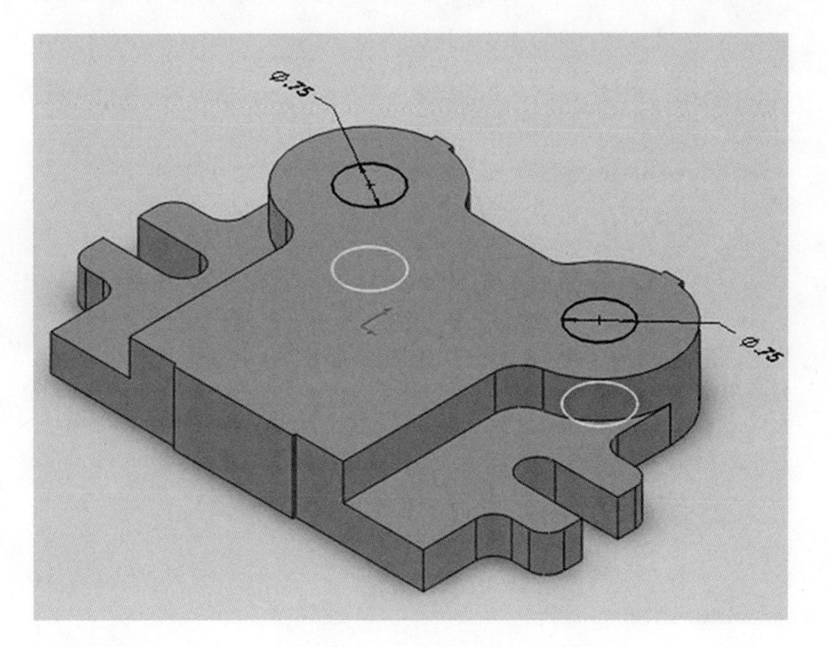

Figure 25.11

A hole created and mirrored about the Right Plane.

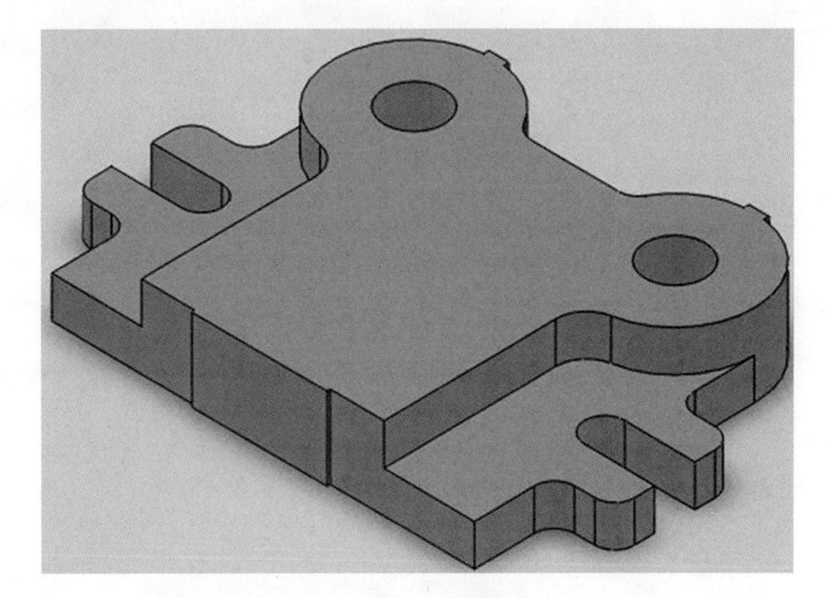

Figure 25.12

Die holder model.

Design of Punch Holder of a Die Set

Designing the punch holder of a die set is another major task in die design. Standard dimensions are available from manufacturers' websites.

1. Create the *basic profile* of a Die Set Punch Holder on the Top Plane, as shown in Figure 25.13.
2. Extrude the basic punch holder sketch through 1.00 in., as shown in Figure 25.14.
3. Create one hole, 1.00 in. in diameter as shown in Figure 25.15.
4. Mirror the hole about the Right Plane as shown in Figure 25.15.
5. Add the Punch Shank as shown in Figure 25.16.

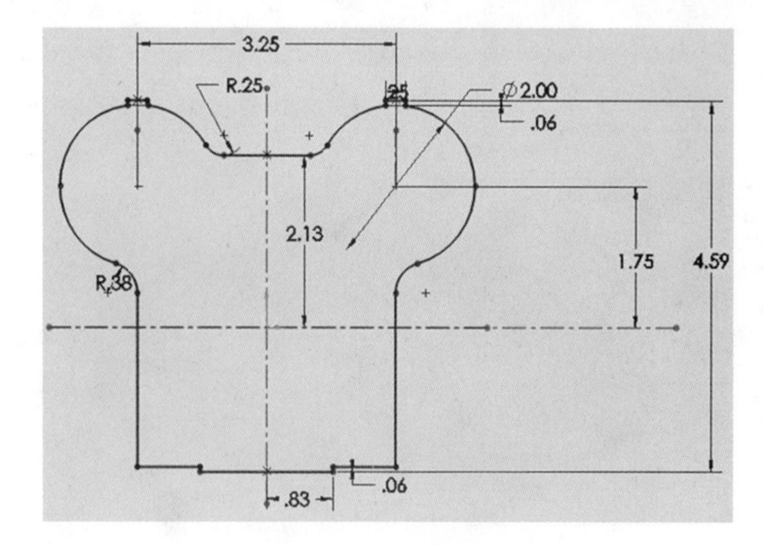

Figure 25.13

Basic profile of a Die Set Punch Holder.

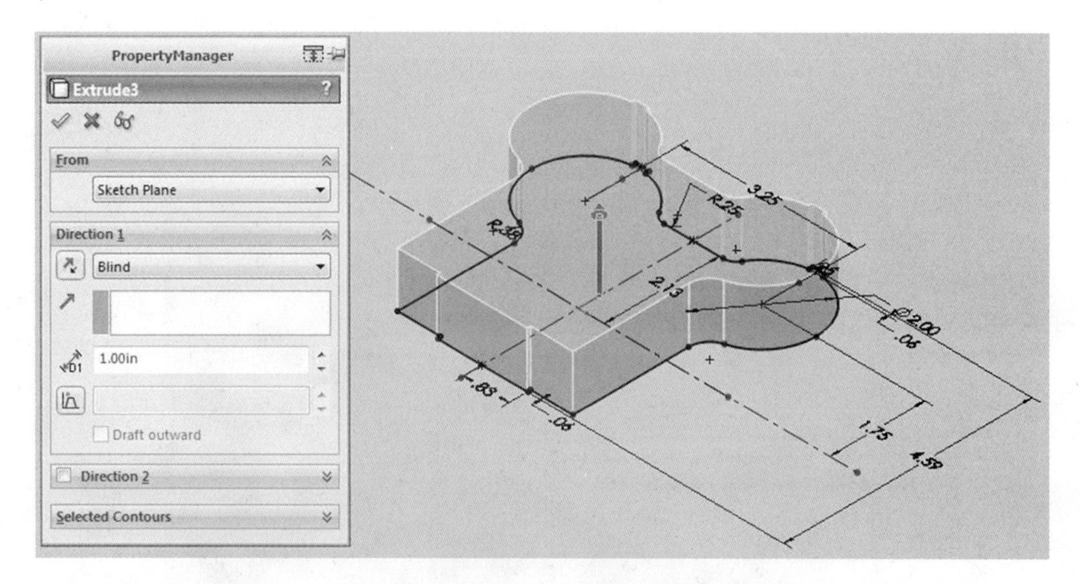

Figure 25.14

Basic punch holder sketch.

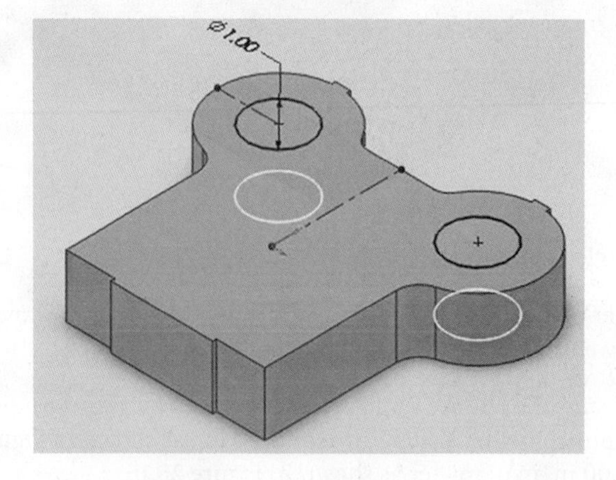

Figure 25.15

Create one hole and mirror about Right Plane.

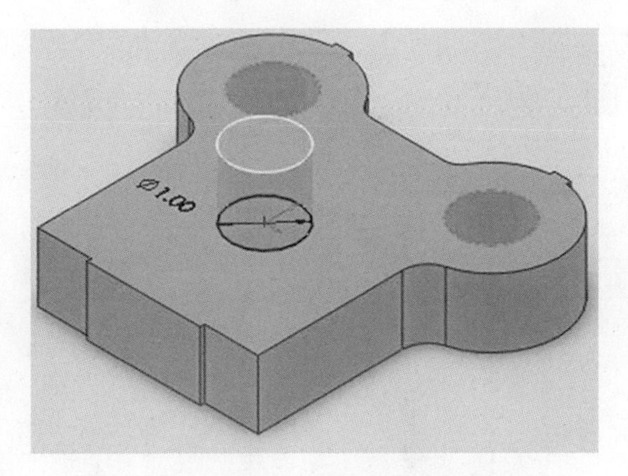

Figure 25.16

Add the Punch Shank.

Design of Guide Post

1. Create a circle, 0.75 in. diameter as shown in Figure 25.17a.
2. Extrude the circle 5.00 in. as shown in Figure 25.17b.

Design of Bushing

1. Create two concentric circles, 0.75 in. and 1.00 in. in diameter, respectively, as shown in Figure 25.18.
2. Extrude the circles through 1.00 in., as shown in Figure 25.18.

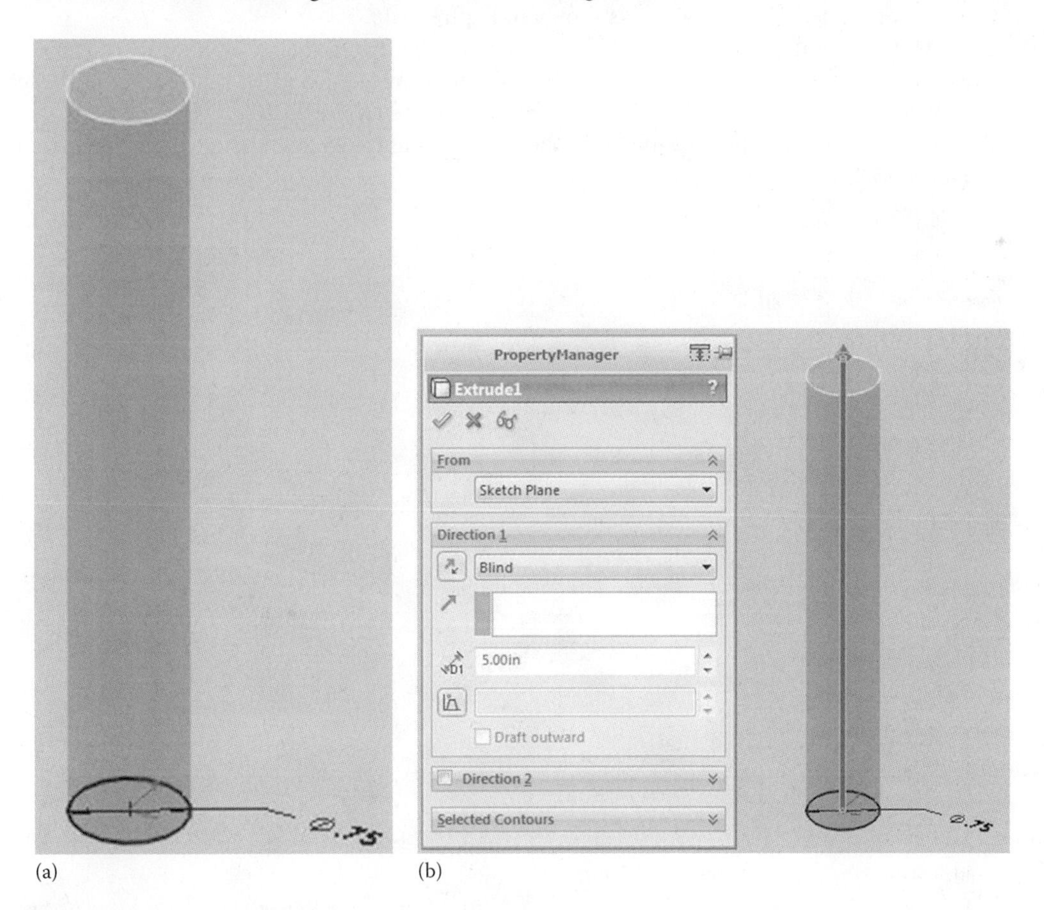

Figure 25.17

Extrude PropertyManager for guide post: (a) circle and (b) extrusion.

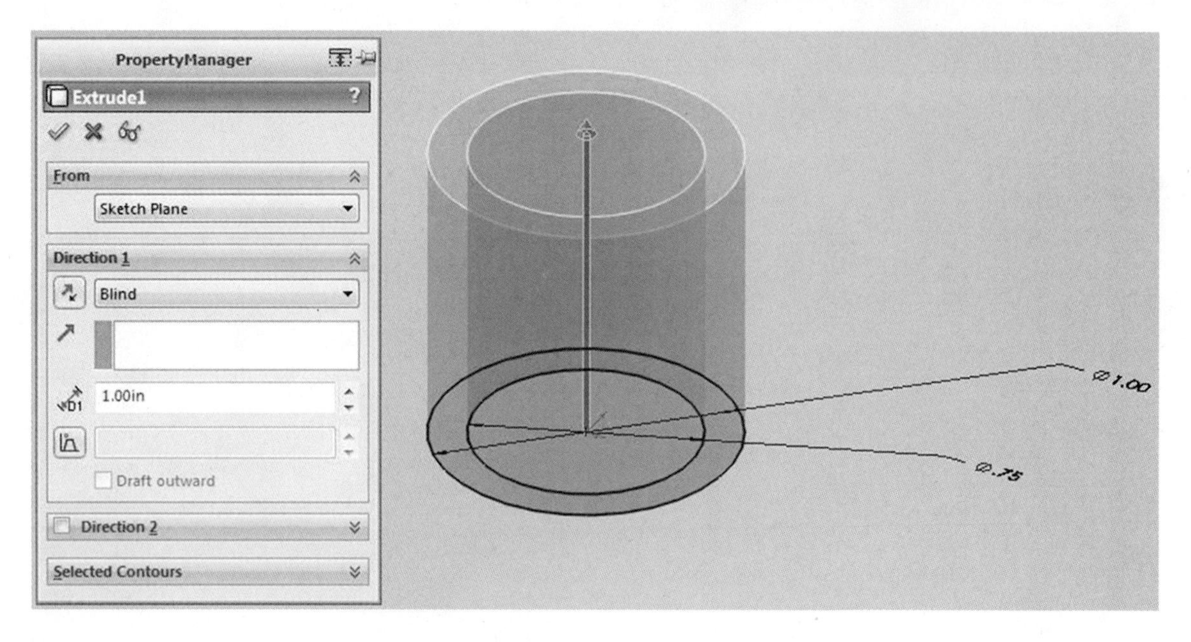

Figure 25.18

Extrude PropertyManager for bushing.

Design of Die Block

1. Create a rectangle, 3.5 in. × 3.0 in., as shown in Figure 25.19.
2. Extrude the rectangle through 0.9375 in., as shown in Figure 25.19.
3. Create a hole using Hole Wizard with center 0.63 in. from each edge near the corner of the feature; choose Hole Size = "I" and All Drill Sizes for the Tool (see Figure 25.20).
4. Linear Pattern/Mirror the hole to obtain the part (see Figure 25.20). The die block model is shown in Figure 25.21.

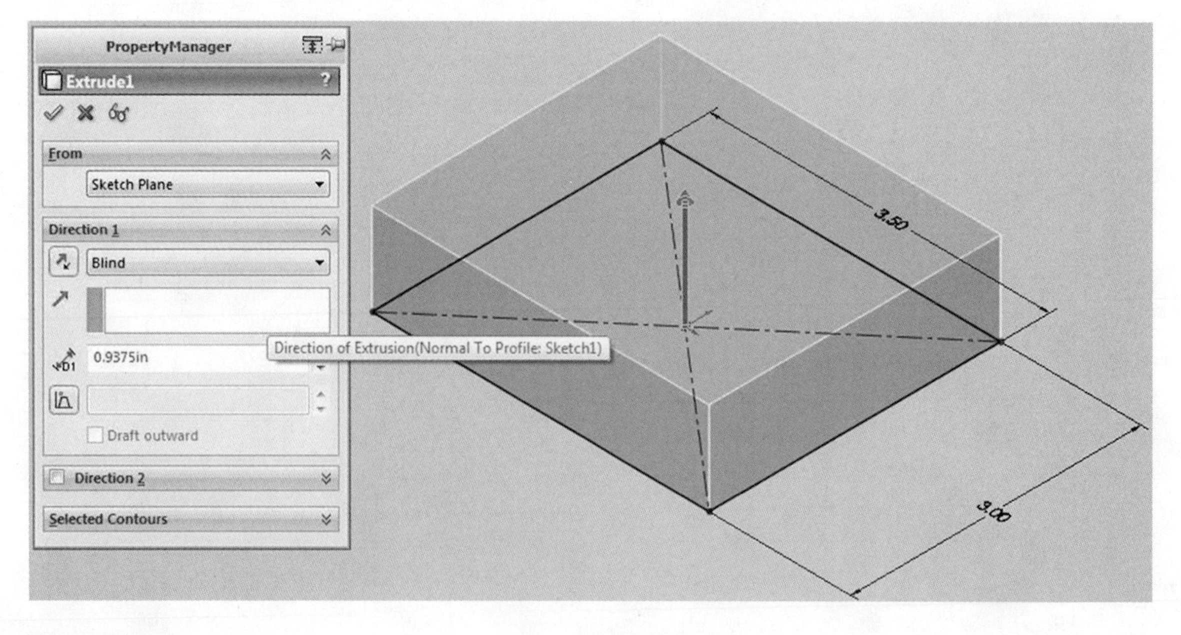

Figure 25.19

Rectangular profile and extrusion.

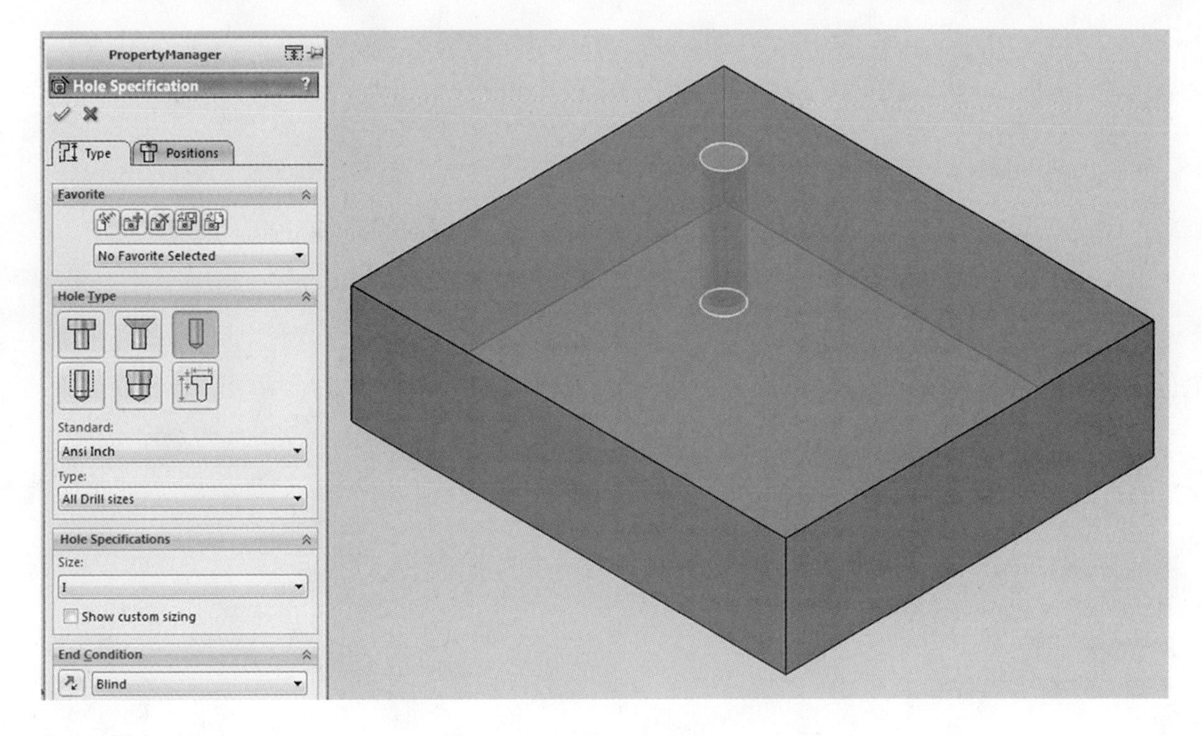

Figure 25.20

Create a hole using Hole Wizard.

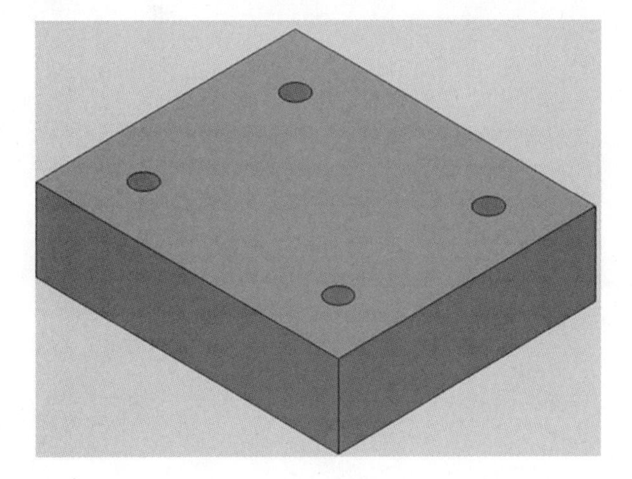

Figure 25.21

Die block model.

Summary

Die design is extremely useful in manufacturing engineering applications. This chapter shows how the design of the different components of a die set can be achieved using SolidWorks.

Exercises

1. Design the Blanking Punch.
2. Design the Punch Plate.
3. Design the Pilot.

4. Design the Stripper.
5. Design the Back Gauge.
6. Design the Front Spacer.
7. Design the Automatic Stop.
8. Assemble the components above with the partial die set and the complete die set.
9. Produce the drawings of the parts and assembly drawing of the die set.

26

Aluminum Extrusion from Manufacturers' Internet Websites

Objectives:

When you complete this chapter, you will have

- Understood how to access manufacturers' Internet websites for AutoCAD two-dimensional (2D) aluminum section profiles
- Understood how convert the AutoCAD 2D aluminum section profiles into SolidWorks three-dimensional (3D) extrusions
- Understood how to use the SolidWorks 3D Aluminum Extrusions to realize structural assemblies

Accessing Manufacturers' AutoCAD 2D Aluminum Section Profiles

Assuming that some engineers belonging to a computer-aided design (CAD) company in Toronto, Canada are working with local customers, and delivering structural machine frames having particular configurations of aluminum sectional profiles, the Canadian CAD company partners with Bosch Rexroth in the United States and needs to download profiles of interest for designing the machine frames that are required by local customers. In the current business environment, manufacturers have several pieces of information relating to their products on their Websites. It is no longer necessary to order a hardcopy catalog. The starting point would be for the engineers to access the Internet Website of Bosch Rexroth for their products.

To download the cross section of a Bosch Rexroth Extrusion, do the following steps:

1. Google: "aluminum structural framing bosch rexroth download CAD"

   To speed up getting to the correct homepage, look for the link having the sentence, "Download Bosch Rexroth CAD Files," as shown here:

   Aluminum Framing—Bosch Rexroth

   www13.boschrexroth-us.com/.../Load_Category.aspx?... - United States

   Download Bosch Rexroth CAD Files · Aluminum Framing · Profiles · Connectors · Fasteners · Door Components · Ecosafe Guarding · Floor To Frame Elements...

2. Click the link to give the following URL:
   http://www13.boschrexroth-us.com/partstream/Load_Category.aspx?category=Aluminum%20Framing&menu=1,0,0.

The link accesses the homepage of Bosch Rexroth, which is shown in Figure 26.1.

1. Click Profiles.
2. Select 40 series (see Figure 26.2).

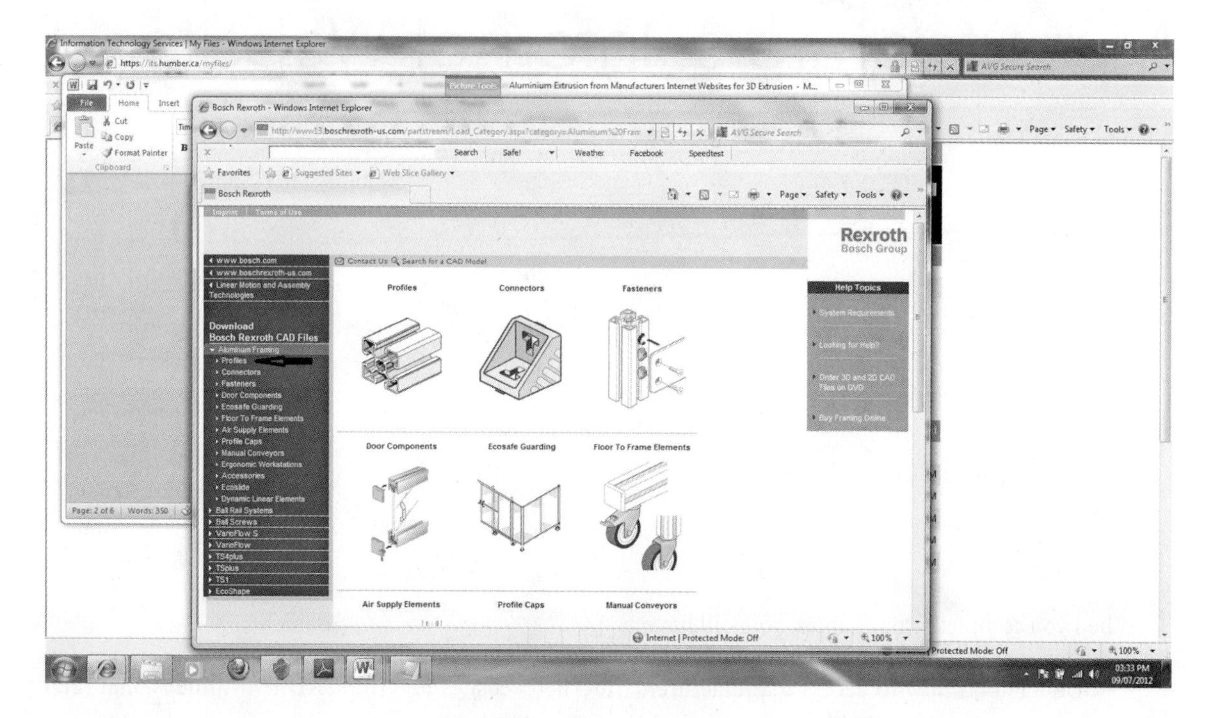

Figure 26.1

Downloadable Bosch Rexroth CAD files (profiles, connectors, etc.).

Figure 26.2

Series selection.

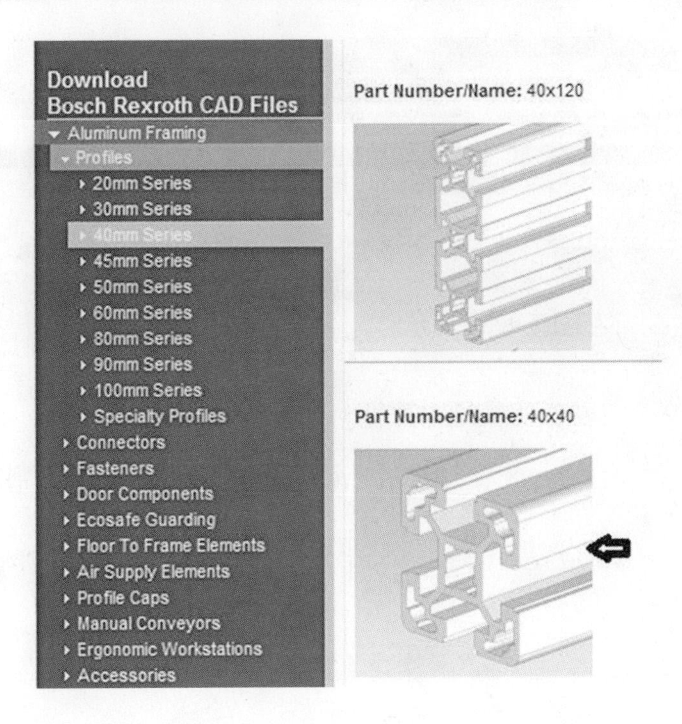

Figure 26.3

Part-number 40 × 40 selection.

3. Click Part Number/Name: 40 × 40 (see Figure 26.3).
   Clicking Part-number 40 × 40 brings up another page, which is shown in Figure 26.4.
4. Click Download (top right of Figure 26.4).
5. Select DWG file(s) (see Figure 26.5).
6. Click Download (bottom right of Figure 26.6).
   When the *Download* button is clicked, a new page shown in Figure 26.7 is displayed. This is the final step of downloading the CAD file(s). An icon in yellow, showing the Part Number, is displayed, as shown in Figure 26.7.
7. Drag and drop this link 40x40 into SolidWorks, or click it to download the file.
8. Save the downloaded file (see Figure 26.8) in a folder of interest.

Creating SolidWorks Sketch from AutoCAD 2D Section Profile

In this section, we will show how to create a SolidWorks Sketch from the AutoCAD 2D Section Profile that is imported. The following steps should be followed:

1. Open a New SolidWorks Part Document.
2. Select Front Plane (marked "1" in Figure 26.9).
3. Click Insert > DXF/DWG (marked "2" in Figure 26.9).
4. Filter DWG files (see Figure 26.10).
5. Select the document marked Rexroth-3842 993 120_200MM (see Figure 26.10).
6. Click Open to open the aluminum profile.
   The DXF/DWG Import page automatically shows up on the screen, as shown in Figure 26.11.
7. Click Next > Millimeters > Next > Finish.
   (See Figure 26.12 for choosing Units ["Inches" chosen] of imported data and preview.)
   When the Finish option is clicked, the views for the aluminum section are automatically displayed. If you do not see the view, hit the "F" button on the keyboard to display the views. Figure 26.13 shows the views.
8. Delete the Top, Side, and Isometric views.

We can confirm using the Measure tool that the width and length of the Rexroth-3842 993 120_200MM profile imported is 40 mm × 40 mm (see Figure 26.14).

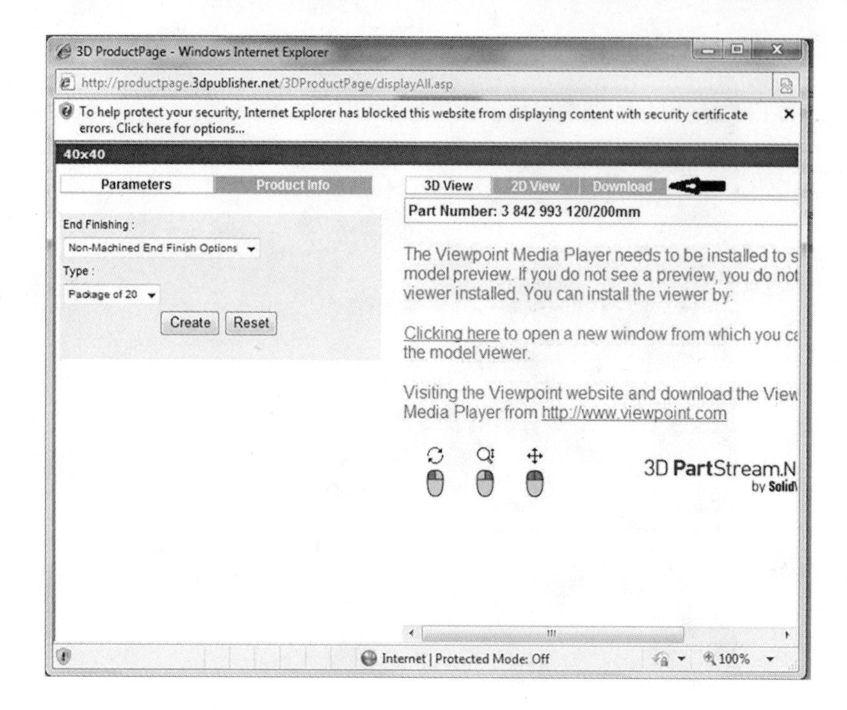

Figure 26.4

Window to create Part Number 40 × 40 aluminum profile.

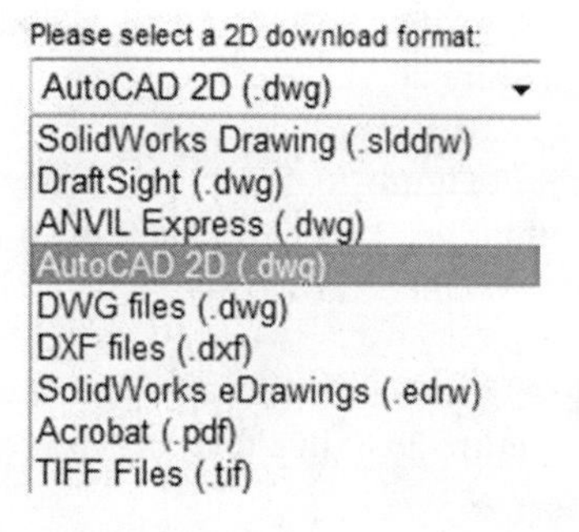

Figure 26.5

Selecting AutoCAD 2D (dwg) files from pull-down menu.

Creating SolidWorks Structural Elements Using AutoCAD 2D Section Profile

Begin your design by extruding Sketch to create features. Our features are extrusions of the profile that is imported from the manufacturer's Website (in this case, having a dimension of 40 mm × 40 mm).

You can now build your structural frame by creating different structural elements of different lengths that are obtained from the initial profile and saving them using names that are associated with their lengths for the ease of future usage.

We will now create four lengths of our imported aluminum profile having the following dimensions:

1. 1200 mm
2. 900 mm
3. 800 mm
4. 300 mm with 45° cuts at each end

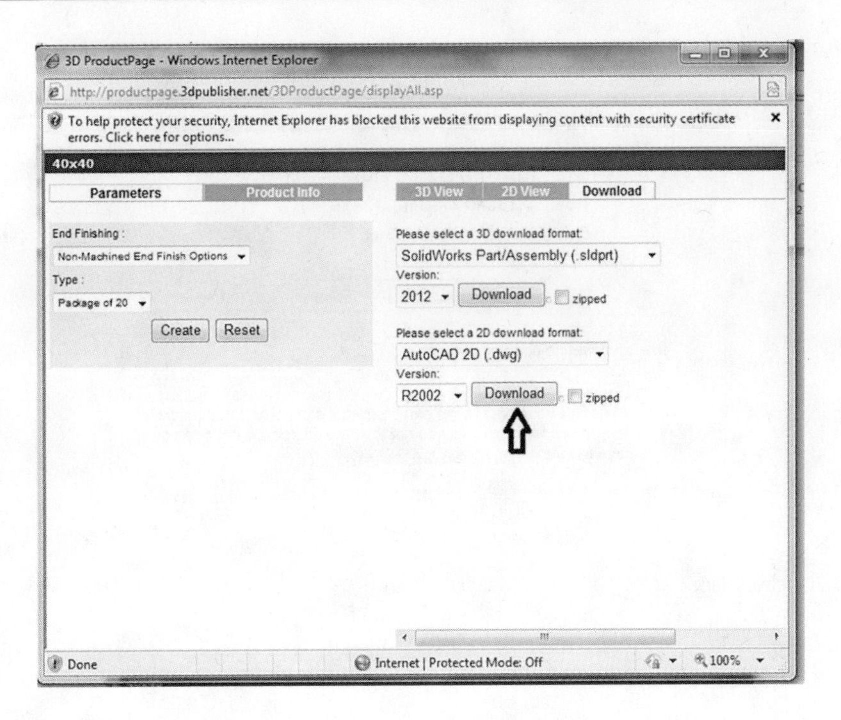

Figure 26.6

Downloading AutoCAD 2D (dwg) file.

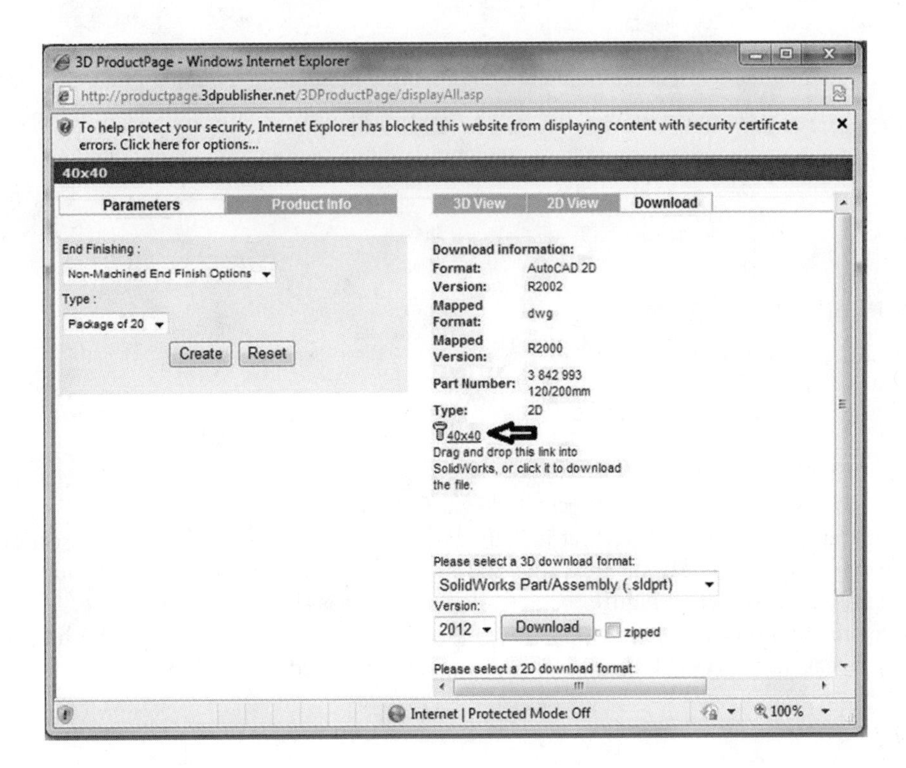

Figure 26.7

Final step in downloading AutoCAD 2D (dwg) file.

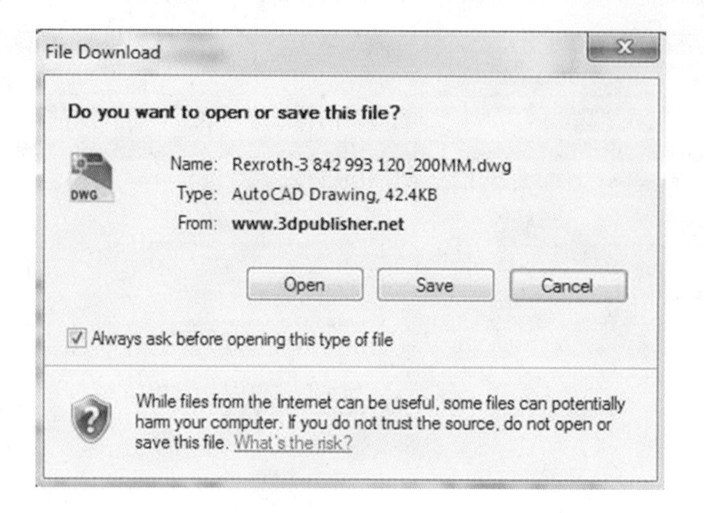

Figure 26.8

Save AutoCAD 2D (dwg) file.

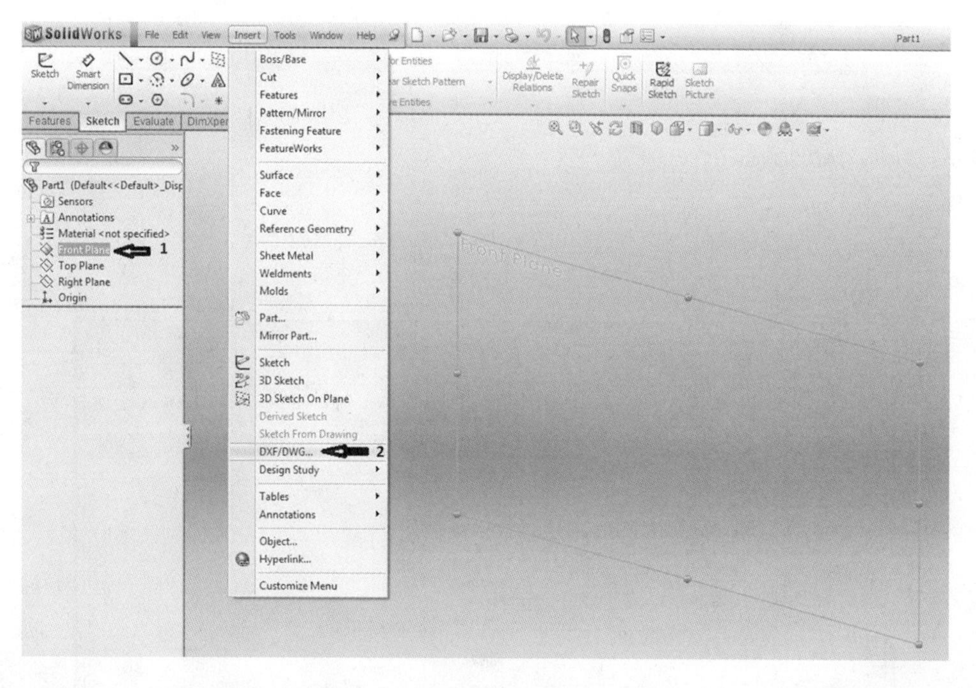

Figure 26.9

Importing DWG file (AutoCAD) to SolidWorks.

Let us create an aluminum feature that is 800-mm long from the 40 mm × 40 mm section. The step involved is to extrude the profile of the Rexroth-3842 993 120_200MM in Figure 26.14 through a distance of 800 mm. The preview is shown in Figure 26.15, while the feature is shown in Figure 26.16.

Save this feature as aluminum_800_40×40 so that we can easily associate the name with the part.

We simply repeat the process for the other features that are named:

aluminum_300_40×40
aluminum_900_40×40
aluminum_1200_40×40

An easy way of creating subsequent features from the first one (aluminum_800_40×40) is to Save As and then change the name to aluminum_300_40×40, for example. Then, edit the feature and change the length to 300. The procedure can then be repeated for lengths 900 and 1200, respectively. This approach is very

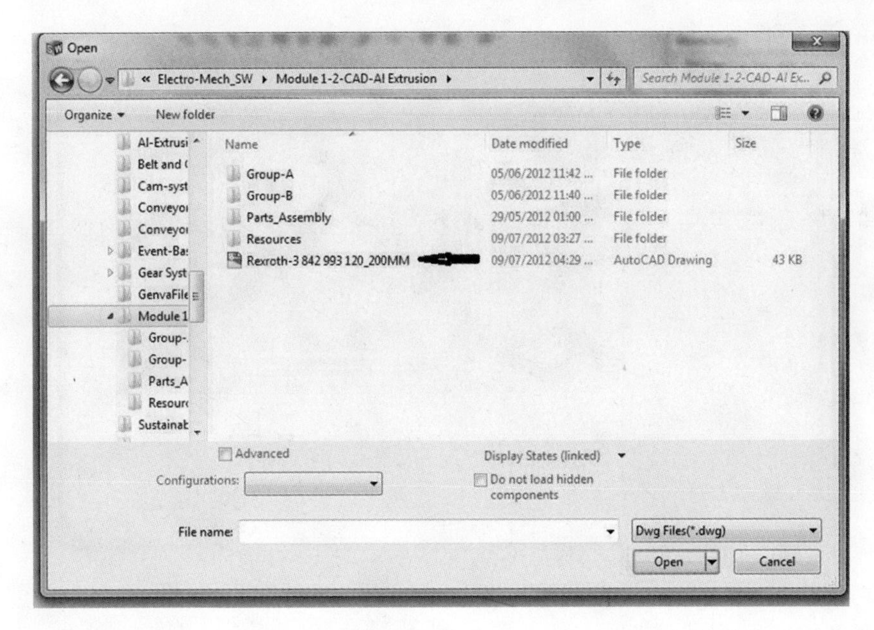

Figure 26.10

Locating the Rexroth file that is already saved.

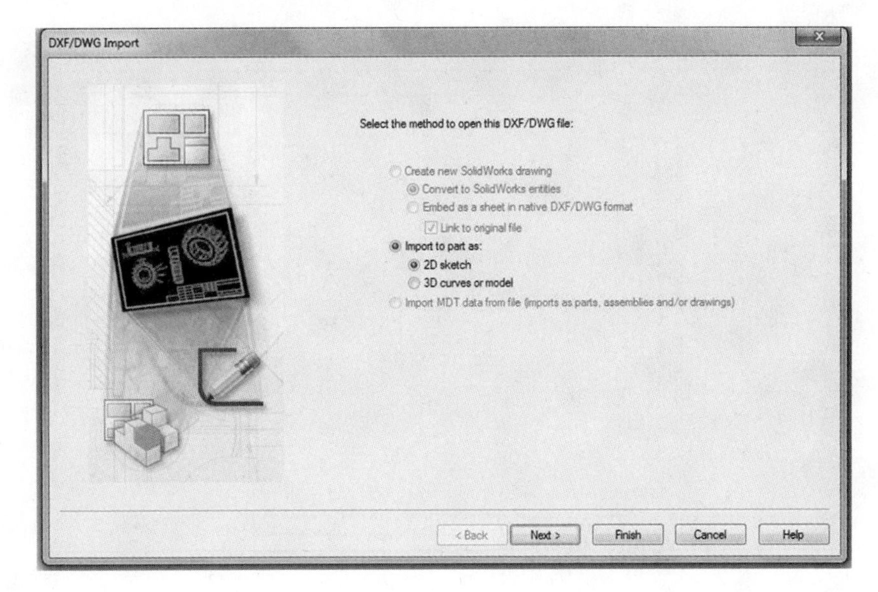

Figure 26.11

DXF/DWG Import tool.

efficient since the cross-sectional dimensions are the same for all the features. At the end of the exercise, the 300-mm long feature is given cut 45° at each end.

Creating SolidWorks Structural Machine Frame Using AutoCAD 2D Section Profile

The task is to use the four structural elements of the following dimensions that we have modeled using the imported Rexroth-3842 993 120_200MM to build an assembly that is shown in Figure 26.17:

1. 1200 mm
2. 900 mm
3. 800 mm
4. 300 mm with 45° cuts at each end

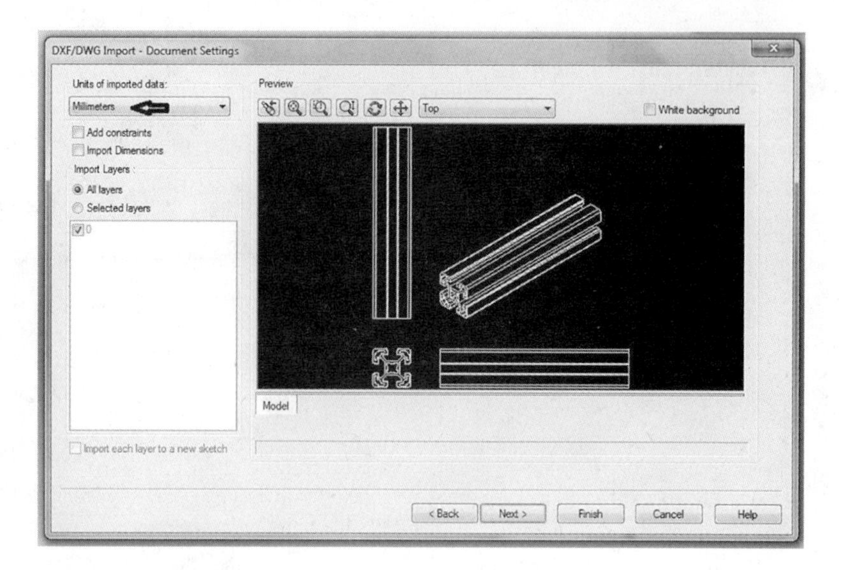

Figure 26.12

Units of imported data and preview.

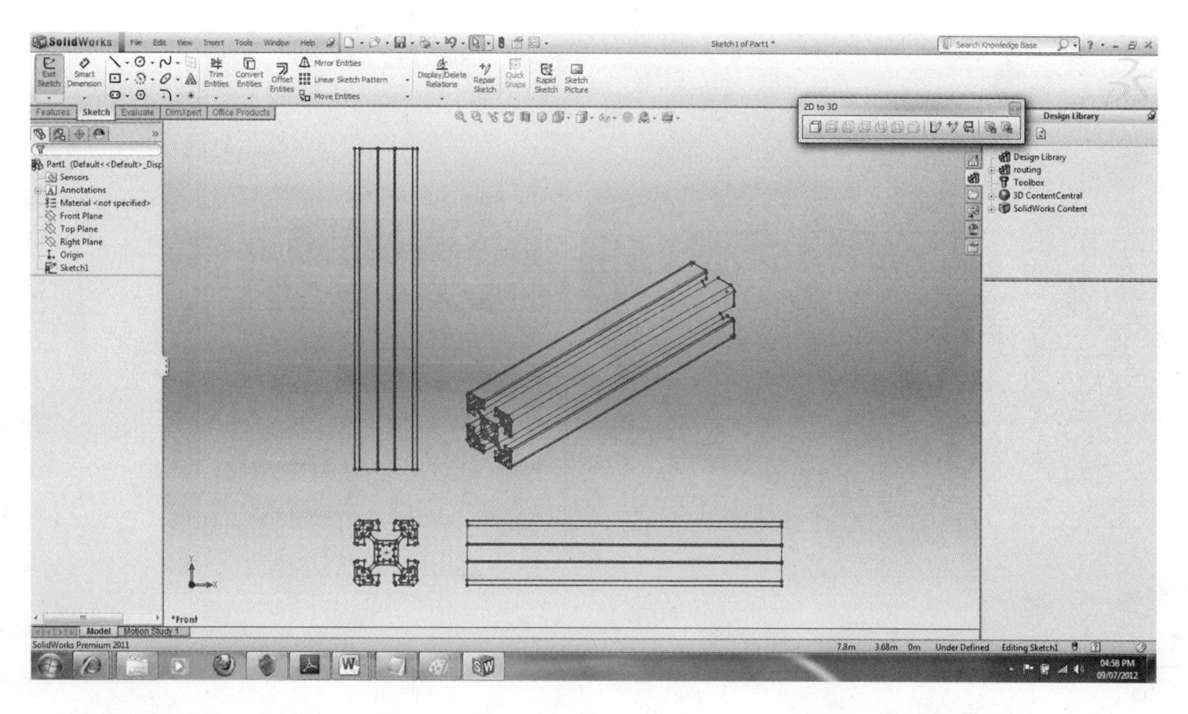

Figure 26.13

Views for the aluminum section.

The remaining task reduces to starting a *New SolidWorks Assembly* document and creating an assembly of parts. A structural frame of overall length = 900 mm, width = 800 mm, and height = 1200 mm is obtained simply by inserting each of these in the assembly, and then we use the Mirror tool to mirror each of them about some convenient Planes. In order to be able to mirror features more efficiently, the way we create the features is important. For example, in Figure 26.15, the Mid Plane option is used, which facilitates mirroring. Figure 26.17 is a structure of 900-mm long, 800-mm wide, 1200-mm high, having three braces at each corner for the purpose of reinforcement.

Different configurations can easily be obtained using the procedure that is presented in this chapter.

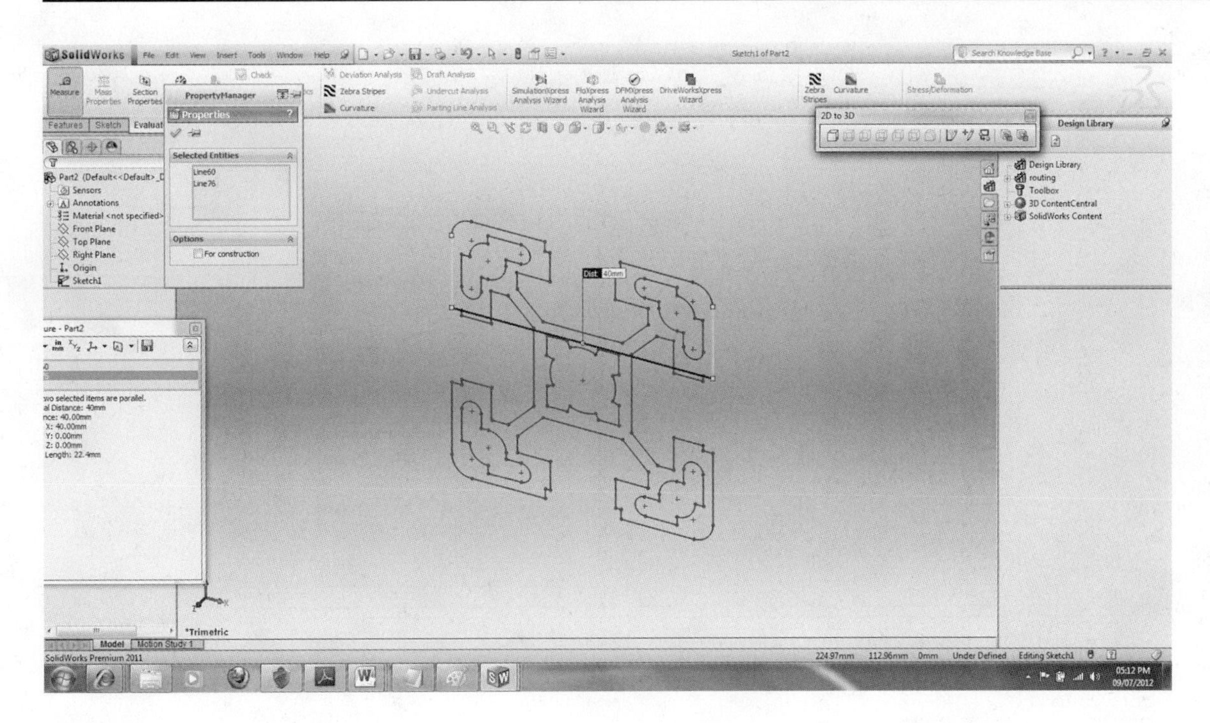

Figure 26.14

Profile of the Rexroth-3842 993 120_200MM imported.

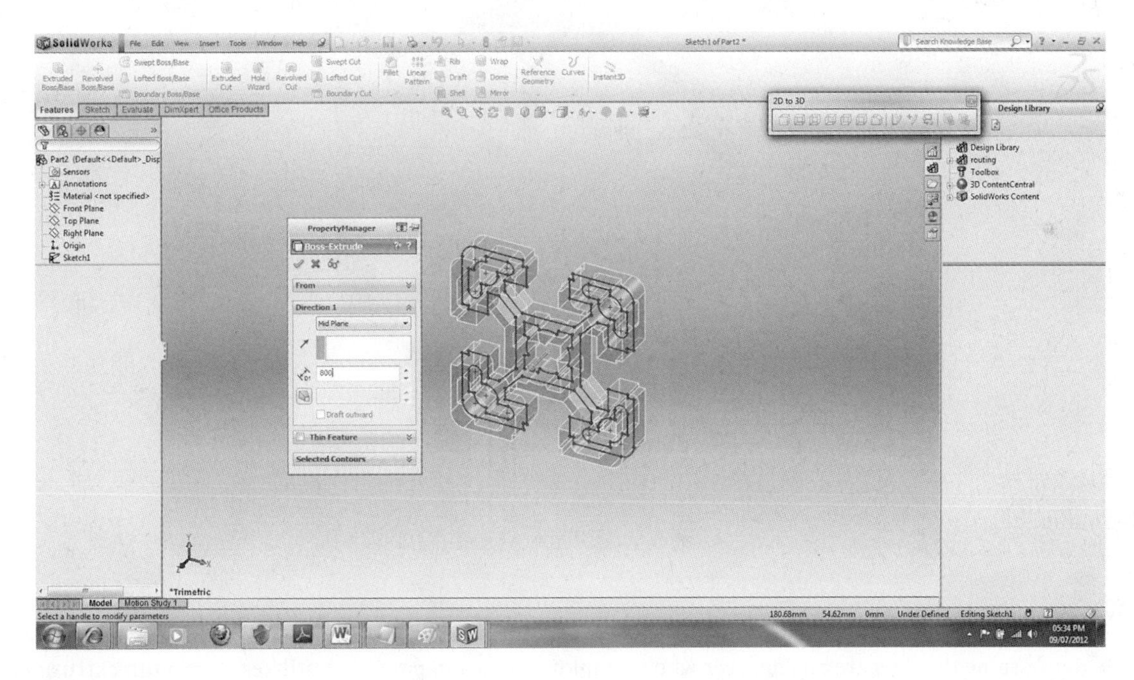

Figure 26.15

3D preview of the aluminum extrusion.

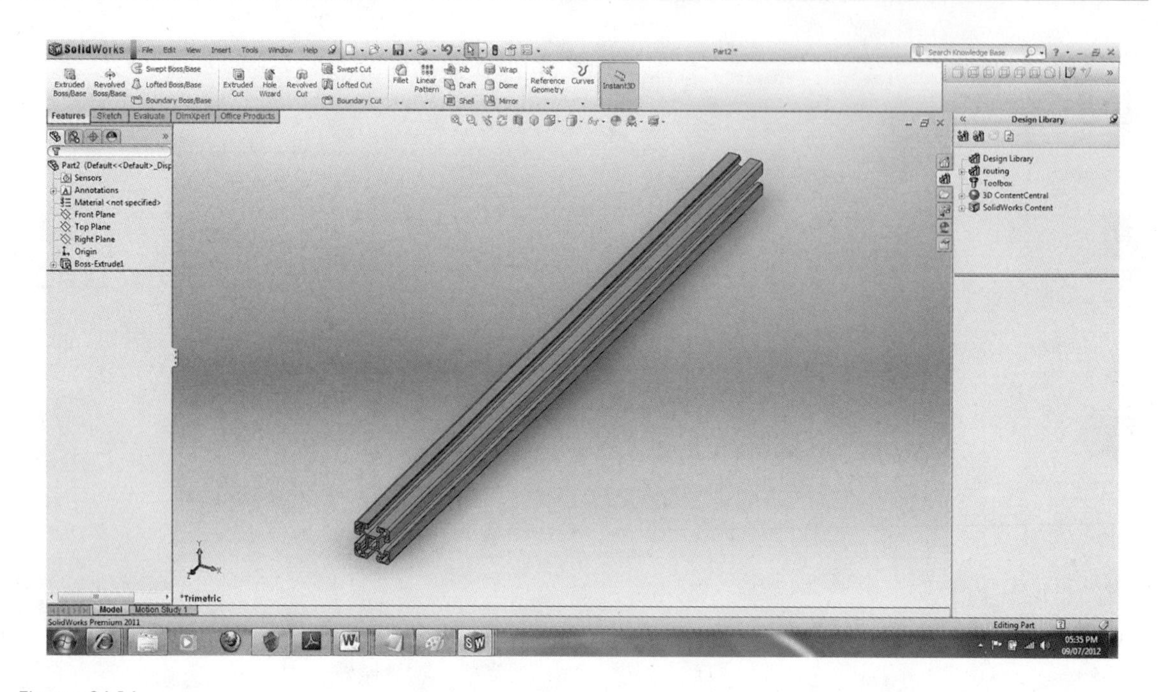

Figure 26.16

A 3D aluminum extrusion realized in SolidWorks environment.

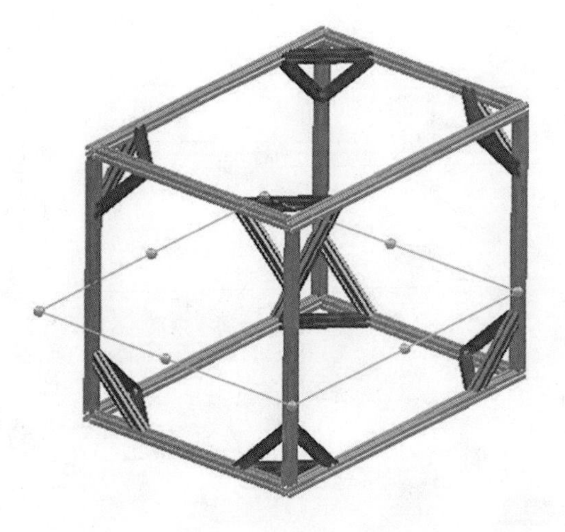

Figure 26.17

Frame structure.

Exercises

1. Assume that you are an engineer who is employed in a company that utilizes aluminum extrusions that are produced by another company, which has AutoCAD drawings of different cross-sectional profiles on their Website. Import a profile (20 mm × 20 mm cross section) from the Bosch Rexroth Website and use it to model a frame structure (see Figure P26.1), using SolidWorks CAD software for your company; maintain the orientation shown. There are three configurations to create: (1) 700 mm, (2) 800 mm (for the base), and (3) 1250 mm (the other dimension). Dimension your 3D structure.

2. Assume that you are an engineer who is employed in a company that utilizes aluminum extrusions that are produced by another company, which has AutoCAD drawings of different cross-sectional profiles on their Website. Import a profile (30 mm × 30 mm cross section) from the Bosch Rexroth

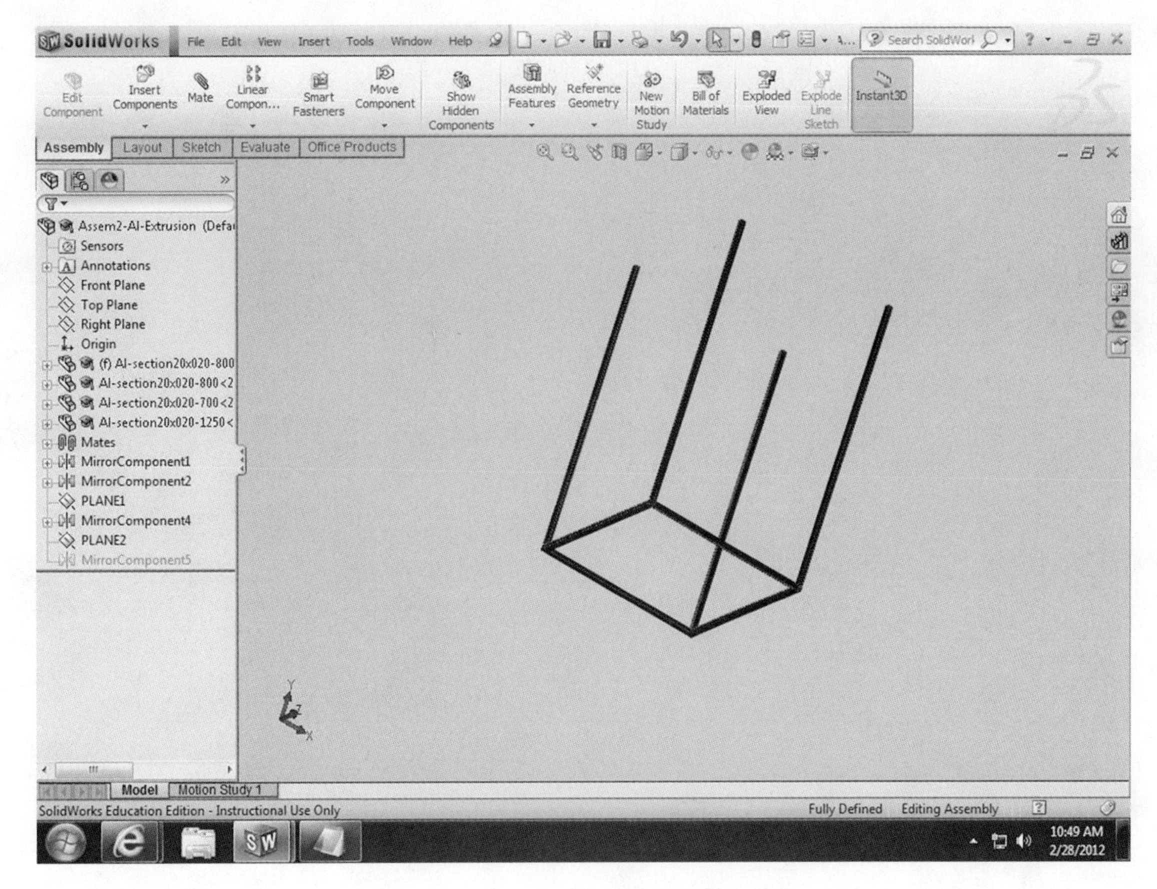

Figure P26.1

Inverted stool.

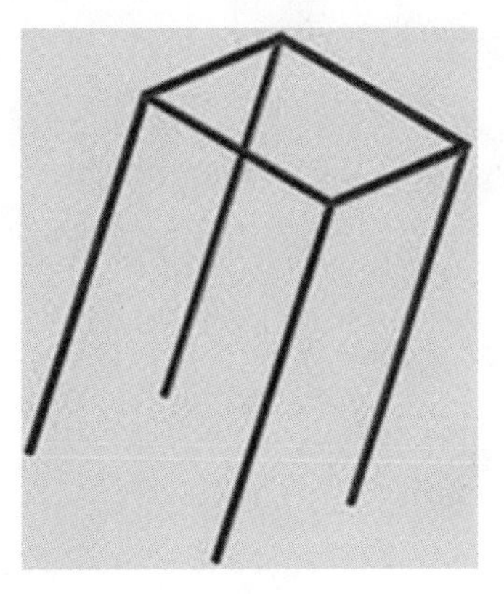

Figure P26.2

Upright stool.

Website and use it to design a frame structure (see Figure P26.2) using SolidWorks CAD software for your company; maintain the orientation that is shown. There are three configurations to create: (1) 500 mm, (2) 600 mm (for the base), and (3) 1000 mm (the other dimension). Dimension your 3D structure.

Bibliography

http://www13.boschrexroth-us.com/.

27

Geneva Wheel Mechanism

Objectives:

When you complete this chapter, you will have

- Understood how to determine the basic design parameters of the Geneva wheel mechanism
- Learned how to use the SolidWorks to model and simulate the Geneva wheel mechanism

Historical Background

The name is derived from the earliest application of the device in *mechanical watches*, especially in Switzerland and Geneva. Geneva mechanism was originally invented by a watchmaker. The watchmaker only put a limited number of slots in one of the rotating disks so that the system could only go through so many rotations. This prevented the spring on the watch from being wound too tight, thus giving the mechanism its other name, the Geneva Stop Clock mechanism in eighteenth-century pocket watch application. Other applications of the Geneva drive include the pen change mechanism in *plotters*, automated sampling devices, and *indexing* tables in assembly lines.

Introduction

Geneva wheel mechanism translates a continuous rotation into an intermittent rotary motion, using an intermittent gear where the drive wheel has a pin that reaches into a slot of the driven wheel and thereby advances it by one step and having a raised circular blocking disc that locks the driven wheel in position between steps.

The Geneva wheel mechanism is a timing device that is used in many counting instruments and in other applications where an intermittent rotary motion is required. Essentially, the Geneva mechanism consists of a rotating disk with a pin and another rotating disk with slots (usually four) into which the pin slides. Once the number of slots and the crank diameter are known, the layout can be constructed with the basic knowledge of geometry.

There are three types of Geneva wheels:

1. External, which is the most popular, and which is covered in this chapter
2. Internal, which is also very common
3. Spherical, which is extremely rare

The Geneva wheel mechanism discussed in this chapter has many applications for the product designers, tool engineers, and others who are involved in the design and manufacture of machinery, tooling, and mechanical devices and assemblies that are used in the industrial context.

Principles of Operation of the Geneva Drive

The Geneva mechanism is a simple mechanism that does not require any complicated design parts. However, the mechanism requires correct dimension and close tolerance for the correct engagements and disengagements. The parts of a Geneva mechanism are (a) crank and (b) Geneva wheel.

- *Crank*

  The crank is the driving member. The main parts of the crank are a pin (roller) and a circular segment. The roller, which rolls or slides in the slots and the circular segment guide, effectively locks the wheel against rotation when the roller is not in engagement and positions the wheel for correct engagement of the roller with the next slot.
- *Geneva wheel*

  The Geneva wheel is a driven member, which contains a number of radial cuts called slots. Slots are the radial path for rolling the rollers. The number of slots depend on the type of drive. The Slot surface should be perfectly smooth. If the slot surface is not smooth, this may cause a wear on the slot or roller. This will change the timing between the driver and the driven wheel.

  In the most common arrangement, the driven wheel has four slots and thus advances for each rotation of the drive wheel by one step of 90°. If the driven wheel has n slots, it advances by $360°/n$ per full rotation of the driver.

  The Geneva mechanism can theoretically have from 3 to any number of slots. In practice, from 4 to 12 is enough to cover most requirements. From a practical standpoint, the follower can have a minimum of 3 slots and a maximum of 18. Most followers are made with four, five, six, or eight slots, which correspond to index lengths of 90, 72, 60, and 45 degrees, respectively; because the mechanism needs to be well lubricated, it is often enclosed in an oil capsule.

Advantages and Disadvantages

Geneva wheels may be the simplest and least expensive of all intermittent motion mechanisms. As mentioned before, they come in a wide variety of sizes, ranging from those that are used in instruments, to those that are used in machine tools to index spindle carriers weighing several tons. They have good motion-curve characteristics compared to ratchets but exhibit more *jerk*, or instantaneous change in acceleration than do better cam systems. (The Geneva, you will remember, is a special type of cam system.)

The Geneva maintains good control of its load at all times, since it is provided with locking ring surfaces, as shown in Figure 27.1, to hold the output during dwell periods. In addition, if properly sized to the load, the Geneva generally exhibits a very long life (about 20 years).

The Geneva is not a versatile mechanism. It can be used to produce no less than three and, usually, no more than 18 dwells per revolution of the output shaft. Furthermore, once the number of dwells has been selected, the designer is well locked into a given set of motion curves. The ratio of dwell period is also established once the number of dwells per revolution has been selected. Also, all Geneva acceleration curves start and end with finite acceleration and deceleration. This means that they produce jerk.

The geometry and kinematics of the external Geneva drive mechanism summarized in the subsequent sections are from Walsh (2000) where they are described in detail.

Geometry of the External Geneva Mechanism

The crank and wheel schematic is shown in Figure 27.1, while the geometry is shown in Figure 27.2.

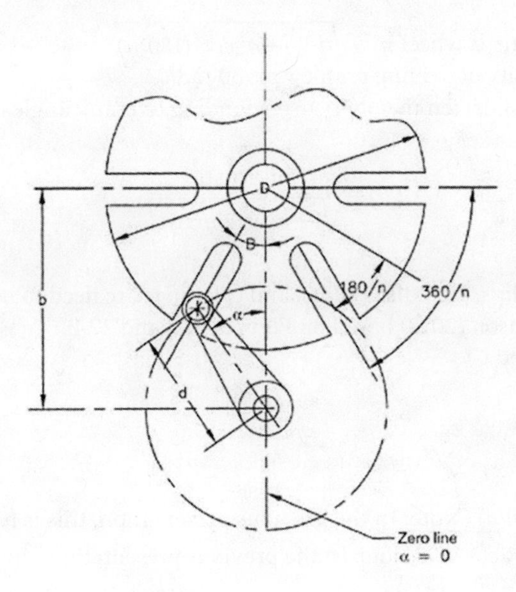

Figure 27.1

External Geneva mechanism.

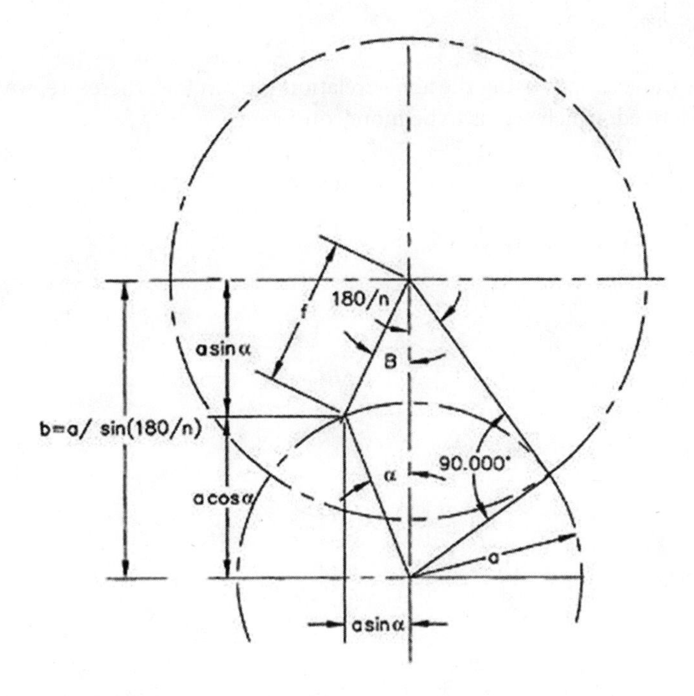

Figure 27.2

External Geneva geometry.

Kinematics of the External Geneva Drive

Based on the external Geneva geometry of Figure 27.2, the kinematics can be better understood.

Assumed or given: a, n, d, and p.

Let $m = 1/\sin(180/n)$

a = crank radius of driving member (This is normally on the inner side of the roller.)

n = number of slots in drive

d_r = roller diameter

p = constant velocity of driving crank

b = center distance = $a \times m$

D = diameter of driven Geneva wheel = $2\sqrt{d_r^2/4+a^2\cot^2(180/n)}$
ω = constant angular velocity of driving crank = $p\pi/30$ rad/s
β = angular displacement of driven member corresponding to crank angle α

$$\beta = \frac{m-\cos\alpha}{\sqrt{1+m^2-2m\cos\alpha}}$$

The slot width, stop arc radius, stop disc radius, and clearance arc need to be defined. To do this, we present other formulas from Johnson (2012) based on Figures 27.3 and 27.4:
where

- a = drive crank radius
- n = driven slot quantity
- p = drive pin diameter
- t = allowed clearance
- c = *center dist.* = $a/\sin(180/n)$ (Note: In the previous presentation, this is b.)
- b = *Geneva wheel radius* = $\sqrt{c^2-a^2}$ (Note: In the previous presentation, this is $D/2$.)
- s = *slot center length* = $(a + b) - c$
- w = *slot width* = $p + t$
- y = *stop arc radius* = $a - (p \times 1.5)$
- z = *stop disc radius* = $y - t$
- v = *clearance arc* = bz/a

The value of the clearance arc, v, using the above relation is often low; therefore, we present another formula that the author derived, which seems to be more consistent:

$$v = \textit{clearance arc} = (r_1 + r_2)\cos(180/n) - \varepsilon; \quad \varepsilon = 4.6\%v$$

where the radii are b and y, respectively.

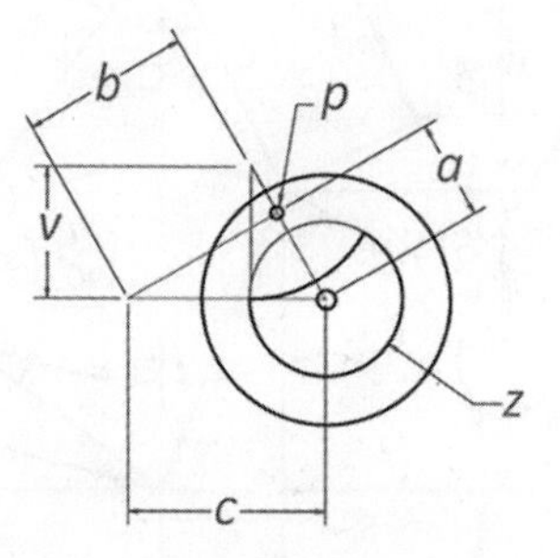

Figure 27.3

Geneva Drive.

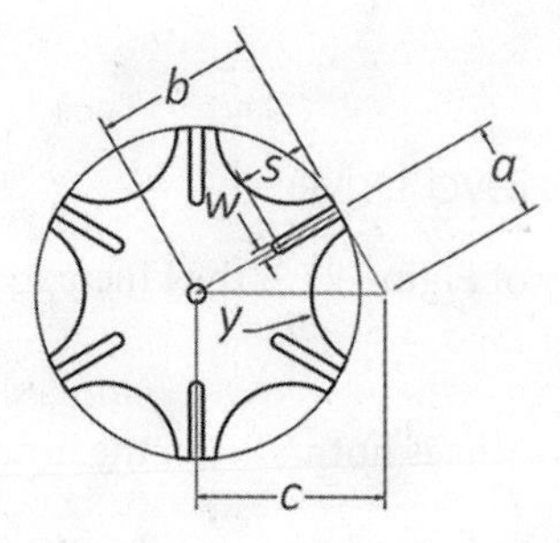

Figure 27.4

Geneva Driven.

SolidWorks Modeling and Simulation of Geneva Wheel Mechanism

This section puts together the geometry and kinematics that are covered so far in modeling the Geneva wheel mechanism that is similar to methods that are presented in Morld Tech Gossips (2011) and by Gupta (2011). For our configuration, the author used the dimensions in Enzo to model the Geneva wheel mechanism and simulated it. The geometric definitions are first discussed to ensure that user understand how the geometric relations are required in defining the mechanism for it to properly function.

For our illustrative example, there are six slots. Therefore,

Let $m = 1/\sin(180/6) = 1/\sin(30)$

$a =$ 1 (diameter of 2/2) (on driver side)

$n =$ 6 (on driven side)

$d_r =$ 0.2 (on driver side)

$\mathrm{b} = 1/\sin(30) = 2$

$D = 2\sqrt{d_r^2/4 + a^2\cot^2(180/n)} = 2\sqrt{(0.1)^2 + (1)^2\cot^2(30)} = 3.47 \approx 3.50$ (on driven side)

$p = d_r/2 = 0.2/2 = 0.1$ (on driver side)

$a = (D/2)\tan 30° = (1.75)\tan 30° = 1.01$ (This is a refined value due to calculated D.)

$y = 1.01 - (0.1 \times 1.5) = 0.86 \approx 0.90$ (on driver side)

$v = (1.75 + 0.9)\cos(30)[1 - 0.046] = 2.18$ (The value of 2.15 is used.) (on driver/driven sides)

Using these design parameters, the driver (wheel) and the driven wheel (Geneva gear) are modeled using the detailed parts descriptions that are given. This is a design phase, and consequently, the estimations should be carefully done to ensure that the assembled parts function correctly.

The step-by-step procedure for using SolidWorks to model the Geneva wheel mechanism is now given. We assume that each part has now been modeled as discussed. The parts are assembled, and then at the motion simulation stage, two contact properties must be defined for the Geneva wheel mechanism to properly function. The definition of the contacts is the key to the successful implementation of the motion simulation. If the pin is not separated from the wheel, there could be a problem to get the assembly to function correctly. Users must define the following contacts:

- Between the Geneva gear and the Wheel
- Between the Geneva gear and the Pin

1. Open the "Assem1Enz" assembly from the downloaded files (see Figure 27.5).
2. Switch to Motion study and set the model orientation as required (see Figure 27.6).
3. Change the Motion study type to "Basic Animation" (see Figure 27.7).
4. Click on "Contact" (see Figure 27.8).
5. Now, in the Contact property manager, click on the pin to keep it visible as we need to use it twice (see Figure 27.9).
6. Now under components selection box, select "GenevaGear_Enz-1" and "GenevaPin_Enz-1" parts (see Figure 27.10).
7. Click OK to apply the contact.
8. With Contact property manager visible, select "GenevaGear_Enz-1" and "GenevaWheel_Enz-2" parts and click OK to apply the contact. You can now close the Contact property manager (see Figure 27.11).
9. Click on "Motor" (see Figure 27.12).
10. Set the motor type to "Rotary motor." Select the cylindrical face or circular edge of "Geneva Wheel" part to define the direction of motion. Set the motion function to "Constant Speed" and RPM to 30 (see Figure 27.13).
11. Click OK and apply the motor.
12. Now click on "Motion Study Properties" (see Figure 27.14).
13. In the Motion Study Properties property manager, under Basic Motion, set the Frames per second to 30 (the larger the number, the smoother the motion) and set the Geometry Accuracy and 3D Contact Resolution settings to High side (move the sliders to right). This will make collision simulation more accurate and smoother motion but requires more time to compute (see Figure 27.15).
14. Click OK to set the properties.
15. Finally click on "Calculate" (see Figure 27.16).
16. And now is the show time. Hit Play to enjoy the show (see Figure 27.17).

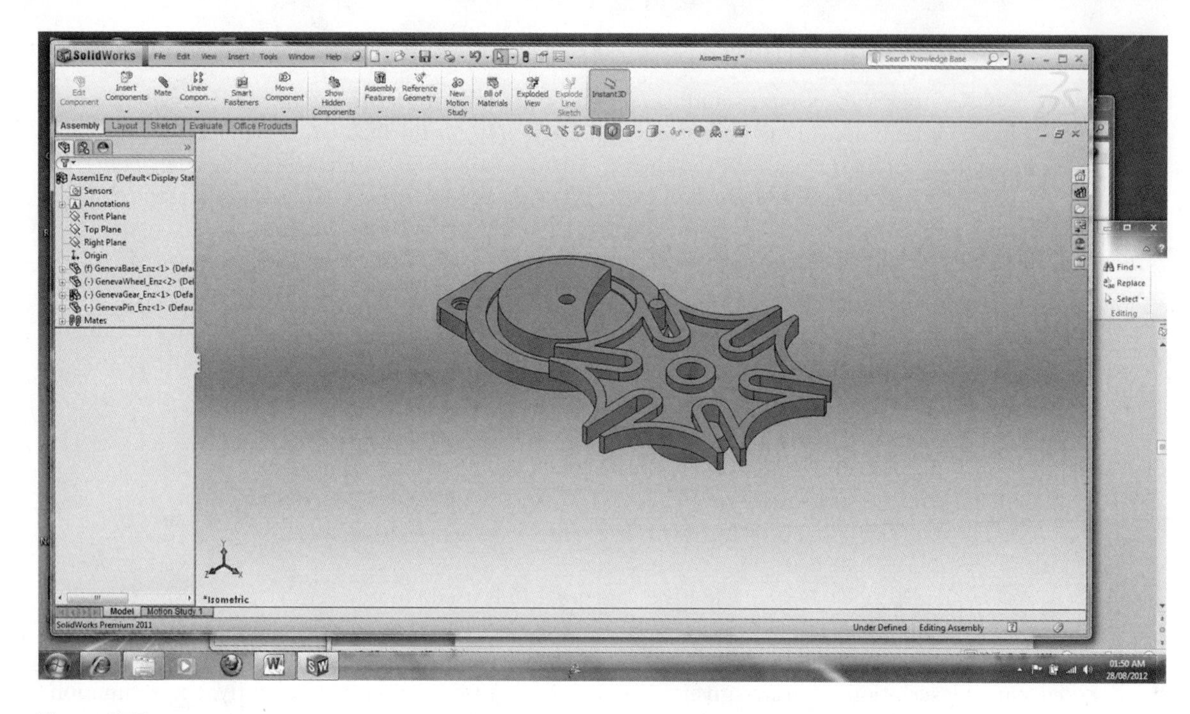

Figure 27.5

Assembly of Geneva mechanism parts.

Figure 27.6

Switch to Motion study.

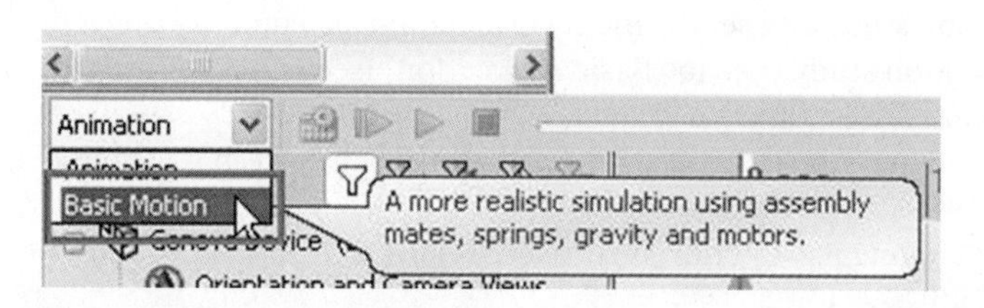

Figure 27.7

Basic Motion mode selected.

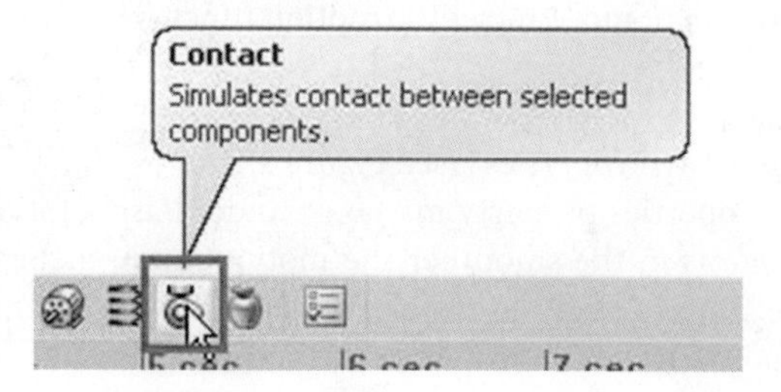

Figure 27.8

Contact tool is accessed.

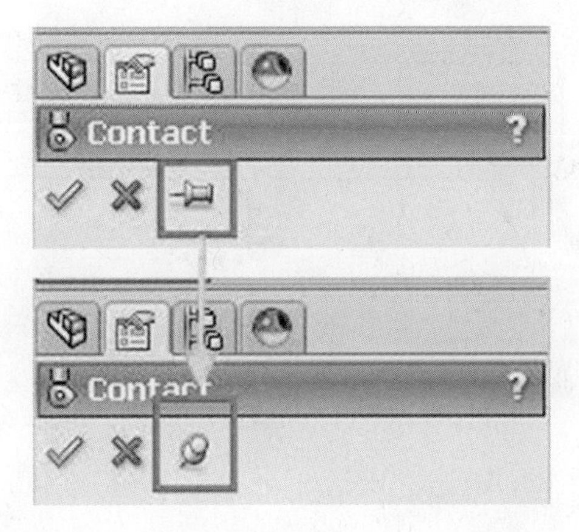

Figure 27.9

Pin the Contact PropertyManager.

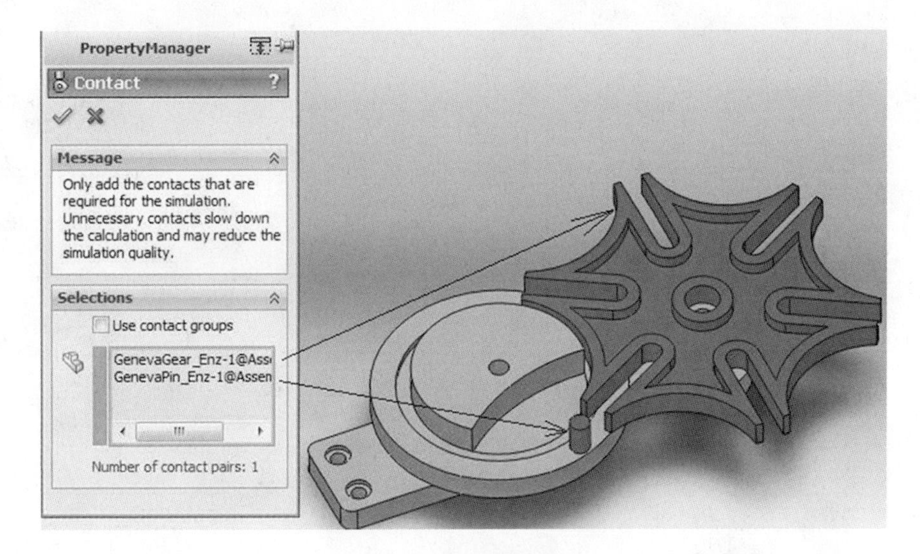

Figure 27.10

First set of Contacts is defined.

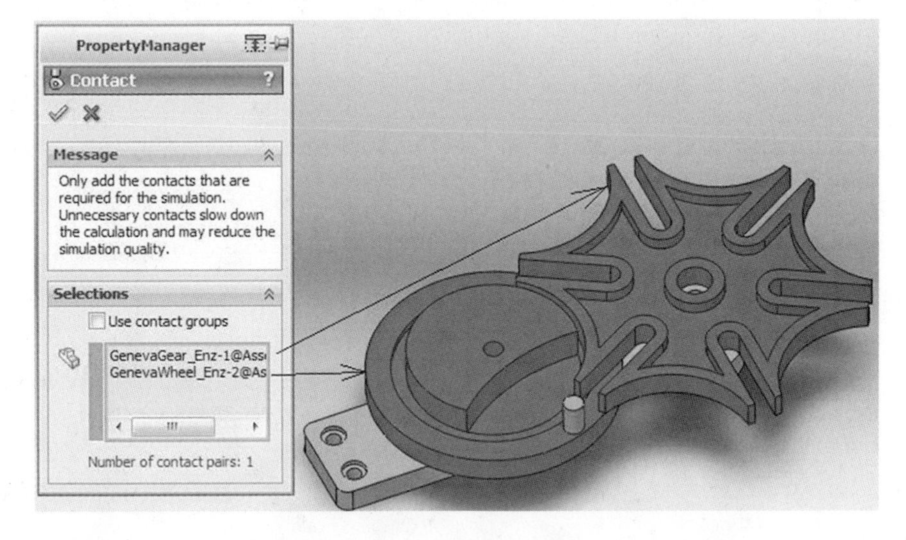

Figure 27.11

Second set of Contacts are defined.

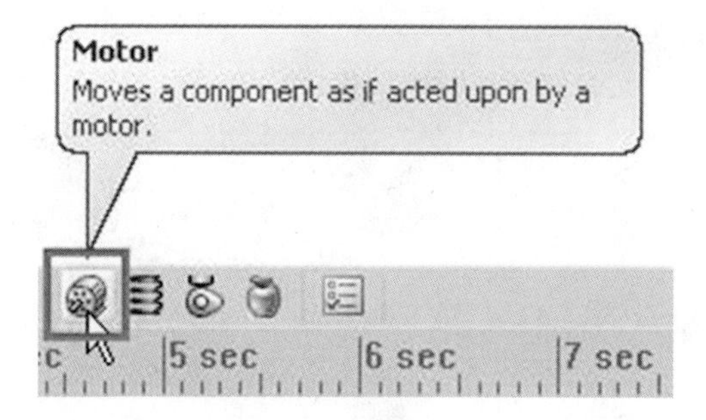

Figure 27.12

Motor definition tool accessed.

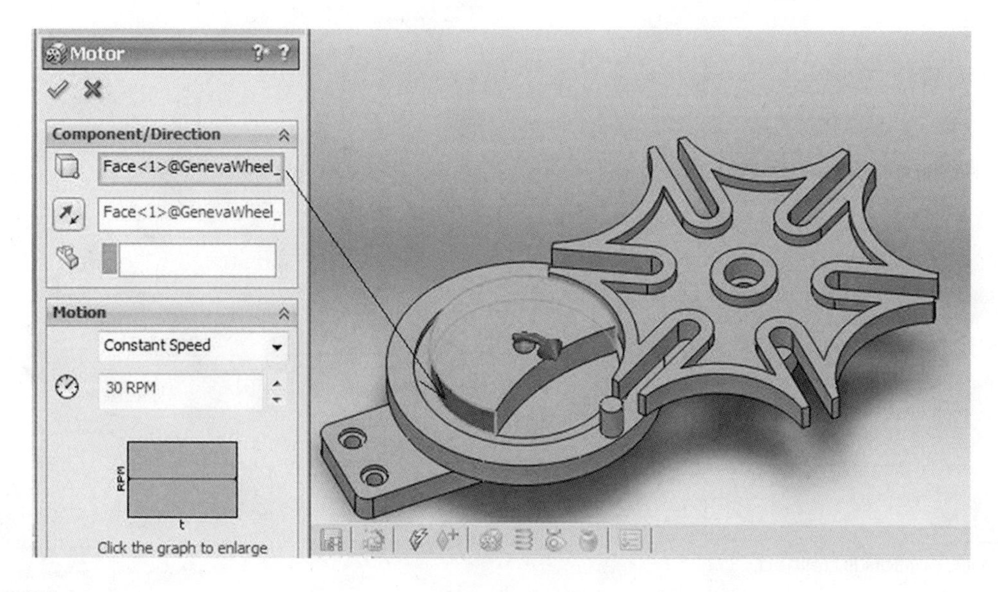

Figure 27.13

Engaging face on the wheel is selected.

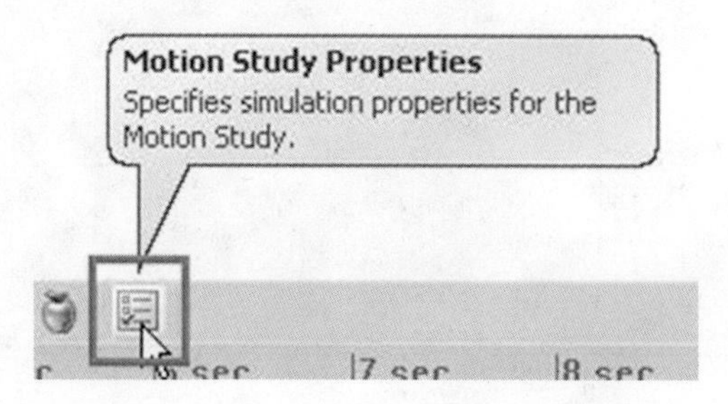

Figure 27.14

Accessing motion study properties.

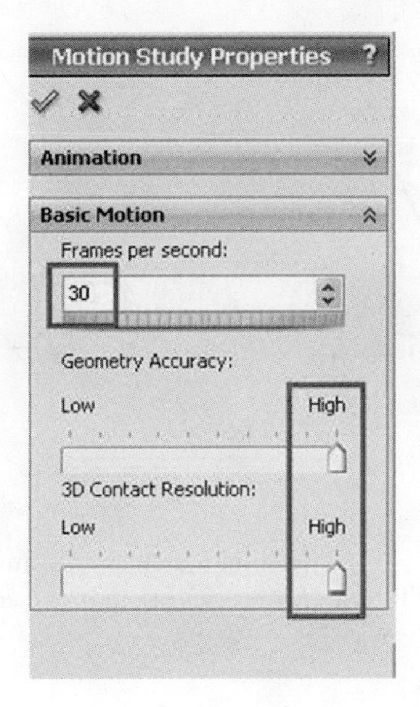

Figure 27.15

Adjusting the Motion Study Properties.

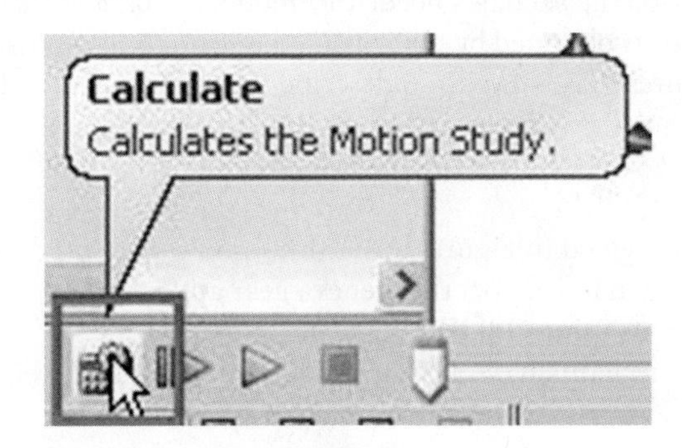

Figure 27.16

Calculate to start simulation.

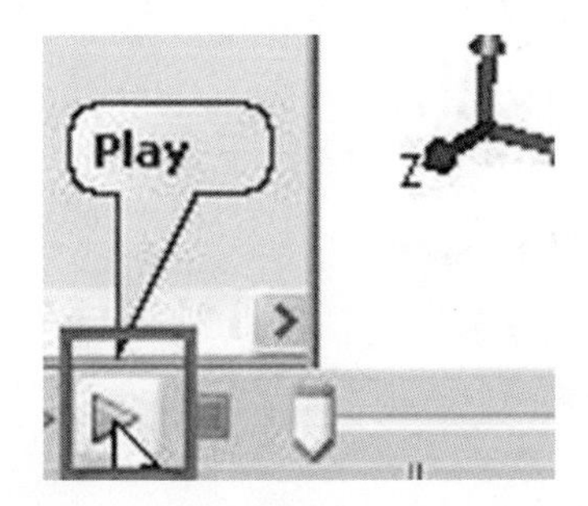

Figure 27.17

Play.

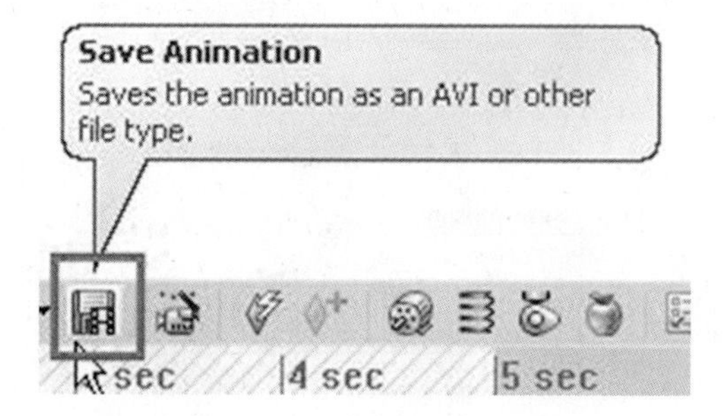

Figure 27.18

Saving the simulation.

You can change and experiments with the settings to get a better animation. Click on Save if you want to export the animation as Audio Video Interleave or series of pictures (see Figure 27.18). You can change other settings in the Save window.

Summary

This chapter has presented the design issues that are related to Geneva wheel mechanism, as well as the modeling and simulation procedure for the mechanism.

Project

Figures P27.1 and P27.2 describe the sketches needed to model a sector of the Geneva gear (Figure P27.3). The full model of the Geneva gear obtained by replicating the sector 5 times (using the circular pattern tool) is shown in Figure P27.4. Figures P27.5 and P27.6 describe the sketches needed to model the Geneva wheel (Figure P27.7). Two through-holes are shown in Figure P27.7; they are 21.796 apart. Figure P27.8 describes the sketch needed to model the pin (Figure P27.9).

1. Use the information provided in Figures P27.1 through P27.8 to model the components
2. Model a base on your own to support the Geneva gear and wheel
3. Assemble all components using SolidWorks
4. Simulate the Geneva mechanism (your final model should look like Figure P27.10)

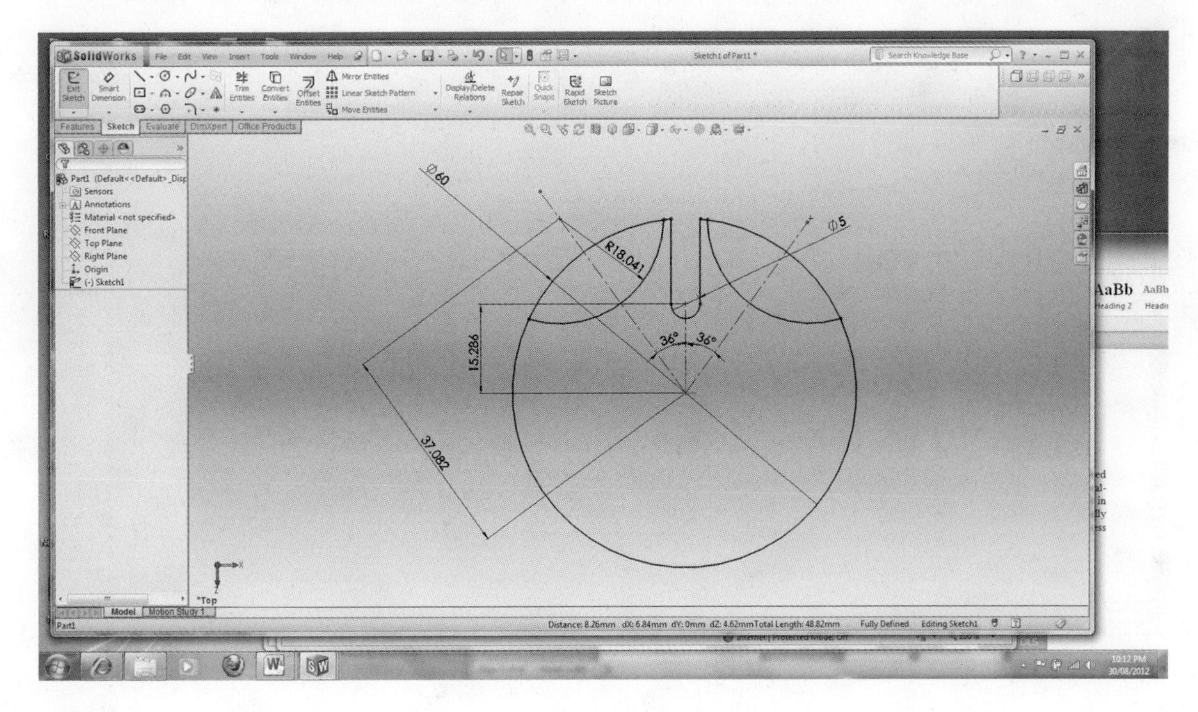

Figure P27.1

Sketch1.

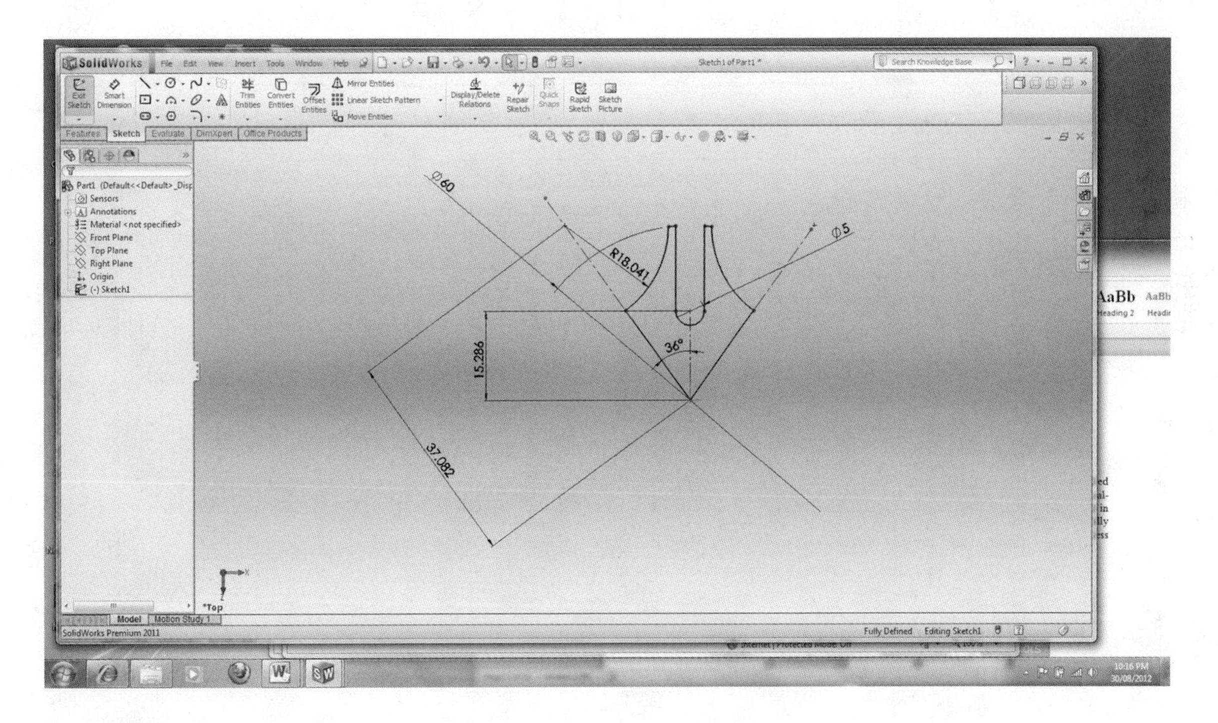

Figure P27.2

Sketch1 trimmed.

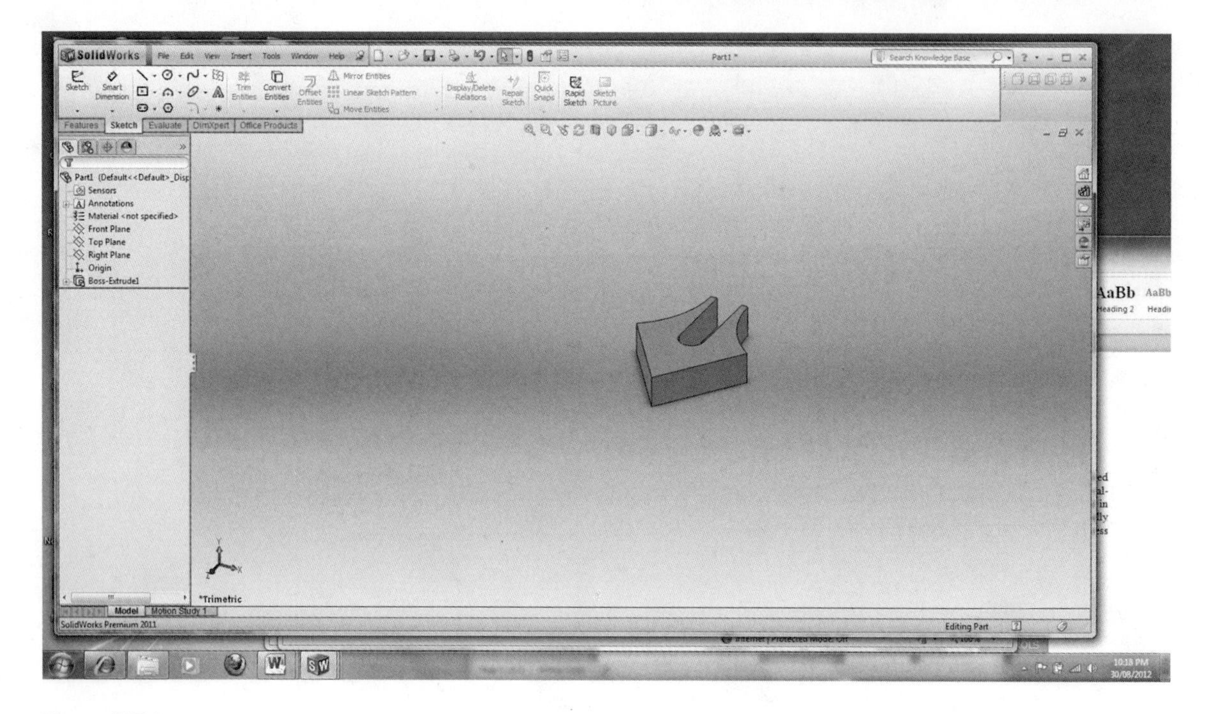

Figure P27.3

Sketch1 extruded.

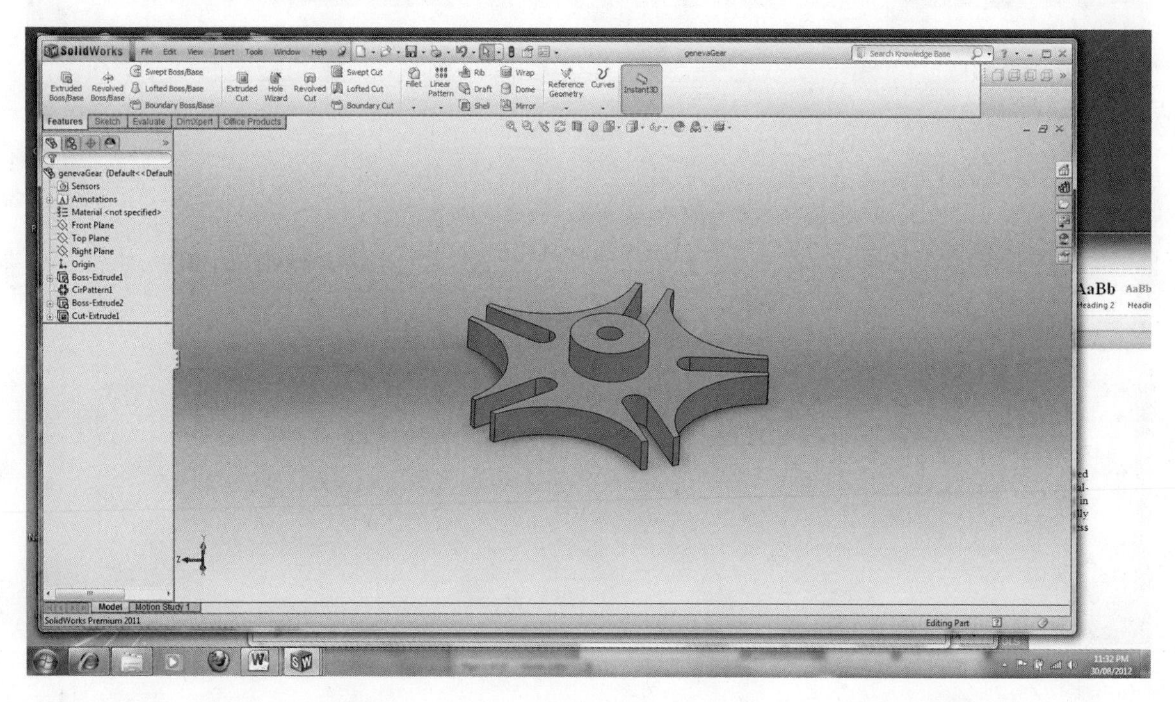

Figure P27.4

Extruded model replication using circular array tool.

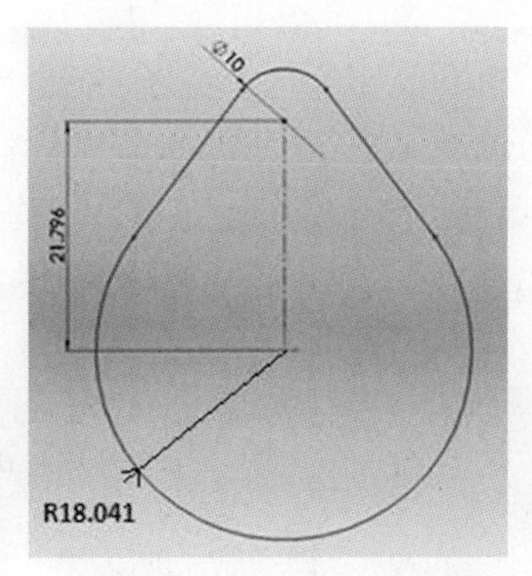

Figure P27.5

Sketch2.

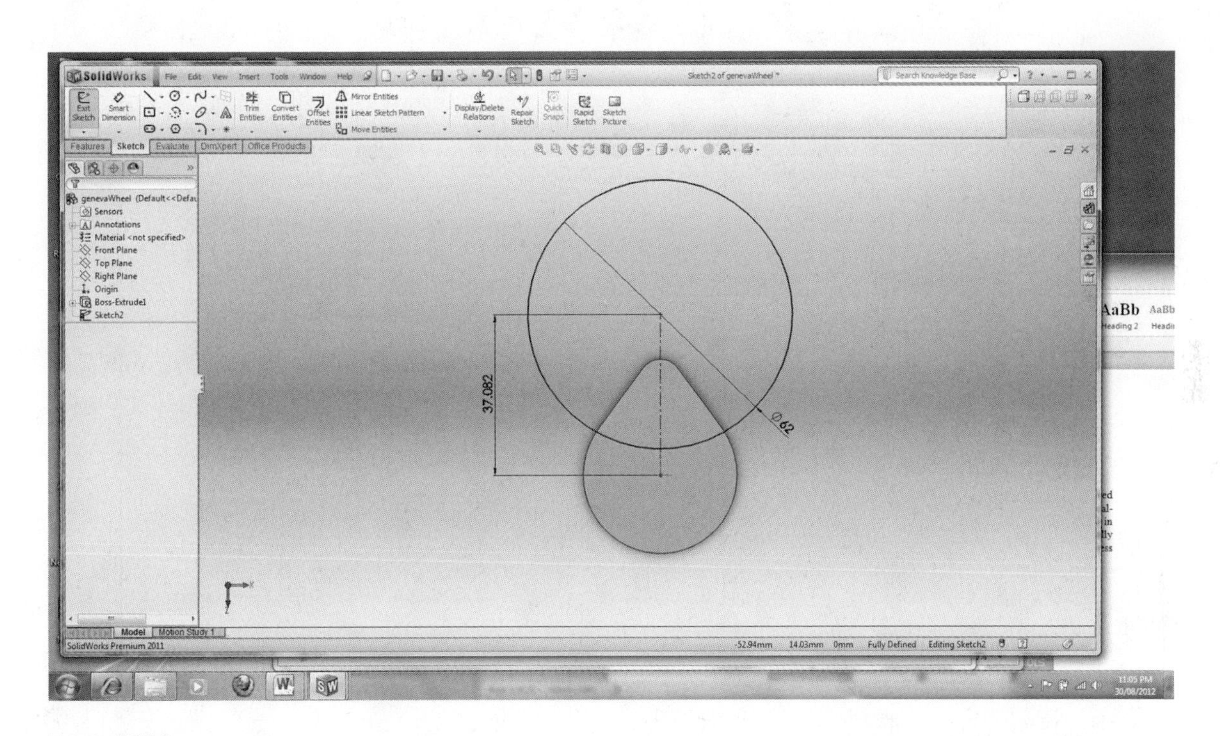

Figure P27.6

Sketch2 extruded with additional feature.

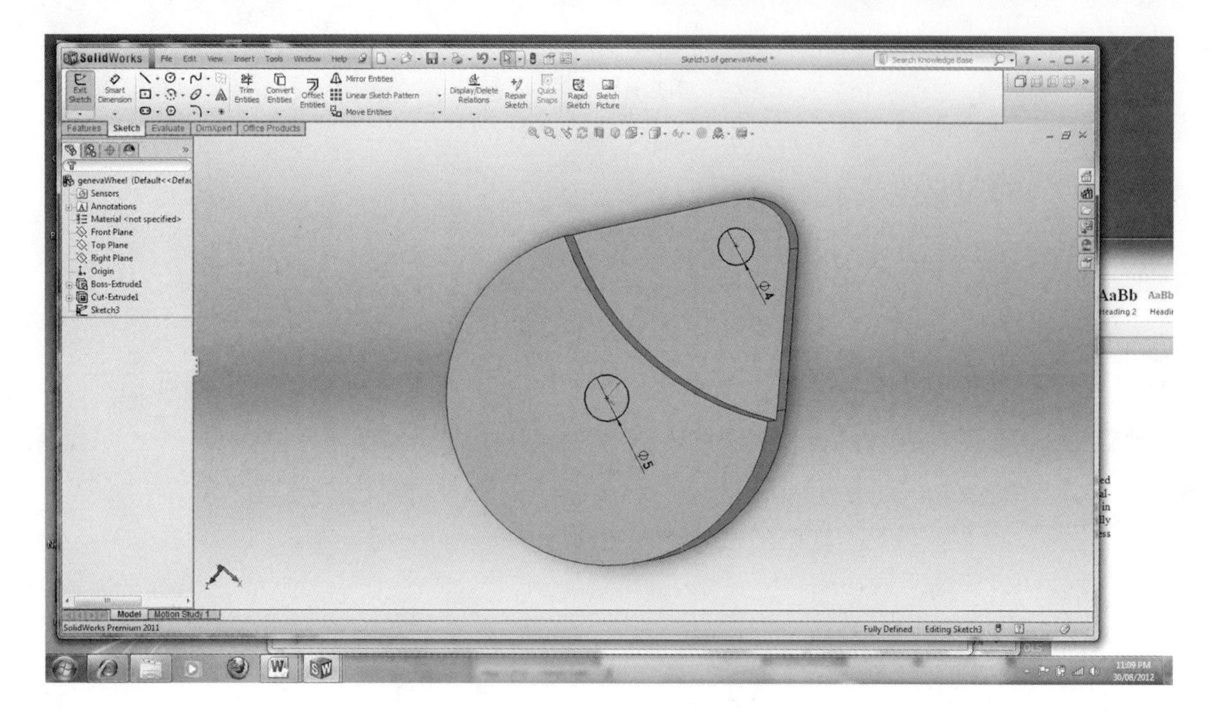

Figure P27.7

Extruded feature with hole insertion.

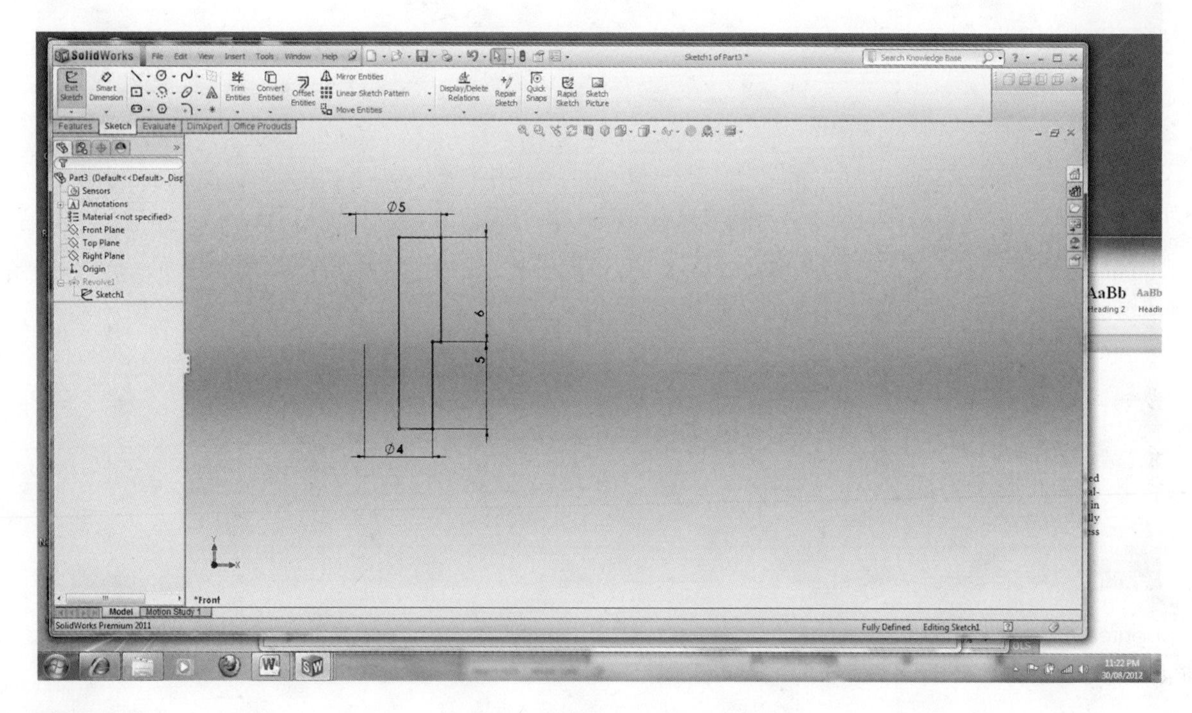

Figure P27.8

Sketch3.

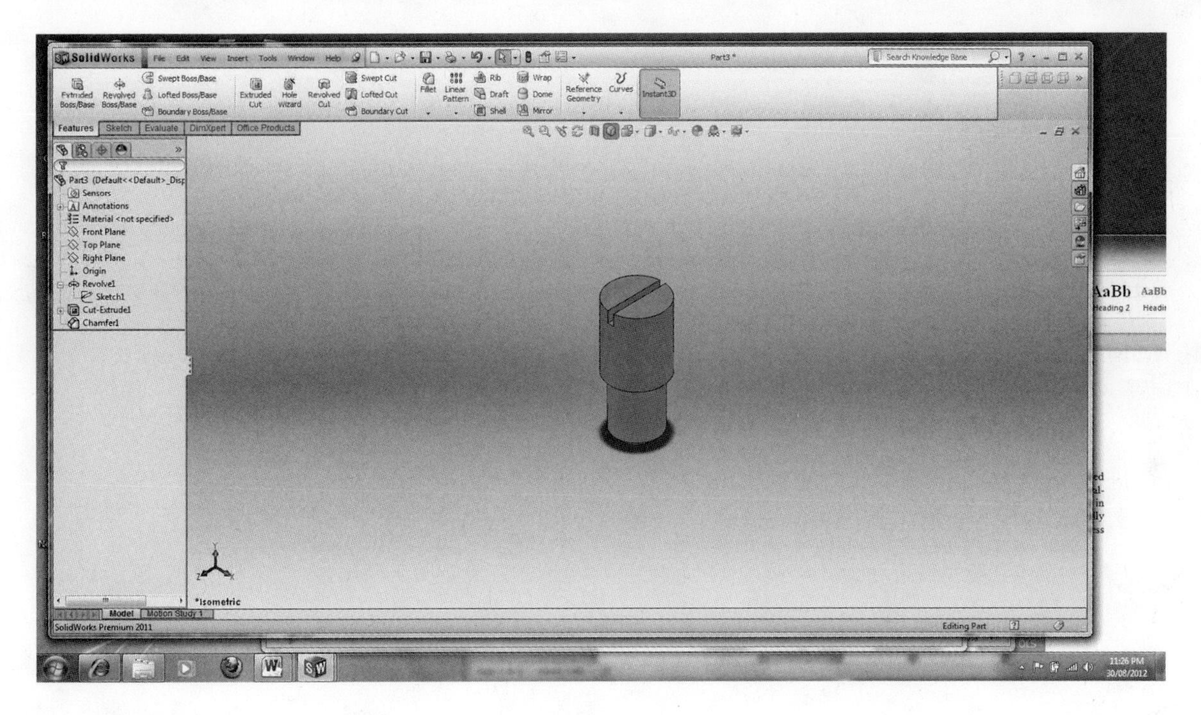

Figure P27.9

Sketch3 revolved.

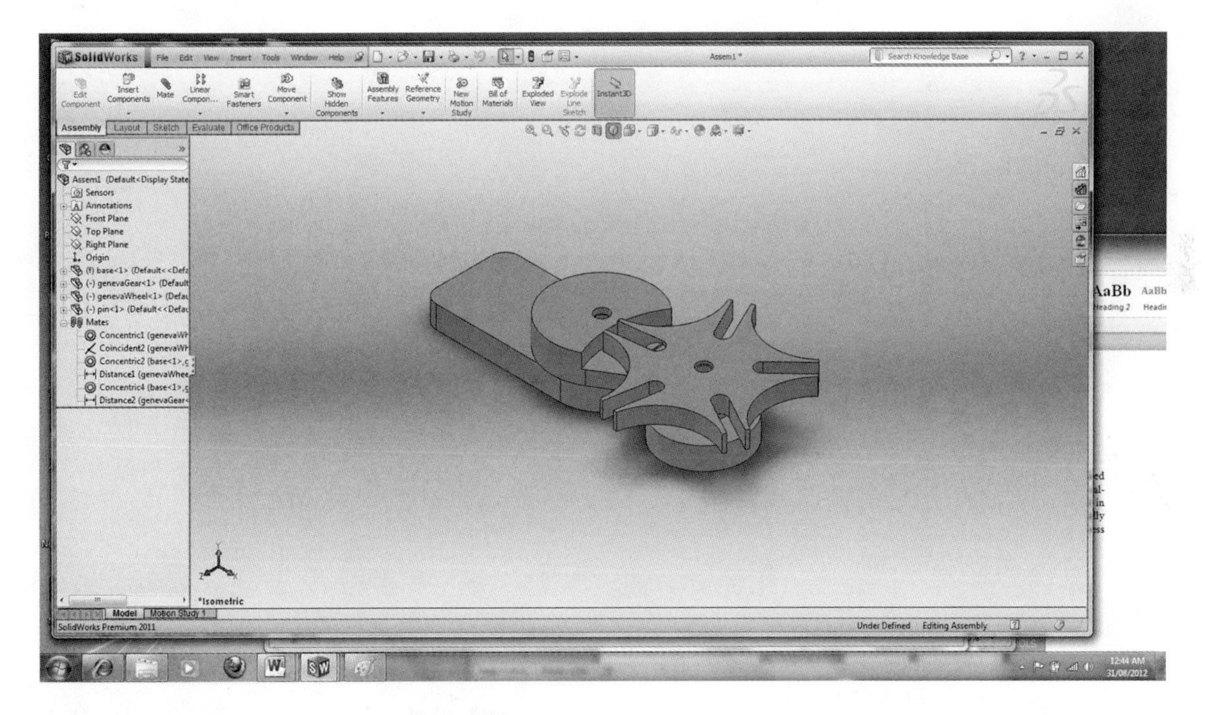

Figure P27.10

Assembly of parts.

Bibliography

Gupta, D. 2011. Animating Geneva Mechanism in SolidWorks. http://gupta9665.wordpress.com/2011/04/22/animating-geneva-mechanism-in-solidworks/.

Johnson, J.E. 2012. Make Geneva Wheels of Any Size. http://newgottland.com/2012/01/08/make-geneva-wheels-of-any-size/#.

Morld Tech Gossips. 2011. Geneva Mechanism. http://morldtechgossips.blogspot.ca/2012/06/geneva-mechanism.html.

Pattnaik, B. Mechanisms, Linkage Geometry, and Calculations. http://www.scribd.com/doc/83874208/Mechanisms-Linkage-Geometry-And-Calculations.

Walsh, R.A. 2000. Chapter 11: Mechanisms, linkage geometry, and calculations. In *Handbook of Machining and Metalworking Calculations* (Ed. R. A. Walsh), McGraw-Hill Companies, Inc. USA.

28

Event-Based Motion Analysis

Objectives:

When you complete this chapter, you will have

- Understood how event-based motion analysis (EBMA) works
- Understood the role of EBMA in the new design paradigm
- Learned about how to simulate EBMA SolidWorks routing

Introduction

SolidWorks event-based motion analysis (EBMA) is the state-of-the-art motion analysis that was launched in 2010 (SolidWorks Corp., Chapter 15—Motion Studies, Reference 1) for handling real-life applications, such as are applicable in the beverage industry, assembly lines, and generally in environments where motions are encountered. Machine designers typically follow the metaprocess that is shown in Figure 28.1, whereas the new machining design process afforded by event-based simulation is shown in Figure 28.2.

Event-Based Motion View

With SolidWorks Simulation® Professional added in, one can use a Motion Analysis study to calculate the motion of an assembly that incorporates event-based motion control. To display the event-based view, from a motion study, select Motion Analysis from the Motion Study Type list (MotionManager toolbar) and click Event-based Motion View (right end of the MotionManager toolbar).

Tasks

Event-based motion requires a set of tasks. The tasks can be sequential or can overlap in time. Each task is defined by a triggering event and its associated task action. Task actions control or define motion during the task. The result of an event-based motion task is an action. Tasks are named and described as shown in Table 28.1.

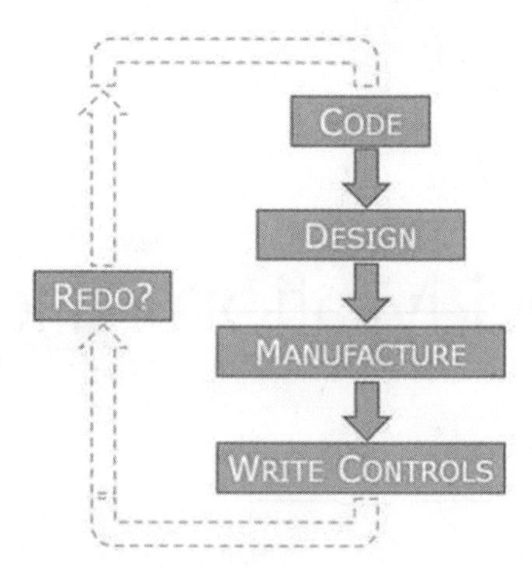

Figure 28.1

Meta-process.

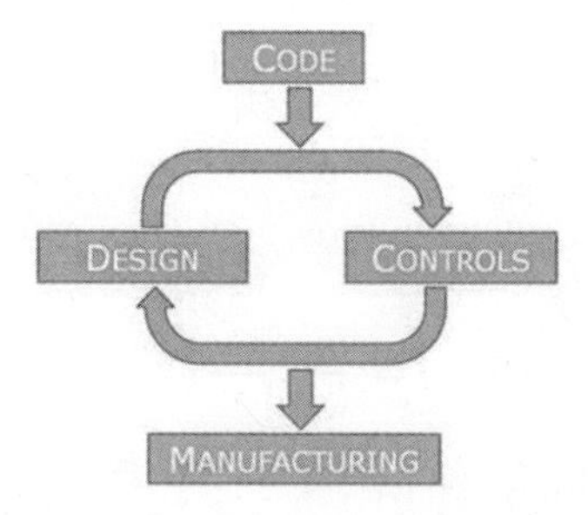

Figure 28.2

New machining design process.

Table 28.1 Description of Tasks.

Name		Displays task names. Click a cell to modify the name.
Description		Displays task descriptions. Click a cell to modify the description.

Triggers

A task trigger (Table 28.2) is the event that drives the motion control action for a task. One can define task triggers based on time, previous tasks, or sensed values, such as component position.

Actions

A task action defines or constrains the motion of one or more components in the assembly. One can define actions to suppress or activate mates, stop motion, or to change values for motors, forces, or torques. One can define a task action to (1) start, stop, or change the value of a constant speed motor, constant force, constant torque, or a servo motor; (2) stop the motion; and (3) suppress a selected mate. Actions are defined by feature, action, value, duration, and profile (Table 28.3), whereas time is defined by start/finish (Table 28.4).

Example 1

To understand how SolidWorks EBMA works, we will consider the motion of a block through a maze that is shown in Figure 28.3. Simulate the EBMA.

Table 28.2 Task Trigger

Trigger		One can create triggers from the following: • Sensors. • Interference detection. • Detects collisions. • Proximity. • Detects the motion of a body crossing a line. • Dimension. • Detects the relative position of components from dimensions. • Previous tasks in the event schedule. • Start and finish times for task actions. • Double-click a cell to modify the trigger.
		Specifies triggering task.
		Specifies triggering sensor. You can include sensors to trigger actions only when you set the Alert condition in the Sensor PropertyManager.
		Specifies a time-based trigger.
Condition		• *Alert On*. Triggers action when sensor is triggered. • *Alert Off*. Triggers action when sensor is off. • *Task Start*. Triggers action at start of a triggering task. • *Task End*. Triggers action at end of a triggering task.
Time/Delay		For triggering tasks or sensors, displays time delay for starting the task action. Click a cell to modify the delay. For time-based triggers, displays start time of the task action. Click a cell to modify the time.

Table 28.3 Actions in EBMA

Feature		Indicates linear motor action.
		Indicates rotary motor action.
	(2)	Indicates multiple motor action.
	(3)	Indicates action by a mixed selection of features.
		Indicates force action.
		Indicates torque action.
		Indicates suppressed or included mate action.
	STOP	Indicates stop motion.
		Double-click a cell to select one or more features. Select only features to which you can apply to the same action.
Action		*On*. Turns on motors, forces, or torques and includes selected mates for the duration of action. *Off*. Turns off motors, forces, or torques and excludes selected mates for the duration of action. *Change*. Changes the value of constant speed motors, servo motors, or constant forces or torques. *Stop*. Stops constant speed or servo motors. Click a cell to change the action.
Value		Specifies changed constant for changed constant speed motors or constant forces or torques.
Duration		Specifies action duration for changed constant speed motors or constant forces or torques.
Profile		Specifies the shape of a constant speed motor profile or a constant force or torque profile. The profile is calculated from the value and duration. Click a cell to change the profile.
		Linear
		Constant Acceleration
		Cycloidal
		Harmonic
		Cubic

Table 28.4 Time

Start	Displays the start time for task action.
End	Displays the end time for task action.

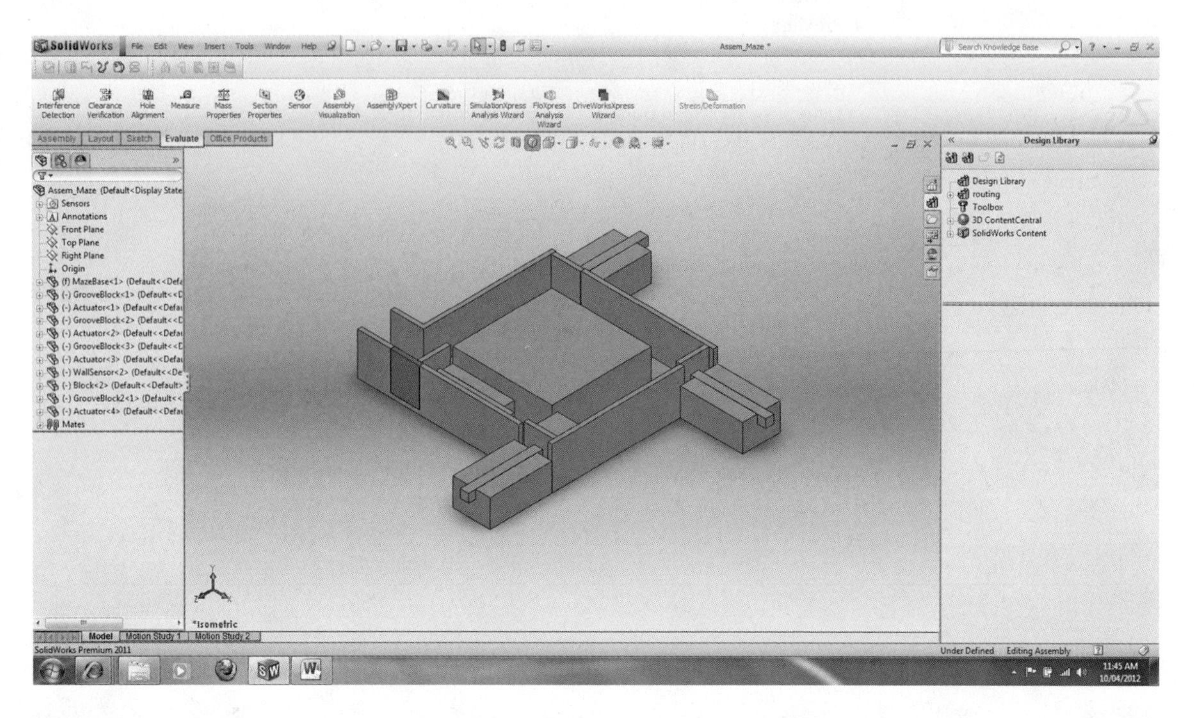

Figure 28.3

Maze configuration.

SOLIDWORKS EBMA SIMULATION SOLUTION

The parts and assembly are first created by the user; then, *Motion Analysis* is *Added-In* before the simulation can proceed. The simulation starts with a *time trigger*, which can be equal to or greater than zero seconds. The subsequent tasks (or actions) are triggered to occur after the previous one finishes. We will now work through how to define the *Tasks*, *Triggers*, and *Actions* before carrying out the required simulation.

Preambles

1. Open the assembly document, Assem_Maze.
2. Click Add-Ins (see Figure 28.4).
3. Select Motion Study 2. (You could start with Study 1 or any other.)
4. Expand the drop-down menu for motion-type at the bottom-left corner of the Graphics Window.
5. Select SolidWorks Motion Analysis (see Figure 28.5).
6. Click OK.

Define Linear Actuators (Motors)

1. Click the Motor icon at the bottom of the *Motion Analysis* window.
   For Motor Type, select Linear Motor (Actuator) (see Figure 28.6).
2. The Motor PropertyManager is automatically displayed.
3. Select Face<1>@Actuator-1 of the actuator as the Motor Location.
4. Motor Direction is automatically assigned to Face<1>@Actuator-1 of the actuator.
5. Reverse if required by clicking the Reverse Direction option.
6. For Motion, select Servo Motor.
7. For second Motion description, select Displacement (see Figure 28.7).

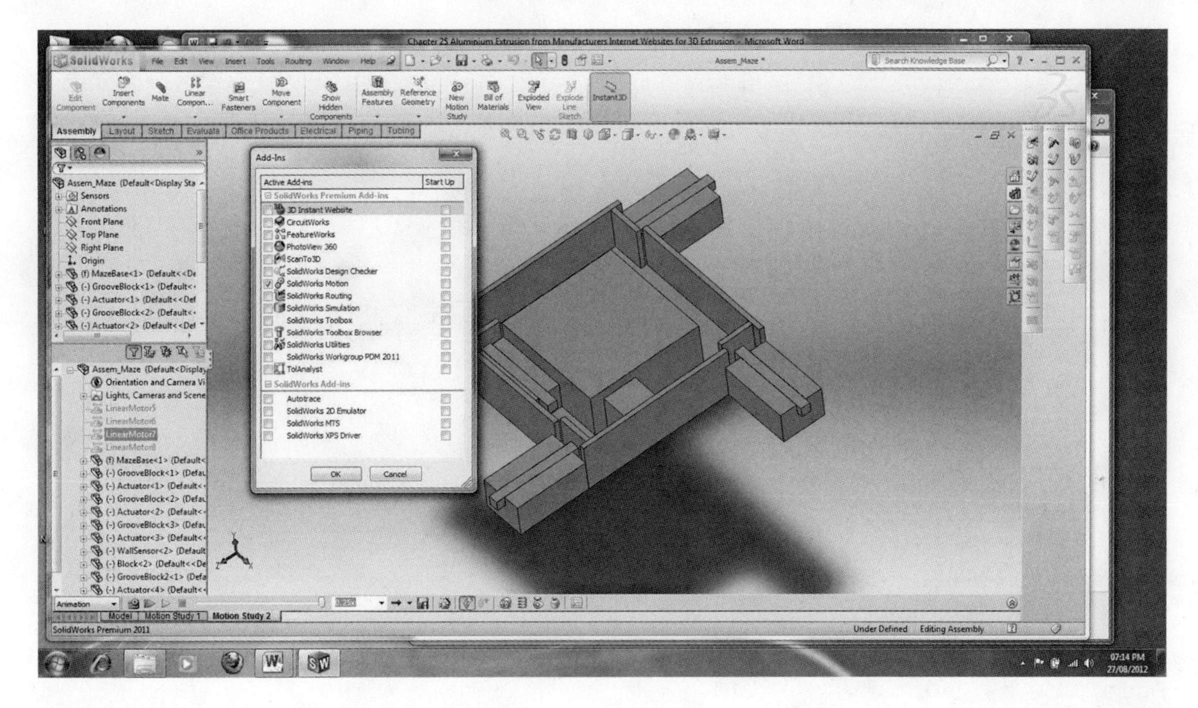

Figure 28.4

Add-Ins.

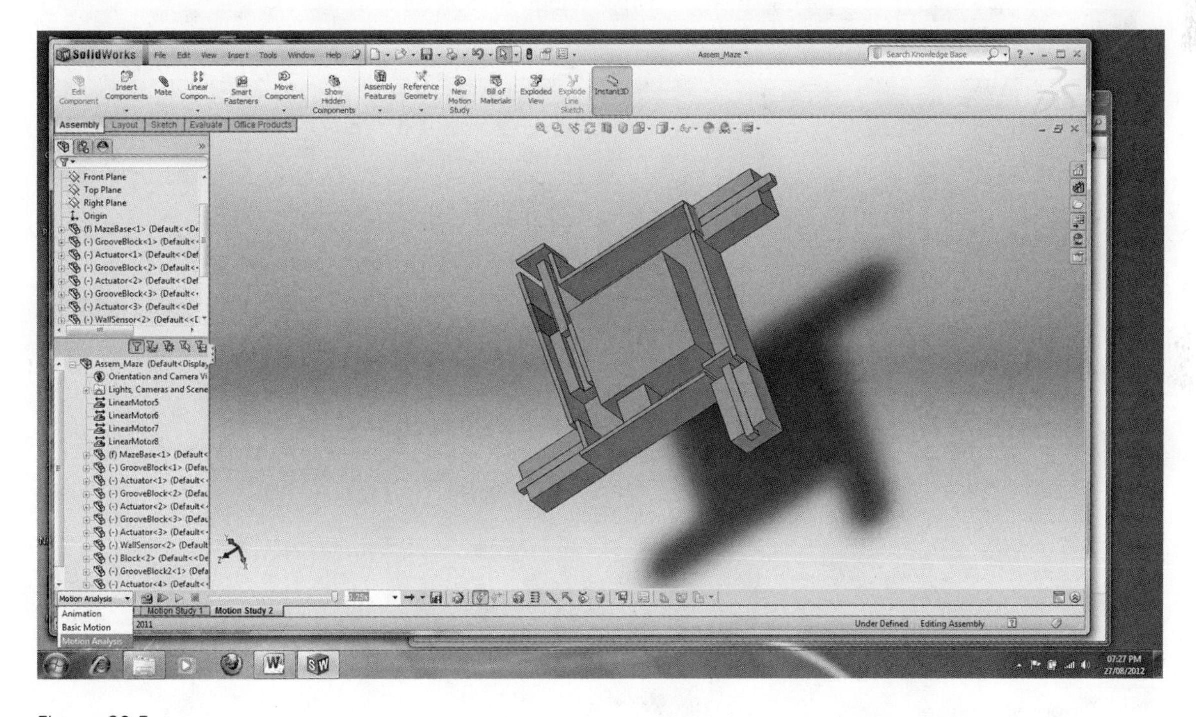

Figure 28.5

Motion Analysis option.

Note that

Motor 5 = Motor 1
Motor 6 = Motor 2
Motor 7 = Motor 3
Motor 8 = Motor 4

The procedure for defining motors is repeated for the other three motors.

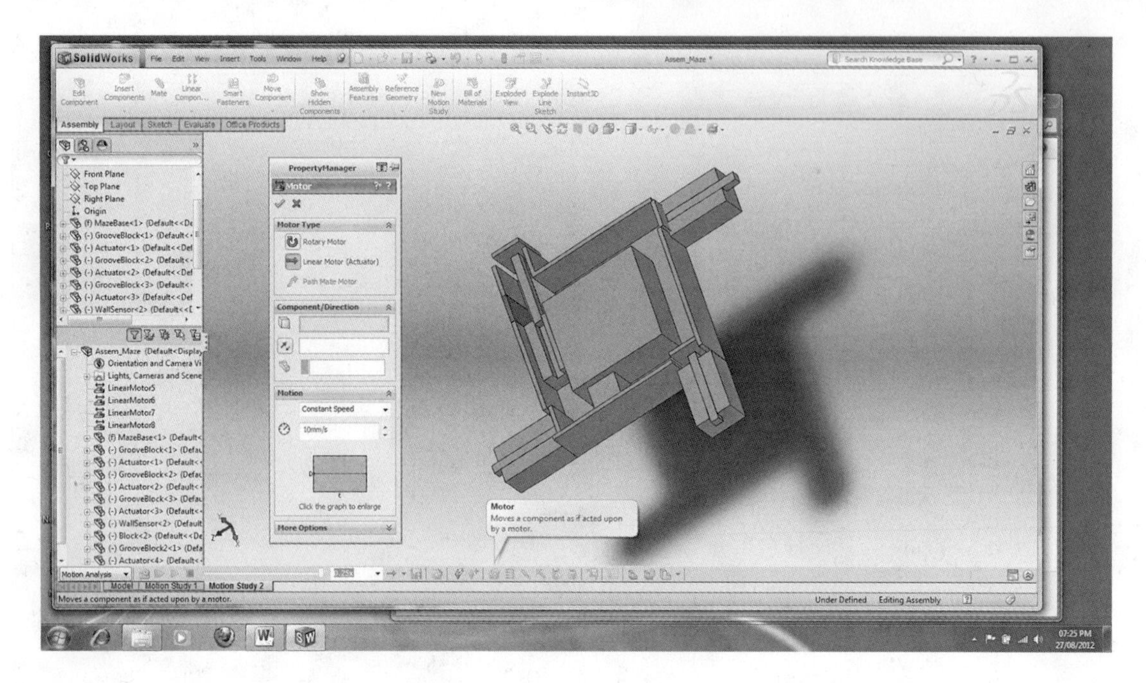

Figure 28.6

Linear Motor (Actuator) definition.

Figure 28.7

Motor PropertyManager.

EBMA Definitions

1. Click the Expand MotionManager at the bottom right of the Graphics Window (see Figure 28.8). (The Event-Based Motion Analysis window is automatically displayed.)
2. Click the + (plus) sign with the description “Click here to add” (see Figure 28.9) to add Tasks to the Task Name (see Figure 28.10).
   Our first step is to define Triggers for Task1.
   In the second column, add descriptions under Task Description.
3. Add the description: *2 last actuators off.*

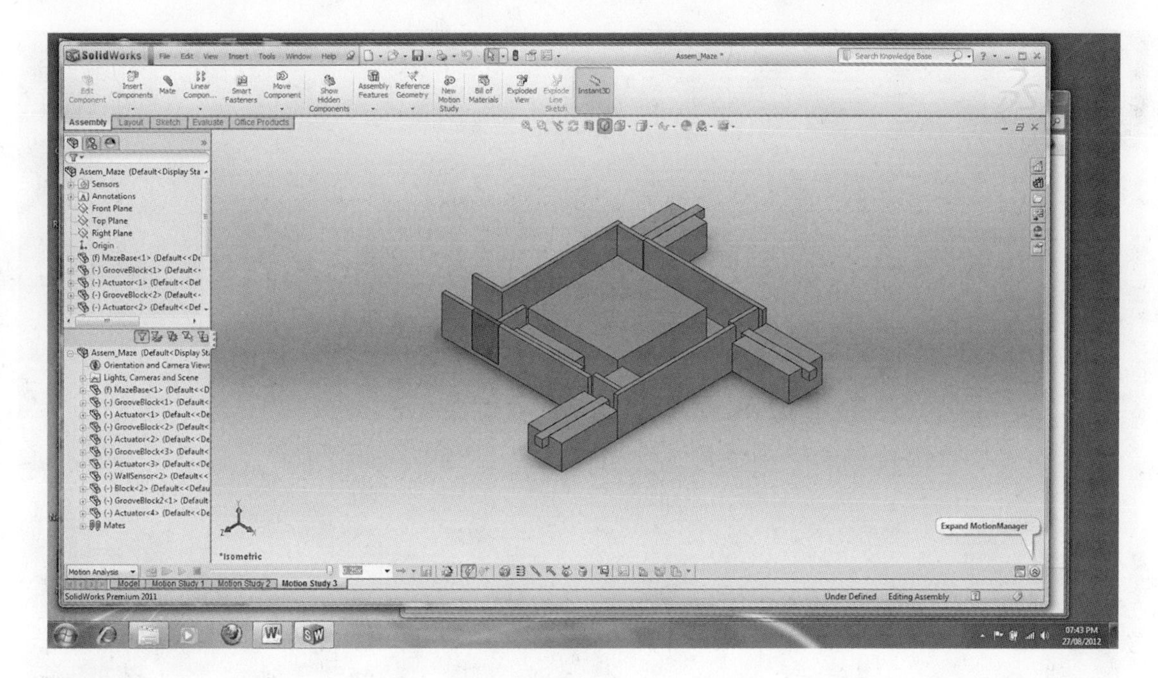

Figure 28.8

Expand MotionManager gateway to EBMA.

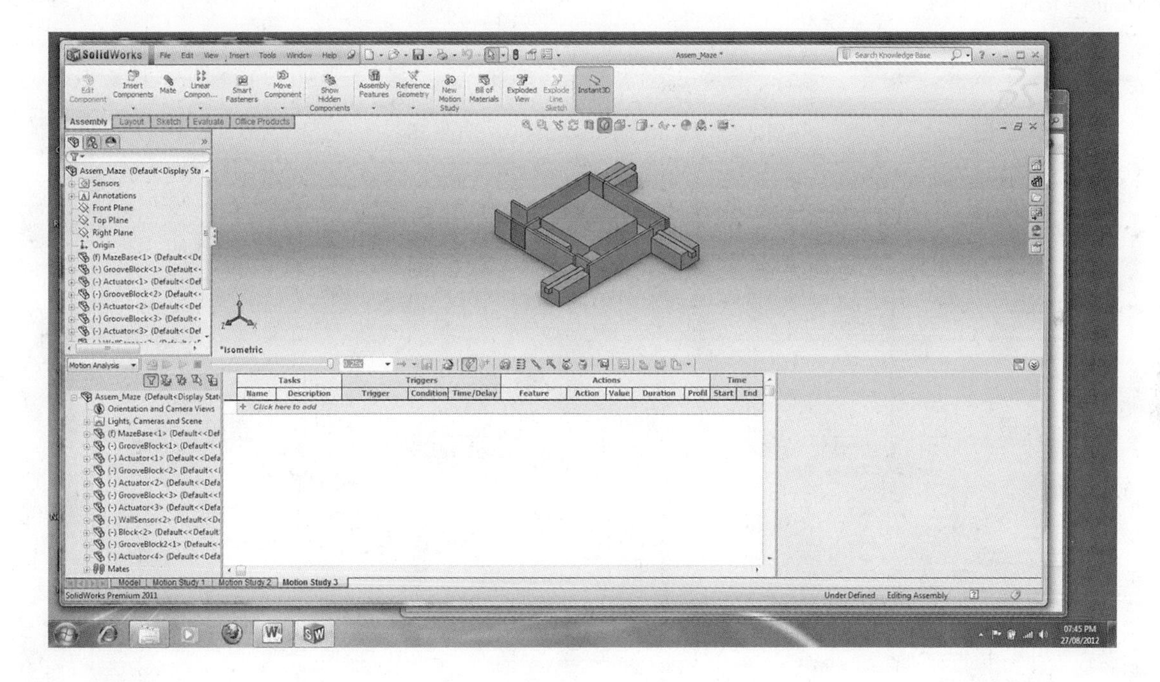

Figure 28.9

Adding Tasks.

Triggers Definition for Task 1

4. Click Trigger—select Time.
5. Condition—None.
6. Time/Delay—0.

Actions Definition for Task 1

7. Click Features—select LinearMotor7 and LinearMotor8 (see Figure 28.11).
8. Click Action—select Off.

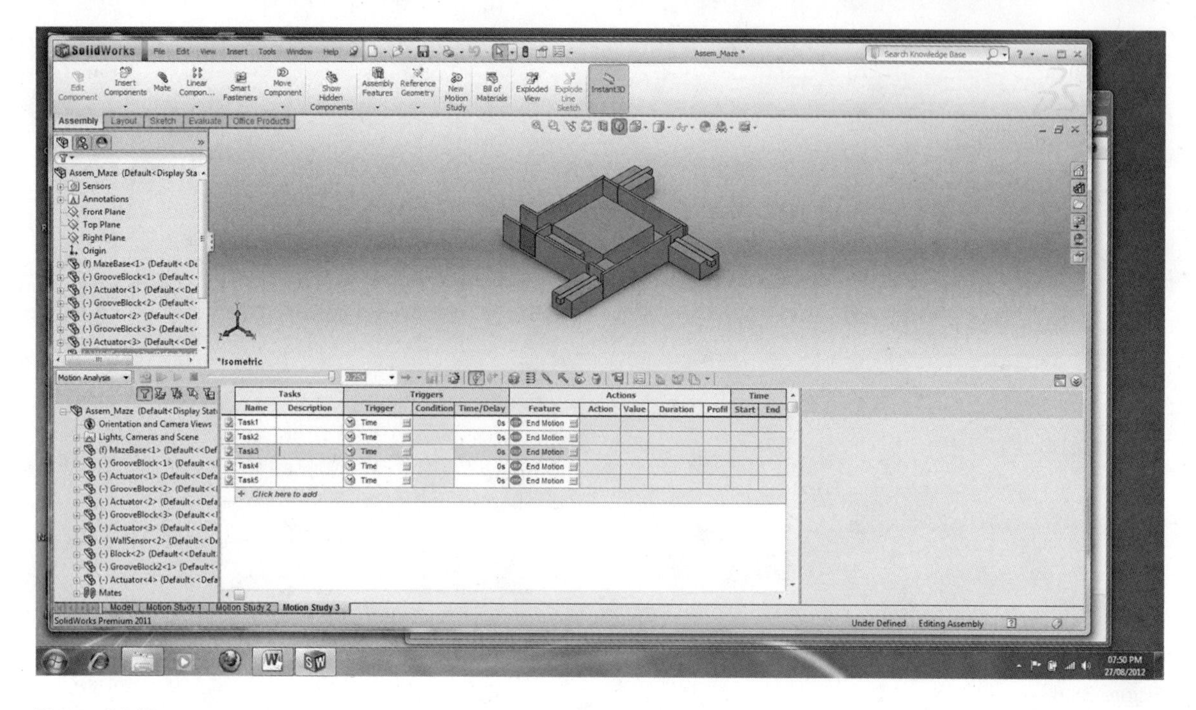

Figure 28.10

Multiple tasks clicked.

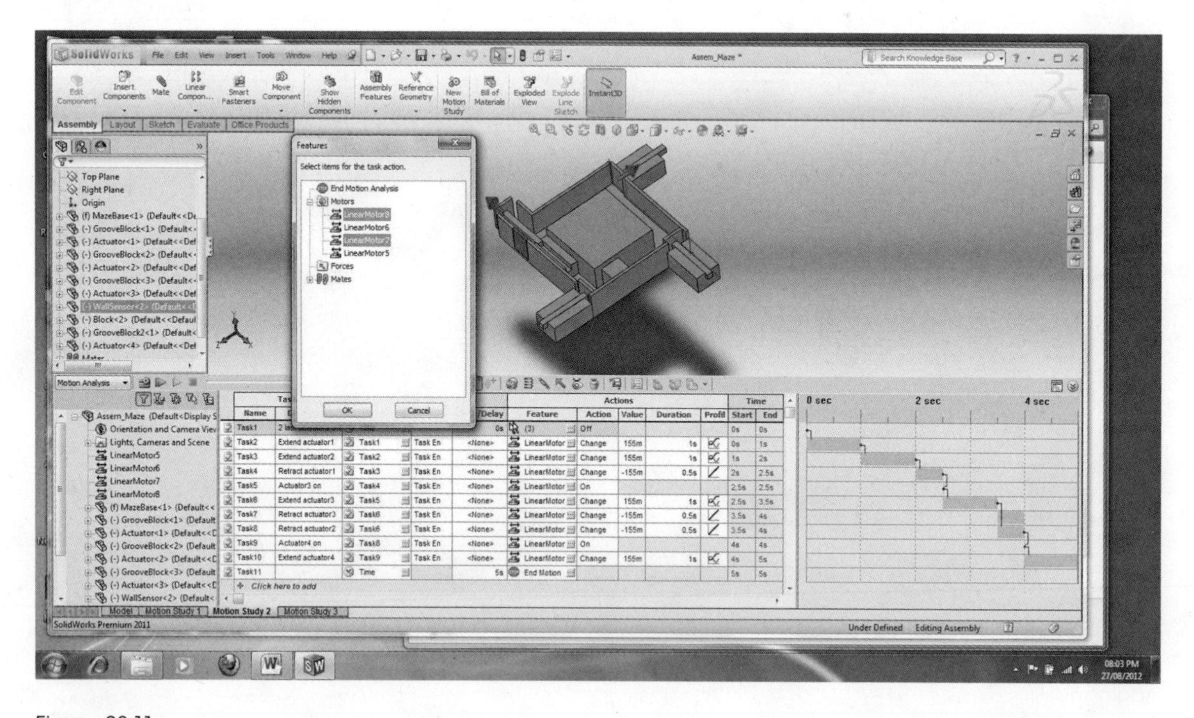

Figure 28.11

Actions definitions using linear motors.

9. Value—None.
10. Duration—None.
11. Profile—None.

    Our second step is to define Triggers for Task2.

    In the second column, add descriptions under Task Description.

12. Add the description: *Extend actuator 1.*

Triggers Definition for Task 2

13. Click Trigger—select Task1 (see Figure 28.12).
14. Condition—select Task End.
15. Time/Delay—<None>.

Actions Definition for Task 2

1. Click Features—select LinearMotor5 (see Figure 28.13).
2. Click Action—select Change (see Figure 28.14).

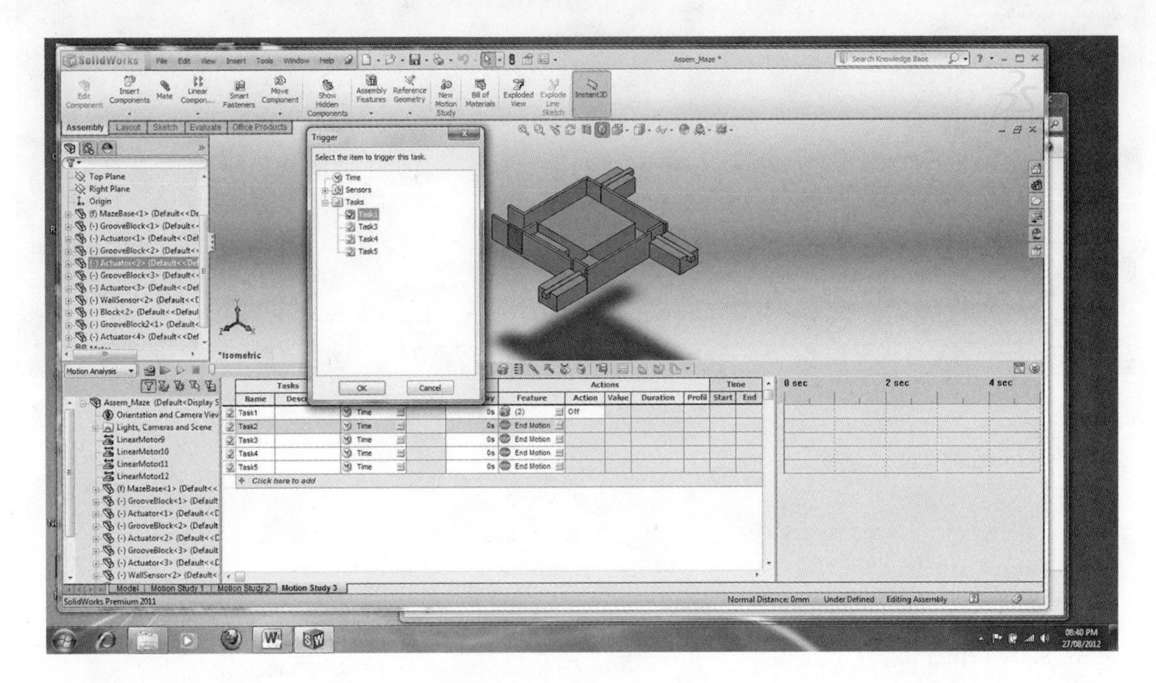

Figure 28.12

Trigger definition using Task 1.

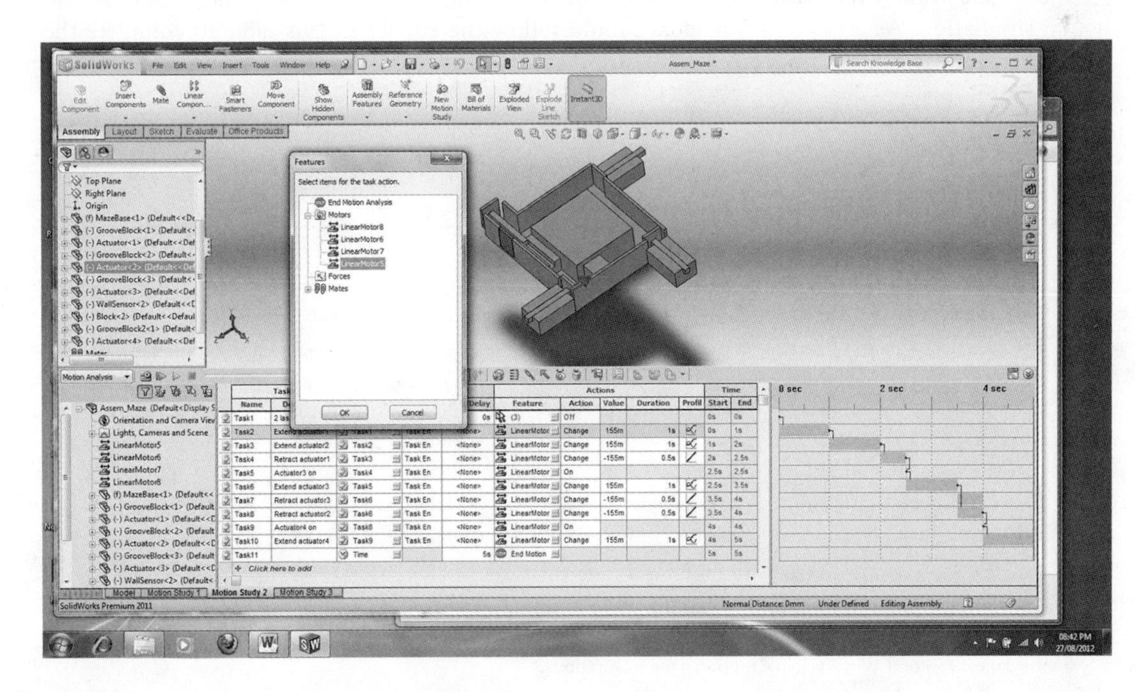

Figure 28.13

Actions-feature definition for Task 2 using first linear motor.

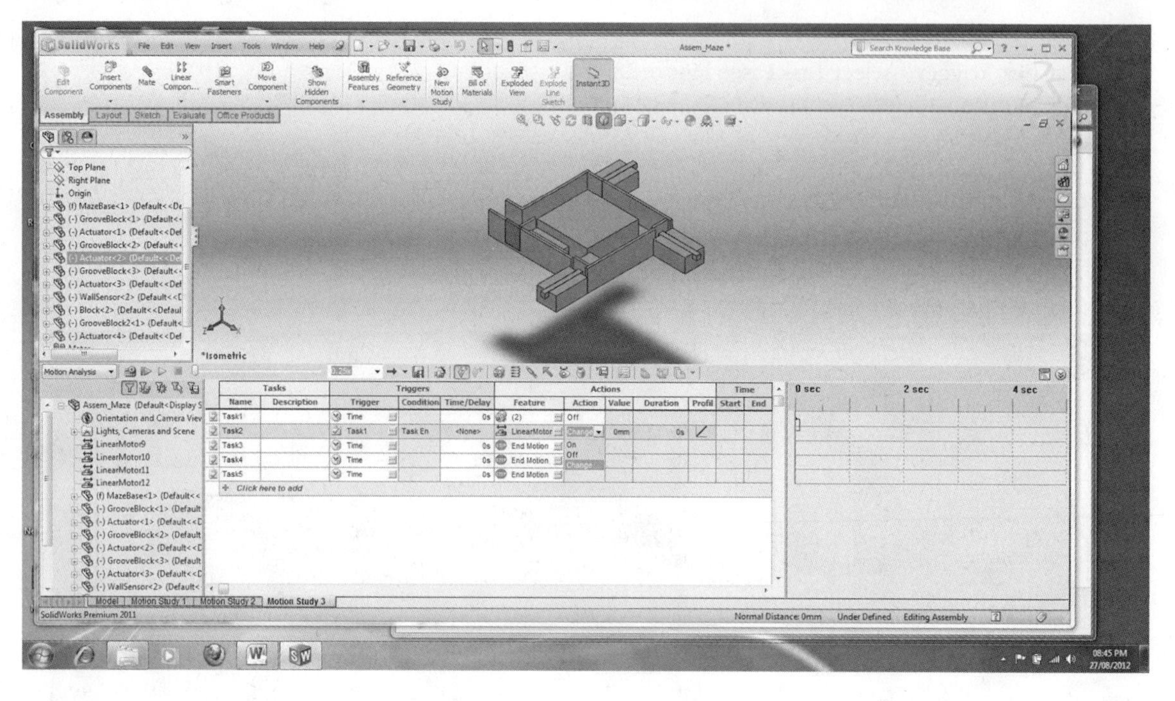

Figure 28.14

Actions–action definition for Task 2 with Change selected.

3. Value—enter 155 mm.
4. Duration—enter 1 s.
5. Profile—select Cycloidal.
6. Start—enter 0.
7. Finish—enter 1 s.

The details of the Event-based Motion View for this example are shown in Table 28.5, based on actual simulation. As an exercise, readers should follow the entries in this table to complete the definitions for each task. The complete definitions for the Event-based Motion View are shown in Figure 28.15.

Table 28.5 Complete Definitions for the Event-Based Motion View

Tasks		Triggers			Actions				
Name	Descriptions	Trigger	Condition	Time/Delay	Feature	Action	Value	Dur	Profile
Task 1	2 last actuators off	Time	–	0 s	LMotor 7,8	Off	–	–	–
Task 2	Extend actuator 1	Task 1	Task End	<None>	LMotor 5	Change	155	1 s	Cyc
Task 3	Extend actuator 2	Task 2	Task End	<None>	LMotor 6	Change	155	1 s	Cyc
Task 4	Retract actuator 1	Task 3	Task End	<None>	LMotor 5	Change	−155	0.5 s	Linear
Task 5	Actuator 3 on	Task 4	Task End	<None>	LMotor 7	On	–	–	–
Task 6	Extend actuator 3	Task 5	Task End	<None>	LMotor 7	Change	155	1 s	Cyc
Task 7	Retract actuator 3	Task 6	Task End	<None>	LMotor 7	Change	−155	0.5 s	Linear
Task 8	Retract actuator 2	Task 6	Task End	<None>	LMotor 6	Change	−155	0.5 s	Linear
Task 9	Actuator 4 on	Task 8	Task End	<None>	LMotor 8	On	–	–	–
Task 10	Extend actuator 4	Task 9	Task End	<None>	LMotor 8	Change	155	1 s	Cyc
Task 11		Time	–	5 s					

Key: LMotor = Linear Motor; Cyc = Cycloidal motion; Dur = Duration.

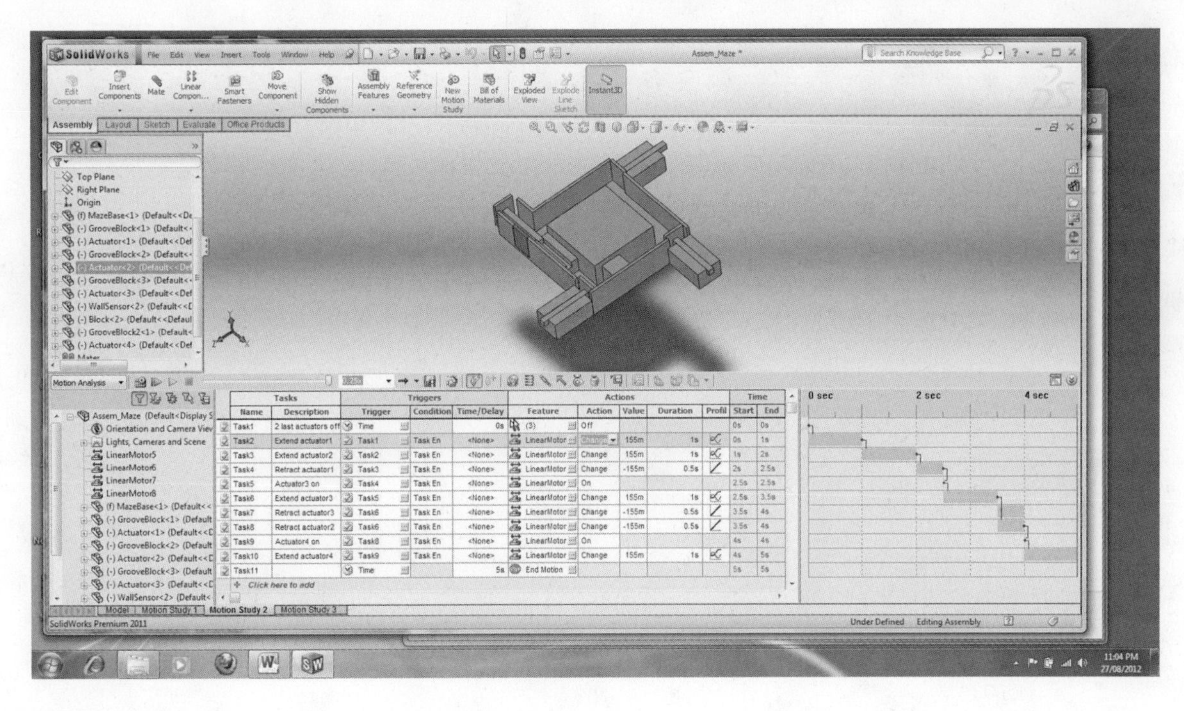

Figure 28.15

Event-based Motion View for the maze problem.

Summary

To date, there is hardly any comprehensive step-by-step documentation on EBMA that users can follow to understand how motion simulations can be realized. There are, however, dozens of YouTube documentations on the subject that are extremely challenging to follow in order to understand the detailed steps that are involved in EBMA. This chapter fills the gap.

Bibliography

SolidWorks Corp. What's New in SolidWorks version 2010? http://www.scribd.com/doc/87413953/167/Servo-Motors-for-Event-based-Motion-Analysis#outer_page_107.

29

Electrical Routing

Objectives:

When you complete this chapter, you will have

- Learned about how to Add Routing to SolidWorks and set routing options
- Learned about how to manually create a route by dragging connectors from the electrical routing Design Library to create a harness

Introduction

When designing wiring systems, the preferred approach is to model the wiring as a wiring harness. A cable harness, also known as a wire harness, cable assembly, wiring assembly or wiring loom, is a string of cables and/or wires that are used to design wiring systems. This method has many advantages over modeling wires and cables individually. It allows SolidWorks to make many calculations for the user and greatly reduces errors. There are two ways to model a wiring harness.

1. *Electrical routing with clips*
   The easiest way to model a harness is to use clips. You can place clips as required in the assembly. The harness is then routed through the clips.
2. *Using a from–to list*
   A *from–to list* automates much of the wiring process. It contains each of the wires and cables in the design and identifies the wire type to use and the required connectivity. Using a from–to list eliminates the need to enter wiring data manually, makes the design process easier, and improves the accuracy of electrical routes.

Creating the Housing

The illustration in this chapter starts with creating a housing (an open container with a number of holes) through which electrical plugs will be positioned. The plugs are connected or wired using routing cables.

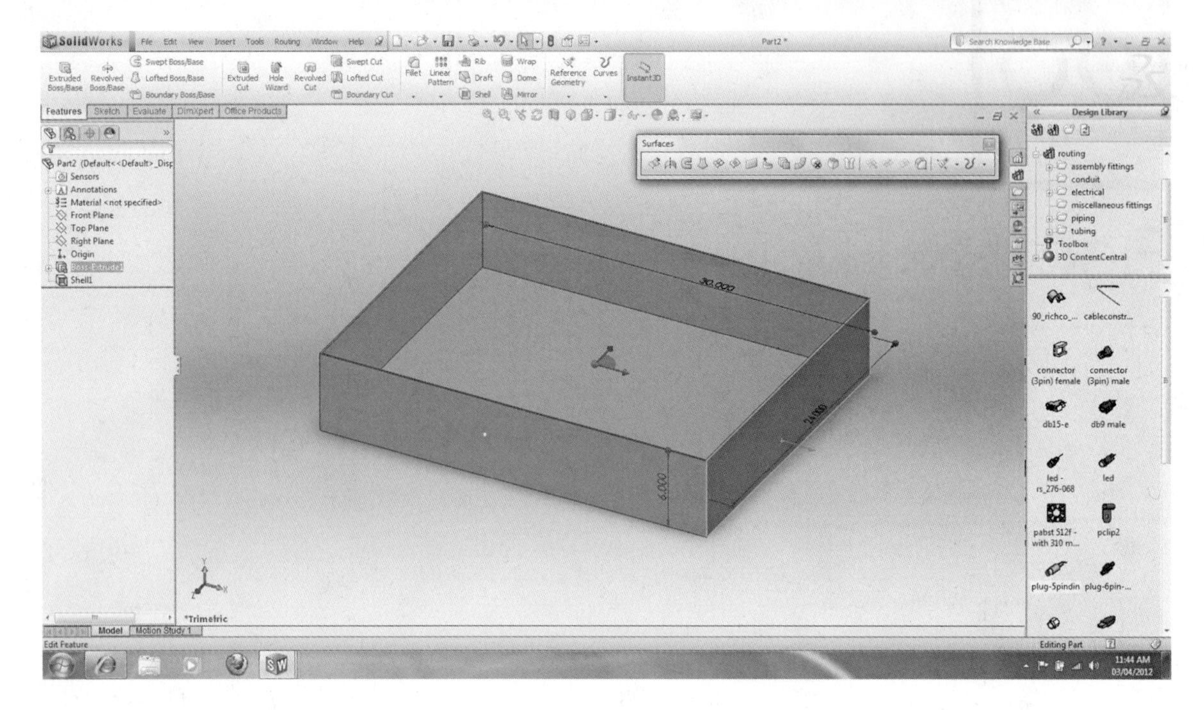

Figure 29.1

Shelled object to create the hollow circuit base.

1. Open a New SolidWorks Part document.
2. Select the Top Plane and be in Sketch Mode.
3. Select the Center Rectangle and dimension it, length = 30 in. and width = 24 in.
4. Extrude Mid Plane through a distance (height) = 6 in.
   Figure 29.1 shows the solid that is obtained.
5. Click the Shell option.
6. Select *top of the object* and apply Thickness = 0.12 in.
   A shelled model is obtained as shown in Figure 29.1.
7. Select the right face and be Normal To (Perpendicular) mode to bring this face perpendicular to your view.
8. Be in Sketch Mode.
9. Sketch a circle and dimension it to be 5-in. diameter, position its center 4 in. from the right edge and 1.75 in. from the top edge (see Figure 29.2).
10. Extrude Cut the circle through the right face of the hollow housing (see Figure 29.3).
11. Linear Pattern 3 replications (see Figure 29.4).
12. Mirror the three patterns about the Right Plane (see Figure 29.5).
13. Linear Pattern the three replications on the Right Plane downward at a distance of 2.5 in. (see Figure 29.6).
14. Save the 3D model as Circuit_Base (name).

Creating the Electrical Harness

So far, we have modeled the housing as a part. We need to open a New SolidWorks Assembly document and insert the housing as the first part in the assembly.

1. Open a New SolidWorks Assembly.
2. Insert the first part as Circuit_Base (already modeled).
   Before creating the electrical harness, first add in SolidWorks Routing.
3. Click Tools > Add-Ins (see Figure 29.7).
4. Select SolidWorks Routing.
   Select SolidWorks Routing in the Start Up column to activate Routing every time you start the SolidWorks application.

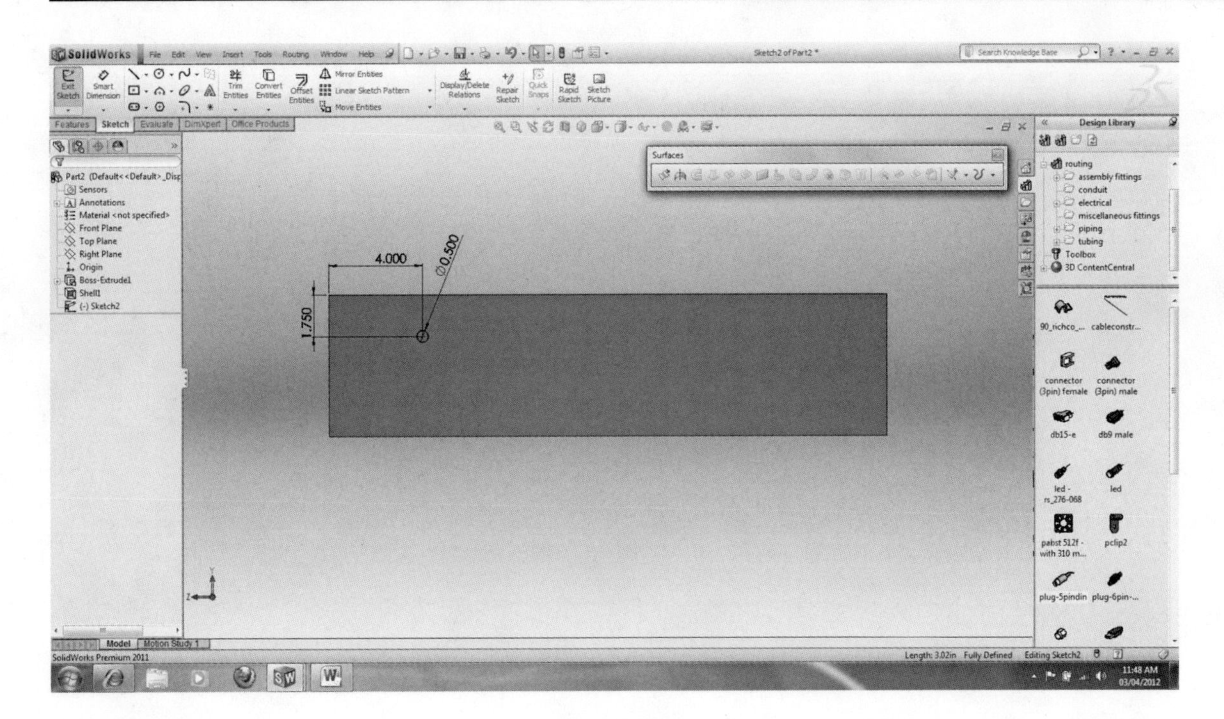

Figure 29.2

First circle is sketched.

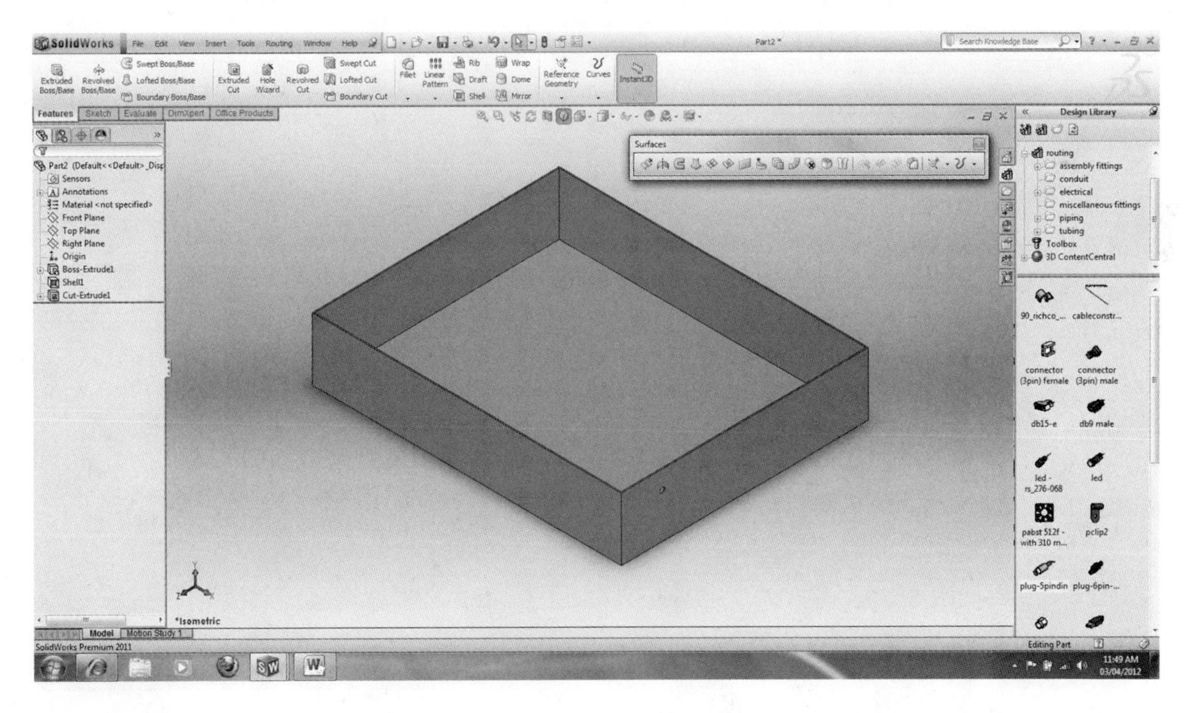

Figure 29.3

Extrude cut for first hole.

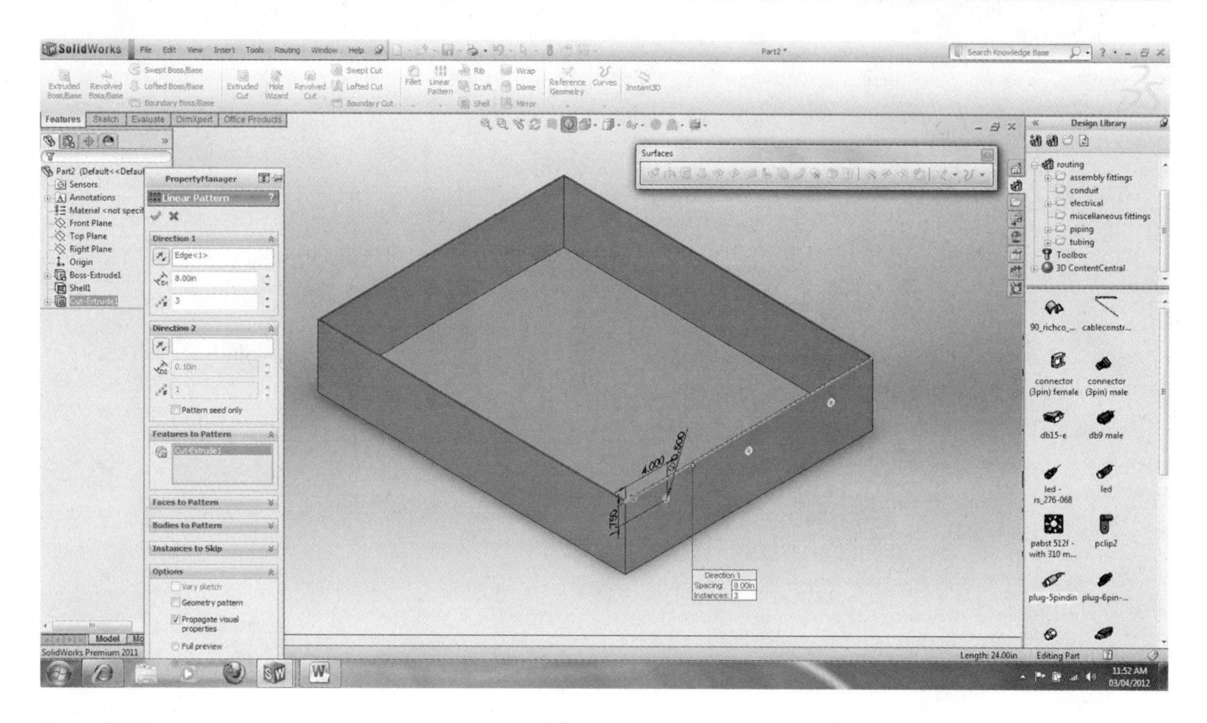

Figure 29.4

Linear patterning to obtain three holes.

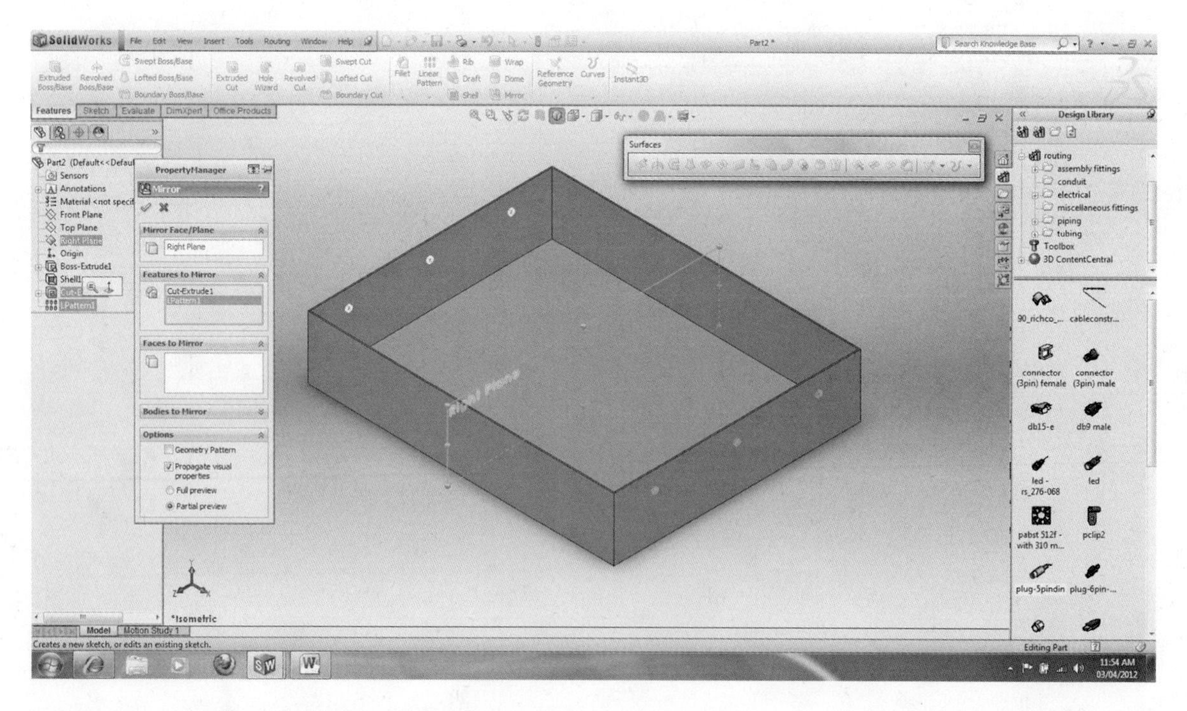

Figure 29.5

Mirrored holes about the Right Plane.

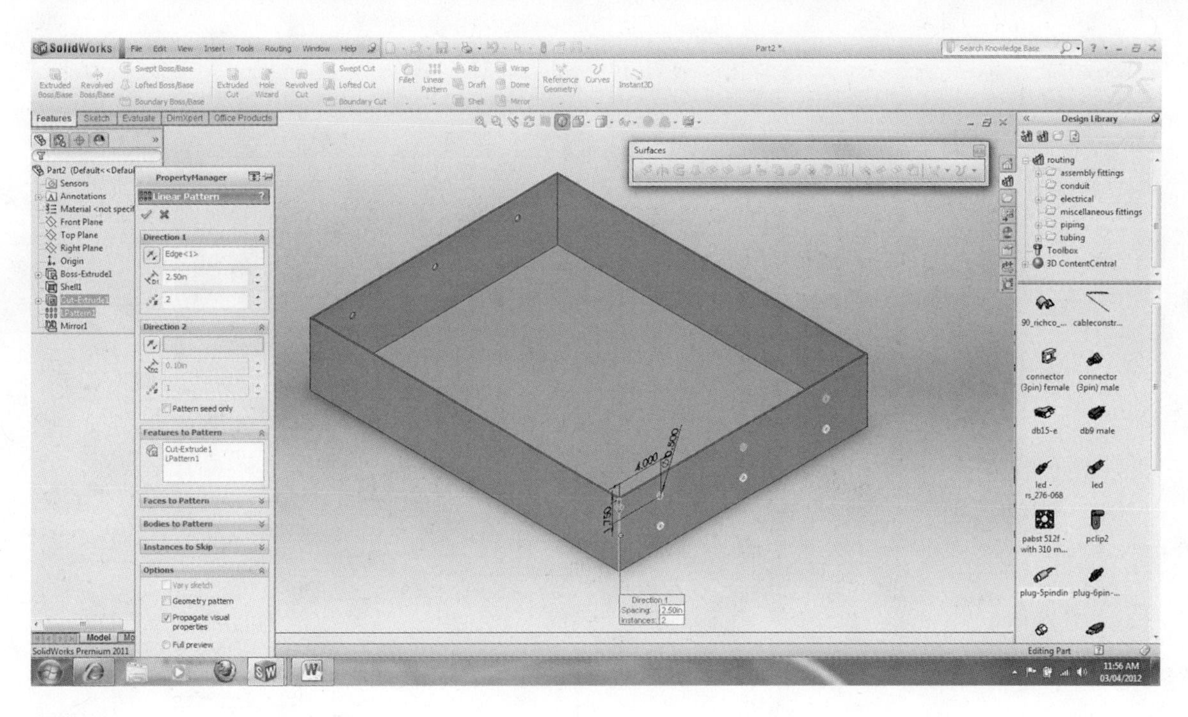

Figure 29.6

Subsequent linear patterns to create six holes.

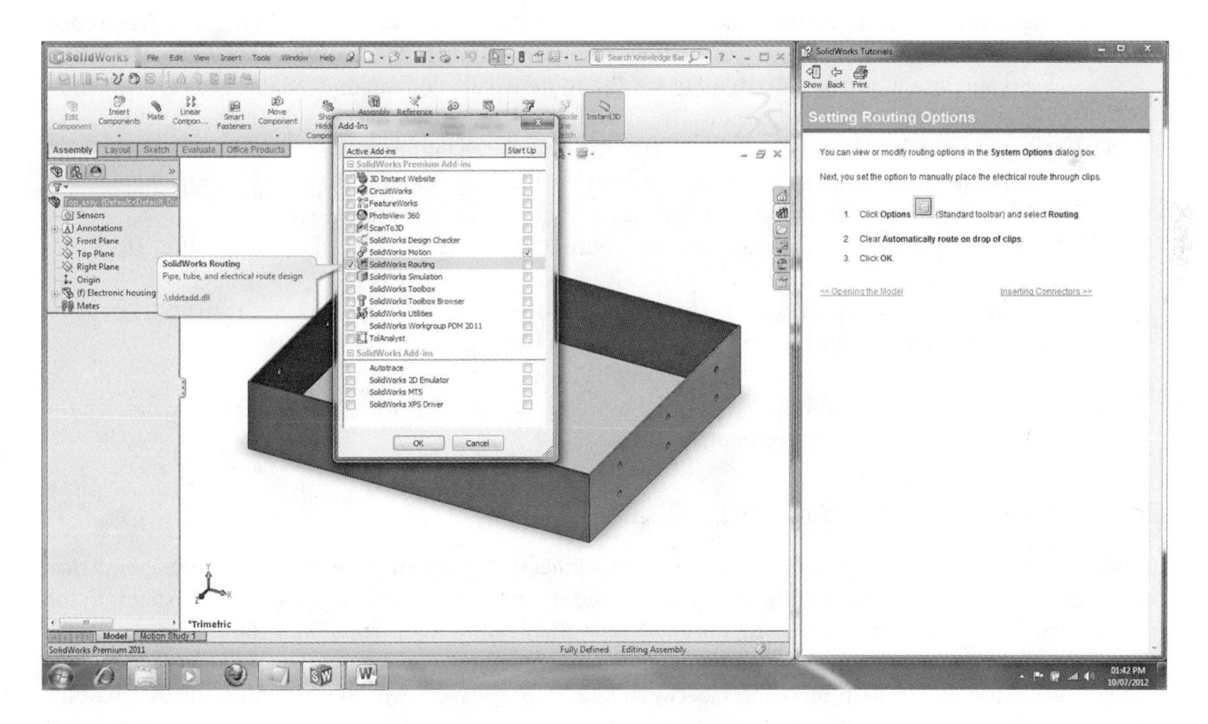

Figure 29.7

Add-In for Routing.

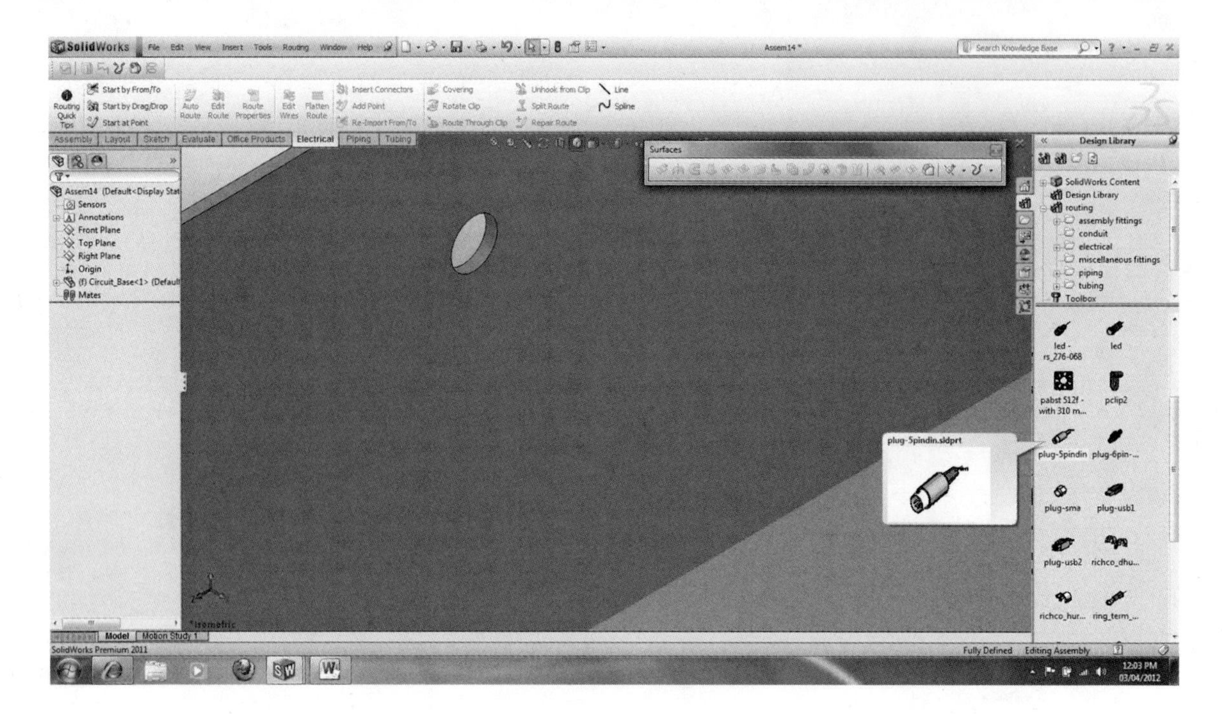

Figure 29.8

Drag a plug from Design Library.

5. Click OK.

   Start the route by inserting route components, which vary by route type, into the assembly. These components define the start and end connection points of the path.

   Next, you insert plug connectors from the Design Library into the assembly. When you drag a component from the Routing Library into the assembly, the Route Properties PropertyManager appears. When you close it, a new subassembly for the route harness opens in 3D sketch mode. Assembly colors appear gray in 3D sketch mode. In addition, the Auto Route PropertyManager appears.
6. Rotate the assembly to view the holes in the inside wall (see Figure 29.8).
7. From the electrical folder in the Routing Library, select plug-5pindin.sldprt.
8. Drag the plug into the assembly and mate it with the *middle hole* on the left side.
9. Click Yes if prompted to set options for routing, and click OK to close the Route Properties PropertyManager if it appears.

   The assembly turns gray, and the Auto Route PropertyManager appears (see Figure 29.9).
10. Click OK to complete the insertion of the first plug (see Figure 29.10).
11. Rotate the assembly to view the six holes opposite the plug.
12. Drag another plug-5pindin.sldprt into the assembly and mate it with the lower middle hole on the side opposite the first plug (see Figure 29.11).

Creating the Route Using the Auto Route

Each route component has a CPoint. CPoints are the connection points from which you connect electrical route segments. When you drag routing components into the assembly, a small length of cable extends a stub from the CPoint. Most electrical routing options are available only when you edit a route. If you exit this mode, select Edit Route (Electrical Routing toolbar) to continue.

Next, use Auto Route to route cables between the stub ends of two components. The Auto Route PropertyManager must be open to perform this procedure. If it is closed, right-click anywhere to the right side of the CommandManager to display a long list of tools. Then, click the Routing Tools as shown in Figure 29.12. This pops up the Auto Route (Routing Tools toolbar), which is shown in Figure 29.12.

1. Zoom in on one of the plugs in the assembly and select the stub at the end of its CPoint.

   The Auto Route PropertyManager displays the selected point in Current Selection.

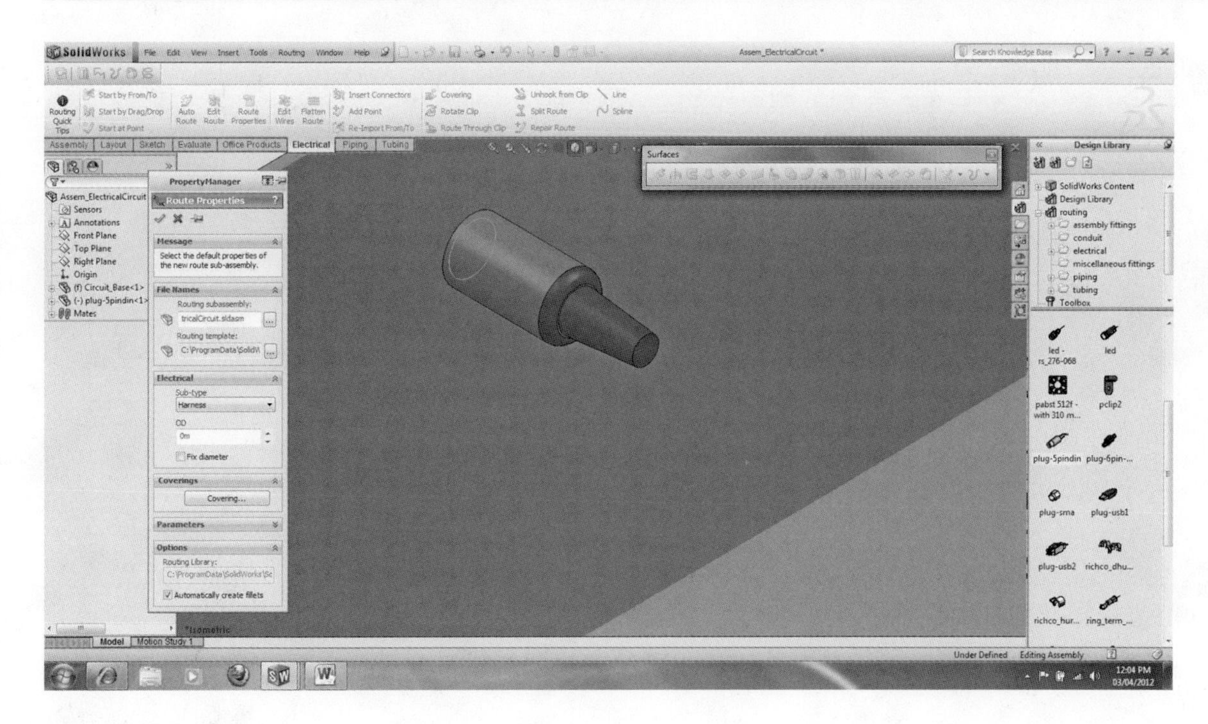

Figure 29.9

AutoRoute PropertyManager.

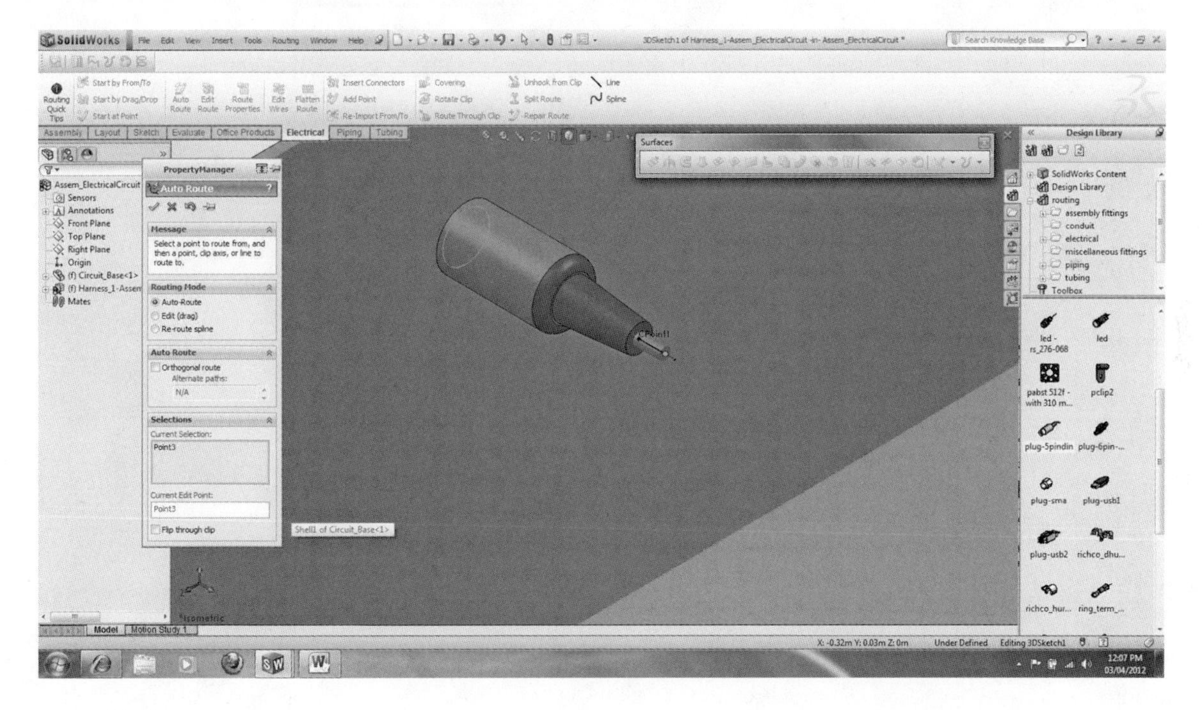

Figure 29.10

First plug is completely inserted.

2. Zoom to the other plug and select its end stub.

   The route connecting the two points appears (see Figure 29.13).

3. Click OK. Routing is complete (see Figure 29.14).

   Next, specify the details of the wires running through the harness.

4. Click Routing > Edit Wires (Electrical Routing toolbar).

Figure 29.11

Second plug is completely inserted.

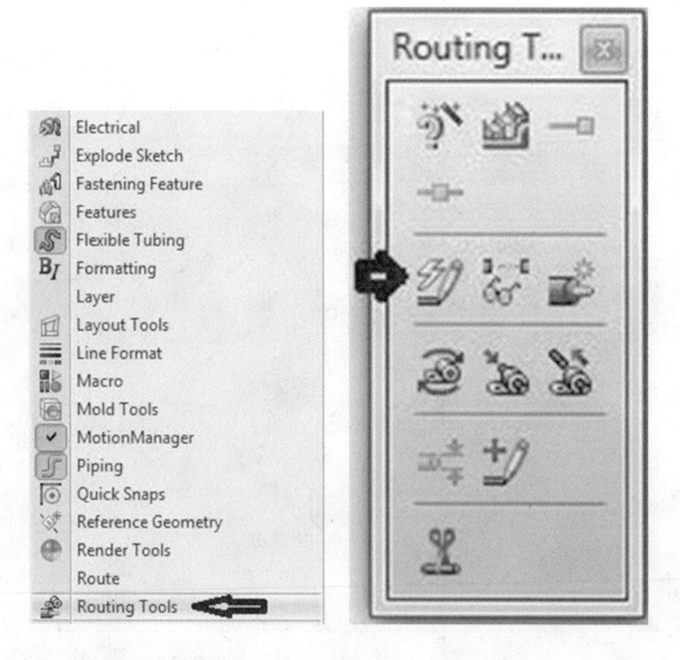

Figure 29.12

Routing Tools for modeling.

5. Click Add Wire in the PropertyManager (see Figure 29.15). The Electrical Library is automatically displayed (see Figure 29.16).
6. Select 20g blue.

   Selecting 20g blue assigns the 20-gauge blue wire part to the internal cable wire when you complete this procedure.

   20g blue appears in Selected Wires.

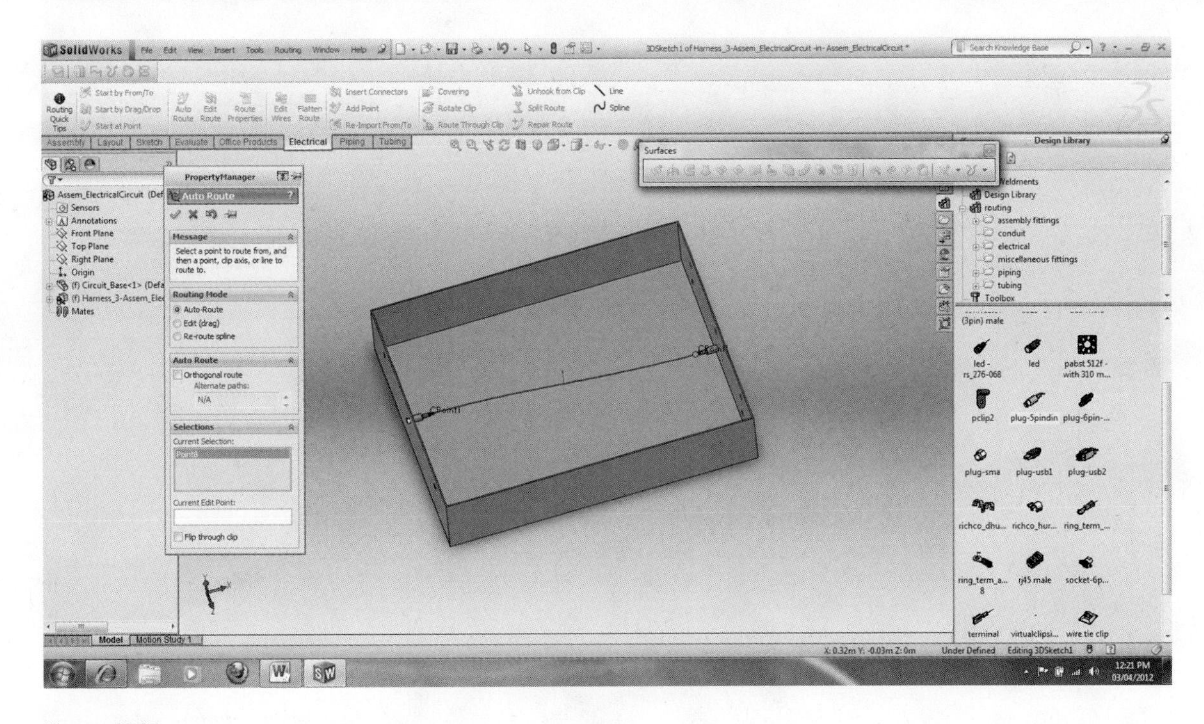

Figure 29.13

AutoRoute PropertyManager for routing the two plugs.

Figure 29.14

Routing of the two plugs.

7. Click OK in the Electrical Library.
8. Click Select Path in the Edit Wires PropertyManager.
9. Select the cable (see Figure 29.17).
10. Click OK.
11. Under From–To Parameters, select 1 for Pin for each plug.
12. Click OK.

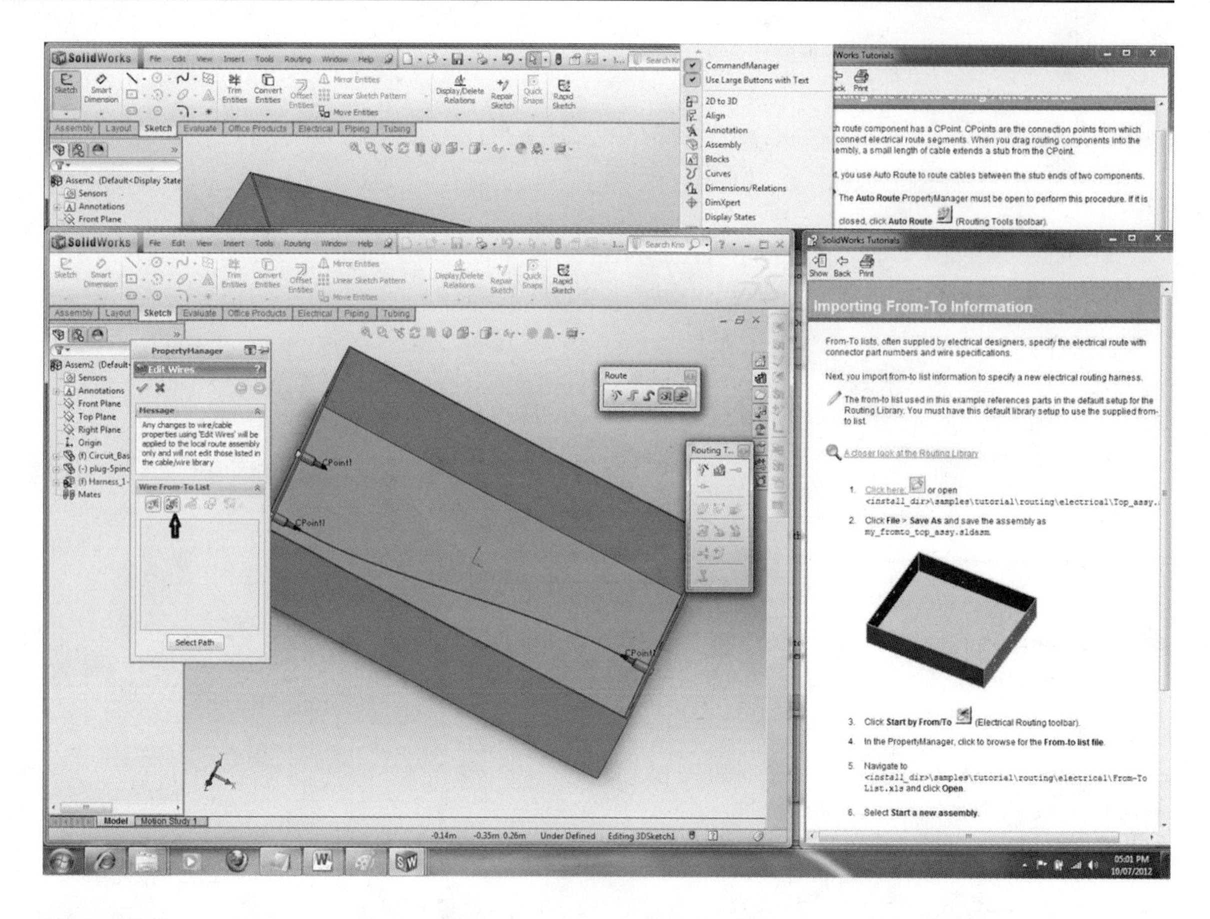

Figure 29.15

Add Wire PropertyManager.

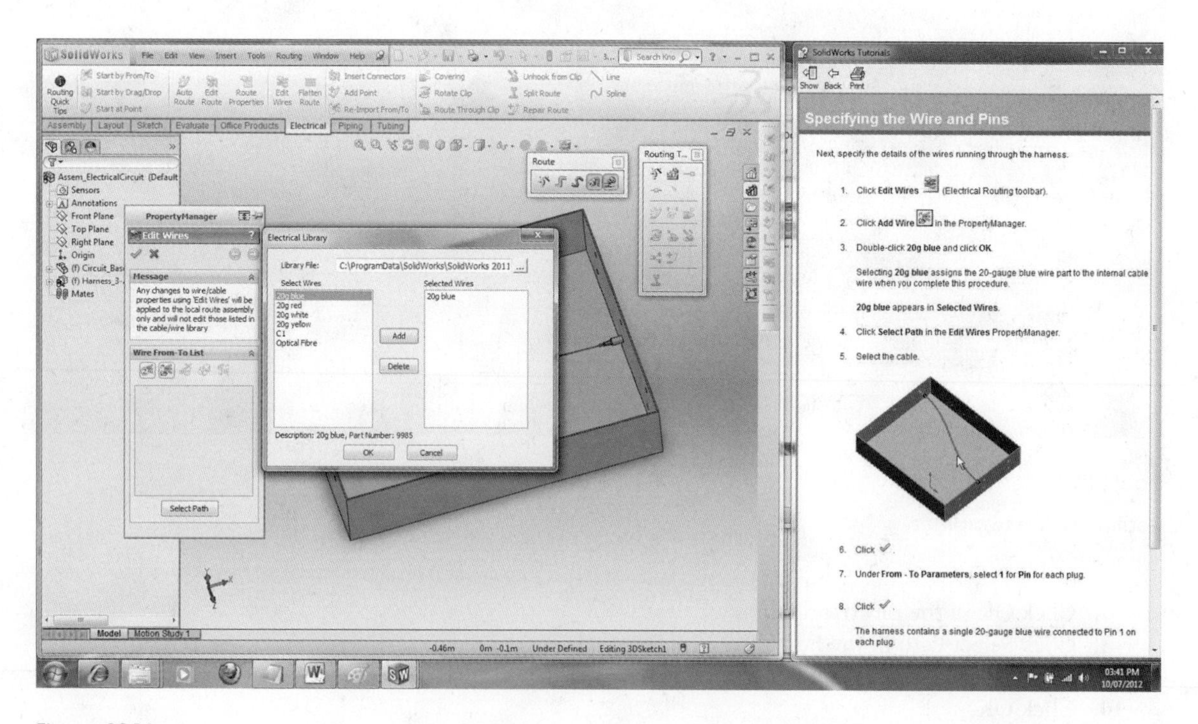

Figure 29.16

Electrical Library.

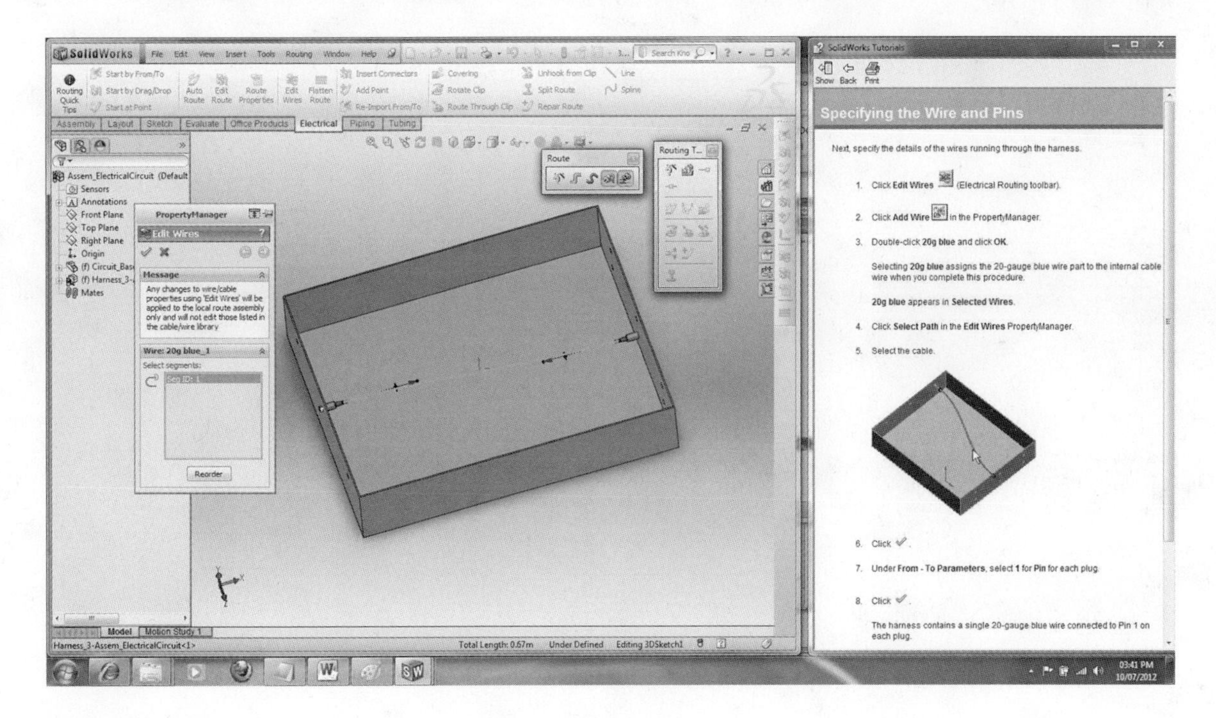

Figure 29.17

Select the cable.

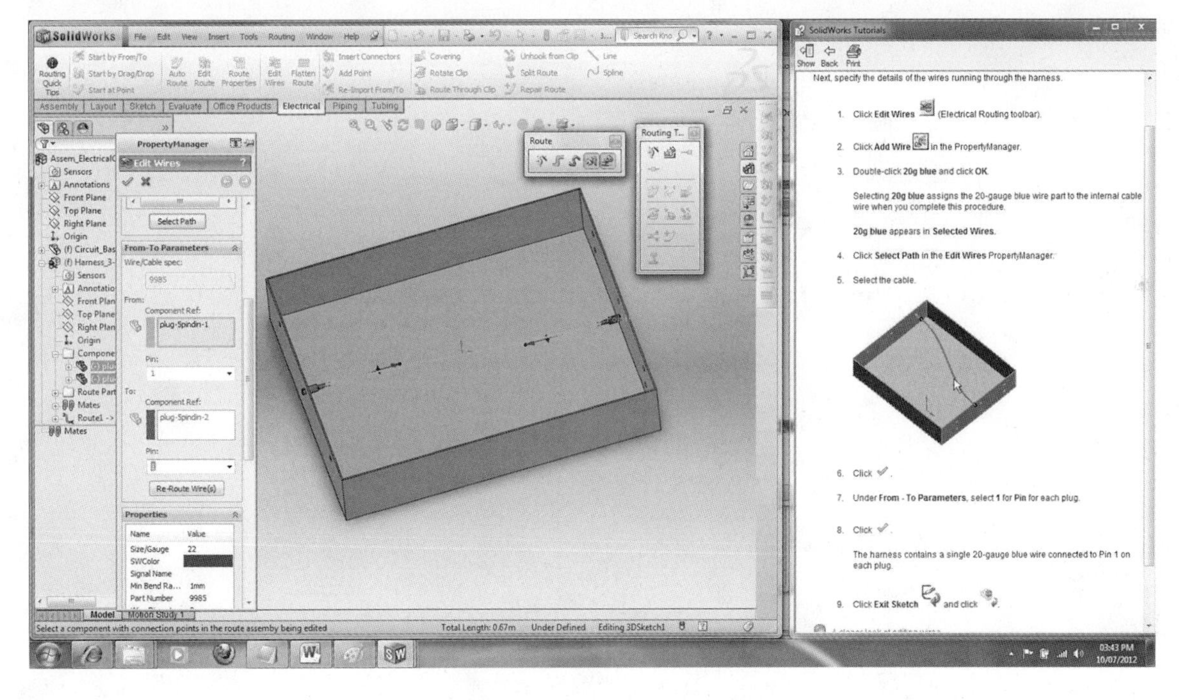

Figure 29.18

Pin assignment.

The harness contains a single 20-gauge blue wire that is connected to Pin 1 on each plug (see Figure 29.18).

13. Click Exit 3D Sketch and click Edit Component to exit assembly mode.

Bibliography

http://help.solidworks.com/2012/English/SolidWorks/sldpiping/c_modeling_electrical_routes.htm.

30

Customized Internal and External Threads

Objectives:

When you complete this chapter, you will have

- Learned how to model customized internal threads
- Learned how to model customized external threads
- Learned how to model internal threads to become external threads

The Hole Wizard tool is preferred for creating standard internal threads (nuts, drilled holes, etc.), whereas Toolbox is preferred for creating standard external threads (bolts, screws, etc.). However, when nonstandard threads (internal/external) are required, it is still recommended to use the Hole Wizard tool for creating internal threads (select the nearest standard thread), whereas manual methods have to be used to create customized external threads.

Customized Internal Threads

As discussed, it is recommended to use the *Hole Wizard* tool for creating internal threads. (Select the nearest standard thread.) For a 2-in.-diameter, 5-in.-long tube, a *1-12-tapped hole* is created using the Hole Wizard tool (see Figure 30.1).

Note that when you first click a surface before invoking the Hole Wizard tool, you will be in 2D Sketching mode, whereas when you do not first click a surface before invoking the Hole Wizard tool, you will be in 3D Sketching mode. This distinction is important because it affects how to go about dimensioning the position of the hole that is created.

The other point worth noting is that the threads created using the Hole Wizard tool could be designated as cosmetic threads and displayed accordingly.

Cosmetic Threads

For a shaded display of cosmetic threads, click Options. On the Document Properties tab, select Detailing. Under Display filter, select or check the box for Shaded cosmetic threads.

Figure 30.1

Internal Threading using Hole Wizard.

Customized External Threads

As discussed, manual methods have to be used to create customized external threads.

Problem Description

Let us consider a lens cap with a base circle diameter of 4.9 in., depth of 1.725 in. and a 5° draft angle. External threads are to be created, 0.45 in. from the top of the larger diameter with a pitch of 0.25 in. and 2.5 revolutions while still maintaining a 5° draft angle.

SolidWorks Solution

As could be observed from the problem description, this part cannot be automatically created using the SolidWorks Toolbox nor the Design Library because the part is not a standard one. To create the threads, we need to follow the principles based on Swept Parts, already covered.

1. Click New from the Menu bar, accepting Imperial units.
2. Right-click Front Plane from the FeatureManager.
3. Click Sketch from the CommandManager or from the Context tool bar.
4. Click the Circle Sketch tool and sketch a circle centered at the origin.
5. Click Smart Dimension and dimension the diameter as 4.9 (see Figure 30.2).

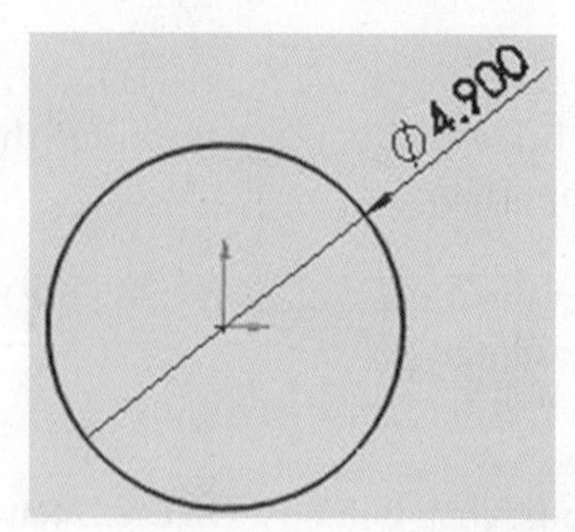

Figure 30.2

Sketch1.

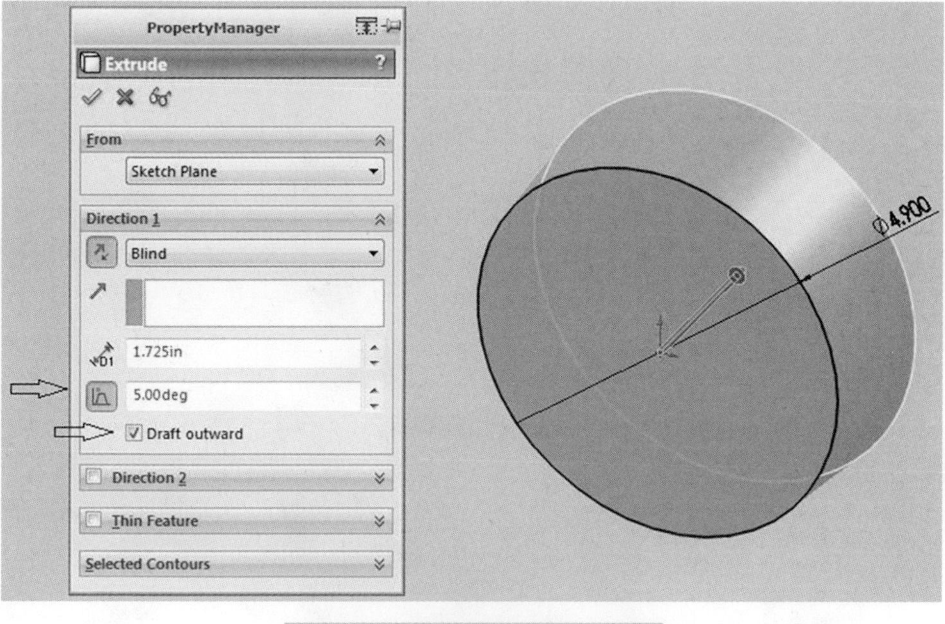

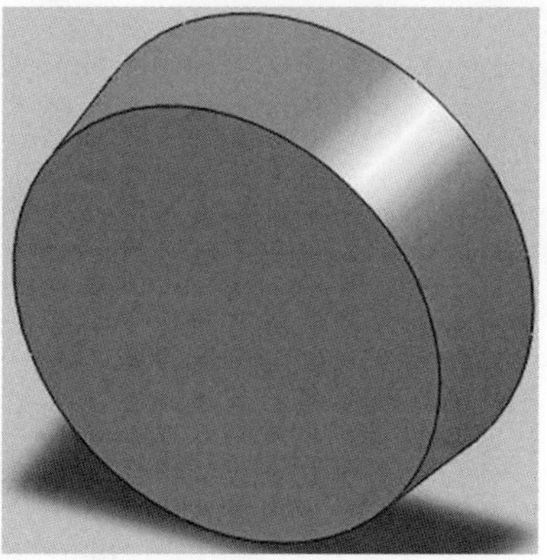

Figure 30.3

Extrusion of Sketch1 to create a draft.

Insert and Extruded Base Feature

6. Click Extruded Boss/Base feature tool. (The Extrude PropertyManager appears as in Figure 30.3.)
7. Accept the default Blind condition and set the Distance as 1.725 (see Figure 30.3).
8. Click the Draft On/Off button to activate it (this makes it possible to model a drafted feature) (see Figure 30.3).
9. Click the Draft Outward box (this makes the draft to be outward) (see Figure 30.3).
10. Enter 5 deg for Angle.
11. Click OK to complete the extrusion.

Creating a Sketch for Extruded Cut Feature

12. Click the *front face* of the smaller diameter.
13. Click Sketch from the CommandManager or from the Context tool bar.
14. Click the Circle Sketch tool and sketch a circle that is centered at the origin.
15. Click Smart Dimension and dimension the diameter as 3.875 (see Figure 30.4).

Insert and Extruded Cut Feature

16. Click Extruded Cut feature tool. (The Extrude PropertyManager appears as in Figure 30.5.)
17. Accept the default Blind condition and set the Distance as 0.275 (see Figure 30.5).

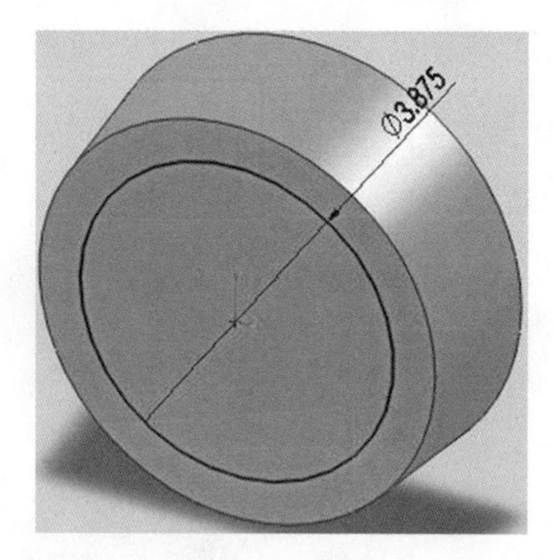

Figure 30.4

Sketch2 for Extrude Cut Feature.

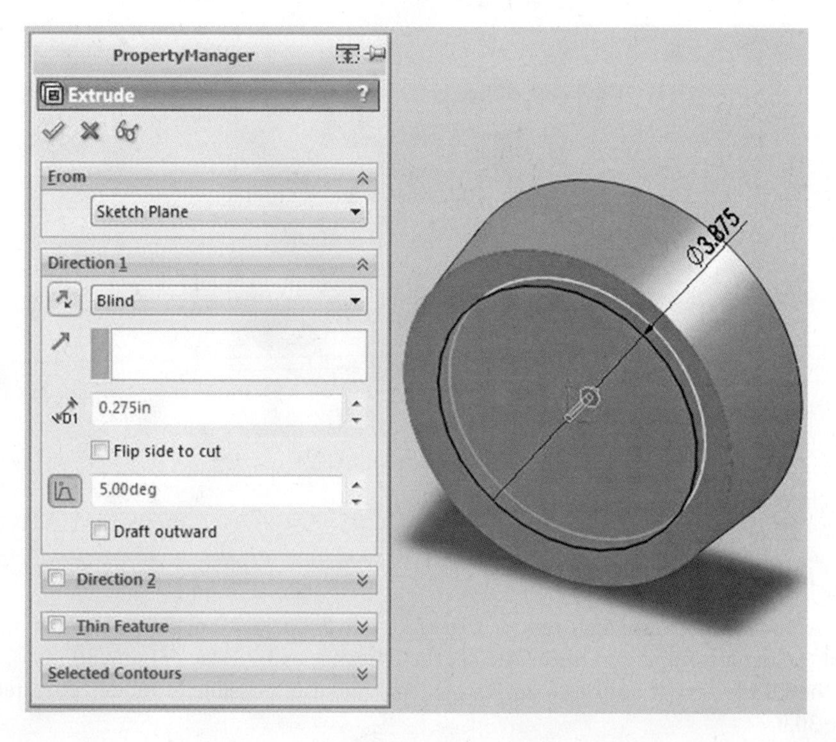

Figure 30.5

Extrude Cut Feature.

18. Click the Draft On/Off button to activate it (this makes it possible to model a drafted feature) (see Figure 30.5; note that the Draft Outward condition of 5 deg is still active).
19. Click OK to complete the extrude cut.

Inserting the Shell Feature

20. Click the Shell Feature tool from the CommandManager. (The Shell PropertyManager appears, as shown in Figure 30.6.)
21. Click *both the inner smaller and larger circular faces* (Face<1> and Face<2>) (see Figure 30.6).
22. Set shelling Thickness as 0.15.
23. Click OK to complete shelling.

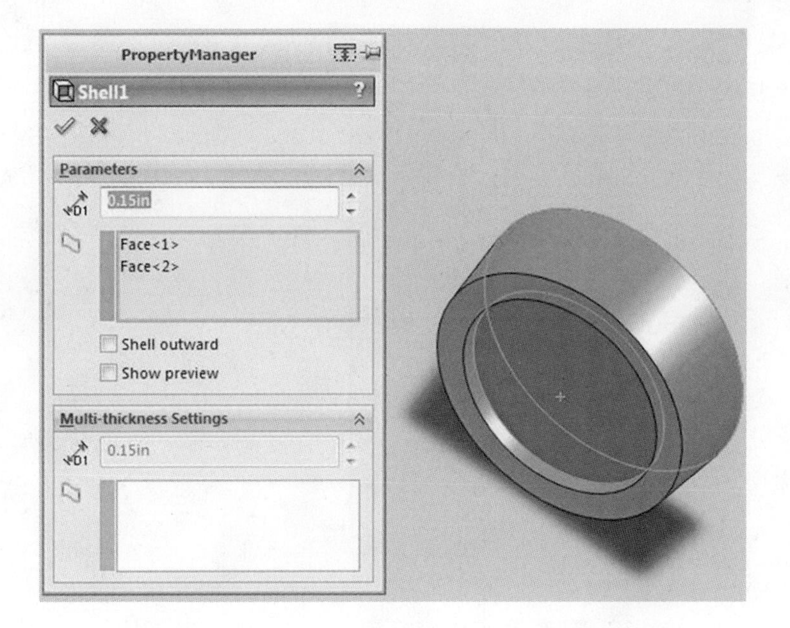

Figure 30.6

Shell1 PropertyManager.

Creating a Part-Revolved Thin Cut Feature

24. Click the Right Plane.
25. Click Hidden Lines Visible from the Heads-up View toolbar (see Figure 30.7).
26. Click Sketch from the CommandManager or from the Context tool bar.
27. Sketch *a horizontal dimension line.*
28. Click the *top slanted edge.*
29. Click Convert Entities to extract the edge.
30. Click the left vertex and drag it toward the right vertex.
31. Click Smart Dimension and dimension the slanting edge as 0.25 (see Figure 30.8).
32. Click Revolved Cut from the CommandManager. (A warning message appears, as shown in Figure 30.9.)
33. Accept the warning by clicking Yes. (See Figure 30.10 for the revolved cut FeatureManager.)
34. Expand the Thin Feature rollout and set the Thickness as 0.05 (see Figure 30.10).

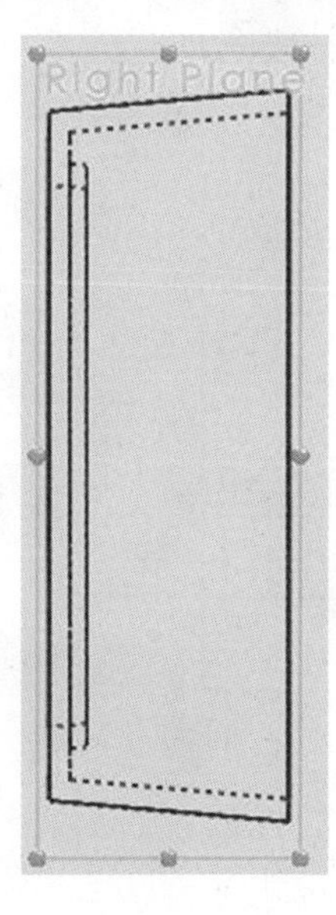

Figure 30.7

Hidden Lines Visible.

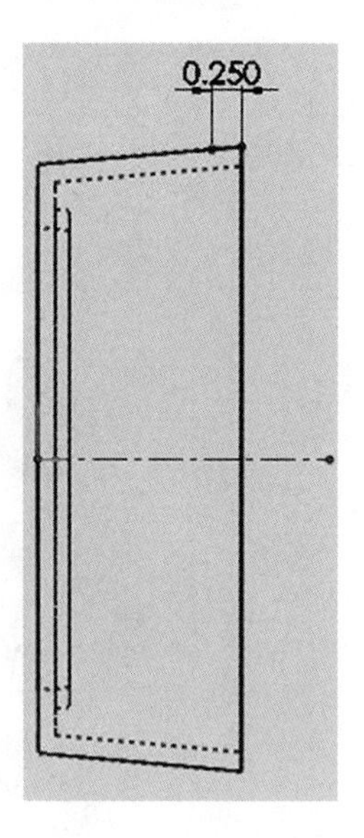

Figure 30.8

Dimensioning the Slanting Edge.

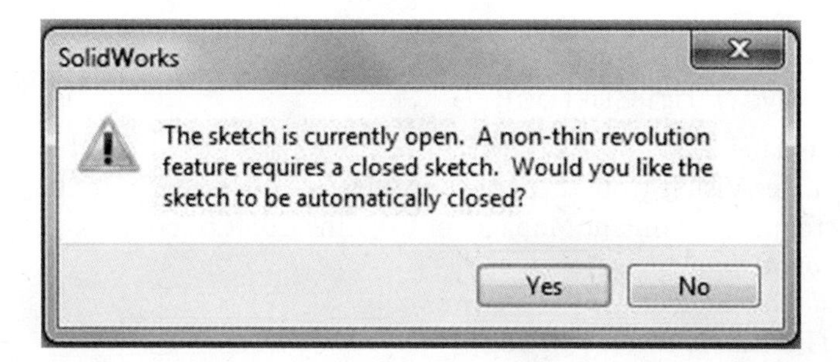

Figure 30.9

Open sketch warning.

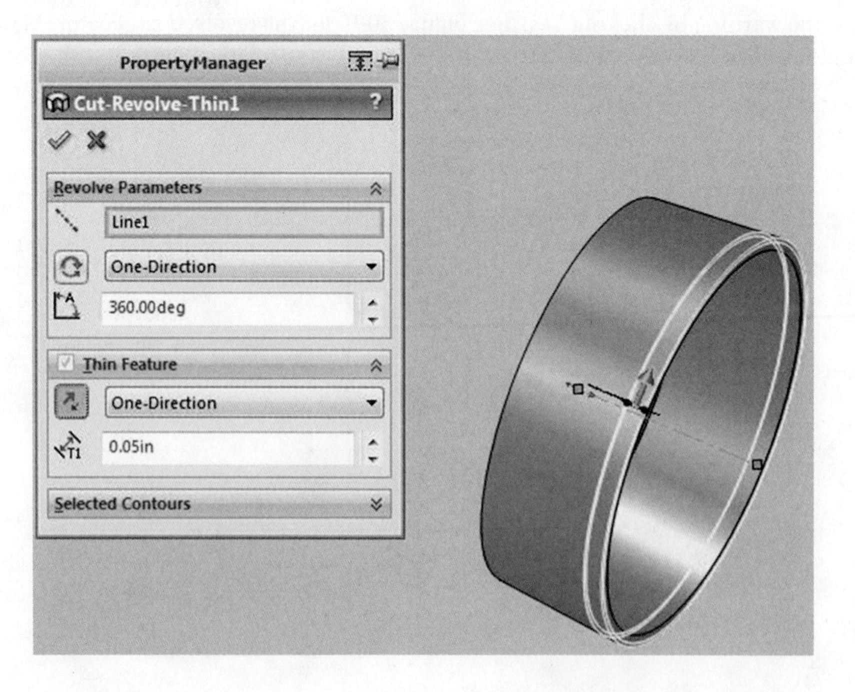

Figure 30.10

Revolved Cut FeatureManager.

Creating Path for External Threads

35. Click Top View from the View Orientation tool.
36. Click *the top narrow strip face of the model* (at the larger diameter) (see Figure 30.11).
37. Click Features > Plane and create a plane Plane1, 0.45 from the Top (see Figure 30.12).
38. Right-click Plane1.
39. Click Sketch from the CommandManager or from the Context tool bar.
40. Click the *inner circle of the narrow strip of the top face of the model.*
41. Click Convert Entities to extract the circle unto Plane1 (see Figure 30.13).
42. Click Features > Curves > Helix and Spiral. (The Helix PropertyManager appears, as in Figure 30.14.)
43. Select Pitch and Revolution for Define By: option (Figure 30.14).
44. Enter the value of Pitch as 0.25 (Figure 30.14).

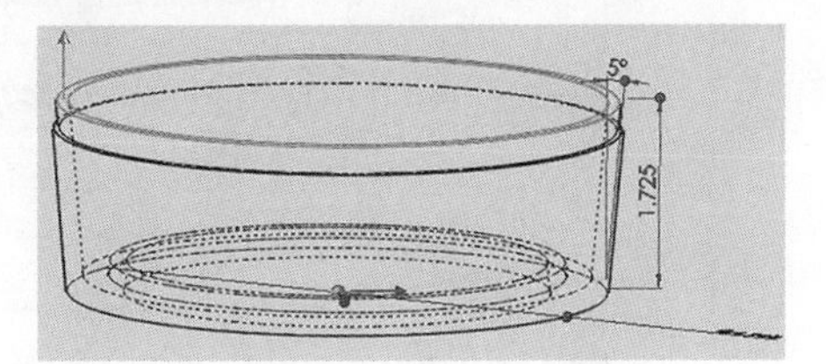

Figure 30.11

Selecting strip of the top face of the model.

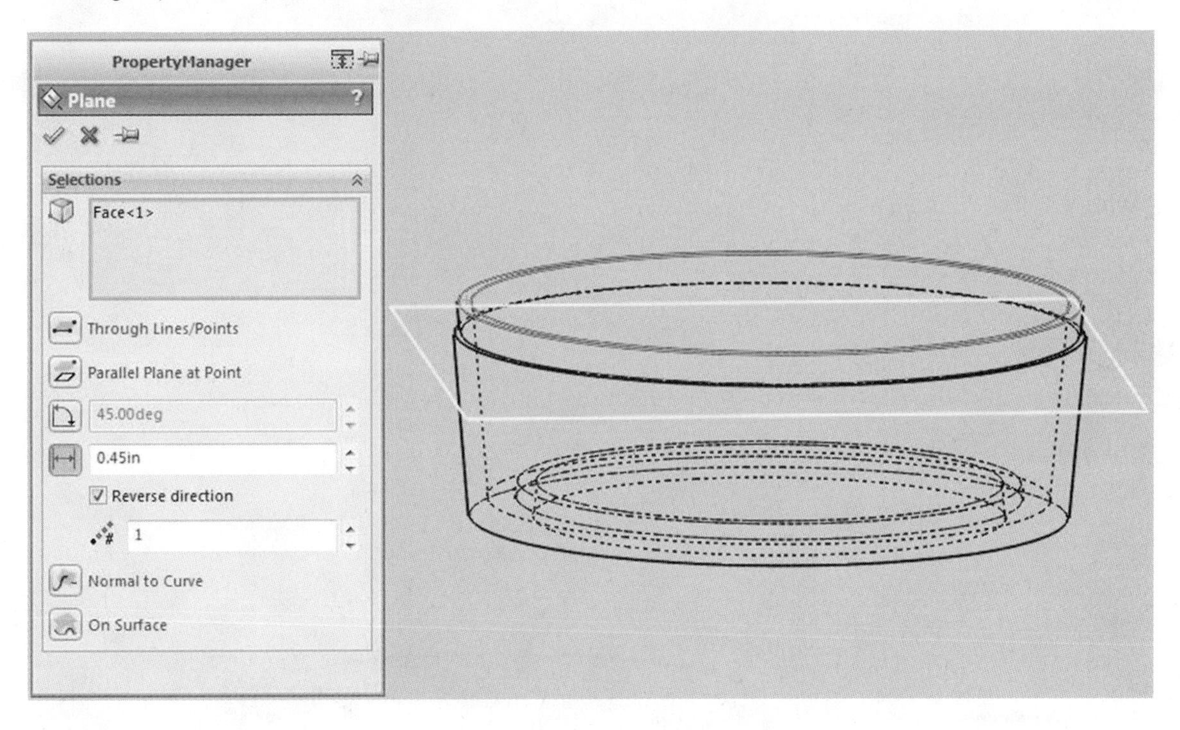

Figure 30.12

Plane1.

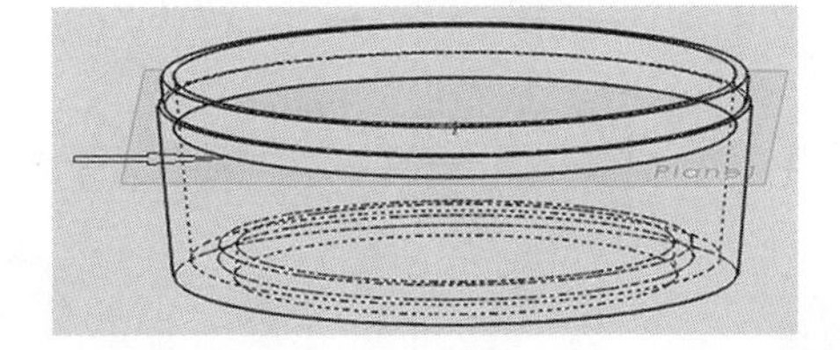

Figure 30.13

Converted Entity on Plane1.

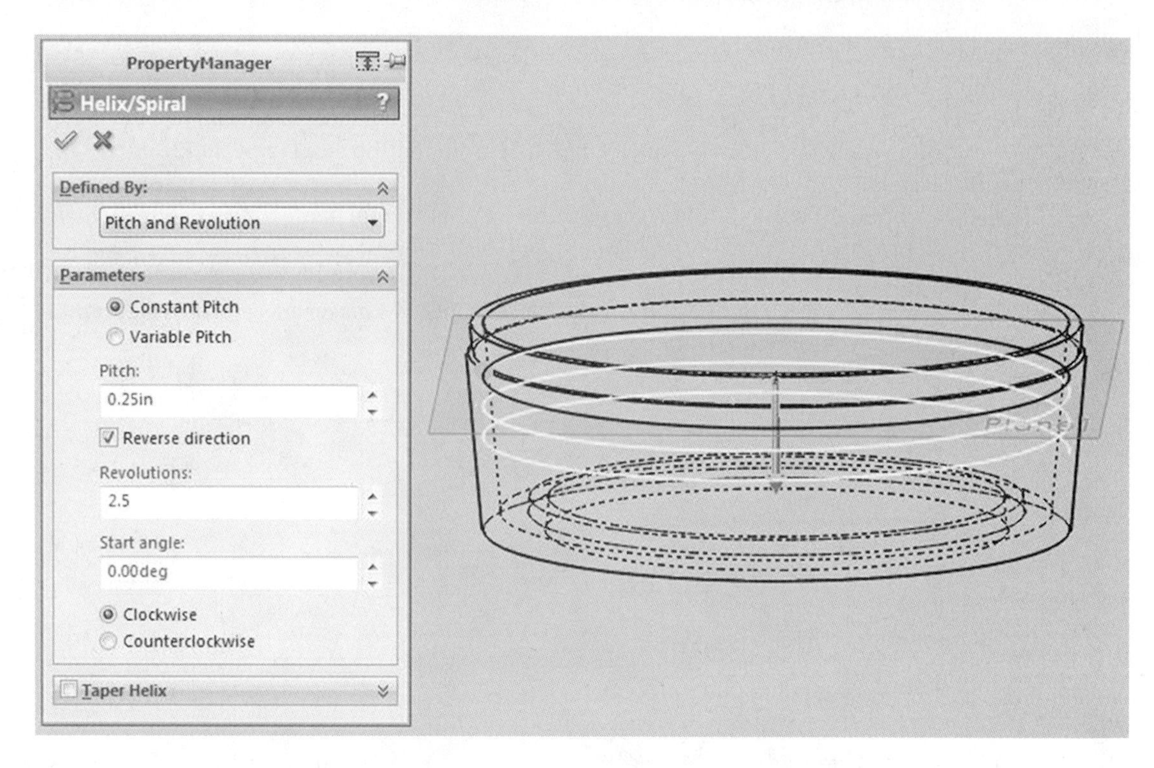

Figure 30.14

Helix definition to complete Path.

45. Enter the Number of Revolution as 2.5 (Figure 30.14).
46. Click OK the complete the Path definition.
47. Exit Sketch. (This is a very important step.)

Creating Profile for External Threads

48. Click Features > Plane. (See Plane PropertyManager in Figure 30.15.)
49. Click the Normal to Curve option (see Figure 30.15).

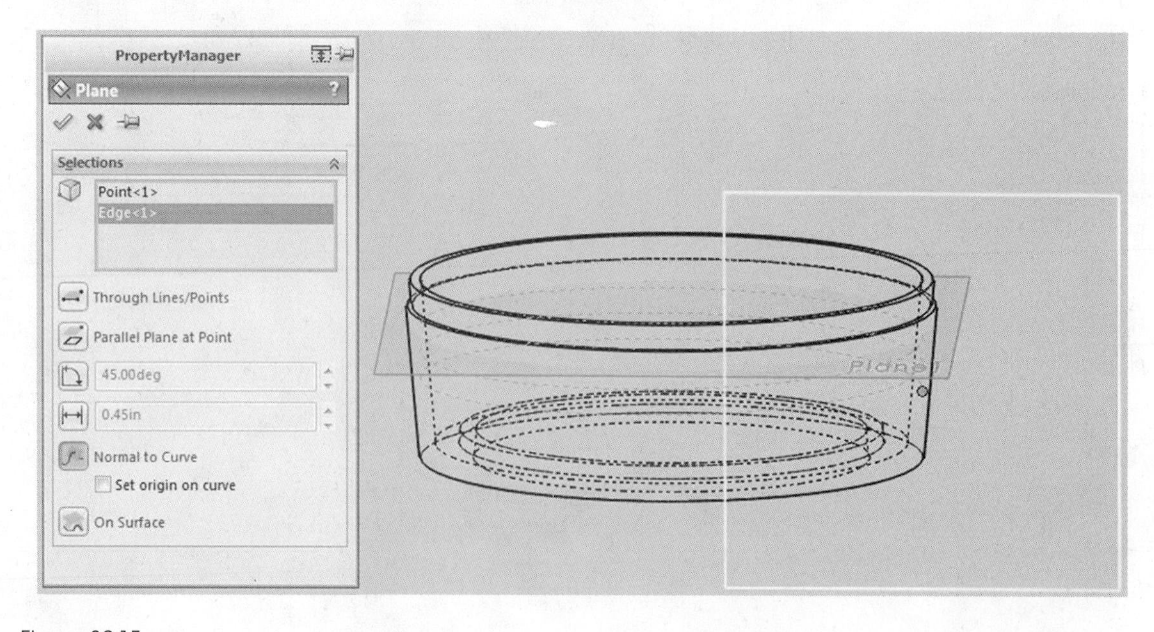

Figure 30.15

Plane2.

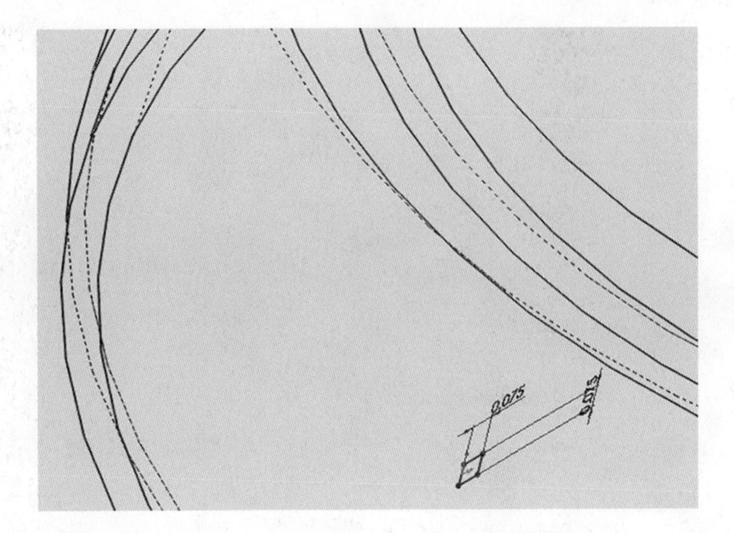

Figure 30.16

Hexagonal center pierces the helix.

50. Click *one endpoint of the Helix* (see Figure 30.15).
51. Click the Helix from the graphics window (see Figure 30.15).
52. Click OK to define Plane2, which is orthogonal to the helix at any point.
53. Right-click Plane2.
54. Click Sketch from the CommandManager or from the Context tool bar.
55. Sketch a Corner Rectangle.
56. Dimension *a side of Corner Rectangle* to a size of 0.075.
57. Use the Relations tool to make the two orthogonal sides to be Equal.
58. Use the Relations tool to make the *Midpoint of the Corner Rectangle* to coincide with the Helix (Edge<1>) by selecting the Pierce condition (see Figure 30.16).
59. Click OK to complete the definition of the profile.
60. Exit Sketch. (This is a very important step.)

Creating External Threads

61. Click Swept Boss/Base. (The Sweep PropertyManager appears as in Figure 30.17.)
62. Click Profile as Sketch5 (Corner Rectangle).

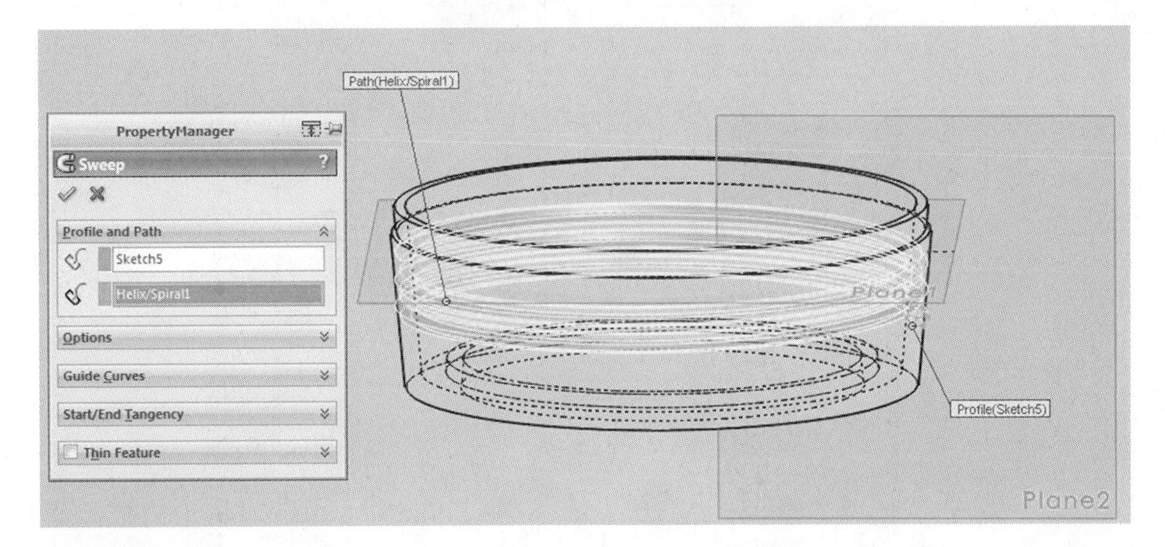

Figure 30.17

Sweep PropertyManager.

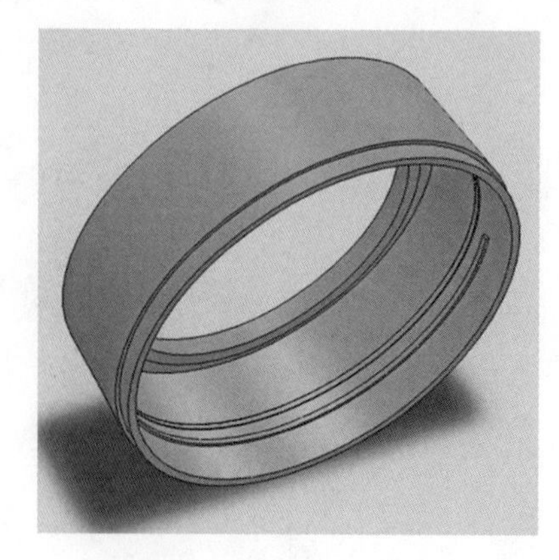

Figure 30.18

Completed part model.

63. Click Path as Helix/Spiral1 (see Figure 30.17).
64. Click OK to complete the part modeling (see Figure 30.18).
65. Save the completed part model.

Editing Features

Editing Features or Modification of Part Model involves a proper understanding of the FeatureManager (see Figure 30.19). Our goal is to change the internal threads to external threads.

66. Suppress the Cut-Revolve-Thin feature (see Figure 30.19).
67. Expand the Sweep1 tree.
68. Right-click the *base sketch*, Sketch4.
69. Click Edit Sketch to be in Sketch mode.
70. Delete the base sketch, Sketch4.
71. Click the *outer circle of the narrow strip of the top face of the model.*
72. Click Convert Entities to extract the circle unto Plane1 (see Figure 30.20).
73. Exit Sketch. (See Figure 30.21 for the modified part model.)

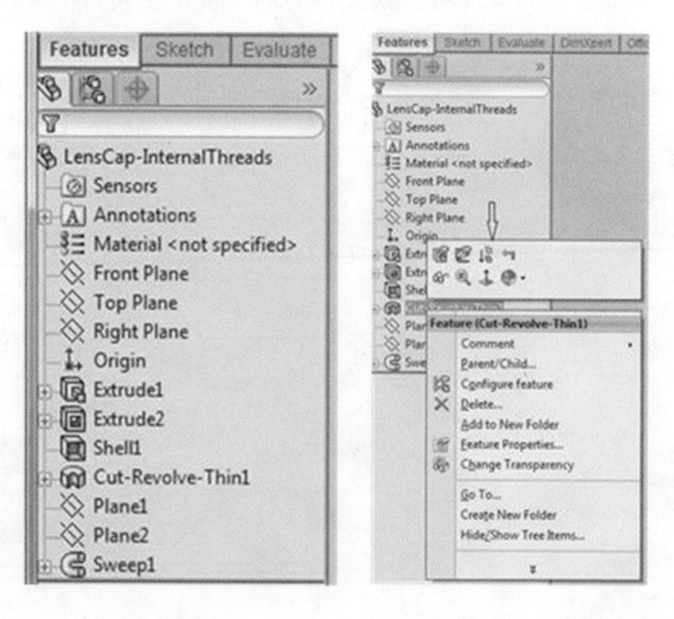

Figure 30.19

FeatureManager of the completed part model.

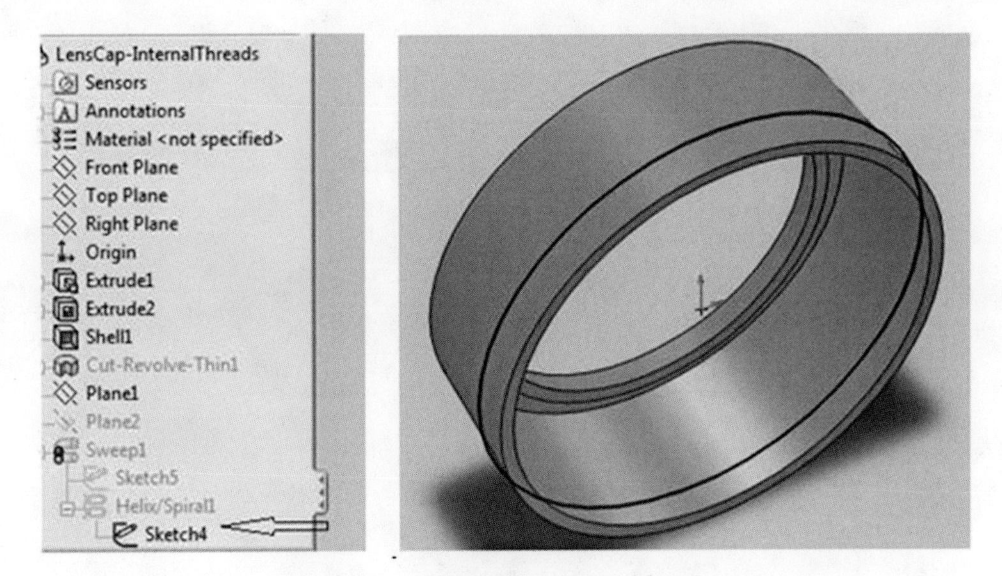

Figure 30.20

Completed part model.

Figure 30.21

Modified part model.

Unified Thread Standard

The *Unified Thread Standard (UTS)* defines a standard thread form and series—along with allowances, tolerances, and designations—for screw threads that are commonly used in the United States and Canada. It has the same 60° profile as the International Organization for Standardization (ISO) metric screw thread that is used in the rest of the world, but the characteristic dimensions of each UTS thread (outer diameter and pitch) were chosen as an inch fraction rather than a round millimeter value. The UTS is currently controlled by the American Society of Mechanical Engineers/American National Standards Institute (ANSI) in the United States.

Basic Profile

Each thread in the series is characterized by its major diameter D_{maj} and its pitch, *P*. UTS threads consist of a symmetric V-shaped thread (see Figure 30.22). In the plane of the thread axis, the flanks of the V have an angle of 60° to each other. The outermost 0.125 and the innermost 0.25 of the height *H* of the V-shape are cut off from the profile.

The pitch *P* is the distance between thread peaks. For UTS threads, which are single-start threads, it is equal to the lead, the axial distance that the screw advances during a 360° rotation. UTS threads do not

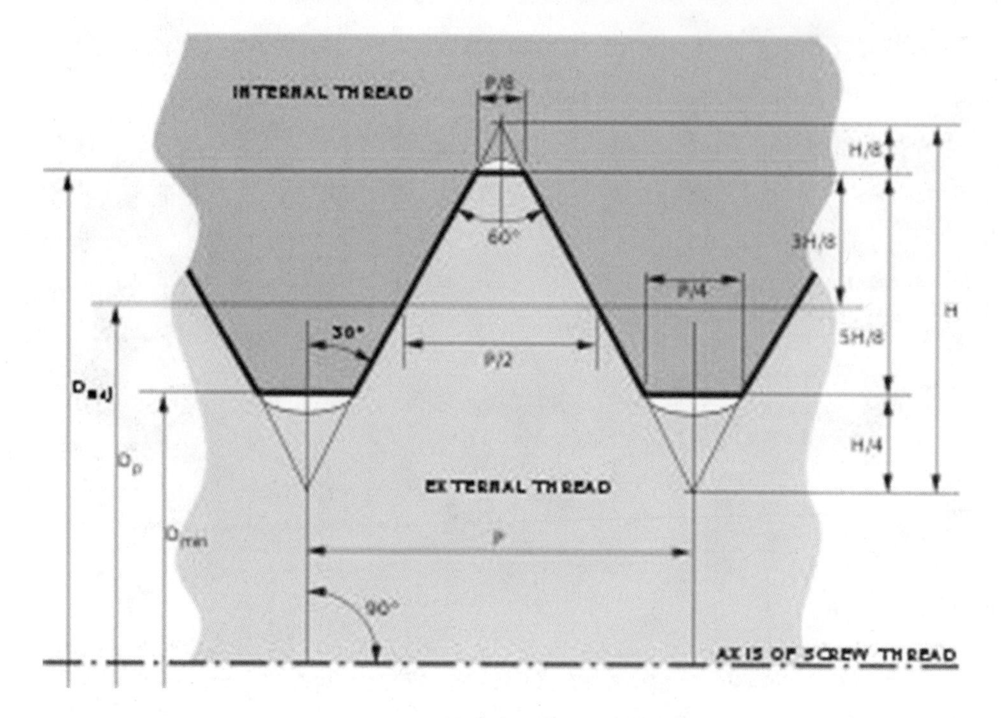

Figure 30.22

Basic profile.

usually use the pitch parameter; instead, a parameter known as threads per inch (TPI) is used, which is the reciprocal of the pitch.

The basic profile of all UTS threads is the same as that of all ISO metric screw threads. Only the commonly used values for D_{maj} and P differ between the two standards.

The relationship between the height H and the pitch P is found using the following equation:

$$H = \cos(30°) \times P = \frac{\sqrt{3}}{2} \times P \approx 0.866 \times P$$

In an external (male) thread (e.g., on a bolt), the major diameter D_{maj} and the minor diameter D_{min} define the *maximum* dimensions of the thread. This means that the external thread must end flat at D_{maj} but can be rounded out below the minor diameter D_{min}. Conversely, in an internal (female) thread (e.g., in a nut), the major and minor diameters are *minimum* dimensions; therefore, the thread profile must end flat at D_{min} but may be rounded out beyond D_{maj}.

The minor diameter D_{min} and the effective pitch diameter D_p are derived from the major diameter and pitch as

$$D_{min} = D_{maj} - 2 \times \frac{5}{8} \times H = D_{maj} - \frac{5\sqrt{3}}{8} \times P = D_{maj} - 1.082532 \times P$$

$$D_p = D_{maj} - 2 \times \frac{3}{8} \times H = D_{maj} - \frac{3\sqrt{3}}{8} \times P = D_{maj} - 0.649519 \times P$$

Designation

The standard designation for a UTS thread is a number indicating the nominal (major) diameter of the thread, followed by the pitch that is measured in TPI. For diameters that are smaller than 0.25 in., the diameter is indicated by an integer number that is defined in the standard; for all other diameters, the value in inches is given. This number pair is optionally followed by the letters UNC, UNF, or UNEF if the diameter–pitch combination is from the *coarse*, *fine*, or *extra fine* series and may also be followed by a tolerance class.

Major Diameter (in mm)	TPI			Tap Drill Size	
	Coarse (UNC)	Fine (UNF)	Extra Fine (UNEF)	Coarse	Fine
#0 = 0.0600 (1.5240)	–	80			3/64 in.
#1 = 0.0730 (1.8542)	64	72		#53	#53
#2 = 0.0860 (2.1844)	56	64		#50	#50
#3 = 0.0990 (2.5146)	48	56		#47	#45
#4 = 0.1120 (2.8448)	40	48		#43	#42
#5 = 0.1250 (3.1750)	40	44		#38	#37
#6 = 0.1380 (3.5052)	32	40		#36	#33
#8 = 0.1640 (4.1656)	32	36		#29	#29
#10 = 0.1900 (4.8260)	24	32		#25	#21
#12 = 0.2160 (5.4864)	24	28	32	#16	#14
1/4 (6.3500)	20	28	32	#7	#3
5/16 (7.9375)	18	24	32	F	I
3/8 (9.5250)	16	24	32	5/16 in.	Q
7/16 (11.1125)	14	20	28	U	25/64 in.
1/2 (12.7000)	13	20	28	27/64 in.	29/64 in.
9/16 (14.2875)	12	18	24	31/64 in.	33/64 in.
5/8 (15.8750)	11	18	24	17/32 in.	37/64 in.
3/4 (19.0500)	10	16	20	21/32 in.	11/16 in.
7/8 (22.2250)	9	14	20	49/64 in.	13/16 in.
1 (25.4000)	8	14[a]	20	7/8 in.	59/64 in.

[a]1–12 was formerly a widespread standard, but 1–14 is the current UNF.

Calculating the Major Diameter of a Numbered Screw Greater than or Equal to 0

The following formula is used to calculate the major diameter of a numbered screw that is greater than or equal to 0:

$$\text{Major diameter} = \text{Screw \#} \times 0.013'' + 0.060''.$$

Example 1

A number 10 is calculated as #10 × 0.013″ + 0.060″ = 0.190″ major diameter.

Example 2

#6-32 UNC 2B (major diameter: #6 × 0.013″ + 0.060″ = 0.1380 in., pitch: 32 TPI)

Unified Miniature Screw Thread Series

A Unified Miniature screw thread series is defined in ANSI standard B1.10, for fasteners of 0.3–1.4-mm (0.0118–0.0551-in.) diameter. These sizes are intended for watches, instruments, and miniature mechanisms and are interchangeable with the threads that are made to ISO Standard 68. These screw sizes are denoted by multiple zeroes, i.e., #000.

Calculating the Major Diameter of a Numbered Screw Greater than or Equal to 0

The formula for number sizes smaller than size #0 is given by

$$\text{Major diameter} = 0.060'' - \text{zero size} \times 0.013''$$

where with the zero size being the number of zeroes after the first.

Example 3

A #00 screw is 0.047″ diameter (0.060″ – (2 – 1) × 0.013″)

Example 4

A #000 screw is 0.034″ diameter (0.060″ – (3 – 1) × 0.013″)

Design Equations for Unified Thread Standard

The *Unified Thread Standard (UTS)* defines a standard thread form and series—along with allowances, tolerances, and designations—for screw threads that are commonly used in the United States and Canada. UTS threads consist of a symmetric V-shaped thread. In the plane of the thread axis, the flanks of the V have an angle of 60° to each other. The outermost 0.125 and the innermost 0.25 of the height H of the V-shape are cut off from the profile.

Each thread in the series is characterized by its major diameter D_{maj} and its pitch, P. The geometry of the threads is fully defined based on dimensions that are based on the height and the minor diameter.

The relationship between the height H and the pitch P is found using the following equation:

$$H = \cos(30°) \times P = \frac{\sqrt{3}}{2} \times P \approx 0.866 \times P$$

The minor diameter D_{min} is derived from the major diameter and pitch as

$$D_{min} = D_{maj} - 2 \times \frac{5}{8} \times H = D_{maj} - \frac{5\sqrt{3}}{8} \times P = D_{maj} - 1.082532 \times P$$

External Threads

The basic profile of an external thread is shown in Figure 30.23. An isosceles triangle is constructed such that its height y is equal to $6H/8$. A horizontal line is drawn at a distance of $H/8$ from the vertical vertex of the triangle to produce a landing of $P/8$ in dimension. Hence, the actual height of the thread is $5H/8$. This height is obtained from subtracting $H/8$ from the height of the isosceles triangle ($6H/8 - H/8$). There are three dimensions that are of interest to us for constructing the profile of external threads: (1) base-dimension, (2) top-dimension, and (3) height.

$$\tan 60° = \frac{(H/8)}{(P/8)} = \frac{(6H/8)}{L}$$

(Note: we have omitted dividing *fall* [$P/8$ and L, respectively] by 2 for both the numerator and the denominator, since this will be cancelled out.)

$$L = \frac{6P}{8} = 0.75 \times P$$

Alternatively, we can obtain the length from the following:

$$\sin 60° = \frac{(6H/8)}{L}$$

$$L = \frac{6H}{8 \sin 60°} = \frac{6 \times 0.866P}{8 \sin 60°} = 0.75 \times P$$

$$h = \frac{5H}{8} = \frac{5 \times 0.866P}{8} = 0.54125 \times P$$

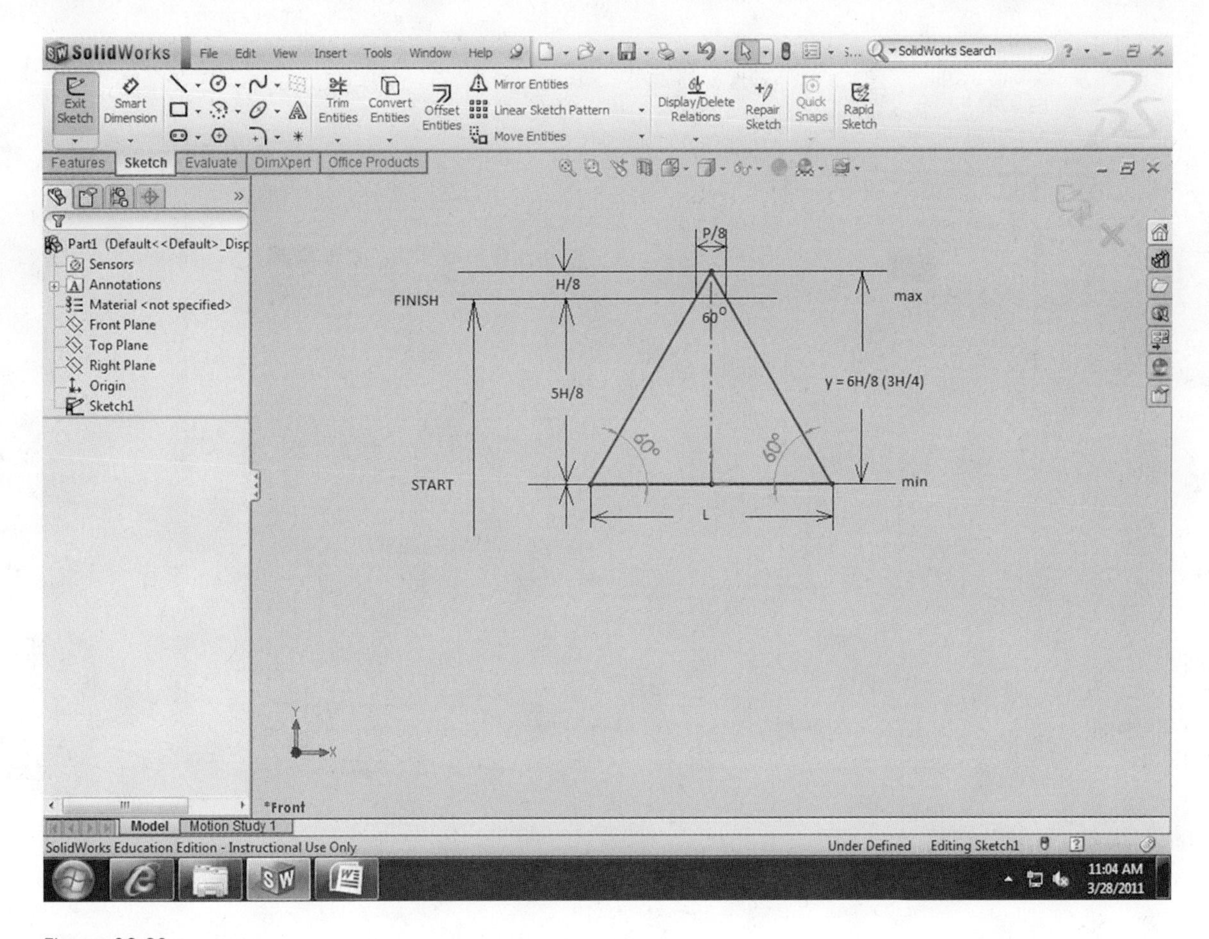

Figure 30.23

Basic profile of an external thread.

Internal Threads

The basic profile of an internal thread is shown in Figure 30.24. An isosceles triangle is constructed such that its height y is equal to $7H/8$. A horizontal line is drawn at a distance of $H/4$ or $2H/8$ from the vertical vertex of the triangle to produce a landing of $P/4$ in dimension. Hence, the actual height of the thread is $5H/8$. This height is obtained from subtracting $2H/8$ from the height of the isosceles triangle ($7H/8 - 2H/8$). There are three dimensions that are of interest to us for constructing the profile of external threads: (1) base-dimension, (2) top-dimension, and (3) height.

$$\tan 60° = \frac{(H/4)}{(P/4)} = \frac{(7H/8)}{L}$$

(Note: we have omitted dividing *fall* [$P/4$ and L, respectively] by 2 for both the numerator and the denominator, since this will cancel out.)

$$L = \frac{7P}{8} = 0.875 \times P$$

Alternatively, we can obtain the length from the following:

$$\sin 60° = \frac{(7H/8)}{L}$$

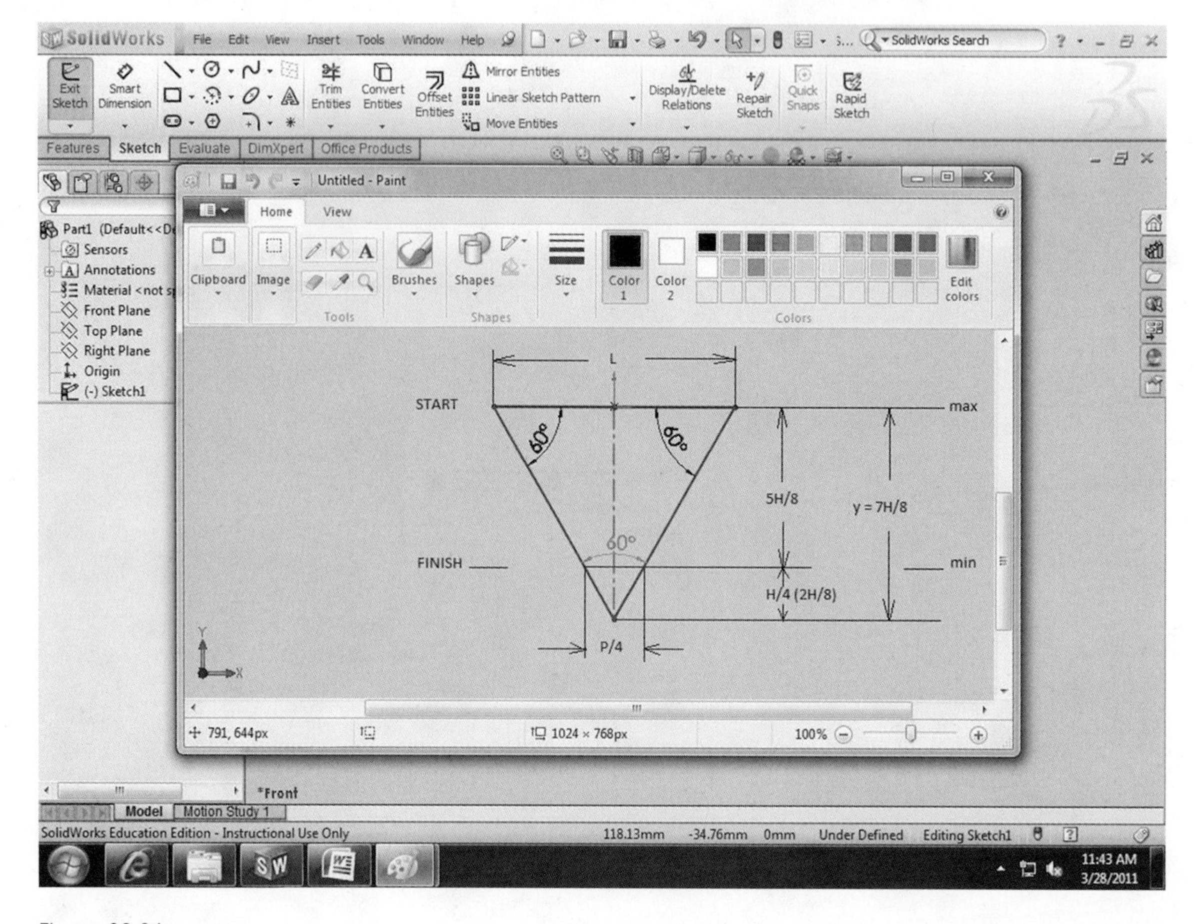

Figure 30.24

Basic profile of an internal thread.

$$L = \frac{7H}{8\sin 60^\circ} = \frac{7 \times 0.866P}{8\sin 60^\circ} = 0.875 \times P$$

$$h = \frac{5H}{8} = \frac{5 \times 0.866P}{8} = 0.54125 \times P$$

We now summarize the quantities that are needed for defining external and internal threads in Table 30.1.

Example 5

Determine the dimensions that are required to model the threads that are designated as 1 1/2 – 14UNF – 3A.

Table 30.1 Parameters for External and Internal Threads

	External Thread	Internal Thread
$L_{bottom} = L$	$0.75 \times P$	$0.875 \times P$
L_{top}	$P/8$	$P/4$
h	$0.54125 \times P$	$0.54125 \times P$
D_{helix}	$D_{min} = D_{maj} - 1.082532 \times P$	D_{max}

SOLUTION

The threads are external since the last letter is A.

The maximum diameter is $D_{max} = 1.5''$.

Teeth per inch is 14; therefore, the pitch, $P = 1/14 = 0.071428$.

$$L_{bottom} = 0.75 \times P = 0.75 \times 0.071428 = 0.05357''$$
$$L_{top} = P/8 = 0.071428/8 = 0.0089''$$
$$h = 0.54125 \times P = 0.54125 \times 0.071428 = 0.03866''$$
$$D_{helix} = D_{min} = D_{maj} - 1.082532 \times P = 1.5 - 1.082532/14 = 1.422676''$$

Example 6

Determine the dimensions that are required to model the threads that are designated as 1 1/2 – 14UNF – 3B.

SOLUTION

The threads are external since the last letter is B.

The maximum diameter is $D_{max} = 1.5''$.

Teeth per inch is 14; therefore, the pitch, $P = 1/14 = 0.071428''$.

$$L_{bottom} = 0.875 \times P = 0.875 \times 0.071428 = 0.0625''$$
$$L_{top} = P/4 = 0.071428/8 = 0.017857''$$
$$h = 0.54125 \times P = 0.54125 \times 0.071428 = 0.03866''$$
$$D_{helix} = D_{maj} = 1.5''$$

Example 7

Create a 3D model for the threads that are designated as 1 1/2 – 14UNF – 3A.

SOLUTION

The threads are external since the last letter is A. Example 5 shows the values of the design parameters for the tooth as follows:

$$D_{max} = 1.5''$$

$$P = 0.071428$$

$$L_{bottom} = 0.05357''$$
$$L_{top} = 0.0089''$$
$$h = 0.03866''$$
$$D_{helix} = 1.422676''$$

1. Open a New SolidWorks Document.
2. Choose the Front Plane.
3. Sketch a Circle of diameter equal to that of the helix diameter = 1.422676″ (see Figure 30.25).
4. Extrude Mid Plane 5″ (see Figure 30.26).
5. Click OK.
6. Click Top of the Cylinder.
7. Be in *Sketch* mode.

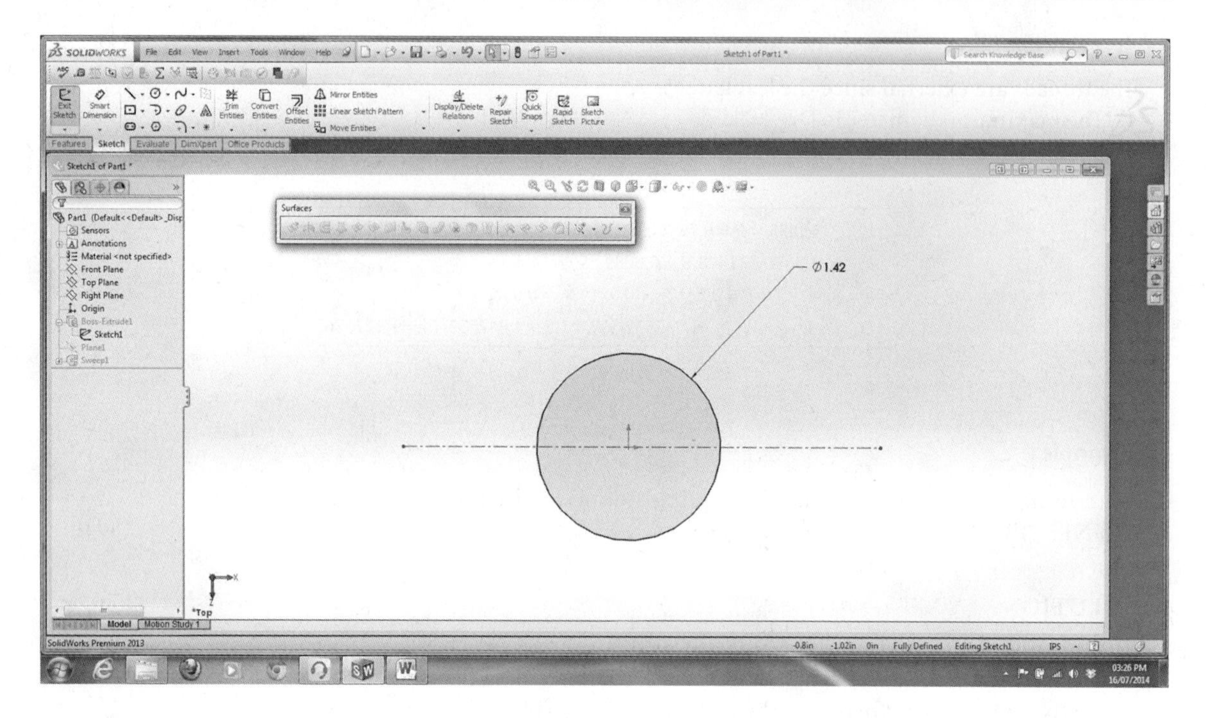

Figure 30.25

Circle of diameter equal to that of the helix diameter.

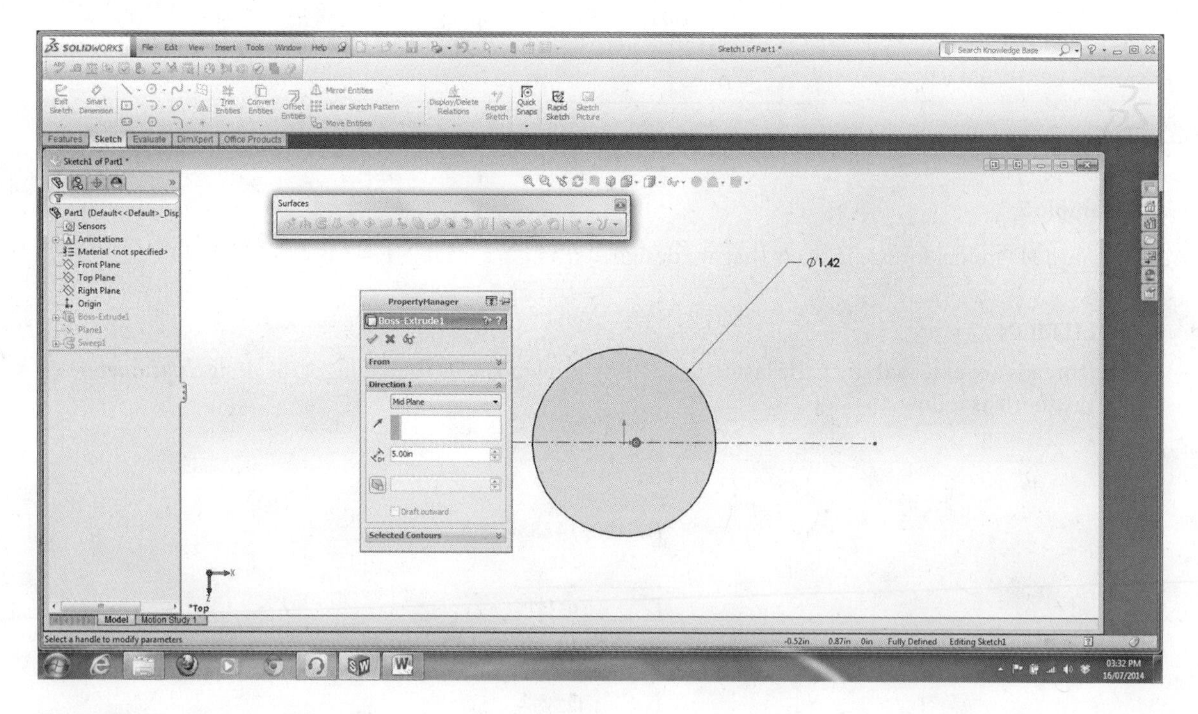

Figure 30.26

Set Mid Plane for Direction and Extrusion Distance = 5.00.

8. Click Convert Entities and select the Edge of the Circular Face to Extract the Circle.
9. Click Features/Curves/Helix and Spiral.
10. In the Helix/Spiral1 PropertyManager, assign Height = 2.00 and Pitch = 0.07142 (see Figure 30.27).
11. Click *endpoint of Helix*.
12. Click Features/Reference Geometry/Plane.
13. Assign First Reference = Point<1>, which is the endpoint of the helix (see Figure 30.28).

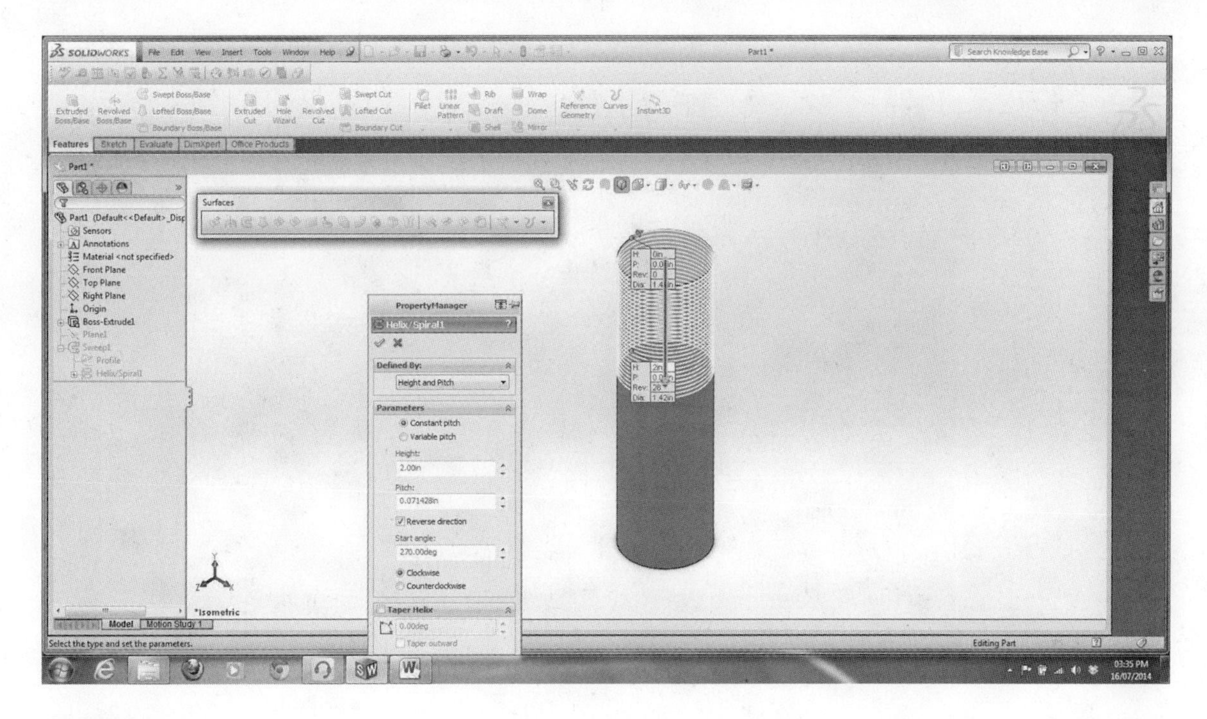

Figure 30.27

Helix/Spiral1 PropertyManager.

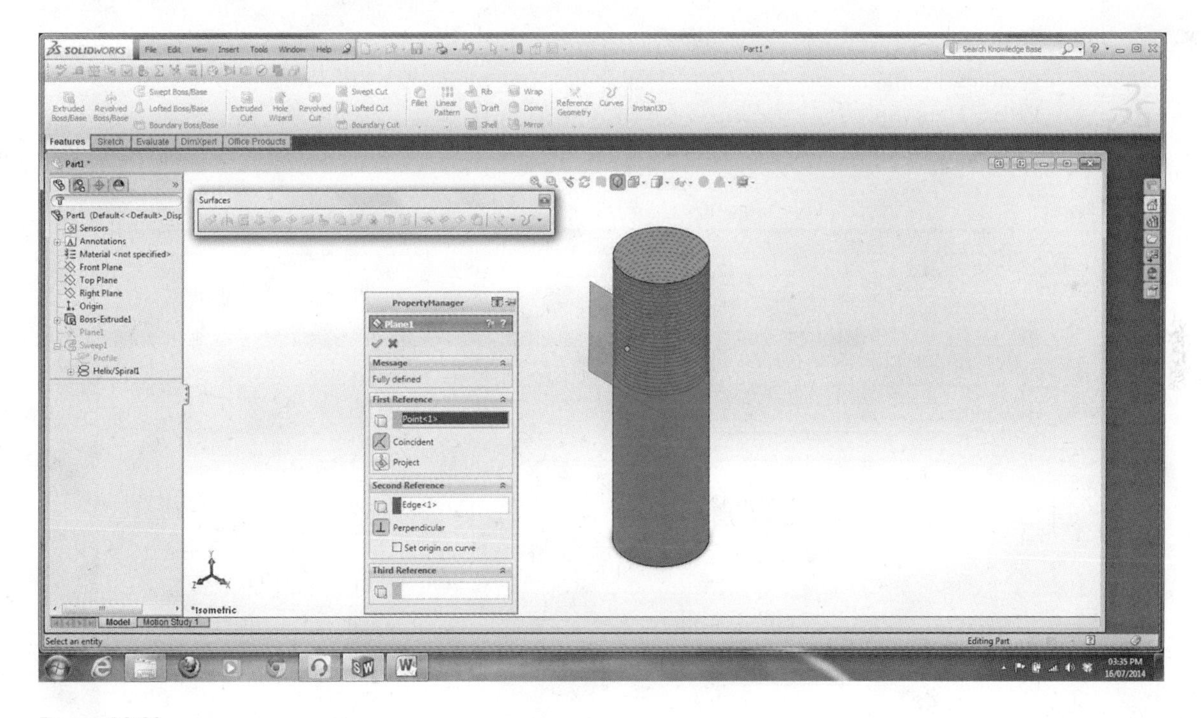

Figure 30.28

Plane defined, which is perpendicular to the helix.

14. Assign Second Reference = Edge<1>, which is obtained by clicking the Helix.
15. Sketch a tooth Profile of the *Thread* using the parameters that are already obtained (Figure 30.29).
16. Click Relations and assign Pierce condition between the *midpoint* of the root of the Thread and the Helix. (Note: this geometry defines the Profile.)
17. Exit Sketch mode.
18. Click Features/Sweep.

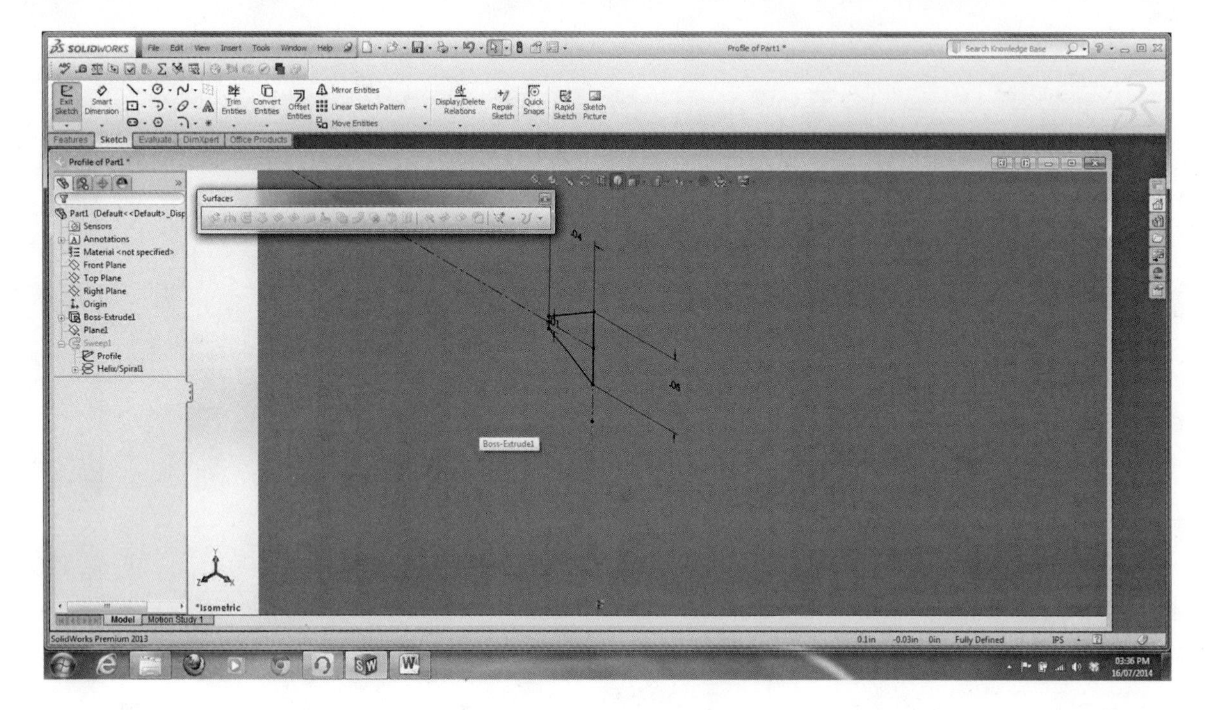

Figure 30.29

Tooth definition for the Plane.

19. In the Helix PropertyManager (see Figure 30.30) set the following:
    Profile = Profile
    Path = Helix/Spiral1
20. Click OK. (See Figure 30.31 for the final model.)

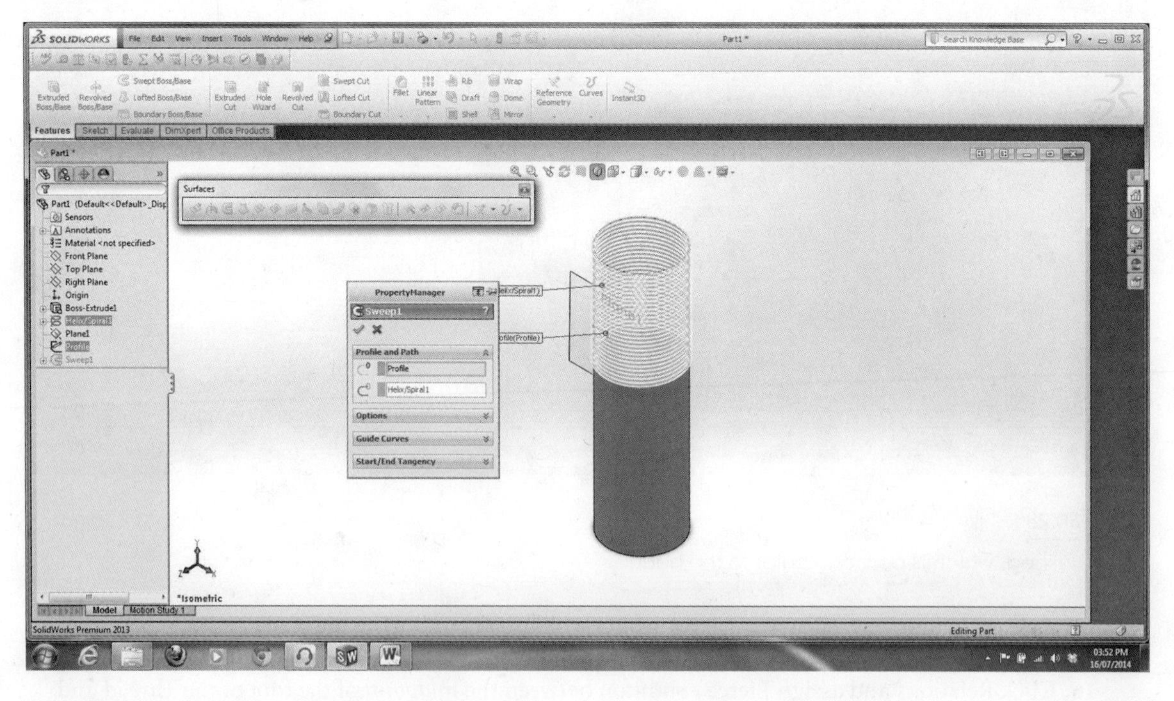

Figure 30.30

Parameter settings for the Sweep operation.

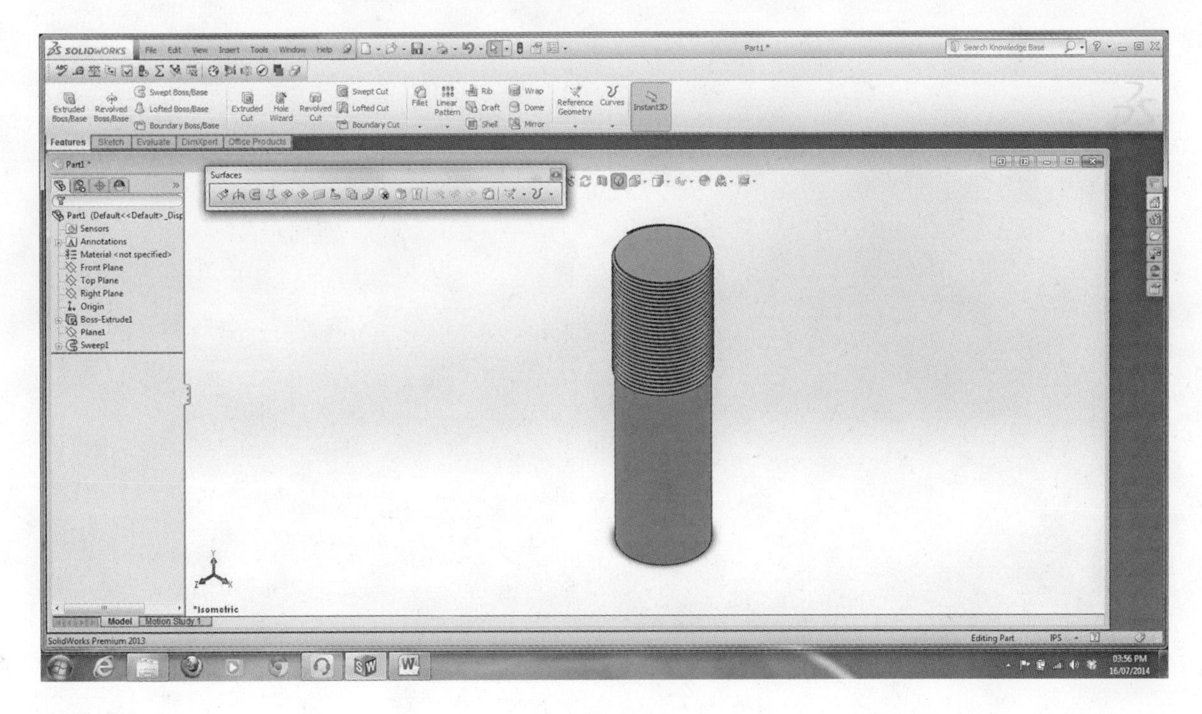

Figure 30.31

Threaded model.

Exercises

1. Use the UTS theory that is presented in this chapter to model external threads for the problem definition of the completed part model of Figure 30.18.
2. Use the UTS theory that is presented in this chapter to model external threads for the problem definition of the completed part model of Figure 30.21.

Bibliography

Noll Incorporated, Lead Screw Manufacturers. http://www.nollinc.com/index.php/contents/Products/Leadscrews/UNV.html

Unified Thread Standard. http://en.wikipedia.org/wiki/Unified_Thread_Standard

31

Sustainability Design for Parts

Objectives:

When you complete this chapter, you will

- Have learned about SolidWorks Sustainability
- Have used the SolidWorks Sustainability tools to analyze the environmental impact of a design throughout the life cycle of a product

Introduction

SolidWorks Sustainability evaluates the environmental impact of a design throughout the life cycle of a product. The engineer or designer can compare results from different designs to ensure a sustainable solution for the product and the environment. Sustainability measures the following areas of environmental impact: (a) carbon footprint, (b) energy consumption, (c) air acidification, and (d) water eutrophication. These are briefly defined.

- *Carbon footprint.* This is a measure of carbon dioxide and equivalents, such as carbon monoxide and methane that are released into the atmosphere primarily by burning fossil fuels.
- *Energy consumption.* All forms of nonrenewable energy that are consumed over the entire life cycle of the product.
- *Air acidification.* Acidic emissions occur, such as sulfur dioxide and nitrous oxides, which eventually lead to acid rain.
- *Water eutrophication.* Contamination of water ecosystems is due to wastewater and fertilizers, resulting in algae blooms and the eventual death of plant and animal life.

SolidWorks measures the environmental impact based on the following parameters:

- Materials used
- Manufacturing process and region
- Transportation and use region
- End of life disposal

SolidWorks Solution Procedure

SolidWorks follows five steps in environmental impact assessment on a part:

1. Activating the sustainability application
2. Selecting a material
3. Setting the manufacturing and use options
4. Setting the transportation and use options
5. Comparing similar materials

Problem Description

This example uses Sustainability to perform an environmental impact analysis of a part. We analyze a common part that is used in holding dish in the kitchen, which is shown in Figure 31.1.

Activate the Sustainability Application

1. Start a New SolidWorks document.
2. Open the part for sustainability study, dish_container.sldprt.
3. Click Sustainability (Tools toolbar) or Tools > Sustainability (see Figure 31.2). The application opens in the Task Pane (see Figure 31.3).
4. Click to keep the Task Pane open (see Figure 31.3).

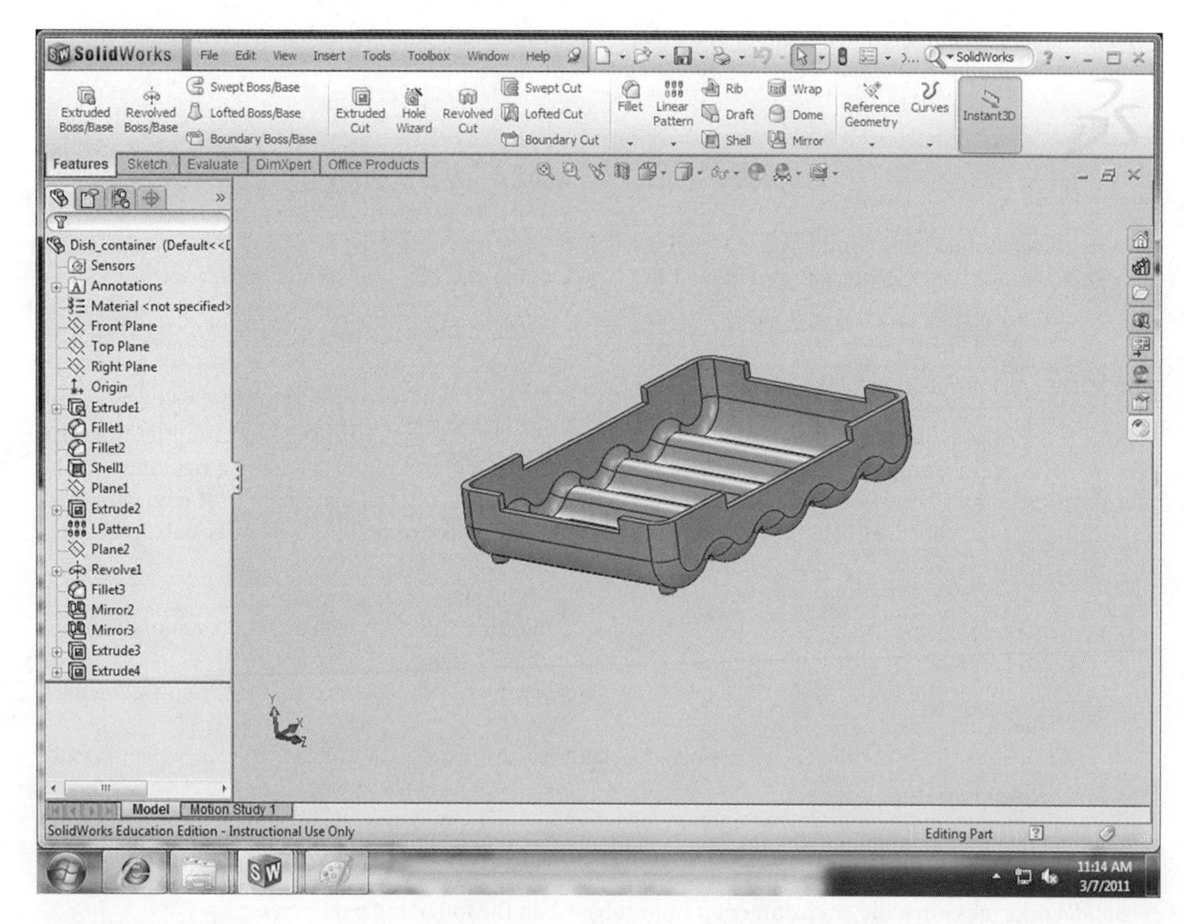

Figure 31.1

Model for study.

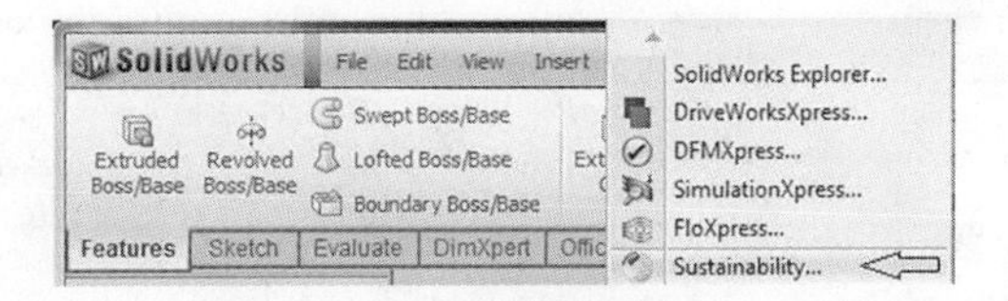

Figure 31.2

Sustainability study gateway.

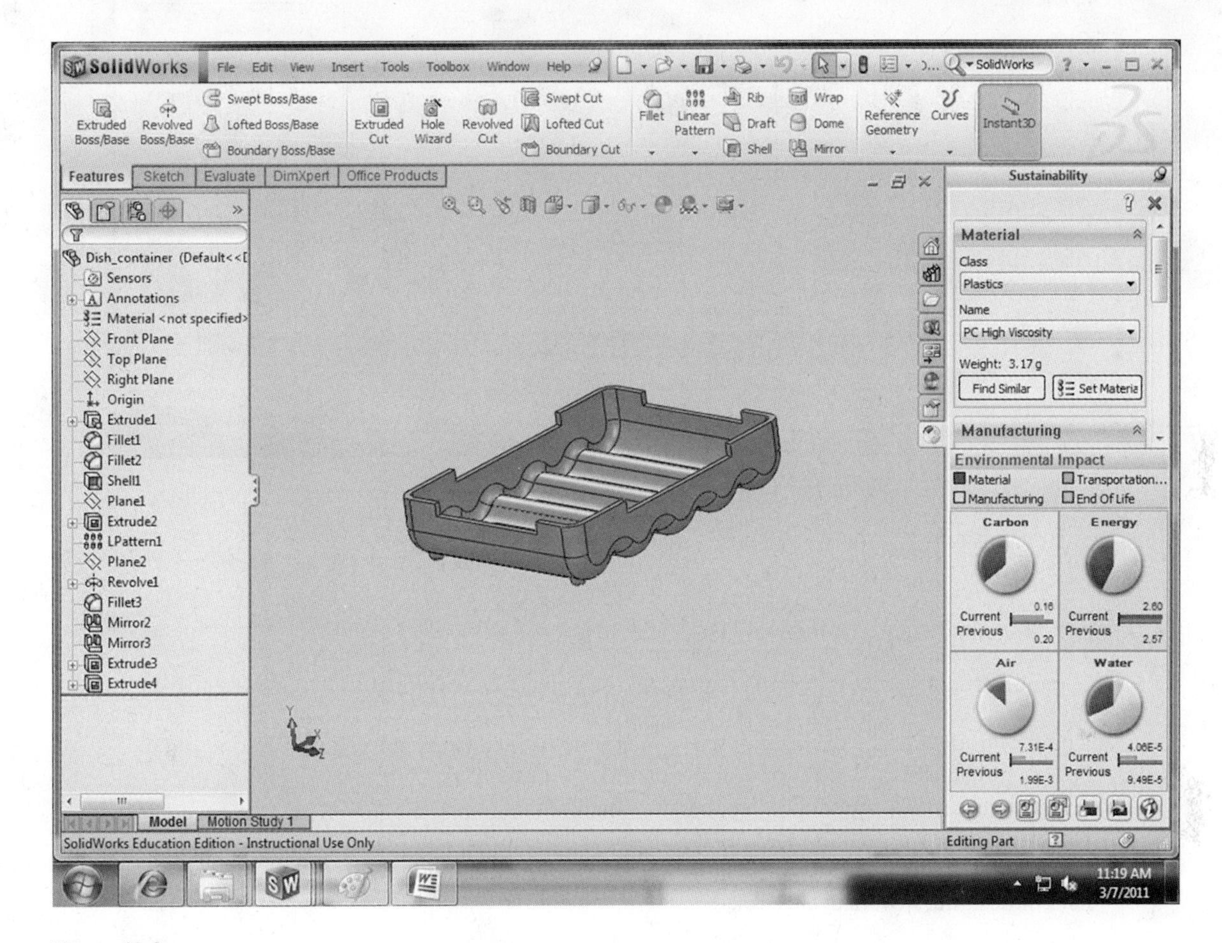

Figure 31.3

Pin the task pane.

Selecting a Material

Select the material that is used for the part.

1. Under Material (see Figure 31.4),
   a. In Class, select Plastics.
   b. In Name, select PC High Viscosity.
   SolidWorks displays the part's weight. The Environmental Impact dashboard at the bottom of the Task Pane provides real-time feedback about the environmental impact of the engineer's or designer's design.

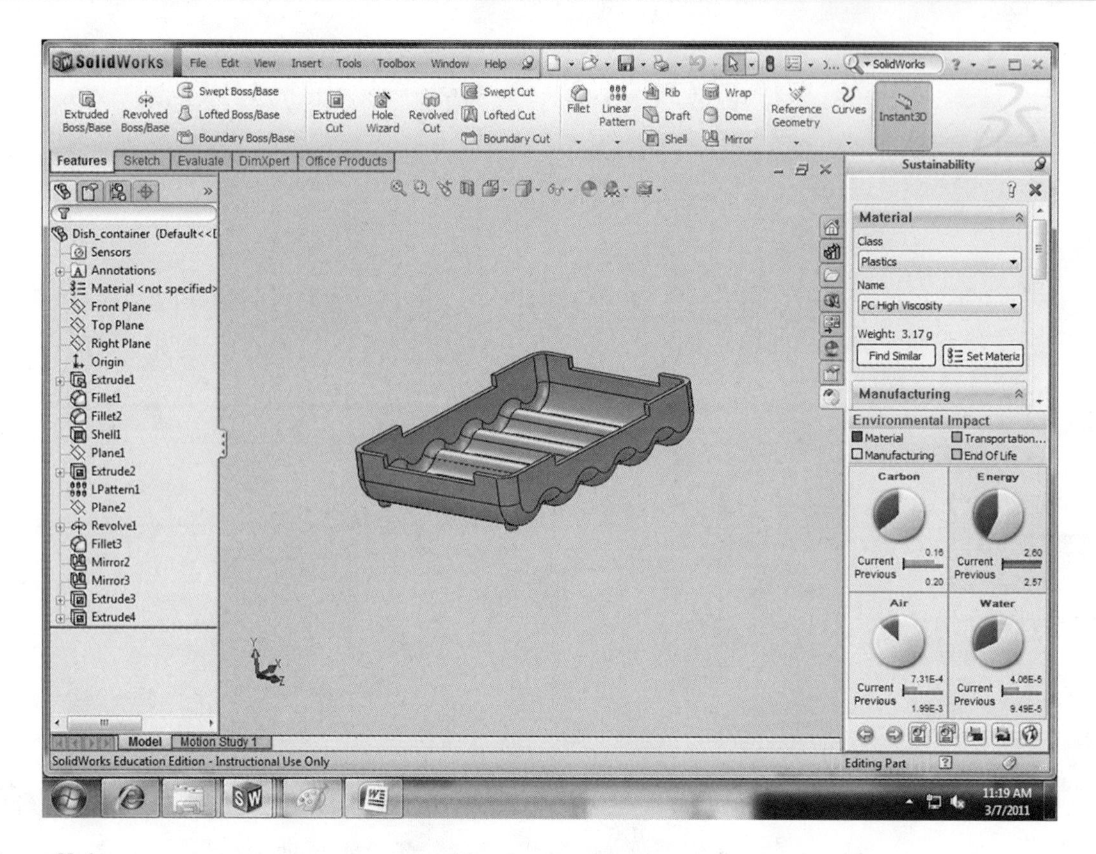

Figure 31.4

Material selection.

Setting the Manufacturing and Use Options

Select the manufacturing process and the regions where the part is manufactured and used.

1. Under Manufacturing, in Process, select Injection Molded (see Figure 31.5).
2. For Region, click North America on the map (see Figure 31.5).
   On each Sustainability map, Japan has its own region.
3. Under Transportation and Use, for Use Region, click North America (see Figure 31.6).
   Data are not available for all regions. Regions that contain data are highlighted when the engineer or designer hovers over them.

Comparing Similar Materials

Now, the engineer or designer sets the baseline material and compare it with other materials, in order to minimize the environmental impact, using the *Environmental Impact* dashboard.

1. At the bottom of the Task Pane, click Set Baseline (see Figure 31.7).
   The Baseline bar for each environmental impact adjusts to show the values for the selected material, PC High Viscosity (see Figure 31.8).
   Next, the engineer or designer tries to find a similar material that is a better environmental choice.
2. Under Material, click Find Similar (see Figure 31.9).
   The dialog box displays the current material with values for multiple parameters.
3. Set these values (see Figure 31.10):

Property	Condition
Density	~ (Approximately)
Tensile strength	> (Greater than)

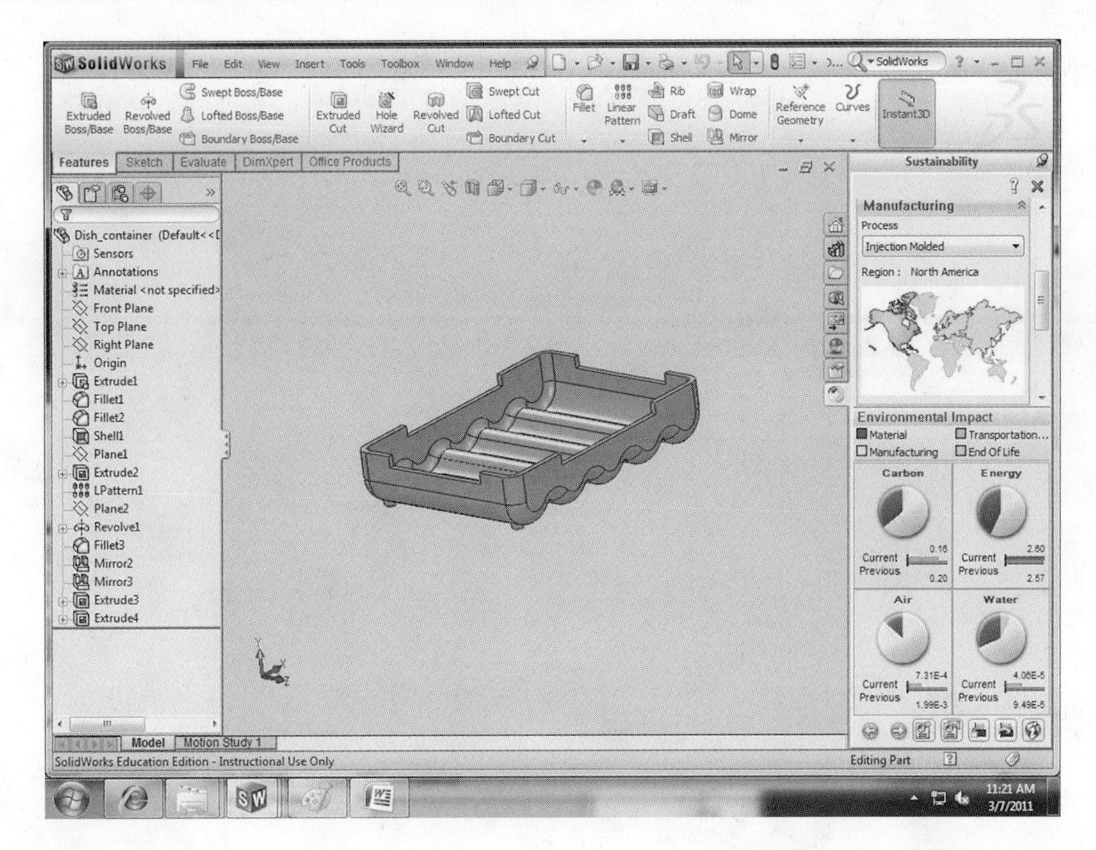

Figure 31.5

Manufacturing process and regions related to part.

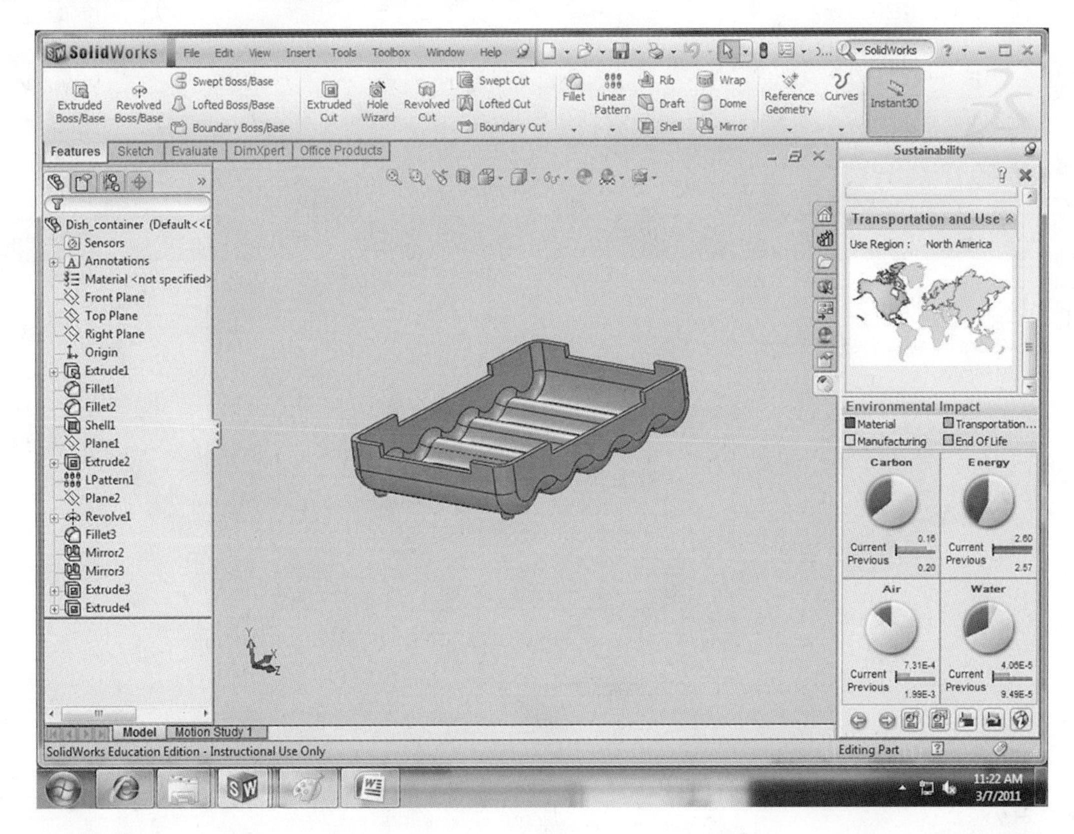

Figure 31.6

Regional identification.

Figure 31.7

Baseline selection.

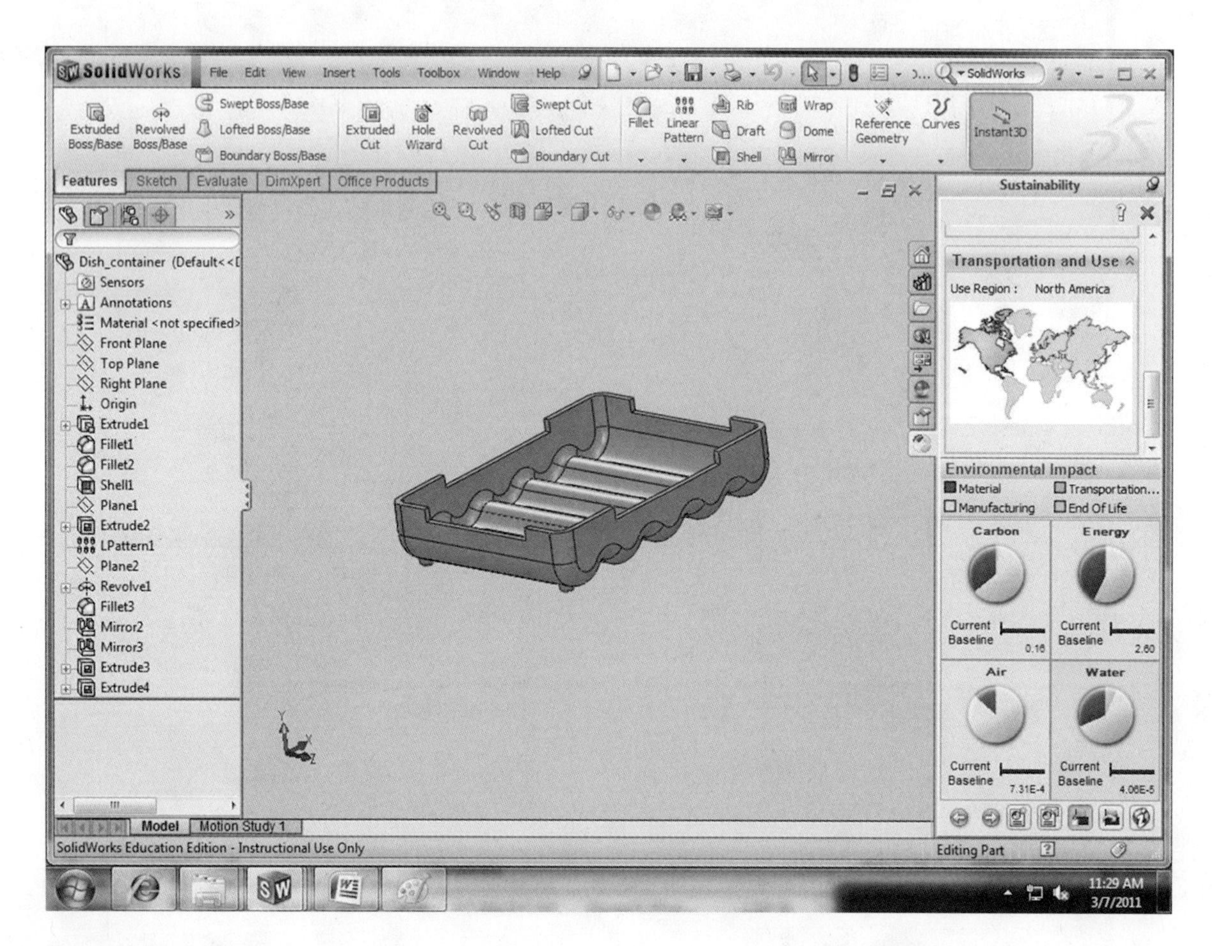

Figure 31.8

PC high viscosity.

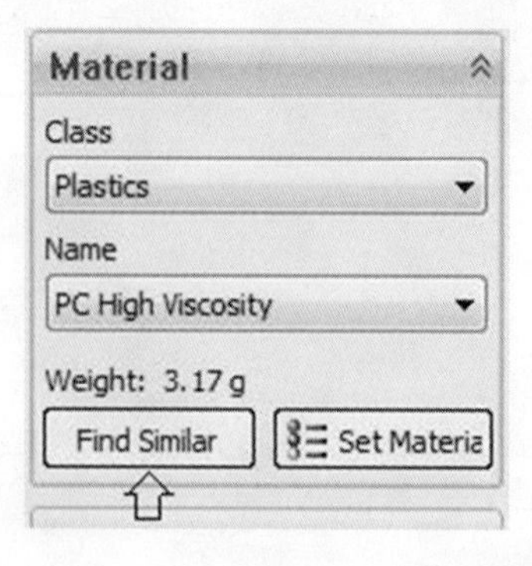

Figure 31.9

Material selection continues.

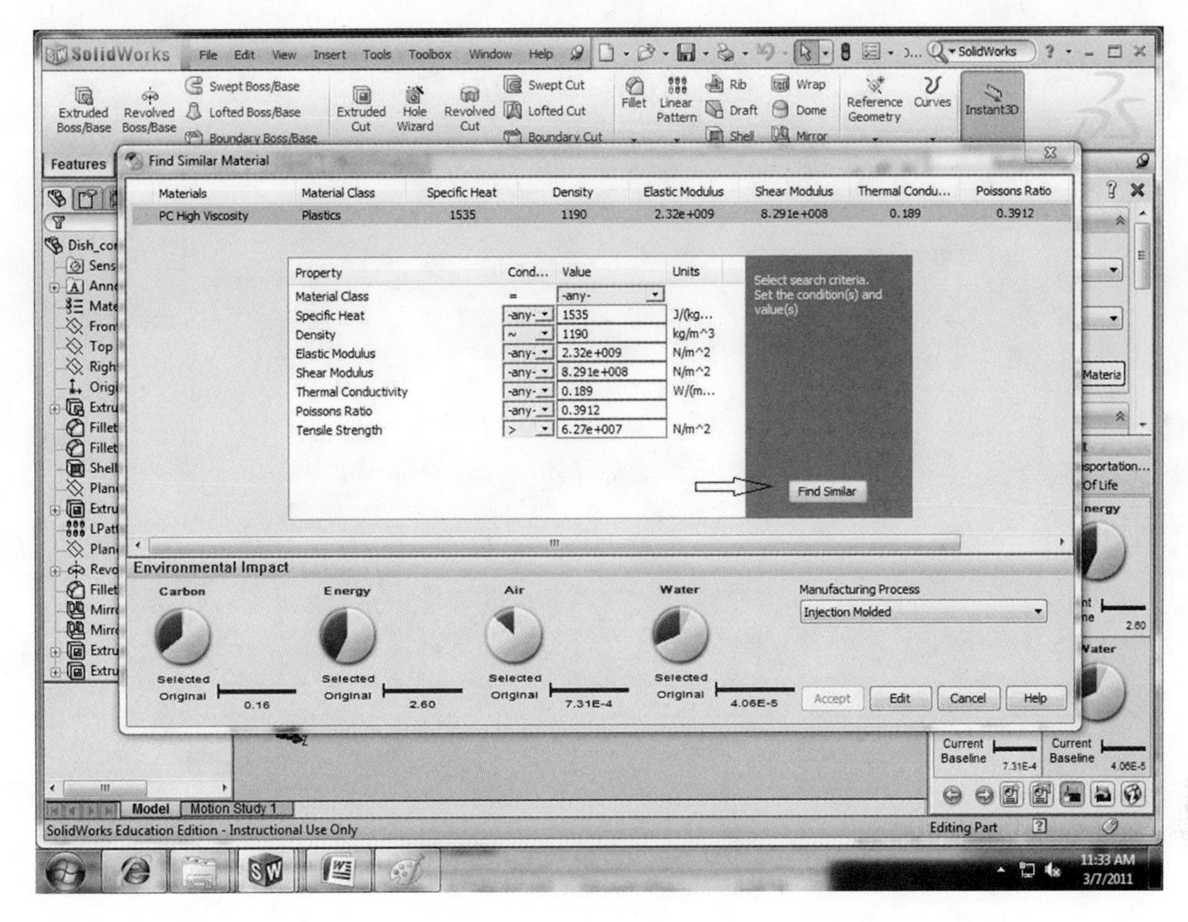

Figure 31.10

Set values.

4. Click Find Similar next to the list in the dialog box (see Figure 31.10).

   A list of similar materials appears. The engineer or designer selects materials from this list to compare them to the original material. The Environmental Impact dashboard at the bottom of the dialog box gives the engineer or designer an intermediate feedback on the selections that are made.

   To filter the list, select the checkbox next to the materials to list and click Show selected only.

5. Under the Materials column, select Acrylic (Medium-high impact).

   Click the *name of the material, not the checkbox*. Click Accept (see Figure 31.11).

   In the dialog box's Environmental Impact dashboard, a green bar for Selected appears above the black bar for Original for all four impact areas. The pie charts are updated.

   The bar's green color and shorter length indicate that the selected material, Acrylic (Medium-high impact), is a better environmental choice than the original material, PC High Viscosity, represented by the black baseline (see Figure 31.11).

6. Select Nylon 101 to see how it compares to the original material. The bars and pie charts are updated. The visual cues indicate that this material is an even better choice than Acrylic (Medium-high impact) (see Figure 31.12). The engineer or designer decides to accept this material.

   The engineer or designer can modify the Manufacturing Process using the menu next to the pie charts.

7. Click Accept (see Figure 31.13).

   The dialog box closes.

   In the Task Pane, under Material, Plastics Nylon 101 is the current material. The Environmental Impact dashboard is updated.

8. In the Task Pane, under Material, click Set Materials (see Figure 31.14).

   The material for the part is updated as Plastics Nylon 101 in the FeatureManager (see Figure 31.14).

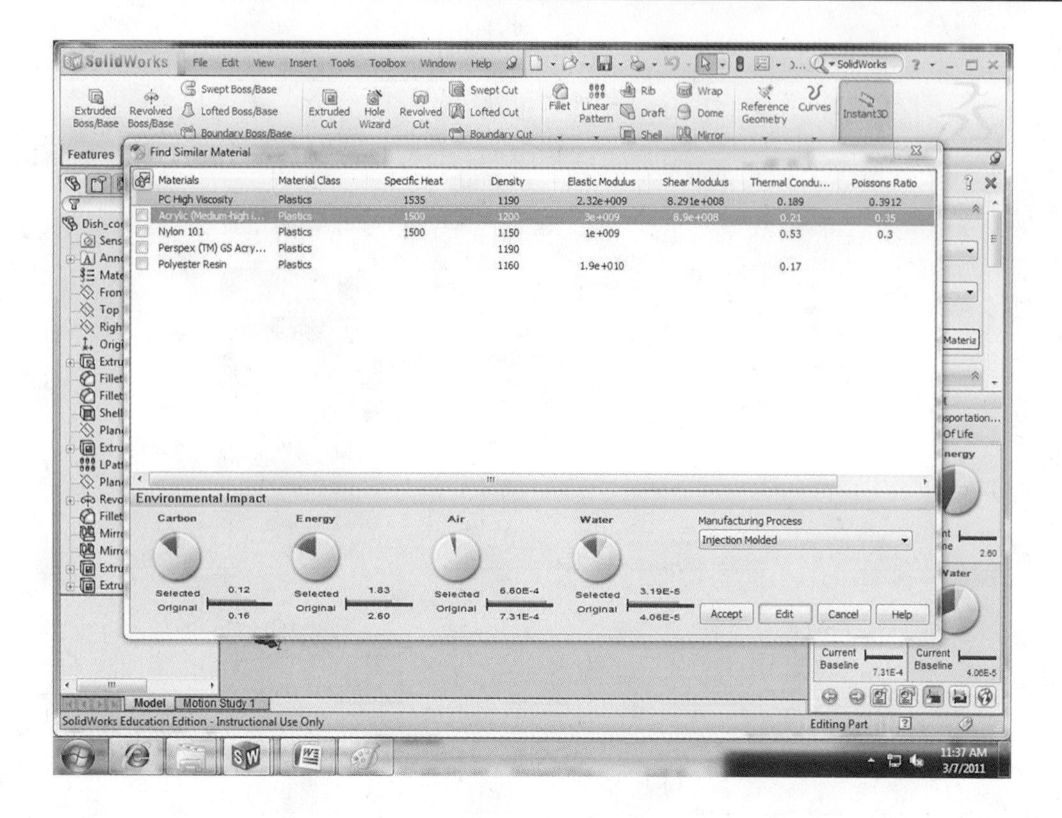

Figure 31.11

Acrylic selection.

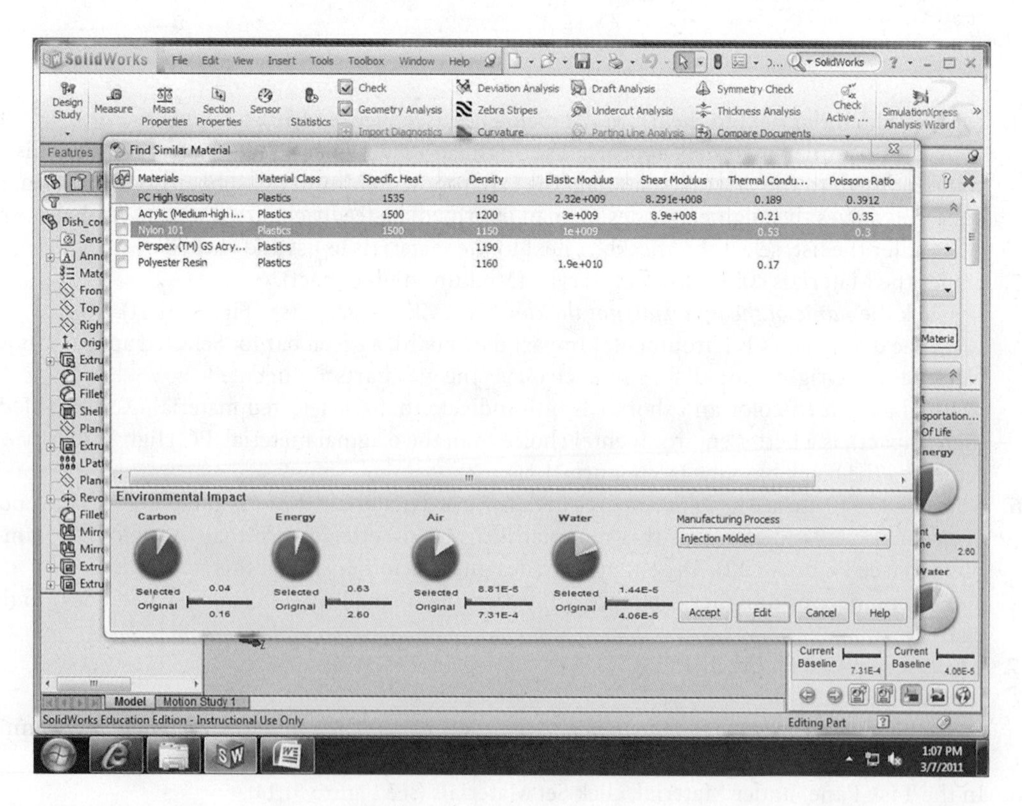

Figure 31.12

Nylon as an option.

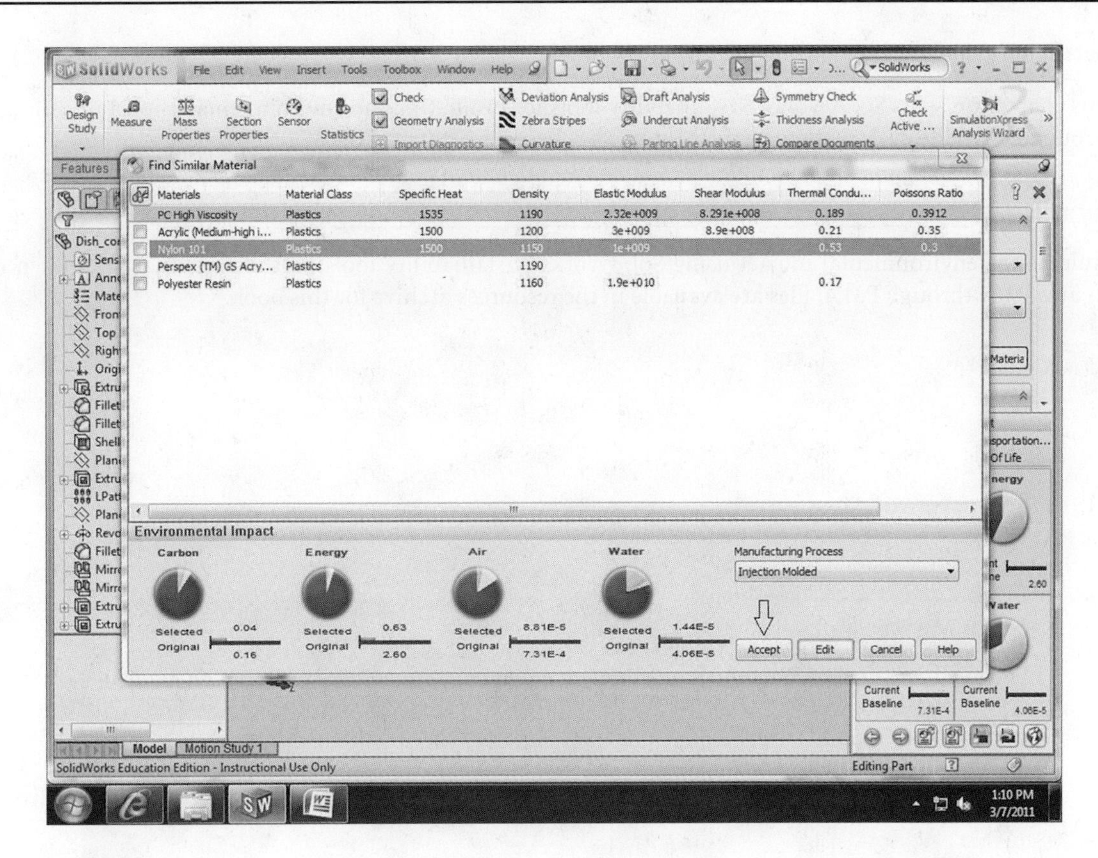

Figure 31.13

Accept options.

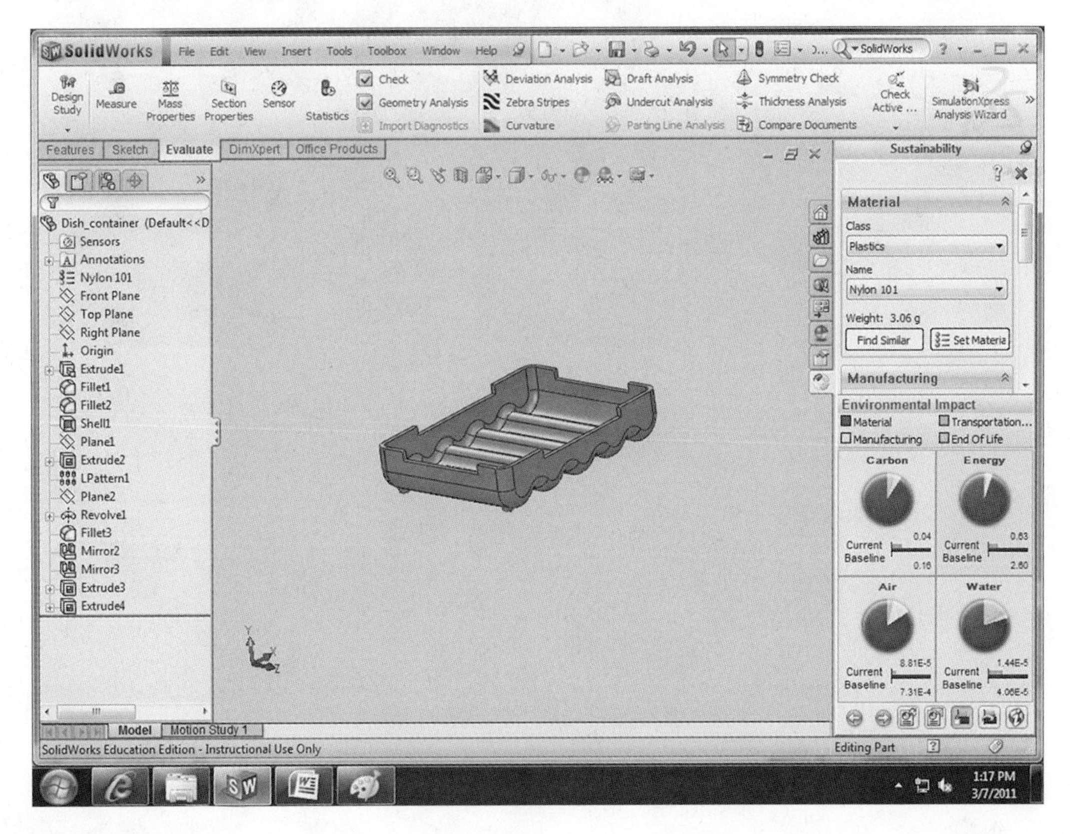

Figure 31.14

Optimal solution.

Summary

This chapter presents SolidWorks Sustainability tools for evaluating the environmental impact of a design throughout the life cycle of a product. This topic has become a hot area of interest.

Exercises

Evaluate the environmental impact using SolidWorks Sustainability tools on the parts that are shown in Figures P31.1 through P31.4; files are available in the resources archive for this book.

1. Pulley
2. Bowl
3. Plastic cover
4. Sump

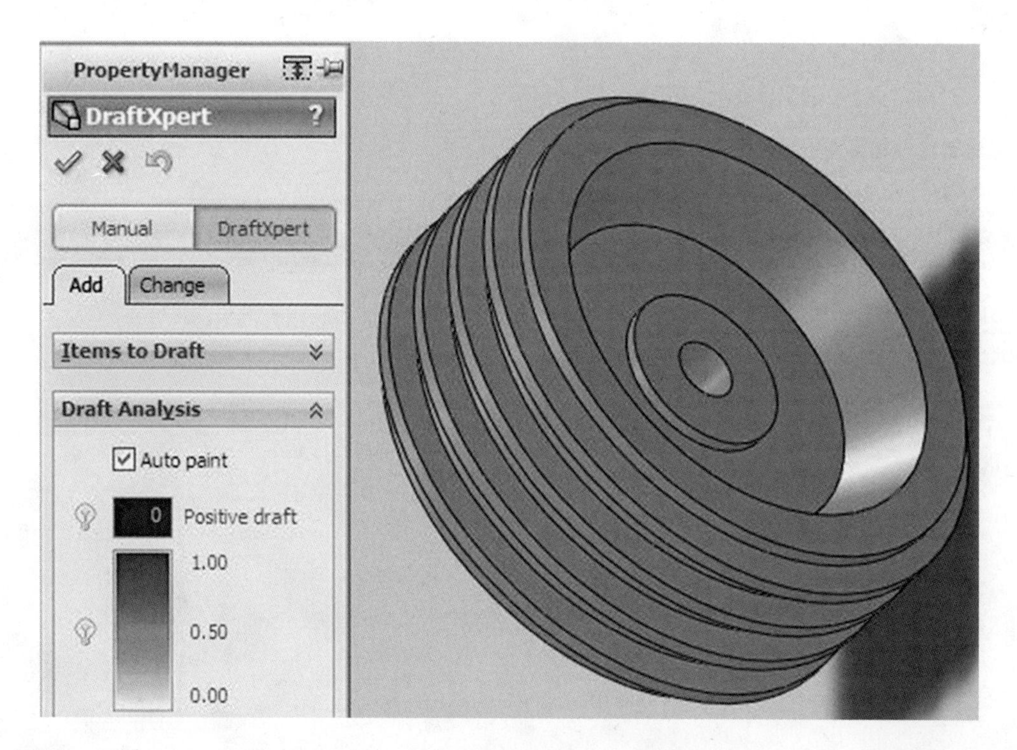

Figure P31.1

Part for which the mold is being designed: pulley.

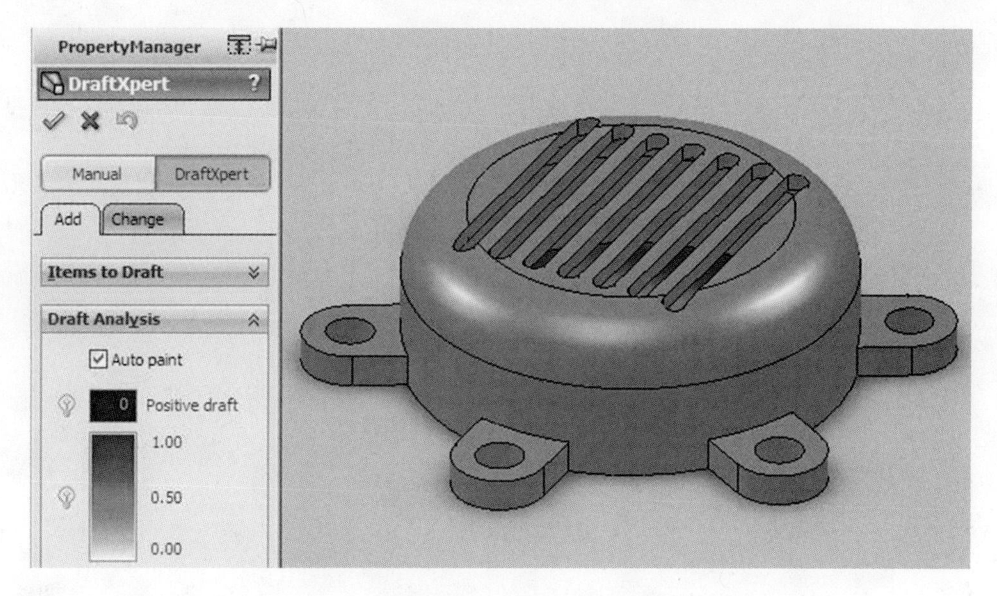

Figure P31.2

Part for which the mold is being designed: bowl.

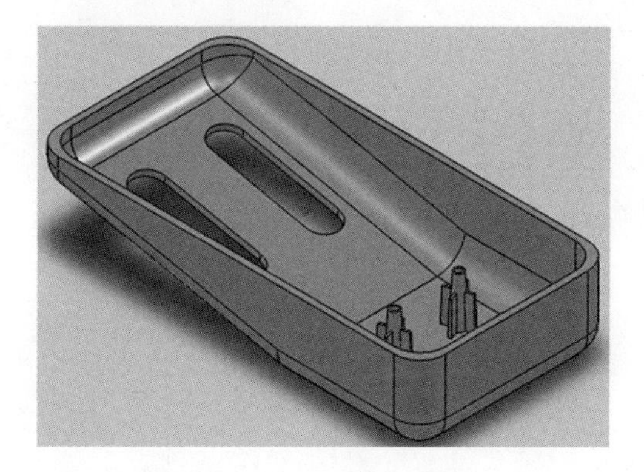

Figure P31.3

Part for which the mold is being designed: plastic cover.

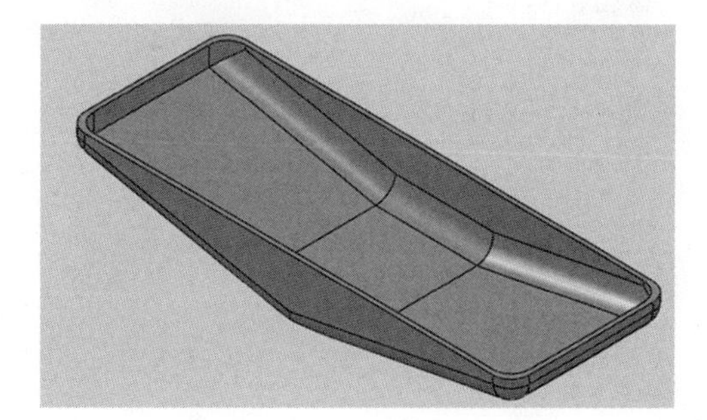

Figure P31.4

Part for which the mold is being designed: sump.

32

Geometric Dimensioning and Tolerancing

Objectives:

When you complete this chapter, you will have

- Learned how to interpret diagram applications specifying the maximum material condition (MMC) and the least material condition (LMC)
- Learned how to determine geometric tolerances for produced sizes at MMC and LMC
- Learned how to insert tolerances to drawings
- Learned how to insert half-, offset-, and aligned-section views to drawings

Introduction

Conventional tolerancing refers to tolerances that are related to dimensioning practices without regard to geometric tolerancing. On the other hand, geometric tolerancing is the dimensioning and tolerancing of individual features of parts in which the permissible variations relate to characteristic form, profile, orientation, runout, or the relationship between features. In this chapter, we will consider conventional tolerancing and then move on to geometric tolerancing.

There are three systems for applying dimensions and tolerances to a drawing: (1) baseline, (2) chain, and (3) direct. Let us illustrate the concept of dimensional tolerance that is built up by using an unspecified tolerance value of ±0.2 for the following examples (Figures 32.1 through 32.3).

Baseline dimensioning is a common method of dimensioning machines parts whereby each feature dimension originates from a common surface, axis, or center plane. Tolerance buildup is less likely to occur than when using chain dimensioning (see Figure 32.1). There are two dimensions between X and Y (16 and 26); hence, the tolerance accumulation is 2(±2) = ±4.

Chain dimensioning, also known as point-to-point dimensioning, is a method of dimensioning from one feature to the next. Tolerance buildup is more likely to occur than when using baseline dimensioning (see Figure 32.2). Each dimension is dependent on the previous dimension or dimensions. There are three dimensions between X and Y (6, 10, and 10); hence, the tolerance accumulation is 3(±2) = ±6.

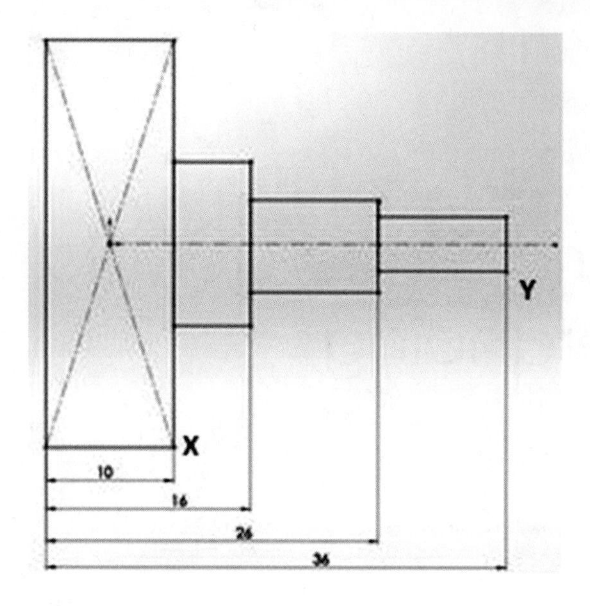

Figure 32.1

Baseline dimensioning.

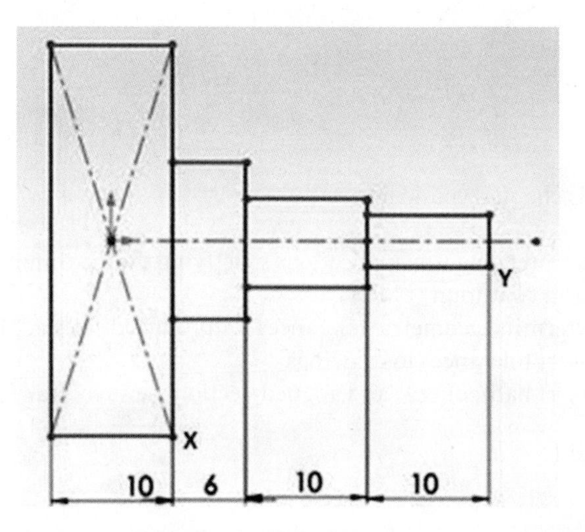

Figure 32.2

Chain dimensioning.

Direct dimensioning is applied to control the size or location of one or more specific features. Tolerance buildup is least likely to occur than when using the other types of dimensioning (see Figure 32.3). There is one dimension between X and Y (26); hence, the tolerance accumulation is ±2.

Tolerance Study Using SolidWorks

When a feature is sketched, and any one side is dimensioned and the value is accepted by clicking the OK sign, the Dimension PropertyManager automatically appears, as shown in Figure 32.4.

Select the Limit option to display the Upper (+) and Lower (–) level options. In this illustration, we dimension the bottom horizontal edge as 75 and accept this value.

1. Click the Tolerance/Precision toolbar.
2. Select Limit option.

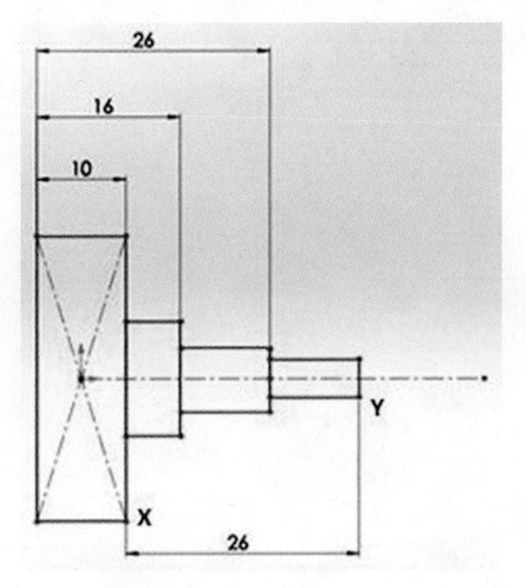

Figure 32.3

Direct dimensioning.

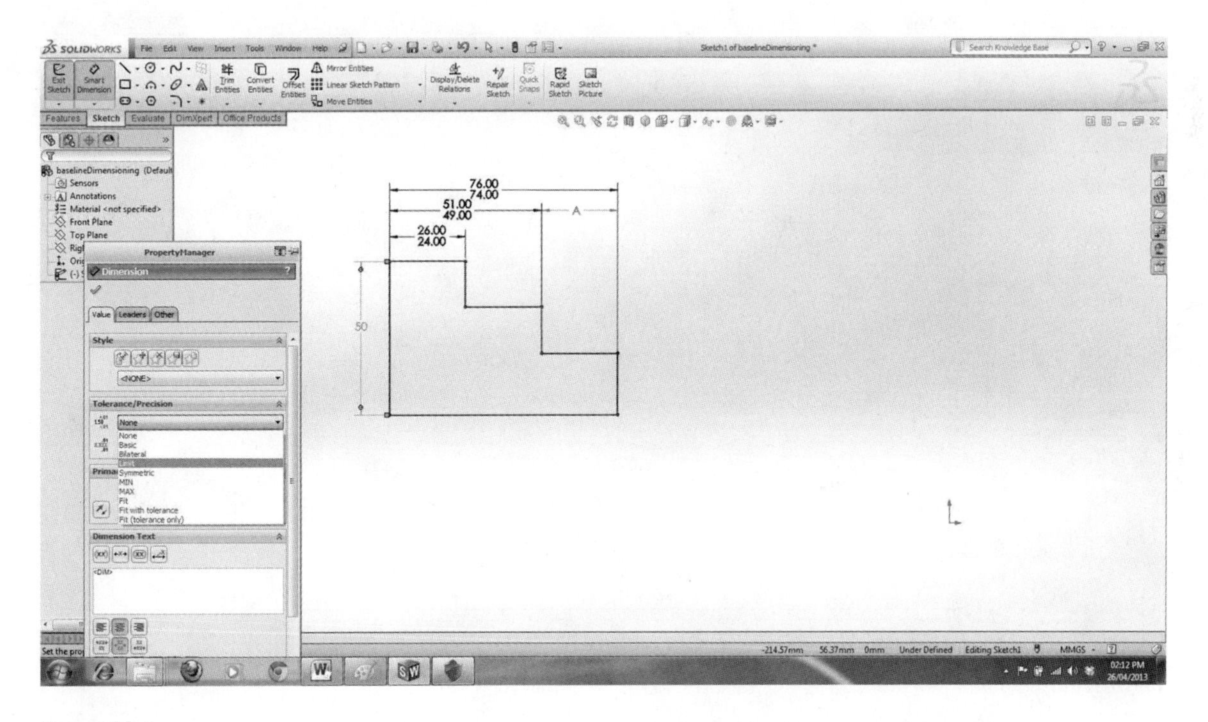

Figure 32.4

Dimensions PropertyManager.

3. Fill the Upper Limit as 1 and Lower Limit as 1 (see Figure 32.5a).
   Maximum dimension becomes 76 and Minimum dimension becomes 74; (see Figure 32.5b).
4. Click OK.

Chain Dimensioning Tolerance Calculation

Repeat for the two top (leftmost) lines starting with the length of 25 and using an upper limit = 1 and a lower limit = −1, as shown in Figure 32.6. The horizontal length A is not dimensioned. Its length is dependent of the tolerances of the two top leftmost horizontal lengths.

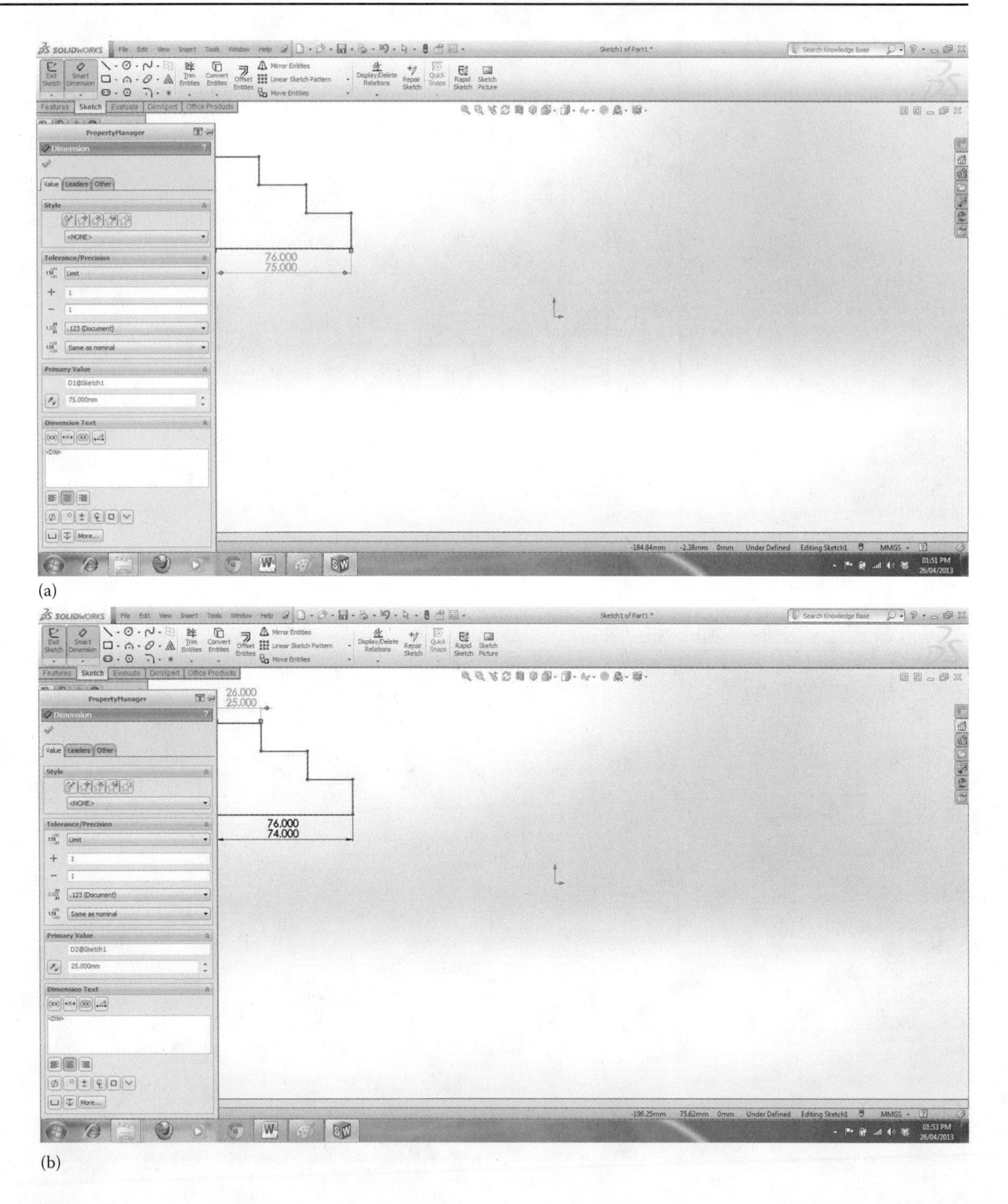

(a)

(b)

Figure 32.5

(a) Upper and (b) lower limits for one of the lines of the feature.

We will now determine the tolerance of length A. The principle used is to determine the maximum length of A and the minimum length of A. The difference will give us the indication of tolerance accumulation for the dimensions that are used in the computation.

Maximum Length of A:

Length A will be maximum when the overall length is at its longest value, and the other distances are at their shortest values.

Minimum Length of A:

Length A will be minimum when the overall length is at its shortest value, and the other distances are at their longest values.

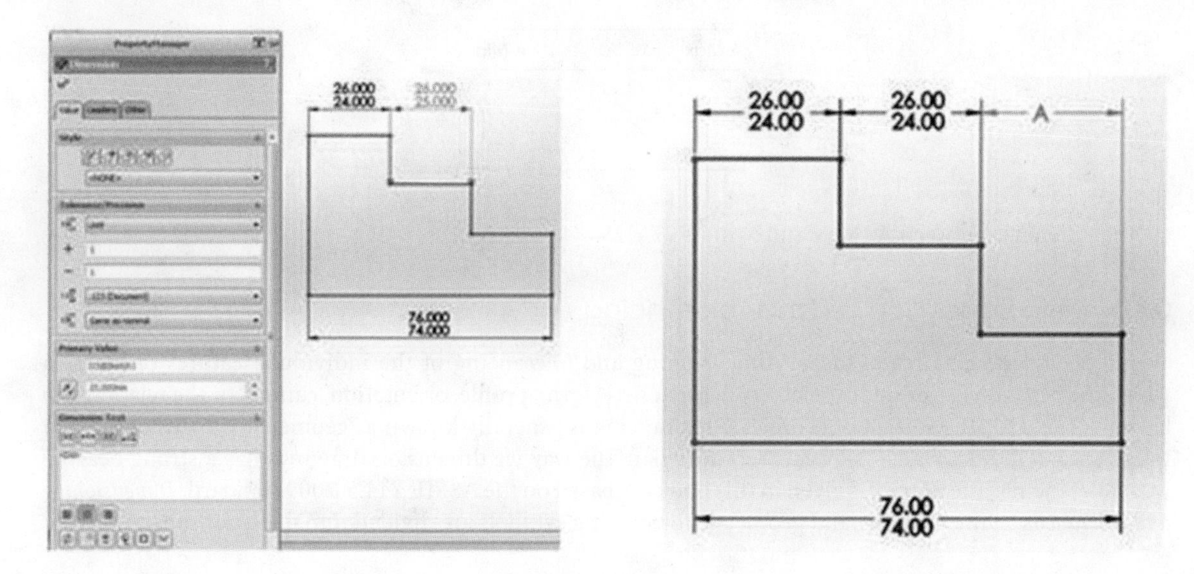

Figure 32.6

Upper and lower limits for one of the lines of the feature: Chain dimensioning.

A Max	A Min
76	74
−24	−26
−24	−26
28	22

The difference between A (Max and Min) is 28 − 22 = 6. This illustration describes the chain dimensioning concept.

Baseline Dimensioning Tolerance Calculation

Figure 32.7 shows the SolidWorks Dimensioning tool for tolerancing. We will now determine the tolerance of length A.

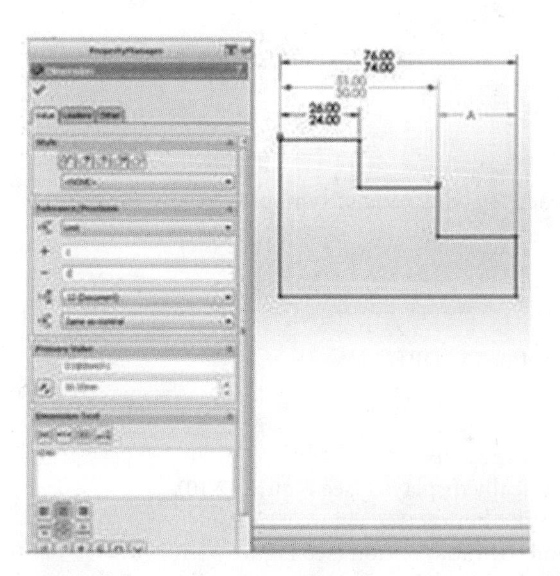

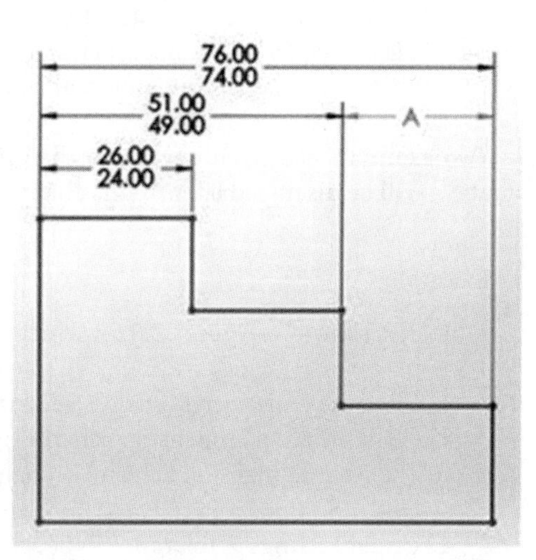

Figure 32.7

Upper and lower limits for one of the lines of the feature: Baseline dimensioning.

A Max	A Min
76	74
−49	−51
27	23

The difference between A (Max and Min) is 27 – 23 = 4.

Geometric Dimensioning and Tolerancing

Geometric tolerancing refers to the dimensioning and tolerancing of the individual features of parts in which the permissible variations relate to characteristic form, profile, orientation, runout, or the relationship between features. The subject of geometric tolerancing is generally known as geometric dimensioning and tolerancing (GD&T), which has been very popular. The way we dimension drawings has a strong bearing with tolerancing. The examples given in this book are based on the ASME Y14.5-2009 standard, *Dimensioning and Tolerancing.* Dimensioning and geometric tolerancing symbols are divided into five basic types:

1. Dimensioning symbols
2. Datum feature and datum target symbols
3. Geometric characteristic symbols
4. Material condition symbols
5. Feature control frame

Datum Reference Frame

Datum features are selected based on the importance to the design of the part. In general, there are three orthogonal datum features to be selected: the datum reference frame (DRF). These three datum features are referred to as the (1) *primary datum*, (2) *secondary datum*, and (3) *tertiary datum*. The primary datum is the most important, followed by the two others according to their placement in the *GD&T Note*.

Geometric Characteristic Symbols

Geometric characteristic symbols are symbols that are used in GD&T to provide specific controls that are related to the form of an object, the orientation of the features, the outline of features, the relationship of features to an axis, or the location of the features. There are normally five types of geometric characteristic symbols: (1) form, (2) profile, (3) location, (4) orientation, and (5) runout.

1. *Form:* straightness, flatness, circularity, and cylindricity
2. *Profile:* line and surface
3. *Location:* position, concentricity, and symmetry
4. *Orientation:* parallelism, perpendicularity, and angularity
5. *Runout:* circular runout and total runout

Two examples are given to explain the different geometric characteristic symbols. The details of these symbols will be discussed later in this chapter.

Example 1

A block is shown in Figure 32.8 for which we want to define the DRF.

1. Click the DimXpertManager (see Figure 32.9).
2. Click the Auto Dimension Scheme option (see Figure 32.9).
   The Auto Dimension Scheme Manager is automatically displayed (see Figure 32.10).
3. Under Settings > Part Type, click the Prismatic radio button (see Figure 32.10).
4. Under Settings > Tolerance Type, click the Geometric radio button (see Figure 32.10).
5. Under Datum Selection, click the *back face* of the block as Primary datum (A).
6. Under Datum Selection, click the *rightmost face* of the block as Secondary datum (B).

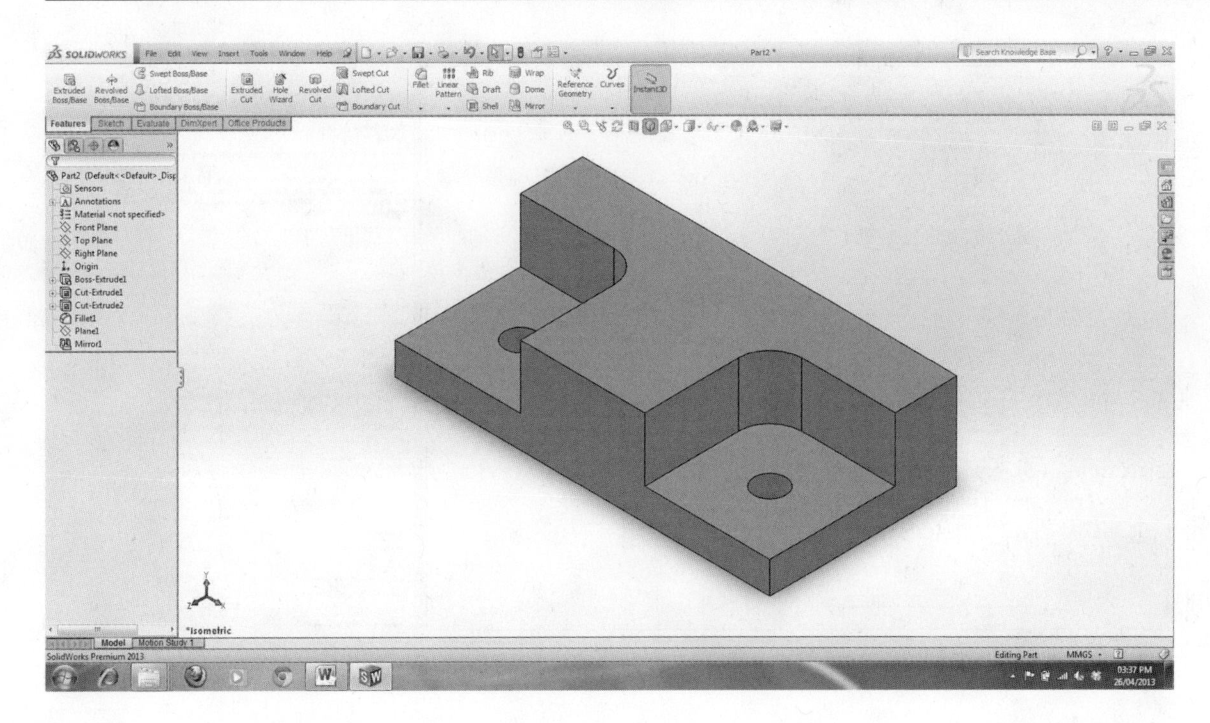

Figure 32.8

A block.

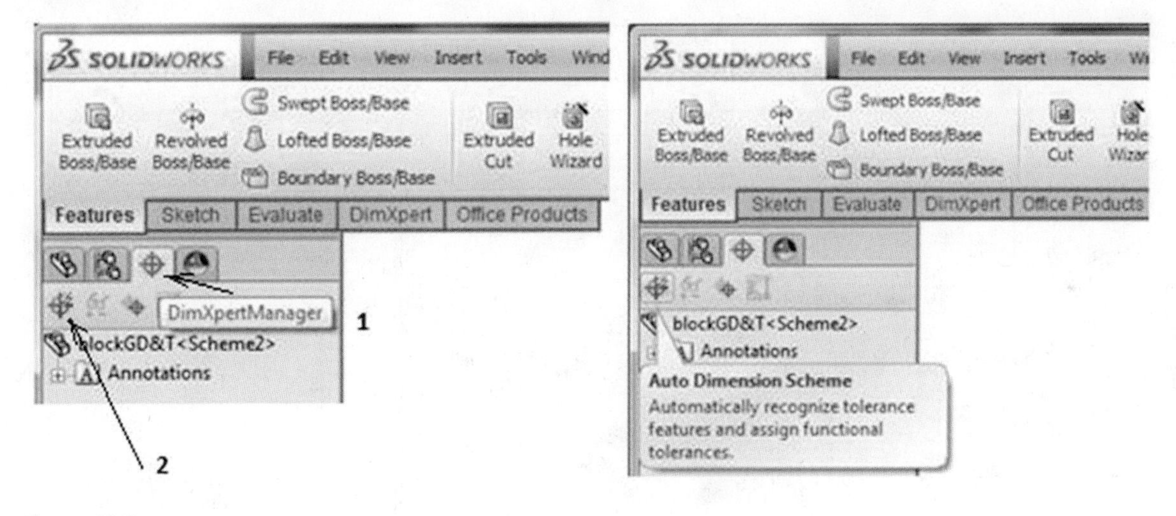

Figure 32.9

DRF for a block.

7. Under Datum Selection, click the *bottom face* of the block as Tertiary datum (C).
8. Under Scope, click the Selected features radio button (see Figure 32.11).
9. Select the two holes (see Figure 32.11).
10. Click OK. (GD&T Notes are automatically displayed; see Figure 32.12.)

Referring to Figure 32.12, we find that

- The back face of the block is designated A; it has a *Form* of *Flatness.*
- The right face of the block is designated B; it has an *Orientation* of *Perpendicularity* with respect to A.
- The bottom face of the block is designated C; it has an *Orientation* of *Perpendicularity* with respect to A.

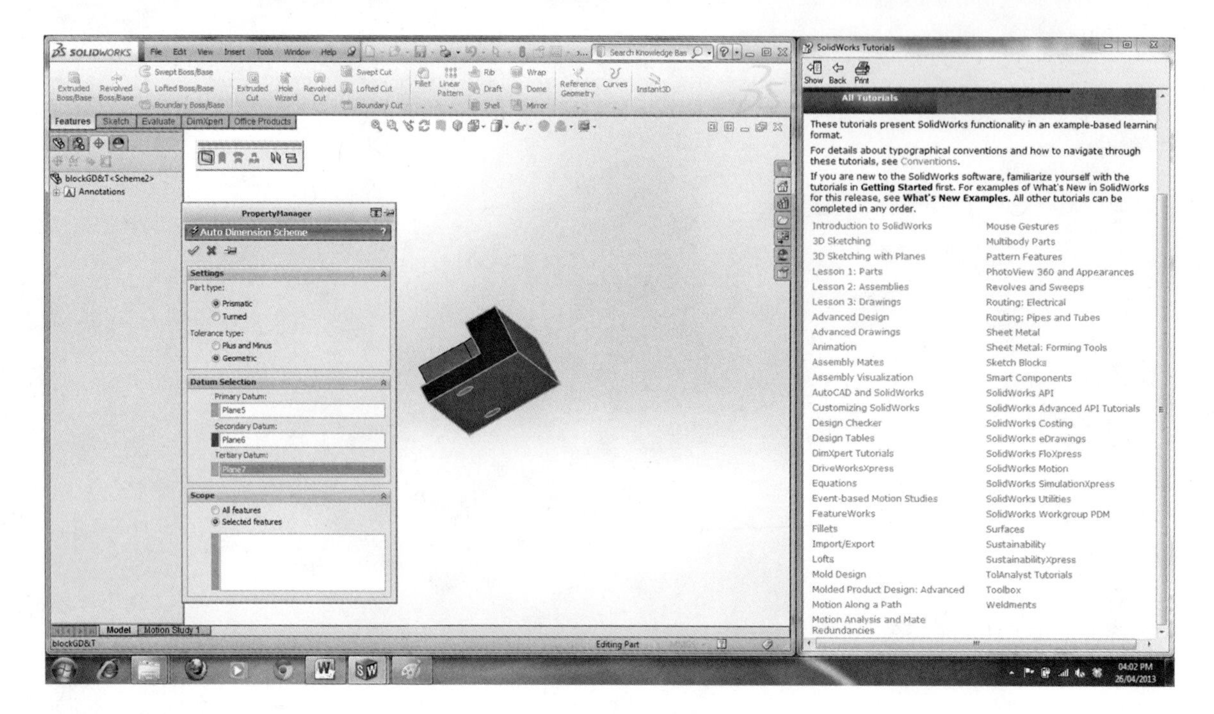

Figure 32.10

The *Auto Dimension Scheme* Manager is automatically displayed.

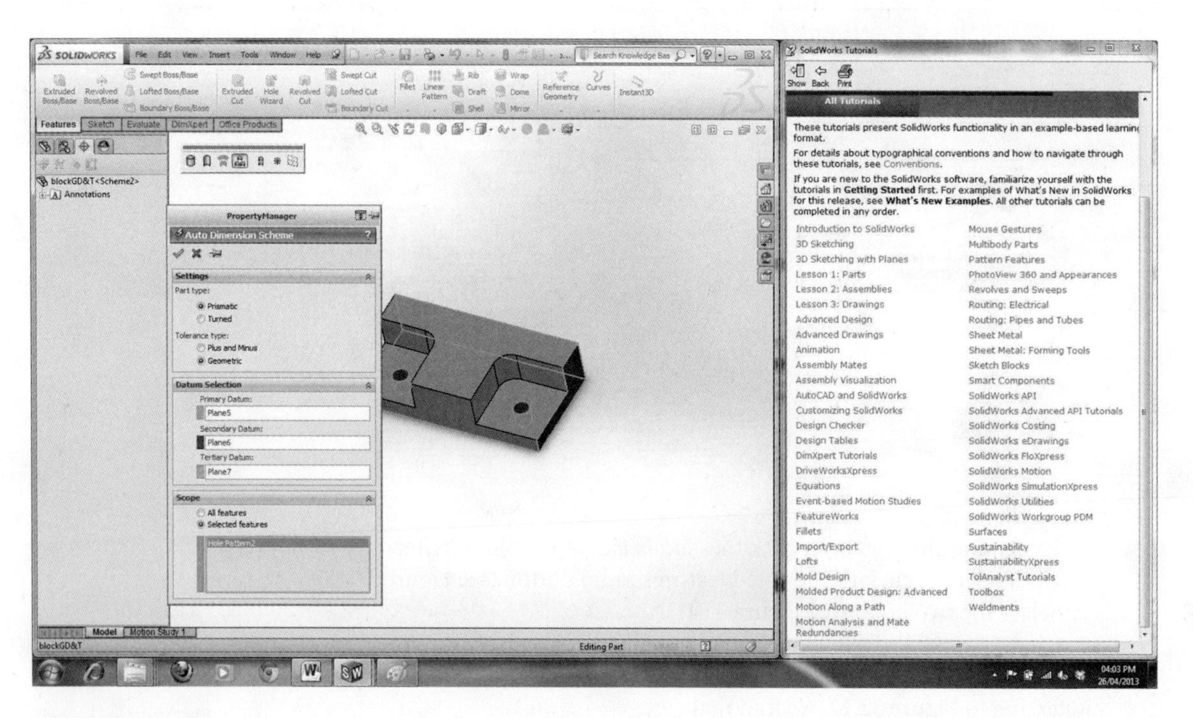

Figure 32.11

The two holes are chosen as the Selected features.

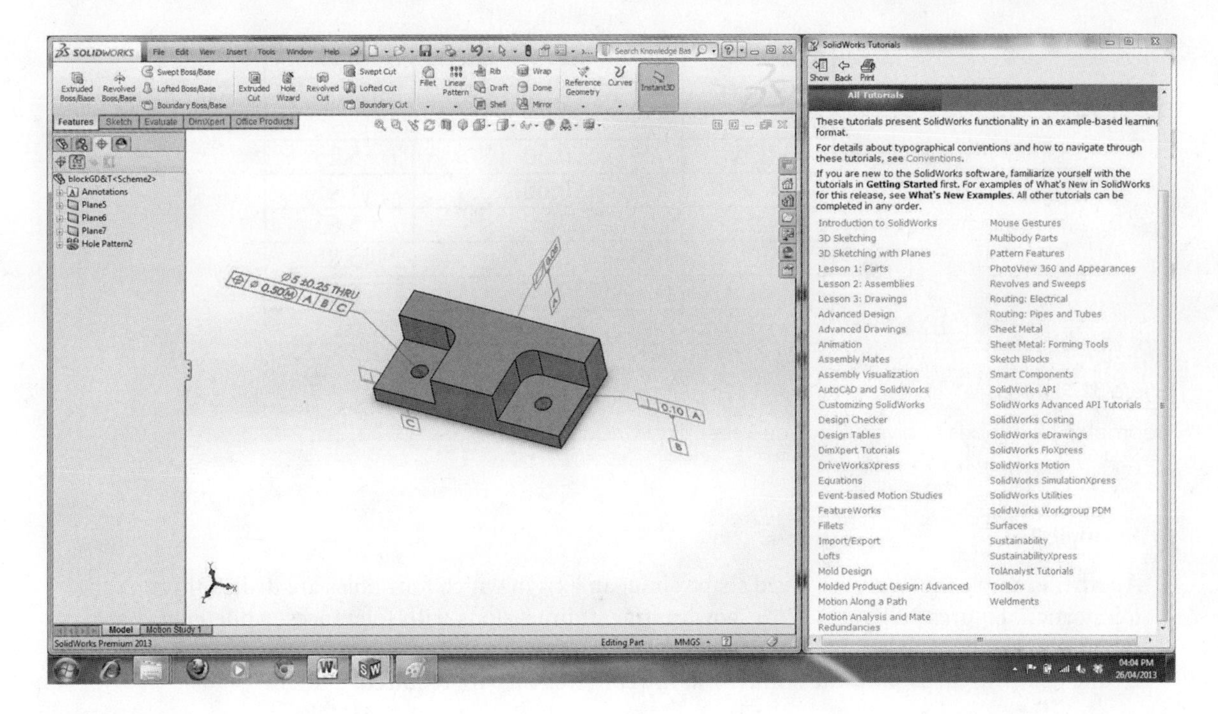

Figure 32.12

GD&T Notes for the block model.

Applying Maximum Material Condition

Maximum material condition (MMC) is the condition in which a feature contains the maximum amount of material within the stated limits of size. When MMC is used in the feature control frame, the given geometric tolerance is maintained when the feature is produced.

> *External Feature*:
> MMC – produced size + given geometric tolerance = applied geometric tolerance
> *Internal Feature*:
> Produced size – MMC + given geometric tolerance = applied geometric tolerance

In Figure 32.12, there are two types of tolerances: (1) feature tolerance and (2) positional tolerance. The feature tolerance is for the hole, while the positional tolerance is for locating the center of the hole. Let us distinguish these tolerances.

> MMC hole = 5 – 0.25 = 4.75
> LMC hole = 5 + 0.25 = 5.25
> Given geometric tolerance = 0.50
> Geometric tolerance at a given produced size of 4.85 = 4.85 – 4.75 + 0.50 = 0.60.
> Virtual condition = 4.85 + 0.60 = 5.45.

The calculations for geometric tolerances at given produced sizes and virtual condition are listed in Figure 32.13.

Expanding the DimXpertManager reveals the DRF that is made up of planes A (Flatness2), B (Perpendicularity1 with respect to A), and C (Perpendicularity1 with respect to A), as well as the two hole features (Simple Hole3 and Simple Hole4) (see Figure 32.14).

Save the object together with the GD&T features (see Figure 32.14).

	Possible produced size	Geometric tolerances at given produced sizes	Virtual condition
MMC →	4.75	0.50	5.25
	4.85	0.60	5.45
	4.95	0.70	5.65
	5.05	0.80	5.85
	5.15	0.90	6.05
LMC →	5.25	1.00	6.25

Figure 32.13

Geometric tolerances at given produced sizes and virtual condition.

Example 2

Another example involves a revolved shape about an axis in which four holes are drilled through the top portion. Figures 32.15 and 32.16 show pertinent dimensions with tolerances, while Figure 32.17 shows the object.

The following steps show the SolidWorks GD&T tools that are required:

1. Click the DimXpertManager.
2. Click the Auto Dimension Scheme option.
   The Auto Dimension Scheme Manager is automatically displayed.
3. Under Settings > Part Type, click the Prismatic radio button.
4. Under Settings > Tolerance Type, click the Geometric radio button.
5. Under Datum Selection, click the *bottom face* of the upper boss as Primary datum (A).
6. Under Datum Selection, click the *outer face* of the lower boss as Secondary datum (B).
   (See Figure 32.17 for the model and the GD&T features and descriptions.)

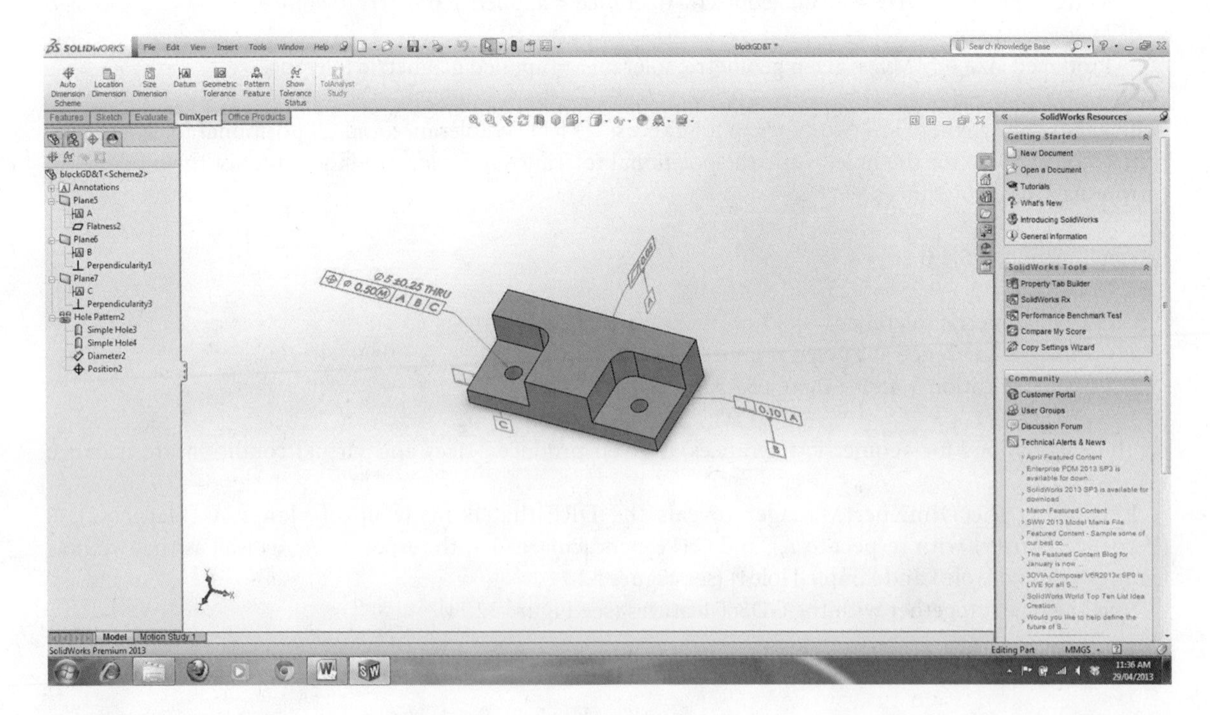

Figure 32.14

GD&T features for the block model.

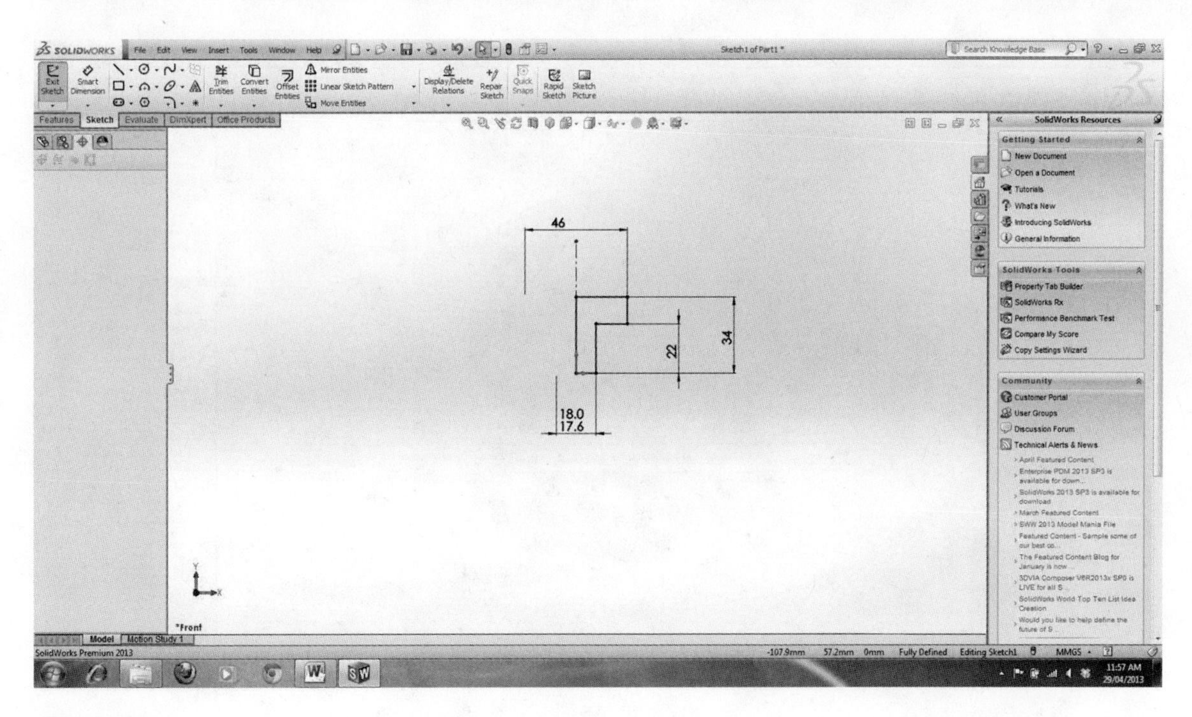

Figure 32.15

Sketch for creating the model.

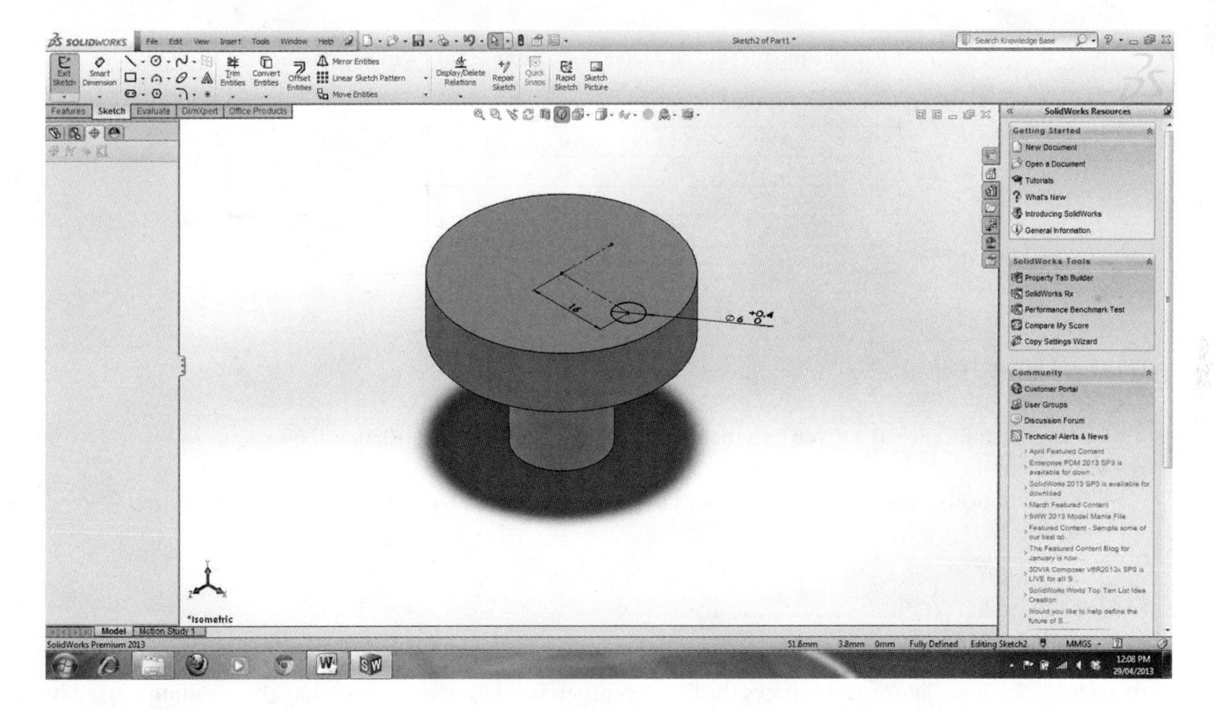

Figure 32.16

Sketch for creating a hole.

As could be observed, the datum A is a flat surface with a tolerance of 0.05, while datum B is perpendicular to A with a tolerance of 0.10 with the MMC related. There are four holes with limits shown. With respect to datum A and datum B (only two in this case), the holes have

MMC hole = 6 − 0.25 = 5.75
LMC hole = 6 + 0.25 = 6.25
Given geometric tolerance = 0.50 (M)

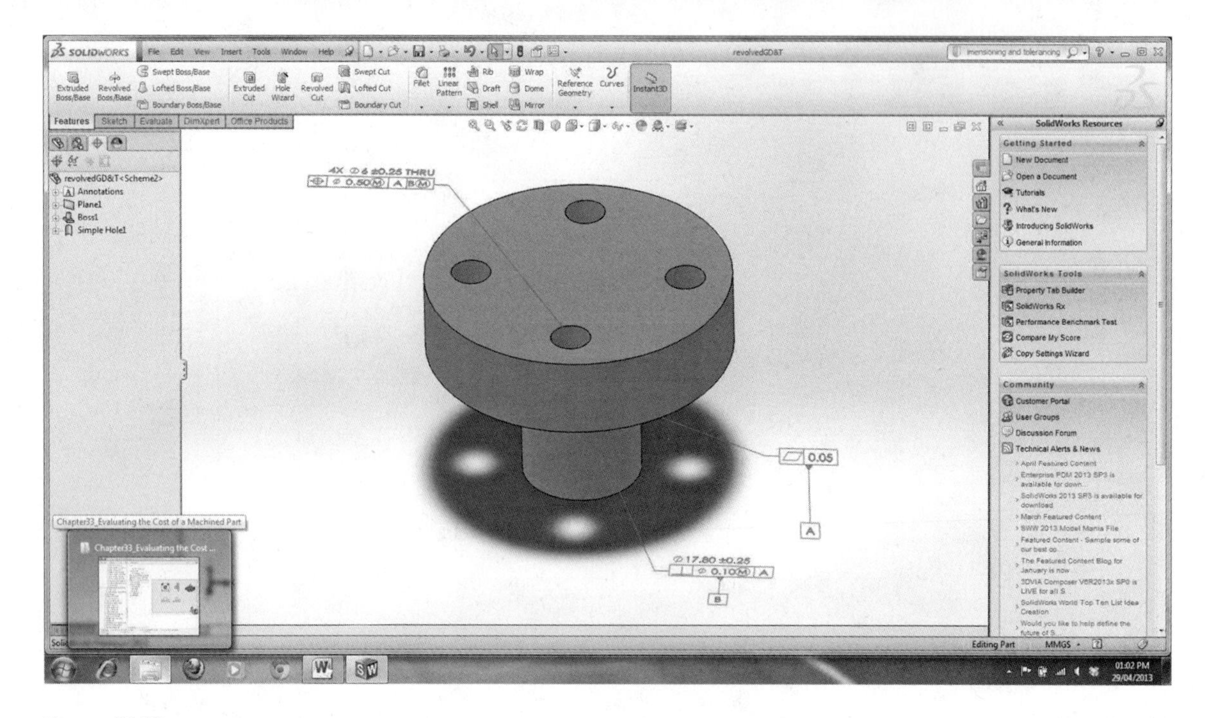

Figure 32.17

GD&T features and descriptions for the model.

	Possible produced size	Geometric tolerances at given produced sizes	Virtual condition
MMC →	5.75	0.50	6.25
	5.85	0.60	6.45
	5.95	0.70	6.65
	6.05	0.80	6.85
	6.15	0.90	7.05
LMC →	6.25	1.00	7.25

Figure 32.18

Geometric tolerances at given produced sizes and virtual condition.

Geometric tolerance at a given produced size of 5.85 = 5.85 − 5.75 + 0.50 = 0.60.

Virtual condition = 5.85 + 0.60 = 6.45.

The calculations for geometric tolerances at given produced sizes and virtual condition are listed in Figure 32.18.

GD&T with SolidWorks

We will now discuss in detail how to use SolidWorks to define GD&T and apply the concepts to views of orthogonal drawings. Figure 32.19 shows the list of geometric tolerance symbols that are commonly used in drawings. The list summarizes the following types of tolerance:

Form: straightness, flatness, circularity, and cylindricity
Profile: line, surface
Location: position, concentricity, and symmetry
Orientation: parallelism, perpendicularity, and angularity
Runout: circular runout, and total runout

Form tolerances are applicable to single (individual) features or elements of single features; therefore, form tolerances are not related to datums.

	Type of Tolerance	Characteristic	Symbol
For individual features	Form	Straightness	—
		Flatness	▱
		Circularity	○
		Cylindricity	⌭
Individual or related features	Profile	Profile of a plane	⌒
		Profile of a surface	⌓
Related features	Orientation	Angularity	∠
		Perpendicularity	⊥
		Parallelism	//
	Location	Position	⌖
		Concentricity	◎
		Symmetry	⌯
	Runout	Circular runout	↗
		Total runout	⌰

Figure 32.19

List of geometric tolerance symbols.

The following subsections discuss the particulars of the form tolerances—(a) straightness, (b) flatness, (c) circularity, and (d) cylindricity.

Straightness

Straightness is a condition where an element of a surface, or an axis, is a straight line. A straightness tolerance specifies a tolerance zone within which the considered element or derived median line must lie. A straightness tolerance is applied in the view where the elements to be controlled are represented by a straight line. Straightness tolerances are most often applied to circular or matching objects to help ensure that the parts are not barreled or warped within the given feature tolerance range and therefore do not fit together well. Figure 32.20 shows a cylindrical object dimensioned and toleranced using an MMC condition that is applied to the tolerance. The surface of the cylinder may vary within the specified range, that is, from 21 to 19.

The allowable tolerance zone (ATZ) is defined as the feature size + flatness size. Figure 32.21 shows how to define the ATZ and the virtual condition.

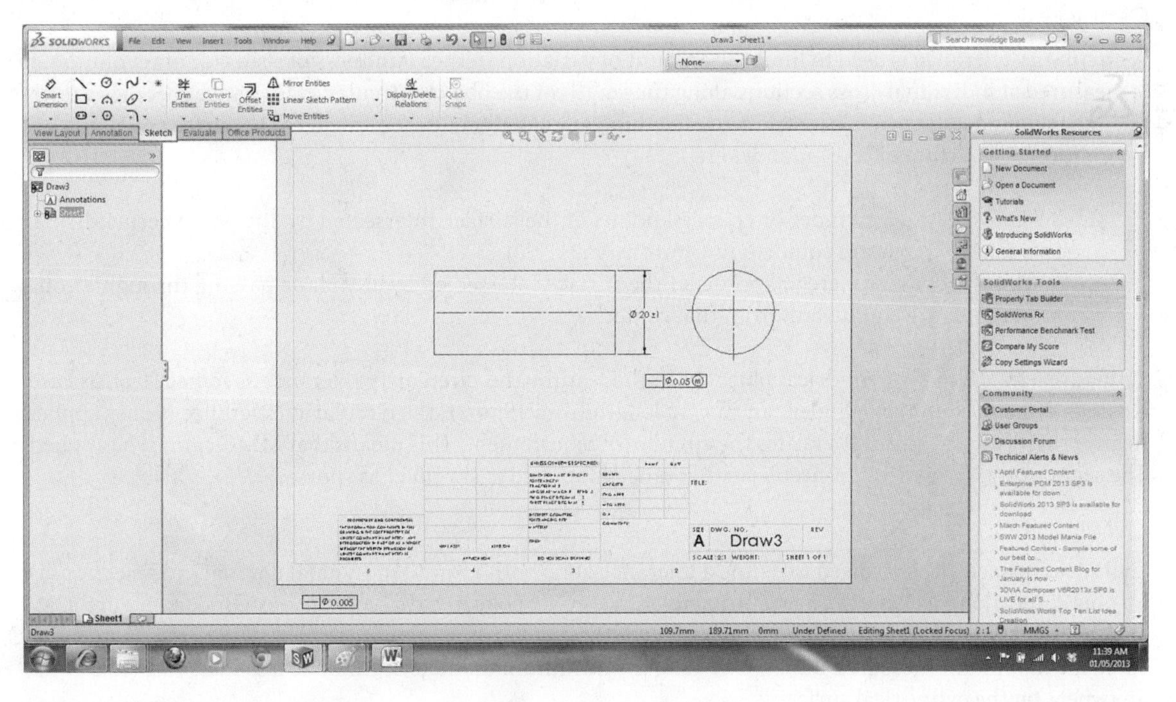

Figure 32.20

Cylindrical object dimensioned and toleranced using an MMC condition.

	Measured size	Allowable Tolerance Zone (ATZ)	Virtual condition
MMC →	21.0	0.0 + 0.05 = 0.05	21.0 + 0.05 = 21.05
	20.9	0.1 + 0.05 = 0.15	21.0 + 0.15 = 21.15
	20.8	0.2 + 0.05 = 0.25	21.0 + 0.25 = 21.25
	20.7	0.3 + 0.05 = 0.35	21.0 + 0.35 = 21.35
	20.6		
	.		
	.		
	19.7	1.3 + 0.05 = 1.35	21.0 + 1.35 = 22.35
	19.6	1.4 + 0.05 = 1.45	21.0 + 1.45 = 22.45
	19.5	1.5 + 0.05 = 1.55	21.0 + 1.55 = 22.55
	.		
	.		
	19.2	1.8 + 0.05 = 1.85	21.0 + 1.85 = 22.85
	19.1	1.9 + 0.05 = 1.95	21.0 + 1.95 = 22.95
LMC →	19.0	2.0 + 0.05 = 2.05	21.0 + 2.05 = 23.05

Figure 32.21

ATZ at given measured sizes and virtual condition.

Flatness

Flatness is the condition of a surface having all elements in one plane. Flatness tolerances are used to define the amount of variation that is permitted in an individual that is surfaced. A flatness tolerance specifies a tolerance zone that is defined by two parallel planes within which the surface must lie. Figure 32.22 shows a rectangular object that varies in height from 25.5 to 24.5. How flat is the top surface? This question can be better answered if an additional flatness tolerance value is added (0.3 in this case). Although the feature can vary based on the tolerance of ±0.5 that is specified, the surface could not vary by more than 0.3. When a flatness tolerance is specified, the feature control frame is attached to a leader that is directed to the surface or to an extension line of the surface. It is placed in a view where the surface elements to be controlled are represented by a line. Where the considered surface is associated with a size dimension, the flatness tolerance must be less than the size tolerance. Notice that the value of the flatness tolerance is normally less than the feature tolerance (0.3 < 0.5).

Circularity

A circularity tolerance is used to limit the amount of variation in the roundness of a surface of revolution. It is measured at individual cross sections along the length of the object. Circularity tolerance specifies a tolerance zone bounded by two concentric circles within which each circular element of the surface must lie. It is a condition of a surface of revolution where

1. With respect to a cylinder or cone, all points of the surface intersected by any plane perpendicular to a common axis are equidistant from that axis.
2. With respect to a sphere, all points of the surface intersected by any plane passing through a common center are equidistant from that center.

Figure 32.23 shows a cylindrical object with the feature and circularity tolerance specified. In this case, we note that although the cylinder can vary in length from 19 to 21, the circularity tolerance is only applied to any cross section and never violates the circularity requirement. This means that MMC cannot be applied. The section A-A for which the circularity tolerance is applied is specified as shown.

Cylindricity

Cylindricity tolerances are used to define a tolerance zone both around individual circular cross sections of an object and along its length. The resulting tolerance zone looks like two concentric cylinders. Figure 32.24 shows a shaft that includes a cylindricity tolerance that establishes a tolerance zone of 0.2. This means that since the maximum measured diameter is 15.3, the minimum diameter cannot be less than 15.1 (15.3 – 0.2) anywhere on the cylindrical surface.

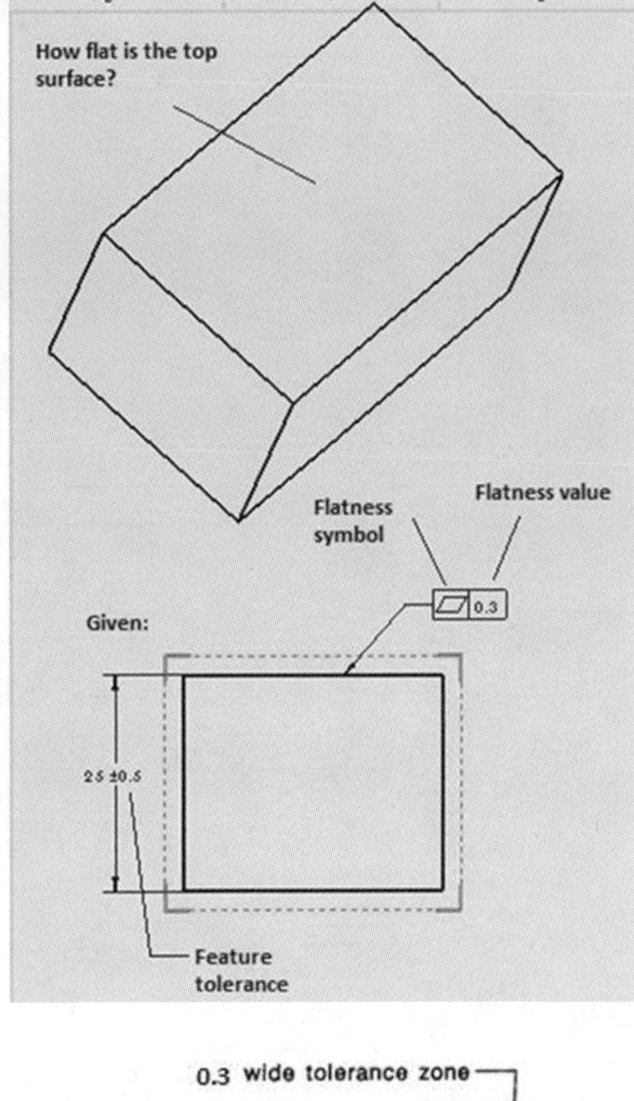

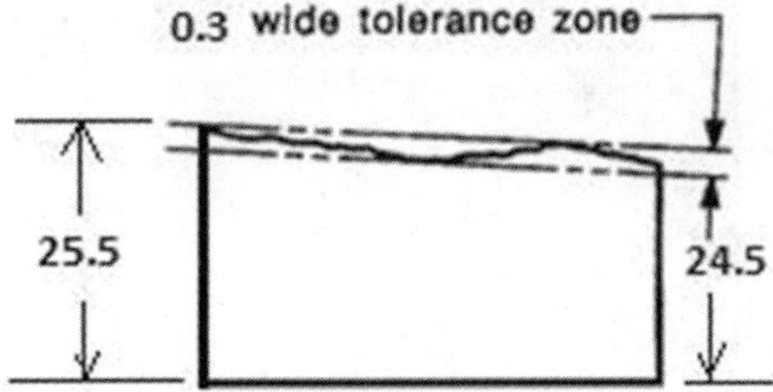

Figure 32.22

A rectangular object with feature and flatness tolerances.

Line Profile

Profiles of line tolerances are specified to irregular surfaces (see Figure 32.25). Profiles of line tolerances are particularly helpful when tolerancing an irregular surface that is consistently changing, such as an airplane wing.

Surface Profile

Profiles of surface tolerances are specified to irregular surfaces (see Figure 32.26). Four cases are identified. In case A, Figure 32.26, a 0.02 bilateral tolerance is specified with no further information. Therefore, the

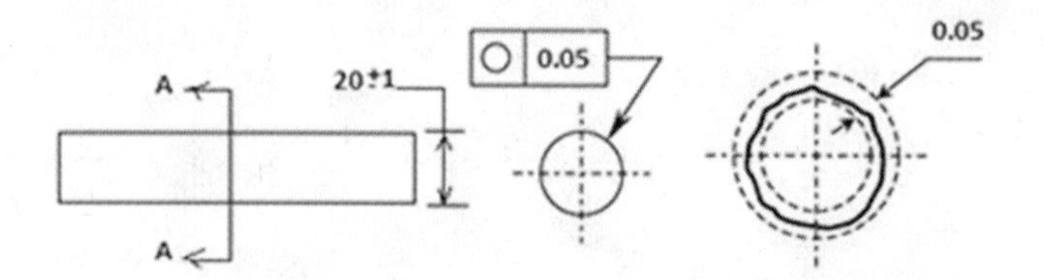

Figure 32.23

A cylindrical object with feature and circularity tolerance.

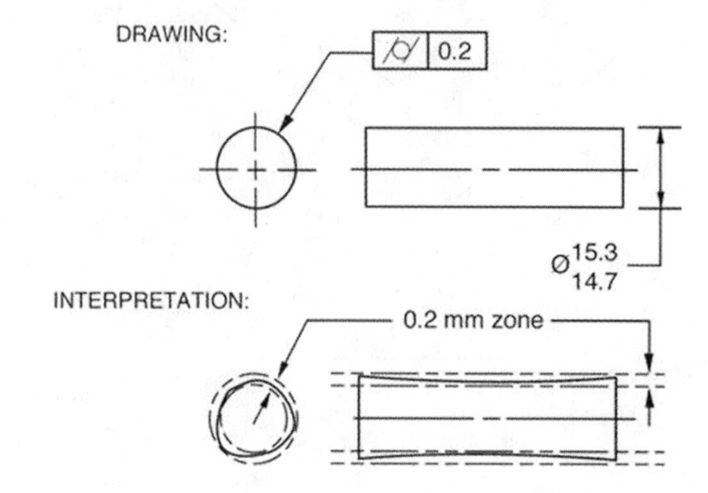

Figure 32.24

A cylindrical object with feature and cylindricity tolerance.

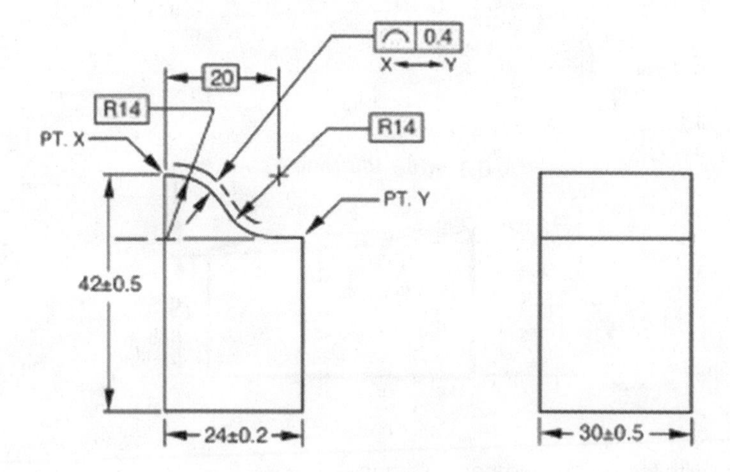

Figure 32.25

Line profile.

outside tolerance is 0.01, and the inside tolerance is 0.01. In case B, Figure 32.26, an unequal distributed bilateral tolerance is specified in which the outside has a tolerance of 0.02 and the inside has a tolerance of 0.005.

In case C, Figure 32.27, a 0.02 unilateral tolerance is specified for the outside tolerance, while in case D, a 0.02 unilateral tolerance is specified for the inside tolerance.

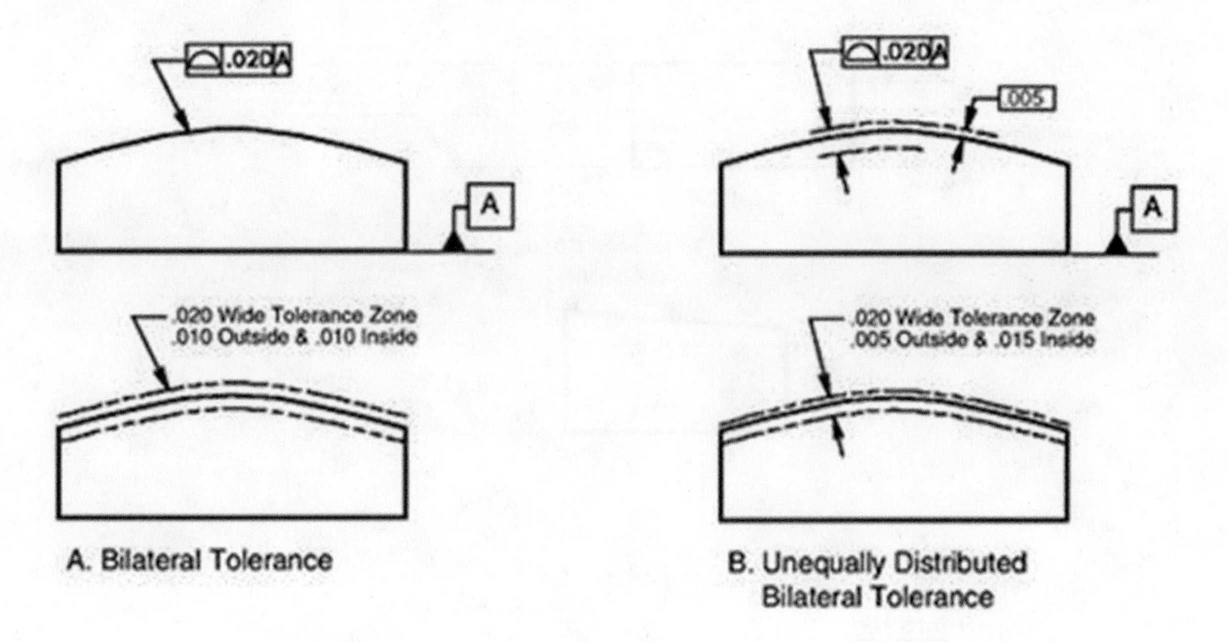

Figure 32.26

Bilateral and unequally distributed bilateral tolerance definitions.

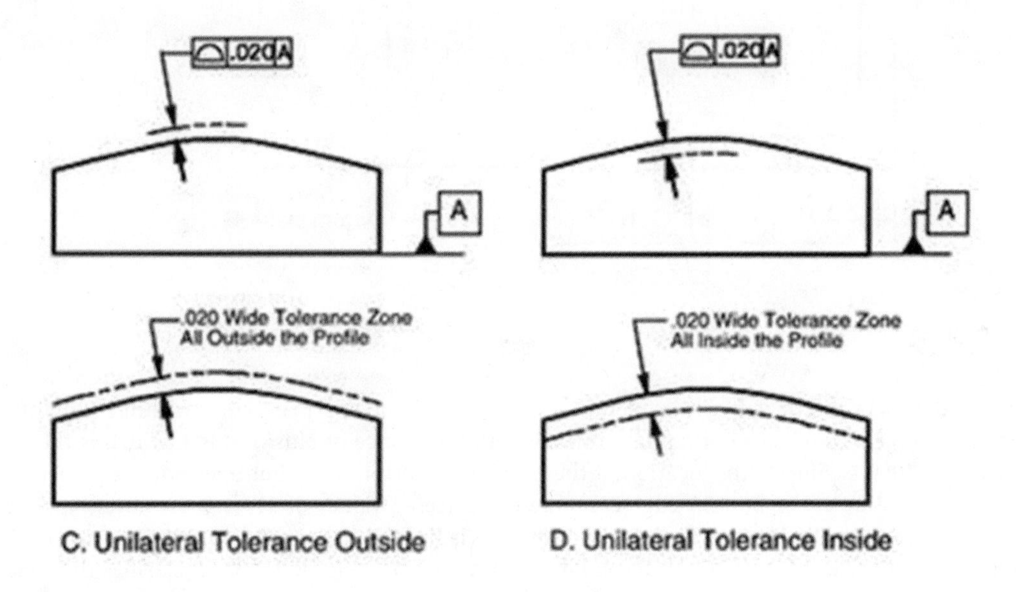

Figure 32.27

Unilateral tolerance definitions.

Parallelism

Parallelism tolerances are used to ensure that all points are within two parallel planes that are parallel to a referenced datum plane (see Figure 32.28).

Perpendicularity

Perpendicularity tolerances are used to limit the amount of variation for a surface or feature within two planes that are parallel to a specified datum. Figure 32.29 shows an L-shape in which the bottom surface is

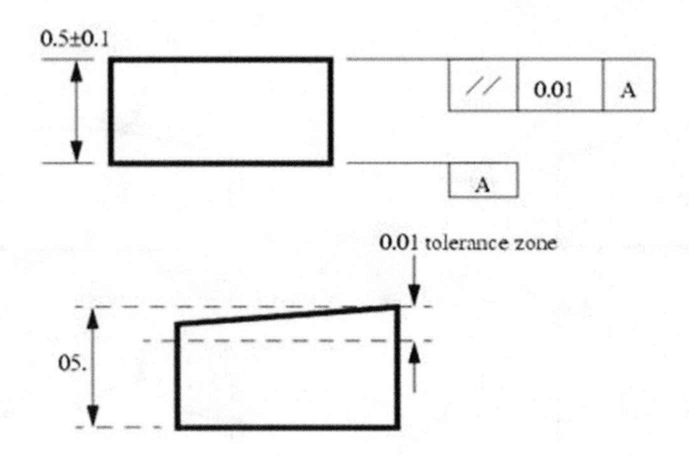

Figure 32.28

Parallelism.

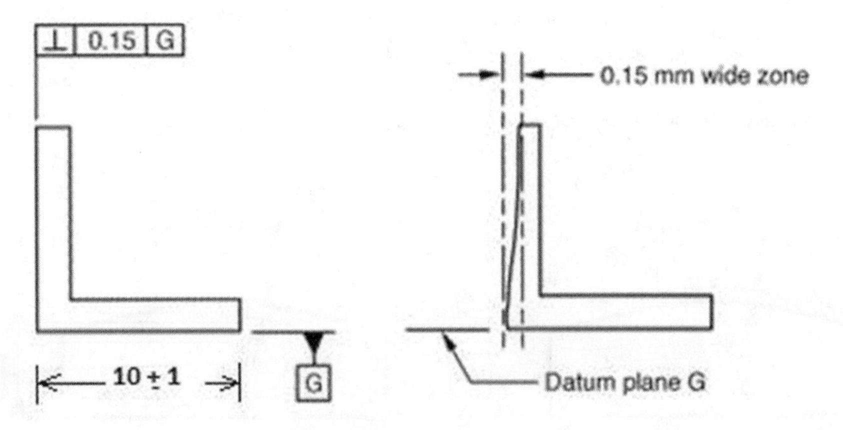

Figure 32.29

Perpendicularity.

assigned as datum G and the left side is toleranced so that it must be perpendicular within a limit of 0.15 to datum G. The object also has a horizontal dimension and tolerance of plus/minus 1, which is the location tolerance. The 10 ± 1 controls the location of the horizontal bottom edge but does not necessarily control the shape. For example, the bottom edge can take the extreme values of 11 or 9. However, applying the perpendicularity tolerances means that the bottom edge is limited to perpendicularity 0.15 within for the upper and lower bounds of 11 or 9. The second diagram to the right in Figure 32.29 clarifies the concept of perpendicularity tolerances.

Angularity

Angularity tolerances are used to limit the variance of surfaces and axes that are at an angle relative to the datum (see Figure 32.30).

Circular Runout

Runout tolerances are used to limit the variations between the features of an object and a datum. They are applied to surfaces around a datum axis such as a cylinder or to a surface that is constructed perpendicular to a datum axis. There are two types: (1) circular and (2) total. Figure 32.31 shows a circular runout tolerance.

Using SolidWorks to Define Tolerances in Drawings

To use SolidWorks to define tolerances in drawings requires that the Drawing Document should be active. Figure 32.32 shows the tools for defining tolerances in SolidWorks. The steps are fairly straightforward:

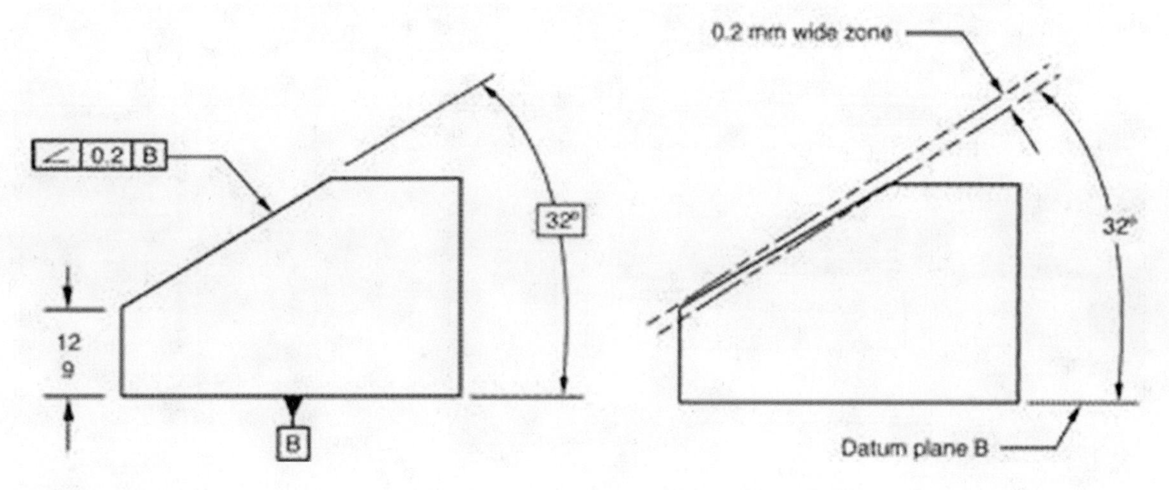

Figure 32.30

Angularity.

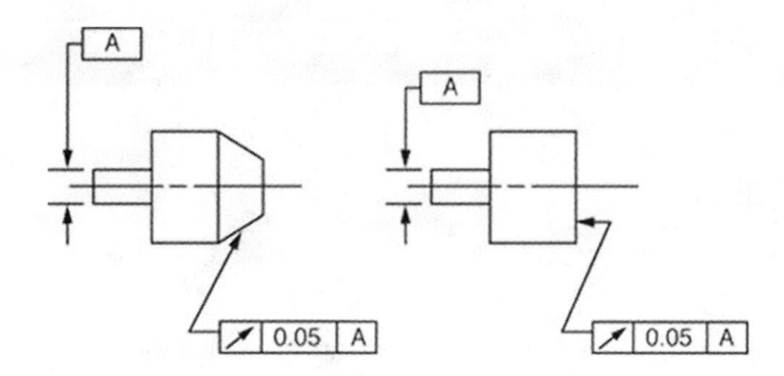

Figure 32.31

Circular runout.

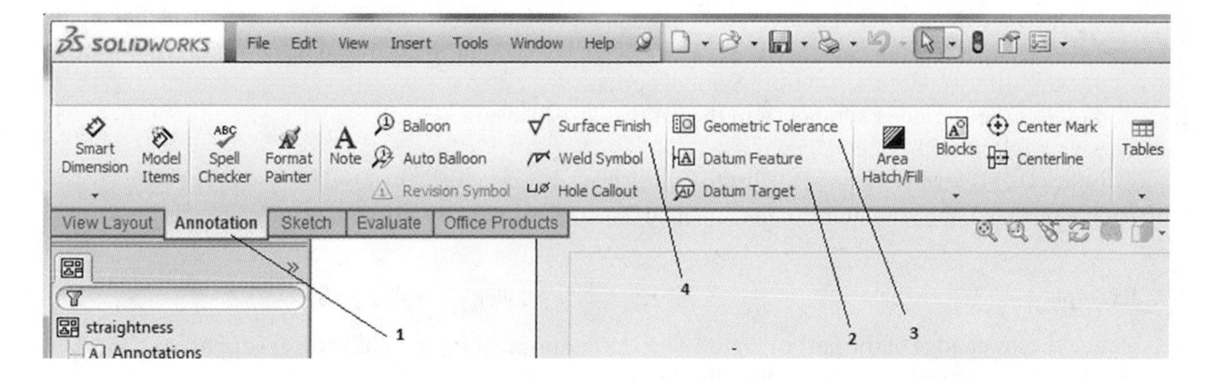

Figure 32.32

Tool for defining tolerances in SolidWorks.

1. Click Annotation (1) to show the ConfigurationManager that is required.
2. Click Datum Feature (2) to define datums.
3. Click Geometric Tolerance (3) to define tolerances.
4. Click Surface Finish (4) if needed for surface finish definition.

When Geometric Tolerance (3) is clicked, the Geometric Tolerance PropertiesManager is automatically displayed, as shown in Figure 32.33. All the tolerance symbols are accessed by clicking the Symbol pull-down button. The box Tolerance 1 is used to show the tolerance value. Then, select Primary as A, or Secondary as B, or Tertiary as C as the case may be; these are the datum surfaces.

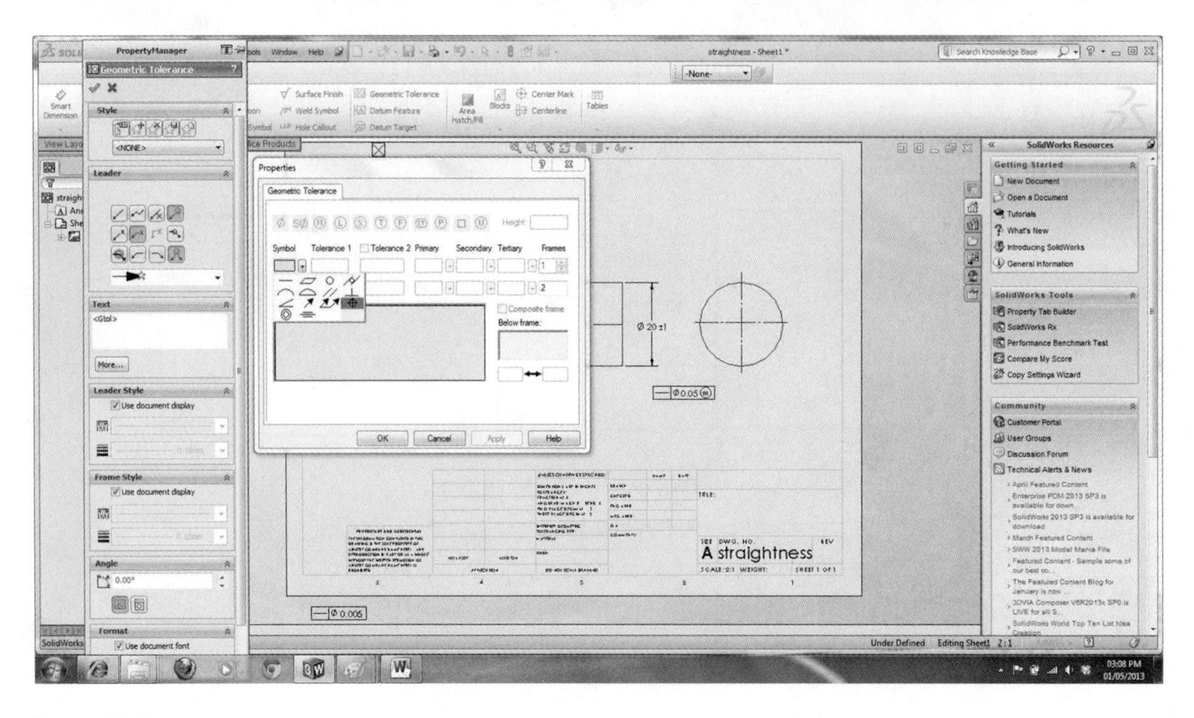

Figure 32.33

Geometric Tolerance PropertiesManager in SolidWorks.

Example 3

Figure 32.34 shows a cylindrical object with the diametral dimension given. Add a straightness tolerance with an MMC condition of 0.05 across the diameter.

SOLIDWORKS SOLUTION

1. Open the Drawing Document.
2. Click Annotation > Geometric Tolerance. (Figure 32.35 automatically appears.)
3. Select the Straightness (–) from Symbol option.
4. Select the Diameter symbol from the list that is displayed.
5. For Tolerance 1, Enter 0.05.
6. Select (M) for MMC from the list that is displayed.
7. Click OK (tolerance automatically displayed on drawing; see Figure 32.36).

Example 4

Define the lower edge of the part in Figure 32.37 as datum A, and the right vertical edge of the part as datum B and perpendicular to datum A within a tolerance of 0.001.

SOLIDWORKS SOLUTION

1. Open the Drawing Document.
2. Click Annotation > Datum Feature. (Figure 32.38 automatically appears.)
3. Select "A" from the Label Settings option.
4. Select the filled triangle from the Leader option.
5. Locate the datum symbol on the drawing.
6. Click OK.
7. Click Annotation > Datum Feature.
8. Select "B" from the Label Settings option.

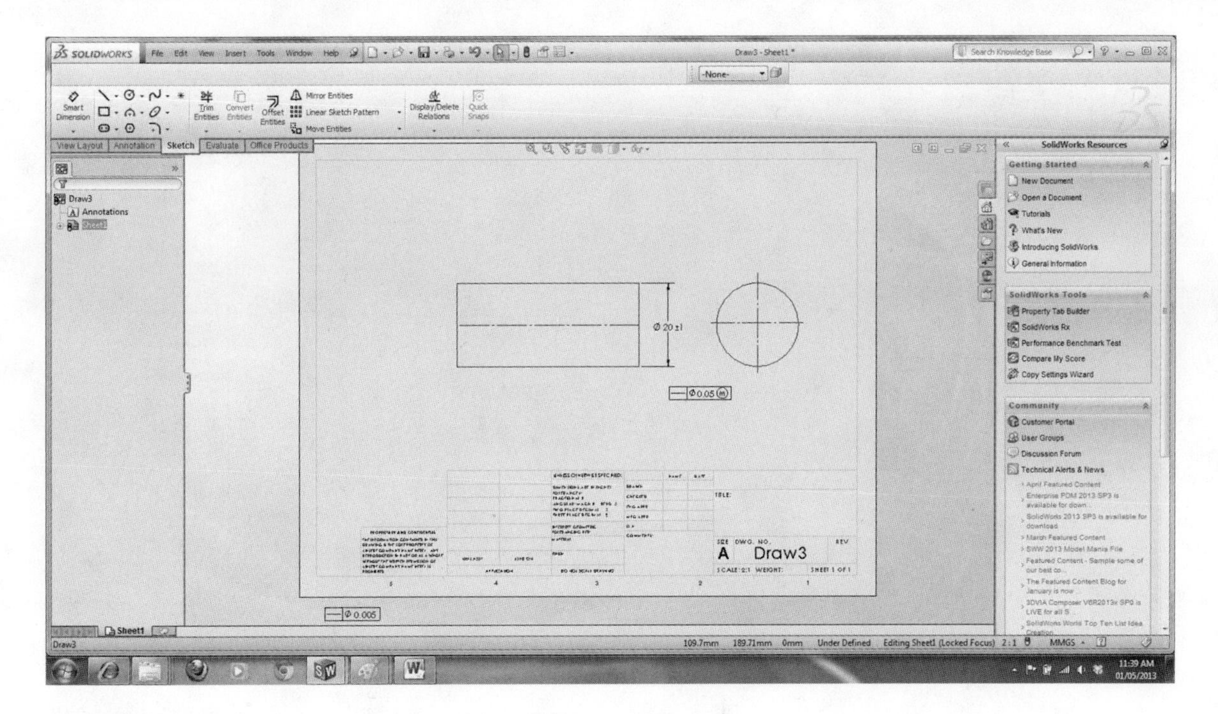

Figure 32.34

A cylindrical object with diametral dimension.

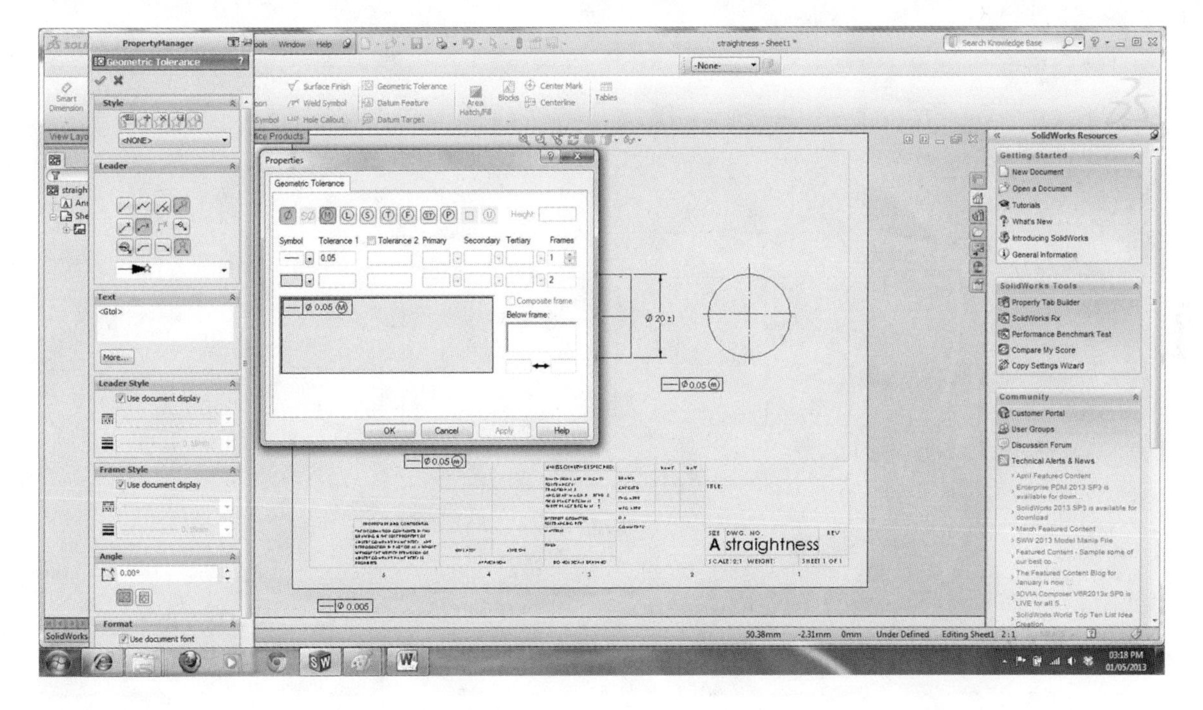

Figure 32.35

Geometric Tolerance PropertiesManager.

9. Select the filled triangle from the Leader option.
10. Click on the "B" label and click OK.
11. Click Annotation > Geometric Tolerance.
12. Select the Perpendicular from Symbol option (see Figure 32.39).
13. For Tolerance 1, enter 0.001 (see Figure 32.39).

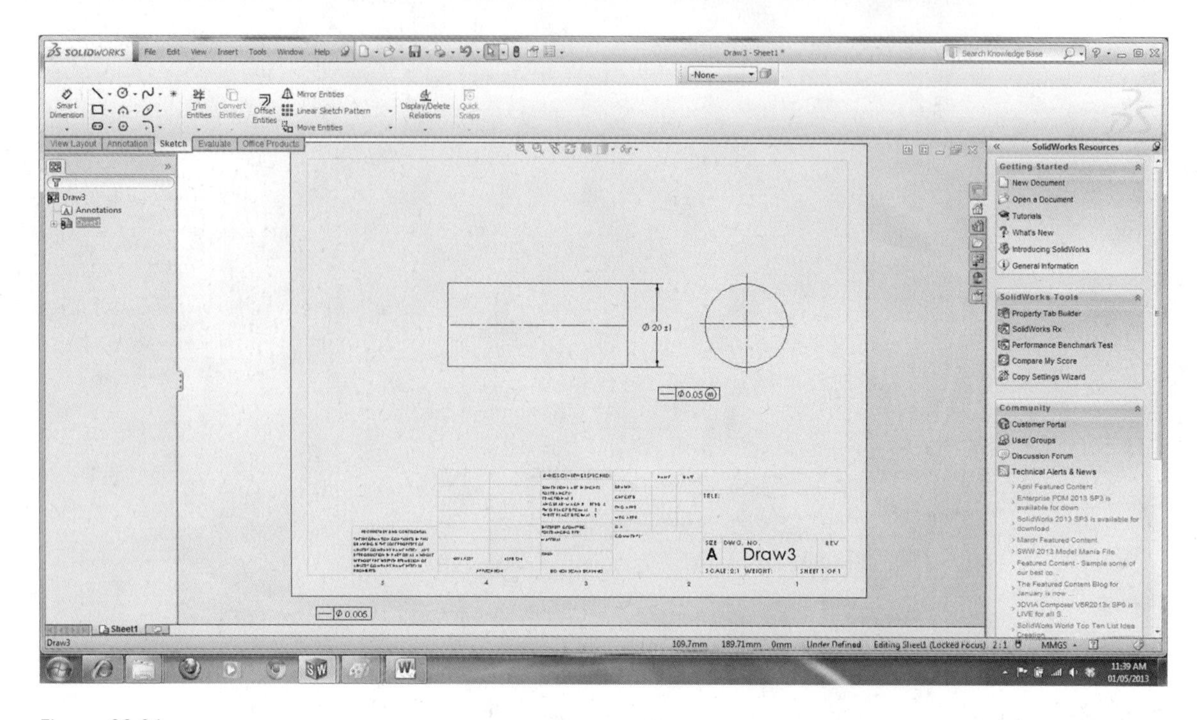

Figure 32.36

Cylindrical object dimensioned and toleranced using an MMC condition.

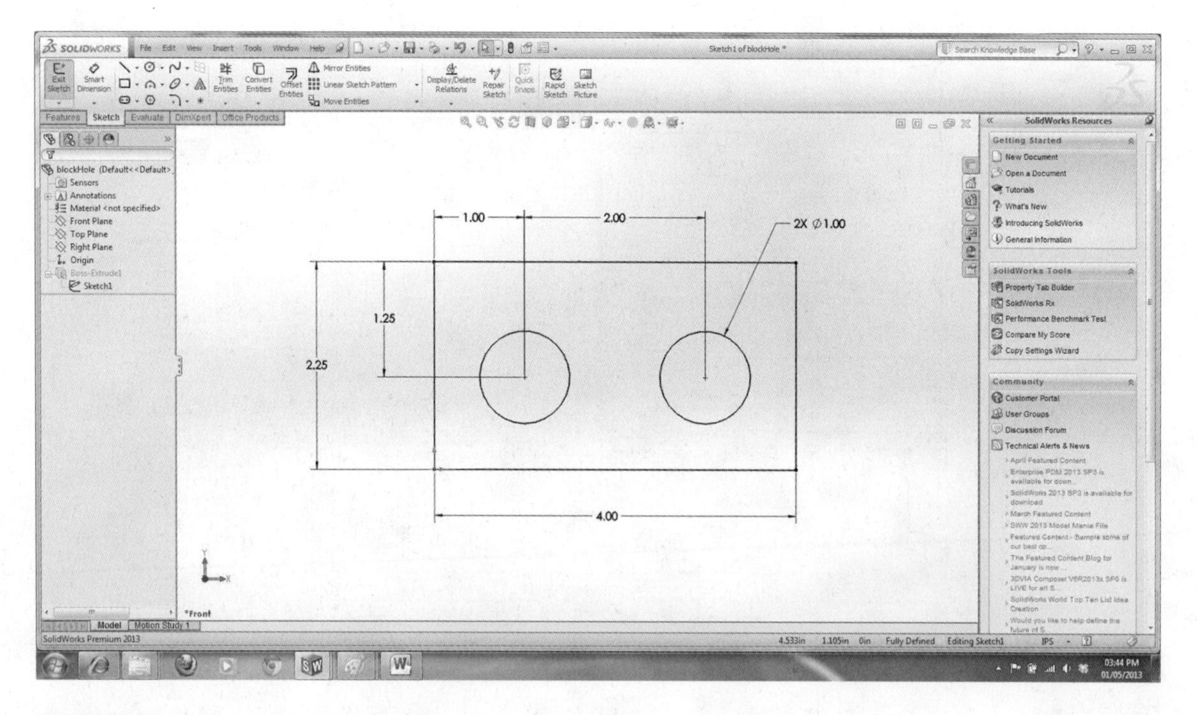

Figure 32.37

Part for tolerance definition using SolidWorks.

14. Select A for Primary Datum (see Figure 32.39).
15. Click OK (tolerance automatically displayed on drawing).
16. Click Annotation > Surface Finish (see Figure 32.40).
17. Select appropriate Symbol and add Surface Finish value = 16.
18. Click OK to complete. (See Figure 32.41 for the final drawing.)

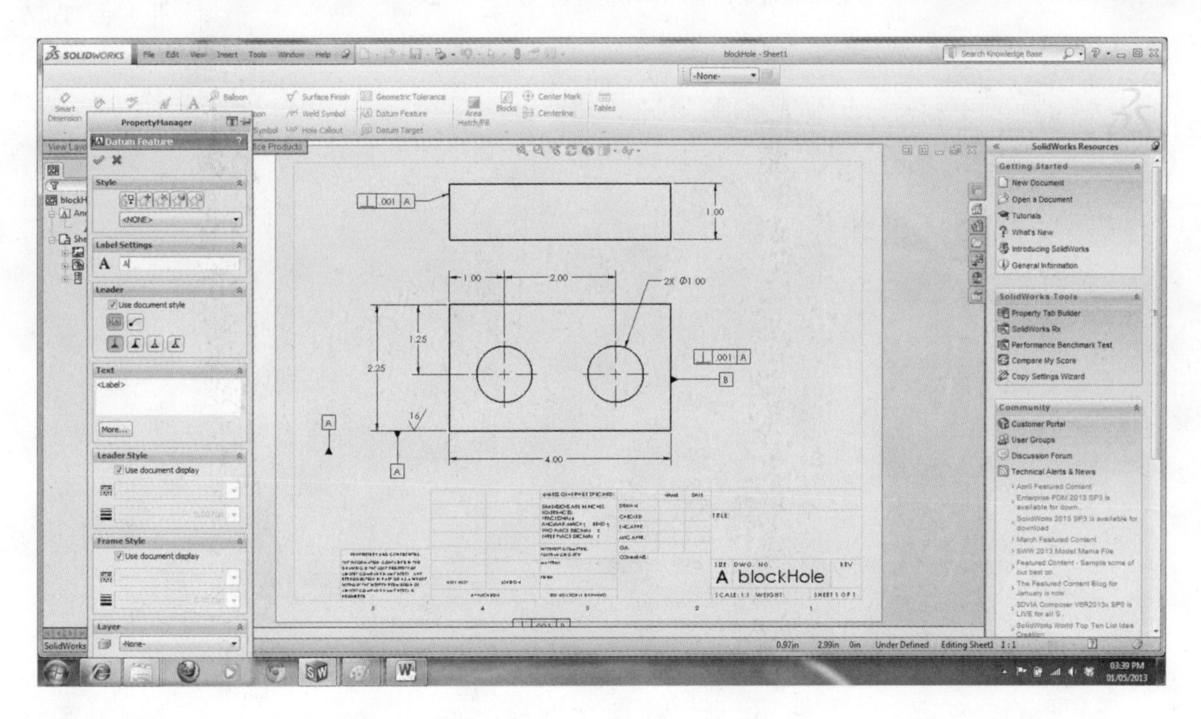

Figure 32.38

Datum Feature PropertiesManager.

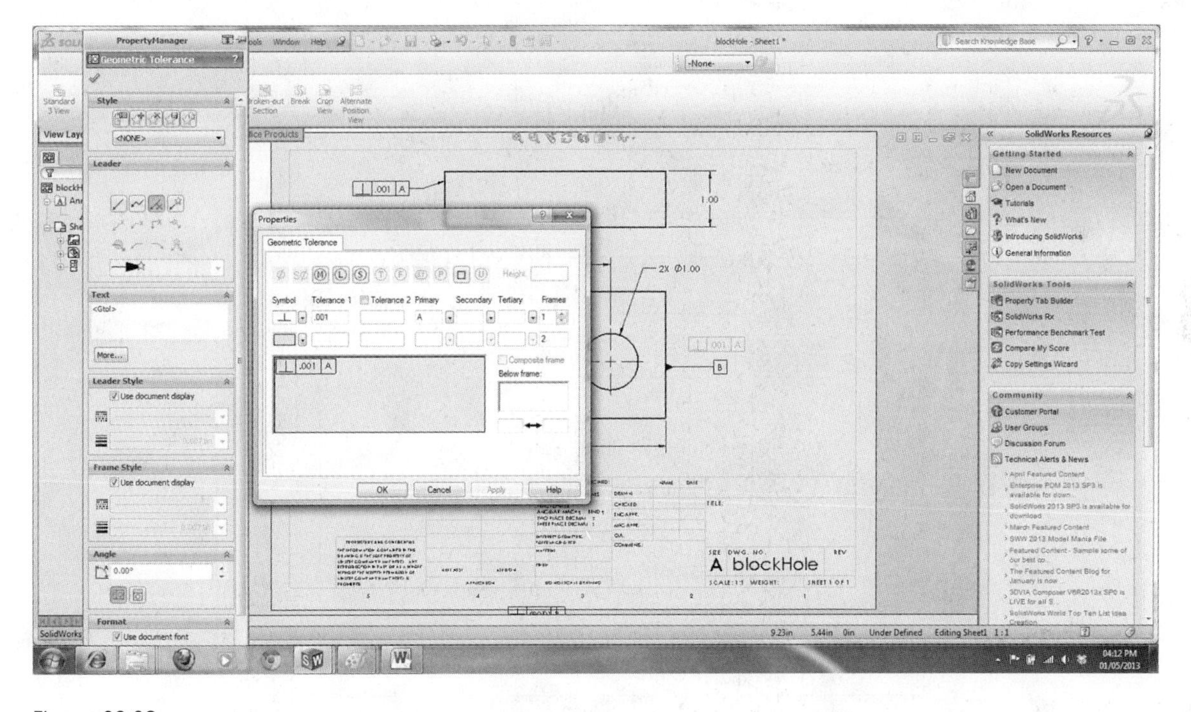

Figure 32.39

Geometric Tolerance PropertiesManager.

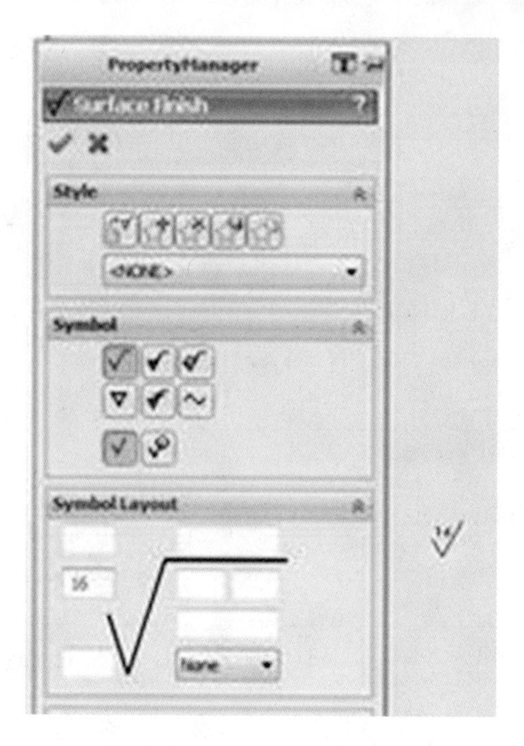

Figure 32.40

Surface Finish PropertiesManager.

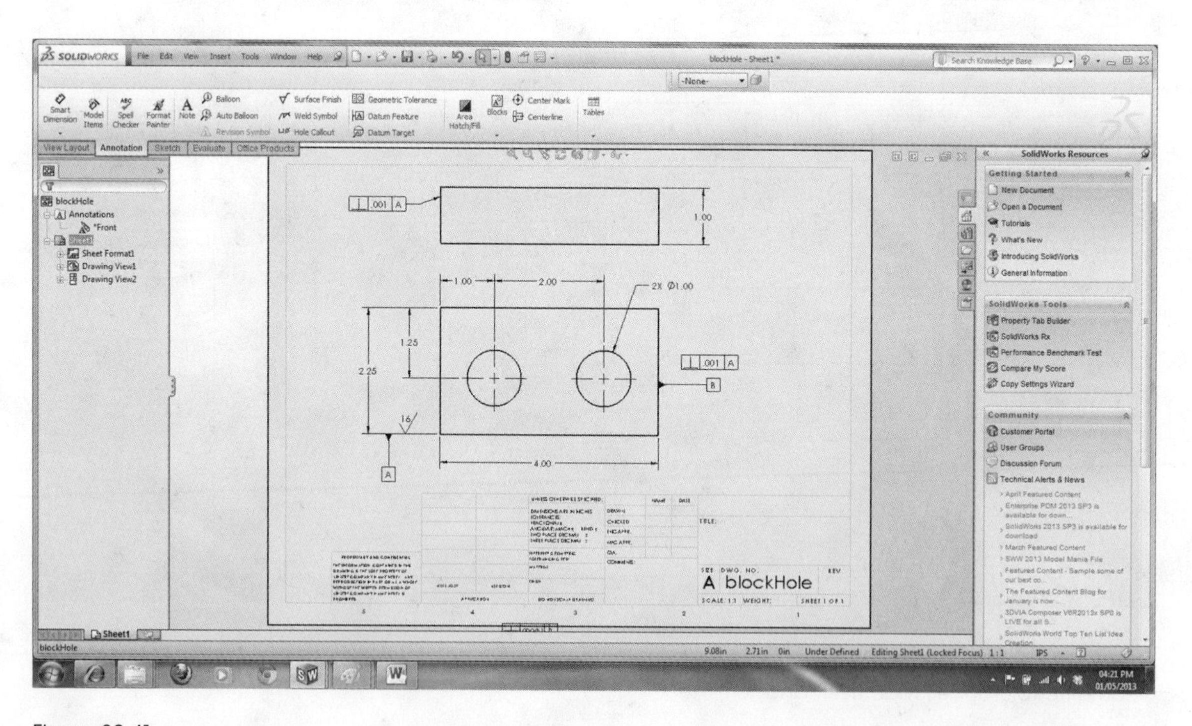

Figure 32.41

Drawing with dimensions surface finishing, datums, and tolerances.

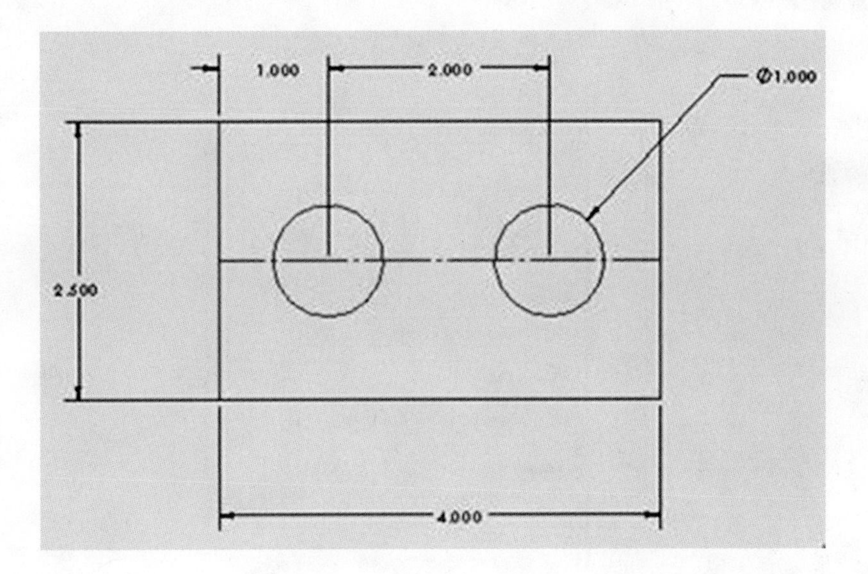

Figure 32.42

Basic drawing.

Example 5

Define a feature tolerance of 0.001 for the hole and apply a 0.001 positional tolerance about the hole's center point at the MMC for the drawing that is shown in Figure 32.42.

SOLIDWORKS SOLUTION

1. Open the Drawing Document.
2. Click Annotation > Geometric Tolerance.
3. Select Position from Symbol option (see Figure 32.43).

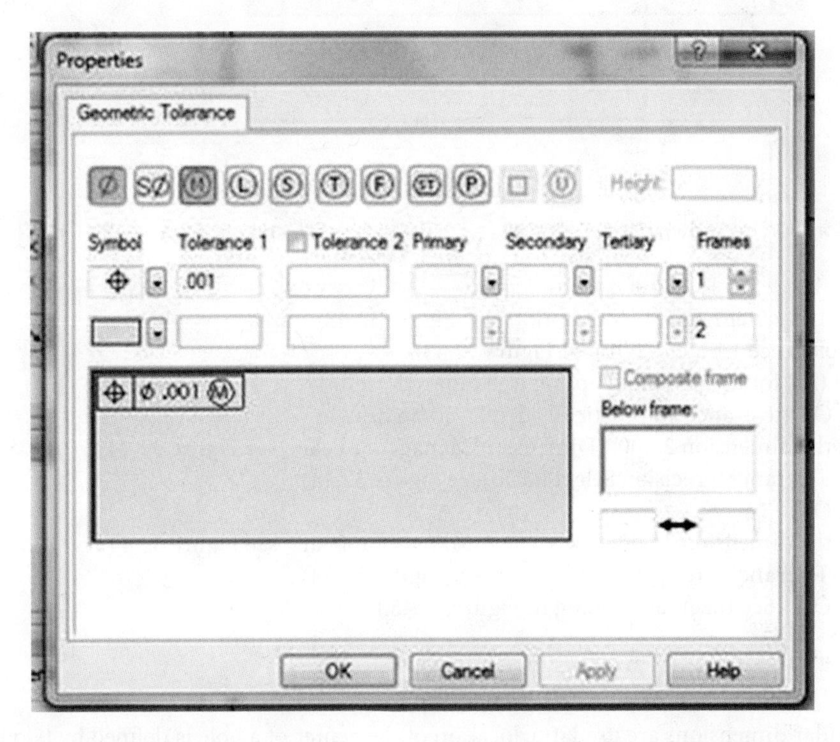

Figure 32.43

Geometric Tolerance PropertiesManager.

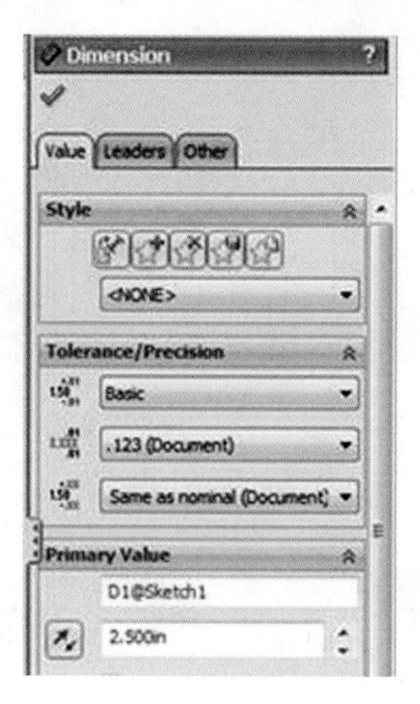

Figure 32.44

Dimension Manager.

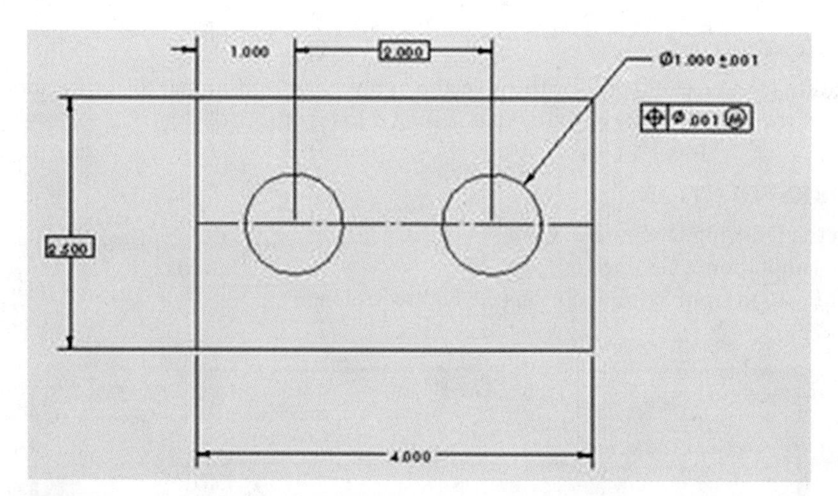

Figure 32.45

Complete annotation on drawing.

4. Select Diameter from Symbol option (see Figure 32.43).
5. For Tolerance 1 Enter 0.001 (see Figure 32.43).
6. Select (M) for from Symbol option (see Figure 32.43).
7. Click OK (tolerance automatically displayed on drawing).
8. Click the dimension 2.000. (Dimension Manager appears; see Figure 32.44.)
9. From Tolerance/Precision Select Basic (see Figure 32.44).
10. Click OK.
11. Click the dimension 2.500. (Dimension Manager appears; see Figure 32.44.)
12. From Tolerance/Precision Select Basic (see Figure 32.44).
13. Click OK. (See the final drawing in Figure 32.45.)

Hole Locations and Joining a Shaft to a Toleranced Hole

When rectangular dimensions are used, the location of the center of a hole is defined by two linear dimensions. The result is a rectangular tolerance zone whose size is based on the linear dimension's tolerances.

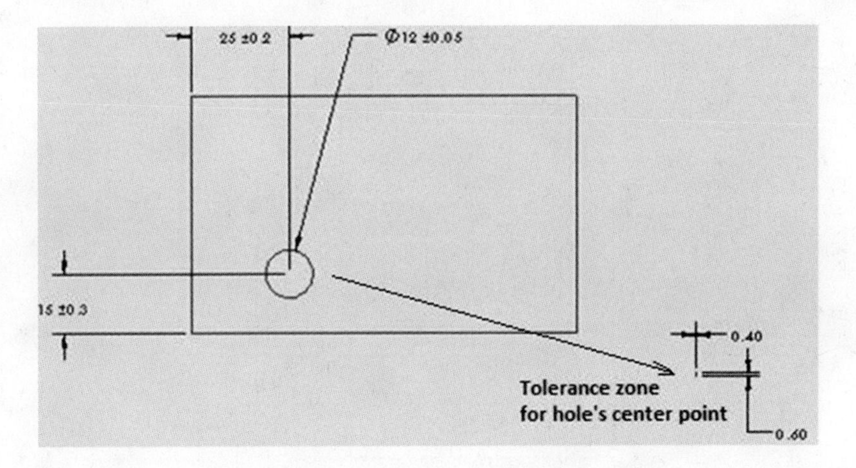

Figure 32.46

Location and size dimensions for a hole, shown.

Figure 32.46 shows the location and size dimensions for a hole, together with the resulting tolerance zone. The dimensions of the tolerance zone are each twice the size of the linear dimension tolerances ($2 \times 0.2 = 0.4$; $2 \times 0.3 = 0.6$).

Figure 32.46 shows the location and size dimensions for a hole.

Example 6

What is the largest diameter shaft that will always fit into the hole that is shown in Figure 32.46?

SOLUTION

For line dimensions and tolerances,

$$S_{max} = H_{min} - DTZ$$

where

S_{max} = maximum shaft diameter
H_{min} = minimum hole diameter
DTZ = diagonal distance across the tolerance zone

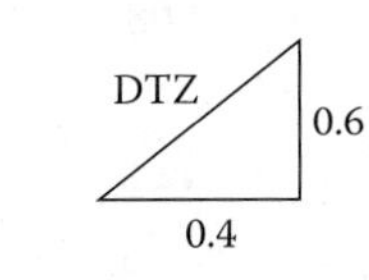

$$DTZ = \sqrt{(0.4)^2 + (0.6)^2} = 0.72$$

$$H_{min} = 12 - 0.05 = 11.95$$

$$S_{max} = 11.95 - 0.72 = 11.23$$

Example 7

Parts A and B in Figure 32.47 are to be joined by a common shaft. The total tolerance for the shaft is to be 0.05. What are the minimum and maximum shaft diameters?

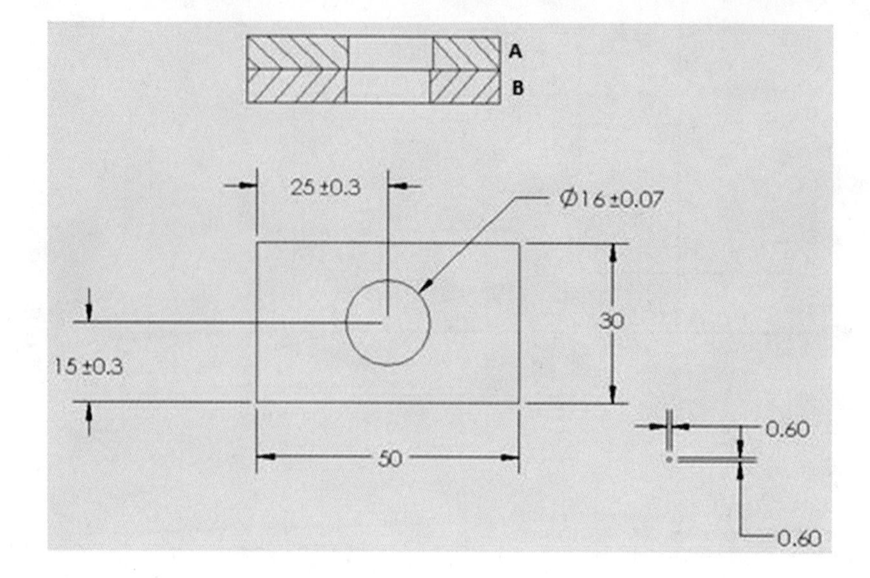

Figure 32.47

Parts A and B to be joined by a common shaft.

SOLUTION

Since both objects A and B have the same dimensions and tolerances and are floating relative to each other,

$$S_{max} = H_{min} - DTZ$$

$$DTZ = \sqrt{(0.6)^2 + (0.6)^2} = 0.85$$

$$H_{min} = 16 - 0.07 = 15.93$$

$$S_{max} = 15.93 - 0.85 = 15.08$$

$$S_{min} = S_{max} - 0.05 = 15.08 - 0.05 = 15.03$$

$$\therefore S_{max} = 15.08$$

$$S_{min} = 15.03$$

Example 8

Parts A and B in Figure 32.48 are to be joined by a common shaft whose maximum diameter is 0.248. What is the minimum hole size for the parts that will always accept the shaft? What is the maximum hole size if the total tolerance for the hole is 0.005?

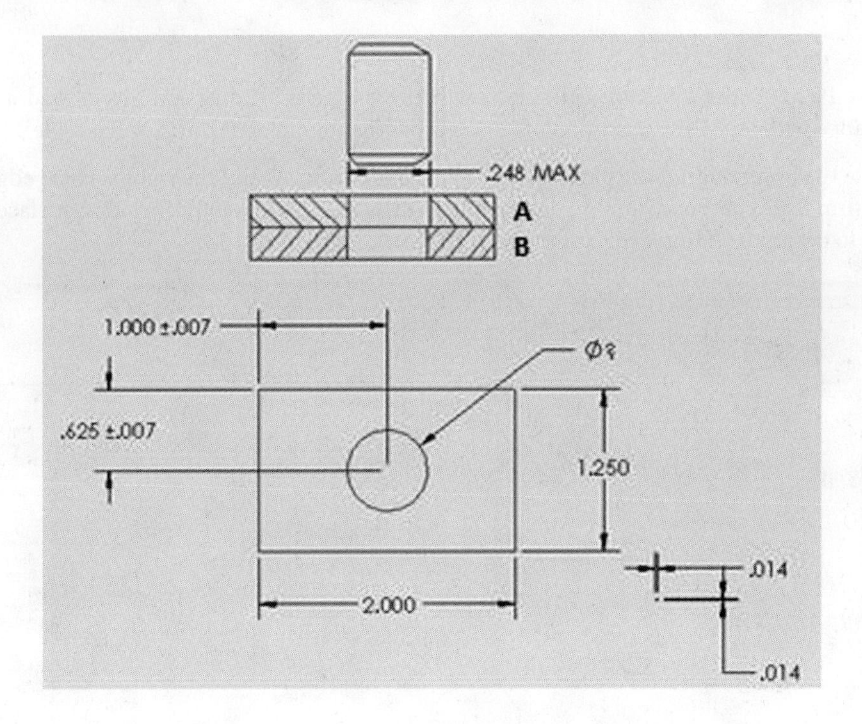

Figure 32.48

Parts A and B to be joined by a common shaft.

SOLUTION

Since both objects A and B have the same dimensions and tolerances and are floating relative to each other,

$$S_{max} = H_{min} - DTZ$$

$$DTZ = \sqrt{(.014)^2 + (.014)^2} = .02$$

$$H_{min} = S_{max} + DTZ$$

$$H_{min} = .248 + .02 = .268$$

$$H_{max} = .268 + .005 = .273$$

$$\therefore H_{max} = .273$$

$$H_{min} = .268$$

Summary

This chapter deals with GD&T, which is very important in the manufacturing industry since production on the shop floor is significantly affected by the proper dimensioning and tolerancing of the machined parts. The discussions and examples given in this chapter are based on the ASME Y14.5-2009 standard, *Dimensioning and Tolerancing*. The materials provided in this chapter are sufficient for beginners to understand the basic concepts that are involved in GD&T.

Exercises

1. Figure P32.1 shows a cylindrical object with the diametral dimension given. Add a straightness tolerance with an MMC condition of 0.025 across the diameter. (Hint: See Example 3.)

2. Define the lower edge of the part in Figure P32.2 as datum A, and the right vertical edge of the part as datum B and perpendicular to datum A within a tolerance of 0.001. Include a surface finish value of 32 to datum A. (Hint: See Example 4.)

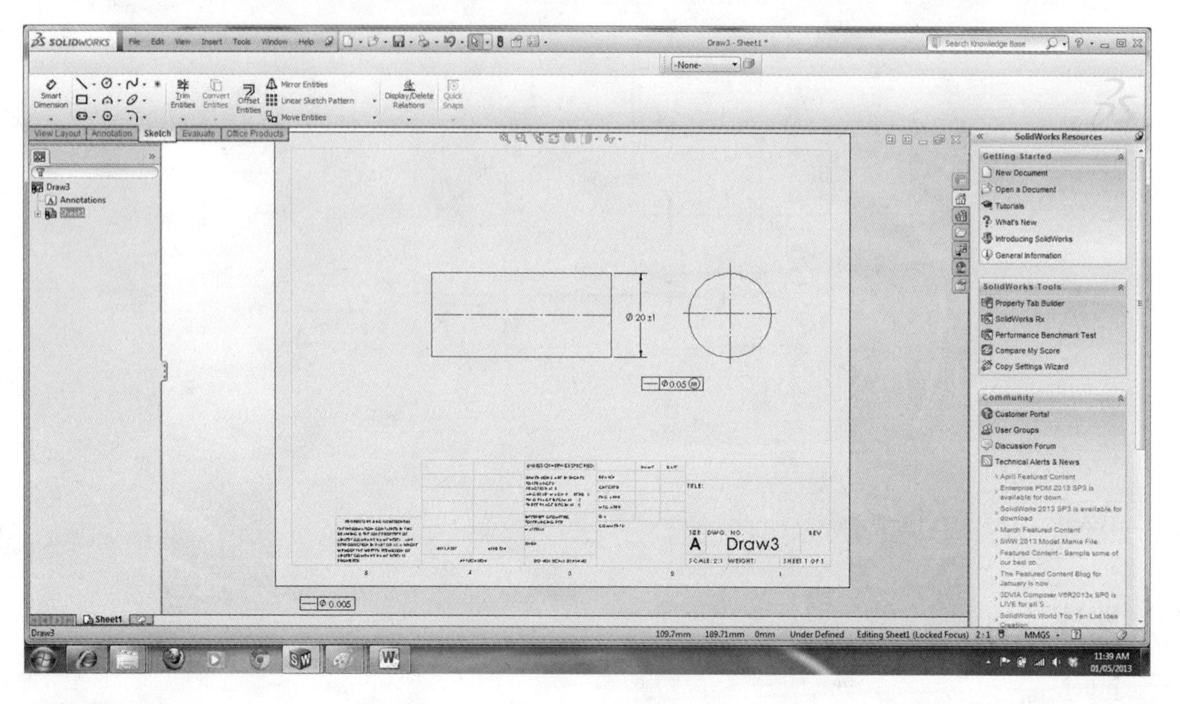

Figure P32.1

Metrology case 1.

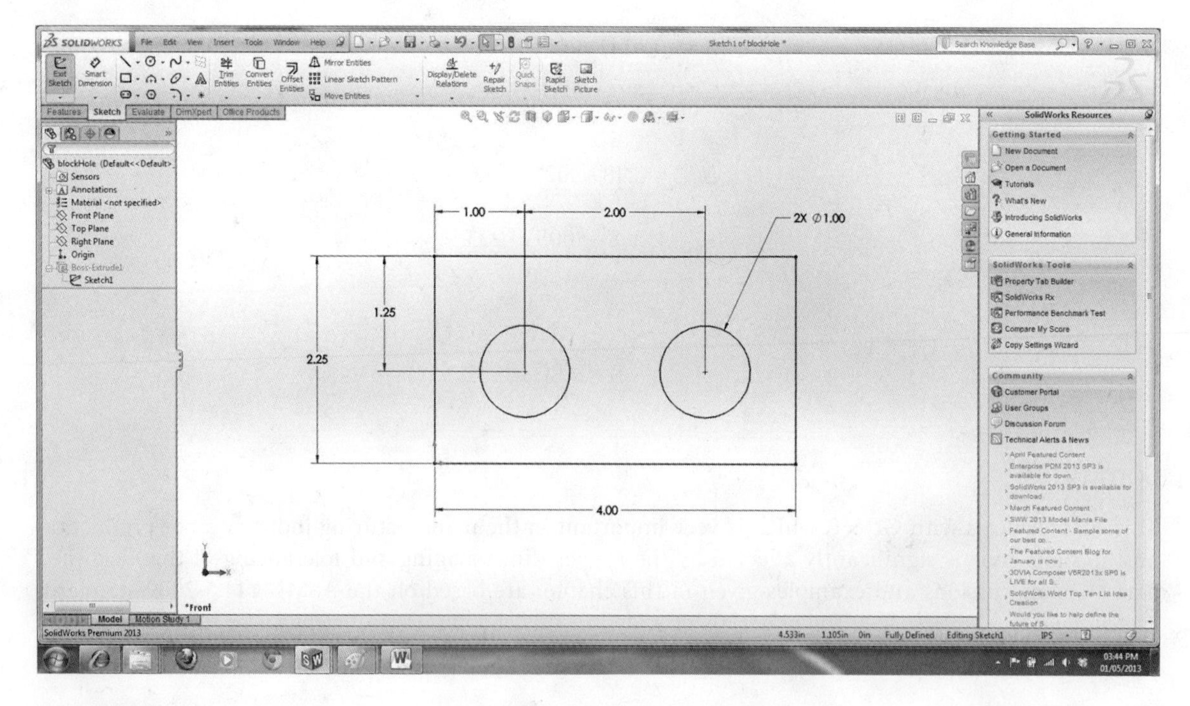

Figure P32.2

Metrology case 2.

3. Define a feature tolerance of 0.002 for the hole and apply a 0.002 positional tolerance about the hole's center point at the MMC for the drawing that is shown in Figure P32.3. (Hint: See Example 5.)
4. What is the largest diameter shaft that will always fit into the hole shown in Figure P32.4? (Hint: See Example 6.)
5. Parts A and B in Figure P32.5 are to be joined by a common shaft. The total tolerance for the shaft is to be 0.05. What are the minimum and maximum shaft diameters? (Hint: See Example 7.)
6. Parts A and B in Figure P32.6 are to be joined by a common shaft whose maximum diameter is 0.248. What is the minimum hole size for the parts that will always accept the shaft? What is the maximum hole size if the total tolerance for the hole is 0.005? (Hint: See Example 8.)

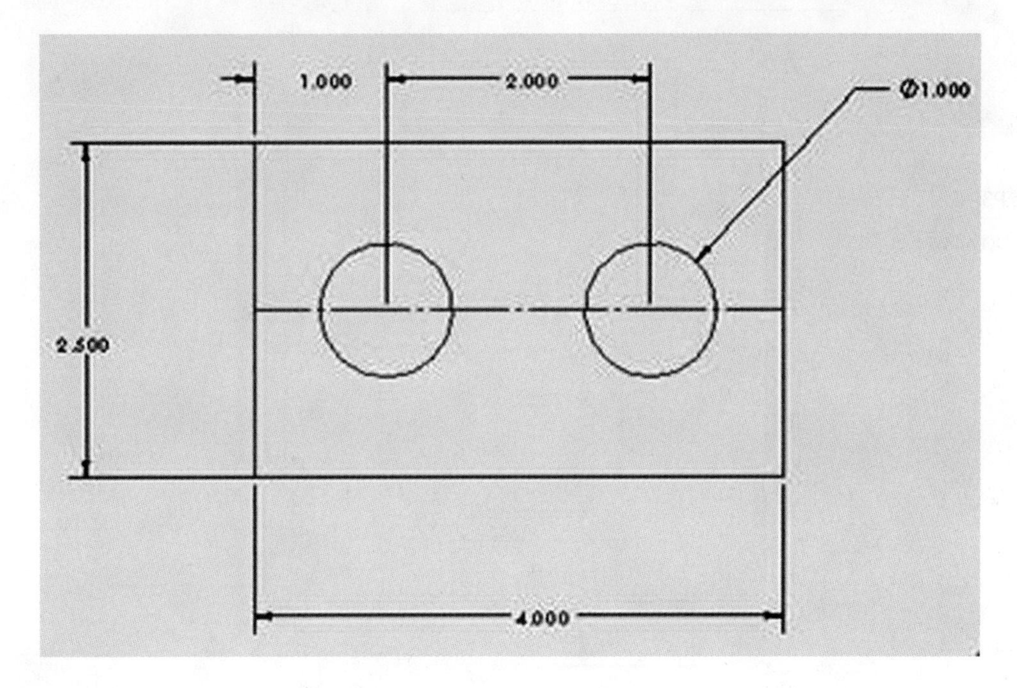

Figure P32.3

Metrology case 3.

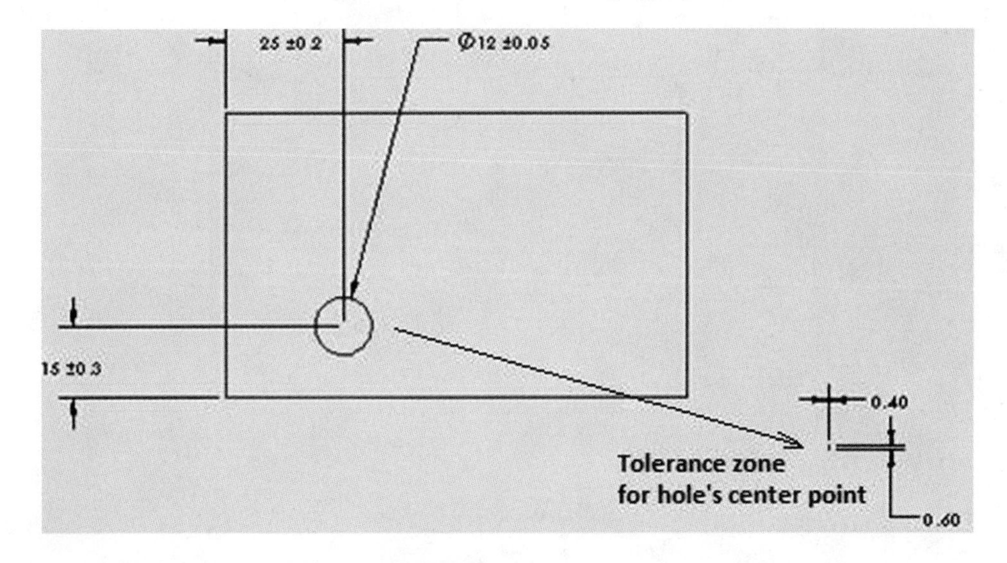

Figure P32.4

Metrology case 4.

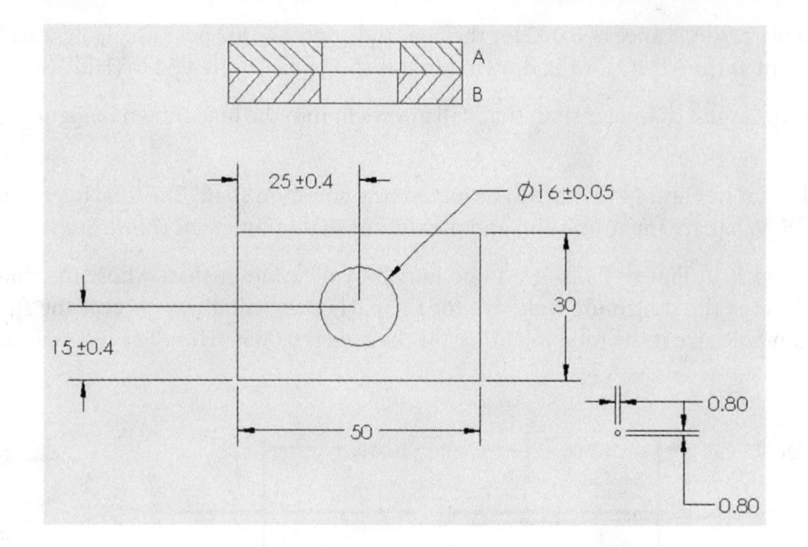

Figure P32.5

Metrology case 5.

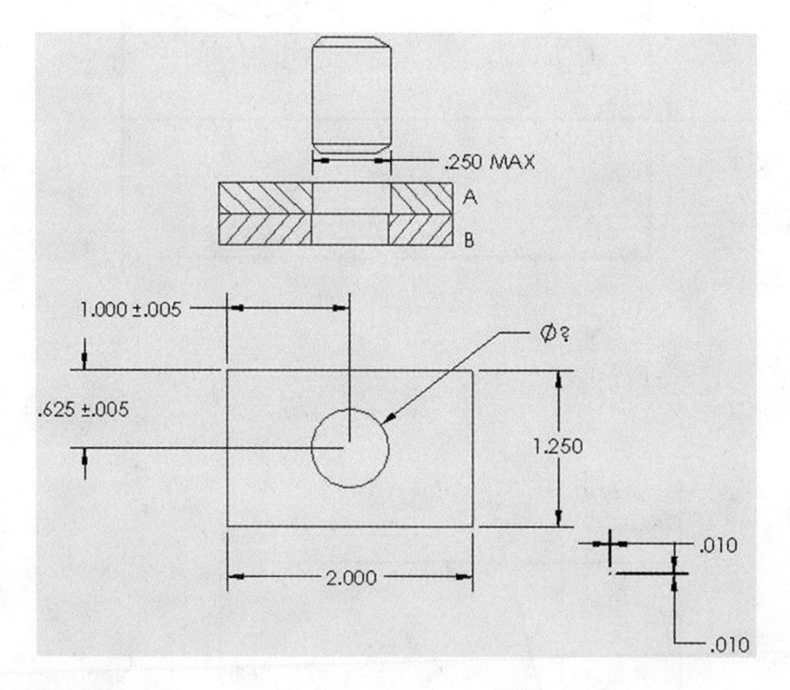

Figure P32.6

Metrology case 6.

33

Evaluating the Cost of Machined Parts

Objectives:

When you complete this chapter, you will have

- Changed Model Geometry and know it affects cost
- Added Material to the Stock Body and know it affects cost
- Examined the CostingManager: Setup Folder
- Examined the CostingManager: Mill Operations
- Changed Material Costs
- Changing the Stock Body
- Changed the Quantity of Parts to Manufacture and known that it affects cost
- Included a Discount
- Applied Custom Operations and know it affects cost
- Compared Material Costs
- Learned how to carry out automatic manufacturing cost estimation for machined parts

Introduction

Manufacturing cost is one of the main factors in machine design process in order to choose the most suitable solution; therefore, accurate estimation in the early design phases is fundamental. Design to cost implies to manage a vast amount of manufacturing knowledge that has to be linked to the design parameters.

Cost management or costing is a process for planning, managing, and controlling the costs of doing business. Ideally, product design projects should have customized costing plans, and companies as a whole should integrate costing into their business models. When properly implemented, costing translates into reduced costs for products and services, as well as increased value being delivered to the customer.

Taking this approach to costing helps a company determine whether they accurately estimated expenses initially and will help to more closely predict expenses in the future. However, costing cannot be used in isolation—projects must be organized and conducted with costing as a vital part of an overall business strategy.

A project that is defined through costing will facilitate the effective management of the costs that it incurs. Effective costing strategies will help deliver a high-quality product within a predetermined budget, as well as making it more valuable to the customer.

Feature-based three-dimensional (3D) computer-aided design (CAD) models contain data that are useful for cost estimation, but, despite the numerous work on features recognition and extraction, no cost estimation software system yet assures reliable results. SolidWorks 3D CAD software now has a tool for rapid manufacturing cost estimation where design features are automatically linked to manufacturing operations. Engineers and designers can now optimize designs, make informed decisions, and save time and development costs using SolidWorks 3D CAD software to automatically generate real-time manufacturing cost estimates for sheet metal and machined parts. The most current SolidWorks Costing functionality has been enhanced to make it more comprehensive than the previous version. Automatic cost estimation estimates part manufacturing costs using built-in cost templates. These manufacturing templates are customizable, allowing the entry of specific manufacturing costs and data, such as material, labor, machine speed and feeds, and setup costs. For the first time, designers can get automatic, real-time estimates of part manufacturing costs for sheet metal and machined parts with SolidWorks Costing, using built-in cost templates that can be customized for your specific operation. Design and manufacturing groups can share knowledge directly through the templates, enabling designers to make more informed product design decisions based on automatic cost estimation.

SolidWorks Costing functionality includes the ability to do the following:

- Design to a cost target using fast, repeatable data analysis
- Automatically track costs for both native SolidWorks or non-native imported models as the design develops
- Automate the quoting process for manufacturing
- Customize standard cost templates as needed to mirror in-house or outsourced manufacturing environments
- Update costs automatically with every design change—no extra work required
- Share manufacturing knowledge directly between design and manufacturing through costing templates

Using SolidWorks Costing, designers can automatically calculate manufacturing cost estimates to ensure that they are within design cost goals, and manufacturers can instantly create detailed quotes that are accurate to their specific manufacturing costs and processes. SolidWorks Costing works directly from 3D models—no preprocessing of the models is required. As a design is created, manufacturing costs are automatically calculated allowing designers to always have a current and accurate cost estimate.

SolidWorks Costing lets more people in your organization become engaged in reducing product cost while maintaining product quality. It provides a method for establishing and monitoring cost priorities. It improves both the bottom line (cost reduction) and the top line (increasing sales through higher-quality products). Costing is part of an overall business solution that maximizes profits and product quality.

SolidWorks Costing is a tool for getting more people to think about both material and process costs from different perspectives. Engineering personnel can leverage costing/cost reduction knowledge and pass it along to others within an organization, so costing is repeatable and becomes a vital aspect of overall corporate culture.

The SolidWorks Costing template contains the procedures that a user or the user's manufacturing supplier would use for manufacturing the part. The user can specify information such as material costs and sizes, the costs of manufacturing operations, and the manufacturing setup costs in the template.

This chapter briefly presents the geometric modeling of the parts for which manufacturing cost estimates is to be carried out, and then the SolidWorks automatic manufacturing Cost Estimation tool is discussed so that users can use it for carrying out the costing task for their product during the design stage. This way, the overall costs from design to manufacturing can be minimized because changes can to be made to the model before deciding on the optimal solution.

Model for Automatic Manufacturing Cost Estimation

1. Open a SolidWorks Part Document.
2. Choose Front Plane.
3. Create the Sketch that is shown in Figure 33.1.

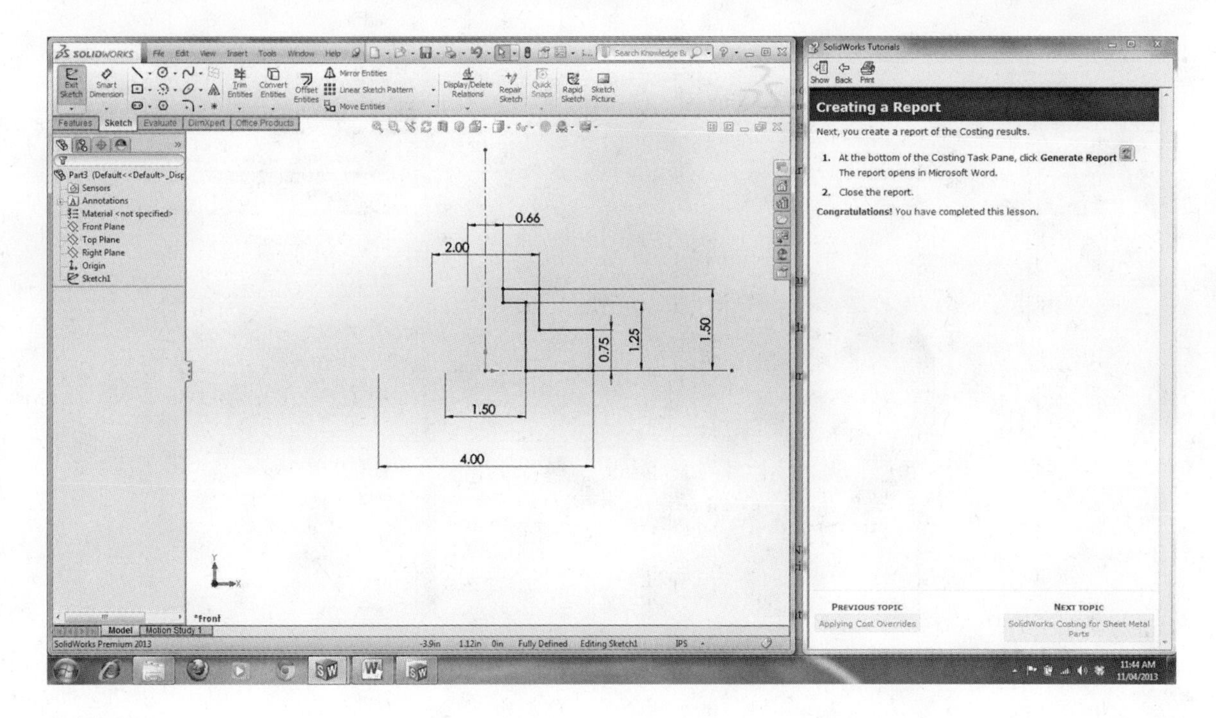

Figure 33.1

Sketch for creating part.

4. Revolve about the vertical centerline. (See Figures 33.2 and 33.3 for part.)
5. Create a construction circle for inserting three holes (see Figure 33.4).
6. Create a circle for inserting a radial hole (see Figure 33.5).
7. Create multiple holes using Circular Pattern to realize the final model (see Figure 33.6).

Now that the model is ready, we will now learn how to use SolidWorks automatic manufacturing cost estimation tool for this product.

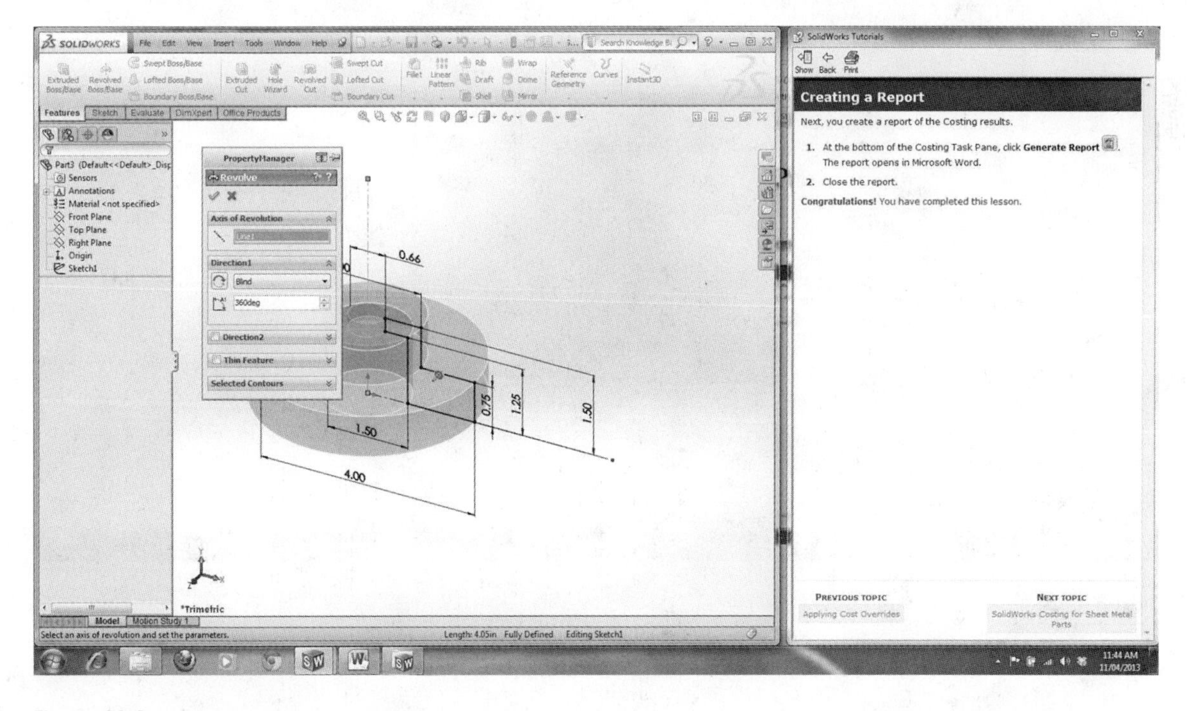

Figure 33.2

Revolve PropertyManager.

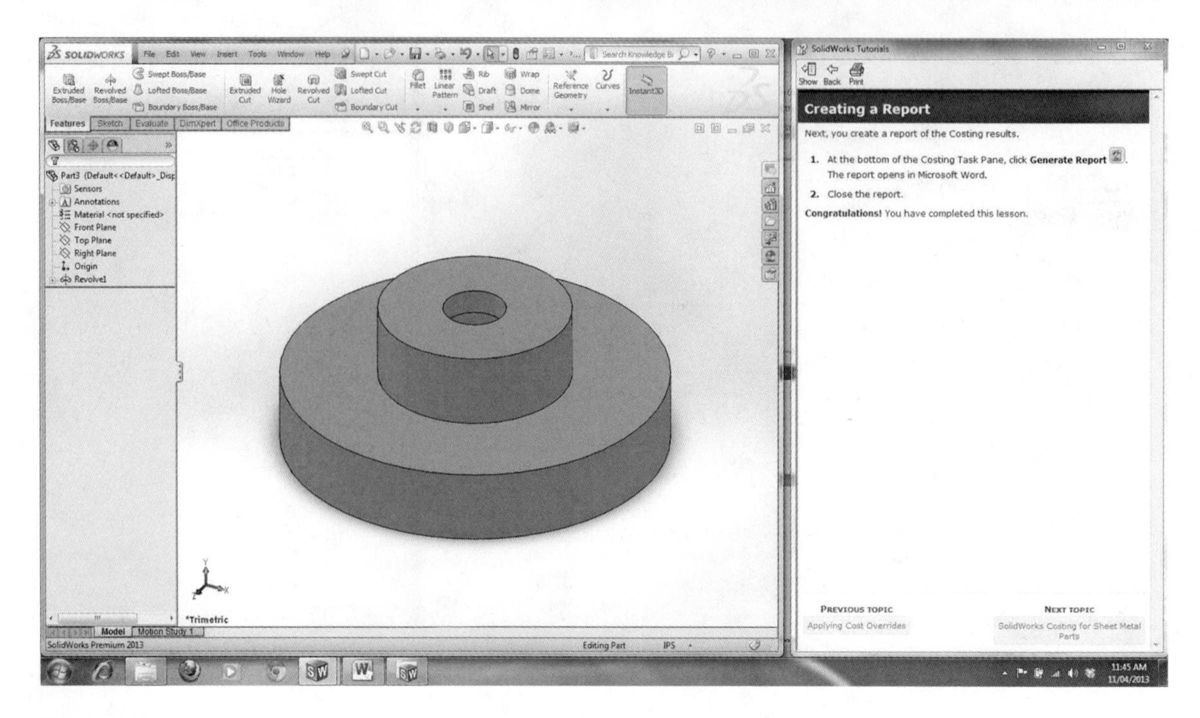

Figure 33.3

Revolved part.

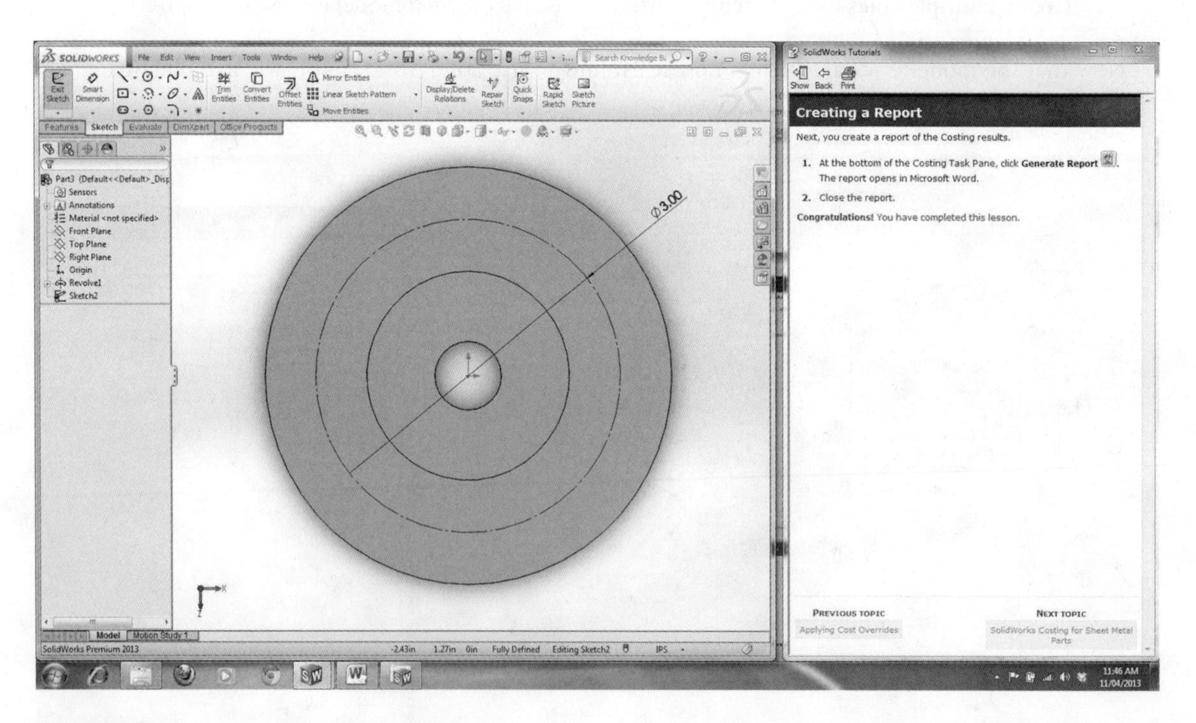

Figure 33.4

Construction circle for creating three holes.

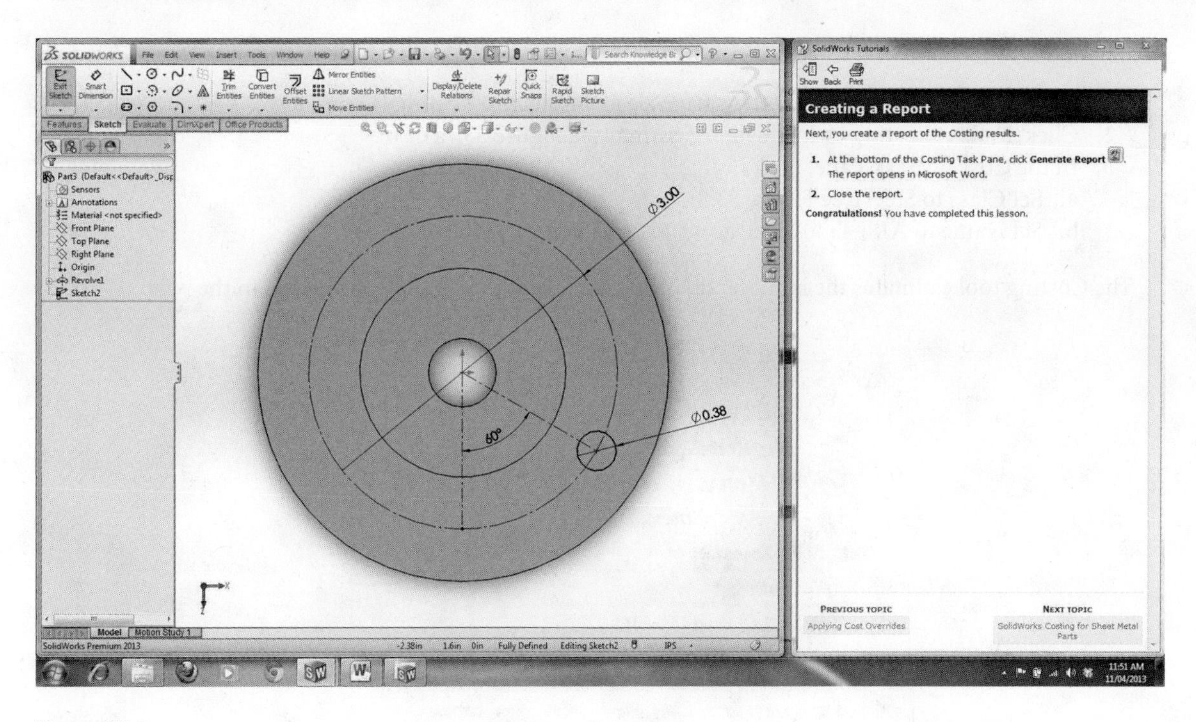

Figure 33.5

Circle for radial hole.

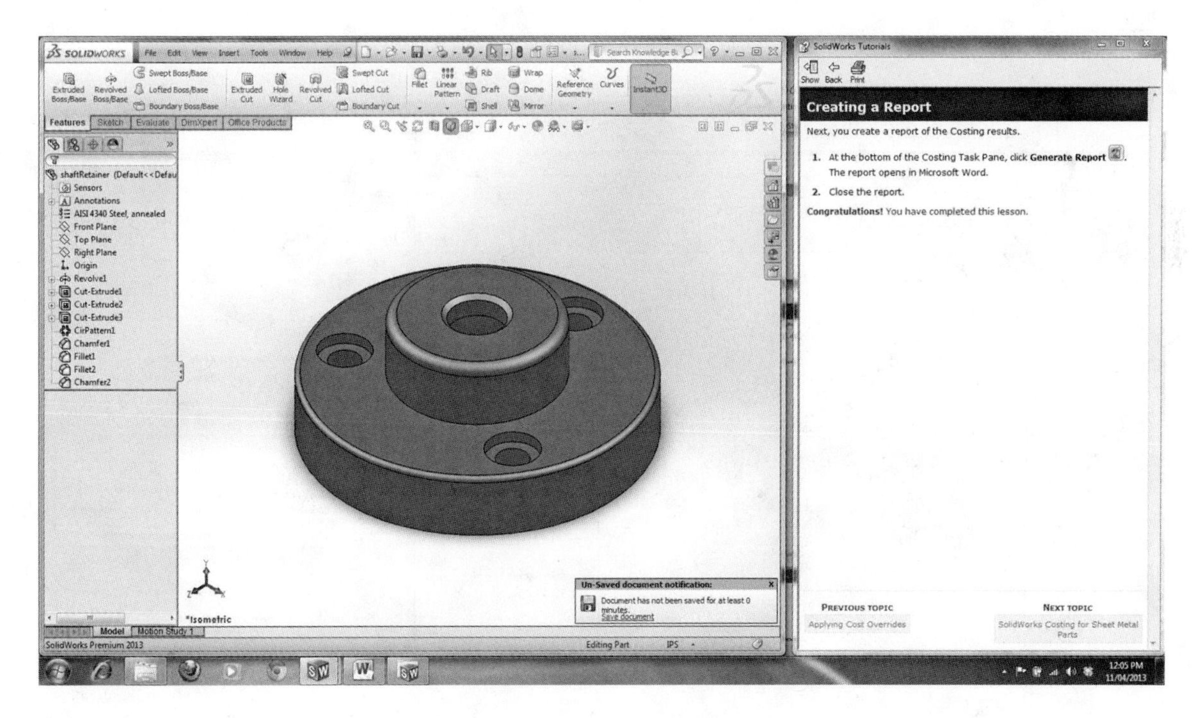

Figure 33.6

Final model.

Automatic Manufacturing Cost Estimation

1. Open the Part Document for the part to be examined for costing (Figure 33.6).
2. Click Tools > Costing or Evaluate > Costing (see Figure 33.7a or b).
3. In the Costing Task Pane, under Material,
   a. Set Class to Steel (see Figure 33.8).
   b. Set Name to AISI 4340 Steel, annealed (see Figure 33.9).

The Costing tool estimates the manufacturing cost using the material information in the part.

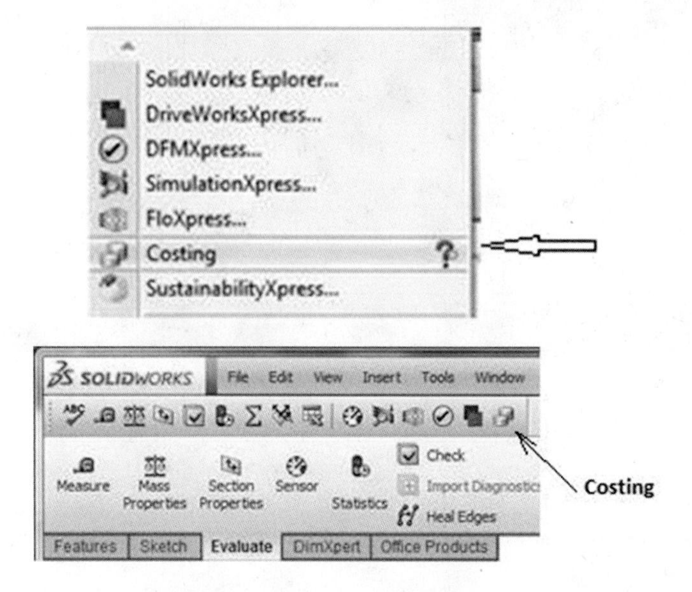

Figure 33.7

Costing tool.

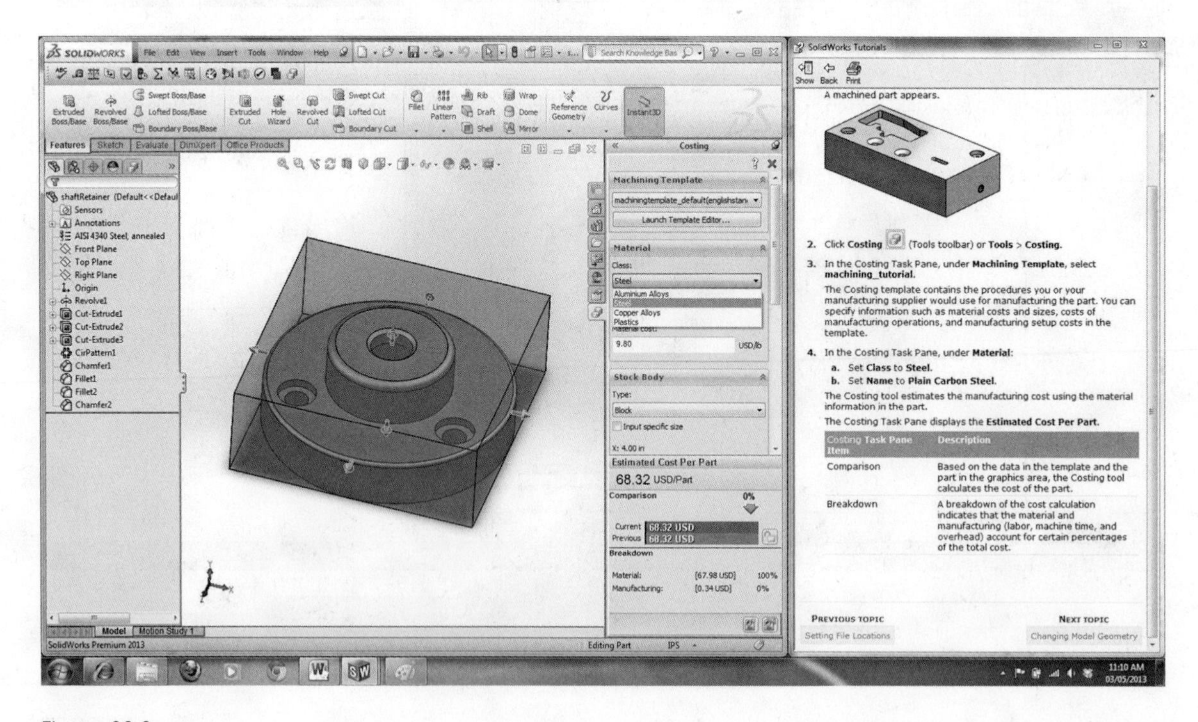

Figure 33.8

Class option in Costing Pane.

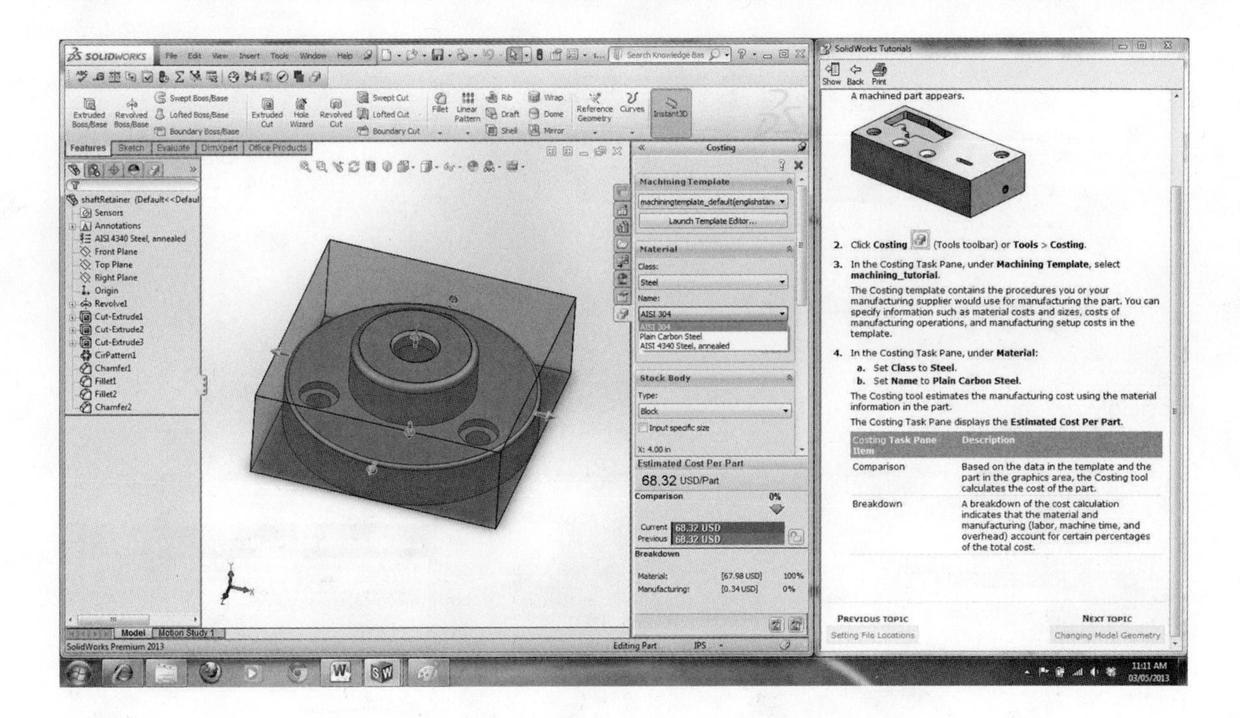

Figure 33.9

Name option in Costing Pane.

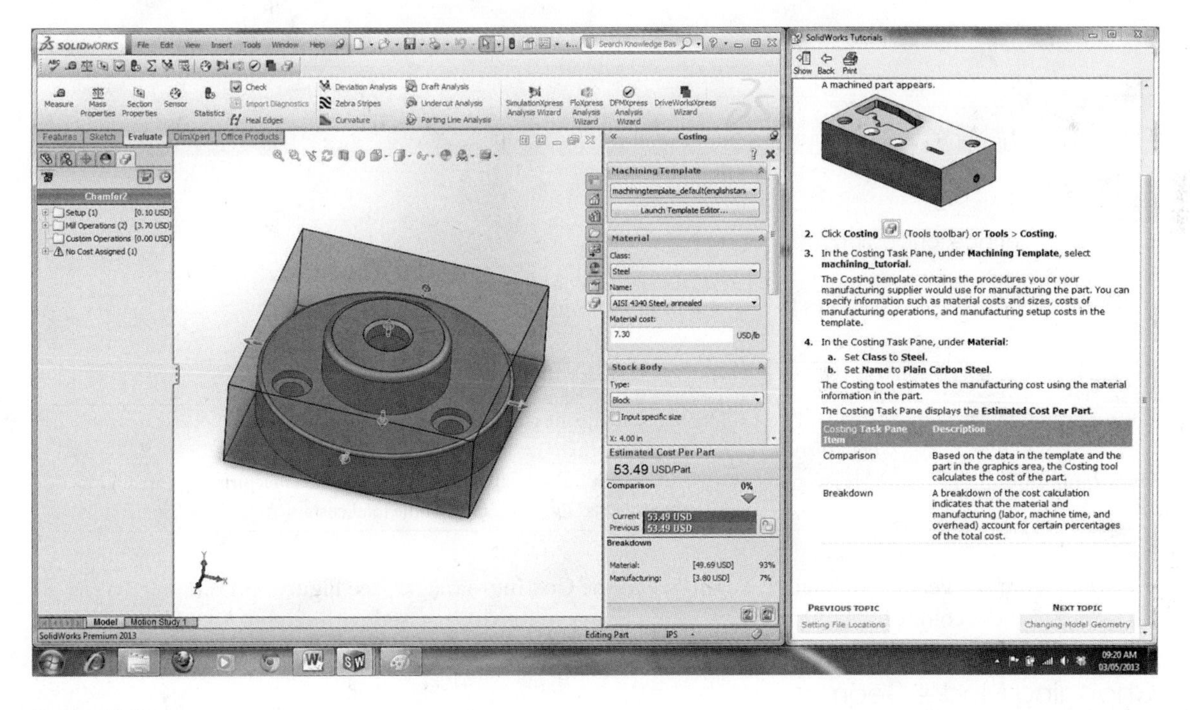

Figure 33.10

Costing Pane.

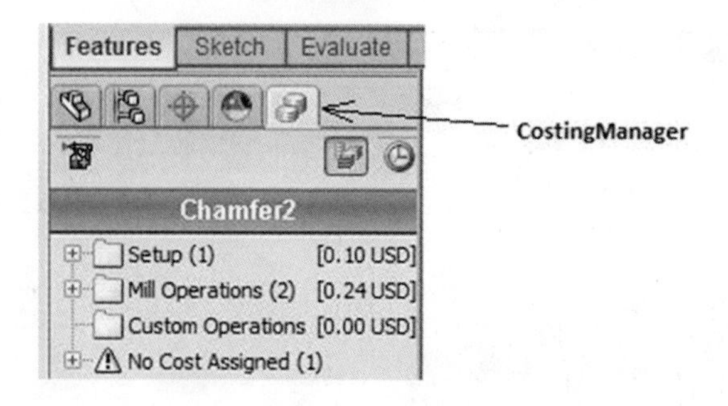

Figure 33.11

CostingManager.

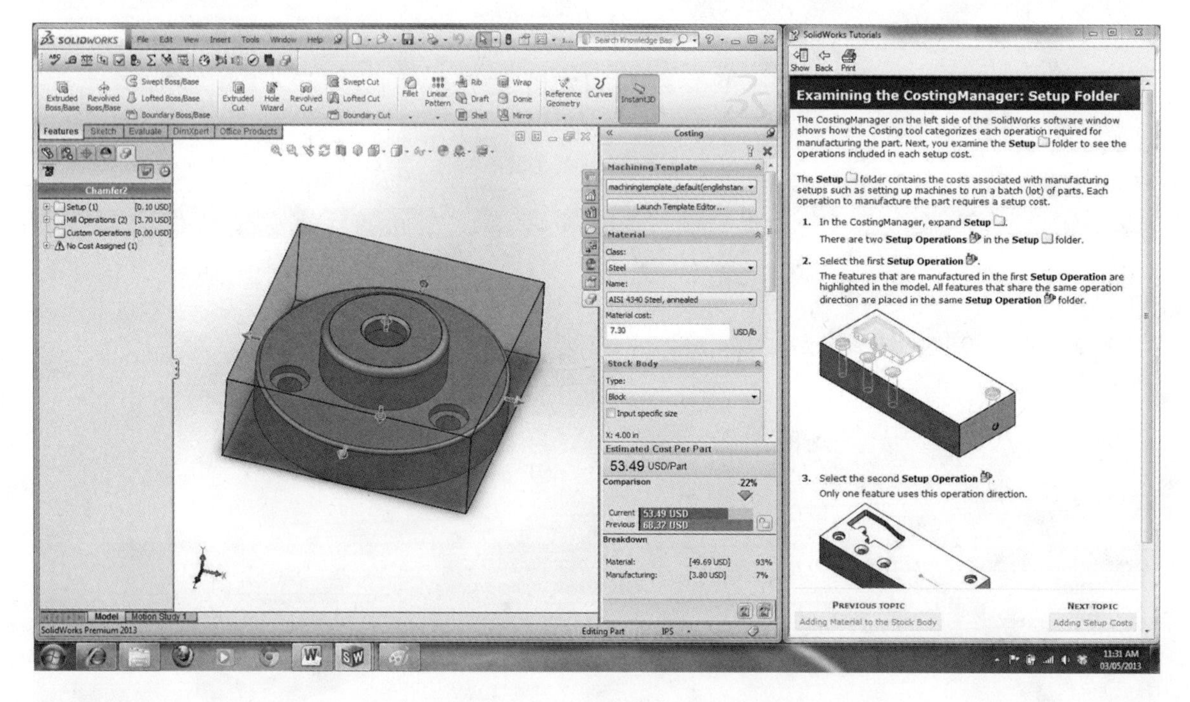

Figure 33.12

SolidWorks interface for Costing.

The *Costing Task Pane* displays the Estimated Cost Per Part (see Figure 33.10).

Costing Task Pane Item	Description
Comparison	Based on the data in the template and the part in the graphics area, the Costing tool calculates the cost of the part.
Breakdown	A breakdown of the cost calculation indicates that the material and manufacturing (labor, machine time, and overhead) account for certain percentages of the total cost.

Switch from the FeatureManager Design Tree to the CostingManager (see Figure 33.11).
Notice that the color of the model changes and is automatically encased within a box (see Figure 33.12).

Changing Model Geometry

Next, you change the model geometry to see how it affects manufacturing costs.

1. In the Costing Task Pane, click Auto Show.
2. In the graphics area, double-click the part.

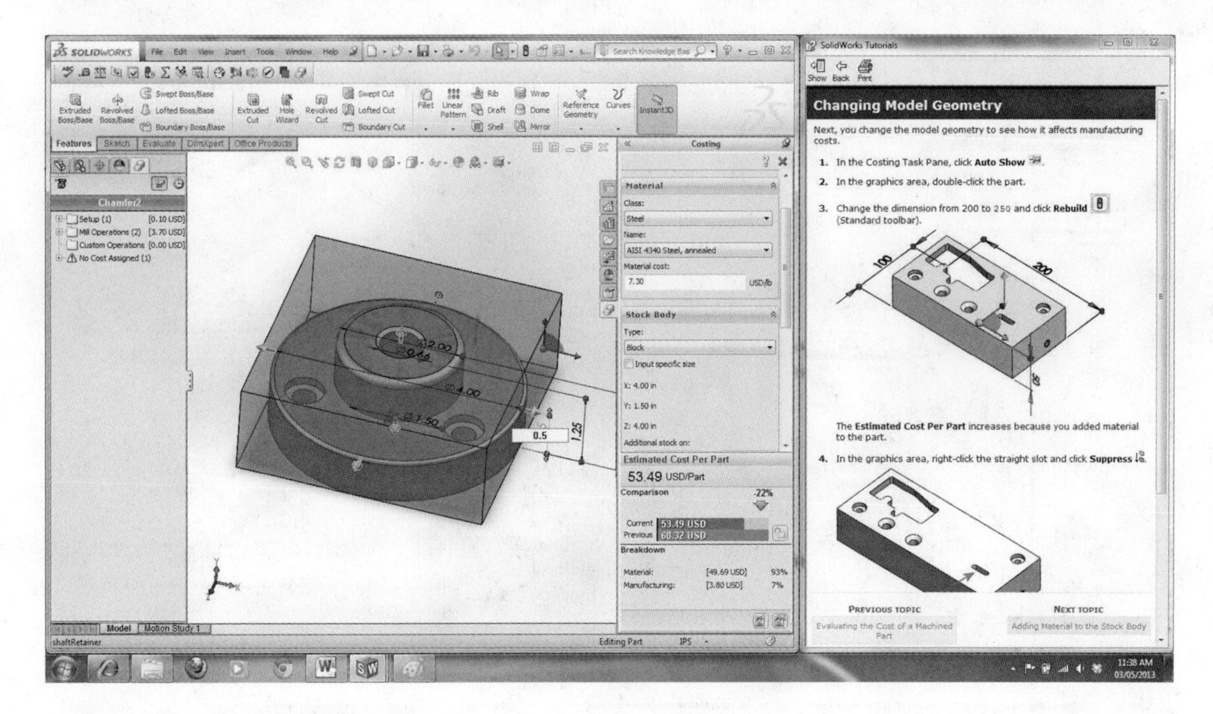

Figure 33.13

Dimension reduced from 0.75 to 0.5 resulting in reduction in cost.

3. Change the dimension from 0.75 to 0.5 and click Click to Update to Rebuild (see Figure 33.13).
   The Estimated Cost Per Part decreases because you added material to the part.
4. In the graphics area, right-click the hole and click Suppress (see Figure 33.14). (Note that you cannot suppress the slot in the CostingManager because the CostingManager shows manufacturing features, not SolidWorks features.)
5. In the Costing Task Pane, under Estimated Cost Per Part, click where it reads Click to Update.
   You manually update the cost estimate because you changed the model geometry.
   The Estimated Cost Per Part decreases because you simplified the part.

Adding Material to the Stock Body

Surfaces on the part sometimes require high tolerance or specific surface finishes. Next, you add more material on the stock body so the part can be precisely machined to meet specific tolerances and surface finishes.

1. In the Costing Task Pane, under Stock Body, for +Y, type 0.5 and press Enter (see Figure 33.15). You must type 0.5, not.5.
2. Click to Update.
3. Change +Y back to 0.
4. Clear Preview stock.
5. Click to Update.

Examining the CostingManager: Setup Folder

The CostingManager on the left side of the SolidWorks software window shows how the Costing tool categorizes each operation that is required for manufacturing the part. Next, you examine the Setup folder to see the operations that are included in each setup cost.

The Setup folder contains the costs that are associated with manufacturing setups such as setting up machines to run a batch (lot) of parts. Each operation to manufacture the part requires a setup cost.

1. In the CostingManager, expand Setup.
   There is one Setup Operation in the Setup folder.

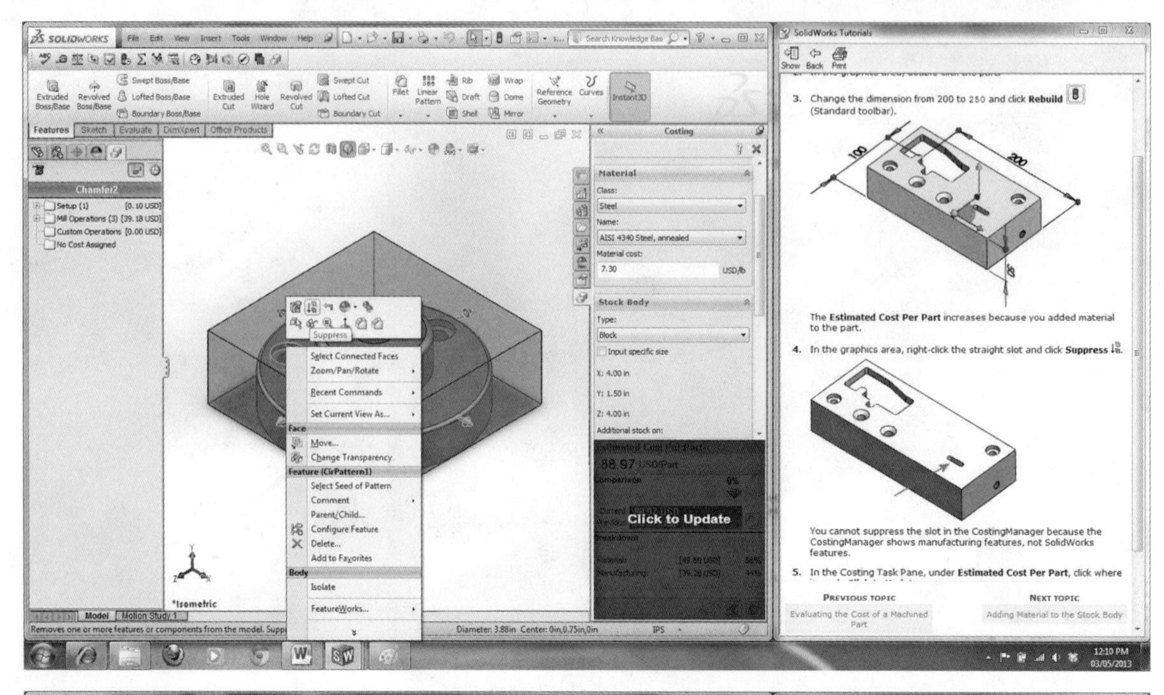

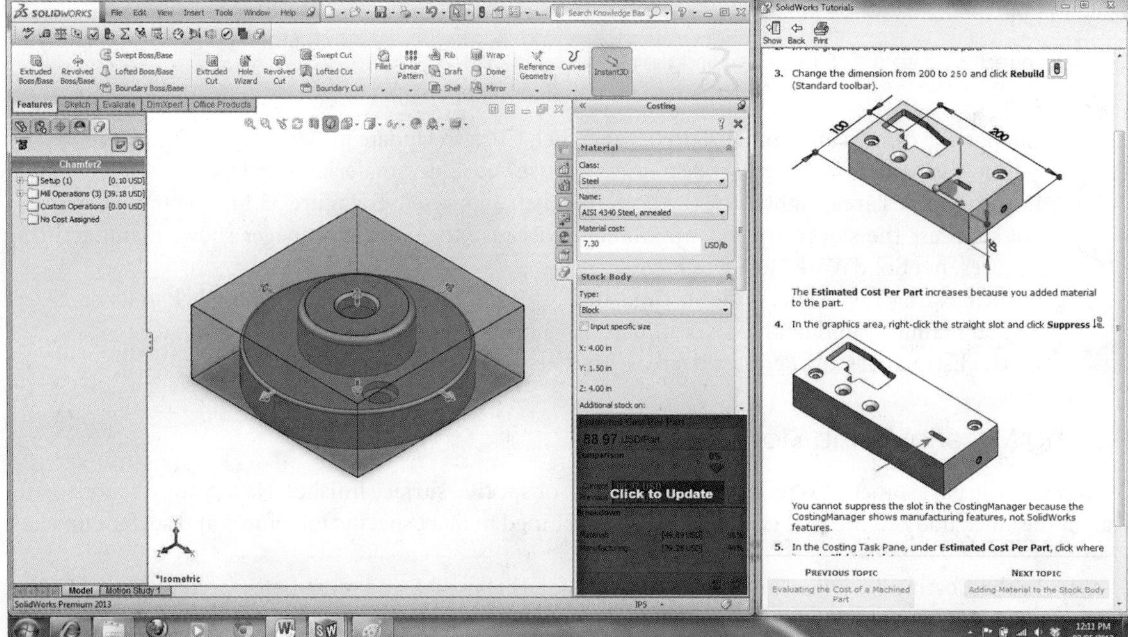

Figure 33.14

Suppression of holes resulting in reduction in cost.

Examining the CostingManager: Mill Operations

The Mill Operations folder contains information on all of the features to be milled in the part. Next, you examine the Mill Operations folder to see the operations that are included in each setup cost.

1. Expand Mill Operations.
   There are three features to be milled.
2. Expand Mill 1.
   This is a chamfer.

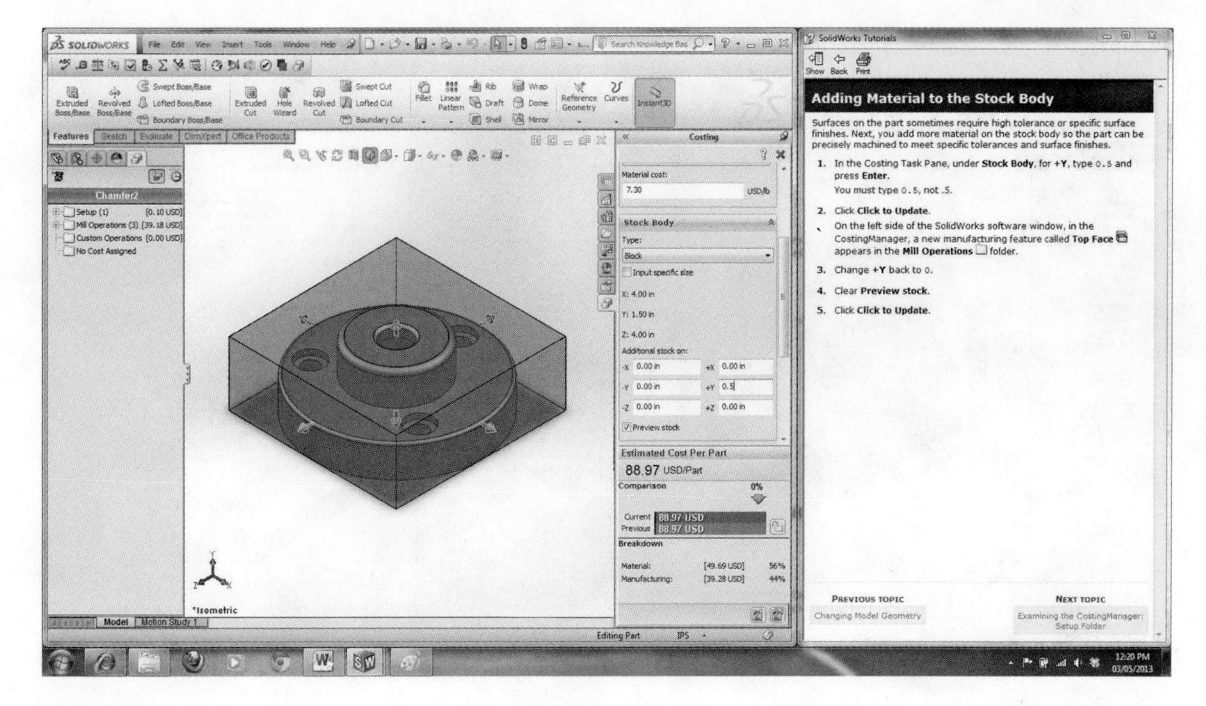

Figure 33.15

Adding material to the stock body.

Changing Material Costs

The Costing Task Pane on the right side of the SolidWorks software window displays the input values that determine the cost of manufacturing the part. Next, you change the material cost to see how it affects the manufacturing cost.

The Costing Task Pane contains a section for Material. All of the material information comes from the Costing template.

1. In the Costing Task Pane, under Material, for Material cost, type **8.0** (see Figure 33.16).
2. Press Enter.

   The text box appears in yellow to indicate that you overrode the cost from the template. Additionally, the Estimated Cost Per Part increases.

Changing the Stock Body

Stock body is the raw material from which the part is manufactured. Next, you change the stock body to see how it affects the manufacturing cost.

1. Under Stock Body, in Type, select Plate (see Figure 33.17).

   There are three types of stock bodies in machining Costing: (1) block, (2) plate, and (3) cylinder. (Note: when you select Plate, the software uses cutting operations such as waterjet and laser to create the through-cuts on the part, rather than milling and drilling. When you select the stock body, you not only control the stock but also the operations that remove material from the part. Choosing Plate costs less than choosing Block.)

   The software selects the plate thickness that is closest to the part thickness. The thickest plate available in the template is 1.0 in., but the part thickness is 1.5 in. The warning icon indicates that the plate stock is not thick enough. Because the part is much thicker than the plate thicknesses that are available in the template, you need to use block stock (see Figure 33.18).
2. Under Stock Body, in Type, select Block.

   Block generally fits all geometries. The software selects the smallest rectangular cuboid to fit the part.
3. Click Update to update.

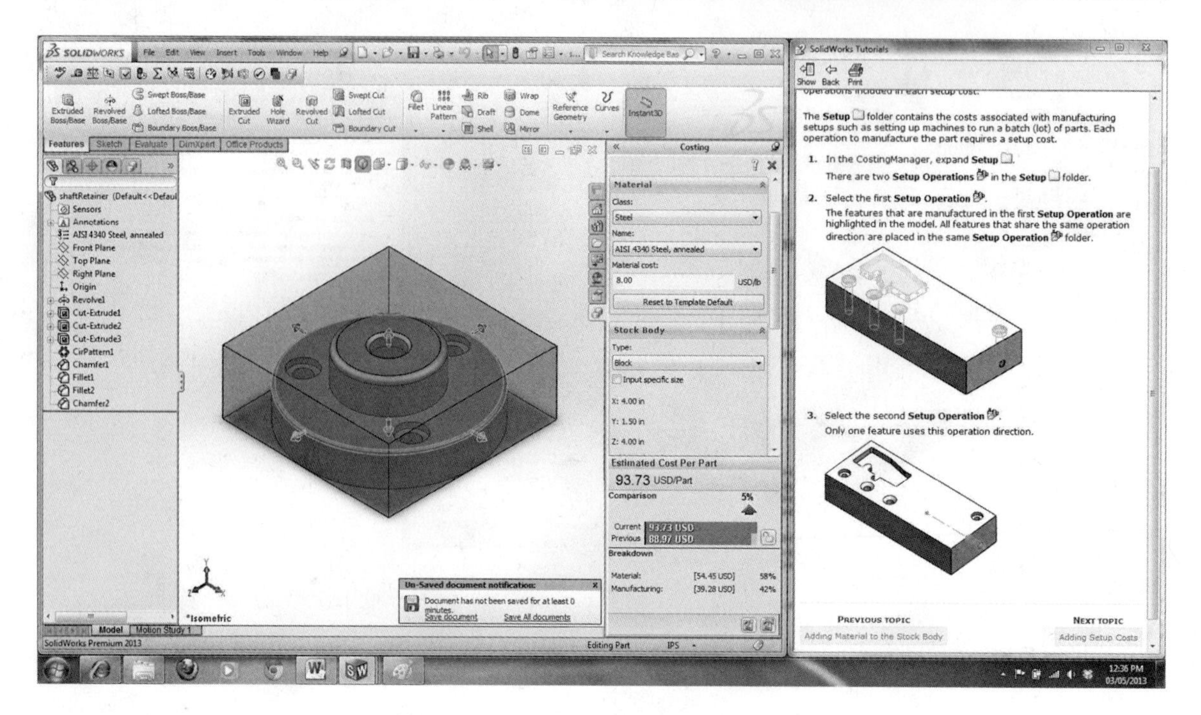

Figure 33.16

Change material cost.

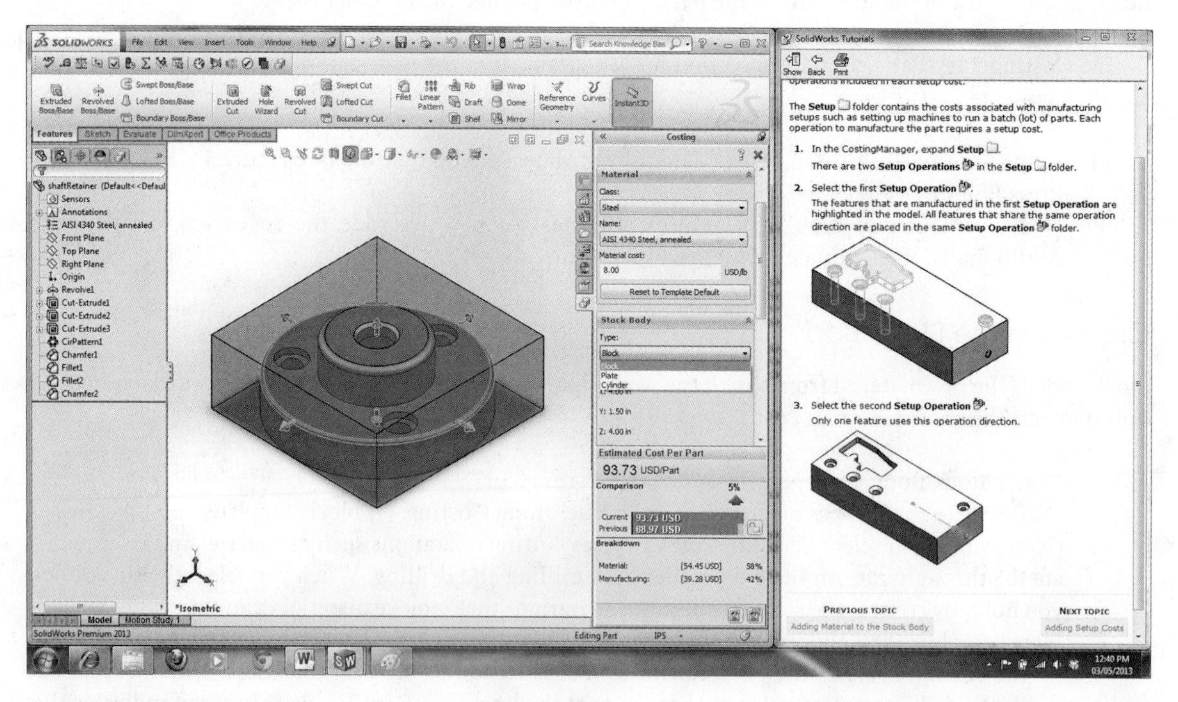

Figure 33.17

Change the stock body.

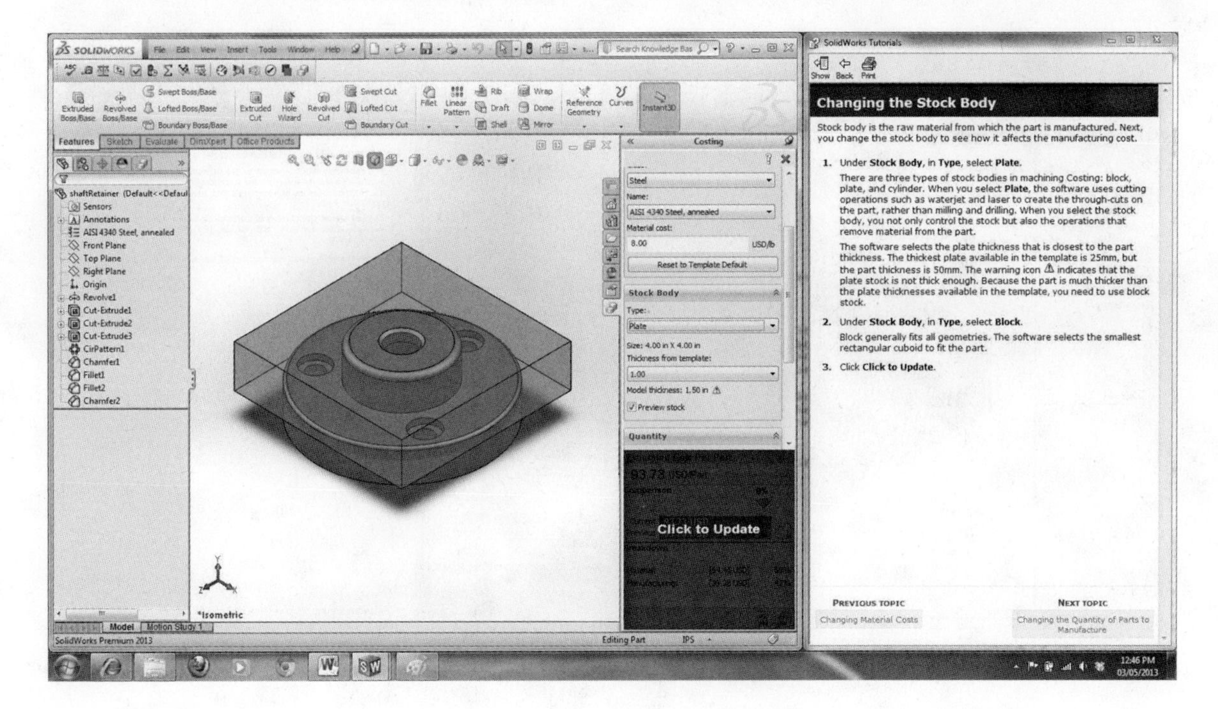

Figure 33.18

Warning icon appears.

Changing the Quantity of Parts to Manufacture

Next, you change the quantity of parts to manufacture to see how it affects the manufacturing cost.

1. Under Quantity, set Total number of parts to 1 and press Enter.
   The Lot size also changes to 1. The Estimated Cost Per Part increases because the machine setup costs are distributed over the lot. Each part carries more of the setup cost when there are fewer parts in the lot.
   The Lot size is the number of parts run in one machine setup. For example, if you run 200 parts and run 100 parts per lot, you need two machine setups (see Figure 33.19).
2. Under Quantity, set Total number of parts to 100 and press Enter.

Including a Discount

Next, you add a discount to the part to see how it affects the manufacturing cost.

1. Under Markup/Discount, select Markup/Discount.
   Under Markup/Discount, you can apply markups or discounts to the part. If you want to add a profit margin, type a positive value. If you get a discount from a vendor, type a negative value.
   Type: **10** for % of Total Cost and press Enter (see Figure 33.20).
   The Estimated Cost Per Part decreases by the **10%** discount.

Applying Custom Operations

Next, you apply a custom operation to the part. Because the part is made out of stainless steel and will not rust, you do not need to apply paint to the part as a custom operation. But, we will merely show the steps that are required in case you need to apply paint.

1. In the CostingManager, click Add Custom Operation (see Figure 33.21).
   Painting is already listed as a custom operation because it is in the machining template. The PropertyManager displays all of the cost information that is associated with the painting operation.

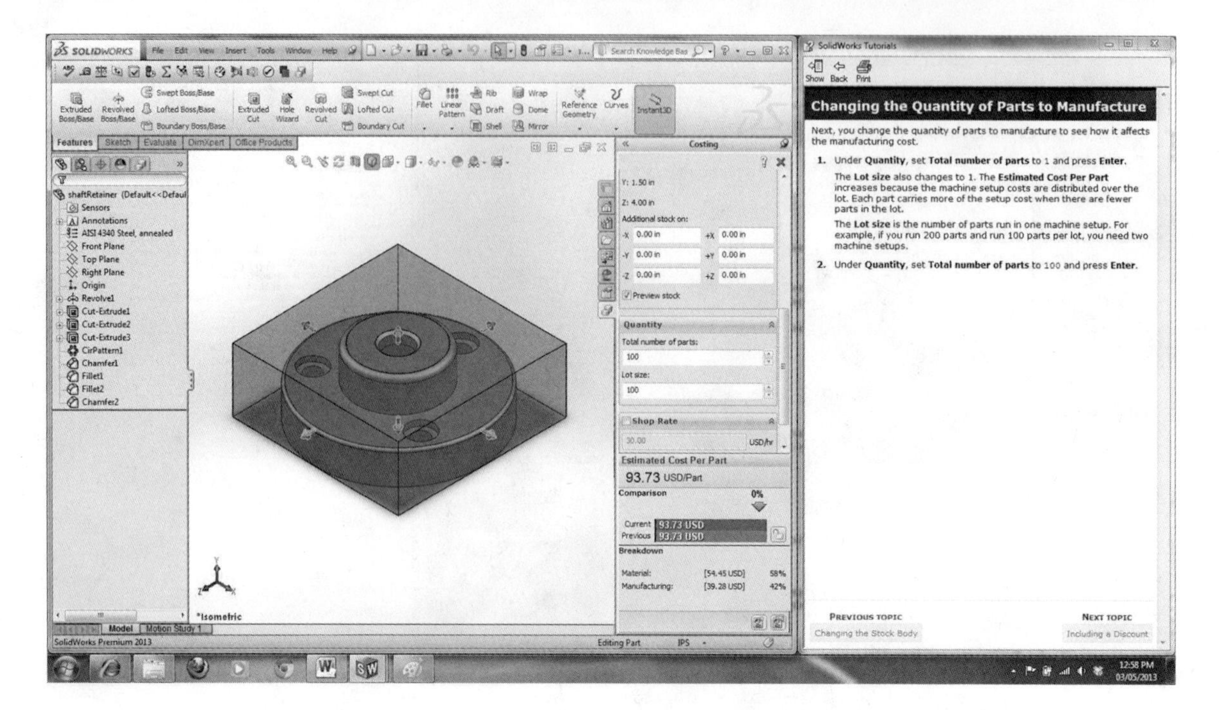

Figure 33.19

Quantity of part to manufacture.

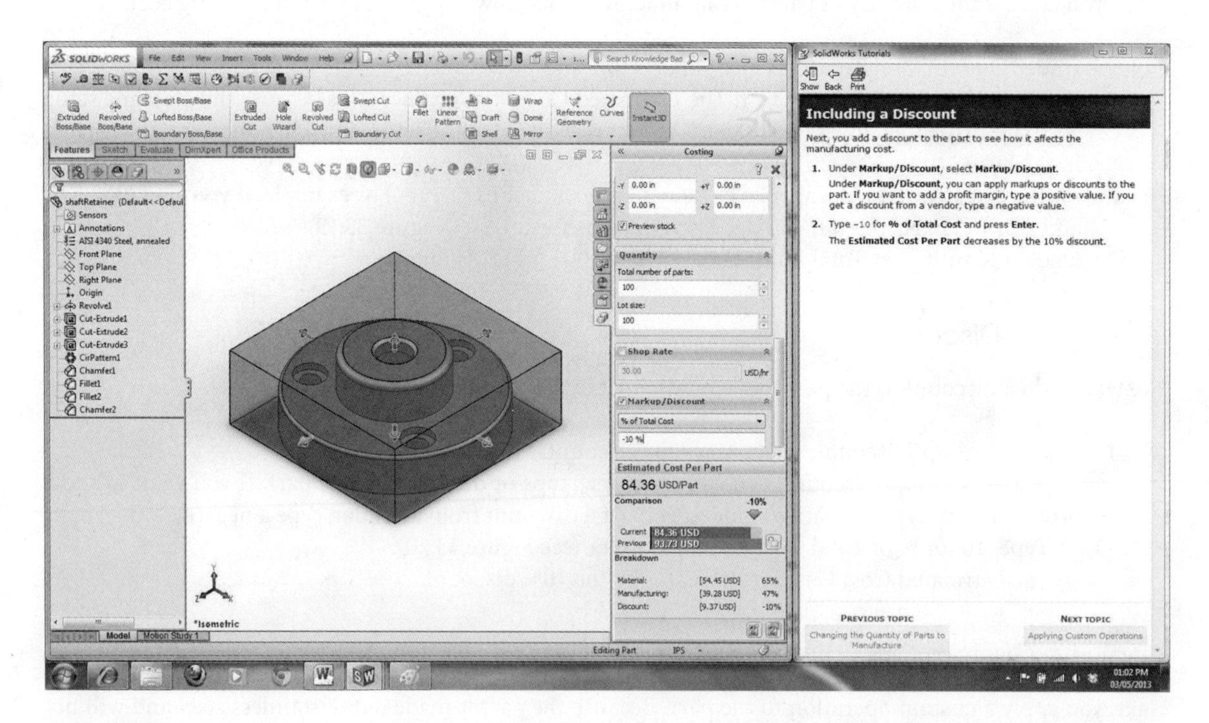

Figure 33.20

Include discount.

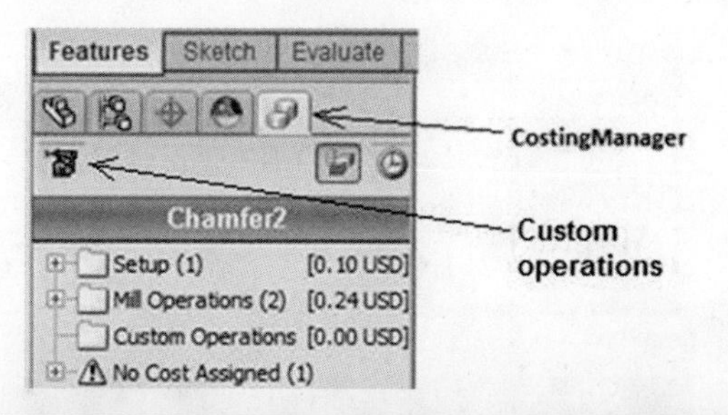

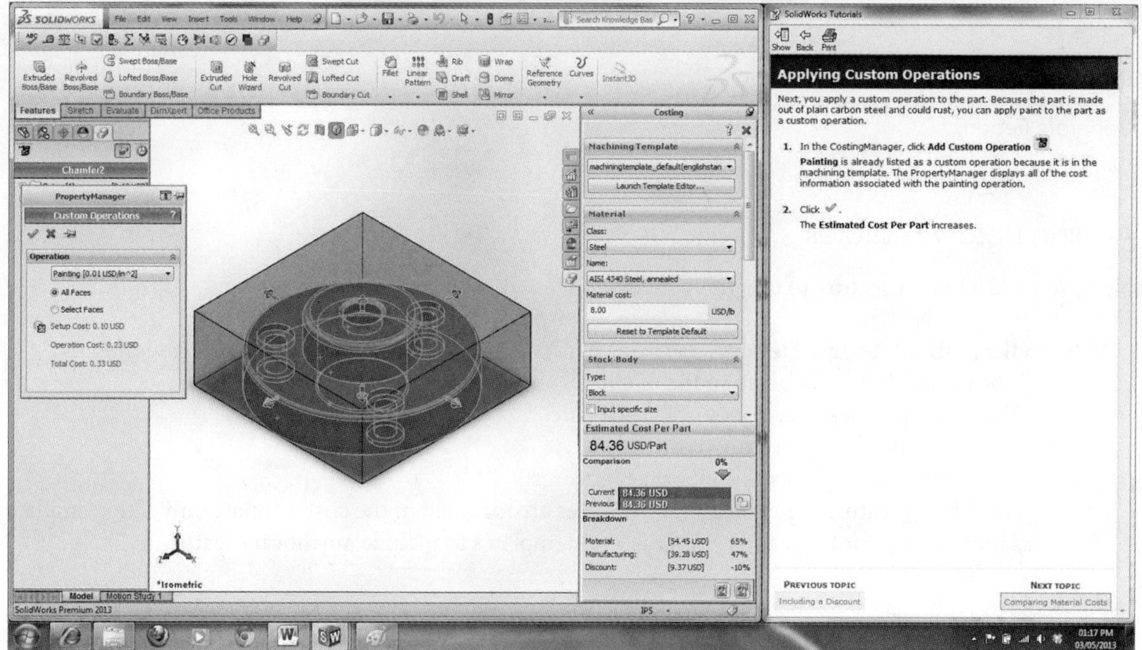

Figure 33.21

Add Custom Operation.

2. Click OK.
   The Estimated Cost Per Part increases.

Note that because American National Standards Institute (ANSI) is stainless steel, painting is not required to prevent rust, and you can remove the painting operation.

Comparing Material Costs

Next, you set a baseline price to compare alternative materials for the part.

1. In the Costing Task Pane, under Estimated Cost Per Part, click Set Baseline.
2. Under Material, in Name, select Plain Carbon Steel.
   The Estimated Cost Per Part decreases.
3. Next, you apply a custom operation to the part because the part is made out of Plain Carbon steel and will rust.
4. In the CostingManager, expand Custom Operations.
5. Right-click Painting <1> - Face and click Remove Custom Operation.
   The Estimated Cost Per Part decreases, but the stainless steel part is still significantly more expensive than the plain carbon steel part.
6. In the Costing Task Pane, under Estimated Cost Per Part, click Set Baseline to remove the baseline price.

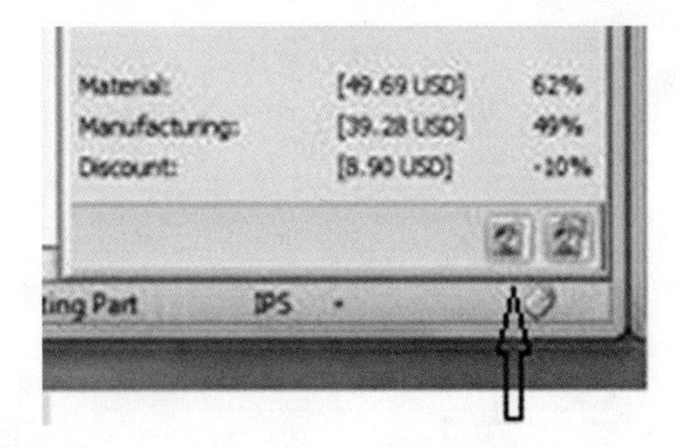

Figure 33.22

Generate Report.

Adding Library Features

Next, you add a library feature to the part to see how it affects the cost; otherwise, skip.

1. In the FeatureManager Design Tree, right-click any library feature that was suppressed.
   The library feature appears in the part.
2. In the Costing Task Pane, click Click to Update.
   The Estimated Cost Per Part increases.
3. In the CostingManager, expand Library Features.
   The library feature appears. Library features are included in the cost estimate only if the template has information for the feature. You can edit templates to include any library feature.

Creating a Report

Next, you create a report of the Costing results.

1. At the bottom of the Costing Task Pane, click Generate Report (see Figure 33.22).
   The report opens in Microsoft Word.
2. Close the report.

Summary

We have completed the discussion on costing, which is a new feature in SolidWorks. In this chapter, we have learned how to specify information such as material costs and sizes in the template, the costs of manufacturing operations in the template, and manufacturing setup costs in the template.

Also, we examined the impact of the following on cost: (a) changing model geometry, (b) adding material to the stock body, (c) changing material costs, (d) changing the stock body, (e) changing the quantity of parts to manufacture, (f) including a discount, and (g) applying custom operations.

Not only that, we also covered the following: (a) examining the CostingManager: Setup Folder, (b) examining the CostingManager: Mill Operations, and (c) comparing Material Costs.

Exercises

1. Create the part in Figure P33.1 and evaluate the cost of the machined part.

   Part name: V-block

   Material: A-steel

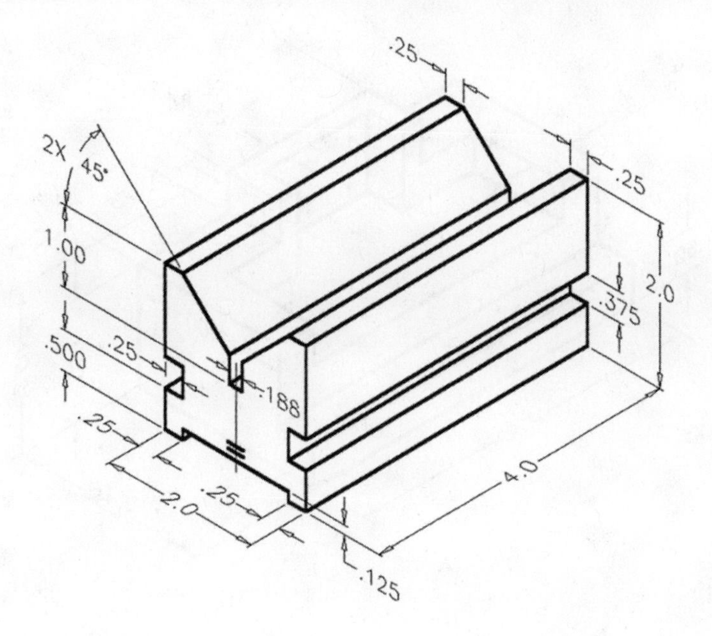

Figure P33.1

V-block.

2. Create the part in Figure P33.2 and evaluate the cost of the machined part.

   Part name: Angle Bracket

   Material: Mild steel

   Specific instruction: Center $2X\ \phi$ 15 holes and provide location dimensions.

3. Create the part in Figure P33.3 and evaluate the cost of the machined part.

   Part name: Support

   Material: Mild steel

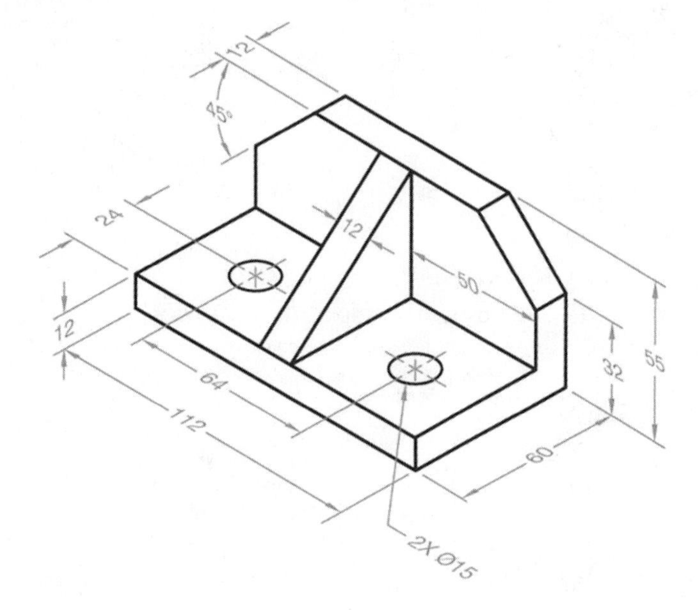

Figure P33.2

Angle bracket.

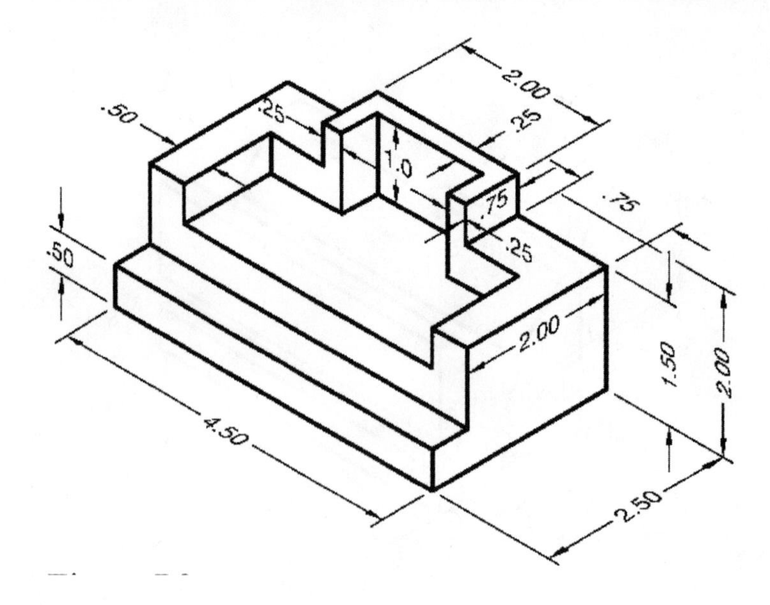

Figure P33.3

Support.

Bibliography

Germani, M., Cicconi, P., Mandolini, M., 2011. Manufacturing cost estimation during early phases of machine design, *Proceedings of the 18th International Conference on Engineering Design (ICED 11)*, Impacting Society through Engineering Design, Vol. 5: Design for X/Design to X, Lyngby/Copenhagen, Denmark, August 15–18, 2011, pp. 198–209.

http://www.solidworks.com/sw/products/3d-cad/manufacturing-cost-estimation.htm.

http://www10.mcadcafe.com/blogs/jeffrowe/2013/02/07/cost-management-moves-up-front-in-design-process-with-solidworks-2013/.

34

Finite Element Analysis Using SolidWorks

Objectives:

When you complete this chapter, you will have

- Understood what COSMOS is used for
- Understood the historical trend leading to SolidWorks Simulation
- Used the SolidWorks Simulation interface
- Understood the fundamental steps that are involved in finite element analysis
- Understood static analysis with solid elements
- Understood the effect of mesh variations on analysis outputs
- Used SolidWorks Simulation to solve stress analysis problems

Introduction to COSMOS/SolidWorks Simulation

What Is SolidWorks Simulation?

The Structural Research and Analysis Corporation (SRAC) developed an engineering analysis software product called COSMOS, based on finite element analysis (FEA). The SRAC was established in 1982, and it made significant contributions toward FEA for engineering analysis.

In 1995, SRAC partnered with the SolidWorks Corporation and developed COSMOS Works, which became the top-selling analysis solution. The commercial success of COSMOS Works integrated with SolidWorks CAD software resulted in Dassault Systemes, the parent of SolidWorks Corporation, acquiring SRAC in 2001. In 2003, SRAC operations merged with SolidWorks Corporation. In the 2009 revision, COSMOS Works was renamed SolidWorks Simulation. This historical perspective is important. SolidWorks Simulation is fully integrated into the SolidWorks Simulation CAD software. SolidWorks Simulation can be used to create and edit model geometry. SolidWorks Simulation is solid-driven, parametric, and feature-driven and runs on Windows.

Table 34.1 SolidWorks Family

Functionality	Until 2008	From 2009	Scope
	COSMOS Works[a]	SolidWorks Simulation[b]	Static, frequency, buckling, fatigue, drop test analysis, linear dynamic, nonlinear thermal analysis: temperature, temperature gradient, heat flow
Stress Analysis: FEA-based	COSMOS FloWorks	SolidWorks Flow Simulation	Fluid flow, heat transfer, forces
Motion Analysis	COSMOS Motion	SolidWorks Motion	Kinematic modeling/analysis of mechanisms
Animation	COSMOS Animation	SolidWorks Animation	Animation of modeled systems

[a] Designer and Professional versions.
[b] SolidWorks SimulationXpress is an introductory version of SolidWorks Simulation.

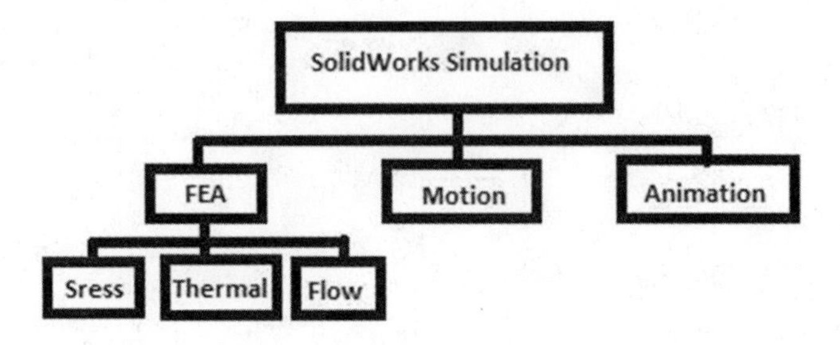

Figure 34.1

SolidWorks Simulation.

There are a number of well-known commercially available FEA packages:

Software	Owner
ANSYS	ANSYS, Inc.
ABACUS	
SolidWorks Simulation/COSMOS Works	Dassault Systemes
I-DEAS	UGS
Pro/MECHANICA	PTC

The functionalities, historical trends, and scope of the SolidWorks family of products are summarized in Table 34.1. A conceptual model of SolidWorks Simulation is shown in Figure 34.1.

Product Development Cycle

In the industry, the product development cycle (PDC) has the following steps:

1. Build model in a CAD system (Inventor, CATIA, Pro/E, SolidWorks, etc.).
2. Prototype the design.
3. Test the prototype in the field.
4. Evaluate the results of the field tests.
5. Modify the design based on the field test results.

The process continues until a satisfactory solution is reached.
FEA could be used to replace field tests in the PDC.
Advantages of Analysis

- Reduced cost by simulation instead of tests
- Reduced development time
- Improved products

What Is Finite Element Analysis?

Finite element analysis is a numerical method, usually modeled on a computer, that analyzes the stresses in a part. The results would otherwise be difficult to obtain. It can be used to predict the failure of a part or structure, due to unknown stresses, by showing problem areas and allowing designers to see all of the theoretical internal stresses. This method of product design and testing is far cheaper than the manufacturing costs of building and testing each sample. For fracture analysis, FEA calculates the stress intensity factors.

FEA, however, has many applications such as for fluid flow and heat transfer. While this range is growing, one thing will remain the same: the theory of how the method works.

FEA is used in new product design and existing product refinement. A company is able to verify that a proposed design will perform to the client's specifications prior to manufacturing or construction. It is used to ensure that a modified product or structure will meet its new specifications. In the case of structural failure, FEA may be used to help determine the design modifications that are necessary to overcome the problem.

There are generally two types of analysis that are used in industry: (1) two-dimensional (2D) modeling and (2) three-dimensional (3D) modeling. While 2D modeling is simple and only requires a relatively normal computer, it tends to yield less accurate results. 3D modeling, however, produces more accurate results while sacrificing the ability to run on all but the fastest computers. For each of these modeling schemes, the programmer can insert numerous algorithms (functions) to make the system behave linearly or nonlinearly. Linear systems are far less complex and generally do not take into account plastic deformation. Nonlinear systems do account for plastic deformation, and many are also capable of testing a material all the way to fracture.

The stiffness of a member can be used to provide a simplified overview of the mathematical basis of FEA. We begin by considering a simple member of original length L subject to an external axial deformation, ΔL. Force and deformation are related by

$$\Delta L = \frac{F.L}{A.E}$$

where E is the modulus of elasticity, F is the force, L is the length of the member, and the cross-sectional area is A.

The strain, which is the change in length divided by the original length, is defined as

$$\varepsilon = \frac{\Delta L}{L}$$

From the classical stress–strain relation, the stress can be determined as

$$\sigma = E\varepsilon$$

In other words, FEA starts from a simple mathematical description of the deformation in a part due to some loading, and then progresses to determine the strain, and finally finds the stress in the part. However, it should be noted that this stress formulation is only valid within the elastic region where stress is proportional to strain.

How Does FEA Work?

In FEA, a part is divided into a number of simple elements:

- Rod
- Beam
- Plate/shell/composite
- Shear panel
- Solid
- Spring
- Mass
- Rigid element
- Viscous damping element

The most commonly used elements are solid, shell, and beam.

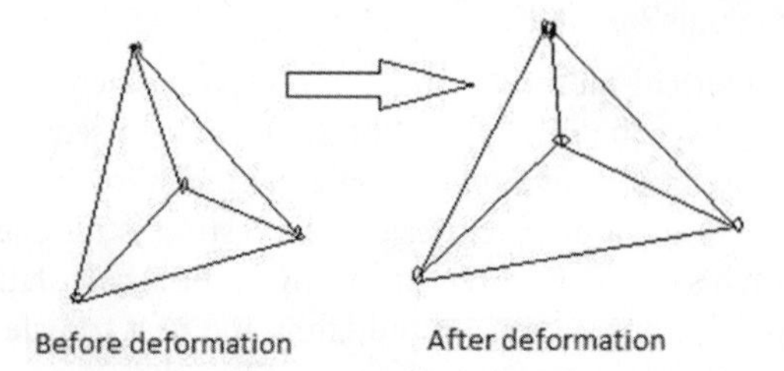

Figure 34.2

First-order tetrahedral element before and after deformation.

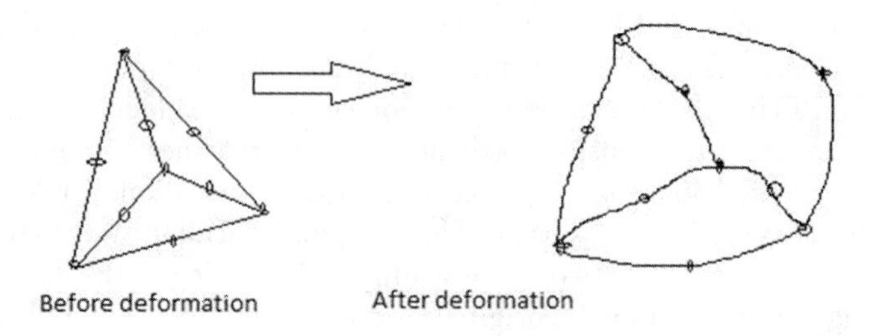

Figure 34.3

Second-order tetrahedral element before and after deformation.

Solid Elements

The majority of parts analyzed with FEA utilize 3D models, based on solid geometry, to define the boundaries of the part or assembly. The solid element is either a first-order tetrahedron (see Figure 34.2), which has 4 flat faces and 4 vertices, or a second-order tetrahedron (see Figure 34.3), which has 4 flat faces and 10 nodes that are the 4 vertices and the midpoints of its edges. Solid elements have three degrees of freedom per node consisting of three deformations.

Shell Elements

Thin-walled parts, such as sheet metal parts, are analyzed using shell elements (see Figure 34.4). Thin-walled parts are commonly found in tanks, beverage containers, plastic parts, thin-walled pressure vessels, etc. Shell elements have six degrees of freedom per node consisting of three deformations and three rotations.

Beam Elements

Beam or truss elements are commonly used in structural members since the number of nodes and elements is greatly reduced (see Figure 34.5). A beam element should be used when the length/height ratio (l/h) is greater than or equal to 20:1.

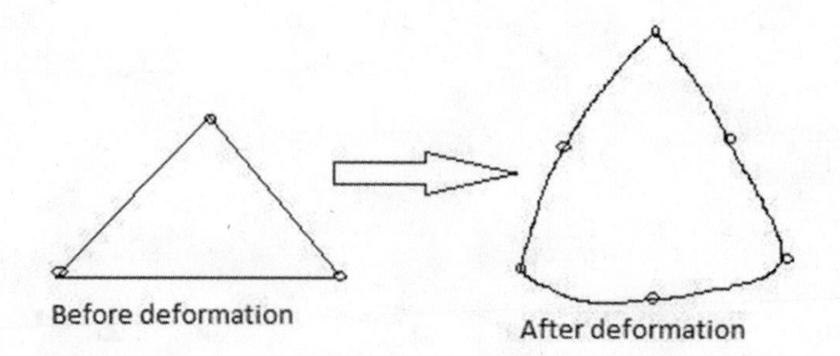

Figure 34.4

Shell element before and after deformation.

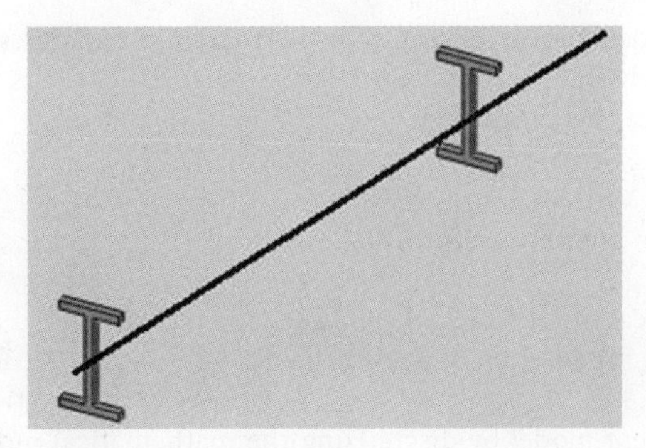

Figure 34.5

Beam element.

FEA uses a complex system of points or nodes to make a grid that is called a mesh. This mesh is programmed to contain the material and structural properties that define how the structure will react under loading. Nodes are assigned at a density throughout the material depending on the anticipated stress levels of a particular area. Regions that will experience a high degree of stress usually have a higher node density than those that experience little or no stress. Points of interest include a fracture point of a previously tested material, fillets, corners, complex detail, and high-stress areas. The mesh acts like a spider's web; from each node, there extends a mesh element to the adjacent nodes. This web of vectors carries the material properties of the object.

A wide range of objective functions (variables within the system) are available for minimization or maximization:

- Mass, volume, temperature
- Strain energy, stress, strain
- Force, displacement, velocity, acceleration

Different loading conditions may be applied to a system:

- Point, pressure, thermal, gravity, and centrifugal static loads
- Thermal loads derived from heat transfer analysis
- Enforced displacements
- Heat flux and convection
- Point, pressure, and gravity dynamic loads

Many FEA programs can use multiple materials within the structure. The structure would be classified as follows:

- Isotropic, identical throughout
- Orthotropic, identical at 90°
- General anisotropic, different throughout

Types of Engineering Analysis

Structural analysis uses linear and nonlinear models. Linear models use simple parameters and assume that the material is not plastically deformed. Nonlinear models stress the material past its elastic capabilities. The stress in the material varies with the amount of deformation.

Vibration analysis is used to test a material against random vibrations, shock, and impact. Each of these may act on the natural vibration frequency of the material, which, in turn, may cause resonance and subsequent failure.

Fatigue analysis helps designers to predict the life of a material or structure by showing the effects of cyclic loading on a specimen. Such analysis can show the areas where crack propagation is most likely to occur. Failure due to fatigue may also show the damage tolerance of the material.

Heat transfer analysis models the conductivity or thermal fluid dynamics of the material or structure. The heat transfer may be steady state or transient. Steady-state transfer refers to constant thermoproperties in the material and yields linear heat diffusion.

Principles of FEA

The methodology for FEA can be summarized as follows:

- Build the mathematical model using the CAD geometry (simplified if required), material properties, loads, restraints, types of analysis, etc. The different types of loads and connectors are shown in Tables 34.2 and 34.3.
- Build the finite element model by discretizing the mathematical model into solid elements, shell elements, beam elements, etc.
- Solve the finite element model. (Use the solver that is provided in SolidWorks Simulation.)
- Analyze the results.

Build the Mathematical Model

The starting point for FEA using SolidWorks Simulation is the availability of a CAD model. If the part is complex, the model may need to be simplified (for example, by removing fillets). Material properties are then assigned. The type of analysis is specified. Define the restraints and loads. This completes the mathematical model, as illustrated in Figure 34.6.

Table 34.2 Types of Loads

Structural Loads	Thermal Loads
Remote loads	Convection
Bearing loads	Radiation
Centrifugal loads	Conduction
Force	Temperature
Gravity	Heat flux
Pressure	Heat power
Shrink fit	

Table 34.3 Types of Connectors

Rigid connectors
Spring connectors
Pin connectors
Elastic support connectors

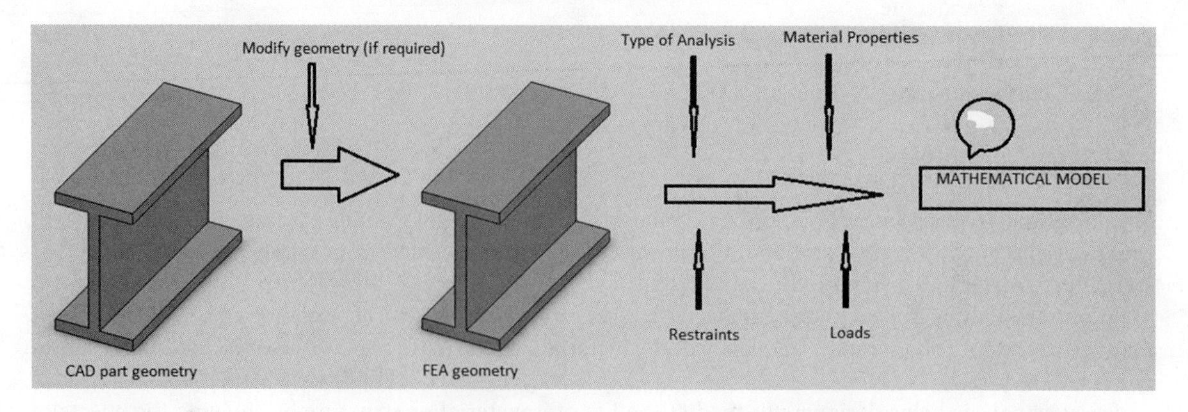

Figure 34.6

Mathematical model for FEA.

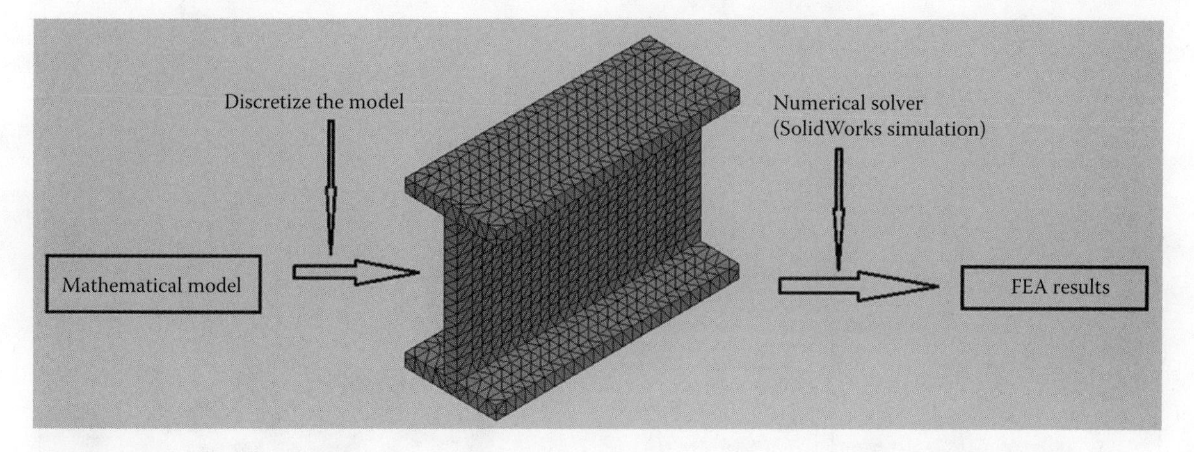

Figure 34.7

FEA model.

Build the Finite Element Model

Discretize the mathematical model using any of the following: (a) solid elements, (b) shell elements, or (c) beam elements (see Figure 34.7). This is also known as meshing the model. The geometry, loads, and restraints are all discretized and applied to the nodes of the elements. The elements are appropriately renumbered.

Solve the Finite Element Model

The finite element model is solved using a solver that is provided in SolidWorks Simulation (see Figure 34.7). Solving FEA problems can take seconds for a simple model or hours for a complex model. Even for small problems, the number of nodes can run into thousands. Coarse elements yield inferior solutions compared to fine elements. The costs, in terms of computation time, are inversely proportional to the quality of the solution.

Analyze the Results

It is not enough to simply accept any results from FEA. The results have to be analyzed to ensure that they are correctly interpreted. There are a number of sources of errors that users of FEA software should understand. Modifying the geometry of a part can be a major source of error if the modification is made without thinking how the solution would be affected. For example, not all fillets should be removed from a part. Some fillets are necessary to reduce corner stresses. Discretizing the model is another area where errors could arise. The mesh size has a significant impact on the quality of the solution as discussed in subsection "Solve the finite element model."

SolidWorks Simulation Add-Ins

SolidWorks Simulation is an add-in that must be enabled:

1. Open SolidWorks.
2. Open a model file.
3. Click Add-Ins (see Figure 34.8).
4. Check SolidWorks Simulation (see Figure 34.9).
5. Click OK. (The Simulation tool is added.)

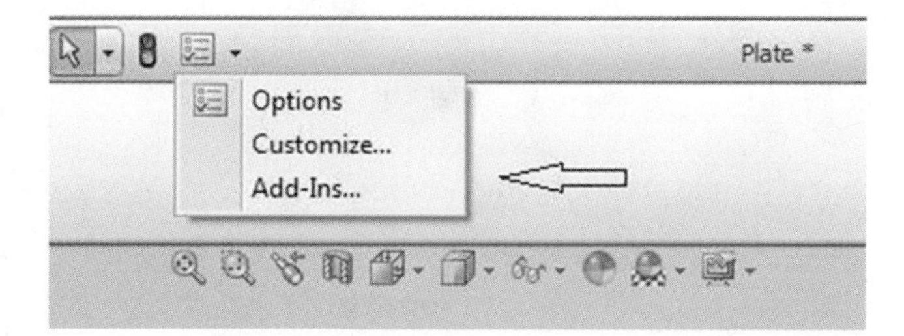

Figure 34.8

Add-Ins option.

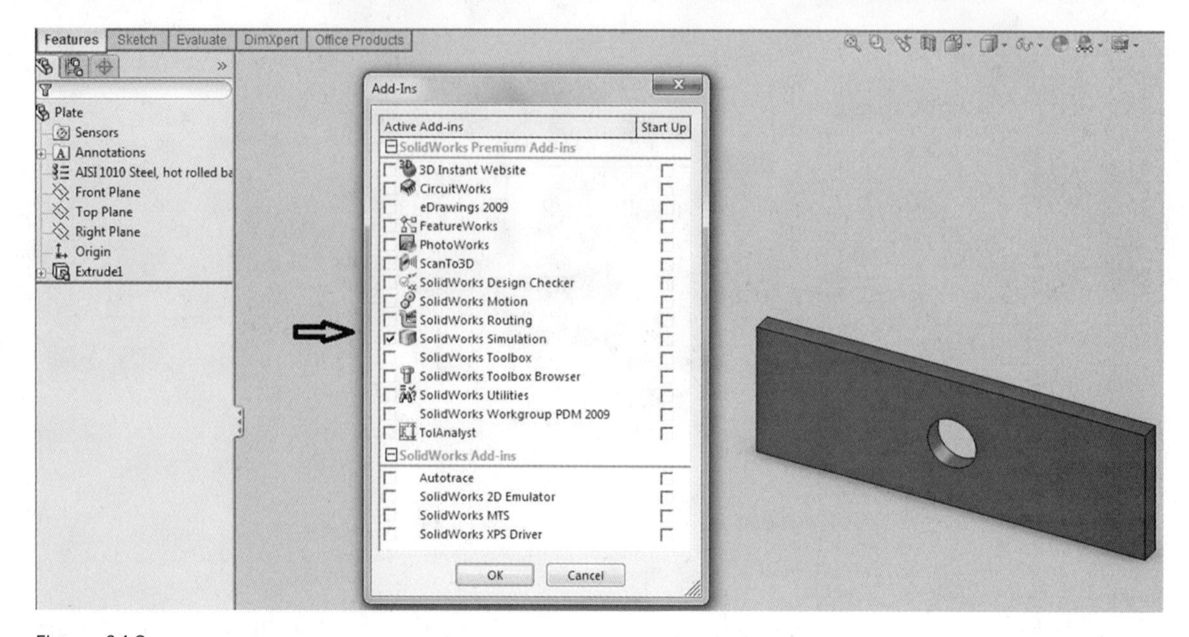

Figure 34.9

SolidWorks add-ins.

SolidWorks Simulation CommandManager

The SolidWorks Simulation CommandManager has a number of advisors. These are found in the menus for Study, Fixtures, External Loads, Connections, Run, and Results, as shown in Figure 34.10. A simulation advisor is a set of tools that guides you through the analysis process. The simulation advisor works with the SolidWorks Simulation interface by starting the appropriate PropertyManager and linking to online help topics for additional information.

Study Advisor

Click Study (from Simulation CommandManager) to access Study Advisor (see Figure 34.11). The simulation advisor tab appears in the task pane. It recommends the study types and outputs to expect. Study Advisor helps you define sensors and creates studies automatically.

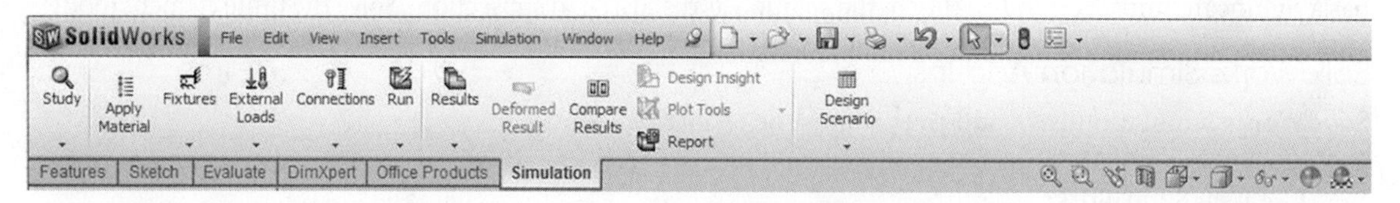

Figure 34.10

SolidWorks Simulation CommandManager.

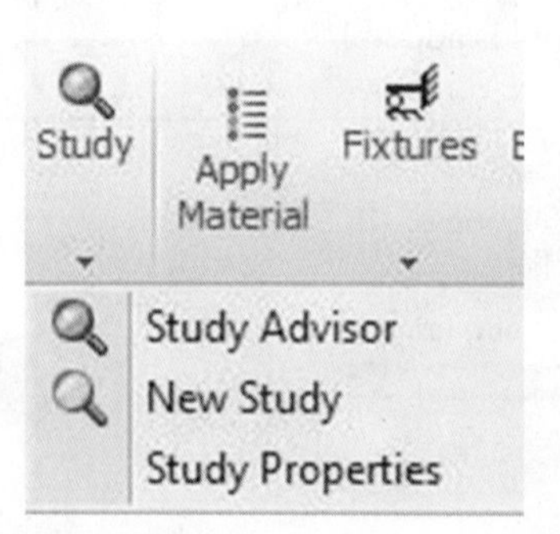

Figure 34.11

Study Advisor.

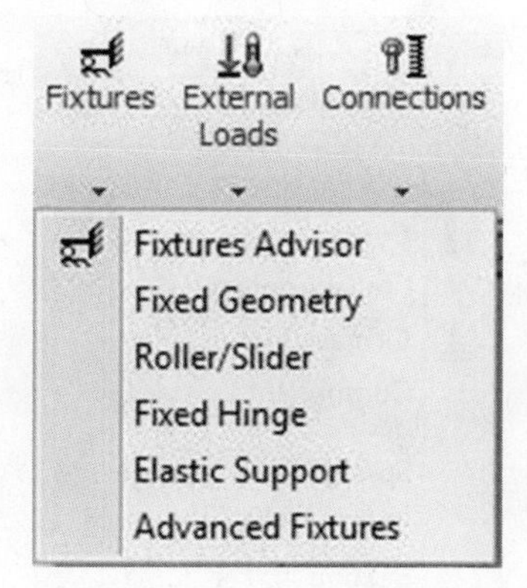

Figure 34.12

Fixtures Advisor.

Fixtures Advisor

Click Fixtures (from Simulation CommandManager) to access Fixtures Advisor (see Figure 34.12). Fixtures Advisor defines internal interactions between bodies in the model. Fixtures are restraints that are applied to the model.

External Loads Advisor

Click External Loads (from Simulation CommandManager) to access External Loads Advisor (see Figure 34.13). External Loads Advisor defines the external interactions between the model and the environment. There are several types of external loads: (a) force/torque, (b) pressure, (c) gravity, (d) centrifugal force, (e) bearing load, (f) remote load/mass, (g) distributed load, (h) temperature, (i) flow effects, (j) thermal effects, etc.

Connections Advisor

Click Connections (from Simulation CommandManager) to access Connections Advisor (see Figure 34.14). Connections Advisor suggests the techniques for connecting components within an assembly model.

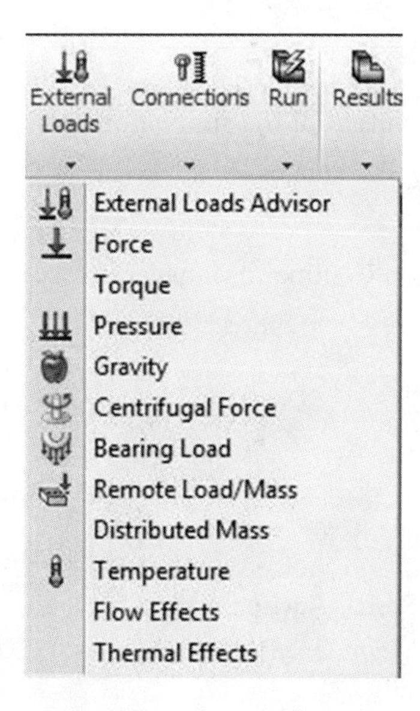

Figure 34.13

External Loads Advisor.

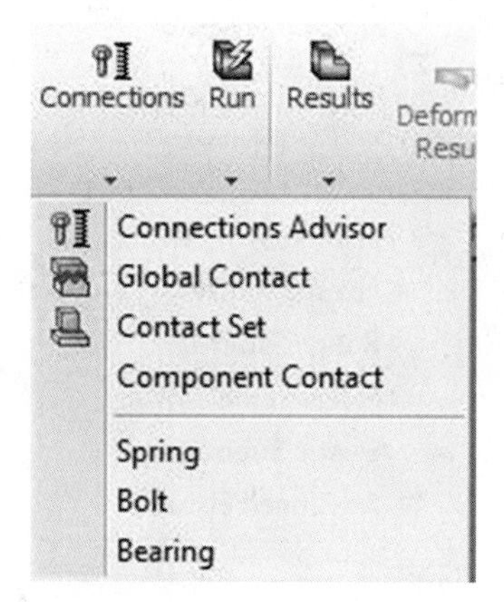

Figure 34.14

Connections Advisor.

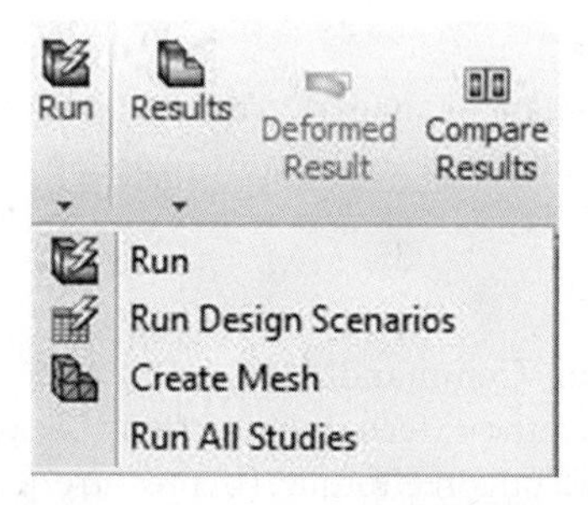

Figure 34.15

Run Advisor.

Run Advisor

Click Run (from Simulation CommandManager) to access Run Advisor (see Figure 34.15). Run Advisor solves the simulation problem.

Results Advisor

Click Results (from Simulation CommandManager) to access Results Advisor (see Figure 34.16). It provides tips for interpreting and viewing the output of the simulation. Also, it helps to determine if frequency or buckling might be areas of concern.

Design Scenario

Click Design Scenario (from Simulation CommandManager) to access Design Scenario (see Figure 34.17).

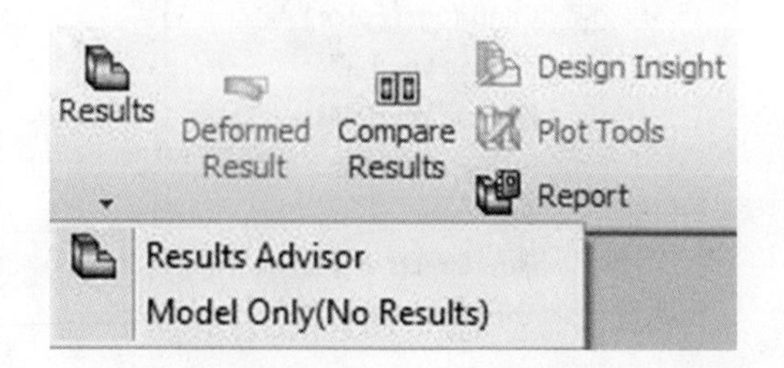

Figure 34.16

Results Advisor.

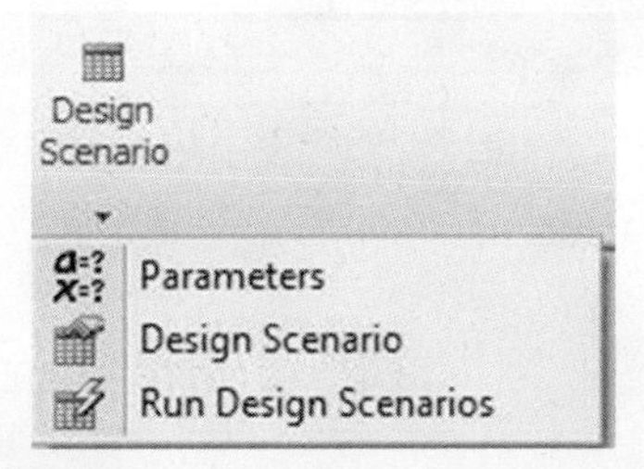

Figure 34.17

Design scenario.

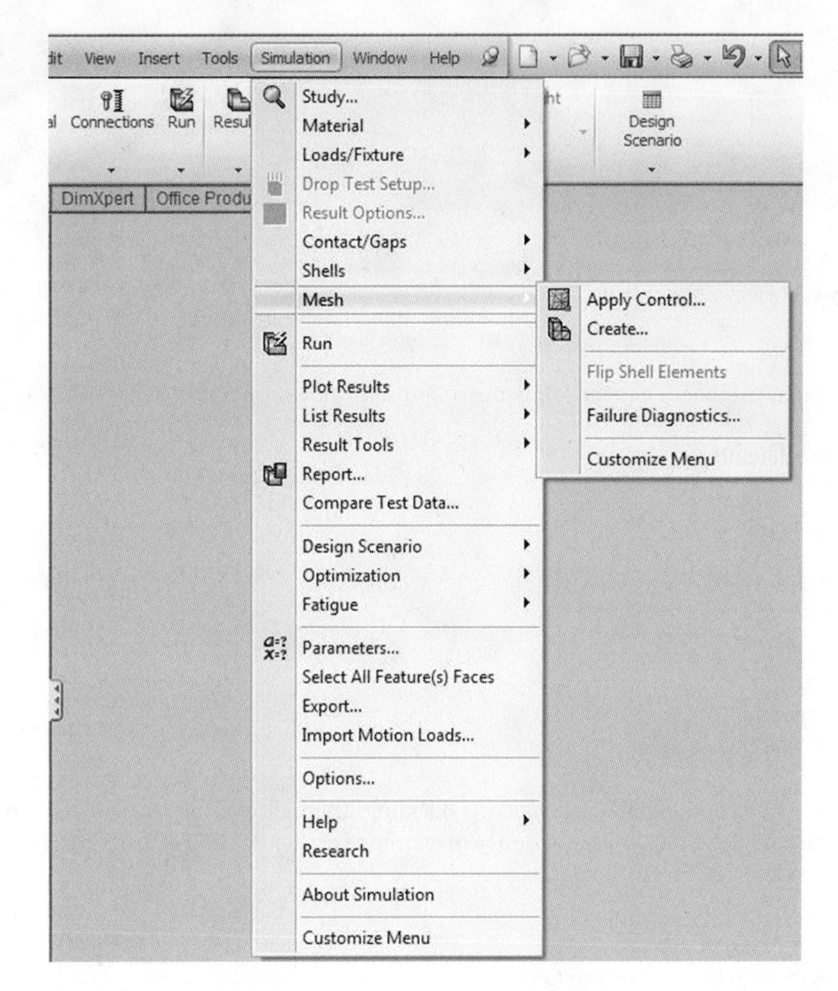

Figure 34.18

SolidWorks Simulation toolbars.

SolidWorks Simulation Toolbars

Another way to access the functions for creating, solving, and analyzing a model is through the SolidWorks Simulation toolbars (see Figure 34.18).

Starting a New Study in SolidWorks Simulation

1. Open a model file.
2. Click Simulation > New Study (see Figure 34.19).

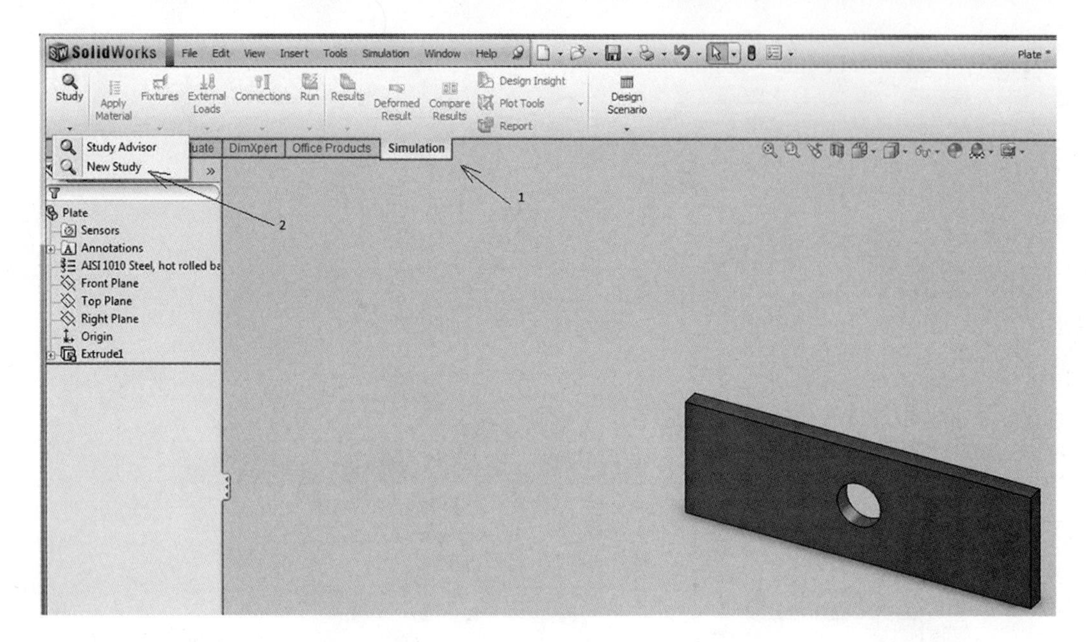

Figure 34.19

Simulation > New Study.

The Study PropertyManager has the following options, as shown in Figure 34.20:

- Static (the default)
- Frequency
- Buckling
- Thermal
- Drop Test
- Fatigue
- Optimization
- Nonlinear
- Linear Dynamic
- Pressure Vessel Design

When any of the study options (Static, Frequency, Buckling, Thermal, etc.) is selected in the PropertyManager and OK is clicked, the SolidWorks SimulationManager appears below the FeatureManager. The following options are the roots of the SimulationManager:

- Connections
- Fixtures
- External Loads
- Mesh

Basic SolidWorks Simulation Steps

The steps involved in SolidWorks Simulation for solving FEA problems (any of the study options) are summarized as follows:

1. Geometric preparation (if required).
2. Apply material to the model.
3. Define connections.
4. Define fixtures.
5. Define external loads.
6. Create the model mesh.

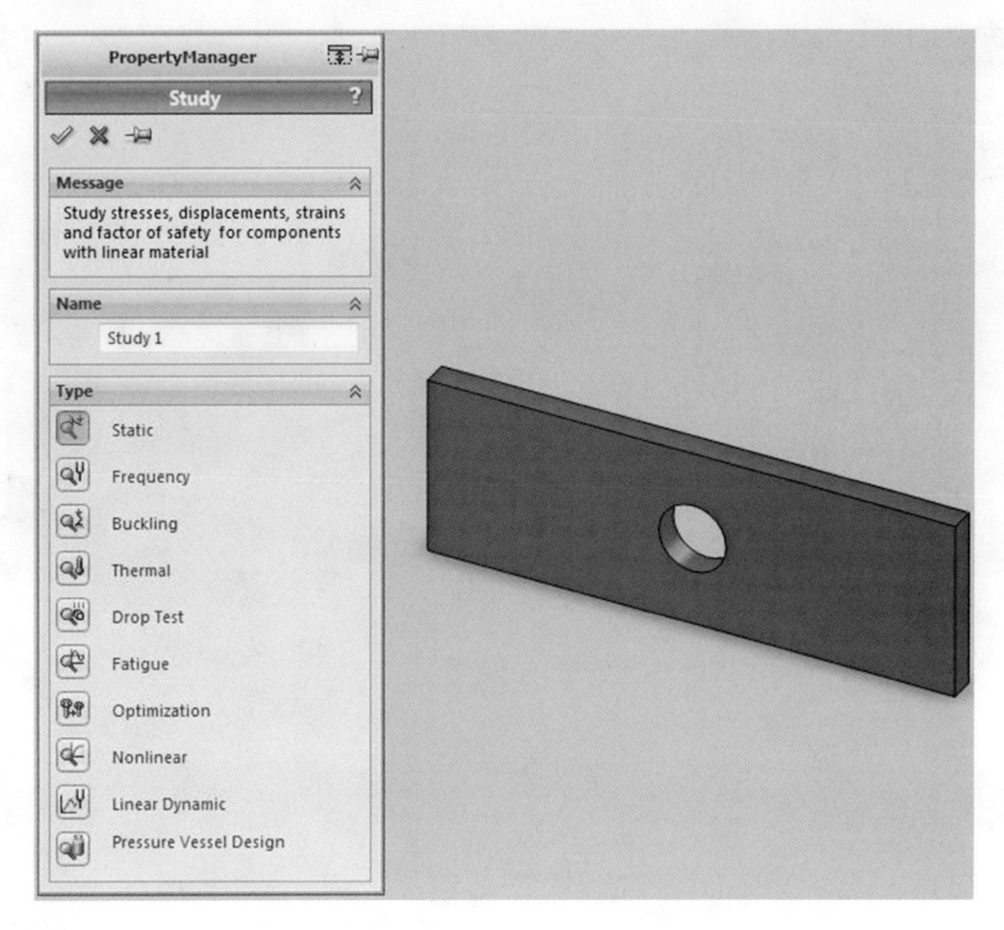

Figure 34.20

Study PropertyManager.

7. Run the model solution.
8. Analyze the results.

FEA of a Sheet Metal Part

The problem that we are solving is one of the sheet metal parts that we designed in Chapter 19.

Start a new study in SolidWorks Simulation:

1. Select Simulation > New Study.
2. Click OK (to use the default study option, Static).

   We are now ready to define the analysis model. The user has to define the connections, fixtures, external loads, and mesh (see Figure 34.21). If material has already been assigned to the part, then Material is not listed as one of the options. Applying materials during part design is the preferred approach, especially in cases where several parts make up an assembly.

Defining Connections

3. Right-click the Connections folder and select the appropriate connections (see Figure 34.22).

Defining Fixtures

4. Right-click the Fixtures folder and select the appropriate fixtures (see Figure 34.23).
5. Click Fixed Geometry and select the faces to fix. (The preview of Figure 34.24 appears.)

Defining External Loads

6. Right-click the External Loads folder and select the appropriate external loads (see Figure 34.25).
7. Click Force; a preview appears (see Figure 34.26).
8. For the force, check Normal.
9. Select the face to apply the pressure, and enter a value of 1.5 kPa. (For reverse pressures, check Reverse direction.)

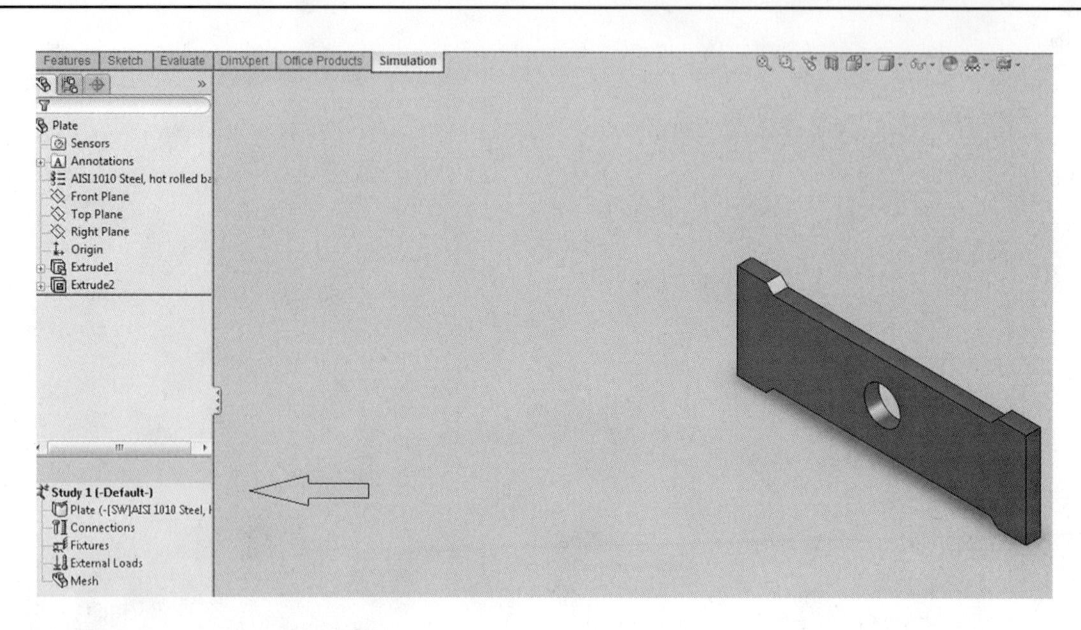

Figure 34.21

Study folders in SimulationManager.

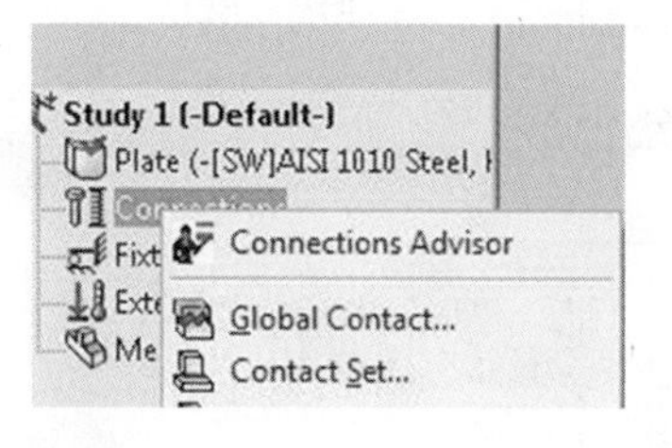

Figure 34.22

Assigning connections.

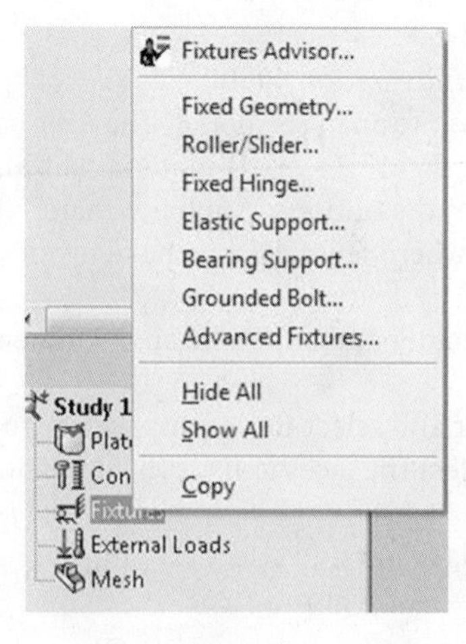

Figure 34.23

Assigning fixtures.

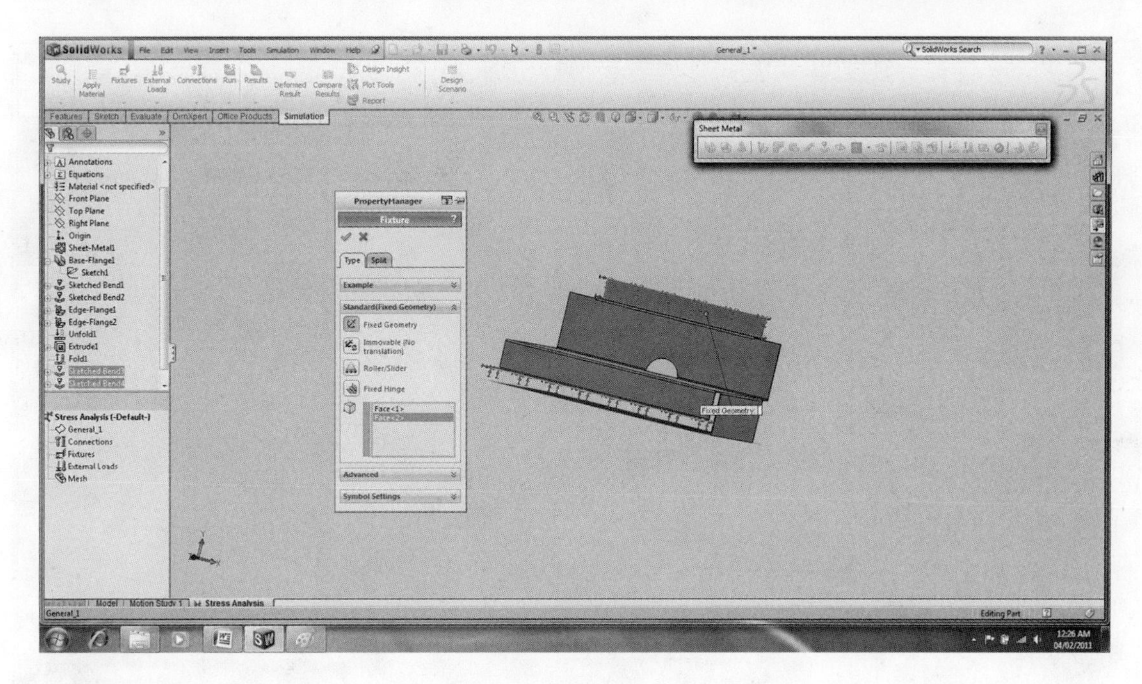

Figure 34.24

Fixtures preview for fixed geometry.

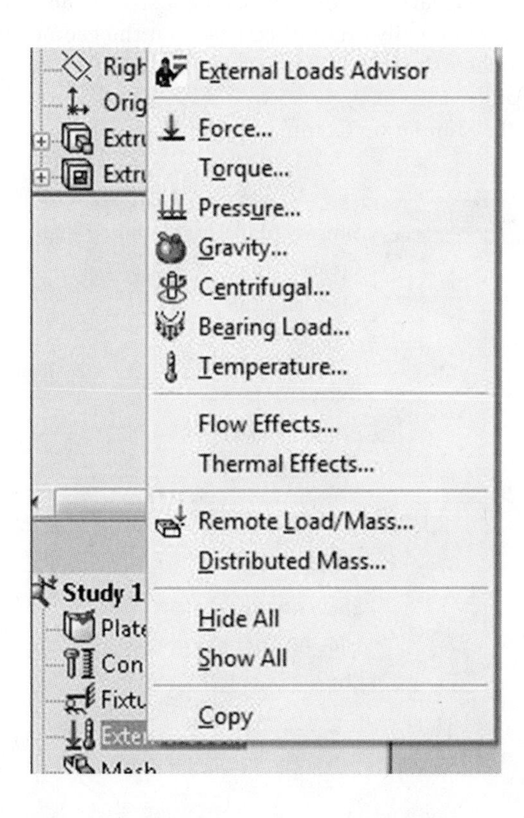

Figure 34.25

Assigning external loads.

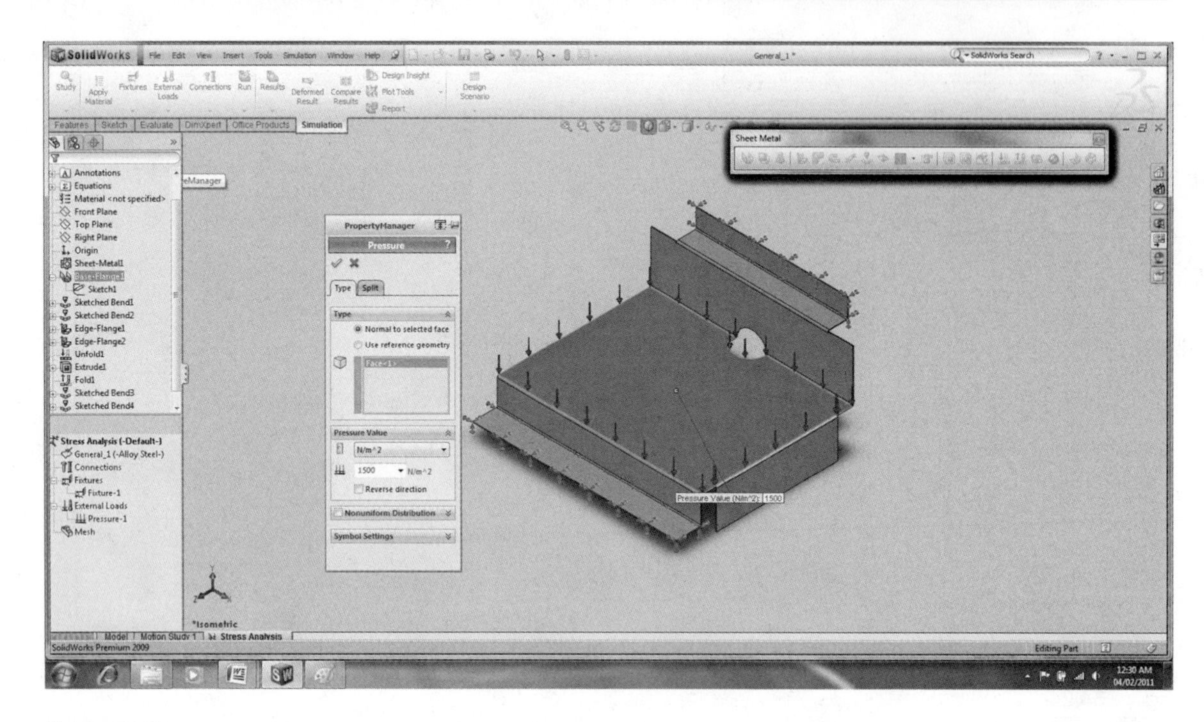

Figure 34.26

External Loads preview for normal pressure.

Defining the mesh

10. Right-click the Mesh folder and select the appropriate mesh (see Figure 34.27).
11. Click Create Mesh. (The preview of Figure 34.28 appears.) Note that we can control the mesh density by moving the slider from Coarse to Fine. The element size (4.786 mm) and the element size tolerance (0.239 mm) are automatically established based on the geometric features of the SolidWorks model. The meshed model is shown in Figure 34.29.

Running the model solution

12. Click Run Advisor in the Simulation CommandManager.

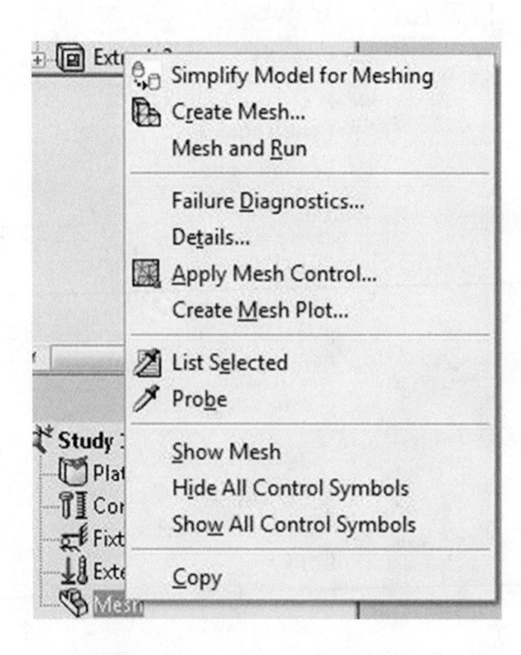

Figure 34.27

Assigning the mesh.

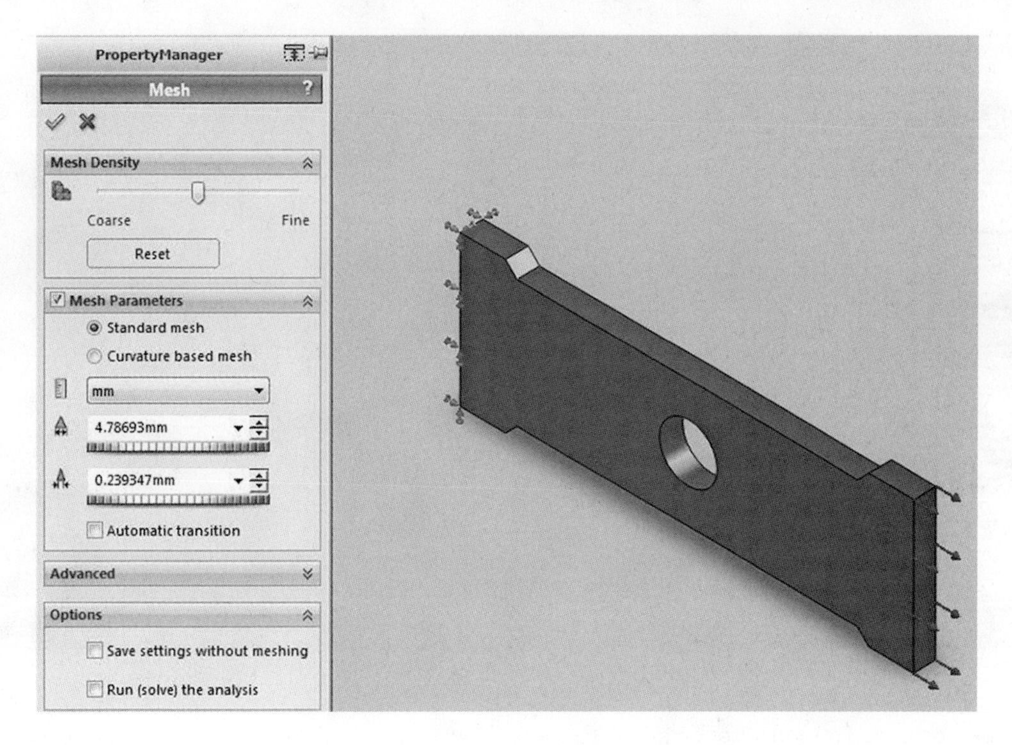

Figure 34.28

Mesh preview. (Create Mesh is selected.)

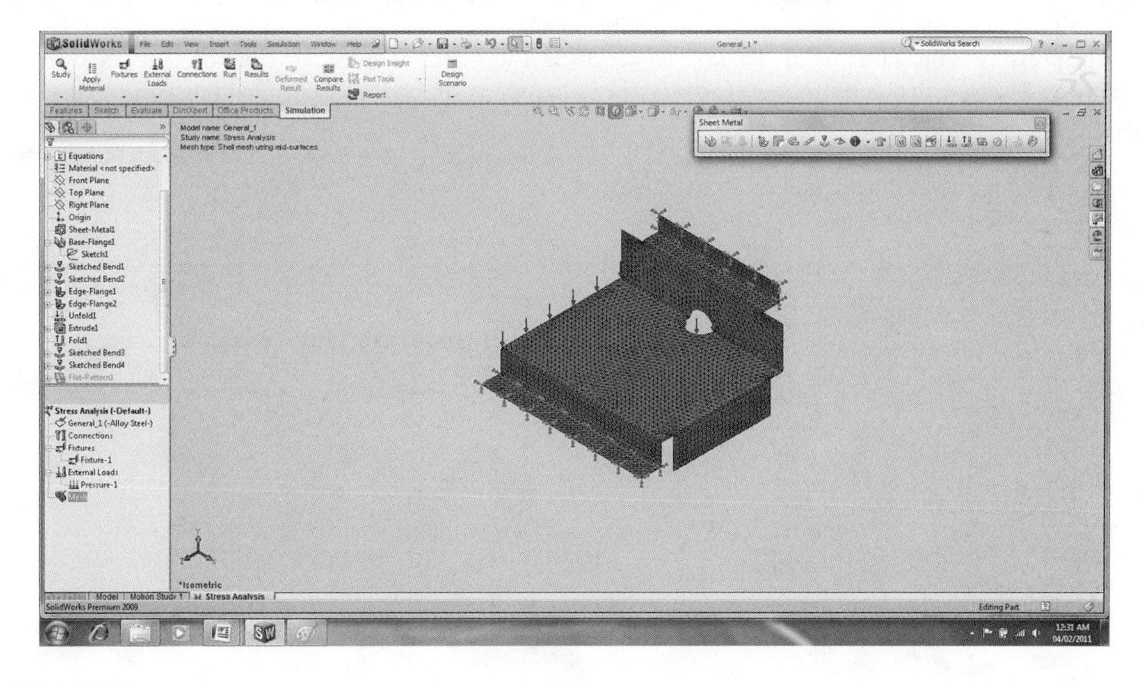

Figure 34.29

Meshed model.

The results are shown (see Figure 34.30), and three plots are automatically created in the Results folder:

1. Stress1: von Mises stresses
2. Displacement1: resultant stresses
3. Strain1: equivalent strain

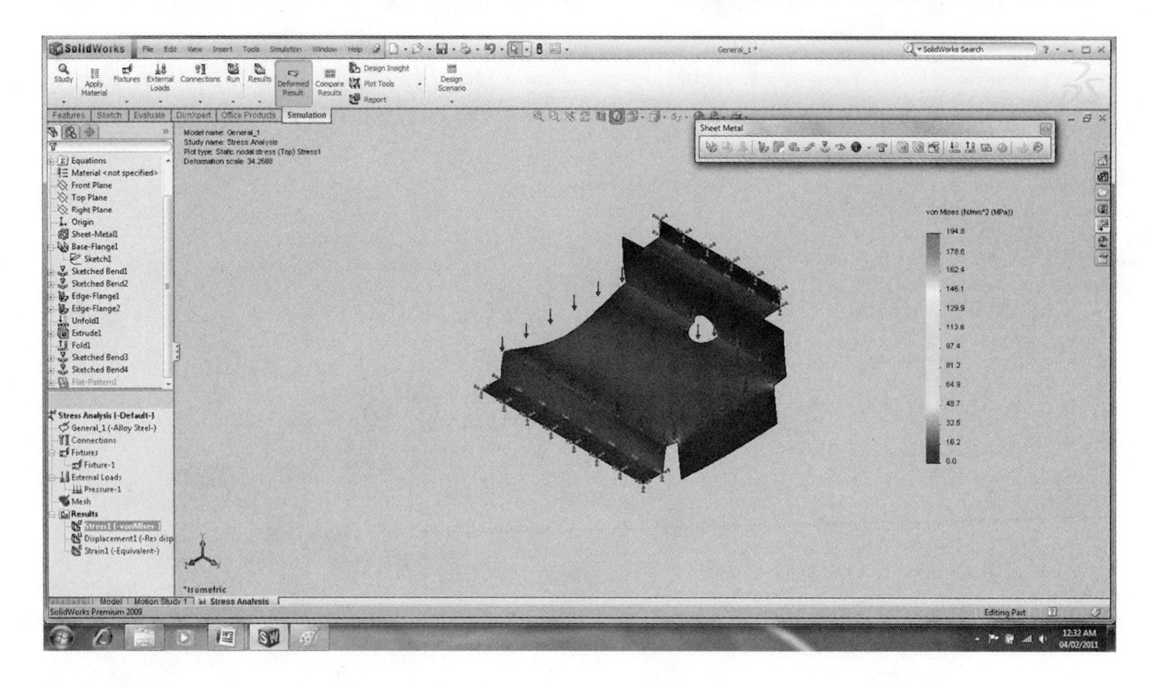

Figure 34.30

Model solution based on von Mises criterion.

The pressure is increased to 3.5 kPa (see Figure 34.31) leading to the stress distribution that is shown in Figure 34.32.

The pressure is increased to 20 kPa. (See Figure 34.33 leading to the stress distribution that is shown in Figure 34.34.) Note the high stress level around the sharp edges, which is consistent with what we expect in practice.

Reverse 20 kPa loading results is shown in Figure 34.35. Beyond 20 kPa loading, the model begins to experience large deflection and becomes unstable.

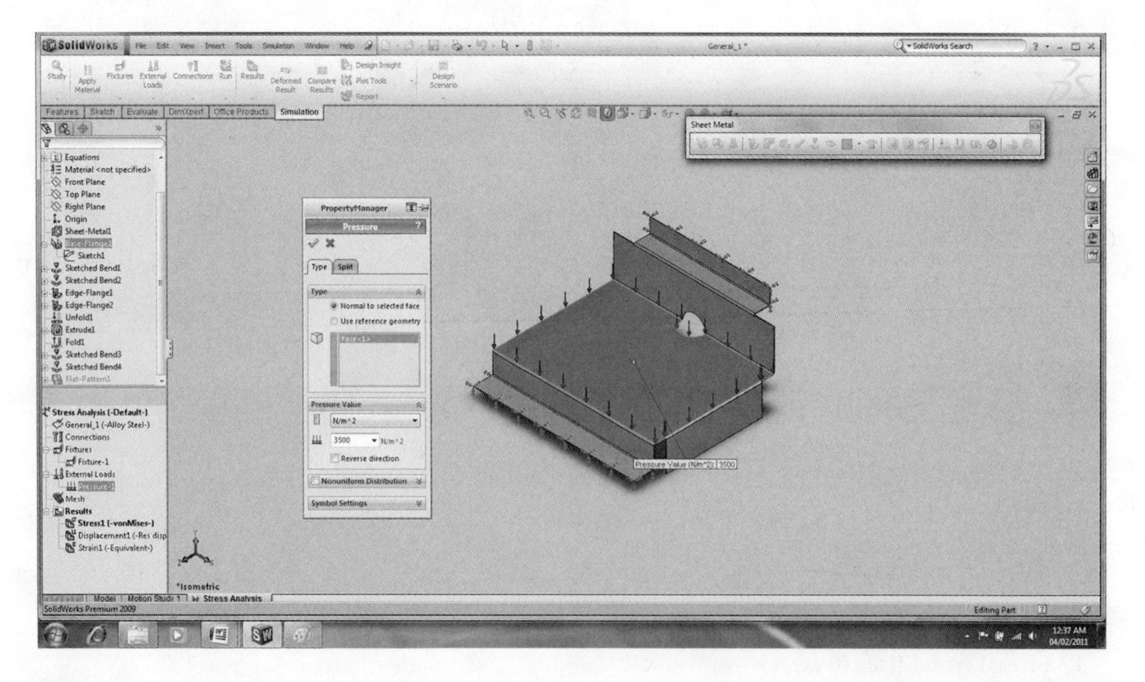

Figure 34.31

Model at 3500 Pa (3.5 kPa) loading.

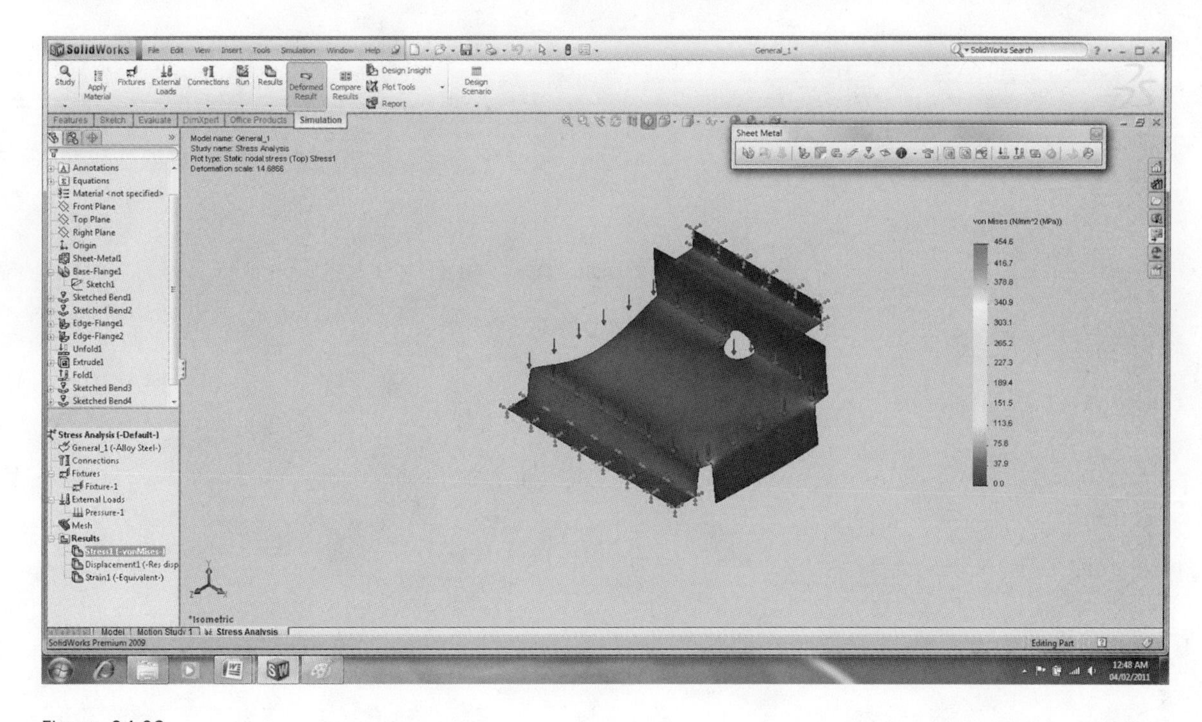

Figure 34.32

Model solution based on von Mises criterion.

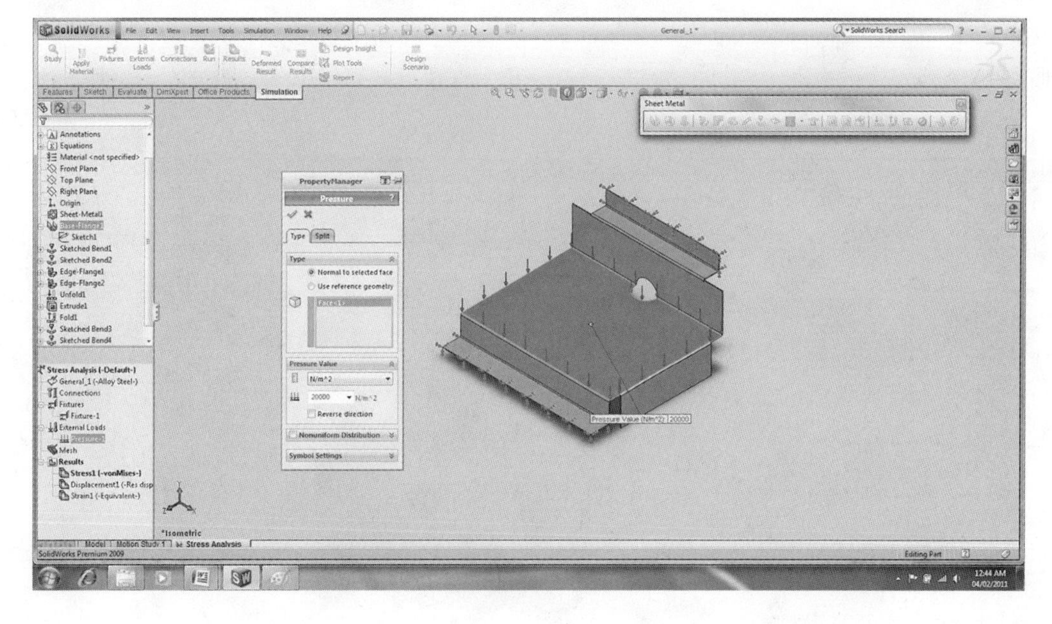

Figure 34.33

Model at 20000 Pa (20 kPa) loading.

Summary

When we run FEA software, we must understand the results. For simple parts with classical solutions, it is a good engineering practice to compare the FEA solution with a manual calculation. Once we have mastered FEA, then we can be more confident of our results when we solve complex, unfamiliar problems. In this chapter, we have applied FEA to analyze the loading of the sheet metal part that we had earlier designed in Chapter 19. By increasing the pressure loading, we are able to track the limit beyond which large deflection occurs. This kind of analysis is important in the design stage in order to anticipate how a model designed should be loaded in practical usage. The Design Scenario tool can be used to realize multiple design solutions from which the optimum is chosen.

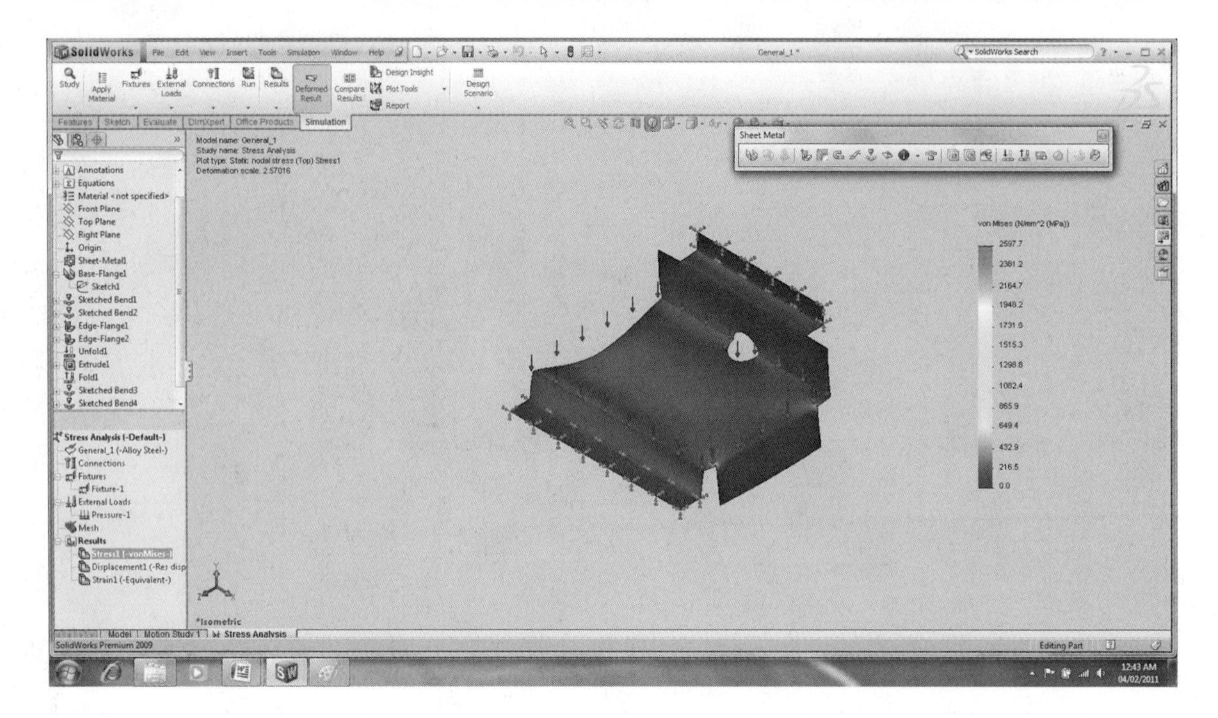

Figure 34.34

Model solution based on von Mises criterion.

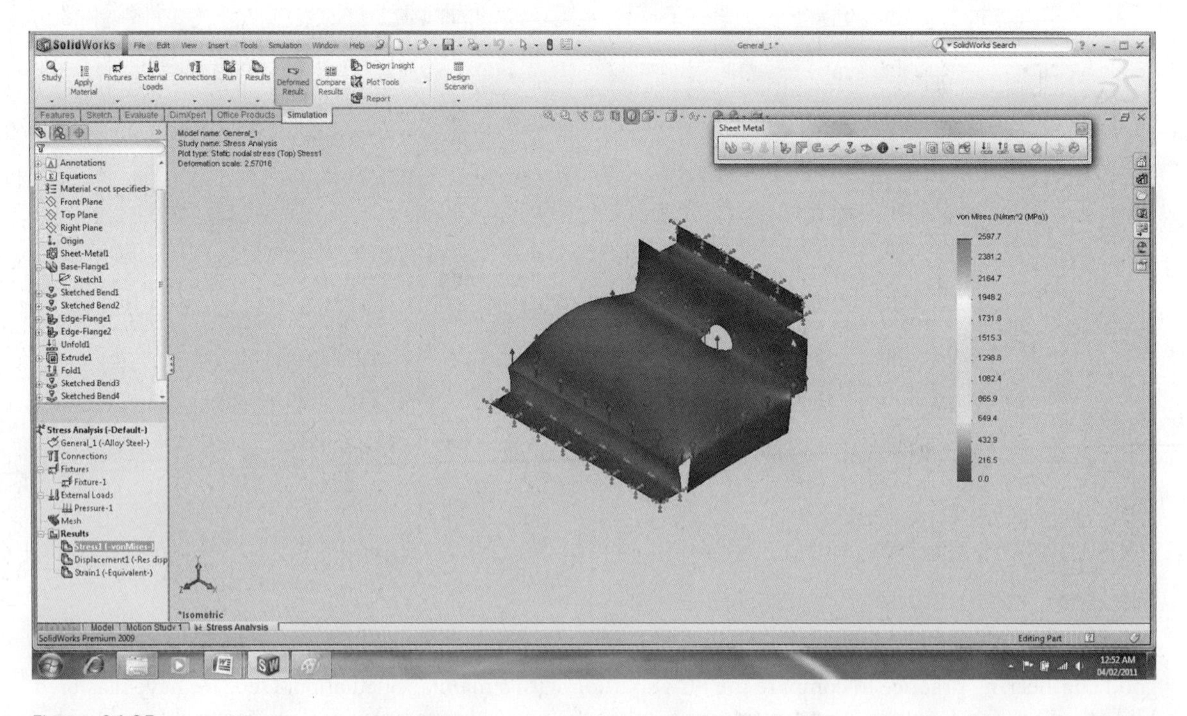

Figure 34.35

Model solution based on von Mises criterion.

Exercises

1. Figure P34.1 shows the model that is presented in this chapter. Change the fixture to only one, as shown. Apply the following loads (similar to the ones that are used in this chapter) and compare the results between the conditions that are shown in this problem and the ones that are discussed earlier in this chapter:
   a. 1.5 kPa loading
   b. 3.5 kPa loading
   c. 20 kPa loading
2. In Figure P34.2, the fixtures are labeled *Fixture 1* and *Fixture 2*. Apply the three different loading conditions of problem 1 to this model on the face that is marked *Face to load*.

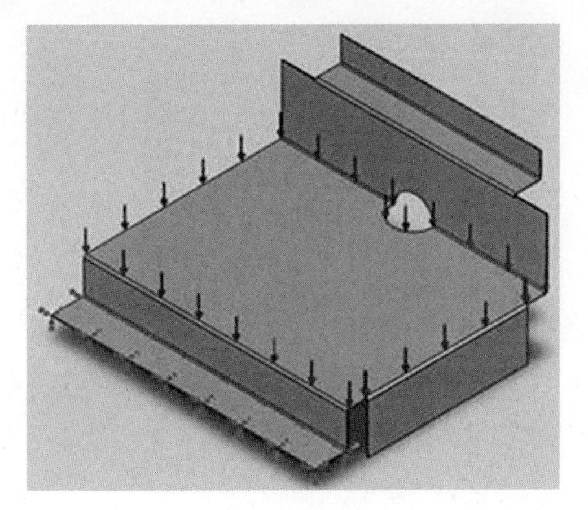

Figure P34.1
Model 1.

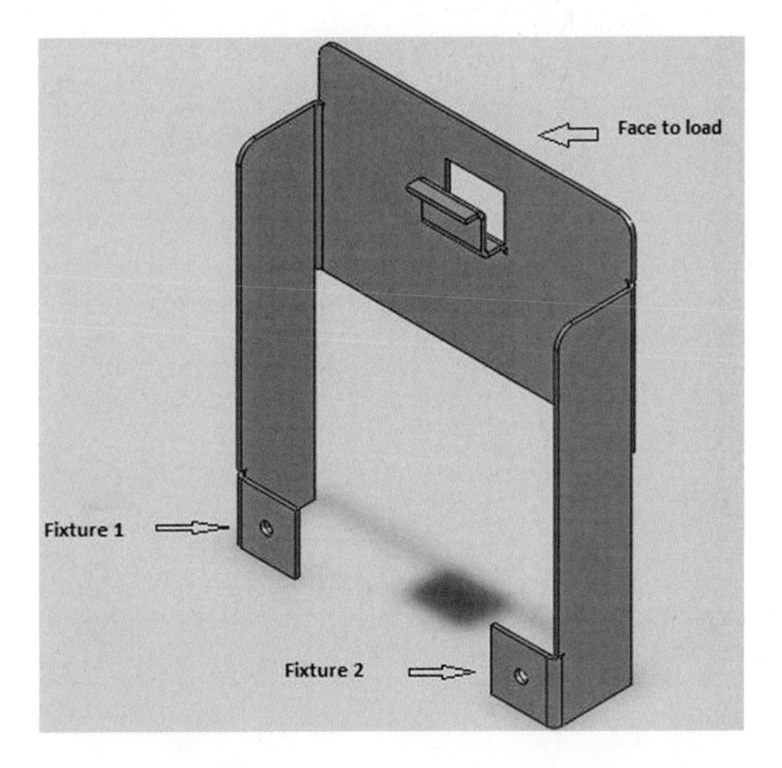

Figure P34.2
Model 2.

Section IV
Introductory 3D Printing

35

Overview of Additive Manufacturing

Objectives:

When you complete this module, you will

- Understand the concept of additive manufacturing (AM)
- Understand the classification and different facets of AM
- Understand the fused deposition modeling process

What Is Additive Manufacturing?

Additive manufacturing (AM) is a relatively new manufacturing technology that is driven by computer-aided design (CAD) in which the solid shape to be produced is first sliced from the Standard Tessellation Language (STL) file that is submitted, and tool paths are generated using an interface software and submitted to the production system. Based on the information that is received, the production system produces the part layer by layer until the part is completely produced, thereby making it possible for companies to significantly cut design and manufacturing cycle times. AM technologies have revolutionized the way that parts are manufactured because they build parts by adding materials, which is opposite to the traditional way of building parts by removing materials.

Classification of AM Processes

Classification of the technologies for AM is based on the raw materials that are used in the process. These categories are as follows:

1. Liquid-based systems
2. Powder-based systems
3. Solid-based systems

1. Liquid-based systems

Stereolithography	Chuk Hull, 1986	3D Systems, 1987
Perfactory	EnvisionTec	EnvisionTec, 2003

2. *Powder-based systems*

Selective laser sintering (SLS; polymer)	Ross Householder, 1979	DTM Corp., 1992
	Carl Deckard, late 1980s	EOS GmbH, 1994
SLS (ceramics and metals)		DTM/EOS, 1990s
Direct laser sintering		EOS, 1990s
Direct metal laser sintering (DMLS)		EOS, 1990s
Three-dimensional (3D) printing		MIT
Electron beam melting		Arcam, 1997
Selective laser melting (SLM)		MCP Group
Selective masking sintering		RP3
Selective inhibition sintering		USC, USA

Note: Desk Top Manufacturing Corporation (DTM); Electro Optical System (EOS); Massachusetts Institute of Technology (MIT); MCP GmbH; University of South Carolina (USC).

3. *Solid-based systems*

Fused deposition modeling (FDM)	Scott Crump, 1991	Stratasys, 1992
Sheet stacking (laminate object manufacturing)	Helysis, 1991	Helysis, 1991
		KIRA, 1990s
		Kinergy, 1990s
		Solidica, 1990s
		Solidimension, 1990s

There are several AM machines that utilize different building methods and materials, such as 3D printing, FDM, laminated object manufacturing, SLS, SLM, and 3D laser cladding. Every AM technique has its advantages and disadvantages, respectively, among which, SLS is an optimal method to directly manufacture metal parts because of its wider material range and better flexibility.

Similar to SLS, DMLS, as a typical rapid prototyping (RP) technique, enables the quick production of complex-shaped 3D parts directly from metal powder. This process uses a laser that is directly exposed to the metal powder in liquid phase sintering and creates parts by selective fusing and consolidating thin layers of loose powder with a scanning laser beam process.

Additionally, due to its flexibility in materials, shapes and control of parameters in the construction may also lead to produce porous metallic components.

Many groups around the world have developed ideas, prototype methods, and commercial systems using various different ways of creating metal parts. In particular, several companies have recently started developing variations of the DMLS method. This process was developed by EOS GmbH of Munich, Germany, and has been available commercially since 1995.

Layer manufacturing techniques are moving from RP and rapid tooling to rapid manufacturing (RM). The production of end-use parts made of metal is one of the most promising applications for these techniques. RM of metal parts is especially suitable for the fabrication of a small number of pieces and mass customization. By this technique, it is possible achieve metal parts with excellent mechanical properties. Aside from that, the advent of the new machines equipped with different lasers may increase the accuracy and mechanical properties of the fabricated parts.

FDM

The FDM process works as shown in Figure 35.1. Beginning with slices of CAD data, which define a tool path in layers, the FDM machine draws thermoplastic material in a filament form from a reel or canister, which then enters a liquefier and extruded in a molten form through the tip unto the built sheet that is placed on the platen. The molten thermoplastic, which is in a semiliquid state, is deposited in layers on the build sheet in the x–y plane to build the part from the bottom up (in z plane), which makes very complex parts easy to build. Two materials, one to make the part, and one to support it, enter the extrusion head. Alternating between part material and support material, the system deposits layers. The FDM machine microcontroller

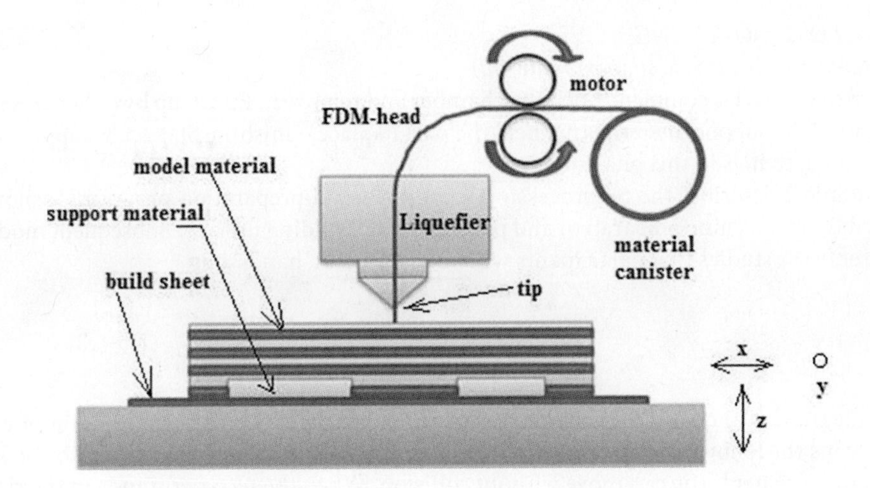

Figure 35.1

FDM process.

coordinates the movements of the platen in x–y–z directions. The FDM head carries a tip that moves in the x–y directions to deposit the molten thermoplastic on the build sheet that is placed on the platen; when one layer is complete, the platen is raised one layer, and the process is continued until the entire model is built, based on the sliced CAD model data. The slicing of the CAD data is carried out in the preprocessor section, which is driven by one of two build-preparation programs, (1) *CatalystEX* or (2) *Insight*. The FDM technology is an environmentally clean, simple-to-use, office-friendly AM process.

1. *Step 1: CAD DESIGN*
   (The design process)
   All FDM parts must commence from a software model that fully describes the external geometry. Any professional CAD solid modeling software such as AutoCAD 3D, Inventor, SolidWorks, CATIA, UNIGRAPHICS, SolidEdge, or Pro-Engineer, can be used for solid modeling. In some cases, reverse engineering equipment, such as laser scanner, can also be used to create 3D solid or surface representation. What is important is that the solid model output must be saved as an STL file format, which has become the *de facto* industry standard for geometric data input into AM machines. The STL file format describes the external closed surface of the original CAD model, and it forms the basis for the computation of the slices that are required for the machine tool paths.
2. *Step 2: PREPROCESSING*
   ("Slicing" or sectioning CAD design into layers)
   The FDM process begins in one of two build-preparation programs: (1) Catalyst EX or (2) Insight.
   In operation, your first step is to import a design file, pick options, and create slices (layers). The preprocessing software calculates sections and *slices* the part design into many layers, ranging from 0.005 in. (0.127 mm) to 0.013 in. (0.3302 mm) in height. Using the sectioning data, the software then generates *tool paths* or building instructions that will drive the extrusion head. This step is automatic when using Catalyst EX. Next, send the job to the 3D printer.
3. *Step 3: PRODUCTION*
   (The layering process)
   Press *print* to start the building process.
   Two materials, one to make the part, and one to support it, enter the extrusion head. Heat is applied to soften the plastics, which are extruded in a ribbon, roughly the size of a human hair. Alternating between part material and support material, the system deposits layers as thin as 0.005 in. (0.13 mm).
   The following hands-on activities will be covered:
   - Material loading
   - Replacement of printing tip
   - Machine calibration
   - Machine setup

4. *Step 4: POSTPROCESSING*

   (Removing disposable support material)

   When the part is complete, open the chamber and remove it. Finish up by either washing or stripping away the support material that held the part in place. Finishing Station & Support Removal are the main activities in this phase.

   Module 2 describes the preprocessing using the build-preparation program, Insight. Module 3 describes the machine preparation and production or building process. Subsequent modules present different case studies that participants will be involved in, hands-on.

DMLS Principle

A schematic diagram of the DMLS system is shown in Figure 35.2. For the construction of any part, the machine performs the following steps. The building and dispenser platform are lowered by one layer thickness; for that, the recoater blade can move without collision. When the recoater stands in the right position, the dispenser platform rises to supply the amount of powder for the next layer. Then, the recoater moves from the right to the left position; in this way, the metal powder is spread from the dispenser to the building area, and the excess metal powder falls into the collector. Then, the heads can move the laser beam through two-dimensional cross section and is precisely switched on and off during the exposure of designated areas.

The absorption of energy by metal powder will generate the cure and sinter of the already solidified areas below. This process proceeds, layer by layer until all the parts in a job are completed. Thus, in a few hours, the machine can produce 3D parts with high complexity and accuracy. In addition, during the building process, sintered parts reach more or less their final properties, but, depending on the application of piece, it is necessary as a postprocessing treatment, like a tempering or surface treatment.

Exposure Strategies

Figure 35.3 shows the laser scanning the top surface of a thin powder layer to form the area that is enclosed by cross sections of the sliced object. Initially, all the contours of the layer structure are exposed with a laser power (L_{pw}) and contour speed (C_{sp}). As the diameter of the sintered zone is usually larger than the laser diameter, the effective laser diameter, or the curing zone, it is necessary to compensate the dimensional

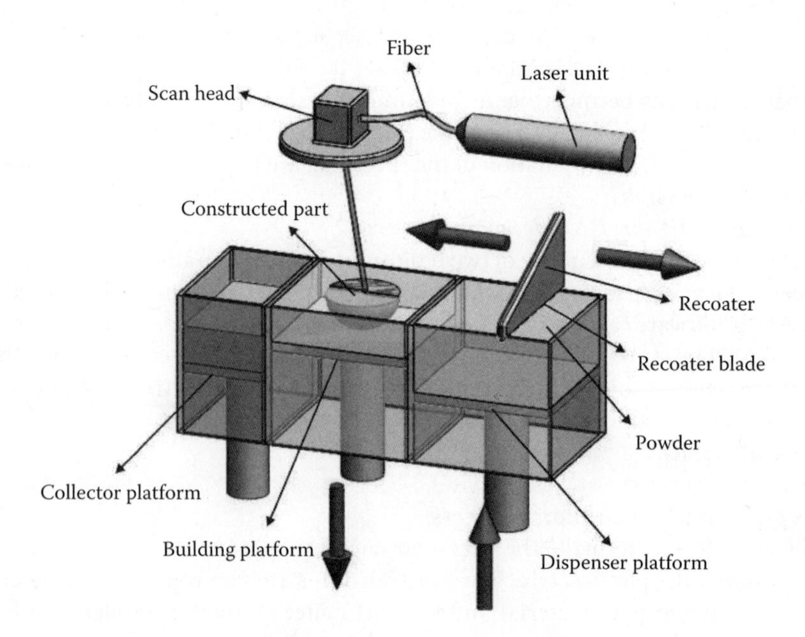

Figure 35.2

Schematic diagram of the DMLS system. (From Romão Bineli et al., Direct Metal Laser Sintering (DMLS): Technology for Design and Construction of Micro-reactors, *6th Brazilian Conference on Manufacturing Engineering*, April 11–15, 2011, Caxias do Sul, RS, Brazil © Associação Brasileira de Engenharia e Ciências Mecânicas, 2011.)

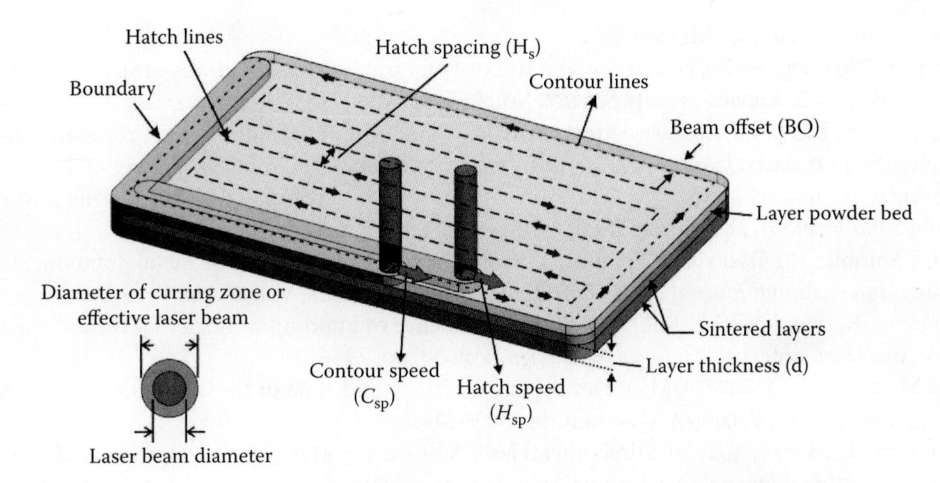

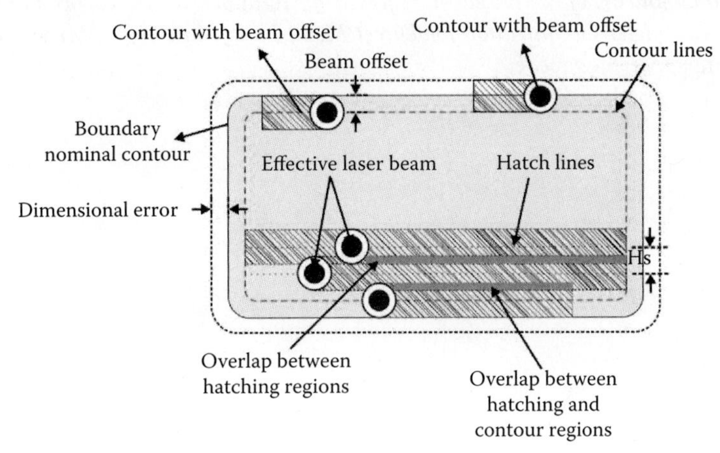

Figure 35.3

Exposure strategies and process parameters: (From Romão Bineli et al., Direct Metal Laser Sintering (DMLS): Technology for Design and Construction of Micro-reactors, *6th Brazilian Conference on Manufacturing Engineering*, April 11–15, 2011, Caxias do Sul, RS, Brazil © Associação Brasileira de Engenharia e Ciências Mecânicas, 2011.)

error, and the laser beam must be shifted by half the curing width from the contour to the inside, to make sure that the contour of the later part will correspond exactly to the original CAD data. This correction of the position is called beam offset (BO).

During hatching, the laser beam moves line after line several times to assure that the sintering process can unroll completely because it maintains the temperature for a long period. The distance between the lines is called hatch spacing (Hs) and is set about one-quarter of the laser beam. Here, the BO value is again defined with respect to the edge of the boundary Figure 35.3, and if this value is higher or lower than the correct value, the particles of the irradiated region may not be sintered or be oversintered.

Another important parameter that can lead to a distorted part or a process interruption is the layer thickness. If the value is too high, no optimal adhesion between the single layers can be realized because the curing depth is not high enough; furthermore, mechanical tension can be generated through this layer, which can lead to the detachment of the layer below. If the selected value is smaller, a tearing-off of a structure can happen during the recoating process, since the sintered particles get struck between it and the recoater blade.

Bibliography

Bineli, A.R.R., Peres, A.P.G., Jardini, L.F., Maciel Filho, R. 2011. Direct metal laser sintering (DMLS): Technology for design and construction of micro-reactors. *6th Brazilian Conference on Manufacturing Engineering*, April 11–15, 2011, Caxias do Sul, RS, Brazil © Associação Brasileira de Engenharia e Ciências Mecânicas.

EOS. 2009. EOSint M 270 User Manual.
Gu, D., Shen, Y. 2008. Processing conditions and microstructural features of porous 316L stainless steel components by DMLS. *Applied Surface Science*, Vol. 255, pp. 1880–1887.
Khaing, M.W., Fuh, J.Y.H., Lu, L. 2001. Direct metal laser sintering for rapid tooling: Processing and characterization of EOS parts. *Journal of Materials Processing Technology*, Vol. 113, pp. 269–272.
Pohl, H., Simchi, A., Issa, M., Dias, H.C. 2001. Thermal stresses in direct metal laser sintering. *Proceedings of Twelfth Solid Freeform Fabrication (SFF) Symposium*, Austin, TX, pp. 366–372.
Santos, E.C., Shiomi, M., Osakada, K., Laoui, T. 2006. Rapid manufacturing of metal components by laser forming. *International Journal of Machine Tools and Manufacture*, Vol. 46, pp. 1459–1468.
Senthilkumaran, K., Pandey, P.M., Rao, P.V.M. 2009. Influence of building strategies on the accuracy of parts in selective laser sintering. *Materials & Design*, Vol. 30, pp. 2946–2954.
Shellabear, M., Nyrhilä, O. 2004. DMLS-Development history and state of the art. *Laser Assisted Net Shape Engineering (LANE)*, Erlangen, Germany, pp. 393–404.
Yang, J., Ouyang, H., Wang, Y. 2010. Direct metal laser fabrication: Machine development and experimental work. *International Journal of Advanced Manufacturing Technology*, Vol. 46, pp. 1133–1143.
Yu, N. 2005. *Process Parameter Optimization for Direct Metal Laser Sintering (DMLS)*. PhD thesis, National University of Singapore, Singapore.

36

Insight Software for Fortus Production Systems

Objectives:

When you complete this module, you will have

- Understood how Insight software works
- Used Insight software to prepare tool paths for additive manufacturing

Step 1: Open Insight software (see Figure 36.1).

Step 2: Specify the configuration of machine by clicking (1) (see Figure 36.2).

Specify the following characteristics of the modeler and the slice height that is used to build the model:

- *Modeler type:* The type of modeler.
- *Modeler name:* The name of a specific modeler that is used to determine the modeler type.
- *Model material:* The type of material that is used to build the part.
- *Model material color:* The color of the model material.
- *Support material:* The type of material that is used to build the disposable supports.
- *Slice height:* The interval at which to slice the Standard Tessellation Language (STL) model.
- *Model tip:* The size of the model-material extrusion tip.
- *Support tip:* The size of the support-material extrusion tip.

Specify the system configuration by choosing the following:

a. Modeler type = Fortus 400mc Large (see (2) in Figures 36.2 and 36.3)
b. Model materials = ABS-M30 (see (3) in Figures 36.2 and 36.4)
c. Model material color (see (4) in Figures 36.2 and 36.5)
d. Support material (see (5) in Figure 36.2) (SR 20 support and SR 30 support)
e. Slice height (Layer thickness) = 0.0050 (see (6) in Figures 36.2 and 36.6)
f. Model tip = T10 (see (7) in Figure 36.2) (not under the control of the user)
g. Support tip = T12 (see (8) in Figure 36.2) (not under the control of the user)

Select the OK button to accept changes and reconfigure system parameters.

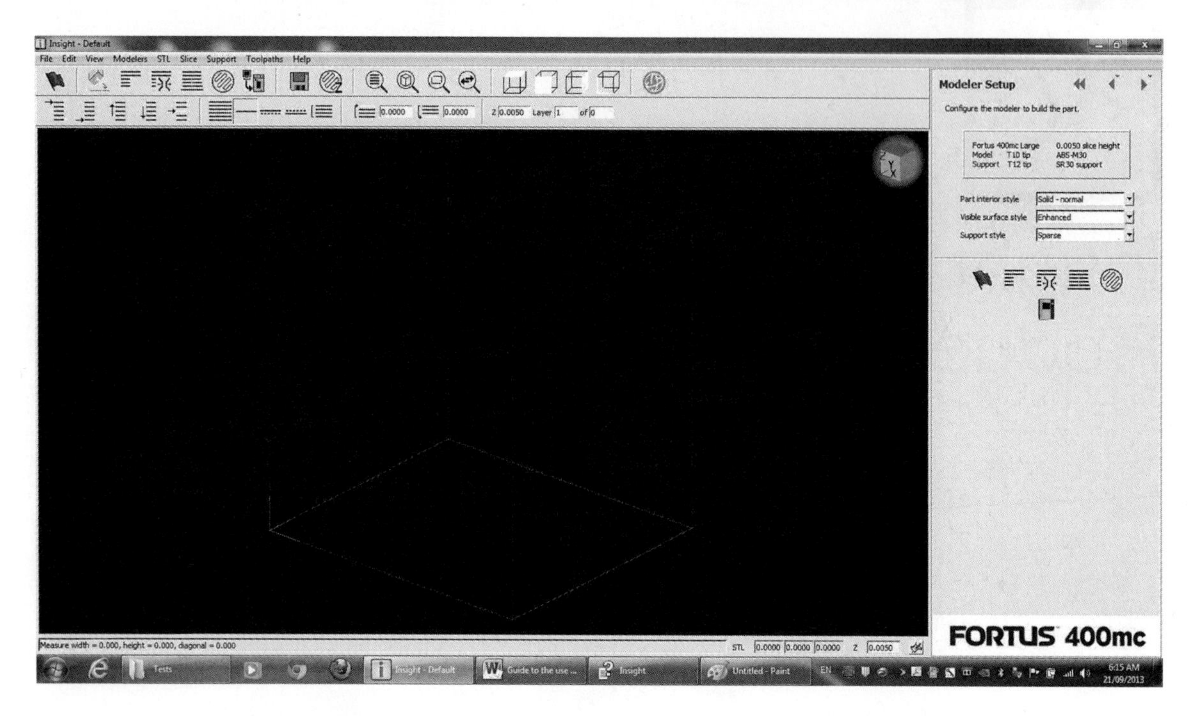

Figure 36.1

Insight Default User Interface.

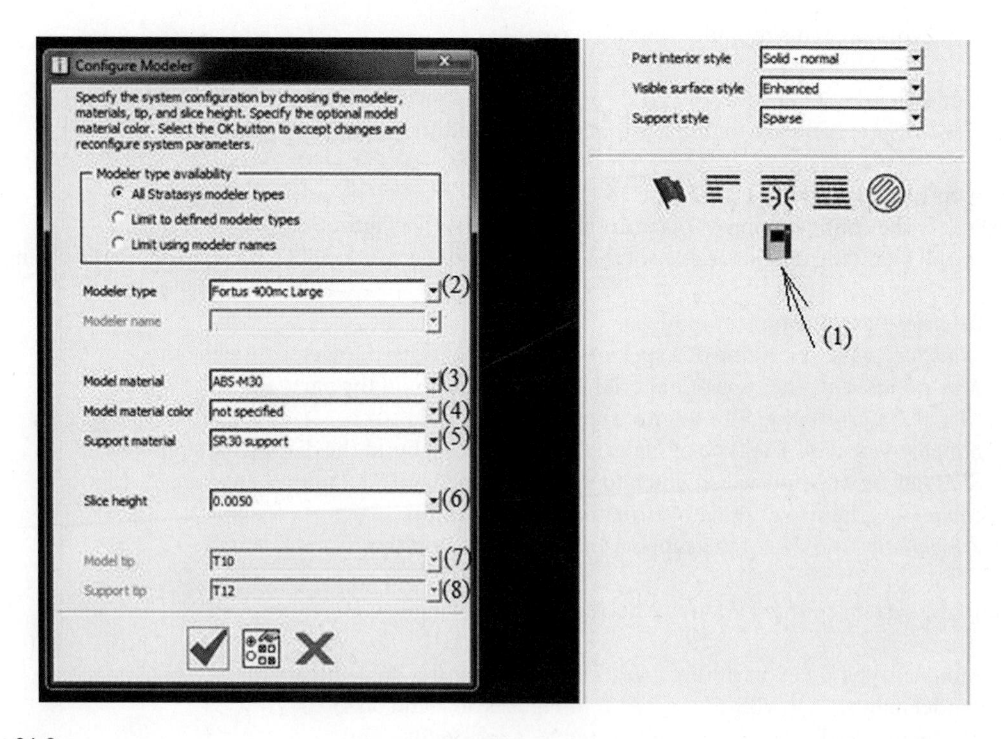

Figure 36.2

Configure Modeler.

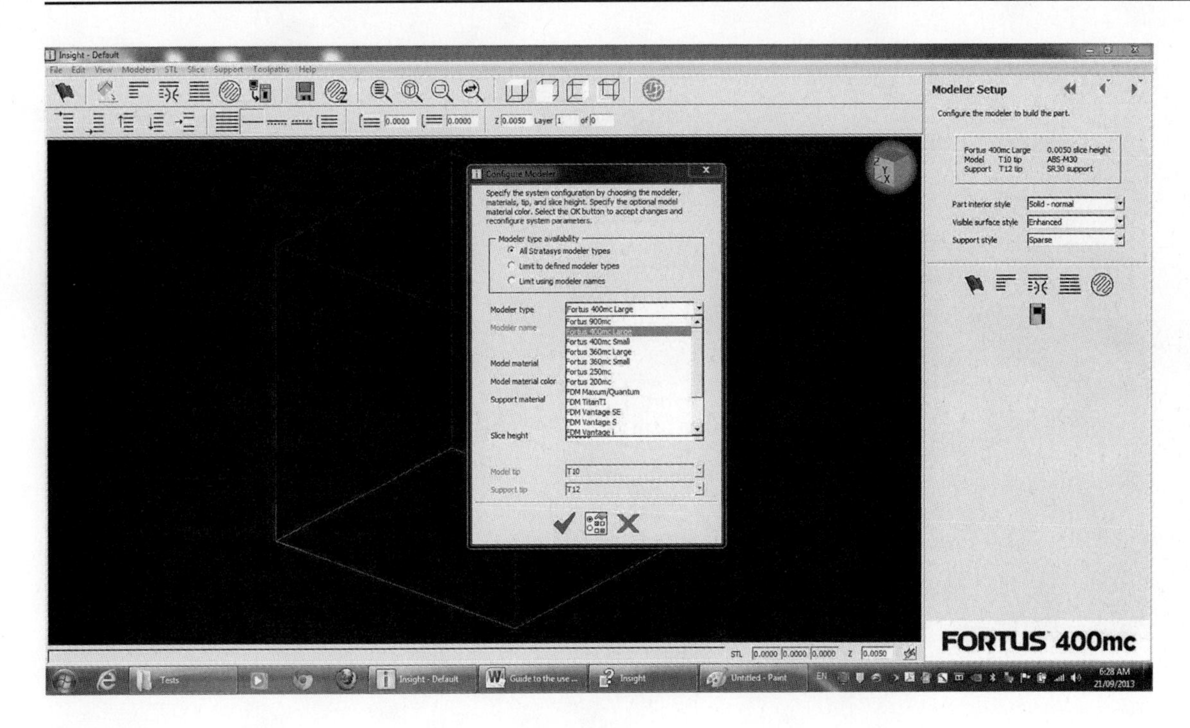

Figure 36.3

Modeler type.

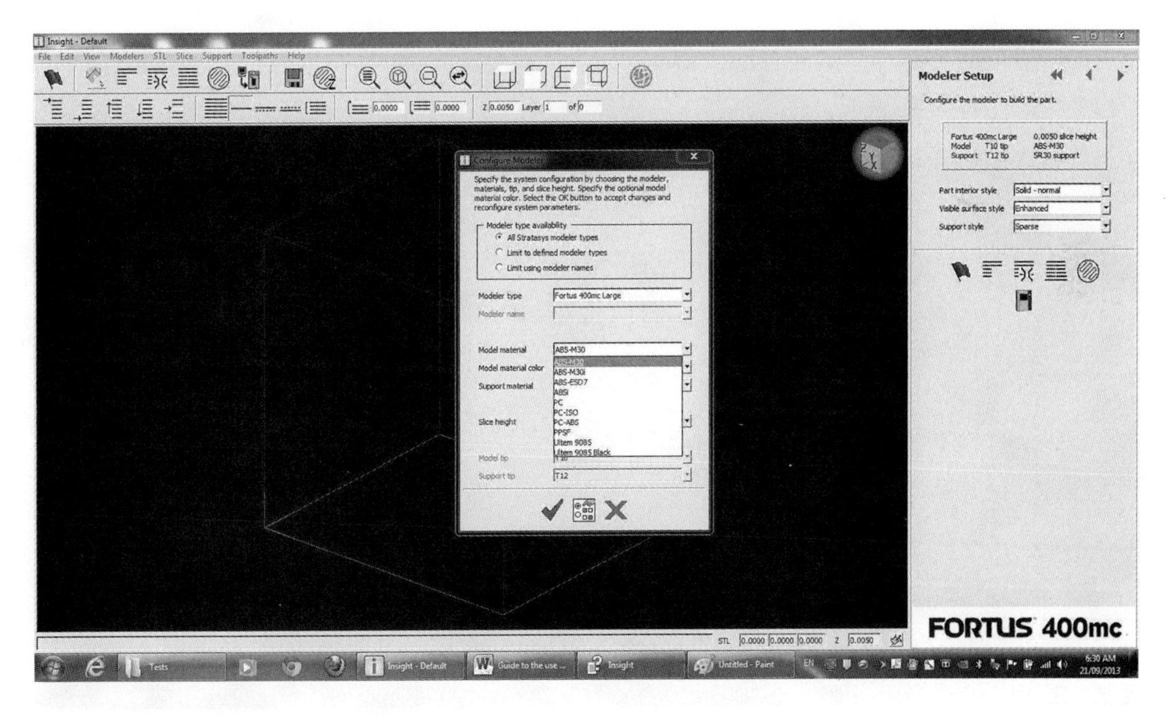

Figure 36.4

Model material.

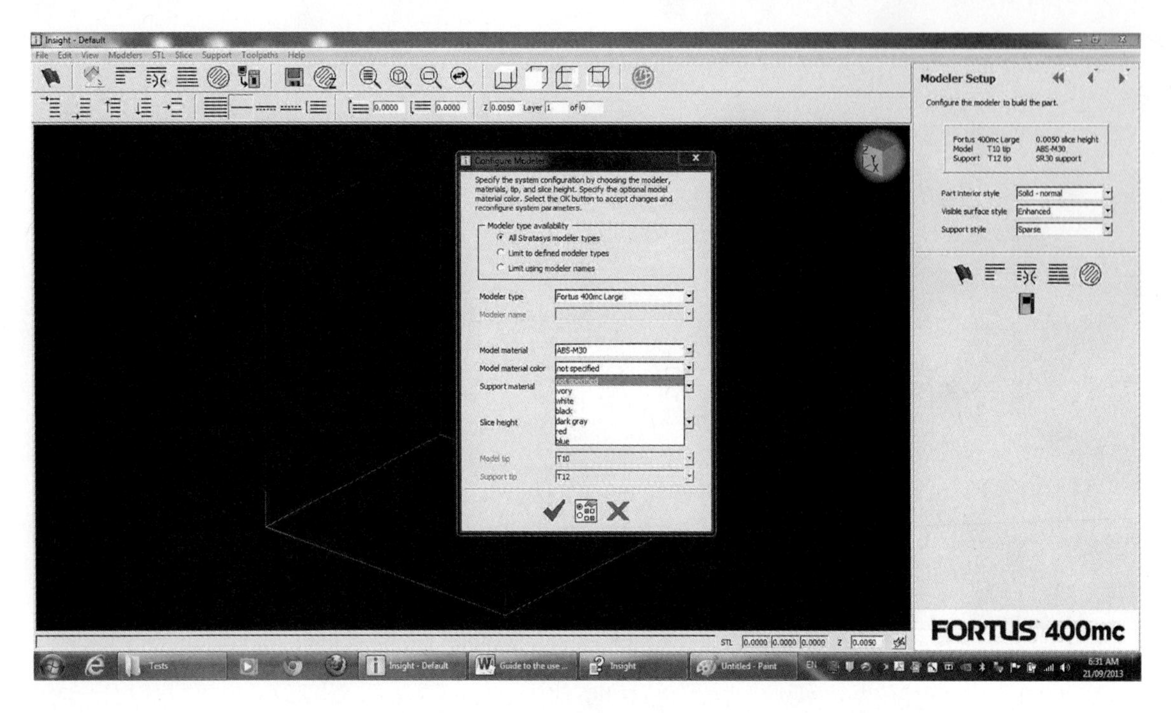

Figure 36.5

Model material color.

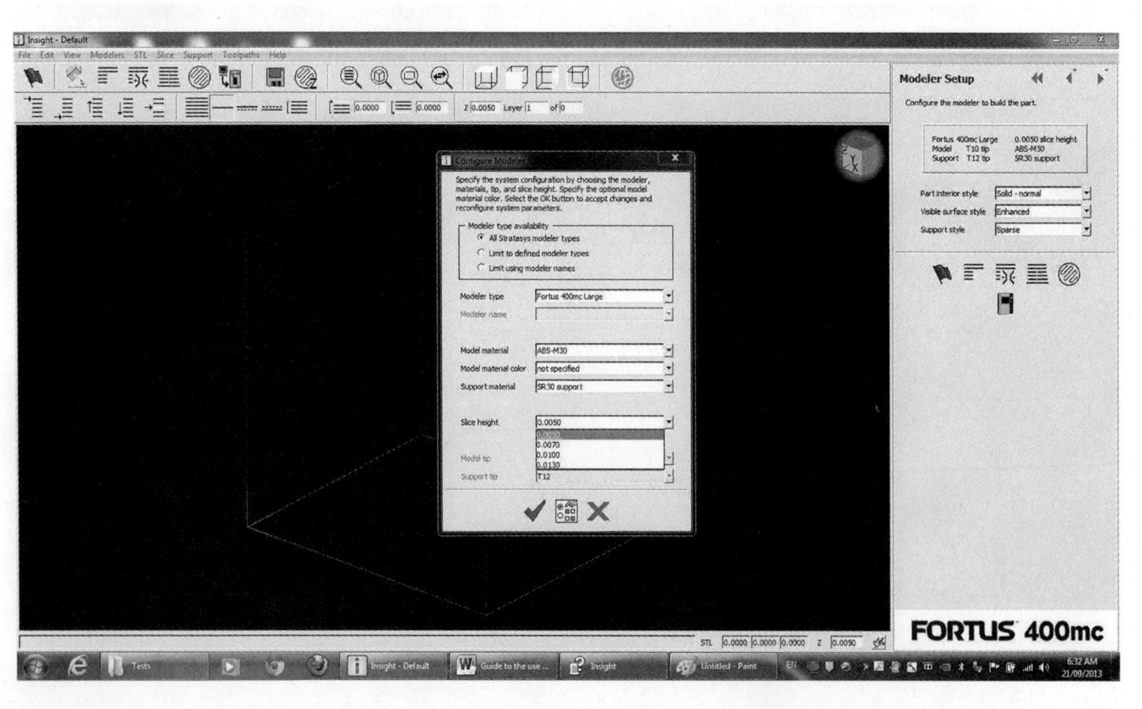

Figure 36.6

Slice height.

Step 3: Open model (see Figures 36.7 through 36.9).

Step 4: Click Orient ("?" is displayed); select Bottom (1) and *Face of model* (2) (see Figure 36.10).
The model is automatically oriented as shown in Figure 36.11.

Step 5: Click Slice the STL model (1); the model is sliced (2) (see Figure 36.12).

Step 6: Click STL > STL display > Clip STL display of current layer (see Figure 36.13).

Step 7: Click Toolpaths > Setup (1) (see Figure 36.14).
Click Access Advanced path generation parameters toolbar (2).

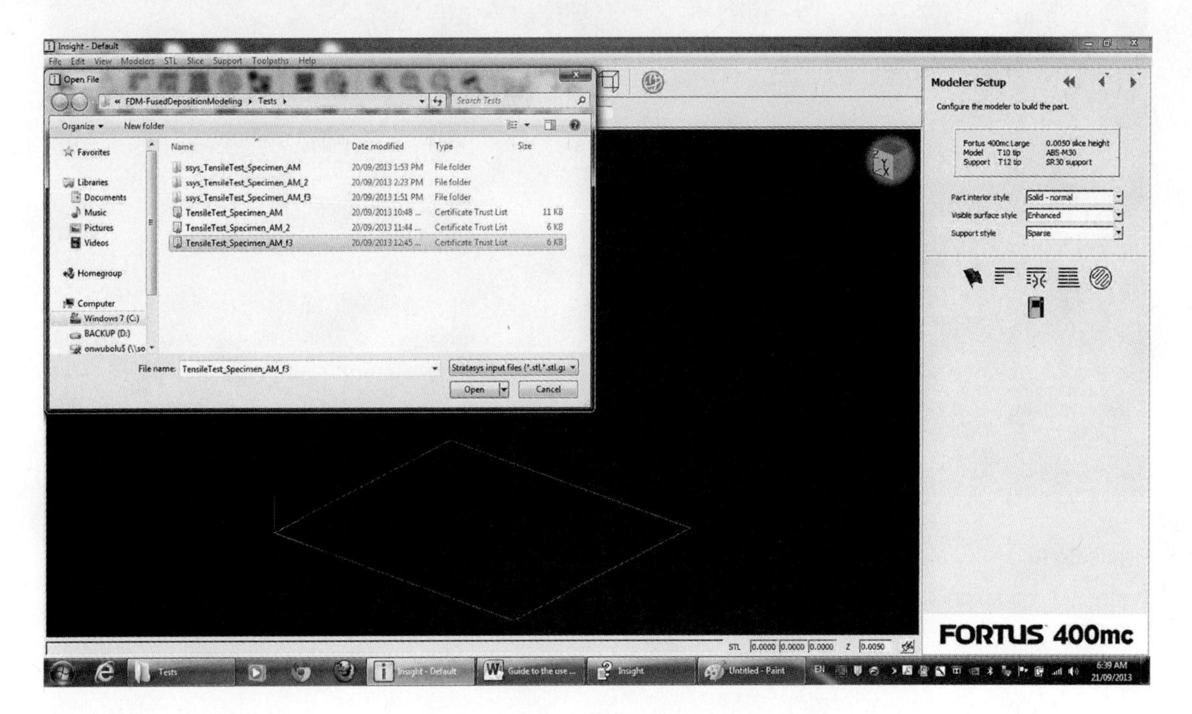

Figure 36.7

Open file.

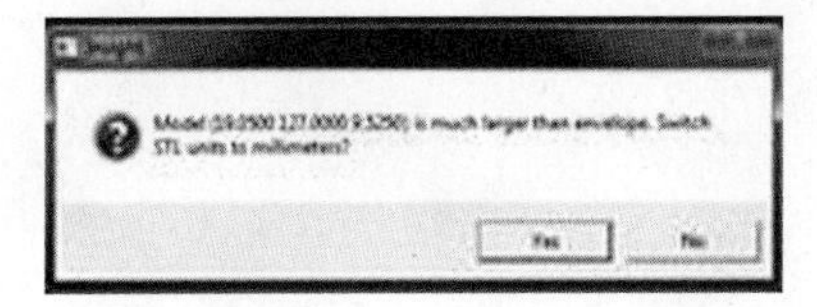

Figure 36.8

Accept "Yes" when message appears on the screen.

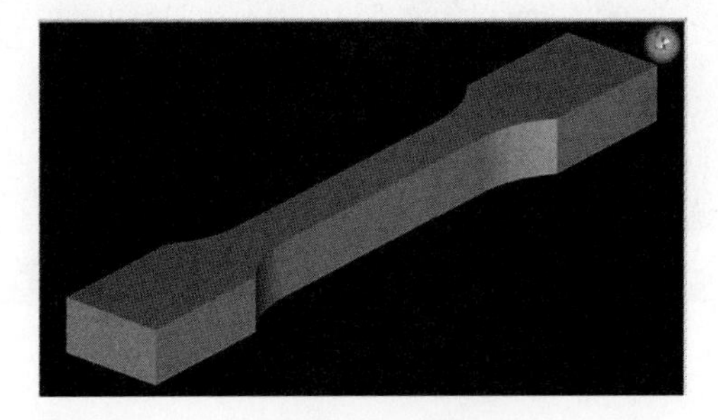

Figure 36.9

STL model uploaded from file.

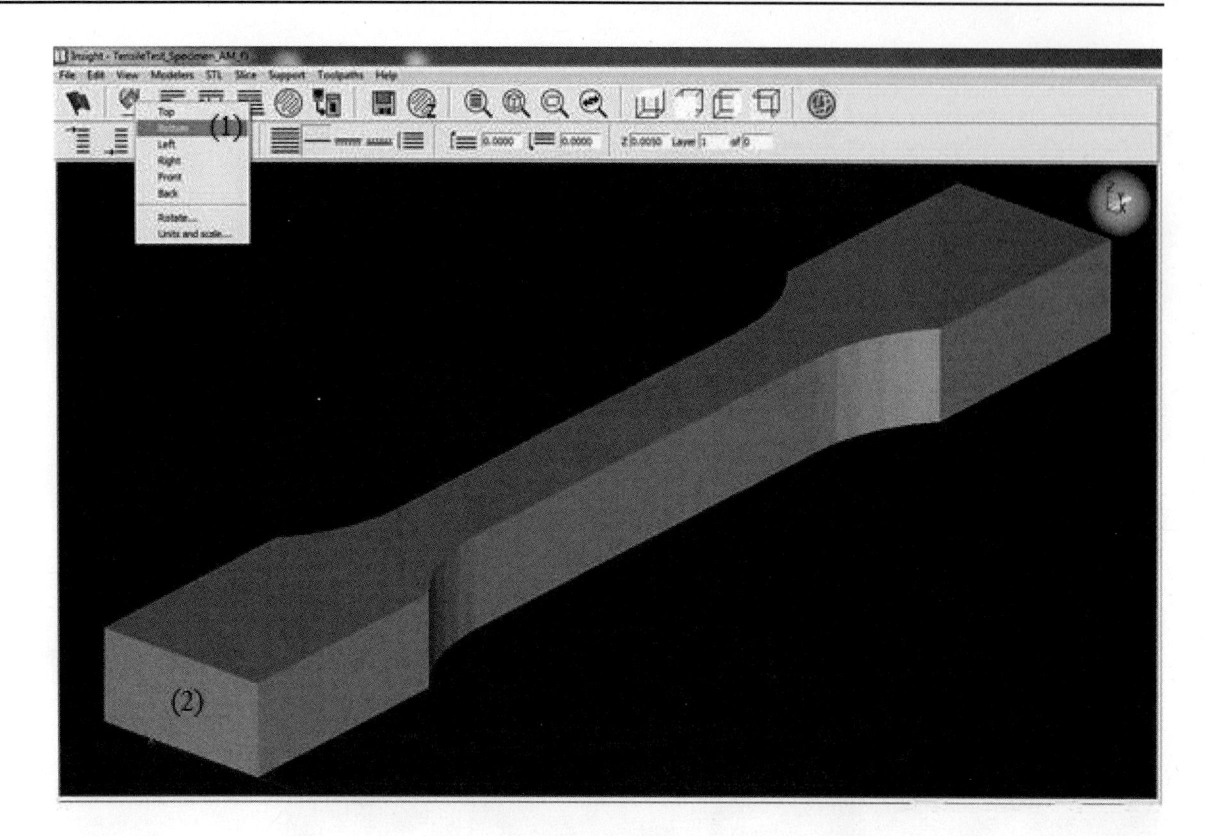

Figure 36.10

Bottom of model chosen.

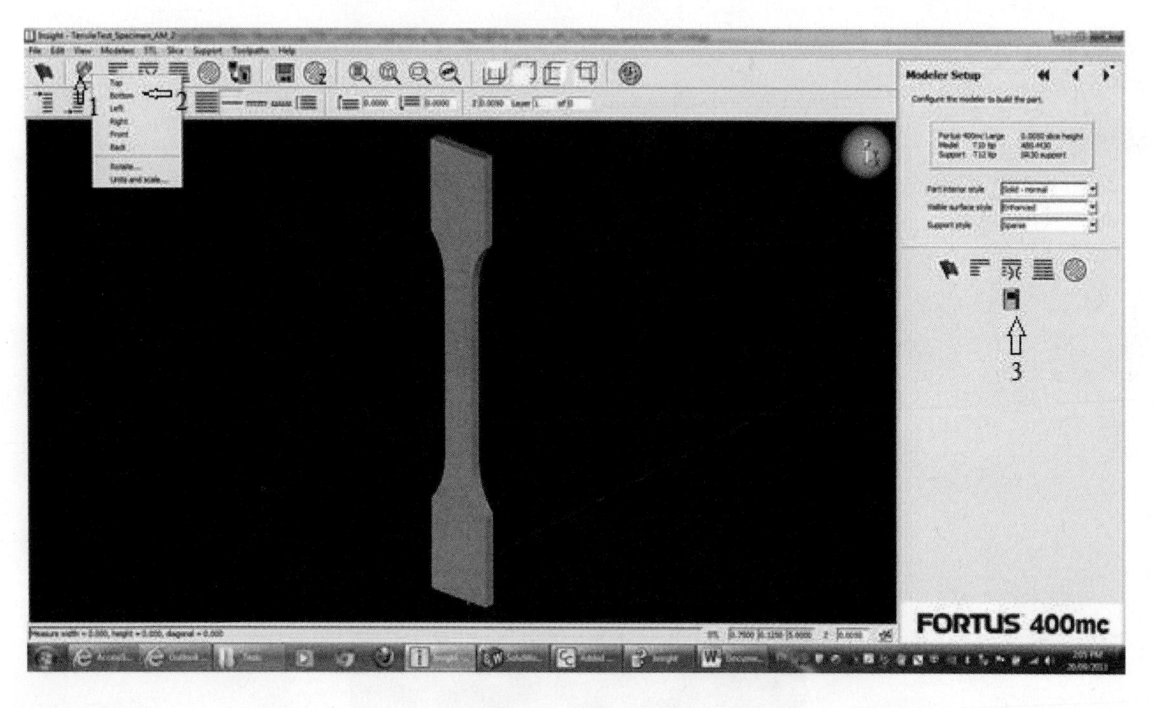

Figure 36.11

Model orientation is changed.

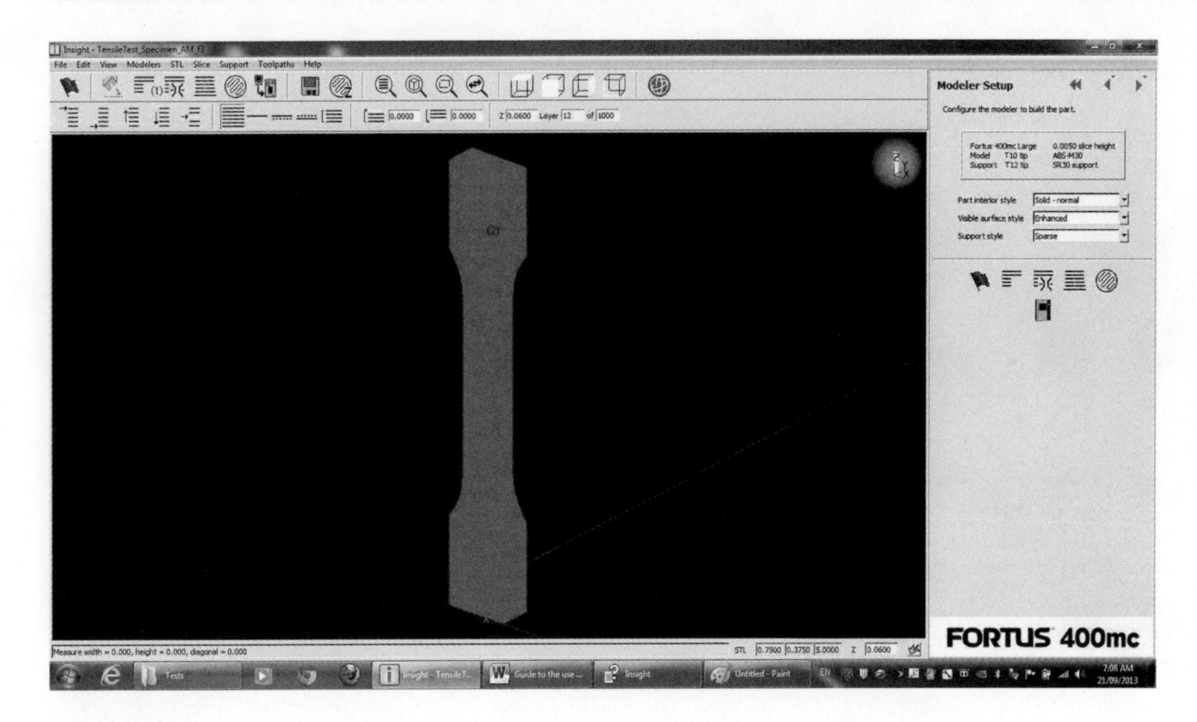

Figure 36.12

Sliced model.

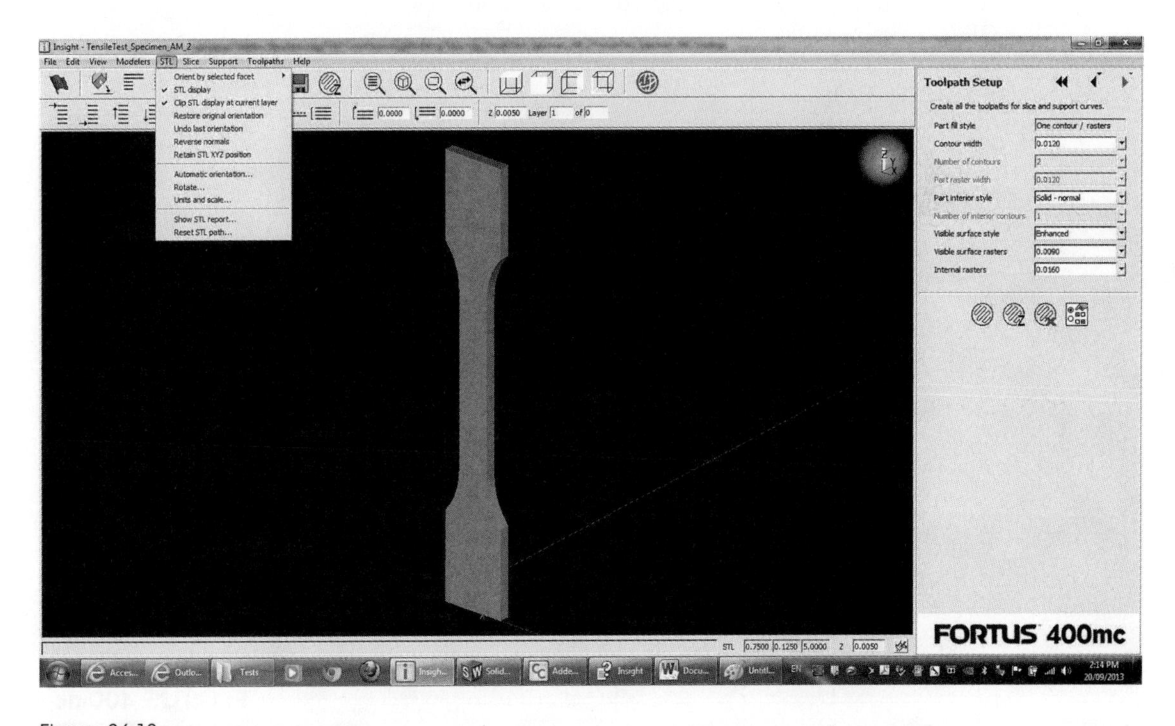

Figure 36.13

Display STL and clipped current layer.

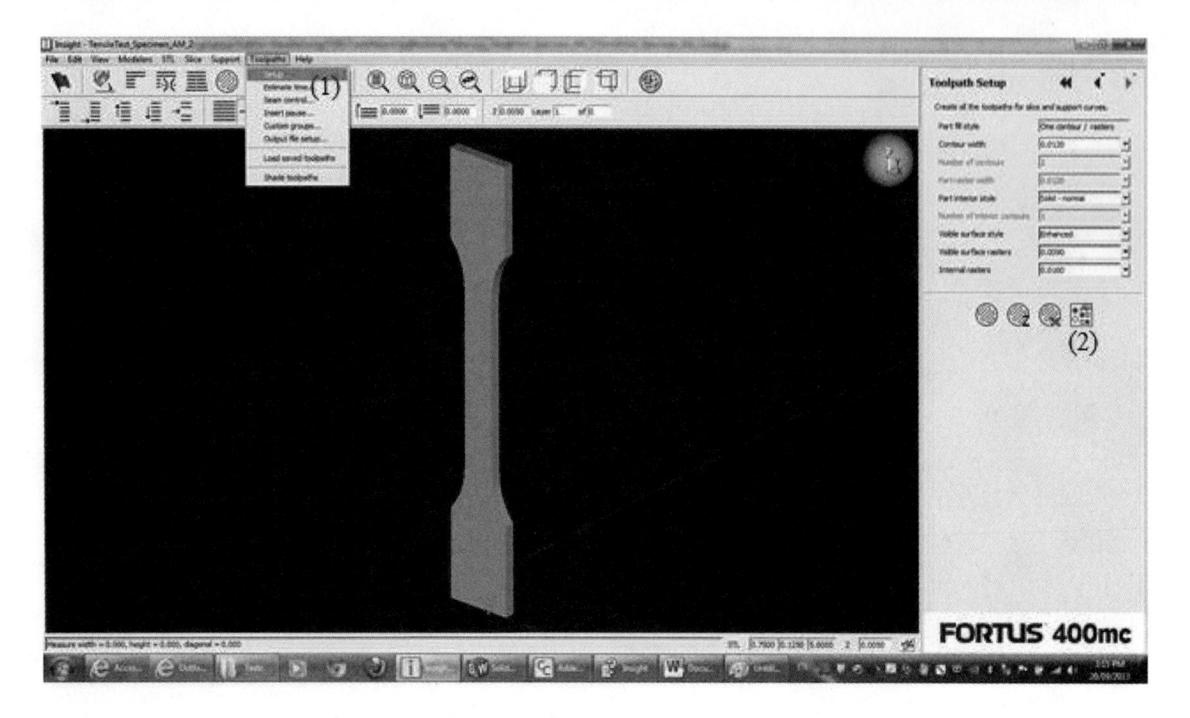

Figure 36.14

Toolpath setup.

Set the parameters for Raster Angle, Raster Width (Internal Rasters), and Air Gap (see Figure 36.15).

- **a.** Raster angle = 0.0000 (see Figure 36.16).
- **b.** Internal rasters (raster width) = 0.0080 (see Figure 36.17).
- **c.** Raster to raster air gap (Air gap) = −0.0010 (see Figure 36.18).
- **d.** Click OK (see Figure 36.18).

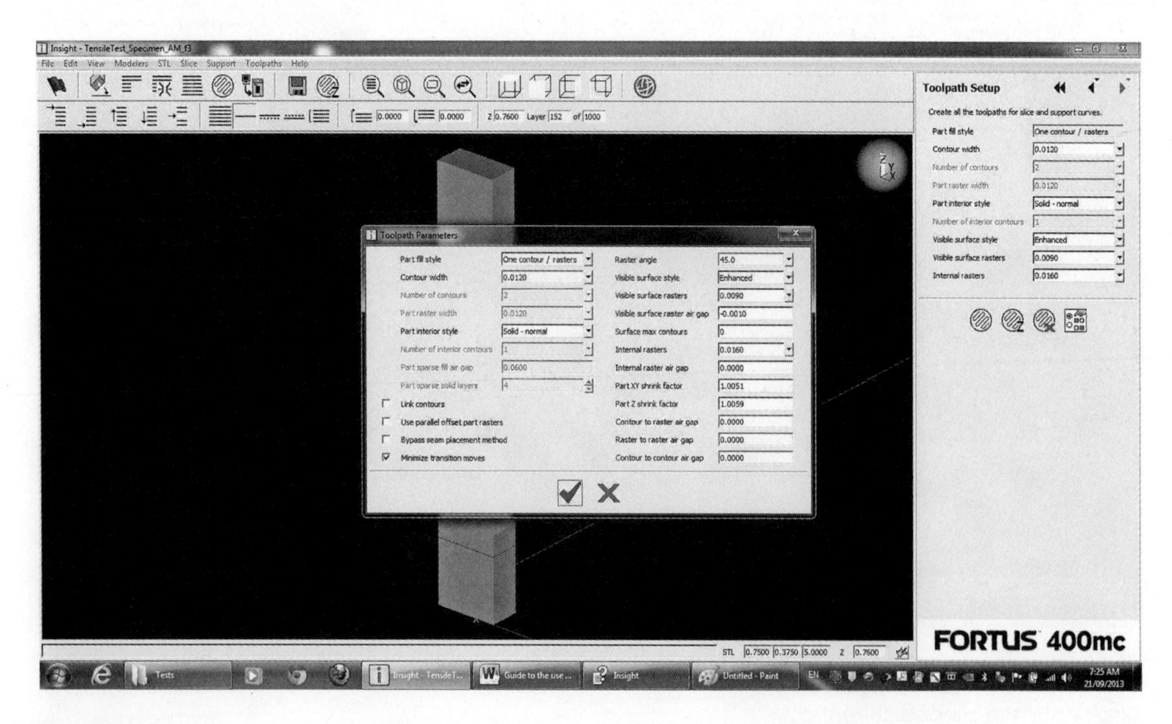

Figure 36.15

Toolpath parameters.

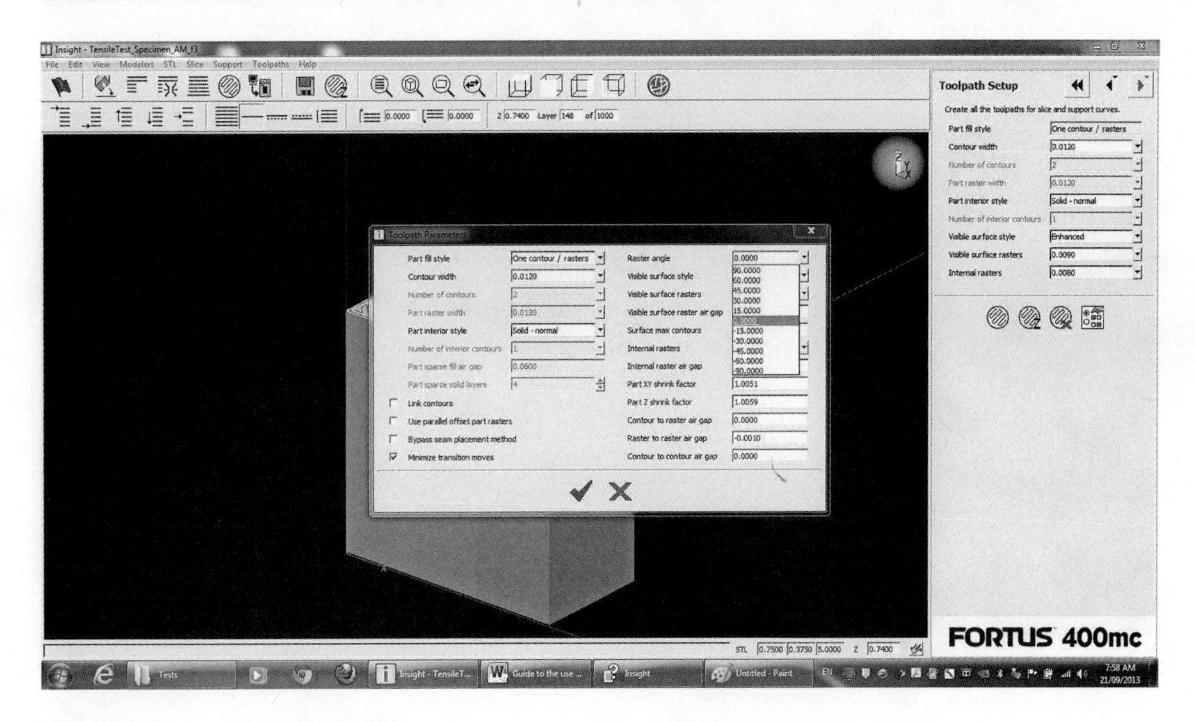

Figure 36.16

Raster angle.

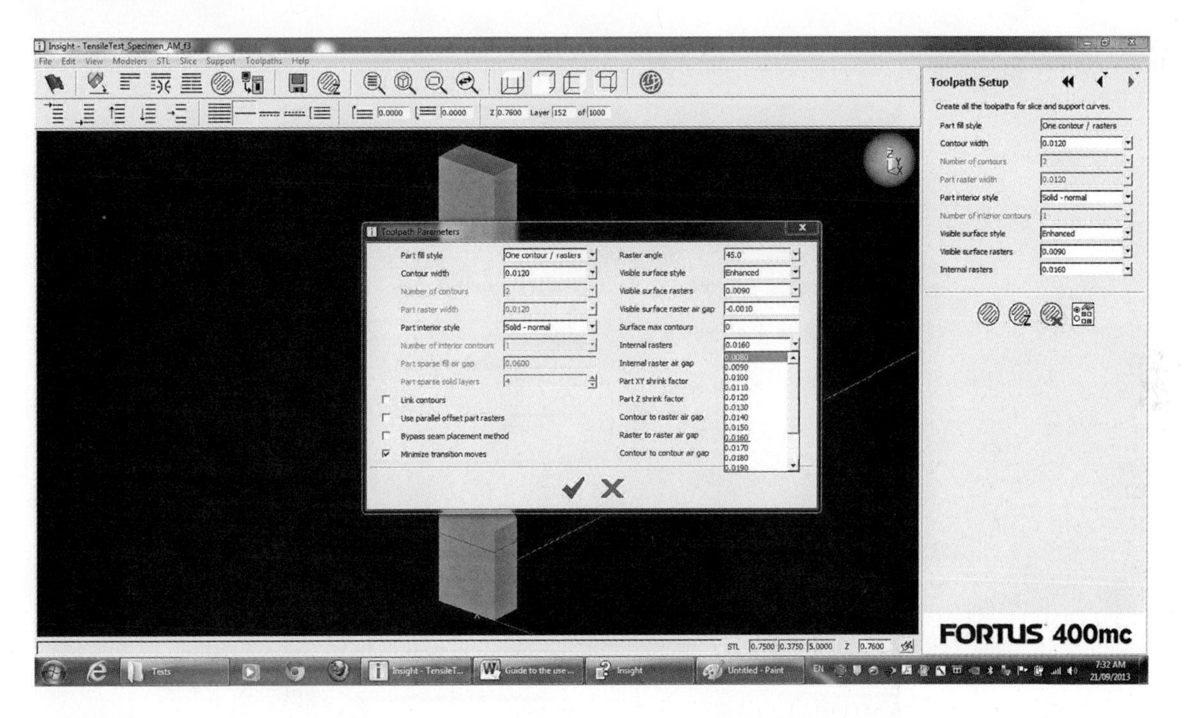

Figure 36.17

Internal rasters (raster width).

Step 8: Click Create toolpaths for all boundaries of layers (1) (see Figure 36.19).

Click Create toolpaths for all boundaries of current layers (2) (see Figure 36.19).

Step 9: Right-click on Graphics Window and select Shade toolpaths (see Figure 36.20).

Use the Page Up (PgUp) and Page Down (PgDn) keys on the keyboard to show toolpaths for different layers (see Figure 36.21).

To print the STL file that is opened, the software must be processed into a CMB file. The CMB file contains specific information that is used by the printer to create a part.

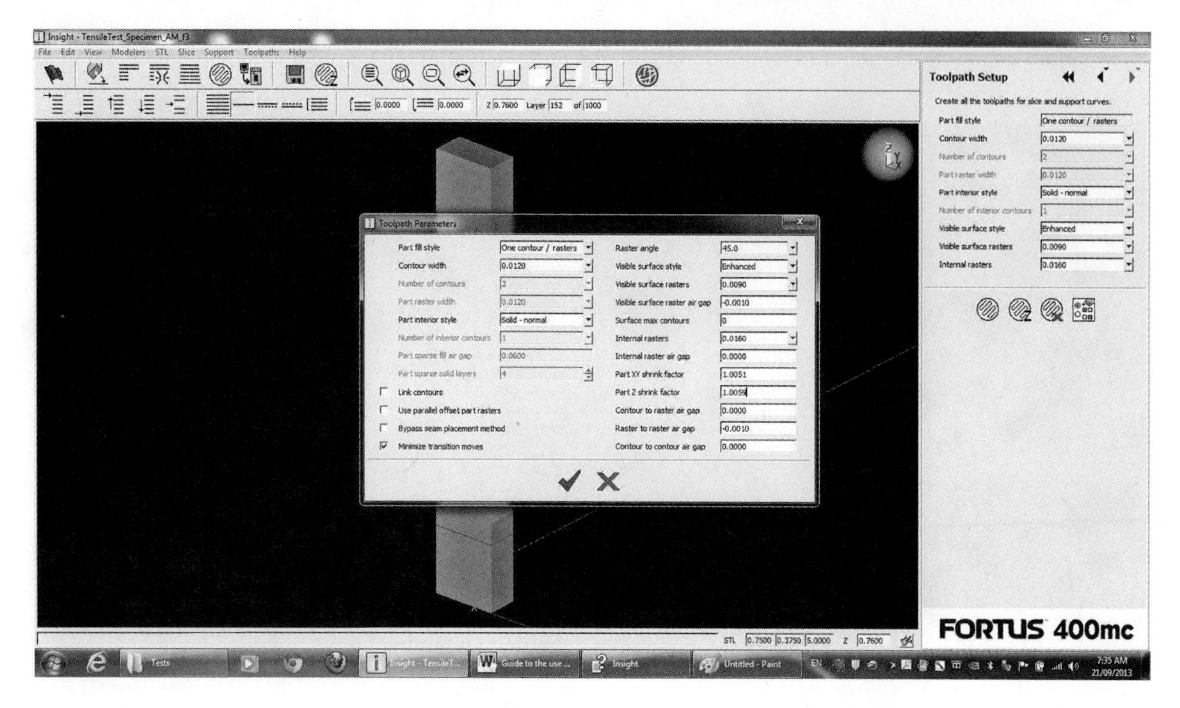

Figure 36.18

Raster to raster air gap (Air gap).

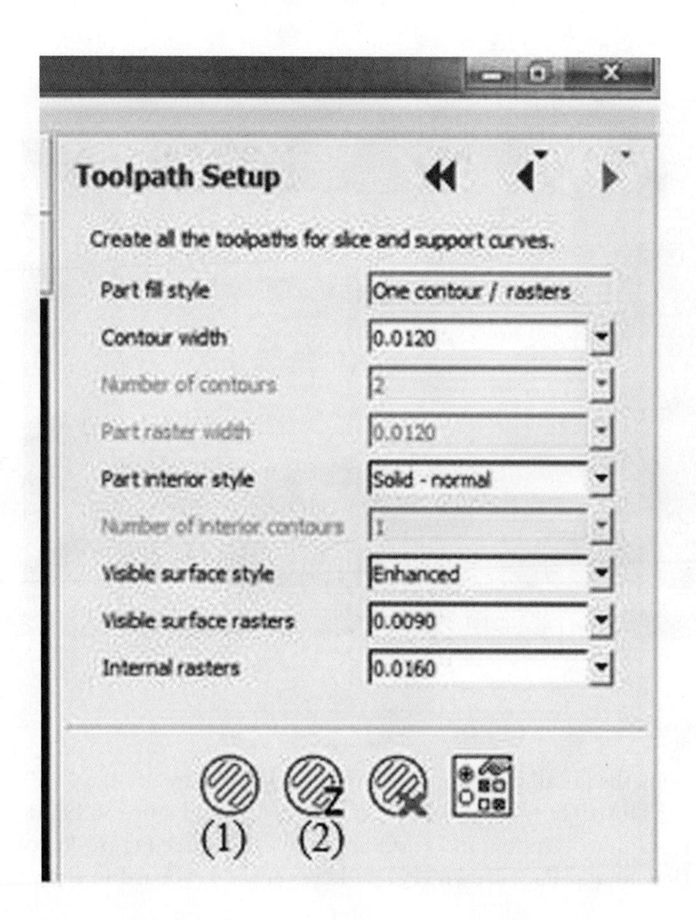

Figure 36.19

Toolpaths for all boundaries on layers/current layers.

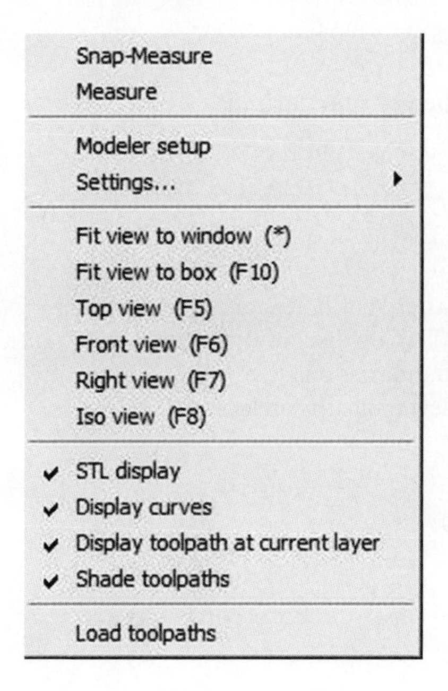

Figure 36.20

Shade toolpaths.

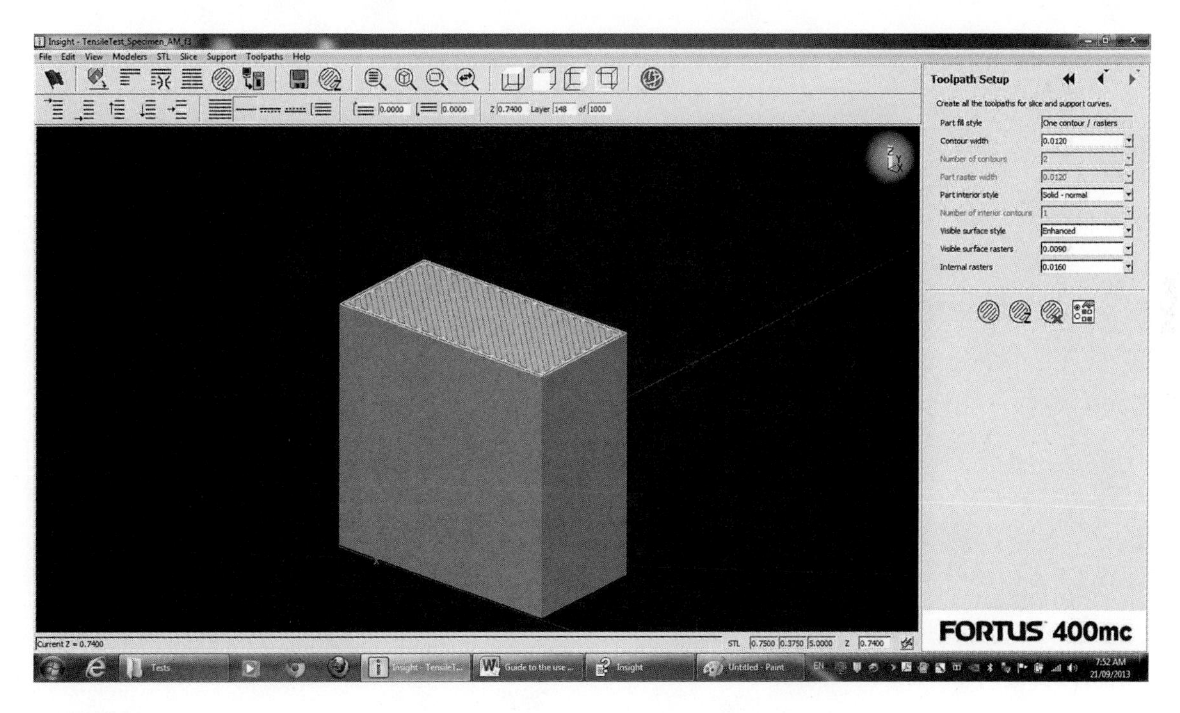

Figure 36.21

Tool paths displayed.

Summary

The steps to create tool paths are hereby summarized:

1. Open CAD model.
2. Orient model.
3. Slice model.
4. Support model.
5. Click Toolpaths/Setup: Raster Width, Raster Angle, Air Gap.
6. Click STL/STL Display: STL/Clip display at current layer; Right Click/Shade toolpath ENTER.
7. Create toolpaths for all boundaries on all layers.
8. Page Up/Page Down to view toolpaths on layers.
9. Finish.
10. Build model.

37

CatalystEX Software User Guide for Dimension sst 1200es

Objectives:

When you complete this module, you will have

- Understood how Catalyst software works
- Used Catalyst software to prepare tool paths for additive manufacturing

CatalystEX is an intuitive, user-friendly application that is designed to interface with Dimension 3D printers. It allows you to quickly and easily open an .stl file that represents a 3D *part*, process the file, and print the part.

Printing Type

Each printer models with ABS plastic—the modeled parts are strong and durable. ABS also ensures that you will be able to drill, tap, sand, and paint your creations.

- *P400 Model Material*—Standard ABS material.
- *P430 Model Material*—Strength-enhanced ABS material.

Some printers differ in the support material that is used.

- *Water-soluble support technology.* This type of support material is removed through a washing process—making it ideal for models that are delicate in design or contain small orifices or chambers. The uPrint, SST 1200es and Dimension Elite use water-soluble support technology.
- *Breakaway support technology.* This type of support material is removed by mechanical means. The BST 1200es uses breakaway support technology.

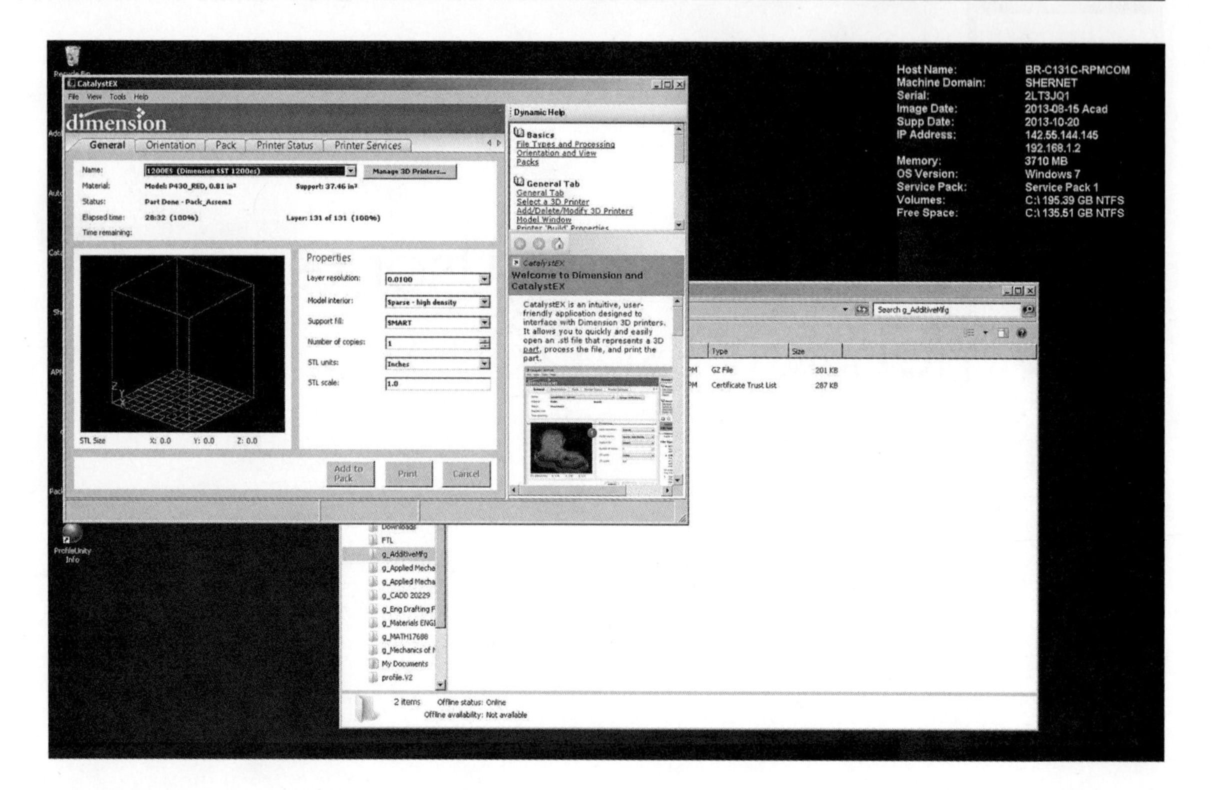

Figure 37.1

Catalyst Default User Interface.

File Types

- *Standard Tessellation Language (STL) file*—A 3D drawing of your part that is created by a CAD software program.
- *CMB file*—This file is created from an STL file through a process that is run by the CatalystEX software. It is the CMB file that is sent to a printer.

In order for you to print a part, you must do one of the following:

1. Open an STL file, process the STL file into a CMB file, and send the CMB file to a printer.
   When the "Print" button is used, CatalystEX will automatically do the following:
   a. Process the STL file into a CMB file.
   b. Save the CMB file.
   c. Print the file.
   or
2. Add an existing CMB file to a *pack* and send the pack to a printer.

Step 1: Open Catalyst software (see Figure 37.1),

Two tabs are useful in defining the general properties, defining the toolpaths, and printing a model: these are (1) *general* and (2) *orientation* tabs.

General Tab

- *Layer Resolution*—The part is sliced parallel to the modeling platform. Each layer, or slice, is equal in height to the *resolution* that is chosen from the properties section of the General Tab. Each slice represents a cross section of the part at a specific level of the model. Available resolutions are based on the printer type and may include the following options: (a) .010 in. (.254 mm) and (b) .013 in. (.330 mm).
- *Model Interior*—Available Model Interior: Solid, Sparse-high density, Sparse-low density.

- *Support Fill*—After the part has been sliced, CatalystEX calculates where to best support the model as it is being built—based upon the choice you made in the properties section of the General Tab. Available Support fill: Basic, Surround, SMART, Sparse.
- *STL Unit*—Available STL units: Inches, Millimeters.
- *STL*—The value of STL scale is set by user.

Step 2: Opening file

Click File > Open STL (Figure 37.2).

Navigate to open the file—in this case, Vacuum Forming.stl (Figure 37.3).

Step 3: Orientation

Orientation refers to how the part is positioned within the modeling envelope—which surface is in contact with the base of the envelope; is the part laying *flat* or at some kind of an angle within the envelope? Changing orientation reorients the part within the modeling envelope. *Orientation will affect printing.*

Click Orientation tab (Figure 37.4).

Select Bottom radio in the Orient Selected Surface option (a flag appears for the user to select bottom surface) (Figure 37.4).

Click surface that is bottom from model. (See Figure 37.5 for the orientation of the model.)

Step 4: Click Process STL (1) (see Figure 37.6).

- *Toolpaths*—The next task in the processing of an STL file is the creation of toolpaths. Using the generated slice and support information—along with the model interior choice that is made in the properties section of the General Tab—CatalystEX precisely determines the paths that are necessary for the material tips to traverse in order for Dimension to accurately create your part. The width of a path is fixed.

Step 5: Click Add to Pack (2) (see Figure 37.6).

- The final task in processing an STL file is the creation of the CMB file. This is done automatically by the software. The CMB file, by default, is saved in the same directory as the STL file—the default name will be *stlfilename*.cmb.gz. (The default CMB file save location can be changed from the *Menu Bar*). If a file of the same name already exists, you will be given the option to

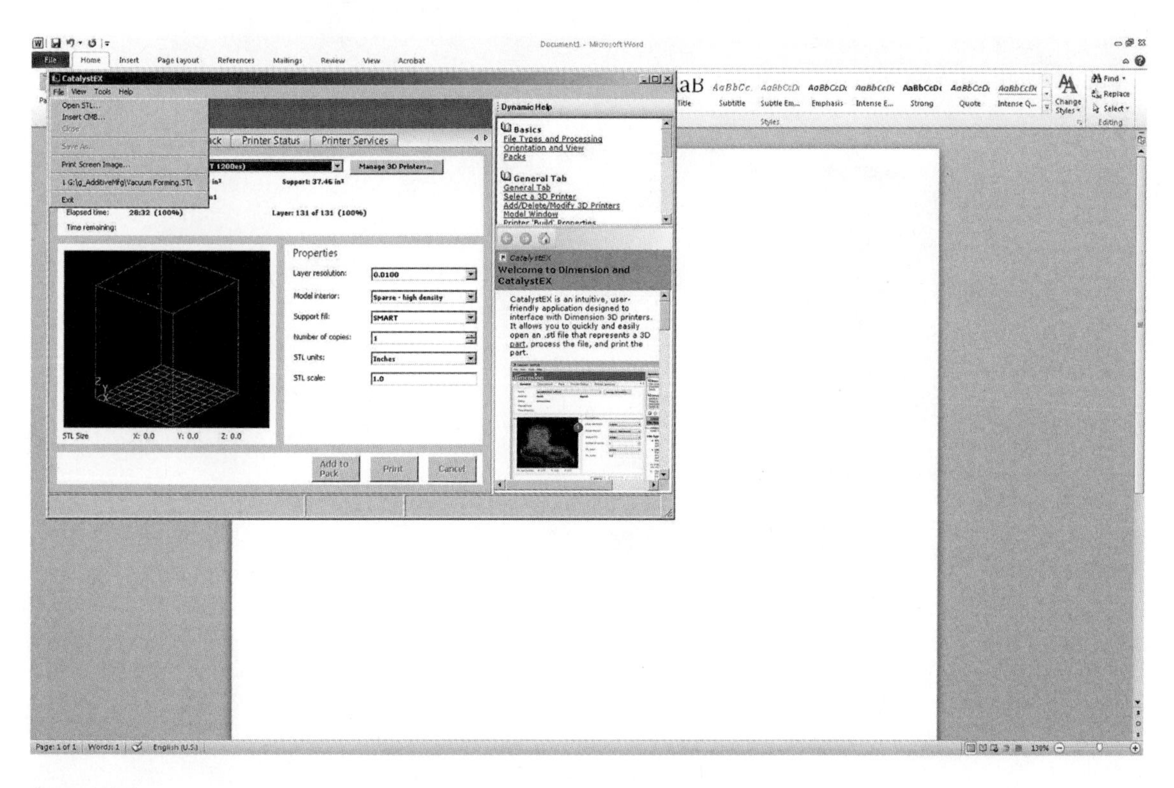

Figure 37.2

Opening an existing STL file.

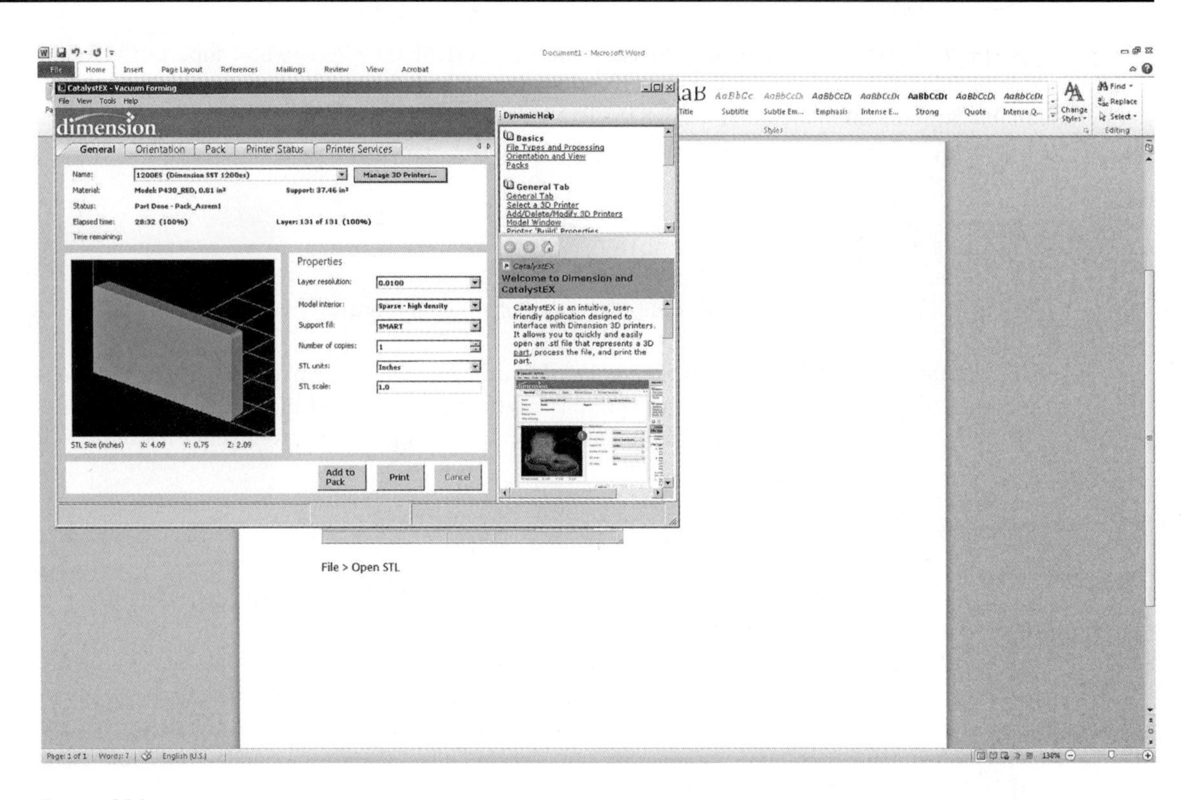

Figure 37.3

Vacuum Forming.stl file is accessed for opening.

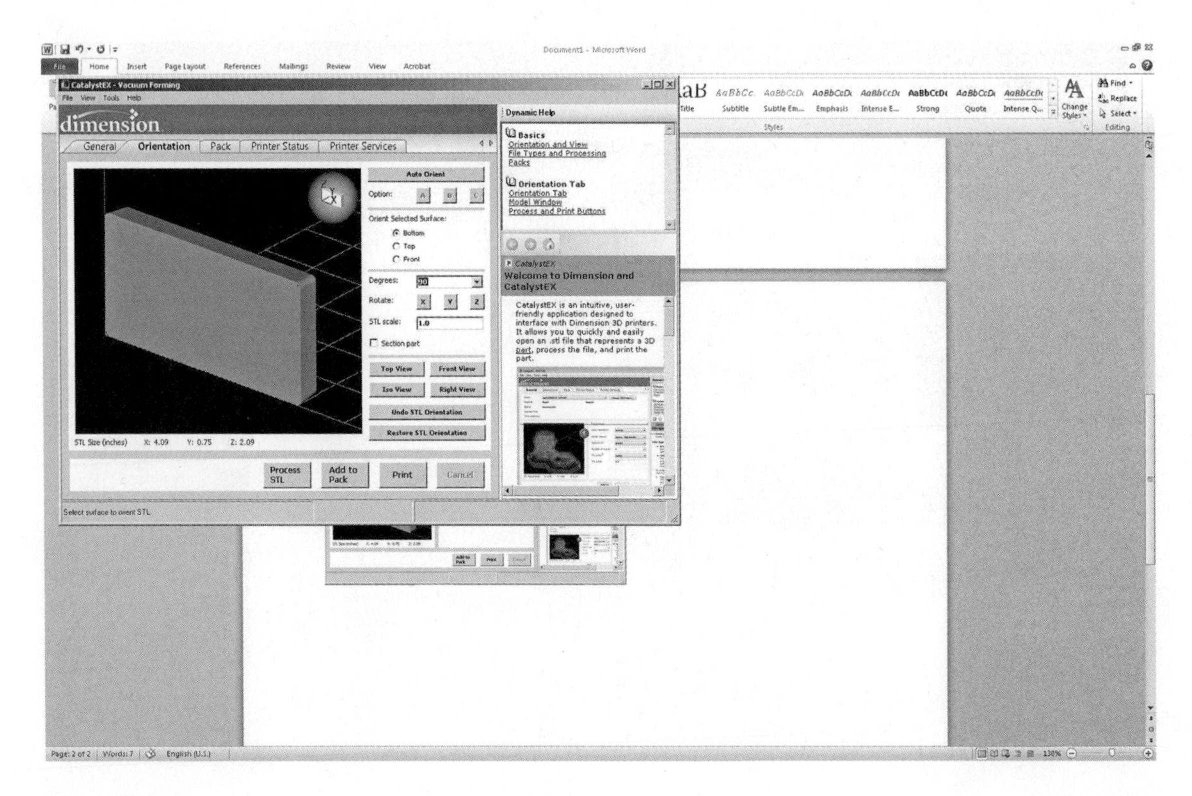

Figure 37.4

Orientation of bottom of the model.

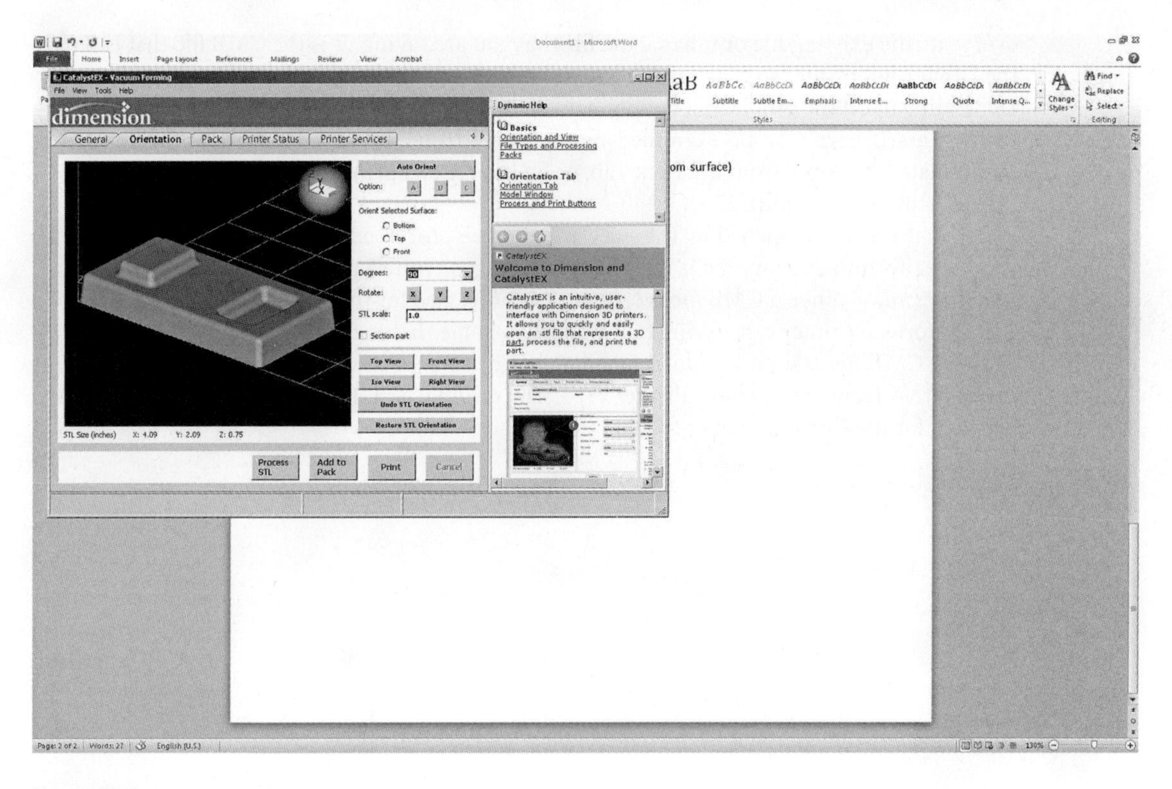

Figure 37.5

Orientation of the model.

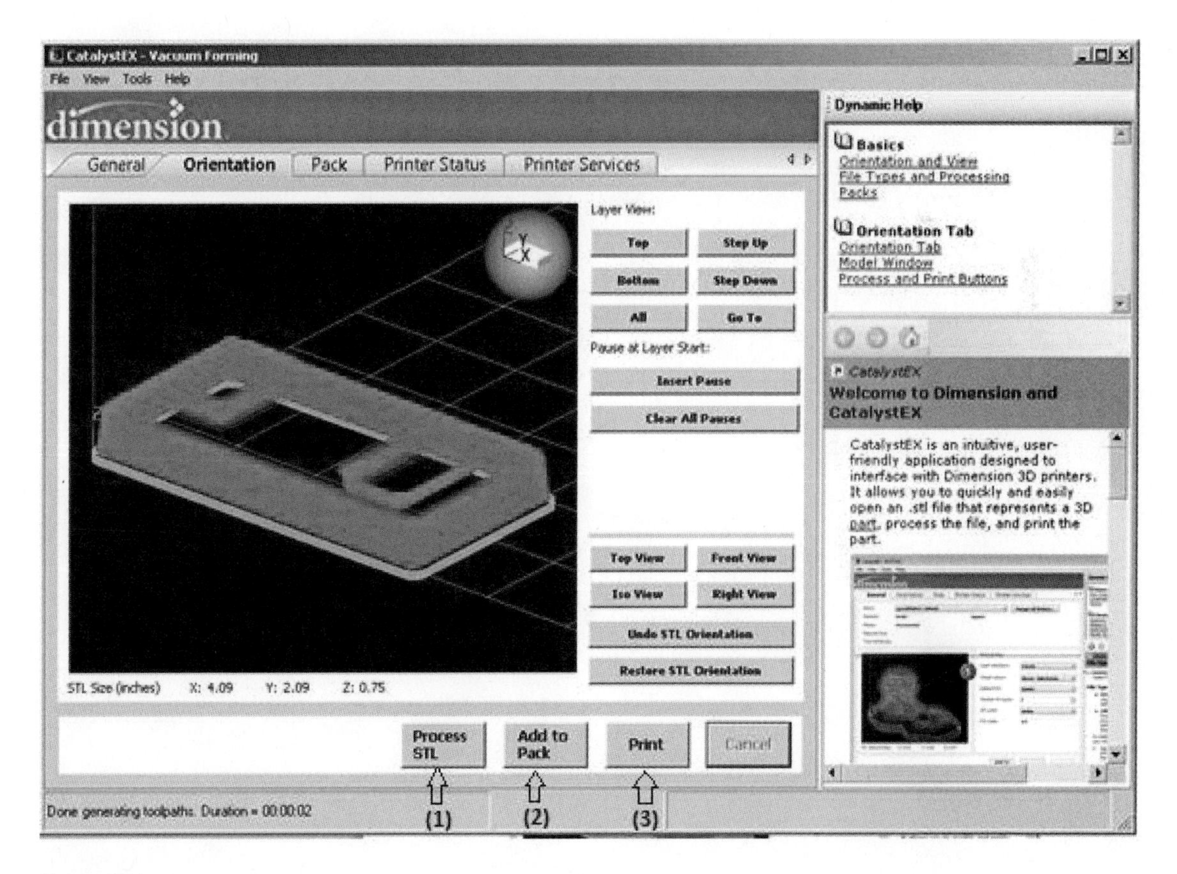

Figure 37.6

Process STL, Add to Pack, and Print.

overwrite the existing file or rename the file that you are saving. It is the CMB file that is sent to the Dimension printer so that a model can be created.

- Depending upon the processing option that you chose (e.g., Process STL, Add to Pack, or Print), the processed part may be examined more closely from the Orientation Tab, positioned or manipulated in a pack from the Pack Tab, or on its way to a printer for build.

Step 6: Click Print (3) (see Figure 37.6).

To Print, the STL file opened by the software must be processed into a CMB file. The CMB file contains specific information that is used by the printer to create a part.

File Processing—When a CMB file is created, it is processed for the Current Printer type. A CMB file created for one printer type is different from a CMB file that is created for another printer type. An existing CMB file cannot be added to a pack (or printed) if it was processed for a printer type that is different from the Current Printer type. The original STL file will need to be opened and reprocessed for the Current Printer type.

38

Bead Design with Orientation Considerations

Objectives:

When you complete this module, you will

- Be able to prepare the toolpaths for model simple parts such as a bead (doughnut)
- Understand the cost implication of orientation consideration
- Be able to 3D Print simple parts using the Fortus machine

In this module, we consider two orientations of a simple part, the bead (otherwise referred to in this module as a doughnut). The module takes you through the steps that are required to prepare the toolpaths and build the part on Fortus machine.

Doughnut in Upright Position (Axis of Hole in Z-Direction)

Load the STL file of doughnut part to Insight software. Orient the *doughnut* in an upright position (axis of hole in Z-direction) (see Figure 38.1).

These steps should be followed:

1. Slice the model (Figure 38.2). (You can use the icons on the *Modeler Setup* tab.)
2. Apply support (Figure 38.3).
3. Prepare the toolpaths (Figure 38.4).
4. If toolpath is to be set up, then click Toolpaths > Setup and make changes.
5. Finish.

To estimate the build time, click Toolpath > Estimate time (Figure 38.5).

Doughnut Lying on Side (Axis of Hole in Y-Direction)

Load the STL file of doughnut part to Insight software. Orient the *doughnut* on its side (axis of hole in Y-direction) (Figure 38.6).

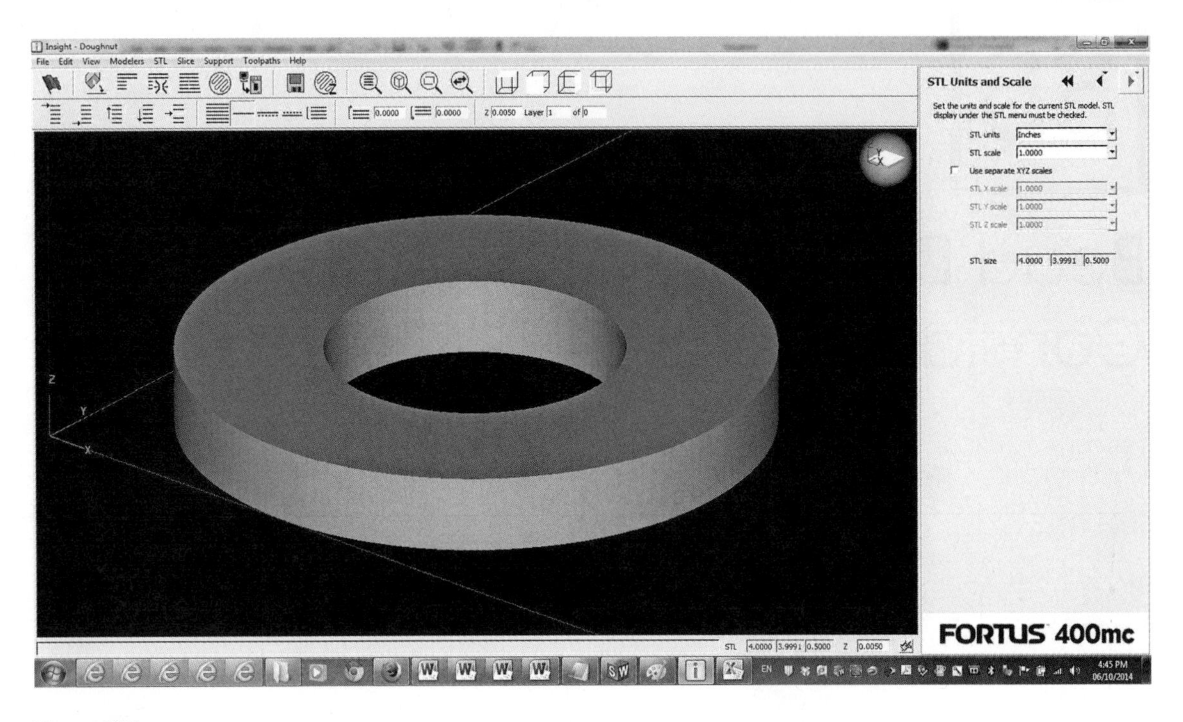

Figure 38.1

Doughnut with flat end on X–Y plane.

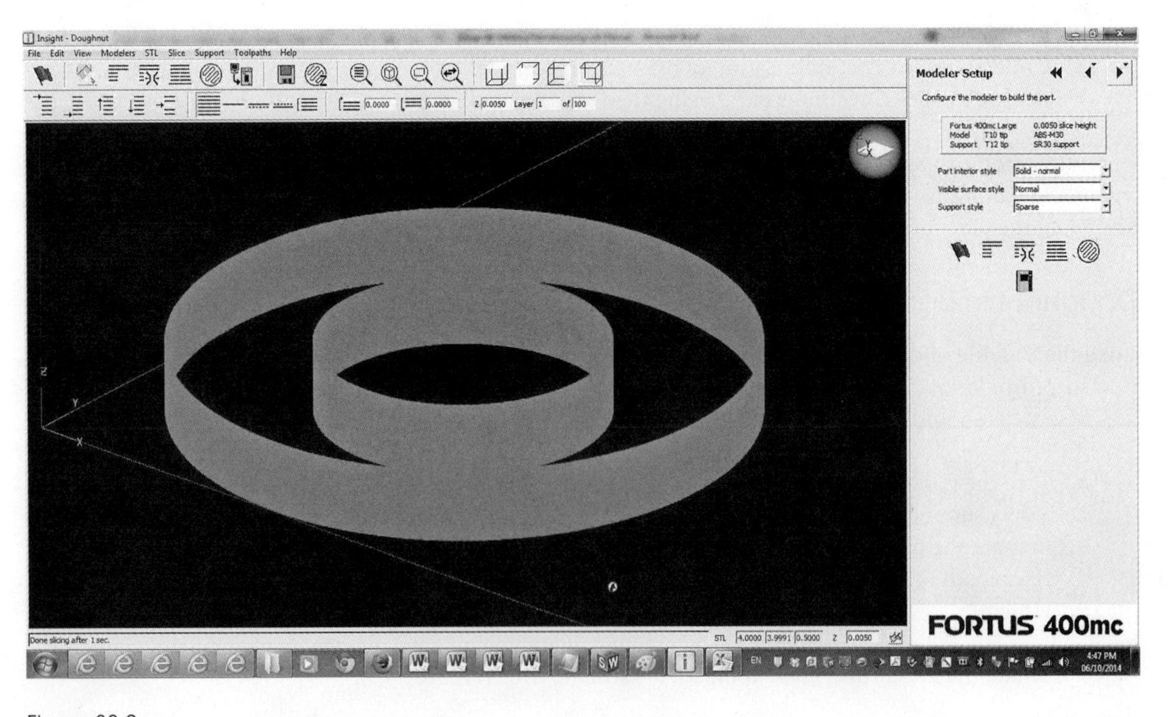

Figure 38.2

Slice the model.

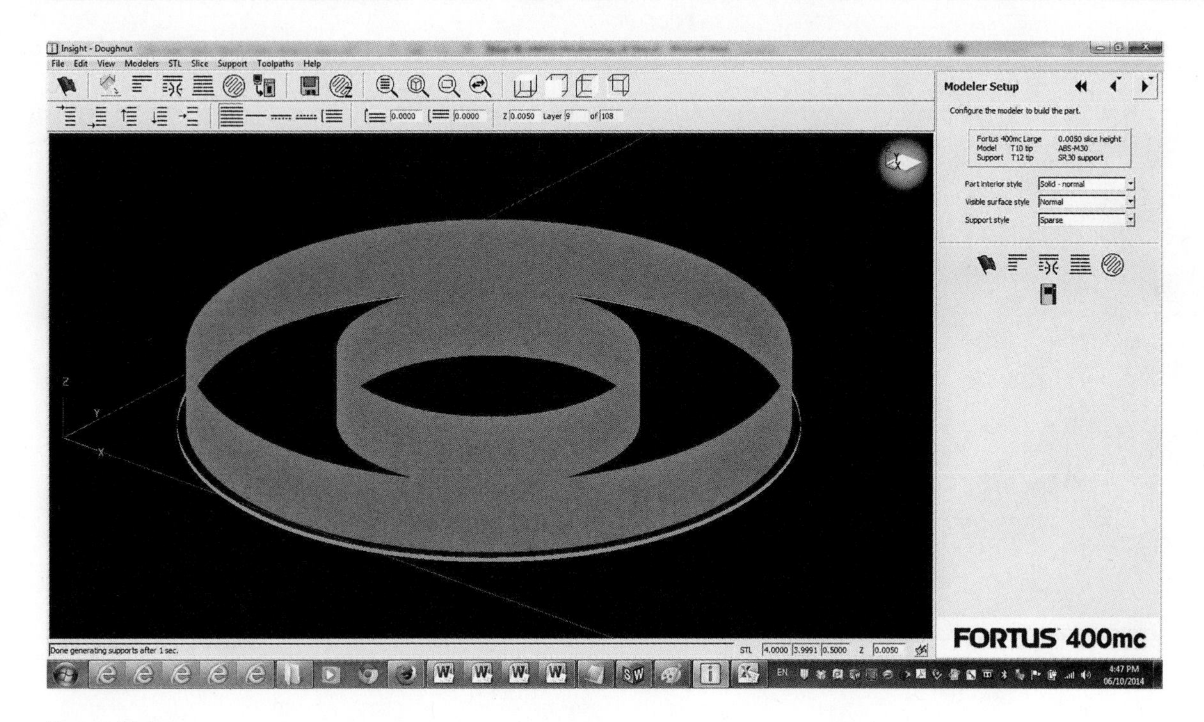

Figure 38.3

Apply support.

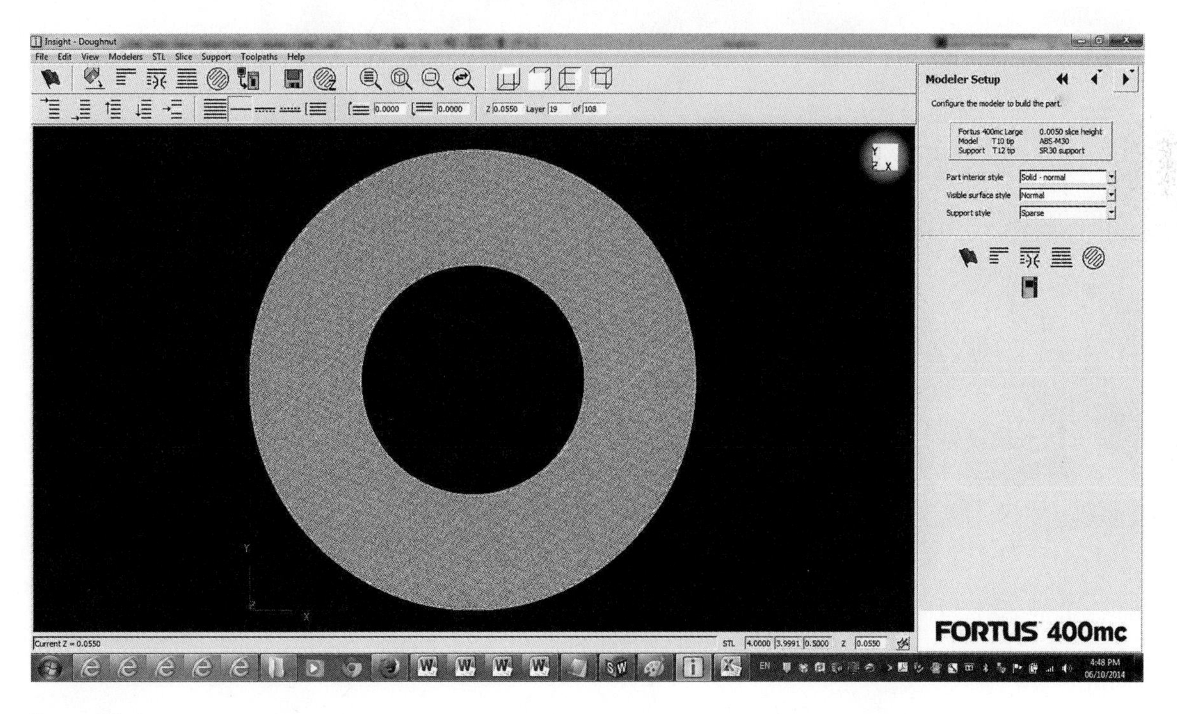

Figure 38.4

Prepare toolpaths.

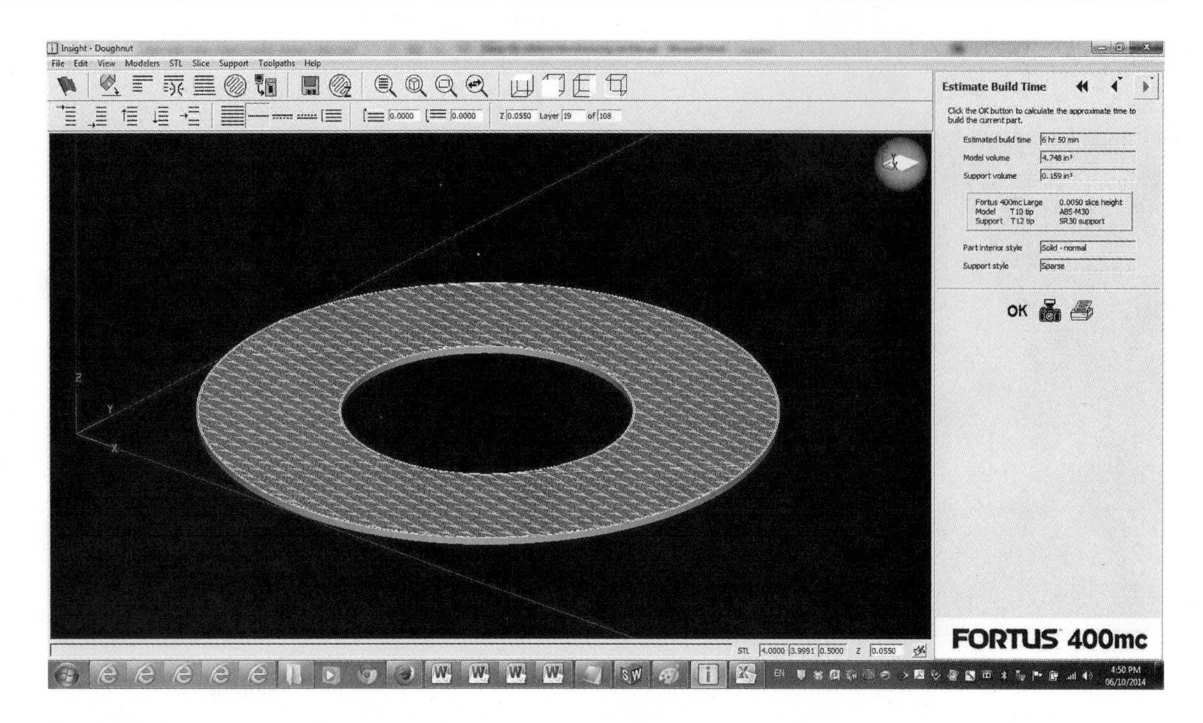

Figure 38.5

Estimate the build time.

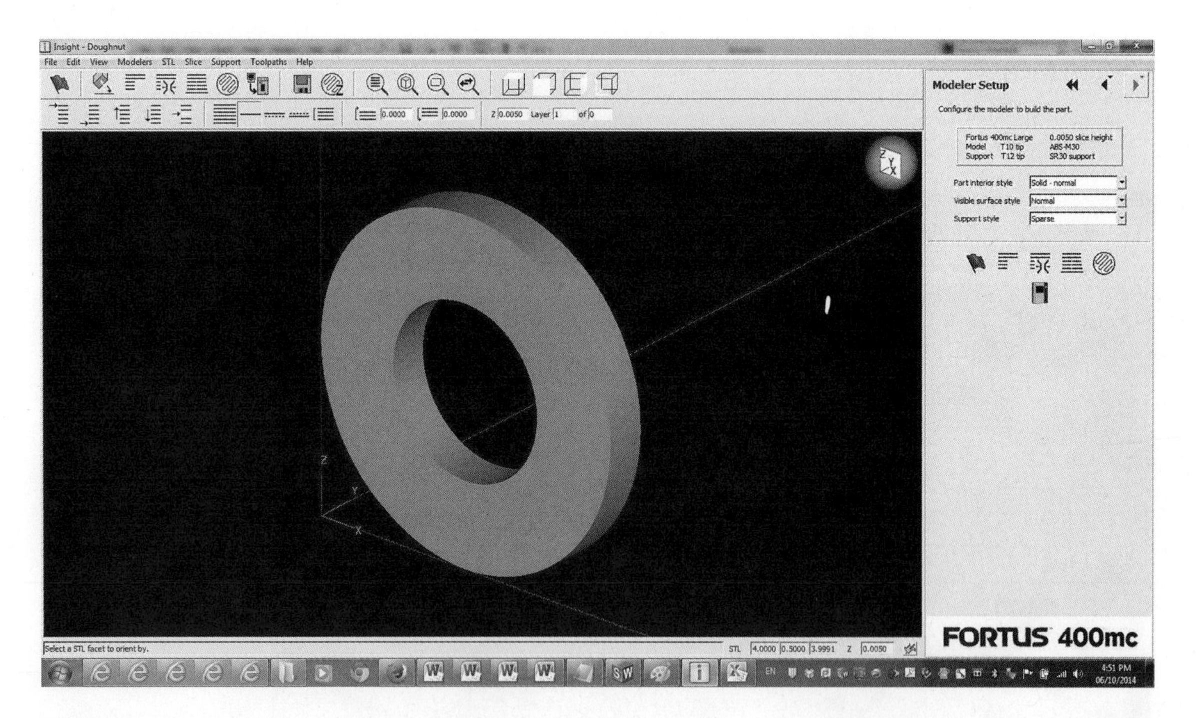

Figure 38.6

Doughnut lying on its side.

The steps to be followed are similar to the previous ones:

1. Slice the model (Figure 38.7). (You can use the icons on the *Modeler Setup* tab.)
2. Apply support (Figure 38.8).
3. Prepare the toolpaths (Figure 38.9).
4. If toolpath is to be set up, then click Toolpaths > Setup and make changes.
5. Finish (Figure 38.10).

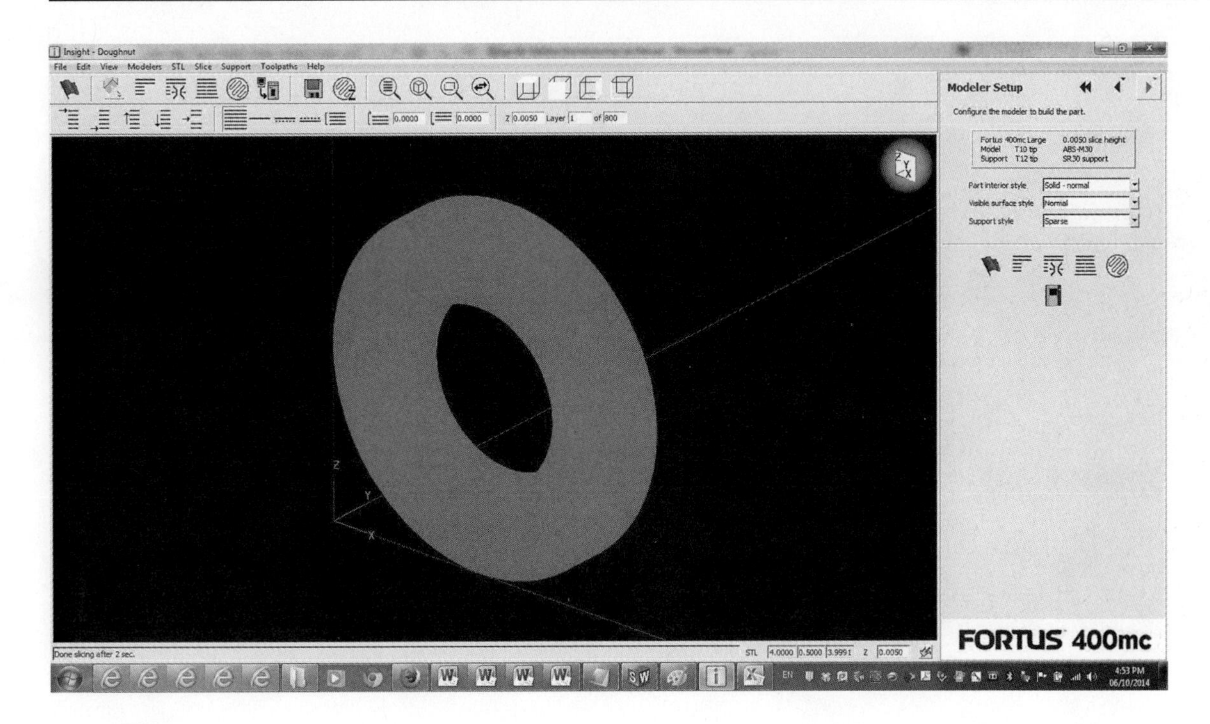

Figure 38.7

Slice the model.

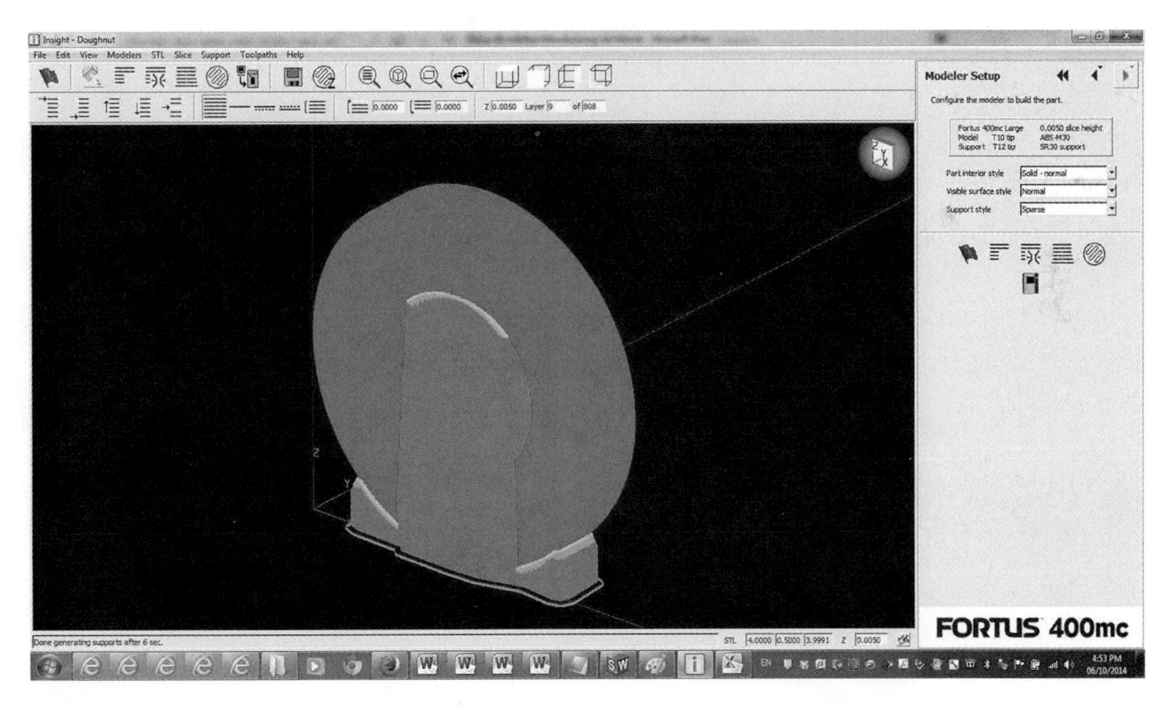

Figure 38.8

Apply support.

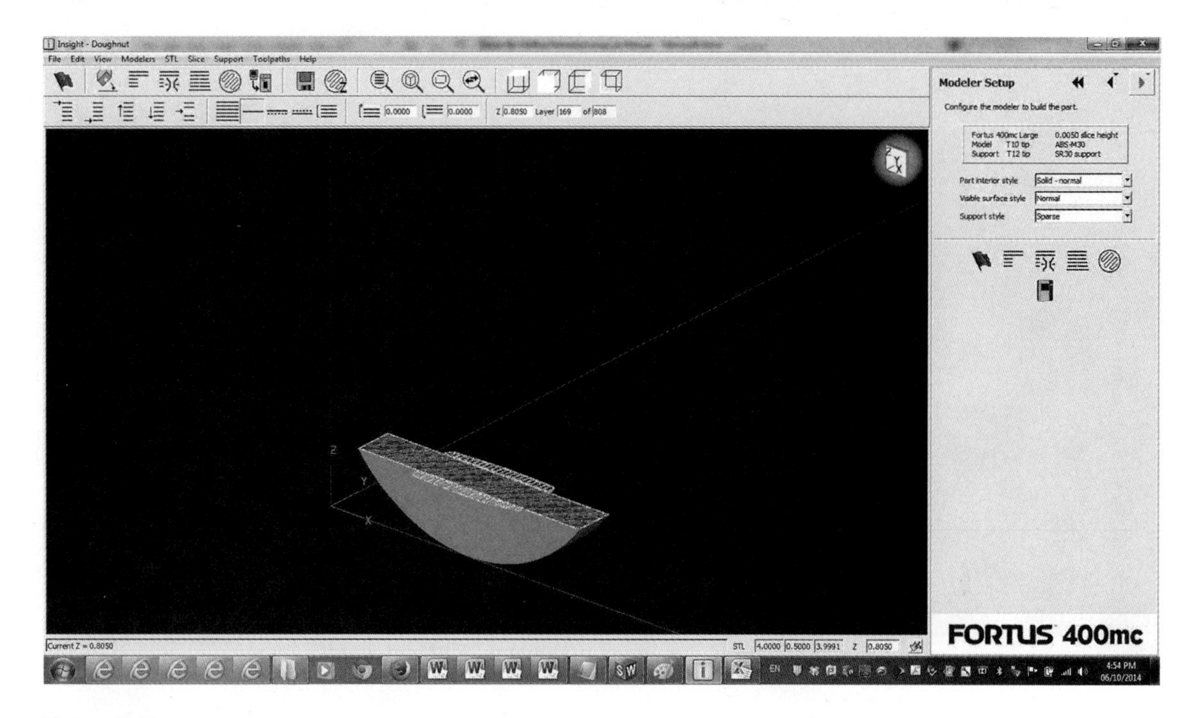

Figure 38.9

Toolpaths are generated.

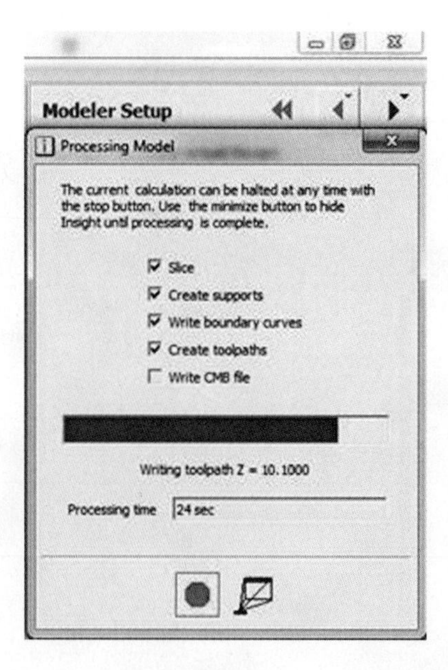

Figure 38.10

Finish operation goes through sequences and generates CMB file.

To estimate the build time, click Toolpath > Estimate time (Figure 38.11).

Table 38.1 shows the comparison between the two orientations. The first orientation is cheaper to build and more aesthetically friendly on the outer curved surface.

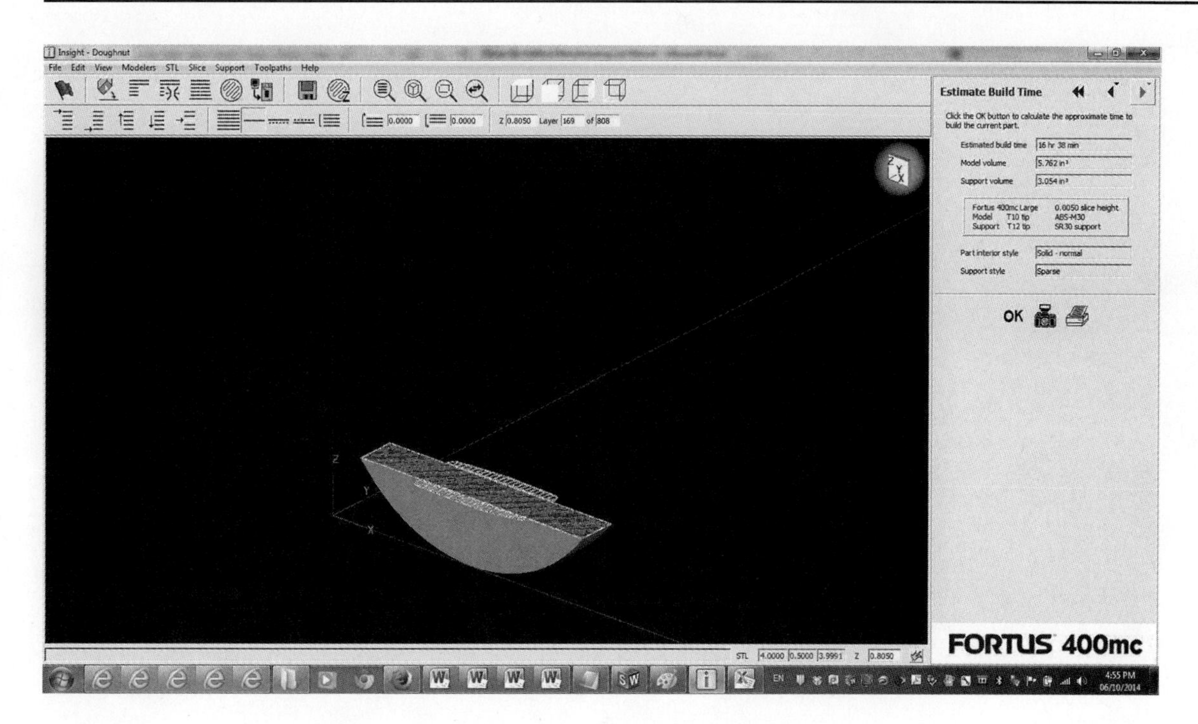

Figure 38.11

Estimate the build time.

Table 38.1 Costs for Different Orientations

Face	Axis of Hole	Build Time (Hours)	Model Volume (in³)	Support Volume (in³)	Cost ($)
X–Y	Z	6.5	4.748	0.159	75
X–Z	Y	16.38	5.762	3.054	176

39

Building Tall Parts Using Surround Support

Objectives:

When you complete this module, you will

- Learn how to prepare the toolpaths for tall parts using the stabilizing wall feature
- Understand the implication of building tall parts using the Fortus machine
- Be able to 3D Print tall parts using the Fortus machine

This module demonstrates how to preprocess *tall parts*, as well as *thin parts*, using the *stabilizing wall* feature.

Open the STL file (tallStructurel) (Figure 39.1).

Orient the model upright by clicking the bottom, which is circular as the bottom of the model (see Figure 39.2) for the model in the upright position.

In the Modeler Setup, change the Support Style to be "SMART" (see Figure 39.3).

Slice model (see Figure 39.4).

Modeler Setup > Slice model.

Modeler setup > Support style > Stabilize wall (see Figure 39.5).

Notice the Stabilizing Wall property manager on the right side of Figure 39.6.

View the single layer at the Z level of the desired top of the stabilizing wall (using Top, or Side View, whichever gives a preferred view of the model to show the curves that are needed for defining the stabilizing wall).

- Select curve at the location to begin stabilization, and confirm with the "+" icon (see Figure 39.7).
- Select the end location (on the same level) and create the stabilizer with the green check (see Figure 39.8).

Notice that the support layout is created, as shown in Figure 39.9.

Now, create the support by carrying out the following operations:

Modeler Setup > Create support for current job (see Figure 39.10).

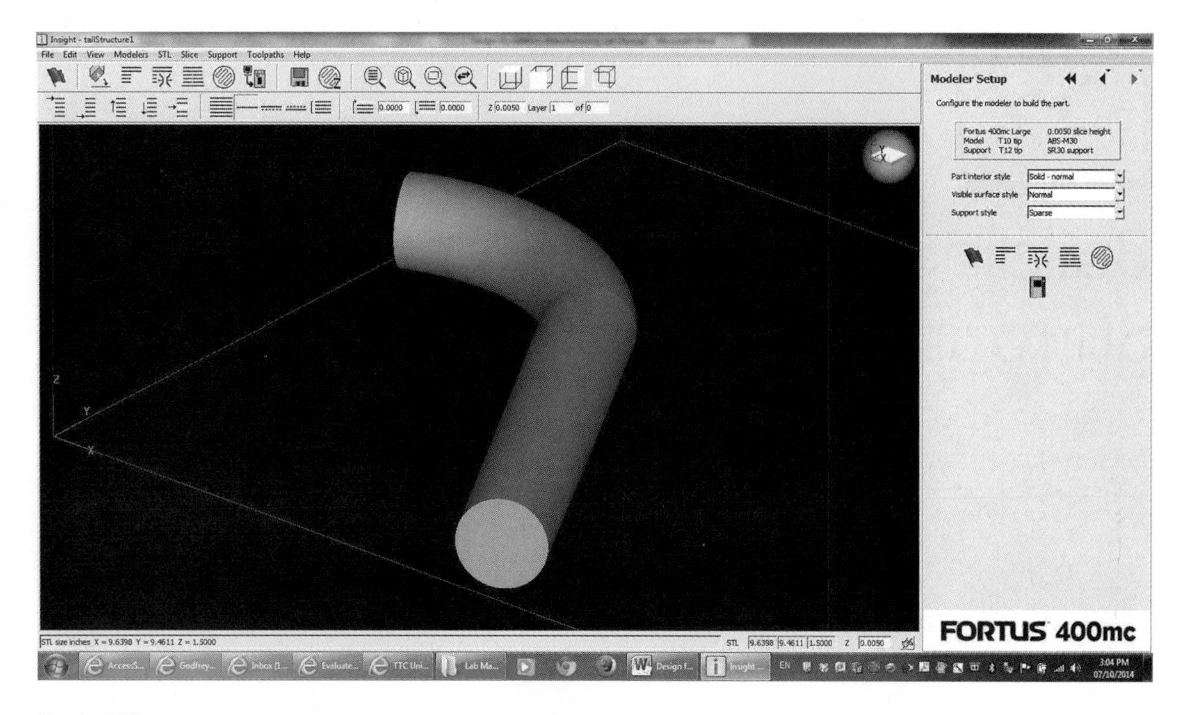

Figure 39.1

STL file opened.

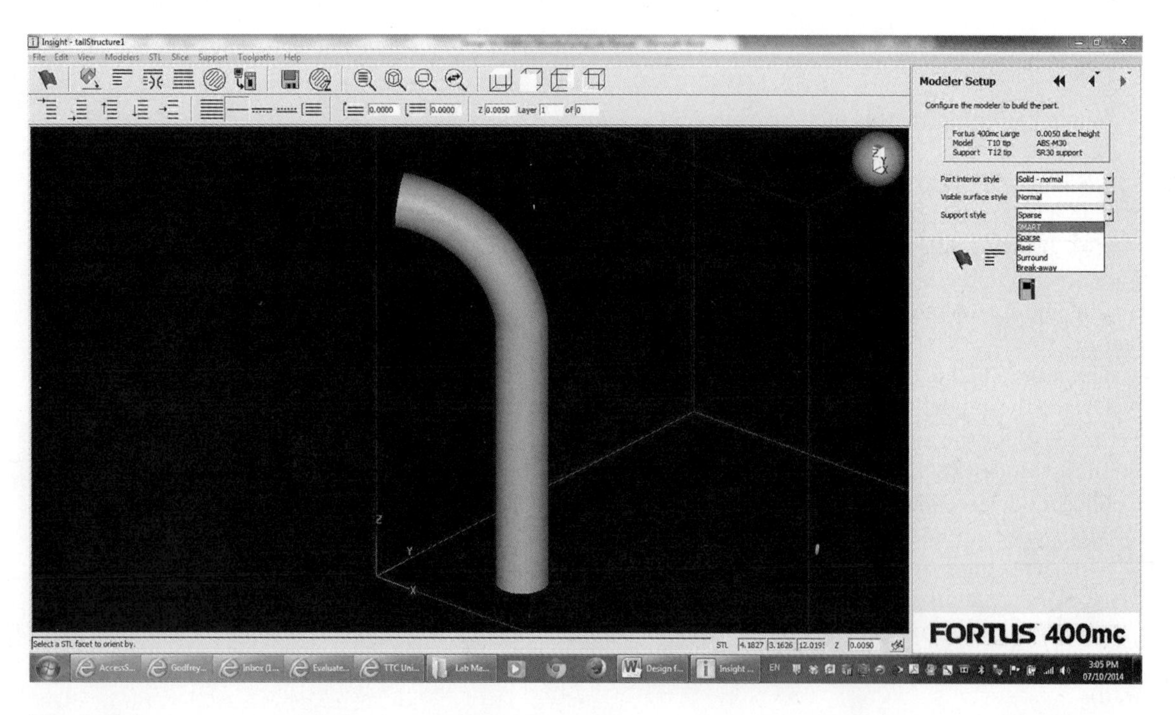

Figure 39.2

Orientation of model.

It is important to design correctly for additive manufacturing. The designer should analyze the support solution that the software provides. In Figure 39.10, there will be a problem when the part is built. Why? The structure is 12-in. tall. During the building session, the part is likely to vibrate and fall before the bent area is printed. To avoid this situation, which we have experienced in the past, we have modified the part to include a very thin feature, 0.01 in. in diameter, so that it can be easily broken without affecting the part (see Figures 39.11 and 39.12). The new support created is now more reasonable to avoid the

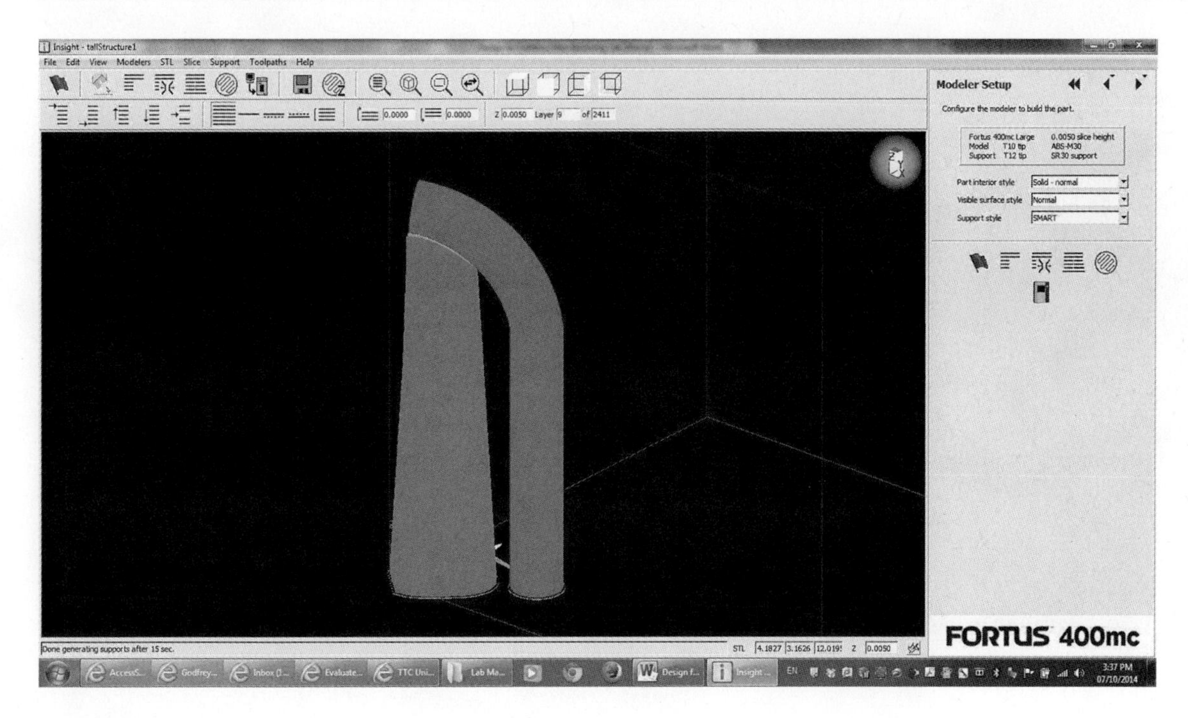

Figure 39.3

Support Style is chosen to be "SMART."

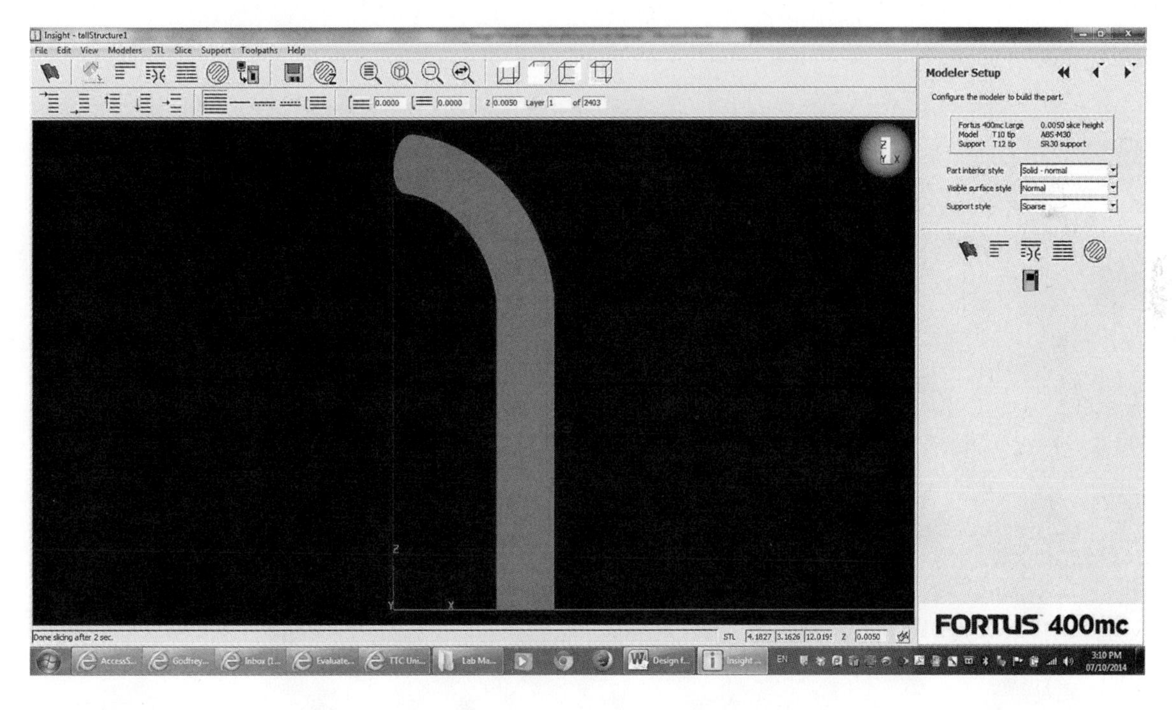

Figure 39.4

Model slicing.

part being printed from tipping off due to vibration of the structure (see Figure 39.13). When the *Finish* tool is clicked, and the estimated build time option is activated, Figure 39.14 shows that the build time is significantly increased to 70 hours and 32 minutes, but we are guaranteed that the part will not fall while being built.

The costs for the different supports are shown in Table 39.1.

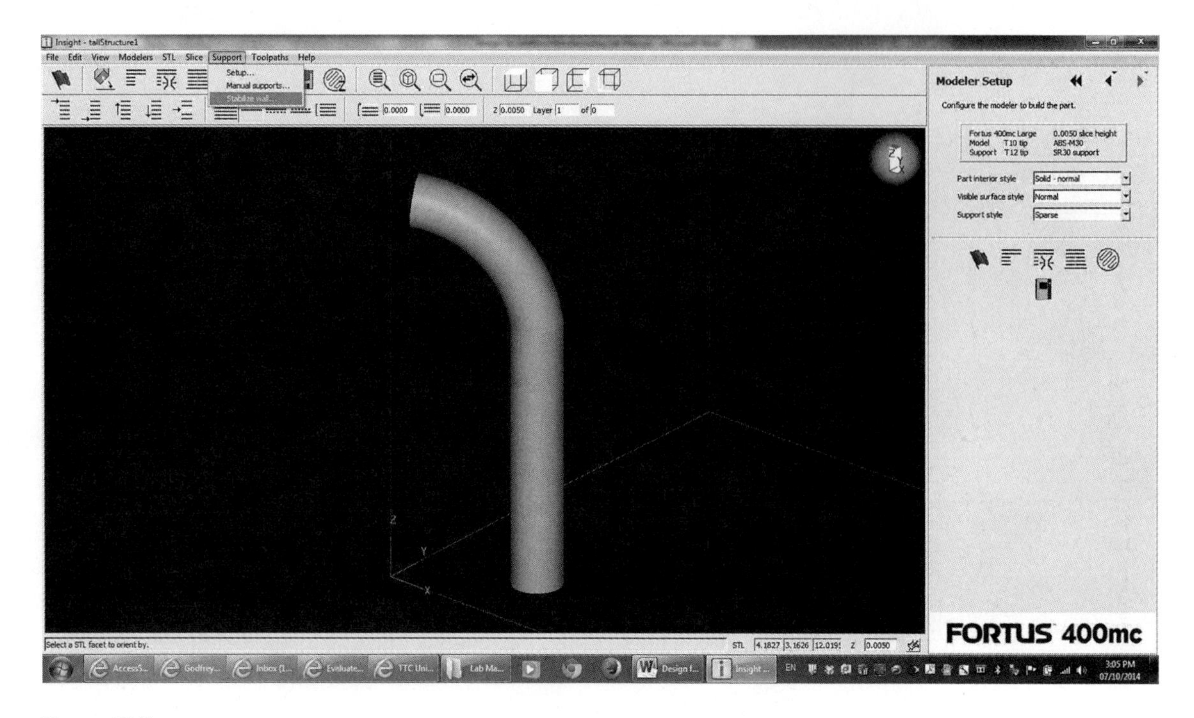

Figure 39.5

Activation of Stabilizing wall for support definition.

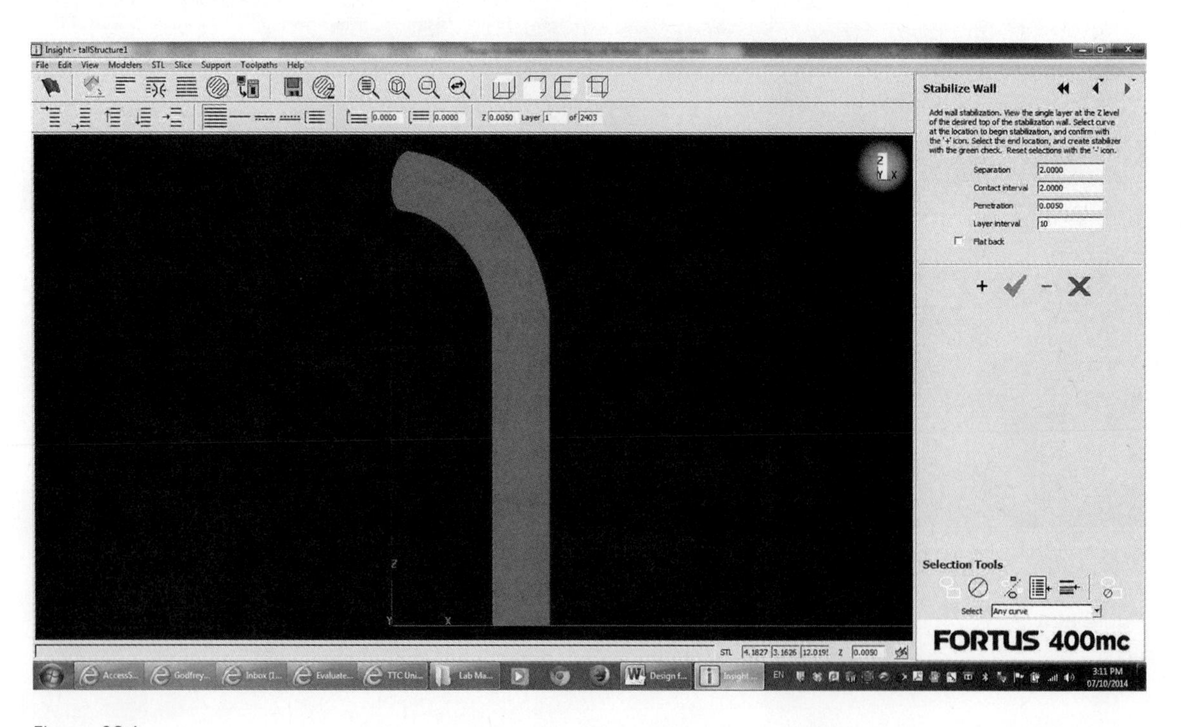

Figure 39.6

Stabilizing Wall property manager pops up on top-right corner.

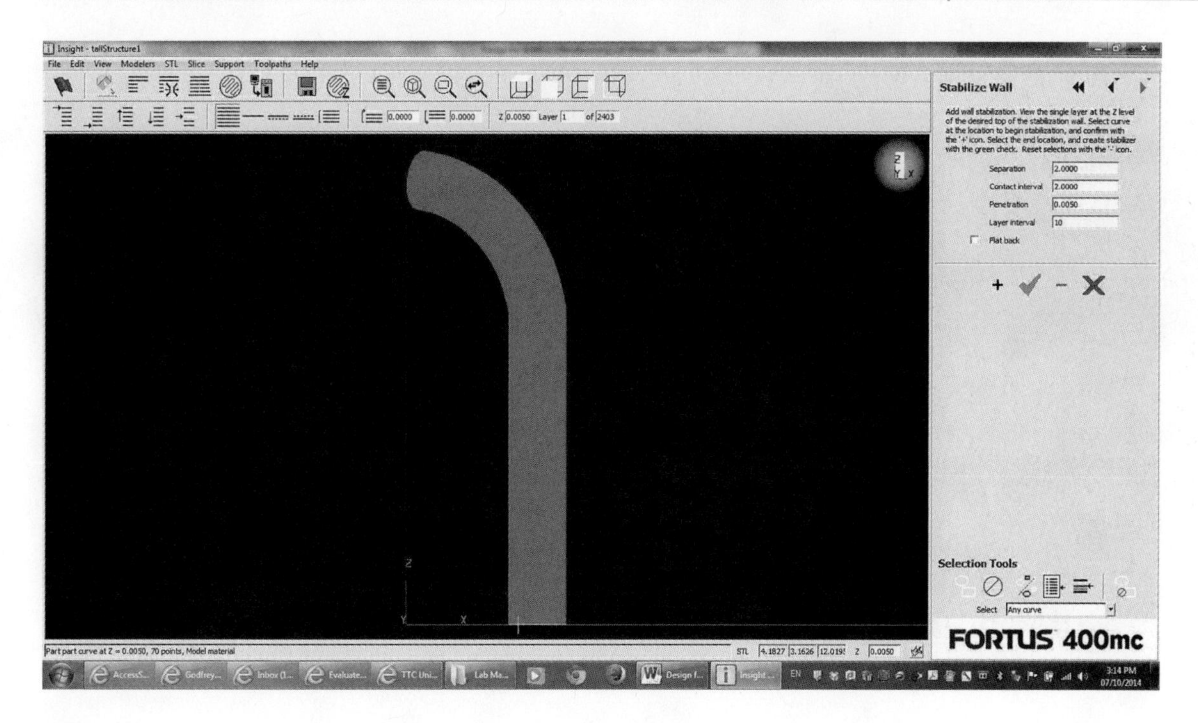

Figure 39.7

First location at the chosen curve (bottom) is clicked and confirmed with "+."

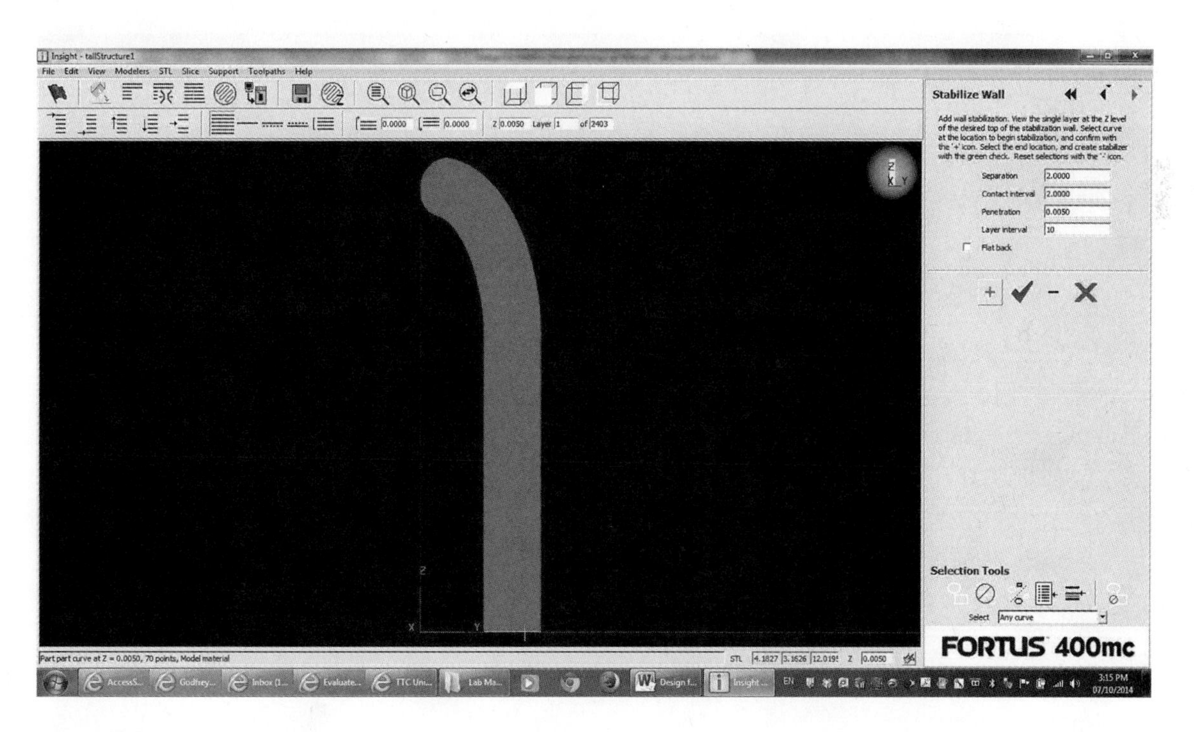

Figure 39.8

Second location on the same curve is selected and confirmed with "√" (check mark).

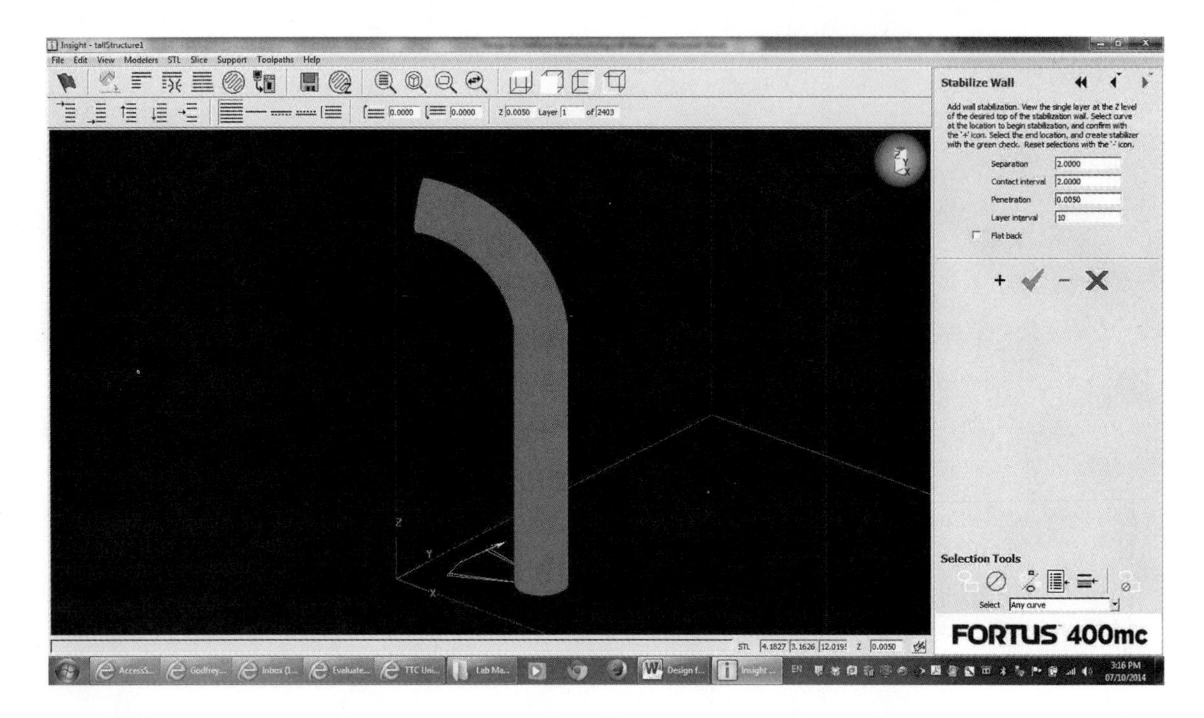

Figure 39.9

Support layout is created as shown at bottom-left position.

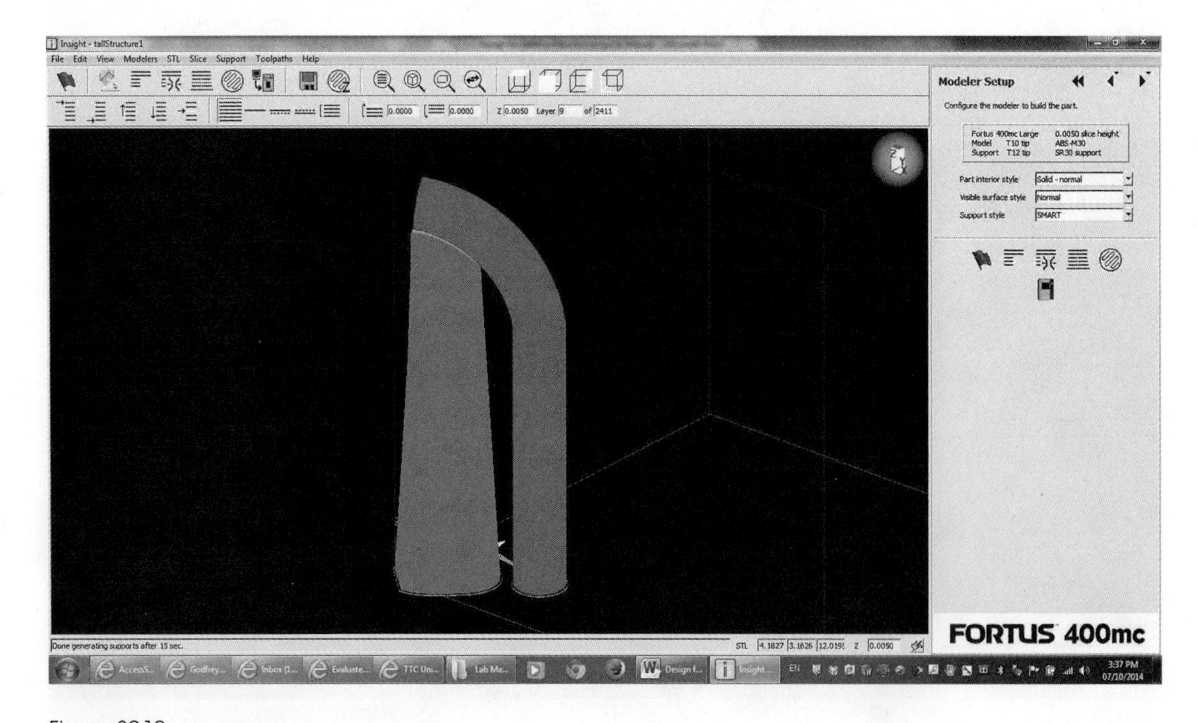

Figure 39.10

Create support for current job.

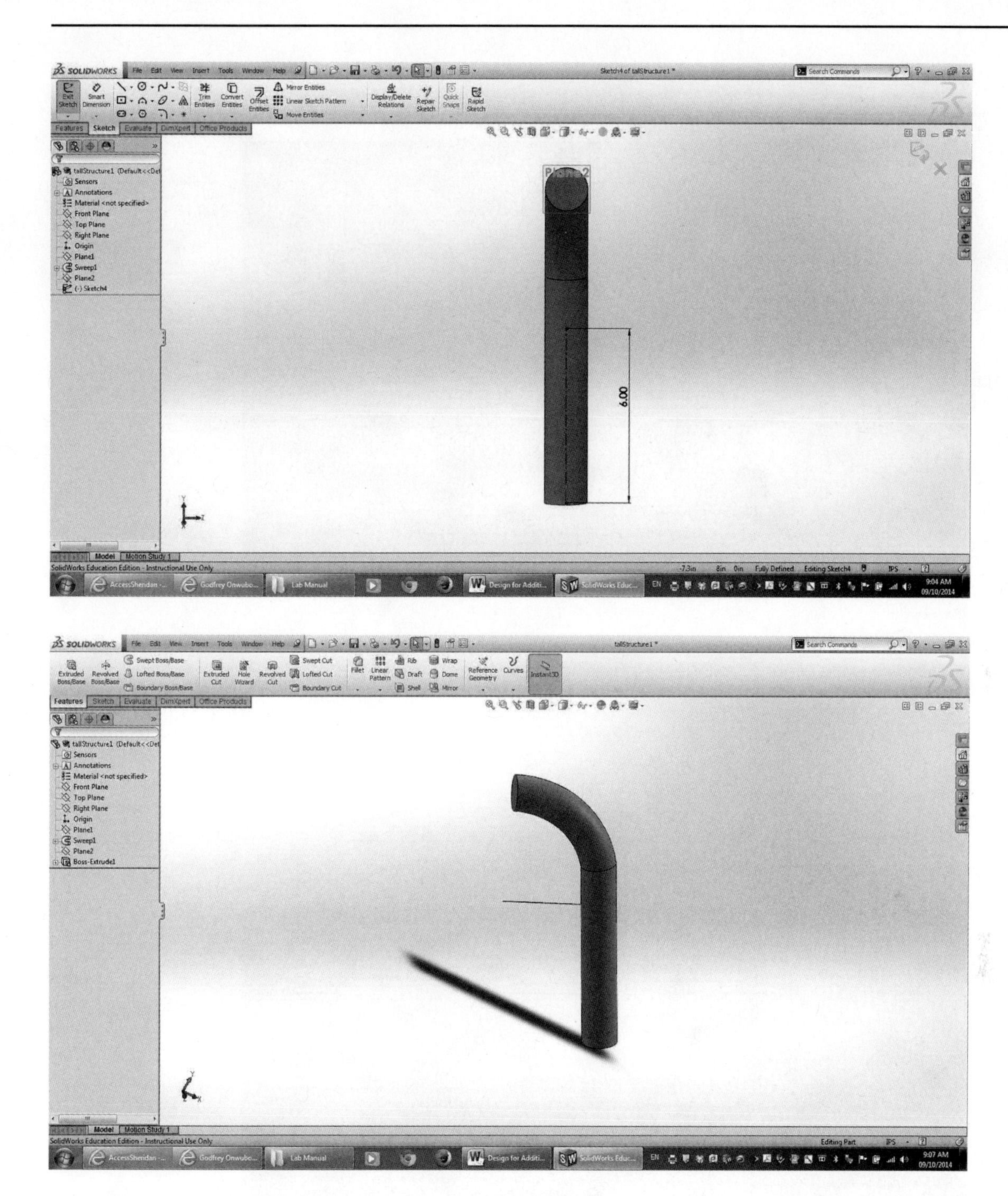

Figure 39.11

Extra feature created to modify support.

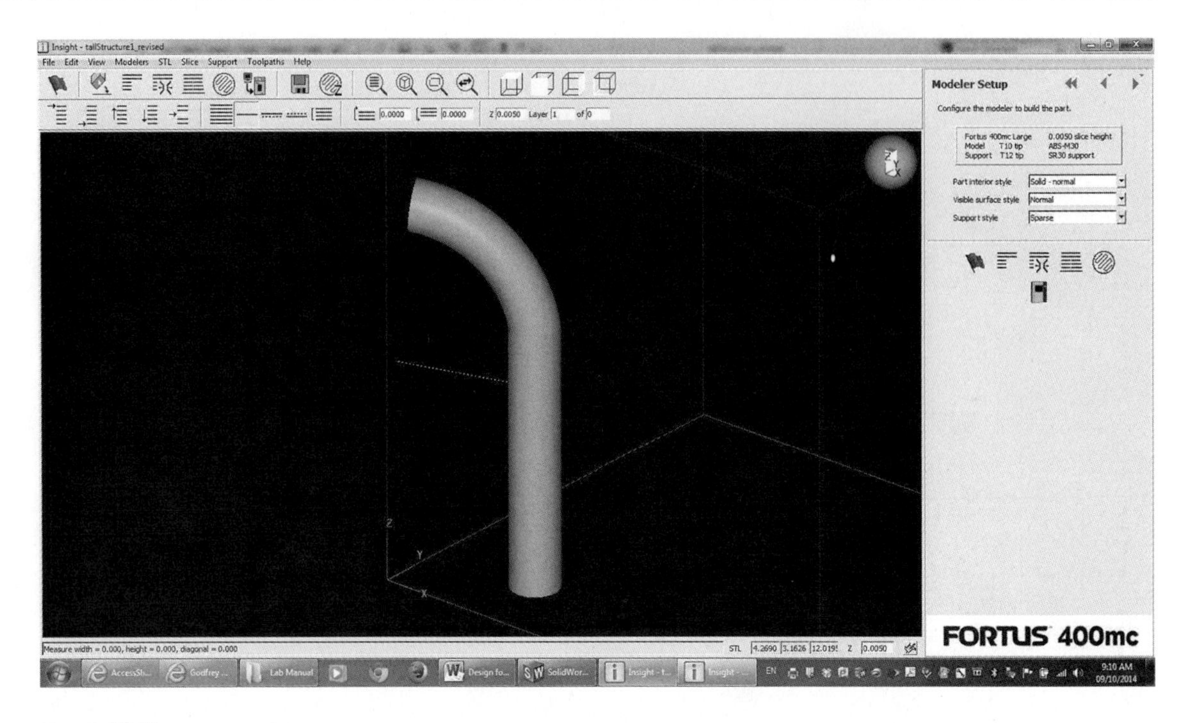

Figure 39.12

Modified model is imported to Insight processing software.

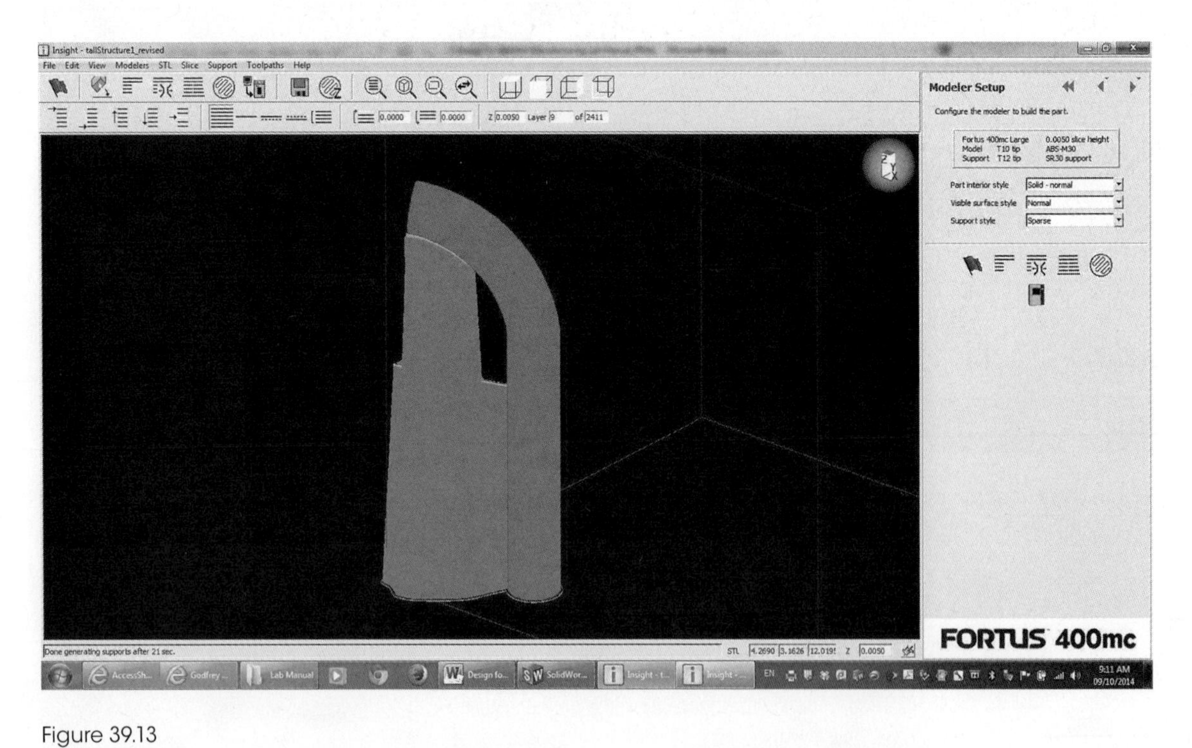

Figure 39.13

New support created for tall structure is more reasonable than the one in Figure 39.10.

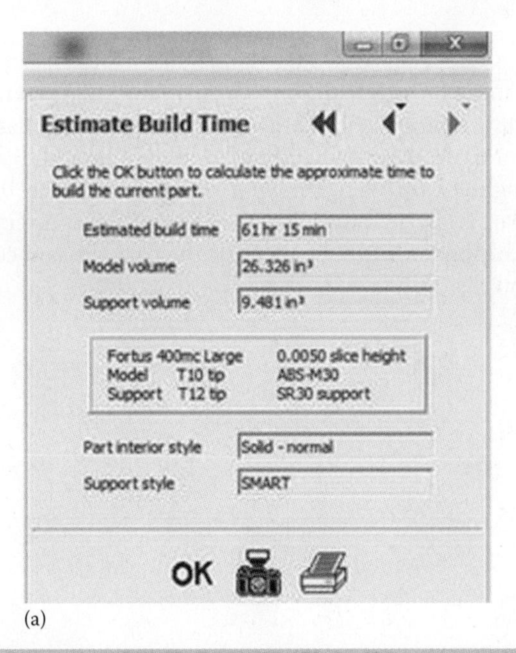

(a)

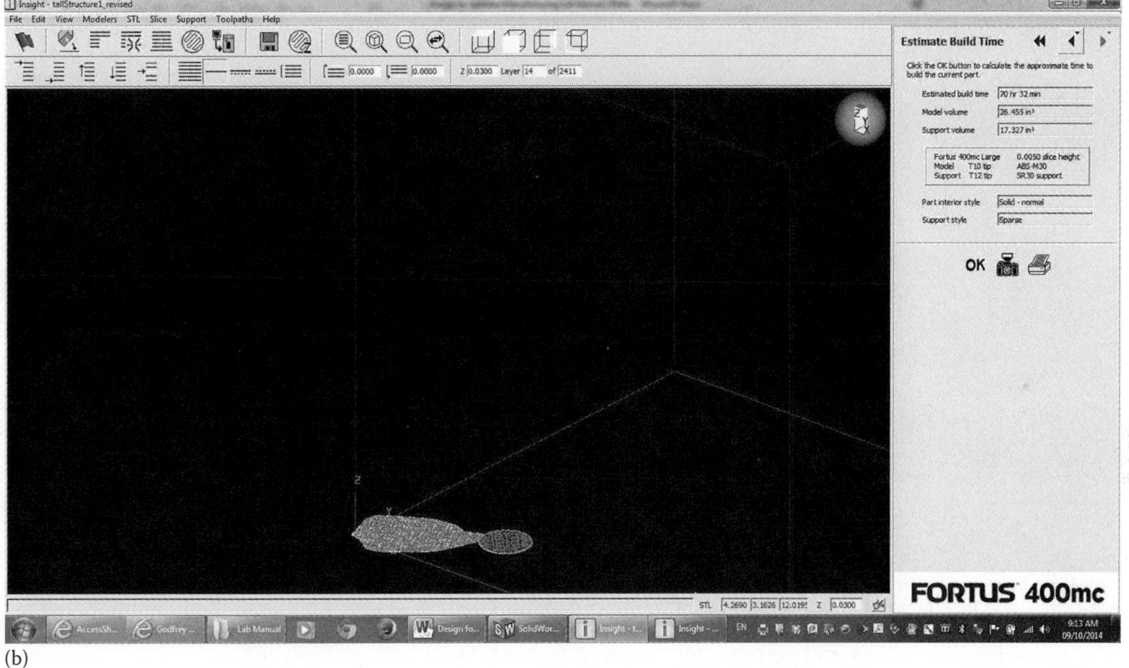

(b)

Figure 39.14

Finishing and checking estimated build time. (a) Estimated build cost for Figure 39.10. (b) Estimated build cost for Figure 39.13.

Table 39.1 Costs for Different Supports

	Build Time (Hours)	Model Volume (in.^{3})	Support Volume (in.^{3})	Cost ($)
Initial support solution	61.15	26.326	9.481	667
Modified support solution	70.32	26.455	17.327	777

Summary

This module has shown how to create stabilizing support for thin walls and tall parts. This is an advanced concept that is treated here. It is a very useful tool for dealing with these classes of models during the 3D Printing of parts. In our tutorial, we have shown how to design for additive manufacturing because the solution that the preprocessing software offers is not optimal; the part being built is very likely going to fall during printing because the part is tall (12" tall and slender). By including a very thin feature that is is 0.01 in. in diameter, which can be easily broken without affecting the part, we save costs because the part will not fall, but completion is guaranteed.

40

Vacuum Forming

Objectives:

When you complete this module, you will

- Understand how to design tool for vacuum forming
- Learn how to use layer feature to define different layers in a part
- Be able to 3D Print the tool for vacuum forming using the Fortus machine

The CAD model of a mold produced using SolidWorks is shown in Figure 40.1, while some of the basic steps involved in using the interface software, Insight for preparing 3D printing data are shown in Figures 40.2 through 40.8.

Modeler Setup

The tools for Modeler Setup are shown in Figure 40.5.

Specify the configuration of the modeler
Click (1) in Modeler Setup (see in Figure 40.5).
Slice
Click (2) in Modeler Setup (see in Figure 40.5).
Create support for the current job
Click (3) in Modeler Setup (see in Figure 40.5).

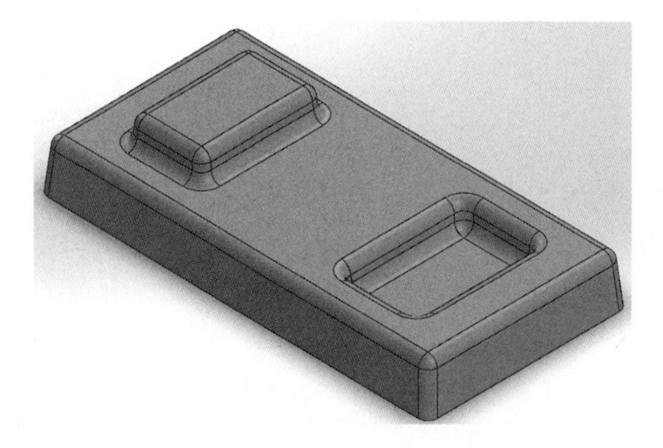

Figure 40.1

Model of mold produced using SolidWorks.

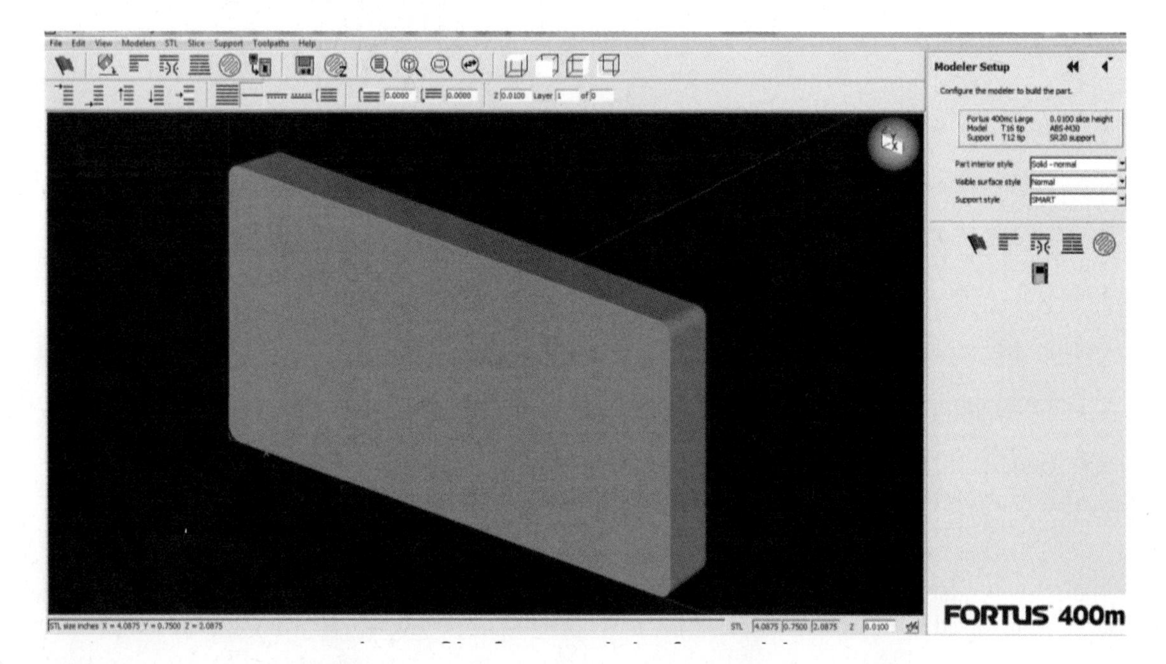

Figure 40.2

Exported STL file for model of mold.

Custom Groups

Menu navigation (see Figure 40.9): *Toolpaths > Custom groups*
Create a New Group

1. Click the New button to display the *Create New Group* window.
2. Enter the *Group name* of the new custom group. Every group must have a name. A default name is provided by Insight.
3. Choose the *Display color* for the custom group.
4. Decide if the group parameters should be based upon a different existing group, i.e., a template group. If so, click the Template button. This will display a small window with a list of all system- and user-created custom groups. Pick one of the groups from the list, and click the OK button. This will update all of the parameters in the *Create New Group* window to match those of the template group.
5. Set the appropriate values for the remaining group parameters.
6. Click the button. The new group will be created.
7. Select the curves to add to the new custom group.
8. Click the Add button to move the selected curves into the new custom group.

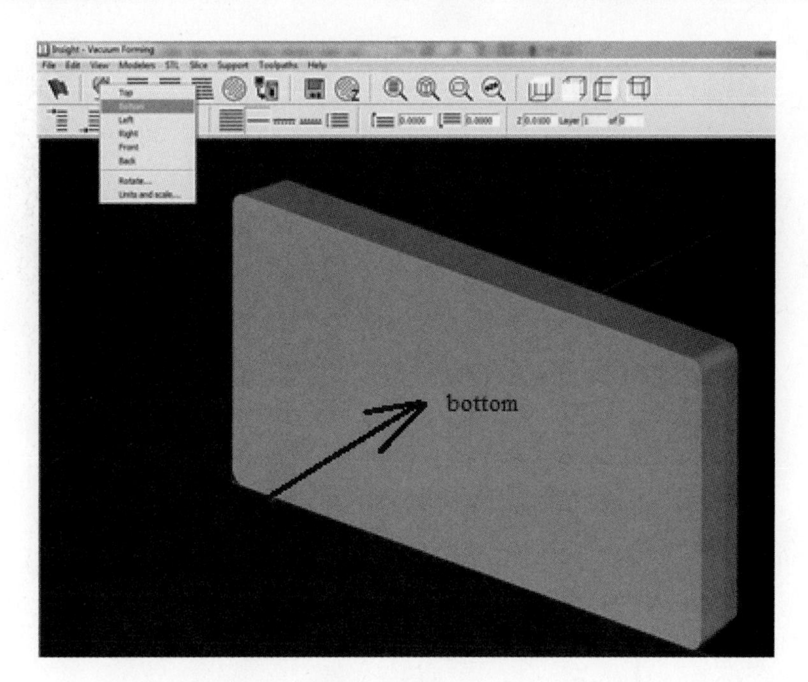

Figure 40.3

Establish bottom of job.

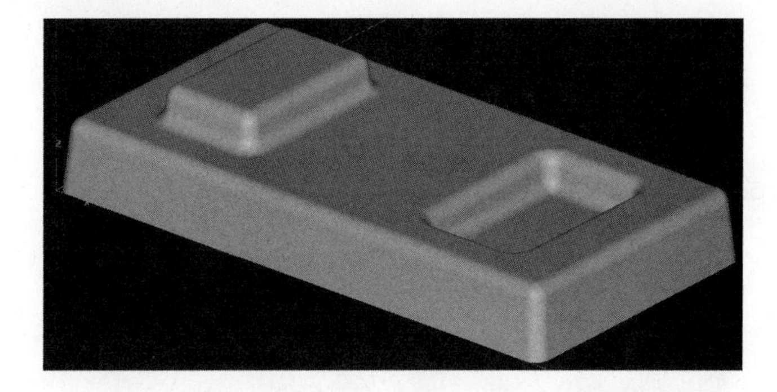

Figure 40.4

Orientation chosen to print the job.

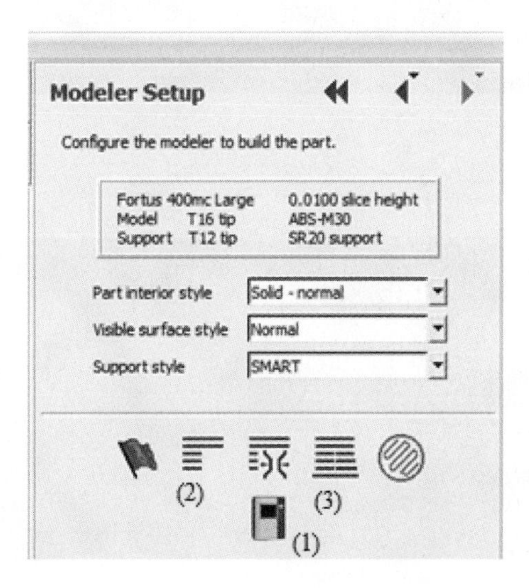

Figure 40.5

Modeler Setup.

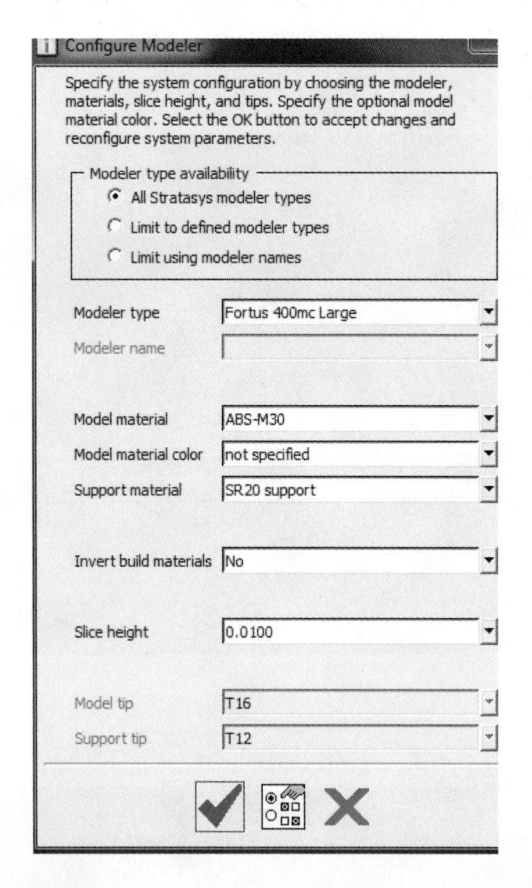

Figure 40.6

Model tip T16 and Support tip T12 are chosen.

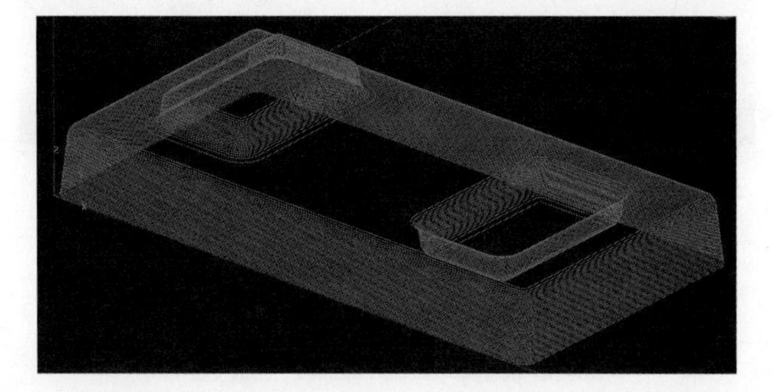

Figure 40.7

Model is sliced.

Figure 40.8

Support is created.

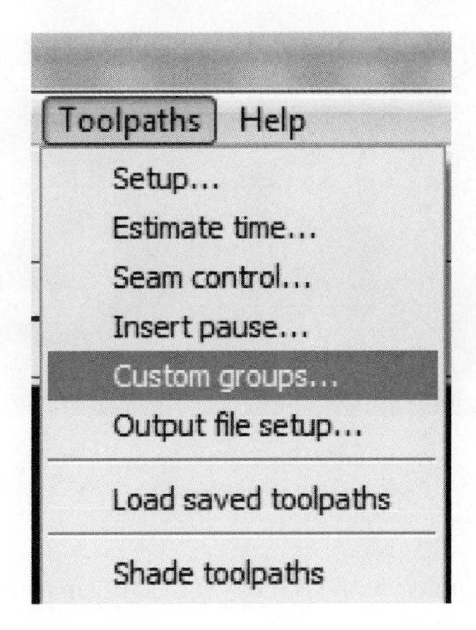

Figure 40.9

Custom Group menu navigation.

Tool Paths Custom Groups

The Tool Paths Custom Group tool is used to control the tool path parameters for the different zones of our product. In the vacuum-forming part, we define five zones using four groups. The following custom groups were created, leading to an optimal solution (see Table 40.1 and Figure 40.10 for clarity):

- Zone 1: is the *crest*, with magenta color; the air gap is medium (0.0100), stopping few layers before the large flat surface.
- Zone 2: is the *large flat surface*, with *cyan* color; the air gap is tight (0.0050), one or two layers on each side of the large flat surface.
- Zone 3: is the *upper base*, with *light red* color; the air gap is coarse (0.0200), stopping a few layers before the valley flat surface.

Table 40.1 Summary of the Tool Paths Custom Group Parameters

Group	Color	Contour size	# Contours	Raster Width	Raster Angle	Air Gap	Levels
1	Magenta	0.0120	2	0.0120	45	0.0100	Crest
2	Cyan	0.0120	2	0.0120	45	0.0050	Surface
3	Light red	0.0120	2	0.0120	45	0.0200	Base
4	Light green	NA	NA	0.0120	45	0.0100	Outlet

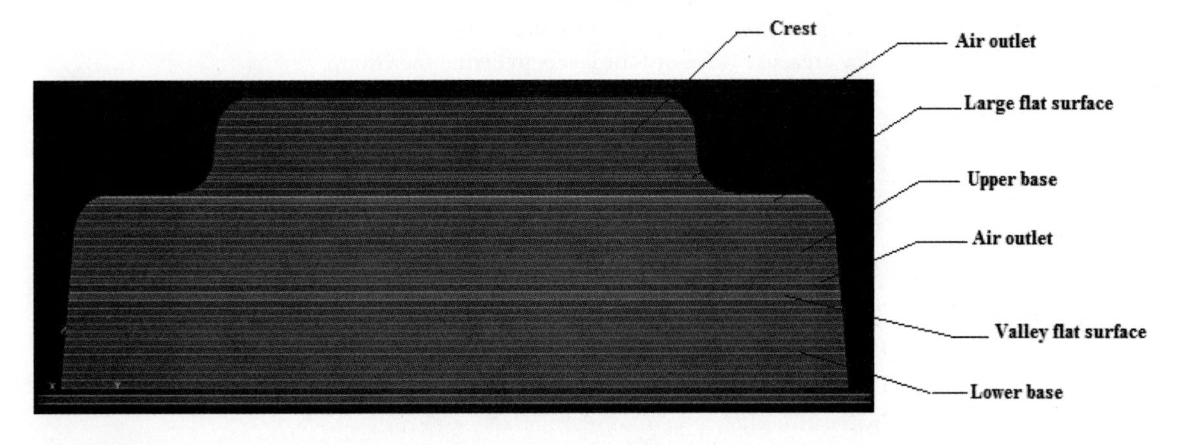

Figure 40.10

Right side view.

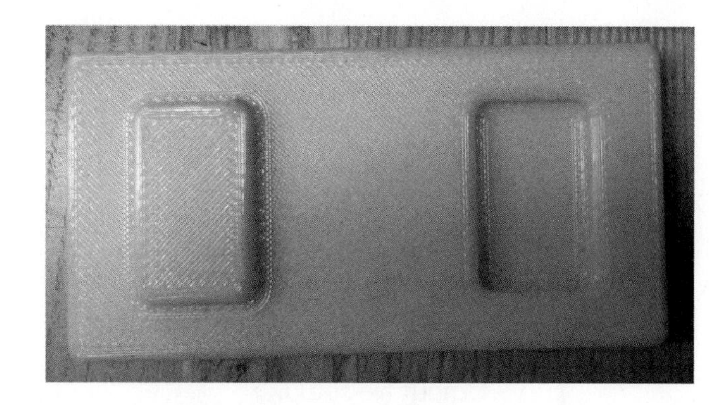

Figure 40.11

Vacuum forming tooling produced using FORTUS 400mc machine.

- Zone 4: is the *valley flat surface*, with *cyan* color; the air gap is tight (0.0050), one or two layers on each side of the valley flat surface.
- Zone 5: is the *lower base*, with *light red* color; the air gap is coarse (0.0200), stopping on top of the base support.
- Zone 6: is the *air outlet*, with *light green* color; the air gap is tight (0.0100), slightly above the large flat surface and valley flat surface; there is no contour in this zone to allow air to exit from the tool when vacuum is formed, thereby assisting in suction effect.

Layer-by-Layer Building of Model

Using the model STL file and processed data for the tool paths, Figure 40.11 shows the Vacuum forming tooling that is produced using the FORTUS 400mc machine.

Summary

1. Select Top layer.
2. Select Bottom layer.
3. Grab the set of layers.
4. Click New.
5. Click Select Tool.
6. Click Add.
7. Click Regenerate.

Important Optimization Issues

1. Use *multiple contours* for all layers (to eliminate the waviness problem).
2. To create suction, use *no contour*—1 layer (for sucking air).
3. In the Top and Valley area, use three or four layers to define the Group.
4. All Enbose (valley) should have porous internal structures so that air can be sucked (air gap = 0.0200; this value is HIGH).

41

Design for Assembly Manufacturing

Objectives:

When you complete this module, you will

- Be able to create an assembly of parts
- Have an experience of the 3D printing of an assembly of parts

Introduction

This module is an advanced module, requiring that participants understand how to deal with tolerance for mating parts and how to 3D-print the assembly of parts. There are two ways to deal with the 3D printing of an assembly of parts: (1) build each part and assemble them or, alternatively, (2) print the entire assembly at one shot. Our experience shows that the 3D printing of an assembly of parts is more troublesome than printing each part and assembling them. However, the 3D printing of an assembly of parts at one shot shows the power of additive manufacturing because it is not possible to use any other manufacturing process such as computer numerically controlled machining to produce an assembly of parts at one shot.

In this module, we will first present the best practice guidelines that are published by Stratasys for building assembly parts using Fortus 3D Production Systems. Then, we will present a methodology that we developed for solving the problem of the assembly of parts based on a research that we carried out on tolerances for mating parts in an assembly of parts.

In our approach, there are essentially five steps:

1. Model the parts using SolidWorks.
2. Assemble the parts while tolerances are built into the parts.
3. Save the assembly as "part."
4. Save the "part" as STL.
5. 3D-print the part.

These steps will be applied to a project, which is described in the next section.

Figure 41.1

The Radial Engine assembly.

Figure 41.2

The exploded view of the assembly.

Project

In this project, we will create the radial engine assembly that is shown in Figures 41.1 and 41.2. The radial engine assembly will be created in two parts: one will be the subassembly, and the other will be the main assembly. The dimensions of the components of the radial engine assembly are shown in Figures 7.3 through 7.6. We will then 3D-print the assembly using the Fortus machine.

Fortus 3D Production Systems Best Practice for Building Assembly Parts

Stratasys has given useful tips that can be utilized on parts that will fit together in an assembly. Since the parts will be assembled, the orientation and offsetting techniques outlined will ensure a proper fit.

Orientation

Step 1

Orient the mating surfaces in the X/Y plane. When orienting the parts in the build envelope, consider which surfaces are to be mated together. This will eliminate stair stepping, producing smooth surfaces that will mate together more precisely. Also, the Z stage is often approximated during the slice function, which can degrade the accuracy of features in the Z plane. The X/Y plane is the most accurate plane and is thus the plane to orient critical mating surfaces.

Step 2

Use the STL rotate menu (located at STL > Rotate) or the Orient by selected facet feature (located at STL > Orient by Selected Facet) to orient your part in the X/Y plane.

Offsetting

If the assembly parts contain features that fit into one another, the EDIT/OFFSET function in the Insight can be used to offset the geometry to allow enough clearance for the mating part to fit together easily. If tolerances were designed into the parts and were part of the STL file, offsetting may not be necessary. Check with the original designer to be sure.

Step 1

After slicing the STL file, select the geometry that should be offset. Go to Edit > Offset, and the offset menu will appear on the right side of the geometry window.

Step 2

Set the direction of the offset for the selected curves: inside or outside the curve. (Inside will shrink the curve; outside will enlarge the curves.)

Step 3

Set the offset distance or clearance that you want to achieve. Typically, 0.002–0.004 in. will work for mating features.

Step 4

Set the Destination Group to "same as Selected." Change the Keep Original flag to No. (If set to yes, the original curves would also appear.) Hit the OK icon, and the selected curves will be offset by the amount that you chose. Hit the undo arrow to undo the offsetting action.

Bottom-Up Solution Approach

An initial research was conducted to study fits and assembly in FDM additive manufacturing. The outcomes are guidelines for designing mating parts:

Material used: ABS
Machine tools: Tool tip: T16 (0.0100" slice height); Support—T12SR30
Configuration investigated: cylindrical mating parts at three orientations

Case I: Printing Assembly of Parts

Maximum bearing load arrangement: In this case, there is internal support material between the mating parts. This leads to welding, and support material must be removed in the cleaning station.

45° bearing load arrangement: The minimum clearance required between mating coaxial parts is 0.010 in.

Coaxial nonsupporting load arrangement: The minimum clearance required between mating coaxial parts is 0.0075 in.

Recommendation: It is not advisable to print an assembly of parts having maximum supporting load.

Notes: We cannot generalize by this test; we need to apply results to real-life cases and check.

As already discussed, in our approach, there are essentially five steps:

1. Model the parts using SolidWorks.
2. Assemble the parts while tolerances are built into the parts.
3. Save the assembly as "part."
4. Save the "part" as STL.
5. 3D-print the part.

Part A: Solid Modeling Using SolidWorks

Step 1

Using the dimensions in Figures 41.3 through 41.6, model each part with the aid of SolidWorks.

Step 2

Assemble the parts, with tolerances designed into the parts (use the guideline that is given from the study that is reported in the preceding section); these tolerances will be a part of the STL file so that offsetting may not be necessary. See Figure 41.6 for the assembly of parts using SolidWorks.

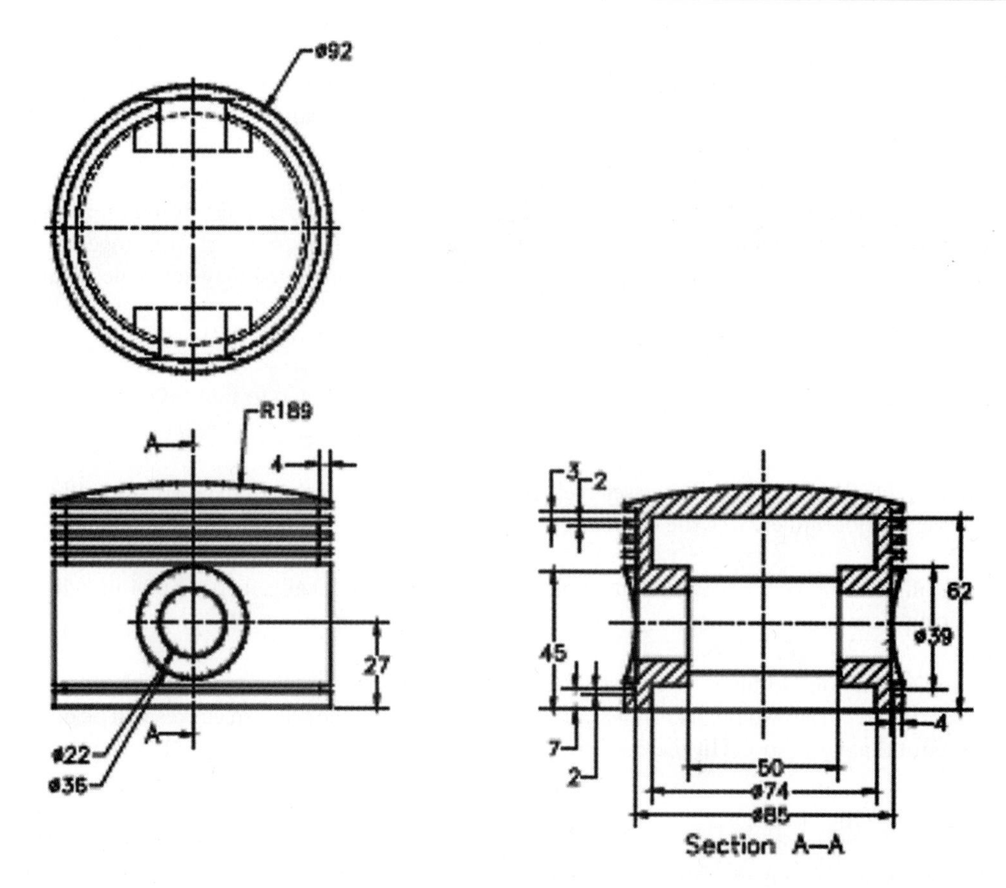

Figure 41.3

Views and dimensions of the piston.

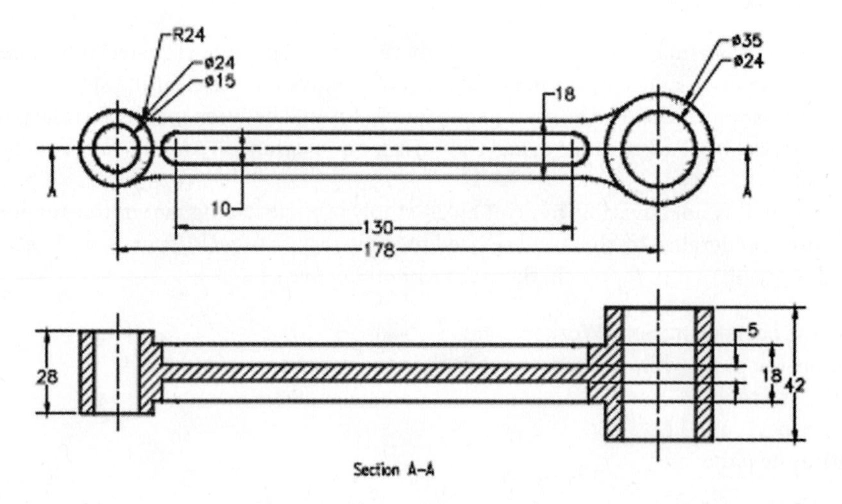

Figure 41.4

Views and dimensions of the articulated rod.

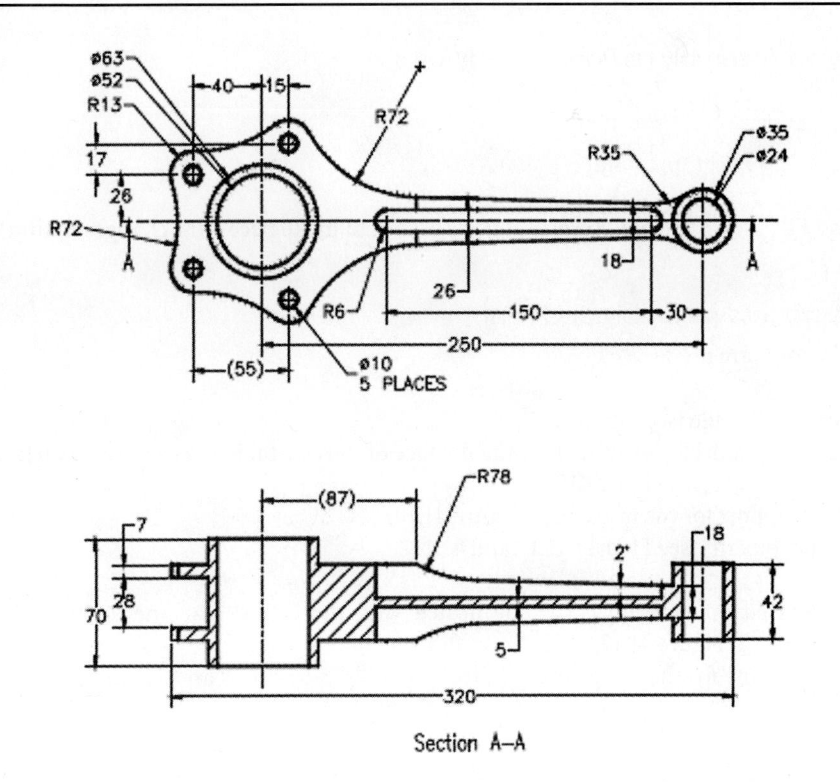

Figure 41.5

Views and dimensions of the master rod.

Figure 41.6

CAD model of assembly of radial engine parts.

Part B: Saving Assembly as Part in SolidWorks

Step 1
Click Save As >
Save as type > Part (*.prt; *.sldprt)
Step 2
Save as STL (*.stl)   (Note: Access Options > Unit to make necessary changes in unit)

Part C: Preprocess Assembly and Print

The following steps are used:

1. Open the STL file (see Figure 41.7).
2. Orient the model for printing (use the flat face of the central hole as a bottom surface) (Figure 41.8).
3. Slice the model (see Figure 41.9).
4. Create support for the model (see Figure 41.10).
5. Create toolpaths (see Figures 41.11 and 41.12).

The preprocessed information is communicated with the Fortus machine for the 3D-printing of the assembly, as shown in Figure 41.13.

The estimated time for the 3D-printing of the assembly is 55 hours and 50 minutes as shown in Figure 41.14.

Case II: Printing Individual Parts

Based on the investigations that we carried out at CAMDT, the following guidelines were used to print individual parts.

Maximum Bearing Load Arrangement: Minimum clearance required between mating coaxial parts is 0.0075 in.

45° Bearing Load Arrangement: Minimum clearance required between mating coaxial parts is 0.0075 in.

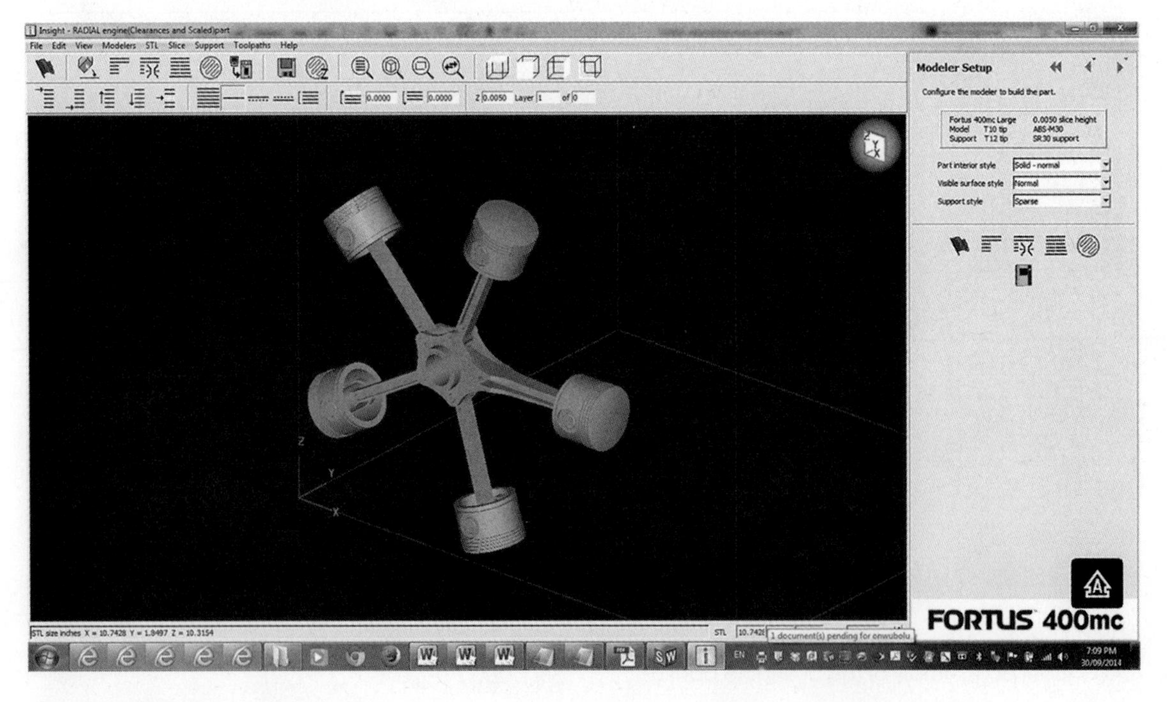

Figure 41.7

STL file of assembly imported.

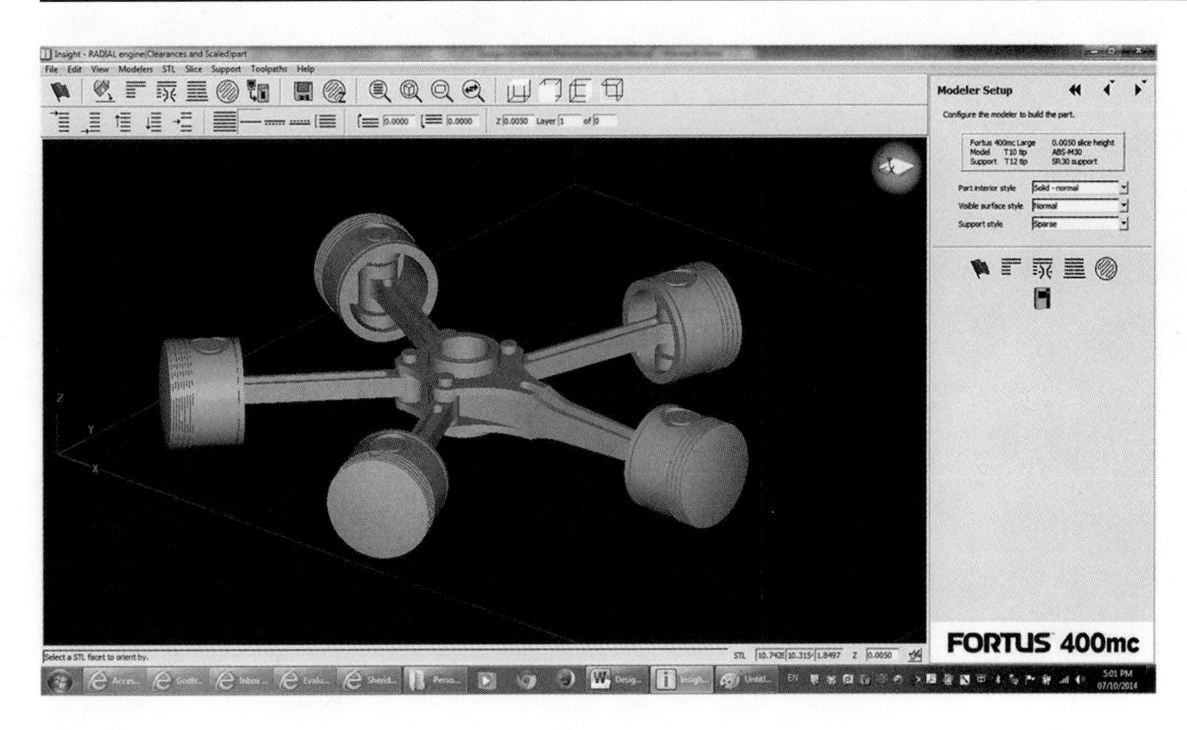

Figure 41.8

Orientation of assembly for printing.

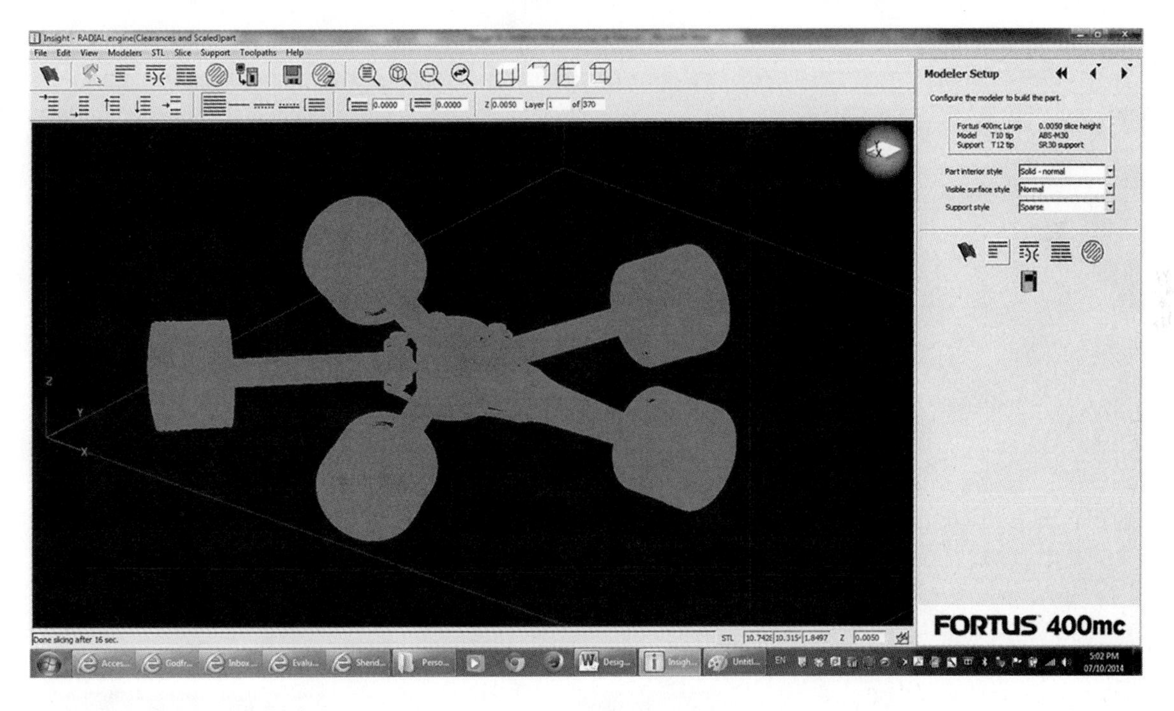

Figure 41.9

Assembly sliced.

Co-axial Non-supporting Load Arrangement: The minimum clearance required between mating coaxial parts is 0.0075 in.

Recommendation: As much as possible, it is advisable to print individual parts before assembling them.

Notes: The maximum clearance indicated in this document may differ from the actual clearance, which must be determined. This is because of the thermal effects on clearance when parts are produced.

Figure 41.15 shows the different parts that are printed and assembled with ease.

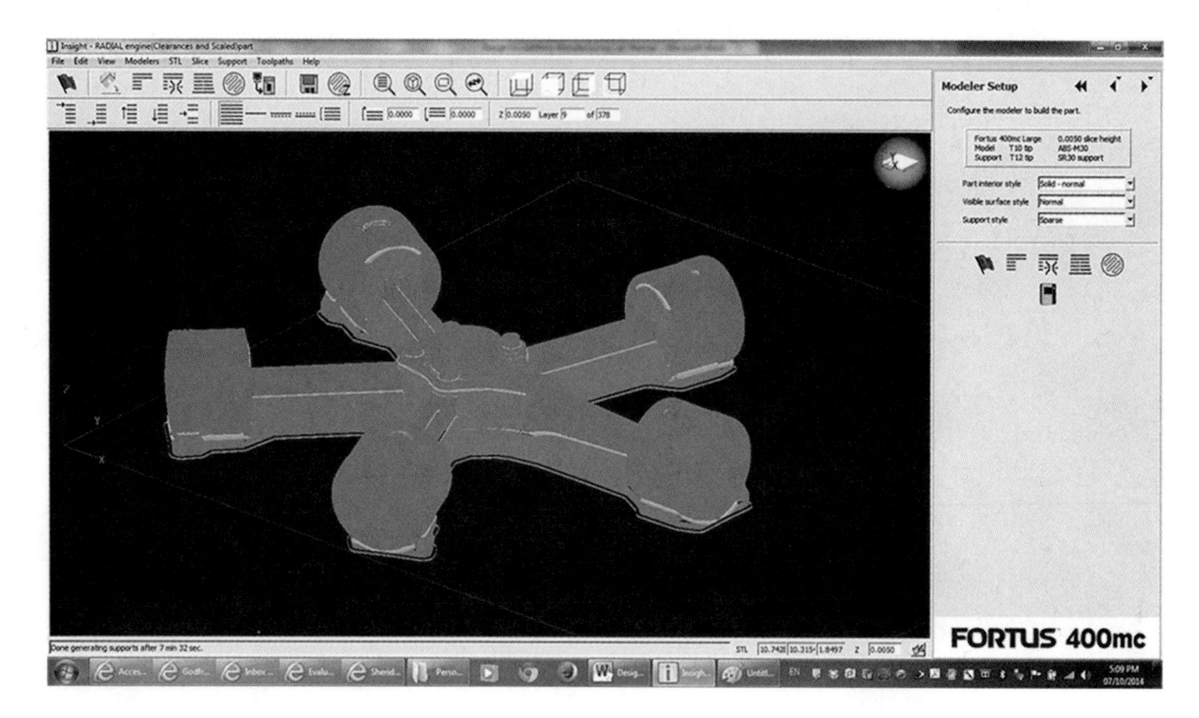

Figure 41.10

Support created for assembly.

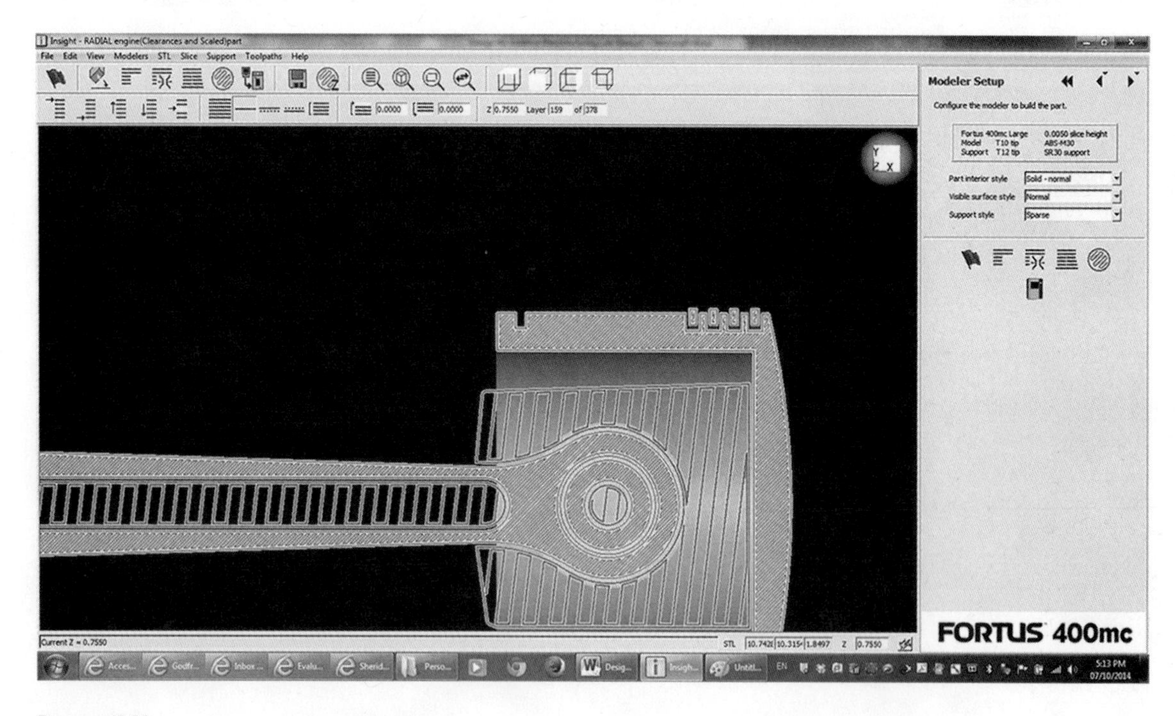

Figure 41.11

Toolpaths for slice exposed in detail (notice support material).

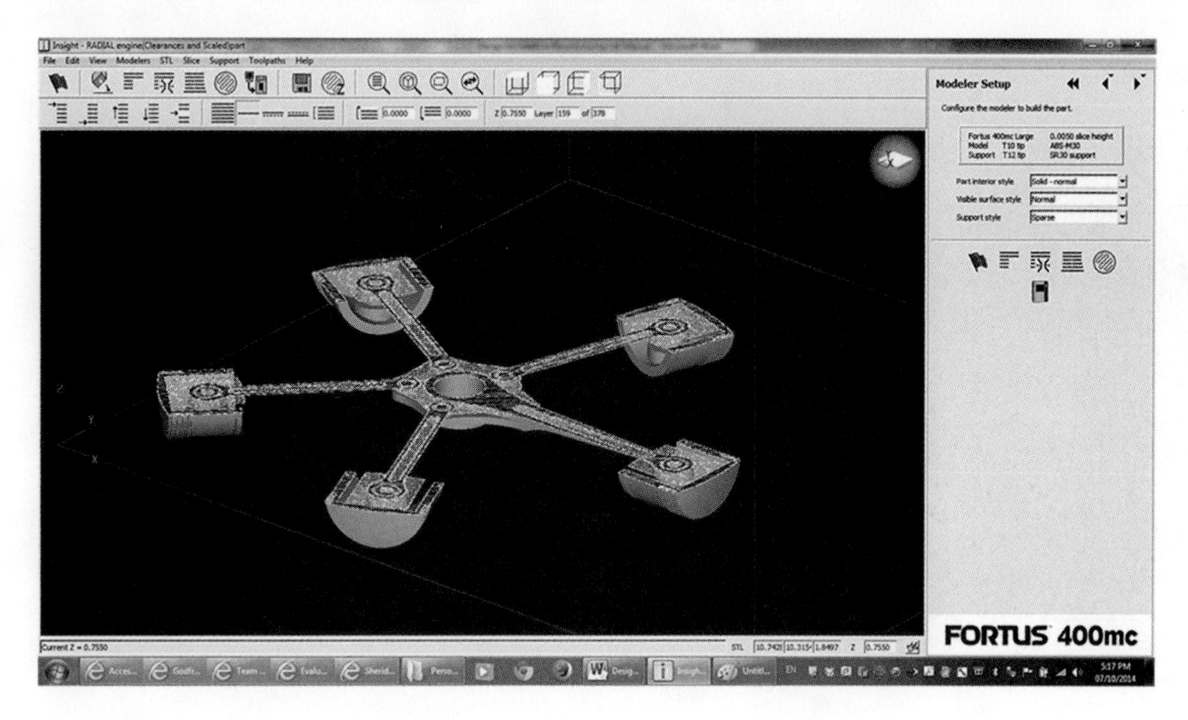

Figure 41.12

Toolpaths for entire assembly for a particular slice height.

Figure 41.13

Radial engine 3D-printed as an assembly of parts.

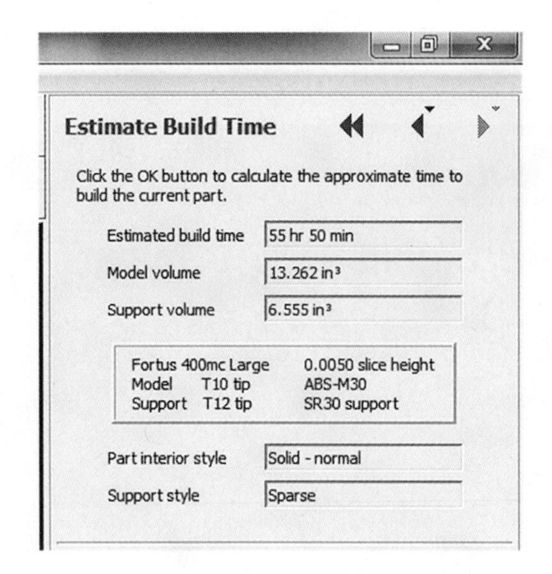

Figure 41.14

Estimated time of printing.

Figure 41.15

Radial engine that is 3D-printed as individual parts.

Summary

While this module presents two approaches for printing parts (in assembly form and in individual form), it is highly recommended that the second approach be used—printing parts as individual parts and then assembling them together. It is suggested that we *keep it simple.*

42

Design Guidelines for Advanced Build Options

Objectives:

When you complete this module, you will

- Understand the guidelines for advanced build options
- Will use some of our designs for bottle parts for toolpath definition
- Be able to 3D Print bottles using Fortus machine

Advanced Build Options

The models in this module were produced in the author's SolidWorks lecture in Summer 2014. We will orient the bottles differently and study the cost implications of 3D printing for different orientations.

The Insight quoting engine now has the ability to calculate the optimal build orientation and layer height for each of the geometry. These build options will be the default choice, *unless specified by the user.* The default orientation is called *Optimal Build*, and the default layer height has an asterisk.

Fused deposition modeling (*FDM*) build orientation and layer height can have a substantial impact on *part quality, build speed* and *price*.

- For FDM, each Standard Tessellation Language (STL) file is analyzed in six different build orientations.
- Support volume and runtime are estimated at each build orientation.
- The optimal build orientation is based on the orientation with the least amount of support volume.
- The default layer height is based on the overall part size.

Considerations for Choosing Build Orientation and Layer Height

Aesthetics: If aesthetics are critical to the final use of the part, you will want to reduce the amount of stepping or build layer lines. In general, smaller layer heights improve the amount of stepping or layer lines. But part geometry and build orientation can also affect aesthetics.

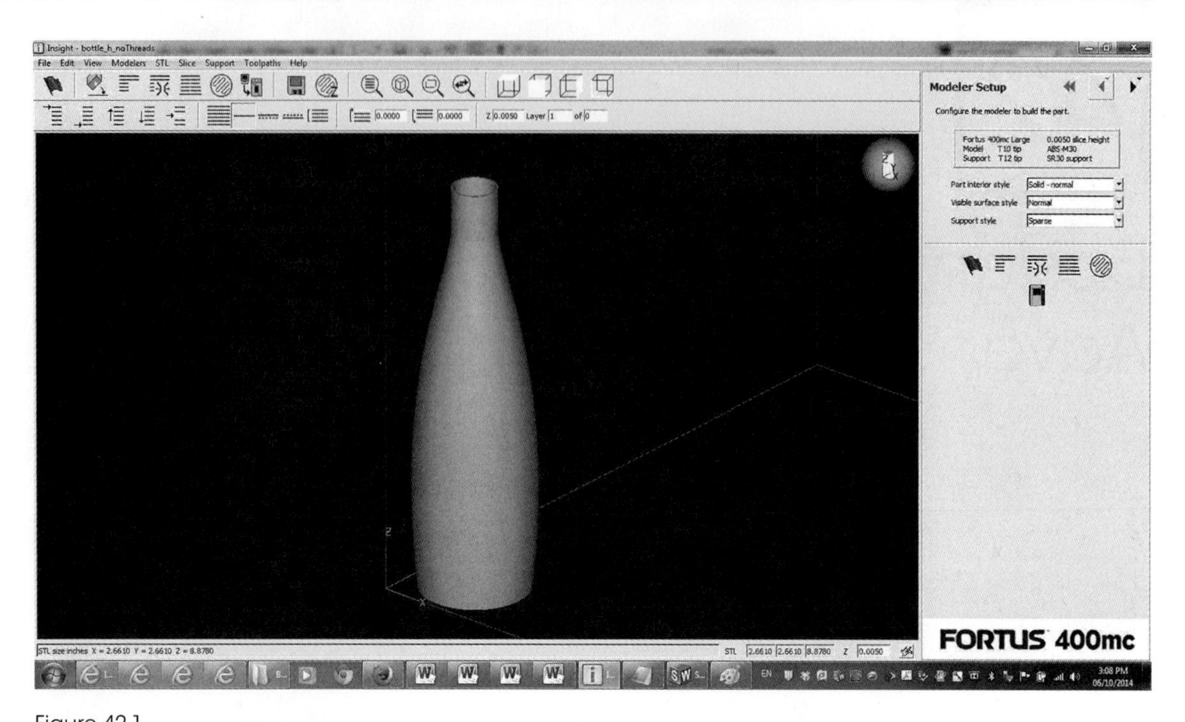

Figure 42.1

Bottle model using SolidWorks.

Case Study 1: Bottle_h-noThreads.Stl in ABS-M30

Open the STL file named 'Bottle_h-noThread' (Figure 42.1), which the instructor will give to you during the hands-on session for this module. You will experiment with six build orientations following the tutorial that is given in this module.

Default layer height for ABS-M30 = 0.005 in. Note: depending on the machine, some materials can be built in the full range of layer heights of 0.005, 0.007, 0.010, and 0.013 in., whereas others are limited to only one or two layer heights.

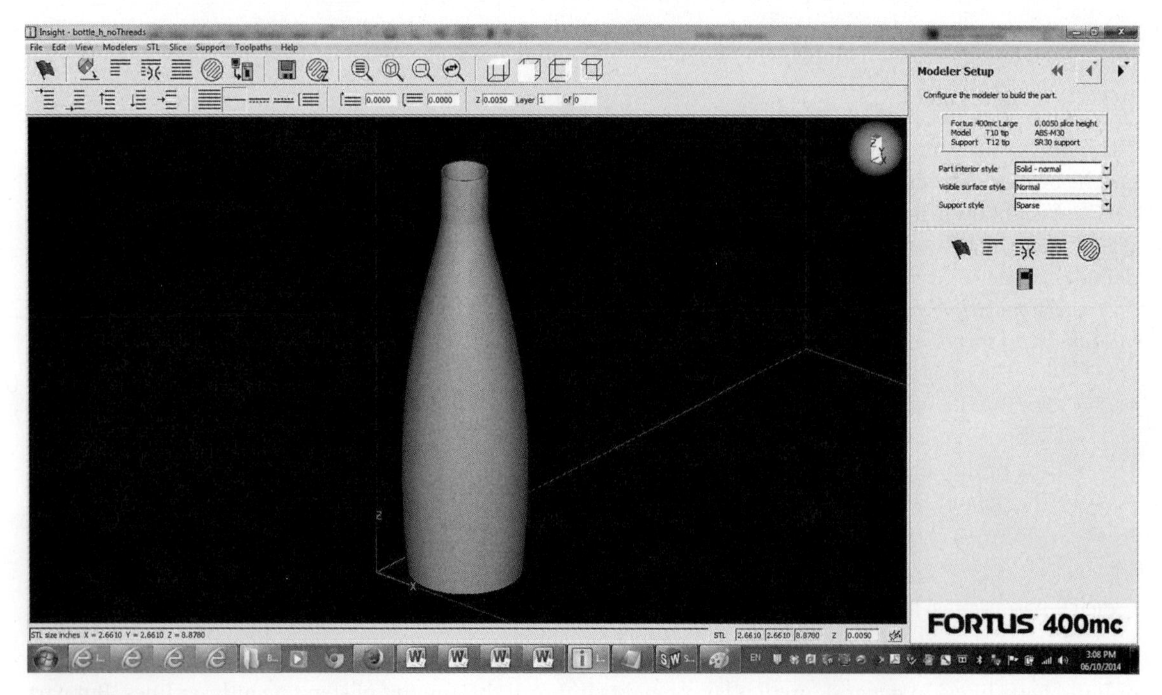

Figure 42.2

Outcomes of our experimentations.

(Continued)

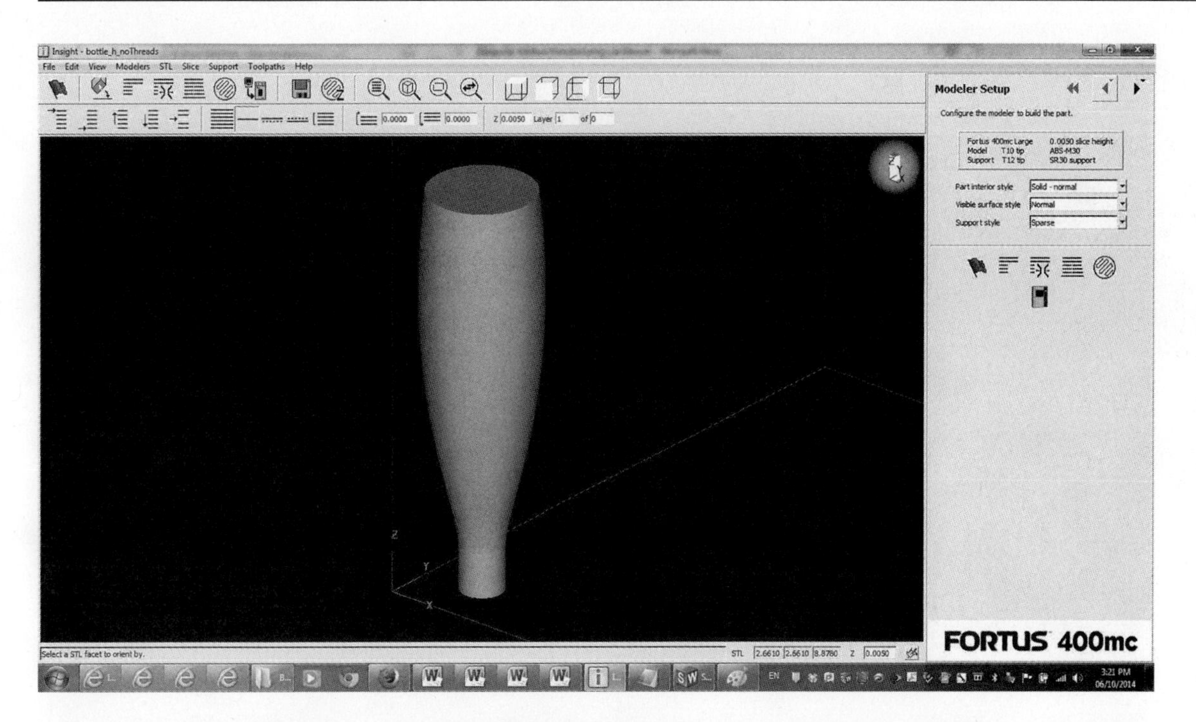

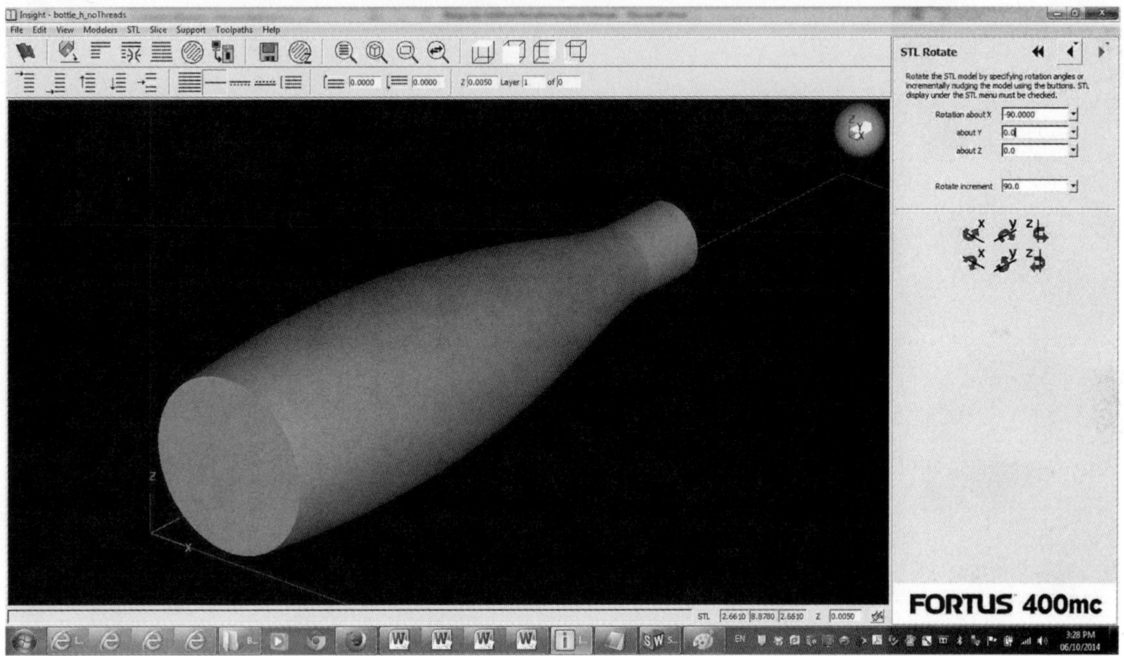

Figure 42.2 (Continued)

Outcomes of our experimentations. *(Continued)*

Six Build Orientations Analyzed and the Cost of Each at Layer Height of 0.005 in.

The outcomes of the experiment for six build orientations are shown in Figure 42.2.

Orientation 1 (Z) $125	Orientation 2 (−Z) $369	Orientation 3 (−90 X) $239
Orientation 4 (−90Y) $243	Orientation 5 (−90 Y) $145	Orientation 6 (90 Y) $134

Orientation Is +Z

Figures 42.3 through 42.6 show the processing for this orientation.

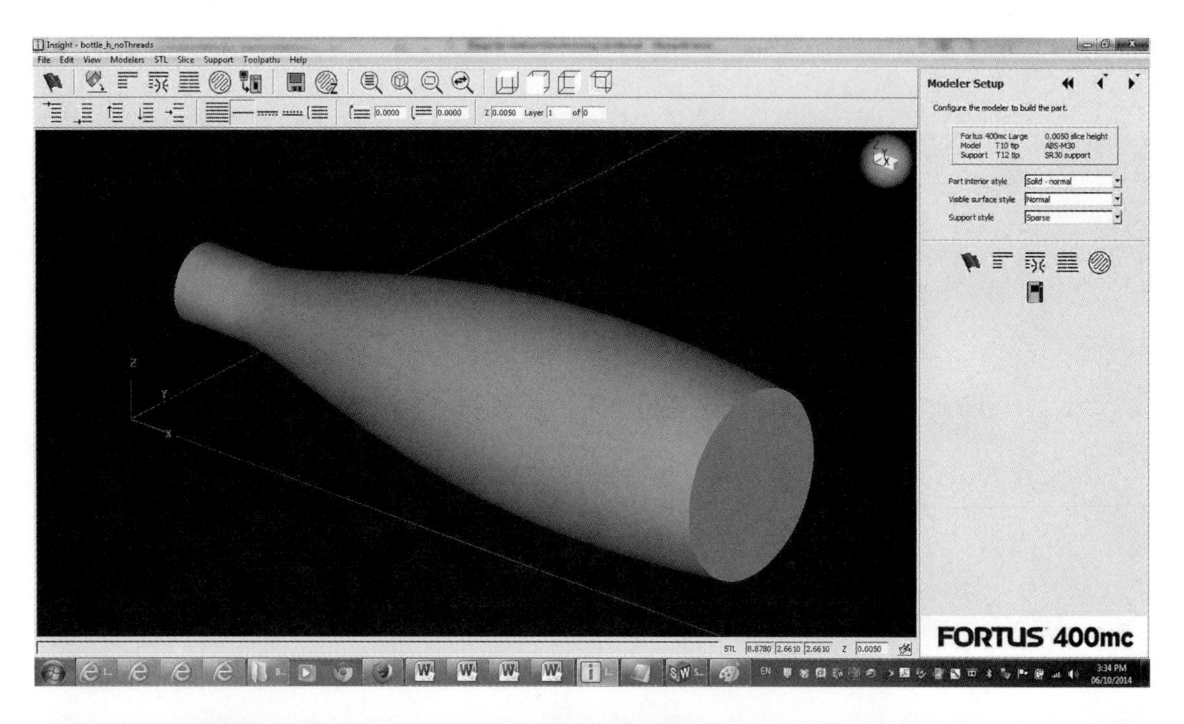

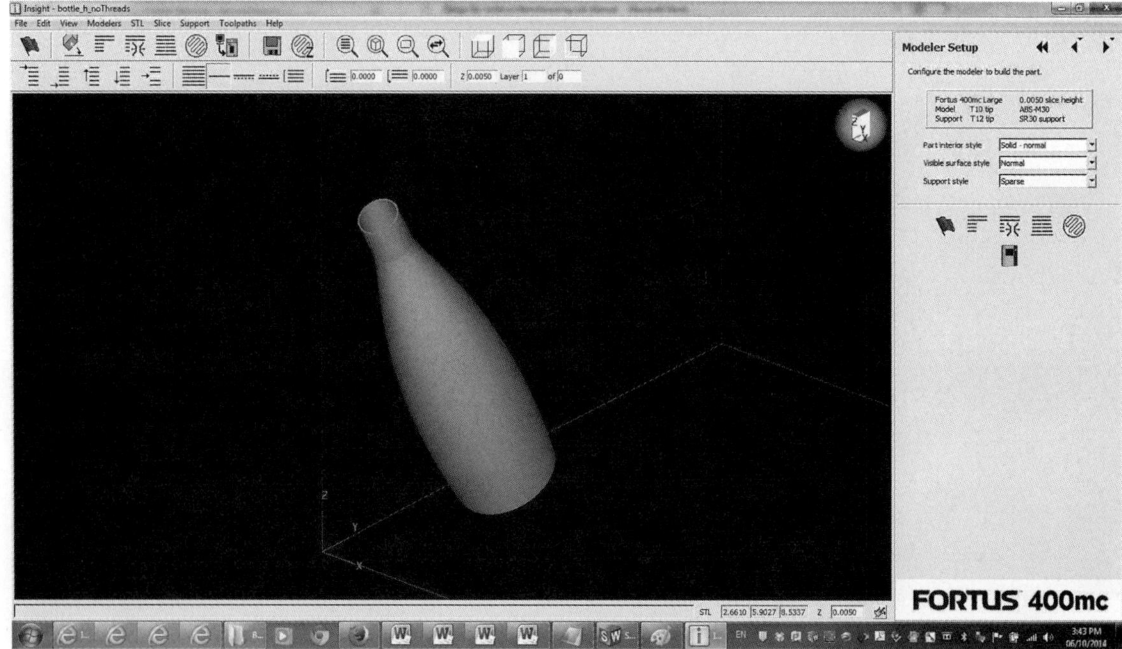

Figure 42.2 (Continued)

Outcomes of our experimentations.

(Continued)

Orientation Is –Z

Figures 42.7 through 42.11 show the processing for this orientation.

Orientation Is –90X

Figures 42.12 through 42.15 show the processing for this orientation.

Orientation Is –90Y

Figures 42.16 through 42.20 show the processing for this orientation.

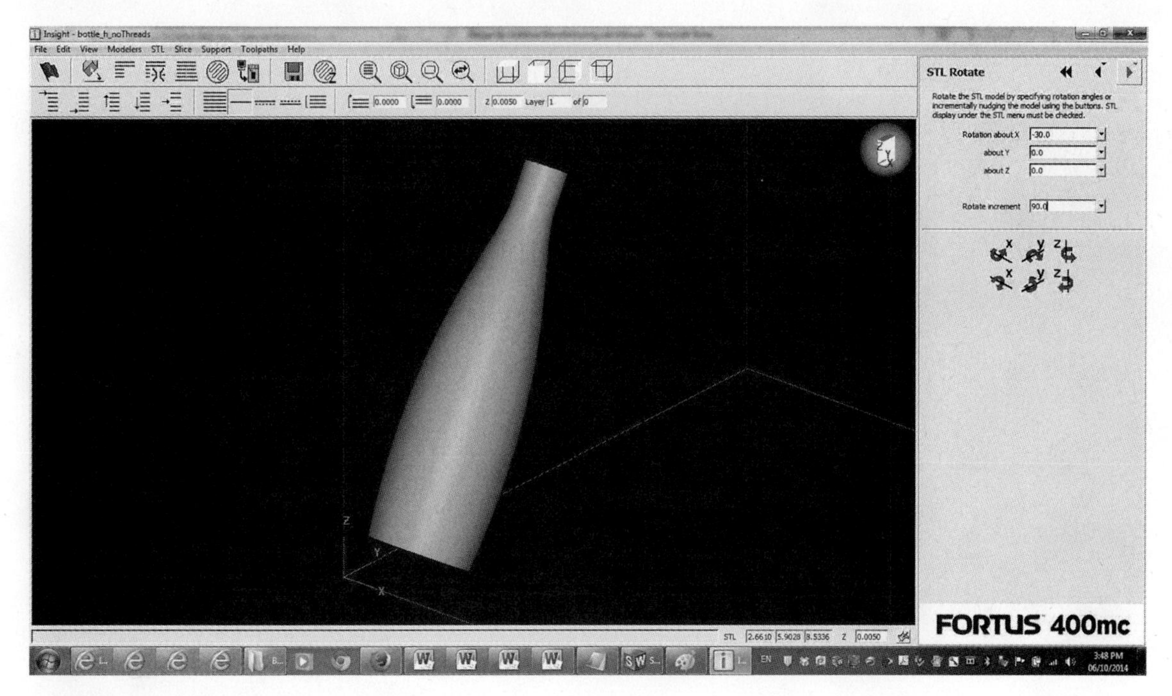

Figure 42.2 (Continued)

Outcomes of our experimentations.

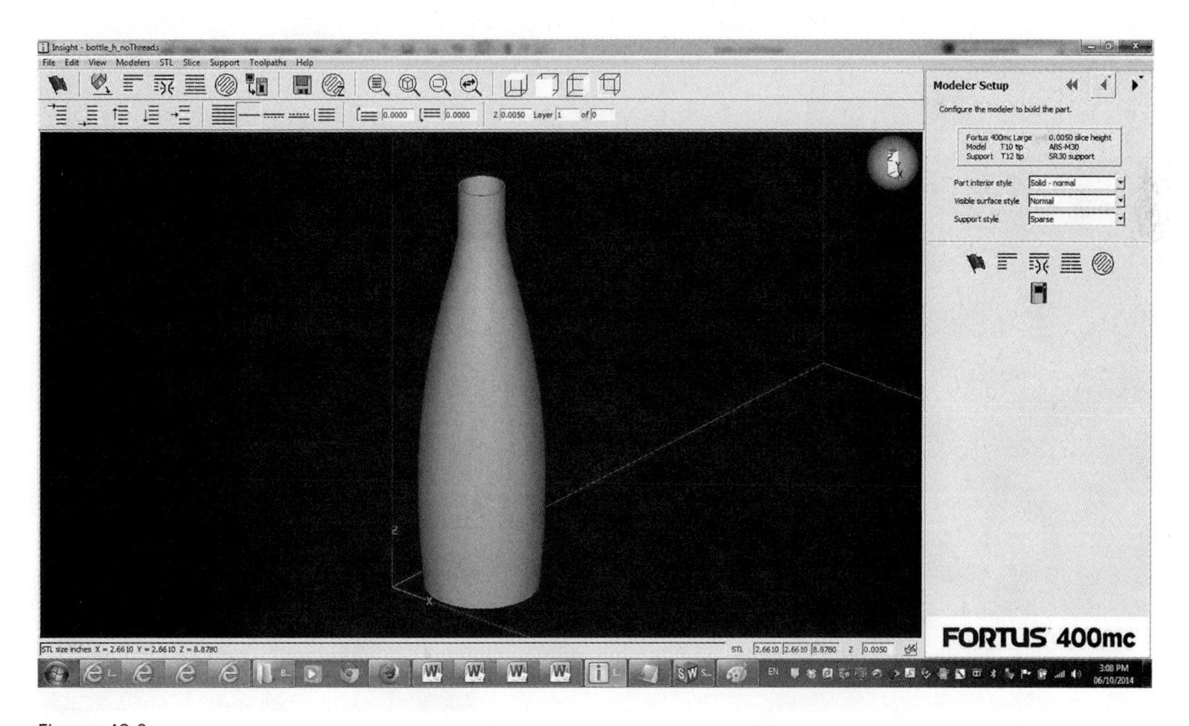

Figure 42.3

Z-orientation.

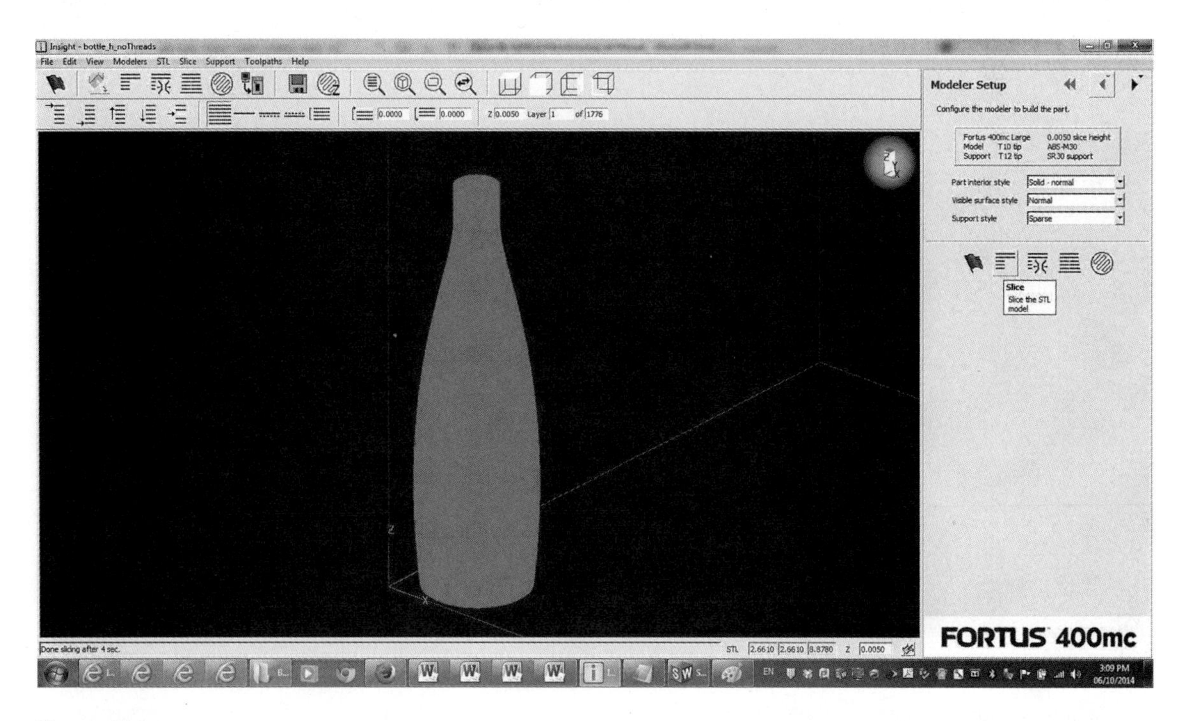

Figure 42.4

Slicing of part.

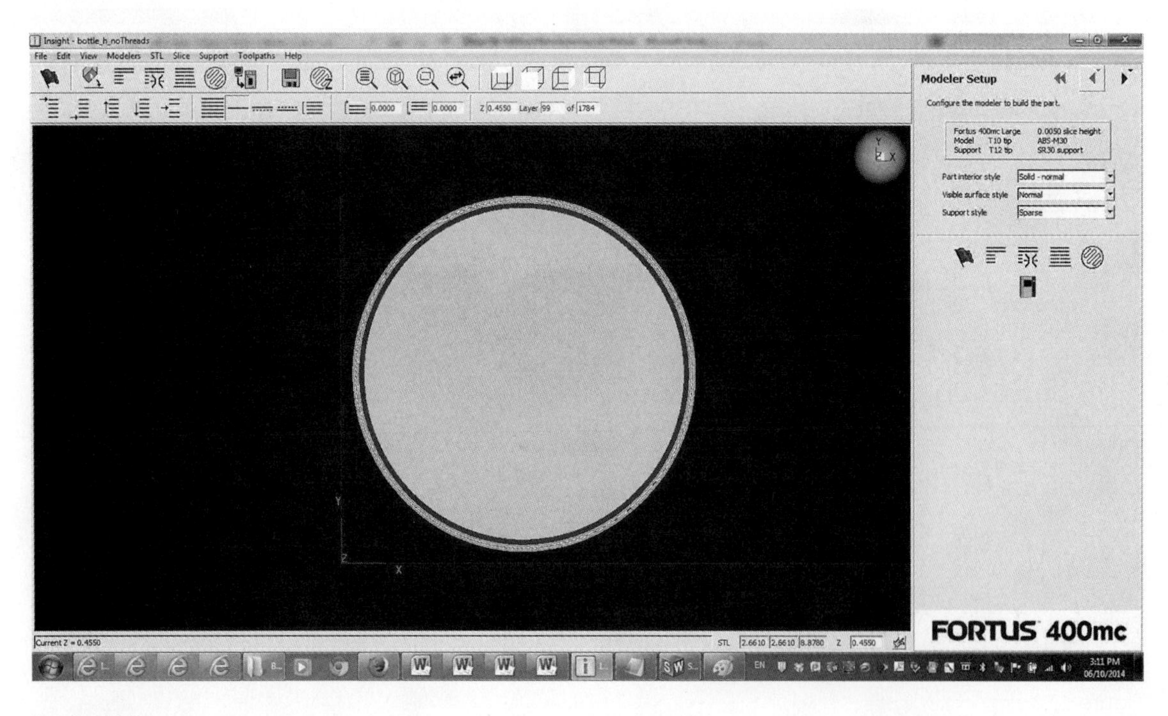

Figure 42.5

Interior toward the bottom of bottle.

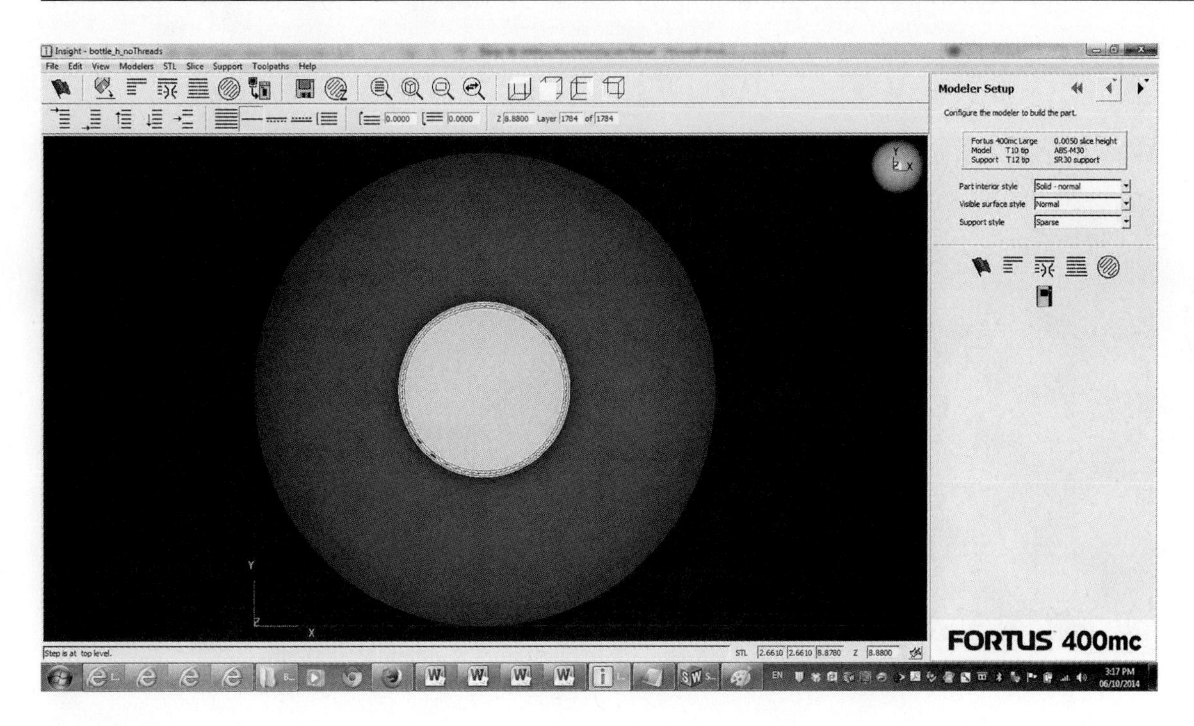

Figure 42.6

Interior toward the top of bottle.

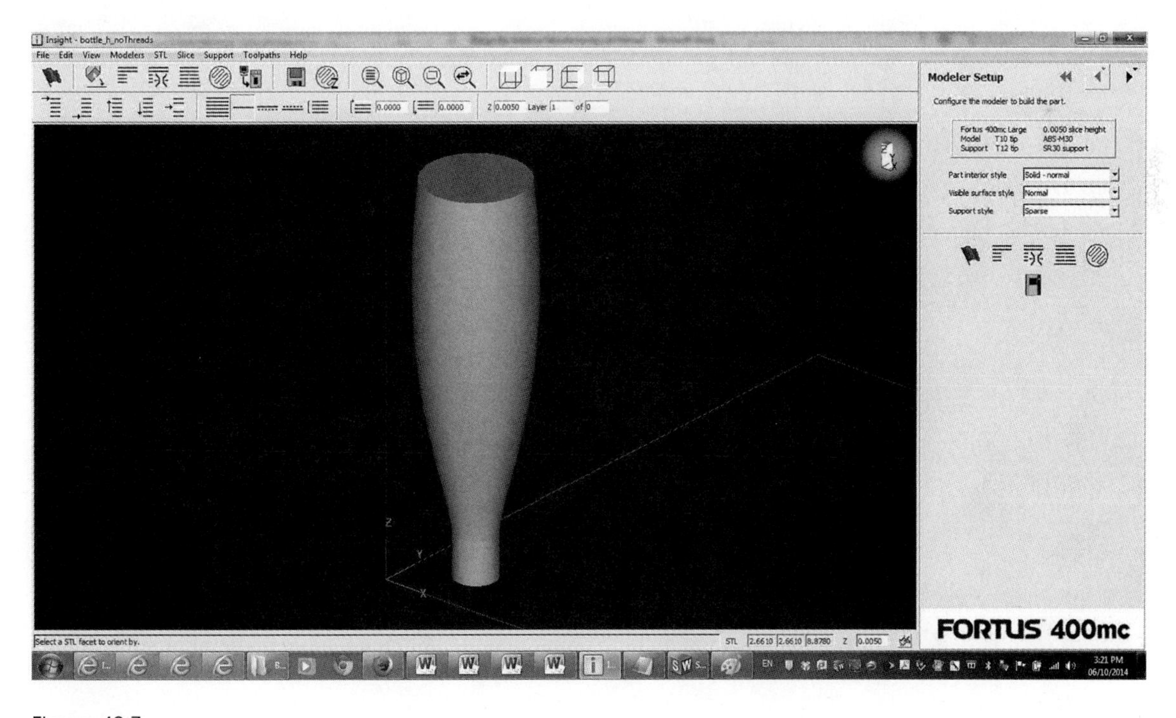

Figure 42.7

Orientation of model.

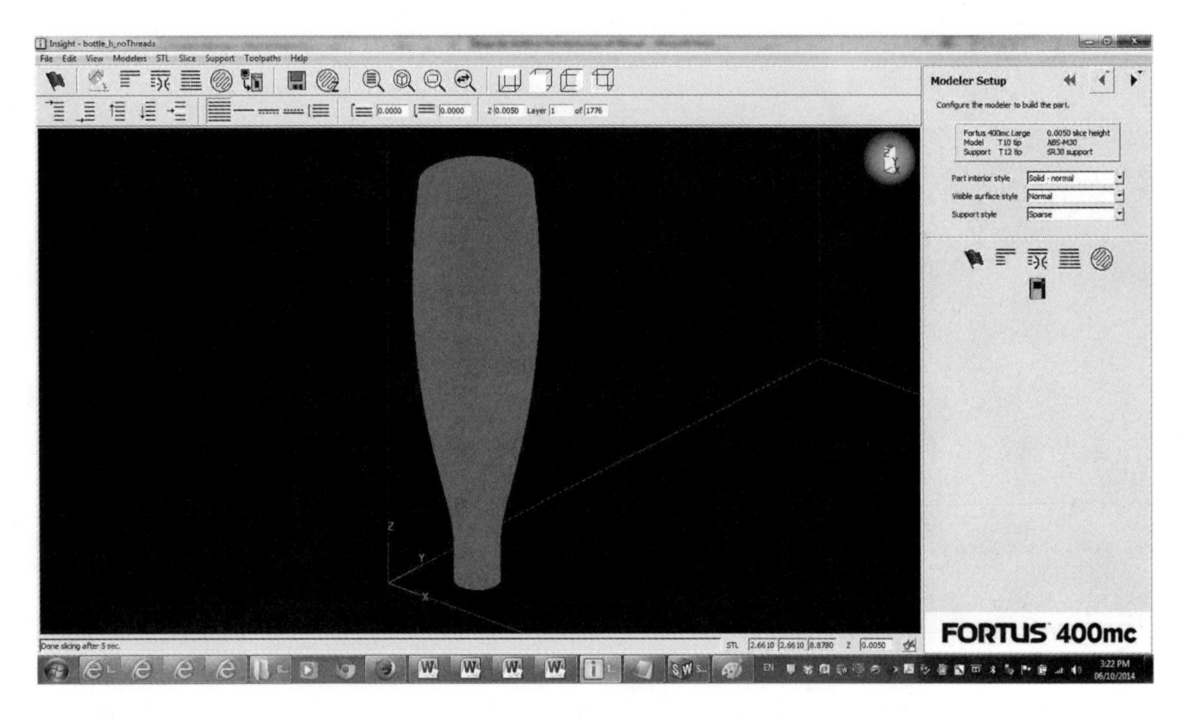

Figure 42.8

Slicing of model.

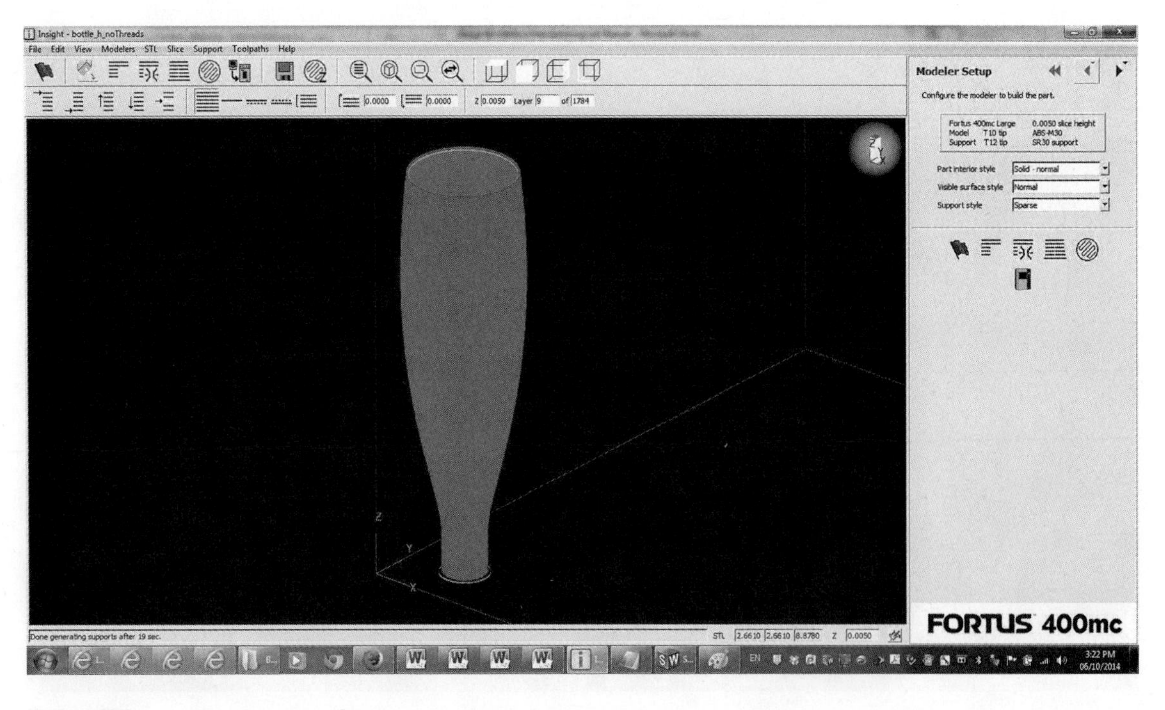

Figure 42.9

Support created.

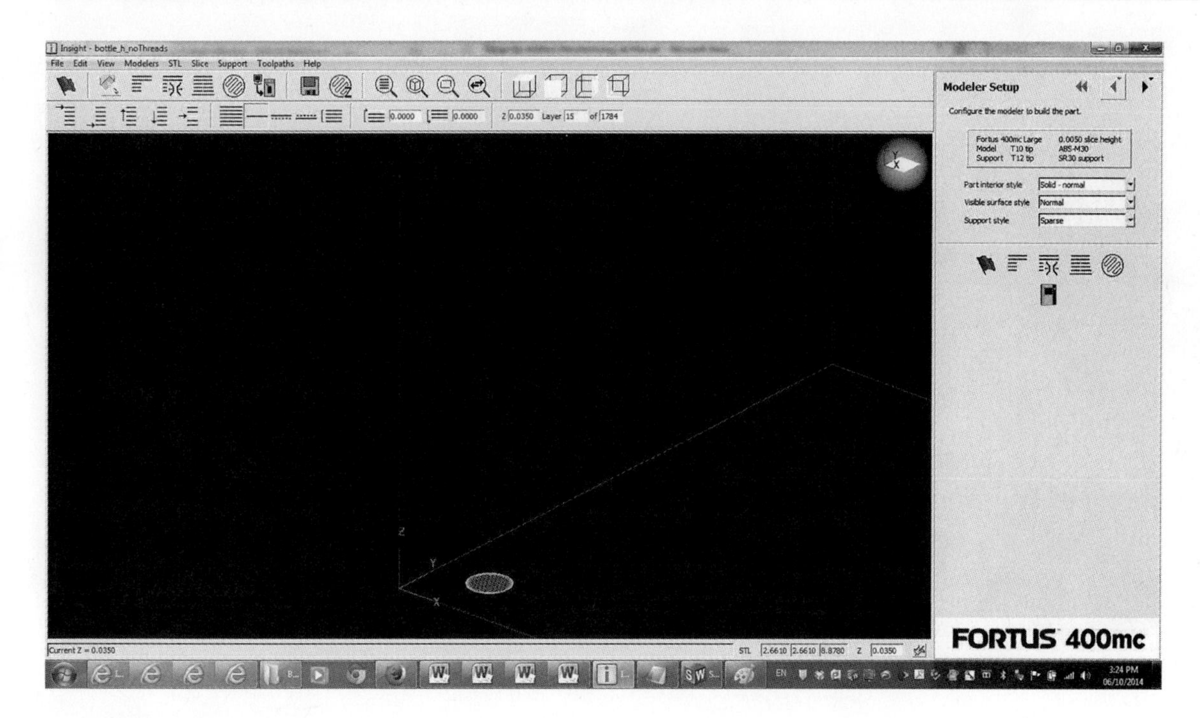

Figure 42.10

Generating toolpath toward bottom.

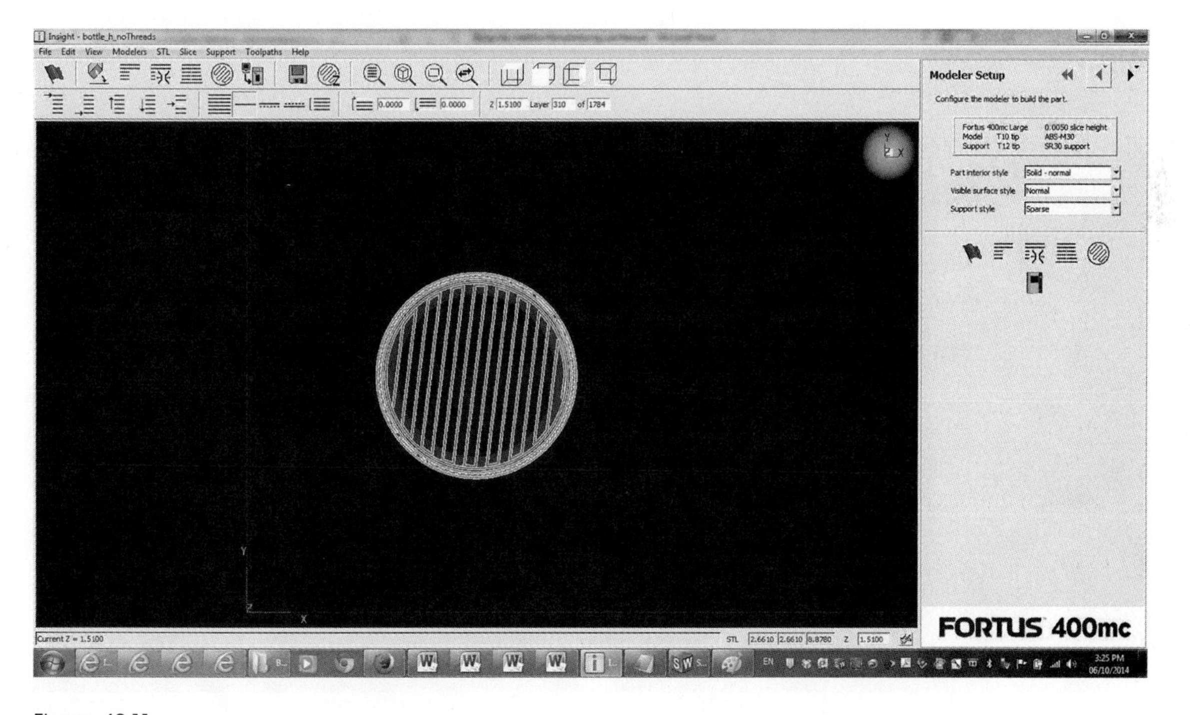

Figure 42.11

Notice that support is everywhere inside.

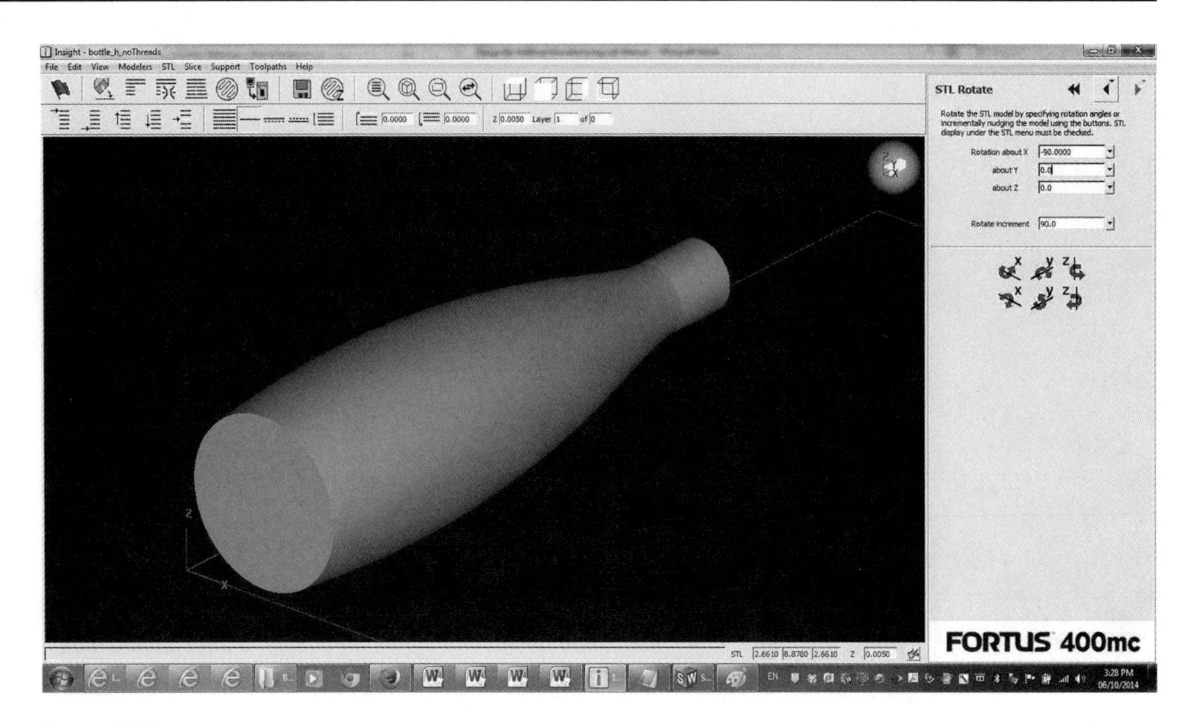

Figure 42.12

Orientation is −90X.

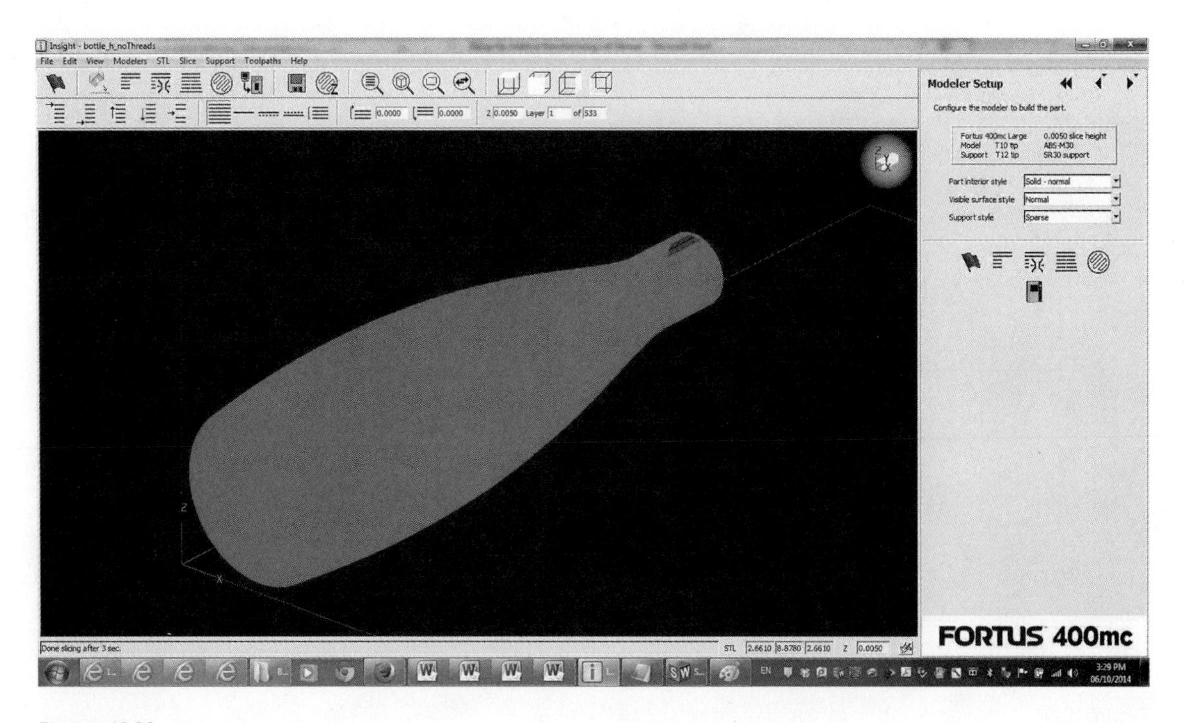

Figure 42.13

Slicing of model.

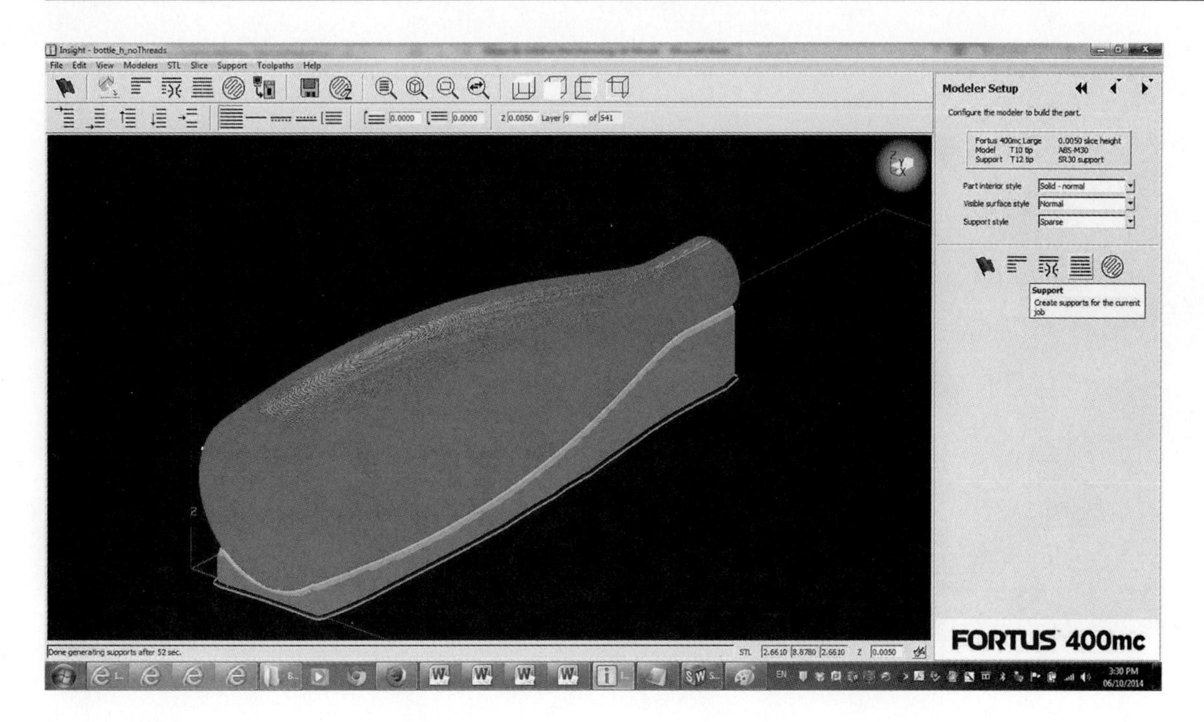

Figure 42.14

Support created.

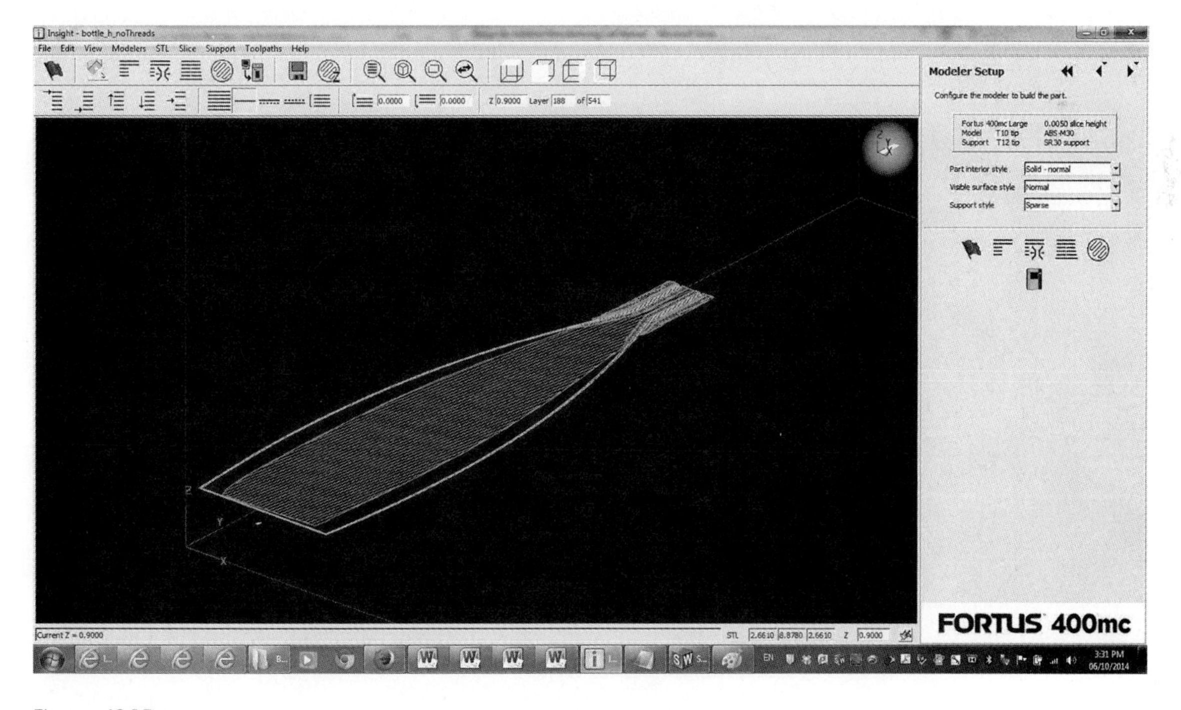

Figure 42.15

Toolpath definition.

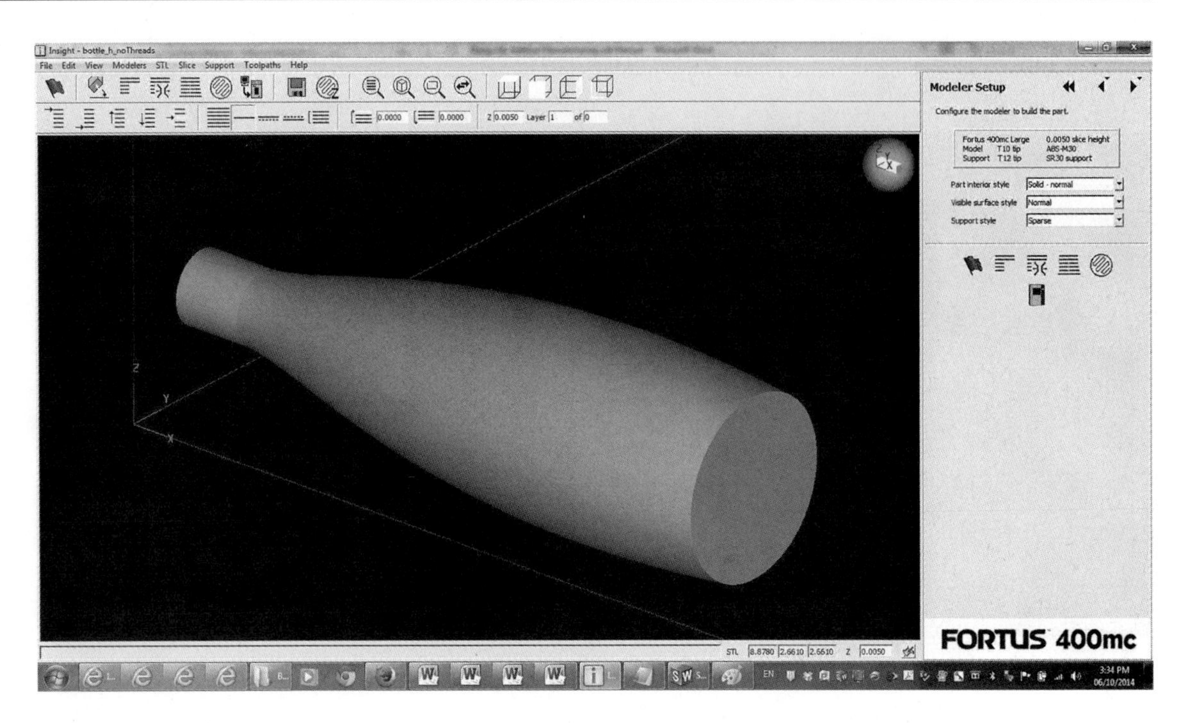

Figure 42.16

STL model.

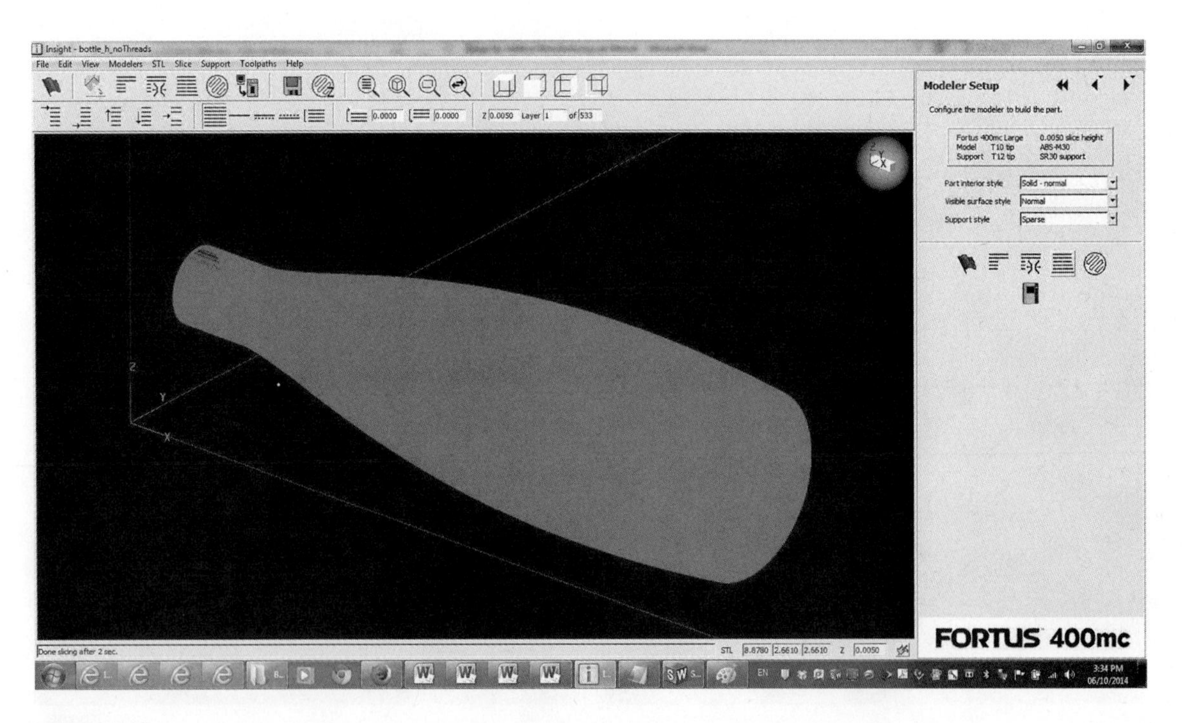

Figure 42.17

Slicing model.

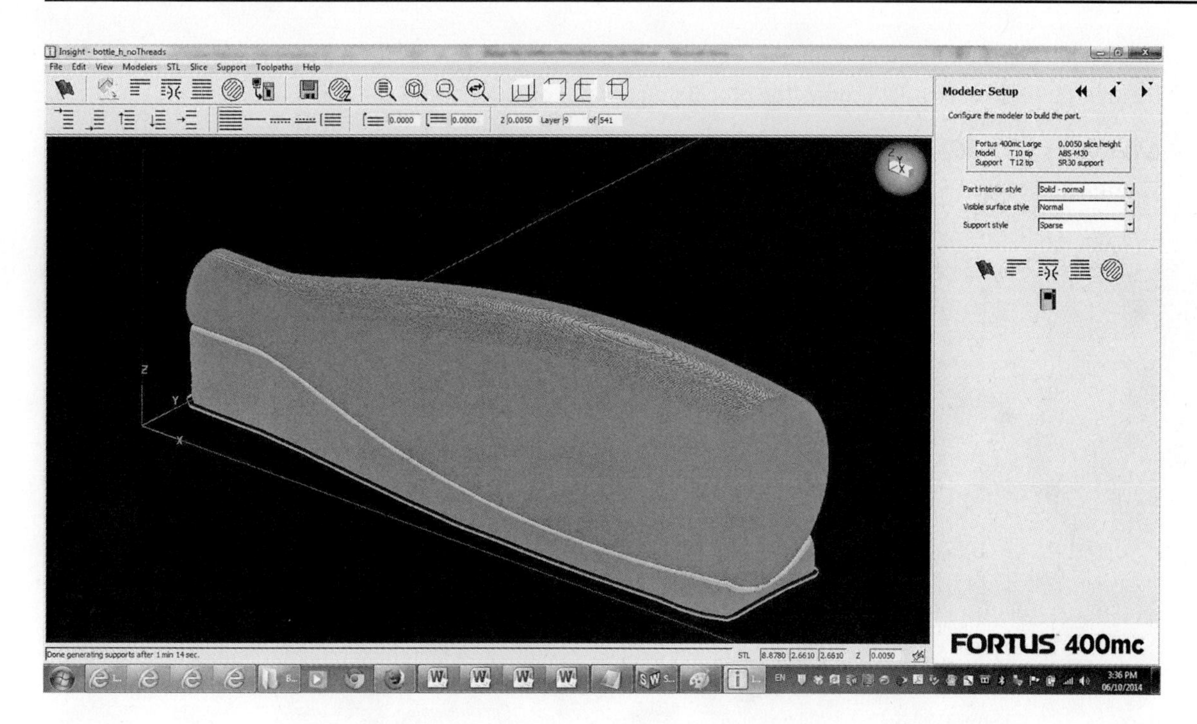

Figure 42.18

Support created.

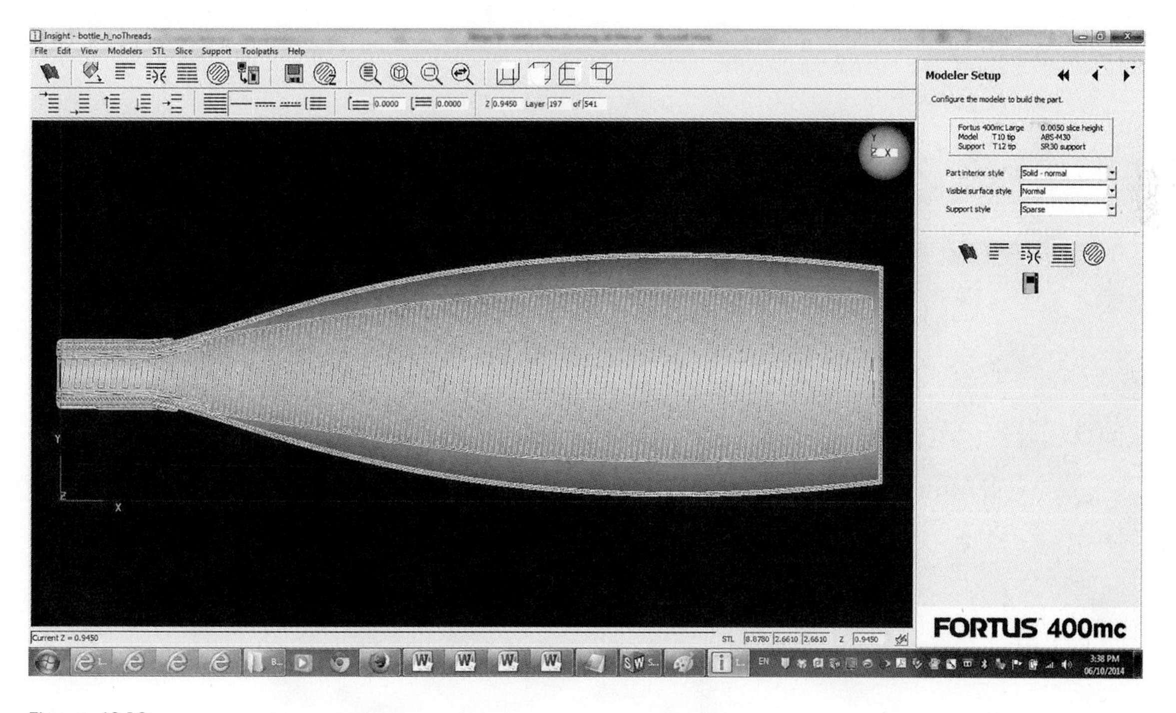

Figure 42.19

Toolpath generation exposed inside of bottle.

Orientation Is 30X

Figures 42.20 through 42.25 shows the processing for this orientation.

Orientation Is −30X

Figures 42.23 through 42.25 show the processing for this orientation.

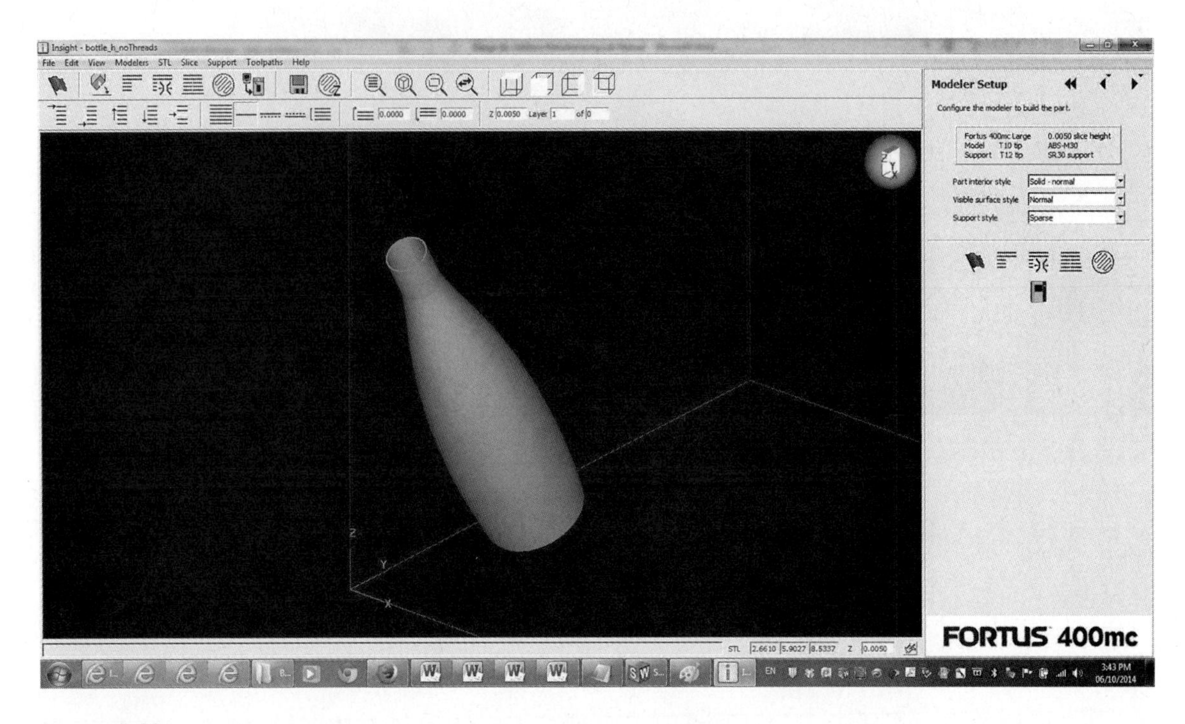

Figure 42.20

STL model opened and oriented in Insight.

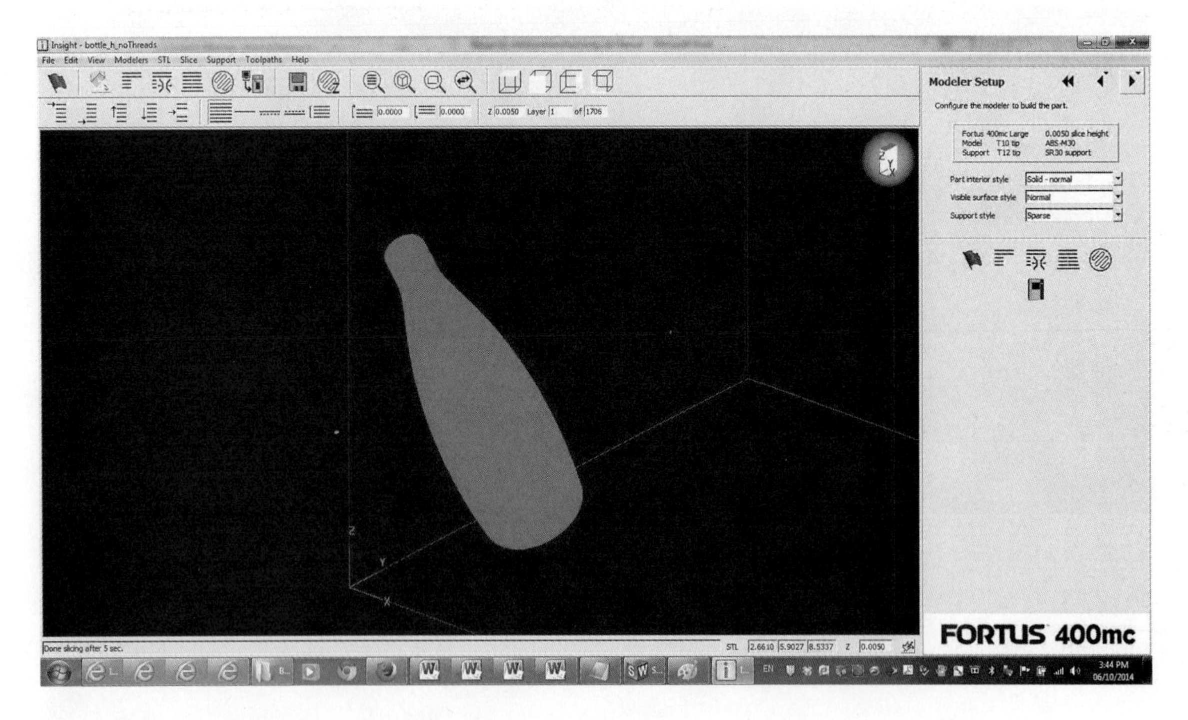

Figure 42.21

Slicing of model.

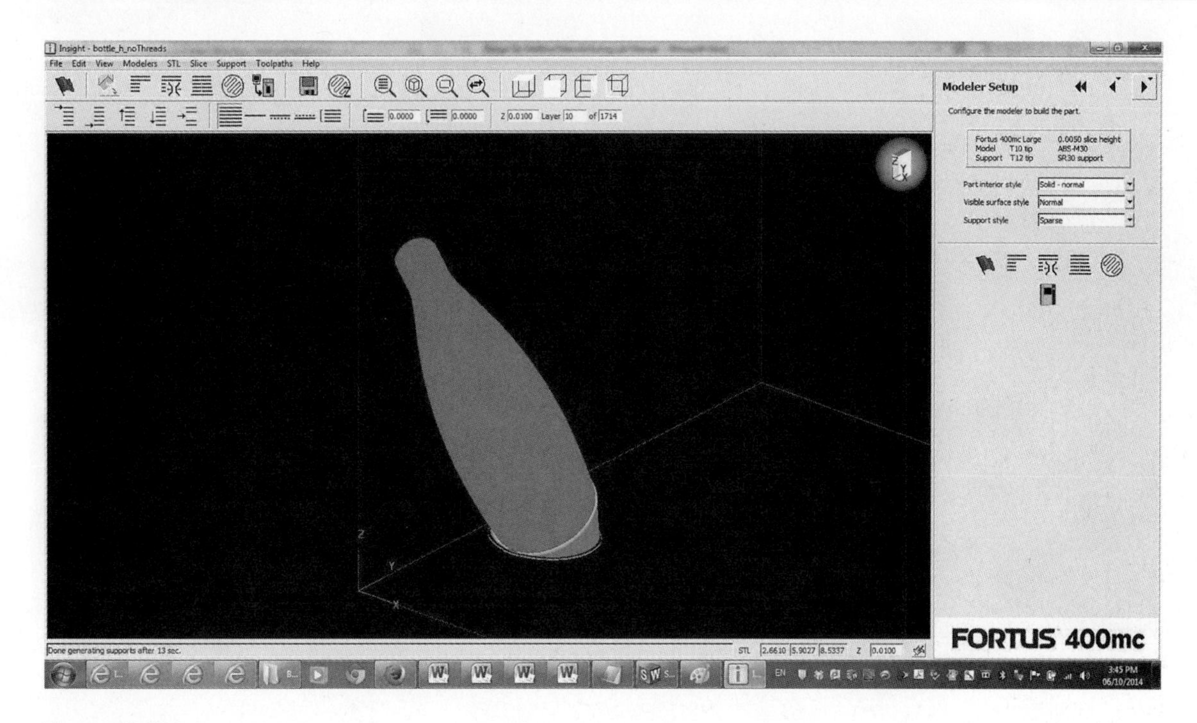

Figure 42.22

Support for model.

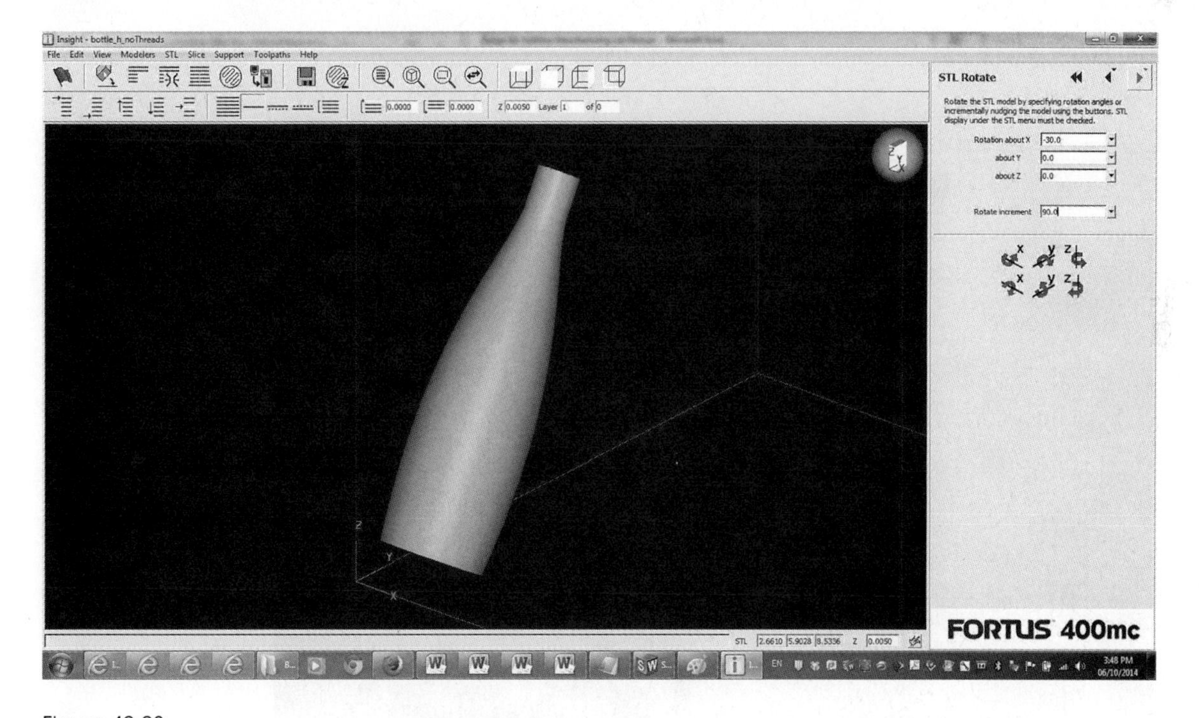

Figure 42.23

STL model opened and oriented in Insight.

The cost implications for the six orientations are summarized in Table 42.1 based on the cost model that the author developed.

Case Study 2

In this case study, another bottle is given to you. Follow the steps that are described in Figures 42.26 through 42.34. For *Printing Upside Down*, see Figures 42.32 through 42.34.

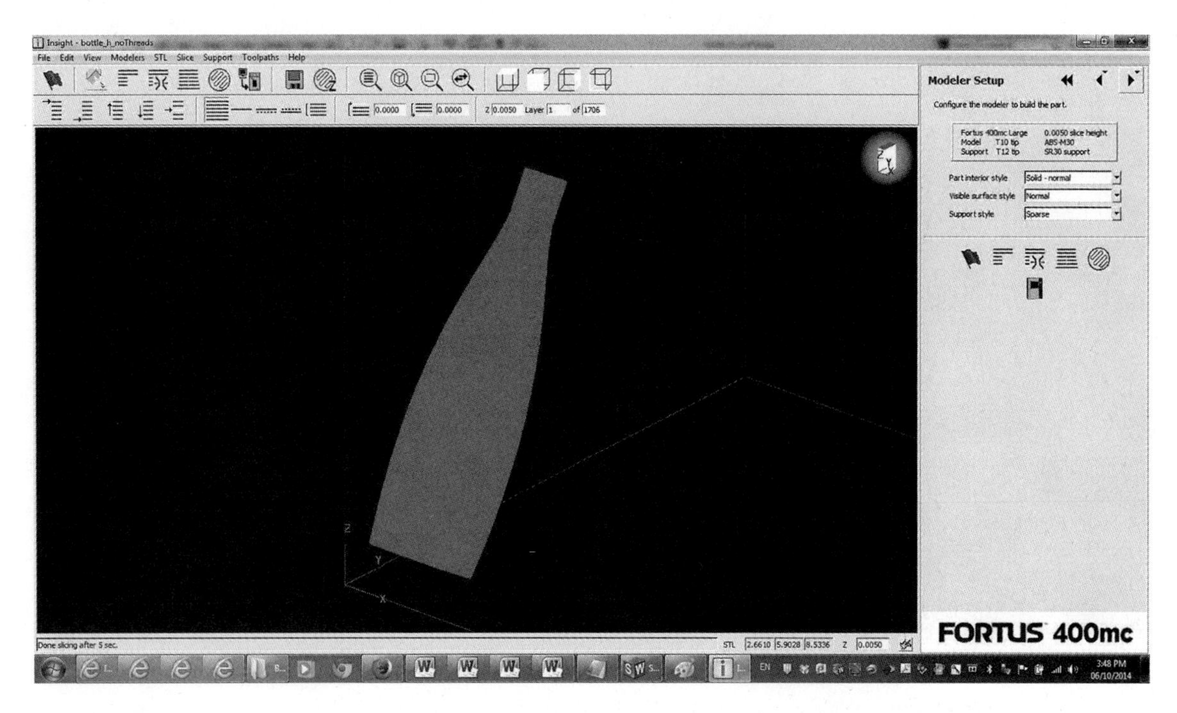

Figure 42.24

Slicing of model.

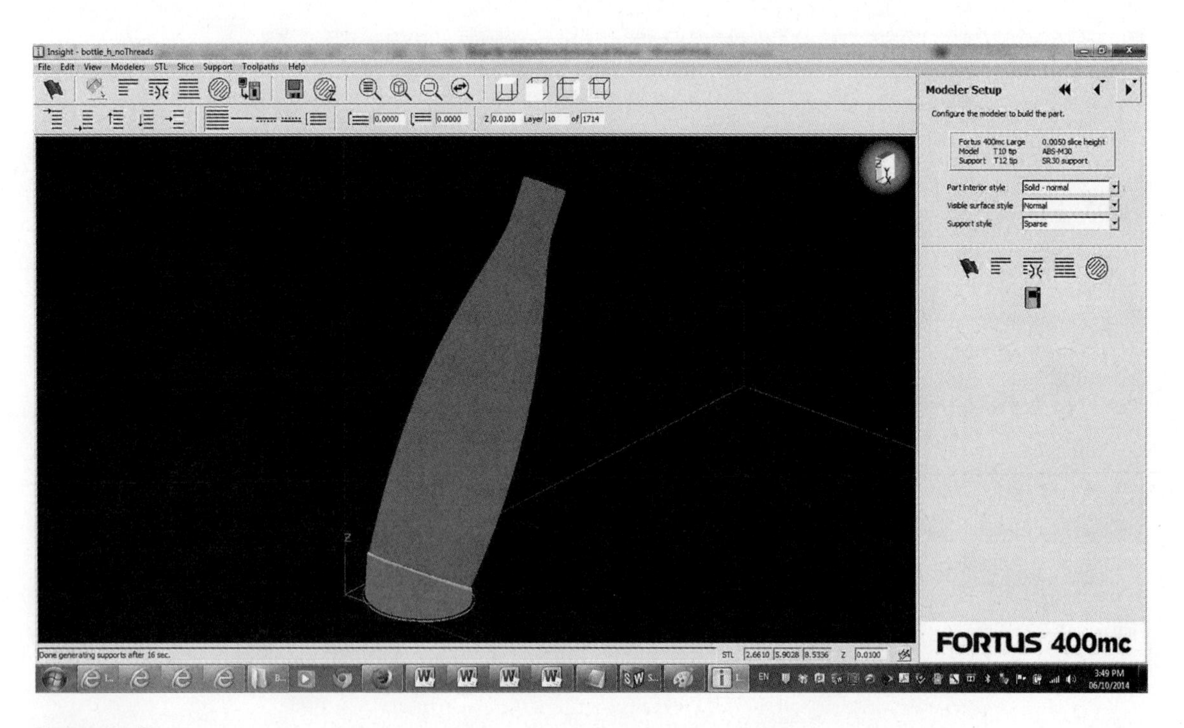

Figure 42.25

Support created for model.

Table 42.1 Cost of Building Different Orientations for the Bottle

	Orientation (Axis)	Build Time (Hours)	Model Volume (in.^{3})	Support Volume (in.^{3})	Cost ($)
Vertical (upright)	Z	10.7	2.445	0.065	**125**
Vertical (downward)	−Z	38.54	5.979	13.211	369
Horizontal	−90X	24.4	3.754	9.5	239
Horizontal	−90Y	25.2	3.754	9.417	243
Inclined	30X	14.21	2.923	1.668	145
Inclined	−30X	14.22	2.923	1.67	134

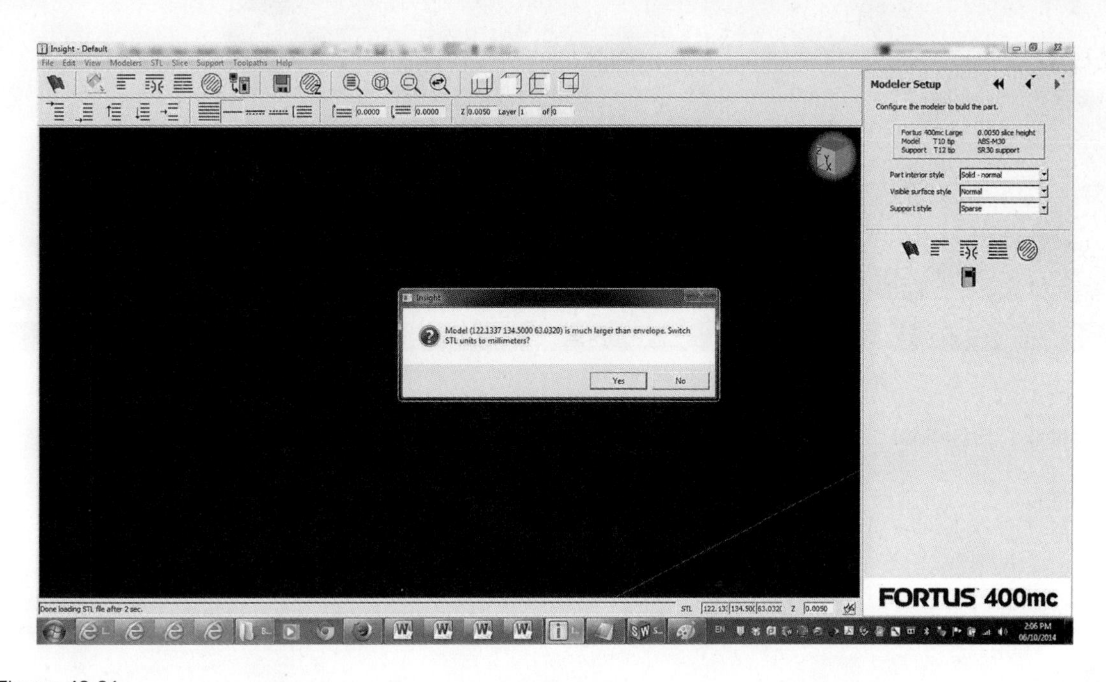

Figure 42.26

Open the STL file.

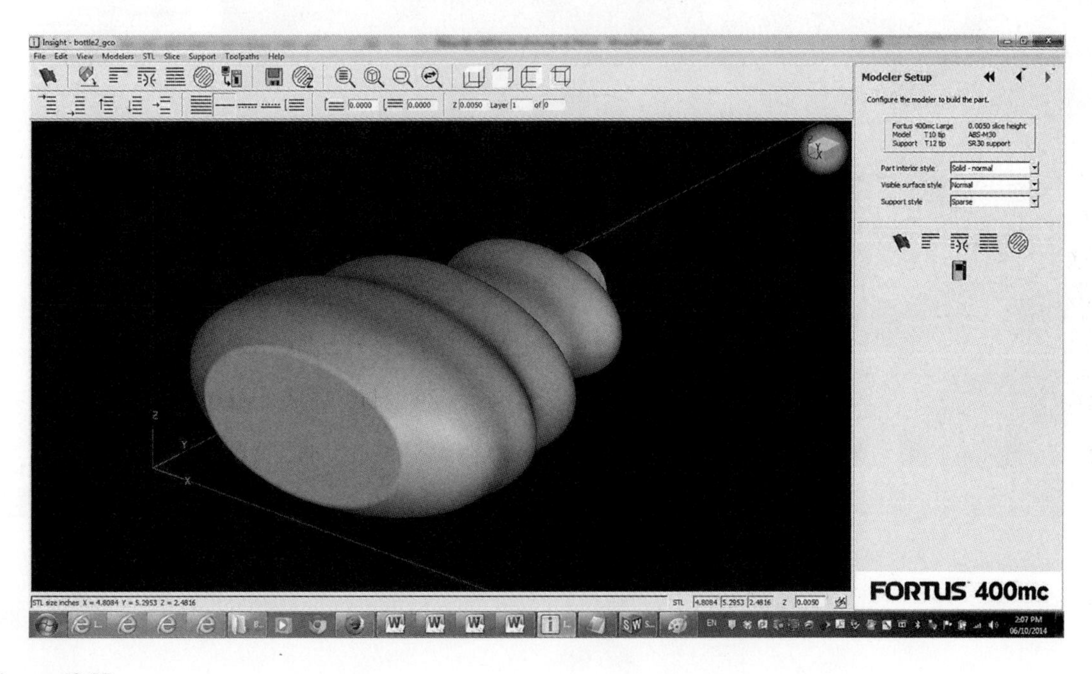

Figure 42.27

STL model.

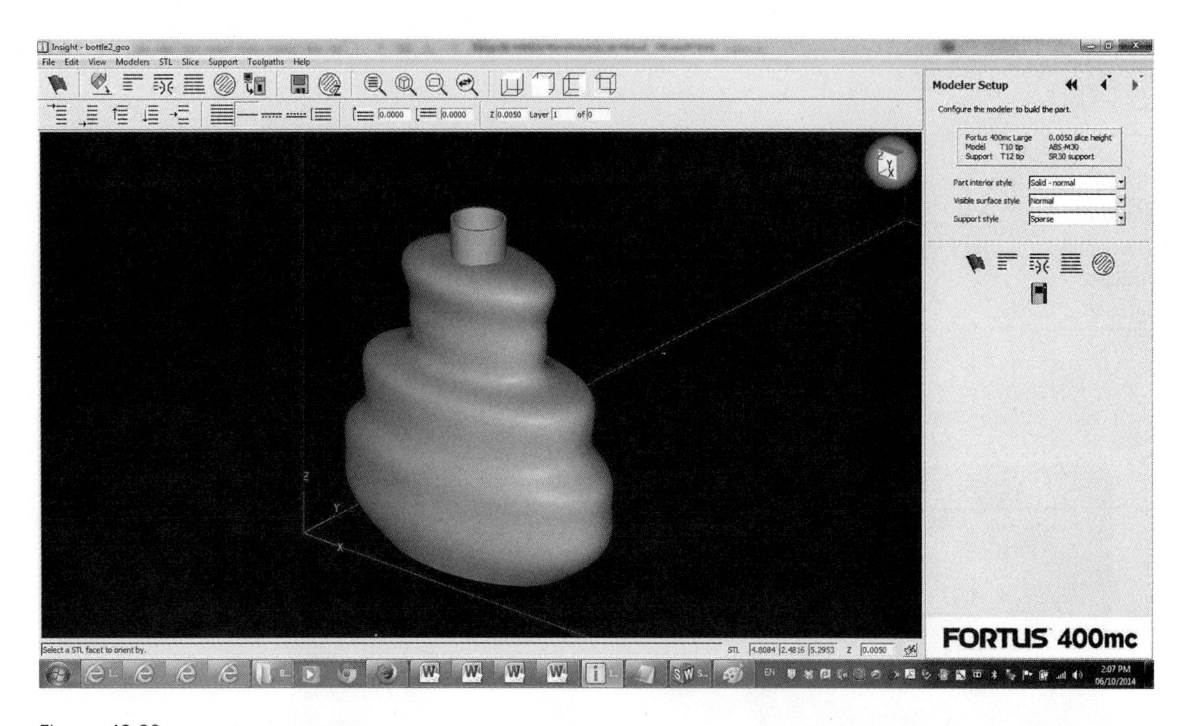

Figure 42.28

Orientation in vertical direction.

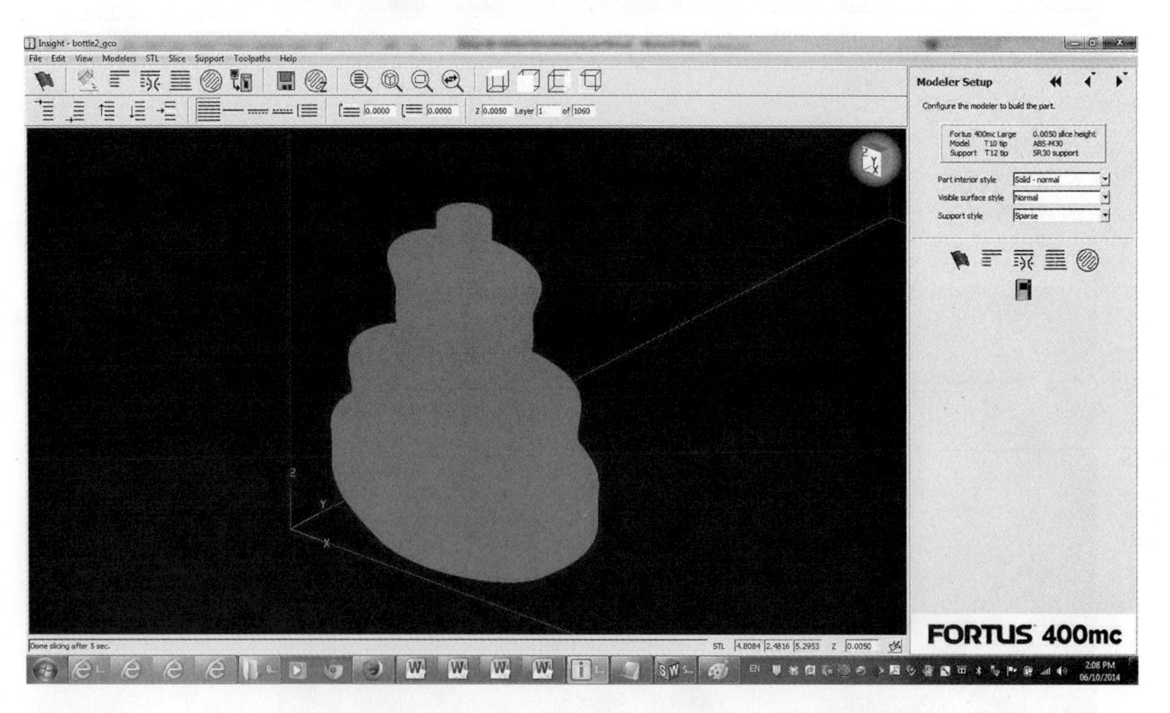

Figure 42.29

Slice model.

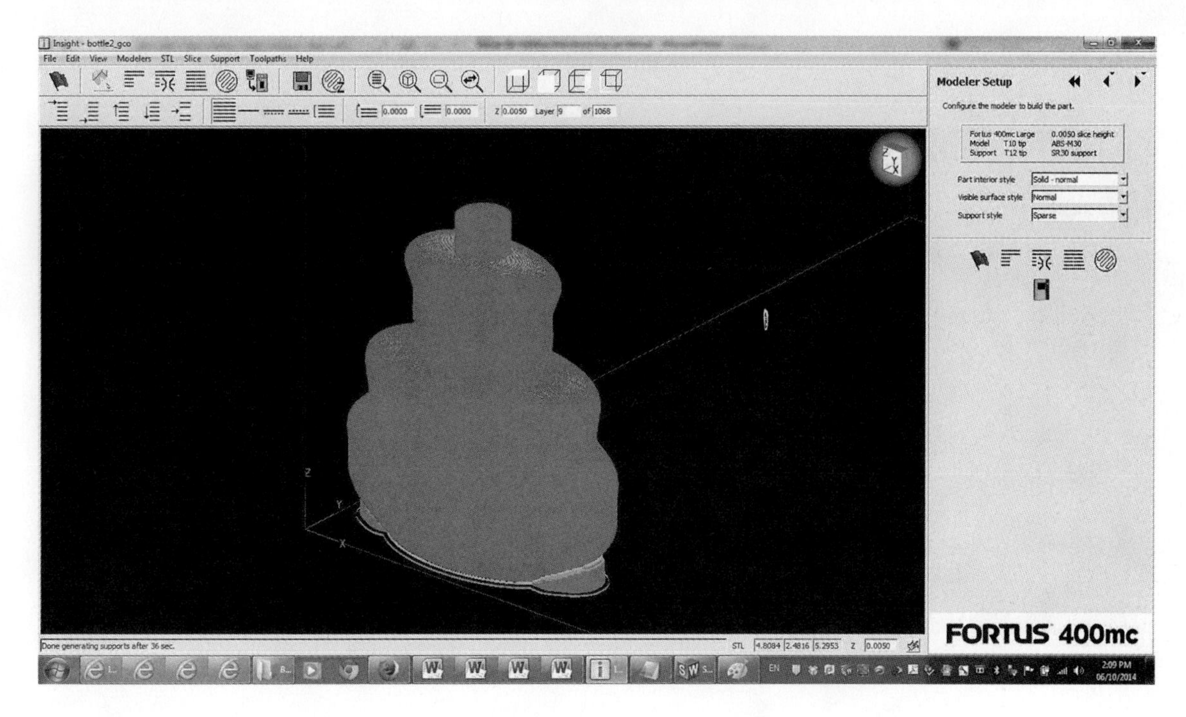

Figure 42.30

Support created.

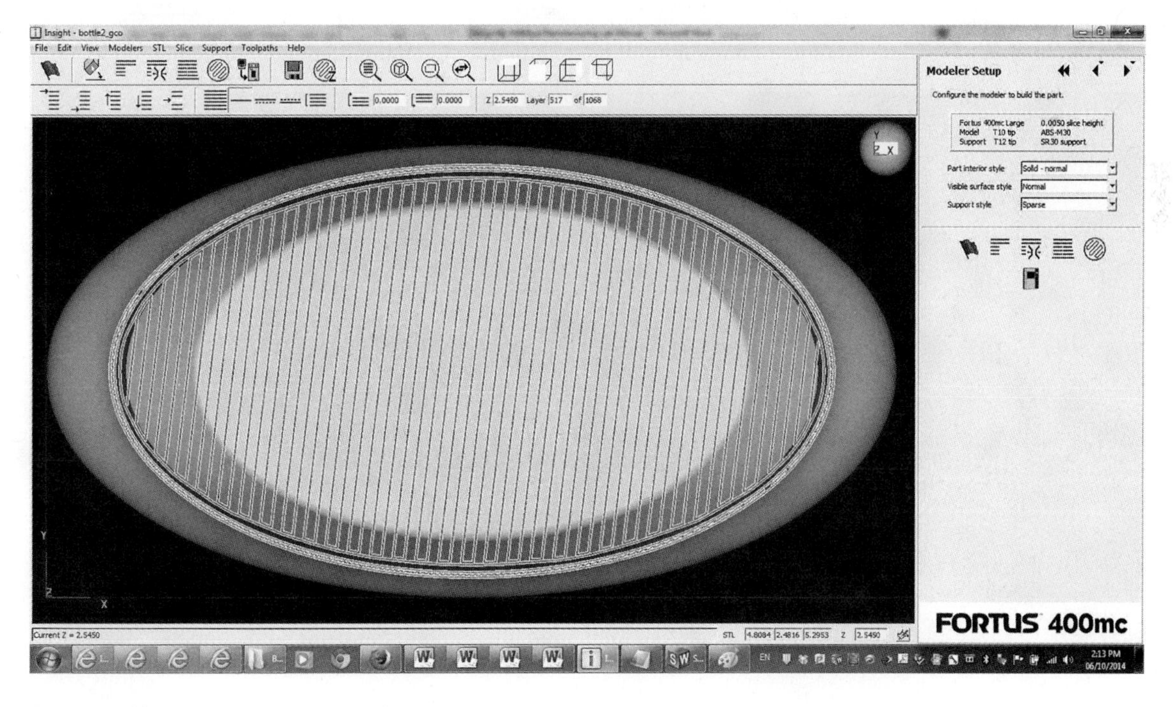

Figure 42.31

Layer 514 out of 1068 with Z = 2.5450 in.

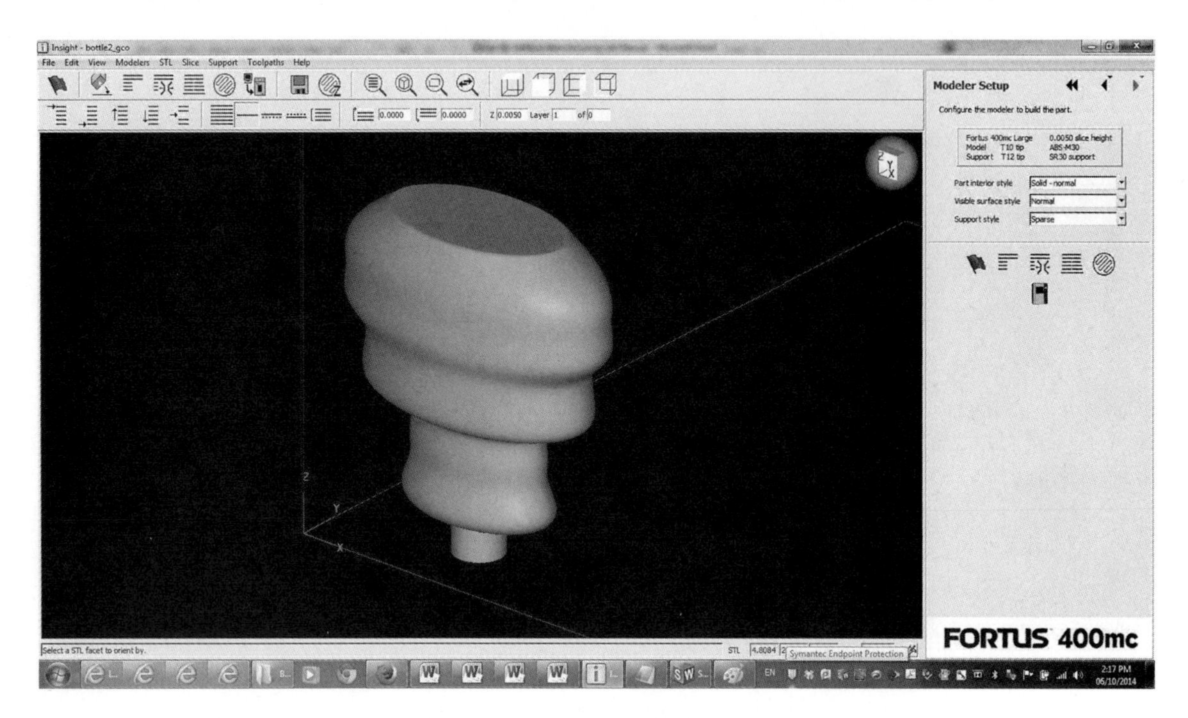

Figure 42.32

Model oriented differently.

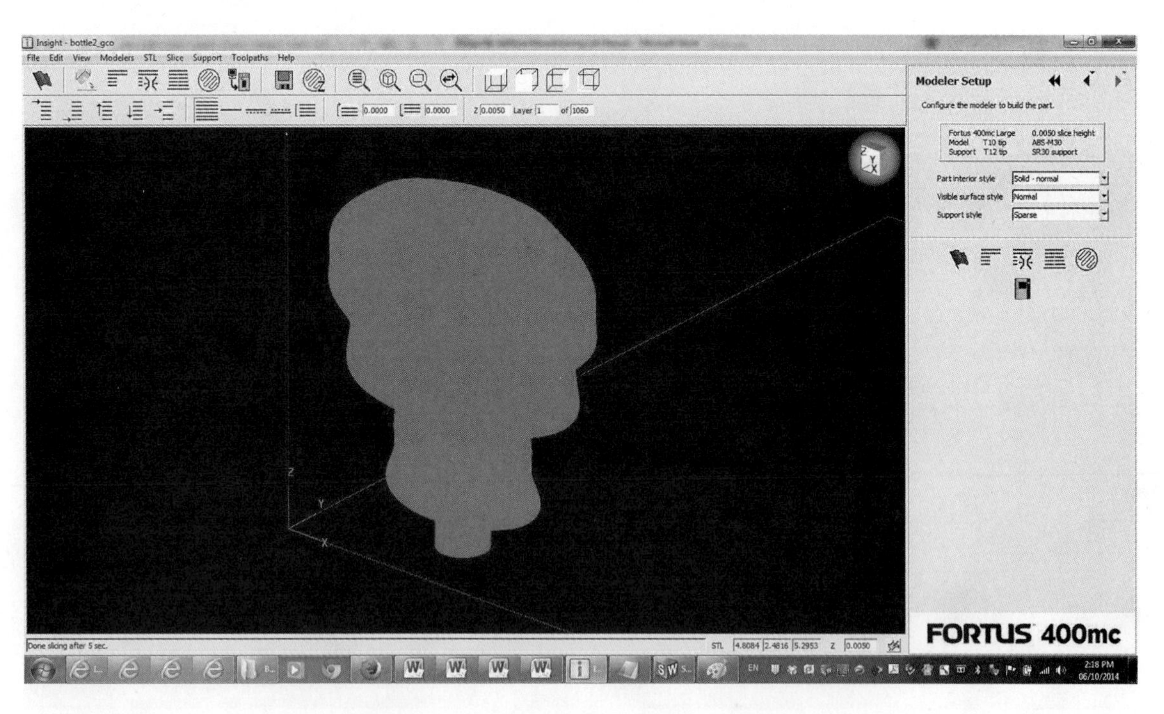

Figure 42.33

Slice model.

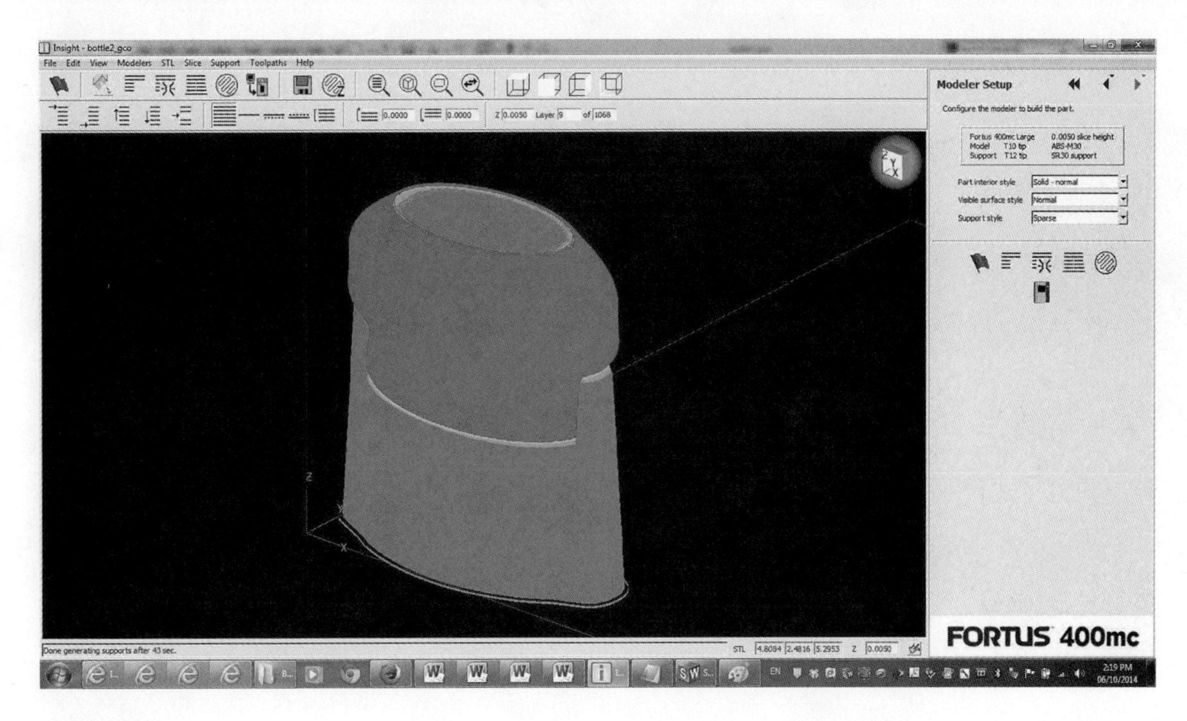

Figure 42.34

Support material created. Note: this model encounters problem for this orientation; experiment on your own and find out why.

Summary

In this module, we have carried out a cost analysis of the 3D printing of bottles in different orientations. A decision should be made regarding whether aesthetics should take preference over building cost. For the first bottle case study, the optimum cost and aesthetic appeal are currently obtained for the +Z orientation. Such analysis is important because we can observe that turning the bottle upside down is most expensive. Why? This is because support material is stuffed inside the bottle. How will we remove the support material after printing? That orientation should be avoided as much as possible.

43

Design for Functionality

Objectives:

When you complete this module, you will

- Understand the guidelines for designing for functionality
- Prepare the toolpath for a 19-mm female Quick-Release Buckle
- 3D-print the 19-mm female Quick-Release Buckle using the Fortus machine

Introduction

Build orientation has the biggest impact on part functionality and feature strength. However, a thinner layer height can improve strength to certain features and geometries. In the examples in this module, you can see how build orientation affects feature strength. In general, when extruding material, the parts will be stronger along the layer lines versus between layers.

Case Study: 19-mm Female Quick-Release Buckle

Printing in Horizontal Orientation

Figures 43.1 through 43.9 show the steps for the preprocessing of the Standard Tessellation Language (STL) file for printing in horizontal orientation.

Open the STL for the 19-mm female Quick Release Buckle, which your instructor will supply. Notice the warning from the software (see Figure 43.1).

Printing in Horizontal Orientation

Click Toolpaths > Setup

Accept settings as follows:

Contour width: 0.0120"

Raster width: 0.0120"

Raster to raster air gap: 0.0000"

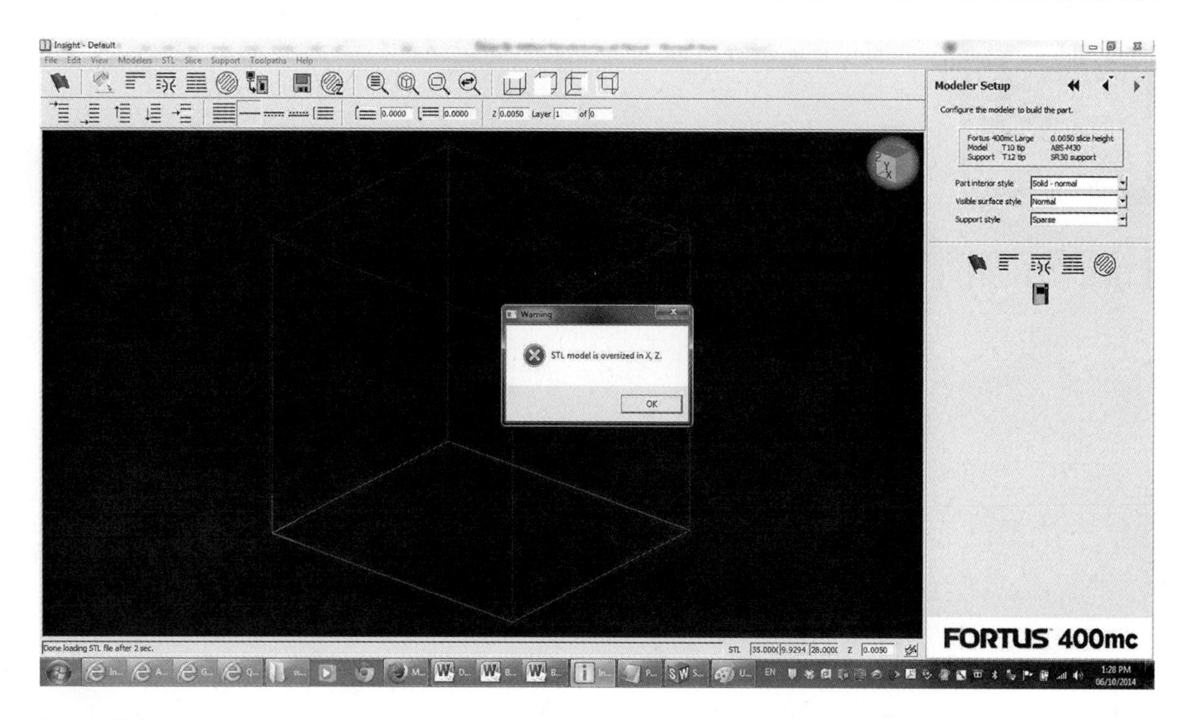

Figure 43.1

Warning when opening STL model.

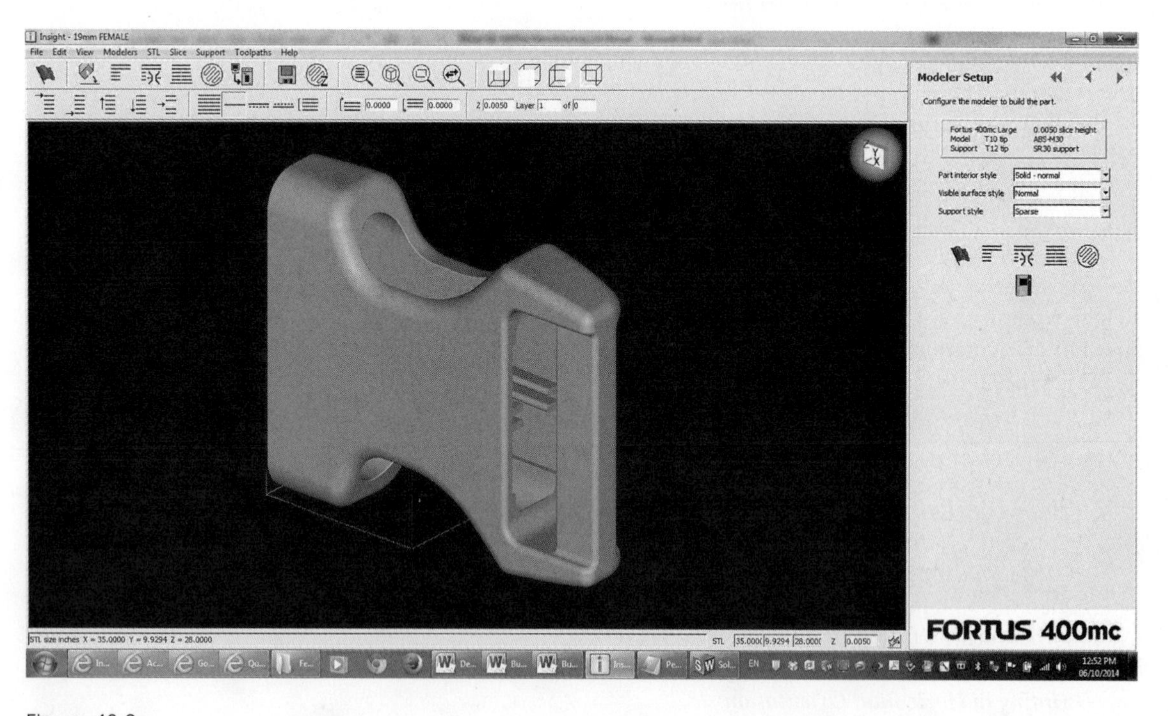

Figure 43.2

STL model.

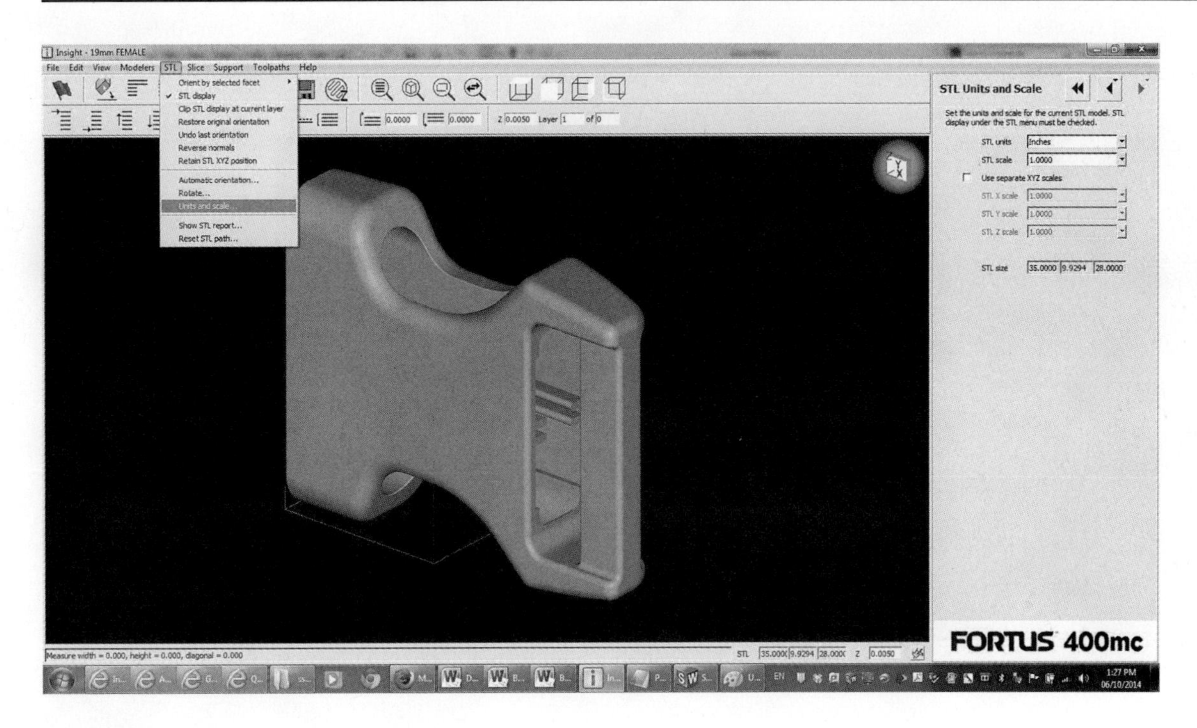

Figure 43.3

Correct unit is chosen: inches in this case.

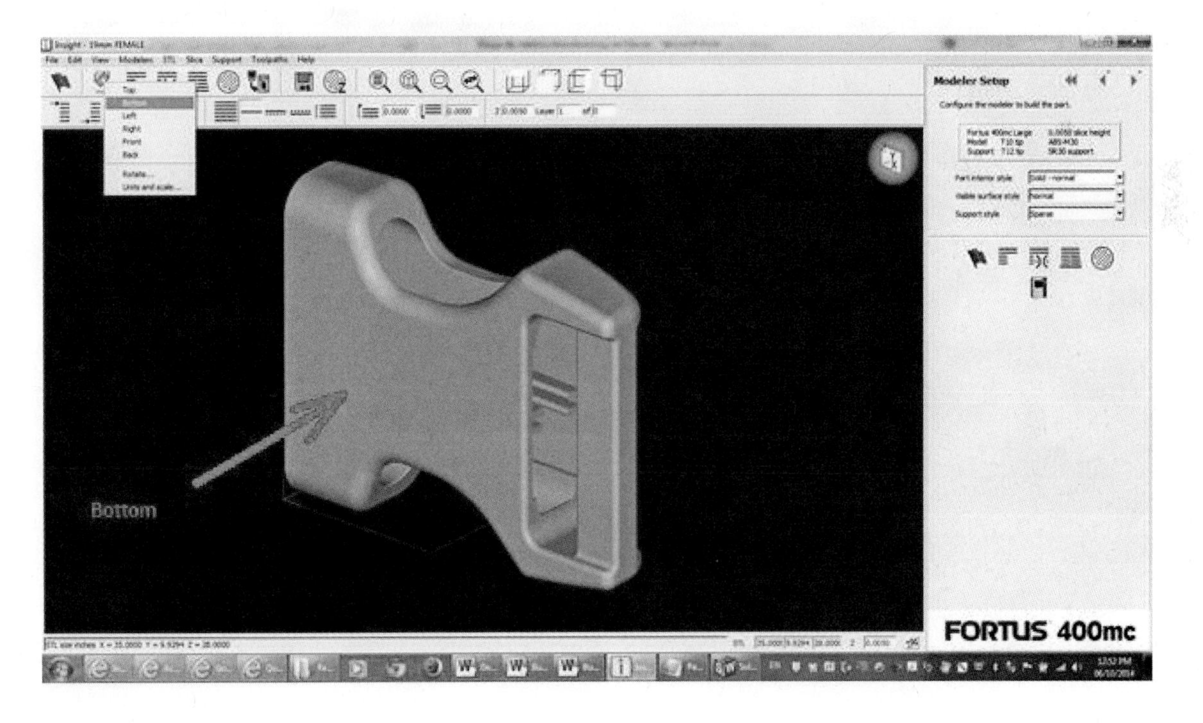

Figure 43.4

Bottom is chosen as shown.

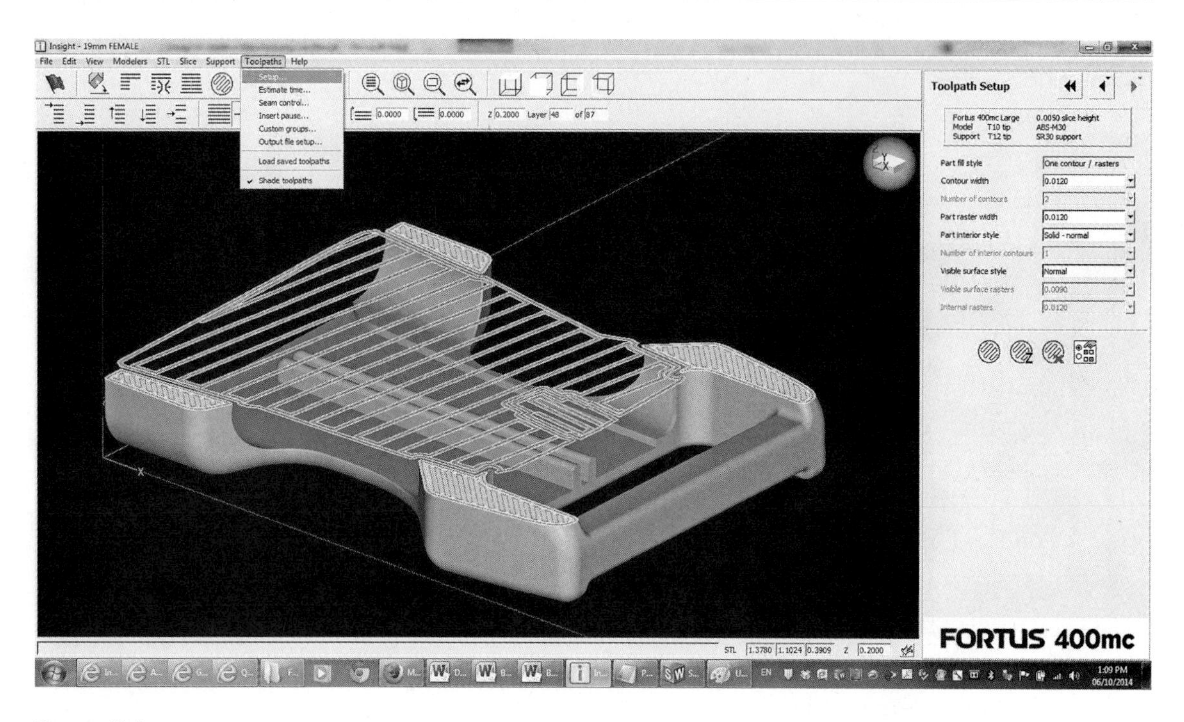

Figure 43.5

Accessing Toolpath setup tool.

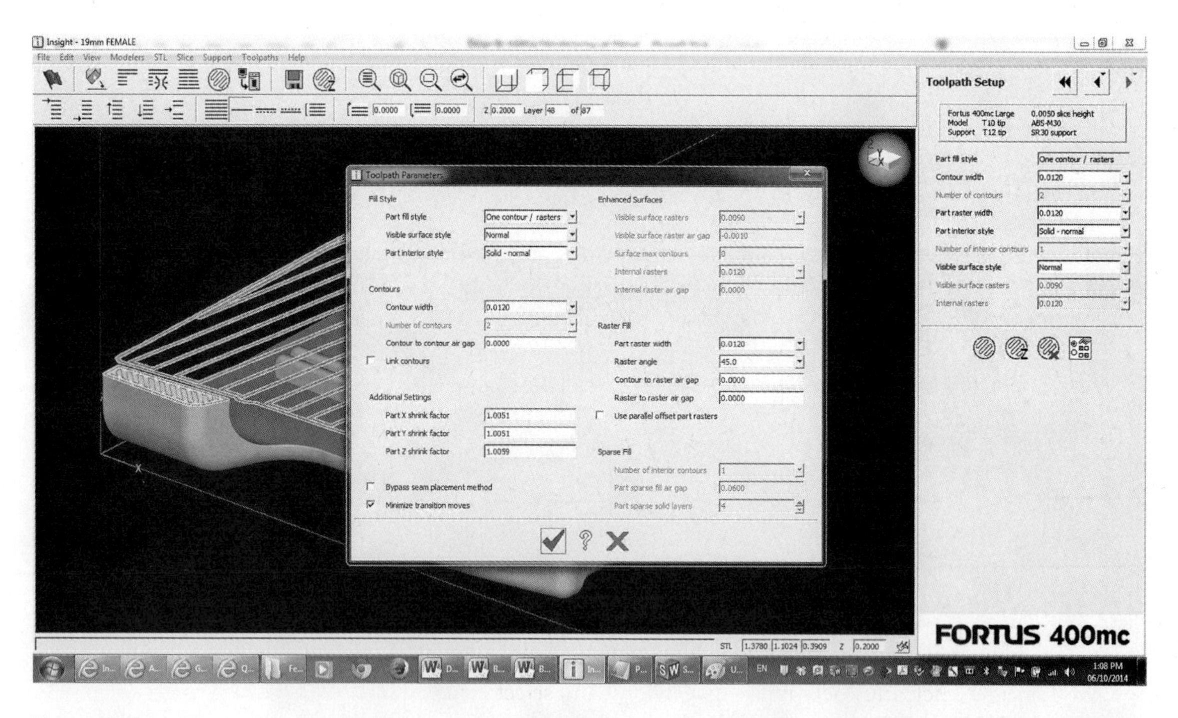

Figure 43.6

Toolpath setup.

Do the following steps (see Figure 43.7):
Slice: Slice the STL (marked [1])
Support: Create Support for the current job (marked [2])
Toolpaths: Create toolpaths for all boundaries on all layers (marked [3])

1. Click STL.
2. Select STL display.

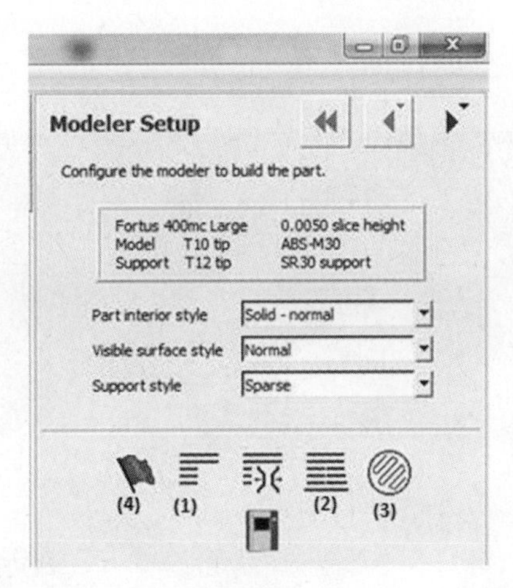

Figure 43.7

Simple sequence to follow for preprocessing.

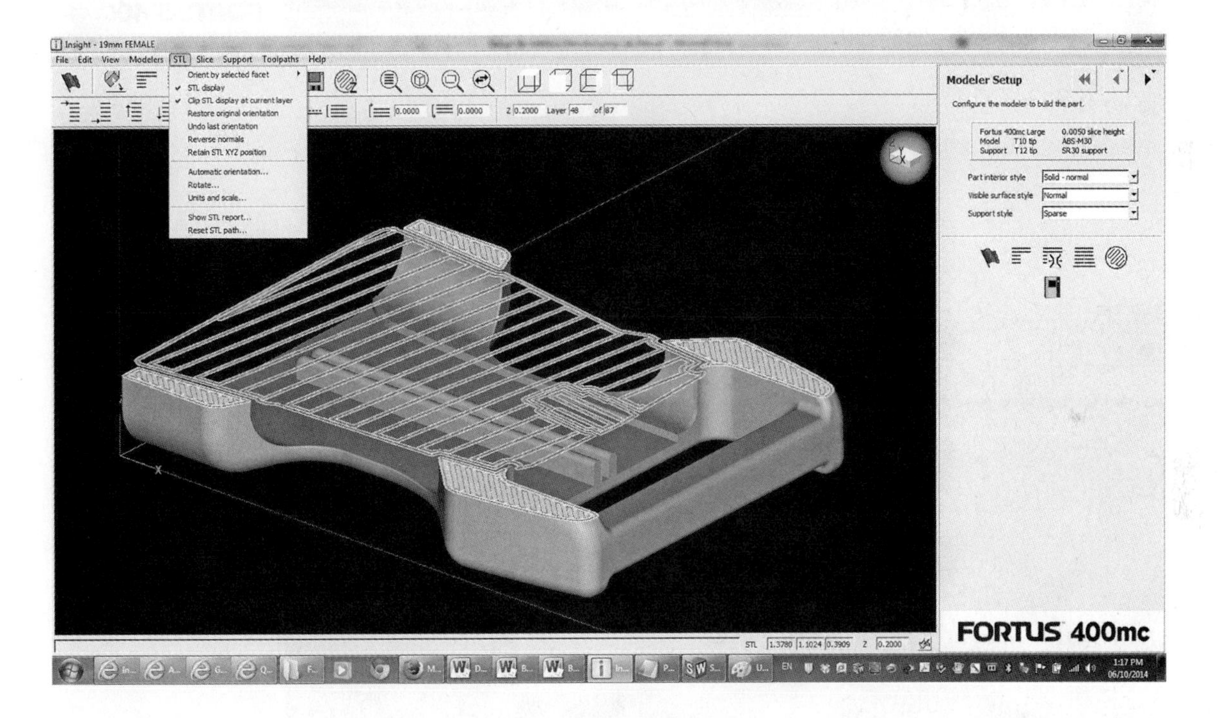

Figure 43.8

Visualization tool activated.

3. Select Clip STL display at current layer.
4. Right-click anywhere in the display unit and select Shade toolpath.

Use the PgUp or PgDn keys on the keyboard to view the layers as shown in Figure 43.9.

Finish: Do all remaining processes (marked [4]) (see Figure 43.7).

Printing in Vertical Orientation

Figures 43.10 through 43.20 show the steps for the preprocessing of the STL file for printing in horizontal orientation.

Open the STL for the 19-mm female Quick Release Buckle, which your instructor will supply. Notice the warning from the software (see Figure 43.10).

To estimate the build time,

Click Toolpaths > Estimate time

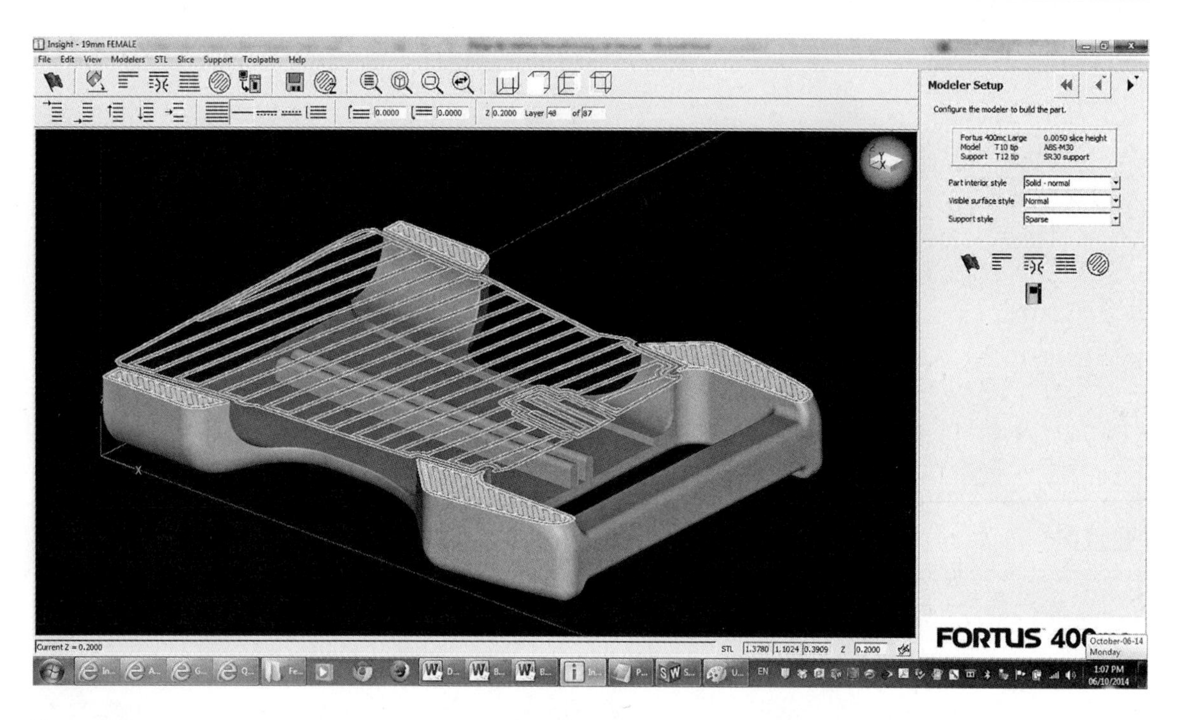

Figure 43.9

Layer exposed for detailed viewing.

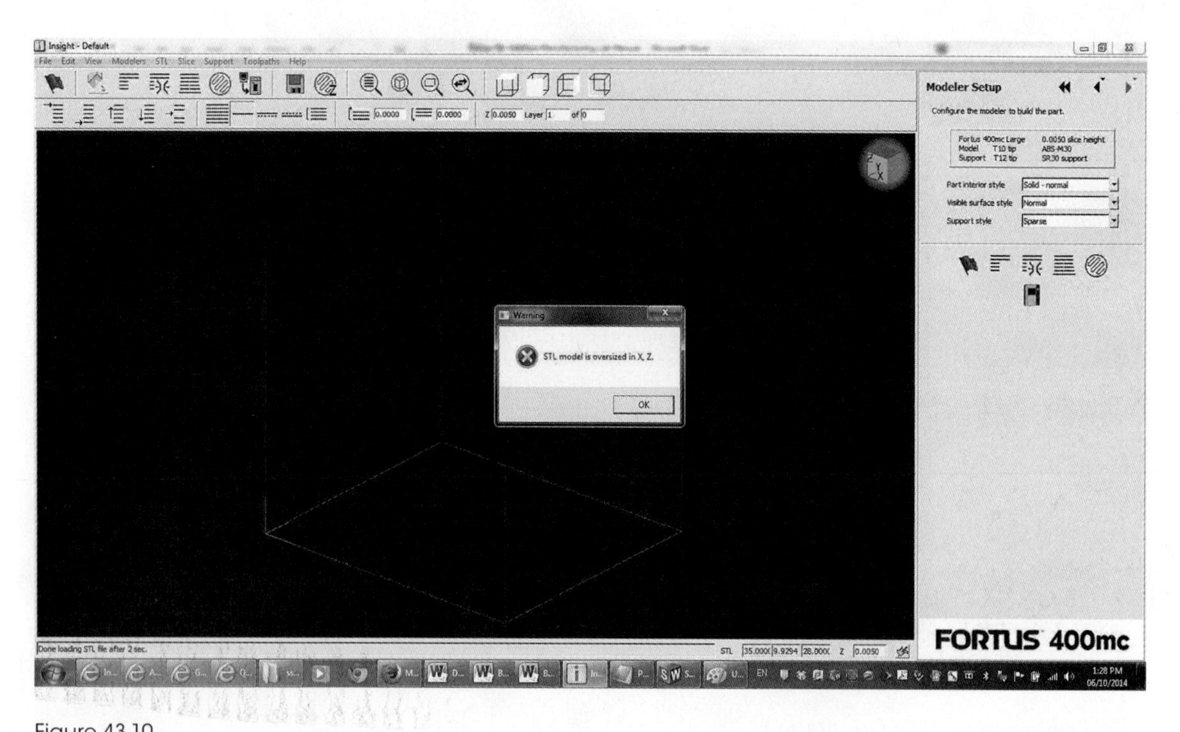

Figure 43.10

Warning when opening STL model.

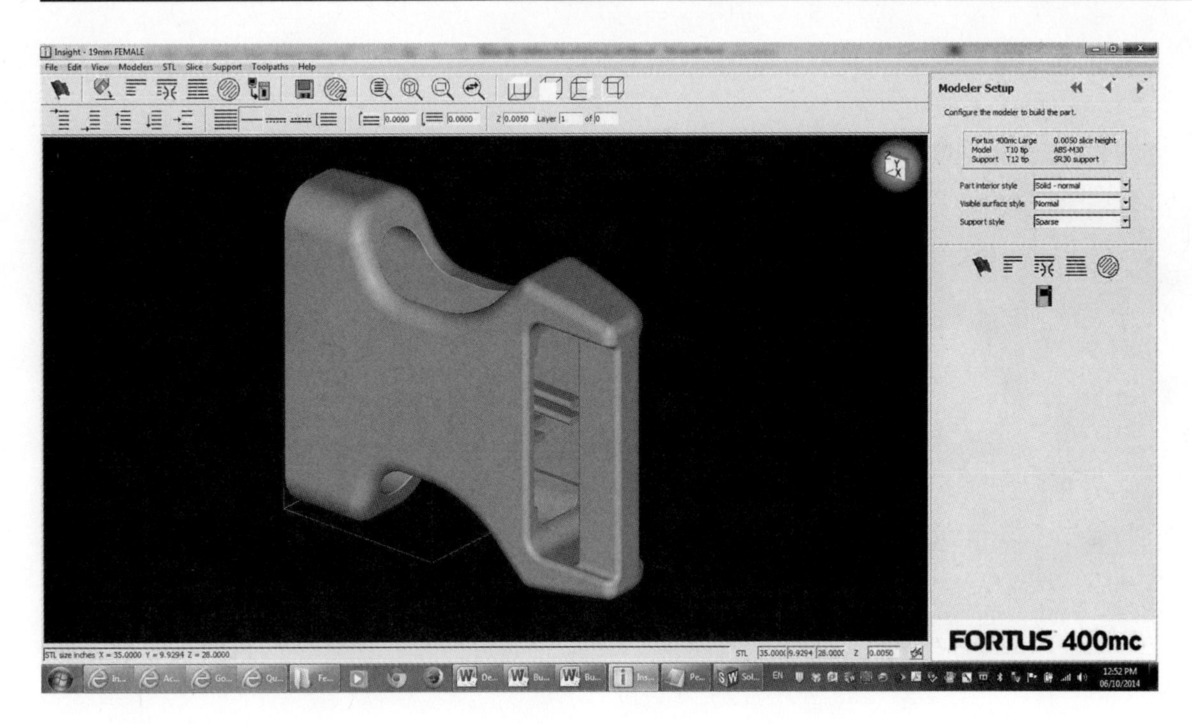

Figure 43.11

STL model.

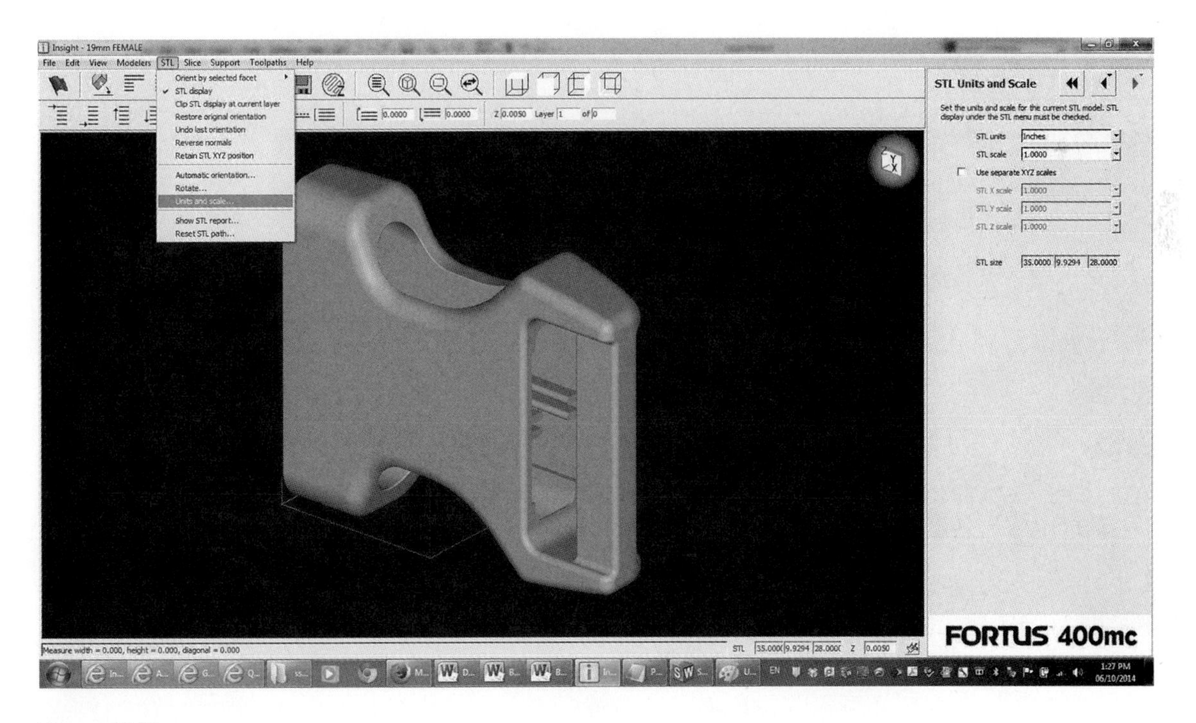

Figure 43.12

Selecting correct unit.

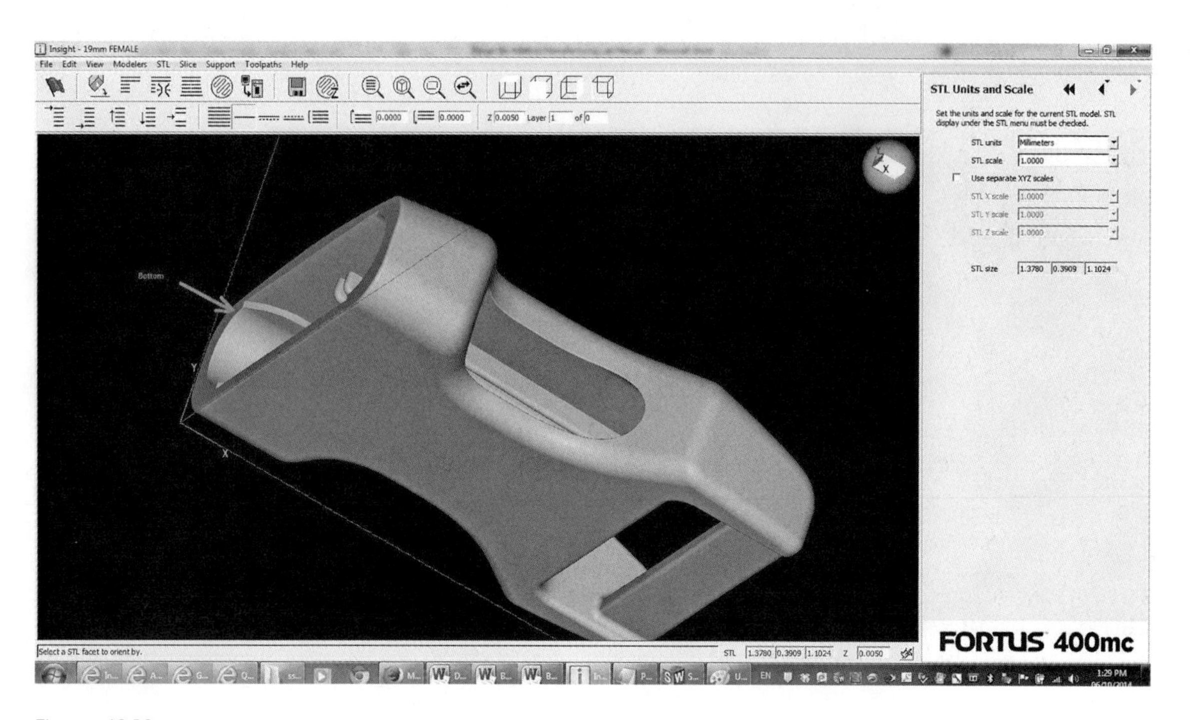

Figure 43.13

Choosing a *Bottom.*

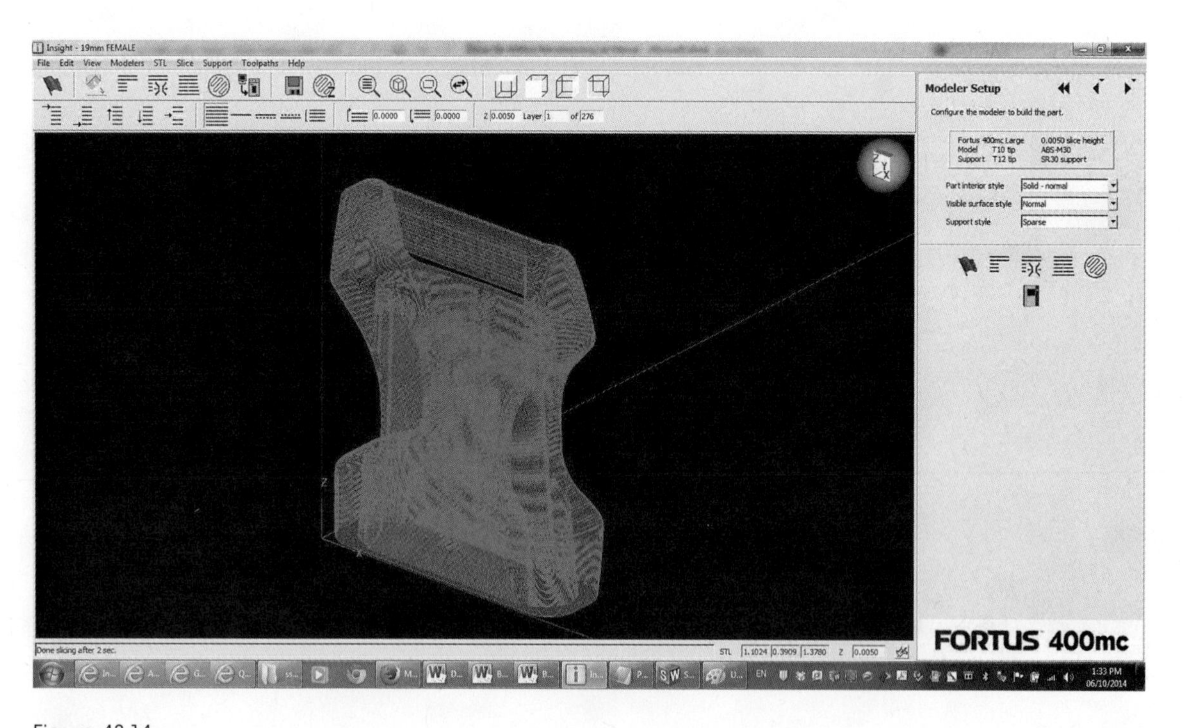

Figure 43.14

Slice model.

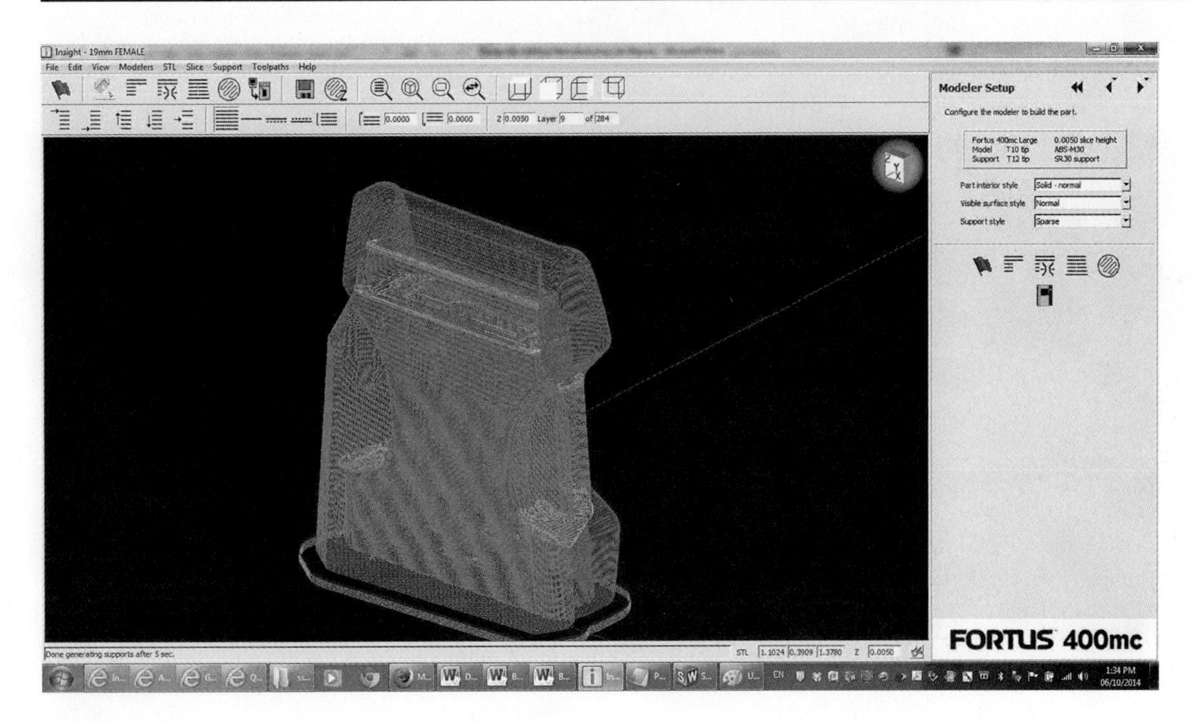

Figure 43.15

Add support.

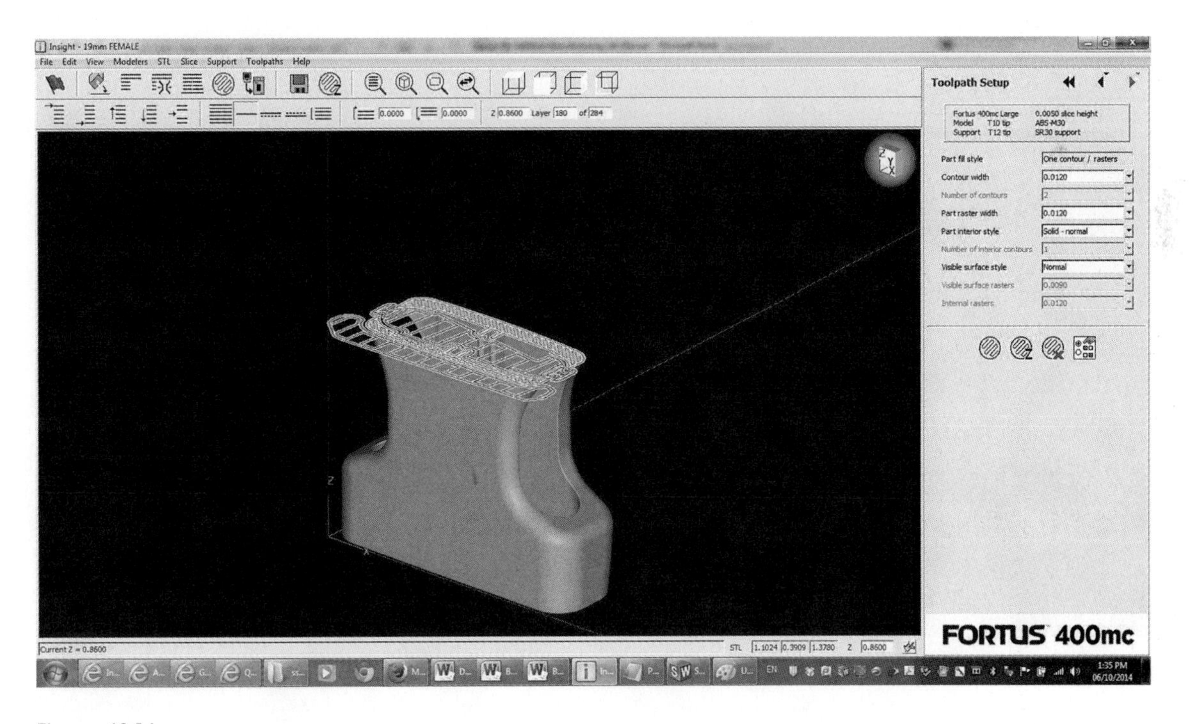

Figure 43.16

Layer exposed for detailed viewing.

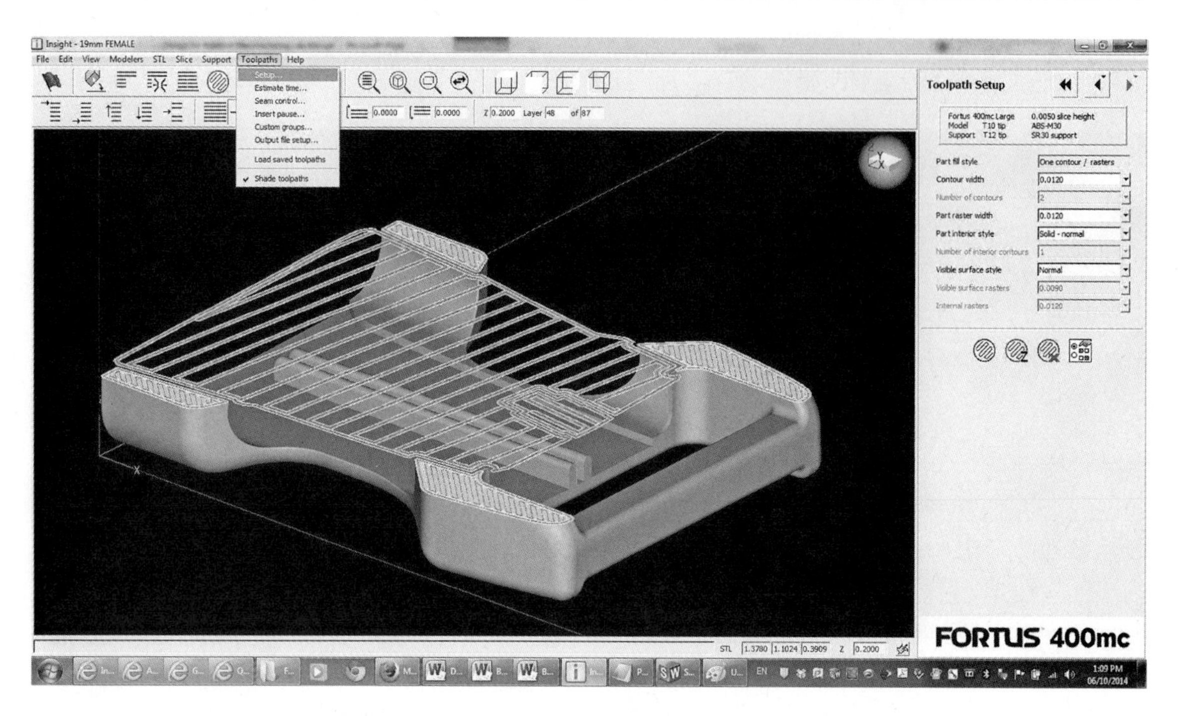

Figure 43.17

Accessing the build time toolbar.

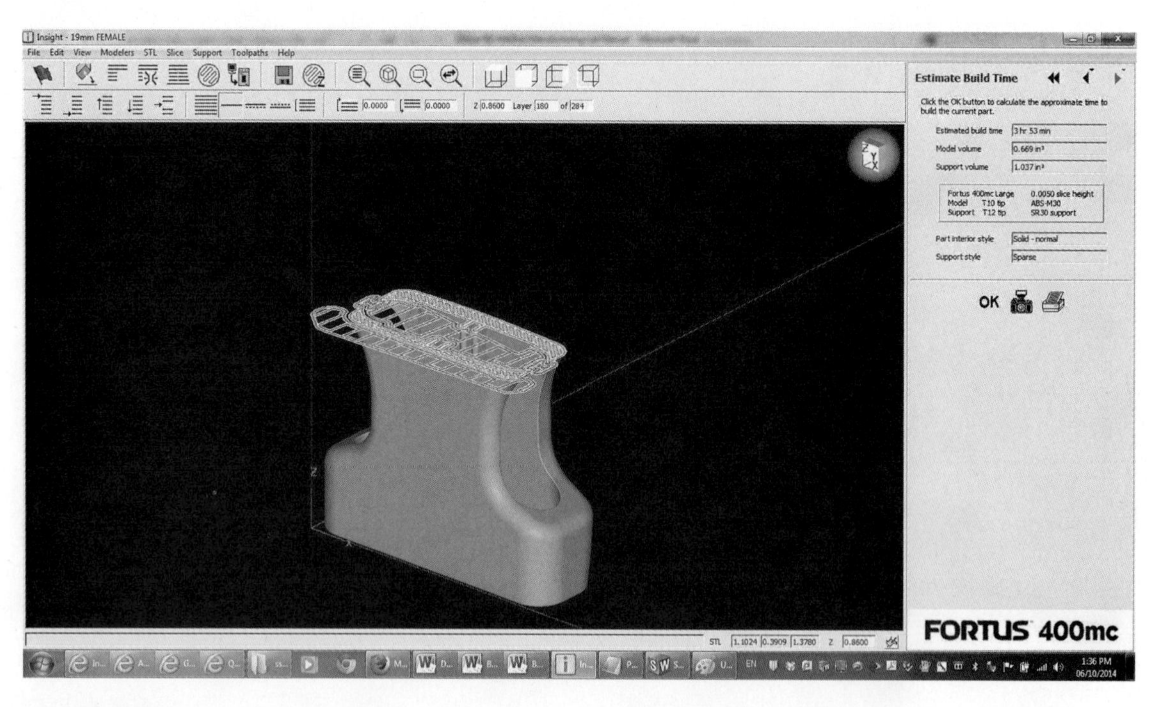

Figure 43.18

Estimating the build time.

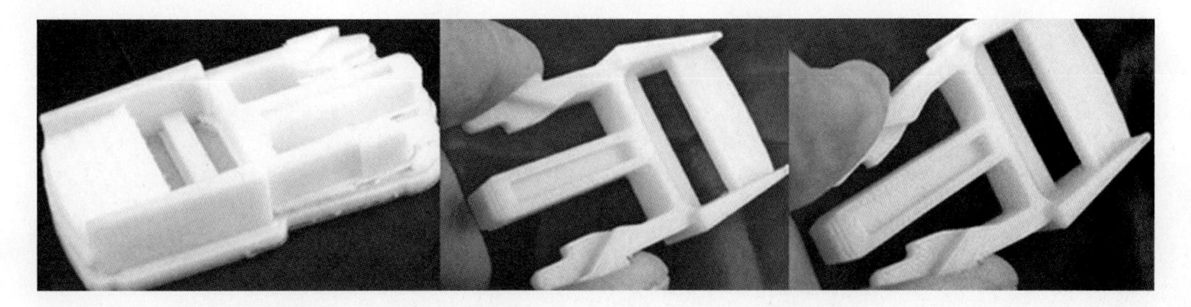

Figure 43.19

Horizontal orientation strengthens the features along layer lines.

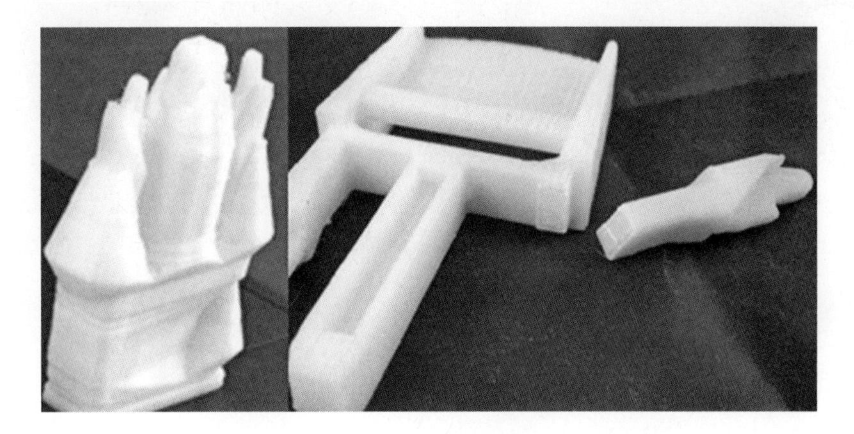

Figure 43.20

Vertical orientation weakens the features layer to layer.

Table 43.1 Comparison of Build Time for the Quick-Release Buckle

Machine	Fortus 400mc	Slice height = 0.0050"	
Model tip	T10	ABS-M30	
Support tip	T12	SR30 support	
Orientation	**Estimated Printing Time (hr.)**	**Model Volume (in.^{3})**	**Support Volume (in.^{3})**
Horizontal	1.29	0.330	0.344
Vertical	3.53	0.669	1.037

Comparing building the part in horizontal and vertical orientations in terms of estimated build time and material usage, it is found that building the part horizontally is more cost-effective (see Table 43.1).

Comparing building the part in horizontal and vertical orientations in terms of strength, it is found that horizontal orientation strengthens the features along layer lines, whereas vertical orientation weakens the features layer to layer.

Mold Tooling Using Z-Corp Spectrum 510 3D Printer

Objectives:

When you complete this module, you will

- Understand how the Z510 3D Printer works
- Be able to use the ZPrint and ZEdit software for preprocessing
- Be able to 3D-print parts using Z510 3D Printer

How Z-Corp Spectrum 510 3D Printer Works

The Spectrum System is based on Powder Bed/Inkjet Head (PB/IH) technology that Massachusetts Institute of Technology patented, which is the basis for three-dimensional printing (3DP) using high-performance composite material (HPCM). HPCM consists of a highly engineered powder with numerous additives that maximize surface finish, feature resolution, and part strength. Each material, paired with its corresponding binder, produces high-definition parts that are fit for the most demanding 3DP application.

The software first converts a three-dimensional (3D) design that is built using 3D computer-aided design (CAD) software into cross section or slices that can be between 0.0035- and 0.004-in. (0.0875–0.1-mm) thick.

The printer then prints these cross sections one after another from the bottom of the part to the top.

Inside the printer, there are two pistons (see Figure 44.1). The feed piston is represented in the diagram on the left and is shown in the *down* position filled with powder. The build piston is the piston on the right, shown in the *up* position. Also represented in the diagrams are the roller (drawn as a circle) and the print assembly (drawn as a square). On the printer, the roller and the print assembly are mounted together on the gantry, which moves horizontally across the build area.

To begin the 3DP process, the printer first spreads a layer of powder in the same thickness as the cross section to be printed. The print heads then apply a binder solution to the powder causing the powder particles to bind to one another and to the printed cross section one level below. The fed piston comes up one

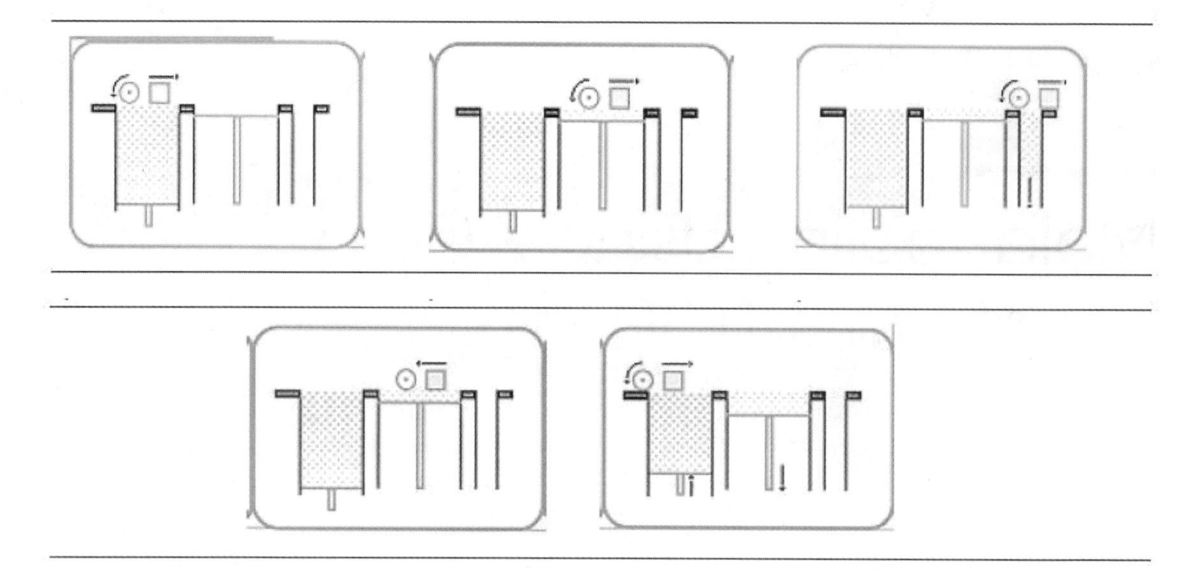

Figure 44.1

Printing Process.

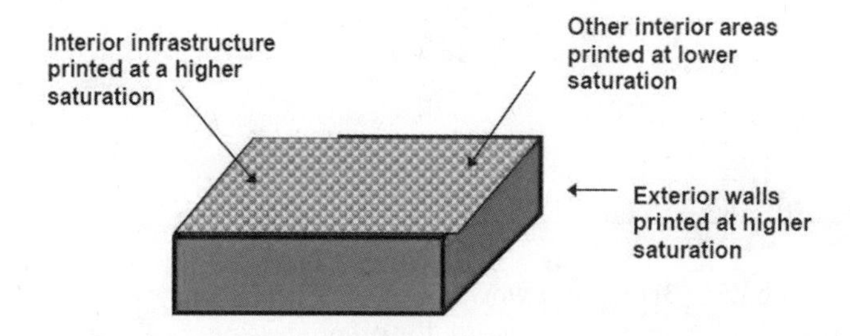

Figure 44.2

The Z-Corp. 3D Printing shelling and infrastructure features.

layer, and the build piston drops one layer. The printer then spreads a new layer of powder and repeats the process, and, in a short time, the entire part is printed.

Step 1: As the gantry traverses left to right, the roller collects powder.	Step 2: The roller spreads a thin layer of powder over the build piston.	Step 3: The roller discharges excess powder down the powder overflow chute.
	Step 4: As the gantry traverses right to left, the print head prints the part cross section.	Step 5: The feed piston moves up one layer, the build piston moves down one layer, and the process is repeated.

The printing employs several techniques to quickly build parts. First, a binder solution is applied in a higher concentration around the edges of the part, creating a strong *shell* around the exterior of the part (see Figure 44.2). Within parts, the printer builds an infrastructure by printing a strong scaffolding within part walls with a higher concentration of binder solution. The remaining interior areas are printed with a lower saturation, which gives them stability, but prevents oversaturation, which can lead to part distortion of the part.

Z-Corp Spectrum 510 3D Printer

The Z-Corp Spectrum 510 3D printer (see Figure 44.3) uses a specialty powder and a weak binder, or glue, to create parts in its 10 in. × 14 in. × 8 in. build envelope. Because the Z-Corp builds the object layer by layer,

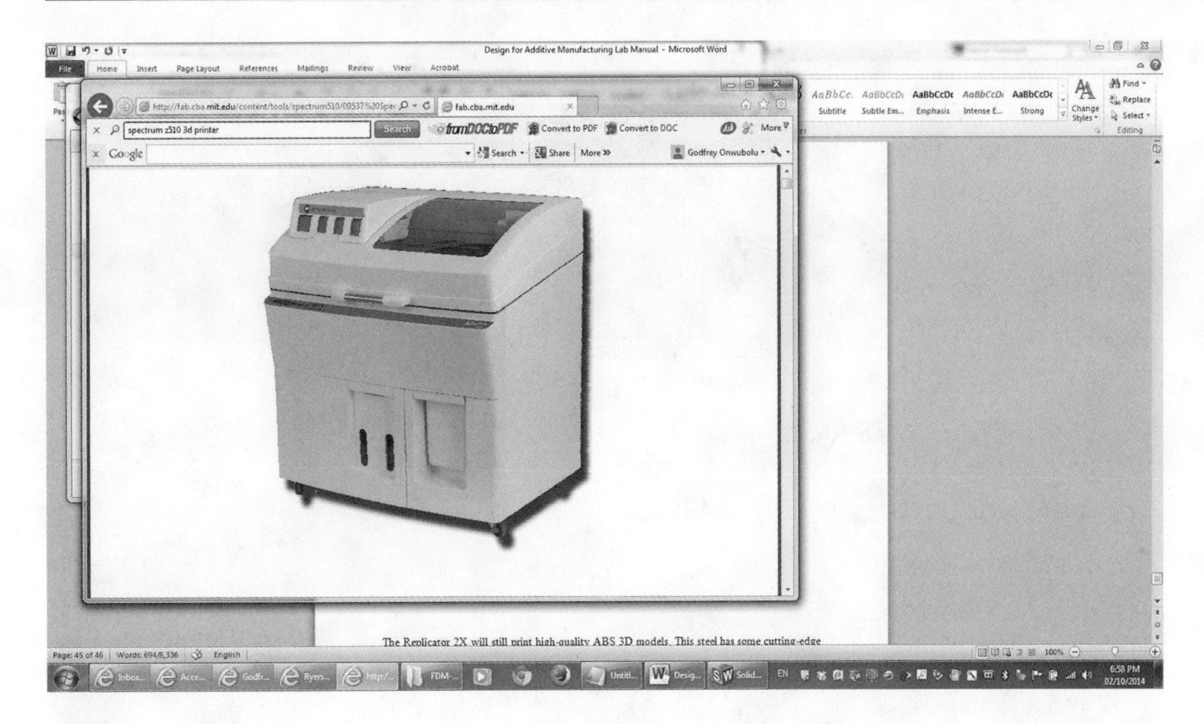

Figure 44.3

Z-Corp Spectrum 510 3D printer.

it is capable of producing complex and organic shapes, as well as linked and fully assembled parts, such as chains, gearboxes, curved tubes, and bottles—essentially any form that is imaginable.

Modeling Tips

- zCast is only good for making molds for casting non-ferrous material such as aluminum, zinc, and silver.
- Different powders require different binders.
- *Layer thickness:* 0.0035–0.0040 in.
- Objects need to be modeled as a solid object—this means that the outer surface of the object needs to be closed and watertight. The object should also be able to support its own weight in the real world—the glue used in the printing process is fairly weak, so thin parts may break before infiltration can be completed. If you have any concerns, contact a staff member for more information.
- *Orientation:* there will be slightly visible lines (cross-section lines) in your completed part as a result of the layer-by-layer printing process. Think about which direction these should run along your part. Generally, parts are stronger in the horizontal direction (the same direction as the lines).

Materials:

The Z-Corp Spectrum uses a plaster-like powder (called ZP-150), which can be dyed to a variety of colors during the printing process. At this time, this is the only material that we will be using with this machine, although others are available.

Mold Tooling Using Z-Corp Spectrum 510 3D Printer

I Module 1

Start Zprint software

Choose Units > Next (see Figure 44.4).

Open the STL file (*coin_mold003_1*) from the Zprint software (see Figure 44.5).

Notice the message that appears automatically; click Next (see Figure 44.6).

II Module 2

Click Settings > General Preferences (see Figure 44.7).

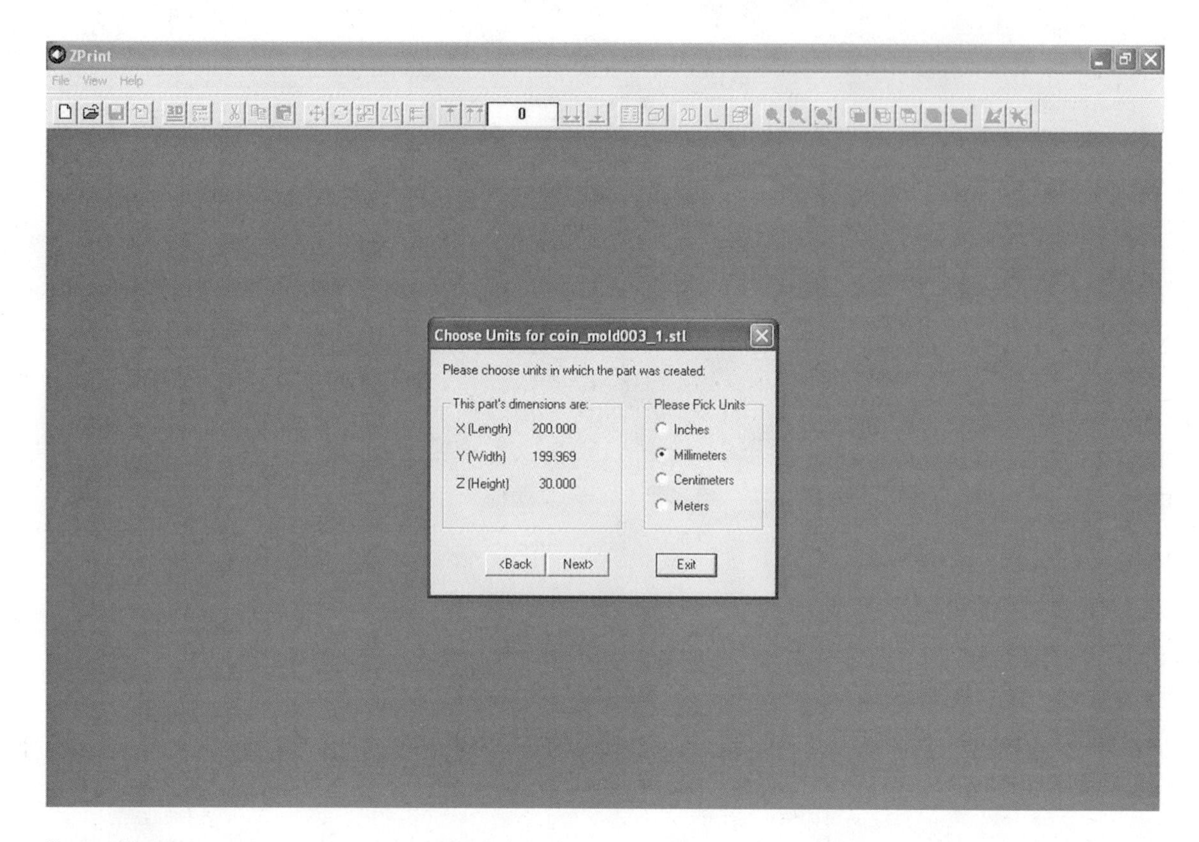

Figure 44.4

Units selection.

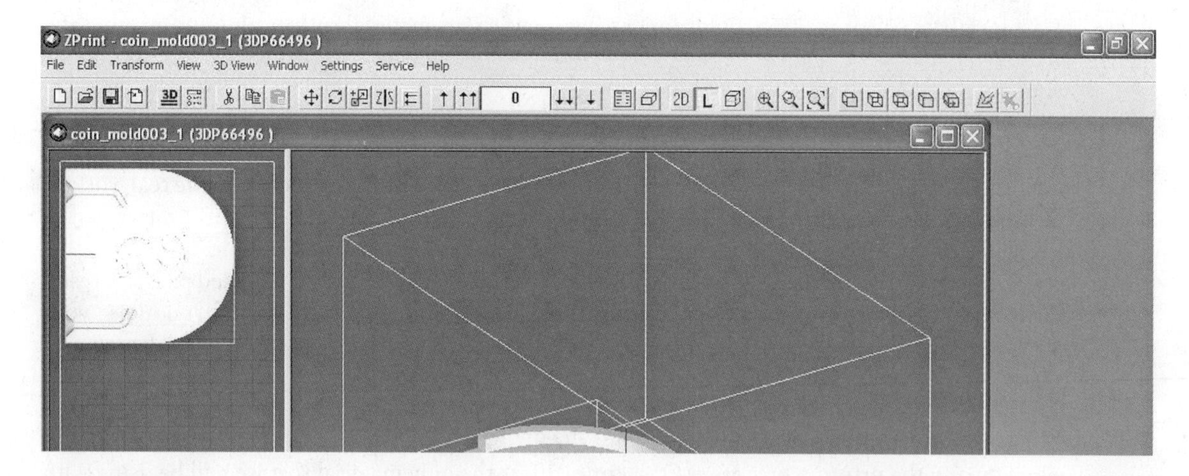

Figure 44.5

STL of model being loaded.

Click Settings > Default Printer Type (see Figure 44.8).
Click Setup > Printer Setup (see Figure 44.9).
Click Setup > Powder Settings (see Figure 44.10).
III Module 3
Click File > 3D Print Setup (see Figure 44.11).
Click File > 3D Print Setup > Powder Type (Select ZP140) (see Figure 44.12).
Click File > 3D Print Setup > Layer Thickness (Select Layer Thickness = 0.004") (see Figure 44.13).

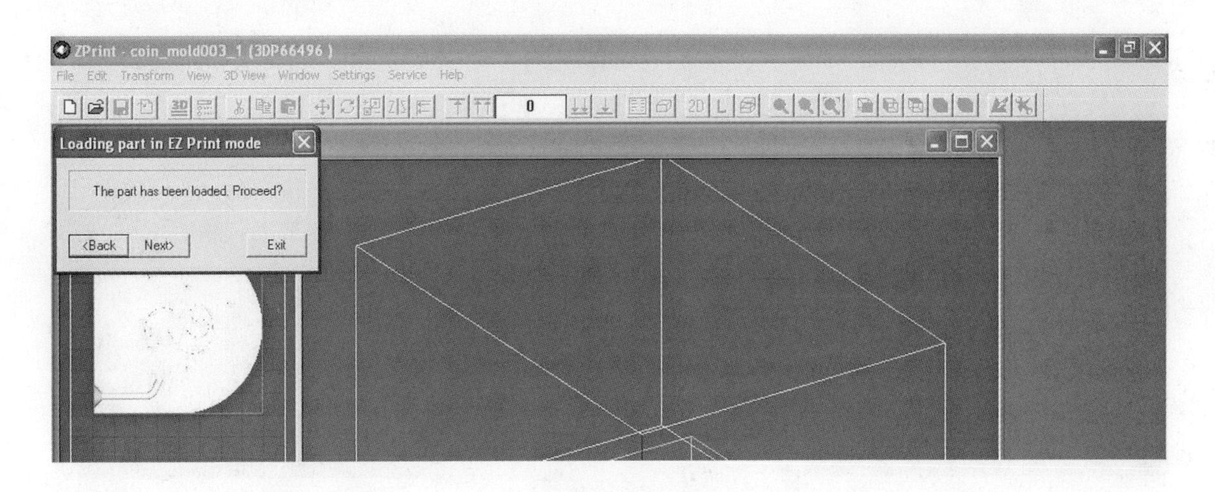

Figure 44.6

Proceeding with loaded model.

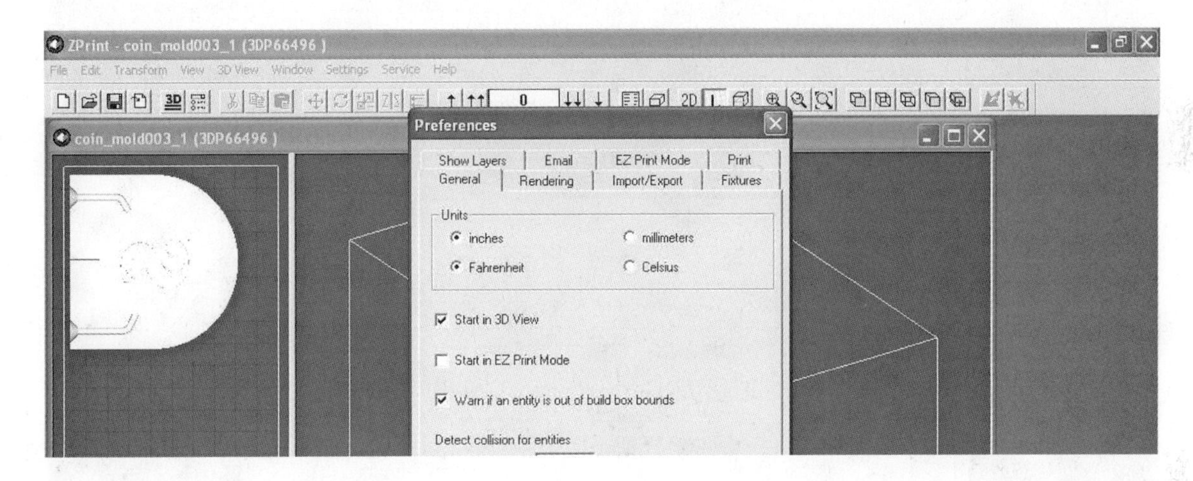

Figure 44.7

General Preferences.

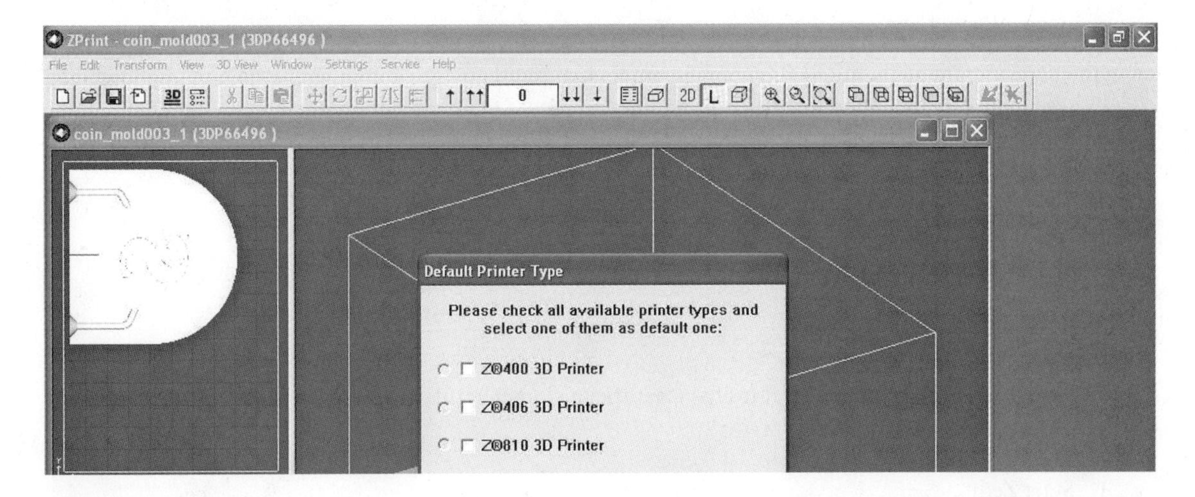

Figure 44.8

Selecting printer type.

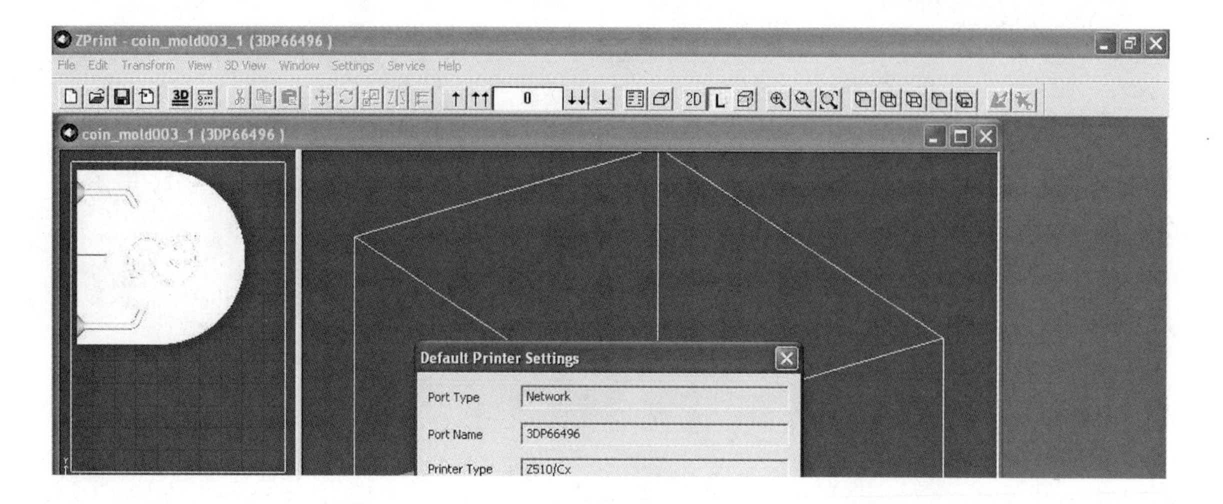

Figure 44.9

Setting up printer.

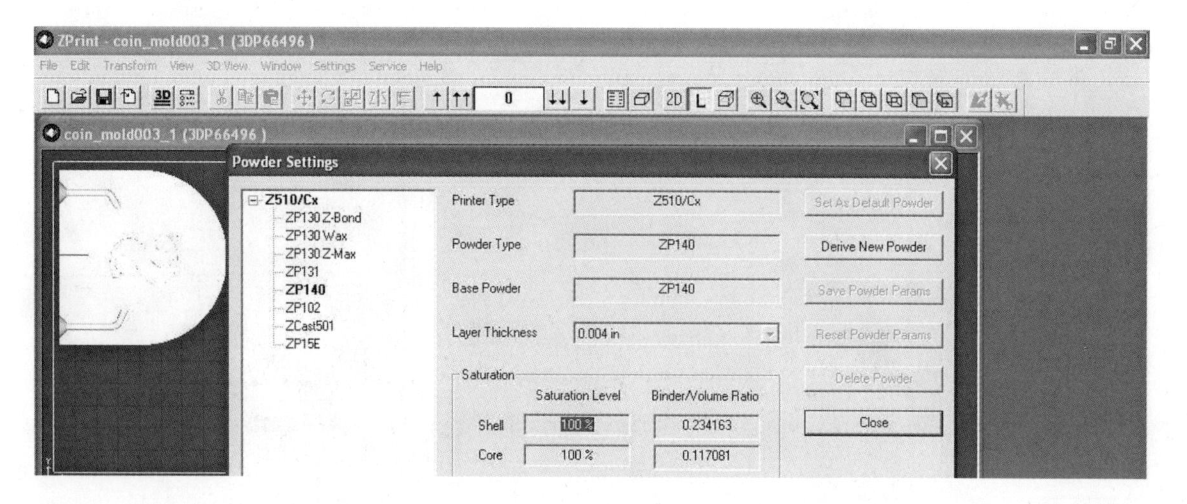

Figure 44.10

Powder default selection.

The top of the mold is shown in Figure 44.14.

The second part of the mold (bottom) is processed exactly as the top part that is already described (see Figure 44.15). Therefore, there is no need to repeat the steps here.

Second Part of Mold: Bottom

IV Module 4: Ready for Printing

When you are ready to print, do the following:

Click the icon 3D with arrow pointing to it, marked (1) (see Figure 44.16).

Click OK to commence printing.

Printing with Colors

Z510 can print with different colors, if that is what the customer wants. There are a few steps to print using colors:

1. Start Zedit.
2. Make Edge Lines.
3. Select Surface.

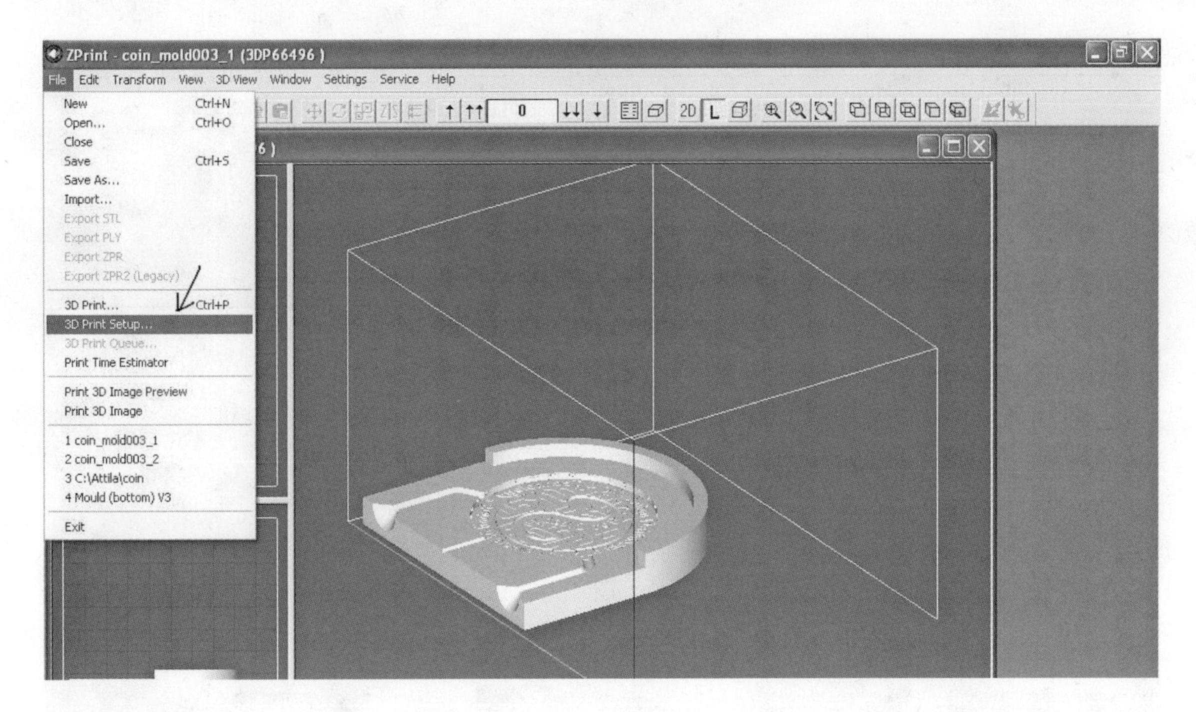

Figure 44.11

3D Print Setup.

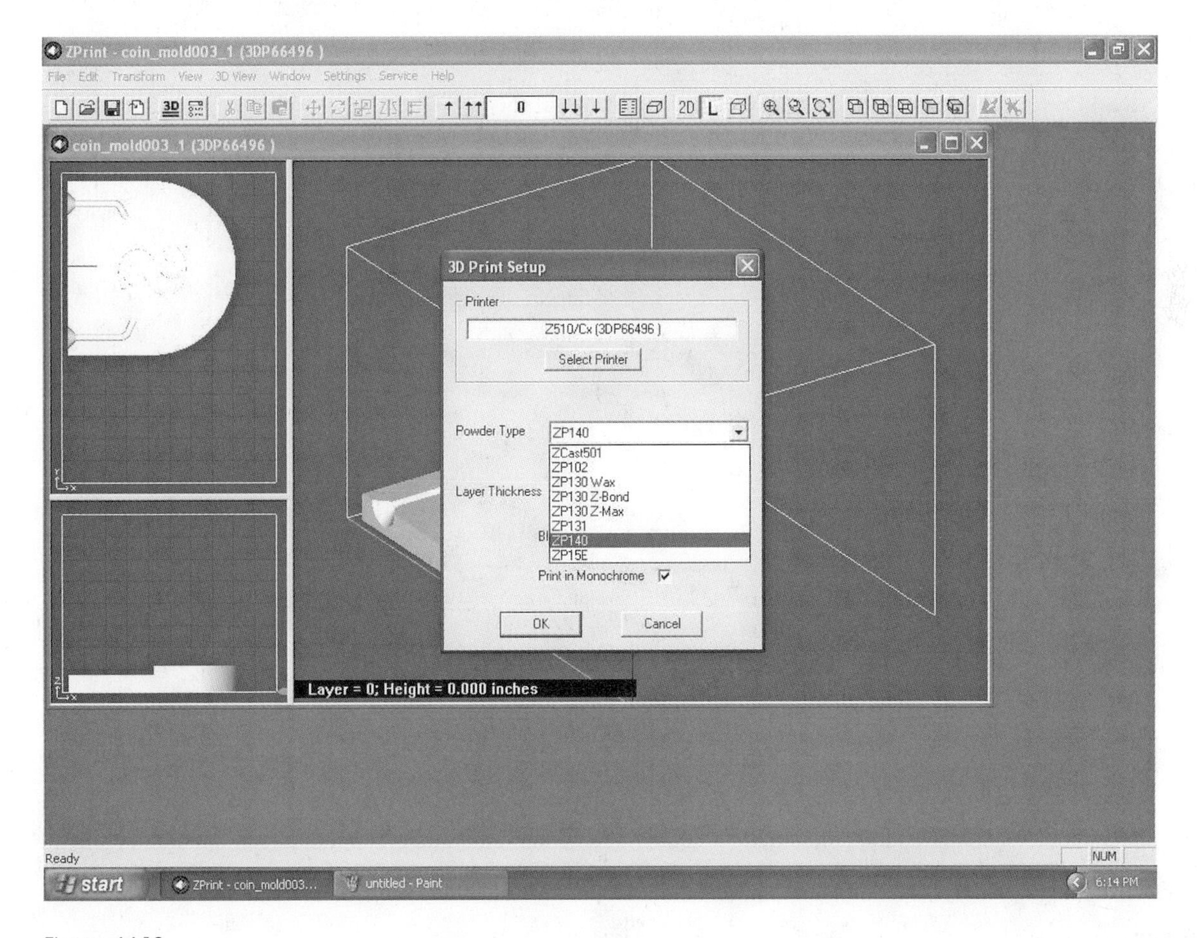

Figure 44.12

Powder Type selection.

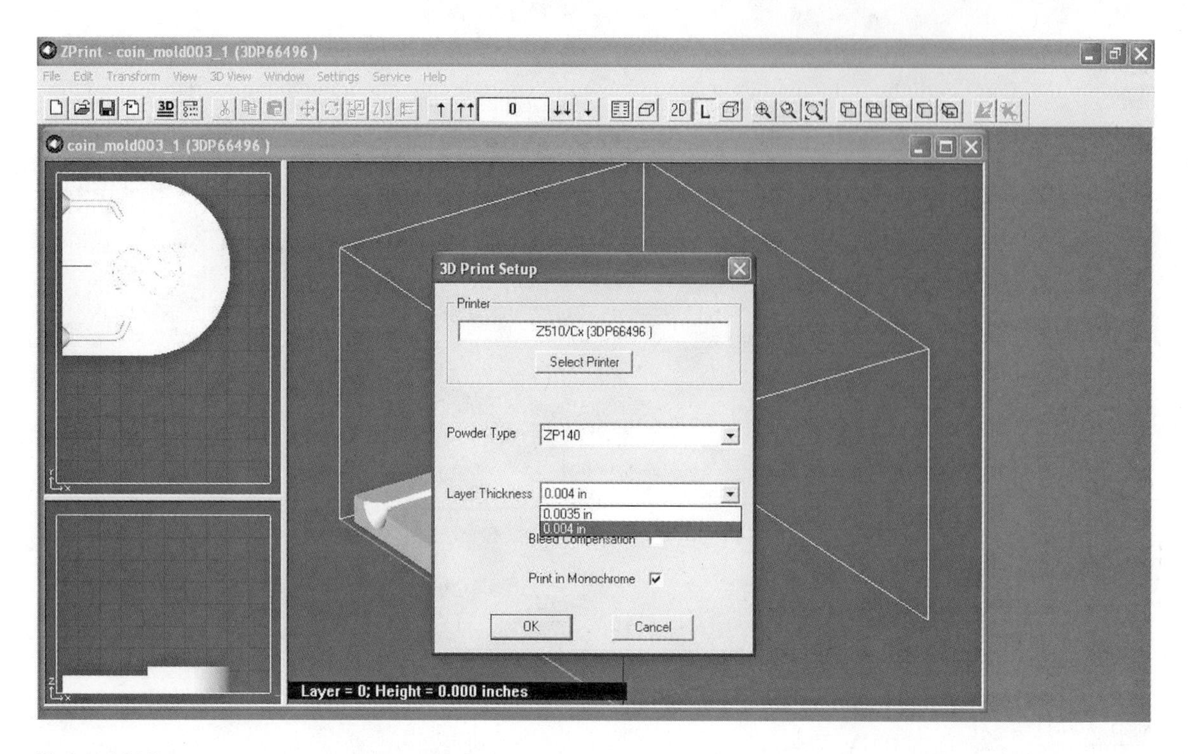

Figure 44.13

Layer Thickness selection.

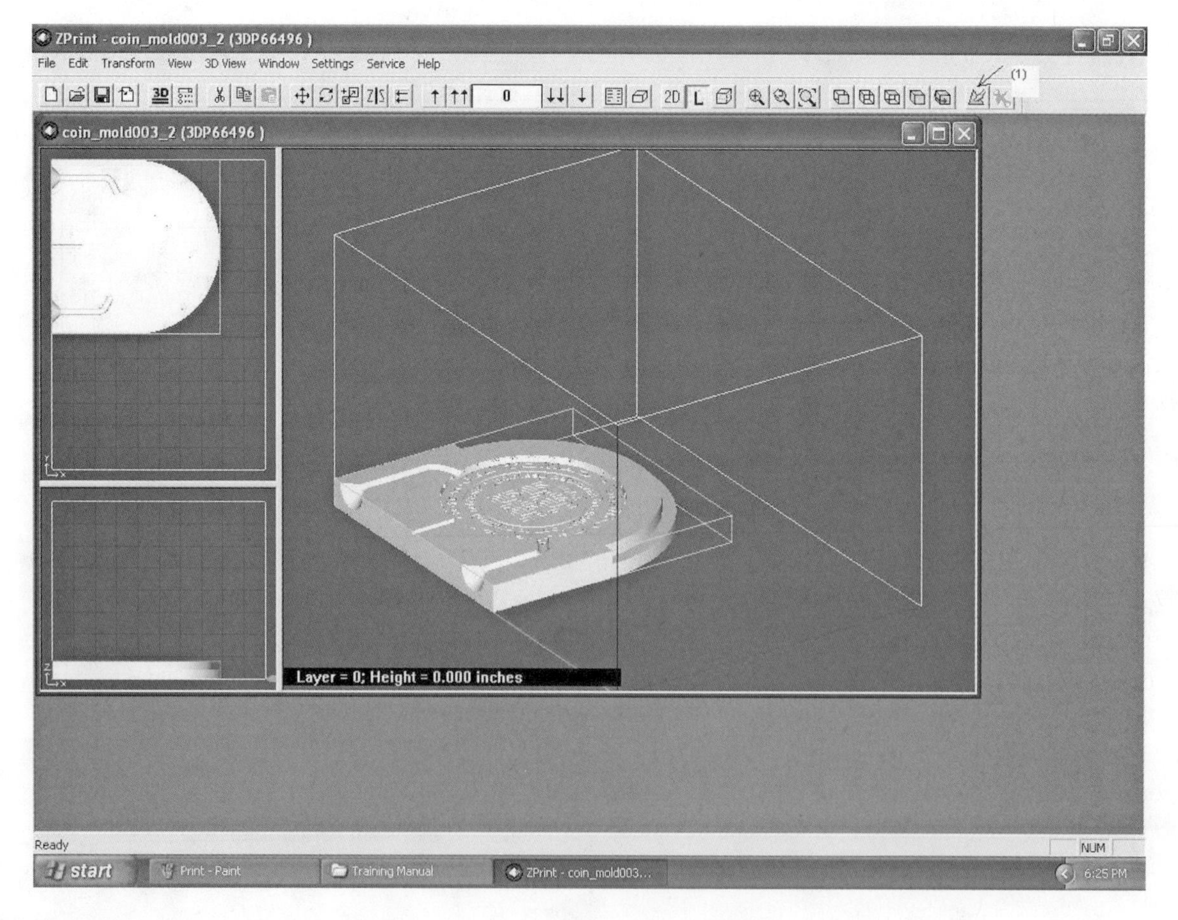

Figure 44.14

Top of Mold ready for printing.

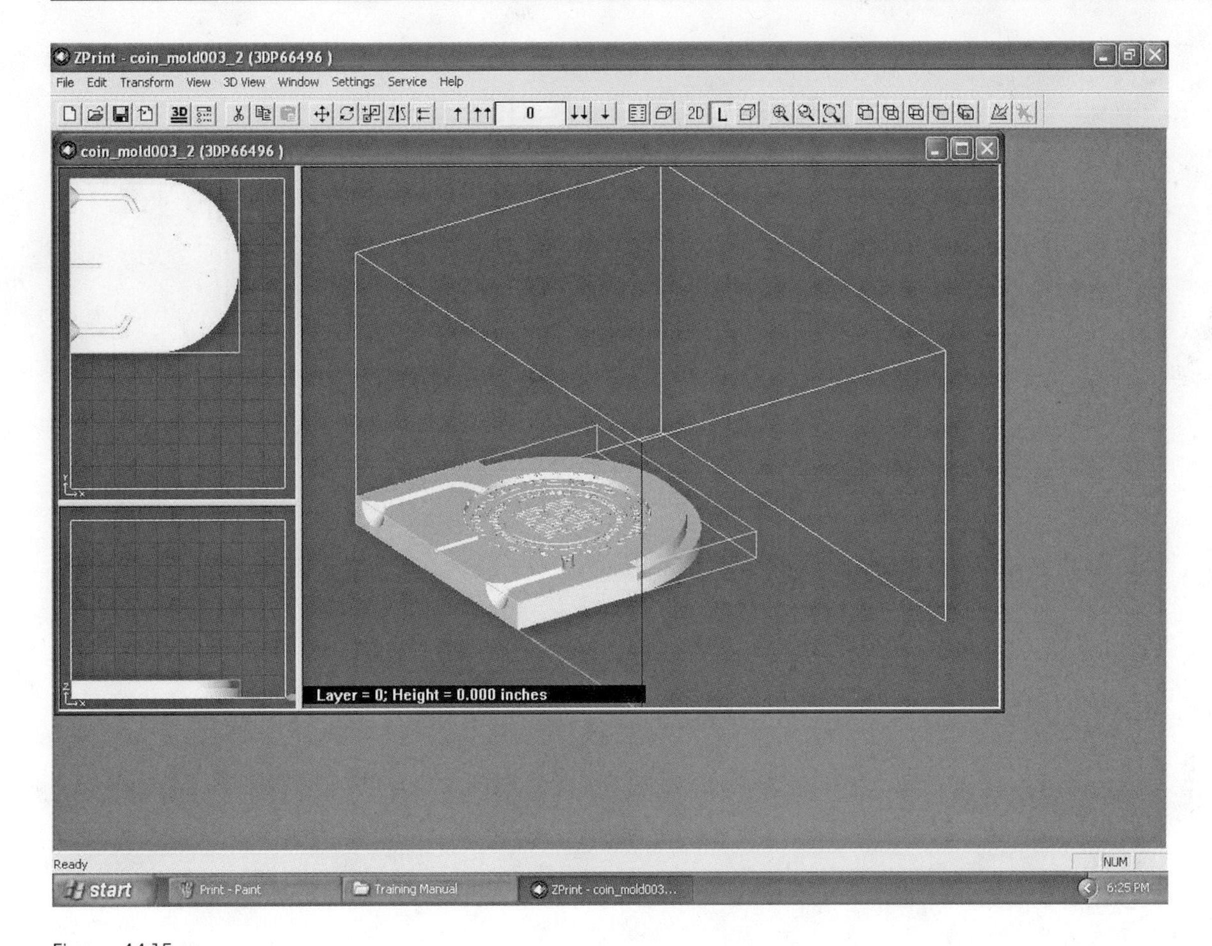

Figure 44.15

Bottom part of mold.

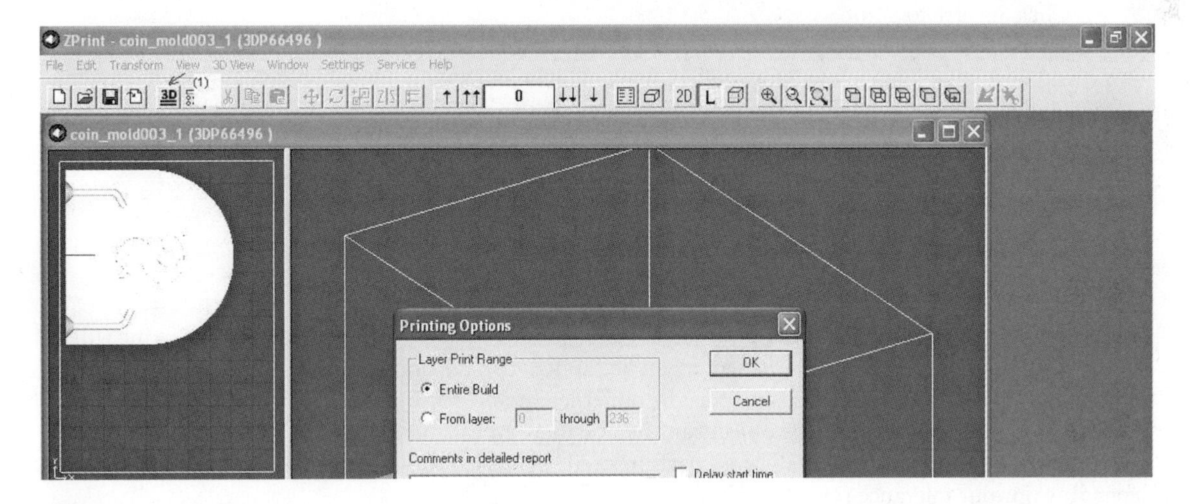

Figure 44.16

To commence printing.

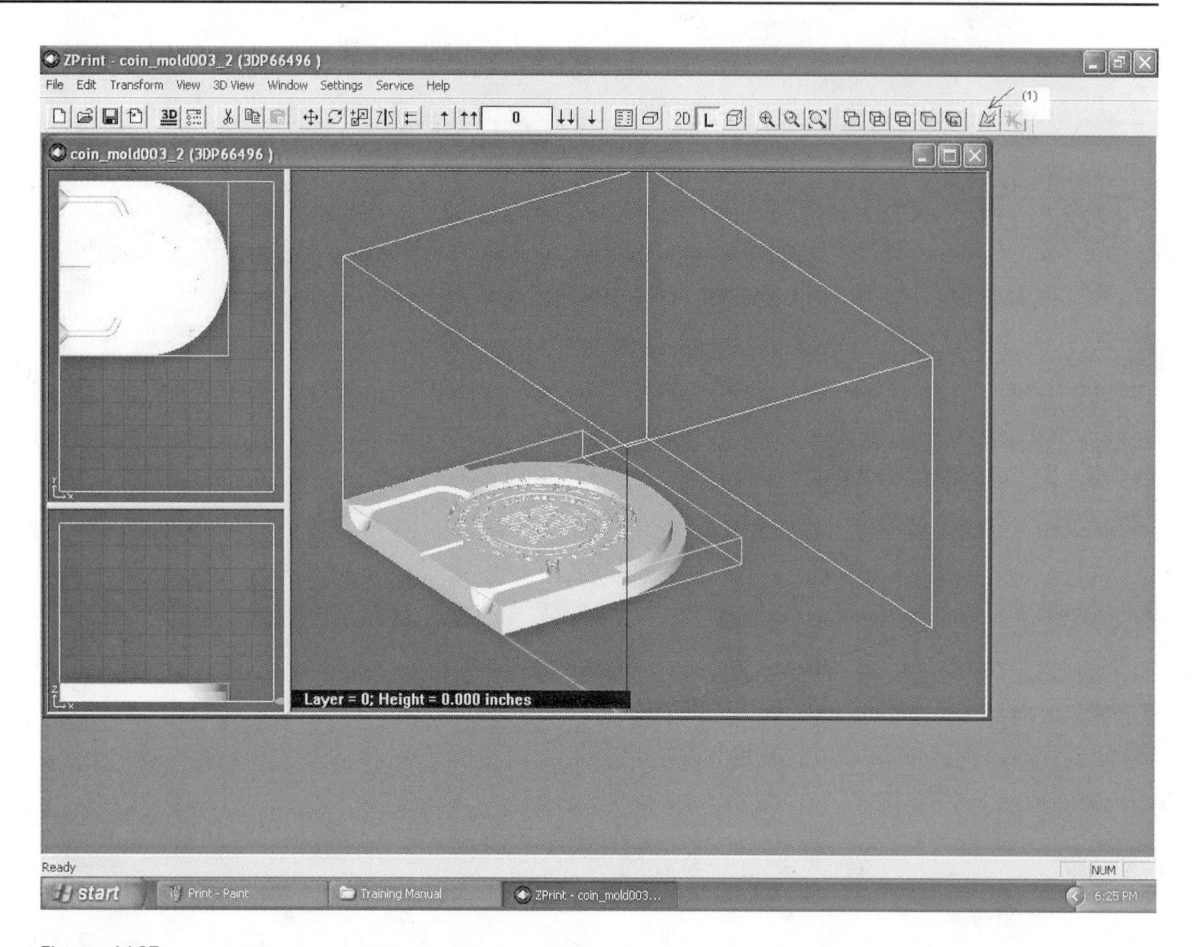

Figure 44.17

Start ZEdit.

4. **Paint.**
5. **Color (choice of color).**
6. **Click Surface to Paint.**

Starting Zprint (see Figure 44.17 for model). (Note: Follow steps as previously described.)
Start ZEdit by click the start button. (See Figure 44.17 with an arrow and marked "(1)".)
Make Edge Line. (Click the Edge Lines tool, marked "(1)"; see Figure 44.18.)
Select Surface. (Click the Selection Mode and choose Surface, marked "(2)"; see Figure 44.19.)
Select tool for painting.

Postprocessing

Once your part is removed from the build bed, excess powder will be brushed off, and the print will be cleaned with a stream of air. Once cleaned, it can be sanded, drilled, and adjusted as you see fit, but it may still be somewhat fragile. The object can then be either left alone or infiltrated. Infiltration will harden the piece for durability and testing. The object can then be touched up and painted to get a large range of finishes. Please note that infiltration can cause parts to swell slightly, so you may need to take this into account, depending on your tolerance requirements.

The cost for 3D-printed parts is based on the volume of material that is used and the quantity of the infiltrant that is needed to harden the part.

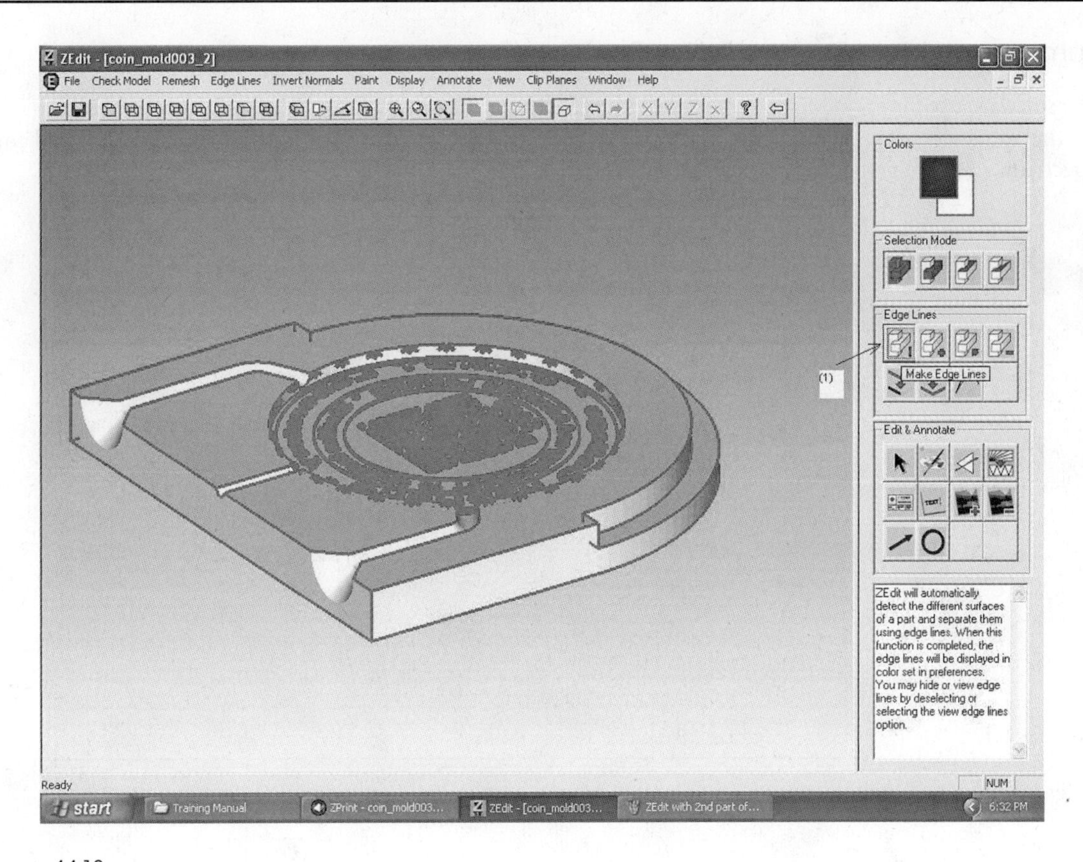

Figure 44.18

Make Edge Lines.

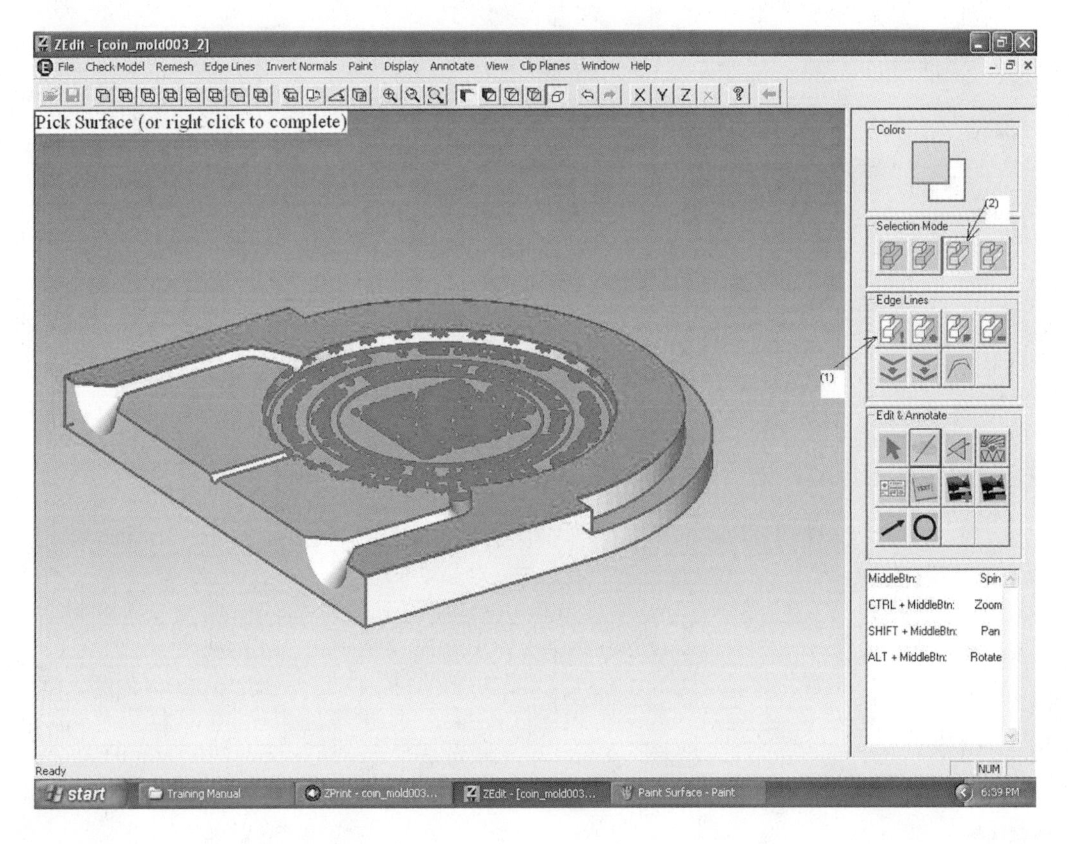

Figure 44.19

Select Surface.

Summary of Mold Design Using zCast

The software used is *zEdit and zPrint Software for Spectrum Z510*

The steps for preprocessing the 3D CAD model for the 3D printing: Powder Bed/Inkjet Head and Spectrum Z510 are given in this section.

I Coloring
Open zEdit software
Make Edge Lines
Surface
Paint > Surface > Color
Save As *.zpr
Close
II Printing
Open zPrint software
Open *.zpr
File > 3D Print

Index

Page numbers followed by f and t indicate figures and tables, respectively.

B

C

D

E

F

G

H

I

J

K

L

N

O

P

R

S

T

U